Fundamental Equations of Dynamics

KINEMATICS

Particle Rectilinear Motion

Variable a	*Constant $a = a_c$*
$a = \dfrac{dv}{dt}$	$v = v_0 + a_c t$
$v = \dfrac{ds}{dt}$	$s = s_0 + v_0 t + \frac{1}{2} a_c t^2$
$v\,dv = a\,ds$	$v^2 = v_0^2 + 2a_c(s - s_0)$

Particle Curvilinear Motion

x, y, z Coordinates		*r, θ, z Coordinates*	
$v_x = \dot{x}$	$a_x = \ddot{x}$	$v_r = \dot{r}$	$a_r = \ddot{r} - r\dot{\theta}^2$
$v_y = \dot{y}$	$a_y = \ddot{y}$	$v_\theta = r\dot{\theta}$	$a_\theta = r\ddot{\theta} + 2\dot{r}\dot{\theta}$
$v_z = \dot{z}$	$a_z = \ddot{z}$	$v_z = \dot{z}$	$a_z = \ddot{z}$

n, t, Coordinates

$v = \dot{s}$ | $a_t = \dot{v} = v\dfrac{dv}{ds}$

$a_n = \dfrac{v^2}{\rho} \qquad \rho = \left|\dfrac{[1 + (dy/dx)^2]^{3/2}}{d^2y/dx^2}\right|$

Relative Motion

$\mathbf{v}_B = \mathbf{v}_A + \mathbf{v}_{B/A} \qquad \mathbf{a}_B = \mathbf{a}_A + \mathbf{a}_{B/A}$

Rigid Body Motion About a Fixed Axis

Variable α	*Constant $\alpha = \alpha_c$*
$\alpha = \dfrac{d\omega}{dt}$	$\omega = \omega_0 + \alpha_c t$
$\omega = \dfrac{d\theta}{dt}$	$\theta = \theta_0 + \omega_0 t + \frac{1}{2}\alpha_c t^2$
$\omega\,d\omega = \alpha\,d\theta$	$\omega^2 = \omega_0^2 + 2\alpha_c(\theta - \theta_0)$

For Point P

$s = \theta r \qquad v = \omega r \qquad a_t = \alpha r \qquad a_n = \omega^2 r$

Relative General Plane Motion-Translating Axes

$\mathbf{v}_B = \mathbf{v}_A + \mathbf{v}_{B/A(\text{pin})} \qquad \mathbf{a}_B = \mathbf{a}_A + \mathbf{a}_{B/A(\text{pin})}$

Relative General Plane Motion-Trans. and Rot. Axis

$\mathbf{v}_B = \mathbf{v}_A + \boldsymbol{\Omega} \times \mathbf{r}_{B/A} + (\mathbf{v}_{B/A})_{\text{rel}}$

$\mathbf{a}_B = \mathbf{a}_A + \dot{\boldsymbol{\Omega}} \times \mathbf{r}_{B/A} + \boldsymbol{\Omega} \times (\boldsymbol{\Omega} \times \mathbf{r}_{B/A}) + 2\boldsymbol{\Omega} \times (\mathbf{v}_{B/A})_{\text{rel}} + (\mathbf{a}_{B/A})_{\text{rel}}$

KINETICS

Mass Moment of Inertia	$I = \int r^2\,dm$
Parallel-Axis Theorem	$I = I_G + md^2$
Radius of Gyration	$k = \sqrt{\dfrac{I}{m}}$

Equations of Motion

Particle	$\Sigma \mathbf{F} = m\mathbf{a}$
Rigid Body (Plane Motion)	$\Sigma F_x = m(a_G)_x$ $\Sigma F_y = m(a_G)_y$ $\Sigma M_G = I_G \alpha \,/\, \Sigma M_P = \Sigma(\mathcal{M}_k)_P$

Principle of Work and Energy

$T_1 + U_{1-2} = T_2$

Kinetic Energy

Particle	$T = \frac{1}{2}mv^2$
Rigid Body (Plane Motion)	$T = \frac{1}{2}mv_G^2 + \frac{1}{2}I_G\omega^2$

Work

Variable force	$U_F = \int F\cos\theta\,ds$
Constant force	$U_{F_c} = (F_c \cos\theta)\,\Delta s$
Weight	$U_W = -W\,\Delta y$
Spring	$U_s = -(\frac{1}{2}ks_2^2 - \frac{1}{2}ks_1^2)$
Couple moment	$U_M = M\,\Delta\theta$

Power and Efficiency

$P = \dfrac{dU}{dt} = \mathbf{F} \cdot \mathbf{v} \qquad \varepsilon = \dfrac{P_{\text{out}}}{P_{\text{in}}} = \dfrac{U_{\text{out}}}{U_{\text{in}}}$

Conservation of Energy Theorem

$T_1 + V_1 = T_2 + V_2$

Potential Energy

$V = V_g + V_e$, *where* $V_g = \pm Wy$, $V_e = +\frac{1}{2}ks^2$

Principle of Linear Impulse and Momentum

Particle	$m\mathbf{v}_1 + \Sigma \int \mathbf{F}\,dt = m\mathbf{v}_2$
Rigid Body	$m(\mathbf{v}_G)_1 + \Sigma \int \mathbf{F}\,dt = m(\mathbf{v}_G)_2$

Conservation of Linear Momentum

$\Sigma(\text{syst. } m\mathbf{v})_1 = \Sigma(\text{syst. } m\mathbf{v})_2$

Coefficient of Restitution

$e = \dfrac{(v_B)_2 - (v_A)_2}{(v_A)_1 - (v_B)_1}$

Principle of Angular Impulse and Momentum

Particle	$(\mathbf{H}_O)_1 + \Sigma \int \mathbf{M}_O\,dt = (\mathbf{H}_O)_2$, *where* $H_O = (d)(mv)$
Rigid Body (Plane Motion)	$(\mathbf{H}_G)_1 + \Sigma \int \mathbf{M}_G\,dt = (\mathbf{H}_G)_2$, *where* $H_G = I_G\omega$ $(\mathbf{H}_O)_1 + \Sigma \int \mathbf{M}_O\,dt = (\mathbf{H}_O)_2$, *where* $H_O = I_G\omega + (d)(mv_G)$

Conservation of Angular Momentum

$\Sigma(\text{syst. } \mathbf{H})_1 = \Sigma(\text{syst. } \mathbf{H})_2$

Statics & Dynamics

ENGINEERING MECHANICS
Statics & Dynamics

SEVENTH EDITION

R. C. HIBBELER

PRENTICE HALL, Upper Saddle River, New Jersey 07458

LIBRARY OF CONGRESS CATALOGING-IN-PUBLICATION DATA

Hibbeler, R. C.
Engineering mechanics—statics and dynamics / R. C. Hibbeler. — 7th ed.
p. cm.
Includes index.
ISBN 0-02-354761-8
1. Mechanics, Applied I. Title.
TA350.H48 1995
620.1—dc20 94-41381
CIP

Acquisition Editor: Bill Stenquist
Editor in Chief: Marcia Horton
Marketing Manager: Frank Nicolazzo
Text Designer: Robert Freese
Cover Designer: Singer Design
Cover Art (Photo): Comstock
Photo Researcher: Julie Tesser
Photo Editor: Melinda Reo
Editorial Assistant: Meg Weist
Production Supervisor: York Production Services
Text composition: York Graphic Services
Art Studio: Precision Graphics

Published by Prentice-Hall, Inc.
A Simon & Schuster / A Viacom Company
Upper Saddle River, New Jersey 07458

PRINTED IN THE UNITED STATES OF AMERICA

10 9 8 7 6 5 4 3

ISBN 0-02-354761-8

Prentice-Hall International (UK) Limited, London
Prentice-Hall of Australia Pty. Limited, Sydney
Prentice-Hall Canada Inc., Toronto
Prentice-Hall Hispanoamericana, S.A., Mexico
Prentice-Hall of India Private Limited, New Delhi
Prentice-Hall of Japan, Inc., Tokyo
Simon & Schuster Asia Pte. Ltd., Singapore
Editora Prentice-Hall do Brasil, Ltda., Rio de Janeiro

TO THE STUDENT

With the hope that this work
will stimulate an interest in Engineering Mechanics
and provide an acceptable guide to its understanding.

Preface

The main purpose of this book is to provide the student with a clear and thorough presentation of the theory and applications of engineering mechanics. To achieve this objective the author has by no means worked alone, for to a large extent this book has been shaped by the comments and suggestions of more than a hundred reviewers in the teaching profession as well as many of the author's students.

Continued improvements have been made to this the seventh edition. Previous users of the book may first notice that the art work has been enhanced in a multi-color presentation in order to provide the reader with a more realistic and understandable sense of the material. Also, the problem sets have been greatly expanded. Often, several problem statements refer to the same drawing, so that the instructor can reinforce concepts discussed in class. The problem sets also provide a wider variation in the degree of difficulty in problem solutions, and instructors can now select problems that focus on design rather than on analysis.

Although the contents of the book have remained in the same order, the details of some topics have been expanded, some examples have been changed and others have been replaced with new ones. Also, the explanation of many topics has been improved by a careful rewording of selected sentences. The hallmarks of the book, however, remain the same: where necessary, a strong emphasis is placed on drawing a free-body diagram, and the importance of selecting an appropriate coordinate system and associated sign convention for vector components is stressed when the equations of mechanics are applied.

Organization and Approach. The contents of each chapter are organized into well-defined sections. Selected groups of sections contain an explanation of specific topics, illustrative example problems, and a set of homework problems. The topics within each section are placed into subgroups defined by boldface titles. The purpose of this is to present a structured method for introducing each new definition or concept, and to make the book convenient for later reference and review.

A ''procedure for analysis'' is given at the end of many sections of the book in order to provide the student with a review or summary of the material and a logical and orderly method to follow when applying the theory. As in the previous editions, the example problems are solved using this outlined method in order to clarify its numerical application. It is to be understood, however, that once the relevant principles have been mastered and enough confidence and judgment have been obtained, the student can then develop his or her own procedures for solving problems. In most cases, it is felt that the first step in any procedure should require drawing a diagram. By doing so, the student forms the habit of tabulating the necessary data while focusing on the physical aspects of the problem and its associated geometry. If this step is correctly performed, applying the relevant equations of mechanics becomes somewhat methodical, since the data can be taken directly from the diagram. This step is particularly important when solving problems involving equilibrium, and for this reason, drawing a free-body diagram is strongly emphasized throughout the book.

Since mathematics provides a systematic means of applying the principles of mechanics, the student is expected to have prior knowledge of algebra, geometry, trigonometry, and, for complete coverage, some calculus. Vector analysis is introduced at points where it is most applicable. Its use often provides a convenient means for presenting concise derivations of the theory, and it makes possible a simple and systematic solution of many complicated three-dimensional problems. Occasionally, the example problems are solved using more than one method of analysis so that the student develops the ability to use mathematics as a tool whereby the solution of any problem may be carried out in the most direct and effective manner.

Problems. Numerous problems in the book depict realistic situations encountered in engineering practice. It is hoped that this realism will both stimulate the student's interest in engineering mechanics and provide a means for developing the skill to reduce any such problem from its physical description to a model or symbolic representation to which the principles of mechanics may be applied. As in the previous edition, an effort has been made to include some problems which may be solved using a numerical procedure executed on either a desktop computer or a programmable pocket calculator. Suitable numerical techniques along with associated computer programs are given in Appendix B. The intent here is to broaden the student's capacity for using other forms of mathematical analysis *without* sacrificing the time needed to focus on the application of the principles of mechanics. Problems of this type which either can or must be solved using numerical procedures are identified by a ‘‘square’’ symbol (■) preceding the problem number.

Throughout the text there is an approximate balance of problems using either SI or FPS units. Furthermore, in any set, an attempt has been made to arrange the problems in order of increasing difficulty.* The answers to all but every fourth problem are listed in the back of the book. To alert the user to a problem without a reported answer, an asterisk (*) is placed before the problem number.

Contents: *Statics*. The subject of statics is presented in 11 chapters, in which the principles introduced are first applied to simple situations. Most often, each principle is applied first to a particle, then to a rigid body subjected to a coplanar system of forces, and finally to the general case of three-dimensional force systems acting on a rigid body.

The text begins in Chapter 1 with an introduction to mechanics and a discussion of units. The notion of a vector and the properties of a concurrent force system are introduced in Chapter 2. This theory is then applied to the equilibrium of particles in Chapter 3. Chapter 4 contains a general discussion of both concentrated and distributed force systems and the methods used to simplify them. The principles of rigid-body equilibrium are developed in Chapter 5 and then applied to specific problems involving the equilibrium of trusses, frames, and machines in Chapter 6, and to the analysis of internal forces in beams and cables in Chapter 7. Applications to problems involving frictional forces are discussed in Chapter 8, and topics related to the center of gravity and centroid are treated in Chapter 9. If time permits, sections concerning more advanced topics, indicated by stars (★), may be covered. Most of these topics are included in Chapter 10 (area and mass moments of inertia) and Chapter 11 (virtual work and potential energy). Note that this material also provides a suitable reference for basic principles when it is discussed in more advanced courses.

At the discretion of the instructor, some of the material may be presented in a different sequence with no loss in continuity. For example, it is possible to introduce the concept of a force and all the necessary methods of vector analysis by first covering Chapter 2 and Sec. 4.1. Then, after covering the rest of Chapter 4 (force and moment systems), the equilibrium methods in Chapters 3 and 5 can be discussed.

Contents: *Dynamics*. The subject of dynamics is presented in the last 11 chapters. In particular, the kinematics of a particle is discussed in Chapter 12, followed by a discussion of particle kinetics in Chapter 13 (equation of motion), Chapter 14 (work and energy), and Chapter 15 (impulse and momentum). The concepts of particle dynamics contained in these four chapters are

*Review problems, wherever they appear, are presented in random order.

then summarized in a ''review'' section and the student is given the chance to identify and solve a variety of different types of problems. A similar sequence of presentation is given for the planar motion of a rigid body: Chapter 16 (planar kinematics), Chapter 17 (equations of motion), Chapter 18 (work and energy), and Chapter 19 (impulse and momentum), followed by a summary and review set of problems for these chapters. If desired, it is possible to cover Chapters 12 through 19 in the following order with no loss in continuity: Chapters 12 and 16 (kinematics), Chapters 13 and 17 (equations of motion), Chapters 14 and 18 (work and energy), and Chapters 15 and 19 (impulse and momentum).

Time permitting, some of the material involving three-dimensional rigid-body motion may be included in the course. The kinematics and kinetics of this motion are discussed in Chapters 20 and 21, respectively. Chapter 22 (vibrations) may be included if the student has the necessary mathematical background. Sections of the book which are considered to be beyond the scope of the basic dynamics course are indicated by a star (★) and may be omitted. As in *Statics,* however, this more advanced material provides a suitable reference for basic principles when it is covered in other courses.

Acknowledgments. I have endeavored to write this book so that it will appeal to both the student and instructor. Through the years many people have helped in its development and I should like to acknowledge their valued suggestions and comments. Specifically, I wish to thank all the reviewers who contributed to this edition. A particular note of thanks is also given to Professor Will Lidell, Jr., Auburn University at Montgomery for his help and support.

Many thanks are also extended to all my students and to members of the teaching profession who have freely taken the time to offer their suggestions and comments. Since the list is too long to mention, it is hoped that those who have given help in this manner will accept this anonymous recognition. Lastly, I should like to acknowledge the assistance of my wife, Conny, during the time it has taken to prepare the manuscript for publication.

Russell Charles Hibbeler

Contents

STATICS

1
General Principles 3

2
Force Vectors 17

3
Equilibrium of a Particle 77

4
Force System Resultants 107

10

Moments of Inertia 469

11

Virtual Work 519

APPENDIXES

A

Mathematical Expressions 550

B

Numerical and Computer Analysis 552

Answers 558

Index 572

DYNAMICS

12

Kinematics of a Particle 3

13

Kinetics of a Particle: Force and Acceleration 91

14

Kinetics of a Particle: Work and Energy 145

15

Kinetics of a Particle: Impulse and Momentum 189

Review 1: Kinematics and Kinetics of a Particle 254

16

Planar Kinematics of a Rigid Body 269

APPENDIXES

Statics

Although computers are often used in engineering, the design and analysis of any structural or mechanical part requires a fundamental understanding of the principles of engineering mechanics.

1

General Principles

This chapter provides an introduction to many of the fundamental concepts in mechanics. It includes a discussion of models or idealizations that are used to apply the theory, a statement of Newton's laws of motion, upon which this subject is based, and a general review of the principles for applying the SI system of units. Standard procedures for performing numerical calculations are then discussed. At the end of the chapter we will present a general guide that should be followed for solving problems.

1.1 Mechanics

Mechanics can be defined as that branch of the physical sciences concerned with the state of rest or motion of bodies that are subjected to the action of forces. In general, this subject is subdivided into three branches: *rigid-body mechanics, deformable-body mechanics,* and *fluid mechanics.* This book treats only rigid-body mechanics, since it forms a suitable basis for the design and analysis of many types of structural, mechanical, or electrical devices encountered in engineering. Also, rigid-body mechanics provides part of the necessary background for the study of the mechanics of deformable bodies and the mechanics of fluids.

Rigid-body mechanics is divided into two areas: statics and dynamics. *Statics* deals with the equilibrium of bodies, that is, those that are either at rest or move with a constant velocity; whereas *dynamics* is concerned with the accelerated motion of bodies. Although statics can be considered as a special case of dynamics, in which the acceleration is zero, statics deserves separate treatment in engineering education, since many objects are designed with the intention that they remain in equilibrium.

Historical Development. The subject of statics developed very early in history, because the principles involved could be formulated simply from measurements of geometry and force. For example, the writings of Archimedes (287–212 B.C.) deal with the principle of the lever. Studies of the pulley, inclined plane, and wrench are also recorded in ancient writings—at times when the requirements of engineering were limited primarily to building construction.

Since the principles of dynamics depend on an accurate measurement of time, this subject developed much later. Galileo Galilei (1564–1642) was one of the first major contributors to this field. His work consisted of experiments using pendulums and falling bodies. The most significant contributions in dynamics, however, were made by Isaac Newton (1642–1727), who is noted for his formulation of the three fundamental laws of motion and the law of universal gravitational attraction. Shortly after these laws were postulated, important techniques for their application were developed by Euler, D'Alembert, Lagrange, and others.

1.2 Fundamental Concepts

Before beginning our study of rigid-body mechanics, it is important to understand the meaning of certain fundamental concepts and principles.

Basic Quantities. The following four quantities are used throughout rigid-body mechanics.

Length. *Length* is needed to locate the position of a point in space and thereby describe the size of a physical system. Once a standard unit of length is defined, one can then quantitatively define distances and geometric properties of a body as multiples of the unit length.

Time. *Time* is conceived as a succession of events. Although the principles of statics are time independent, this quantity does play an important role in the study of dynamics.

Mass. *Mass* is a property of matter by which we can compare the action of one body with that of another. This property manifests itself as a gravitational attraction between two bodies and provides a quantitative measure of the resistance of matter to a change in velocity.

Force. In general, *force* is considered as a ''push'' or ''pull'' exerted by one body on another. This interaction can occur when there is direct contact between the bodies, such as a person pushing on a wall, or it can occur through a distance when the bodies are physically separated. Examples of the latter type include gravitational, electrical, and magnetic forces. In any case, a force is completely characterized by its magnitude, direction, and point of application.

Idealizations. Models or idealizations are used in mechanics in order to simplify application of the theory. A few of the more important idealizations will now be defined. Others that are noteworthy will be discussed at points where they are needed.

Particle. A *particle* has a mass but a size that can be neglected. For example, the size of the earth is insignificant compared to the size of its orbit, and therefore the earth can be modeled as a particle when studying its orbital motion. When a body is idealized as a particle, the principles of mechanics reduce to a rather simplified form, since the geometry of the body will not be involved in the analysis of the problem.

Rigid Body. A *rigid body* can be considered as a combination of a large number of particles in which all the particles remain at a fixed distance from one another both before and after applying a load. As a result, the material properties of any body that is assumed to be rigid will not have to be considered when analyzing the forces acting on the body. In most cases the actual deformations occurring in structures, machines, mechanisms, and the like are relatively small, and the rigid-body assumption is suitable for analysis.

Concentrated Force. A *concentrated force* represents the effect of a loading which is assumed to act at a point on a body. We can represent this effect by a concentrated force, provided the area over which the load is applied is very small compared to the overall size of the body.

Newton's Three Laws of Motion. The entire subject of rigid-body mechanics is formulated on the basis of Newton's three laws of motion, the validity of which is based on experimental observation. They apply to the motion of a particle as measured from a nonaccelerating reference frame and may be briefly stated as follows.

First Law. A particle originally at rest, or moving in a straight line with constant velocity, will remain in this state provided the particle is *not* subjected to an unbalanced force.

Second Law. A particle acted upon by an *unbalanced force* **F** experiences an acceleration **a** that has the same direction as the force and a magnitude that is directly proportional to the force.* If **F** is applied to a particle of mass m, this law may be expressed mathematically as

$$\mathbf{F} = m\mathbf{a} \qquad (1\text{–}1)$$

Third Law. The mutual forces of action and reaction between two particles are equal, opposite, and collinear.

*Stated another way, the unbalanced force acting on the particle is proportional to the time rate of change of the particle's linear momentum.

Newton's Law of Gravitational Attraction. Shortly after formulating his three laws of motion, Newton postulated a law governing the gravitational attraction between any two particles. Stated mathematically,

$$F = G\frac{m_1 m_2}{r^2} \tag{1-2}$$

where F = force of gravitation between the two particles
G = universal constant of gravitation; according to experimental evidence, $G = 66.73(10^{-12})\ \text{m}^3/(\text{kg} \cdot \text{s}^2)$
m_1, m_2 = mass of each of the two particles
r = distance between the two particles

Weight. According to Eq. 1–2, any two particles or bodies have a mutual attractive (gravitational) force acting between them. In the case of a particle located at or near the surface of the earth, however, the only gravitational force having any sizable magnitude is that between the earth and the particle. Consequently, this force, termed the *weight,* will be the only gravitational force considered in our study of mechanics.

From Eq. 1–2, we can develop an approximate expression for finding the weight W of a particle having a mass $m_1 = m$. If we assume the earth to be a nonrotating sphere of constant density and having a mass m_2, then if r is the distance between the earth's center and the particle, we have

$$W = G\frac{m m_2}{r^2}$$

Letting $g = Gm_2/r^2$ yields

$$W = mg \tag{1-3}$$

By comparison with Eq. 1–1, we term g the acceleration due to gravity. Since it depends on r, it can be seen that the weight of a body is *not* an absolute quantity. Instead, its magnitude is determined from where the measurement was made. For most engineering calculations, however, g is determined at sea level and at a latitude of 45°, which is considered the "standard location."

1.3 Units of Measurement

The four basic quantities—length, time, mass, and force—are not all independent from one another; in fact, they are *related* by Newton's second law of motion, $\mathbf{F} = m\mathbf{a}$. Hence, the *units* used to define force, mass, length, and time cannot *all* be selected arbitrarily. The equality $\mathbf{F} = m\mathbf{a}$ is maintained only if three of the four units, called *base units,* are *arbitrarily defined* and the fourth unit is *derived* from the equation.

SI Units. The International System of units, abbreviated SI after the French "Système International d'Unités," is a modern version of the metric system which has received worldwide recognition. As shown in Table 1–1, the SI system specifies length in meters (m), time in seconds (s), and mass in kilograms (kg). The unit of force, called a newton (N), is *derived* from $\mathbf{F} = m\mathbf{a}$. Thus, 1 newton is equal to a force required to give 1 kilogram of mass an acceleration of 1 m/s^2 (N = kg · m/s^2).

If the weight of a body located at the "standard location" is to be determined in newtons, then Eq. 1–3 must be applied. Here $g = 9.806\,65$ m/s^2; however, for calculations, the value $g = 9.81$ m/s^2 will be used. Thus,

$$W = mg \qquad (g = 9.81 \text{ m/s}^2) \tag{1–4}$$

Therefore, a body of mass 1 kg has a weight of 9.81 N, a 2-kg body weighs 19.62 N, and so on.

U.S. Customary. In the U.S. Customary system of units (FPS) length is measured in feet (ft), force in pounds (lb), and time in seconds (s), Table 1–1. The unit of mass, called a *slug,* is *derived* from $\mathbf{F} = m\mathbf{a}$. Hence, 1 slug is equal to the amount of matter accelerated at 1 ft/s^2 when acted upon by a force of 1 lb (slug = lb · s^2/ft).

In order to determine the mass of a body having a weight measured in pounds, we must apply Eq. 1–3. If the measurements are made at the "standard location," then $g = 32.2$ ft/s^2 will be used for calculations. Therefore,

$$m = \frac{W}{g} \qquad (g = 32.2 \text{ ft/s}^2) \tag{1–5}$$

And so a body weighing 32.2 lb has a mass of 1 slug, a 64.4-lb body has a mass of 2 slugs, and so on.

Table 1–1 Systems of Units

Name	*Length*	*Time*	*Mass*	*Force*
International System of Units (SI)	meter (m)	second (s)	kilogram (kg)	newton* (N) $\left(\frac{\text{kg} \cdot \text{m}}{\text{s}^2}\right)$
U.S. Customary (FPS)	foot (ft)	second (s)	slug* $\left(\frac{\text{lb} \cdot \text{s}^2}{\text{ft}}\right)$	pound (lb)

*Derived unit.

Conversion of Units. In some cases it may be necessary to convert from one system of units to another. In this regard, Table 1–2 provides a set of direct conversion factors between FPS and SI units for the basic quantities. Also, in the FPS system, recall that 1 ft = 12 in. (inches), 5280 ft = 1 mi (mile), 1000 lb = 1 kip (kilo-pound), and 2000 lb = 1 ton.

Table 1–2 Conversion Factors

Quantity	*Unit of Measurement (FPS)*	*Equals*	*Unit of Measurement (SI)*
Force	lb		4.448 2 N
Mass	slug		14.593 8 kg
Length	ft		0.304 8 m

1.4 The International System of Units

The SI system of units is used extensively in this book since it is intended to become the worldwide standard for measurement. Consequently, the rules for its use and some of its terminology relevant to mechanics will now be presented.

Prefixes. When a numerical quantity is either very large or very small, the units used to define its size may be modified by using a prefix. Some of the prefixes used in the SI system are shown in Table 1–3. Each represents a multiple or submultiple of a unit which, if applied successively, moves the decimal point of a numerical quantity to every third place.* For example, 4 000 000 N = 4 000 kN (kilo-newton) = 4 MN (mega-newton), or 0.005 m = 5 mm (milli-meter). Notice that the SI system does not include the multiple deca (10) or the submultiple centi (0.01), which form part of the metric system. Except for some volume and area measurements, the use of these prefixes is to be avoided in science and engineering.

Table 1–3 Prefixes

	Exponential Form	*Prefix*	*SI Symbol*
Multiple			
1 000 000 000	10^9	giga	G
1 000 000	10^6	mega	M
1 000	10^3	kilo	k
Submultiple			
0.001	10^{-3}	milli	m
0.000 001	10^{-6}	micro	μ
0.000 000 001	10^{-9}	nano	n

*The kilogram is the only base unit that is defined with a prefix.

Rules for Use. The following rules are given for the proper use of the various SI symbols:

1. A symbol is *never* written with a plural "s," since it may be confused with the unit for second (s).

2. Symbols are always written in lowercase letters, with the following exceptions: symbols for the two largest prefixes shown in Table 1–3, giga and mega, are capitalized as G and M, respectively; and symbols named after an individual are also capitalized, e.g., N.

3. Quantities defined by several units which are multiples of one another are separated by a *dot* to avoid confusion with prefix notation, as indicated by $N = kg \cdot m/s^2 = kg \cdot m \cdot s^{-2}$. Also, $m \cdot s$ (meter-second), whereas ms (milli-second).

4. The exponential power represented for a unit having a prefix refers to both the unit *and* its prefix. For example, $\mu N^2 = (\mu N)^2 = \mu N \cdot \mu N$. Likewise, mm^2 represents $(mm)^2 = mm \cdot mm$.

5. Physical constants or numbers having several digits on either side of the decimal point should be reported with a *space* between every three digits rather than with a comma; e.g., 73 569.213 427. In the case of four digits on either side of the decimal, the spacing is optional; e.g., 8537 or 8 537. Furthermore, always try to use decimals and avoid fractions; that is, write 15.25, *not* $15\frac{1}{4}$.

6. When performing calculations, represent the numbers in terms of their *base or derived units* by converting all prefixes to powers of 10. The final result should then be expressed using a *single prefix.* Also, after calculation, it is best to keep numerical values between 0.1 and 1000; otherwise, a suitable prefix should be chosen. For example,

$$
\begin{aligned}
(50\ kN)(60\ nm) &= [50(10^3)\ N][60(10^{-9})\ m] \\
&= 3000(10^{-6})\ N \cdot m = 3(10^{-3})\ N \cdot m = 3\ mN \cdot m
\end{aligned}
$$

7. Compound prefixes should not be used; e.g., $k\mu s$ (kilo-micro-second) should be expressed as ms (milli-second) since $1\ k\mu s = 1(10^3)(10^{-6})\ s = 1(10^{-3})\ s = 1\ ms$.

8. With the exception of the base unit the kilogram, in general avoid the use of a prefix in the denominator of composite units. For example, do not write N/mm, but rather kN/m; also, m/mg should be written as Mm/kg.

9. Although not expressed in multiples of 10, the minute, hour, etc., are retained for practical purposes as multiples of the second. Furthermore, plane angular measurement is made using radians (rad). In this book, however, degrees will often be used, where $180° = \pi$ rad.

1.5 Numerical Calculations

Numerical work in engineering practice is most often performed by using hand-held calculators and computers. It is important, however, that the answers to any problem be reported with both justifiable accuracy and appropriate significant figures. In this section we will discuss these topics together with some other important aspects involved in all engineering calculations.

Dimensional Homogeneity. The terms of any equation used to describe a physical process must be *dimensionally homogeneous;* that is, each term must be expressed in the same units. Provided this is the case, all the terms of an equation can then be combined if numerical values are substituted for the variables. Consider, for example, the equation $s = vt + \frac{1}{2}at^2$, where, in SI units, s is the position in meters, m, t is time in seconds, s, v is velocity in m/s, and a is acceleration in m/s^2. Regardless of how this equation is evaluated, it maintains its dimensional homogeneity. In the form stated each of the three terms is expressed in meters [m, $(\text{m}/\cancel{\text{s}})\cancel{\text{s}}$, $(\text{m}/\cancel{\text{s}^2})\cancel{\text{s}^2}$], or solving for a, $a = 2s/t^2 - 2v/t$, the terms are each expressed in units of m/s^2 [m/s^2, m/s^2, (m/s)/s].

Since problems in mechanics involve the solution of dimensionally homogeneous equations, the fact that all terms of an equation are represented by a consistent set of units can be used as a partial check for algebraic manipulations of an equation.

Significant Figures. The accuracy of a number is specified by the number of significant figures it contains. A *significant figure* is any digit, including a zero, provided it is not used to specify the location of the decimal point for the number. For example, the numbers 5604 and 34.52 each have four significant figures. When numbers begin or end with zeros, however, it is difficult to tell how many significant figures are in the number. Consider the number 40. Does it have one (4), or perhaps two (40) significant figures? In order to clarify this situation, the number should be reported using powers of 10. There are two ways of doing this. The format for *scientific notation* specifies one digit to the left of the decimal point, with the remaining digits to the right; for example, 40 expressed to one significant figure would be $4(10^1)$. Using *engineering notation,* which is preferred here, the exponent is displayed in multiples of three in order to facilitate conversion of SI units to those having an appropriate prefix. Thus, 40 expressed to one significant figure would be $0.04(10^3)$. Likewise, 2500 and 0.00546 expressed to three significant figures would be $2.50(10^3)$ and $5.46(10^{-3})$.

Rounding Off Numbers. For numerical calculations, the accuracy obtained from the solution of a problem generally can never be better than the accuracy of the problem data. This is what is to be expected, but often hand-held calculators or computers involve more figures in the answer than the number of significant figures used for the data. For this reason, a calculated result should always be ''rounded off'' to an appropriate number of significant figures.

To ensure accuracy, the following rules for rounding off a number to n significant figures apply:

1. If the $n + 1$ digit is *less than 5,* the $n + 1$ digit and others following it are dropped. For example, 2.326 and 0.451 rounded off to $n = 2$ significant figures would be 2.3 and 0.45.

2. If the $n + 1$ digit is equal to 5 with zeros following it, then round off the nth digit to an *even number.* For example, 1245 and 0.8655 rounded off to $n = 3$ significant figures become 1240 and 0.866.

3. If the $n + 1$ digit is *greater than 5* or equal to 5 with any nonzero digits following it, then increase the nth digit by 1 and drop the $n + 1$ digit and others following it. For example, 0.723 87 and 565.500 3 rounded off to $n = 3$ significant figures become 0.724 and 566.

Calculations. As a general rule, to ensure accuracy of a final result when performing calculations with numbers of unequal accuracy, always retain one extra significant figure in the more accurate numbers than in the least accurate number *before* beginning the computations. Then round off the final result so that it has the same number of significant figures as the least accurate number. If possible, try to work out the computations so that numbers which are approximately equal are not subtracted, since accuracy is often lost from this calculation.

In engineering we generally round off final answers to *three* significant figures since the data for geometry, loads, and other measurements are often reported with this accuracy.* Consequently, in this book the intermediate calculations for the examples are often worked out to four significant figures and the answers are generally reported to *three* significant figures.

The following examples illustrate application of the principles just discussed as related to the proper use and conversion of units.

*Of course, some numbers, such as π, e, or numbers used in derived formulas, are exact and are therefore accurate to an infinite number of significant figures.

Example 1–1

Convert 2 km/h to m/s. How many ft/s is this?

SOLUTION

Since 1 km = 1000 m and 1 h = 3600 s, the factors of conversion are arranged in the following order, so that a cancellation of the units can be applied:

$$2\text{ km/h} = \frac{2\ \cancel{\text{km}}}{\cancel{\text{h}}}\left(\frac{1000\text{ m}}{\cancel{\text{km}}}\right)\left(\frac{1\ \cancel{\text{h}}}{3600\text{ s}}\right)$$

$$= \frac{2000\text{ m}}{3600\text{ s}} = 0.556\text{ m/s} \qquad \textit{Ans.}$$

From Table 1–2, 1 ft = 0.304 8 m. Thus

$$0.556\text{ m/s} = \frac{0.556\ \cancel{\text{m}}}{\text{s}}\ \frac{1\text{ ft}}{0.304\,8\ \cancel{\text{m}}}$$

$$= 1.82\text{ ft/s} \qquad \textit{Ans.}$$

Example 1–2

Convert the quantity 300 lb · s to appropriate SI units.

SOLUTION

Using Table 1–2, 1 lb = 4.448 2 N.

$$300\text{ lb}\cdot\text{s} = 300\ \cancel{\text{lb}}\cdot\text{s}\left(\frac{4.448\,2\text{ N}}{\cancel{\text{lb}}}\right)$$

$$= 1334.5\text{ N}\cdot\text{s} = 1.33\text{ kN}\cdot\text{s} \qquad \textit{Ans.}$$

Note that the calculations for both of these examples were performed to several significant figures and then rounded off to three significant figures for the answer.

Example 1–3

Evaluate each of the following and express with SI units having an appropriate prefix: (a) (50 mN)(6 GN), (b) $(400\text{ mm})(0.6\text{ MN})^2$, (c) $45\text{ MN}^3/900\text{ Gg}$.

SOLUTION

First convert each number to base units, perform the indicated operations, then choose an appropriate prefix (see Rule 6 on p. 9).

Part (a)

$$\begin{aligned}(50\text{ mN})(6\text{ GN}) &= [50(10^{-3})\text{ N}][6(10^{9})\text{ N}]\\ &= 300(10^{6})\text{ N}^2\\ &= 300(10^{6})\,\cancel{\text{N}^2}\left(\frac{1\text{ kN}}{10^3\,\cancel{\text{N}}}\right)\left(\frac{1\text{ kN}}{10^3\,\cancel{\text{N}}}\right)\\ &= 300\text{ kN}^2 \qquad \textit{Ans.}\end{aligned}$$

Note carefully the convention $\text{kN}^2 = (\text{kN})^2 = 10^6\text{ N}^2$ (Rule 4 on p. 9).

Part (b)

$$\begin{aligned}(400\text{ mm})(0.6\text{ MN})^2 &= [400(10^{-3})\text{ m}][0.6(10^{6})\text{ N}]^2\\ &= [400(10^{-3})\text{ m}][0.36(10^{12})\text{ N}^2]\\ &= 144(10^{9})\text{ m}\cdot\text{N}^2\\ &= 144\text{ Gm}\cdot\text{N}^2 \qquad \textit{Ans.}\end{aligned}$$

We can also write

$$\begin{aligned}144(10^{9})\text{ m}\cdot\text{N}^2 &= 144(10^{9})\text{ m}\cdot\cancel{\text{N}^2}\left(\frac{1\text{ MN}}{10^6\,\cancel{\text{N}}}\right)\left(\frac{1\text{ MN}}{10^6\,\cancel{\text{N}}}\right)\\ &= 0.144\text{ m}\cdot\text{MN}^2\end{aligned}$$

Part (c)

$$\begin{aligned}45\text{ MN}^3/900\text{ Gg} &= \frac{45(10^6\text{ N})^3}{900(10^6)\text{ kg}}\\ &= 0.05(10^{12})\text{ N}^3/\text{kg}\\ &= 0.05(10^{12})\,\cancel{\text{N}^3}\left(\frac{1\text{ kN}}{10^3\,\cancel{\text{N}}}\right)^3\frac{1}{\text{kg}}\\ &= 0.05(10^{3})\text{ kN}^3/\text{kg}\\ &= 50\text{ kN}^3/\text{kg} \qquad \textit{Ans.}\end{aligned}$$

Here we have used Rules 4 and 8 on p. 9.

1.6 General Procedure for Analysis

The most effective way of learning the principles of engineering mechanics is to *solve problems.* To be successful at this, it is important always to present the work in a *logical* and *orderly manner,* as suggested by the following sequence of steps:

1. Read the problem carefully and try to correlate the actual physical situation with the theory studied.
2. Draw any necessary diagrams and tabulate the problem data.
3. Apply the relevant principles, generally in mathematical form.
4. Solve the necessary equations algebraically as far as practical, then, making sure they are dimensionally homogeneous, use a consistent set of units and complete the solution numerically. Report the answer with no more significant figures than the accuracy of the given data.
5. Study the answer with technical judgment and common sense to determine whether or not it seems reasonable.
6. Once the solution has been completed, review the problem. Try to think of other ways of obtaining the same solution.

In applying this general procedure, do the work as neatly as possible. Being neat generally stimulates clear and orderly thinking, and vice versa.

PROBLEMS

1–1. What is the weight in newtons of an object that has a mass of (a) 8 kg, (b) 0.04 g, and (c) 760 Mg?

1–2. Wood has a density of 4.70 slug/ft^3. What is its density expressed in SI units?

1–3. Using Table 1–3, determine your own mass in kilograms, your weight in newtons, and your height in meters.

***1–4.** Represent each of the following combinations of units in the correct SI form using an appropriate prefix: (a) m/ms, (b) μkm, (c) ks/mg, and (d) km · μN.

1–5. Represent each of the following as a number between 0.1 and 1000 using an appropriate prefix: (a) 45 320 kN, (b) 568(10^5) mm, and (c) 0.00563 mg.

1–6. Evaluate each of the following and express with an appropriate prefix: (a) $(430\text{ kg})^2$, (b) $(0.002\text{ mg})^2$, and (c) $(230\text{ m})^3$.

1–7. Represent each of the following combinations of units in the correct SI form: (a) GN · μm, (b) kg/μm, (c) N/ks^2, and (d) kN/μs.

***1–8.** Represent each of the following combinations of units in the correct SI form: (a) kN/μs, (b) Mg/mN, and (c) MN/(kg · ms).

1–9. The *pascal* (Pa) is actually a very small unit of pressure. To show this, convert 1 Pa = 1 N/m^2 to lb/ft^2. Atmospheric pressure at sea level is 14.7 lb/in^2. How many pascals is this?

1–10. Convert: (a) 20 lb · ft to N · m, (b) 450 lb/ft^3 to kN/m^3, and (c) 15 ft/h to mm/s.

1–11. Determine the number of cubic millimeters contained in one cubic inch.

***1–12.** Convert each of the following and express the answer using an appropriate prefix: (a) 175 lb/ft^3 to kN/m^3, (b) 6 ft/h to mm/s, and (c) 835 lb · ft to kN · m.

1–13. A steel disk has a diameter of 500 mm and a thickness of 70 mm. If the density of steel is 7850 kg/m^3, determine the weight of the disk in pounds.

1–14. If an object has a mass of 40 slugs, determine its mass in kilograms.

1–15. Using the base units of the SI system, show that Eq. 1–2 is a dimensionally homogeneous equation which gives F in newtons. Compute the gravitational force acting between two identical spheres that are touching each other. The mass of each sphere is 150 kg and the radius is 275 mm.

***1–16.** Two particles have a mass of 8 kg and 12 kg, respectively. If they are 800 mm apart, determine the force of gravity acting between them. Compare this result with the weight of each particle.

1–17. Evaluate each of the following to three significant figures and express each answer in SI units using an appropriate prefix: (a) $(212\text{ mN})^2$, (b) $(52\,800\text{ ms})^2$, and (c) $[548(10^6)]^{1/2}$ ms.

1–18. If a man weighs 155 lb on earth, specify (a) his mass in slugs, (b) his mass in kilograms, and (c) his weight in newtons. If the man is on the moon, where the acceleration due to gravity is $g_m = 5.30\text{ ft/s}^2$, determine (d) his weight in pounds, and (e) his mass in kilograms.

1–19. Evaluate each of the following and express each answer in SI units using an appropriate prefix: (a) (684 μm)/43 ms, (b) (28 ms)(0.0458 Mm)/(348 mg), and (c) (2.68 mm)(426 Mg).

***1–20.** Determine the mass in kilograms of an object that has a weight of (a) 20 mN, (b) 150 kN, and (c) 60 MN. Express each answer using an appropriate prefix.

This communications tower is stabilized by cables that exert resultant forces at the three points of connection. In this chapter we will show how to determine the magnitudes and directions of these resultant forces.

2

Force Vectors

In this chapter we will introduce the concept of a concentrated force and give the procedures for adding forces, resolving them into components, and projecting them along an axis. Since force is a vector quantity, we must use the rules of vector algebra whenever forces are considered. We will begin our study by defining scalar and vector quantities and then develop some of the basic rules of vector algebra.

2.1 Scalars and Vectors

Most of the physical quantities in mechanics can be expressed mathematically by means of scalars and vectors.

Scalar. A quantity characterized by a positive or negative number is called a *scalar.* Mass, volume, and length are scalar quantities often used in statics. In this book, scalars are indicated by letters in italic type, such as the scalar A. The mathematical operations involving scalars follow the same rules as those of elementary algebra.

Vector. A *vector* is a quantity that has both a magnitude and a direction. In statics the vector quantities frequently encountered are position, force, and moment. For handwritten work, a vector is generally represented by a letter with an arrow written over it, such as $\vec{\mathrm{A}}$. The magnitude is designated $|\vec{\mathrm{A}}|$ or simply A. In this book vectors will be symbolized in boldface type; for example, **A** is used to designate the vector "A". Its magnitude, which is always a positive quantity, is symbolized in italic type, written as $|A|$, or simply A when it is understood that A is a positive scalar.

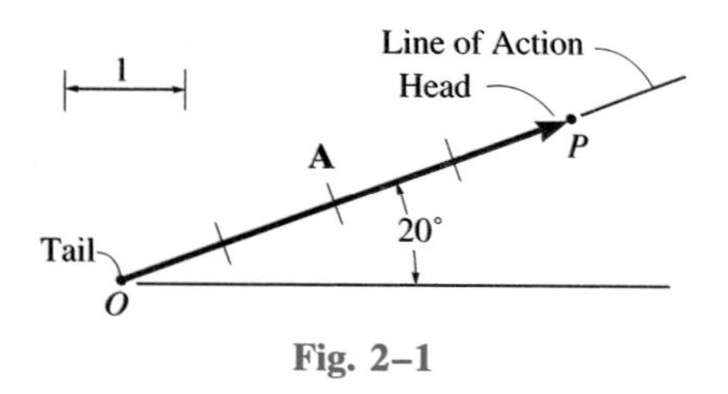

Fig. 2–1

A vector is represented graphically by an arrow, which is used to define its magnitude, direction, and sense. The *magnitude* of the vector is indicated by the length of the arrow, the *direction* is defined by the angle between a reference axis and the arrow's line of action, and the *sense* is indicated by the arrowhead. For example, the vector **A** shown in Fig. 2–1 has a magnitude of 4 units, a direction which is 20° measured counterclockwise from the horizontal axis, and a sense which is upward and to the right. The point O is called the *tail* of the vector, the point P is the *tip* or *head.*

2.2 Vector Operations

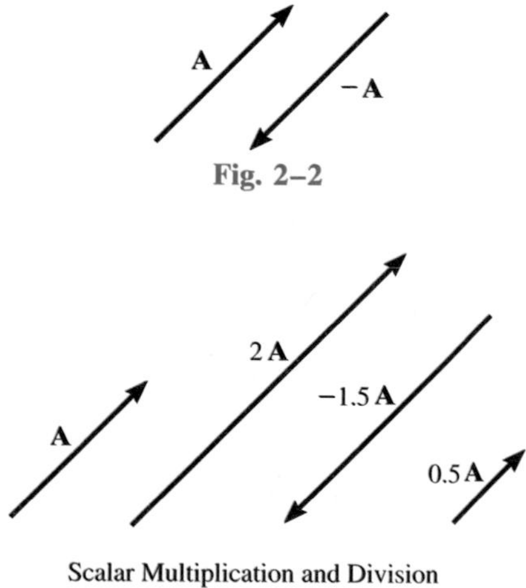

Fig. 2–2

Scalar Multiplication and Division

Fig. 2–3

Multiplication and Division of a Vector by a Scalar. The product of vector **A** and scalar a, yielding $a\mathbf{A}$, is defined as a vector having a magnitude $|aA|$. The *sense* of $a\mathbf{A}$ is the *same* as **A** provided a is *positive;* it is *opposite* to **A** if a is *negative.* Consequently, the negative of a vector is formed by multiplying the vector by the scalar (-1), Fig. 2–2. Division of a vector by a scalar can be defined using the laws of multiplication, since $\mathbf{A}/a = (1/a)\mathbf{A}$, $a \neq 0$. Graphic examples of these operations are shown in Fig. 2–3.

Vector Addition. Two vectors **A** and **B** of the same type, Fig. 2–4*a*, may be added to form a "resultant" vector $\mathbf{R} = \mathbf{A} + \mathbf{B}$ by using the *parallelogram law.* To do this, **A** and **B** are joined at their tails, Fig. 2–4*b*. Parallel lines drawn from the head of each vector intersect at a common point, thereby forming the adjacent sides of a parallelogram. As shown, the resultant **R** is the diagonal of the parallelogram, which extends from the tails of **A** and **B** to the intersection of the lines.

We can also add **B** to **A** using a *triangle construction,* which is a special case of the parallelogram law, whereby vector **B** is added to vector **A** in a "head-to-tail" fashion, i.e., by connecting the head of **A** to the tail of **B,** Fig. 2–4*c*. The resultant **R** extends from the tail of **A** to the head of **B.** In a similar manner, **R** can also be obtained by adding **A** to **B,** Fig. 2–4*d*. By

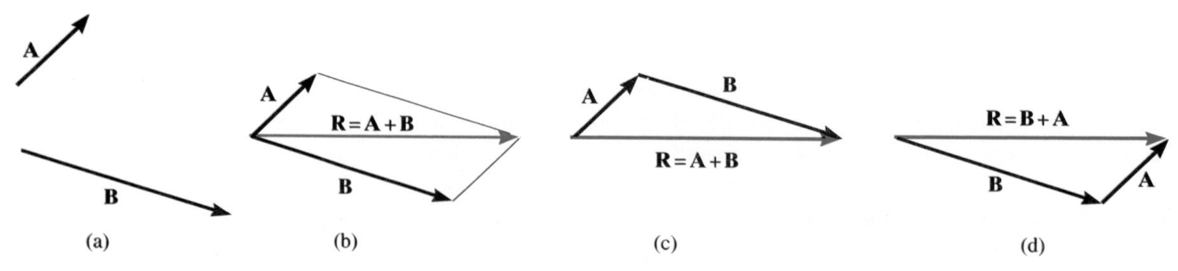

Vector Addition

Fig. 2–4

comparison, it is seen that vector addition is commutative; in other words, the vectors can be added in either order, i.e., $\mathbf{R} = \mathbf{A} + \mathbf{B} = \mathbf{B} + \mathbf{A}$.

As a special case, if the two vectors **A** and **B** are *collinear,* i.e., both have the same line of action, the parallelogram law reduces to an *algebraic* or *scalar addition* $R = A + B$, as shown in Fig. 2–5.

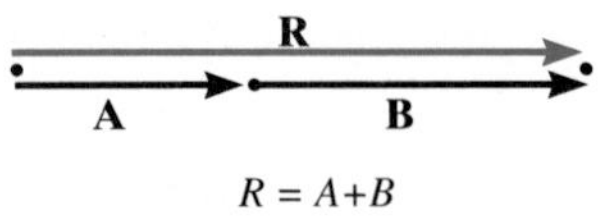

R

A B

$R = A+B$

Addition of Collinear Vectors

Fig. 2–5

Vector Subtraction. The resultant *difference* between two vectors **A** and **B** of the same type may be expressed as

$$\mathbf{R}' = \mathbf{A} - \mathbf{B} = \mathbf{A} + (-\mathbf{B})$$

This vector sum is shown graphically in Fig. 2–6. Subtraction is therefore defined as a special case of addition, so the rules of vector addition also apply to vector subtraction.

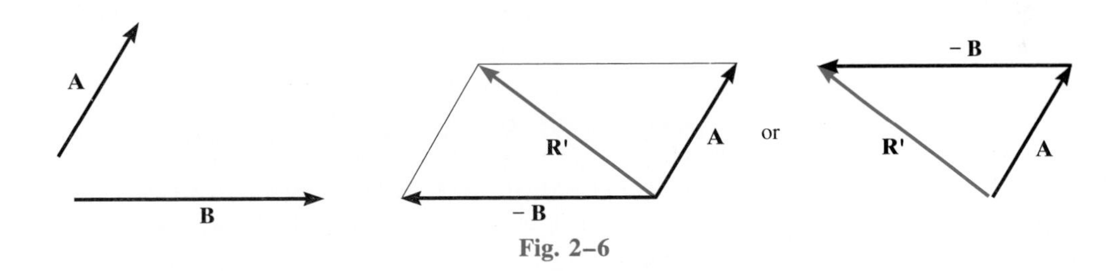

Fig. 2–6

Resolution of a Vector. A vector may be resolved into two "components" having known lines of action by using the parallelogram law. For example, if **R** in Fig. 2–7*a* is to be resolved into components acting along the lines *a* and *b*, one starts at the *head* of **R** and extends a line *parallel* to *a* until it intersects *b*. Likewise, a line parallel to *b* is drawn from the *head* of **R** to the point of intersection with *a*, Fig. 2–7*a*. The two components **A** and **B** are then drawn such that they extend from the tail of **R** to the points of intersection, as shown in Fig. 2–7*b*.

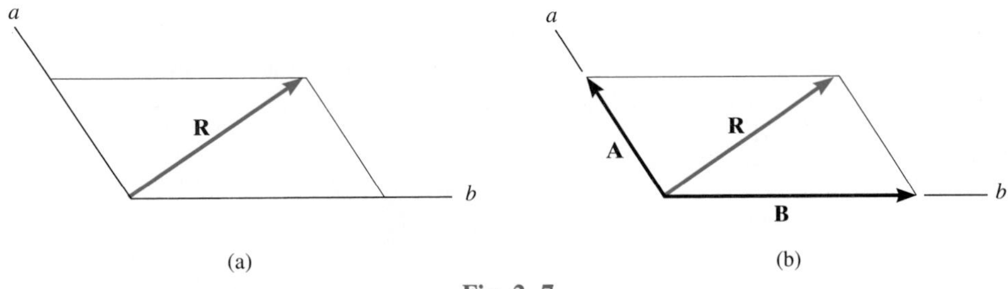

Fig. 2–7

2.3 Vector Addition of Forces

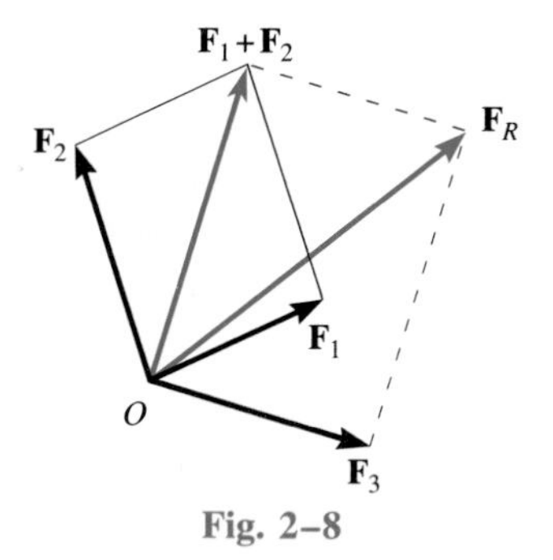

Fig. 2–8

Experimental evidence has shown that a force is a vector quantity since it has a specified magnitude, direction, and sense and it adds according to the parallelogram law. Two common problems in statics involve either finding the resultant force, knowing its components, or resolving a known force into two components. As described in Sec. 2–2, both of these problems require application of the parallelogram law.

If more than two forces are to be added, successive applications of the parallelogram law can be carried out in order to obtain the resultant force. For example, if three forces $\mathbf{F}_1$, $\mathbf{F}_2$, $\mathbf{F}_3$ act at point O, Fig. 2–8, the resultant of any two of the forces is found—say, $\mathbf{F}_1 + \mathbf{F}_2$—and then this resultant is added to the third force, yielding the resultant of all three forces; i.e., $\mathbf{F}_R = (\mathbf{F}_1 + \mathbf{F}_2) + \mathbf{F}_3$. Using the parallelogram law to add more than two forces, as shown here, often requires extensive geometric and trigonometric calculation to determine the numerical values for the magnitude and direction of the resultant. Instead, problems of this type are easily solved by using the "rectangular-component method," which is explained in Sec. 2.4.

PROCEDURE FOR ANALYSIS

Problems that involve the addition of two forces and contain at most *two unknowns* can be solved by using the following procedure:

Parallelogram Law. Make a sketch showing the vector addition using the parallelogram law. If possible, determine the interior angles of the parallelogram from the geometry of the problem. Recall that the sum total of these angles is 360°. Unknown angles, along with known and unknown force magnitudes, should be clearly labeled on this sketch. Redraw a half portion of the constructed parallelogram to illustrate the triangular head-to-tail addition of the components.

Trigonometry. By using trigonometry, the two unknowns can be determined from the data listed on the triangle. If the triangle does *not* contain a 90° angle, the law of sines and/or the law of cosines may be used for the solution. These formulas are given in Fig. 2–9 for the triangle shown.

The following examples illustrate this method numerically.

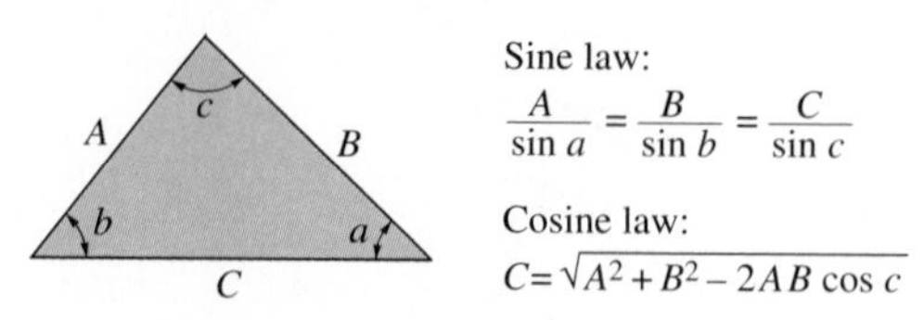

Fig. 2–9

Example 2–1

The screw eye in Fig. 2–10*a* is subjected to two forces, $\mathbf{F}_1$ and $\mathbf{F}_2$. Determine the magnitude and direction of the resultant force.

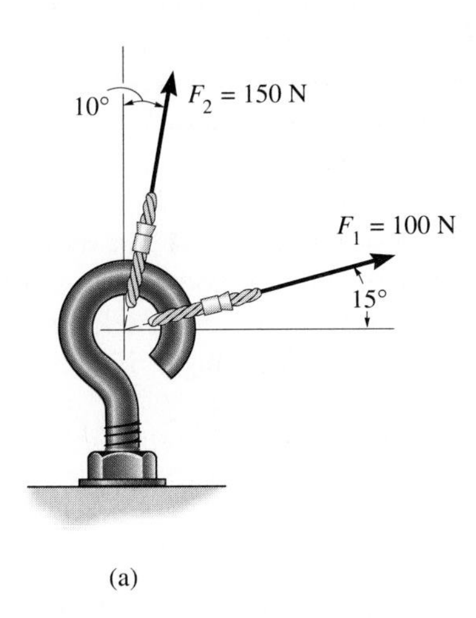

(a)

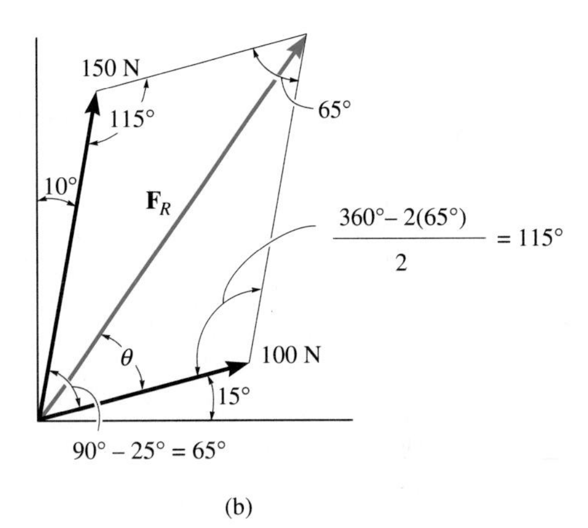

(b)

SOLUTION

The resultant force is formed from the parallelogram law.

Parallelogram Law. The addition is shown in Fig. 2–10*b*. The two unknowns are the magnitude of $\mathbf{F}_R$ and the angle θ (theta). From Fig. 2–10*b*, the vector triangle, Fig. 2–10*c*, is constructed.

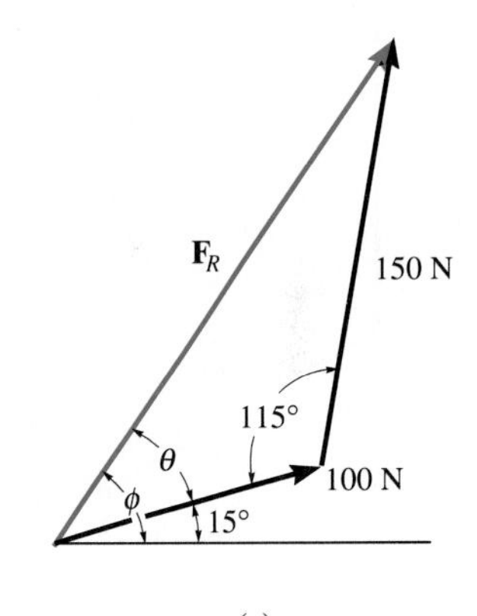

(c)

Fig. 2–10

Trigonometry. F_R is determined by using the law of cosines:

$$F_R = \sqrt{(100\text{ N})^2 + (150\text{ N})^2 - 2(100\text{ N})(150\text{ N})\cos 115^\circ}$$
$$= \sqrt{10\,000 + 22\,500 - 30\,000(-0.4226)} = 212.6\text{ N}$$
$$= 213\text{ N} \qquad \textit{Ans.}$$

The angle θ is determined by applying the law of sines, using the computed value of F_R.

$$\frac{150\text{ N}}{\sin\theta} = \frac{212.6\text{ N}}{\sin 115^\circ}$$
$$\sin\theta = \frac{150\text{ N}}{212.6\text{ N}}(0.9063)$$
$$\theta = 39.8^\circ$$

Thus, the direction ϕ (phi) of $\mathbf{F}_R$, measured from the horizontal, is

$$\phi = 39.8^\circ + 15.0^\circ = 54.8^\circ \quad \measuredangle^{\phi} \qquad \textit{Ans.}$$

Example 2–2

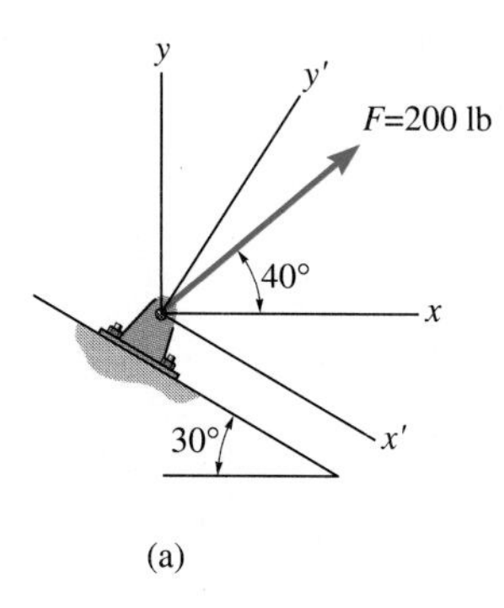

Resolve the 200-lb force shown acting on the pin, Fig. 2–11*a*, into components in the (a) x and y directions, and (b) x' and y directions.

SOLUTION

In each case the parallelogram law is used to resolve **F** into its two components, and then the vector triangle is constructed to determine the numerical results by trigonometry.

Part (a). The vector addition $\mathbf{F} = \mathbf{F}_x + \mathbf{F}_y$ is shown in Fig. 2–11*b*. In particular, note that the length of the components is scaled along the x and y axes by first constructing lines parallel to the axes in accordance with the parallelogram law. From the vector triangle, Fig. 2–11*c*,

$$F_x = 200 \text{ lb} \cos 40° = 153 \text{ lb} \qquad \textbf{\textit{Ans.}}$$
$$F_y = 200 \text{ lb} \sin 40° = 129 \text{ lb} \qquad \textbf{\textit{Ans.}}$$

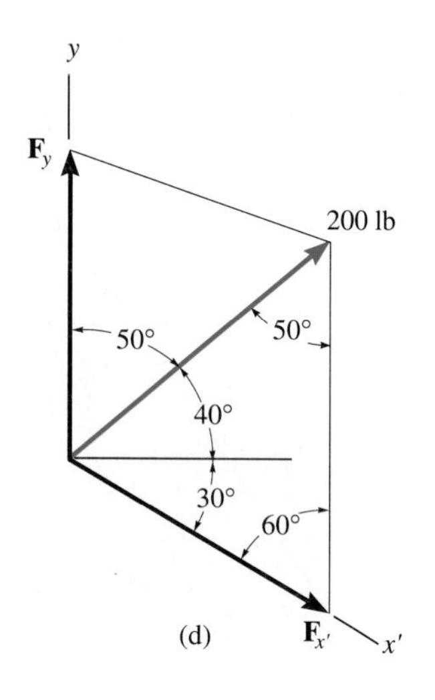

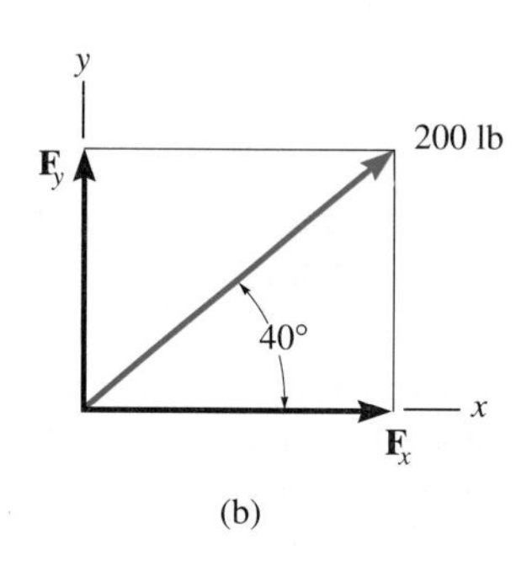

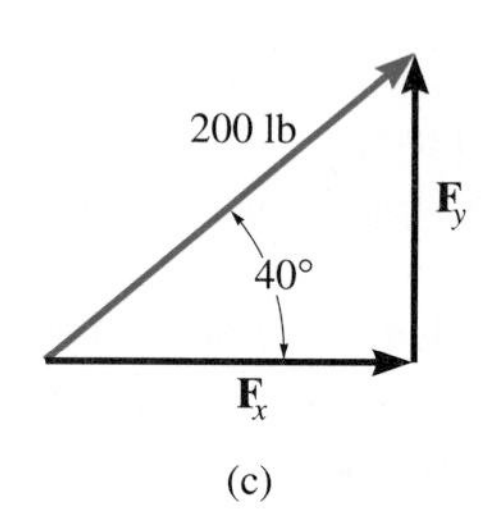

Fig. 2–11

Part (b). The vector addition $\mathbf{F} = \mathbf{F}_{x'} + \mathbf{F}_y$ is shown in Fig. 2–11*d*. Note carefully how the parallelogram is constructed. Applying the law of sines and using the data listed on the vector triangle, Fig. 2–11*e*, yields

$$\frac{F_{x'}}{\sin 50°} = \frac{200 \text{ lb}}{\sin 60°}$$

$$F_{x'} = 200 \text{ lb}\left(\frac{\sin 50°}{\sin 60°}\right) = 177 \text{ lb} \qquad \textbf{\textit{Ans.}}$$

$$\frac{F_y}{\sin 70°} = \frac{200 \text{ lb}}{\sin 60°}$$

$$F_y = 200 \text{ lb}\left(\frac{\sin 70°}{\sin 60°}\right) = 217 \text{ lb} \qquad \textbf{\textit{Ans.}}$$

(e)

Example 2–3

The force **F** acting on the frame shown in Fig. 2–12a has a magnitude of 500 N and is to be resolved into two components acting along struts AB and AC. Determine the angle θ, measured *below* the horizontal, so that the component $\mathbf{F}_{AC}$ is directed from A toward C and has a magnitude of 400 N.

Fig. 2–12

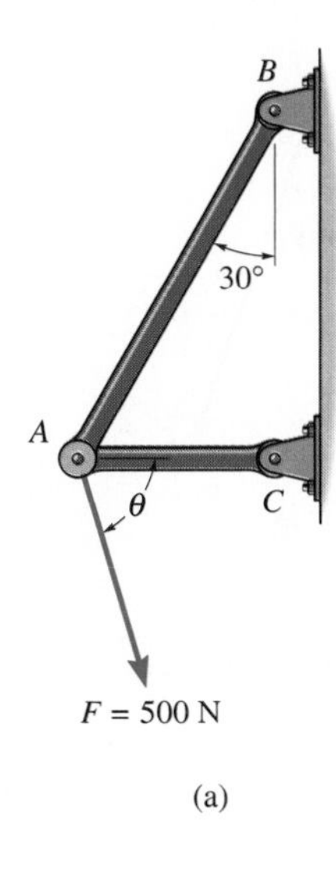

(a)

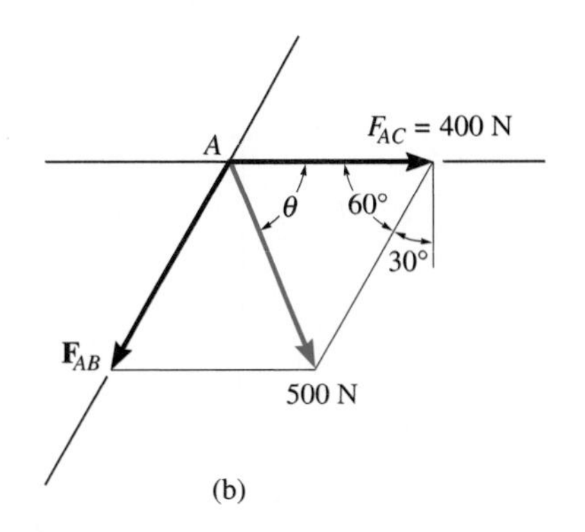

(b)

SOLUTION

By using the parallelogram law, the vector addition of the two components yielding the resultant is shown in Fig. 2–12b. Note carefully how the resultant force is resolved into the two components $\mathbf{F}_{AB}$ and $\mathbf{F}_{AC}$, which have specified lines of action. The corresponding vector triangle is shown in Fig. 2–12c. The angle ϕ can be determined by using the law of sines:

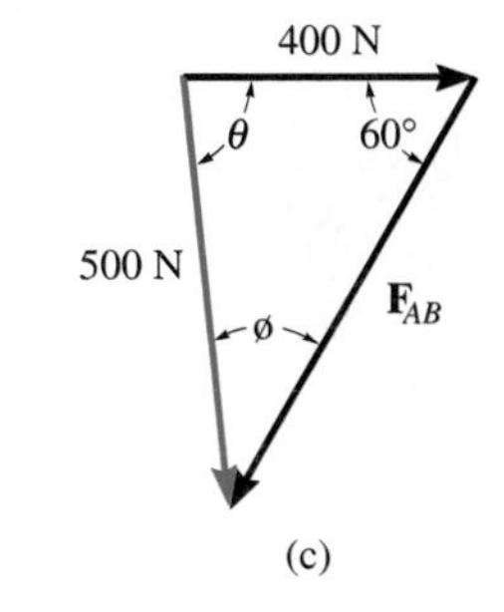

(c)

$$\frac{400 \text{ N}}{\sin \phi} = \frac{500 \text{ N}}{\sin 60^\circ}$$

$$\sin \phi = \left(\frac{400 \text{ N}}{500 \text{ N}}\right) \sin 60^\circ = 0.6928$$

$$\phi = 43.9^\circ$$

Hence,

$$\theta = 180^\circ - 60^\circ - 43.9^\circ = 76.1^\circ \quad \text{Ans.}$$

Using this value for θ, apply the law of cosines and show that $\mathbf{F}_{AB}$ has a magnitude of 561 N.

Notice that **F** can also be directed at an angle θ *above* the horizontal, as shown in Fig. 2–12d, and still produce the required component $\mathbf{F}_{AC}$. Show that in this case $\theta = 16.1^\circ$ and $F_{AB} = 161$ N.

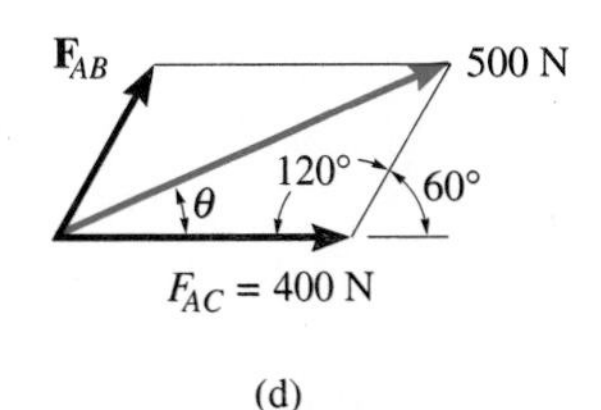

(d)

Example 2–4

The ring shown in Fig. 2–13*a* is subjected to two forces, $\mathbf{F}_1$ and $\mathbf{F}_2$. If it is required that the resultant force have a magnitude of 1 kN and be directed vertically downward, determine (a) the magnitudes of $\mathbf{F}_1$ and $\mathbf{F}_2$ provided $\theta = 30°$, and (b) the magnitudes of $\mathbf{F}_1$ and $\mathbf{F}_2$ if F_2 is to be a minimum.

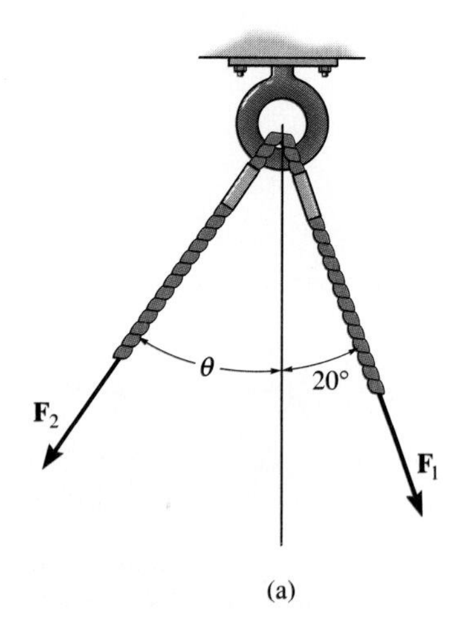

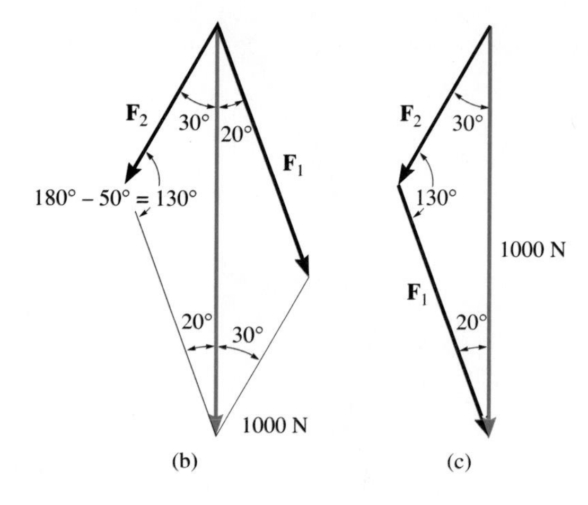

SOLUTION

Part (a). A sketch of the vector addition according to the parallelogram law is shown in Fig. 2–13*b*. From the vector triangle constructed in Fig. 2–13*c*, the unknown magnitudes F_1 and F_2 can be determined by using the law of sines.

$$\frac{F_1}{\sin 30°} = \frac{1000\text{ N}}{\sin 130°}$$

$$F_1 = 653\text{ N} \qquad \textit{Ans.}$$

$$\frac{F_2}{\sin 20°} = \frac{1000\text{ N}}{\sin 130°}$$

$$F_2 = 446\text{ N} \qquad \textit{Ans.}$$

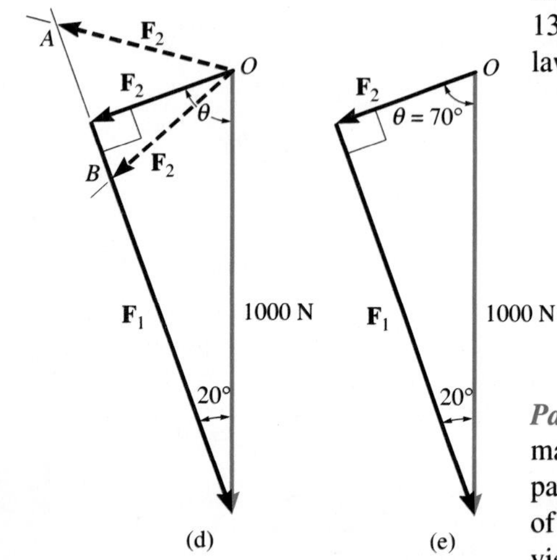

Fig. 2–13

Part (b). If θ is not specified, then by the vector triangle, Fig. 2–13*d*, $\mathbf{F}_2$ may be added to $\mathbf{F}_1$ in various ways to yield the resultant 1000-N force. In particular, the *minimum* length or magnitude of $\mathbf{F}_2$ will occur when its line of action is *perpendicular to* $\mathbf{F}_1$. Any other direction, such as *OA* or *OB*, yields a larger value for F_2. Hence, when $\theta = 90° - 20° = 70°$, F_2 is minimum. From the triangle shown in Fig. 2–13*e*, it is seen that

$$F_1 = 1000 \sin 70°\text{ N} = 940\text{ N} \qquad \textit{Ans.}$$

$$F_2 = 1000 \sin 20°\text{ N} = 342\text{ N} \qquad \textit{Ans.}$$

PROBLEMS

2–1. Determine the magnitude of the resultant force $\mathbf{F}_R = \mathbf{F}_1 + \mathbf{F}_2$ and its direction, measured counterclockwise from the positive x axis.

2–2. Determine the magnitude of the resultant force $\mathbf{F}_R = \mathbf{F}_1 + \mathbf{F}_3$ and its direction, measured counterclockwise from the positive x axis.

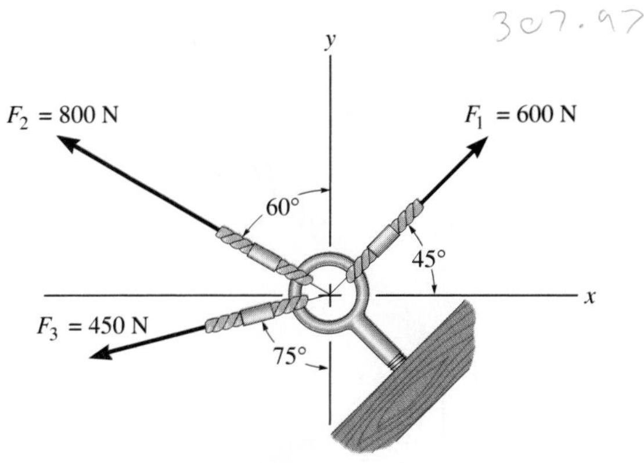

Probs. 2–1/2–2

2–3. Determine the magnitude of the resultant force $\mathbf{F}_R = \mathbf{F}_1 + \mathbf{F}_2$ and its direction, measured counterclockwise from the positive x axis.

***2–4.** Determine the magnitude of the resultant force $\mathbf{F}_R = \mathbf{F}_1 - \mathbf{F}_2$ and its direction, measured counterclockwise from the positive x axis.

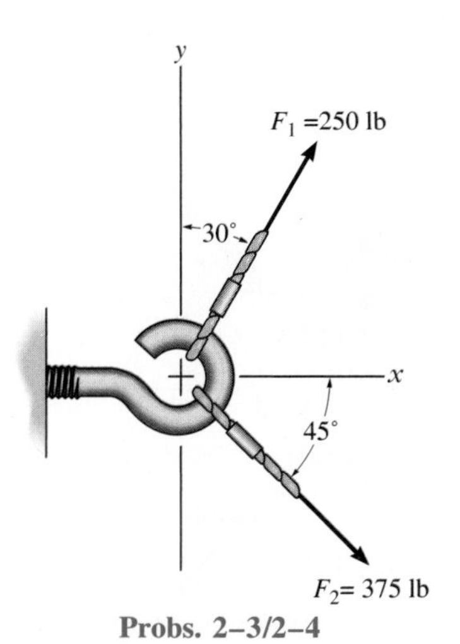

Probs. 2–3/2–4

2–5. Determine the magnitude of the resultant force and its direction, measured counterclockwise from the positive x axis.

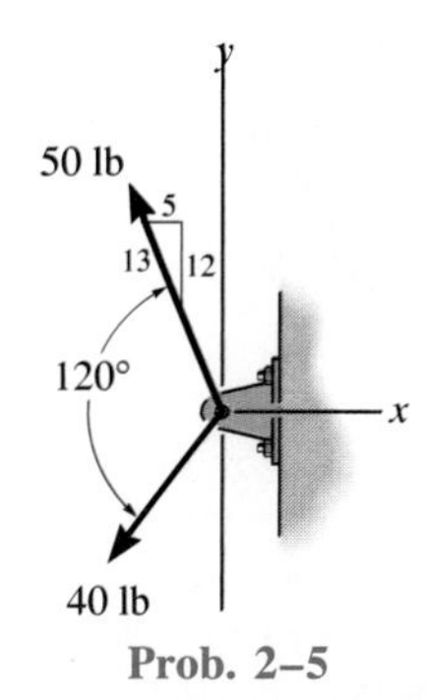

Prob. 2–5

2–6. Determine the magnitude of the resultant force $\mathbf{F}_R = \mathbf{F}_1 + \mathbf{F}_2$ and its direction, measured clockwise from the positive u axis.

2–7. Resolve the force $\mathbf{F}_1$ into components acting along the u and v axes and determine the magnitudes of the components.

***2–8.** Resolve the force $\mathbf{F}_2$ into components acting along the u and v axes and determine the magnitudes of the components.

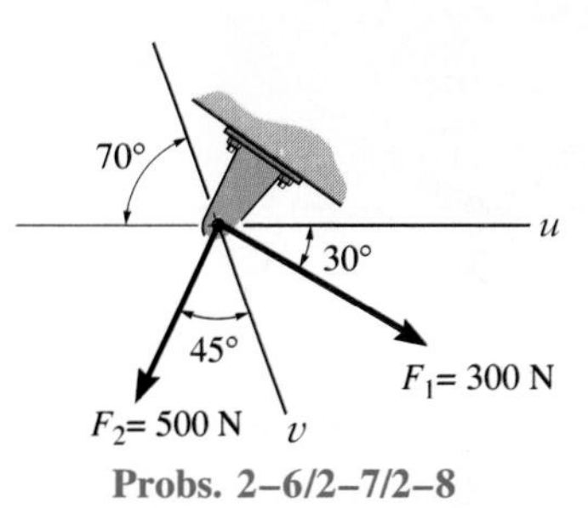

Probs. 2–6/2–7/2–8

2–9. The V-grooved wheel is used to run along the track. If the track exerts a vertical force of 200 lb on the wheel, determine the components of this force acting along the a and b axes, which are perpendicular to the sides of the groove.

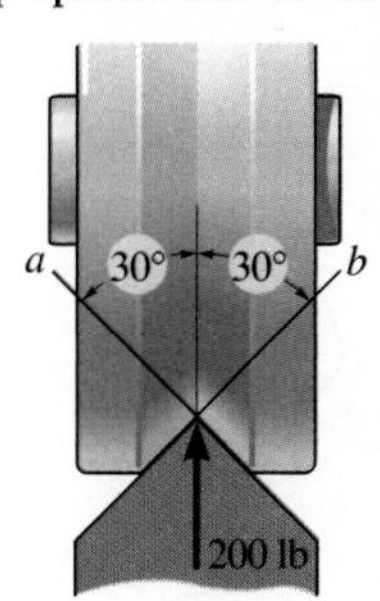

Prob. 2–9

2–10. Resolve the 60-lb force into components acting along the u and v axes and determine the magnitudes of the components.

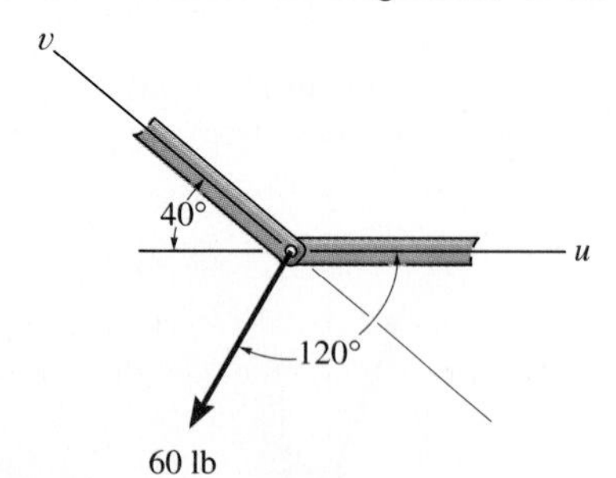

Prob. 2–10

2–11. The wind is deflected by the sail of a boat such that it exerts a resultant force of $F = 110$ lb perpendicular to the sail. Resolve this force into two components, one parallel and one perpendicular to the keel aa of the boat. *Note:* The ability to sail into the wind is known as tacking, made possible by the force parallel to the boat's keel. The perpendicular component tends to tip the boat or push it over.

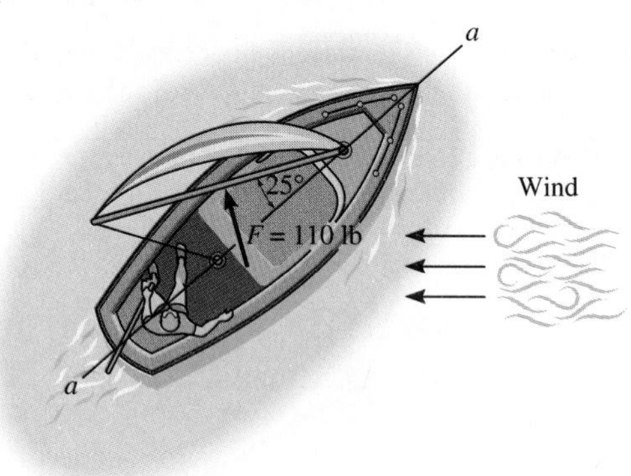

Prob. 2–11

***2–12.** The hook supports the two cable forces $F_1 = 500$ N and $F_2 = 300$ N. If the resultant of these forces acts vertically downward and has a magnitude of $F_R = 750$ N, determine the angles θ and ϕ of the cables.

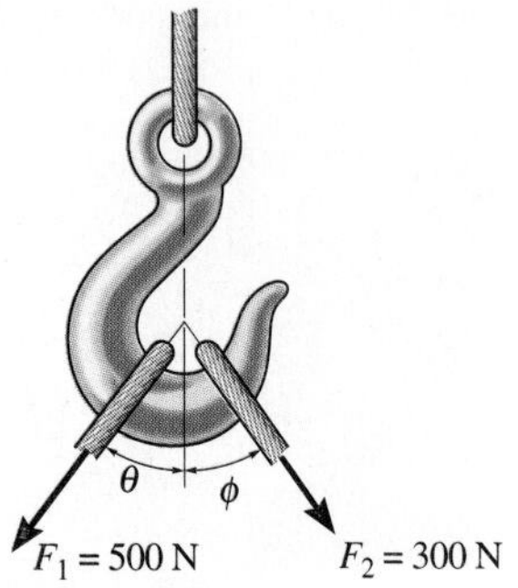

Prob. 2–12

2–13. The vertical force of $F = 60$ lb acts downward at A on the two-member frame. Determine the magnitudes of the two components of **F** directed along the axes of members AB and AC. Set $\theta = 45°$.

2–14. The vertical force of $F = 60$ lb acts downward at A on the two-member frame. Determine the angle θ $(0° \leq \theta \leq 90°)$ of member AB so that the component of **F** acting along the axis of AB is 80 lb. What is the magnitude of the force component acting along the axis of member AC?

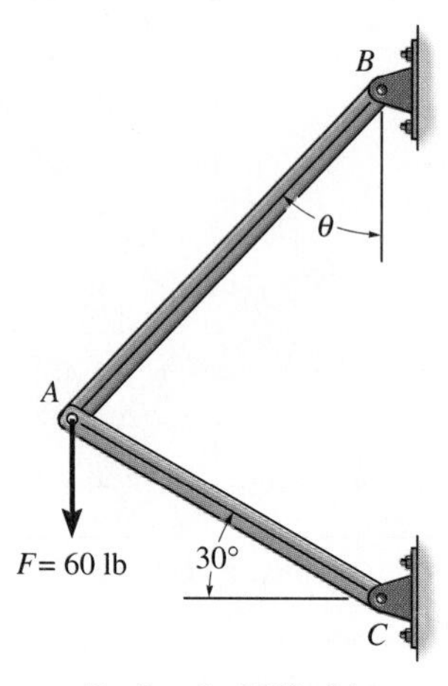

Probs. 2–13/2–14

2–15. The plate is subjected to the two forces at A and B as shown. If $\theta = 60°$, determine the magnitude of the resultant of these two forces and its direction measured clockwise from the positive x axis.

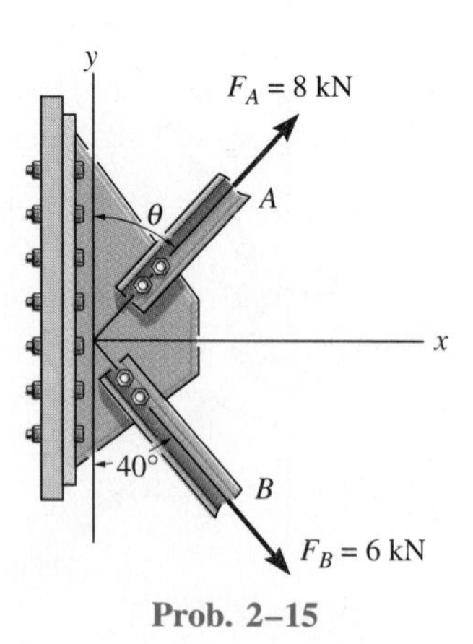

Prob. 2–15

*2–16. Resolve the 50-lb force into components acting along (a) the x and y axes, and (b) the x and y' axes.

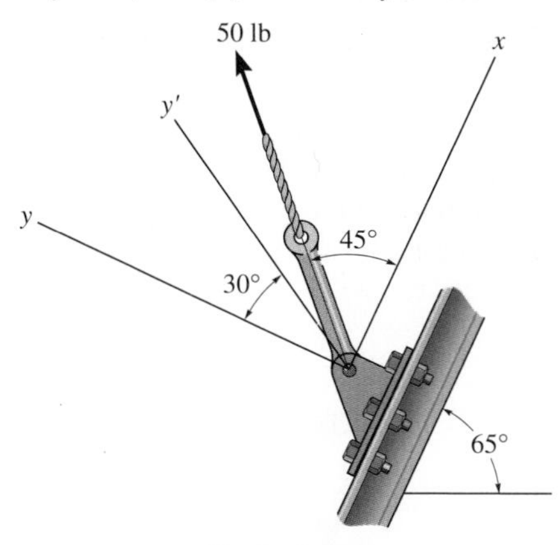

Prob. 2–16

2–17. The force acting on the gear tooth is $F = 20$ lb. Resolve this force into two components acting along the lines aa and bb.

2–18. The component of force **F** acting along line aa is required to be 30 lb. Determine the magnitude of **F** and its component along line bb.

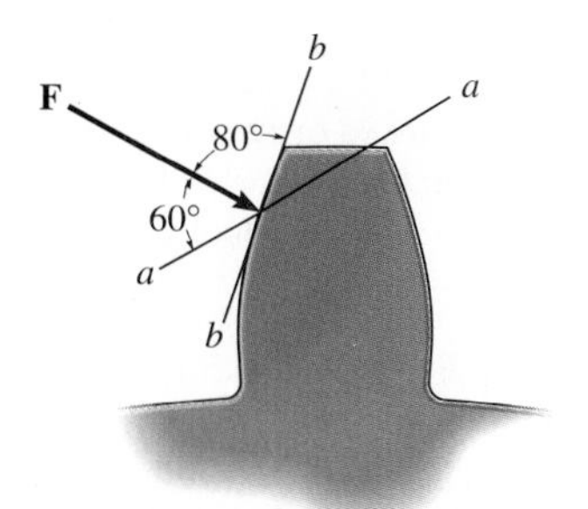

Probs. 2–17/2–18

2–19. Two forces having a magnitude of 10 lb and 6 lb act on the ring. If the largest magnitude of the resultant force the ring can support is 14 lb, determine the angle θ between the forces.

*2–20. Determine the angle θ ($0° \leq \theta \leq 90°$) between the two forces so that the magnitude of the resultant force acting on the ring is a minimum. What is the magnitude of the resultant force?

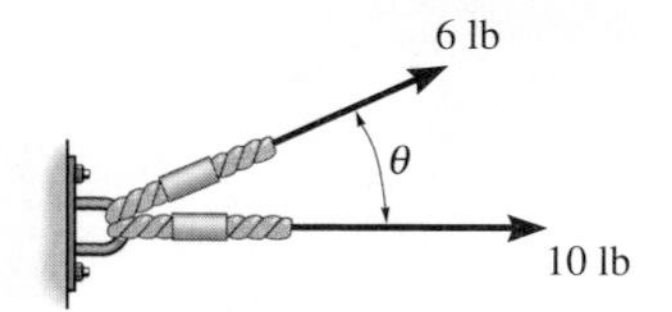

Probs. 2–19/2–20

2–21. The post is to be pulled out of the ground using two ropes A and B. Rope A is subjected to a force of 600 lb and is directed at 60° from the horizontal. Determine the force T in rope B if the post starts to lift out when $\theta = 20°$. For this to occur, the resultant force on the post is to be directed vertically upward. Also calculate the magnitude of the resultant force.

2–22. The post is to be pulled out of the ground using two ropes A and B. Rope A is subjected to a force of 600 lb and is directed at 60° from the horizontal. If the resultant force acting on the post is to be 1200 lb, vertically upward, determine the force T in rope B and the corresponding angle θ.

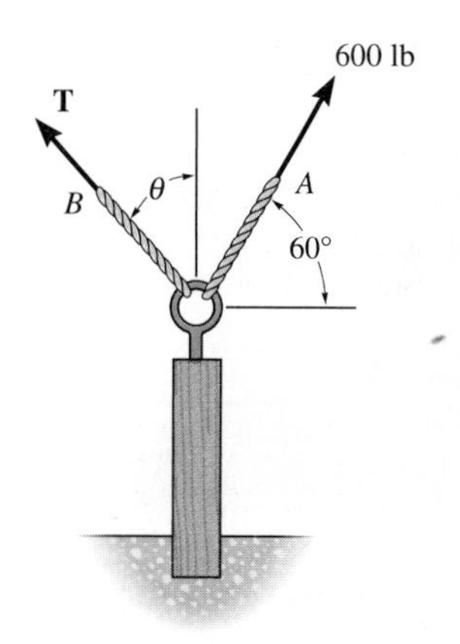

Probs. 2–21/2–22

2–23. The chisel exerts a force of 20 lb on the wood dowel rod which is turning in a lathe. Resolve this force into components acting (a) along the n and t axes and (b) along the x and y axes.

*2–24. The chisel exerts a force of 20 lb on the wood dowel rod which is turning in a lathe. Resolve this force into components acting (a) along the n and y axes and (b) along the x and t axes.

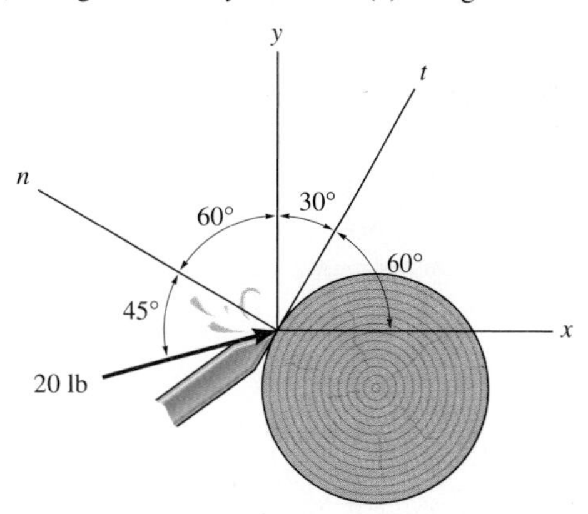

Probs. 2–23/2–24

2–25. If $\theta = 20°$ and $\phi = 35°$, determine the magnitudes of $\mathbf{F}_1$ and $\mathbf{F}_2$ so that the resultant force has a magnitude of 20 lb and is directed along the positive x axis.

2–26. If $F_1 = F_2 = 30$ lb, determine the angles θ and ϕ so that the resultant force is directed along the positive x axis and has a magnitude of $F_R = 20$ lb.

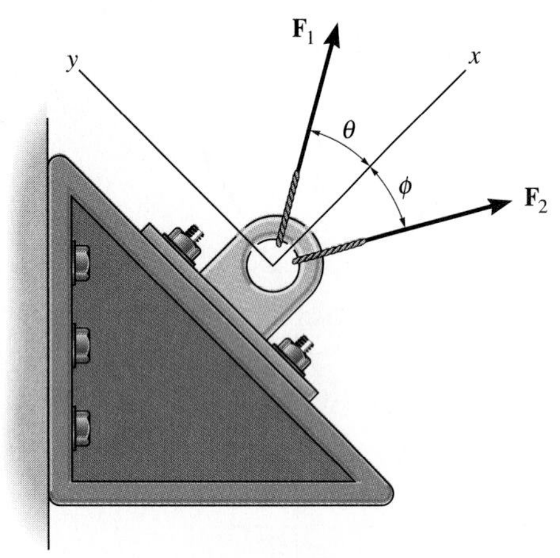

Probs. 2–25/2–26

2–27. Determine the magnitude and direction of the resultant $\mathbf{F}_R = \mathbf{F}_1 + \mathbf{F}_2 + \mathbf{F}_3$ of the three forces by first finding the resultant $\mathbf{F}' = \mathbf{F}_1 + \mathbf{F}_2$ and then forming $\mathbf{F}_R = \mathbf{F}' + \mathbf{F}_3$.

***2–28.** Determine the magnitude and direction of the resultant $\mathbf{F}_R = \mathbf{F}_1 + \mathbf{F}_2 + \mathbf{F}_3$ of the three forces by first finding the resultant $\mathbf{F}' = \mathbf{F}_2 + \mathbf{F}_3$ and then forming $\mathbf{F}_R = \mathbf{F}' + \mathbf{F}_1$.

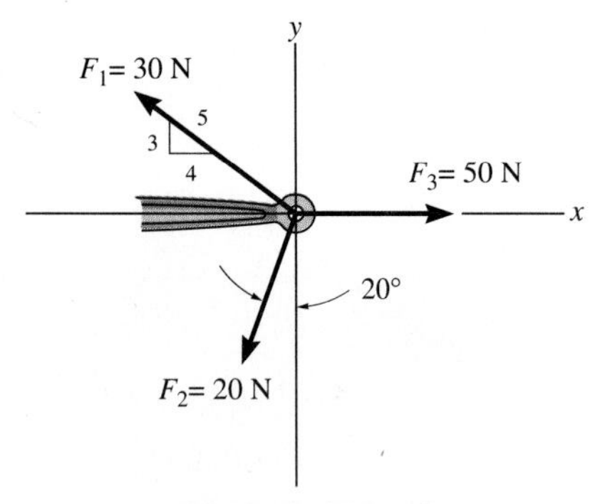

Probs. 2–27/2–28

2–29. Determine the design angle θ ($0° \leq \theta \leq 90°$) for strut AB so that the 400-lb horizontal force has a component of 500 lb directed from A towards C. What is the component of force acting along member AB? Take $\phi = 40°$.

2–30. Determine the design angle ϕ ($0° \leq \phi \leq 90°$) between struts AB and AC so that the 400-lb horizontal force has a component of 600 lb which acts up to the left, in the same direction as from B towards A. Take $\theta = 30°$.

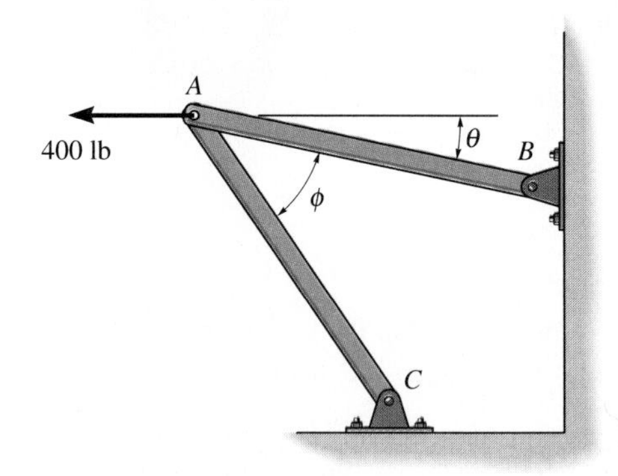

Probs. 2–29/2–30

2–31. The log is being towed by two tractors A and B. Determine the magnitudes of the two towing forces $\mathbf{F}_A$ and $\mathbf{F}_B$ if it is required that the resultant force have a magnitude $F_R = 10$ kN and be directed along the x axis. Set $\theta = 15°$.

***2–32.** If the resultant $\mathbf{F}_R$ of the two forces acting on the log is to be directed along the positive x axis and have a magnitude of 10 kN, determine the angle θ of the cable attached to B such that the force $\mathbf{F}_B$ in this cable is a *minimum*. What is the magnitude of force in each cable for this situation?

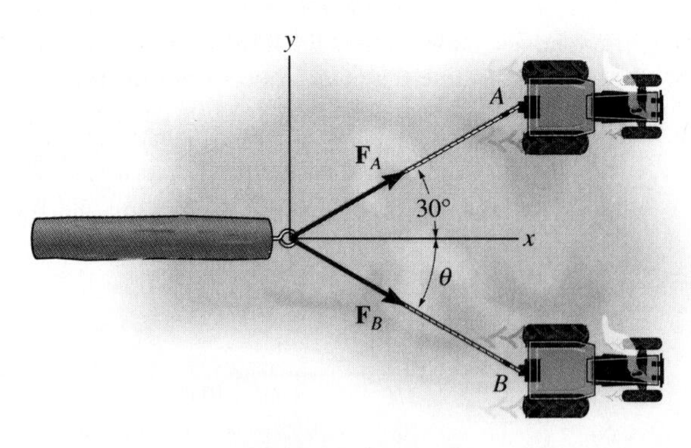

Probs. 2–31/2–32

2.4 Addition of a System of Coplanar Forces

When the resultant of more than two forces has to be obtained, it is easier to find the components of each force along specified axes, add these components algebraically, and then form the resultant, rather than form the resultant of the forces by successive application of the parallelogram law as discussed in Sec. 2.3. In this section we will resolve each force into its rectangular components $\mathbf{F}_x$ and $\mathbf{F}_y$, which lie along the x and y axes, respectively, Fig. 2–14*a*. Although the axes are shown here to be horizontal and vertical, they may in general be directed at any inclination, as long as they remain perpendicular to one another, Fig. 2–14*b*. In either case, by the parallelogram law, we require:

$$\mathbf{F} = \mathbf{F}_x + \mathbf{F}_y$$

and

$$\mathbf{F}' = \mathbf{F}'_x + \mathbf{F}'_y$$

(a)

As shown in Fig. 2–14, the sense of direction of each component is represented *graphically* by the *arrowhead.* For *analytical* work, however, we must establish a notation for representing the directional sense of the rectangular components for each coplanar vector. This can be done in one of two ways.

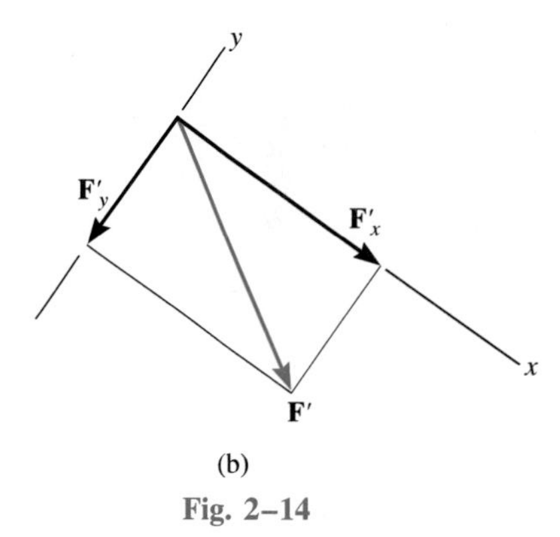

(b)

Fig. 2–14

Scalar Notation. Since the x and y axes have designated positive and negative directions, the magnitude and directional sense of the rectangular components of a force can be expressed in terms of *algebraic scalars.* For example, the components of $\mathbf{F}$ in Fig. 2–14*a* can be represented by positive scalars F_x and F_y since their sense of direction is along the *positive* x and y axes, respectively. In a similar manner, the components of $\mathbf{F}'$ in Fig. 2–14*b* are F'_x and $-F'_y$. Here the y component is negative, since $\mathbf{F}'_y$ is directed along the negative y axis. It is important to keep in mind that this scalar notation is to be used only for computational purposes, not for graphical representations in figures. Throughout the text, the *head of a vector arrow* in any figure indicates the sense of the vector *graphically;* algebraic signs are not used for this purpose. Thus, the vectors in Figs. 2–14*a* and 2–14*b* are designated by using boldface (vector) notation.* Whenever italic symbols are written near vector arrows in figures, they indicate the *magnitude* of the vector, which is *always* a *positive* quantity.

*Negative signs are used only in figures with boldface notation when showing equal but opposite pairs of vectors as in Fig. 2–2.

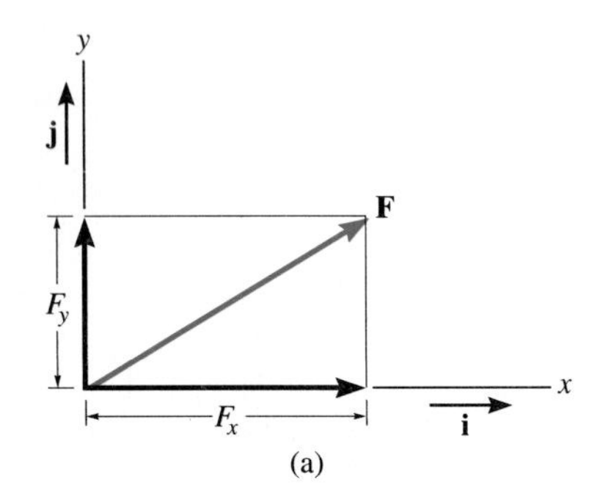

(a)

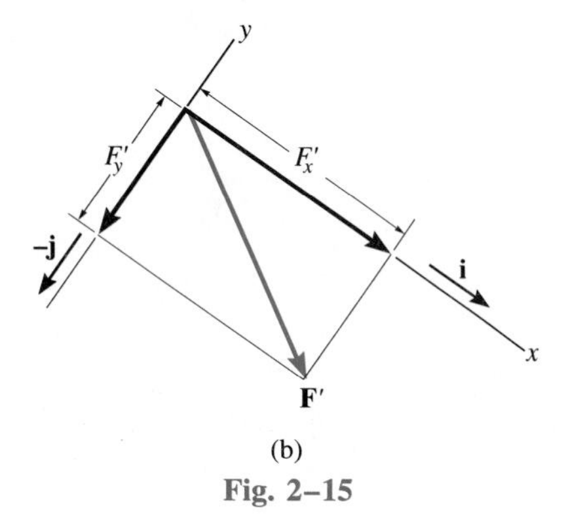

(b)
Fig. 2–15

Cartesian Vector Notation. It is also possible to represent the components of a force in terms of Cartesian unit vectors. By doing this the methods of vector algebra are easier to apply, and we will see that this becomes particularly advantageous for solving problems in three dimensions. In two dimensions the *Cartesian unit vectors* **i** and **j** are used to designate the *directions* of the x and y axes, respectively, Fig. 2–15*a*.* These vectors have a dimensionless magnitude of unity, and their sense (or arrowhead) will be described analytically by a plus or minus sign, depending on whether they are pointing along the positive or negative x or y axis.

As shown in Fig. 2–15*a*, the *magnitude* of each component of **F** is *always a positive quantity,* which is represented by the (positive) scalars F_x and F_y. Therefore, having established notation to represent the magnitude and the direction of each component, we can express **F** in Fig. 2–15*a* as a *Cartesian vector,* i.e.,

$$\mathbf{F} = F_x\mathbf{i} + F_y\mathbf{j}$$

And in the same way, **F**′ in Fig. 2–15*b* can be expressed as

$$\mathbf{F}' = F'_x\mathbf{i} + F'_y(-\mathbf{j})$$

or simply

$$\mathbf{F}' = F'_x\mathbf{i} - F'_y\mathbf{j}$$

Coplanar Force Resultants. Either of the two methods just described for representing the rectangular components of a force can be used to determine the resultant of several *coplanar forces.* To do this, each force is first resolved into its x and y components and then the respective components are added using *scalar algebra* since they are collinear. The resultant force is then formed by adding the resultants of the x and y components using the parallelogram law. For example, consider the three forces in Fig. 2–16*a*, which have x and y components as shown in Fig. 2–16*b*. To solve this problem using *Cartesian vector notation,* each force is first represented as a Cartesian vector, i.e.,

$$\mathbf{F}_1 = F_{1x}\mathbf{i} + F_{1y}\mathbf{j}$$
$$\mathbf{F}_2 = -F_{2x}\mathbf{i} + F_{2y}\mathbf{j}$$
$$\mathbf{F}_3 = F_{3x}\mathbf{i} - F_{3y}\mathbf{j}$$

*For handwritten work, unit vectors are usually indicated using a circumflex, e.g., $\hat{\mathbf{i}}$ and $\hat{\mathbf{j}}$.

The vector resultant is therefore

$$\begin{aligned}\mathbf{F}_R &= \mathbf{F}_1 + \mathbf{F}_2 + \mathbf{F}_3 \\ &= F_{1x}\mathbf{i} + F_{1y}\mathbf{j} - F_{2x}\mathbf{i} + F_{2y}\mathbf{j} + F_{3x}\mathbf{i} - F_{3y}\mathbf{j} \\ &= (F_{1x} - F_{2x} + F_{3x})\mathbf{i} + (F_{1y} + F_{2y} - F_{3y})\mathbf{j} \\ &= (F_{Rx})\mathbf{i} + (F_{Ry})\mathbf{j}\end{aligned}$$

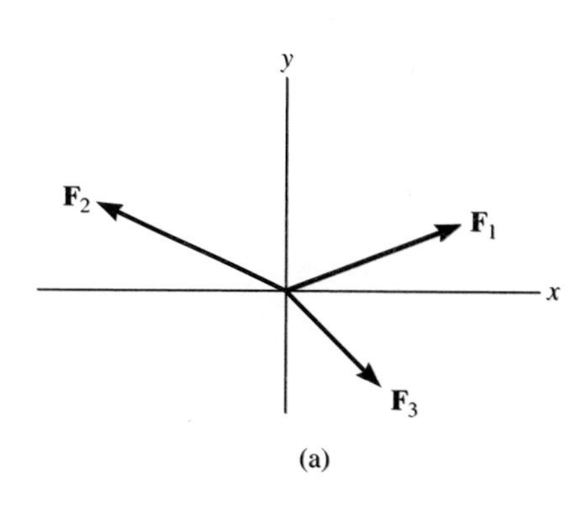

(a)

If *scalar notation* is used, then, from Fig. 2–16*b*, since *x* is positive to the right and *y* is positive upward, we have

$$(\overset{+}{\rightarrow}) \qquad F_{Rx} = F_{1x} - F_{2x} + F_{3x}$$
$$(+\uparrow) \qquad F_{Ry} = F_{1y} + F_{2y} - F_{3y}$$

These results are the *same* as the **i** and **j** components of $\mathbf{F}_R$ determined above.

In the general case, the *x* and *y* components of the resultant of any number of coplanar forces can be represented symbolically by the algebraic sum of the *x* and *y* components of all the forces, i.e.,

$$\begin{aligned}F_{Rx} &= \Sigma F_x \\ F_{Ry} &= \Sigma F_y\end{aligned} \qquad (2\text{–}1)$$

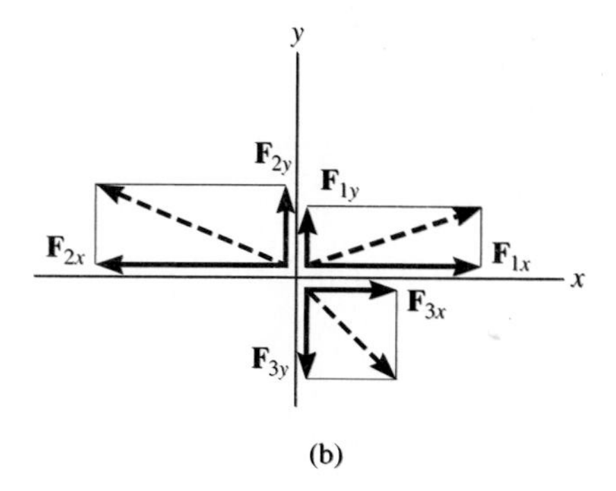

(b)

When applying these equations it is important to use the *sign convention* established for the components; and that is, components having a directional sense along the positive coordinate axes are considered positive scalars, whereas those having a directional sense along the negative coordinate axes are considered negative scalars. If this convention is followed, then the signs of the resultant components will specify the sense of these components. For example, a positive result indicates that the component has a directional sense which is in the positive coordinate direction.

Once the resultant components are determined, they may be sketched along the *x* and *y* axes in their proper directions, and the resultant force can be determined from vector addition, as shown in Fig. 2–16*c*. From this sketch, the magnitude of $\mathbf{F}_R$ is then found from the Pythagorean theorem; that is,

$$F_R = \sqrt{F_{Rx}^2 + F_{Ry}^2}$$

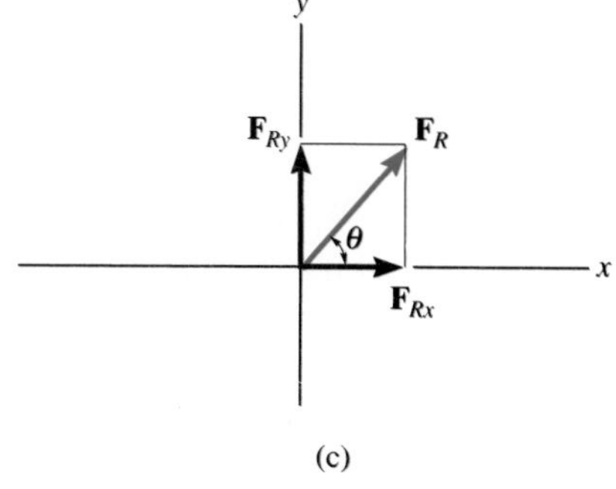

(c)

Fig. 2–16

Also, the direction angle θ, which specifies the orientation of the force, is determined from trigonometry.

$$\theta = \tan^{-1}\left|\frac{F_{Ry}}{F_{Rx}}\right|$$

The above concepts are illustrated numerically in the following examples.

Example 2–5

Determine the x and y components of $\mathbf{F}_1$ and $\mathbf{F}_2$ shown in Fig. 2–17*a*. Express each force as a Cartesian vector.

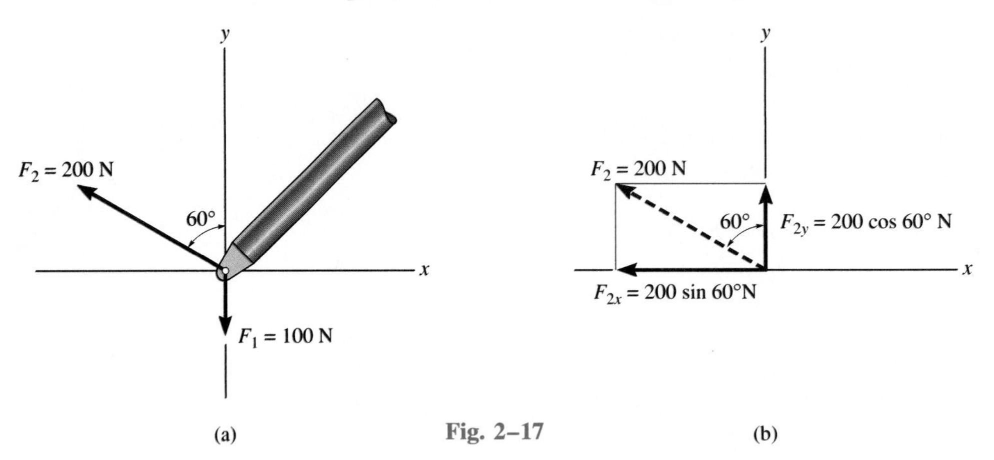

Fig. 2–17

SOLUTION

Scalar Notation. Since $\mathbf{F}_1$ acts along the negative y axis, and the magnitude of $\mathbf{F}_1$ is 100 N, the components written in scalar form are

$$F_{1x} = 0, \qquad F_{1y} = -100 \text{ N} \qquad \textit{Ans.}$$

or, alternatively,

$$F_{1x} = 0, \qquad F_{1y} = 100 \text{ N} \downarrow \qquad \textit{Ans.}$$

By the parallelogram law, $\mathbf{F}_2$ is resolved into x and y components, Fig. 2–17*b*. The magnitude of each component is determined by trigonometry. Since $\mathbf{F}_{2x}$ acts in the $-x$ direction, and $\mathbf{F}_{2y}$ acts in the $+y$ direction, we have

$$F_{2x} = -200 \sin 60° \text{ N} = -173 \text{ N} = 173 \text{ N} \leftarrow \qquad \textit{Ans.}$$
$$F_{2y} = 200 \cos 60° \text{ N} = 100 \text{ N} = 100 \text{ N} \uparrow \qquad \textit{Ans.}$$

Cartesian Vector Notation. Having computed the magnitudes of the components of $\mathbf{F}_2$, Fig. 2–17*b*, we can express each force as a Cartesian vector.

$$\mathbf{F}_1 = 0\mathbf{i} + 100 \text{ N}(-\mathbf{j})$$
$$= \{-100\mathbf{j}\} \text{ N} \qquad \textit{Ans.}$$

and

$$\mathbf{F}_2 = 200 \sin 60° \text{ N}(-\mathbf{i}) + 200 \cos 60° \text{ N}(\mathbf{j})$$
$$= \{-173\mathbf{i} + 100\mathbf{j}\} \text{ N} \qquad \textit{Ans.}$$

Example 2–6

Determine the x and y components of the force **F** shown in Fig. 2–18*a*.

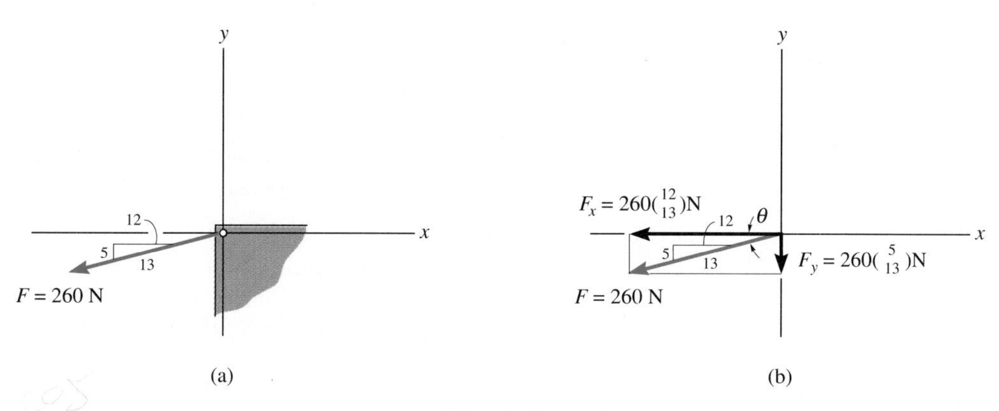

Fig. 2–18

SOLUTION

The force is resolved into its x and y components as shown in Fig. 2–18*b*. Here the *slope* of the line of action for the force is indicated. From this "slope triangle" we could obtain the direction angle θ, e.g., $\theta = \tan^{-1}(\frac{5}{12})$, and then proceed to determine the magnitudes of the components in the same manner as for $\mathbf{F}_2$ in Example 2–5. An easier method, however, consists of using proportional parts of similar triangles, i.e.,

$$\frac{F_x}{260 \text{ N}} = \frac{12}{13} \qquad F_x = 260 \text{ N}\left(\frac{12}{13}\right) = 240 \text{ N}$$

Similarly,

$$F_y = 260 \text{ N}\left(\frac{5}{13}\right) = 100 \text{ N}$$

Notice that the magnitude of the *horizontal component, F_x*, was obtained by multiplying the force magnitude by the ratio of the *horizontal leg* of the slope triangle divided by the hypotenuse; whereas the magnitude of the *vertical component, F_y*, was obtained by multiplying the force magnitude by the ratio of the *vertical leg* divided by the hypotenuse. Hence, using scalar notation,

$$F_x = -240 \text{ N} = 240 \text{ N} \leftarrow \qquad \textit{Ans.}$$
$$F_y = -100 \text{ N} = 100 \text{ N} \downarrow \qquad \textit{Ans.}$$

If **F** is expressed as a Cartesian vector, we have

$$\mathbf{F} = \{-240\mathbf{i} - 100\mathbf{j}\} \text{ N} \qquad \textit{Ans.}$$

Example 2–7

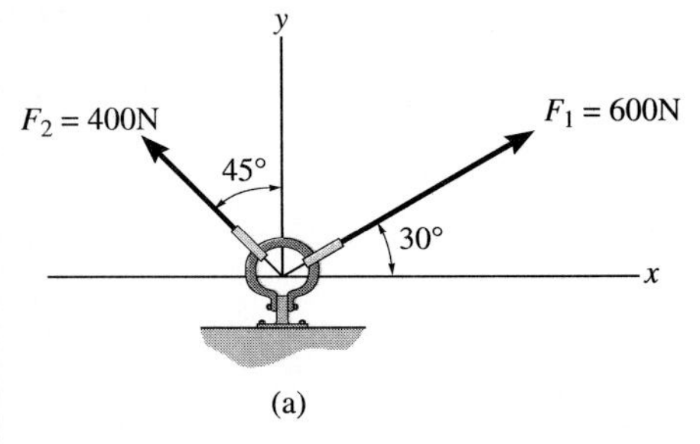

(a)

(b)

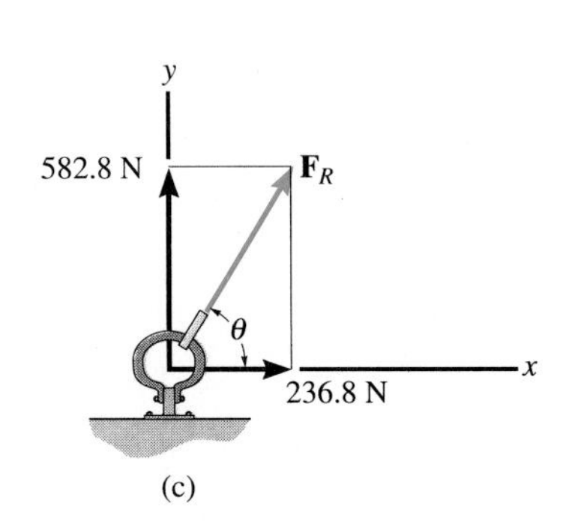

(c)

Fig. 2–19

The link in Fig. 2–19*a* is subjected to two forces $\mathbf{F}_1$ and $\mathbf{F}_2$. Determine the magnitude and orientation of the resultant force.

SOLUTION I

Scalar Notation. This problem can be solved by using the parallelogram law; however, here we will resolve each force into its x and y components, Fig. 2–19*b*, and sum these components algebraically. Indicating the "positive" sense of the x and y force components alongside Eqs. 2–1, we have

$$\overset{+}{\rightarrow} F_{Rx} = \Sigma F_x; \qquad F_{Rx} = 600 \cos 30° \text{ N} - 400 \sin 45° \text{ N} = 236.8 \text{ N} \rightarrow$$

$$+\uparrow F_{Ry} = \Sigma F_y; \qquad F_{Ry} = 600 \sin 30° \text{ N} + 400 \cos 45° \text{ N} = 582.8 \text{ N} \uparrow$$

The resultant force, shown in Fig. 2–19*c*, has a *magnitude* of

$$F_R = \sqrt{(236.8 \text{ N})^2 + (582.8 \text{ N})^2} = 629 \text{ N} \qquad \textit{Ans.}$$

From the vector addition, Fig. 2–19*c*, the direction angle θ is

$$\theta = \tan^{-1}\left(\frac{582.8 \text{ N}}{236.8 \text{ N}}\right) = 67.9° \qquad \textit{Ans.}$$

SOLUTION II

Cartesian Vector Notation. From Fig. 2–19*b*, each force expressed as a Cartesian vector is

$$\mathbf{F}_1 = \{600 \cos 30°\mathbf{i} + 600 \sin 30°\mathbf{j}\} \text{ N}$$
$$\mathbf{F}_2 = \{-400 \sin 45°\mathbf{i} + 400 \cos 45°\mathbf{j}\} \text{ N}$$

Thus

$$\mathbf{F}_R = \mathbf{F}_1 + \mathbf{F}_2 = (600 \cos 30° \text{ N} - 400 \sin 45° \text{ N})\mathbf{i} + (600 \sin 30° \text{ N} + 400 \cos 45° \text{ N})\mathbf{j} = \{236.8\mathbf{i} + 582.8\mathbf{j}\} \text{ N}$$

The magnitude and direction of $\mathbf{F}_R$ are determined in the same manner as shown above.

Comparing the two methods of solution, it is seen that use of scalar notation is more efficient, since the scalar components can be found *directly,* without first having to express each force as a Cartesian vector before adding the components. Cartesian vector analysis, however, will later be shown to be more advantageous for solving three-dimensional problems.

Example 2–8

The end of the boom O in Fig. 2–20a is subjected to three concurrent and coplanar forces. Determine the magnitude and orientation of the resultant force.

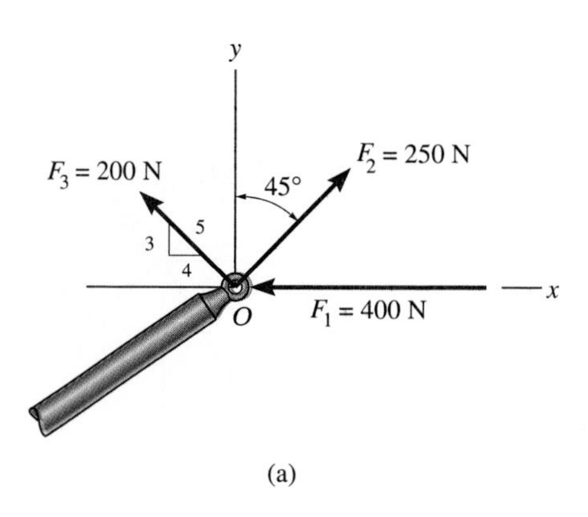

(a)

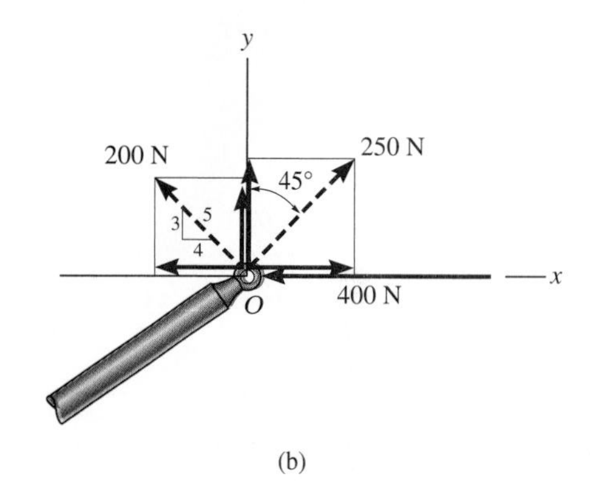

(b)

SOLUTION

Each force is resolved into its x and y components as shown in Fig. 2–20b. Summing the x components, we have

$$\overset{+}{\rightarrow} F_{Rx} = \Sigma F_x; \quad F_{Rx} = -400\text{ N} + 250 \sin 45^\circ \text{ N} - 200(\tfrac{4}{5})\text{ N}$$
$$= -383.2\text{ N} = 383.2\text{ N} \leftarrow$$

The negative sign indicates that F_{Rx} acts to the left, i.e., in the negative x direction as noted by the small arrow. Summing the y components yields

$$+\uparrow F_{Ry} = \Sigma F_y; \quad F_{Ry} = 250 \cos 45^\circ \text{ N} + 200(\tfrac{3}{5})\text{ N}$$
$$= 296.8\text{ N} \uparrow$$

The resultant force, shown in Fig. 2–20c, has a *magnitude* of

$$F_R = \sqrt{(-383.2)^2 + (296.8)^2}$$
$$= 485\text{ N} \qquad \textit{Ans.}$$

From the vector addition in Fig. 2–20c, the direction angle θ is

$$\theta = \tan^{-1}\left(\frac{296.8}{383.2}\right) = 37.8^\circ \qquad \textit{Ans.}$$

Realize that the single force $\mathbf{F}_R$ shown in Fig. 2–20c creates the *same effect* on the boom as the three forces in Fig. 2–20a.

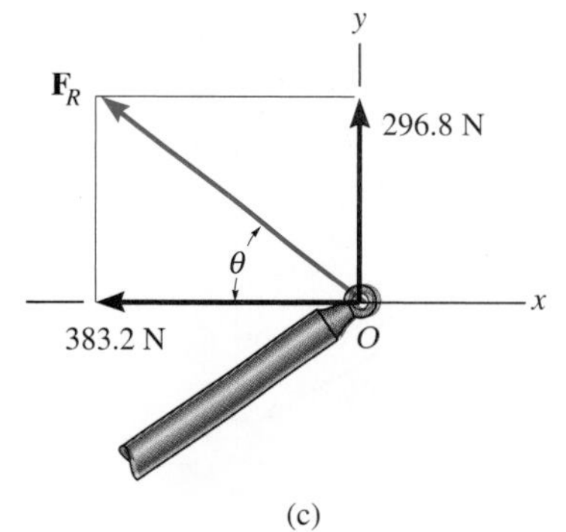

(c)

Fig. 2–20

PROBLEMS

2–33. Determine the x and y components of the 800-lb force.

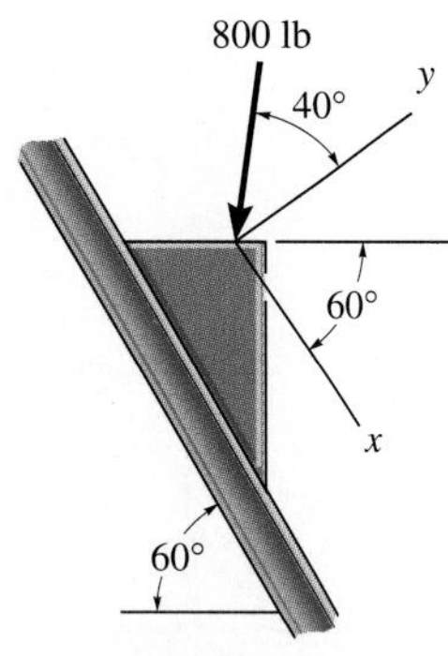

Prob. 2–33

2–34. Determine the magnitude of the resultant force and its direction, measured counterclockwise from the positive x axis.

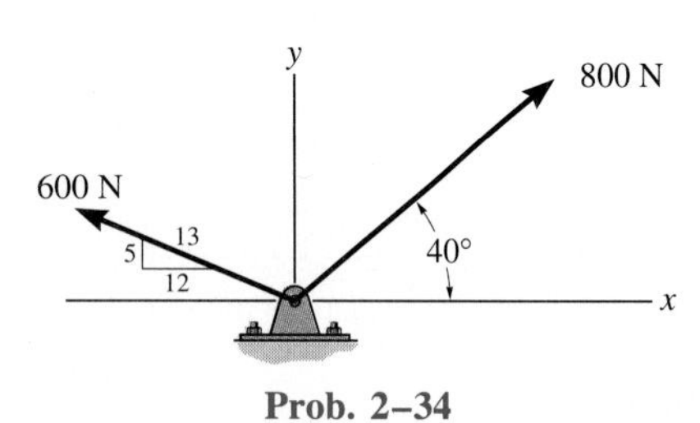

Prob. 2–34

2–35. Determine the magnitude of the resultant force and its direction, measured clockwise from the positive x axis.

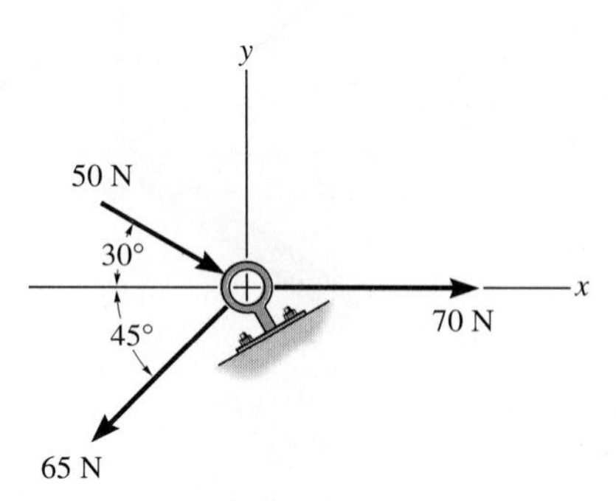

Prob. 2–35

***2–36.** Express $\mathbf{F}_1$, $\mathbf{F}_2$, and $\mathbf{F}_3$ as Cartesian vectors.

2–37. Determine the magnitude of the resultant force and its direction, measured counterclockwise from the positive x axis.

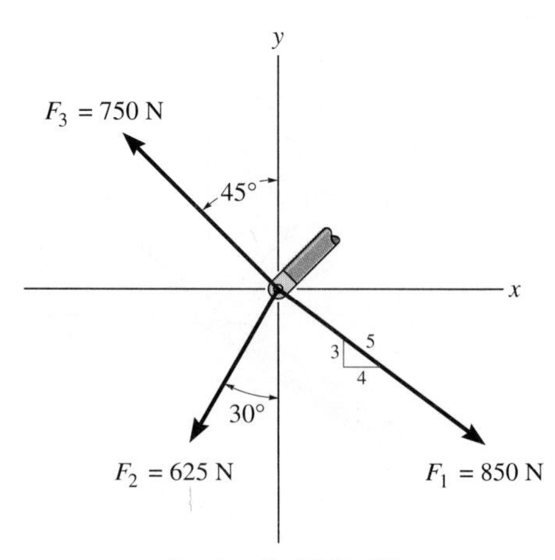

Probs. 2–36/2–37

2–38. Express $\mathbf{F}_1$ and $\mathbf{F}_2$ as Cartesian vectors.

2–39. Determine the magnitude of the resultant force and its direction, measured counterclockwise from the positive x axis.

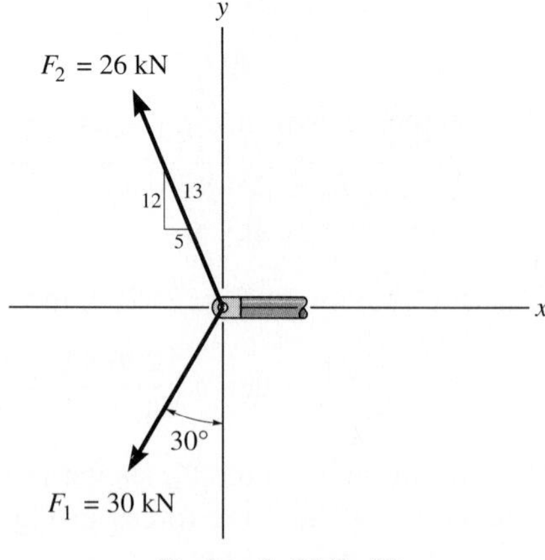

Probs. 2–38/2–39

***2–40.** Determine the x and y components of $\mathbf{F}_1$ and $\mathbf{F}_2$.

2–41. Determine the magnitude of the resultant force and its direction, measured counterclockwise from the positive x axis.

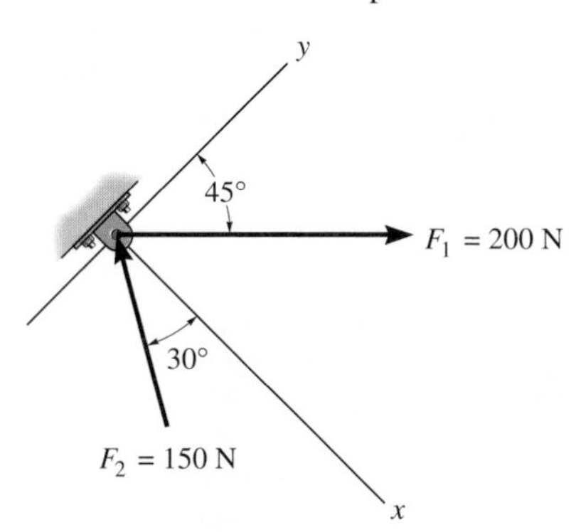

Probs. 2–40/2–41

2–42. Solve Prob. 2–1 by summing the rectangular or x, y components of the forces to obtain the resultant force.

2–43. Solve Prob. 2–2 by summing the rectangular or x, y components of the forces to obtain the resultant force.

***2–44.** Solve Prob. 2–3 by summing the rectangular or x, y components of the forces to obtain the resultant force.

2–45. Solve Prob. 2–15 by summing the rectangular or x, y components of the forces to obtain the resultant force.

2–46. Solve Prob. 2–27 by summing the rectangular or x, y components of the forces to obtain the resultant force.

2–47. Determine the x and y components of each force acting on the *gusset plate* of the bridge truss. Show that the resultant force is zero.

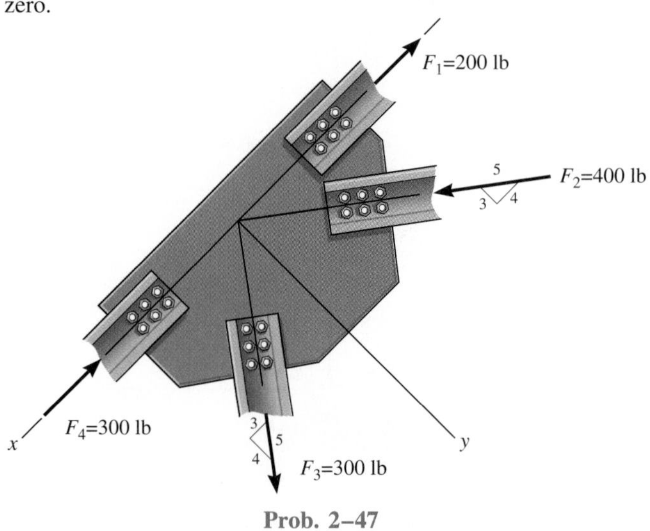

Prob. 2–47

***2–48.** If $\theta = 60°$ and $F = 20$ kN, determine the magnitude of the resultant force and its direction measured clockwise from the positive x axis.

2–49. Determine the magnitude F and direction θ of force $\mathbf{F}$ so that the resultant of the three forces acting on the hook is zero.

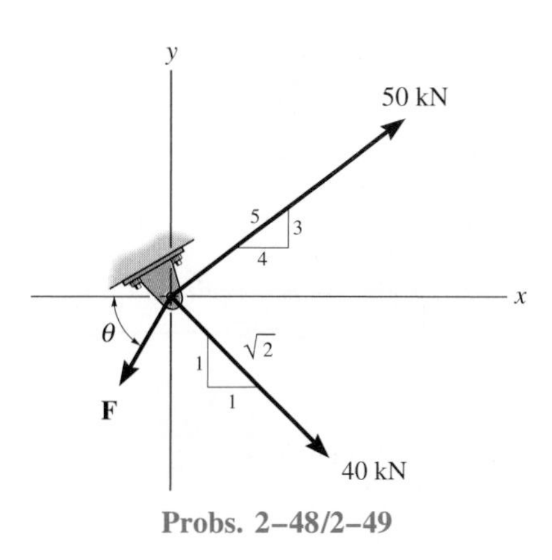

Probs. 2–48/2–49

2–50. Express each of the three forces acting on the column in Cartesian vector form and compute the magnitude of the resultant force.

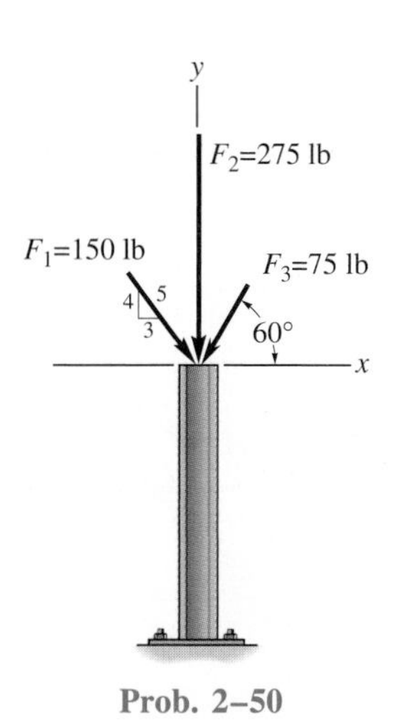

Prob. 2–50

2–51. Three forces act on the bracket. Determine the magnitude and direction θ of $\mathbf{F}_1$ so that the resultant force is directed along the positive x' axis and has a magnitude of 1 kN.

***2–52.** If $F_1 = 300$ N and $\theta = 20°$, determine the magnitude and direction, measured counterclockwise from the x' axis, of the resultant force of the three forces acting on the bracket.

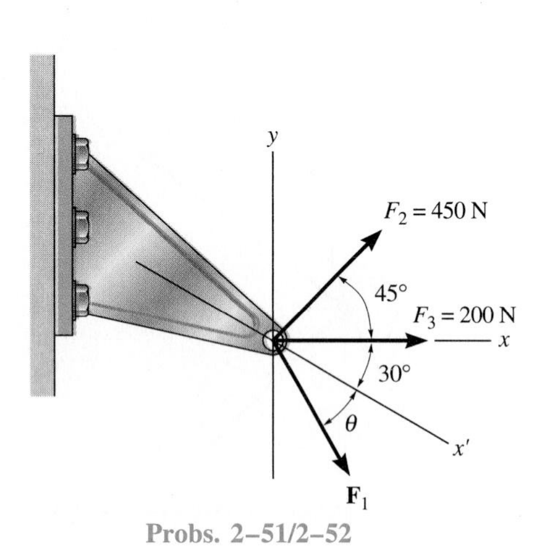

Probs. 2–51/2–52

2–53. Three forces act on the ring. Determine the range of values for the magnitude of **P** so that the magnitude of the resultant force does not exceed 2500 N. Force **P** is always directed to the right.

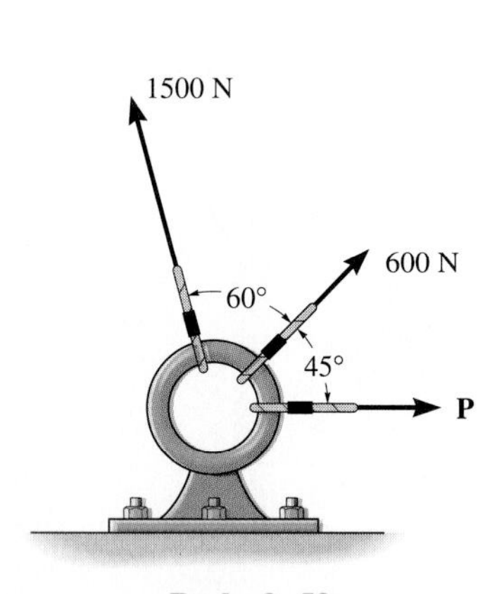

Prob. 2–53

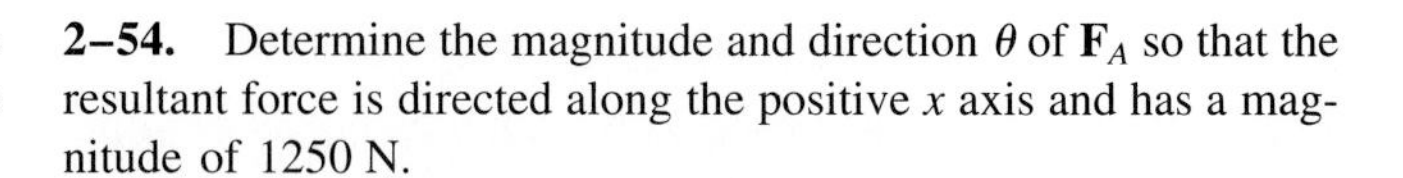

2–54. Determine the magnitude and direction θ of $\mathbf{F}_A$ so that the resultant force is directed along the positive x axis and has a magnitude of 1250 N.

2–55. If $F_A = 750$ N and $\theta = 45°$, determine the magnitude and direction, measured counterclockwise from the positive x axis, of the resultant force acting on the ring at O.

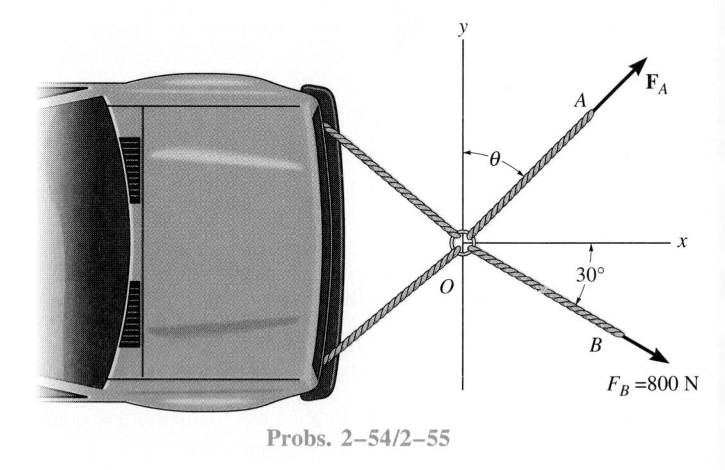

Probs. 2–54/2–55

***2–56.** Three forces act on the bracket. Determine the magnitude and direction θ of $\mathbf{F}_1$ so that the resultant force is directed along the positive x' axis and has a magnitude of 800 N.

2–57. If $F_1 = 300$ N and $\theta = 10°$, determine the magnitude and direction, measured counterclockwise from the positive x' axis, of the resultant force acting on the bracket.

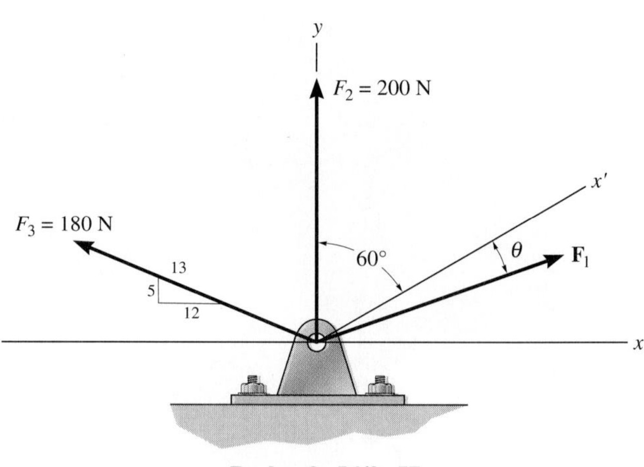

Probs. 2–56/2–57

2–58. Express each of the three forces acting on the bracket in Cartesian vector form with respect to the x and y axes. Determine the magnitude and direction θ of $\mathbf{F}_1$ so that the resultant force is directed along the positive x' axis and has a magnitude of $F_R = 600$ N.

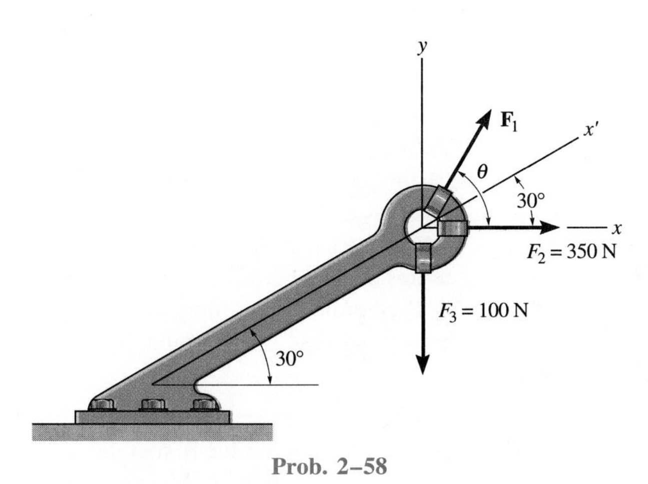

Prob. 2–58

2–59. The three concurrent forces acting on the post produce a resultant force $\mathbf{F}_R = \mathbf{0}$. If $F_2 = \frac{1}{2}F_1$, and $\mathbf{F}_1$ is to be 90° from $\mathbf{F}_2$ as shown, determine the required magnitude F_3 expressed in terms of F_1 and the angle θ.

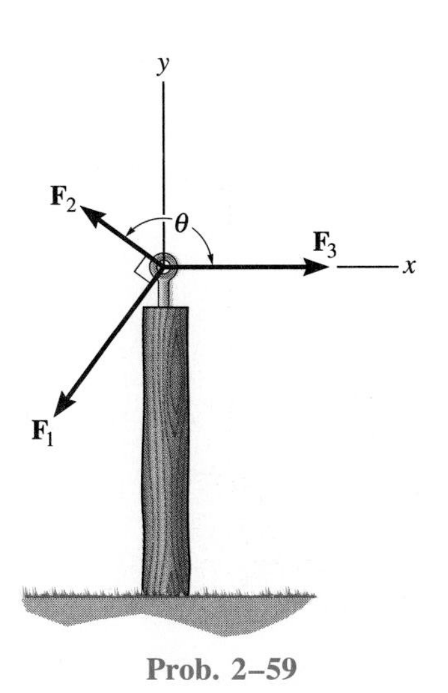

Prob. 2–59

***2–60.** Determine the direction θ of the cable and the required tension F_1 so that the resultant force is directed vertically upward and has a magnitude of 800 N.

2–61. Determine the magnitude and direction of the resultant force of the three forces acting on the ring A. Take $F_1 = 500$ N and $\theta = 20°$.

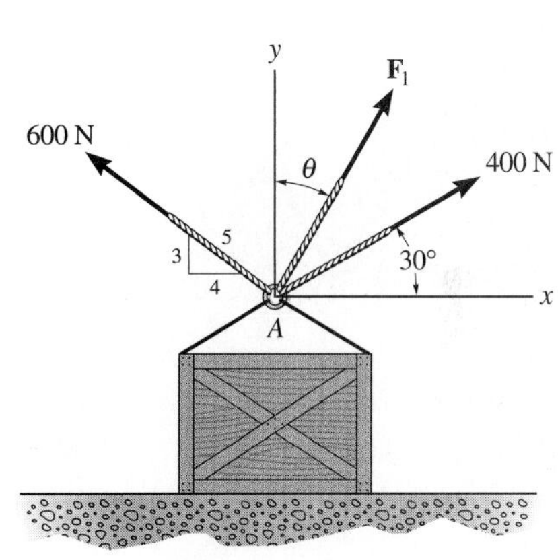

Probs. 2–60/2–61

2–62. Determine the magnitude of force $\mathbf{F}$ so that the magnitude of the resultant $\mathbf{F}_R$ of the three forces is as small as possible. What is the minimum magnitude of $\mathbf{F}_R$?

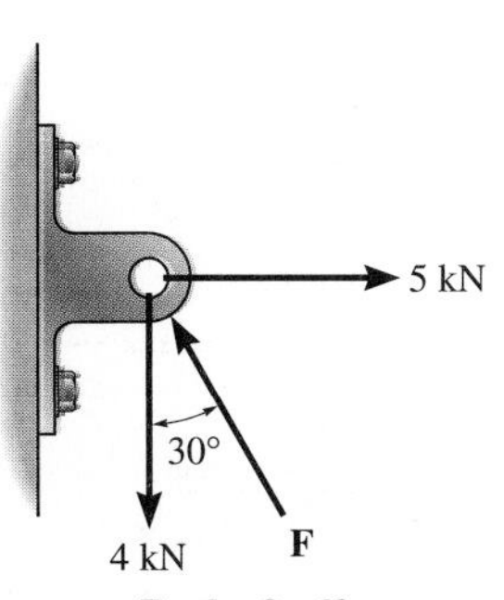

Prob. 2–62

2.5 Cartesian Vectors

The operations of vector algebra, when applied to solving problems in *three dimensions,* are greatly simplified if the vectors are first represented in Cartesian vector form. In this section we will present a general method for doing this, then in Sec. 2.6 we will apply this method to solving problems involving the addition of forces. Similar applications will be illustrated for the position and moment vectors given in later sections of the text.

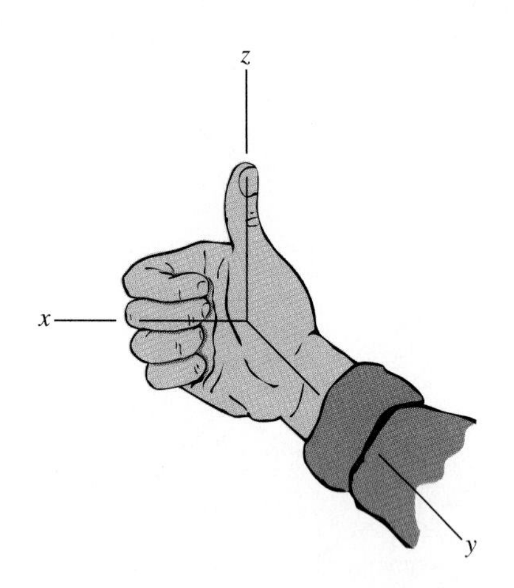

Right-handed coordinate system

Fig. 2–21

Right-Handed Coordinate System. A right-handed coordinate system will be used for developing the theory of vector algebra that follows. A rectangular or Cartesian coordinate system is said to be *right-handed* provided the thumb of the right hand points in the direction of the positive z axis when the right-hand fingers are curled about this axis and directed from the positive x toward the positive y axis, Fig. 2–21. Furthermore, according to this rule, the z axis for a two-dimensional problem as in Fig. 2–20 would be directed outward, perpendicular to the page.

Rectangular Components of a Vector. A vector $\mathbf{A}$ may have one, two, or three rectangular components along the x, y, z coordinate axes, depending on how the vector is oriented relative to the axes. In general, though, when $\mathbf{A}$ is directed within an octant of the x, y, z frame, Fig. 2–22, then by two successive applications of the parallelogram law, we may resolve the vector into components as $\mathbf{A} = \mathbf{A}' + \mathbf{A}_z$ and then $\mathbf{A}' = \mathbf{A}_x + \mathbf{A}_y$. Combining these equations, $\mathbf{A}$ is represented by the vector sum of its *three* rectangular components,

$$\mathbf{A} = \mathbf{A}_x + \mathbf{A}_y + \mathbf{A}_z \quad (2\text{–}2)$$

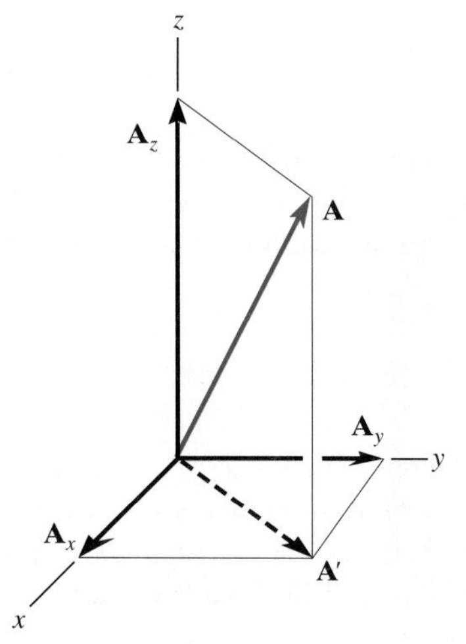

Fig. 2–22

Unit Vector. In general, a *unit vector* is a vector having a magnitude of 1. If **A** is a vector having a magnitude $A \neq 0$, then a unit vector having the *same direction* as **A** is represented by

$$\mathbf{u}_A = \frac{\mathbf{A}}{A} \tag{2–3}$$

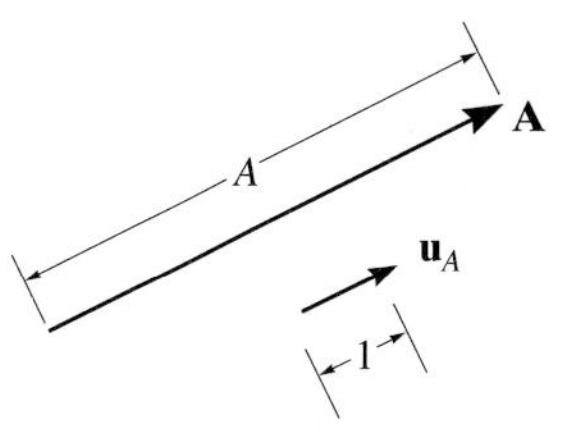

Fig. 2–23

Rewriting this equation gives

$$\mathbf{A} = A\mathbf{u}_A \tag{2–4}$$

Since vector **A** is of a certain type, e.g., a force vector, it is customary to use the proper set of units for its description. The magnitude A also has this same set of units; hence, from Eq. 2–3, the *unit vector will be dimensionless* since the units will cancel out. Equation 2–4 therefore indicates that vector **A** may be expressed in terms of both its magnitude and direction *separately;* i.e., A (a positive scalar) defines the *magnitude* of **A,** and $\mathbf{u}_A$ (a dimensionless vector) defines the *direction* and sense of **A,** Fig. 2–23.

Cartesian Unit Vectors. In three dimensions, the set of Cartesian unit vectors, **i, j, k,** is used to designate the directions of the *x, y, z* axes respectively. As stated in Sec. 2–4, the *sense* (or arrowhead) of these vectors will be described analytically by a plus or minus sign, depending on whether they are pointing along the positive or negative *x, y,* or *z* axis. Thus the positive unit vectors are shown in Fig. 2–24.

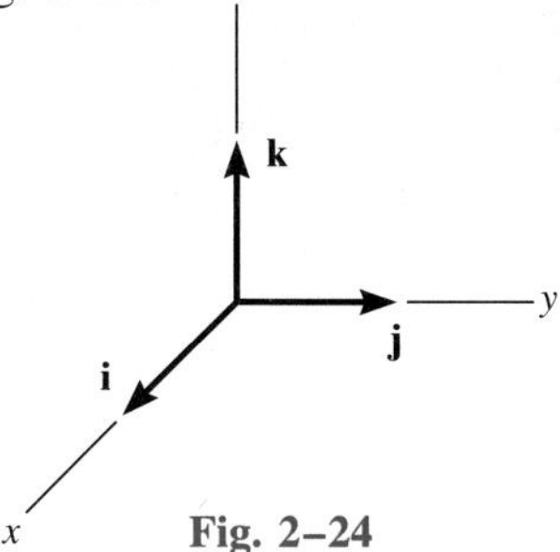

Fig. 2–24

Cartesian Vector Representation. Using Cartesian unit vectors, the three vector components of Eq. 2–2 may be written in "Cartesian vector form." Since the components act in the positive **i, j,** and **k** directions, Fig. 2–25, we have

$$\mathbf{A} = A_x\mathbf{i} + A_y\mathbf{j} + A_z\mathbf{k} \tag{2–5}$$

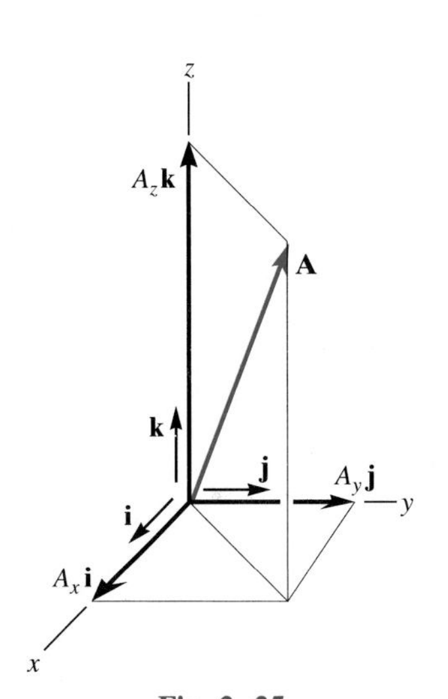

Fig. 2–25

There is a distinct advantage to writing vectors in terms of their Cartesian components. Since each of these components has the same form as Eq. 2–4, the *magnitude* and *direction* of each *component vector* are *separated,* and it will be shown that this will simplify the operations of vector algebra, particularly in three dimensions.

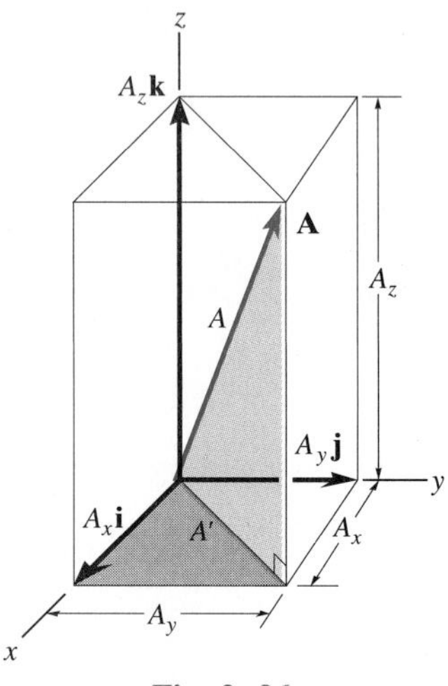

Fig. 2–26

Magnitude of a Cartesian Vector. It is always possible to obtain the magnitude of vector **A** provided the vector is expressed in Cartesian vector form. As shown in Fig. 2–26, from the colored right triangle, $A = \sqrt{A'^2 + A_z^2}$, and from the shaded right triangle, $A' = \sqrt{A_x^2 + A_y^2}$. Combining these equations yields

$$A = \sqrt{A_x^2 + A_y^2 + A_z^2} \tag{2–6}$$

Hence, the magnitude of **A** *is equal to the positive square root of the sum of the squares of its components.*

Direction of a Cartesian Vector. The *orientation* of vector **A** is defined by the *coordinate direction angles* α (alpha), β (beta), and γ (gamma), measured between the *tail* of **A** and the *positive* x, y, z axes located at the tail of **A,** Fig. 2–27. Note that regardless of where **A** is directed, each of these angles will be between 0° and 180°. To determine α, β, and γ, consider the projection of **A** onto the x, y, z axes, Fig. 2–28. Referring to the colored right triangles shown in each figure, we have

$$\cos\alpha = \frac{A_x}{A} \qquad \cos\beta = \frac{A_y}{A} \qquad \cos\gamma = \frac{A_z}{A} \tag{2–7}$$

These numbers are known as the *direction cosines* of **A.** Once they have been obtained, the coordinate direction angles α, β, γ can then be determined from the inverse cosines.

An easy way of obtaining the direction cosines of **A** is to form a unit vector in the direction of **A,** Eq. 2–3. Provided **A** is expressed in Cartesian

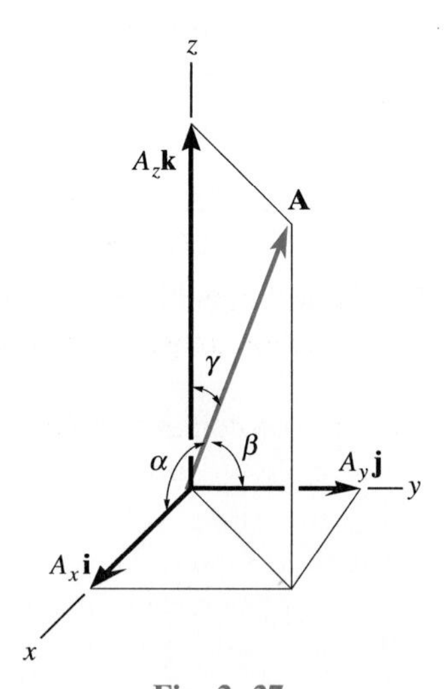

Fig. 2–27

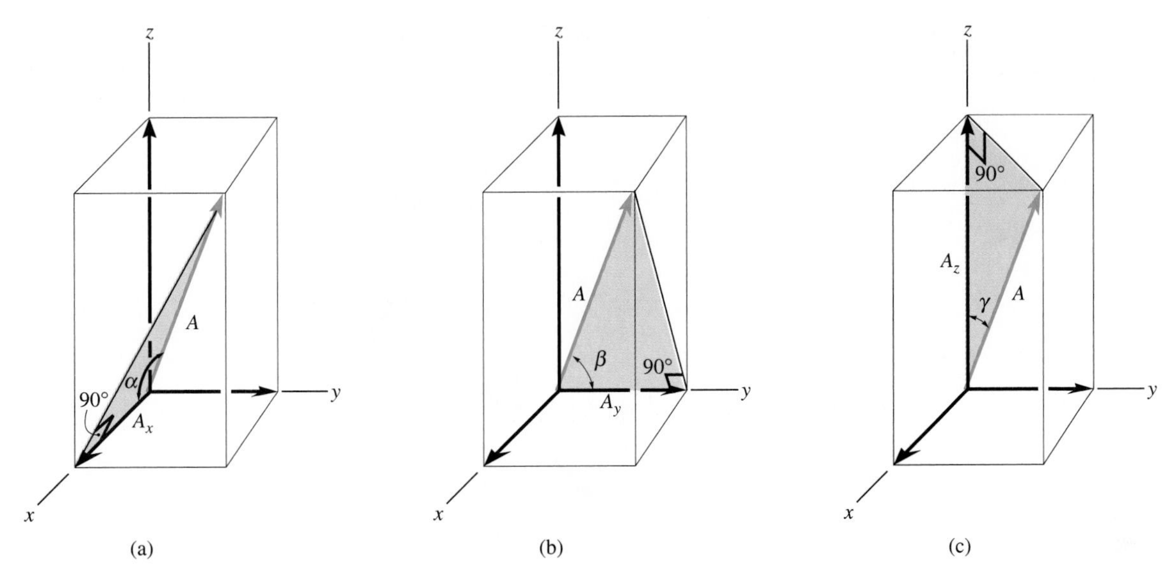

Fig. 2–28

vector form as $\mathbf{A} = A_x\mathbf{i} + A_y\mathbf{j} + A_z\mathbf{k}$ (Eq. 2–5), we have

$$\mathbf{u}_A = \frac{\mathbf{A}}{A} = \frac{A_x}{A}\mathbf{i} + \frac{A_y}{A}\mathbf{j} + \frac{A_z}{A}\mathbf{k} \qquad (2\text{–}8)$$

where $A = \sqrt{(A_x)^2 + (A_y)^2 + (A_z)^2}$ (Eq. 2–6). By comparison with Eqs. 2–7, it is seen that *the* **i, j,** *and* **k** *components of* $\mathbf{u}_A$ *represent the direction cosines of* **A,** i.e.,

$$\mathbf{u}_A = \cos\alpha\mathbf{i} + \cos\beta\mathbf{j} + \cos\gamma\mathbf{k} \qquad (2\text{–}9)$$

Since the magnitude of a vector is equal to the positive square root of the sum of the squares of the magnitudes of its components, and $\mathbf{u}_A$ has a magnitude of 1, then from Eq. 2–9 an important relation between the direction cosines can be formulated as

$$\cos^2\alpha + \cos^2\beta + \cos^2\gamma = 1 \qquad (2\text{–}10)$$

Provided vector **A** lies in a known octant, this equation can be used to determine one of the coordinate direction angles if the other two are known. (See Example 2–10.)

Finally, if the magnitude and coordinate direction angles of **A** are given, **A** may be expressed in Cartesian vector form as

$$\begin{aligned} \mathbf{A} &= A\mathbf{u}_A \\ &= A\cos\alpha\mathbf{i} + A\cos\beta\mathbf{j} + A\cos\gamma\mathbf{k} \\ &= A_x\mathbf{i} + A_y\mathbf{j} + A_z\mathbf{k} \end{aligned} \qquad (2\text{–}11)$$

2.6 Addition and Subtraction of Cartesian Vectors

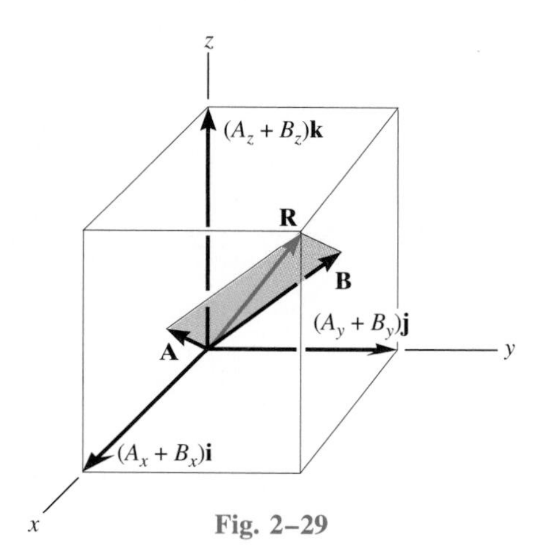

Fig. 2–29

The vector operations of addition and subtraction of two or more vectors are greatly simplified if the vectors are expressed in terms of their Cartesian components. For example, consider the two vectors **A** and **B,** both of which are directed within the positive octant of the x, y, z frame, Fig. 2–29. If $\mathbf{A} = A_x\mathbf{i} + A_y\mathbf{j} + A_z\mathbf{k}$ and $\mathbf{B} = B_x\mathbf{i} + B_y\mathbf{j} + B_z\mathbf{k}$, then the resultant vector, **R,** has components which represent the scalar sums of the **i, j,** and **k** components of **A** and **B,** i.e.,

$$\mathbf{R} = \mathbf{A} + \mathbf{B} = (A_x + B_x)\mathbf{i} + (A_y + B_y)\mathbf{j} + (A_z + B_z)\mathbf{k}$$

Vector subtraction, being a special case of vector addition, simply requires a scalar subtraction of the respective **i, j,** and **k** components of either **A** or **B.** For example,

$$\mathbf{R}' = \mathbf{A} - \mathbf{B} = (A_x - B_x)\mathbf{i} + (A_y - B_y)\mathbf{j} + (A_z - B_z)\mathbf{k}$$

Concurrent Force Systems. In particular, the above concept of vector addition may be generalized and applied to a system of several concurrent forces. In this case, the force resultant is the vector sum of all the forces in the system and can be written as

$$\mathbf{F}_R = \Sigma\mathbf{F} = \Sigma F_x\mathbf{i} + \Sigma F_y\mathbf{j} + \Sigma F_z\mathbf{k} \qquad (2\text{–}12)$$

Here ΣF_x, ΣF_y, and ΣF_z represent the algebraic sums of the respective x, y, z, or **i, j, k** components of each force in the system.

The following examples illustrate numerically the methods used to apply the above theory to the solution of problems involving force as a vector quantity.

Example 2–9

Determine the magnitude and the coordinate direction angles of the resultant force acting on the ring in Fig. 2–30a.

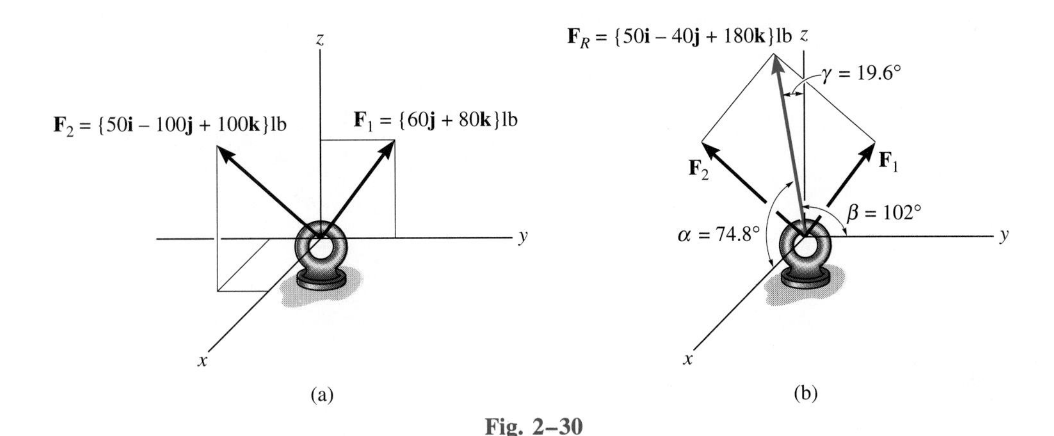

Fig. 2–30

SOLUTION

Since each force is represented in Cartesian vector form, the resultant force, shown in Fig. 2–30b, is

$$\mathbf{F}_R = \Sigma\mathbf{F} = \mathbf{F}_1 + \mathbf{F}_2 = \{60\mathbf{j} + 80\mathbf{k}\}\text{ lb} + \{50\mathbf{i} - 100\mathbf{j} + 100\mathbf{k}\}\text{ lb}$$
$$= \{50\mathbf{i} - 40\mathbf{j} + 180\mathbf{k}\}\text{ lb}$$

The magnitude of $\mathbf{F}_R$ is found from Eq. 2–6, i.e.,

$$F_R = \sqrt{(50)^2 + (-40)^2 + (180)^2}$$
$$= 191.0\text{ lb} \qquad \textit{Ans.}$$

The coordinate direction angles α, β, γ are determined from the components of the unit vector acting in the direction of $\mathbf{F}_R$.

$$\mathbf{u}_{F_R} = \frac{\mathbf{F}_R}{F_R} = \frac{50}{191.0}\mathbf{i} - \frac{40}{191.0}\mathbf{j} + \frac{180}{191.0}\mathbf{k}$$
$$= 0.2617\mathbf{i} - 0.2094\mathbf{j} + 0.9422\mathbf{k}$$

so that

$$\cos\alpha = 0.2617 \qquad \alpha = 74.8° \qquad \textit{Ans.}$$
$$\cos\beta = -0.2094 \qquad \beta = 102° \qquad \textit{Ans.}$$
$$\cos\gamma = 0.9422 \qquad \gamma = 19.6° \qquad \textit{Ans.}$$

These angles are shown in Fig. 2–30b. In particular, note that $\beta > 90°$ since the $\mathbf{j}$ component of $\mathbf{u}_{F_R}$ is negative.

Example 2–10

Express the force **F** shown in Fig. 2–31 as a Cartesian vector.

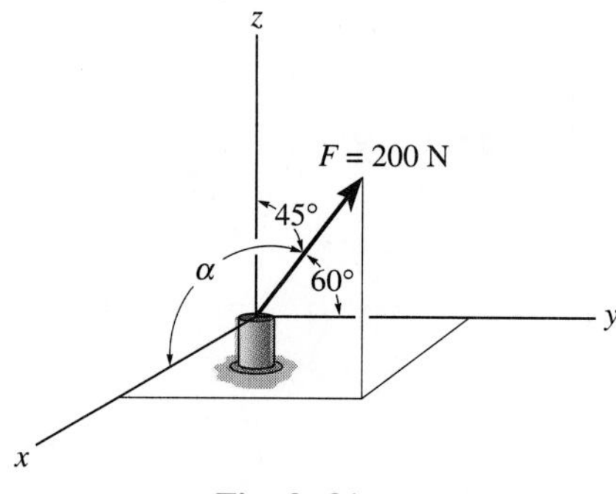

Fig. 2–31

SOLUTION

Since only two coordinate direction angles are specified, the third angle α is determined from Eq. 2–10; i.e.,

$$\cos^2\alpha + \cos^2\beta + \cos^2\gamma = 1$$
$$\cos^2\alpha + \cos^2 60° + \cos^2 45° = 1$$
$$\cos\alpha = \sqrt{1 - (0.707)^2 - (0.5)^2} = \pm 0.5$$

Hence,

$$\alpha = \cos^{-1}(0.5) = 60° \qquad \text{or} \qquad \alpha = \cos^{-1}(-0.5) = 120°$$

By inspection of Fig. 2–31, however, it is necessary that $\alpha = 60°$, since $\mathbf{F}_x$ is in the $+x$ direction.

Using Eq. 2–11, with $F = 200$ N, we have

$$\mathbf{F} = F\cos\alpha\mathbf{i} + F\cos\beta\mathbf{j} + F\cos\gamma\mathbf{k}$$
$$= 200\cos 60°\ \text{N}\mathbf{i} + 200\cos 60°\ \text{N}\mathbf{j} + 200\cos 45°\ \text{N}\mathbf{k}$$
$$= \{100.0\mathbf{i} + 100.0\mathbf{j} + 141.4\mathbf{k}\}\ \text{N} \qquad \textit{Ans.}$$

By applying Eq. 2–6, note that indeed the magnitude of $F = 200$ N.

$$F = \sqrt{F_x^2 + F_y^2 + F_z^2}$$
$$= \sqrt{(100.0)^2 + (100.0)^2 + (141.4)^2} = 200\ \text{N}$$

Example 2–11

Express the force **F** shown acting on the hook in Fig. 2–32*a* as a Cartesian vector.

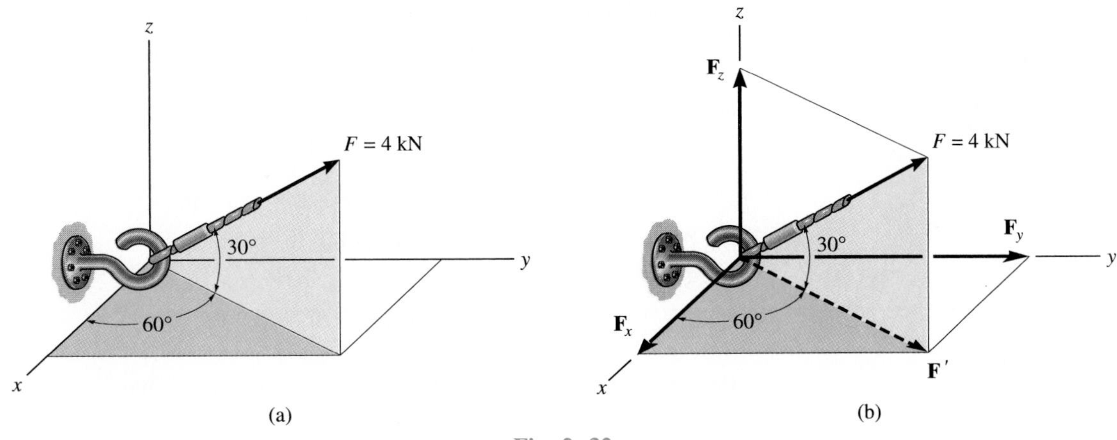

Fig. 2–32

SOLUTION

In this case the angles 60° and 30° defining the direction of **F** are *not* coordinate direction angles. Why? By two successive applications of the parallelogram law, however, **F** can be resolved into its *x, y, z* components as shown in Fig. 2–32*b*. First, from the colored triangle,

$$F' = 4 \cos 30° \text{ kN} = 3.46 \text{ kN}$$
$$F_z = 4 \sin 30° \text{ kN} = 2.00 \text{ kN}$$

Next, using **F′** and the shaded triangle,

$$F_x = 3.46 \cos 60° \text{ kN} = 1.73 \text{ kN}$$
$$F_y = 3.46 \sin 60° \text{ kN} = 3.00 \text{ kN}$$

Thus,

$$\mathbf{F} = \{1.73\mathbf{i} + 3.00\mathbf{j} + 2.00\mathbf{k}\} \text{ kN} \qquad \textit{Ans.}$$

As an exercise, show that the magnitude of **F** is indeed 4 kN, and that the coordinate direction angle $\alpha = 64.3°$.

Example 2–12

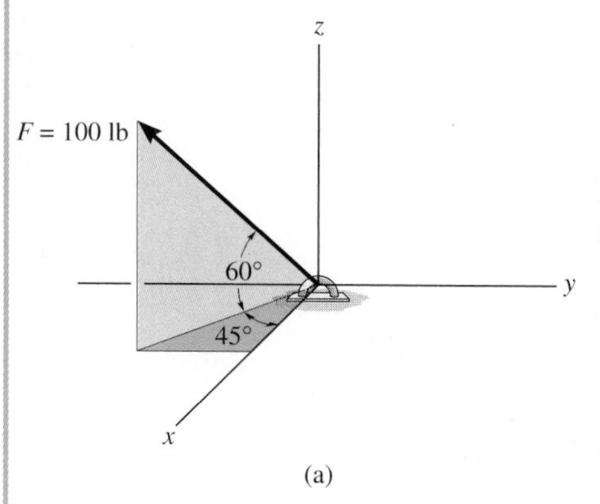

(a)

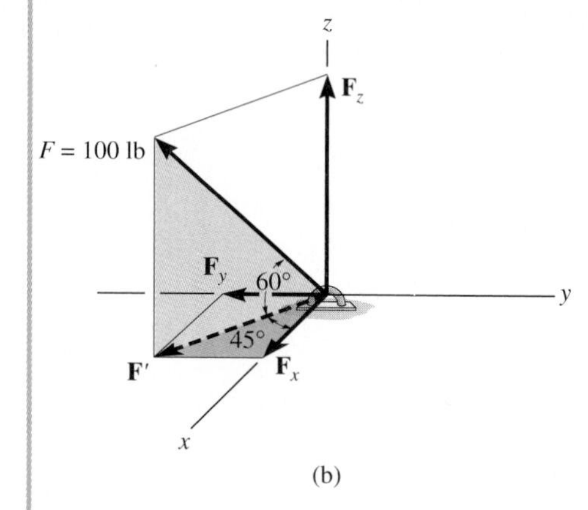

(b)

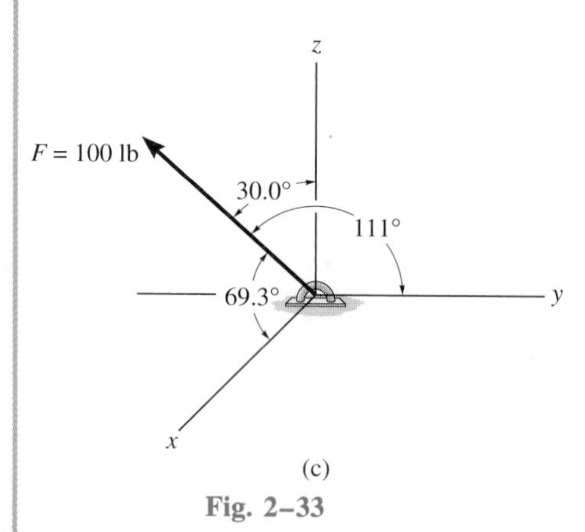

(c)

Fig. 2–33

Express force **F** shown in Fig. 2–33*a* as a Cartesian vector.

SOLUTION

As in Example 2–11, the angles of 60° and 45° defining the direction of **F** are *not* coordinate direction angles. The two successive applications of the parallelogram law needed to resolve **F** into its *x, y, z* components are shown in Fig. 2–33*b*. By trigonometry, the magnitudes of the components are

$$F_z = 100 \sin 60° \text{ lb} = 86.6 \text{ lb}$$
$$F' = 100 \cos 60° \text{ lb} = 50 \text{ lb}$$
$$F_x = 50 \cos 45° \text{ lb} = 35.4 \text{ lb}$$
$$F_y = 50 \sin 45° \text{ lb} = 35.4 \text{ lb}$$

Realizing that $\mathbf{F}_y$ has a direction defined by $-\mathbf{j}$, we have

$$\mathbf{F} = F_x\mathbf{i} + F_y\mathbf{j} + F_z\mathbf{k}$$
$$\mathbf{F} = \{35.4\mathbf{i} - 35.4\mathbf{j} + 86.6\mathbf{k}\} \text{ lb} \qquad \textit{Ans.}$$

To show that the magnitude of this vector is indeed 100 lb, apply Eq. 2–6,

$$F = \sqrt{F_x^2 + F_y^2 + F_z^2}$$
$$= \sqrt{(35.4)^2 + (-35.4)^2 + (86.6)^2} = 100 \text{ lb}$$

If needed, the coordinate direction angles of **F** can be determined from the components of the unit vector acting in the direction of **F.** Hence,

$$\mathbf{u} = \frac{\mathbf{F}}{F} = \frac{F_x}{F}\mathbf{i} + \frac{F_y}{F}\mathbf{j} + \frac{F_z}{F}\mathbf{k}$$
$$= \frac{35.4}{100}\mathbf{i} - \frac{35.4}{100}\mathbf{j} + \frac{86.6}{100}\mathbf{k}$$
$$= 0.354\mathbf{i} - 0.354\mathbf{j} + 0.866\mathbf{k}$$

so that

$$\alpha = \cos^{-1}(0.354) = 69.3°$$
$$\beta = \cos^{-1}(-0.354) = 111°$$
$$\gamma = \cos^{-1}(0.866) = 30.0°$$

These results are shown in Fig. 2–33*c*.

Example 2–13

Two forces act on the hook shown in Fig. 2–34*a*. Specify the coordinate direction angles of $\mathbf{F}_2$ so that the resultant force $\mathbf{F}_R$ acts along the positive y axis and has a magnitude of 800 N.

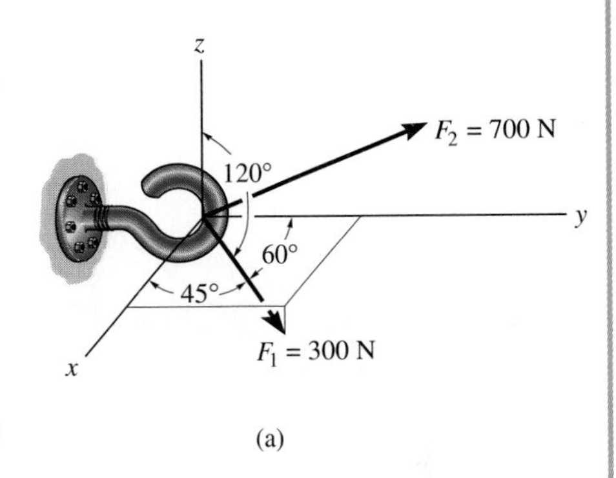

(a)

SOLUTION

To solve this problem, the resultant force and its two components, $\mathbf{F}_1$ and $\mathbf{F}_2$, will each be expressed in Cartesian vector form. Then, as shown in Fig. 2–34*b*, it is necessary that $\mathbf{F}_R = \mathbf{F}_1 + \mathbf{F}_2$.

Applying Eq. 2–11,

$$\begin{aligned} \mathbf{F}_1 = F_1\mathbf{u}_{F_1} &= F_1 \cos \alpha_1 \mathbf{i} + F_1 \cos \beta_1 \mathbf{j} + F_1 \cos \gamma_1 \mathbf{k} \\ &= 300 \cos 45^\circ \text{ N}\mathbf{i} + 300 \cos 60^\circ \text{ N}\mathbf{j} + 300 \cos 120^\circ 300 \text{ N}\mathbf{k} \\ &= \{212.1\mathbf{i} + 150\mathbf{j} - 150\mathbf{k}\} \text{ N} \\ \mathbf{F}_2 = F_2\mathbf{u}_{F_2} &= F_{2x}\mathbf{i} + F_{2y}\mathbf{j} + F_{2z}\mathbf{k} \end{aligned}$$

According to the problem statement, the resultant force $\mathbf{F}_R$ has a magnitude of 800 N and acts in the $+\mathbf{j}$ direction. Hence,

$$\mathbf{F}_R = (800 \text{ N})(+\mathbf{j}) = \{800\mathbf{j}\} \text{ N}$$

We require

$$\begin{aligned} \mathbf{F}_R &= \mathbf{F}_1 + \mathbf{F}_2 \\ 800\mathbf{j} &= 212.1\mathbf{i} + 150\mathbf{j} - 150\mathbf{k} + F_{2x}\mathbf{i} + F_{2y}\mathbf{j} + F_{2z}\mathbf{k} \\ 800\mathbf{j} &= (212.1 + F_{2x})\mathbf{i} + (150 + F_{2y})\mathbf{j} + (-150 + F_{2z})\mathbf{k} \end{aligned}$$

To satisfy this equation, the corresponding **i, j,** and **k** components on the left and right sides must be equal. This is equivalent to stating that the x, y, z components of $\mathbf{F}_R$ be equal to the corresponding x, y, z components of $(\mathbf{F}_1 + \mathbf{F}_2)$. Hence,

$$\begin{aligned} 0 &= 212.1 + F_{2x} & F_{2x} &= -212.1 \text{ N} \\ 800 &= 150 + F_{2y} & F_{2y} &= 650 \text{ N} \\ 0 &= -150 + F_{2z} & F_{2z} &= 150 \text{ N} \end{aligned}$$

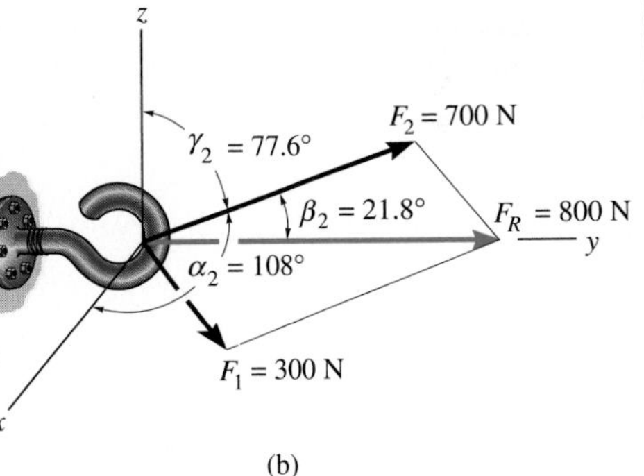

(b)

Since the magnitudes of $\mathbf{F}_2$ and its components are known, we can use Eq. 2–11 to determine α, β, γ.

$$-212.1 = 700 \cos \alpha_2; \qquad \alpha_2 = \cos^{-1}\left(\frac{-212.1}{700}\right) = 108^\circ \qquad \textit{Ans.}$$

$$650 = 700 \cos \beta_2; \qquad \beta_2 = \cos^{-1}\left(\frac{650}{700}\right) = 21.8^\circ \qquad \textit{Ans.}$$

$$150 = 700 \cos \gamma_2; \qquad \gamma_2 = \cos^{-1}\left(\frac{150}{700}\right) = 77.6^\circ \qquad \textit{Ans.}$$

These results are shown in Fig. 2–34*b*.

Fig. 2–34

PROBLEMS

2–63. The cable at the end of the crane boom exerts a force of $F = 250$ lb on the boom as shown. Express **F** as a Cartesian vector.

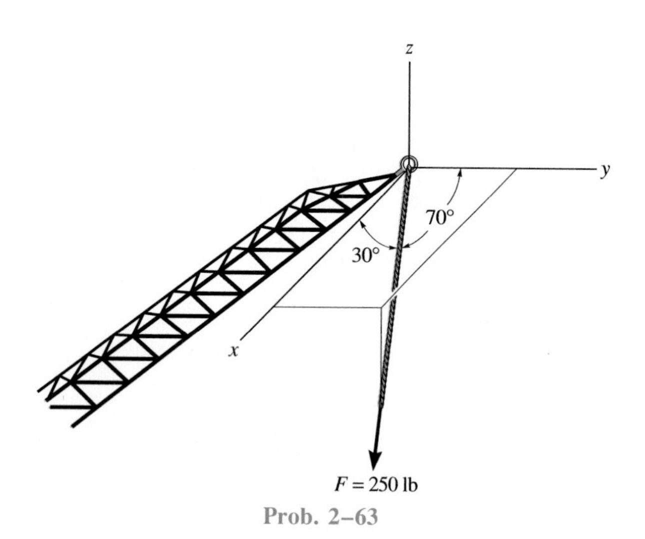

Prob. 2–63

***2–64.** The force **F** acting on the stake has a component of 40 N acting in the *x-y* plane as shown. Express **F** as a Cartesian vector.

2–65. Determine the magnitude and coordinate direction angles of the force **F** acting on the stake.

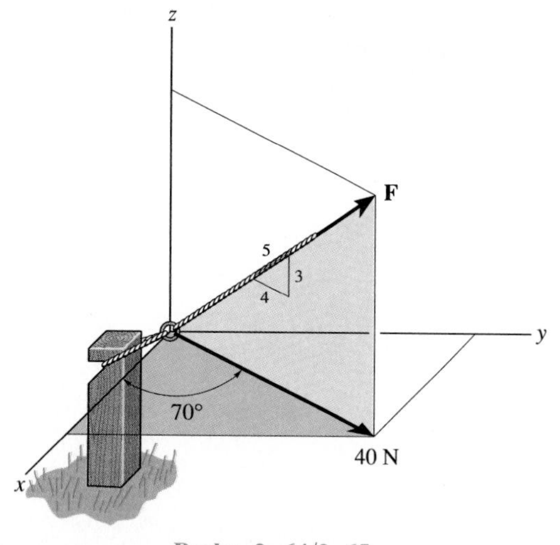

Probs. 2–64/2–65

2–66. The stock *S* mounted on the lathe is subjected to a force of 60 N, which is caused by a die. Determine the coordinate direction angle β and express the force as a Cartesian vector.

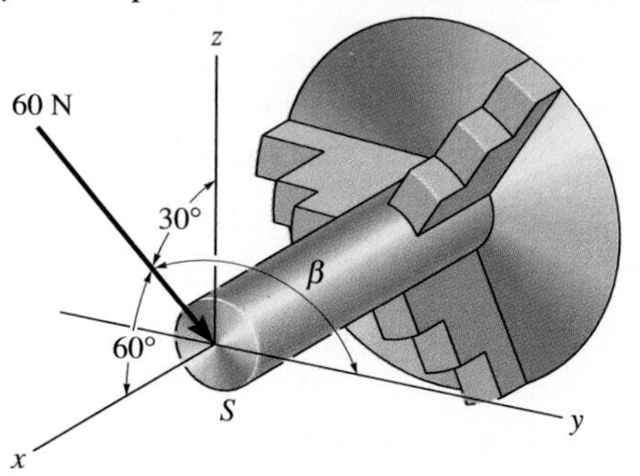

Prob. 2–66

2–67. Express each force as a Cartesian vector and then determine the resultant force $\mathbf{F}_R$. Find the magnitude and coordinate direction angles and sketch this vector on the coordinate system.

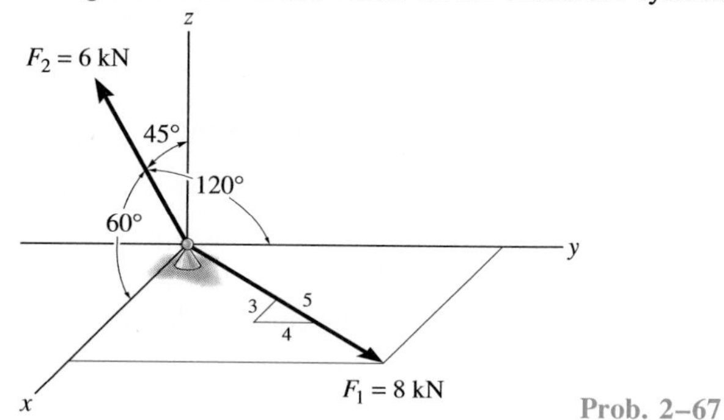

Prob. 2–67

***2–68.** Determine the magnitude and coordinate direction angles of the resultant force and sketch this vector on the coordinate system.

2–69. Specify the coordinate direction angles of $\mathbf{F}_1$ and $\mathbf{F}_2$ and express each force as a Cartesian vector.

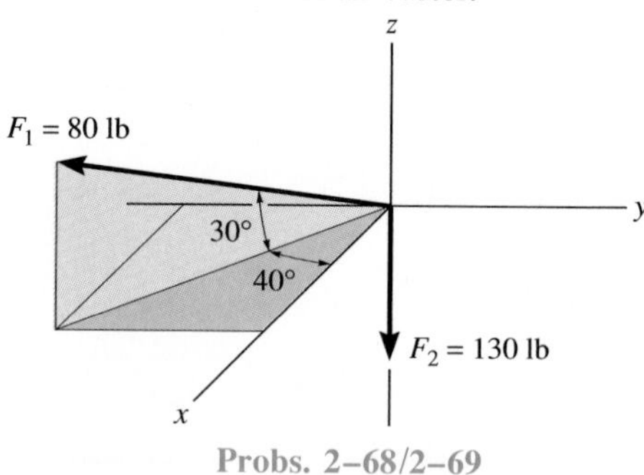

Probs. 2–68/2–69

2–70. Express each force as a Cartesian vector.

2–71. Determine the magnitude and coordinate direction angles of the resultant force and sketch this vector on the coordinate system.

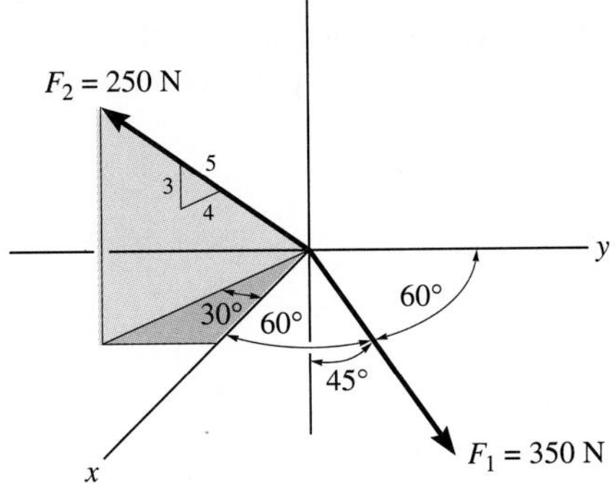

Probs. 2–70/2–71

***2–72.** Determine the magnitude and coordinate direction angles of the resultant force.

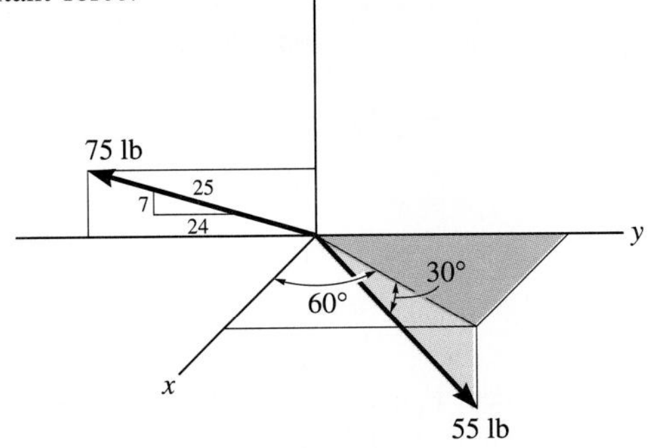

Prob. 2–72

2–73. The beam is subjected to the two forces shown. Express each force in Cartesian vector form and determine the magnitude and coordinate direction angles of the resultant force.

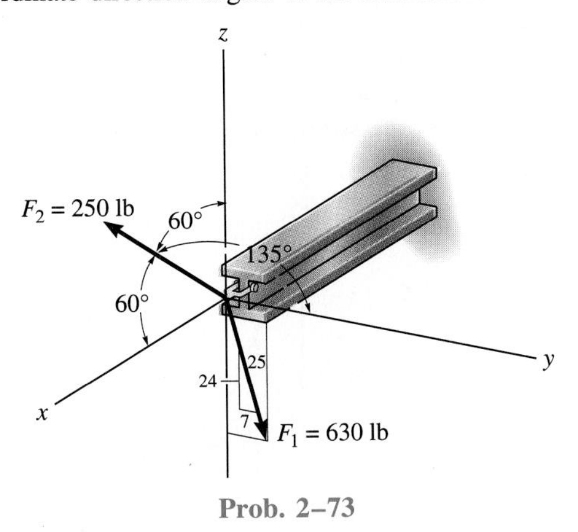

Prob. 2–73

2–74. The mast is subjected to the three forces shown. Determine the coordinate direction angles α_1, β_1, γ_1 of $\mathbf{F}_1$ so that the resultant force acting on the mast is $\mathbf{F}_R = \{350\mathbf{i}\}$ N.

2–75. The mast is subjected to the three forces shown. Determine the coordinate direction angles α_1, β_1, γ_1 of $\mathbf{F}_1$ so that the resultant force acting on the mast is zero.

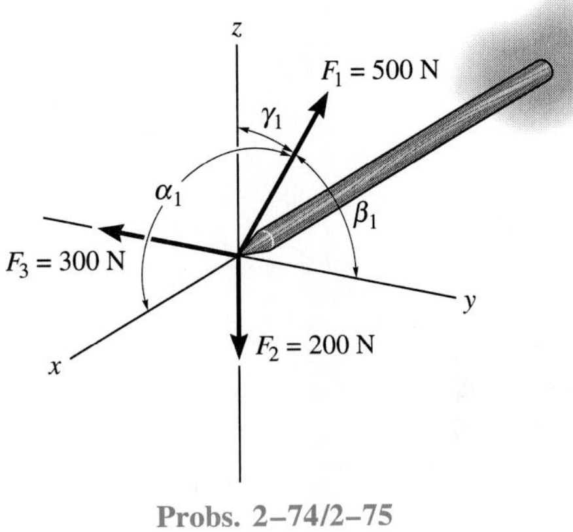

Probs. 2–74/2–75

***2–76.** The two forces $\mathbf{F}_1$ and $\mathbf{F}_2$ acting at A have a resultant force of $\mathbf{F}_R = \{-100\mathbf{k}\}$ lb. Determine the magnitude and coordinate direction angles of $\mathbf{F}_2$.

2–77. Determine the coordinate direction angles of the force $\mathbf{F}_1$ and indicate them on the figure.

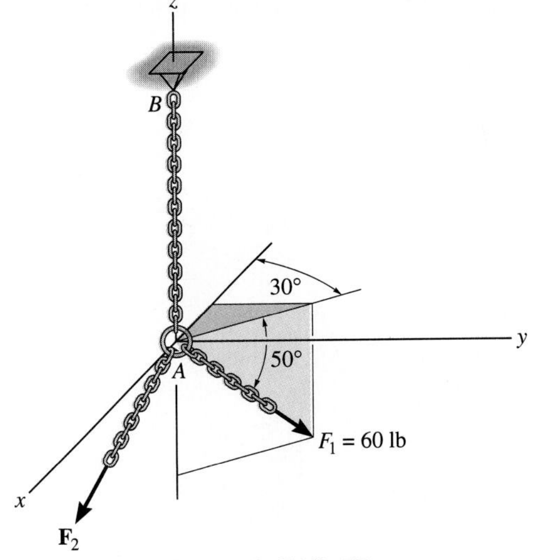

Probs. 2–76/2–77

2–78. The pole is subjected to the force **F**, which has components acting along the x, y, z axes as shown. If the magnitude of **F** is 3 kN, and $\beta = 30°$ and $\gamma = 75°$, determine the magnitudes of its three components.

2–79. The pole is subjected to the force **F** which has components $F_x = 1.5$ kN and $F_z = 1.25$ kN. If $\beta = 75°$, determine the magnitudes of **F** and $\mathbf{F}_y$.

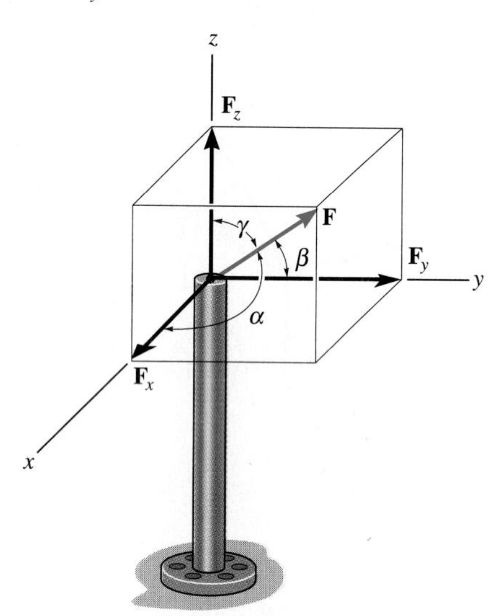

Probs. 2–78/2–79

2–81. The bolt is subjected to the force **F**, which has components acting along the x, y, z axes as shown. If the magnitude of **F** is 80 N, and $\alpha = 60°$ and $\gamma = 45°$, determine the magnitudes of its components.

2–82. The bolt is subjected to the force **F** which has components $F_x = 20$ N, $F_z = 20$ N. If $\beta = 120°$, determine the magnitudes of **F** and $\mathbf{F}_y$.

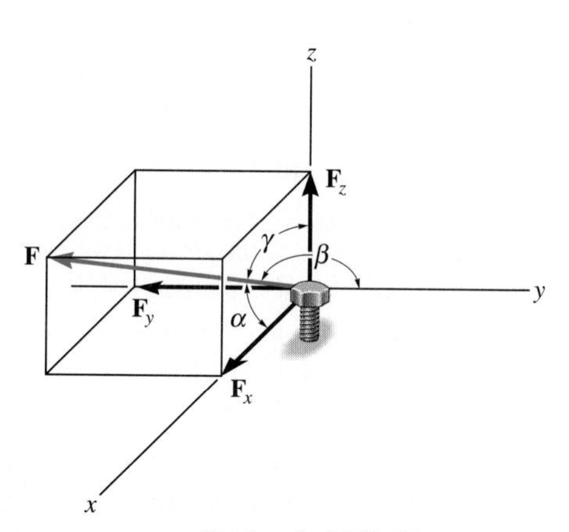

Probs. 2–81/2–82

***2–80.** A force **F** is applied at the top of the tower at A. If it acts in the direction shown, such that one of its components lying in the shaded y-z plane has a magnitude of 80 lb, determine the magnitude of **F** and its coordinate direction angles α, β, γ.

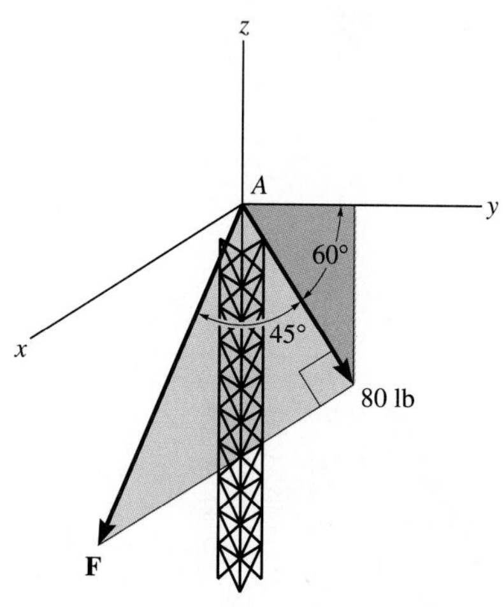

Prob. 2–80

2–83. Two forces $\mathbf{F}_1$ and $\mathbf{F}_2$ act on the bolt. If the resultant force $\mathbf{F}_R$ has a magnitude of 50 lb and coordinate direction angles $\alpha = 110°$ and $\beta = 80°$, as shown, determine the magnitude of $\mathbf{F}_2$ and its coordinate direction angles.

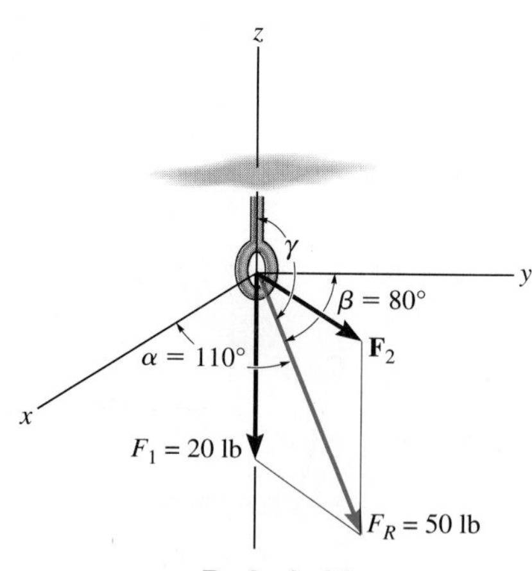

Prob. 2–83

2.7 Position Vectors

In this section we will introduce the concept of a position vector. It will be shown in Sec. 2.8 that this vector is of importance in formulating a Cartesian force vector directed between any two points in space, and later, in Chapter 4, we will use it for finding the moment of a force.

***x, y, z* Coordinates.** Throughout this text we will use a *right-handed* coordinate system to reference the location of points in space. Furthermore, we will use the convention followed in many technical books, and that is to require the positive z axis to be directed *upward* (the zenith direction) so that it measures the height of an object or the altitude of a point. The x, y axes then lie in the horizontal plane, Fig. 2–35. Points in space are located relative to the origin of coordinates, O, by successive measurements along the x, y, z axes. For example, in Fig. 2–35 the coordinates of point A are obtained by starting at O and measuring $x_A = +4$ m along the x axis, $y_A = +2$ m along the y axis, and $z_A = -6$ m along the z axis. Thus, $A(4, 2, -6)$. In a similar manner, measurements along the x, y, z axes from O to B yield the coordinates of B, i.e., $B(0, 2, 0)$. Also notice that $C(6, -1, 4)$.

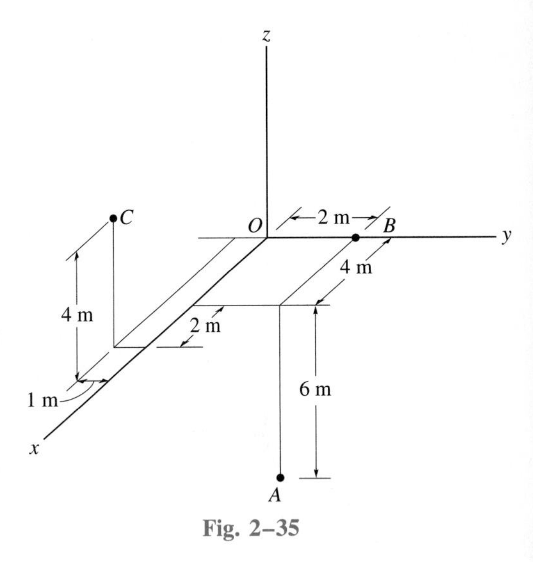

Fig. 2–35

Position Vector. The *position vector* **r** is defined as a fixed vector which locates a point in space relative to another point. For example, if **r** extends from the origin of coordinates, O, to point $P(x, y, z)$, Fig. 2–36*a*, then **r** can be expressed in Cartesian vector form as

$$\mathbf{r} = x\mathbf{i} + y\mathbf{j} + z\mathbf{k}$$

In particular, note how the head-to-tail vector addition of the three components yields vector **r**, Fig. 2–36*b*. Starting at the origin O, one travels x in the $+\mathbf{i}$ direction, then y in the $+\mathbf{j}$ direction, and finally z in the $+\mathbf{k}$ direction to arrive at point $P(x, y, z)$.

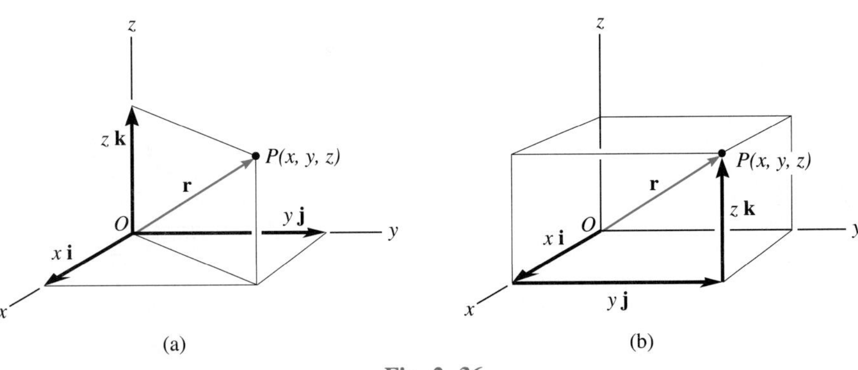

Fig. 2–36

In the more general case, the position vector may be directed from point A to point B in space, Fig. 2–37*a*. As noted, this vector is also designated by the symbol $\mathbf{r}$. As a matter of convention, however, we will *sometimes* refer to this vector with *two subscripts* to indicate from and to the point where it is directed, thus, $\mathbf{r}$ can also be designated as $\mathbf{r}_{AB}$. Also, note that $\mathbf{r}_A$ and $\mathbf{r}_B$ in Fig. 2–37*a* are referenced with only one subscript since they extend from the origin of coordinates.

From Fig. 2–37*a*, by the head-to-tail vector addition, we require

$$\mathbf{r}_A + \mathbf{r} = \mathbf{r}_B$$

Solving for $\mathbf{r}$ and expressing $\mathbf{r}_A$ and $\mathbf{r}_B$ in Cartesian vector form yields

$$\mathbf{r} = \mathbf{r}_B - \mathbf{r}_A = (x_B\mathbf{i} + y_B\mathbf{j} + z_B\mathbf{k}) - (x_A\mathbf{i} + y_A\mathbf{j} + z_A\mathbf{k})$$

or

$$\mathbf{r} = (x_B - x_A)\mathbf{i} + (y_B - y_A)\mathbf{j} + (z_B - z_A)\mathbf{k} \qquad (2\text{–}13)$$

Thus, the **i, j, k** *components of the position vector* $\mathbf{r}$ *may be formed by taking the coordinates of the tail of the vector,* $A(x_A, y_A, z_A)$, *and subtracting them from the corresponding coordinates of the head,* $B(x_B, y_B, z_B)$. Again note how the head-to-tail addition of these three components yields $\mathbf{r}$, i.e., going from A to B, Fig. 2–37*b*, one first travels $(x_B - x_A)$ in the $+\mathbf{i}$ direction, then $(y_B - y_A)$ in the $+\mathbf{j}$ direction, and finally $(z_B - z_A)$ in the $+\mathbf{k}$ direction.

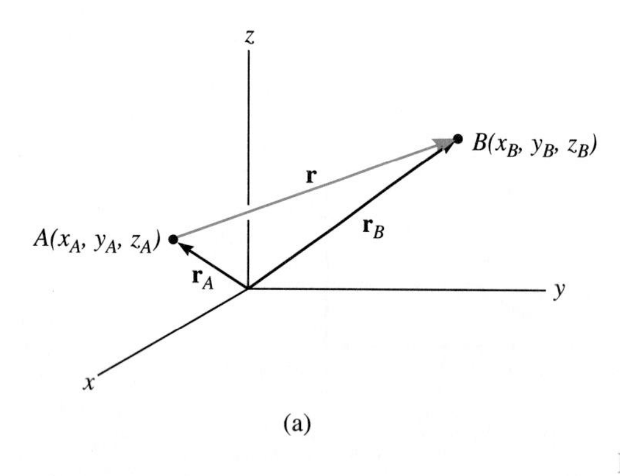

(a)

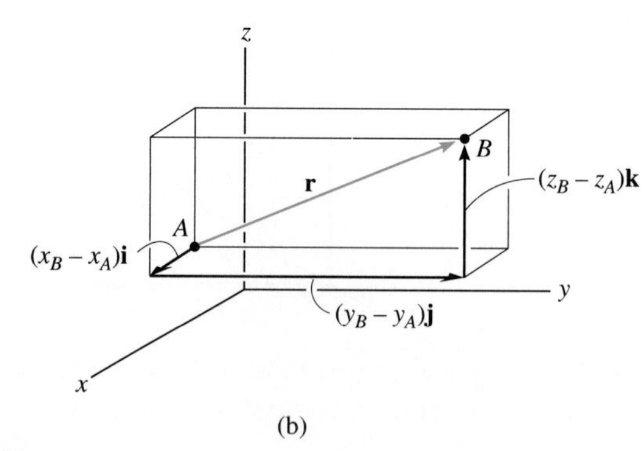

(b)

Fig. 2–37

Example 2–14

An elastic rubber band is attached to points A and B as shown in Fig. 2–38a. Determine its length and its direction measured from A toward B.

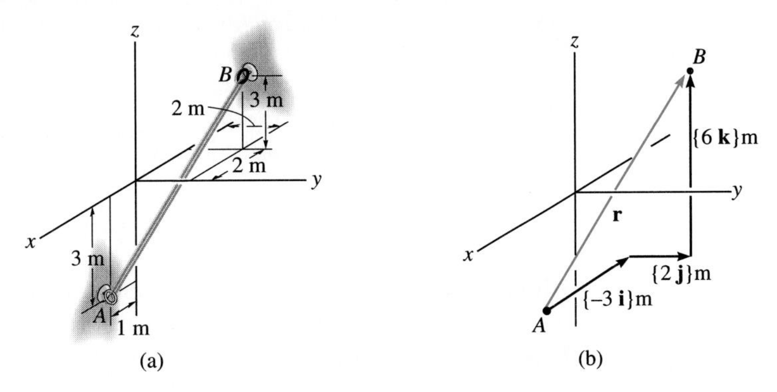

SOLUTION

We first establish a position vector from A to B, Fig. 2–38b. In accordance with Eq. 2–13, the coordinates of the tail $A(1 \text{ m}, 0, -3 \text{ m})$ are subtracted from the coordinates of the head $B(-2 \text{ m}, 2 \text{ m}, 3 \text{ m})$, which yields

$$\mathbf{r} = (-2\text{ m} - 1\text{ m})\mathbf{i} + (2\text{ m} - 0)\mathbf{j} + [3\text{ m} - (-3\text{ m})]\mathbf{k}$$
$$= \{-3\mathbf{i} + 2\mathbf{j} + 6\mathbf{k}\}\text{ m}$$

As shown, the three components of $\mathbf{r}$ represent the direction and distance one must go along each axis in order to move from A to B, i.e., along the x axis $\{-3\mathbf{i}\}$ m, along the y axis $\{2\mathbf{j}\}$ m, and finally along the z axis $\{6\mathbf{k}\}$ m.

The magnitude of $\mathbf{r}$ represents the length of the rubber band.

$$r = \sqrt{(-3)^2 + (2)^2 + (6)^2} = 7 \text{ m} \qquad \textit{Ans.}$$

Formulating a unit vector in the direction of $\mathbf{r}$, we have

$$\mathbf{u} = \frac{\mathbf{r}}{r} = \frac{-3}{7}\mathbf{i} + \frac{2}{7}\mathbf{j} + \frac{6}{7}\mathbf{k}$$

The components of this unit vector yield the coordinate direction angles

$$\alpha = \cos^{-1}\left(\frac{-3}{7}\right) = 115° \qquad \textit{Ans.}$$

$$\beta = \cos^{-1}\left(\frac{2}{7}\right) = 73.4° \qquad \textit{Ans.}$$

$$\gamma = \cos^{-1}\left(\frac{6}{7}\right) = 31.0° \qquad \textit{Ans.}$$

These angles are measured from the *positive axes* of a localized coordinate system placed at the tail of $\mathbf{r}$, point A, as shown in Fig. 2–38c.

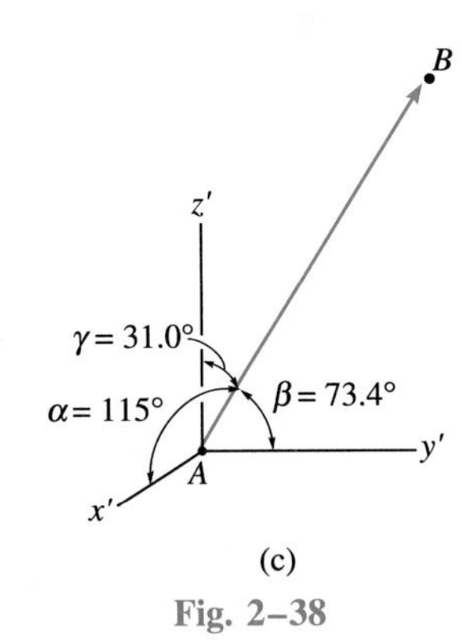

Fig. 2–38

2.8 Force Vector Directed Along a Line

Quite often in three-dimensional statics problems, the direction of a force is specified by two points through which its line of action passes. Such a situation is shown in Fig. 2–39, where the force **F** is directed along the cord AB. We can formulate **F** as a Cartesian vector by realizing that it has the *same direction* and *sense* as the position vector **r** directed from point A to point B on the cord. This common direction is specified by the *unit vector* $\mathbf{u} = \mathbf{r}/r$. Hence,

$$\mathbf{F} = F\mathbf{u} = F\left(\frac{\mathbf{r}}{r}\right)$$

Although we have represented **F** symbolically in Fig. 2–39, note that it has units of force, and unlike **r**, or coordinates x, y, z, which have units of length, **F** cannot be scaled along the coordinate axes.

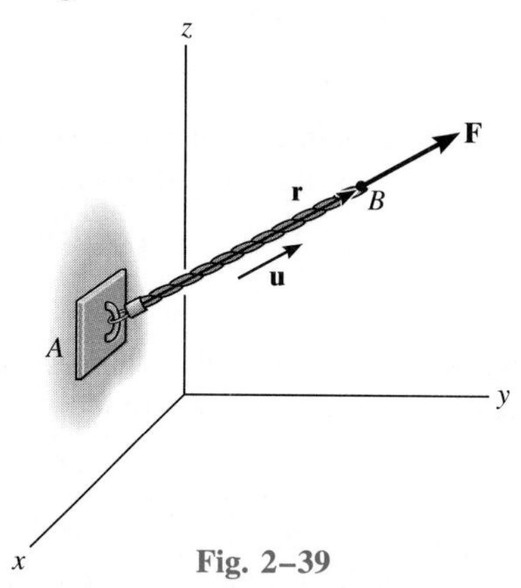

Fig. 2–39

PROCEDURE FOR ANALYSIS

When **F** is directed along a line which extends from point A to point B, then **F** can be expressed in Cartesian vector form as follows:

Position Vector. Determine the position vector **r** directed from A to B, and compute its magnitude r.

Unit Vector. Determine the unit vector $\mathbf{u} = \mathbf{r}/r$ which defines the *direction* and *sense* of *both* **r** and **F**.

Force Vector. Determine **F** by combining its magnitude F and direction **u**, i.e., $\mathbf{F} = F\mathbf{u}$.

This procedure is illustrated numerically in the following example problems.

Example 2–15

The man shown in Fig. 2–40*a* pulls on the cord with a force of 70 lb. Represent this force, acting on the support *A*, as a Cartesian vector and determine its direction.

SOLUTION

Force **F** is shown in Fig. 2–40*b*. The *direction* of this vector, **u,** is determined from the position vector **r,** which extends from *A* to *B*, Fig. 2–40*b*. To formulate **F** as a Cartesian vector we use the following procedure.

Position Vector. The coordinates of the end points of the cord are *A*(0, 0, 30 ft) and *B*(12 ft, −8 ft, 6 ft). Forming the position vector by subtracting the corresponding *x*, *y*, and *z* coordinates of *A* from those of *B*, we have

$$\mathbf{r} = (12\text{ ft} - 0)\mathbf{i} + (-8\text{ ft} - 0)\mathbf{j} + (6\text{ ft} - 30\text{ ft})\mathbf{k}$$
$$= \{12\mathbf{i} - 8\mathbf{j} - 24\mathbf{k}\}\text{ ft}$$

Show on Fig. 2–40*a* how one can write **r** *directly* by going from *A* $\{12\mathbf{i}\}$ ft, then $\{-8\mathbf{j}\}$ ft, and finally $\{-24\mathbf{k}\}$ ft to get to *B*.

The magnitude of **r,** which represents the *length* of cord *AB*, is

$$r = \sqrt{(12)^2 + (-8)^2 + (-24)^2} = 28\text{ ft}$$

Unit Vector. Forming the unit vector that defines the direction and sense of both **r** and **F** yields

$$\mathbf{u} = \frac{\mathbf{r}}{r} = \frac{12}{28}\mathbf{i} - \frac{8}{28}\mathbf{j} - \frac{24}{28}\mathbf{k}$$

Force Vector. Since **F** has a *magnitude* of 70 lb and a *direction* specified by **u,** then

$$\mathbf{F} = F\mathbf{u} = 70\text{ lb}\left(\frac{12}{28}\mathbf{i} - \frac{8}{28}\mathbf{j} - \frac{24}{28}\mathbf{k}\right)$$
$$= \{30\mathbf{i} - 20\mathbf{j} - 60\mathbf{k}\}\text{ lb} \qquad \textit{Ans.}$$

As shown in Fig. 2–40*b*, the coordinate direction angles are measured between **r** (or **F**) and the *positive axes* of a localized coordinate system with origin placed at *A*. From the components of the unit vector:

$$\alpha = \cos^{-1}\left(\frac{12}{28}\right) = 64.6° \qquad \textit{Ans.}$$

$$\beta = \cos^{-1}\left(\frac{-8}{28}\right) = 107° \qquad \textit{Ans.}$$

$$\gamma = \cos^{-1}\left(\frac{-24}{28}\right) = 149° \qquad \textit{Ans.}$$

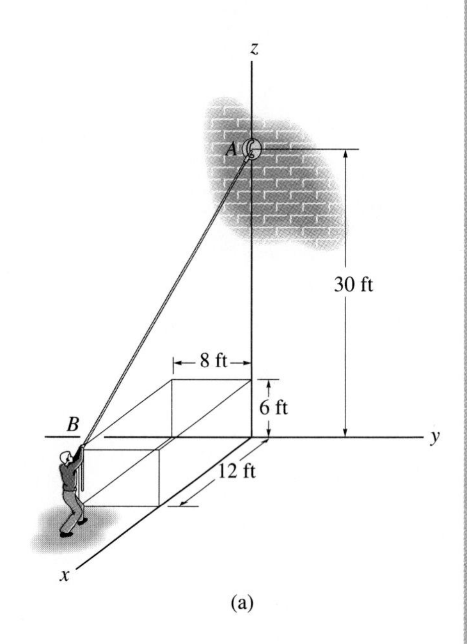

(a)

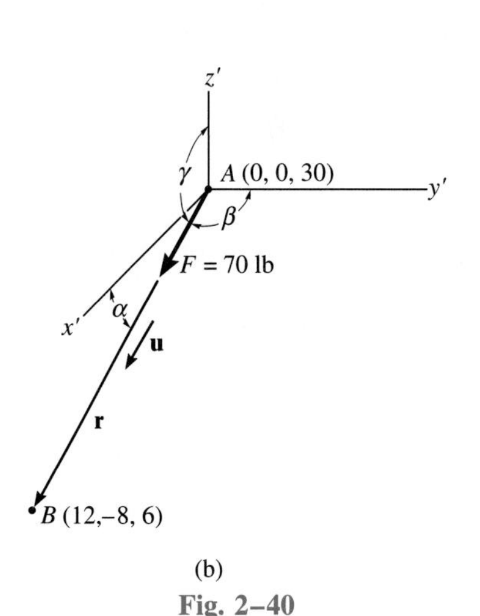

(b)

Fig. 2–40

Example 2–16

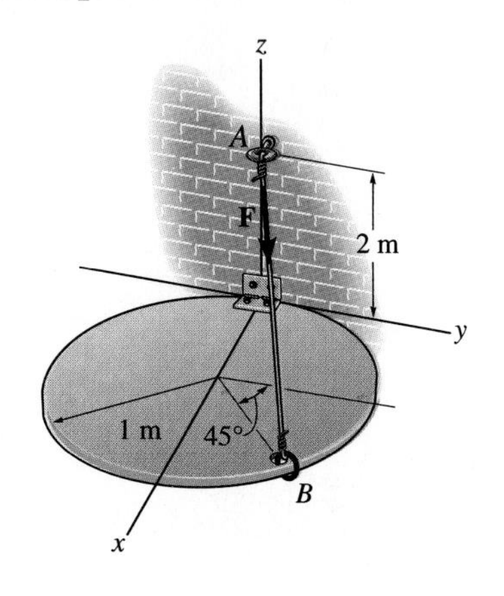

(a)

The circular plate shown in Fig. 2–41*a* is partially supported by the cable *AB*. If the force of the cable on the hook at *A* is $F = 500$ N, express **F** as a Cartesian vector.

SOLUTION

As shown in Fig. 2–41*b*, **F** has the same direction and sense as the position vector **r**, which extends from *A* to *B*.

Position Vector. The coordinates of the end points of the cable are $A(0, 0, 2\text{ m})$ and $B(1.707\text{ m}, 0.707\text{ m}, 0)$, as indicated in the figure. Thus,

$$\mathbf{r} = (1.707\text{ m} - 0)\mathbf{i} + (0.707\text{ m} - 0)\mathbf{j} + (0 - 2\text{ m})\mathbf{k}$$
$$= \{1.707\mathbf{i} + 0.707\mathbf{j} - 2\mathbf{k}\}\text{ m}$$

Note how one can calculate these components *directly* by going from *A*, $\{-2\mathbf{k}\}$ m along the z axis, then $\{1.707\mathbf{i}\}$ m along the x axis, and finally $\{0.707\mathbf{j}\}$ m along the y axis to get to *B*.

The magnitude of **r** is

$$r = \sqrt{(1.707)^2 + (0.707)^2 + (-2)^2} = 2.72\text{ m}$$

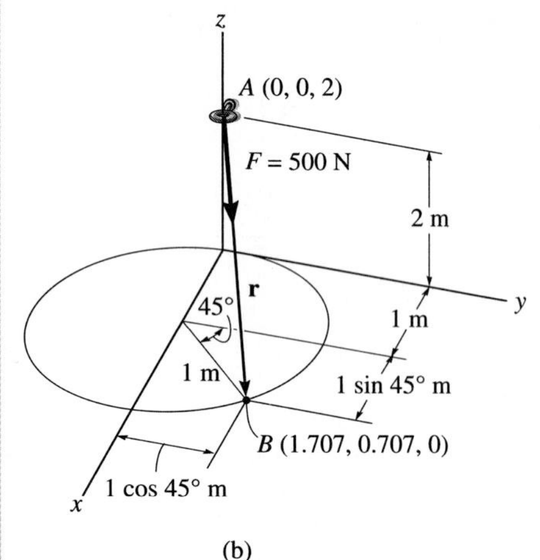

(b)

Fig. 2–41

Unit Vector. Thus,

$$\mathbf{u} = \frac{\mathbf{r}}{r} = \frac{1.707}{2.72}\mathbf{i} + \frac{0.707}{2.72}\mathbf{j} - \frac{2}{2.72}\mathbf{k}$$
$$= 0.627\mathbf{i} + 0.260\mathbf{j} - 0.735\mathbf{k}$$

Force Vector. Since $F = 500$ N and **F** has the direction **u**, we have

$$\mathbf{F} = F\mathbf{u} = 500\text{ N}(0.627\mathbf{i} + 0.260\mathbf{j} - 0.735\mathbf{k})$$
$$= \{314\mathbf{i} + 130\mathbf{j} - 368\mathbf{k}\}\text{ N} \qquad \textit{Ans.}$$

Using these components, notice that indeed the magnitude of **F** is 500 N; i.e.,

$$F = \sqrt{(314)^2 + (130)^2 + (-368)^2} = 500\text{ N}$$

Show that the coordinate direction angle $\gamma = 137°$, and indicate this angle on the figure.

Example 2–17

The cables exert forces $F_{AB} = 100$ N and $F_{AC} = 120$ N on the ring at A as shown in Fig. 2–42*a*. Determine the magnitude of the resultant force acting at A.

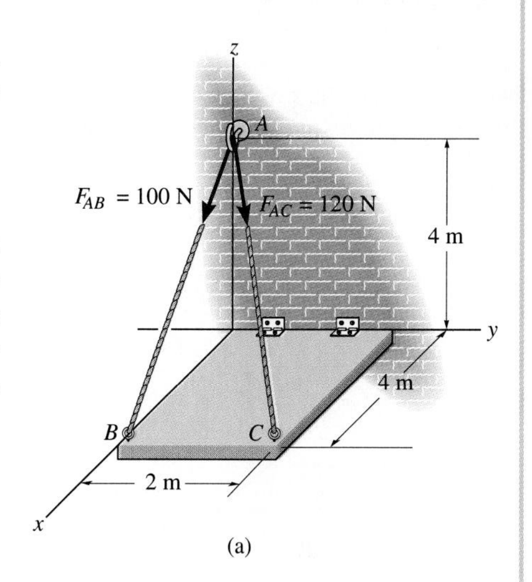

(a)

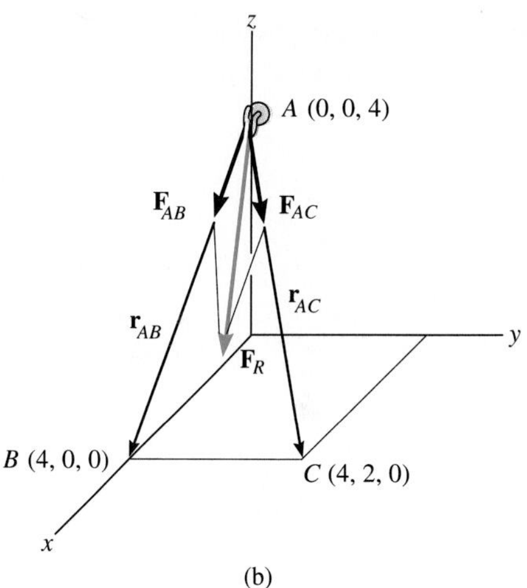

(b)

Fig. 2–42

SOLUTION

The resultant force $\mathbf{F}_R$ is shown graphically in Fig. 2–42*b*. We can express this force as a Cartesian vector by first formulating $\mathbf{F}_{AB}$ and $\mathbf{F}_{AC}$ as Cartesian vectors and then adding their components. The directions of $\mathbf{F}_{AB}$ and $\mathbf{F}_{AC}$ are specified by forming unit vectors $\mathbf{u}_{AB}$ and $\mathbf{u}_{AC}$ along the cables. These unit vectors are obtained from the associated position vectors $\mathbf{r}_{AB}$ and $\mathbf{r}_{AC}$. With reference to Fig. 2–42*b* for $\mathbf{F}_{AB}$ we have

$$\mathbf{r}_{AB} = (4\text{ m} - 0)\mathbf{i} + (0 - 0)\mathbf{j} + (0 - 4\text{ m})\mathbf{k}$$
$$= \{4\mathbf{i} - 4\mathbf{k}\}\text{ m}$$
$$r_{AB} = \sqrt{(4)^2 + (-4)^2} = 5.66\text{ m}$$
$$\mathbf{F}_{AB} = 100\text{ N}\left(\frac{\mathbf{r}_{AB}}{r_{AB}}\right) = 100\text{ N}\left(\frac{4}{5.66}\mathbf{i} - \frac{4}{5.66}\mathbf{k}\right)$$
$$\mathbf{F}_{AB} = \{70.7\mathbf{i} - 70.7\mathbf{k}\}\text{ N}$$

For $\mathbf{F}_{AC}$ we have

$$\mathbf{r}_{AC} = (4\text{ m} - 0)\mathbf{i} + (2\text{ m} - 0)\mathbf{j} + (0 - 4\text{ m})\mathbf{k}$$
$$= \{4\mathbf{i} + 2\mathbf{j} - 4\mathbf{k}\}\text{ m}$$
$$r_{AC} = \sqrt{(4)^2 + (2)^2 + (-4)^2} = 6\text{ m}$$
$$\mathbf{F}_{AC} = 120\text{ N}\left(\frac{\mathbf{r}_{AC}}{r_{AC}}\right) = 120\text{ N}\left(\frac{4}{6}\mathbf{i} + \frac{2}{6}\mathbf{j} - \frac{4}{6}\mathbf{k}\right)$$
$$= \{80\mathbf{i} + 40\mathbf{j} - 80\mathbf{k}\}\text{ N}$$

The resultant force is therefore

$$\mathbf{F}_R = \mathbf{F}_{AB} + \mathbf{F}_{AC} = \{70.7\mathbf{i} - 70.7\mathbf{k}\}\text{ N} + \{80\mathbf{i} + 40\mathbf{j} - 80\mathbf{k}\}\text{ N}$$
$$= \{150.7\mathbf{i} + 40\mathbf{j} - 150.7\mathbf{k}\}\text{ N}$$

The magnitude of $\mathbf{F}_R$ is thus

$$F_R = \sqrt{(150.7)^2 + (40)^2 + (-150.7)^2}$$
$$= 217\text{ N} \qquad \textit{Ans.}$$

PROBLEMS

*2–84. Express the position vector **r** in Cartesian vector form; then determine its magnitude and coordinate direction angles.

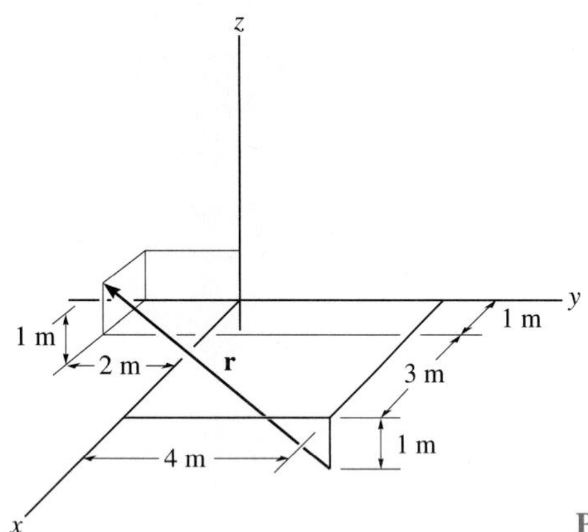

Prob. 2–84

2–85. Express the position vector **r** in Cartesian vector form; then determine its magnitude and coordinate direction angles.

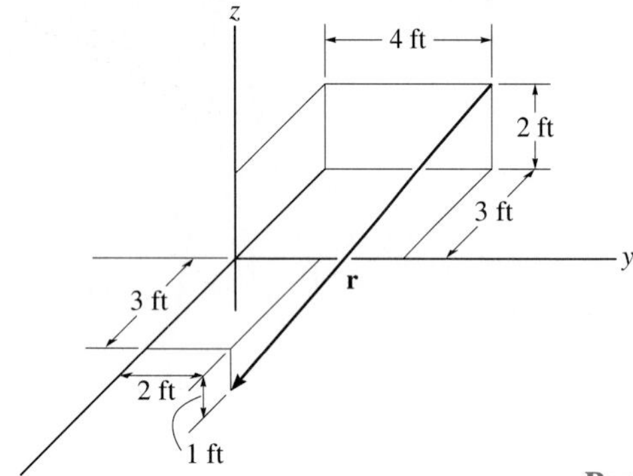

Prob. 2–85

2–86. Express the position vector **r** in Cartesian vector form; then determine its magnitude and coordinate direction angles.

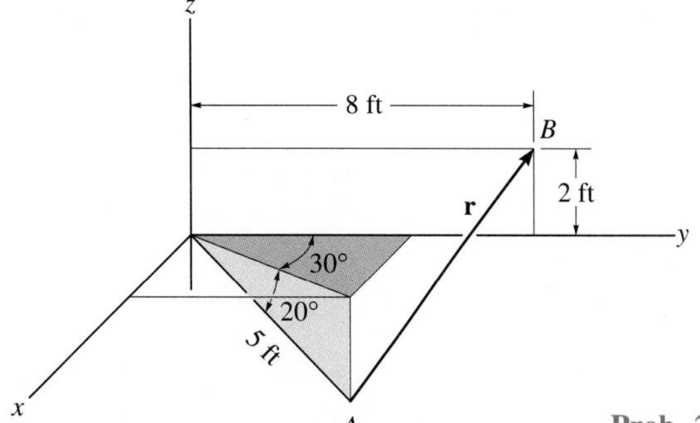

Prob. 2–86

2–87. Determine the length of member *AB* of the truss by first establishing a Cartesian position vector from *A* to *B* and then determining its magnitude.

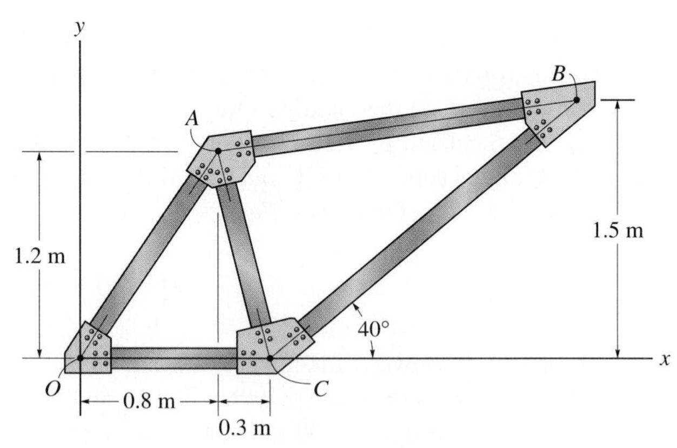

Prob. 2–87

*2–88. The 8-m-long cable is anchored to the ground at *A*. If $x = 4$ m and $y = 2$ m, determine the coordinate z to the highest point of attachment along the column.

2–89. The 8-m-long cable is anchored to the ground at *A*. If $z = 5$ m, determine the location $+x$, $+y$ of point *A*. Choose a value such that $x = y$.

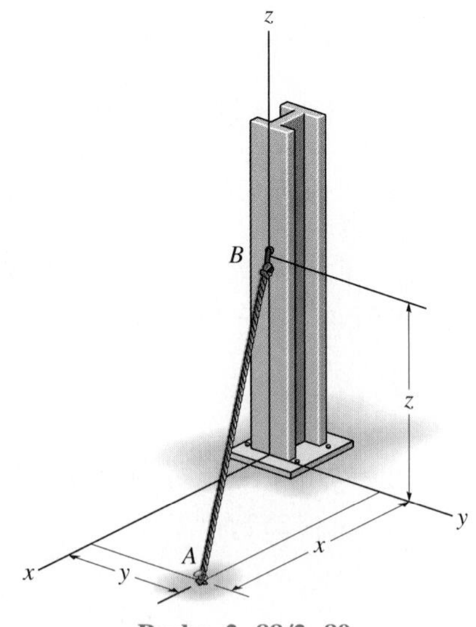

Probs. 2–88/2–89

2–90. Determine the length of the crankshaft AB by first formulating a Cartesian position vector from A to B and then determining its magnitude.

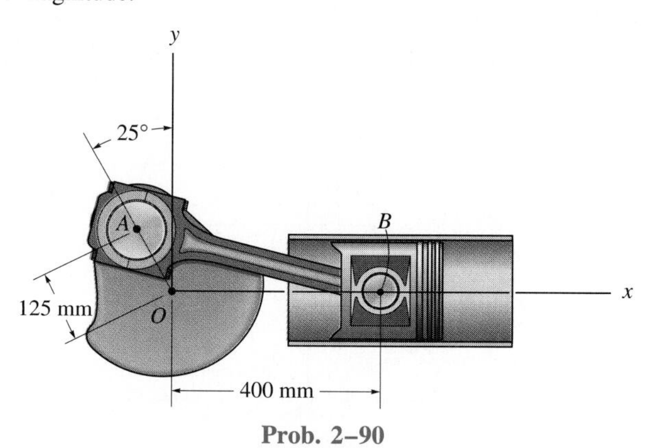

Prob. 2–90

2–91. At the instant shown, position vectors along the robotic arm from O to B and B to A are $\mathbf{r}_{OB} = \{100\mathbf{i} + 300\mathbf{j} + 400\mathbf{k}\}$ mm and $\mathbf{r}_{BA} = \{350\mathbf{i} + 225\mathbf{j} - 640\mathbf{k}\}$ mm, respectively. Determine the distance from O to the grip at A.

***2–92.** If $\mathbf{r}_{OA} = \{0.5\mathbf{i} + 4\mathbf{j} + 0.25\mathbf{k}\}$ m and $\mathbf{r}_{OB} = \{0.3\mathbf{i} + 2\mathbf{j} + 2\mathbf{k}\}$ m, express $\mathbf{r}_{BA}$ as a Cartesian vector.

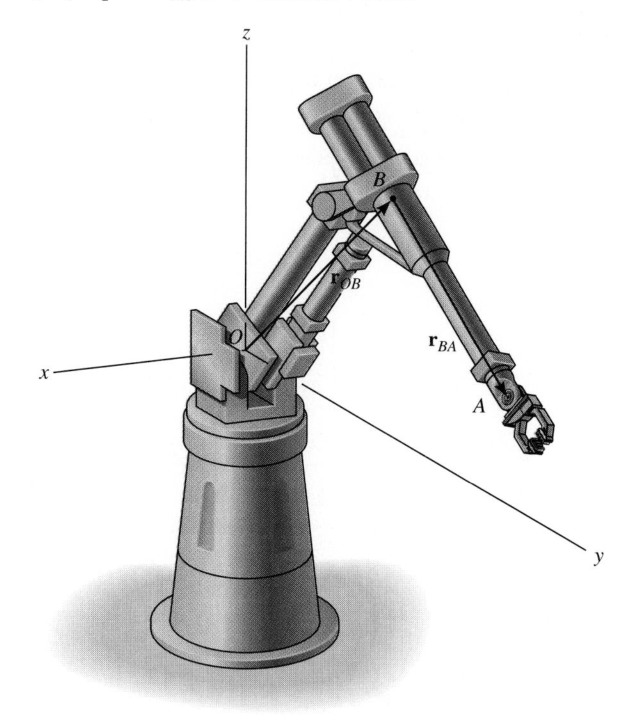

Probs. 2–91/2–92

2–93. At a given instant, the positions of a plane at A and a train at B are measured relative to a radar antenna at O. Determine the distance d between A and B at this instant. To solve the problem, formulate a position vector, directed from A to B, and then determine its magnitude.

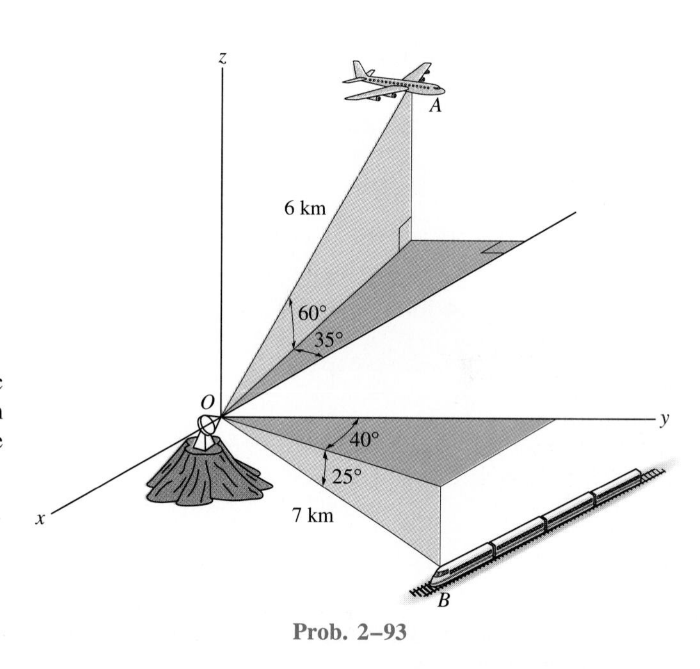

Prob. 2–93

2–94. Determine the lengths of wires AD, BD, and CD. The ring at D is midway between A and B.

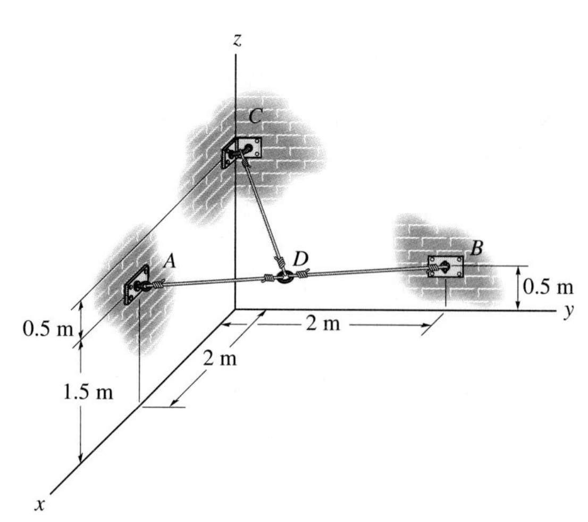

Prob. 2–94

2–95. Express force **F** as a Cartesian vector, then determine its coordinate direction angles.

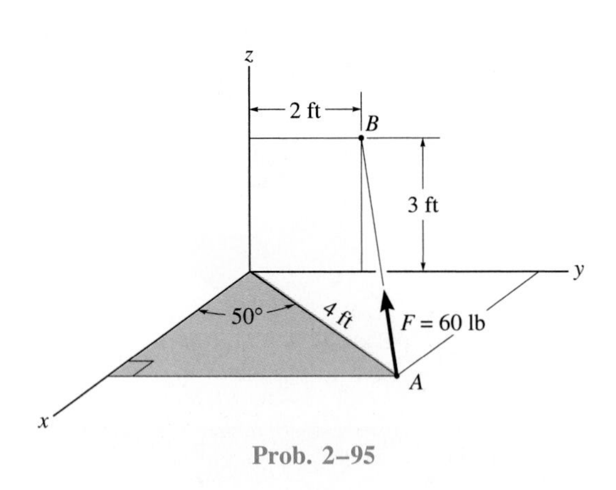

Prob. 2–95

***2–96.** Express force **F** as a Cartesian vector; then determine its coordinate direction angles.

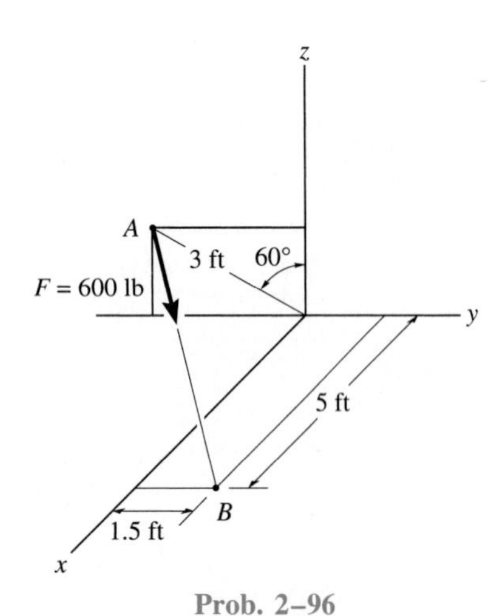

Prob. 2–96

2–97. Express each of the two forces in Cartesian vector form.

2–98. Determine the magnitude and coordinate direction angles of the resultant force acting at point A.

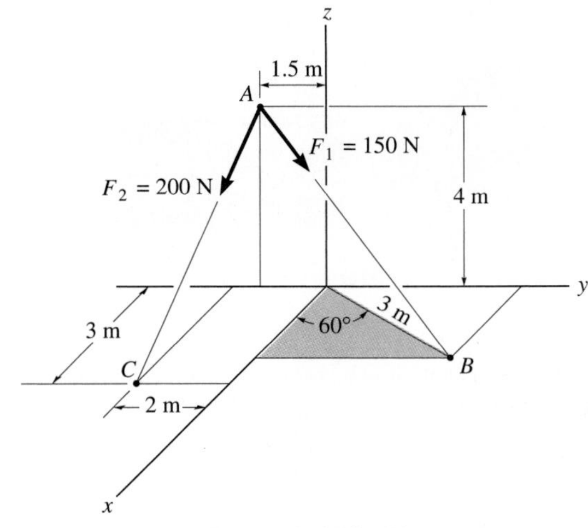

Probs. 2–97/2–98

2–99. Express each of the two forces in Cartesian vector form.

***2–100.** Determine the magnitude and coordinate direction angles of the resultant force acting at point A.

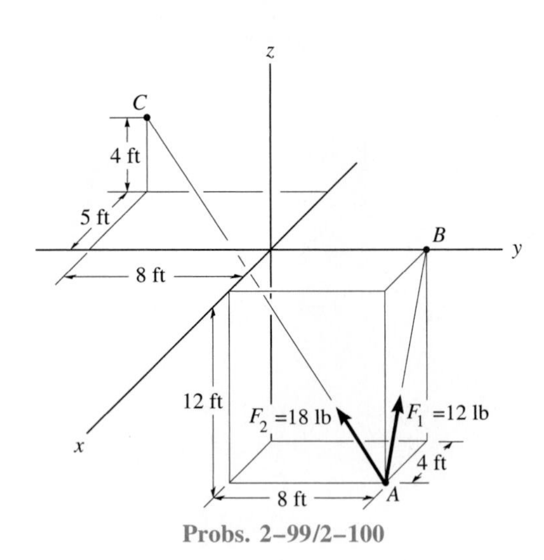

Probs. 2–99/2–100

2–101. The cord exerts a force of $\mathbf{F} = \{12\mathbf{i} + 9\mathbf{j} - 8\mathbf{k}\}$ lb on the hook. If the cord is 8 ft long, determine the location x, y of the point of attachment B, and the height z of the hook.

2–102. The cord exerts a force of $F = 30$ lb on the hook. If the cord is 8 ft long, $z = 4$ ft, and the x component of the force is $F_x = 25$ lb, determine the location x, y of the point of attachment B of the cord to the ground.

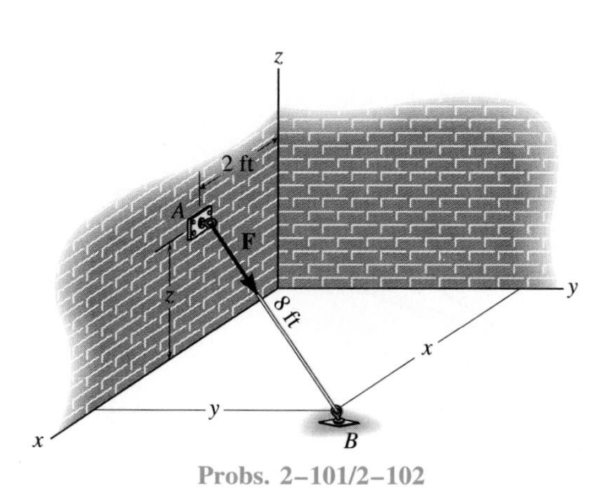

Probs. 2–101/2–102

2–103. The force **F** has a magnitude of 80 lb and acts at the midpoint C of the thin rod. Express the force as a Cartesian vector.

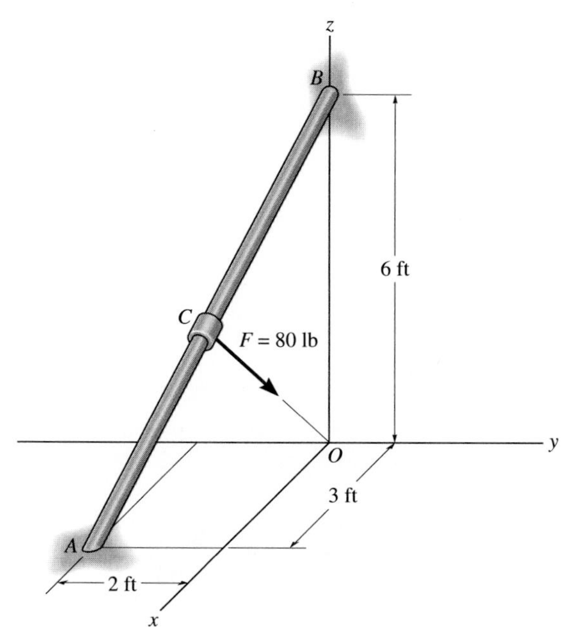

Prob. 2–103

***2–104.** The window is held open by chain AB. Determine the length of the chain, and express the 50-lb force acting at A along the chain as a Cartesian vector. Determine its coordinate direction angles.

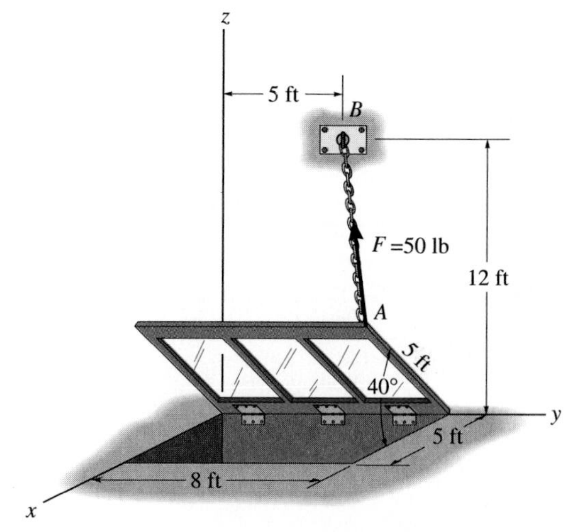

Prob. 2–104

2–105. The cable attached to the tractor at B exerts a force of 350 lb on the framework. Express this force as a Cartesian vector.

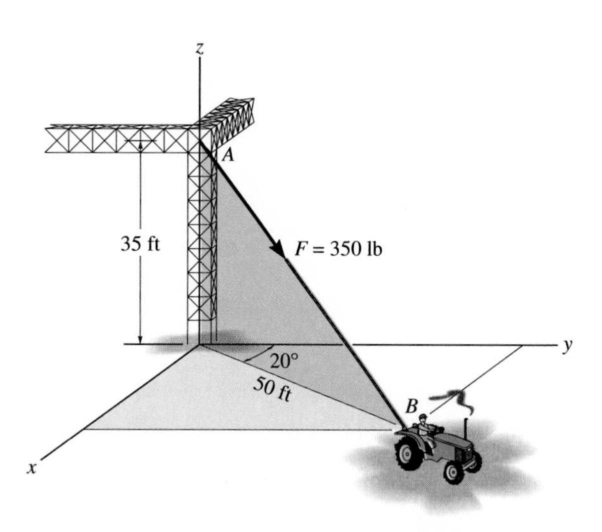

Prob. 2–105

2–106. Express force **F** as a Cartesian vector; then determine its coordinate direction angles.

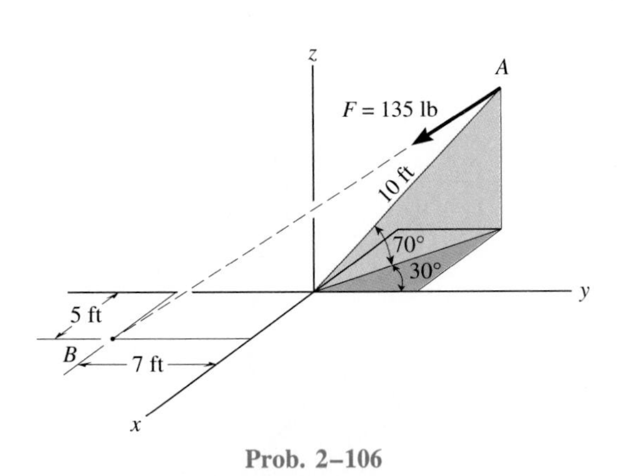

Prob. 2–106

2–107. Express each of the forces in Cartesian vector form and determine the magnitude and coordinate direction angles of the resultant force.

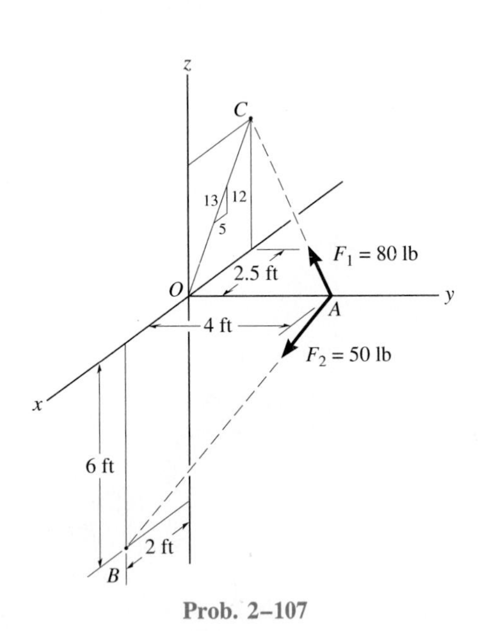

Prob. 2–107

***2–108.** The three supporting cables exert the forces shown on the sign. Represent each force as a Cartesian vector.

2–109. Determine the magnitude and coordinate direction angles of the resultant force of the two forces acting on the sign at point A.

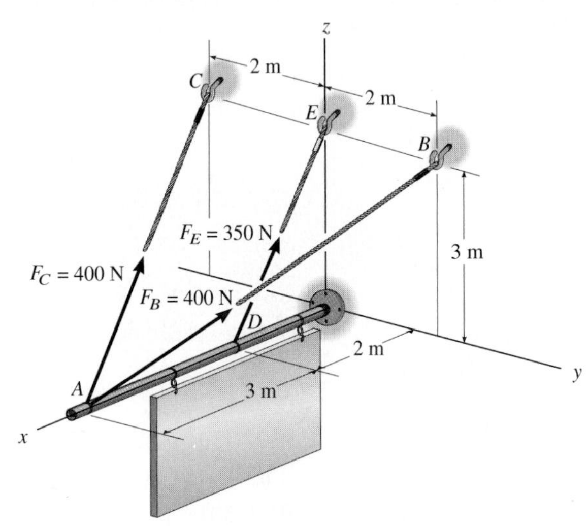

Probs. 2–108/2–109

2–110. The window is held open by chain AB. Determine the length of the chain and express the 30-N force acting at A along the chain as a Cartesian vector.

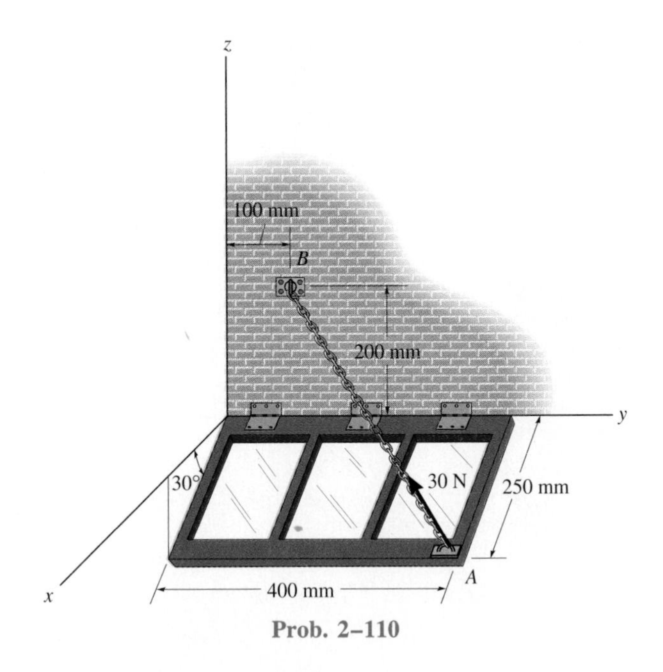

Prob. 2–110

2–111. Each of the four forces acting at E has a magnitude of 28 kN. Express each force as a Cartesian vector and determine the resultant force.

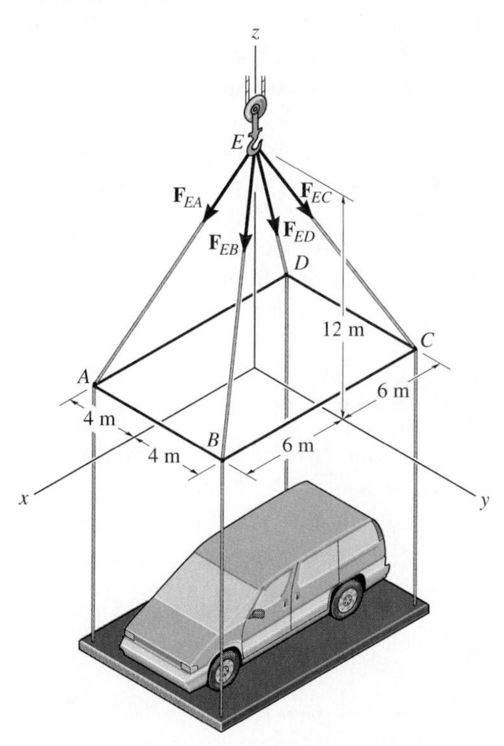

Prob. 2–111

***2–112.** Express force $\mathbf{F}$ in Cartesian vector form if its acts at the midpoint B of the rod.

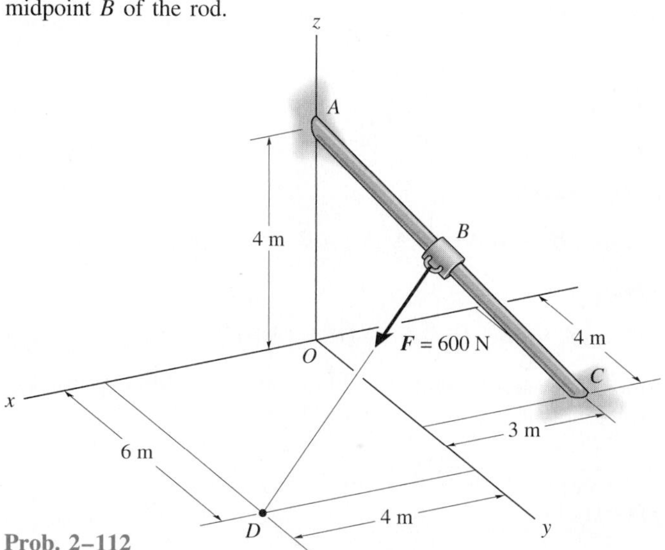

Prob. 2–112

2–113. Express force $\mathbf{F}$ in Cartesian vector form if point B is located 3 m along the rod from end C.

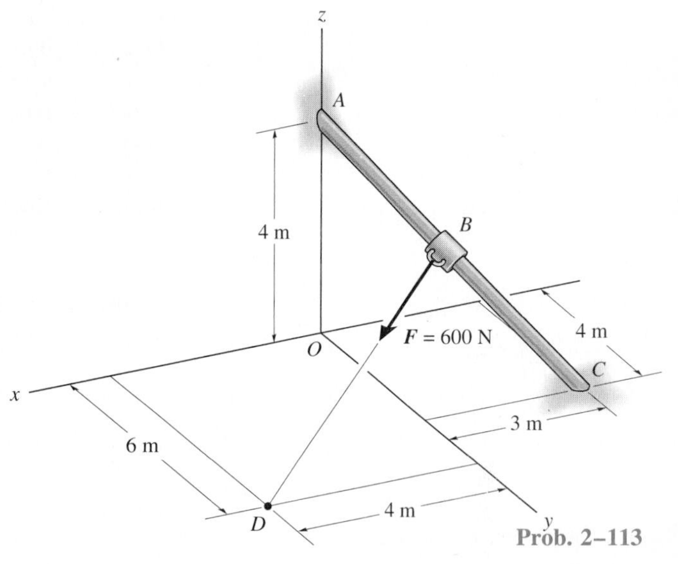

Prob. 2–113

2–114. The tower is held in place by three cables. If the force of each cable acting on the tower is shown, determine the position (x, y) for fixing cable DA so that the resultant force exerted on the tower is directed along its axis, from D toward O.

2–115. The tower is held in place by three cables. If the force of each cable acting on the tower is shown, determine the magnitude and coordinate direction angles α, β, γ of the resultant force. Take $x = 20$ m, $y = 15$ m.

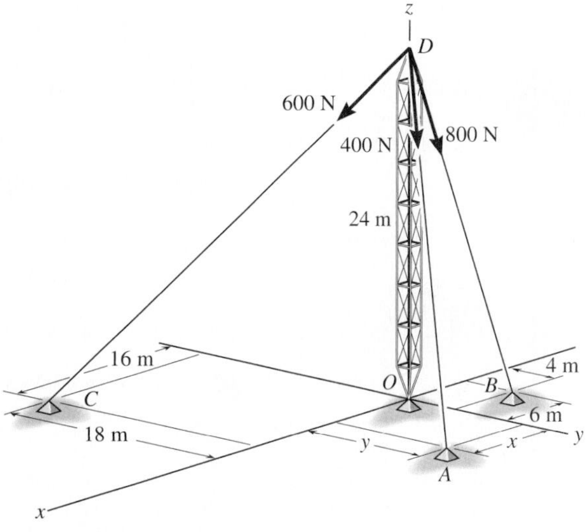

Probs. 2–114/2–115

2.9 Dot Product

Occasionally in statics one has to find the angle between two lines or the components of a force parallel and perpendicular to a line. In two dimensions, these problems can readily be solved by trigonometry since the geometry is easy to visualize. In three dimensions, however, this is often difficult, and consequently vector methods should be employed for the solution. The dot product defines a particular method for ''multiplying'' two vectors and is used to solve the above-mentioned problems.

The *dot product* of vectors **A** and **B**, written $\mathbf{A} \cdot \mathbf{B}$, and read **"A** dot **B,"** is defined as the product of the magnitudes of **A** and **B** and the cosine of the angle θ between their tails, Fig. 2–43. Expressed in equation form,

$$\mathbf{A} \cdot \mathbf{B} = AB \cos \theta \qquad (2\text{–}14)$$

where $0° \leq \theta \leq 180°$. The dot product is often referred to as the *scalar product* of vectors, since the result is a *scalar* and not a vector.

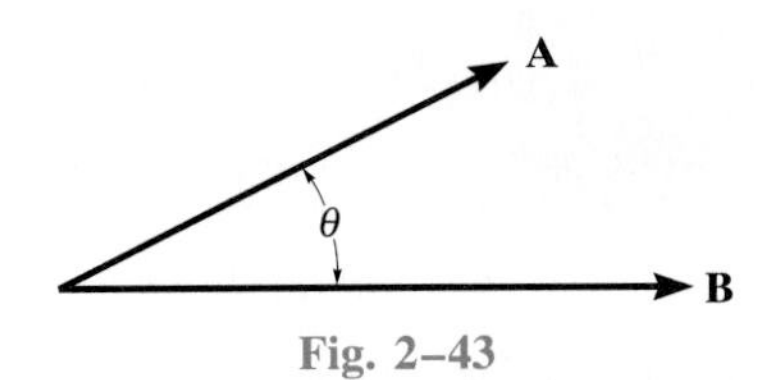

Fig. 2–43

Laws of Operation

1. Commutative law:

$$\mathbf{A} \cdot \mathbf{B} = \mathbf{B} \cdot \mathbf{A}$$

2. Multiplication by a scalar:

$$a(\mathbf{A} \cdot \mathbf{B}) = (a\mathbf{A}) \cdot \mathbf{B} = \mathbf{A} \cdot (a\mathbf{B}) = (\mathbf{A} \cdot \mathbf{B})a$$

3. Distributive law:

$$\mathbf{A} \cdot (\mathbf{B} + \mathbf{D}) = (\mathbf{A} \cdot \mathbf{B}) + (\mathbf{A} \cdot \mathbf{D})$$

It is easy to prove the first and second laws by using Eq. 2–14. The proof of the distributive law is left as an exercise (see Prob. 2–116).

Cartesian Vector Formulation. Equation 2–14 may be used to find the dot product for each of the Cartesian unit vectors. For example, $\mathbf{i} \cdot \mathbf{i} = (1)(1) \cos 0° = 1$ and $\mathbf{i} \cdot \mathbf{j} = (1)(1) \cos 90° = 0$. In a similar manner,

$$\begin{array}{lll} \mathbf{i} \cdot \mathbf{i} = 1 & \mathbf{j} \cdot \mathbf{j} = 1 & \mathbf{k} \cdot \mathbf{k} = 1 \\ \mathbf{i} \cdot \mathbf{j} = 0 & \mathbf{i} \cdot \mathbf{k} = 0 & \mathbf{k} \cdot \mathbf{j} = 0 \end{array}$$

These results should not be memorized; rather, it should be clearly understood how each is obtained.

Consider now the dot product of two general vectors **A** and **B** which are expressed in Cartesian vector form. We have

$$\begin{aligned} \mathbf{A} \cdot \mathbf{B} &= (A_x\mathbf{i} + A_y\mathbf{j} + A_z\mathbf{k}) \cdot (B_x\mathbf{i} + B_y\mathbf{j} + B_z\mathbf{k}) \\ &= A_xB_x(\mathbf{i} \cdot \mathbf{i}) + A_xB_y(\mathbf{i} \cdot \mathbf{j}) + A_xB_z(\mathbf{i} \cdot \mathbf{k}) \\ &\quad + A_yB_x(\mathbf{j} \cdot \mathbf{i}) + A_yB_y(\mathbf{j} \cdot \mathbf{j}) + A_yB_z(\mathbf{j} \cdot \mathbf{k}) \\ &\quad + A_zB_x(\mathbf{k} \cdot \mathbf{i}) + A_zB_y(\mathbf{k} \cdot \mathbf{j}) + A_zB_z(\mathbf{k} \cdot \mathbf{k}) \end{aligned}$$

Carrying out the dot-product operations, the final result becomes

$$\mathbf{A} \cdot \mathbf{B} = A_xB_x + A_yB_y + A_zB_z \qquad (2\text{–}15)$$

Thus, to determine the dot product of two Cartesian vectors, multiply their corresponding x, y, z components and sum their products algebraically. Since the result is a scalar, be careful *not* to include any unit vectors in the final result.

Applications. The dot product has two important applications in mechanics.

1. *The angle formed between two vectors or intersecting lines.* The angle θ between the tails of vectors **A** and **B** in Fig. 2–43 can be determined from Eq. 2–14 and written as

$$\theta = \cos^{-1}\left(\frac{\mathbf{A} \cdot \mathbf{B}}{AB}\right) \qquad 0° \le \theta \le 180°$$

Here $\mathbf{A} \cdot \mathbf{B}$ is computed from Eq. 2–15. In particular, notice that if $\mathbf{A} \cdot \mathbf{B} = 0$, $\theta = \cos^{-1} 0 = 90°$, so that **A** will be *perpendicular* to **B**.

2. *The components of a vector parallel and perpendicular to a line.* The component of vector **A** parallel to or collinear with the line aa' in Fig. 2–44 is defined by $\mathbf{A}_{\|}$, where $A_{\|} = A \cos \theta$. This component is sometimes referred to as the *projection* of **A** onto the line, since a right angle is formed in the construction. If the *direction* of the line is specified by the unit vector **u,** then, since $u = 1$, we can determine $A_{\|}$ directly from the dot product (Eq. 2–14); i.e.,

$$A_{\|} = A \cos \theta = \mathbf{A} \cdot \mathbf{u}$$

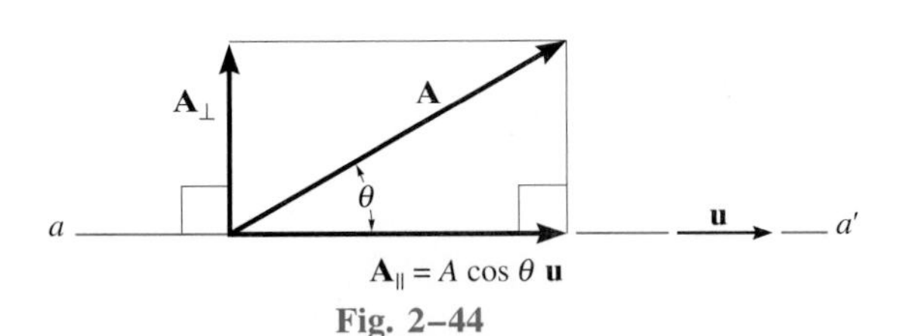

Fig. 2–44

Hence, the scalar projection of **A** *along a line is determined from the dot product of* **A** *and the unit vector* **u** *which defines the direction of the line.* Notice that if this result is positive, then $\mathbf{A}_{\|}$ has a directional sense which is the same as **u,** whereas if $A_{\|}$ is a negative scalar, then $\mathbf{A}_{\|}$ has the opposite sense of direction to **u.** The component $\mathbf{A}_{\|}$ represented as a *vector* is therefore

$$\mathbf{A}_{\|} = A \cos \theta \mathbf{u} = (\mathbf{A} \cdot \mathbf{u})\mathbf{u}$$

Note that the component of **A** which is *perpendicular* to line aa' can also be obtained, Fig. 2–44. Since $\mathbf{A} = \mathbf{A}_{\|} + \mathbf{A}_{\perp}$, then $\mathbf{A}_{\perp} = \mathbf{A} - \mathbf{A}_{\|}$. There are two possible ways of obtaining $A_{\perp}$. The first would be to determine θ from the dot product, $\theta = \cos^{-1} (\mathbf{A} \cdot \mathbf{u}/A)$, then $A_{\perp} = A \sin \theta$. Alternatively, if $A_{\|}$ is known, then by the Pythagorean theorem we can also write $A_{\perp} = \sqrt{A^2 - A_{\|}^2}$.

The above two applications are illustrated numerically in the following example problems.

Example 2–18

The frame shown in Fig. 2–45*a* is subjected to a horizontal force $\mathbf{F} = \{300\mathbf{j}\}$ N acting at its corner. Determine the magnitude of the components of this force parallel and perpendicular to member *AB*.

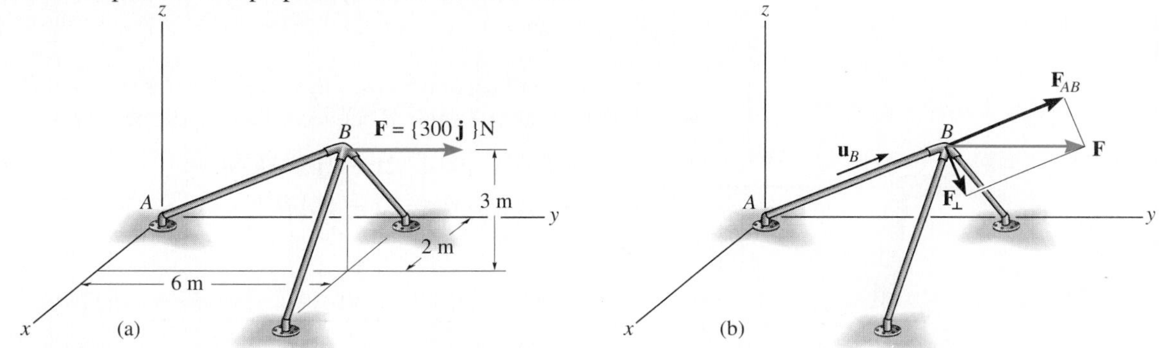

Fig. 2–45

SOLUTION

The magnitude of the component of **F** along *AB* is equal to the dot product of **F** and the unit vector $\mathbf{u}_B$ which defines the direction of *AB*, Fig. 2–45*b*. Since

$$\mathbf{u}_B = \frac{\mathbf{r}_B}{r_B} = \frac{2\mathbf{i} + 6\mathbf{j} + 3\mathbf{k}}{\sqrt{(2)^2 + (6)^2 + (3)^2}} = 0.286\mathbf{i} + 0.857\mathbf{j} + 0.429\mathbf{k}$$

Then

$$\begin{aligned} F_{AB} &= F\cos\theta = \mathbf{F}\cdot\mathbf{u}_B = (300\mathbf{j})\cdot(0.286\mathbf{i} + 0.857\mathbf{j} + 0.429\mathbf{k}) \\ &= (0)(0.286) + (300)(0.857) + (0)(0.429) \\ &= 257.1 \text{ N} \end{aligned} \qquad \textit{Ans.}$$

Since the result is a positive scalar, $\mathbf{F}_{AB}$ has the same sense of direction as $\mathbf{u}_B$, Fig. 2–45*b*.

Expressing $\mathbf{F}_{AB}$ in Cartesian vector form, we have

$$\begin{aligned} \mathbf{F}_{AB} &= F_{AB}\mathbf{u}_B = 257.1 \text{ N}(0.286\mathbf{i} + 0.857\mathbf{j} + 0.429\mathbf{k}) \\ &= \{73.5\mathbf{i} + 220\mathbf{j} + 110\mathbf{k}\} \text{ N} \end{aligned} \qquad \textit{Ans.}$$

The perpendicular component, Fig. 2–45*b*, is therefore

$$\begin{aligned} \mathbf{F}_\perp &= \mathbf{F} - \mathbf{F}_{AB} = 300\mathbf{j} - (73.5\mathbf{i} + 220\mathbf{j} + 110\mathbf{k}) \\ &= \{-73.5\mathbf{i} + 80\mathbf{j} - 110\mathbf{k}\} \text{ N} \end{aligned}$$

Its magnitude can be determined either from this vector or from the Pythagorean theorem, Fig. 2–45*b*:

$$\begin{aligned} F_\perp &= \sqrt{F^2 - F_{AB}^2} \\ &= \sqrt{(300)^2 - (257.1)^2} \\ &= 155 \text{ N} \end{aligned} \qquad \textit{Ans.}$$

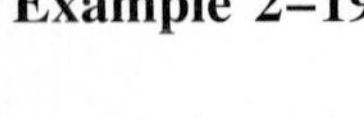

Example 2–19

The pipe in Fig. 2–46*a* is subjected to the force $F = 80$ lb at its end B. Determine the angle θ between $\mathbf{F}$ and the pipe segment BA, and the magnitudes of the components of $\mathbf{F}$, which are parallel and perpendicular to BA.

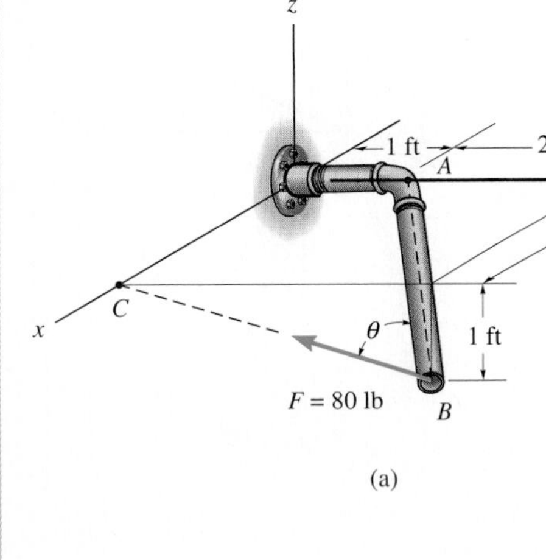

(a)

SOLUTION

Angle θ. First we will establish position vectors from B to A and B to C, then we will determine the angle θ between the tails of these two vectors.

$$\mathbf{r}_{BA} = \{-2\mathbf{i} - 2\mathbf{j} + 1\mathbf{k}\} \text{ ft}$$
$$\mathbf{r}_{BC} = \{-3\mathbf{j} + 1\mathbf{k}\} \text{ ft}$$

Thus,

$$\cos\theta = \frac{\mathbf{r}_{BA} \cdot \mathbf{r}_{BC}}{r_{BA}\, r_{BC}} = \frac{(-2)(0) + (-2)(-3) + (1)(1)}{3\sqrt{10}}$$
$$= 0.7379$$
$$\theta = 42.5^\circ \qquad \textit{Ans.}$$

***Components of* F.** The force $\mathbf{F}$ is resolved into components as shown in Fig. 2–46*b*. Since $F_{BA} = \mathbf{F} \cdot \mathbf{u}_{BA}$, we must first formulate the unit vector along BA and force $\mathbf{F}$ as Cartesian vectors.

$$\mathbf{u}_{BA} = \frac{\mathbf{r}_{BA}}{r_{BA}} = \frac{(-2\mathbf{i} - 2\mathbf{j} + 1\mathbf{k})}{3} = -\frac{2}{3}\mathbf{i} - \frac{2}{3}\mathbf{j} + \frac{1}{3}\mathbf{k}$$

$$\mathbf{F} = 80 \text{ lb}\left(\frac{\mathbf{r}_{BC}}{r_{BC}}\right) = 80\left(\frac{-3\mathbf{j} + 1\mathbf{k}}{\sqrt{10}}\right) = -75.89\mathbf{j} + 25.30\mathbf{k}$$

Thus,

$$F_{BA} = \mathbf{F} \cdot \mathbf{u}_{BA} = (-75.89\mathbf{j} + 25.30\mathbf{k}) \cdot \left(-\frac{2}{3}\mathbf{i} - \frac{2}{3}\mathbf{j} + \frac{1}{3}\mathbf{k}\right)$$
$$= 0 + 50.60 + 8.43$$
$$= 59.0 \text{ lb} \qquad \textit{Ans.}$$

Since θ was calculated in Fig. 2–46*b*, this same result can also be obtained directly from trigonometry.

$$F_{BA} = 80 \cos 42.5^\circ \text{ lb} = 59.0 \text{ lb} \qquad \textit{Ans.}$$

The perpendicular component can be obtained by trigonometry,

$$F_{\perp} = F \sin\theta$$
$$= 80 \sin 42.5^\circ \text{ lb}$$
$$= 54.0 \text{ lb} \qquad \textit{Ans.}$$

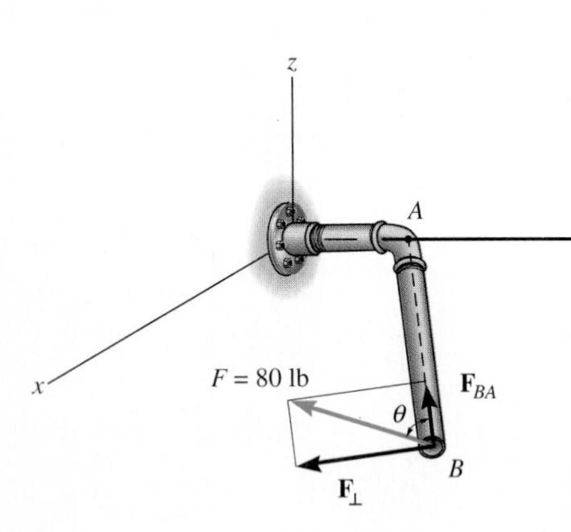

(b)

Fig. 2–46

Or, by the Pythagorean theorem,

$$F_{\perp} = \sqrt{F^2 - F_{BA}^2} = \sqrt{(80)^2 - (59.0)^2}$$
$$= 54.0 \text{ lb} \qquad \textit{Ans.}$$

PROBLEMS

*2–116. Given the three vectors **A**, **B**, and **D**, show that $\mathbf{A} \cdot (\mathbf{B} + \mathbf{D}) = (\mathbf{A} \cdot \mathbf{B}) + (\mathbf{A} \cdot \mathbf{D})$.

2–117. Determine the angle θ between the tails of the two vectors.

2–118. Determine the magnitude of the projection of $\mathbf{r}_1$ along $\mathbf{r}_2$, and the projected component of $\mathbf{r}_2$ along $\mathbf{r}_1$.

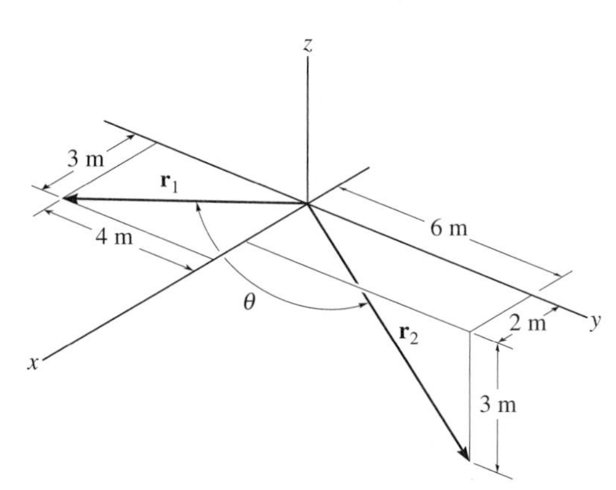

Probs. 2–117/2–118

2–119. Determine the angle θ between the tails of the two vectors.

*2–120. Determine the magnitude of the projected component of $\mathbf{r}_1$ along $\mathbf{r}_2$, and the projection of $\mathbf{r}_2$ along $\mathbf{r}_1$.

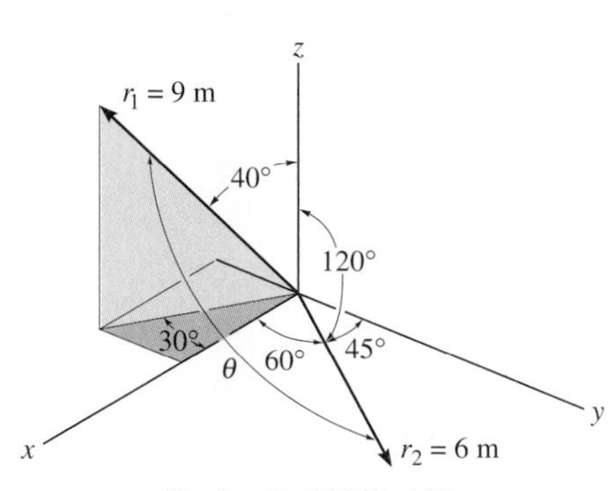

Probs. 2–119/2–120

2–121. Determine the two components of the force **F** along the lines Oa and Ob such that $\mathbf{F} = \mathbf{F}_A + \mathbf{F}_B$. Also find the projected component of **F** along Oa and Ob. Show graphically how the components and projections are constructed.

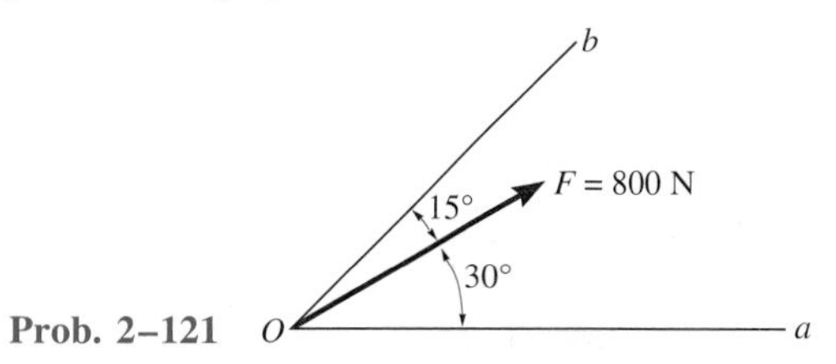

Prob. 2–121

2–122. Determine the angle θ between the edges of the sheet-metal bracket.

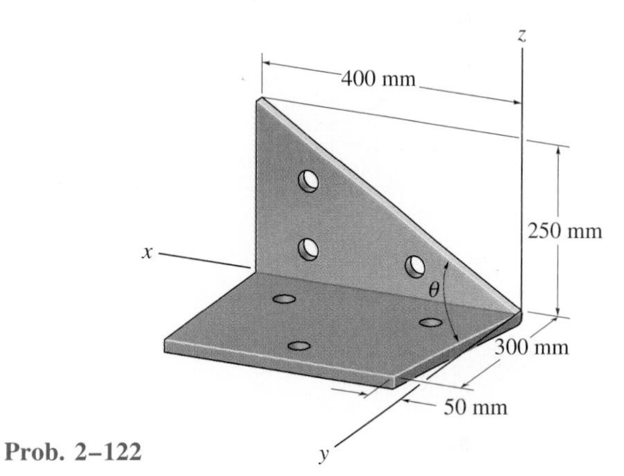

Prob. 2–122

2–123. Determine the magnitude of the projected component of the position vector **r** along the Oa axis.

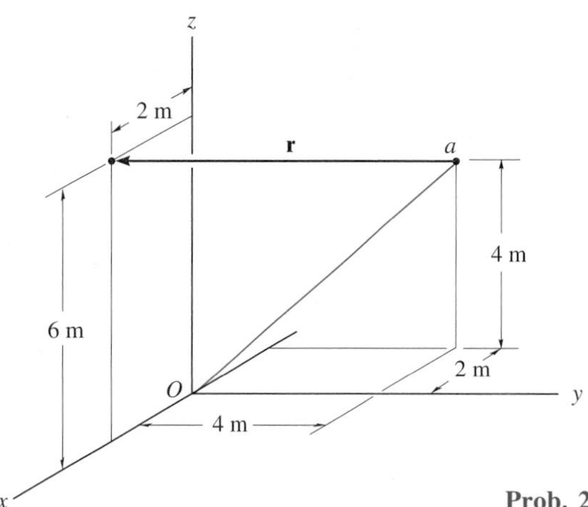

Prob. 2–123

***2–124.** Determine the projected component of the 80-N force acting along the axis *AB* of the pipe.

2–125. Determine the angle θ between pipe segments *BA* and *BC*.

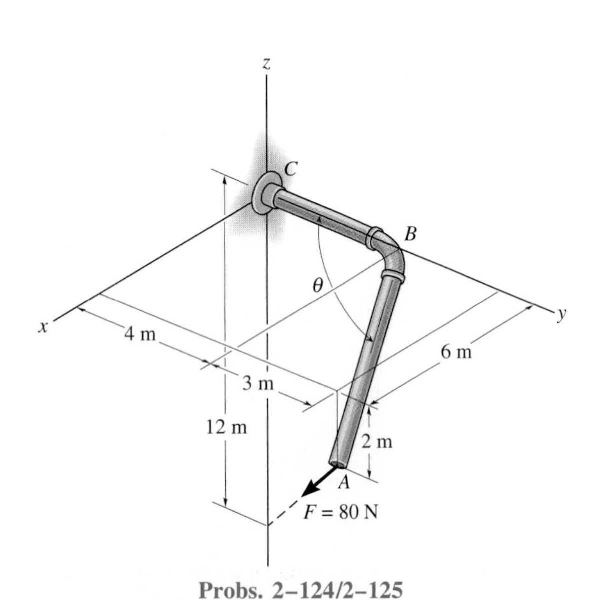

Probs. 2–124/2–125

2–126. The force **F** acts at the end *A* of the pipe assembly. Determine the magnitudes of the components $\mathbf{F}_1$ and $\mathbf{F}_2$ which act along the axis of *AB* and perpendicular to it.

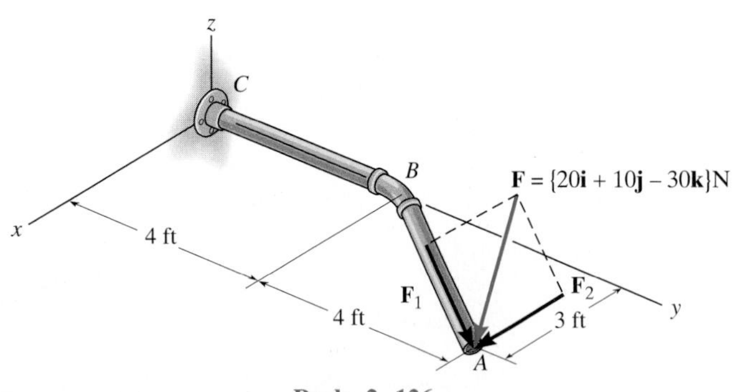

Prob. 2–126

2–127. The clamp is used on a jig. If the vertical force acting on the bolt is $\mathbf{F} = \{-500\mathbf{k}\}$ N, determine the magnitudes of the components $\mathbf{F}_1$ and $\mathbf{F}_2$ which act along the *OA* axis and perpendicular to it.

***2–128.** The clamp is used on a jig. Determine the angle θ between the line of action of **F** and the clamp axis *OA*.

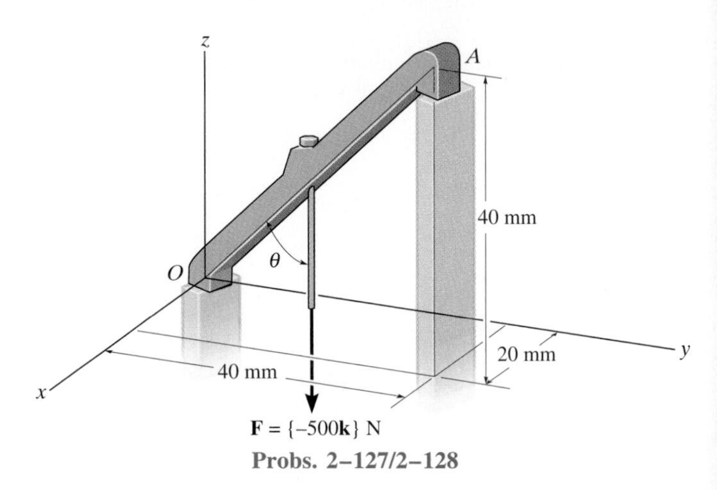

Probs. 2–127/2–128

2–129. The cables each exert a force of 400 N on the post. Determine the magnitude of the projected component of $\mathbf{F}_1$ along the line of action of $\mathbf{F}_2$.

2–130. Determine the angle θ between the two cables.

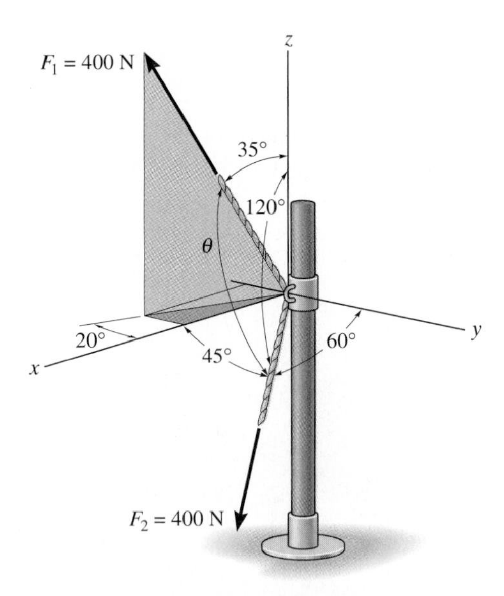

Probs. 2–129/2–130

2–131. Determine the components of **F** that act along rod AC and perpendicular to it. Point B is located at the midpoint of the rod.

***2–132.** Determine the components of **F** that act along rod AC and perpendicular to it. Point B is located 3 m along the rod from end C.

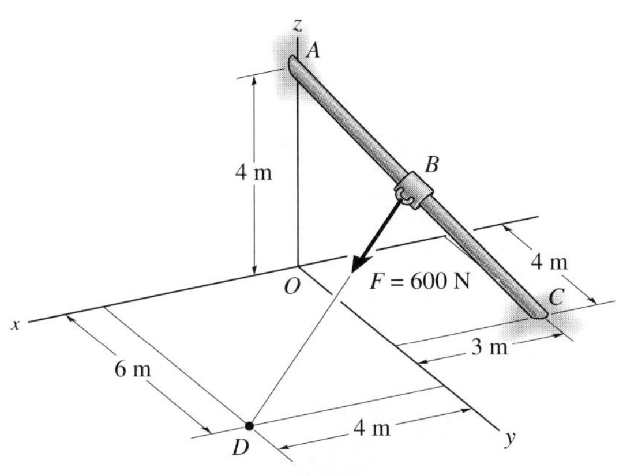

Probs. 2–131/2–132

2–133. Determine the angles θ and ϕ made between the axes OA of the flag pole and AB and AC, respectively, of each cable.

2–134. The two supporting cables exert the forces shown on the flag pole. Determine the projected component of each force acting along the axis OA of the pole.

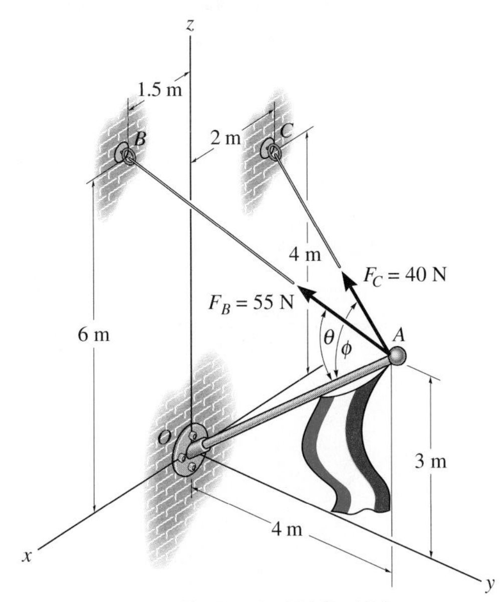

Probs. 2–133/2–134

2–135. Determine the angle θ cable OA makes with beam OC.

***2–136.** Determine the angle ϕ cable OA makes with beam OD.

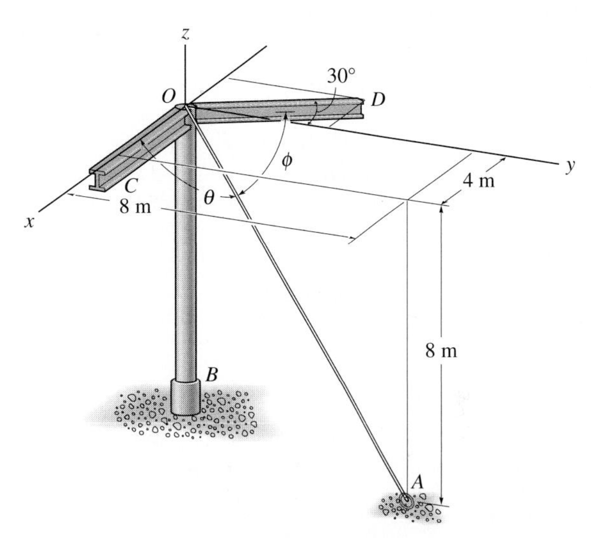

Probs. 2–135/2–136

2–137. Determine the magnitude of the projected component of the 100-lb force acting along the axis BC of the pipe.

2–138. Determine the angle θ between pipe segments BA and BC.

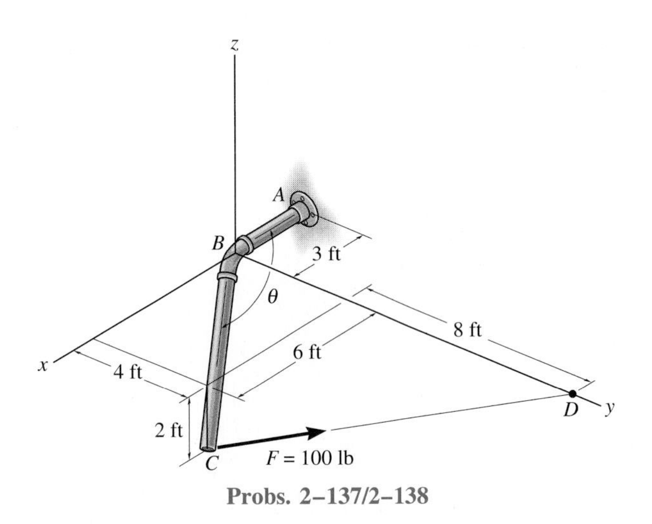

Probs. 2–137/2–138

REVIEW PROBLEMS

2–139. The boat is to be pulled onto the shore using two ropes. Determine the magnitudes of forces **T** and **P** acting in each rope in order to develop a resultant force of 80 lb, directed along the *aa* keel as shown. Take $\theta = 40°$.

***2–140.** The boat is to be pulled onto the shore using two ropes. If the resultant force is to be 80 lb, directed along the keel *aa*, as shown, determine the magnitudes of forces **T** and **P** acting in each rope and the angle θ of **P** so that the magnitude of **P** is a *minimum*. **T** acts at 30° from the keel as shown.

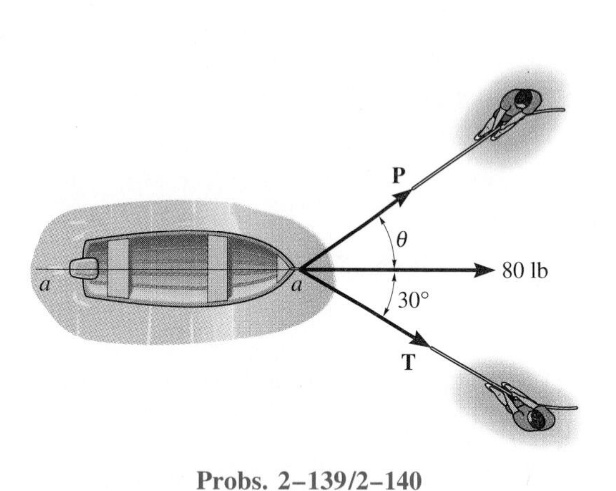

Probs. 2–139/2–140

2–141. Determine the components of the 250-N force acting along the *u* and *v* axes.

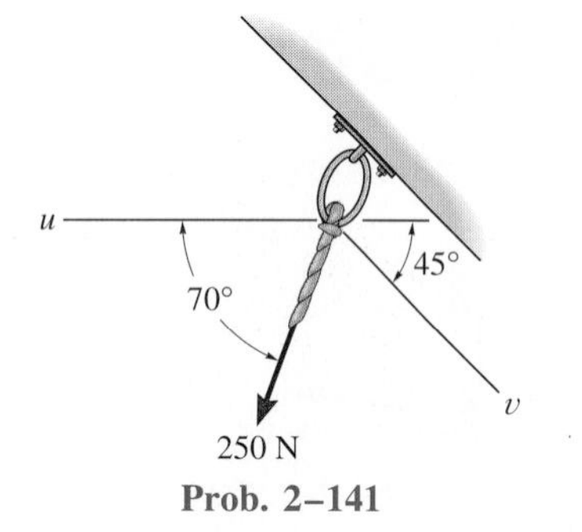

Prob. 2–141

2–142. Determine the magnitude and coordinate direction angles of $\mathbf{F}_3$ so that the resultant of the three forces acts along the positive *y* axis and has a magnitude of 600 lb.

2–143. Determine the magnitude and coordinate direction angles of $\mathbf{F}_3$ so that the resultant of the three forces is zero.

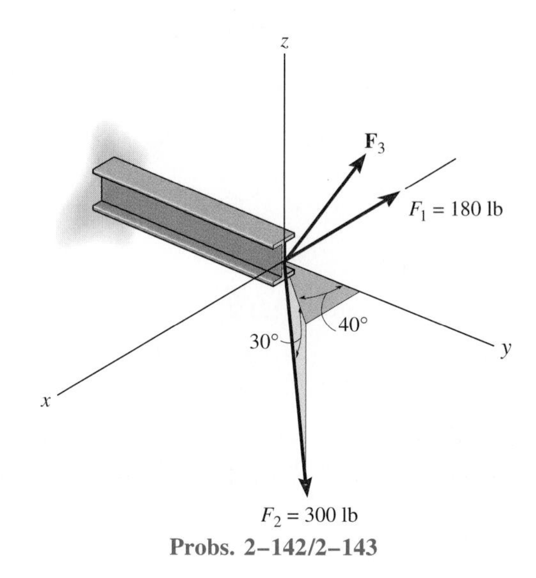

Probs. 2–142/2–143

***2–144.** Two forces $\mathbf{F}_1$ and $\mathbf{F}_2$ act on the hook. If their lines of action are at an angle θ apart and the magnitude of each force is $F_1 = F_2 = F$, determine the magnitude of the resultant force $\mathbf{F}_R$ and the angle between $\mathbf{F}_R$ and $\mathbf{F}_1$.

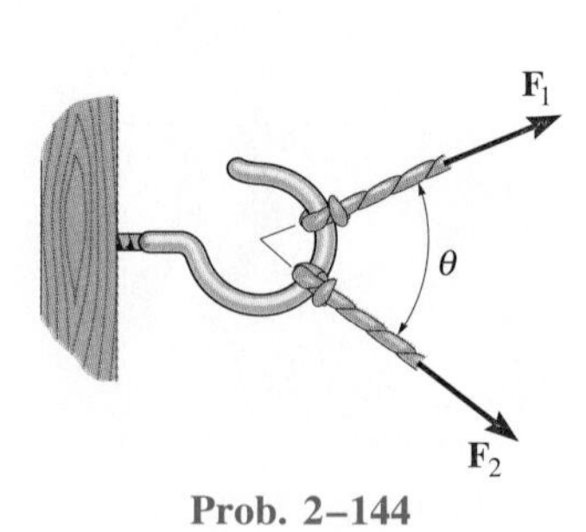

Prob. 2–144

2–145. Express $\mathbf{F}_1$ and $\mathbf{F}_2$ as Cartesian vectors.

2–146. Determine the magnitude of the resultant force and its direction, measured counterclockwise from the positive x axis.

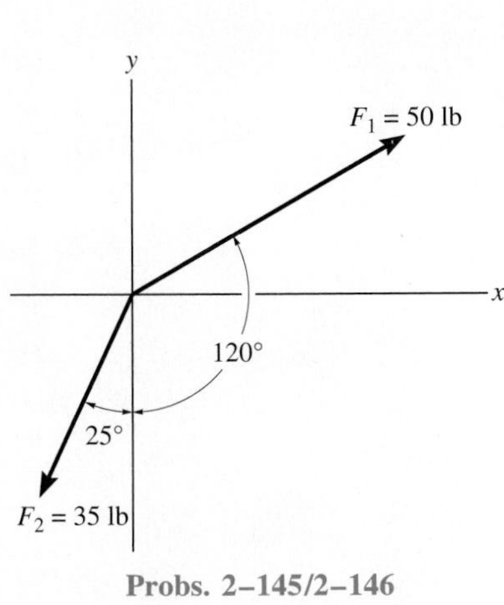

Probs. 2–145/2–146

2–147. Determine the angles θ and ϕ between the wire segments.

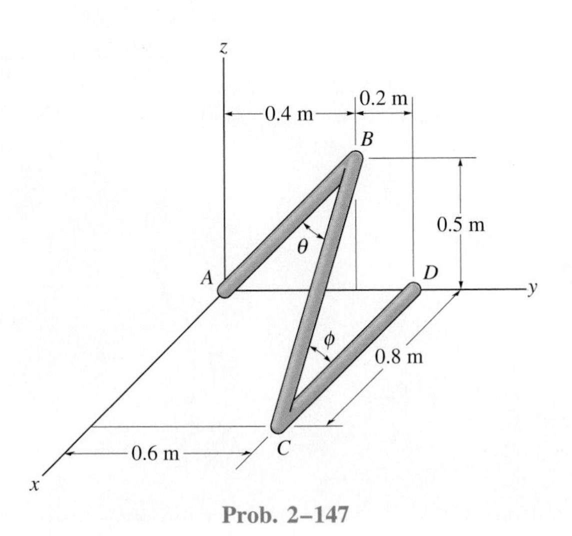

Prob. 2–147

***2–148.** Determine the magnitudes of the projected components of the force $\mathbf{F} = \{60\mathbf{i} + 12\mathbf{j} - 40\mathbf{k}\}$ N in the direction of the cables AB and AC.

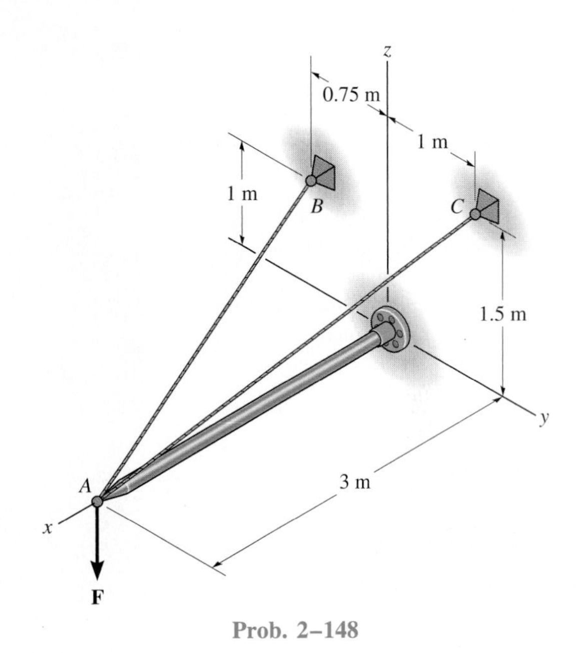

Prob. 2–148

2–149. A force of 23 kN is developed by the main rotor of a helicopter while it is flying forward. Resolve this force into its x and y components and explain what physical effects on the helicopter are caused by each of these components.

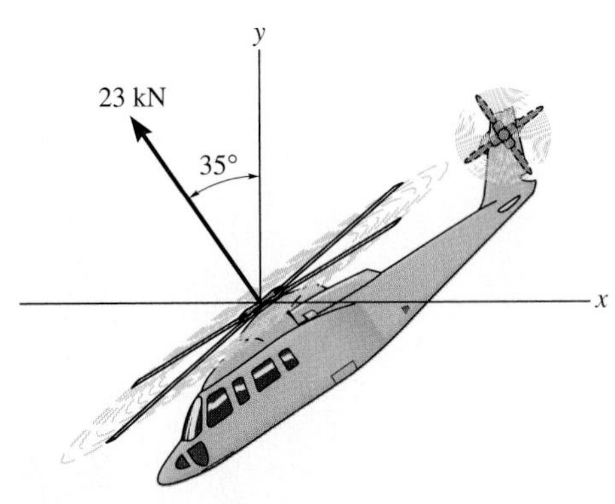

Prob. 2–149

Whenever cables are used for hoisting structural members they must be selected so that they do not fail when they are placed at their points of attachment. In this chapter we will show how to calculate cable loadings for such cases.

3

Equilibrium of a Particle

In this chapter the methods for resolving a force into components and expressing a force as a Cartesian vector will be used to solve problems involving the equilibrium of a particle. To simplify the discussion, particle equilibrium for a concurrent coplanar force system will be considered first. Then, in the last part of the chapter, equilibrium problems involving concurrent three-dimensional force systems will be considered.

3.1 Condition for the Equilibrium of a Particle

A particle is in *equilibrium* provided it is at rest if originally at rest, or has a constant velocity if originally in motion. Most often, however, the term "equilibrium" or, more specifically, "static equilibrium" is used to describe an object at rest. To maintain a state of equilibrium, it is *necessary* to satisfy Newton's first law of motion, which states that if the *resultant force* acting on a particle is *zero,* then the particle is in equilibrium. This condition may be stated mathematically as

$$\Sigma \mathbf{F} = \mathbf{0} \tag{3–1}$$

where $\Sigma \mathbf{F}$ is the vector *sum of all the forces* acting on the particle.

Not only is Eq. 3–1 a necessary condition for equilibrium, it is also a *sufficient* condition. This follows from Newton's second law of motion, which can be written as $\Sigma \mathbf{F} = m\mathbf{a}$. Since the force system satisfies Eq. 3–1, then $m\mathbf{a} = \mathbf{0}$, and therefore the particle's acceleration $\mathbf{a} = \mathbf{0}$, and consequently the particle indeed moves with constant velocity or remains at rest.

3.2 The Free-Body Diagram

To apply the equation of equilibrium correctly, we must account for *all* the known and unknown forces ($\Sigma\mathbf{F}$) which act *on* the particle. The best way to do this is to draw the particle's *free-body diagram.* This diagram is a sketch of the particle which represents it as being isolated or "free" from its surroundings. On this sketch it is necessary to show *all* the forces that act *on* the particle. Once this diagram is drawn, it will then be easy to apply Eq. 3–1.

Before presenting a formal procedure as to how to draw a free-body diagram, we will first discuss two types of connections often encountered in particle equilibrium problems.

Springs. If a *linear elastic spring* is used for support, the length of the spring will change in direct proportion to the force acting on it. A characteristic that defines the "elasticity" of a spring is the *spring constant* or *stiffness k.* Specifically, the magnitude of force developed by a linear elastic spring which has a stiffness k, and is deformed (elongated or compressed) a distance s measured from its *unloaded* position, is

$$F = ks \tag{3–2}$$

Note that s is determined from the difference in the spring's deformed length l and its undeformed length l_o, i.e., $s = l - l_o$. Thus, if s is positive, $\mathbf{F}$ "pulls" on the spring; whereas if s is negative, $\mathbf{F}$ must "push" on it. For example, the spring shown in Fig. 3–1 has an undeformed length $l_o = 0.4$ m and stiffness $k = 500$ N/m. To stretch it so that $l = 0.6$ m, a force $F = ks = (500 \text{ N/m})(0.6 \text{ m} - 0.4 \text{ m}) = 100$ N is needed. Likewise, to compress it to a length $l = 0.2$ m, a force $F = ks = (500 \text{ N/m})(0.2 \text{ m} - 0.4 \text{ m}) = -100$ N is required, Fig. 3–1.

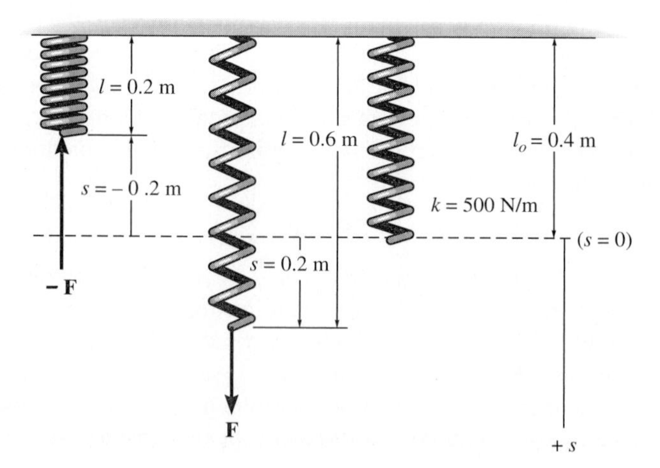

Fig. 3–1

Cables and Pulleys. Throughout this book, except in Sec. 7.4, all cables (or cords) are assumed to have negligible weight and they cannot be stretched. A cable can support *only* a tension or "pulling" force, and this force always acts in the direction of the cable. In Chapter 5 it will be shown that the tension force developed in a *continuous cable* which passes over a frictionless pulley must have a *constant* magnitude to keep the cable in equilibrium. Hence, for any angle θ, shown in Fig. 3–2, the cable is subjected to a constant tension T throughout its length.

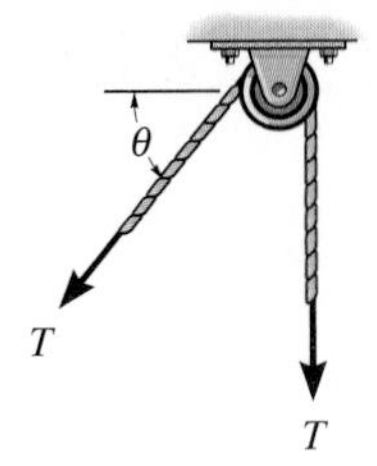

Cable is in tension

Fig. 3–2

PROCEDURE FOR DRAWING A FREE-BODY DIAGRAM

Since we must account for all the forces acting on the particle, the importance of drawing a free-body diagram before applying the equation of equilibrium to the solution of a problem cannot be overemphasized. To construct a free-body diagram, the following three steps are necessary.

Step 1. Imagine the particle to be *isolated* or cut "free" from its surroundings. Hence the name "free-body" diagram. Draw or sketch its outlined shape.

Step 2. Indicate on this sketch *all* the forces that act *on the particle.* These forces can be *active forces,* which tend to set the particle in motion, such as those caused by attached cables, weight, or magnetic and electrostatic interaction. Also, *reactive forces* will occur, such as those caused by the constraints or supports that tend to prevent motion. To account for all these forces, it may help to trace around the particle's boundary, carefully noting each force acting on it.

Step 3. The forces that are *known* should be labeled with their proper magnitudes and directions. Letters are used to represent the magnitudes and directions of forces that are unknown. In particular, if a force has a known line of action but unknown magnitude, the "arrowhead," which defines the sense of the force, can be *assumed.* The correct sense will become apparent after solving for the unknown magnitude. By definition, the *magnitude* of a force is *always positive* so that, if the solution yields a "negative" scalar, the *minus sign* indicates that the arrowhead or sense of the force is opposite to that which was originally assumed.

Application of the above steps is illustrated in the following two examples.

Example 3–1

The crate in Fig. 3–3*a* has a weight of 20 lb. Draw a free-body diagram of the crate, the cord *BD*, and the ring at *B*.

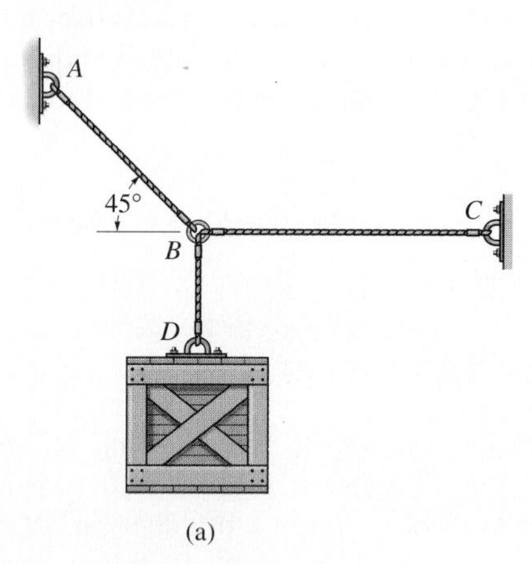

(a)

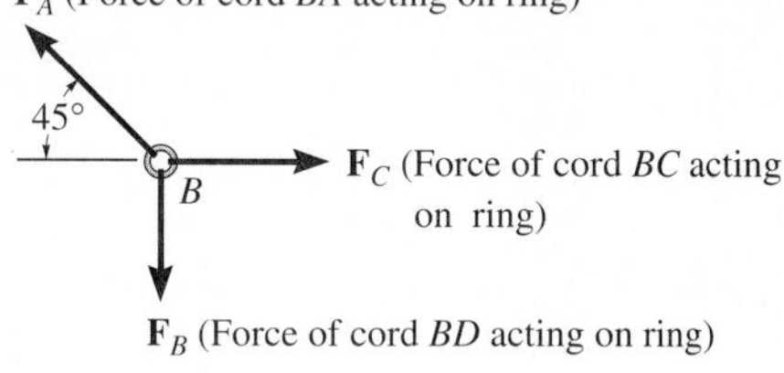

(b)

SOLUTION

If we imagine the crate to be *isolated from its surroundings,* then by inspection there are only two forces acting on it, namely, the gravitational force or weight of 20 lb, and the force of the cord *BD*. Thus the free-body diagram is shown in Fig. 3–3*d*.

If the cord *BD* is isolated from its surroundings, then there are only two forces acting on it, Fig. 3–3*c*, namely, the force of the crate, $\mathbf{F}_D$, and the force $\mathbf{F}_B$ caused by the ring. Since these forces tend to pull on the cord, we can state that the cord is in *tension.* (This must be the case, since compressive, or pushing, forces would cause the cord to collapse.)

When the ring at *B* is isolated from its surroundings, it should be noted that three forces act on it. All these forces are caused by the attached cords, Fig. 3–3*b*. Notice that $\mathbf{F}_B$ shown here is equal but opposite to that shown in Fig. 3–3*c*, a consequence of Newton's third law.

$\mathbf{F}_B$ (Force of ring acting on cord)

B

D

$\mathbf{F}_D$ (Force of crate acting on cord)

(c)

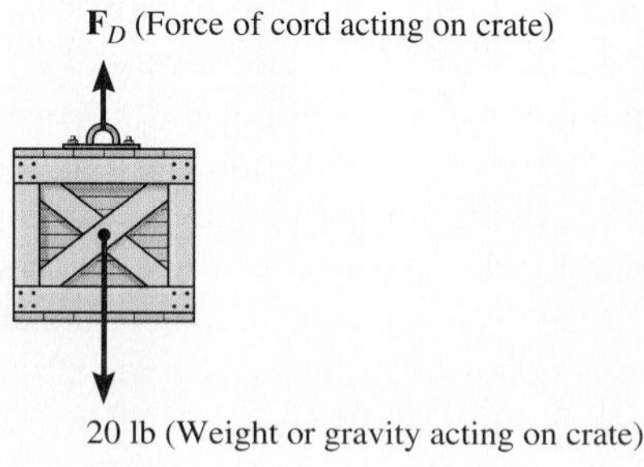

(d)

Fig. 3–3

Example 3–2

The sphere in Fig. 3–4*a* has a mass of 6 kg and is supported as shown. Draw a free-body diagram of the sphere and the knot at *C*.

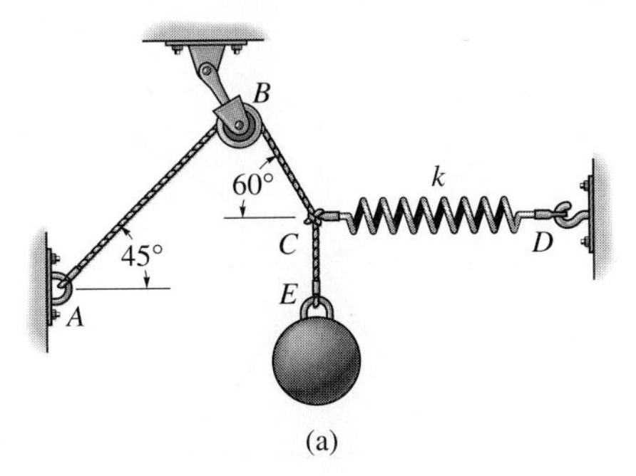

(a)

SOLUTION

There are two forces acting on the sphere, namely, its weight and the force **P** of the cord *CE*. The sphere has a weight of (6 kg)(9.81 m/s^2) = 58.9 N. Its free-body diagram is shown in Fig. 3–4*b*.

By inspection, three forces act on the knot at *C* when it is isolated. They are caused by cords *CBA* and *CE*, and the spring *CD*. Thus, the free-body diagram of the knot is shown in Fig. 3–4*c*.

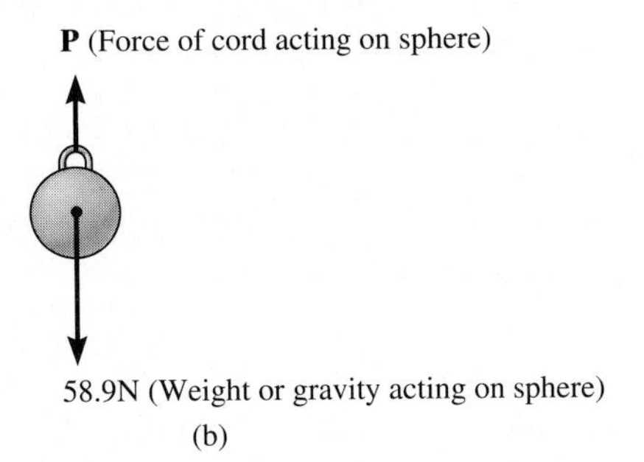

(b)

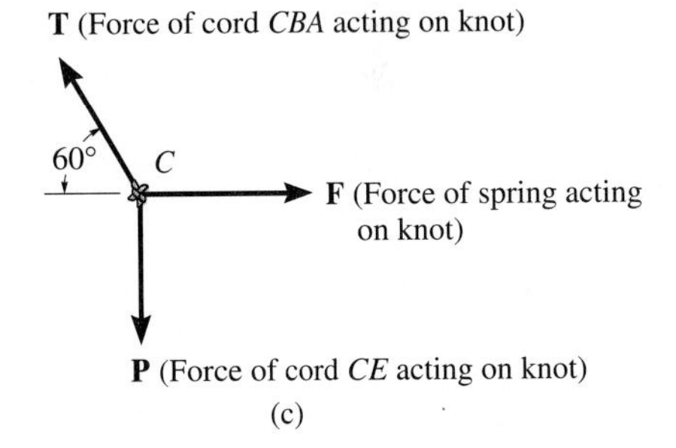

(c)

Fig. 3–4

3.3 Coplanar Force Systems

Many particle equilibrium problems involve a coplanar force system, Fig. 3–5. If the forces lie in the x–y plane, they can each be resolved into their

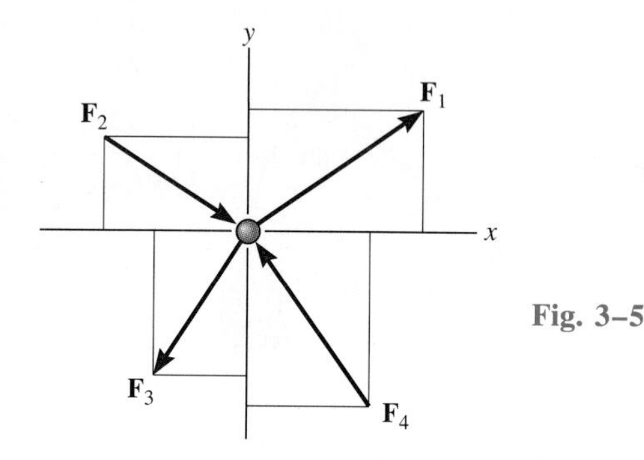

Fig. 3–5

respective **i** and **j** components and Eq. 3–1 can be written as

$$\Sigma \mathbf{F} = \mathbf{0}$$

$$\Sigma F_x \mathbf{i} + \Sigma F_y \mathbf{j} = \mathbf{0}$$

For this vector equation to be satisfied, both the x and y components must equal zero, otherwise $\Sigma \mathbf{F} \neq \mathbf{0}$. Hence, we require

$$\begin{aligned} \Sigma F_x &= 0 \\ \Sigma F_y &= 0 \end{aligned} \qquad (3\text{–}3)$$

These scalar equilibrium equations state that the algebraic sum of the x and y components of all the forces acting on the particle be equal to zero. As a result, Eqs. 3–3 can be solved for at most two unknowns, generally represented as angles and magnitudes of forces shown on the particle's free-body diagram.

Scalar Notation. Since each of the two equilibrium equations requires the resolution of vector components along a specified axis (x or y), we will use scalar notation to represent the components when applying these equations. By doing this, the sense of direction for each component is accounted for by an *algebraic sign* which corresponds to the arrowhead direction of the component as indicated graphically on the free-body diagram. In particular, if a force component has an unknown magnitude, then the arrowhead sense of the force on the free-body diagram can be *assumed.* Since the magnitude of a force is *always positive,* then if the *solution* yields a *negative scalar,* it indicates that the sense of the force as shown on the free-body diagram is opposite to that which was assumed.

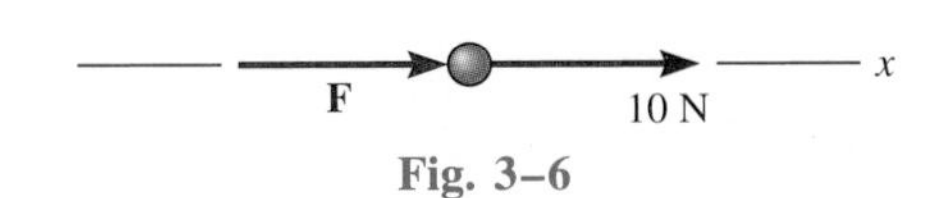

Fig. 3–6

For example, consider the free-body diagram of the particle subjected to the two forces as shown in Fig. 3–6. For the sake of discussion, we have *assumed* that the *unknown force* **F** acts to the right to maintain equilibrium. Application of the equation of equilibrium along the x axis yields

$$\overset{+}{\rightarrow}\Sigma F_x = 0; \qquad +F + 10\text{ N} = 0$$

Both terms are "positive" since both forces act in the positive x direction, as indicated by the arrow placed alongside the equation. When this equation is solved, $F = -10$ N. Here the *negative sign* refers to the fact that **F** in Fig. 3–6 is shown in the opposite sense of the actual direction. In other words, **F** must act to the left to hold the particle in equilibrium. Notice that if the $+x$ axis in Fig. 3–6 was directed to the left both terms in the above equation would be negative, but again $F = -10$ N, indicating F would be directed to the left.

PROCEDURE FOR ANALYSIS

The following procedure provides a method for solving coplanar force problems involving particle equilibrium:

Free-Body Diagram. Draw a free-body diagram of the particle. As outlined in Sec. 3.2, this requires that all the known and unknown force magnitudes and angles be labeled on the diagram. The sense of a force having an unknown magnitude can be assumed.

Equations of Equilibrium. Establish the x, y axes in *any* suitable direction and apply the two equations of equilibrium, $\Sigma F_x = 0$, and $\Sigma F_y = 0$. For application, components are positive if they are directed along the positive axes, and negative if they are directed along the negative axes. If more than two unknowns exist and the problem involves a spring, apply $F = ks$ (Eq. 3–2) to relate the spring force to the deformation s of the spring.

The following example problems illustrate this solution procedure numerically.

Example 3–3

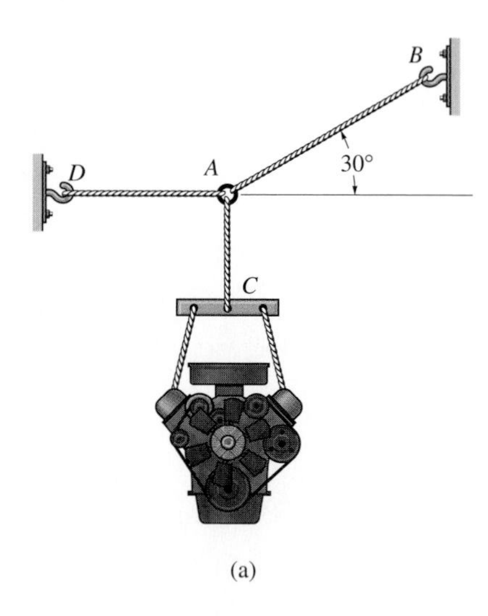

(a)

Determine the tension in cables AB and AD for equilibrium of the 250-kg engine shown in Fig. 3–7a.

SOLUTION

Free-Body Diagram. To solve this problem we will investigate the equilibrium of the ring at A, because this "particle" is subjected to the forces of both cables AB and AD. First, however, note from the free-body diagram of the engine, Fig. 3–7c, that its weight $(250\ \text{kg})(9.81\ \text{m/s}^2) = 2.452\ \text{kN}$ is balanced by the 2.452-kN force of cable CA on the spreader bar, a consequence of equilibrium. Furthermore, by Newton's third law, the spreader bar exerts an equal but opposite force of 2.452 kN on the cable at C, Fig. 3–7b. And again, for equilibrium, the force of the ring at A on the cable must be 2.452 kN. Finally, as shown in Fig. 3–7d, there are three concurrent forces *acting on the ring*. The forces $\mathbf{T}_B$ and $\mathbf{T}_D$ have unknown magnitudes but known directions, and cable AC exerts a downward force on A equal to 2.452 kN. Why?

2.452 kN (Force of ring A acting on cable)

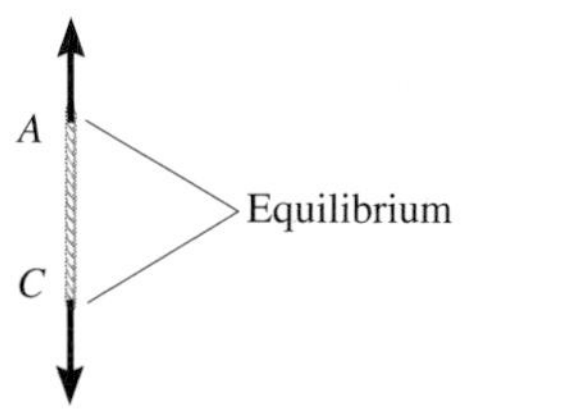

2.452 kN (Force of engine support acting on cable)

(b)

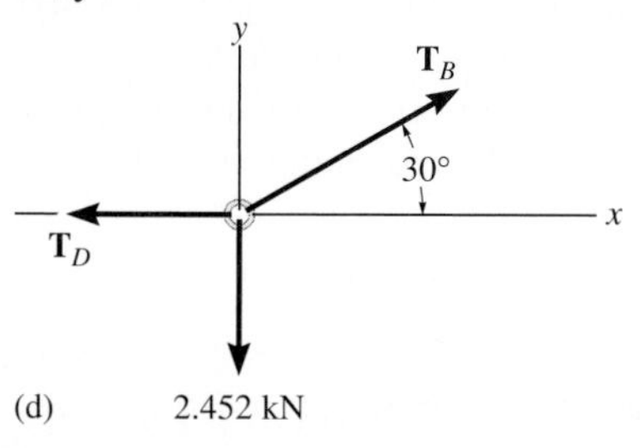

(d) 2.452 kN

Equations of Equilibrium. The two unknown magnitudes T_B and T_D can be obtained from the two scalar equations of equilibrium, $\Sigma F_x = 0$ and $\Sigma F_y = 0$. To apply these equations, the x, y axes are established on the free-body diagram and $\mathbf{T}_B$ is resolved into its dashed x and y components. Thus,

$$\overset{+}{\rightarrow}\Sigma F_x = 0; \qquad T_B \cos 30° - T_D = 0 \qquad (1)$$

$$+\uparrow \Sigma F_y = 0; \qquad T_B \sin 30° - 2.452\ \text{kN} = 0 \qquad (2)$$

Solving Eq. (2) for T_B and substituting into Eq. (1) to obtain T_D yields

$$T_B = 4.91\ \text{kN} \qquad \textit{Ans.}$$

$$T_D = 4.25\ \text{kN} \qquad \textit{Ans.}$$

The accuracy of these results, of course, depends on the accuracy of the data, i.e., measurements of geometry and loads. For most engineering work involving a problem such as this, the data as measured to three significant figures would be sufficient. Also, note that here we have neglected the weights of the cables, a reasonable assumption since they would be small in comparison with the weight of the engine.

2.452 kN (Force of cable acting on engine spreader bar)

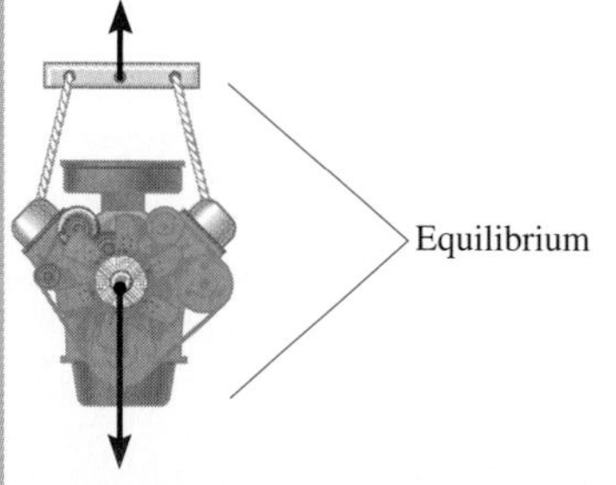

2.452 kN (Weight or gravity acting on engine)

(c)

Fig. 3–7

Example 3–4

If the sack at A in Fig. 3–8*a* has a weight of 20 lb, determine the weight of the sack at B and the force in each cord needed to hold the system in the equilibrium position shown.

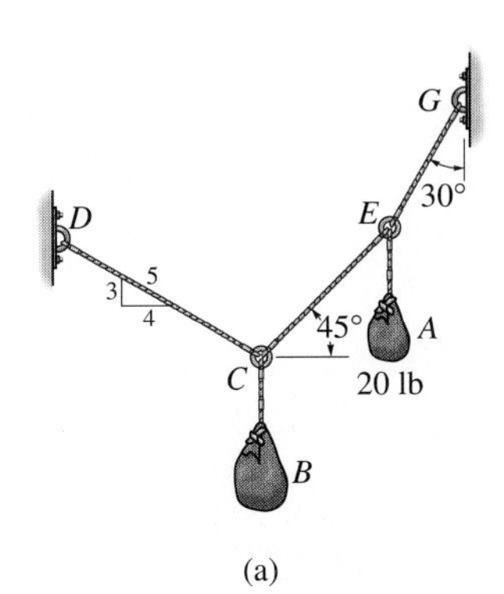

(a)

SOLUTION

Since the weight of A is known, the unknown tension in the two cords EG and EC can be determined by investigating the equilibrium of the ring at E. Why?

Free-Body Diagram. There are three forces acting on E, as shown in Fig. 3–8*b*.

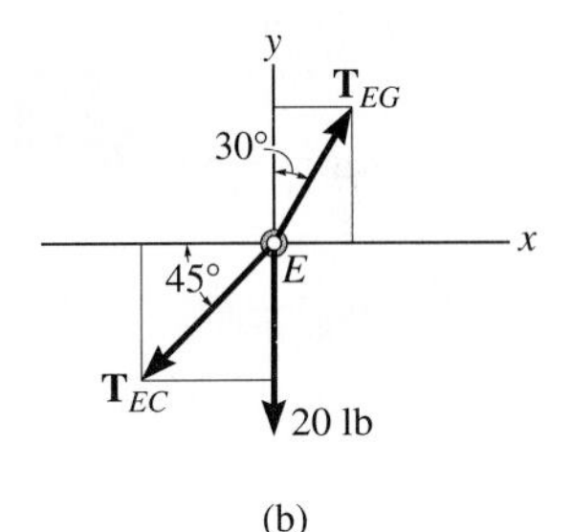

(b)

Equations of Equilibrium. Establishing the x, y axes and resolving each force into its x and y components using trigonometry, we have

$$\overset{+}{\rightarrow}\Sigma F_x = 0; \qquad T_{EG}\sin 30^\circ - T_{EC}\cos 45^\circ = 0 \qquad (1)$$

$$+\uparrow \Sigma F_y = 0; \qquad T_{EG}\cos 30^\circ - T_{EC}\sin 45^\circ - 20\text{ lb} = 0 \qquad (2)$$

Solving Eq. 1 for T_{EG} in terms of T_{EC} and substituting the result into Eq. 2 allows a solution for T_{EC}. One then obtains T_{EG} from Eq. 1. The results are

$$T_{EC} = 38.6\text{ lb} \qquad \textit{Ans.}$$

$$T_{EG} = 54.6\text{ lb} \qquad \textit{Ans.}$$

Using the calculated result for T_{EC}, the equilibrium of the ring at C can now be investigated to determine the tension in CD and the weight of B.

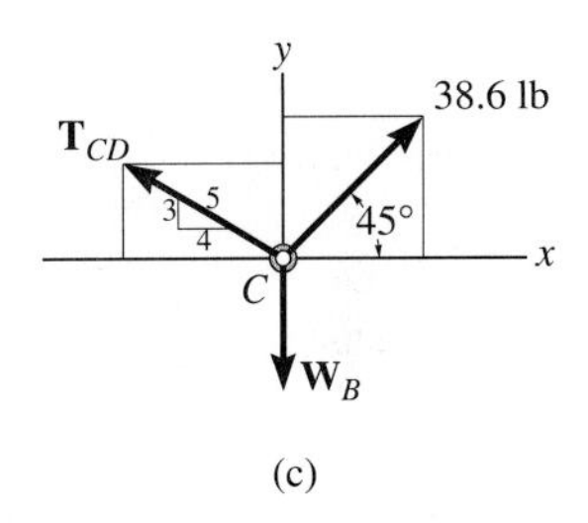

(c)

Free-Body Diagram. As shown in Fig. 3–8*c*, $T_{EC} = 38.6$ lb "pulls" on C. The reason for this becomes clear when one draws the free-body diagram of cord CE and applies both equilibrium and the principle of action, equal but opposite force reaction (Newton's third law), Fig. 3–8*d*.

Equations of Equilibrium. Establishing the x, y axes and noting the components of $\mathbf{T}_{CD}$ are proportional to the slope of the cord as defined by the 3–4–5 triangle, we have

$$\overset{+}{\rightarrow}\Sigma F_x = 0; \qquad 38.6\cos 45^\circ\text{ lb} - \left(\tfrac{4}{5}\right)T_{CD} = 0 \qquad (3)$$

$$+\uparrow \Sigma F_y = 0; \qquad \left(\tfrac{3}{5}\right)T_{CD} + 38.6\sin 45^\circ\text{ lb} - W_B = 0 \qquad (4)$$

Solving Eq. 3 and substituting the result into Eq. 4 yields

$$T_{CD} = 34.2\text{ lb} \qquad \textit{Ans.}$$

$$W_B = 47.8\text{ lb} \qquad \textit{Ans.}$$

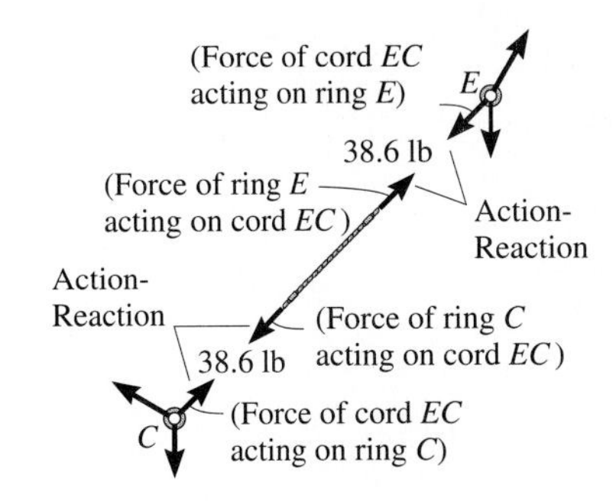

(d)

Fig. 3–8

Example 3–5

Determine the required length of cord AC in Fig. 3–9a so that the 8-kg lamp is suspended in the position shown. The *undeformed* length of the spring AB is $l'_{AB} = 0.4$ m, and the spring has a stiffness of $k_{AB} = 300$ N/m.

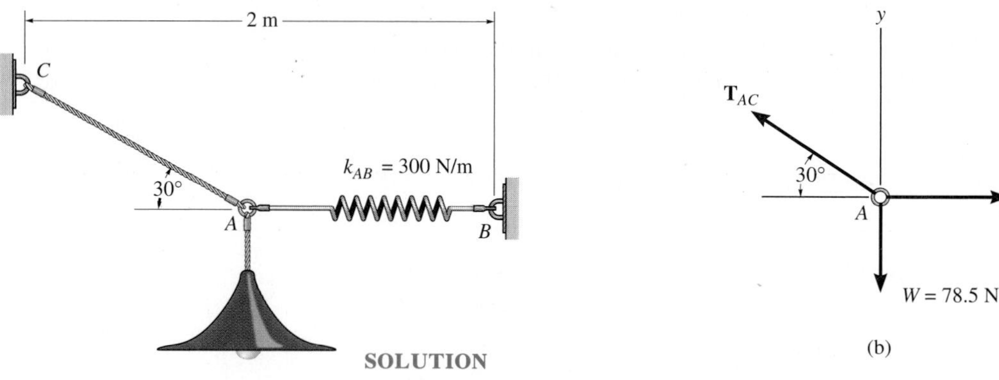

(a)

(b)

Fig. 3–9

SOLUTION

If the force in spring AB is known, the stretch of the spring can be found ($F = ks$). Using the problem geometry, it is then possible to calculate the required length of AC.

Free-Body Diagram. The lamp has a weight $W = 8(9.81) = 78.5$ N. The free-body diagram of the ring at A is shown in Fig. 3–9b.

Equations of Equilibrium Using the x, y axes,

$$\overset{+}{\rightarrow}\Sigma F_x = 0; \qquad T_{AB} - T_{AC}\cos 30° = 0$$

$$+\uparrow \Sigma F_y = 0; \qquad T_{AC}\sin 30° - 78.5\text{ N} = 0$$

Solving, we obtain

$$T_{AC} = 157.0\text{ N}$$

$$T_{AB} = 136.0\text{ N}$$

The stretch of spring AB is therefore

$$T_{AB} = k_{AB}s_{AB}; \qquad 136.0\text{ N} = 300\text{ N/m}(s_{AB})$$

$$s_{AB} = 0.453\text{ m}$$

so the stretched length is

$$l_{AB} = l'_{AB} + s_{AB}$$

$$l_{AB} = 0.4\text{ m} + 0.453\text{ m} = 0.853\text{ m}$$

The horizontal distance from C to B, Fig. 3–9a, requires

$$2\text{ m} = l_{AC}\cos 30° + 0.853\text{ m}$$

$$l_{AC} = 1.32\text{ m}$$

Ans.

PROBLEMS

3–1. Determine the magnitudes of $\mathbf{F}_1$ and $\mathbf{F}_2$ so that the particle is in equilibrium.

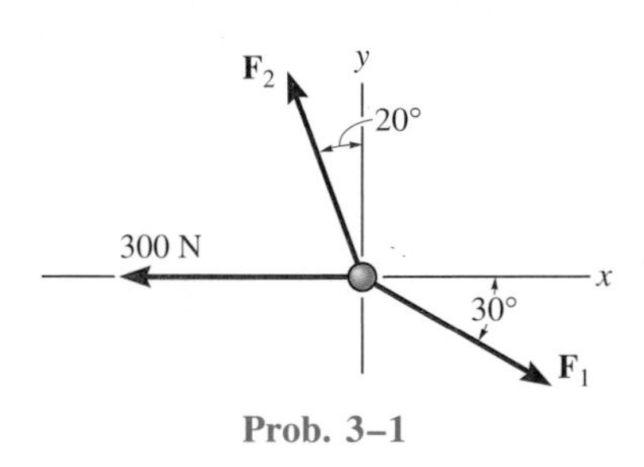

Prob. 3–1

3–2. Determine the magnitude and direction θ of $\mathbf{F}_1$ so that the particle is in equilibrium.

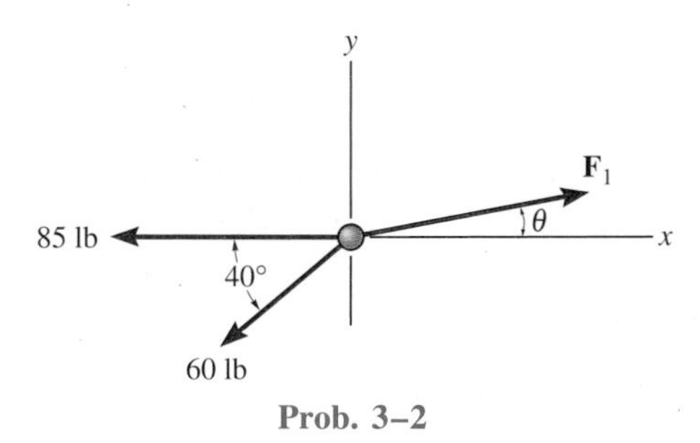

Prob. 3–2

3–3. Determine the magnitude and direction θ of $\mathbf{F}$ so that the particle is in equilibrium.

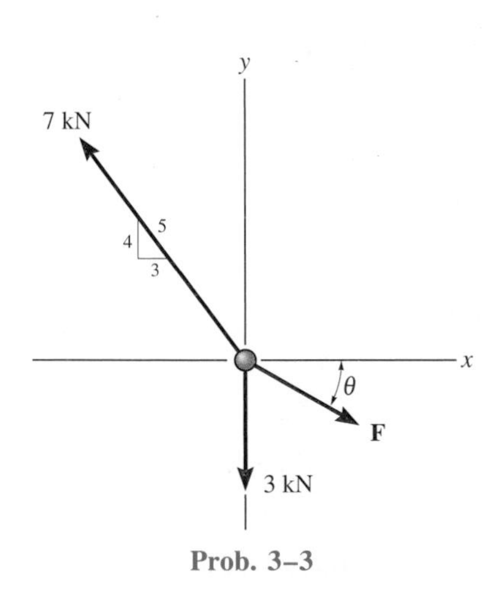

Prob. 3–3

***3–4.** Determine the magnitude and angle θ of $\mathbf{F}$ so that the particle is in equilibrium.

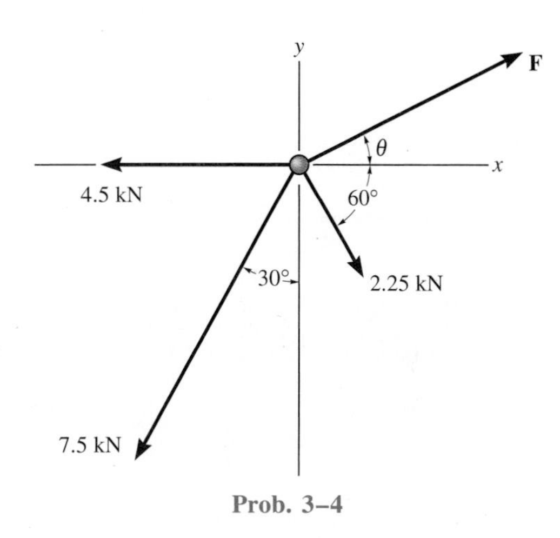

Prob. 3–4

3–5. The members of a truss are pin-connected at joint O. Determine the magnitudes of $\mathbf{F}_1$ and $\mathbf{F}_2$ for equilibrium. Set $\theta = 60°$.

3–6. The members of a truss are pin-connected at joint O. Determine the magnitude of $\mathbf{F}_1$ and its angle θ for equilibrium. Set $F_2 = 6$ kN.

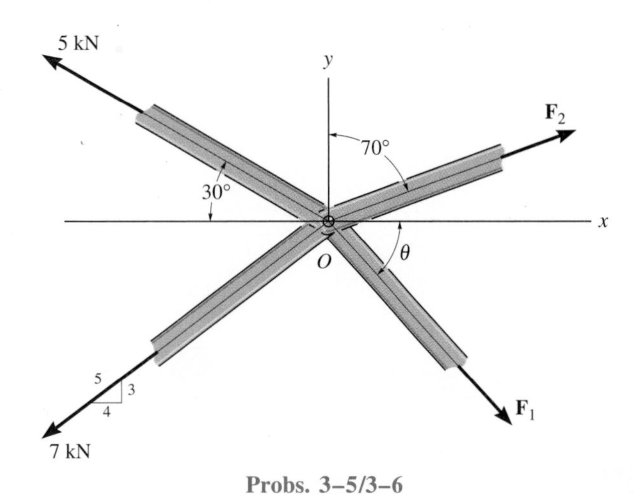

Probs. 3–5/3–6

3–7. The sling is used to support a drum having a weight of 900 lb. Determine the force in cords AB and AC for equilibrium. Take $\theta = 20°$.

***3–8.** Cords AB and AC can each sustain a maximum tension of 800 lb. If the drum has a weight of 900 lb, determine the smallest angle θ at which they can be attached to the drum.

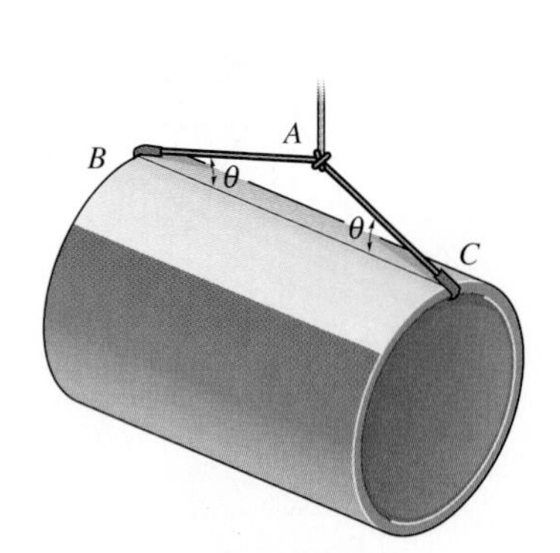

Probs. 3–7/3–8

3–9. Two electrically charged pith balls, each having a mass of 0.2 g, are suspended from light threads of equal length. Determine the resultant horizontal force of repulsion, F, acting on each ball if the measured distance between them is $r = 200$ mm.

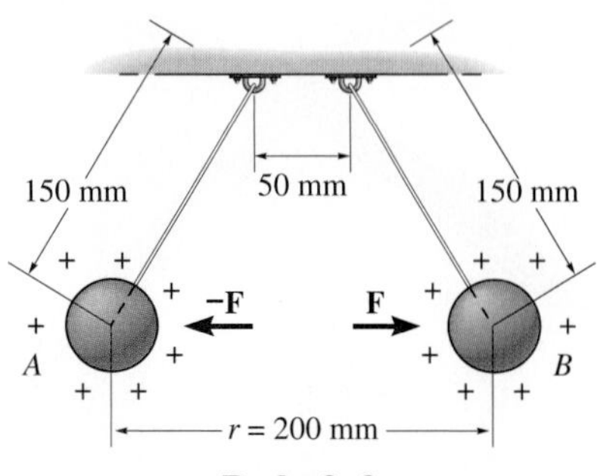

Prob. 3–9

3–10. The spring ABC has a stiffness of 500 N/m and an unstretched length of 6 m. Determine the horizontal force **F** applied to the cord which is attached to the *small* pulley B so that the displacement of the pulley from the wall is $d = 1.5$ m.

■**3–11.** The spring ABC has a stiffness of 500 N/m and an unstretched length of 6 m. Determine the displacement d of the cord from the wall when a force $F = 175$ N is applied to the cord.

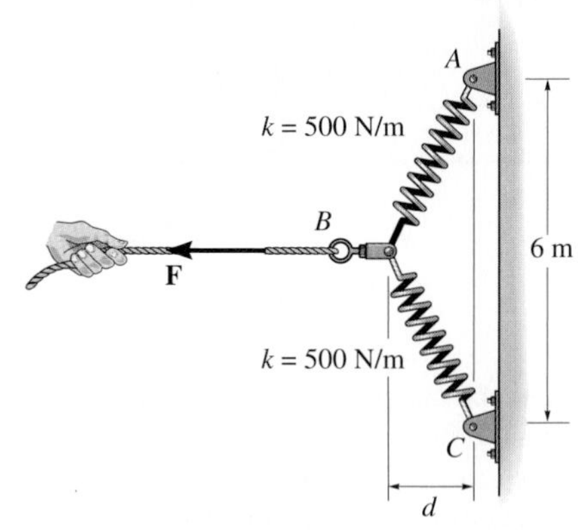

Probs. 3–10/3–11

***3–12.** Determine the stretch in each spring for equilibrium of the 2-kg block. The springs are shown in the equilibrium position.

3–13. The unstretched length of spring AB is 2 m. If the block is held in the equilibrium position shown, determine the mass of the block at D.

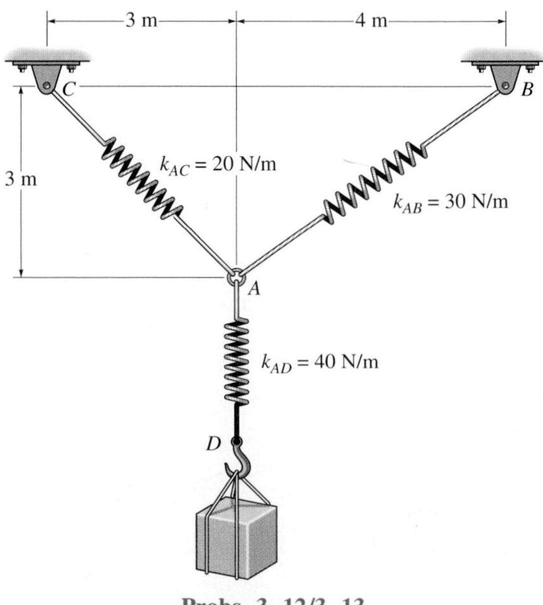

Probs. 3–12/3–13

3–14. Determine the stiffness k_T of the single spring such that the force **F** will stretch it by the same amount s as the force **F** stretches the two springs. Express k_T in terms of the stiffness k_1 and k_2 of the two springs.

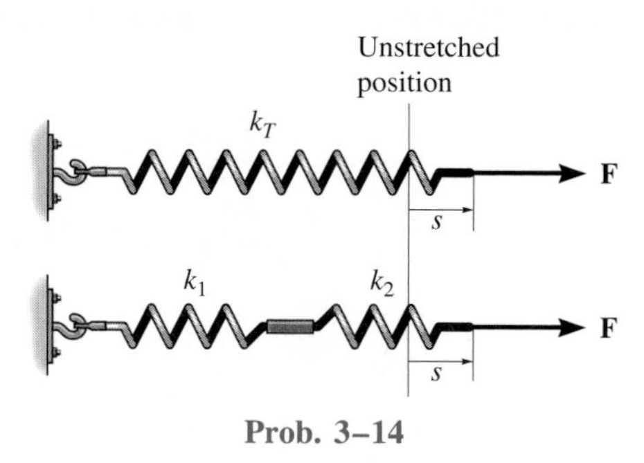

Prob. 3–14

3–15. The motor at B winds up the cord attached to the 65-lb crate with a constant speed. Determine the force in cord CD supporting the pulley and the angle θ for equilibrium. Neglect the size of the pulley at C.

***3–16.** The cords BCA and CD can each support a maximum load of 100 lb. Determine the maximum weight of the crate that can be hoisted at constant velocity, and the angle θ for equilibrium.

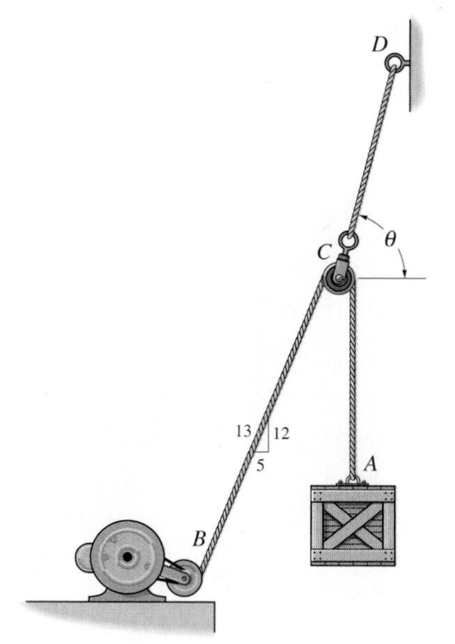

Probs. 3–15/3–16

3–17. Determine the mass that must be supported at A and the angle θ of the connecting cord in order to hold the system in equilibrium.

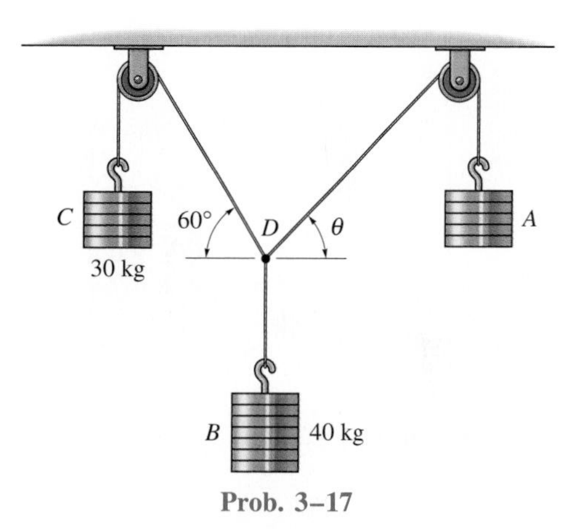

Prob. 3–17

3–18. The 500-lb crate is hoisted using the ropes AB and AC. Each rope can withstand a maximum tension of 2500 lb before it breaks. If AB always remains horizontal, determine the smallest angle θ to which the crate can be hoisted.

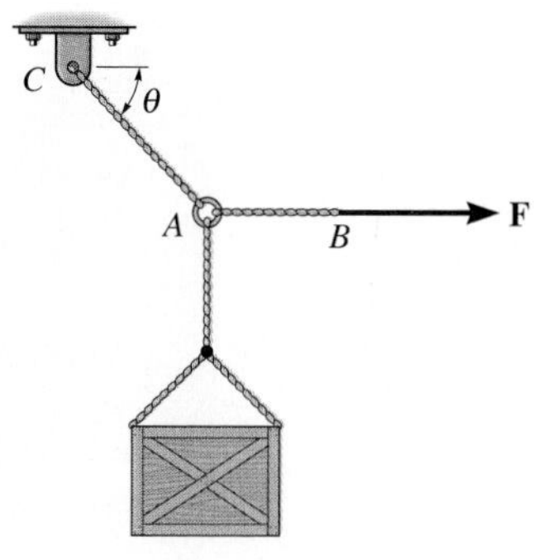

Prob. 3–18

3–19. The street-lights at A and B are suspended from the two poles as shown. If each light has a weight of 50 lb, determine the tension in each of the three supporting cables and the required height h of the pole DE so that cable AB is horizontal.

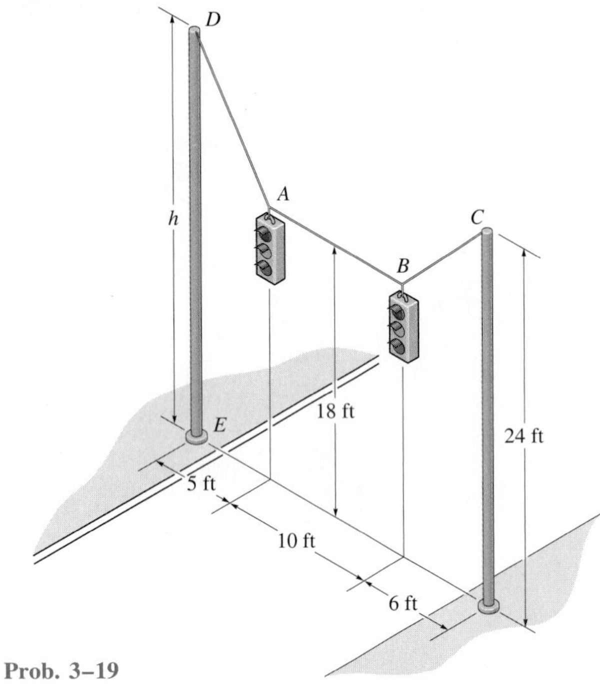

Prob. 3–19

***3–20.** A nuclear-reactor vessel has a weight of $500(10^3)$ lb. Determine the horizontal compressive force that the spreader bar AB exerts on point A and the force that each cable segment CA and AD exert on this point while the vessel is hoisted upward at constant velocity.

Prob. 3–20

3–21. The block has a weight of 20 lb and is being hoisted at uniform velocity. Determine the angle θ for equilibrium and the required force in each cord.

3–22. Determine the maximum weight W of the block that can be suspended in the position shown if each cord can support a maximum tension of 80 lb. Also, what is the angle θ for equilibrium?

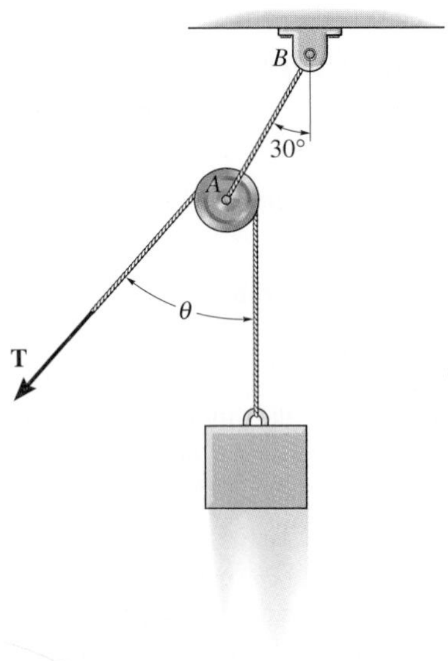

Probs. 3–21/3–22

3–23. The pipe is held in place by the vice. If the bolt exerts a force of 50 lb on the pipe in the direction shown, determine the forces F_A and F_B that the smooth contacts at A and B exert on the pipe.

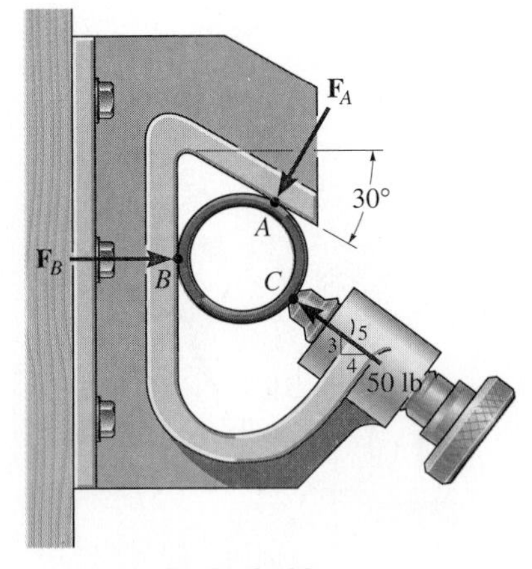

Prob. 3–23

***3–24.** Blocks D and F weigh 5 lb each and block E weighs 8 lb. Determine the sag s for equilibrium. Neglect the size of the pulleys.

3–25. If blocks D and F weigh 5 lb each, determine the weight of block E if the sag $s = 3$ ft. Neglect the size of the pulleys.

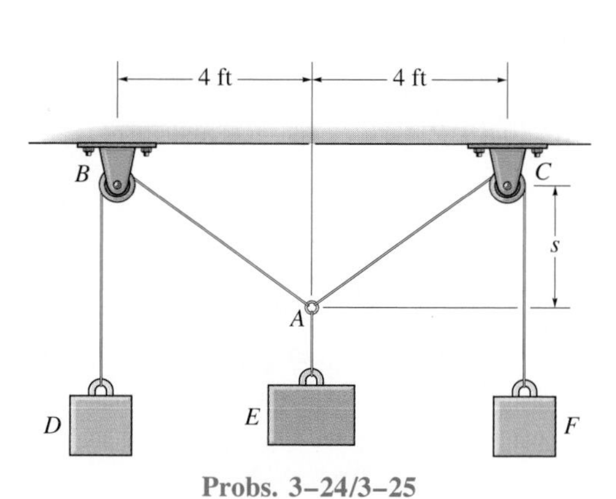

Probs. 3–24/3–25

■**3–26.** A vertical force $P = 10$ lb is applied to the ends of the 2-ft cord AB and spring AC. If the spring has an unstretched length of 2 ft, determine the angle θ for equilibrium. Take $k = 15$ lb/ft.

3–27. Determine the unstretched length of spring AC if a force $P = 80$ lb causes the angle $\theta = 60°$ for equilibrium. Cord AB is 2 ft long. Take $k = 50$ lb/ft.

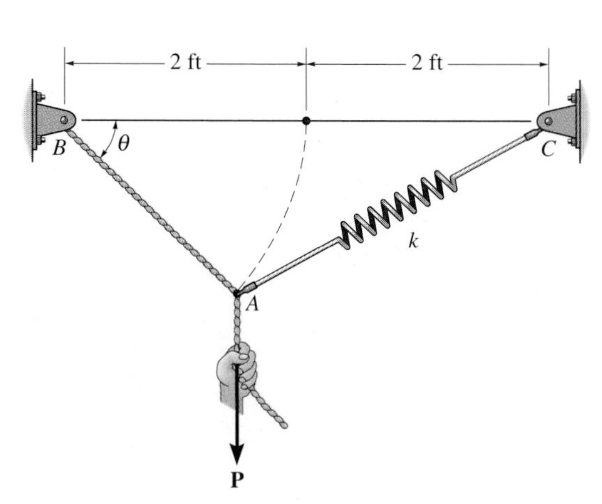

Probs. 3–26/3–27

*■**3–28.** A car is to be towed using the rope arrangement shown. The towing force required is 600 lb. Determine the minimum length l of rope AB so that the tension in either rope AB or AC does not exceed 750 lb. *Hint:* Use the equilibrium condition at point A to determine the required angle θ for attachment, then determine l using trigonometry applied to triangle ABC.

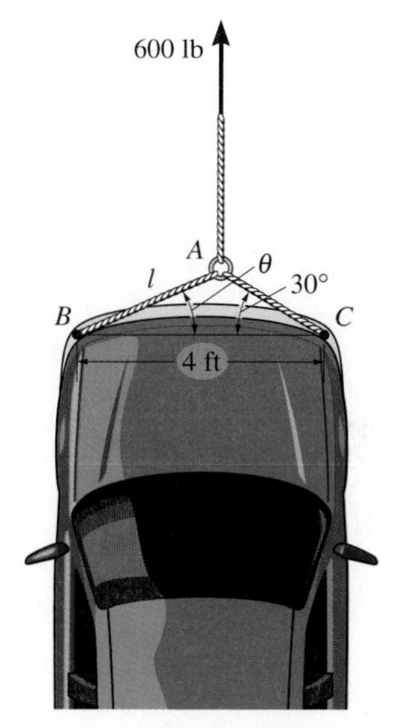

Prob. 3–28

3–29. The sling BAC is used to lift the 100-lb load with constant velocity. Determine the force in the sling and plot its value T (ordinate) as a function of its orientation θ, where $0 \leq \theta \leq 90°$.

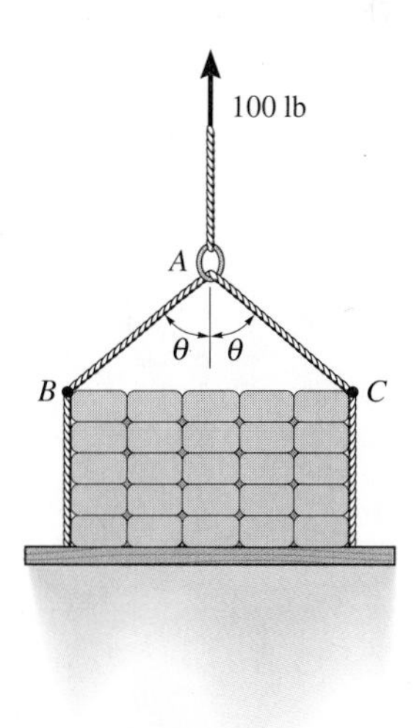

Prob. 3–29

3–30. When y is zero, the springs sustain a force of 60 lb. Determine the magnitude of the applied vertical forces **F** and −**F** required to pull point A away from point B a distance of $y = 2$ ft. The ends of cords CAD and CBD are attached to rings at C and D.

■**3–31.** When y is zero, the springs are each stretched 1.5 ft. Determine the distance y if a force of $F = 60$ lb is applied to points A and B as shown. The ends of cords CAD and CBD are attached to rings at C and D.

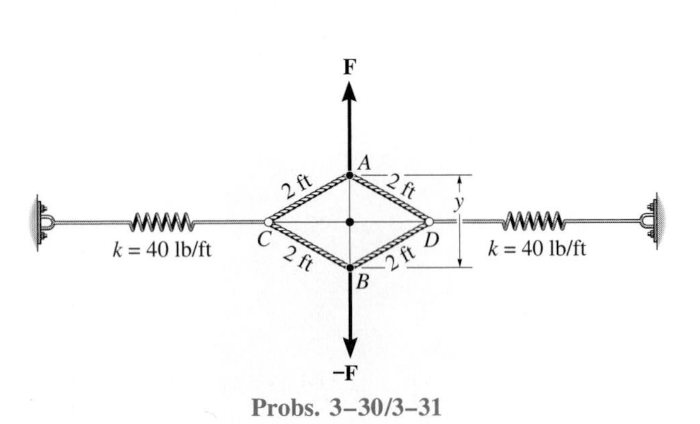

Probs. 3–30/3–31

■**3–33.** The 10-lb lamp fixture is suspended from two springs, each having an unstretched length of 4 ft and stiffness of $k = 5$ lb/ft. Determine the angle θ for equilibrium.

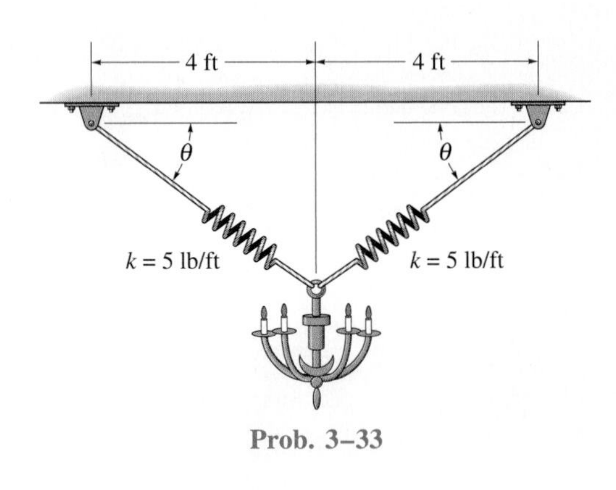

Prob. 3–33

*__3–32.__ Determine the maximum weight W that can be supported in the position shown if each cable AC and AB can support a maximum tension of 600 lb before it fails.

Prob. 3–32

3–34. If the cords suspend the two buckets in the equilibrium position shown, determine the weight of bucket B. Bucket A has a weight of 60 lb.

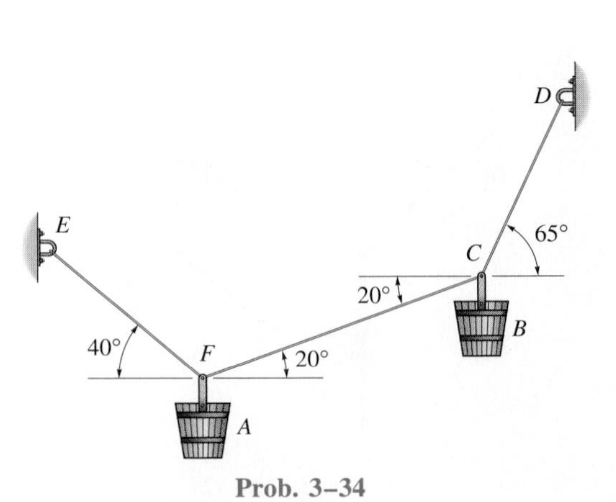

Prob. 3–34

3–35. The 30-kg pipe is supported at A by a system of five cords. Determine the force in each cord for equilibrium.

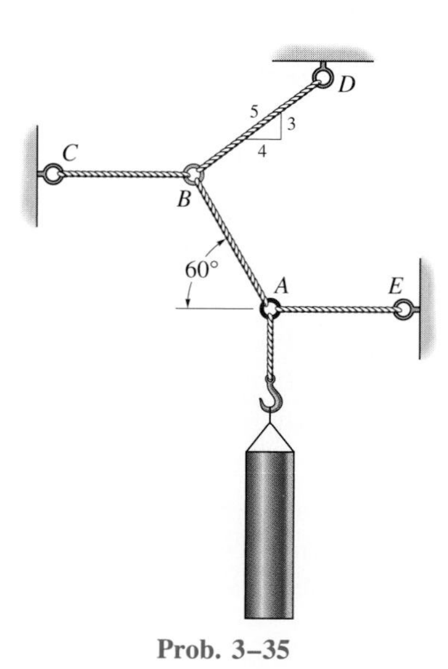

Prob. 3–35

***■3–36.** The ring of negligible size is subjected to a vertical force of 200 lb. Determine the required length l of cord AC such that the tension acting in AC is 160 lb. Also, what is the force in cord AB? *Hint:* Use the equilibrium condition to determine the required angle θ for attachment, then determine l using trigonometry applied to triangle ABC.

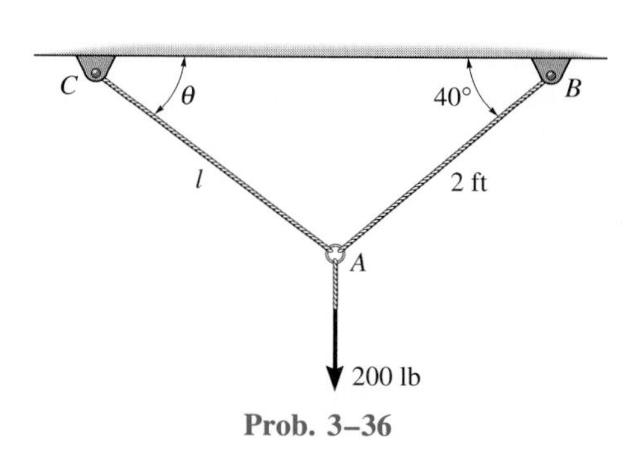

Prob. 3–36

■3–37. The cord AB of length 5 ft is attached to the end B of a spring having a stiffness $k = 10$ lb/ft and an unstretched length of 5 ft. The other end of the spring is attached to a roller C so that the spring remains horizontal as it stretches. If a 10-lb weight is suspended from B, determine the angle θ of cord AB for equilibrium.

3–38. The cord AB has a length of 5 ft and is attached to the end B of the spring having a stiffness $k = 10$ lb/ft. The other end of the spring is attached to a roller C so that the spring remains horizontal as it stretches. If a 10-lb weight is suspended from B, determine the necessary unstretched length of the spring, so that $\theta = 40°$ for equilibrium.

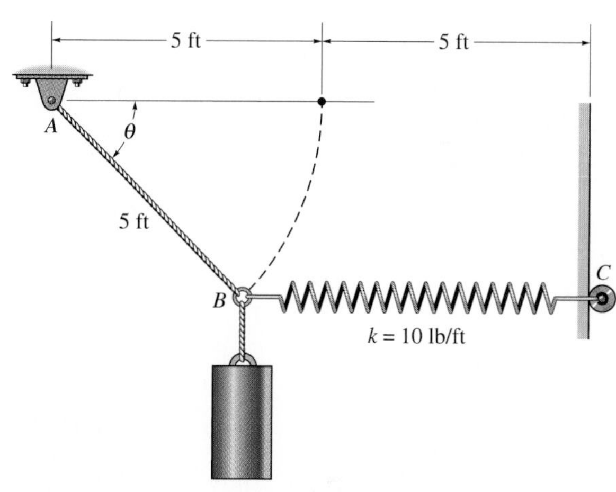

Probs. 3–37/3–38

3–39. The pail and its contents have a mass of 60 kg. If the cable is 15 m long, determine the elevation y of the pulley for equilibrium. Neglect the size of the pulley at A.

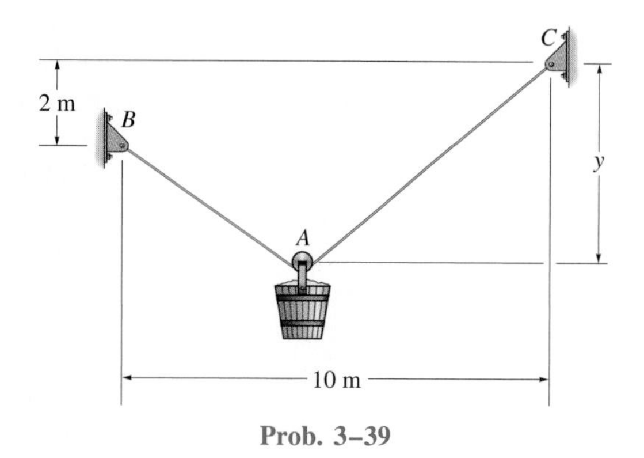

Prob. 3–39

3.4 Three-Dimensional Force Systems

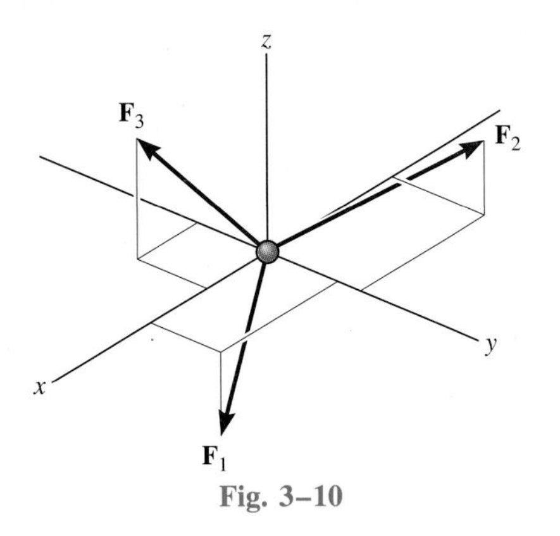

Fig. 3–10

It was shown in Sec. 3.1 that particle equilibrium requires

$$\Sigma \mathbf{F} = \mathbf{0} \tag{3–4}$$

If the forces acting on the particle are resolved into their respective **i, j, k** components, Fig. 3–10, we can then write

$$\Sigma F_x \mathbf{i} + \Sigma F_y \mathbf{j} + \Sigma F_z \mathbf{k} = \mathbf{0}$$

To ensure that Eq. 3–4 is satisfied, we must therefore require that the following three scalar component equations be satisfied:

$$\begin{aligned} \Sigma F_x &= 0 \\ \Sigma F_y &= 0 \\ \Sigma F_z &= 0 \end{aligned} \tag{3–5}$$

These equations represent the *algebraic sums* of the x, y, z force components acting on the particle. Using them we can solve for at most three unknowns, generally represented as angles or magnitudes of forces shown on the particle's free-body diagram.

PROCEDURE FOR ANALYSIS

The following procedure provides a method for solving three-dimensional force equilibrium problems.

Free-Body Diagram. Draw a free-body diagram of the particle and label all the known and unknown forces on this diagram.

Equations of Equilibrium. Establish the x, y, z coordinate axes with origin located at the particle and apply the equations of equilibrium. Use the three scalar Eqs. 3–5 in cases where it is easy to resolve each force acting on the particle into its x, y, z components. If this appears difficult, first express each force acting on the particle in Cartesian vector form, and then substitute these vectors into Eq. 3–4. By setting the respective **i, j, k** components equal to zero, the three scalar Eqs. 3–5 can be generated. If more than three unknowns exist and the problem involves a spring, consider using $F = ks$ to relate the spring force to the deformation s of the spring.

The following example problems numerically illustrate this solution procedure.

Example 3–6

A 90-lb load is suspended from the hook shown in Fig. 3–11*a*. The load is supported by two cables and a spring having a stiffness $k =$ 500 lb/ft. Determine the force in the cables and the stretch of the spring for equilibrium. Cable *AD* lies in the *x*–*y* plane and cable *AC* lies in the *x*–*z* plane.

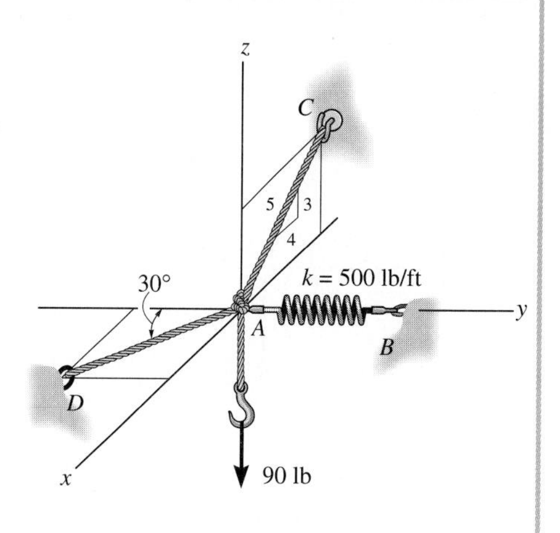

(a)

SOLUTION

The stretch of the spring can be determined once the force in the spring is determined.

Free-Body Diagram. The connection at *A* is chosen for the equilibrium analysis since the cable forces are concurrent at this point. The free-body diagram is shown in Fig. 3–11*b*.

Equations of Equilibrium. By inspection, each force can easily be resolved into its *x*, *y*, *z* components, and therefore the three scalar equations of equilibrium can be directly applied. Considering components directed along the positive axes as "positive," we have

$$\Sigma F_x = 0; \qquad F_D \sin 30^\circ - \tfrac{4}{5}F_C = 0 \qquad (1)$$

$$\Sigma F_y = 0; \qquad -F_D \cos 30^\circ + F_B = 0 \qquad (2)$$

$$\Sigma F_z = 0; \qquad \tfrac{3}{5}F_C - 90 \text{ lb} = 0 \qquad (3)$$

Solving Eq. 3 for F_C, then Eq. 1 for F_D, and finally Eq. 2 for F_B, we get

$$F_C = 150 \text{ lb} \qquad \textit{Ans.}$$

$$F_D = 240 \text{ lb} \qquad \textit{Ans.}$$

$$F_B = 208 \text{ lb} \qquad \textit{Ans.}$$

The stretch of the spring is therefore

$$F_B = ks_{AB}$$

$$208 \text{ lb} = 500 \text{ lb/ft } (s_{AB})$$

$$s_{AB} = 0.416 \text{ ft} \qquad \textit{Ans.}$$

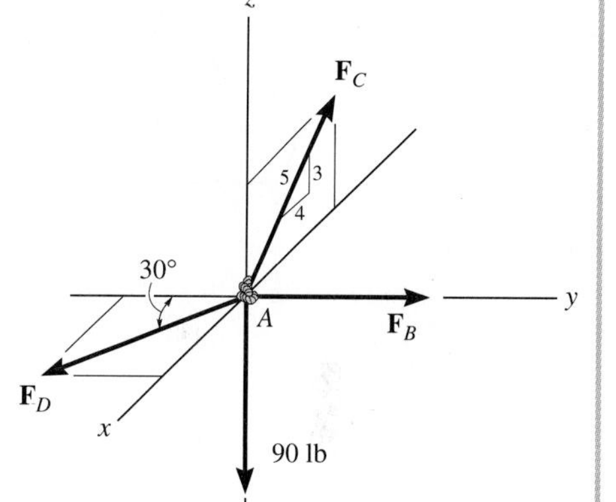

(b)

Fig. 3–11

Example 3–7

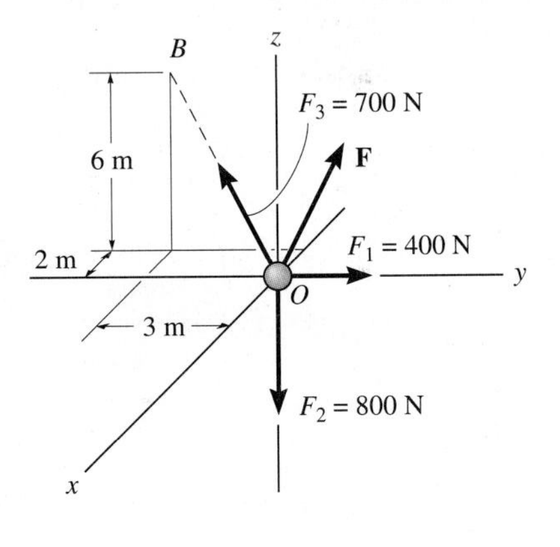

(a)

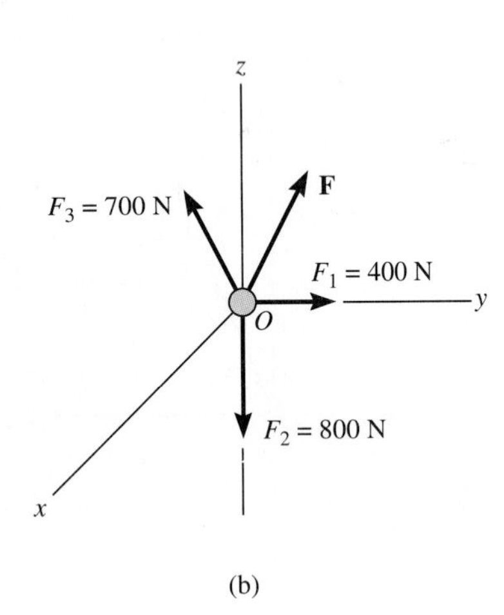

(b)

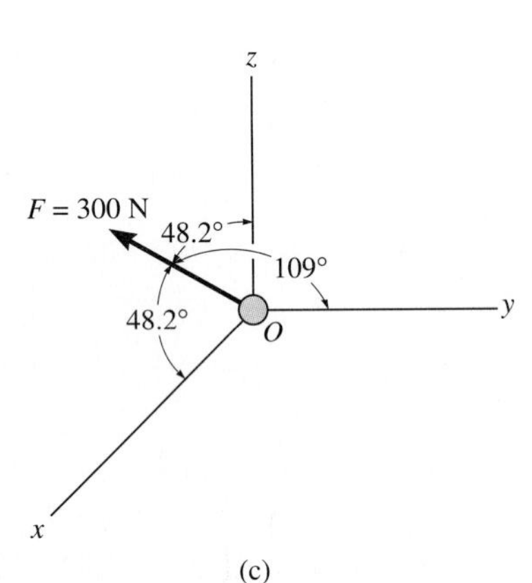

(c)

Fig. 3–12

Determine the magnitude and coordinate direction angles of force **F** in Fig. 3–12*a* that are required for equilibrium of particle *O*.

SOLUTION

Free-Body Diagram. Four forces act on particle *O*, Fig. 3–12*b*.

Equations of Equilibrium. Each of the forces can be expressed in Cartesian vector form and the equations of equilibrium can be applied to determine the *x*, *y*, *z* components of **F**. Hence, noting that the coordinates of *B* are $B(-2\text{ m}, -3\text{ m}, 6\text{ m})$, we have

$$\mathbf{F}_1 = \{400\mathbf{j}\}\text{ N}$$

$$\mathbf{F}_2 = \{-800\mathbf{k}\}\text{ N}$$

$$\mathbf{F}_3 = F_3\mathbf{u}_B = F_3\left(\frac{\mathbf{r}_B}{r_B}\right) = 700\text{ N}\left[\frac{-2\mathbf{i} - 3\mathbf{j} + 6\mathbf{k}}{\sqrt{(-2)^2 + (-3)^2 + (6)^2}}\right]$$

$$= \{-200\mathbf{i} - 300\mathbf{j} + 600\mathbf{k}\}\text{ N}$$

$$\mathbf{F} = F_x\mathbf{i} + F_y\mathbf{j} + F_z\mathbf{k}$$

For equilibrium

$$\Sigma\mathbf{F} = \mathbf{0};\ \mathbf{F}_1 + \mathbf{F}_2 + \mathbf{F}_3 + \mathbf{F} = \mathbf{0}$$

$$400\mathbf{j} - 800\mathbf{k} - 200\mathbf{i} - 300\mathbf{j} + 600\mathbf{k} + F_x\mathbf{i} + F_y\mathbf{j} + F_z\mathbf{k} = \mathbf{0}$$

Equating the respective **i, j, k** components to zero, we have

$$\Sigma F_x = 0; \qquad -200 + F_x = 0 \qquad F_x = 200\text{ N}$$

$$\Sigma F_y = 0; \qquad 400 - 300 + F_y = 0 \qquad F_y = -100\text{ N}$$

$$\Sigma F_z = 0; \qquad -800 + 600 + F_z = 0 \qquad F_z = 200\text{ N}$$

Thus,

$$\mathbf{F} = \{200\mathbf{i} - 100\mathbf{j} + 200\mathbf{k}\}\text{ N}$$

$$F = \sqrt{(200)^2 + (-100)^2 + (200)^2} = 300\text{ N} \qquad \textit{Ans.}$$

$$\mathbf{u}_F = \frac{\mathbf{F}}{F} = \frac{200}{300}\mathbf{i} - \frac{100}{300}\mathbf{j} + \frac{200}{300}\mathbf{k}$$

$$\alpha = \cos^{-1}\left(\frac{200}{300}\right) = 48.2° \qquad \textit{Ans.}$$

$$\beta = \cos^{-1}\left(\frac{-100}{300}\right) = 109° \qquad \textit{Ans.}$$

$$\gamma = \cos^{-1}\left(\frac{200}{300}\right) = 48.2° \qquad \textit{Ans.}$$

The magnitude and correct direction of **F** are shown in Fig. 3–12*c*.

Example 3–8

Determine the force developed in each cable used to support the 40-lb crate shown in Fig. 3–13*a*.

SOLUTION

Free-Body Diagram. As shown in Fig. 3–13*b*, the free-body diagram of point *A* is considered in order to "expose" the three unknown forces in the cables, and by applying the condition for equilibrium we can obtain their magnitudes.

Equations of Equilibrium. First we will express each force in Cartesian vector form. Since the coordinates of points *B* and *C* are $B(-3 \text{ ft}, -4 \text{ ft}, 8 \text{ ft})$ and $C(-3 \text{ ft}, 4 \text{ ft}, 8 \text{ ft})$, we have

$$\mathbf{F}_B = F_B\left[\frac{-3\mathbf{i} - 4\mathbf{j} + 8\mathbf{k}}{\sqrt{(-3)^2 + (-4)^2 + (8)^2}}\right]$$
$$= -0.318F_B\mathbf{i} - 0.424F_B\mathbf{j} + 0.848F_B\mathbf{k}$$
$$\mathbf{F}_C = F_C\left[\frac{-3\mathbf{i} + 4\mathbf{j} + 8\mathbf{k}}{\sqrt{(-3)^2 + (4)^2 + (8)^2}}\right]$$
$$= -0.318F_C\mathbf{i} + 0.424F_C\mathbf{j} + 0.848F_C\mathbf{k}$$
$$\mathbf{F}_D = F_D\mathbf{i}$$
$$\mathbf{W} = \{-40\mathbf{k}\} \text{ lb}$$

Equilibrium requires

$$\Sigma\mathbf{F} = \mathbf{0}; \qquad \mathbf{F}_B + \mathbf{F}_C + \mathbf{F}_D + \mathbf{W} = \mathbf{0}$$
$$-0.318F_B\mathbf{i} - 0.424F_B\mathbf{j} + 0.848F_B\mathbf{k} - 0.318F_C\mathbf{i} + 0.424F_C\mathbf{j} + 0.848F_C\mathbf{k} + F_D\mathbf{i} - 40\mathbf{k} = \mathbf{0}$$

Equating the respective **i, j, k** components to zero yields

$$\Sigma F_x = 0; \qquad -0.318F_B - 0.318F_C + F_D = 0 \qquad (1)$$
$$\Sigma F_y = 0; \qquad -0.424F_B + 0.424F_C = 0 \qquad (2)$$
$$\Sigma F_z = 0; \qquad 0.848F_B + 0.848F_C - 40 = 0 \qquad (3)$$

Equation 2 states that $F_B = F_C$. Thus, solving Eq. 3 for F_B and F_C and substituting the result into Eq. 1 to obtain F_D, we have

$$F_B = F_C = 23.6 \text{ lb} \qquad \textit{Ans.}$$
$$F_D = 15.0 \text{ lb} \qquad \textit{Ans.}$$

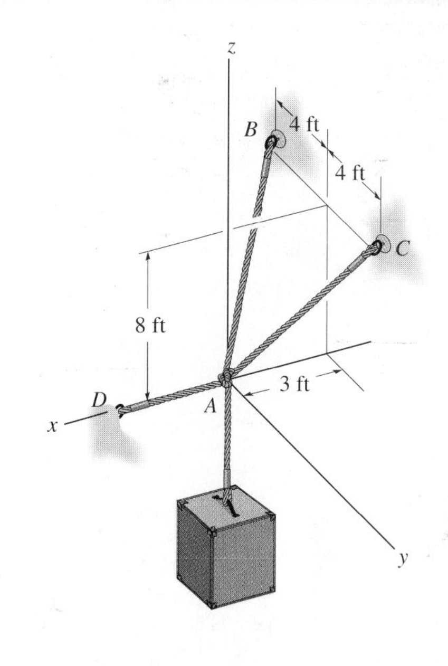

(a)

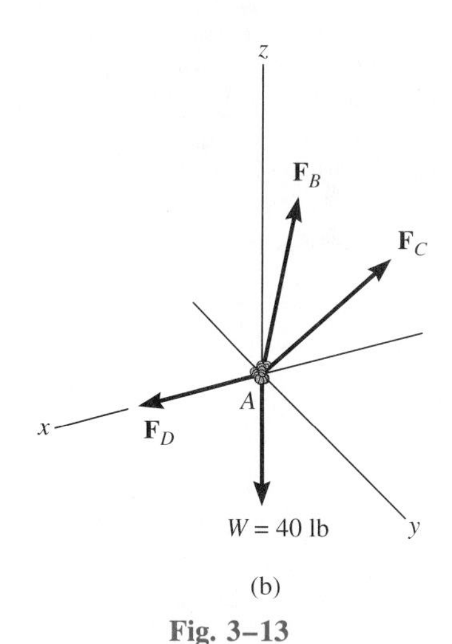

(b)

Fig. 3–13

Example 3–9

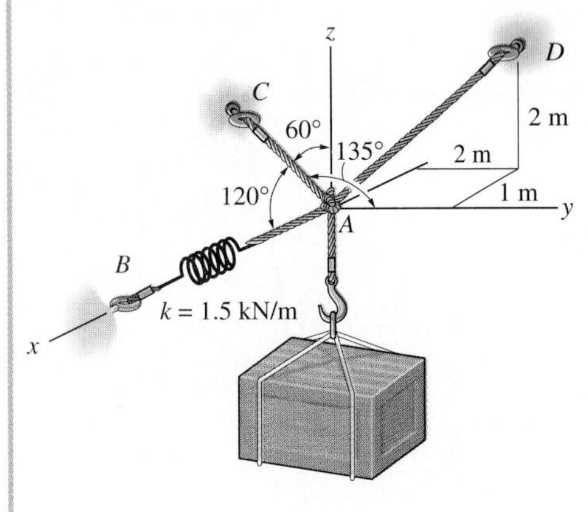

(a)

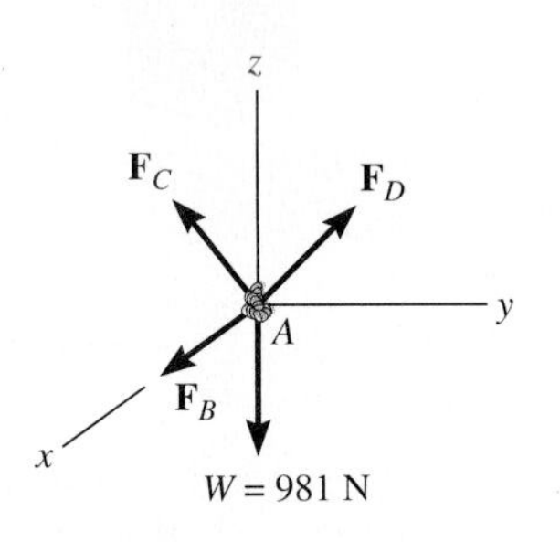

(b)

Fig. 3–14

The 100-kg box shown in Fig. 3–14*a* is supported by three cords, one of which is connected to a spring. Determine the tension in each cord and the stretch of the spring.

SOLUTION

Free-Body Diagram. The force in each of the cords can be determined by investigating the equilibrium of point A. The free-body diagram is shown in Fig. 3–14*b*. The weight of the cylinder is $W = 100(9.81) = 981$ N.

Equations of Equilibrium. Each vector on the free-body diagram is first expressed in Cartesian vector form. Using Eq. 2–11 for $\mathbf{F}_C$, and noting point $D(-1\text{ m}, 2\text{ m}, 2\text{ m})$ for $\mathbf{F}_D$, we have

$$\mathbf{F}_B = F_B\mathbf{i}$$

$$\mathbf{F}_C = F_C\cos 120°\mathbf{i} + F_C\cos 135°\mathbf{j} + F_C\cos 60°\mathbf{k}$$
$$= -0.5F_C\mathbf{i} - 0.707F_C\mathbf{j} + 0.5F_C\mathbf{k}$$

$$\mathbf{F}_D = F_D\left[\frac{-1\mathbf{i} + 2\mathbf{j} + 2\mathbf{k}}{\sqrt{(-1)^2 + (2)^2 + (2)^2}}\right]$$
$$= -0.333F_D\mathbf{i} + 0.667F_D\mathbf{j} + 0.667F_D\mathbf{k}$$

$$\mathbf{W} = \{-981\mathbf{k}\}\text{ N}$$

Equilibrium requires

$$\Sigma\mathbf{F} = \mathbf{0}; \qquad \mathbf{F}_B + \mathbf{F}_C + \mathbf{F}_D + \mathbf{W} = \mathbf{0}$$

$$F_B\mathbf{i} - 0.5F_C\mathbf{i} - 0.707F_C\mathbf{j} + 0.5F_C\mathbf{k} - 0.333F_D\mathbf{i} + 0.667F_D\mathbf{j} + 0.667F_D\mathbf{k} - 981\mathbf{k} = \mathbf{0}$$

Equating the respective **i, j, k** components to zero,

$$\Sigma F_x = 0; \qquad F_B - 0.5F_C - 0.333F_D = 0 \qquad (1)$$

$$\Sigma F_y = 0; \qquad -0.707F_C + 0.667F_D = 0 \qquad (2)$$

$$\Sigma F_z = 0; \qquad 0.5F_C + 0.667F_D - 981 = 0 \qquad (3)$$

Solving Eq. 2 for F_D in terms of F_C and substituting into Eq. 3 yields F_C. F_D is determined from Eq. 2. Finally, substituting the results into Eq. 1 yields F_B. Hence,

$$F_C = 813\text{ N} \qquad \textit{Ans.}$$

$$F_D = 862\text{ N} \qquad \textit{Ans.}$$

$$F_B = 693.7\text{ N} \qquad \textit{Ans.}$$

The stretch of the spring is therefore

$$F = ks; \qquad 693.7 = 1500s$$

$$s = 0.462\text{ m} \qquad \textit{Ans.}$$

PROBLEMS

*3–40. Determine the magnitudes of $\mathbf{F}_1$, $\mathbf{F}_2$, and $\mathbf{F}_3$ for equilibrium of the particle.

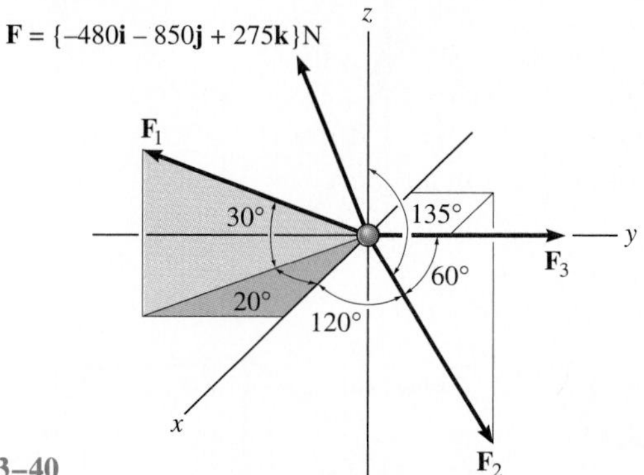

Prob. 3–40

3–41. Determine the magnitudes of $\mathbf{F}_1$, $\mathbf{F}_2$, and $\mathbf{F}_3$ for equilibrium of the particle.

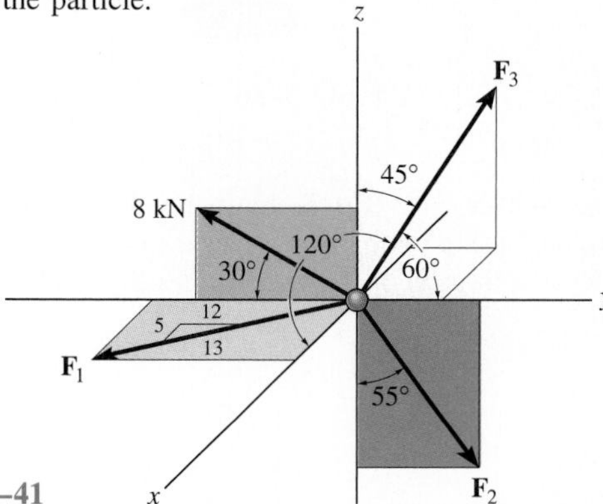

Prob. 3–41

3–42. Determine the magnitudes of $\mathbf{F}_1$, $\mathbf{F}_2$, and $\mathbf{F}_3$ for equilibrium of the particle.

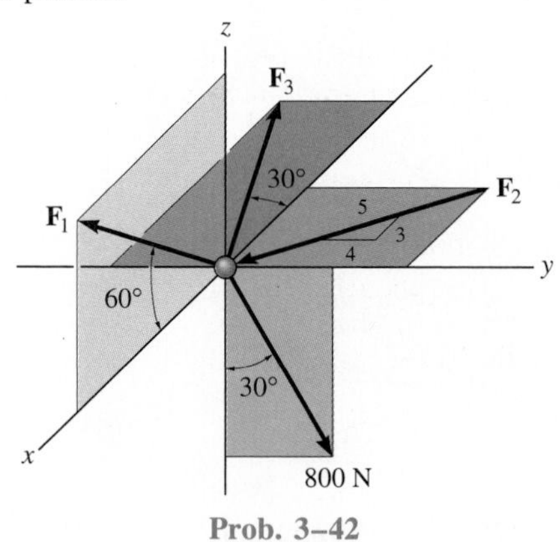

Prob. 3–42

3–43. Determine the magnitudes of $\mathbf{F}_1$, $\mathbf{F}_2$, and $\mathbf{F}_3$ for equilibrium of the particle.

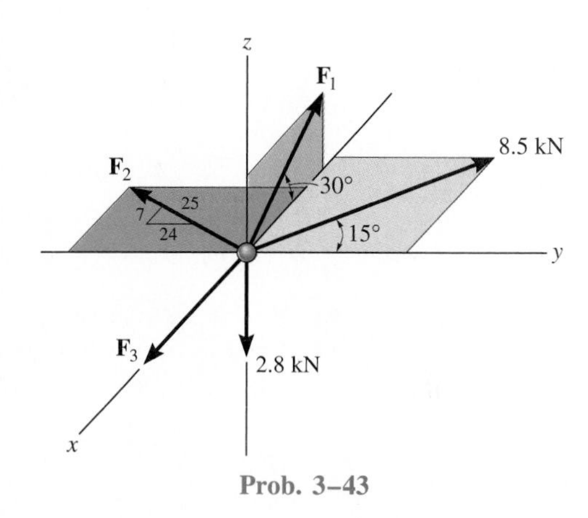

Prob. 3–43

*3–44. The 25-kg flowerpot is supported at *A* by the three cords. Determine the force acting in each cord for equilibrium.

3–45. If each cord can sustain a maximum tension of 50 N before it fails, determine the greatest weight of the flowerpot the cords can support.

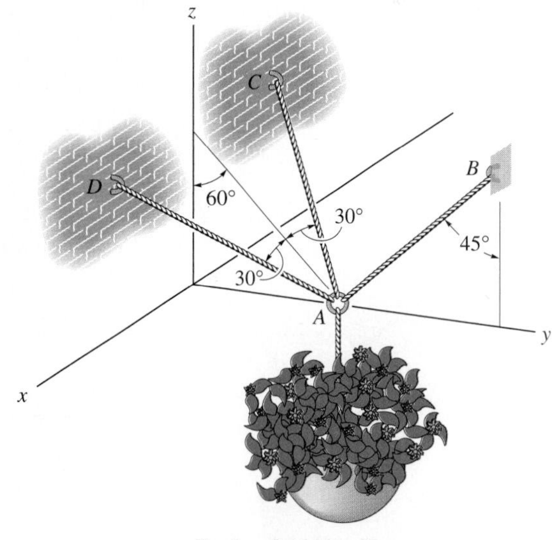

Probs. 3–44/3–45

3–46. Determine the tension developed in the three cables required to support the traffic light, which has a mass of 15 kg. Take $h = 4$ m.

3–47. Determine the tension developed in the three cables required to support the traffic light, which has a mass of 20 kg. Take $h = 3.5$ m.

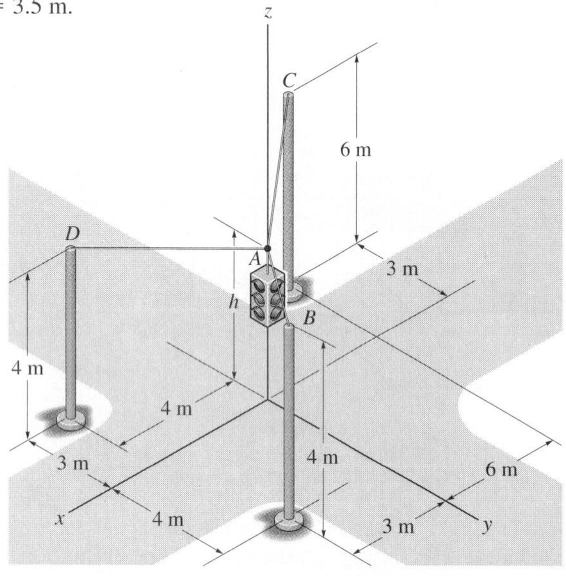

Probs. 3–46/3–47

***3–48.** If the bucket and its contents have a total weight of 20 lb, determine the force in the supporting cables *DA*, *DB*, and *DC*.

3–49. If each cable can sustain a maximum tension of 600 lb, determine the greatest weight of the bucket and its contents that can be supported.

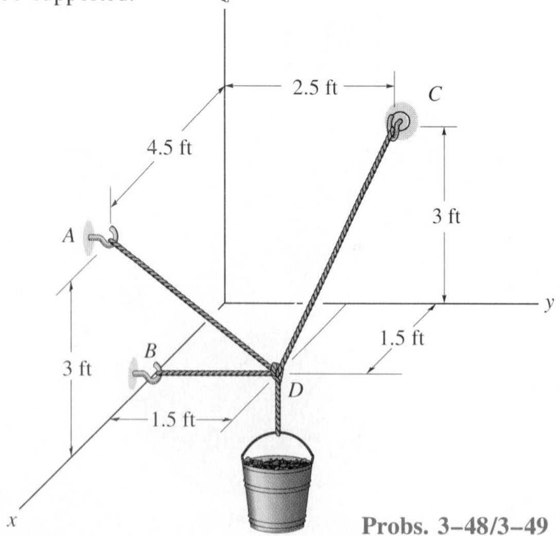

Probs. 3–48/3–49

3–50. The three cables are used to support the 800-N lamp. Determine the force developed in each cable for equilibrium.

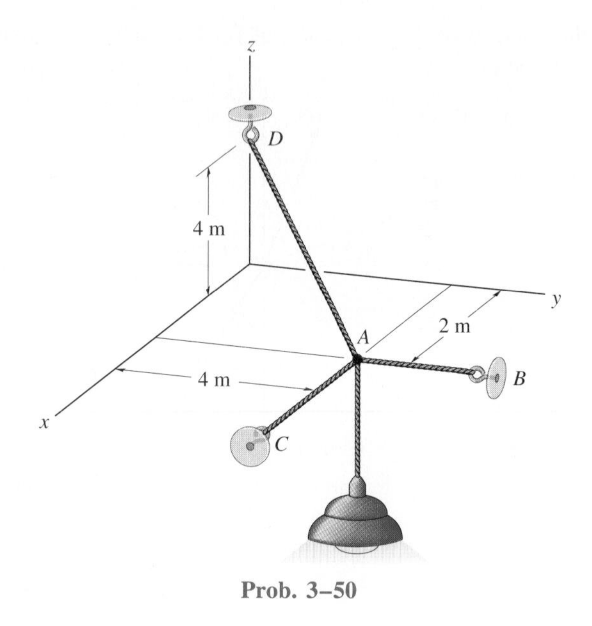

Prob. 3–50

■**3–51.** The 2500-N crate is to be hoisted with constant velocity from the hold of a ship using the cable arrangement shown. Determine the tension in each of the three cables for equilibrium.

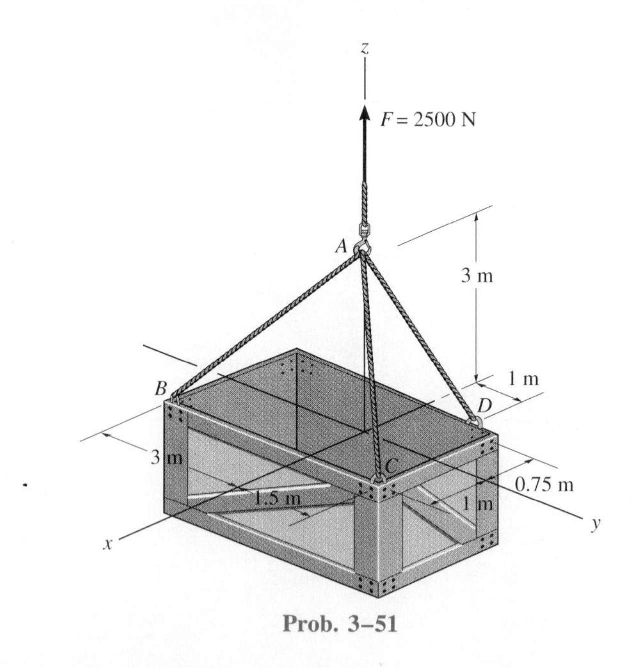

Prob. 3–51

*■3–52. The lamp has a mass of 15 kg and is supported by a pole AO and cables AB and AC. If the force in the pole acts along its axis, determine the forces in AO, AB, and AC for equilibrium.

3–53. Cables AB and AC can sustain a maximum tension of 500 N, and the pole can support a maximum compression of 300 N. Determine the maximum weight of the lamp that can be supported in the position shown. The force in the pole acts along the axis of the pole.

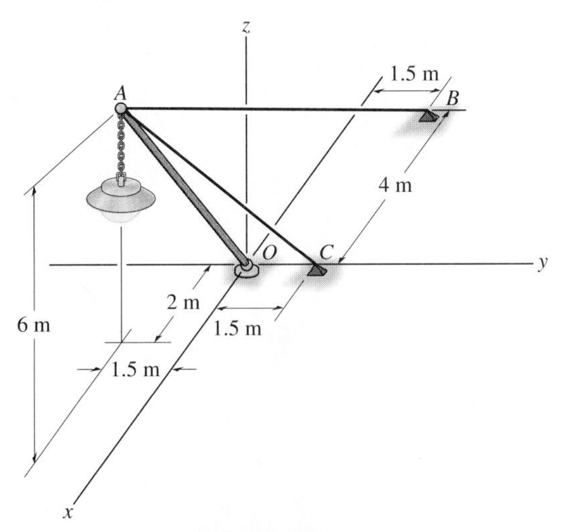

Probs. 3–52/3–53

3–54. The mast OA is supported by three cables. If cable AB is subjected to a tension of 500 N, determine the tension in cables AC and AD and the vertical force F_{AO} which the mast exerts along its axis at A.

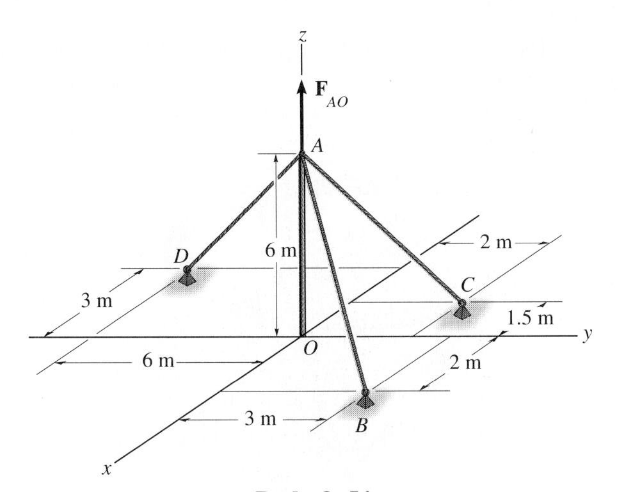

Prob. 3–54

3–55. The 500-lb crate is suspended from the cable system shown. Determine the force in each segment of the cable, i.e., AB, AC, CD, CE, and CF. *Hint:* First analyze the equilibrium of point A, then using the result for AC, analyze the equilibrium of point C.

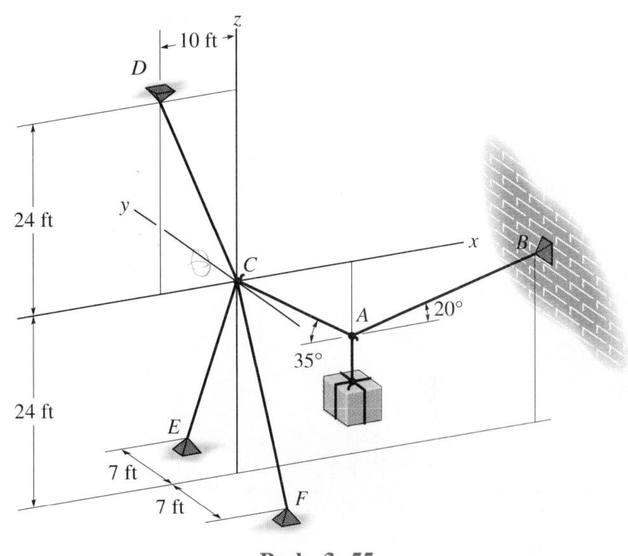

Prob. 3–55

*3–56. The 9500-lb crucible is supported by three cables. Determine the force in each cable for equilibrium of the hook at A.

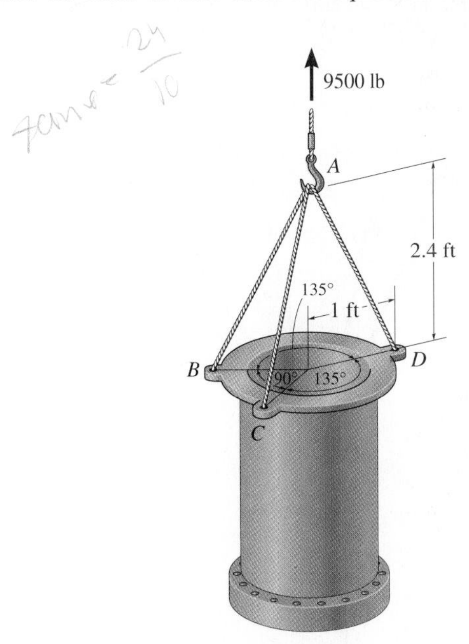

Prob. 3–56

3–57. Determine the force in each cable needed to support the 3500-lb platform. Set $d = 2$ ft.

3–58. Determine the force in each cable needed to support the 3500-lb platform. Set $d = 4$ ft.

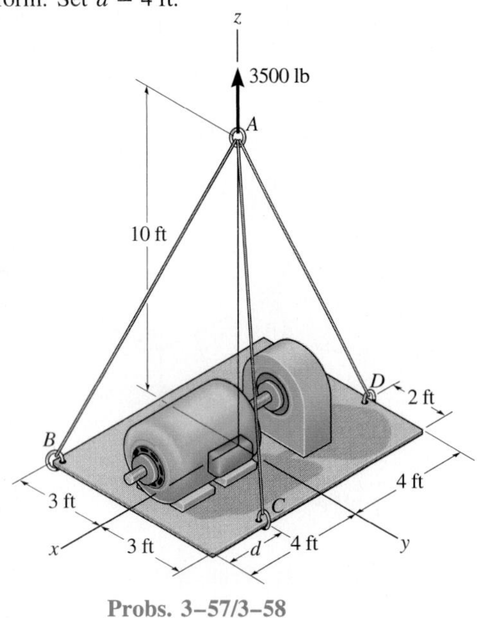

Probs. 3–57/3–58

3–59. The 800-lb cylinder is supported by three chains as shown. Determine the force in each chain for equilibrium. Take $d = 1$ ft.

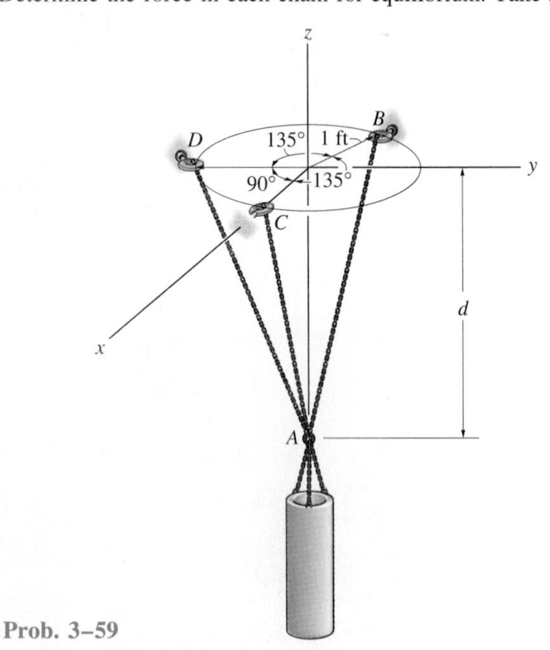

Prob. 3–59

***3–60.** The 80-lb chandelier is supported by three wires as shown. Determine the force in each wire for equilibrium.

3–61. If each wire can sustain a maximum tension of 120 lb before it fails, determine the greatest weight of the chandelier the wires will support in the position shown.

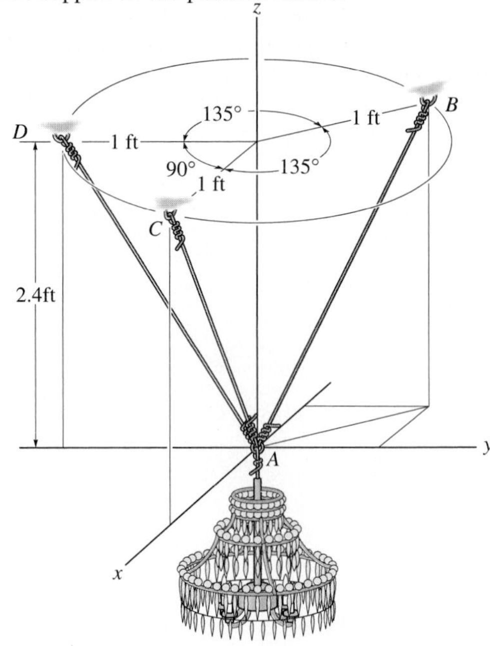

Probs. 3–60/3–61

■**3–62.** The 80-lb ball is suspended from the horizontal ring using three springs each having an unstretched length of 1.5 ft and stiffness of $k = 50$ lb/ft. Determine the vertical distance h from the ring to point A for equilibrium.

3–63. The ball is suspended from the horizontal ring using three springs each having a stiffness of $k = 50$ lb/ft and an unstretched length of 1.5 ft. If $h = 2$ ft, determine the weight of the ball.

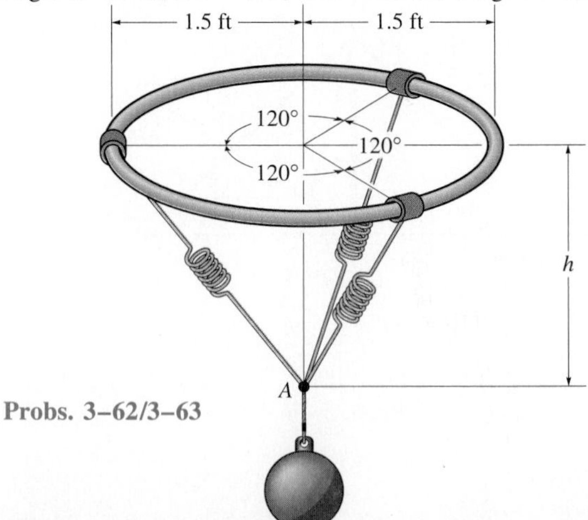

Probs. 3–62/3–63

REVIEW PROBLEMS

***3–64.** The man attempts to pull the log at C by using the three ropes. Determine the direction θ in which he should pull on his rope with a force of 80 lb, so that he exerts a maximum force on the log. What is the force on the log for this case? Also, determine the direction in which he should pull in order to maximize the force in the rope attached to B. What is this maximum force?

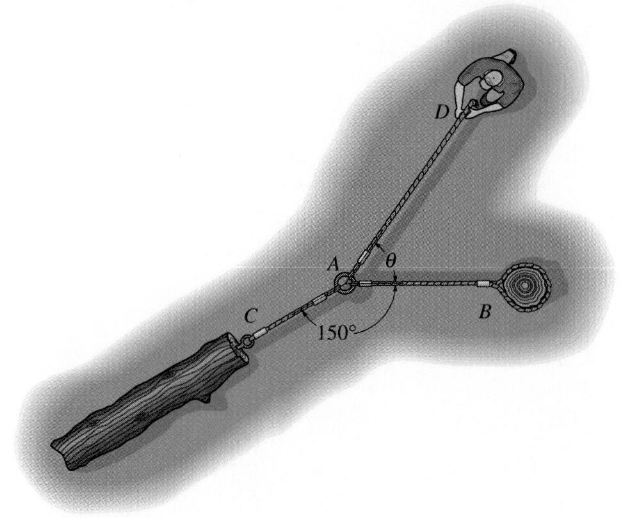

Prob. 3–64

3–65. Determine the tension in cables AB, AC, and AD, required to hold the 60-lb crate in equilibrium.

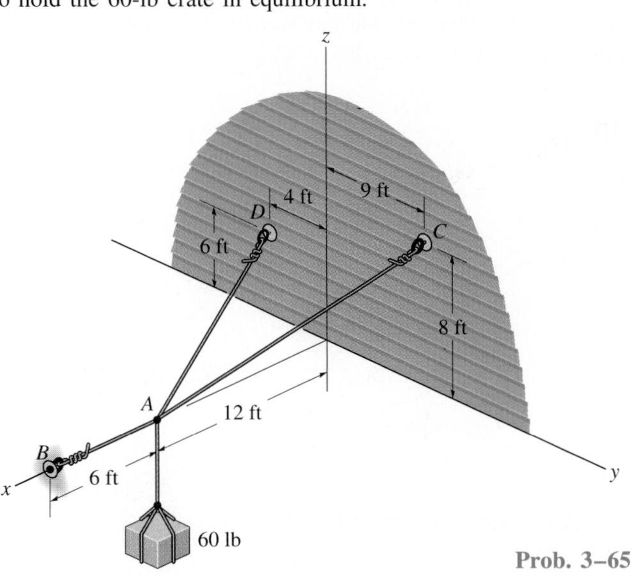

Prob. 3–65

3–66. Romeo tries to reach Juliet by climbing with constant velocity up a rope which is knotted at point A. Any of the three segments of the rope can sustain a maximum force of 2 kN before it breaks. Determine if Romeo, who has a mass of 65 kg, can climb the rope, and if so, can he along with his Juliet, who has a mass of 60 kg, climb down with constant velocity?

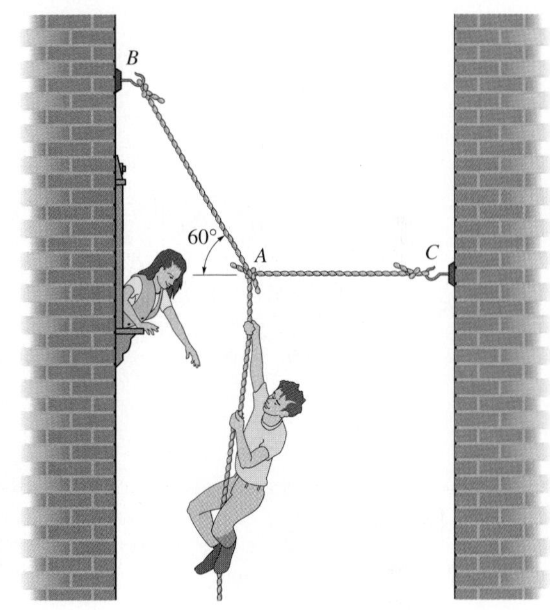

Prob. 3–66

3–67. The 30-kg block is supported by two springs having the stiffness shown. Determine the unstretched length of each spring after the block is removed.

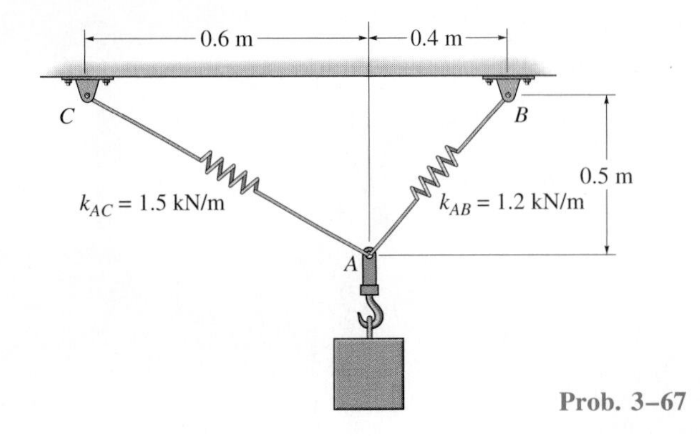

Prob. 3–67

*■**3–68.** Determine the magnitudes of forces $\mathbf{F}_1$, $\mathbf{F}_2$, and $\mathbf{F}_3$ necessary to hold the force $\mathbf{F} = \{-9\mathbf{i} - 8\mathbf{j} - 5\mathbf{k}\}$ kN in equilibrium.

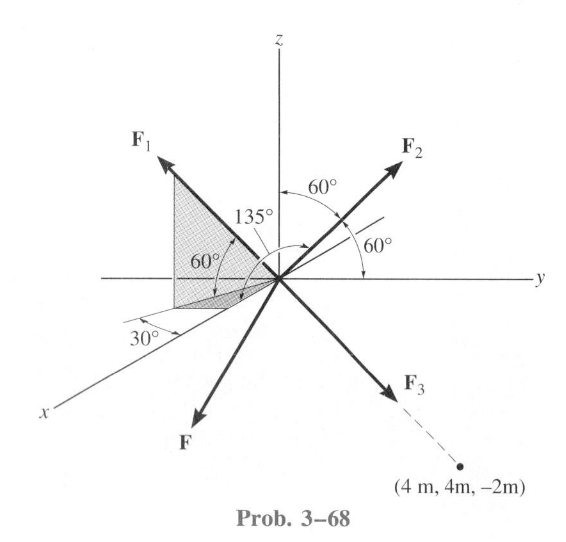

Prob. 3–68

3–69. The boy has a mass of 60 kg and attempts to cross the creek by using the pulley and 15-m-long rope shown. If he leaves the shore at A, determine how close s he comes to the shore at B once he reaches a state of equilibrium.

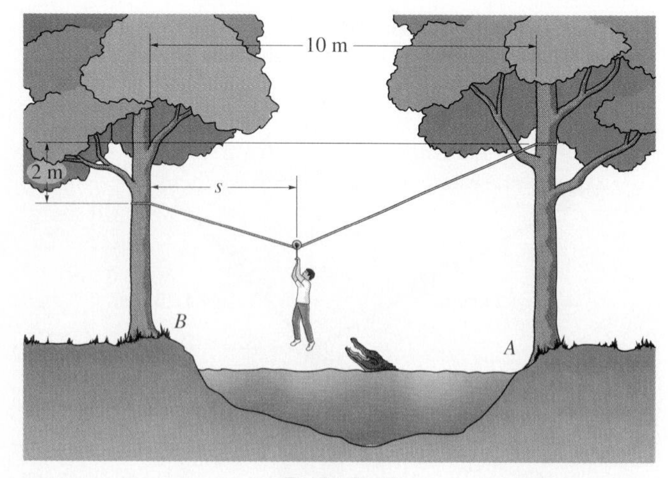

Prob. 3–69

3–70. The 2-m-long cord AB is attached to a spring BC having an unstretched length of 2 m. If the cord sags downward an amount $\theta = 30°$ as shown, determine the vertical force F applied. The spring has a stiffness of $k = 50$ N/m.

Prob. 3–70

3–71. The bulldozer attempts to pull down the chimney using the cable and *small* pulley arrangement shown. If the tension in AB is 600 lb, determine the tension in cable CAD and the angle θ which the cable makes at the pulley.

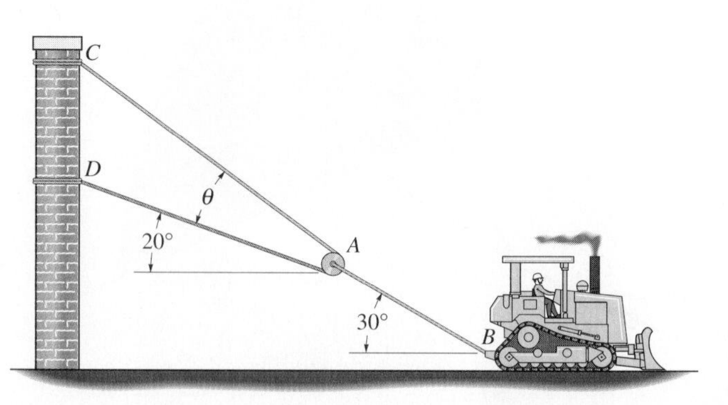

Prob. 3–71

*3–72. Determine the magnitudes of $\mathbf{F}_1$, $\mathbf{F}_2$, and $\mathbf{F}_3$ for equilibrium.

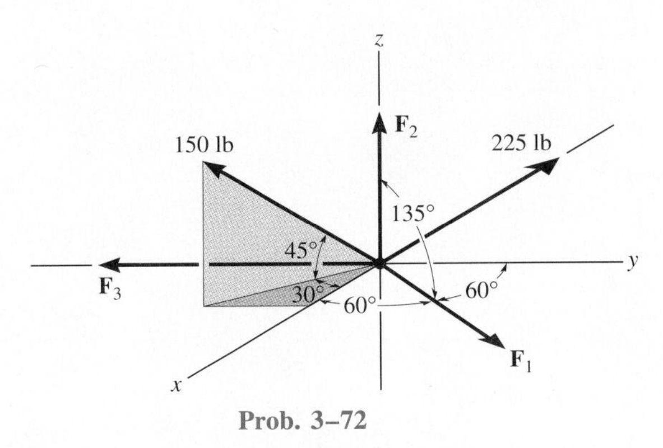

Prob. 3–72

3–73. Determine the maximum weight of the engine that can be supported without exceeding a tension of 450 lb in chain AB and 480 lb in chain AC.

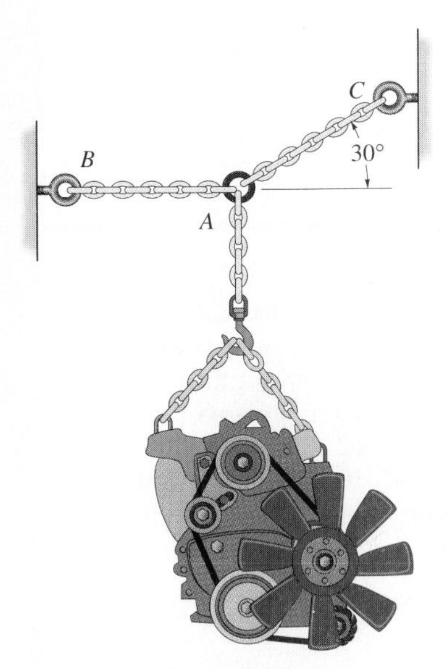

Prob. 3–73

Oftentimes a structure is subjected to a system of forces. Here the many forces of the cable-stayed bridge support the deck; the cables of the crane exert forces on the yellow spreader beam, which in turn supports the metal spool. In this chapter we will study ways of representing these force systems by their resultants.

4

Force System Resultants

In Chapter 3 it was shown that the condition for the equilibrium of a particle or a concurrent force system simply requires that the resultant of the force system be equal to zero, i.e., $\Sigma \mathbf{F} = \mathbf{0}$. In Chapter 5 it will be shown that such a restriction is necessary but not sufficient for the equilibrium of a rigid body. Since a body has a physical size, a further restriction must be made with regard to the nonconcurrency of the applied force system, giving rise to the concept of *moment*. A moment tends to turn a body, and equilibrium requires the body to have no rotation.

In this chapter a formal definition of a moment will be presented and ways of finding the moment of a force about a point or axis will be discussed. We will also present methods for determining the resultants of nonconcurrent force systems. This development is important since application of the equations for force-system simplification is similar to applying the equations of equilibrium for a rigid body. Furthermore, the resultants of a force system will influence the state of equilibrium or motion of a rigid body in the same way as the force system, and therefore we can study rigid-body behavior in a simpler manner by using the resultants.

4.1 Cross Product

The moment of a force will be formulated using Cartesian vectors in the next section. Before doing this, however, it is first necessary to expand our knowledge of vector algebra and introduce the cross-product method of vector multiplication.

The *cross product* of two vectors **A** and **B** yields the vector **C**, which is written

$$\mathbf{C} = \mathbf{A} \times \mathbf{B}$$

and is read "**C** equals **A** cross **B**."

Magnitude. The *magnitude* of **C** is defined as the product of the magnitudes of **A** and **B** and the sine of the angle θ between their tails ($0° \leq \theta \leq 180°$). Thus, $C = AB \sin \theta$.

Direction. Vector **C** has a *direction* that is perpendicular to the plane containing **A** and **B** such that the direction of **C** is specified by the right-hand rule; i.e., curling the fingers of the right hand from vector **A** (cross) to vector **B,** the thumb then points in the direction of **C,** as shown in Fig. 4–1.

Knowing both the magnitude and direction of **C,** we can write

$$\mathbf{C} = \mathbf{A} \times \mathbf{B} = (AB \sin \theta)\mathbf{u}_C \qquad (4\text{–}1)$$

where the scalar $AB \sin \theta$ defines the *magnitude* of **C** and the unit vector $\mathbf{u}_C$ defines the *direction* of **C.** The terms of Eq. 4–1 are illustrated graphically in Fig. 4–2.

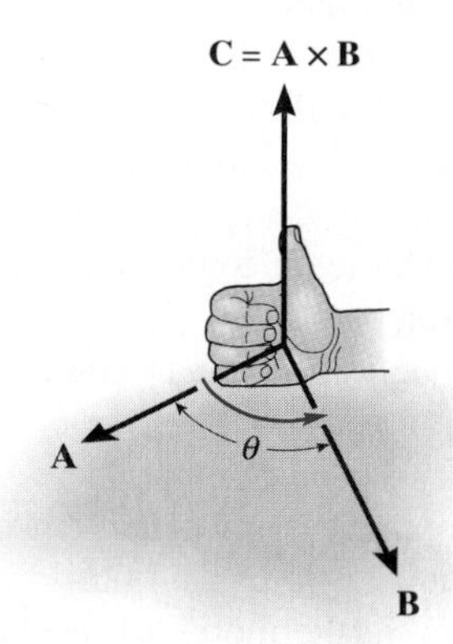

Fig. 4–1

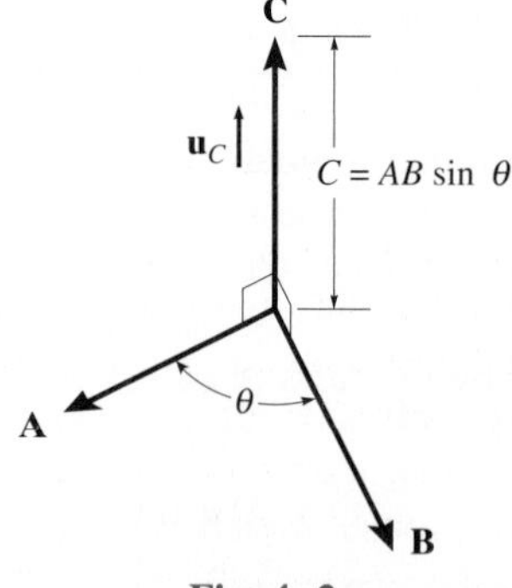

Fig. 4–2

Laws of Operation

1. The commutative law is *not* valid; i.e.,

$$\mathbf{A} \times \mathbf{B} \neq \mathbf{B} \times \mathbf{A}$$

Rather,

$$\mathbf{A} \times \mathbf{B} = -\mathbf{B} \times \mathbf{A}$$

This is shown in Fig. 4–3 by using the right-hand rule. The cross product $\mathbf{B} \times \mathbf{A}$ yields a vector that acts in the opposite direction to $\mathbf{C}$; i.e., $\mathbf{B} \times \mathbf{A} = -\mathbf{C}$.

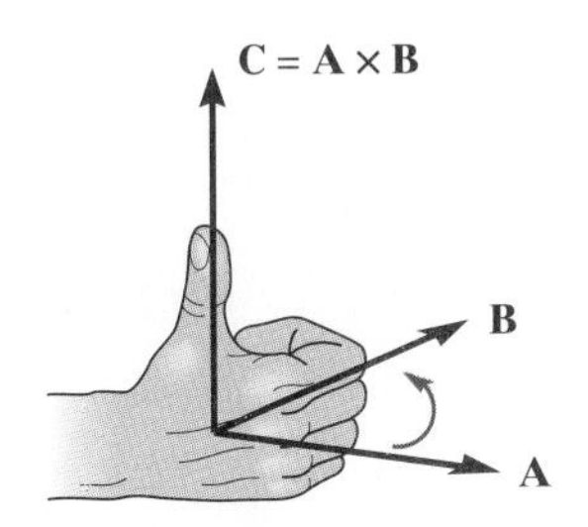

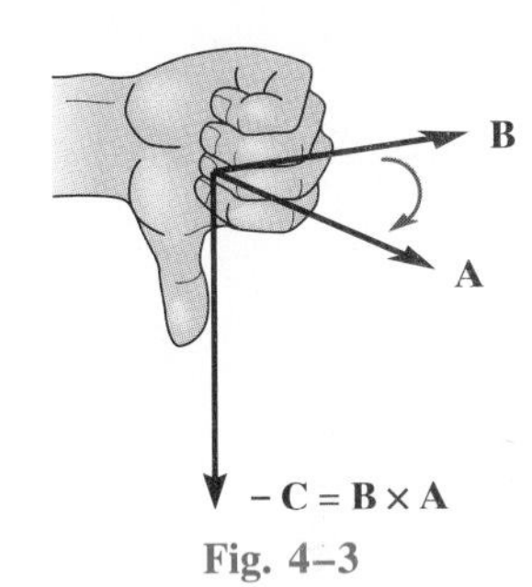

Fig. 4–3

2. Multiplication by a scalar:

$$a(\mathbf{A} \times \mathbf{B}) = (a\mathbf{A}) \times \mathbf{B} = \mathbf{A} \times (a\mathbf{B}) = (\mathbf{A} \times \mathbf{B})a$$

This property is easily shown, since the magnitude of the resultant vector ($|a|AB \sin \theta$) and its direction are the same in each case.

3. The distributive law:

$$\mathbf{A} \times (\mathbf{B} + \mathbf{D}) = (\mathbf{A} \times \mathbf{B}) + (\mathbf{A} \times \mathbf{D})$$

The proof of this identity is left as an exercise (see Prob. 4–1). It is important to note that *proper order* of the cross products must be maintained, since they are not commutative.

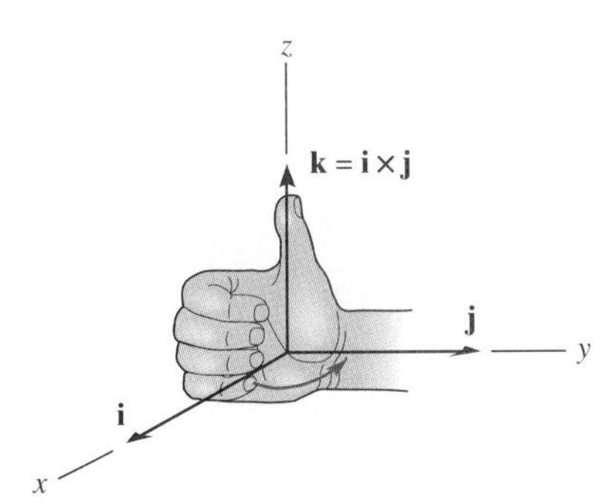

Fig. 4–4

Cartesian Vector Formulation. Equation 4–1 may be used to find the cross product of a pair of Cartesian unit vectors. For example, to find $\mathbf{i} \times \mathbf{j}$, the *magnitude* of the resultant vector is $(i)(j)(\sin 90°) = (1)(1)(1) = 1$, and its *direction* is determined using the right-hand rule. As shown in Fig. 4–4, the resultant vector points in the $+\mathbf{k}$ direction. Thus, $\mathbf{i} \times \mathbf{j} = (1)\mathbf{k}$. In a similar manner,

$$\begin{array}{lll} \mathbf{i} \times \mathbf{j} = \mathbf{k} & \mathbf{i} \times \mathbf{k} = -\mathbf{j} & \mathbf{i} \times \mathbf{i} = \mathbf{0} \\ \mathbf{j} \times \mathbf{k} = \mathbf{i} & \mathbf{j} \times \mathbf{i} = -\mathbf{k} & \mathbf{j} \times \mathbf{j} = \mathbf{0} \\ \mathbf{k} \times \mathbf{i} = \mathbf{j} & \mathbf{k} \times \mathbf{j} = -\mathbf{i} & \mathbf{k} \times \mathbf{k} = \mathbf{0} \end{array}$$

These results should *not* be memorized; rather, it should be clearly understood how each is obtained by using the right-hand rule and the definition of the cross product. A simple scheme shown in Fig. 4–5 is helpful for obtaining the same results when the need arises. If the circle is constructed as shown, then "crossing" two unit vectors in a *counterclockwise* fashion around the circle yields the *positive* third unit vector; e.g., $\mathbf{k} \times \mathbf{i} = \mathbf{j}$. Moving *clockwise*, a *negative* unit vector is obtained; e.g., $\mathbf{i} \times \mathbf{k} = -\mathbf{j}$.

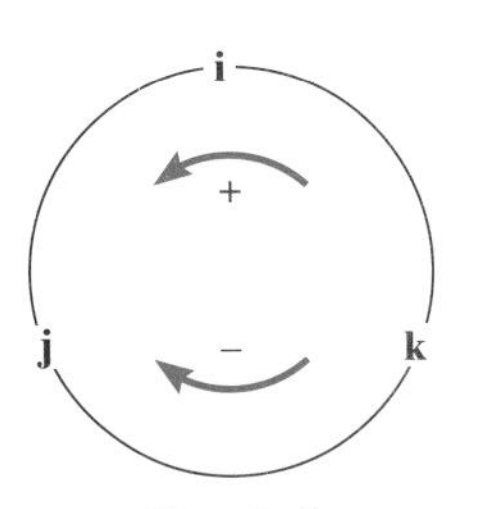

Fig. 4–5

Consider now the cross product of two general vectors **A** and **B** which are expressed in Cartesian vector form. We have

$$\begin{aligned}\mathbf{A} \times \mathbf{B} &= (A_x\mathbf{i} + A_y\mathbf{j} + A_z\mathbf{k}) \times (B_x\mathbf{i} + B_y\mathbf{j} + B_z\mathbf{k}) \\ &= A_xB_x(\mathbf{i} \times \mathbf{i}) + A_xB_y(\mathbf{i} \times \mathbf{j}) + A_xB_z(\mathbf{i} \times \mathbf{k}) \\ &\qquad + A_yB_x(\mathbf{j} \times \mathbf{i}) + A_yB_y(\mathbf{j} \times \mathbf{j}) + A_yB_z(\mathbf{j} \times \mathbf{k}) \\ &\qquad\qquad + A_zB_x(\mathbf{k} \times \mathbf{i}) + A_zB_y(\mathbf{k} \times \mathbf{j}) + A_zB_z(\mathbf{k} \times \mathbf{k})\end{aligned}$$

Carrying out the cross-product operations and combining terms yields

$$\mathbf{A} \times \mathbf{B} = (A_yB_z - A_zB_y)\mathbf{i} - (A_xB_z - A_zB_x)\mathbf{j} + (A_xB_y - A_yB_x)\mathbf{k} \tag{4–2}$$

This equation may also be written in a more compact determinant form as

$$\mathbf{A} \times \mathbf{B} = \begin{vmatrix} \mathbf{i} & \mathbf{j} & \mathbf{k} \\ A_x & A_y & A_z \\ B_x & B_y & B_z \end{vmatrix} \tag{4–3}$$

Thus, to find the cross product of any two Cartesian vectors **A** and **B,** it is necessary to expand a determinant whose first row of elements consists of the unit vectors **i, j,** and **k** and whose second and third rows represent the x, y, z components of the two vectors **A** and **B,** respectively.*

*A determinant having three rows and three columns can be expanded using three minors, each of which is multiplied by one of the three terms in the first row. There are four elements in each minor, e.g.,

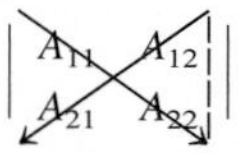

$$\begin{vmatrix} A_{11} & A_{12} \\ A_{21} & A_{22} \end{vmatrix}$$

By *definition,* this notation represents the terms $(A_{11}A_{22} - A_{12}A_{21})$, which is simply the product of the two elements of the arrow slanting downward to the right $(A_{11}A_{22})$ *minus* the product of the two elements intersected by the arrow slanting downward to the left $(A_{12}A_{21})$. For a 3×3 determinant, such as Eq. 4–3, the three minors can be generated in accordance with the following scheme:

For element **i:** $\begin{vmatrix} \mathbf{i} & \mathbf{j} & \mathbf{k} \\ A_x & A_y & A_z \\ B_x & B_y & B_z \end{vmatrix} = \mathbf{i}(A_yB_z - A_zB_y)$

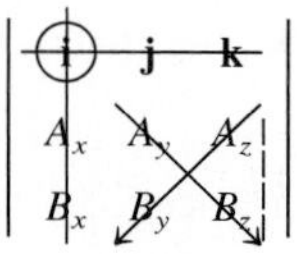

For element **j:** $\begin{vmatrix} \mathbf{i} & \mathbf{j} & \mathbf{k} \\ A_x & A_y & A_z \\ B_x & B_y & B_z \end{vmatrix} = -\mathbf{j}(A_xB_z - A_zB_x)$

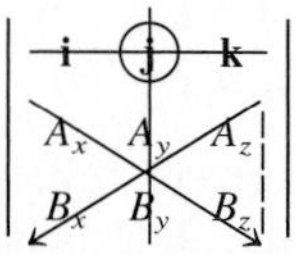

For element **k:** $\begin{vmatrix} \mathbf{i} & \mathbf{j} & \mathbf{k} \\ A_x & A_y & A_z \\ B_x & B_y & B_z \end{vmatrix} = \mathbf{k}(A_xB_y - A_yB_x)$

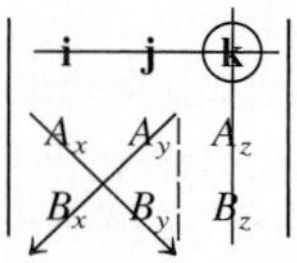

Adding the results and noting that the **j** element *must include the minus sign* yields the expanded form of **A × B** given by Eq. 4–2.

4.2 Moment of a Force—Scalar Formulation

The *moment* of a force about a point or axis provides a measure of the tendency of the force to cause a body to rotate about the point or axis. For example, consider the horizontal force $\mathbf{F}_x$, which acts perpendicular to the handle of the wrench and is located a distance d_y from point O, Fig. 4–6*a*. It is seen that this force tends to cause the pipe to turn about the z axis. The larger the force or the length d_y, the greater the turning effect. This tendency for rotation caused by $\mathbf{F}_x$ is sometimes called a *torque,* but most often it is called the *moment of a force* or simply the *moment* $(\mathbf{M}_O)_z$. In particular, note that the *moment axis* (z) is perpendicular to the shaded plane (x–y) which contains both $\mathbf{F}_x$ and d_y and that this axis intersects the plane at point O. Now consider applying the force $\mathbf{F}_z$ to the wrench, Fig. 4–6*b*. This force will *not* rotate the pipe about the z axis. Instead, it tends to rotate it about the x axis. Keep in mind that although it may not be possible actually to "rotate" or turn the pipe in this manner, $\mathbf{F}_z$ still creates the *tendency* for rotation and so the moment $(\mathbf{M}_O)_x$ is produced. As before, the force and distance d_y lie in the shaded plane (y–z) which is perpendicular to the moment axis (x). Lastly, if a force $\mathbf{F}_y$ is applied to the wrench, Fig. 4–6*c*, no moment is produced about point O. This lack of turning effect results, since the line of action of the force passes through O and therefore no tendency for rotation is possible.

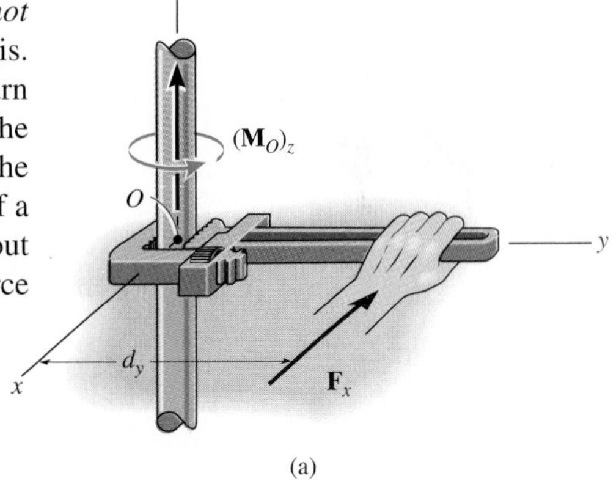

(a)

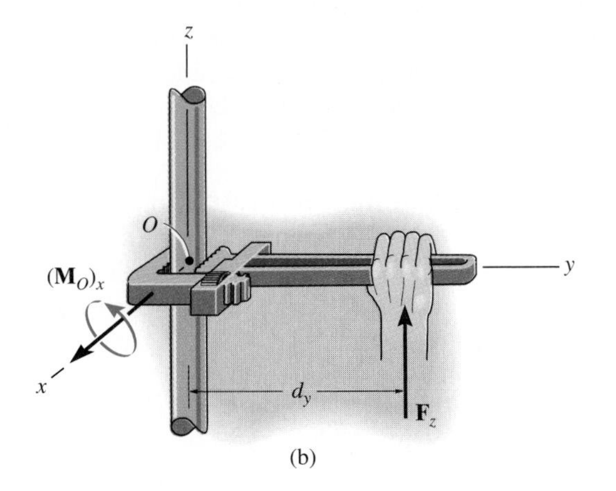

(b)

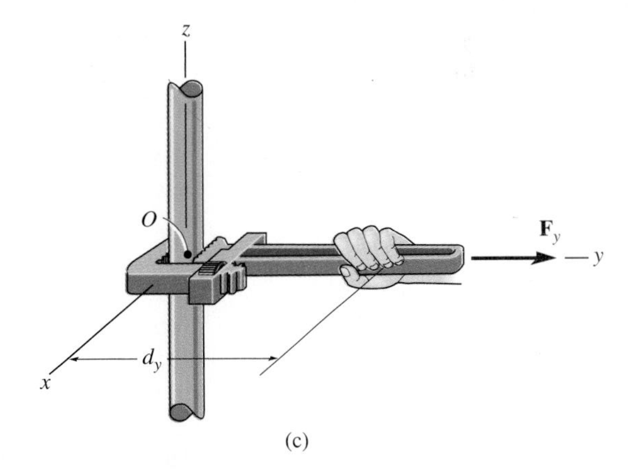

(c)

Fig. 4–6

We will now generalize the above discussion and consider the force $\mathbf{F}$ and point O which lie in a shaded plane as shown in Fig. 4–7*a*. The moment $\mathbf{M}_O$ about point O, or about an axis passing through O and perpendicular to the plane, is a *vector quantity* since it has a specified magnitude and direction.

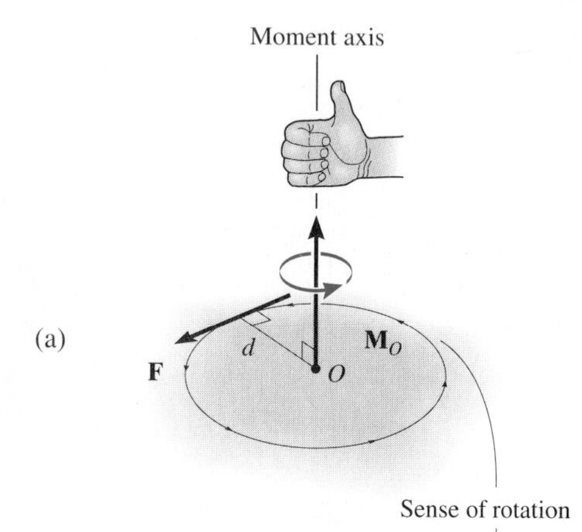

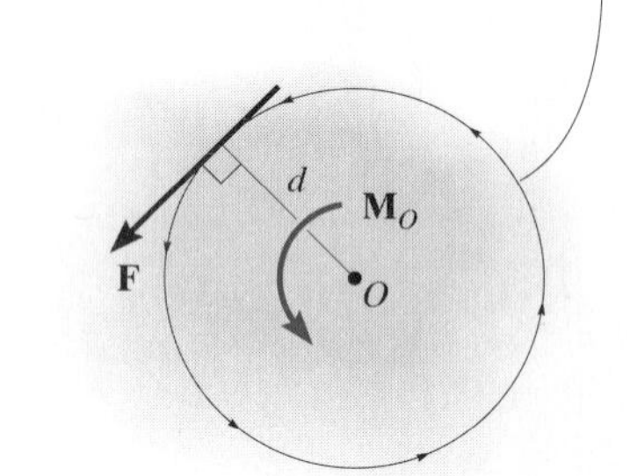

Fig. 4–7

Magnitude. The magnitude of $\mathbf{M}_O$ is

$$M_O = Fd \tag{4–4}$$

where d is referred to as the *moment arm* or perpendicular distance from the axis at point O to the line of action of the force. Units of moment magnitude consist of force times distance, e.g., N · m or lb · ft.

Direction. The direction of $\mathbf{M}_O$ will be specified by using the ''right-hand rule.'' To do this, the fingers of the right hand are curled such that they follow the sense of rotation, which would occur if the force could rotate about point O, Fig. 4–7*a*. The *thumb* then *points* along the *moment axis* so that it gives the direction and sense of the moment vector, which is *upward* and *perpendicular* to the shaded plane containing **F** and d. By this definition, the moment $\mathbf{M}_O$ can be considered as a *sliding vector* and therefore acts at any point along the moment axis.

In three dimensions, $\mathbf{M}_O$ is illustrated by a vector arrow with a curl on it to *distinguish* it from a force vector, Fig. 4–7*a*. Many problems in mechanics, however, involve coplanar force systems that may be conveniently viewed in two dimensions. For example, a two-dimensional view of Fig. 4–7*a* is given in Fig. 4–7*b*. Here $\mathbf{M}_O$ is simply represented by the (counterclockwise) curl, which indicates the action of **F.** The arrowhead on this curl is used to show the *sense of rotation* caused by **F.** Using the right-hand rule, however, realize that the direction and sense of the moment vector in Fig. 4–7*b* are specified by the thumb, which points *out* of the page, since the fingers follow the curl. In particular, notice that *this curl or sense of rotation can always be determined by observing in which direction the force would ''orbit'' about point O* (counterclockwise in Fig. 4–7*b*). In two dimensions we will often refer to finding the moment of a force ''about a point'' (O). Keep in mind, however, that the moment *always acts about an axis* which is perpendicular to the plane containing **F** and d, and this axis intersects the plane at the point (O), Fig. 4–7*a*.

Resultant Moment of a System of Coplanar Forces. If a system of forces all lie in an x–y plane, then the moment produced by each force about point O will be directed along the z axis, Fig. 4–8. Consequently, the resultant moment $\mathbf{M}_{R_O}$ of the system can be determined by simply adding the moments of all forces *algebraically,* since all the moment vectors are collinear. We can write this vector sum symbolically as

$$\circlearrowleft + M_{R_O} = \Sigma Fd \tag{4–5}$$

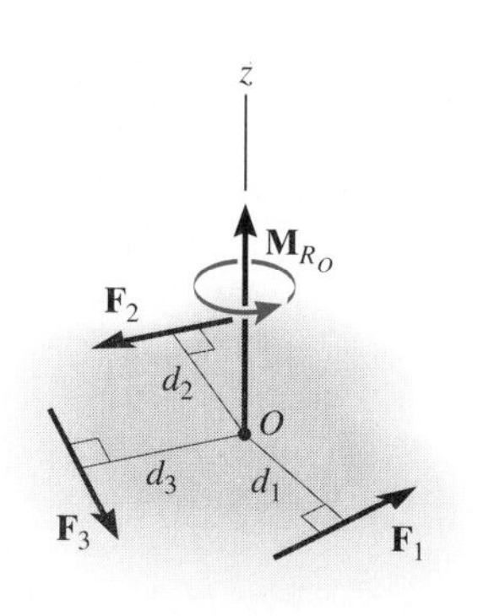

Fig. 4–8

Here the counterclockwise curl written alongside the equation indicates that by the scalar sign convention, the moment of any force will be positive if it is directed along the $+z$ axis, whereas a negative moment is directed along the $-z$ axis.

The following examples illustrate numerical application of Eqs. 4–4 and 4–5.

Example 4–1

Determine the moment of the 800-N force acting on the frame in Fig. 4–9 about points A, B, C, and D.

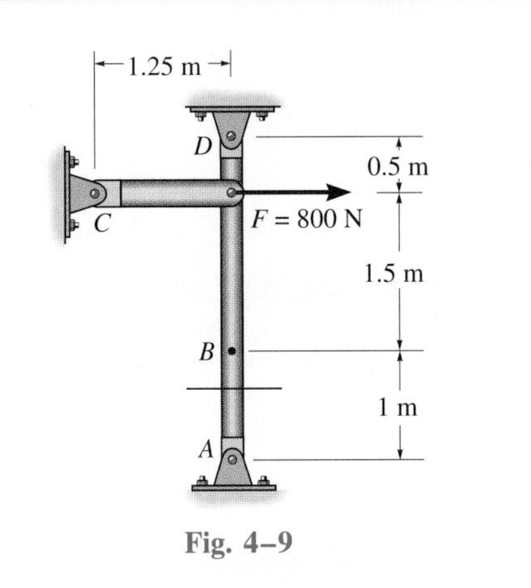

Fig. 4–9

SOLUTION *(SCALAR ANALYSIS)*

In general, $M = Fd$, where d is the moment arm or *perpendicular distance* from the *point* on the moment axis to the *line of action* of the force. Hence,

$M_A = 800\text{ N}(2.5\text{ m}) = 2000\text{ N}\cdot\text{m} \curvearrowright$ ***Ans.***

$M_B = 800\text{ N}(1.5\text{ m}) = 1200\text{ N}\cdot\text{m} \curvearrowright$ ***Ans.***

$M_C = 800\text{ N}(0) = 0$ (line of action of $\mathbf{F}$ passes through C) ***Ans.***

$M_D = 800\text{ N}(0.5\text{ m}) = 400\text{ N}\cdot\text{m} \curvearrowleft$ ***Ans.***

The curls indicate the sense of rotation of the moment, which is defined by the direction the force orbits about each point.

Example 4–2

Determine the location of the point of application P and the direction of a 20-lb force that lies in the plane of the square plate shown in Fig. 4–10a, so that this force creates the greatest counterclockwise moment about point O. What is this moment?

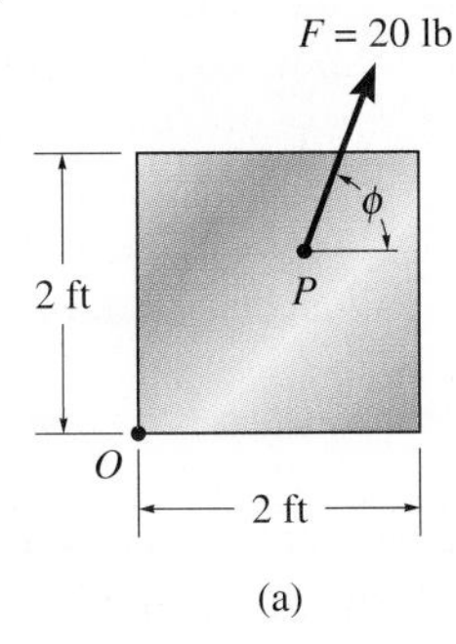

(a)

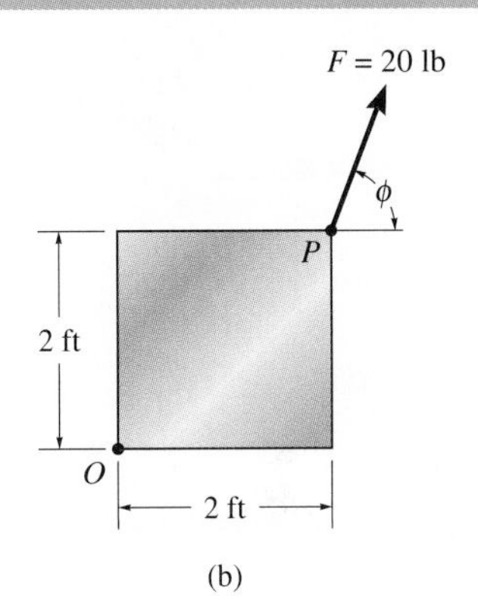

(b)

SOLUTION *(SCALAR ANALYSIS)*

Since the maximum moment created by the force is required, the force must act on the plate at a distance *farthest* from point O. As shown in Fig. 4–10b, the point of application must therefore be at the diagonal corner. In order to produce *counterclockwise* rotation of the plate about O, $\mathbf{F}$ must act at an angle $45° < \phi < 225°$. The greatest moment is produced when the line of action of $\mathbf{F}$ is *perpendicular* to d, i.e., $\phi = 135°$, Fig. 4–10c. The maximum moment is therefore

$$M_O = Fd = (20\text{ lb})(2\sqrt{2}\text{ ft}) = 56.6\text{ lb}\cdot\text{ft} \curvearrowleft \qquad \textit{Ans.}$$

By the right-hand rule, $\mathbf{M}_O$ is directed out of the page.

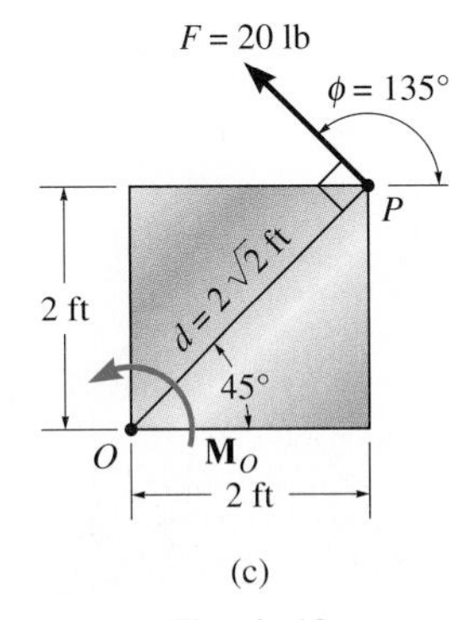

(c)

Fig. 4–10

Example 4–3

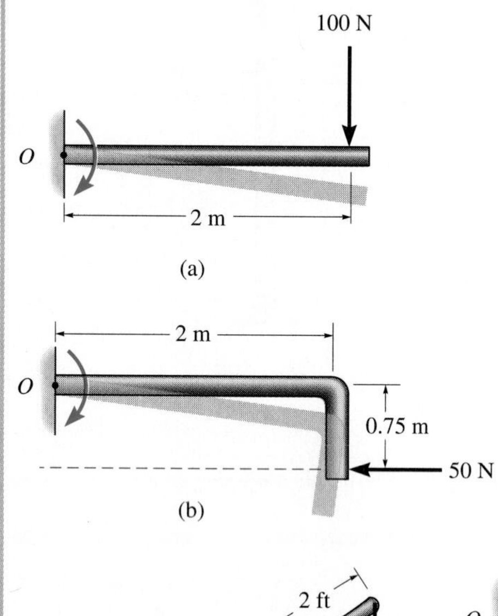

For each case illustrated in Fig. 4–11, determine the moment of the force about point O.

SOLUTION (*SCALAR ANALYSIS*)

The line of action of each force is extended as a dashed line in order to establish the moment arm d. Also illustrated is the tendency of rotation of the member as caused by the force. Furthermore, the orbit of the force is shown as a colored curl. Thus,

Fig. 4–11*a*, $M_O = (100 \text{ N})(2 \text{ m}) = 200 \text{ N} \cdot \text{m} \circlearrowright$ *Ans.*

Fig. 4–11*b*, $M_O = (50 \text{ N})(0.75 \text{ m}) = 37.5 \text{ N} \cdot \text{m} \circlearrowright$ *Ans.*

Fig. 4–11*c*, $M_O = (40 \text{ lb})(4 \text{ ft} + 2 \cos 30° \text{ ft}) = 229 \text{ lb} \cdot \text{ft} \circlearrowright$ *Ans.*

Fig. 4–11*d*, $M_O = (60 \text{ lb})(1 \sin 45° \text{ ft}) = 42.4 \text{ lb} \cdot \text{ft} \circlearrowleft$ *Ans.*

Fig. 4–11*e*, $M_O = (7 \text{ kN})(4 \text{ m} - 1 \text{ m}) = 21.0 \text{ kN} \cdot \text{m} \circlearrowleft$ *Ans.*

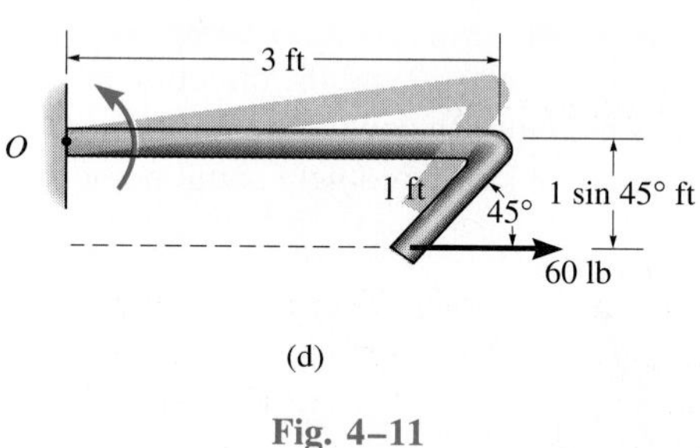

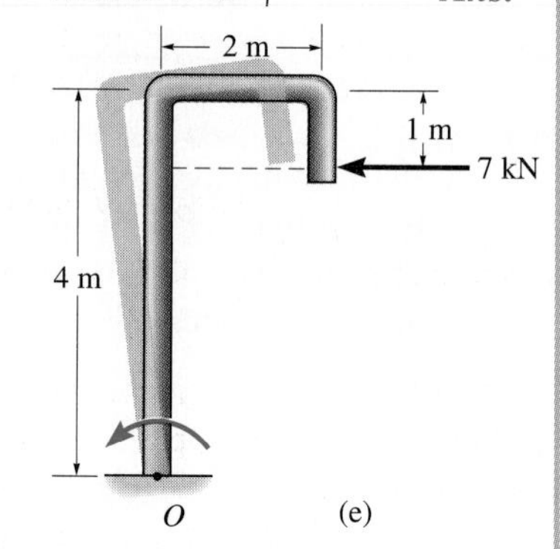

Fig. 4–11

Example 4–4

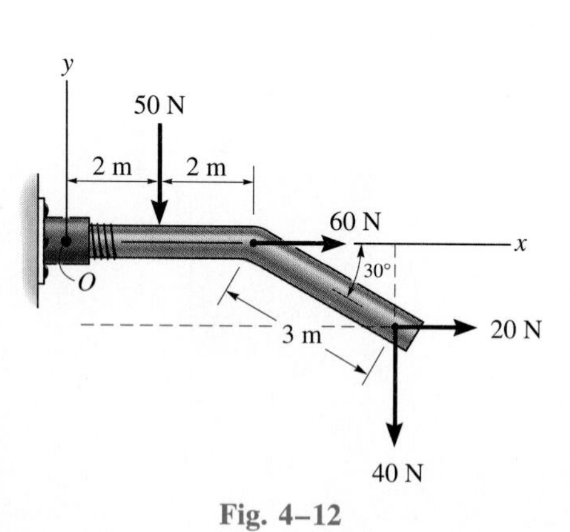

Fig. 4–12

Determine the resultant moment of the four forces acting on the rod shown in Fig. 4–12 about point O.

SOLUTION

Here it is necessary to apply Eq. 4–5. Assuming that positive moments act in the $+\mathbf{k}$ direction, i.e., counterclockwise, we have

$$\circlearrowleft + M_{R_O} = \Sigma Fd;$$

$$M_{R_O} = -50 \text{ N}(2 \text{ m}) + 60 \text{ N}(0) + 20 \text{ N}(3 \sin 30° \text{ m}) - 40 \text{ N}(4 \text{ m} + 3 \cos 30° \text{ m})$$

$$M_{R_O} = -334 \text{ N} \cdot \text{m} = 334 \text{ N} \cdot \text{m} \circlearrowright$$ *Ans.*

For this calculation, note how the moment-arm distances for the 20-N and 40-N forces are established from the extended (dashed) lines of action of each of these forces.

4.3 Moment of a Force—Vector Formulation

The moment of a force **F** about point O, or actually about the moment axis passing through O and perpendicular to the plane containing O and **F**, Fig. 4–13*a*, can also be expressed using the vector cross product, namely,

$$\mathbf{M}_O = \mathbf{r} \times \mathbf{F} \tag{4–6}$$

Here **r** represents a position vector drawn from O to *any point* lying on the line of action of **F**. It will now be shown that indeed the moment $\mathbf{M}_O$, when determined by this cross product, has the proper magnitude and direction.

Magnitude. The magnitude of the above cross product is defined from Eq. 4–1 as $M_O = rF \sin \theta$. Here, the angle θ is measured between the *tails* of **r** and **F**. To establish this angle, **r** must be treated as a sliding vector so that θ can be constructed properly, Fig. 4–13*b*. Since the moment arm $d = r \sin \theta$, then

$$M_O = rF \sin \theta = F\,(r \sin \theta) = Fd$$

which agrees with Eq. 4–4.

Direction. The direction and sense of $\mathbf{M}_O$ in Eq. 4–6 are determined by the right-hand rule as it applies to the cross product. Thus, extending **r** to the dashed position and curling the right-hand fingers from **r** toward **F**, "**r** cross **F**," the thumb is directed upward or perpendicular to the plane containing **r** and **F** and this is in the *same direction* as $\mathbf{M}_O$, the moment of the force about point O, Fig. 4–13*b*. Note that the "curl" of the fingers, like the curl around the moment vector, indicates the sense of rotation caused by the force. Since the cross product is not commutative, it is important that the *proper order* of **r** and **F** be maintained in Eq. 4–6.

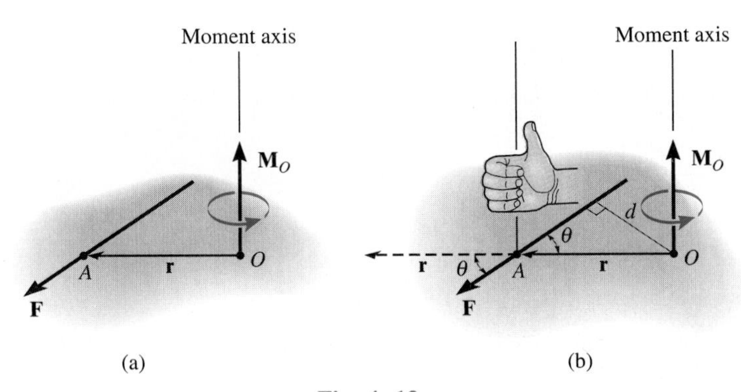

Fig. 4–13

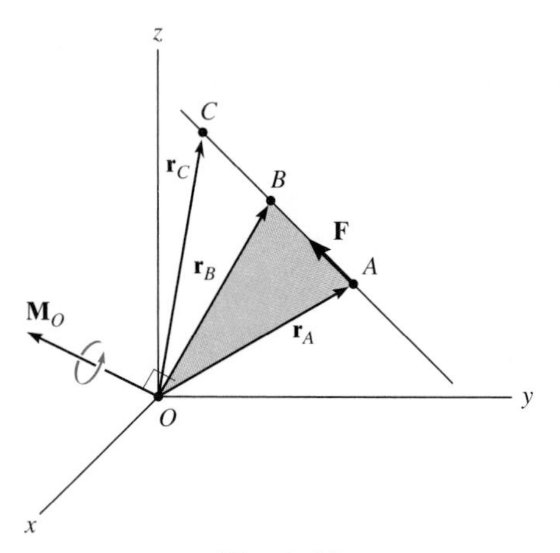

Fig. 4–14

Transmissibility of a Force. Consider the force $\mathbf{F}$ applied at point A in Fig. 4–14. The moment created by $\mathbf{F}$ about point O is $\mathbf{M}_O = \mathbf{r}_A \times \mathbf{F}$; however, it was shown that the position vector "$\mathbf{r}$" can extend from O to *any point* on the line of action of $\mathbf{F}$. Consequently, $\mathbf{F}$ may be applied at point B or C and the same moment $\mathbf{M}_O = \mathbf{r}_B \times \mathbf{F} = \mathbf{r}_C \times \mathbf{F}$ will be computed. As a result, $\mathbf{F}$ has the properties of a *sliding vector* and can therefore act at *any point along its line of action and still create the same moment about point O.* We refer to $\mathbf{F}$ in this regard as being "transmissible," and we will discuss this property further in Sec. 4.7.

Cartesian Vector Formulation. If we establish x, y, z coordinate axes, then the position vector $\mathbf{r}$ and force $\mathbf{F}$ can be expressed as Cartesian vectors, Fig. 4–15. Applying Eq. 4–6 we have

$$\mathbf{M}_O = \mathbf{r} \times \mathbf{F} = \begin{vmatrix} \mathbf{i} & \mathbf{j} & \mathbf{k} \\ r_x & r_y & r_z \\ F_x & F_y & F_z \end{vmatrix} \tag{4–7}$$

where r_x, r_y, r_z represent the x, y, z components of the position vector drawn from point O to *any point* on the line of action of the force

F_x, F_y, F_z represent the x, y, z components of the force vector

If the determinant is expanded, then like Eq. 4–2 we have

$$\mathbf{M}_O = (r_yF_z - r_zF_y)\mathbf{i} - (r_xF_z - r_zF_x)\mathbf{j} + (r_xF_y - r_yF_x)\mathbf{k} \tag{4–8}$$

The physical meaning of these three moment components becomes evident by studying Fig. 4–15*a*. For example, the $\mathbf{i}$ component of $\mathbf{M}_O$ is determined from the moments of $\mathbf{F}_x$, $\mathbf{F}_y$, and $\mathbf{F}_z$ about the x axis. In particular, note that $\mathbf{F}_x$ does *not* create a moment or tendency to cause turning about the x axis, since this

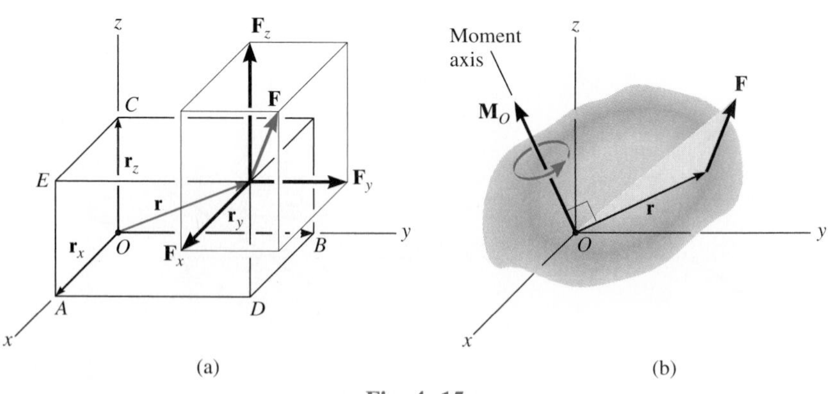

Fig. 4–15

force is *parallel* to the x axis. The line of action of $\mathbf{F}_y$ passes through point E, and so the magnitude of the moment of $\mathbf{F}_y$ about point A on the x axis is $r_z F_y$. By the right-hand rule this component acts in the negative $\mathbf{i}$ direction. Likewise, $\mathbf{F}_z$ contributes a moment component of $r_y F_z \mathbf{i}$. Thus, $(M_O)_x = (r_y F_z - r_z F_y)$ as shown in Eq. 4–8. As an exercise, establish the $\mathbf{j}$ and $\mathbf{k}$ components of $\mathbf{M}_O$ in this manner, and show that indeed the expanded form of the determinant, Eq. 4–8, represents the moment of $\mathbf{F}$ about point O. Once determined, realize that $\mathbf{M}_O$ will always be *perpendicular* to the shaded plane containing vectors $\mathbf{r}$ and $\mathbf{F}$, Fig. 4–15*b*.

It will be shown in Example 4–5 that the computation of the moment using the cross product has a distinct advantage over the scalar formulation when solving problems in *three dimensions*. This is because it is generally easier to establish the position vector $\mathbf{r}$ to the force, rather than determining the moment-arm distance d that must be directed *perpendicular* to the line of action of the force.

Resultant Moment of a System of Forces. The resultant moment of a system of forces about point O can be determined by vector addition resulting from successive applications of Eq. 4–7. This resultant can be written symbolically as

$$\mathbf{M}_{R_O} = \Sigma(\mathbf{r} \times \mathbf{F}) \qquad (4\text{–}9)$$

and is shown in Fig. 4–16.

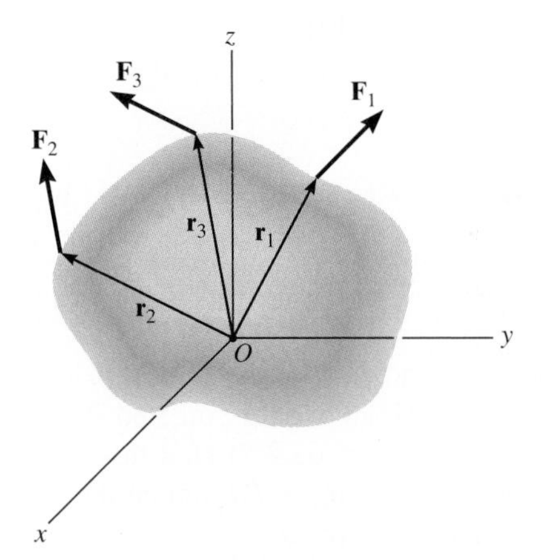

Fig. 4–16

Example 4–5

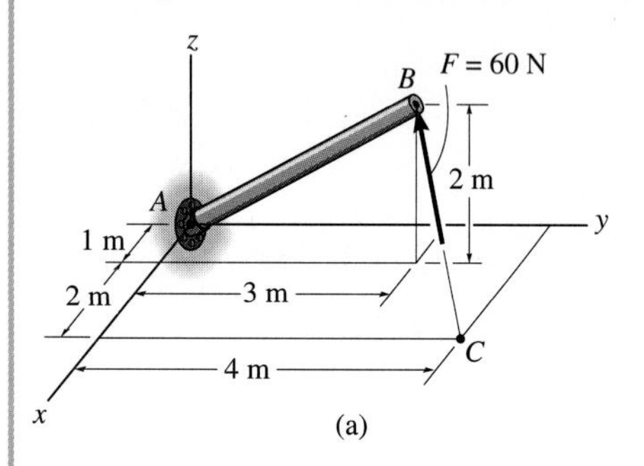

(a)

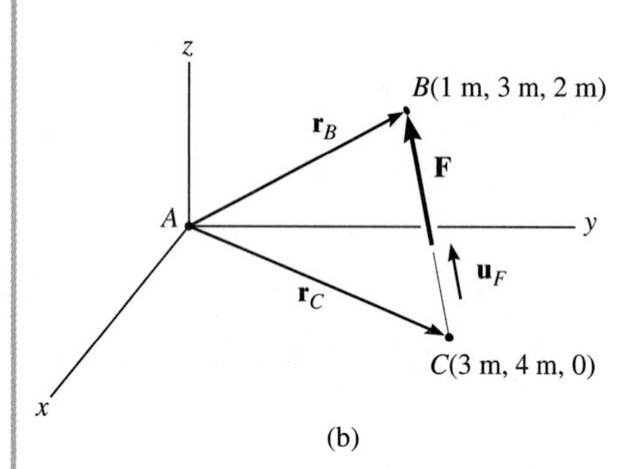

(b)

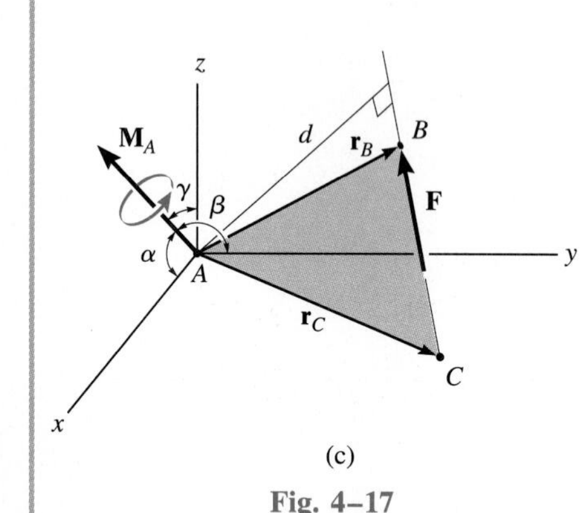

(c)

Fig. 4–17

The pole in Fig. 4–17*a* is subjected to a 60-N force that is directed from *C* to *B*. Determine the magnitude of the moment created by this force about the support at *A*.

SOLUTION *(VECTOR ANALYSIS)*

As shown in Fig. 4–17*b*, either one of two position vectors can be used for the solution, since $\mathbf{M}_A = \mathbf{r}_B \times \mathbf{F}$ or $\mathbf{M}_A = \mathbf{r}_C \times \mathbf{F}$. The position vectors are represented as

$$\mathbf{r}_B = \{1\mathbf{i} + 3\mathbf{j} + 2\mathbf{k}\}\text{ m} \quad \text{and} \quad \mathbf{r}_C = \{3\mathbf{i} + 4\mathbf{j}\}\text{ m}$$

The force has a magnitude of 60 N and a direction specified by the unit vector $\mathbf{u}_F$, directed from C to B. Thus,

$$\mathbf{F} = (60\text{ N})\mathbf{u}_F = (60\text{ N})\left[\frac{(1-3)\mathbf{i} + (3-4)\mathbf{j} + (2-0)\mathbf{k}}{\sqrt{(-2)^2 + (-1)^2 + (2)^2}}\right]$$

$$= \{-40\mathbf{i} - 20\mathbf{j} + 40\mathbf{k}\}\text{ N}$$

Substituting into the determinant formulation, Eq. 4–7, and following the scheme for determinant expansion as stated in the footnote on page 110, we have

$$\mathbf{M}_A = \mathbf{r}_B \times \mathbf{F} = \begin{vmatrix} \mathbf{i} & \mathbf{j} & \mathbf{k} \\ 1 & 3 & 2 \\ -40 & -20 & 40 \end{vmatrix}$$

$$= [3(40) - 2(-20)]\mathbf{i} - [1(40) - 2(-40)]\mathbf{j} + [1(-20) - 3(-40)]\mathbf{k}$$

or

$$\mathbf{M}_A = \mathbf{r}_C \times \mathbf{F} = \begin{vmatrix} \mathbf{i} & \mathbf{j} & \mathbf{k} \\ 3 & 4 & 0 \\ -40 & -20 & 40 \end{vmatrix}$$

$$= [4(40) - 0(-20)]\mathbf{i} - [3(40) - 0(-40)]\mathbf{j} + [3(-20) - 4(-40)]\mathbf{k}$$

In both cases,

$$\mathbf{M}_A = \{160\mathbf{i} - 120\mathbf{j} + 100\mathbf{k}\}\text{ N}\cdot\text{m}$$

The *magnitude* of $\mathbf{M}_A$ is therefore

$$M_A = \sqrt{(160)^2 + (-120)^2 + (100)^2} = 224\text{ N}\cdot\text{m} \qquad \textit{Ans.}$$

As expected, $\mathbf{M}_A$ acts perpendicular to the shaded plane containing vectors $\mathbf{F}$, $\mathbf{r}_B$, and $\mathbf{r}_C$, Fig. 4–17*c*. (How would you find its coordinate direction angles $\alpha = 44.3°$, $\beta = 122°$, $\gamma = 63.4°$?) Had this problem been worked using a scalar approach, where $M_A = Fd$, notice the difficulty that might arise in obtaining the moment arm d.

Example 4–6

Three forces act on the rod shown in Fig. 4–18*a*. Determine the resultant moment they create about the flange at *O*, and determine the direction of the moment axis.

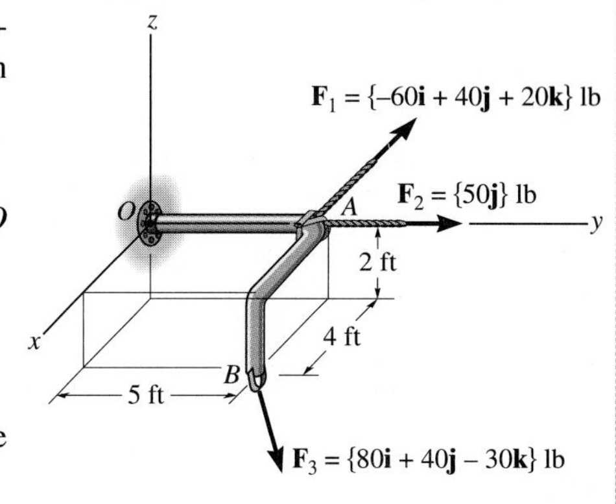

(a)

SOLUTION

Here we must apply Eq. 4–9. Position vectors are directed from point *O* to each force as shown in Fig. 4–18*b*. These vectors are

$$\mathbf{r}_A = \{5\mathbf{j}\}\text{ ft}$$
$$\mathbf{r}_B = \{4\mathbf{i} + 5\mathbf{j} - 2\mathbf{k}\}\text{ ft}$$

Since $\mathbf{F}_2 = \{50\mathbf{j}\}$ lb, and the Cartesian components of the other forces are given, the resultant moment about *O* is therefore

$$\begin{aligned}\mathbf{M}_{R_O} &= \Sigma(\mathbf{r} \times \mathbf{F})\\ &= \mathbf{r}_A \times \mathbf{F}_1 + \mathbf{r}_A \times \mathbf{F}_2 + \mathbf{r}_B \times \mathbf{F}_3\\ &= \begin{vmatrix}\mathbf{i} & \mathbf{j} & \mathbf{k}\\ 0 & 5 & 0\\ -60 & 40 & 20\end{vmatrix} + \begin{vmatrix}\mathbf{i} & \mathbf{j} & \mathbf{k}\\ 0 & 5 & 0\\ 0 & 50 & 0\end{vmatrix} + \begin{vmatrix}\mathbf{i} & \mathbf{j} & \mathbf{k}\\ 4 & 5 & -2\\ 80 & 40 & -30\end{vmatrix}\\ &= [5(20) - 40(0)]\mathbf{i} - [0\mathbf{j}] + [0(40) - (-60)(5)]\mathbf{k} + [0\mathbf{i} - 0\mathbf{j} + 0\mathbf{k}]\\ &\quad + [5(-30) - (40)(-2)]\mathbf{i} - [4(-30) - 80(-2)]\mathbf{j} + [4(40) - 80(5)]\mathbf{k}\\ &= \{30\mathbf{i} - 40\mathbf{j} + 60\mathbf{k}\}\text{ lb}\cdot\text{ft} \qquad \textit{Ans.}\end{aligned}$$

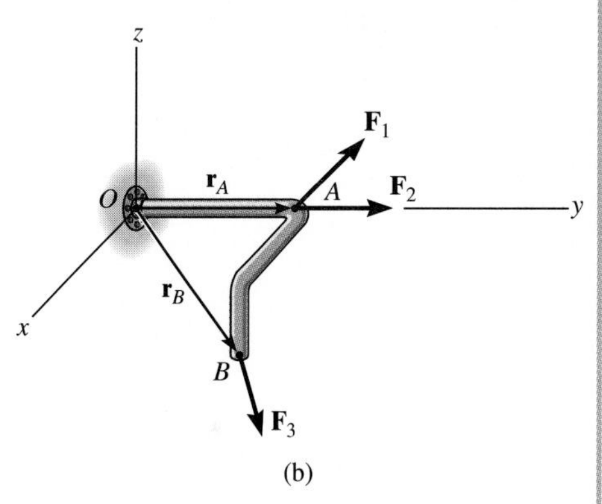

(b)

The moment axis is directed along the line of action of $\mathbf{M}_{R_O}$. Since the magnitude of this moment is

$$M_{R_O} = \sqrt{(30)^2 + (-40)^2 + (60)^2} = 78.10\text{ lb}\cdot\text{ft}$$

the unit vector which defines the direction of the moment axis is

$$\mathbf{u} = \frac{\mathbf{M}_{R_O}}{M_{R_O}} = \frac{30\mathbf{i} - 40\mathbf{j} + 60\mathbf{k}}{78.10} = 0.3841\mathbf{i} - 0.5121\mathbf{j} + 0.7682\mathbf{k}$$

Therefore, the coordinate direction angles of the moment axis are

$$\cos\alpha = 0.3841; \qquad \alpha = 67.4° \qquad \textit{Ans.}$$
$$\cos\beta = -0.5121; \qquad \beta = 121° \qquad \textit{Ans.}$$
$$\cos\gamma = 0.7682; \qquad \gamma = 39.8° \qquad \textit{Ans.}$$

These results are shown in Fig. 4–18*c*. Realize that the three forces tend to cause the rod to rotate about this axis in the manner shown by the curl indicated on the moment vector.

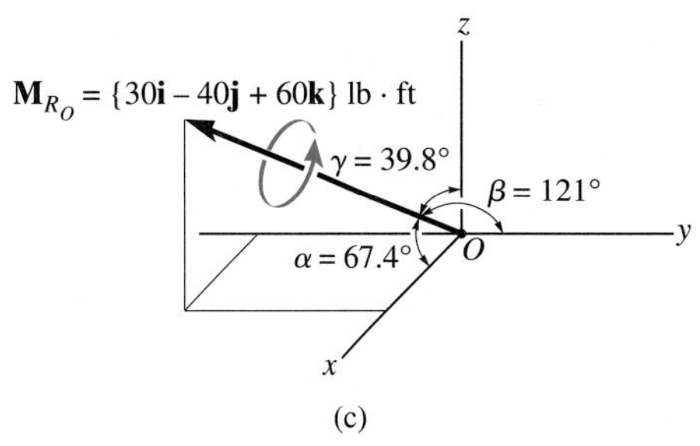

(c)

Fig. 4–18

4.4 Principle of Moments

A concept often used in mechanics is the *principle of moments,* which is sometimes referred to as *Varignon's theorem* since it was originally developed by the French mathematician Varignon (1654–1722). It states that *the moment of a force about a point is equal to the sum of the moments of the force's components about the point.* The proof follows directly from the distributive law of the vector cross product. To show this, consider the force **F** and two of its components, where $\mathbf{F} = \mathbf{F}_1 + \mathbf{F}_2$, Fig. 4–19. We have

$$\mathbf{M}_O = \mathbf{r} \times \mathbf{F}_1 + \mathbf{r} \times \mathbf{F}_2 = \mathbf{r} \times (\mathbf{F}_1 + \mathbf{F}_2) = \mathbf{r} \times \mathbf{F}$$

This concept has important applications to the solution of problems and proofs of theorems that follow, since it is often easier to determine the moments of a force's components rather than the moment of the force itself.

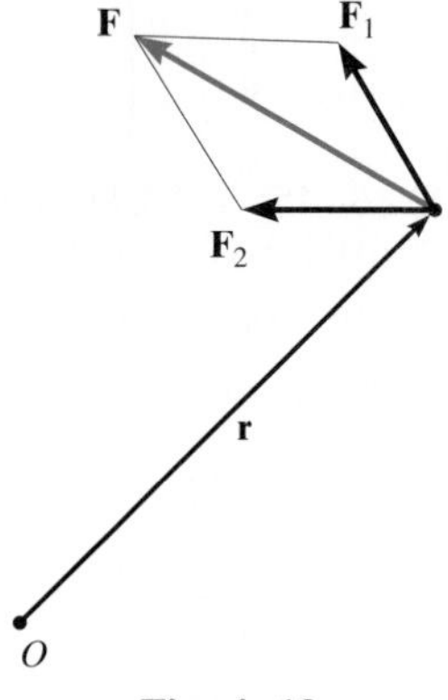

Fig. 4–19

Example 4–7

A 200-N force acts on the bracket shown in Fig. 4–20*a*. Determine the moment of the force about point *A*.

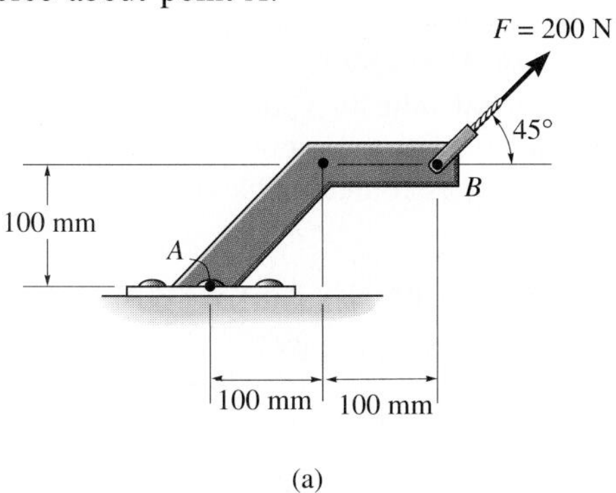

(a)

SOLUTION I

The moment arm *d* can be found by trigonometry, using the construction shown in Fig. 4–20*b*. From the right triangle *BCD*,

$$CB = d = 100 \cos 45^\circ = 70.71 \text{ mm} = 0.070\,71 \text{ m}$$

Thus,

$$M_A = Fd = 200 \text{ N}(0.070\,71 \text{ m}) = 14.1 \text{ N} \cdot \text{m} \circlearrowleft$$

According to the right-hand rule, $\mathbf{M}_A$ is directed in the $+\mathbf{k}$ direction since the force tends to rotate or orbit *counterclockwise* about point *A*. Hence, reporting the moment as a Cartesian vector, we have

$$\mathbf{M}_A = \{14.1\mathbf{k}\} \text{ N} \cdot \text{m} \qquad \textit{Ans.}$$

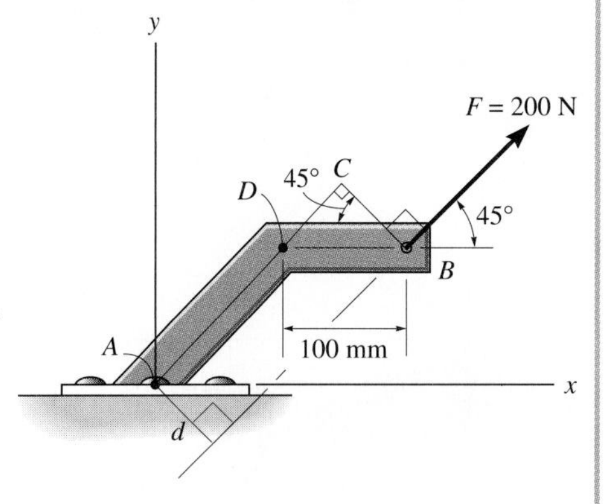

(b)

SOLUTION II

The 200-N force may be resolved into *x* and *y* components, as shown in Fig. 4–20*c*. In accordance with the principle of moments, the moment of **F** computed about point *A* is equivalent to the sum of the moments produced by the two force components. Assuming counterclockwise rotation as positive, i.e., in the $+\mathbf{k}$ direction, we can apply Eq. 4–5 ($M_A = \Sigma\, Fd$), in which case

$$\circlearrowleft + M_A = (200 \sin 45^\circ \text{ N})(0.20 \text{ m}) - (200 \cos 45^\circ \text{ N})(0.10 \text{ m})$$
$$= 14.1 \text{ N} \cdot \text{m} \circlearrowleft$$

Thus

$$\mathbf{M}_A = \{14.1\mathbf{k}\} \text{ N} \cdot \text{m} \qquad \textit{Ans.}$$

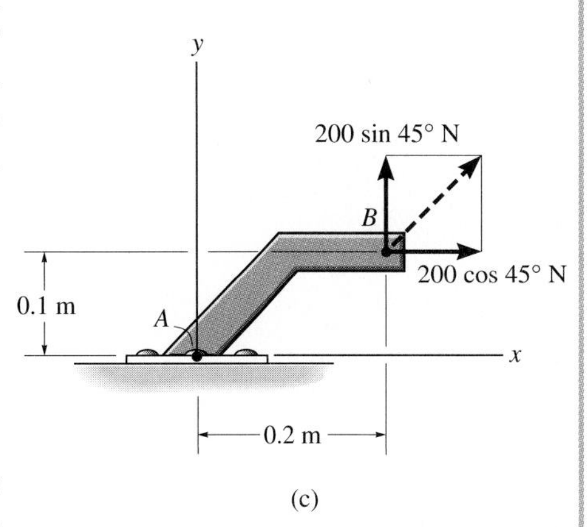

(c)

Fig. 4–20

By comparison, it is seen that Solution II provides a more *convenient method* for analysis than Solution I, since the moment arm for each component force is easier to establish.

Example 4–8

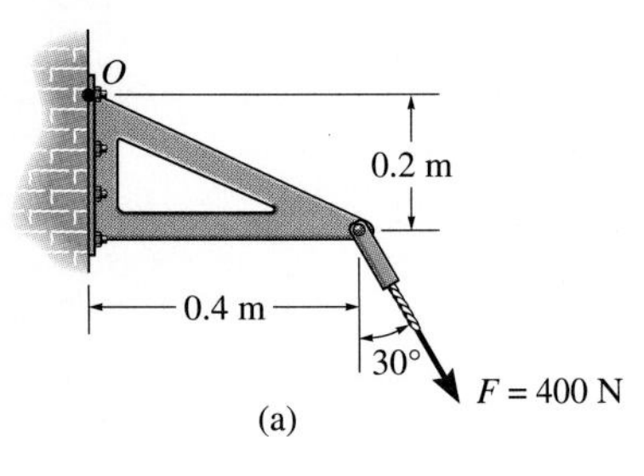

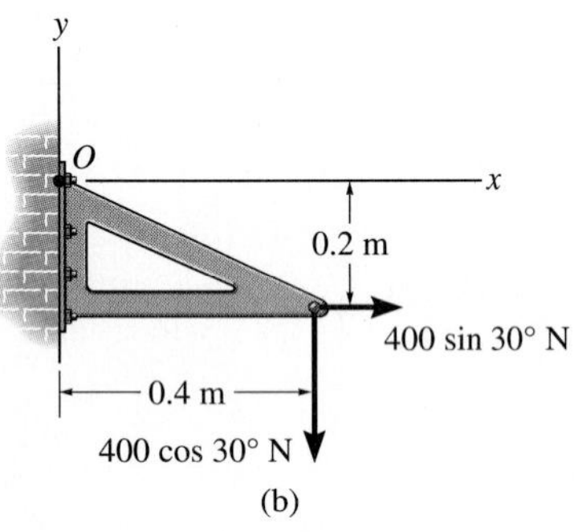

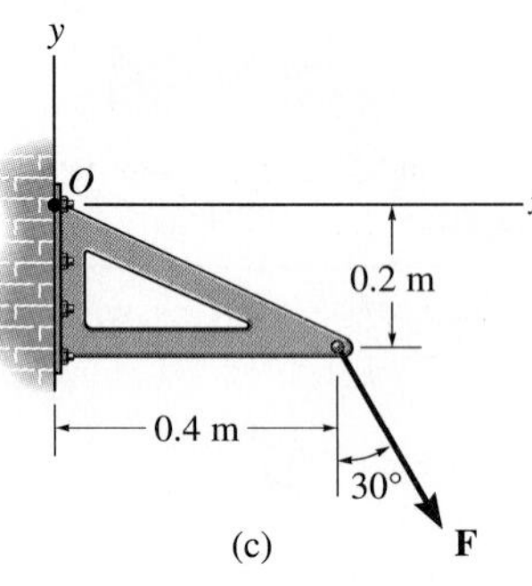

Fig. 4–21

The force **F** acts at the end of the angle bracket shown in Fig. 4–21*a*. Determine the moment of the force about point *O*.

SOLUTION I (*SCALAR ANALYSIS*)

The force is resolved into its *x* and *y* components as shown in Fig. 4–21*b* and the moments of the components are computed about point *O*. Taking positive moments as counterclockwise, i.e., in the $+\mathbf{k}$ direction, we have

$$\circlearrowleft + M_O = 400 \sin 30° \text{ N}(0.2 \text{ m}) - 400 \cos 30° \text{ N}(0.4 \text{ m})$$
$$= -98.6 \text{ N} \cdot \text{m} = 98.6 \text{ N} \cdot \text{m} \circlearrowright$$

or

$$\mathbf{M}_O = \{-98.6\mathbf{k}\} \text{ N} \cdot \text{m} \qquad \textit{Ans.}$$

SOLUTION II (*VECTOR ANALYSIS*)

Using a Cartesian vector approach, the force and position vectors shown in Fig. 4–21*c* can be represented as

$$\mathbf{r} = \{0.4\mathbf{i} - 0.2\mathbf{j}\} \text{ m}$$
$$\mathbf{F} = \{400 \sin 30°\mathbf{i} - 400 \cos 30°\mathbf{j}\} \text{ N}$$
$$= \{200.0\mathbf{i} - 346.4\mathbf{j}\} \text{ N}$$

The moment is therefore

$$\mathbf{M}_O = \mathbf{r} \times \mathbf{F} = \begin{vmatrix} \mathbf{i} & \mathbf{j} & \mathbf{k} \\ 0.4 & -0.2 & 0 \\ 200.0 & -346.4 & 0 \end{vmatrix}$$
$$= 0\mathbf{i} - 0\mathbf{j} + [0.4(-346.4) - (-0.2)(200.0)]\mathbf{k}$$
$$= \{-98.6\mathbf{k}\} \text{ N} \cdot \text{m} \qquad \textit{Ans.}$$

By comparison, it is seen that the scalar analysis (Solution I) provides a more *convenient method* for analysis than Solution II, since the direction of the moment and the moment arm for each component force are easy to establish. Hence, this method is generally recommended for solving problems displayed in two dimensions. On the other hand, Cartesian vector analysis is generally recommended only for solving three-dimensional problems, where the moment arms and force components are often more difficult to determine.

PROBLEMS

4–1 If **A, B,** and **D** are given vectors, prove the distributive law for the vector cross product, i.e., $\mathbf{A} \times (\mathbf{B} + \mathbf{D}) = (\mathbf{A} \times \mathbf{B}) + (\mathbf{A} \times \mathbf{D})$.

4–2. Prove the triple scalar product identity $\mathbf{A} \cdot \mathbf{B} \times \mathbf{C} = \mathbf{A} \times \mathbf{B} \cdot \mathbf{C}$.

4–3. Given the three nonzero vectors **A**, **B**, and **C**, show that if $\mathbf{A} \cdot (\mathbf{B} \times \mathbf{C}) = 0$, the three vectors *must* lie in the same plane.

***4–4.** Determine the magnitude and directional sense of the moment of the force at *A* about point *O*.

4–5. Determine the magnitude and directional sense of the moment of the force at *A* about point *P*.

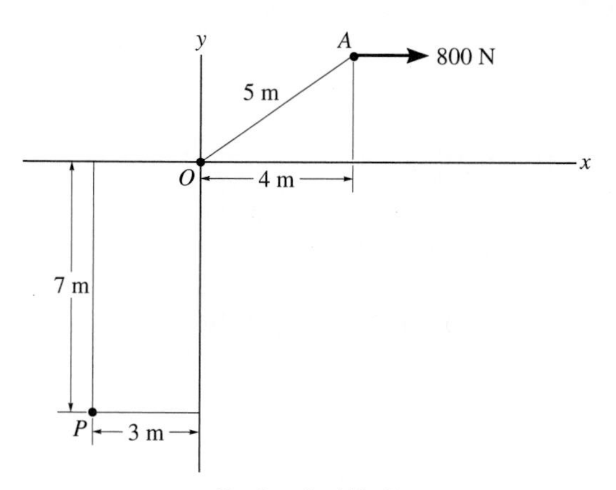

Probs. 4–4/4–5

4–6. Determine the magnitude and directional sense of the moment of the force at *A* about point *P*.

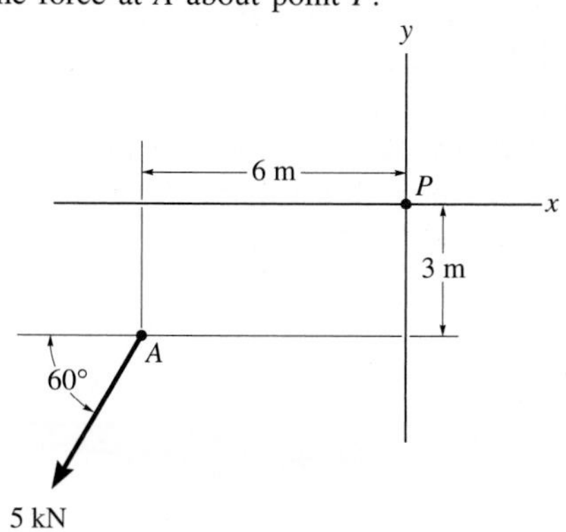

Prob. 4–6

4–7. Determine the magnitude and directional sense of the resultant moment of the forces about point *O*.

***4–8.** Determine the magnitude and directional sense of the resultant moment of the forces about point *P*.

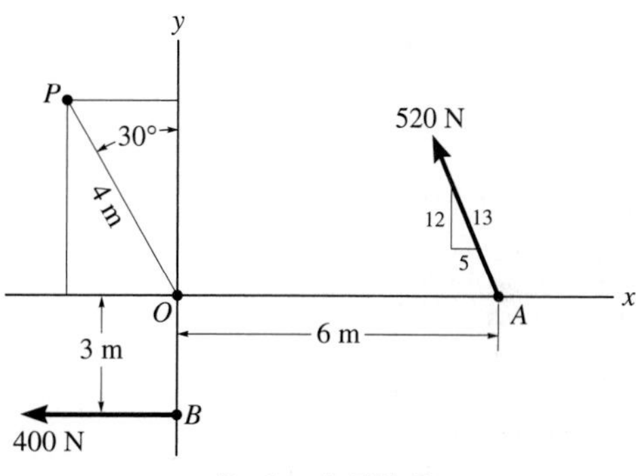

Probs. 4–7/4–8

4–9. Determine the magnitude and directional sense of the resultant moment of the forces about point *O*.

4–10. Determine the magnitude and directional sense of the resultant moment of the forces about point *P*.

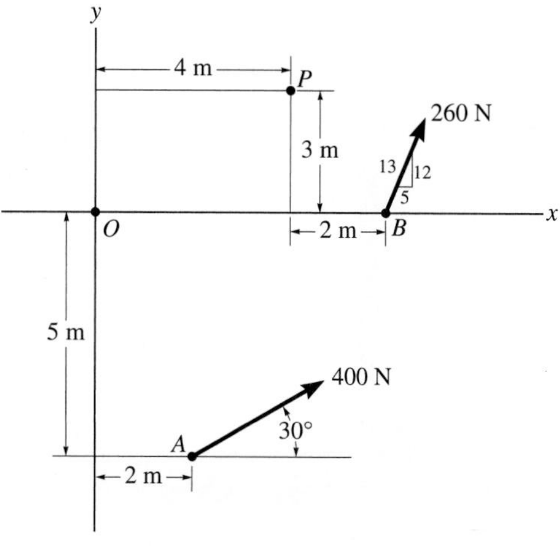

Probs. 4–9/4–10

4–11. Determine the magnitude and directional sense of the resultant moment of the forces about point O.

***4–12.** Determine the magnitude and directional sense of the resultant moment of the forces about point P.

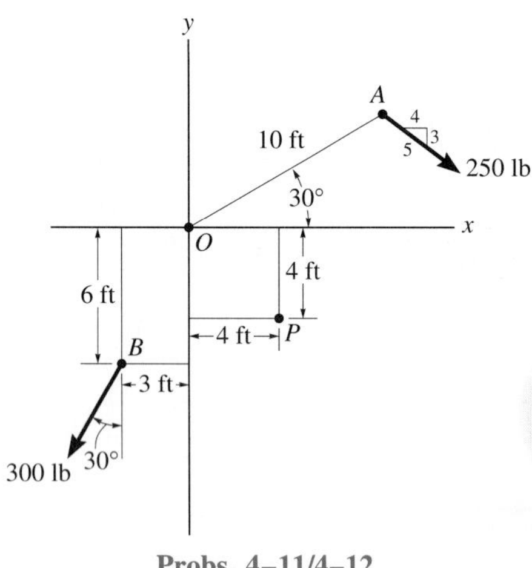

Probs. 4–11/4–12

4–13. The 70-N force acts on the end of the pipe at B. Determine (a) the moment of this force about point A, and (b) the magnitude and direction of a horizontal force, applied at C, which produces the same moment. Take $\theta = 60°$.

4–14. The 70-N force acts on the end of the pipe at B. Determine the angles θ ($0° \leq \theta \leq 180°$) of the force that will produce maximum and minimum moments about point A. What are the magnitudes of these moments?

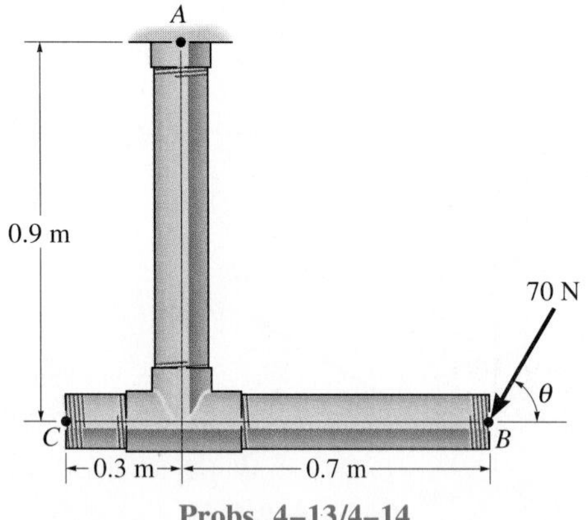

Probs. 4–13/4–14

4–15. Determine the moment of each force about the bolt located at A. Take $F_B = 40$ lb, $F_C = 50$ lb.

***4–16.** If $F_B = 30$ lb and $F_C = 45$ lb, determine the resultant moment about the bolt located at A.

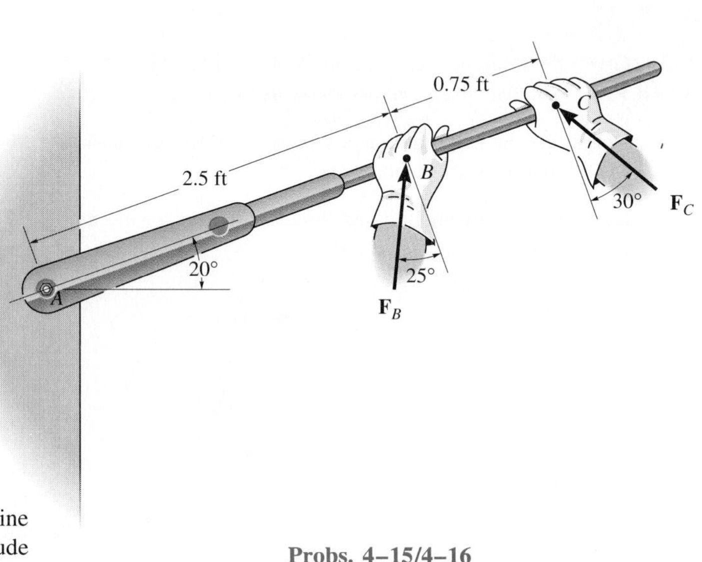

Probs. 4–15/4–16

4–17. The torque wrench ABC is used to measure the moment or torque applied to a bolt when the bolt is located at A and a force is applied to the handle at C. The mechanic reads the torque on the scale at B. If an extension AO of length d is used on the wrench, determine the required scale reading if the desired torque on the bolt at O is to be M.

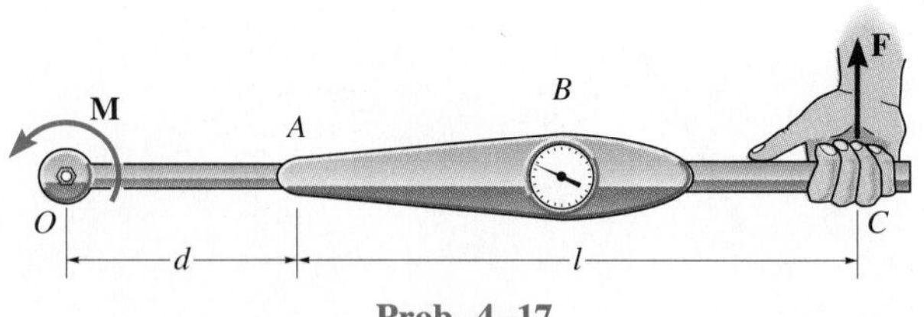

Prob. 4–17

4–18. Determine the resultant moment of the forces about point A. Solve the problem first by considering each force as a whole, and then by using the principle of moments. Take $F_1 = 250$ N, $F_2 = 300$ N, $F_3 = 500$ N.

4–19. If the resultant moment about point A is 4800 N · m clockwise, determine the magnitude of $\mathbf{F}_3$ if $F_1 = 300$ N and $F_2 = 400$ N.

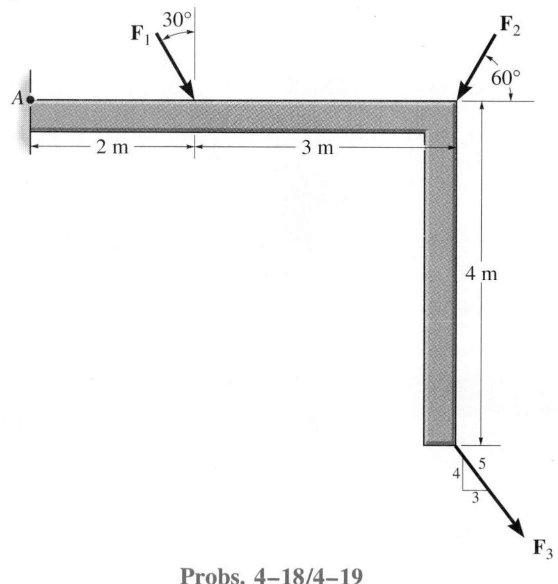

Probs. 4–18/4–19

*4–20. The lift has a boom that has a length of 30 ft, a weight of 800 lb, and mass center at G. If the bucket is designed to hold $W = 350$ lb, with mass center at G', determine the moment $\mathbf{M}$ that must be supplied by the motor at A to overcome the moment of the two forces of 800 lb and 350 lb. Take $\theta = 30°$.

4–21. The boom has a length of 30 ft, a weight of 800 lb, and mass center at G. If the maximum moment that can be developed by the motor at A is $M = 20(10^3)$ lb · ft, determine the maximum load W, having a mass center at G', that can be lifted. Take $\theta = 30°$.

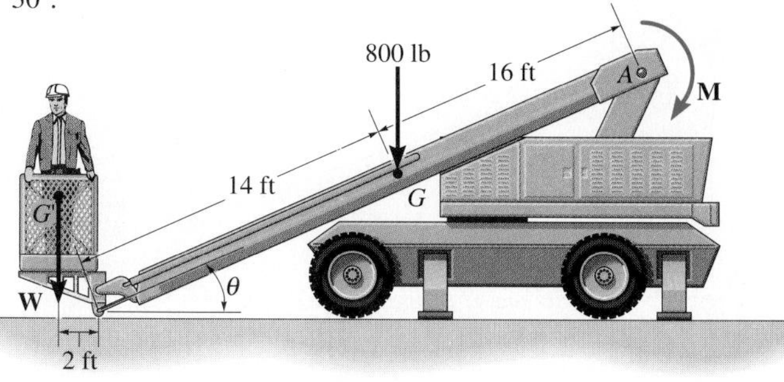

Probs. 4–20/4–21

4–22. Determine the direction $\theta(0° \leq \theta \leq 180°)$ of the force $F = 40$ lb so that it produces (a) the maximum moment about point A and (b) the minimum moment about point A. Compute the moment in each case.

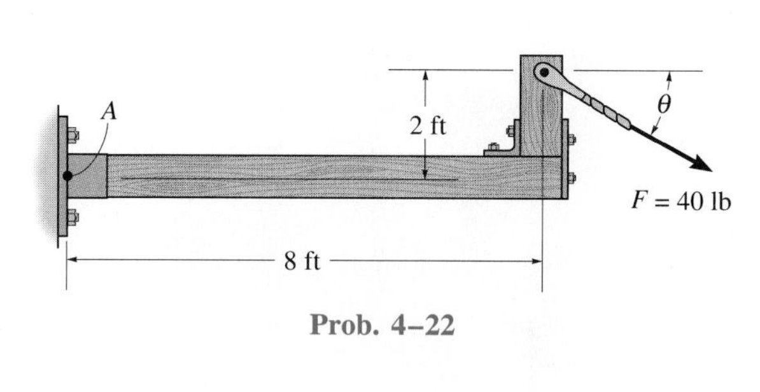

Prob. 4–22

4–23. The towline exerts a force of $P = 4$ kN at the end of the 20-m-long crane boom. If $\theta = 30°$, determine the placement x of the hook at A so that this force creates a maximum moment about point O. What is this moment?

*4–24. The towline exerts a force of $P = 4$ kN at the end of the 20-m-long crane boom. If $x = 25$ m, determine the position θ of the boom so that this force creates a maximum moment about point O. What is this moment?

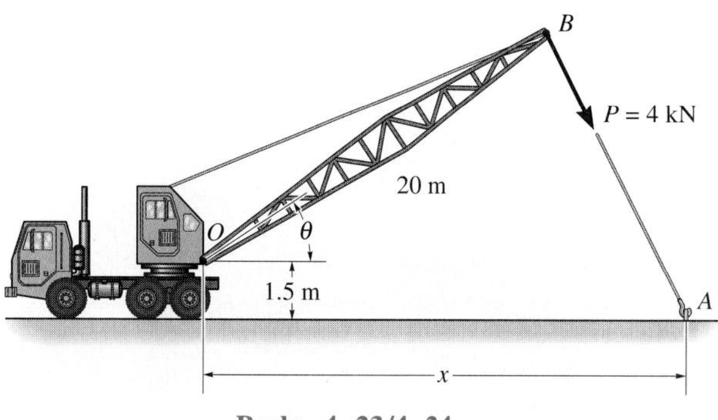

Probs. 4–23/4–24

4–25. The tool at A is used to hold a power lawnmower blade stationary while the nut is being loosened with the wrench. If a force of 50 N is applied to the wrench at B in the direction shown, determine the moment it creates about the nut at C. What is the magnitude of force **F** at A so that it creates the opposite moment about C?

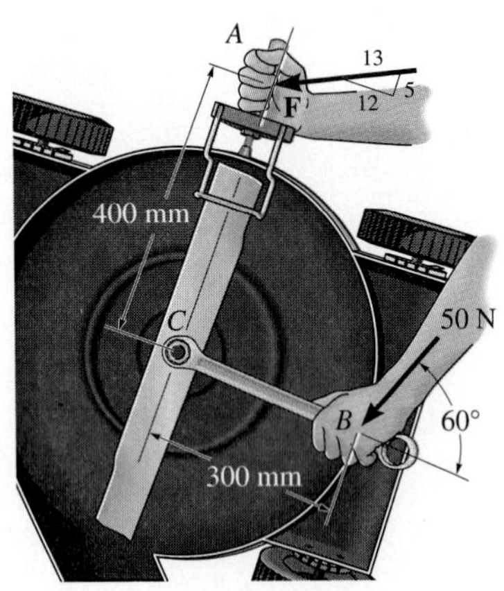

Prob. 4–25

4–26. The bucket boom carries a worker who has a weight of 230 lb and mass center at G. Determine the moment of this force about (a) point A and (b) point B.

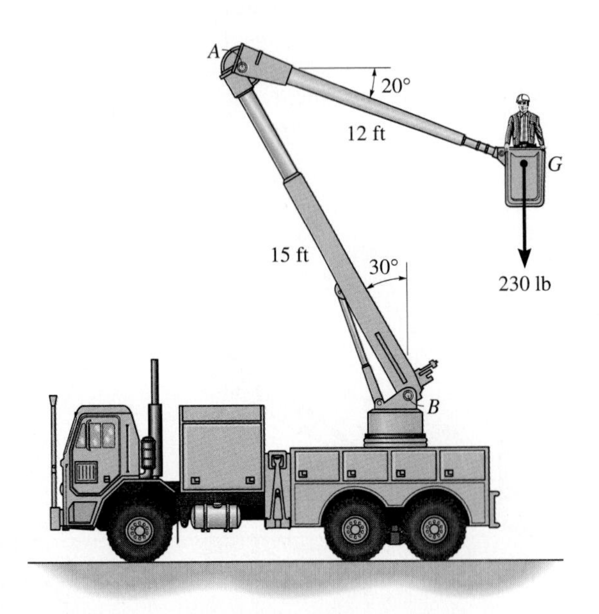

Prob. 4–26

4–27. The worker is using the bar to pull two pipes together in order to complete the connection. If he applies a horizontal force of 80 lb to the handle of the lever, determine the moment of this force about the end A. What would be the tension T in the cable needed to cause the opposite moment about point A?

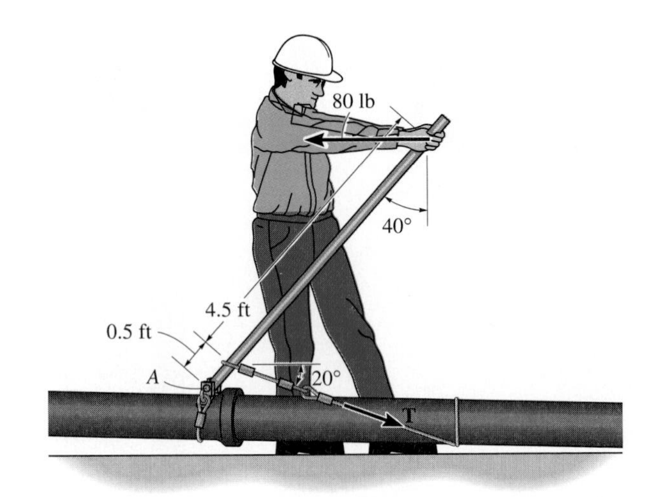

Prob. 4–27

***4–28.** The crowbar is subjected to a vertical force of $P = 25$ lb at the grip, whereas it takes a force of $F = 155$ lb at the claw to pull the nail out. Find the moment of each force about point A and determine if **P** is sufficient to pull out the nail. The crowbar contacts the board at point A.

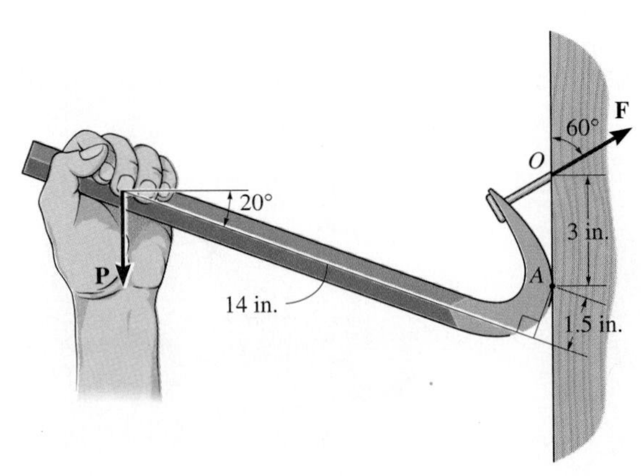

Prob. 4–28

4–29. If it takes a force of $F = 125$ lb to pull the nail out, determine the smallest vertical force **P** that must be applied to the handle of the crowbar. *Hint:* This requires the moment of **F** about point A to be equal to the moment of **P** about A. Why?

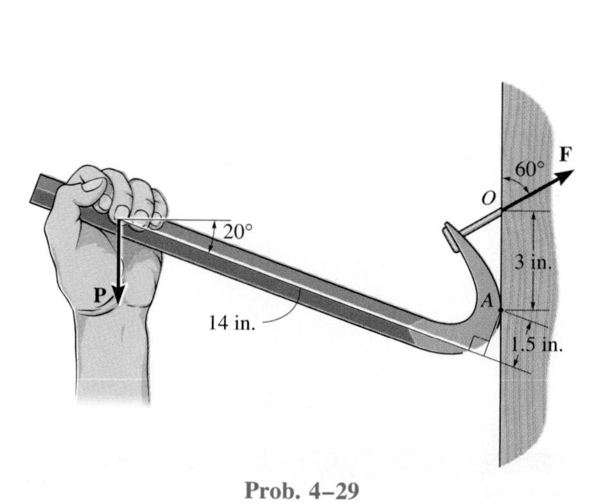

Prob. 4–29

4–30. Two forces act on the skew caster. Determine the resultant moment of these forces about point A and about point B.

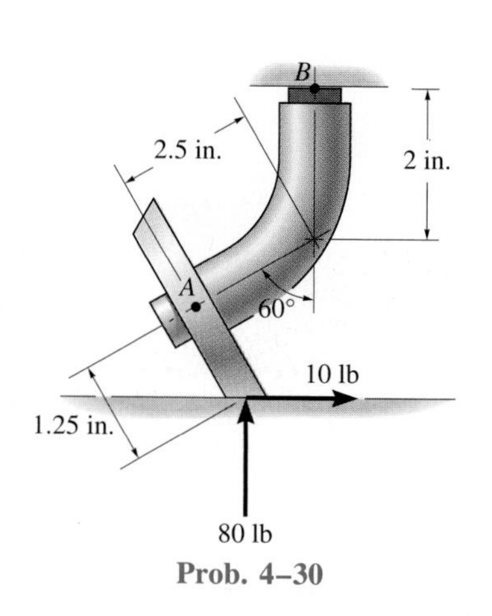

Prob. 4–30

4–31. Determine the moment of the force **F** at A about point O. Express the result as a Cartesian vector.

***4–32.** Determine the moment of the force **F** at A about point P. Express the result as a Cartesian vector.

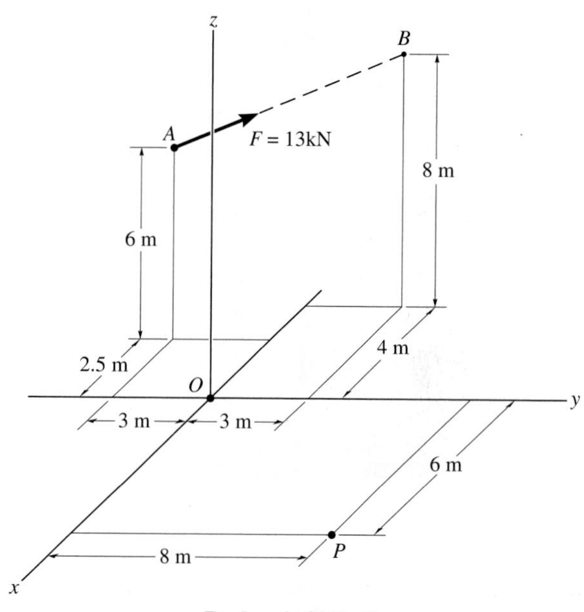

Probs. 4–31/4–32

4–33. Determine the moment of the force at A about point O. Express the result as a Cartesian vector.

4–34. Determine the moment of the force at A about point P. Express the result as a Cartesian vector.

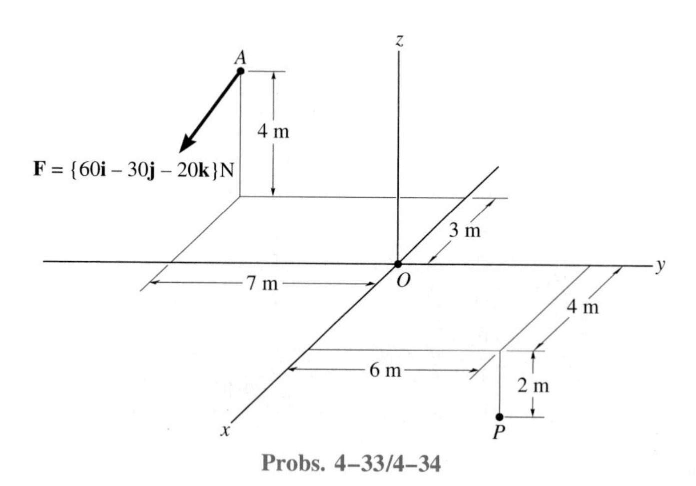

Probs. 4–33/4–34

4–35. The pole supports a 22-lb traffic light. Using Cartesian vectors, determine the moment of the weight of the traffic light about the base of the pole at A.

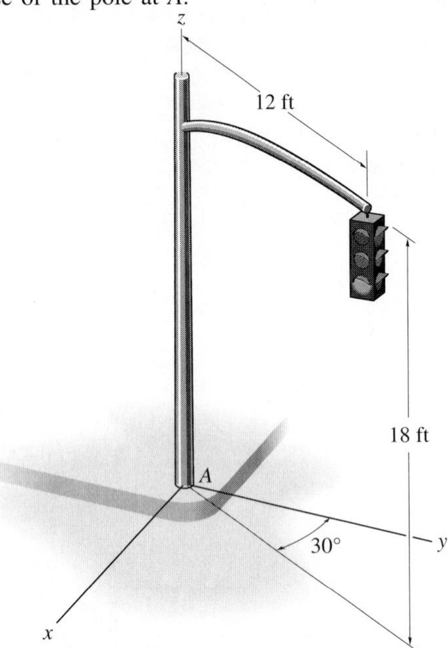

Prob. 4–35

*4–36. Using Cartesian vector analysis, determine the moment of each of the three forces acting on the column about point A. Take $\mathbf{F}_1 = \{300\mathbf{i} + 200\mathbf{j} + 80\mathbf{k}\}$ N.

4–37. Using Cartesian vector analysis, determine the resultant moment of the three forces about the base of the column at A. Take $\mathbf{F}_1 = \{400\mathbf{i} + 300\mathbf{j} + 120\mathbf{k}\}$ N.

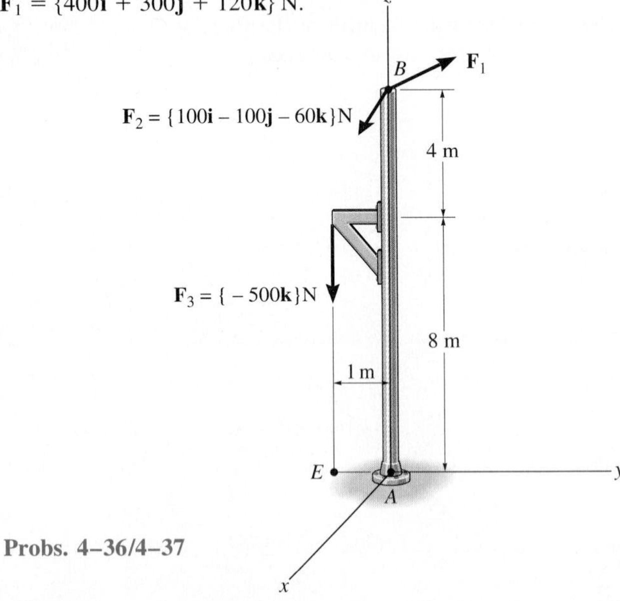

Probs. 4–36/4–37

4–38. The man pulls on the rope with a force of $F = 20$ N. Determine the moment that this force exerts about the base of the pole at O. Solve the problem two ways, i.e., by using a position vector from O to A, then O to B.

4–39. Determine the smallest force F that must be applied to the rope in order to cause the pole to break at its base O. This requires a moment of $M = 900$ N · m to be developed at O.

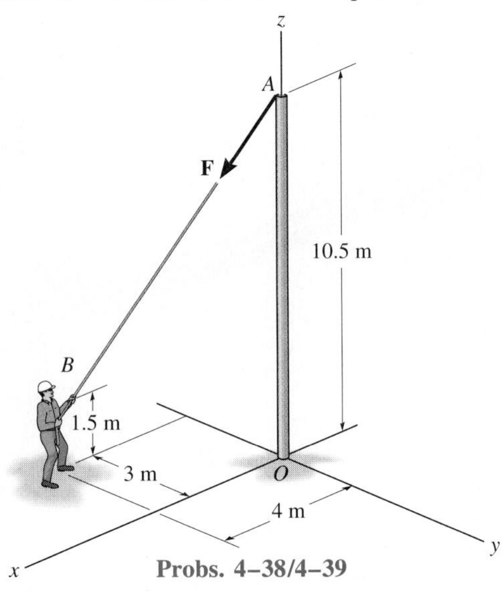

Probs. 4–38/4–39

*4–40. The curved rod lies in the x-y plane and has a radius of 3 m. If a force of $F = 80$ N acts at its end as shown, determine the moment of this force about point O.

4–41. The curved rod lies in the x-y plane and has a radius of 3 m. If a force of $F = 80$ N acts at its end as shown, determine the moment of this force about point B.

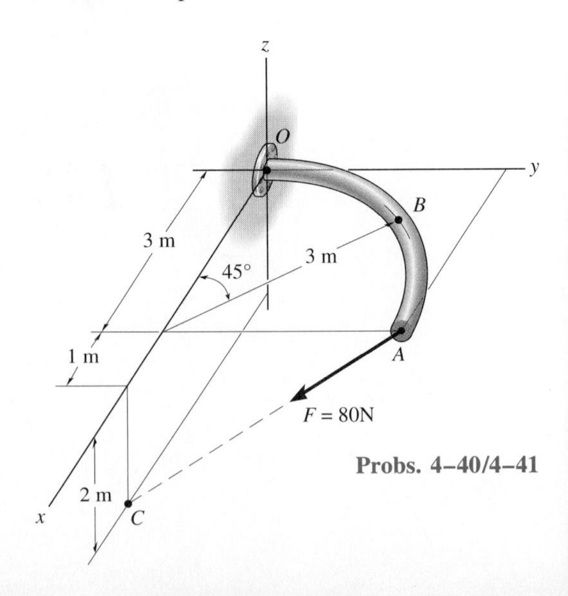

Probs. 4–40/4–41

4–42. The force $\mathbf{F} = \{600\mathbf{i} + 300\mathbf{j} - 600\mathbf{k}\}$ N acts at the end B of the beam. Determine the moment of this force about point O.

4–43. The force $\mathbf{F} = \{600\mathbf{i} + 300\mathbf{j} - 600\mathbf{k}\}$ N acts at the end of the beam. Determine the moment of the force about point A.

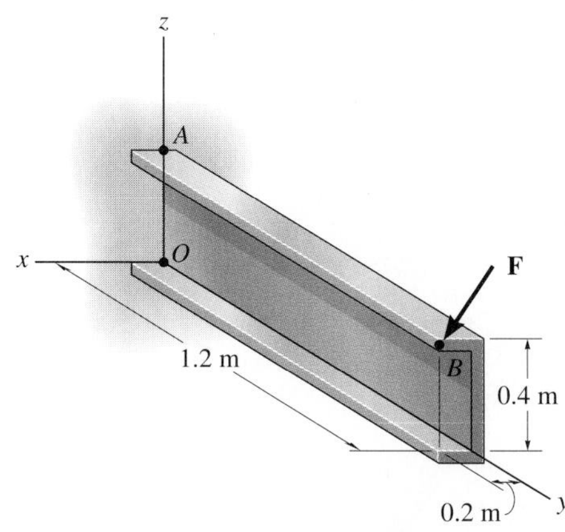

Probs. 4–42/4–43

*__4–44.__ The curved rod has a radius of 5 ft. If a force of $F = 60$ lb acts at its end as shown, determine the moment of this force about point C.

4–45. Determine the smallest force F that must be applied along the rope in order to cause the curved rod, which has a radius of 5 ft, to fail at the support C. This requires a moment of $M = 80$ lb · ft to be developed at C.

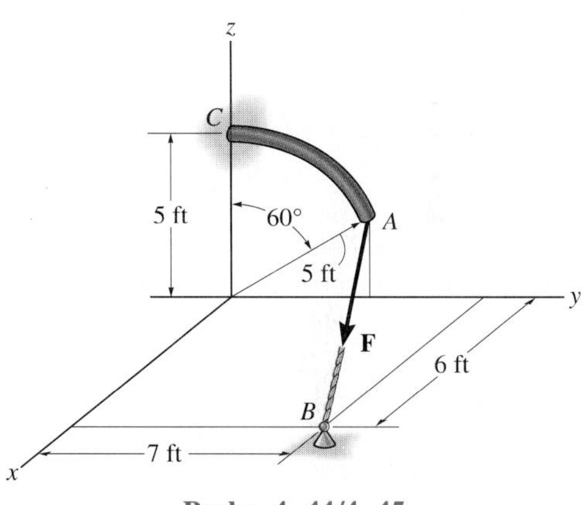

Probs. 4–44/4–45

4–46. A force of $\mathbf{F} = \{6\mathbf{i} - 2\mathbf{j} + 1\mathbf{k}\}$ kN produces a moment of $\mathbf{M}_O = \{4\mathbf{i} + 5\mathbf{j} - 14\mathbf{k}\}$ kN · m about the origin of coordinates, point O. If the force acts at a point having an x coordinate of $x = 1$ m, determine the y and z coordinates.

4–47. The force $\mathbf{F} = \{6\mathbf{i} + 8\mathbf{j} + 10\mathbf{k}\}$ N creates a moment about point O of $\mathbf{M}_O = \{-14\mathbf{i} + 8\mathbf{j} + 2\mathbf{k}\}$ N · m. If the force passes through a point having an x coordinate of 1 m, determine the y and z coordinates of the point. Also, realizing that $M_O = Fd$, determine the perpendicular distance d from point O to the line of action of $\mathbf{F}$.

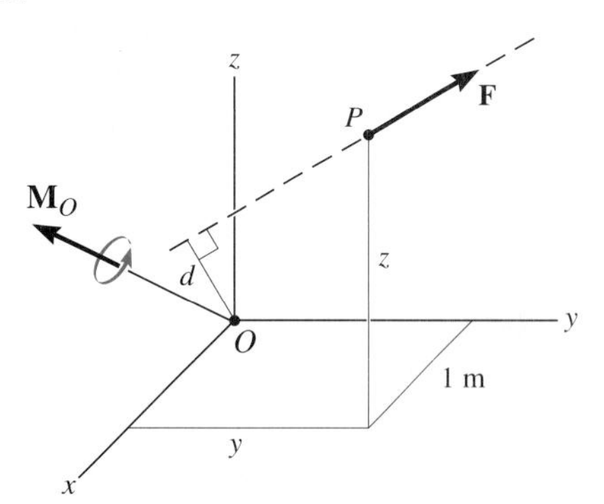

Probs. 4–46/4–47

*__4–48.__ Segments of drill pipe D for an oil well are tightened a prescribed amount by using a set of tongs T, which grip the pipe, and a hydraulic cylinder (not shown) to regulate the force $\mathbf{F}$ applied to the tongs. This force acts along the cable which passes around the small pulley P. If the cable is originally perpendicular to the tongs as shown, determine the magnitude of force $\mathbf{F}$ which must be applied so that the moment about the pipe is $M =$ 2000 lb · ft. In order to maintain this same moment what magnitude of $\mathbf{F}$ is required when the tongs rotate 30° to the dashed position?

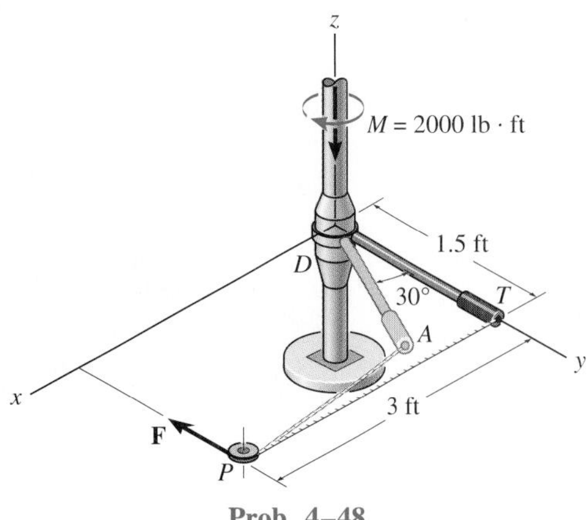

Prob. 4–48

4.5 Moment of a Force About a Specified Axis

Recall that when the moment of a force is computed about a point, the moment and its axis are *always* perpendicular to the plane containing the force and the moment arm. In some problems it is important to find the *component* of this moment along a *specified axis* that passes through the point. To solve this problem either a scalar or vector analysis can be used.

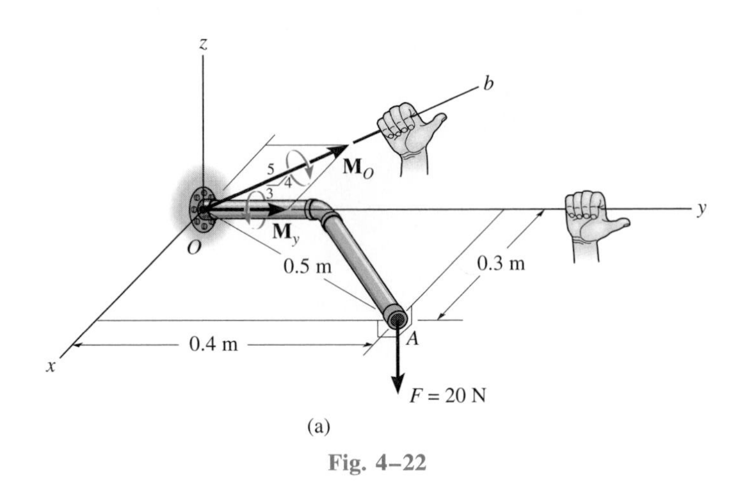

Fig. 4–22

Scalar Analysis. As a numerical example of this problem, consider the pipe assembly shown in Fig. 4–22*a*, which lies in the horizontal plane and is subjected to the vertical force of $F = 20$ N applied at point A. The moment of this force about point O has a *magnitude* of $M_O = (20 \text{ N})(0.5 \text{ m}) = 10 \text{ N} \cdot \text{m}$ and a *direction* defined by the right-hand rule, as shown in Fig. 4–22*a*. This moment tends to turn the pipe about the Ob axis. For practical reasons, however, it may be necessary to determine the *component* of $\mathbf{M}_O$ about the y axis, $\mathbf{M}_y$, since this component tends to unscrew the pipe from the flange at O. From Fig. 4–22*a*, $\mathbf{M}_y$ has a magnitude of $M_y = \frac{3}{5}(10 \text{ N} \cdot \text{m}) = 6 \text{ N} \cdot \text{m}$ and a sense of direction shown by the vector resolution. Rather than performing this *two-step* process of first finding the moment of the force about point O and then resolving the moment along the y axis, it is also possible to solve this problem *directly*. To do so, it is necessary to determine the perpendicular or moment-arm distance from the line of action of $\mathbf{F}$ to the y axis. From Fig. 4–22*a* this distance is 0.3 m. Thus the *magnitude* of the moment of the force about the y axis is again $M_y = 0.3(20 \text{ N}) = 6 \text{ N} \cdot \text{m}$, and the *direction* is determined by the right-hand rule as shown.

In general, then, *if the line of action of a force* **F** *is perpendicular to any specified axis aa*, the magnitude of the moment of **F** about the axis can be determined from the equation

$$M_a = Fd_a \tag{4–10}$$

Here d_a is the *perpendicular or shortest distance* from the force line of action to the axis. The direction is determined from the thumb of the right hand when the fingers are curled in accordance with the direction of rotation as produced by the force. In particular, realize that a *force will not contribute a moment about a specified axis if the force line of action is parallel to the axis or its line of action passes through the axis.*

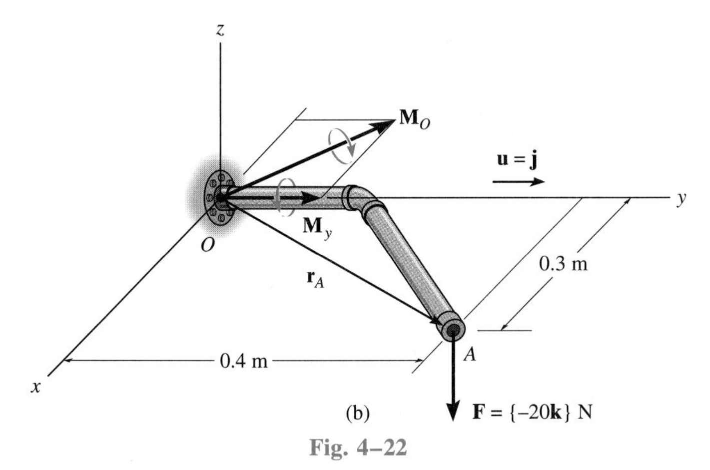

(b)

Fig. 4–22

Vector Analysis. The previous two-step solution of first finding the moment of the force about a point on the axis and then finding the projected component of the moment about the axis can also be performed using a vector analysis, Fig. 4–22*b*. Here the moment about point O is first determined from $\mathbf{M}_O = \mathbf{r}_A \times \mathbf{F} = (0.3\mathbf{i} + 0.4\mathbf{j}) \times (-20\mathbf{k}) = \{-8\mathbf{i} + 6\mathbf{j}\}$ N · m. The component or projection of this moment along the y axis is then determined from the dot product (Sec. 2.9). Since the unit vector for this axis (or line) is $\mathbf{u} = \mathbf{j}$, then $M_y = \mathbf{M}_O \cdot \mathbf{u} = (-8\mathbf{i} + 6\mathbf{j}) \cdot \mathbf{j} = 6$ N · m. This result, of course, is to be expected, since it represents the **j** component of $\mathbf{M}_O$.

A vector analysis such as this is particularly advantageous for finding the moment of a force about an axis when the force components or the appropriate moment arms are difficult to determine. For this reason, the above two-step process will now be generalized and applied to a body of arbitrary shape. To do so, consider the body in Fig. 4–23, which is subjected to the force **F** acting at point A. Here we wish to determine the effect of **F** in tending to rotate the body about the aa' axis. This tendency for rotation is measured by the moment component $\mathbf{M}_a$. To determine $\mathbf{M}_a$ we first compute the moment of **F** about any *arbitrary point O* that lies on the aa' axis. In this case, $\mathbf{M}_O$ is expressed by the

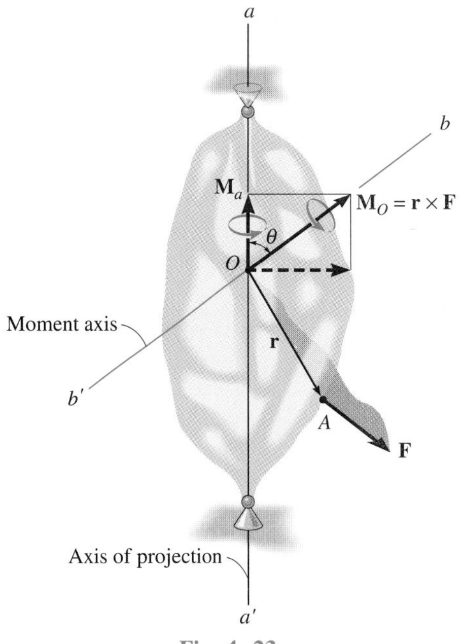

Fig. 4–23

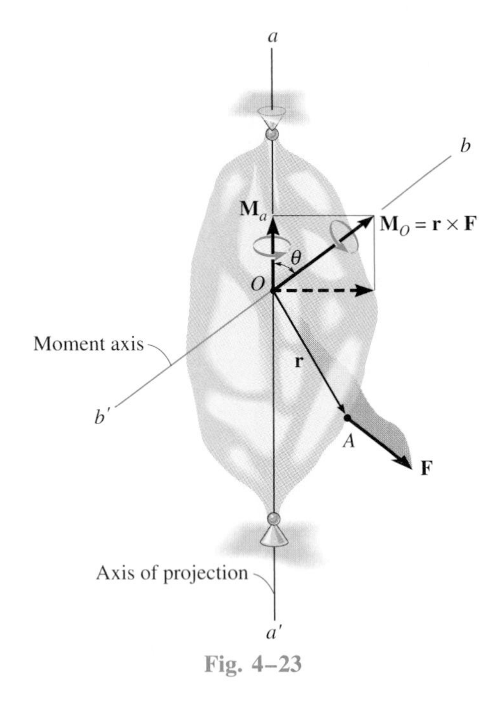

Fig. 4–23

cross product $\mathbf{M}_O = \mathbf{r} \times \mathbf{F}$, where $\mathbf{r}$ is directed from O to A. Since $\mathbf{M}_O$ acts along the moment axis bb', which is perpendicular to the plane containing $\mathbf{r}$ and $\mathbf{F}$, the component or projection of $\mathbf{M}_O$ onto the aa' axis is then represented by $\mathbf{M}_a$. The *magnitude* of $\mathbf{M}_a$ is determined by the dot product, $M_a = M_O \cos \theta = \mathbf{M}_O \cdot \mathbf{u}_a$, where $\mathbf{u}_a$ is a unit vector that defines the direction of the aa' axis. Combining these two steps as a general expression, we have $M_a = (\mathbf{r} \times \mathbf{F}) \cdot \mathbf{u}_a$. Since the dot product is commutative, we can also write

$$M_a = \mathbf{u}_a \cdot (\mathbf{r} \times \mathbf{F})$$

In vector algebra, this combination of dot and cross product yielding the scalar M_a is called the *triple scalar product*. Provided x, y, z axes are established and the Cartesian components of each of the vectors can be determined, then the triple scalar product may be written in determinant form as

$$M_a = (u_{a_x}\mathbf{i} + u_{a_y}\mathbf{j} + u_{a_z}\mathbf{k}) \cdot \begin{vmatrix} \mathbf{i} & \mathbf{j} & \mathbf{k} \\ r_x & r_y & r_z \\ F_x & F_y & F_z \end{vmatrix}$$

or simply

$$M_a = \mathbf{u}_a \cdot (\mathbf{r} \times \mathbf{F}) = \begin{vmatrix} u_{a_x} & u_{a_y} & u_{a_z} \\ r_x & r_y & r_z \\ F_x & F_y & F_z \end{vmatrix} \qquad (4\text{–}11)$$

where u_{a_x}, u_{a_y}, u_{a_z} represent the x, y, z components of the unit vector defining the direction of the aa' axis

r_x, r_y, r_z represent the x, y, z components of the position vector drawn from any point O on the aa' axis to any point A on the line of action of the force

F_x, F_y, F_z represent the x, y, z components of the force vector.

When M_a is evaluated from Eq. 4–11, it will yield a positive or negative scalar. The sign of this scalar indicates the sense of direction of $\mathbf{M}_a$ along the aa' axis. If it is positive, then $\mathbf{M}_a$ will have the same sense as $\mathbf{u}_a$, whereas if it is negative, then $\mathbf{M}_a$ will act opposite to $\mathbf{u}_a$.

Once M_a is determined, we can then express $\mathbf{M}_a$ as a Cartesian vector, namely,

$$\mathbf{M}_a = M_a\mathbf{u}_a = [\mathbf{u}_a \cdot (\mathbf{r} \times \mathbf{F})]\mathbf{u}_a \qquad (4\text{–}12)$$

Finally, if the resultant moment of a series of forces is to be computed about the axis, then the moment components of each force are added together *algebraically,* since each component lies along the same axis, i.e., its magnitude is

$$M_A = \Sigma\, [\mathbf{u}_a \cdot (\mathbf{r} \times \mathbf{F})] = \mathbf{u}_a \cdot \Sigma\, (\mathbf{r} \times \mathbf{F})$$

The following examples illustrate a numerical application of the above concepts.

Example 4–9

The force $\mathbf{F} = \{-40\mathbf{i} + 20\mathbf{j} + 10\mathbf{k}\}$ N acts at point A shown in Fig. 4–24*a*. Determine the moments of this force about the x and Oa axes.

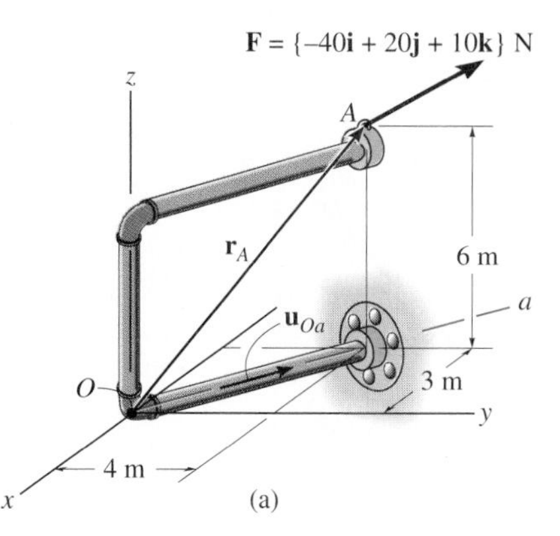

(a)

SOLUTION I *(VECTOR ANALYSIS)*

We can solve this problem by using the position vector $\mathbf{r}_A$. Why? Since $\mathbf{r}_A = \{-3\mathbf{i} + 4\mathbf{j} + 6\mathbf{k}\}$ m, and $\mathbf{u}_x = \mathbf{i}$, then applying Eq. 4–11,

$$M_x = \mathbf{i} \cdot (\mathbf{r}_A \times \mathbf{F}) = \begin{vmatrix} 1 & 0 & 0 \\ -3 & 4 & 6 \\ -40 & 20 & 10 \end{vmatrix}$$

$$= 1[4(10) - 6(20)] - 0[(-3)(10) - 6(-40)] + 0[(-3)(20) - 4(-40)]$$

$$= -80 \text{ N} \cdot \text{m} \qquad \textit{Ans.}$$

The negative sign indicates that the sense of $\mathbf{M}_x$ is opposite to **i**.

We can compute M_{Oa} also using $\mathbf{r}_A$ because $\mathbf{r}_A$ extends from a point on the Oa axis to the force. Also, $\mathbf{u}_{Oa} = -\frac{3}{5}\mathbf{i} + \frac{4}{5}\mathbf{j}$. Thus,

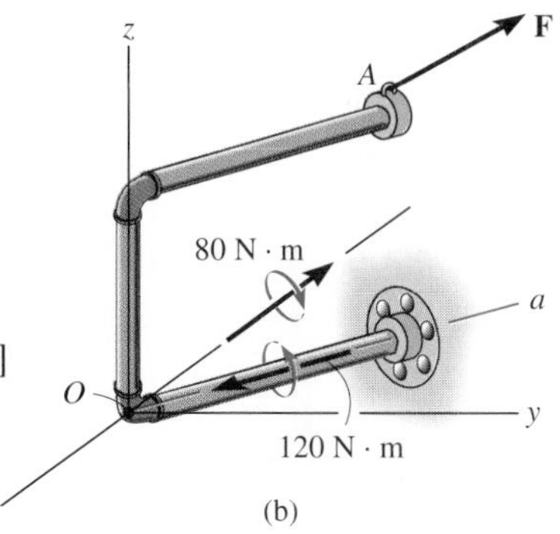

(b)

$$M_{Oa} = \mathbf{u}_{Oa} \cdot (\mathbf{r}_A \times \mathbf{F}) = \begin{vmatrix} -\frac{3}{5} & \frac{4}{5} & 0 \\ -3 & 4 & 6 \\ -40 & 20 & 10 \end{vmatrix}$$

$$= -\tfrac{3}{5}[4(10) - 6(20)] - \tfrac{4}{5}[(-3)(10) - 6(-40)] + 0[(-3)(20) - 4(-40)]$$

$$= -120 \text{ N} \cdot \text{m} \qquad \textit{Ans.}$$

What does the negative sign indicate?

The moment components are shown in Fig. 4–24*b*.

SOLUTION II *(SCALAR ANALYSIS)*

Since the force components and moment arms are easy to determine for computing M_x a scalar analysis can be used to solve this problem. Referring to Fig. 4–24*c*, only the 10-N and 20-N forces contribute moments about the x axis. (The line of action of the 40-N force is *parallel* to this axis and hence its moment about the x axis is zero.) Using the right-hand rule, the algebraic sum of the moment components about the x axis is therefore

$$M_x = (10 \text{ N})(4 \text{ m}) - (20 \text{ N})(6 \text{ m}) = -80 \text{ N} \cdot \text{m} \qquad \textit{Ans.}$$

Although not required here, note also that

$$M_y = (10 \text{ N})(3 \text{ m}) - (40 \text{ N})(6 \text{ m}) = -210 \text{ N} \cdot \text{m}$$
$$M_z = (40 \text{ N})(4 \text{ m}) - (20 \text{ N})(3 \text{ m}) = 100 \text{ N} \cdot \text{m}$$

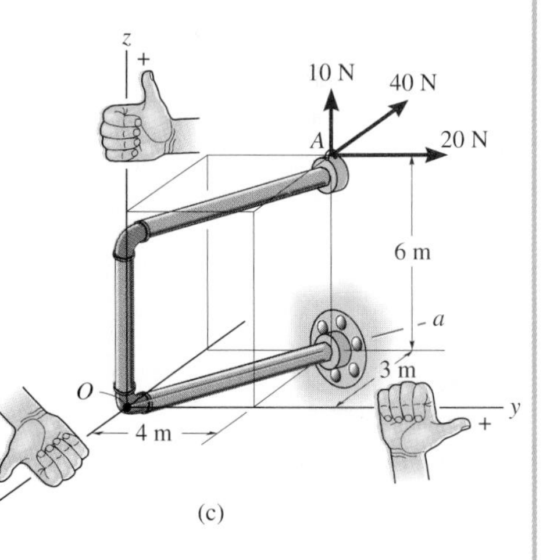

(c)

Fig. 4–24

If we were to determine M_{Oa} by this scalar method it would require much more effort, since the force components of 40 N and 20 N are *not perpendicular* to the direction of Oa. The vector analysis yields a more direct solution.

Example 4–10

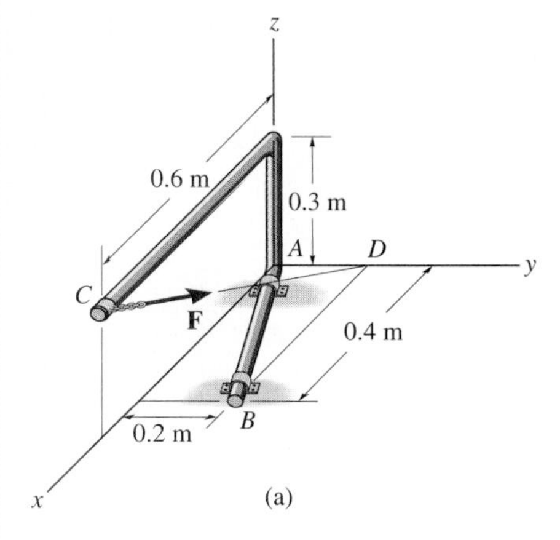

(a)

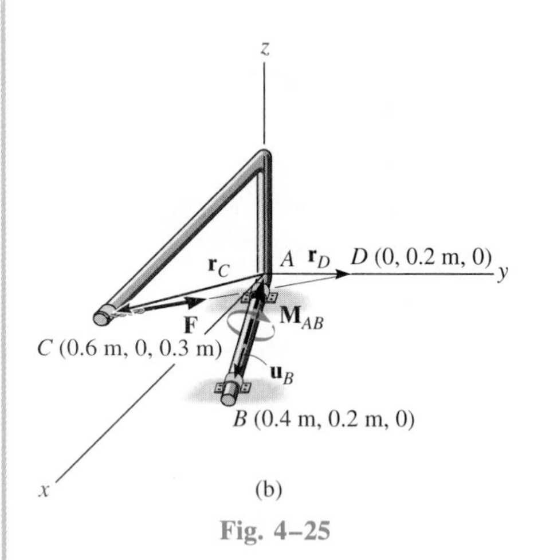

(b)

Fig. 4–25

The rod shown in Fig. 4–25*a* is supported by two brackets at *A* and *B*. Determine the moment $\mathbf{M}_{AB}$ produced by $\mathbf{F} = \{-600\mathbf{i} + 200\mathbf{j} - 300\mathbf{k}\}$ N, which tends to rotate the rod about the *AB* axis.

SOLUTION

A vector analysis using $M_{AB} = \mathbf{u}_B \cdot (\mathbf{r} \times \mathbf{F})$ will be considered for the solution since the moment arm or perpendicular distance from the line of action of **F** to the *AB* axis is difficult to determine. Each of the terms in the equation will now be identified.

Unit vector $\mathbf{u}_B$ defines the direction of the *AB* axis of the rod, Fig. 4–25*b*, where

$$\mathbf{u}_B = \frac{\mathbf{r}_B}{r_B} = \frac{0.4\mathbf{i} + 0.2\mathbf{j}}{\sqrt{(0.4)^2 + (0.2)^2}} = 0.894\mathbf{i} + 0.447\mathbf{j}$$

Vector **r** is directed from *any point* on the *AB* axis to *any point* on the line of action of the force. For example, position vectors $\mathbf{r}_C$ and $\mathbf{r}_D$ are suitable, Fig. 4–25*b*. (Although not shown, $\mathbf{r}_{BC}$ or $\mathbf{r}_{BD}$ can also be used.) For simplicity, we choose $\mathbf{r}_D$, where

$$\mathbf{r}_D = \{0.2\mathbf{j}\} \text{ m}$$

The force is

$$\mathbf{F} = \{-600\mathbf{i} + 200\mathbf{j} - 300\mathbf{k}\} \text{ N}$$

Substituting these vectors into the determinant form and expanding, we have

$$M_{AB} = \mathbf{u}_B \cdot (\mathbf{r}_D \times \mathbf{F}) = \begin{vmatrix} 0.894 & 0.447 & 0 \\ 0 & 0.2 & 0 \\ -600 & 200 & -300 \end{vmatrix}$$

$$= 0.894[0.2(-300) - 0(200)] - 0.447[0(-300) - 0(-600)] + 0[0(200) - 0.2(-600)]$$

$$= -53.67 \text{ N} \cdot \text{m}$$

The negative sign indicates that the sense of $\mathbf{M}_{AB}$ is opposite to that of $\mathbf{u}_B$.

Expressing $\mathbf{M}_{AB}$ as a Cartesian vector yields

$$\mathbf{M}_{AB} = M_{AB}\mathbf{u}_B = (-53.67 \text{ N} \cdot \text{m})(0.894\mathbf{i} + 0.447\mathbf{j})$$

$$= \{-48.0\mathbf{i} - 24.0\mathbf{j}\} \text{ N} \cdot \text{m} \qquad \textit{Ans.}$$

The result is shown in Fig. 4–25*b*.

Note that if axis *AB* is defined using a unit vector directed from *B* toward *A*, then in the above formulation $-\mathbf{u}_B$ would have to be used. This would lead to $M_{AB} = +53.67$ N · m. Consequently, $\mathbf{M}_{AB} = M_{AB}(-\mathbf{u}_B)$, and the above result would again be determined.

PROBLEMS

4–49. Determine the moment of the force **F** about the *Oa* axis. Express the result as a Cartesian vector.

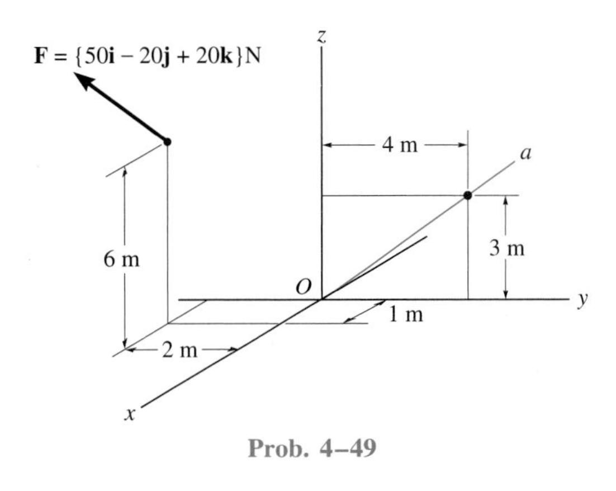

Prob. 4–49

4–50. Determine the moment of the force **F** about the *Aa* axis. Express the result as a Cartesian vector.

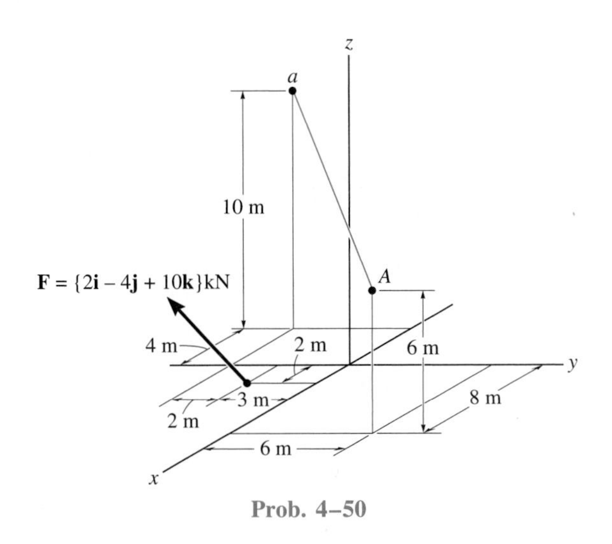

Prob. 4–50

4–51. Determine the resultant moment of the two forces about the *Oa* axis. Express the result as a Cartesian vector.

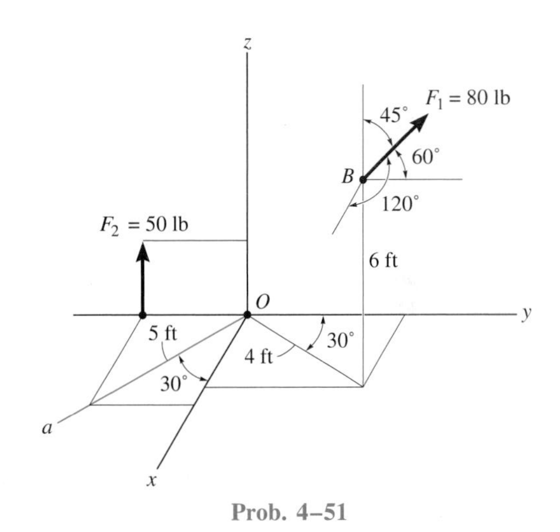

Prob. 4–51

***4–52.** Determine the resultant moment of the two forces about the *aa* axis. Express the result as a Cartesian vector.

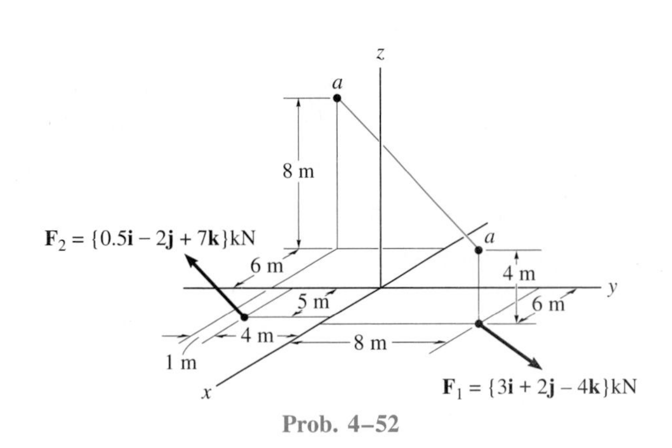

Prob. 4–52

4–53. The cutting tool on the lathe exerts a force **F** on the shaft in the direction shown. Determine the moment of this force about the y axis of the shaft.

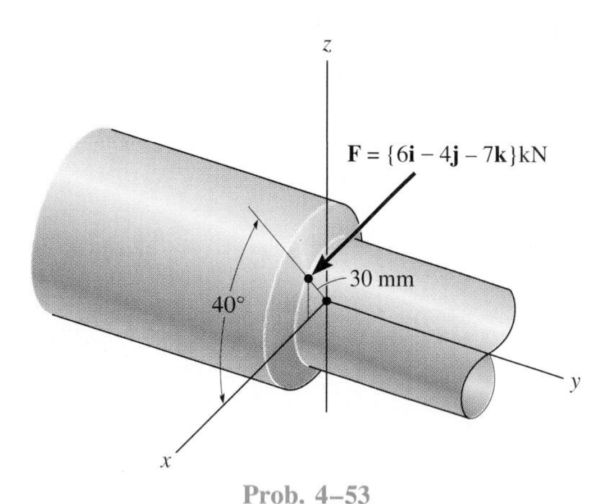

Prob. 4–53

4–54. The hood of the automobile is supported by the strut AB, which exerts a force of $F = 24$ lb on the hood. Determine the moment of this force about the hinged axis y.

4–55. When the strut AB is removed, a moment of required about the y axis to hold the hood of the automobile in the position shown. Determine the magnitude of the *smallest force* $\mathbf{F}'$ that should be applied at B to do this.

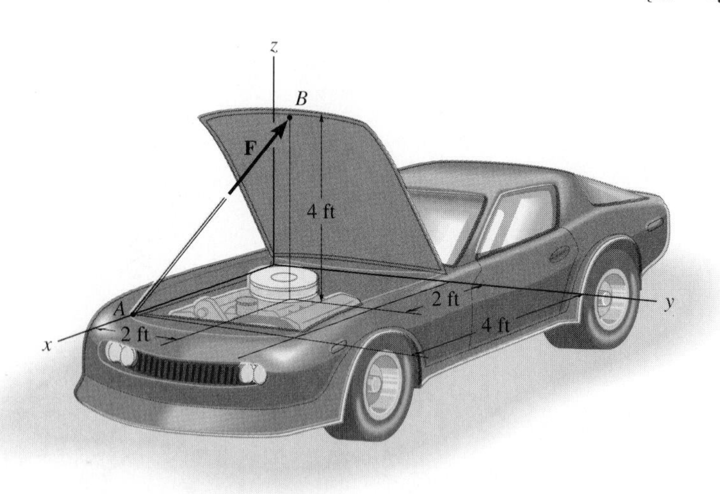

Probs. 4–54/4–55

***4–56.** Determine the magnitude of the moment of the force $\mathbf{F} = \{50\mathbf{i} - 20\mathbf{j} - 80\mathbf{k}\}$ N about the base line AB of the tripod.

4–57. Determine the magnitude of the moment of the force $\mathbf{F} = \{50\mathbf{i} - 20\mathbf{j} - 80\mathbf{k}\}$ N about the base line BC of the tripod.

4–58. Determine the magnitude of the moment of the force $\mathbf{F} = \{50\mathbf{i} - 20\mathbf{j} - 80\mathbf{k}\}$ N about the base line CA of the tripod.

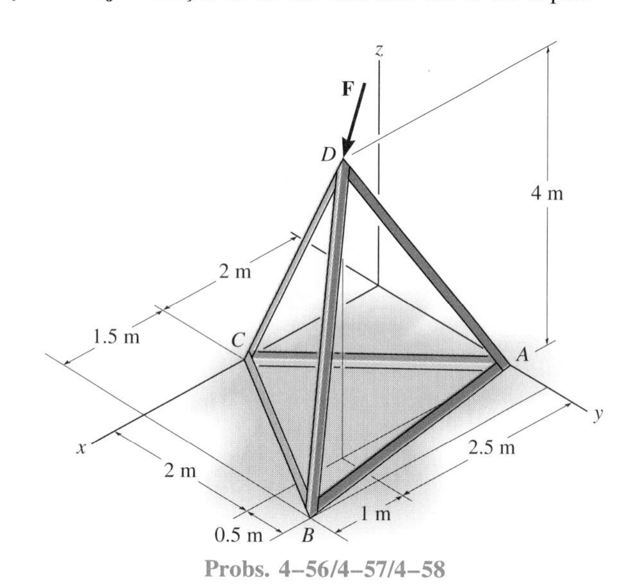

Probs. 4–56/4–57/4–58

4–59. The showerhead H is installed into the shower arm using a wrench. Determine the moment that the force $\mathbf{F}_C = \{2\mathbf{i} - 6\mathbf{j} + 4\mathbf{k}\}$ lb applied to the wrench develops about the y axis.

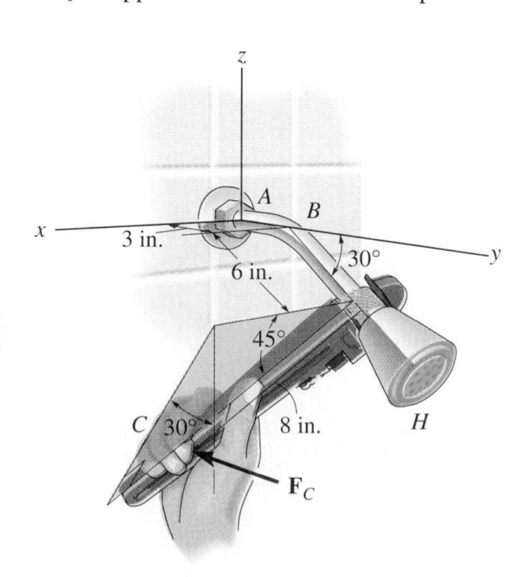

Prob. 4–59

***4–60.** The flex-headed ratchet wrench is subjected to a force of $P = 16$ lb, applied perpendicular to the handle as shown. Determine the moment or torque this imparts along the vertical axis of the bolt at A.

4–61. If a torque or moment of 80 lb · in. is required to loosen the bolt at A, determine the force P that must be applied perpendicular to the handle of the flex-headed ratchet wrench.

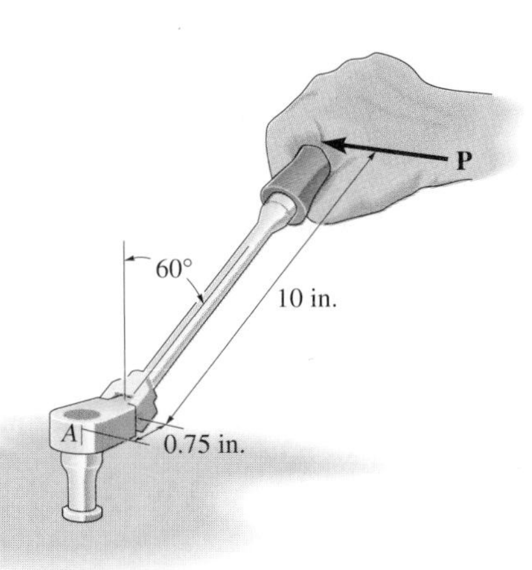

Probs. 4–60/4–61

4–62. A vertical force of $F = 60$ N is applied to the handle of the pipe wrench. Determine the moment that this force exerts along the axis AB (x axis) of the pipe assembly. Both the wrench and pipe assembly ABC lie in the x-y plane. *Suggestion:* Use a scalar analysis.

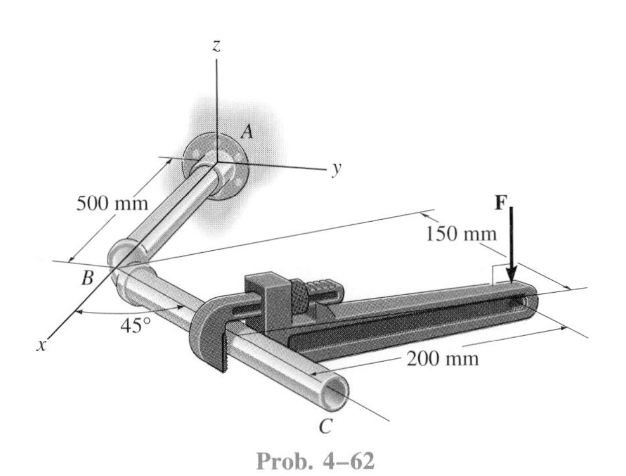

Prob. 4–62

4–63. Determine the magnitude of the vertical force **F** acting on the handle of the wrench so that this force produces a component of moment along the AB axis (x axis) of the pipe assembly of $(\mathbf{M}_A)_x = \{-5\mathbf{i}\}$ N · m. Both the pipe assembly ABC and the wrench lie in the x-y plane. *Suggestion:* Use a scalar analysis.

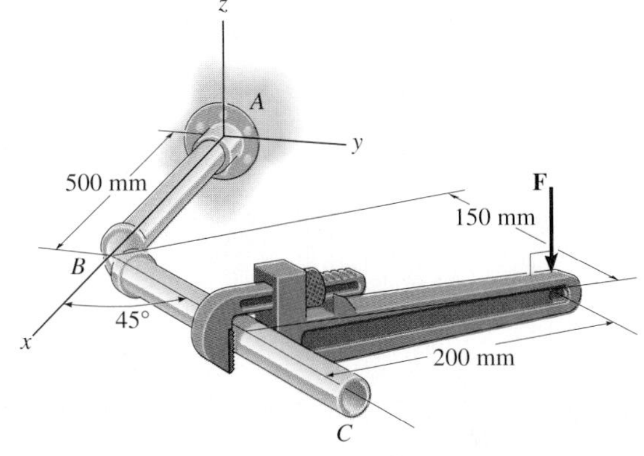

Prob. 4–63

***4–64.** A horizontal force of $\mathbf{F} = \{-50\mathbf{i}\}$ N is applied perpendicular to the handle of the pipe wrench. Determine the moment that this force exerts along the axis OA (z axis) of the pipe assembly. Both the wrench and pipe assembly, $OABC$, lie in the y-z plane. *Suggestion:* Use a scalar analysis.

4–65. Determine the magnitude of the horizontal force $\mathbf{F} = -F\mathbf{i}$ acting on the handle of the wrench so that this force produces a component of moment along the OA axis (z axis) of the pipe assembly of $\mathbf{M}_z = \{4\mathbf{k}\}$ N · m. Both the wrench and the pipe assembly, $OABC$, lie in the y-z plane. *Suggestion:* Use a scalar analysis.

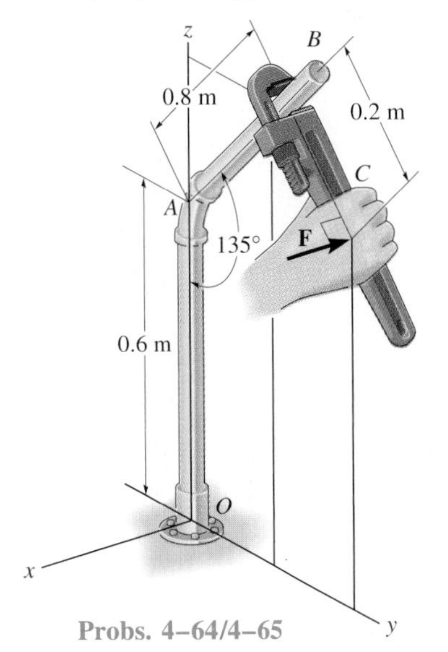

Probs. 4–64/4–65

4.6 Moment of a Couple

A *couple* is defined as two parallel forces that have the same magnitude, opposite directions, and are separated by a perpendicular distance d, Fig. 4–26. Since the resultant force of the two forces composing the couple is zero, the only effect of a couple is to produce a rotation or tendency of rotation in a specified direction. As a practical example, a couple is produced on the steering wheel of an automobile when turning the wheels.

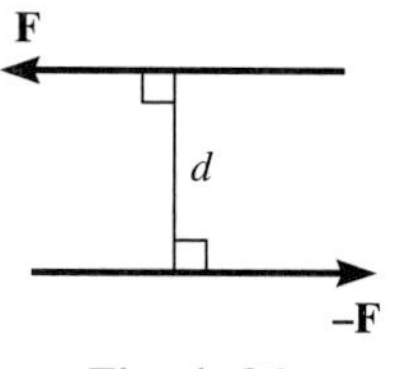

Fig. 4–26

The moment produced by a couple, called a *couple moment,* is equivalent to the sum of the moments of both couple forces, determined about *any* arbitrary point O in space. To show this, consider position vectors $\mathbf{r}_A$ and $\mathbf{r}_B$, directed from O to points A and B lying on the line of action of $-\mathbf{F}$ and $\mathbf{F}$, Fig. 4–27. The couple moment computed about O is therefore

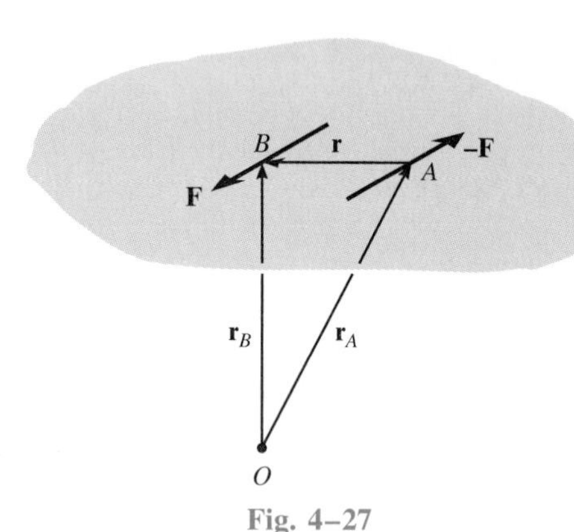

Fig. 4–27

$$\begin{aligned}\mathbf{M} &= \mathbf{r}_A \times (-\mathbf{F}) + \mathbf{r}_B \times (\mathbf{F}) \\ &= (\mathbf{r}_B - \mathbf{r}_A) \times \mathbf{F}\end{aligned}$$

By the triangle law of vector addition, $\mathbf{r}_A + \mathbf{r} = \mathbf{r}_B$ or $\mathbf{r} = \mathbf{r}_B - \mathbf{r}_A$, so that

$$\mathbf{M} = \mathbf{r} \times \mathbf{F} \qquad (4\text{–}13)$$

This result indicates that a couple moment is a *free vector,* i.e., it can act at *any point,* since $\mathbf{M}$ depends *only* upon the position vector directed *between* the forces and *not* the position vectors $\mathbf{r}_A$ and $\mathbf{r}_B$, directed from point O to the forces. This concept is therefore unlike the moment of a force, which requires a definite point (or axis) about which moments are determined.

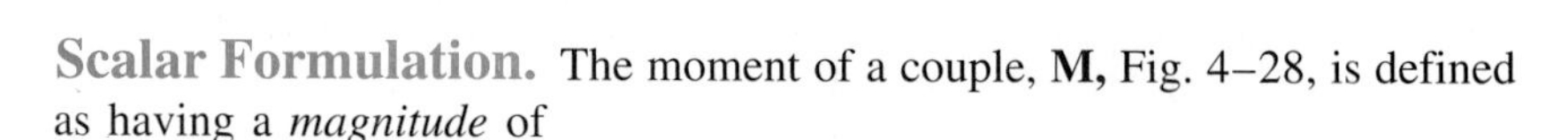

Scalar Formulation. The moment of a couple, **M,** Fig. 4–28, is defined as having a *magnitude* of

$$M = Fd \qquad (4\text{–}14)$$

Fig. 4–28

where F is the magnitude of one of the forces and d is the perpendicular distance or moment arm between the forces. The *direction* and sense of the couple moment are determined by the right-hand rule, where the thumb indicates the direction when the fingers are curled with the sense of rotation caused by the two forces. In all cases, **M** acts perpendicular to the plane containing these forces.

Vector Formulation. The moment of a couple can also be expressed by the vector cross product using Eq. 4–13, i.e.,

$$\mathbf{M} = \mathbf{r} \times \mathbf{F} \qquad (4\text{–}15)$$

Application of this equation is easily remembered if one thinks of taking the moments of both forces about a point lying on the line of action of one of the forces. For example, if moments are taken about point A in Fig. 4–27 the moment of $-\mathbf{F}$ is zero about this point and the moment of $\mathbf{F}$ is defined from Eq. 4–15. Therefore, in the formulation $\mathbf{r}$ is crossed with the force $\mathbf{F}$ to which it is directed.

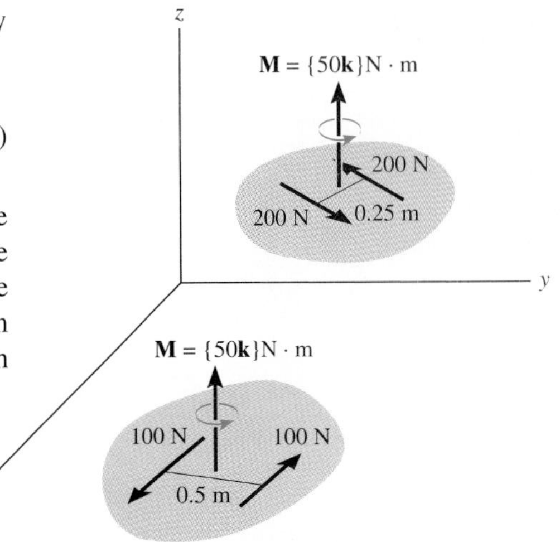

Fig. 4–29

Equivalent Couples. Two couples are said to be equivalent if they produce the same moment. Since the moment produced by a couple is always perpendicular to the plane containing the couple forces, it is therefore necessary that the forces of equal couples lie either in the same plane or in planes that are *parallel* to one another. In this way, the direction of each couple moment will be the same, that is, perpendicular to the parallel planes. For example, the two couples shown in Fig. 4–29 are equivalent. One couple is produced by a pair of 100-N forces separated by a distance of $d = 0.5$ m, and the other is produced by a pair of 200-N forces separated by a distance of 0.25 m. Since the planes in which the forces act are parallel to the x–y plane, the moment produced by each of the couples may be expressed as $\mathbf{M} = \{50\mathbf{k}\}$ N · m.

Resultant Couple Moment. Since couple moments are free vectors, they may be applied at any point P on a body and added vectorially. For example, the two couples acting on different planes of the rigid body in Fig. 4–30a may be replaced by their corresponding couple moments $\mathbf{M}_1$ and $\mathbf{M}_2$, Fig. 4–30b, and then these free vectors may be moved to the *arbitrary point P* and added to obtain the resultant couple moment $\mathbf{M}_R = \mathbf{M}_1 + \mathbf{M}_2$, shown in Fig. 4–30$c$.

If more than two couple moments act on the body, we may generalize this concept and write the vector resultant as

$$\mathbf{M}_R = \Sigma\,(\mathbf{r} \times \mathbf{F}) \qquad (4\text{–}16)$$

where each couple moment is computed in accordance with Eq. 4–15.

The following examples illustrate these concepts numerically. In general, problems projected in two dimensions should be solved using a scalar analysis, since the moment arms and force components are easy to compute.

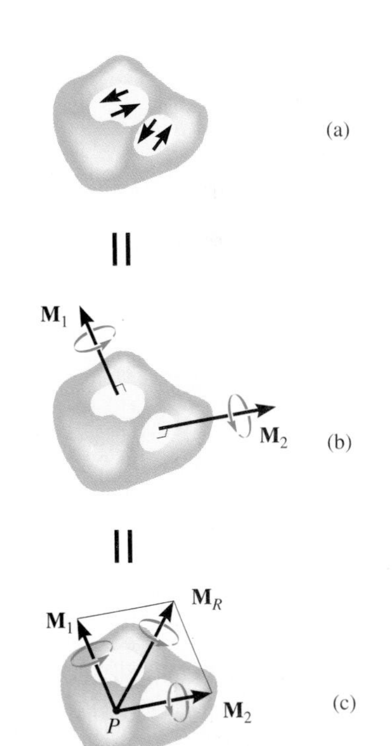

Fig. 4–30

Example 4–11

A couple acts on the gear teeth as shown in Fig. 4–31*a*. Replace it by an equivalent couple having a pair of forces that act through (a) points A and B, and (b) points D and E.

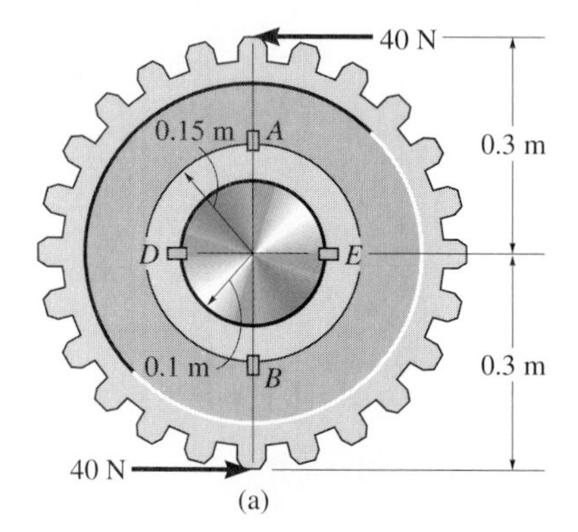

(a)

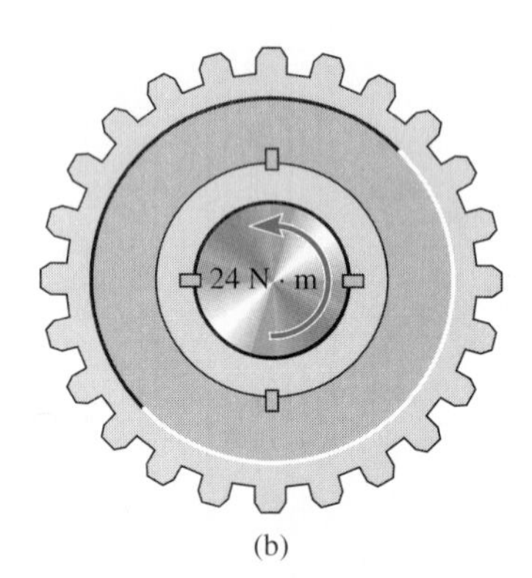

(b)

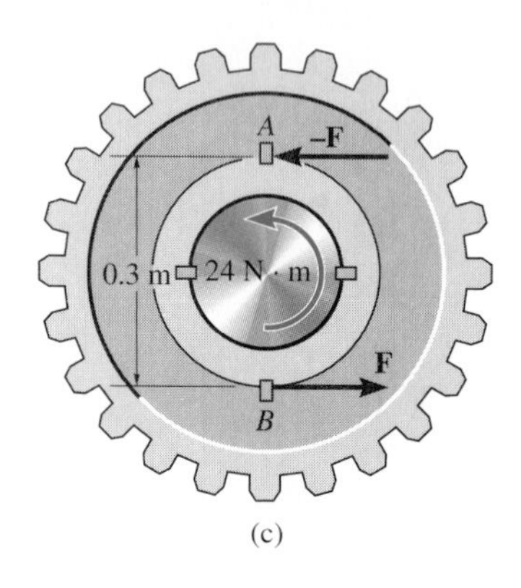

(c)

Fig. 4–31

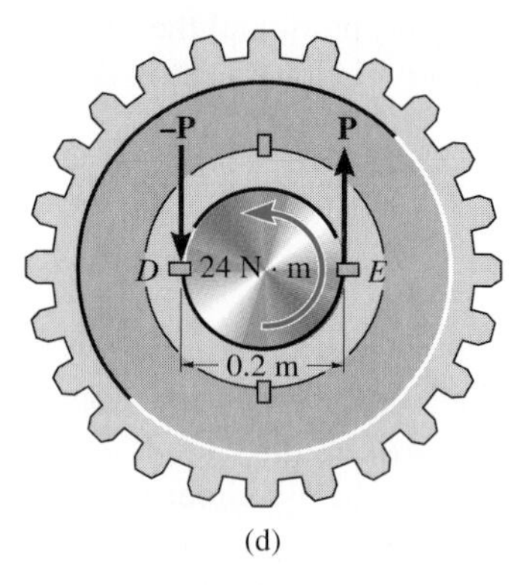

(d)

SOLUTION *(SCALAR ANALYSIS)*

The couple has a magnitude of $M = Fd = 40(0.6) = 24\text{ N}\cdot\text{m}$ and a direction that is out of the page since the forces tend to rotate counterclockwise. **M** is a free vector so that it can be placed at any point on the gear, Fig. 4–31*b*.

Part (a). To preserve the counterclockwise rotation of **M,** *horizontal* forces acting through points A and B must be directed as shown in Fig. 4–31*c*. The magnitude of each force is

$$M = Fd$$

$$24\text{ N}\cdot\text{m} = F(0.3\text{ m})$$

$$F = 80\text{ N} \qquad \textit{Ans.}$$

Part (b). Likewise, for counterclockwise rotation, forces acting through points D and E must be *vertical* and directed as shown in Fig. 4–31*d*. The magnitude of each force is

$$M = Pd$$

$$24\text{ N}\cdot\text{m} = P(0.2\text{ m})$$

$$P = 120\text{ N} \qquad \textit{Ans.}$$

Example 4–12

Determine the moment of the couple acting on the machine member shown in Fig. 4–32*a*.

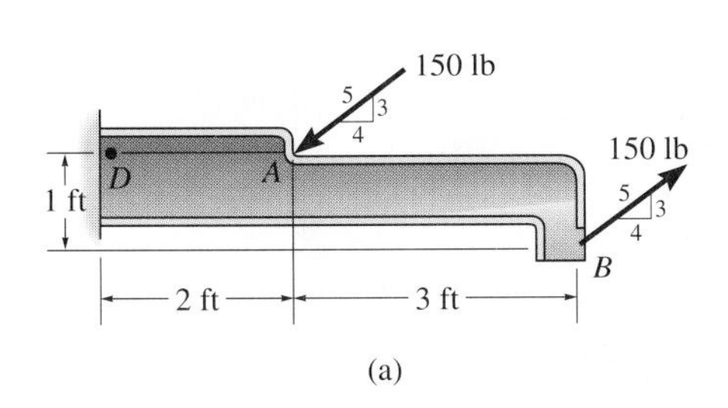

(a)

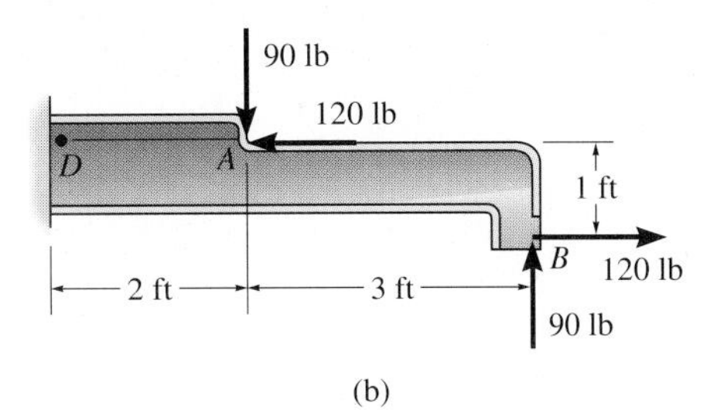

(b)

SOLUTION (*SCALAR ANALYSIS*)

Here it is somewhat difficult to determine the perpendicular distance between the forces and compute the couple moment as $M = Fd$. Instead, we can resolve each force into its horizontal and vertical components, $F_x = \frac{4}{5}(150 \text{ lb}) = 120 \text{ lb}$ and $F_y = \frac{3}{5}(150 \text{ lb}) = 90 \text{ lb}$, Fig. 4–32*b*, and then use the principle of moments. The couple moment can be determined about *any point*. For example, if point D is chosen, we have for all four forces,

$$\circlearrowleft + M = 120 \text{ lb}(0 \text{ ft}) - 90 \text{ lb}(2 \text{ ft}) + 90 \text{ lb}(5 \text{ ft}) + 120 \text{ lb}(1 \text{ ft})$$
$$= 390 \text{ lb} \cdot \text{ft} \circlearrowleft \qquad \textit{Ans.}$$

It is easier, however, to determine the moments about point A or B in order to *eliminate* the moment of the forces acting at the moment point. For point A, Fig. 4–32*b*, we have

$$\circlearrowleft + M = 90 \text{ lb}(3 \text{ ft}) + 120 \text{ lb}(1 \text{ ft})$$
$$= 390 \text{ lb} \cdot \text{ft} \circlearrowleft \qquad \textit{Ans.}$$

Show that one obtains this same result if moments are summed about point B. Notice also that the couple in Fig. 4–32*a* can be replaced by *two* couples in Fig. 4–32*b*. Using $M = Fd$, one couple has a moment of $M_1 = 90 \text{ lb}(3 \text{ ft}) = 270 \text{ lb} \cdot \text{ft}$ and the other has a moment of $M_2 = 120 \text{ lb}(1 \text{ ft}) = 120 \text{ lb} \cdot \text{ft}$. By the right-hand rule, both couple moments are counterclockwise and are therefore directed out of the page. Since these couples are free vectors, they can be moved to any point and added, which yields $M = 270 \text{ lb} \cdot \text{ft} + 120 \text{ lb} \cdot \text{ft} = 390 \text{ lb} \cdot \text{ft} \circlearrowleft$, the same result determined above. **M** is a free vector and can therefore act at any point on the member, Fig. 4–32*c*. Also, realize that the external effect, such as the support reactions on the member, will be the *same* if the member supports the couple, Fig. 4–32*a*, or the couple moment, Fig. 4–32*c*.

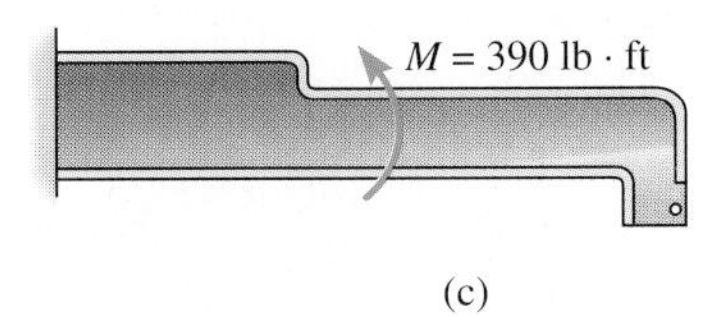

(c)

Fig. 4–32

Example 4–13

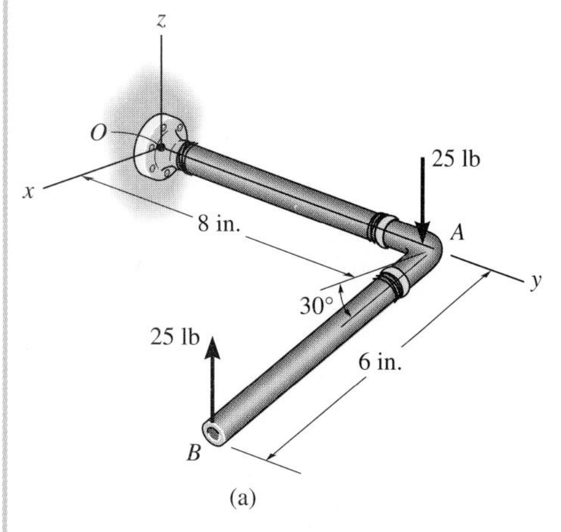

(a)

Determine the couple moment acting on the pipe shown in Fig. 4–33*a*. Segment *AB* is directed 30° below the *x–y* plane.

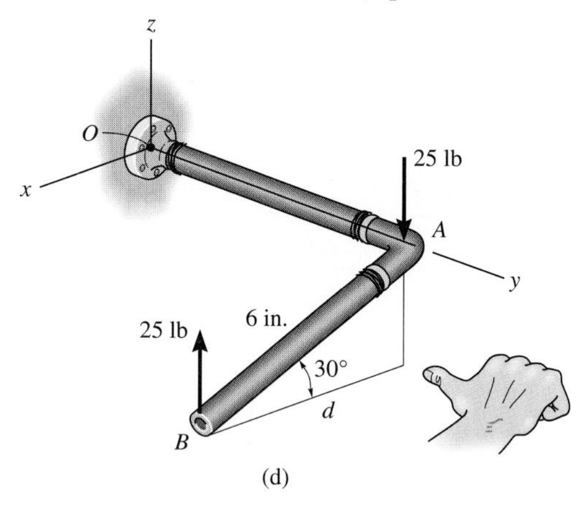

(d)

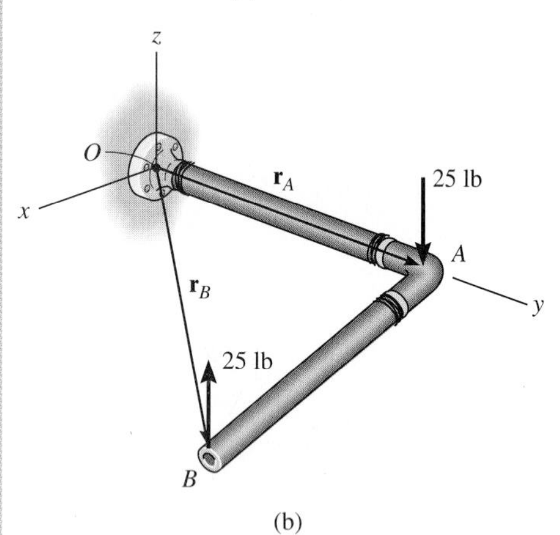

(b)

SOLUTION I (*VECTOR ANALYSIS*)

The moment of the two couple forces can be found about *any point*. If point *O* is considered, Fig. 4–33*b*, we have

$$\begin{aligned}\mathbf{M} &= \mathbf{r}_A \times (-25\mathbf{k}) + \mathbf{r}_B \times (25\mathbf{k}) \\ &= (8\mathbf{j}) \times (-25\mathbf{k}) + (6 \cos 30°\mathbf{i} + 8\mathbf{j} - 6 \sin 30°\mathbf{k}) \times (25\mathbf{k}) \\ &= -200\mathbf{i} - 129.9\mathbf{j} + 200\mathbf{i} \\ &= \{-130\mathbf{j}\} \text{ lb} \cdot \text{in.}\end{aligned}$$ *Ans.*

It is *easier* to take moments of the couple forces about a point lying on the line of action of one of the forces, e.g., point *A*, Fig. 4–33*c*. In this case the moment of the force at *A* is zero, so that

$$\begin{aligned}\mathbf{M} &= \mathbf{r}_{AB} \times (25\mathbf{k}) \\ &= (6 \cos 30°\mathbf{i} - 6 \sin 30°\mathbf{k}) \times (25\mathbf{k}) \\ &= \{-130\mathbf{j}\} \text{ lb} \cdot \text{in.}\end{aligned}$$ *Ans.*

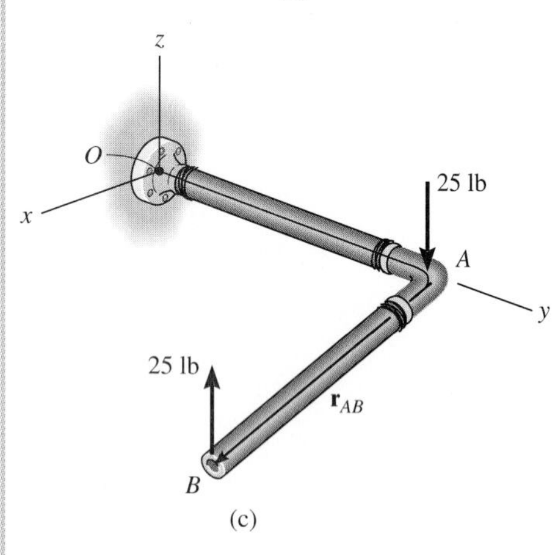

(c)

Fig. 4–33

SOLUTION II (*SCALAR ANALYSIS*)

Although this problem is shown in three dimensions, the geometry is simple enough to use the scalar equation $M = Fd$. The perpendicular distance between the lines of action of the forces is $d = 6 \cos 30° = 5.20$ in., Fig. 4–33*d*. Hence, taking moments of the forces about either point *A* or *B* yields

$$M = Fd = 25 \text{ lb}(5.20 \text{ in.}) = 129.9 \text{ lb} \cdot \text{in.}$$

Applying the right-hand rule, **M** acts in the $-\mathbf{j}$ direction. Thus,

$$\mathbf{M} = \{-130\mathbf{j}\} \text{ lb} \cdot \text{in.}$$ *Ans.*

Example 4–14

Replace the two couples acting on the pipe column in Fig. 4–34*a* by a resultant couple moment.

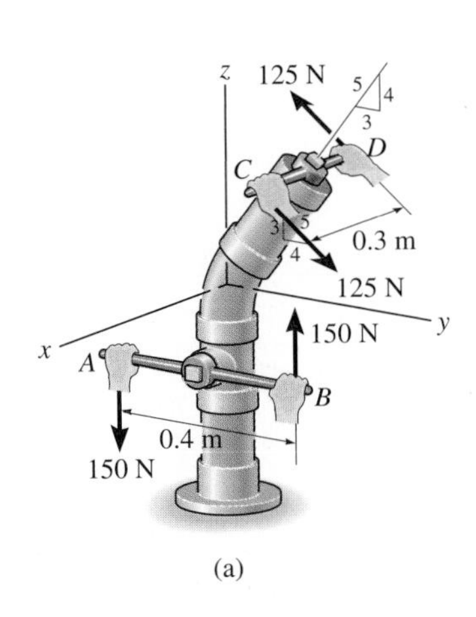

(a)

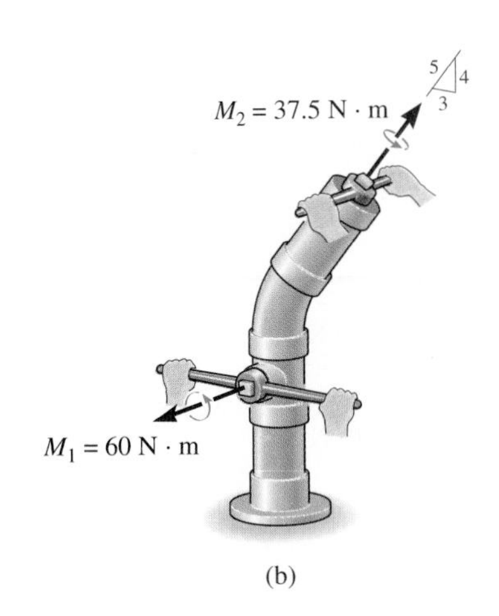

(b)

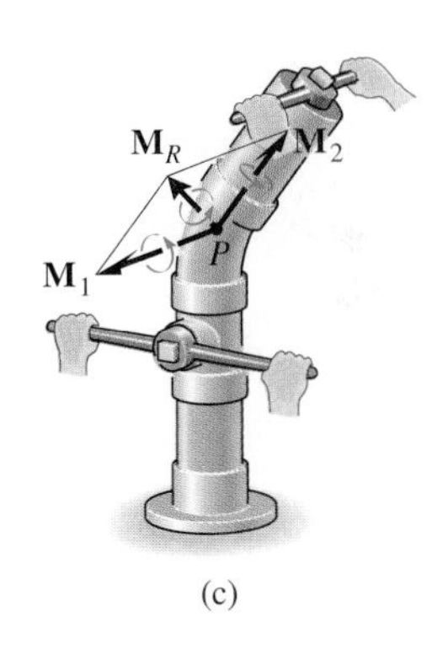

(c)

Fig. 4–34

SOLUTION *(VECTOR ANALYSIS)*

The couple moment $\mathbf{M}_1$, developed by the forces at A and B, can easily be determined from a scalar formulation.

$$M_1 = Fd = 150\text{ N}(0.4\text{ m}) = 60\text{ N}\cdot\text{m}$$

By the right-hand rule, $\mathbf{M}_1$ acts in the $+\mathbf{i}$ direction, Fig. 4–34*b*. Hence,

$$\mathbf{M}_1 = \{60\mathbf{i}\}\text{ N}\cdot\text{m}$$

Vector analysis will be used to determine $\mathbf{M}_2$, caused by forces at C and D. If moments are computed about point D, Fig. 4–34*a*, $\mathbf{M}_2 = \mathbf{r}_{DC} \times \mathbf{F}_C$, then

$$\begin{aligned}\mathbf{M}_2 = \mathbf{r}_{DC} \times \mathbf{F}_C &= (0.3\mathbf{i}) \times [125(\tfrac{4}{5})\mathbf{j} - 125(\tfrac{3}{5})\mathbf{k}]\\ &= (0.3\mathbf{i}) \times [100\mathbf{j} - 75\mathbf{k}] = 30(\mathbf{i}\times\mathbf{j}) - 22.5(\mathbf{i}\times\mathbf{k})\\ &= \{22.5\mathbf{j} + 30\mathbf{k}\}\text{ N}\cdot\text{m}\end{aligned}$$

Try to establish $\mathbf{M}_2$ by using a scalar formulation, Fig. 4–34*b*.

Since $\mathbf{M}_1$ and $\mathbf{M}_2$ are free vectors, they may be moved to some arbitrary point P and added vectorially, Fig. 4–34*c*. The resultant couple moment becomes

$$\mathbf{M}_R = \mathbf{M}_1 + \mathbf{M}_2 = \{60\mathbf{i} + 22.5\mathbf{j} + 30\mathbf{k}\}\text{ N}\cdot\text{m} \qquad \textit{Ans.}$$

PROBLEMS

4–66. Determine the magnitude and sense of the couple moment.

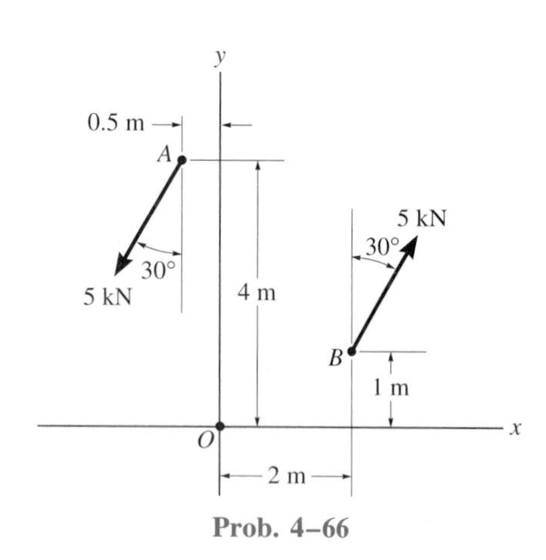

Prob. 4–66

4–67. Determine the magnitude and sense of the couple moment. Each force has a magnitude of $F = 8$ kN.

***4–68.** If the couple moment has a magnitude of $250\ \text{N} \cdot \text{m}$, determine the magnitude F of the couple forces.

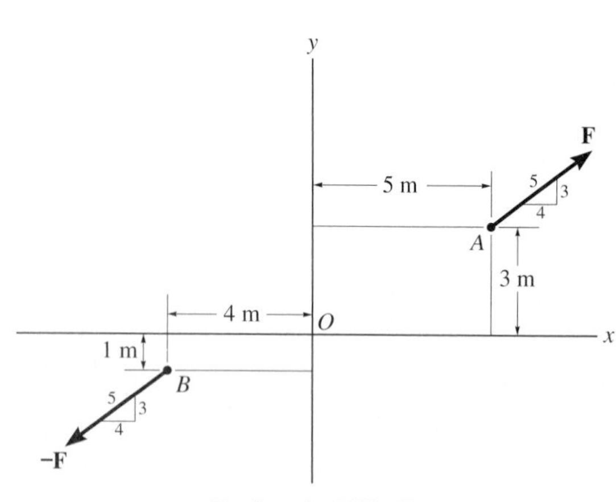

Probs. 4–67/4–68

4–69. Determine the magnitude and sense of the couple moment. Take $F = 260$ lb.

4–70. If the couple moment has a magnitude of $300\ \text{lb} \cdot \text{ft}$, determine the magnitude F of the couple forces.

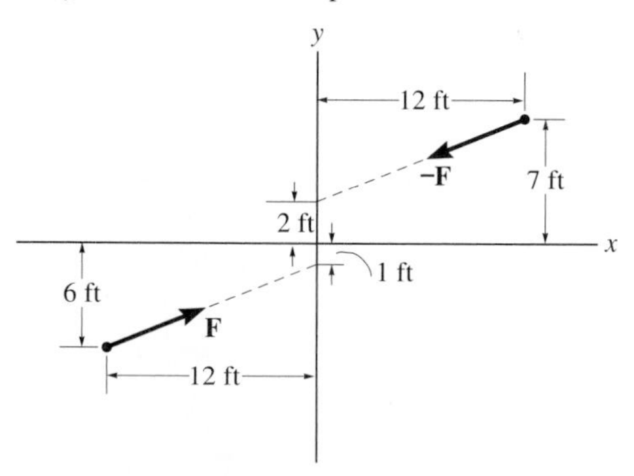

Probs. 4–69/4–70

4–71. Three couple moments act on the pipe assembly. Determine the magnitude of the resultant couple moment if $M_2 = 50\ \text{N} \cdot \text{m}$ and $M_3 = 35\ \text{N} \cdot \text{m}$.

***4–72.** Three couple moments act on the pipe assembly. Determine the magnitudes of $\mathbf{M}_2$ and $\mathbf{M}_3$ so that the resultant couple moment is zero.

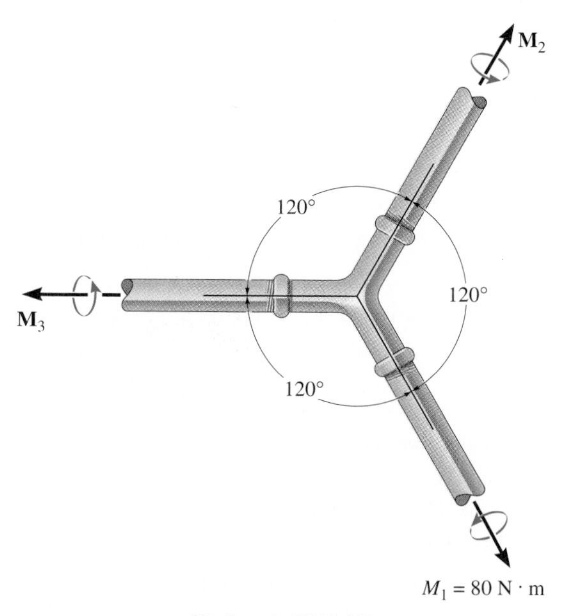

Probs. 4–71/4–72

4–73. Two couples act on the beam as shown. Determine the magnitude of **F** so that the resultant couple moment is 300 lb · ft counterclockwise. Where on the beam does the resultant couple act?

4–74. If $F = 180$ lb, determine the magnitude and sense of the resultant couple moment. Where on the beam does the resultant couple act?

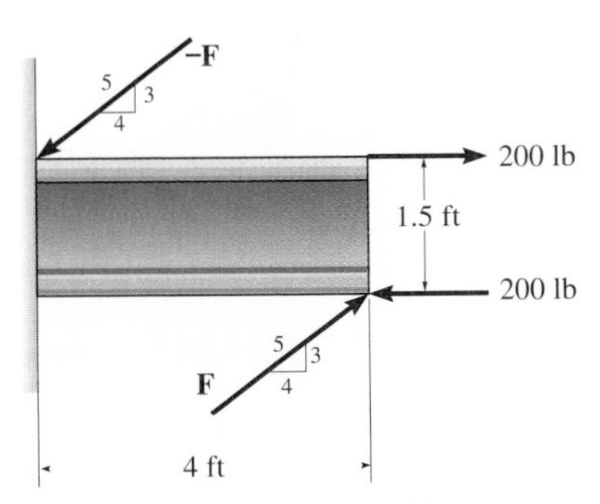

Probs. 4–73/4–74

4–75. Two couples act on the frame. If the resultant couple moment is to be zero, determine the distance d between the 80-lb couple forces.

***4–76.** Two couples act on the frame. If $d = 4$ ft, determine the resultant couple moment. Compute the result by resolving each force into x and y components and (a) finding the moment of each couple (Eq. 4–13) and (b) summing the moments of all the force components about point A.

4–77. Two couples act on the frame. If $d = 4$ ft, determine the resultant couple moment. Compute the result by resolving each force into x and y components and (a) finding the moment of each couple (Eq. 4–13) and (b) summing the moments of all the force components about point B.

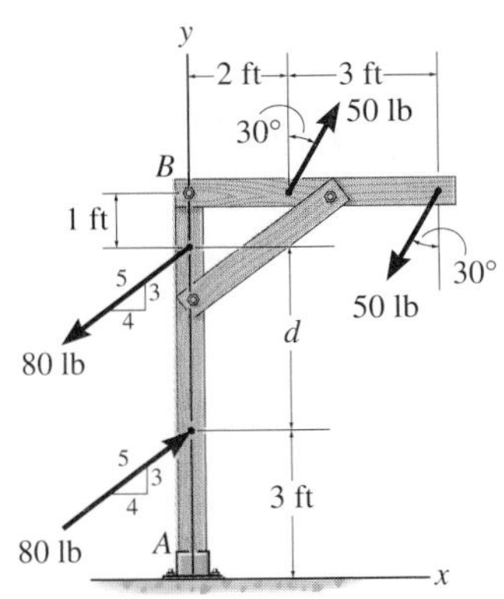

Probs. 4–75/4–76/4–77

4–78. The resultant couple moment created by the two couples acting on the disk is $\mathbf{M}_R = \{10\mathbf{k}\}$ kip · in. Determine the magnitude of force **T**.

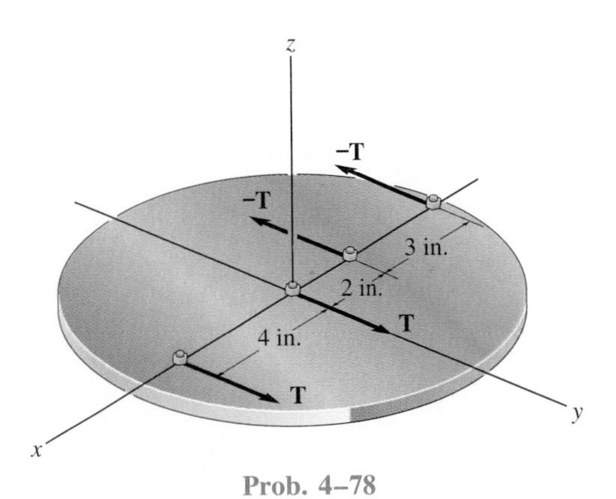

Prob. 4–78

4–79. Two couples act on the beam. Determine the magnitude of **F** so that the resultant couple moment is 450 lb · ft, counterclockwise. Where on the beam does the resultant couple moment act?

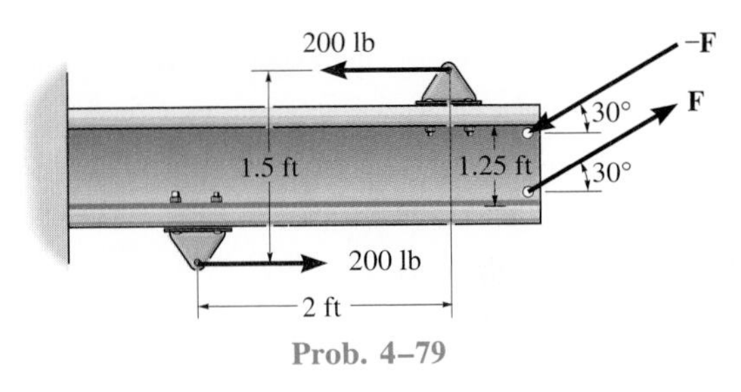

Prob. 4–79

***4–80.** Determine the couple moment. Express the result as a Cartesian vector.

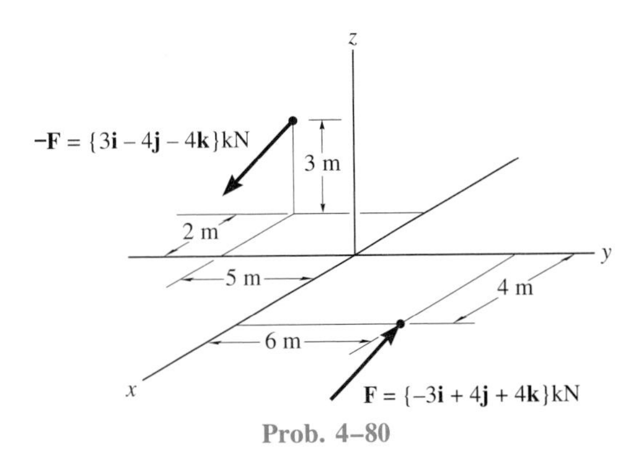

Prob. 4–80

4–81. Determine the couple moment. Express the result as a Cartesian vector.

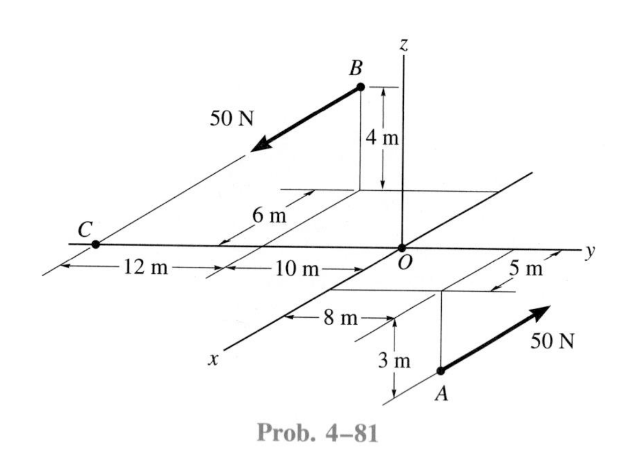

Prob. 4–81

***4–84.** A couple acts on each of the handles of the minidual valve. Determine the magnitude and coordinate direction angles of the resultant couple moment.

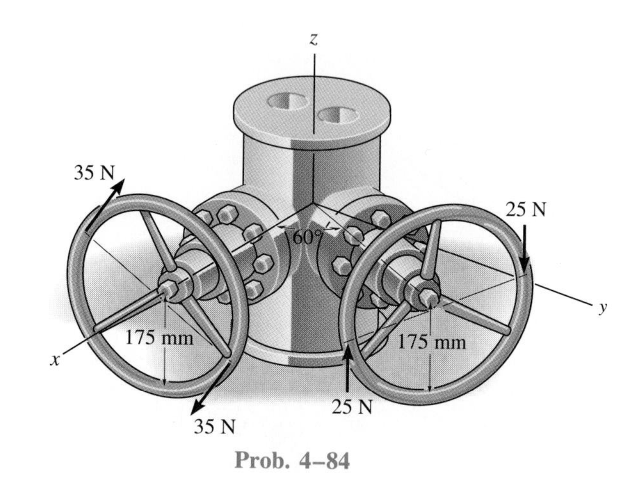

Prob. 4–84

4–82. Express the moment of the couple acting on the pipe in Cartesian vector form. What is the magnitude of the couple moment? Take $F = 125$ N.

4–83. If the couple moment acting on the pipe has a magnitude of 300 N · m, determine the magnitude F of the forces applied to the wrenches.

4–85. Express the moment of the couple acting on the pipe assembly in Cartesian vector form. Solve the problem (a) using Eq. 4–13, and (b) summing the moment of each force about point O. Take $\mathbf{F} = \{25\mathbf{k}\}$ N.

4–86. If the couple moment acting on the pipe has a magnitude of 400 N · m, determine the magnitude F of the vertical force applied to each wrench.

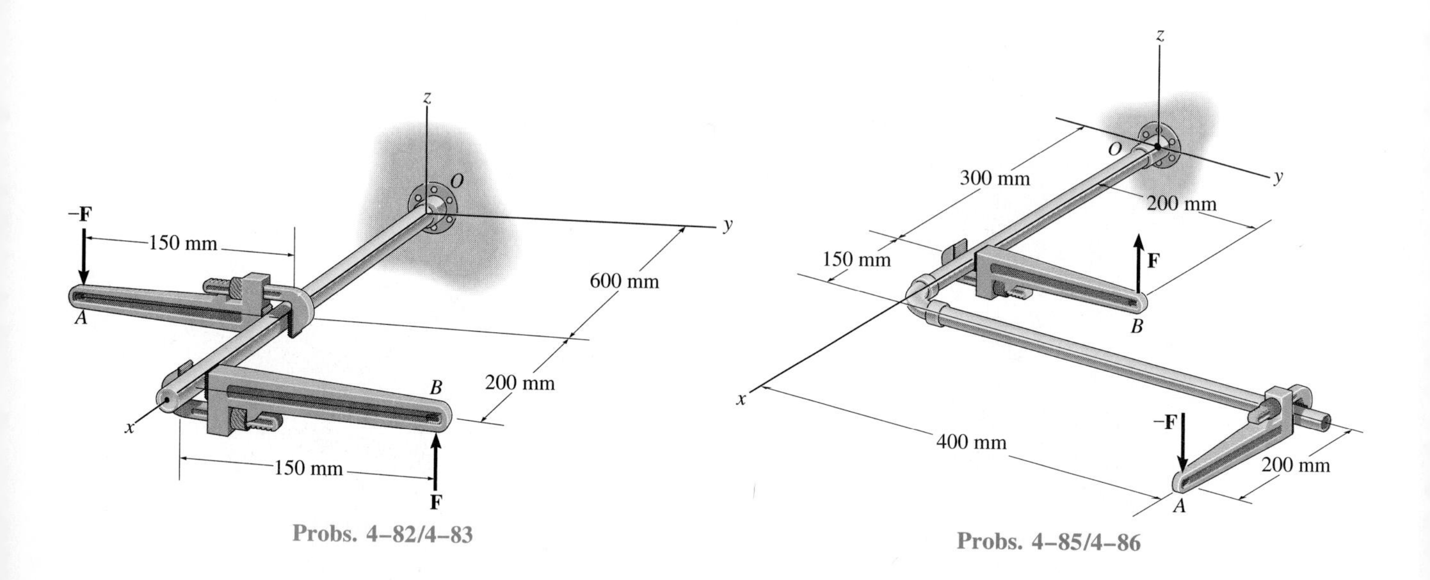

Probs. 4–82/4–83

Probs. 4–85/4–86

4–87. The meshed gears are subjected to the couple moments shown. Determine the magnitude and coordinate direction angles of the resultant couple moment.

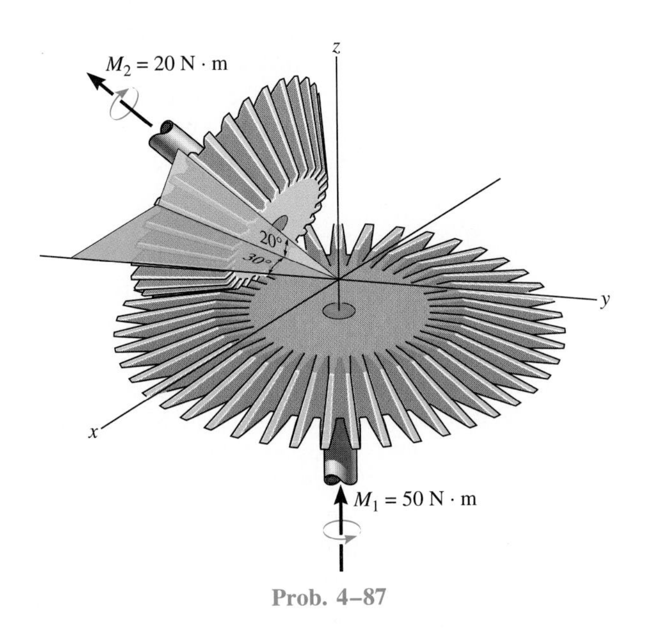

Prob. 4–87

4–88. Determine the resultant couple moment of the two couples that act on the assembly. Member *OB* lies in the *x-z* plane.

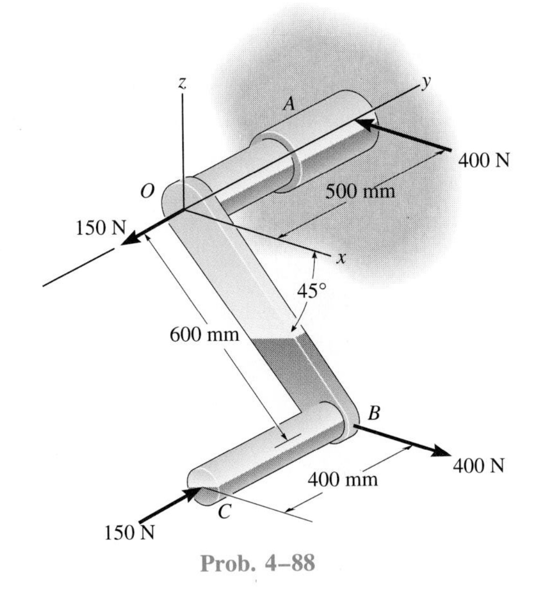

Prob. 4–88

4–89. Express the moment of the couple acting on the pipe assembly in Cartesian vector form. What is the magnitude of the couple moment?

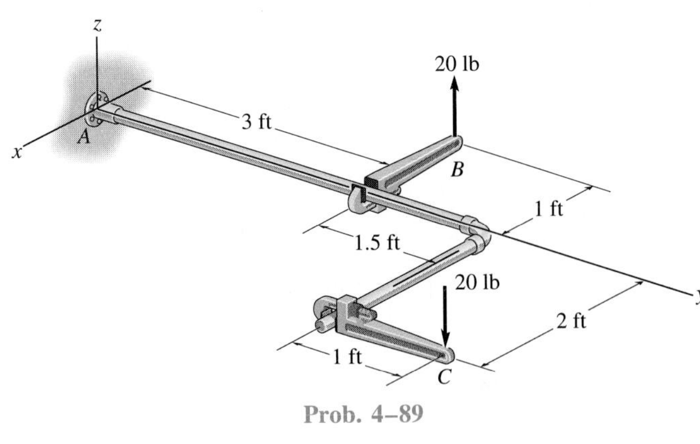

Prob. 4–89

4–90. If $\mathbf{F} = \{100\ \mathbf{k}\}$ N, determine the couple moment that acts on the assembly. Express the result as a Cartesian vector. Member *BA* lies in the *x-y* plane.

4–91. If the magnitude of the resultant couple moment is 15 N · m, determine the magnitude *F* of the forces applied to the wrenches.

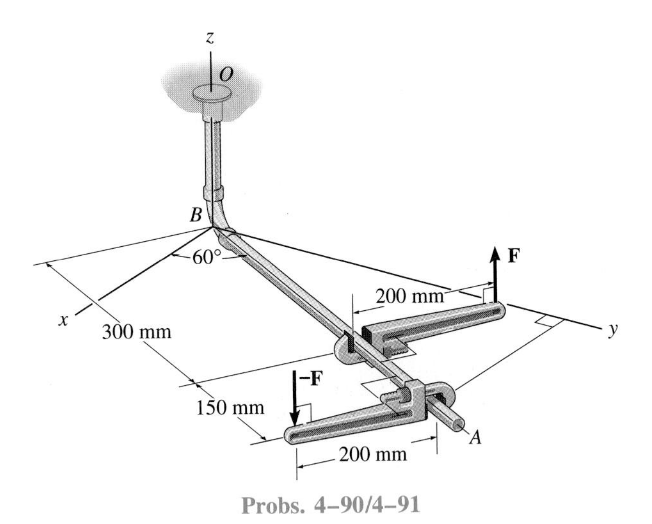

Probs. 4–90/4–91

4.7 Movement of a Force on a Rigid Body

Many problems in statics, including the reduction of a force system to its simplest possible form, require moving a force from one point to another on a rigid body. Since a force tends to both *translate* and *rotate* a body, it is important that these two ''external'' effects remain the same if the force is moved from one point to another on the body. Two cases for the location of the point O to which the force is moved will now be considered.

Point O Is On the Line of Action of the Force. Consider the rigid body shown in Fig. 4–35*a*, which is subjected to the force **F** applied to point A. In order to move the force to point O without altering the external effects on the body, we will first apply equal but opposite forces **F** and $-$**F** at O, as shown in Fig. 4–35*b*. The two forces indicated by the slash across them can be canceled, leaving the force at point O as required, Fig. 4–35*c*. By using this construction procedure, an *equivalent system* has been maintained between each of the diagrams, as shown by the equal sign. Note, however, that the force has simply been ''transmitted'' along its line of action, from point A, Fig. 4–35*a*, to point O, Fig. 4–35*c*. In other words, the force can be considered as a *sliding vector* since it can act at any point O along its line of action. In Sec. 4.3 we referred to this concept as the *principle of transmissibility*. It can be formally stated as follows: *The external effects on a rigid body remain unchanged when a force, acting at a given point on the body, is applied to another point lying on the line of action of the force.* It is important to realize that only the *external effects,* such as the body's motion or the forces needed to support the body if it is stationary, remain *unchanged* after **F** is moved. Certainly the *internal effects* depend on where **F** is located. For example, when **F** acts at A, the internal forces in the body have a high intensity around A; whereas movement of **F** away from this point will cause these internal forces to decrease.

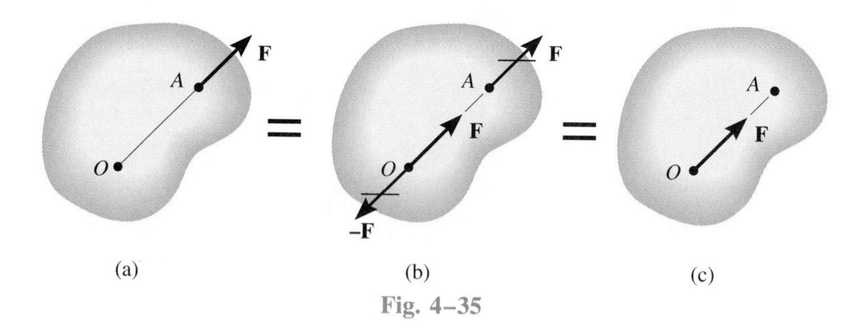

Fig. 4–35

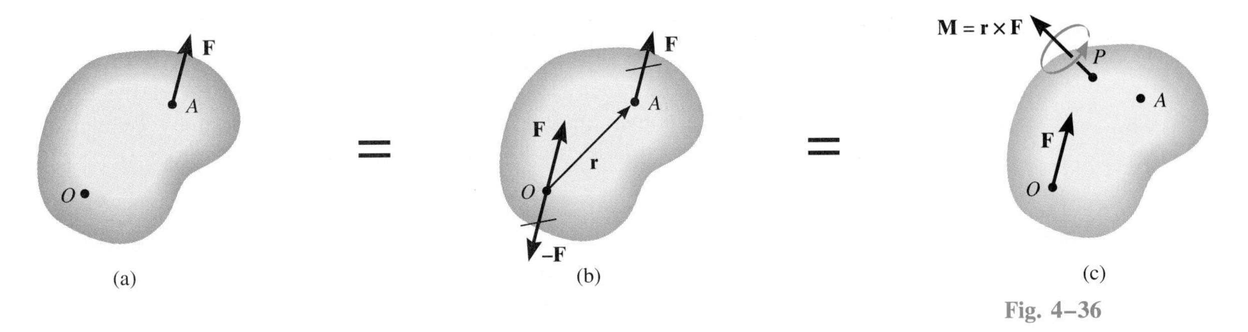

Fig. 4–36

Point O Is Not On the Line of Action of the Force. This case is shown in Fig. 4−36*a*, where **F** is to be moved to point O without altering the external effects on the body. Following the same procedure as before, we first apply equal but opposite forces **F** and −**F** at point O, Fig. 4–36*b*. Here the two forces indicated by a slash across them form a couple which has a moment that is perpendicular to **F** and is defined by the cross product $\mathbf{M} = \mathbf{r} \times \mathbf{F}$. Since the couple moment is a *free vector,* it may be applied at *any point* P on the body as shown in Fig. 4–36*c*. In addition to this couple moment, **F** now acts at point O as required.

As a physical illustration of the above two cases, consider the effect on the hand when holding the end O of a stick of negligible weight. If a vertical force **F** is applied at the other end A and the stick is held in the vertical position, Fig. 4–37*a*, then only the force is felt at the grip, regardless of where **F** is applied along its line of action OA. This is a consequence of the principle of transmissibility. When the stick is held in the horizontal position, Fig. 4–37*b*, the force at A has the effect of producing *both* a downward force at the grip O and a clockwise twist. Conversely, these same effects are felt at the grip if the force **F** is applied at the grip and the couple moment $M = Fd$ is applied to the stick.

Fig. 4–37

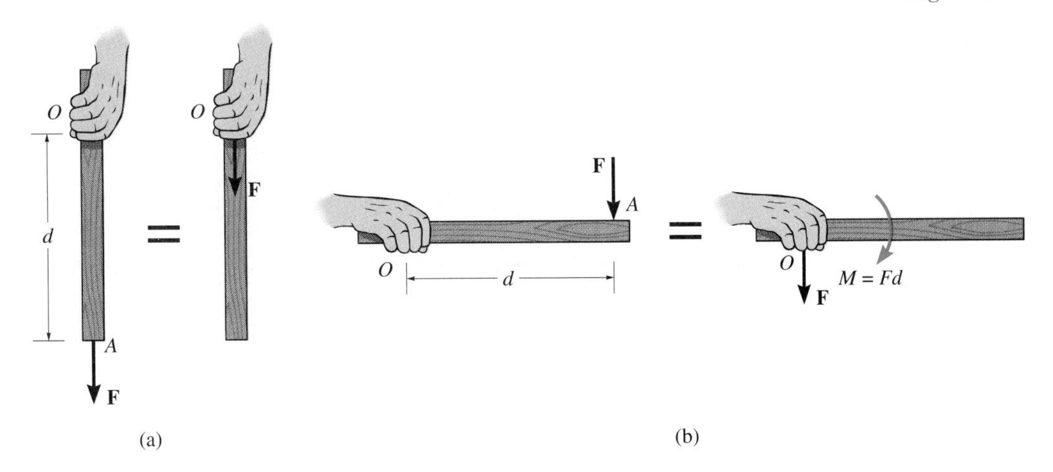

Example 4–15

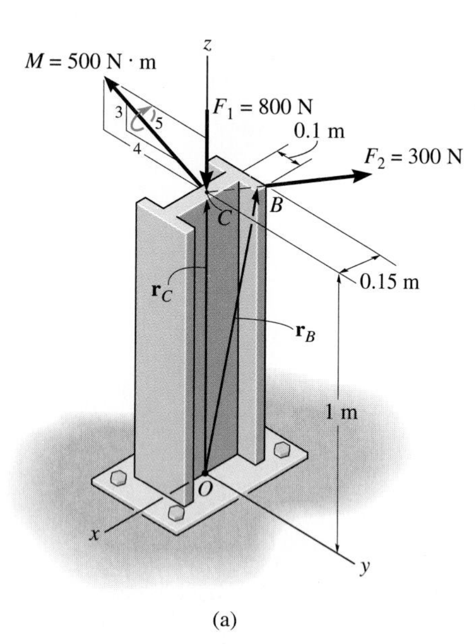

(a)

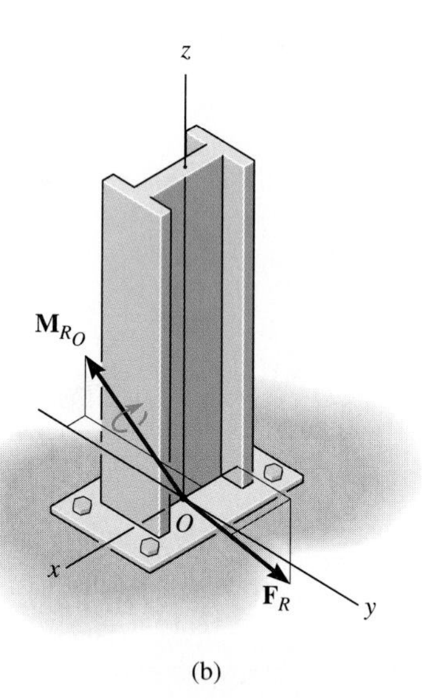

(b)

Fig. 4–38

A structural member is subjected to a couple moment **M** and forces $\mathbf{F}_1$ and $\mathbf{F}_2$ as shown in Fig. 4–38*a*. Replace this system by an equivalent resultant force and couple moment acting at its base, point *O*.

SOLUTION *(VECTOR ANALYSIS)*

The three-dimensional aspects of the problem can be simplified by using a Cartesian vector analysis. Expressing the forces and couple moment as Cartesian vectors, we have

$$\mathbf{F}_1 = \{-800\mathbf{k}\}\text{ N}$$

$$\mathbf{F}_2 = (300\text{ N})\mathbf{u}_{CB} = (300\text{ N})\left(\frac{\mathbf{r}_{CB}}{r_{CB}}\right)$$

$$= 300\left[\frac{-0.15\mathbf{i} + 0.1\mathbf{j}}{\sqrt{(-0.15)^2 + (0.1)^2}}\right] = \{-249.6\mathbf{i} + 166.4\mathbf{j}\}\text{ N}$$

$$\mathbf{M} = -500(\tfrac{4}{5})\mathbf{j} + 500(\tfrac{3}{5})\mathbf{k} = \{-400\mathbf{j} + 300\mathbf{k}\}\text{ N}\cdot\text{m}$$

Force Summation

$$\mathbf{F}_R = \Sigma\mathbf{F}; \qquad \mathbf{F}_R = \mathbf{F}_1 + \mathbf{F}_2 = -800\mathbf{k} - 249.6\mathbf{i} + 166.4\mathbf{j}$$

$$= \{-249.6\mathbf{i} + 166.4\mathbf{j} - 800\mathbf{k}\}\text{ N} \qquad \textit{Ans.}$$

Moment Summation

$$\mathbf{M}_{R_O} = \Sigma\mathbf{M}_O;$$

$$\mathbf{M}_{R_O} = \mathbf{M} + \mathbf{r}_C \times \mathbf{F}_1 + \mathbf{r}_B \times \mathbf{F}_2$$

$$= (-400\mathbf{j} + 300\mathbf{k}) + (1\mathbf{k}) \times (-800\mathbf{k}) + \begin{vmatrix} \mathbf{i} & \mathbf{j} & \mathbf{k} \\ -0.15 & 0.1 & 1 \\ -249.6 & 166.4 & 0 \end{vmatrix}$$

$$= (-400\mathbf{j} + 300\mathbf{k}) + (\mathbf{0}) + (-166.4\mathbf{i} - 249.6\mathbf{j})$$

$$= \{-166.4\mathbf{i} - 649.6\mathbf{j} + 300\mathbf{k}\}\text{ N}\cdot\text{m} \qquad \textit{Ans.}$$

The results are shown in Fig. 4–38*b*.

4.8 Resultants of a Force and Couple System

When a rigid body is subjected to a system of forces and couple moments, it is often simpler to study the external effects on the body by using the force and couple moment resultants, rather than the force and couple moment system. To show how to simplify a system of forces and couple moments to their resultants, consider the rigid body in Fig. 4–39*a*. The force and couple moment system acting on it will be simplified by moving the forces and couple moments to the arbitrary point O. In this regard, the couple moment $\mathbf{M}$ is simply moved to O, since it is a free vector. Forces $\mathbf{F}_1$ and $\mathbf{F}_2$ are sliding vectors, and since O does not lie on the line of action of these forces, each must be moved to O in accordance with the procedure stated in Sec. 4.7. For example, when $\mathbf{F}_1$ is applied at O, a corresponding couple moment $\mathbf{M}_1 = \mathbf{r}_1 \times \mathbf{F}_1$ must also be applied to the body, Fig. 4–39*b*. By vector addition, the force and couple moment system shown in Fig. 4–39*b* can now be reduced to an *equivalent* resultant force $\mathbf{F}_R = \mathbf{F}_1 + \mathbf{F}_2$ and resultant couple moment $\mathbf{M}_{R_O} = \mathbf{M} + \mathbf{M}_1 + \mathbf{M}_2$ as shown in Fig. 4–39*c*. Note that both the magnitude and direction of $\mathbf{F}_R$ are independent of the location of point O; however, $\mathbf{M}_{R_O}$ depends upon this location, since the moments $\mathbf{M}_1$ and $\mathbf{M}_2$ are computed using the position vectors $\mathbf{r}_1$ and $\mathbf{r}_2$. Realize also that $\mathbf{M}_{R_O}$ is a free vector and can act at *any point* on the body, although point O is generally chosen as its point of application.

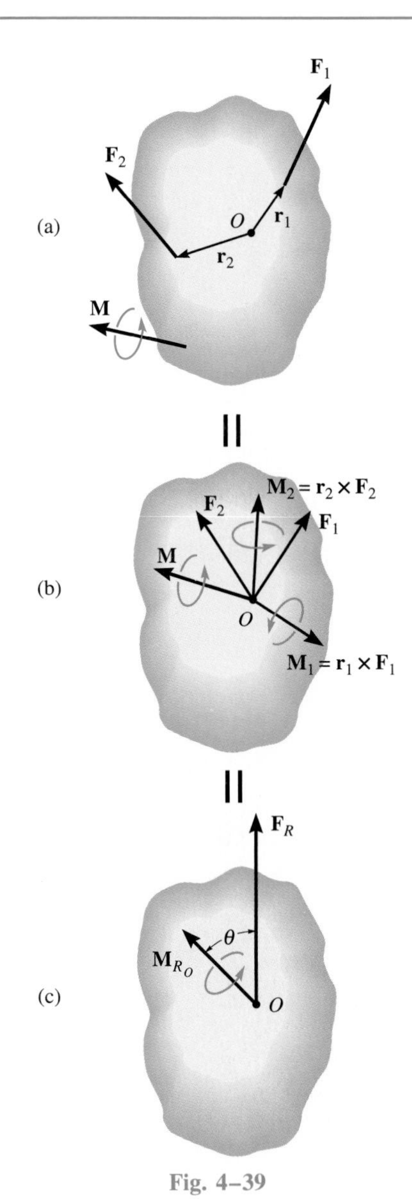

Fig. 4–39

PROCEDURE FOR ANALYSIS

The above method for simplifying any force and couple moment system to a resultant force acting at point O and a resultant couple moment will now be stated in general terms. To apply this method it is first necessary to establish an x, y, z coordinate system.

Three-Dimensional Systems. A Cartesian vector analysis is generally used to solve problems involving three-dimensional force and couple systems for which the force components and moment arms are difficult to determine.

Force Summation. The resultant force is equivalent to the vector sum of all the forces in the system; i.e.,

$$\mathbf{F}_R = \Sigma\mathbf{F} \tag{4–17}$$

Moment Summation. The resultant couple moment is equivalent to the vector sum of all the couple moments in the system plus the moments about point O of all the forces in the system; i.e.,

$$\mathbf{M}_{R_O} = \Sigma\mathbf{M} + \Sigma\mathbf{M}_O \qquad (4\text{–}18)$$

Coplanar Force Systems. Since force components and moment arms are easy to determine in two dimensions, a scalar analysis provides the most convenient solution to problems involving coplanar force systems. Assuming that the forces lie in the x–y plane and any couple moments are perpendicular to this plane (along the z axis), then the resultants are determined as follows:

Force Summation. The *resultant force* $\mathbf{F}_R$ is equivalent to the vector sum of its two components $\mathbf{F}_{R_x}$ and $\mathbf{F}_{R_y}$. Each component is found from the scalar (algebraic) sum of the components of all the forces in the system that act in the same direction; i.e.,

$$\begin{aligned} F_{R_x} &= \Sigma F_x \\ F_{R_y} &= \Sigma F_y \end{aligned} \qquad (4\text{–}19)$$

Moment Summation. The *resultant couple moment* $\mathbf{M}_{R_O}$ is perpendicular to the plane containing the forces and is equivalent to the scalar (algebraic) sum of all the couple moments in the system *plus* the moments about point O of all the forces in the system; i.e.,

$$M_{R_O} = \Sigma M + \Sigma M_O \qquad (4\text{–}20)$$

When determining the moments of the forces about O, it is generally advantageous to use the *principle of moments;* i.e., determine the moments of the *components* of each force rather than the moment of the force itself.

It is important to remember that, when applying any of these equations, attention should be paid to the sense of direction of the force components and the moments of the forces. If they are along the positive coordinate axes, they represent positive scalars; whereas if these components have a directional sense along the negative coordinate axes, they are negative scalars. By following this convention, a positive result, for example, indicates that the resultant vector has a sense of direction along the positive coordinate axis.

The following example illustrates these procedures numerically.

Example 4–16

Replace the forces acting on the brace shown in Fig. 4–40*a* by an equivalent resultant force and couple moment acting at point *A*.

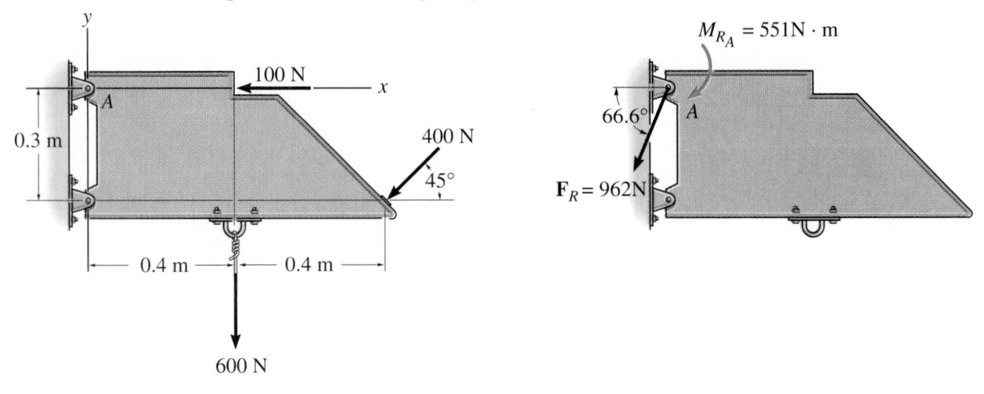

(a) **Fig. 4–40** (b)

SOLUTION (*SCALAR ANALYSIS*)

The principle of moments will be applied to the 400-N force, whereby the moments of its two rectangular components will be considered.

Force Summation. The resultant force has *x* and *y* components of

$$\overset{+}{\rightarrow} F_{R_x} = \Sigma F_x; \quad F_{R_x} = -100 \text{ N} - 400 \cos 45° \text{ N} = -382.8 \text{ N} = 382.8 \text{ N} \leftarrow$$

$$+\uparrow F_{R_y} = \Sigma F_y; \quad F_{R_y} = -600 \text{ N} - 400 \sin 45° \text{ N} = -882.8 \text{ N} = 882.8 \text{ N} \downarrow$$

As shown in Fig. 4–40*b*, $\mathbf{F}_R$ has a magnitude of

$$F_R = \sqrt{(F_{R_x})^2 + (F_{R_y})^2} = \sqrt{(382.8)^2 + (882.8)^2} = 962 \text{ N} \qquad \textit{Ans.}$$

and a direction defined from the vector sketch of

$$\theta = \tan^{-1}\left(\frac{F_{R_y}}{F_{R_x}}\right) = \tan^{-1}\left(\frac{882.8}{382.8}\right) = 66.6° \quad \theta \swarrow \qquad \textit{Ans.}$$

Moment Summation. The resultant couple moment $\mathbf{M}_{R_A}$ is determined by summing the moments of the forces about point *A*. Assuming that positive moments act counterclockwise, i.e., in the $+\mathbf{k}$ direction, we have

$$\circlearrowleft + M_{R_A} = \Sigma M_A;$$

$$M_{R_A} = 100 \text{ N}(0) - 600 \text{ N}(0.4 \text{ m}) - (400 \sin 45° \text{ N})(0.8 \text{ m}) - (400 \cos 45° \text{ N})(0.3 \text{ m})$$

$$= -551 \text{ N} \cdot \text{m} = 551 \text{ N} \cdot \text{m} \circlearrowright \qquad \textit{Ans.}$$

In conclusion, when $\mathbf{M}_{R_A}$ and $\mathbf{F}_R$ act on the brace at point *A*, Fig. 4–40*b*, they will produce the *same* external effect or reactions at the supports as that produced by the force system in Fig. 4–40*a*.

4.9 Further Reduction of a Force and Couple System

Simplification to a Single Resultant Force. Consider now a special case for which the system of forces and couple moments acting on a rigid body, Fig. 4–41*a*, reduces at point O to a resultant force $\mathbf{F}_R = \Sigma\mathbf{F}$ and resultant couple moment $\mathbf{M}_{R_O} = \Sigma\mathbf{M}_O$, which are *perpendicular* to one another, Fig. 4–41*b*. Whenever this occurs, we can further simplify the force and couple moment system by moving $\mathbf{F}_R$ to another point P, located either on or off the body so that no resultant couple moment has to be applied to the body, Fig. 4–41*c*. In other words, if the force and couple moment system in Fig. 4–41*a* is reduced to a resultant system at point P, only the force resultant will have to be applied to the body, Fig. 4–41*c*.

The location of point P, measured from point O, can always be determined provided $\mathbf{F}_R$ and $\mathbf{M}_{R_O}$ are known, Fig. 4–41*b*. As shown in Fig. 4–41*c*, P must lie on the *bb* axis, which is perpendicular to the line of action of $\mathbf{F}_R$ and the *aa* axis. This point is chosen such that the distance d satisfies the scalar equation $M_{R_O} = F_R d$ or $d = M_{R_O}/F_R$. With $\mathbf{F}_R$ so located, it will produce the same external effects on the body as the force and couple moment system in Fig. 4–41*a*, or the force and couple moment resultants in Fig. 4–41*b*. We refer to the force and couple moment system in Fig. 4–41*a* as being *equivalent or equipollent* to the single force "system" in Fig. 4–41*c*, because each system produces the *same* resultant force and resultant moment when replaced at point O.

If a system of forces is either concurrent, coplanar, or parallel, it can always be reduced, as in the above case, to a single resultant force $\mathbf{F}_R$ acting through a unique point P. This is because in each of these cases $\mathbf{F}_R$ and $\mathbf{M}_{R_O}$ will always be perpendicular to each other when the force system is simplified at *any* point O.

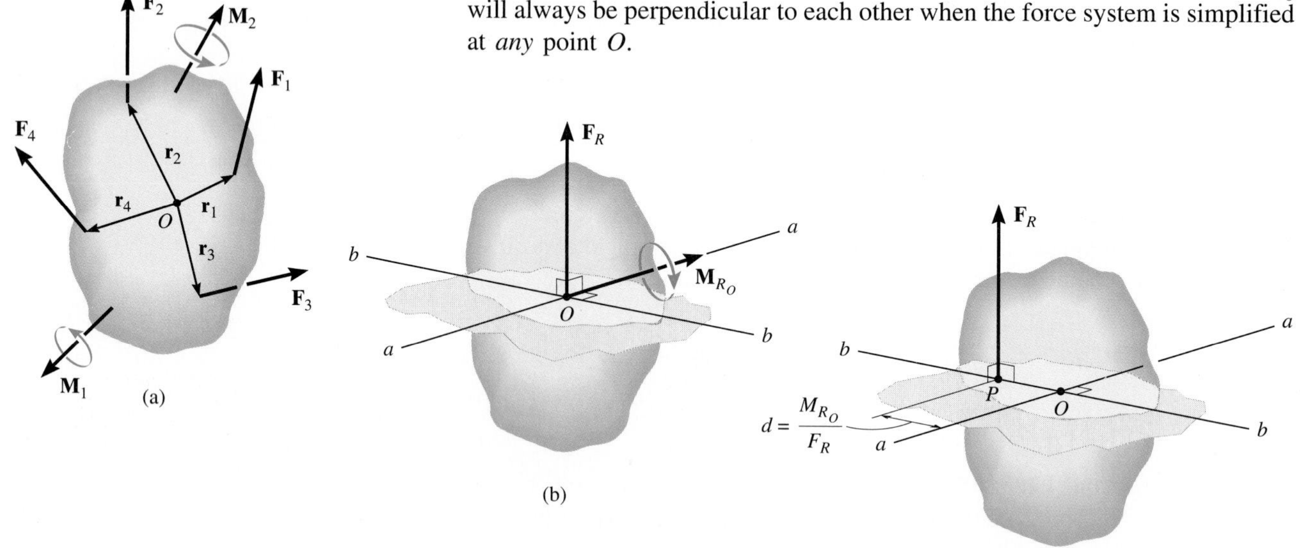

Fig. 4–41

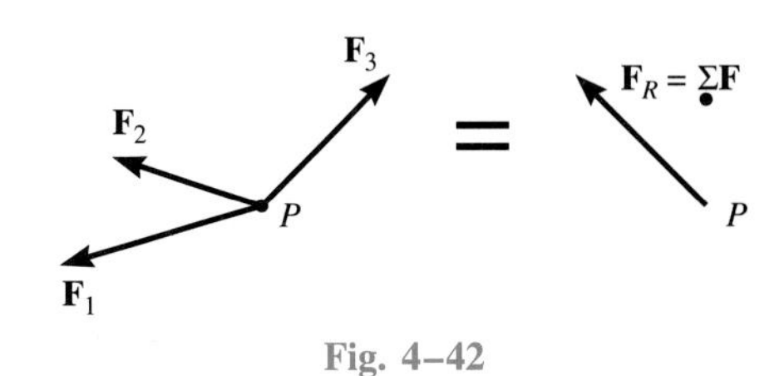

Fig. 4–42

Concurrent Force Systems. A concurrent force system has been treated in detail in Chapter 2. Obviously, all the forces act at a point for which there is no resultant couple moment, so the point P is automatically specified, Fig. 4–42.

Coplanar Force Systems. Coplanar force systems, which may include couple moments directed perpendicular to the plane of the forces, as shown in Fig. 4–43*a*, can be reduced to a single resultant force, because when each force in the system is moved to any point O in the x–y plane, it produces a couple moment that is *perpendicular* to the plane, i.e., in the $\pm\mathbf{k}$ direction. The resultant moment $\mathbf{M}_{R_O} = \Sigma\mathbf{M} + \Sigma\mathbf{r} \times \mathbf{F}$ is thus perpendicular to the resultant force $\mathbf{F}_R$, Fig. 4–43*b*; and so $\mathbf{F}_R$ can be positioned a distance d from O so as to create this same moment $\mathbf{M}_{R_O}$ about O, Fig. 4–43*c*.

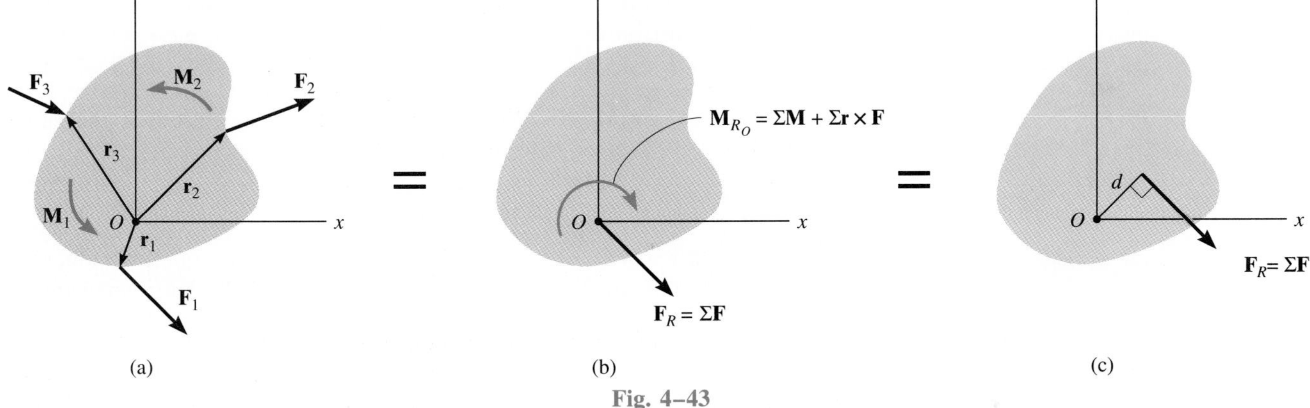
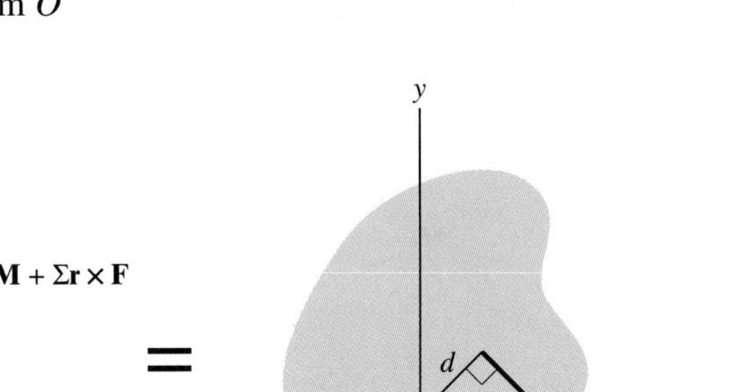

Fig. 4–43

Parallel Force Systems. Parallel force systems, which can include couple moments that are perpendicular to the forces, as shown in Fig. 4–44*a*, can be reduced to a single resultant force, because when each force is moved to any point O in the x–y plane, it produces a couple moment that has components only about the x and y axes. The resultant moment $\mathbf{M}_{R_O} = \Sigma\mathbf{M} + \Sigma\mathbf{r} \times \mathbf{F}$ is thus perpendicular to the resultant force $\mathbf{F}_R$, Fig. 4–44*b*; and so $\mathbf{F}_R$ can be moved to a point a distance d away so that it produces the same moment about O.

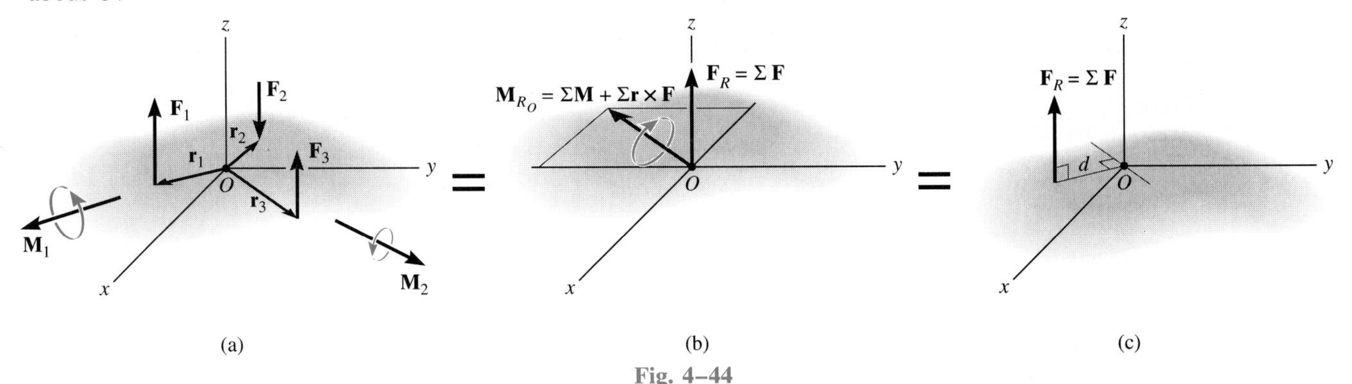

Fig. 4–44

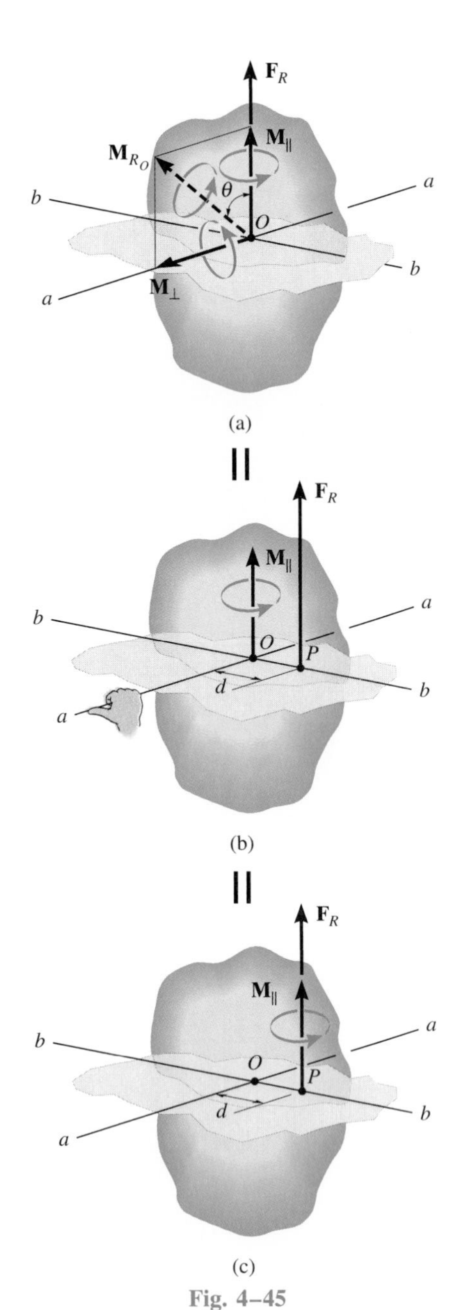

Fig. 4–45

PROCEDURE FOR ANALYSIS

The technique used to reduce a coplanar or parallel force system to a single resultant force follows the general procedure outlined in the previous section. First establish an x, y, z coordinate system. Then simplification requires the following two steps:

Force Summation. The resultant force $\mathbf{F}_R$ equals the sum of all the forces of the system, Fig. 4–41*a* and 4–41*c*; i.e.,

$$\mathbf{F}_R = \Sigma \mathbf{F} \tag{4–21}$$

Moment Summation. The distance d from the arbitrary point O to the line of action of $\mathbf{F}_R$ is determined by equating the moment of $\mathbf{F}_R$ about O, $\mathbf{M}_{R_O}$, Fig. 4–41*c*, to the sum of the moments about point O of all the couple moments and forces in the system, $\Sigma \mathbf{M}_O$, Fig. 4–41*a*; i.e.,

$$\mathbf{M}_{R_O} = \Sigma \mathbf{M}_O \tag{4–22}$$

Most often a scalar analysis can be used to apply these equations, since the force components and the moment arms are easily determined for either coplanar or parallel force systems.

Reduction to a Wrench. In the general case, the force and couple moment system acting on a body, Fig. 4–39*a*, will reduce to a single resultant force $\mathbf{F}_R$ and couple moment $\mathbf{M}_{R_O}$ at O which are *not* perpendicular. Instead, $\mathbf{F}_R$ will act at an angle θ from $\mathbf{M}_{R_O}$, Fig. 4–39*c*. As shown in Fig. 4–45*a*, however, $\mathbf{M}_{R_O}$ may be resolved into two components: one perpendicular, $\mathbf{M}_\perp$, and the other parallel, $\mathbf{M}_\|$, to the line of action of $\mathbf{F}_R$. As in the previous discussion, the perpendicular component $\mathbf{M}_\perp$ may be *eliminated* by moving $\mathbf{F}_R$ to point P, as shown in Fig. 4–45*b*. This point lies on axis *bb*, which is perpendicular to both $\mathbf{M}_{R_O}$ and $\mathbf{F}_R$. In order to maintain an equivalency of loading, the distance from O to P is $d = M_\perp/F_R$. Furthermore, when $\mathbf{F}_R$ is applied at P, the moment of $\mathbf{F}_R$ tending to cause rotation of the body *about O* is in the *same direction* as $\mathbf{M}_\perp$, Fig. 4–45*a*. Finally, since $\mathbf{M}_\|$ is a free vector, it may be moved to P so that it is collinear with $\mathbf{F}_R$, Fig. 4–45*c*. This combination of a collinear force and couple moment is called a *wrench* or *screw*. The *axis of the wrench* has the same line of action as the force. Hence, the wrench tends to cause both a translation along and a rotation about this axis. Comparing Fig. 4–45*a* to Fig. 4–45*c*, it is seen that a general force and couple moment system acting on a body can be reduced to a wrench. The axis of the wrench and a point through which this axis passes are unique and can always be determined.

Example 4–17

Replace the system of forces acting on the aircraft wing shown in Fig. 4–46*a* by an equivalent resultant force. Specify the distance the force acts from point *A* on the fuselage.

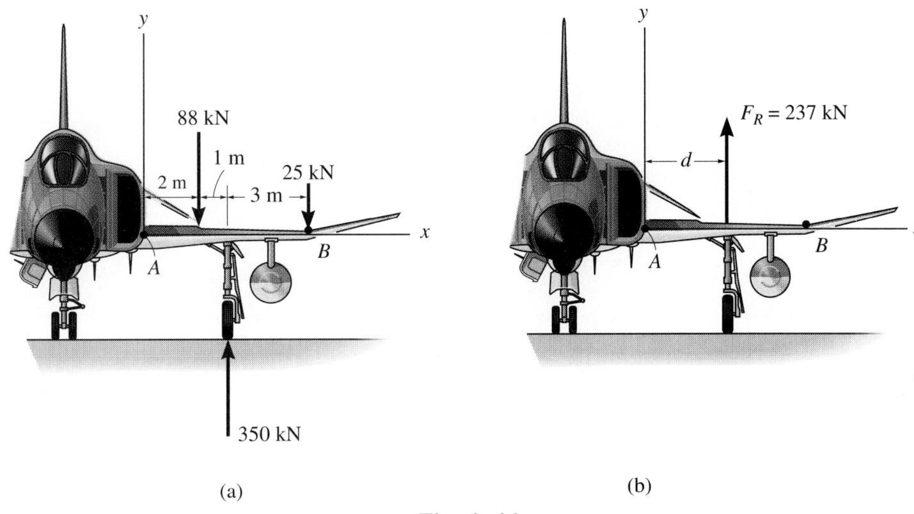

Fig. 4–46

SOLUTION

Here the force system is parallel and coplanar. We will use the established *x*, *y*, *z* axes to solve this problem.

Force Summation. From Fig. 4–46*a* the force resultant $\mathbf{F}_R$ is

$$+\uparrow F_R = \Sigma F; \quad F_R = -88 \text{ kN} + 350 \text{ kN} - 25 \text{ kN} = 237 \text{ kN} \uparrow \qquad \textit{Ans.}$$

Moment Summation. Moments will be summed about point *A*. Considering counterclockwise rotations as positive, i.e., positive moment vectors act in the $+\mathbf{k}$ direction, then from Figs. 4–46*a* and 4–46*b* we require the moment of $\mathbf{F}_R$ about *A* to equal the moments of the force system about *A*; i.e.,

$$\circlearrowleft + M_{R_A} = \Sigma M_A;$$

$$237 \text{ kN}(d) = -(88 \text{ kN})(2 \text{ m}) + (350 \text{ kN})(3 \text{ m}) - (25 \text{ kN})(6 \text{ m})$$

$$237 \text{ kN}(d) = 724 \text{ kN} \cdot \text{m}$$

$$d = 3.05 \text{ m} \qquad \textit{Ans.}$$

Note that using a clockwise sign convention would yield the same result. Since *d* is *positive*, $\mathbf{F}_R$ acts to the right of *A* as shown in Fig. 4–46*b*. Try and solve this problem by summing moments about point *B* and show that $d' = 2.95$ m, measured to the left of *B*.

Example 4–18

The beam *AE* in Fig. 4–47*a* is subjected to a system of coplanar forces. Determine the magnitude, direction, and location on the beam of a resultant force which is equivalent to the given system of forces.

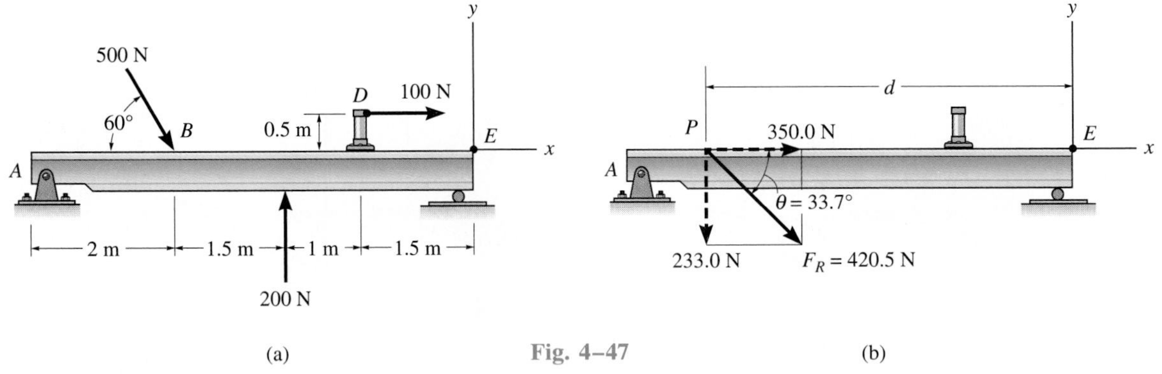

Fig. 4–47

SOLUTION

The origin of coordinates is arbitrarily located at point *E* as shown in Fig. 4–47*a*.

Force Summation. Resolving the 500-N force into *x* and *y* components, and summing the force components, yields

$$\overset{+}{\rightarrow} F_{R_x} = \Sigma F_x; \qquad F_{R_x} = 500 \cos 60° \text{ N} + 100 \text{ N} = 350.0 \text{ N} \rightarrow$$

$$+\uparrow F_{R_y} = \Sigma F_y; \qquad F_{R_y} = -500 \sin 60° \text{ N} + 200 \text{ N} = -233.0 \text{ N}$$

$$= 233.0 \text{ N} \downarrow$$

The magnitude and direction of the resultant force are established from the vector addition shown in Fig. 4–47*b*. We have

$$F_R = \sqrt{(350.0)^2 + (233.0)^2} = 420.5 \text{ N} \qquad \textit{Ans.}$$

$$\theta = \tan^{-1}\left(\frac{233.0}{350.0}\right) = 33.7° \ \measuredangle_\theta \qquad \textit{Ans.}$$

Moment Summation. Moments will be summed about point *E*. Hence, from Figs. 4–47*a* and 4–47*b*, we require the moments of the components of $\mathbf{F}_R$ (or the moment of $\mathbf{F}_R$) about point *E* to equal the moments of the force system about *E*; i.e.,

$$\circlearrowright + M_{R_E} = \Sigma M_E;$$

$$233.0 \text{ N}(d) + 350.0 \text{ N}(0) = (500 \sin 60° \text{ N})(4 \text{ m}) + (500 \cos 60° \text{ N})(0)$$
$$-(100 \text{ N})(0.5 \text{ m}) - (200 \text{ N})(2.5 \text{ m})$$

$$d = \frac{1182.1}{233.0} = 5.07 \text{ m} \qquad \textit{Ans.}$$

Example 4–19

The jib crane shown in Fig. 4–48a is subjected to three coplanar forces. Replace this loading by an equivalent resultant force and specify where the resultant's line of action intersects the column AB and boom BC.

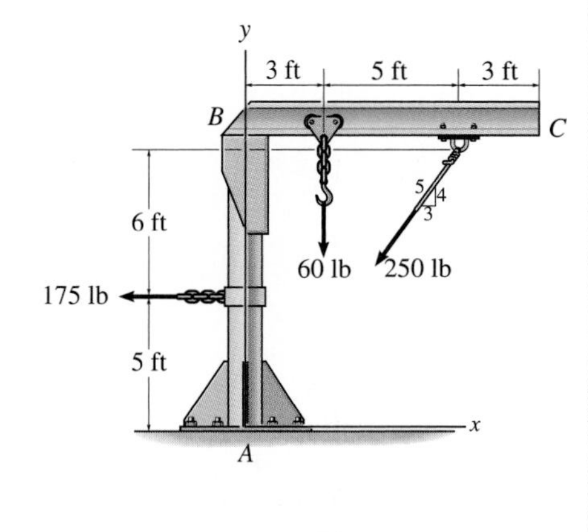

(a)

SOLUTION

Force Summation. Resolving the 250-lb force into x and y components and summing the force components yields

$\overset{+}{\rightarrow} F_{R_x} = \Sigma F_x;$ $\quad F_{R_x} = -250 \text{ lb}(\frac{3}{5}) - 175 \text{ lb} = -325 \text{ lb} = 325 \text{ lb} \leftarrow$

$+\uparrow F_{R_y} = \Sigma F_y;$ $\quad F_{R_y} = -250 \text{ lb}(\frac{4}{5}) - 60 \text{ lb} = -260 \text{ lb} = 260 \text{ lb} \downarrow$

As shown by the vector addition in Fig. 4–48b,

$$F_R = \sqrt{(325)^2 + (260)^2} = 416 \text{ lb} \qquad \textit{Ans.}$$

$$\theta = \tan^{-1}\left(\frac{260}{325}\right) = 38.7^\circ \quad \theta\swarrow \qquad \textit{Ans.}$$

Moment Summation. Moments will be summed about point A. If the line of action of $\mathbf{F}_R$ *intersects AB*, Fig. 4–48b, we require the moment of the components of $\mathbf{F}_R$ in Fig. 4–48b about A to equal the moments of the force system in Fig. 4–48a about A; i.e.,

$$\circlearrowleft + M_{R_A} = \Sigma M_A; \quad 325 \text{ lb}(y) + 260 \text{ lb}(0)$$
$$= 175 \text{ lb}(5 \text{ ft}) - 60 \text{ lb}(3 \text{ ft}) + 250 \text{ lb}(\tfrac{3}{5})(11 \text{ ft}) - 250 \text{ lb}(\tfrac{4}{5})(8 \text{ ft})$$
$$y = 2.29 \text{ ft} \qquad \textit{Ans.}$$

By the principle of transmissibility, $\mathbf{F}_R$ can also intersect BC, Fig. 4–48b, in which case we have

$$\circlearrowleft + M_{R_A} = \Sigma M_A; \quad 325 \text{ lb}(11 \text{ ft}) - 260 \text{ lb}(x)$$
$$= 175 \text{ lb}(5 \text{ ft}) - 60 \text{ lb}(3 \text{ ft}) + 250 \text{ lb}(\tfrac{3}{5})(11 \text{ ft}) - 250 \text{ lb}(\tfrac{4}{5})(8 \text{ ft})$$
$$x = 10.9 \text{ ft} \qquad \textit{Ans.}$$

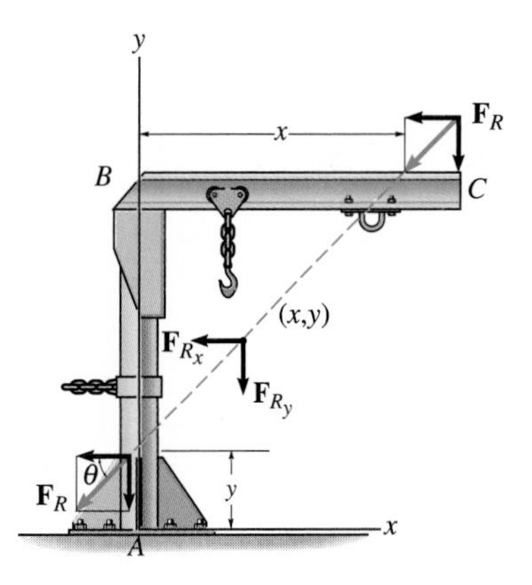

(b)

Fig. 4–48

We can also solve for these positions by assuming $\mathbf{F}_R$ acts at the arbitrary point (x, y) on its line of action, Fig. 4–48b. Summing moments about point A yields

$$\circlearrowleft + M_{R_A} = \Sigma M_A; \quad 325 \text{ lb}(y) - 260 \text{ lb}(x)$$
$$= 175 \text{ lb}(5 \text{ ft}) - 60 \text{ lb}(3 \text{ ft}) + 250 \text{ lb}(\tfrac{3}{5})(11 \text{ ft}) - 250 \text{ lb}(\tfrac{4}{5})(8 \text{ ft})$$
$$325y - 260x = 745$$

which is the equation of the colored dashed line in Fig. 4–48b. To find the points of intersection with the crane, set $x = 0$, then $y = 2.29$ ft, and set $y = 11$ ft, then $x = 10.9$ ft.

Example 4–20

The slab in Fig. 4–49*a* is subjected to four parallel forces. Determine the magnitude and direction of a resultant force equivalent to the given force system, and locate its point of application on the slab.

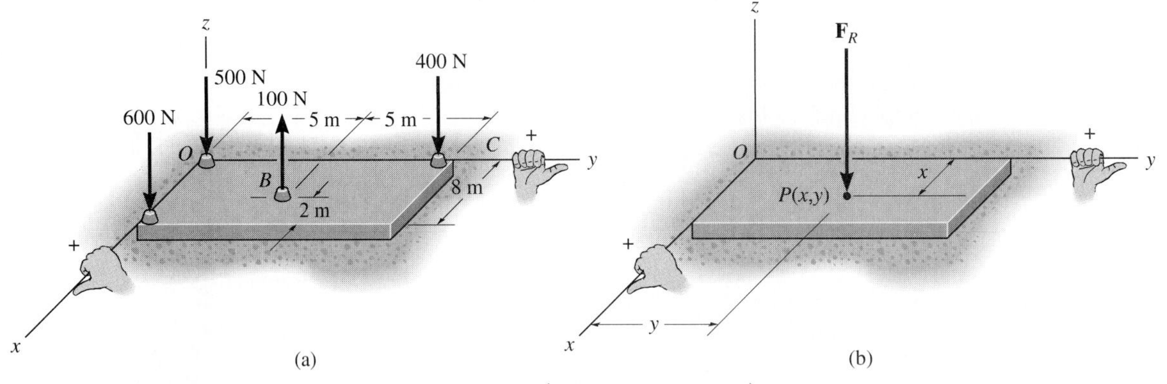

Fig. 4–49

SOLUTION *(SCALAR ANALYSIS)*

Force Summation. From Fig. 4–49*a*, the resultant force is

$$+\uparrow F_R = \Sigma F; \quad F_R = -600\text{ N} + 100\text{ N} - 400\text{ N} - 500\text{ N}$$
$$= -1400\text{ N} = 1400\text{ N}\downarrow \qquad \textit{Ans.}$$

Moment Summation. We require the moment about the x axis of the resultant force, Fig. 4–49*b*, to be equal to the sum of the moments about the x axis of all the forces in the system, Fig. 4–49*a*. The moment arms are determined from the y coordinates since these coordinates represent the *perpendicular distances* from the x axis to the lines of action of the forces. Using the right-hand rule, where positive moments act in the $+\mathbf{i}$ direction, we have

$$M_{R_x} = \Sigma M_x;$$
$$-(1400\text{ N})y = 600\text{ N}(0) + 100\text{ N}(5\text{ m}) - 400\text{ N}(10\text{ m}) + 500\text{ N}(0)$$
$$-1400y = -3500 \qquad y = 2.50\text{ m} \qquad \textit{Ans.}$$

In a similar manner, assuming that positive moments act in the $+\mathbf{j}$ direction, a moment equation can be written about the y axis using moment arms defined by the x coordinates of each force.

$$M_{R_y} = \Sigma M_y;$$
$$(1400\text{ N})x = 600\text{ N}(8\text{ m}) - 100\text{ N}(6\text{ m}) + 400\text{ N}(0) + 500\text{ N}(0)$$
$$1400x = 4200 \qquad x = 3.00\text{ m} \qquad \textit{Ans.}$$

Hence, a force of $F_R = 1400$ N placed at point $P(3.00\text{ m}, 2.50\text{ m})$ on the slab, Fig. 4–49*b*, is equivalent to the parallel force system acting on the slab in Fig. 4–49*a*.

Example 4–21

Three parallel bolting forces act on the rim of the circular cover plate in Fig. 4–50*a*. Determine the magnitude and direction of a resultant force equivalent to the given force system and locate its point of application, *P*, on the cover plate.

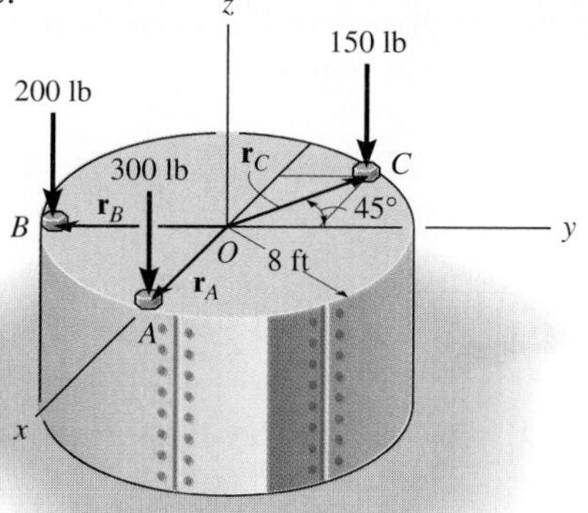

(a)

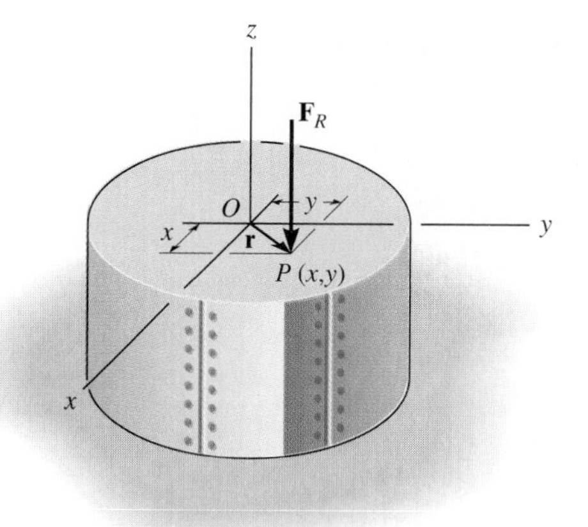

(b)

Fig. 4–50

SOLUTION ***(VECTOR ANALYSIS)***

Force Summation. From Fig. 4–50*a*, the force resultant $\mathbf{F}_R$ is

$$\mathbf{F}_R = \Sigma\mathbf{F}; \qquad \mathbf{F}_R = -300\mathbf{k} - 200\mathbf{k} - 150\mathbf{k}$$
$$= \{-650\mathbf{k}\} \text{ lb} \qquad \textit{Ans.}$$

Moment Summation. Choosing point *O* as a reference for computing moments and assuming that $\mathbf{F}_R$ acts at a point $P(x, y)$, Fig. 4–50*b*, we require

$$\mathbf{M}_{R_O} = \Sigma\mathbf{M}_O; \quad \mathbf{r} \times \mathbf{F}_R = \mathbf{r}_A \times (-300\mathbf{k}) + \mathbf{r}_B \times (-200\mathbf{k}) + \mathbf{r}_C \times (-150\mathbf{k})$$
$$(x\mathbf{i} + y\mathbf{j}) \times (-650\mathbf{k}) = (8\mathbf{i}) \times (-300\mathbf{k}) + (-8\mathbf{j}) \times (-200\mathbf{k})$$
$$+ (-8 \sin 45°\mathbf{i} + 8 \cos 45°\mathbf{j}) \times (-150\mathbf{k})$$
$$650x\mathbf{j} - 650y\mathbf{i} = 2400\mathbf{j} + 1600\mathbf{i} - 848.5\mathbf{j} - 848.5\mathbf{i}$$

Equating the corresponding **j** and **i** components yields

$$650x = 2400 - 848.5 \qquad (1)$$
$$-650y = 1600 - 848.5 \qquad (2)$$

Solving these equations, we obtain the coordinates of point *P*,

$$x = 2.39 \text{ ft} \qquad y = -1.16 \text{ ft} \qquad \textit{Ans.}$$

The negative sign indicates that it was wrong to have assumed a $+y$ position for $\mathbf{F}_R$ as shown in Fig. 4–50*b*.

As a review, try to establish Eqs. 1 and 2 by using a scalar analysis; i.e., apply the sum of moments about the *x* and *y* axes, respectively.

PROBLEMS

***4–92.** Replace the force at A by an equivalent force and couple moment at point O.

4–93. Replace the force at A by an equivalent force and couple moment at point P.

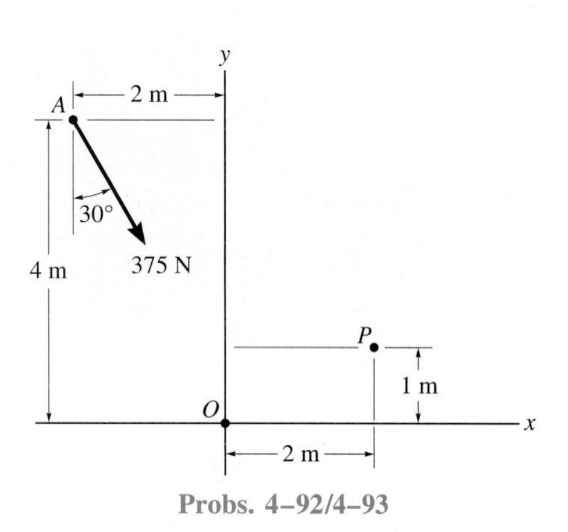

Probs. 4–92/4–93

***4–96.** Replace the force system by an equivalent force and couple moment at point O.

4–97. Replace the force system by an equivalent force and couple moment at point P.

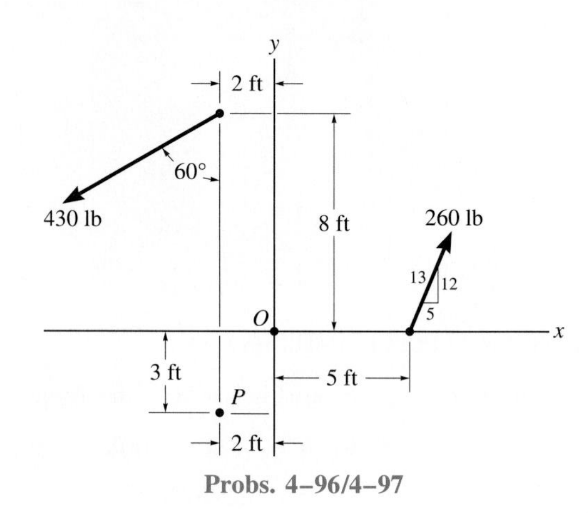

Probs. 4–96/4–97

4–94. Replace the force at A by an equivalent force and couple moment at point O.

4–95. Replace the force at A by an equivalent force and couple moment at point P.

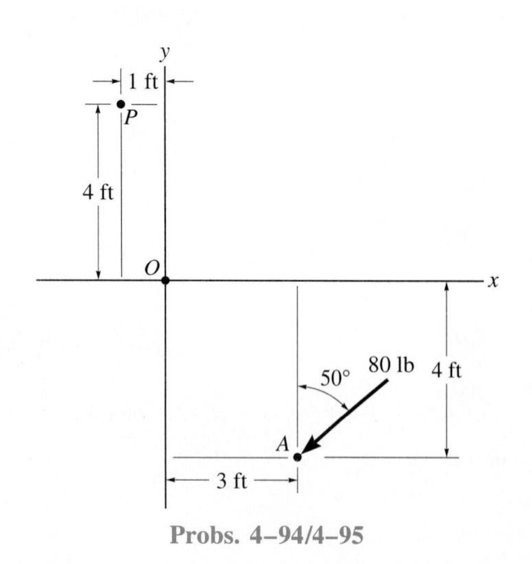

Probs. 4–94/4–95

4–98. Replace the force and couple system by an equivalent force and couple moment at point O.

4–99. Replace the force and couple system by an equivalent force and couple moment at point P.

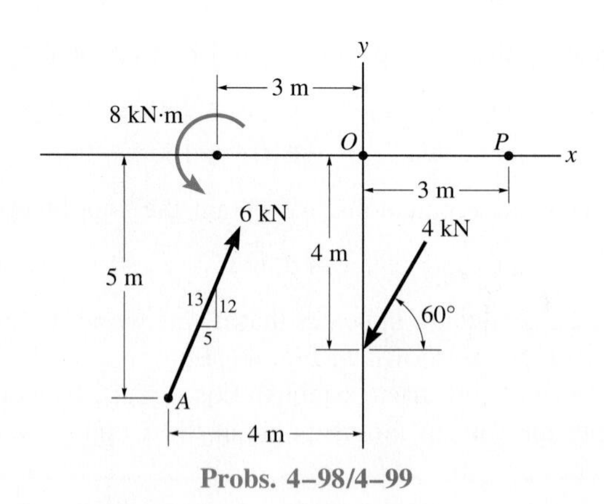

Probs. 4–98/4–99

***4–100.** Replace the force system by a single force resultant and specify its point of application, measured along the x axis from point O.

4–101. Replace the force system by a single force resultant and specify its point of application, measured along the x axis from point P.

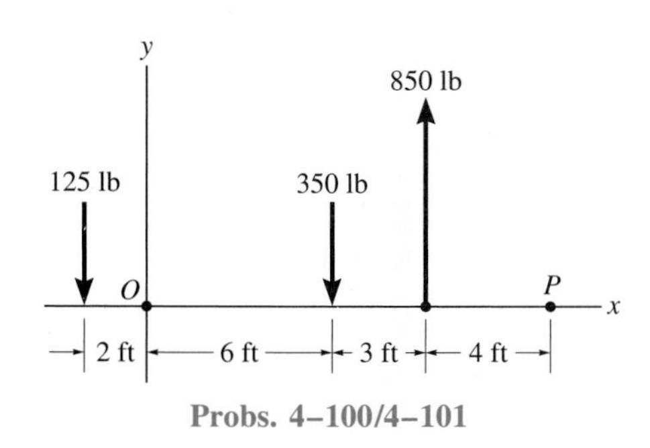

Probs. 4–100/4–101

4–102. The forces $\mathbf{F}_1 = \{-4\mathbf{i} + 2\mathbf{j} - 3\mathbf{k}\}$ kN and $\mathbf{F}_2 = \{3\mathbf{i} - 4\mathbf{j} - 2\mathbf{k}\}$ kN act on the end of the beam. Replace these forces by an equivalent force and couple moment acting at point O.

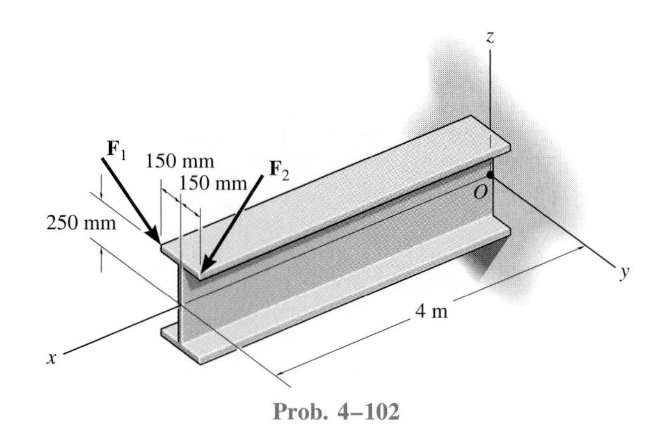

Prob. 4–102

4–103. Replace the three forces acting on the shaft by a single resultant force. Specify where the force acts, measured from end A.

***4–104.** Replace the three forces acting on the shaft by a single resultant force. Specify where the force acts, measured from end B.

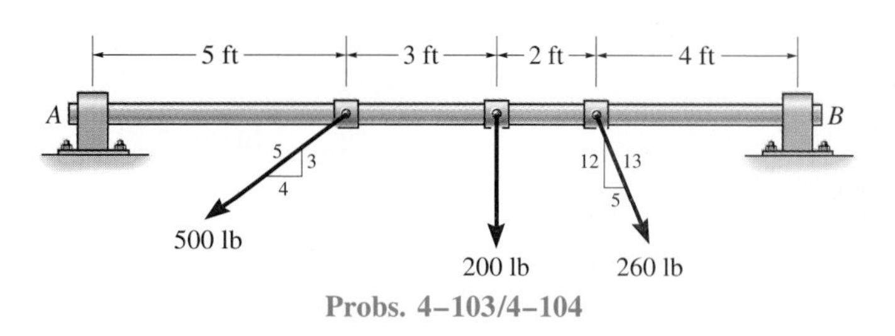

Probs. 4–103/4–104

4–105. Replace the three forces acting on the beam by a single resultant force. Specify where the force acts, measured from end A.

4–106. Replace the three forces acting on the beam by a single resultant force. Specify where the force acts, measured from B.

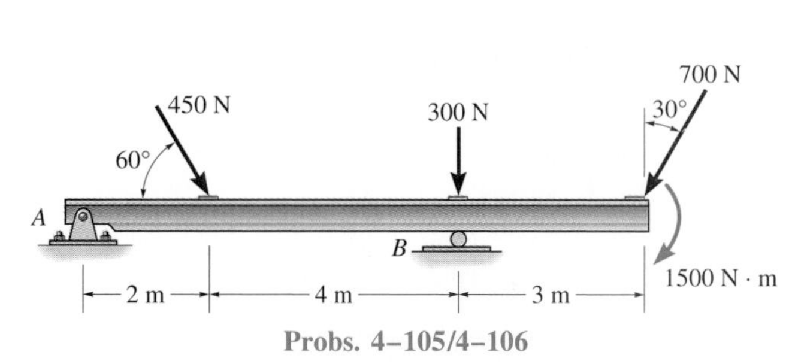

Probs. 4–105/4–106

4–107. Replace the loading on the frame by a single resultant force. Specify where its line of action intersects member AB, measured from A.

***4–108.** Replace the loading on the frame by a single resultant force. Specify where its line of action intersects member AB, measured from B.

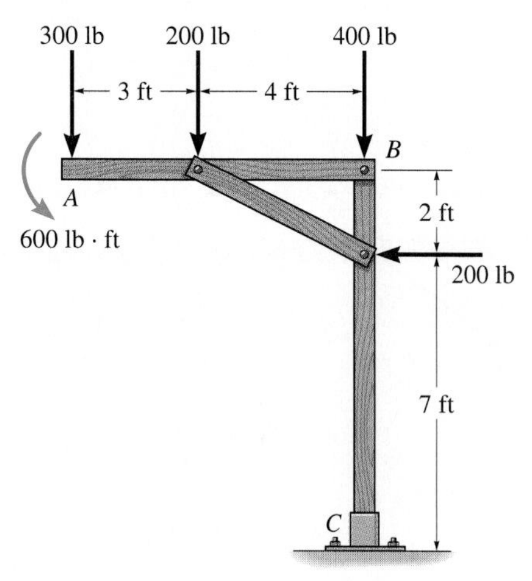

Probs. 4–107/4–108

4–109. Replace the loading on the frame by a single resultant force. Specify where its line of action intersects member AB, measured from A.

4–110. Replace the loading on the frame by a single resultant force. Specify where its line of action intersects member CD, measured from end C.

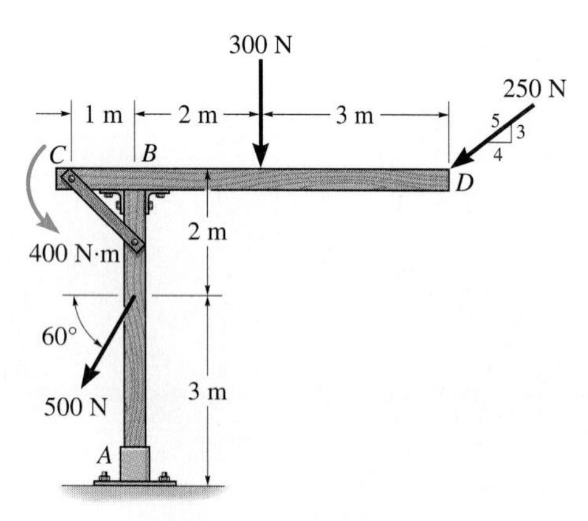

Probs. 4–109/4–110

4–111. The resultant force of the wind and the weights of the various components act on the sign. Determine the equivalent force and moment acting at its base, A.

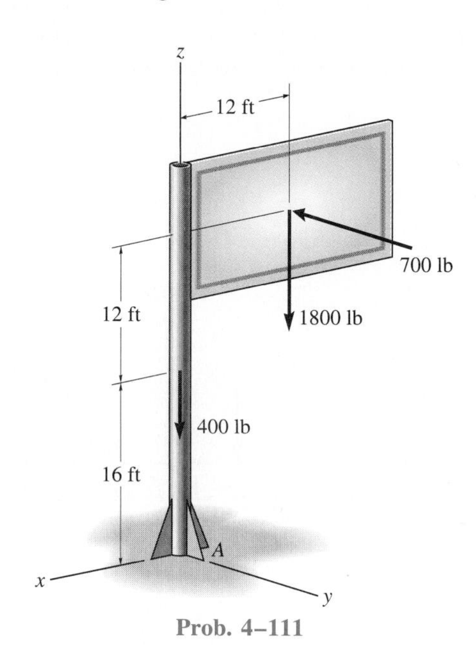

Prob. 4–111

***4–112.** The forces and couple moments which are exerted on the toe and heel plates of a snow ski are $\mathbf{F}_t = \{-50\mathbf{i} + 80\mathbf{j} - 158\mathbf{k}\}$ N, $\mathbf{M}_t = \{-6\mathbf{i} + 4\mathbf{j} + 2\mathbf{k}\}$ N · m, and $\mathbf{F}_h = \{-20\mathbf{i} + 60\mathbf{j} - 250\mathbf{k}\}$ N, $\mathbf{M}_h = \{-20\mathbf{i} + 8\mathbf{j} + 3\mathbf{k}\}$ N · m, respectively. Replace this system by an equivalent force and couple moment acting at point O. Express the results in Cartesian vector form.

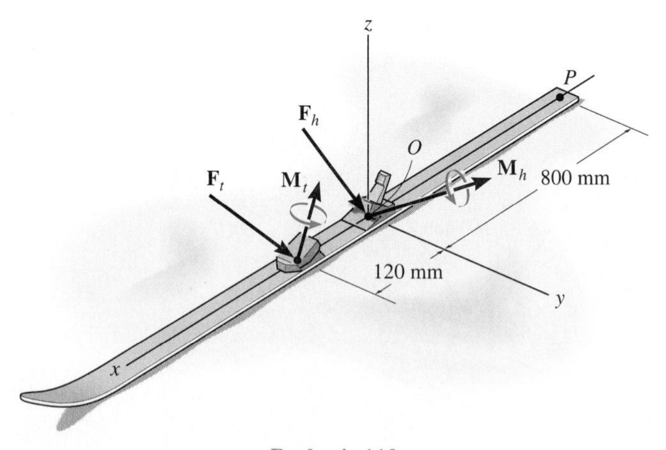

Prob. 4–112

4–113. The forces and couple moments that are exerted on the toe and heel plates of a snow ski are $\mathbf{F}_t = \{-50\mathbf{i} + 80\mathbf{j} - 158\mathbf{k}\}$ N, $\mathbf{M}_t = \{-6\mathbf{i} + 4\mathbf{j} + 2\mathbf{k}\}$ N · m, and $\mathbf{F}_h = \{-20\mathbf{i} + 60\mathbf{j} - 250\mathbf{k}\}$ N, $\mathbf{M}_h = \{-20\mathbf{i} + 8\mathbf{j} + 3\mathbf{k}\}$ N · m, respectively. Replace this system by an equivalent force and couple moment acting at point P. Express the results in Cartesian vector form.

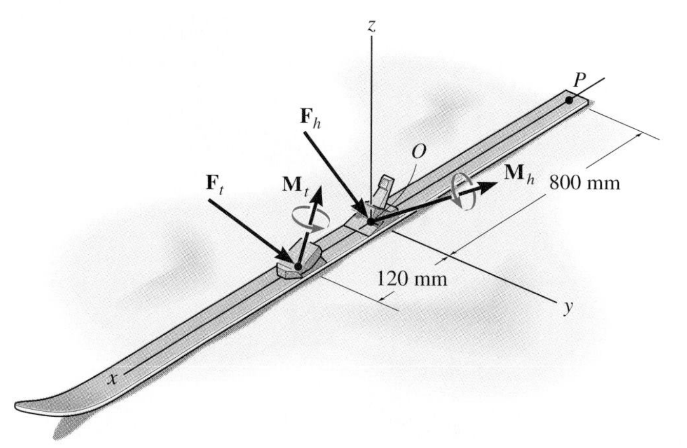

Prob. 4–113

4–114. The belt passing over the pulley is subjected to forces $\mathbf{F}_1$ and $\mathbf{F}_2$, each having a magnitude of 40 N. $\mathbf{F}_1$ acts in the $-\mathbf{k}$ direction. Replace these forces by an equivalent force and couple moment at point A. Express the result in Cartesian vector form. Set $\theta = 0°$ so that $\mathbf{F}_2$ acts in the $-\mathbf{j}$ direction.

4–115. The belt passing over the pulley is subjected to two forces $\mathbf{F}_1$ and $\mathbf{F}_2$, each having a magnitude of 40 N. $\mathbf{F}_1$ acts in the $-\mathbf{k}$ direction. Replace these forces by an equivalent force and couple moment at point A. Express the result in Cartesian vector form. Take $\theta = 45°$.

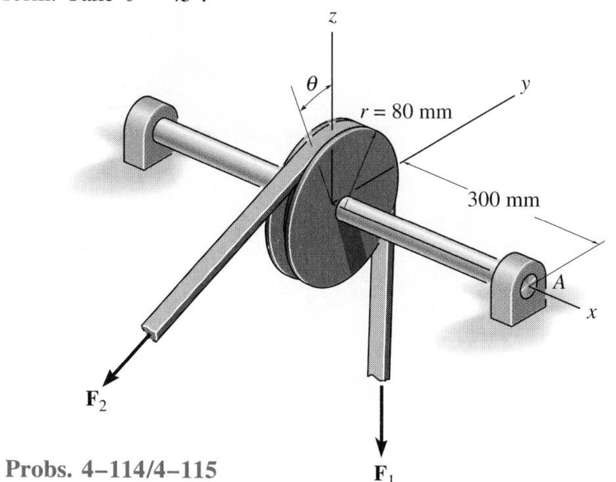

Probs. 4–114/4–115

***4–116.** Three parallel bolting forces act on the circular plate. Determine the resultant force, and specify its location (x, z) on the plate. $F_A = 200$ lb, $F_B = 100$ lb, and $F_C = 400$ lb.

4–117. The three parallel bolting forces act on the circular plate. If the force at A has a magnitude of $F_A = 200$ lb, determine the magnitudes of $\mathbf{F}_B$ and $\mathbf{F}_C$ so that the resultant force $\mathbf{F}_R$ of the system has a line of action that coincides with the y axis. *Hint:* This requires $\Sigma M_x = 0$ and $\Sigma M_z = 0$. Why?

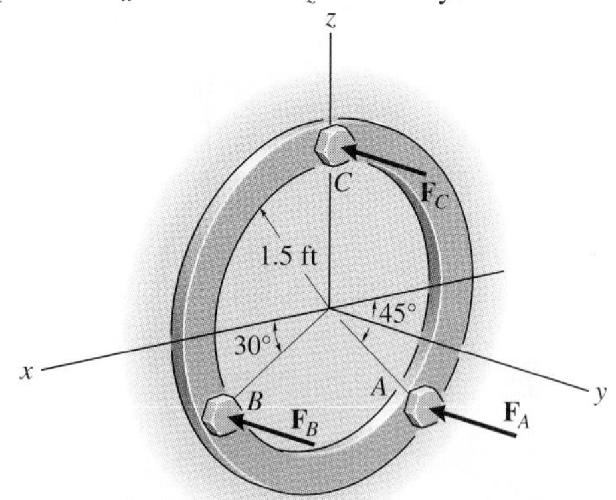

Probs. 4–116/4–117

4–118. A biomechanical model of the lumbar region of the human trunk is shown. The forces acting in the four muscle groups consist of $F_R = 35$ N for the rectus, $F_O = 45$ N for the oblique, $F_L = 23$ N for the lumbar latissimus dorsi, and $F_E = 32$ N for the erector spinae. These loadings are symmetric with respect to the y-z plane. Replace this system of parallel forces by an equivalent force and couple moment acting at the spine, point O. Express the results in Cartesian vector form.

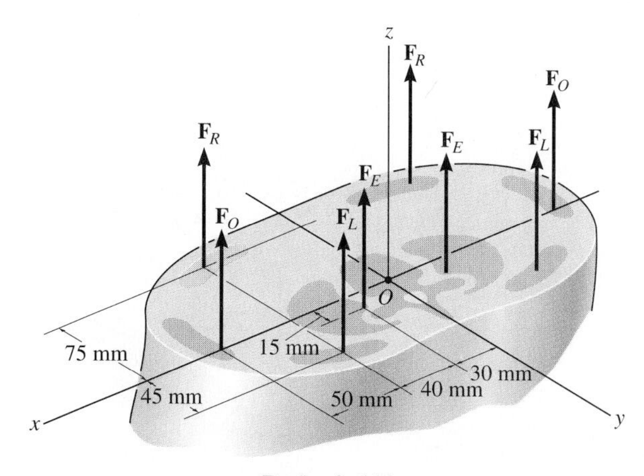

Prob. 4–118

4–119. A biomechanical model of the lumbar region of the human trunk is shown. The forces acting in the four muscle groups consist of $F_R = 35$ N for the rectus, $F_O = 45$ N for the oblique, $F_L = 23$ N for the lumbar latissimus dorsi, and $F_E = 32$ N for the erector spinae. These loadings are symmetric with respect to the y-z plane. Determine the resultant force equivalent to the given parallel force system, and locate its point of application (x, y) on the trunk.

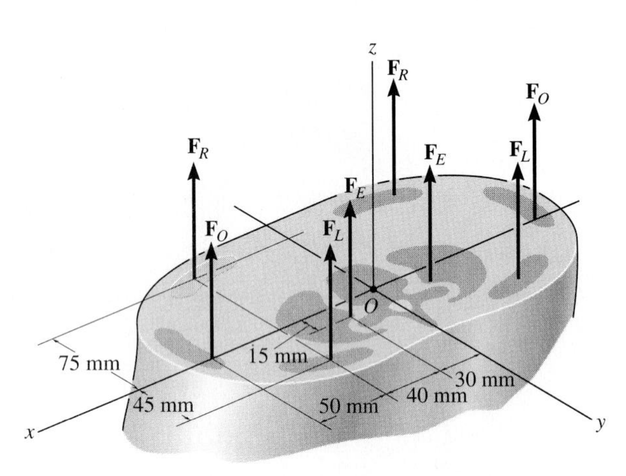

Prob. 4–119

*4–120. The building slab is subjected to four parallel column loadings. Determine the equivalent resultant force and specify its location (x, y) on the slab. Take $F_1 = 30$ kN, $F_2 = 40$ kN.

4–121. The building slab is subjected to four parallel column loadings. Determine the equivalent resultant force and specify its location (x, y) on the slab. Take $F_1 = 20$ kN, $F_2 = 50$ kN.

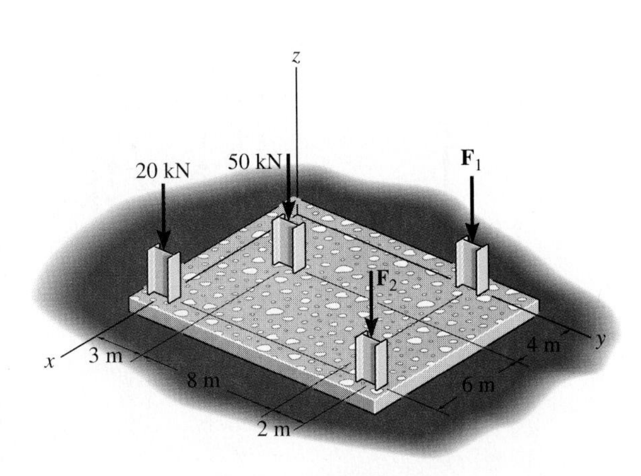

Probs. 4–120/4–121

4–122. A force and couple act on the pipe assembly. Replace this system by an equivalent resultant force. Specify the point where the line of action of the resultant force intersects the x axis. The pipe lies in the x-y plane. Take $F_1 = F_2 = 45$ N.

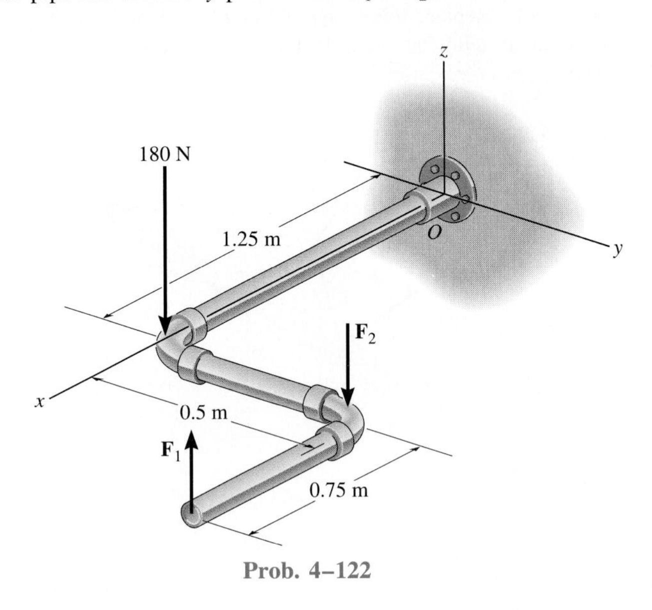

Prob. 4–122

4–123. A force and couple act on the pipe assembly. If $F_1 = 50$ N and $F_2 = 80$ N, replace this system by an equivalent resultant force and couple moment acting at O. Express the results in Cartesian vector form.

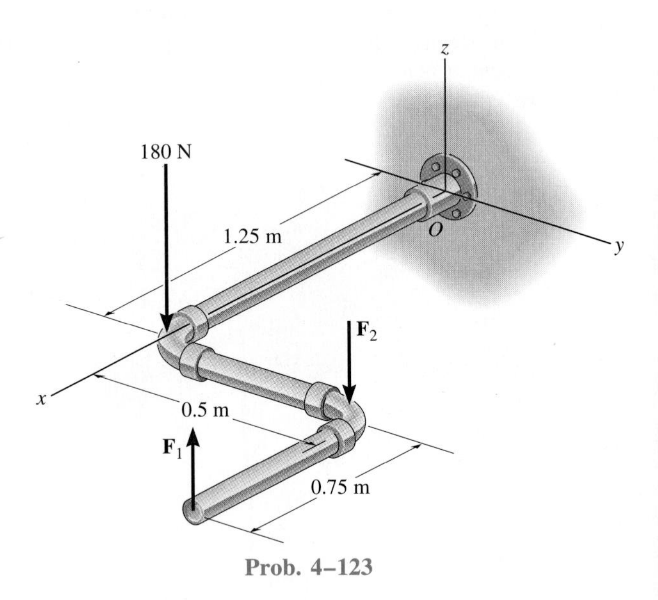

Prob. 4–123

***4–124.** The three forces acting on the block each have a magnitude of 10 lb. Replace this system by a wrench and specify the point where the wrench intersects the z axis, measured from point O.

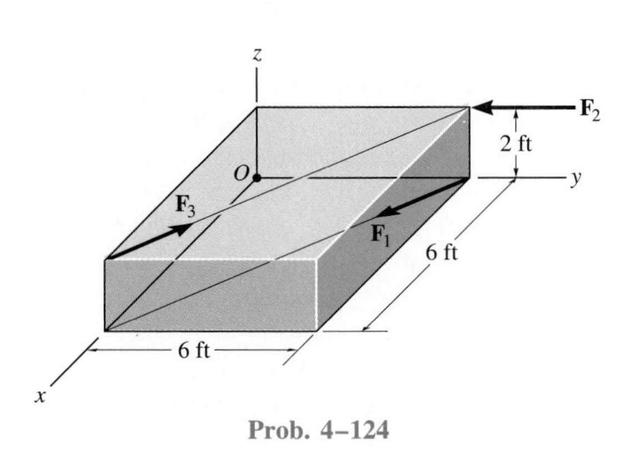

Prob. 4–124

4–125. The pipe assembly is subjected to the action of a wrench at B and a couple at A. Simplify this system to a resultant wrench and specify the location of the wrench along the axis of pipe CD, measured from point C. Set $F = 40$ N.

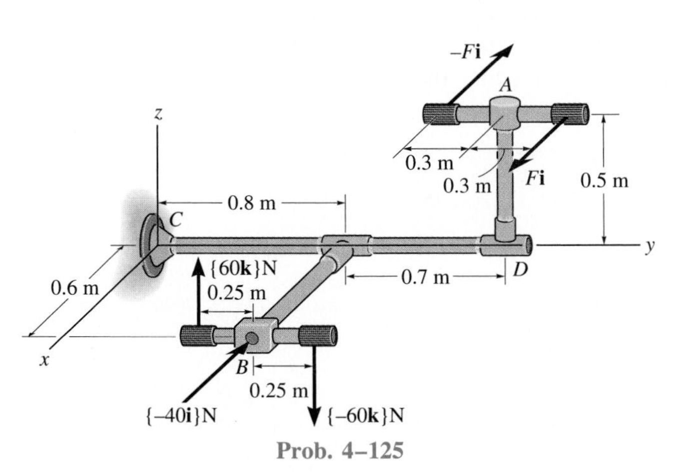

Prob. 4–125

4–126. The pipe assembly is subjected to the action of a wrench at B and a couple at A. Determine the magnitude F of the couple forces so that the system can be simplified to a wrench acting at point C.

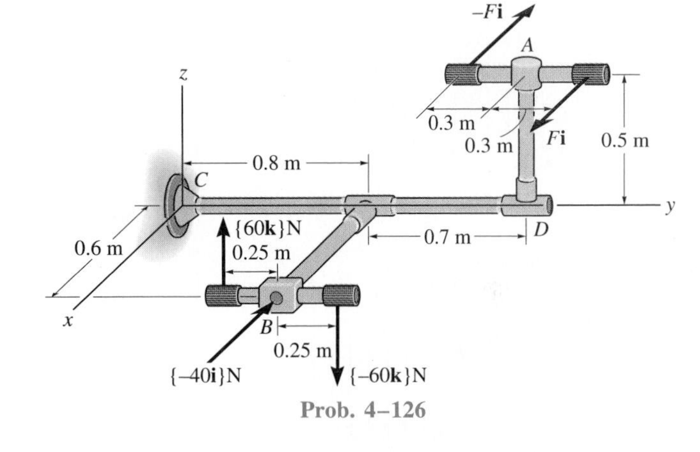

Prob. 4–126

4–127. Replace the three forces acting on the plate by a wrench. Specify the magnitude of the force and couple moment for the wrench and the point $P(x, y)$ where its line of action intersects the plate.

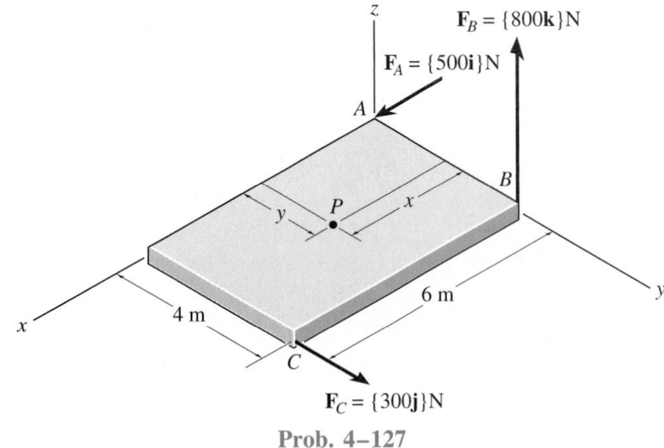

Prob. 4–127

4.10 Reduction of a Simple Distributed Loading

In many situations a very large surface area of a body may be subjected to *distributed loadings* such as those caused by wind, fluids, or simply the weight of material supported over the body's surface. The *intensity* of these loadings at each point on the surface is defined as the *pressure* p (force per unit area), which can be measured in units of lb/ft^2 or pascals (Pa), where 1 Pa = 1 N/m^2.

In this section we will consider the most common case of a distributed pressure loading, which is *uniform* along one axis of a flat rectangular body upon which the loading is applied.* An example of such a loading is shown in Fig. 4–51a. The direction of the intensity of the pressure load is indicated by arrows shown on the *load-intensity diagram*. The entire loading on the plate is therefore a system of parallel forces, infinite in number and each acting on a separate differential area of the plate. Here the *loading function*, $p = p(x)$ Pa, is only a function of x since the pressure is uniform along the y axis. If we multiply $p = p(x)$ by the *width* a m of the plate, we obtain $w = [p(x)\ \text{N/m}^2]a\ \text{m} = w(x)$ N/m. This loading function, shown in Fig. 4–51b, is a measure of load distribution along the line $y = 0$ which is in the plane of symmetry of the loading, Fig. 4–51a. As noted, it is measured as a force per unit length, rather than a force per unit area. Consequently, the load-intensity diagram for $w = w(x)$ can be represented by a system of *coplanar* parallel forces, shown in two dimensions in Fig. 4–51b. Using the methods of Sec. 4.9, this system of forces can be simplified to a single resultant force $\mathbf{F}_R$ and its location $\bar{x}$ specified, Fig. 4–51c.

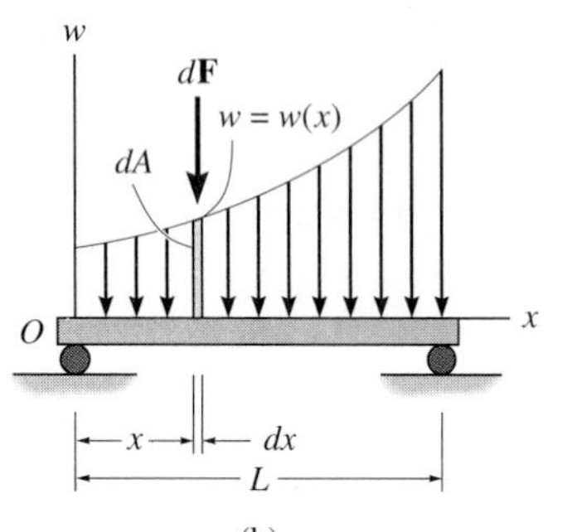

(b)

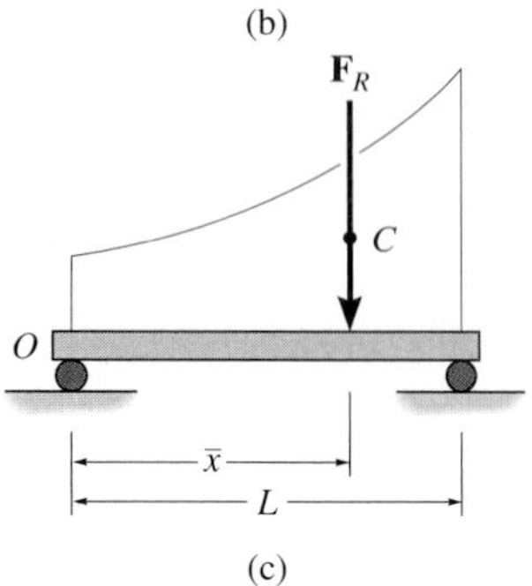

(c)

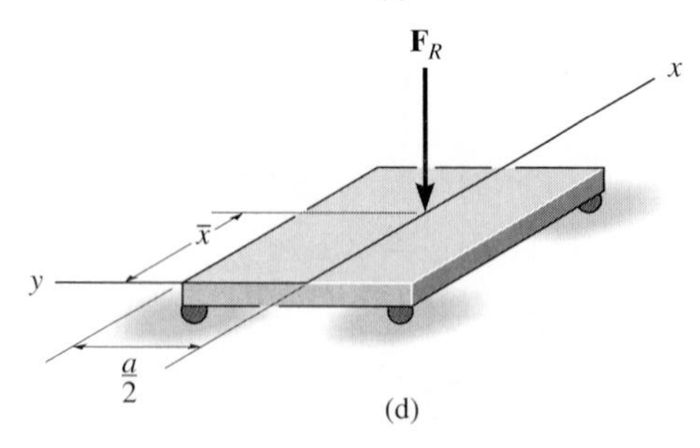

(d)

Fig. 4–51

Magnitude of Resultant Force. From Eq. 4–21 ($F_R = \Sigma F$), the magnitude of $\mathbf{F}_R$ is equivalent to the sum of all the forces in the system. In this case integration must be used, since there is an infinite number of parallel forces $d\mathbf{F}$ acting along the plate, Fig. 4–51b. Since $d\mathbf{F}$ is acting on an element of length dx, and $w(x)$ is a force per unit length, then at the location x, $dF = w(x)\,dx = dA$. In other words, the magnitude of $d\mathbf{F}$ is determined from the colored differential *area* dA under the loading curve. For the entire plate length,

$$+\downarrow F_R = \Sigma F; \qquad F_R = \int_L w(x)\,dx = \int_A dA = A \tag{4–23}$$

Hence, the magnitude of the resultant force is equal to the total area under the loading diagram $w = w(x)$.

*The more general case of a nonuniform surface loading acting on a body is considered in Sec. 9.5.

Location of Resultant Force. Applying Eq. 4–22 ($M_{R_O} = \Sigma M_O$), the location $\bar{x}$ of the line of action of $\mathbf{F}_R$ can be determined by equating the moments of the force resultant and the force distribution about point O (the y axis). Since $d\mathbf{F}$ produces a moment of $x\,dF = x\,w(x)\,dx$ about O, Fig. 4–51b, then for the entire plate, Fig. 4–51c,

$$\curvearrowleft + M_{R_O} = \Sigma M_O; \qquad \bar{x}F_R = \int_L x\,w(x)\,dx$$

Solving for $\bar{x}$, using Eq. 4–23, we can write

$$\bar{x} = \frac{\int_L x\,w(x)\,dx}{\int_L w(x)\,dx} = \frac{\int_A x\,dA}{\int_A dA} \tag{4–24}$$

This equation represents the x coordinate for the geometric center or *centroid* of the *area* under the distributed-loading diagram $w(x)$. *Therefore, the resultant force has a line of action which passes through the centroid C (geometric center) of the area defined by the distributed-loading diagram* $w(x)$, Fig. 4–51c.

Once $\bar{x}$ is determined, $\mathbf{F}_R$ by symmetry passes through point ($\bar{x}$, 0) on the surface of the plate, Fig. 4–51d. If we now consider the three-dimensional pressure loading $p(x)$, Fig. 4–51a, we can therefore conclude that *the resultant force has a magnitude equal to the volume under the distributed-loading curve* $p = p(x)$ *and a line of action which passes through the centroid (geometric center) of this volume.* Detailed treatment of the integration techniques for computing the centroids of volumes or areas is given in Chapter 9. In many cases, however, the distributed-loading diagram is in the shape of a rectangle, triangle, or other simple geometric form. The centroids for such common shapes do not have to be determined from Eq. 4–24; rather, they can be obtained directly from the tabulation given on the inside back cover.

Example 4–22

In each case, determine the magnitude and location of the resultant of the distributed load acting on the beams in Fig. 4–52.

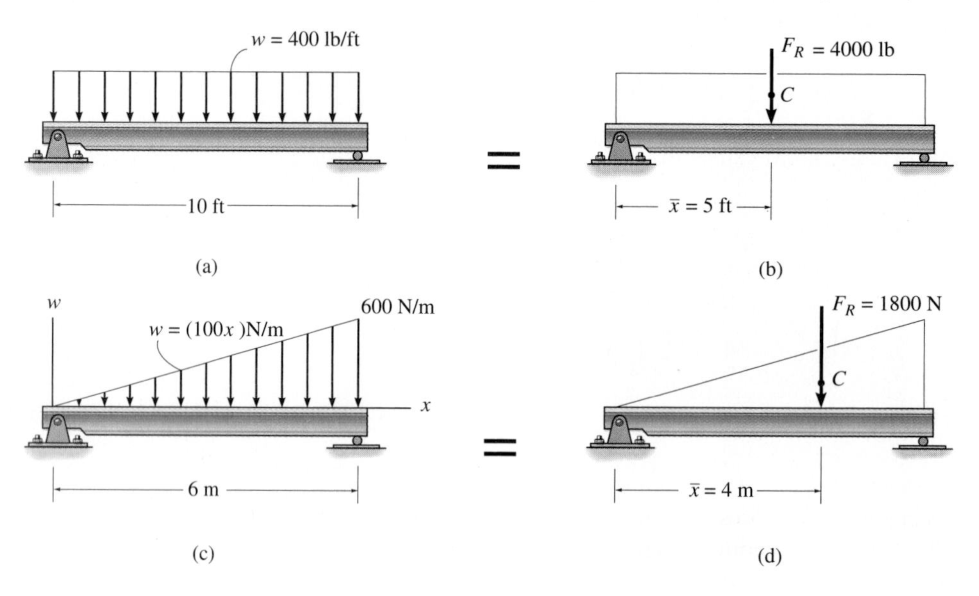

Fig. 4–52

SOLUTION

Uniform Loading. As indicated $w = 400$ lb/ft, which is constant over the entire beam, Fig. 4–52*a*. This loading forms a rectangle, the area of which is equal to the resultant force, Fig. 4–52*b*; i.e.,

$$F_R = (400 \text{ lb/ft})(10 \text{ ft}) = 4000 \text{ lb} \qquad \textit{Ans.}$$

The location of $\mathbf{F}_R$ passes through the geometric center or centroid C of this rectangular area, so

$$\bar{x} = 5 \text{ ft} \qquad \textit{Ans.}$$

Triangular Loading. Here the loading varies uniformly in intensity from 0 to 600 N/m, Fig. 4–52*c*. These values can be verified by substitution of $x = 0$ and $x = 6$ m into the loading function $w = 100x$ N/m. The area of this triangular loading is equal to $\mathbf{F}_R$, Fig. 4–52*d*. From the table on the inside back cover, $A = \frac{1}{2}bh$, so that

$$F_R = \tfrac{1}{2}(6 \text{ m})(600 \text{ N/m}) = 1800 \text{ N} \qquad \textit{Ans.}$$

The line of action of $\mathbf{F}_R$ passes through the centroid C of the triangle. Using the table on the inside back cover, this point lies at a distance of one third the length of the beam, measured from the right side. Hence,

$$\bar{x} = 6 \text{ m} - \tfrac{1}{3}(6 \text{ m}) = 4 \text{ m} \qquad \textit{Ans.}$$

Example 4–23

The granular material exerts the distributed loading on the beam as shown in Fig. 4–53*a*. Determine the magnitude and location of the resultant of this load.

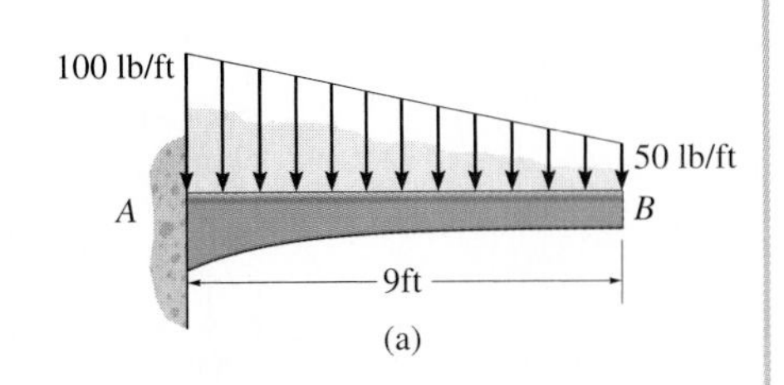

(a)

SOLUTION

The area of the loading diagram is a *trapezoid,* and therefore the solution can be obtained directly from the area and centroid formulas for a trapezoid listed on the inside back cover. Since these formulas are not easily remembered, instead we will solve this problem by using "composite" areas. In this regard, we can divide the trapezoidal loading into a rectangular and triangular loading as shown in Fig. 4–53*b*. The magnitude of the force represented by each of these loadings is equal to its associated *area,*

$$F_1 = \tfrac{1}{2}(9\text{ ft})(50\text{ lb/ft}) = 225\text{ lb}$$
$$F_2 = (9\text{ ft})(50\text{ lb/ft}) = 450\text{ lb}$$

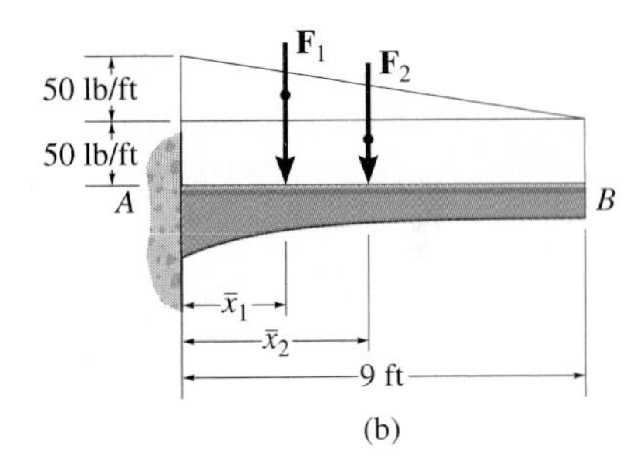

(b)

The lines of action of these parallel forces act through the *centroid* of their associated areas and therefore intersect the beam at

$$\bar{x}_1 = \tfrac{1}{3}(9\text{ ft}) = 3\text{ ft}$$
$$\bar{x}_2 = \tfrac{1}{2}(9\text{ ft}) = 4.5\text{ ft}$$

The two parallel forces $\mathbf{F}_1$ and $\mathbf{F}_2$ can be reduced to a single resultant $\mathbf{F}_R$. The magnitude of $\mathbf{F}_R$ is

$+\downarrow F_R = \Sigma F;$ $\qquad F_R = 225 + 450 = 675\text{ lb}$ ***Ans.***

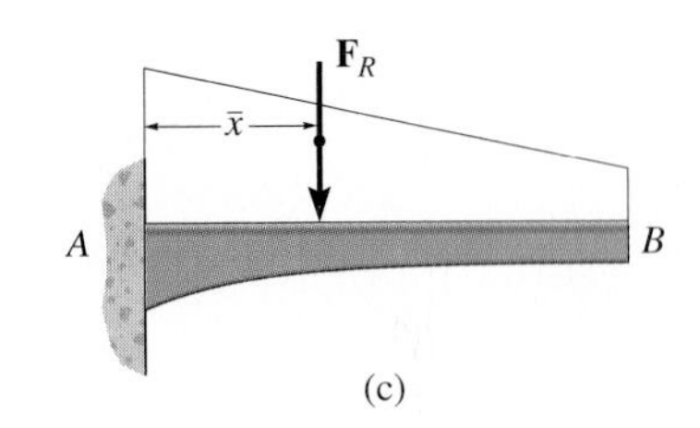

(c)

With reference to point *A*, Fig. 4–53*b* and *c*, we can define the location of $\mathbf{F}_R$. We require

$\curvearrowleft + M_{R_A} = \Sigma M_A;$ $\qquad \bar{x}(675) = 3(225) + 4.5(450)$

$$\bar{x} = 4\text{ ft}$$ ***Ans.***

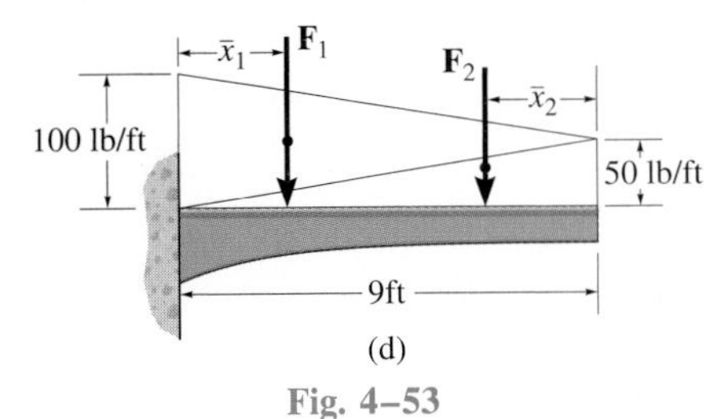

(d)

Fig. 4–53

Note: The trapezoidal area in Fig. 4–53*a* can also be divided into two triangular areas as shown in Fig. 4–53*d*. In this case

$$F_1 = \tfrac{1}{2}(9\text{ ft})(100\text{ lb/ft}) = 450\text{ lb}$$
$$F_2 = \tfrac{1}{2}(9\text{ ft})(50\text{ lb/ft}) = 225\text{ lb}$$

and

$$\bar{x}_1 = \tfrac{1}{3}(9\text{ ft}) = 3\text{ ft}$$
$$\bar{x}_2 = \tfrac{1}{3}(9\text{ ft}) = 3\text{ ft}$$

Using these results, show that again $F_R = 675$ lb and $\bar{x} = 4$ ft.

Example 4–24

Determine the magnitude and location of the resultant force acting on the shaft in Fig. 4–54*a*.

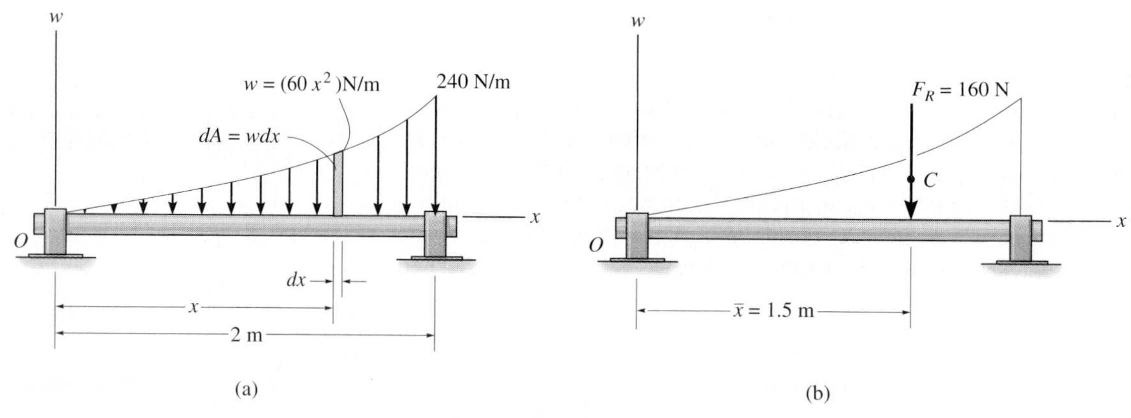

Fig. 4–54

SOLUTION

Since $w = w(x)$ is given, this problem will be solved by integration. The colored differential area element $dA = w\,dx = 60x^2\,dx$. Applying Eq. 4–23, by summing these elements from $x = 0$ to $x = 2$ m, we obtain the resultant force $\mathbf{F}_R$,

$F_R = \Sigma F;$

$$F_R = \int_A dA = \int_0^2 60x^2\,dx = 60\left[\frac{x^3}{3}\right]_0^2 = 60\left[\frac{2^3}{3} - \frac{0^3}{3}\right]$$
$$= 160 \text{ N} \qquad \textit{Ans.}$$

Since the element of area dA is located an arbitrary distance x from O, the location $\bar{x}$ of $\mathbf{F}_R$ *measured from O,* Fig. 4–54*b*, is determined from Eq. 4–24.

$$\bar{x} = \frac{\int_A x\,dA}{\int_A dA} = \frac{\int_0^2 x(60x^2)\,dx}{160} = \frac{60\left[\frac{x^4}{4}\right]_0^2}{160} = \frac{60\left[\frac{2^4}{4} - \frac{0^4}{4}\right]}{160}$$
$$= 1.5 \text{ m} \qquad \textit{Ans.}$$

These results may be checked by using the table on the inside back cover, where it is shown that for an exparabolic area of length a, height b, and shape shown in Fig. 4–54*a*,

$$A = \frac{ab}{3} = \frac{2(240)}{3} = 160 \text{ N} \quad \text{and} \quad \bar{x} = \frac{3}{4}a = \frac{3}{4}(2) = 1.5 \text{ m}$$

Example 4–25

A distributed loading of $p = 800x$ Pa acts over the top surface of the beam shown in Fig. 4–55*a*. Determine the magnitude and location of the resultant force.

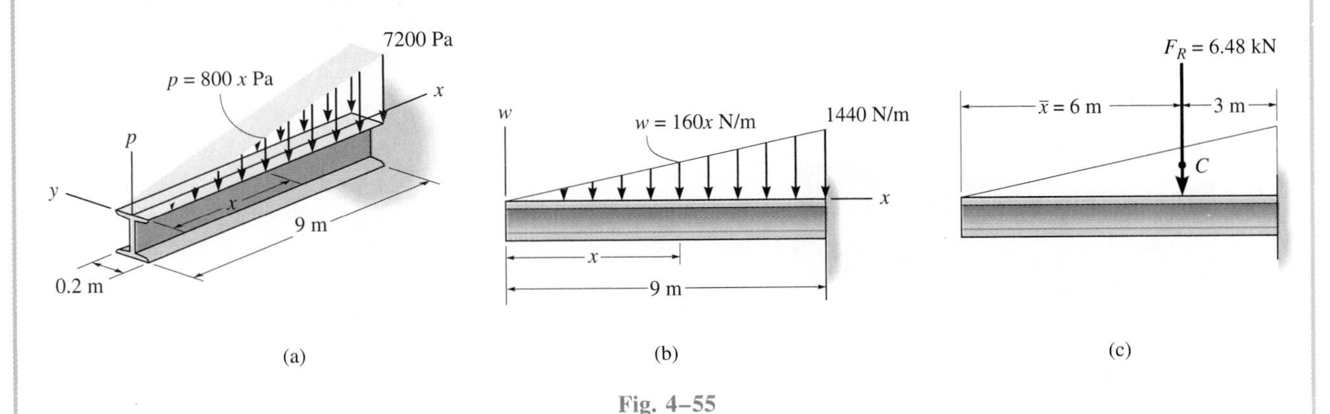

Fig. 4–55

SOLUTION

The loading function $p = 800x$ Pa indicates that the load intensity varies uniformly from $p = 0$ at $x = 0$ to $p = 7200$ Pa at $x = 9$ m. Since the intensity is uniform along the width of the beam (the y axis), the loading may be viewed in two dimensions as shown in Fig. 4–55*b*. Here

$$\begin{aligned} w &= (800x \text{ N/m}^2)(0.2 \text{ m}) \\ &= (160x) \text{ N/m} \end{aligned}$$

At $x = 9$ m, note that $w = 1440$ N/m. Although we may again apply Eqs. 4–23 and 4–24 as in Example 4–24, it is simpler to use the table on the inside back cover.

The magnitude of the resultant force is

$$F_R = \tfrac{1}{2}(9 \text{ m})(1440 \text{ N/m}) = 6480 \text{ N} = 6.48 \text{ kN} \qquad \textit{Ans.}$$

The line of action of $\mathbf{F}_R$ passes through the *centroid* C of the triangle. Hence,

$$\bar{x} = 9 \text{ m} - \tfrac{1}{3}(9 \text{ m}) = 6 \text{ m} \qquad \textit{Ans.}$$

The results are shown in Fig. 4–55*c*.

We may also view the resultant $\mathbf{F}_R$ as *acting* through the *centroid* of the *volume* of the loading diagram $p = p(x)$ in Fig. 4–55*a*. Hence $\mathbf{F}_R$ intersects the x–y plane at the point (6 m, 0). Furthermore, the *magnitude* of $\mathbf{F}_R$ is equal to the *volume* under the loading diagram; i.e.,

$$F_R = V = \tfrac{1}{2}(7200 \text{ N/m}^2)(9 \text{ m})(0.2 \text{ m}) = 6.48 \text{ kN} \qquad \textit{Ans.}$$

PROBLEMS

***4–128.** The loading on the bookshelf is distributed as shown. Determine the magnitude of the equivalent resultant force and its location, measured from point O.

4–129. The loading on the bookshelf is distributed as shown. Determine the equivalent resultant force and its location, measured from point A.

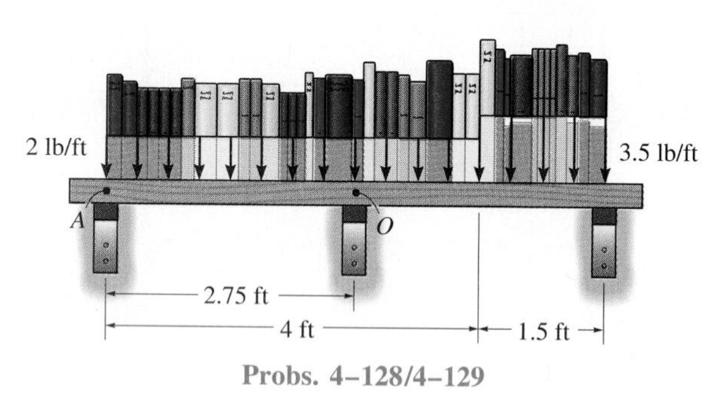

Probs. 4–128/4–129

4–130. Replace the loading by an equivalent resultant force and couple moment acting at point O.

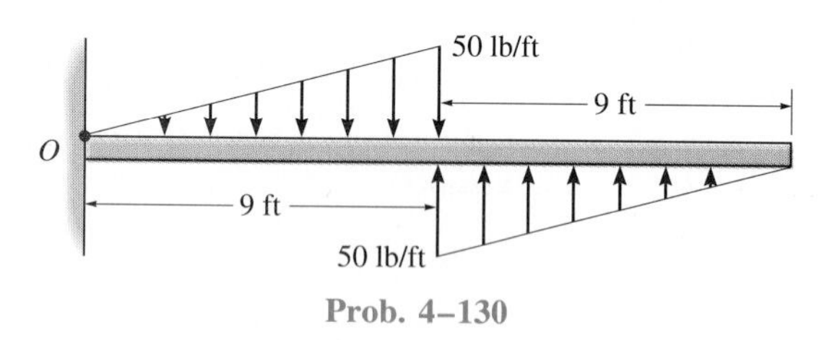

Prob. 4–130

4–131. Replace the distributed loading by an equivalent resultant force, and specify its location on the beam, measured from the pin at C.

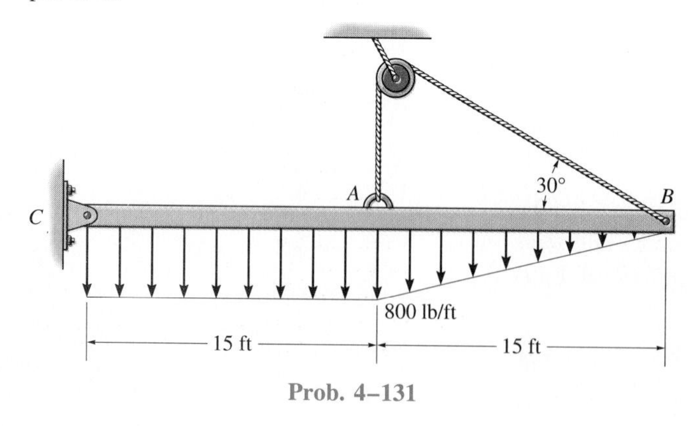

Prob. 4–131

***4–132.** The column is used to support the floor which exerts a force of 3000 lb on the top of the column. The effect of soil pressure along its side is distributed as shown. Replace this loading by an equivalent resultant force and specify where it acts along the column, measured from its base A.

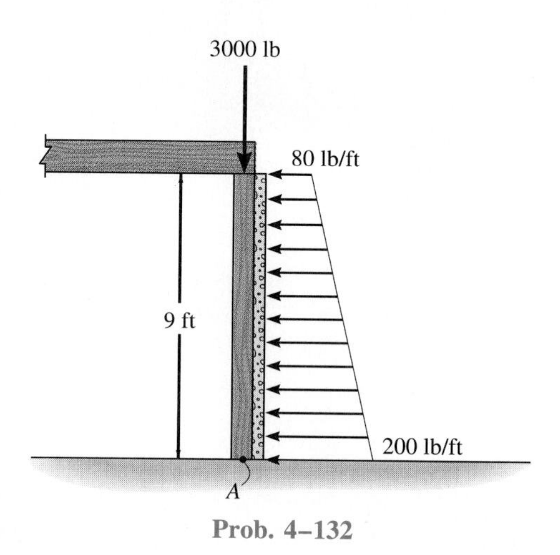

Prob. 4–132

4–133. Replace the loading by an equivalent force and couple moment acting at point O.

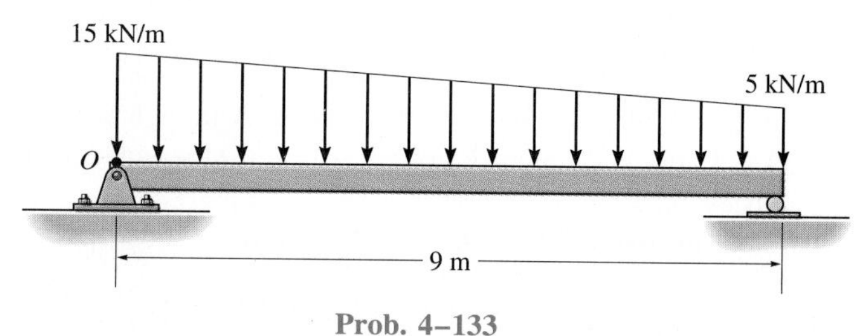

Prob. 4–133

4–134. The masonry support creates the loading distribution acting on the end of the beam. Replace this load by an equivalent resultant force and specify its location, measured from point O.

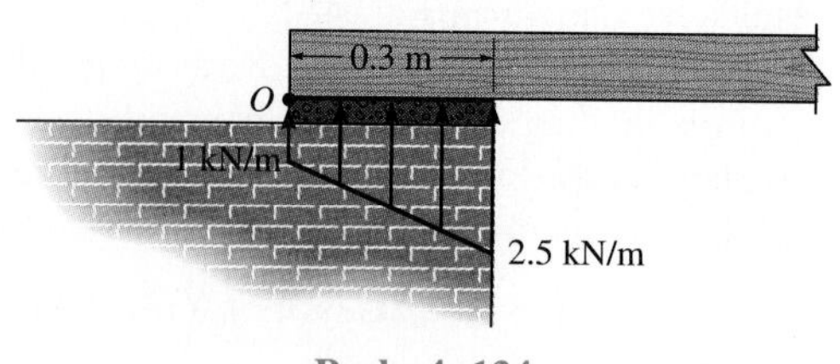

Prob. 4–134

4–135. Replace the loading by an equivalent resultant force and specify its location on the beam, measured from point *B*.

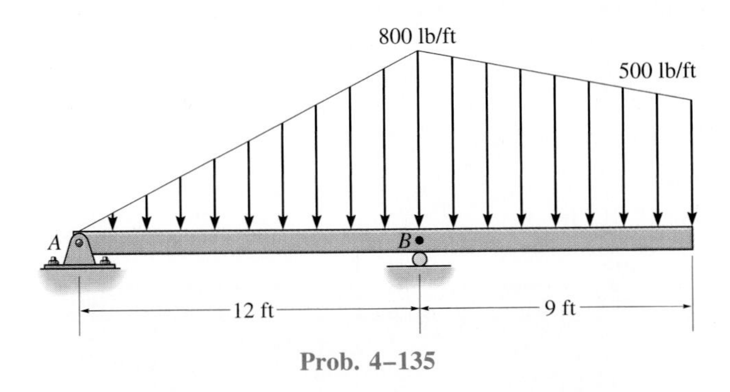

Prob. 4–135

***4–136.** Replace the distributed loading by an equivalent resultant force and specify its location, measured from point *A*.

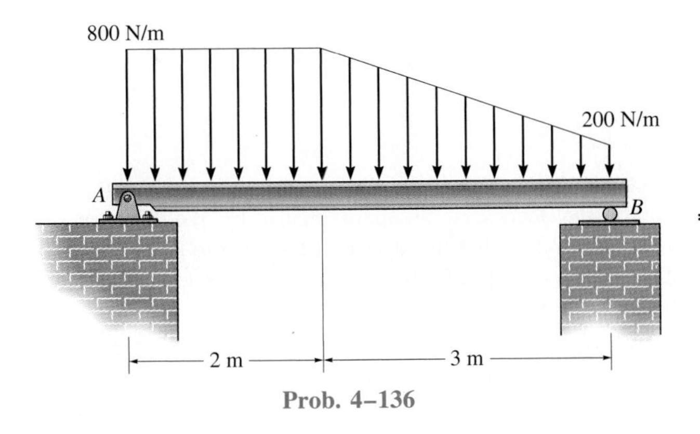

Prob. 4–136

4–137. The beam is subjected to the parabolic loading. Replace this loading by an equivalent force and couple moment at point *O*.

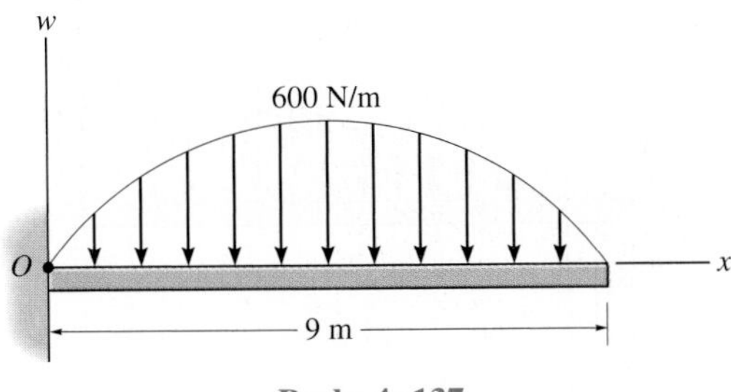

Prob. 4–137

4–138. Replace the distributed loading by an equivalent resultant force and specify where its line of action intersects member *AB*, measured from *A*.

4–139. Replace the distributed loading by an equivalent resultant force and specify where its line of action intersects member *BC*, measured from *C*.

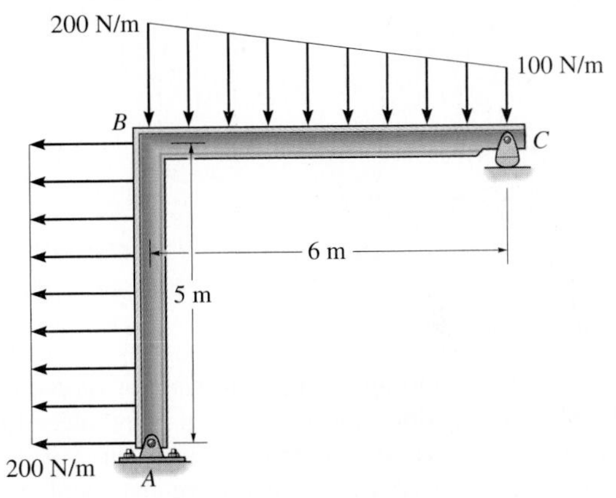

Probs. 4–138/4–139

***4–140.** Replace the loading by an equivalent resultant force and couple moment acting at point *O*.

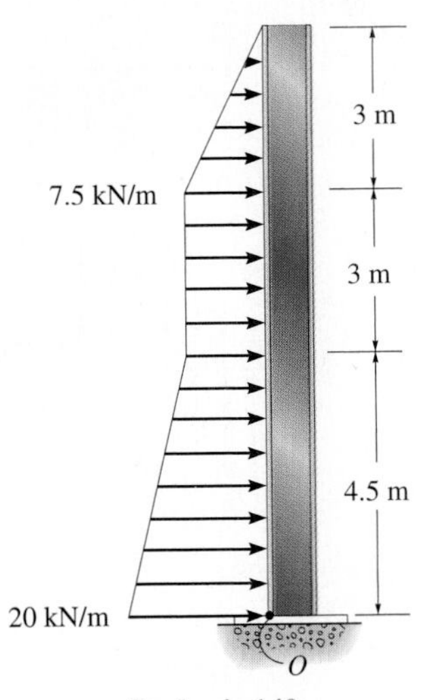

Prob. 4–140

4–141. The distribution of soil loading on the bottom of a building slab is shown. Replace this loading by an equivalent resultant force and specify its location, measured from point O.

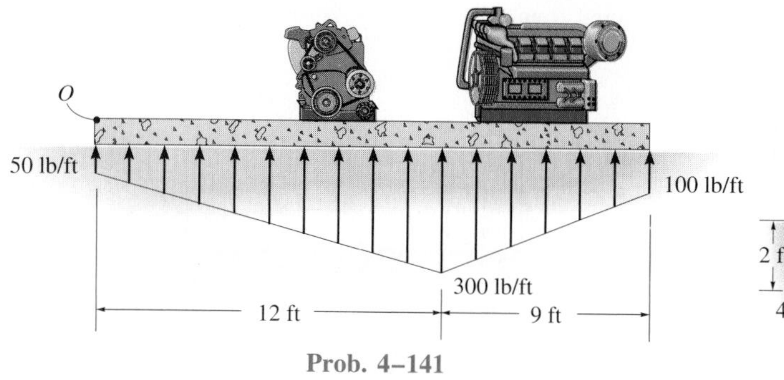

Prob. 4–141

4–142. The bricks on top of the beam and the supports at the bottom create the distributed loading shown in the second figure. Determine the required intensity w and dimension d of the right support so that the equivalent resultant force and couple moment about point A of the system are both zero.

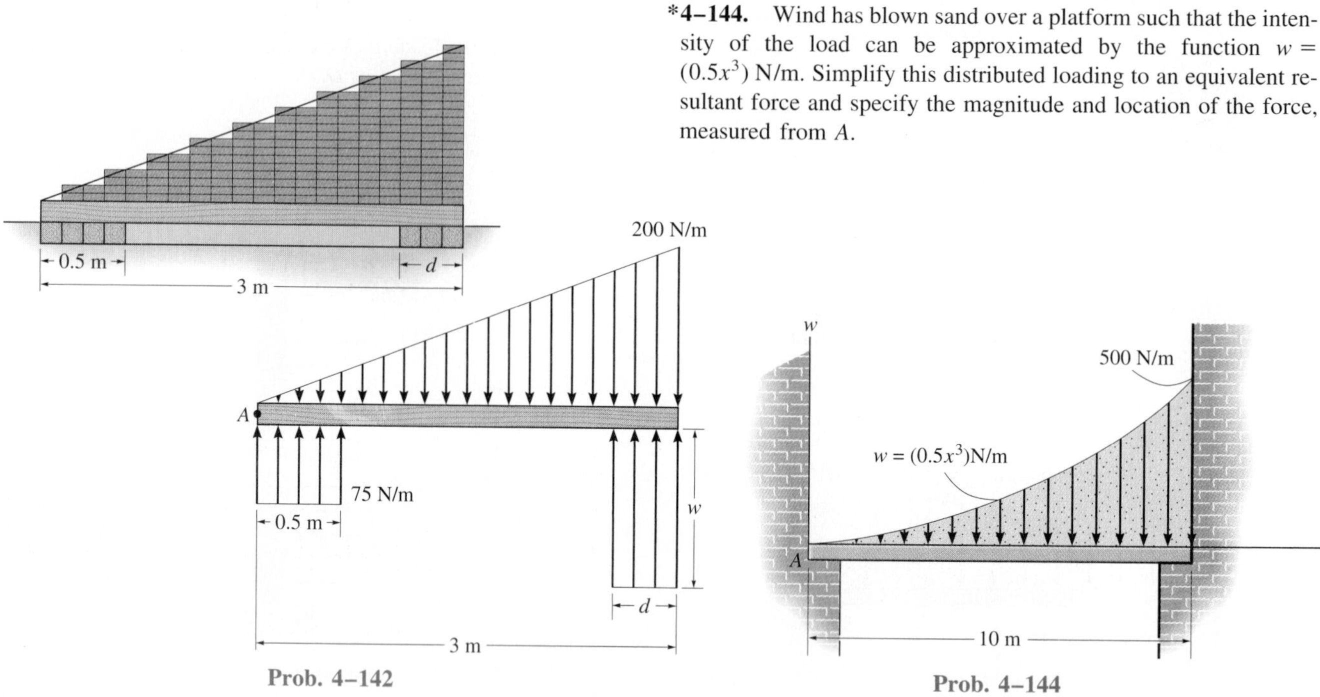

Prob. 4–142

4–143. The distribution of soil loading on the bottom of a building slab is shown. The center portion of the loading is parabolic. Simplify this loading to an equivalent resultant force and specify its location, measured from point O.

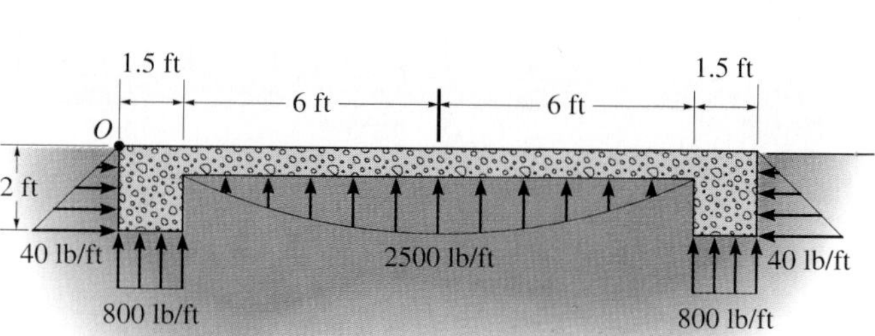

Prob. 4–143

***4–144.** Wind has blown sand over a platform such that the intensity of the load can be approximated by the function $w = (0.5x^3)$ N/m. Simplify this distributed loading to an equivalent resultant force and specify the magnitude and location of the force, measured from A.

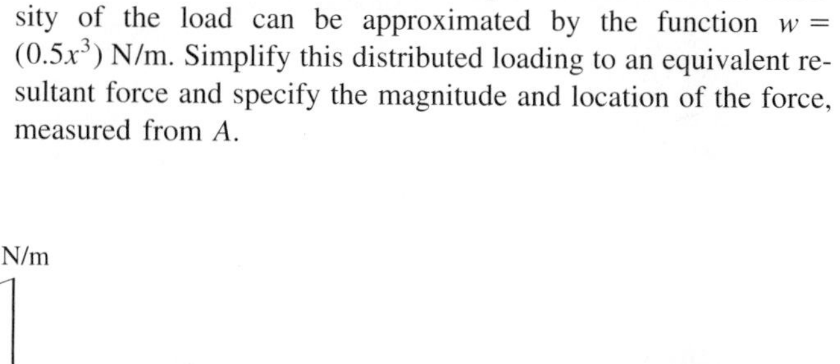

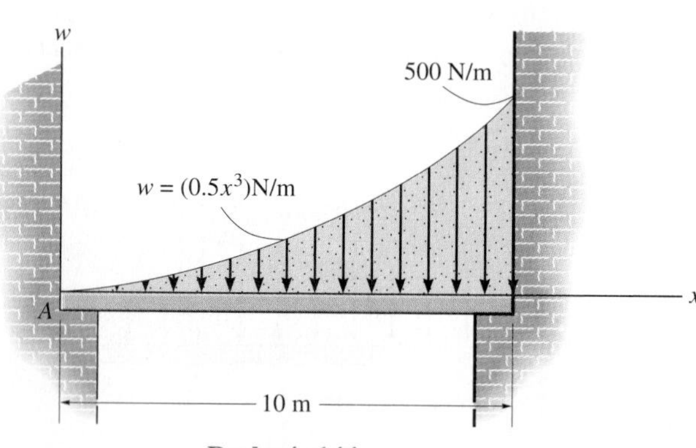

Prob. 4–144

4–145. The form is used to cast a concrete wall having a width of 5 m. Determine the equivalent resultant force the wet concrete exerts on the form *AB* if the pressure distribution due to the concrete can be approximated as shown. Specify the location of the resultant force, measured from point *B*.

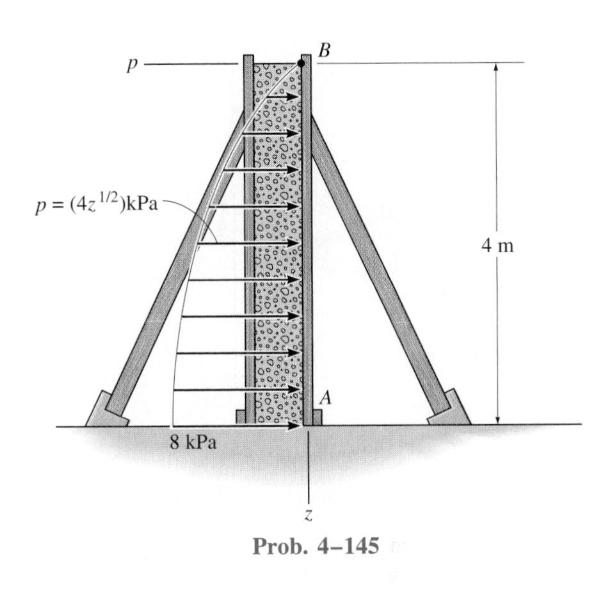

Prob. 4–145

■**4–146.** Determine the equivalent resultant force of the distributed loading and its location, measured from point *A*. Evaluate the integrals using Simpson's rule.

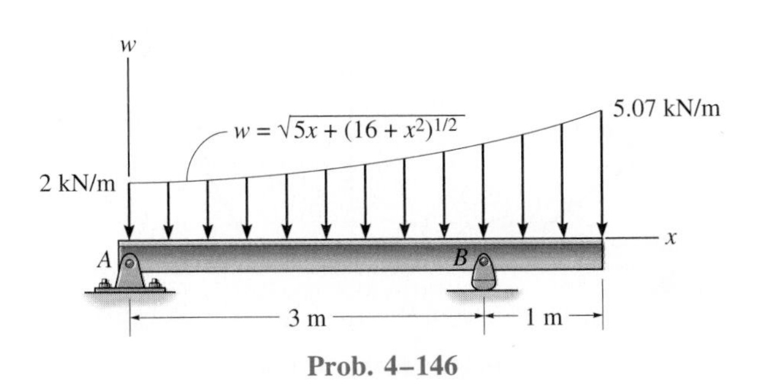

Prob. 4–146

4–147. Determine the magnitude of the equivalent resultant force and its location, measured from point *O*.

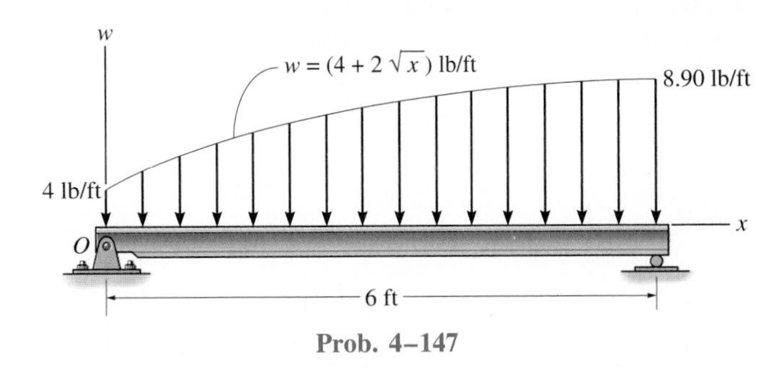

Prob. 4–147

***4–148.** The distributed load acts on the beam as shown. Determine the magnitude of the equivalent resultant force and specify where it acts, measured from the support, *A*.

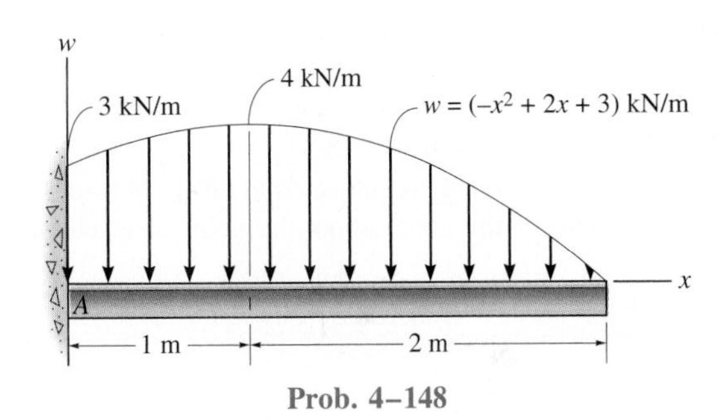

Prob. 4–148

4–149. The distributed load acts on the shaft as shown. Determine the magnitude of the equivalent resultant force and specify its location, measured from the support, *A*.

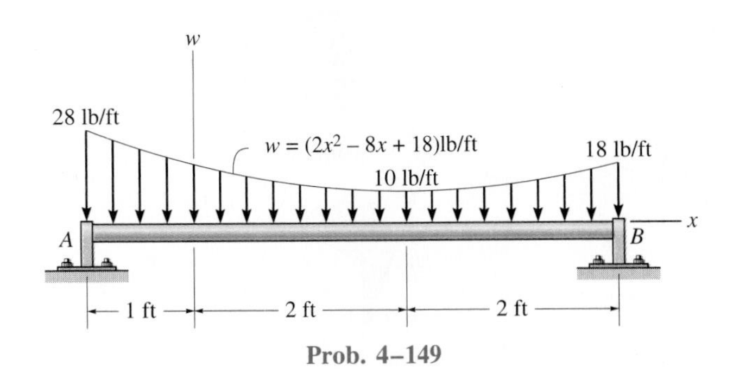

Prob. 4–149

REVIEW PROBLEMS

4–150. Determine the equivalent resultant force of the distributed loading and its location, measured from point A.

***4–152.** Replace the force **F** having a magnitude of $F = 50$ lb and acting at point A by an equivalent force and couple moment at point C.

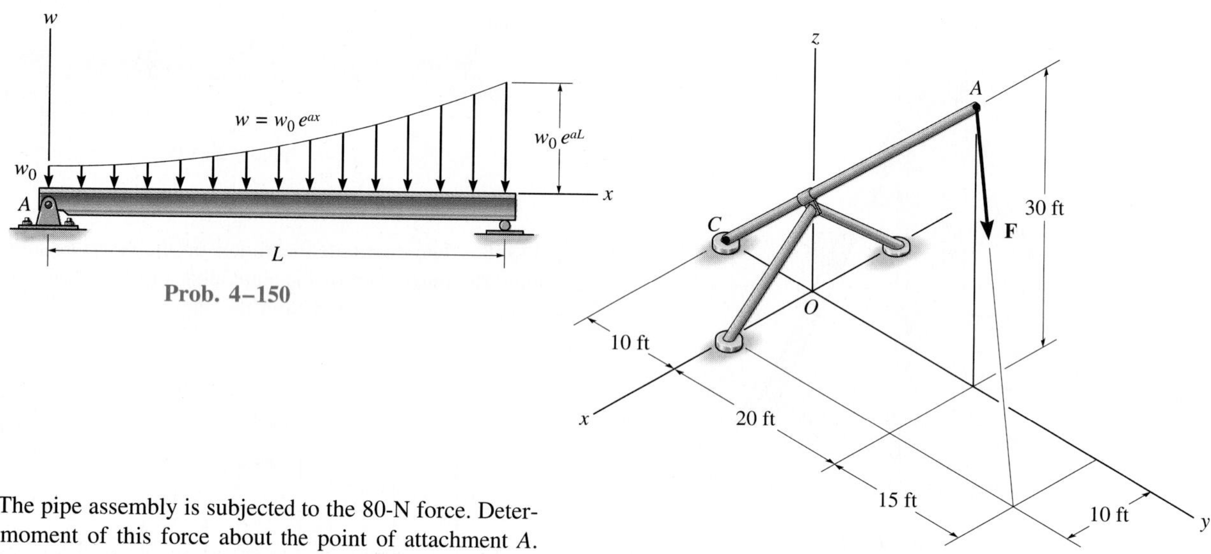

Prob. 4–150

Prob. 4–152

4–151. The pipe assembly is subjected to the 80-N force. Determine the moment of this force about the point of attachment A.

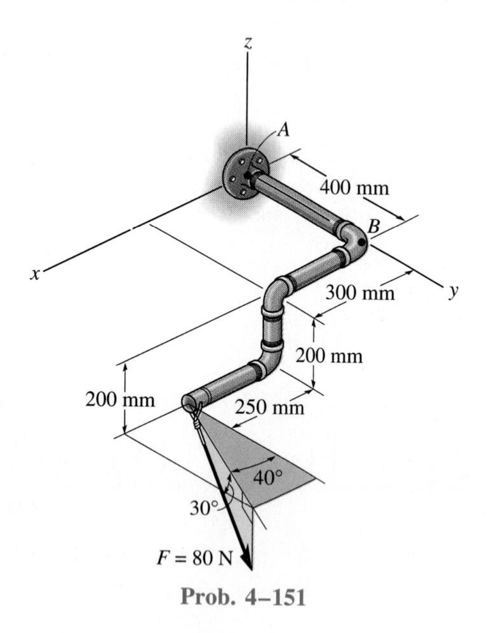

Prob. 4–151

4–153. Determine the resultant moment about point A of the forces acting on the beam.

4–154. Determine the resultant moment about point B of the forces acting on the beam.

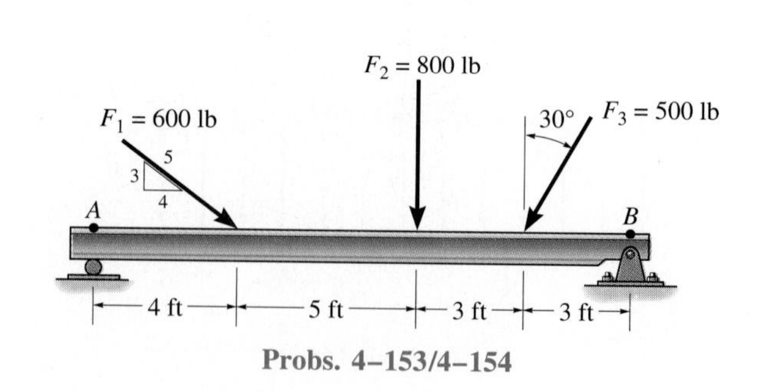

Probs. 4–153/4–154

4–155. The gear reducer is subjected to the two couple moments shown. Express the resultant couple moment in Cartesian vector form and specify its magnitude and coordinate direction angles. The coordinate direction angles for $\mathbf{M}_1$ are $\alpha_1 = 120°$, $\beta_1 = 45°$, $\gamma_1 = 60°$.

***4–156.** The gear reducer is subjected to the two couple moments shown. Determine the coordinate direction angles of $\mathbf{M}_1$ so that the resultant couple moment acts in the $+\mathbf{k}$ direction. What is the magnitude of the resultant couple moment?

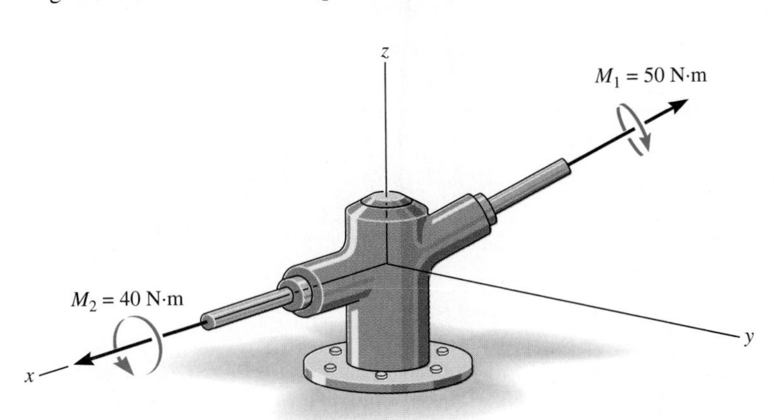

Probs. 4–155/4–156

4–157. The force of $F = 80$ lb acts along the edge DB of the tetrahedron. Determine the magnitude of the moment of this force about the edge AC.

4–158. If the moment of the force $\mathbf{F}$ about the edge AC of the tetrahedron has a magnitude of $M = 200$ lb · ft and is directed from C towards A, determine the magnitude of $\mathbf{F}$.

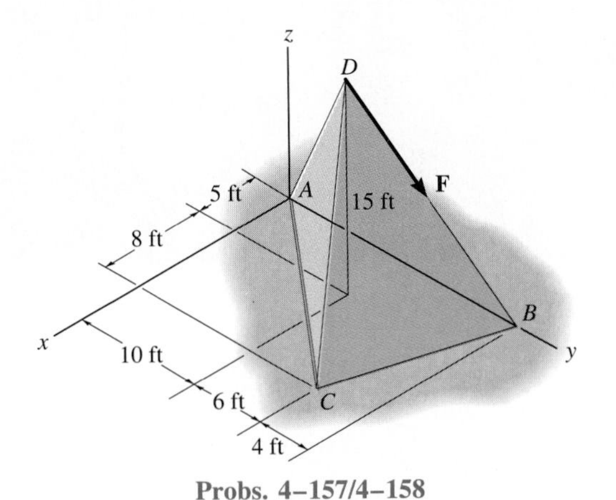

Probs. 4–157/4–158

4–159. Determine the moments of the force $\mathbf{F}$ about the x, y, and z axes. Solve the problem (a) using a Cartesian vector approach and (b) using a scalar approach. Express each result as a Cartesian vector.

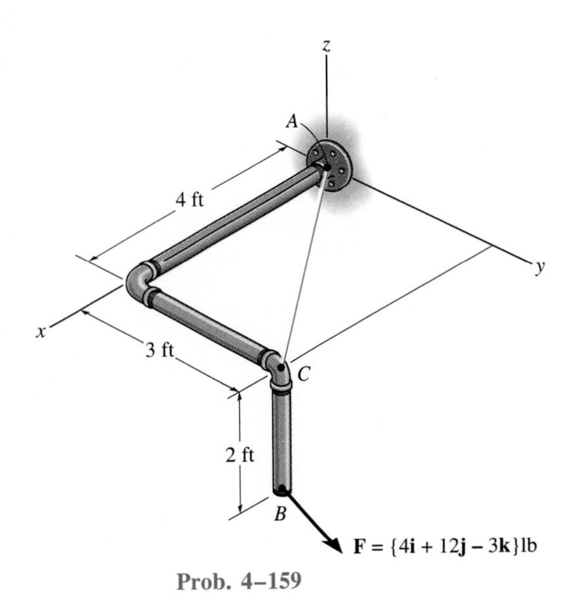

Prob. 4–159

***4–160.** Determine the moment of the force $\mathbf{F}$ about an axis extending between A and C. Express the result as a Cartesian vector.

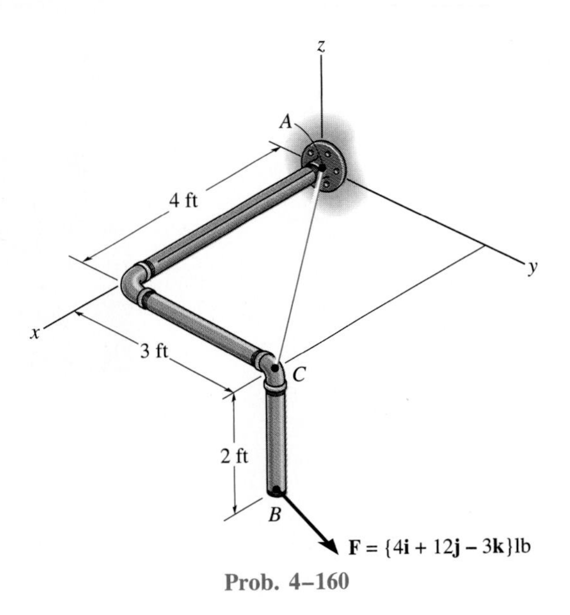

Prob. 4–160

Utility poles are often subjected to the forces of cables and the weight of transformers. The equilibrium analysis, as explained in this chapter, must account for these loadings.

5

Equilibrium of a Rigid Body

In this chapter the fundamental concepts of rigid-body equilibrium will be discussed. It will be shown that equilibrium requires both a *balance of forces,* to prevent the body from translating with accelerated motion, and a *balance of moments,* to prevent the body from rotating.

Many types of engineering problems involve symmetric loadings and can be solved by projecting all the forces acting on a body onto a single plane. Hence, in the first part of this chapter, the equilibrium of a body subjected to a *coplanar* or *two-dimensional force system* will be considered. Ordinarily the geometry of such problems is not very complex, so a scalar solution is suitable for analysis. The more general discussion of rigid bodies subjected to *three-dimensional force systems* is given in the second part of this chapter. It will be seen that many of these types of problems can best be solved by using vector analysis.

5.1 Conditions for Rigid-Body Equilibrium

In Chapter 3 it was stated that a particle is in equilibrium if it remains at rest or moves with constant velocity. For this to be the case, it is both necessary and sufficient to require the resultant force acting on the particle to be equal to zero. Using this fact, we will now develop the conditions required to maintain equilibrium for a rigid body. To do this, consider the rigid body in Fig. 5–1*a*, which is fixed in the x, y, z reference and is either at rest or moves with the reference at constant velocity. A free-body diagram of the arbitrary ith particle of the body is shown in Fig. 5–1*b*. There are two types of forces which act on

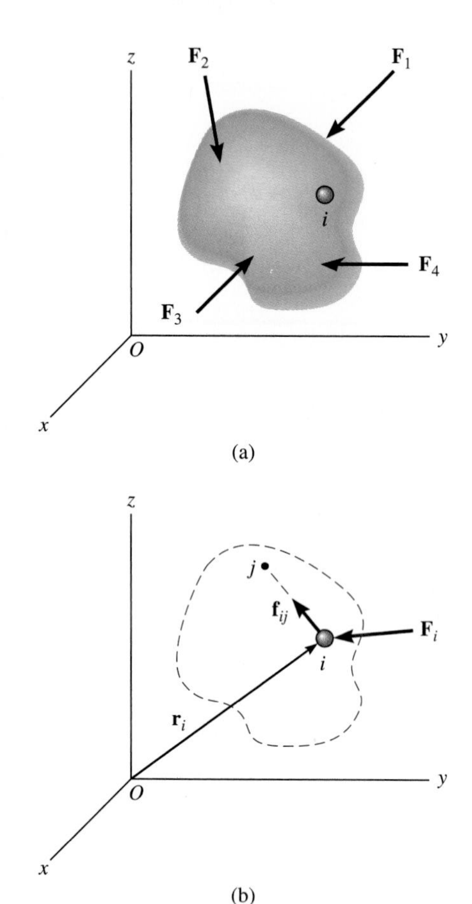

Fig. 5–1

it. The *internal forces*, represented symbolically as

$$\sum_{\substack{j=1 \\ (j \neq 1)}}^{n} \mathbf{f}_{ij} = \mathbf{f}_i$$

are forces which all the other particles exert on the ith particle and produce the resultant $\mathbf{f}_i$. Although only one of these particles is shown in Fig. 5–1b, the summation extends over all n particles composing the body. For this summation note that it is meaningless for $i = j$ since the ith particle cannot exert a force on itself. The resultant *external force* $\mathbf{F}_i$ represents, for example, the effects of gravitational, electrical, magnetic, or contact forces between the ith particle and adjacent bodies or particles *not* included within the body. If the particle is in equilibrium, then applying Newton's first law we have

$$\mathbf{F}_i + \mathbf{f}_i = \mathbf{0}$$

When the equation of equilibrium is applied to each of the other particles of the body, similar equations will result. If all these equations are added together *vectorially*, we obtain

$$\Sigma \mathbf{F}_i + \Sigma \mathbf{f}_i = \mathbf{0}$$

The summation of the internal forces if carried out will equal zero since the internal forces between particles within the body will occur in equal but opposite collinear pairs, Newton's third law. Consequently, only the sum of the *external forces* will remain; and therefore, letting $\Sigma \mathbf{F}_i = \Sigma \mathbf{F}$, the above equation can be written as

$$\Sigma \mathbf{F} = \mathbf{0}$$

Let us now consider the moments of the forces acting on the ith particle about the arbitrary point O, Fig. 5–1b. Using the particle equilibrium equation and the distributive law of the vector cross product yields

$$\mathbf{r}_i \times (\mathbf{F}_i + \mathbf{f}_i) = \mathbf{r}_i \times \mathbf{F}_i + \mathbf{r}_i \times \mathbf{f}_i = \mathbf{0}$$

Similar equations can be written for the other particles of the body, and adding them together vectorially, we obtain

$$\Sigma \mathbf{r}_i \times \mathbf{F}_i + \Sigma \mathbf{r}_i \times \mathbf{f}_i = \mathbf{0}$$

The second term is zero since, as stated above, the internal forces occur in equal but opposite collinear pairs, and by the transmissibility of a force, as discussed in Sec. 4.4, the moment of each pair of forces about point O is therefore zero. Hence, using the notation $\Sigma \mathbf{M}_O = \Sigma \mathbf{r}_i \times \mathbf{F}_i$, we can write the previous equation as

$$\Sigma \mathbf{M}_O = \mathbf{0}$$

Hence the *equations of equilibrium* for a rigid body can be summarized as follows:

$$\Sigma \mathbf{F} = \mathbf{0}$$
$$\Sigma \mathbf{M}_O = \mathbf{0} \qquad (5\text{–}1)$$

These equations require that a rigid body will be in equilibrium provided the sum of all the *external forces* acting on the body is equal to zero and the sum of the moments of the external forces about a point is equal to zero. The fact that these conditions are *necessary* for equilibrium has now been proven. They are also *sufficient* conditions. To show this, let us assume that the body is *not* in equilibrium, and yet the force system acting on it satisfies Eqs. 5–1. Suppose that an *additional force* $\mathbf{F}'$ is required to hold the body in equilibrium. As a result, the equilibrium equations become

$$\Sigma \mathbf{F} + \mathbf{F}' = \mathbf{0}$$
$$\Sigma \mathbf{M}_O + \mathbf{M}'_O = \mathbf{0}$$

where $\mathbf{M}'_O$ is the moment of $\mathbf{F}'$ about O. Since $\Sigma \mathbf{F} = \mathbf{0}$ and $\Sigma \mathbf{M}_O = \mathbf{0}$, then we require $\mathbf{F}' = \mathbf{0}$ (also $\mathbf{M}'_O = \mathbf{0}$). Consequently, the additional force $\mathbf{F}'$ is not required for holding the body, and indeed Eqs. 5–1 are also sufficient conditions for equilibrium.

Equilibrium in Two Dimensions

5.2 Free-Body Diagrams

Successful application of the equations of equilibrium requires a complete specification of *all* the known and unknown external forces that act *on* the body. The best way to account for these forces is to draw the body's free-body diagram. This diagram is a sketch of the outlined shape of the body, which represents it as being *isolated* or "free" from its surroundings, i.e., a "free body". On this sketch it is necessary to show *all* the forces and couple moments that the surroundings exert *on the body.* By using this diagram the effects of all the applied forces and couple moments acting on the body can be accounted for when the equations of equilibrium are applied. For this reason, *a thorough understanding of how to draw a free-body diagram is of primary importance for solving problems in mechanics.*

Table 5–1 **Supports for Rigid Bodies Subjected to Two-Dimensional Force Systems**

Types of Connection	*Reaction*	*Number of Unknowns*
(1) 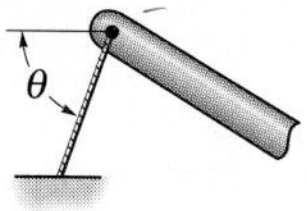 cable	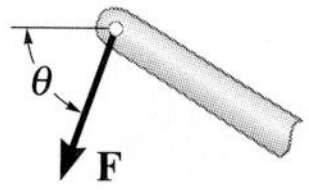	One unknown. The reaction is a tension force which acts away from the member in the direction of the cable.
(2) 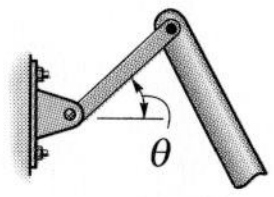 weightless link	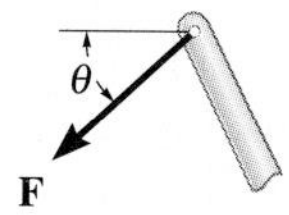 or 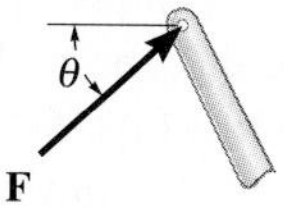	One unknown. The reaction is a force which acts along the axis of the link.
(3) 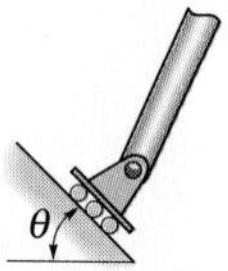 roller	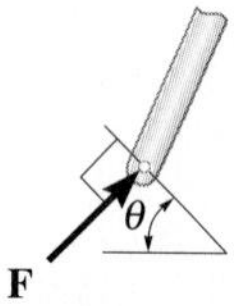	One unknown. The reaction is a force which acts perpendicular to the surface at the point of contact.
(4) 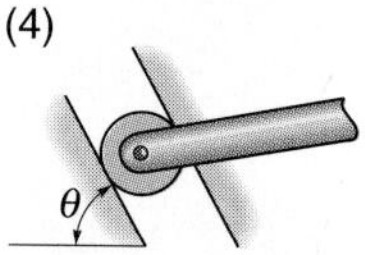 roller or pin in confined smooth slot	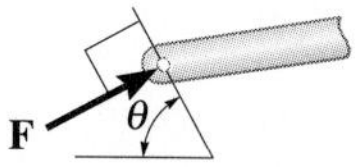 or 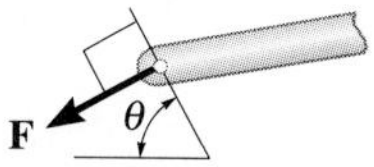	One unknown. The reaction is a force which acts perpendicular to the slot.
(5) 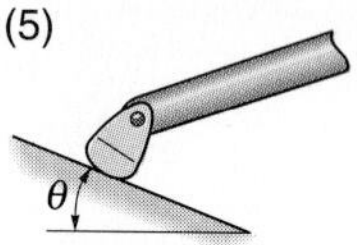 rocker	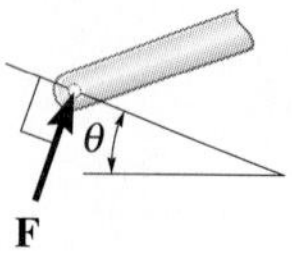	One unknown. The reaction is a force which acts perpendicular to the surface at the point of contact.
(6) 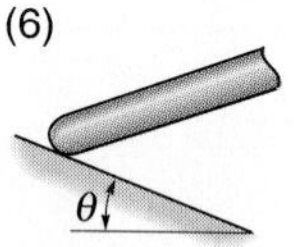 smooth contacting surface	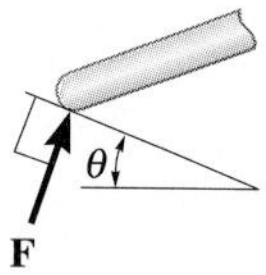	One unknown. The reaction is a force which acts perpendicular to the surface at the point of contact.
(7) 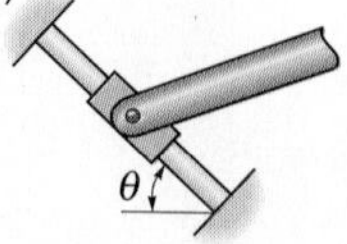 member pin connected to collar on smooth rod	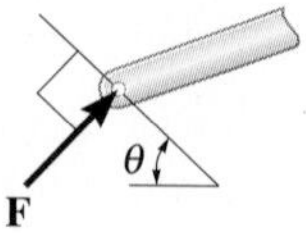 or 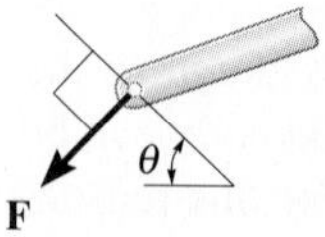	One unknown. The reaction is a force which acts perpendicular to the rod.

Table 5–1 (Contd.)

Types of Connection	*Reaction*	*Number of Unknowns*
(8) θ smooth pin or hinge	$\mathbf{F}_y$, $\mathbf{F}_x$ or $\mathbf{F}$, ϕ, $\mathbf{F}_x$	Two unknowns. The reactions are two components of force, or the magnitude and direction ϕ of the resultant force. Note that ϕ and θ are not necessarily equal [usually not, unless the rod shown is a link as in (2)].
(9) member fixed connected to collar on smooth rod	$\mathbf{F}$, $\mathbf{M}$	Two unknowns. The reactions are the couple moment and the force which acts perpendicular to the rod.
(10) fixed support	$\mathbf{F}_y$, $\mathbf{F}_x$, $\mathbf{M}$ or $\mathbf{F}$, ϕ, $\mathbf{M}$	Three unknowns. The reactions are the couple moment and the two force components, or the couple moment and the magnitude and direction ϕ of the resultant force.

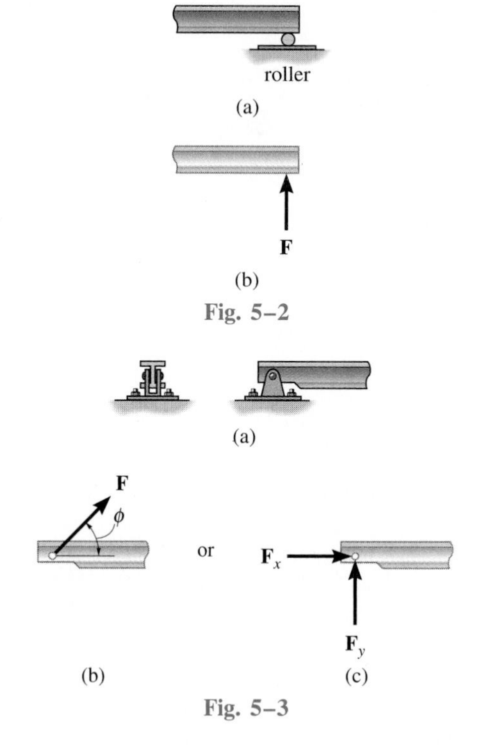

Fig. 5–2

Fig. 5–3

Support Reactions.

Before presenting a formal procedure as to how to draw a free-body diagram, we will first consider the various types of reactions that occur at supports and points of support between bodies subjected to coplanar force systems. *As a general rule, if a support prevents the translation of a body in a given direction, then a force is developed on the body in that direction. Likewise, if rotation is prevented, a couple moment is exerted on the body.*

For example, let us consider three ways in which a horizontal member, such as a beam, is commonly supported at its end. The first method of support consists of a *roller* or cylinder, Fig. 5–2*a*. Since this type of support only prevents the beam from translating in the vertical direction, the roller can only exert a force on the beam in this direction, Fig. 5–2*b*.

The beam can be supported in a more restrictive manner by using a *pin* as shown in Fig. 5–3*a*. The pin passes through holes in the beam and two leaves which are fixed to the ground. Here the pin will prevent translation of the beam in *any direction* ϕ, Fig. 5–3*b*, and so the pin must exert a force **F** on the beam in this direction. For purposes of analysis, it is generally easier to represent this effect by its two components $\mathbf{F}_x$ and $\mathbf{F}_y$, Fig. 5–3*c*. If F_x and F_y are known, then F and ϕ can be calculated.

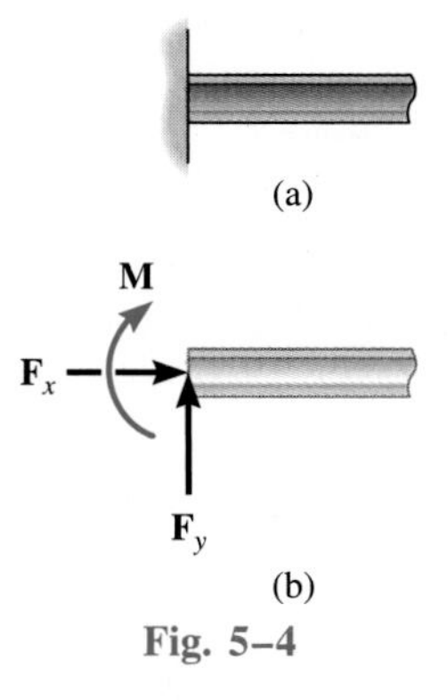

Fig. 5–4

The most restrictive way to support the beam would be to use a *fixed support* as shown in Fig. 5–4*a*. This support will prevent both translation and rotation of the beam, and so to do this a force and couple moment must be developed on the beam at its point of connection, Fig. 5–4*b*. As in the case of the pin, the force is usually represented by its components $\mathbf{F}_x$ and $\mathbf{F}_y$.

Table 5–1 lists other common types of supports for bodies subjected to coplanar force systems. (In all cases the angle θ is assumed to be known.) Carefully study each of the symbols used to represent these supports and the types of reactions they exert on their contacting members. Although concentrated forces and couple moments are shown in this table, they actually represent the *resultants* of *distributed surface loads* that exist between each support and its contacting member. It is these *resultants* which will be determined from the equations of equilibrium. It is generally not important to determine the actual distribution of the load, since the surface area over which it acts is considerably *smaller* than the *total surface area* of the connected member.

External and Internal Forces. Since a rigid body is a composition of particles, both *external* and *internal* loadings may act on it. It is important to realize, however, that if the free-body diagram for the body is drawn, the forces that are *internal* to the body are *not represented* on the free-body diagram. As discussed in Sec. 5.1, these forces always occur in equal but opposite collinear pairs, and therefore their *net effect* on the body is zero.

In some problems, a free-body diagram for a ''system'' of connected bodies may be used for an analysis. An example would be the free-body diagram of an entire automobile (system) composed of its many parts. Obviously, the connecting forces between its parts would represent *internal forces* which would *not* be included on the free-body diagram of the automobile. To summarize, then, internal forces act between particles which are located *within* a specified system which is contained within the boundary of the free-body diagram. Particles or bodies outside this boundary exert external forces on the system, and these alone must be shown on the free-body diagram.

Weight and the Center of Gravity. When a body is subjected to a gravitational field, each of its particles has a specified weight as defined by Newton's law of gravitation, $F = Gm_1m_2/r^2$, Eq. 1–2. If we assume the size of the body to be ''small'' in relation to the size of the earth, then it is appropriate to consider these gravitational forces to be represented as a *system of parallel forces* acting on the particles contained within the boundary of the body. It was shown in Sec. 4.9 that such a system can be reduced to a single resultant force acting through a specified point. We refer to this force resultant as the *weight* $\mathbf{W}$ of the body, and to the location of its point of application as the *center of gravity* G. The methods used for its calculation will be developed in Chapter 9.

In the examples and problems that follow, if the weight of the body is important for the analysis, this force will then be reported in the problem statement. Also, when the body is *uniform* or made of homogeneous material, the center of gravity will be located at the body's *geometric center* or *centroid;* however, if the body is nonhomogeneous or has an unusual shape, then its center of gravity will be given.

PROCEDURE FOR DRAWING A FREE-BODY DIAGRAM

To construct a free-body diagram for a rigid body or group of bodies considered as a single system, the following steps should be performed:

Step 1. Imagine the body to be *isolated* or cut "free" from its constraints and connections, and draw (sketch) its outlined shape.

Step 2. Identify all the external forces and couple moments that act on the body. Those generally encountered are due to (1) applied loadings, (2) reactions occurring at the supports or at points of contact with other bodies (see Table 5–1), and (3) the weight of the body. To account for all these effects, it may help to trace over the boundary, carefully noting each force or couple moment acting on it.

Step 3. Indicate the dimensions of the body necessary for computing the moments of forces. The forces and couple moments that are known should be labeled with their proper magnitudes and directions. Letters are used to represent the magnitudes and direction angles of forces and couple moments that are *unknown.* Establish an x, y coordinate system so that these unknowns, A_x, B_y, etc., can be identified. In particular, if a force or couple moment has a known line of action but unknown magnitude, the arrowhead which defines the sense of the vector can be assumed. The correctness of the assumed sense will become apparent after solving the equilibrium equations for the unknown magnitude. By definition, the *magnitude* of a vector is *always positive,* so that if the solution yields a "negative" scalar, the *minus sign* indicates that the vector's sense is *opposite* to that which was originally assumed.

Before proceeding, review this section; then carefully study the following examples. Afterward, attempt to draw the free-body diagrams for the objects in Figs. 5–5 through 5–9 without "looking" at the solutions. Further practice in drawing free-body diagrams should be gained by solving *all* the problems given at the end of this section.

Example 5–1

Draw the free-body diagram of the uniform beam shown in Fig. 5–5*a*. The beam has a mass of 100 kg.

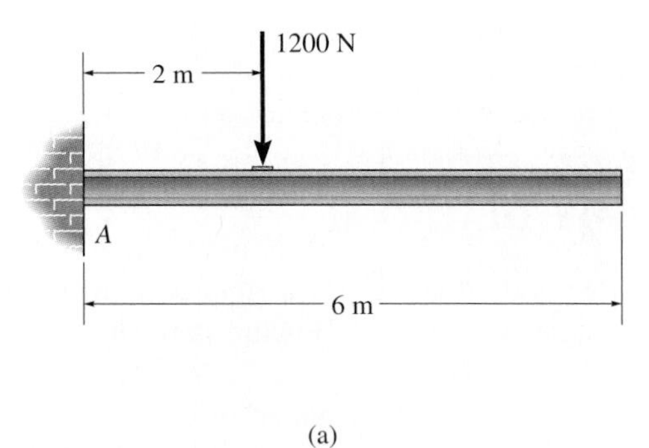

(a)

SOLUTION

The free-body diagram of the beam is shown in Fig. 5–5*b*. Since the support at *A* is a fixed wall, there are three reactions acting *on the beam* at *A*, denoted as $\mathbf{A}_x$, $\mathbf{A}_y$, and $\mathbf{M}_A$. The magnitudes of these vectors are *unknown,* and their sense has been *assumed.* (How does one obtain the *correct* sense of these vectors?) The weight of the beam, $W = 100(9.81) = 981$ N, acts through the beam's center of gravity *G*, 3 m from *A* since the beam is uniform.

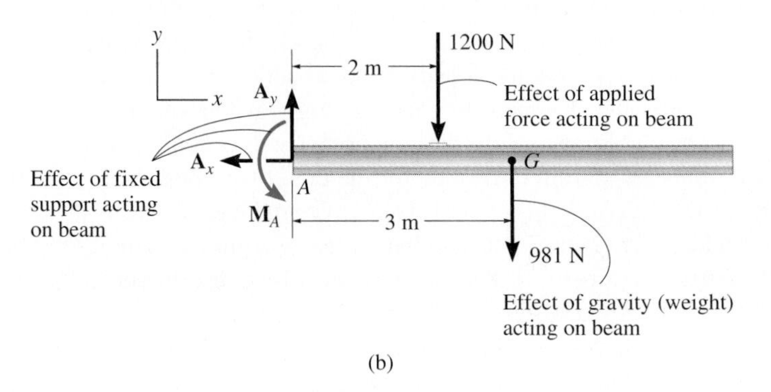

(b)

Fig. 5–5

Example 5–2

Draw the free-body diagram for the bell crank *ABC* shown in Fig. 5–6*a*.

SOLUTION

The free-body diagram is shown in Fig. 5–6*b*. The pin support at *B* exerts force components $\mathbf{B}_x$ and $\mathbf{B}_y$ *on the bell crank,* each having a known line of action but unknown magnitude. The link at *C* exerts a force $\mathbf{F}_C$ acting in the direction of the link and having an unknown magnitude. The dimensions of the crank are also labeled on the free-body diagram, since this information will be useful in computing the moments of the forces. As usual, the sense of the three unknown forces has been assumed. The correct sense will become apparent after solving the equilibrium equations.

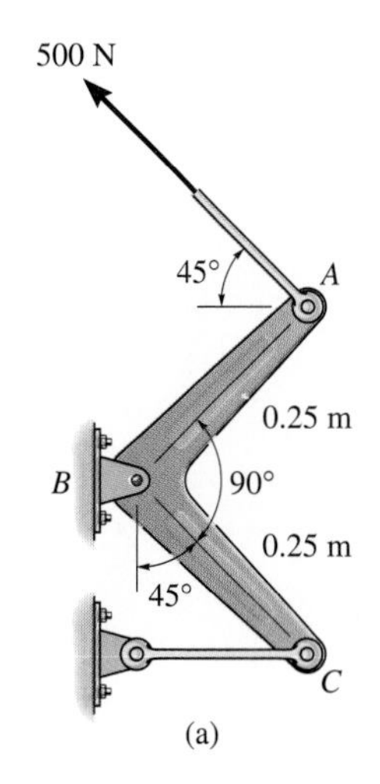

(a)

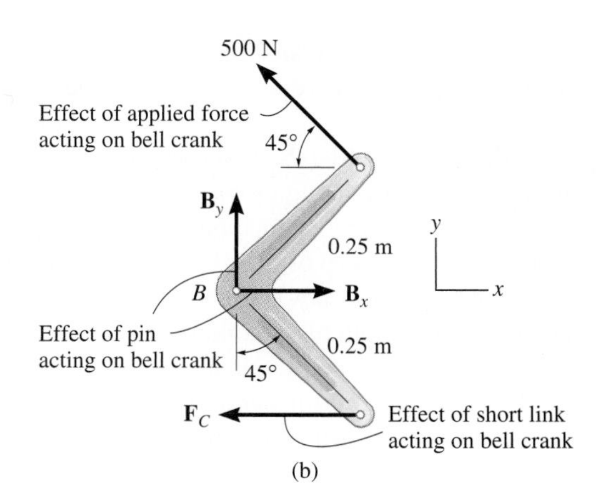

(b)

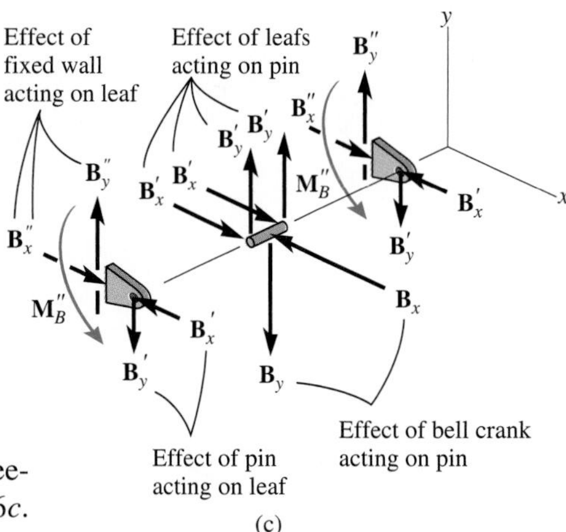

(c)

Fig. 5–6

Although not part of this problem, three-dimensional views of the free-body diagrams of the pin and two pin leaves at *B* are shown in Fig. 5–6*c*. Since the leaves are *fixed-connected* to the wall, there are three unknowns that the wall exerts on each leaf; namely, $\mathbf{B}''_x$, $\mathbf{B}''_y$, $\mathbf{M}''_B$. These reactions are shown to be equal in magnitude and direction on each leaf due to the symmetry of the loading and geometry. Note carefully how the principle of action—equal but opposite collinear reaction—is used when applying the forces $\mathbf{B}'_x$ and $\mathbf{B}'_y$ to each leaf and the pin. All of these unknowns can be obtained from the equations of equilibrium once $\mathbf{B}_x$ and $\mathbf{B}_y$ are obtained.

Example 5–3

Two smooth tubes A and B, each having a mass of 2 kg, rest between the inclined plates shown in Fig. 5–7*a*. Draw the free-body diagrams for tube A, tube B, and tubes A and B together.

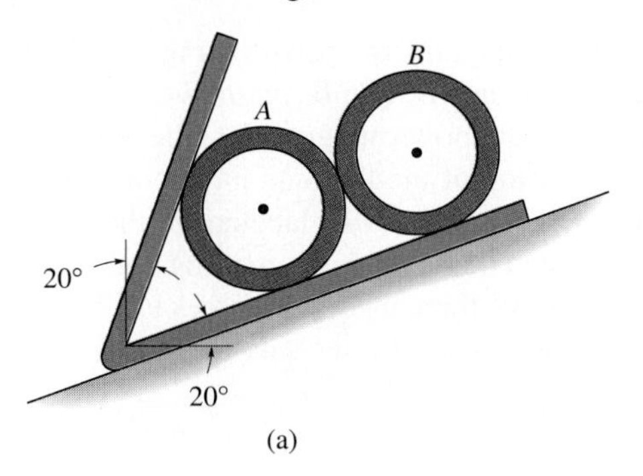

(a)

SOLUTION

The free-body diagram for tube A is shown in Fig. 5–7*b*. Its weight is $W = 2(9.81) = 19.62$ N. Since all contacting surfaces are *smooth,* the reactive forces **T, F, R** act in a direction *normal* to the tangent at their surfaces of contact.

The free-body diagram of tube B is shown in Fig. 5–7*c*. Can you identify each of the three forces acting *on the tube?* In particular, note that **R,** representing the force of tube A on tube B, Fig. 5–7*c*, is equal and opposite to **R** representing the force of tube B on tube A, Fig. 5–7*b*. This is a consequence of Newton's third law of motion.

The free-body diagram of both tubes combined (''system'') is shown in Fig. 5–7*d*. Here the contact force **R,** which acts between A and B, is considered as an *internal* force and hence is not shown on the free-body diagram. That is, it represents a pair of equal but opposite collinear forces which cancel each other.

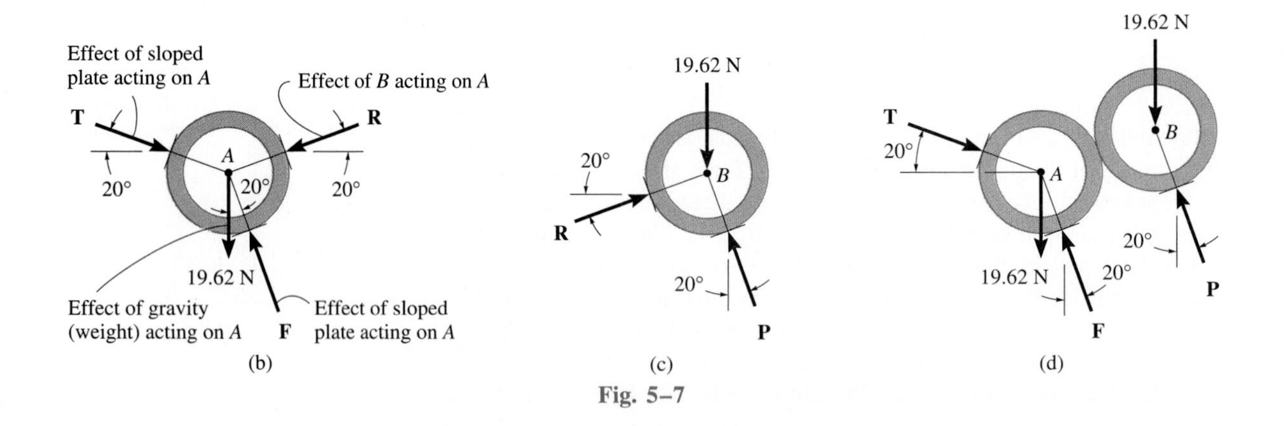

Fig. 5–7

Example 5–4

The free-body diagram of each object in Fig. 5–8 is drawn. Carefully study each solution and identify what each loading represents, as was done in Fig. 5–7*b*. The weight of the objects is neglected except where indicated.

SOLUTION

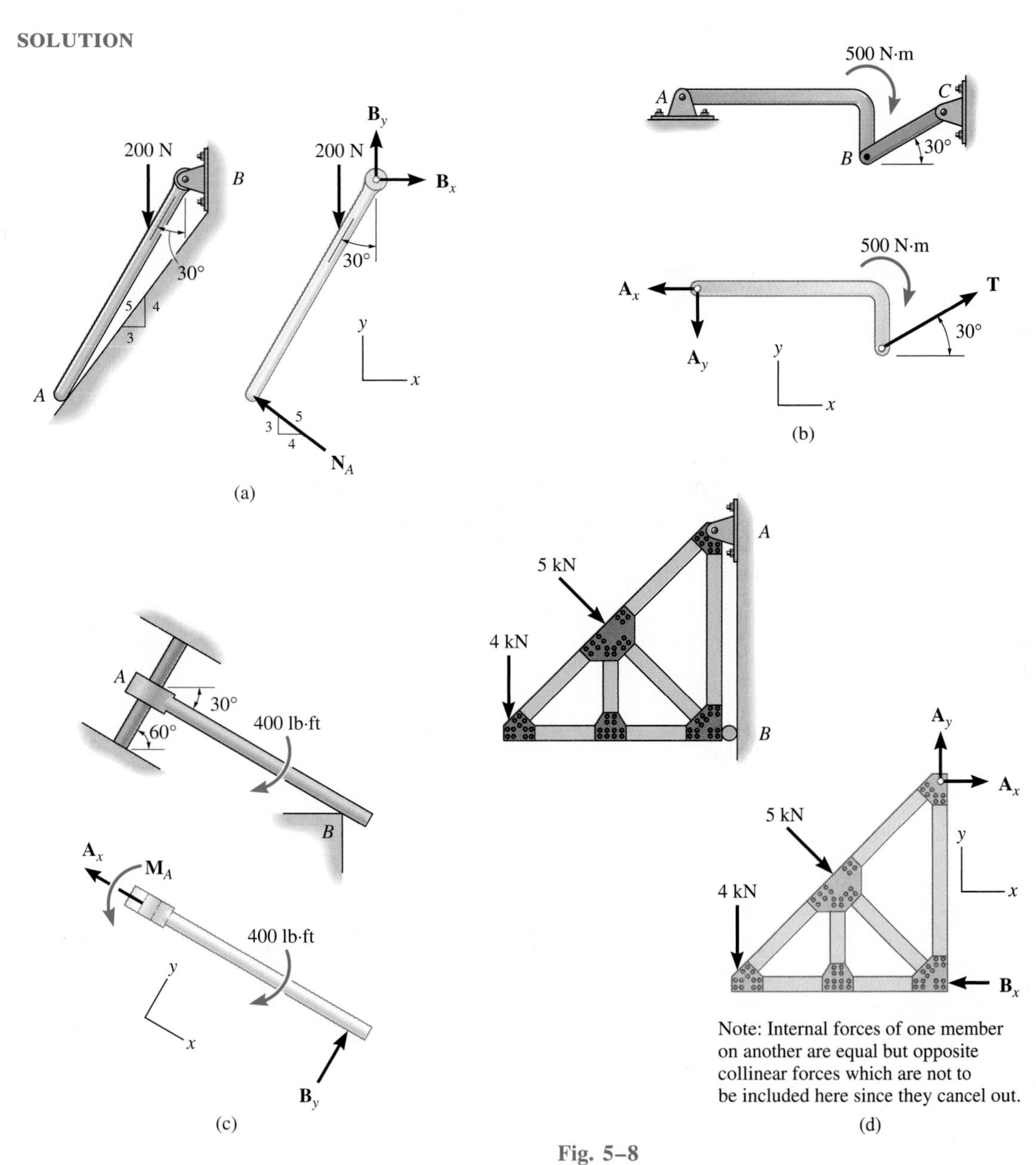

Fig. 5–8

Example 5–5

The highway sign shown in Fig. 5–9*a* has a mass of 100 kg with a center of gravity at *G*. It is supported by pins at *C* and *D* and a cable *AB*. Draw a free-body diagram of the sign and the supporting frame. Neglect the weight of the frame.

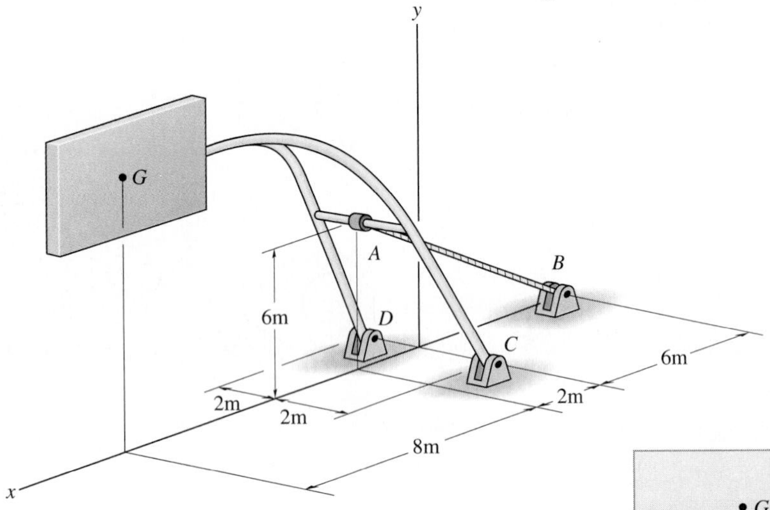

(a)

Fig. 5–9

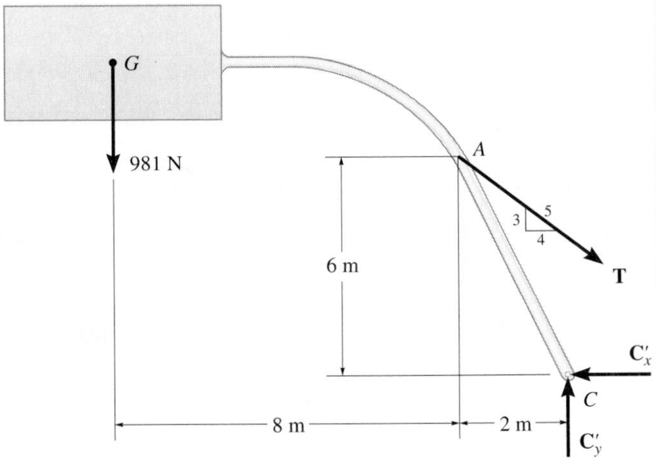

(b)

SOLUTION

By observation, the frame, sign, and the loading are all symmetrical about the vertical x–y plane, hence the problem may be analyzed using a system of *coplanar forces*. The free-body diagram is shown in Fig. 5–9*b*. Note that the force **T** that the cable exerts on the frame has a known line of action indicated by the 3–4–5 slope triangle. The force components $\mathbf{C}'_x$ and $\mathbf{C}'_y$ represent the horizontal and vertical reactions that *both* pins *C* and *D* exert on the frame. Consequently, after the solution for these reactions is obtained, *half* their magnitude is developed at *C* and half at *D*.

PROBLEMS

5–1. Draw the free-body diagram of the 50-kg uniform pipe, which is supported by the smooth contacts at A and B.

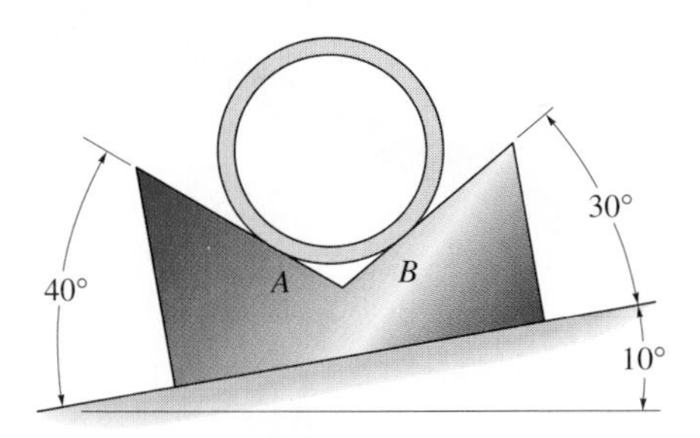

Prob. 5–1

5–2. Draw the free-body diagram of the hand punch, which is pinned at A and bears down on the smooth surface at B.

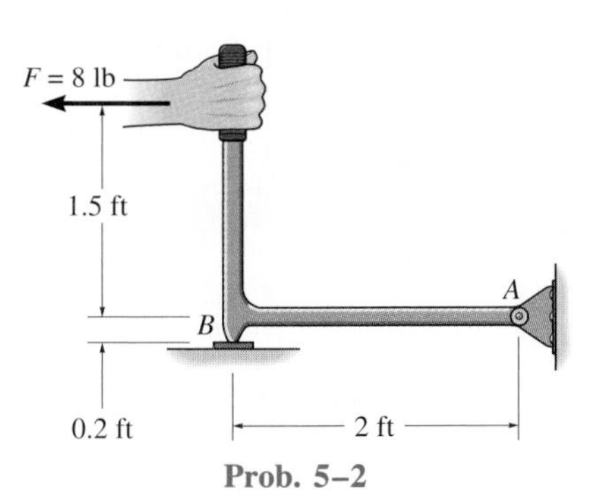

Prob. 5–2

5–3. Draw the free-body diagram of the smooth bar, which has points of contact at A, B, and C.

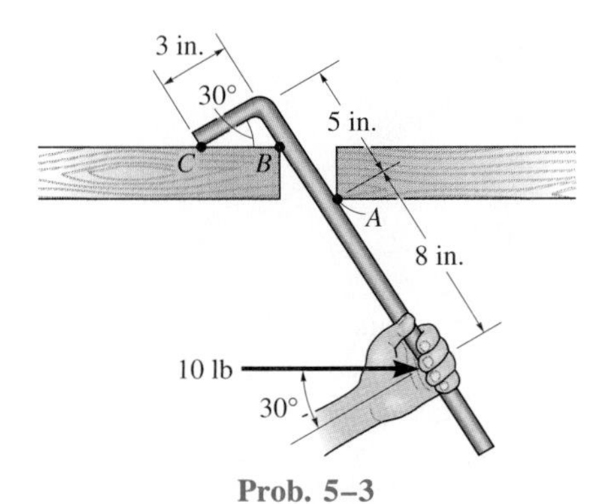

Prob. 5–3

***5–4.** Draw the free-body diagram of the jib crane AB, which is pin-connected at A and supported by member (link) BC.

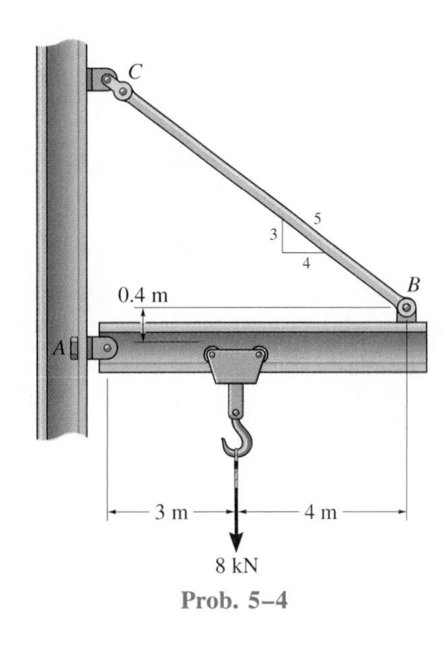

Prob. 5–4

5–5. Draw the free-body diagram of the dumpster D of the truck, which has a weight of 5000 lb and a center of gravity at G. It is supported by a pin at A and a pin-connected hydraulic cylinder BC (short link).

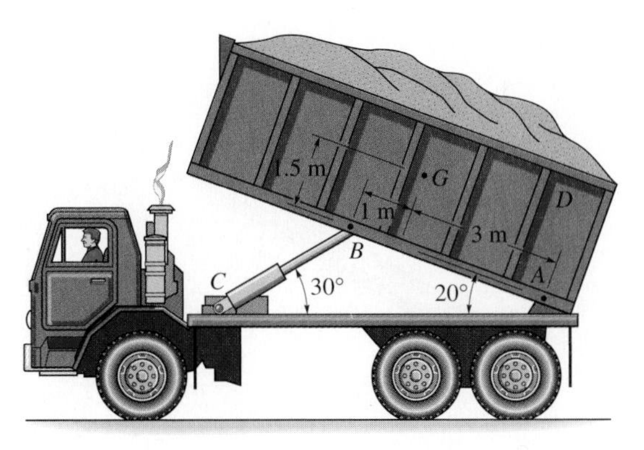

Prob. 5–5

5–6. Draw the free-body diagram of the link *CAB*, which is pin-connected at *A* and rests on the smooth cam at *B*.

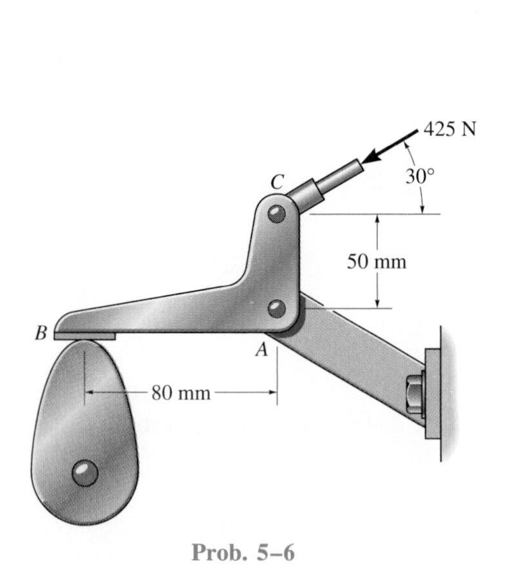

Prob. 5–6

5–7. Draw the free-body diagram of the uniform pipe, which has a mass of 100 kg and a center of mass at *G*. The supports *A*, *B*, and *C* are smooth.

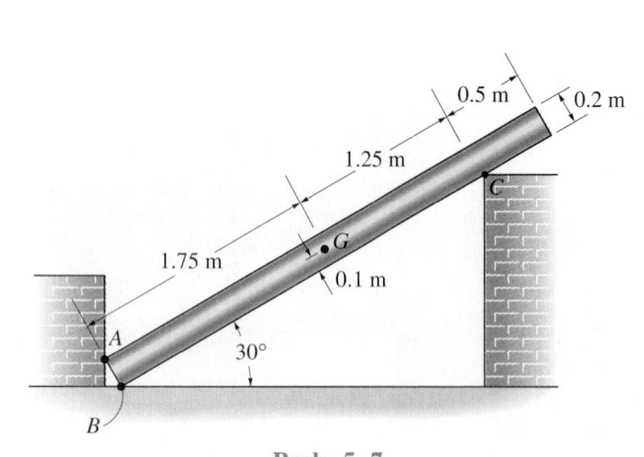

Prob. 5–7

***5–8.** Draw the free-body diagram of the beam, which is pin-connected at *A* and rocker-supported at *B*.

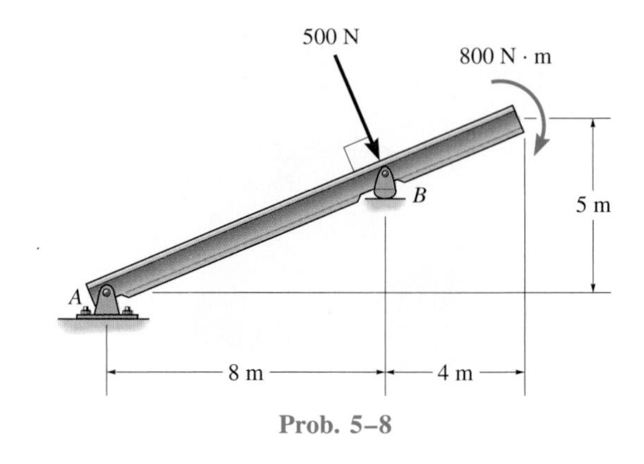

Prob. 5–8

5–9. Draw the free-body diagram of the beam, which is pin-supported at *A* and rests on the smooth incline at *B*.

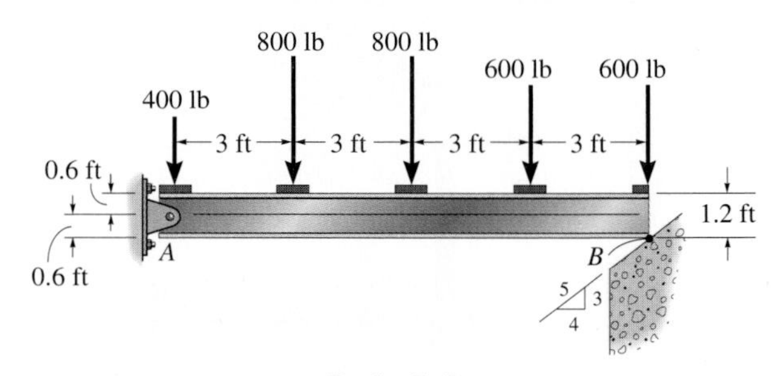

Prob. 5–9

5–10. Draw the free-body diagram of member *ABC*, which is supported by a pin at *A* and a horizontal short link *BD*.

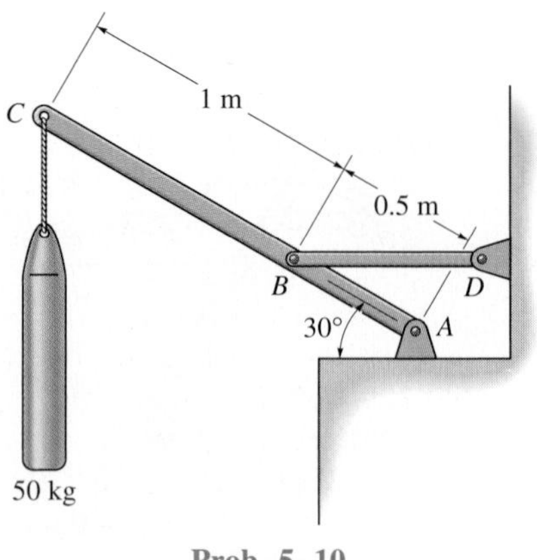

Prob. 5–10

5.3 Equations of Equilibrium

In Sec. 5.1 we developed the two equations which are both necessary and sufficient for the equilibrium of a rigid body, namely, $\Sigma \mathbf{F} = \mathbf{0}$ and $\Sigma \mathbf{M}_O = \mathbf{0}$. When the body is subjected to a system of forces, which all lie in the x–y plane, then the forces can be resolved into their x and y components. Consequently, the conditions for equilibrium in two dimensions are

$$\Sigma F_x = 0$$
$$\Sigma F_y = 0 \qquad (5\text{–}2)$$
$$\Sigma M_O = 0$$

Here ΣF_x and ΣF_y represent, respectively, the algebraic sums of the x and y components of all the forces acting on the body, and ΣM_O represents the algebraic sum of the couple moments and the moments of all the force components about an axis perpendicular to the x–y plane and passing through the arbitrary point O, which may lie either on or off the body.

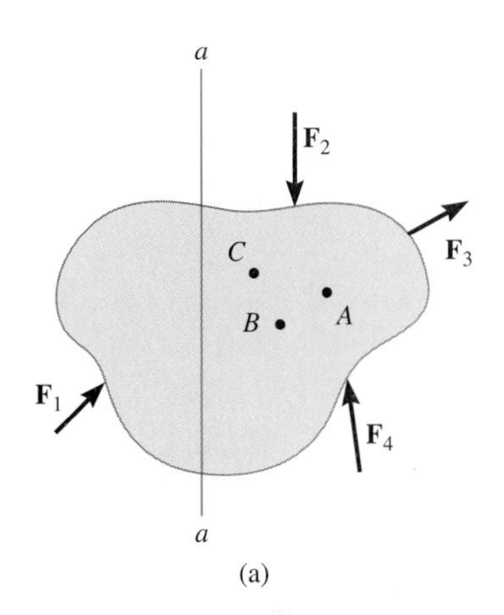

(a)

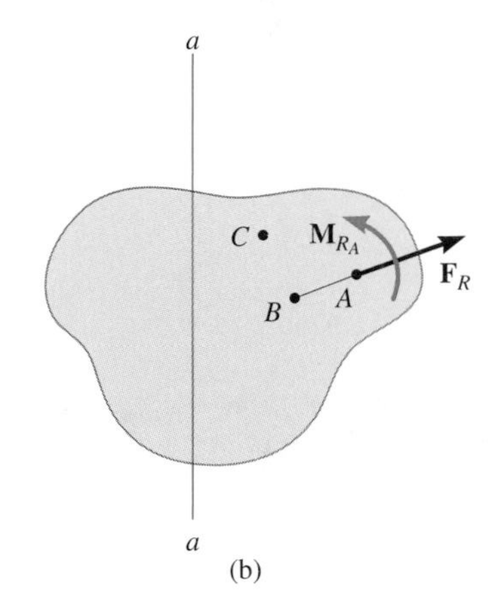

(b)

Alternative Sets of Equilibrium Equations. Although Eqs. 5–2 are *most often* used for solving equilibrium problems involving coplanar force systems, two *alternative* sets of three independent equilibrium equations may also be used. One such set is

$$\Sigma F_a = 0$$
$$\Sigma M_A = 0 \qquad (5\text{–}3)$$
$$\Sigma M_B = 0$$

When using these equations it is required that the moment points A and B do *not* lie on a line that is *perpendicular* to the a axis. To prove that Eqs. 5–3 provide the *conditions* for equilibrium, consider the free-body diagram of an arbitrarily shaped body shown in Fig. 5–10a. Using the methods of Sec. 4.8, the loading on the free-body diagram may be replaced by a single resultant force $\mathbf{F}_R = \Sigma \mathbf{F}$, acting at point A, and a resultant couple moment $\mathbf{M}_{R_A} = \Sigma \mathbf{M}_A$, Fig. 5–10$b$. If $\Sigma M_A = 0$ is satisfied, it is necessary that $\mathbf{M}_{R_A} = \mathbf{0}$. Furthermore, in order that $\mathbf{F}_R$ satisfy $\Sigma F_a = 0$, it must have *no component* along the a axis, and therefore its line of action must be perpendicular to the a axis, Fig. 5–10c. Finally, if it is required that $\Sigma M_B = 0$, where B does not lie on the line of action of $\mathbf{F}_R$, then $\mathbf{F}_R = \mathbf{0}$, and indeed the body shown in Fig. 5–10a must be in equilibrium.

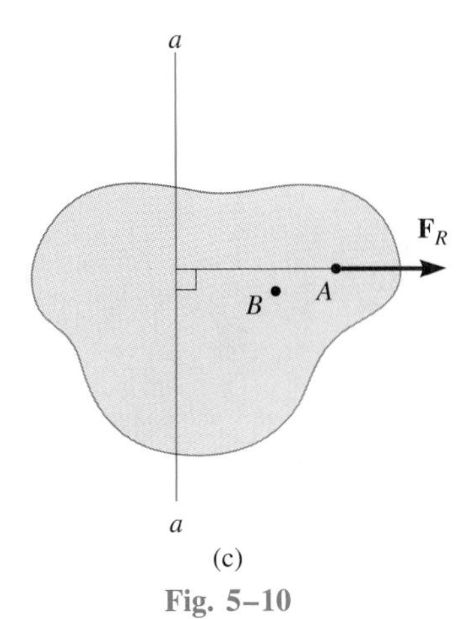

(c)

Fig. 5–10

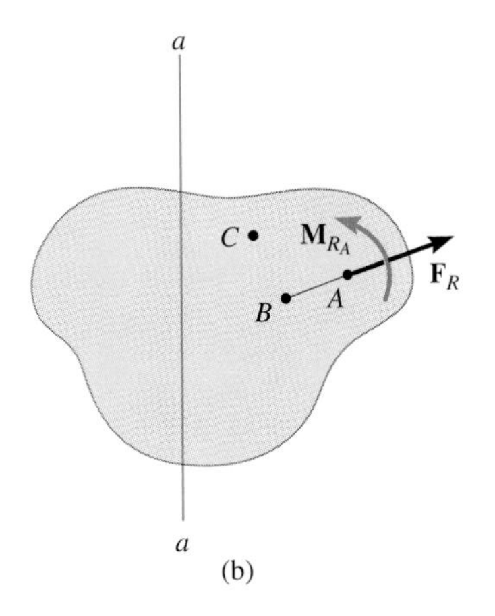

(b)

A second alternative set of equilibrium equations is

$$\Sigma M_A = 0$$
$$\Sigma M_B = 0 \qquad (5\text{–}4)$$
$$\Sigma M_C = 0$$

Here it is necessary that points A, B, and C do not lie on the same line. To prove that these equations when satisfied ensure equilibrium, consider again the free-body diagram in Fig. 5–10*b*. If $\Sigma M_A = 0$ is to be satisfied, then $\mathbf{M}_{R_A} = \mathbf{0}$. $\Sigma M_B = 0$ is satisfied if the line of action of $\mathbf{F}_R$ passes through point B as shown, and finally, if we require $\Sigma M_C = 0$, where C does not lie on line AB, Fig. 5–10*b*, it is necessary that $\mathbf{F}_R = \mathbf{0}$, and the body in Fig. 5–10*a* must then be in equilibrium.

PROCEDURE FOR ANALYSIS

The following procedure provides a method for solving coplanar force equilibrium problems:

Free-Body Diagram. Draw a free-body diagram of the body as discussed in Sec. 5.2. Briefly, this requires showing all the external forces and couple moments acting *on the body*. The magnitudes of these vectors must be labeled and their directions specified relative to an established set of x, y axes. Dimensions of the body, necessary for computing the moments of forces, are also included on the free-body diagram. Identify the unknowns. The sense of a force or couple moment having an *unknown* magnitude but known line of action can be *assumed.*

Equations of Equilibrium. Using the established x, y axes, apply the equations of equilibrium: $\Sigma F_x = 0$, $\Sigma F_y = 0$, $\Sigma M_O = 0$ (or the alternative sets of Eqs. 5–3 or 5–4). To *avoid* having to solve simultaneous equations, apply the moment equation $\Sigma M_O = 0$ about a point (O) *that lies at the intersection of the lines of action of two unknown forces.* In this way, the moments of these unknowns are *zero* about O, and one can obtain a *direct solution* for the third unknown. When applying the force equations $\Sigma F_x = 0$ and $\Sigma F_y = 0$, orient the x and y axes along lines that will provide the simplest resolution of the forces into their x and y components. If the solution of the equilibrium equations yields a *negative* scalar for an unknown force or couple moment, it indicates that the sense is *opposite* to that which was assumed on the free-body diagram.

The following example problems illustrate this procedure numerically.

Example 5–6

Determine the horizontal and vertical components of reaction for the beam loaded as shown in Fig. 5–11*a*. Neglect the weight of the beam in the calculations.

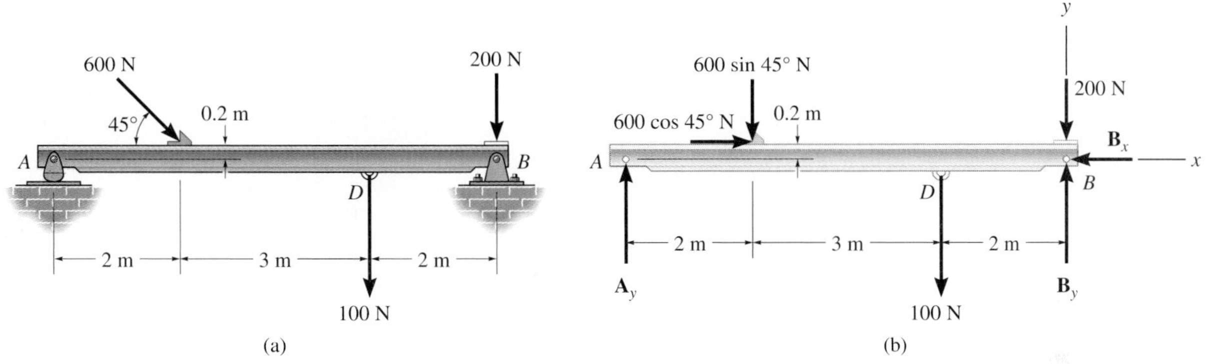

Fig. 5–11

SOLUTION

Free-Body Diagram. Can you identify each of the forces shown on the free-body diagram of the beam, Fig. 5–11*b*? For simplicity in applying the equilibrium equations, the 600-N force is represented by its x and y components as shown. Also, note that a 200-N force acts on the beam at B, and is independent of the force components $\mathbf{B}_x$ and $\mathbf{B}_y$ which represent the effect of the pin on the beam.

Equations of Equilibrium. Summing forces in the x direction yields

$$\overset{+}{\rightarrow}\Sigma F_x = 0; \qquad 600 \cos 45° \text{ N} - B_x = 0$$

$$B_x = 424 \text{ N} \qquad \textit{Ans.}$$

A direct solution for $\mathbf{A}_y$ can be obtained by applying the moment equation $\Sigma M_B = 0$ about point B. For the calculation, it should be apparent that forces 200 N, $\mathbf{B}_x$, and $\mathbf{B}_y$ all create zero moment about B. Assuming counterclockwise rotation about B to be positive (in the $+\mathbf{k}$ direction), Fig. 5–11*b*, we have

$$\circlearrowleft + \Sigma M_B = 0; \quad 100 \text{ N}(2 \text{ m}) + (600 \sin 45° \text{ N})(5 \text{ m}) - (600 \cos 45° \text{ N})(0.2 \text{ m}) - A_y(7 \text{ m}) = 0$$

$$A_y = 319 \text{ N} \qquad \textit{Ans.}$$

Summing forces in the y direction, using this result, gives

$$+\uparrow \Sigma F_y = 0; \quad 319 \text{ N} - 600 \sin 45° \text{ N} - 100 \text{ N} - 200 \text{ N} + B_y = 0$$

$$B_y = 405 \text{ N} \qquad \textit{Ans.}$$

Example 5–7

The cord shown in Fig. 5–12*a* supports a force of 100 lb and wraps over the frictionless pulley. Determine the tension in the cord at *C* and the horizontal and vertical components of reaction at pin *A*.

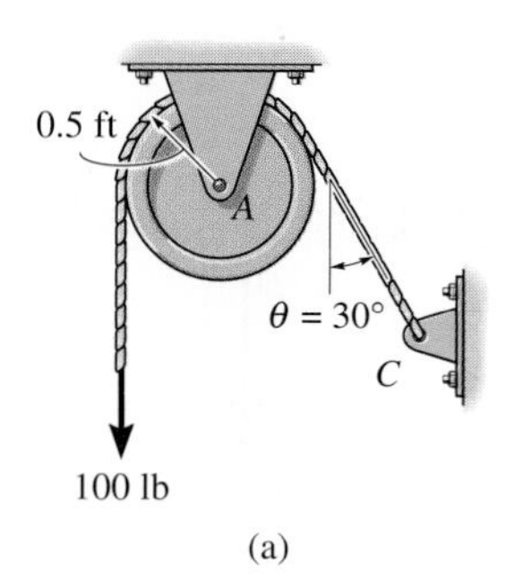

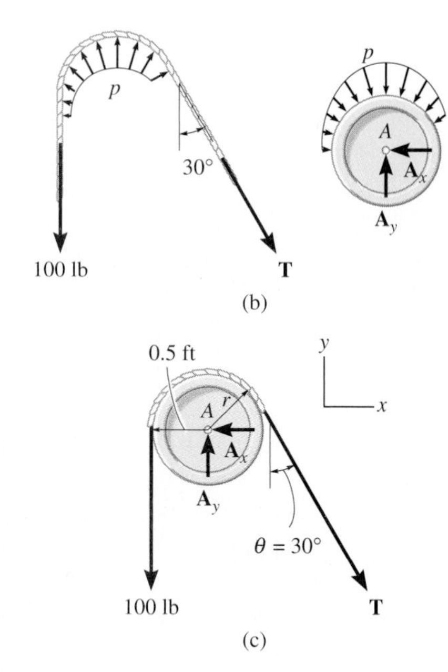

Fig. 5–12

SOLUTION

Free-Body Diagrams. The free-body diagrams of the cord and pulley are shown in Fig. 5–12*b*. Note that the principle of action, equal but opposite reaction, must be carefully observed when drawing each of these diagrams: the cord exerts an unknown load distribution p along part of the pulley's surface, whereas the pulley exerts an equal but opposite effect on the cord. For the solution, however, it is simpler to *combine* the free-body diagrams of the pulley and the contacting portion of the cord, so that the distributed load becomes *internal* to the system and is therefore eliminated from the analysis, Fig. 5–12*c*.

Equations of Equilibrium. Summing moments about point A to eliminate $\mathbf{A}_x$ and $\mathbf{A}_y$, Fig. 5–12*c*, we have

$$\circlearrowleft + \Sigma M_A = 0; \qquad 100\text{ lb}(0.5\text{ ft}) - T(0.5\text{ ft}) = 0$$

$$T = 100\text{ lb} \qquad \textit{Ans.}$$

It is seen that the tension remains *constant* as the cord passes over the pulley. (This of course is true for *any angle* θ at which the cord is directed and for *any radius* r of the pulley.) Using the result for T, a force summation is applied to determine the components of reaction at pin A.

$$\overset{+}{\rightarrow}\Sigma F_x = 0; \qquad -A_x + 100 \sin 30° \text{ lb} = 0$$

$$A_x = 50.0\text{ lb} \qquad \textit{Ans.}$$

$$+\uparrow \Sigma F_y = 0; \qquad A_y - 100\text{ lb} - 100 \cos 30° \text{ lb} = 0$$

$$A_y = 187\text{ lb} \qquad \textit{Ans.}$$

Example 5–8

The open-ended box wrench in Fig. 5–13*a* is used to tighten the bolt at *A*. If the wrench does not turn when the load is applied to the handle, determine the torque or moment applied to the bolt and the force of the wrench on the bolt.

SOLUTION

Free-Body Diagram. The free-body diagram for the wrench is shown in Fig. 5–13*b*. Since the bolt acts as a "fixed support," it exerts force components $\mathbf{A}_x$ and $\mathbf{A}_y$ and a torque $\mathbf{M}_A$ on the wrench at *A*.

Equations of Equilibrium

$$\overset{+}{\rightarrow}\Sigma F_x = 0; \qquad A_x - 52\left(\tfrac{5}{13}\right)\text{ N} + 30\cos 60^\circ\text{ N} = 0$$

$$A_x = 5.00\text{ N} \qquad \textit{Ans.}$$

$$+\uparrow \Sigma F_y = 0; \qquad A_y - 52\left(\tfrac{12}{13}\right)\text{ N} - 30\sin 60^\circ\text{ N} = 0$$

$$A_y = 74.0\text{ N} \qquad \textit{Ans.}$$

$$\circlearrowleft+\Sigma M_A = 0; \quad M_A - 52\left(\tfrac{12}{13}\right)\text{ N }(0.3\text{ m}) - (30\sin 60^\circ\text{ N})(0.7\text{ m}) = 0$$

$$M_A = 32.6\text{ N}\cdot\text{m} \qquad \textit{Ans.}$$

Point *A* was chosen for summing moments because the lines of action of the *unknown* forces $\mathbf{A}_x$ and $\mathbf{A}_y$ pass through this point, and therefore these forces were not included in the moment summation. Realize, however, that $\mathbf{M}_A$ must be *included* in this moment summation. This couple moment is a free vector and represents the twisting resistance of the bolt on the wrench. By Newton's third law, the wrench exerts an equal but opposite moment or torque on the bolt. Furthermore, the resultant force on the wrench or bolt is

$$F_A = \sqrt{(5.00)^2 + (74.0)^2} = 74.1\text{ N} \qquad \textit{Ans.}$$

Because the force components A_x and A_y were calculated as positive quantities, their directional sense is shown correctly on the free-body diagram in Fig. 5–13*b*. Hence

$$\theta = \tan^{-1}\frac{74.0\text{ N}}{5.00\text{ N}} = 86.1^\circ \quad \measuredangle$$

Realize that $\mathbf{F}_A$ acts in the opposite direction on the bolt. Why?

Although only *three* independent equilibrium equations can be written for a rigid body, it is a good practice to *check* the calculations using a fourth equilibrium equation. For example, the above computations may be verified in part by summing moments about point *C*:

$$\circlearrowleft+\Sigma M_C = 0; \quad 52\left(\tfrac{12}{13}\right)\text{ N }(0.4\text{ m}) + 32.6\text{ N}\cdot\text{m} - 74.0\text{ N}(0.7\text{ m}) = 0$$

$$19.2\text{ N}\cdot\text{m} + 32.6\text{ N}\cdot\text{m} - 51.8\text{ N}\cdot\text{m} = 0$$

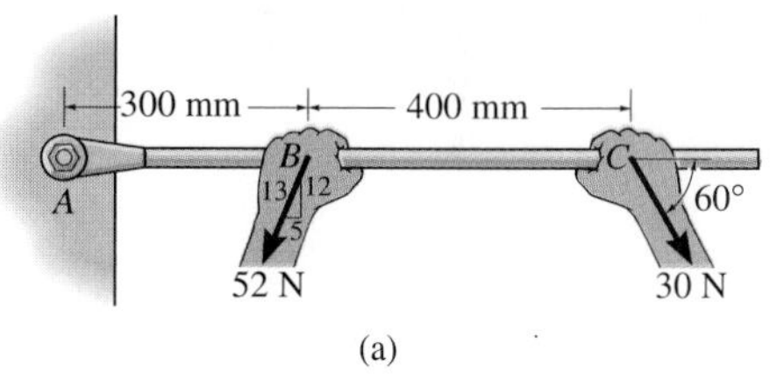

(a)

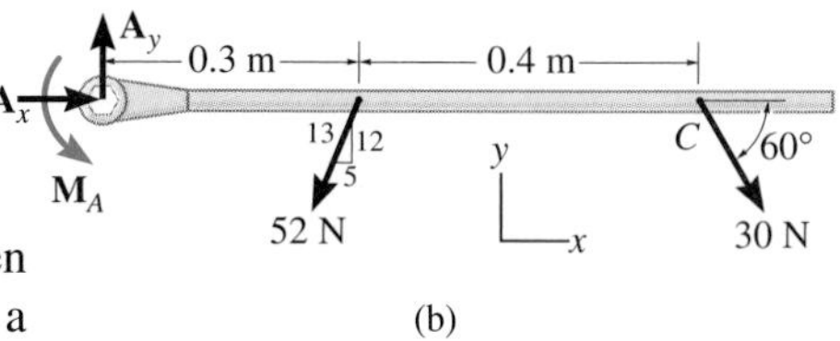

(b)

Fig. 5–13

Example 5-9

The uniform smooth rod shown in Fig. 5-14a is subjected to a force and couple moment. If the rod is supported at A by a smooth wall and at B and C either at the top or bottom by rollers, determine the reactions at these supports. Neglect the weight of the rod.

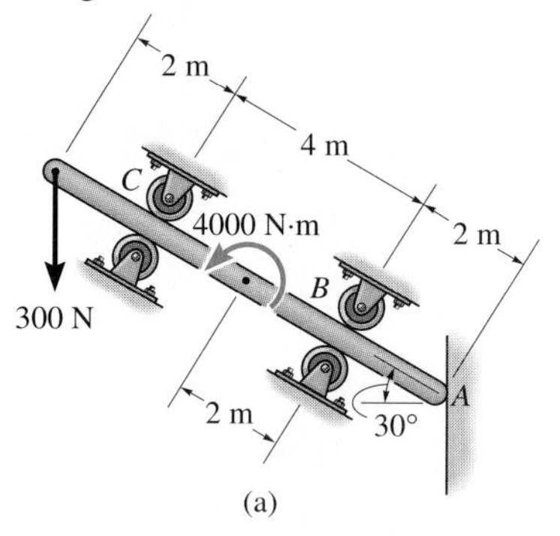

(a)

SOLUTION

Free-Body Diagram. As shown in Fig. 5-14b, all the support reactions act normal to the surface of contact since the contacting surfaces are smooth. The reactions at B and C are shown acting in the positive y' direction. This assumes that only the rollers located on the bottom of the rod are used for support.

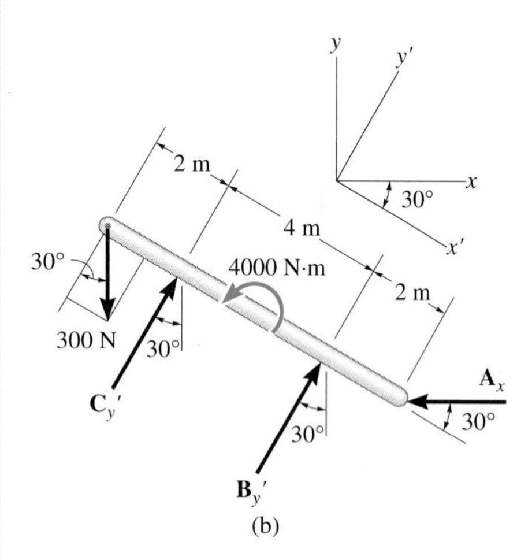

(b)

Fig. 5-14

Equations of Equilibrium. Using the x, y coordinate system in Fig. 5-14b, we have

$$\overset{+}{\rightarrow}\Sigma F_x = 0; \qquad C_{y'} \sin 30° + B_{y'} \sin 30° - A_x = 0 \qquad (1)$$

$$+\uparrow \Sigma F_y = 0; \qquad -300\text{ N} + C_{y'} \cos 30° + B_{y'} \cos 30° = 0 \qquad (2)$$

$$\circlearrowleft + \Sigma M_A = 0; \qquad -B_{y'}(2\text{ m}) + 4000\text{ N}\cdot\text{m} - C_{y'}(6\text{ m}) + (300 \cos 30°\text{ N})(8\text{ m}) = 0 \qquad (3)$$

When writing the moment equation, it should be noticed that the line of action of the force component 300 sin 30° N passes through point A, and therefore this force is not included in the moment equation.

Solving Eqs. 2 and 3 simultaneously, we obtain

$$B_{y'} = -1000.0\text{ N} \qquad \textit{Ans.}$$

$$C_{y'} = 1346.4\text{ N} \qquad \textit{Ans.}$$

Since $B_{y'}$ is a negative scalar, the sense of $\mathbf{B}_{y'}$ is opposite to that shown on the free-body diagram in Fig. 5-14b. Therefore, the top roller at B serves as the support rather than the bottom one. Retaining the negative sign for $B_{y'}$ (Why?) and substituting the results into Eq. 1, we obtain

$$1346.4 \sin 30°\text{ N} - 1000.0 \sin 30°\text{ N} - A_x = 0$$

$$A_x = 173.2\text{ N} \qquad \textit{Ans.}$$

Example 5–10

The link shown in Fig. 5–15*a* is pin-connected at *A* and rests against a smooth support at *B*. Compute the horizontal and vertical components of reaction at the pin *A*.

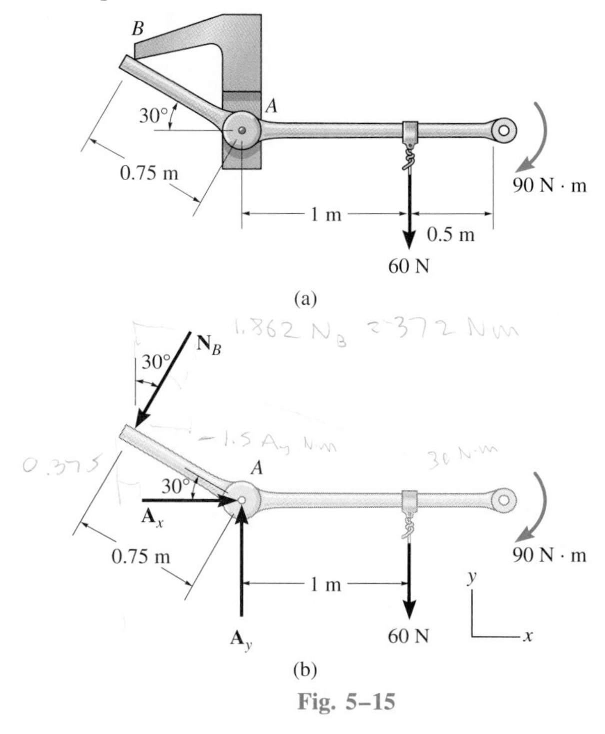

Fig. 5–15

SOLUTION

Free-Body Diagram. As shown in Fig. 5–15*b*, the reaction $\mathbf{N}_B$ is perpendicular to the link at *B*. Also, horizontal and vertical components of reaction are represented at *A*, even though the base of the pin support is tilted.

Equations of Equilibrium. Summing moments about *A*, we obtain a direct solution for N_B,

$$\circlearrowleft + \Sigma M_A = 0; \quad -90\ \text{N}\cdot\text{m} - 60\ \text{N}(1\ \text{m}) + N_B(0.75\ \text{m}) = 0$$

$$N_B = 200\ \text{N}$$

Using this result,

$$\overset{+}{\rightarrow}\Sigma F_x = 0; \qquad A_x - 200 \sin 30^\circ\ \text{N} = 0$$

$$A_x = 100\ \text{N} \qquad \textit{Ans.}$$

$$+\uparrow \Sigma F_y = 0; \qquad A_y - 200 \cos 30^\circ\ \text{N} - 60\ \text{N} = 0$$

$$A_y = 233\ \text{N} \qquad \textit{Ans.}$$

Example 5–11

A force of 150 lb acts on the end of the beam shown in Fig. 5–16*a*. Determine the magnitude and direction of the reaction at the pin *A* and the tension in the cable.

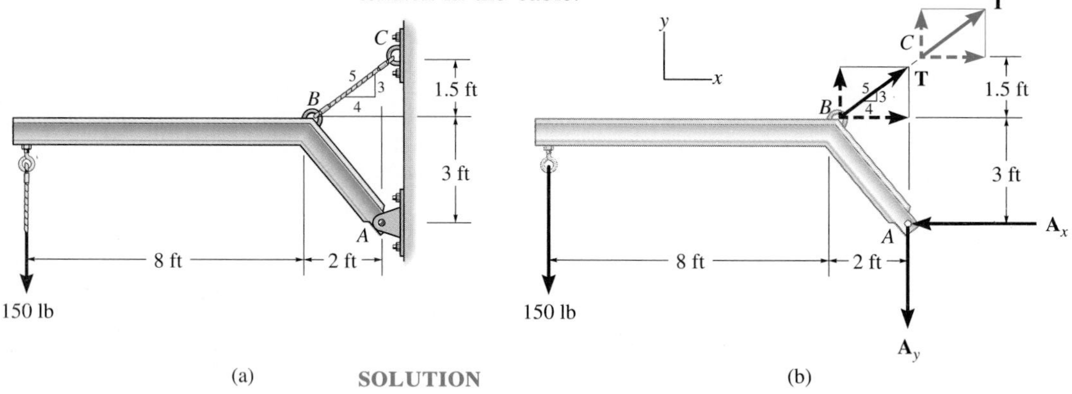

Fig. 5–16

SOLUTION

Free-Body Diagram. The forces acting on the beam are shown in Fig. 5–16*b*.

Equations of Equilibrium. Summing moments about point *A* to obtain a direct solution for the cable tension yields

$$\circlearrowleft + \Sigma M_A = 0; \quad -(\tfrac{3}{5}T)(2 \text{ ft}) - (\tfrac{4}{5}T)(3 \text{ ft}) + 150 \text{ lb}(10 \text{ ft}) = 0$$

$$-3.6T + 150 \text{ lb}(10 \text{ ft}) = 0 \qquad (1)$$

$$T = 416.7 \text{ lb} \qquad \textit{Ans.}$$

Using the principle of transmissibility it is also possible to locate **T** at *C*, even though this point is not on the beam, Fig. 5–16*b*. In this case, the vertical component of **T** creates *zero moment* about *A* and the moment arm of the horizontal component ($\tfrac{4}{5}T$) becomes 4.5 ft. Hence, $\Sigma M_A = 0$ yields Eq. 1 directly since $(\tfrac{4}{5}T)(4.5) = 3.6T$.

Summing forces to obtain A_x and A_y, using the result for T, we have

$$\overset{+}{\rightarrow}\Sigma F_x = 0; \qquad -A_x + (\tfrac{4}{5})(416.7 \text{ lb}) = 0$$

$$A_x = 333.3 \text{ lb} \leftarrow$$

$$+\uparrow \Sigma F_y = 0; \qquad (\tfrac{3}{5})416.7 \text{ lb} - 150 \text{ lb} - A_y = 0$$

$$A_y = 100 \text{ lb} \downarrow$$

Thus,

$$F_A = \sqrt{(333.3 \text{ lb})^2 + (100 \text{ lb})^2}$$

$$= 348.0 \text{ lb} \qquad \textit{Ans.}$$

$$\theta = \tan^{-1}\frac{100 \text{ lb}}{333.3 \text{ lb}} = 16.7^\circ \qquad \textit{Ans.}$$

Example 5–12

The oil-drilling rig shown in Fig. 5–17*a* has a mass of 24 Mg and mass center at *G*. If the rig is pin-connected at its base, determine the tension in the hoisting cable and the magnitude of the resultant force at *A* when the rig is in the position shown.

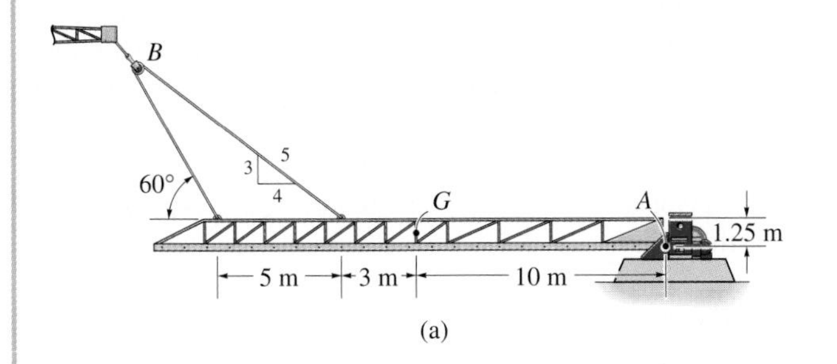

(a)

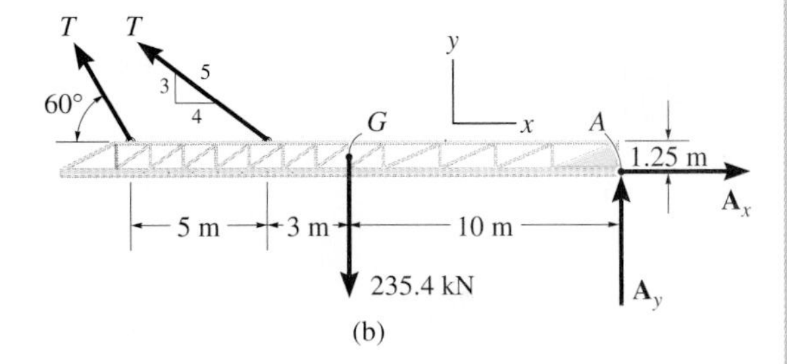

(b)

Fig. 5–17

SOLUTION

Free-Body Diagram. Because the hoisting cable is continuous and passes over the pulley, the cable is subjected to the same tension *T* throughout its length. Hence the cable exerts a force *T* on the rig at its points of attachment, Fig. 5–17*b*. The weight of the rig is $(24(10^3)\text{ kg})(9.81\text{ m/s}^2) = 235.4\text{ kN}$.

Equations of Equilibrium. By summing moments about point *A*, it is possible to obtain a direct solution for *T*. Why?

$$\circlearrowleft + \Sigma M_A = 0; \quad (235.4\text{ kN})(10\text{ m}) - T(\tfrac{3}{5})(13\text{ m}) + T(\tfrac{4}{5})(1.25\text{ m})$$
$$-(T\sin 60°)(18\text{ m}) + (T\cos 60°)(1.25\text{ m}) = 0$$
$$T = 108.2\text{ kN} \qquad \textit{Ans.}$$

$$\overset{+}{\rightarrow}\Sigma F_x = 0; \quad A_x - 108.2(\tfrac{4}{5})\text{ kN} - 108.2\cos 60°\text{ kN} = 0$$
$$A_x = 140.6\text{ kN}$$

$$+\uparrow \Sigma F_y = 0; \quad A_y - 235.4\text{ kN} + 108.2(\tfrac{3}{5})\text{ kN} + 108.2\sin 60°\text{ kN} = 0$$
$$A_y = 76.8\text{ kN}$$

Thus,

$$F_A = \sqrt{(140.6)^2 + (76.8)^2} = 160\text{ kN} \qquad \textit{Ans.}$$

5.4 Two- and Three-Force Members

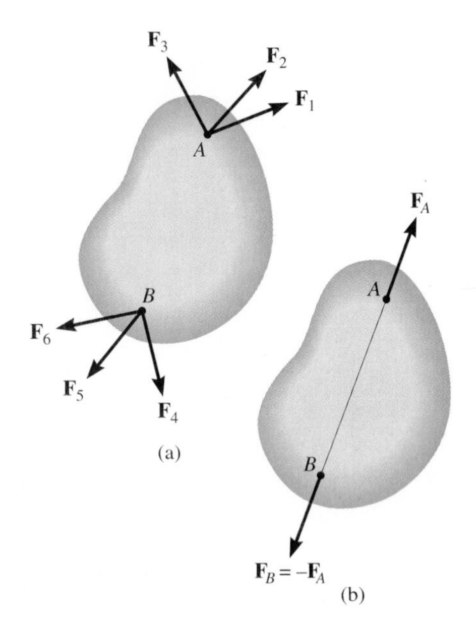

Two-force member
Fig. 5–18

The solution to some equilibrium problems can be simplified if one is able to recognize members that are subjected to only two or three forces.

Two-Force Members. When a member is subject to *no couple moments* and forces are applied at only two points on a member, the member is called a *two-force member.* An example of this situation is shown in Fig. 5–18*a*. The forces at A and B are first summed to obtain their respective *resultants* $\mathbf{F}_A$ and $\mathbf{F}_B$, Fig. 5–18*b*. These two forces will maintain *translational or force equilibrium* ($\Sigma\mathbf{F} = \mathbf{0}$) provided $\mathbf{F}_A$ is of equal magnitude and opposite direction to $\mathbf{F}_B$. Furthermore, *rotational or moment equilibrium* ($\Sigma\mathbf{M}_O = \mathbf{0}$) is satisfied if $\mathbf{F}_A$ is *collinear* with $\mathbf{F}_B$. As a result, the line of action of both forces is known, since it always passes through A and B. Hence, only the force magnitude must be determined or stated. Other examples of two-force members held in equilibrium are shown in Fig. 5–19.

Three-Force Members. If a member is subjected to only three forces, then it is necessary that the forces be either *concurrent* or *parallel* if the member is to be in equilibrium. To show the concurrency requirement, consider the body in Fig. 5–20*a* and suppose that any two of the three forces acting on the body have lines of action that intersect at point O. To satisfy moment equilibrium about O, i.e., $\Sigma M_O = 0$, the third force must also pass through O, which then makes the force system *concurrent.* If two of the three forces are parallel, Fig. 5–20*b*, the point of concurrency, O, is considered to be at ''infinity'' and the third force must be parallel to the other two forces to intersect at this ''point.''

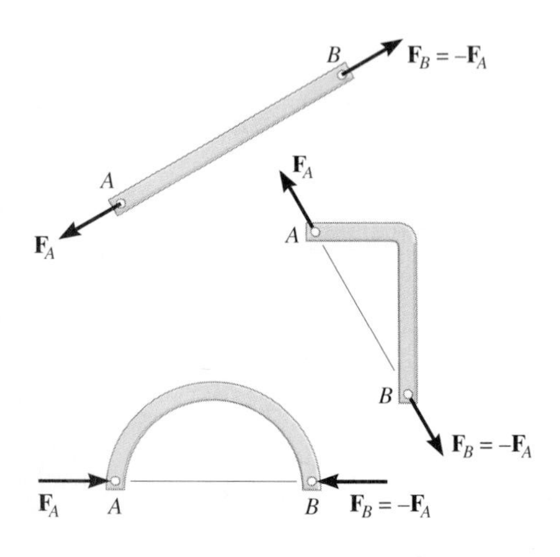

Two-force members
Fig. 5–19

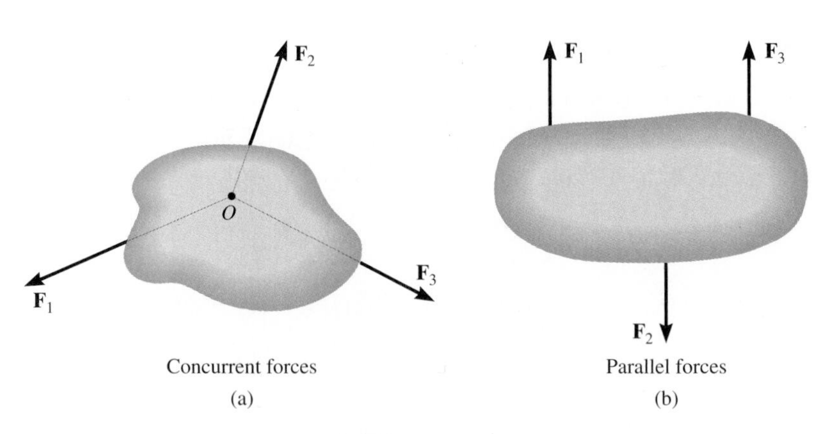

Three-force members
Fig. 5–20

Example 5–13

The lever ABC is pin-supported at A and connected to a short link BD as shown in Fig. 5–21*a*. If the weight of the members is negligible, determine the force of the pin on the lever at A.

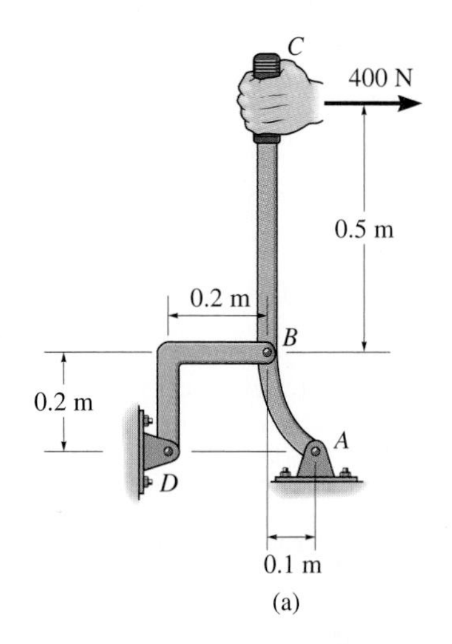

(a)

SOLUTION

Free-Body Diagrams. As shown by the free-body diagram, Fig. 5–21*b*, the short link BD is a *two-force member,* so the *resultant forces* at pins D and B must be equal, opposite, and collinear. Although the magnitude of the force is unknown, the line of action is known, since it passes through B and D.

Lever ABC is a *three-force member,* and therefore, in order to satisfy moment equilibrium, the three nonparallel forces acting on it must be concurrent at O, Fig. 5–21*c*. In particular, note that the force $\mathbf{F}$ on the lever is equal but opposite to $\mathbf{F}$ acting at B on the link. Why? The distance CO must be 0.5 m, since the lines of action of $\mathbf{F}$ and the 400-N force are known.

Equations of Equilibrium. By requiring the force system to be concurrent at O, so that $\Sigma M_O = 0$, the angle θ which defines the line of action of $\mathbf{F}_A$ can be determined from trigonometry,

$$\theta = \tan^{-1}\left(\frac{0.7}{0.4}\right) = 60.3^\circ \quad \measuredangle^{\theta} \qquad \textit{Ans.}$$

Using the x, y axes and applying the force equilibrium equations, we can obtain F and F_A.

$$\overset{+}{\rightarrow}\Sigma F_x = 0; \qquad F_A \cos 60.3^\circ - F \cos 45^\circ + 400 \text{ N} = 0$$

$$+\uparrow \Sigma F_y = 0; \qquad F_A \sin 60.3^\circ - F \sin 45^\circ = 0$$

Solving, we get

$$F_A = 1075 \text{ N} \qquad \textit{Ans.}$$

$$F = 1320 \text{ N}$$

Note: We can also solve this problem by representing the force at A by its two components $\mathbf{A}_x$ and $\mathbf{A}_y$ and applying $\Sigma M_A = 0$, $\Sigma F_x = 0$, $\Sigma F_y = 0$ to the lever. Once A_x and A_y are determined, how would you find F_A and θ?

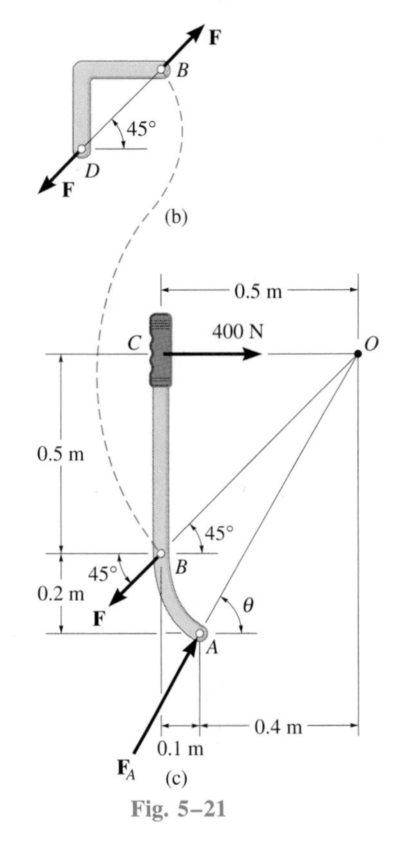

(c)

Fig. 5–21

PROBLEMS

5–11. Determine the magnitude of the resultant force acting at pin A of the hand punch in Prob. 5–2.

***5–12.** Determine the reactions at the supports A and B of the jib crane in Prob. 5–4.

5–13. Determine the reactions at the supports for the beam in Prob. 5–8.

5–14. Determine the reactions at the pin A and the force on the hydraulic cylinder of the truck dumpster in Prob. 5–5.

5–15. Determine the support reactions on the beam in Prob. 5–9.

***5–16.** Determine the reactions on the uniform pipe at A, B, and C in Prob. 5–7.

5–17. Determine the reactions at the points of contact at A, B, and C of the bar in Prob. 5–3.

5–18. Determine the magnitude of the reactions on the beam at A and B. Neglect the thickness of the beam.

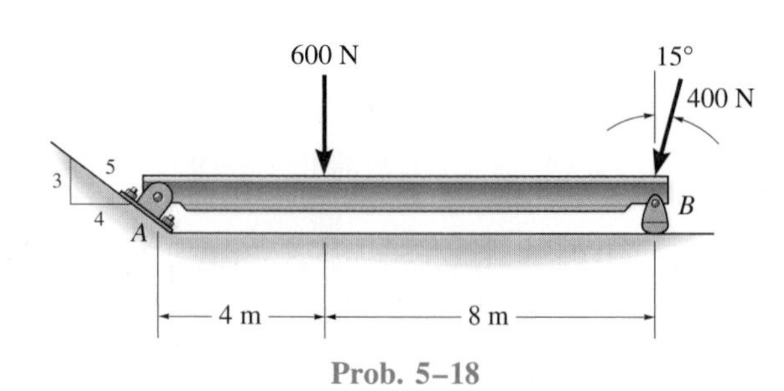

Prob. 5–18

5–19. The forces acting on the plane while it is flying at constant velocity are shown. If the engine thrust is $F_T = 110$ kip and the plane's weight is $W = 170$ kip, determine the atmospheric drag $\mathbf{F}_D$ and the wing lift $\mathbf{F}_L$. Also, determine the distance s to the line of action of the drag force.

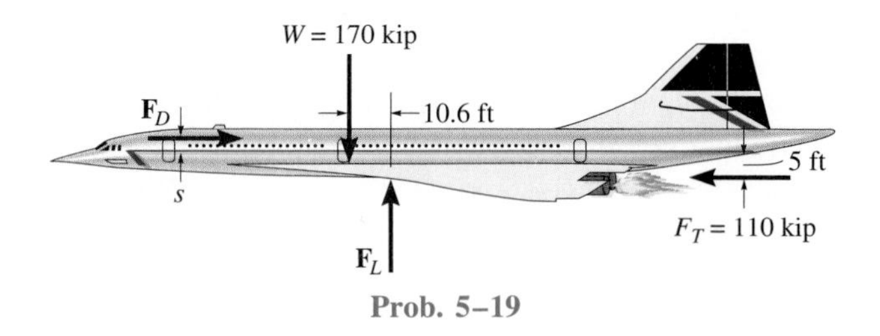

Prob. 5–19

***5–20.** When holding the 5-lb stone in equilibrium, the humerus H, assumed to be smooth, exerts normal forces $\mathbf{F}_C$ and $\mathbf{F}_A$ on the radius C and ulna A as shown. Determine these forces and the force $\mathbf{F}_B$ that the biceps B exerts on the radius for equilibrium. The stone has a center of mass at G. Neglect the weight of the arm.

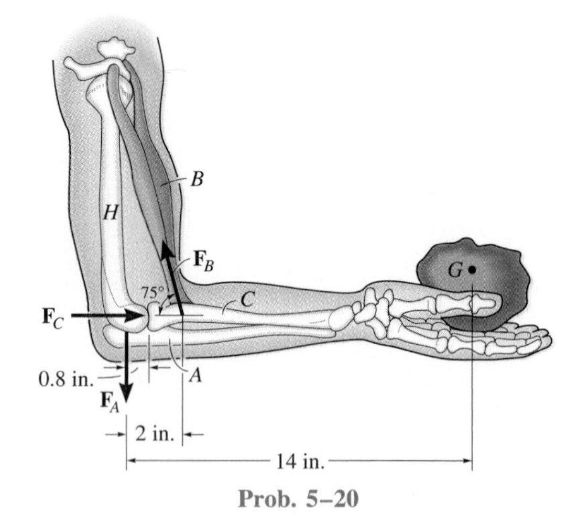

Prob. 5–20

5–21. Determine the reactions at the roller B, the rocker C, and where the beam contacts the smooth plane at A. Neglect the thickness of the beam.

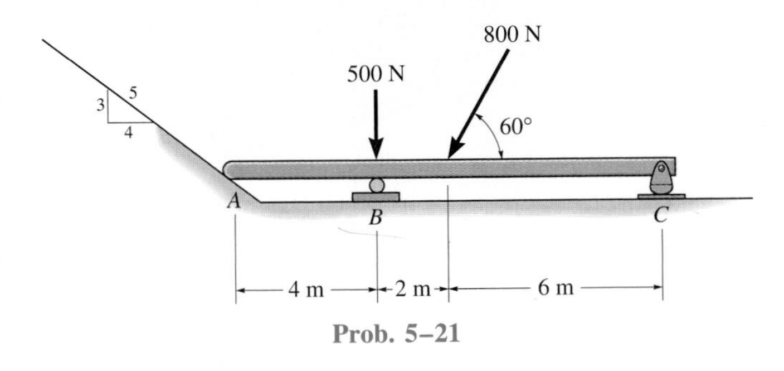

Prob. 5–21

5–22. The ramp of a ship has a weight of 200 lb and a center of gravity at G. Determine the cable force in CD needed to just start lifting the ramp, i.e., so the reaction at B becomes zero. Also, determine the horizontal and vertical components of force at the hinge (pin) at A.

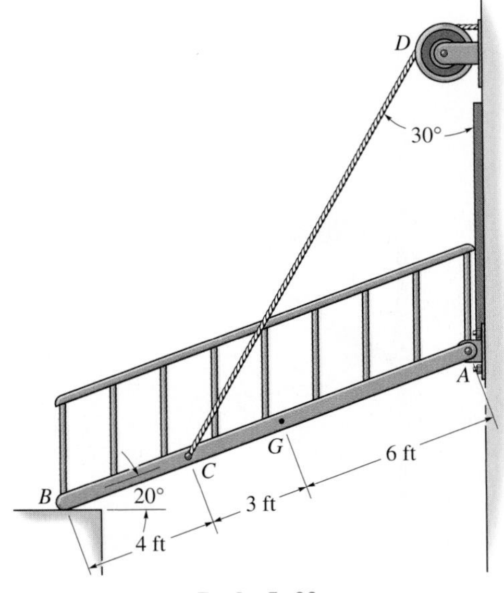

Prob. 5–22

5–23. The man is pulling a load of 8 lb with one arm held as shown. Determine the force $\mathbf{F}_H$ this exerts on the humerus bone H, and the tension developed in the biceps muscle B. Neglect the weight of the man's arm.

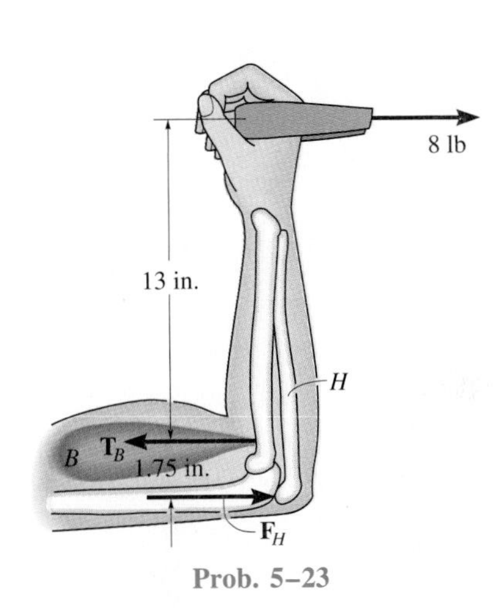

Prob. 5–23

***5–24.** Determine the reactions at the pin A and the tension in cord BC. Set $F = 40$ kN. Neglect the thickness of the beam.

5–25. If rope BC will fail when the tension becomes 50 kN, determine the greatest vertical load F that can be applied to the beam at B. What is the magnitude of the reaction at A for this loading? Neglect the thickness of the beam.

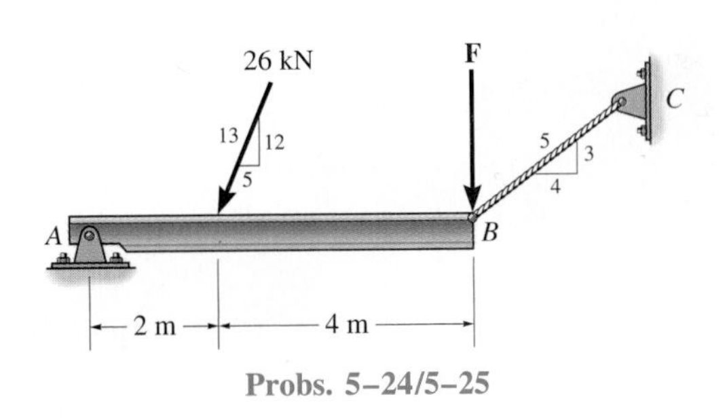

Probs. 5–24/5–25

5–26. The platform assembly has a weight of 250 lb and center of gravity at G_1. If it is intended to support a maximum load of 400 lb placed at point G_2, determine the smallest counterweight W that should be placed at B in order to prevent the platform from tipping over.

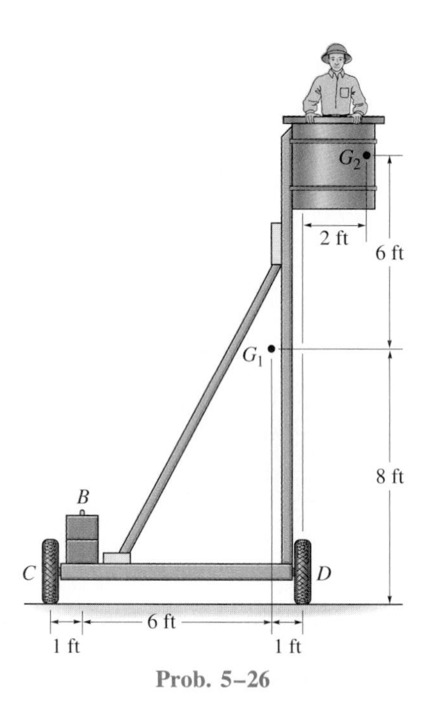

Prob. 5–26

5–27. The device is used to hold an elevator door open. If the spring has a stiffness of $k = 40$ N/m and it is compressed 0.2 m, determine the horizontal and vertical components of reaction at the pin A and the resultant force at the wheel bearing B.

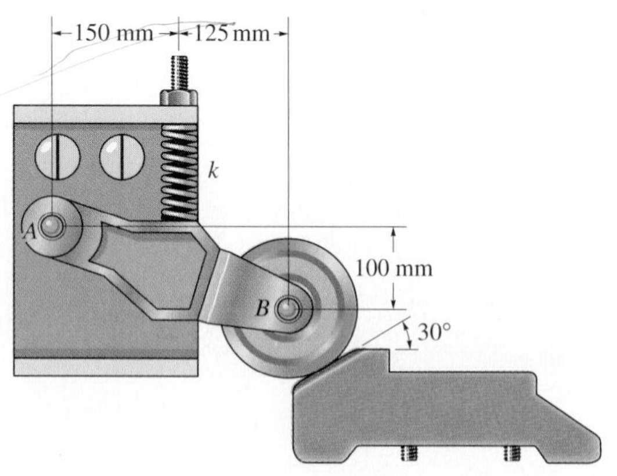

Prob. 5–27

***5–28.** The linkage rides along the top and bottom flanges of the crane rail. If the load it supports is 500 lb, determine the force of each roller on the flange.

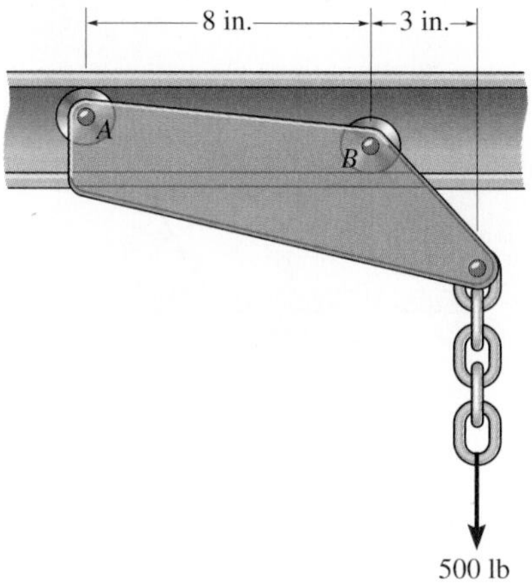

Prob. 5–28

5–29. The wall crane is supported by the journal bearing (smooth collar) at B and thrust bearing at A. Determine the horizontal and vertical components of force at A and the horizontal force at B if $P = 8$ kN.

5–30. The wall crane is supported by the journal bearing (smooth collar) at B and thrust bearing at A. If the journal bearing can support a force of 12 kN before it fails, determine the maximum load P that can be suspended from the crane. The thrust bearing at A can support both horizontal and vertical components of force.

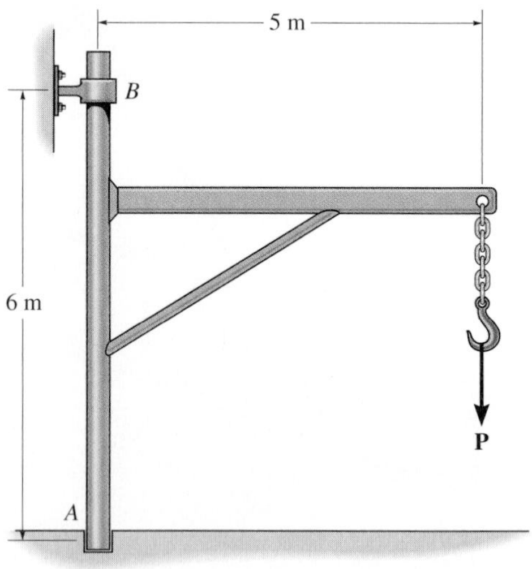

Probs. 5–29/5–30

5–31. The cantilevered jib crane is used to support the load of 780 lb. If the trolley T can be placed anywhere between $1.5 \text{ ft} \leq x \leq 7.5 \text{ ft}$, determine the maximum magnitude of reaction at the supports A and B. Note that the supports are collars that allow the crane to rotate freely about the vertical axis. The collar at B supports a force in the vertical direction, whereas the one at A does not.

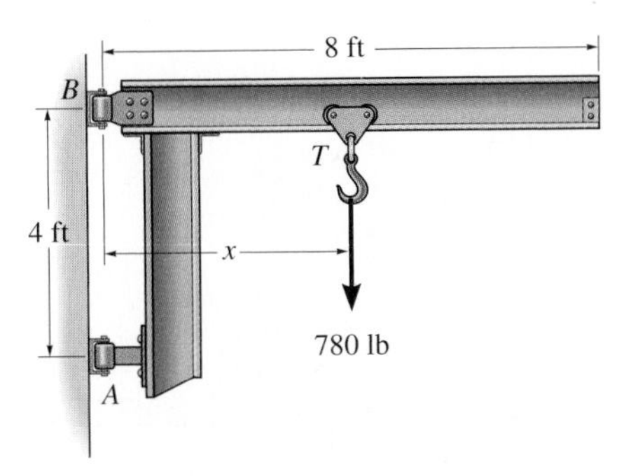

Prob. 5–31

5–33. The power pole supports the three lines, each line exerting a vertical force on the pole due to its weight as shown. Determine the reactions at the fixed support D. If it is possible for wind or ice to snap the lines, determine which line(s) when removed create(s) a condition for the greatest moment reaction at D.

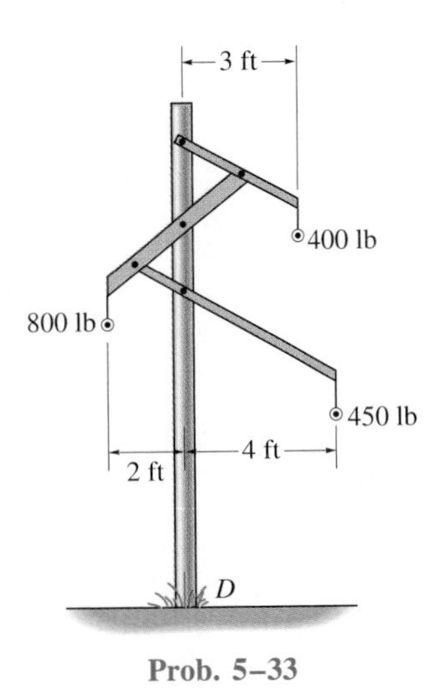

Prob. 5–33

5–34. The framework is supported by the member AB which rests on the smooth floor. When loaded, the pressure distribution on AB is linear as shown. Determine the smallest size d of member AB so that it will not cause the frame to tip over. What is the intensity w for this case?

***5–32.** Determine the reactions on the beam at A and B.

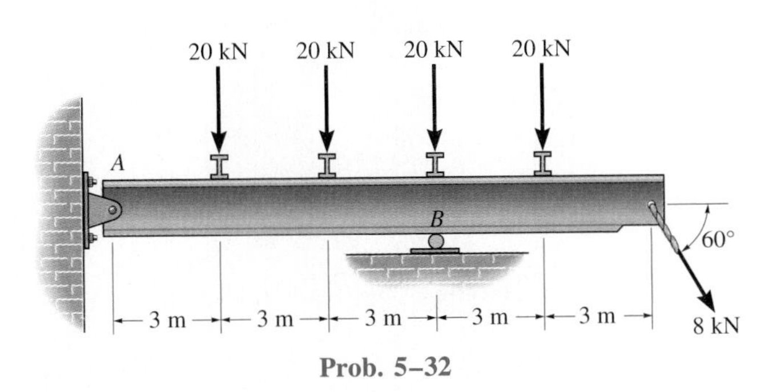

Prob. 5–32

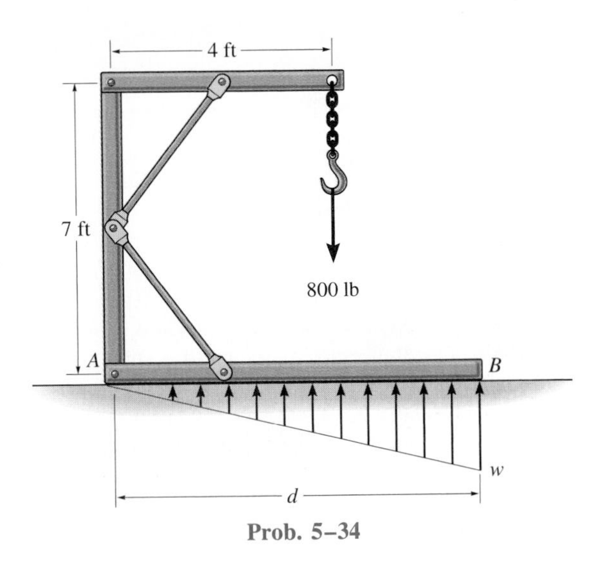

Prob. 5–34

5–35. If the wheelbarrow and its contents have a mass of 60 kg and center of mass at G, determine the magnitude of the resultant force which the man must exert on *each* of the two handles in order to hold the wheelbarrow in equilibrium.

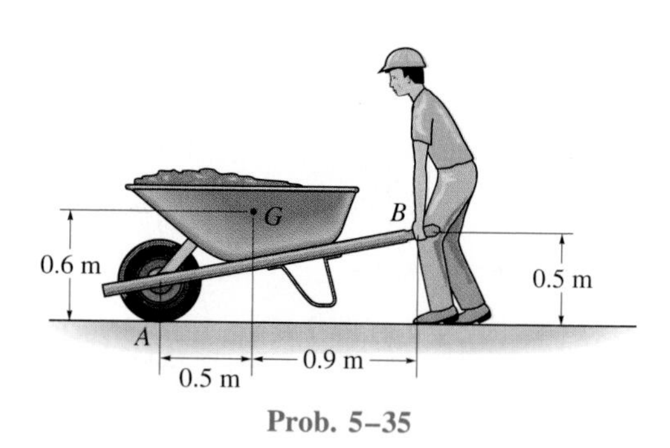

Prob. 5–35

***5–36.** Determine the resultant normal force acting on *each* set of wheels of the airplane. There is a set of wheels in the front, A, and a set of wheels under each wing, B. Both wings have a total weight of 50 kip and center of gravity at G_w, the fuselage has a weight of 180 kip and center of gravity at G_f, and both engines (one on each side) have a total weight of 22 kip and center of gravity at G_e.

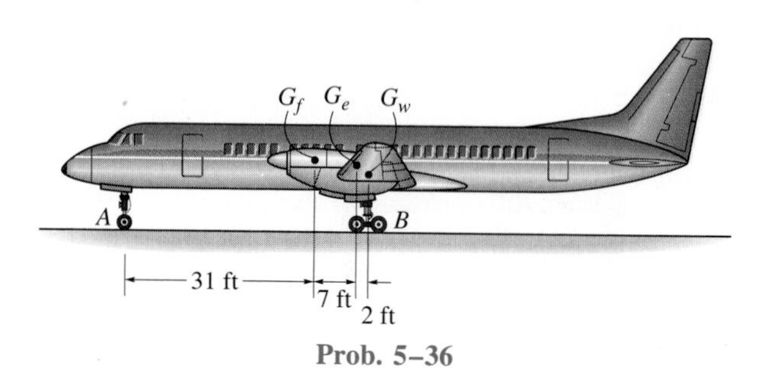

Prob. 5–36

5–37. The telephone pole of negligible thickness is subjected to the cable force of 80 lb directed as shown. It is supported by the cable BCD and the pole can be assumed pinned at its base A. In order to provide clearance for a sidewalk right of way, where D is located, a strut CE is attached to the pole at its midheight $h = 15$ ft and two cables, BC and CD', are attached to the strut at C, as shown by the dashed line (cable segment CD is removed). Determine the tension in cable CD' and the resultant force at A.

5–38. The telephone pole of negligible thickness is subjected to the force of 80 lb directed as shown. It is supported by the cable BCD and can be assumed pinned at its base A. In order to provide clearance for a sidewalk right of way, where D is located, a strut tached to the strut at C, as shown by the dashed lines (cable segment CD is removed). If the tension in CD' is to be twice the tension in BCD, determine the height h for placement of the strut CE.

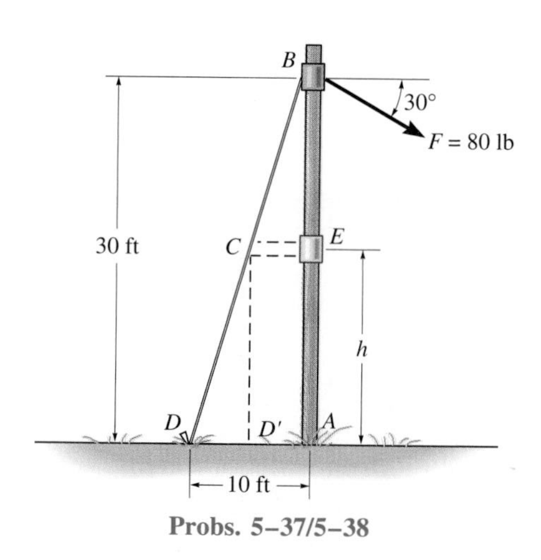

Probs. 5–37/5–38

5–39. The mechanism shown was thought by its inventor to be a perpetual-motion machine. It consists of the stand A, two smooth idler wheels B and C, and in between a uniform hollow cylindrical ring D suspended in the manner shown. The ring has a weight W and it was expected to revolve in the direction indicated by the arrow. Draw a free-body diagram of the ring and using an appropriate equation of equilibrium show that it will not rotate.

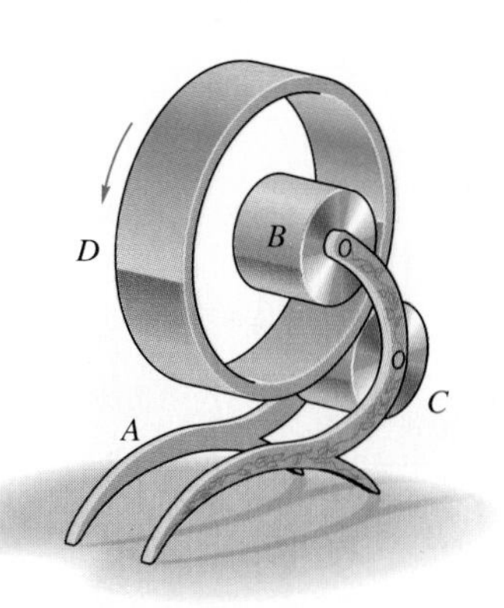

Prob. 5–39

***5–40** The shelf supports the electric motor which has a mass of 15 kg and mass center at G_m. The platform upon which it rests has a mass of 4 kg and mass center at G_p. Assuming that a single bolt B holds the shelf up and the bracket bears against the smooth wall at A, determine this normal force at A and the horizontal and vertical components of reaction of the bolt on the bracket.

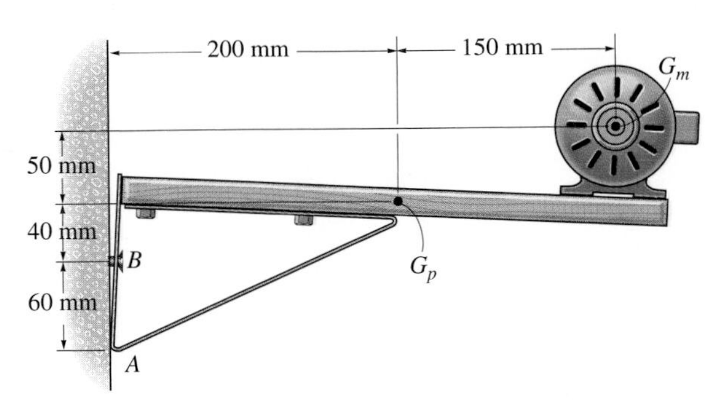

Prob. 5–40

5–43. Determine the reactions at the smooth collar A, the rocker B, and the short link CD.

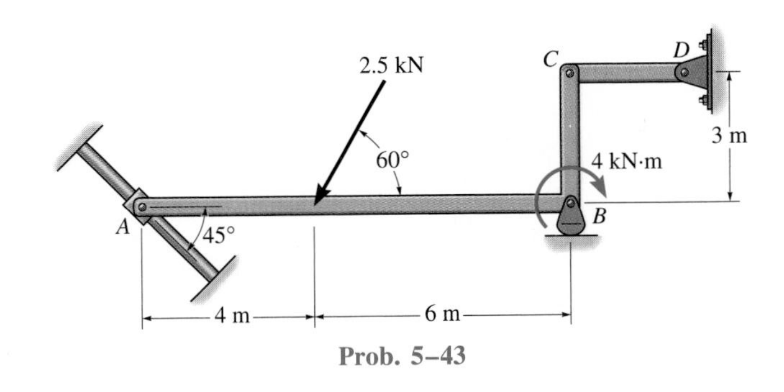

Prob. 5–43

5–41. The boom supports the two vertical loads. Neglect the size of the collars at D and B and the thickness of the boom, and compute the horizontal and vertical components of force at the pin A and the force in cable CB. Set $F_1 = 800$ N and $F_2 = 350$ N.

5–42. The boom is intended to support two vertical loads, $\mathbf{F}_1$ and $\mathbf{F}_2$. If the cable CB can sustain a maximum load of 1500 lb before it fails, determine the critical loads if $F_1 = 2F_2$. Also, what is the magnitude of the maximum reaction at pin A?

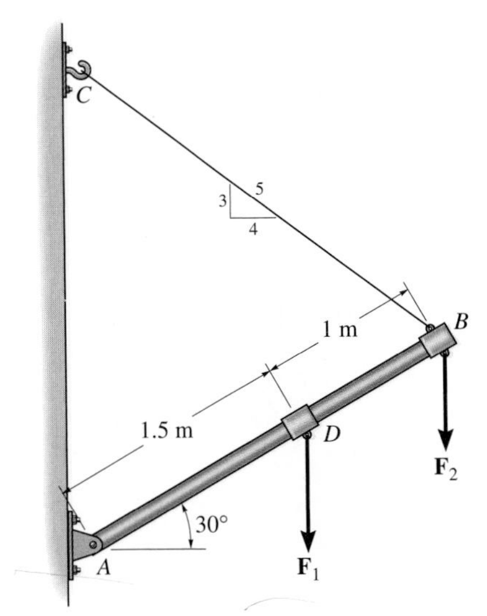

Probs. 5–41/5–42

***5–44.** The worker uses the hand truck to move material down the ramp. If the truck and its contents are held in the position shown and have a weight of 100 lb with center of gravity at G, determine the resultant normal force of both wheels on the ground A and the magnitude of the force required at the grip B.

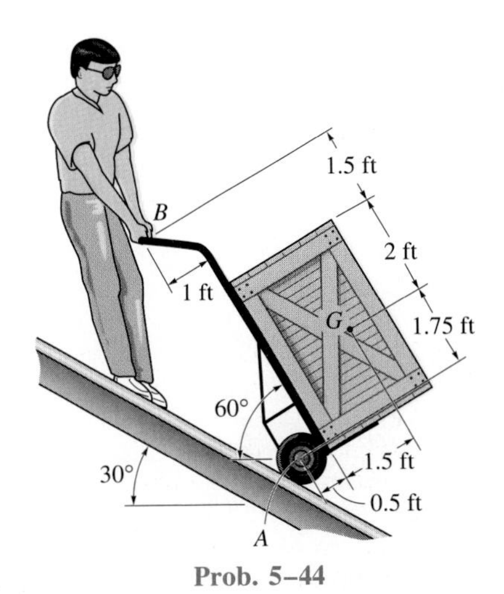

Prob. 5–44

5–45. The upper portion of the crane boom consists of the jib AB, which is supported by the pin at A, the guy line BC, and the backstay CD, each cable being separately attached to the mast at C. If the 5-kN load is supported by the hoist line, which passes over the pulley at B, determine the magnitude of the resultant force the pin exerts on the jib at A for equilibrium, the tension in the guy line BC, and the tension T in the hoist line. Neglect the weight of the jib. The pulley at B has a radius of 0.1 m.

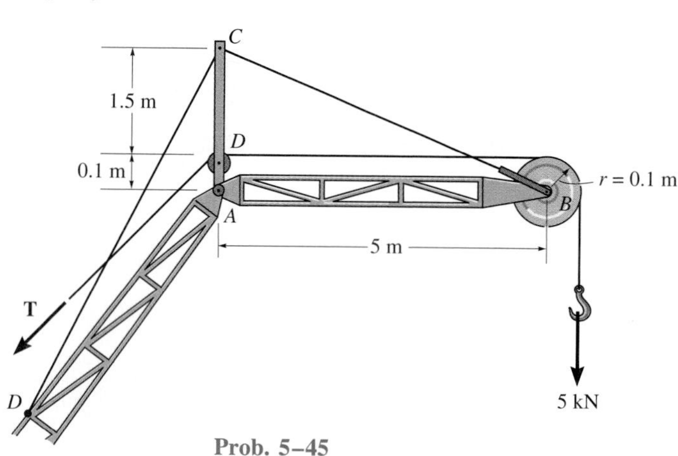

Prob. 5–45

5–46. The mobile crane has a weight of 120,000 lb and center of gravity at G_1; the boom has a weight of 30,000 lb and center of gravity at G_2. Determine the smallest angle of tilt θ of the boom, without causing the crane to overturn if the suspended load is $W = 40{,}000$ lb. Neglect the thickness of the tracks at A and B.

5–47. The mobile crane has a weight of 120,000 lb and center of gravity at G_1; the boom has a weight of 30,000 lb and center of gravity at G_2. If the suspended load has a weight of $W = 16{,}000$ lb, determine the normal reactions at the tracks A and B. For the calculation, neglect the thickness of the tracks and take $\theta = 30°$.

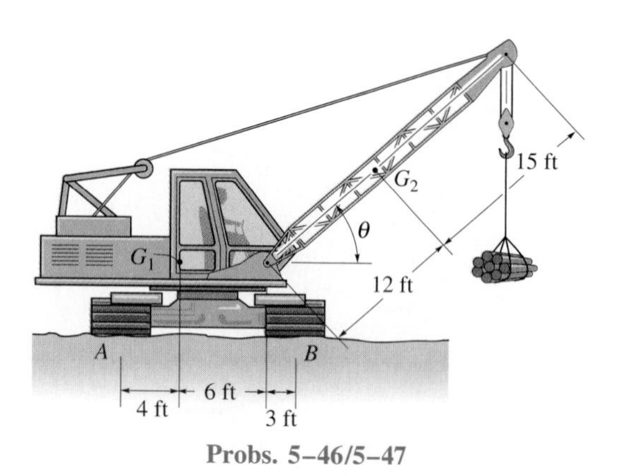

Probs. 5–46/5–47

***5–48.** The toggle switch consists of a cocking lever that is pinned to a fixed frame at A and held in place by the spring which has an unstretched length of 200 mm. Determine the magnitude of the resultant force at A and the normal force on the peg at B when the lever is in the position shown.

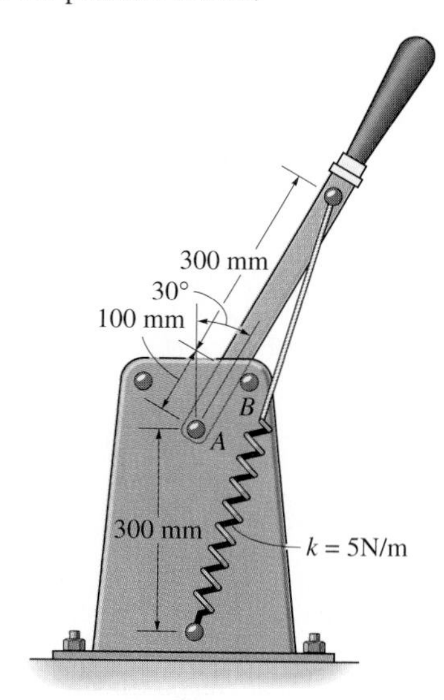

Prob. 5–48

5–49. The lineman has a weight of 175 lb, mass center at G, and stands in the position shown. If he lets go of the pole with his hands, determine the magnitude of the resultant force that both his feet must exert on the pole at B and the horizontal force on the ring at A. Assume the pole and his waist have the same diameter, so the sides of the belt are parallel.

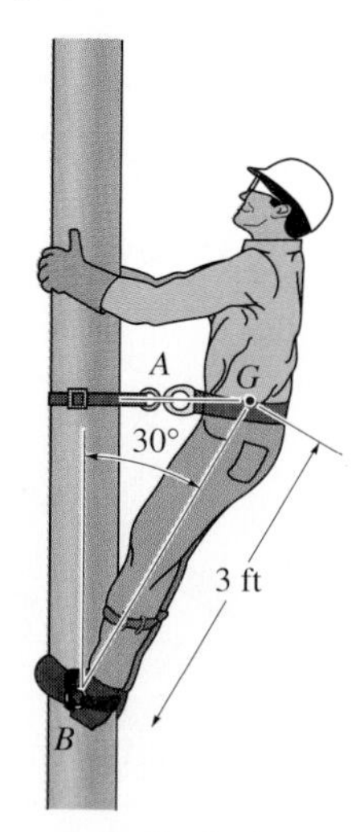

Prob. 5–49

5–50. The wheel support on a cart is attached to the frame of the cart by a smooth collar. Since it is not a snug fit, the shaft on the wheel support bears on the collar at the smooth points A, B, and C. If the wheel loading is 900 N, determine the reactive forces on the shaft at its points of contact.

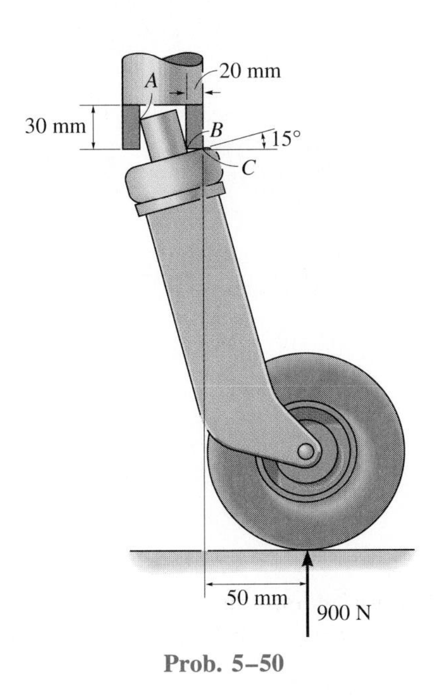

Prob. 5–50

5–51. The rigid beam of negligible weight is supported horizontally by two springs and a pin. If the springs are uncompressed when the load is removed, determine the force in each spring when the load **P** is applied. Also, compute the vertical deflection of end C. Assume the spring stiffness k is large enough so that only small deflections occur. *Hint:* The beam rotates about A so the deflections in the springs can be related.

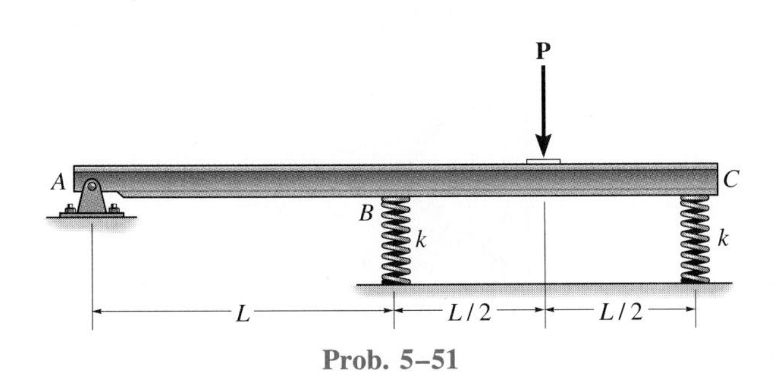

Prob. 5–51

***5–52.** The file cabinet contains four uniform drawers, each 2.5 ft long and weighing 30 lb/ft. If the empty cabinet (without drawers) has a weight of 40 lb and a center of mass at G, determine how many of the drawers can be fully pulled out and the extension x of the last drawer that will cause the assembly to be on the verge of tipping over.

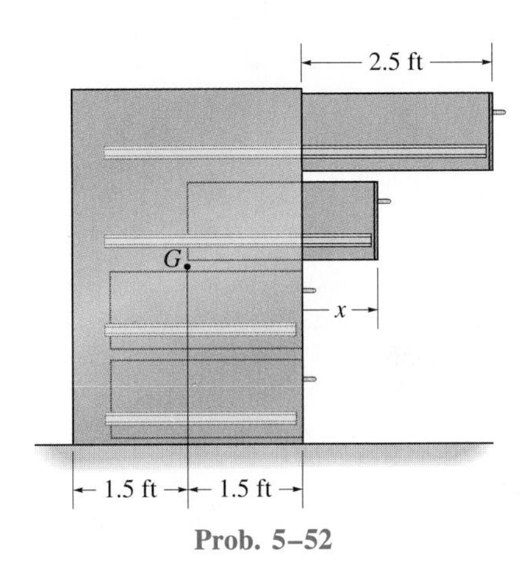

Prob. 5–52

5–53. The smooth pipe rests against the wall at the points of contact A, B, and C. Determine the reactions at these points needed to support the vertical force of 45 lb. Neglect the pipe's thickness in the calculation.

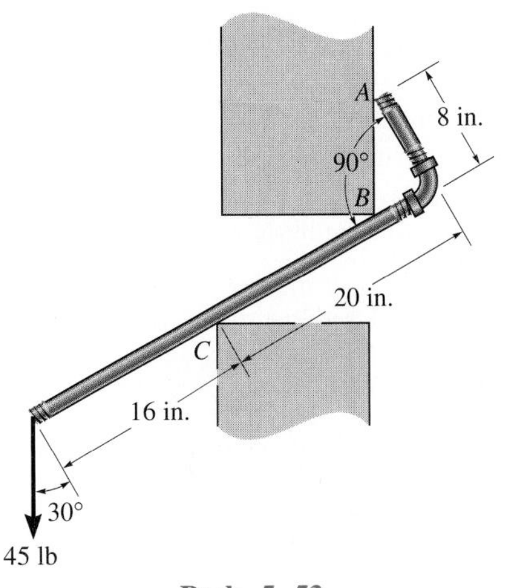

Prob. 5–53

5–54. Determine the distance d for placement of the load $\mathbf{P}$ for equilibrium of the smooth bar in the position θ as shown. Neglect the weight of the bar.

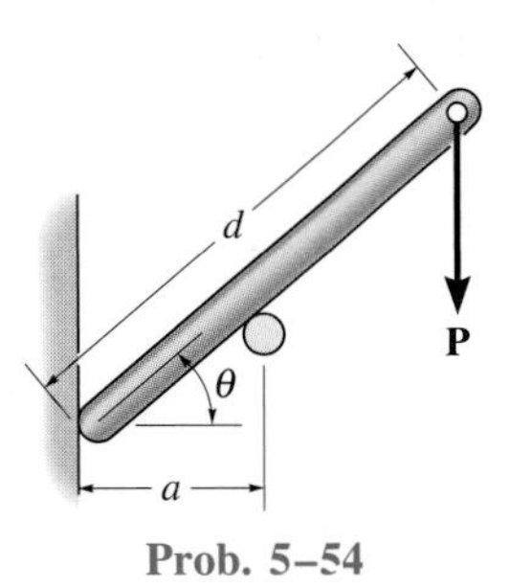

Prob. 5–54

5–55. The assembly is made from two boards. The board on the left has a weight of 10 lb and center of gravity at G_1, and the board on the right has a weight of 7 lb and center of gravity at G_2. Determine the force that the two smooth pipes exert on it at A, B, and C.

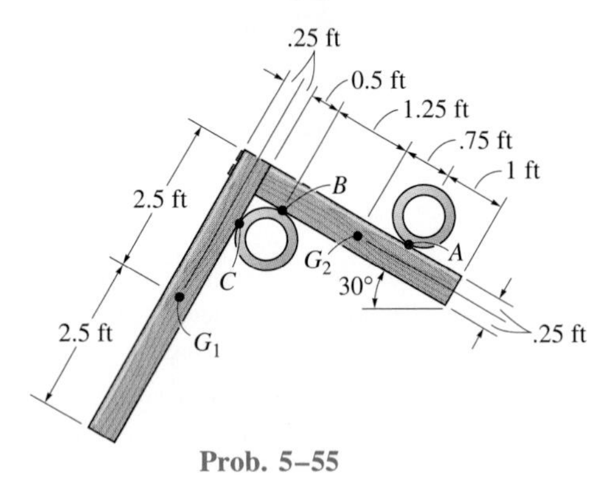

Prob. 5–55

***5–56.** The disk has a mass of 20 kg and is supported on the smooth cylindrical surface by a spring having a stiffness of $k = 400$ N/m and unstretched length of $l_0 = 1$ m. The spring remains in the horizontal position since its end A is attached to the small roller guide which has negligible weight. Determine the angle θ to the nearest degree for equilibrium of the roller.

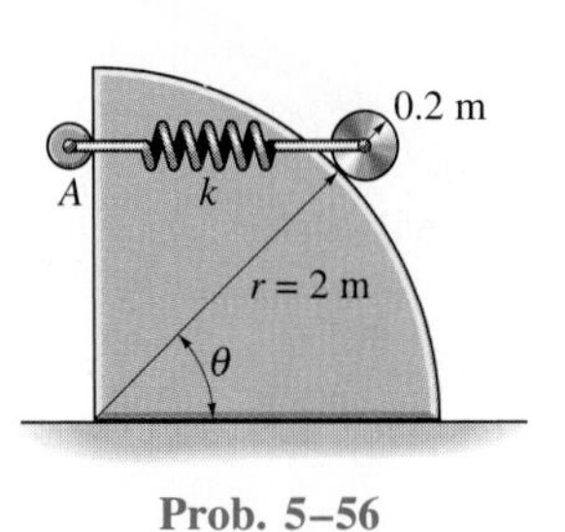

Prob. 5–56

5–57. The wheelbarrow and its contents have a mass m and center of mass at G. Determine the greatest angle of tilt θ without causing the wheelbarrow to tip over.

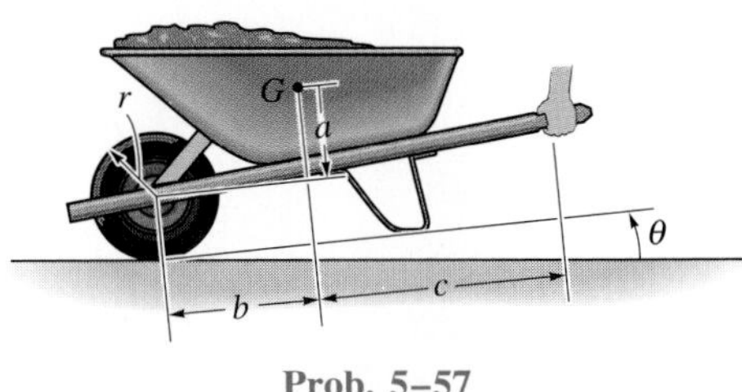

Prob. 5–57

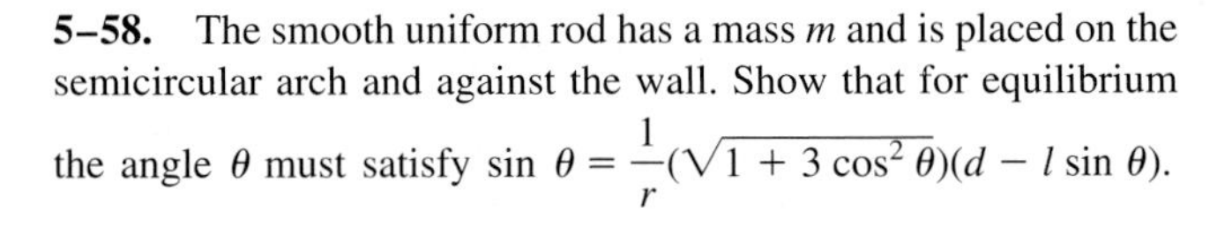

5–58. The smooth uniform rod has a mass m and is placed on the semicircular arch and against the wall. Show that for equilibrium the angle θ must satisfy $\sin\theta = \dfrac{1}{r}(\sqrt{1 + 3\cos^2\theta})(d - l\sin\theta)$.

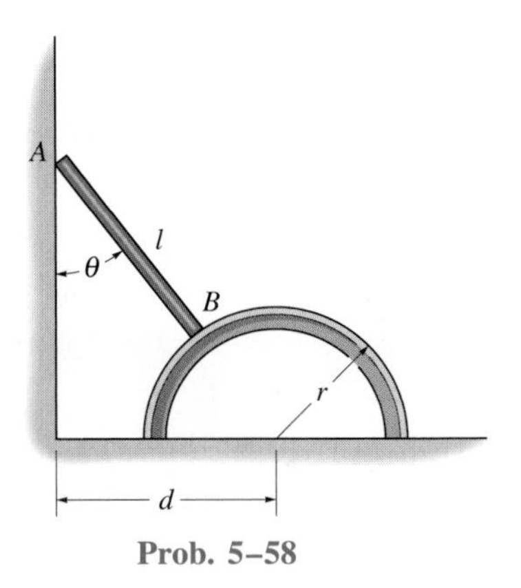

Prob. 5–58

5–59. The uniform ladder has a mass of 60 kg and is placed against the smooth step A. It is lowered to the horizontal position by a man who applies a normal force to it always from a height of 2.5 m. Determine the largest length L at which it can be so that he can let it down slowly without causing the ladder to slip at A.

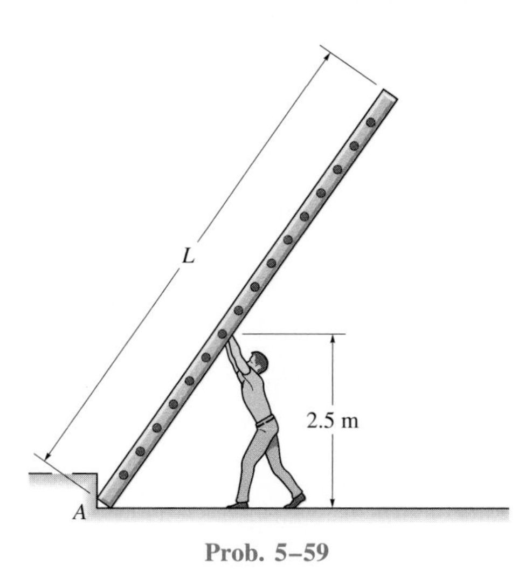

Prob. 5–59

***5–60.** The rod BC is supported by two cords, each of length a, which are attached to the pin at A. If weights W and $2W$ are suspended from the ends of the rod, determine the angle θ for equilibrium, measured from the horizontal. Express the answer in terms of a and l. Neglect the weight of the rod.

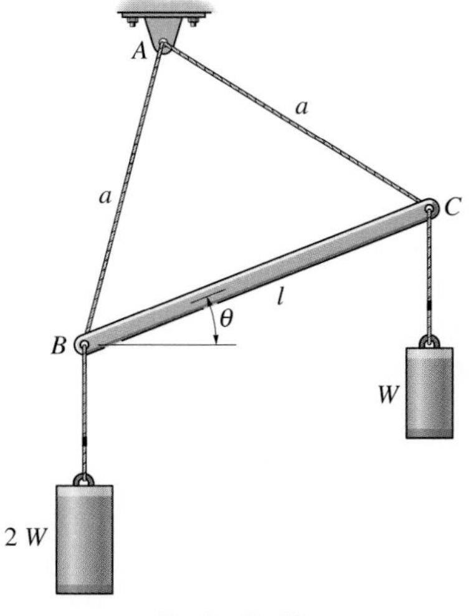

Prob. 5–60

5–61. The disk has a mass of 20 kg and rests on the smooth inclined surface. One end of a spring is attached to the center of the disk and the other end is attached to a roller at A. Consequently, the spring remains in the horizontal position when the disk is in equilibrium. If the unstretched length of the spring is 200 mm, determine its stretched length when the disk is in equilibrium. Neglect the size and weight of the roller.

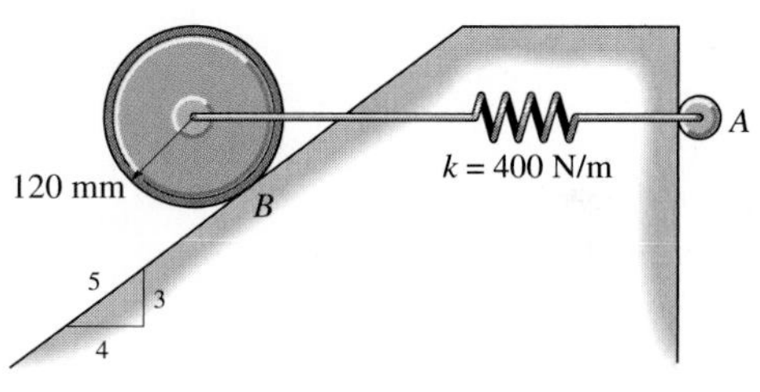

Prob. 5–61

▪5–62. A linear *torsional spring* deforms such that an applied couple moment M is related to the spring's rotation θ in radians by the equation $M = (20\ \theta)\ \text{N}\cdot\text{m}$. If such a spring is attached to the end of a pin-connected uniform 10-kg rod, determine the angle θ for equilibrium. The spring is undeformed when $\theta = 0°$.

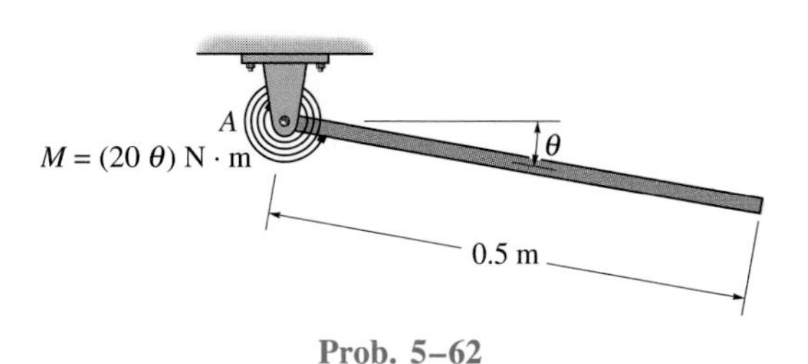

Prob. 5–62

Equilibrium in Three Dimensions

5.5 Free-Body Diagrams

The first step in solving three-dimensional equilibrium problems, as in the case of two dimensions, is to draw a free-body diagram of the body (or group of bodies considered as a system). Before we show this, however, it is necessary to discuss the types of reactions that can occur at the supports.

Support Reactions. The reactive forces and couple moments acting at various types of supports and connections, when the members are viewed in three dimensions, are listed in Table 5–2. It is important to recognize the symbols used to represent each of these supports and to understand clearly how the forces and couple moments are developed by each support. As in the two-dimensional case, *a force is developed by a support that restricts the translation of the attached member, whereas a couple moment is developed when rotation of the attached member is prevented.* For example, in Table 5–2, the ball-and-socket joint (4) prevents any translation of the connecting member; therefore, a force must act on the member at the point of connection. This force has three components having unknown magnitudes, F_x, F_y, F_z. Provided these components are known, one can obtain the magnitude of force, $F = \sqrt{F_x^2 + F_y^2 + F_z^2}$, and the force's orientation defined by the coordinate direction angles α, β, γ, Eqs. 2–7.* Since the connecting member is allowed to rotate freely about *any* axis, no couple moment is resisted by a ball-and-socket joint.

It should be noted that the *single* bearing supports (5) and (7), the *single* pin (8), and the *single* hinge (9) are shown to support both force and couple-moment components. If, however, these supports are used in conjunction with *other* bearings, pins, or hinges to hold the body in equilibrium, and provided the physical body maintains its *rigidity* when loaded and the supports are *properly aligned* when connected to the body, then the *force reactions* at these supports may *alone* be adequate for supporting the body. In other words, the couple moments become redundant and may be neglected on the free-body diagram. The reason for this will be clear after studying the examples which follow, but essentially the couple moments will not be developed at these supports since the rotation of the body is prevented by the reactions developed at the other supports and not by the supporting couple moments.

*The three unknowns may also be represented as an unknown force magnitude F and two unknown coordinate direction angles. The third direction angle is obtained using the identity $\cos^2\alpha + \cos^2\beta + \cos^2\gamma = 1$, Eq. 2–10.

Table 5–2 Supports for Rigid Bodies Subjected to Three-Dimensional Force Systems

Types of Connection	*Reaction*	*Number of Unknowns*
(1) 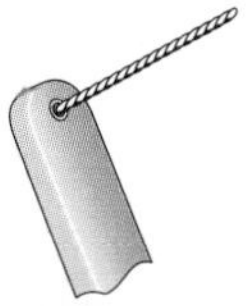 cable	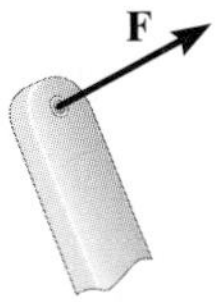 $\mathbf{F}$	One unknown. The reaction is a force which acts away from the member in the direction of the cable.
(2) 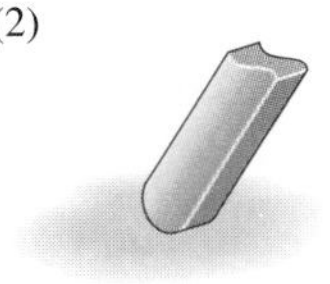 smooth surface support	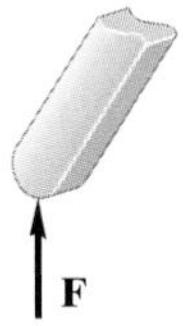 $\mathbf{F}$	One unknown. The reaction is a force which acts perpendicular to the surface at the point of contact.
(3) 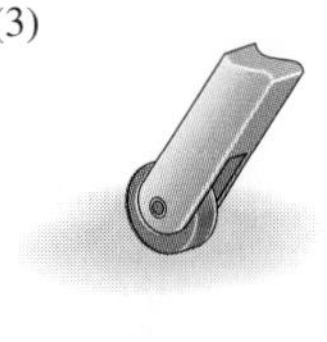 roller	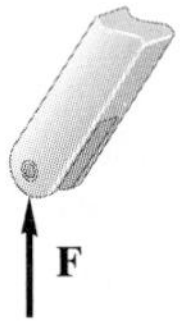 $\mathbf{F}$	One unknown. The reaction is a force which acts perpendicular to the surface at the point of contact.
(4) 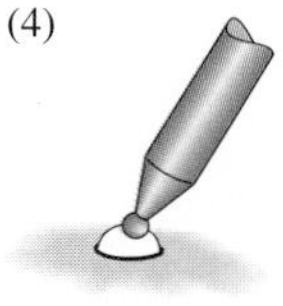 ball and socket	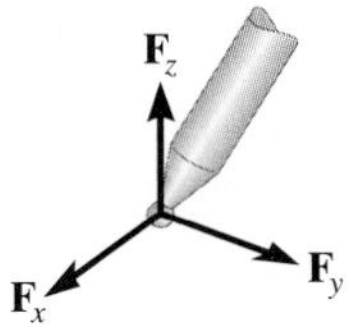 $\mathbf{F}_z$, $\mathbf{F}_x$, $\mathbf{F}_y$	Three unknowns. The reactions are three rectangular force components.
(5) 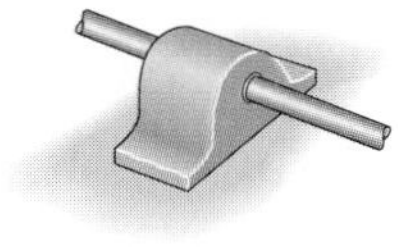 single journal bearing	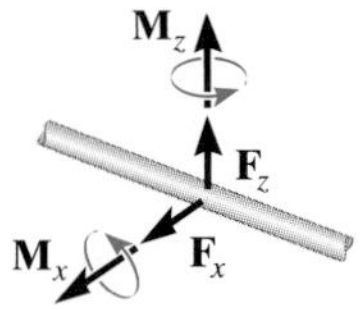 $\mathbf{M}_z$, $\mathbf{F}_z$, $\mathbf{M}_x$, $\mathbf{F}_x$	Four unknowns. The reactions are two force and two couple-moment components which act perpendicular to the shaft.
(6)  single journal bearing with square shaft	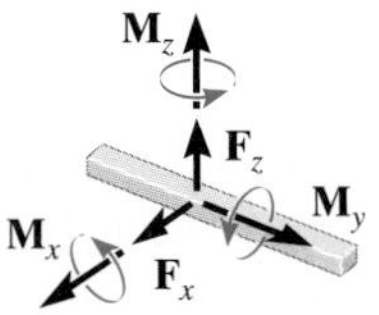 $\mathbf{M}_z$, $\mathbf{F}_z$, $\mathbf{M}_y$, $\mathbf{M}_x$, $\mathbf{F}_x$	Five unknowns. The reactions are two force and three couple-moment components.

Table 5–2 (Contd.)

Types of Connection	*Reaction*	*Number of Unknowns*
(7) single thrust bearing	$\mathbf{M}_z$, $\mathbf{F}_y$, $\mathbf{F}_z$, $\mathbf{M}_x$, $\mathbf{F}_x$	Five unknowns. The reactions are three force and two couple-moment components.
(8) single smooth pin	$\mathbf{M}_z$, $\mathbf{F}_z$, $\mathbf{F}_y$, $\mathbf{M}_y$, $\mathbf{F}_x$	Five unknowns. The reactions are three force and two couple-moment components.
(9) single hinge	$\mathbf{M}_z$, $\mathbf{F}_z$, $\mathbf{F}_y$, $\mathbf{F}_x$, $\mathbf{M}_x$	Five unknowns. The reactions are three force and two couple-moment components.
(10) fixed support	$\mathbf{M}_z$, $\mathbf{F}_z$, $\mathbf{F}_x$, $\mathbf{M}_x$, $\mathbf{F}_y$, $\mathbf{M}_y$	Six unknowns. The reactions are three force and three couple-moment components.

Free-Body Diagrams. The general procedure for establishing the free-body diagram of a rigid body has been outlined in Sec. 5.2. Essentially it requires first "isolating" the body by drawing its outlined shape. This is followed by a careful *labeling* of *all* the forces and couple moments in reference to an established x, y, z coordinate system. As a general rule, *components of reaction* having an *unknown magnitude* are shown acting on the free-body diagram in the *positive sense.* In this way, if any negative values are obtained, they will indicate that the components act in the negative coordinate directions.

Example 5–14

Several examples of objects along with their associated free-body diagrams are shown in Fig. 5–22. In all cases, the x, y, z axes are established and the unknown reaction components are indicated in the positive sense. The weight of the objects is neglected.

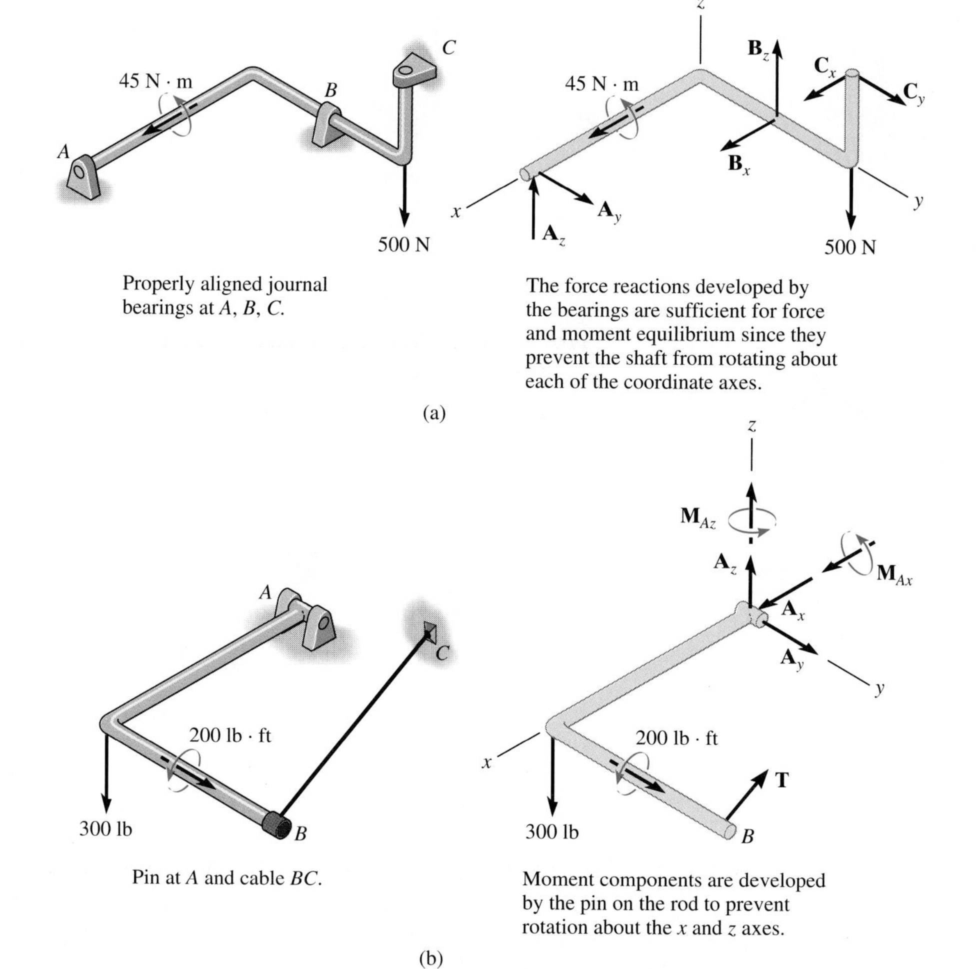

Fig. 5–22*a* and *b*

Fig. 5–22 (*continued*)

Fig. 5–22 (*continued*)

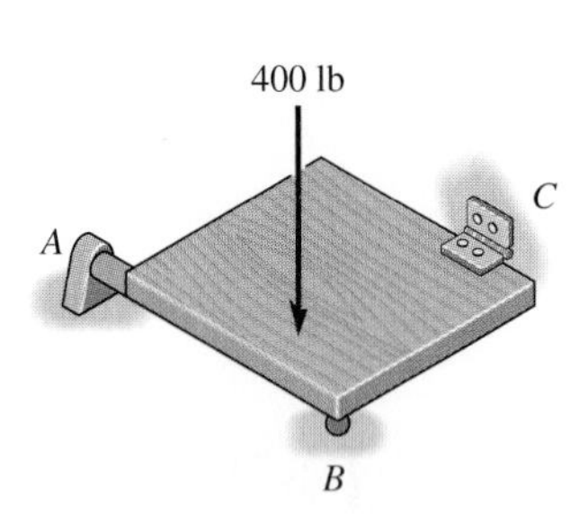

Properly aligned journal bearing at A and hinge at C. Roller at B.

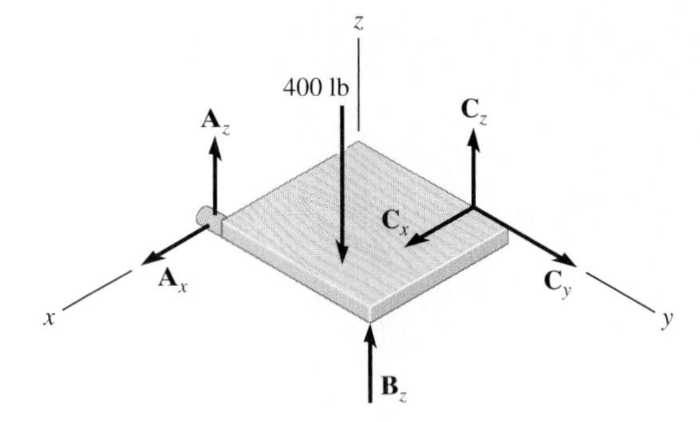

Only force reactions are developed by the bearing and hinge on the plate to prevent rotation about each coordinate axis. No moments at the hinge are developed.

(c)

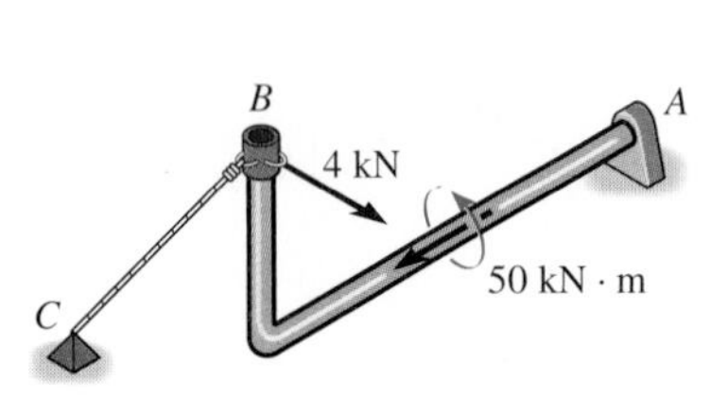

Thrust bearing at A and cable BC

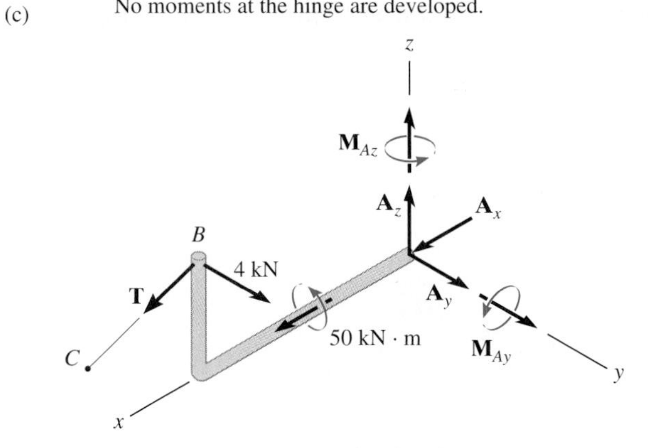

Moment components are developed by the bearing on the rod in order to prevent rotation about the y and z axes.

(d)

Fig. 5–22*c* and *d*

5.6 Equations of Equilibrium

As stated in Sec. 5.1, the conditions for equilibrium of a rigid body subjected to a three-dimensional force system require that both the *resultant* force and *resultant* couple moment acting on the body be equal to *zero*.

Vector Equations of Equilibrium. The two conditions for equilibrium of a rigid body may be expressed mathematically in vector form as

$$\Sigma \mathbf{F} = \mathbf{0}$$
$$\Sigma \mathbf{M}_O = \mathbf{0} \qquad (5\text{–}5)$$

where $\Sigma \mathbf{F}$ is the vector sum of all the external forces acting on the body and $\Sigma \mathbf{M}_O$ is the sum of the couple moments and the moments of all the forces about any point O located either on or off the body.

Scalar Equations of Equilibrium. If all the applied external forces and couple moments are expressed in Cartesian vector form and substituted into Eqs. 5–5, we have

$$\Sigma \mathbf{F} = \Sigma F_x \mathbf{i} + \Sigma F_y \mathbf{j} + \Sigma F_z \mathbf{k} = \mathbf{0}$$
$$\Sigma \mathbf{M}_O = \Sigma M_x \mathbf{i} + \Sigma M_y \mathbf{j} + \Sigma M_z \mathbf{k} = \mathbf{0}$$

Since the **i, j,** and **k** components are independent from one another, the above equations are satisfied provided

$$\Sigma F_x = 0$$
$$\Sigma F_y = 0 \qquad (5\text{–}6a)$$
$$\Sigma F_z = 0$$

and

$$\Sigma M_x = 0$$
$$\Sigma M_y = 0 \qquad (5\text{–}6b)$$
$$\Sigma M_z = 0$$

These *six scalar equilibrium equations* may be used to solve for at most six unknowns shown on the free-body diagram. Equations 5–6*a* express the fact that the sum of the external force components acting in the x, y, and z directions must be zero, and Eqs. 5–6*b* require the sum of the moment components about the x, y, and z axes to be zero.

PROCEDURE FOR ANALYSIS

The following procedure provides a method for solving three-dimensional equilibrium problems.

Free-Body Diagram. Construct the free-body diagram for the body. Be sure to include *all* the forces and couple moments that act *on* the body. These interactions are commonly caused by the externally applied loadings, contact forces exerted by adjacent bodies, support reactions, and the weight of the body if it is significant compared to the magnitudes of the other applied forces. Establish the origin of the x, y, z axes at a convenient point, and orient the axes so that they are parallel to as many of the external forces and moments as possible. Identify the unknowns, and in general show all the unknown components having a positive sense if the sense cannot be determined. Dimensions of the body, necessary for calculating the moments of forces, are also included on the free-body diagram.

Equations of Equilibrium. Apply the equations of equilibrium. In many cases, problems can be solved by *direct application* of the six scalar equations $\Sigma F_x = 0$, $\Sigma F_y = 0$, $\Sigma F_z = 0$, $\Sigma M_x = 0$, $\Sigma M_y = 0$, $\Sigma M_z = 0$, Eqs. 5–6; however, if the force components or moment arms seem difficult to determine, it is recommended that the solution be obtained by using vector equations: $\Sigma \mathbf{F} = \mathbf{0}$, $\Sigma \mathbf{M}_O = \mathbf{0}$, Eqs. 5–5. In any case, it is *not necessary* that the set of axes chosen for force summation *coincide* with the set of axes chosen for moment summation. Instead, it is recommended that one *choose the direction of an axis for moment summation such that it intersects the lines of action of as many unknown forces as possible.* The moments of forces passing through points on this axis or forces which are parallel to the axis will then be zero. Furthermore, *any set of three nonorthogonal axes* may be chosen for either the force or moment summations. By the proper choice of axes, it may be possible to solve directly for an unknown quantity, or at least reduce the need for solving a large number of simultaneous equations for the unknowns.

5.7 Constraints for a Rigid Body

To ensure the equilibrium of a rigid body, it is not only necessary to satisfy the equations of equilibrium, but the body must also be properly held or constrained by its supports. Some bodies may have more supports than are necessary for equilibrium, whereas others may not have enough or the supports may be arranged in a particular manner that could cause the body to collapse. Each of these cases will now be discussed.

Redundant Constraints. When a body has redundant supports, that is, more supports than are necessary to hold it in equilibrium, it becomes statically indeterminate. *Statically indeterminate* means that there will be more unknown loadings on the body than equations of equilibrium available for their solution. For example, the two-dimensional problem, Fig. 5–23*a*, and the three-dimensional problem, Fig. 5–23*b*, shown together with their free-body diagrams, are both statically indeterminate because of additional support reactions. In the two-dimensional case, there are five unknowns, that is, M_A, A_x, A_y, B_y, and C_y, for which only three equilibrium equations can be written ($\Sigma F_x = 0$, $\Sigma F_y = 0$, and $\Sigma M_O = 0$, Eqs. 5–2). The three-dimensional problem has eight unknowns, for which only six equilibrium equations can be written, Eqs. 5–6. The additional equations needed to solve indeterminate problems of the type shown in Fig. 5–23 are generally obtained from the deformation conditions at the points of support. These equations involve the physical properties of the body which are studied in subjects dealing with the mechanics of deformation, such as "mechanics of materials."*

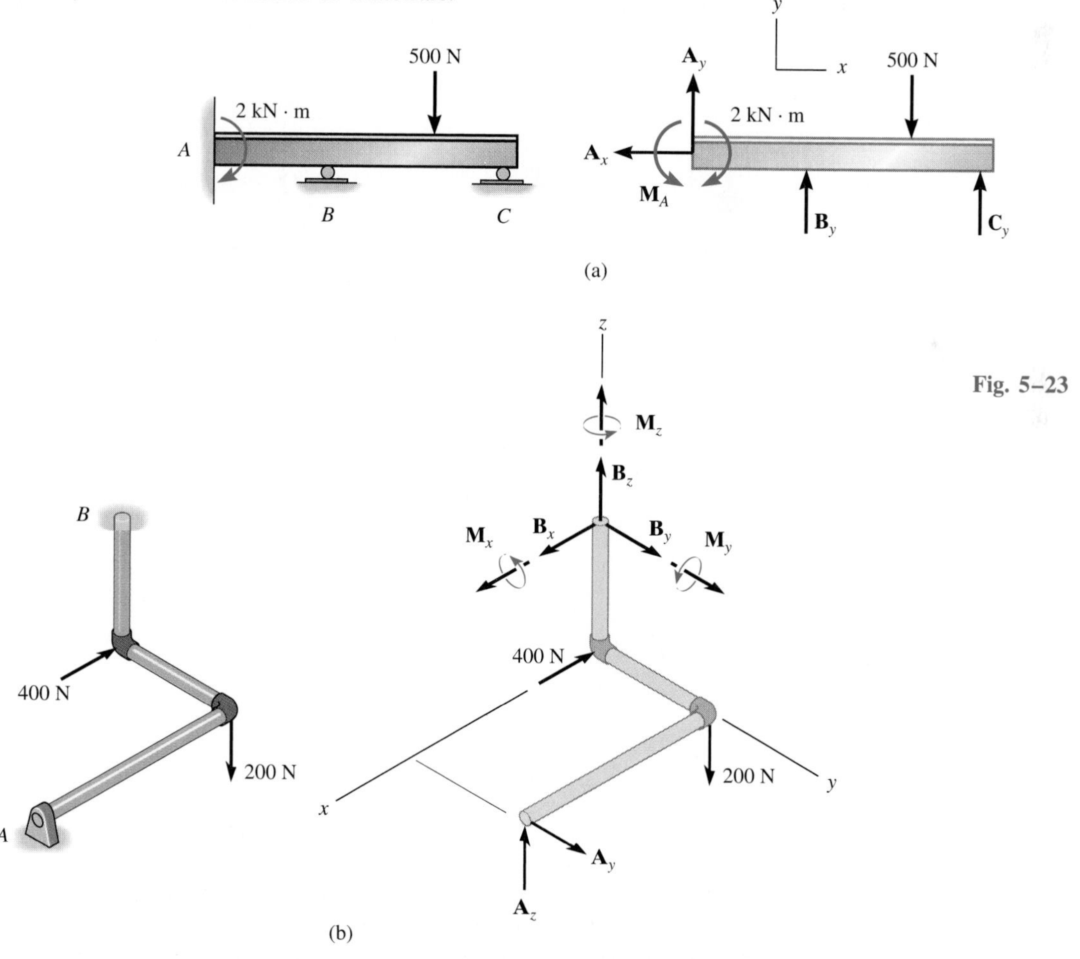

Fig. 5–23

*See R. C. Hibbeler, *Mechanics of Materials*, 2nd edition (New York: Macmillan, 1994).

Improper Constraints. In some cases, there may be as many unknown forces on the body as there are equations of equilibrium; however, *instability* of the body can develop because of *improper constraining* by the supports. In the case of three-dimensional problems, the body is improperly constrained if the support reactions *all intersect a common axis.* For two-dimensional problems, this axis is *perpendicular* to the plane of the forces and therefore appears as a point. Hence, when all the reactive forces are *concurrent* at this point, the body is improperly constrained. Examples of both cases are given in Fig. 5–24. From the free-body diagrams it is seen that the summation of moments about the x axis, Fig. 5–24a, or point O, Fig. 5–24b, will *not* be equal to zero; thus rotation about the x axis or point O will take place.* Furthermore, in both cases, it becomes *impossible* to solve *completely* for all the unknowns, since one can write a moment equation that *does not* involve any of the unknown support reactions, and as a result, this reduces the number of available equilibrium equations by one.

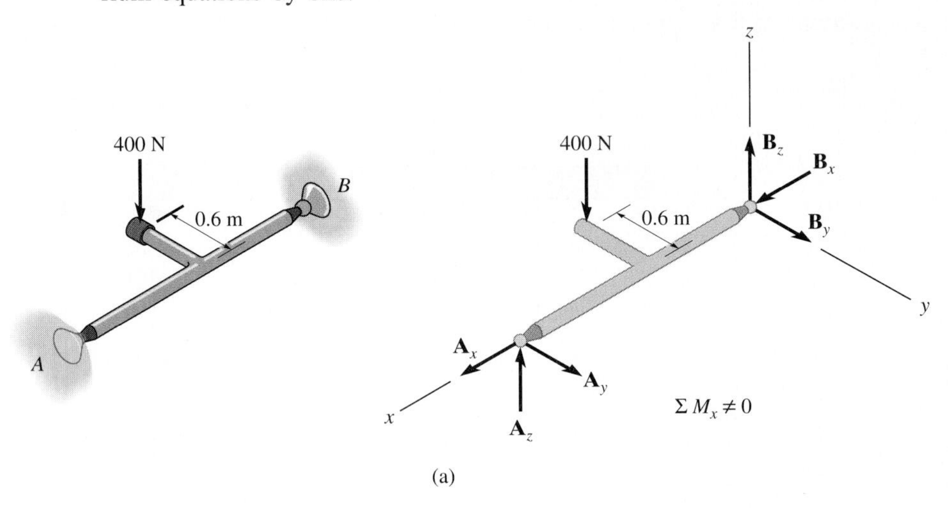

(a)

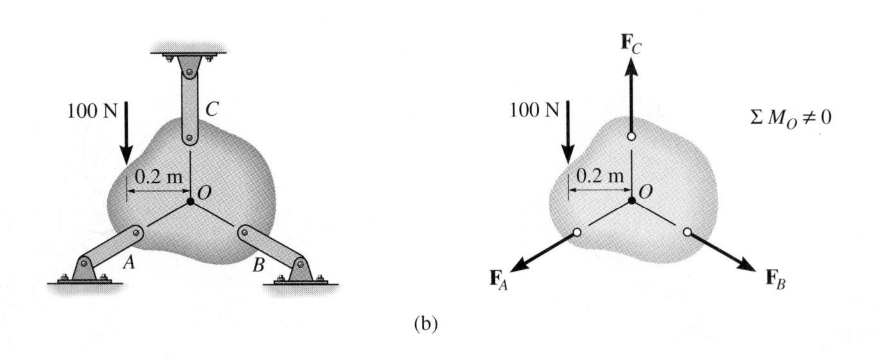

(b)

Fig. 5–24

*For the three-dimensional problem, $\Sigma M_x = (400\text{ N})(0.6\text{ m}) \neq 0$, and for the two-dimensional problem, $\Sigma M_O = (100\text{ N})(0.2\text{ m}) \neq 0$.

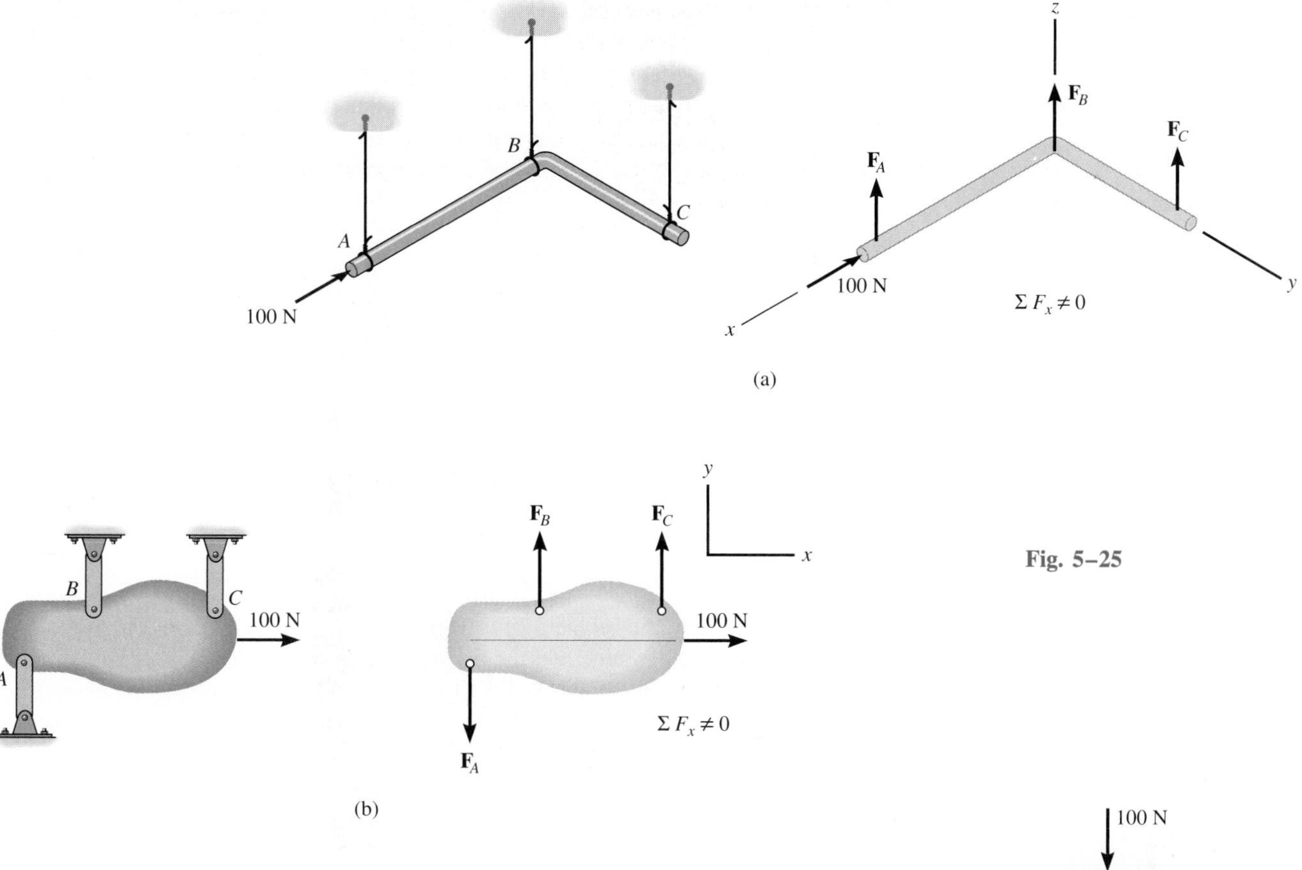

Fig. 5–25

Another way in which improper constraining leads to instability occurs when the *reactive forces* are all *parallel.* Three- and two-dimensional examples of this are shown in Fig. 5–25. In both cases, the summation of forces along the x axis will not equal zero.

In some cases, a body may have *fewer* reactive forces than equations of equilibrium that must be satisfied. The body then becomes only *partially constrained.* For example, consider the body shown in Fig. 5–26*a* with its corresponding free-body diagram in Fig. 5–26*b*. If O is a point not located on the line AB, the equations $\Sigma F_x = 0$ and $\Sigma M_O = 0$ will be satisfied by proper choice of the reactions $\mathbf{F}_A$ and $\mathbf{F}_B$. The equation $\Sigma F_y = 0$, however, will not be satisfied for the loading conditions and therefore equilibrium will not be maintained.

Proper constraining therefore requires that (1) the lines of action of the reactive forces do not intersect points on a common axis, and (2) the reactive forces must not all be parallel to one another. When the number of reactive forces needed to properly constrain the body in question is a *minimum,* the problem will be statically determinate, and therefore the equations of equilibrium can be used to determine *all* the reactive forces.

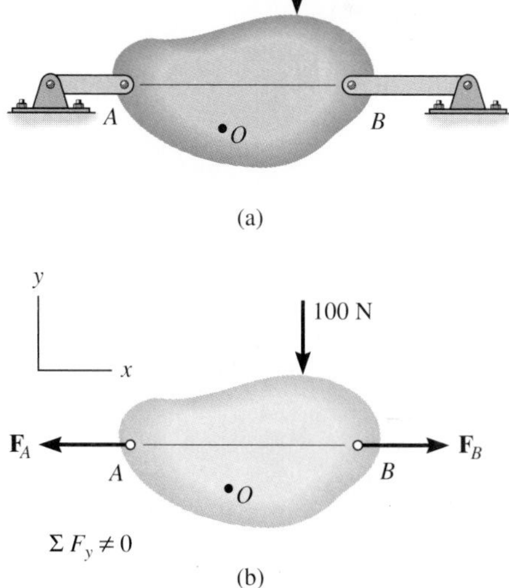

Fig. 5–26

Example 5–15

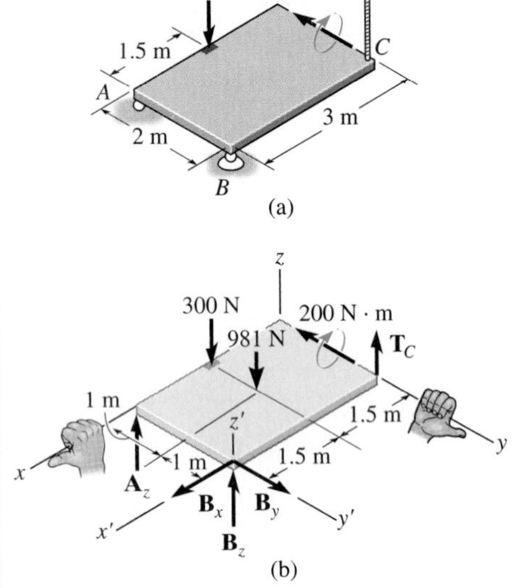

Fig. 5–27

The homogeneous plate shown in Fig. 5–27*a* has a mass of 100 kg and is subjected to a force and couple moment along its edges. If it is supported in the horizontal plane by means of a roller at *A*, a ball-and-socket joint at *B*, and a cord at *C*, determine the components of reaction at the supports.

SOLUTION (*SCALAR ANALYSIS*)

Free-Body Diagram. There are five unknown reactions acting on the plate, as shown in Fig. 5-27*b*. Each of these reactions is assumed to act in a positive coordinate direction.

Equations of Equilibrium. Since the three-dimensional geometry is rather simple, a *scalar analysis* provides a *direct solution* to this problem. A force summation along each axis yields

$\Sigma F_x = 0;\qquad B_x = 0$ *Ans.*

$\Sigma F_y = 0;\qquad B_y = 0$ *Ans.*

$\Sigma F_z = 0;\qquad A_z + B_z + T_C - 300\text{ N} - 981\text{ N} = 0 \qquad (1)$

Recall that the moment of a force about an axis is equal to the product of the force magnitude and the perpendicular distance (moment arm) from the line of action of the force to the axis. The sense of the moment is determined by the right-hand rule. Hence, summing moments of the forces on the free-body diagram, with positive moments acting along the positive *x* or *y* axis, we have

$\Sigma M_x = 0;\qquad T_C(2\text{ m}) - 981\text{ N}(1\text{ m}) + B_z(2\text{ m}) = 0 \qquad (2)$

$\Sigma M_y = 0;$

$300\text{ N}(1.5\text{ m}) + 981\text{ N}(1.5\text{ m}) - B_z(3\text{ m}) - A_z(3\text{ m}) - 200\text{ N}\cdot\text{m} = 0 \qquad (3)$

The components of force at *B* can be eliminated if the x', y', z' axes are used. We obtain

$\Sigma M_{x'} = 0;\qquad 981\text{ N}(1\text{ m}) + 300\text{ N}(2\text{ m}) - A_z(2\text{ m}) = 0 \qquad (4)$

$\Sigma M_{y'} = 0;$

$-300\text{ N}(1.5\text{ m}) - 981\text{ N}(1.5\text{ m}) - 200\text{ N}\cdot\text{m} + T_C(3\text{ m}) = 0 \qquad (5)$

Solving Eqs. 1 through 3 or the more convenient Eqs. 1, 4, and 5 yields

$A_z = 790\text{ N}\qquad B_z = -217\text{ N}\qquad T_C = 707\text{ N}$ *Ans.*

The negative sign indicates that $\mathbf{B}_z$ acts downward.

Note that the solution of this problem does not require the use of a summation of moments about the *z* axis. The plate is partially constrained since the supports will not prevent it from turning about the *z* axis if a force is applied to it in the *x–y* plane.

Example 5–16

The windlass shown in Fig. 5–28*a* is supported by a thrust bearing at *A* and a smooth journal bearing at *B*, which are properly aligned on the shaft. Determine the magnitude of the vertical force **P** that must be applied to the handle to maintain equilibrium of the 100-kg crate. Also calculate the reactions at the bearings.

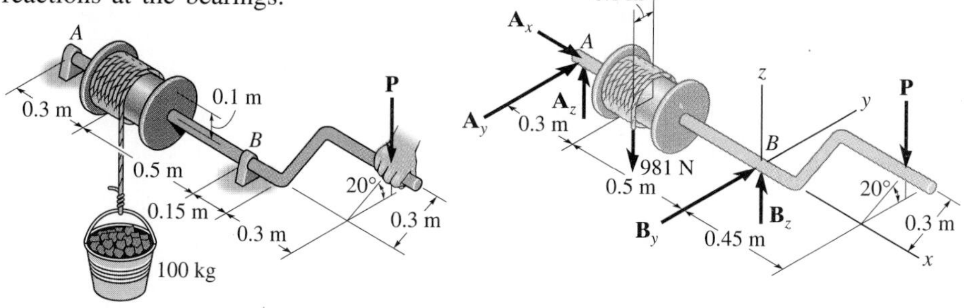

Fig. 5–28

SOLUTION (*SCALAR ANALYSIS*)

Free-Body Diagram. Since the bearings at *A* and *B* are aligned correctly, *only* force reactions occur at these supports, Fig. 5–28*b*. Why are there no moment reactions?

Equations of Equilibrium. Summing moments about the *x* axis yields a direct solution for **P.** Why? For a scalar moment summation, it is necessary to determine the moment of each force as the product of the force magnitude and the *perpendicular distance* from the *x* axis to the line of action of the force. Using the right-hand rule and assuming positive moments act in the $+\mathbf{i}$ direction, we have

$$\Sigma M_x = 0; \qquad 981\text{ N}(0.1\text{ m}) - P(0.3\cos 20°\text{ m}) = 0$$

$$P = 348.0\text{ N} \qquad \textit{Ans.}$$

Using this result and summing moments about the *y* and *z* axes yields

$\Sigma M_y = 0;$

$$-981\text{ N}(0.5\text{ m}) + A_z(0.8\text{ m}) + (348.0\text{ N})(0.45\text{ m}) = 0$$

$$A_z = 417.4\text{ N} \qquad \textit{Ans.}$$

$$\Sigma M_z = 0; \qquad -A_y(0.8\text{ m}) = 0 \qquad A_y = 0 \qquad \textit{Ans.}$$

The reactions at *B* are determined by a force summation, using the results obtained above.

$$\Sigma F_x = 0; \qquad A_x = 0 \qquad \textit{Ans.}$$

$$\Sigma F_y = 0; \qquad 0 + B_y = 0 \qquad B_y = 0 \qquad \textit{Ans.}$$

$$\Sigma F_z = 0; \qquad 417.4 - 981 + B_z - 348.0 = 0 \qquad B_z = 911.6\text{ N} \qquad \textit{Ans.}$$

Example 5–17

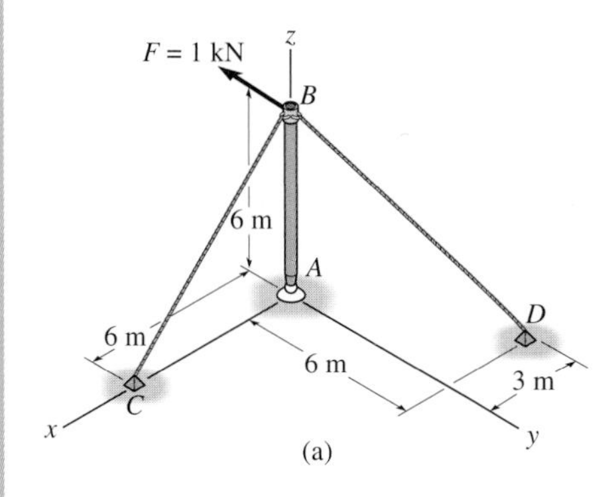

(a)

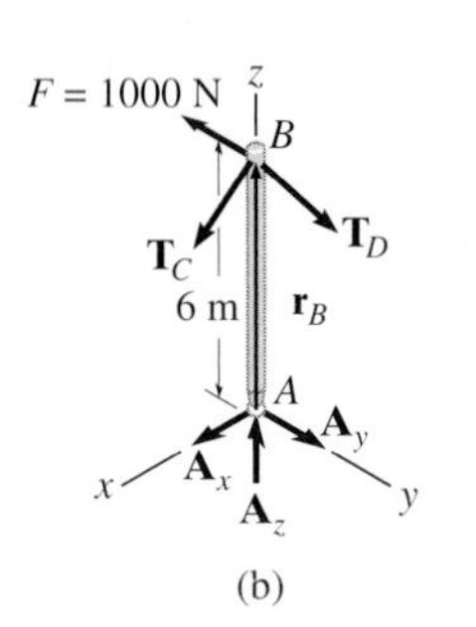

(b)

Fig. 5–29

Determine the tension in cables BC and BD and the reactions at the ball-and-socket joint A for the mast shown in Fig. 5–29a.

SOLUTION (*VECTOR ANALYSIS*)

Free-Body Diagram. There are five unknown force magnitudes shown on the free-body diagram, Fig. 5–29b.

Equations of Equilibrium. Expressing each force in Cartesian vector form, we have

$$\mathbf{F} = \{-1000\mathbf{j}\}\text{ N}$$
$$\mathbf{F}_A = A_x\mathbf{i} + A_y\mathbf{j} + A_z\mathbf{k}$$
$$\mathbf{T}_C = 0.707T_C\mathbf{i} - 0.707T_C\mathbf{k}$$
$$\mathbf{T}_D = T_D\left(\frac{\mathbf{r}_{BD}}{r_{BD}}\right) = -0.333T_D\mathbf{i} + 0.667T_D\mathbf{j} - 0.667T_D\mathbf{k}$$

Applying the force equation of equilibrium gives

$$\Sigma\mathbf{F} = \mathbf{0}; \qquad \mathbf{F} + \mathbf{F}_A + \mathbf{T}_C + \mathbf{T}_D = \mathbf{0}$$
$$(A_x + 0.707T_C - 0.333T_D)\mathbf{i} + (-1000 + A_y + 0.667T_D)\mathbf{j} + (A_z - 0.707T_C - 0.667T_D)\mathbf{k} = \mathbf{0}$$

$$\Sigma F_x = 0; \qquad A_x + 0.707T_C - 0.333T_D = 0 \qquad (1)$$
$$\Sigma F_y = 0; \qquad A_y + 0.667T_D - 1000 = 0 \qquad (2)$$
$$\Sigma F_z = 0; \qquad A_z - 0.707T_C - 0.667T_D = 0 \qquad (3)$$

Summing moments about point A, we have

$$\Sigma\mathbf{M}_A = \mathbf{0}; \qquad \mathbf{r}_B \times (\mathbf{F} + \mathbf{T}_C + \mathbf{T}_D) = \mathbf{0}$$
$$6\mathbf{k} \times (-1000\mathbf{j} + 0.707T_C\mathbf{i} - 0.707T_C\mathbf{k} - 0.333T_D\mathbf{i} + 0.667T_D\mathbf{j} - 0.667T_D\mathbf{k}) = \mathbf{0}$$

Evaluating the cross product and combining terms yields

$$(-4T_D + 6000)\mathbf{i} + (4.24T_C - 2T_D)\mathbf{j} = \mathbf{0}$$
$$\Sigma M_x = 0; \qquad -4T_D + 6000 = 0 \qquad (4)$$
$$\Sigma M_y = 0; \qquad 4.24T_C - 2T_D = 0 \qquad (5)$$

The moment equation about the z axis, $\Sigma M_z = 0$, is automatically satisfied. Why? Solving Eqs. 1 through 5 we have

$$T_C = 707\text{ N} \qquad T_D = 1500\text{ N} \qquad \textit{Ans.}$$
$$A_x = 0\text{ N} \qquad A_y = 0\text{ N} \qquad A_z = 1500\text{ N} \qquad \textit{Ans.}$$

Since the mast is a two-force member, note that the value $A_x = A_y = 0$ could have been determined *by inspection*.

Example 5–18

Rod AB shown in Fig. 5–30a is subjected to the 200-N force. Determine the reactions at the ball-and-socket joint A and the tension in cables BD and BE.

SOLUTION (*VECTOR ANALYSIS*)

Free-Body Diagram. Fig. 5–30b.

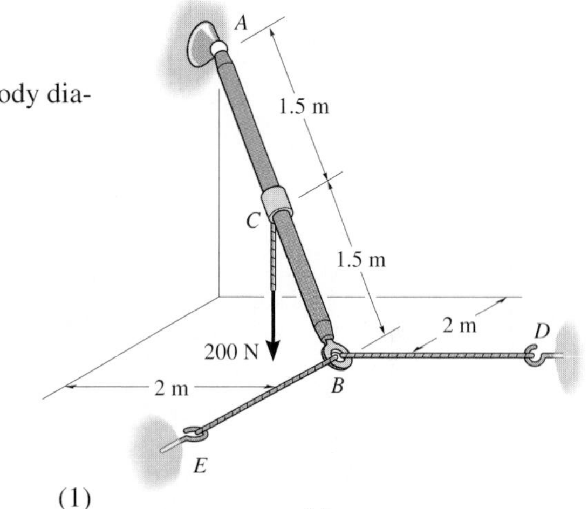

(a)

Equations of Equilibrium. Representing each force on the free-body diagram in Cartesian vector form, we have

$$\mathbf{F}_A = A_x\mathbf{i} + A_y\mathbf{j} + A_z\mathbf{k}$$
$$\mathbf{T}_E = T_E\mathbf{i}$$
$$\mathbf{T}_D = T_D\mathbf{j}$$
$$\mathbf{F} = \{-200\mathbf{k}\}\ \text{N}$$

Applying the force equation of equilibrium,

$\Sigma\mathbf{F} = \mathbf{0};$ $$\mathbf{F}_A + \mathbf{T}_E + \mathbf{T}_D + \mathbf{F} = \mathbf{0}$$

$$(A_x + T_E)\mathbf{i} + (A_y + T_D)\mathbf{j} + (A_z - 200)\mathbf{k} = \mathbf{0}$$

$\Sigma F_x = 0;$ $$A_x + T_E = 0 \qquad (1)$$

$\Sigma F_y = 0;$ $$A_y + T_D = 0 \qquad (2)$$

$\Sigma F_z = 0;$ $$A_z - 200 = 0 \qquad (3)$$

Summing moments about point A yields

$\Sigma\mathbf{M}_A = \mathbf{0};$ $$\mathbf{r}_C \times \mathbf{F} + \mathbf{r}_B \times (\mathbf{T}_E + \mathbf{T}_D) = \mathbf{0}$$

Since $\mathbf{r}_C = \frac{1}{2}\mathbf{r}_B$, then

$$(1\mathbf{i} + 1\mathbf{j} - 0.5\mathbf{k}) \times (-200\mathbf{k}) + (2\mathbf{i} + 2\mathbf{j} - 1\mathbf{k}) \times (T_E\mathbf{i} + T_D\mathbf{j}) = \mathbf{0}$$

Expanding and rearranging terms gives

$$(T_D - 200)\mathbf{i} + (-T_E + 200)\mathbf{j} + (2T_D - 2T_E)\mathbf{k} = \mathbf{0}$$

$\Sigma M_x = 0;$ $$T_D - 200 = 0 \qquad (4)$$

$\Sigma M_y = 0;$ $$-T_E + 200 = 0 \qquad (5)$$

$\Sigma M_z = 0;$ $$2T_D - 2T_E = 0 \qquad (6)$$

Solving Eqs. 1 through 6, we get

$$A_x = A_y = -200\ \text{N} \qquad \textit{Ans.}$$
$$A_z = T_E = T_D = 200\ \text{N} \qquad \textit{Ans.}$$

The negative sign indicates that $\mathbf{A}_x$ and $\mathbf{A}_y$ have a sense which is opposite to that shown on the free-body diagram, Fig. 5–30b.

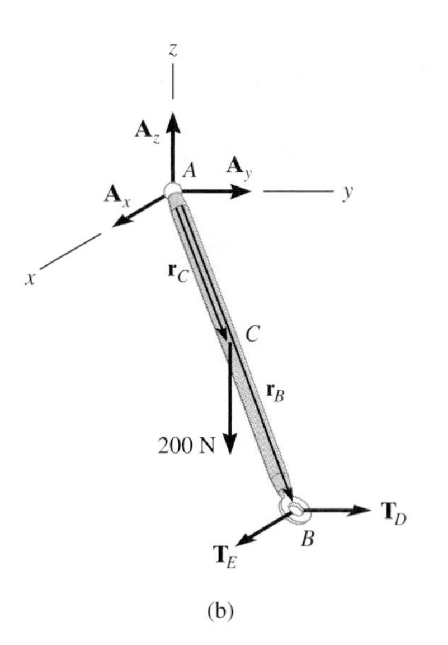

(b)

Fig. 5–30

Example 5–19

The bent rod in Fig. 5–31*a* is supported at *A* by a journal bearing, at *D* by a ball-and-socket joint, and at *B* by means of cable *BC*. Using only *one equilibrium equation,* obtain a direct solution for the tension in cable *BC*. The bearing at *A* is capable of exerting force components only in the *z* and *y* directions, since it is properly aligned on the shaft.

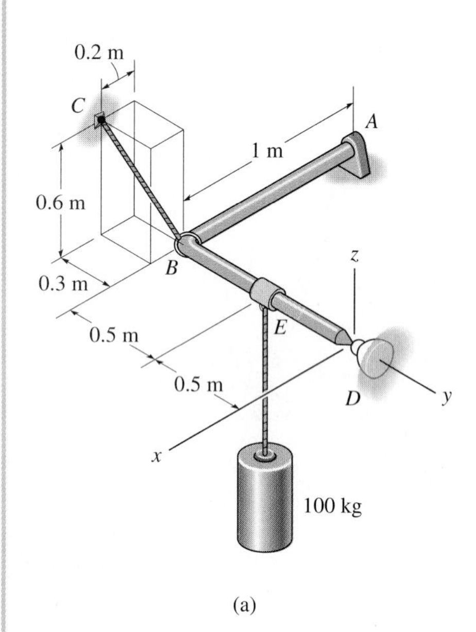

(a)

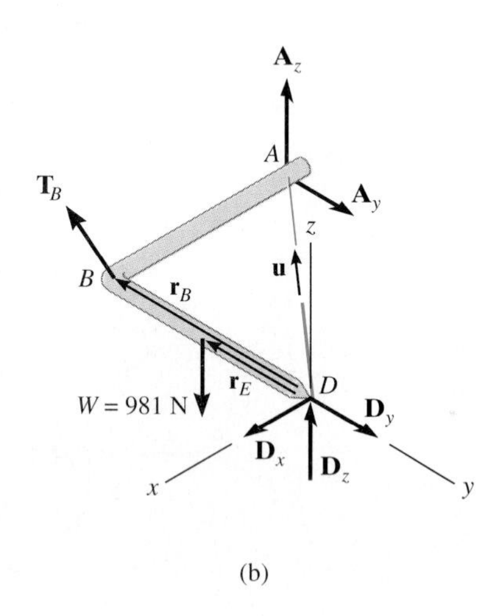

(b)

Fig. 5–31

SOLUTION (*VECTOR ANALYSIS*)

Free-Body Diagram. As shown in Fig. 5–31*b*, there are six unknowns: three force components caused by the ball-and-socket joint, two caused by the bearing, and one caused by the cable.

Equations of Equilibrium. The cable tension $\mathbf{T}_B$ may be obtained *directly* by summing moments about an axis passing through points *D* and *A*. Why? The direction of the axis is defined by the unit vector **u,** where

$$\mathbf{u} = \frac{\mathbf{r}_{DA}}{r_{DA}} = -\frac{1}{\sqrt{2}}\mathbf{i} - \frac{1}{\sqrt{2}}\mathbf{j}$$
$$= -0.707\mathbf{i} - 0.707\mathbf{j}$$

Hence, the sum of the moments about this axis is zero provided

$$\Sigma M_{DA} = \mathbf{u} \cdot \Sigma(\mathbf{r} \times \mathbf{F}) = 0$$

Here **r** represents a position vector drawn from *any point* on the axis *DA* to any point on the line of action of force **F** (see Eq. 4–11). With reference to Fig. 5–31*b*, we can therefore write

$$\mathbf{u} \cdot (\mathbf{r}_B \times \mathbf{T}_B + \mathbf{r}_E \times \mathbf{W}) = \mathbf{0}$$

$$(-0.707\mathbf{i} - 0.707\mathbf{j}) \cdot [(-1\mathbf{j}) \times \left(\tfrac{0.2}{0.7}T_B\mathbf{i} - \tfrac{0.3}{0.7}T_B\mathbf{j} + \tfrac{0.6}{0.7}T_B\mathbf{k}\right)$$
$$+ (-0.5\mathbf{j}) \times (-981\mathbf{k})] = \mathbf{0}$$

$$(-0.707\mathbf{i} - 0.707\mathbf{j}) \cdot [(-0.857T_B + 490.5)\mathbf{i} + 0.286T_B\mathbf{k}] = \mathbf{0}$$

$$-0.707(-0.857T_B + 490.5) + 0 + 0 = 0$$

$$T_B = \frac{490.5}{0.857} = 572 \text{ N} \qquad \textit{Ans.}$$

The advantage of using Cartesian vectors for this solution should be noted. It would be especially tedious to determine the perpendicular distance from the *DA* axis to the line of action of $\mathbf{T}_B$ using scalar methods.

Note: In Example 5–17, a direct solution for A_z is possible by summing moments about an axis passing through the supports at *C* and *D*, Fig. 5–29*a*. If this is done only the moment of **F** and $\mathbf{A}_z$ must be considered. Go back to that example and try to apply the above technique to determine the result $A_z = 1500$ N.

PROBLEMS

5–63. Determine the x, y, z components of reaction at the fixed wall A. The 150-N force is parallel to the z axis and the 200-N force is parallel to the y axis.

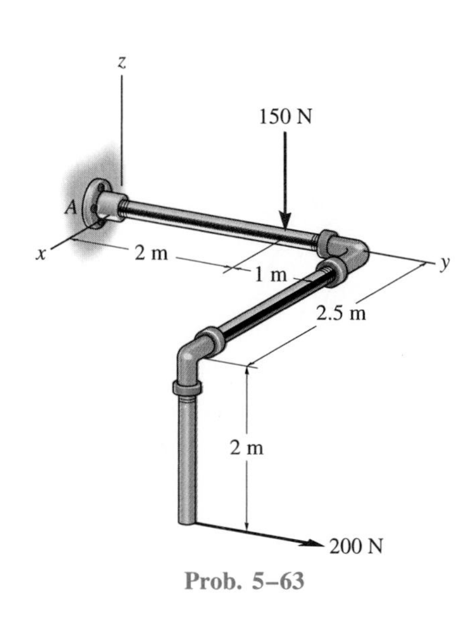

Prob. 5–63

***5–64.** The wing of the jet aircraft is subjected to a thrust of $T = 8$ kN from its engine and the resultant lift force $L = 45$ kN. If the mass of the wing is 2.1 Mg and the mass center is at G, determine the x, y, z components of reaction where the wing is fixed to the fuselage at A.

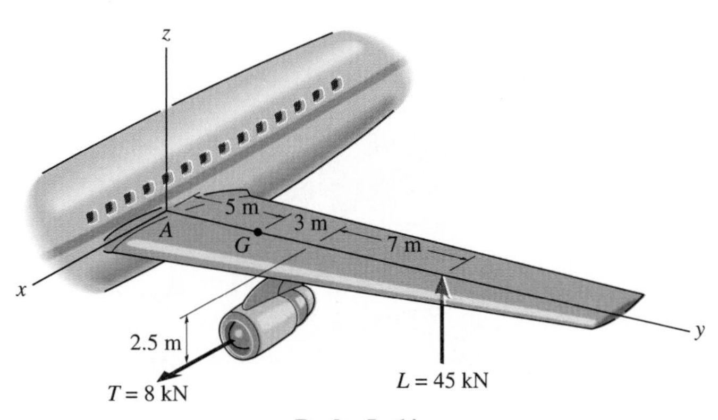

Prob. 5–64

5–65. The uniform concrete slab has a weight of 5500 lb. Determine the tension in each of the three parallel supporting cables when the slab is held in the horizontal plane as shown.

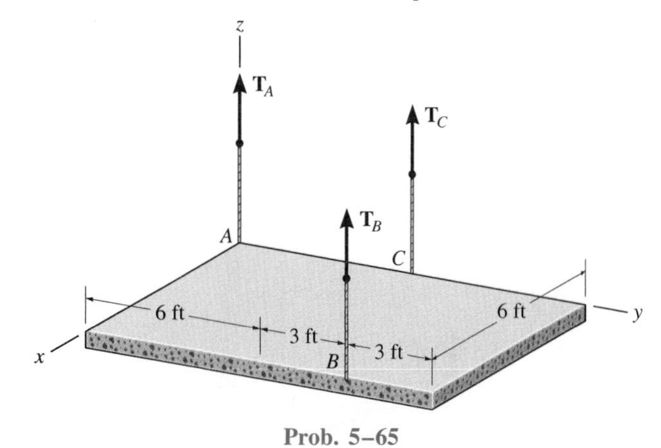

Prob. 5–65

5–66. The air-conditioning unit is hoisted to the roof of a building using the three cables. If the unit has a weight of 900 lb and a center of gravity G, located at $x = 5.5$ ft, $y = 6$ ft, determine the tension in each of the cables for equilibrium.

5–67. The air-conditioning unit is hoisted to the roof of a building using the three cables. If the tensions in the cables are $T_A = 250$ lb, $T_B = 300$ lb, and $T_C = 200$ lb, determine the weight of the unit and the location (x, y) of its center of gravity G.

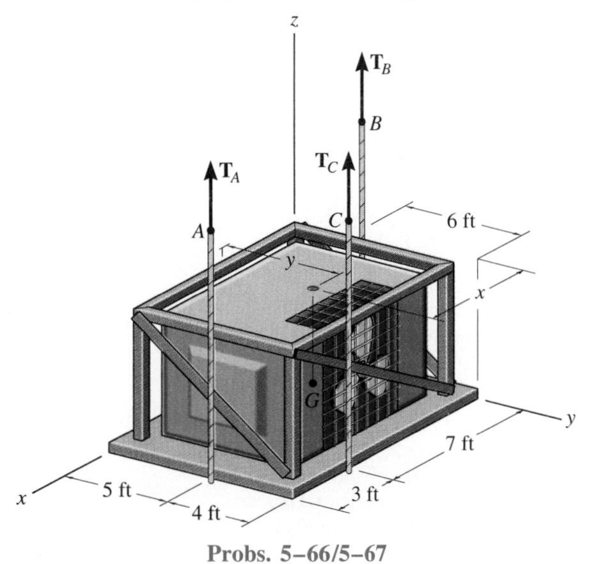

Probs. 5–66/5–67

*5–68. The platform truck supports the three loadings shown. Determine the normal reactions on each of its three wheels.

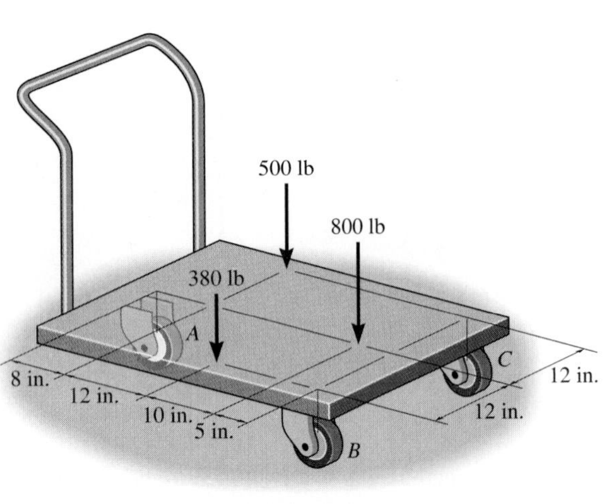
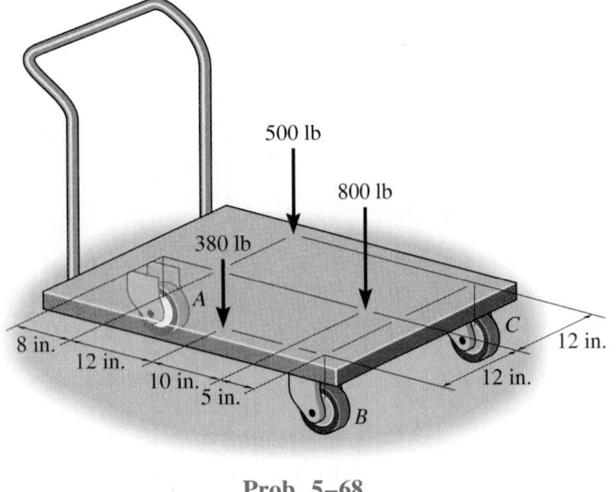

Prob. 5–68

5–69. Determine the force components acting on the ball-and-socket at A, the reaction at the roller B and the tension in the cord CD needed for equilibrium of the quarter circular plate.

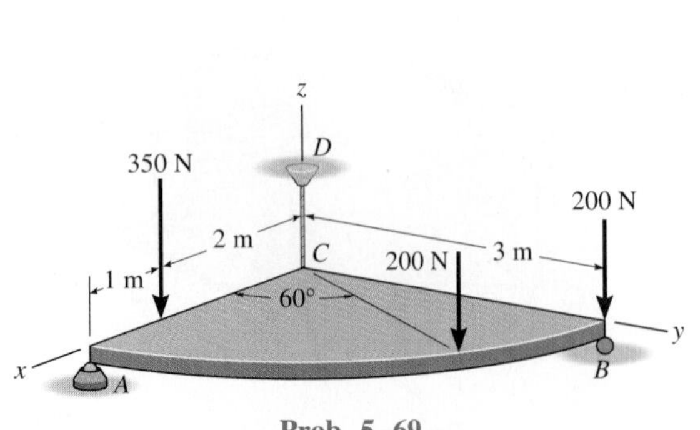

Prob. 5–69

5–70. The pole for a power line is subjected to the two cable forces of 60 lb, each force lying in a plane parallel to the x-y plane. If the tension in the guy wire AB is 80 lb, determine the x, y, z components of reaction at the base of the pole, O

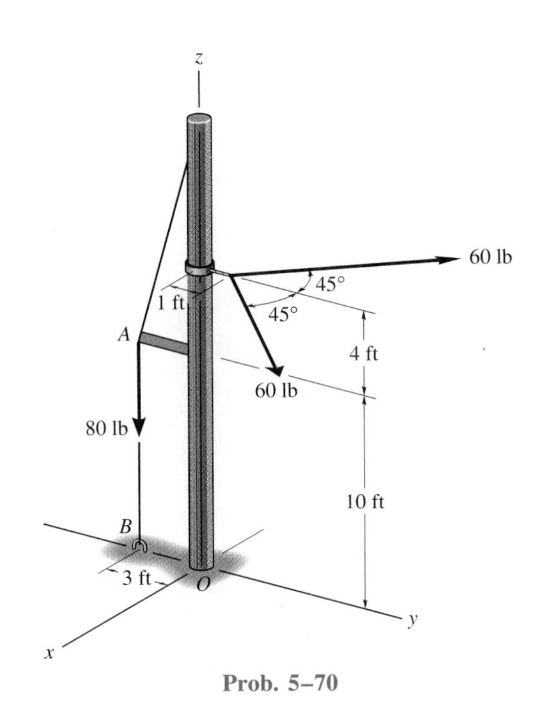

Prob. 5–70

5–71. The windlass is subjected to a load of 150 lb. Determine the horizontal force **P** needed to hold the handle in the position shown, and the components of reaction at the ball-and-socket joint A and the smooth journal bearing B. The bearing at B is in proper alignment and exerts only force reactions on the windlass.

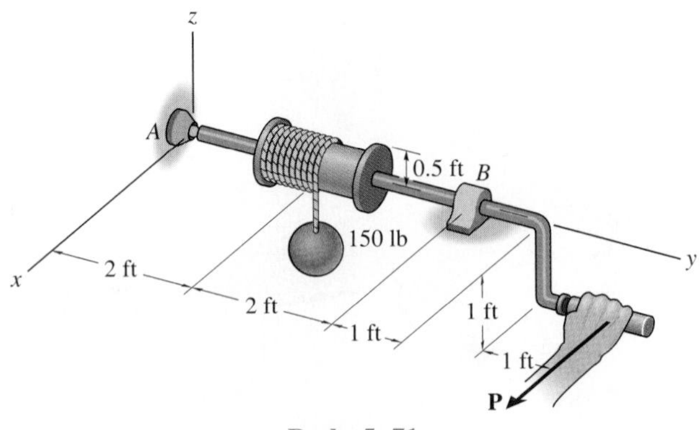

Prob. 5–71

***5–72.** The stiff-leg derrick used on ships is supported by a ball-and-socket joint at D and two cables BA and BC. The cables are attached to a smooth collar ring at B, which allows rotation of the derrick about the z axis. If the derrick supports a crate having a mass of 200 kg, determine the tension in the cables and the x, y, z components of reaction at D.

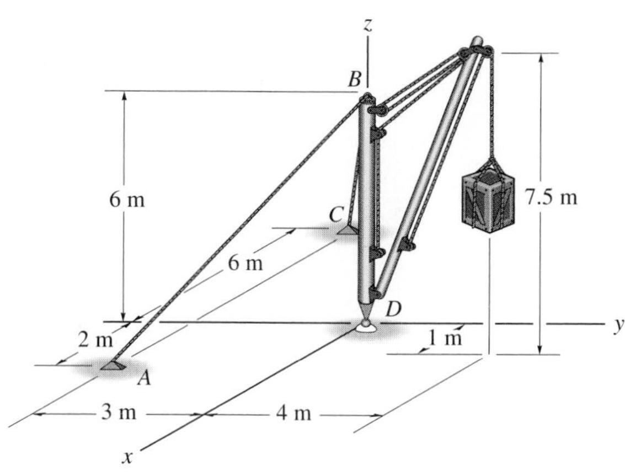

Prob. 5–72

5–73. The pole is subjected to the two forces shown. Determine the components of reaction at A assuming it to be a ball-and-socket joint. Also, compute the tension in each of the guy wires, BC and ED.

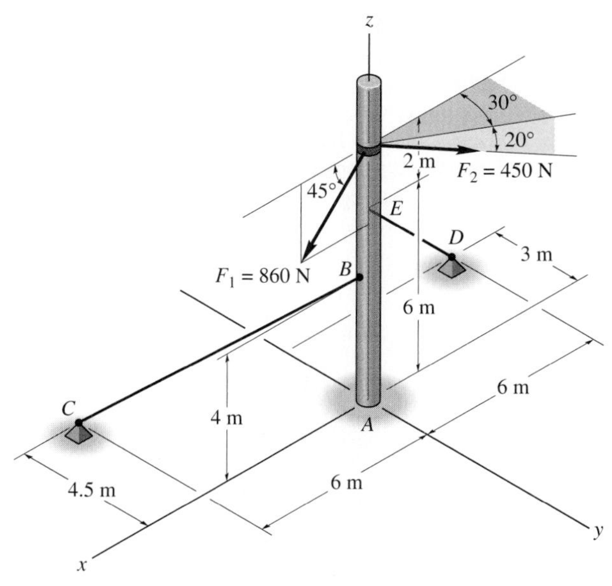

Prob. 5–73

5–74. The boom AB is held in equilibrium by a ball-and-socket joint A and a pulley and cord system as shown. Determine the x, y, z components of reaction at A and the tension in cable DEC if $\mathbf{F} = \{-1500\mathbf{k}\}$ lb.

5–75. The cable CED can sustain a maximum tension of 800 lb before it fails. Determine the greatest vertical force F that can be applied to the boom. Also, what are the x, y, z components of reaction at the ball-and-socket joint A?

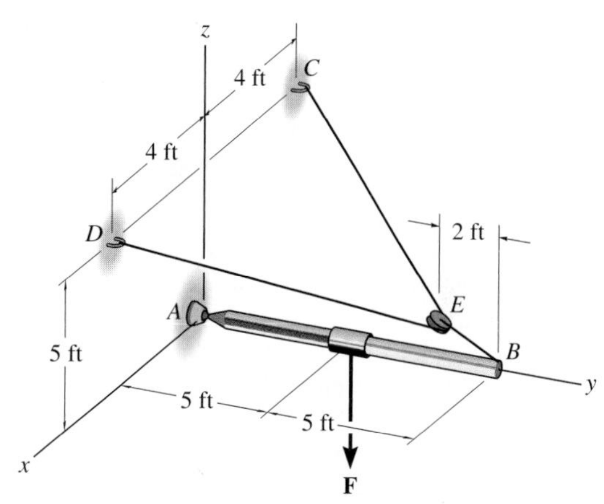

Probs. 5–74/5–75

***5–76.** The power line is subjected to a tension of 1700 lb. If the insulator AB weighs 50 lb with center of gravity at G, determine the angle θ it makes with the pole.

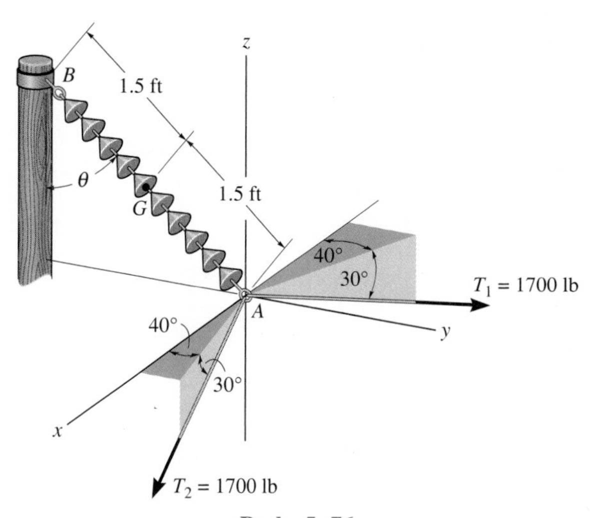

Prob. 5–76

5–77. The boom is supported by a ball-and-socket joint at A and a guy wire at B. If the loads in the cables are each 5 kN and they lie in a plane which is parallel to the x-z plane, determine the components of reaction at A for equilibrium.

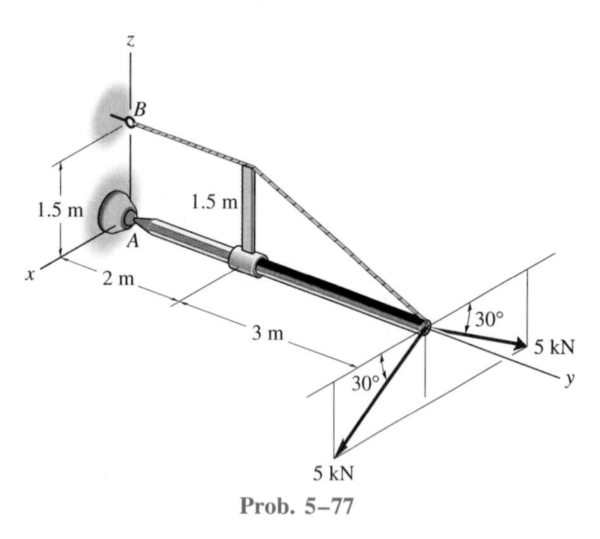

Prob. 5–77

5–78. The pipe assembly supports the vertical loads shown. Determine the components of reaction at the ball-and-socket joint A and the tension in the supporting cables BC and BD.

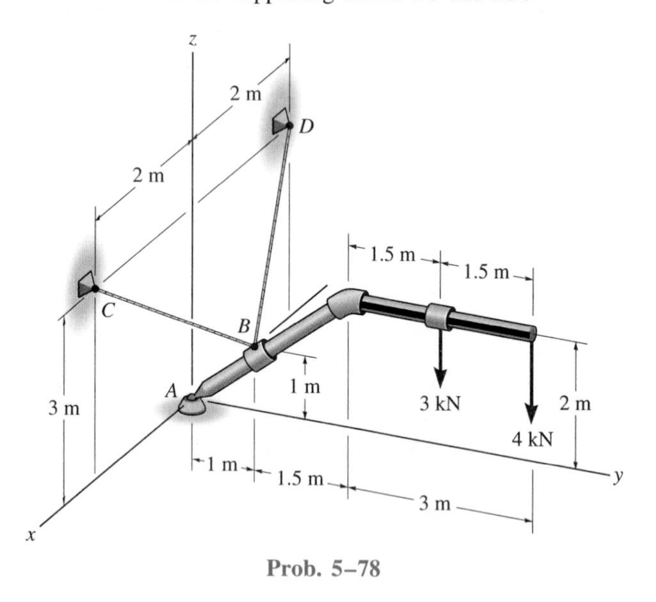

Prob. 5–78

5–79. The boom supports a crate having a weight of 850 lb. Determine the x, y, z components of reaction at the ball-and-socket joint A and the tension in cables BC and DE.

***5–80.** Cable BC or DE can support a maximum tension of 900 lb before it fails. Determine the greatest weight W of the crate that can be suspended from the end of the boom. Also, determine the x, y, z components of reaction at the ball-and-socket joint A.

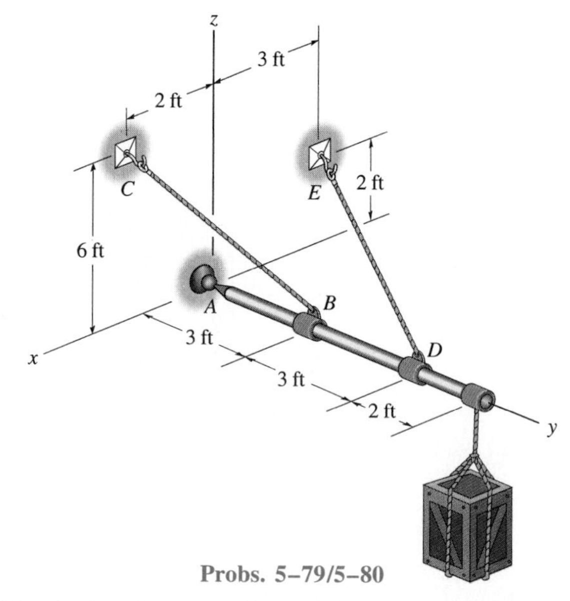

Probs. 5–79/5–80

5–81. Both pulleys are fixed to the shaft and as the shaft turns with constant angular velocity, the power of pulley A is transmitted to pulley B. Determine the horizontal tension **T** in the belt on pulley B and the x, y, z components of reaction at the journal bearing C and thrust bearing D if $\theta = 0°$. The bearings are in proper alignment and exert only force reactions on the shaft.

5–82. Both pulleys are fixed to the shaft and as the shaft turns with constant angular velocity, the power of pulley A is transmitted to pulley B. Determine the horizontal tension **T** in the belt on pulley B and the x, y, z components of reaction at the journal bearing C and thrust bearing D if $\theta = 45°$. The bearings are in proper alignment and exert only force reactions on the shaft.

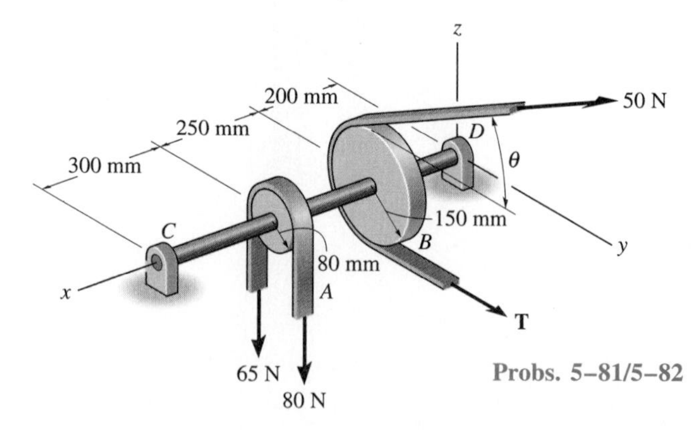

Probs. 5–81/5–82

5–83. The platform has a mass of 3 Mg and center of mass located at G. If it is lifted with constant velocity using the three cables, determine the force in each of the cables.

***5–84.** The platform has a mass of 2 Mg and center of mass located at G. If it is lifted using the three cables, determine the force in each of these cables. Solve for each force by using a single moment equation of equilibrium.

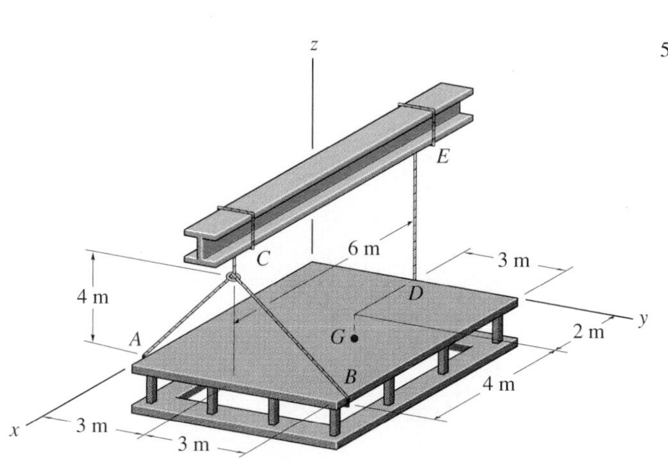

Probs. 5–83/5–84

5–85. The cables exert the forces shown on the pole. Assuming the pole is supported by a ball-and-socket joint at its base, determine the components of reaction at A. The forces of 140 lb and 75 lb lie in a horizontal plane.

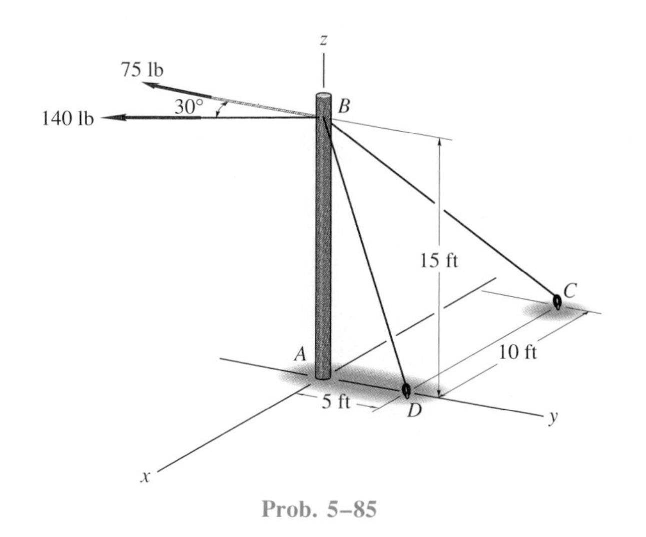

Prob. 5–85

5–86. The silo has a weight of 3500 lb and a center of gravity at G. Determine the vertical component of force that each of the three struts at A, B, and C exerts on the silo if it is subjected to a resultant wind loading of 250 lb which acts in the direction shown.

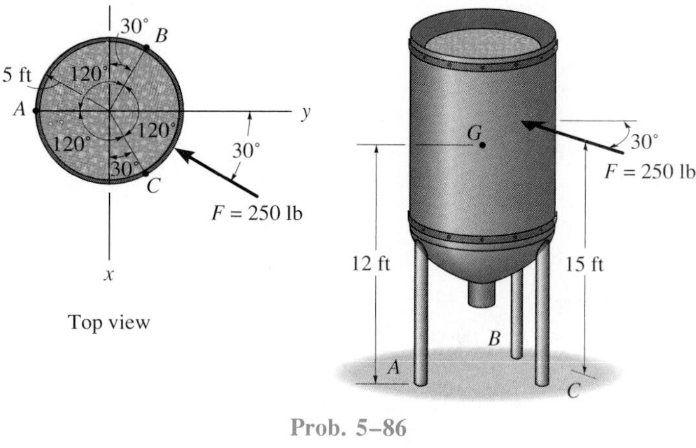

Prob. 5–86

5–87. Member AB is supported by a cable BC and at A by a *square* rod which fits loosely through the square hole at the end joint of the member as shown. Determine the components of reaction at A and the tension in the cable needed to hold the 800-lb cylinder in equilibrium.

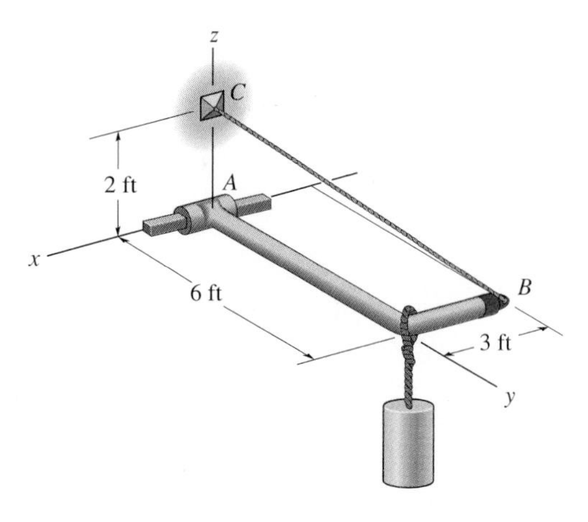

Prob. 5–87

*5–88. Determine the tensions in the cables and the components of reaction acting on the smooth collar at A necessary to hold the 50-lb sign in equilibrium. The center of gravity for the sign is at G.

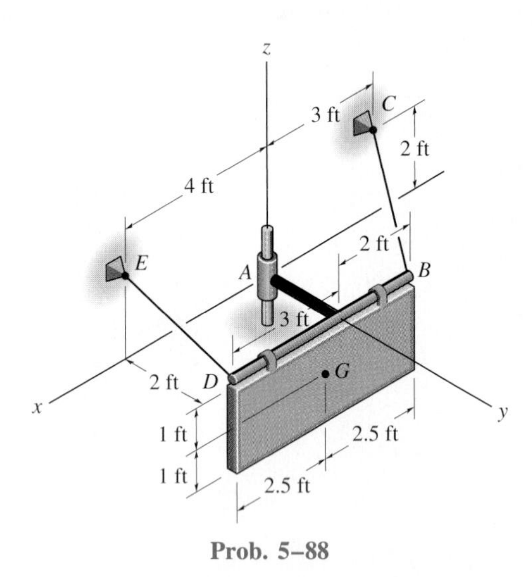

Prob. 5–88

5–89. The boom AC is supported at A by a ball-and-socket joint and by two cables BDC and CE. Cable BDC is continuous and passes over a pulley at D. Calculate the tension in the cables and the x, y, z components of reaction at A if a crate has a weight of 80 lb.

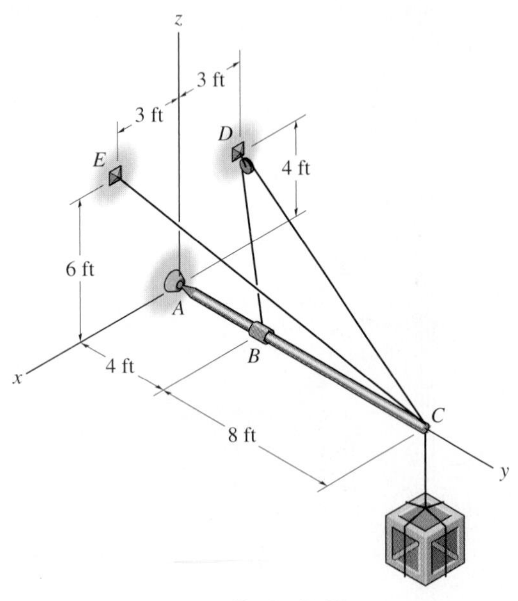

Prob. 5–89

5–90. The bent rod is supported at A, B, and C by smooth journal bearings. Compute the x, y, z components of reaction at the bearings if the rod is subjected to forces $F_1 = 300$ lb and $F_2 = 250$ lb. F_1 lies in the y-z plane. The bearings are in proper alignment and exert only force reactions on the rod.

5–91. The bent rod is supported at A, B, and C by smooth journal bearings. Determine the magnitude of $\mathbf{F}_2$ which will cause the reaction $\mathbf{C}_y$ at the bearing C to be equal to zero. The bearings are in proper alignment and exert only force reactions on the rod. Set $F_1 = 300$ lb.

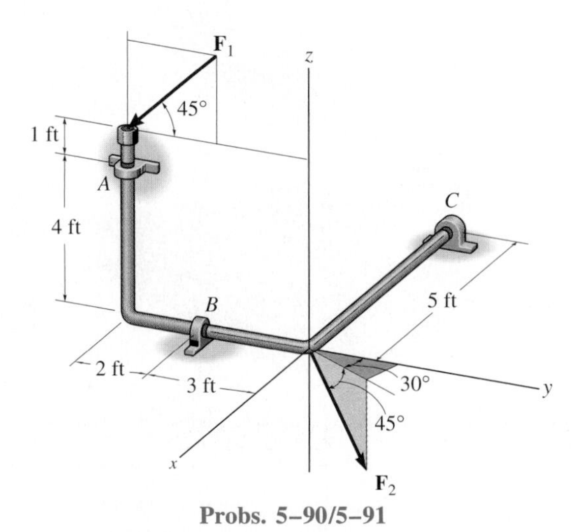

Probs. 5–90/5–91

*5–92. The bar AB is supported by two smooth collars. At A the connection is with a ball-and-socket joint and at B it is a rigid attachment. If a 50-lb load is applied to the bar, determine the x, y, z components of reaction at A and B.

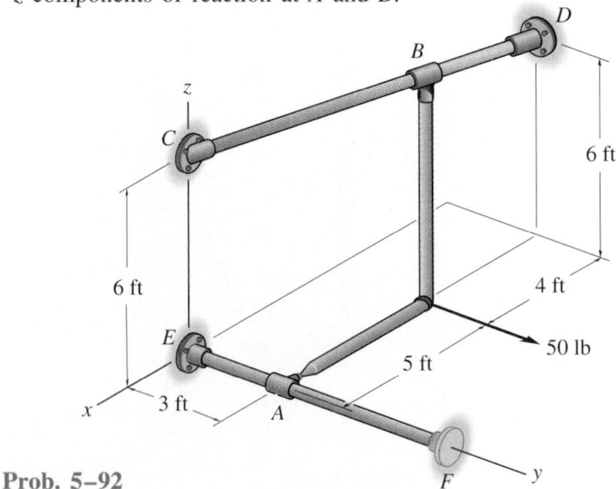

Prob. 5–92

5–93. The left shaft of the universal joint (or Hooke's joint) is subjected to a torque (or couple moment) of 50 N · m. Determine the required equilibrium couple moment M' on the connected shaft when the shafts are in the position $\theta = 30°$, $\phi = 60°$ as shown. Axles AB and CD are perpendicular to one another and are free to turn in their bearings.

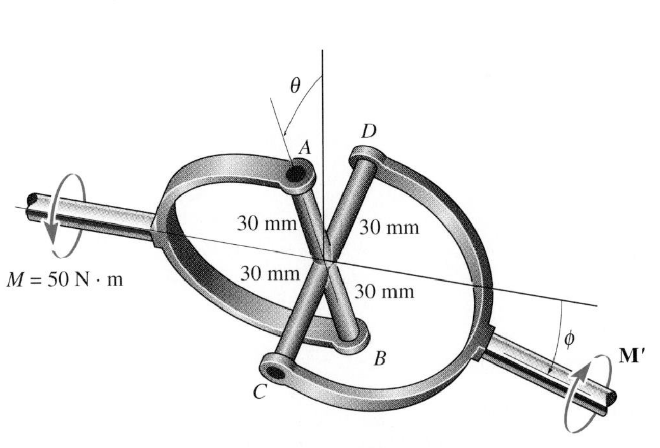

Prob. 5–93

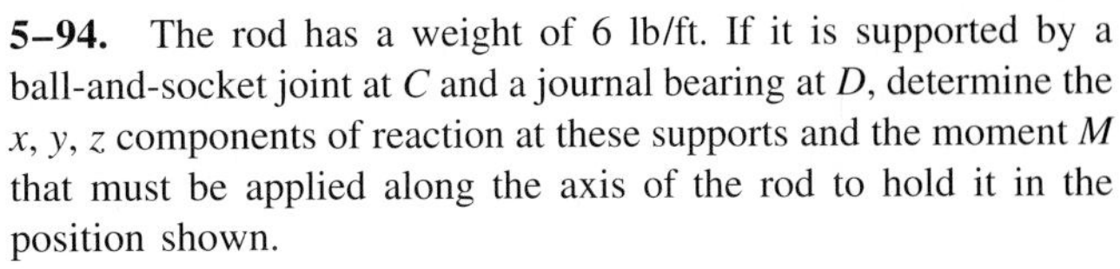

5–94. The rod has a weight of 6 lb/ft. If it is supported by a ball-and-socket joint at C and a journal bearing at D, determine the x, y, z components of reaction at these supports and the moment M that must be applied along the axis of the rod to hold it in the position shown.

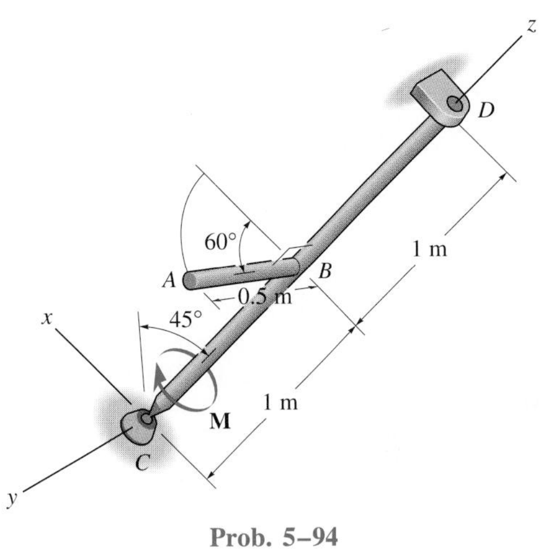

Prob. 5–94

REVIEW PROBLEMS

5–95. Determine the x and z components of reaction at the journal bearing A and the tension in cords BC and BD necessary for equilibrium of the rod.

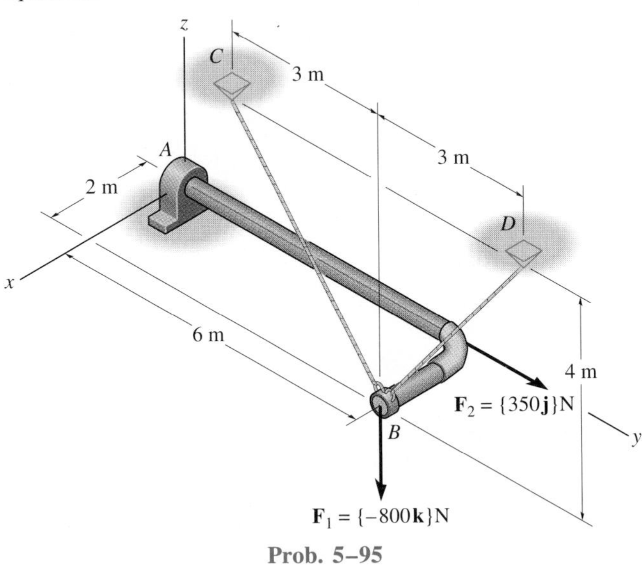

Prob. 5–95

***5–96.** The shaft assembly is supported by two smooth journal bearings A and B and a short link DC. If a couple moment is applied to the shaft as shown, determine the components of force reaction at the bearings and the force in the link. The link lies in a plane parallel to the y-z plane and the bearings are properly aligned on the shaft.

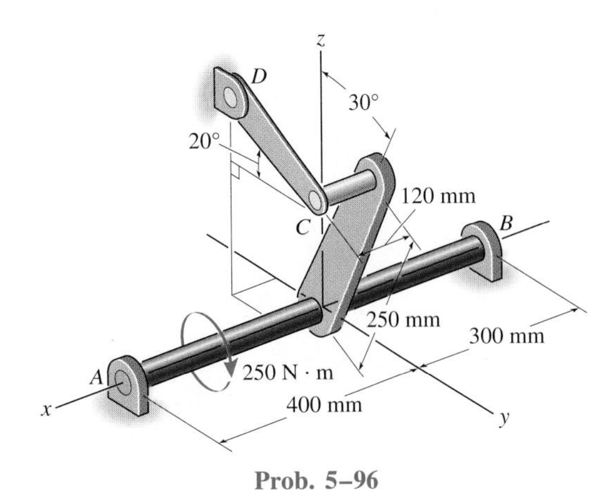

Prob. 5–96

5–97. Determine the reactions at the roller A and pin B.

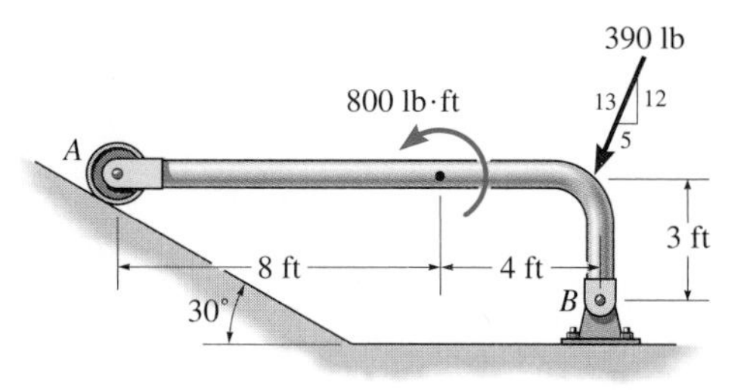

Prob. 5–97

5–98. The wheel of radius R has a spring attached to its central hub. If the spring has a stiffness k and unstretched length R, show that the *horizontal* force **F** needed to pull the wheel forward so that the spring makes an angle θ with the horizontal is $F = kR(\cot\theta - \cos\theta)$.

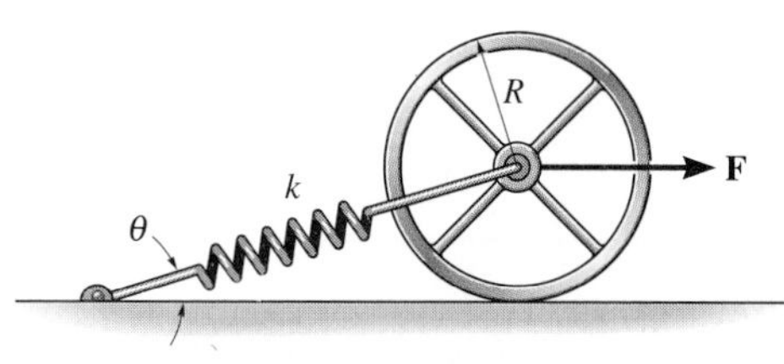

Prob. 5–98

5–99. Determine the reactions at the roller A and pin B.

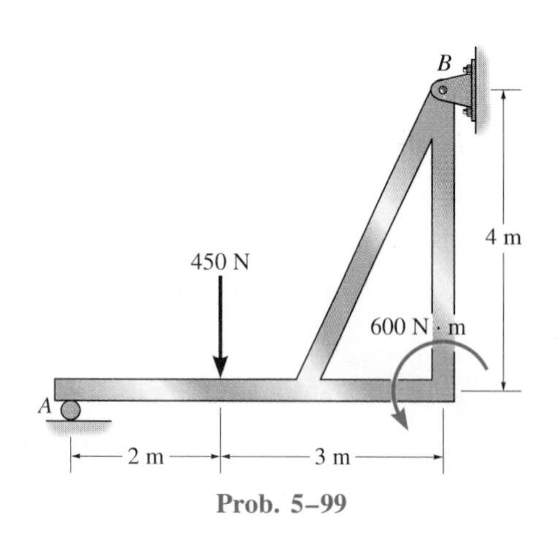

Prob. 5–99

***5–100.** The member is supported by cable BC and at A by a smooth fixed *square* rod which fits loosely through the square hole of the collar. Determine the x, y, z components of reaction at A and the tension in the cable needed to hold the member in equilibrium.

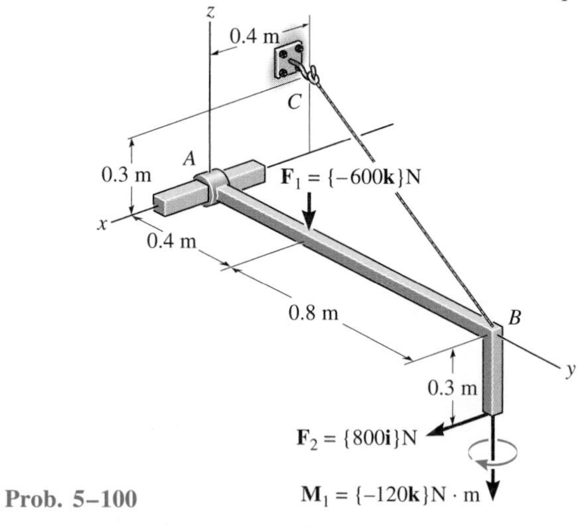

Prob. 5–100

5–101. Member AB is supported at B by a cable and at A by a smooth fixed *square* rod which fits loosely through the square hole of the collar. If $\mathbf{F} = \{20\mathbf{i} - 40\mathbf{j} - 75\mathbf{k}\}$ lb, determine the x, y, z components of reaction at A and the tension in the cable.

5–102. Member AB is supported at B by a cable and at A by a smooth fixed *square* rod which fits loosely through the square hole of the collar. Determine the tension in cable BC if the force $\mathbf{F} = \{-45\mathbf{k}\}$ lb.

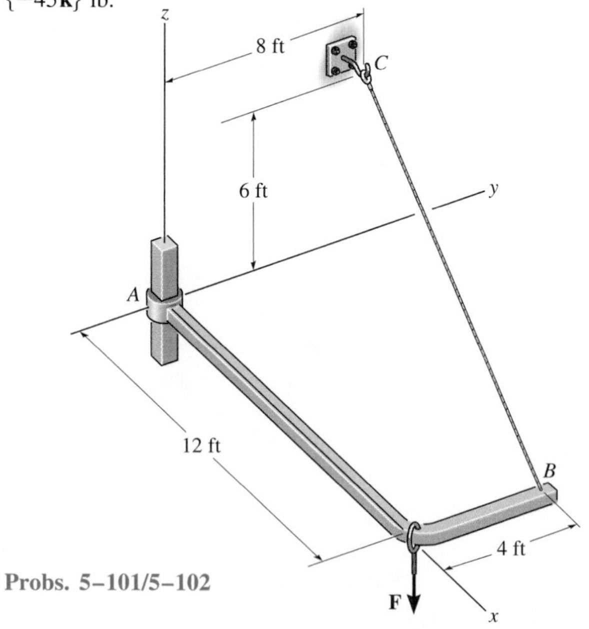

Probs. 5–101/5–102

5–103. The wall footing is used to support the column load of 12,000 lb. Determine the intensities w_1 and w_2 of the distributed loading acting on the base of the footing for equilibrium.

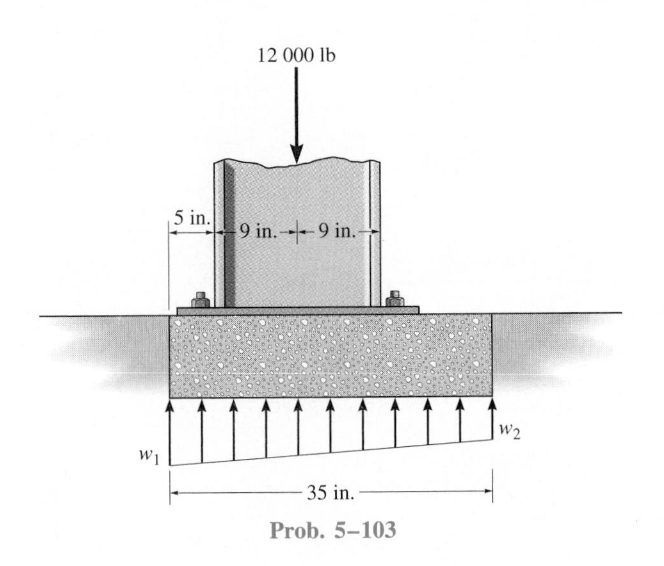

Prob. 5–103

5–105. The uniform rod AB has a mass of 5 kg and is supported by a ball-and-socket joint at A, a cord BC, and a smooth wall at B. Determine the x, y, z components of reaction at the supports.

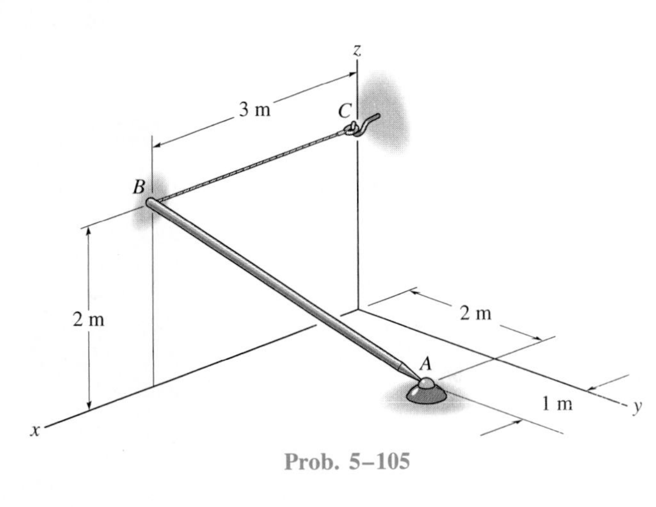

Prob. 5–105

***5–104.** Compute the horizontal and vertical components of force at pin B. The belt is subjected to a tension of $T = 100$ N and passes over each of the three pulleys.

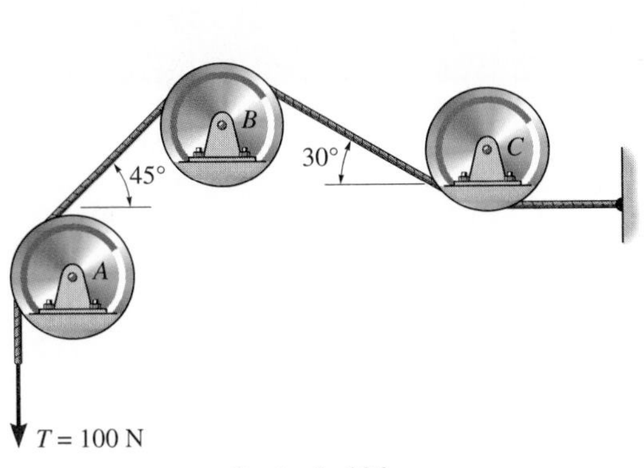

Prob. 5–104

5–106. Determine the reactions at roller A and pin B for equilibrium of the member.

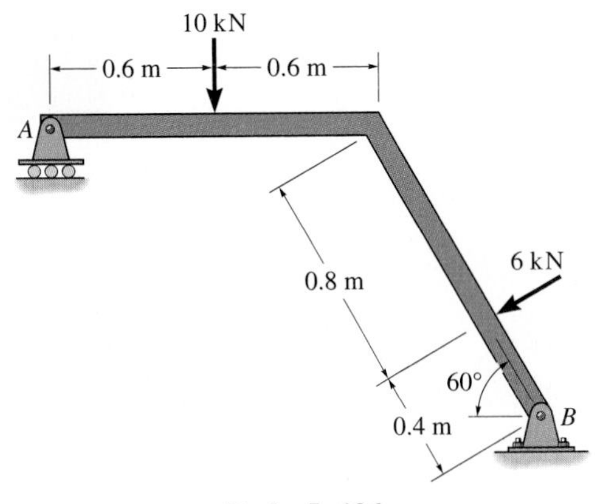

Prob. 5–106

The forces in the truss members of the booms and in the supporting cables of these tower cranes can be determined using the principles discussed in this chapter.

6

Structural Analysis

In this chapter we will use the equations of equilibrium to analyze structures composed of pin-connected members. The analysis is based on the principle that if a structure is in equilibrium, then each of its members is also in equilibrium. By applying the equations of equilibrium to the various parts of a simple truss, frame, or machine, we will be able to determine all the forces acting at the connections.

The topics in this chapter are very important since they provide practice in drawing free-body diagrams, using the principle of action, equal but opposite collinear force reaction, and applying the equations of equilibrium.

6.1 Simple Trusses

A *truss* is a structure composed of slender members joined together at their end points. The members commonly used in construction consist of wooden struts or metal bars. The joint connections are usually formed by bolting or welding the ends of the members to a common plate, called a *gusset plate,* as shown in Fig. 6–1*a*, or by simply passing a large bolt or pin through each of the members, Fig. 6–1*b*.

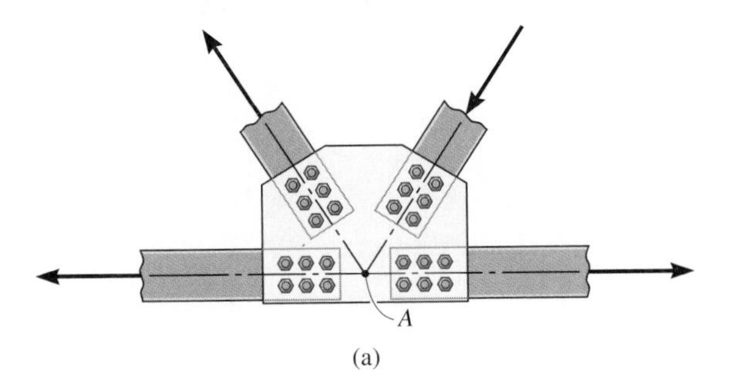

(a)

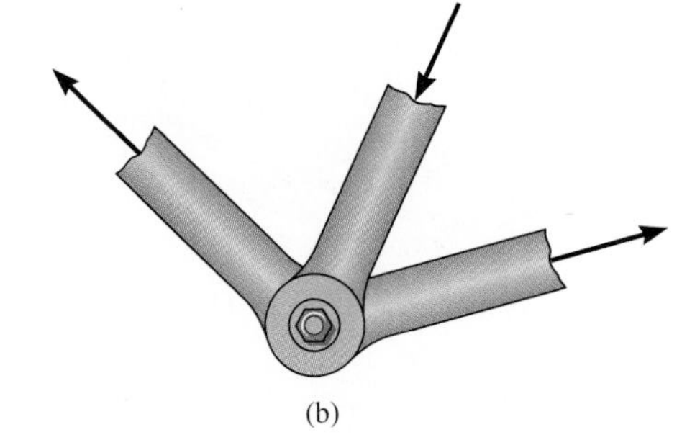
(b)

Fig. 6–1

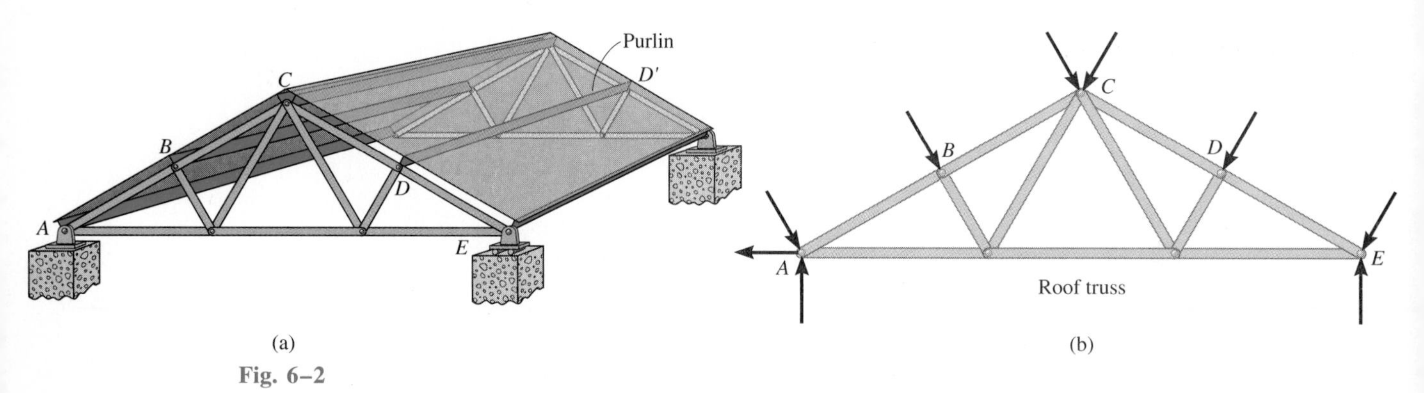

Fig. 6–2

Planar Trusses. *Planar* trusses lie in a single plane and are often used to support roofs and bridges. The truss *ABCDE,* shown in Fig. 6–2*a*, is an example of a typical roof-supporting truss. In this figure, the roof load is transmitted to the truss *at the joints* by means of a series of *purlins,* such as *DD'*. Since the imposed loading acts in the same plane as the truss, Fig. 6–2*b*, the analysis of the forces developed in the truss members is two-dimensional.

In the case of a bridge, such as shown in Fig. 6–3*a*, the load on the deck is first transmitted to *stringers,* then to *floor beams,* and finally to the *joints B*, *C*, and *D* of the two supporting side trusses. Like the roof truss, the bridge truss loading is also coplanar, Fig. 6–3*b*.

When bridge or roof trusses extend over large distances, a rocker or roller is commonly used for supporting one end, joint *E* in Figs. 6–2*a* and 6–3*a*. This type of support allows freedom for expansion or contraction of the members due to temperature or application of loads.

Fig. 6–3

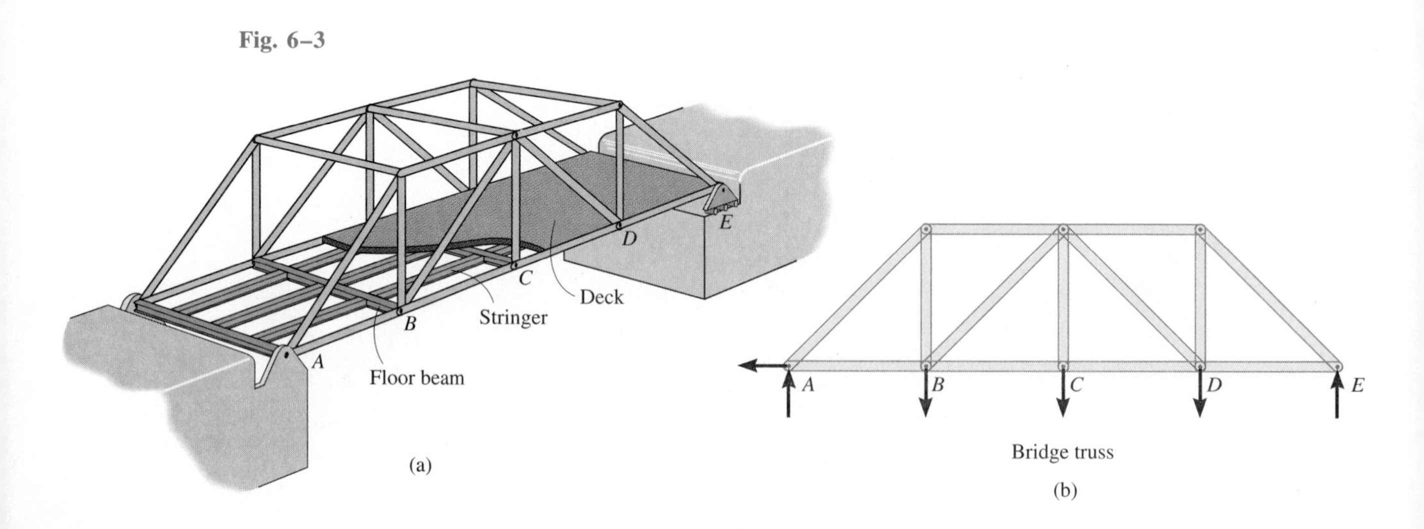

Assumptions for Design. To design both the members and the connections of a truss, it is first necessary to determine the *force* developed in each member when the truss is subjected to a given loading. In this regard, two important assumptions will be made:

1. *All loadings are applied at the joints.* In most situations, such as for bridge and roof trusses, this assumption is true. Frequently in the force analysis the weights of the members are neglected, since the forces supported by the members are usually large in comparison with their weights. If the member's weight is to be included in the analysis, it is generally satisfactory to apply it as a vertical force, half of its magnitude applied at each end of the member.
2. *The members are joined together by smooth pins.* In cases where bolted or welded joint connections are used, this assumption is satisfactory provided the center lines of the joining members are *concurrent,* as in the case of point A in Fig. 6–1a.

Because of these two assumptions, *each truss member acts as a two-force member,* and therefore the forces at the ends of the member must be directed along the axis of the member. If the force tends to *elongate* the member, it is a *tensile force* (**T**), Fig. 6–4a; whereas if it tends to *shorten* the member, it is a *compressive force* (**C**), Fig. 6–4b. In the actual design of a truss it is important to state whether the nature of the force is tensile or compressive. Often, compression members must be made *thicker* than tension members, because of the buckling or column effect that occurs when a member is in compression.

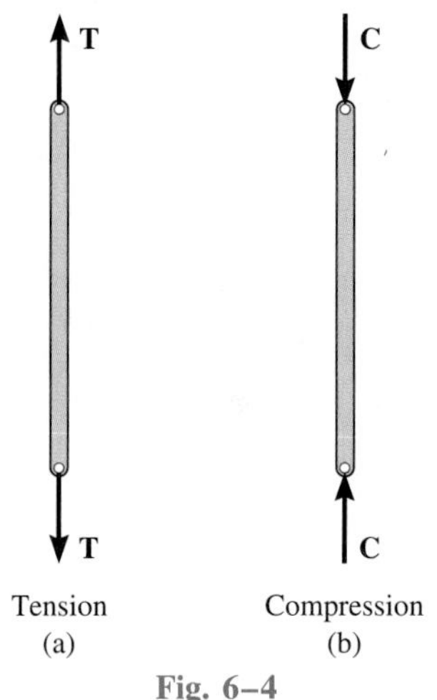

Fig. 6–4

Simple Truss. To prevent collapse, the form of a truss must be rigid. Obviously, the four-bar shape $ABCD$ in Fig. 6–5 will collapse unless a diagonal, such as AC, is added for support. The simplest form that is rigid or stable is a *triangle.* Consequently, a *simple truss* is constructed by *starting* with a basic triangular element, such as ABC in Fig. 6–6, and connecting two members (AD and BD) to form an additional element. Thus it is seen that as each additional element of two members is placed on the truss, the number of joints for a simple truss is increased by one.

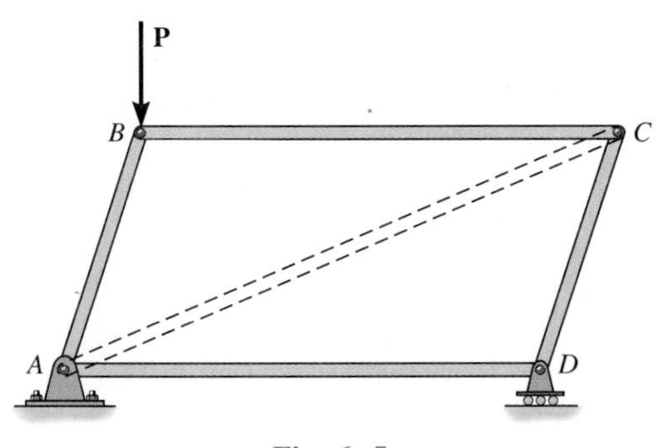

Fig. 6–5

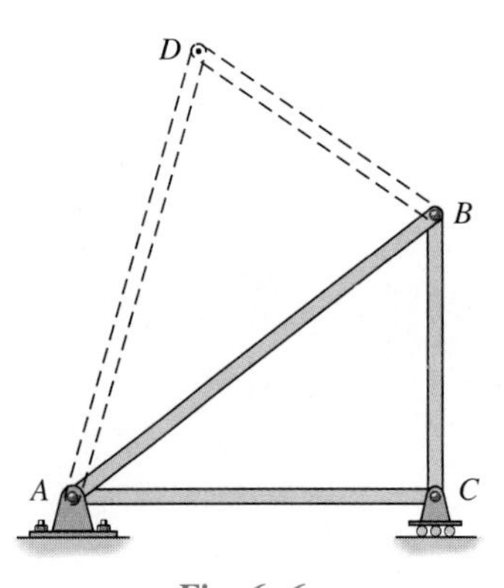

Fig. 6–6

6.2 The Method of Joints

If a truss is in equilibrium, then each of its joints must also be in equilibrium. The method of joints is based on this fact, since it consists of satisfying the equilibrium conditions for the forces exerted *on the pin* at each joint of the truss. Because the truss members are all straight two-force members lying in the same plane, the force system acting at each pin is *coplanar and concurrent.* Consequently, rotational or moment equilibrium is automatically satisfied at the joint (or pin), and it is only necessary to satisfy $\Sigma F_x = 0$ and $\Sigma F_y = 0$ to ensure translational or force equilibrium.

When using the method of joints, it is *first* necessary to draw the joint's free-body diagram before applying the equilibrium equations. To do this, recall that the *line of action* of each member force acting on the joint is *specified* from the geometry of the truss, since the force in a member passes along the axis of the member. As an example, consider the pin at joint B of the truss in Fig. 6–7*a*. Three forces act on the pin, namely, the 500-N force and the forces exerted by members BA and BC. The free-body diagram is shown in Fig. 6–7*b*. As shown, $\mathbf{F}_{BA}$ is "pulling" on the pin, which means that member BA is in *tension;* whereas $\mathbf{F}_{BC}$ is "pushing" on the pin, and consequently member BC is in *compression.* These effects are clearly demonstrated by isolating the joint with small segments of the member connected to the pin, Fig. 6–7*c*. Notice that pushing or pulling on these small segments indicates the effect of the member being either in compression or tension.

In all cases, the analysis should start at a joint having at least one known force and at most two unknown forces, as in Fig. 6–7*b*. In this way, application of $\Sigma F_x = 0$ and $\Sigma F_y = 0$ yields two algebraic equations which can be solved for the two unknowns. When applying these equations, the correct sense of an unknown member force can be determined using one of two possible methods.

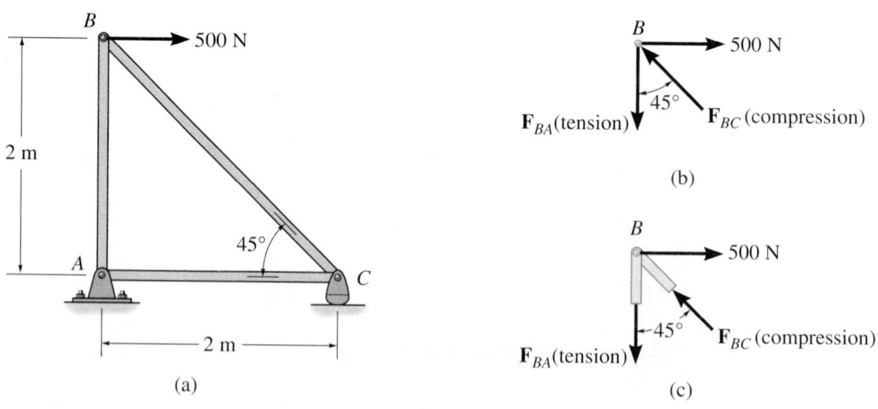

Fig. 6–7

1. *Always assume* the *unknown member forces* acting on the joint's free-body diagram to be in *tension,* i.e., ''pulling'' on the pin. If this is done, then numerical solution of the equilibrium equations will yield *positive scalars for members in tension and negative scalars for members in compression.* Once an unknown member force is found, use its *correct* magnitude and sense (T or C) on subsequent joint free-body diagrams.
2. The *correct* sense of direction of an unknown member force can, in many cases, be determined ''by inspection.'' For example, $\mathbf{F}_{BC}$ in Fig. 6–7*b* must push on the pin (compression) since its horizontal component, $F_{BC} \sin 45°$, must balance the 500-N force ($\Sigma F_x = 0$). Likewise, $\mathbf{F}_{BA}$ is a tensile force since it balances the vertical component, $F_{BC} \cos 45°$ ($\Sigma F_y = 0$). In more complicated cases, the sense of an unknown member force can be *assumed;* then, after applying the equilibrium equations, the assumed sense can be verified from the numerical results. A *positive* answer indicates that the sense is *correct,* whereas a *negative* answer indicates that the sense shown on the free-body diagram must be *reversed.* This is the method we will use in the example problems which follow.

PROCEDURE FOR ANALYSIS

The following procedure provides a typical means for analyzing a truss using the method of joints.

Draw the free-body diagram of a joint having at least one known force and at most two unknown forces. (If this joint is at one of the supports, it generally will be necessary to know the external reactions at the truss support.) Use one of the two methods described above for establishing the sense of an unknown force. Orient the x and y axes such that the forces on the free-body diagram can be easily resolved into their x and y components and then apply the two force equilibrium equations $\Sigma F_x = 0$ and $\Sigma F_y = 0$. Solve for the two unknown member forces and verify their correct sense.

Continue to analyze each of the other joints, where again it is necessary to choose a joint having at most two unknowns and at least one known force. Realize that once the force in a member is found from the analysis of a joint at one of its ends, the result can be used to analyze the forces acting on the joint at its other end. Strict adherence to the principle of action, equal but opposite reaction must, of course, be observed. Remember, a member in *compression* ''pushes'' on the joint and a member in *tension* ''pulls'' on the joint.

Once the force analysis of the truss has been completed, the size of the members and their connections can be determined using the theory of mechanics of materials along with information given in engineering design codes.

Example 6–1

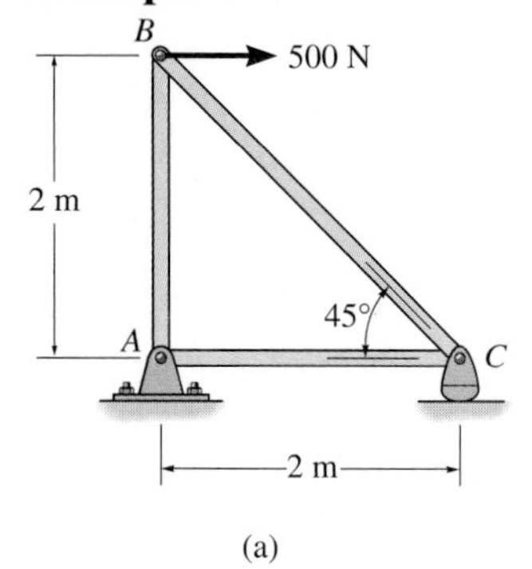

(a)

Determine the force in each member of the truss shown in Fig. 6–8*a* and indicate whether the members are in tension or compression.

SOLUTION

By inspection of Fig. 6–8*a*, there are two unknown member forces at joint *B*, two unknown member forces and an unknown reaction force at joint *C*, and two unknown member forces and two unknown reaction forces at joint *A*. Since we must have no more than two unknowns at the joint and at least one known force acting there, we must begin the analysis at joint *B*.

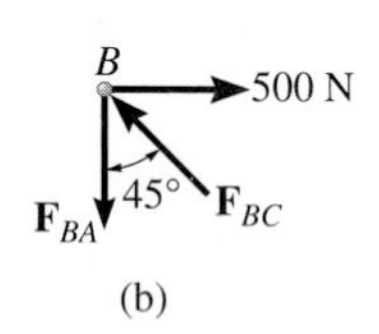

(b)

Joint B. The free-body diagram of the pin at *B* is shown in Fig. 6–8*b*. Three forces act on the pin: the external force of 500 N and the *two* unknown forces developed by members *BA* and *BC*. Applying the equations of joint equilibrium, we have

$$\xrightarrow{+}\Sigma F_x = 0; \quad 500\text{ N} - F_{BC}\sin 45^\circ = 0 \qquad F_{BC} = 707.1\text{ N (C)} \quad \textit{Ans.}$$

$$+\uparrow \Sigma F_y = 0; \quad F_{BC}\cos 45^\circ - F_{BA} = 0 \qquad F_{BA} = 500\text{ N (T)} \quad \textit{Ans.}$$

Since the force in member *BC* has been calculated, we can proceed to analyze joint *C* in order to determine the force in member *CA* and the support reaction at the rocker.

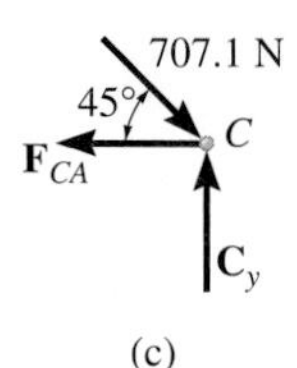

(c)

Joint C. From the free-body diagram of joint *C*, Fig. 6–8*c*, we have

$$\xrightarrow{+}\Sigma F_x = 0; \quad -F_{CA} + 707.1\cos 45^\circ\text{ N} = 0 \qquad F_{CA} = 500\text{ N (T)} \quad \textit{Ans.}$$

$$+\uparrow \Sigma F_y = 0; \quad C_y - 707.1\sin 45^\circ\text{ N} = 0 \qquad C_y = 500\text{ N} \quad \textit{Ans.}$$

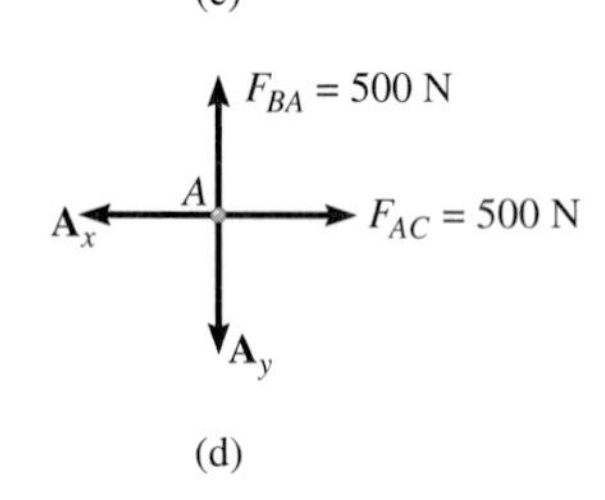

(d)

Joint A. Although not necessary, we can determine the support reactions at joint *A* using the results of $F_{AC} = 500$ N and $F_{AB} = 500$ N. From the free-body diagram, Fig. 6–8*d*, we have

$$\xrightarrow{+}\Sigma F_x = 0; \quad 500\text{ N} - A_x = 0 \qquad A_x = 500\text{ N}$$

$$+\uparrow \Sigma F_y = 0; \quad 500\text{ N} - A_y = 0 \qquad A_y = 500\text{ N}$$

The results of the analysis are summarized in Fig. 6–8*e*. Note that the free-body diagram of each pin shows the effects of all the connected members and external forces applied to the pin, whereas the free-body diagram of each member shows only the effects of the end pins on the member.

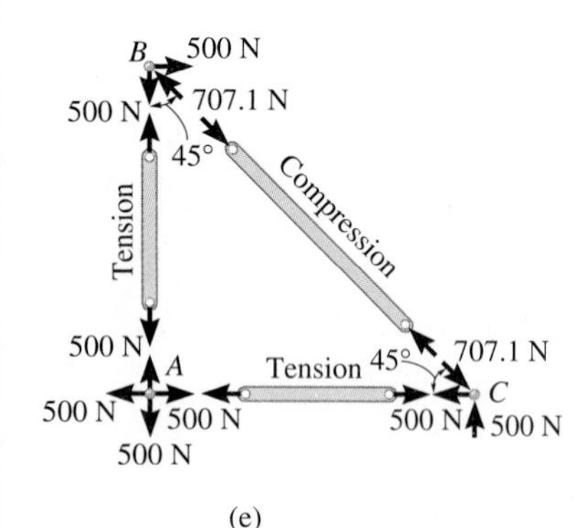

(e)

Fig. 6–8

Example 6–2

Determine the forces acting in all the members of the roof truss shown in Fig. 6–9*a*.

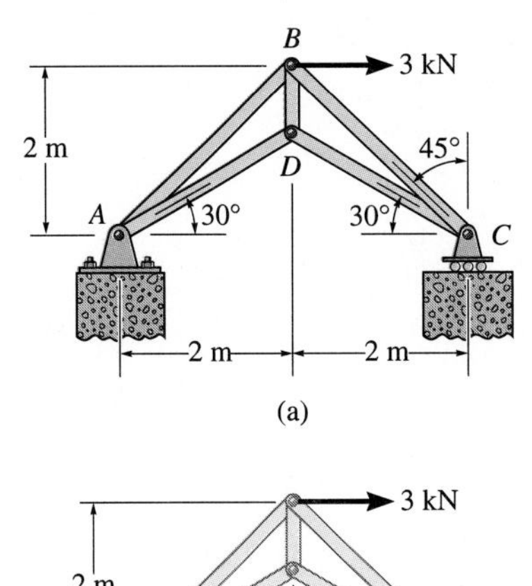

(a)

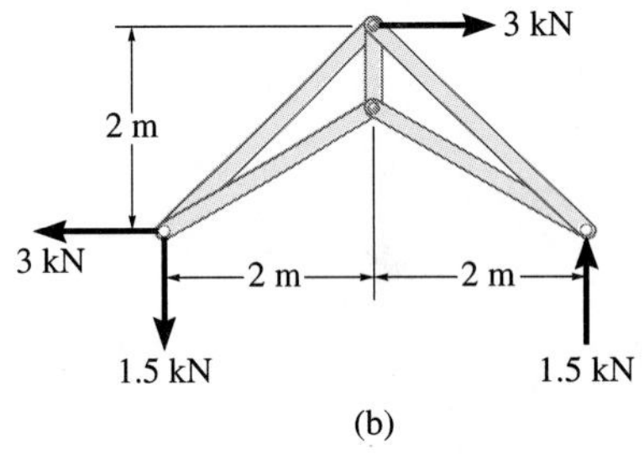

(b)

SOLUTION

By inspection, there are more than two unknowns at each joint. Consequently, the support reactions on the truss must first be determined. Show that they have been correctly calculated on the free-body diagram in Fig. 6–9*b*. We can now begin the analysis at joint *C*. Why?

Joint C. From the free-body diagram, Fig. 6–9*c*,

$$\overset{+}{\rightarrow}\Sigma F_x = 0; \qquad -F_{CD}\cos 30^\circ + F_{CB}\sin 45^\circ = 0$$

$$+\uparrow \Sigma F_y = 0; \qquad 1.5\text{ kN} + F_{CD}\sin 30^\circ - F_{CB}\cos 45^\circ = 0$$

These two equations must be solved *simultaneously* for each of the two unknowns. Note, however, that a *direct solution* for one of the unknown forces may be obtained by applying a force summation along an axis that is *perpendicular* to the direction of the other unknown force. For example, summing forces along the y' axis, which is perpendicular to the direction of $\mathbf{F}_{CD}$, Fig. 6–9*d*, yields a direct solution for F_{CB}.

$$+\nearrow\Sigma F_{y'} = 0;$$

$$1.5\cos 30^\circ\text{ kN} - F_{CB}\sin 15^\circ = 0 \qquad F_{CB} = 5.02\text{ kN (C)} \qquad \textit{Ans.}$$

In a similar fashion, summing forces along the y'' axis, Fig. 6–9*e*, yields a direct solution for F_{CD}.

$$+\nearrow\Sigma F_{y''} = 0;$$

$$1.5\cos 45^\circ\text{ kN} - F_{CD}\sin 15^\circ = 0 \qquad F_{CD} = 4.10\text{ kN (T)} \qquad \textit{Ans.}$$

(c) (d)

Joint D. We can now proceed to analyze joint *D*. The free-body diagram is shown in Fig. 6–9*f*.

$$\overset{+}{\rightarrow}\Sigma F_x = 0; \qquad -F_{DA}\cos 30^\circ + 4.10\cos 30^\circ\text{ kN} = 0$$

$$F_{DA} = 4.10\text{ kN (T)} \qquad \textit{Ans.}$$

$$+\uparrow \Sigma F_y = 0; \qquad F_{DB} - 2(4.10\sin 30^\circ\text{ kN}) = 0$$

$$F_{DB} = 4.10\text{ kN (T)} \qquad \textit{Ans.}$$

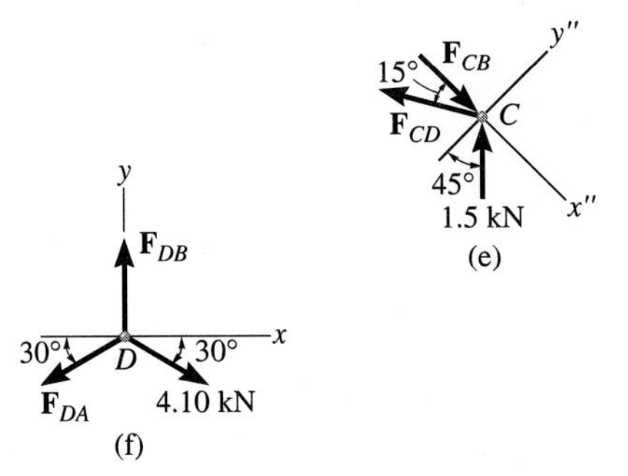

(e) (f)

The force in the last member, *BA*, can be obtained from joint *B* or joint *A*. As an exercise, draw the free-body diagram of joint *B*, sum the forces in the horizontal direction, and show that $F_{BA} = 0.776$ kN (C).

Fig. 6–9

Example 6–3

Determine the force in each member of the truss shown in Fig. 6–10*a*. Indicate whether the members are in tension or compression.

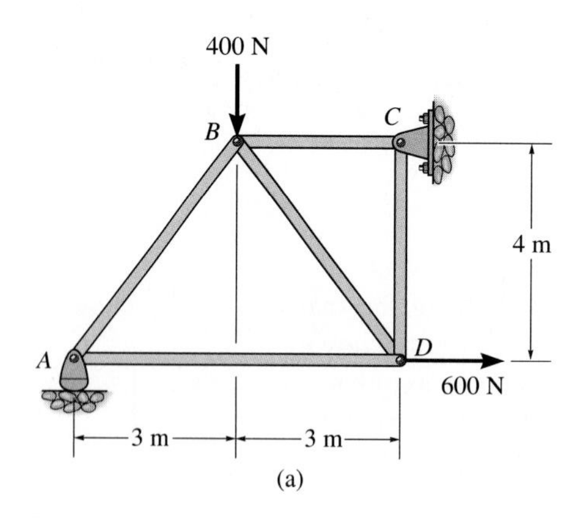

(a)

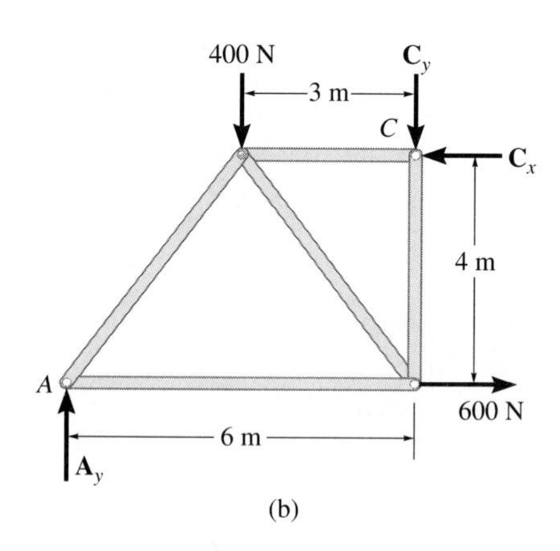

(b)

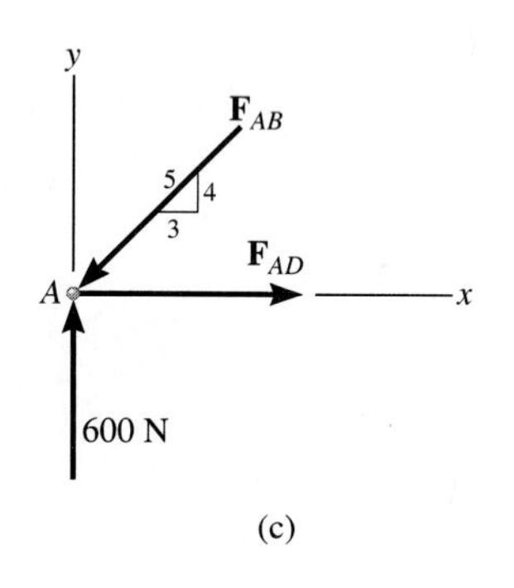

(c)

Fig. 6–10(a–f)

SOLUTION

Support Reactions. No joint can be analyzed until the support reactions are determined. Why? A free-body diagram of the entire truss is given in Fig. 6–10*b*. Applying the equations of equilibrium, we have

$$\overset{+}{\rightarrow}\Sigma F_x = 0; \qquad 600\text{ N} - C_x = 0 \qquad C_x = 600\text{ N}$$

$$\circlearrowleft + \Sigma M_C = 0; \qquad -A_y(6\text{ m}) + 400\text{ N}(3\text{ m}) + 600\text{ N}(4\text{ m}) = 0 \qquad A_y = 600\text{ N}$$

$$+\uparrow \Sigma F_y = 0; \qquad 600\text{ N} - 400\text{ N} - C_y = 0 \qquad C_y = 200\text{ N}$$

The analysis can now start at either joint *A* or *C*. The choice is arbitrary, since there are one known and two unknown member forces acting on the pin at each of these joints.

Joint A (Fig. 6–10*c*). As shown on the free-body diagram, there are three forces that act on the pin at joint *A*. The inclination of $\mathbf{F}_{AB}$ is determined from the geometry of the truss. By inspection, can you see why this force is assumed to be compressive and $\mathbf{F}_{AD}$ tensile? Applying the equations of equilibrium, we have

$$+\uparrow \Sigma F_y = 0; \qquad 600\text{ N} - \tfrac{4}{5}F_{AB} = 0 \qquad F_{AB} = 750\text{ N (C)} \qquad \textit{Ans.}$$

$$\overset{+}{\rightarrow}\Sigma F_x = 0; \qquad F_{AD} - \tfrac{3}{5}(750\text{ N}) = 0 \qquad F_{AD} = 450\text{ N (T)} \qquad \textit{Ans.}$$

Joint D (Fig. 6–10*d*). The pin at this joint is chosen next since, by inspection of Fig. 6–10*a*, the force in *AD* is known and the unknown forces in *DB* and *DC* can be determined. Summing forces in the horizontal direction, Fig. 6–10*d*, we have

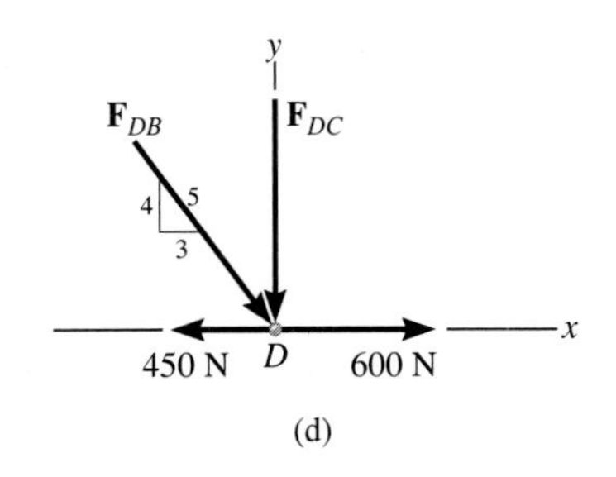

(d)

$$\overset{+}{\rightarrow}\Sigma F_x = 0; \quad -450\text{ N} + \tfrac{3}{5}F_{DB} + 600\text{ N} = 0 \qquad F_{DB} = -250\text{ N}$$

The negative sign indicates that $\mathbf{F}_{DB}$ acts in the *opposite sense* to that shown in Fig. 6–10*d*.* Hence,

$$F_{DB} = 250\text{ N} \quad (\text{T}) \qquad \textit{Ans.}$$

To determine $\mathbf{F}_{DC}$, we can either correct the sense of $\mathbf{F}_{DB}$ and then apply $\Sigma F_y = 0$, or apply this equation and retain the negative sign for F_{DB}, i.e.,

$$+\uparrow \Sigma F_y = 0; \quad -F_{DC} - \tfrac{4}{5}(-250\text{ N}) = 0 \qquad F_{DC} = 200\text{ N} \quad (\text{C}) \qquad \textit{Ans.}$$

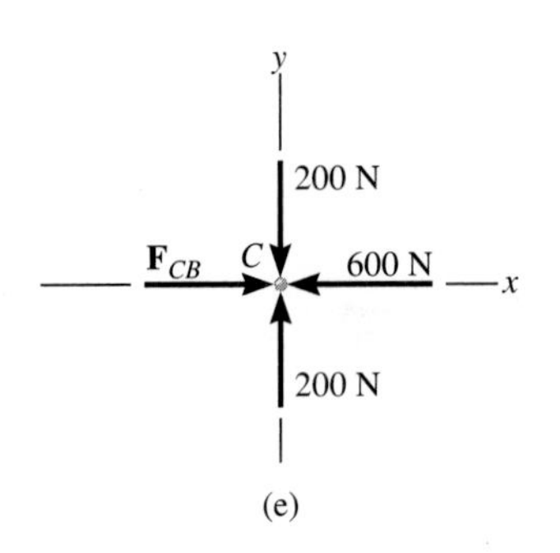

(e)

Joint C (Fig. 6–10*e*)

$$\overset{+}{\rightarrow}\Sigma F_x = 0; \qquad F_{CB} - 600\text{ N} = 0 \qquad F_{CB} = 600\text{ N} \quad (\text{C}) \qquad \textit{Ans.}$$

$$+\uparrow \Sigma F_y = 0; \qquad 200\text{ N} - 200\text{ N} \equiv 0 \quad (\text{check})$$

The analysis is summarized in Fig. 6–10*f*, which shows the correct free-body diagram for each pin and member.

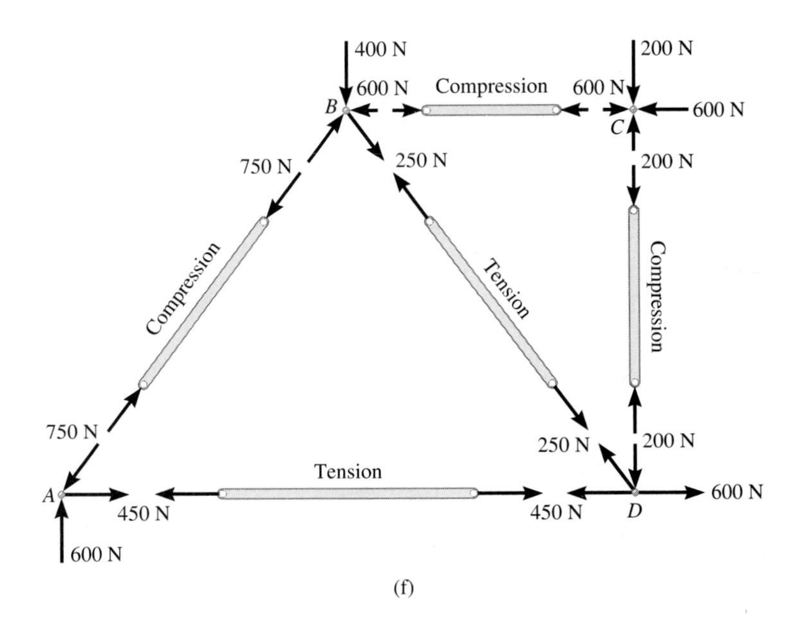

(f)

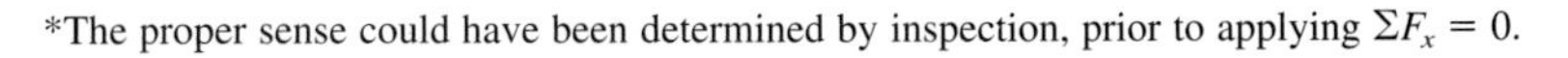
*The proper sense could have been determined by inspection, prior to applying $\Sigma F_x = 0$.

6.3 Zero-Force Members

Truss analysis using the method of joints is greatly simplified if one is able to first determine those members which support *no loading*. These *zero-force members* are used to increase the stability of the truss during construction and to provide support if the applied loading is changed.

The zero-force members of a truss can generally be determined *by inspection* of each of its joints. For example, consider the truss shown in Fig. 6–11*a*. If a free-body diagram of the pin at joint *A* is drawn, Fig. 6–11*b*, it is seen that members *AB* and *AF* are zero-force members. On the other hand, notice that we could not have come to this conclusion if we had considered the free-body diagrams of joints *F* or *B*, simply because there are five unknowns at each of these joints. In a similar manner, consider the free-body diagram of joint *D*, Fig. 6–11*c*. Here again it is seen that *DC* and *DE* are zero-force members. As a general rule, then, *if only two members form a truss joint and no external load or support reaction is applied to the joint, the members must be zero-force members.* The load on the truss in Fig. 6–11*a* is therefore supported by only five members as shown in Fig. 6–11*d*.

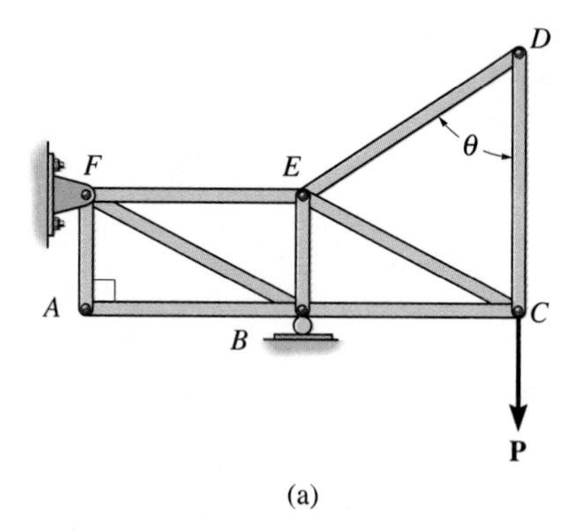

(a)

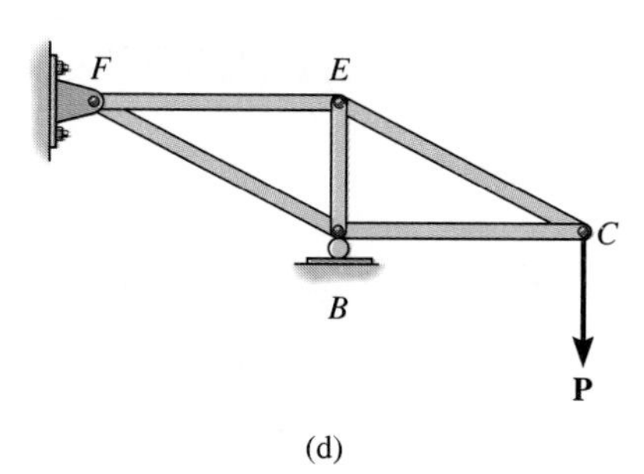

(d)

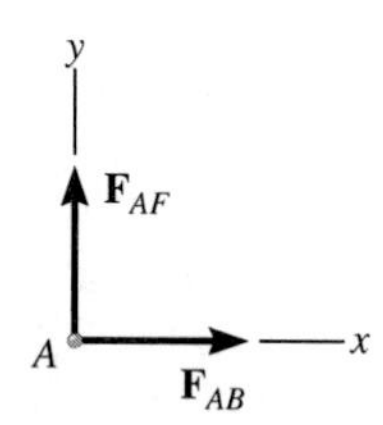

$\overset{+}{\rightarrow} \Sigma F_x = 0; \; F_{AB} = 0$

$+\uparrow \Sigma F_y = 0; \; F_{AF} = 0$

(b)

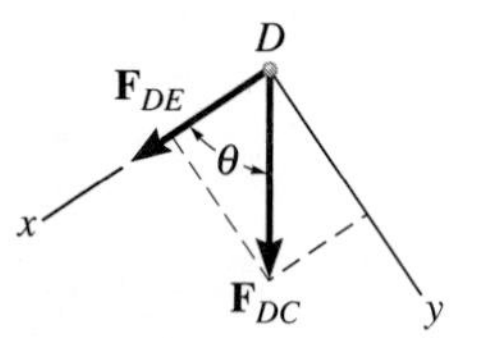

$+\searrow \Sigma F_y = 0; \; F_{DC} \sin \theta = 0; \; F_{DC} = 0 \text{ since } \sin \theta \neq 0$

$+\swarrow \Sigma F_x = 0; \; F_{DE} + 0 = 0; \; F_{DE} = 0$

(c)

Fig. 6–11

Now consider the truss shown in Fig. 6–12*a*. The free-body diagram of the pin at joint D is shown in Fig. 6–12*b*. By orienting the y axis along members DC and DE and the x axis along member DA, it is seen that DA is a zero-force member. Note that this is also the case for member CA, Fig. 6–12*c*. In general, then, *if three members form a truss joint for which two of the members are collinear, the third member is a zero-force member provided no external force or support reaction is applied to the joint.* The truss shown in Fig. 6–12*d* is therefore suitable for supporting the load **P**.

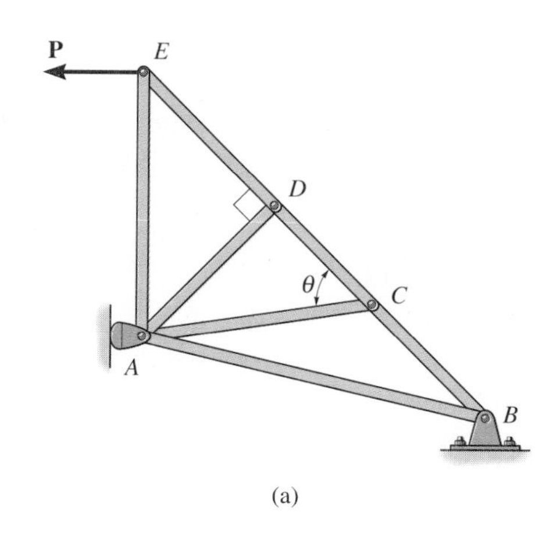

(a)

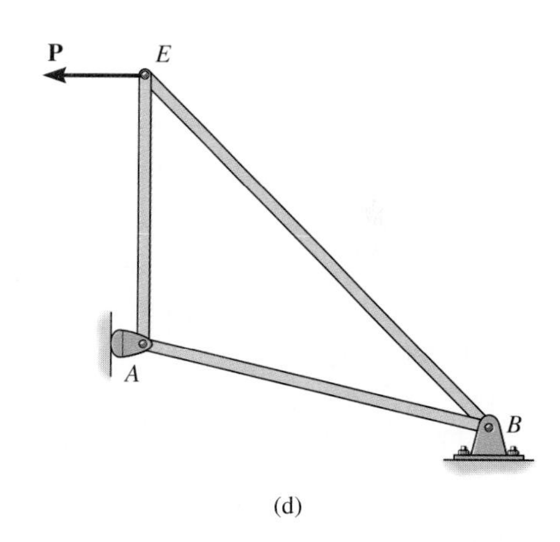

(d)

Fig. 6–12

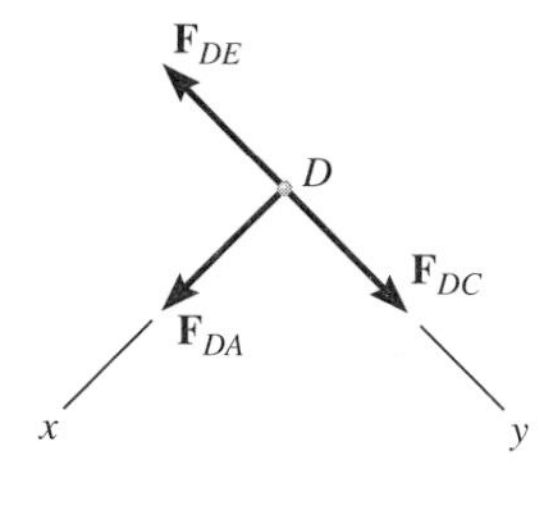

$+\swarrow \Sigma F_x = 0; \quad F_{DA} = 0$

$+\searrow \Sigma F_y = 0; \quad F_{DC} = F_{DE}$

(b)

$+\swarrow \Sigma F_x = 0; \quad F_{CA} \sin\theta = 0; \quad F_{CA} = 0 \text{ since } \sin\theta \neq 0;$

$+\searrow \Sigma F_y = 0; \quad F_{CB} = F_{CD}$

(c)

Example 6–4

Using the method of joints, determine all the zero-force members of the *Fink roof truss* shown in Fig. 6–13*a*. Assume all joints are pin-connected.

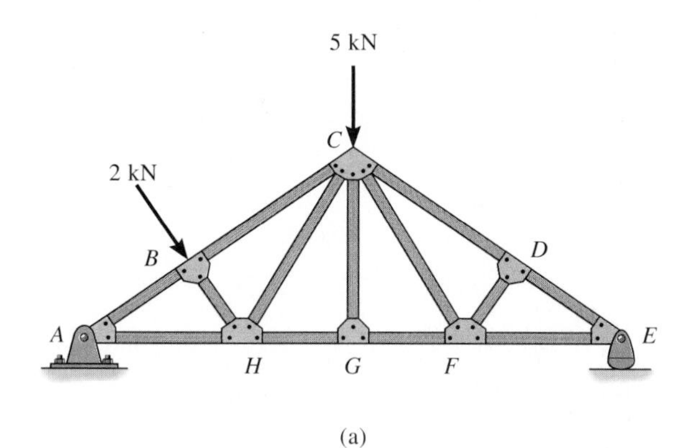

(a)

Fig. 6–13

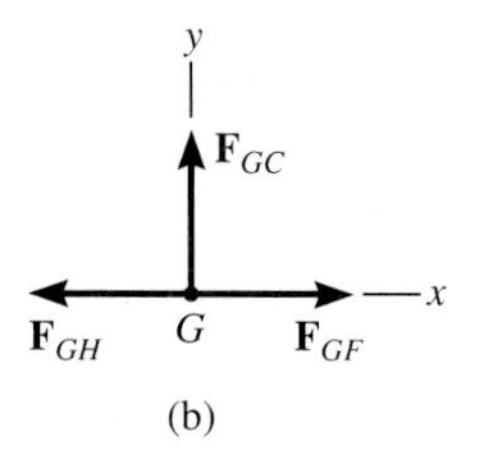

(b)

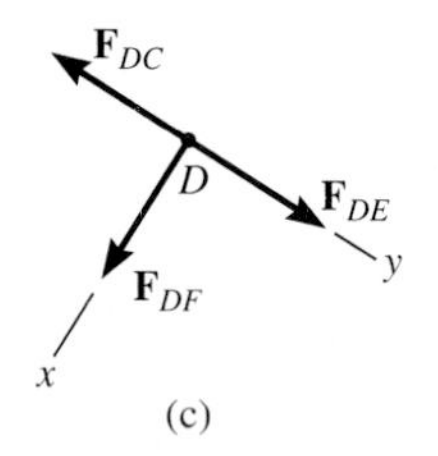

(c)

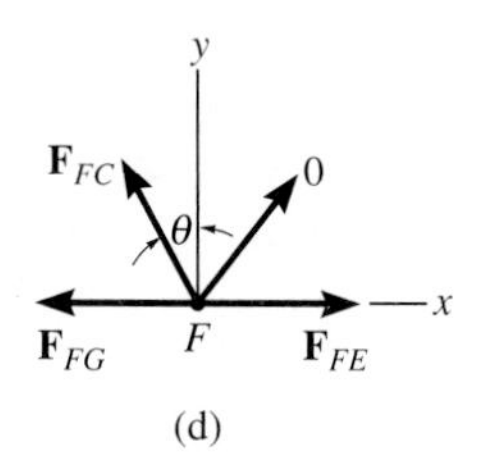

(d)

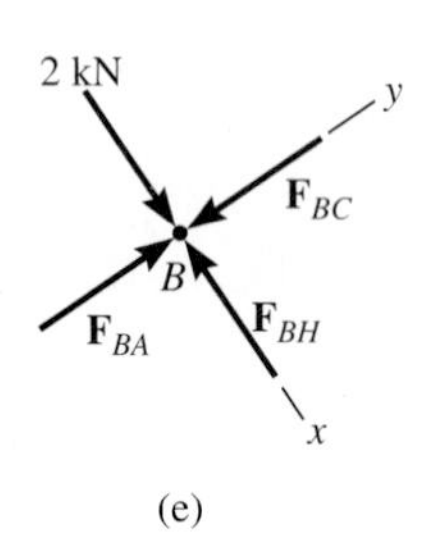

(e)

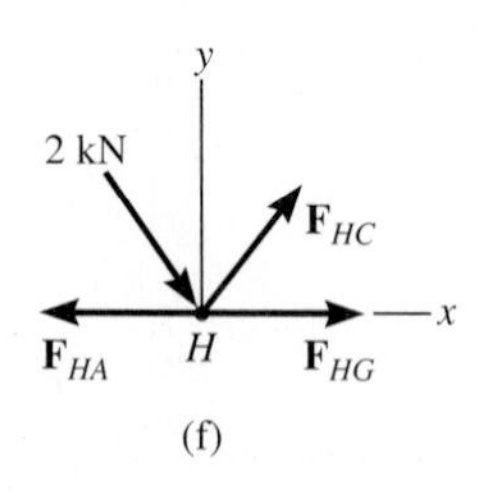

(f)

SOLUTION

Looking for joint geometries that are similar to those outlined in Figs. 6–11 and 6–12, we have

Joint G (Fig. 6–13*b*)

$+\uparrow \Sigma F_y = 0;$ $\qquad F_{GC} = 0$ $\qquad$ ***Ans.***

Realize that we could not conclude that GC is a zero-force member by considering joint C, where there are five unknowns. The fact that GC is a zero-force member means that the 5-kN load at C must be supported by members CB, CH, CF, and CD.

Joint D (Fig. 6–13*c*)

$+\swarrow \Sigma F_x = 0;$ $\qquad F_{DF} = 0$ $\qquad$ ***Ans.***

Joint F (Fig. 6–13*d*)

$+\uparrow \Sigma F_y = 0; \quad F_{FC} \cos \theta = 0 \qquad \text{Since } \theta \neq 90°, \qquad F_{FC} = 0$ $\qquad$ ***Ans.***

Note that if joint B is analyzed, Fig. 6–13*e*,

$+\searrow \Sigma F_x = 0; \quad 2 \text{ kN} - F_{BH} = 0 \qquad F_{BH} = 2 \text{ kN (C)}$

Consequently, the numerical value of F_{HC} must satisfy $\Sigma F_y = 0$, Fig. 6–13*f*, and therefore HC is *not* a zero-force member.

PROBLEMS

6–1. The truss, used to support a balcony, is subjected to the loading shown. Approximate each joint as a pin and determine the force in each member. State whether the members are in tension or compression. Set $P_1 = 600$ lb, $P_2 = 400$ lb.

6–2. The truss, used to support a balcony, is subjected to the loading shown. Approximate each joint as a pin and determine the force in each member. State whether the members are in tension or compression. Set $P_1 = 800$ lb, $P_2 = 0$.

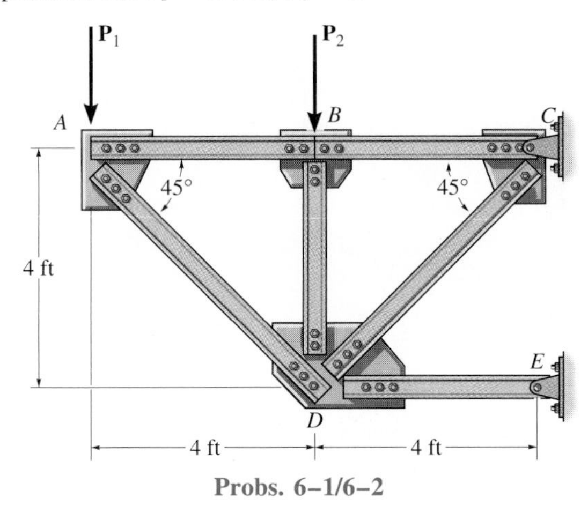

Probs. 6–1/6–2

6–3. Determine the force in each member of the truss and state if the members are in tension or compression. Set $P_1 = 7$ kN, $P_2 = 7$ kN.

***6–4.** Determine the force in each member of the truss and state if the members are in tension or compression. Set $P_1 = 8$ kN, $P_2 = 10$ kN.

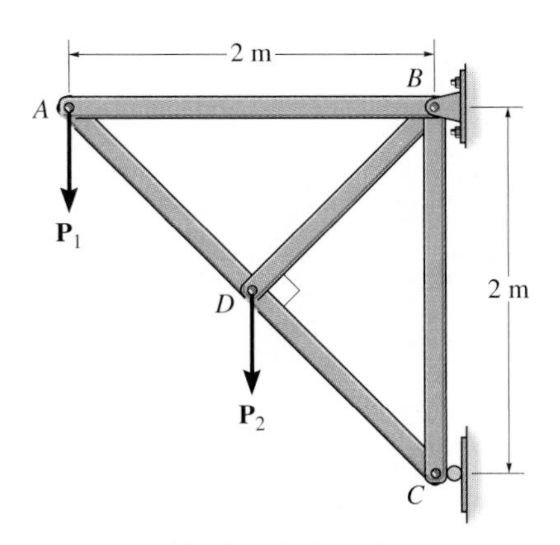

Probs. 6–3/6–4

6–5. Determine the force in each member of the truss and state if the members are in tension or compression. Set $P_1 = 0$, $P_2 = 1000$ lb.

6–6. Determine the force in each member of the truss and state if the members are in tension or compression. Set $P_1 = 500$ lb, $P_2 = 1500$ lb.

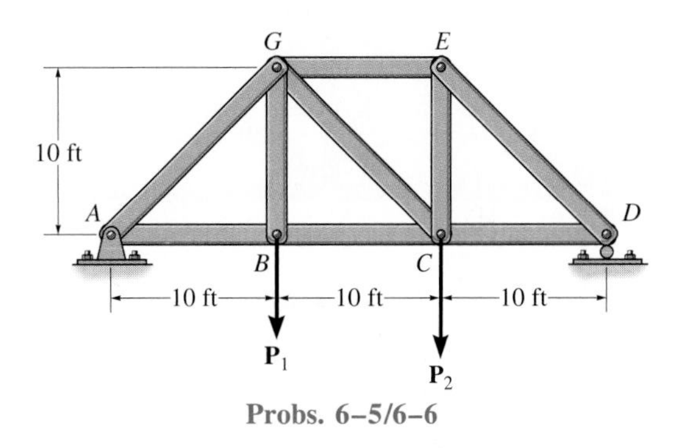

Probs. 6–5/6–6

6–7. Determine the force in each member of the truss and state if the members are in tension or compression. Set $P_1 = 10$ kN, $P_2 = 15$ kN.

***6–8.** Determine the force in each member of the truss and state if the members are in tension or compression. Set $P_1 = 0$, $P_2 = 20$ kN.

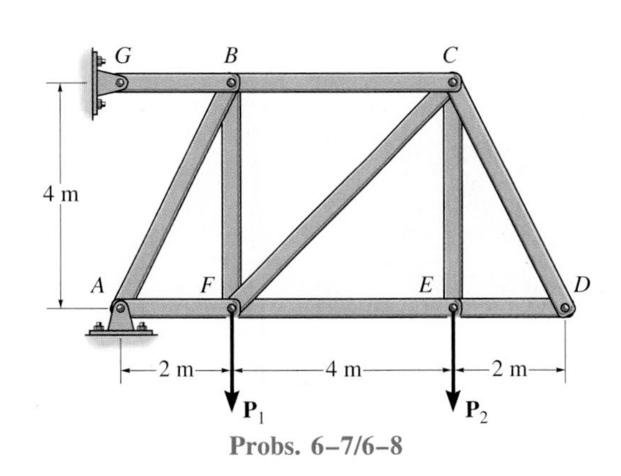

Probs. 6–7/6–8

6–9. Determine the force in each member of the truss and state if the members are in tension or compression.

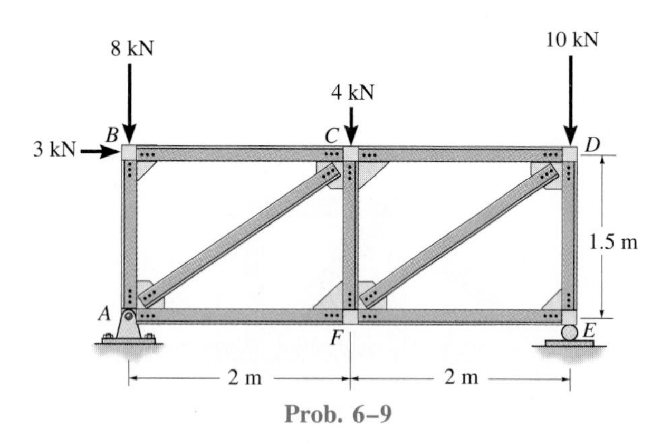

Prob. 6–9

***6–12.** Determine the force in each member of the truss in terms of the load P and state if the members are in tension or compression.

6–13. Members AB and BC can support a maximum compressive force of 800 lb, and members AD, DC, and BD can support a maximum tensile force of 1500 lb. If $a = 10$ ft, determine the greatest load P the truss can support.

6–14. Members AB and BC can support a maximum compressive force of 800 lb, and members AD, DC, and BD can support a maximum tensile force of 2000 lb. If $a = 6$ ft, determine the greatest load P the truss can support.

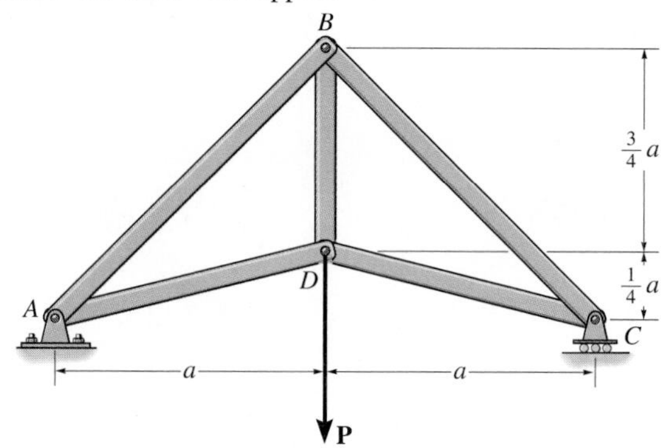

Probs. 6–12/6–13/6–14

6–10. Determine the force in each member of the truss and state if the members are in tension or compression. Set $P_1 = 100$ lb, $P_2 = 200$ lb, $P_3 = 300$ lb.

6–11. Determine the force in each member of the truss and state if the members are in tension or compression. Set $P_1 = 400$ lb, $P_2 = 400$ lb, $P_3 = 0$.

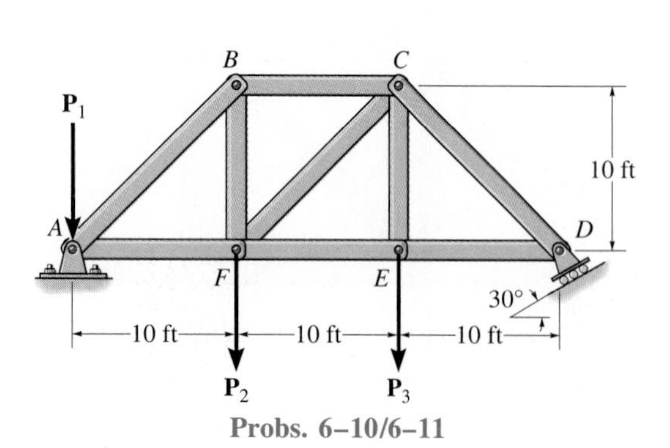

Probs. 6–10/6–11

6–15. Determine the force in each member of the truss and state if the members are in tension or compression. Approximate each joint as a pin. Set $P = 4$ kN.

***6–16.** Assume that each member of the truss is made of steel having a mass per length of 4 kg/m. Set $P = 0$, determine the force in each member, and state if the members are in tension or compression. Neglect the weight of the gusset plates and approximate each joint as a pin. Solve the problem by *assuming* the weight of each member can be represented as a vertical force, half of which is applied at each end of the member.

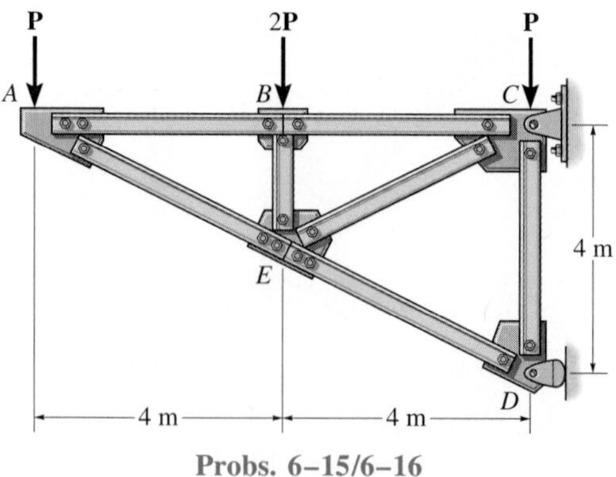

Probs. 6–15/6–16

6–17. Determine the force in each member of the truss and state if the members are in tension or compression. *Hint:* The vertical component of force at C must equal zero. Why?

6–18. Each member of the truss is uniform and has a mass of 8 kg/m. Remove the external loads of 6 kN and 8 kN and determine the approximate force in each member due to the weight of the truss. State if the members are in tension or compression. Solve the problem by *assuming* the weight of each member can be represented as a vertical force, half of which is applied at each end of the member.

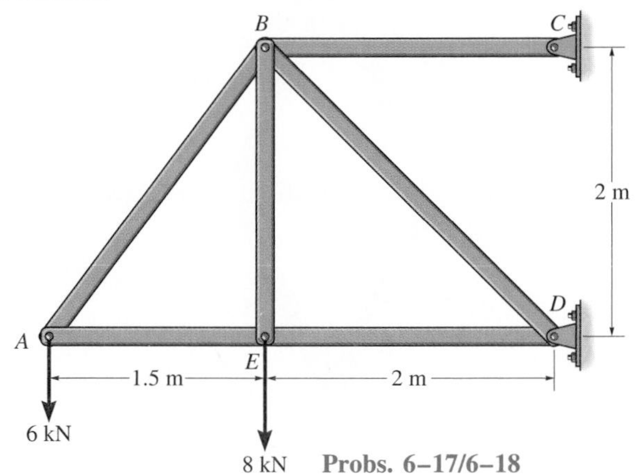

Probs. 6–17/6–18

6–19. Determine the force in each member of the truss and state if the members are in tension or compression. *Hint:* The resultant force at the pin E acts along member ED. Why?

***6–20.** Each member of the truss is uniform and has a mass of 8 kg/m. Remove the external loads of 3 kN and 2 kN and determine the approximate force in each member due to the weight of the truss. State if the members are in tension or compression. Solve the problem by *assuming* the weight of each member can be represented as a vertical force, half of which is applied at each end of the member.

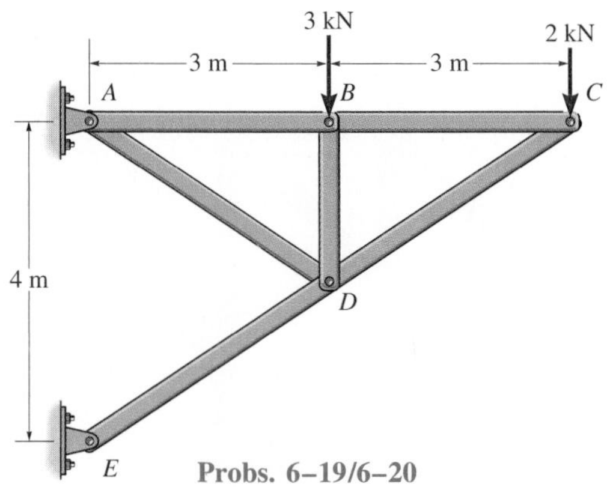

Probs. 6–19/6–20

6–21. Determine the force in each member of the truss and state if the members are in tension or compression. *Hint:* The horizontal force component at A must be zero. Why?

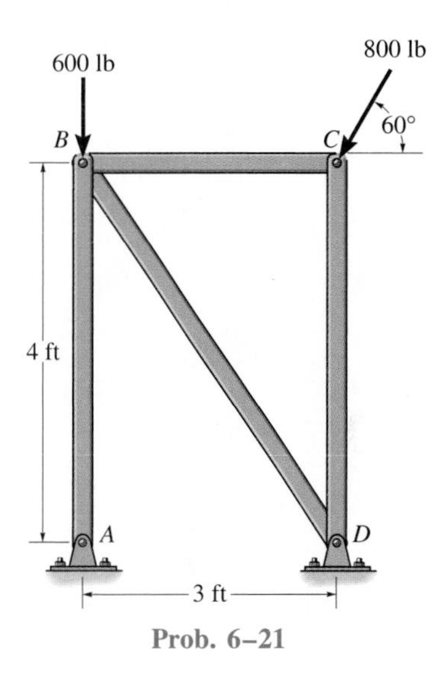

Prob. 6–21

6–22. Determine the force in each member of the double scissors truss in terms of the load P and state if the members are in tension or compression.

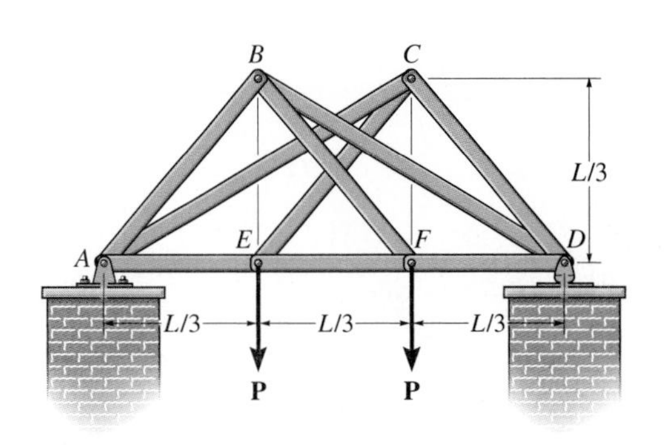

Prob. 6–22

6–23. Determine the force in each member of the truss in terms of the external loading and state if the members are in tension or compression.

***6–24.** The maximum allowable tensile force in the members of the truss is $(F_t)_{max} = 1500$ lb, and the maximum allowable compressive force is $(F_c)_{max} = 800$ lb. Determine the maximum magnitude P of the two loads that can be applied to the truss. Take $a = 8$ ft.

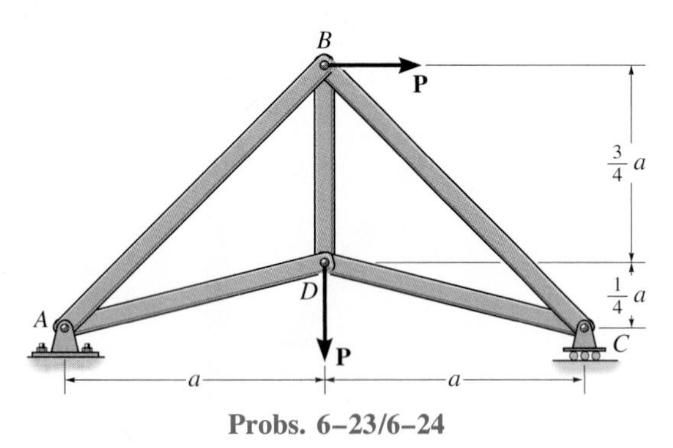

Probs. 6–23/6–24

6–25. Determine the force in each member of the truss in terms of the external loading and state if the members are in tension or compression.

6–26. The maximum allowable tensile force in the members of the truss is $(F_t)_{max} = 2$ kN, and the maximum allowable compressive force is $(F_c)_{max} = 1.2$ kN. Determine the maximum magnitude P of the two loads that can be applied to the truss. Take $L = 2$ m and $\theta = 30°$.

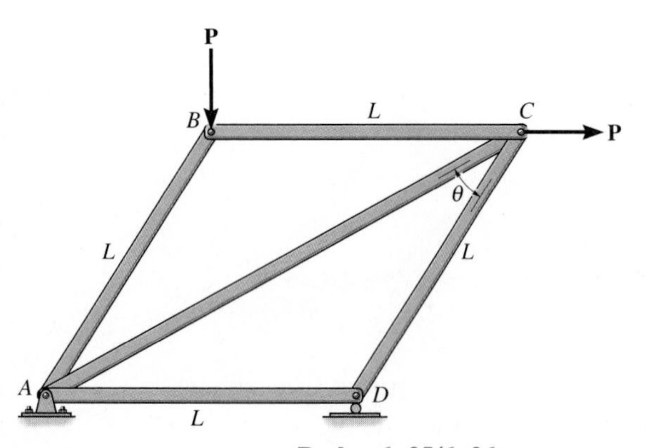

Probs. 6–25/6–26

6–27. Determine the force in each member of the truss in terms of the load P and state if the members are in tension or compression.

***6–28.** The maximum allowable tensile force in the members of the truss is $(F_t)_{max} = 3$ kN, and the maximum allowable compressive force is $(F_c)_{max} = 5$ kN. Determine the maximum magnitude of the load **P** that can be applied to the truss. Take $d = 2$ m.

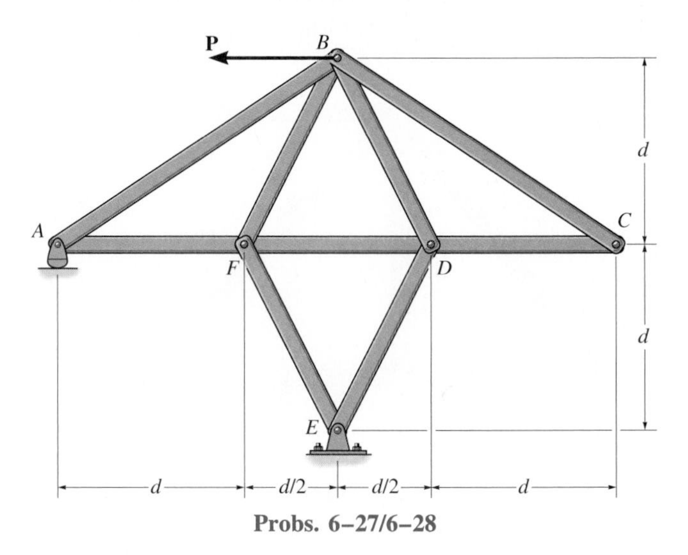

Probs. 6–27/6–28

6–29. Determine the force in each member of the truss and state if the members are in tension or compression. Set $P_1 = 4$ kN, $P_2 = 5$ kN.

6–30. Determine the force in each member of the truss and state if the members are in tension or compression. Set $P_1 = 0$, $P_2 = 8$ kN.

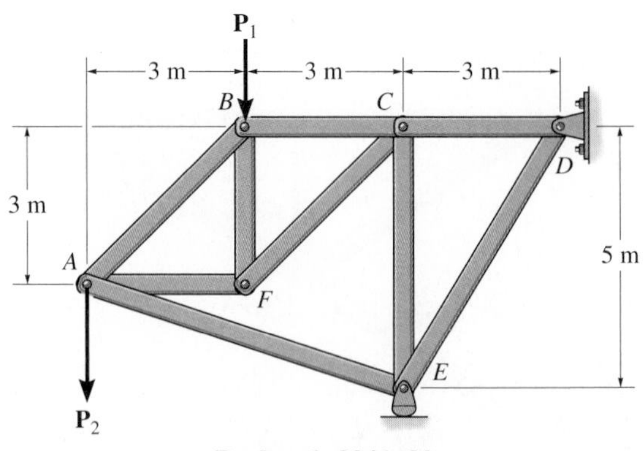

Probs. 6–29/6–30

6.4 The Method of Sections

The *method of sections* is used to determine the loadings acting within a body. It is based on the principle that if a body is in equilibrium, then any part of the body is also in equilibrium. To apply this method, one passes an *imaginary section* through the body, thus cutting it into two parts. When a free-body diagram of one of the parts is drawn, the loads acting at the section must be *included* on the free-body diagram. One then applies the equations of equilibrium to the part in order to determine the loading at the section. For example, consider the two truss members shown colored in Fig. 6–14. The internal loads at the section indicated by the blue line can be obtained using one of the free-body diagrams shown on the right. Clearly, it can be seen that equilibrium requires that the member in tension be subjected to a "pull" **T** at the section, whereas the member in compression is subjected to a "push" **C**.

The method of sections can also be used to "cut" or section several members of an entire truss. If either of the two parts of the truss is isolated as a free-body diagram, we can then apply the equations of equilibrium to that part to determine the member forces at the "cut section." Since only *three* independent equilibrium equations ($\Sigma F_x = 0$, $\Sigma F_y = 0$, $\Sigma M_O = 0$) can be applied to the isolated part of the truss, one should try to select a section that, in general, passes through not more than *three* members in which the forces are unknown. For example, consider the truss in Fig. 6–15*a*. If the force in member *GC* is to be determined, section *aa* would be appropriate. The free-body diagrams of the two parts are shown in Figs. 6–15*b* and 6–15*c*. In particular, note that the line of action of each cut member force is specified from the *geometry* of the truss, since the force in a member passes along its axis. Also, the member forces acting on one part of the truss are equal but opposite to those acting on the other part—Newton's third law. As noted above, members assumed to be in *tension* (*BC* and *GC*) are subjected to a "pull," whereas the member in *compression* (*GF*) is subjected to a "push."

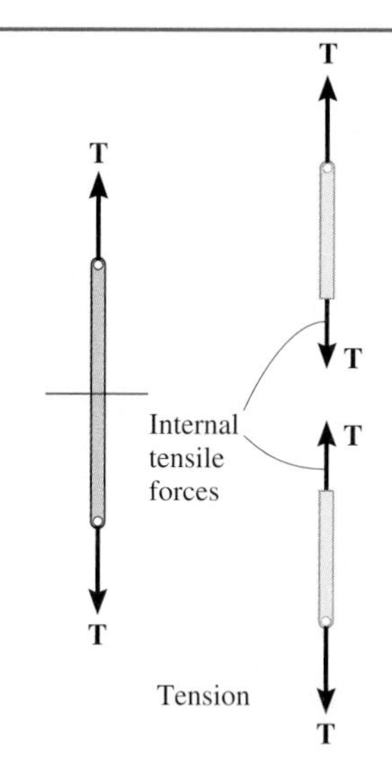

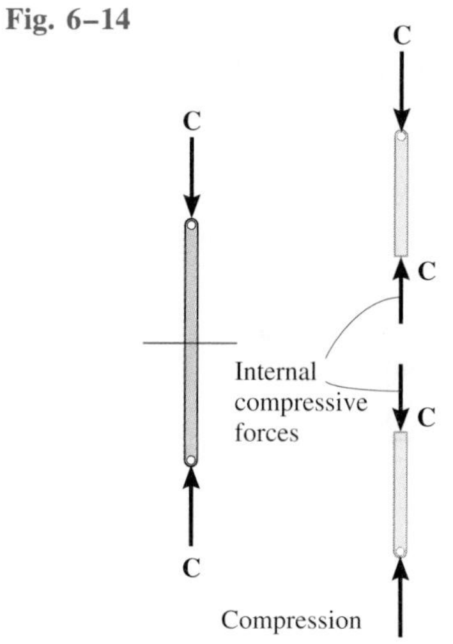

Fig. 6–14

Fig. 6–15

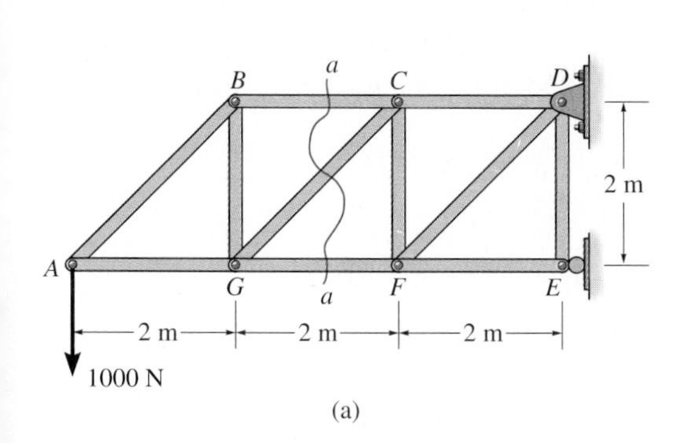

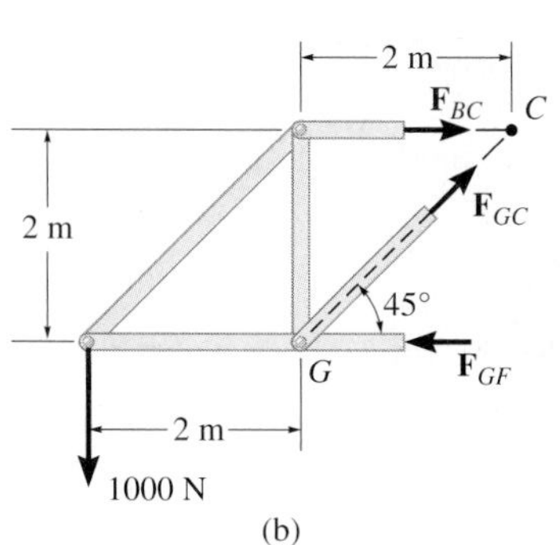

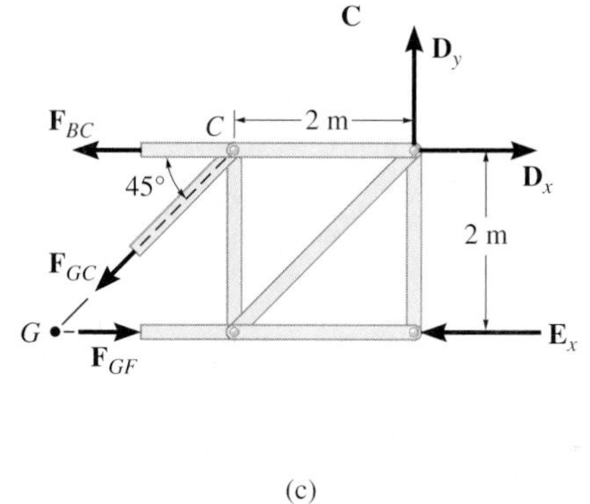

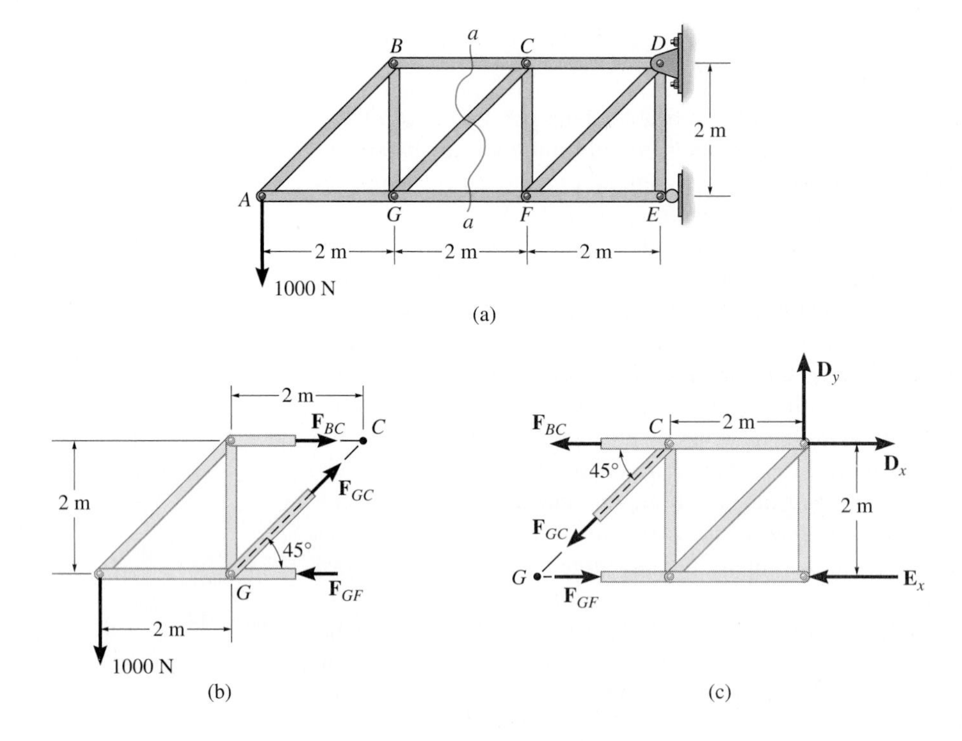

Fig. 6–15 ***(Repeated)***

The three unknown member forces $\mathbf{F}_{BC}$, $\mathbf{F}_{GC}$, and $\mathbf{F}_{GF}$ can be obtained by applying the three equilibrium equations to the free-body diagram in Fig. 6–15*b*. If, however, the free-body diagram in Fig. 6–15*c* is considered, the three support reactions $\mathbf{D}_x$, $\mathbf{D}_y$, and $\mathbf{E}_x$ will have to be determined *first.* Why? (This, of course, is done in the usual manner by considering a free-body diagram of the *entire truss.*) When applying the equilibrium equations, one should consider ways of writing the equations so as to yield a *direct solution* for each of the unknowns, rather than having to solve simultaneous equations. For example, summing moments about C in Fig. 6–15*b* would yield a direct solution for $\mathbf{F}_{GF}$ since $\mathbf{F}_{BC}$ and $\mathbf{F}_{GC}$ create zero moment about C. Likewise, $\mathbf{F}_{BC}$ can be directly obtained by summing moments about G. Finally, $\mathbf{F}_{GC}$ can be found directly from a force summation in the vertical direction since $\mathbf{F}_{GF}$ and $\mathbf{F}_{BC}$ have no vertical components. This ability to *determine directly* the force in a particular truss member is one of the main advantages of using the method of sections.*

*By comparison, if the method of joints were used to determine, say, the force in member GC, it would be necessary to analyze joints A, B, and G in sequence.

As in the method of joints, there are two ways in which one can determine the correct sense of an unknown member force.

1. *Always assume* that the unknown member forces at the cut section are in *tension,* i.e., "pulling" on the member. By doing this, the numerical solution of the equilibrium equations will yield *positive scalars for members in tension and negative scalars for members in compression.*
2. The correct sense of an unknown member force can in many cases be determined "by inspection." For example, $\mathbf{F}_{BC}$ is a tensile force as represented in Fig. 6–15*b*, since moment equilibrium about G requires that $\mathbf{F}_{BC}$ create a moment opposite to that of the 1000-N force. Also, $\mathbf{F}_{GC}$ is tensile since its vertical component must balance the 1000-N force acting downward. In more complicated cases, the sense of an unknown member force may be *assumed.* If the solution yields a *negative* scalar, it indicates that the force's sense is *opposite* to that shown on the free-body diagram. This is the method we will use in the example problems which follow.

PROCEDURE FOR ANALYSIS

The following procedure provides a means for applying the method of sections to determine the forces in the members of a truss.

Free-Body Diagram. Make a decision as to how to "cut" or section the truss through the members where forces are to be determined. Before isolating the appropriate section, it may first be necessary to determine the truss's *external* reactions, so that the three equilibrium equations are used *only* to solve for member forces at the cut section. Draw the free-body diagram of that part of the sectioned truss which has the least number of forces acting on it. Use one of the two methods described above for establishing the sense of an unknown member force.

Equations of Equilibrium. Try to apply the three equations of equilibrium such that simultaneous solution of equations is avoided. In this regard, moments should be summed about a point that lies at the intersection of the lines of action of two unknown forces, so that the third unknown force is determined directly from the moment equation. If two of the unknown forces are *parallel,* forces may be summed *perpendicular* to the direction of these unknowns to determine *directly* the third unknown force.

The following examples illustrate these concepts numerically.

Example 6–5

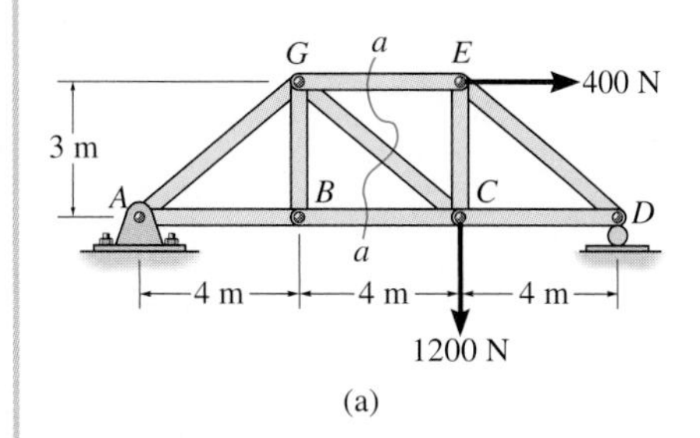

(a)

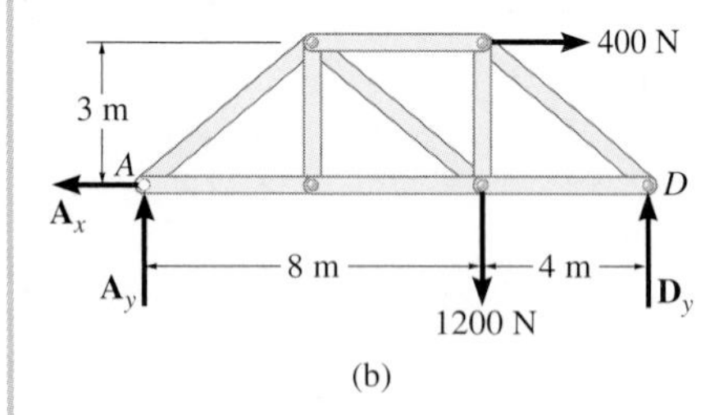

(b)

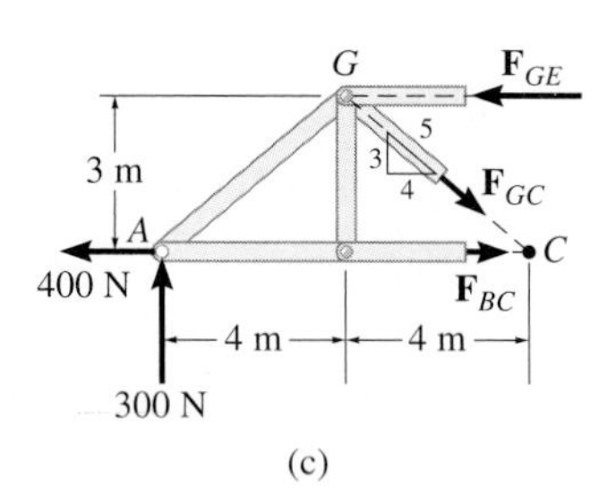

(c)

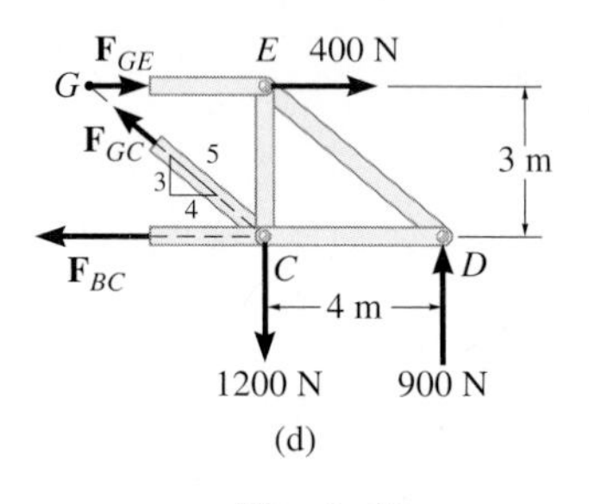

(d)

Fig. 6–16

Determine the force in members *GE*, *GC*, and *BC* of the truss shown in Fig. 6–16*a*. Indicate whether the members are in tension or compression.

SOLUTION

Section *aa* in Fig. 6–16*a* has been chosen since it cuts through the *three* members whose forces are to be determined. In order to use the method of sections, however, it is *first* necessary to determine the external reactions at *A* or *D*. Why? A free-body diagram of the entire truss is shown in Fig. 6–16*b*. Applying the equations of equilibrium, we have

$$\overset{+}{\rightarrow}\Sigma F_x = 0; \qquad 400\text{ N} - A_x = 0 \qquad A_x = 400\text{ N}$$

$$\curvearrowleft + \Sigma M_A = 0; \qquad -1200\text{ N}(8\text{ m}) - 400\text{ N}(3\text{ m}) + D_y(12\text{ m}) = 0$$

$$D_y = 900\text{ N}$$

$$+\uparrow \Sigma F_y = 0; \qquad A_y - 1200\text{ N} + 900\text{ N} = 0 \qquad A_y = 300\text{ N}$$

Free-Body Diagrams. The free-body diagrams of the sectioned truss are shown in Figs. 6–16*c* and 6–16*d*. For the analysis the free-body diagram in Fig. 6–16*c* will be used since it involves the least number of forces.

Equations of Equilibrium. Summing moments about point *G* eliminates $\mathbf{F}_{GE}$ and $\mathbf{F}_{GC}$ and yields a direct solution for F_{BC}.

$$\curvearrowleft + \Sigma M_G = 0; \qquad -300\text{ N}(4\text{ m}) - 400\text{ N}(3\text{ m}) + F_{BC}(3\text{ m}) = 0$$

$$F_{BC} = 800\text{ N} \quad \text{(T)} \qquad \textit{Ans.}$$

In the same manner, by summing moments about point *C* we obtain a direct solution for F_{GE}.

$$\curvearrowleft + \Sigma M_C = 0; \qquad -300\text{ N}(8\text{ m}) + F_{GE}(3\text{ m}) = 0$$

$$F_{GE} = 800\text{ N} \quad \text{(C)} \qquad \textit{Ans.}$$

Since $\mathbf{F}_{BC}$ and $\mathbf{F}_{GE}$ have no vertical components, summing forces in the *y* direction directly yields F_{GC}, i.e.,

$$+\uparrow \Sigma F_y = 0; \qquad 300\text{ N} - \tfrac{3}{5}F_{GC} = 0$$

$$F_{GC} = 500\text{ N} \quad \text{(T)} \qquad \textit{Ans.}$$

Obtain these results by applying the equations of equilibrium to the free-body diagram shown in Fig. 6–16*d*.

Example 6–6

Determine the force in member *CF* of the bridge truss shown in Fig. 6–17*a*. Indicate whether the member is in tension or compression. Assume each member is pin-connected.

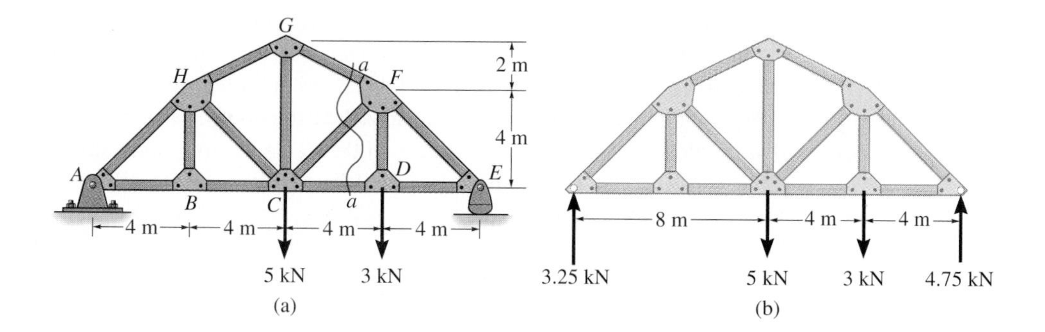

SOLUTION

Free-Body Diagram. Section *aa* in Fig. 6–17*a* will be used since this section will "expose" the internal force in member *CF* as "external" on the free-body diagram of either the right or left portion of the truss. It is first necessary, however, to determine the external reactions on either the left or right side of the truss. Verify the results shown on the free-body diagram in Fig. 6–17*b*.

The free-body diagram of the right portion of the truss, which is the easiest to analyze, is shown in Fig. 6–17*c*. There are three unknowns, F_{FG}, F_{CF}, and F_{CD}.

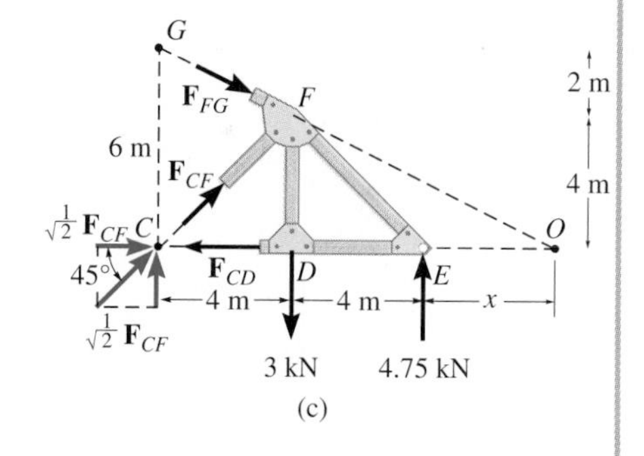

Fig. 6–17

Equations of Equilibrium. The most direct method for solving this problem requires application of the moment equation about a point that eliminates two of the unknown forces. Hence, to obtain $\mathbf{F}_{CF}$, we will eliminate $\mathbf{F}_{FG}$ and $\mathbf{F}_{CD}$ by summing moments about point *O*, Fig. 6–17*c*. Note that the location of point *O* measured from *E* is determined from proportional triangles, i.e., $4/(4 + x) = 6/(8 + x)$, $x = 4$ m. Or, stated in another manner, the slope of member *GF* has a drop of 2 m to a horizontal distance of $CD = 4$ m. Since *FD* is 4 m, Fig. 6–17*c*, then from *D* to *O* the distance must be 8 m.

An easy way to determine the moment of $\mathbf{F}_{CF}$ about point *O* is to resolve $\mathbf{F}_{CF}$ into its two rectangular components and then use the principle of transmissibility to move $\mathbf{F}_{CF}$ to point *C*. We have

$$\circlearrowleft + \Sigma M_O = 0; \quad -\frac{1}{\sqrt{2}}F_{CF}(12 \text{ m}) + (3 \text{ kN})(8 \text{ m}) - (4.75 \text{ kN})(4 \text{ m}) = 0$$

$$F_{CF} = 0.589 \text{ kN} \quad \text{(C)} \qquad \textit{Ans.}$$

Example 6–7

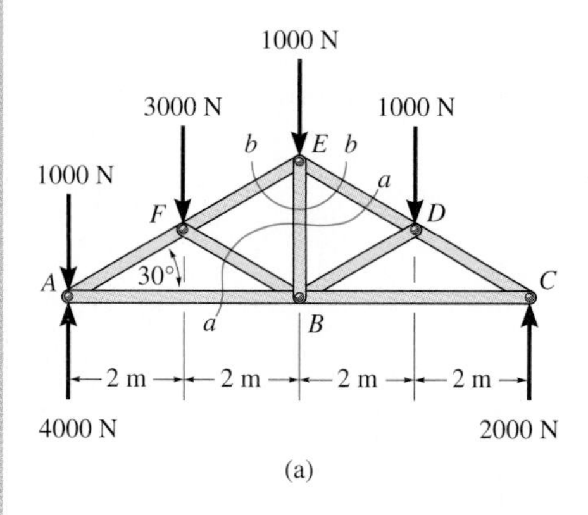

(a)

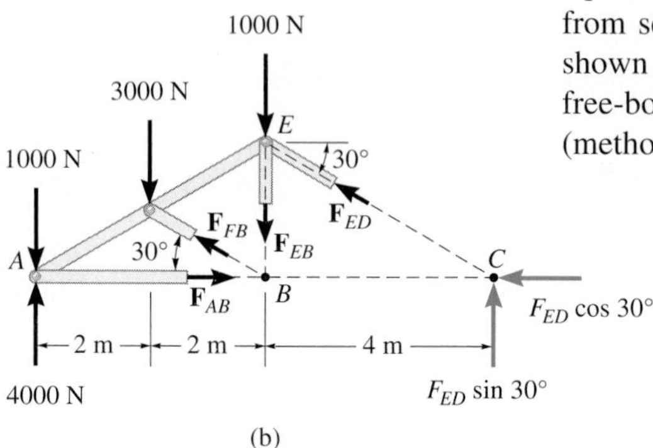

(b)

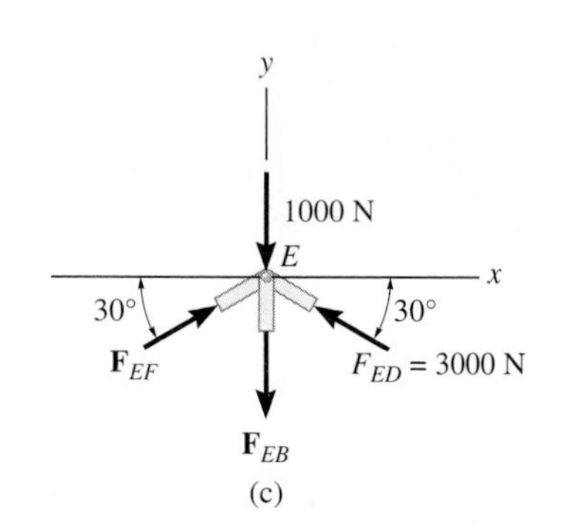

(c)

Fig. 6–18

Determine the force in member EB of the roof truss shown in Fig. 6–18a. Indicate whether the member is in tension or compression.

SOLUTION

Free-Body Diagrams. By the method of sections, any imaginary vertical section that cuts through EB, Fig. 6–18a, will also have to cut through three other members for which the forces are unknown. For example, section aa cuts through ED, EB, FB, and AB. If the components of reaction at A are calculated first ($A_x = 0$, $A_y = 4000$ N) and a free-body diagram of the left side of this section is considered, Fig. 6–18b, it is possible to obtain $\mathbf{F}_{ED}$ by summing moments about B to eliminate the other three unknowns; however, $\mathbf{F}_{EB}$ cannot be determined from the remaining two equilibrium equations. One possible way of obtaining $\mathbf{F}_{EB}$ is first to determine $\mathbf{F}_{ED}$ from section aa, then use this result on section bb, Fig. 6–18a, which is shown in Fig. 6–18c. Here the force system is concurrent and our sectioned free-body diagram is the same as the free-body diagram for the pin at E (method of joints).

Equations of Equilibrium. In order to determine the moment of $\mathbf{F}_{ED}$ about point B, Fig. 6–18b, we will resolve the force into its rectangular components and, by the principle of transmissibility, extend it to point C as shown. The moments of 1000 N, F_{AB}, F_{FB}, F_{EB}, and F_{ED} cos 30° are all zero about B. Therefore,

$$\circlearrowleft + \Sigma M_B = 0; \quad 1000\text{ N}(4\text{ m}) + 3000\text{ N}(2\text{ m}) - 4000\text{ N}(4\text{ m}) + F_{ED}\sin 30^\circ(4) = 0$$

$$F_{ED} = 3000\text{ N} \quad \text{(C)}$$

Considering now the free-body diagram of section bb, Fig. 6–18c, we have

$$\xrightarrow{+} \Sigma F_x = 0; \quad F_{EF}\cos 30^\circ - 3000\cos 30^\circ\text{ N} = 0$$

$$F_{EF} = 3000\text{ N} \quad \text{(C)}$$

$$+\uparrow \Sigma F_y = 0; \quad 2(3000\sin 30^\circ\text{ N}) - 1000\text{ N} - F_{EB} = 0$$

$$F_{EB} = 2000\text{ N} \quad \text{(T)} \qquad \textit{Ans.}$$

PROBLEMS

6–31. Determine the force in members *BC*, *HC*, and *HG* of the bridge truss and state if these members are in tension or compression.

***6–32.** Determine the force in members *GF*, *CF*, and *CD* of the bridge truss and state if these members are in tension or compression.

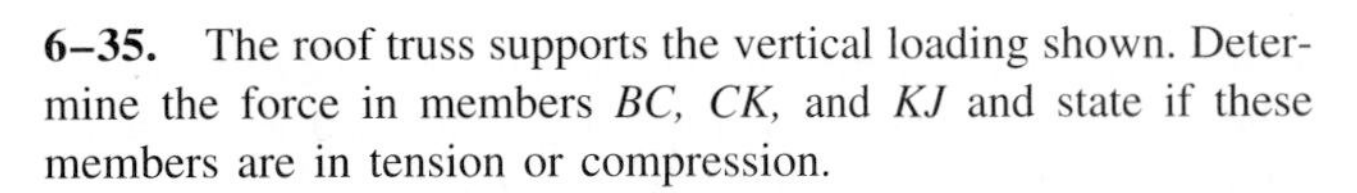

6–35. The roof truss supports the vertical loading shown. Determine the force in members *BC*, *CK*, and *KJ* and state if these members are in tension or compression.

***6–36.** The roof truss supports the vertical loading shown. Determine the force in members *DE* and *DJ* and state if these members are in tension or compression.

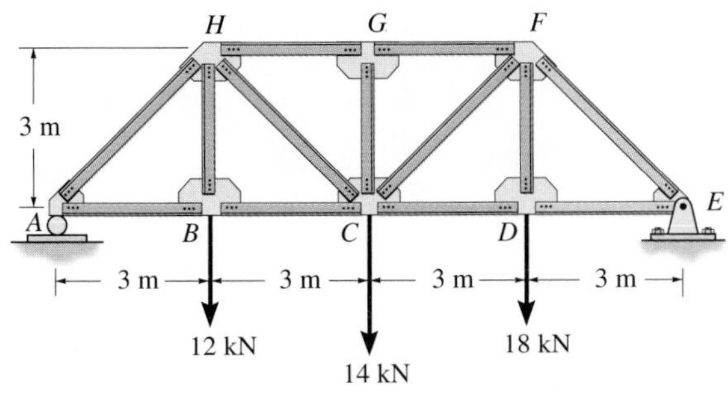

Probs. 6–31/6–32

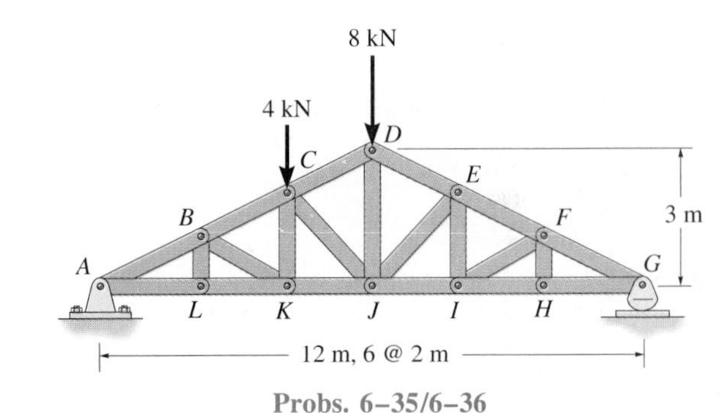

Probs. 6–35/6–36

6–33. The *Howe bridge truss* is subjected to the loading shown. Determine the force in members *HD*, *CD*, and *GD* and state if these members are in tension or compression.

6–34. The *Howe bridge truss* is subjected to the loading shown. Determine the force in members *HI*, *HB*, and *BC* and state if these members are in tension or compression.

6–37. Determine the force in members *CD*, *CJ*, *KJ*, and *DJ* of the truss which serves to support the deck of a bridge. State if these members are in tension or compression.

6–38. Determine the force in members *EI* and *JI* of the truss which serves to support the deck of a bridge. State if these members are in tension or compression.

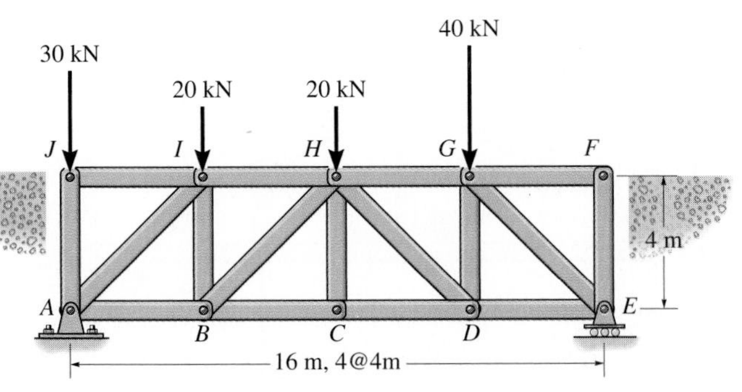

Probs. 6–33/6–34

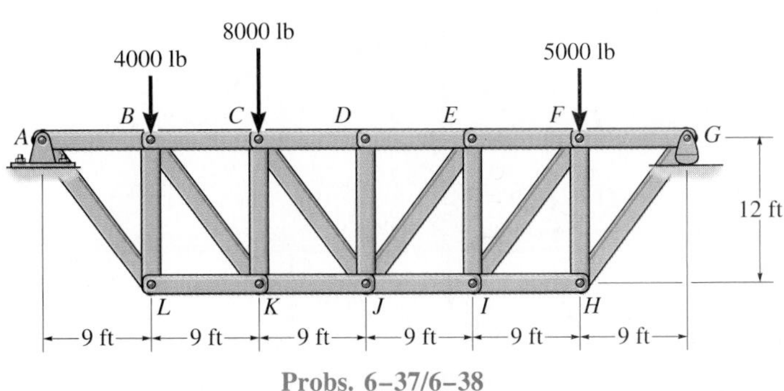

Probs. 6–37/6–38

6–39. Determine the force in members *CE*, *FE*, and *CD* and state if these members are in tension or compression. *Hint:* The force acting at the pin *G* is directed along bar *GD*. Why?

***6–40.** Determine the force in members *BC*, *FC*, and *FE* and state if these members are in tension or compression. *Hint:* The force acting at the pin *G* is directed along bar *GD*. Why?

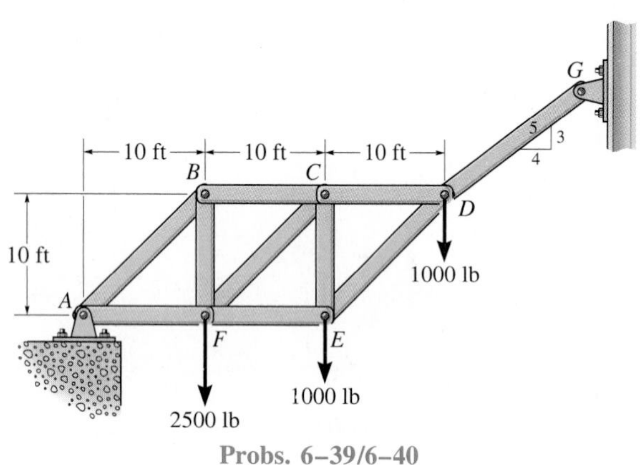

Probs. 6–39/6–40

6–41. Determine the force developed in members *GB* and *GF* of the bridge truss and state if these members are in tension or compression.

6–42. Determine the force developed in members *FC* and *BC* of the bridge truss and state if these members are in tension or compression.

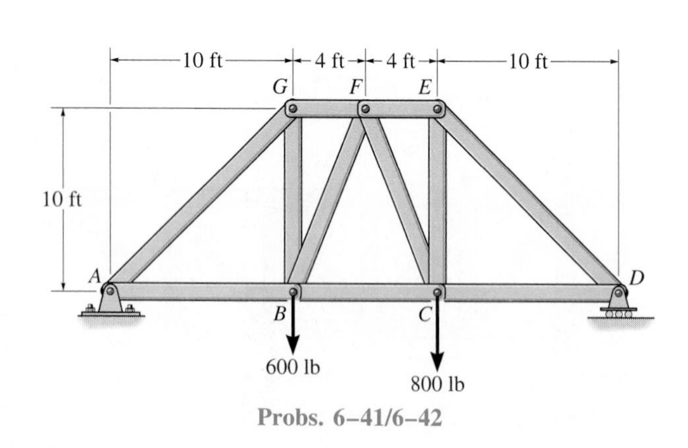

Probs. 6–41/6–42

6–43. Determine the force in members *BC*, *HC*, and *HG*. After the truss is sectioned use a single equation of equilibrium for the calculation of each force. State if these members are in tension or compression.

***6–44.** Determine the force in members *CD*, *CF*, and *CG* and state if these members are in tension or compression.

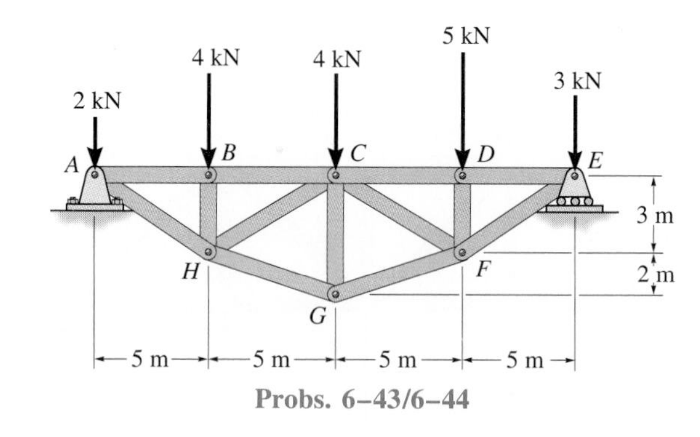

Probs. 6–43/6–44

6–45. Determine the force in member *BC* of the truss and state if this member is in tension or compression.

6–46. Determine the force in member *GJ* of the truss and state if this member is in tension or compression.

6–47. Determine the force in member *GC* of the truss and state if this member is in tension or compression.

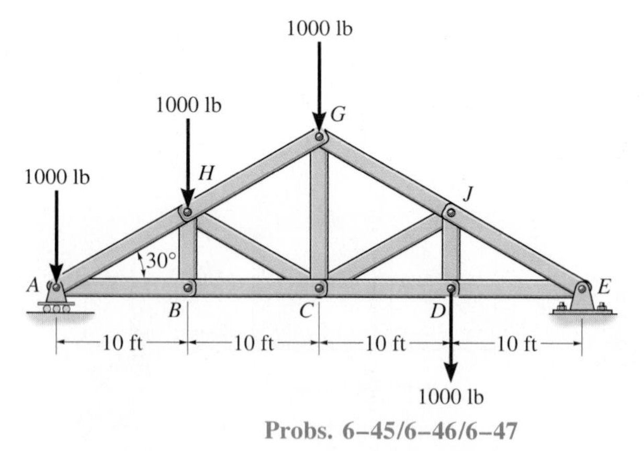

Probs. 6–45/6–46/6–47

***6–48.** The truss is used to support two electrical power lines that exert the forces shown on the structure. Determine the force developed in members *BC*, *BD*, and *DE* and state if these members are in tension or compression.

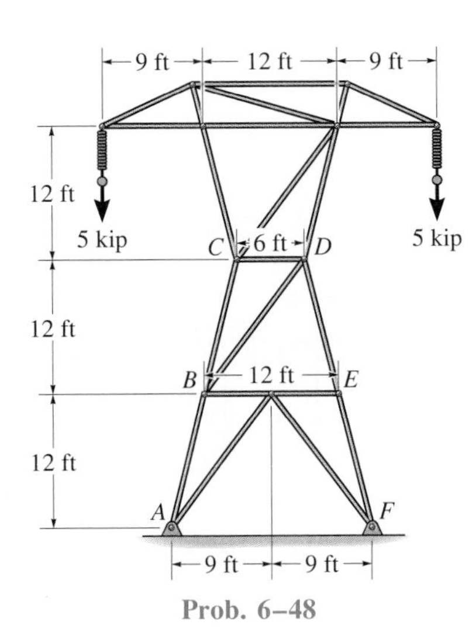

Prob. 6–48

6–49. Determine the force in members *IC* and *CG* of the truss and state if these members are in tension or compression. Also, indicate all zero-force members.

6–50. Determine the force in members *JE* and *GF* of the truss and state if these members are in tension or compression. Also, indicate all zero-force members.

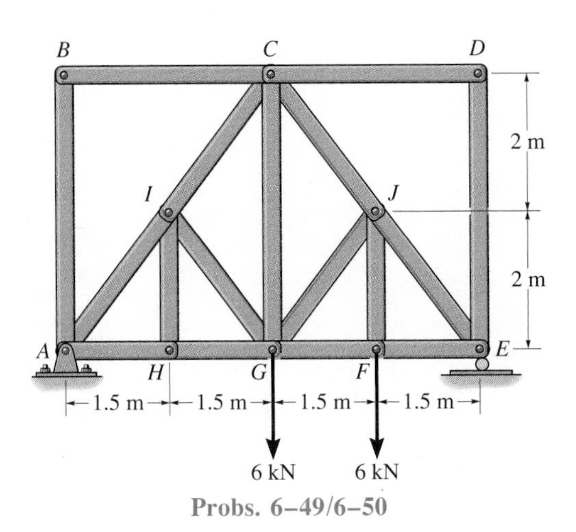

Probs. 6–49/6–50

6–51. The skewed truss carries the load shown. Determine the force in members *CB*, *BE*, and *EF* and state if these members are in tension or compression. Assume that all joints are pinned.

***6–52.** The skewed truss carries the load shown. Determine the force in members *AB*, *BF*, and *EF* and state if these members are in tension or compression. Assume that all joints are pinned.

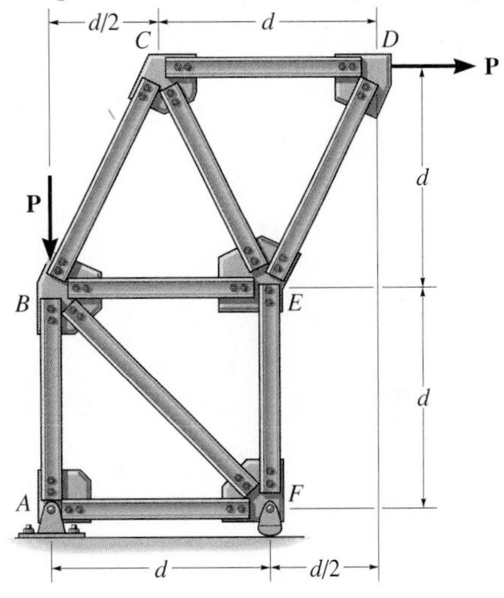

Probs. 6–51/6–52

6–53. The suspension tower consists of an overhead truss which supports the 2-kN cable weights at *A*, *C*, *E*, and *G*. If the truss can be assumed pin-supported at *A* and roller-supported at *G*, determine the force in members *EH*, *EF*, and *IH* and state if these members are in tension or compression.

6–54. The suspension tower consists of an overhead truss which supports the 2-kN cable weights at *A*, *C*, *E*, and *G*. If the truss can be assumed pin-supported at *A* and roller-supported at *G*, determine the force in members *KD*, *CD*, and *KJ* and state if these members are in tension or compression.

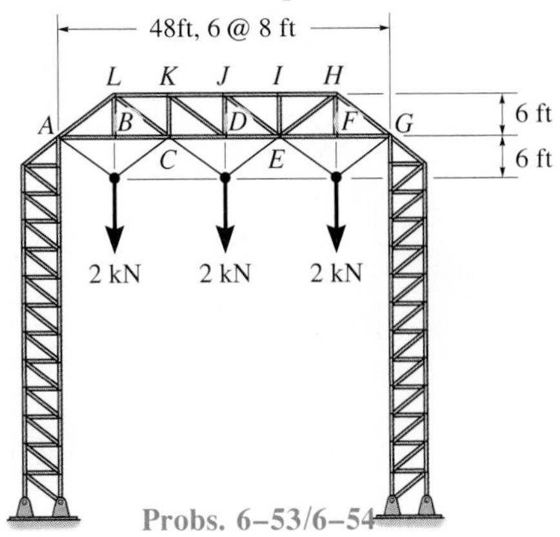

Probs. 6–53/6–54

6–55. Determine the force in members *HG, HC,* and *BC* of the truss and state if these members are in tension or compression. After the truss is sectioned, use a single equation of equilibrium for the calculation of each force.

***6–56.** Determine the force in members *GF, CF,* and *CD* of the truss and state if these members are in tension or compression. After the truss is sectioned, use a single equation of equilibrium for the calculation of each force.

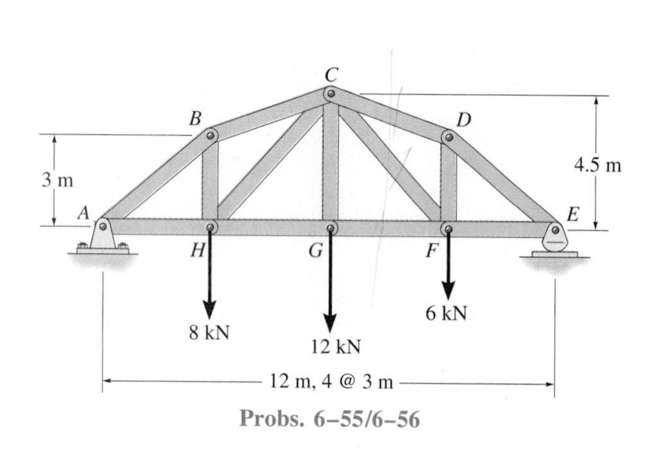

Probs. 6–55/6–56

6–57. Determine the force in members *CB, CG,* and *GF* of the symmetrical truss and state if these members are in tension or compression.

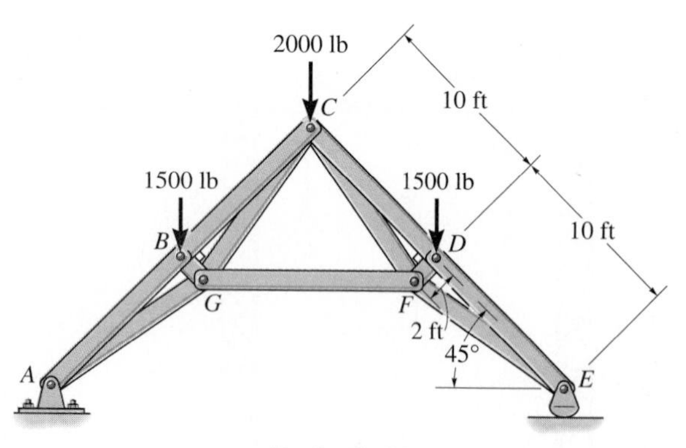

Prob. 6–57

6–58. Determine the force in members *DE, JI,* and *DO* of the *K truss* and state if these members are in tension or compression. *Hint:* Use sections *aa* and *bb*.

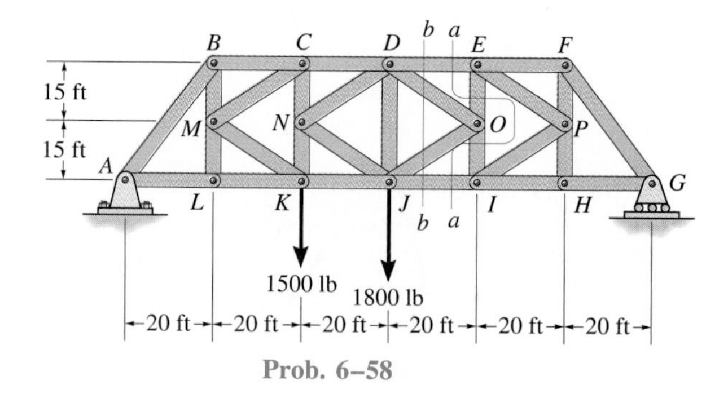

Prob. 6–58

6–59. Determine the force in members *CD* and *KJ* of the *K truss* and state if these members are in tension or compression.

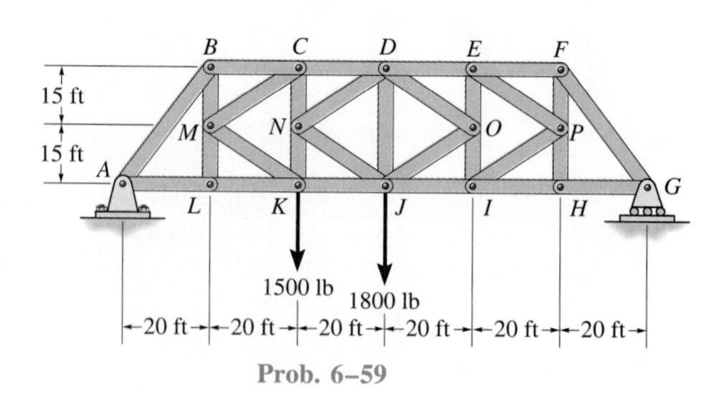

Prob. 6–59

★6.5 Space Trusses

A *space truss* consists of members joined together at their ends to form a stable three-dimensional structure. The simplest element of a space truss is a *tetrahedron,* formed by connecting six members together, as shown in Fig. 6–19. Any additional members added to this basic element would be redundant in supporting the force **P.** A *simple space truss* can be built from this basic tetrahedral element by adding three additional members and a joint forming a system of multiconnected tetrahedrons.

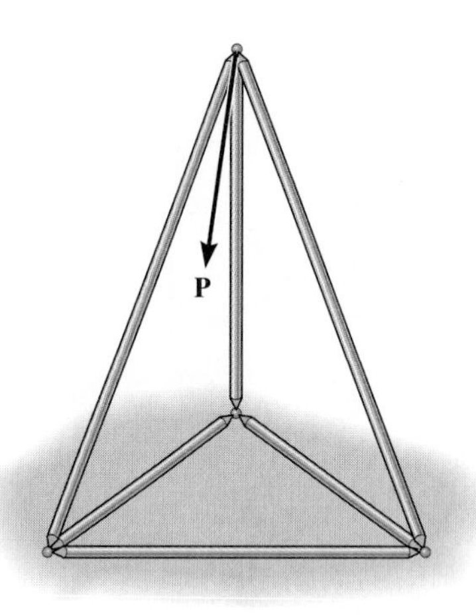

Fig. 6–19

Assumptions for Design. The members of a space truss may be treated as two-force members provided the external loading is applied at the joints and the joints consist of ball-and-socket connections. These assumptions are justified if the welded or bolted connections of the joined members intersect at a common point and the weight of the members can be neglected. In cases where the weight of a member is to be included in the analysis, it is generally satisfactory to apply it as a vertical force, half of its magnitude applied at each end of the member.

PROCEDURE FOR ANALYSIS

Either the method of joints or the method of sections can be used to determine the forces developed in the members of a simple space truss.

Method of Joints. Generally, if the forces in *all* the members of the truss must be determined, the method of joints is most suitable for the analysis. When using the method of joints, it is necessary to solve the three scalar equilibrium equations $\Sigma F_x = 0$, $\Sigma F_y = 0$, $\Sigma F_z = 0$ at each joint. The solution of many simultaneous equations can be avoided if the force analysis begins at a joint having at least one known force and at most three unknown forces. If the three-dimensional geometry of the force system at the joint is hard to visualize, it is recommended that a Cartesian vector analysis be used for the solution.

Method of Sections. If only a *few* member forces are to be determined, the method of sections may be used. When an imaginary section is passed through a truss, and the truss is separated into two parts, the force system acting on one of the parts must satisfy the *six* scalar equilibrium equations: $\Sigma F_x = 0$, $\Sigma F_y = 0$, $\Sigma F_z = 0$, $\Sigma M_x = 0$, $\Sigma M_y = 0$, $\Sigma M_z = 0$ (Eqs. 5–6). By proper choice of the section and axes for summing forces and moments, many of the unknown member forces in a space truss can be computed *directly,* using a single equilibrium equation.

Example 6–8

Determine the forces acting in the members of the space truss shown in Fig. 6–20*a*. Indicate whether the members are in tension or compression.

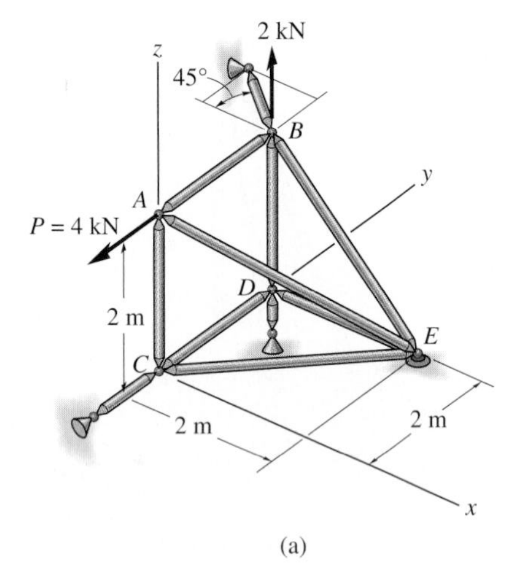

(a)

SOLUTION

Since there are one known force and three unknown forces acting at joint *A*, the force analysis of the truss will begin at this joint.

Joint A (Fig. 6–20*b*). Expressing each force that acts on the free-body diagram of joint *A* in vector notation, we have

$$\mathbf{P} = \{-4\mathbf{j}\}\text{ kN}, \qquad \mathbf{F}_{AB} = F_{AB}\mathbf{j}, \qquad \mathbf{F}_{AC} = -F_{AC}\mathbf{k},$$

$$\mathbf{F}_{AE} = F_{AE}\left(\frac{\mathbf{r}_{AE}}{r_{AE}}\right) = F_{AE}(0.577\mathbf{i} + 0.577\mathbf{j} - 0.577\mathbf{k})$$

For equilibrium,

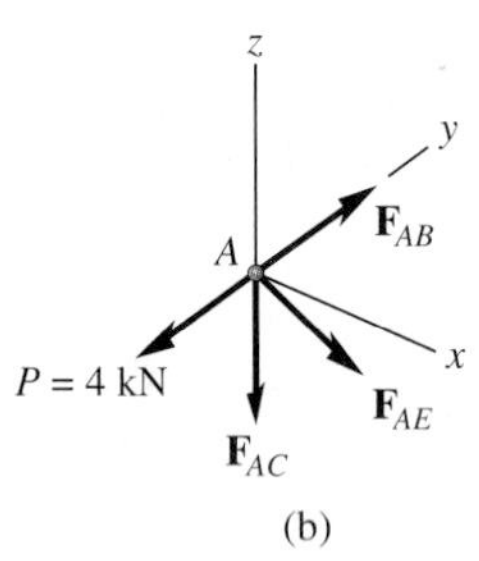

(b)

$$\Sigma\mathbf{F} = \mathbf{0}; \qquad \mathbf{P} + \mathbf{F}_{AB} + \mathbf{F}_{AC} + \mathbf{F}_{AE} = \mathbf{0}$$

$$-4\mathbf{j} + F_{AB}\mathbf{j} - F_{AC}\mathbf{k} + 0.577F_{AE}\mathbf{i} + 0.577F_{AE}\mathbf{j} - 0.577F_{AE}\mathbf{k} = \mathbf{0}$$

$$\Sigma F_x = 0; \qquad 0.577F_{AE} = 0$$

$$\Sigma F_y = 0; \qquad -4 + F_{AB} + 0.577F_{AE} = 0$$

$$\Sigma F_z = 0; \qquad -F_{AC} - 0.577F_{AE} = 0$$

$$F_{AC} = F_{AE} = 0 \qquad \textit{Ans.}$$

$$F_{AB} = 4\text{ kN} \quad (\text{T}) \qquad \textit{Ans.}$$

Since F_{AB} is known, joint *B* may be analyzed next.

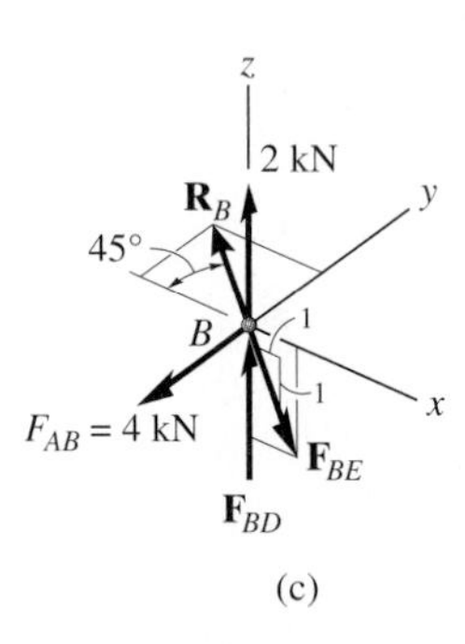

(c)

Joint B (Fig. 6–20*c*)

$$\Sigma F_x = 0; \qquad -R_B \cos 45° + 0.707F_{BE} = 0$$

$$\Sigma F_y = 0; \qquad -4 + R_B \sin 45° = 0$$

$$\Sigma F_z = 0; \qquad 2 + F_{BD} - 0.707F_{BE} = 0$$

$$R_B = F_{BE} = 5.66\text{ kN} \quad (\text{T}), \qquad F_{BD} = 2\text{ kN} \quad (\text{C}) \qquad \textit{Ans.}$$

The *scalar* equations of equilibrium may also be applied directly to the force systems on the free-body diagrams of joints *D* and *C*, since the force components are easily determined. Show that

$$F_{DE} = F_{DC} = F_{CE} = 0 \qquad \textit{Ans.}$$

Fig. 6–20

PROBLEMS

*6–60. Determine the force in each member of the space truss and state if the members are in tension or compression. The truss is supported by ball-and-socket joints at *D, C,* and *E*. *Hint:* The support reaction at *E* acts along member *EB*. Why?

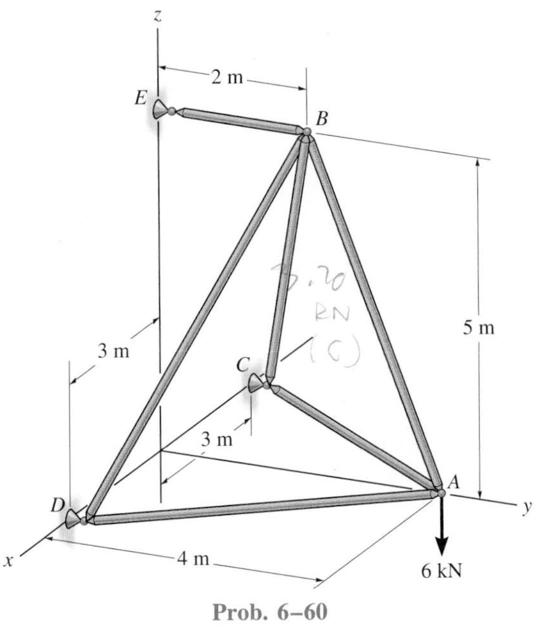

Prob. 6–60

6–61. The tetrahedral truss rests on roller supports at points *A, B,* and *C*. Determine the force in each member and state if the members are in tension or compression.

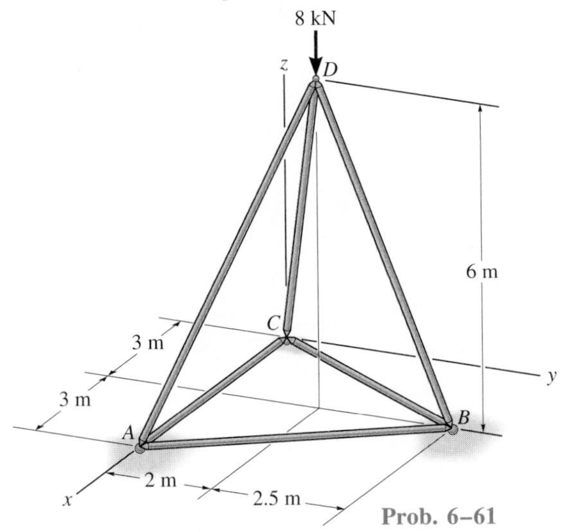

Prob. 6–61

6–62. Determine the force developed in each member of the space truss and state if the members are in tension or compression. The crate has a weight of 150 lb.

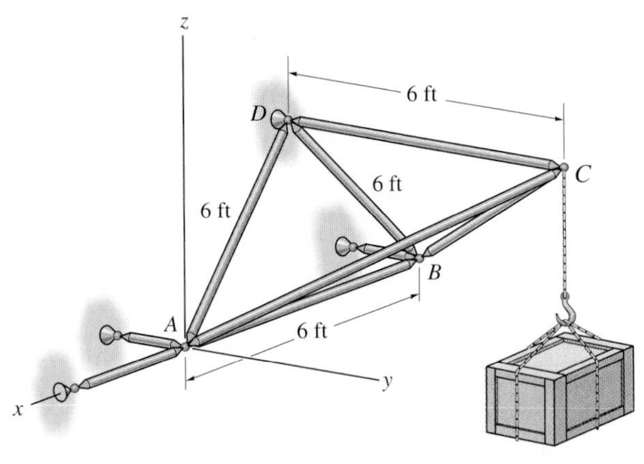

Prob. 6–62

6–63. The space truss is used to support vertical forces at joints *B, C,* and *D*. Determine the force in each member and state if the members are in tension or compression.

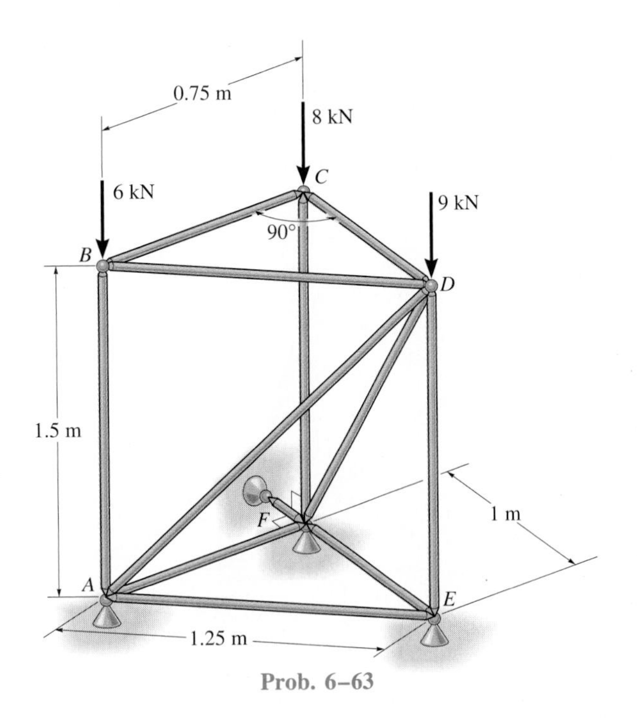

Prob. 6–63

*6–64. The space truss is supported by a ball-and-socket joint at D and short links at C and E. Determine the force in each member and state if the members are in tension or compression. Take $\mathbf{F}_1 = \{-500\mathbf{k}\}$ lb and $\mathbf{F}_2 = \{400\mathbf{j}\}$ lb.

6–65. The space truss is supported by a ball-and-socket joint at D and short links at C and E. Determine the force in each member and state if the members are in tension or compression. Take $\mathbf{F}_1 = \{200\mathbf{i} + 300\mathbf{j} - 500\mathbf{k}\}$ lb and $\mathbf{F}_2 = \{400\mathbf{j}\}$ lb.

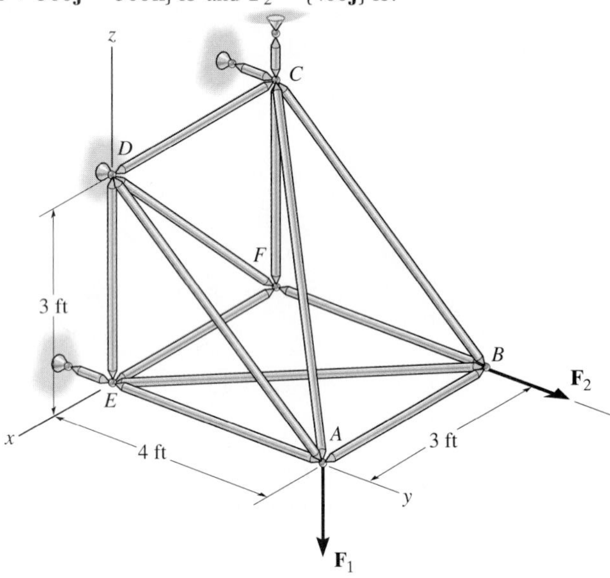

Probs. 6–64/6–65

6–66. Determine the force in each member of the space truss and state if the members are in tension or compression. The truss is supported by a ball-and-socket joint at A and short links at B and C.

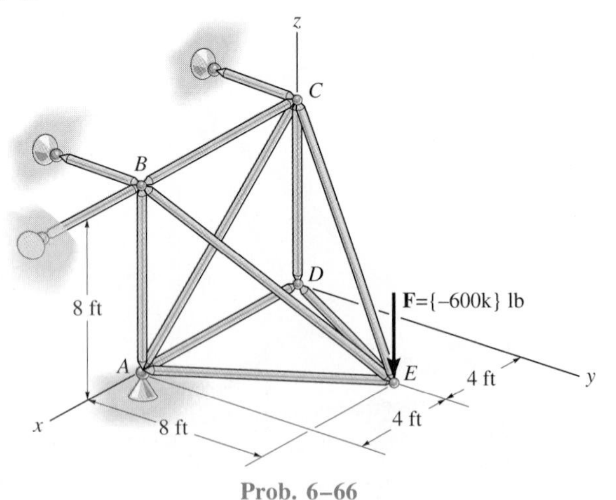

Prob. 6–66

6–67. Determine the force in each member of the space truss and state if the members are in tension or compression. The truss is supported by ball-and-socket joints at A, B, and E. Set $\mathbf{F} = \{800\mathbf{j}\}$ N. *Hint:* The support reaction at E acts along member EC. Why?

*6–68. Determine the force in each member of the space truss and state if the members are in tension or compression. The truss is supported by ball-and-socket joints at A, B, and E. Set $\mathbf{F} = \{-200\mathbf{i} + 400\mathbf{j}\}$ N. *Hint:* The support reaction at E acts along member EC. Why?

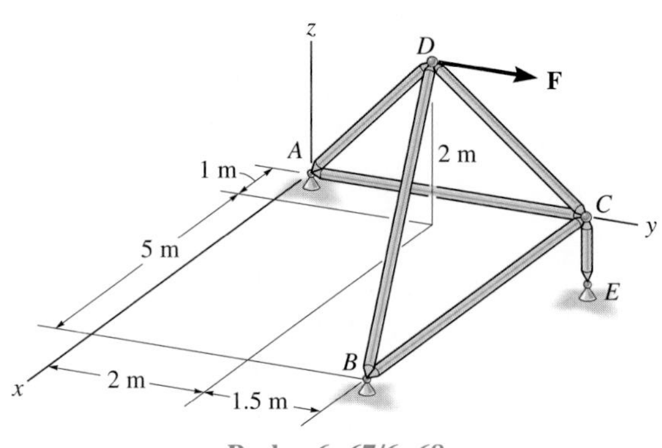

Probs. 6–67/6–68

6–69. Determine the force in each member of the space truss and state if the members are in tension or compression. The truss is supported by ball-and-socket joints at C, D, E, and G.

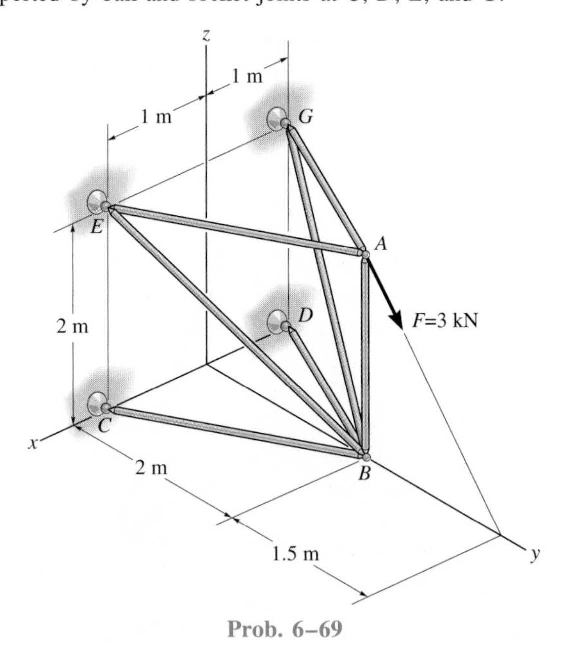

Prob. 6–69

6.6 Frames and Machines

Frames and machines are two common types of structures which are often composed of pin-connected *multiforce members,* i.e., members that are subjected to more than two forces. *Frames* are generally stationary and are used to support loads, whereas *machines* contain moving parts and are designed to transmit and alter the effect of forces. Provided a frame or machine is properly constrained and contains no more supports or members than are necessary to prevent collapse, the forces acting at the joints and supports can be determined by applying the equations of equilibrium to each member. Once the forces at the joints are obtained, it is then possible to *design* the size of the members, connections, and supports using the theory of mechanics of materials and an appropriate engineering design code.

Free-Body Diagrams. In order to determine the forces acting at the joints and supports of a frame or machine, the structure must be disassembled and the free-body diagrams of its parts must be drawn. In this regard, the following important points *must* be observed:

1. Isolate each part by drawing its *outlined shape.* Then show all the forces and/or couple moments that act on the part. Make sure to *label* or *identify* each known and unknown force and couple moment with reference to an established x, y coordinate system. Also, indicate any dimensions used for taking moments. Most often the equations of equilibrium are easier to apply if the forces are represented by their rectangular components. As usual, the sense of an unknown force or couple moment can be assumed.
2. Identify all the two-force members in the structure, and represent their free-body diagrams as having two equal but opposite forces acting at their points of application. The line of action of the forces is defined by the line joining the two points where the forces act (see Sec. 5.4). By recognizing the two-force members, we can avoid solving an unnecessary number of equilibrium equations. (See Example 6–14.)
3. Forces common to any two *contacting* members act with equal magnitudes but opposite sense on the respective members. If the two members are treated as a *''system'' of connected members,* then these forces are *''internal''* and are *not shown* on the *free-body diagram of the system;* however, if the free-body diagram of *each member* is drawn, the forces are *''external''* and *must* be shown on each of the free-body diagrams.

The following examples graphically illustrate application of these points in drawing the free-body diagrams of a dismembered frame or machine. In all cases, the weight of the members is neglected.

Example 6–9

For the frame shown in Fig. 6–21*a*, draw the free-body diagram of (a) each member, (b) the pin at *B*, and (c) the two members connected together.

(a)

(b)

SOLUTION

Part (a). By inspection, the members are *not* two-force members. Instead, as shown on the free-body diagrams, Fig. 6–21*b*, *BC* is subjected to *not* five but *three forces,* namely, the *resultants* of the two components of reaction at pins *B* and *C* and the external force **P.** Likewise, *AB* is subjected to the *resultant* pin-reactive forces at *A* and *B* and the external couple moment **M.**

Part (b). It can be seen in Fig. 6–21*a* that the pin at *B* is subjected to only *two forces,* i.e., the force of member *BC* on the pin and the force of member *AB* on the pin. For *equilibrium* these forces and therefore their respective components must be equal but opposite, Fig. 6–21*c*. Notice carefully how Newton's third law is applied between the pin and its contacting members, i.e., the effect of the pin on the two members, Fig. 6–21*b*, and the equal but opposite effect of the two members on the pin, Fig. 6–21*c*. Also note that $\mathbf{B}_x$ and $\mathbf{B}_y$ shown equal but opposite in Fig. 6–21*b* on members *AB* and *BC* is *not* the effect of Newton's third law; instead, this results from the *equilibrium* analysis of the pin, Fig. 6–21*c*.

Part (c). The free-body diagram of both members connected together, yet removed from the supporting pins at *A* and *C*, is shown in Fig. 6–21*d*. The force components $\mathbf{B}_x$ and $\mathbf{B}_y$ are *not shown* on this diagram since they form equal but opposite collinear pairs of *internal* forces (Fig. 6–21*b*) and therefore cancel out.* Also, to be consistent when later applying the equilibrium equations, the unknown force components at *A* and *C* must act in the *same sense* as those shown in Fig. 6–21*b*.

*This is similar to not including internal forces exerted between adjacent particles of a rigid body when drawing the free-body diagram of the entire rigid body.

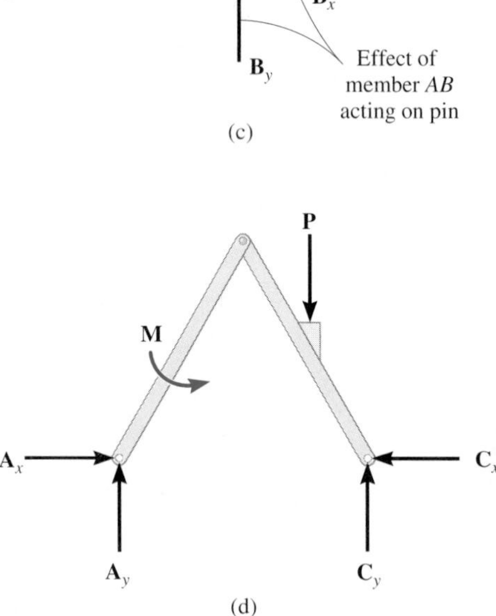

(c)

(d)

Fig. 6–21

Example 6–10

For the frame shown in Fig. 6–22*a*, draw the free-body diagrams of (a) each of the three members, and (b) members *ABC* and *BD* together.

SOLUTION

Part (a). By inspection, none of the three members of the frame are two-force members. Instead, each is subjected to *three* forces. The components of these forces are shown on the free-body diagrams in Fig. 6–22*b*. Notice that equal but opposite force reactions occur at *B*, *C*, and *D*. Draw a free-body diagram of one of the pins at *B*, *C*, or *D* and show why this is so.

Part (b). The free-body diagram of *ABC* and *BD* together is shown in Fig. 6–22*c*. Since the entire frame is in equilibrium, the force system on these two members also satisfies the equilibrium equations. Why not show the force components $\mathbf{B}_x$ and $\mathbf{B}_y$ on this diagram?

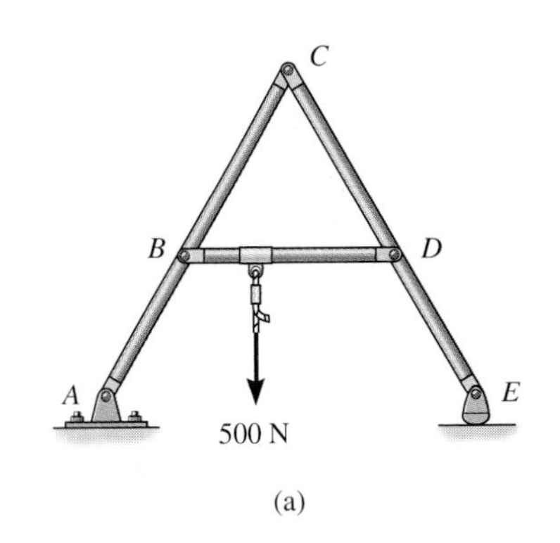

(a)

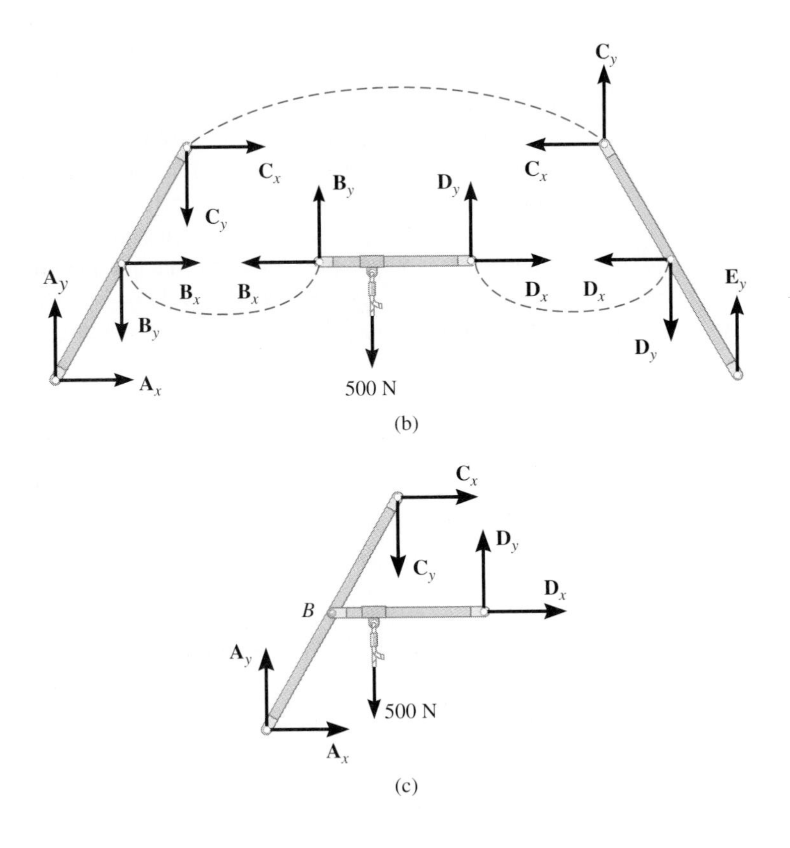

(b)

(c)

Fig. 6–22

Example 6–11

Draw the free-body diagram of each part of the smooth piston and link mechanism used to recycle crushed cans, which is shown in Fig. 6–23*a*.

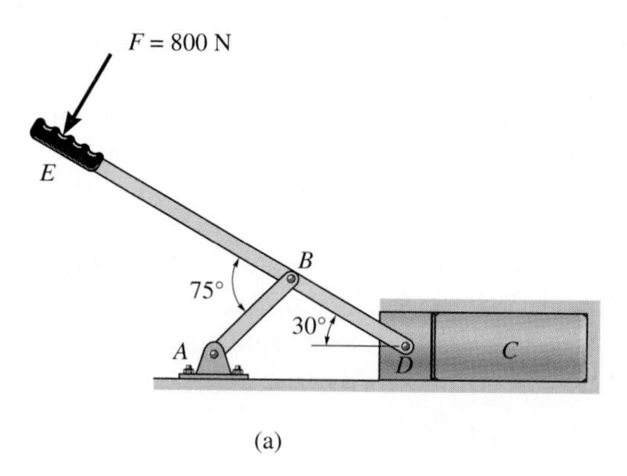

(a)

SOLUTION

By inspection, member *AB* is a two-force member. The free-body diagrams of the parts are shown in Fig. 6–23*b*. Since the pins at *B* and *D* *connect only two parts together,* the forces there are shown as equal but opposite on the separate free-body diagrams of their connected members. In particular, four components of force act on the piston: $\mathbf{D}_x$ and $\mathbf{D}_y$ represent the effect of the pin (or lever *EBD*), $\mathbf{N}_w$ is the *resultant force* of the cylinder's wall, and $\mathbf{P}$ is the resultant compressive force caused by the can *C*.

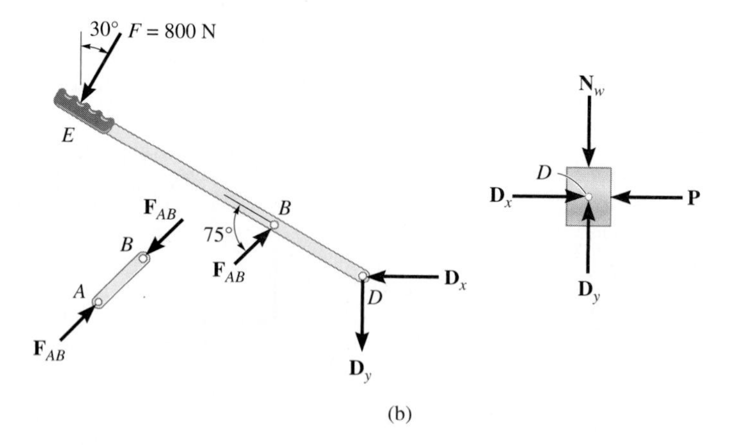

(b)

Fig. 6–23

Example 6–12

For the frame shown in Fig. 6–24*a*, draw the free-body diagrams of (a) the entire frame including the pulleys and cords, (b) the frame without the pulleys and cords, and (c) each of the pulleys.

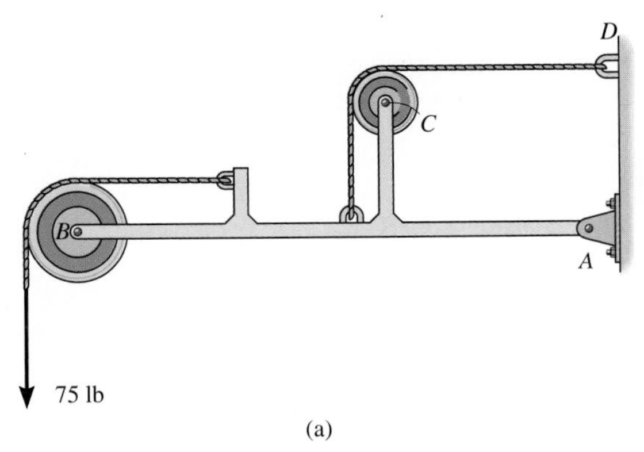

(a)

SOLUTION

Part (a): When the entire frame including the pulleys and cords is considered, the interactions at the points where the pulleys and cords are connected to the frame become pairs of *internal forces* which cancel each other and therefore are not shown on the free-body diagram, Fig. 6–24*b*.

Part (b): When the cords and pulleys are removed, their effect *on the frame* must be shown, Fig. 6–24*c*.

Part (c): The force components $\mathbf{B}_x$, $\mathbf{B}_y$, $\mathbf{C}_x$, $\mathbf{C}_y$ of the pins on the pulleys, Fig. 6–24*d*, are equal but opposite to the force components exerted by the pins on the frame, Fig. 6–24*c*. Why?

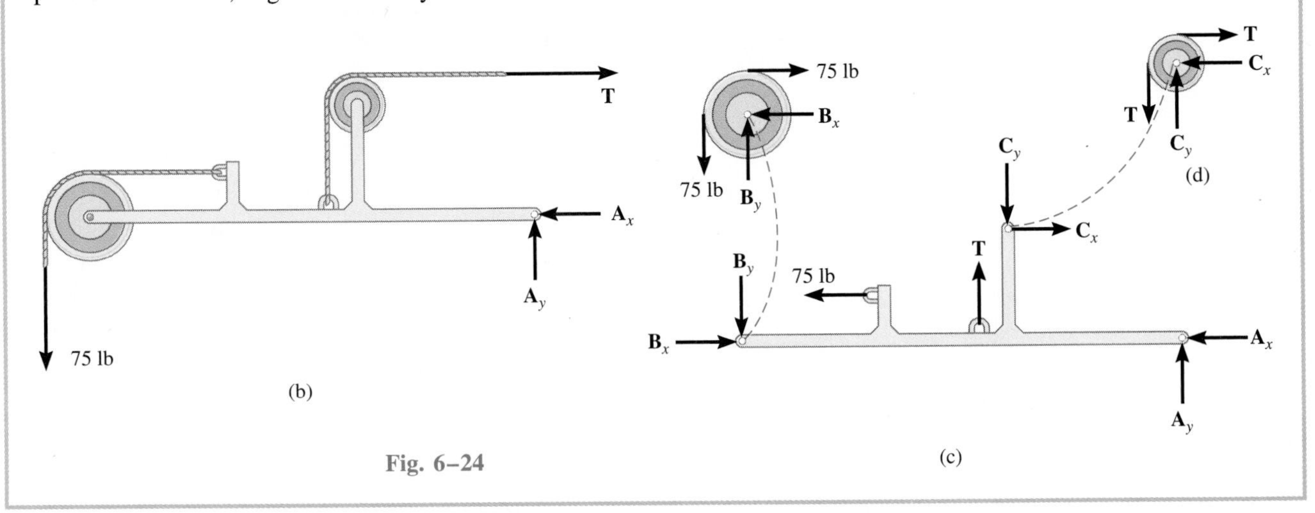

Fig. 6–24

Example 6–13

The hydraulic truck-mounted crane shown in Fig. 6–25*a* is used to lift a beam that has a mass of 1 Mg. Draw the free-body diagrams of each of its parts, including the pins at *A* and *C*.

Fig. 6–25

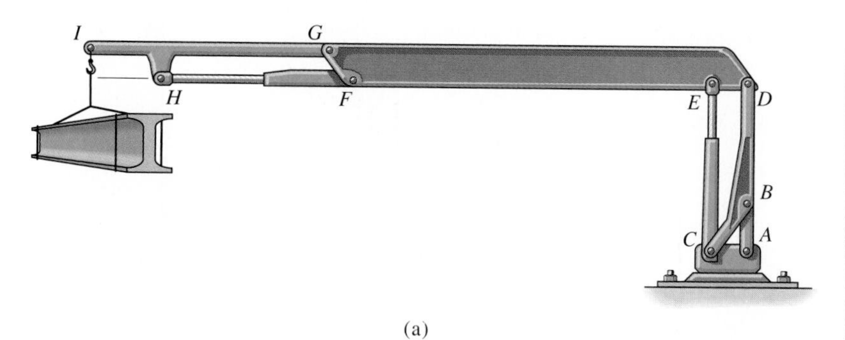

(a)

SOLUTION

By inspection, *HF*, *EC*, and *AB* are all two-force members. The free-body diagrams are shown in Fig. 6–25*b*. The pin at *A* is subjected to only *two* forces, namely, the force of the link *AB* and the force of the support. For equilibrium, these forces must be equal in magnitude but opposite in direction. The pin at *C*, however, is subjected to *three* forces. The force $\mathbf{F}_{EC}$ is caused by the hydraulic cylinder, the force components $\mathbf{C}_x$ and $\mathbf{C}_y$ are caused by member *CBD*, and finally, $\mathbf{C}'_x$ and $\mathbf{C}'_y$ are caused by the support. These components can be related by the equations of force equilibrium applied to the pin.

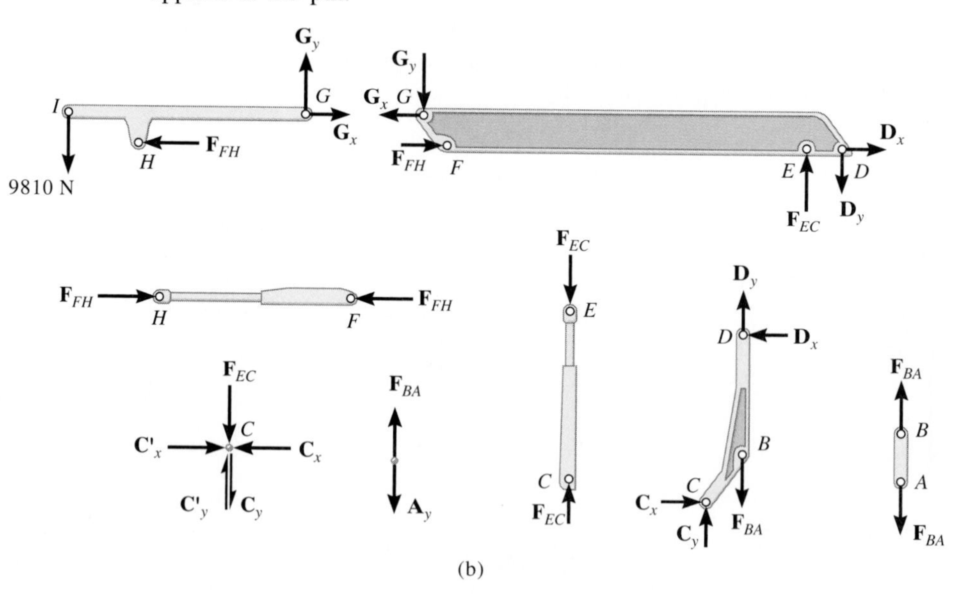

(b)

Before proceeding, it is recommended to cover the solutions to the previous examples and attempt to draw the requested free-body diagrams. When doing so, make sure the work is neat and that all the forces and couple moments are properly labeled.

Equations of Equilibrium. Provided the structure (frame or machine) is properly supported and contains no more supports or members than are necessary to prevent its collapse, then the unknown forces at the supports and connections can be determined from the equations of equilibrium. If the structure lies in the x–y plane, then for *each* free-body diagram drawn the loading must satisfy $\Sigma F_x = 0$, $\Sigma F_y = 0$, and $\Sigma M_O = 0$. The selection of the free-body diagrams used for the analysis is *completely arbitrary*. They may represent each of the members of the structure, a portion of the structure, or its entirety. For example, consider finding the six components of the pin reactions at A, B, and C for the frame shown in Fig. 6–26a. If the frame is dismembered, Fig. 6–26b, these unknowns can be determined by applying the three equations of equilibrium to each of the two members (total of six equations). The free-body diagram of the *entire frame* can also be used for part of the analysis, Fig. 6–26c. Hence, if so desired, all six unknowns can be determined by applying the three equilibrium equations to the entire frame, Fig. 6–26c, and also to either one of its members. Furthermore, the answers can be checked in part by applying the three equations of equilibrium to the remaining ''second'' member. In general, then, this problem can be solved by writing *at most* six equilibrium equations using free-body diagrams of the members and/or the combination of connected members. Any more than six equations written would *not* be unique from the original six and would serve only to check the results.

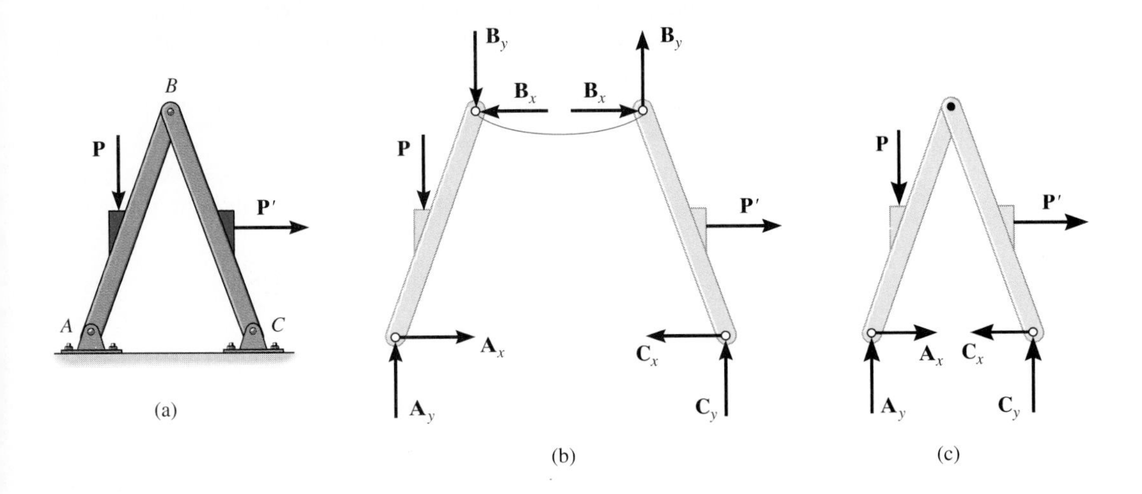

Fig. 6–26

PROCEDURE FOR ANALYSIS

The following procedure provides a method for determining the *joint reactions* of frames and machines (structures) composed of multiforce members.

Free-Body Diagrams. Draw the free-body diagram of the entire structure, a portion of the structure, or each of its members. The choice should be made so that it leads to the most direct solution to the problem.

Forces common to two members which are in contact act with equal magnitude but opposite sense on the respective free-body diagrams of the members. Recall that all *two-force members,* regardless of their shape, have equal but opposite collinear forces acting at the ends of the member. The unknown forces acting at the joints of multiforce members should be represented by their rectangular components. In many cases it is possible to tell by inspection the proper sense of the unknown forces; however, if this seems difficult, the sense can be assumed.

Equations of Equilibrium. Count the total number of unknowns to make sure that an equivalent number of equilibrium equations can be written for solution. Recall that in general three equilibrium equations can be written for each rigid body represented in two dimensions. Many times, the solution for the unknowns will be straightforward if moments are summed about a point that lies at the intersection of the lines of action of as many unknown forces as possible. If after obtaining the solution an unknown force magnitude is found to be negative, it means the sense of the force is the reverse of that shown on the free-body diagrams.

The examples that follow illustrate this procedure. All these examples should be *thoroughly understood* before proceeding to solve the problems.

Example 6–14

Determine the horizontal and vertical components of force which the pin at C exerts on member CB of the frame in Fig. 6–27*a*.

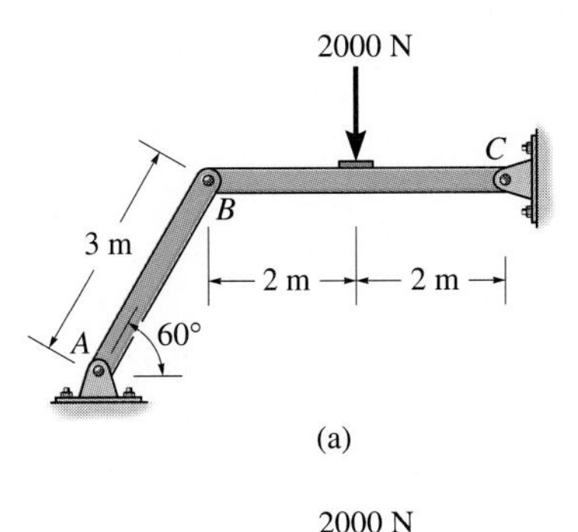

(a)

SOLUTION I

Free-Body Diagrams. By inspection it can be seen that AB is a two-force member. The free-body diagrams are shown in Fig. 6–27*b*.

Equations of Equilibrium. The *three unknowns,* C_x, C_y, and F_{AB}, can be determined by applying the three equations of equilibrium to member CB.

$$\circlearrowleft + \Sigma M_C = 0; \quad 2000\text{ N}(2\text{ m}) - (F_{AB} \sin 60°)(4\text{ m}) = 0 \quad F_{AB} = 1154.7\text{ N}$$

$$\xrightarrow{+} \Sigma F_x = 0; \quad 1154.7 \cos 60°\text{ N} - C_x = 0 \quad C_x = 577\text{ N} \quad \textit{Ans.}$$

$$+\uparrow \Sigma F_y = 0; \quad 1154.7 \sin 60°\text{ N} - 2000\text{ N} + C_y = 0 \quad C_y = 1000\text{ N} \quad \textit{Ans.}$$

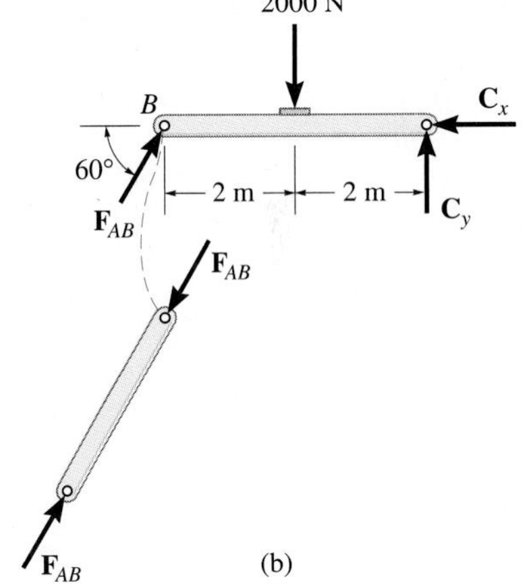

(b)

SOLUTION II

Free-Body Diagrams. If one does not recognize that AB is a two-force member, then more work is involved in solving this problem. The free-body diagrams are shown in Fig. 6–27*c*.

Equations of Equilibrium. The *six unknowns,* A_x, A_y, B_x, B_y, C_x, C_y, are determined by applying the three equations of equilibrium to each member.

Member AB

$$\circlearrowleft + \Sigma M_A = 0; \quad B_x(3 \sin 60°\text{ m}) - B_y(3 \cos 60°\text{ m}) = 0 \quad (1)$$

$$\xrightarrow{+} \Sigma F_x = 0; \quad A_x - B_x = 0 \quad (2)$$

$$+\uparrow \Sigma F_y = 0; \quad A_y - B_y = 0 \quad (3)$$

Member BC

$$\circlearrowleft + \Sigma M_C = 0; \quad 2000\text{ N}(2\text{ m}) - B_y(4\text{ m}) = 0 \quad (4)$$

$$\xrightarrow{+} \Sigma F_x = 0; \quad B_x - C_x = 0 \quad (5)$$

$$+\uparrow \Sigma F_y = 0; \quad B_y - 2000\text{ N} + C_y = 0 \quad (6)$$

The results for C_x and C_y can be determined by solving these equations in the following sequence: 4, 1, 5, then 6. The results are

$$B_y = 1000\text{ N}$$

$$B_x = 577\text{ N}$$

$$C_x = 577\text{ N} \quad \textit{Ans.}$$

$$C_y = 1000\text{ N} \quad \textit{Ans.}$$

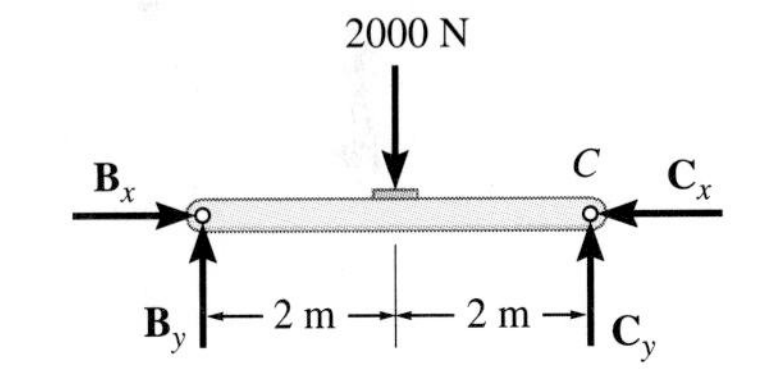

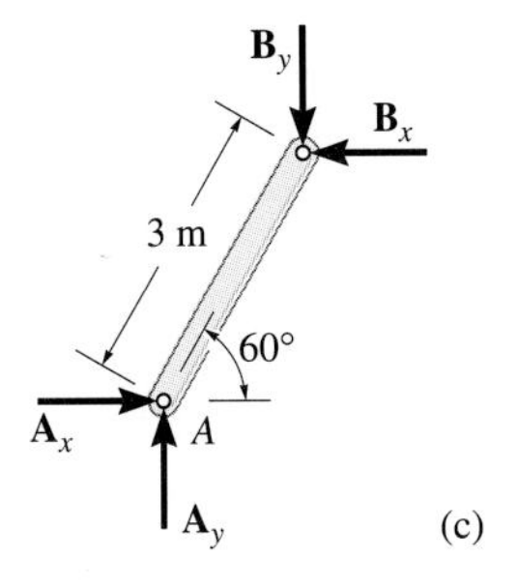

(c)

Fig. 6–27

By comparison, Solution I is simpler since the requirement that $\mathbf{F}_{AB}$ in Fig. 6–27*b* be equal, opposite, and collinear at the ends of member AB automatically satisfies Eqs. 1, 2, and 3 above and therefore eliminates the need to write these equations. *As a result, always identify the two-force members before starting the analysis!*

Example 6–15

The compound beam shown in Fig. 6–28*a* is pin-connected at *B*. Determine the reactions at its supports. Neglect its weight and thickness.

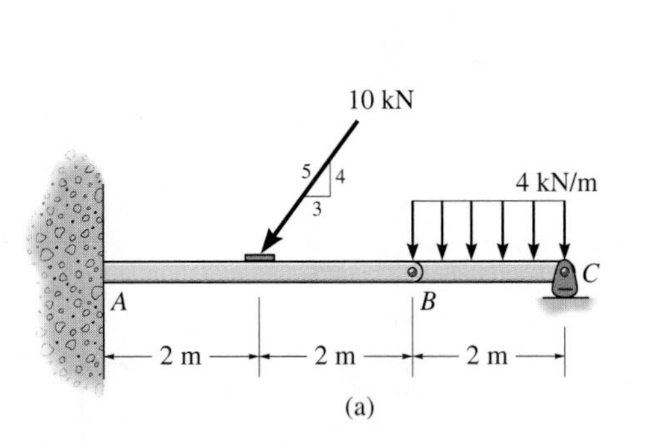

(a)

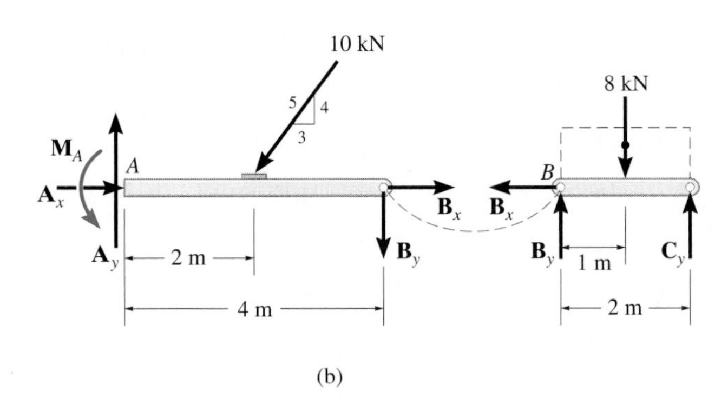

(b)

Fig. 6–28

SOLUTION

Free-Body Diagrams. By inspection, if we consider a free-body diagram of the entire beam *ABC*, there will be three unknown reactions at *A* and one at *C*. These four unknowns cannot all be obtained from the three equations of equilibrium, and so it will become necessary to dismember the beam into its two segments as shown in Fig. 6–28*b*.

Equations of Equilibrium. The six unknowns are determined as follows:

Segment BC

$\overset{+}{\rightarrow}\Sigma F_x = 0; \qquad B_x = 0$

$\circlearrowleft + \Sigma M_B = 0; \qquad -8\text{ kN}(1\text{ m}) + C_y(2\text{ m}) = 0$

$+\uparrow \Sigma F_y = 0; \qquad B_y - 8\text{ kN} + C_y = 0$

Segment AB

$\overset{+}{\rightarrow}\Sigma F_x = 0; \qquad A_x - (10\text{ kN})(\tfrac{3}{5}) + B_x = 0$

$\circlearrowleft + \Sigma M_A = 0; \qquad M_A - (10\text{ kN})(\tfrac{4}{5})(2\text{ m}) - B_y(4\text{ m}) = 0$

$+\uparrow \Sigma F_y = 0; \qquad A_y - (10\text{ kN})(\tfrac{4}{5}) - B_y = 0$

Solving each of these equations successively, using previously calculated results, we obtain

$A_x = 6\text{ kN} \qquad A_y = 12\text{ kN} \qquad M_A = 32\text{ kN}\cdot\text{m}$ ***Ans.***

$B_x = 0 \qquad B_y = 4\text{ kN}$

$C_y = 4\text{ kN}$ ***Ans.***

Example 6–16

Determine the horizontal and vertical components of force which the pin at C exerts on member $ABCD$ of the frame shown in Fig. 6–29a.

SOLUTION

Free-Body Diagrams. By inspection, the three components of reaction that the supports exert on $ABCD$ can be determined from a free-body diagram of the entire frame, Fig. 6–29b. Also, the free-body diagram of each frame member is shown in Fig. 6–29c. Notice that member BE is a two-force member. As shown by the colored dashed lines, the forces at B, C, and E have equal magnitudes but opposite directions on the separate free-body diagrams.

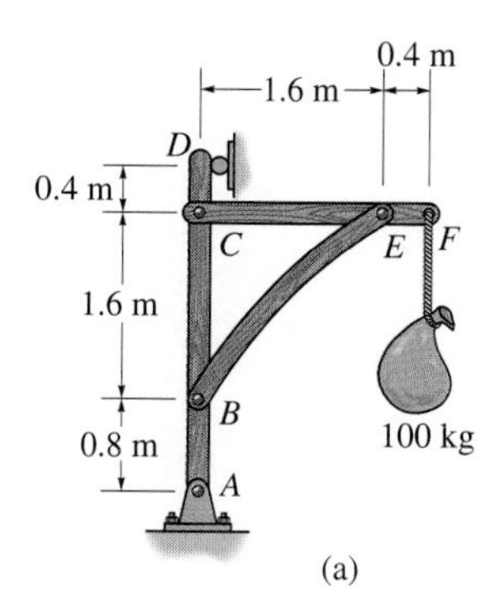

(a)

Equations of Equilibrium. The six unknowns A_x, A_y, F_B, C_x, C_y, and D_x will be determined from the equations of equilibrium applied to the entire frame and then to member CEF. We have

Entire Frame

$\circlearrowleft+\Sigma M_A = 0;\quad -981\text{ N}(2\text{ m}) + D_x(2.8\text{ m}) = 0 \qquad D_x = 700.7\text{ N}$

$\xrightarrow{+}\Sigma F_x = 0;\quad A_x - 700.7\text{ N} = 0 \qquad A_x = 700.7\text{ N}$

$+\uparrow \Sigma F_y = 0;\quad A_y - 981\text{ N} = 0 \qquad A_y = 981\text{ N}$

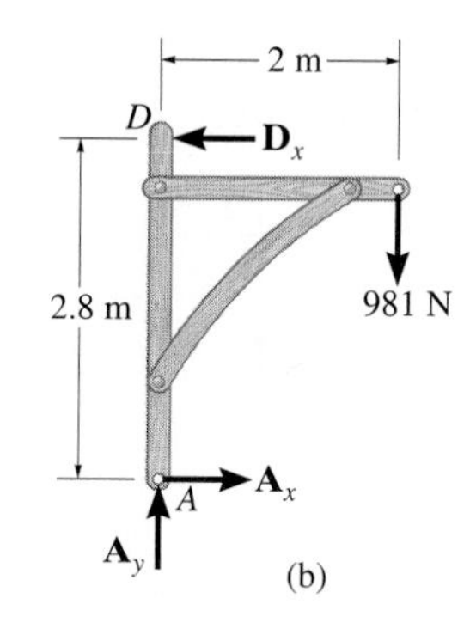

(b)

Member CEF

$\circlearrowleft+\Sigma M_C = 0;\quad -981\text{ N}(2\text{ m}) - (F_B \sin 45°)(1.6\text{ m}) = 0$

$$F_B = -1734.2\text{ N}$$

$\xrightarrow{+}\Sigma F_x = 0;\quad -C_x - (-1734.2 \cos 45°\text{ N}) = 0$

$$C_x = 1230\text{ N} \qquad \textit{Ans.}$$

$+\uparrow \Sigma F_y = 0;\quad C_y - (-1734.2 \sin 45°\text{ N}) - 981\text{ N} = 0$

$$C_y = -245\text{ N} \qquad \textit{Ans.}$$

Since the magnitudes of forces $\mathbf{F}_B$ and $\mathbf{C}_y$ were calculated as negative quantities, they were assumed to be acting in the wrong sense on the free-body diagrams, Fig. 6–29c. The correct sense of these forces might have been determined "by inspection" *before* applying the equations of equilibrium to member CEF. As shown in Fig. 6–29c, moment equilibrium about point E on member CEF indicates that $\mathbf{C}_y$ must actually act *downward* to counteract the moment created by the 981-N force about point E. Similarly, summing moments about point C, it is seen that the vertical component of force $\mathbf{F}_B$ must actually act *upward,* and so $\mathbf{F}_B$ must act upward to the right.

The above calculations can be checked by applying the three equilibrium equations to member $ABCD$, Fig. 6–29c.

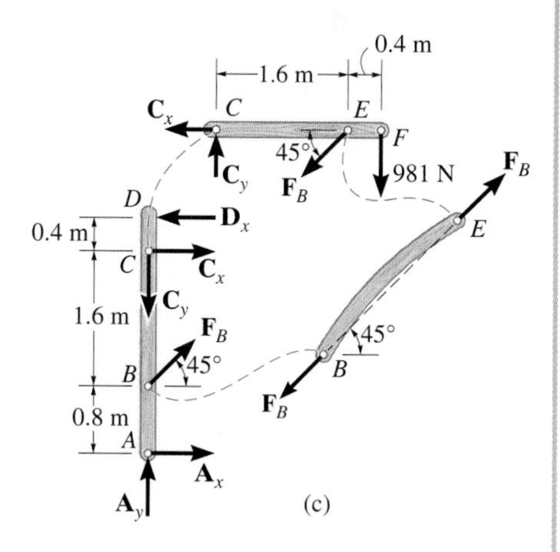

(c)

Fig. 6–29

Example 6–17

The smooth disk shown in Fig. 6–30*a* is pinned at D and has a weight of 20 lb. Neglecting the weights of the other members, determine the horizontal and vertical components of reaction at pins B and D.

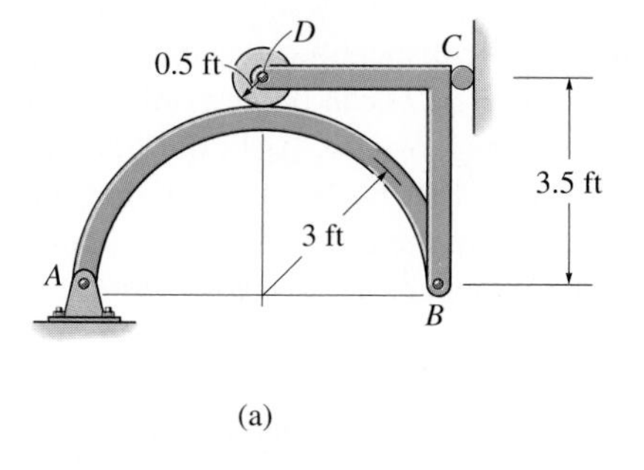

(a)

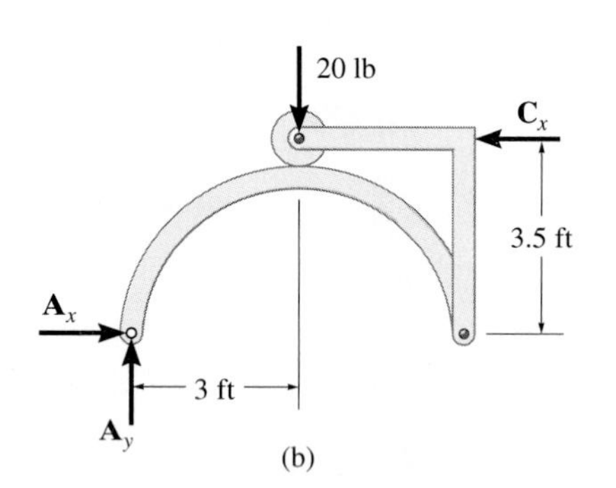

(b)

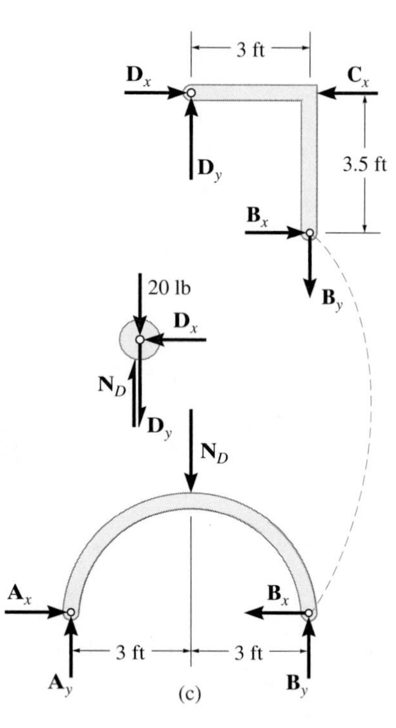

(c)

Fig. 6–30

SOLUTION

Free-Body Diagrams. By inspection, the three components of reaction at the supports can be determined from a free-body diagram of the entire frame, Fig. 6–30*b*. Also, free-body diagrams of the members are shown in Fig. 6–30*c*.

Equations of Equilibrium. The eight unknowns can of course be obtained by applying the eight equilibrium equations to each member—three to member AB, three to member BCD, and two to the disk. (Moment equilibrium is automatically satisfied for the disk.) If this is done, however, all the results can be obtained only from a simultaneous solution of some of the equations. (Try it and find out.) To avoid this situation, it is best to first determine the three support reactions on the *entire* frame; then, using these results, the remaining five equilibrium equations can be applied to two other parts in order to solve successively for the other unknowns.

Entire Frame

$$\circlearrowright+\Sigma M_A = 0; \quad -20\text{ lb}(3\text{ ft}) + C_x(3.5\text{ ft}) = 0 \quad C_x = 17.1\text{ lb}$$

$$\xrightarrow{+}\Sigma F_x = 0; \quad A_x - 17.1\text{ lb} = 0 \quad A_x = 17.1\text{ lb}$$

$$+\uparrow\Sigma F_y = 0; \quad A_y - 20\text{ lb} = 0 \quad A_y = 20\text{ lb}$$

Member AB

$$\xrightarrow{+}\Sigma F_x = 0; \quad 17.1\text{ lb} - B_x = 0 \quad B_x = 17.1\text{ lb} \qquad \textit{Ans.}$$

$$\circlearrowright+\Sigma M_B = 0; \quad -20\text{ lb}(6\text{ ft}) + N_D(3\text{ ft}) = 0 \quad N_D = 40\text{ lb}$$

$$+\uparrow\Sigma F_y = 0; \quad 20\text{ lb} - 40\text{ lb} + B_y = 0 \quad B_y = 20\text{ lb} \qquad \textit{Ans.}$$

Disk

$$\xrightarrow{+}\Sigma F_x = 0; \quad D_x = 0 \qquad \textit{Ans.}$$

$$+\uparrow\Sigma F_y = 0; \quad 40\text{ lb} - 20\text{ lb} - D_y = 0 \quad D_y = 20\text{ lb} \qquad \textit{Ans.}$$

Example 6–18

Determine the tension in the cables and also the force **P** required to support the 600-N force using the frictionless pulley system shown in Fig. 6–31*a*.

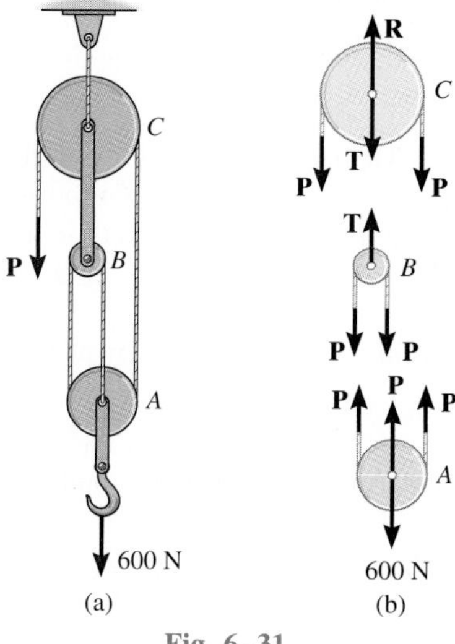

Fig. 6–31

SOLUTION

Free-Body Diagrams. A free-body diagram of each pulley *including* its pin and a portion of the contacting cable is shown in Fig. 6–31*b*. Since the cable is *continuous* and the pulleys are frictionless, the cable has a *constant tension P* acting throughout its length (see Example 5–7). The link connection between pulleys *B* and *C* is a two-force member and therefore it has an unknown tension *T* acting on it. Notice that the *principle of action, equal but opposite reaction* must be carefully observed for forces **P** and **T** when the *separate* free-body diagrams are drawn.

Equations of Equilibrium. The three unknowns are obtained as follows:

Pulley A

$+\uparrow \Sigma F_y = 0;\qquad 3P - 600\text{ N} = 0\qquad P = 200\text{ N}$ *Ans.*

Pulley B

$+\uparrow \Sigma F_y = 0;\qquad T - 2P = 0\qquad T = 400\text{ N}$ *Ans.*

Pulley C

$+\uparrow \Sigma F_y = 0;\qquad R - 2P - T = 0\qquad R = 800\text{ N}$ *Ans.*

Example 6–19

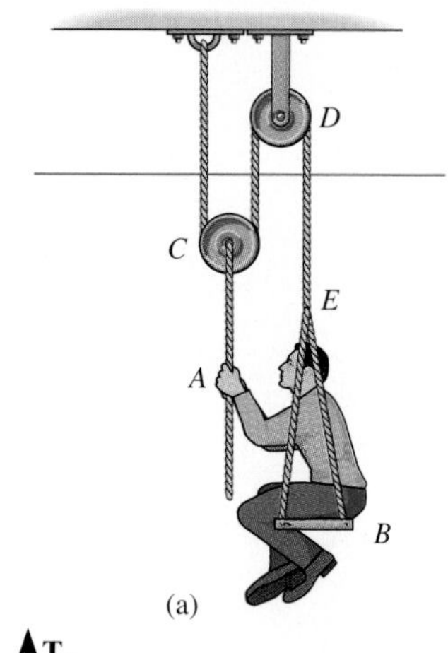

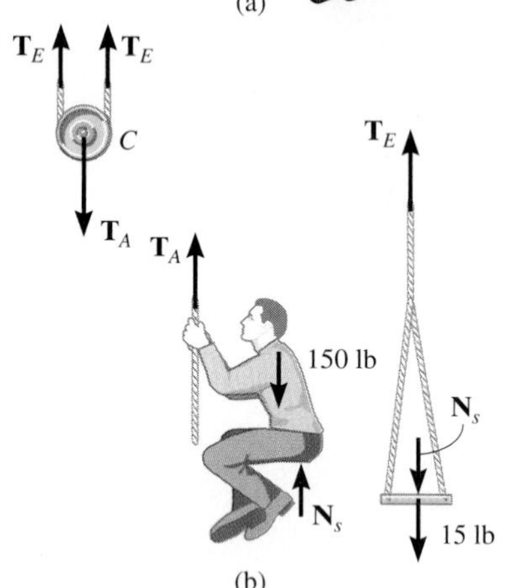

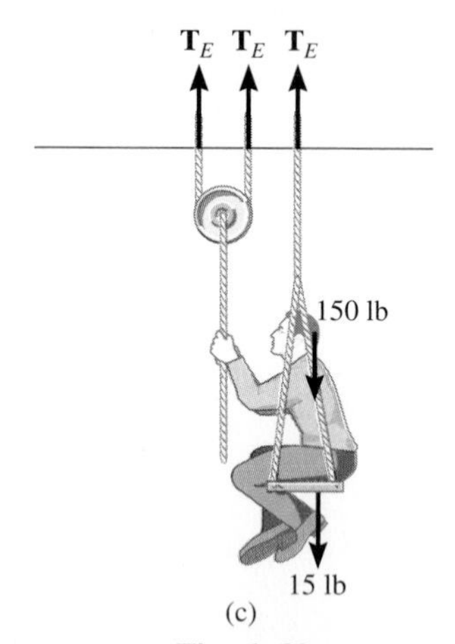

Fig. 6–32

A man having a weight of 150 lb supports himself by means of the cable and pulley system shown in Fig. 6–32*a*. If the seat has a weight of 15 lb, determine the equilibrium force that he must exert on the cable at *A* and the force he exerts on the seat. Neglect the weight of the cables and pulleys.

SOLUTION I

Free-Body Diagrams. The free-body diagrams of the man, seat, and pulley *C* are shown in Fig. 6–32*b*. The *two* cables are subjected to tensions $\mathbf{T}_A$ and $\mathbf{T}_E$, respectively. The man is subjected to three forces: his weight, the tension $\mathbf{T}_A$ of cable *AC*, and the reaction $\mathbf{N}_s$ of the seat.

Equations of Equilibrium. The three unknowns are obtained as follows:

Man

$$+\uparrow \Sigma F_y = 0; \qquad T_A + N_s - 150 \text{ lb} = 0 \qquad (1)$$

Seat

$$+\uparrow \Sigma F_y = 0; \qquad T_E - N_s - 15 \text{ lb} = 0 \qquad (2)$$

Pulley C

$$+\uparrow \Sigma F_y = 0; \qquad 2T_E - T_A = 0 \qquad (3)$$

The magnitude of force $\mathbf{T}_E$ can be determined by adding Eqs. 1 and 2 to eliminate N_s and then using Eq. 3. The other unknowns are then obtained by resubstitution of T_E.

$$T_A = 110 \text{ lb} \qquad \textit{Ans.}$$
$$T_E = 55 \text{ lb}$$
$$N_s = 40 \text{ lb} \qquad \textit{Ans.}$$

SOLUTION II

Free-Body Diagrams. By using the blue section shown in Fig. 6–32*a*, the man, pulley, and seat can be considered as a *single system,* Fig. 6–32*c*. Here $\mathbf{N}_s$ and $\mathbf{T}_A$ are *internal* forces and hence are not included on the "combined" free-body diagram.

Equations of Equilibrium. Applying $\Sigma F_y = 0$ yields a *direct* solution for T_E.

$$+\uparrow \Sigma F_y = 0; \qquad 3T_E - 15 \text{ lb} - 150 \text{ lb} = 0 \qquad T_E = 55 \text{ lb}$$

The other unknowns can be obtained from Eqs. 2 and 3.

Example 6–20

The hand exerts a force of 8 lb on the grip of the spring compressor shown in Fig. 6–33*a*. Determine the force in the spring needed to maintain equilibrium of the mechanism in the position shown.

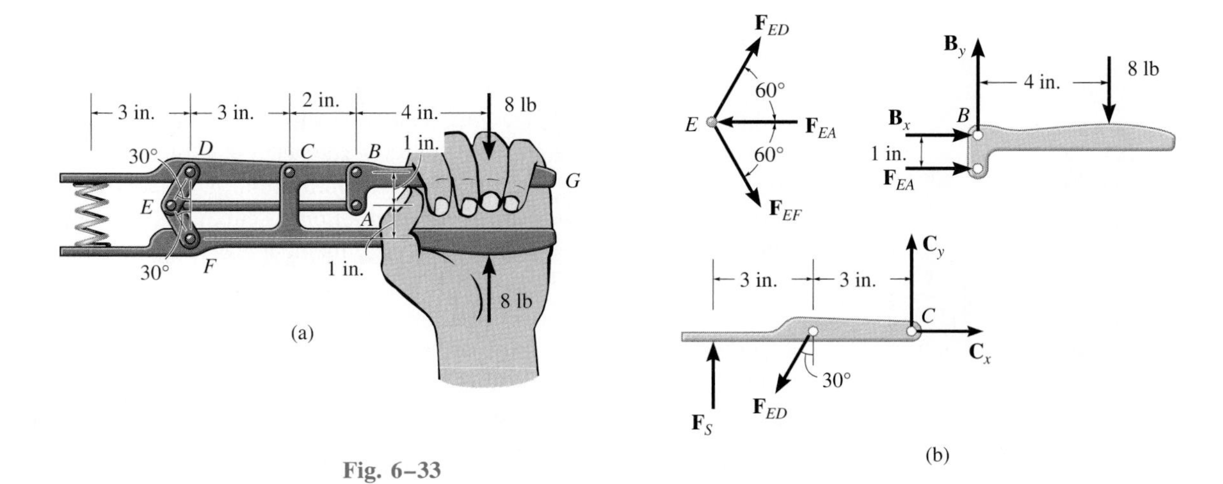

Fig. 6–33

SOLUTION

Free-Body Diagrams. By inspection, members *EA*, *ED*, and *EF* are all two-force members. The free-body diagrams for parts *DC* and *ABG* are shown in Fig. 6–33*b*. The pin at *E* has also been included here since *three* force interactions occur on this pin. They represent the effects of members *ED*, *EA*, and *EF*. Note carefully how equal and opposite force reactions occur between each of the parts.

Equations of Equilibrium. By studying the free-body diagrams, the most direct way to obtain the spring force is to apply the equations of equilibrium in the following sequence:

Lever ABG

$\circlearrowleft + \Sigma M_B = 0; \quad F_{EA}(1 \text{ in.}) - 8 \text{ lb}(4 \text{ in.}) = 0 \qquad F_{EA} = 32 \text{ lb}$

Pin E

$+\uparrow \Sigma F_y = 0; \quad F_{ED} \sin 60° - F_{EF} \sin 60° = 0 \quad F_{ED} = F_{EF} = F$

$\xrightarrow{+} \Sigma F_x = 0; \quad 2F \cos 60° - 32 \text{ lb} = 0 \qquad F = 32 \text{ lb}$

Arm DC

$\circlearrowleft + \Sigma M_C = 0; \quad -F_s(6 \text{ in.}) + 32 \cos 30° \text{ lb}(3 \text{ in.}) = 0$

$$F_s = 13.9 \text{ lb} \qquad \textit{Ans.}$$

Example 6–21

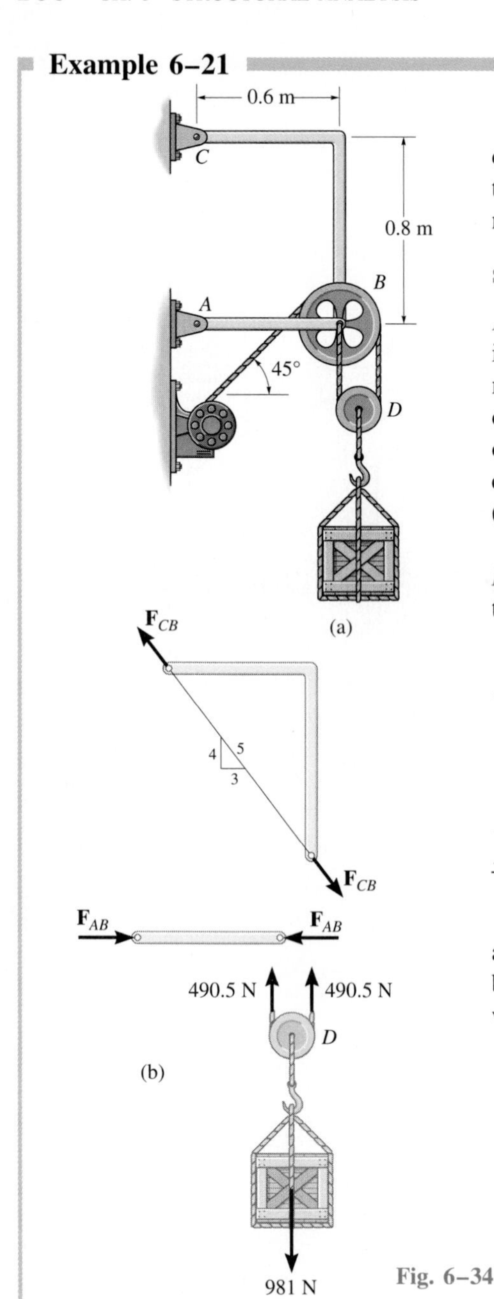

The 100-kg block is held in equilibrium by means of the pulley and continuous cable system shown in Fig. 6–34*a*. If the cable is attached to the pin at *B*, compute the forces which this pin exerts on each of its connecting members.

SOLUTION

Free-Body Diagrams. A free-body diagram of each member of the frame is shown in Fig. 6–34*b*. By inspection, members *AB* and *CB* are two-force members. Furthermore, the cable must be subjected to a force of 490.5 N in order to hold pulley *D* and the block in equilibrium. A free-body diagram of the pin at *B* is needed, since *four interactions* occur at this pin. These are caused by the attached cable (490.5 N), member *AB* ($\mathbf{F}_{AB}$), member *CB* ($\mathbf{F}_{CB}$), and pulley *B* ($\mathbf{B}_x$ and $\mathbf{B}_y$).

Equations of Equilibrium. Applying the equations of force equilibrium to pulley *B*, we have

$\overset{+}{\rightarrow}\Sigma F_x = 0;\quad B_x - 490.5 \cos 45^\circ \text{ N} = 0 \quad B_x = 346.8 \text{ N}$ *Ans.*

$+\uparrow \Sigma F_y = 0;\quad B_y - 490.5 \sin 45^\circ \text{ N} - 490.5 \text{ N} = 0$

$B_y = 837.3 \text{ N}$ *Ans.*

Using these results, equilibrium of the pin requires that

$+\uparrow \Sigma F_y = 0;\quad \frac{4}{5}F_{CB} - 837.3 \text{ N} - 490.5 \text{ N} = 0 \quad F_{CB} = 1660 \text{ N}$ *Ans.*

$\overset{+}{\rightarrow}\Sigma F_x = 0;\quad F_{AB} - \frac{3}{5}(1660 \text{ N}) - 346.8 \text{ N} = 0 \quad F_{AB} = 1343 \text{ N}$ *Ans.*

It may be noted that the two-force member *CB* is subjected to bending as caused by the force $\mathbf{F}_{CB}$. From the standpoint of design, it would be better to make this member *straight* (from *C* to *B*) so that the force $\mathbf{F}_{CB}$ would only create tension in the member.

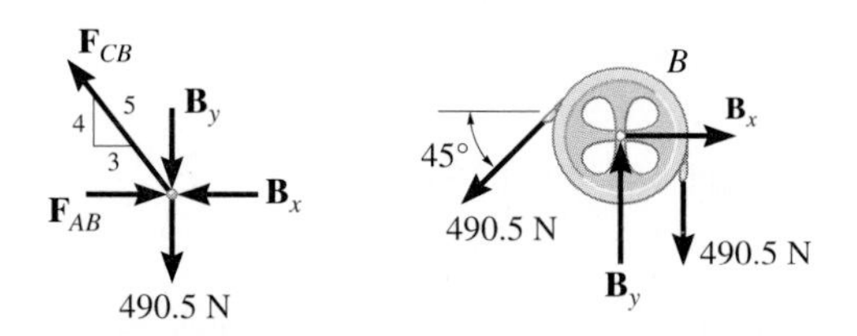

Fig. 6–34

Before solving the following problems, it is suggested that a brief review be made of all the previous examples. This may be done by covering each solution and trying to locate the two-force members, drawing the free-body diagrams, and conceiving ways of applying the equations of equilibrium to obtain the solution.

PROBLEMS

6–70. A force of $P = 8$ lb is applied to the handles of the pliers. Determine the force developed on the smooth bolt B and the reaction that pin A exerts on its attached members.

6–71. Determine the force P that must be applied to the handles of the pliers so that it develops a force of 100 lb on the smooth bolt at B. Also, what is the magnitude of the resultant force acting on the pin at A?

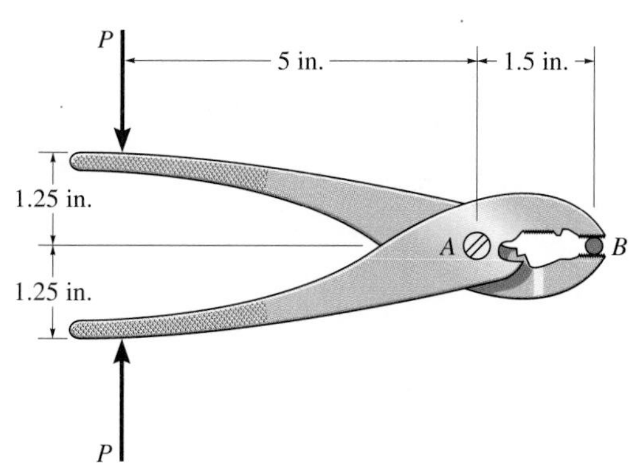

Probs. 6–70/6–71

***6–72.** The eye hook has a positive locking latch when it supports the load because its two parts are pin-connected at A and they bear against one another along the smooth surface at B. Determine the resultant force at the pin and the normal force at B when the eye hook supports a load of 800 lb.

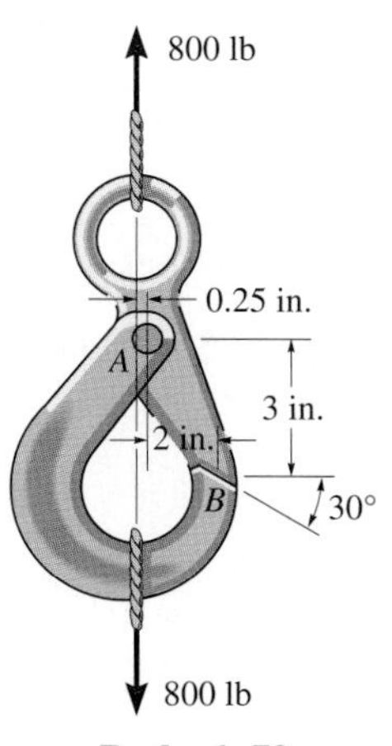

Prob. 6–72

6–73. The link is used to hold the rod in place. Determine the required axial force on the screw at E if the largest force to be exerted on the rod at B, C, or D is to be 100 lb. Also, find the magnitude of the force reaction at pin A. Assume all surfaces of contact are smooth.

6–74. The link is used to hold the rod in place. Determine the force on the rod at B, C, and D and the magnitude of the reaction at pin A if the axial load on the screw E is 200 N.

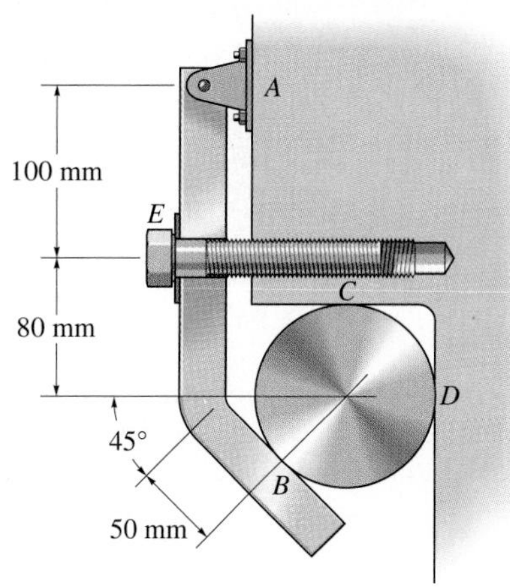

Probs. 6–73/6–74

6–75. Determine the horizontal and vertical components of force at pins A, B, and C, and the reactions at the fixed support D of the three-member frame.

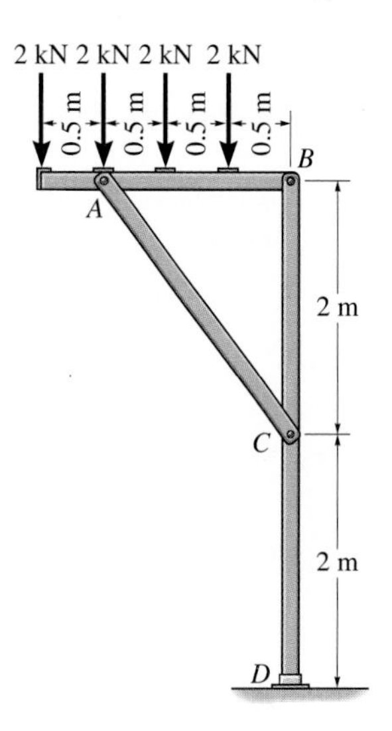

Prob. 6–75

*6–76. The smooth block is held in place using the vice clamp. If the screw exerts a force of 500 N on the block, determine the magnitude of the resultant force on the pin at A.

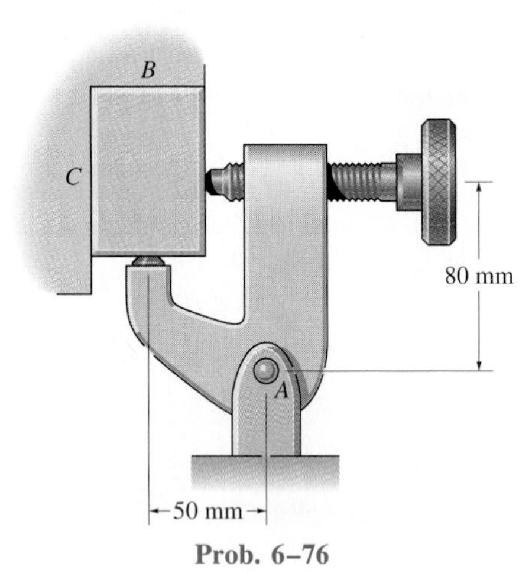

Prob. 6–76

6–77. The two ends of the spanner wrench fit loosely into the smooth slots of the bolt head. Determine the required force P on the handle in order to develop a torque of $M = 50$ N · m on the bolt. Also, what is the resultant force on the pin at B?

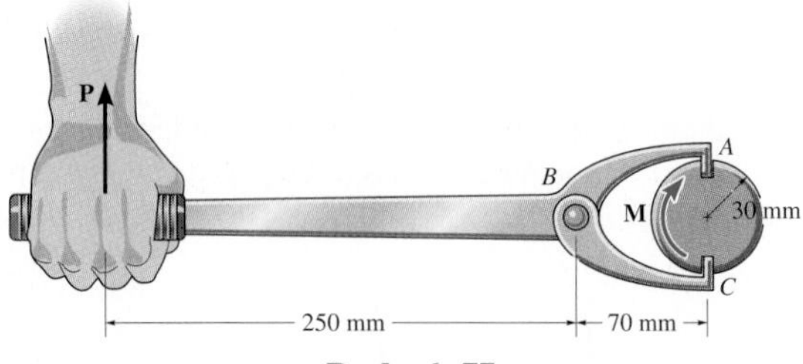

Prob. 6–77

6–78. The two ends of the spanner wrench fit loosely into the smooth slots of the bolt head. Determine the torque M on the bolt and the resultant force on the pin at B when a force of $P = 80$ N is applied to the handle.

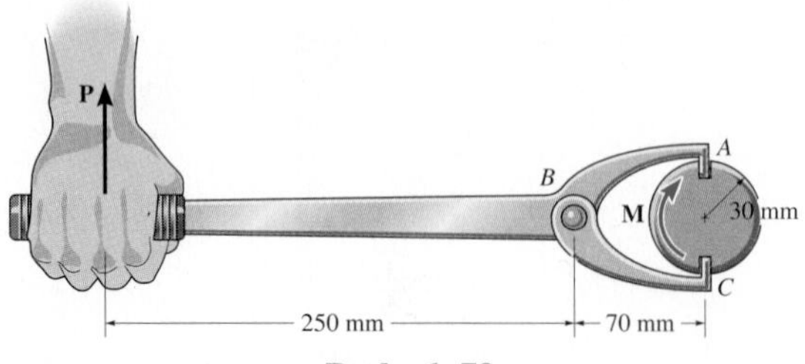

Prob. 6–78

6–79. The three-hinged arch supports the loads $F_1 = 8$ kN and $F_2 = 5$ kN. Determine the horizontal and vertical components of reaction at the pin supports A and B. Take $h = 2$ m.

*6–80. The three-hinged arch supports the loads $F_1 = 4$ kN and $F_2 = 7$ kN. Determine the horizontal and vertical components of reaction at the pin supports A and B. Take $h = 0$.

6–81. The three-hinged arch supports the loads $F_1 = 8$ kN and $F_2 = 0$. Determine the horizontal and vertical components of reaction at the pin supports A and B. Take $h = 3$ m.

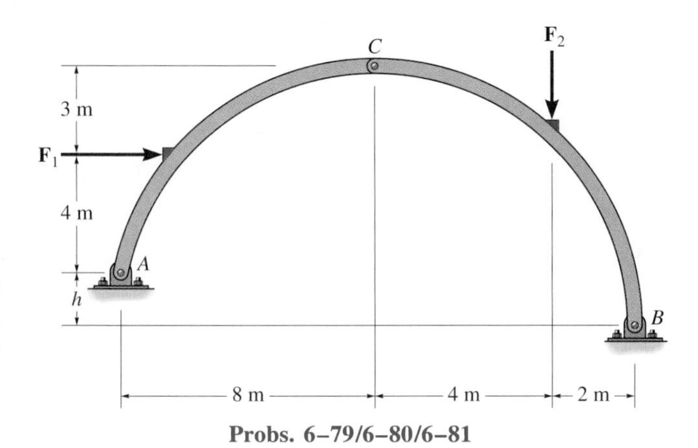

Probs. 6–79/6–80/6–81

6–82. Determine the horizontal and vertical components of force at pin A. Take $P = 600$ N.

6–83. Determine the greatest force P that can be applied to the frame if the largest force resultant acting at A can have a magnitude of 2 kN.

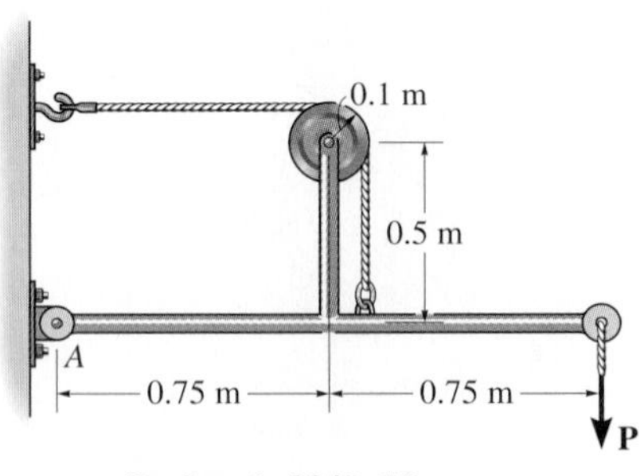

Probs. 6–82/6–83

***6–84.** Determine the horizontal and vertical components of force that the pins at A, B, and C exert on their connecting members.

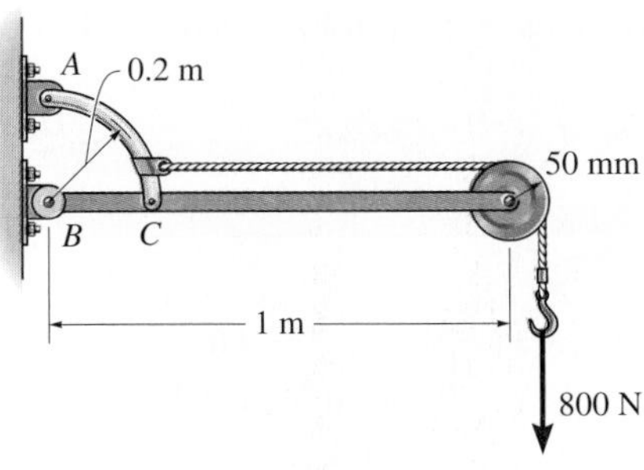
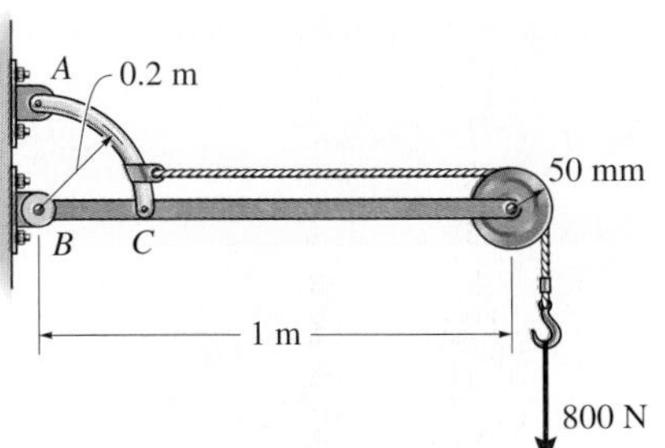

Prob. 6–84

6–85. Determine the force P needed to support the 20-kg mass using the *Spanish Burton rig.* Also, what are the reactions at the supporting hooks A, B, and C?

6–86. Determine the force P needed to support the 50-kg mass using the *Spanish Burton rig.* The pulleys have a mass of $m_D = 10$ kg, $m_E = m_F = 5$ kg, and $m_G = m_H = 2$ kg. Also, what are the reactions at the supporting hooks A, B, and C?

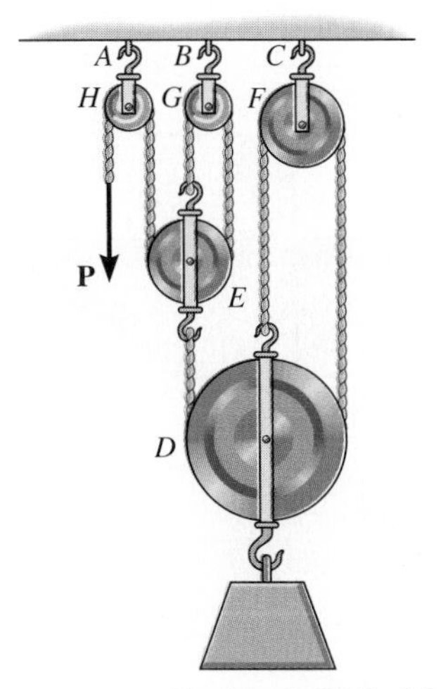

Probs. 6–85/6–86

6–87. Determine the force P needed to suspend the 100-lb weight. Each pulley has a weight of 10 lb. Also, what are the cord reactions at A and B?

***6–88.** If each cord can support a maximum tension of 500 lb, determine the largest weight that can be supported by the pulley system. Each pulley has a weight of 10 lb.

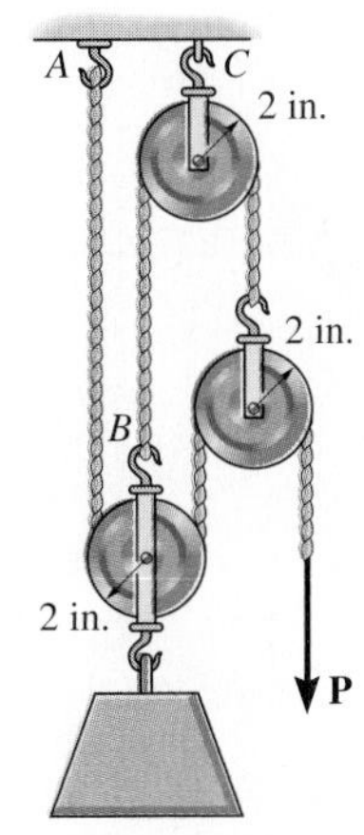

Probs. 6–87/6–88

6–89. Determine the force P on the cord, and the angle θ that the pulley-supporting link AB makes with the vertical. Neglect the mass of the pulleys and the link. The block has a weight of 200 lb and the cord is attached to the pin at B. The pulleys have radii of $r_1 = 2$ in. and $r_2 = 1$ in.

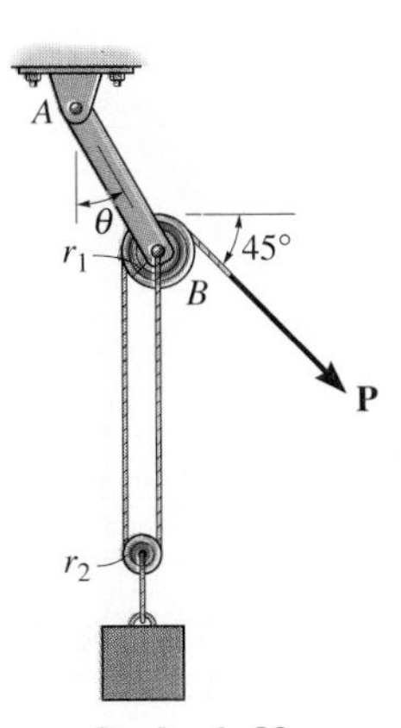

Prob. 6–89

6–90. Determine the horizontal and vertical components of force at C which member ABC exerts on member CEF.

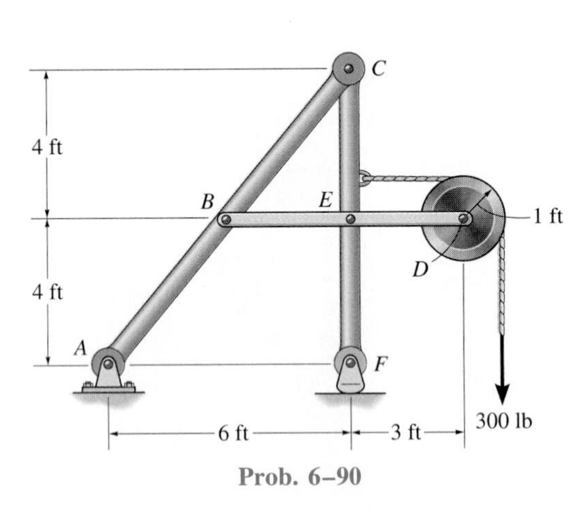

Prob. 6–90

6–91. Determine the horizontal and vertical components of force which the connecting pins at B, E, and D exert on member BED.

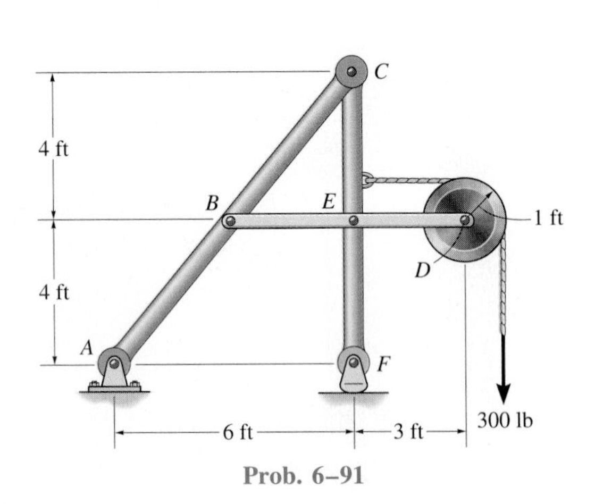

Prob. 6–91

***6–92.** Determine the force that the smooth roller C exerts on beam AB. Also, what are the horizontal and vertical components of reaction at pin A? Neglect the weight of the frame and roller.

6–93. Solve Prob. 6–92 if roller C has a weight of 20 lb.

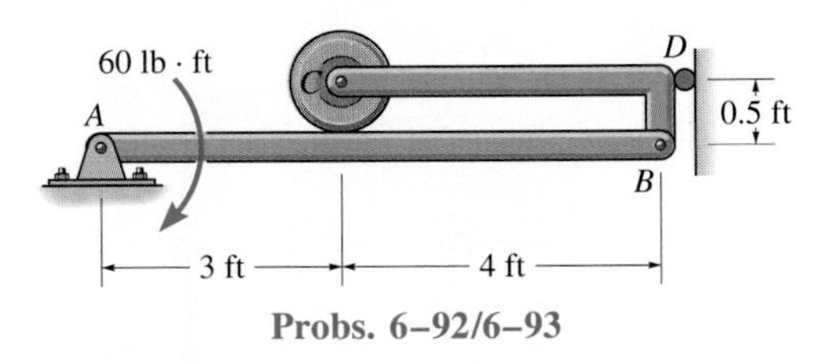

Probs. 6–92/6–93

6–94. Determine the horizontal and vertical components of force which the pins exert on member ABC.

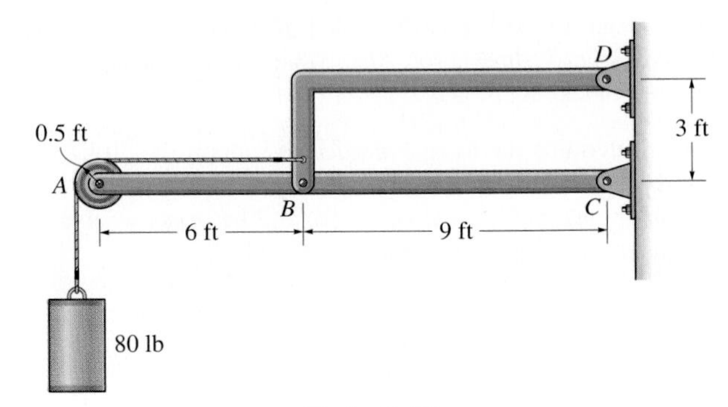

Prob. 6–94

6–95. Determine the horizontal and vertical components of force at pins B and C.

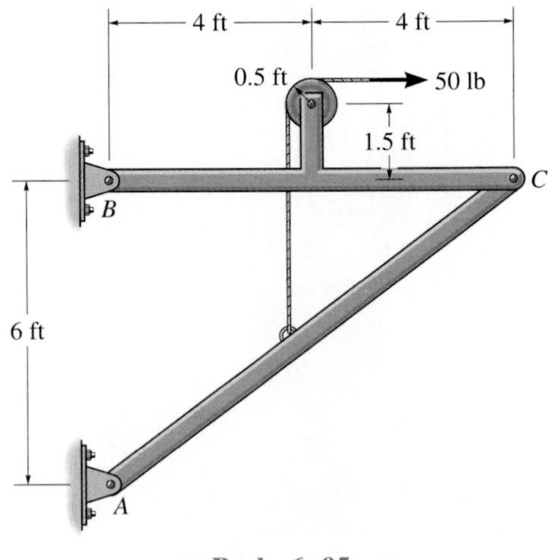

Prob. 6–95

***6–96.** Determine the horizontal and vertical components of force at each pin. The suspended cylinder has a weight of 80 lb.

6–97. Determine the largest weight of the cylinder if the maximum reaction at the roller A is 150 lb.

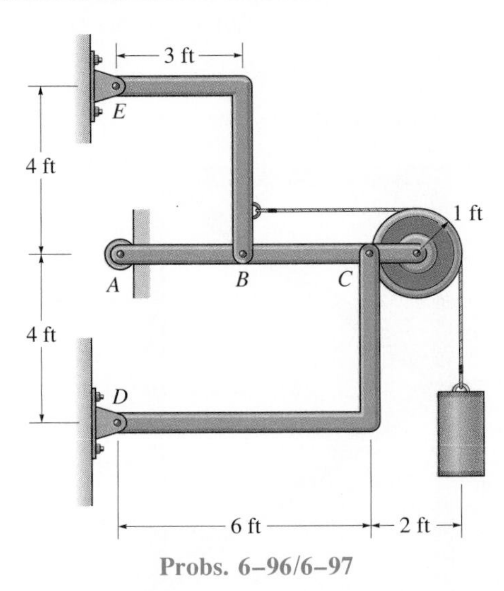

Probs. 6–96/6–97

6–98. Determine the force P on the cable if the spring is compressed 0.5 in. when the mechanism is in the position shown. The spring has a stiffness of $k = 800$ lb/ft.

6–99. Determine the compression of the spring if the cable force is to be $P = 80$ lb. The spring has a stiffness of $k = 800$ lb/ft and the mechanism is to be held in the position shown when the spring is compressed.

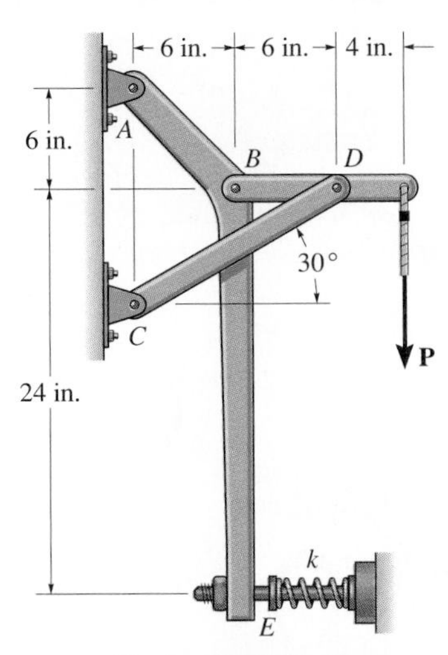

Probs. 6–98/6–99

***6–100.** If a force of $P = 6$ lb is applied perpendicular to the handle of the mechanism, determine the magnitude of force $\mathbf{F}$ for equilibrium. The members are pin-connected at A, B, C, and D.

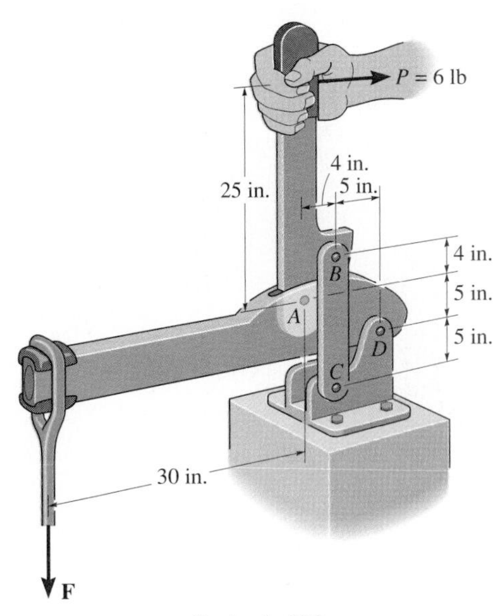

Prob. 6–100

6–101. The clamp is used to hold the smooth strut S in place. If the tensile force in the bolt GH is 300 N, determine the force exerted at points A and B.

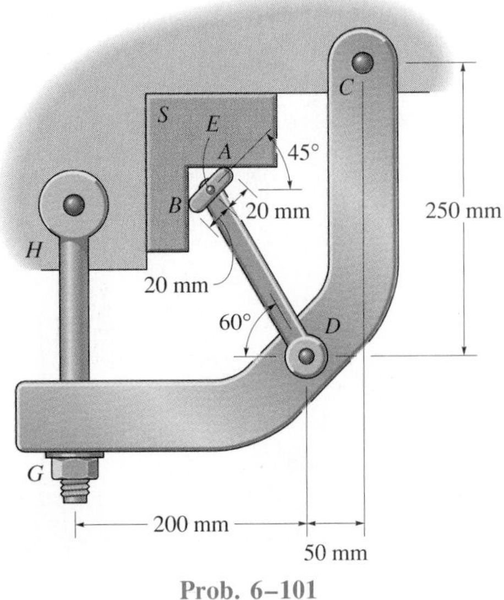

Prob. 6–101

6–102. Determine the required mass of the suspended cylinder if the tension in the chain wrapped around the freely turning gear is to be 2 kN. Also, what is the magnitude of the resultant force on pin A?

Prob. 6–102

6–103. The derrick is pin-connected to the pivot at A. Determine the largest mass that can be supported by the derrick if the maximum force that can be sustained by the pin at A is 18 kN.

***6–104.** The derrick is pin-connected to the pivot at A. Determine the force in the cable at C and in the hoisting cable at D if the suspended crate is 900 kg. Also, what is the resultant force acting on the pin at A?

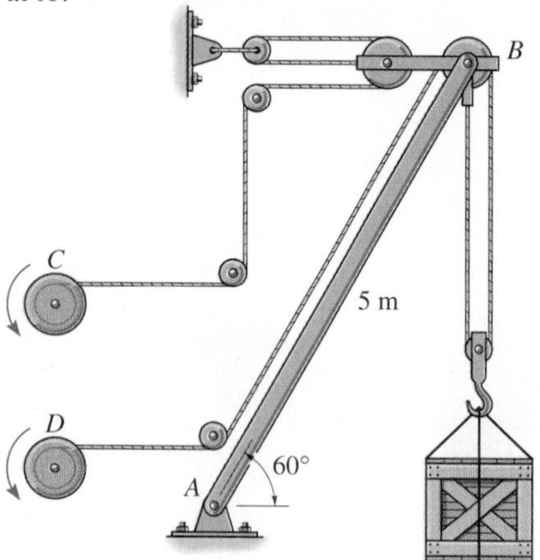

Probs. 6–103/6–104

6–105. Determine the horizontal and vertical components of force which the pins at A, B, and C exert on member ABC of the frame.

6–106. Determine the horizontal and vertical components of force which the pins at D and E exert on member DE of the frame.

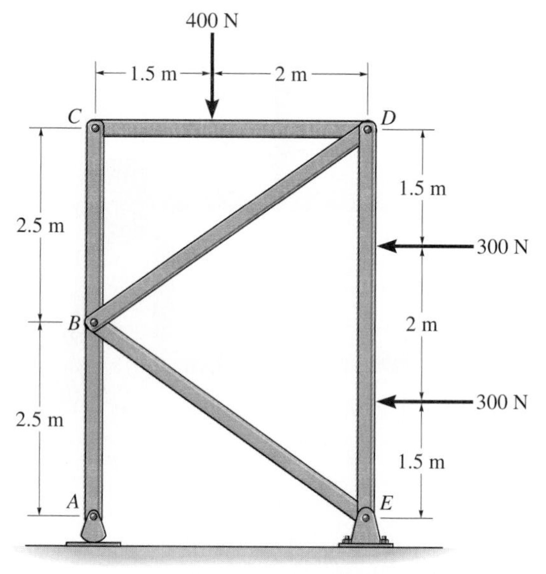

Probs. 6–105/6–106

6–107. By squeezing on the hand brake of the bicycle, the rider subjects the brake cable to a tension of 50 lb. If the caliper mechanism is pin-connected to the bicycle frame at B, determine the normal force each brake pad exerts on the rim of the wheel. Is this the force that stops the wheel from turning? Explain.

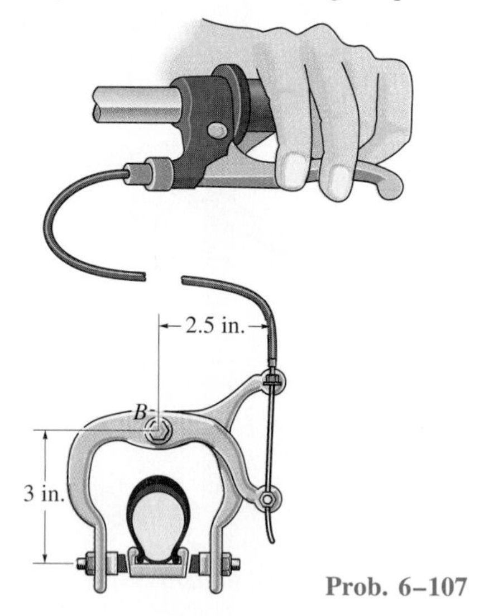

Prob. 6–107

*6–108. A man having a weight of 175 lb attempts to lift himself using one of the two methods shown. Determine the total force he must exert on bar AB in each case and the normal reaction he exerts on the platform at C. Neglect the weight of the platform.

6–109. A man having a weight of 175 lb attempts to lift himself using one of the two methods shown. Determine the total force he must exert on bar AB in each case and the normal reaction he exerts on the platform at C. The platform has a weight of 30 lb.

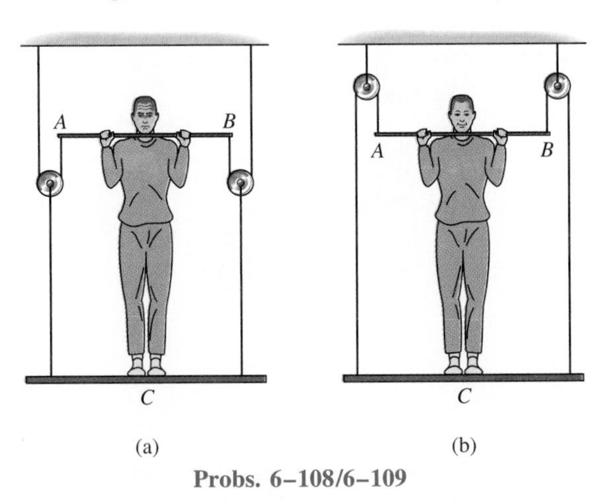

Probs. 6–108/6–109

6–110. Multiple levers can be used in a *compound arrangement* such as shown for the pan scale. If the mass on the pan is 4 kg, determine the reactions at pins A, B, and C and the distance x of the 25-g mass to keep the scale in balance.

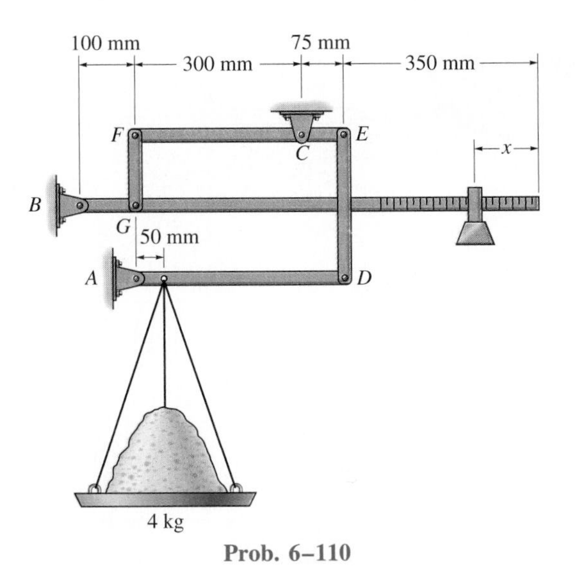

Prob. 6–110

6–111. Determine the horizontal and vertical components of force at pin B and the normal force the pin at C exerts on the smooth slot. Also, determine the moment and horizontal and vertical reactions of force at A. There is a pulley at E.

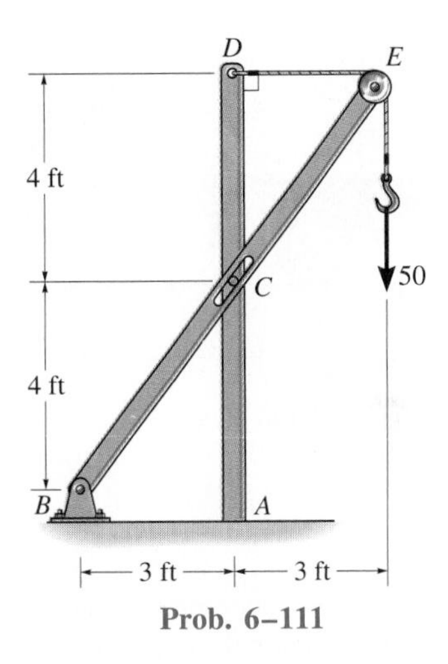

Prob. 6–111

*6–112. The flat-bed trailer has a weight of 7000 lb and center of gravity at G_T. It is pin-connected to the cab at D. The cab has a weight of 6000 lb and center of gravity at G_C. Determine the range of values x for the position of the 2000-lb load L so that when it is placed over the rear axle, no axle is subjected to more than 5500 lb. The load has a center of gravity at G_L.

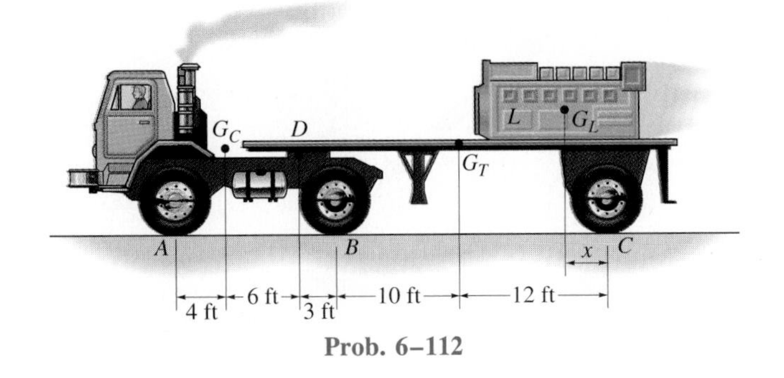

Prob. 6–112

6–113. The flat-bed trailer has a weight of 7000 lb and center of gravity at G_T. It is pin-connected to the cab at D. The cab has a weight of 6000 lb and center of gravity at G_C. Determine the largest load L that can be placed on the bed if no axle is to be subjected to more than 5500 lb. The center of gravity G_L of the load is at $x = 4$ ft.

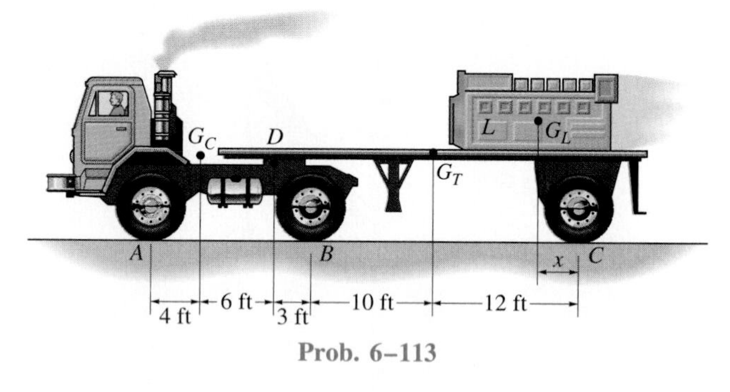

Prob. 6–113

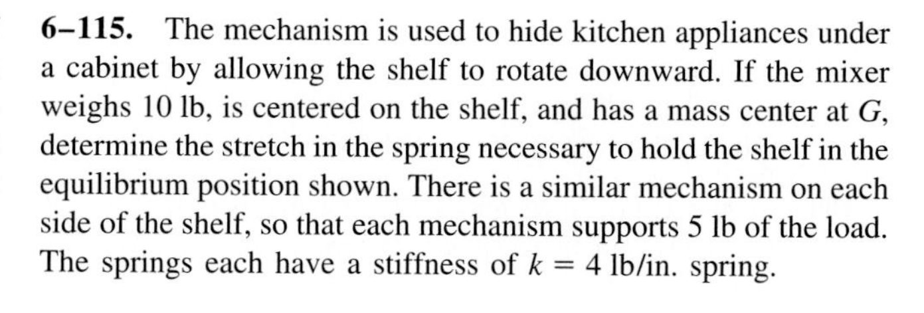

6–115. The mechanism is used to hide kitchen appliances under a cabinet by allowing the shelf to rotate downward. If the mixer weighs 10 lb, is centered on the shelf, and has a mass center at G, determine the stretch in the spring necessary to hold the shelf in the equilibrium position shown. There is a similar mechanism on each side of the shelf, so that each mechanism supports 5 lb of the load. The springs each have a stiffness of $k = 4$ lb/in. spring.

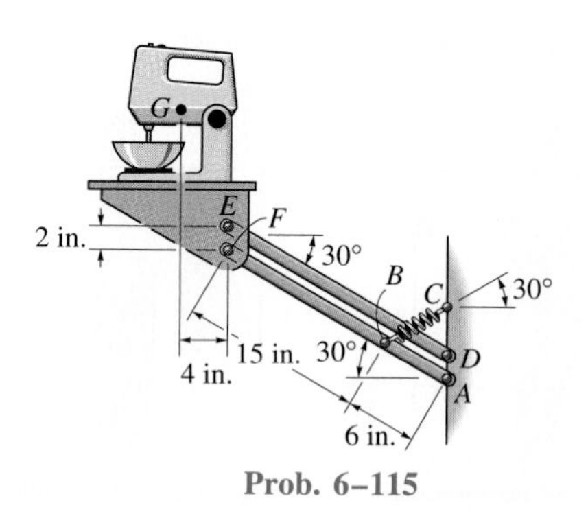

Prob. 6–115

6–114. The aircraft-hangar door opens and closes slowly by means of a motor which draws in the cable AB. If the door is made in two sections (bifold) and each section has a uniform weight W and length L, determine the force in the cable as a function of the door's position θ. The sections are pin-connected at C and D and the bottom is attached to a roller that travels along the vertical track.

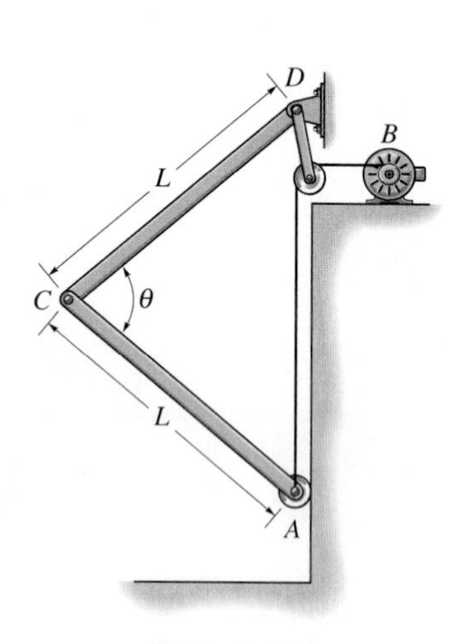

Prob. 6–114

***6–116.** The tractor boom supports the uniform mass of 500 kg in the bucket which has a center of mass at G. Determine the force in each hydraulic cylinder AB and CD and the resultant force at pins E and F. The load is supported equally on each side of the tractor by a similar mechanism.

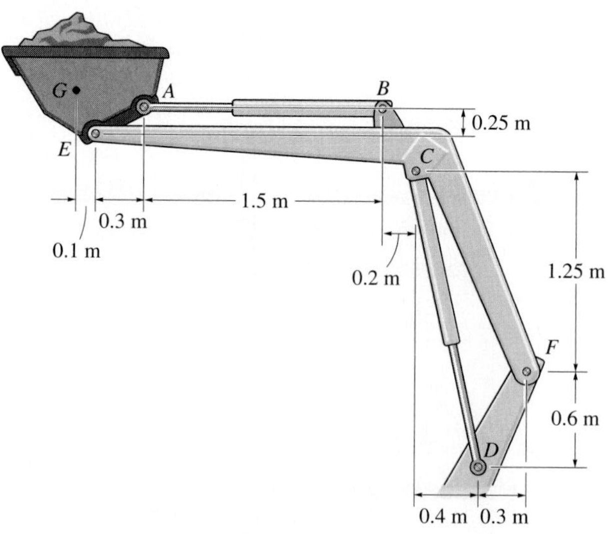

Prob. 6–116

6–117. The pruning shears are subjected to a squeezing force of $P = 8$ lb at the grip. Determine the normal force developed on the twig at the blade E.

6–118. Determine the required force P that must be applied at the blade of the pruning shears so that the blade exerts a normal force of 20 lb on the twig at E.

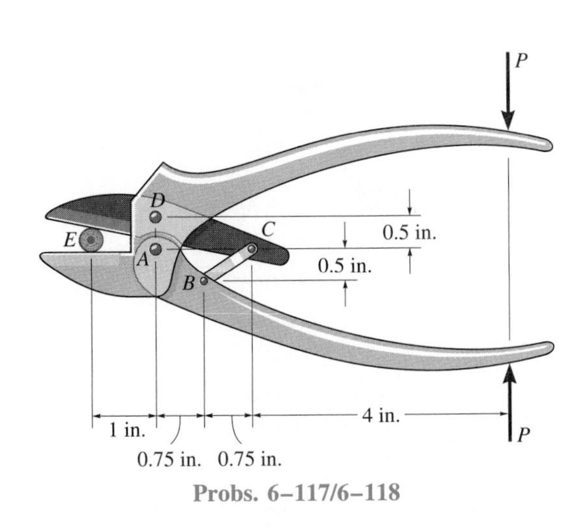

Probs. 6–117/6–118

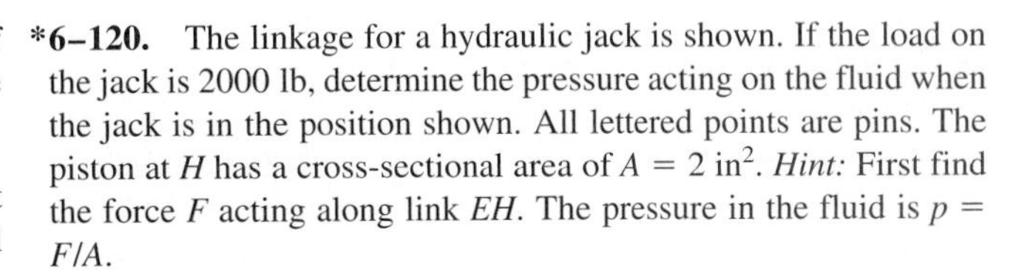

***6–120.** The linkage for a hydraulic jack is shown. If the load on the jack is 2000 lb, determine the pressure acting on the fluid when the jack is in the position shown. All lettered points are pins. The piston at H has a cross-sectional area of $A = 2\text{ in}^2$. *Hint:* First find the force F acting along link EH. The pressure in the fluid is $p = F/A$.

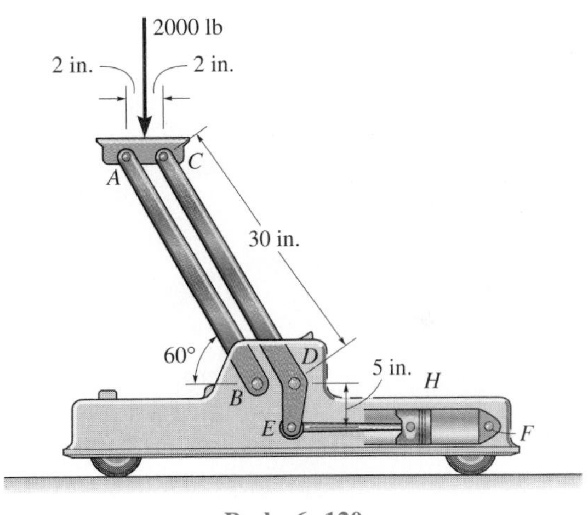

Prob. 6–120

6–119. The three power lines exert the forces shown on the truss joints, which in turn are pin-connected to the poles AH and EG. Determine the force in the guy cable AI and the pin reaction at the support H.

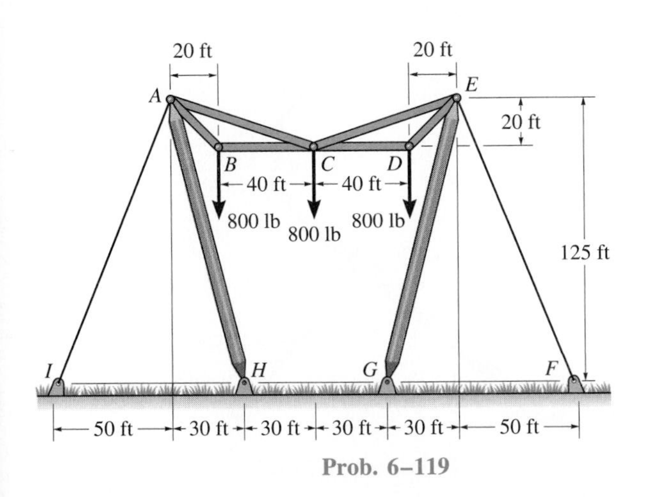

Prob. 6–119

6–121. The hydraulic crane is used to lift the 1400-lb load. Determine the force in the hydraulic cylinder AB and the force in links AC and AD when the load is held in the position shown.

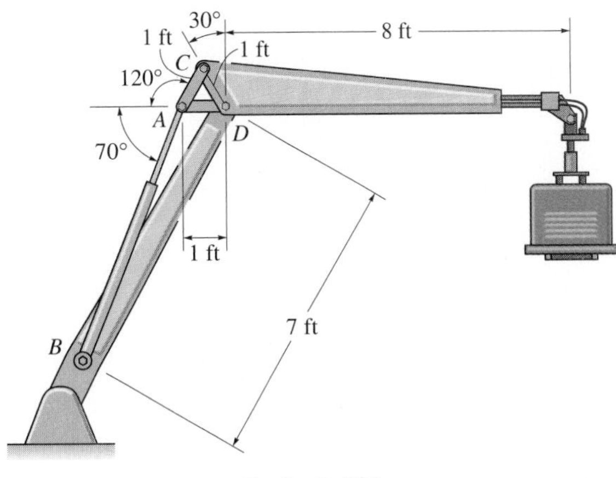

Prob. 6–121

6–122. The kinetic sculpture requires that each of the three pinned beams be in perfect balance at all times during its slow motion. If each member has a uniform weight of 2 lb/ft and length of 3 ft, determine the necessary counterweights W_1, W_2, and W_3 which must be added to the ends of each member to keep the system in balance for any position. Neglect the size of the counterweights.

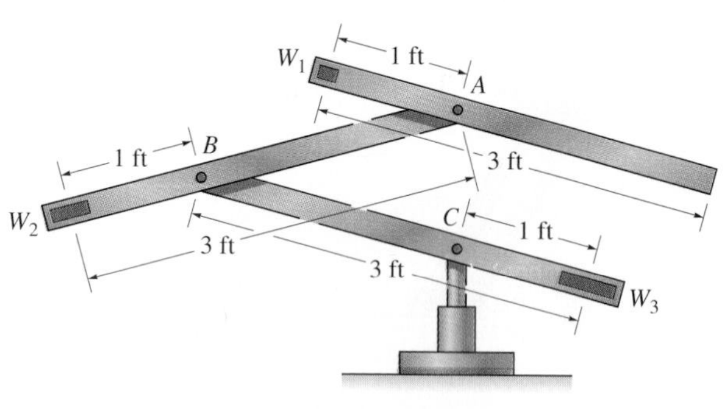

Prob. 6–122

■**6–123.** The spring has an unstretched length of 0.3 m. Determine the angle θ for equilibrium if the uniform links each have a mass of 5 kg.

***6–124.** The spring has an unstretched length of 0.3 m. Determine the mass m of each uniform link if the angle $\theta = 15°$ for equilibrium.

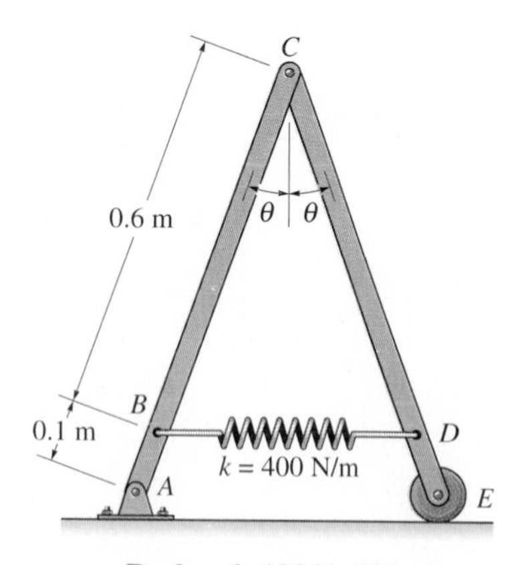

Probs. 6–123/6–124

6–125. Determine the horizontal force F required to maintain equilibrium of the slider mechanism when $\theta = 60°$.

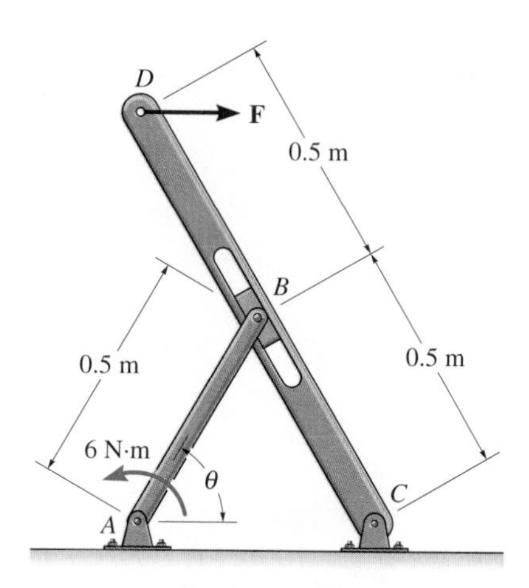

Prob. 6–125

6–126. The two-bar mechanism consists of a lever arm AB and smooth link CD, which has a fixed collar at its end C and a roller at the other end D. Determine the force P needed to hold the lever in the position θ. The spring has a stiffness k and unstretched length of $2L$. The roller contacts either the top or bottom portion of the horizontal guide.

6–127. The two-bar mechanism consists of a lever arm AB and smooth link CD, which has a fixed collar at its end C and a roller at the other end D. If a force **P** is applied to the lever, determine the required stiffness of the spring so that the lever will reach the equilibrium position when $\theta = 45°$. The unstretched length of the spring is $2L$, and the roller contacts either the top or bottom portion of the horizontal guide. Express k in terms of P and L.

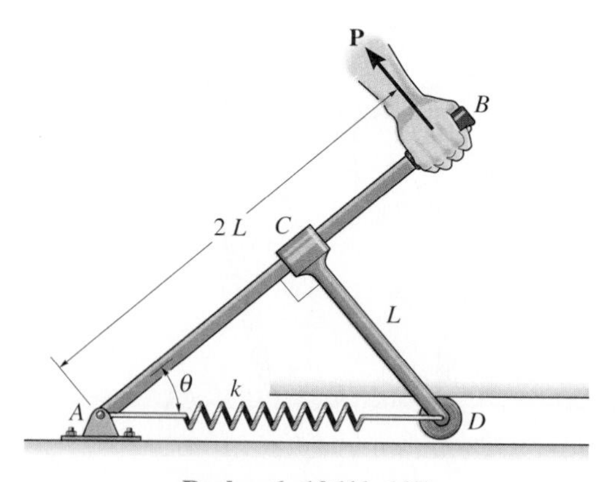

Probs. 6–126/6–127

*6–128. The spring has an unstretched length of 0.3 m. Determine the angle θ for equilibrium if the uniform links each have a mass of 5 kg.

6–129. The spring has an unstretched length of 0.3 m. Determine the mass m of each uniform link if the angle $\theta = 20°$ for equilibrium.

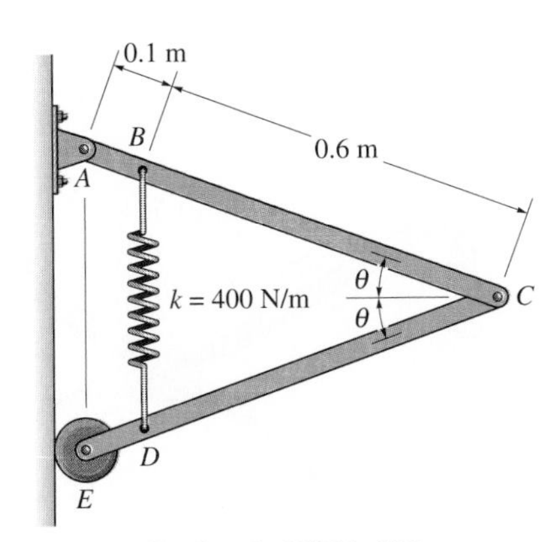

Probs. 6–128/6–129

6–130. The three-member frame is connected at its ends using ball-and-socket joints. Determine the x, y, z components of reaction at B and the tension in member ED. The force acting at D is $\mathbf{F} = \{250\mathbf{i} - 350\mathbf{k}\}$ lb.

6–131. The three-member frame is connected at its ends using ball-and-socket joints. Determine the x, y, z components of reaction at B and the tension in member ED. The force acting at D is $\mathbf{F} = \{135\mathbf{i} + 200\mathbf{j} - 180\mathbf{k}\}$ lb.

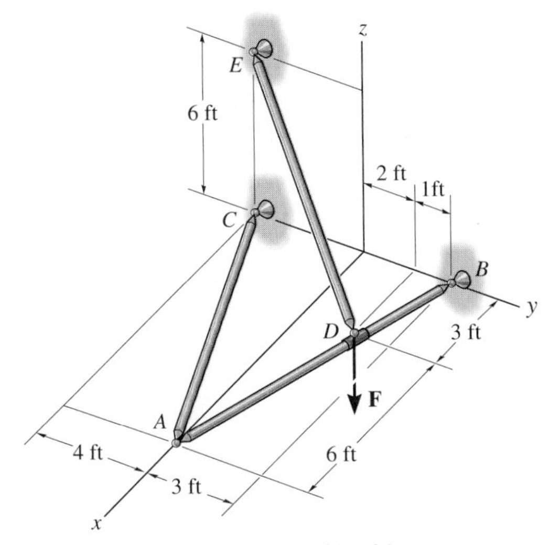

Probs. 6–130/6–131

*6–132. The four-member "A" frame supports a vertical force of $P = 600$ N. If it is assumed that the supports at A and E are smooth collars, G is a pin, and all other joints are ball-and-sockets, determine the x, y, z force components which member BD exerts on members EDC and FG. The collars at A and E and the pin at G only exert force components on the frame.

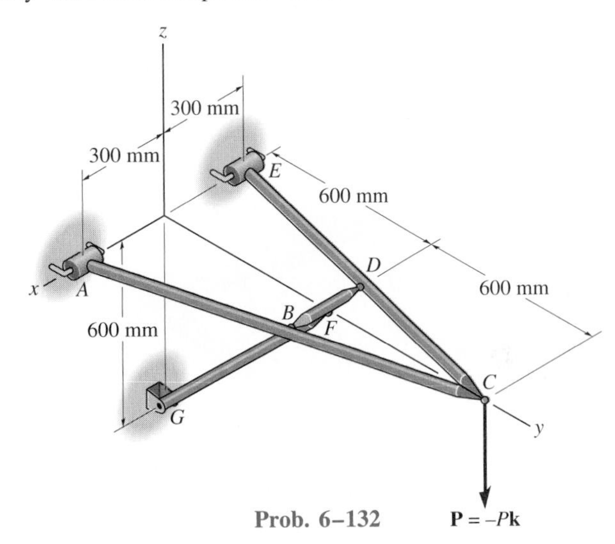

Prob. 6–132 $\mathbf{P} = -P\mathbf{k}$

6–133. The four-member "A" frame is supported at A and E by smooth collars and at G by a pin. All the other joints are ball-and-sockets. If the pin at G will fail when the resultant force there is 800 N, determine the largest vertical force P that can be supported by the frame. Also, what are the x, y, z force components which member BD exerts on members EDC and ABC? The collars at A and E and the pin at G only exert force components on the frame.

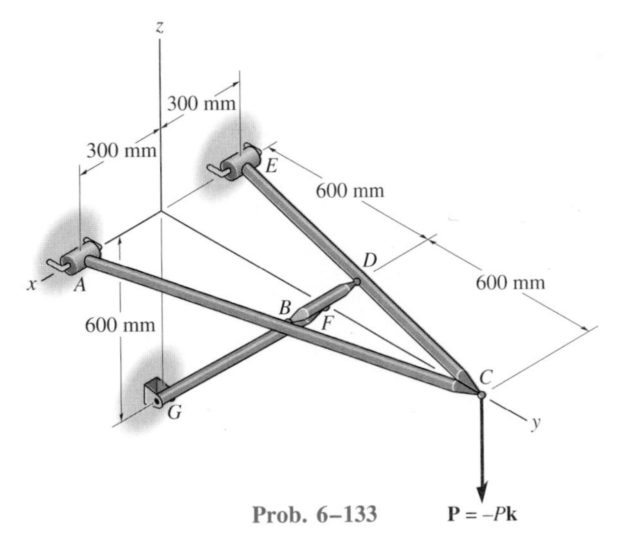

Prob. 6–133 $\mathbf{P} = -P\mathbf{k}$

6–134. The structure is subjected to the loading shown. Member AD is supported by a cable AB and roller at C and fits through a smooth circular hole at D. Member ED is supported by a roller at D and a pole that fits in a smooth snug circular hole at E. Determine the x, y, z components of reaction at E and the tension in cable AB.

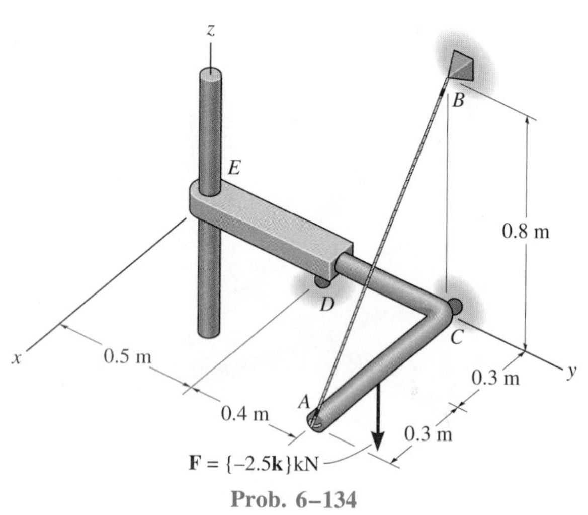

Prob. 6–134

6–135. The structure is subjected to the force of 450 lb which lies in a plane parallel to the y-z plane. Member AB is supported by a ball-and-socket joint at A and fits through a snug hole at B. Member CD is supported by a pin at C. Determine the x, y, z components of reaction at A and C.

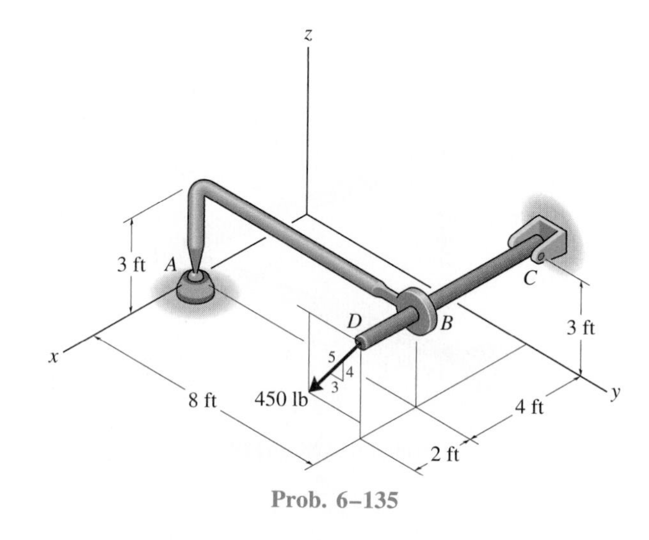

Prob. 6–135

REVIEW PROBLEMS

***6–136.** Determine the horizontal and vertical components of force that the pins at A and C exert on the two-bar mechanism.

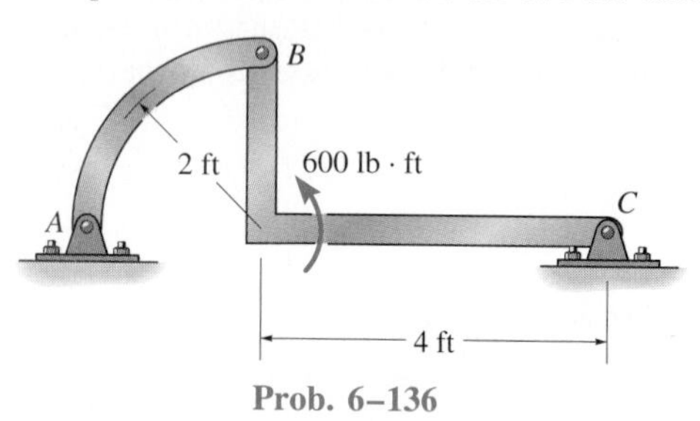

Prob. 6–136

6–137. Determine the horizontal and vertical components of force at pins A and C of the two-member frame.

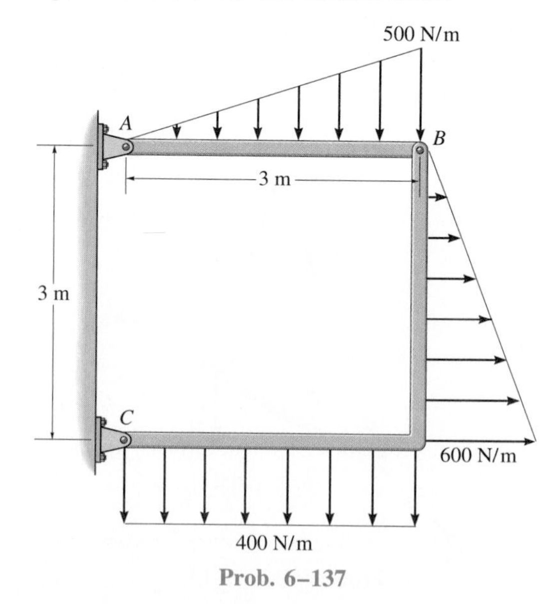

Prob. 6–137

6–138. Determine the force in each member of the truss and state if the members are in tension or compression.

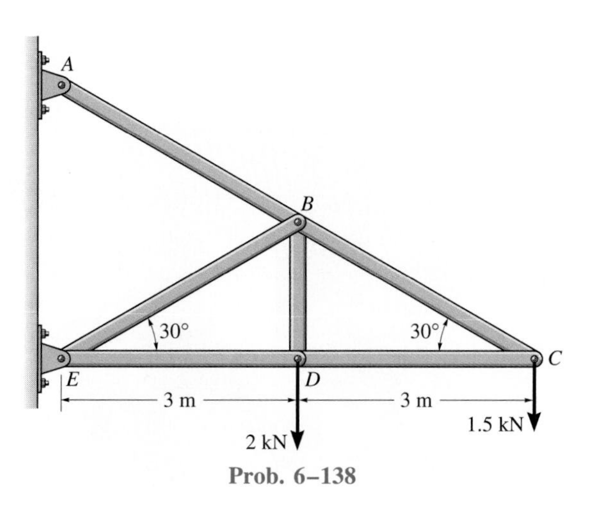

Prob. 6–138

6–139. The compound beam is fixed at A and supported by a rocker at B and C. There are pins at D and E. Determine the reactions at the supports.

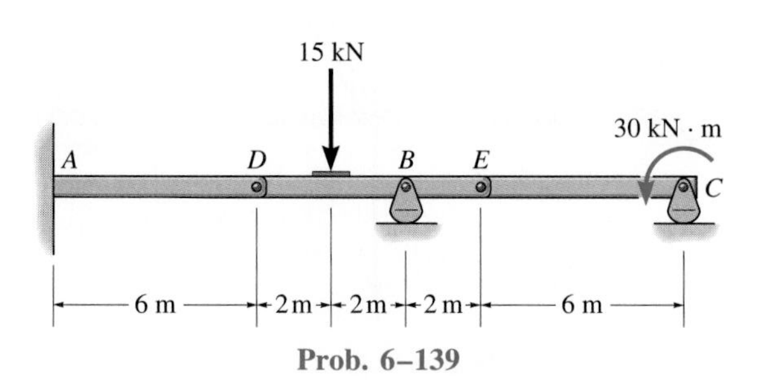

Prob. 6–139

***6–140.** Determine the force in each member of the truss and state if the members are in tension or compression.

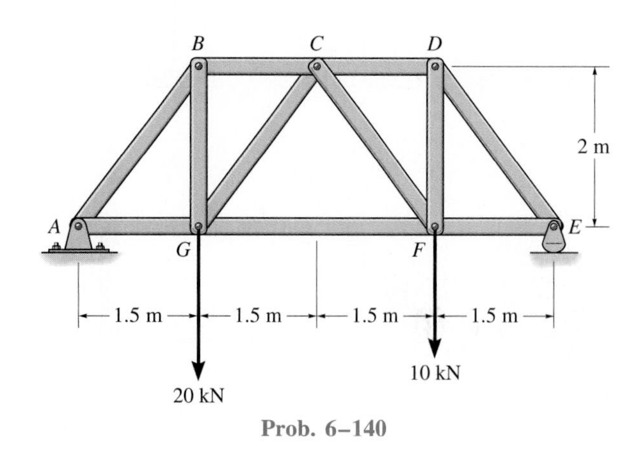

Prob. 6–140

6–141. The pipe cutter is clamped around the pipe P. If the cutting wheel at A exerts a normal force of $F_A = 80$ N on the pipe, determine the normal forces of wheels B and C on the pipe. Also compute the pin reaction on the wheel at C. The three wheels each have a radius of 7 mm and the pipe has an outer radius of 10 mm.

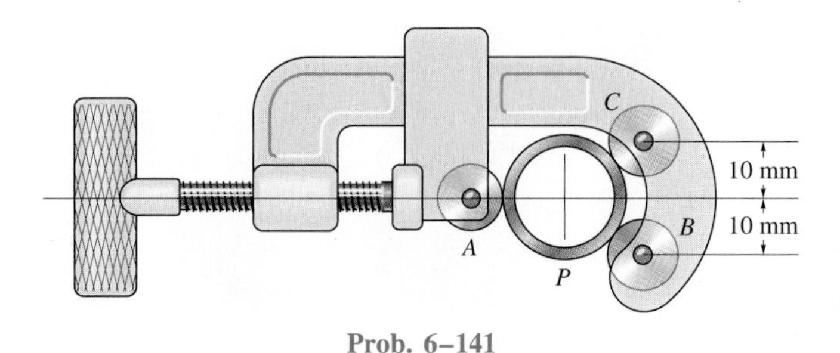

Prob. 6–141

6–142. Determine the resultant forces at pins B and C on member ABC of the four-member frame.

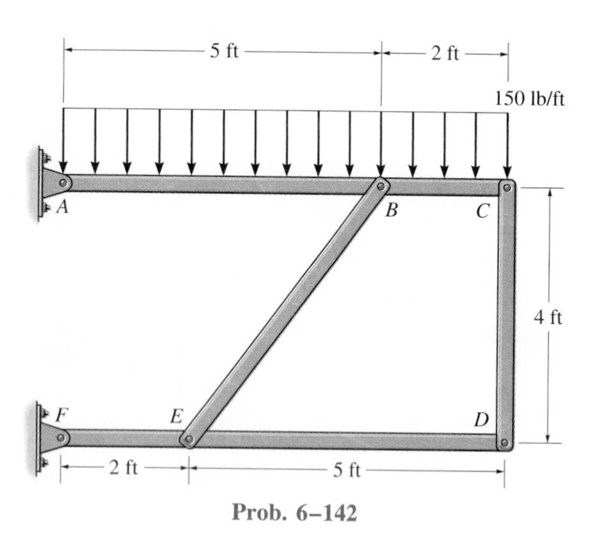

Prob. 6–142

6–143. The mechanism consists of identical meshed gears A and B and arms which are fixed to the gears. The spring attached to the ends of the arms has an unstretched length of 100 mm and a stiffness of $k = 250$ N/m. If a torque of $M = 6$ N · m is applied to gear A, determine the angle θ through which each arm rotates. The gears are each pinned to fixed supports at their centers.

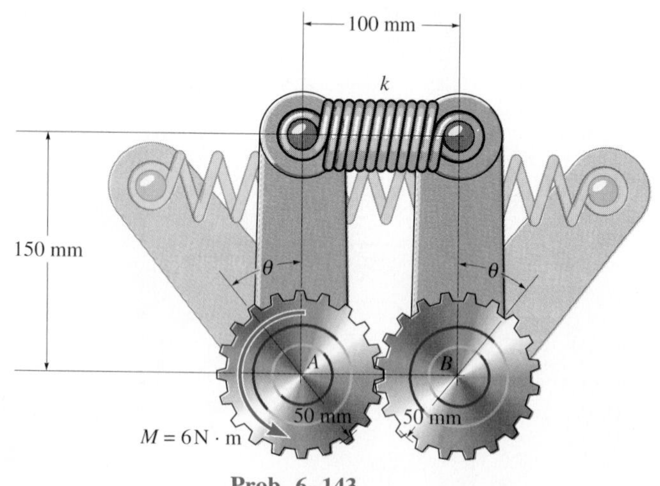

Prob. 6–143

***6–144.** The mechanism consists of identical meshed gears A and B and arms which are fixed to the gears. If a torque of $M = 6$ N · m is applied to gear A as shown, determine the required stiffness k of the spring so that each arm rotates $\theta = 30°$. The gears are each pinned to fixed supports at their centers. The spring has an unstretched length of 100 mm.

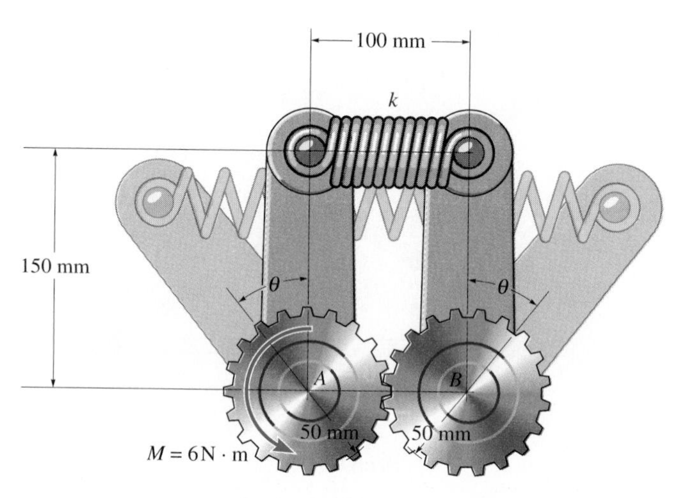

Prob. 6–144

6–145. Determine the force in each member of the truss in terms of the load P and state if the members are in tension or compression.

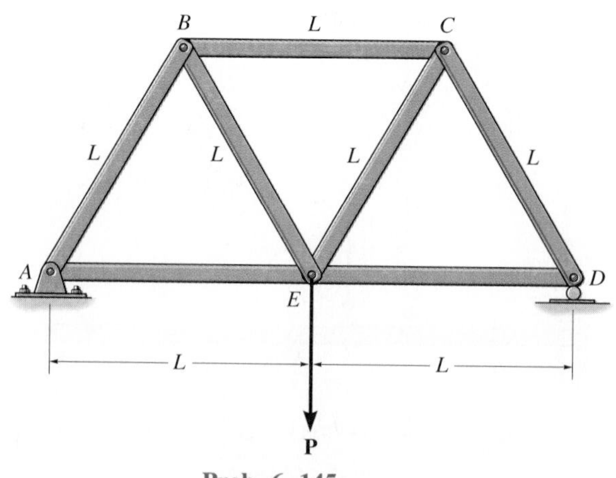

Prob. 6–145

6–146. Each member of the truss is uniform and has a weight W. Remove the external force $\mathbf{P}$ and determine the approximate force in each member due to the weight of the truss. State if the members are in tension or compression. Solve the problem by *assuming* the weight of each member can be represented as a vertical force, half of which is applied at each end of the member.

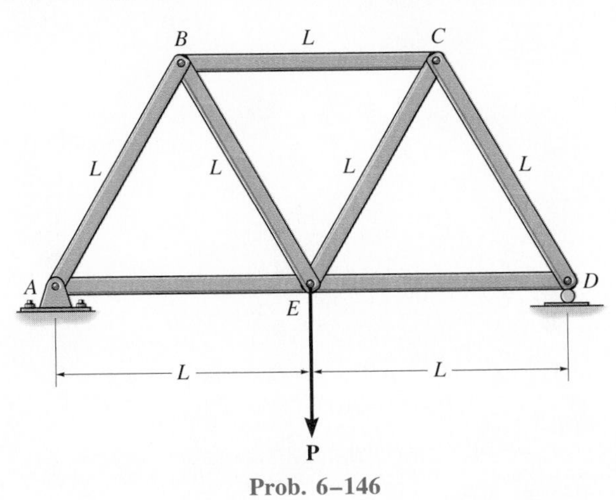

Prob. 6–146

***6–148.** The man has a weight of 150 lb and stands on the uniform plank having a weight of 40 lb. Determine the force he exerts on the plank if he pulls with just enough force to lift the plank off the support at B. The plank rests on the smooth surface at A.

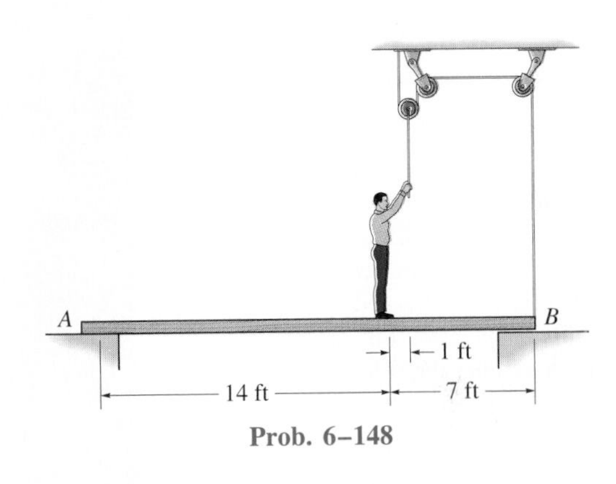

Prob. 6–148

6–147. The spring mechanism is used as a shock absorber for a load applied to the drawbar AB. Determine the equilibrium length of each spring when the 80-N force is applied. Each spring has an unstretched length of 200 mm, and the drawbar slides along the smooth guideposts CG and EF. The bottom springs pass around the guideposts, and the ends of all springs are attached to their respective members.

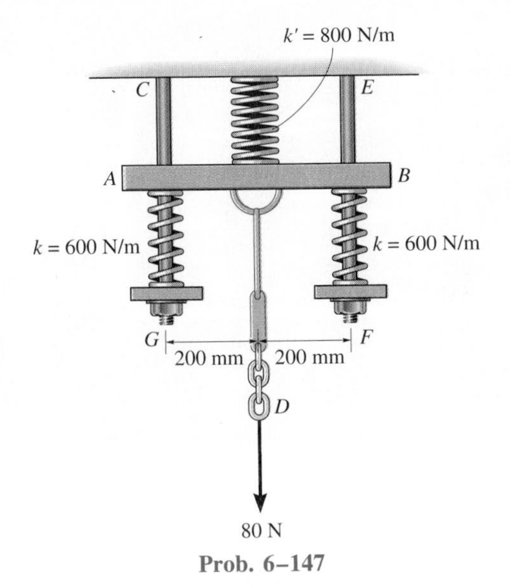

Prob. 6–147

6–149. The bucket of the backhoe and its contents have a weight of 1200 lb and a center of gravity at G. Determine the forces in the hydraulic cylinder AB and in links AC and AD in order to hold the load in the position shown. The bucket is pinned at E. *Hint:* AB, AC, and AD are all two-force members.

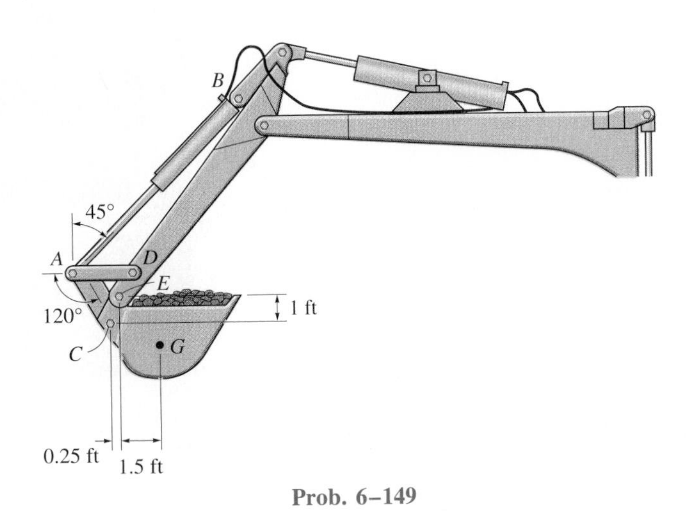

Prob. 6–149

The design and analysis of any structural member requires knowledge of the internal loadings acting within it, not only when it is in place and subjected to service loads, but also when it is being hoisted as shown here. In this chapter we will discuss how engineers determine these loadings.

7

Internal Forces

In this chapter we will develop a technique for finding the internal loading at specific *points* within a structural member, and then we will generalize this method to find the point-to-point variation of loading along the axis of the member. A graph showing this variation of internal load will allow us to find the critical points where the *maximum* internal loading occurs. The analysis of cables will be treated in the last part of the chapter.

7.1 Internal Forces Developed in Structural Members

The design of a structural member requires an investigation of the loadings acting *within* the member which are necessary to balance the loadings acting external to it. The *method of sections* can be used for this purpose. In order to illustrate the procedure, consider the "simply supported" beam shown in Fig. 7–1*a*, which is subjected to the forces $\mathbf{F}_1$ and $\mathbf{F}_2$. The *support reactions* $\mathbf{A}_x$, $\mathbf{A}_y$, and $\mathbf{B}_y$ can be determined by applying the equations of equilibrium, using the free-body diagram of the *entire beam,* Fig. 7–1*b*. If the *internal loadings* at point C are to be determined, it is necessary to pass an imaginary section through the beam, cutting it into two segments at that point, Fig. 7–1*a*. Doing this "exposes" the internal loadings as *external* on the free-body diagram of

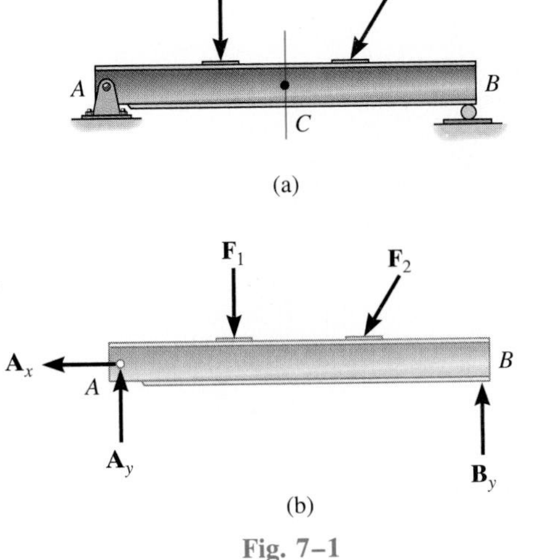

Fig. 7–1

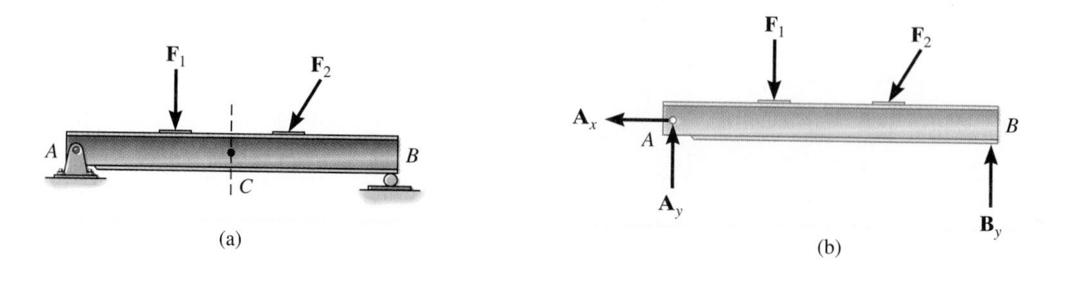

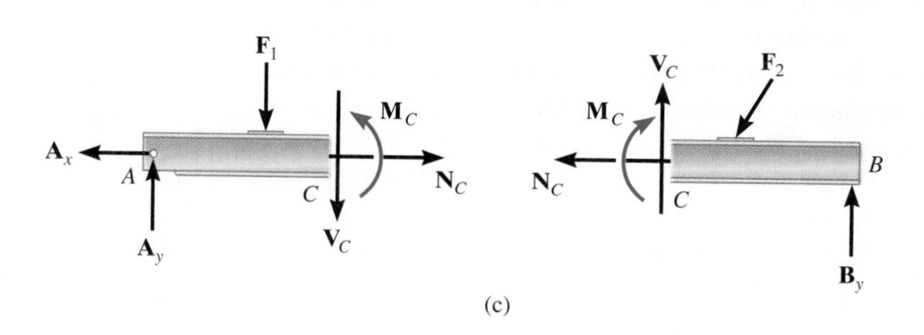

Fig. 7–1

each segment, Fig. 7–1*c*. Since both segments (*AC* and *CB*) were prevented from translating and rotating *before* the beam was sectioned, equilibrium of each segment is maintained if rectangular force components $\mathbf{N}_C$ and $\mathbf{V}_C$ and a resultant couple moment $\mathbf{M}_C$ are developed at the cut section. Note that these loadings must be equal in magnitude and opposite in direction on each of the segments (Newton's third law). The magnitude of each unknown can now be determined by applying the three equations of equilibrium to either segment *AC* or *CB*. A *direct solution* for $\mathbf{N}_C$ is obtained by applying $\Sigma F_x = 0$; $\mathbf{V}_C$ is obtained directly from $\Sigma F_y = 0$; and $\mathbf{M}_C$ is determined by summing moments about point *C*, $\Sigma M_C = 0$, in order to eliminate the moments of the unknowns $\mathbf{N}_C$ and $\mathbf{V}_C$.

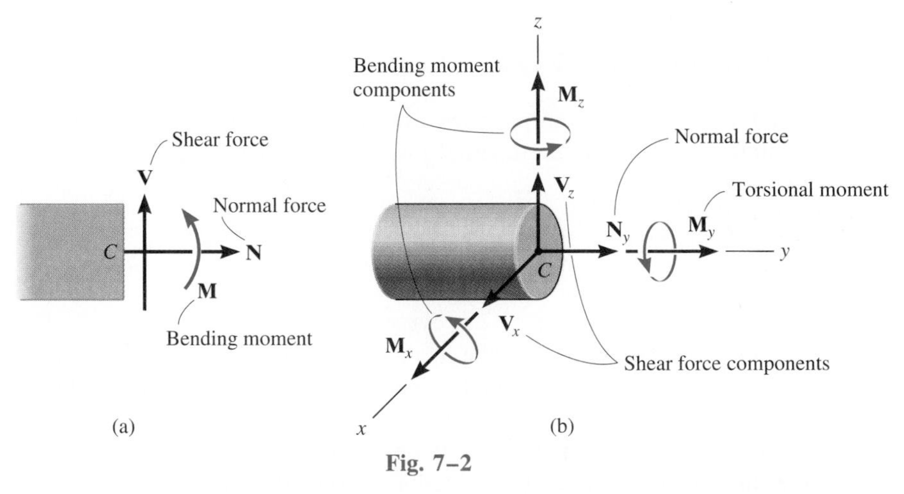

Fig. 7–2

In mechanics, the force components **N,** acting normal to the beam at the cut section, and **V,** acting tangent to the section, are termed the *normal or axial force* and the *shear force,* respectively. The couple moment **M** is referred to as the *bending moment,* Fig. 7–2*a*. In three dimensions, a general internal force and couple moment resultant will act at the section. The *x, y, z* components of these loadings are shown in Fig. 7–2*b*. Here $\mathbf{N}_y$ is the *normal force,* $\mathbf{V}_x$ and $\mathbf{V}_z$ are *shear force components,* $\mathbf{M}_y$ is a *torsional or twisting moment,* and $\mathbf{M}_x$ and $\mathbf{M}_z$ are *bending moment components.* For most applications, these *resultant loadings* will act at the geometric center or centroid (C) of the section's cross-sectional area. Although the magnitude for each loading generally will be different at various points along the axis of the member, the method of sections can always be used to determine their values.

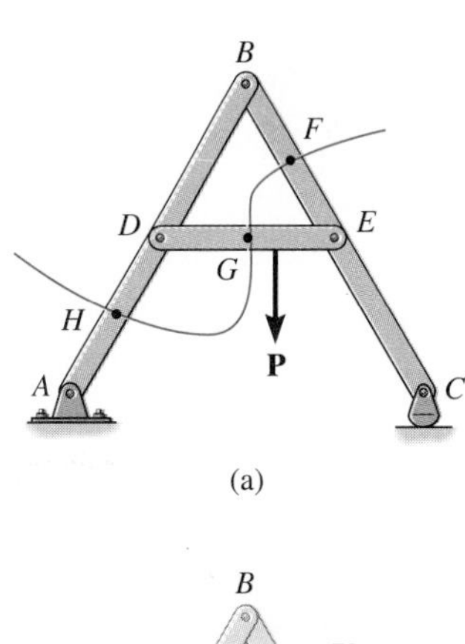

Free-Body Diagrams. Since frames and machines are composed of *multiforce members,* each of these members will generally be subjected to internal normal, shear, and bending loadings. For example, consider the frame shown in Fig. 7–3*a*. If the blue section is passed through the frame to determine the internal loadings at points *H*, *G*, and *F*, the resulting free-body diagram of the top portion of this section is shown in Fig. 7–3*b*. At each point where a member is sectioned there is an unknown normal force, shear force, and bending moment. As a result, we cannot apply the *three* equations of equilibrium to this section in order to obtain these *nine unknowns.** Instead, to solve this problem we must *first dismember* the frame and determine the reactions at the connections of the members using the techniques of Sec. 6.6. Once this is done, *each member* may then be sectioned at its appropriate point and the three equations of equilibrium can be applied to determine **N, V,** and **M.** For example, the free-body diagram of segment *DG*, Fig. 7–3*c*, can be used to determine the internal loadings at *G* provided the reactions of the pin, $\mathbf{D}_x$ and $\mathbf{D}_y$, are known.

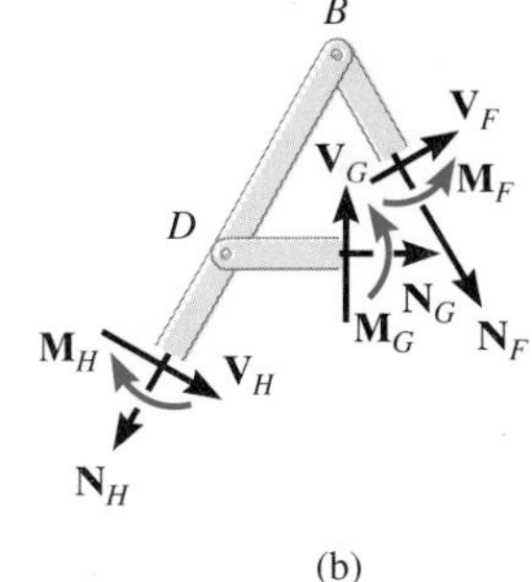

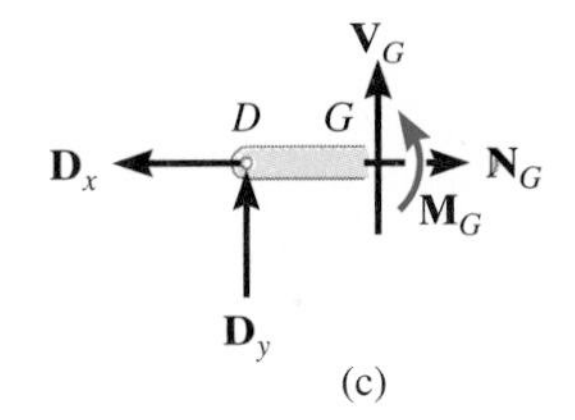

Fig. 7–3

*Recall that this method of analysis worked well for trusses, since truss members are *straight two-force members* which support only an axial or normal load.

PROCEDURE FOR ANALYSIS

The following procedure provides a means for applying the method of sections to determine the internal loadings at a specific location in a member.

Support Reactions. Before the member is ''cut'' or sectioned, it may first be necessary to determine the member's support reactions, so that the equilibrium equations are used only to solve for the internal loadings when the member is sectioned. If the member is part of a frame or machine, the reactions at its connections are determined using the methods of Sec. 6.6.

Free-Body Diagram. Keep all distributed loadings, couple moments, and forces acting on the member in their *exact location,* then pass an imaginary section through the member, perpendicular to its axis at the point where the internal loading is to be determined. Draw a free-body diagram of one of the ''cut'' segments on either side of the section and indicate the x, y, z components of the force and couple moment resultants at the section. In particular, if the member is subjected to a *coplanar* system of forces, only **N, V,** and **M** act at the section. In many cases it may be possible to tell by inspection the proper sense of the unknown loadings; however, if this seems difficult, the sense can be assumed.

Equations of Equilibrium. Apply the equations of equilibrium to obtain the unknown internal loadings. Generally, moments should be summed at the section about axes passing through the *centroid* or geometric center of the member's cross-sectional area, in order to eliminate the unknown normal and shear forces and thereby obtain direct solutions for the moment components. If the solution of the equilibrium equations yields a negative scalar, the assumed sense of the quantity is opposite to that shown on the free-body diagram.

The following examples numerically illustrate this procedure.

Example 7–1

The bar is fixed at its end and is loaded as shown in Fig. 7–4*a*. Determine the internal normal force at points *B* and *C*.

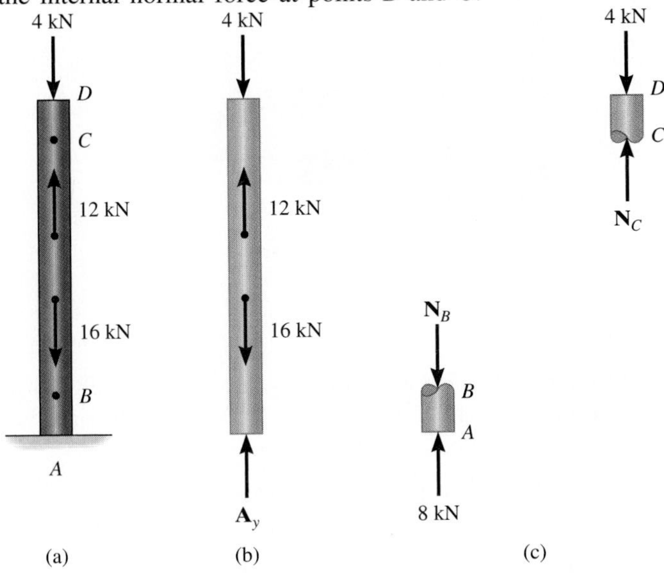

Fig. 7–4

SOLUTION

Support Reactions. A free-body diagram of the entire bar is shown in Fig. 7–4*b*. By inspection, only a normal force $\mathbf{A}_y$ acts at the fixed support, since the loads are applied symmetrically along the bar's axis.

$$+\uparrow \Sigma F_y = 0; \qquad A_y - 16\text{ kN} + 12\text{ kN} - 4\text{ kN} = 0 \qquad A_y = 8\text{ kN}$$

Free-Body Diagrams. The internal forces at *B* and *C* will be found using the free-body diagrams of the sectioned bar shown in Fig. 7–4*c*. In particular, segments *AB* and *DC* will be chosen here, since they contain the *least* number of forces.

Equations of Equilibrium

Segment AB

$$+\uparrow \Sigma F_y = 0; \qquad 8\text{ kN} - N_B = 0 \qquad N_B = 8\text{ kN} \qquad \textit{Ans.}$$

Segment DC

$$+\uparrow \Sigma F_y = 0; \qquad N_C - 4\text{ kN} = 0 \qquad N_C = 4\text{ kN} \qquad \textit{Ans.}$$

Try working this problem in the following manner: Determine N_B from segment *BD*. (Note that this approach does *not require* solution for the support reaction at *A*.) Using the result for N_B, isolate segment *BC* to determine N_C.

Example 7–2

The circular shaft is subjected to three concentrated torques as shown in Fig. 7–5*a*. Determine the internal torques at points B and C.

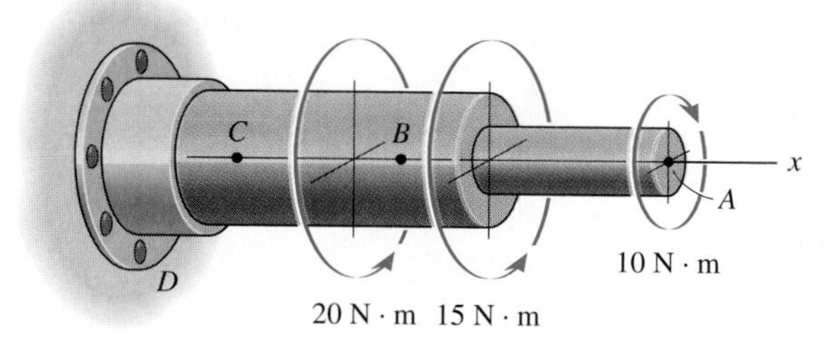

(a)

SOLUTION

Support Reactions. Since the shaft is subjected only to collinear torques, a torque reaction occurs at the support, Fig. 7–5*b*. Using the right-hand rule to define the positive directions of the torques, we require

$$\Sigma M_x = 0; \qquad -10\text{ N}\cdot\text{m} + 15\text{ N}\cdot\text{m} + 20\text{ N}\cdot\text{m} - T_D = 0$$
$$T_D = 25\text{ N}\cdot\text{m}$$

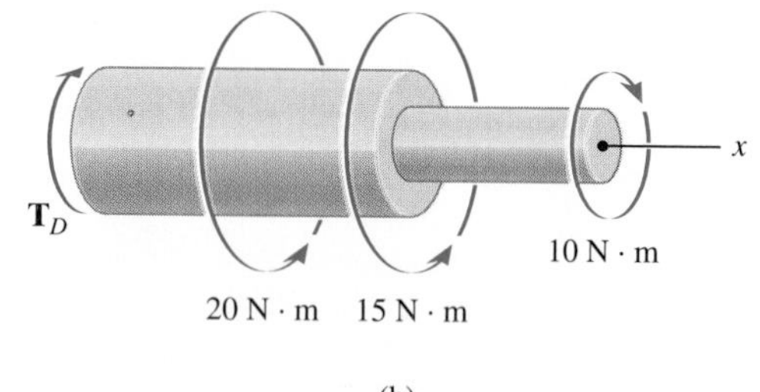

(b)

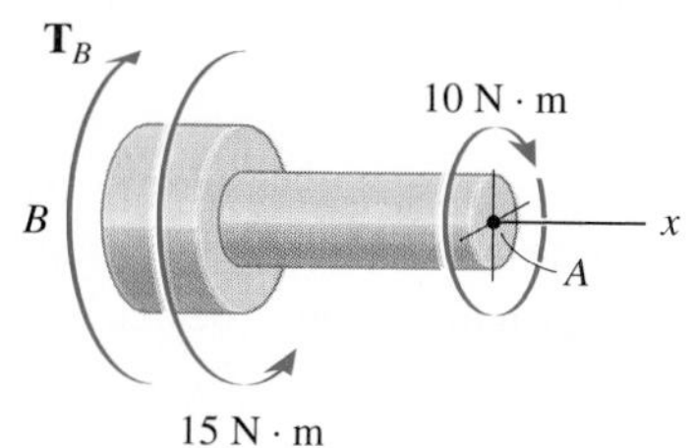

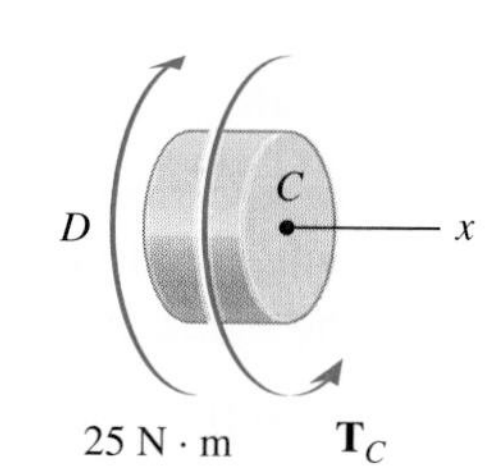

(c)

Fig. 7–5

Free-Body Diagrams. The internal torques at B and C will be found using the free-body diagrams of the shaft segments AB and CD shown in Fig. 7–5*c*.

Equations of Equilibrium. Applying the equation of moment equilibrium along the shaft's axis, we have

Segment AB

$$\Sigma M_x = 0; \quad -10\text{ N}\cdot\text{m} + 15\text{ N}\cdot\text{m} - T_B = 0 \qquad T_B = 5\text{ N}\cdot\text{m} \qquad \textit{Ans.}$$

Segment CD

$$\Sigma M_x = 0; \qquad T_C - 25\text{ N}\cdot\text{m} = 0 \qquad T_C = 25\text{ N}\cdot\text{m} \qquad \textit{Ans.}$$

Try to solve for T_C by using segment CA. Note that this approach does not require a solution for the support reaction at D.

Example 7–3

The beam supports the loading shown in Fig. 7–6*a*. Determine the internal normal force, shear force, and bending moment acting just to the left, point *B*, and just to the right, point *C*, of the 6-kN force.

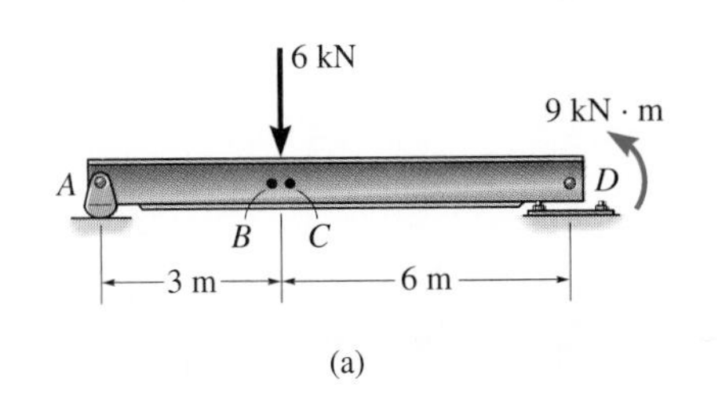

(a)

SOLUTION

Support Reactions. The free-body diagram of the beam is shown in Fig. 7–6*b*. When determining the *external reactions,* realize that the 9-kN · m couple moment is a free vector and therefore it can be placed *anywhere* on the free-body diagram of the entire beam. Here we will only determine $\mathbf{A}_y$, since segments *AB* and *AC* will be used for the analysis.

$$\circlearrowleft + \Sigma M_D = 0; \quad 9\text{ kN}\cdot\text{m} + (6\text{ kN})(6\text{ m}) - A_y(9\text{ m}) = 0$$
$$A_y = 5\text{ kN}$$

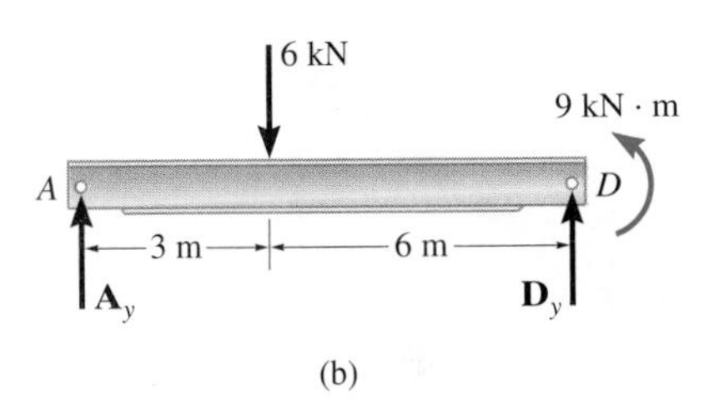

(b)

Free-Body Diagrams. The free-body diagrams of the left segments *AB* and *AC* of the beam are shown in Figs. 7–6*c* and 7–6*d*. In this case the 9-kN · m couple moment is *not included* on these diagrams, since it must be kept in its *original position* until *after* the section is made and the appropriate body isolated. In other words, the internal loadings, when determined from the left segments of the beam, are not influenced by the effect of the couple moment since this moment does not actually act on these segments.

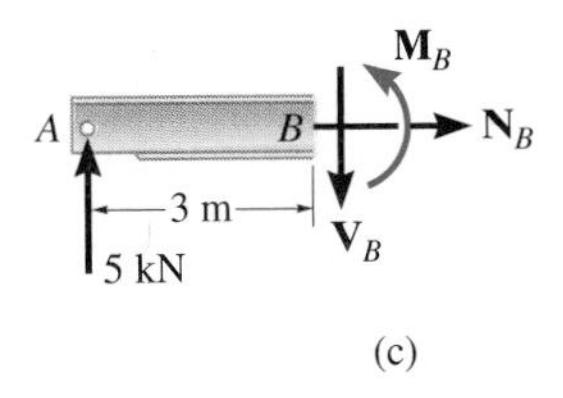

(c)

Equations of Equilibrium

Segment AB

$\overset{+}{\rightarrow}\Sigma F_x = 0; \quad N_B = 0$ **Ans.**

$+\uparrow \Sigma F_y = 0; \quad 5\text{ kN} - V_B = 0 \quad V_B = 5\text{ kN}$ **Ans.**

$\circlearrowleft + \Sigma M_B = 0; \quad -(5\text{ kN})(3\text{ m}) + M_B = 0 \quad M_B = 15\text{ kN}\cdot\text{m}$ **Ans.**

Segment AC

$\overset{+}{\rightarrow}\Sigma F_x = 0; \quad N_C = 0$ **Ans.**

$+\uparrow \Sigma F_y = 0; \quad 5\text{ kN} - 6\text{ kN} + V_C = 0 \quad V_C = 1\text{ kN}$ **Ans.**

$\circlearrowleft + \Sigma M_C = 0; \quad -(5\text{ kN})(3\text{ m}) + M_C = 0 \quad M_C = 15\text{ kN}\cdot\text{m}$ **Ans.**

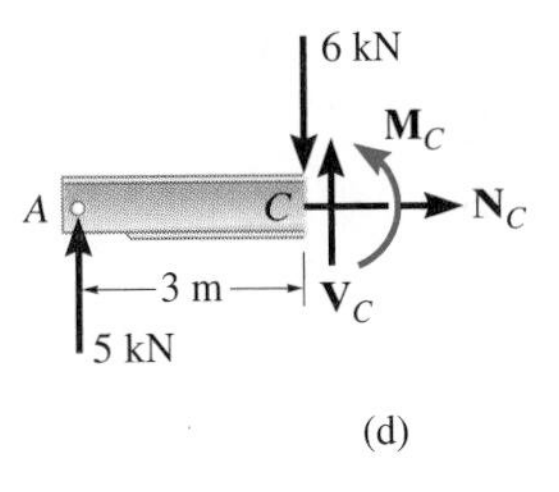

(d)

Fig. 7–6

Here the moment arm for the 5-kN force in both cases is approximately 3 m, since *B* and *C* are "almost" coincident.

Example 7–4

Determine the internal normal force, shear force, and bending moment acting at point B of the two-member frame shown in Fig. 7–7a.

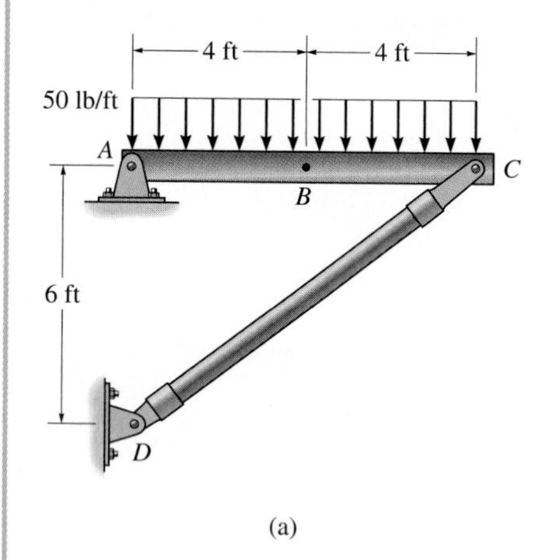

(a)

SOLUTION

Support Reactions. A free-body diagram of each member is shown in Fig. 7–7b. Since CD is a two-force member, the equations of equilibrium need to be applied only to member AC.

$$\circlearrowleft+\Sigma M_A = 0; \quad -400 \text{ lb}(4 \text{ ft}) + (\tfrac{3}{5})F_{DC}(8 \text{ ft}) = 0 \quad F_{DC} = 333.3 \text{ lb}$$

$$\xrightarrow{+}\Sigma F_x = 0; \quad -A_x + (\tfrac{4}{5})(333.3 \text{ lb}) = 0 \quad A_x = 266.7 \text{ lb}$$

$$+\uparrow \Sigma F_y = 0; \quad A_y - 400 \text{ lb} + \tfrac{3}{5}(333.3 \text{ lb}) = 0 \quad A_y = 200 \text{ lb}$$

Free-Body Diagrams. Passing an imaginary section perpendicular to the axis of member AC through point B yields the free-body diagrams of segments AB and BC shown in Fig. 7–7c. When constructing these diagrams it is important to keep the distributed loading exactly as it is until *after* the section is made. Only then can it be replaced by a single resultant force. Why? Also, notice that $\mathbf{N}_B$, $\mathbf{V}_B$, and $\mathbf{M}_B$ act with equal magnitude but opposite direction on each segment—Newton's third law.

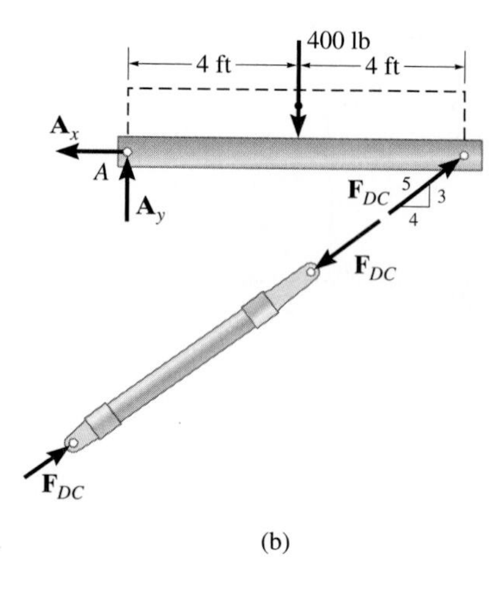

(b)

Fig. 7–7

Equations of Equilibrium. Applying the equations of equilibrium to segment AB, we have

$$\xrightarrow{+}\Sigma F_x = 0; \quad N_B - 266.7 \text{ lb} = 0 \quad N_B = 267 \text{ lb} \quad \textit{Ans.}$$

$$+\uparrow \Sigma F_y = 0; \quad 200 \text{ lb} - 200 \text{ lb} - V_B = 0 \quad V_B = 0 \quad \textit{Ans.}$$

$$\circlearrowleft+\Sigma M_B = 0; \quad M_B - 200 \text{ lb}(4 \text{ ft}) + 200 \text{ lb}(2 \text{ ft}) = 0$$

$$M_B = 400 \text{ lb} \cdot \text{ft} \quad \textit{Ans.}$$

As an exercise, try to obtain these same results using segment BC.

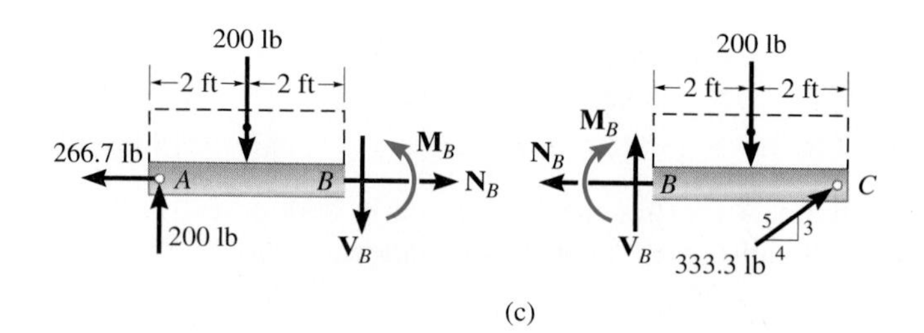

(c)

Example 7–5

Determine the normal force, shear force, and bending moment acting at point E of the frame loaded as shown in Fig. 7–8a.

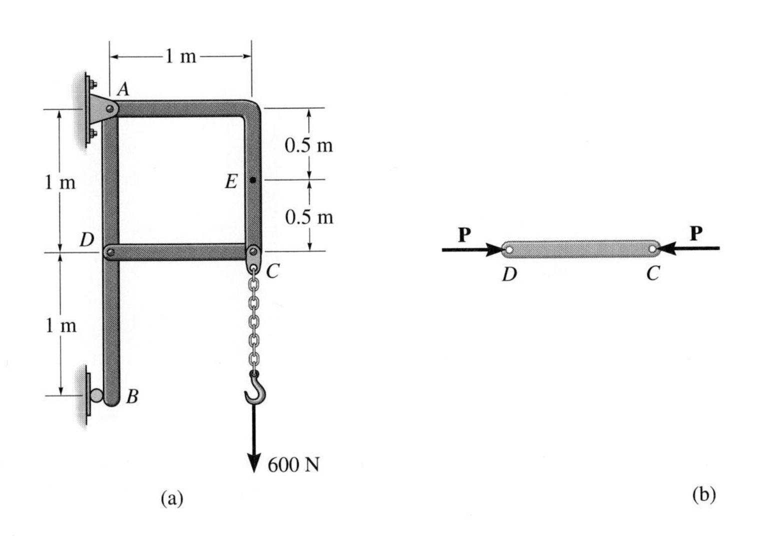

Fig. 7–8

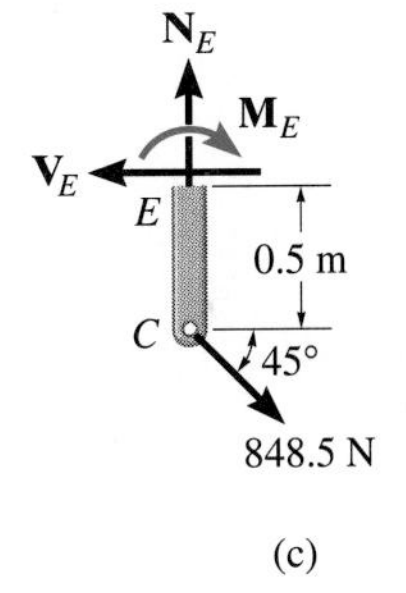

SOLUTION

Support Reactions. By inspection, members AC and CD are two-force members, Fig. 7–8b. In order to determine the internal loadings at E, we must first determine the force $\mathbf{R}$ at the end of member AC. To do this we must analyze the equilibrium of the pin at C. Why?

Summing forces in the vertical direction on the pin, Fig. 7–8b, we have

$$+\uparrow \Sigma F_y = 0; \qquad R \sin 45^\circ - 600\text{ N} = 0 \qquad R = 848.5\text{ N}$$

Free-Body Diagram. The free-body diagram of segment CE is shown in Fig. 7–8c.

Equations of Equilibrium

$$\overset{+}{\rightarrow}\Sigma F_x = 0; \quad 848.5 \cos 45^\circ\text{ N} - V_E = 0 \qquad V_E = 600\text{ N} \qquad \textit{Ans.}$$

$$+\uparrow \Sigma F_y = 0; \quad -848.5 \sin 45^\circ\text{ N} + N_E = 0 \qquad N_E = 600\text{ N} \qquad \textit{Ans.}$$

$$\circlearrowright + \Sigma M_E = 0; \quad 848.5 \cos 45^\circ\text{ N}(0.5\text{ m}) - M_E = 0 \quad M_E = 300\text{ N} \cdot \text{m} \qquad \textit{Ans.}$$

As in Example 6–21, member AC should be *straight* (from A to C) so that bending within the member is eliminated, and the internal force would only create tension in the member.

Example 7–6

A force of $\mathbf{F} = \{-3\mathbf{i} + 7\mathbf{j} - 4\mathbf{k}\}$ kN acts at the corner of a beam extended from a fixed wall as shown in Fig. 7–9*a*. Determine the internal loadings at *A*.

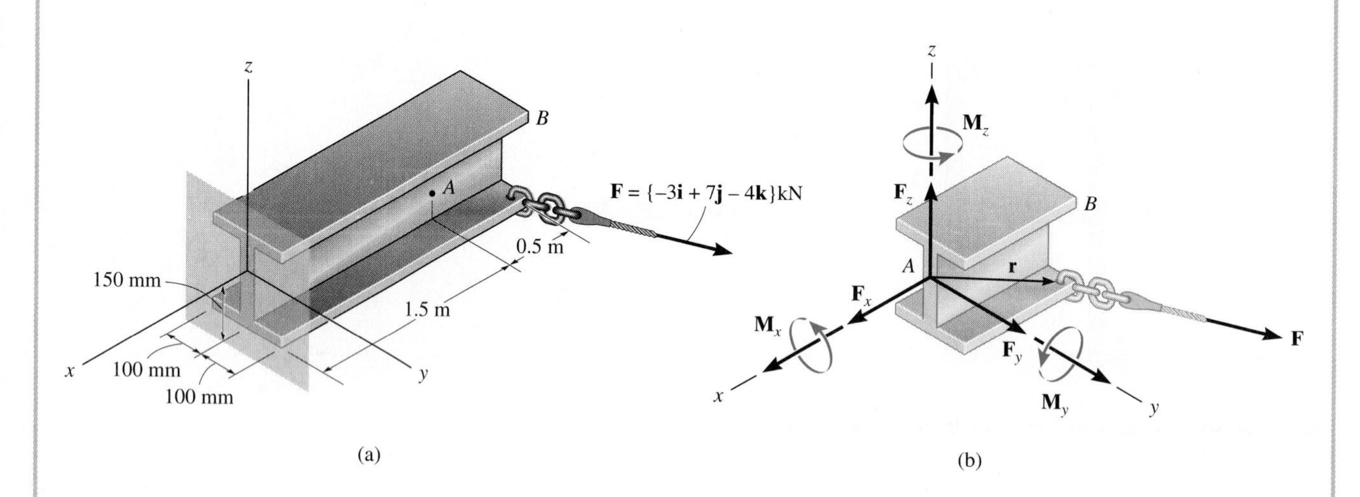

SOLUTION **Fig. 7–9**

This problem will be solved by considering segment *AB* of the beam, since it does *not* involve the support reactions.

Free-Body Diagram. The free-body diagram of segment *AB* is shown in Fig. 7–9*b*. The components of the resultant force $\mathbf{F}_A$ and moment $\mathbf{M}_A$ pass through the *centroid* or geometric center of the cross-sectional area at *A*.

Equations of Equilibrium

$\Sigma\mathbf{F} = \mathbf{0};\ \mathbf{F}_A + \mathbf{F} = \mathbf{0} \qquad \mathbf{F}_A - 3\mathbf{i} + 7\mathbf{j} - 4\mathbf{k} = \mathbf{0}$

$$\mathbf{F}_A = \{3\mathbf{i} - 7\mathbf{j} + 4\mathbf{k}\}\text{ kN} \qquad \textit{Ans.}$$

$$\Sigma\mathbf{M}_A = \mathbf{0};\quad \mathbf{M}_A + \mathbf{r}\times\mathbf{F} = \mathbf{M}_A + \begin{vmatrix} \mathbf{i} & \mathbf{j} & \mathbf{k} \\ -0.5 & 0.1 & -0.15 \\ -3 & 7 & -4 \end{vmatrix} = \mathbf{0}$$

$$\mathbf{M}_A = \{-0.650\mathbf{i} + 1.55\mathbf{j} + 3.20\mathbf{k}\}\text{ kN}\cdot\text{m} \qquad \textit{Ans.}$$

Here $\mathbf{F}_x = \{3\mathbf{i}\}$ kN represents the normal force N, whereas $\mathbf{F}_y = \{-7\mathbf{j}\}$ kN and $\mathbf{F}_z = \{4\mathbf{k}\}$ kN are components of the shear force $V = \sqrt{F_y^2 + F_z^2}$. Also, the torsional moment is $\mathbf{M}_x = \{-0.65\mathbf{i}\}$ kN · m, and the bending moment is determined from its components $\mathbf{M}_y = \{1.55\mathbf{j}\}$ kN · m and $\mathbf{M}_z = \{3.20\mathbf{k}\}$ kN · m; i.e., $M_b = \sqrt{M_y^2 + M_z^2}$.

PROBLEMS

7–1. Three torques act on the shaft. Determine the torques at sections passing through points *A*, *B*, *C*, and *D*.

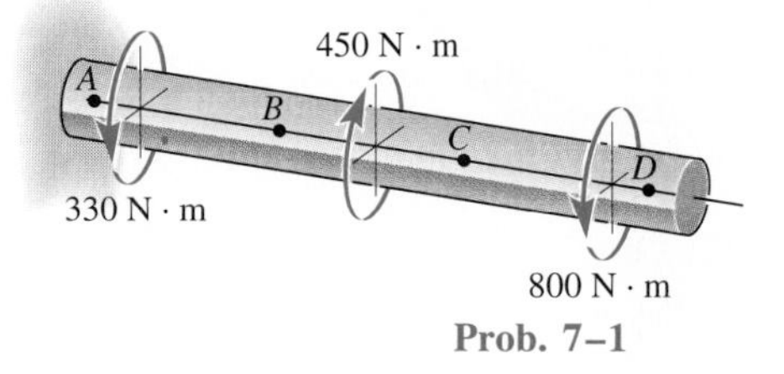

Prob. 7–1

7–2. The shaft is supported by smooth bearings at *A* and *B* and subjected to the torques shown. Determine the torques at sections passing through points *C*, *D*, and *E*.

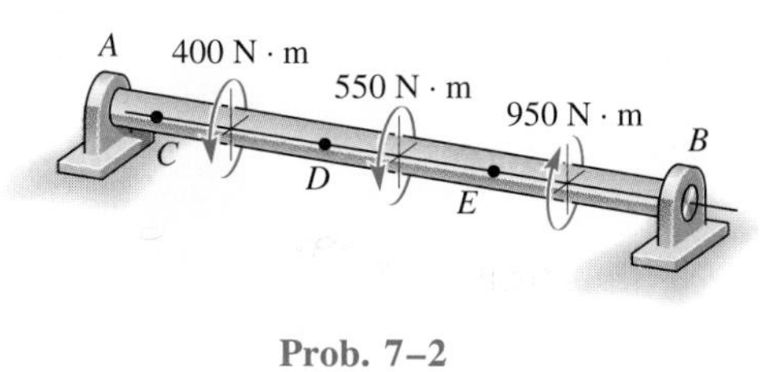

Prob. 7–2

7–3. Three torques act on the shaft as shown. Determine the torques at sections passing through points *A*, *B*, *C*, and *D*.

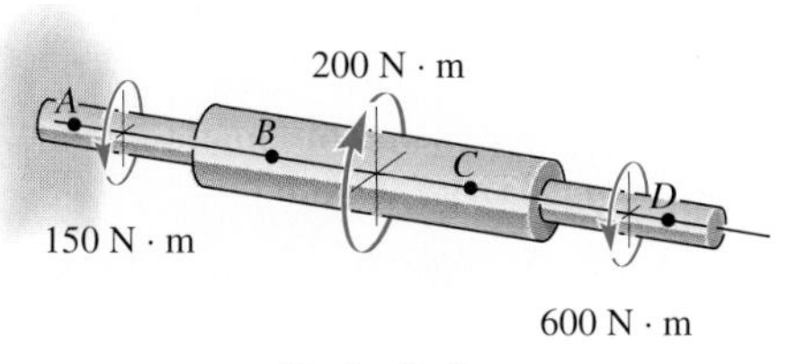

Prob. 7–3

***7–4.** The shaft is supported by a journal bearing at *A* and a thrust bearing at *B*. Determine the normal force, shear force, and moment at a section passing through (a) point *C*, which is just to the right of the bearing at *A*, and (b) point *D*, which is just to the left of the 3000-lb force.

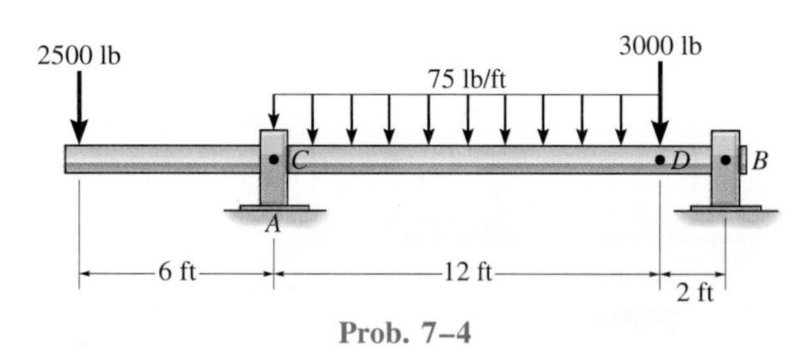

Prob. 7–4

7–5. Determine the normal force, shear force, and moment at a section passing through point *C*. Assume the support at *A* can be approximated by a pin and *B* as a roller.

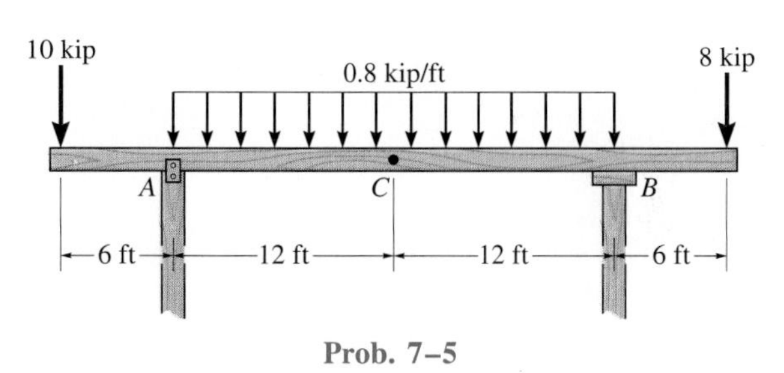

Prob. 7–5

7–6. Determine the normal force, shear force, and moment at a section passing through point *C*.

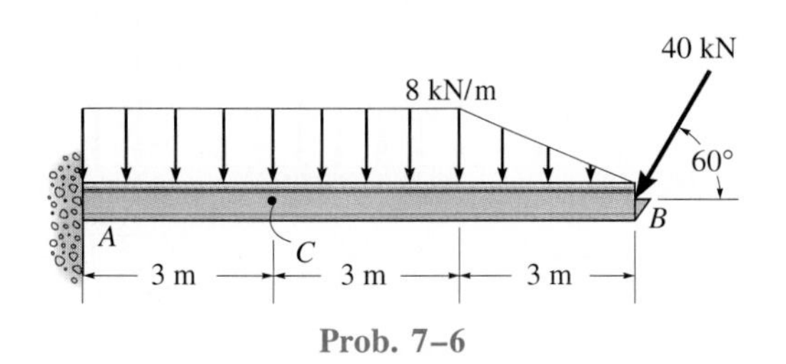

Prob. 7–6

7–7. Determine the normal force, shear force, and moment at a section passing through point D. Take $w = 150$ N/m.

***7–8.** The beam AB will fail if the maximum internal moment at D reaches 800 N · m or the normal force in member BC becomes 1500 N. Determine the largest load w it can support.

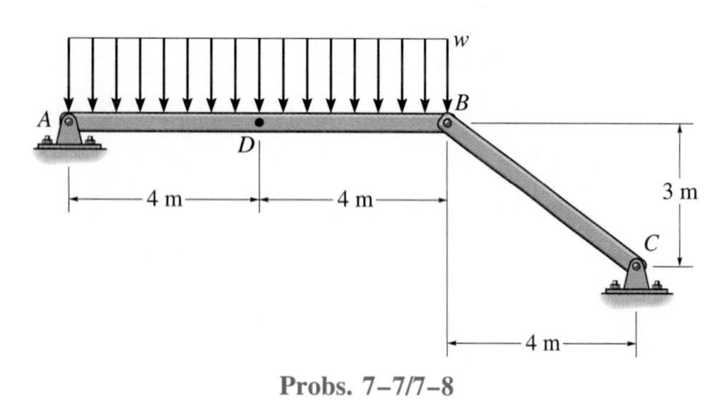

Probs. 7–7/7–8

7–9. Determine the shear force and moment acting at a section passing through point C in the beam.

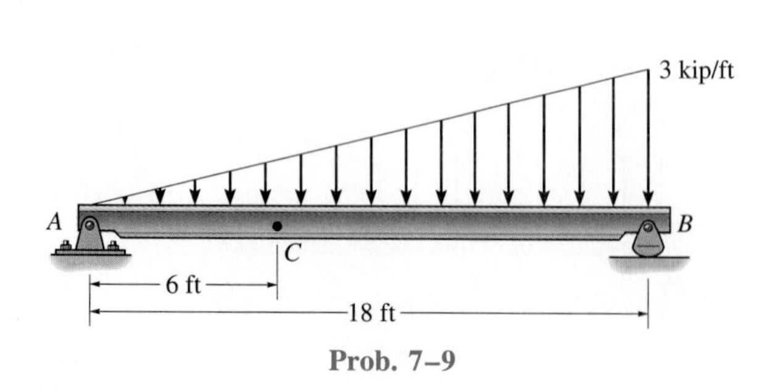

Prob. 7–9

7–10. The cantilevered rack is used to support each end of a smooth pipe that has a total weight of 300 lb. Determine the normal force, shear force, and moment that act in the arm at its fixed support A along a vertical section.

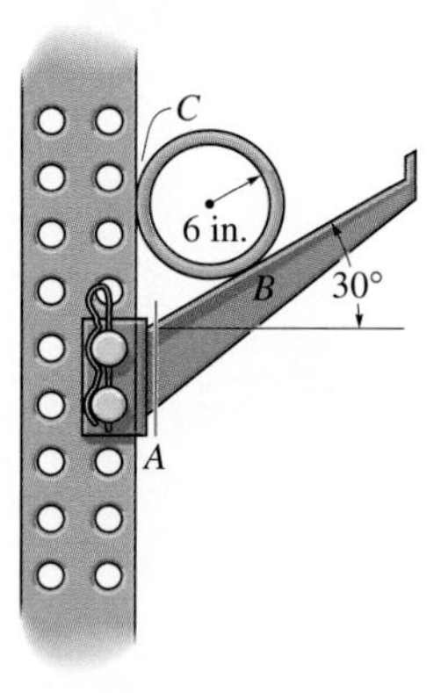

Prob. 7–10

7–11. The A-frame ladder rests on the smooth floor. If a 175-lb man stands uniformly on it at D, determine the moment each of the two lock joints at C must resist to hold the ladder stationary. Assume the man exerts only a vertical reaction at D and no other force on the ladder.

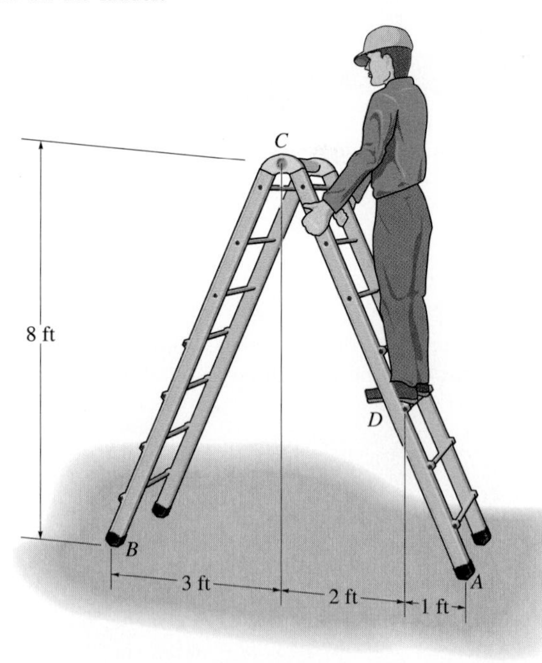

Prob. 7–11

***7–12.** Determine the normal force, shear force, and moment at a section passing through point D of the two-member frame.

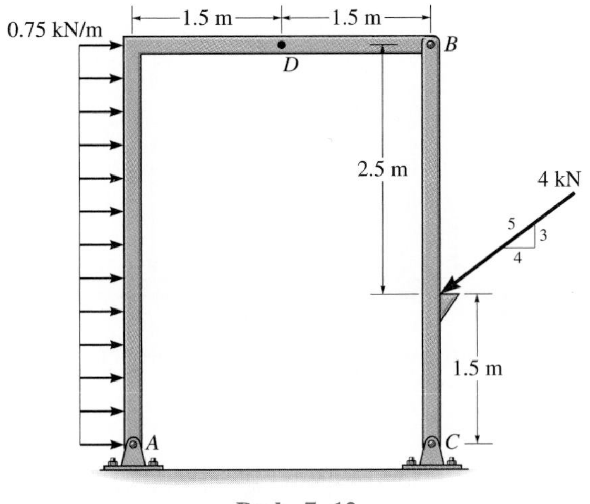

Prob. 7–12

7–13. The boom DF of the jib crane and the column DE have a uniform weight of 50 lb/ft. If the hoist and load weigh 300 lb, determine the normal force, shear force, and moment in the crane at sections passing through points A, B, and C.

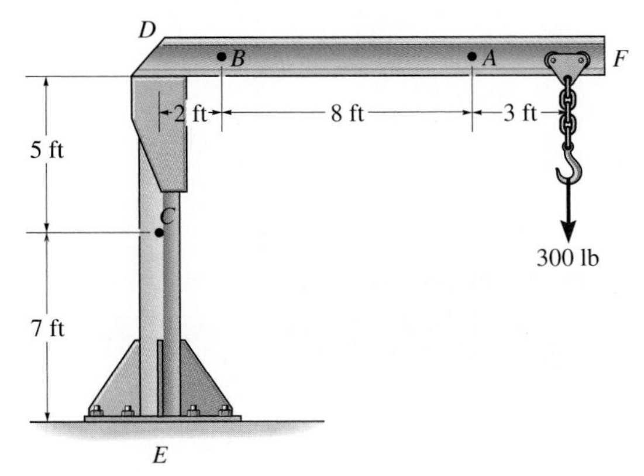

Prob. 7–13

7–14. Determine the normal force, shear force, and moment at a section passing through point D of the two-member frame.

7–15. Determine the normal force, shear force, and moment at a section passing through point E of the two-member frame.

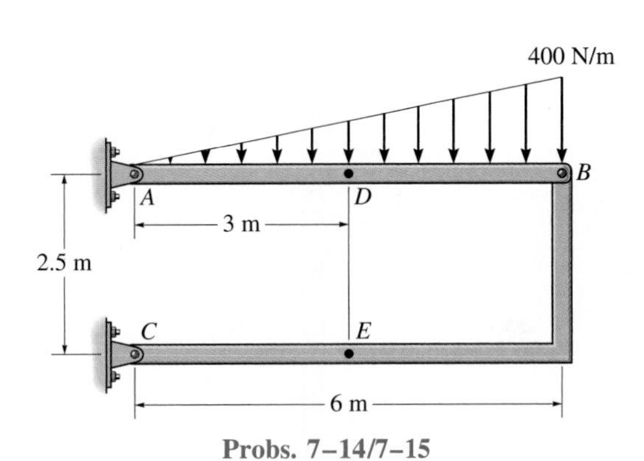

Probs. 7–14/7–15

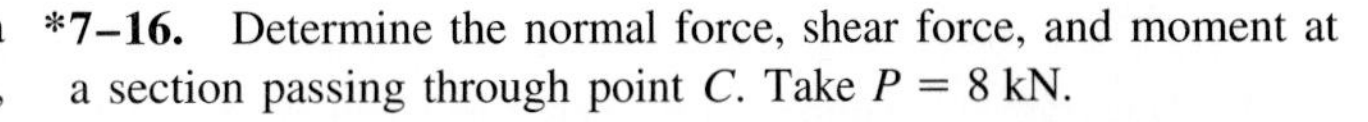

***7–16.** Determine the normal force, shear force, and moment at a section passing through point C. Take $P = 8$ kN.

7–17. The cable will fail when subjected to a tension of 2 kN. Determine the largest vertical load P the frame will support and calculate the internal normal force, shear force, and moment at a section passing through point C for this loading.

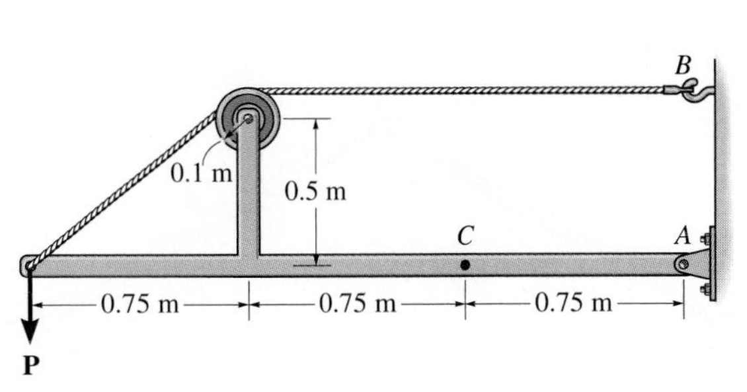

Probs. 7–16/7–17

7–18. The wishbone construction of the power pole supports the three lines, each exerting a force of 800 lb on the bracing struts. If the struts are pin-connected at *A*, *B*, and *C*, determine the normal force, shear force, and moment at sections passing through points *D*, *E*, and *F*.

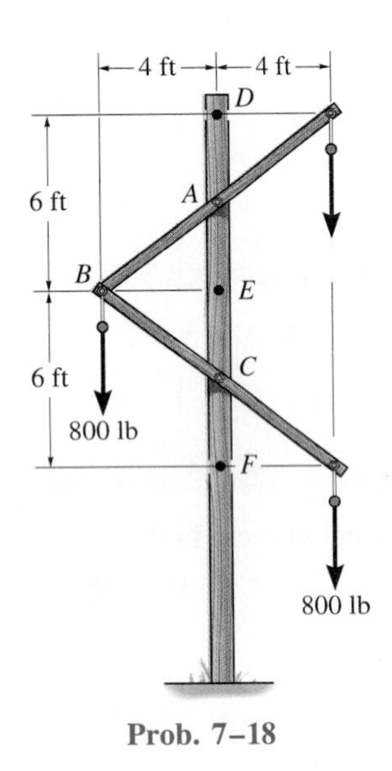

Prob. 7–18

7–19. Determine the normal force, shear force, and moment acting at a section passing through point *C*.

***7–20.** Determine the normal force, shear force, and moment acting at a section passing through point *D*.

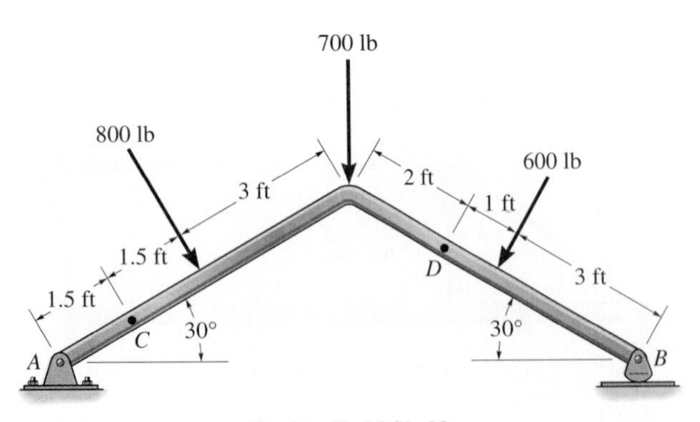

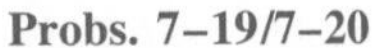
Probs. 7–19/7–20

7–21. Determine the normal force, shear force, and moment in the beam at sections passing through points *D* and *E*. Point *E* is just to the right of the 4-kip load.

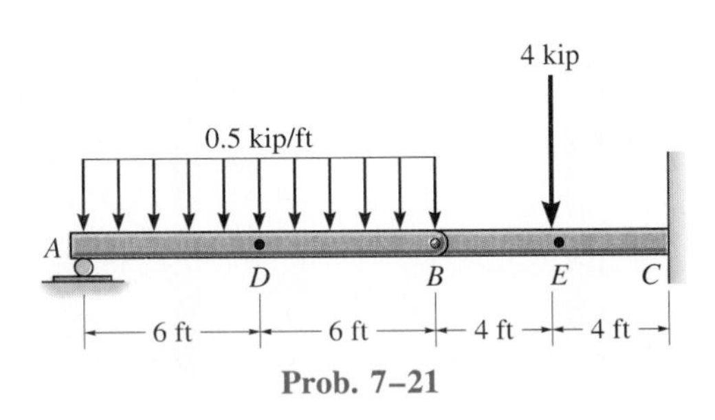

Prob. 7–21

7–22. The 9-kip force is supported by the floor panel *DE*, which in turn is simply supported at its ends by floor beams. These beams transmit their loads to the girder *AB*. Determine the shear and moment acting at a section passing through point *C* in the girder.

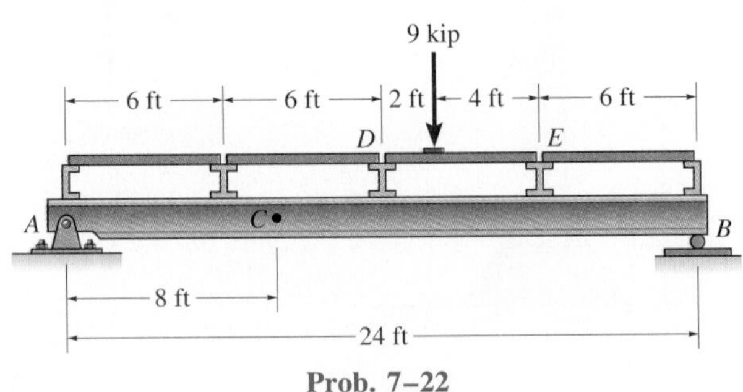

Prob. 7–22

7–23. Determine the normal force, shear force, and moment in the beam at sections passing through points D and E. Point E is just to the right of the 3-kip load.

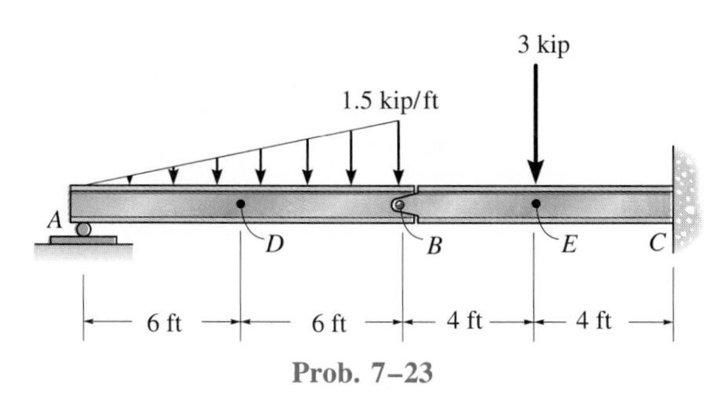

Prob. 7–23

*7–24. Determine the normal force, shear force, and moment at a section passing through point D of the two-member frame.

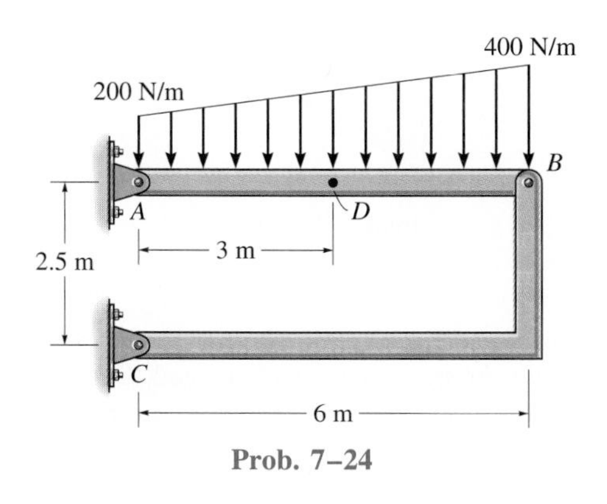

Prob. 7–24

7–25. Determine the normal force, shear force, and moment at sections passing through points E and F. Member BC is pinned at B and there is a smooth slot in it at C. The pin at C is fixed to member CD.

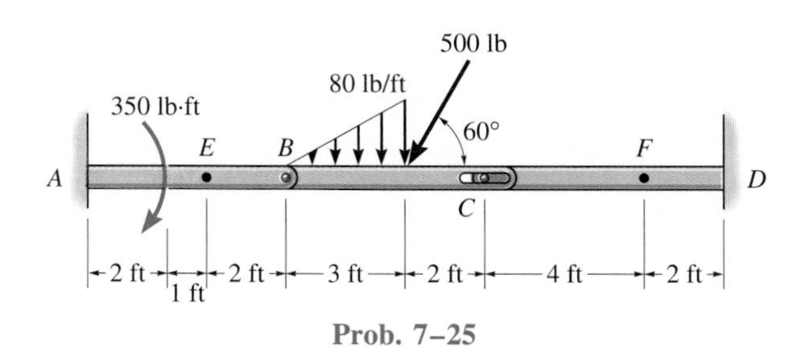

Prob. 7–25

7–26. Determine the normal force, shear force, and moment acting at sections passing through points B and C on the curved rod.

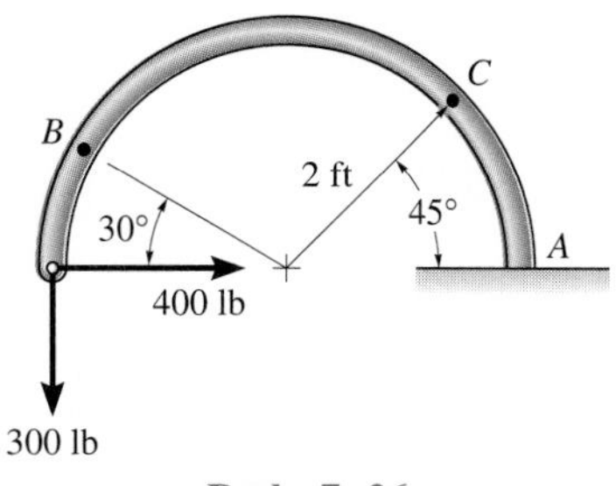

Prob. 7–26

7–27. The semicircular arch is subjected to a uniform distributed load along its axis of w_0 per unit length. Determine the normal force, shear force, and moment in the arch at $\theta = 45°$.

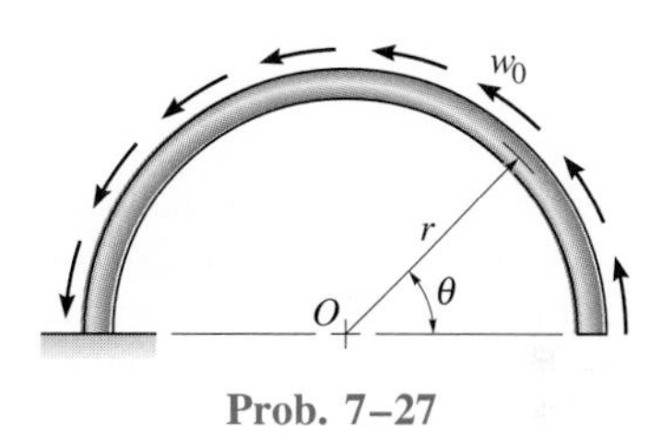

Prob. 7–27

*7–28. The distributed loading $w = w_0 \sin \theta$, measured per unit length, acts on the curved rod. Determine the normal force, shear force, and moment in the rod at $\theta = 45°$.

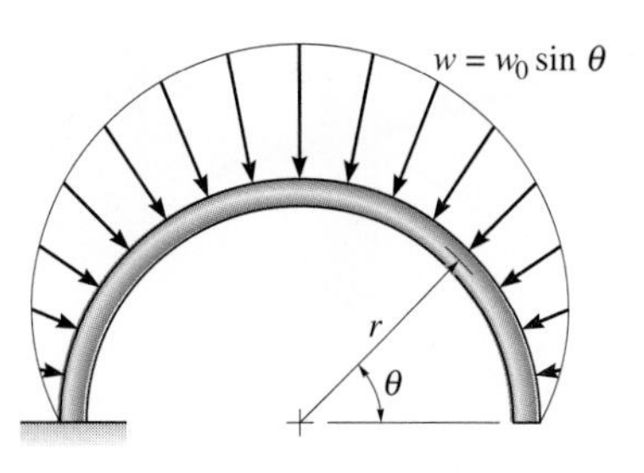

Prob. 7–28

■*7–29. Determine the shear force and moment acting at a section through point C of the beam. For the calculation use Simpson's rule to evaluate the integrals.

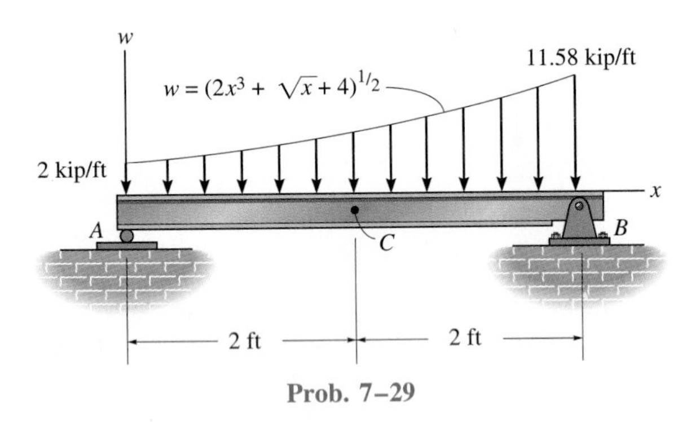

Prob. 7–29

7–30. The man has a weight of 250 lb and a center of gravity at G. If the bench upon which he is sitting is fix-connected to the center support, determine the internal x, y, z components of internal loading at sections passing through point A and the base B.

Prob. 7–30

7–31. Determine the x, y, z components of internal loading at a section passing through point C in the pipe assembly. Neglect the weight of the pipe. Take $\mathbf{F}_1 = \{350\mathbf{j} - 400\mathbf{k}\}$ lb and $\mathbf{F}_2 = \{150\mathbf{i} - 300\mathbf{k}\}$ lb.

*7–32. Determine the x, y, z components of internal loading at a section passing through point C in the pipe assembly. Neglect the weight of the pipe. Take $\mathbf{F}_1 = \{-80\mathbf{i} + 200\mathbf{j} - 300\mathbf{k}\}$ lb and $\mathbf{F}_2 = \{250\mathbf{i} - 150\mathbf{j} - 200\mathbf{k}\}$ lb.

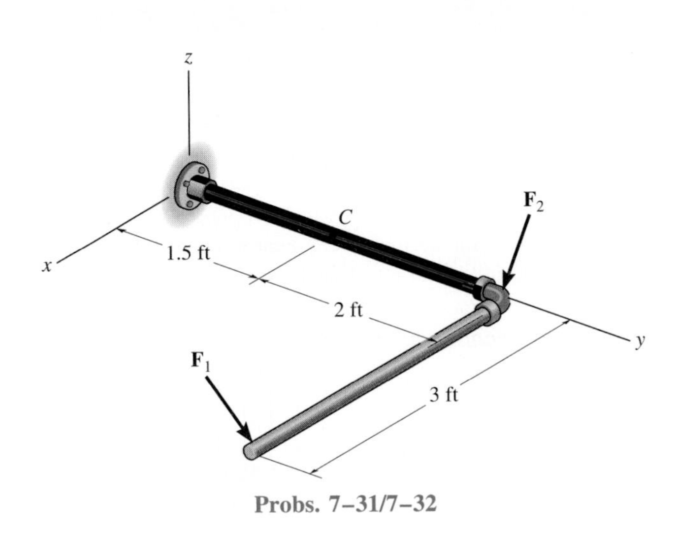

Probs. 7–31/7–32

7–33. Determine the x, y, z components of internal loading at a section passing through point D of the rod. Take $\mathbf{F} = \{-7\mathbf{i} + 12\mathbf{j} - 5\mathbf{k}\}$ kN. The supports at A, B, and C are journal bearings.

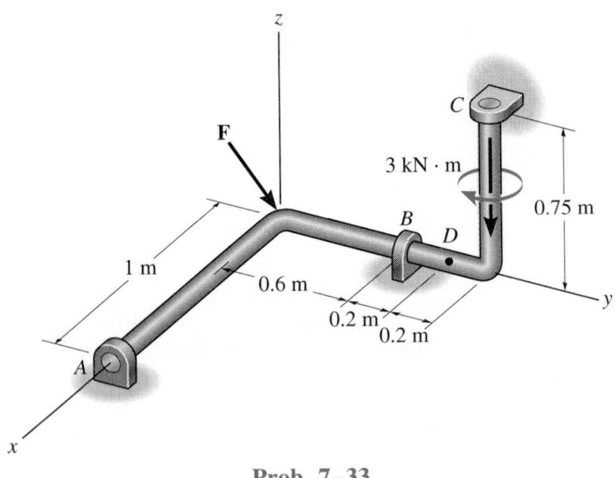

Prob. 7–33

*7.2 Shear and Moment Equations and Diagrams

Beams are structural members which are designed to support loadings applied perpendicular to their axes. In general, beams are long, straight bars having a constant cross-sectional area. Often they are classified as to how they are supported. For example, a *simply supported beam* is pinned at one end and roller-supported at the other, Fig. 7–10, whereas a *cantilevered beam* is fixed at one end and free at the other. The actual design of a beam requires a detailed knowledge of the *variation* of the internal shear force V and bending moment M acting at *each point* along the axis of the beam. After this force and bending-moment analysis is complete, one can then use the theory of mechanics of materials and an appropriate engineering design code to determine the beam's required cross-sectional area.

The *variations* of V and M as a function of the position x along the beam's axis can be obtained by using the method of sections discussed in Sec. 7.1. Here, however, it is necessary to section the beam at an arbitrary distance x from one end rather than at a specified point. If the results are plotted, the graphical variations of V and M as a function of x are termed the *shear diagram* and *bending-moment diagram,* respectively.

In general, the internal shear and bending-moment functions will be discontinuous, or their slope will be discontinuous at points where a distributed load changes or where concentrated forces or couple moments are applied. Because of this, these functions must be determined for *each segment* of the beam located between any two discontinuities of loading. For example, sections located at x_1, x_2, and x_3 will have to be used to describe the variation of V and M throughout the length of the beam in Fig. 7–10. These functions will be valid *only* within regions from O to a for x_1, from a to b for x_2, and from b to L for x_3.

The internal normal force will not be considered in the following discussion for two reasons. In most cases, the loads applied to a beam act perpendicular to the beam's axis and hence produce only an internal shear force and bending moment. For design purposes, the beam's resistance to shear, and particularly to bending, is more important than its ability to resist a normal force.

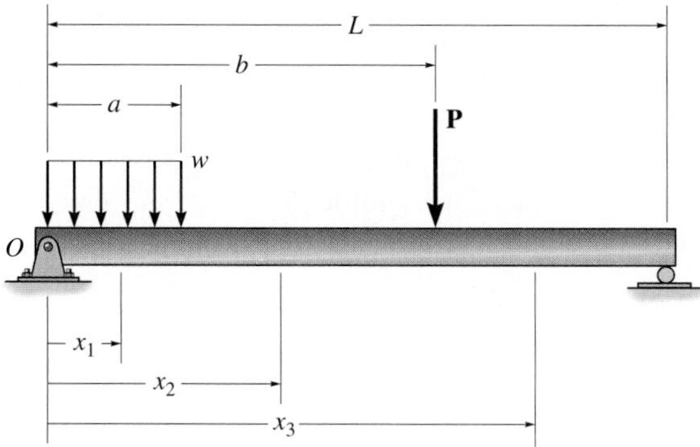

Fig. 7–10

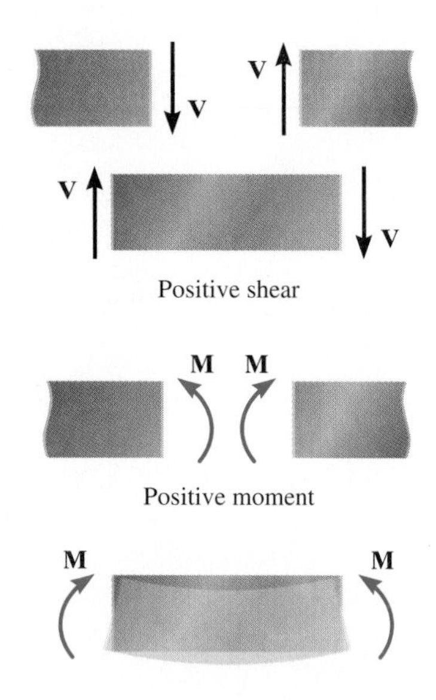

Beam sign convention
Fig. 7–11

Sign Convention. Before presenting a method for determining the shear and bending moment as functions of x and later plotting these functions (shear and bending-moment diagrams), it is first necessary to establish a *sign convention* so as to define a ''positive'' and ''negative'' shear force and bending moment acting in the beam. [This is analogous to assigning coordinate directions x positive to the right and y positive upward when plotting a function $y = f(x)$.] Although the choice of a sign convention is arbitrary, here we will choose the one used for the majority of engineering applications. It is illustrated in Fig. 7–11. Here the positive directions are denoted by an internal *shear force* that causes *clockwise rotation* of the member on which it acts, and an internal *moment* that causes *compression, or pushing on the upper part* of the member. Also, positive moment would tend to bend the member if it were elastic, concave upward. Loadings that are opposite to these are considered negative.

PROCEDURE FOR ANALYSIS

The following procedure provides a method for constructing the shear and bending-moment diagrams for a beam.

Support Reactions. Determine all the reactive forces and couple moments acting on the beam, and resolve all the forces into components acting perpendicular and parallel to the beam's axis.

Shear and Moment Functions. Specify separate coordinates x having an origin at the beam's *left end* and extending to regions of the beam *between* concentrated forces and/or couple moments, or where there is no discontinuity of distributed loading. Section the beam perpendicular to its axis at each distance x. On the free-body diagram, be sure **V** and **M** are shown acting in their *positive sense,* in accordance with the sign convention given in Fig. 7–11. V is obtained by summing forces perpendicular to the beam's axis, and M is obtained by summing moments about the sectioned end of the segment.

Shear and Moment Diagrams. Plot the shear diagram (V versus x) and the moment diagram (M versus x). If computed values of the functions describing V and M are *positive,* the values are plotted above the x axis, whereas *negative* values are plotted below the x axis. Generally, it is convenient to plot the shear and bending-moment diagrams directly below the free-body diagram of the beam.

The following examples illustrate this procedure numerically.

Example 7–7

Draw the shear and bending-moment diagrams for the shaft shown in Fig. 7–12*a*. The support at *A* is a thrust bearing and at *C* a journal bearing.

SOLUTION

Support Reactions. The support reactions have been computed, as shown on the shaft's free-body diagram, Fig. 7–12*d*.

Shear and Moment Functions. The shaft is sectioned at an arbitrary distance *x* from point *A*, extending within the region *AB*, and the free-body diagram of the left segment is shown in Fig. 7–12*b*. The unknowns **V** and **M** are assumed to act in the *positive sense* on the right-hand face of the segment according to the established sign convention. Why? Applying the equilibrium equations yields

$$+\uparrow \Sigma F_y = 0; \qquad V = 2.5 \text{ kN} \qquad (1)$$

$$\circlearrowleft + \Sigma M = 0; \qquad M = 2.5x \text{ kN} \cdot \text{m} \qquad (2)$$

A free-body diagram for a left segment of the shaft extending a distance *x* within the region *BC* is shown in Fig. 7–12*c*. As always, **V** and **M** are shown acting in the positive sense. Hence,

$$+\uparrow \Sigma F_y = 0; \qquad 2.5 \text{ kN} - 5 \text{ kN} - V = 0$$

$$V = -2.5 \text{ kN} \qquad (3)$$

$$\circlearrowleft + \Sigma M = 0; \qquad M + 5 \text{ kN}(x - 2 \text{ m}) - 2.5 \text{ kN}(x) = 0$$

$$M = (10 - 2.5x) \text{ kN} \cdot \text{m} \qquad (4)$$

Shear and Moment Diagrams. When Eqs. 1 through 4 are plotted within the regions in which they are valid, the shear and bending-moment diagrams shown in Fig. 7–12*d* are obtained. The shear diagram indicates that the internal shear force is always 2.5 kN (positive) within shaft segment *AB*. Just to the right of point *B*, the shear force changes sign and remains at a constant value of −2.5 kN for segment *BC*. The moment diagram starts at zero, increases linearly to point *B* at $x = 2$ m, where $M_{max} = 2.5 \text{ kN}(2 \text{ m}) = 5 \text{ kN} \cdot \text{m}$, and thereafter decreases back to zero.

It is seen in Fig. 7–12*d* that the graph of the shear and moment diagrams is discontinuous at points of concentrated force, i.e., points *A*, *B*, and *C*. For this reason, as stated earlier, it is necessary to express both the shear and bending-moment functions separately for regions between concentrated loads. It should be realized, however, that all loading discontinuities are mathematical, arising from the *idealization of a concentrated force and couple moment.* Physically, loads are always applied over a finite area, and if this load variation could be accounted for, the shear and bending-moment diagrams would actually be continuous over the shaft's entire length.

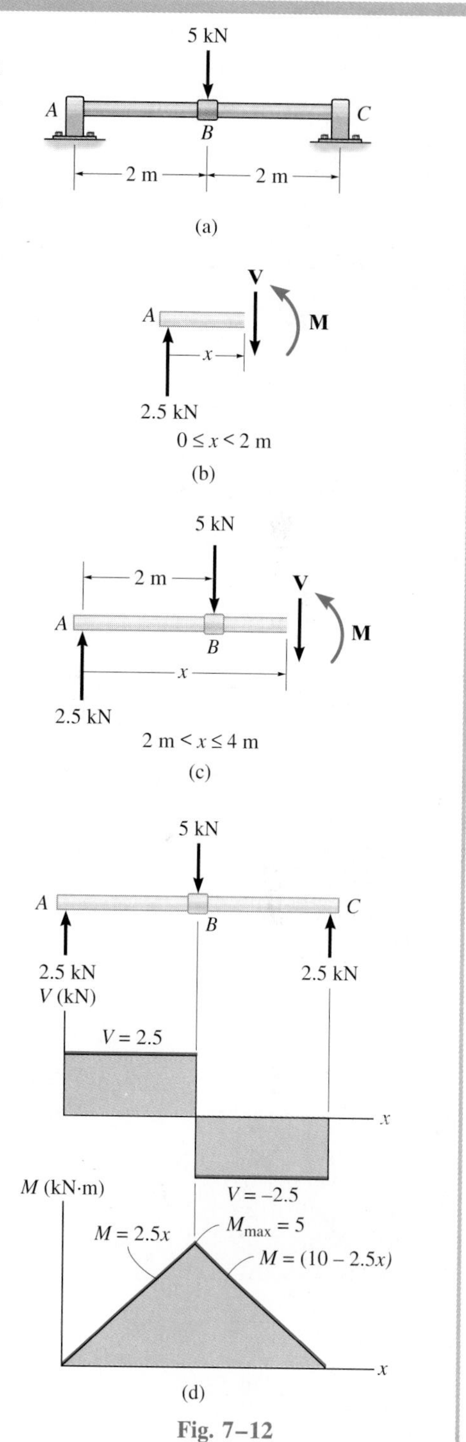

Fig. 7–12

Example 7–8

Draw the shear and bending-moment diagrams for the beam shown in Fig. 7–13*a*.

6 kN/m

9 m

(a)

SOLUTION

Support Reactions. The support reactions have been computed as shown on the beam's free-body diagram, Fig. 7–13*c*.

Shear and Moment Functions. A free-body diagram for a left segment of the beam having a length x is shown in Fig. 7–13*b*. The distributed loading acting on this segment has an intensity of $\frac{2}{3}x$ at its end and is replaced by a resultant force *after* the segment is isolated as a free-body diagram. The *magnitude* of the resultant force is equal to $\frac{1}{2}(x)(\frac{2}{3}x) = \frac{1}{3}x^2$. This force *acts through the centroid* of the distributed loading area, a distance $\frac{1}{3}x$ from the right end. Applying the two equations of equilibrium yields

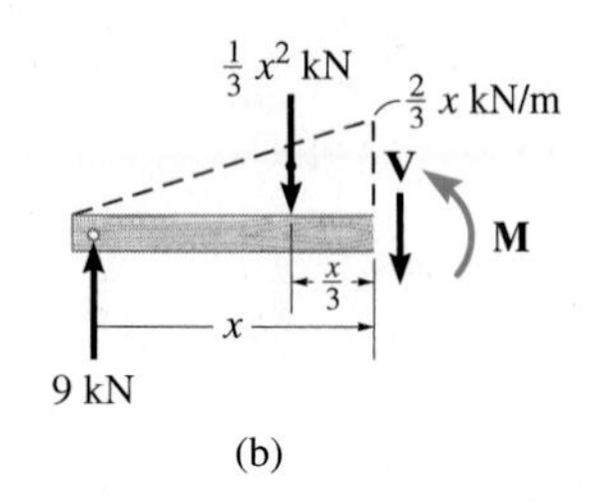

(b)

$$+\uparrow \Sigma F_y = 0; \qquad 9 - \frac{1}{3}x^2 - V = 0$$

$$V = \left(9 - \frac{x^2}{3}\right) \text{ kN} \tag{1}$$

$$\circlearrowleft + \Sigma M = 0; \qquad M + \frac{1}{3}x^2\left(\frac{x}{3}\right) - 9x = 0$$

$$M = \left(9x - \frac{x^3}{9}\right) \text{ kN} \cdot \text{m} \tag{2}$$

Shear and Moment Diagrams. The shear and bending-moment diagrams shown in Fig. 7–13*c* are obtained by plotting Eqs. 1 and 2.

The point of *zero shear* can be found using Eq. 1:

$$V = 9 - \frac{x^2}{3} = 0$$

$$x = 5.20 \text{ m}$$

This value of x happens to represent the point on the beam where the *maximum moment* occurs (see Sec. 7.3). Using Eq. (2), we have

$$M_{\text{max}} = \left(9(5.20) - \frac{(5.20)^3}{9}\right) \text{ kN} \cdot \text{m}$$

$$= 31.2 \text{ kN} \cdot \text{m}$$

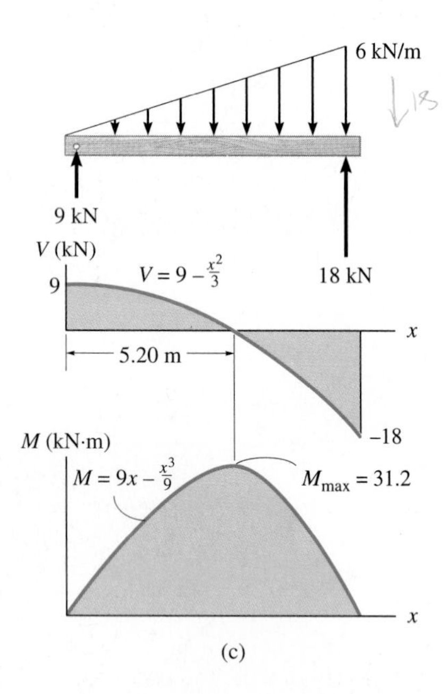

(c)

Fig. 7–13

PROBLEMS

For each of the following problems, establish the x axis with the origin at the left side of the beam, and obtain the internal shear and moment as a function of x. Use these results to plot the shear and moment diagrams.

7–34. Draw the shear and moment diagrams for the shaft (a) in terms of the parameters shown; (b) set $P = 9$ kN, $a = 2$ m, $L = 6$ m. There is a thrust bearing at A and a journal bearing at B.

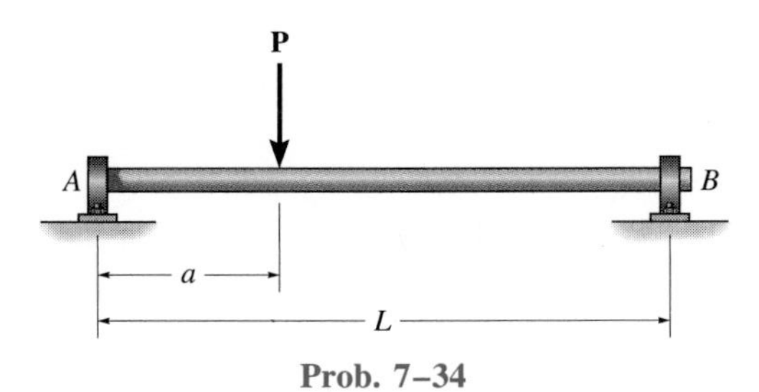

Prob. 7–34

7–35. Draw the shear and moment diagrams for the beam (a) in terms of the parameters shown; (b) set $P = 800$ lb, $a = 5$ ft, $L = 12$ ft.

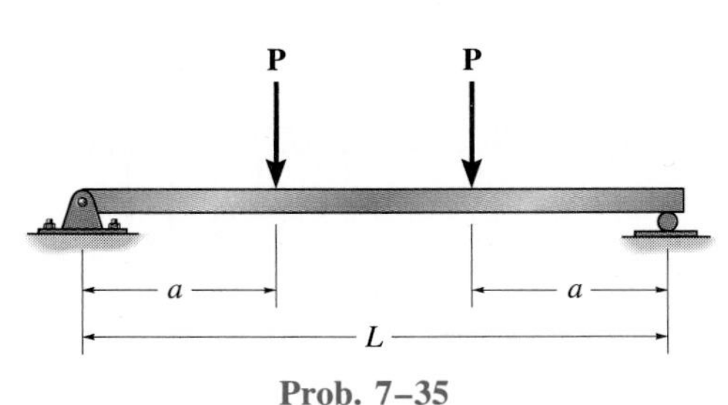

Prob. 7–35

***7–36.** Draw the shear and moment diagrams for the beam (a) in terms of the parameters shown; (b) set $M_0 = 500$ N · m, $L = 8$ m.

7–37. If $L = 9$ m, the beam will fail when the maximum shear force is $V_{max} = 5$ kN or the maximum bending moment is $M_{max} = 2$ kN · m. Determine the magnitude M_0 of the largest couple moments it will support.

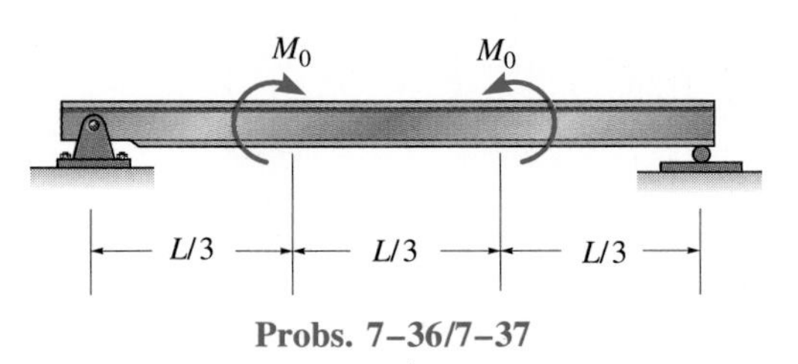

Probs. 7–36/7–37

7–38. The shaft is supported by a thrust bearing at A and a journal bearing at B. Draw the shear and moment diagrams for the shaft (a) in terms of the parameters shown; (b) set $w = 500$ lb/ft, $L = 10$ ft.

7–39. The shaft is supported by a thrust bearing at A and a journal bearing at B. If $L = 10$ ft, the shaft will fail when the maximum moment is $M_{max} = 5$ kip · ft. Determine the largest uniform distributed load w the shaft will support.

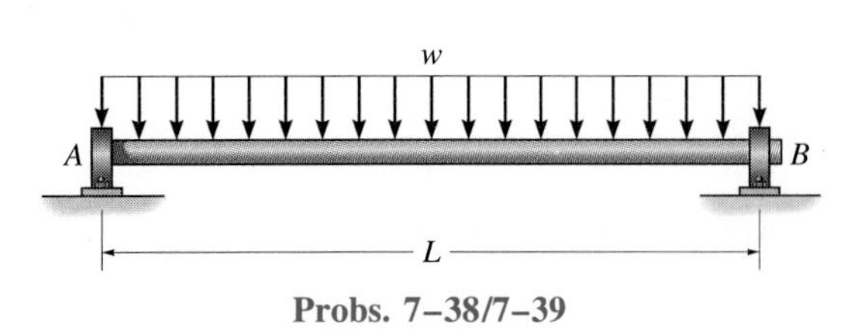

Probs. 7–38/7–39

***7–40.** The shaft is supported by a journal bearing at A and a thrust bearing at B. Draw the shear and moment diagrams for the shaft.

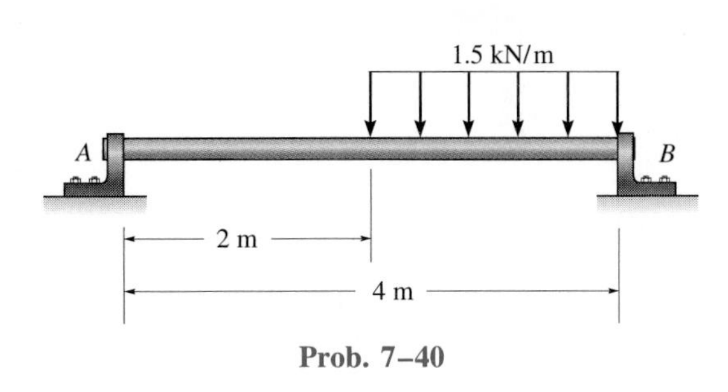

Prob. 7–40

7–41. Draw the shear and moment diagrams for the beam.

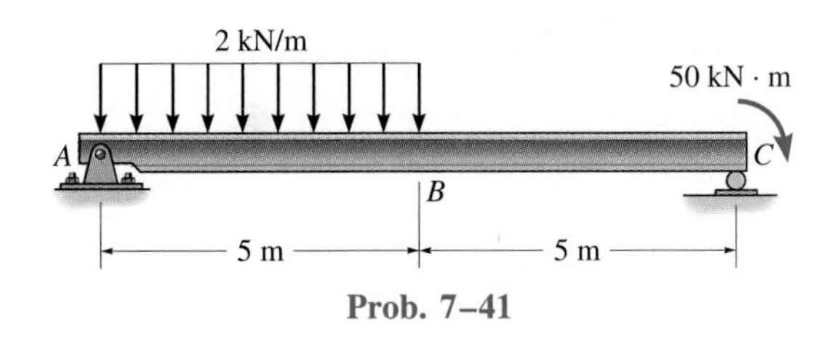

Prob. 7–41

7–42. Draw the shear and moment diagrams for the beam.

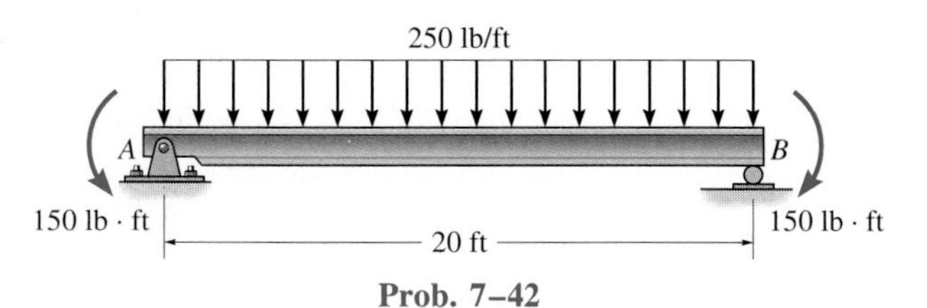

Prob. 7–42

7–43. Draw the shear and moment diagrams for the beam.

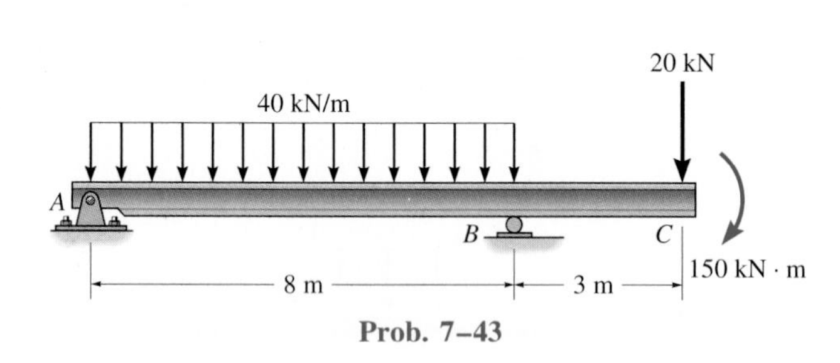

Prob. 7–43

***7–44.** Draw the shear and moment diagrams for the beam.

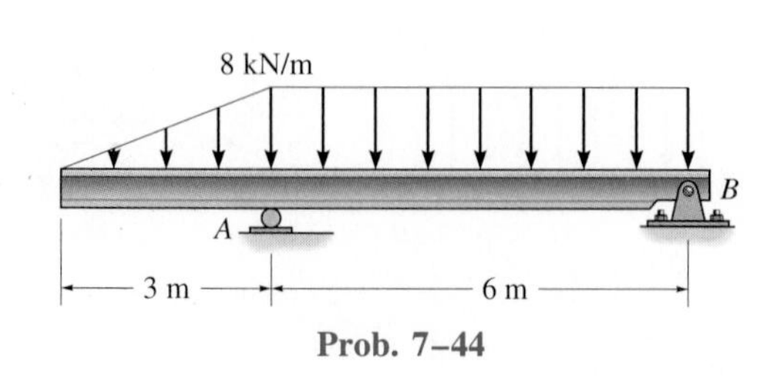

Prob. 7–44

7–45. Draw the shear and moment diagrams for beam *ABC*. Note that there is a pin at *B*. Solve the problem (a) in terms of the parameters shown; (b) set $w = 5$ kN/m, $L = 12$ m.

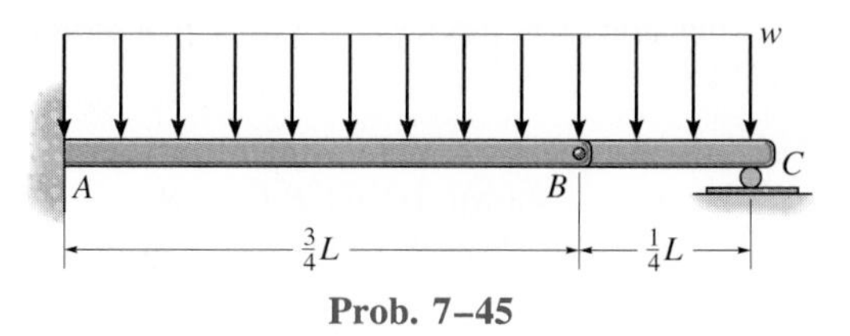

Prob. 7–45

7–46. Draw the shear and moment diagrams for the beam.

Prob. 7–46

7–47. Determine the distance *a* between the bearings in terms of the shaft's length *L* so that the moment in the *symmetric* shaft is zero at its center.

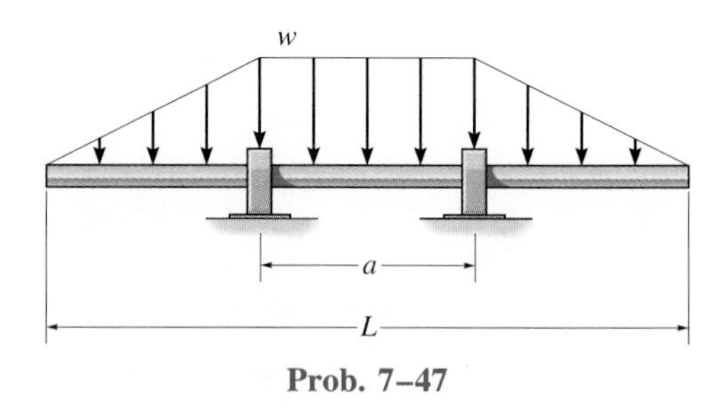

Prob. 7–47

***7–48.** Draw the shear and moment diagrams for the beam (a) in terms of the parameters shown; (b) set $w = 250$ lb/ft, $L = 12$ ft.

7–49. If $L = 18$ ft, the beam will fail when the maximum shear force is $V_{max} = 800$ lb, or the maximum moment is $M_{max} = 1200$ lb · ft. Determine the largest intensity w of the distributed loading it will support.

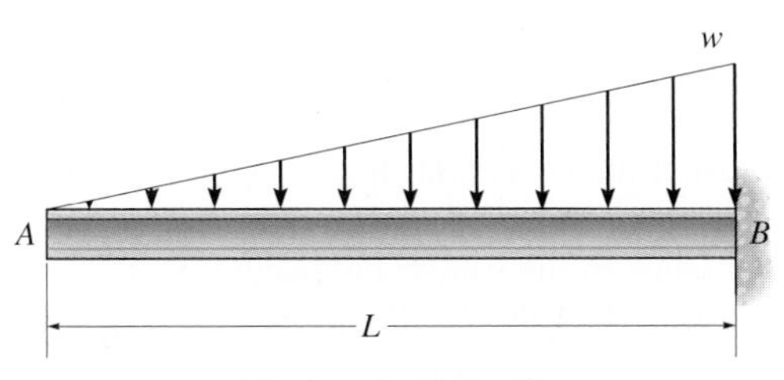

Probs. 7–48/7–49

7–50. The beam will fail when the maximum internal moment is M_{max}. Determine the position x of the concentrated force **P** and its smallest magnitude that will cause failure.

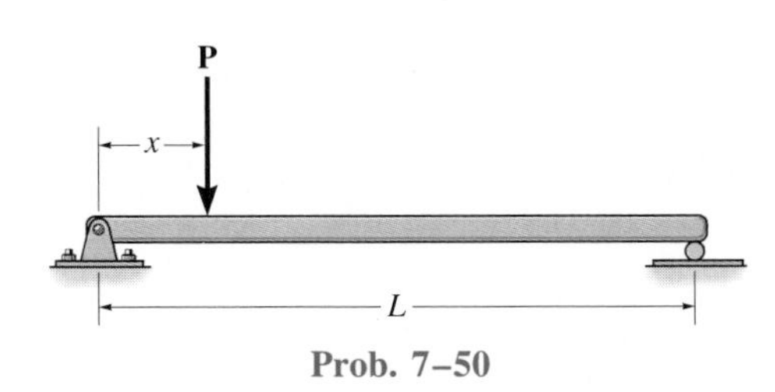

Prob. 7–50

7–51. Determine the distance a as a fraction of the beam's length L for locating the roller support so that the moment in the beam at B is zero.

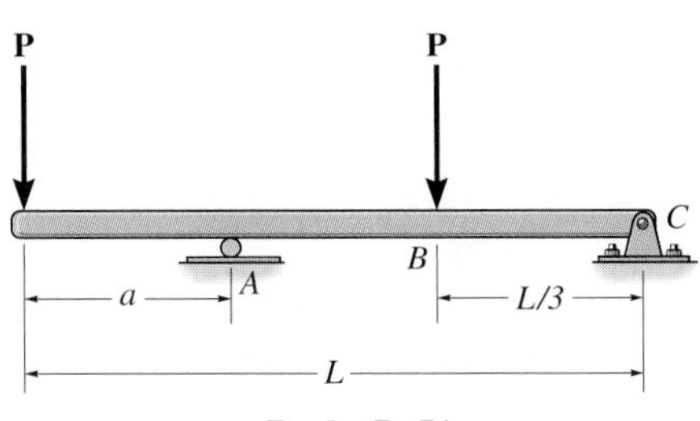

Prob. 7–51

***7–52.** Express the x, y, z components of internal loading in the rod as a function of y, where $0 \leq y \leq 4$ ft.

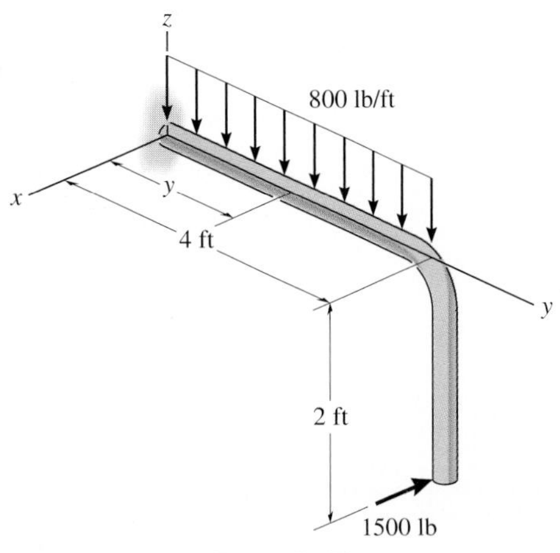

Prob. 7–52

7–53. Determine the normal force, shear force, and moment in the curved rod as a function of θ.

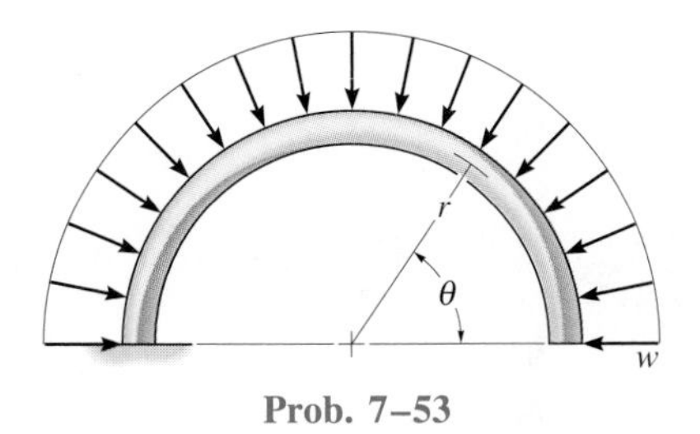

Prob. 7–53

7–54. Determine the normal force, shear force, and moment in the curved rod as a function of θ.

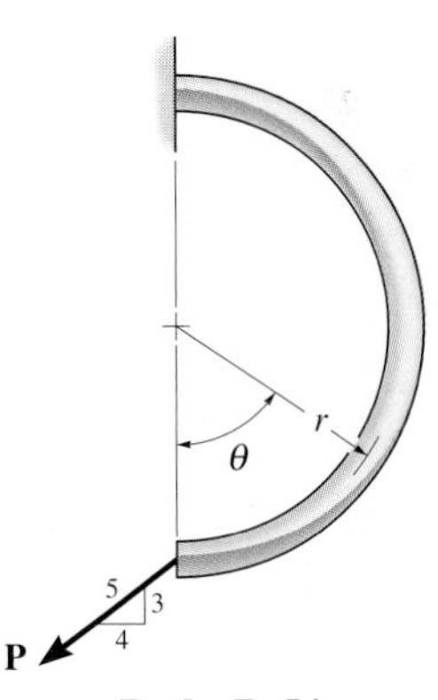

Prob. 7–54

*7.3 Relations Between Distributed Load, Shear, and Moment

In cases where a beam is subjected to several concentrated forces, couple moments, and distributed loads, the method of constructing the shear and bending-moment diagrams discussed in Sec. 7.2 may become quite tedious. In this section a simpler method for constructing these diagrams is discussed—a method based on differential relations that exist between the load, shear, and bending moment.

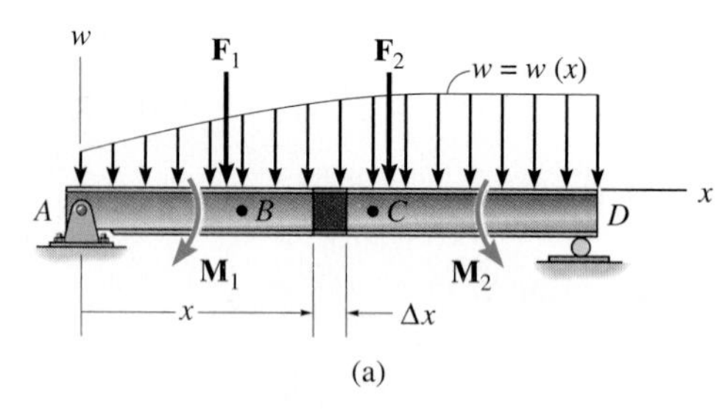

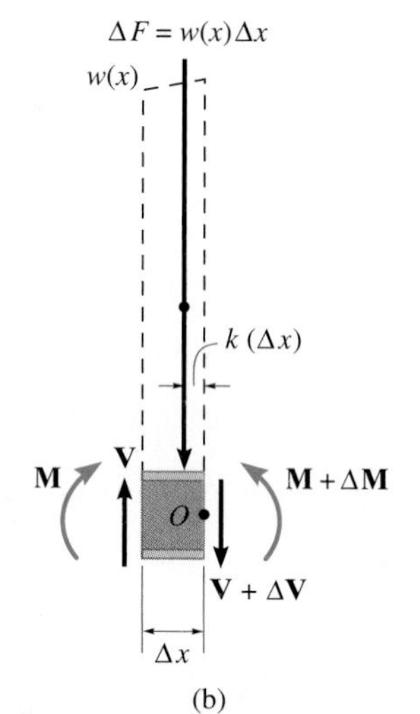

Fig. 7–14

Distributed Load. Consider the beam *AD* shown in Fig. 7–14*a*, which is subjected to an arbitrary distributed load $w = w(x)$ and a series of concentrated forces and couple moments. In the following discussion, the *distributed load* will be considered *positive* when the *loading acts downward* as shown. The free-body diagram for a small segment of the beam having a length Δx is shown in Fig. 7–14*b*. Since this segment has been chosen at a point x along the beam which is *not* subjected to a concentrated force or couple moment, any results obtained will not apply at points of concentrated loading. The internal shear force and bending moment shown on the free-body diagram are assumed to act in the *positive sense* according to the established sign convention, Fig. 7–14*b*. Note that both the shear force and moment acting on the right-hand face must be increased by a small, finite amount in order to keep the segment in equilibrium. The distributed loading has been replaced by a resultant force $\Delta F = w(x)\,\Delta x$ that acts at a fractional distance $k\,(\Delta x)$ from the right end, where $0 < k < 1$ [for example, if $w(x)$ is *uniform*, $k = \frac{1}{2}$]. Applying the equations of equilibrium, we have

$$+\uparrow \Sigma F_y = 0; \qquad V - w(x)\,\Delta x - (V + \Delta V) = 0$$
$$\Delta V = -w(x)\,\Delta x$$

$$\circlearrowright\!+\Sigma M_O = 0; \quad -V\,\Delta x - M + w(x)\,\Delta x[k(\Delta x)] + (M + \Delta M) = 0$$
$$\Delta M = V\,\Delta x - w(x)k(\Delta x)^2$$

Dividing by Δx and taking the limit as $\Delta x \rightarrow 0$, these two equations become

$$\frac{dV}{dx} = -w(x) \tag{7–1}$$

Slope of shear diagram = Negative of distributed load intensity

$$\frac{dM}{dx} = V$$

$$\text{Slope of moment diagram} = \text{Shear} \tag{7-2}$$

These two equations provide a convenient means for plotting the shear and moment diagrams for a beam. At a specific point in a beam, Eq. 7–1 states that the *slope of the shear diagram is equal to the negative of the intensity of the distributed load,* while Eq. 7–2 states that the *slope of the moment diagram is equal to the shear.* In particular, if the shear is equal to zero, $dM/dx = 0$, and therefore *a point of zero shear corresponds to a point of maximum (or possibly minimum) moment.*

Equations 7–1 and 7–2 may also be rewritten in the form $dV = -w(x)\,dx$ and $dM = V\,dx$. Noting that $w(x)\,dx$ and $V\,dx$ represent differential areas under the distributed-loading and shear diagrams, respectively, we can integrate these areas between two points B and C along the beam, Fig. 7–14a, and write

$$\Delta V_{BC} = -\int w(x)\,dx$$

$$\text{Change in shear} = \text{Negative of area under loading curve} \tag{7-3}$$

and

$$\Delta M_{BC} = \int V\,dx$$

$$\text{Change in moment} = \text{Area under shear diagram} \tag{7-4}$$

Equation 7–3 states that the *change in shear between points B and C is equal to the negative of the area under the distributed-loading curve between these points.* Similarly, from Eq. 7–4, the *change in moment between B and C is equal to the area under the shear diagram within region BC.* Because two integrations are involved, first to determine the change in shear, Eq. 7–3, then to determine the change in moment, Eq. 7–4, we can state that if the loading curve $w = w(x)$ is a polynomial of degree n, then $V = V(x)$ will be a curve of degree $n + 1$, and $M = M(x)$ will be a curve of degree $n + 2$.

As stated previously, the above equations do not apply at points where a *concentrated* force or couple moment acts. These two special cases create *discontinuities* in the shear and moment diagrams, and as a result, each deserves separate treatment.

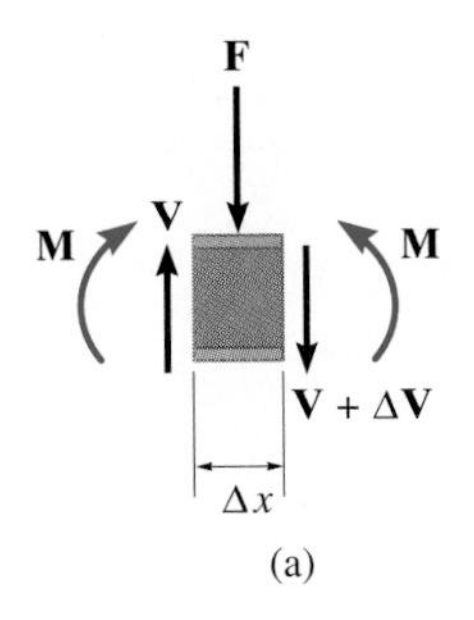

(a)

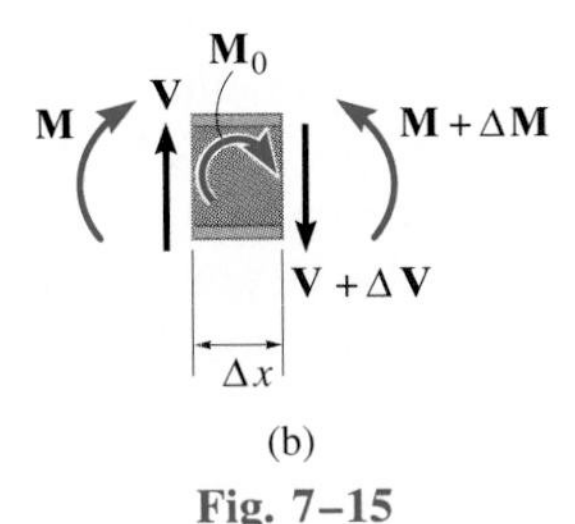

(b)

Fig. 7–15

Force. A free-body diagram of a small segment of the beam in Fig. 7–14*a*, taken from under one of the forces, is shown in Fig. 7–15*a*. Here it can be seen that force equilibrium requires

$$+\uparrow \Sigma F_y = 0; \qquad \Delta V = -F \qquad (7\text{–}5)$$

Thus, the *change in shear is negative,* so that on the shear diagram the shear "jumps" *downward when* **F** *acts downward* on the beam. Likewise, the jump in shear (ΔV) is upward when **F** acts upward.

Couple Moment. If we remove a segment of the beam in Fig. 7–14*a* that is located at the couple moment, the free-body diagram shown in Fig. 7–15*b* results. In this case letting $\Delta x \rightarrow 0$, moment equilibrium requires

$$\circlearrowright + \Sigma M = 0; \qquad \Delta M = M_0 \qquad (7\text{–}6)$$

Thus, the *change in moment is positive,* or the moment diagram "jumps" *upward if* $\mathbf{M}_0$ *is clockwise.* Likewise, the jump ΔM is downward when $\mathbf{M}_0$ is counterclockwise.

The following examples illustrate application of the above equations for the construction of the shear and moment diagrams. After working through these examples, it is recommended that Examples 7–7 and 7–8 be solved using this method.

Example 7–9

Draw the shear and bending-moment diagrams for the beam shown in Fig. 7–16*a*.

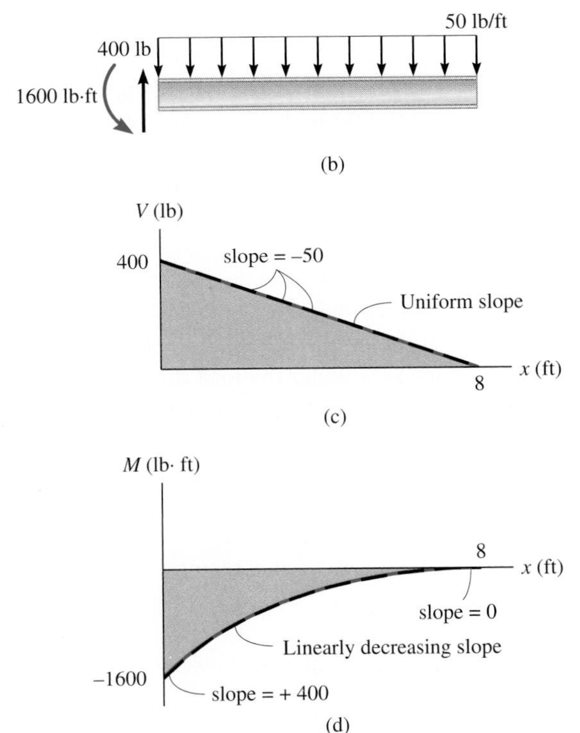

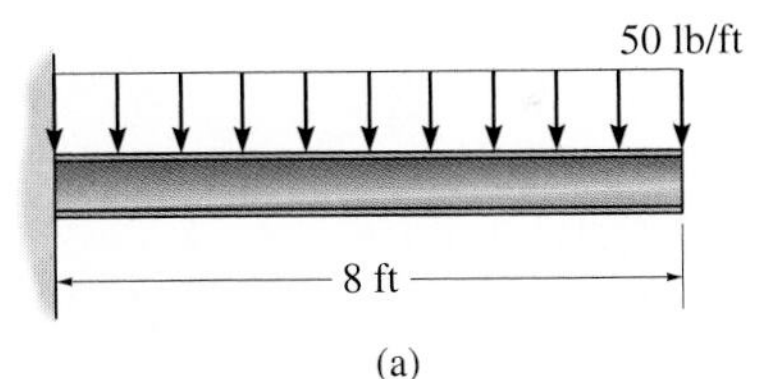

Fig. 7–16

SOLUTION

Support Reactions. The reactions at the fixed support have been calculated and are shown on the free-body diagram of the beam, Fig. 7–16*b*.

Shear Diagram. The shear at the end points is plotted first, Fig. 7–16*c*. From the sign convention, Fig. 7–11, $V = +400$ at $x = 0$ and $V = 0$ at $x = 8$. Since $dV/dx = -w = -50$, a straight, *negative* sloping line connects the end points.

Moment Diagram. From our sign convention, Fig. 7–11, the moments at the beam's end points, $M = -1600$ at $x = 0$ and $M = 0$ at $x = 8$, are plotted first, Fig. 7–16*d*. Successive values of shear taken from the shear diagram, Fig. 7–16*c*, indicate that the *slope* $dM/dx = V$ of the moment diagram, Fig. 7–16*d*, is always positive yet *linearly decreasing* from $dM/dx = 400$ at $x = 0$ to $dM/dx = 0$ at $x = 8$. Thus, due to the integrations, w a constant becomes V a sloping line (first-degree curve) and M a parabola (second-degree curve).

Example 7–10

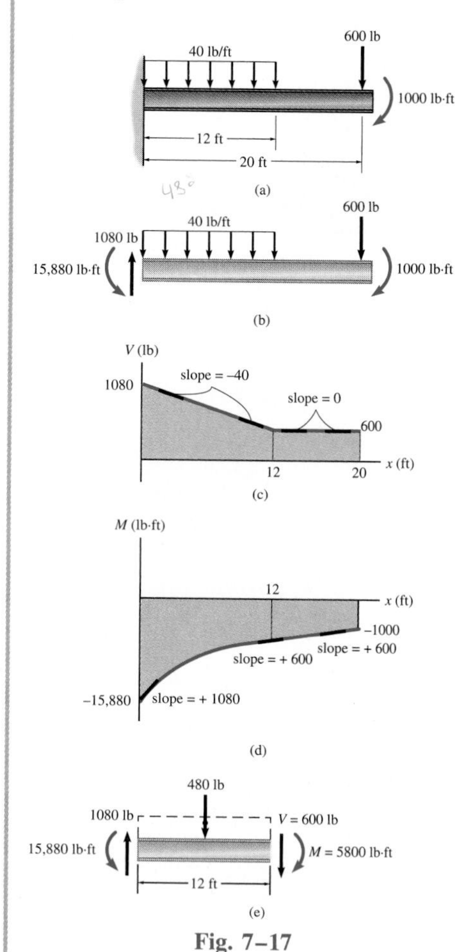

Fig. 7–17

Draw the shear and bending-moment diagrams for the beam shown in Fig. 7–17*a*.

SOLUTION

Support Reactions. The reactions at the fixed support have been calculated and are shown on the free-body diagram of the beam, Fig. 7–17*b*.

Shear Diagram. Using the established sign convention, Fig. 7–11, the shear at the ends of the beam is plotted first; i.e., $x = 0$, $V = +1080$; $x = 20$, $V = +600$, Fig. 7–17*c*.

Since the uniform distributed load is downward and *constant,* the slope of the shear diagram is $dV/dx = -w = -40$ for $0 \leq x < 12$ as indicated.

The magnitude of shear at $x = 12$ is $V = +600$. This can be determined by first finding the area under the load diagram between $x = 0$ and $x = 12$. This represents the change in shear. That is, $\Delta V = -\int w(x)\,dx = -40(12) = -480$. Thus $V|_{x=12} = V|_{x=0} + (-480) = 1080 - 480 = 600$. Also, we can obtain this value by using the method of sections, Fig. 7–17*e*, where for equilibrium $V = +600$.

Since the load between $12 < x \leq 20$ is $w = 0$, the slope $dV/dx = 0$ as indicated. This brings the shear to the required value of $V = +600$ at $x = 20$.

Moment Diagram. Again, using the established sign convention, Fig. 7–11, the moments at the ends of the beam are plotted first; i.e., $x = 0$, $M = -15{,}880$; $x = 20$, $M = -1000$, Fig. 7–17*d*.

Each value of shear gives the slope of the moment diagram since $dM/dx = V$. As indicated, at $x = 0$, $dM/dx = +1080$; and at $x = 12$, $dM/dx = +600$. For $0 \leq x < 12$, specific values of the shear diagram are positive but linearly decreasing. Hence, the moment diagram is parabolic with a linear, decreasing, positive slope.

The magnitude of moment at $x = 12$ is -5800. This can be found by first determining the trapezoidal area under the shear diagram, which represents the change in moment, $\Delta M = \int V dx = 600(12) + \frac{1}{2}(1080 - 600)(12) = +10{,}080$. Thus, $M|_{x=12} = M|_{x=0} + 10{,}080 = -15{,}880 + 10{,}080 = -5800$. The more "basic" method of sections can also be used, where equilibrium at $x = 12$ requires $M = -5800$, Fig. 7–17*e*.

The moment diagram has a constant slope for $12 < x \leq 20$ since, from the shear diagram, $dM/dx = V = +600$. This brings the value of $M = -1000$ at $x = 20$, as required.

Example 7–11

Draw the shear and moment diagrams for the shaft in Fig. 7–18*a*. The support at *A* is a thrust bearing and at *B* a journal bearing.

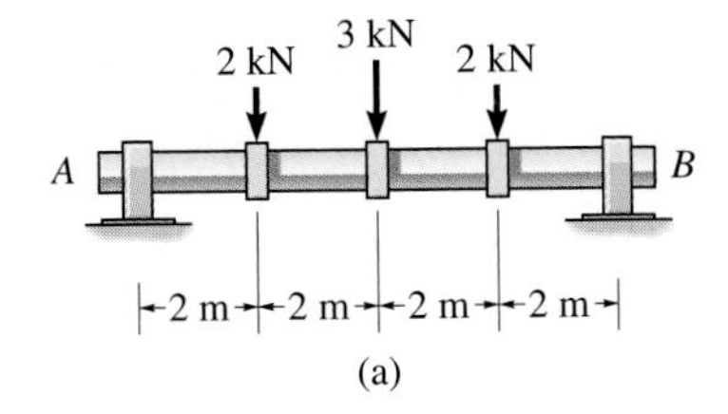

(a)

SOLUTION

Support Reactions. The reactions at the supports are shown on the free-body diagram in Fig. 7–18*b*.

Shear Diagram. The end points $x = 0$, $V = +3.5$ and $x = 8$, $V = -3.5$ are plotted first, as shown in Fig. 7–18*c*.

Since there is no distributed load on the shaft, the slope of the shear diagram throughout the shaft's length is zero; i.e., $dV/dx = -w = 0$. There is a discontinuity or "jump" of the shear diagram, however, at each concentrated force. From Eq. 7–5, $\Delta V = -F$, the change in shear is negative when the force acts downward and positive when the force acts upward. Stated another way, the "jump" follows the force, i.e., a downward force causes a downward jump, and vice versa. Thus, the 2-kN force at $x = 2$ m changes the shear from 3.5 kN to 1.5 kN; the 3-kN force at $x = 4$ m changes the shear from 1.5 kN to -1.5 kN, etc. We can *also* obtain numerical values for the shear at a specified point in the shaft by using the method of sections, as for example, $x = 2^+$ m, $V = 1.5$ kN in Fig. 7–18*e*.

Moment Diagram. The end points $x = 0$, $M = 0$ and $x = 8$, $M = 0$ are plotted first, as shown in Fig. 7–18*d*.

Since the shear is constant in each region of the shaft, the moment diagram has a corresponding constant positive or negative slope as indicated on the diagram. Numerical values for the change in moment at any point can be computed from the *area* under the shear diagram. For example, at $x = 2$ m, $\Delta M = \int V\,dx = 3.5(2) = 7$. Thus, $M|_{x=2} = M|_{x=0} + 7 = 0 + 7 = 7$. Also, by the method of sections, we can determine the moment at a specified point, as for example, $x = 2^+$ m, $M = 7$ kN · m, Fig. 7–18*e*.

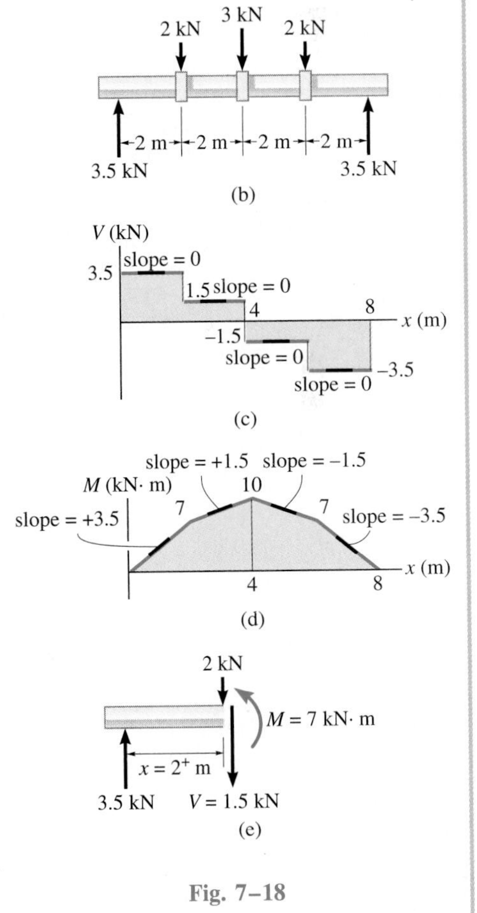

Fig. 7–18

Example 7–12

Sketch the shear and bending-moment diagrams for the beam shown in Fig. 7–19*a*.

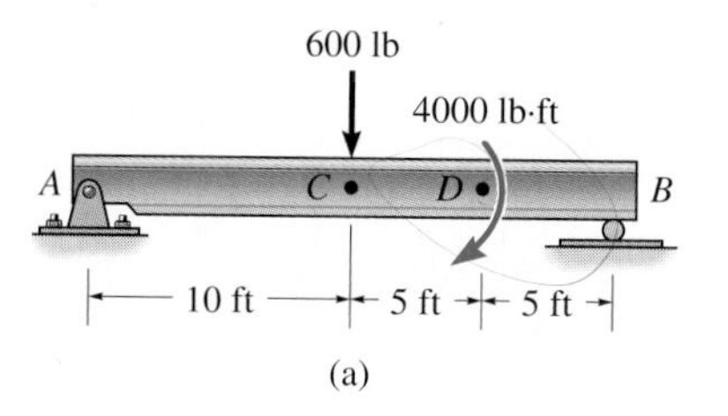

(a)

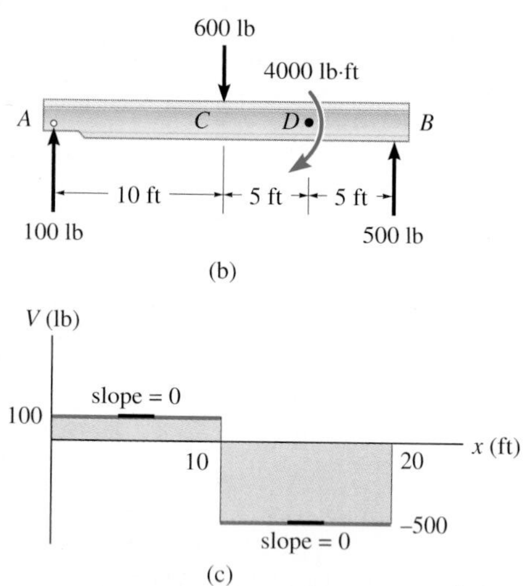

(b)

(c)

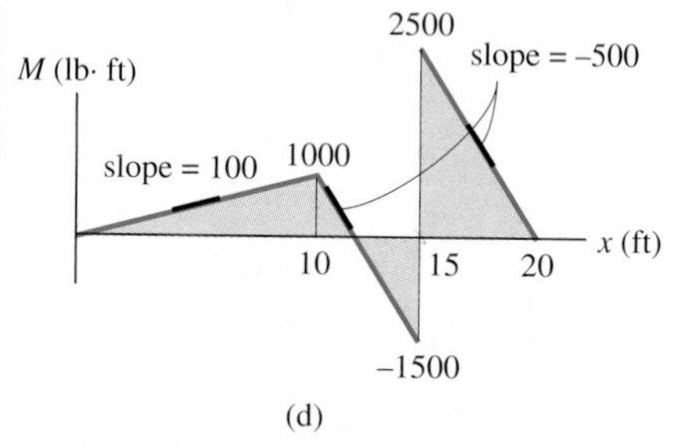

(d)

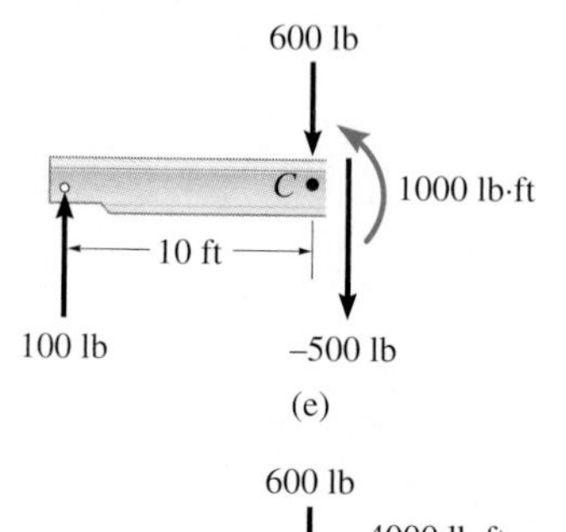

(e)

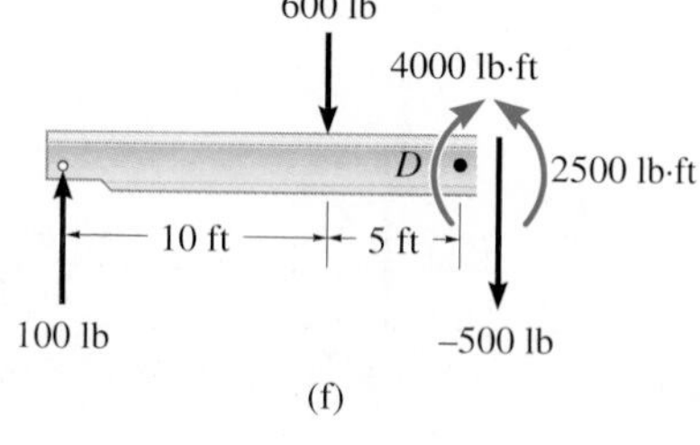

(f)

Fig. 7–19

SOLUTION

Support Reactions. The reactions are calculated and indicated on the free-body diagram, Fig. 7–19*b*.

Shear Diagram. As in Example 7–11, the shear diagram can be constructed by "following the load" on the free-body diagram. In this regard, beginning at A, $V_A = +100$ lb. No load acts between A and C, so the shear remains constant; i.e., $dV/dx = -w(x) = 0$, Fig. 7–19*c*. At C the 600-lb force acts downward, so the shear jumps down 600 lb, from 100 lb to -500 lb. Again the shear is constant (no load) and ends at -500 lb, point B. Notice that no jump or discontinuity in shear occurs at D, the point where the 4000-lb · ft couple moment is applied, Fig. 7–19*b*. This is because, for force equilibrium, $\Delta V = 0$ in Fig. 7–15*b*.

Moment Diagram. The moment at each end of the beam is zero. These two points are plotted first, Fig. 7–19*d*. The slope of the moment diagram from A to C is constant since $dM/dx = V = +100$. The value of the moment at C can be determined by the method of sections, Fig. 7–19*e* where $M_C = +1000$ lb · ft; or by first computing the rectangular area under the shear diagram between A and C to obtain the change in moment $\Delta M_{AC} = (100 \text{ lb})(10 \text{ ft}) = 1000$ lb · ft. Since $M_A = 0$, then $M_C = 0 + 1000$ lb · ft $= 1000$ lb · ft. From C to D the slope of the moment diagram is $dM/dx = V = -500$, Fig. 7–19*c*. The area under the shear diagram between points C and D is $\Delta M_{CD} = M_D - M_C = (-500 \text{ lb})(5 \text{ ft}) = -2500$ lb · ft, so that $M_D = 1000 - 2500 = -1500$ lb · ft. A jump in the moment diagram occurs at point D, which is caused by the concentrated couple moment of 4000 lb · ft. From Eq. 7–6, the jump is *positive* since the couple moment is *clockwise*. Thus, at $x = 15^+$ ft, the moment is $M_D = -1500 + 4000 = 2500$ lb · ft. This value can *also* be determined by the method of sections, Fig. 7–19*f*. From point D the slope of $dM/dx = -500$ is maintained until the diagram closes to zero at B, Fig. 7–19*d*.

PROBLEMS

7–55. Draw the shear and moment diagrams for the beam in Prob. 7–35.

***7–56.** Draw the shear and moment diagrams for the beam in Prob. 7–40.

7–57. Draw the shear and moment diagrams for the beam in Prob. 7–38.

7–58. Draw the shear and moment diagrams for the beam in Prob. 7–42.

7–59. Draw the shear and moment diagrams for the beam in Prob. 7–46.

***7–60.** Draw the shear and moment diagrams for the beam in Prob. 7–48.

7–61. Draw the shear and moment diagrams for the beam.

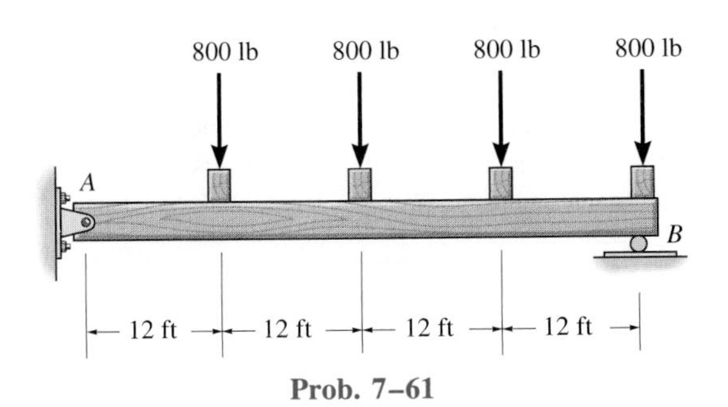

Prob. 7–61

7–62. Draw the shear and moment diagrams for the beam.

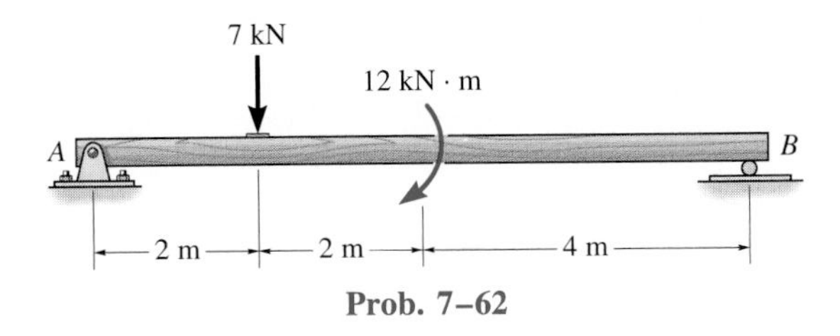

Prob. 7–62

7–63. Draw the shear and moment diagrams for the shaft. The support at A is a thrust bearing and at B it is a journal bearing.

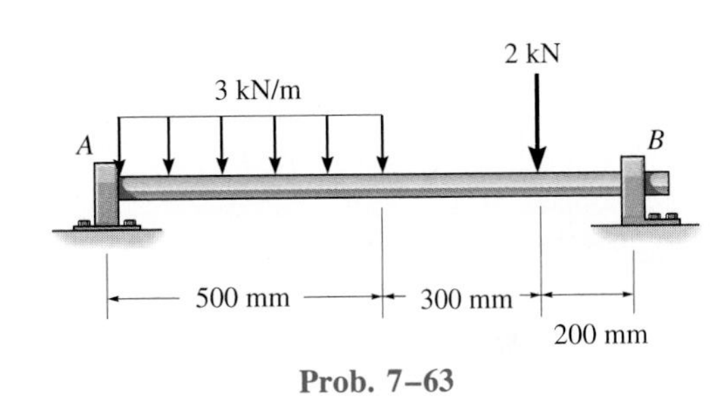

Prob. 7–63

***7–64.** Draw the shear and moment diagrams for the beam.

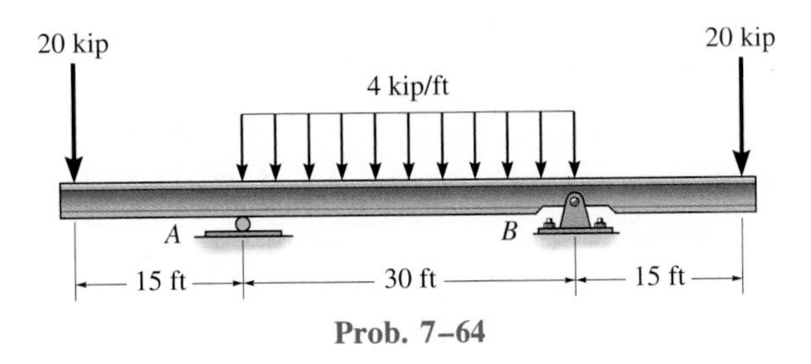

Prob. 7–64

7–65. Draw the shear and moment diagrams for the shaft. The support at A is a journal bearing and at B it is a thrust bearing.

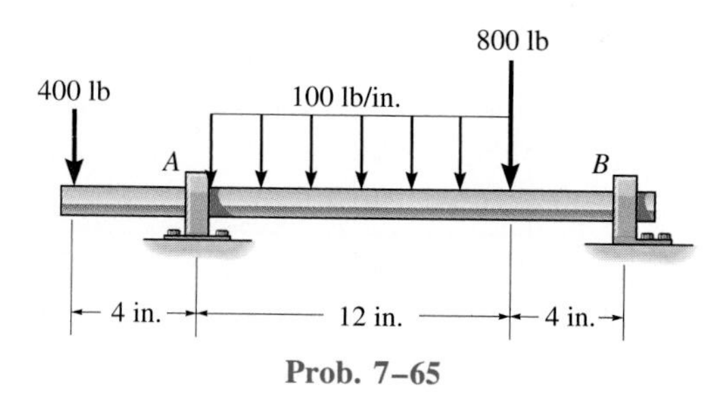

Prob. 7–65

7–66. Draw the shear and moment diagrams for the shaft. The support at A is a journal bearing and at B it is a thrust bearing.

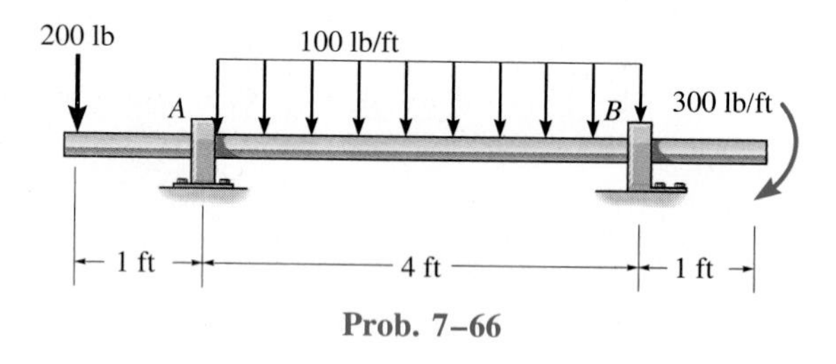

Prob. 7–66

7–67. Draw the shear and moment diagrams for the beam.

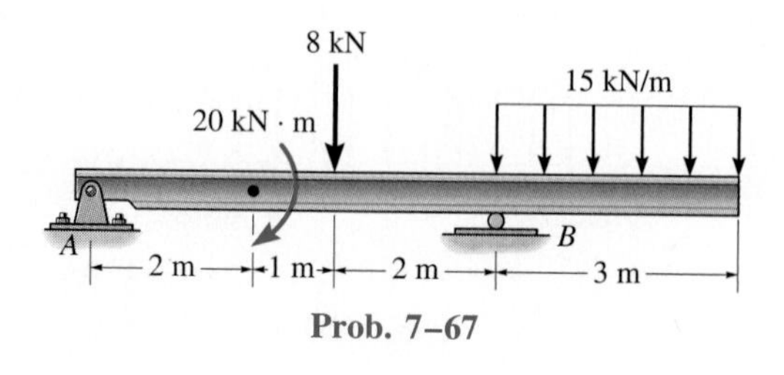

Prob. 7–67

***7–68.** Draw the shear and moment diagrams for the beam (a) in terms of the parameters shown; (b) set $w = 500$ N/m, $L = 3$ m.

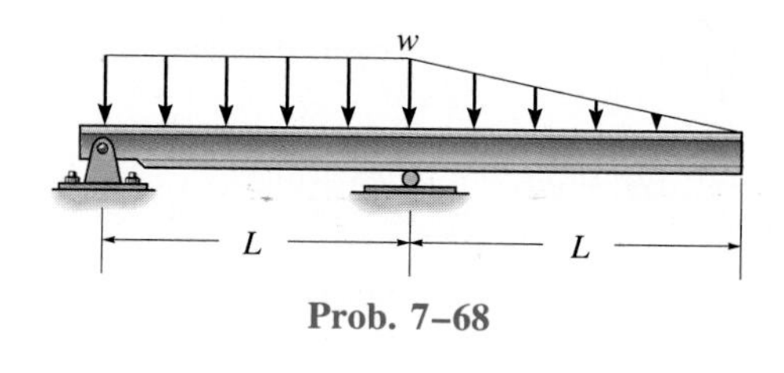

Prob. 7–68

7–69. Draw the shear and moment diagrams for the beam. Take $w = 200$ lb/ft.

7–70. The beam will fail when the maximum moment is $M_{max} =$ 30 kip · ft or the maximum shear is $V_{max} = 8$ kip. Determine the largest distributed load w the beam will support.

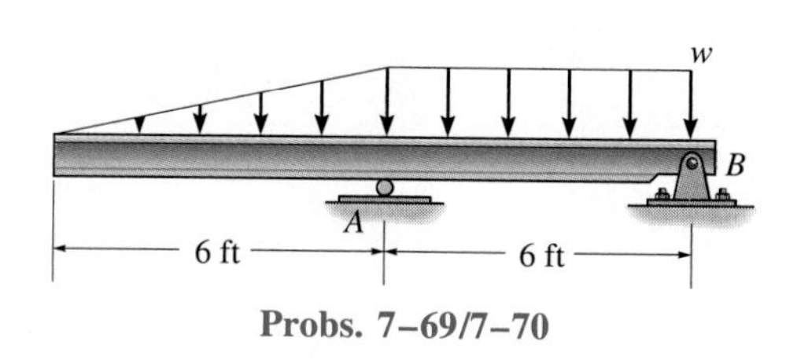

Probs. 7–69/7–70

7–71. Draw the shear and moment diagrams for the beam.

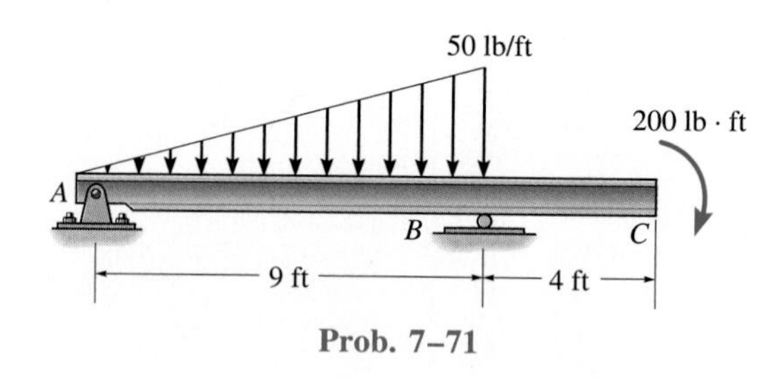

Prob. 7–71

***7–72.** Draw the shear and moment diagrams for the beam.

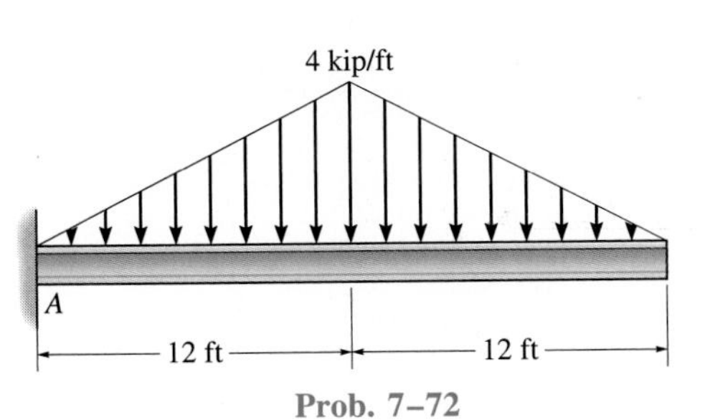

Prob. 7–72

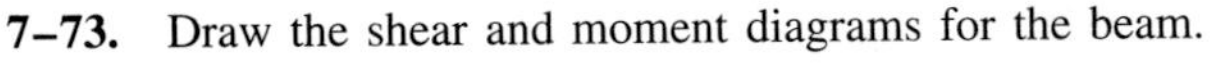
7–73. Draw the shear and moment diagrams for the beam.

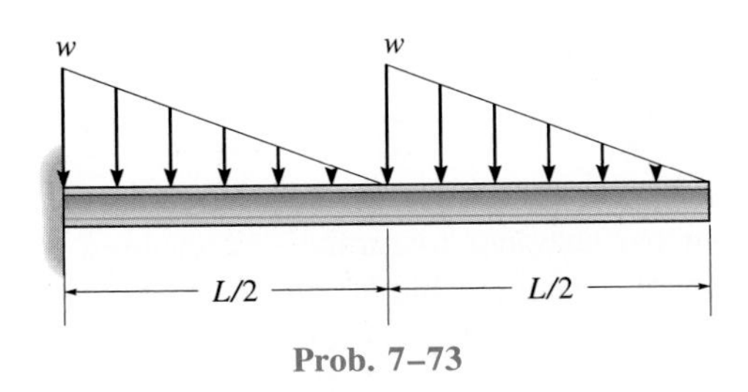

Prob. 7–73

7–74. Draw the shear and moment diagrams for the beam.

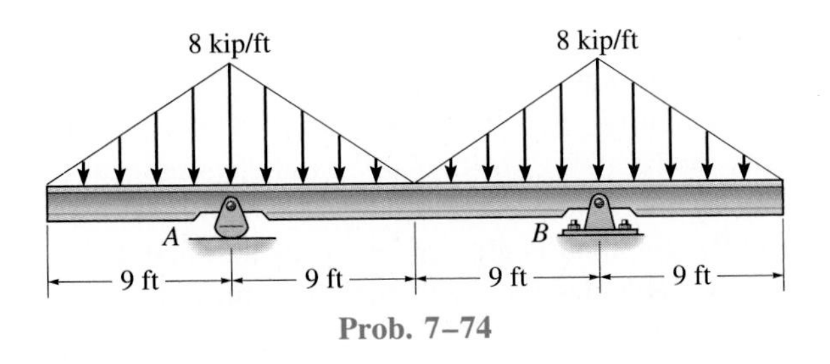

Prob. 7–74

7–75. Draw the shear and moment diagrams for the beam.

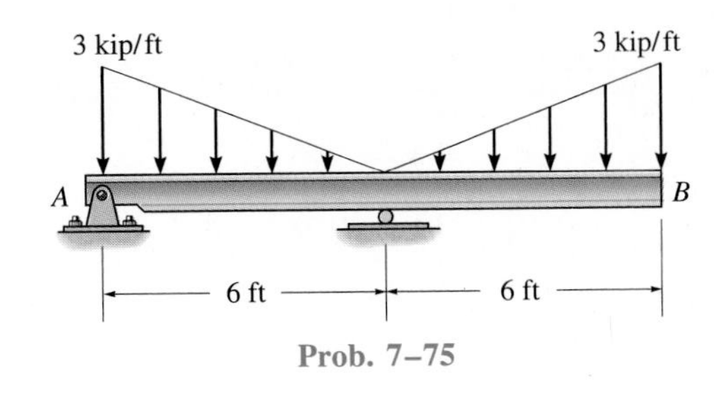

Prob. 7–75

***7–76.** Draw the shear and moment diagrams for the beam.

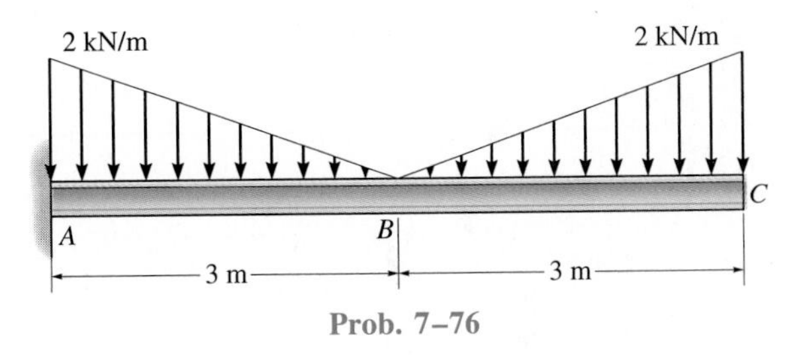

Prob. 7–76

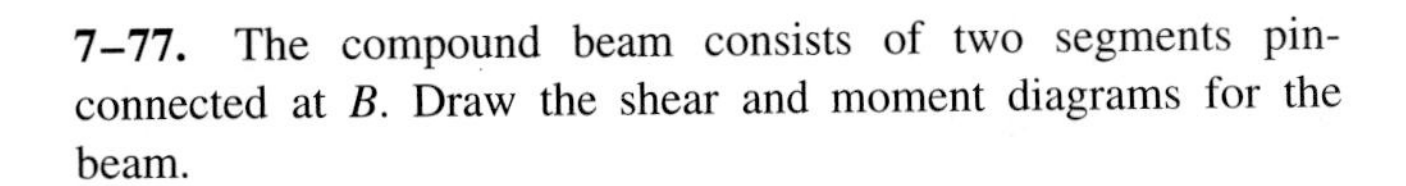
7–77. The compound beam consists of two segments pin-connected at *B*. Draw the shear and moment diagrams for the beam.

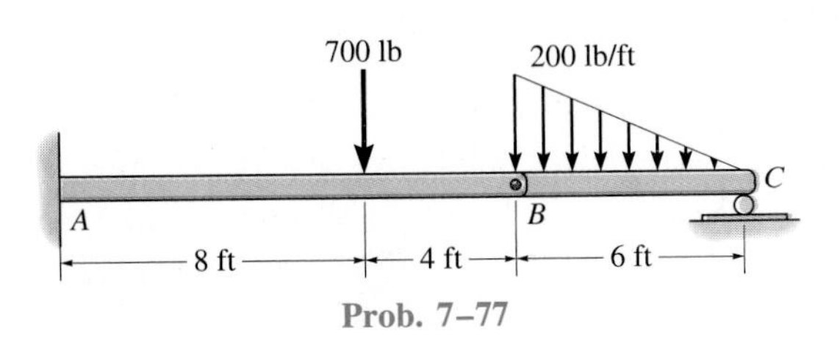

Prob. 7–77

7–78. Draw the shear and moment diagrams for the compound beam. The beam is pin-connected at E and F.

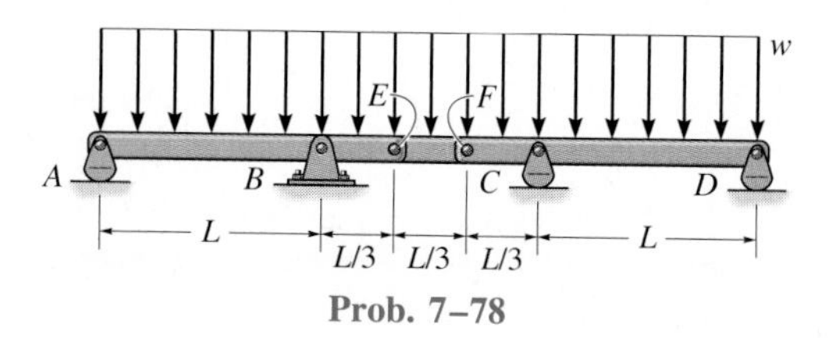

Prob. 7–78

7–79. Draw the shear and moment diagrams for the beam.

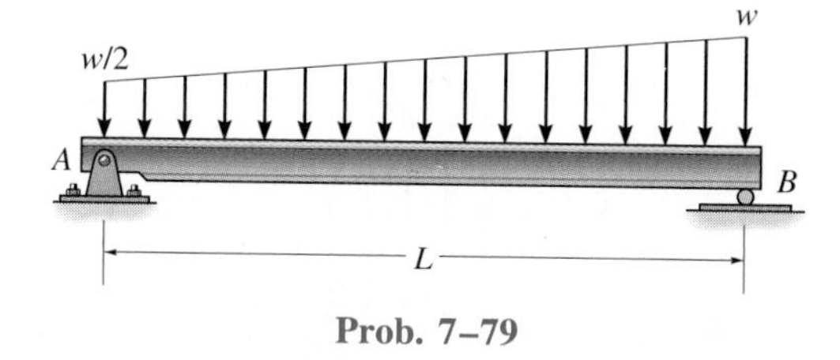

Prob. 7–79

*7.4 Cables

Flexible cables and chains are often used in engineering structures for support and to transmit loads from one member to another. When used to support suspension bridges and trolley wheels, cables form the main load-carrying element of the structure. In the force analysis of such systems, the weight of the cable itself may be neglected because it is often small compared to the load it carries. On the other hand, when cables are used as transmission lines and guys for radio antennas and derricks, the cable weight may become important and must be included in the structural analysis. Three cases will be considered in the analysis that follows: (1) a cable subjected to concentrated loads; (2) a cable subjected to a distributed load; and (3) a cable subjected to its own weight. Regardless of which loading conditions are present, provided the loading is coplanar with the cable, the requirements for equilibrium are formulated in an identical manner.

When deriving the necessary relations between the force in the cable and its slope, we will make the assumption that the cable is *perfectly flexible* and *inextensible*. Due to its flexibility, the cable offers no resistance to bending, and therefore, the tensile force acting in the cable is always tangent to the cable at points along its length. Being inextensible, the cable has a constant length both before and after the load is applied. As a result, once the load is applied, the geometry of the cable remains fixed, and the cable or a segment of it can be treated as a rigid body.

Cable Subjected to Concentrated Loads. When a cable of negligible weight supports several concentrated loads, the cable takes the form of several straight-line segments, each of which is subjected to a constant tensile force. Consider, for example, the cable shown in Fig. 7–20, where the distances h, L_1, L_2, and L_3 and the loads $\mathbf{P}_1$ and $\mathbf{P}_2$ are known. The problem here is to determine the *nine unknowns* consisting of the tension in each of the *three* segments, the *four* components of reaction at A and B, and the sags y_C and y_D at the *two* points C and D. For the solution we can write *two* equations of force equilibrium at each of points A, B, C, and D. This results in a total of *eight equations.** To complete the solution, it will be necessary to know something about the geometry of the cable in order to obtain the necessary ninth equation. For example, if the cable's total *length* L is specified, then the Pythagorean theorem can be used to relate each of the three segmental lengths, written in terms of h, y_C, y_D, L_1, L_2, and L_3, to the total length L. Unfortunately, this type of problem cannot be solved easily by hand. Another possibility, however, is to specify one of the sags, either y_C or y_D, instead of the cable length. By doing this, the equilibrium equations are then sufficient for obtaining the unknown forces and the remaining sag. Once the sag at each point of loading is obtained, the length of the cable can be determined by trigonometry. The following example illustrates a procedure for performing the equilibrium analysis for a problem of this type.

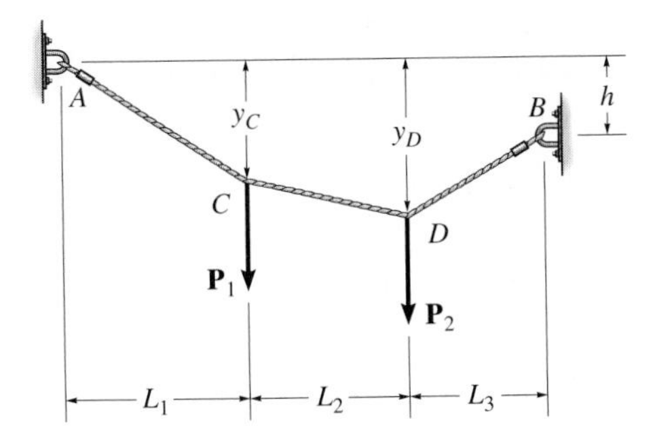

Fig. 7–20

*As will be shown in the following example, the eight equilibrium equations can *also* be written for the entire cable, or any part thereof. But *no more* than *eight* equations are available.

Example 7–13

Determine the tension in each segment of the cable shown in Fig. 7–21*a*.

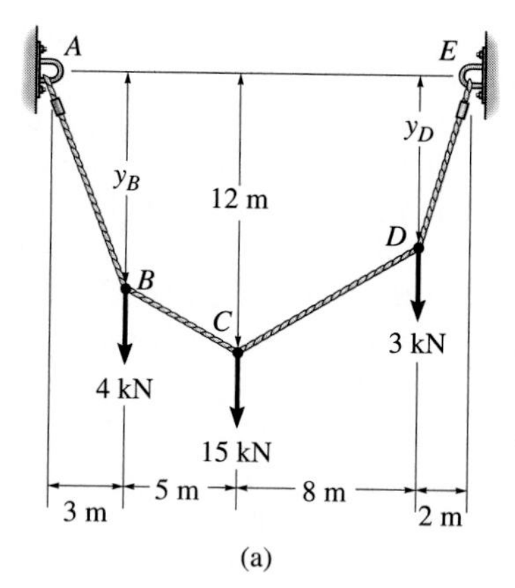

(a)

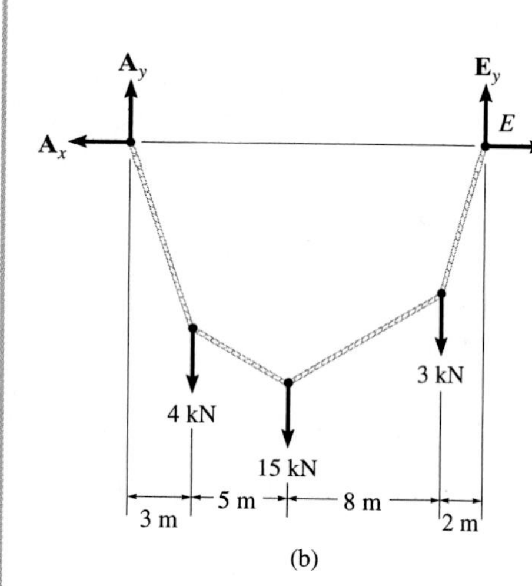

(b)

SOLUTION

By inspection, there are four unknown external reactions (A_x, A_y, E_x, and E_y) and four unknown cable tensions, one in each cable segment. These eight unknowns along with the two unknown sags y_B and y_D can be determined from *ten* available equilibrium equations. One method is to apply these equations as force equilibrium ($\Sigma F_x = 0$, $\Sigma F_y = 0$) to each of the five points A through E. Here, however, we will take a more direct approach.

Consider the free-body diagram for the entire cable, Fig. 7–21*b*. Thus,

$\overset{+}{\rightarrow}\Sigma F_x = 0;$ $\quad -A_x + E_x = 0$

$\circlearrowleft+\Sigma M_E = 0;$ $\quad -A_y(18 \text{ m}) + 4 \text{ kN}(15 \text{ m}) + 15 \text{ kN}(10 \text{ m}) + 3 \text{ kN}(2 \text{ m}) = 0$

$$A_y = 12 \text{ kN}$$

$+\uparrow\Sigma F_y = 0;$ $\quad 12 \text{ kN} - 4 \text{ kN} - 15 \text{ kN} - 3 \text{ kN} + E_y = 0$

$$E_y = 10 \text{ kN}$$

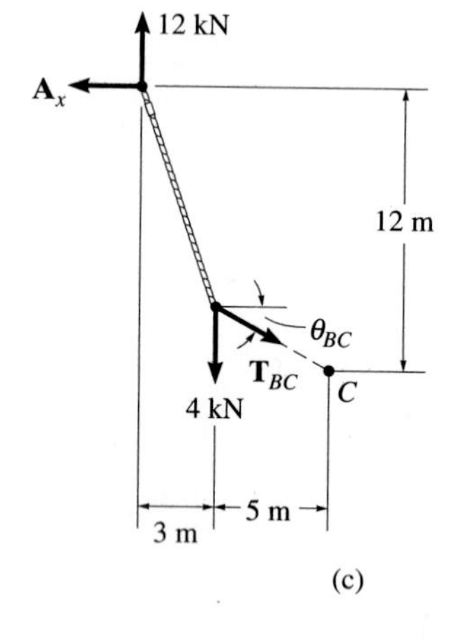

(c)

Fig. 7–21

Since the sag $y_C = 12$ m is known, we will now consider the leftmost section, which cuts cable BC, Fig. 7–21*c*.

$\circlearrowleft+\Sigma M_C = 0;$ $\quad A_x(12 \text{ m}) - 12 \text{ kN}(8 \text{ m}) + 4 \text{ kN}(5 \text{ m}) = 0$

$$A_x = E_x = 6.33 \text{ kN}$$

$\overset{+}{\rightarrow}\Sigma F_x = 0;$ $\quad T_{BC} \cos \theta_{BC} - 6.33 \text{ kN} = 0$

$+\uparrow\Sigma F_y = 0;$ $\quad 12 \text{ kN} - 4 \text{ kN} - T_{BC} \sin \theta_{BC} = 0$

Thus,

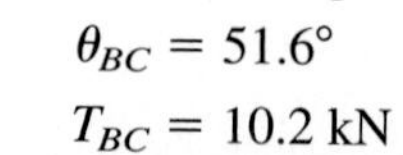

$$\theta_{BC} = 51.6^\circ$$

$$T_{BC} = 10.2 \text{ kN} \qquad \textit{Ans.}$$

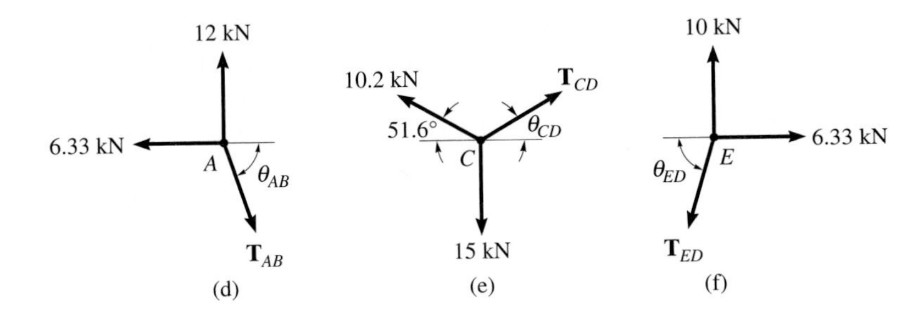

Proceeding now to analyze the equilibrium of points A, C, and E in sequence, we have

Point A (Fig. 7–21*d*)

$\xrightarrow{+}\Sigma F_x = 0;$ $\quad T_{AB} \cos \theta_{AB} - 6.33 \text{ kN} = 0$

$+\uparrow \Sigma F_y = 0;$ $\quad -T_{AB} \sin \theta_{AB} + 12 \text{ kN} = 0$

$$\theta_{AB} = 62.2°$$

$$T_{AB} = 13.6 \text{ kN}$$ ***Ans.***

Point C (Fig. 7–21*e*)

$\xrightarrow{+}\Sigma F_x = 0;$ $\quad T_{CD} \cos \theta_{CD} - 10.2 \cos 51.6° \text{ kN} = 0$

$+\uparrow \Sigma F_y = 0;$ $\quad T_{CD} \sin \theta_{CD} + 10.2 \sin 51.6° \text{ kN} - 15 \text{ kN} = 0$

$$\theta_{CD} = 47.9°$$

$$T_{CD} = 9.44 \text{ kN}$$ ***Ans.***

Point E (Fig. 7–21*f*)

$\xrightarrow{+}\Sigma F_x = 0;$ $\quad 6.33 \text{ kN} - T_{ED} \cos \theta_{ED} = 0$

$+\uparrow \Sigma F_y = 0;$ $\quad 10 \text{ kN} - T_{ED} \sin \theta_{ED} = 0$

$$\theta_{ED} = 57.7°$$

$$T_{ED} = 11.8 \text{ kN}$$ ***Ans.***

By comparison, the maximum cable tension is in segment AB, since this segment has the greatest slope (θ) and it is required that for any left-hand segment the horizontal component $T \cos \theta = A_x$ (a constant). Also, since the slope angles that the cable segments make with the horizontal have now been determined, it is possible to determine the sags y_B and y_D, Fig. 7–21*a*, using trigonometry.

Cable Subjected to a Distributed Load. Consider the weightless cable shown in Fig. 7–22*a*, which is subjected to a loading function $w = w(x)$ *as measured in the x direction.* The free-body diagram of a small segment of the cable having a length Δs is shown in Fig. 7–22*b*. Since the tensile force in the cable changes continuously in both magnitude and direction along the cable's length, this change is denoted on the free-body diagram by ΔT. The distributed load is represented by its resultant force $w(x)(\Delta x)$, which acts at a fractional distance $k(\Delta x)$ from point O, where $0 < k < 1$. Applying the equations of equilibrium yields

$$\overset{+}{\rightarrow}\Sigma F_x = 0; \quad -T\cos\theta + (T + \Delta T)\cos(\theta + \Delta\theta) = 0$$

$$+\uparrow\Sigma F_y = 0; \quad -T\sin\theta - w(x)(\Delta x) + (T + \Delta T)\sin(\theta + \Delta\theta) = 0$$

$$\circlearrowleft+\Sigma M_O = 0; \quad w(x)(\Delta x)k(\Delta x) - T\cos\theta\,\Delta y + T\sin\theta\,\Delta x = 0$$

Dividing each of these equations by Δx and taking the limit as $\Delta x \rightarrow 0$, and hence $\Delta y \rightarrow 0$, $\Delta\theta \rightarrow 0$, and $\Delta T \rightarrow 0$, we obtain

$$\frac{d(T\cos\theta)}{dx} = 0 \tag{7–7}$$

$$\frac{d(T\sin\theta)}{dx} - w(x) = 0 \tag{7–8}$$

$$\frac{dy}{dx} = \tan\theta \tag{7–9}$$

Integrating Eq. 7–7, we have

$$T\cos\theta = \text{constant} = F_H \tag{7–10}$$

Here F_H represents the horizontal component of tensile force at *any point* along the cable.

Integrating Eq. 7–8 gives

$$T\sin\theta = \int w(x)\,dx \tag{7–11}$$

Dividing Eq. 7–11 by Eq. 7–10 eliminates T. Then, using Eq. 7–9, we can obtain the slope

$$\tan\theta = \frac{dy}{dx} = \frac{1}{F_H}\int w(x)\,dx$$

Performing a second integration yields

$$y = \frac{1}{F_H} \int \left(\int w(x)\, dx \right) dx \qquad (7\text{–}12)$$

This equation is used to determine the curve for the cable, $y = f(x)$. The horizontal force component F_H and the two constants, say C_1 and C_2, resulting from the integration are determined by applying the boundary conditions for the cable.

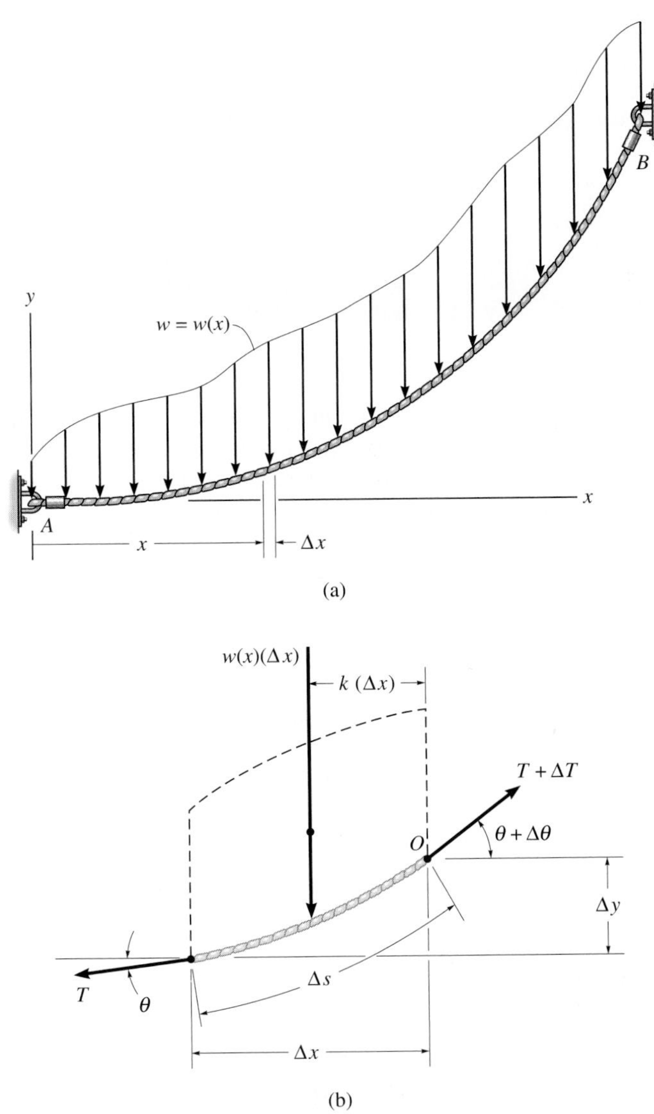

Fig. 7–22

Example 7–14

The cable of a suspension bridge supports half of the uniform road surface between the two columns at A and B, as shown in Fig. 7–23a. If this distributed loading is w_o, determine the maximum force developed in the cable and the cable's required length. The span length L and sag h are known.

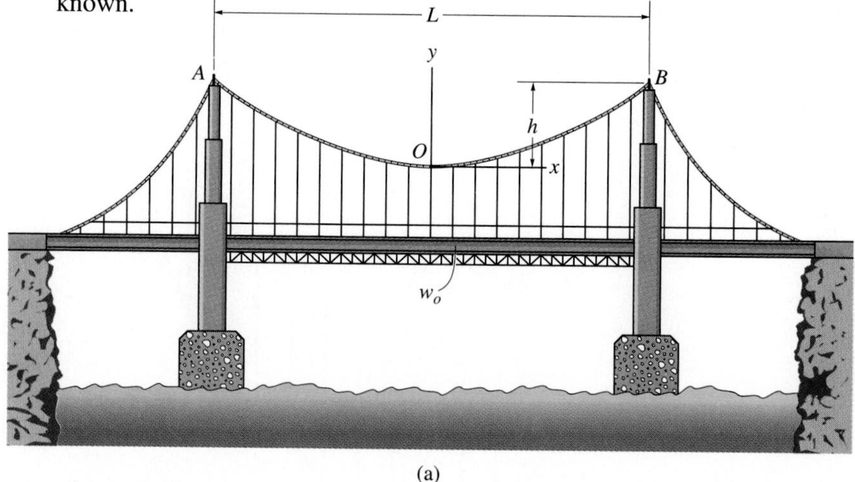

(a)

Fig. 7–23

SOLUTION

We can determine the unknowns in the problem by first finding the curve that defines the shape of the cable by using Eq. 7–12. For reasons of symmetry, the origin of coordinates has been placed at the cable's center. Noting that $w(x) = w_o$, we have

$$y = \frac{1}{F_H} \int \left(\int w_o \, dx \right) dx$$

Integrating this equation twice gives

$$y = \frac{1}{F_H}\left(\frac{w_o x^2}{2} + C_1 x + C_2 \right) \tag{1}$$

The constants of integration may be determined by using the boundary conditions $y = 0$ at $x = 0$ and $dy/dx = 0$ at $x = 0$. Substituting into Eq. 1 yields $C_1 = C_2 = 0$. The curve then becomes

$$y = \frac{w_o}{2F_H} x^2 \tag{2}$$

This is the equation of a *parabola*. The constant F_H may be obtained by using the boundary condition $y = h$ at $x = L/2$. Thus,

$$F_H = \frac{w_o L^2}{8h} \tag{3}$$

Therefore, Eq. 2 becomes

$$y = \frac{4h}{L^2}x^2 \tag{4}$$

Since F_H is known, the tension in the cable may be determined using Eq. 7–10, written as $T = F_H/\cos\theta$. For $0 \leq \theta < \pi/2$, the maximum tension will occur when θ is *maximum,* i.e., at point B, Fig. 7–23*a*. From Eq. 2, the slope at this point is

$$\left.\frac{dy}{dx}\right|_{x=L/2} = \tan\theta_{\max} = \left.\frac{w_o}{F_H}x\right|_{x=L/2}$$

or

$$\theta_{\max} = \tan^{-1}\left(\frac{w_o L}{2F_H}\right) \tag{5}$$

Therefore,

$$T_{\max} = \frac{F_H}{\cos(\theta_{\max})} \tag{6}$$

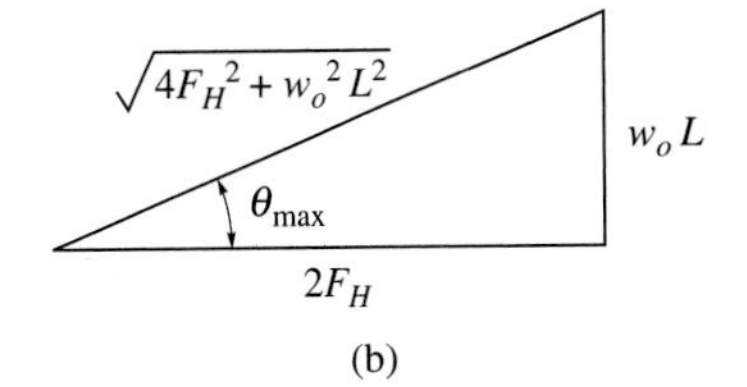

(b)

Using the triangular relationship shown in Fig. 7–23*b*, which is based on Eq. 5, Eq. 6 may be written as

$$T_{\max} = \frac{\sqrt{4F_H^2 + w_o^2 L^2}}{2}$$

Substituting Eq. 3 into the above equation yields

$$T_{\max} = \frac{w_o L}{2}\sqrt{1 + \left(\frac{L}{4h}\right)^2} \qquad \textit{Ans.}$$

For a differential segment of cable length ds, we can write

$$ds = \sqrt{(dx)^2 + (dy)^2} = \sqrt{1 + \left(\frac{dy}{dx}\right)^2}\,dx$$

Hence, the total length of the cable, $\mathcal{L}$, can be determined by integration. Using Eq. 4, we have

$$\mathcal{L} = \int ds = 2\int_0^{L/2} \sqrt{1 + \left(\frac{8h}{L^2}x\right)^2}\,dx \tag{7}$$

Integrating and substituting the limits yields

$$\mathcal{L} = \frac{L}{2}\left[\sqrt{1 + \left(\frac{4h}{L}\right)^2} + \frac{L}{4h}\sinh^{-1}\left(\frac{4h}{L}\right)\right] \qquad \textit{Ans.}$$

Cable Subjected to Its Own Weight. When the weight of the cable becomes important in the force analysis, the loading function along the cable becomes a function of the arc length s rather than the projected length x. A generalized loading function $w = w(s)$ acting along the cable is shown in Fig. 7–24*a*. The free-body diagram for a segment of the cable is shown in Fig. 7–24*b*. Applying the equilibrium equations to the force system on this diagram, one obtains relationships identical to those given by Eqs. 7–7 through 7–9, but with ds replacing dx. Therefore, it may be shown that

$$T \cos \theta = F_H \qquad (7\text{–}13)$$

$$T \sin \theta = \int w(s)\, ds$$

$$\frac{dy}{dx} = \frac{1}{F_H} \int w(s)\, ds \qquad (7\text{–}14)$$

To perform a direct integration of Eq. 7–14, it is necessary to replace dy/dx by ds/dx. Since

$$ds = \sqrt{dx^2 + dy^2}$$

then

$$\frac{dy}{dx} = \sqrt{\left(\frac{ds}{dx}\right)^2 - 1}$$

Therefore,

$$\frac{ds}{dx} = \left\{1 + \frac{1}{F_H^2}\left(\int w(s)\, ds\right)^2\right\}^{1/2}$$

Separating the variables and integrating yields

$$x = \int \frac{ds}{\left\{1 + \frac{1}{F_H^2}\left(\int w(s)\, ds\right)^2\right\}^{1/2}} \qquad (7\text{–}15)$$

The two constants of integration, say C_1 and C_2, are found using the boundary conditions for the cable.

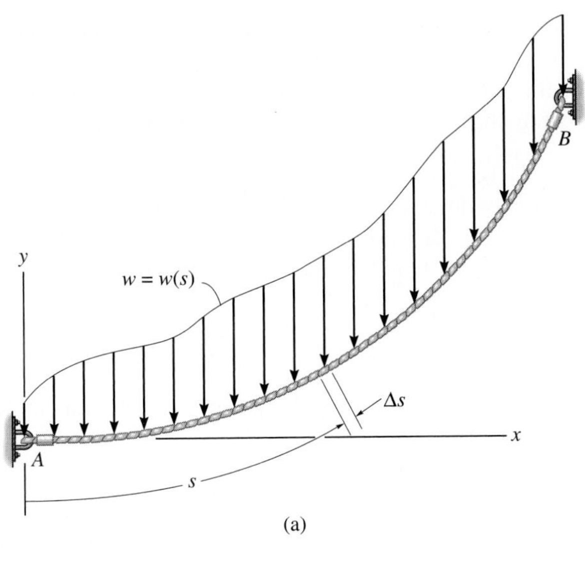

(a)

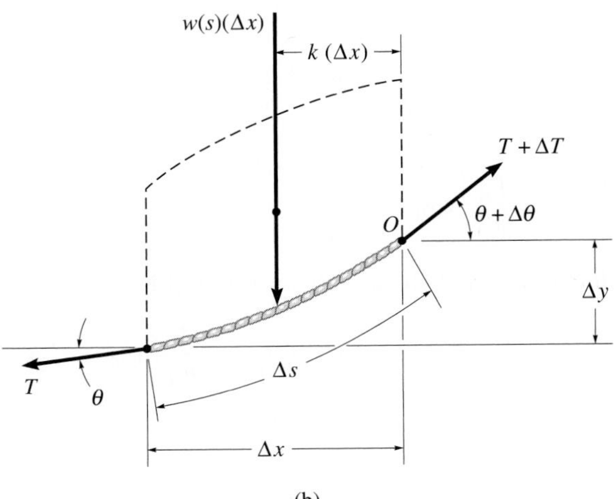

(b)

Fig. 7–24

Example 7–15

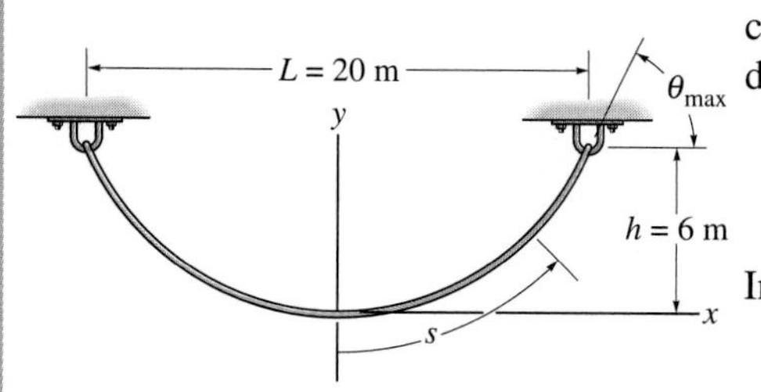

Fig. 7–25

Determine the deflection curve, the length, and the maximum tension in the uniform cable shown in Fig. 7–25. The cable weighs $w_o = 5$ N/m.

SOLUTION

By reasons of symmetry, the origin of coordinates is located at the center of the cable. The deflection curve is expressed as $y = f(x)$. We can determine it by first applying Eq. 7–15, where $w(s) = w_o$.

$$x = \int \frac{ds}{[1 + (1/F_H^2)(\int w_o\, ds)^2]^{1/2}}$$

Integrating the term under the integral sign in the denominator, we have

$$x = \int \frac{ds}{[1 + (1/F_H^2)(w_o s + C_1)^2]^{1/2}}$$

Substituting $u = (1/F_H)(w_o s + C_1)$ so that $du = (w_o/F_H)\, ds$, a second integration yields

$$x = \frac{F_H}{w_o}(\sinh^{-1} u + C_2)$$

or

$$x = \frac{F_H}{w_o}\left\{\sinh^{-1}\left[\frac{1}{F_H}(w_o s + C_1)\right] + C_2\right\} \tag{1}$$

To evaluate the constants note that, from Eq. 7–14,

$$\frac{dy}{dx} = \frac{1}{F_H}\int w_o\, ds \qquad \text{or} \qquad \frac{dy}{dx} = \frac{1}{F_H}(w_o s + C_1)$$

Since $dy/dx = 0$ at $s = 0$, then $C_1 = 0$. Thus,

$$\frac{dy}{dx} = \frac{w_o s}{F_H} \tag{2}$$

The constant C_2 may be evaluated by using the condition $s = 0$ at $x = 0$ in Eq. 1, in which case $C_2 = 0$. To obtain the deflection curve, solve for s in Eq. 1, which yields

$$s = \frac{F_H}{w_o}\sinh\left(\frac{w_o}{F_H}x\right) \tag{3}$$

Now substitute into Eq. 2, in which case

$$\frac{dy}{dx} = \sinh\left(\frac{w_o}{F_H}x\right)$$

Hence

$$y = \frac{F_H}{w_o} \cosh\left(\frac{w_o}{F_H}x\right) + C_3 \tag{4}$$

If the boundary condition $y = 0$ at $x = 0$ is applied, the constant $C_3 = -F_H/w_o$, and therefore the deflection curve becomes

$$y = \frac{F_H}{w_o}\left[\cosh\left(\frac{w_o}{F_H}x\right) - 1\right]$$

This equation defines the shape of a *catenary curve.* The constant F_H is obtained by using the boundary condition that $y = h$ at $x = L/2$, in which case

$$h = \frac{F_H}{w_o}\left[\cosh\left(\frac{w_o L}{2F_H}\right) - 1\right] \tag{5}$$

Since $w_o = 5$ N/m, $h = 6$ m, and $L = 20$ m, Eqs. 4 and 5 become

$$y = \frac{F_H}{5 \text{ N/m}}\left[\cosh\left(\frac{5 \text{ N/m}}{F_H}x\right) - 1\right] \tag{6}$$

$$6 \text{ m} = \frac{F_H}{5 \text{ N/m}}\left[\cosh\left(\frac{50 \text{ N}}{F_H}\right) - 1\right] \tag{7}$$

Equation 7 can be solved for F_H by using a trial-and-error procedure. The result is

$$F_H = 45.8 \text{ N}$$

and therefore the deflection curve, Eq. 6, becomes

$$y = 9.16[\cosh(0.109x) - 1] \text{ m} \qquad \textit{Ans.}$$

Using Eq. 3, with $x = 10$ m, the half-length of the cable is

$$\frac{\mathscr{L}}{2} = \frac{45.8 \text{ N}}{5 \text{ N/m}} \sinh\left[\frac{5 \text{ N/m}}{45.8 \text{ N}}(10 \text{ m})\right] = 12.1 \text{ m}$$

Hence,

$$\mathscr{L} = 24.2 \text{ m} \qquad \textit{Ans.}$$

Since $T = F_H/\cos\theta$, Eq. 7–13, the maximum tension occurs when θ is maximum, i.e., at $s = \mathscr{L}/2 = 12.1$ m. Using Eq. 2 yields

$$\left.\frac{dy}{dx}\right|_{s=12.1 \text{ m}} = \tan\theta_{max} = \frac{5 \text{ N/m}(12.1 \text{ m})}{45.8 \text{ N}} = 1.32$$

$$\theta_{max} = 52.9^\circ$$

Thus,

$$T_{max} = \frac{F_H}{\cos\theta_{max}} = \frac{45.8 \text{ N}}{\cos 52.9^\circ} = 75.9 \text{ N} \qquad \textit{Ans.}$$

PROBLEMS

Neglect the weight of the cable in the following problems, *unless* specified.

***7–80.** Determine the tension in each segment of the cable and the cable's total length.

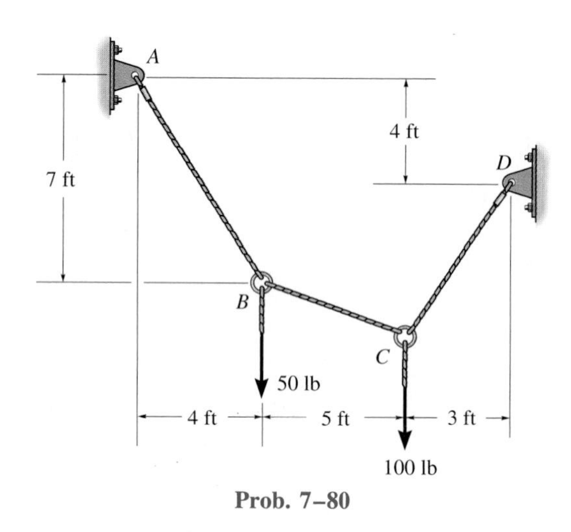

Prob. 7–80

7–81. Cable $ABCD$ supports the 4-kg flowerpot E and 6-kg flowerpot F. Determine the maximum tension in the cable and the sag of point B.

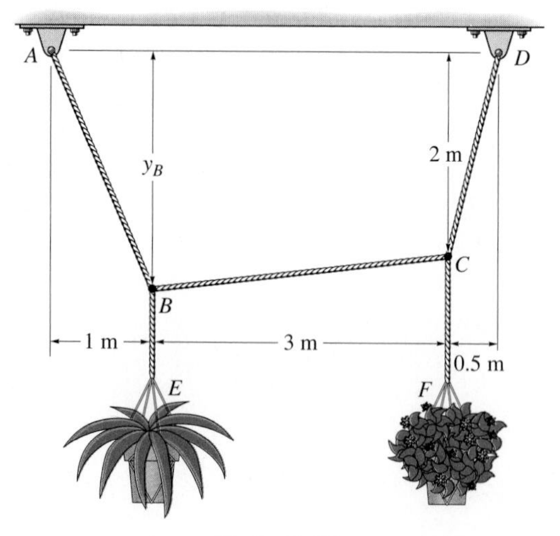

Prob. 7–81

7–82. The cable supports the three loads shown. Determine the sags y_B and y_D of points B and D. Take $P_1 = 400$ lb, $P_2 = 250$ lb.

7–83. The cable supports the three loads shown. Determine the magnitude of $\mathbf{P}_1$ if $P_2 = 300$ lb and $y_B = 8$ ft. Also find the sag y_D.

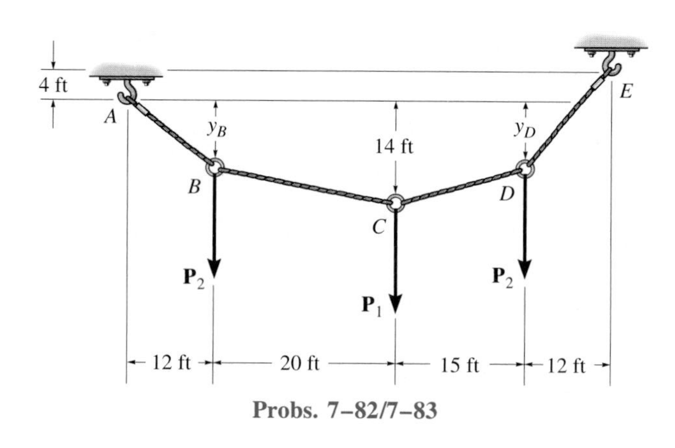

Probs. 7–82/7–83

***7–84.** The cable supports the loading shown. Determine the distance x_B the force at point B acts from A. Set $P = 40$ lb.

7–85. The cable supports the loading shown. Determine the magnitude of the horizontal force **P** so that $x_B = 6$ ft.

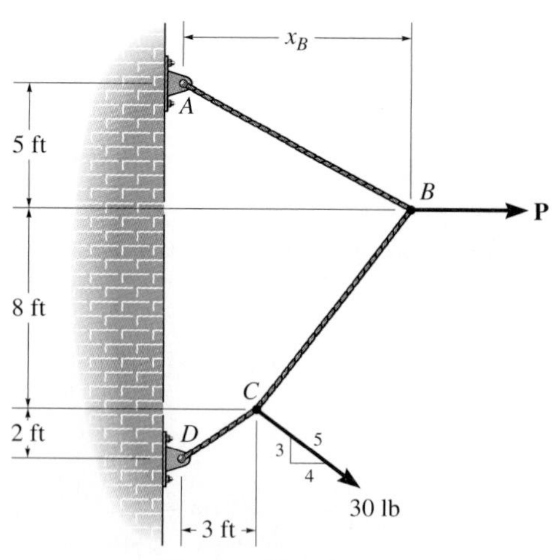

Probs. 7–84/7–85

7–86. Determine the tension in each cable segment and the cable's total length.

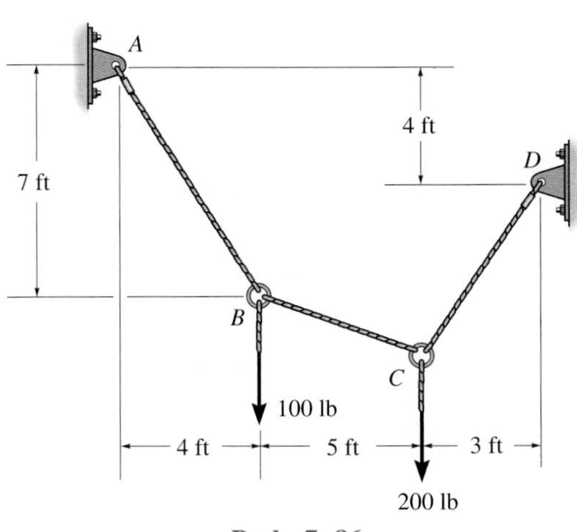

Prob. 7–86

7–87. The cable segments support the loading shown. Determine the distance x_B from the force at B to point A. Set $P = 40$ lb.

***7–88.** The cable segments support the loading shown. Determine the magnitude of the horizontal force **P** so that $x_B = 6$ ft.

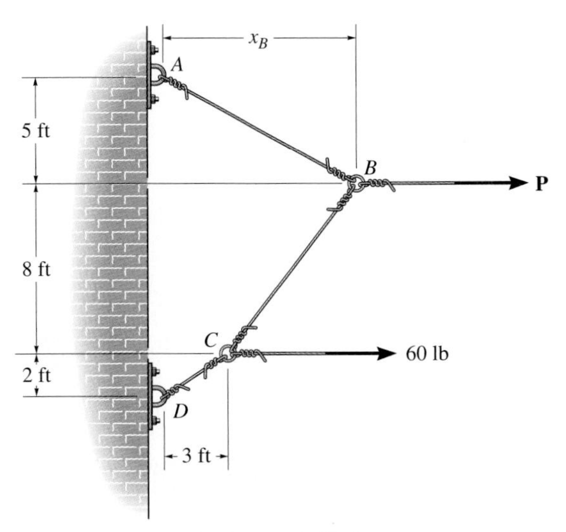

Probs. 7–87/7–88

7–89. The cable supports a girder which weighs 850 lb/ft. Determine the tension in the cable at points A, B, and C.

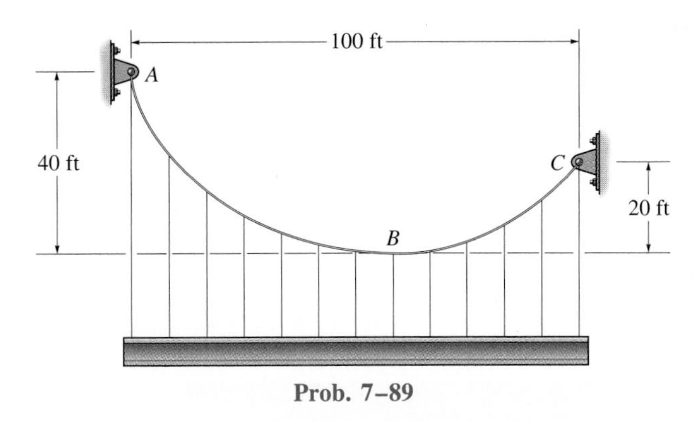

Prob. 7–89

7–90. Determine the maximum uniform loading w lb/ft that the cable can support if it is capable of sustaining a maximum tension of 3000 lb before it will break.

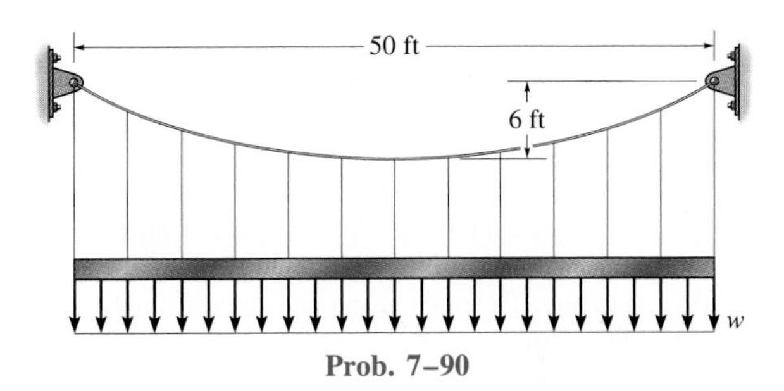

Prob. 7–90

7–91. The cable is subjected to the parabolic loading $w = 150(1 - (x/50)^2)$ lb/ft, where x is in ft. Determine the equation $y = f(x)$ which defines the cable shape AB and the maximum tension in the cable.

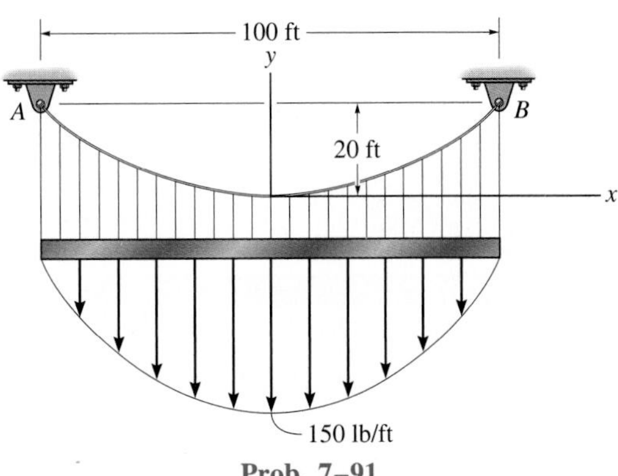

Prob. 7–91

*7–92. The cable is subjected to the triangular loading. If the slope of the cable at point O is zero, determine the equation of the curve $y = f(x)$ which defines the cable shape OB, and the maximum tension developed in the cable.

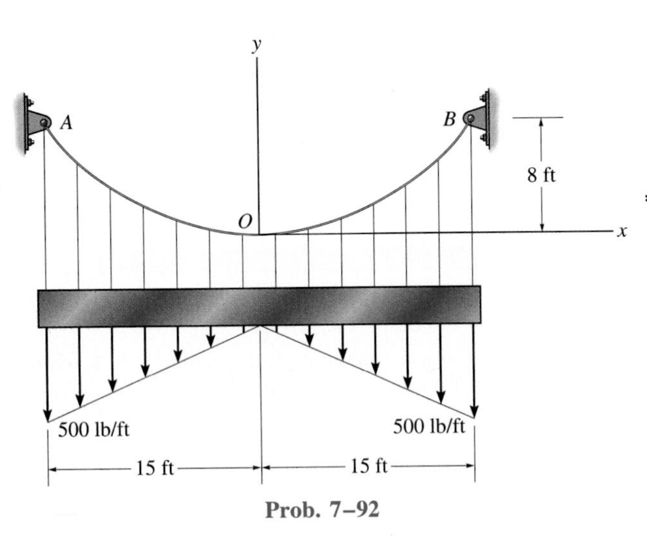

Prob. 7–92

7–93. The cable AB is subjected to a uniform loading of 200 N/m. If the weight of the cable is neglected and the slope angles at points A and B are 30° and 60°, respectively, determine the curve that defines the cable shape and the maximum tension developed in the cable.

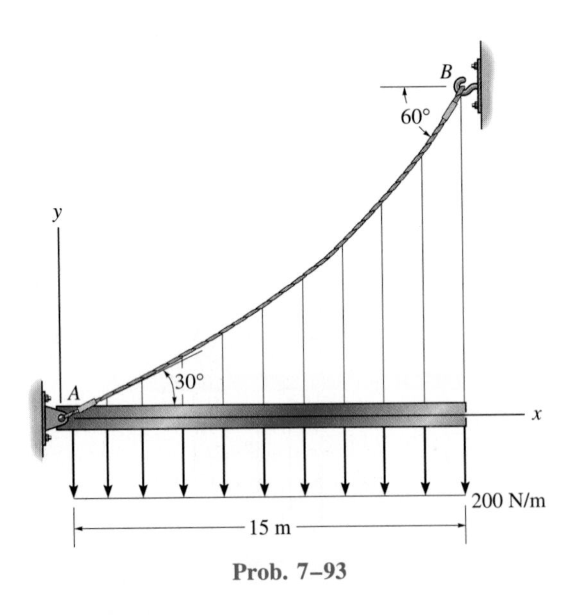

Prob. 7–93

7–94. A uniform cord is suspended between two points having the same elevation. Determine the sag-to-span ratio so that the maximum tension in the cord equals the cord's total weight.

7–95. Show that the deflection curve of the cable discussed in Example 7–15 reduces to Eq. (4) in Example 7–14 when the *hyperbolic cosine function* is expanded in terms of a series and only the first two terms are retained. (The answer indicates that the *catenary* may be replaced by a *parabola* in the analysis of problems in which the sag is small. In this case, the cable weight is assumed to be uniformly distributed along the horizontal.)

*■7–96. A cable has a weight of 3 lb/ft and is supported at points that are 500 ft apart and at the same elevation. If it has a length of 600 ft, determine the sag.

■7–97. A cable has a weight of 5 lb/ft. If it can span 300 ft and has a sag of 15 ft, determine the length of the cable. The ends of the cable are supported at the same elevation.

7–98. A 200-lb cable is attached between two points that are 75 ft apart and at the same elevation. If the maximum tension developed in the cable is 120 lb, determine the length of the cable and the sag.

7–99. The power line is supported at A by the tower. If the cable weighs 0.75 lb/ft, and the sag $s = 3$ ft, determine the resultant horizontal force the cable exerts at A.

*7–100. The power line is supported at A by the tower. If the cable weighs 0.75 lb/ft, determine the total length of the cable, BAC. Set $s = 3$ ft.

7–101. The power line is supported at A by the tower. If the cable weighs 0.75 lb/ft, determine the required sag s so that the resultant horizontal force the cable exerts at A is zero.

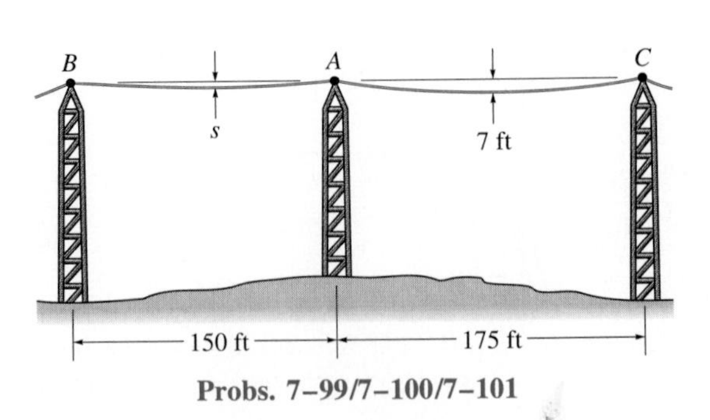

Probs. 7–99/7–100/7–101

7–102. The transmission cable having a weight of 20 lb/ft is strung across the river as shown. Determine the required force that must be applied to the cable at its points of attachment to the towers at B and C.

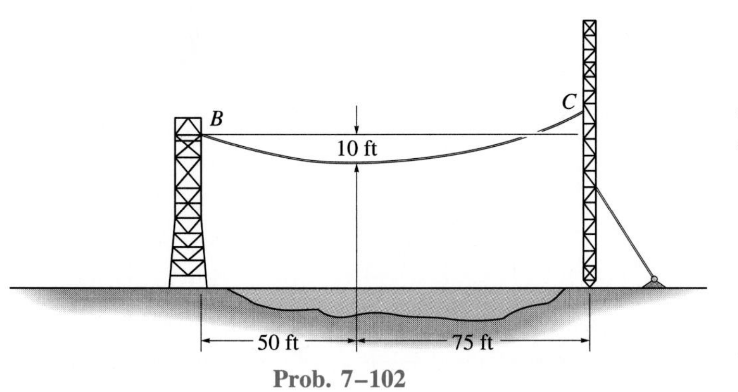

Prob. 7–102

7–103. The uniform beam weighs 800 lb/ft and is held in the horizontal position by means of the cable AB, which has a weight of 10 lb/ft. If the slope angle of the cable at A is 15°, determine the length of the cable.

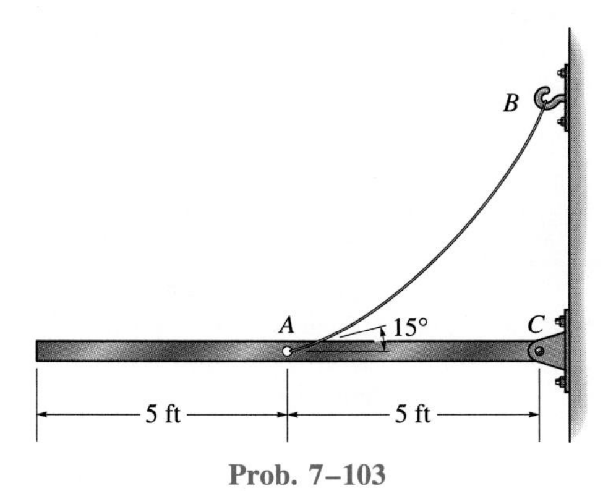

Prob. 7–103

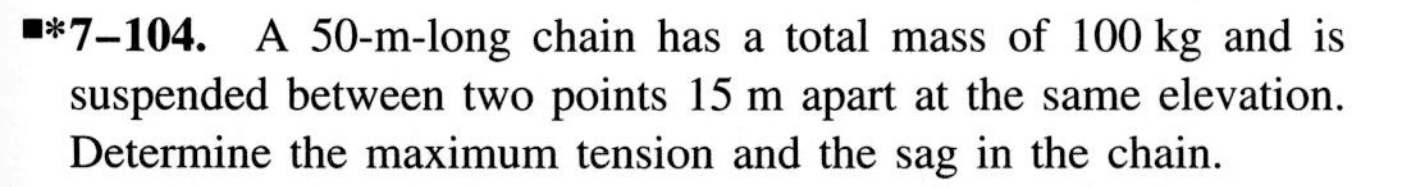

■***7–104.** A 50-m-long chain has a total mass of 100 kg and is suspended between two points 15 m apart at the same elevation. Determine the maximum tension and the sag in the chain.

■**7–105.** A chain has a mass of 3 kg/m and is supported at points which are 3 m apart and at the same elevation. If the sag in the chain is 1 m, determine the maximum tension in the chain.

■**7–106.** The telephone wire has a mass of 500 g/m. If the cable has a length of 32 m between the poles, determine the maximum tension in the cable and its sag.

29 m

Prob. 7–106

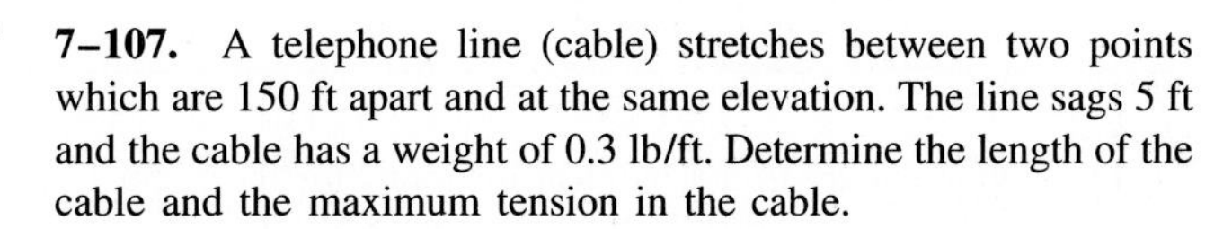

7–107. A telephone line (cable) stretches between two points which are 150 ft apart and at the same elevation. The line sags 5 ft and the cable has a weight of 0.3 lb/ft. Determine the length of the cable and the maximum tension in the cable.

■***7–108.** The cable has a mass of 0.5 kg/m and is 25 m long. Determine the vertical and horizontal components of force it exerts on the top of the tower.

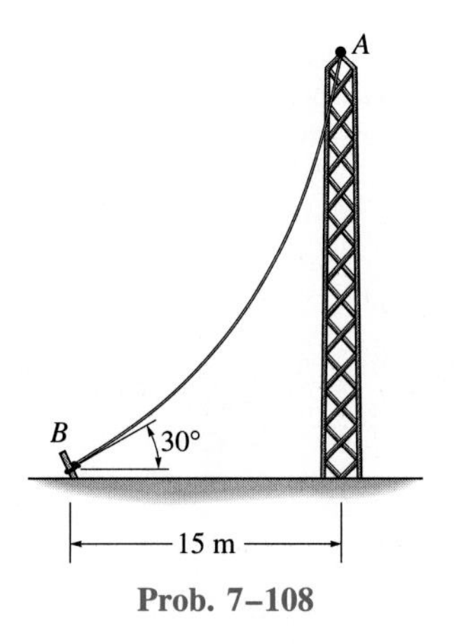

Prob. 7–108

7–109. The buoyant (or vertical) component of force acting on the weather balloon is 600 N. If cable OB is 70 m in length and has a mass of 500 g/m, determine the altitude h of the balloon.

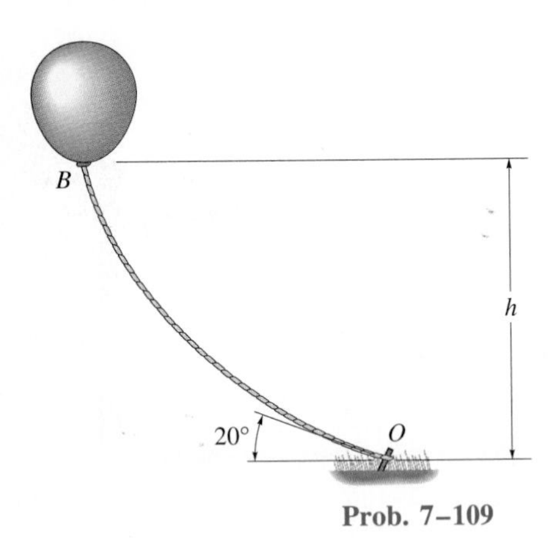

Prob. 7–109

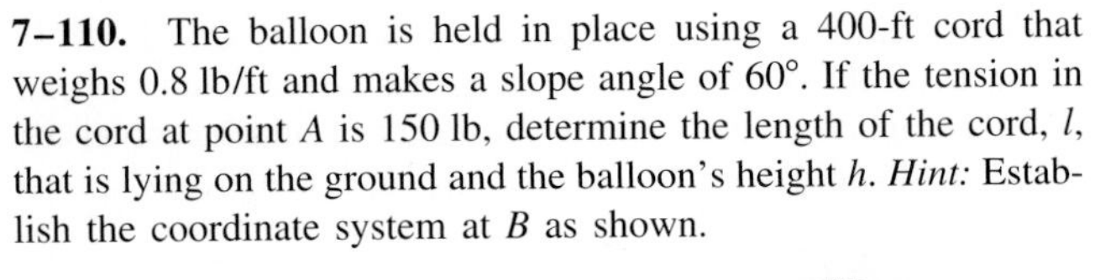

7–110. The balloon is held in place using a 400-ft cord that weighs 0.8 lb/ft and makes a slope angle of 60°. If the tension in the cord at point A is 150 lb, determine the length of the cord, l, that is lying on the ground and the balloon's height h. *Hint:* Establish the coordinate system at B as shown.

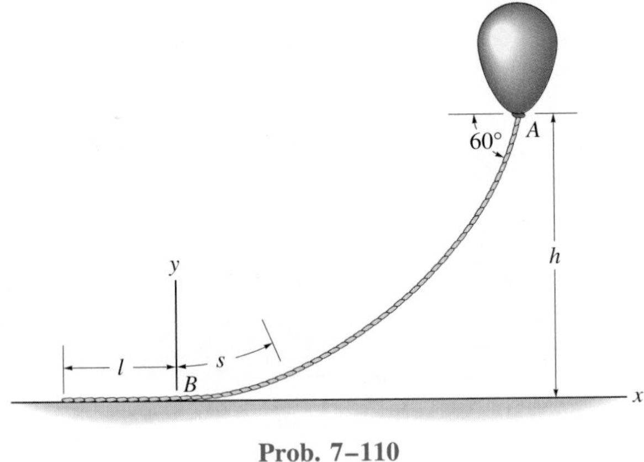

Prob. 7–110

REVIEW PROBLEMS

7–111. Draw the shear and moment diagrams for the beam.

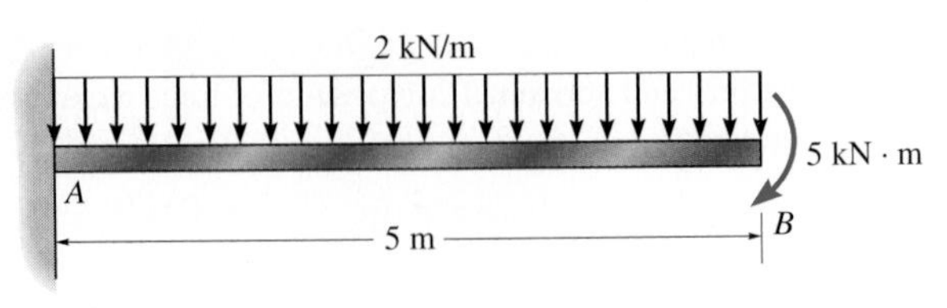

Prob. 7–111

***7–112.** The chain is suspended between points A and B. If it has a weight of 0.5 lb/ft and the sag is 3 ft, determine the maximum tension in the chain.

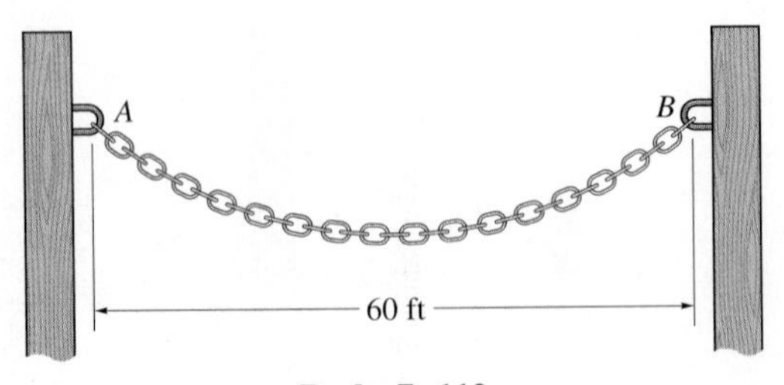

Prob. 7–112

7–113. Draw the shear and moment diagrams for the shaft. The supports at A and B are journal bearings.

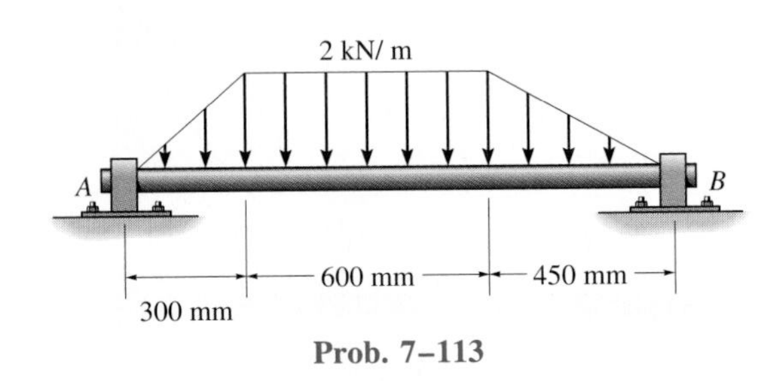

Prob. 7–113

7–114. The 80-ft-long chain is fixed at its ends and hoisted at its midpoint B using a crane. If the chain has a weight of 0.5 lb/ft, determine the minimum height h of the hook in order to lift the chain *completely* off the ground. What is the horizontal force at pin A or C when the chain is in this position? *Hint:* When h is a minimum, the slope at A and C is zero.

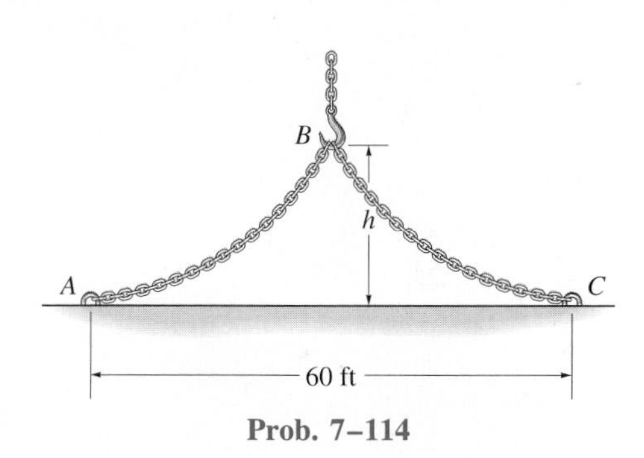

Prob. 7–114

7–115. The two segments of the girder are pin-connected together by a short vertical link BC. Draw the shear and moment diagrams for the girder.

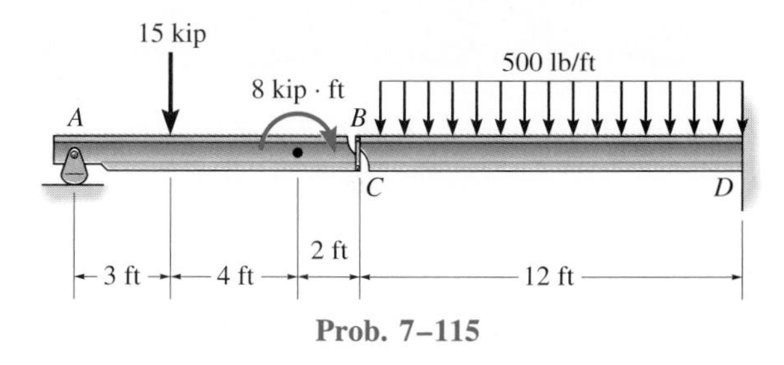

Prob. 7–115

***7–116.** Determine the ratio of a/b for which the shear force will be zero at the midpoint C of the beam.

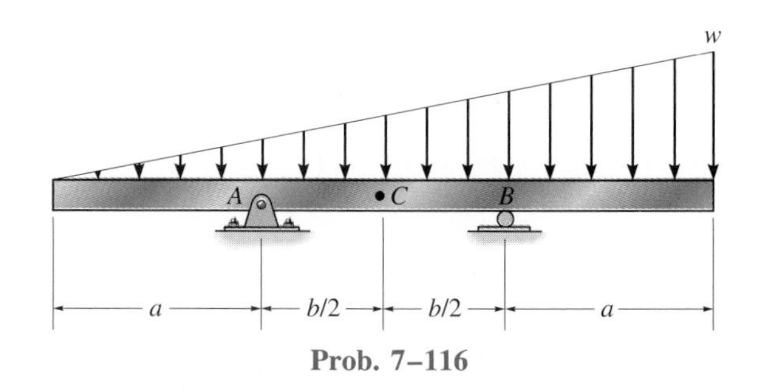

Prob. 7–116

7–117. Draw the shear and moment diagrams for the beam.

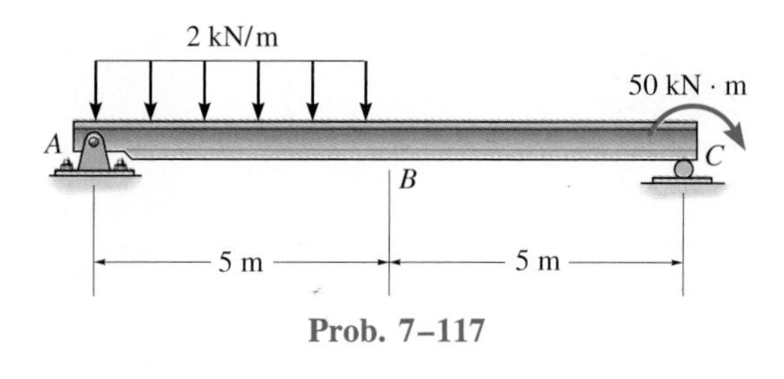

Prob. 7–117

7–118. Determine the normal force, shear force, and moment at sections through points D and E of the frame.

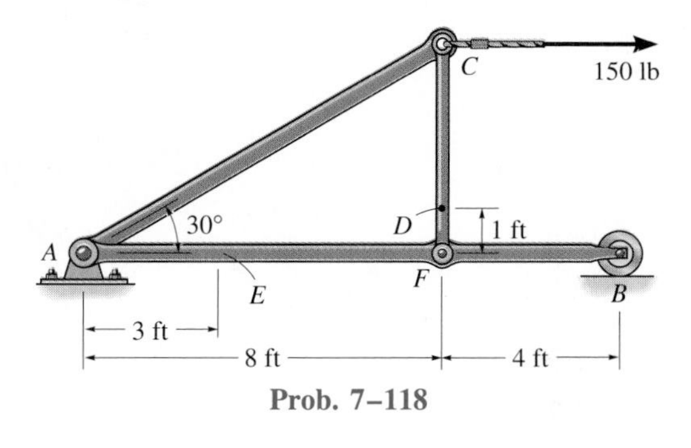

Prob. 7–118

7–119. Express the shear and moment acting in the pipe as a function of y, where $0 \le y \le 4$ ft.

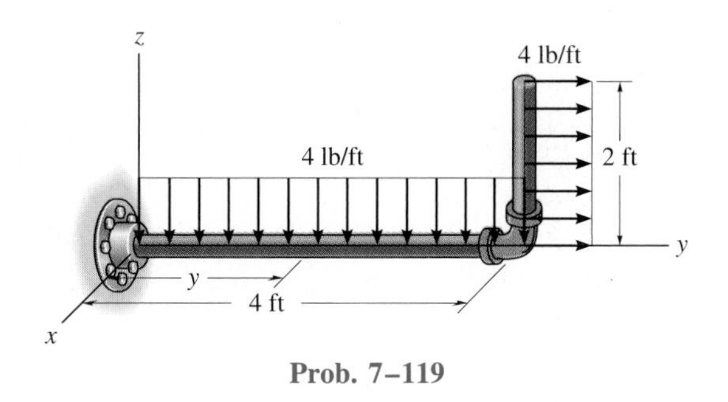

Prob. 7–119

The effective design of a brake system, such as the one for this bicycle, requires efficient capacity for the mechanism to resist frictional forces. In this chapter we will study the nature of friction and show how these forces are considered in engineering analysis.

8

Friction

In the previous chapters the surfaces of contact between two bodies were considered to be perfectly *smooth.* Because of this, the force of interaction between the bodies always acts *normal* to the surface at points of contact. In reality, however, all surfaces are *rough,* and depending on the nature of the problem, the ability of a body to support a *tangential* as well as a *normal* force at its contacting surface must be considered. The tangential force is caused by friction, and in this chapter we will show how to analyze problems involving frictional forces. Specific application will include frictional forces on screws, bearings, disks, and belts. The analysis of rolling resistance is given in the last part of the chapter.

8.1 Characteristics of Dry Friction

Friction may be defined as a force of resistance acting on a body which prevents or retards slipping of the body relative to a second body or surface with which it is in contact. This force always acts *tangent* to the surface at points of contact with other bodies and is directed so as to oppose the possible or existing motion of the body relative to these points.

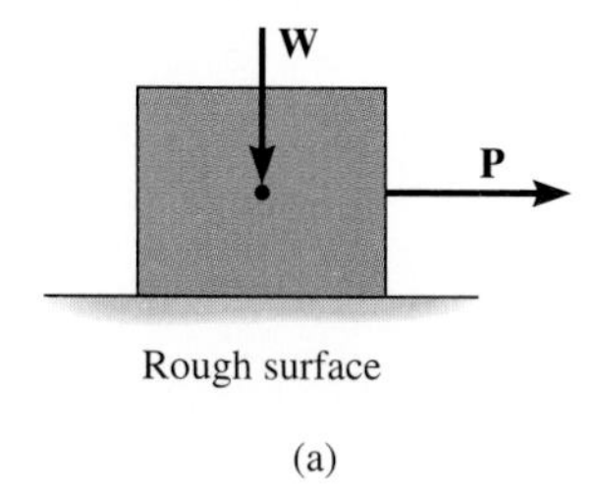

(a)

In general, two types of friction can occur between surfaces. *Fluid friction* exists when the contacting surfaces are separated by a film of fluid (gas or liquid). The nature of fluid friction is studied in fluid mechanics, since it depends on knowledge of the velocity of the fluid and the fluid's ability to resist shear force. In this book only the effects of *dry friction* will be presented. This type of friction is often called *Coulomb friction,* since its characteristics were studied extensively by C. A. Coulomb in 1781. Specifically, dry friction occurs between the contacting surfaces of bodies in the absence of a lubricating fluid.

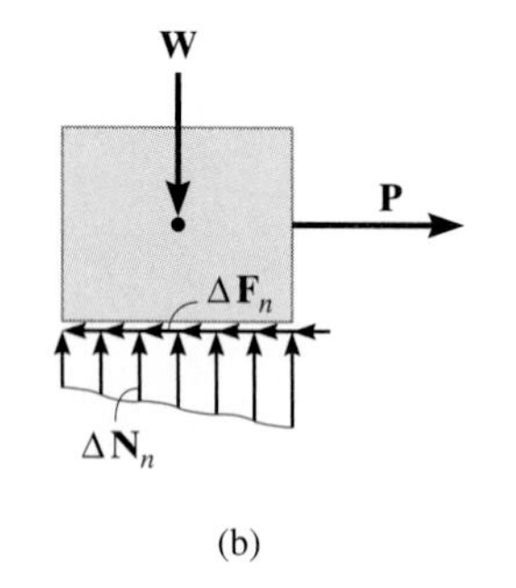

(b)

Theory of Dry Friction. The theory of dry friction can best be explained by considering what effects are caused by pulling horizontally on a block of uniform weight **W** which is resting on a rough horizontal surface, Fig. 8–1*a*. To properly develop a full understanding of the nature of friction, it is necessary to consider the surfaces of contact to be *nonrigid or deformable.* The other portion of the block, however, will be considered rigid. As shown on the free-body diagram of the block, Fig. 8–1*b*, the floor exerts a *distribution* of both *normal force* $\Delta \mathbf{N}_n$ and *frictional force* $\Delta \mathbf{F}_n$ along the contacting surface. For equilibrium, the normal forces must act *upward* to balance the block's weight **W,** and the frictional forces act to the left to prevent the applied force **P** from moving the block to the right. Close examination of the contacting surfaces between the floor and block reveals how these frictional and normal forces develop, Fig. 8–1*c*. It can be seen that many microscopic irregularities exist between the two surfaces and, as a result, reactive forces $\Delta \mathbf{R}_n$ are developed at each of the protuberances.* These forces act at all points of contact and, as shown, each reactive force contributes both a frictional component $\Delta \mathbf{F}_n$ and a normal component $\Delta \mathbf{N}_n$.

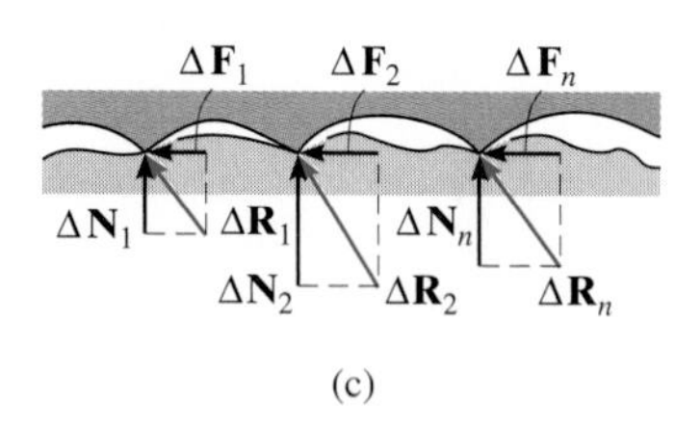

(c)

Equilibrium. For simplicity in the following analysis, the effect of the *distributed* normal and frictional loadings will be indicated by their *resultants* **N** and **F,** which are represented on the free-body diagram of the block as shown in Fig. 8–1*d*. Clearly, the distribution of $\Delta \mathbf{F}_n$ in Fig. 8–1*b* indicates that **F** always acts *tangent to the contacting surface, opposite* to the direction of **P.** On the other hand, the normal force **N** is determined from the distribution of $\Delta \mathbf{N}_n$ in Fig. 8–1*b* and is directed upward to balance the block's weight **W.** Notice that **N** acts a distance x to the right of the line of action of **W,** Fig. 8–1*d*. This location, which coincides with the centroid of the loading diagram in Fig. 8–1*b*, is necessary in order to balance the "tipping effect" caused by **P.** For example, if **P** is applied at a height h from the surface, Fig. 8–1*d*, then moment equilibrium about point O is satisfied if $Wx = Ph$ or $x = Ph/W$. In particular, note that the block will be on the verge of *tipping* if **N** acts at the right corner of the block, $x = a/2$.

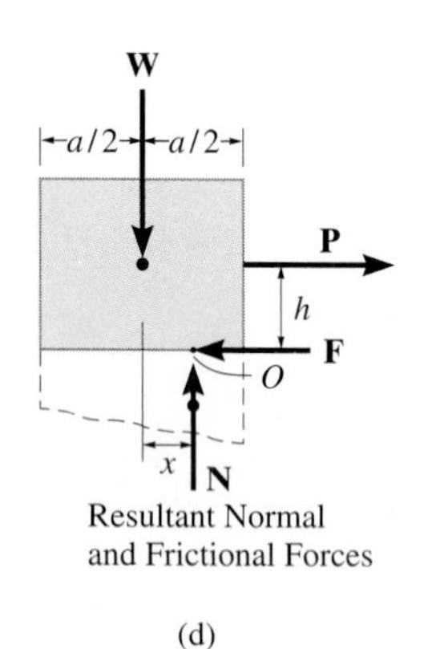

Resultant Normal and Frictional Forces

(d)

Fig. 8–1a–d

*Besides mechanical interactions as explained here, a detailed treatment of the nature of frictional forces must also include the effects of temperature, density, cleanliness, and atomic or molecular attraction between the contacting surfaces. See D. Tabor, *Journal of Lubrication Technology,* 103, 169, 1981.

Impending Motion. In cases where h is small or the surfaces of contact are rather "slippery," the frictional force **F** may *not* be great enough to balance the magnitude of **P,** and consequently the block will tend to slip *before* it can tip. In other words, as the magnitude of **P** is slowly increased, the magnitude of **F** correspondingly increases until it attains a certain *maximum value* F_s, called the *limiting static frictional force,* Fig. 8–1*e*. When this value is reached, the block is in *unstable equilibrium,* since any further increase in P will cause deformations and fractures at the points of surface contact and consequently the block will begin to move. Experimentally, it has been determined that the magnitude of the limiting static frictional force $\mathbf{F}_s$ is *directly proportional* to the magnitude of the resultant normal force **N.** This may be expressed mathematically as

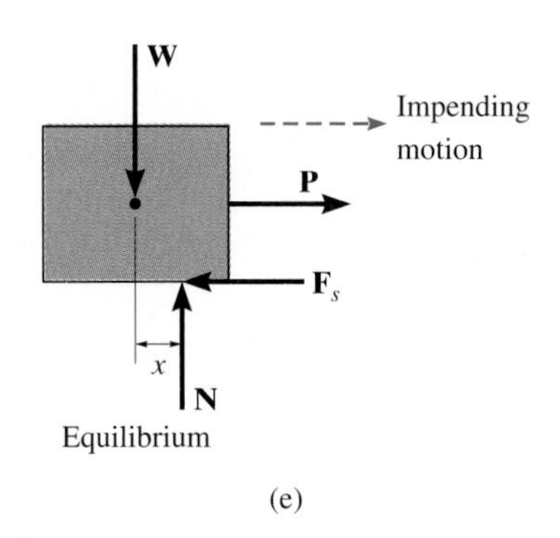

(e)

$$F_s = \mu_s N \tag{8–1}$$

where the constant of proportionality, μ_s (mu "sub" s), is called the *coefficient of static friction.*

Typical values for μ_s, found in many engineering handbooks, are given in Table 8–1. Although this coefficient is generally less than 1, be aware that in some cases it is possible, as in the case of aluminum on aluminum, for μ_s to be greater than 1. Physically this means, of course, that in this case the frictional force is greater than the corresponding normal force. Furthermore, it should be noted that μ_s is dimensionless and depends only on the characteristics of the two surfaces in contact. A wide range of values is given for each value of μ_s, since experimental testing was done under variable conditions of roughness and cleanliness of the contacting surfaces. For applications, therefore, it is important that both caution and judgment be exercised when selecting a coefficient of friction for a given set of conditions. When an exact calculation of F_s is required, the coefficient of friction should be determined directly by an experiment that involves the two materials to be used.

Table 8–1 Typical Values for μ_s

Contact Materials	*Coefficient of Static Friction (μ_s)*
Metal on ice	0.03–0.05
Wood on wood	0.30–0.70
Leather on wood	0.20–0.50
Leather on metal	0.30–0.60
Aluminum on aluminum	1.10–1.70

Motion. If the magnitude of **P** acting on the block is increased so that it becomes greater than F_s, the frictional force at the contacting surfaces drops slightly to a smaller value F_k, called the *kinetic frictional force.* The block will *not* be held in equilibrium ($P > F_k$); instead, it will begin to slide with increasing speed, Fig. 8–1*f*. The drop made in the frictional force magnitude, from F_s (static) to F_k (kinetic), can be explained by again examining the surfaces of contact, Fig. 8–1*g*. Here it is seen that when $P > F_s$, then P has the capacity to shear off the peaks at the contact surfaces and cause the block to "lift" somewhat out of its settled position and "ride" on top of the peaks. Once the block begins to slide, high local temperatures at the points of contact cause momentary adhesion (welding) of these points. The continued shearing of these welds is the dominant mechanism creating friction. Since the resultant contact forces $\Delta\mathbf{R}_n$ are aligned slightly more in the vertical direction than before, Fig. 8–1*c*, they thereby contribute *smaller* frictional components, $\Delta\mathbf{F}_n$, as when the irregularities are meshed.

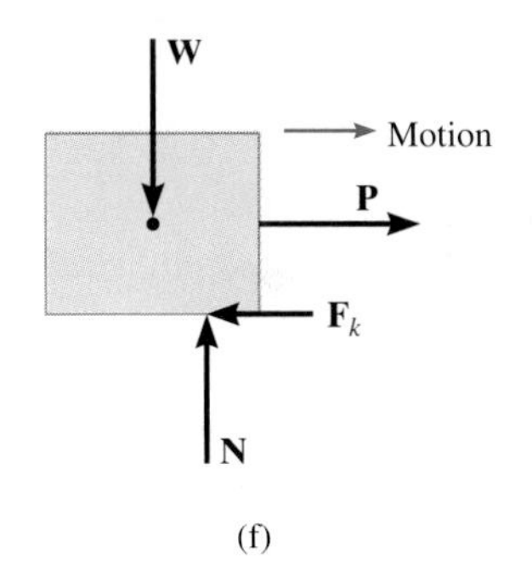

(f)

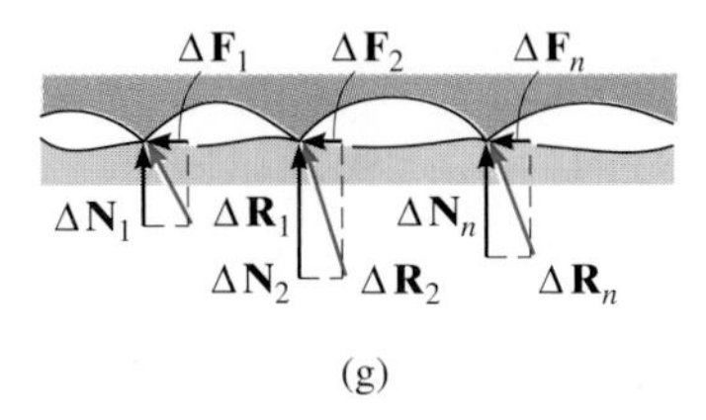

(g)

Fig. 8–1e–g

Experiments with sliding blocks indicate that the magnitude of the resultant frictional force $\mathbf{F}_k$ is directly proportional to the magnitude of the resultant normal force $\mathbf{N}$. This may be expressed mathematically as

$$F_k = \mu_k N \tag{8-2}$$

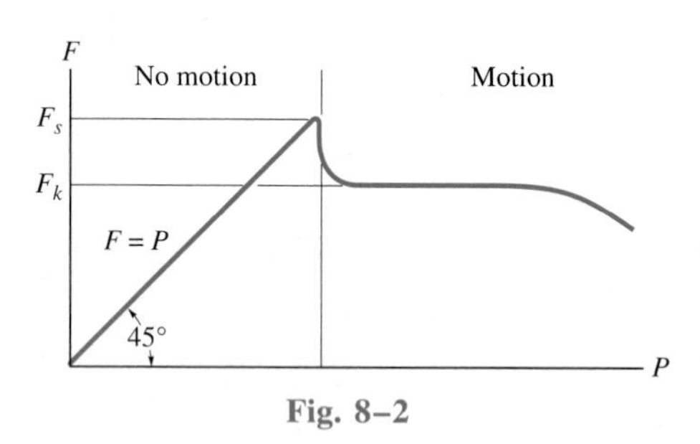

Fig. 8–2

where the constant of proportionality, μ_k, is called the *coefficient of kinetic friction.* Typical values for μ_k are approximately 25 percent *smaller* than those listed in Table 8–1 for μ_s.

the graph in Fig. 8–2, which shows the variation of the frictional force F versus the applied load P. Here the frictional force is categorized in three different ways: namely, F is a *static-frictional force* if equilibrium is maintained; F is a *limiting static-frictional force* $\mathbf{F}_s$ when its magnitude reaches a maximum value needed to maintain equilibrium; and finally, F is termed a *kinetic-frictional force* F_k when sliding occurs at the contacting surface. Notice also from the graph that for very large values of P or for high speeds, because of aerodynamic effects, F_k and likewise μ_k begin to decrease.

Characteristics of Dry Friction. As a result of *experiments* that pertain to the foregoing discussion, the following rules which apply to bodies subjected to dry friction may be stated.

1. The frictional force acts *tangent* to the contacting surfaces in a direction *opposed* to the *relative motion* or tendency for motion of one surface against another.
2. The magnitude of the maximum static frictional force $\mathbf{F}_s$ that can be developed is independent of the area of contact, provided the normal pressure is not very low nor great enough to severely deform or crush the contacting surfaces of the bodies.
3. The magnitude of the maximum static frictional force is generally greater than the magnitude of the kinetic frictional force for any two surfaces of contact. However, if one of the bodies is moving with a *very low velocity* over the surface of another, F_k becomes approximately equal to F_s, i.e., $\mu_s \approx \mu_k$.
4. When *slipping* at the surface of contact is *about to occur,* the magnitude of the maximum static frictional force is proportional to the magnitude of the normal force, such that $F_s = \mu_s N$, Eq. 8–1.
5. When *slipping* at the surface of contact is *occurring,* the magnitude of the kinetic frictional force is proportional to the magnitude of the normal force, such that $F_k = \mu_k N$, Eq. 8–2.

Angle of Friction. It should be observed that Eqs. 8–1 and 8–2 have a specific yet *limited* use in the solution of friction problems. In particular, the frictional force acting at a contacting surface is determined from $F_k = \mu_k N$ *only* if *relative motion* is occurring between the two surfaces. Furthermore, if two bodies are *stationary,* the magnitude of the frictional force, F, *does not necessarily* equal $\mu_s N$; instead, F must satisfy the inequality $F \leq \mu_s N$. Only when *impending motion* occurs does F reach its upper limit, $F = F_s = \mu_s N$. This situation may be better understood by considering the block shown in Fig. 8–3*a*, which is acted upon by a force **P.** In this case consider $P = F_s$, so that the block is on the *verge of sliding*. For equilibrium, the normal force **N** and frictional force $\mathbf{F}_s$ combine to create a resultant $\mathbf{R}_s$. The angle ϕ_s that $\mathbf{R}_s$ makes with **N** is called the *angle of static friction.* From the figure,

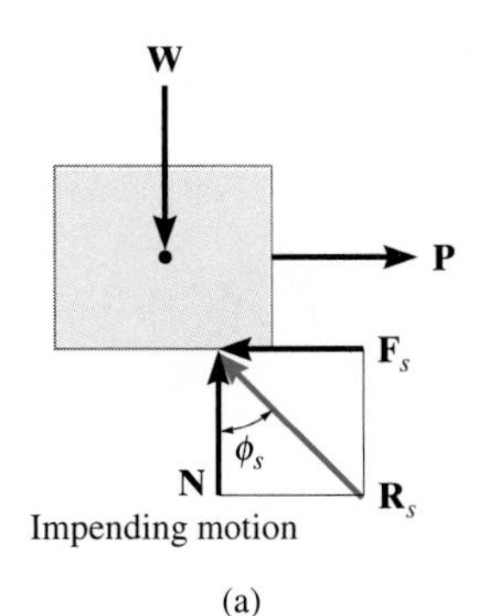

(a)

$$\phi_s = \tan^{-1}\left(\frac{F_s}{N}\right) = \tan^{-1}\left(\frac{\mu_s N}{N}\right) = \tan^{-1}\mu_s$$

Provided the block is *not in motion,* any horizontal force $P < F_s$ causes a resultant **R** which has a line of action directed at an angle ϕ from the vertical such that $\phi \leq \phi_s$. If **P** creates uniform *motion* of the block, then $P = F_k$. In this case, the resultant $\mathbf{R}_k$ has a line of action defined by ϕ_k, Fig. 8–3*b*. This angle is referred to as the *angle of kinetic friction,* where

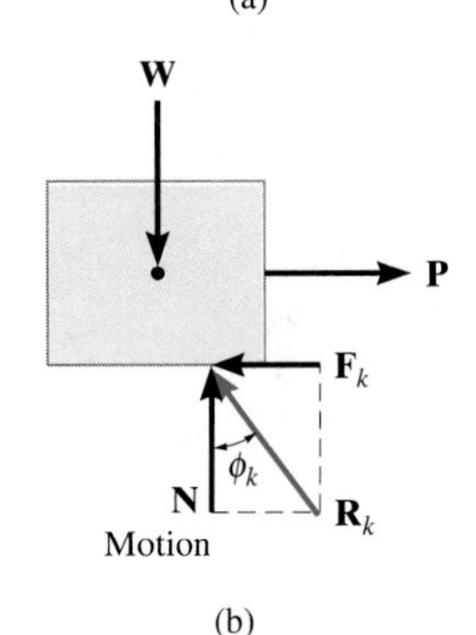

(b)

Fig. 8–3

$$\phi_k = \tan^{-1}\left(\frac{F_k}{N}\right) = \tan^{-1}\left(\frac{\mu_k N}{N}\right) = \tan^{-1}\mu_k$$

By comparison, $\phi_s \geq \phi_k$.

Angle of Repose. An experimental method which can be used to measure the coefficient of friction between two contacting surfaces consists of placing a block of one material having a weight W on a plane made of another material, Fig. 8–4*a*. The plane is inclined to the angle θ_s, at which point the block is on the *verge of sliding* and therefore $F_s = \mu_s N$. The free-body diagram of the block at this instant is shown in Fig. 8–4*b*. Applying the force equations of equilibrium the normal force $N = W\cos\theta_s$, and the frictional force $F_s = W\sin\theta_s$. Since $F_s = \mu_s N$, then $W\sin\theta_s = \mu_s(W\cos\theta_s)$ or

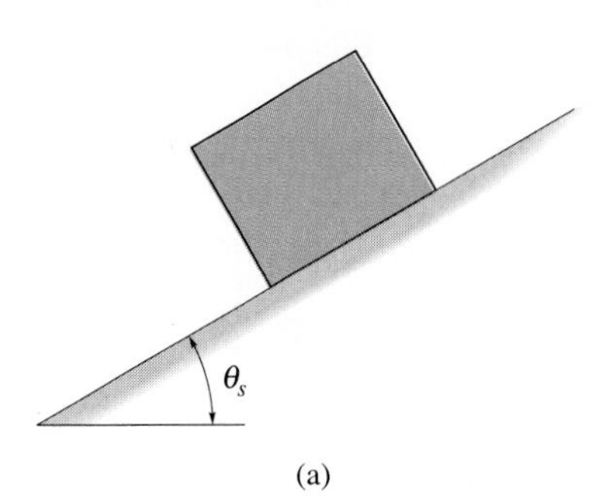

(a)

$$\theta_s = \tan^{-1}\mu_s$$

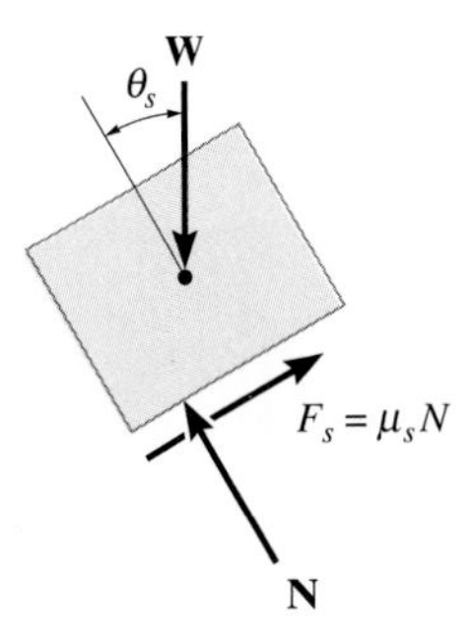

(b)

Fig. 8–4

The angle θ_s is referred to as the *angle of repose,* and by comparison it is equal to the angle of static friction ϕ_s. Once it is measured, the coefficient of static friction is obtained from $\mu_s = \tan\theta_s$. Note that this calculation is independent of the weight of the block, and so for the experiment W does not have to be known.

8.2 Problems Involving Dry Friction

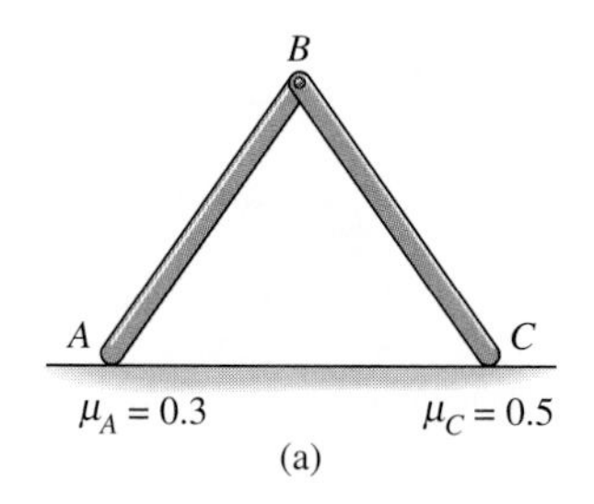

(a)

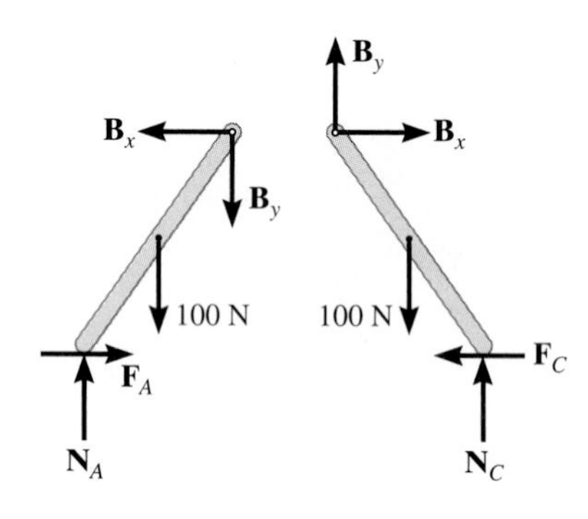

(b)

Fig. 8–5

If a rigid body is in equilibrium when it is subjected to a system of forces that includes the effect of friction, the force system must satisfy not only the equations of equilibrium but *also* the laws that govern the frictional forces.

Types of Friction Problems. In general, there are three types of mechanics problems involving dry friction. They can easily be classified once the free-body diagrams are drawn and the total number of unknowns are identified and compared with the total number of available equilibrium equations. Each type of problem will now be explained and illustrated graphically by examples. In all these cases the geometry and dimensions for the problem are assumed to be known.

Equilibrium. Problems in this category are strictly equilibrium problems which require *the total number of unknowns to be equal to the total number of available equilibrium equations.* Once the frictional forces are determined from the solution, however, their numerical values must be checked to be sure they satisfy the inequality $F \leq \mu_s N$; otherwise, slipping will occur and the body will not remain in equilibrium. A problem of this type is shown in Fig. 8–5*a*. Here we must determine the frictional forces at A and C to check if the equilibrium position of the bars can be maintained. If the bars are uniform and have known weights of 100 N each, then the free-body diagrams are as shown in Fig. 8–5*b*. There are six unknown force components which can be determined *strictly* from the six equilibrium equations (three for each member). Once F_A, N_A, F_C, and N_C are determined, then the bars will remain in equilibrium provided $F_A \leq 0.3N_A$ and $F_C \leq 0.5N_C$ are satisfied.

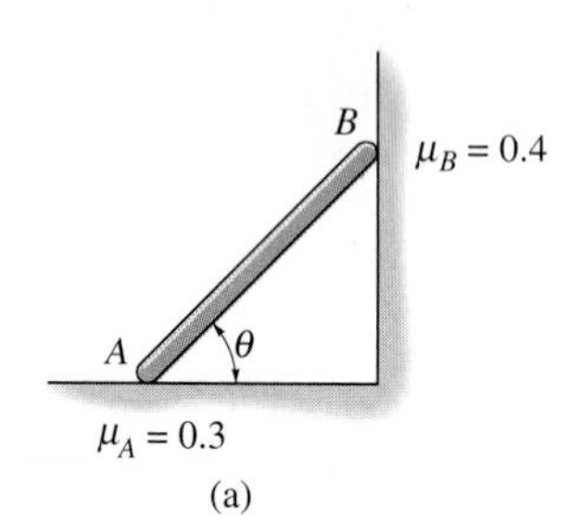

(a)

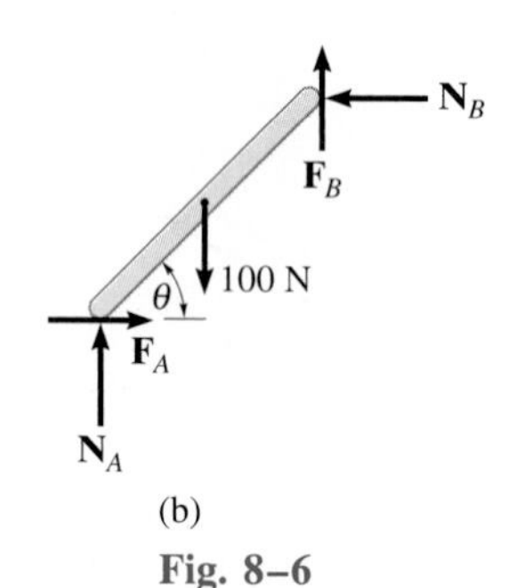

(b)

Fig. 8–6

Impending Motion at All Points. In this case *the total number of unknowns will equal the total number of available equilibrium equations plus the total number of available frictional equations,* $F = \mu N$. In particular, if *motion is impending* at the points of contact, then $F_s = \mu_s N$; whereas if the body is *slipping,* then $F_k = \mu_k N$. For example, consider the problem of finding the smallest angle θ at which the 100-N bar in Fig. 8–6*a* can be placed against the wall without slipping. The free-body diagram is shown in Fig. 8–6*b*. Here there are *five* unknowns: F_A, N_A, F_B, N_B, θ. For the solution there are *three* equilibrium equations and *two* static frictional equations which apply at *both* points of contact, so that $F_A = 0.3N_A$ and $F_B = 0.4N_B$. (It should also be noted that the bar will not be in a state where motion impends *unless* the bar slips at *both* points A and B simultaneously.)

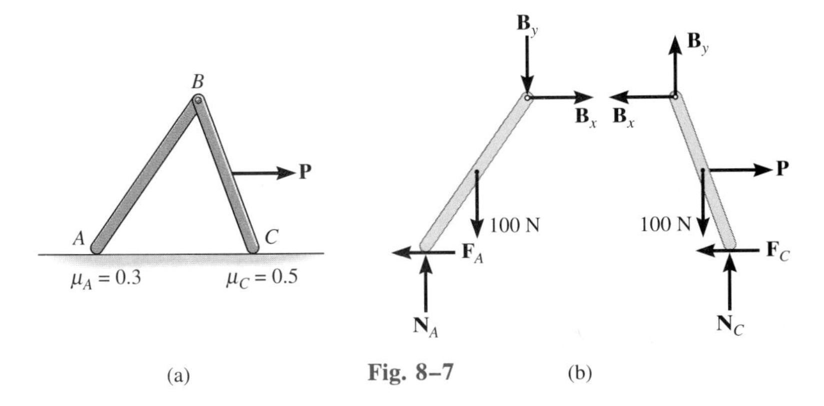

(a) **Fig. 8–7** (b)

Tipping or Impending Motion at Some Points. Here *the total number of unknowns will be less than the number of available equilibrium equations plus the total number of frictional equations or conditional equations for tipping.* As a result, several possibilities for motion or impending motion will exist and the problem will involve a determination of the kind of motion which actually occurs. For example, consider the two-member frame shown in Fig. 8–7*a*. In this problem we wish to determine the horizontal force **P** needed to cause movement of the frame. If each member has a weight of 100 N, then the free-body diagrams are as shown in Fig. 8–7*b*. There are *seven* unknowns: N_A, F_A, N_C, F_C, B_x, B_y, P. For a unique solution we must satisfy the *six* equilibrium equations (three for each member) and only *one* of two possible static frictional equations. This means that as **P** increases its magnitude it will either cause slipping at A and no slipping at C, so that $F_A = 0.3N_A$ and $F_C \leq 0.5N_C$; or slipping occurs at C and no slipping at A, in which case $F_C = 0.5N_C$ and $F_A \leq 0.3N_A$. The actual situation can be determined by calculating P for each case and then choosing the case for which P is *smallest.* If in both cases the *same value* for P is calculated, which in practice would be highly improbable, then slipping at both points occurs simultaneously; i.e., the *seven unknowns* will satisfy *eight equations.* As a second example, consider a block having a width b, height h, and weight W which is resting on a rough surface, Fig. 8–8*a*. The force **P** needed to cause motion is to be determined. Inspection of the free-body diagram, Fig. 8–8*b*, indicates that there are *four unknowns,* namely, P, F, N, and x. For a unique solution, however, we must satisfy the *three* equilibrium equations and either *one* static friction equation or *one* conditional equation which requires the block not to tip. Hence two possibilities of motion exist. Either the block will *slip,* Fig. 8–8*b*, in which case $F = \mu_s N$ and the value obtained for x must satisfy $0 \leq x \leq b/2$; or the block will *tip,* Fig. 8–8*c*, in which case $x = b/2$ and the frictional force will satisfy the inequality $F \leq \mu_s N$. The solution yielding the *smallest* value of P will define the type of motion the block undergoes. If it happens that the same value of P is calculated for both cases, although this would be very improbable, then slipping and tipping will occur simultaneously; i.e., the *four unknowns* will satisfy *five equations.*

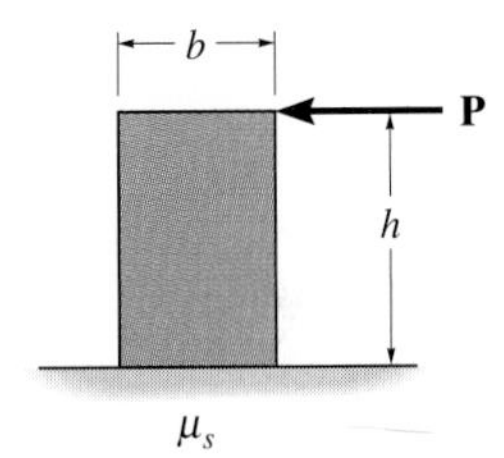

(a)

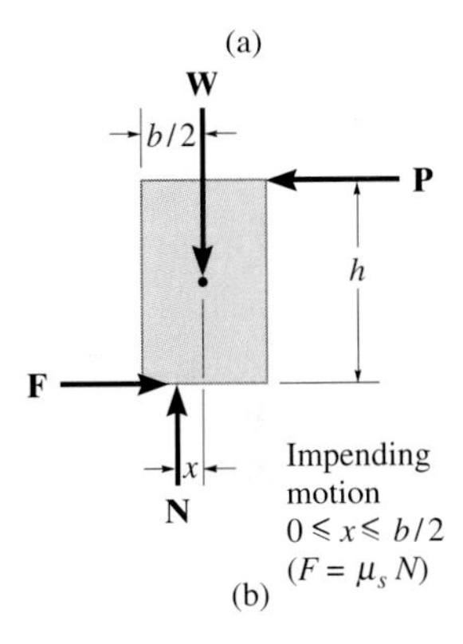

(b)

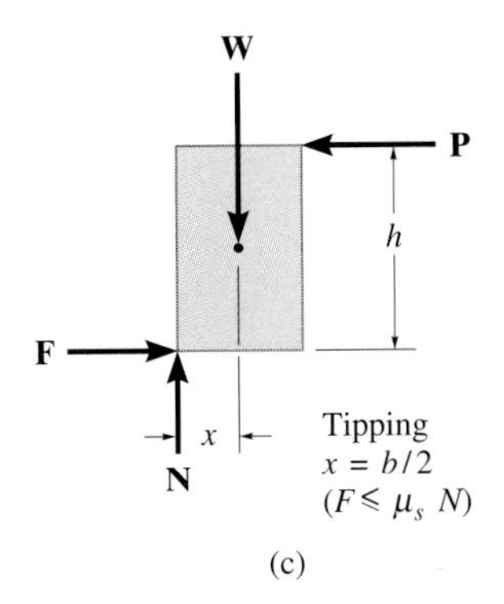

(c)

Fig. 8–8

Equilibrium Versus Frictional Equations. It was stated earlier that the frictional force *always* acts so as to either oppose the relative motion or impede the motion of a body over its contacting surface. Realize, however, that we can *assume* the sense of the frictional force in problems which require F to be an "equilibrium force" and satisfy the inequality $F < \mu_s N$. The correct sense is made known *after* solving the equations of equilibrium for F. For example, if F is a negative scalar, the sense of $\mathbf{F}$ is the reverse of that which was assumed. This convenience of *assuming* the sense of $\mathbf{F}$ is possible because the equilibrium equations equate to zero the *components of vectors* acting in the *same direction.* In cases where the frictional equation $F = \mu N$ is used in the solution of a problem, however, the convenience of *assuming* the sense of $\mathbf{F}$ is *lost,* since the frictional equation relates only the *magnitudes* of two perpendicular vectors. Consequently, $\mathbf{F}$ *must always* be shown acting with its *correct sense* on the free-body diagram whenever the frictional equation is used for the solution of a problem.

PROCEDURE FOR ANALYSIS

The following procedure provides a method for solving equilibrium problems involving dry friction.

Free-Body Diagrams. Draw the necessary free-body diagrams and determine the number of unknowns or equations required for a complete solution. Unless stated in the problem, *always* show the frictional forces as *unknowns;* i.e., *do not assume that* $F = \mu N$. Recall that only three equations of coplanar equilibrium can be written for each body. Consequently, if there are more unknowns than equations of equilibrium, it will be necessary to apply the frictional equation at some, if not all, points of contact to obtain the extra equations needed for a complete solution.

Equations of Friction and Equilibrium. Apply the equations of equilibrium and the necessary frictional equations (or conditional equations if tipping is involved) and solve for the unknowns. If the problem involves a three-dimensional force system such that it becomes difficult to obtain the force components or the necessary moment arms, apply the equations of equilibrium using Cartesian vectors.

The following example problems illustrate this procedure numerically.

Example 8–1

The uniform crate shown in Fig. 8–9*a* has a mass of 20 kg. If a force $P = 80$ N is applied to the crate, determine if it remains in equilibrium. The coefficient of static friction is $\mu_s = 0.3$.

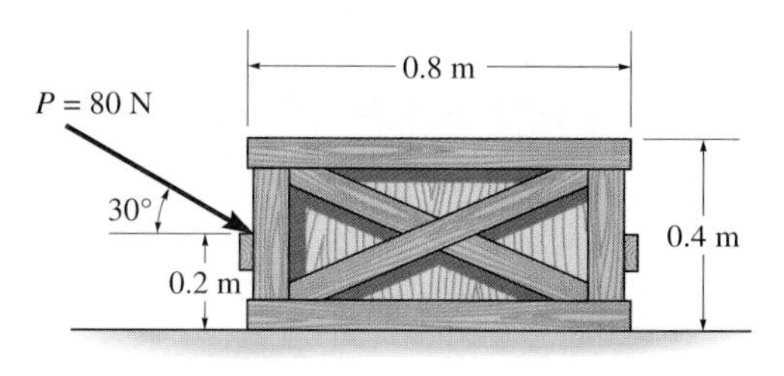

(a)

Fig. 8–9

SOLUTION

Free-Body Diagram. As shown in Fig. 8–9*b*, the *resultant* normal force $\mathbf{N}_C$ must act a distance x from the crate's center line in order to counteract the tipping effect caused by **P**. There are *three unknowns: F, N_C,* and x, which can be determined strictly from the *three* equations of equilibrium.

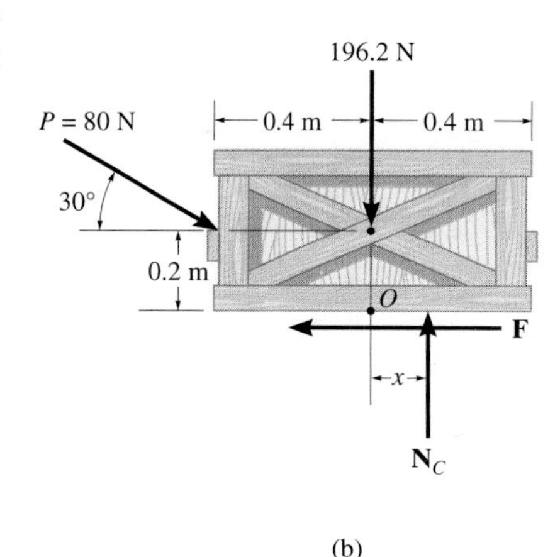

(b)

Equations of Equilibrium

$$\xrightarrow{+}\Sigma F_x = 0; \qquad 80 \cos 30° \text{ N} - F = 0$$

$$+\uparrow\Sigma F_y = 0; \qquad -80 \sin 30° \text{ N} + N_C - 196.2 \text{ N} = 0$$

$$\circlearrowright+\Sigma M_O = 0; \quad 80 \sin 30° \text{ N}(0.4 \text{ m}) - 80 \cos 30° \text{ N}(0.2 \text{ m}) + N_C(x) = 0$$

Solving,

$$F = 69.3 \text{ N}$$

$$N_C = 236 \text{ N}$$

$$x = -0.00908 \text{ m} = -9.08 \text{ mm}$$

Since x is negative it indicates the *resultant* normal force acts (slightly) to the *left* of the crate's center line. No tipping will occur since $x \leq 0.4$ m. Also, the *maximum* frictional force which can be developed at the surface of contact is $F_{\text{max}} = \mu_s N_C = 0.3(236 \text{ N}) = 70.8$ N. Since $F = 69.3 \text{ N} < 70.8$ N, the crate will *not slip,* although it is very close to doing so.

Example 8–2

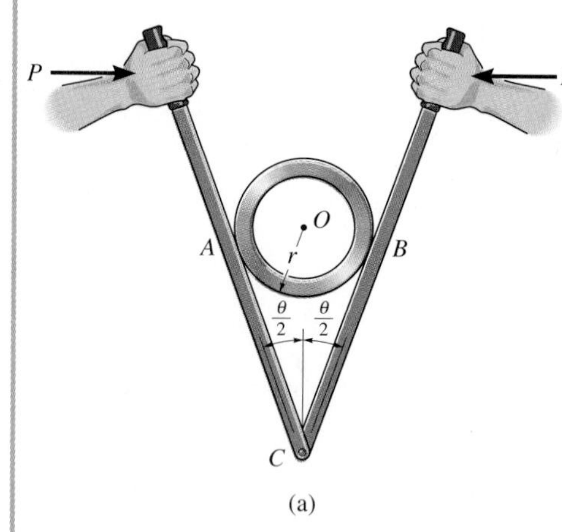

(a)

The pipe shown in Fig. 8–10*a* is gripped between two levers that are pinned together at *C*. If the coefficient of static friction between the levers and the pipe is $\mu = 0.3$, determine the maximum angle θ at which the pipe can be gripped without slipping. Neglect the weight of the pipe.

SOLUTION

Free-Body Diagram. As shown in Fig. 8–10*b*, there are five unknowns: N_A, F_A, N_B, F_B, and θ. The *three* equations of equilibrium and *two* frictional equations at *A* and *B* apply. The frictional forces act toward *C* to prevent upward motion of the pipe.

Equations of Friction and Equilibrium. The frictional equations are

$F_s = \mu_s N;$ $$F_A = \mu N_A$$ $$F_B = \mu N_B$$

Using these results, and applying the equations of equilibrium, yields

$\overset{+}{\rightarrow}\Sigma F_x = 0;$

$$N_A \cos\left(\frac{\theta}{2}\right) + \mu N_A \sin\left(\frac{\theta}{2}\right) - N_B \cos\left(\frac{\theta}{2}\right) - \mu N_B \sin\left(\frac{\theta}{2}\right) = 0 \quad (1)$$

$\curvearrowright + \Sigma M_O = 0;$

$$-\mu N_B(r) + \mu N_A(r) = 0 \quad (2)$$

$+\uparrow \Sigma F_y = 0;$

$$N_A \sin\left(\frac{\theta}{2}\right) - \mu N_A \cos\left(\frac{\theta}{2}\right) + N_B \sin\left(\frac{\theta}{2}\right) - \mu N_B \cos\left(\frac{\theta}{2}\right) = 0 \quad (3)$$

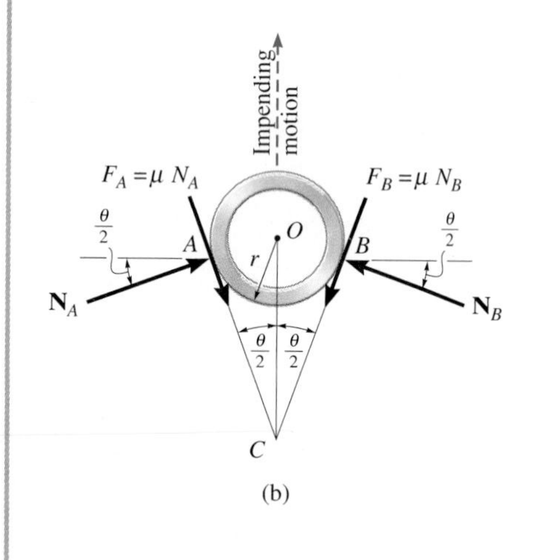

(b)

Fig. 8–10

From either Eq. 1 or 2 it is seen that $N_A = N_B$. This could also have been determined directly from the symmetry of *both* geometry and loading. Substituting the result into Eq. 3, we obtain

$$\sin\left(\frac{\theta}{2}\right) - \mu \cos\left(\frac{\theta}{2}\right) = 0$$

so that

$$\tan\left(\frac{\theta}{2}\right) = \frac{\sin(\theta/2)}{\cos(\theta/2)} = \mu = 0.3$$

$$\theta = 2 \tan^{-1} 0.3 = 33.4^\circ$$ ***Ans.***

Example 8–3

The uniform rod having a weight W and length l is supported at its ends A and B, where the coefficient of static friction is μ, Fig. 8–11a. Determine the greatest angle θ so the rod does not slip. Neglect the thickness of the rod for the calculation.

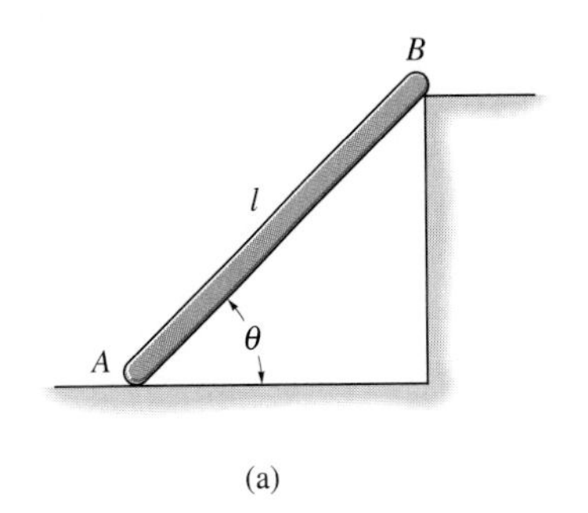

(a)

SOLUTION

Free-Body Diagram. As shown in Fig. 8–11b, there are *five* unknowns: F_A, N_A, F_B, N_B, and θ. These can be determined from the *three* equilibrium equations and *two* frictional equations applied at points A and B. The frictional forces must be drawn with their correct sense so that they oppose the tendency for motion of the rod. Why?

Equations of Friction and Equilibrium. Writing the frictional equations,

$F = \mu_s N;$
$$F_A = \mu N_A$$
$$F_B = \mu N_B$$

Using these results and applying the equations of equilibrium yields

$\xrightarrow{+} \Sigma F_x = 0;$
$$\mu N_A + \mu N_B \cos\theta - N_B \sin\theta = 0 \qquad (1)$$

$+\uparrow \Sigma F_y = 0;$
$$N_A - W + N_B \cos\theta + \mu N_B \sin\theta = 0 \qquad (2)$$

$\circlearrowleft + \Sigma M_G = 0;$
$$-N_A\left(\frac{l}{2}\cos\theta\right) + \mu N_A\left(\frac{l}{2}\sin\theta\right) + N_B\left(\frac{l}{2}\right) = 0 \qquad (3)$$

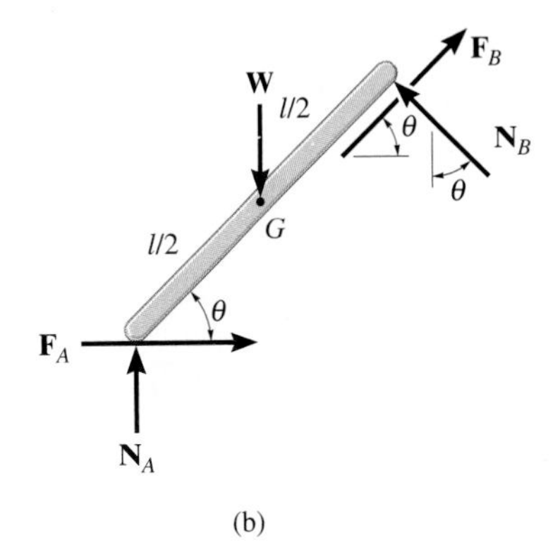

(b)

Fig. 8–11

Moments were summed about the center of the rod G in order to eliminate W. We can solve Eqs. 1 and 3, which reduce to

$$\mu N_A = N_B(\sin\theta - \mu\cos\theta)$$
$$N_B = N_A(\cos\theta - \mu\sin\theta)$$

Thus

$$\mu N_A = N_A(\cos\theta - \mu\sin\theta)(\sin\theta - \mu\cos\theta)$$
$$\mu = \sin\theta\cos\theta - \mu\cos^2\theta - \mu\sin^2\theta + \mu^2\sin\theta\cos\theta$$
$$\mu = (1+\mu^2)\sin\theta\cos\theta - \mu(\sin^2\theta + \cos^2\theta)$$

Since $\sin^2\theta + \cos^2\theta = 1$ and $\sin 2\theta = 2\sin\theta\cos\theta$, then

$$2\mu = \left(\frac{1+\mu^2}{2}\right)\sin 2\theta$$

Solving for θ, we have

$$\theta = \frac{1}{2}\sin^{-1}\left(\frac{4\mu}{1+\mu^2}\right) \qquad \textit{Ans.}$$

Example 8–4

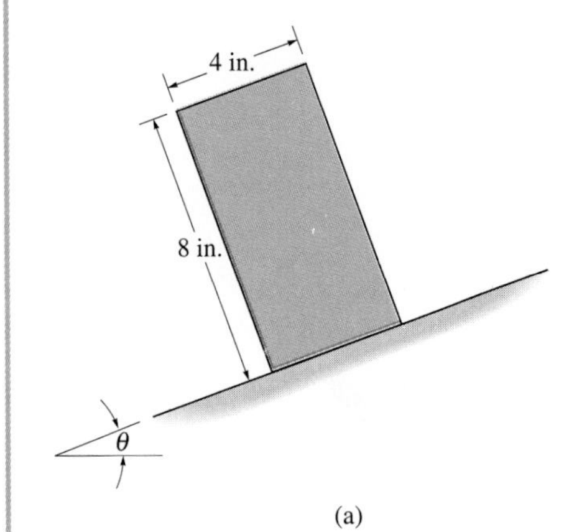

(a)

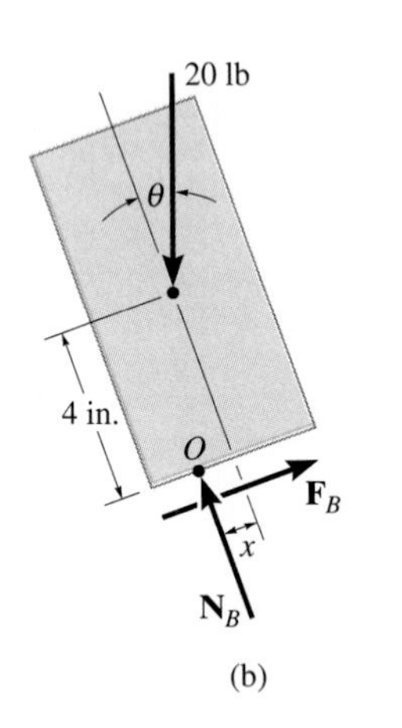

(b)

Fig. 8–12

The homogeneous block shown in Fig. 8–12*a* has a weight of 20 lb and rests on the incline for which $\mu_s = 0.55$. Determine the largest angle of tilt, θ, of the plane before the block moves.

SOLUTION

Free-Body Diagram. As shown in Fig. 8–12*b*, the dimension x is used to locate the position of the resultant normal force $\mathbf{N}_B$ under the block. There are *four* unknowns, θ, N_B, F_B, and x. *Three* equations of equilibrium are available. The *fourth* equation is obtained by investigating the conditions for tipping or sliding of the block.

Equations of Equilibrium. Applying the equations of equilibrium yields

$$+\swarrow \Sigma F_x = 0; \qquad 20 \sin \theta \text{ lb} - F_B = 0 \qquad (1)$$

$$+\nwarrow \Sigma F_y = 0; \qquad N_B - 20 \cos \theta \text{ lb} = 0 \qquad (2)$$

$$\circlearrowright + \Sigma M_O = 0; \qquad 20 \sin \theta \text{ lb}(4 \text{ in.}) - 20 \cos \theta \text{ lb}(x) = 0 \qquad (3)$$

(Impending Motion of Block.) This requires use of the frictional equation

$$F_s = \mu_s N; \qquad F_B = 0.55 N_B \qquad (4)$$

Solving Eqs. 1 through 4 yields

$$N_B = 17.5 \text{ lb} \qquad F_B = 9.64 \text{ lb} \qquad \theta = 28.8^\circ \qquad x = 2.2 \text{ in.}$$

Since $x = 2.2 \text{ in.} > 2 \text{ in.}$, the block will tip *before* sliding.

(Tipping of Block.) This requires

$$x = 2 \text{ in.} \qquad (5)$$

Solving Eqs. 1 through 3 using Eq. 5 yields

$$N_B = 17.9 \text{ lb} \qquad F_B = 8.94 \text{ lb}$$

$$\theta = 26.6^\circ \qquad \textit{Ans.}$$

Note: If we *first* assumed that the block tips, then the results for F_B would have to be checked with the maximum *possible* static frictional force; i.e.,

$$F_B = 8.94 \text{ lb} \overset{?}{<} (0.55)(17.9 \text{ lb}) = 9.84 \text{ lb}$$

Since the inequality holds, indeed the block will tip before it slips.

Example 8–5

Beam AB is subjected to a uniform load of 200 N/m and is supported at B by a post BC, Fig. 8–13a. If the coefficients of static friction at B and C are $\mu_B = 0.2$ and $\mu_C = 0.5$, determine the force $\mathbf{P}$ needed to pull the post out from under the beam. Neglect the weight of the members and the thickness of the post.

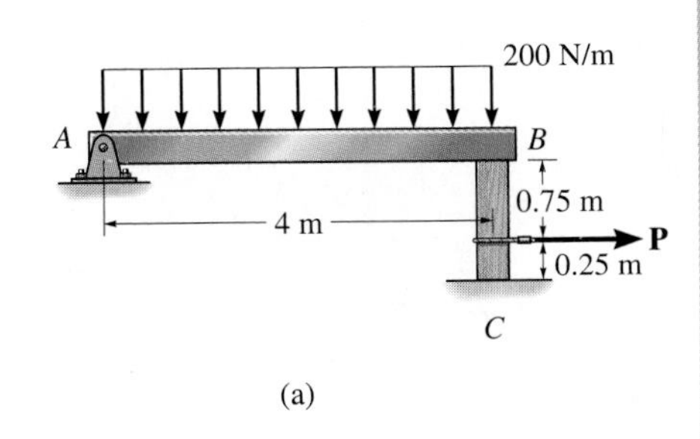

(a)

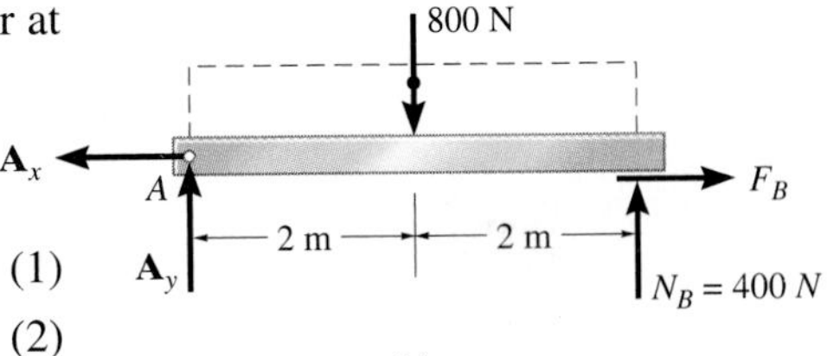

(b)

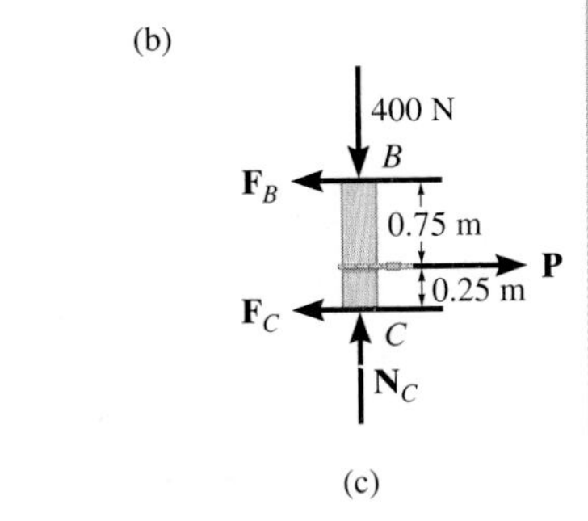

(c)

Fig. 8–13

SOLUTION

Free-Body Diagrams. The free-body diagram of beam AB is shown in Fig. 8–13b. Applying $\Sigma M_A = 0$, we obtain $N_B = 400$ N. This result is shown on the free-body diagram of the post, Fig. 8–13c. Referring to this member, the *four* unknowns F_B, P, F_C, and N_C are determined from the *three* equations of equilibrium and *one* frictional equation applied either at B or C.

Equations of Equilibrium and Friction.

$\overset{+}{\rightarrow}\Sigma F_x = 0;$ $\quad P - F_B - F_C = 0 \quad (1)$

$+\uparrow\Sigma F_y = 0;$ $\quad N_C - 400\text{ N} = 0 \quad (2)$

$\circlearrowleft+\Sigma M_C = 0;$ $\quad -P(0.25\text{ m}) + F_B(1\text{ m}) = 0 \quad (3)$

(Post Slips Only at B.) This requires $F_C \leq \mu N_C$ and

$F_B = \mu_B N_B;$ $\quad F_B = 0.2(400\text{ N}) = 80\text{ N}$

Using this result and solving Eqs. 1 through 3, we obtain

$$P = 320\text{ N}$$
$$F_C = 240\text{ N}$$
$$N_C = 400\text{ N}$$

Since $F_C = 240\text{ N} > \mu_C N_C = 0.5(400\text{ N}) = 200\text{ N}$, the other case of movement must be investigated.

(Post Slips Only at C.) Here $F_B \leq \mu_B N_B$ and

$F_C = \mu_C N_C;$ $\quad F_C = 0.5 N_C \quad (4)$

Solving Eqs. 1 through 4 yields

$$P = 267\text{ N} \qquad \textit{Ans.}$$
$$N_C = 400\text{ N}$$
$$F_C = 200\text{ N}$$
$$F_B = 66.7\text{ N}$$

Obviously, this case occurs first, since it requires a *smaller* value for P.

Example 8–6

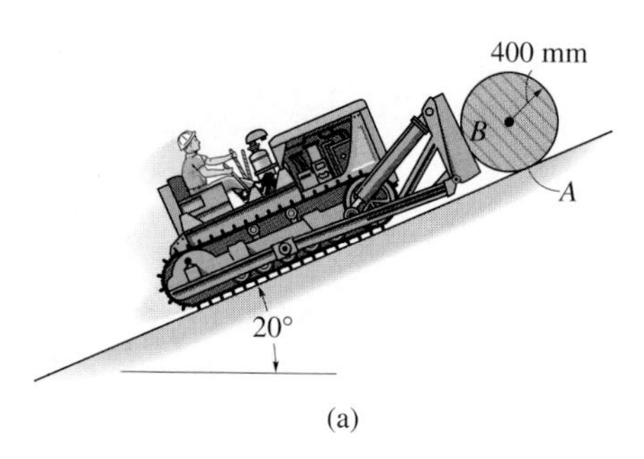

(a)

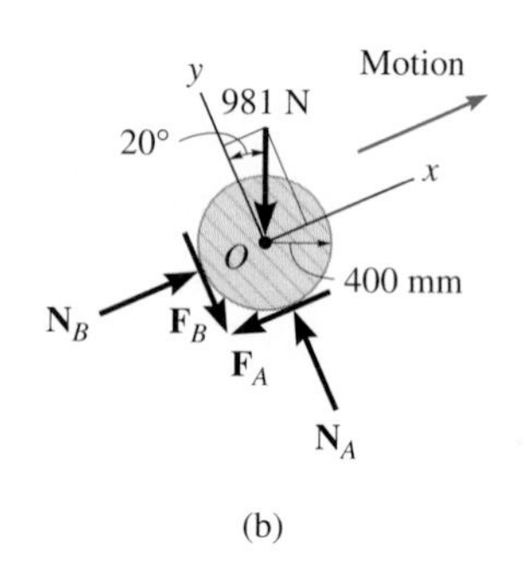

(b)

Fig. 8–14

Determine the normal force that must be exerted on the 100-kg spool shown in Fig. 8–14*a* to push it up the 20° incline at constant velocity. The coefficients of static and kinetic friction at the points of contact are $(\mu_s)_A = 0.18$, $(\mu_k)_A = 0.15$ and $(\mu_s)_B = 0.45$, $(\mu_k)_B = 0.4$.

SOLUTION

Free-Body Diagram. As shown in Fig. 8–14*b*, there are four unknowns N_A, F_A, N_B, and F_B acting on the spool. These can be determined from the *three* equations of equilibrium and *one* frictional equation, which applies either at *A* or *B*. If slipping only occurs at *B*, the spool *rolls* up the incline; whereas if slipping only occurs at *A*, the spool will *slide* up the incline. Here we must calculate N_B.

Equations of Equilibrium and Friction

$$+\nearrow\Sigma F_x = 0; \qquad -F_A + N_B - 981 \sin 20° \text{ N} = 0 \qquad (1)$$

$$+\nwarrow\Sigma F_y = 0; \qquad N_A - F_B - 981 \cos 20° \text{ N} = 0 \qquad (2)$$

$$\circlearrowright+\Sigma M_O = 0; \qquad F_B(400 \text{ mm}) - F_A(400 \text{ mm}) = 0 \qquad (3)$$

(Spool Rolls up Incline.) In this case $F_A \leq 0.18N_A$ and

$$(F_k)_B = (\mu_k)_B N_B; \qquad F_B = 0.40N_B \qquad (4)$$

The direction of the frictional force at *B* must be specified correctly. Why? Since the spool is being forced up the plane, $\mathbf{F}_B$ acts downward to prevent the clockwise rolling motion of the spool, Fig. 8–14*b*. Solving Eqs. 1 through 4, we have

$$N_A = 1146 \text{ N} \qquad F_A = 224 \text{ N} \qquad N_B = 559 \text{ N} \qquad F_B = 224 \text{ N}$$

The assumption regarding no slipping at *A* should be checked.

$$F_A \leq (\mu_s)_A N_A; \qquad 224 \text{ N} \overset{?}{\leq} 0.18(1146 \text{ N}) = 206 \text{ N}$$

The inequality does *not apply,* and therefore slipping occurs at *A* and not at *B*. Hence, the other case of motion must be investigated.

(Spool Slides up Incline.) In this case, $F_B \leq 0.45N_B$ and

$$(F_k)_A = (\mu_k)_A N_A; \qquad F_A = 0.15N_A \qquad (5)$$

Solving Eqs. 1 through 3 and 5 yields

$$N_A = 1084 \text{ N} \qquad F_A = 163 \text{ N} \qquad N_B = 498 \text{ N} \qquad F_B = 163 \text{ N}$$

The validity of the solution ($N_B = 498$ N) can be checked by testing the assumption that indeed no slipping occurs at *B*.

$$F_B \leq (\mu_s)_B N_B; \qquad 163 \text{ N} < 0.45(498 \text{ N}) = 224 \text{ N} \qquad \text{(check)}$$

PROBLEMS

8–1. Determine the horizontal force P needed to just start moving the 300-lb crate up the plane. Take $\mu_s = 0.3$.

8–2. Determine the range of values for which the horizontal force **P** will prevent the 300-lb crate from slipping down or up the inclined plane. Take $\mu_s = 0.1$.

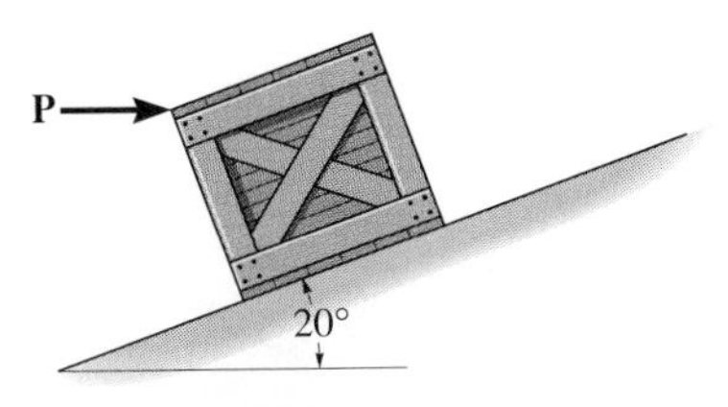

Probs. 8–1/8–2

8–3. If the horizontal force $P = 80$ lb, determine the normal and frictional forces acting on the 300-lb crate. Take $\mu_s = 0.3$, $\mu_k = 0.2$.

***8–4.** If the horizontal force $P = 140$ lb, determine the normal and frictional forces acting on the 300-lb crate. Take $\mu_s = 0.3$, $\mu_k = 0.2$.

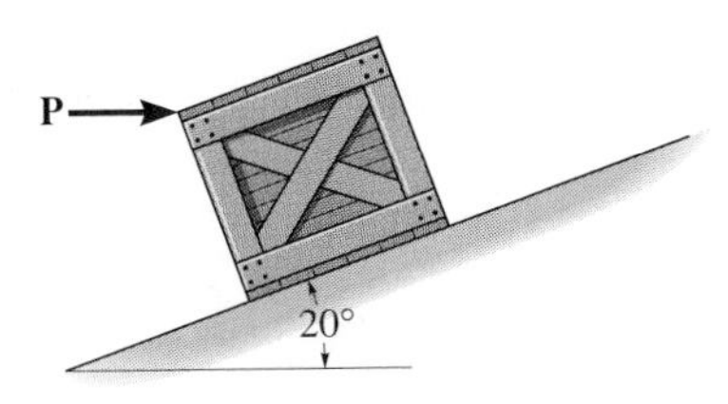

Probs. 8–3/8–4

8–5. Determine the magnitude of force **P** needed to start towing the 40-kg crate. Also determine the location of the resultant normal force acting on the crate, measured from point A. Take $\mu_s = 0.3$.

8–6. Determine the friction force on the 40-kg crate, and the resultant normal force if the force $P = 300$ N. Take $\mu_s = 0.5$ and $\mu_k = 0.2$.

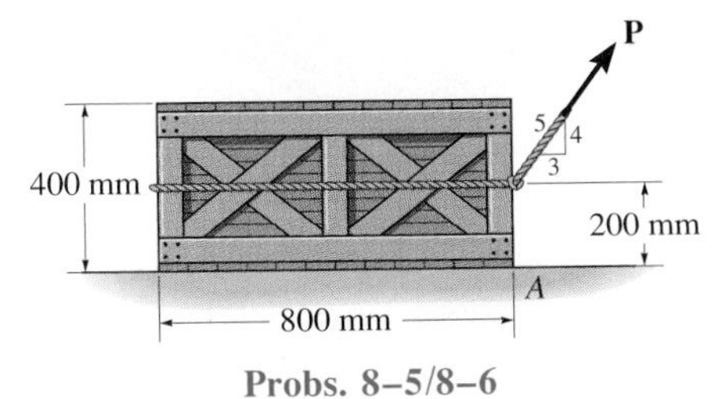

Probs. 8–5/8–6

8–7. The loose-fitting collar is supported by the pipe for which the coefficient of static friction at the points of contact A and B is $\mu_s = 0.2$. Determine the smallest dimension d so the rod will not slip when the load **P** is applied.

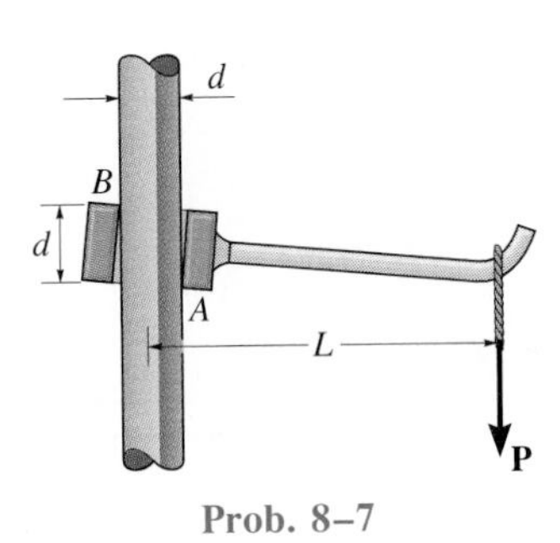

Prob. 8–7

***8–8.** An axial force of $T = 800$ lb is applied to the bar. If the coefficient of static friction at the jaws C and D is $\mu_s = 0.5$, determine the smallest normal force that the screw at A must exert on the smooth surface of the links at B and C in order to hold the bar stationary. The links are pin-connected at F and G.

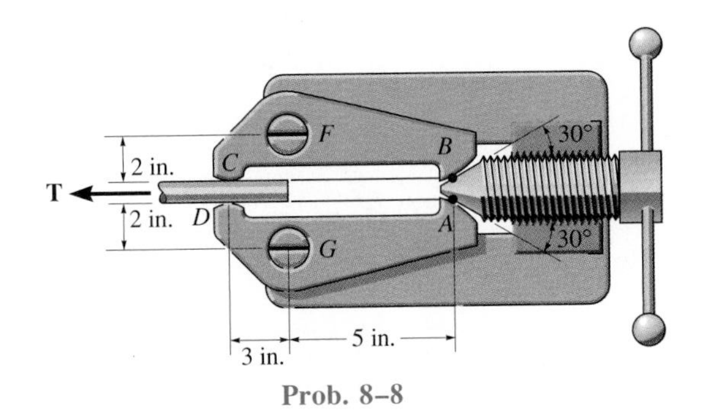

Prob. 8–8

8–9. The block brake consists of a pin-connected lever and friction block at B. The coefficient of static friction between the wheel and the lever is $\mu_s = 0.3$, and a torque of 5 N · m is applied to the wheel. Determine if the brake can hold the wheel stationary when the force applied to the lever is (a) $P = 30$ N, (b) $P = 70$ N.

8–10. Solve Prob. 8–9 if the 5-N · m torque is applied counterclockwise.

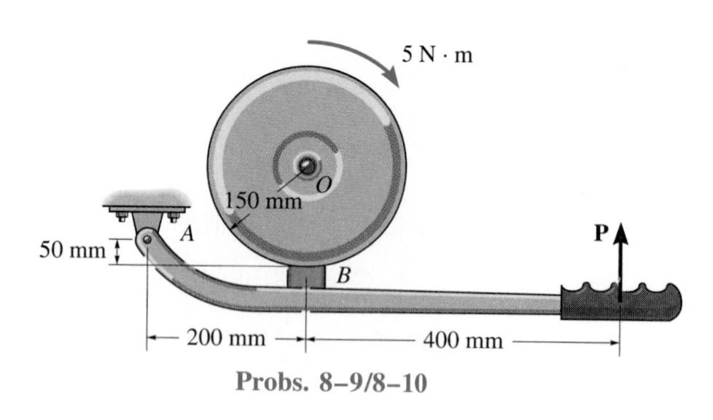

Probs. 8–9/8–10

8–11. The 15-ft ladder has a uniform weight of 80 lb and rests against the smooth wall at B. If the coefficient of static friction at A is $\mu_A = 0.4$, determine if the ladder will slip. Take $\theta = 60°$.

***8–12.** Determine the smallest angle θ at which the ladder in Prob. 8–11 can be placed against the side of the smooth wall without having it slip.

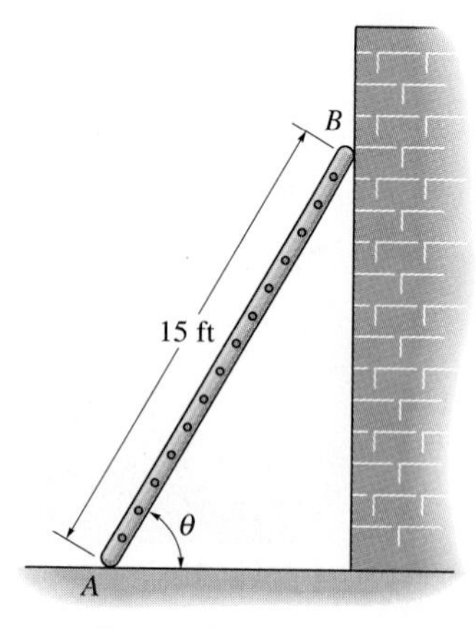

Probs. 8–11/8–12

8–13. The winch on the truck is used to hoist the garbage bin onto the bed of the truck. If the loaded bin has a weight of 8500 lb and center of gravity at G, determine the force in the cable needed to begin the lift. The coefficients of static friction at A and B are $\mu_A = 0.3$ and $\mu_B = 0.2$, respectively. Neglect the height of the support at A.

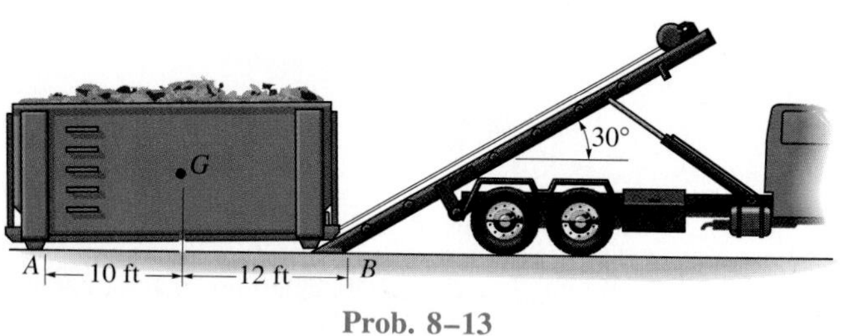

Prob. 8–13

8–14. The car has a mass of 1.6 Mg and center of mass at G. If the coefficient of static friction between the shoulder of the road and the tires is $\mu_s = 0.4$, determine the greatest slope θ the shoulder can have without causing the car to slip or tip over if the car travels along the shoulder at constant velocity.

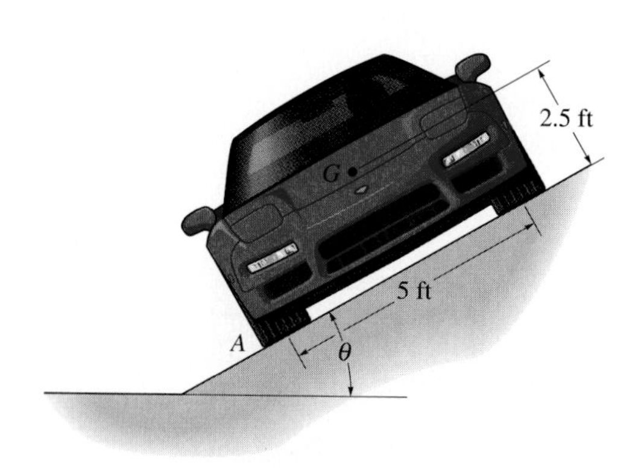

Prob. 8–14

8–15. The man pushes against the stack of four uniform boxes each weighing 20 lb. If the coefficient of static friction between each box is $\mu_s = 0.6$ and between the floor and the bottom box $\mu_s' = 0.4$, determine the greatest horizontal force P that can be applied without causing any slipping or tipping.

***8–16.** The man pushes against the stack of four uniform boxes each weighing 20 lb. If $P = 10$ lb, determine if the stack will slip or tip. The coefficient of static friction between each box is $\mu_s = 0.6$ and between the floor and the bottom box $\mu_s' = 0.4$.

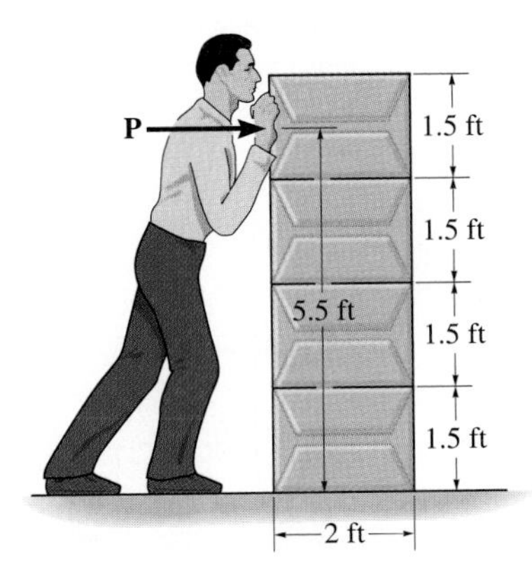

Probs. 8–15/8–16

8–17. The uniform hoop of weight W is suspended from the peg at A and a horizontal force **P** is slowly applied at B. If the hoop begins to slip at A when $\theta = 30°$, determine the coefficient of static friction between the hoop and the peg.

8–18. The uniform hoop of weight W is suspended from the peg at A and a horizontal force **P** is slowly applied at B. If the coefficient of static friction between the hoop and peg is $\mu_s = 0.2$, determine if it is possible for the angle $\theta = 30°$ before the hoop begins to slip.

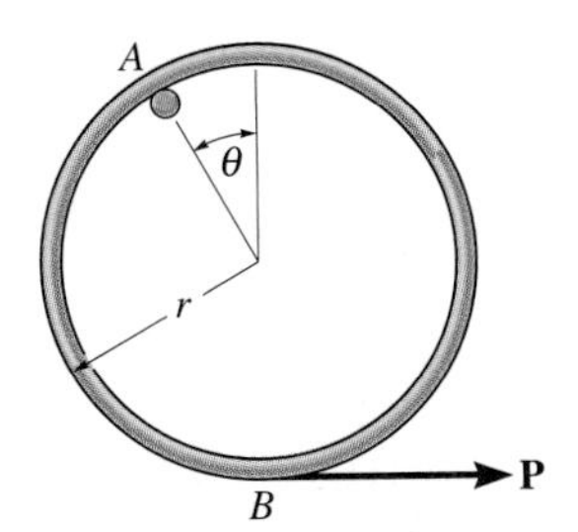

Probs. 8–17/8–18

8–19. The tractor has a weight of 4500 lb with center of gravity at G. The driving traction is developed at the rear wheels B, while the front wheels at A are free to roll. If the coefficient of static friction between the wheels at B and the ground is $\mu_s = 0.5$, determine the largest force P at which it can pull without causing the wheels at B to slip or the front wheels at A to lift off the ground.

***8–20.** The tractor has a weight of 4500 lb with center of gravity at G. The driving traction is developed at the rear wheels B, while the front wheels at A are free to roll. If the coefficient of static friction between the wheels at B and the ground is $\mu_s = 0.5$, determine if it is possible to pull at $P = 1200$ lb without causing the wheels at B to slip or the front wheels at A to lift off the ground.

Probs. 8–19/8–20

8–21. The friction tongs are used to drag the 100-kg pallet and load along the floor. Determine the tension in the chain and the minimum coefficient of static friction at the shoes A and B of the tongs so slipping of the tongs does not occur. The coefficient of static friction between the pallet and the floor is $\mu_s = 0.4$.

8–22. The coefficient of static friction between the shoes at A and B of the tongs and the pallet is $\mu_s' = 0.5$, and between the pallet and the floor $\mu_s = 0.4$. If a horizontal towing force of $P = 300$ N is applied to the tongs, determine the largest mass that can be towed.

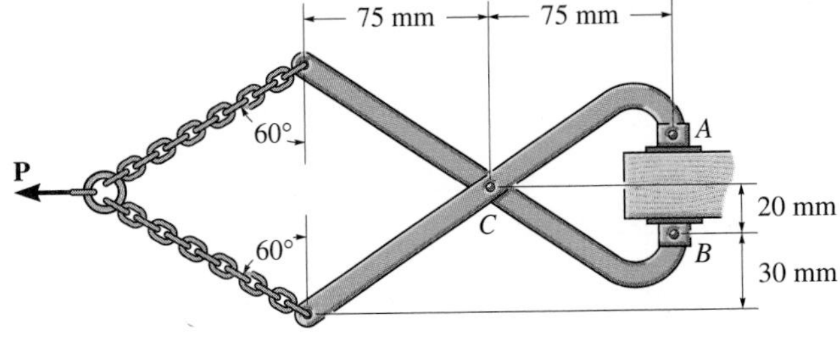

Probs. 8–21/8–22

8–23. Determine the maximum weight W the man can lift with constant velocity using the pulley system, without and then with the "leading block" or pulley at A. The man has a weight of 200 lb and the coefficient of static friction between his feet and the ground is $\mu_s = 0.6$.

***8–24.** If the weight of the load is $W = 80$ lb, determine the normal and frictional forces acting on the 200-lb man needed to support the load in each case. The coefficient of static friction between his feet and the ground is $\mu_s = 0.6$.

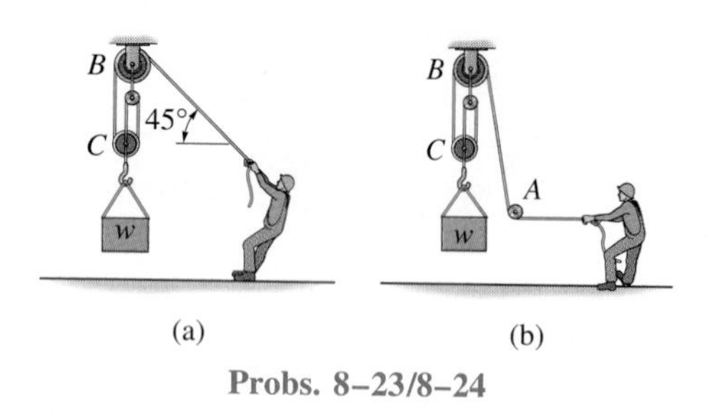

Probs. 8–23/8–24

8–26. The uniform dresser has a weight of 90 lb and rests on a tile floor for which $\mu_s = 0.25$. If the man pushes on it in the horizontal direction $\theta = 0°$, determine the smallest magnitude of force **F** needed to move the dresser. Also, if the man has a weight of 150 lb, determine the smallest coefficient of static friction between his shoes and the floor so that he does not slip.

8–27. The uniform dresser has a weight of 90 lb and rests on a tile floor for which $\mu_s = 0.25$. If the man pushes on it in the direction $\theta = 30°$, determine the smallest magnitude of force **F** needed to move the dresser. Also, if the man has a weight of 150 lb, determine the smallest coefficient of static friction between his shoes and the floor so that he does not slip.

Probs. 8–26/8–27

8–25. The pipe is hoisted using the tongs. If the coefficient of static friction at A and B is μ_s, determine the smallest dimension b so that any pipe of inner diameter d can be lifted.

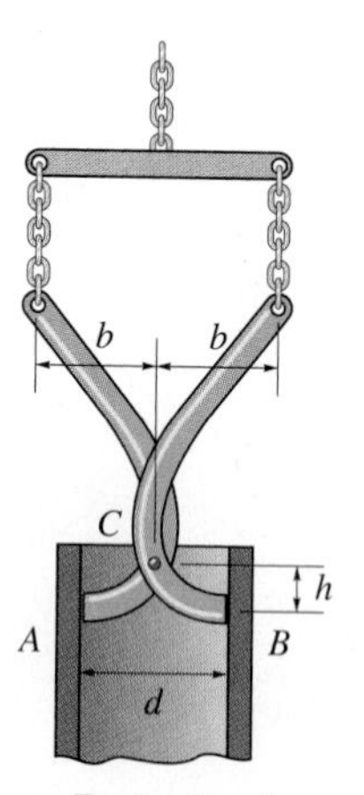

Prob. 8–25

***8–28.** The spool has a mass of 200 kg and rests against the wall and on the beam. If the coefficient of static friction at A and B is $\mu_A = 0.4$ and $\mu_B = 0.5$, respectively, determine the smallest vertical force P that must be applied to the cable that will cause the spool to turn.

8–29. The spool has a mass of 200 kg and rests against the wall and on the beam. If the coefficient of static friction at B is $\mu_B = 0.3$, and the wall is smooth, determine the friction force developed at B when the vertical force applied to the cable is $P = 800$ N.

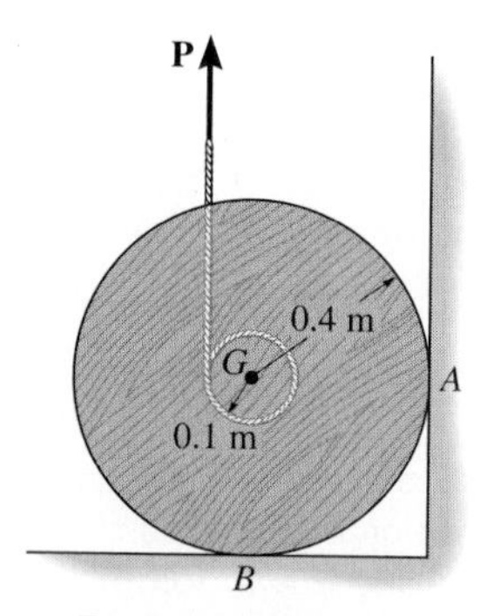

Probs. 8–28/8–29

8–30. Car A has a mass of 1.4 Mg and mass center at G. If car B exerts a horizontal force on A of 2 kN, determine if this force is great enough to move car A. The coefficients of static and kinetic friction between the tires and the road are $\mu_s = 0.5$ and $\mu_k = 0.35$. Assume B's bumper is smooth and B does not slip.

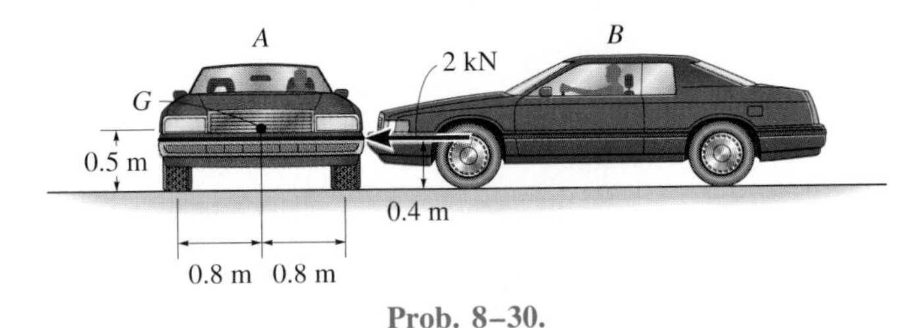

Prob. 8–30.

8–31. The board can be adjusted vertically by tilting it up and sliding the smooth pin A along the vertical guide G. When placed horizontally, the bottom C then bears along the edge of the guide, where $\mu_s = 0.4$. Determine the largest dimension d which will support any applied force $\mathbf{F}$ without causing the board to slip downward.

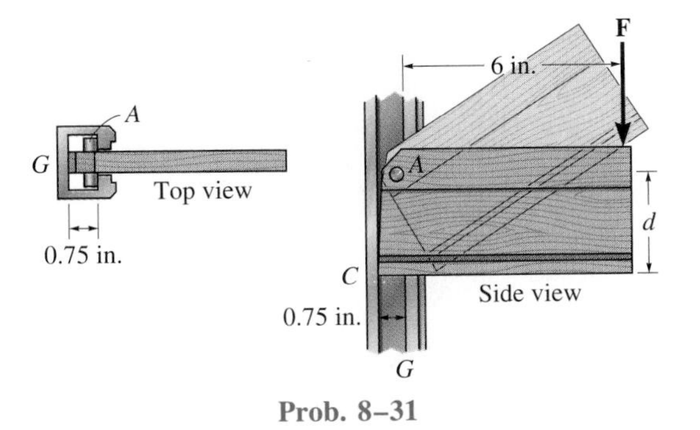

Prob. 8–31

***8–32.** Determine how far d the man can walk slowly up the plank without causing the plank to slip. The coefficient of static friction at A and B is $\mu_s = 0.3$. The man has a weight of 200 lb and a center of gravity at G. Neglect the thickness and weight of the plank.

8–33. Determine how far d the man can walk slowly up the 25-lb plank without causing it to slip. The coefficient of static friction at A is $\mu_A = 0.5$, and the surface at B is smooth. The man has a weight of 200 lb and a center of gravity at G. Neglect the thickness of the plank.

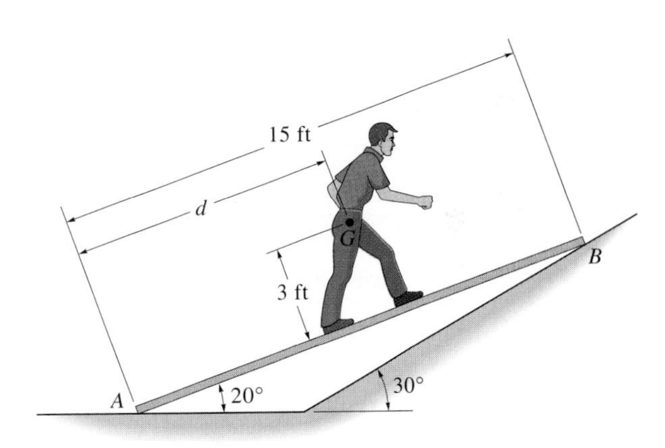

Prob. 8–32/8–33

8–34. The homogeneous semicylinder has a mass m and mass center at G. Determine the largest angle θ of the inclined plane upon which it rests so that it does not slip down the plane. The coefficient of static friction between the plane and the cylinder is $\mu_s = 0.3$. Also, what is the angle ϕ for this case?

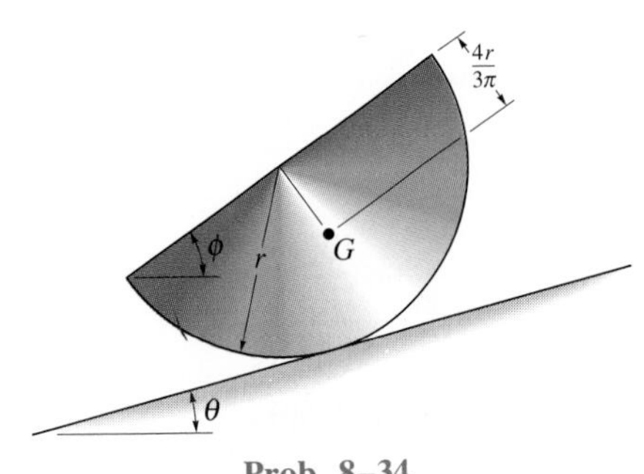

Prob. 8–34.

8–35. The 800-lb concrete pipe is being lowered from the truck bed when it is in the position shown. If the coefficient of static friction at the points of support A and B is $\mu_s = 0.4$, determine where it begins to slip first; at A or B, or both at A and B.

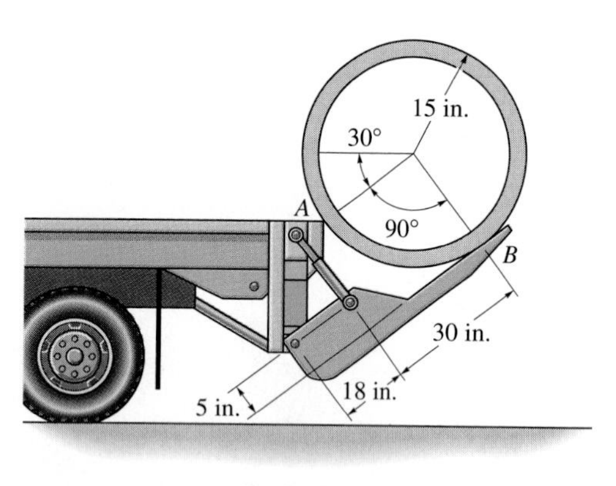

Prob. 8–35

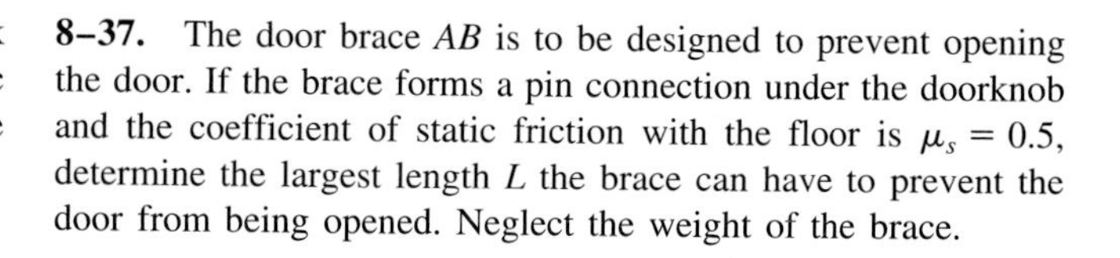

8–37. The door brace AB is to be designed to prevent opening the door. If the brace forms a pin connection under the doorknob and the coefficient of static friction with the floor is $\mu_s = 0.5$, determine the largest length L the brace can have to prevent the door from being opened. Neglect the weight of the brace.

Prob. 8–37

***8–36.** The uniform crate resting on the dolly has a mass of 500 kg and mass center at G. If the front casters contact a high step, and the coefficient of static friction between the crate and the dolly is $\mu_s = 0.45$, determine the greatest force P that can be applied without causing motion of the crate. The dolly does not move.

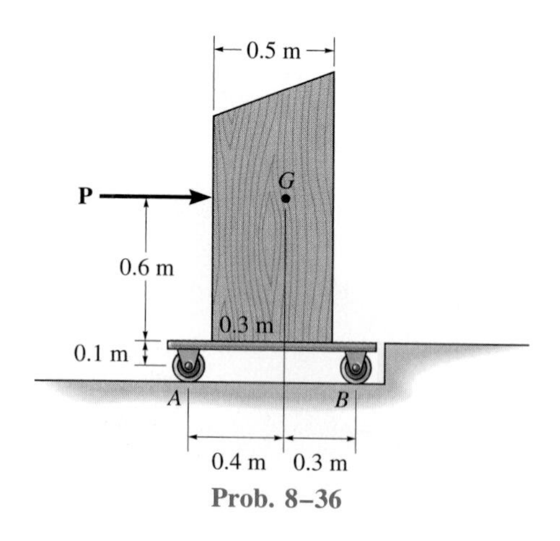

Prob. 8–36

8–38. The double-block brake mechanism is used to prevent the wheel from turning when the wheel is subjected to the torque of $M = 5\ \text{N} \cdot \text{m}$. If the coefficient of static friction between the blocks and the wheel is $\mu_s = 0.8$, determine the smallest vertical force P applied to the handle needed to stop the wheel.

8–39. Solve Prob. 8–38 if the torque is $M = 5\ \text{N} \cdot \text{m}$ clockwise.

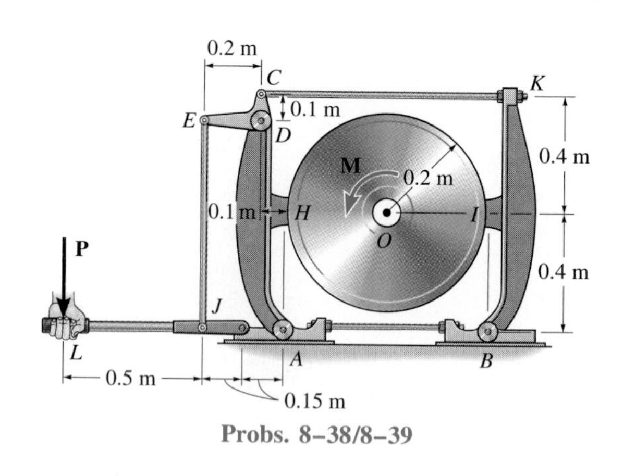

Probs. 8–38/8–39

***8–40.** The carpenter slowly pushes the uniform board horizontally over the top of the saw horse. The board has a uniform weight of 3 lb/ft, and the saw horse has a weight of 15 lb and a center of gravity at G. Determine if the saw horse will stay in position, slip, or tip if the board is pushed forward when $d = 10$ ft. The coefficients of static friction are shown in the figure.

8–41. The carpenter slowly pushes the uniform board horizontally over the top of the saw horse. The board has a uniform weight of 3 lb/ft, and the saw horse has a weight of 15 lb and a center of gravity at G. Determine if the saw horse will stay in position, slip, or tip if the board is pushed forward when $d = 14$ ft. The coefficients of static friction are shown in the figure.

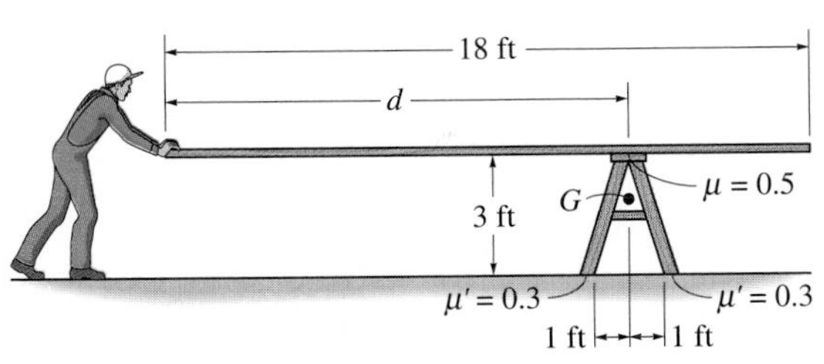

Probs. 8–40/8–41

8–42. The man has a weight of 200 lb, and the coefficient of static friction between his shoes and the floor is $\mu_s = 0.5$. Determine where he should position his center of gravity G at d in order to exert the maximum horizontal force on the door. What is this force?

Prob. 8–42

8–43. The smooth barrel has a weight W and is to be held on the incline using the chock at A. If the coefficient of static friction between the chock and the incline is $\mu_s = 0.5$, determine the design angle θ of the chock so the barrel will not move. Neglect the weight of the chock.

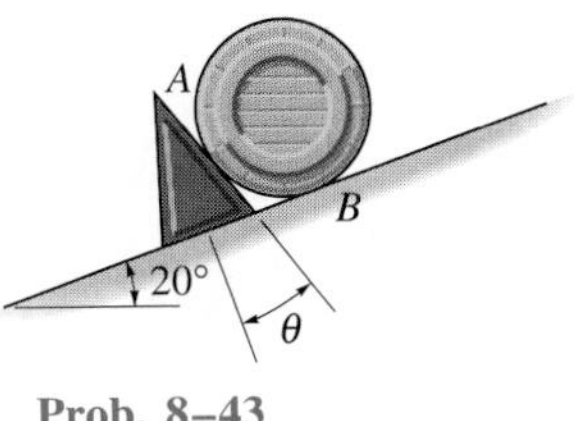

Prob. 8–43

***8–44.** Determine the smallest force the man must exert on the rope in order to move the 80-kg crate. Also, what is the angle θ at this moment? The coefficient of static friction between the crate and the floor is $\mu_s = 0.3$.

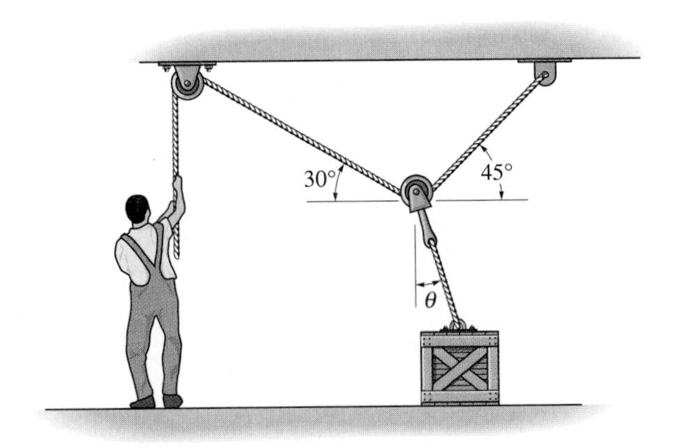

Prob. 8–44

8–45. The friction hook is made from a fixed frame which is shown colored and a cylinder of negligible weight. A piece of paper is placed between the wall and the cylinder. If $\mu_s = 0.3$ at all points of contact, determine the design angle θ so that any weight W of paper p can be held.

8–46. The friction hook is made from a fixed frame which is shown colored and a cylinder of negligible weight. A piece of paper is placed between the smooth wall and the cylinder. If $\theta = 20°$, determine the smallest coefficient of static friction μ at all points of contact so that any weight W of paper p can be held.

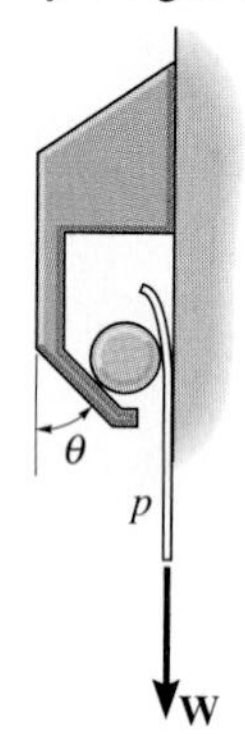

Probs. 8–45/8–46

8–47. The beam AB has a negligible mass and thickness and is subjected to a triangular distributed loading. It is supported at one end by a pin and at the other end by a post having a mass of 50 kg and negligible thickness. Determine the minimum force P needed to move the post. The coefficients of static friction at B and C are $\mu_B = 0.4$ and $\mu_C = 0.2$, respectively.

***8–48.** The beam AB has a negligible mass and thickness and is subjected to a triangular distributed loading. It is supported at one end by a pin and at the other end by a post having a mass of 50 kg and negligible thickness. Determine the two coefficients of static friction at B and at C so that when the magnitude of the applied force is increased to $P = 150$ N, the post slips at both B and C simultaneously.

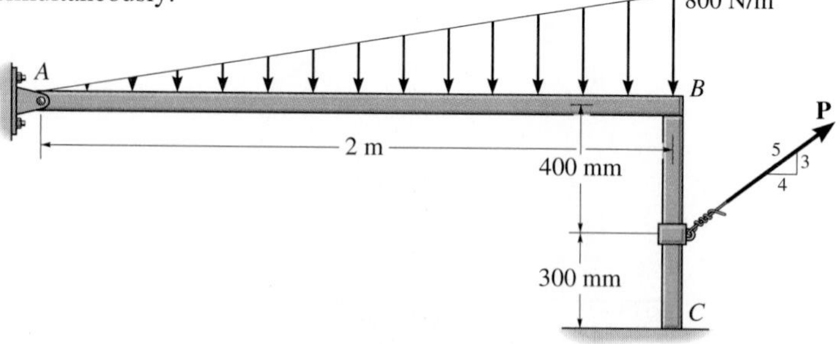

Probs. 8–47/8–48

8–49. The 45-kg disk rests on the surface for which the coefficient of static friction is $\mu_A = 0.2$. Determine the largest couple moment M that can be applied to the bar without causing motion.

8–50. The 45-kg disk rests on the surface for which the coefficient of static friction is $\mu_A = 0.15$. If $M = 50$ N · m, determine the friction force at A.

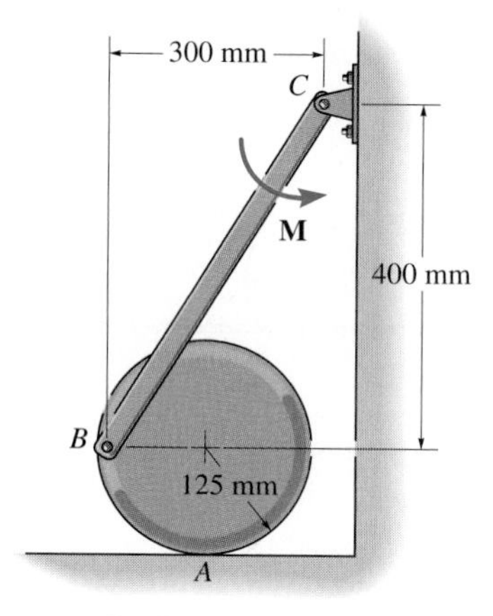

Probs. 8–49/8–50

8–51. The wheel weighs 20 lb and rests on a surface for which $\mu_B = 0.2$. A cord wrapped around it is attached to the top of the 30-lb homogeneous block. If the coefficient of static friction at D is $\mu_D = 0.3$, determine the smallest vertical force that can be applied tangentially to the wheel which will cause motion to impend.

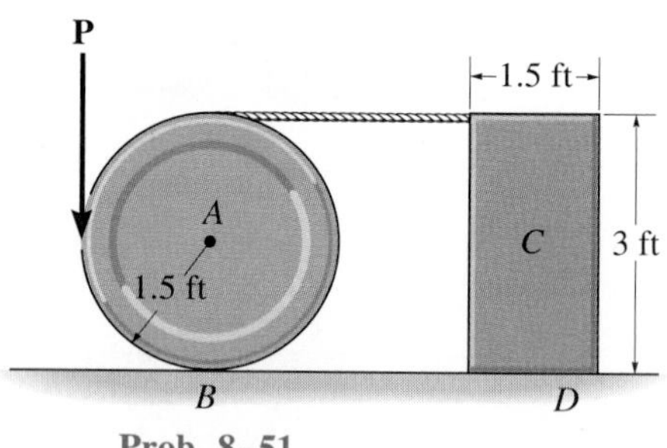

Prob. 8–51

*8–52. The ring has a mass of 0.5 kg and is resting on the surface of the table. In an effort to move the ring a normal force **P** from the finger is exerted on it. If this force is directed towards the ring's center O as shown, determine its magnitude when the ring is on the verge of slipping at A. The coefficient of static friction at A is $\mu_A = 0.2$ and at B, $\mu_B = 0.3$.

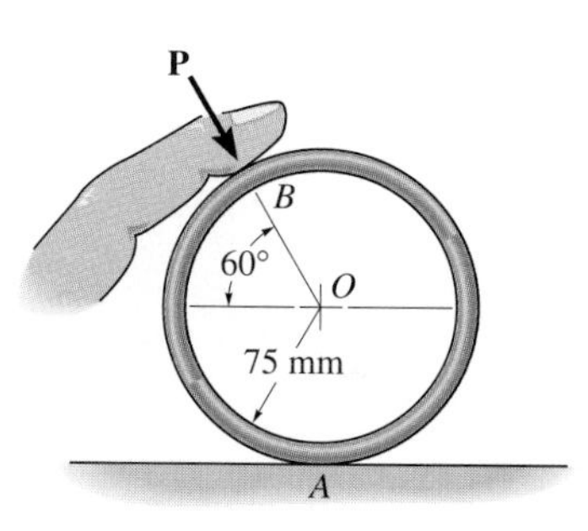

Prob. 8–52

8–53. Two blocks A and B, each having a mass of 6 kg, are connected by the linkage shown. If the coefficients of static friction at the contacting surfaces are $\mu_B = 0.8$ and $\mu_A = 0.2$, determine the largest vertical force P that may be applied to pin C without causing the blocks to slip. Neglect the weight of the links.

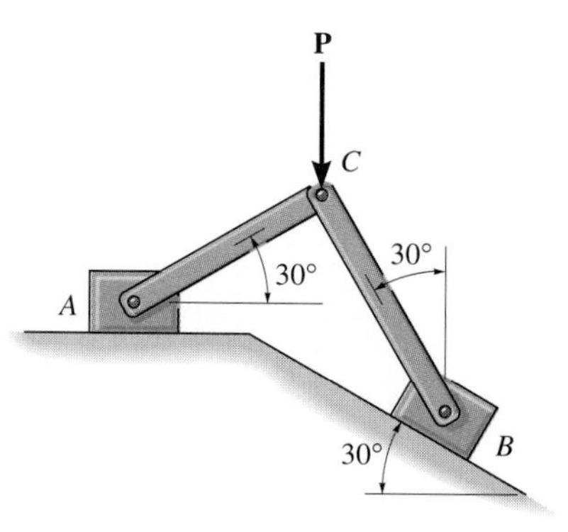

Prob. 8–53

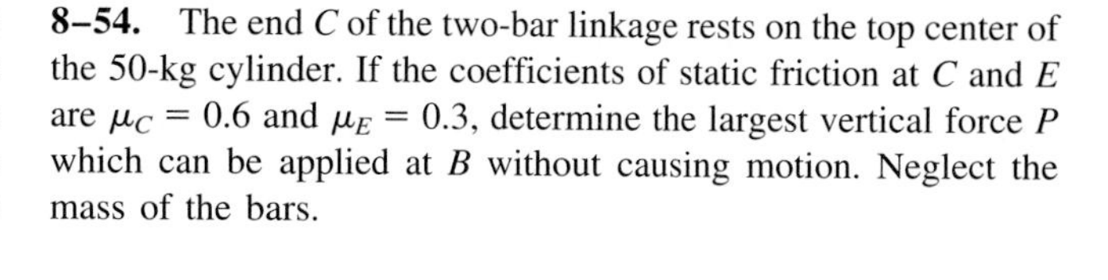
8–54. The end C of the two-bar linkage rests on the top center of the 50-kg cylinder. If the coefficients of static friction at C and E are $\mu_C = 0.6$ and $\mu_E = 0.3$, determine the largest vertical force P which can be applied at B without causing motion. Neglect the mass of the bars.

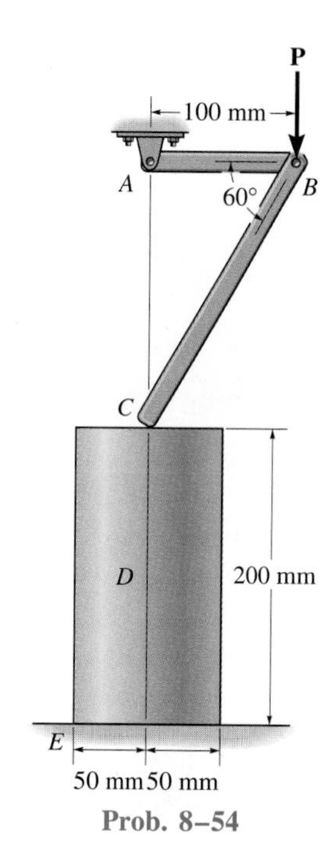

Prob. 8–54

8–55. Block C has a mass of 50 kg and is confined between two walls by smooth rollers. If the block rests on top of the 40-kg spool, determine the minimum cable force P needed to move the spool. The cable is wrapped around the spool's inner core. The coefficients of static friction at A and B are $\mu_A = 0.3$ and $\mu_B = 0.6$.

Prob. 8–55

***8–56.** Block C has a mass of 50 kg and is confined between two walls by smooth rollers. If the block rests on top of the 40-kg spool, determine the required coefficients of static friction at A and B so that the spool slips at A and B when the magnitude of the applied force is increased to $P = 300$ N.

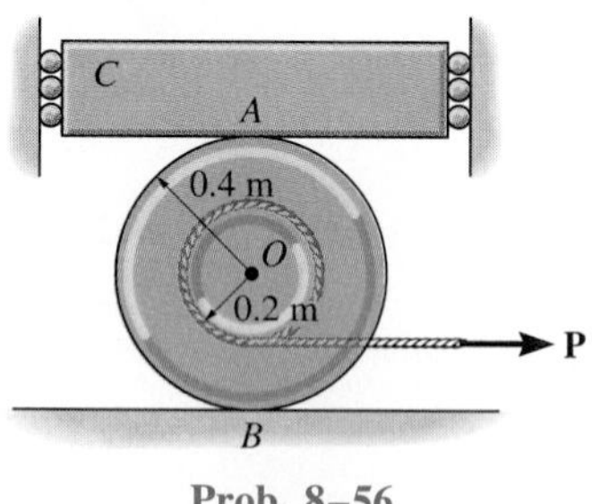

Prob. 8–56

8–57. The block of weight W is being pulled up the inclined plane of slope α using a force $\mathbf{P}$. If $\mathbf{P}$ acts at the angle ϕ as shown, show that for slipping to occur, $P = W \sin(\alpha + \theta)/\cos(\phi - \theta)$, where θ is the angle of friction; $\theta = \tan^{-1} \mu$.

8–58. Determine the angle ϕ at which $\mathbf{P}$ should act on the block so that the magnitude of $\mathbf{P}$ is as small as possible. What is the corresponding value of P? The block weighs W and the slope α is known.

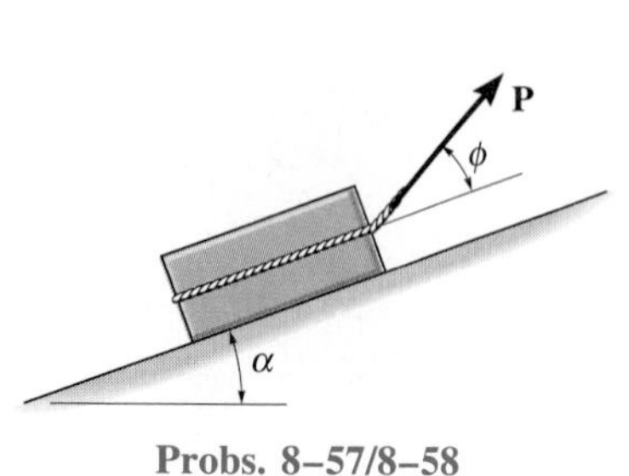

Probs. 8–57/8–58

8–59. The uniform rod has a length l, a mass m, and is supported by a ball-and-socket joint at A. If it is located a distance a from the wall, determine the smallest height h for placement against the wall which will not allow the rod to slip. The coefficient of static friction at B is μ.

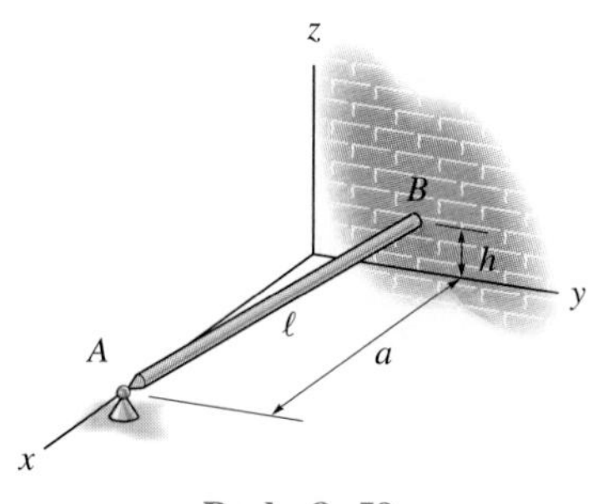

Prob. 8–59

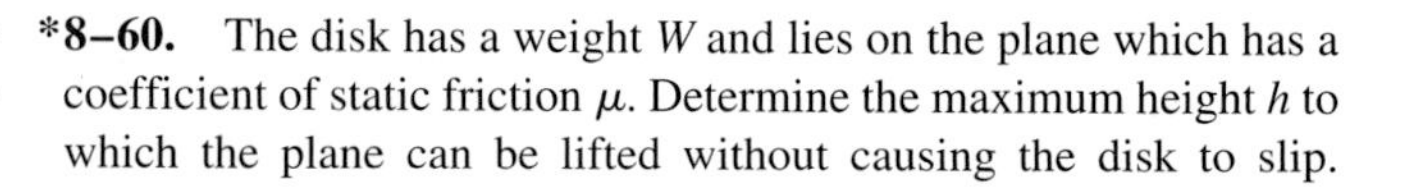

***8–60.** The disk has a weight W and lies on the plane which has a coefficient of static friction μ. Determine the maximum height h to which the plane can be lifted without causing the disk to slip.

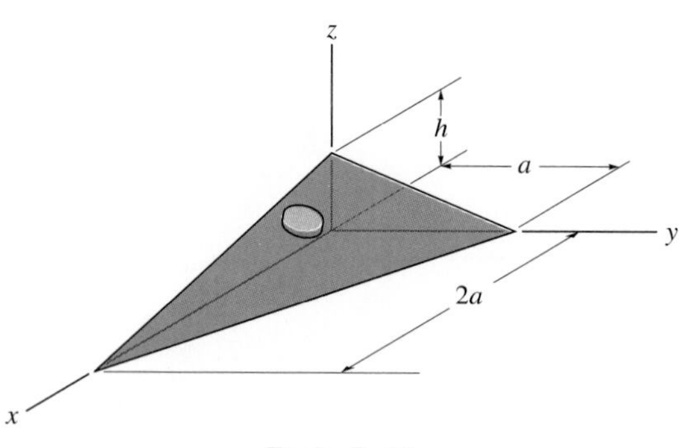

Prob. 8–60

8.3 Wedges

A *wedge* is a simple machine which is often used to transform an applied force into much larger forces, directed at approximately right angles to the applied force. Also, wedges can be used to give small displacements or adjustments to heavy loads.

Consider, for example, the wedge shown in Fig. 8–15*a*, which is used to *lift* a block of weight **W** by applying a force **P** to the wedge. Free-body diagrams of the block and wedge are shown in Fig. 8–15*b*. Here we have excluded the weight of the wedge since it is usually *small* compared to the weight of the block. Also, note that the frictional forces $\mathbf{F}_1$ and $\mathbf{F}_2$ must oppose the motion of the wedge. Likewise, the frictional force $\mathbf{F}_3$ of the wall on the block must act downward so as to oppose the block's upward motion. The locations of the resultant normal forces are not important in the force analysis, since neither the block nor wedge will "tip." Hence the moment equilibrium equations will not be considered. There are seven unknowns consisting of the applied force **P**, needed to cause motion of the wedge, and six normal and frictional forces. The seven available equations consist of two force equilibrium equations ($\Sigma F_x = 0$, $\Sigma F_y = 0$) applied to the wedge and block (four equations total) and the frictional equation $F = \mu N$ applied at each surface of contact (three equations total).

If the block is to be *lowered,* the frictional forces will all act in a sense opposite to that shown in Fig. 8–15*b*. The applied force **P** will act to the right as shown if the coefficient of friction is very *small* or the wedge angle θ is *large*. Otherwise, **P** may have the reverse sense of direction in order to *pull* on the wedge to remove it. If **P** is *removed,* or **P** = **0,** and friction forces hold the block in place, then the wedge is referred to as *self-locking.*

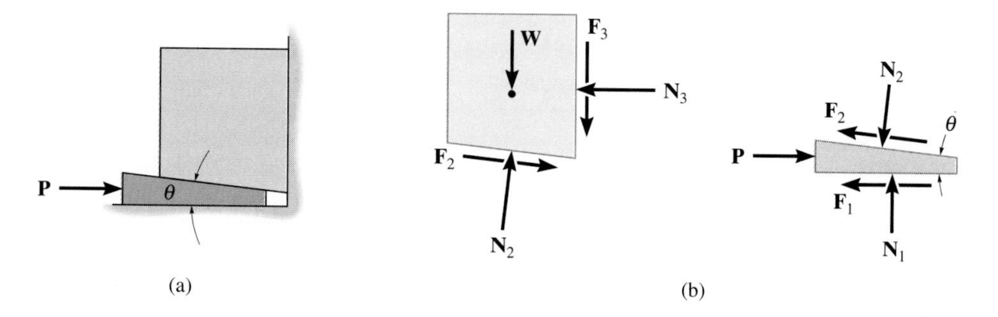

Fig. 8–15

Example 8–7

The uniform stone has a mass of 500 kg and is held in the horizontal position using a wedge at B as shown in Fig. 8–16a. If the coefficient of static friction is $\mu_s = 0.3$ at the surface in contact with the wedge, determine the force **P** needed to remove the wedge. Is the wedge self-locking? Assume that the stone does not slip at A.

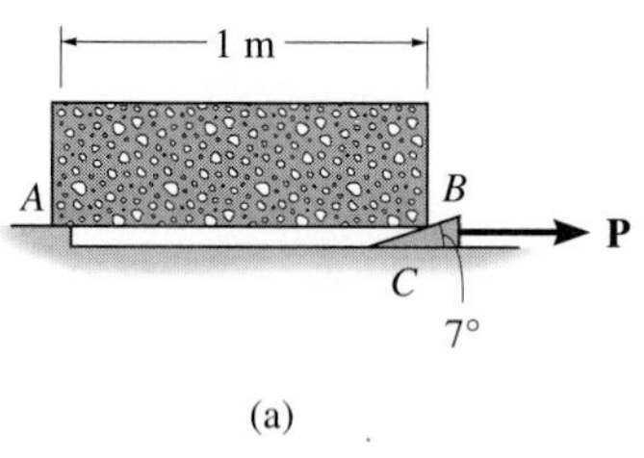

(a)

(b)

Fig. 8–16

SOLUTION

Since the wedge is to be removed, slipping is about to occur at the surfaces of contact. Thus, $F = \mu_s N$ on the wedge, and the free-body diagrams are shown in Fig. 8–16b. Note that on the wedge the friction force opposes the motion and on the stone at A, $F_A \leq \mu_s N_A$, since slipping does not occur there. There are five unknowns F_A, N_A, N_B, N_C, and P. Three equilibrium equations for the stone and two for the wedge are available for solution. From the free-body diagram of the stone,

$$\circlearrowright+\Sigma M_A = 0; \quad -4905\text{ N}(0.5\text{ m}) + (N_B \cos 7°\text{ N})(1\text{ m}) + (0.3N_B \sin 7°\text{ N})(1\text{ m}) = 0$$

$$N_B = 2383.1\text{ N}$$

Using this result for the wedge, we have

$$\xrightarrow{+}\Sigma F_x = 0; \quad 2383.1 \sin 7°\text{ N} - 0.3(2383.1 \cos 7°\text{ N}) + P - 0.3N_C = 0$$

$$+\uparrow\Sigma F_y = 0; \quad N_C - 2383.1 \cos 7°\text{ N} - 0.3(2383.1 \sin 7°\text{ N}) = 0$$

$$N_C = 2452.5\text{ N}$$

$$P = 1154.9\text{ N} = 1.15\text{ kN} \qquad \textit{Ans.}$$

Since P is positive, indeed the wedge must be pulled out. Obviously, if P is zero, the wedge will remain in place (self-locking) and the frictional forces $\mathbf{F}_C$ and $\mathbf{F}_B$ developed at the points of contact will satisfy $F_B < \mu_s N_B$ and $F_C < \mu_s N_C$.

*8.4 Frictional Forces on Screws

In most cases screws are used as fasteners; however, in many types of machines they are incorporated to transmit power or motion from one part of the machine to another. A *square-threaded screw* is commonly used for the latter purpose, especially when large forces are applied along its axis. In this section we will analyze the forces acting on square-threaded screws. The analysis of other types of screws, such as the V-thread, is based on the same principles.

A *screw* may be thought of simply as an inclined plane or wedge wrapped around a cylinder. A nut initially at position A on the screw shown in Fig. 8–17a will move up to B when rotated 360° around the screw. This rotation is equivalent to translating the nut up an inclined plane of height l and length $2\pi r$, where r is the mean radius of the thread, Fig. 8–17b. The rise l for a single revolution is referred to as the *lead* of the screw, where the *lead angle* is given by $\theta = \tan^{-1}(l/2\pi r)$.

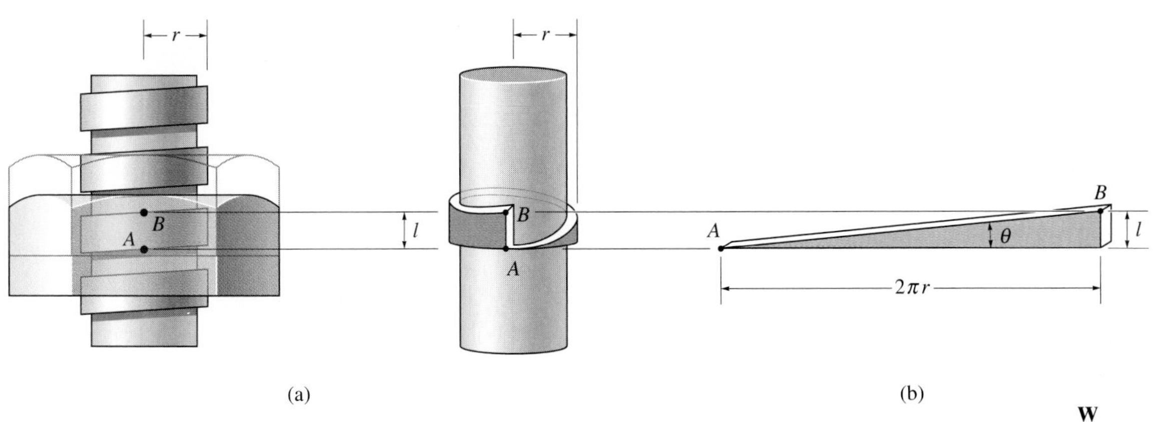

Fig. 8–17

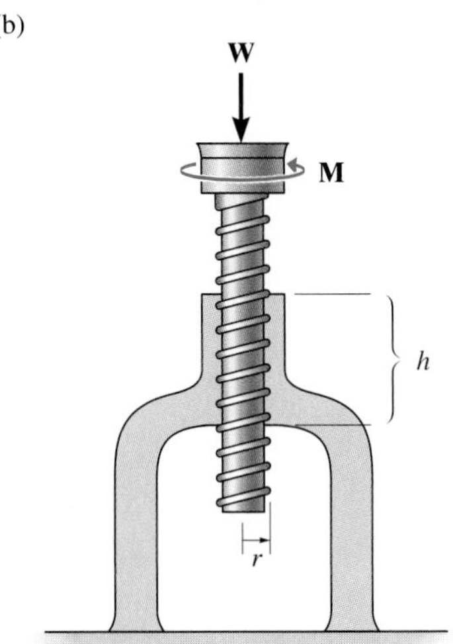

Fig. 8–18

Frictional Analysis. When a screw is subjected to large axial loads, the frictional forces developed on the thread become important if we are to determine the moment **M*** needed to turn the screw. Consider, for example, the square-threaded jack screw shown in Fig. 8–18, which supports the vertical load **W.** The reactive forces of the jack to this load are actually distributed over the circumference of the screw thread in contact with the screw hole in the jack, that is, within region h shown in Fig. 8–18. For simplicity, this

*For applications, **M** is developed by applying a horizontal force **P** at a right angle to the end of a lever that would be fixed to the screw.

portion of thread can be imagined as being unwound from the screw and represented as a simple block resting on an inclined plane having the screw's lead angle θ, Fig. 8–19*a*. Here the inclined plane represents the inside *supporting thread* of the jack base. Three forces act on the block or screw. The force **W** is the total axial load applied to the screw. The horizontal force **S** is caused by the applied moment **M**, such that the magnitudes of these loads can be related by summing moments about the axis of the screw. We require $M = Sr$, where r is the screw's mean radius. As a result of **W** and **S**, the inclined plane exerts a resultant force **R** on the block, which is shown to have components acting normal, **N**, and tangent, **F**, to the contacting surfaces.

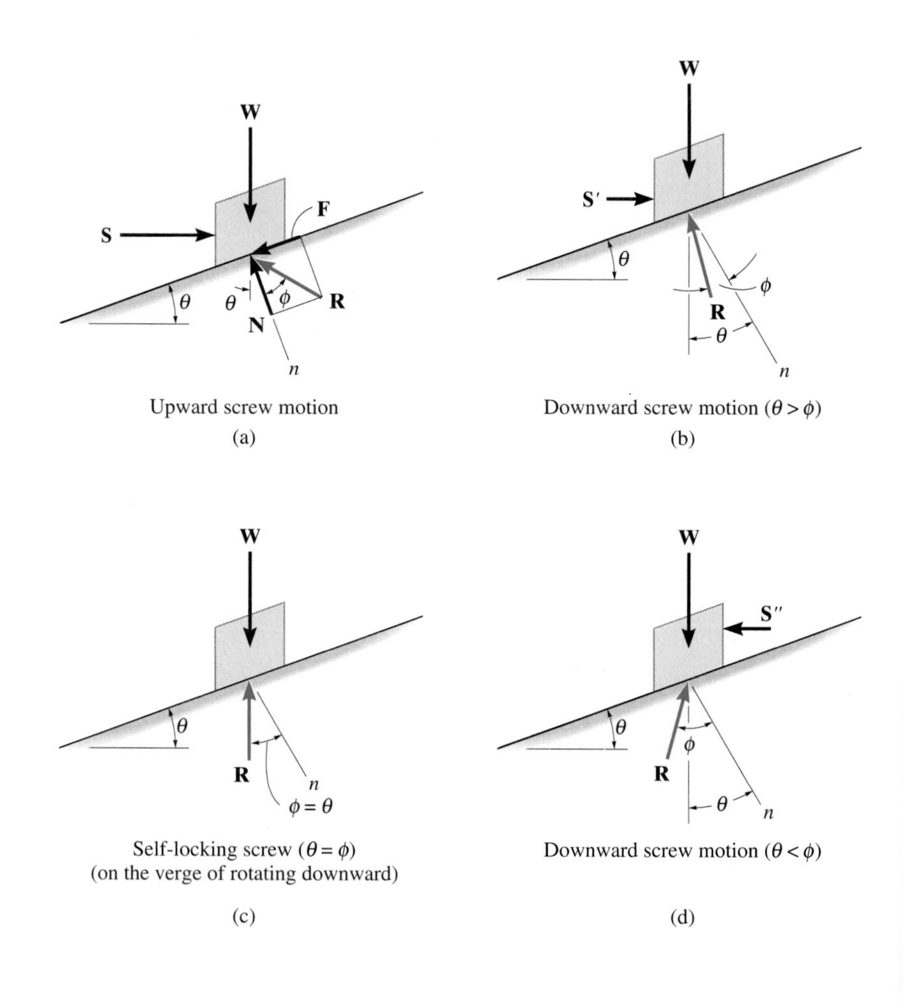

Fig. 8–19

Upward Screw Motion. Provided M is great enough, the screw (and hence the block) can either be brought to the verge of upward impending motion or motion can be occurring. Under these conditions, **R** acts at an angle $(\theta + \phi)$ from the vertical as shown in Fig. 8–19*a*, where $\phi = \tan^{-1}(F/N) = \tan^{-1}(\mu N/N) = \tan^{-1}\mu$. Applying the two force equations of equilibrium to the block, we obtain

$$\stackrel{+}{\rightarrow}\Sigma F_x = 0; \qquad S - R\sin(\theta + \phi) = 0$$

$$+\uparrow \Sigma F_y = 0; \qquad R\cos(\theta + \phi) - W = 0$$

Eliminating R and solving for S, then substituting this value into the equation $M = Sr$, yields

$$M = Wr\tan(\theta + \phi) \tag{8–3}$$

As indicated, M is the moment necessary to cause upward impending motion of the screw, provided $\phi = \phi_s = \tan^{-1}\mu_s$ (the angle of static friction). If ϕ is replaced by $\phi_k = \tan^{-1}\mu_k$ (the angle of kinetic friction), Eq. 8–3 will give a smaller value M necessary to maintain uniform upward motion of the screw.

Downward Screw Motion ($\theta > \phi$). If the surface of the screw is very *slippery,* it may be possible for the screw to rotate downward if the magnitude, *not the direction,* of the moment is reduced to, say, $M' < M$. As shown in Fig. 8–19*b*, this causes the effect of $\mathbf{M}'$ to become $\mathbf{S}'$, and it requires the angle ϕ (ϕ_s or ϕ_k) to lie on the opposite side of the normal n to the plane supporting the block, such that $\theta > \phi$. For this case, Eq. 8–3 becomes

$$M' = Wr\tan(\theta - \phi) \tag{8–4}$$

Self-Locking Screw. If the moment **M** (or its effect **S**) is *removed,* the screw will remain *self-locking;* i.e., it will support the load **W** by *friction forces alone* provided $\phi \geq \theta$. To show this, consider the necessary limiting case when $\phi = \theta$, Fig. 8–19*c*. Here vertical equilibrium is maintained since **R** is vertical and thus balances **W.**

Downward Screw Motion ($\theta < \phi$). When the surface of the screw is *very rough,* the screw will not rotate downward as stated above. Instead, the direction of the applied moment must be *reversed* in order to cause the motion. The free-body diagram shown in Fig. 8–19*d* is representative of this case. Here $\mathbf{S}''$ is caused by the applied (reverse) moment $\mathbf{M}''$. Hence Eq. 8–3 becomes

$$M'' = Wr\tan(\phi - \theta) \tag{8–5}$$

Each of the above cases should be thoroughly understood before proceeding to solve problems.

Example 8–8

The turnbuckle shown in Fig. 8–20 has a square thread with a mean radius of 5 mm and a lead of 2 mm. If the coefficient of static friction between the screw and the turnbuckle is $\mu_s = 0.25$, determine the moment **M** that must be applied to draw the end screws closer together. Is the turnbuckle self-locking?

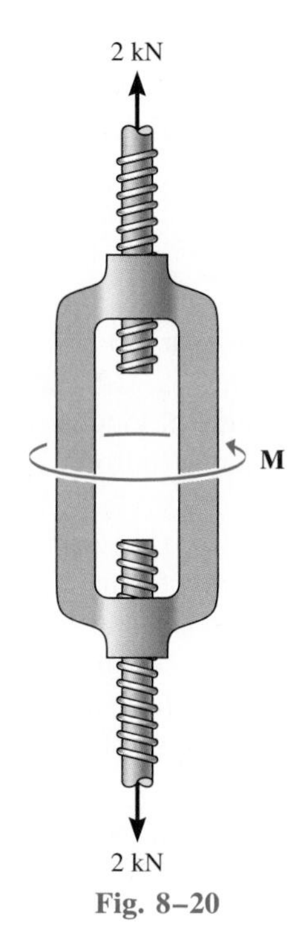

Fig. 8–20

SOLUTION

The moment may be obtained by using Eq. 8–3. Why? Since friction at *two screws* must be overcome, this requires

$$M = 2[Wr\tan(\theta + \phi)] \qquad (1)$$

Here $W = 2000$ N, $r = 5$ mm, $\phi_s = \tan^{-1}\mu_s = \tan^{-1}(0.25) = 14.04°$, and $\theta = \tan^{-1}(l/2\pi r) = \tan^{-1}(2\text{ mm}/[2\pi(5\text{ mm})]) = 3.64°$. Substituting these values into Eq. 1 and solving gives

$$M = 2[(2000\text{ N})(5\text{ mm})\tan(14.04° + 3.64°)]$$
$$= 6375.1\text{ N}\cdot\text{mm} = 6.38\text{ N}\cdot\text{m} \qquad \textit{Ans.}$$

When the moment is *removed,* the turnbuckle will be self-locking; i.e., it will not unscrew, since $\phi_s > \theta$.

PROBLEMS

8–61. The blocks each have a weight of 50 lb. If the coefficient of static friction at A is $\mu_s = 0.2$ and between each block $\mu_s' = 0.4$, determine how many blocks can be stacked as shown before they begin to topple.

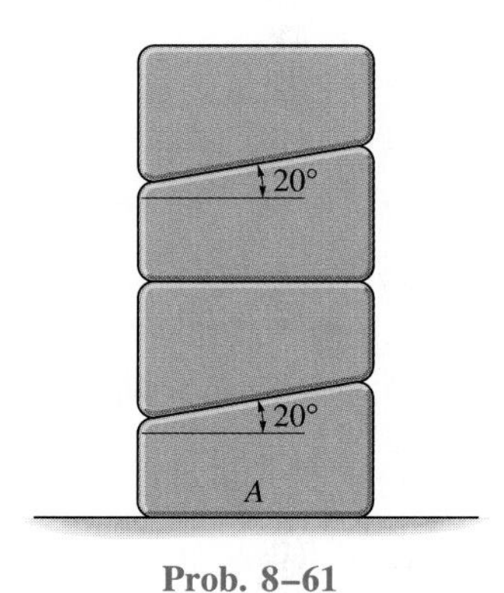

Prob. 8–61

8–62. Each block has a weight of 400 lb. Determine how far the force **P** can compress the spring until block B slips on block A. What is the magnitude of **P** for this to occur?

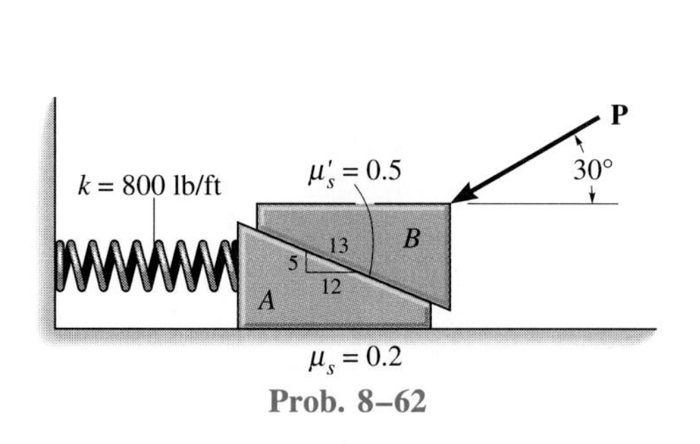

Prob. 8–62

8–63. Column D is subjected to a vertical load of 8000 lb. It is supported on two identical wedges A and B for which the coefficient of static friction at the contacting surfaces between A and B and B and C is $\mu_s = 0.4$. Determine the force P needed to raise wedge B and the equilibrium force P' needed to hold wedge A stationary. The contacting surface between A and D is smooth.

***8–64.** Column D is subjected to a vertical load of 8000 lb. It is supported on two identical wedges A and B for which the coefficient of static friction at the contacting surfaces between A and B and B and C is $\mu_s = 0.4$. If the forces **P** and **P′** are removed are the wedges self-locking? The contacting surface between A and D is smooth.

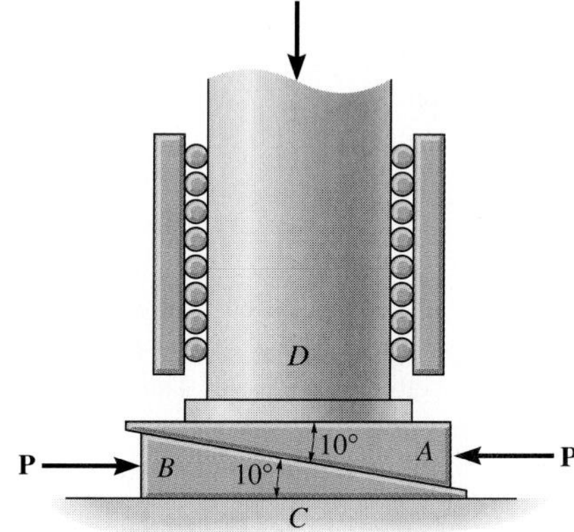

Probs. 8–63/8–64

8–65. If the spring is compressed 60 mm and the coefficient of static friction between the tapered stub S and the slider A is $\mu_{SA} = 0.5$, determine the horizontal force P needed to move the slider forward. The stub is free to move without friction within the fixed collar C. The coefficient of static friction between A and surface B is $\mu_{AB} = 0.4$. Neglect the weights of the slider and stub.

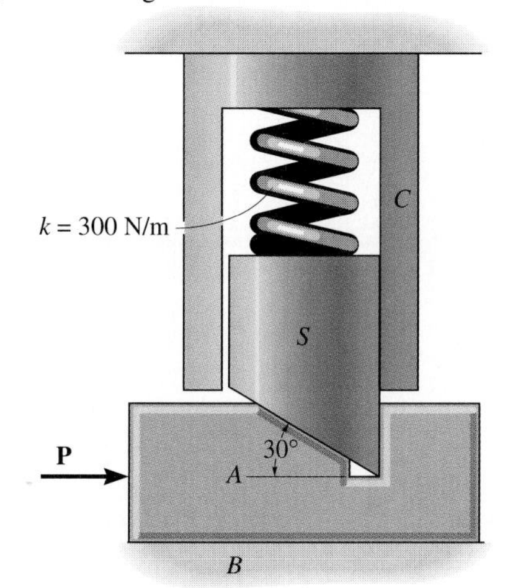

Prob. 8–65

8–66. If the beam AD is loaded as shown, determine the horizontal force P which must be applied to the wedge in order to remove it from under the beam. The coefficients of static friction at the wedge's top and bottom surfaces are $\mu_{CA} = 0.25$ and $\mu_{CB} = 0.35$, respectively. If $P = 0$, is the wedge self-locking?

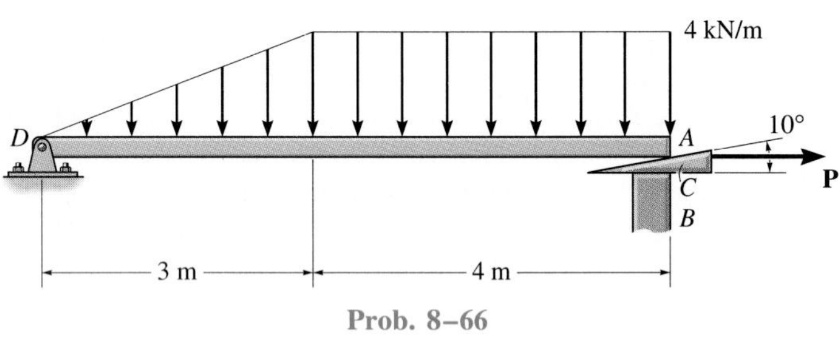

Prob. 8–66

8–67. The wedge is used to level the member. For the loading shown, determine the horizontal force P that must be applied to move the wedge to the right. The coefficient of static friction between the wedge and its two surfaces of contact is $\mu_s = 0.25$. Neglect the size and weight of the wedge.

***8–68.** The wedge is used to level the member. For the loading shown, determine the reversed horizontal force $-\mathbf{P}$ that must be applied to pull the wedge out to the left. The coefficient of static friction between the wedge and its two surfaces of contact is $\mu_s = 0.15$. Neglect the weight of the wedge.

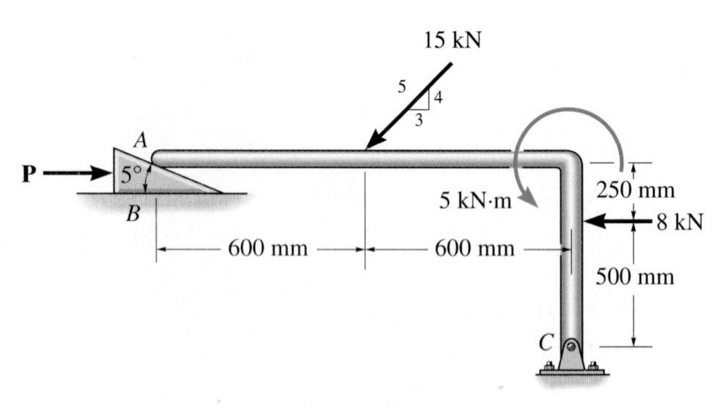

Probs. 8–67/8–68

8–69. The coefficient of static friction between wedges B and C is $\mu_s = 0.6$ and between the surfaces of contact B and A and C and D, $\mu_s' = 0.4$. If the spring is compressed 200 mm when in the position shown, determine the smallest force P needed to move wedge C to the left. Neglect the weight of the wedges.

8–70. The coefficient of static friction between the wedges B and C is $\mu_s = 0.6$ and between the surfaces of contact B and A and C and D, $\mu_s' = 0.4$. If $P = 50$ N, determine the largest allowable compression of the spring without causing wedge C to move to the left. Neglect the weight of the wedges.

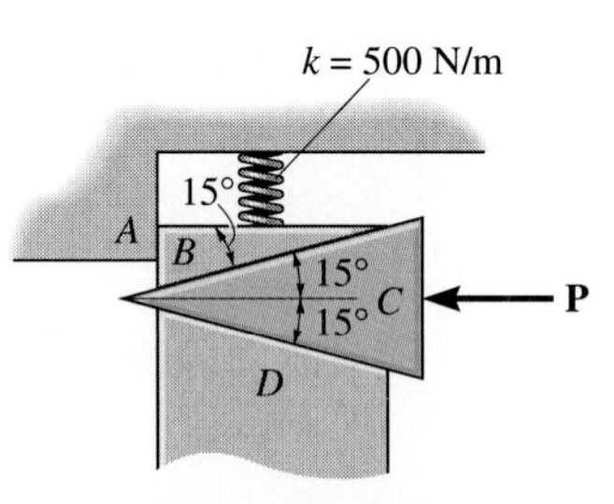

Probs. 8–69/8–70

8–71. The wedge blocks are used to hold the specimen in a tension testing machine. Determine the design angle θ of the wedges so that the specimen will not slip regardless of the applied load. The coefficients of static friction are $\mu_A = 0.1$ at A and $\mu_B = 0.6$ at B. Neglect the weight of the blocks.

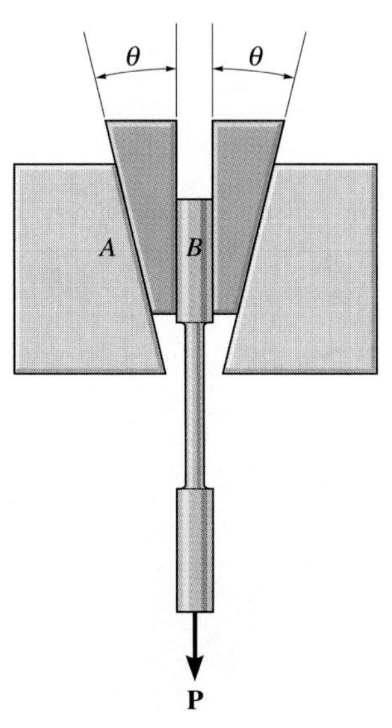

Prob. 8–71

*8–72. The two blocks have weights of $W_A = 600$ lb and $W_B = 500$ lb. Determine the smallest horizontal force **P** that must be applied to block A in order to move it. The coefficient of static friction between the blocks is $\mu_s = 0.3$ and between the floor and each block $\mu_s' = 0.5$.

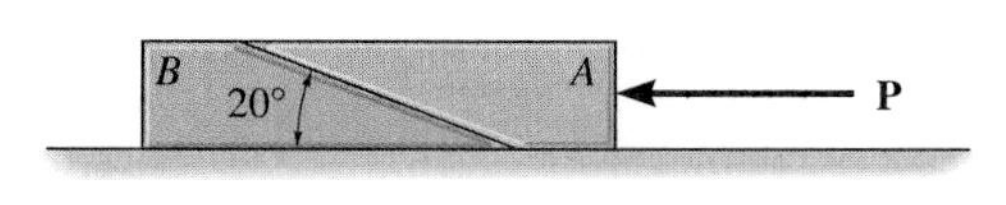

Prob. 8–72

8–73. The device is used to pull the battery-cable terminal C from the post of a battery. If the required pulling force is 85 lb, determine the torque M that must be applied to the handle on the screw to tighten it. The screw has square threads, a mean diameter of 0.2 in., a lead of 0.08 in., and a coefficient of static friction of $\mu_s = 0.5$.

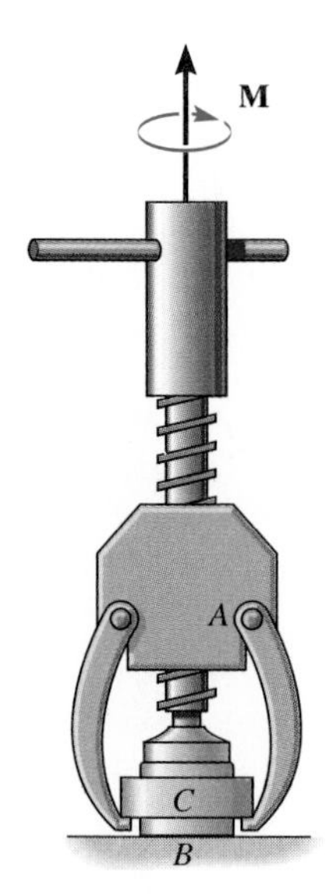

Prob. 8–73

8–74. The column is used to support the upper floor. If a force $F = 80$ N is applied perpendicular to the handle to tighten the screw, determine the compressive force in the column. The square-threaded screw on the jack has a coefficient of static friction of $\mu_s = 0.4$, mean diameter of 25 mm, and a lead of 3 mm.

8–75. If the force **F** is removed from the handle of the jack in Prob. 8–74, determine if the screw is self-locking.

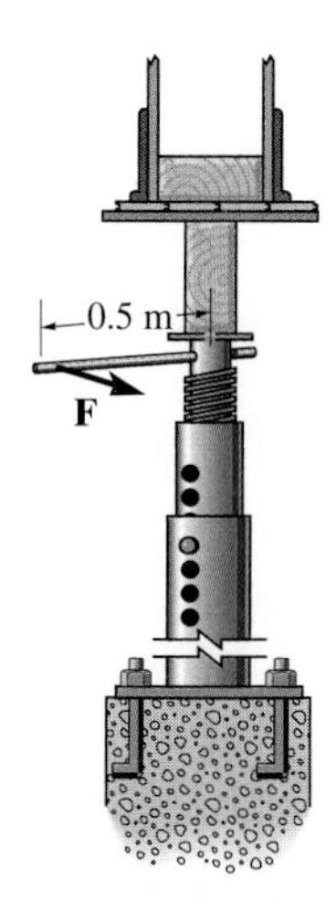

Probs. 8–74/8–75

*8–76. The clamp provides pressure from several directions on the edges of the board. If the square-threaded screw has a lead of 3 mm, radius of 10 mm, and the coefficient of static friction is $\mu_s = 0.4$, determine the horizontal force developed on the board at A and the vertical forces developed at B and C if a torque of $M = 1.5$ N · m is applied to the handle to tighten it further. The blocks at B and C are pin-connected to the board.

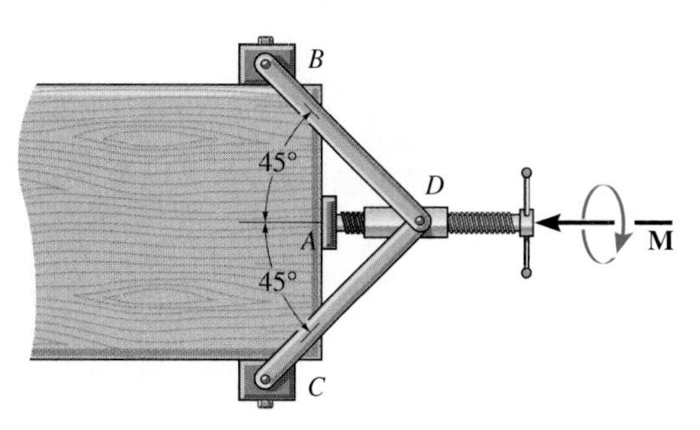

Prob. 8–76

8–77. Determine the clamping force on the board A if the screw of the "C" clamp is tightened with a twist of $M = 8\ \text{N} \cdot \text{m}$. The single square-threaded screw has a mean radius of 10 mm, a lead of 3 mm, and the coefficient of static friction is $\mu_s = 0.35$.

8–78. If the required clamping force at the board A is to be 50 N, determine the torque M that must be applied to the handle of the "C" clamp to tighten it down. The single square-threaded screw has a mean radius of 10 mm, a lead of 3 mm, and the coefficient of static friction is $\mu_s = 0.35$.

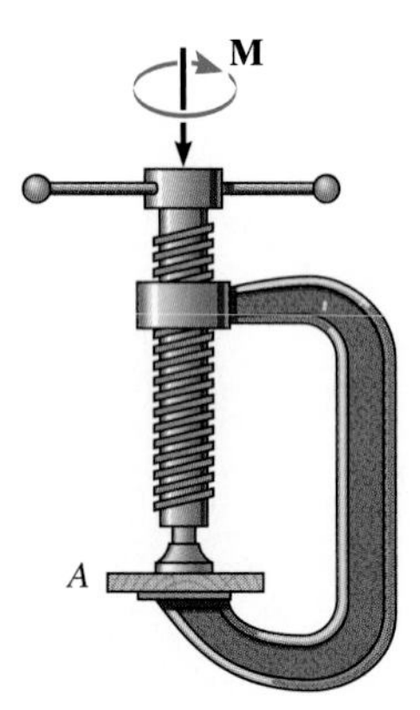

Probs. 8–77/8–78

8–79. The square-threaded screw has a mean diameter of 20 mm and a lead of 4 mm. If the weight of the plate A is 5 lb, determine the smallest coefficient of static friction between the screw and the plate so that the plate does not travel down the screw when the plate is suspended as shown.

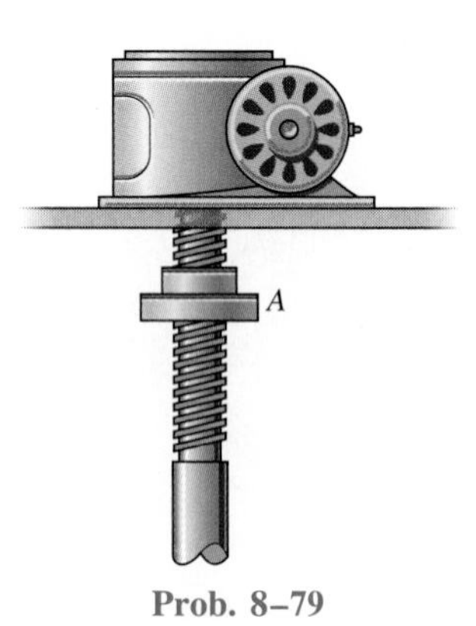

Prob. 8–79

***8–80.** Determine the horizontal force P applied perpendicular to the handle of the jack screw necessary to start lifting the 3-kN load. The square-threaded screw has a lead of 5 mm and a mean diameter of 60 mm. The coefficient of static friction for the screw is $\mu_s = 0.2$.

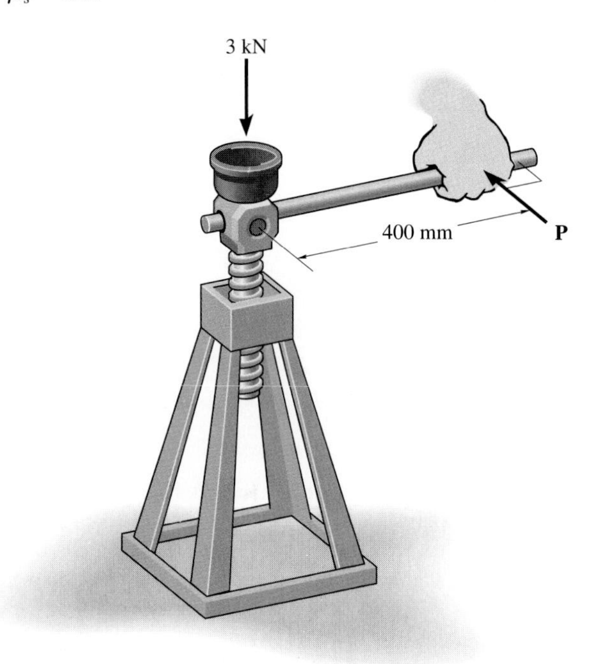

Prob. 8–80

8–81. The shaft has a square-threaded screw with a lead of 9 mm and a mean radius of 15 mm. If it is in contact with a plate gear having a mean radius of 20 mm, determine the resisting torque M on the plate gear which can be overcome if a torque of $7\ \text{N} \cdot \text{m}$ is applied to the shaft. The coefficient of static friction between the gear and the screw is $\mu_s = 0.2$. Neglect friction of the bearings located at A and B.

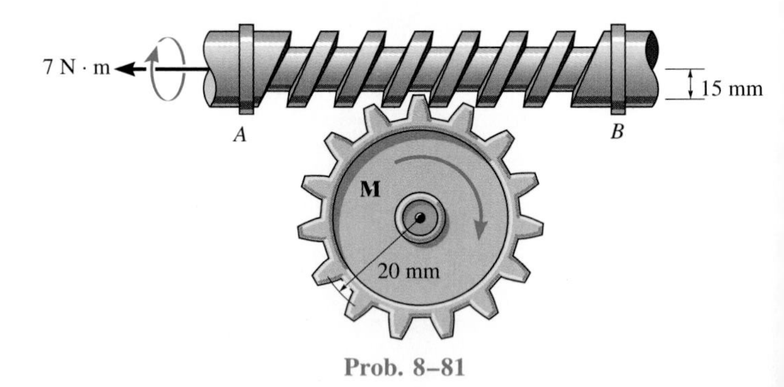

Prob. 8–81

8–82. The fixture clamp consists of a square-threaded screw having a coefficient of static friction of $\mu_s = 0.3$, mean diameter of 3 mm, and a lead of 1 mm. The five points indicated are pin connections. Determine the clamping force at the smooth blocks D and E when a torque of $M = 0.08\ \text{N} \cdot \text{m}$ is applied to the handle of the screw.

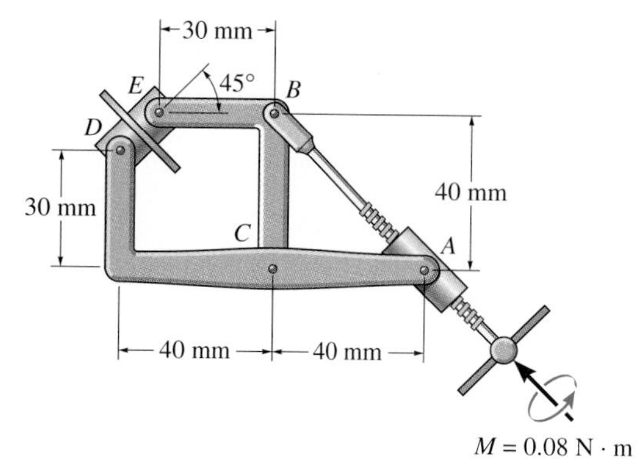

Prob. 8–82

*8.5 Frictional Forces on Flat Belts

Whenever belt drives or band brakes are designed, it is necessary to determine the frictional forces developed between the belt and its contacting surface. In this section we will analyze the frictional forces acting on a flat belt, although the analysis of other types of belts, such as the V-belt, is based on similar principles.

Here we will consider the flat belt shown in Fig. 8–21*a*, which passes over a fixed curved surface, such that the total angle of belt to surface contact in radians is β and the coefficient of friction between the two surfaces is μ. We will determine the tension T_2 in the belt which is needed to pull the belt counterclockwise over the surface and thereby overcome both the frictional forces at the surface of contact and the known tension T_1. Obviously, $T_2 > T_1$.

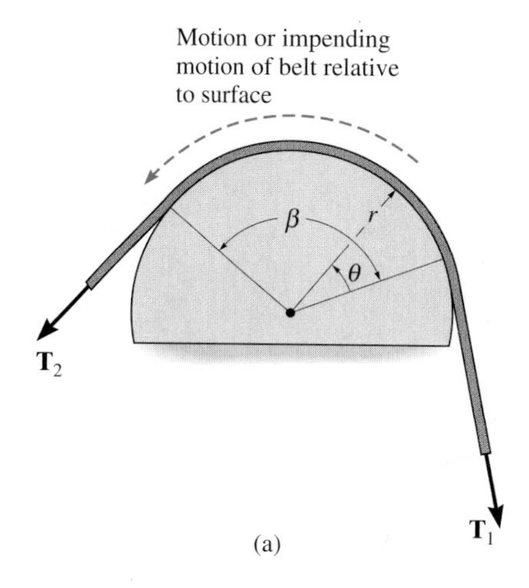

(a)

Frictional Analysis. A free-body diagram of the belt segment in contact with the surface is shown in Fig. 8–21*b*. Here the normal force **N** and the frictional force **F,** acting at different points along the belt, will vary both in magnitude and direction. Due to this *unknown* force distribution, the analysis of the problem will proceed on the basis of initially studying the forces acting on a differential element of the belt.

A free-body diagram of an element having a length ds is shown in Fig. 8–21*c*. Assuming either impending motion or motion of the belt, the magnitude of the frictional force $dF = \mu\, dN$. This force opposes the sliding motion of the belt and thereby increases the magnitude of the tensile force acting in the belt by dT. Applying the two force equations of equilibrium, we have

$$\overset{+}{\rightarrow}\Sigma F_x = 0; \qquad T\cos\left(\frac{d\theta}{2}\right) + \mu\, dN - (T + dT)\cos\left(\frac{d\theta}{2}\right) = 0$$

$$+\uparrow\Sigma F_y = 0; \qquad dN - (T + dT)\sin\left(\frac{d\theta}{2}\right) - T\sin\left(\frac{d\theta}{2}\right) = 0$$

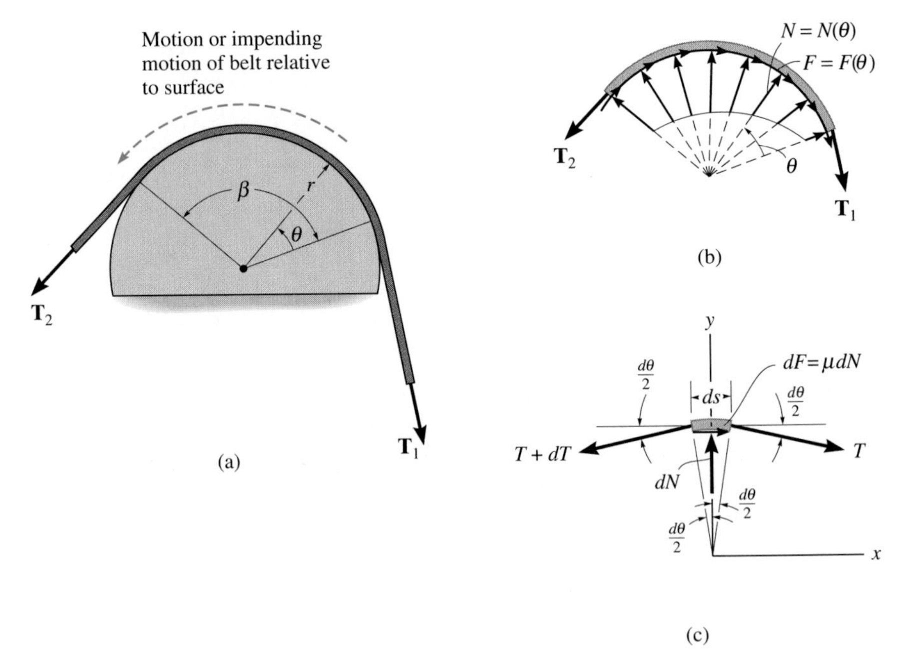

Fig. 8–21

Since $d\theta$ is of *infinitesimal size,* $\sin(d\theta/2)$ and $\cos(d\theta/2)$ can be replaced by $d\theta/2$ and 1, respectively. Also, the *product* of the two infinitesimals dT and $d\theta/2$ may be neglected when compared to infinitesimals of the first order. The above two equations therefore reduce to

$$\mu\, dN = dT$$

and

$$dN = T\, d\theta$$

Eliminating dN yields

$$\frac{dT}{T} = \mu\, d\theta$$

Integrating this equation between all the points of contact that the belt makes with the drum, and noting that $T = T_1$ at $\theta = 0$ and $T = T_2$ at $\theta = \beta$, yields

$$\int_{T_1}^{T_2} \frac{dT}{T} = \mu \int_0^{\beta} d\theta$$

$$\ln \frac{T_2}{T_1} = \mu\beta$$

Solving for T_2, we obtain

$$T_2 = T_1 e^{\mu\beta} \quad (8\text{–}6)$$

where T_2, T_1 = belt tensions; $\mathbf{T}_1$ opposes the direction of motion (or impending motion) of the belt measured relative to the surface, while $\mathbf{T}_2$ acts in the direction of the relative belt motion (or impending motion); because of friction, $T_2 > T_1$

μ = coefficient of static or kinetic friction between the belt and the surface of contact

β = angle of belt to surface contact, measured in radians

e = 2.718. . . , base of the natural logarithm

Note that Eq. 8–6 is *independent* of the *radius* of the drum and instead depends on the angle of belt to surface contact, β. Furthermore, as indicated by the integration, this equation is valid for flat belts placed on *any shape* of contacting surface. For application, however, keep in mind that Eq. 8–6 is valid only when *impending motion* occurs.

Example 8–9

The maximum tension that can be developed in the belt shown in Fig. 8–22*a* is 500 N. If the pulley at *A* is free to rotate and the coefficient of static friction at the fixed drums *B* and *C* is $\mu_s = 0.25$, determine the largest mass of the cylinder that can be lifted by the belt. Assume that the force **T** applied at the end of the belt is directed vertically downward, as shown.

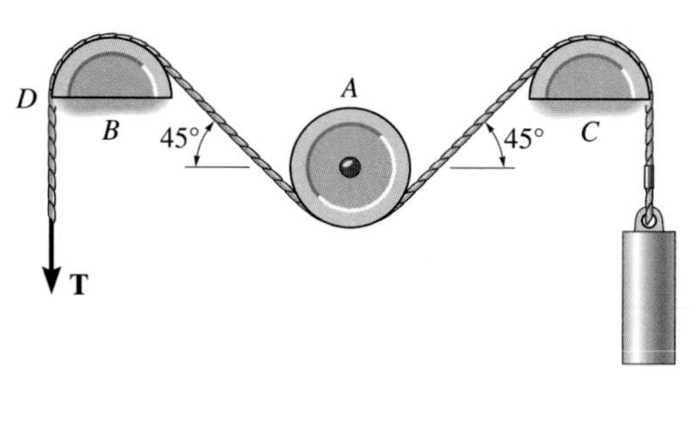

(a)

SOLUTION

Lifting the cylinder, which has a weight $W = mg$, causes the belt to move counterclockwise over the drums at *B* and *C*; hence, the maximum tension T_2 in the belt occurs at *D*. Thus, $T_2 = 500$ N. A section of the belt passing over the drum at *B* is shown in Fig. 8–22*b*. Since $180° = \pi$ rad, the angle of contact between the drum and the belt is $\beta = (135°/180°)\pi = 3\pi/4$ rad. Using Eq. 8–6, we have

$$T_2 = T_1 e^{\mu_s \beta}; \qquad 500 \text{ N} = T_1 e^{0.25[(3/4)\pi]}$$

Hence,

$$T_1 = \frac{500 \text{ N}}{e^{0.25[(3/4)\pi]}} = \frac{500 \text{ N}}{1.80} = 277.4 \text{ N}$$

Since the pulley at *A* is free to rotate, equilibrium requires that the tension in the belt remains the *same* on both sides of the pulley.

The section of the belt passing over the drum at *C* is shown in Fig. 8–22*c*. The weight $W < 277.4$ N. Why? Applying Eq. 8–6, we obtain

$$T_2 = T_1 e^{\mu_s \beta}; \qquad 277.4 \text{ N} = W e^{0.25[(3/4)\pi]}$$

$$W = 153.9 \text{ N}$$

so that

$$m = \frac{W}{g} = \frac{153.9 \text{ N}}{9.81 \text{ m/s}^2}$$

$$= 15.7 \text{ kg} \qquad \textit{Ans.}$$

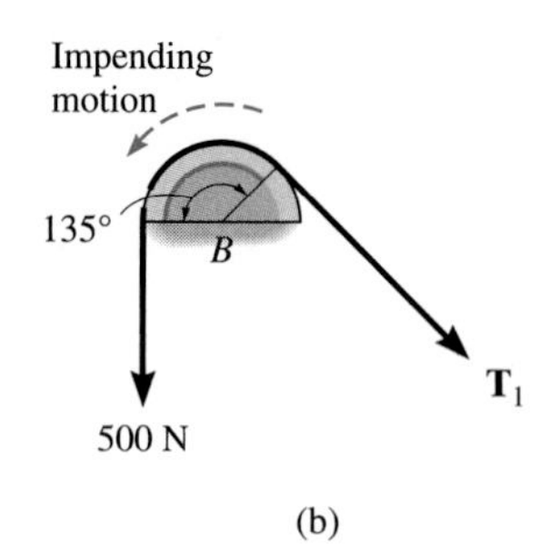

(b)

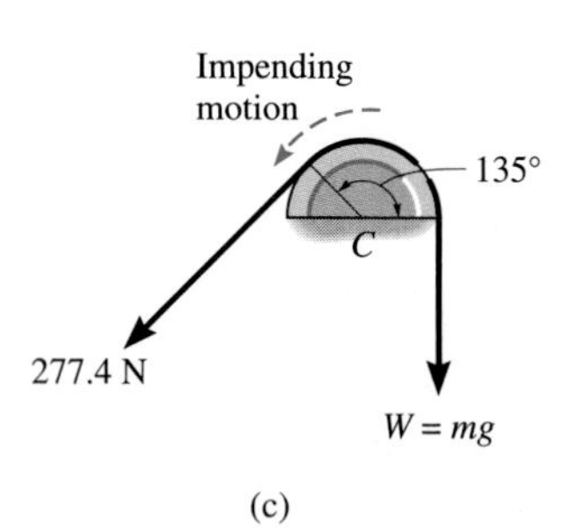

(c)

Fig. 8–22

PROBLEMS

8–83. Determine the *minimum* tension in the rope at points A and B that is necessary to maintain equilibrium. Take $\mu_s = 0.3$ between the rope and the fixed post D.

***8–84.** Determine the *maximum* tension in the rope at points A and B that is necessary to maintain equilibrium. Take $\mu_s = 0.3$ between the rope and the fixed post D.

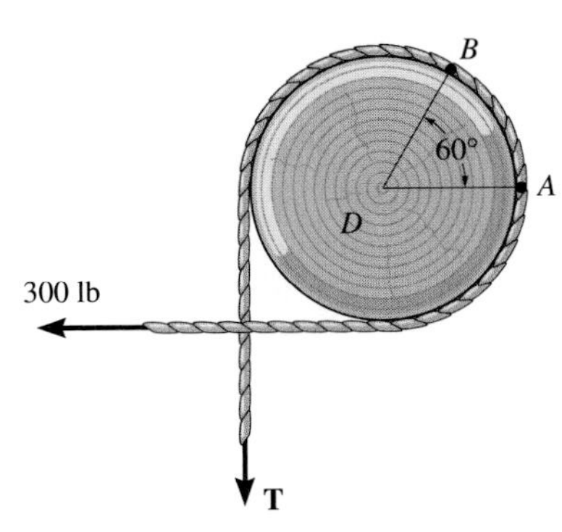

Probs. 8–83/8–84

8–85. A cylinder having a mass of 250 kg is to be supported by the cord which wraps over the pipe. Determine the *smallest* vertical force F needed to support the load if the cord passes (a) once over the pipe, $\beta = 180°$, and (b) two times over the pipe, $\beta = 540°$. Take $\mu_s = 0.2$.

8–86. A cylinder having a mass of 250 kg is to be supported by the cord which wraps over the pipe. Determine the *largest* vertical force F that can be applied to the cord without moving the cylinder. The cord passes (a) once over the pipe, $\beta = 180°$, and (b) two times over the pipe, $\beta = 540°$. Take $\mu_s = 0.2$.

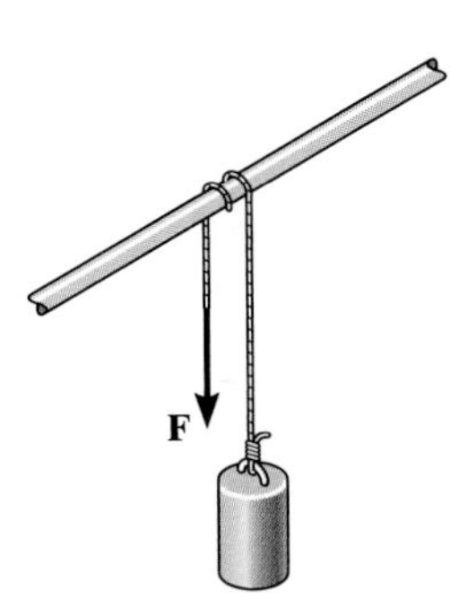

Probs. 8–85/8–86

8–87. The truck, which has a mass of 3.4 Mg, is to be lowered down the slope by a rope that is wrapped around a tree. If the wheels are free to roll and the man at A can resist a pull of 300 N, determine the minimum number of turns the rope should be wrapped around the tree to lower the truck at a constant speed. The coefficient of kinetic friction between the tree and rope is $\mu_k = 0.3$.

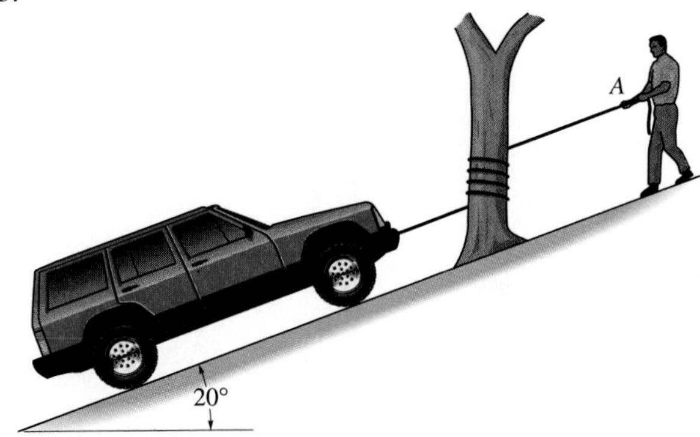

Prob. 8–87

***8–88.** The boat has a weight of 500 lb and is held in position off the side of a ship by the spars at A and B. A sailor having a weight of 130 lb gets in the boat, wraps a rope around an overhead boom at C, and ties it to the ends of the boat as shown. If the boat is disconnected from the spars, determine the *minimum number* of *half turns* the rope must make around the boom so that the boat can be safely lowered into the water. The coefficient of kinetic friction between the rope and the boom is $\mu_k = 0.15$. *Hint:* The problem requires that the normal force between the man's feet and the boat be as small as possible.

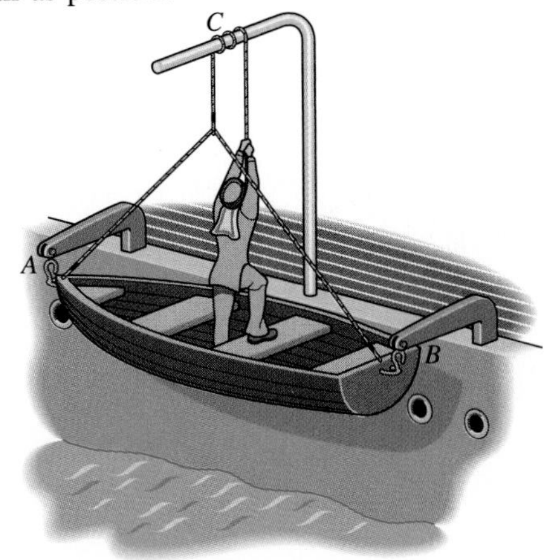

Prob. 8–88

8–89. A cord having a weight of 0.5 lb/ft and a total length of 10 ft is suspended over a peg P as shown. If the coefficient of static friction between the peg and cord is $\mu_s = 0.5$, determine the longest length h which one side of the suspended cord can have without causing motion. Neglect the size of the peg and the length of cord draped over it.

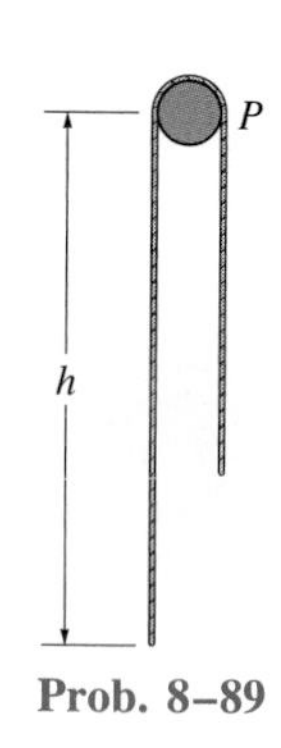

Prob. 8–89

8–90. The uniform concrete pipe has a weight of 800 lb and is unloaded slowly from the truck bed using the rope and skids shown. If the coefficient of kinetic friction between the rope and pipe is $\mu_k = 0.3$, determine the force the worker must exert on the rope to lower the pipe at constant speed. There is a pulley at B, and the pipe does not slip on the skids. The lower portion of the rope is parallel to the skids.

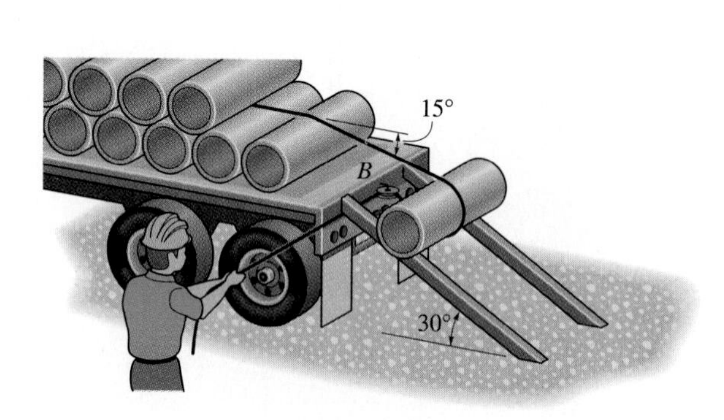

Prob. 8–90

■**8–91.** The choker sling is used to lift the smooth pipe that has a mass of 600 kg. If the coefficient of static friction between the loop at the end A of the sling and the rope is $\mu_s = 0.3$, determine the angle θ at the connection.

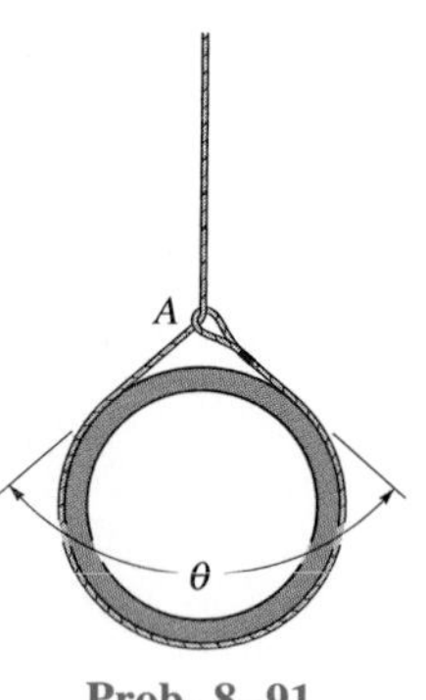

Prob. 8–91

***8–92.** The belt on the portable dryer wraps around the drum D, idler pulley A, and motor pulley B. If the motor can develop a maximum torque of $M = 0.80 \text{ N} \cdot \text{m}$, determine the smallest spring tension required to hold the belt from slipping. The coefficient of static friction between the belt and the drum and motor pulley is $\mu_s = 0.3$.

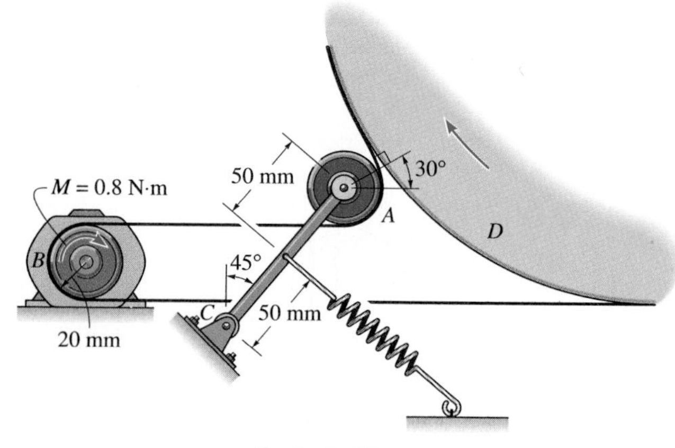

Prob. 8–92

8–93. Blocks A and B weigh 50 lb and 30 lb, respectively. Using the coefficients of static friction indicated, determine the greatest weight of block D without causing motion.

8–94. Blocks A, B, and D weigh 50 lb, 30 lb, and 12 lb, respectively. Using the coefficients of static friction indicated, determine the frictional force between blocks A and B and between block A and the floor C.

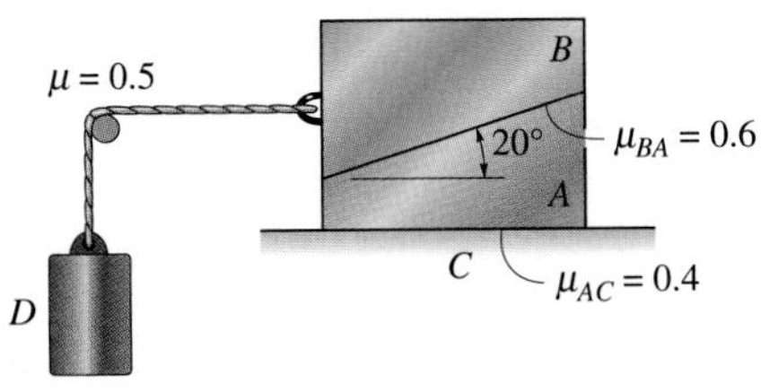

Probs. 8–93/8–94

8–95. Determine the smallest lever force P needed to prevent the wheel from rotating if it is subjected to a torque of $M = 250\ \text{N} \cdot \text{m}$. The coefficient of static friction between the belt and the wheel is $\mu_s = 0.3$. The wheel is pin-connected at its center, B.

***8–96.** Determine the torque M that can be resisted by the band brake if a force of $P = 30$ N is applied to the handle of the lever. The coefficient of static friction between the belt and the wheel is $\mu_s = 0.3$. The wheel is pin-connected at its center, B.

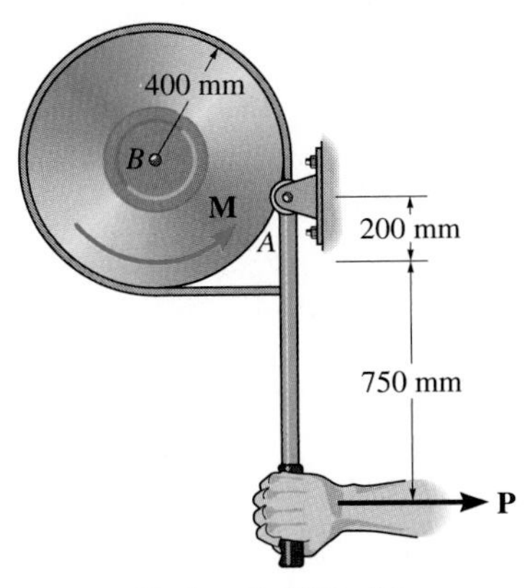

Probs. 8–95/8–96

8–97. The simple band brake is constructed so that the ends of the friction strap are connected to the pin at A and the lever arm at B. If the wheel is subjected to a torque of $M = 80\ \text{lb} \cdot \text{ft}$, determine the smallest force P applied to the lever that is required to hold the wheel stationary. The coefficient of static friction between the strap and wheel is $\mu_s = 0.5$.

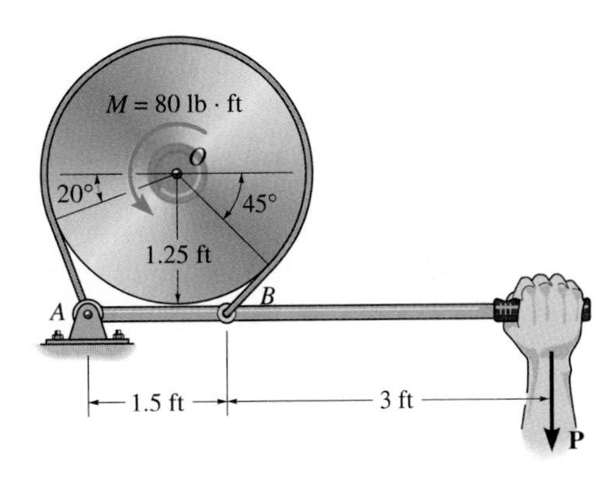

Prob. 8–97

8–98. Show that the frictional relationship between the belt tensions, the coefficient of friction μ, and the angular contacts α and β for the V-belt is $T_2 = T_1 e^{\mu\beta/\sin(\alpha/2)}$.

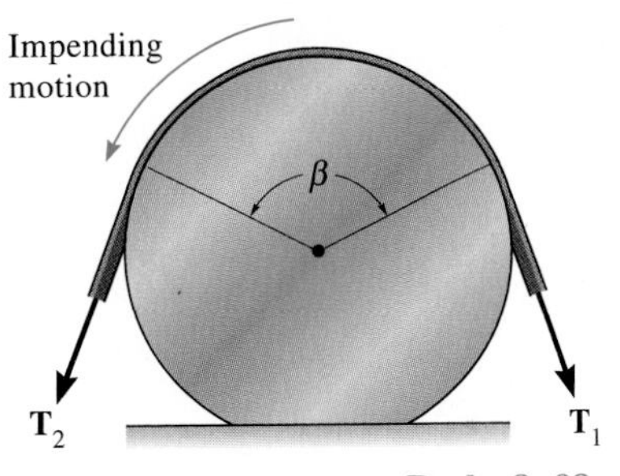

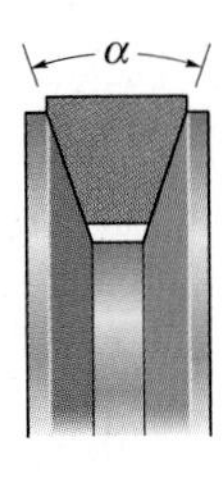

Prob. 8–98

8–99. Block A has a mass of 50 kg and rests on surface B for which $\mu_s = 0.25$. If the coefficient of static friction between the cord and the fixed peg at C is $\mu_s' = 0.3$, determine the greatest mass of the suspended cylinder D without causing motion.

***8–100.** Block A has a mass of 50 kg and rests on surface B for which $\mu_s = 0.25$. If the mass of the suspended cylinder D is 4 kg, determine the frictional force of the surface on A. The coefficient of static friction between the cord and the fixed peg at C is $\mu_s' = 0.3$.

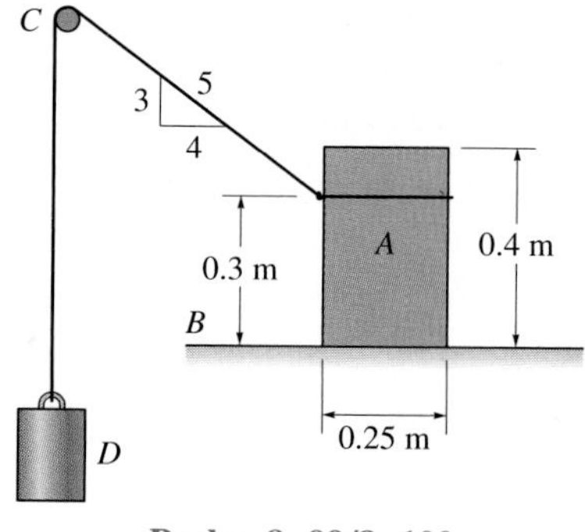

Probs. 8–99/8–100

*8.6 Frictional Forces on Collar Bearings, Pivot Bearings, and Disks

Pivot and *collar bearings* are commonly used in machines to support an *axial load* on a rotating shaft. These two types of support are shown in Fig. 8–23. Provided the bearings are not lubricated, or are only partially lubricated, the laws of dry friction may be applied to determine the moment **M** needed to turn the shaft when it supports an axial force **P**.

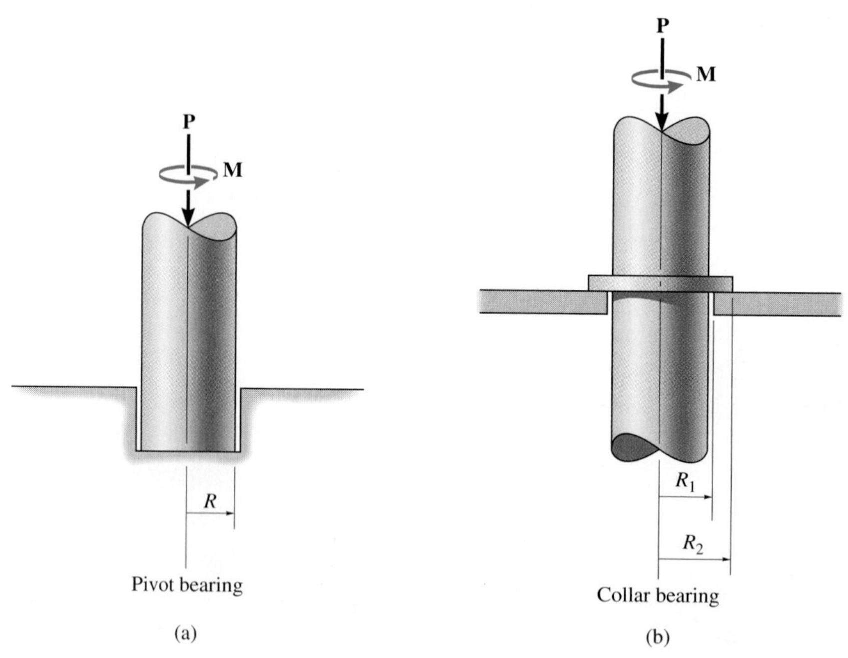

Fig. 8–23

Frictional Analysis. The collar bearing on the shaft shown in Fig. 8–24 is subjected to an axial force **P** and has a total bearing or contact area $\pi(R_2^2 - R_1^2)$. In the following analysis, the normal pressure p is considered to be *uniformly distributed* over this area—a reasonable assumption provided the bearing is new and evenly supported. Since $\Sigma F_z = 0$, p, measured as a force per unit area, is $p = P/\pi(R_2^2 - R_1^2)$.

The moment needed to cause impending rotation of the shaft can be determined from moment equilibrium of the frictional forces $d\mathbf{F}$ developed at the bearing surface by applying $\Sigma M_z = 0$. A small area element $dA = (r\,d\theta)(dr)$, shown in Fig. 8–24, is subjected to both a normal force $dN = p\,dA$ and an associated frictional force,

$$dF = \mu_s\,dN = \mu_s\,p\,dA = \frac{\mu_s P}{\pi(R_2^2 - R_1^2)}\,dA$$

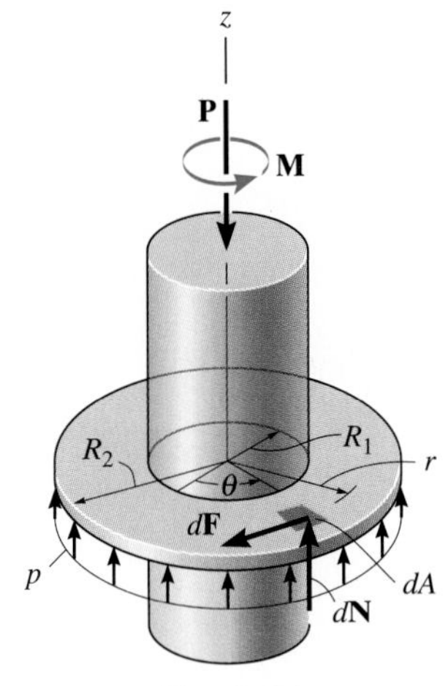

Fig. 8–24

The normal force does not create a moment about the z axis of the shaft; however, the frictional force does: namely, $dM = r\,dF$. Integration is needed to compute the total moment created by all the frictional forces acting on differential areas dA. Therefore, for impending rotational motion,

$$\Sigma M_z = 0; \qquad M - \int_A r\,dF = 0$$

Substituting for dF and dA and integrating over the entire bearing area yields

$$M = \int_{R_1}^{R_2}\int_0^{2\pi} r\left[\frac{\mu_s P}{\pi(R_2^2 - R_1^2)}\right](r\,d\theta\,dr) = \frac{\mu_s P}{\pi(R_2^2 - R_1^2)}\int_{R_1}^{R_2} r^2\,dr\int_0^{2\pi} d\theta$$

or

$$M = \tfrac{2}{3}\mu_s P\left(\frac{R_2^3 - R_1^3}{R_2^2 - R_1^2}\right) \qquad (8\text{–}7)$$

This equation gives the magnitude of moment required for impending rotation of the shaft. The frictional moment developed at the end of the shaft, when it is *rotating* at constant speed, can be found by substituting μ_k for μ_s in Eq. 8–7.

When $R_2 = R$ and $R_1 = 0$, as in the case of a pivot bearing, Fig. 8–23*a*, Eq. 8–7 reduces to

$$M = \tfrac{2}{3}\mu_s PR \qquad (8\text{–}8)$$

Recall from the initial assumption that both Eqs. 8–7 and 8–8 apply only for bearing surfaces subjected to *constant pressure*. If the pressure is not uniform, a variation of the pressure as a function of the bearing area must be determined before integrating to obtain the moment. The following example illustrates this concept.

Example 8–10

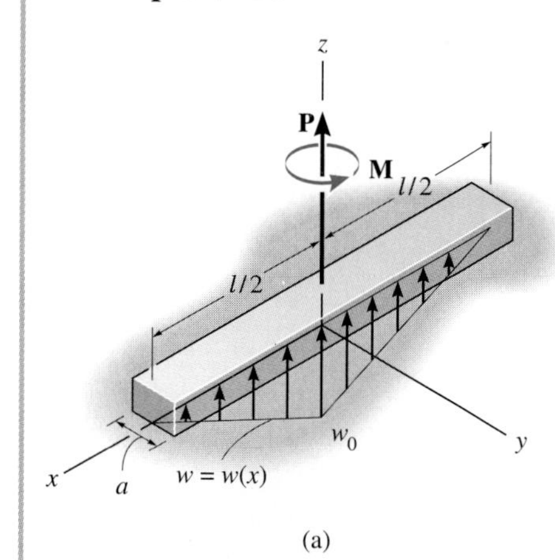

(a)

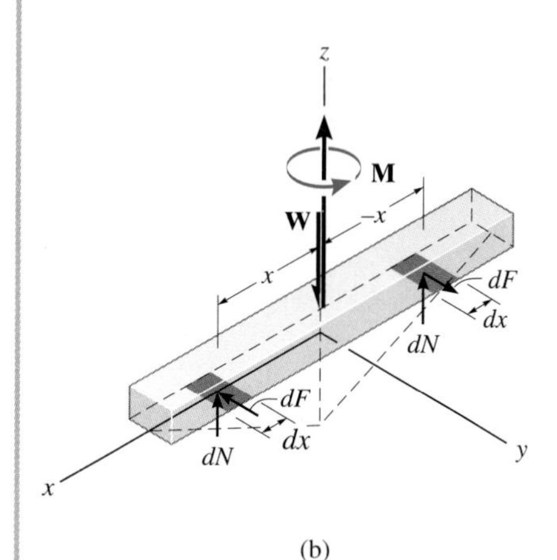

(b)

Fig. 8–25

The uniform bar shown in Fig. 8–25*a* has a total mass *m*. If it is assumed that the normal pressure acting at the contacting surface varies linearly along the length of the bar as shown, determine the couple moment **M** required to rotate the bar. Assume that the bar's width *a* is negligible in comparison to its length *l*. The coefficient of static friction is equal to μ.

SOLUTION

A free-body diagram of the bar is shown in Fig. 8–25*b*. Since the bar has a total weight of $W = mg$, the intensity w_0 of the distributed load at the center ($x = 0$) is determined from vertical force equilibrium, Fig. 8–25*a*.

$$+\uparrow \Sigma F_z = 0; \quad -mg + 2\left[\frac{1}{2}\left(\frac{l}{2}\right)w_0\right] = 0 \qquad w_0 = \frac{2mg}{l}$$

Since $w = 0$ at $x = l/2$, the distributed load expressed as a function of x is

$$w = w_0\left(1 - \frac{2x}{l}\right) = \frac{2mg}{l}\left(1 - \frac{2x}{l}\right)$$

The magnitude of the normal force acting on a segment of area having a length dx is therefore

$$dN = w\,dx = \frac{2mg}{l}\left(1 - \frac{2x}{l}\right)dx$$

The magnitude of the frictional force acting on the same element of area is

$$dF = \mu\,dN = \frac{2\mu mg}{l}\left(1 - \frac{2x}{l}\right)dx$$

Hence, the moment created by this force about the z axis is

$$dM = x\,dF = \frac{2\mu mg}{l}x\left(1 - \frac{2x}{l}\right)dx$$

The summation of moments about the z axis of the bar is determined by integration, which yields

$$\Sigma M_z = 0; \qquad M - 2\int_0^{l/2}\frac{2\mu mg}{l}x\left(1 - \frac{2x}{l}\right)dx = 0$$

$$M = \frac{4\mu mg}{l}\left(\frac{x^2}{2} - \frac{2x^3}{3l}\right)\Bigg|_0^{l/2}$$

$$M = \frac{\mu mgl}{6} \qquad \textit{Ans.}$$

*8.7 Frictional Forces on Journal Bearings

When a shaft or axle is subjected to lateral loads, a *journal bearing* is commonly used for support. Well-lubricated journal bearings are subjected to the laws of fluid mechanics, in which the viscosity of the lubricant, the speed of rotation, and the amount of clearance between the shaft and bearing are needed to determine the frictional resistance of the bearing. When the bearing is not lubricated or is only partially lubricated, however, a reasonable analysis of the frictional resistance can be based on the laws of dry friction.

Frictional Analysis. A typical journal-bearing support is shown in Fig. 8–26*a*. As the shaft rotates in the direction shown in the figure, it rolls up against the wall of the bearing to some point A where slipping occurs. If the lateral load acting at the end of the shaft is **W**, it is necessary that the bearing reactive force **R** acting at A be equal and opposite to **W**, Fig. 8–26*b*. The moment needed to maintain constant rotation of the shaft can be found by summing moments about the z axis of the shaft; i.e.,

$$\Sigma M_z = 0; \qquad -M + (R \sin \phi_k)r = 0$$

or

$$M = Rr \sin \phi_k \tag{8–9}$$

where ϕ_k is the angle of kinetic friction defined by $\tan \phi_k = F/N = \mu_k N/N = \mu_k$. In Fig. 8–26*c*, it is seen that $r \sin \phi_k = r_f$. The dashed circle with radius r_f is called the *friction circle*, and as the shaft rotates, the reaction **R** will always be tangent to it. If the bearing is partially lubricated, μ_k is small, and therefore $\mu_k = \tan \phi_k \approx \sin \phi_k \approx \phi_k$. Under these conditions, a reasonable *approximation* to the moment needed to overcome the frictional resistance becomes

$$M \approx Rr\mu_k \tag{8–10}$$

The following example illustrates a common application of this analysis.

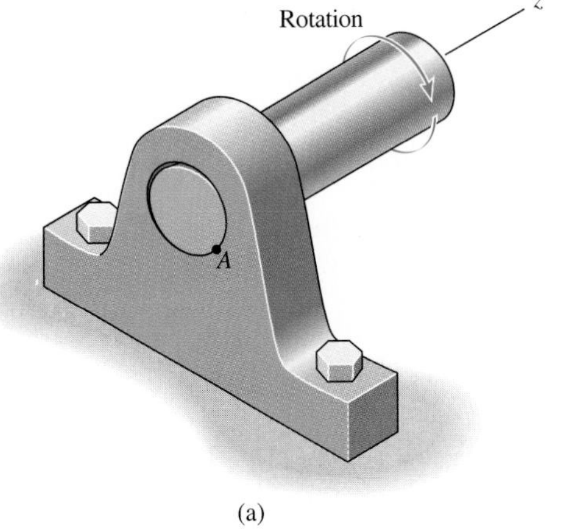

(a)

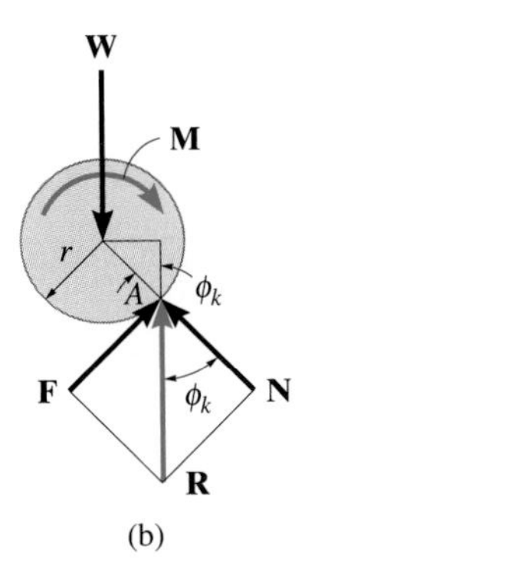

(b)

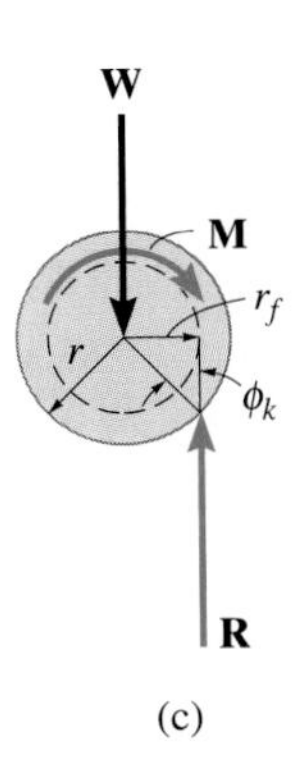

(c)

Fig. 8–26

Example 8–11

The 100-mm-diameter pulley shown in Fig. 8–27*a* fits loosely on a 10-mm-diameter shaft for which the coefficient of static friction is $\mu_s = 0.4$. Determine the minimum tension T in the belt needed to (a) raise the 100-kg block and (b) lower the block. Assume that no slipping occurs between the belt and pulley and neglect the weight of the pulley.

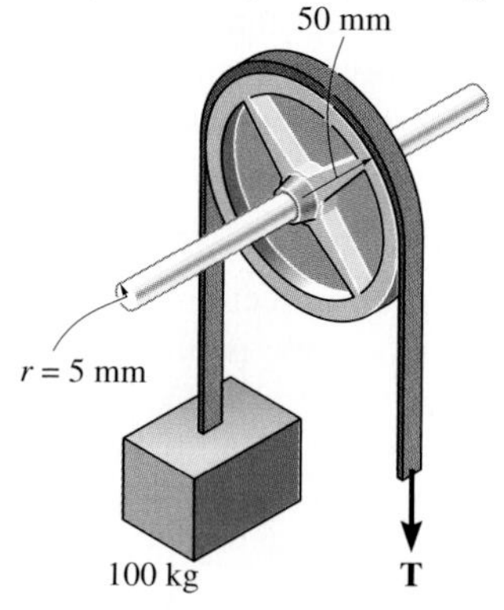

(a)

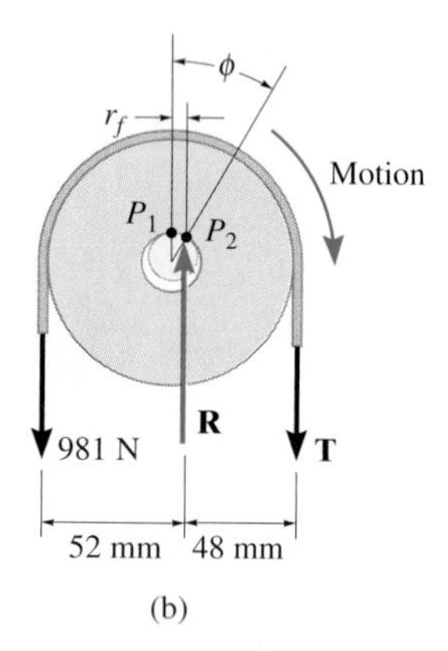

(b)

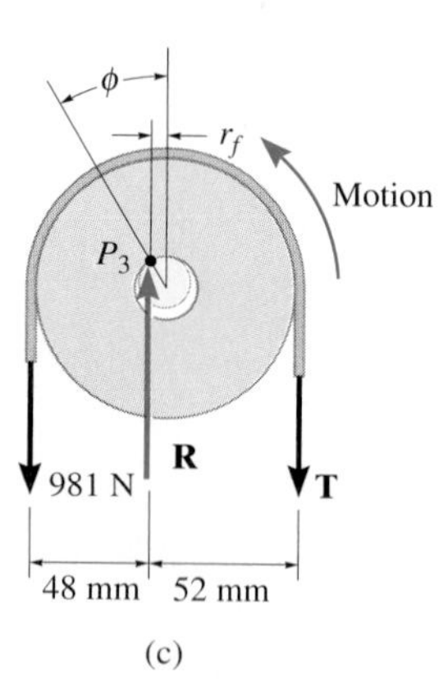

(c)

Fig. 8–27

SOLUTION

Part (a). A free-body diagram of the pulley is shown in Fig. 8–27*b*. When the pulley is subjected to belt tensions of 981 N each, it makes contact with the shaft at point P_1. As the tension T is *increased*, the pulley will roll around the shaft to point P_2 before motion impends. From the figure, the friction circle has a radius $r_f = r \sin \phi$. Using the simplification $\sin \phi \approx \phi$, $r_f \approx r\mu_s = (5 \text{ mm})(0.4) = 2 \text{ mm}$, so that summing moments about P_2 gives

$$\circlearrowright +\Sigma M_{P_2} = 0; \qquad 981 \text{ N}(52 \text{ mm}) - T(48 \text{ mm}) = 0$$

$$T = 1063 \text{ N} = 1.06 \text{ kN} \qquad \textit{Ans.}$$

If a more exact analysis is used, then $\phi = \tan^{-1} 0.4 = 21.8°$. Thus, the radius of the friction circle would be $r_f = r \sin \phi = 5 \sin 21.8° = 1.86 \text{ mm}$. Therefore,

$$\circlearrowright +\Sigma M_{P_2} = 0;$$

$$981 \text{ N}(50 \text{ mm} + 1.86 \text{ mm}) - T(50 \text{ mm} - 1.86 \text{ mm}) = 0$$

$$T = 1057 \text{ N} = 1.06 \text{ kN} \qquad \textit{Ans.}$$

Part (b). When the block is lowered, the resultant force **R** acting on the shaft passes through point P_3, as shown in Fig. 8–27*c*. Summing moments about this point yields

$$\circlearrowright +\Sigma M_{P_3} = 0; \qquad 981 \text{ N}(48 \text{ mm}) - T(52 \text{ mm}) = 0$$

$$T = 906 \text{ N} \qquad \textit{Ans.}$$

★8.8 Rolling Resistance

If a *rigid* cylinder of weight **W** rolls at constant velocity along a *rigid* surface, the normal force exerted by the surface on the cylinder acts at the tangent point of contact, as shown in Fig. 8–28*a*. Under these conditions, provided the cylinder does not encounter frictional resistance from the air, motion will continue indefinitely. Actually, however, no materials are perfectly rigid, and therefore the reaction of the surface on the cylinder consists of a distribution of normal pressure. For example, consider the cylinder to be made of a very hard material, and the surface on which it rolls to be soft. Due to its weight, the cylinder compresses the surface underneath it, Fig. 8–28*b*. As the cylinder rolls, the surface material in front of the cylinder *retards* the motion since it is being *deformed*, whereas the material in the rear is *restored* from the deformed state and therefore tends to *push* the cylinder forward. The normal pressures acting on the cylinder in this manner are represented in Fig. 8–28*b* by their resultant forces $\mathbf{N}_d$ and $\mathbf{N}_r$. Unfortunately, the magnitude of the force of *deformation* and its horizontal component is *always greater* than that of *restoration,* and consequently a horizontal driving force **P** must be applied to the cylinder to maintain the motion, Fig. 8–28*b*.*

Rolling resistance is caused primarily by this effect, although it is also, to a smaller degree, the result of surface adhesion and relative micro-sliding between the surfaces of contact. Because the actual force **P** needed to overcome these effects is difficult to determine, a simplified method will be developed here to explain one way engineers have analyzed this phenomenon. To do this, we will consider the resultant of the *entire* normal pressure, $\mathbf{N} = \mathbf{N}_d + \mathbf{N}_r$, acting on the cylinder, Fig. 8–28*c*. As shown in Fig. 8–28*d*, this force acts at an angle θ with the vertical. To keep the cylinder in equilibrium, i.e., rolling at a constant rate, it is necessary that **N** be *concurrent* with the driving force **P** and the weight **W.** Summing moments about point A gives $Wa = P(r\cos\theta)$. Since the deformations are generally very small in relation to the cylinder's radius, $\cos\theta \approx 1$; hence,

$$Wa \approx Pr$$

or

$$P \approx \frac{Wa}{r} \qquad (8\text{–}11)$$

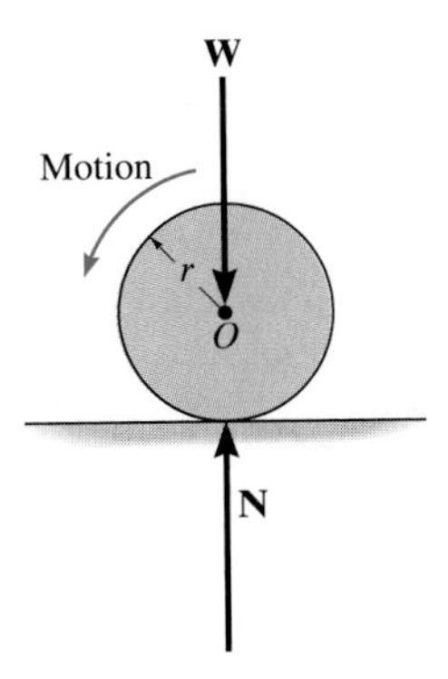

Rigid surface of contact

(a)

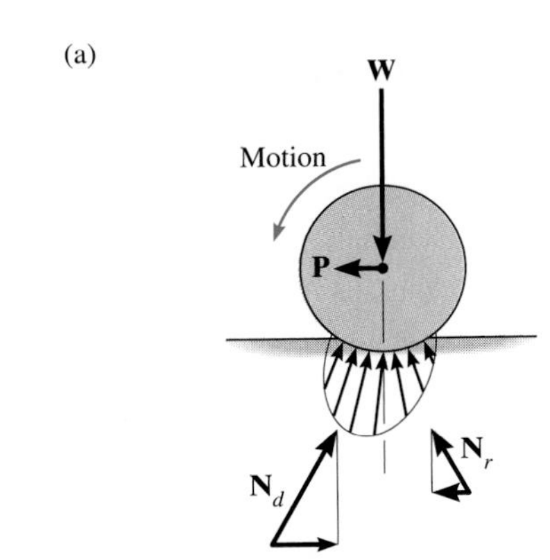

Soft surface of contact

(b)

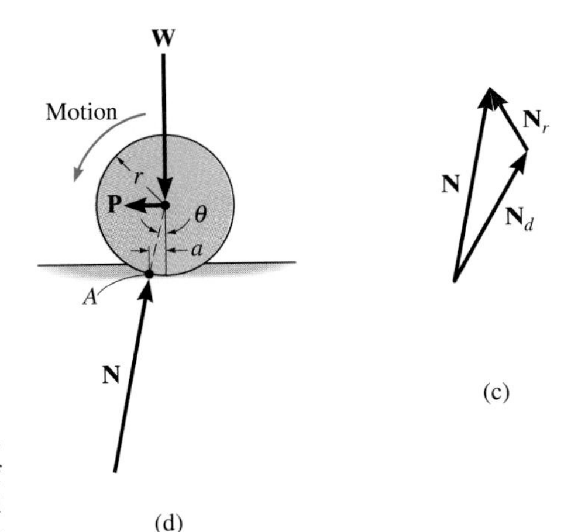

(d)

(c)

Fig. 8–28

*Actually, the deformation force $\mathbf{N}_d$ causes *energy* to be stored in the material as its magnitude is increased, whereas the restoration force $\mathbf{N}_r$, as its magnitude is decreased, allows some of this energy to be released. The remaining energy is *lost* since it is used to heat up the surface, and if the cylinder's weight is very large, it accounts for permanent deformation of the surface. Work must be done by the horizontal force **P** to make up for this loss.

The distance a is termed the *coefficient of rolling resistance*, which has the dimension of length. For instance, $a \approx 0.5$ mm for a wheel rolling on a rail, both of which are made of mild steel. For hardened steel ball bearings on steel, $a \approx 0.1$ mm. Experimentally, though, this factor is difficult to measure, since it depends on such parameters as the rate of rotation of the cylinder, the elastic properties of the contacting surfaces, and the surface finish. For this reason, little reliance is placed on the data for determining a. The analysis presented here does, however, indicate why a heavy load (W) offers greater resistance to motion (P) than a light load under the same conditions. Furthermore, since the force needed to *roll* the cylinder over the surface will be much less than that needed to *slide* the cylinder across the surface, the analysis indicates why roller or ball bearings are often used to minimize the frictional resistance between moving parts.

Example 8–12

A 10-kg steel wheel shown in Fig. 8–29*a* has a radius of 100 mm and rests on an inclined plane made of wood. If θ is increased so that the wheel begins to roll down the incline with constant velocity when $\theta = 1.2°$, determine the coefficient of rolling resistance.

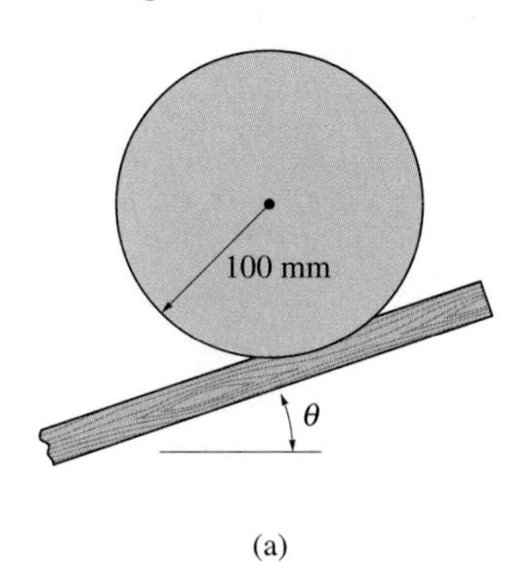

(a)

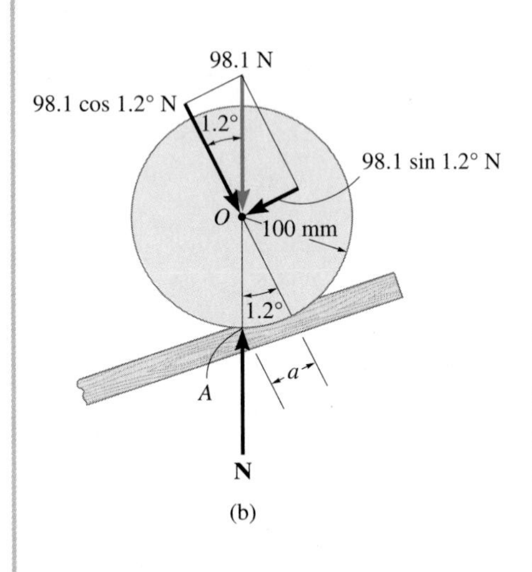

(b)

Fig. 8–29

SOLUTION

As shown on the free-body diagram, Fig. 8–29*b*, when the wheel has impending motion, the normal reaction **N** acts at point A defined by the dimension a. Resolving the weight into components parallel and perpendicular to the incline, and summing moments about point A, yields (approximately)

$$\circlearrowleft + \Sigma M_A = 0; \quad 98.1 \cos 1.2°(a) - 98.1 \sin 1.2°(100) = 0$$

Solving, we obtain

$$a = 2.1 \text{ mm} \qquad \textit{Ans.}$$

PROBLEMS

8–101. The collar bearing uniformly supports an axial force of $P = 800$ lb. If the coefficient of static friction is $\mu_s = 0.3$, determine the torque M required to overcome friction.

8–102. The collar bearing uniformly supports an axial force of $P = 500$ lb. If a torque of $M = 3$ lb · ft is applied to the shaft and causes it to rotate at constant angular velocity, determine the coefficient of kinetic friction at the surface of contact.

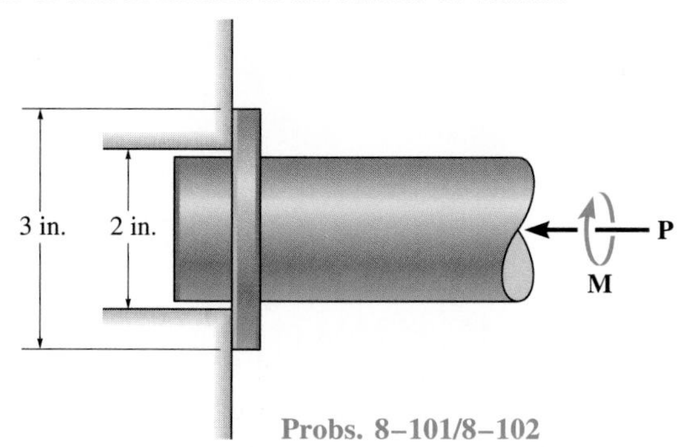

Probs. 8–101/8–102

8–103. The annular ring bearing is subjected to a thrust of 800 lb. If $\mu_s = 0.35$, determine the torque M that must be applied to overcome friction.

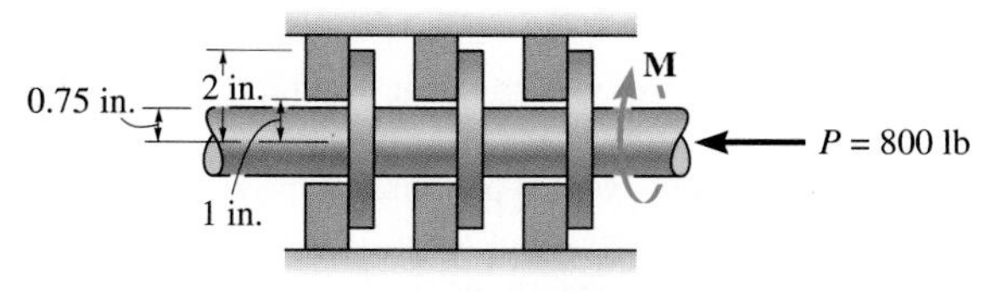

Prob. 8–103

***8–104.** Assuming the pressure of the sanding disk D on the surface to be uniform, determine the vertical couple forces developed at the handles which are necessary to hold it in equilibrium. The disk is pushed against the surface with a total horizontal force of 18 lb. Neglect the weight of the sander and take $\mu_k = 0.59$.

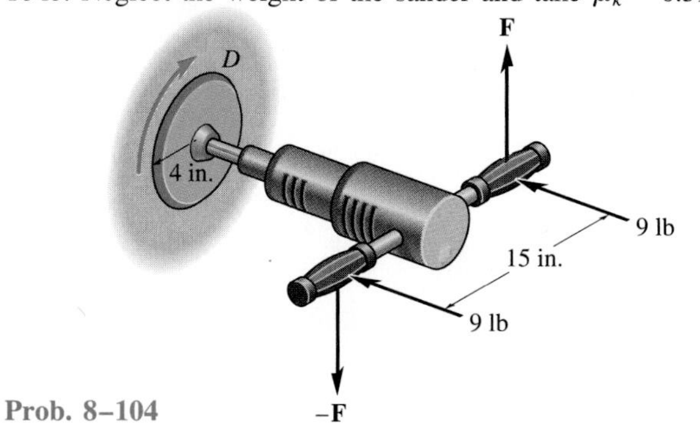

Prob. 8–104

8–105. The floor-polishing machine rotates at a constant angular velocity. If it has a weight of 80 lb, determine the couple forces F the operator must apply to the handles to hold the machine stationary. The coefficient of kinetic friction between the floor and brush is $\mu_k = 0.3$. Assume the brush exerts a uniform pressure on the floor.

Prob. 8–105

8–106. The plate clutch consists of a flat plate A that slides over the rotating shaft S. The shaft is fixed to the driving plate gear B. If the gear C, which is in mesh with B, is subjected to a torque of $M = 0.8$ N · m, determine the smallest force P, that must be applied via the control arm, to stop the rotation. The coefficient of static friction between the plates A and D is $\mu_s = 0.4$. Assume the bearing pressure between A and D to be uniform.

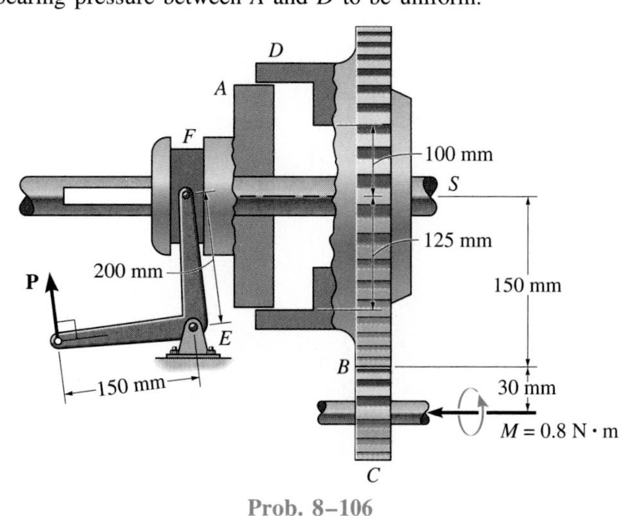

Prob. 8–106

8–107. Because of wearing at the edges, the pivot bearing is subjected to a conical pressure distribution at its surface of contact. Determine the torque M required to overcome friction and turn the shaft, which supports an axial force $\mathbf{P}$. The coefficient of static friction is μ. For the solution, it is necessary to determine the peak pressure p_0 in terms of P and the bearing radius R.

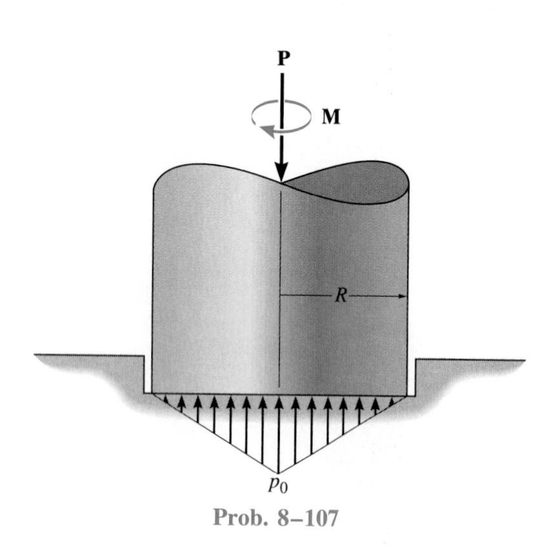

Prob. 8–107

8–109. The pivot bearing is subjected to a pressure distribution at its surface of contact which varies as shown. If the coefficient of static friction is μ, determine the torque M required to overcome friction if the shaft supports an axial force $\mathbf{P}$.

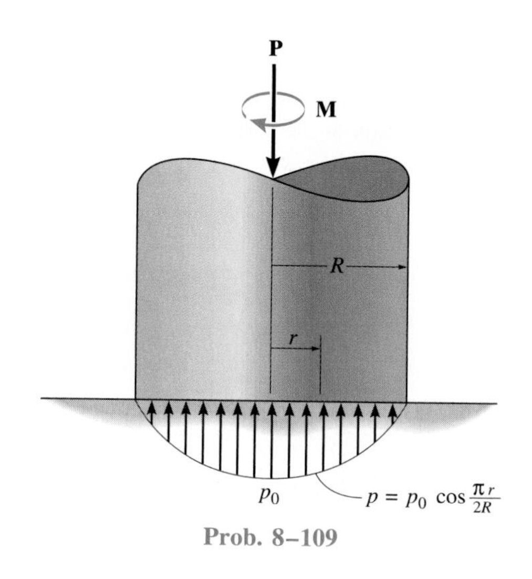

Prob. 8–109

***8–108.** The pivot bearing is subjected to a parabolic pressure distribution at its surface of contact. If the coefficient of static friction is μ, determine the torque M required to overcome friction and turn the shaft if it supports an axial force $\mathbf{P}$.

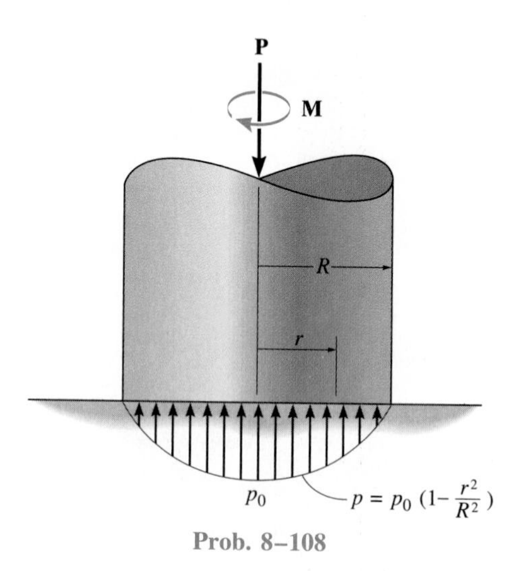

Prob. 8–108

8–110. The tractor is used to push the 1500-lb pipe. To do this it must overcome the frictional forces at the ground, caused by sand. Assuming that the sand exerts a pressure on the bottom of the pipe as shown, and the coefficient of static friction between the pipe and the sand is $\mu_s = 0.3$, determine the force required to push the pipe forward. Also, determine the peak pressure p_0.

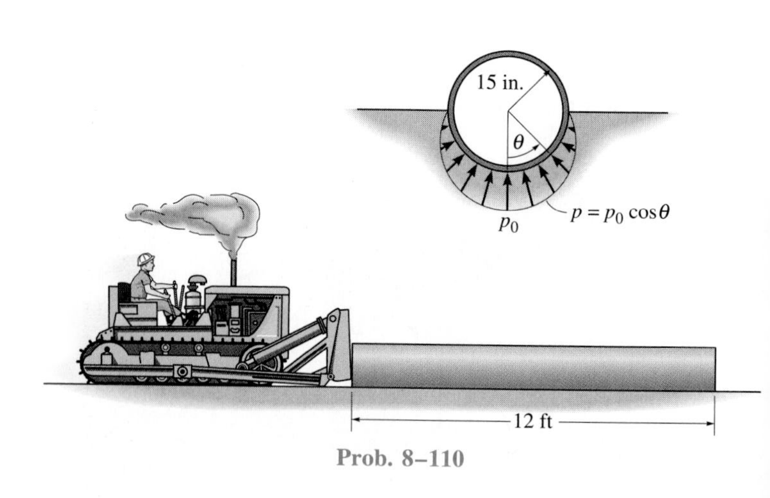

Prob. 8–110

8–111. The corkscrew is used to remove the 15-mm-diameter cork from the bottle. Determine the smallest vertical force P that must be applied to the handle if the gauge pressure in the bottle is $p = 175$ kPa and the cork pushes against the sides of the bottle's neck with a uniform pressure of 90 kPa. The coefficient of static friction between the bottle and the cork is $\mu_s = 0.4$. *Hint:* The force exerted on the bottom of the cork is $F = pA$, where A is the surface area of the cork's bottom and p is the gauge pressure.

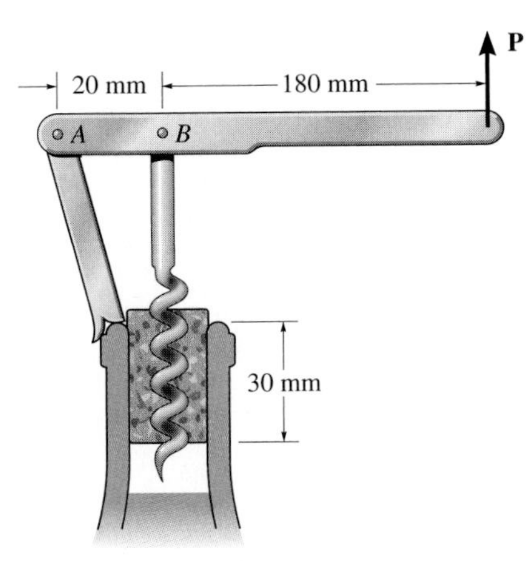

Prob. 8–111

*8–112. If the smallest tension force $T_A = 500$ N is required to pull the belt downward at A over the shaft S, determine the coefficient of kinetic friction between the loosely fitting collar bushing B and the shaft. Assume that the belt does not slip on the collar; rather, the collar slips on the shaft.

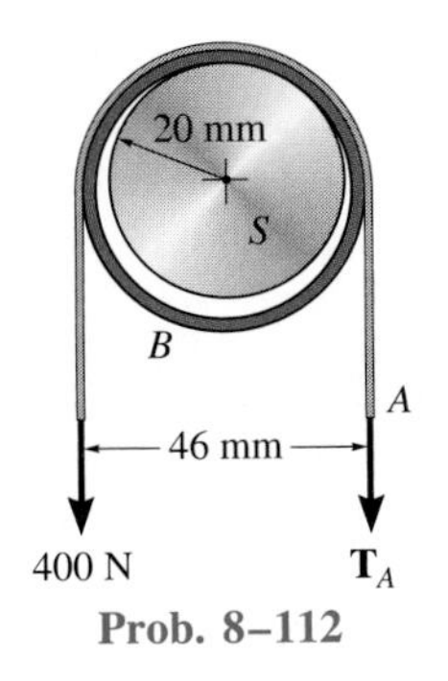

Prob. 8–112

8–113. The 5-kg pulley has a diameter of 240 mm and the axle has a diameter of 40 mm. If the coefficient of kinetic friction between the axle and the pulley is $\mu_k = 0.15$, determine the vertical force P on the rope required to lift the 80-kg block at constant velocity.

8–114. Solve Prob. 8–113 if the force **P** is applied horizontally to the right.

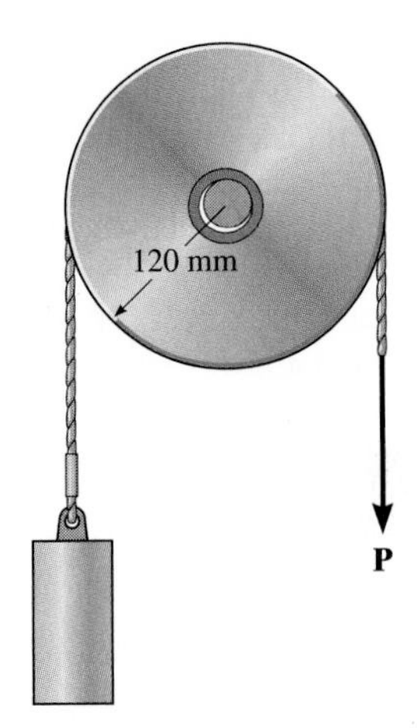

Probs. 8–113/8–114

8–115. The pulley has a radius of 3 in. and fits loosely on the 0.5-in.-diameter shaft. If the loadings acting on the belt cause the pulley to rotate with constant angular velocity, determine the frictional force between the shaft and the pulley and compute the coefficient of kinetic friction. The pulley weighs 18 lb.

*8–116. The pulley has a radius of 3 in. and fits loosely on the 0.5-in.-diameter shaft. If the loadings acting on the belt cause the pulley to rotate with constant angular velocity, determine the frictional force between the shaft and the pulley and compute the coefficient of kinetic friction. Neglect the weight of the pulley.

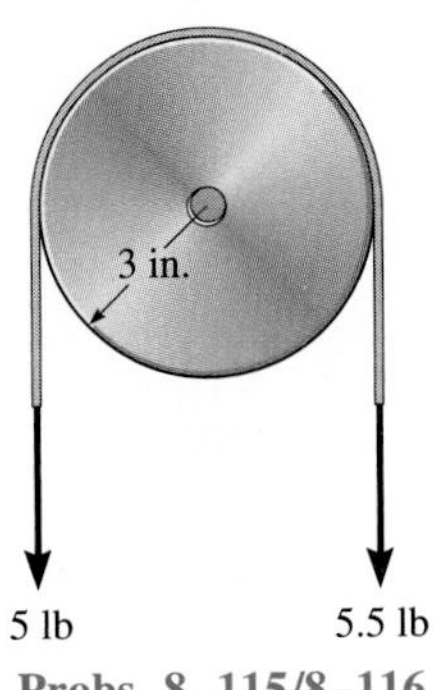

Probs. 8–115/8–116

8–117. The collar fits *loosely* around a fixed shaft that has a radius of 2 in. If the coefficient of kinetic friction between the shaft and the collar is $\mu_k = 0.3$, determine the force P on the horizontal segment of the belt so that the collar rotates counterclockwise with a constant angular velocity. Assume that the belt does not slip on the collar; rather, the collar slips on the shaft. Neglect the weight and thickness of the belt and collar. The radius, measured from the center of the collar to the mean thickness of the belt, is 2.25 in.

8–118. The collar fits *loosely* around a fixed shaft that has a radius of 2 in. If the coefficient of kinetic friction between the shaft and the collar is $\mu_k = 0.3$, determine the force P on the horizontal segment of the belt so that the collar rotates clockwise with a constant angular velocity. Assume that the belt does not slip on the collar; rather, the collar slips on the shaft. Neglect the weight and thickness of the belt and collar. The radius, measured from the center of the collar to the mean thickness of the belt, is 2.25 in.

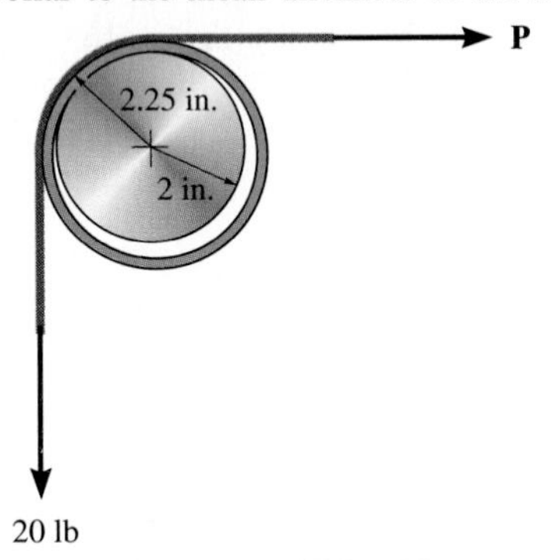

Probs. 8–117/8–118

8–119. A pulley having a diameter of 80 mm and a mass of 1.25 kg is supported loosely on a shaft having a diameter of 20 mm. Determine the torque M that must be applied to the pulley to cause it to rotate with constant angular velocity. The coefficient of kinetic friction between the shaft and pulley is $\mu_k = 0.4$. Also calculate the angle θ which the normal force at the point of contact makes with the horizontal. The shaft itself cannot rotate.

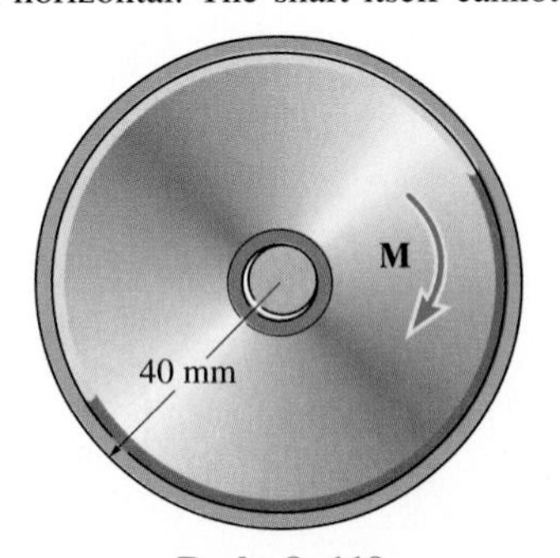

Prob. 8–119

***8–120.** The weight of the body on the tibiotalar joint J is 125 lb. If the radius of curvature of the talus surface of the ankle is 1.40 in., and the coefficient of static friction between the bones is $\mu_s = 0.1$, determine the force T developed in the Achilles tendon necessary to rotate the joint.

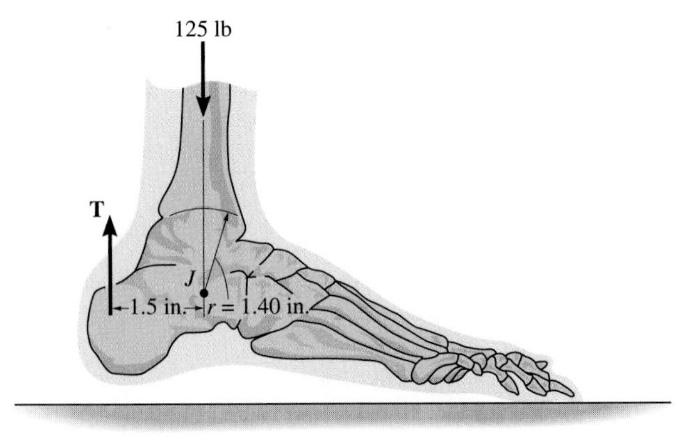

Prob. 8–120

8–121. A car has a mass of 1.5 Mg. Determine the horizontal force P that must be applied to overcome the rolling resistance of the tires if the coefficient of rolling resistance is 0.05 in. The tires have a diameter of 2.75 ft.

8–122. The car has a weight of 2600 lb and center of gravity at G. Determine the horizontal force P that must be applied to overcome the rolling resistance of the wheels. The coefficient of rolling resistance is 0.5 in. The tires have a diameter of 2.75 ft.

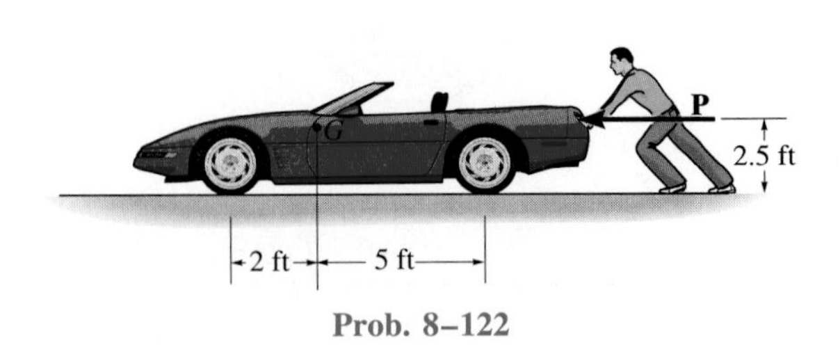

Prob. 8–122

8–123. The hand cart has wheels with a diameter of 80 mm. If a crate having a mass of 500 kg is placed on the cart so that each wheel carries an equal load, determine the horizontal force P that must be applied to the handle to overcome the rolling resistance. The coefficient of rolling resistance is 2 mm. Neglect the mass of the cart.

Prob. 8–123

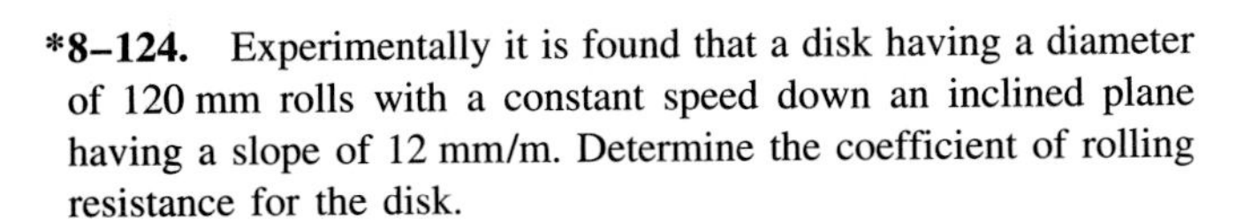

***8–124.** Experimentally it is found that a disk having a diameter of 120 mm rolls with a constant speed down an inclined plane having a slope of 12 mm/m. Determine the coefficient of rolling resistance for the disk.

8–125. The cylinder is subjected to a load that has a weight W. If the coefficients of rolling resistance for the cylinder's top and bottom surfaces are a_A and a_B, respectively, show that a force having a magnitude of $P = [W(a_A + a_B)]/2r$ is required to move the load and thereby roll the cylinder forward. Neglect the weight of the cylinder.

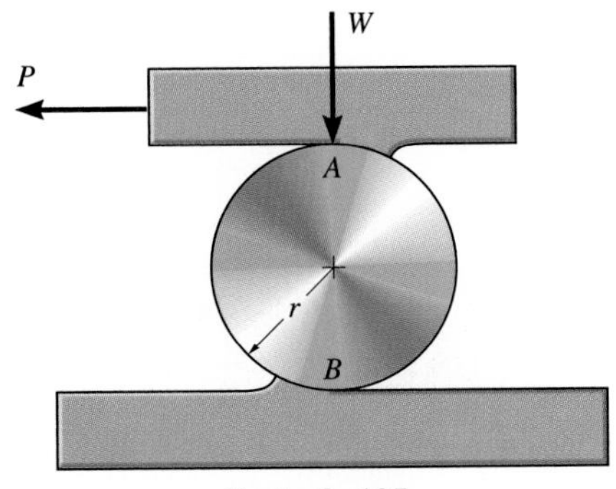

Prob. 8–125

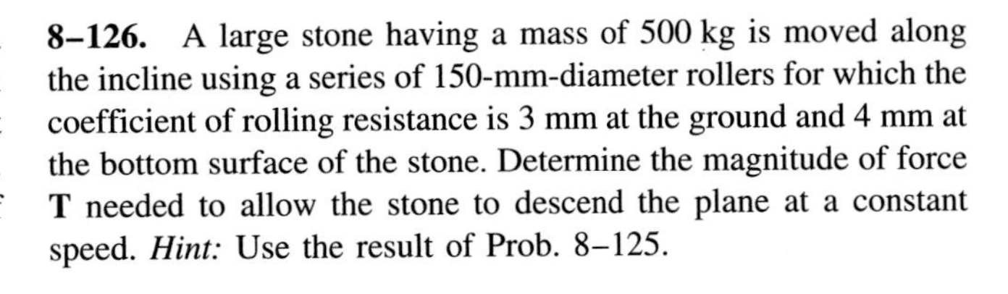

8–126. A large stone having a mass of 500 kg is moved along the incline using a series of 150-mm-diameter rollers for which the coefficient of rolling resistance is 3 mm at the ground and 4 mm at the bottom surface of the stone. Determine the magnitude of force **T** needed to allow the stone to descend the plane at a constant speed. *Hint:* Use the result of Prob. 8–125.

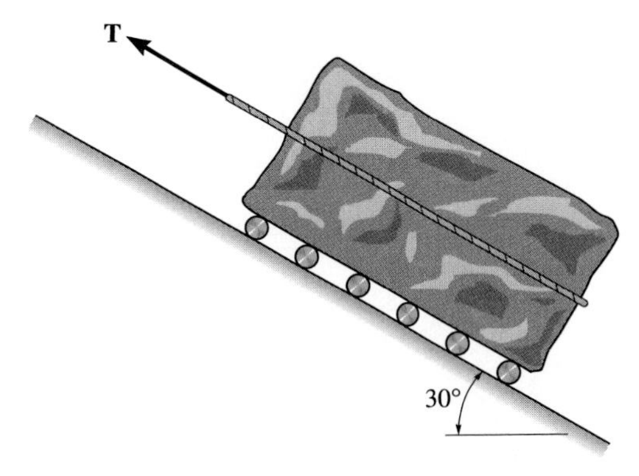

Prob. 8–126

8–127. The carriage is used to support a crane load of $F = 1.25(10^6)$ lb. The nine axles shown in the arrangement serve to equalize the load on each of the eight wheels. (By comparison note that a single supporting beam would *deform* and the wheels would share an unequal amount of load.) Determine the horizontal force that must be applied to the carriage in order to overcome the rolling resistance. The coefficient of rolling resistance is 0.007 in. and each wheel has a diameter of 36 in.

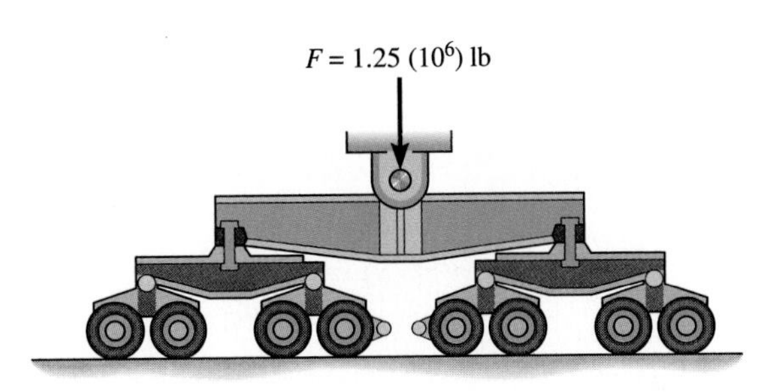

Prob. 8–127

REVIEW PROBLEMS

*8–128. The man attempts to pull open the door, which requires a force of 190 lb. If he has a weight of 200 lb and the coefficient of static friction between his shoes and the floor is $\mu_s = 0.5$, determine if he can do it.

Prob. 8–128

8–129. The uniform board having a length of 10 ft and weight W is placed within the opening of the concrete pipe having an inner diameter of 3 ft. Determine the coefficient of static friction μ at A and B if the board is on the verge of slipping when $\theta = 30°$.

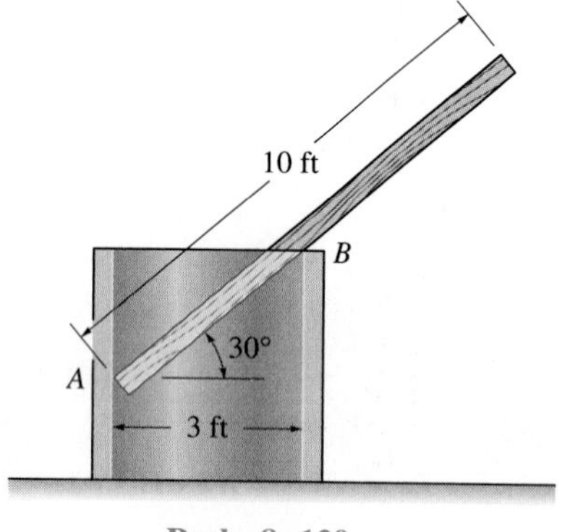

Prob. 8–129

8–130. The cam or short link is pinned at A and is used to hold mops or brooms against a wall. If the coefficient of static friction between the broomstick and the cam is $\mu_s = 0.2$, determine if it is possible to support the broom having a weight W. The surface at B is smooth. Neglect the weight of the cam.

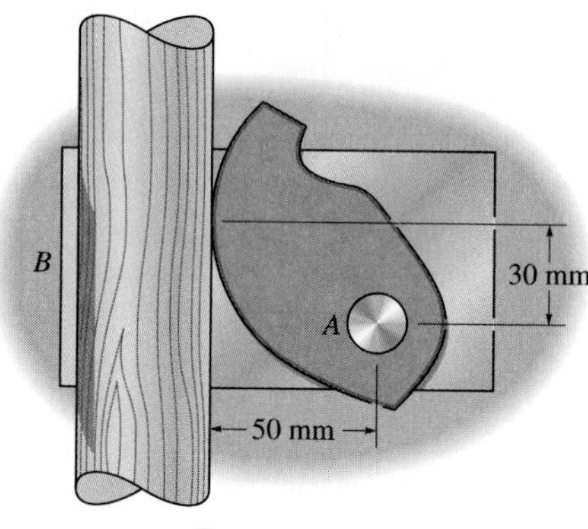

Prob. 8–130

8–131. The 1.4-Mg machine is to be moved over a level surface using a series of rollers for which the coefficient of rolling resistance is 0.5 mm at the ground and 0.2 mm at the bottom surface of the machine. Determine the appropriate diameter of the rollers so that the machine can be pushed forward with a horizontal force of $P = 250$ N. *Hint:* Use the result of Prob. 8–125.

Prob. 8–131

***8–132.** The carton clamp on the forklift has a coefficient of static friction of $\mu_s = 0.5$ with any cardboard carton, whereas a cardboard carton has a coefficient of static friction of $\mu_s' = 0.4$ with any other cardboard carton. Compute the smallest horizontal force P the clamp must exert on the sides of a carton so that two cartons A and B each weighing 30 lb can be lifted. What smallest clamping force P' is required to lift three 30-lb cartons? The third carton C is placed between A and B.

Prob. 8–132

8–133. The differential band brake is constructed so that the ends of the friction strap are connected to the pin at A and the lever at B. If the wheel is subjected to a torque of $M = 100 \text{ lb} \cdot \text{ft}$, determine the vertical force P applied to the lever required to hold the wheel stationary. The coefficient of static friction between the strap and wheel is $\mu_s = 0.5$.

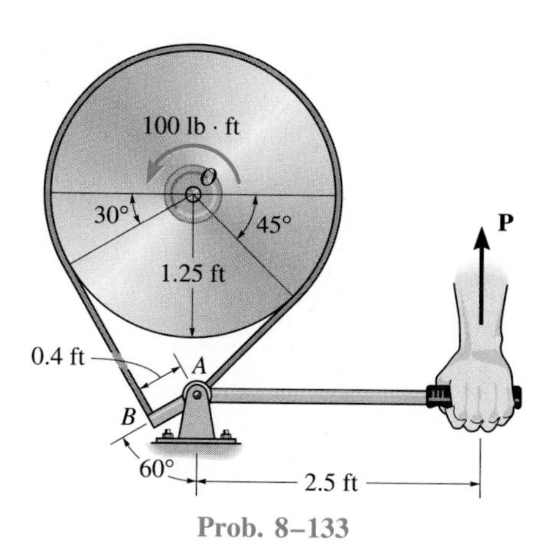

Prob. 8–133

8–134. A turnbuckle is used to tension member BC of the truss. The coefficient of static friction between the square-threaded screws and the turnbuckle is $\mu_s = 0.5$. The screws have a mean radius of 6 mm and a lead of 3 mm. If a torque of $M = 10 \text{ N} \cdot \text{m}$ is applied to the turnbuckle to draw the screws closer together, determine the force in each member of the truss. No external forces act on the truss.

8–135. A turnbuckle is used to tension member BC of the truss. The coefficient of static friction between the square-threaded screws and the turnbuckle is $\mu_s = 0.5$. The screws have a mean radius of 6 mm and a lead of 3 mm. Determine the torque M that must be applied to the turnbuckle to draw the screws closer together, so that a tension force of 500 N is developed in member BC.

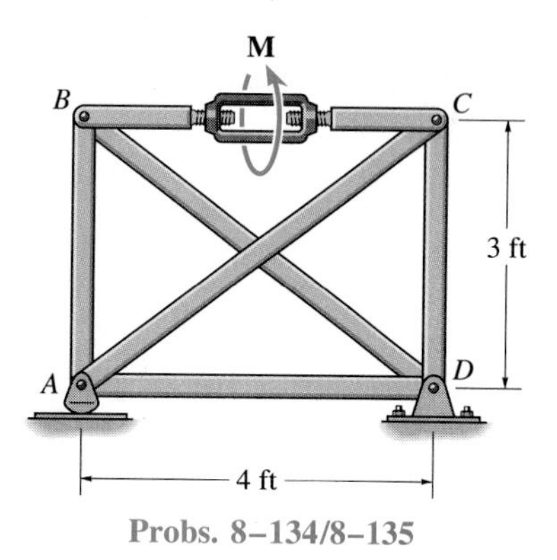

Probs. 8–134/8–135

***8–136.** A cable is attached to a 60-lb plate B, passes over a fixed disk at C, and is attached to the block at A. Using the coefficients of static friction shown in the figure, determine the smallest weight of block A that will prevent sliding motion of B down the plane.

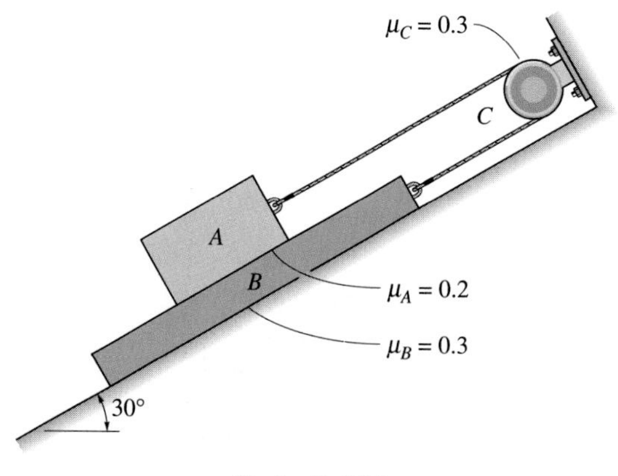

Prob. 8–136

When a pressure vessel is designed it is important to be able to determine the center of gravity of its component parts, calculate its volume and surface area, and reduce three-dimensional distributed loadings to their resultants. These topics are discussed in this chapter.

9

Center of Gravity and Centroid

In this chapter we will discuss the method used to determine the location of the center of gravity and center of mass for a system of discrete particles, and then we will expand its application to include a body of arbitrary shape. The same method of analysis will also be used to determine the geometric center, or centroid, of lines, areas, and volumes. Once the centroid has been located, we will then show how to obtain the area and volume of a surface of revolution and determine the resultants of various types of distributed loadings.

9.1 Center of Gravity and Center of Mass for a System of Particles

Center of Gravity. Consider the system of n particles fixed within a region of space as shown in Fig. 9–1a. The weights of the particles comprise a system of parallel forces* which can be replaced by a single (equivalent) resultant weight and a defined point of application. This point is called the *center of gravity G*. To find its $\bar{x}$, $\bar{y}$, $\bar{z}$ coordinates, we must use the principles outlined in Sec. 4.9. This requires that the resultant weight be equal to the total weight of all n particles; that is,

$$W_R = \Sigma W$$

The sum of the moments of the weights of all the particles about the x, y, and z axes is then equal to the moment of the resultant weight about these axes.

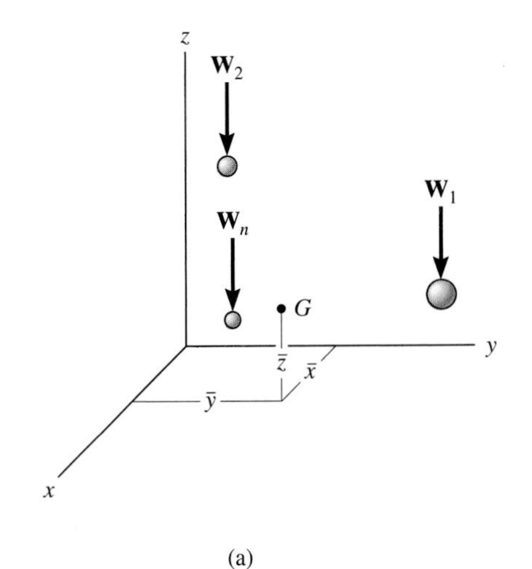

(a)

*This is not true in the exact sense, since the weights are not parallel to each other; rather they are all *concurrent* at the earth's center. Furthermore, the acceleration of gravity g is actually different for each particle, since it depends on the distance from the earth's center to the particle. For all practical purposes, however, both of these effects can be neglected.

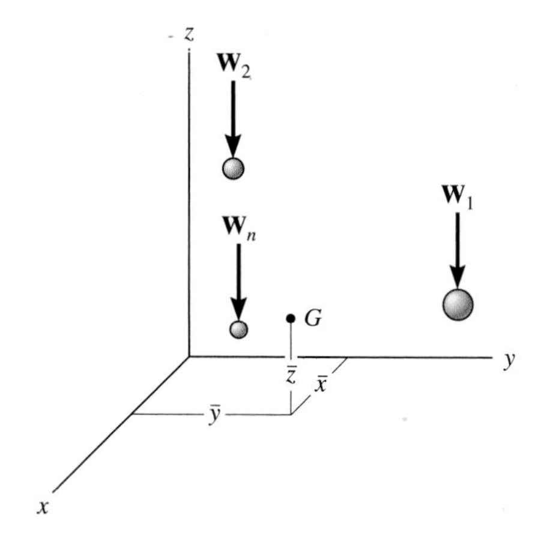

(a)

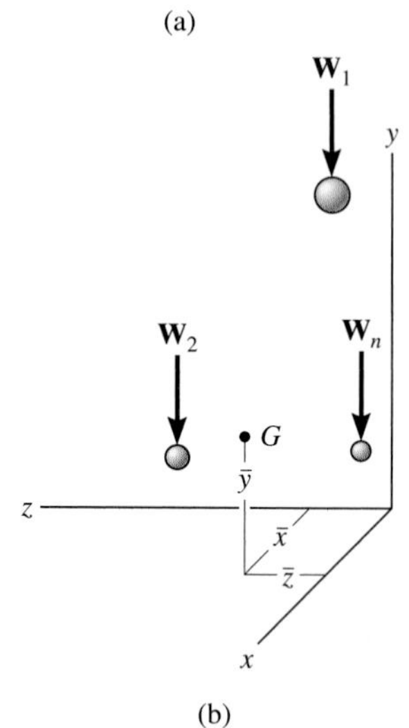

(b)

Fig. 9–1

Thus, to determine the $\bar{x}$ coordinate of G, we can sum moments about the y axis. This yields

$$\bar{x}W_R = \tilde{x}_1 W_1 + \tilde{x}_2 W_2 + \cdots + \tilde{x}_n W_n$$

Likewise, summing moments about the x axis, we can obtain the $\bar{y}$ coordinate; i.e.,

$$\bar{y}W_R = \tilde{y}_1 W_1 + \tilde{y}_2 W_2 + \cdots + \tilde{y}_n W_n$$

Although the weights do not produce a moment about the z axis, we can obtain the z coordinate of G by imagining the coordinate system, with the particles fixed in it, as being rotated 90° about the x (or y) axis, Fig. 9–1b. Summing moments about the x axis, we have

$$\bar{z}W_R = \tilde{z}_1 W_1 + \tilde{z}_2 W_2 + \cdots + \tilde{z}_n W_n$$

We can generalize these formulas, and write them symbolically in the form

$$\bar{x} = \frac{\Sigma \tilde{x} W}{\Sigma W} \qquad \bar{y} = \frac{\Sigma \tilde{y} W}{\Sigma W} \qquad \bar{z} = \frac{\Sigma \tilde{z} W}{\Sigma W} \tag{9–1}$$

Here

$\bar{x}, \bar{y}, \bar{z}$ represent the coordinates of the center of gravity G of the system of particles, regardless of the orientation of the x, y, z axes.

$\tilde{x}, \tilde{y}, \tilde{z}$ represent the coordinates of each particle in the system.

$W_R = \Sigma W$ is the total sum of the weights of all the particles in the system.

Formulas having this same form will be presented throughout this chapter to represent other ''quantities'' for a system. In all cases, however, keep in mind that they simply represent a balance between the sum of the moments of the ''quantity'' for *each part* of the system and the moment of the *resultant* ''quantity'' for the system.

Center of Mass. To study problems concerning the motion of *matter* under the influence of force, i.e., dynamics, it is necessary to locate a point called the *center of mass.* Provided the acceleration of gravity g for every particle is constant, then $W = mg$. Substituting into Eqs. 9–1 and canceling g from both the numerator and denominator yields

$$\bar{x} = \frac{\Sigma \tilde{x} m}{\Sigma m} \qquad \bar{y} = \frac{\Sigma \tilde{y} m}{\Sigma m} \qquad \bar{z} = \frac{\Sigma \tilde{z} m}{\Sigma m} \tag{9–2}$$

By comparison, then, the location of the center of gravity *coincides* with that of the center of mass. Recall, however, that particles have "weight" only when under the influence of a gravitational attraction, whereas the center of mass is independent of gravity. For example, it would be meaningless to define the center of gravity of a system of particles representing the planets of our solar system, while the center of mass of this system is important.

9.2 Center of Gravity, Center of Mass, and Centroid for a Body

Center of Gravity. When the principles used to determine Eqs. 9–1 are applied to a system of particles composing a rigid body having a total weight W, one obtains the same form as these equations except that each particle located at $(\tilde{x}, \tilde{y}, \tilde{z})$ is thought to have a *differential weight dW,* Fig. 9–2. As a result, *integration* is required rather than a discrete summation of the terms. The resulting equations are

$$\bar{x} = \frac{\int \tilde{x}\, dW}{\int dW} \qquad \bar{y} = \frac{\int \tilde{y}\, dW}{\int dW} \qquad \bar{z} = \frac{\int \tilde{z}\, dW}{\int dW} \qquad (9\text{–}3)$$

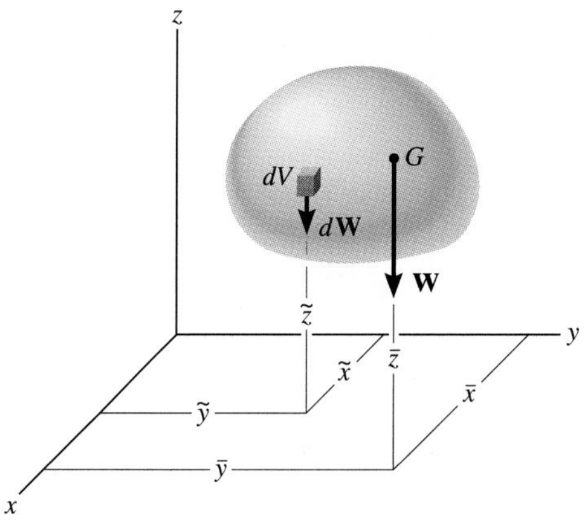

Fig. 9–2

In order to use these equations properly, the differential weight dW must be expressed in terms of its associated volume dV. If γ represents the *specific weight* of the body, measured as a weight per unit volume, then $dW = \gamma\, dV$ and therefore

$$\bar{x} = \frac{\int_V \tilde{x}\gamma\, dV}{\int_V \gamma\, dV} \qquad \bar{y} = \frac{\int_V \tilde{y}\gamma\, dV}{\int_V \gamma\, dV} \qquad \bar{z} = \frac{\int_V \tilde{z}\gamma\, dV}{\int_V \gamma\, dV} \qquad (9\text{–}4)$$

Here integration must be performed throughout the entire volume of the body.

Center of Mass. The *density* ρ, or mass per unit volume, is related to γ by the equation $\gamma = \rho g$, where g is the acceleration of gravity. Substituting this relationship into Eqs. 9–4 and canceling g from both the numerators and denominators yields similar equations (with ρ replacing γ) that can be used to determine the body's *center of mass.*

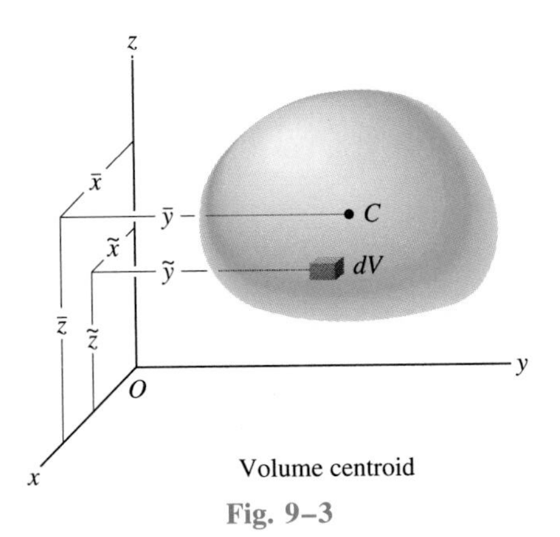

Fig. 9–3

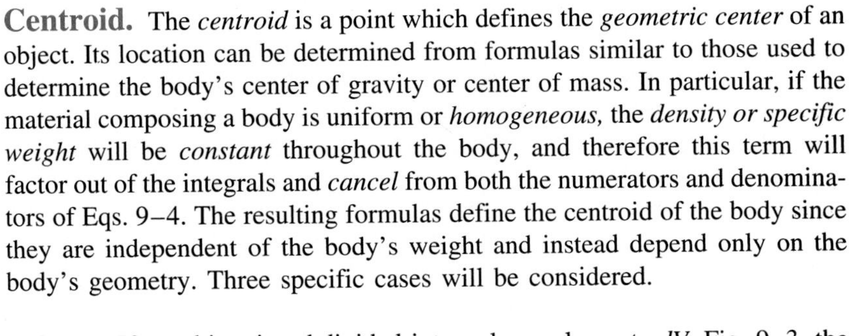

Centroid. The *centroid* is a point which defines the *geometric center* of an object. Its location can be determined from formulas similar to those used to determine the body's center of gravity or center of mass. In particular, if the material composing a body is uniform or *homogeneous,* the *density or specific weight* will be *constant* throughout the body, and therefore this term will factor out of the integrals and *cancel* from both the numerators and denominators of Eqs. 9–4. The resulting formulas define the centroid of the body since they are independent of the body's weight and instead depend only on the body's geometry. Three specific cases will be considered.

Volume. If an object is subdivided into volume elements dV, Fig. 9–3, the location of the centroid $C\ (\bar{x}, \bar{y}, \bar{z})$ for the volume of the object can be determined by computing the "moments" of the elements about the coordinate axes. The resulting formulas are

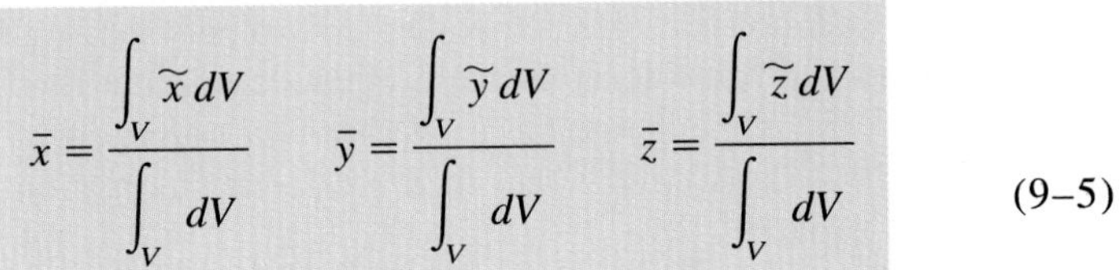

$$\bar{x} = \frac{\int_V \tilde{x}\, dV}{\int_V dV} \qquad \bar{y} = \frac{\int_V \tilde{y}\, dV}{\int_V dV} \qquad \bar{z} = \frac{\int_V \tilde{z}\, dV}{\int_V dV} \tag{9–5}$$

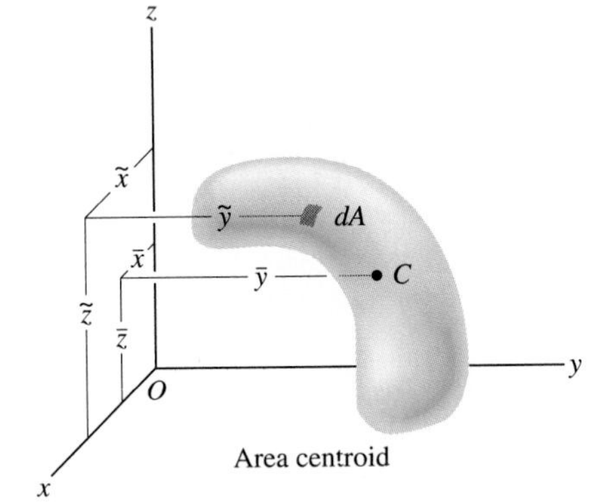

Fig. 9–4

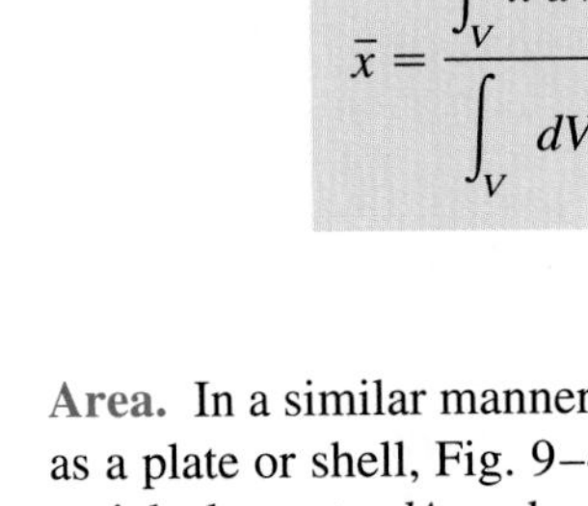

Area. In a similar manner, the centroid for the surface area of an object, such as a plate or shell, Fig. 9–4, can be found by subdividing the area into differential elements dA and computing the "moments" of these area elements about the coordinate axes, namely,

$$\bar{x} = \frac{\int_A \tilde{x}\, dA}{\int_A dA} \qquad \bar{y} = \frac{\int_A \tilde{y}\, dA}{\int_A dA} \qquad \bar{z} = \frac{\int_A \tilde{z}\, dA}{\int_A dA} \tag{9–6}$$

Line. If the geometry of the object, such as a thin rod or wire, takes the form of a line, Fig. 9–5, the manner of finding its centroid is identical to the procedure outlined above. The results are

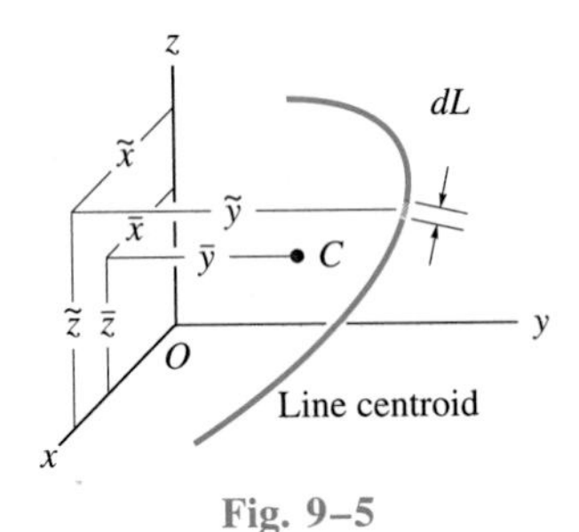

Fig. 9–5

$$\bar{x} = \frac{\int_L \tilde{x}\, dL}{\int_L dL} \qquad \bar{y} = \frac{\int_L \tilde{y}\, dL}{\int_L dL} \qquad \bar{z} = \frac{\int_L \tilde{z}\, dL}{\int_L dL} \tag{9–7}$$

It is important to remember that when applying Eqs. 9–4 through 9–7 it is best to choose a coordinate system that simplifies as much as possible the equation used to describe the object's boundary. For example, polar coordinates are generally appropriate for objects having circular boundaries. Also, if a rectangular coordinate system is used, the terms $\tilde{x}$, $\tilde{y}$, $\tilde{z}$ in the equations refer to the "moment arms" or coordinates of the *center of gravity or centroid for the differential element* used. If possible, this differential element should be chosen such that it has a differential size or thickness in only *one direction*. When this is done, only a single integration is required to cover the entire region.

Symmetry. Notice that in all of the above cases the location of C does not necessarily have to be within the object; rather, it can be located off the object in space. Also, the *centroids* of some shapes may be partially or completely specified by using conditions of *symmetry*. In cases where the shape has an axis of symmetry, the centroid of the shape will lie along that axis. For example, the centroid C for the line shown in Fig. 9–6 must lie along the y axis, since for every elemental length dL at a distance $+\tilde{x}$ to the right of the y axis, there is an identical element at a distance $-\tilde{x}$ to the left. The total moment for all the elements about the axis of symmetry will therefore cancel; i.e., $\int \tilde{x}\, dL = 0$ (Eq. 9–7), so that $\bar{x} = 0$. In cases where a shape has two or three axes of symmetry, it follows that the centroid lies at the intersection of these axes, Fig. 9–7 and Fig. 9–8.

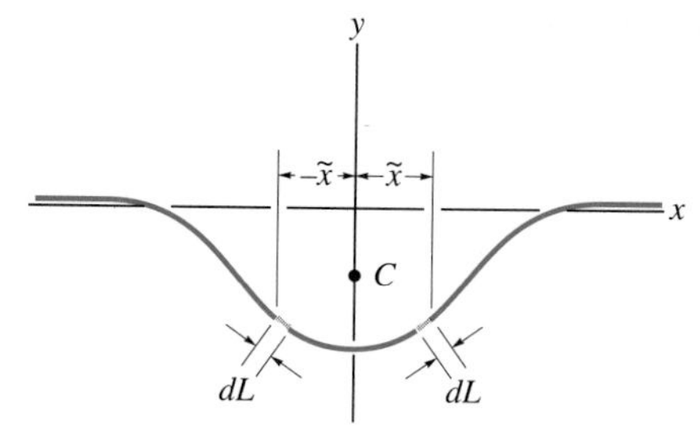

Fig. 9–6

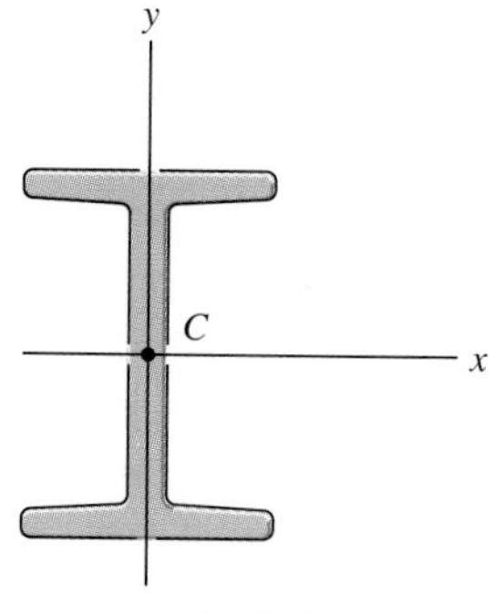

Fig. 9–7

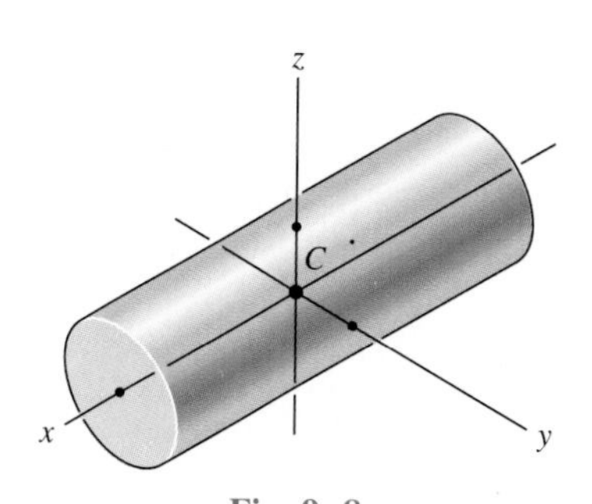

Fig. 9–8

PROCEDURE FOR ANALYSIS

The following procedure provides a method for determining the center of gravity or centroid of an object or shape using a single integration.

Differential Element. Select an appropriate coordinate system and specify the coordinate axes. Then choose an appropriate differential element for integration. If x, y, z axes are used, then for lines this element dL is represented as a differential line segment; for areas the element dA is generally a rectangle having a finite length and differential width; and for volumes the element dV is either a circular disk having a finite radius and differential thickness, or a shell having a finite length and radius and differential thickness. Locate the element so that it intersects the *boundary* of the shape at an *arbitrary point* (x, y, z).

Size and Moment Arms. Express the length dL, area dA, or volume dV of the element in terms of the coordinates used to define the boundary of the shape. Determine the coordinates or moment arms $\tilde{x}$, $\tilde{y}$, $\tilde{z}$ for the centroid or center of gravity of the element.

Integrations. Substitute the data computed above into the appropriate equations (Eqs. 9–4 through 9–7) and perform the integrations.* Note that integration can be accomplished only when the function in the integrand is expressed in terms of the *same variable as the differential thickness of the element.* The limits of the integral are then defined from the two extreme locations of the element's differential thickness, so that when the elements are ''summed'' or the integration performed, the entire region is covered.

*Formulas for integration are given in Appendix A.

The following examples illustrate this procedure numerically.

Example 9–1

Locate the centroid of the rod bent into the shape of a parabolic arc, shown in Fig. 9–9.

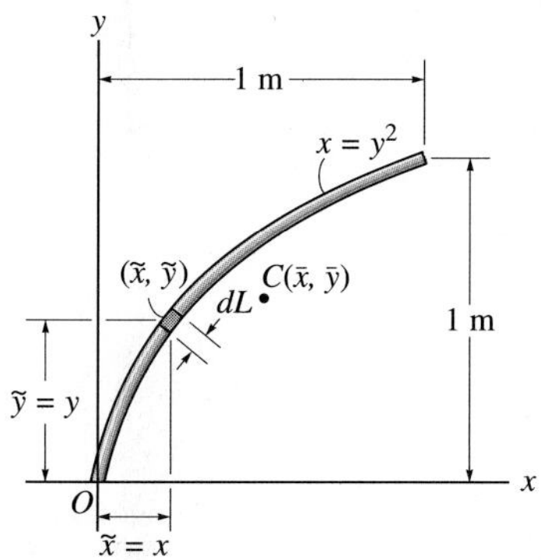

Fig. 9–9

SOLUTION

Differential Element. The differential element is shown in Fig. 9–9. It is located on the curve at the *arbitrary point* (x, y).

Length and Moment Arms. The differential length of the element dL can be expressed in terms of the differentials dx and dy by using the Pythagorean theorem.

$$dL = \sqrt{(dx)^2 + (dy)^2} = \sqrt{\left(\frac{dx}{dy}\right)^2 + 1}\, dy$$

Since $x = y^2$, then $dx/dy = 2y$. Therefore, expressing dL in terms of y and dy, we have

$$dL = \sqrt{(2y)^2 + 1}\, dy$$

The centroid is located at $\tilde{x} = x$, $\tilde{y} = y$.

Integrations. Applying Eqs. 9–7 and integrating with respect to y using the formulas in Appendix A, we have

$$\bar{x} = \frac{\int_L \tilde{x}\, dL}{\int_L dL} = \frac{\int_0^1 x\sqrt{4y^2 + 1}\, dy}{\int_0^1 \sqrt{4y^2 + 1}\, dy} = \frac{\int_0^1 y^2\sqrt{4y^2 + 1}\, dy}{\int_0^1 \sqrt{4y^2 + 1}\, dy}$$

$$= \frac{0.746}{1.479} = 0.504 \text{ m} \qquad \textit{Ans.}$$

$$\bar{y} = \frac{\int_L \tilde{y}\, dL}{\int_L dL} = \frac{\int_0^1 y\sqrt{4y^2 + 1}\, dy}{\int_0^1 \sqrt{4y^2 + 1}\, dy} = \frac{0.848}{1.479} = 0.573 \text{ m} \qquad \textit{Ans.}$$

Example 9–2

Locate the centroid of the circular wire segment shown in Fig. 9–10.

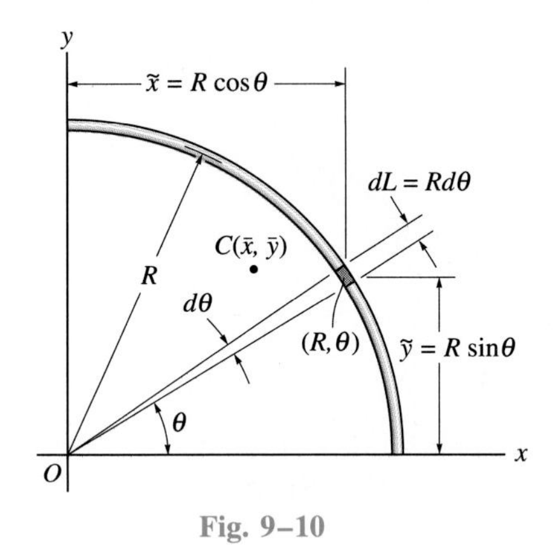

Fig. 9–10

SOLUTION

Polar coordinates will be used to solve this problem since the arc is circular.

Differential Element. A differential circular arc is selected as shown in the figure. This element intersects the curve at (R, θ).

Length and Moment Arm. The differential length of the element is $dL = R\,d\theta$, and its centroid is located at $\widetilde{x} = R\cos\theta$ and $\widetilde{y} = R\sin\theta$.

Integrations. Applying Eqs. 9–7 and integrating with respect to θ, we obtain

$$\bar{x} = \frac{\int_L \widetilde{x}\,dL}{\int_L dL} = \frac{\int_0^{\pi/2} (R\cos\theta)R\,d\theta}{\int_0^{\pi/2} R\,d\theta} = \frac{R^2\int_0^{\pi/2} \cos\theta\,d\theta}{R\int_0^{\pi/2} d\theta} = \frac{2R}{\pi} \qquad \textit{Ans.}$$

$$\bar{y} = \frac{\int_L \widetilde{y}\,dL}{\int_L dL} = \frac{\int_0^{\pi/2} (R\sin\theta)R\,d\theta}{\int_0^{\pi/2} R\,d\theta} = \frac{R^2\int_0^{\pi/2} \sin d\theta}{R\int_0^{\pi/2} d\theta} = \frac{2R}{\pi} \qquad \textit{Ans.}$$

Example 9–3

Determine the distance $\bar{y}$ to the centroid of the area of the triangle shown in Fig. 9–11.

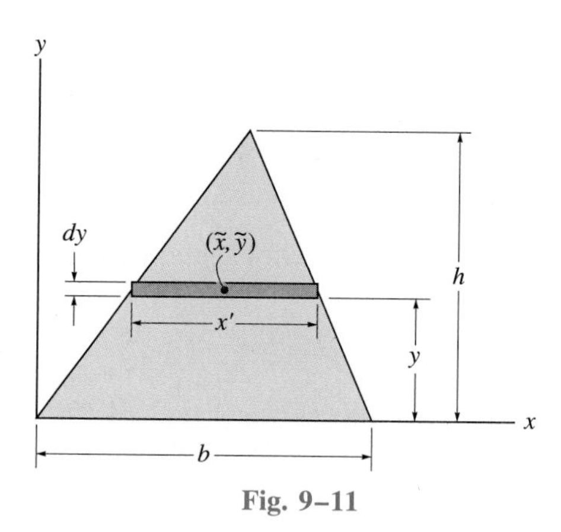

Fig. 9–11

SOLUTION

Differential Element. Consider a rectangular element having thickness dy and *variable length* x', Fig. 9–11. By similar triangles, $b/h = x'/(h-y)$ or $x' = \dfrac{b}{h}(h-y)$. The element intersects the sides of the triangle at a height y above the x axis.

Area and Moment Arms. The area of the element is $dA = x'\,dy = \dfrac{b}{h}(h-y)\,dy$, and its centroid is located a distance $\tilde{y} = y$ from the x axis.

Integrations. Applying the second of Eqs. 9–6, and integrating with respect to y, yields

$$\bar{y} = \frac{\int_A \tilde{y}\,dA}{\int_A dA} = \frac{\int_0^h y\dfrac{b}{h}(h-y)\,dy}{\int_0^h \dfrac{b}{h}(h-y)\,dy} = \frac{\frac{1}{6}bh^2}{\frac{1}{2}bh}$$

$$= \frac{h}{3} \qquad \textit{Ans.}$$

Example 9–4

Locate the centroid for the area of a quarter circle shown in Fig. 9–12*a*.

SOLUTION I

Differential Element. Polar coordinates will be used since the boundary is circular. We choose the element in the shape of a *triangle,* Fig. 9–12*a*. (Actually the shape is a circular sector; however, neglecting higher-order differentials, the element becomes triangular.) The element intersects the curve at point (R, θ).

Area and Moment Arms. The area of the element is

$$dA = \tfrac{1}{2}(R)(R\, d\theta) = \frac{R^2}{2} d\theta$$

and using the results of Example 9–3, the centroid of the (triangular) element is located at $\widetilde{x} = \frac{2}{3}R\cos\theta$, $\widetilde{y} = \frac{2}{3}R\sin\theta$.

Integrations. Applying Eqs. 9–6, and integrating with respect to θ, we obtain

$$\bar{x} = \frac{\int_A \widetilde{x}\, dA}{\int_A dA} = \frac{\int_0^{\pi/2} \left(\frac{2}{3}R\cos\theta\right)\frac{R^2}{2} d\theta}{\int_0^{\pi/2} \frac{R^2}{2} d\theta}$$

$$= \frac{\left(\frac{2}{3}R\right)\int_0^{\pi/2} \cos\theta\, d\theta}{\int_0^{\pi/2} d\theta} = \frac{4R}{3\pi} \qquad \textit{Ans.}$$

$$\bar{y} = \frac{\int_A \widetilde{y}\, dA}{\int_A dA} = \frac{\int_0^{\pi/2} \left(\frac{2}{3}R\sin\theta\right)\frac{R^2}{2} d\theta}{\int_0^{\pi/2} \frac{R^2}{2} d\theta}$$

$$= \frac{\left(\frac{2}{3}R\right)\int_0^{\pi/2} \sin\theta\, d\theta}{\int_0^{\pi/2} d\theta} = \frac{4R}{3\pi} \qquad \textit{Ans.}$$

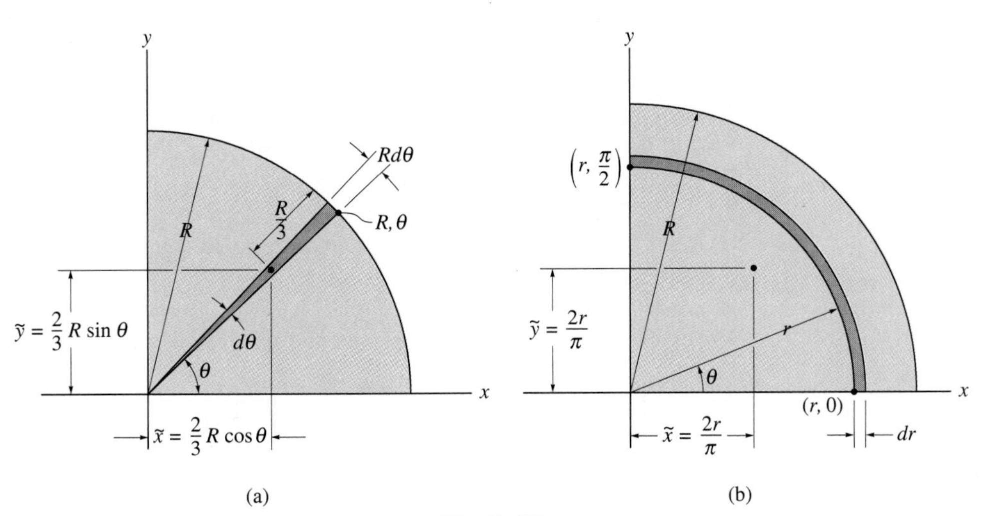

Fig. 9–12

SOLUTION II

Differential Element. The differential element may be chosen in the form of a *circular arc* having a thickness dr as shown in Fig. 9–12*b*. The element intersects the axes at points $(r, 0)$ and $(r, \pi/2)$.

Area and Moment Arms. The area of the element is $dA = (2\pi r/4)\,dr$. Since the centroid of a 90° circular arc was determined in Example 9–2, then for the element $\tilde{x} = 2r/\pi$, $\tilde{y} = 2r/\pi$.

Integrations. Using Eqs. 9–6, and integrating with respect to r, we obtain

$$\bar{x} = \frac{\int_A \tilde{x}\,dA}{\int_A dA} = \frac{\int_0 \frac{2r}{\pi}\left(\frac{2\pi r}{4}\right) dr}{\int_0^R \frac{2\pi r}{4} dr} = \frac{\int_0 r^2\,dr}{\frac{\pi}{2}\int_0^R r\,dr} = \frac{4}{3}\frac{R}{\pi} \qquad \textit{Ans.}$$

$$\bar{y} = \frac{\int_A \tilde{y}\,dA}{\int_A dA} = \frac{\int_0^R \frac{2r}{\pi}\left(\frac{2\pi r}{4}\right) dr}{\int_0^R \frac{2\pi r}{4}\,dr} = \frac{\int_0^R r^2\,dr}{\frac{\pi}{2}\int_0^R r\,dr} = \frac{4}{3}\frac{R}{\pi} \qquad \textit{Ans.}$$

Example 9–5

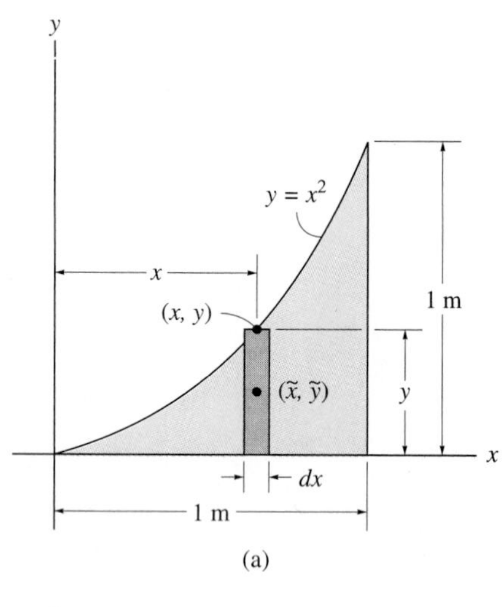

(a)

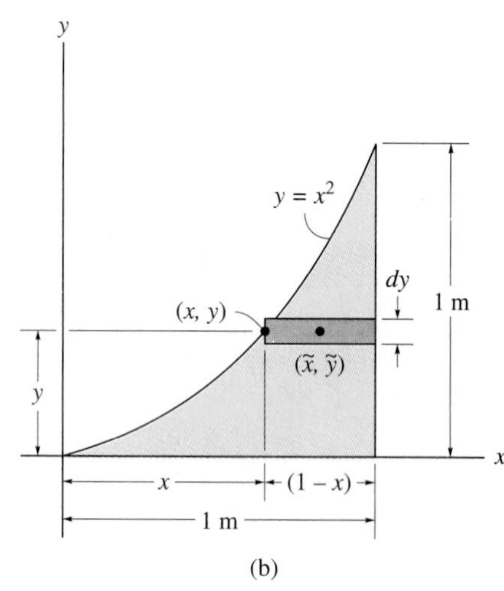

(b)

Fig. 9–13

Locate the centroid of the area shown in Fig. 9–13*a*.

SOLUTION I

Differential Element. A differential element of thickness dx is shown in Fig. 9–13*a*. The element intersects the curve at the *arbitrary point* (x, y), and so it has a height y.

Area and Moment Arms. The area of the element is $dA = y\,dx$, and its centroid is located at $\tilde{x} = x$, $\tilde{y} = y/2$.

Integrations. Applying Eqs. 9–6 and integrating with respect to x yields

$$\bar{x} = \frac{\int_A \tilde{x}\,dA}{\int_A dA} = \frac{\int_0^1 xy\,dx}{\int_0^1 y\,dx} = \frac{\int_0^1 x^3\,dx}{\int_0^1 x^2\,dx} = \frac{0.250}{0.333} = 0.75\text{ m} \qquad \textit{Ans.}$$

$$\bar{y} = \frac{\int_A \tilde{y}\,dA}{\int_A dA} = \frac{\int_0^1 (y/2)y\,dx}{\int_0^1 y\,dx} = \frac{\int_0^1 (x^2/2)x^2\,dx}{\int_0^1 x^2\,dx} = \frac{0.100}{0.333} = 0.3\text{ m} \qquad \textit{Ans.}$$

SOLUTION II

Differential Element. The differential element of thickness dy is shown in Fig. 9–13*b*. The element intersects the curve at the *arbitrary point* (x, y), and so it has a length $(1 - x)$.

Area and Moment Arms. The area of the element is $dA = (1 - x)\,dy$, and its centroid is located at

$$\tilde{x} = x + \left(\frac{1 - x}{2}\right) = \frac{1 + x}{2}, \; \tilde{y} = y$$

Integrations. Applying Eqs. 9–6 and integrating with respect to y, we obtain

$$\bar{x} = \frac{\int_A \tilde{x}\,dA}{\int_A dA} = \frac{\int_0^1 [(1 + x)/2](1 - x)\,dy}{\int_0^1 (1 - x)\,dy} = \frac{\frac{1}{2}\int_0^1 (1 - y)\,dy}{\int_0^1 (1 - \sqrt{y})\,dy} = \frac{0.250}{0.333} = 0.75\text{ m} \qquad \textit{Ans.}$$

$$\bar{y} = \frac{\int_A \tilde{y}\,dA}{\int_A dA} = \frac{\int_0^1 y(1 - x)\,dy}{\int_0^1 (1 - x)\,dy} = \frac{\int_0^1 (y - y^{3/2})\,dy}{\int_0^1 (1 - \sqrt{y})\,dy} = \frac{0.100}{0.333} = 0.3\text{ m} \qquad \textit{Ans.}$$

Example 9–6

Locate the $\bar{x}$ centroid of the shaded area bounded by the two curves $y = x$ and $y = x^2$, Fig. 9–14.

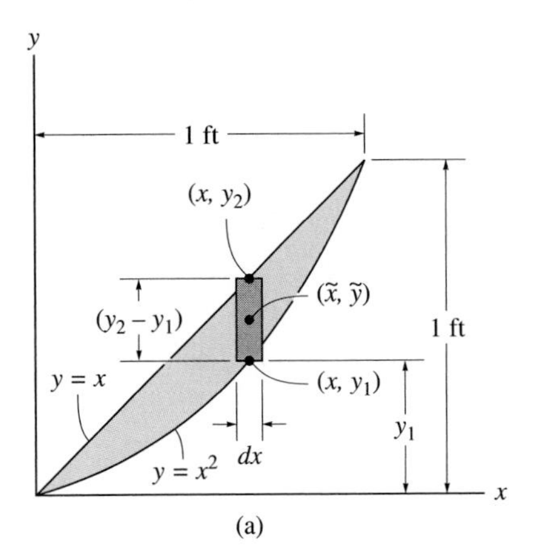

(a)

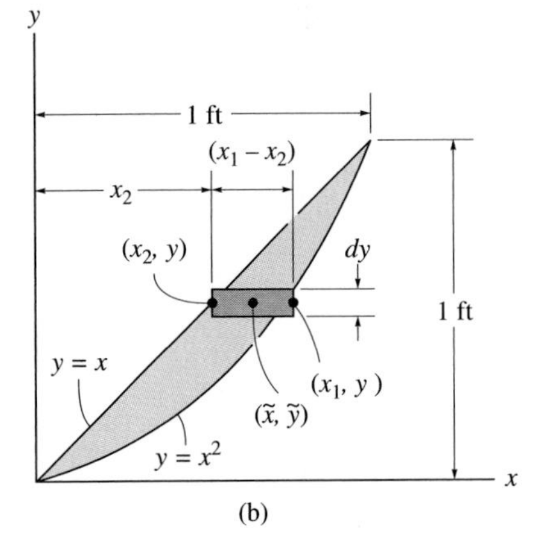

(b)

Fig. 9–14

SOLUTION I

Differential Element. A differential element of thickness dx is shown in Fig. 9–14*a*. The element intersects the curves at *arbitrary points* (x, y_1) and (x, y_2), and so it has a height $(y_2 - y_1)$.

Area and Moment Arm. The area of the element is $dA = (y_2 - y_1)\,dx$, and its centroid is located at $\tilde{x} = x$.

Integrations. Applying Eq. 9–6, we have

$$\bar{x} = \frac{\int_A \tilde{x}\,dA}{\int_A dA} = \frac{\int_0^1 x(y_2 - y_1)\,dx}{\int_0^1 (y_2 - y_1)\,dx} = \frac{\int_0^1 x(x - x^2)\,dx}{\int_0^1 (x - x^2)\,dx} = \frac{\frac{1}{12}}{\frac{1}{6}} = 0.5 \text{ ft} \quad \textit{Ans.}$$

SOLUTION II

Differential Element. A differential element having a thickness dy is shown in Fig. 9–14*b*. The element intersects the curves at the *arbitrary points* (x_2, y) and (x_1, y), and so it has a length $(x_1 - x_2)$.

Area and Moment Arm. The area of the element is $dA = (x_1 - x_2)\,dy$, and its centroid is located at

$$\tilde{x} = x_2 + \frac{x_1 - x_2}{2} = \frac{x_1 + x_2}{2}$$

Integrations. Applying Eq. 9–6, we have

$$\bar{x} = \frac{\int_A \tilde{x}\,dA}{\int_A dA} = \frac{\int_0^1 [(x_1 + x_2)/2](x_1 - x_2)\,dy}{\int_0^1 (x_1 - x_2)\,dy} = \frac{\int_0^1 [(\sqrt{y} + y)/2](\sqrt{y} - y)\,dy}{\int_0^1 (\sqrt{y} - y)\,dy}$$

$$= \frac{\frac{1}{2}\int_0^1 (y - y^2)\,dy}{\int_0^1 (\sqrt{y} - y)\,dy} = \frac{\frac{1}{12}}{\frac{1}{6}} = 0.5 \text{ ft} \quad \textit{Ans.}$$

Example 9–7

Locate the $\bar{y}$ centroid for the paraboloid of revolution, which is generated by revolving the shaded area shown in Fig. 9–15*a* about the *y* axis.

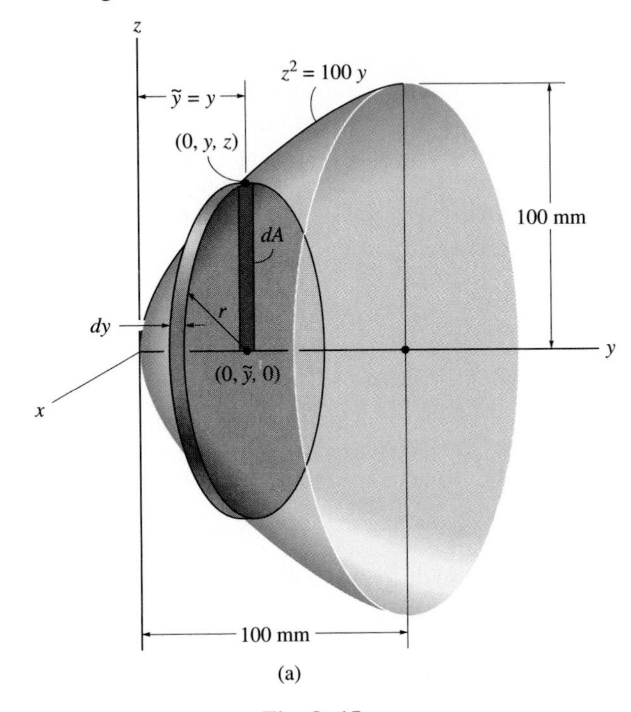

Fig. 9–15

SOLUTION I

Differential Element. An element having the shape of a *thin disk* is chosen, Fig. 9–15*a*. This element has a thickness *dy*. In this ''disk'' method of analysis, the element of planar area, *dA*, is always taken *perpendicular* to the axis of revolution. Here the element intersects the generating curve at the *arbitrary point* (0, *y*, *z*), and so its radius is $r = z$.

Volume and Moment Arm. The volume of the element is $dV = (\pi z^2)\,dy$, and its centroid is located at $\tilde{y} = y$.

Integrations. Applying the second of Eqs. 9–5 and integrating with respect to *y* yields

$$\bar{y} = \frac{\int_V \tilde{y}\,dV}{\int_V dV} = \frac{\int_0^{100} y(\pi z^2)\,dy}{\int_0^{100} (\pi z^2)\,dy} = \frac{100\pi \int_0^{100} y^2\,dy}{100\pi \int_0^{100} y\,dy} = 66.7\text{ mm} \qquad \textit{Ans.}$$

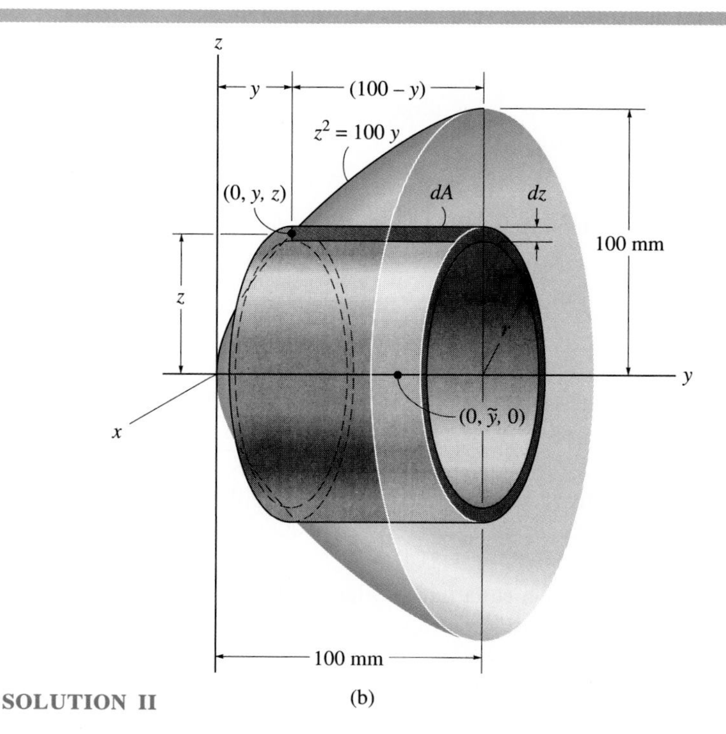

(b)

SOLUTION II

Differential Element. As shown in Fig. 9–15*b*, the volume element can be chosen in the form of a *thin cylindrical shell,* where the shell's thickness is dz. In this "shell" method of analysis, the element of planar area, dA, is always taken *parallel* to the axis of revolution. Here the element intersects the generating curve at point $(0, y, z)$, and so the radius of the shell is $r = z$.

Volume and Moment Arm. The volume of the element is $dV = 2\pi r\, dA = 2\pi z(100 - y)\, dz$, and its centroid is located at $\tilde{y} = y + (100 - y)/2 = (100 + y)/2$.

Integrations. Applying the second of Eqs. 9–5 and integrating with respect to z yields

$$\bar{y} = \frac{\int_V \tilde{y}\, dV}{\int_V dV} = \frac{\int_0^{100} [(100 + y)/2]2\pi z(100 - y)\, dz}{\int_0^{100} 2\pi z(100 - y)\, dz}$$

$$= \frac{\pi \int_0^{100} z(10^4 - 10^{-4}z^4)\, dz}{2\pi \int_0^{100} z(100 - 10^{-2}z^2)\, dz} = 66.7 \text{ mm} \qquad \textit{Ans.}$$

Example 9–8

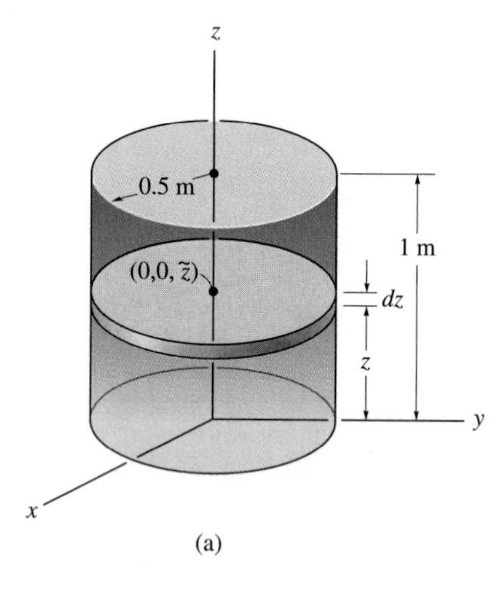

(a)

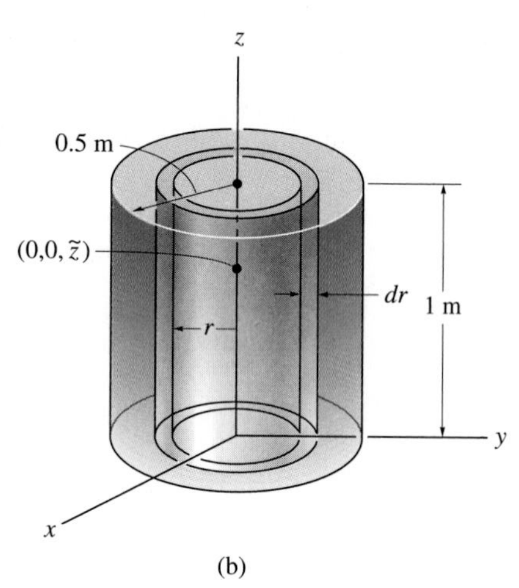

(b)

Fig. 9–16

Determine the location of the center of mass of the cylinder shown in Fig. 9–16*a* if its density varies directly with its distance from the base, such that $\rho = 200z\ \text{kg/m}^3$.

SOLUTION

For reasons of material symmetry,

$$\bar{x} = \bar{y} = 0 \qquad \textit{Ans.}$$

Differential Element. A disk element of radius 0.5 m and thickness dz is chosen for integration, Fig. 9–16*a*, since the *density of the entire element is constant* for a given value of z. The element is located along the z axis at the *arbitrary point* $(0, 0, z)$.

Volume and Moment Arm. The volume of the element is $dV = \pi(0.5)^2\,dz$, and its centroid is located at $\tilde{z} = z$.

Integrations. Using an equation similar to the third of Eqs. 9–4 and integrating with respect to z, noting that $\rho = 200z$, we have

$$\bar{z} = \frac{\int_V \tilde{z}\rho\,dV}{\int_V \rho\,dV} = \frac{\int_0^1 z(200z)\pi(0.5)^2\,dz}{\int_0^1 (200z)\pi(0.5)^2\,dz}$$

$$= \frac{\int_0^1 z^2\,dz}{\int_0^1 z\,dz} = 0.667 \text{ ft} \qquad \textit{Ans.}$$

Note: It is not possible to use a shell element for integration such as shown in Fig. 9–16*b*, since the density of the material composing the shell would *vary* along the shell's height, and hence the location of $\tilde{z}$ for the element cannot be specified.

PROBLEMS

9–1. Locate the center of gravity $\bar{x}$ of the homogeneous rod bent in the form of a semicircular arc. The rod has a weight per unit length of 0.5 lb/ft. Also, determine the horizontal reaction at the smooth support B and the x and y components of reaction at the pin A.

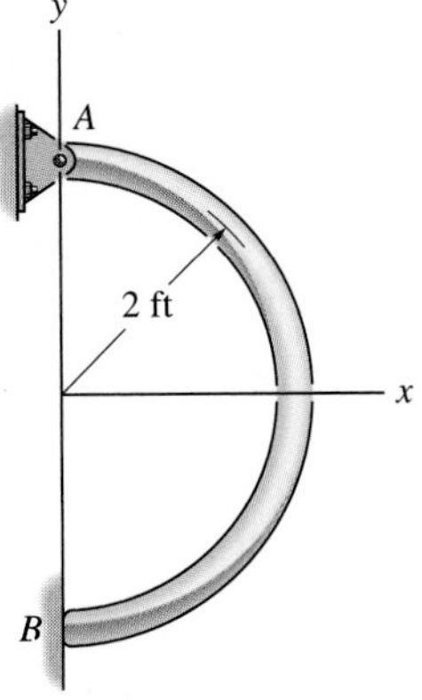

Prob. 9–1

9–2. Locate the center of gravity of the rod having a constant cross-sectional area if its density varies according to $\rho = kx^2$, where k is a constant.

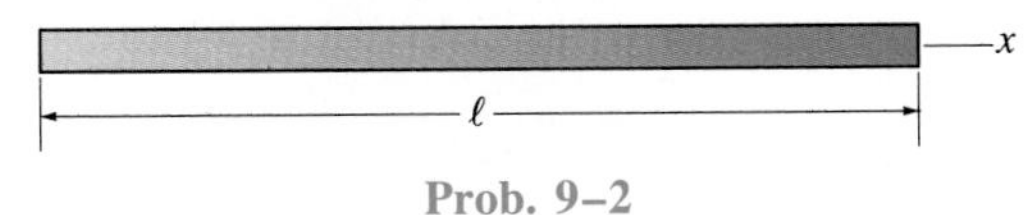

Prob. 9–2

9–3. Locate the centroid $\bar{x}$ of the circular rod. Express the answer in terms of the radius r and semiarc angle α.

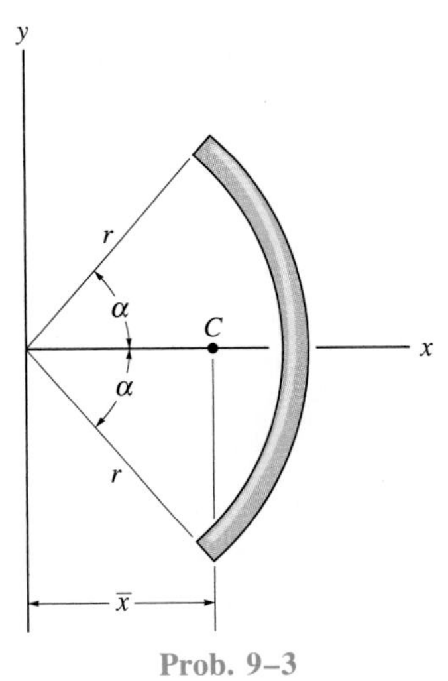

Prob. 9–3

***9–4.** Locate the center of mass of the homogeneous rod bent into the shape of a circular arc.

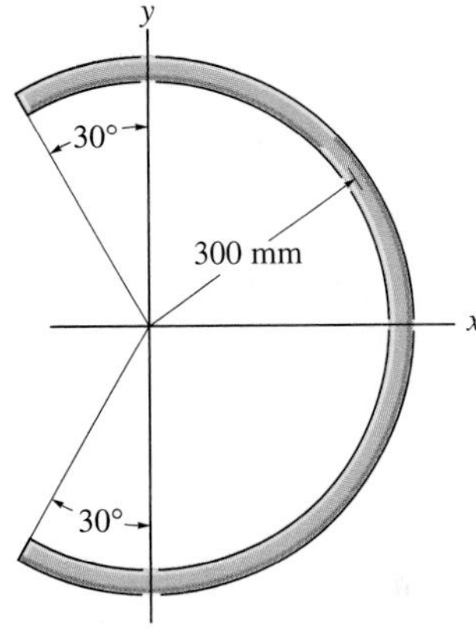

Prob. 9–4

9–5. Determine the distance $\bar{x}$ to the center of gravity of the homogeneous rod bent into the parabolic shape. If the rod has a weight per unit length of 0.5 lb/ft, determine the reactions at the fixed support O.

9–6. Determine the distance $\bar{y}$ to the center of gravity of the homogeneous rod bent into the parabolic shape.

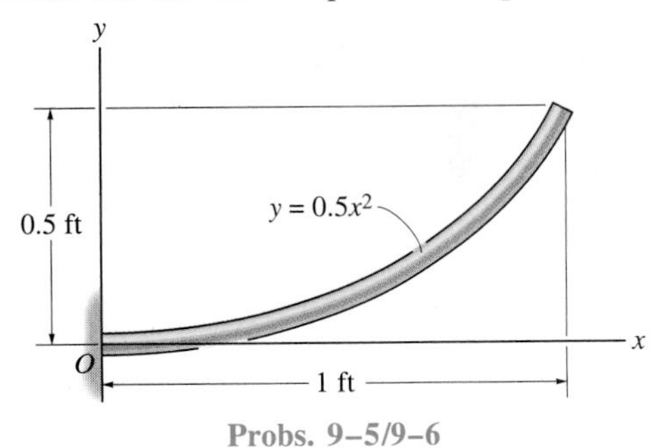

Probs. 9–5/9–6

9–7. Locate the centroid of the shaded area.

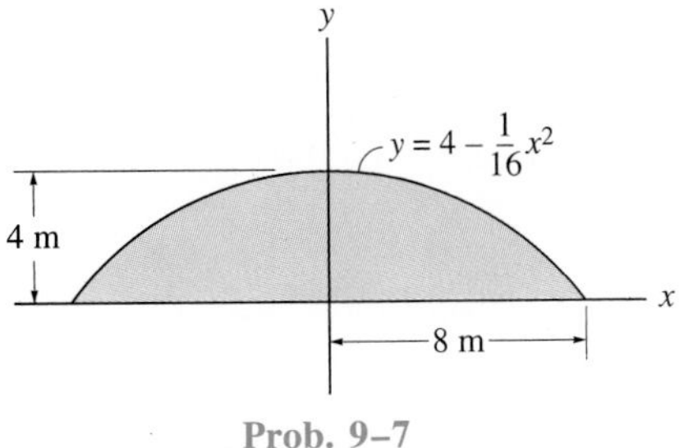

Prob. 9–7

***9–8.** Locate the centroid of the shaded area.

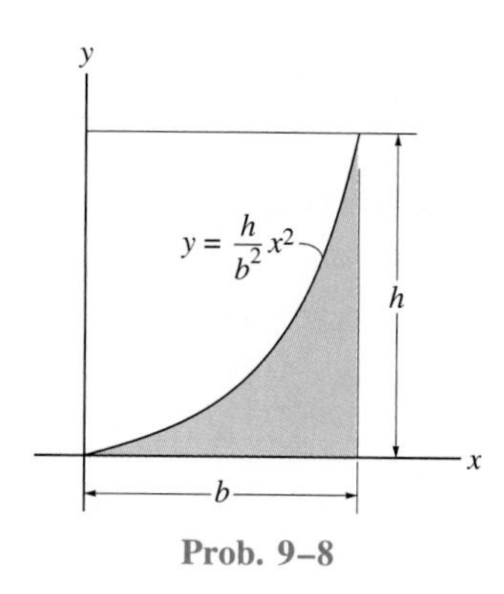

Prob. 9–8

9–9. Locate the centroid of the parabolic area.

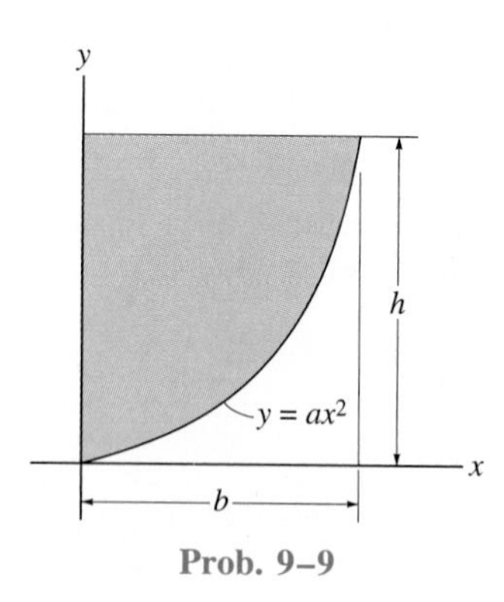

Prob. 9–9

9–10. Locate the centroid of the shaded area.

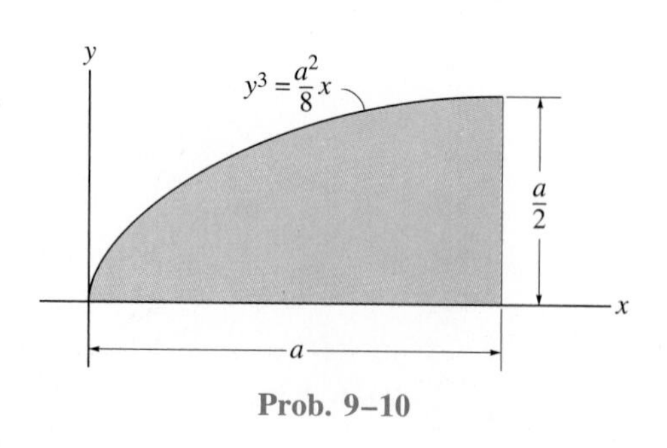

Prob. 9–10

9–11. Locate the centroid of the shaded area.

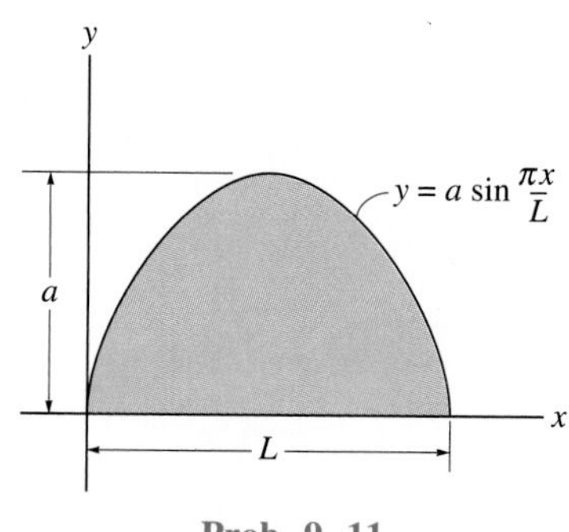

Prob. 9–11

***9–12.** Locate the centroid of the shaded area.

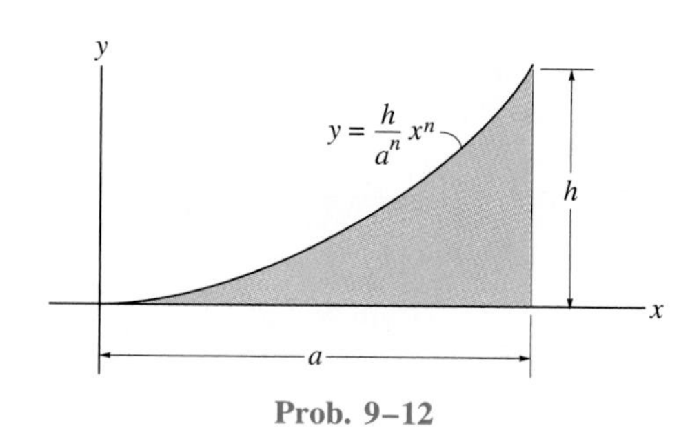

Prob. 9–12

9–13. Locate the centroid of the shaded area.

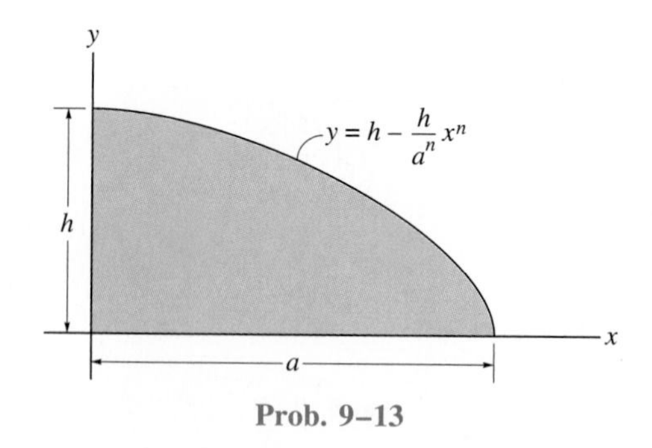

Prob. 9–13

9–14. Locate the centroid of the shaded area bounded by the parabola and the line $y = a$.

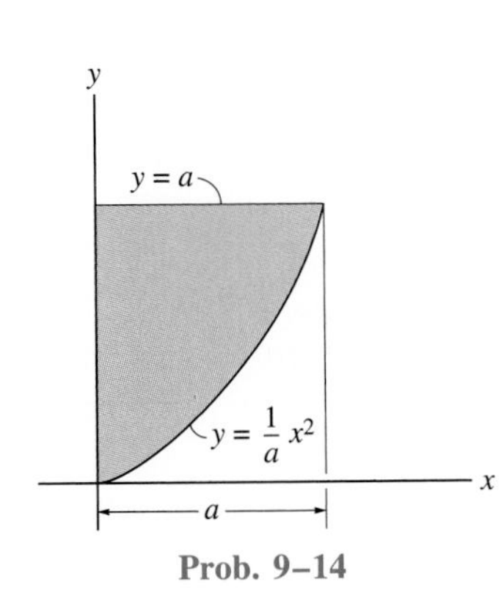

Prob. 9–14

9–15. Locate the centroid of the quarter elliptical area.

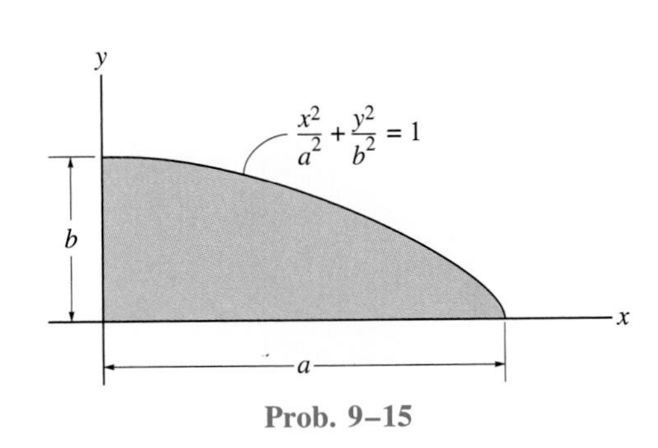

Prob. 9–15

***9–16.** Locate the centroid of the shaded area.

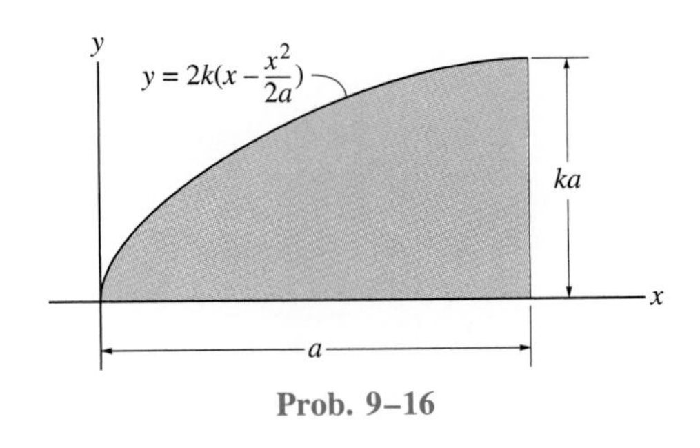

Prob. 9–16

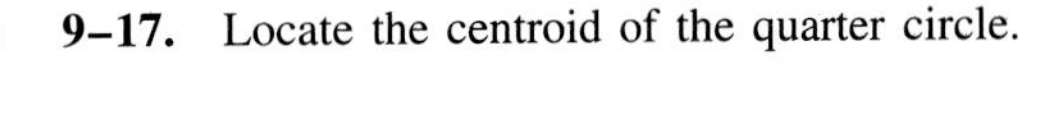

9–17. Locate the centroid of the quarter circle.

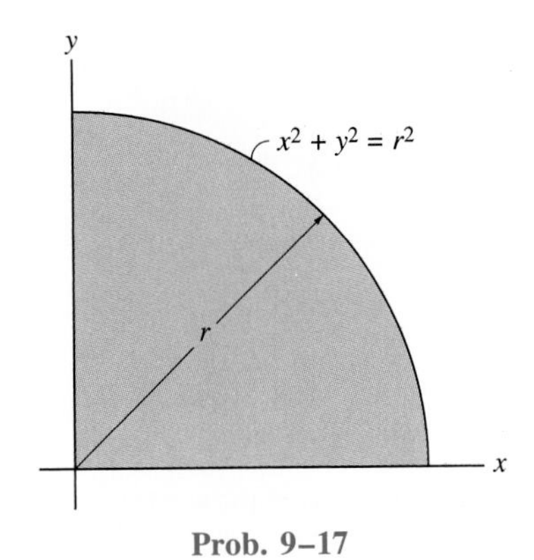

Prob. 9–17

9–18. Locate the centroid of the spandrel area.

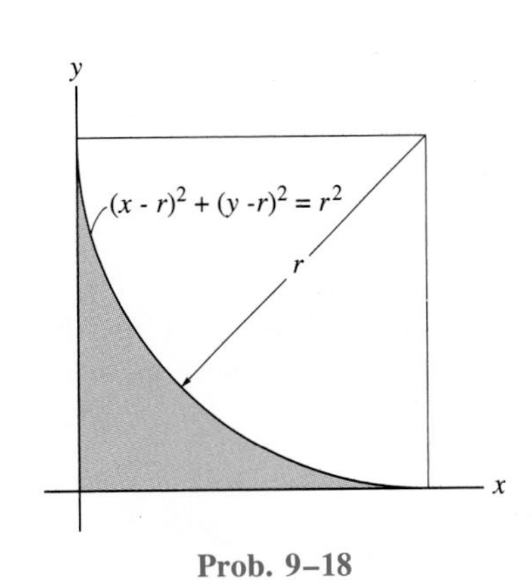

Prob. 9–18

9–19. Locate the centroid of the semicircular area.

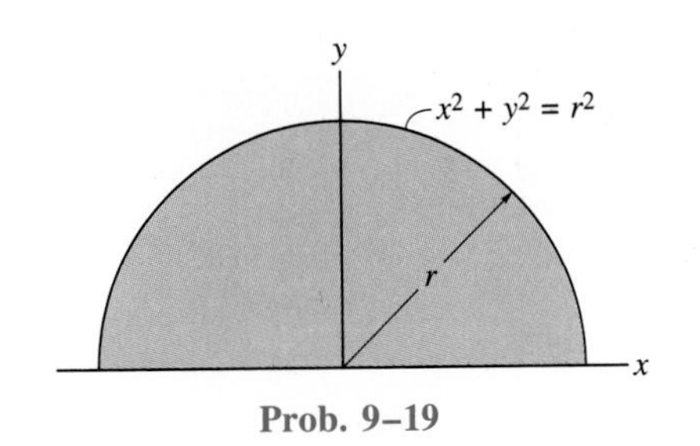

Prob. 9–19

***9–20.** Locate the centroid of the shaded area.

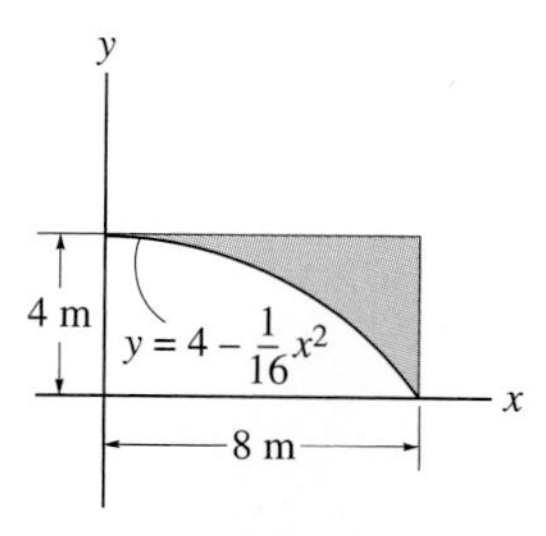

Prob. 9–20

■9–21. Locate the centroid $\bar{x}$ of the shaded area. Solve the problem by evaluating the integrals using Simpson's rule.

■9–22. Locate the centroid $\bar{y}$ of the shaded area. Solve the problem by evaluating the integrals using Simpson's rule.

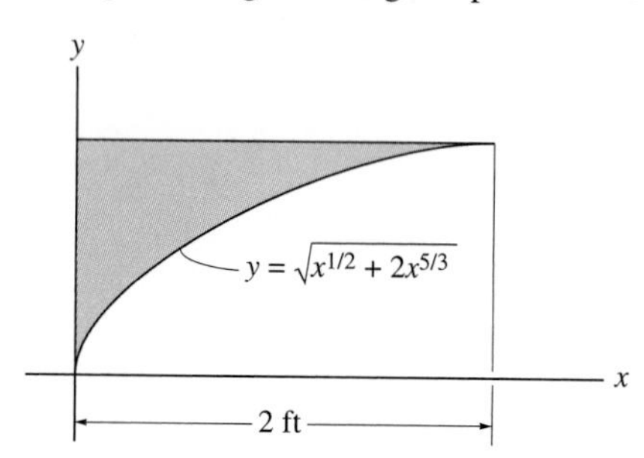

Probs. 9–21/9–22

9–23. The steel plate is 0.3 m thick and has a density of 7850 kg/m^3. Determine the location of its center of mass. Also compute the reactions at the pin and roller support.

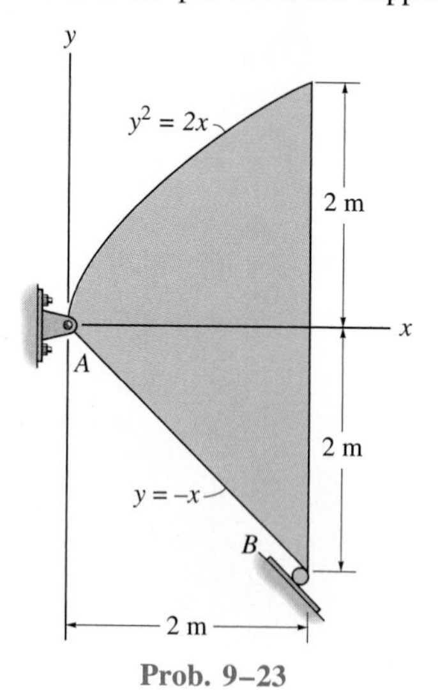

Prob. 9–23

***■9–24.** Locate the centroid $\bar{x}$ of the shaded area. Solve the problem by evaluating the integrals using Simpson's rule.

■9–25. Locate the centroid $\bar{y}$ of the shaded area. Solve the problem by evaluating the integrals using Simpson's rule.

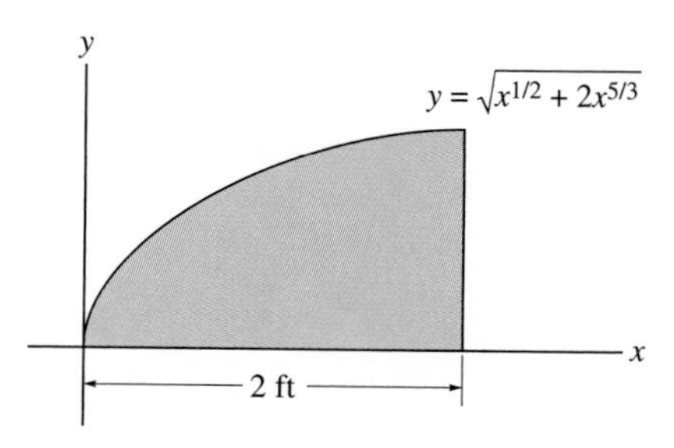

Probs. 9–24/9–25

■9–26. Locate the centroid $\bar{x}$ of the shaded area. Solve the problem by evaluating the integrals using Simpson's rule.

■9–27. Locate the centroid $\bar{y}$ of the shaded area. Solve the problem by evaluating the integrals using Simpson's rule.

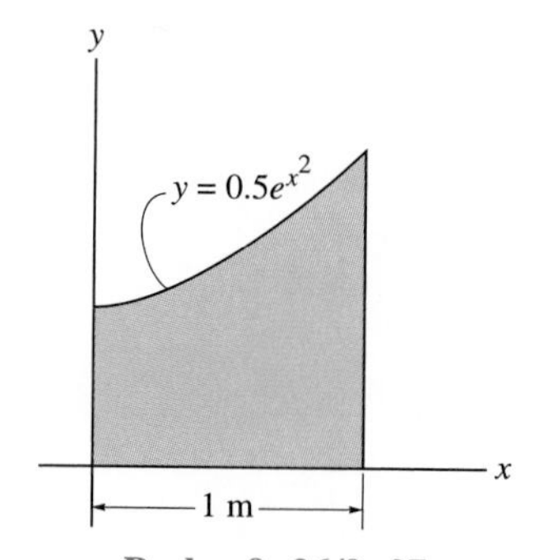

Probs. 9–26/9–27

***■9–28.** Determine the location $\bar{r}$ of the centroid C for the loop of the lemniscate. $r^2 = 2a^2 \cos 2\theta$, $(-45° \le \theta \le 45°)$.

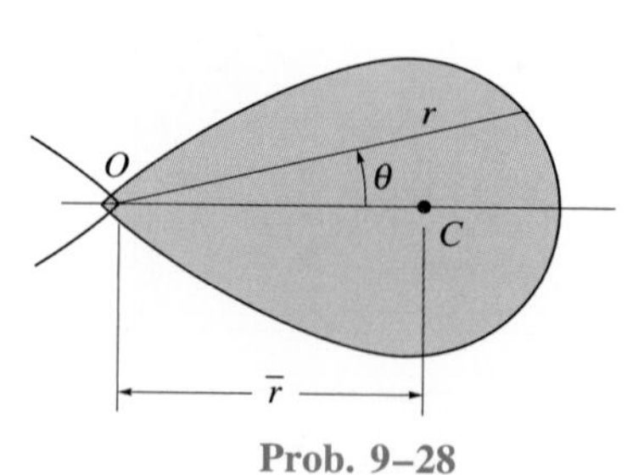

Prob. 9–28

9–29. Determine the location $\bar{r}$ of the centroid C of the upper portion of the cardioid, $r = a(1 - \cos\theta)$.

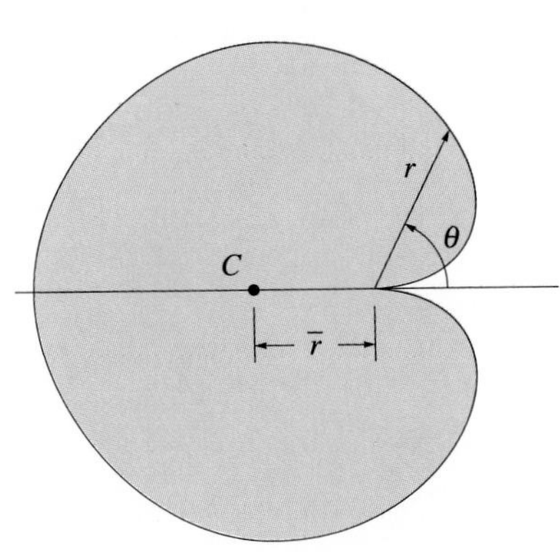

Prob. 9–29

9–30. Locate the centroid of the ellipsoid of revolution.

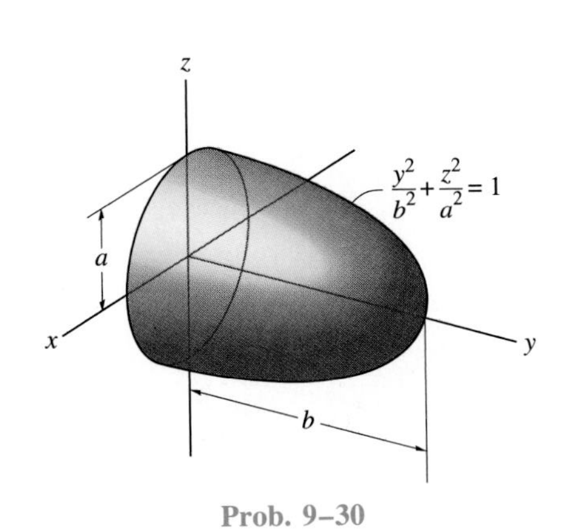

Prob. 9–30

9–31. Locate the centroid of the solid.

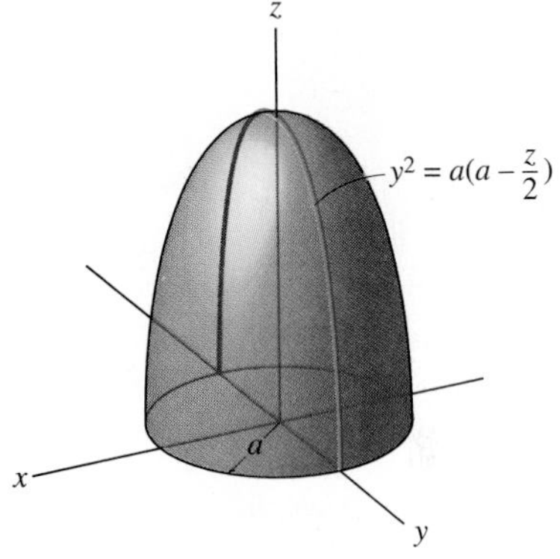

Prob. 9–31

***9–32.** Locate the centroid of the solid.

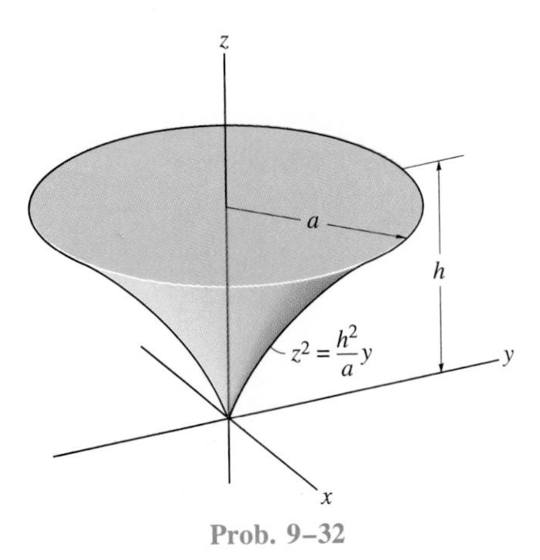

Prob. 9–32

9–33. Locate the centroid $\bar{x}$ of the thin homogeneous hemispherical shell. *Suggestion:* Choose a ring element having a center at $(x, 0, 0)$, radius z, and thickness $dL = \sqrt{(dx)^2 + (dz)^2}$.

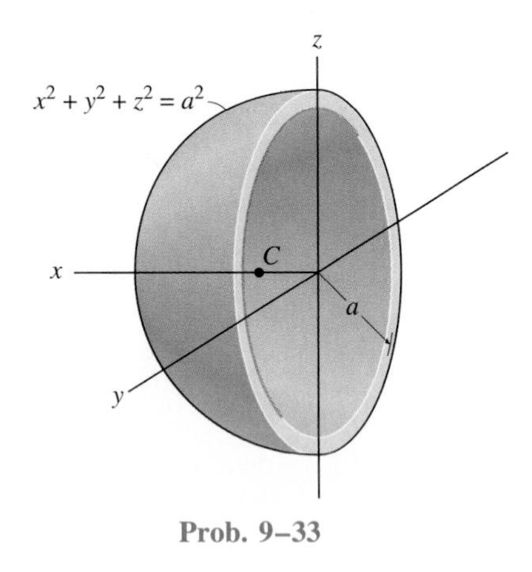

Prob. 9–33

9–34 Locate the centroid $\bar{z}$ of the thin conical shell. *Suggestion:* Use ring elements having a center at $(0, 0\ z)$, radius y, and thickness $dL = \sqrt{(dy)^2 + (dz)^2}$.

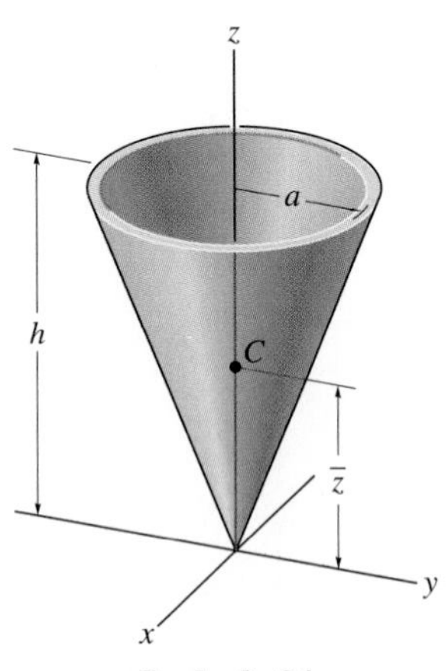

Prob. 9–34

9–35. The king's chamber of the Great Pyramid of Gîza is located at its centroid. Assuming the pyramid to be a solid, prove that this point is at $\bar{z} = \frac{1}{4}h$. *Suggestion:* Use a rectangular differential plate element having a thickness dz and area $(2x)(2y)$.

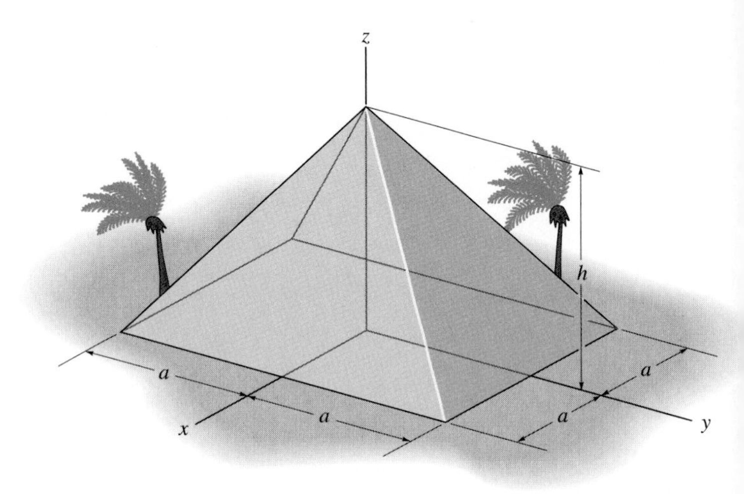

Prob. 9–35

***9–36.** Determine the location $\bar{z}$ of the centroid for the tetrahedron. *Suggestion:* Use a triangular ''plate'' element parallel to the x-y plane and of thickness dz.

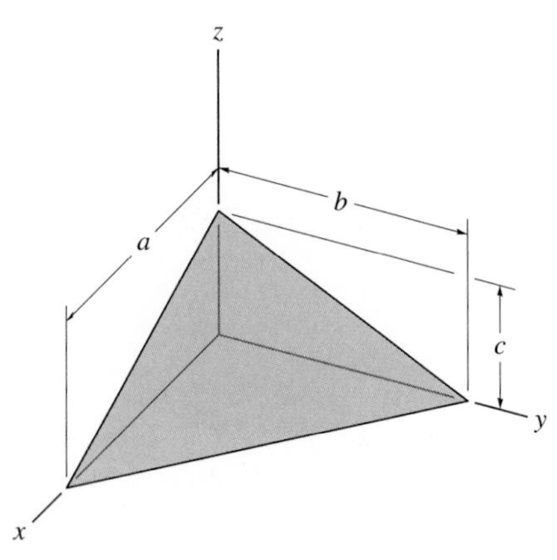

Prob. 9–36

9–37. Locate the centroid of the quarter-cone.

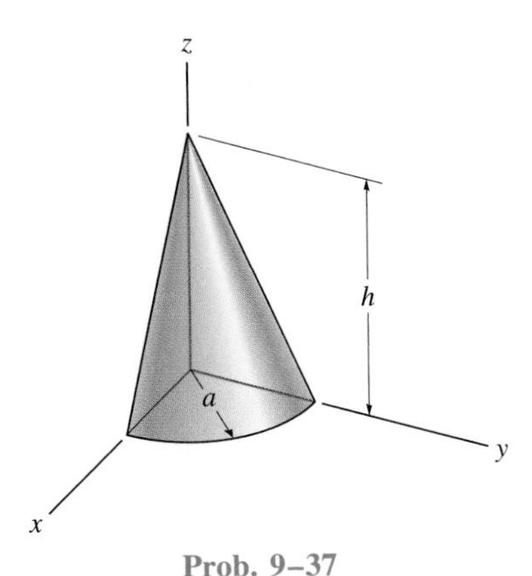

Prob. 9–37

9–38. Locate the center of gravity $\bar{x}$ of the hemisphere. The density of the material varies linearly from zero at the origin O to ρ_0 at the surface. *Suggestion:* Choose a hemispherical shell element for integration and use the result of Prob. 9–33.

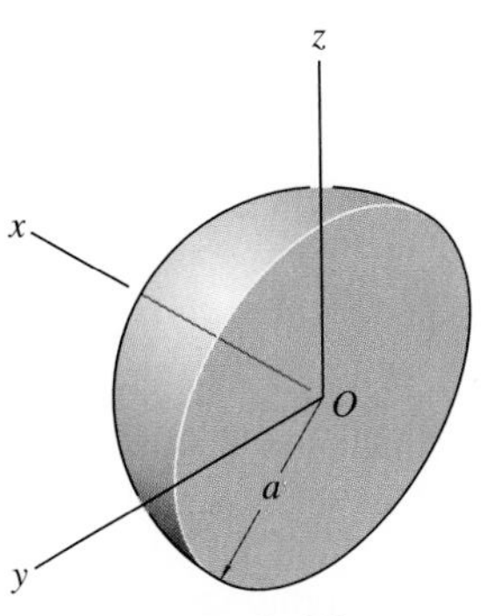

Prob. 9–38

9–39. Locate the centroid $\bar{z}$ of the frustum of the right-circular cone.

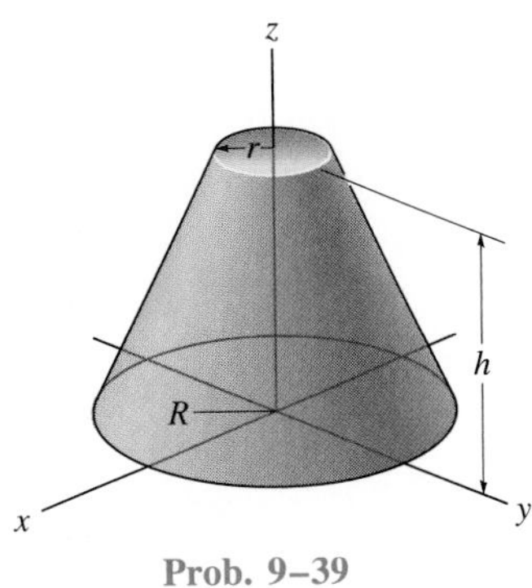

Prob. 9–39

***9–40.** The hemisphere of radius r is made from a stack of very thin plates such that the density varies with height $\rho = kz$, where k is a constant. Determine its mass and the distance $\bar{z}$ to the location of its center of mass G.

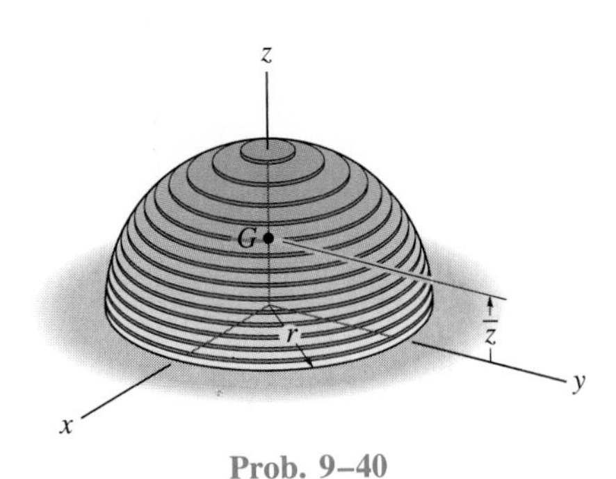

Prob. 9–40

9.3 Composite Bodies

A *composite body* consists of a series of connected "simpler" shaped bodies, which may be rectangular, triangular, semicircular, etc. Such a body can often be sectioned or divided into its composite parts, and provided the *weight* and location of the center of gravity of each of these parts are known, we can eliminate the need for integration to determine the center of gravity for the entire body. The method for doing this requires treating each composite part as a particle and following the procedure outlined in Sec. 9.1. Formulas analogous to Eqs. 9–1 result, since we must account for a finite number of weights. Rewriting these formulas, we have

$$\bar{x} = \frac{\Sigma \tilde{x} W}{\Sigma W} \qquad \bar{y} = \frac{\Sigma \tilde{y} W}{\Sigma W} \qquad \bar{z} = \frac{\Sigma \tilde{z} W}{\Sigma W} \tag{9–8}$$

Here

$\bar{x}, \bar{y}, \bar{z}$ represent the coordinates of the center of gravity G of the composite body.

$\tilde{x}, \tilde{y}, \tilde{z}$ represent the coordinates of the center of gravity of each composite part of the body.

ΣW is the sum of the weights of all the composite parts of the body, or simply the total weight of the body.

When the body has a *constant density or specific weight,* the center of gravity *coincides* with the centroid of the body. The centroid for composite lines, areas, and volumes can be found using relations analogous to Eqs. 9–8; however, the W's are replaced by L's, A's, and V's, respectively. Centroids for common shapes of lines, areas, shells, and volumes are given in the table on the inside back cover.

PROCEDURE FOR ANALYSIS

The following procedure provides a method for determining the center of gravity of a body or the centroid of a composite geometrical object represented by a line, area, or volume.

Composite Parts. Using a sketch, divide the body or object into a finite number of composite parts that have simpler shapes. If a composite part has a *hole,* or geometric region having no material, then consider the composite part without the hole, and the hole as an *additional* composite part having *negative* weight or size.

Moment Arms. Establish the coordinate axes on the sketch and determine the coordinates $\tilde{x}, \tilde{y}, \tilde{z}$ of the center of gravity or centroid of each part.

Summations. Determine $\bar{x}, \bar{y}, \bar{z}$ by applying the center of gravity equations, Eqs. 9–8, or the analogous centroid equations. If an object is *symmetrical* about an axis, recall that the centroid of the object lies on this axis.

If desired, the calculations can be arranged in tabular form, as indicated in the following three examples.

Example 9–9

Locate the centroid of the wire shown in Fig. 9–17*a*.

SOLUTION

Composite Parts. The wire is divided into three segments as shown in Fig. 9–17*b*.

Moment Arms. The location of the centroid for each piece is determined and indicated in the figure. In particular, the centroid of segment (1) is determined either by integration or using the table on the inside back cover.

Summations. The calculations are tabulated as follows:

Segment	L (mm)	$\tilde{x}$ (mm)	$\tilde{y}$ (mm)	$\tilde{z}$ (mm)	$\tilde{x}L$ (mm^2)	$\tilde{y}L$ (mm^2)	$\tilde{z}L$ (mm^2)
1	$\pi(60) = 188.5$	60	−38.2	0	11 310	−7200	0
2	40	0	20	0	0	800	0
3	20	0	40	−10	0	800	−200
	$\Sigma L = 248.5$				$\Sigma\tilde{x}L = 11\,310$	$\Sigma\tilde{y}L = -5600$	$\Sigma\tilde{z}L = -200$

Thus,

$$\bar{x} = \frac{\Sigma\tilde{x}L}{\Sigma L} = \frac{11\,310}{248.5} = 45.5 \text{ mm} \qquad \textit{Ans.}$$

$$\bar{y} = \frac{\Sigma\tilde{y}L}{\Sigma L} = \frac{-5600}{248.5} = -22.5 \text{ mm} \qquad \textit{Ans.}$$

$$\bar{z} = \frac{\Sigma\tilde{z}L}{\Sigma L} = \frac{-200}{248.5} = -0.805 \text{ mm} \qquad \textit{Ans.}$$

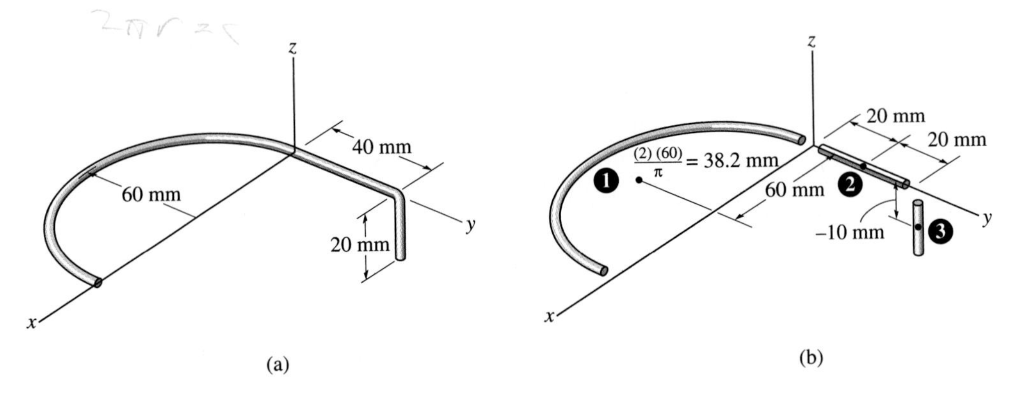

Fig. 9–17

Example 9–10

Locate the centroid of the plate area shown in Fig. 9–18*a*.

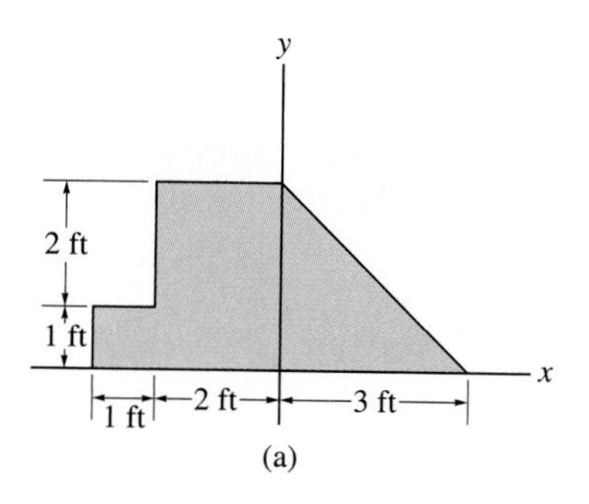

(a)

SOLUTION

Composite Parts. The plate is divided into three segments as shown in Fig. 9–18*b*. Here the area of the small rectangle ③ is considered "negative" since it must be subtracted from the larger one ②.

Moment Arms. The centroid of each segment is located as indicated in the figure. Note that the $\tilde{x}$ coordinates of ② and ③ are *negative.*

Summations. Taking the data from Fig. 9–18*b*, the calculations are tabulated as follows:

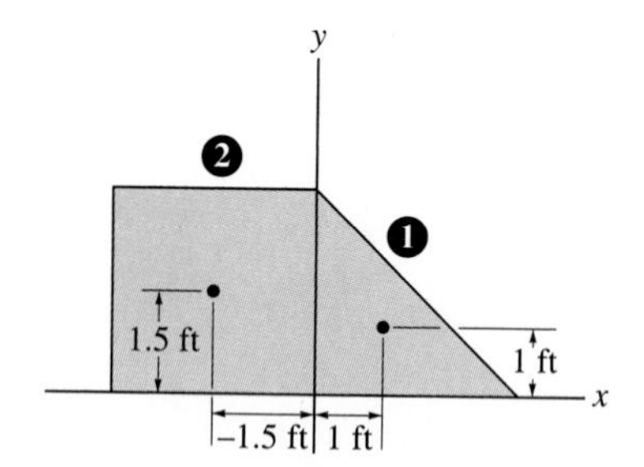

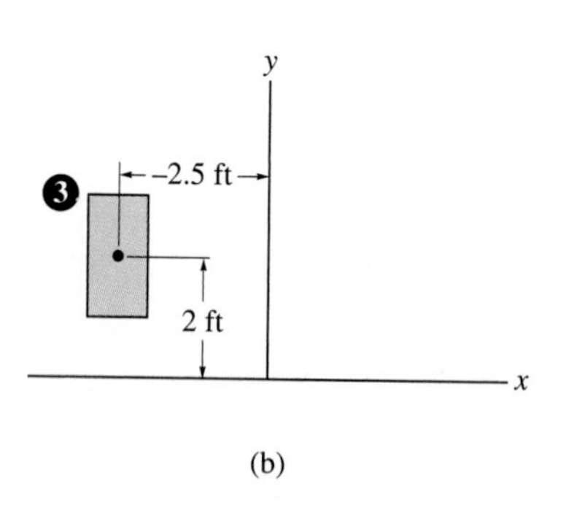

(b)

Fig. 9–18

Segment	A (ft^2)	$\tilde{x}$ (ft)	$\tilde{y}$ (ft)	$\tilde{x}A$ (ft^3)	$\tilde{y}A$ (ft^3)
1	$\frac{1}{2}(3)(3) = 4.5$	1	1	4.5	4.5
2	$(3)(3) = 9$	−1.5	1.5	−13.5	13.5
3	$-(2)(1) = -2$	−2.5	2	5	−4
	$\Sigma A = 11.5$			$\Sigma\tilde{x}A = -4$	$\Sigma\tilde{y}A = 14$

Thus,

$$\bar{x} = \frac{\Sigma\tilde{x}A}{\Sigma A} = \frac{-4}{11.5} = -0.348 \text{ ft} \qquad \textit{Ans.}$$

$$\bar{y} = \frac{\Sigma\tilde{y}A}{\Sigma A} = \frac{14}{11.5} = 1.22 \text{ ft} \qquad \textit{Ans.}$$

Example 9–11

Locate the center of mass of the composite assembly shown in Fig. 9–19*a*. The conical frustum has a density of $\rho_c = 8\ \text{Mg/m}^3$, and the hemisphere has a density of $\rho_h = 4\ \text{Mg/m}^3$.

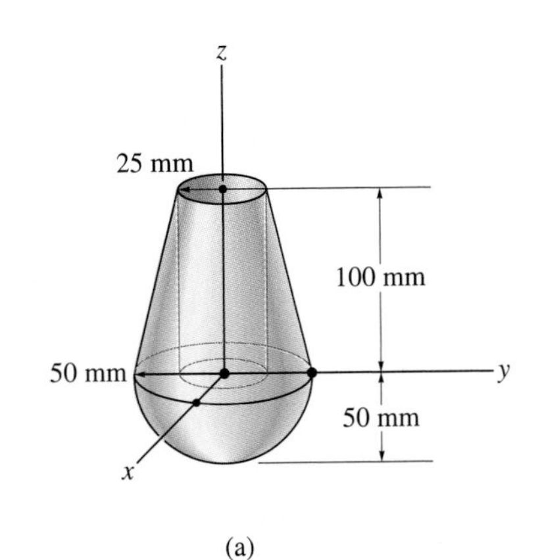

(a)

Fig. 9–19

SOLUTION

Composite Parts. The assembly can be thought of as consisting of four segments as shown in Fig. 9–19*b*. For the calculations, ③ and ④ must be considered as "negative" volumes in order that the four segments, when added together, yield the total composite shape shown in Fig. 9–19*a*.

Moment Arm. Using the table on the inside back cover, the computations for the centroid $\tilde{z}$ of each piece are shown in the figure.

Summations. Because of *symmetry,* note that

$$\bar{x} = \bar{y} = 0 \qquad \textit{Ans.}$$

Since $W = mg$ and g is constant, the third of Eqs. 9–8 becomes $\bar{z} = \Sigma\tilde{z}m/\Sigma m$. The mass of each piece can be computed from $m = \rho V$ and used for the calculations. Also, $1\ \text{Mg/m}^3 = 10^{-6}\ \text{kg/mm}^3$, so that

Segment	m (kg)	$\tilde{z}$ (mm)	$\tilde{z}m$ (kg · mm)
1	$8(10^{-6})(\frac{1}{3})\pi(50)^2(200) = 4.189$	50	209.440
2	$4(10^{-6})(\frac{2}{3})\pi(50)^3 = 1.047$	−18.75	−19.635
3	$-8(10^{-6})(\frac{1}{3})\pi(25)^2(100) = -0.524$	100 + 25 = 125	−65.450
4	$-8(10^{-6})\pi(25)^2(100) = -1.571$	50	−78.540
	$\Sigma m = 3.141$		$\Sigma\tilde{z}m = 45.815$

Thus,

$$\bar{z} = \frac{\Sigma\tilde{z}m}{\Sigma m} = \frac{45.815}{3.141} = 14.6\ \text{mm} \qquad \textit{Ans.}$$

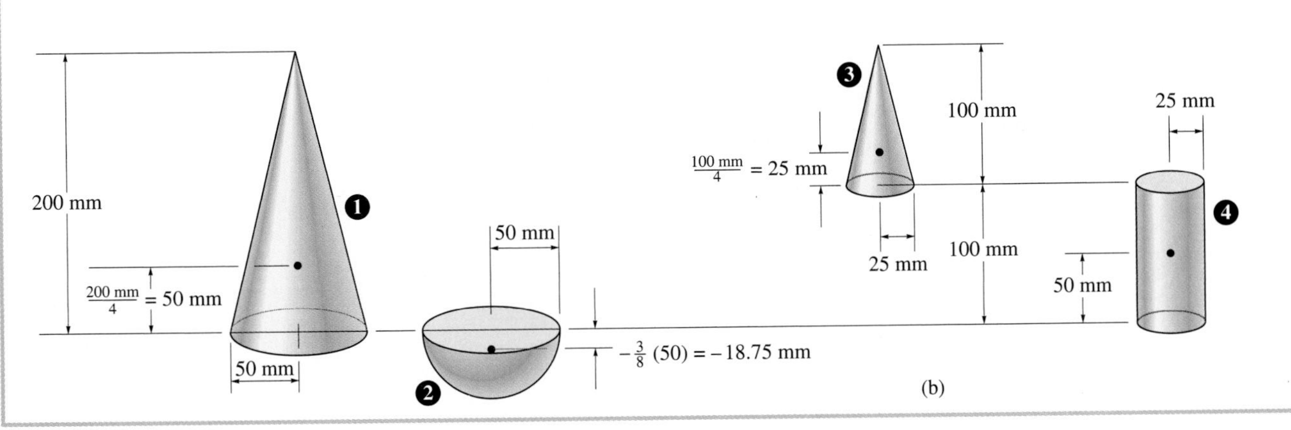

(b)

PROBLEMS

9–41. If the four particles can be replaced by a single 10-kg particle acting at a distance of 2 m to the left of the origin, determine the position $\tilde{x}_p$ and the mass m_p of particle P.

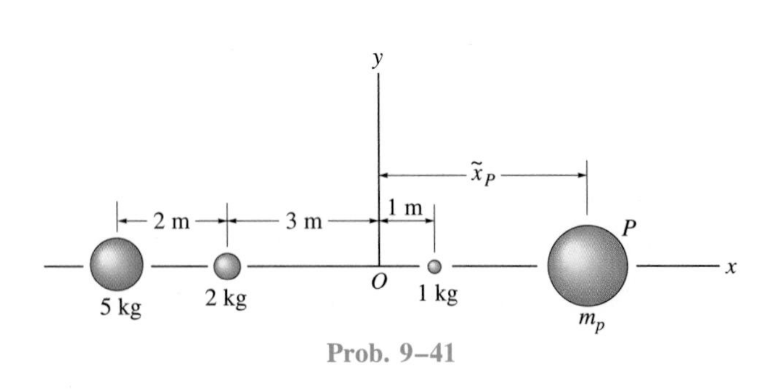

Prob. 9–41

9–42. Determine the location of the center of mass $(\bar{x}, \bar{y}, \bar{z})$ of the three particles.

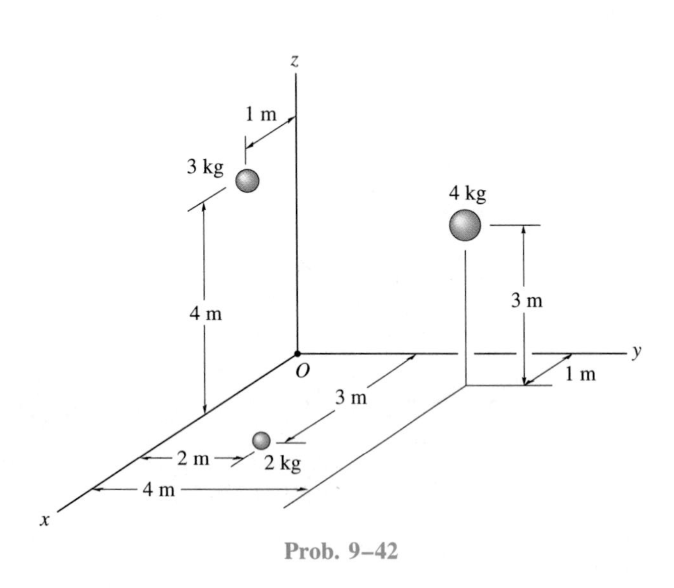

Prob. 9–42

9–43. A "roll-formed" member has the cross section shown. Determine the location $\bar{y}$ of the centroid C. Neglect the thickness of the material and any slight bends at the corners.

Prob. 9–43

***9–44.** The steel and aluminum plate assembly is bolted together and fastened to the wall. Each plate has a constant width in the z direction of 200 mm and thickness of 20 mm. If the density of A and B is $\rho_s = 7.85\ \text{Mg/m}^3$, and for C, $\rho_{al} = 2.71\ \text{Mg/m}^3$, determine the location $\bar{x}$ of the center of mass. Neglect the size of the bolts.

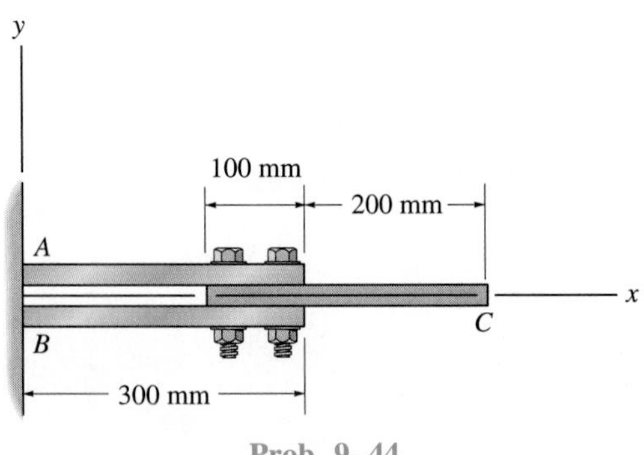

Prob. 9–44

9–45. The truss is made from five members, each having a length of 4 m and a mass of 7 kg/m. If the mass of the gusset plates at the joints and the thickness of the members can be neglected, determine the distance d to where the hoisting cable must be attached, so that the truss does not tip (rotate) when it is lifted.

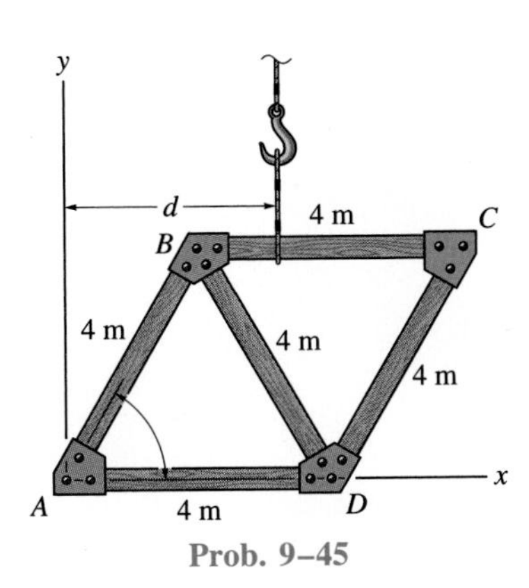

Prob. 9–45

9–46. The three members of the frame each have a weight per unit length of 4 lb/ft. Locate the position $(\bar{x}, \bar{y})$ of the center of gravity. Neglect the size of the pins at the joints and the thickness of the members. Also, calculate the reactions at the fixed support A.

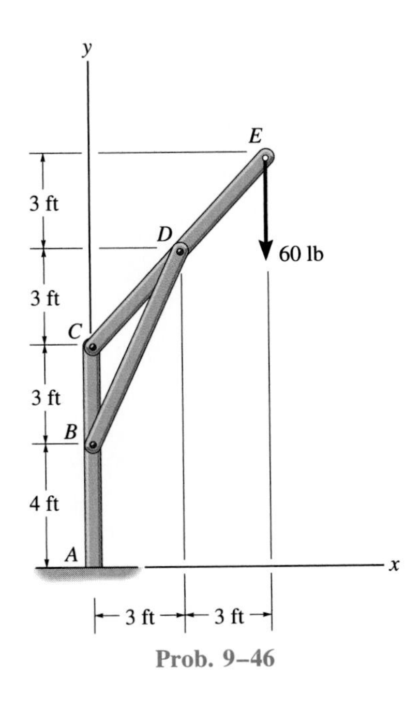

Prob. 9–46

9–47. Determine the location $(\bar{x}, \bar{y})$ of the centroid of the area.

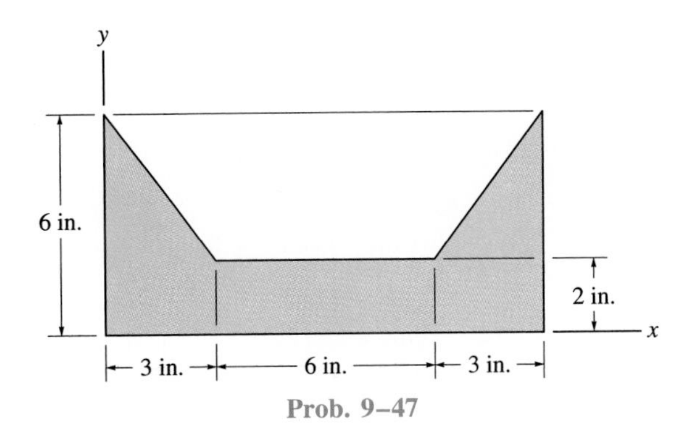

Prob. 9–47

***9–48.** Determine the location $\bar{y}$ of the centroid of the area.

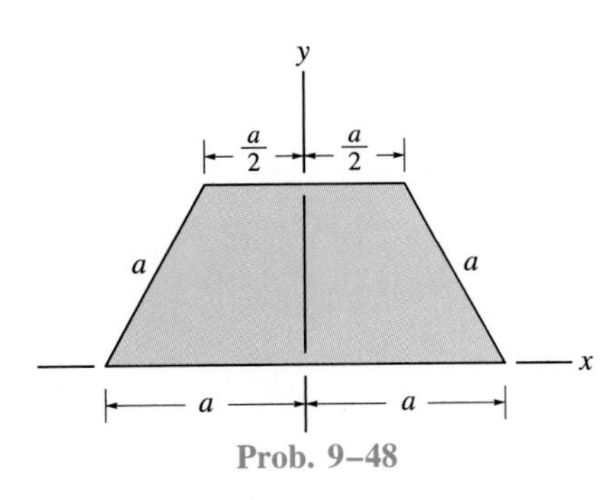

Prob. 9–48

9–49. Determine the weight and location $(\bar{x}, \bar{y})$ of the center of gravity G of the concrete retaining wall. The wall has a length of 10 ft, and concrete has a specific gravity of $\gamma = 150\ \text{lb/ft}^3$.

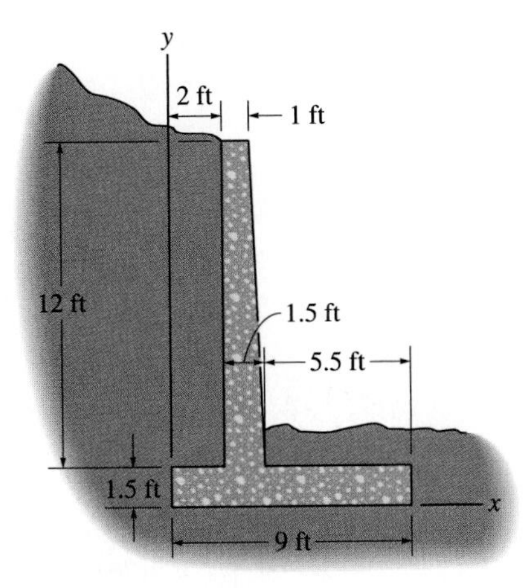

Prob. 9–49

9–50. Locate the centroid $\bar{y}$ of the concrete beam having the tapered cross section shown.

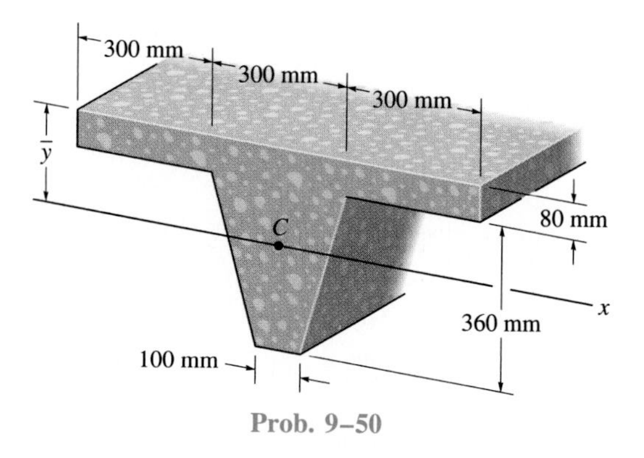

Prob. 9–50

9–51. Determine the distance $\bar{y}$ to the centroid of the trapezoidal area in terms of the dimensions shown.

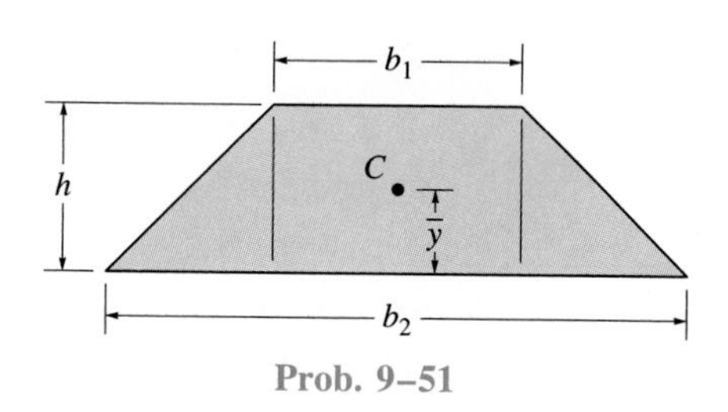

Prob. 9–51

***9–52.** Determine the location $\bar{y}$ of the centroid of the beam's cross section built up from a channel and a wide-flange beam.

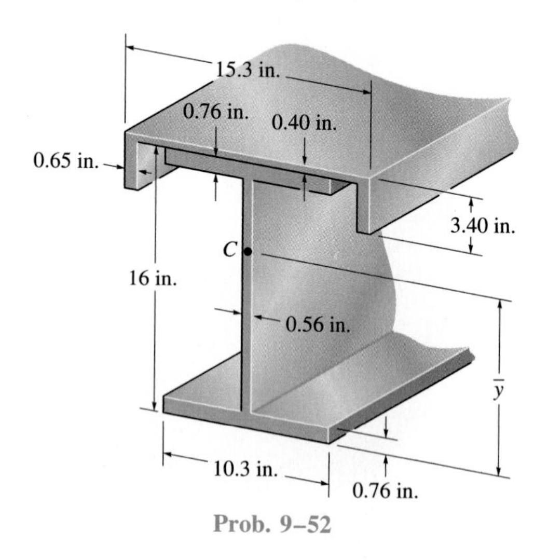

Prob. 9–52

9–53. Determine the location $(\bar{x}, \bar{y})$ of the centroid C for the angle's cross-sectional area.

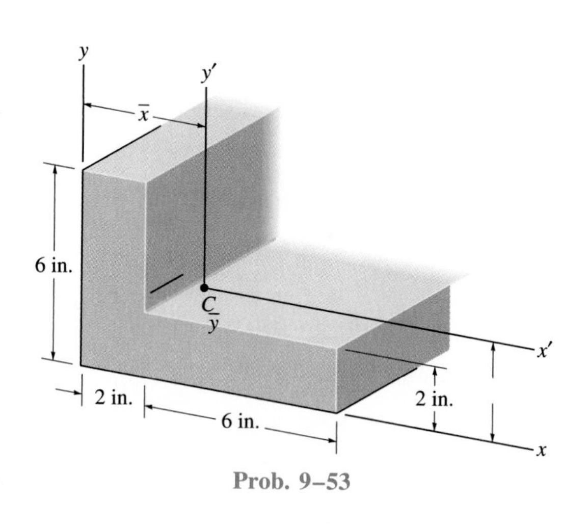

Prob. 9–53

9–54. Determine the location $\bar{y}$ of the centroid of the beam's cross-sectional area.

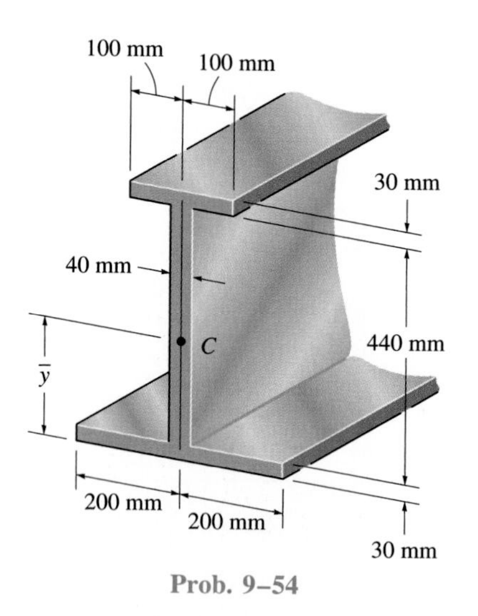

Prob. 9–54

9–55. Determine the location $\bar{y}$ of the centroidal axis $\bar{x}\bar{x}$ of the beam's cross-sectional area. Neglect the size of the corner welds at A and B for the calculation.

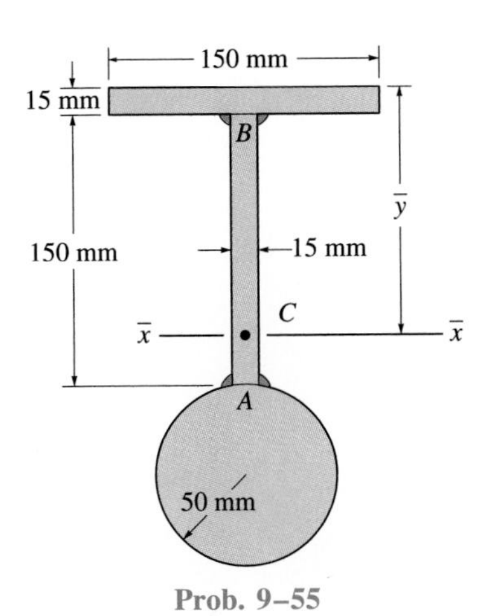

Prob. 9–55

***9–56.** Determine the location $(\bar{x}, \bar{y})$ of the centroid C of the cross-sectional area for the structural member constructed from two equal-sized channels welded together as shown. Assume all corners are square. Neglect the size of the welds.

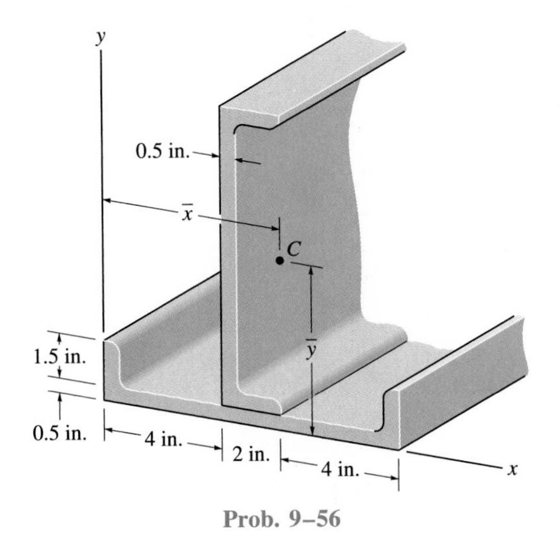

Prob. 9–56

9–57. Determine the location $(\bar{x}, \bar{y})$ of the centroid C of the area.

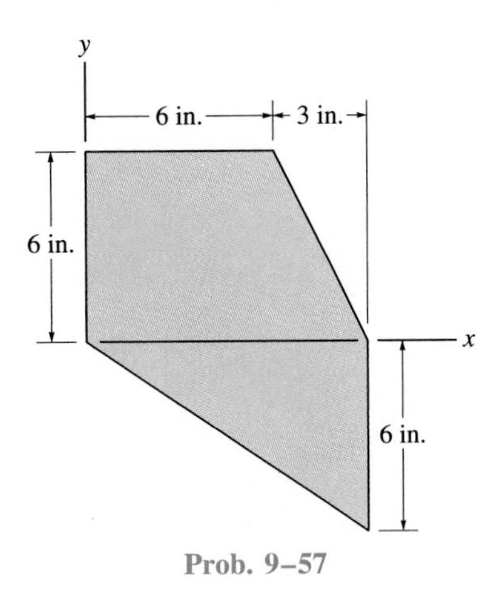

Prob. 9–57

9–58. Determine the location $\bar{y}$ of the centroid for the cross-sectional area.

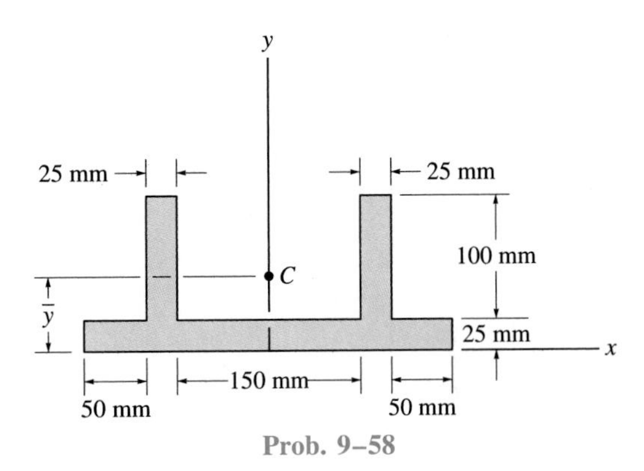

Prob. 9–58

9–59. Determine the location $(\bar{x}, \bar{y})$ of the centroid of the cross-sectional area.

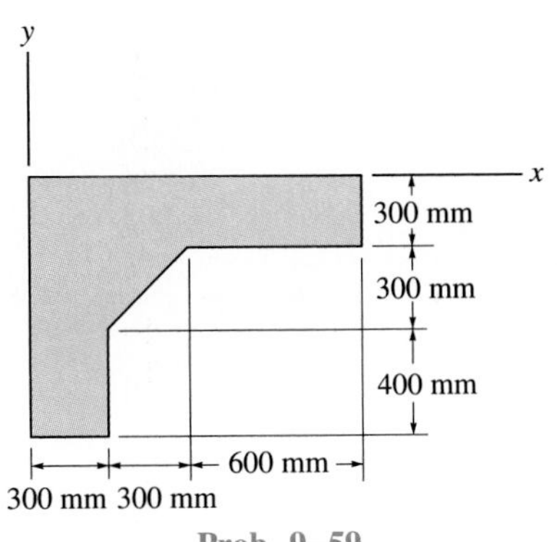

Prob. 9–59

***9–60.** Determine the location $\bar{y}$ of the centroid of the cross-sectional area.

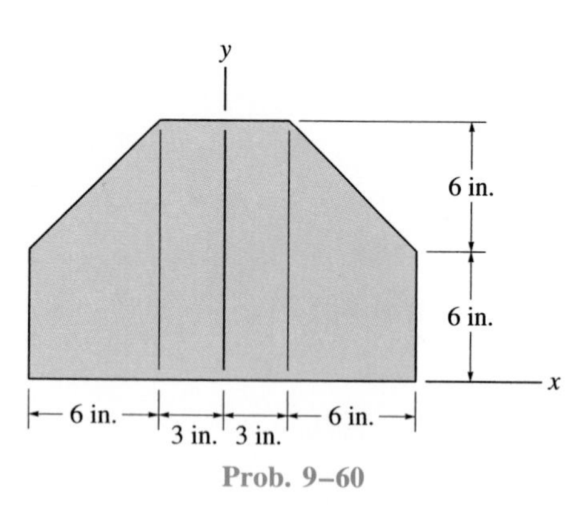

Prob. 9–60

9–61. The tank and compressor have a mass of 15 kg and mass center at G_T, and the motor has a mass of 70 kg and a mass center at G_M. Determine the angle of tilt, θ, of the tank so that the unit will be on the verge of tipping over.

Prob. 9–61

9–62. The wooden table is made from a square board having a weight of 15 lb. Each of the legs weighs 2 lb and is 3 ft long. Determine how high its center of gravity is from the floor. Also, what is the angle, measured from the horizontal, through which its top surface can be tilted on two of its legs before it begins to overturn? Neglect the thickness of each leg.

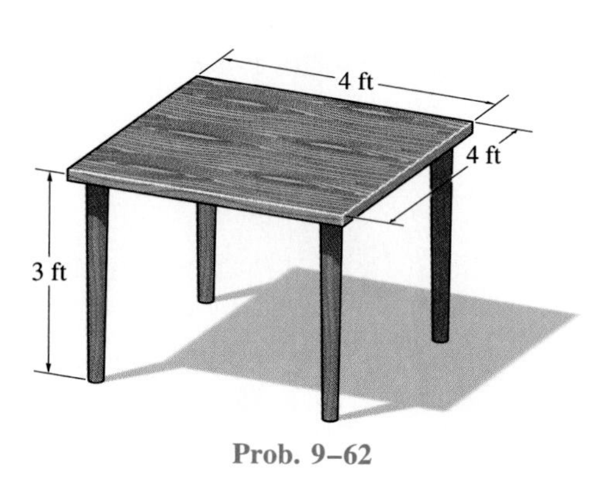

Prob. 9–62

9–63. Determine the location $\bar{x}$ of the centroid C of the shaded area which is part of a circle having a radius r.

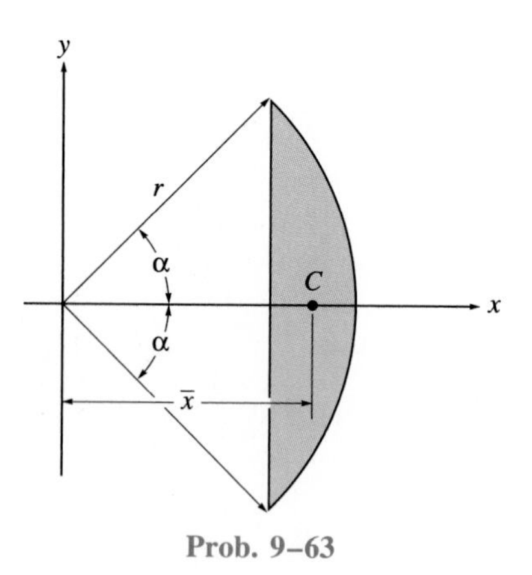

Prob. 9–63

***9–64.** A triangular plate made of homogeneous material has a constant thickness which is very small. If it is folded over as shown, determine the location $\bar{y}$ of the plate's center of gravity G.

9–65. A triangular plate made of homogeneous material has a constant thickness which is very small. If it is folded over as shown, determine the location $\bar{z}$ of the plate's center of gravity G.

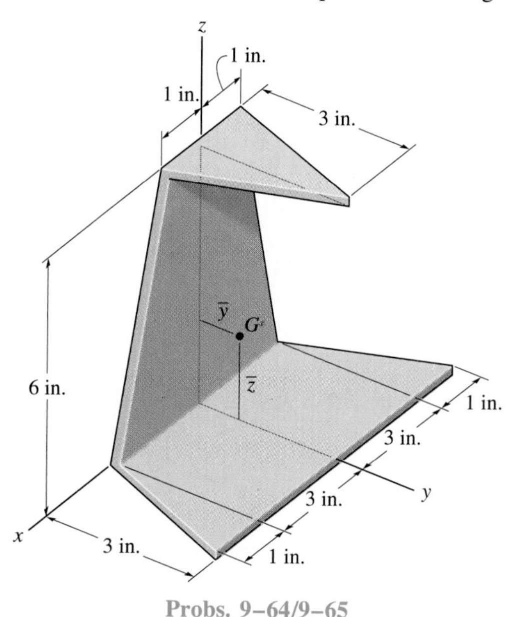

Probs. 9–64/9–65

9–66. Determine the location $(\bar{x}, \bar{y})$ of the center of gravity of the three-wheeler. The location of the center of gravity of each component and its weight are tabulated in the figure. If the three-wheeler is symmetrical with respect to the x-y plane, determine the normal reactions each of its wheels exerts on the ground.

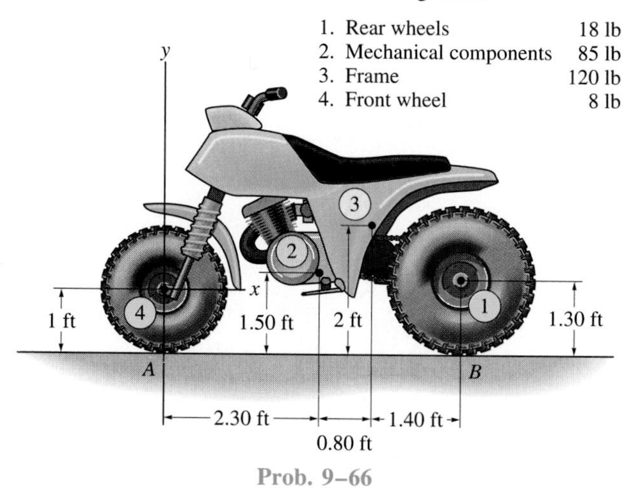

Prob. 9–66

9–67. Locate the center of mass of the two-block assembly. The densities of materials A and B are $\rho_A = 150\ \text{lb/ft}^3$ and $\rho_B = 400\ \text{lb/ft}^3$, respectively.

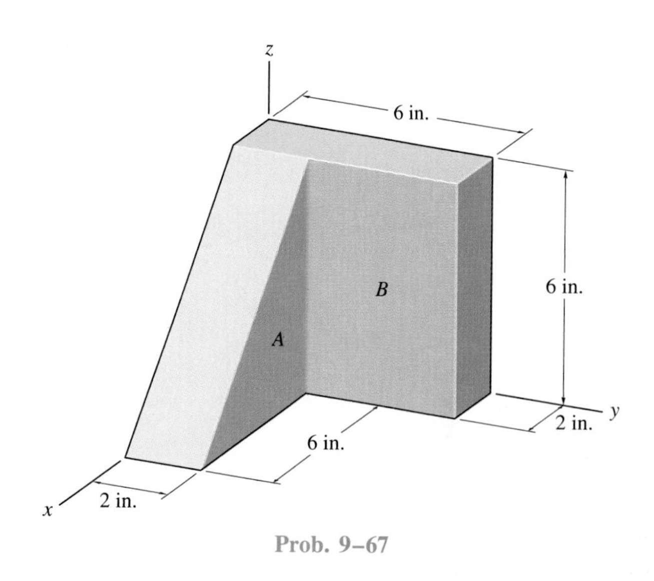

Prob. 9–67

***9–68.** The buoy is made from two homogeneous cones each having a radius of 1.5 ft. If $h = 1.2$ ft, find the distance $\bar{z}$ to the buoy's center of gravity G.

9–69. The buoy is made from two homogeneous cones each having a radius of 1.5 ft. If it is required that the buoy's center of gravity G be located at $\bar{z} = 0.5$ ft, determine the height h of the top cone.

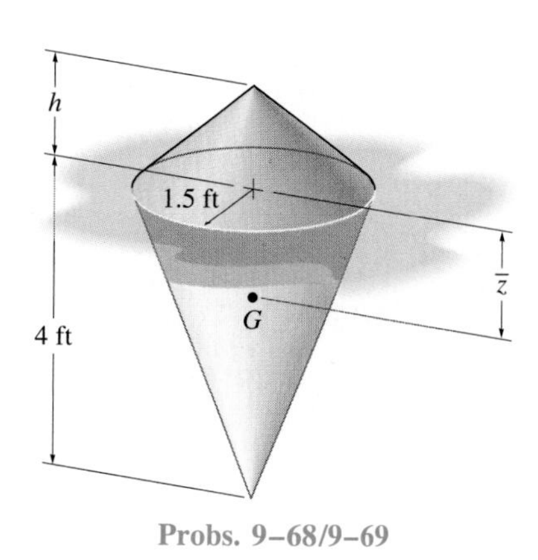

Probs. 9–68/9–69

9–70. The assembly consists of a 20-in. wooden dowel rod and a tight-fitting steel collar. Determine the distance $\bar{x}$ to its center of gravity if the specific weights of the materials are $\gamma_w = 150 \text{ lb/ft}^3$ and $\gamma_{st} = 490 \text{ lb/ft}^3$.

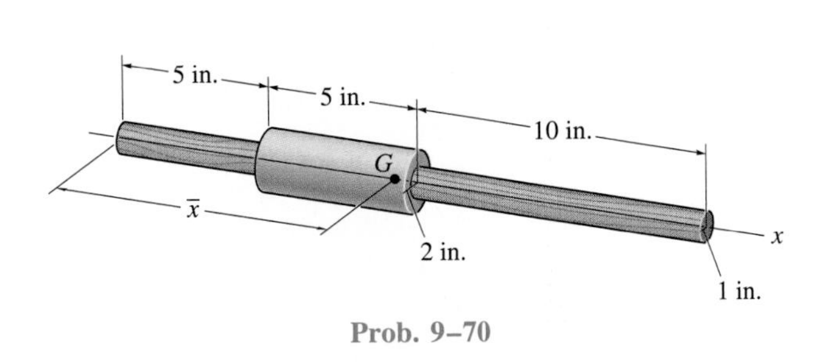

Prob. 9–70

9–71. Determine the distance h to which a 100-mm-diameter hole must be bored into the base of the cone so that the center of mass of the resulting shape is located at $\bar{z} = 115$ mm. The material has a density of 8 Mg/m^3.

***9–72.** Determine the distance $\bar{z}$ to the centroid of the shape which consists of a cone with a hole of height $h = 50$ mm bored into its base.

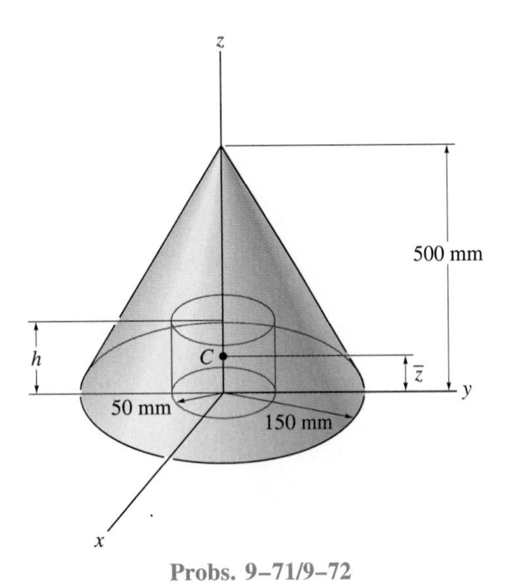

Probs. 9–71/9–72

9–73. Determine the location $\bar{z}$ of the centroid of the top made from a hemisphere and a cone.

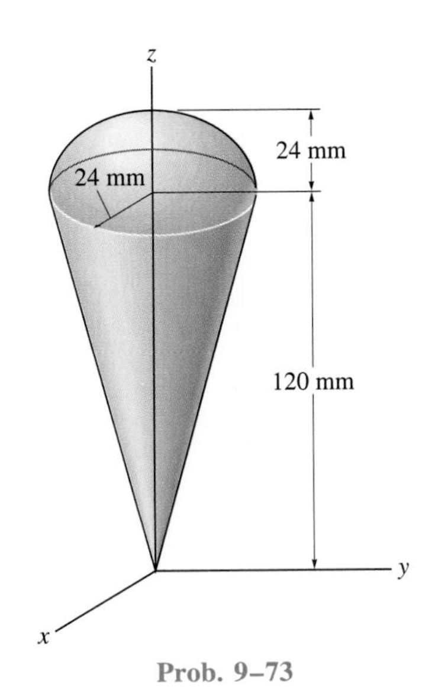

Prob. 9–73

9–74. The assembly is made from a steel hemisphere, $\rho_{st} = 7.80 \text{ Mg/m}^3$, and an aluminum cylinder, $\rho_{al} = 2.70 \text{ Mg/m}^3$. Determine the height h of the cylinder so that the mass center of the assembly is located at $\bar{z} = 160$ mm.

9–75. The assembly is made from a steel hemisphere, $\rho_{st} = 7.80 \text{ Mg/m}^3$, and an aluminum cylinder, $\rho_{al} = 2.70 \text{ Mg/m}^3$. Determine the mass center of the assembly if the height of the cylinder is $h = 200$ mm.

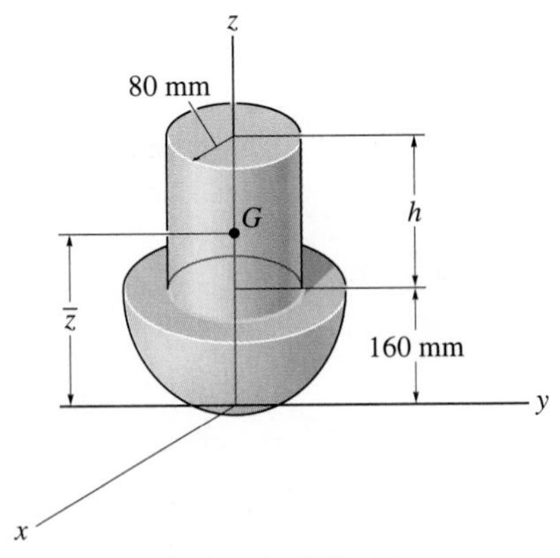

Probs. 9–74/9–75

*9.4 Theorems of Pappus and Guldinus

The two *theorems of Pappus and Guldinus,* which were first developed by Pappus of Alexandria during the third century A.D. and then restated at a later time by the Swiss mathematician Paul Guldin or Guldinus (1577–1643), are used to find the surface area and volume of any object of revolution.

A *surface area of revolution* is generated by revolving a *plane curve* about a nonintersecting fixed axis in the plane of the curve; whereas a *volume of revolution* is generated by revolving a *plane area* about a nonintersecting fixed axis in the plane of the area. For example, if the *line AB* shown in Fig. 9–20 is rotated about a fixed axis, it generates the *surface area* of a cone; if the triangular *area ABC* shown in Fig. 9–21 is rotated about the axis, it generates the *volume* of a cone.

The statements and proofs of the theorems of Pappus and Guldinus follow. The proofs require that the generating curves and areas do *not* cross the axis about which they are rotated; otherwise, two sections on either side of the axis would generate areas or volumes having opposite signs and hence cancel each other.

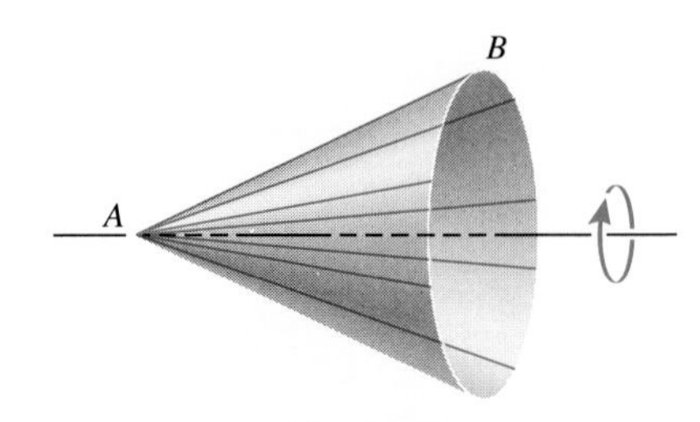

Fig. 9–20

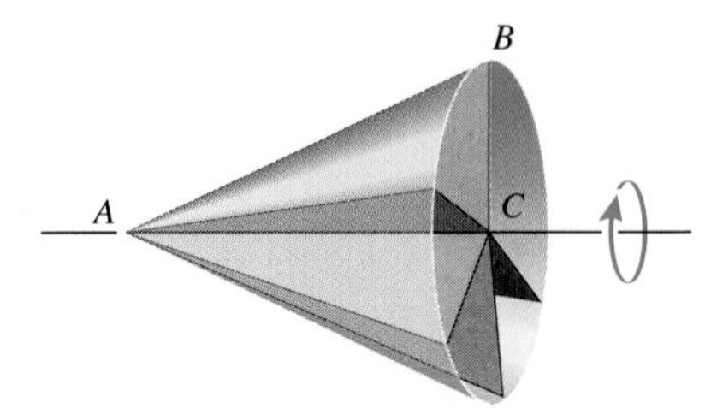

Fig. 9–21

Surface Area

The area of a surface of revolution equals the product of the length of the generating curve and the distance traveled by the centroid of the curve in generating the surface area.

Proof. When a differential length dL of the curve shown in Fig. 9–22 is revolved about an axis through a distance $2\pi r$, it generates a ring having a surface area $dA = 2\pi r\, dL$. The entire surface area, generated by revolving the entire curve about the axis, is therefore $A = 2\pi \int_L r\, dL$. This equation may be simplified, however, by noting that the location r of the centroid for the line of total length L can be determined from an equation having the form of Eqs. 9–7, namely, $\bar{r} = \int r\, dL/L$. Thus the total surface area becomes $A = 2\pi \bar{r} L$. In general, though, if the line does not undergo a complete revolution, then

$$A = \theta \bar{r} L \qquad (9\text{–}9)$$

where A = surface area of revolution
θ = angle of revolution measured in radians, $\theta \leq 2\pi$
$\bar{r}$ = perpendicular distance from the axis of revolution to the centroid of the generating curve
L = length of the generating curve

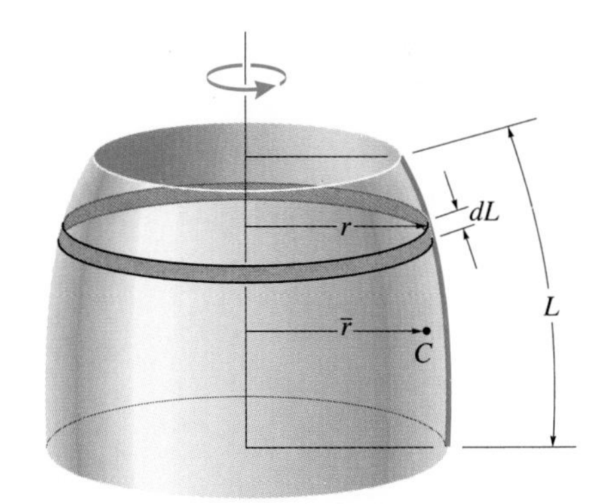

Fig. 9–22

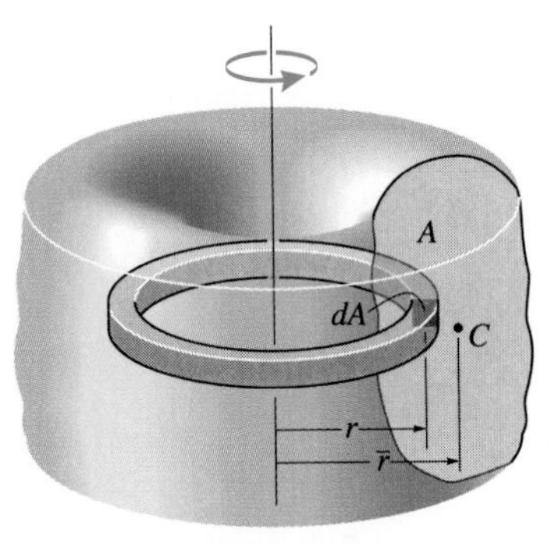

Fig. 9–23

Volume

The volume of a body of revolution equals the product of the generating area and the distance traveled by the centroid of the area in generating the volume.

Proof. When the differential area dA shown in Fig. 9–23 is revolved about an axis through a distance $2\pi r$, it generates a ring having a volume $dV = 2\pi r\, dA$. The entire volume, generated by revolving A about the axis, is therefore $V = 2\pi \int_A r\, dA$. Here the integral can be eliminated by using an equation analogous to Eqs. 9–6, $\int_A r\, dA = \bar{r}A$, where $\bar{r}$ locates the centroid C of the generating area A, and the volume becomes $V = 2\pi\bar{r}A$. In general, though,

$$V = \theta\bar{r}A \qquad (9\text{–}10)$$

where V = volume of revolution
θ = angle of revolution measured in radians, $\theta \leq 2\pi$
$\bar{r}$ = perpendicular distance from the axis of revolution to the centroid of the generating area
A = generating area

Composite Shapes. We may also apply the above two theorems to lines or areas that may be composed of a series of composite parts. In this case the total surface area or volume generated is the addition of the surface areas or volumes generated by each of the composite parts. Since each part undergoes the *same* angle of revolution, θ, and the distance from the axis of revolution to the centroid of each composite part is $\tilde{r}$, then

$$A = \theta\Sigma\tilde{r}L \qquad (9\text{–}11)$$

and

$$V = \theta\Sigma\tilde{r}A \qquad (9\text{–}12)$$

Application of the above theorems is illustrated numerically in the following example.

Example 9–12

Show that the surface area of a sphere is $A = 4\pi R^2$ and its volume is $V = \frac{4}{3}\pi R^3$.

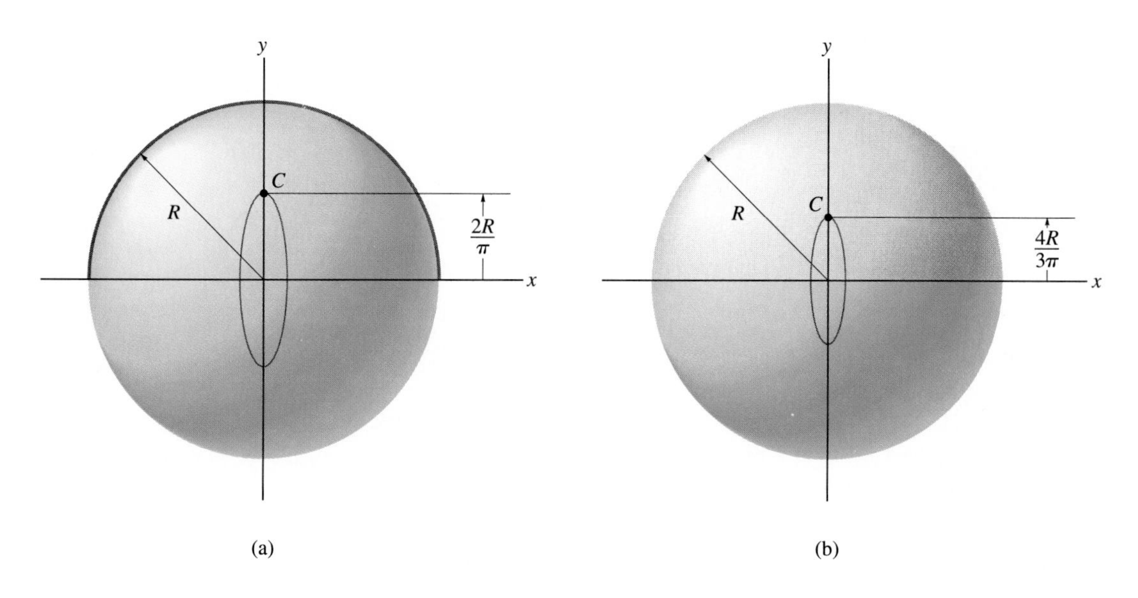

Fig. 9–24

SOLUTION

Surface Area. The surface area of the sphere in Fig. 9–24*a* is generated by rotating a semicircular *arc* about the *x* axis. Using the table on the inside back cover, it is seen that the centroid of this arc is located at a distance $\bar{r} = 2R/\pi$ from the x axis of rotation. Since the centroid moves through an angle of $\theta = 2\pi$ rad in generating the sphere, by applying Eq. 9–9 we have

$$A = \theta\bar{r}L; \qquad A = 2\pi\left(\frac{2R}{\pi}\right)\pi R = 4\pi R^2 \qquad \textit{Ans.}$$

Volume. The volume of the sphere is generated by rotating the semicircular *area* in Fig. 9–24*b* about the *x* axis. Using the table on the inside back cover to locate the centroid and applying Eq. 9–10, we have

$$V = \theta\bar{r}A; \qquad V = 2\pi\left(\frac{4R}{3\pi}\right)\left(\frac{1}{2}\pi R^2\right) = \frac{4}{3}\pi R^3 \qquad \textit{Ans.}$$

PROBLEMS

*9–76. Using integration, determine both the area and the centroidal distance $\bar{x}$ of the shaded area. Then, using the second theorem of Pappus–Guldinus, determine the volume of the solid generated by revolving the area about the y axis.

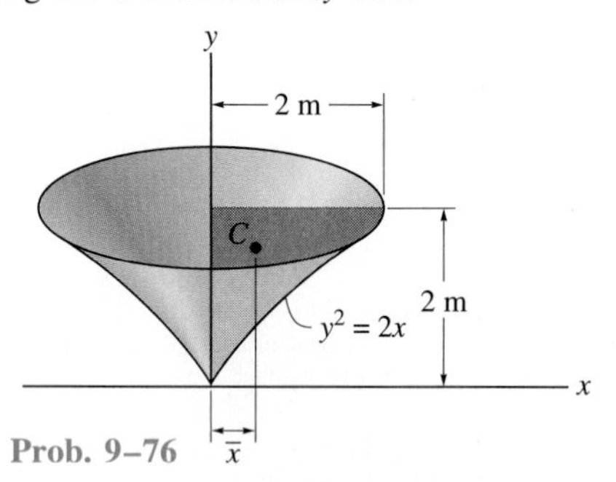

Prob. 9–76

9–77. Using integration, determine the area and the centroidal distance $\bar{y}$ of the shaded area. Then, using the second theorem of Pappus–Guldinus, determine the volume of a paraboloid formed by revolving the area about the x axis.

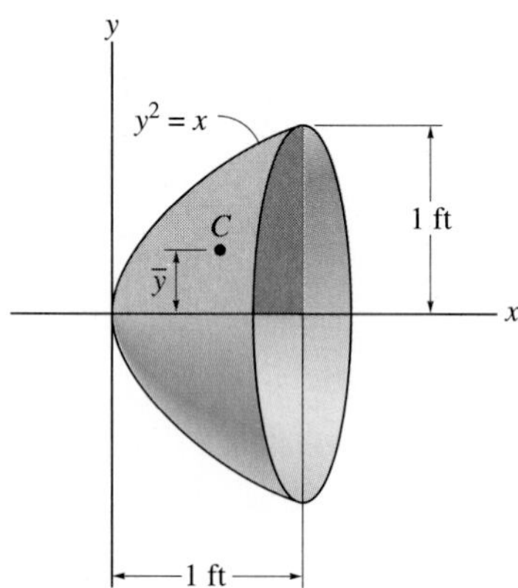

Prob. 9–77

9–78. Using integration, determine the area and the centroidal distance $\bar{x}$ of the shaded area. Then, using the second theorem of Pappus-Guldinus, determine the volume of a solid formed by revolving the area about the y axis.

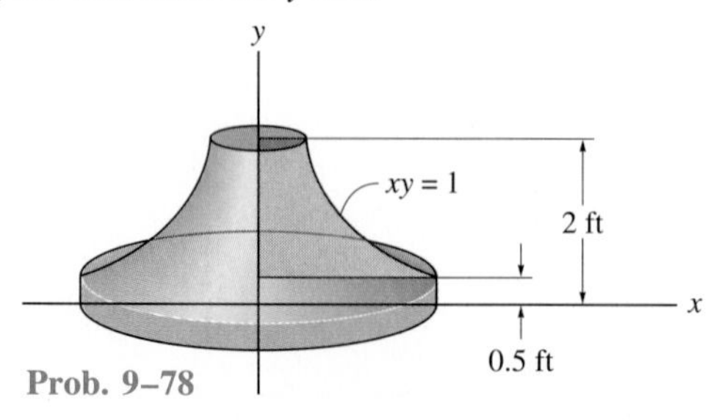

Prob. 9–78

9–79. Determine the surface area of the casting.

*9–80. Determine the volume of material needed to make the casting.

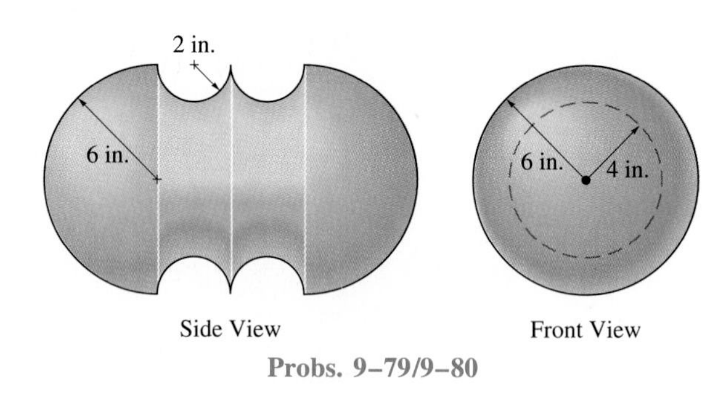

Probs. 9–79/9–80

9–81. Using integration, determine both the area and the distance $\bar{y}$ to the centroid of the shaded area. Then using the second theorem of Pappus—Guldinus, determine the volume of the solid generated by revolving the shaded area about the x axis.

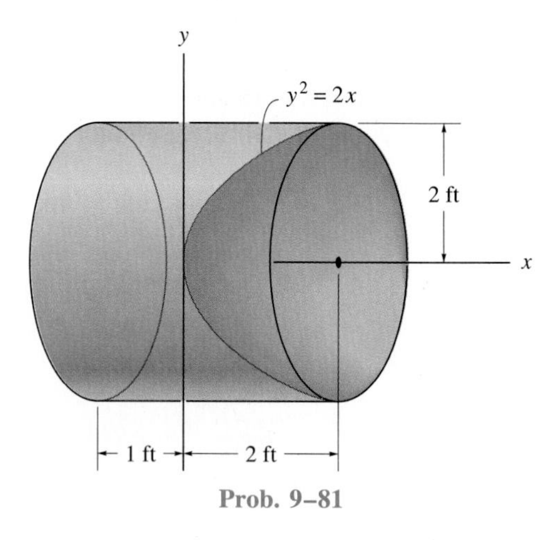

Prob. 9–81

9–82. Using integration, determine both the area and the centroidal distance $\bar{x}$ of the shaded area. Then, using the second theorem of Pappus—Guldinus, determine the volume of the solid generated by revolving the shaded area about the y axis.

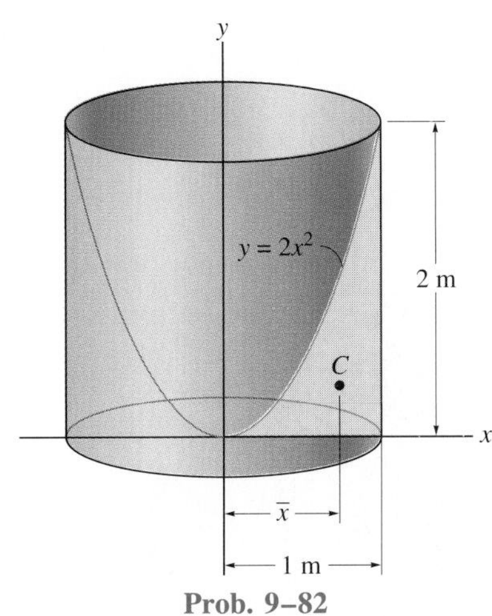
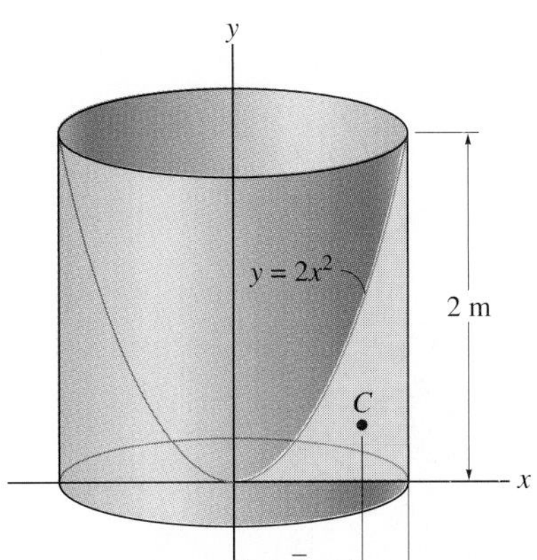

Prob. 9–82

9–83. Sand is piled between two walls as shown. Assume the pile to be a quarter section of a cone and that 26 percent of this volume is voids (air space). Use the second theorem of Pappus–Guldinus to determine the volume of sand.

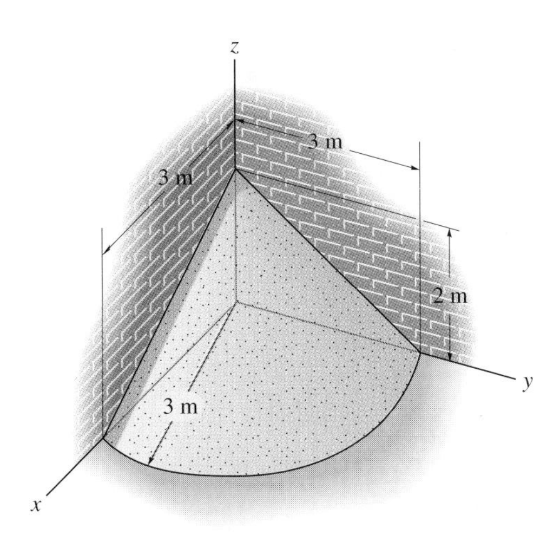

Prob. 9–83

***9–84.** Determine the volume of concrete needed to construct the curb.

9–85. Determine the surface area of the curb. Do not include the area of the ends in the calculation.

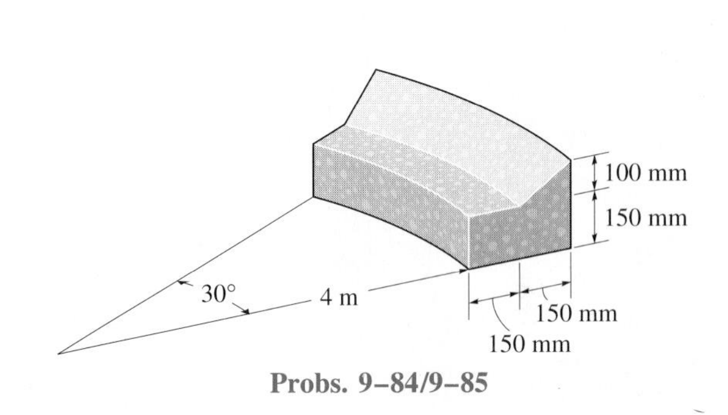

Probs. 9–84/9–85

9–86. The *rim* of a flywheel has the cross section *A-A* shown. Determine the volume of material needed for its construction.

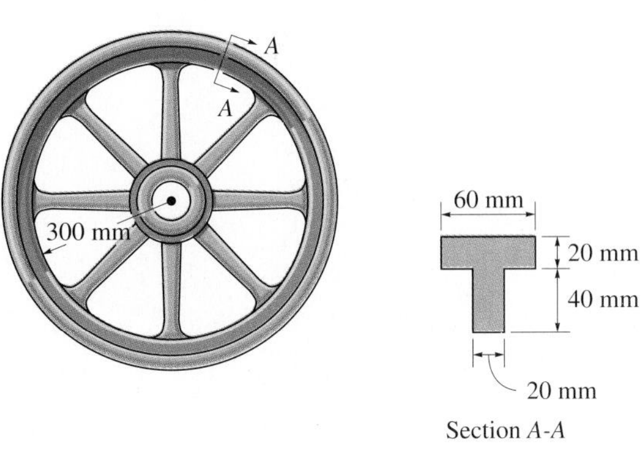

Prob. 9–86

9–87. The anchor ring is made of steel having a specific weight of $\gamma_{st} = 490 \text{ lb/ft}^3$. Determine its weight. The cross section is circular as shown.

***9–88.** The anchor ring is made of steel having a specific weight of $\gamma_{st} = 490 \text{ lb/ft}^3$. Determine the surface area of the ring. The cross section is circular as shown.

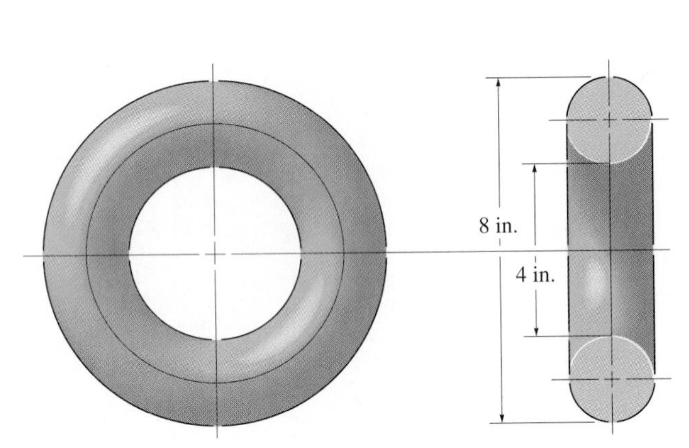

Probs. 9–87/9–88

9–89. A circular sea wall is made of concrete. Determine the total weight of the wall if the concrete has a specific weight of $\gamma_c = 150 \text{ lb/ft}^3$.

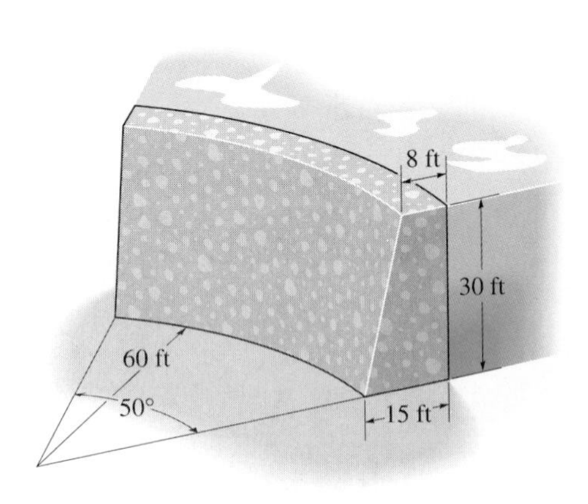

Prob. 9–89

9–90. The water-supply tank has a hemispherical bottom and cylindrical sides. Determine the weight of water in the tank when it is filled to the top at C. Take $\gamma_w = 62.4 \text{ lb/ft}^3$.

9–91. Determine the number of gallons of paint needed to paint the outside surface of the water-supply tank, which consists of a hemispherical bottom, cylindrical sides, and conical top. Each gallon of paint can cover 250 ft^2.

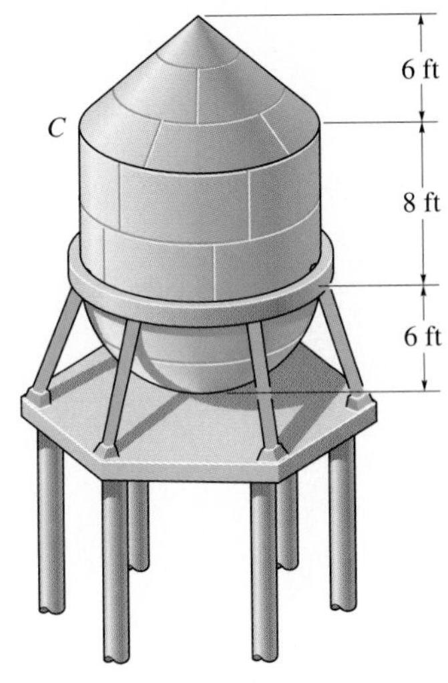

Probs. 9–90/9–91

***9–92.** The hopper is filled to its top with coal. Determine the volume of coal if the voids (air space) are 35 percent of the volume of the hopper.

9–93. If the hopper is made of steel plate having a thickness of 10 mm, determine its weight when empty. Steel has a density of $\rho_{st} = 7.85 \text{ Mg/m}^3$.

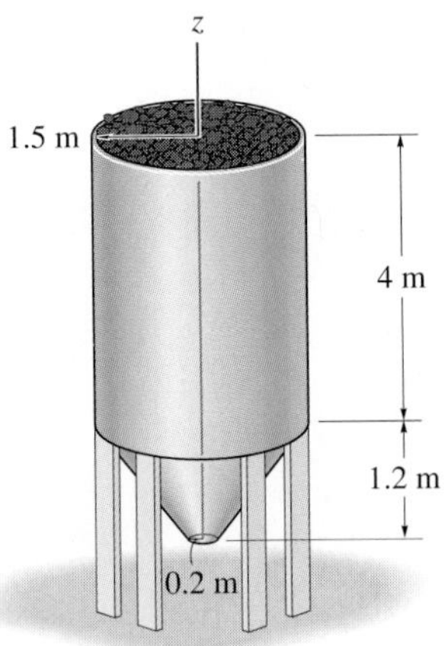

Probs. 9–92/9–93

9–94. The full circular aluminum housing is used in an automotive brake system. Half the cross section is shown in the figure. Determine its weight if aluminum has a specific weight of $\gamma_{al} = 160\ \text{lb/ft}^3$.

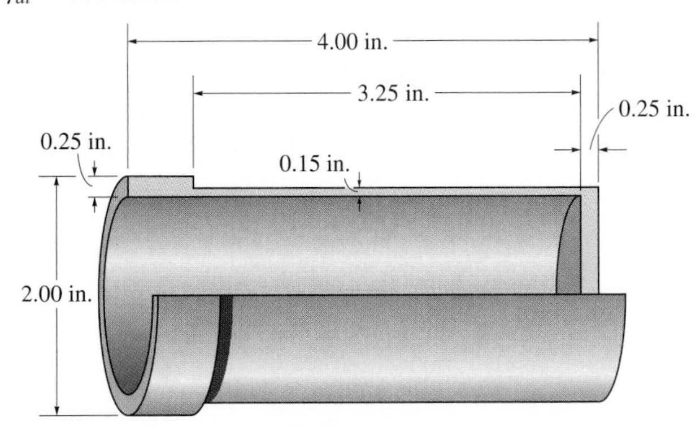

Prob. 9–94

9–95. Determine the approximate amount of aluminum necessary to make the funnel. It consists of a full circular part having a thickness of 2 mm. Its cross section is shown in the figure.

***9–96.** Determine the approximate outer surface area of the funnel. It consists of a full circular part of negligible thickness.

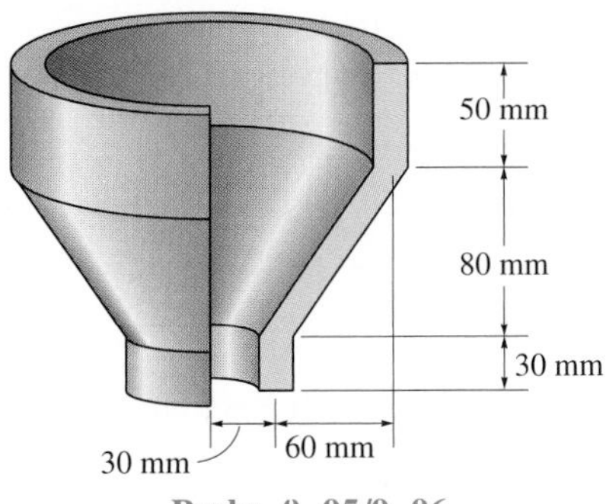

Probs. 9–95/9–96

9–97. Determine the height h to which liquid should be poured into the cup so that it contacts half the surface area on the inside of the cup. Neglect the cup's thickness for the calculation.

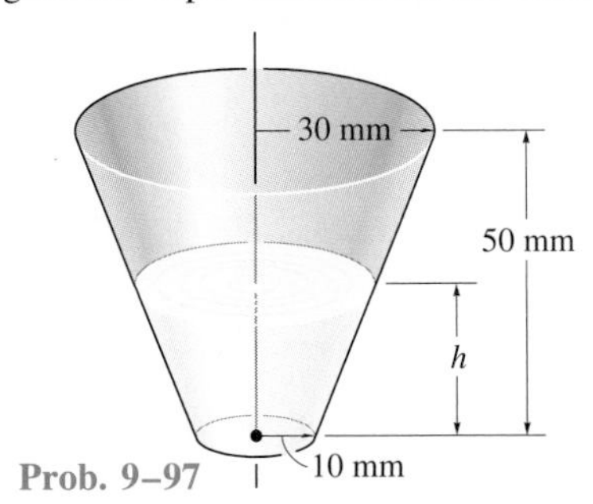

Prob. 9–97

9–98. The process tank is used to store liquids during manufacturing. Estimate both the volume of the tank and its surface area. The tank has a flat top and the plates from which the tank is made have negligible thickness.

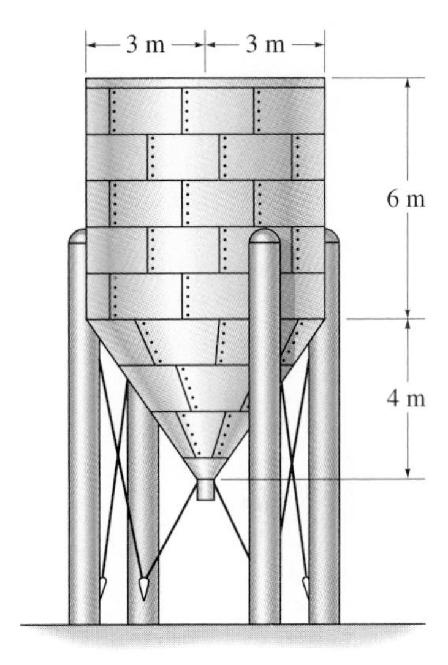

Prob. 9–98

9–99. Determine the volume of steel needed to make the brake piston. It consists of a full circular part. Its cross section is shown in the figure.

***9–100.** Determine the volume of oil that can be contained in the brake piston. It consists of a full circular part. Its cross section is shown in the figure.

9–101. Determine the interior surface area of the brake piston. It consists of a full circular part. Its cross section is shown in the figure.

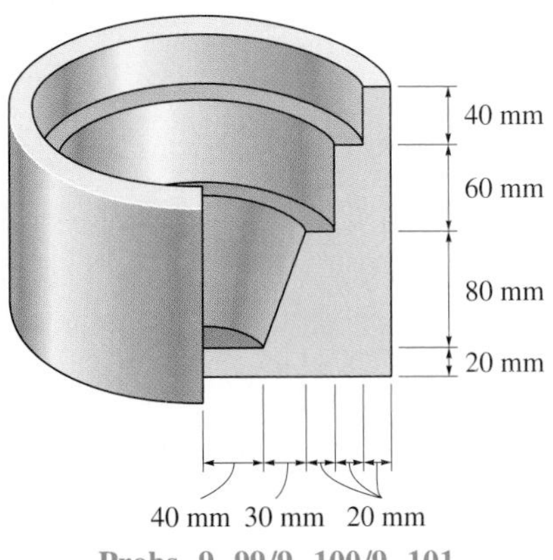

Probs. 9–99/9–100/9–101

★9.5 Resultant of a General Distributed Force System

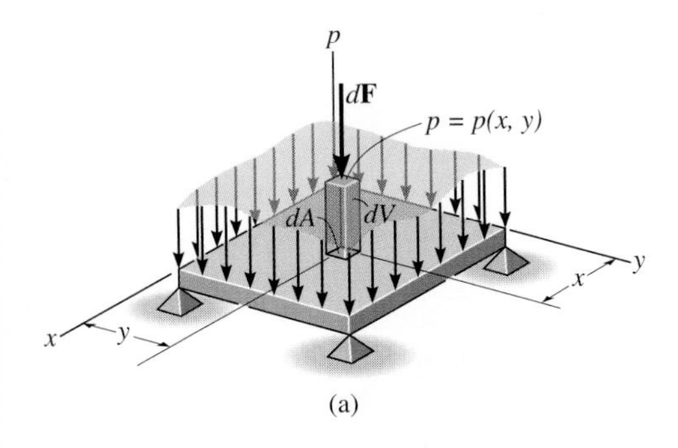

(a)

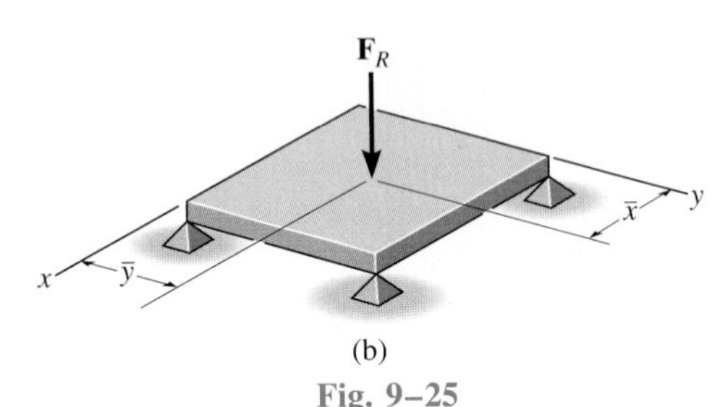

(b)

Fig. 9–25

In Sec. 4.10 we discussed the method used to simplify a distributed loading that is uniform along an axis of a rectangular surface. In this section we will generalize this method to include surfaces that have an arbitrary shape and are subjected to a variable load distribution. As a specific application, in Sec. 9.6 we will find the resultant loading acting on the surface of a body that is submerged in a fluid.

Pressure Distribution over a Surface. Consider the flat plate shown in Fig. 9–25*a*, which is subjected to the loading function $p = p(x, y)$ Pa, where 1 pascal, Pa = 1 N/m^2. Knowing this function, we can determine the *magnitude* of the infinitesimal force $d\mathbf{F}$ acting on the differential area dA m^2 of the plate, located at the arbitrary point (x, y). This force magnitude is simply $dF = [p(x, y)\text{ N/m}^2](dA\text{ m}^2) = [p(x, y)\,dA]$ N. The entire loading on the plate is therefore represented as a system of *parallel forces* infinite in number and each acting on separate differential areas dA. This system of parallel forces will now be simplified to a single resultant force $\mathbf{F}_R$ acting through a unique point $(\bar{x}, \bar{y})$ on the plate, Fig. 9–25*b*.

Magnitude of Resultant Force. To determine the *magnitude* of $\mathbf{F}_R$, it is necessary to sum each of the differential forces $d\mathbf{F}$ acting over the plate's *entire surface area A*. This sum may be expressed mathematically as an integral:

$$F_R = \Sigma F; \qquad F_R = \int_A p(x, y)\,dA = \int_V dV \qquad (9\text{–}13)$$

Note that $p(x, y)\,dA = dV$, the colored differential *volume element* shown in Fig. 9–25*a*. Therefore, the result indicates that the *magnitude of the resultant force is equal to the total volume under the distributed-loading diagram.*

Location of Resultant Force. The location $(\bar{x}, \bar{y})$ of $\mathbf{F}_R$ is determined by setting the moments of $\mathbf{F}_R$ equal to the moments of all the forces $d\mathbf{F}$ about the respective y and x axes. From Fig. 9–25*a* and 9–25*b*, using Eq. 9–13, we have

$$\bar{x} = \frac{\int_A x p(x, y)\,dA}{\int_A p(x, y)\,dA} = \frac{\int_V x\,dV}{\int_V dV} \qquad \bar{y} = \frac{\int_A y p(x, y)\,dA}{\int_A p(x, y)\,dA} = \frac{\int_V y\,dV}{\int_V dV} \qquad (9\text{–}14)$$

Hence, it can be seen that the *line of action of the resultant force passes through the geometric center or centroid of the volume under the distributed loading diagram.*

*9.6 Fluid Pressure

According to Pascal's law, a fluid at rest creates a pressure p at a point that is the *same* in *all* directions. The magnitude of p, measured as a force per unit area, depends on the specific weight γ or mass density ρ of the fluid and the depth z of the point from the fluid surface.* The relationship can be expressed mathematically as

$$p = \gamma z = \rho g z \qquad (9\text{–}15)$$

where g is the acceleration of gravity. Equation 9–15 is valid only for fluids that are assumed *incompressible,* as in the case of most liquids. Gases are compressible fluids, and since their density changes significantly with both pressure and temperature, Eq. 9–15 cannot be used.

To illustrate how Eq. 9–15 is applied, consider the submerged plate shown in Fig. 9–26. Three points on the plate have been specified. Since points A and B are both at depth z_2 from the liquid surface, the *pressure* at these points has a magnitude $p_2 = \gamma z_2$. Likewise, point C is at depth z_1; hence, $p_1 = \gamma z_1$. In all cases, the pressure acts *normal* to the surface area dA located at the specified point, Fig. 9–26. Using Eq. 9–15 and the results of Sec. 9.5, it is possible to determine the resultant force caused by a liquid pressure distribution, and specify its location on the surface of a submerged plate. Three different shapes of plates will now be considered.

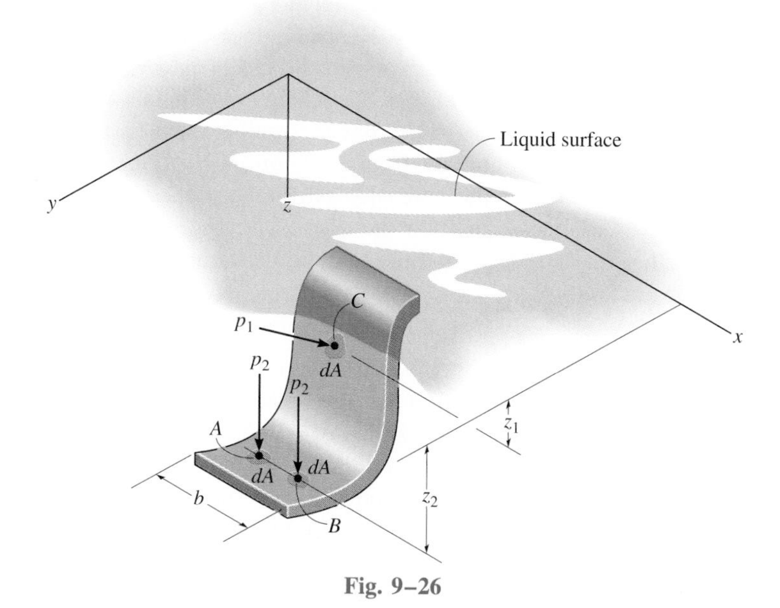

Fig. 9–26

*In particular, for water $\gamma = 62.4\ \text{lb/ft}^3$, or $\gamma = 9810\ \text{N/m}^3$, since $\rho = 1000\ \text{kg/m}^3$ and $g = 9.81\ \text{m/s}^2$.

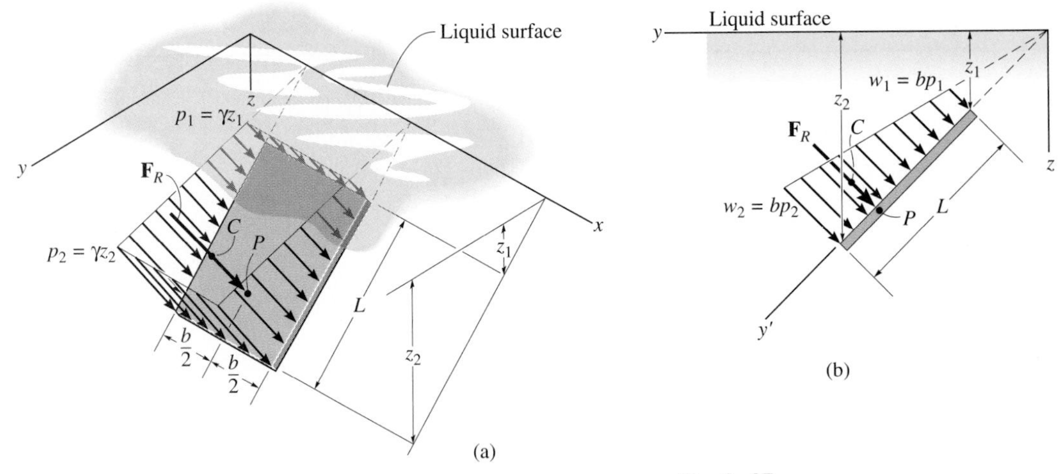

Fig. 9–27

Flat Plate of Constant Width. A flat rectangular plate of constant width, which is submerged in a liquid having a specific weight γ, is shown in Fig. 9–27*a*. The plane of the plate makes an angle with the horizontal, such that its top edge is located at a depth z_1 from the liquid surface and its bottom edge is located at a depth z_2. Since pressure varies linearly with depth, Eq. 9–15, the distribution of pressure over the plate's surface is represented by a trapezoidal volume having an intensity of $p_1 = \gamma z_1$ at depth z_1 and $p_2 = \gamma z_2$ at depth z_2. As noted in Sec. 9.5, the magnitude of the *resultant force* $\mathbf{F}_R$ is equal to the *volume* of this loading diagram and $\mathbf{F}_R$ has a *line of action* that passes through the volume's centroid *C*. Hence $\mathbf{F}_R$ does *not* act at the centroid of the plate; rather, it acts at point *P*, called the *center of pressure.*

Since the plate has a *constant width,* the loading distribution may also be viewed in two dimensions, Fig. 9–27*b*. Here the loading intensity is measured as force/length and varies linearly from $w_1 = bp_1 = b\gamma z_1$ to $w_2 = bp_2 = b\gamma z_2$. The magnitude of $\mathbf{F}_R$ in this case equals the trapezoidal *area,* and $\mathbf{F}_R$ has a *line of action* that passes through the area's *centroid C.* For numerical applications, the area and location of the centroid for a trapezoid are tabulated on the inside back cover.

Curved Plate of Constant Width. When the submerged plate is curved, the pressure acting normal to the plate continually changes direction, and therefore calculation of the magnitude of $\mathbf{F}_R$ and its location *P* is more difficult than for a flat plate. Three- and two-dimensional views of the loading distribution are shown in Figs. 9–28*a* and 9–28*b*, respectively. Here integration can be used to determine both F_R and the location of the centroid *C* or center of pressure *P*.

A simpler method exists, however, for calculating the magnitude of $\mathbf{F}_R$ and its location along a curved (or flat) plate having a *constant width.* This

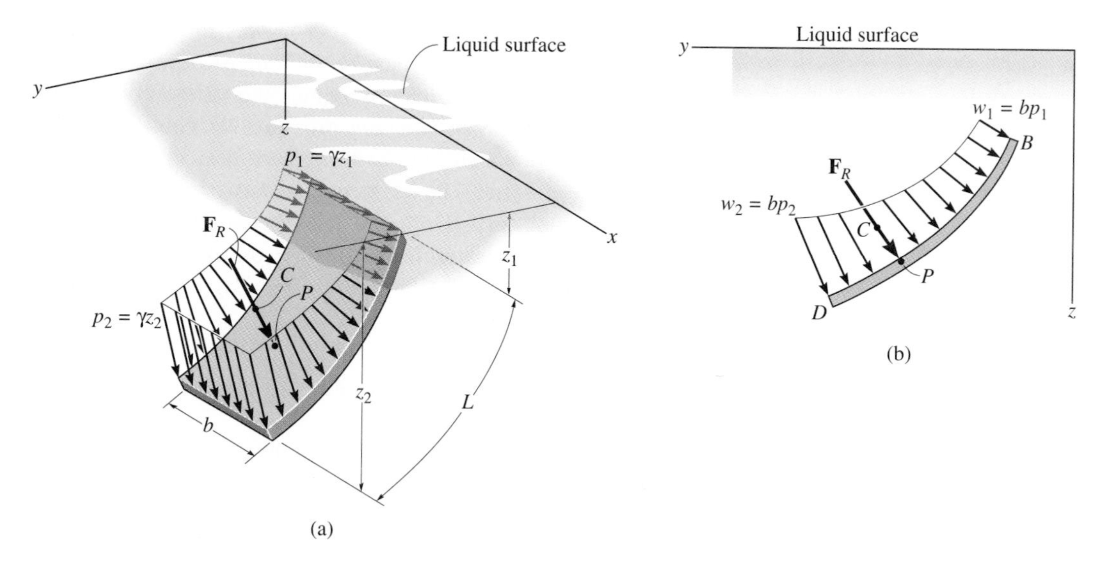

Fig. 9–28

method requires separate calculations for the horizontal and vertical *components* of $\mathbf{F}_R$. For example, the distributed loading acting on the curved plate DB in Fig. 9–28*b* can be represented by the *equivalent loading* shown in Fig. 9–29. Here the plate supports the weight of liquid W_f contained within the block BDA. This force has a magnitude $W_f = (\gamma b)(\text{area}_{BDA})$ and acts through the centroid of BDA. In addition, there are the pressure distributions caused by the liquid acting along the vertical and horizontal sides of the block. Along the vertical side AD, the force $\mathbf{F}_{AD}$ has a magnitude that equals the area under the trapezoid and acts through the centroid C_{AD} of this area. The distributed loading along the horizontal side AB is constant, since all points lying in this plane are at the same depth from the surface of the liquid. The magnitude of $\mathbf{F}_{AB}$ is simply the area of the rectangle. This force acts through the area's centroid C_{AB} or the midpoint of AB. Summing the three forces in Fig. 9–29 yields $\mathbf{F}_R = \Sigma\mathbf{F} = \mathbf{F}_{AD} + \mathbf{F}_{AB} + \mathbf{W}_f$, which is shown in Fig. 9–28. Finally, the location of the center of pressure P on the plate is determined by applying the equation $M_{R_O} = \Sigma M_O$, which states that the moment of the resultant force about a convenient reference point, Fig. 9–28, is equal to the sum of the moments of the three forces in Fig. 9–29 about the same point.

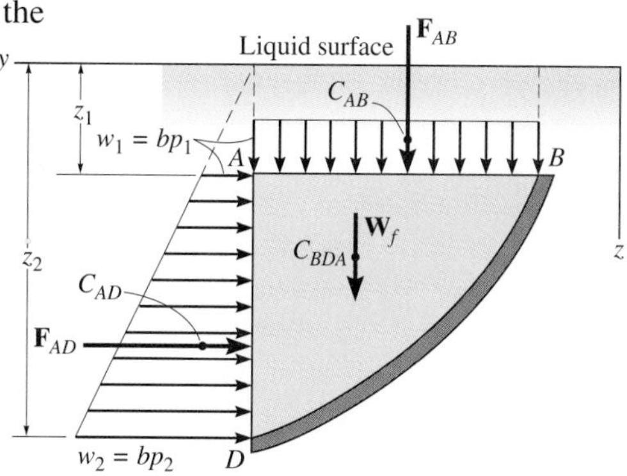

Fig. 9–29

Flat Plate of Variable Width. The pressure distribution acting on the surface of a submerged plate having a variable width is shown in Fig. 9–30. The resultant force of this loading equals the volume described by the plate area as its base and linear varying pressure distribution as its altitude. The shaded element shown in Fig. 9–30 may be used if integration is chosen to determine this volume. The element consists of a rectangular strip of area $dA = x\,dy'$ located at a depth z below the liquid surface. Since a uniform pressure $p = \gamma z$ (force/area) acts on dA, the magnitude of the differential force $d\mathbf{F}$ is equal to $dF = dV = p\,dA = \gamma z(x\,dy')$. Integrating over the entire volume yields Eq. 9–13; i.e.,

$$F_R = \int_A p\,dA = \int_V dV = V$$

From Eq. 9–14, the centroid of V defines the point through which $\mathbf{F}_R$ acts. The center of pressure, which lies on the surface of the plate just below C, has coordinates $P(\bar{x}, \bar{y}')$ defined by the equations

$$\bar{x} = \frac{\int_V \tilde{x}\,dV}{\int_V dV} \qquad \bar{y}' = \frac{\int_V \tilde{y}'\,dV}{\int_V dV}$$

This point should *not* be mistaken for the centroid of the plate's *area.*

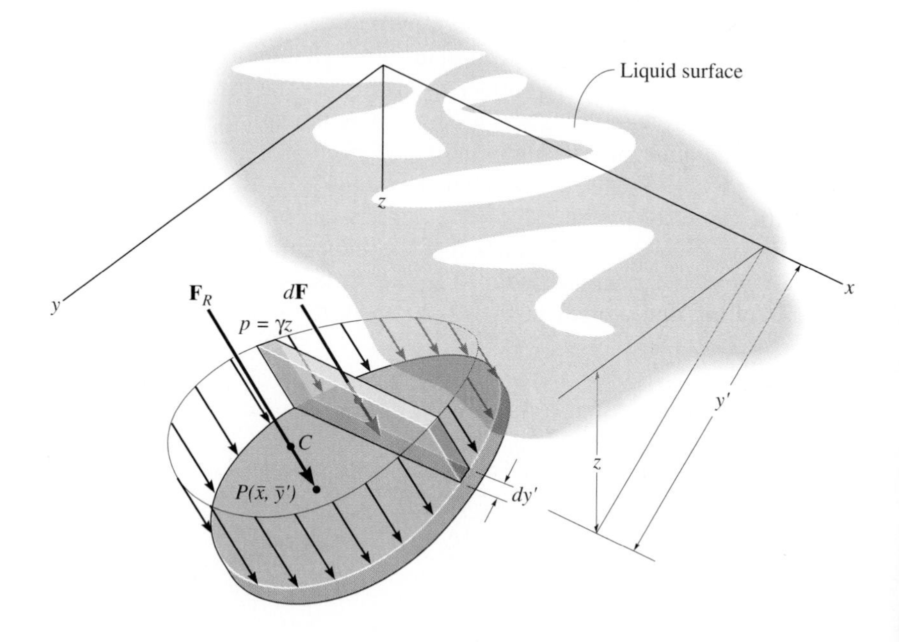

Fig. 9–30

Example 9–13

Determine the magnitude and location of the resultant hydrostatic force acting on the submerged rectangular plate AB shown in Fig. 9–31a. The plate has a width of 1.5 m; $\rho_w = 1000\ \text{kg/m}^3$.

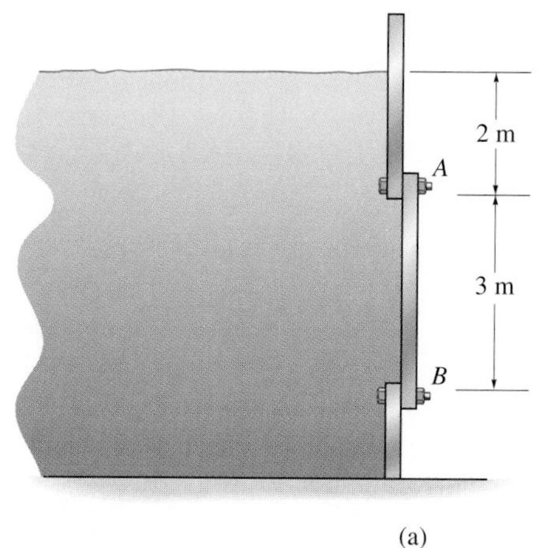

(a)

SOLUTION

The water pressures at depths A and B are

$$p_A = \rho_w g z_A = (1000\ \text{kg/m}^3)(9.81\ \text{m/s}^2)(2\ \text{m}) = 19.62\ \text{kPa}$$
$$p_B = \rho_w g z_B = (1000\ \text{kg/m}^3)(9.81\ \text{m/s}^2)(5\ \text{m}) = 49.05\ \text{kPa}$$

Since the plate has a constant width, the distributed loading can be viewed in two dimensions as shown in Fig. 9–31b. The intensity of the load at A and B is

$$w_A = bp_A = (1.5\ \text{m})(19.62\ \text{kPa}) = 29.4\ \text{kN/m}$$
$$w_B = bp_B = (1.5\ \text{m})(49.05\ \text{kPa}) = 73.6\ \text{kN/m}$$

From the table on the inside back cover, the magnitude of the resultant force $\mathbf{F}_R$ created by the distributed load is

$$F_R = \text{area of trapezoid}$$
$$= \tfrac{1}{2}(3)(29.4 + 73.6) = 154.5\ \text{kN} \qquad \textit{Ans.}$$

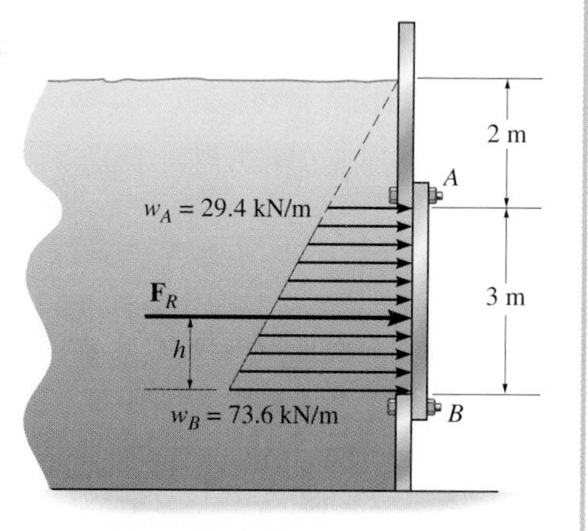

(b)

This force acts through the centroid of the area,

$$h = \frac{1}{3}\left(\frac{2(29.4) + 73.6}{29.4 + 73.6}\right)(3) = 1.29\ \text{m} \qquad \textit{Ans.}$$

measured upward from B, Fig. 9–31b.

The same results can be obtained by considering two components of $\mathbf{F}_R$ defined by the triangle and rectangle shown in Fig. 9–31c. Each force acts through its associated centroid and has a magnitude of

$$F_{Re} = (29.4\ \text{kN/m})(3\ \text{m}) = 88.2\ \text{kN}$$
$$F_t = \tfrac{1}{2}(44.2\ \text{kN/m})(3\ \text{m}) = 66.3\ \text{kN}$$

Hence,

$$F_R = F_{Re} + F_t = 88.2 + 66.3 = 154.5\ \text{kN} \qquad \textit{Ans.}$$

The location of $\mathbf{F}_R$ is determined by summing moments about B, Fig. 9–31b and c, i.e.,

$$\curvearrowleft+(M_R)_B = \Sigma M_B; \quad (154.5)h = 88.2(1.5) + 66.3(1)$$
$$h = 1.29\ \text{m} \qquad \textit{Ans.}$$

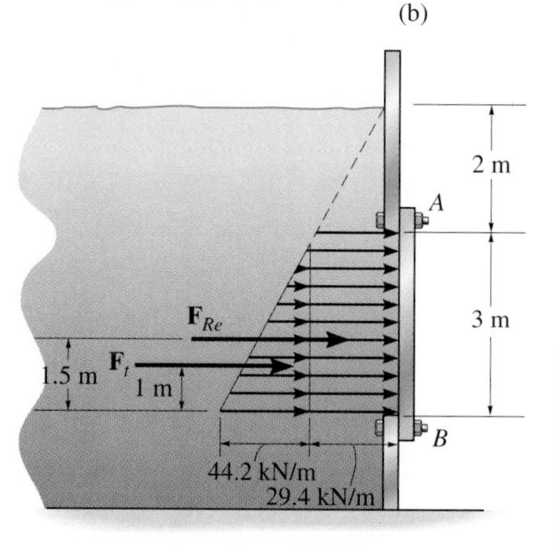

Fig. 9–31 (c)

Example 9–14

Determine the magnitude of the resultant hydrostatic force acting on the surface of a seawall shaped in the form of a parabola as shown in Fig. 9–32*a*. The wall is 5 m long; $\rho_w = 1020\ \text{kg/m}^3$.

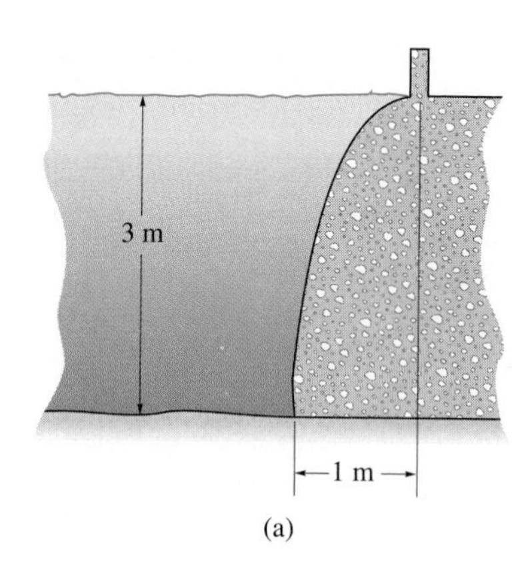

(a)

SOLUTION

The horizontal and vertical components of the resultant force will be calculated, Fig. 9–32*b*. Since

$$p_B = \rho_w g z_B = (1020\ \text{kg/m}^3)(9.81\ \text{m/s}^2)(3\ \text{m}) = 30.02\ \text{kPa}$$

then

$$w_B = b p_B = 5\ \text{m}(30.02\ \text{kPa}) = 150.1\ \text{kN/m}$$

Thus,

$$F_x = \tfrac{1}{2}(3\ \text{m})(150.1\ \text{kN/m}) = 225.2\ \text{kN}$$

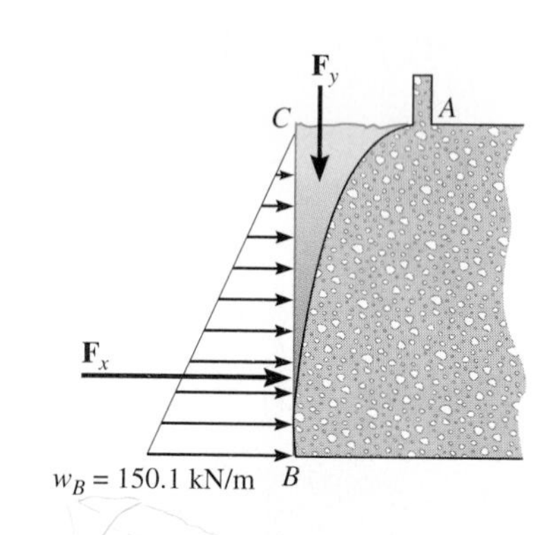

(b)

Fig. 9–32

The area of the parabolic sector *ABC* can be determined using the table on the inside back cover. Hence, the weight of water within this region is

$$F_y = (\rho_w g b)(\text{area}_{ABC})$$
$$= (1020\ \text{kg/m}^3)(9.81\ \text{m/s}^2)(5\ \text{m})[\tfrac{1}{3}(1\ \text{m})(3\ \text{m})] = 50.0\ \text{kN}$$

The resultant force is therefore

$$F_R = \sqrt{F_x^2 + F_y^2} = \sqrt{(225.2)^2 + (50.0)^2}$$
$$= 231\ \text{kN}$$

Ans.

Example 9–15

Determine the magnitude and location of the resultant force acting on the triangular end plates of the water trough shown in Fig. 9–33*a*; $\rho_w = 1000\ \text{kg/m}^3$.

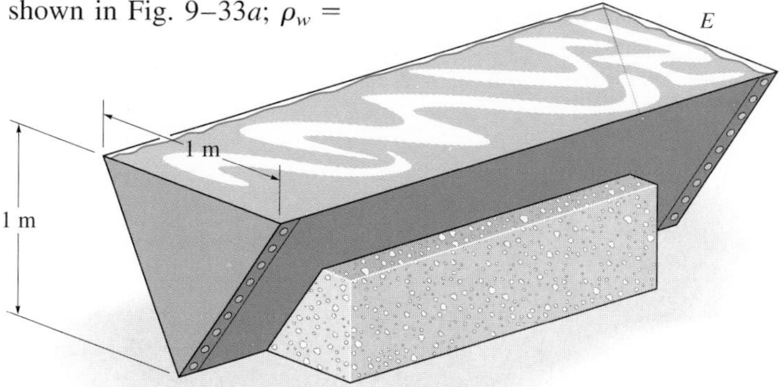

(a)

SOLUTION

The pressure distribution acting on the end plate *E* is shown in Fig. 9–33*b*. The magnitude of the resultant force **F** is equal to the volume of this loading distribution. We will solve the problem by integration. Choosing the differential volume element shown in the figure, we have

$$dF = dV = p\,dA = \rho_w g z(2x\,dz) = 19\,620zx\,dz$$

The equation of line *AB* is

$$x = 0.5(1 - z)$$

Hence, substituting and integrating with respect to z from $z = 0$ to $z = 1$ m yields

$$F = V = \int_V dV = \int_0^1 (19\,620)z[0.5(1 - z)]\,dz$$

$$= 9810 \int_0^1 (z - z^2)\,dz = 1635\ \text{N} = 1.64\ \text{kN} \qquad \textbf{Ans.}$$

This resultant passes through the *centroid of the volume*. Because of symmetry,

$$\bar{x} = 0 \qquad \textbf{Ans.}$$

Since $\tilde{z} = z$ for the volume element in Fig. 9–33*b*, then

$$\bar{z} = \frac{\int_V \tilde{z}\,dV}{\int_V dV} = \frac{\int_0^1 z(19\,620)z[0.5(1 - z)]\,dz}{1635} = \frac{9810\int_0^1 (z^2 - z^3)\,dz}{1635}$$

$$= 0.5\ \text{m} \qquad \textbf{Ans.}$$

(b)

Fig. 9–33

PROBLEMS

9–102. Determine the magnitude of the resultant hydrostatic force acting on the dam and its location, measured from the top surface of the water. The width of the dam is 8 m; $\rho_w = 1.0\ \text{Mg/m}^3$.

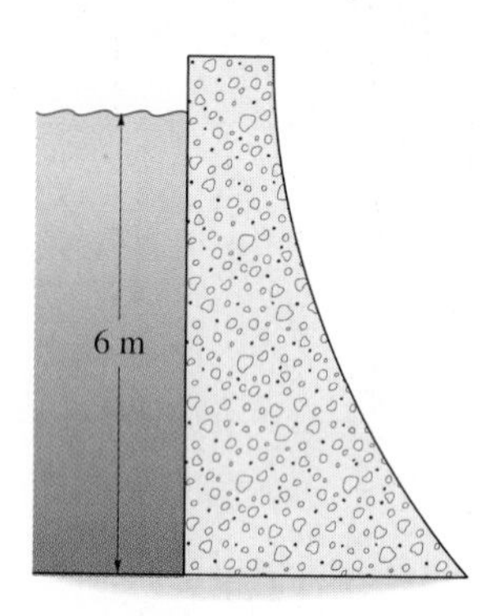

Prob. 9–102

9–103. The tank is filled with water to a depth of $d = 4$ m. Determine the resultant force the water exerts on side A and side B of the tank. If oil instead of water is placed in the tank, to what depth d should it reach so that it creates the same resultant forces? $\rho_o = 900\ \text{kg/m}^3$ and $\rho_w = 1000\ \text{kg/m}^3$.

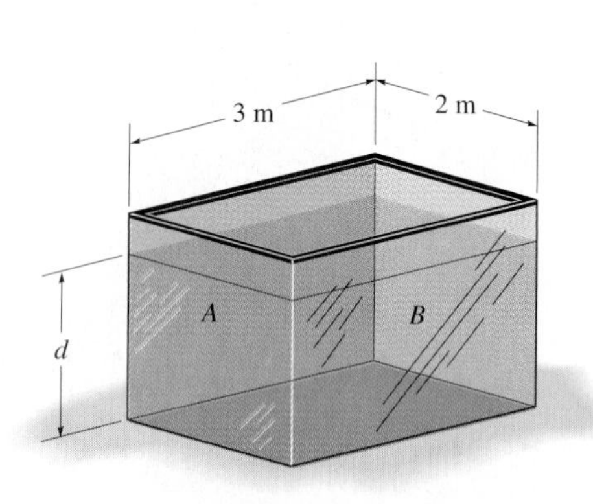

Prob. 9–103

***9–104.** The concrete dam is designed so that its face AB has a gradual slope into the water as shown. Because of this, the frictional force at the base BD of the dam is increased due to the hydrostatic force of the water acting on the dam. Calculate the hydrostatic force acting on the face AB of the dam. The dam is 60 ft wide. $\gamma_w = 62.4\ \text{lb/ft}^3$.

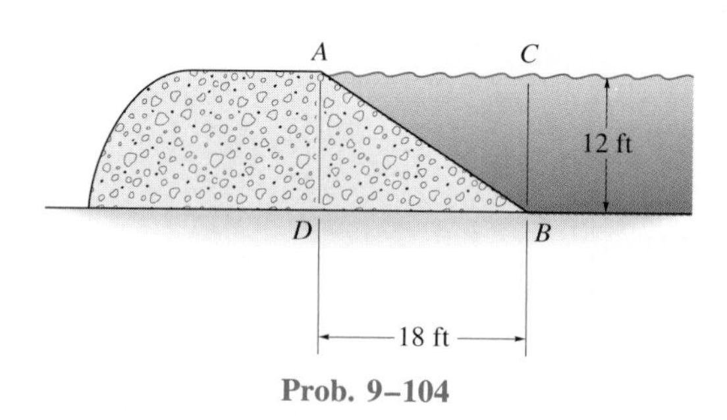

Prob. 9–104

9–105. The storage tank contains oil having a specific weight of $\gamma_o = 56\ \text{lb/ft}^3$. If the tank is 6 ft wide, calculate the resultant force acting on the inclined side BC of the tank, caused by the oil, and specify its location along BC, measured from B. Also compute the total resultant force acting on the bottom of the tank.

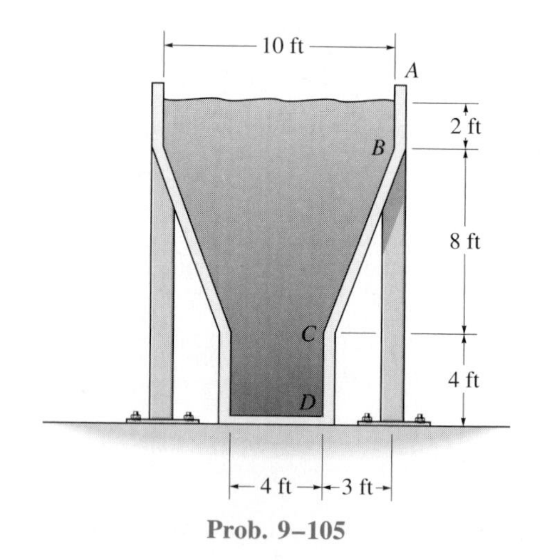

Prob. 9–105

9–106. The semicircular tunnel passes under a river which is 9 m deep. Determine the vertical resultant hydrostatic force acting per meter of length along the length of the tunnel. The tunnel is 6 m wide; $\rho_w = 1.0 \text{ Mg/m}^3$.

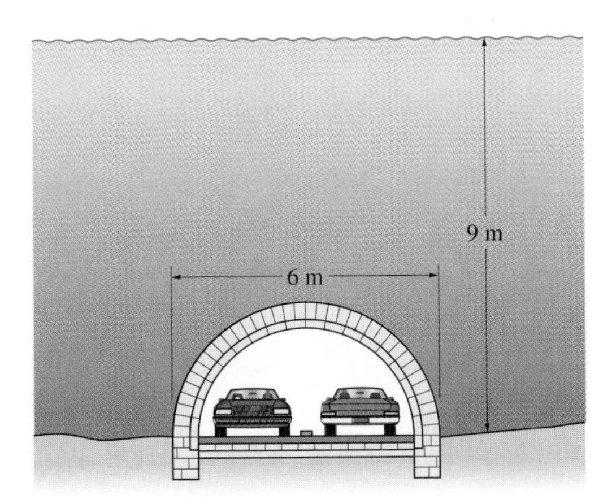

Prob. 9–106

9–107. The storage tank contains oil having a density of $\rho_o = 0.90 \text{ Mg/m}^3$. If the tank is 1.5 m wide, calculate the resultant force acting on the inclined side AB of the tank, caused by the oil, and specify its location along AB, measured from A.

***9–108.** The storage tank contains oil having a density of $\rho_o = 0.90 \text{ Mg/m}^3$. If the tank is 1.5 m wide, calculate the resultant force acting on side BC of the tank, caused by the oil, and specify its location along BC, measured from C.

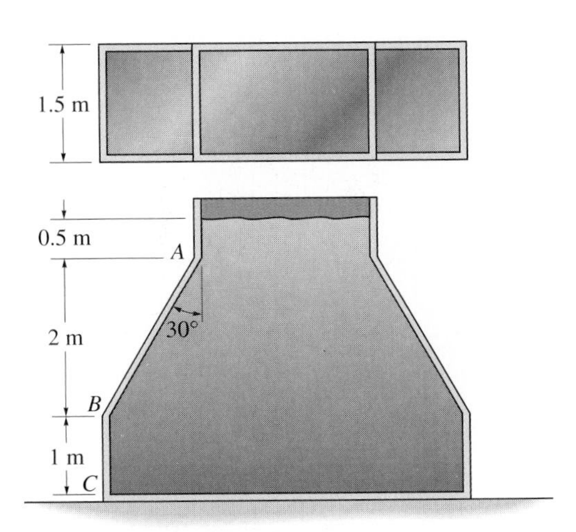

Probs. 9–107/9–108

9–109. The gate AB is 8 m wide. Determine the horizontal and vertical components of force acting on the pin at B and the vertical reaction at the smooth support A. $\rho_w = 1.0 \text{ Mg/m}^3$.

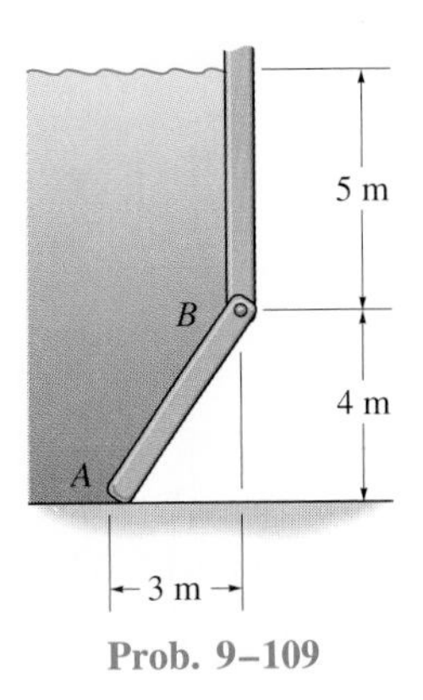

Prob. 9–109

9–110. The structure shown is used for temporary storage of oil at sea for later loading into ships. When it is empty the water level is at A (sea level). As oil is loaded, the water is displaced through exit ports at D. If the riser EC is filled with oil, i.e., to a depth of C, determine the height h to B of the oil level above sea level. $\rho_o = 900 \text{ kg/m}^3$ and $\rho_w = 1020 \text{ kg/m}^3$.

9–111. If the structure in Prob. 9–110 is totally filled with oil, i.e., until it reaches a depth of 58 m below sea level, how high h will the oil level extend above sea level?

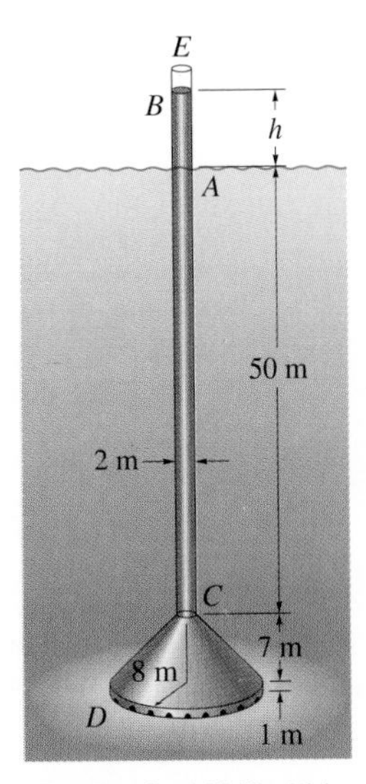

Probs. 9–110/9–111

***9–112.** The arched surface *AB* is shaped in the form of a quarter circle. If it is 8 m long, determine the horizontal and vertical components of the resultant force caused by the water acting on the surface. $\rho_w = 1.0\ \text{Mg/m}^3$.

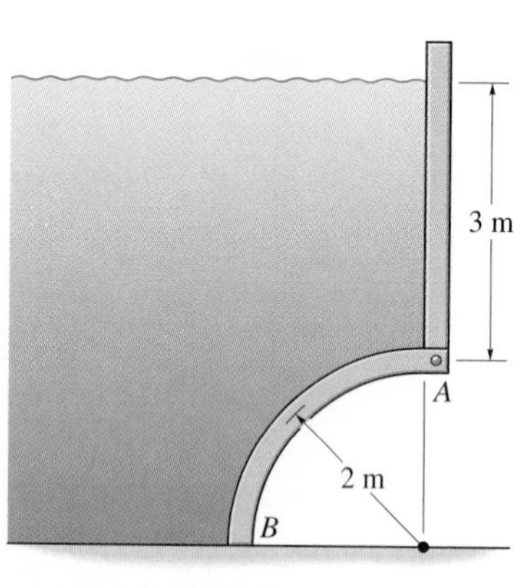

Prob. 9–112

9–113. Determine the magnitude and location of the resultant hydrostatic force acting on each of the cover plates *A* and *B*. $\rho_w = 1.0\ \text{Mg/m}^3$.

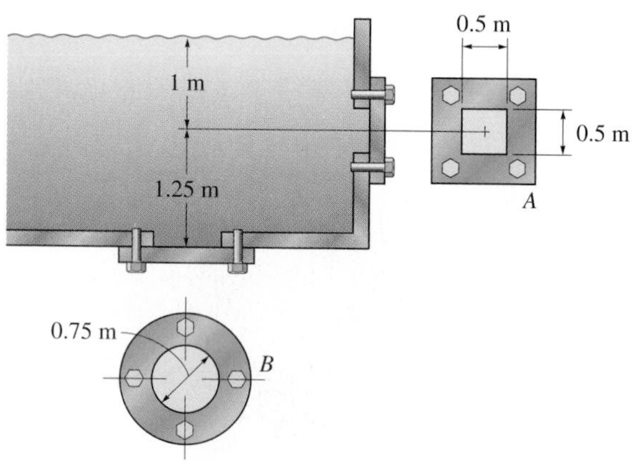

Prob. 9–113

9–114. Determine the magnitude of the resultant hydrostatic force acting per meter of length on the sea wall. $\rho_w = 1.0\ \text{Mg/m}^3$.

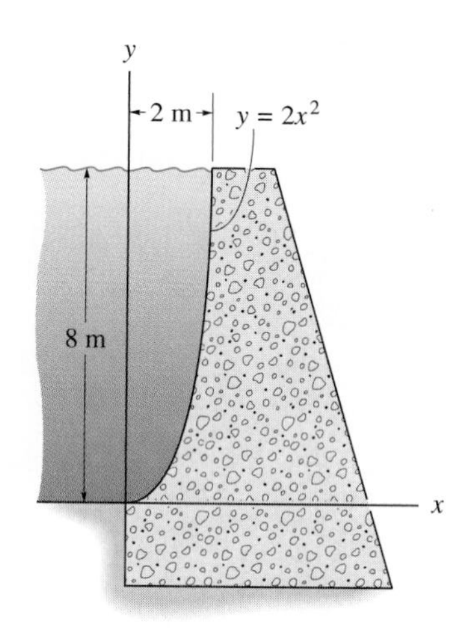

Prob. 9–114

9–115. Determine the magnitude of the resultant hydrostatic force acting per meter of length on the sea wall; $\gamma_w = 62.4\ \text{lb/ft}^3$.

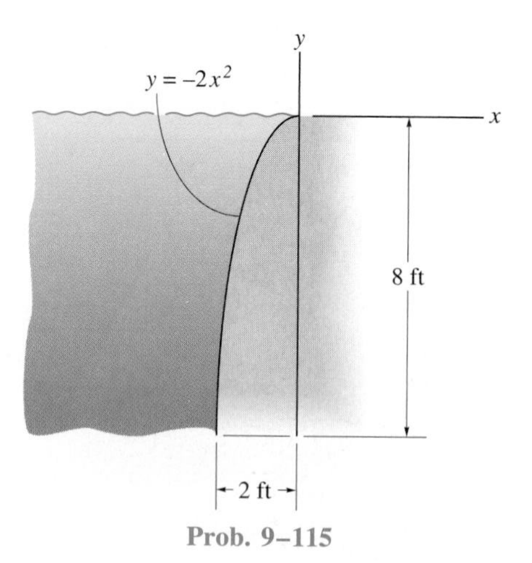

Prob. 9–115

***9–116.** The rear end of the boat has a trapezoidal form with the dimensions shown. If the water on the outside of the boat is 3 in. from the top, determine the resultant force and its location, measured from the top of the boat. $\gamma_w = 62.4\ \text{lb/ft}^3$.

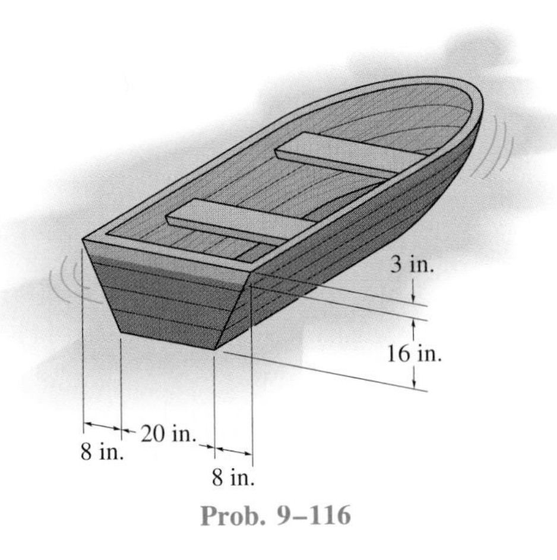

Prob. 9–116

9–117. The tank is filled to the top ($y = 0.5$ m) with water having a density of $\rho_w = 1.0\ \text{Mg/m}^3$. Determine the resultant force of the water pressure acting on the flat end plate C of the tank, and its location, measured from the top of the tank.

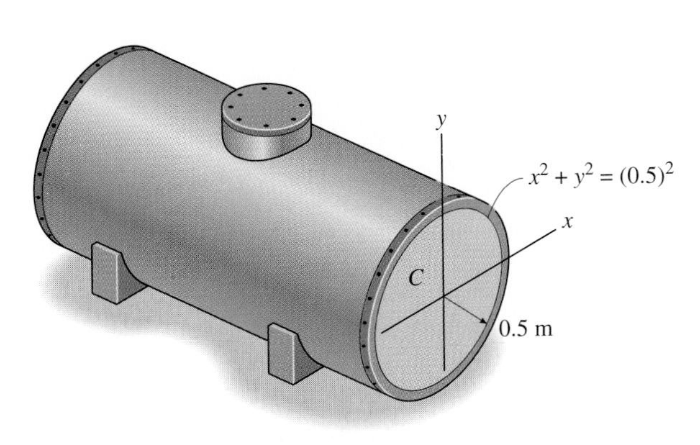

Prob. 9–117

9–118. The end plates of the trough are in the form of a semiellipse. If it is filled to the top with water, determine the resultant force the water exerts on these plates and the location of the center of pressure, measured from the top of the trough. $\gamma_w = 62.4\ \text{lb/ft}^3$.

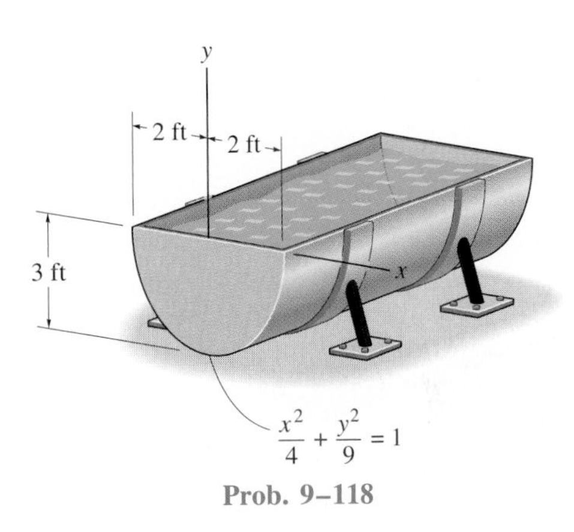

Prob. 9–118

9–119. The wind blows uniformly on the front surface of the metal building with a pressure of $30\ \text{lb/ft}^2$. Determine the resultant force it exerts on the surface and the position of this resultant.

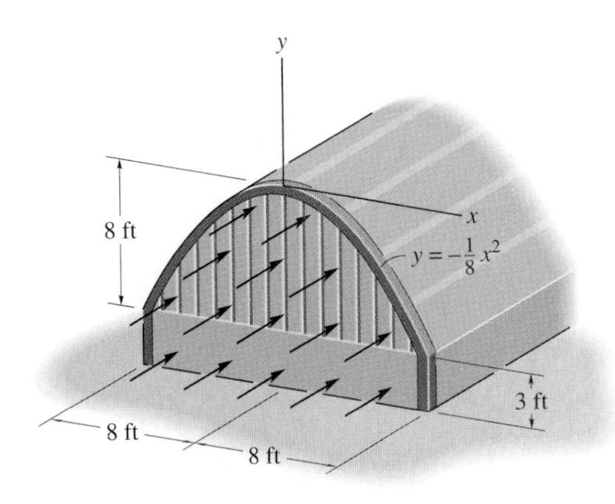

Prob. 9–119

*9–120. The pressure loading on the plate is described by the function $p = [-240/(x + 1) + 340]$ Pa. Determine the magnitude of the resultant force and the coordinates $(\bar{x}, \bar{y})$ of the point where the line of action of the force intersects the plate.

9–122. The loading acting on a square plate is represented by a parabolic pressure distribution. Determine the magnitude of the resultant force and the coordinates $(\bar{x}, \bar{y})$ of the point where the line of action of the force intersects the plate. Also, what are the reactions at the rollers B and C and the ball-and-socket joint A? Neglect the weight of the plate.

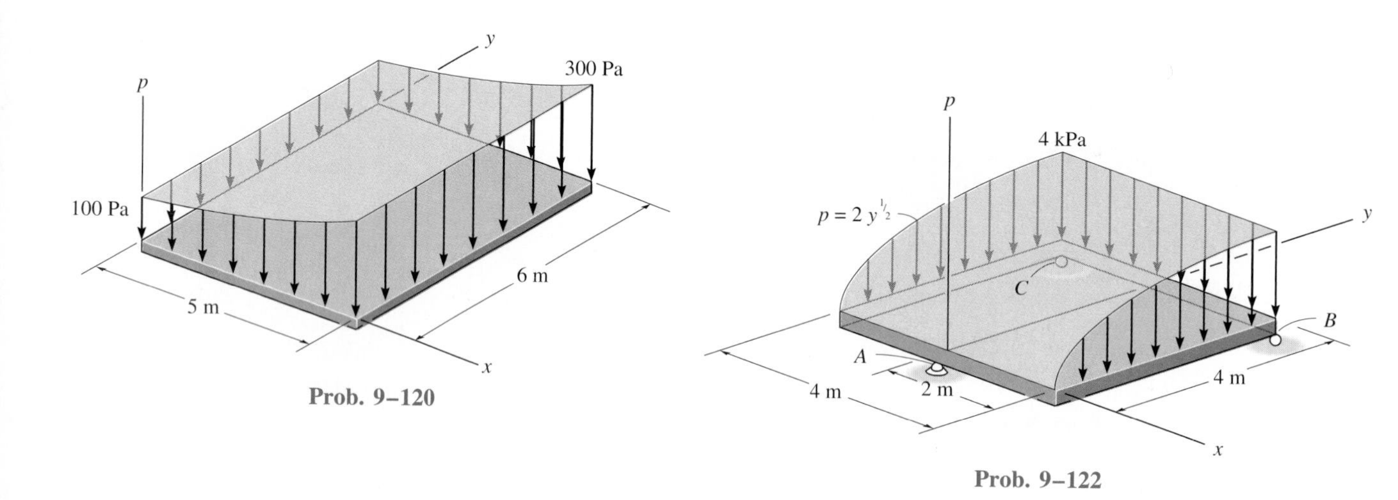

Prob. 9–120

Prob. 9–122

9-121. The pressure loading on the plate varies uniformly along each of its edges. Determine the magnitude of the resultant force and the coordinates $(\bar{x}, \bar{y})$ of the point where the line of action of the force intersects the plate. *Hint:* The equation defining the boundary of the load has the form $p = ax + by + c$, where the constants a, b, and c have to be determined.

9–123. The load over the plate varies linearly along the sides of the plate such that $p = \frac{2}{3}[x(4 - y)]$ kPa. Determine the magnitude of the resultant force and the coordinates $(\bar{x}, \bar{y})$ of the point where the line of action of the force intersects the plate.

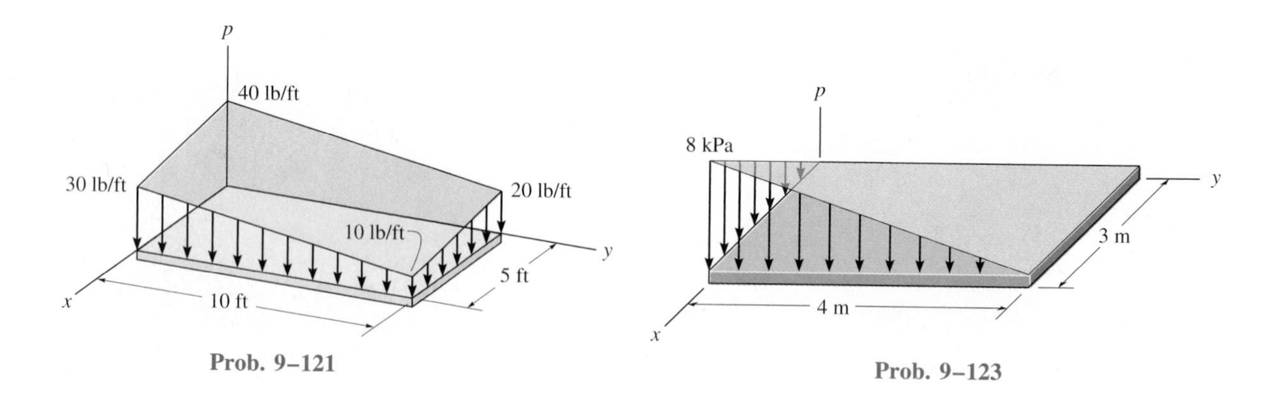

Prob. 9–121

Prob. 9–123

REVIEW PROBLEMS

*9–124. A circular V-belt has an inner radius of 600 mm and a cross-sectional area as shown. Determine the volume of material required to make the belt.

9–125. A circular V-belt has an inner radius of 600 mm and a cross-sectional area as shown. Determine the surface area of the belt.

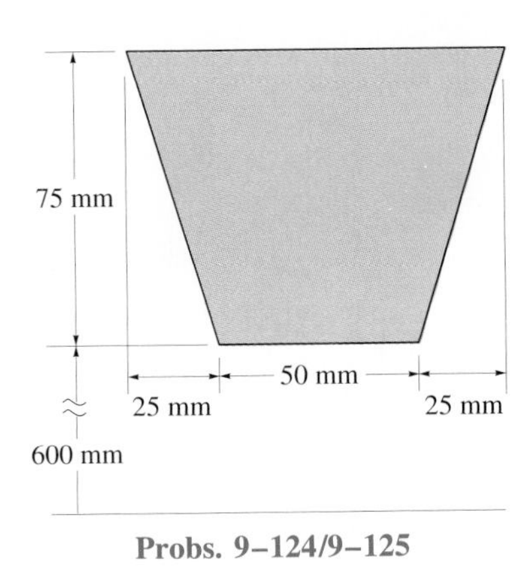

Probs. 9–124/9–125

9–126. The starter for an electric motor has the cross-sectional area shown. If copper wiring has a density of $\rho_{cu} = 8.90\ \text{Mg/m}^3$ and the steel frame has a density of $\rho_{st} = 7.80\ \text{Mg/m}^3$, estimate the total mass of the starter. Neglect the voids within the copper wiring.

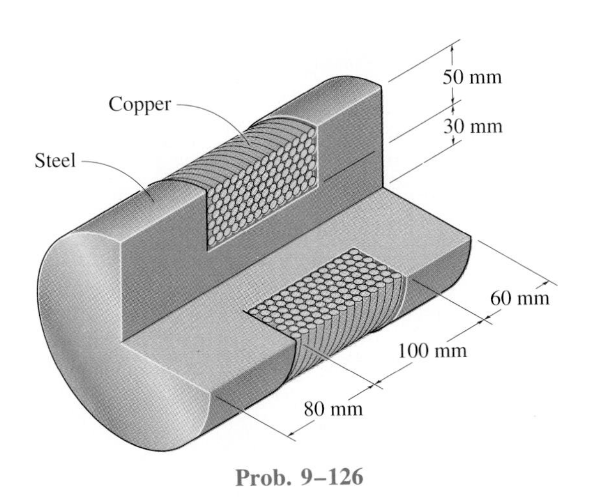

Prob. 9–126

9–127. Determine the distance $\bar{y}$ of the centroid of the cross-sectional area.

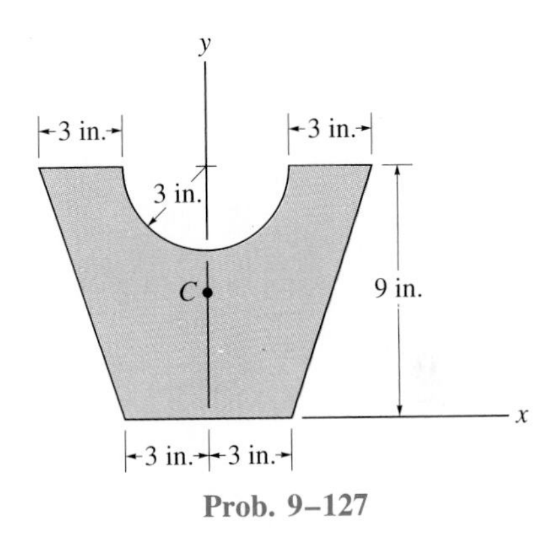

Prob. 9–127

*9–128. Locate the centroid for the cold-formed metal strut having the cross section shown. Neglect the thickness of the material and slight bends at the corners.

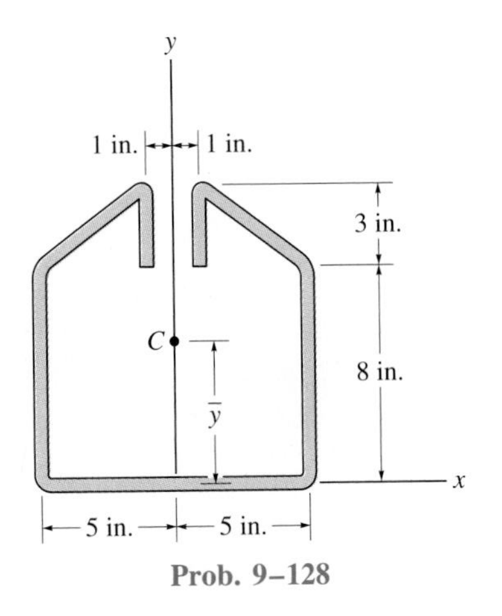

Prob. 9–128

9–129. Locate the center of gravity of the homogeneous rod. The rod has a weight of 2 lb/ft. Also, compute the x, y, z components of reaction at the fixed support A.

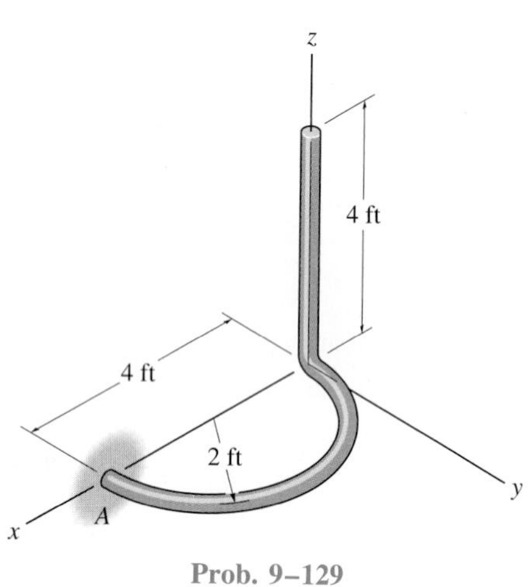

Prob. 9–129

9–130. The thin-walled channel and stiffener have the cross section shown. If the material has a constant thickness, determine the location $\bar{y}$ of its centroid. The dimensions are indicated to the center of each segment.

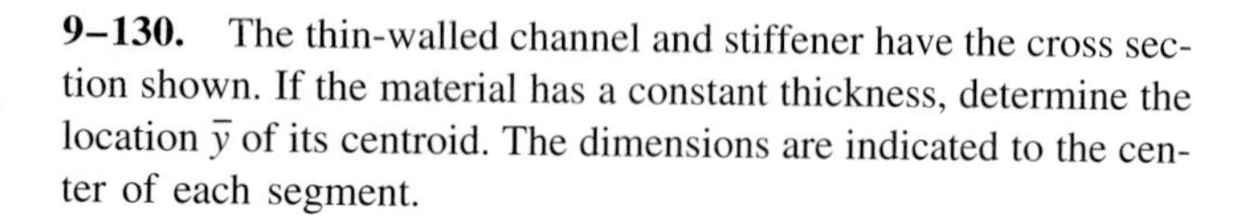

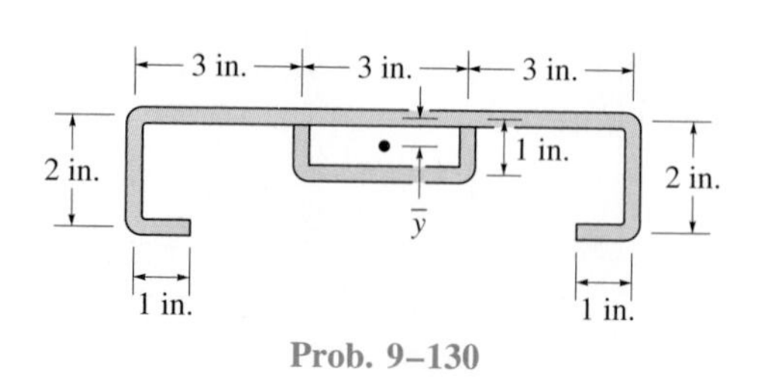

Prob. 9–130

9–131. Determine the volume of steel needed to produce the tapered part. The cross section is shown, although the part is 360° around. Also, compute the outside surface area of the part, excluding its ends.

Prob. 9–131

***9–132.** The rectangular bin is filled with coal, which creates a pressure distribution along wall A that varies as shown, i.e., $p = 4z^3$ lb/ft^2, where z is in feet. Compute the resultant force created by the coal, and its location, measured from the top surface of the coal.

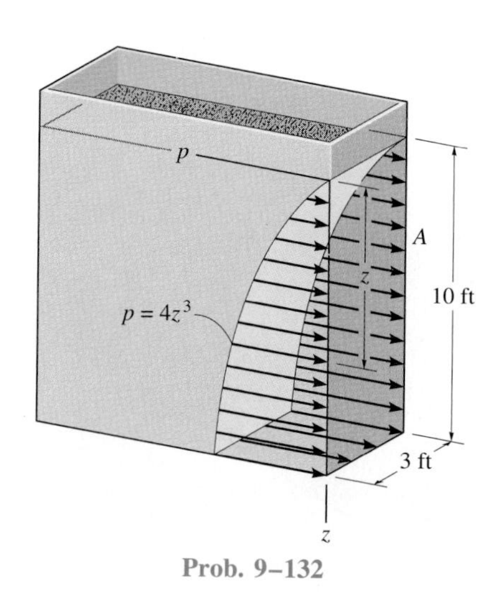

Prob. 9–132

9–133. Determine the location $(\bar{x}, \bar{y})$ of the center of mass for the compressor assembly. The locations of the centers of mass of the various components and their masses are indicated and tabulated in the figure. What are the vertical reactions at blocks A and B needed to support the platform?

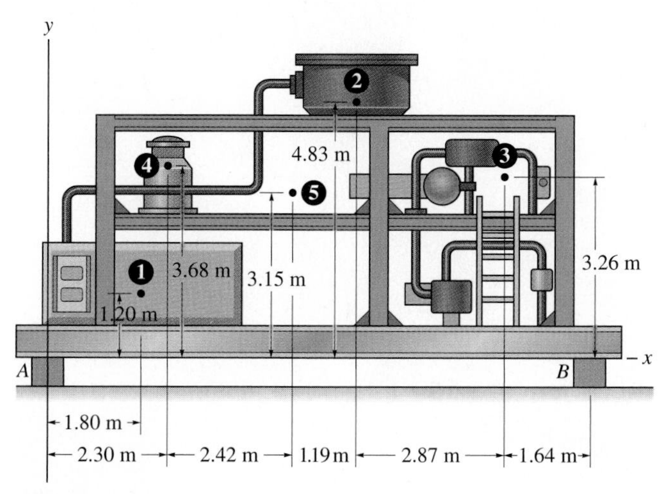

1	Instrument panel	230 kg
2	Filter system	183 kg
3	Piping assembly	120 kg
4	Liquid storage	85 kg
5	Structural framework	468 kg

Prob. 9–133

9–134. Locate the centroid of the channel's cross-sectional area.

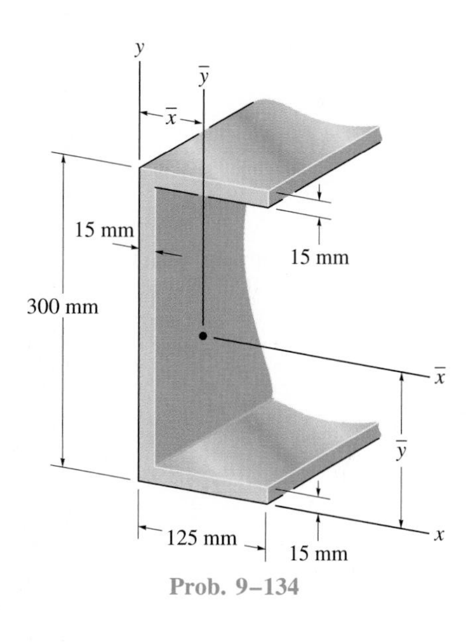

Prob. 9–134

The design of a structural member, such as a beam or column, requires calculation of its cross-sectional moment of inertia. In this chapter we will discuss how this is done.

10

Moments of Inertia

In this chapter we will develop a method for determining the moment of inertia for both an area and a body having a specified mass. The moment of inertia for an area is an important property in engineering, since it must be determined or specified if one is to analyze or design a structural member or a mechanical part. On the other hand, a body's mass moment of inertia must be known if one is studying the motion of the body.

10.1 Definition of Moments of Inertia for Areas

In the last chapter we determined the centroid for an area by considering the first moment of the area about an axis; that is, for the computation we had to evaluate an integral of the form $\int x\,dA$. Integrals of the second moment of an area, such as $\int x^2\,dA$, are referred to as the *moment of inertia* for the area. The terminology ''moment of inertia'' as used here is actually a misnomer; however, it has been adopted because of the similarity with integrals of the same form related to mass.

The moment of inertia of an area originates whenever one has to compute the moment of a distributed load that varies linearly from the moment axis. A typical example of this kind of loading occurs due to the pressure of a liquid acting on the surface of a submerged plate. It was pointed out in Sec. 9.6 that the pressure, or force per unit area, exerted at a point located a distance z below the surface of a liquid is $p = \gamma z$, Eq. 9–15, where γ is the specific weight of the liquid. Thus, the magnitude of force exerted by a liquid on the

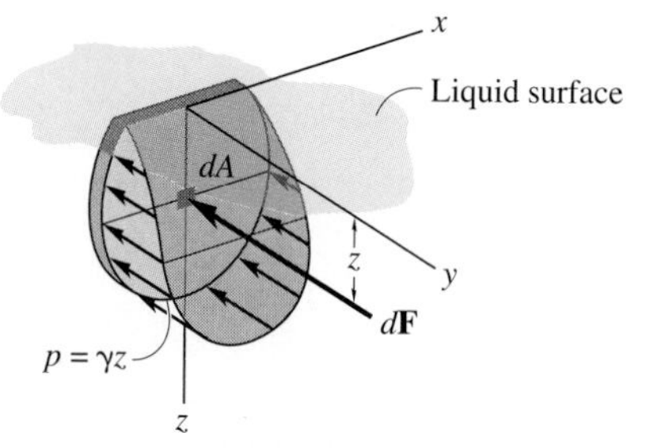

Fig. 10–1

area dA of the submerged plate shown in Fig. 10–1 is $dF = p\,dA = \gamma z\,dA$. The moment of this force about the x axis of the plate is $dM = z\,dF = \gamma z^2\,dA$, and therefore the moment created by the *entire* pressure distribution is $M = \gamma \int z^2\,dA$. Here the integral represents the moment of inertia of the area of the plate about the x axis. Since integrals of this form often arise in formulas used in fluid mechanics, mechanics of materials, structural mechanics, and machine design, the engineer should become familiar with the methods used for their computation.

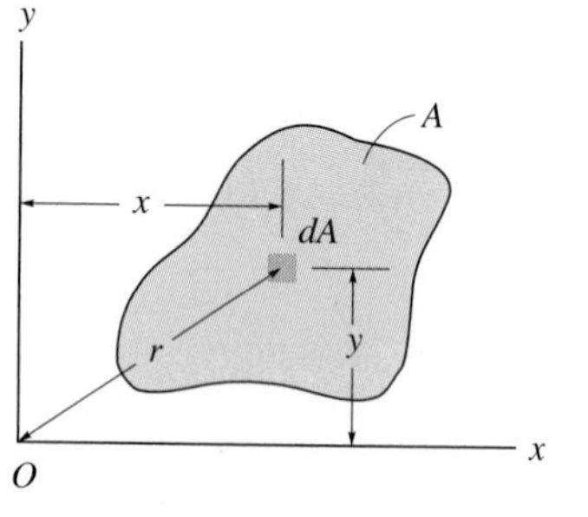

Fig. 10–2

Moment of Inertia. Consider the area A, shown in Fig. 10–2, which lies in the x–y plane. By definition, the moments of inertia of the differential planar area dA about the x and y axes are $dI_x = y^2\,dA$ and $dI_y = x^2\,dA$, respectively. For the entire area the *moments of inertia* are determined by integration; i.e.,

$$I_x = \int_A y^2\,dA$$
$$I_y = \int_A x^2\,dA \qquad (10\text{–}1)$$

We can also formulate the second moment of the differential area dA about the pole O or z axis, Fig. 10–2. This is referred to as the polar moment of inertia, $dJ_O = r^2\,dA$. Here r is the perpendicular distance from the pole (z axis) to the element dA. For the entire area the *polar moment of inertia* is

$$J_O = \int_A r^2\,dA = I_x + I_y \qquad (10\text{–}2)$$

The relationship between J_O and I_x, I_y is possible since $r^2 = x^2 + y^2$, Fig. 10–2.

From the above formulations it is seen that I_x, I_y, and J_O will *always* be *positive,* since they involve the product of distance squared and area. Furthermore, the units for moment of inertia involve length raised to the fourth power, e.g., m^4, mm^4, or ft^4, in^4.

10.2 Parallel-Axis Theorem for an Area

If the moment of inertia for an area is known about an axis passing through its centroid, it is convenient to determine the moment of inertia of the area about a corresponding parallel axis using the *parallel-axis theorem.* To derive this theorem, consider finding the moment of inertia of the shaded area shown in Fig. 10–3 about the x axis. In this case, a differential element dA is located at an arbitrary distance y' from the *centroidal* x' axis, whereas the *fixed distance* between the parallel x and x' axes is defined as d_y. Since the moment of inertia of dA about the x axis is $dI_x = (y' + d_y)^2\, dA$, then for the entire area,

$$I_x = \int_A (y' + d_y)^2\, dA$$

$$= \int_A y'^2\, dA + 2d_y \int_A y'\, dA + d_y^2 \int_A dA$$

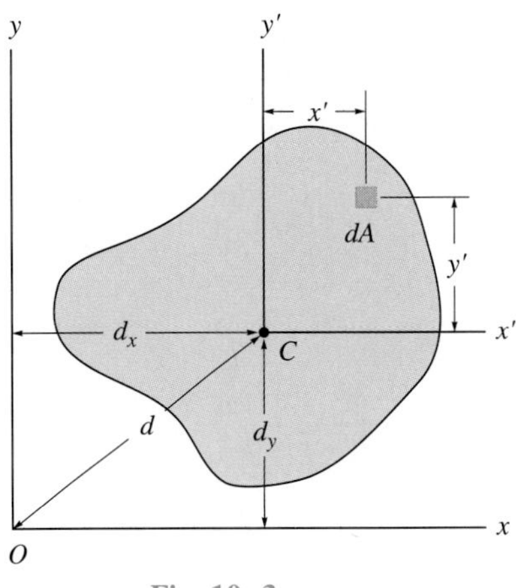

Fig. 10–3

The first integral represents the moment of inertia of the area about the centroidal axis, $\bar{I}_{x'}$. The second integral is zero since the x' axis passes through the area's centroid C; i.e., $\int y'\, dA = \bar{y} \int dA = 0$ since $\bar{y} = 0$. Realizing that the third integral represents the total area A, the final result is therefore

$$I_x = \bar{I}_{x'} + Ad_y^2 \qquad (10\text{–}3)$$

A similar expression can be written for I_y; i.e.,

$$I_y = \bar{I}_{y'} + Ad_x^2 \qquad (10\text{–}4)$$

And finally, for the polar moment of inertia about an axis perpendicular to the x–y plane and passing through the pole O (z axis), Fig. 10–3, we have

$$J_O = \bar{J}_C + Ad^2 \qquad (10\text{–}5)$$

The form of each of these equations states that *the moment of inertia of an area about an axis is equal to the moment of inertia of the area about a parallel axis passing through the area's centroid plus the product of the area and the square of the perpendicular distance between the axes.*

10.3 Radius of Gyration of an Area

The *radius of gyration* of a planar area has units of length and is a quantity that is often used for the design of columns in structural mechanics. Provided the areas and moments of inertia are *known,* the radii of gyration are determined from the formulas

$$k_x = \sqrt{\frac{I_x}{A}} \qquad k_y = \sqrt{\frac{I_y}{A}} \qquad k_O = \sqrt{\frac{J_O}{A}} \qquad (10\text{–}6)$$

The form of these equations is easily remembered, since it is similar to that for finding the moment of inertia of a differential area about an axis. For example, $I_x = k_x^2 A$; whereas for a differential area, $dI_x = y^2\, dA$.

10.4 Moments of Inertia for an Area by Integration

When the boundaries for a planar area are expressed by mathematical functions, Eqs. 10–1 may be integrated to determine the moments of inertia for the area. If the element of area chosen for integration has a differential size in two directions as shown in Fig. 10–2, a double integration must be performed to evaluate the moment of inertia. Most often, however, it is easier to perform only a single integration by choosing an element having a differential size or thickness in only one direction.

PROCEDURE FOR ANALYSIS

If a single integration is performed to determine the moment of inertia of an area about an axis, it will first be necessary to specify the differential element dA. Most often this element will be rectangular, such that it will have a finite length and differential width. The element should be located so that it intersects the boundary of the area at the *arbitrary point* (x, y). There are two possible ways to orient the element with respect to the axis about which the moment of inertia is to be determined.

Case 1. The *length* of the element can be oriented *parallel* to the axis. This situation occurs when the rectangular element shown in Fig. 10–4 is used to determine I_y for the area. Direct application of Eq. 10–1, i.e., $I_y = \int x^2\, dA$, can be made in this case, since the element has an infinitesimal thickness dx and therefore *all parts* of the element lie at the *same* moment-arm distance x from the y axis.*

Case 2. The *length* of the element can be oriented *perpendicular* to the axis. Here Eq. 10–1 *does not apply,* since all parts of the element will *not* lie at the same moment-arm distance from the axis. For example, if the rectangular element in Fig. 10–4 is used for determining I_x for the area, it will first be necessary to calculate the moment of inertia of the *element* about a horizontal axis passing through the element's centroid and then determine the moment of inertia of the *element* about the x axis by using the parallel-axis theorem. Integration of this result will yield I_x.

*In the case of the element $dA = dx\, dy$, Fig. 10–2, the moment arms y and x are appropriate for the formulation of I_x and I_y (Eq. 10–1) since the *entire* element, because of its infinitesimal size, lies at the specified y and x perpendicular distances from the x and y axes.

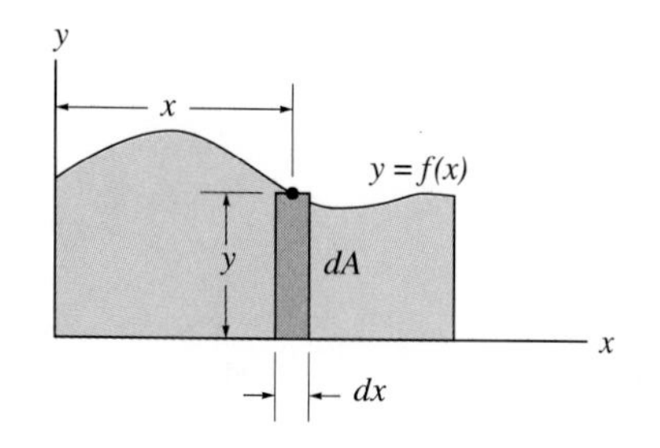

Fig. 10–4

Application of these cases is illustrated in the following examples.

Example 10–1

Determine the moment of inertia for the rectangular area shown in Fig. 10–5 with respect to (a) the centroidal x' axis, (b) the axis x_b passing through the base of the rectangle, and (c) the pole or z' axis perpendicular to the x'–y' plane and passing through the centroid C.

SOLUTION *(CASE 1)*

Part (a). The differential element shown in Fig. 10–5 is chosen for integration. Because of its location and orientation, the *entire element* is at a distance y' from the x' axis. Here it is necessary to integrate from $y' = -h/2$ to $y' = h/2$. Since $dA = b\,dy'$, then

$$\bar{I}_{x'} = \int_A y'^2\,dA = \int_{-h/2}^{h/2} y'^2(b\,dy') = b\int_{-h/2}^{h/2} y'^2\,dy'$$

$$= \frac{1}{12}bh^3 \qquad \textit{Ans.}$$

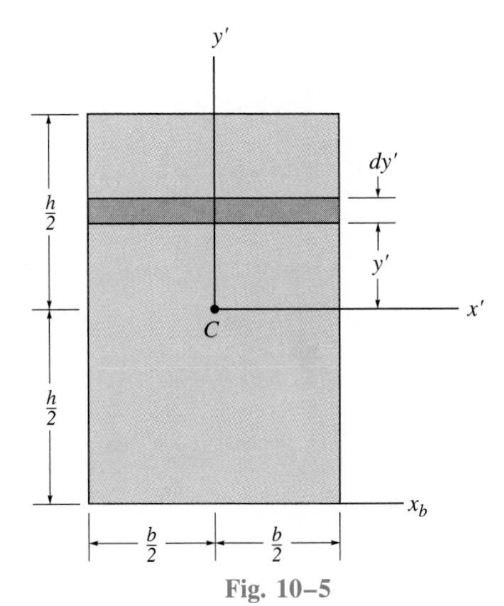

Fig. 10–5

Part (b). The moment of inertia about an axis passing through the base of the rectangle can be obtained by using the result of part (a) and applying the parallel-axis theorem, Eq. 10–3.

$$I_{x_b} = \bar{I}_{x'} + Ad_y^2$$

$$= \frac{1}{12}bh^3 + bh\left(\frac{h}{2}\right)^2 = \frac{1}{3}bh^3 \qquad \textit{Ans.}$$

Part (c). To obtain the polar moment of inertia about point C, we must first obtain $\bar{I}_{y'}$, which may be found by interchanging the dimensions b and h in the result of part (a), i.e.,

$$\bar{I}_{y'} = \frac{1}{12}hb^3$$

Using Eq. 10–2, the polar moment of inertia about C is therefore

$$\bar{J}_C = \bar{I}_{x'} + \bar{I}_{y'} = \frac{1}{12}bh(h^2 + b^2) \qquad \textit{Ans.}$$

Example 10–2

Determine the moment of inertia of the shaded area shown in Fig. 10–6*a* about the *x* axis.

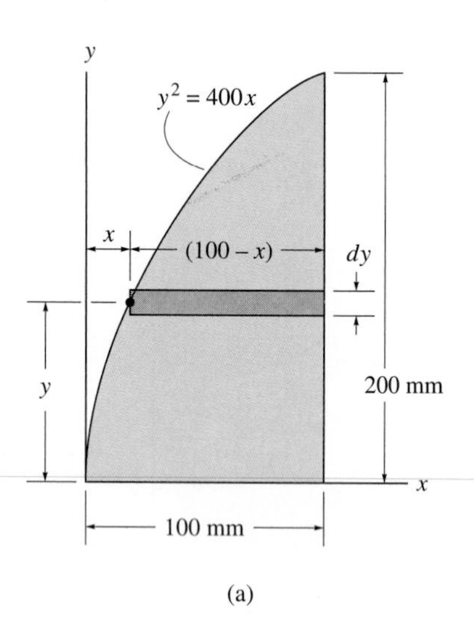

(a)

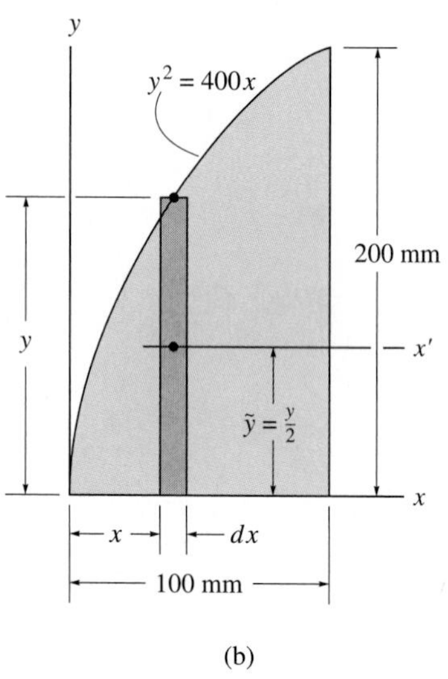

(b)

Fig. 10–6

SOLUTION I *(CASE 1)*

A differential element of area that is *parallel* to the x axis, as shown in Fig. 10–6*a*, is chosen for integration. Since the element has a thickness dy and intersects the curve at the *arbitrary point* (x, y), the area is $dA = (100 - x)\,dy$. Furthermore, all parts of the element lie at the same distance y from the x axis. Hence, integrating with respect to y, from $y = 0$ to $y = 200$ mm, yields

$$I_x = \int_A y^2\,dA = \int_A y^2(100 - x)\,dy$$

$$= \int_0^{200} y^2\left(100 - \frac{y^2}{400}\right) dy = 100 \int_0^{200} y^2\,dy - \frac{1}{400}\int_0^{200} y^4\,dy$$

$$= 107(10^6)\text{ mm}^4$$ ***Ans.***

SOLUTION II *(CASE 2)*

A differential element *parallel* to the y axis, as shown in Fig. 10–6*b*, is chosen for integration. It intersects the curve at the *arbitrary point* (x, y). In this case, all parts of the element do *not* lie at the same distance from the x axis, and therefore the parallel-axis theorem must be used to determine the *moment of inertia of the element* with respect to this axis. For a rectangle having a base b and height h, the moment of inertia about its centroidal axis has been determined in part (a) of Example 10–1. There it was found that $\bar{I}_{x'} = \frac{1}{12}bh^3$. For the differential element shown in Fig. 10–6*b*, $b = dx$ and $h = y$, and thus $d\bar{I}_{x'} = \frac{1}{12}\,dx\,y^3$. Since the centroid of the element is at $\tilde{y} = y/2$ from the x axis, the moment of inertia of the element about this axis is

$$dI_x = d\bar{I}_{x'} + dA\,\tilde{y}^2 = \frac{1}{12}\,dx\,y^3 + y\,dx\left(\frac{y}{2}\right)^2 = \frac{1}{3}y^3\,dx$$

[This result can also be concluded from part (b) of Example 10–1.]

Integrating with respect to x, from $x = 0$ to $x = 100$ mm, yields

$$I_x = \int dI_x = \int_A \frac{1}{3}y^3\,dx = \int_0^{100} \frac{1}{3}(400x)^{3/2}\,dx$$

$$= 107(10^6)\text{ mm}^4$$ ***Ans.***

Example 10–3

Determine the moment of inertia with respect to the x axis of the circular area shown in Fig. 10–7a.

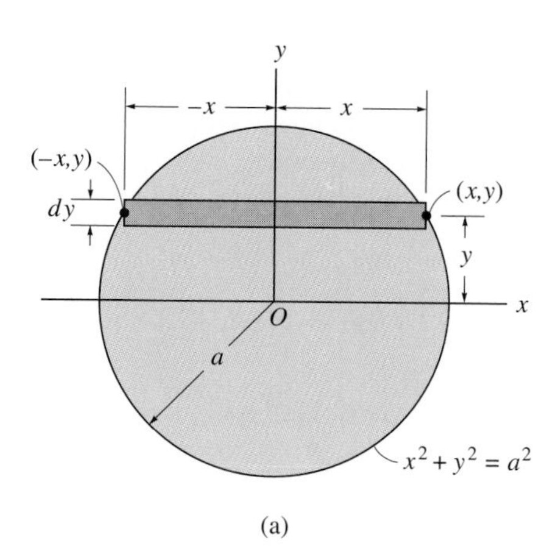

(a)

SOLUTION I *(CASE 1)*

Using the differential element shown in Fig. 10–7a, since $dA = 2x\,dy$, we have

$$I_x = \int_A y^2\,dA = \int_A y^2(2x)\,dy$$

$$= \int_{-a}^{a} y^2(2\sqrt{a^2 - y^2})\,dy = \frac{\pi a^4}{4} \qquad \textit{Ans.}$$

SOLUTION II *(CASE 2)*

When the differential element is chosen as shown in Fig. 10–7b, the centroid for the element happens to lie on the x axis, and so, applying Eq. 10–3, noting that $d_y = 0$, we have

$$dI_x = \frac{1}{12}\,dx\,(2y)^3$$

$$= \frac{2}{3}y^3\,dx$$

Integrating with respect to x yields

$$I_x = \int_{-a}^{a} \frac{2}{3}(a^2 - x^2)^{3/2}\,dx = \frac{\pi a^4}{4} \qquad \textit{Ans.}$$

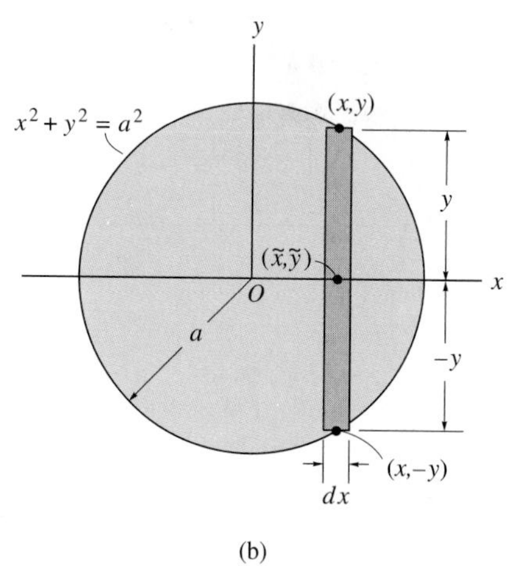

(b)

Fig. 10–7

Example 10–4

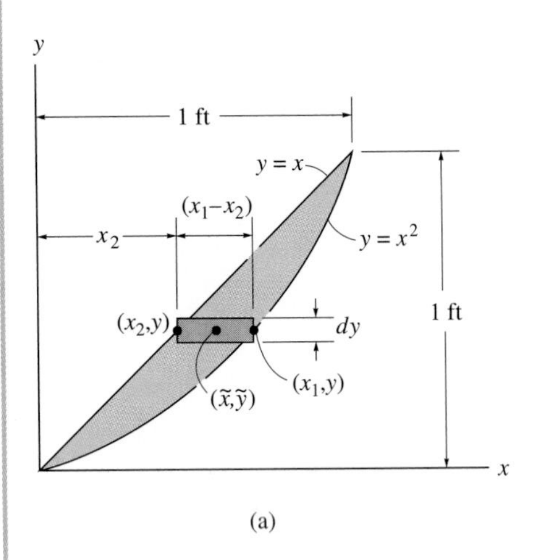

(a)

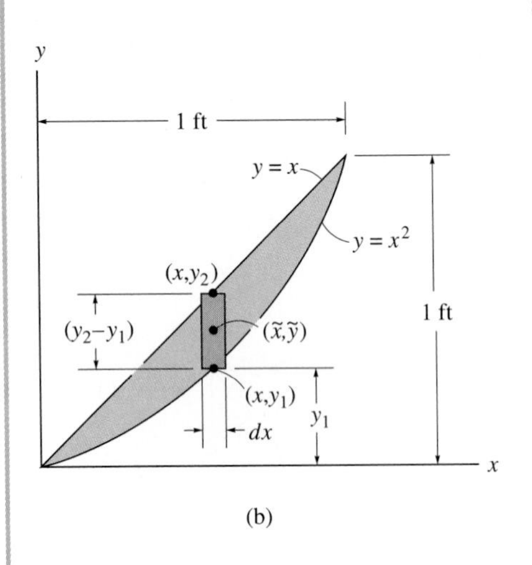

(b)

Fig. 10–8

Determine the moment of inertia of the shaded area shown in Fig. 10–8*a* about the x axis.

SOLUTION I (*CASE 1*)

The differential element of area parallel to the x axis is chosen for integration, Fig. 10–8*a*. The element intersects the curve at the *arbitrary points* (x_2, y) and (x_1, y). Consequently, its area is $dA = (x_1 - x_2)\,dy$. Since all parts of the element lie at the same distance y from the x axis, we have

$$I_x = \int_A y^2\,dA = \int_0^1 y^2(x_1 - x_2)\,dy = \int_0^1 y^2(\sqrt{y} - y)\,dy$$

$$I_x = \frac{2}{7}y^{7/2} - \frac{1}{4}y^4 \Big|_0^1 = 0.0357\text{ ft}^4 \qquad \textit{Ans.}$$

SOLUTION II (*CASE 2*)

The differential element of area parallel to the y axis is shown in Fig. 10–8*b*. It intersects the curves at the *arbitrary points* (x, y_2) and (x, y_1). Since all parts of its entirety do *not* lie at the same distance from the x axis, we must first use the parallel-axis theorem to find the element's moment of inertia about the x axis, then integrate this result to determine I_x. Thus,

$$dI_x = d\bar{I}_{x'} + dA\,\tilde{y}^2 = \frac{1}{12}\,dx\,(y_2 - y_1)^3 + (y_2 - y_1)\,dx\left(y_1 + \frac{y_2 - y_1}{2}\right)^2$$

$$= \frac{1}{3}(y_2^3 - y_1^3)\,dx = \frac{1}{3}(x^3 - x^6)\,dx$$

$$I_x = \frac{1}{3}\int_0^1 (x^3 - x^6)\,dx = \frac{1}{12}x^4 - \frac{1}{21}x^7 \Big|_0^1 = 0.0357\text{ ft}^4 \qquad \textit{Ans.}$$

By comparison, Solution I requires much less computation. If an integral using a particular element appears difficult to evaluate, try solving the problem using an element oriented in the other direction.

PROBLEMS

10–1. The 28-in^2 area has a moment of inertia about the xx axis of 3325 in^4. Determine its moment of inertia about the $x'x'$ axis. The $\bar{x}\bar{x}$ axis passes through the centroid C of the area.

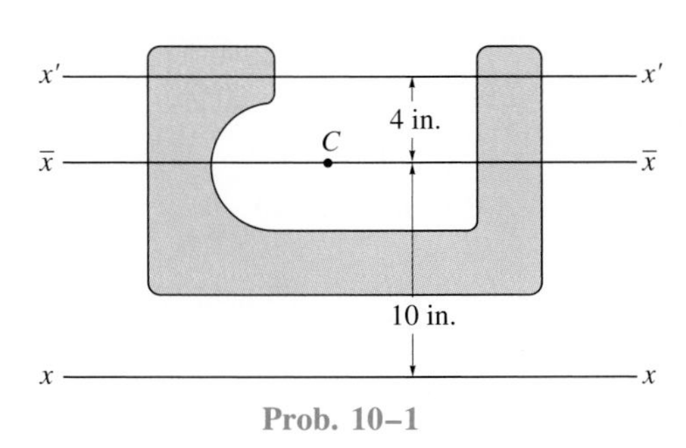

Prob. 10–1

10–2. The $15(10^3)$-mm^2 area has a moment of inertia about the yy axis of $25(10^6)$ mm^4. Determine its moment of inertia about the $y'y'$ axis. The $\bar{y}\bar{y}$ axis passes through the centroid C of the area.

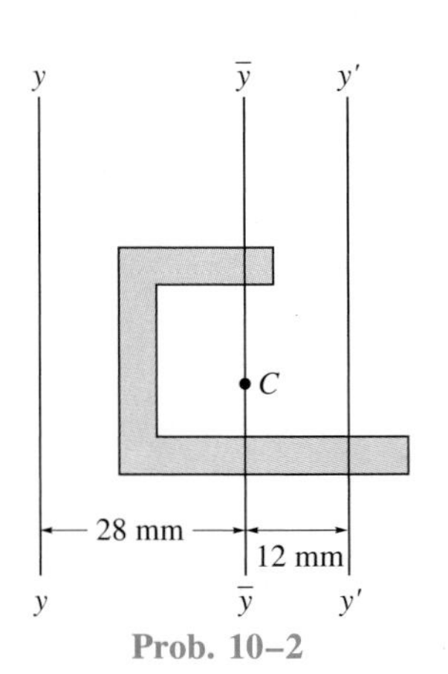

Prob. 10–2

10–3. Determine the moment of inertia of the shaded area about the x axis.

***10–4.** Determine the moment of inertia of the shaded area about the y axis.

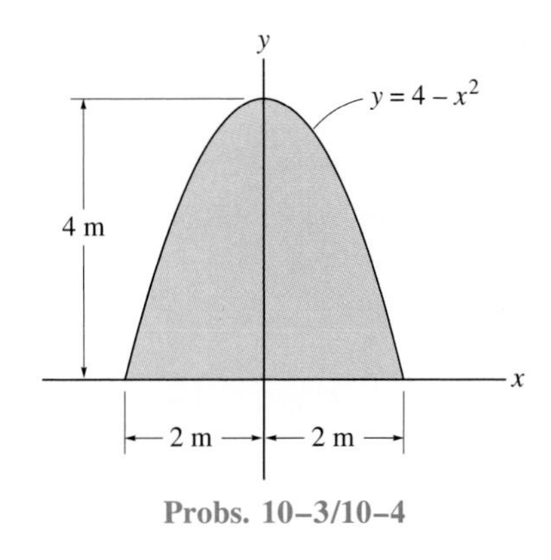

Probs. 10–3/10–4

10–5. Determine the moment of inertia of the shaded area about the x axis.

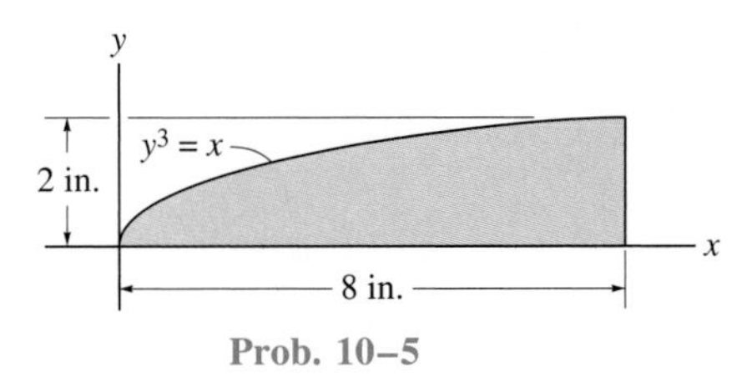

Prob. 10–5

10–6. Determine the moment of inertia of the shaded area about the y axis.

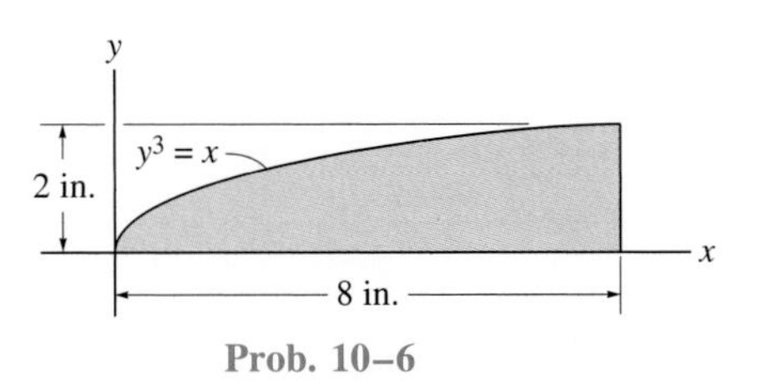

Prob. 10–6

10–7. Determine the moment of inertia of the shaded area about the x axis.

***10–8.** Determine the moment of inertia of the shaded area about the y axis.

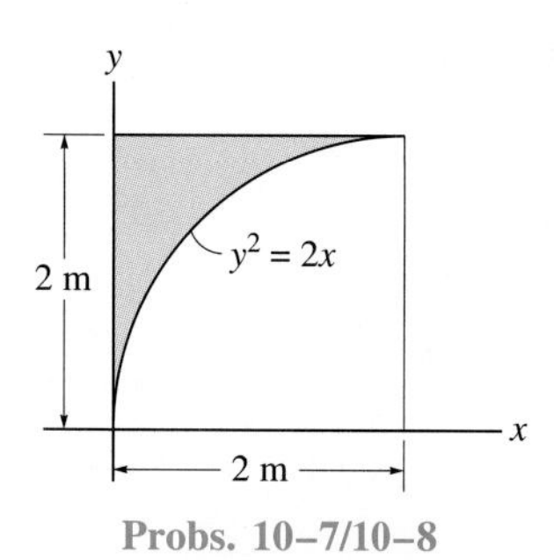

Probs. 10–7/10–8

10–9. Determine the moment of inertia of the shaded area about the x axis.

10–10. Determine the moment of inertia of the shaded area about the y axis.

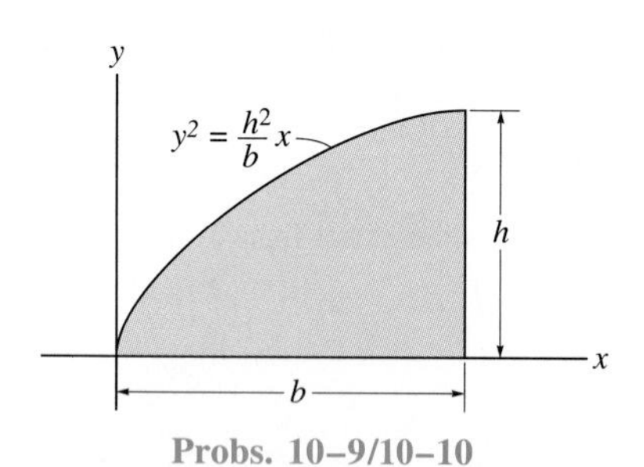

Probs. 10–9/10–10

10–11. Determine the moment of inertia of the area about the x axis. Solve the problem in two ways, using rectangular differential elements: (a) having a thickness dx, and (b) having a thickness dy.

***10–12.** Determine the moment of inertia of the area about the y axis. Solve the problem in two ways, using rectangular differential elements: (a) having a thickness dx, and (b) having a thickness dy.

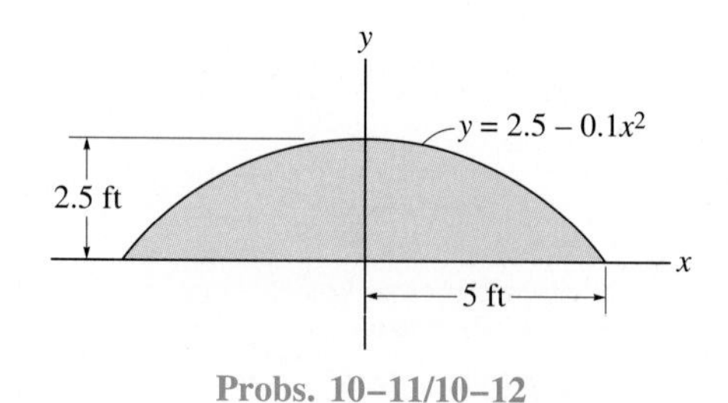

Probs. 10–11/10–12

10–13. Determine the moment of inertia of the shaded area about the y axis.

10–14. Determine the moment of inertia of the shaded area about the x axis.

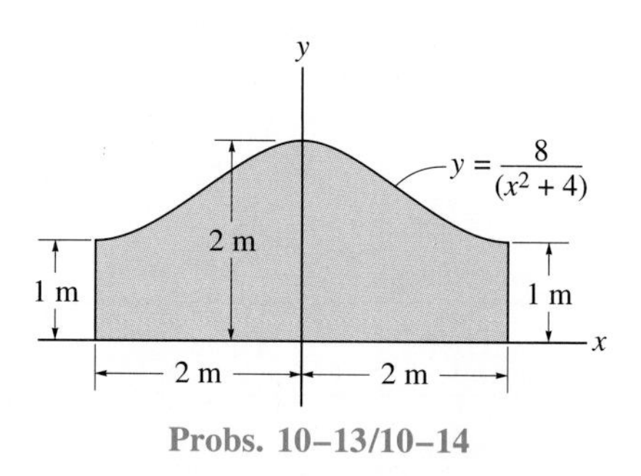

Probs. 10–13/10–14

10–15. Determine the moment of inertia of the shaded area about the x axis.

***10–16.** Determine the moment of inertia of the shaded area about the y axis.

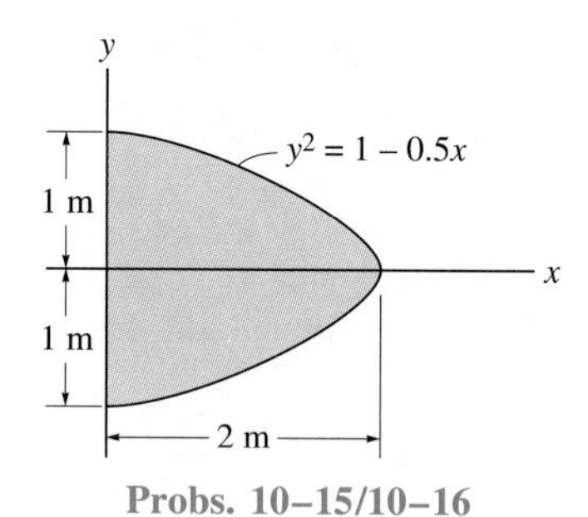

Probs. 10–15/10–16

10–17. Determine the moment of inertia of the shaded area about the x axis.

10–18. Determine the moment of inertia of the shaded area about the y axis.

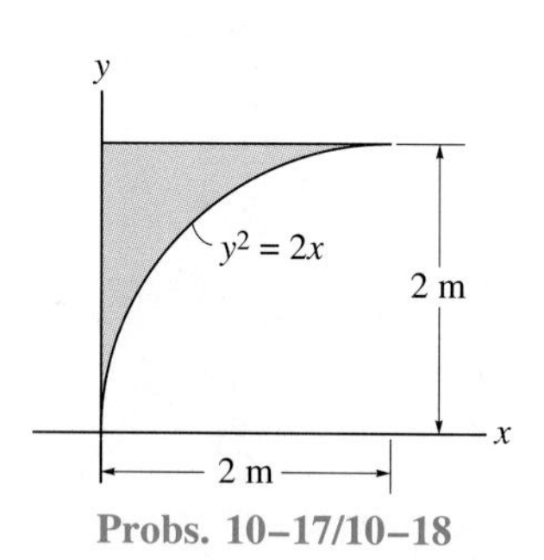

Probs. 10–17/10–18

10–19. Determine the moment of inertia of the equilateral triangle about the x' axis passing through its centroid.

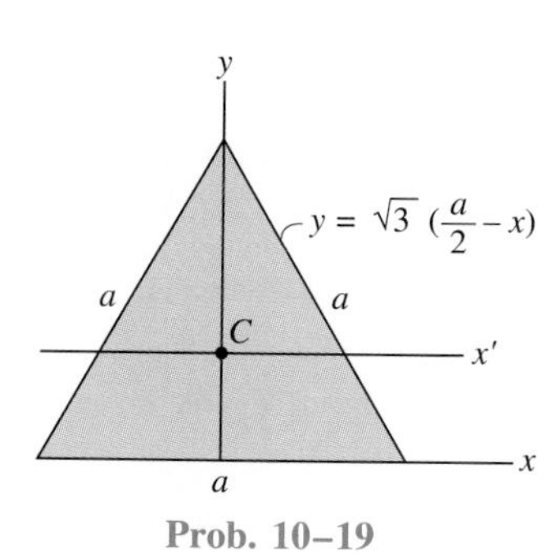

Prob. 10–19

***10–20.** Determine the moment of inertia of the area enclosed by the arch of the curve $y = \sin x$ about the x axis.

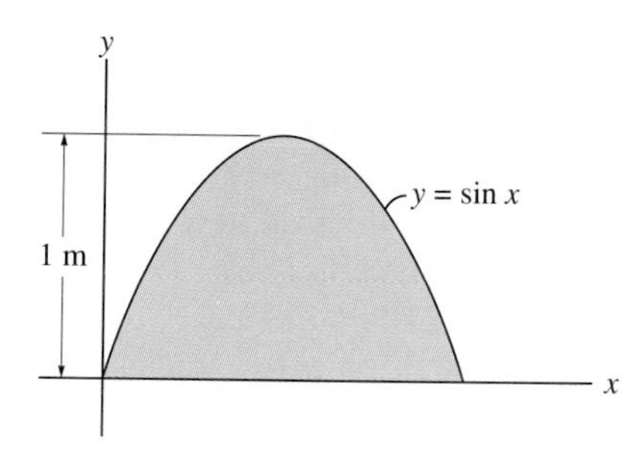

Prob. 10–20

■**10–21.** Determine the moment of inertia of the area enclosed by the arch of the curve $y = \sin x$ about a vertical axis passing through its centroid.

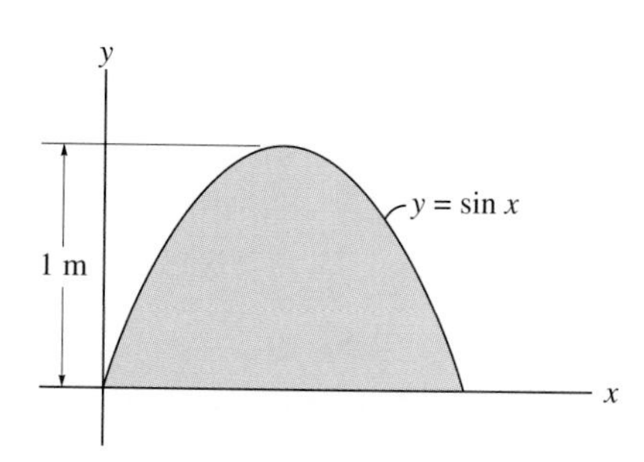

Prob. 10–21

10–22. Determine the moment of inertia of the shaded area about the x axis.

10–23. Determine the moment of inertia of the shaded area about the y axis.

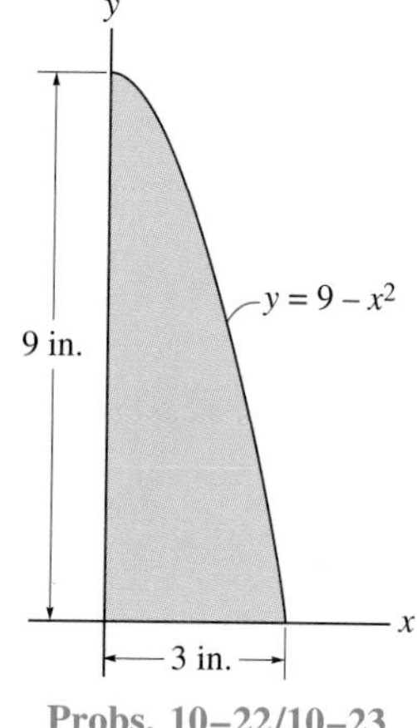

Probs. 10–22/10–23

*■**10–24.** Determine the moment of inertia of the area about the y axis. Use Simpson's rule to evaluate the integral.

*■**10–25.** Determine the moment of inertia of the area about the x axis. Use Simpson's rule to evaluate the integral.

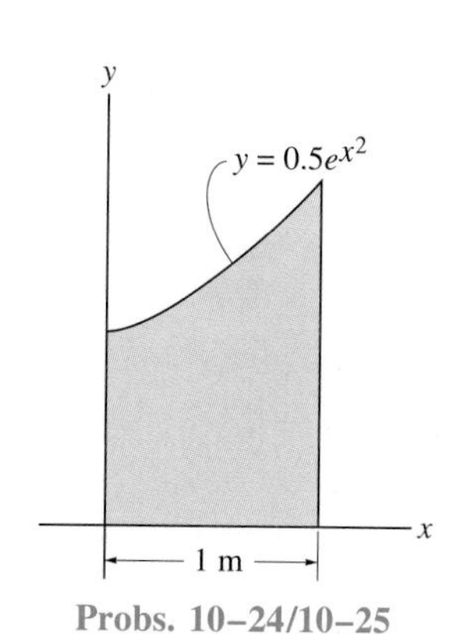

Probs. 10–24/10–25

10.5 Moments of Inertia for Composite Areas

A composite area consists of a series of connected ''simpler'' parts or shapes, such as semicircles, rectangles, and triangles. Provided the moment of inertia of each of these parts is known or can be determined about a common axis, then the moment of inertia of the composite area equals the *algebraic sum* of the moments of inertia of all its parts.

PROCEDURE FOR ANALYSIS

The following procedure provides a method for determining the moment of inertia of a composite area about a reference axis.

Composite Parts. Using a sketch, divide the area into its composite parts and indicate the perpendicular distance from the *centroid* of each part to the reference axis.

Parallel-Axis Theorem. The moment of inertia of each part should be determined about its centroidal axis, which is parallel to the reference axis. For the calculation use the table given on the inside back cover. If the centroidal axis does not coincide with the reference axis, the parallel-axis theorem, $I = \bar{I} + Ad^2$, should be used to determine the moment of inertia of the part about the reference axis.

Summation. The moment of inertia of the entire area about the reference axis is determined by summing the results of its composite parts. In particular, if a composite part has a ''hole,'' its moment of inertia is found by ''subtracting'' the moment of inertia for the hole from the moment of inertia of the entire part including the hole.

Example 10–5

Compute the moment of inertia of the composite area shown in Fig. 10–9*a* about the x axis.

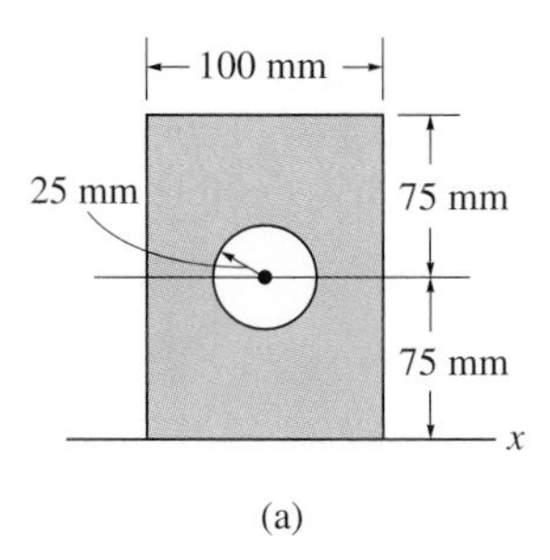

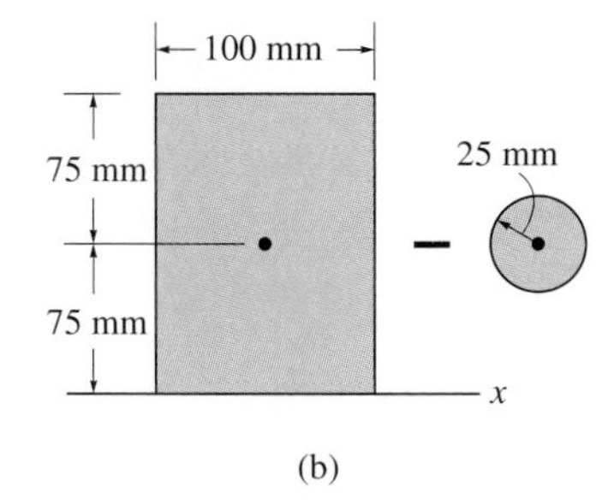

Fig. 10–9

SOLUTION

Composite Parts. The composite area is obtained by *subtracting* the circle from the rectangle as shown in Fig. 10–9*b*. The centroid of each area is located in the figure.

Parallel-Axis Theorem. The moments of inertia about the x axis are determined using the parallel-axis theorem and the data in the table on the inside back cover.

Circle

$$I_x = \bar{I}_{x'} + Ad_y^2$$
$$= \frac{1}{4}\pi(25)^4 + \pi(25)^2(75)^2 = 11.4(10^6)\text{ mm}^4$$

Rectangle

$$I_x = \bar{I}_{x'} + Ad_y^2$$
$$= \frac{1}{12}(100)(150)^3 + (100)(150)(75)^2 = 112.5(10^6)\text{ mm}^4$$

Summation. The moment of inertia for the composite area is thus

$$I_x = -11.4(10^6) + 112.5(10^6)$$
$$= 101(10^6)\text{ mm}^4 \qquad \textit{Ans.}$$

Example 10–6

Determine the moments of inertia of the beam's cross-sectional area shown in Fig. 10–10*a* about the *x* and *y* centroidal axes.

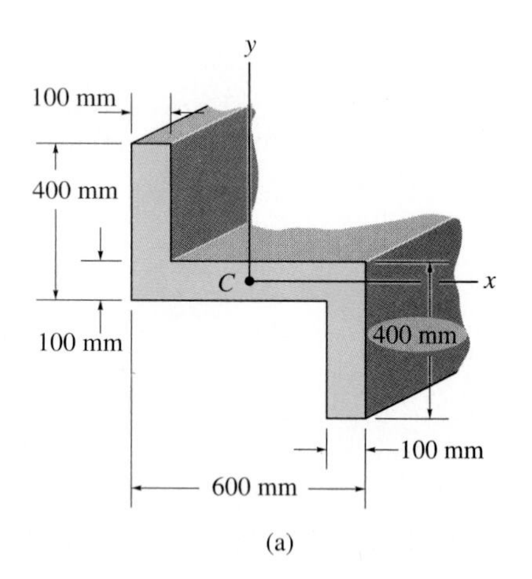

(a)

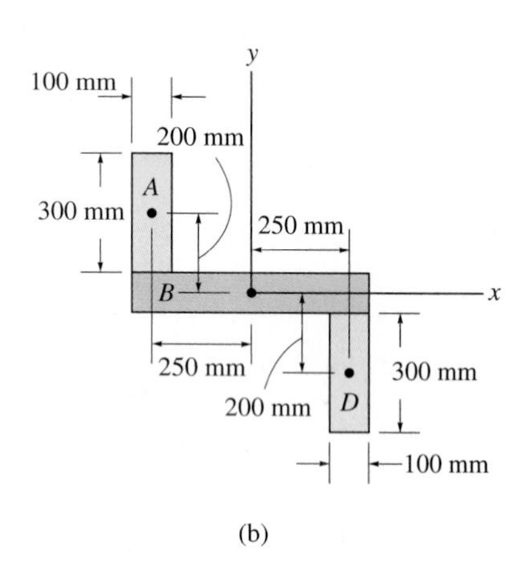

(b)

Fig. 10–10

SOLUTION

Composite Parts. The cross section can be considered as three composite rectangular areas *A*, *B*, and *D* shown in Fig. 10–10*b*. For the calculation, the centroid of each of these rectangles is located in the figure.

Parallel-Axis Theorem. From the table on the inside back cover, or Example 10–1, the moment of inertia of a rectangle about its centroidal axis is $\bar{I} = \frac{1}{12}bh^3$. Hence, using the parallel-axis theorem for rectangles *A* and *D*, the calculations are as follows:

Rectangle A

$$I_x = \bar{I}_{x'} + Ad_y^2 = \frac{1}{12}(100)(300)^3 + (100)(300)(200)^2$$
$$= 1.425(10^9)\text{ mm}^4$$
$$I_y = \bar{I}_{y'} + Ad_x^2 = \frac{1}{12}(300)(100)^3 + (100)(300)(250)^2$$
$$= 1.90(10^9)\text{ mm}^4$$

Rectangle B

$$I_x = \frac{1}{12}(600)(100)^3 = 0.05(10^9)\text{ mm}^4$$
$$I_y = \frac{1}{12}(100)(600)^3 = 1.80(10^9)\text{ mm}^4$$

Rectangle D

$$I_x = \bar{I}_{x'} + Ad_y^2 = \frac{1}{12}(100)(300)^3 + (100)(300)(200)^2$$
$$= 1.425(10^9)\text{ mm}^4$$
$$I_y = \bar{I}_{y'} + Ad_x^2 = \frac{1}{12}(300)(100)^3 + (100)(300)(250)^2$$
$$= 1.90(10^9)\text{ mm}^4$$

Summation. The moments of inertia for the entire cross section are thus

$$I_x = 1.425(10^9) + 0.05(10^9) + 1.425(10^9)$$
$$= 2.90(10^9)\text{ mm}^4 \qquad \textbf{\textit{Ans.}}$$
$$I_y = 1.90(10^9) + 1.80(10^9) + 1.90(10^9)$$
$$= 5.60(10^9)\text{ mm}^4 \qquad \textbf{\textit{Ans.}}$$

PROBLEMS

10–26. Determine the moment of inertia of the beam's cross-sectional area about the x axis.

10–27. Determine the moment of inertia of the beam's cross-sectional area about the y axis.

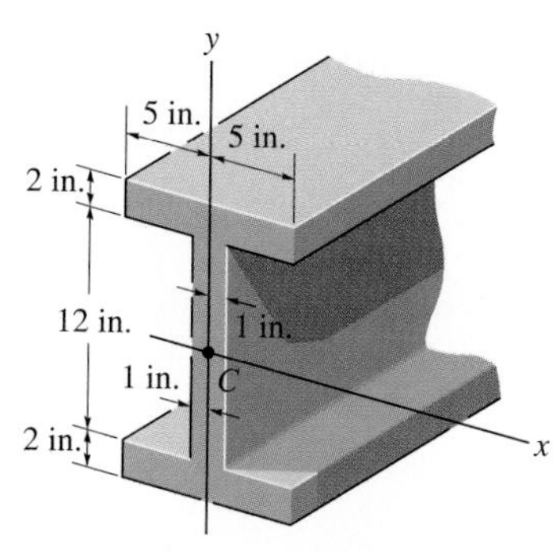

Probs. 10–26/10–27

***10–28.** The beam is constructed from the two channels and two cover plates. If each channel has a cross-sectional area of $A_c = 11.8\ \text{in}^2$ and a moment of inertia about a horizontal axis passing through its own centroid, C_c, of $(\bar{I}_{\bar{x}})_{C_c} = 349\ \text{in}^4$, determine the moment of inertia of the beam about the x axis.

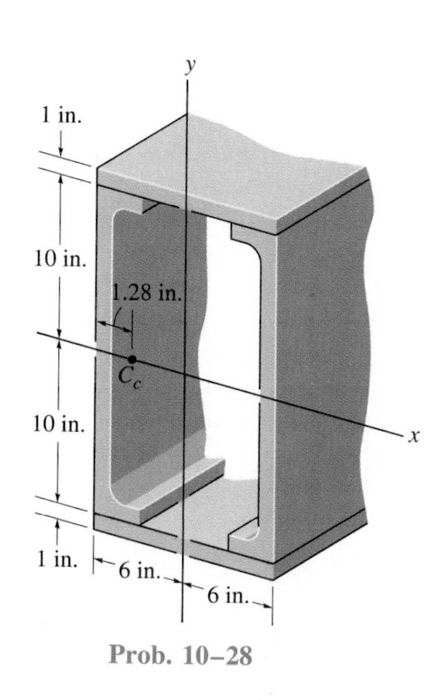

Prob. 10–28

10–29. The beam is constructed from the two channels and two cover plates. If each channel has a cross-sectional area of $A_c = 11.8\ \text{in}^2$ and a moment of inertia about a vertical axis passing through its own centroid, C_c, of $(\bar{I}_{\bar{y}})_{C_c} = 9.23\ \text{in}^4$, determine the moment of inertia of the beam about the y axis.

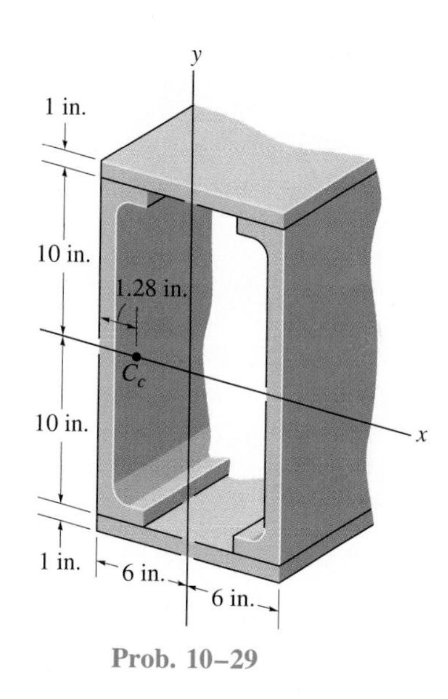

Prob. 10–29

10–30. Determine the distance $\bar{x}$ to the centroid for the beam's cross-sectional area, then find $\bar{I}_{y'}$.

10–31. Determine the moment of inertia of the beam's cross-sectional area about the x axis.

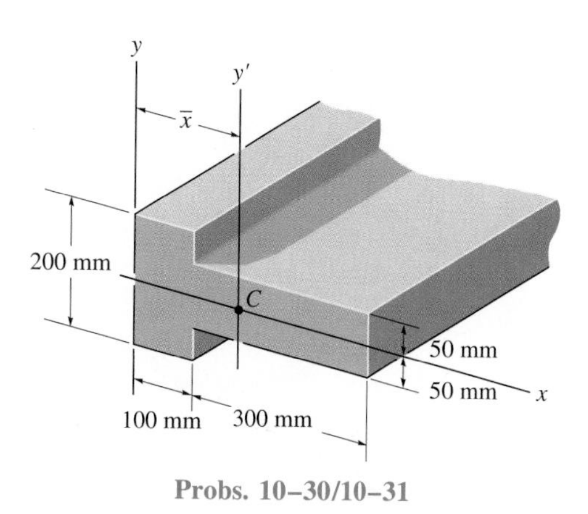

Probs. 10–30/10–31

***10–32.** Determine the distance $\bar{x}$ to the centroid for the beam's cross-sectional area, then find $\bar{I}_{y'}$.

10–33. Determine the moment of inertia of the beams cross-sectional area about the x axis.

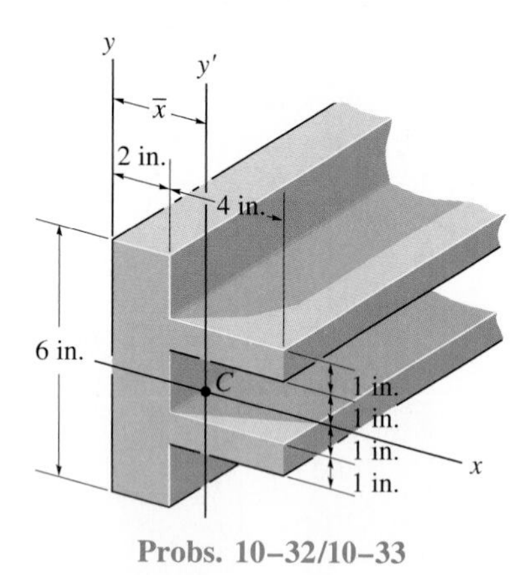

Probs. 10–32/10–33

***10–36.** Determine the distance $\bar{y}$ to the centroid for the beam's cross-sectional area; then find $\bar{I}_{x'}$.

10–37. Determine the moment of inertia of the beam's cross-sectional area about the y axis.

25 mm
25 mm
100 mm
x'
C
ȳ
25 mm
x
50 mm
75 mm
75 mm
50 mm
100 mm
25 mm

Probs. 10–36/10–37

10–34. Determine $\bar{y}$, which locates the centroid C of the wing channel, and then determine the moment of inertia $\bar{I}_{x'}$ about the centroidal x' axis. Neglect the effect of rounded corners. The material has a uniform thickness of 0.5 in.

10–35. Determine the moment of inertia of the wing channel about the y axis.

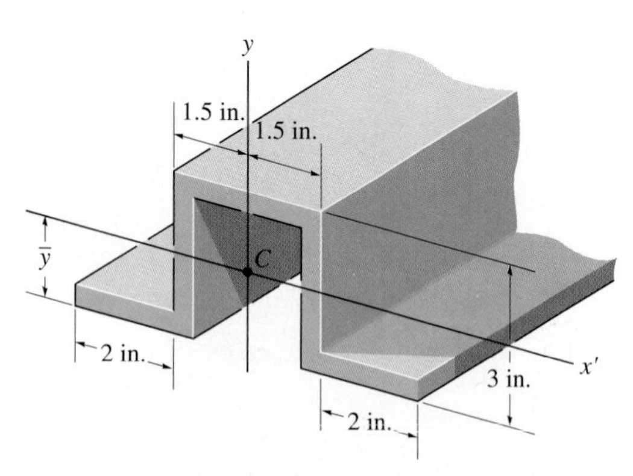

Probs. 10–34/10–35

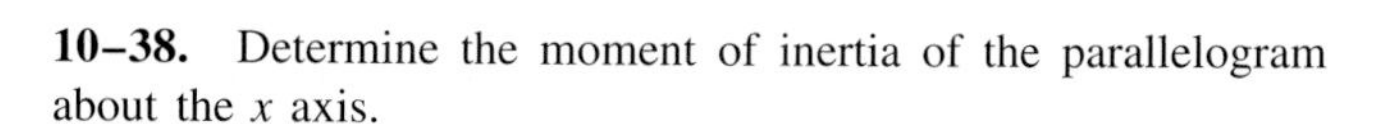

10–38. Determine the moment of inertia of the parallelogram about the x axis.

10–39. Determine the moment of inertia of the parallelogram about the y axis.

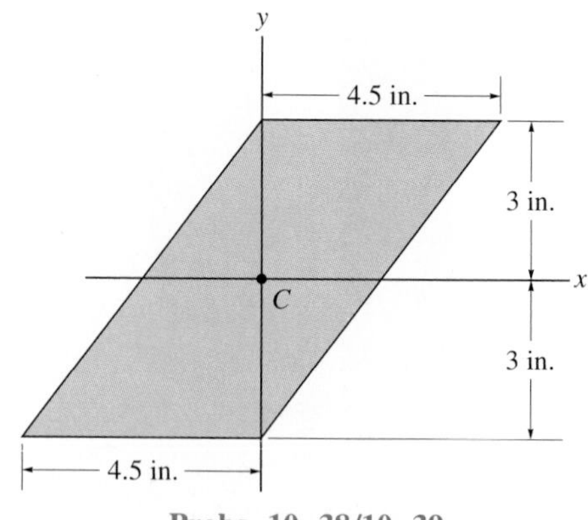

Probs. 10–38/10–39

*10–40. Locate the centroid $\bar{y}$ of the channel's cross-sectional area, and then determine the moment of inertia with respect to the x' axis passing through the centroid.

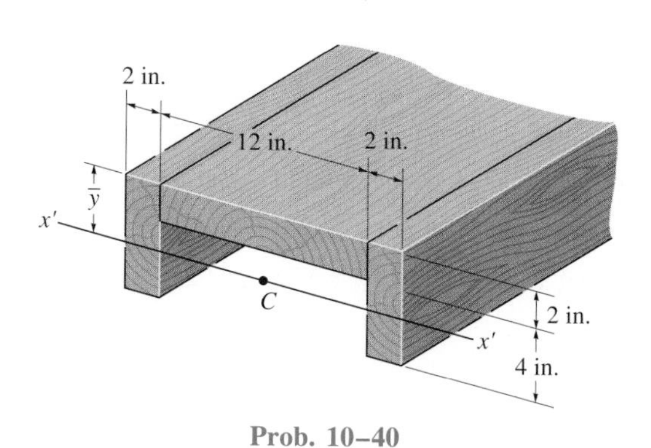

Prob. 10–40

10–41. Determine y, which locates the centroidal axis x for the cross-sectional area of the T-beam, and then find the moments of inertia $\bar{I}_{x'}$ and $\bar{I}_{y'}$.

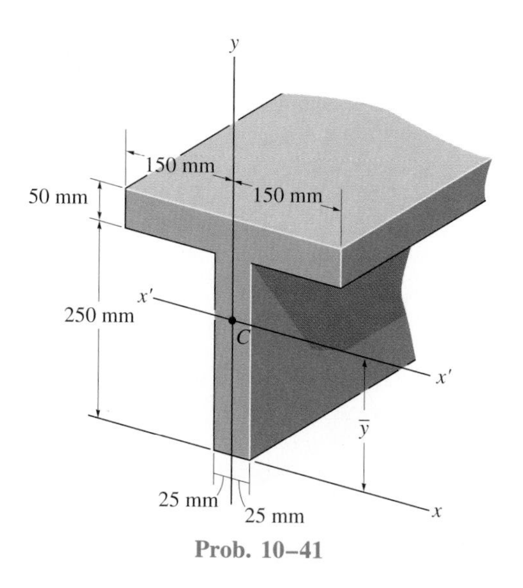

Prob. 10–41

10–42. Determine the moment of inertia I_x of the shaded area about the x axis.

10–43. Determine the moment of inertia I_y of the shaded area about the y axis.

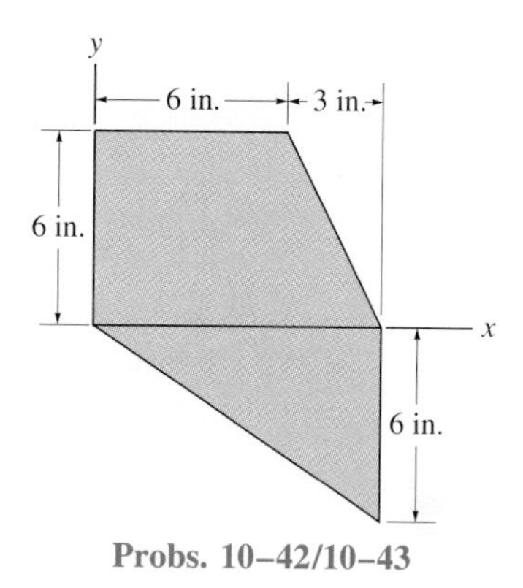

Probs. 10–42/10–43

*10–44. Compute the moments of inertia I_x and I_y for the shaded area about the x and y axes.

10–45. Determine the distance $\bar{y}$ to the centroid C of the beam's cross-sectional area and then compute the moment of inertia $\bar{I}_{x'}$ about the x' axis.

10–46. Determine the distance $\bar{x}$ to the centroid C of the beam's cross-sectional area and then compute the moment of inertia $\bar{I}_{y'}$ about the y' axis.

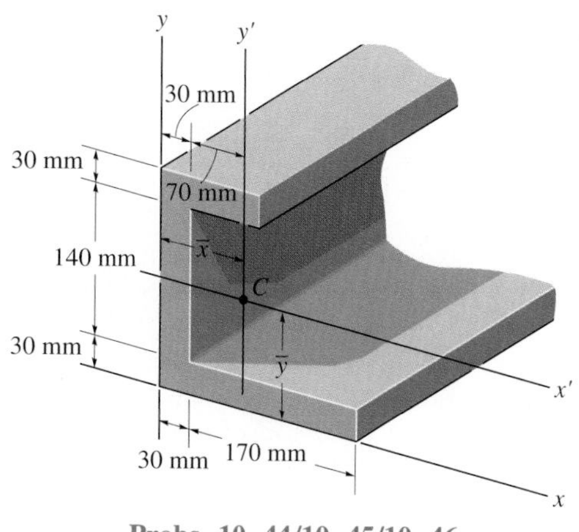

Probs. 10–44/10–45/10–46

10–47. Determine the moments of inertia I_x and I_y of the shaded area.

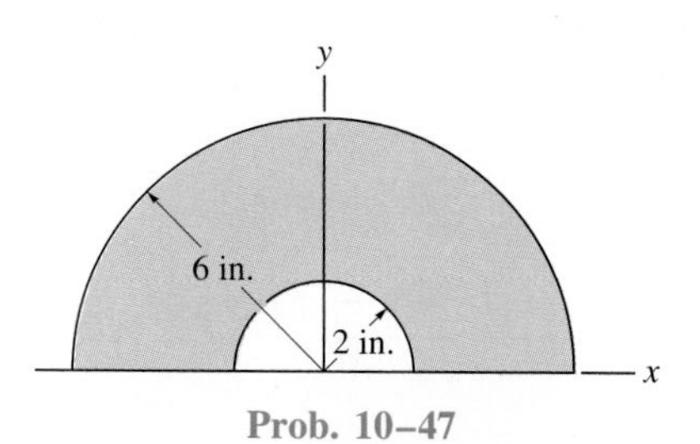

Prob. 10–47

***10–48.** Determine the moment of inertia of the beam's cross-sectional area about the x axis, which passes through the centroid C.

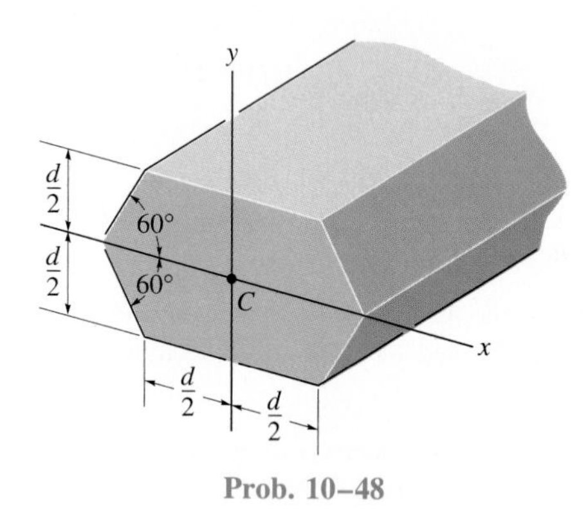

Prob. 10–48

10–49. Determine the moment of inertia of the beam's cross-sectional area about the y axis, which passes through the centroid C.

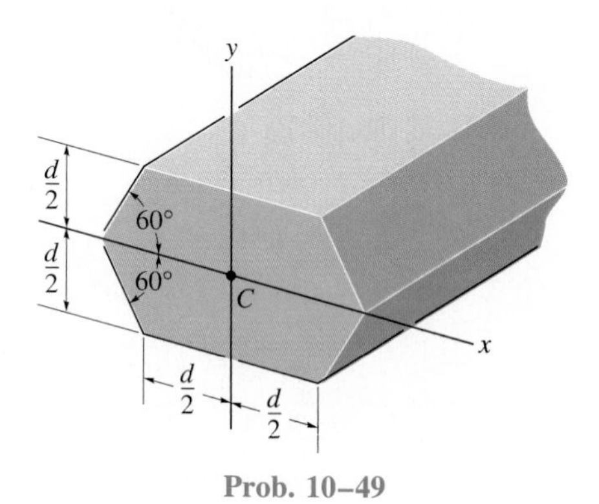

Prob. 10–49

10–50. Compute the polar moments of inertia J_O for the cross-sectional area of the solid shaft and tube. What percentage of J_O is contributed by the tube to that of the solid shaft?

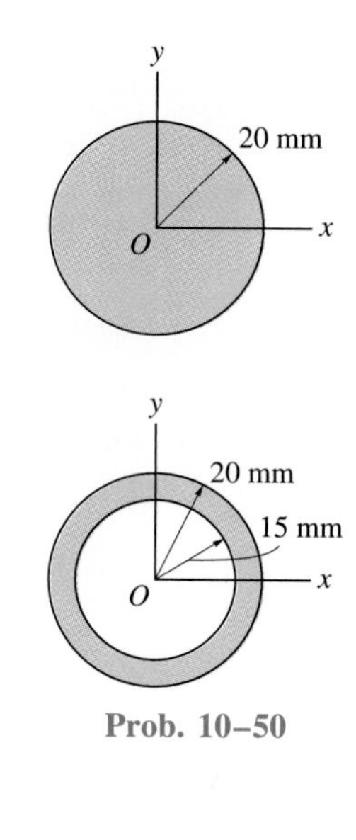

Prob. 10–50

10–51. Determine the moment of inertia of the rectangle about its diagonal axis xx.

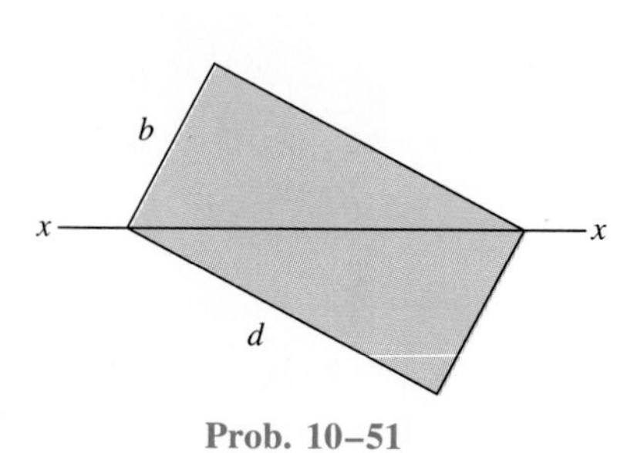

Prob. 10–51

***10–52.** Determine the polar moment of inertia of the shaft's cross-sectional area about the center O.

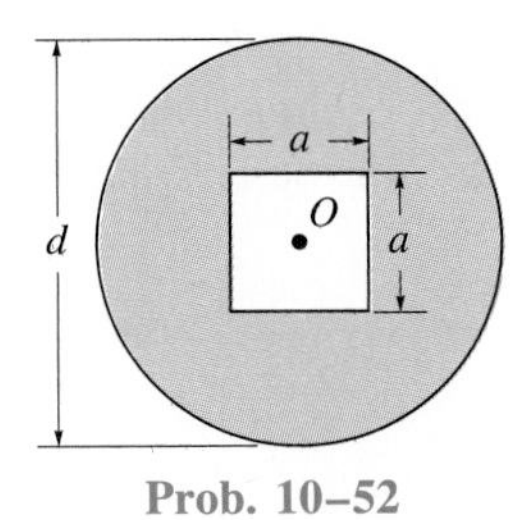

Prob. 10–52

10–53. Determine the radius of gyration k_x for the column's cross-sectional area.

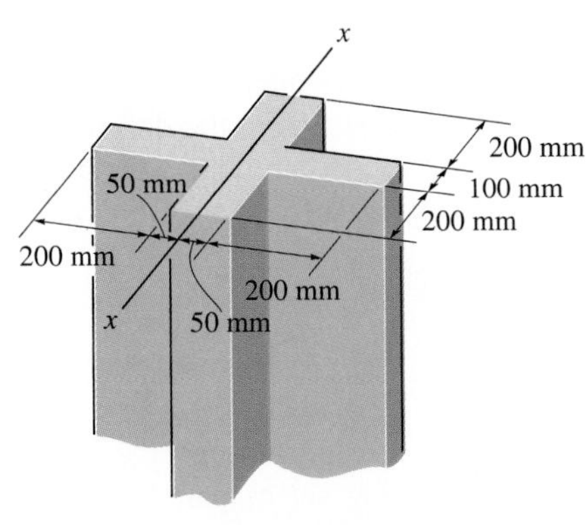

Prob. 10–53

10–54. Determine the moment of inertia of the parallelogram about the x' axis, which passes through the centroid C of the area.

10–55. Determine the moment of inertia of the parallelogram about the y' axis, which passes through the centroid C of the area.

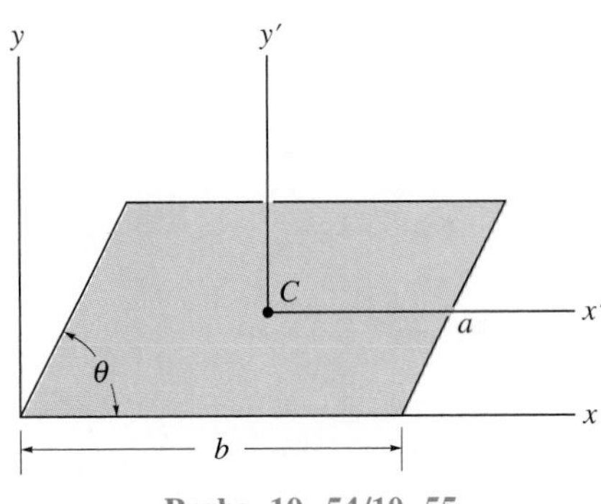

Probs. 10–54/10–55

***10–56.** Determine the moments of inertia of the triangular area about the x' and y' axes, which pass through the centroid C of the area.

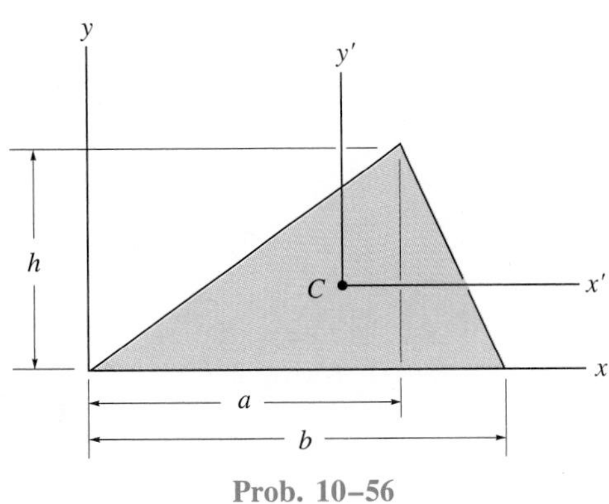

Prob. 10–56

*10.6 Product of Inertia for an Area

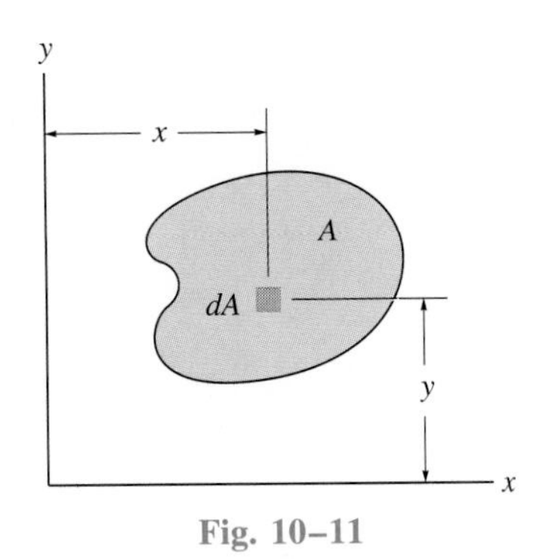

Fig. 10–11

In general, the moment of inertia for an area is different for every axis about which it is computed. In some applications of structural or mechanical design it is necessary to know the orientation of those axes which give, respectively, the maximum and minimum moments of inertia for the area. The method for determining this is discussed in Sec. 10.7. To use this method, however, one must first compute the product of inertia for the area as well as its moments of inertia for given x, y axes.

The product of inertia for an element of area located at point (x, y), Fig. 10–11, is defined as $dI_{xy} = xy\,dA$. Thus, for the entire area A, the *product of inertia* is

$$I_{xy} = \int_A xy\,dA \tag{10–7}$$

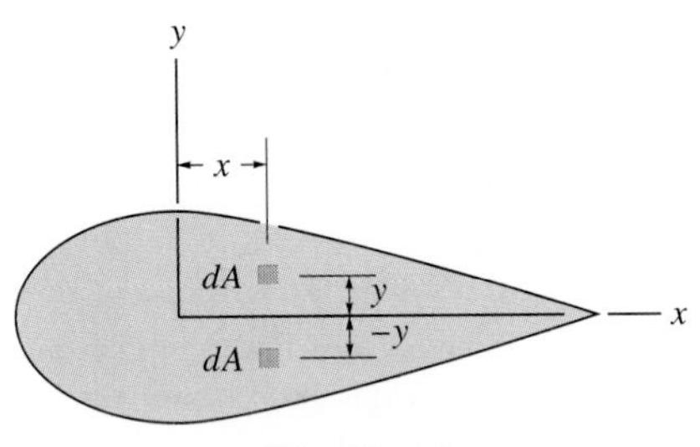

Fig. 10–12

If the element of area chosen has a differential size in two directions, as shown in Fig. 10–11, a double integration must be performed to evaluate I_{xy}. Most often, however, it is easier to choose an element having a differential size or thickness in only one direction, in which case the evaluation requires only a single integration (see Example 10–7).

Like the moment of inertia, the product of inertia has units of length raised to the fourth power, e.g., m^4, mm^4 or ft^4, in^4. However, since x or y may be a negative quantity, while the element of area is always positive, the product of inertia may be positive, negative, or zero, depending on the location and orientation of the coordinate axes. For example, the product of inertia I_{xy} for an area will be *zero* if either the x or y axis is an axis of *symmetry* for the area. To show this, consider the shaded area in Fig. 10–12, where for every element dA located at point (x, y) there is a corresponding element dA located at $(x, -y)$. Since the products of inertia for these elements are, respectively, $xy\,dA$ and $-xy\,dA$, the algebraic sum or integration of all the elements that are chosen in this way will cancel each other. Consequently, the product of inertia for the total area becomes zero. It also follows from the definition of I_{xy} that the "sign" of this quantity depends on the quadrant where the area is located. As shown in Fig. 10–13, if the area is rotated from one quadrant to another one, the sign of I_{xy} will change.

Parallel-Axis Theorem. Consider the shaded area shown in Fig. 10–14, where x' and y' represent a set of axes passing through the *centroid* of the area, and x and y represent a corresponding set of parallel axes. Since the product of inertia of dA with respect to the x and y axes is $dI_{xy} = (x' + d_x)$

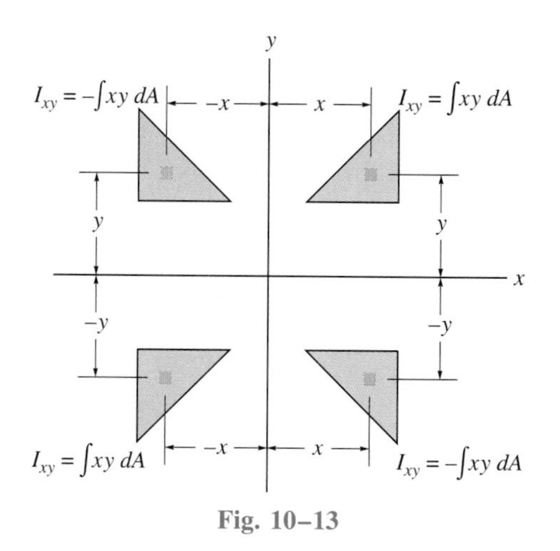

Fig. 10–13

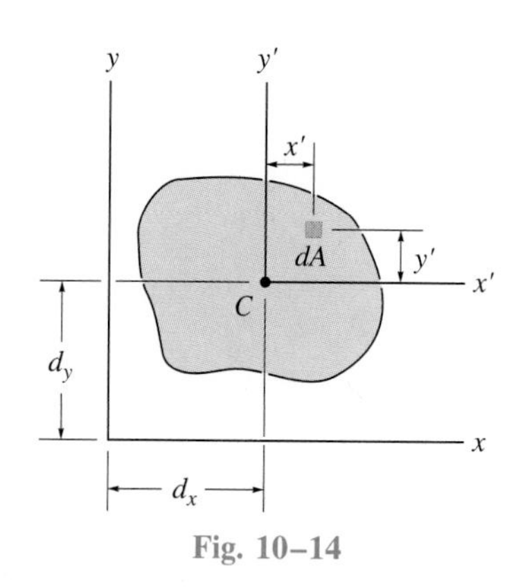

Fig. 10–14

$(y' + d_y)\, dA$, then for the entire area,

$$
\begin{aligned}
I_{xy} &= \int_A (x' + d_x)(y' + d_y)\, dA \\
&= \int_A x'y'\, dA + d_x \int_A y'\, dA + d_y \int_A x'\, dA + d_x d_y \int_A dA
\end{aligned}
$$

The first term on the right represents the product of inertia of the area with respect to the centroidal axis, $\bar{I}_{x'y'}$. The integrals in the second and third terms are zero since the moments of the area are taken about the centroidal axis. Realizing that the fourth integral represents the total area A, the final result is therefore

$$
I_{xy} = \bar{I}_{x'y'} + A d_x\, d_y \qquad (10\text{–}8)
$$

The similarity between this equation and the parallel-axis theorem for moments of inertia should be noted. In particular, it is important that the *algebraic signs* for d_x and d_y be maintained when applying Eq. 10–8. As illustrated in Example 10–8, the parallel-axis theorem finds important application in determining the product of inertia of a *composite area* with respect to a set of x, y axes.

Example 10–7

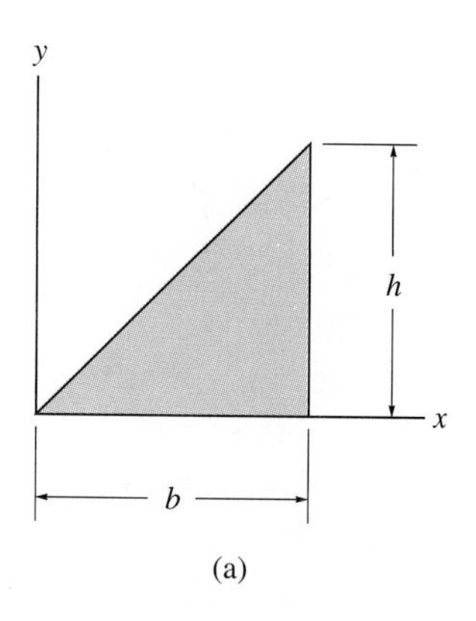

(a)

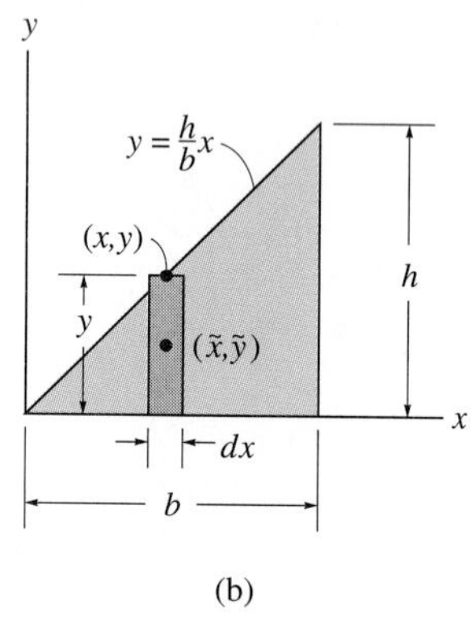

(b)

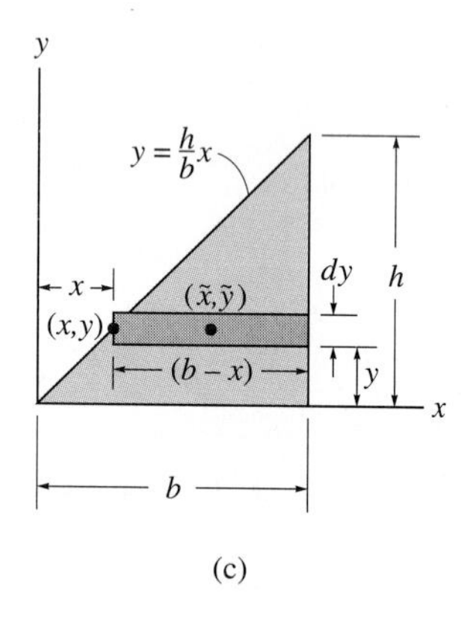

(c)

Fig. 10–15

Determine the product of inertia I_{xy} of the triangle shown in Fig. 10–15*a*.

SOLUTION I

A differential element that has a thickness dx, Fig. 10–15*b*, has an area $dA = y\,dx$. The product of inertia of the element about the x, y axes is determined using the parallel-axis theorem.

$$dI_{xy} = d\bar{I}_{x'y'} + dA\,\tilde{x}\,\tilde{y}$$

where $(\tilde{x}, \tilde{y})$ locates the *centroid* of the element or the origin of the x', y' axes. Since $d\bar{I}_{x'y'} = 0$, due to symmetry, and $\tilde{x} = x$, $\tilde{y} = y/2$, then

$$dI_{xy} = 0 + (y\,dx)x\left(\frac{y}{2}\right) = \left(\frac{h}{b}x\,dx\right)x\left(\frac{h}{2b}x\right)$$

$$= \frac{h^2}{2b^2}x^3\,dx$$

Integrating with respect to x from $x = 0$ to $x = b$ yields

$$I_{xy} = \frac{h^2}{2b^2}\int_0^b x^3\,dx = \frac{b^2h^2}{8} \qquad \textit{Ans.}$$

SOLUTION II

The differential element that has a thickness dy, Fig. 10–15*c*, and area $dA = (b - x)\,dy$ can also be used. The *centroid* is located at point $\tilde{x} = x + (b - x)/2 = (b + x)/2$, $\tilde{y} = y$, so the product of inertia of the element becomes

$$dI_{xy} = d\bar{I}_{x'y'} + dA\,\tilde{x}\,\tilde{y}$$

$$= 0 + (b - x)\,dy\left(\frac{b + x}{2}\right)y$$

$$= \left(b - \frac{b}{h}y\right)dy\left[\frac{b + (b/h)\,y}{2}\right]y = \frac{1}{2}y\left(b^2 - \frac{b^2}{h^2}y^2\right)dy$$

Integrating with respect to y from $y = 0$ to $y = h$ yields

$$I_{xy} = \frac{1}{2}\int_0^h y\left(b^2 - \frac{b^2}{h^2}y^2\right)dy = \frac{b^2h^2}{8} \qquad \textit{Ans.}$$

Example 10–8

Compute the product of inertia of the beam's cross-sectional area, shown in Fig. 10–16*a*, about the x and y centroidal axes.

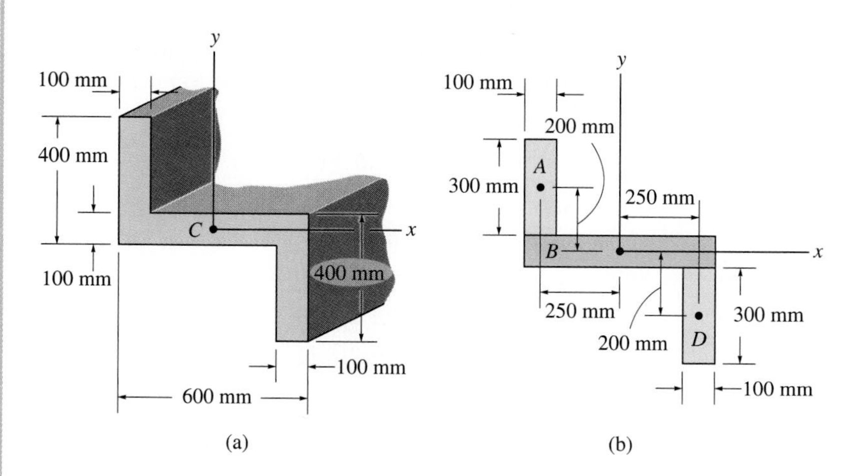

Fig. 10–16

SOLUTION

As in Example 10–6, the cross section can be considered as three composite rectangular areas A, B, and D, Fig. 10–16*b*. The coordinates for the centroid of each of these rectangles are shown in the figure. Due to symmetry, the product of inertia of *each rectangle* is *zero* about a set of x', y' axes that pass through the rectangle's centroid. Hence, application of the parallel-axis theorem to each of the rectangles yields

Rectangle A

$$\begin{aligned} I_{xy} &= \bar{I}_{x'y'} + A d_x d_y \\ &= 0 + (300)(100)(-250)(200) \\ &= -1.50(10^9)\text{ mm}^4 \end{aligned}$$

Rectangle B

$$\begin{aligned} I_{xy} &= \bar{I}_{x'y'} + A d_x d_y \\ &= 0 + 0 \\ &= 0 \end{aligned}$$

Rectangle D

$$\begin{aligned} I_{xy} &= \bar{I}_{x'y'} + A d_x d_y \\ &= 0 + (300)(100)(250)(-200) \\ &= -1.50(10^9)\text{ mm}^4 \end{aligned}$$

The product of inertia for the entire cross section is therefore

$$I_{xy} = -1.50(10^9) + 0 - 1.50(10^9) = -3.00(10^9)\text{ mm}^4 \qquad \textit{Ans.}$$

*10.7 Moments of Inertia for an Area About Inclined Axes

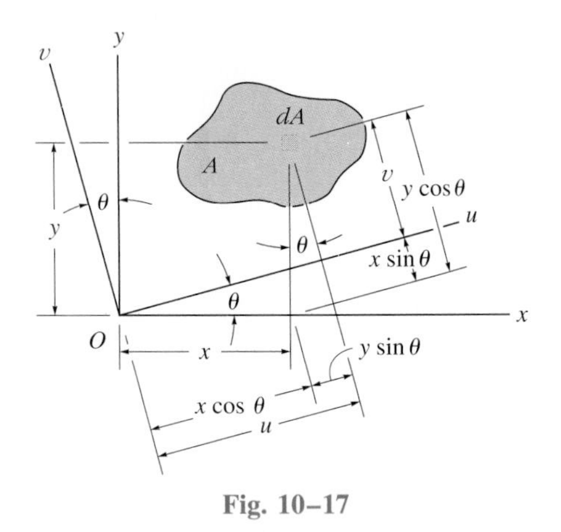

Fig. 10–17

In structural and mechanical design, it is sometimes necessary to calculate the moments and product of inertia I_u, I_v, and I_{uv} for an area with respect to a set of inclined u and v axes when the values for θ, I_x, I_y, and I_{xy} are *known.* To do this we will use *transformation equations* which relate the x, y and u, v coordinates. From Fig. 10–17, these equations are

$$u = x\cos\theta + y\sin\theta$$
$$v = y\cos\theta - x\sin\theta$$

Using these equations, the moments and product of inertia of dA about the u and v axes become

$$dI_u = v^2\,dA = (y\cos\theta - x\sin\theta)^2\,dA$$
$$dI_v = u^2\,dA = (x\cos\theta + y\sin\theta)^2\,dA$$
$$dI_{uv} = uv\,dA = (x\cos\theta + y\sin\theta)(y\cos\theta - x\sin\theta)\,dA$$

Expanding each expression and integrating, realizing that $I_x = \int y^2\,dA$, $I_y = \int x^2\,dA$, and $I_{xy} = \int xy\,dA$, we obtain

$$I_u = I_x\cos^2\theta + I_y\sin^2\theta - 2I_{xy}\sin\theta\cos\theta$$
$$I_v = I_x\sin^2\theta + I_y\cos^2\theta + 2I_{xy}\sin\theta\cos\theta$$
$$I_{uv} = I_x\sin\theta\cos\theta - I_y\sin\theta\cos\theta + I_{xy}(\cos^2\theta - \sin^2\theta)$$

These equations may be simplified by using the trigonometric identities $\sin 2\theta = 2\sin\theta\cos\theta$ and $\cos 2\theta = \cos^2\theta - \sin^2\theta$, in which case

$$I_u = \frac{I_x + I_y}{2} + \frac{I_x - I_y}{2}\cos 2\theta - I_{xy}\sin 2\theta$$
$$I_v = \frac{I_x + I_y}{2} - \frac{I_x - I_y}{2}\cos 2\theta + I_{xy}\sin 2\theta \qquad (10\text{–}9)$$
$$I_{uv} = \frac{I_x - I_y}{2}\sin 2\theta + I_{xy}\cos 2\theta$$

If the first and second equations are added together, we can show that the polar moment of inertia about the z axis passing through point O is *independent* of the orientation of the u and v axes; i.e.,

$$J_O = I_u + I_v = I_x + I_y$$

Principal Moments of Inertia. From Eqs. 10–9, it may be seen that I_u, I_v, and I_{uv} depend on the angle of inclination, θ, of the u, v axes. We will now determine the orientation of the u, v axes about which the moments of inertia for the area, I_u and I_v, are maximum and minimum. This particular set of axes is called the *principal axes* of the area, and the corresponding moments of inertia with respect to these axes are called the *principal moments of inertia.*

In general, there is a set of principal axes for every chosen origin O, although in structural and mechanical design, the area's centroid is an important location for O.

The angle $\theta = \theta_p$, which defines the orientation of the principal axes for the area, may be found by differentiating the first of Eqs. 10–9 with respect to θ and setting the result equal to zero. Thus,

$$\frac{dI_u}{d\theta} = -2\left(\frac{I_x - I_y}{2}\right)\sin 2\theta - 2I_{xy}\cos 2\theta = 0$$

Therefore, at $\theta = \theta_p$,

$$\tan 2\theta_p = \frac{-I_{xy}}{(I_x - I_y)/2} \qquad (10\text{–}10)$$

This equation has two roots, θ_{p_1} and θ_{p_2}, which are 90° apart and so specify the inclination of the principal axes. In order to substitute them into Eq. 10–9, we must first find the sine and cosine of $2\theta_{p_1}$ and $2\theta_{p_2}$. This can be done using the triangles shown in Fig. 10–18, which are based on Eq. 10–10.

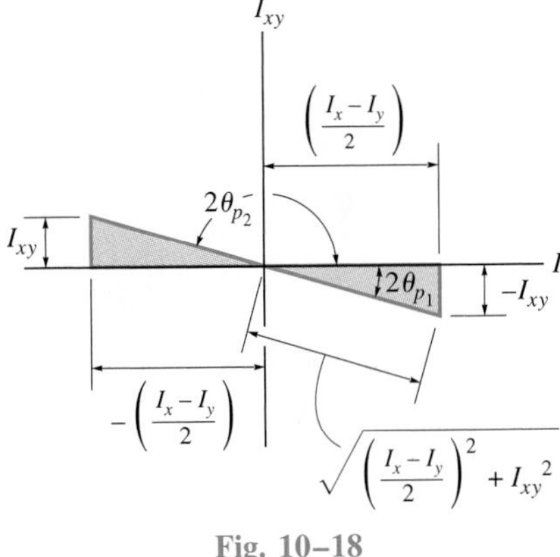

Fig. 10–18

For θ_{p_1},

$$\sin 2\theta_{p_1} = -I_{xy} \Big/ \sqrt{\left(\frac{I_x - I_y}{2}\right)^2 + I_{xy}^2}$$

$$\cos 2\theta_{p_1} = \left(\frac{I_x - I_y}{2}\right) \Big/ \sqrt{\left(\frac{I_x - I_y}{2}\right)^2 + I_{xy}^2}$$

For θ_{p_2},

$$\sin 2\theta_{p_2} = I_{xy} \Big/ \sqrt{\left(\frac{I_x - I_y}{2}\right)^2 + I_{xy}^2}$$

$$\cos 2\theta_{p_2} = -\left(\frac{I_x - I_y}{2}\right) \Big/ \sqrt{\left(\frac{I_x - I_y}{2}\right)^2 + I_{xy}^2}$$

Substituting these two sets of trigonometric relations into the first or second of Eqs. 10–9 and simplifying, we obtain

$$I_{\substack{\max \\ \min}} = \frac{I_x + I_y}{2} \pm \sqrt{\left(\frac{I_x - I_y}{2}\right)^2 + I_{xy}^2} \qquad (10\text{–}11)$$

Depending on the sign chosen, this result gives the maximum or minimum moment of inertia for the area. Furthermore, if the above trigonometric relations for θ_{p_1} and θ_{p_2} are substituted into the third of Eqs. 10–9, it can be shown that $I_{uv} = 0$; that is, the *product of inertia with respect to the principal axes is zero.* Since it was indicated in Sec. 10.6 that the product of inertia is zero with respect to any symmetrical axis, it therefore follows that *any symmetrical axis represents a principal axis of inertia for the area.*

Example 10–9

Determine the principal moments of inertia for the beam's cross-sectional area shown in Fig. 10–19*a* with respect to an axis passing through the centroid.

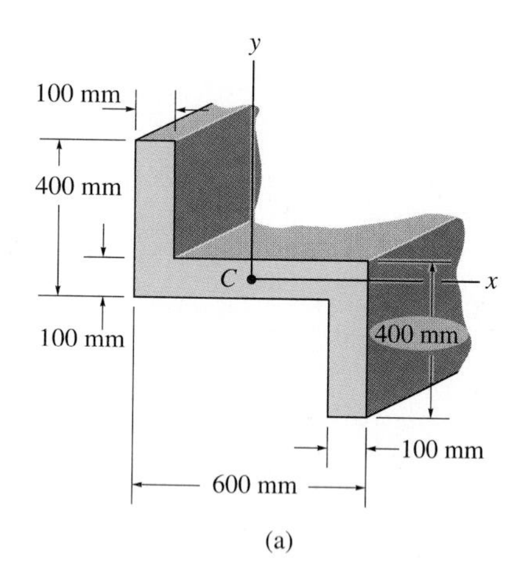

(a)

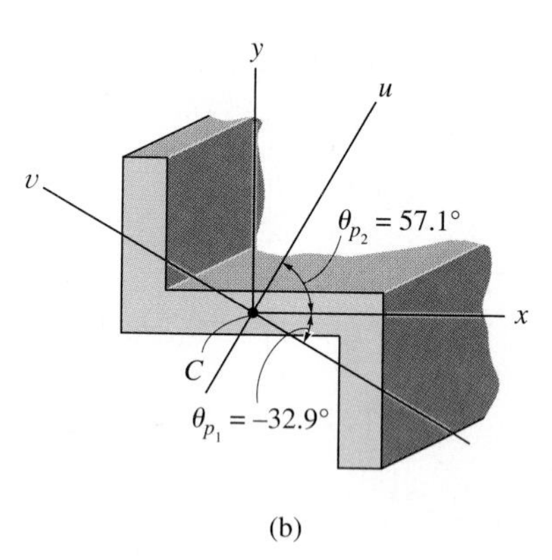

(b)

Fig. 10–19

SOLUTION

The moments and product of inertia of the cross section with respect to the *x*, *y* axes have been computed in Examples 10–6 and 10–8. The results are

$$I_x = 2.90(10^9)\ \text{mm}^4 \qquad I_y = 5.60(10^9)\ \text{mm}^4 \qquad I_{xy} = -3.00(10^9)\ \text{mm}^4$$

Using Eq. 10–10, the angles of inclination of the principal axes *u* and *v* are

$$\tan 2\theta_p = \frac{-I_{xy}}{(I_x - I_y)/2} = \frac{3.00(10^9)}{[2.90(10^9) - 5.60(10^9)]/2} = -2.22$$

$$2\theta_{p_1} = -65.8° \quad \text{and} \quad 2\theta_{p_2} = 114.2°$$

Thus, as shown in Fig. 10–19*b*,

$$\theta_{p_1} = -32.9° \quad \text{and} \quad \theta_{p_2} = 57.1°$$

The principal moments of inertia with respect to the *u* and *v* axes are determined from Eq. 10–11. Hence,

$$I_{\substack{\max\\ \min}} = \frac{I_x + I_y}{2} \pm \sqrt{\left(\frac{I_x - I_y}{2}\right)^2 + I_{xy}^2}$$

$$= \frac{2.90(10^9) + 5.60(10^9)}{2} \pm \sqrt{\left[\frac{2.90(10^9) - 5.60(10^9)}{2}\right]^2 + [-3.00(10^9)]^2}$$

$$I_{\substack{\max\\ \min}} = 4.25(10^9) \pm 3.29(10^9)$$

or

$$I_{\max} = 7.54(10^9)\ \text{mm}^4 \qquad I_{\min} = 0.960(10^9)\ \text{mm}^4 \qquad \textit{Ans.}$$

Specifically, the maximum moment of inertia, $I_{\max} = 7.54(10^9)\ \text{mm}^4$, occurs with respect to the selected *u* axis, since *by inspection* most of the cross-sectional area is farthest away from this axis. Or, stated in another manner, $I_{\max}$ occurs about the *u* axis since it is located within ±45° of the *y* axis, which has the largest value of I ($I_y > I_x$). Also, this may be concluded mathematically by substituting the data with $\theta = 57.1°$ into the first of Eqs. 10–9.

*10.8 Mohr's Circle for Moments of Inertia

Equations 10–9 to 10–11 have a graphical solution that is convenient to use and generally easy to remember. Squaring the first and third of Eqs. 10–9 and adding, it is found that

$$\left(I_u - \frac{I_x + I_y}{2}\right)^2 + I_{uv}^2 = \left(\frac{I_x - I_y}{2}\right)^2 + I_{xy}^2 \qquad (10\text{–}12)$$

In a given problem, I_u and I_{uv} are *variables,* and I_x, I_y, and I_{xy} are *known constants*. Thus, Eq. 10–12 may be written in compact form as

$$(I_u - a)^2 + I_{uv}^2 = R^2$$

When this equation is plotted on a set of axes that represent the respective moment of inertia and the product of inertia, Fig. 10–20, the resulting graph represents a *circle* of radius

$$R = \sqrt{\left(\frac{I_x - I_y}{2}\right)^2 + I_{xy}^2}$$

having its center located at point $(a, 0)$, where $a = (I_x + I_y)/2$. The circle so constructed is called *Mohr's circle,* named after the German engineer Otto Mohr (1835–1918).

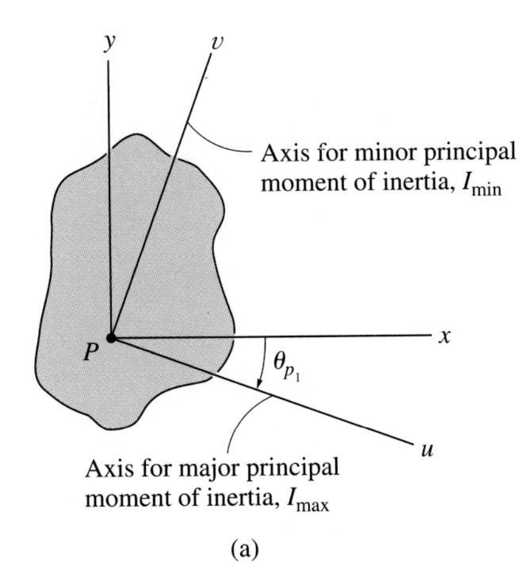

(a)

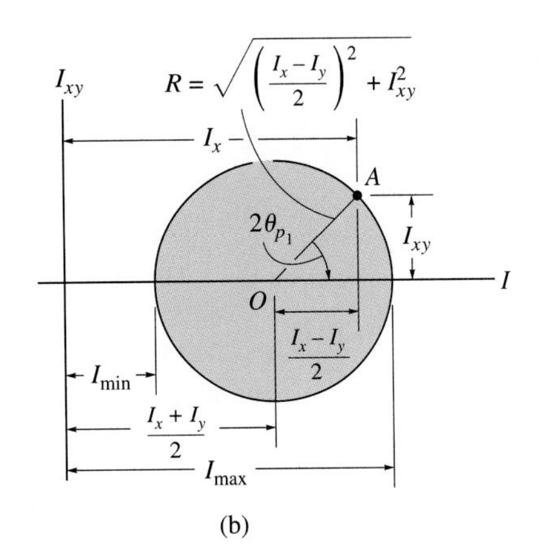

(b)

Fig. 10–20

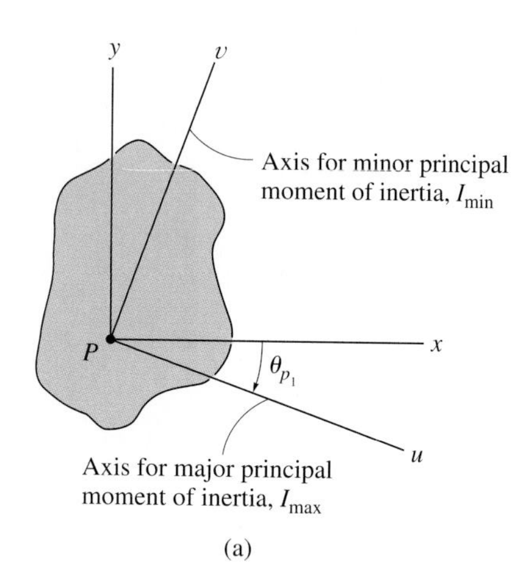

(a)

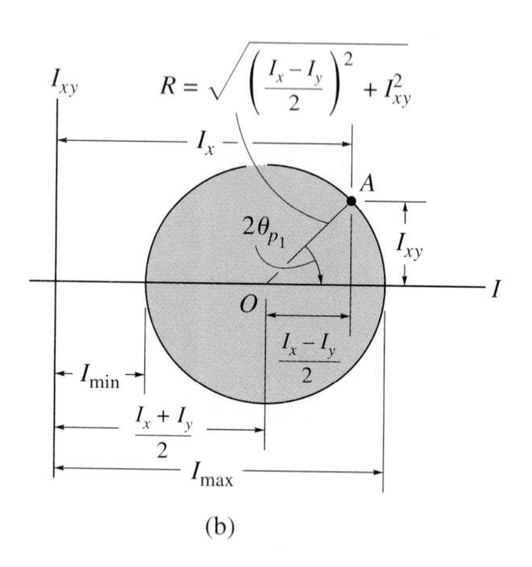

(b)

Fig. 10–20

PROCEDURE FOR ANALYSIS

The main purpose in using Mohr's circle here is to have a convenient means for transforming I_x, I_y, and I_{xy} into the principal moments of inertia. The following procedure provides a method for doing this.

Determine I_x, I_y, I_{xy}. Establish the x, y axes for the area, with the origin located at the point P of interest, and determine I_x, I_y, and I_{xy}, Fig. 10–20*a*.

Construct the Circle. Construct a rectangular coordinate system such that the abscissa represents the moment of inertia I, and the ordinate represents the product of inertia I_{xy}, Fig. 10–20*b*. Determine the center of the circle, O, which is located at a distance $(I_x + I_y)/2$ from the origin, and plot the reference point A having coordinates (I_x, I_{xy}). By definition, I_x is always positive, whereas I_{xy} will be either positive or negative. Connect the reference point A with the center of the circle, and determine the distance OA by trigonometry. This distance represents the radius of the circle, Fig. 10–20*b*. Finally, draw the circle.

Principal Moments of Inertia. The points where the circle intersects the abscissa give the values of the principal moments of inertia I_{min} and I_{max}. Notice that the *product of inertia will be zero at these points,* Fig. 10–20*b*.

Principal Axes. To find the direction of the major principal axis, determine by trigonometry the angle $2\theta_{p_1}$, *measured from the radius OA to the positive I axis,* Fig. 10–20*b*. This angle represents *twice* the angle from the x axis of the area in question to the axis of maximum moment of inertia I_{max}, Fig. 10–20*a*. Both the angle on the circle, $2\theta_{p_1}$, and the angle to the axis on the area, θ_{p_1}, *must be measured in the same sense,* as shown in Fig. 10–20. The axis for minimum moment of inertia I_{min} is perpendicular to the axis for I_{max}.

Using trigonometry, the above procedure may be verified to be in accordance with the equations developed in Sec. 10.7.

Example 10–10

Using Mohr's circle, determine the principal moments of inertia for the beam's cross-sectional area, shown in Fig. 10–21*a*, with respect to an axis passing through the centroid.

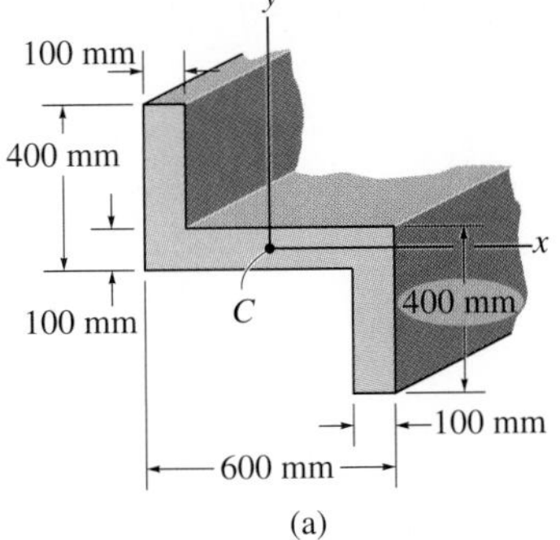

(a)

SOLUTION

Determine I_x, I_y, I_{xy}. The moments of inertia and the product of inertia have been determined in Examples 10–6 and 10–8 with respect to the x, y axes shown in Fig. 10–21*a*. The results are $I_x = 2.90(10^9)\text{ mm}^4$, $I_y = 5.60(10^9)\text{ mm}^4$, and $I_{xy} = -3.00(10^9)\text{ mm}^4$.

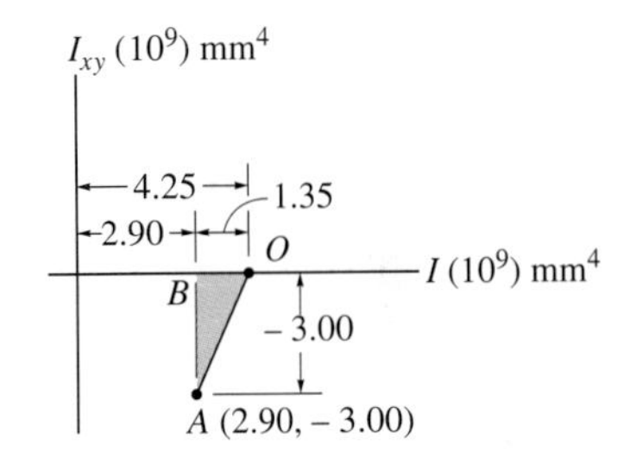

(b)

Construct the Circle. The I and I_{xy} axes are shown in Fig. 10–21*b*. The center of the circle, O, lies at a distance $(I_x + I_y)/2 = (2.90 + 5.60)/2 = 4.25$ from the origin. When the reference point $A(2.90, -3.00)$ is connected to point O, the radius OA is determined from the triangle OBA using the Pythagorean theorem.

$$OA = \sqrt{(1.35)^2 + (-3.00)^2} = 3.29$$

The circle is constructed in Fig. 10–21*c*.

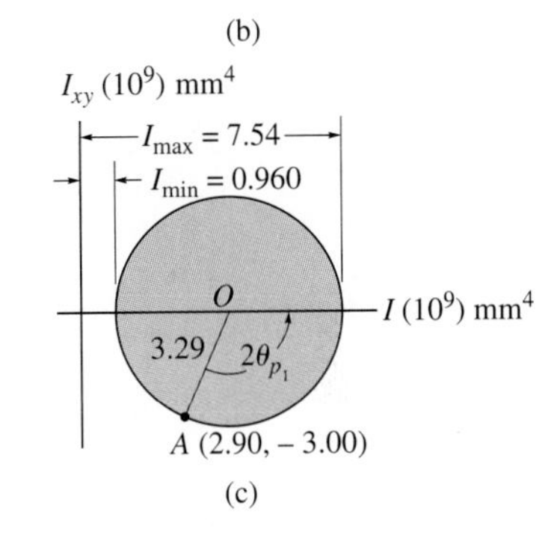

(c)

Principal Moments of Inertia. The circle intersects the I axis at points (7.54, 0) and (0.960, 0). Hence,

$$I_{max} = 7.54(10^9)\text{ mm}^4 \qquad \textit{Ans.}$$
$$I_{min} = 0.960(10^9)\text{ mm}^4 \qquad \textit{Ans.}$$

Principal Axes. As shown in Fig. 10–21*c*, the angle $2\theta_{p_1}$ is determined from the circle by measuring counterclockwise from OA to the direction of the *positive* I axis. Hence,

$$2\theta_{p_1} = 180° - \sin^{-1}\left(\frac{|BA|}{|OA|}\right) = 180° - \sin^{-1}\left(\frac{3.00}{3.29}\right) = 114.2°$$

The principal axis for $I_{max} = 7.54(10^9)\text{ mm}^4$ is therefore oriented at an angle $\theta_{p_1} = 57.1°$, measured *counterclockwise,* from the *positive* x axis to the *positive* u axis. The v axis is perpendicular to this axis. The results are shown in Fig. 10–21*d*.

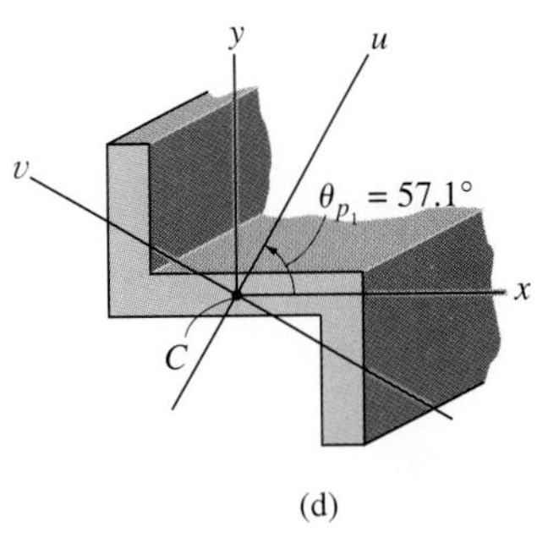

(d)

Fig. 10–21

PROBLEMS

10–57. Determine the product of inertia of the shaded area with respect to the x and y axes.

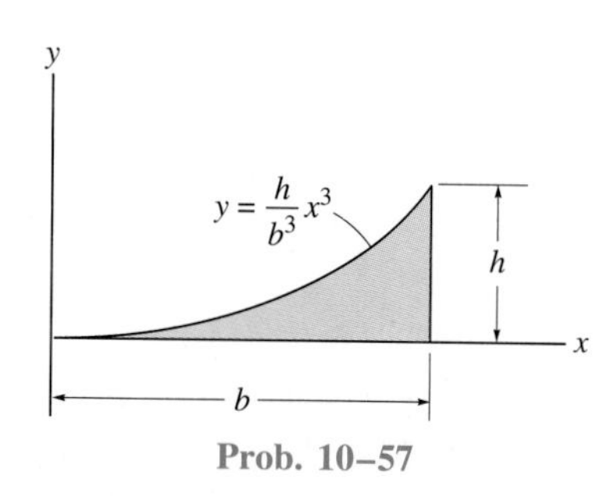

Prob. 10–57

10–58. Determine the product of inertia of the shaded area of the ellipse with respect to the x and y axes.

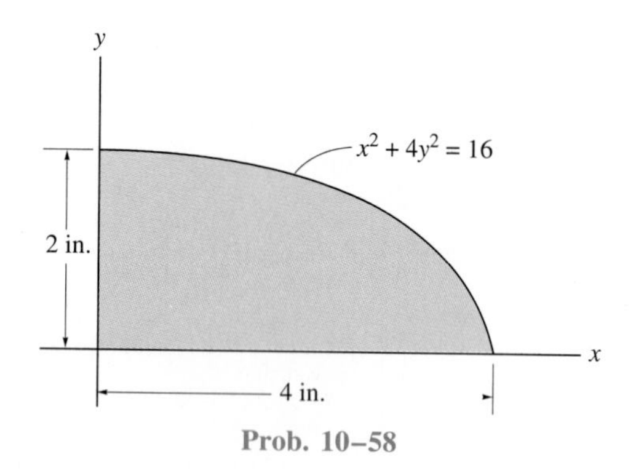

Prob. 10–58

10–59. Determine the product of inertia of the shaded area with respect to the x and y axes.

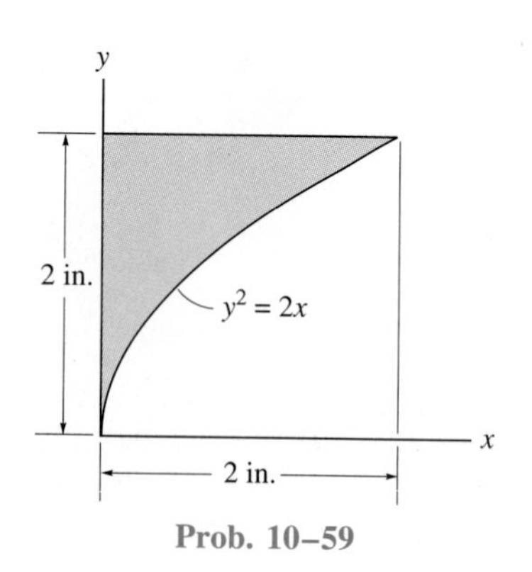

Prob. 10–59

***10–60.** Determine the product of inertia of the parabolic area with respect to the x and y axes.

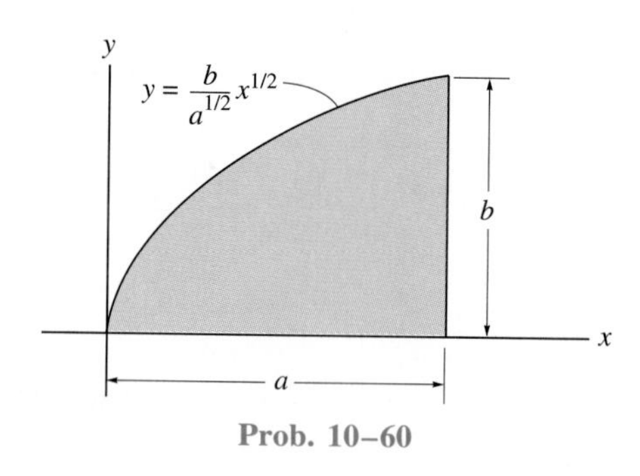

Prob. 10–60

10–61. Determine the product of inertia of the shaded area with respect to the x and y axes.

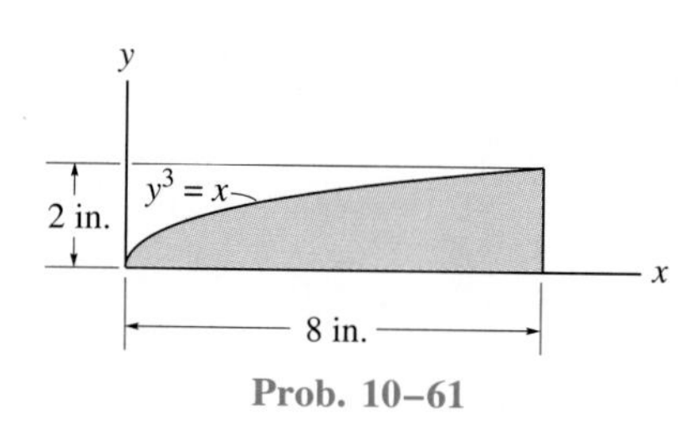

Prob. 10–61

10–62. Determine the product of inertia of the shaded area with respect to the x and y axes.

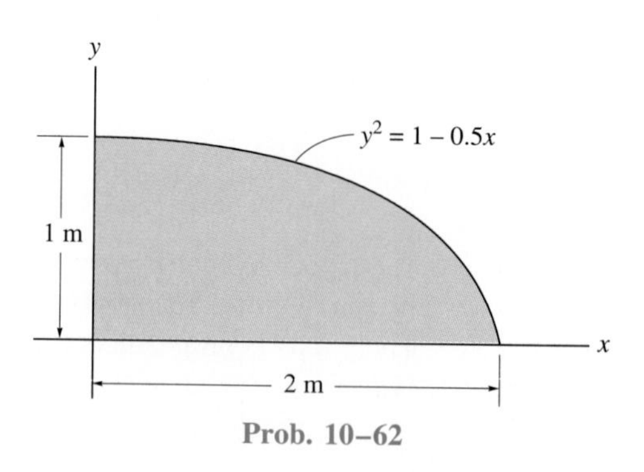

Prob. 10–62

10–63. Determine the product of inertia of the shaded area with respect to the x and y axes.

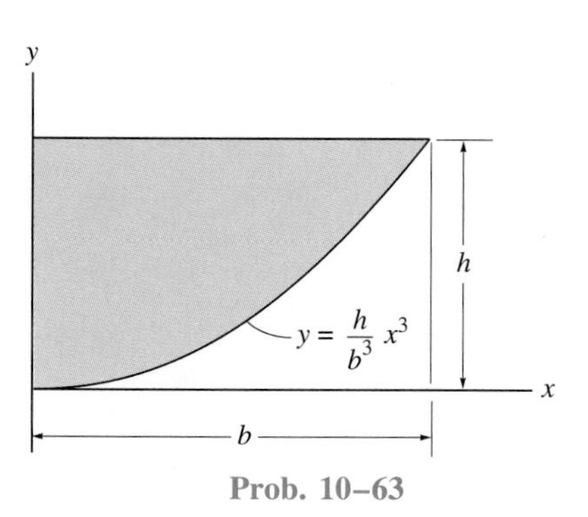

Prob. 10–63

***10–64.** Determine the product of inertia of the shaded area with respect to the x and y axes.

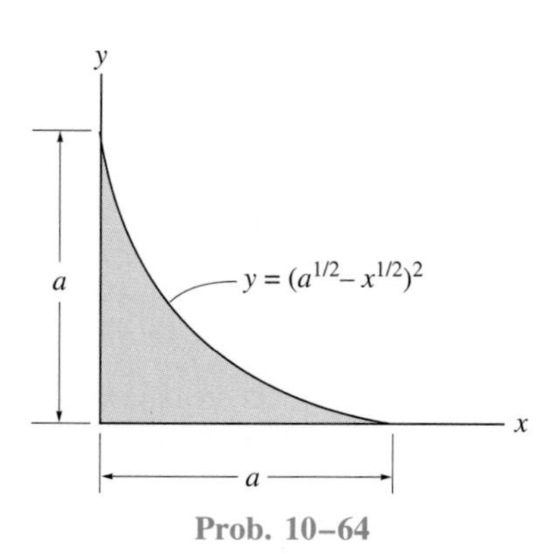

Prob. 10–64

■**10–65.** Determine the product of inertia of the shaded area with respect to the x and y axes. Use Simpson's rule to evaluate the integral.

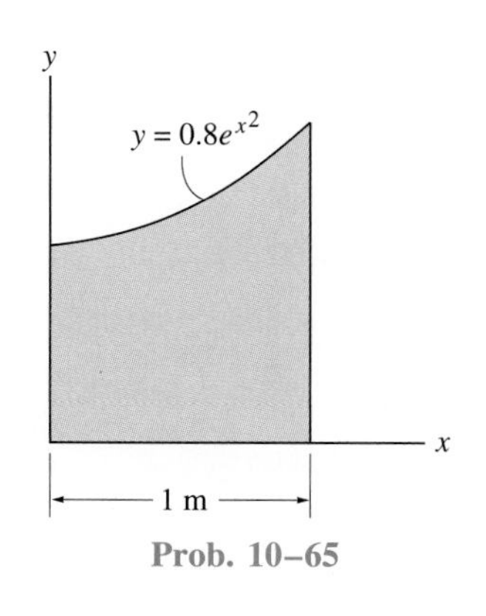

Prob. 10–65

10–66. Determine the product of inertia of the beam's cross-sectional area with respect to the x and y axes.

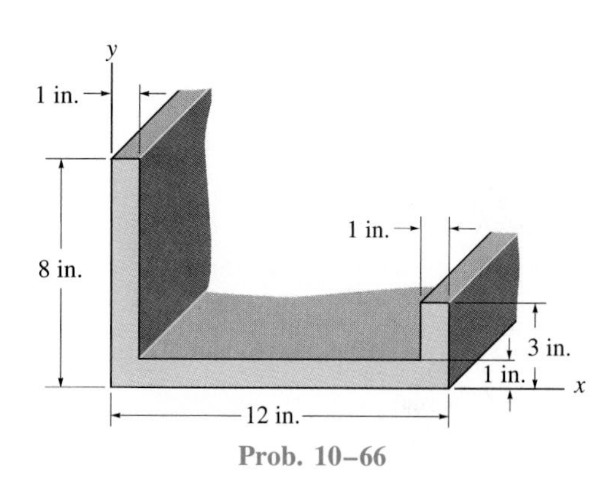

Prob. 10–66

10–67. Locate the position $\bar{x}$, $\bar{y}$ for the centroid C of the beam's cross-sectional area, and then compute the product of inertia with respect to the x' and y' axes.

***10–68.** Determine the product of inertia of the beam's cross-sectional area with respect to the x and y axes.

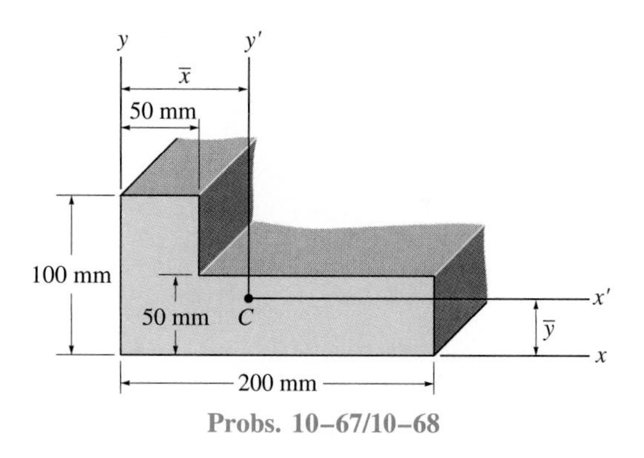

Probs. 10–67/10–68

10–69. Determine the product of inertia of the beam's cross-sectional area with respect to the x and y axes that have their origin located at the centroid C.

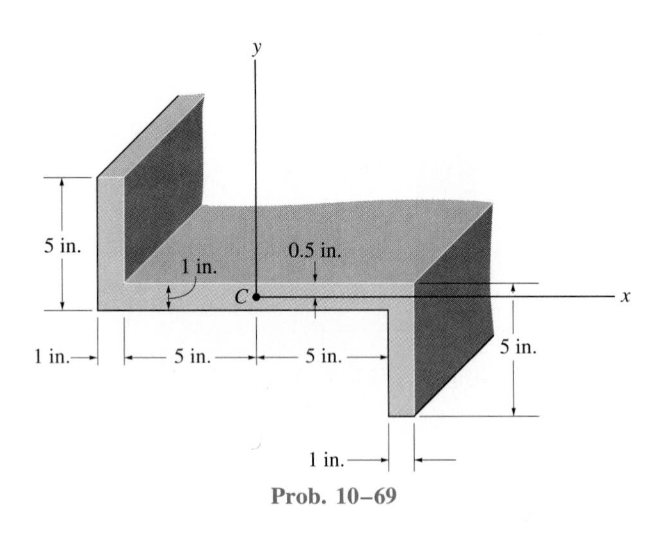

Prob. 10–69

10–70. Determine the product of inertia of the beam's cross-sectional area with respect to the x and y axes that have their origin located at the centroid C.

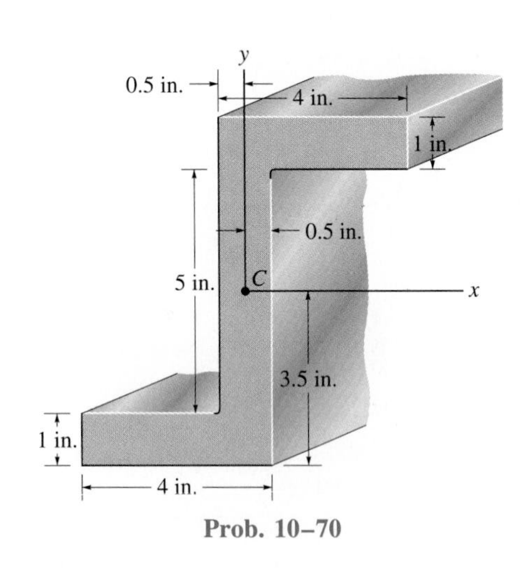

Prob. 10–70

10–71. Determine the product of inertia for the angle with respect to the x and y axes passing through the centroid C. Assume all corners to be square.

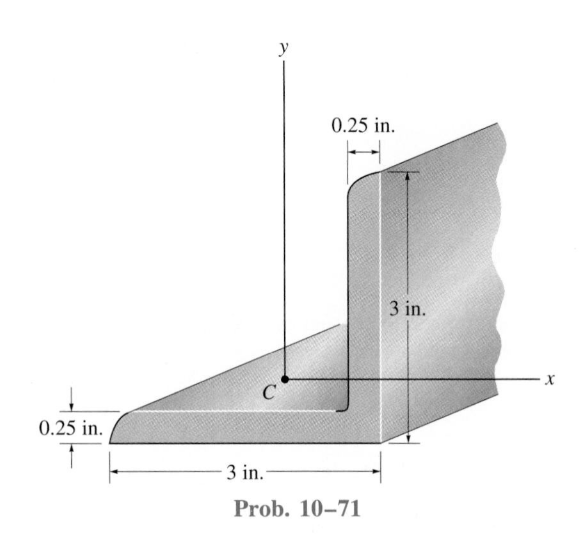

Prob. 10–71

***10–72.** Determine the product of inertia of the beam's cross-sectional area with respect to the x and y axes.

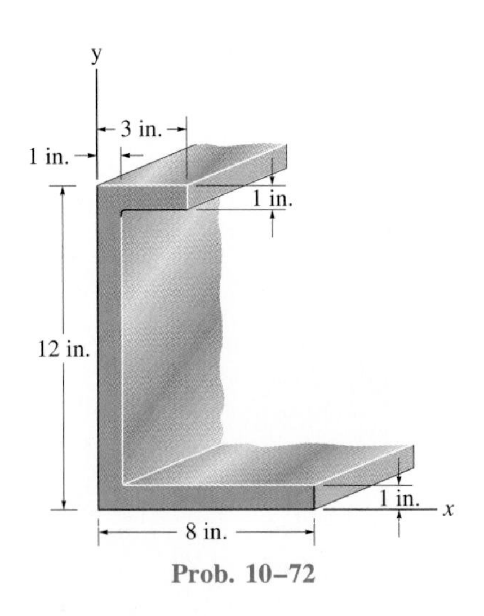

Prob. 10–72

10–73. Determine the product of inertia for the beam's cross-sectional area with respect to the u and v axes.

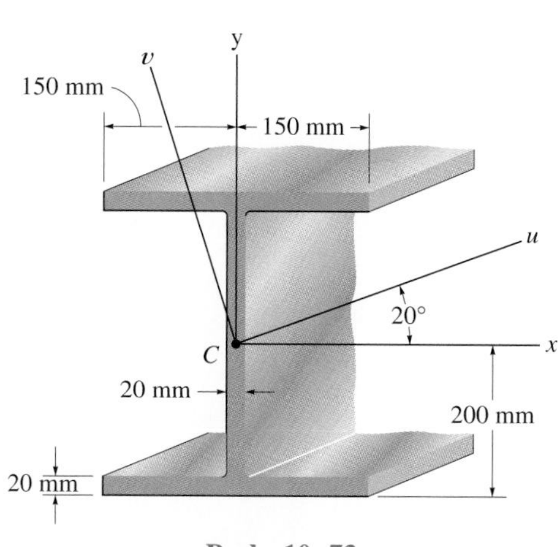

Prob. 10–73

10–74. Determine the moments of inertia I_u and I_v and the product of inertia I_{uv} for the rectangular area. The u and v axes pass through the centroid C. Take $\theta = 30°$.

10–75. Determine the moments of inertia I_u and I_v and the product of inertia I_{uv} for the rectangular area. The u and v axes pass through the centroid C. Take $\theta = 20°$.

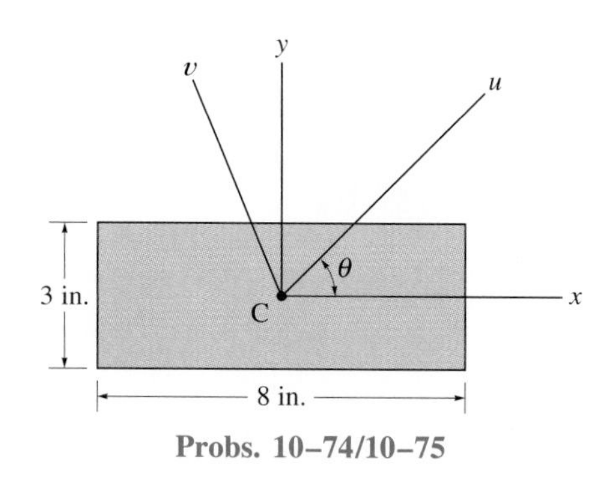

Probs. 10–74/10–75

***10–76.** Determine the distance $\bar{y}$ to the centroid of the area and then calculate the moments of inertia I_u and I_v of the channel's cross-sectional area. The u and v axes have their origin at the centroid C. For the calculation, assume all corners to be square.

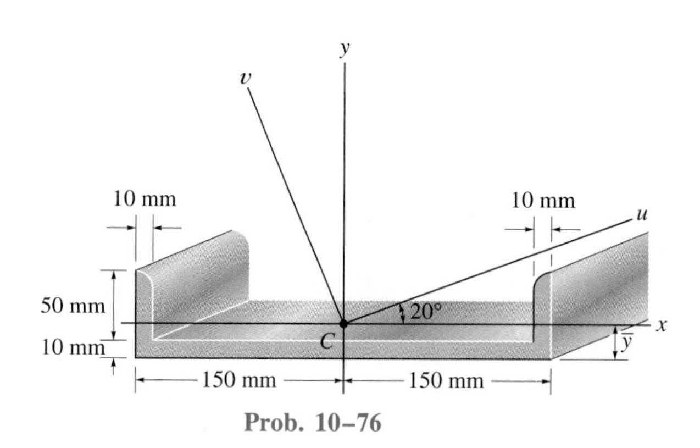

Prob. 10–76

10–77. The area of the cross section of an airplane wing has the following properties about the x and y axes passing through the centroid C: $\bar{I}_x = 450 \text{ in}^4$, $\bar{I}_y = 1730 \text{ in}^4$, $\bar{I}_{xy} = 138 \text{ in}^4$. Determine the orientation of the principal axes and the principal moments of inertia.

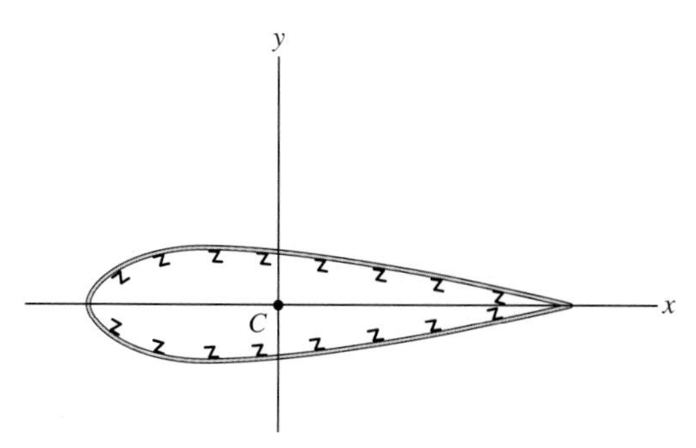

Prob. 10–77

10–78. Determine the moments of inertia I_u, I_v and the product of inertia I_{uv} of the beam's cross-sectional area. Take $\theta = 45°$.

10–79. Determine the moments of inertia I_u, I_v and the product of inertia I_{uv} of the beam's cross-sectional area. Take $\theta = 60°$.

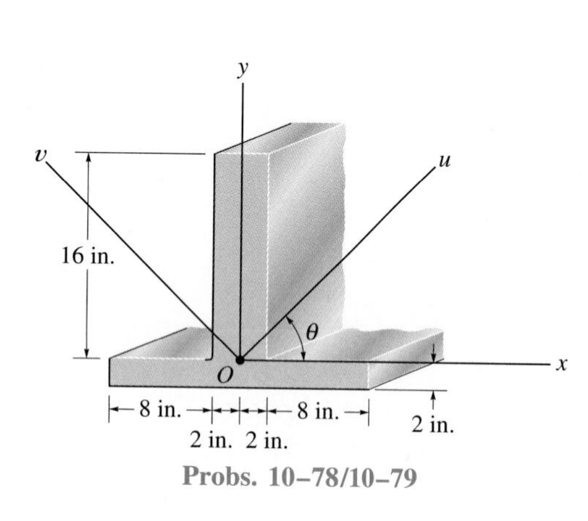

Probs. 10–78/10–79

***10–80.** Determine the moments of inertia I_u and I_v of the beam's cross-sectional area.

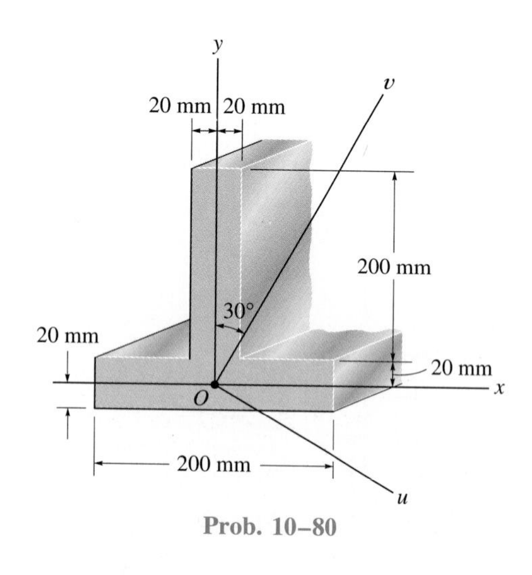

Prob. 10–80

10–81. Determine the principal moments of inertia of the beam's cross-sectional area about the principal axes that have their origin located at the centroid C. Use the equations developed in Section 10–7. For the calculation, assume all corners to be square.

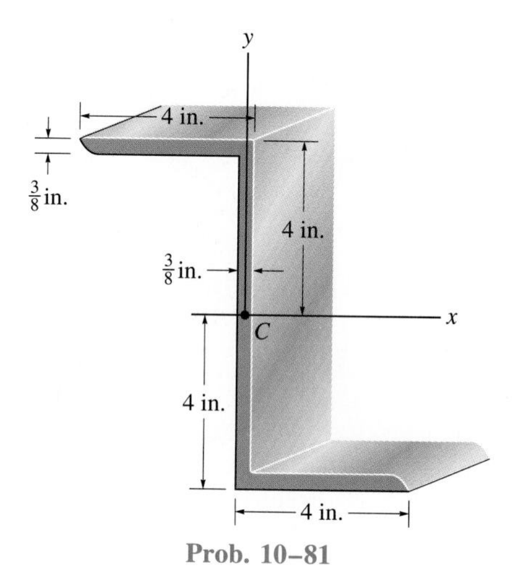

Prob. 10–81

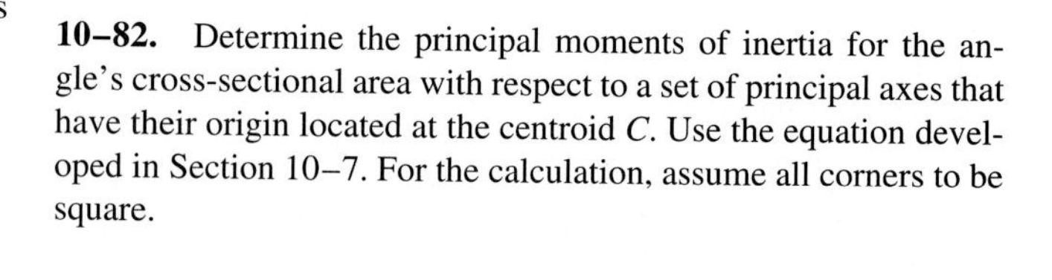

10–82. Determine the principal moments of inertia for the angle's cross-sectional area with respect to a set of principal axes that have their origin located at the centroid C. Use the equation developed in Section 10–7. For the calculation, assume all corners to be square.

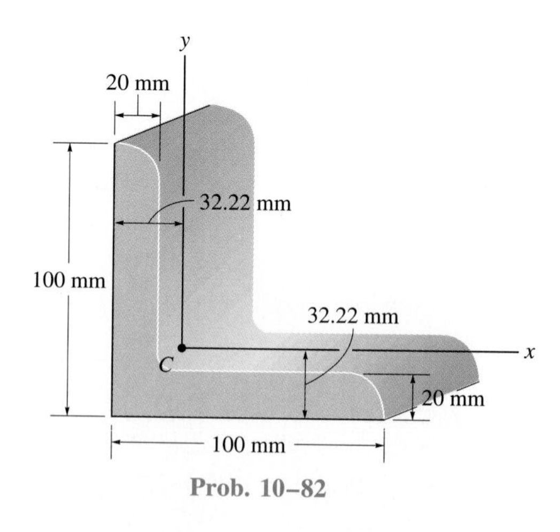

Prob. 10–82

10–83. Locate the centroid, $\bar{y}$, and determine the orientation of the principal centroidal axes for the composite area. What are the moments of inertia with respect to these axes?

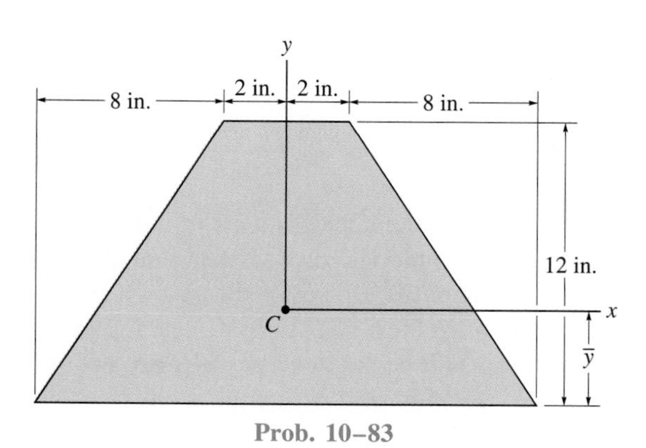

Prob. 10–83

*10–84. Determine the directions of the principal axes with origin located at point O, and the principal moments of inertia for the quarter-circular area about these axes.

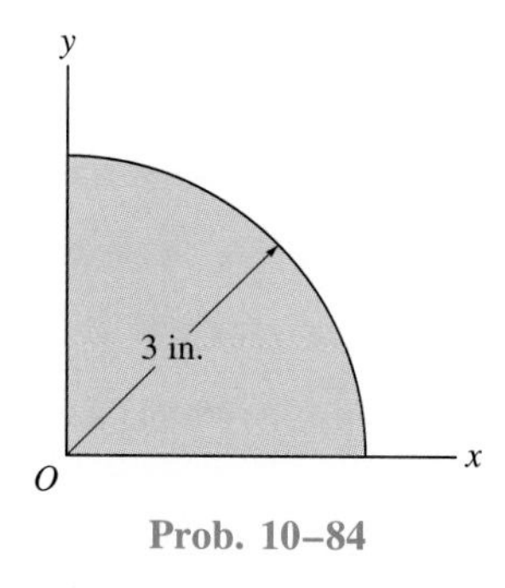

Prob. 10–84

10–85. Determine the directions of the principal axes with origin located at point O, and the principal moments of inertia for the rectangular area about these axes.

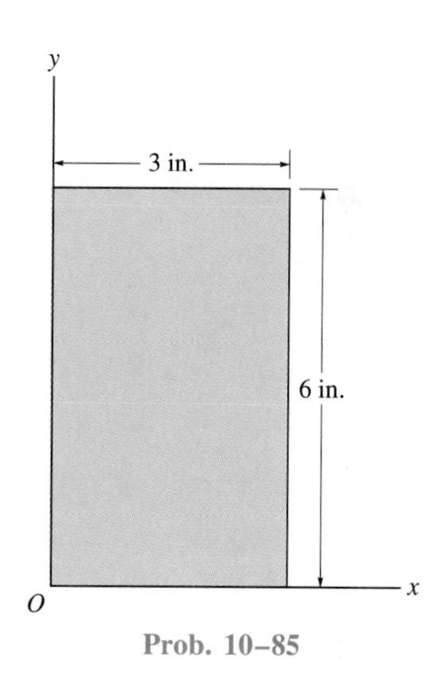

Prob. 10–85

10–86. Solve Prob. 10–77 using Mohr's circle.

10–87. Solve Prob. 10–81 using Mohr's circle.

***10–88.** Solve Prob. 10–84 using Mohr's circle.

10–89. Solve Prob. 10–82 using Mohr's circle.

10–90. Solve Prob. 10–85 using Mohr's circle.

10.9 Mass Moment of Inertia

The mass moment of inertia of a body is a property that measures the resistance of the body to angular acceleration. Since it is used in dynamics to study rotational motion, methods for its calculation will now be discussed.

We define the *mass moment of inertia* as the integral of the "second moment" about an axis of all the elements of mass dm which compose the body.* For example, consider the rigid body shown in Fig. 10–22. The body's moment of inertia about the z axis is

$$I = \int_m r^2\, dm \tag{10–13}$$

Here the "moment arm" r is the perpendicular distance from the axis to the arbitrary element dm. Since the formulation involves r, the value of I is *unique* for each axis z about which it is computed. However, the axis which is generally chosen for analysis passes through the body's mass center G. The moment of inertia computed about this axis will be defined as I_G. Realize that because r is squared in Eq. 10–13, the mass moment of inertia is always a *positive quantity*. Common units used for its measurement are $kg \cdot m^2$ or $slug \cdot ft^2$.

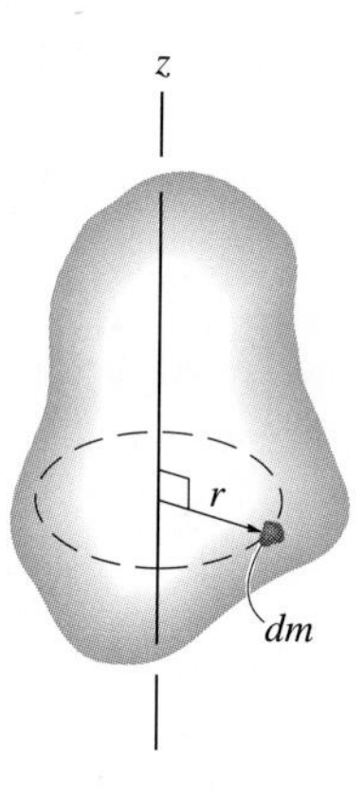

Fig. 10–22

*Another property of the body which measures the symmetry of the body's mass with respect to a coordinate system is the mass product of inertia. This property most often applies to the three-dimensional motion of a body and is discussed in *Engineering Mechanics: Dynamics* (Chapter 21).

PROCEDURE FOR ANALYSIS

For integration, we will consider only symmetric bodies having surfaces which are generated by revolving a curve about an axis. An example of such a body which is generated about the z axis is shown in Fig. 10–23.

If the body consists of material having a variable density, $\rho = \rho(x, y, z)$, the elemental mass dm of the body may be expressed in terms of its density and volume as $dm = \rho\, dV$. Substituting dm into Eq. 10–13, the body's moment of inertia is then computed using *volume elements* for integration; i.e.,

$$I = \int_V r^2 \rho\, dV \tag{10–14}$$

In the special case of ρ being a *constant,* this term may be factored out of the integral and the integration is then purely a function of geometry:

$$I = \rho \int_V r^2\, dV \tag{10–15}$$

When the elemental volume chosen for integration has differential sizes in all three directions, e.g., $dV = dx\, dy\, dz$, Fig. 10–23*a*, the moment of inertia of the body must be determined using "triple integration." The integration process can, however, be simplified to a *single integration* provided the chosen elemental volume has a differential size or thickness in only *one direction.* Shell or disk elements are often used for this purpose.

Shell Element. If a *shell element* having a height z, radius y, and thickness dy is chosen for integration, Fig. 10–23*b*, then the volume $dV = (2\pi y)(z)\, dy$. This element may be used in Eq. 10–14 or 10–15 for determining the moment of inertia I_z of the body about the z axis, since the *entire element,* due to its "thinness," lies at the *same* perpendicular distance $r = y$ from the z axis (see Example 10–11).

Disk Element. If a disk element having a radius y and a thickness dz is chosen for integration, Fig. 10–23*c*, then the volume $dV = (\pi y^2)\, dz$. In this case, however, the element is *finite* in the radial direction, and consequently its parts *do not* all lie at the *same radial distance* r from the z axis. As a result, Eq. 10–14 or 10–15 *cannot* be used to determine I_z. Instead, to perform the integration using this element, it is first necessary to determine the moment of inertia *of the element* about the z axis and then integrate this result (see Example 10–12).

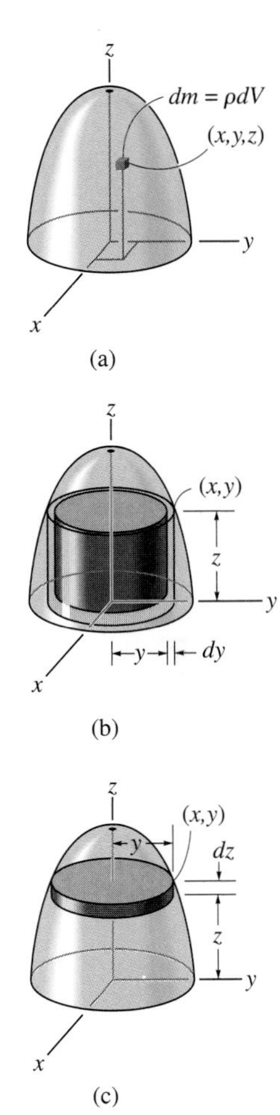

Fig. 10–23

Example 10–11

Determine the moment of inertia of the cylinder shown in Fig. 10–24*a* about the z axis. The density ρ of the material is constant.

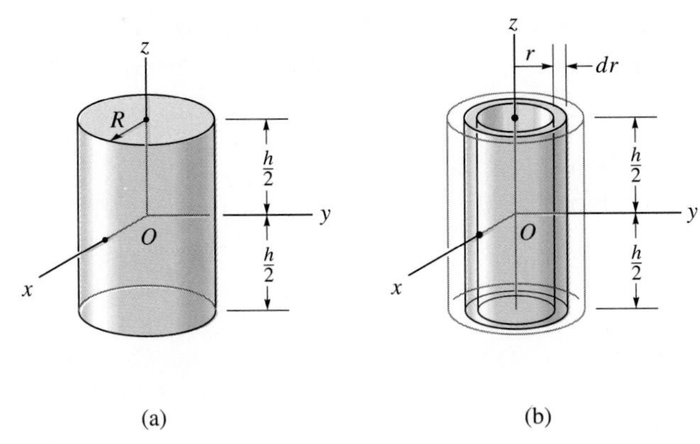

Fig. 10–24

SOLUTION

Shell Element. This problem may be solved using the *shell element* in Fig. 10–24*b* and single integration. The volume of the element is $dV = (2\pi r)(h)\,dr$, so that its mass is $dm = \rho\,dV = \rho(2\pi hr\,dr)$. Since the *entire element* lies at the same distance r from the z axis, the moment of inertia *of the element* is

$$dI_z = r^2\,dm = \rho 2\pi h r^3\,dr$$

Integrating over the entire region of the cylinder yields

$$I_z = \int_m r^2\,dm = \rho 2\pi h \int_0^R r^3\,dr = \frac{\rho\pi}{2} R^4 h$$

The mass of the cylinder is

$$m = \int_m dm = \rho 2\pi h \int_0^R r\,dr = \rho\pi h R^2$$

so that

$$I_z = \frac{1}{2} m R^2 \qquad \textit{Ans.}$$

Example 10–12

A solid is formed by revolving the shaded area shown in Fig. 10–25*a* about the *y* axis. If the density of the material is 5 slug/ft^3, determine the moment of inertia about the *y* axis.

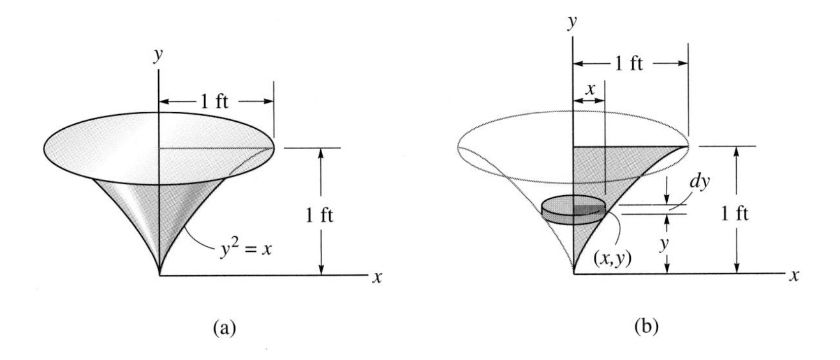

Fig. 10–25

SOLUTION

Disk Element. The moment of inertia will be determined using a *disk element,* as shown in Fig. 10–25*b*. Here the element intersects the curve at the arbitrary point (x, y) and has a mass

$$dm = \rho\, dV = \rho(\pi x^2)\, dy$$

Although all portions of the element are *not* located at the same distance from the *y* axis, it is still possible to determine the moment of inertia dI_y *of the element* about the *y* axis. In Example 10–11 it was shown that the moment of inertia of a cylinder about its longitudinal axis is $I = \frac{1}{2}mR^2$, where m and R are the mass and radius of the cylinder. Since the height of the cylinder is not involved in this formula, we can also use it for a disk. Thus, for the disk element in Fig. 10–25*b*, we have

$$dI_y = \frac{1}{2}(dm)x^2 = \frac{1}{2}[\rho(\pi x^2)\, dy]x^2$$

Substituting $x = y^2$, $\rho = 5$ slug/ft^3, and integrating with respect to y, from $y = 0$ to $y = 1$ ft, yields the moment of inertia for the entire solid:

$$I_y = \frac{5\pi}{2}\int_0^1 x^4\, dy = \frac{5\pi}{2}\int_0^1 y^8\, dy = 0.873 \text{ slug} \cdot \text{ft}^2 \qquad \textit{Ans.}$$

Parallel-Axis Theorem. If the moment of inertia of the body about an axis passing through the body's mass center is known, then the moment of inertia about any other *parallel axis* may be determined by using the *parallel-axis theorem.* This theorem can be derived by considering the body shown in Fig. 10–26. The z' axis passes through the mass center G, whereas the corresponding *parallel* z *axis* lies at a constant distance d away. Selecting the differential element of mass dm which is located at point (x', y') and using the Pythagorean theorem, $r^2 = (d + x')^2 + y'^2$, we can express the moment of inertia of the body about the z axis as

$$I = \int_m r^2\,dm = \int_m [(d + x')^2 + y'^2]\,dm$$

$$= \int_m (x'^2 + y'^2)\,dm + 2d\int_m x'\,dm + d^2\int_m dm$$

Since $r'^2 = x'^2 + y'^2$, the first integral represents I_G. The second integral equals *zero,* since the z' axis passes through the body's mass center, i.e., $\int x'\,dm = \bar{x}\int dm = 0$ since $\bar{x} = 0$. Finally, the third integral represents the total mass m of the body. Hence, the moment of inertia about the z axis can be written as

$$I = I_G + md^2 \qquad (10\text{–}16)$$

where I_G = moment of inertia about the z' axis passing through the mass center G

m = mass of the body

d = perpendicular distance between the parallel axes

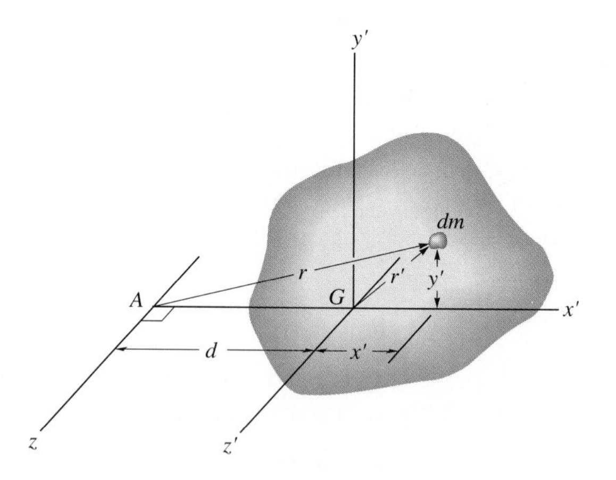

Fig. 10–26

Radius of Gyration. Occasionally, the moment of inertia of a body about a specified axis is reported in handbooks using the *radius of gyration, k*. This value has units of length, and when it and the body's mass m are known, the moment of inertia is determined from the equation

$$I = mk^2 \quad \text{or} \quad k = \sqrt{\frac{I}{m}} \tag{10–17}$$

Note the *similarity* between the definition of k in this formula and r in the equation $dI = r^2\,dm$, which defines the moment of inertia of an elemental mass dm of the body about an axis.

Composite Bodies. If a body is constructed from a number of simple shapes such as disks, spheres, and rods, the moment of inertia of the body about any axis z can be determined by adding algebraically the moments of inertia of all the composite shapes computed about the z axis. Algebraic addition is necessary since a composite part must be considered as a negative quantity if it has already been included within another part—for example, a "hole" subtracted from a solid plate. The parallel-axis theorem is needed for the calculations if the center of mass of each composite part does not lie on the z axis. For the calculation, then, $I = \Sigma(I_G + md^2)$, where I_G for each of the composite parts is computed by integration or can be determined from a table, such as the one given on the inside back cover.

Example 10–13

If the plate shown in Fig. 10–27*a* has a density of 8000 kg/m^3 and a thickness of 10 mm, compute its moment of inertia about an axis directed perpendicular to the page and passing through point *O*.

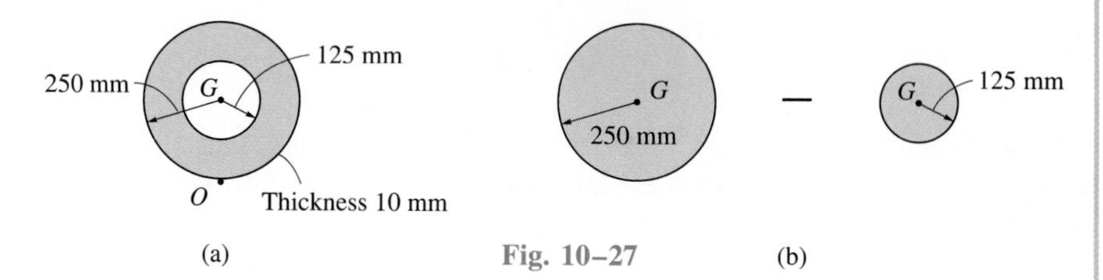

Fig. 10–27

SOLUTION

The plate consists of two composite parts, the 250-mm-radius disk *minus* a 125-mm-radius disk, Fig. 10–27*b*. The moment of inertia about *O* can be determined by computing the moment of inertia of each of these parts about *O* and then *algebraically* adding the results. The computations are performed by using the parallel-axis theorem in conjunction with the data listed in the table on the inside back cover.

Disk. The moment of inertia of a disk about an axis perpendicular to the plane of the disk is $I_G = \frac{1}{2}mr^2$. The mass center of the disk is located at a distance of 0.25 m from point *O*. Thus,

$$m_d = \rho_d V_d = 8000 \text{ kg/m}^3[\pi(0.25 \text{ m})^2(0.01 \text{ m})] = 15.71 \text{ kg}$$

$$\begin{aligned}(I_O)_d &= \tfrac{1}{2}m_d r_d^2 + m_d d^2 \\ &= \frac{1}{2}(15.71 \text{ kg})(0.25 \text{ m})^2 + (15.71 \text{ kg})(0.25 \text{ m})^2 \\ &= 1.473 \text{ kg} \cdot \text{m}^2\end{aligned}$$

Hole. For the 125-mm-radius disk (hole), we have

$$m_h = \rho_h V_h = 8000 \text{ kg/m}^3[\pi(0.125 \text{ m})^2(0.01 \text{ m})] = 3.93 \text{ kg}$$

$$\begin{aligned}(I_O)_h &= \tfrac{1}{2}m_h r_h^2 + m_h d^2 \\ &= \frac{1}{2}(3.93 \text{ kg})(0.125 \text{ m})^2 + (3.93 \text{ kg})(0.25 \text{ m})^2 \\ &= 0.276 \text{ kg} \cdot \text{m}^2\end{aligned}$$

The moment of inertia of the plate about point *O* is therefore

$$\begin{aligned}I_O &= (I_O)_d - (I_O)_h \\ &= 1.473 \text{ kg} \cdot \text{m}^2 - 0.276 \text{ kg} \cdot \text{m}^2 \\ &= 1.20 \text{ kg} \cdot \text{m}^2\end{aligned}$$

Ans.

Example 10–14

The pendulum consists of two thin rods each having a weight of 10 lb and suspended from point O as shown in Fig. 10–28. Determine the pendulum's moment of inertia about an axis passing through (a) the pin at O, and (b) the mass center G of the pendulum.

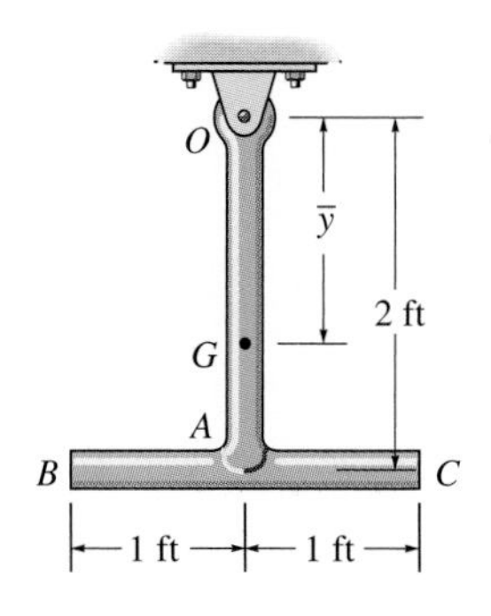

Fig. 10–28

SOLUTION

Part (a). Using the table on the inside back cover, the moment of inertia of rod OA about an axis perpendicular to the page and passing through the end point O of the rod is $I_O = \frac{1}{3}ml^2$. Hence,

$$(I_{OA})_O = \frac{1}{3}ml^2 = \frac{1}{3}\left(\frac{10}{32.2}\right)(2)^2 = 0.414 \text{ slug} \cdot \text{ft}^2$$

The same value may be computed using $I_G = \frac{1}{12}ml^2$ and the parallel-axis theorem; i.e.,

$$(I_{OA})_O = \frac{1}{12}ml^2 + md^2 = \frac{1}{12}\left(\frac{10}{32.2}\right)(2)^2 + \frac{10}{32.2}(1)^2$$
$$= 0.414 \text{ slug} \cdot \text{ft}^2$$

For rod BC we have

$$(I_{BC})_O = \frac{1}{12}ml^2 + md^2 = \frac{1}{12}\left(\frac{10}{32.2}\right)(2)^2 + \frac{10}{32.2}(2)^2$$
$$= 1.346 \text{ slug} \cdot \text{ft}^2$$

The moment of inertia of the pendulum about O is therefore

$$I_O = 0.414 + 1.346 = 1.76 \text{ slug} \cdot \text{ft}^2 \qquad \textit{Ans.}$$

Part (b). The mass center G will be located relative to the pin at O. Assuming this distance to be $\bar{y}$, Fig. 10–28, and using the formula for determining the mass center, we have

$$\bar{y} = \frac{\Sigma \tilde{y}m}{\Sigma m} = \frac{1(10/32.2) + 2(10/32.2)}{(10/32.2) + (10/32.2)} = 1.50 \text{ ft}$$

The moment of inertia I_G may be computed in the same manner as I_O, which requires successive applications of the parallel-axis theorem in order to transfer the moments of inertia of rods OA and BC to G. A more direct solution, however, involves applying the parallel-axis theorem using the result for I_O determined above; i.e.,

$$I_O = I_G + md^2; \qquad 1.76 = I_G + \left(\frac{20}{32.2}\right)(1.50)^2$$

$$I_G = 0.362 \text{ slug} \cdot \text{ft}^2 \qquad \textit{Ans.}$$

PROBLEMS

10–91. Determine the moment of inertia I_y for the slender rod. The rod's density ρ and cross-sectional area A are constant. Express the result in terms of the rod's total mass m.

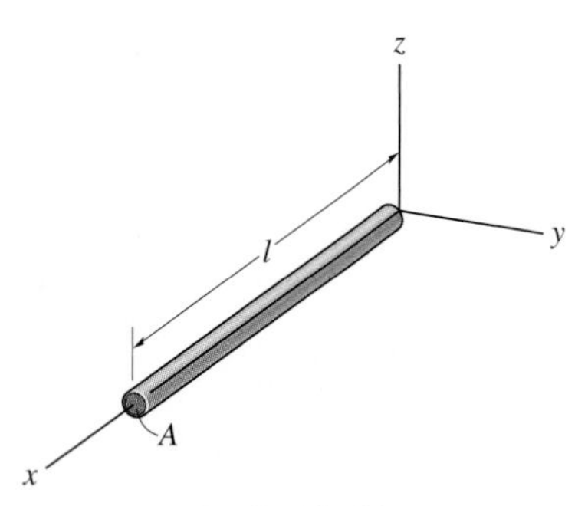

Prob. 10–91

*10–92. Determine the moment of inertia of the thin ring about the z axis. The ring has a mass m.

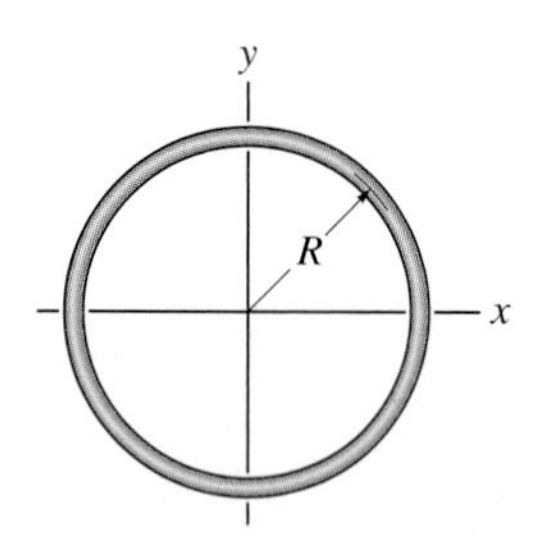

Prob. 10–92

10–93. The right circular cone is formed by revolving the shaded area around the x axis. Determine the moment of inertia I_x and express the result in terms of the total mass m of the cone. The cone has a constant density ρ.

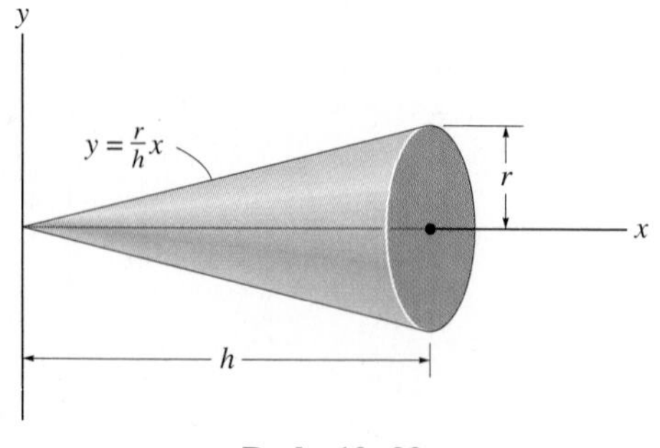

Prob. 10–93

10–94. The sphere is formed by revolving the shaded area around the x axis. Determine the moment of inertia I_x and express the result in terms of the total mass m of the sphere. The sphere has a constant density ρ.

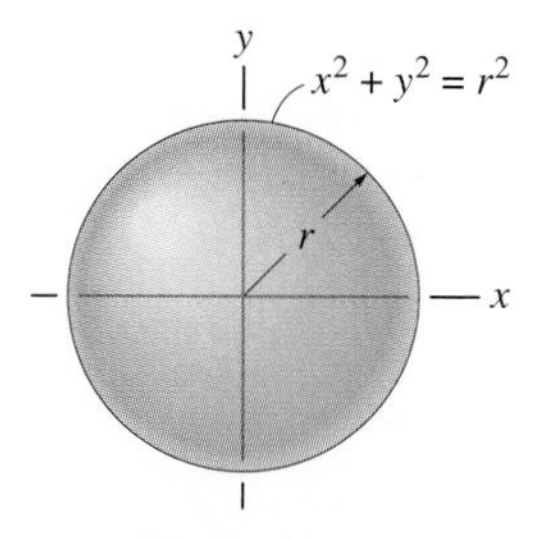

Prob. 10–94

10–95. The paraboloid is formed by revolving the shaded area around the x axis. Determine the radius of gyration k_x. The material has a constant density ρ.

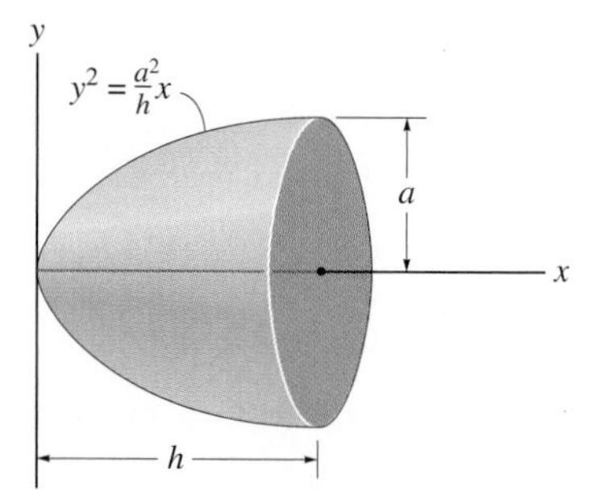

Prob. 10–95

*10–96. A semiellipsoid is formed by rotating the shaded area about the x axis. Determine the moment of inertia with respect to the x axis and express the result in terms of the mass m of the semiellipsoid. The material has constant density ρ.

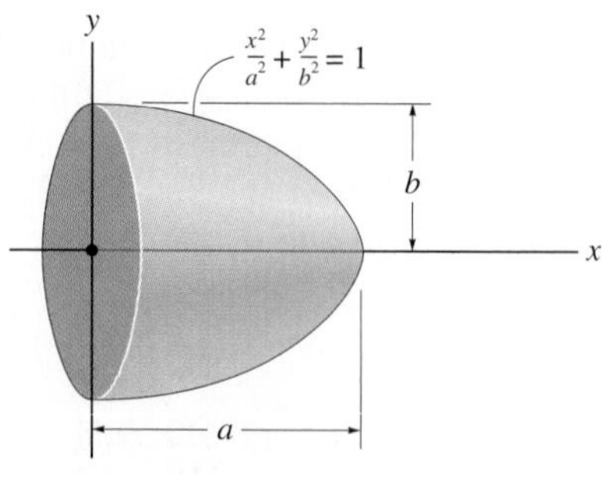

Prob. 10–96

10–97. An ellipsoid is formed by rotating the shaded area about the x axis. Determine the moment of inertia with respect to the x axis and express the result in terms of the mass m of the ellipsoid. The material has a constant density ρ.

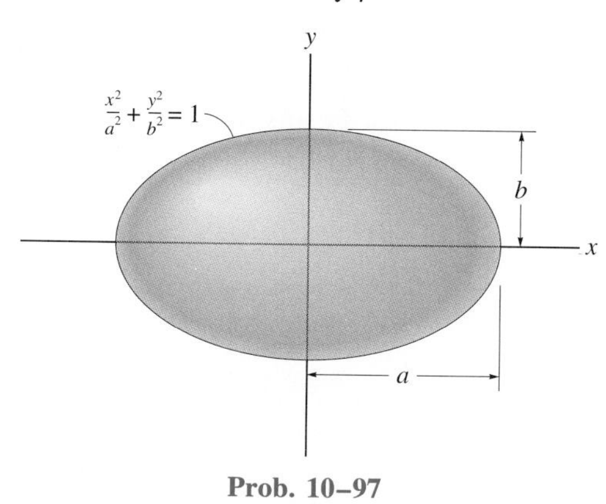

Prob. 10–97

10–98. The concrete shape is formed by rotating the shaded area about the y axis. Determine the moment of inertia I_y. The specific weight of concrete is $\gamma = 150\ \text{lb/ft}^3$.

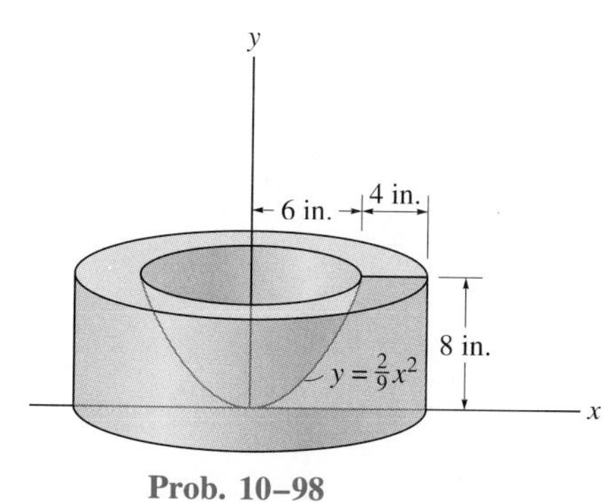

Prob. 10–98

10–99. The frustum is formed by rotating the shaded area around the x axis. Determine the moment of inertia I_x and express the result in terms of the total mass m of the frustum. The frustum has a constant density.

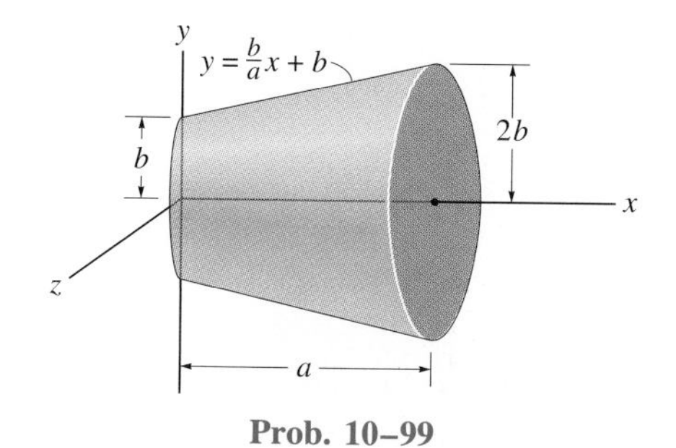

Prob. 10–99

***10–100.** The body is formed by revolving the shaded area around the x axis. Determine the radius of gyration k_x. The specific weight of the material is $\gamma = 380\ \text{lb/ft}^3$.

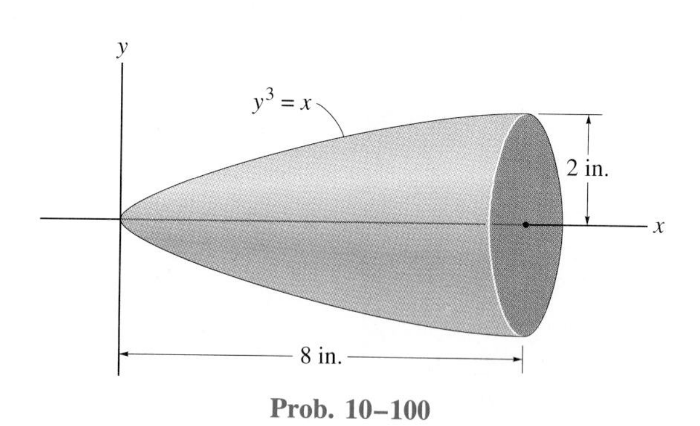

Prob. 10–100

10–101. Determine the moment of inertia of the homogeneous triangular prism with respect to the y axis. Express the result in terms of the mass m of the prism. *Hint:* For integration, use thin plate elements parallel to the x-y plane and having a thickness dz.

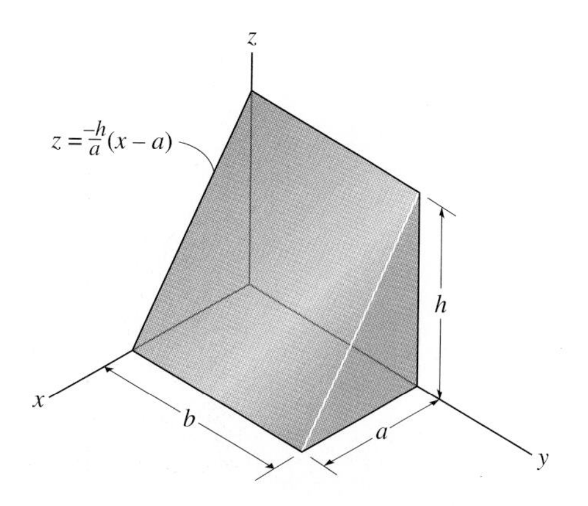

Prob. 10–101

10–102. The slender rods have a weight of 3 lb/ft. Determine the moment of inertia of the assembly about an axis perpendicular to the page and passing through point A.

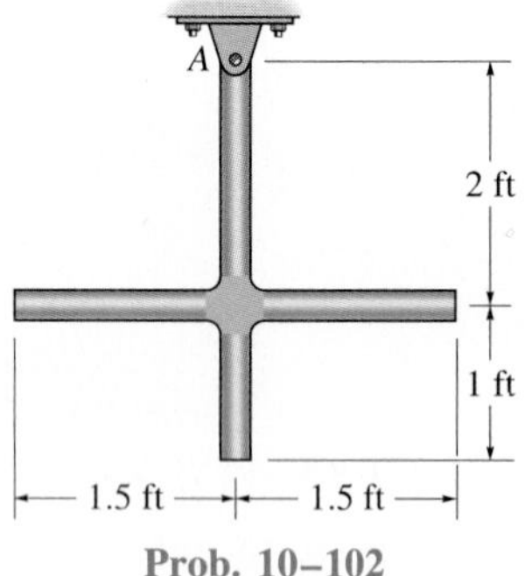
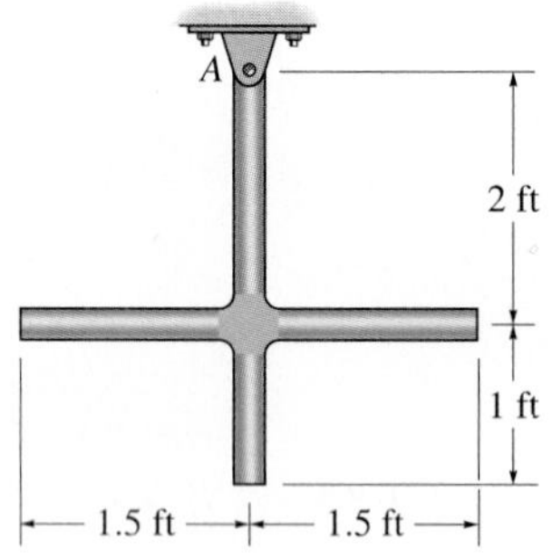

Prob. 10–102

10–103. Determine the moment of inertia I_z of the frustum of the cone which has a conical depression. The material has a density of 200 kg/m^3.

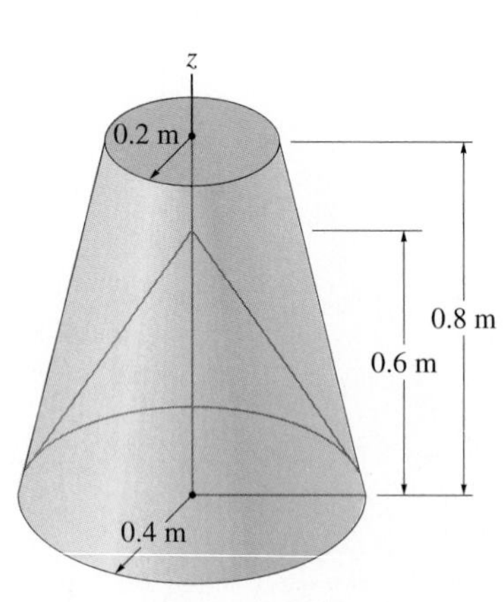

Prob. 10–103

***10–104.** Determine the moment of inertia of the wheel about an axis which is perpendicular to the page and passes through the center of mass G. The material has a specific weight of $\gamma = 90$ lb/ft^3.

10–105. Determine the moment of inertia of the wheel about an axis which is perpendicular to the page and passes through point O. The material has a specific weight of $\gamma = 90$ lb/ft^3.

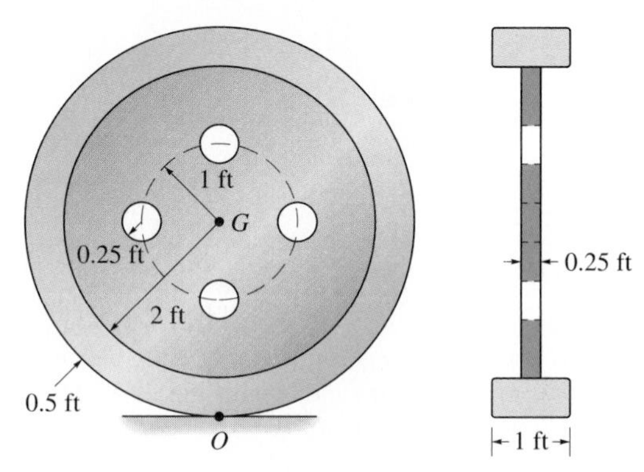

Probs. 10–104/10–105

10–106. The pendulum consists of two slender rods AB and OC which have a mass of 3 kg/m. The thin plate has a mass of 12 kg/m^2. Determine the location $\bar{y}$ of the center of mass G of the pendulum, then calculate the moment of inertia of the pendulum about an axis perpendicular to the page and passing through G.

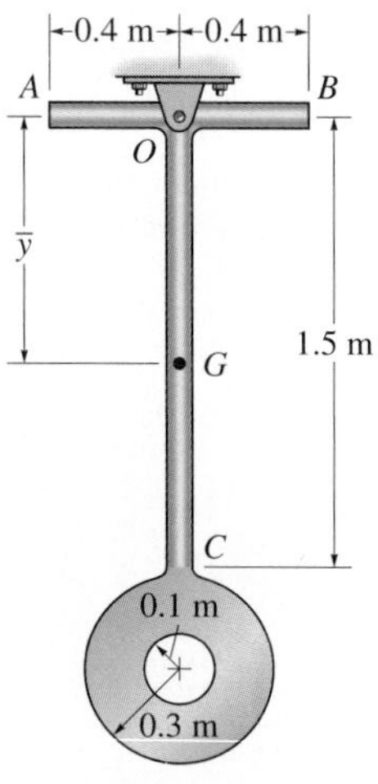

Prob. 10–106

10–107. The pendulum consists of a disk having a mass of 6 kg and slender rods AB and DC which have a mass of 2 kg/m. Determine the length L of DC so that the center of mass is at the bearing O. What is the moment of inertia of the assembly about an axis perpendicular to the page and passing through point O?

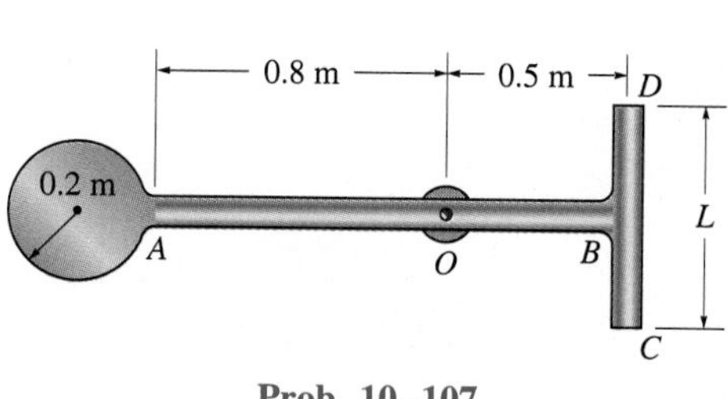

Prob. 10–107

***10–108.** Determine the moment of inertia of the wire triangle about an axis perpendicular to the page and passing through point O. Also, locate the mass center G and determine the moment of inertia about an axis perpendicular to the page and passing through point G. The wire has a mass of 0.3 kg/m. Neglect the size of the ring at O.

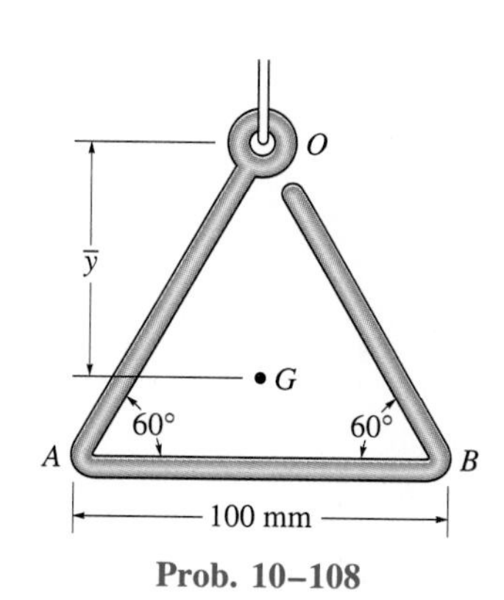

Prob. 10–108

10–109. Determine the location $\bar{y}$ of the center of mass G of the assembly and then calculate the moment of inertia about an axis perpendicular to the page and passing through G. The block has a mass of 3 kg and the mass of the semicylinder is 5 kg.

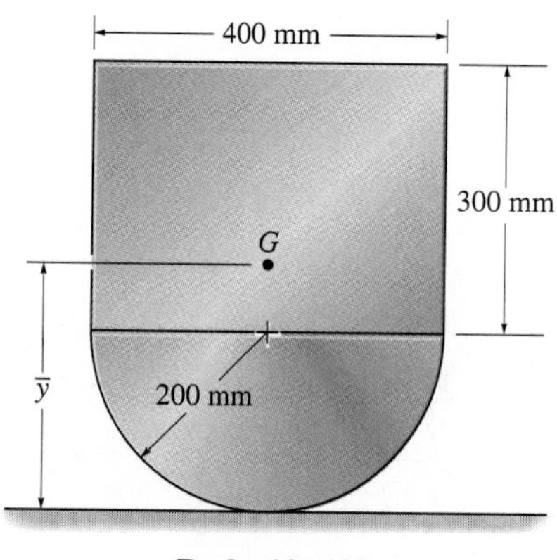

Prob. 10–109

10–110. Determine the moment of inertia I_z of the frustum of the cone which has a conical depression. The material has a density of 200 kg/m^3.

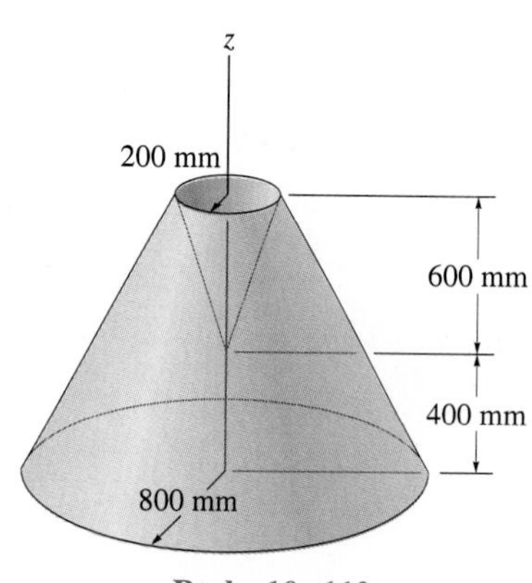

Prob. 10–110

REVIEW PROBLEMS

10–111. Determine the moments of inertia I_u and I_v and the product of inertia I_{uv} for the semicircular area.

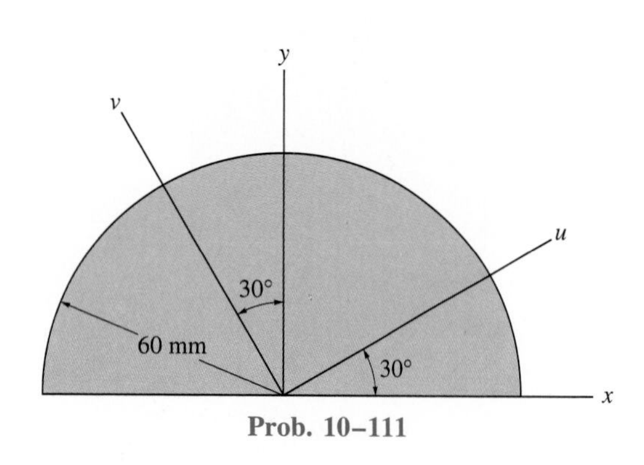

Prob. 10–111

***10–112.** The paraboloid is formed by revolving the shaded area around the x axis. Determine the radius of gyration k_x. The density of the material is $\rho = 5\ \text{Mg/m}^3$.

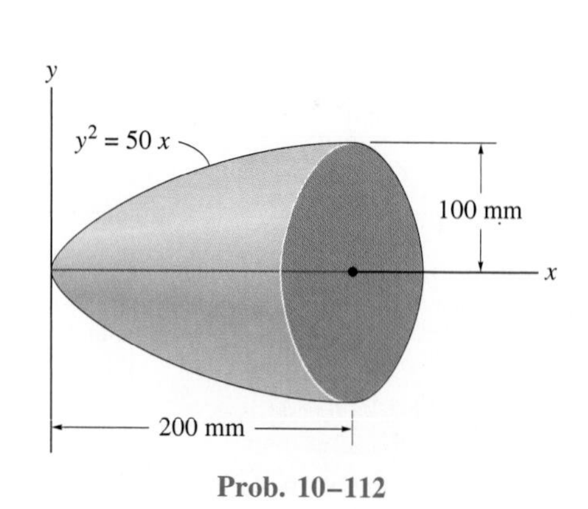

Prob. 10–112

10–113. Determine the moments of inertia I_x and I_y of the shaded area.

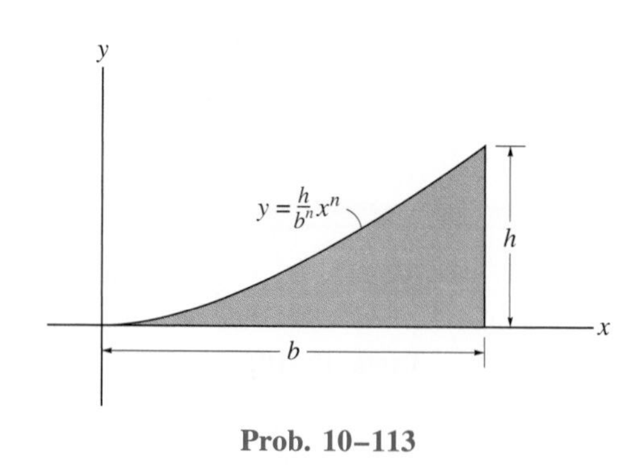

Prob. 10–113

10–114. Determine the moment of inertia of the shaded area about the x axis.

10–115. Determine the moment of inertia of the shaded area about the y axis.

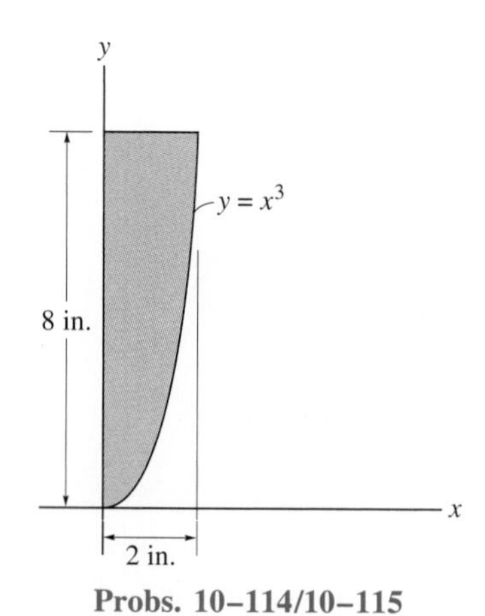

Probs. 10–114/10–115

***10–116.** Determine the moment of inertia of the shaded area about the x axis.

10–117. Determine the moment of inertia of the shaded area about the y axis.

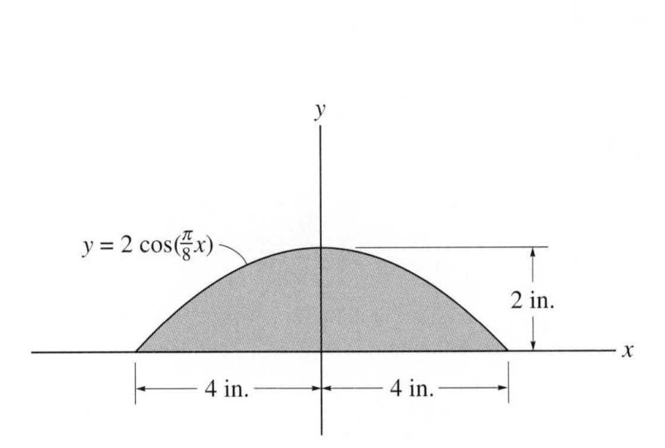

Probs. 10–116/10–117

10–118. Determine the moment of inertia of the shaded area about the x axis.

10–119. Determine the moment of inertia of the shaded area about the y axis.

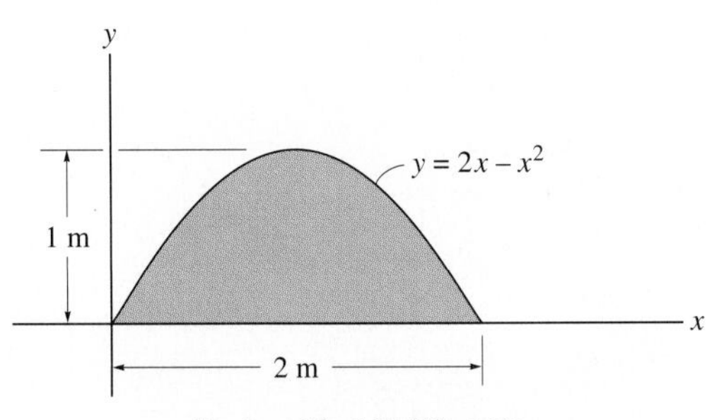

Probs. 10–118/10–119

***10–120.** Determine the moment of inertia of the triangular area about (a) the x axis, and (b) the centroidal x' axis.

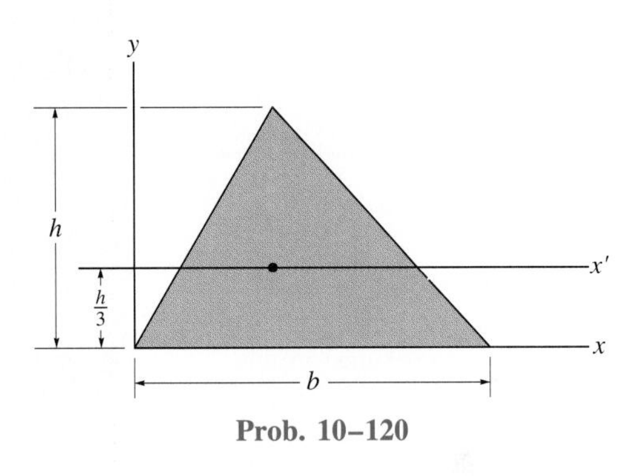

Prob. 10–120

10–121. Determine the moments of inertia I_x and I_y of the shaded area.

10–122. Determine the product of inertia I_{xy} of the shaded area with respect to the x and y axes.

10–123. Determine the directions of the principal axes with origin located at point O, and the principal moments of inertia for the shaded area about these axes. Here $I_x = 0.1667 \text{ in}^4$, $I_y = 0.0333 \text{ in}^4$, $I_{xy} = 0.0625 \text{ in}^4$.

***10–124.** Determine the principal moments of inertia for the shaded area using Mohr's circle. Here $I_x = 0.1667 \text{ in}^4$, $I_y = 0.0333 \text{ in}^4$, $I_{xy} = 0.0625 \text{ in}^4$.

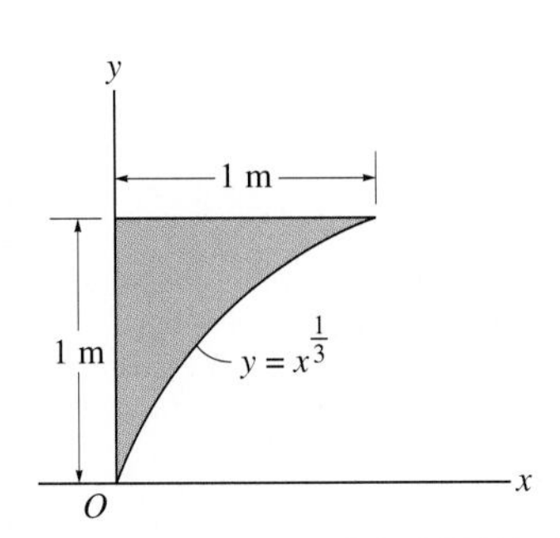

Probs. 10–121/10–122/10–123/10–124

Equilibrium and stability of this articulated crane boom as a function of its position can be analyzed using methods based on work and energy, which are explained in this chapter.

11

Virtual Work

In this chapter we will use the principle of virtual work and the potential-energy method to determine the equilibrium position of a series of connected rigid bodies. Although application requires more mathematical sophistication than using the equations of equilibrium, it will be shown that, once the equation of virtual work or the potential-energy function is established, the solution may be obtained *directly,* without having to dismember the system in order to obtain relationships between forces occurring at the connections. Furthermore, by using the potential-energy method, we will be able to investigate the "type" of equilibrium or the stability of the configuration.

11.1 Definition of Work and Virtual Work

Work of a Force. In mechanics a force $\mathbf{F}$ does work only when it undergoes a displacement in the direction of the force. For example, consider the force $\mathbf{F}$ in Fig. 11–1, which is located on the path s specified by the position vector $\mathbf{r}$. If the force moves along the path to a new position $\mathbf{r}' = \mathbf{r} + d\mathbf{r}$, the displacement is $d\mathbf{r}$ and therefore the work dU is a *scalar quantity,* defined by the dot product

$$dU = \mathbf{F} \cdot d\mathbf{r}$$

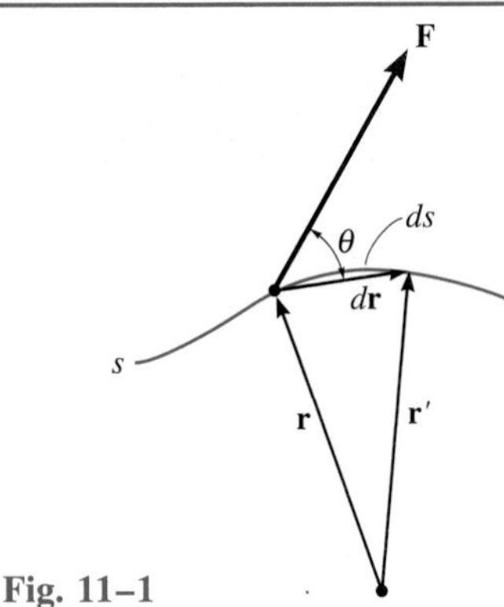

Fig. 11–1

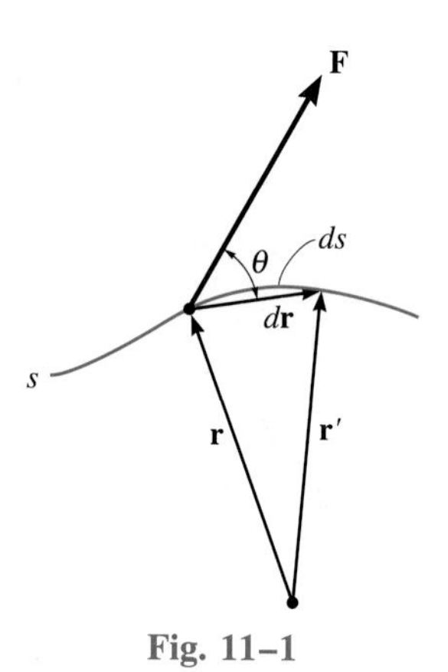

Fig. 11–1

Because $d\mathbf{r}$ is infinitesimal, the magnitude of $d\mathbf{r}$ can be represented by ds, the differential arc segment along the path. If the angle between the tails of $d\mathbf{r}$ and $\mathbf{F}$ is θ, Fig. 11–1, then by definition of the dot product, the above equation may also be written as

$$dU = F\,ds\cos\theta$$

Work expressed by this equation may be interpreted in one of two ways: either as the product of $\mathbf{F}$ and the component of displacement in the direction of the force, i.e., $ds\cos\theta$; or as the product of ds and the component of force in the direction of displacement, i.e., $F\cos\theta$. Note that if $0° \leq \theta < 90°$, then the force component and the displacement have the *same sense,* so that the work is *positive;* whereas if $90° < \theta \leq 180°$, these vectors have an *opposite sense,* and therefore the work is *negative.* Also, $dU = 0$ if the force is *perpendicular* to displacement, since $\cos 90° = 0$, or if the force is applied at a *fixed point,* in which case the displacement $ds = 0$.

The basic unit for work combines the units of force and displacement. In the SI system a *joule* (J) is equivalent to the work done by a force of 1 newton which moves 1 meter in the direction of the force ($1\text{ J} = 1\text{ N}\cdot\text{m}$). In the FPS system, work is defined in units of ft · lb. The moment of a force has the same combination of units; however, the concepts of moment and work are in no way related. A moment is a vector quantity, whereas work is a scalar.

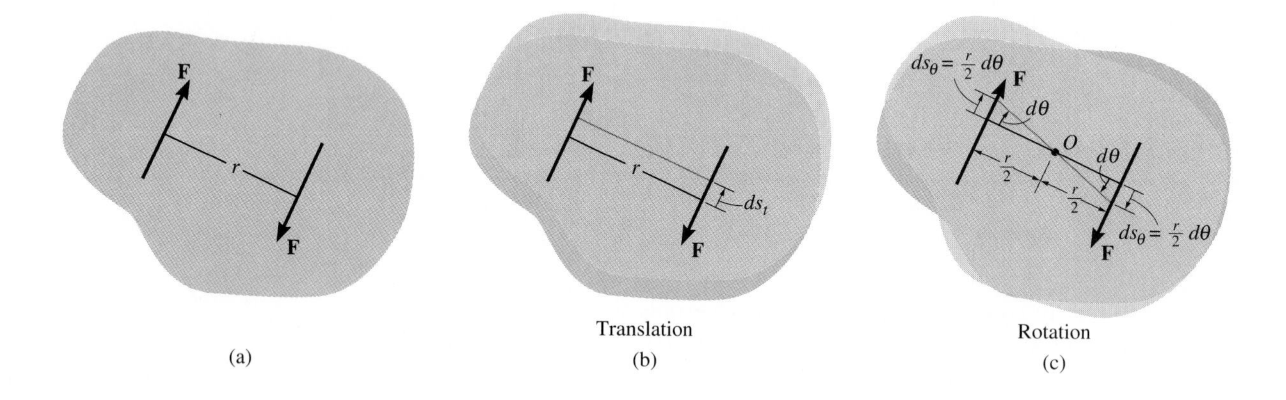

Fig. 11–2

Work of a Couple. The two forces of a couple do work when the couple *rotates* about an axis perpendicular to the plane of the couple. To show this, consider the body in Fig. 11–2*a*, which is subjected to a couple whose moment has a magnitude $M = Fr$. Any general differential displacement of the body can be considered as a combination of a translation and rotation. When the body *translates* such that the *component of displacement* along the line of action of each force is ds_t, clearly the "positive" work of one force ($F\,ds_t$) *cancels* the "negative" work of the other ($-F\,ds_t$), Fig. 11–2*b*. Consider now a differential *rotation* $d\theta$ of the body about an axis perpendicular to the plane of the couple, which intersects the plane at point O, Fig. 11–2*c*. (For the derivation, any other point in the plane may also be considered.) As shown, each force undergoes a displacement $ds_\theta = (r/2)\,d\theta$ in the direction of the force; hence, the work of both forces is

$$dU = F\left(\frac{r}{2}\,d\theta\right) + F\left(\frac{r}{2}\,d\theta\right) = (Fr)\,d\theta$$

or

$$dU = M\,d\theta$$

The resultant work is *positive* when the sense of $\mathbf{M}$ is the *same* as that of $d\boldsymbol{\theta}$, and negative when they have an opposite sense. As in the case of the moment vector, the *direction and sense* of $d\boldsymbol{\theta}$ are defined by the right-hand rule, where the fingers of the right hand follow the rotation or "curl" and the thumb indicates the direction of $d\boldsymbol{\theta}$. Hence, the line of action of $d\boldsymbol{\theta}$ will be *parallel* to the line of action of $\mathbf{M}$ if movement of the body occurs in the *same plane*. If the body rotates in space, however, the *component* of $d\boldsymbol{\theta}$ in the direction of $\mathbf{M}$ is required. Thus, in general, the work done by a couple is defined by the dot product, $dU = \mathbf{M} \cdot d\boldsymbol{\theta}$.

Virtual Work. The definitions of the work of a force and a couple have been presented in terms of *actual movements* expressed by differential displacements having magnitudes of ds and $d\theta$. Consider now an *imaginary* or *virtual movement*, which indicates a displacement or rotation that is *assumed* and *does not actually exist*. These movements are first-order differential quantities and will be denoted by the symbols δs and $\delta\theta$ (delta s and delta θ), respectively. The *virtual work* done by a force undergoing a virtual displacement δs is

$$\delta U = F\cos\theta\,\delta s \qquad (11\text{–}1)$$

Similarly, when a couple undergoes a virtual rotation $\delta\theta$ in the plane of the couple forces, the *virtual work* is

$$\delta U = M\,\delta\theta \qquad (11\text{–}2)$$

11.2 Principle of Virtual Work for a Particle and a Rigid Body

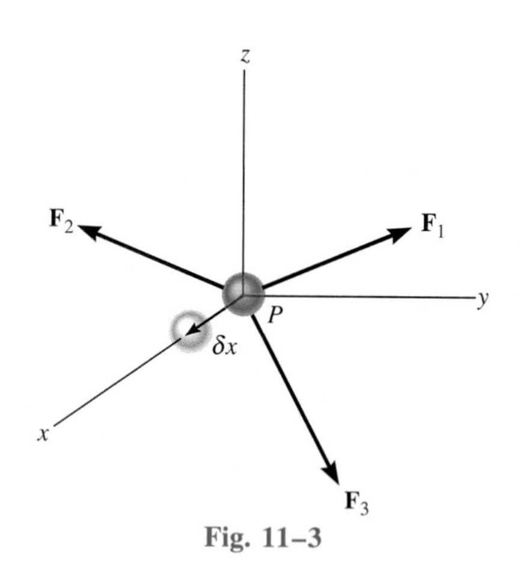

Fig. 11–3

If a particle is in equilibrium, the resultant of the force system acting on it must be equal to zero. Hence, if the particle undergoes an imaginary or virtual displacement in the x, y, or z direction, the virtual work (δU) done by the force system must be equal to zero since the components $\Sigma F_x = 0, \Sigma F_y = 0, \Sigma F_z = 0$. Alternatively, this may be expressed as

$$\delta U = 0$$

For example, if the particle in Fig. 11–3 is given a virtual displacement $\delta\mathbf{x}$, only the x components of the forces acting on the particle do work. (No work is done by the y and z components, since they are perpendicular to the displacement.) The virtual-work equation is therefore

$$\delta U = 0; \qquad F_{1x}\,\delta x + F_{2x}\,\delta x + F_{3x}\,\delta x = 0$$

Factoring out δx, which is common to every term, yields

$$(F_{1x} + F_{2x} + F_{3x})\,\delta x = 0$$

Since $\delta x \neq 0$, this equation is satisfied only if the sum of the force components in the x direction is equal to zero, i.e., $\Sigma F_x = 0$. Two other virtual work equations can be written by assuming virtual displacements $\delta\mathbf{y}$ and $\delta\mathbf{z}$ in the y and z directions, respectively. Doing this, however, amounts to satisfying the equilibrium equations $\Sigma F_y = 0$ and $\Sigma F_z = 0$ for the particle.

In a similar manner, a rigid body that is subjected to a coplanar force system will be in equilibrium provided $\Sigma F_x = 0$, $\Sigma F_y = 0$, and $\Sigma M_O = 0$. We can also write a set of three virtual work equations for the body, each of which requires $\delta U = 0$. If these equations involve separate virtual translations in the x and y directions and a virtual rotation about an axis perpendicular to the x–y plane and passing through point O, then it can be shown that they will correspond to the above-mentioned three equilibrium equations. When writing these equations, it is *not necessary* to include the work done by the *internal forces* acting within the body, since a rigid body *does not deform* when subjected to an external loading, and furthermore, when the body moves through a virtual displacement, the internal forces occur in equal but opposite collinear pairs, so that the corresponding work done by each pair of forces *cancels.*

As in the case of a particle, however, no added advantage would be gained by solving rigid-body equilibrium problems using the principle of virtual work. This is because for each application of the virtual-work equation the virtual displacement, common to every term, factors out, leaving an equation that could have been obtained in a more *direct manner* by applying the equations of equilibrium.

11.3 Principle of Virtual Work for a System of Connected Rigid Bodies

The method of virtual work is most suitable for solving equilibrium problems that involve a system of several *connected* rigid bodies such as the ones shown in Fig. 11–4. Before we can apply the principle of virtual work to these systems, however, we must first specify the number of degrees of freedom for a system and establish coordinates that define the position of the system.

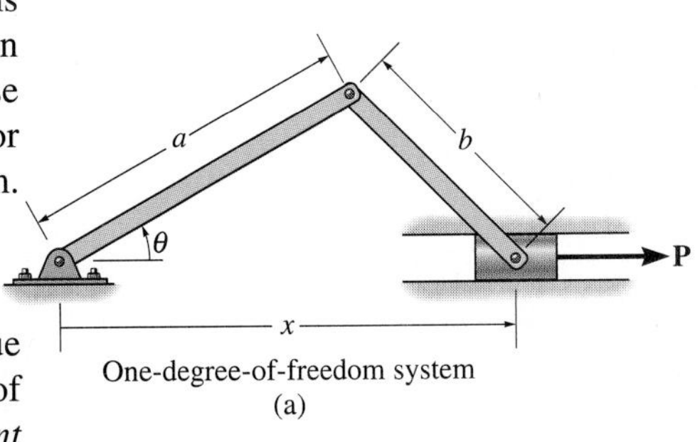

One-degree-of-freedom system
(a)

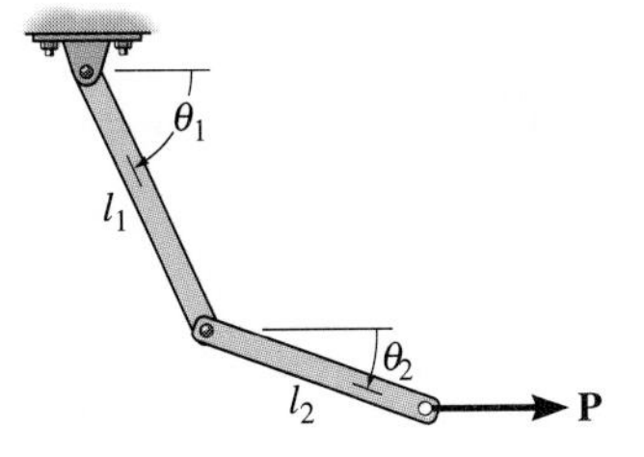

Two-degree-of-freedom system
(b)

Fig. 11–4

Degrees of Freedom. A system of connected bodies takes on a unique shape that can be specified provided we know the position of a number of specific points on the system. These positions are defined using *independent coordinates q,* which are measured from fixed reference points. For every coordinate established, the system will have a *degree of freedom* for displacement along the coordinate axis such that it is consistent with the constraining action of the supports. Thus, an n-degree-of-freedom system requires n independent coordinates q_n to specify the location of all its members. For example, the link and sliding-block arrangement shown in Fig. 11–4*a* is an example of a one-degree-of-freedom system. The independent coordinate $q = \theta$ may be used to specify the location of the two connecting links and the block. The coordinate x could also be used as the independent coordinate. However, since the block is constrained to move within the slot, x is not independent of θ; rather, it can be related to θ using the cosine law, $b^2 = a^2 + x^2 - 2ax \cos \theta$. The double-link arrangement, shown in Fig. 11–4*b*, is an example of a two-degree-of-freedom system. To specify the location of each link, the coordinate angles θ_1 and θ_2 must be known, since a rotation of one link is independent of a rotation of the other.

Principle of Virtual Work. The principle of virtual work for a system of rigid bodies whose connections are *frictionless* may be stated as follows: *A system of connected rigid bodies is in equilibrium provided the virtual work done by all the external forces and couples acting on the system is zero for each independent virtual displacement of the system.* Mathematically, this may be expressed as

$$\delta U = 0 \tag{11–3}$$

where δU represents the virtual work of all the external forces (and couples) acting on the system during any independent virtual displacement.

As stated above, if a system has n degrees of freedom it takes n independent coordinates q_n to completely specify the location of the system. Hence, for the system it is possible to write n independent virtual-work equations, one for every virtual displacement taken along each of the independent coordinate axes, while the remaining $n - 1$ independent coordinates are held *fixed.**

PROCEDURE FOR ANALYSIS

The following procedure provides a method for applying the equation of virtual work to solve problems involving a system of frictionless connected rigid bodies having a single degree of freedom.

Free-Body Diagram. Draw the free-body diagram of the entire system of connected bodies and define the *independent coordinate* q. Sketch the "deflected position" of the system on the free-body diagram when the system undergoes a *positive* virtual displacement δq. From this, specify the "active" forces and couples, that is, those that do work.

Virtual Displacements. Indicate *position coordinates* s_i, measured from a *fixed point* on the free-body diagram to each of the i number of "active" forces and couples. Each coordinate axis should be parallel to the line of action of the "active" force to which it is directed, so that the virtual work along the coordinate axis can be calculated.

Relate each of the position coordinates s_i to the independent coordinate q; then *differentiate* these expressions in order to express the virtual displacements δs_i in terms of δq.

Virtual-Work Equation. Write the *virtual-work equation* for the system assuming that, whether possible or not, all the position coordinates s_i undergo *positive* virtual displacements δs_i. Using the relations for δs_i, express the work of *each* "active" force and couple in the equation in terms of the single independent virtual displacement δq. By factoring out this common displacement, one is left with an equation that generally can be solved for an unknown force, couple, or equilibrium position.

If the system contains n degrees of freedom, n independent coordinates q_n must be specified. In this case, follow the above procedure and let *only one* of the independent coordinates undergo a virtual displacement, while the remaining $n - 1$ coordinates are held fixed. In this way, n virtual-work equations can be written, one for each independent coordinate.

The following examples should help to clarify application of this procedure.

*This method of applying the principle of virtual work is sometimes called the *method of virtual displacements,* since a virtual displacement is applied, resulting in the calculation of a real force. Although it is not to be used here, realize that we can also apply the principle of virtual work as a method of virtual forces. This method is often used to determine the displacements of points on deformable bodies. See R. C. Hibbeler, *Mechanics of Materials,* 2nd edition, Macmillan Publishing Company, New York, 1994.

Example 11–1

Determine the angle θ for equilibrium of the two-member linkage shown in Fig. 11–5*a*. Each member has a mass of 10 kg.

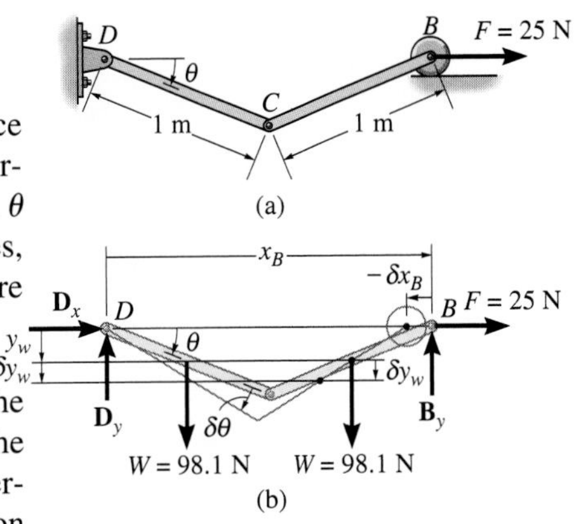

Fig. 11–5

SOLUTION

Free-Body Diagram. The system has only one degree of freedom, since the location of both links may be specified by the single independent coordinate ($q =$) θ. As shown on the free-body diagram in Fig. 11–5*b*, when θ undergoes a *positive* (clockwise) virtual rotation $\delta\theta$, only the active forces, **F** and the two 98.1-N weights, do work. (The reactive forces $\mathbf{D}_x$ and $\mathbf{D}_y$ are fixed, and $\mathbf{B}_y$ does not move along its line of action.)

Virtual Displacements. If the origin of coordinates is established at the *fixed* pin support D, the location of **F** and **W** may be specified by the *position coordinates* x_B and y_w, as shown in the figure. In order to determine the work, note that these coordinates are parallel to the lines of action of their associated forces.

Expressing the position coordinates in terms of the independent coordinate θ and taking the derivatives yields

$$x_B = 2(1 \cos\theta)\text{ m} \qquad \delta x_B = -2\sin\theta\,\delta\theta\text{ m} \qquad (1)$$

$$y_w = \tfrac{1}{2}(1\sin\theta)\text{ m} \qquad \delta y_w = 0.5\cos\theta\,\delta\theta\text{ m} \qquad (2)$$

It is seen by the *signs* of these equations, and indicated in Fig. 11–5*b*, that an *increase* in θ (i.e., $\delta\theta$) causes a *decrease* in x_B and an *increase* in y_w.

Virtual-Work Equation. If the virtual displacements δx_B and δy_w were *both positive,* then the forces **W** and **F** would do positive work since the forces and their corresponding displacements would have the same sense. Hence, the virtual-work equation for the displacement $\delta\theta$ is

$$\delta U = 0; \qquad W\,\delta y_w + W\,\delta y_w + F\,\delta x_B = 0 \qquad (3)$$

Substituting Eqs. 1 and 2 into Eq. 3 in order to relate the virtual displacements to the common virtual displacement $\delta\theta$ yields

$$98.1(0.5\cos\theta\,\delta\theta) + 98.1(0.5\cos\theta\,\delta\theta) + 25(-2\sin\theta\,\delta\theta) = 0$$

Notice that the "negative work" done by **F** (force in the opposite sense to displacement) has been *accounted for* in the above equation by the "negative sign" of Eq. 1. Factoring out the *common displacement* $\delta\theta$ and solving for θ, noting that $\delta\theta \neq 0$, yields

$$(98.1\cos\theta - 50\sin\theta)\,\delta\theta = 0$$

$$\theta = \tan^{-1}\frac{98.1}{50} = 63.0^\circ \qquad \textit{Ans.}$$

If this problem had been solved using the equations of equilibrium, it would have been necessary to dismember the links and apply three scalar equations to *each* link. The principle of virtual work, by means of calculus, has eliminated this task so that the answer is obtained directly.

Example 11–2

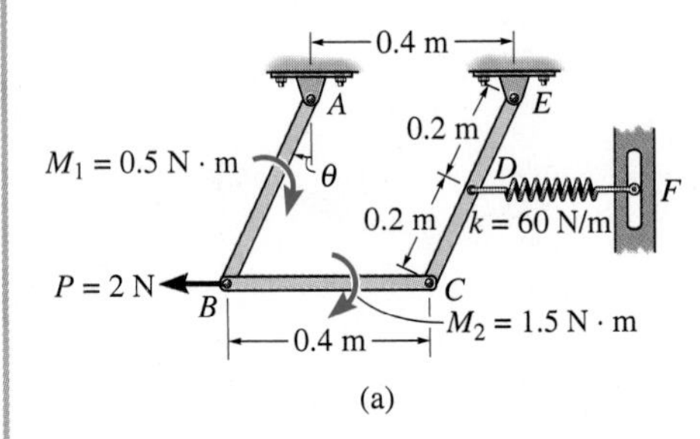

(a)

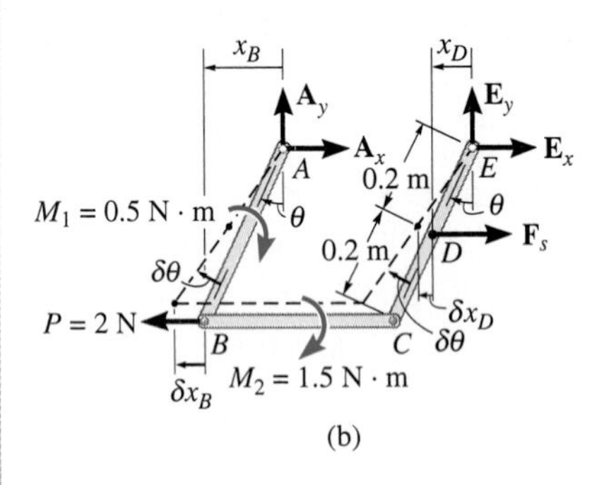

(b)

Fig. 11–6

Using the principle of virtual work, determine the angle θ required to maintain equilibrium of the mechanism shown in Fig. 11–6*a*. Neglect the weight of the links. The spring is unstretched when $\theta = 0°$ and it maintains a horizontal position due to the roller.

SOLUTION

Free-Body Diagram. The mechanism has one degree of freedom, and therefore the location of each member may be specified using the independent coordinate θ. When θ undergoes a *positive* virtual displacement $\delta\theta$, as shown on the free-body diagram in Fig. 11–6*b*, links AB and EC rotate by the same amount since they have the same length, and link BC only translates. Since a couple moment does work *only* when it rotates, the work done by $\mathbf{M}_2$ is zero. The reactive forces at A and E do no work. Why?

Virtual Displacements. The position coordinates x_B and x_D are *parallel* to the lines of action of $\mathbf{P}$ and $\mathbf{F}_s$, and these coordinates locate these forces with respect to the *fixed points* A and E. From Fig. 11–6*b*,

$$x_B = 0.4 \sin\theta \text{ m}$$
$$x_D = 0.2 \sin\theta \text{ m}$$

Thus,

$$\delta x_B = 0.4 \cos\theta\, \delta\theta \text{ m}$$
$$\delta x_D = 0.2 \cos\theta\, \delta\theta \text{ m}$$

Virtual-Work Equation. Applying the equation of virtual work, noting that, for positive virtual displacements, $\mathbf{F}_s$ is opposite to $\delta\mathbf{x}_D$ and hence does negative work, we obtain

$$\delta U = 0; \qquad M_1\,\delta\theta + P\,\delta x_B - F_s\,\delta x_D = 0$$

Relating each of the virtual displacements to the *common* virtual displacement $\delta\theta$ yields

$$0.5\,\delta\theta + 2(0.4\cos\theta\,\delta\theta) - F_s(0.2\cos\theta\,\delta\theta) = 0$$
$$(0.5 + 0.8\cos\theta - 0.2F_s\cos\theta)\,\delta\theta = 0 \qquad (1)$$

For the arbitrary angle θ, the spring is stretched a distance of $x_D = (0.2\sin\theta)$ m; and therefore, $F_s = 60$ N/m$(0.2\sin\theta)$ m $= (12\sin\theta)$ N. Substituting into Eq. 1 and noting that $\delta\theta \neq 0$, we have

$$0.5 + 0.8\cos\theta - 0.2(12\sin\theta)\cos\theta = 0$$

Since $\sin 2\theta = 2\sin\theta\cos\theta$, then

$$1 = 2.4\sin 2\theta - 1.6\cos\theta$$

Solving for θ by trial and error yields

$$\theta = 36.3° \qquad \textit{Ans.}$$

Example 11–3

Using the principle of virtual work, determine the horizontal force that the pin at C must exert in order to hold the mechanism shown in Fig. 11–7*a* in equilibrium when $\theta = 45°$. Neglect the weight of the members.

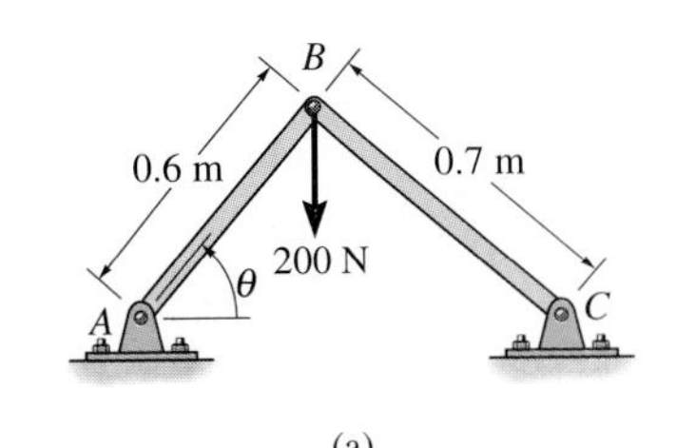

(a)

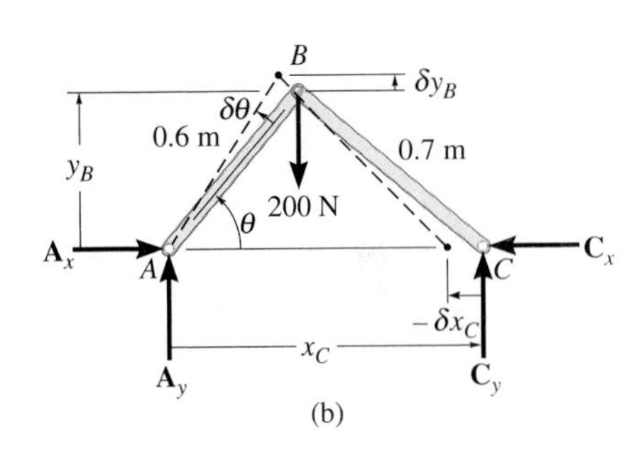

(b)

Fig. 11–7

SOLUTION

Free-Body Diagram. The reaction $\mathbf{C}_x$ can be obtained by *releasing* the pin constraint at C in the x direction and allowing the frame to be displaced in this direction. The system then has only one degree of freedom, defined by the independent coordinate θ, Fig. 11–7*b*. When θ undergoes a *positive* virtual displacement $\delta\theta$, only $\mathbf{C}_x$ and the 200-N force do work.

Virtual Displacements. Forces $\mathbf{C}_x$ and 200 N are located from the fixed origin A using position coordinates y_B and x_C. From Fig. 11–7*b*, x_C can be related to θ by the ‘‘law of cosines.’’ Hence,

$$(0.7)^2 = (0.6)^2 + x_C^2 - 2(0.6)x_C \cos\theta \tag{1}$$

$$0 = 0 + 2x_C\,\delta x_C - 1.2\,\delta x_C \cos\theta + 1.2x_C \sin\theta\,\delta\theta$$

$$\delta x_C = \frac{1.2x_C \sin\theta}{1.2\cos\theta - 2x_C}\,\delta\theta \tag{2}$$

Also,

$$y_B = 0.6\sin\theta$$

$$\delta y_B = 0.6\cos\theta\,\delta\theta \tag{3}$$

Virtual-Work Equation. When y_B and x_C undergo *positive* virtual displacements δy_B and δx_C, $\mathbf{C}_x$ and 200 N do *negative work,* since they both act in the opposite sense to $\delta\mathbf{y}_B$ and $\delta\mathbf{x}_C$. Hence,

$$\delta U = 0; \qquad -200\,\delta y_B - C_x\,\delta x_C = 0$$

Substituting Eqs. 2 and 3 into this equation, factoring out $\delta\theta$, and solving for C_x yields

$$-200(0.6\cos\theta\,\delta\theta) - C_x \frac{1.2x_C\sin\theta}{1.2\cos\theta - 2x_C}\,\delta\theta = 0$$

$$C_x = \frac{-120\cos\theta(1.2\cos\theta - 2x_C)}{1.2x_C\sin\theta} \tag{4}$$

At the required equilibrium position $\theta = 45°$, the corresponding value of x_C can be found by using Eq. 1, in which case

$$x_C^2 - 1.2\cos 45°\,x_C - 0.13 = 0$$

Solving for the positive root yields

$$x_C = 0.981 \text{ m}$$

Thus, from Eq. 4,

$$C_x = 114 \text{ N} \qquad \textit{Ans.}$$

Example 11–4

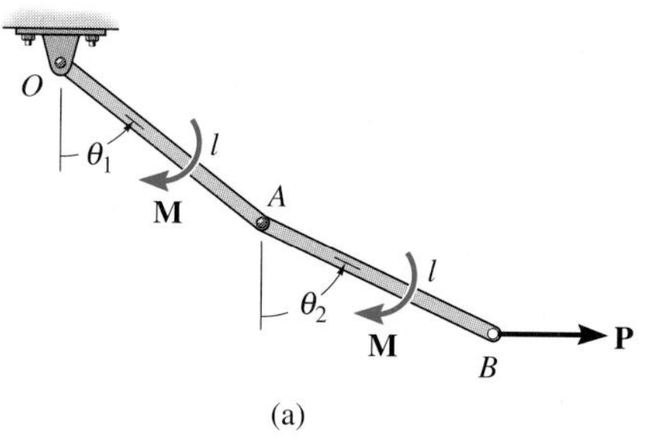

(a)

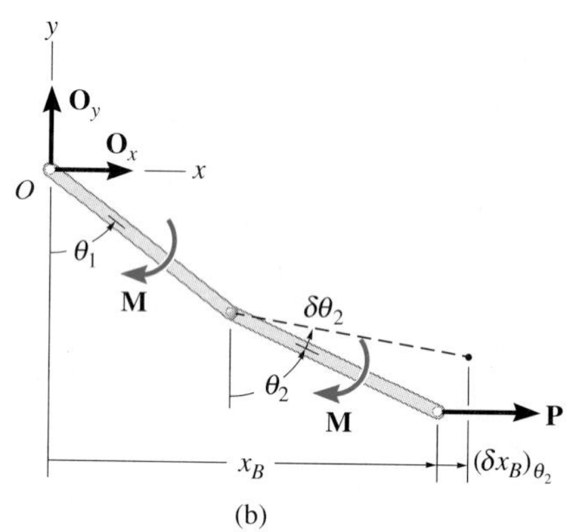

(b)

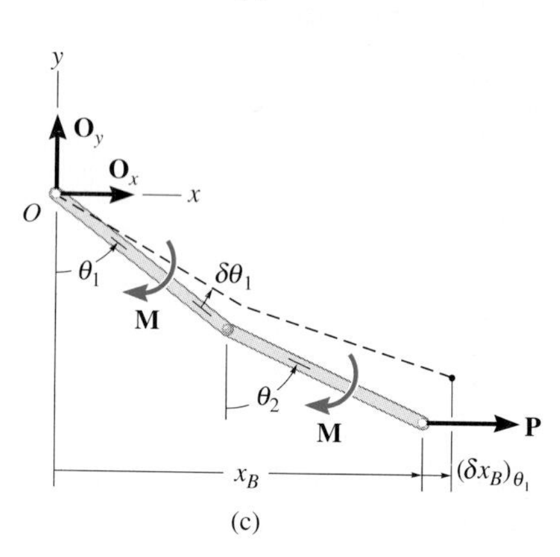

(c)

Fig. 11–8

Using the principle of virtual work, determine the equilibrium position of the two-bar linkage shown in Fig. 11–8*a*. Neglect the weight of the links.

SOLUTION

The system has two degrees of freedom, since the *independent coordinates* θ_1 and θ_2 must be known to locate the position of both links. The position coordinate x_B, measured from the fixed point O, is used to specify the location of **P**, Fig. 11–8*b* and *c*.

If θ_1 is held *fixed* and θ_2 varies by an amount $\delta\theta_2$, as shown in Fig. 11–8*b*, the virtual-work equation becomes

$[\delta U = 0]_{\theta_2};$
$$P(\delta x_B)_{\theta_2} - M\,\delta\theta_2 = 0 \tag{1}$$

Here P and M represent the magnitudes of the applied force and couple moment acting on link AB.

When θ_2 is held *fixed* and θ_1 varies by an amount $\delta\theta_1$, as shown in Fig. 11–8*c*, the virtual-work equation becomes

$[\delta U = 0]_{\theta_1};$
$$P(\delta x_B)_{\theta_1} - M\,\delta\theta_1 - M\,\delta\theta_1 = 0 \tag{2}$$

The *position coordinate* x_B may be related to the independent coordinates θ_1 and θ_2 by the equation

$$x_B = l\sin\theta_1 + l\sin\theta_2 \tag{3}$$

To obtain the variation of δx_B in terms of $\delta\theta_2$, it is necessary to take the *partial derivative* of x_B with respect to θ_2 since x_B is a function of both θ_1 and θ_2. Hence,

$$\frac{\partial x_B}{\partial\theta_2} = l\cos\theta_2 \qquad (\delta x_B)_{\theta_2} = l\cos\theta_2\,\delta\theta_2$$

Substituting into Eq. 1, we have

$$(Pl\cos\theta_2 - M)\,\delta\theta_2 = 0$$

Since $\delta\theta_2 \neq 0$, then

$$\theta_2 = \cos^{-1}\left(\frac{2M}{Pl}\right) \qquad \textit{Ans.}$$

Using Eq. 3 to obtain the variation of x_B with θ_1 yields

$$\frac{\partial x_B}{\partial\theta_1} = l\cos\theta_1 \qquad (\delta x_B)_{\theta_1} = l\cos\theta_1\,\delta\theta_1$$

Substituting into Eq. 2, we have

$$(Pl\cos\theta_1 - M)\,\delta\theta_1 = 0$$

Since $\delta\theta_1 \neq 0$, then

$$\theta_1 = \cos^{-1}\left(\frac{2M}{Pl}\right) \qquad \textit{Ans.}$$

PROBLEMS

11–1. The crankshaft is subjected to a torque of $M = 50\ \text{N}\cdot\text{m}$. Determine the horizontal compressive force F applied to the piston for equilibrium when $\theta = 60^\circ$.

11–2. The crankshaft is subjected to a torque of $M = 50\ \text{N}\cdot\text{m}$. Determine the horizontal compressive force F and plot the result of F (ordinate) versus θ (abscissa) for $0^\circ \leq \theta \leq 90^\circ$.

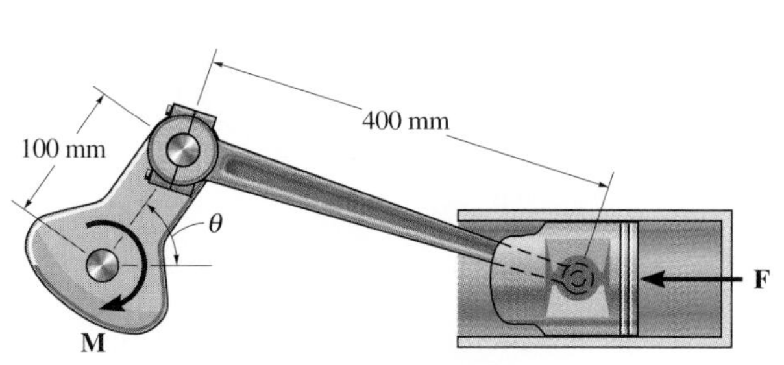

Probs. 11–1/11–2

11–3. The toggle joint is subjected to the load **P**. Determine the compressive force F it creates on the cylinder at A as a function of θ.

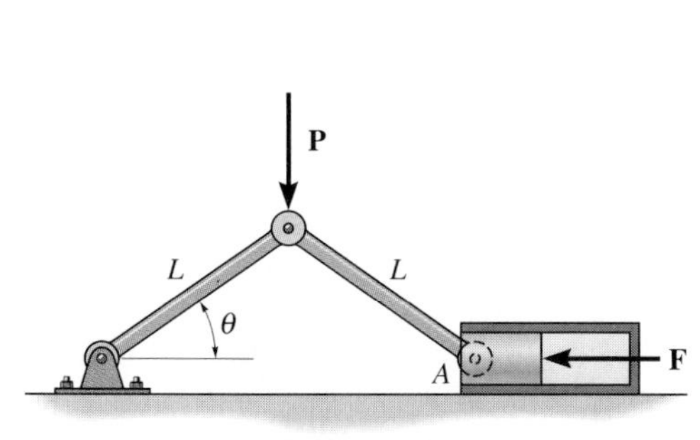

Prob. 11–3

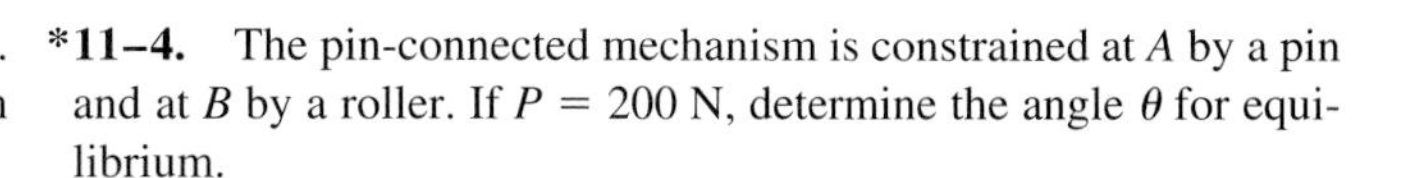

***11–4.** The pin-connected mechanism is constrained at A by a pin and at B by a roller. If $P = 200$ N, determine the angle θ for equilibrium.

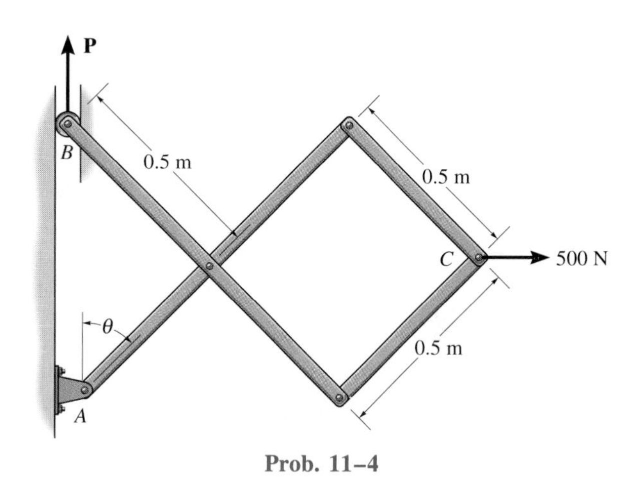

Prob. 11–4

■**11–5.** Each member of the pin-connected mechanism has a mass of 8 kg. If the spring is unstretched when $\theta = 0^\circ$, determine the angle θ for equilibrium. Set $k = 2500$ N/m and $M = 50\ \text{N}\cdot\text{m}$.

11–6. Each member of the pin-connected mechanism has a mass of 8 kg. If the spring is unstretched when $\theta = 0^\circ$, determine the required stiffness k so that the mechanism is in equilibrium when $\theta = 30^\circ$. Set $\mathbf{M} = \mathbf{0}$.

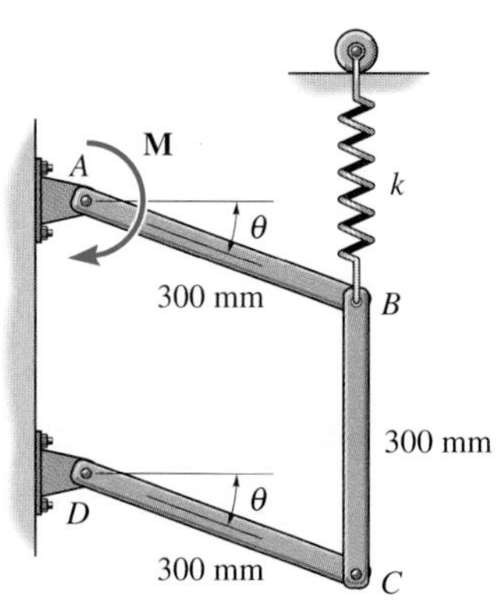

Probs. 11–5/11–6

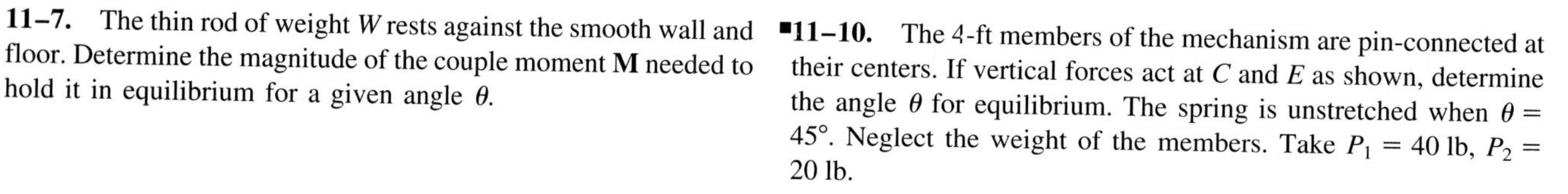

11–7. The thin rod of weight W rests against the smooth wall and floor. Determine the magnitude of the couple moment **M** needed to hold it in equilibrium for a given angle θ.

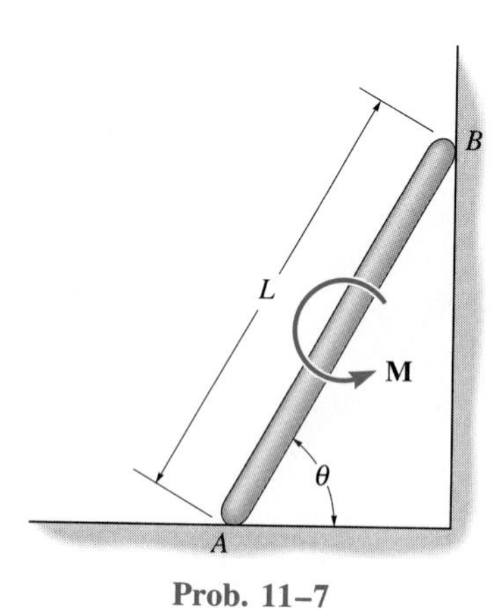

Prob. 11–7

***■11–8.** Determine the angle θ for equilibrium of the uniform 7-kg bar AC. Due to the roller guide at B, the spring remains vertical and is unstretched when $\theta = 0°$. Set $k = 200$ N/m.

11–9. Determine the required stiffness k so that the uniform 7-kg bar AC is in equilibrium when $\theta = 30°$. Due to the roller guide at B, the spring remains vertical and is unstretched when $\theta = 0°$.

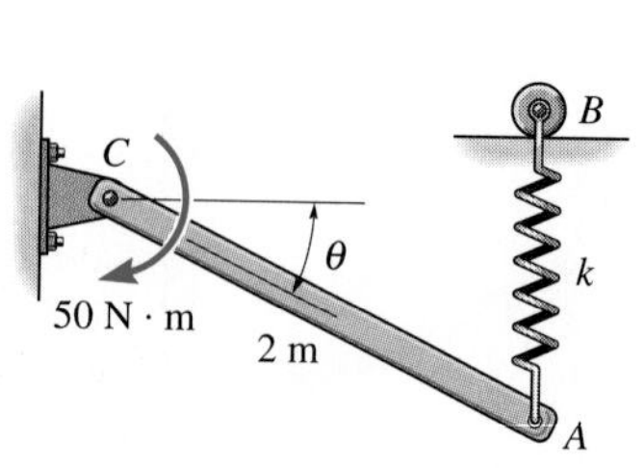

Probs. 11–8/11–9

■11–10. The 4-ft members of the mechanism are pin-connected at their centers. If vertical forces act at C and E as shown, determine the angle θ for equilibrium. The spring is unstretched when $\theta = 45°$. Neglect the weight of the members. Take $P_1 = 40$ lb, $P_2 = 20$ lb.

■11–11. The 4-ft members of the mechanism are pin-connected at their centers. If vertical forces $P_1 = P_2 = 30$ lb act at C and E as shown, determine the angle θ for equilibrium. The spring is unstretched when $\theta = 45°$. Neglect the weight of the members.

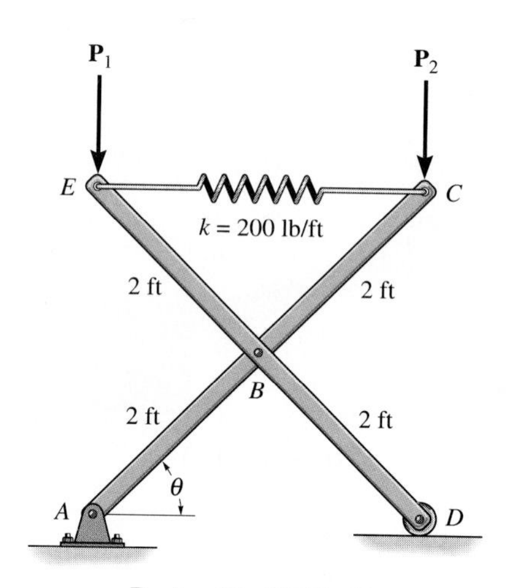

Probs. 11–10/11–11

***11–12.** If each of the three links of the mechanism has a weight of 20 lb, determine the angle θ for equilibrium of the spring, which, due to the roller guide, always remains horizontal and is unstretched when $\theta = 0°$.

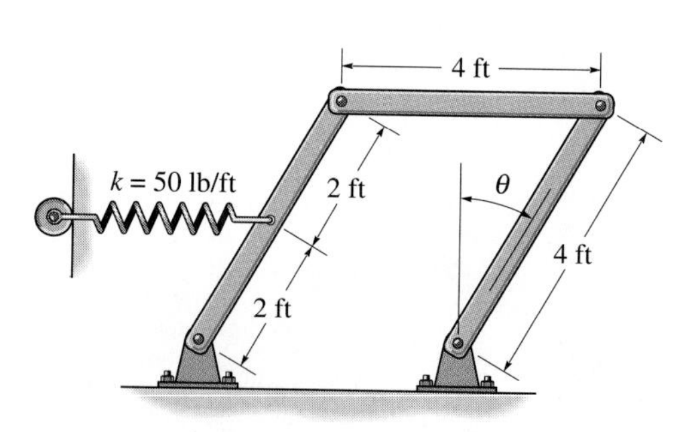

Prob. 11–12

11–13. If each of the three links of the mechanism has a weight of 20 lb, determine the angle θ for equilibrium. The spring, which always remains vertical due to the roller guide, is unstretched when $\theta = 0°$. Set $\mathbf{P} = \mathbf{0}$.

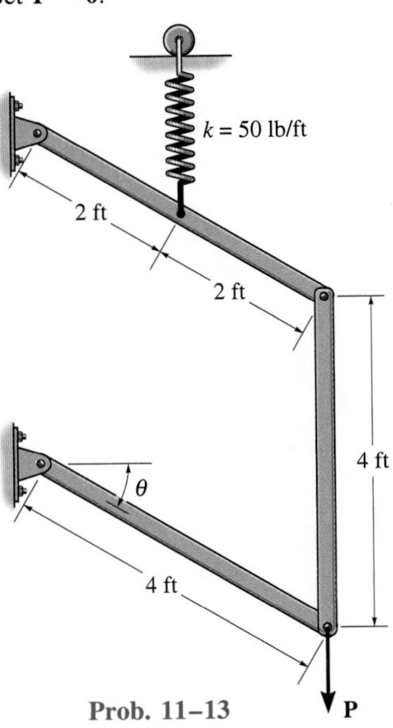

Prob. 11–13

11–14. The spring is unstretched when $\theta = 0°$. If $P = 8$ lb, determine the angle θ for equilibrium. Due to the roller guide, the spring always remains vertical. Neglect the weight of the links.

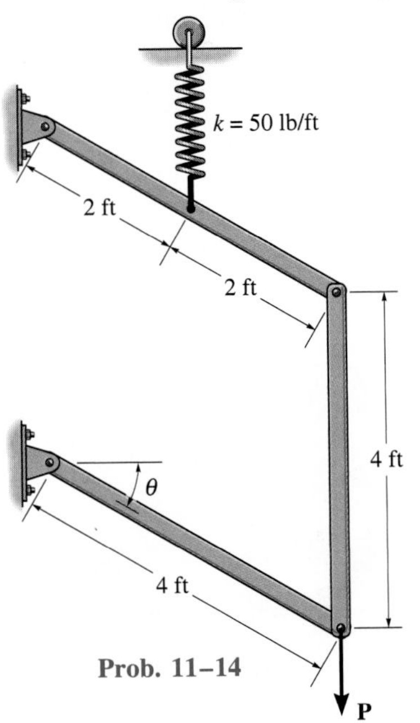

Prob. 11–14

11–15. The two-bar linkage is subjected to a couple moment $M = 100\ \text{N}\cdot\text{m}$ and vertical force $P = 500$ N. Determine the angle θ for equilibrium. The spring is unstretched when $\theta = 45°$. Neglect the mass of each bar.

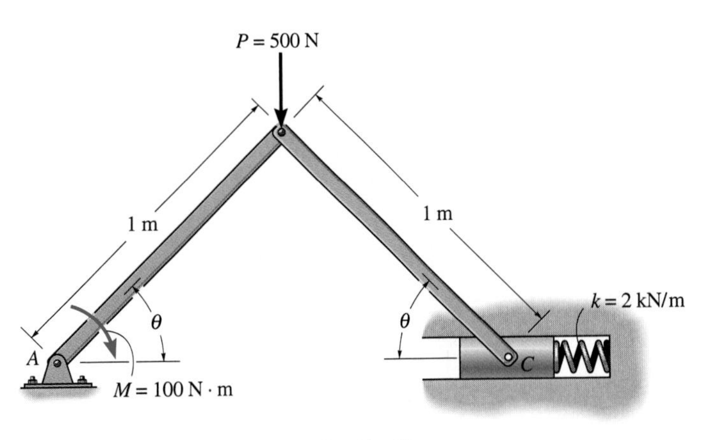

Prob. 11–15

***11–16.** The two-bar linkage is subjected to a couple moment $M = 100\ \text{N}\cdot\text{m}$ and vertical force $P = 500$ N. Also, each bar is uniform and has a mass of 10 kg. Determine the angle θ for equilibrium. The spring is unstretched when $\theta = 45°$.

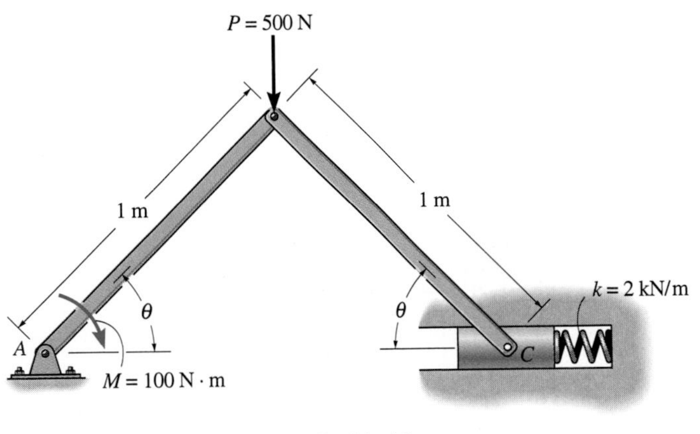

Prob. 11–16

11–17. Determine the force F needed to lift the bucket having a weight of 100 lb. *Hint:* Note that the coordinates s_A and s_B can be related to the *constant* vertical length l of the cord.

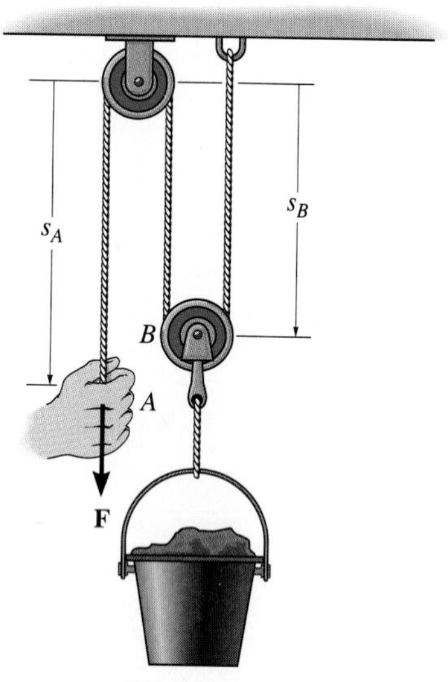

Prob. 11–17

11–18. Determine the force F needed to lift the block having a weight of 100 lb. *Hint:* Note that the coordinates s_A and s_B can be related to the *constant* vertical length l of the cord.

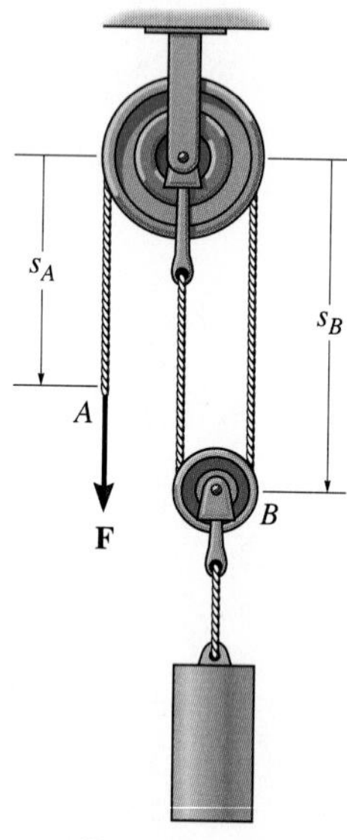

Prob. 11–18

11–19. The assembly is used for exercise. It consists of four pin-connected bars, each of length L, and a spring of stiffness k and unstretched length a ($<2L$). If horizontal forces **P** and $-$**P** are applied to the handles so that θ is slowly decreased, determine the angle θ at which the magnitude of **P** becomes a maximum.

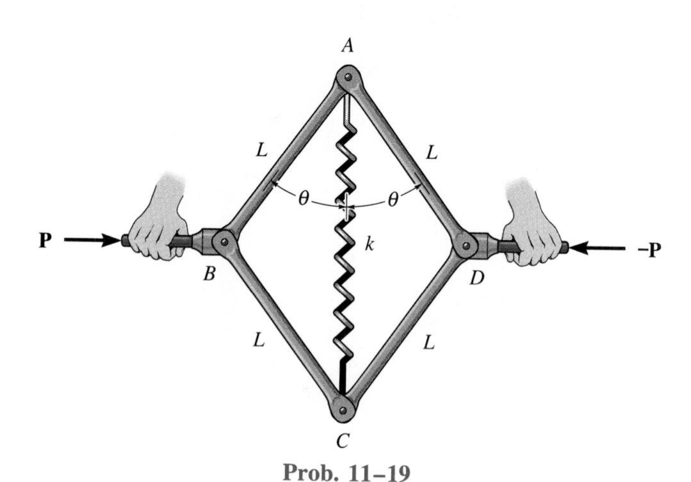

Prob. 11–19

***11–20.** The scissors jack supports a load **P.** Determine the axial force in the screw necessary for equilibrium when the jack is in the position θ. Each of the four links has a length L and is pin-connected at its center. Points B and D can move horizontally.

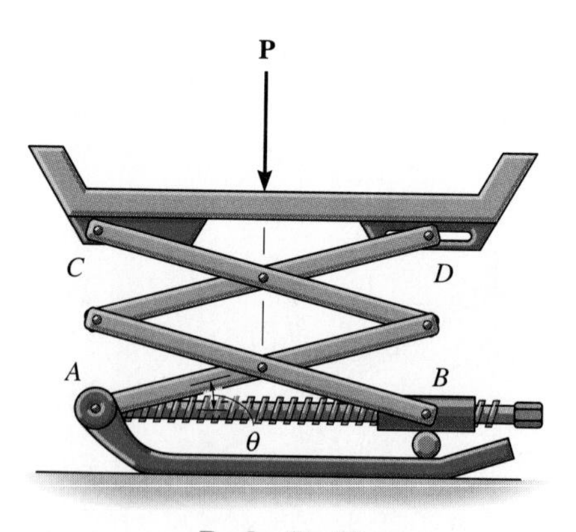

Prob. 11–20

11–21. Determine the horizontal force F required to maintain equilibrium of the slider mechanism when $\theta = 60°$. Set $M = 6\text{ N}\cdot\text{m}$.

11–22. Determine the moment M that must be applied to the slider mechanism in order to maintain the equilibrium position $\theta = 60°$ when the horizontal force $\mathbf{F} = 100\text{ N}$ is applied at D.

11–23. Solve Prob. 11-22 if the force $\mathbf{F}$ acts vertically upward.

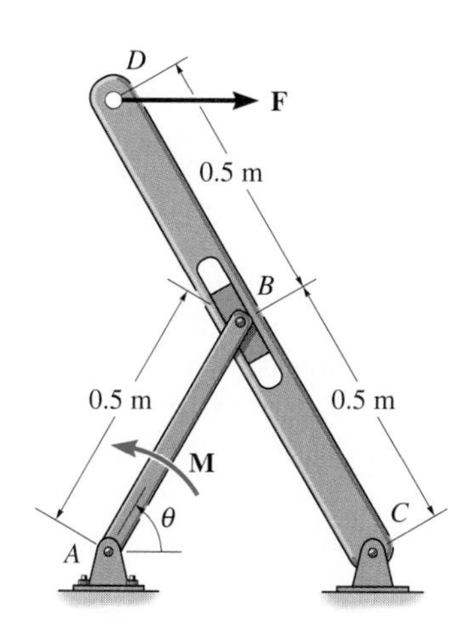

Probs. 11–21/11–22/11–23

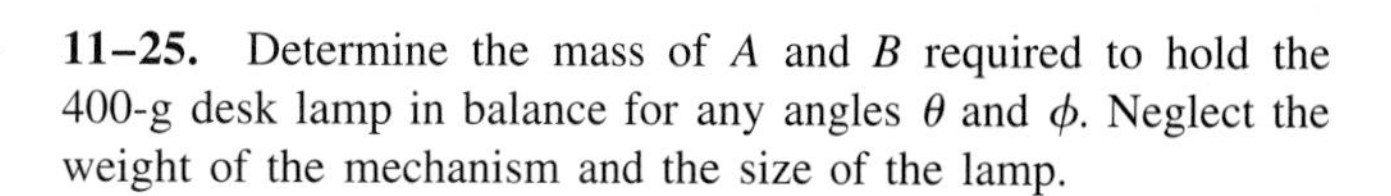

11–25. Determine the mass of A and B required to hold the 400-g desk lamp in balance for any angles θ and ϕ. Neglect the weight of the mechanism and the size of the lamp.

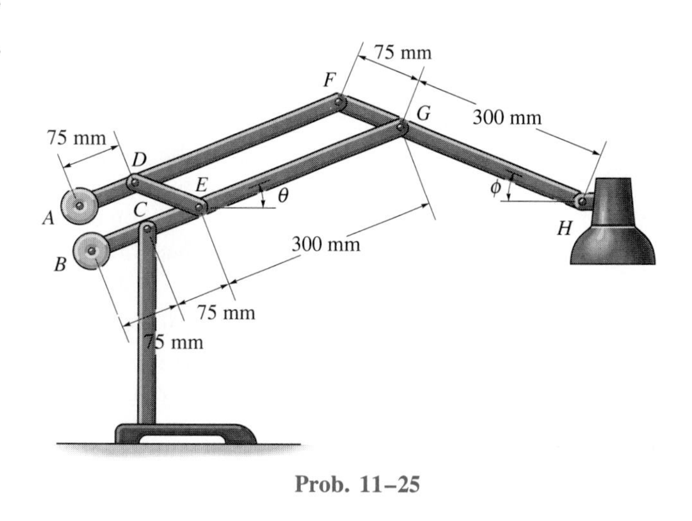

Prob. 11–25

***11–24.** Rods AB and BC have a center of mass located at their midpoints. If all contacting surfaces are smooth and BC has a mass of 100 kg, determine the appropriate mass of AB required for equilibrium.

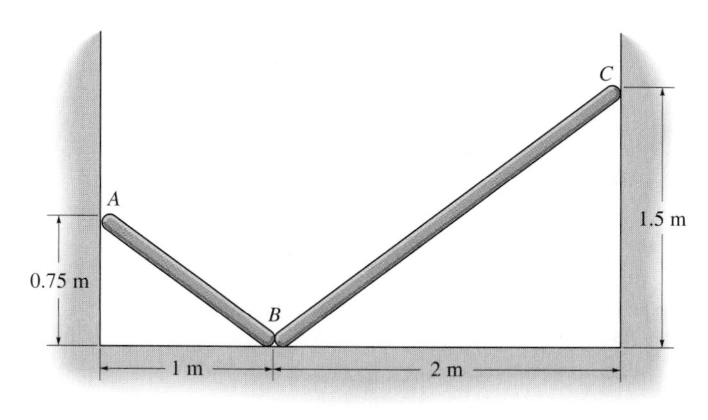

Prob. 11–24

11–26. If a force $P = 40\text{ N}$ is applied perpendicular to the handle of the toggle press, determine the vertical compressive force developed at C; $\theta = 30°$.

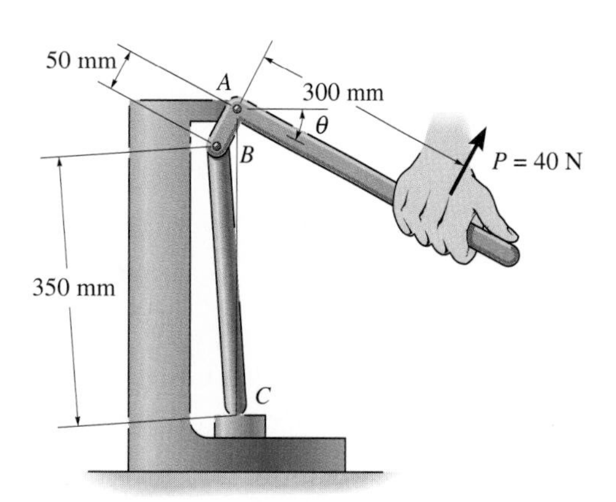

Prob. 11–26

*11.4 Conservative Forces

The work done by a force when it undergoes a *differential displacement* has been defined as $dU = F \cos \theta \, ds$. If the force is displaced over a path that has a *finite length s*, the work is determined by integrating over the path; i.e.,

$$U = \int_s F \cos \theta \, ds$$

To evaluate the integral, it is necessary to obtain a relationship between F and the component of displacement $ds \cos \theta$. In some instances, however, the work done by a force will be *independent* of its path and, instead, will depend only on the initial and final locations of the force along the path. A force that has this property is called a *conservative force.*

Weight. Consider the body in Fig. 11–9, which is initially at P'. If the body is moved *down* along the *arbitrary path* A to the dashed position, then, for a given displacement ds along the path, the displacement component in the direction of $\mathbf{W}$ has a magnitude of $dy = ds \cos \theta$, as shown. Since both the force and displacement are in the same direction, the work is positive; hence,

$$U = \int_s W \cos \theta \, ds = \int_0^y W \, dy$$

or

$$U = Wy$$

In a similar manner, the work done by the weight when the body moves up a distance y back to P', along the arbitrary path A', is

$$U = -Wy$$

Why is the work negative?

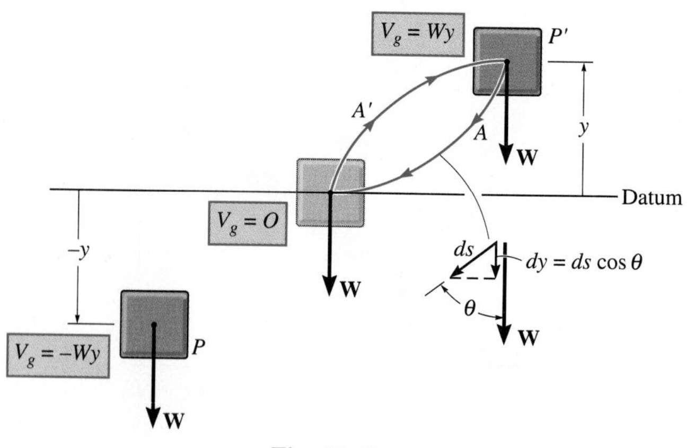

Fig. 11–9

The weight of a body is therefore a conservative force, since the work done by the weight depends *only* on the body's *vertical displacement* and is independent of the path along which the body moves.

Elastic Spring. The force developed by an elastic spring ($F_s = ks$) is also a conservative force. If the spring is attached to a body and the body is displaced along *any path,* such that it causes the spring to elongate or compress from a position s_1 to a further position s_2, the work will be negative, since the spring exerts a force $\mathbf{F}_s$ *on the body* that is opposite to the body's displacement ds, Fig. 11–10. For either extension or compression, the work is independent of the path and is simply

$$U = \int_{s_1}^{s_2} F_s\, ds = \int_{s_1}^{s_2} (-ks)\, ds$$
$$= -(\tfrac{1}{2}ks_2^2 - \tfrac{1}{2}ks_1^2)$$

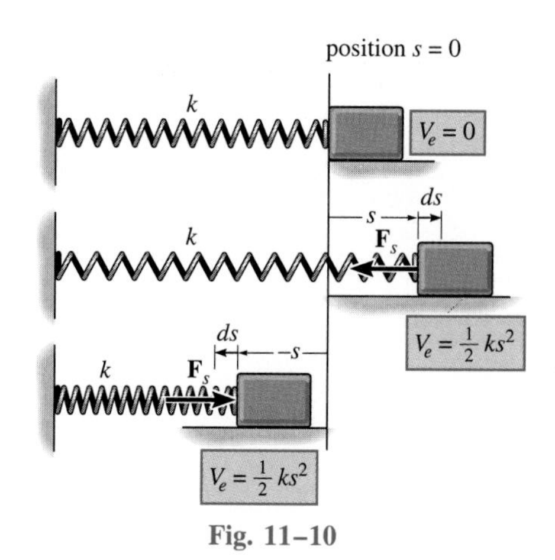

Fig. 11–10

Friction. In contrast to a conservative force, consider the force of *friction* exerted on a moving body by a fixed surface. The work done by the frictional force depends on the path; the longer the path, the greater the work. Consequently, frictional forces are *nonconservative,* and the work done is dissipated from the body in the form of heat.

*11.5 Potential Energy

When a conservative force acts on a body, it gives the body the capacity to do work. This capacity, measured as *potential energy,* depends on the location of the body.

Gravitational Potential Energy. If a body is located a distance y *above* a fixed horizontal reference or datum, Fig. 11–9, the weight of the body has *positive* gravitational potential energy V_g since $\mathbf{W}$ has the capacity of doing positive work when the body is moved back down to the datum. Likewise, if the body is located a distance y *below* the datum, V_g is *negative* since the weight does negative work when the body is moved back up to the datum. At the datum, $V_g = 0$.

Measuring y as *positive upward,* the gravitational potential energy of the body's weight $\mathbf{W}$ is thus

$$V_g = Wy \qquad (11\text{–}4)$$

Elastic Potential Energy. The elastic potential energy V_e that a spring produces on an attached body, when the spring is elongated or compressed from an undeformed position ($s = 0$) to a final position s, is

$$V_e = \tfrac{1}{2}ks^2 \tag{11–5}$$

Here V_e is *always positive,* since in the deformed position the spring has the capacity of doing *positive work* in *returning* the body back to the spring's undeformed position, Fig. 11–10.

Potential Function. In the general case, if a body is subjected to *both* gravitational and elastic forces, the *potential energy or potential function* V of the body can be expressed as the algebraic sum

$$V = V_g + V_e \tag{11–6}$$

where measurement of V depends on the location of the body with respect to a selected datum in accordance with Eqs. 11–4 and 11–5.

In general, if a system of frictionless connected rigid bodies has a *single degree of freedom* such that its position from the datum is defined by the independent coordinate q, then the potential function for the system can be expressed as $V = V(q)$. The work done by all the conservative forces acting on the system in moving it from q_1 to q_2 is measured by the *difference* in V; i.e.,

$$U_{1-2} = V(q_1) - V(q_2) \tag{11–7}$$

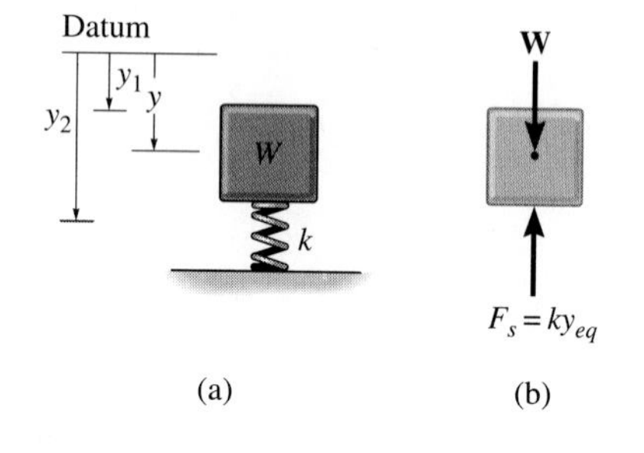

Fig. 11–11

For example, the potential function for a system consisting of a block of weight **W** supported by a spring, Fig. 11–11*a*, can be expressed in terms of its independent coordinate ($q =$) y, measured from a fixed datum located at the unstretched length of the spring; we have

$$\begin{aligned} V &= V_g + V_e \\ &= -Wy + \tfrac{1}{2}ky^2 \end{aligned} \tag{11–8}$$

If the block moves from y_1 to a farther downward position y_2, then the work of **W** and $\mathbf{F}_s$ is

$$U_{1-2} = V(y_1) - V(y_2) = -W[y_1 - y_2] + \tfrac{1}{2}ky_1^2 - \tfrac{1}{2}ky_2^2$$

*11.6 Potential-Energy Criterion for Equilibrium

System Having One Degree of Freedom. When the displacement of a frictionless connected system is *infinitesimal,* i.e., from q to $q + dq$, Eq. 11–7 becomes

$$dU = V(q) - V(q + dq)$$

or

$$dU = -dV$$

Furthermore, if the system undergoes a *virtual displacement* δq, rather than an actual displacement dq, then $\delta U = -\delta V$. For equilibrium, the principle of virtual work requires that $\delta U = 0$ and therefore, provided the potential function for the system is known, this also requires that $\delta V = 0$. We can also express this requirement as

$$\frac{dV}{dq} = 0 \qquad (11\text{–}9)$$

Hence, *when a frictionless connected system of rigid bodies is in equilibrium, the first variation or change in V is zero.* This change is determined by taking the *first derivative* of the potential function and setting it equal to zero. For example, using Eq. 11–8 to determine the equilibrium position for the spring and block in Fig. 11–11*a*, we have

$$\frac{dV}{dy} = W - ky = 0$$

Hence, the equilibrium position $y = y_{eq}$ is

$$y_{eq} = \frac{W}{k}$$

Of course, the *same result* is obtained by applying $\Sigma F_y = 0$ to the forces acting on the free-body diagram of the block, Fig. 11–11*b*.

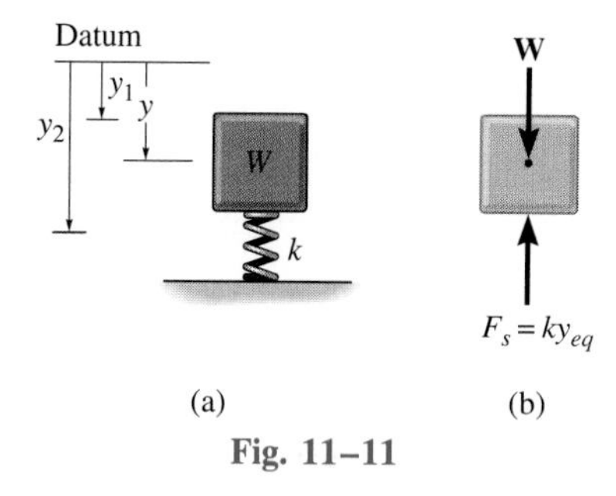

Fig. 11–11

System Having *n* Degrees of Freedom. When the system of connected bodies has n degrees of freedom, the total potential energy stored in the system will be a function of n independent coordinates q_n; i.e., $V = V(q_1, q_2, \ldots, q_n)$. In order to apply the equilibrium criterion $\delta V = 0$, it is necessary to determine the change in potential energy δV by using the "chain rule" of differential calculus; i.e.,

$$\delta V = \frac{\partial V}{\partial q_1}\delta q_1 + \frac{\partial V}{\partial q_2}\delta q_2 + \cdots + \frac{\partial V}{\partial q_n}\delta q_n = 0$$

Since the virtual displacements $\delta q_1, \delta q_2, \ldots, \delta q_n$ are independent of one another, the equation is satisfied provided

$$\frac{\partial V}{\partial q_1} = 0, \quad \frac{\partial V}{\partial q_2} = 0, \quad \ldots, \quad \frac{\partial V}{\partial q_n} = 0$$

Hence *it is possible to write n independent equations for a system having n degrees of freedom.*

★11.7 Stability of Equilibrium

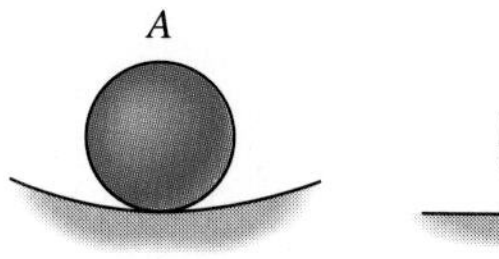

Stable equilibrium Neutral equilibrium

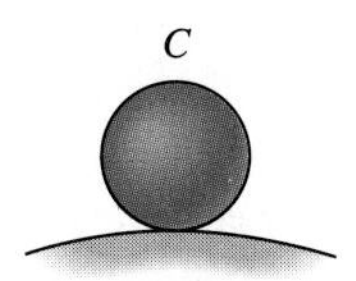

Unstable equilibrium

Fig. 11–12

Once the equilibrium configuration for a body or system of connected bodies is defined, it is sometimes important to investigate the "type" of equilibrium or the stability of the configuration. For example, consider the position of a ball resting at a point on each of the three paths shown in Fig. 11–12. Each situation represents an equilibrium state for the ball. When the ball is at A, it is said to be in *stable equilibrium* because if it is given a small displacement up the hill, it will always *return* to its original, lowest, position. At A, its total potential energy is a *minimum.* When the ball is at B, it is in *neutral equilibrium.* A small displacement to either the left or right of B will not alter this condition. The ball *remains* in equilibrium in the displaced position, and therefore its potential energy is *constant.* When the ball is at C, it is in *unstable equilibrium.* Here a small displacement will cause the ball's potential energy to be *decreased,* and so it will roll farther *away* from its original, highest position. At C, the potential energy of the ball is a *maximum.*

Type of Equilibrium. The example just presented illustrates that one of three types of equilibrium positions can be specified for a body or system of connected bodies.

1. *Stable equilibrium* occurs when a small displacement of the system causes the system to return to its original position. In this case the original potential energy of the system is a minimum.
2. *Neutral equilibrium* occurs when a small displacement of the system causes the system to remain in its displaced state. In this case the potential energy of the system remains constant.
3. *Unstable equilibrium* occurs when a small displacement of the system causes the system to move farther away from its original position. In this case the original potential energy of the system is a maximum.

System Having One Degree of Freedom. For *equilibrium* of a system having a single degree of freedom, defined by the independent coordinate q, it has been shown that the first derivative of the potential function for the system must be equal to zero; i.e., $dV/dq = 0$. If the potential function $V = V(q)$ is plotted, Fig. 11–13, the first derivative (equilibrium position) is represented as the slope dV/dq, which is zero when the function is maximum, minimum, or an inflection point.

If the *stability* of a body at the equilibrium position is to be investigated, it is necessary to determine the *second derivative* of V and evaluate it at the equilibrium position $q = q_{eq}$. As shown in Fig. 11–13*a*, if $V = V(q)$ is a *minimum*, then

$$\frac{dV}{dq} = 0, \quad \frac{d^2V}{dq^2} > 0 \quad \text{stable equilibrium} \tag{11–10}$$

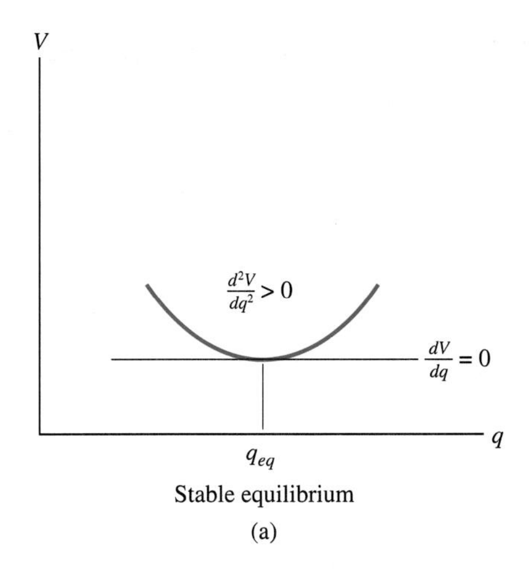

Stable equilibrium
(a)

If $V = V(q)$ is a *maximum,* Fig. 11–13*b*, then

$$\frac{dV}{dq} = 0, \quad \frac{d^2V}{dq^2} < 0 \quad \text{unstable equilibrium} \tag{11–11}$$

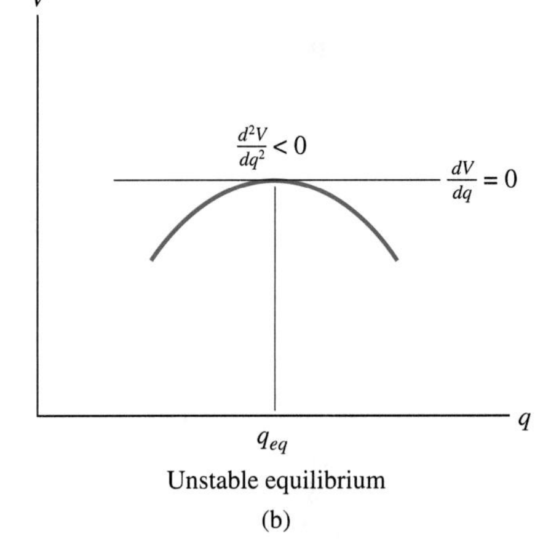

Unstable equilibrium
(b)

If the second derivative is zero, it will be necessary to investigate *higher-order* derivatives to determine the stability. In particular, stable equilibrium will occur if the order of the lowest remaining nonzero derivative is *even* and the sign of this nonzero derivative is positive when it is evaluated at $q = q_{eq}$; otherwise, it is unstable.

If the system is in neutral equilibrium, Fig. 11–13*c*, it is required that

$$\frac{dV}{dq} = \frac{d^2V}{dq^2} = \frac{d^3V}{dq^3} = \cdots = 0 \quad \text{neutral equilibrium} \tag{11–12}$$

since then V must be constant at and around the ''neighborhood'' of q_{eq}.

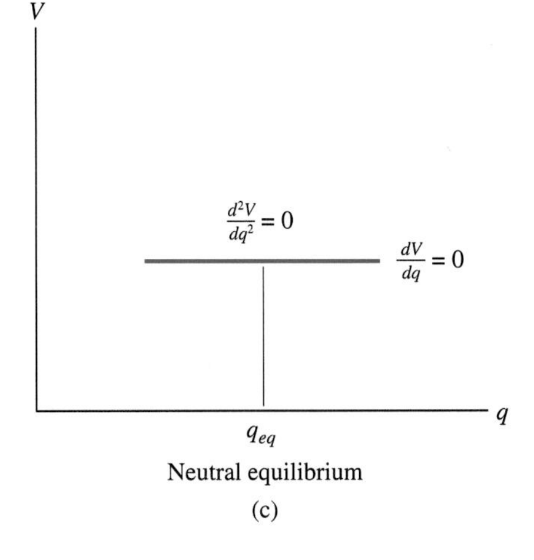

Neutral equilibrium
(c)

Fig. 11–13

System Having Two Degrees of Freedom. A criterion for investigating stability becomes increasingly complex as the number of degrees of freedom for the system increases. For a system having two degrees of freedom, defined by independent coordinates (q_1, q_2), it may be verified (using the calculus of functions of two variables) that equilibrium and stability occur at a point (q_{1eq}, q_{2eq}) when

$$\frac{\partial V}{\partial q_1} = \frac{\partial V}{\partial q_2} = 0$$

$$\left[\left(\frac{\partial^2 V}{\partial q_1\, \partial q_2}\right)^2 - \left(\frac{\partial^2 V}{\partial q_1^2}\right)\left(\frac{\partial^2 V}{\partial q_2^2}\right)\right] < 0$$

$$\left(\frac{\partial^2 V}{\partial q_1^2} + \frac{\partial^2 V}{\partial q_2^2}\right) > 0$$

Both equilibrium and instability occur when

$$\frac{\partial V}{\partial q_1} = \frac{\partial V}{\partial q_2} = 0$$

$$\left[\left(\frac{\partial^2 V}{\partial q_1\, \partial q_2}\right)^2 - \left(\frac{\partial^2 V}{\partial q_1^2}\right)\left(\frac{\partial^2 V}{\partial q_2^2}\right)\right] < 0$$

$$\left(\frac{\partial^2 V}{\partial q_1^2} + \frac{\partial^2 V}{\partial q_2^2}\right) < 0$$

PROCEDURE FOR ANALYSIS

Using potential-energy methods, the equilibrium positions and the stability of a body or a system of connected bodies having a single degree of freedom can be obtained by applying the following procedure.

Potential Function. Formulate the potential function $V = V_g + V_e$ for the system. To do this, sketch the system so that it is located at some *arbitrary position* specified by the independent coordinate q. A horizontal *datum* is established through a *fixed point,** and the *gravitational potential energy* V_g is expressed in terms of the weight W of each member and its vertical distance y from the datum, $V_g = Wy$, Eq. 11–4. The elastic potential energy V_e of the system is expressed in terms of the stretch or compression, s, of any connecting spring and the spring's stiffness k, $V_e = \frac{1}{2}ks^2$, Eq. 11–5. Once V has been established, express the *position coordinates* y and s in terms of the independent coordinate q.

Equilibrium Position. The equilibrium position is determined by taking the first derivative of V and setting it equal to zero, $\delta V = 0$, Eq. 11–9.

Stability. Stability at the equilibrium position is determined by evaluating the second or higher-order derivatives of V as indicated by Eqs. 11–10 to 11–12.

*The location of the datum is *arbitrary* since only the *changes* or differentials of V are required for investigation of the equilibrium position and its stability.

The following examples illustrate this procedure numerically.

Example 11–5

The uniform link shown in Fig. 11–14*a* has a mass of 10 kg. The spring is unstretched when $\theta = 0°$. Determine the angle θ for equilibrium and investigate the stability at the equilibrium position.

SOLUTION

Potential Function. The datum is established at the top of the link when the *spring is unstretched,* Fig. 11–14*b*. When the link is located at the arbitrary position θ, the spring increases its potential energy by stretching and the weight decreases its potential energy. Hence,

$$V = V_e + V_g = \frac{1}{2}ks^2 - W\left(s + \frac{l}{2}\cos\theta - \frac{l}{2}\right)$$

Since $l = s + l\cos\theta$ or $s = l(1 - \cos\theta)$, then

$$V = \frac{1}{2}kl^2(1 - \cos\theta)^2 - \frac{Wl}{2}(1 - \cos\theta)$$

Equilibrium Position. The first derivative of V gives

$$\frac{dV}{d\theta} = kl^2(1 - \cos\theta)\sin\theta - \frac{Wl}{2}\sin\theta = 0$$

or

$$l\left[kl(1 - \cos\theta) - \frac{W}{2}\right]\sin\theta = 0$$

This equation is satisfied provided

$$\sin\theta = 0 \qquad \theta = 0° \qquad \textit{Ans.}$$

$$\theta = \cos^{-1}\left(1 - \frac{W}{2kl}\right) = \cos^{-1}\left[1 - \frac{10(9.81)}{2(200)(0.6)}\right] = 53.8° \qquad \textit{Ans.}$$

Stability. Determining the second derivative of V gives

$$\frac{d^2V}{d\theta^2} = kl^2(1 - \cos\theta)\cos\theta + kl^2\sin\theta\sin\theta - \frac{Wl}{2}\cos\theta$$

$$= kl^2(\cos\theta - \cos 2\theta) - \frac{Wl}{2}\cos\theta$$

Substituting values for the constants, with $\theta = 0°$ and $\theta = 53.8°$, yields

$$\left.\frac{d^2V}{d\theta^2}\right|_{\theta=0°} = 200(0.6)^2(\cos 0° - \cos 0°) - \frac{10(9.81)(0.6)}{2}\cos 0°$$

$$= -29.4 < 0 \qquad \text{(unstable equilibrium at } \theta = 0°) \qquad \textit{Ans.}$$

$$\left.\frac{d^2V}{d\theta^2}\right|_{\theta=53.8°} = 200(0.6)^2(\cos 53.8° - \cos 107.6°) - \frac{10(9.81)(0.6)}{2}\cos 53.8°$$

$$= 46.9 > 0 \qquad \text{(stable equilibrium at } \theta = 53.8°) \qquad \textit{Ans.}$$

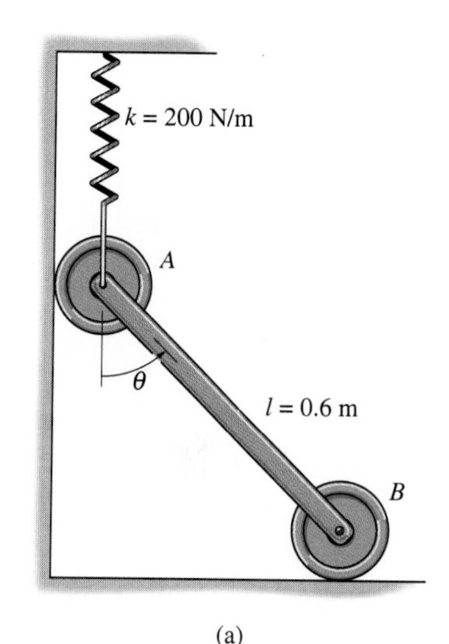

(a)

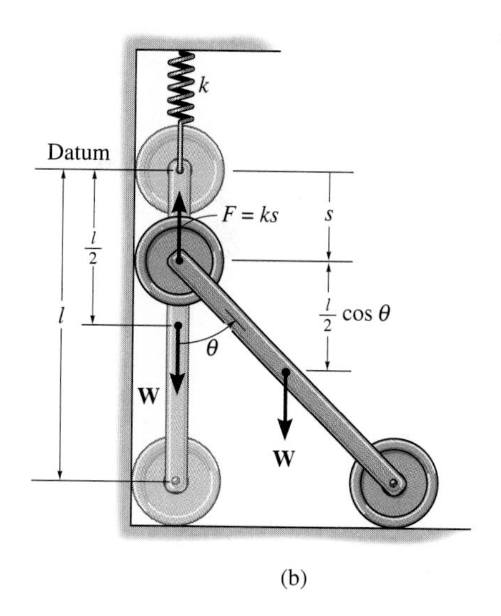

(b)

Fig. 11–14

Example 11–6

Determine the mass m of the block required for equilibrium of the uniform 10-kg rod shown in Fig. 11–15a when $\theta = 20°$. Investigate the stability at the equilibrium position.

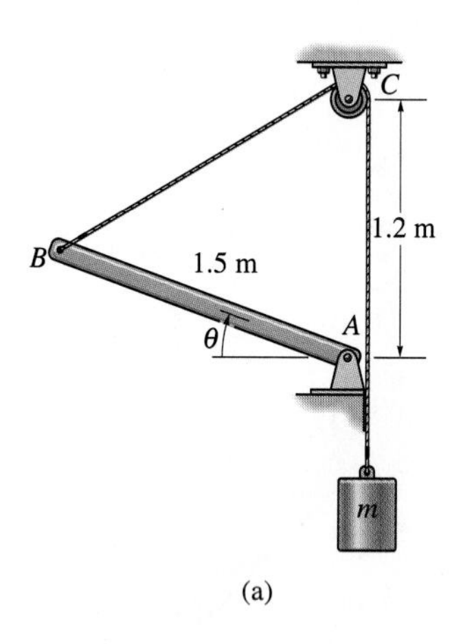

(a)

SOLUTION

Potential Function. The datum is established through point A, Fig. 11–15b. When $\theta = 0°$, the block is assumed to be suspended $(y_W)_1$ below the datum. Hence, in the position θ,

$$V = V_e + V_g = 98.1\left(\frac{1.5 \sin \theta}{2}\right) - m(9.81)(\Delta y) \tag{1}$$

The distance $\Delta y = (y_W)_2 - (y_W)_1$ may be related to the independent coordinate θ by measuring the difference in cord lengths $B'C$ and BC. Since

$$B'C = \sqrt{(1.5)^2 + (1.2)^2} = 1.92$$

$$BC = \sqrt{(1.5 \cos \theta)^2 + (1.2 - 1.5 \sin \theta)^2} = \sqrt{3.69 - 3.60 \sin \theta}$$

then

$$\Delta y = B'C - BC = 1.92 - \sqrt{3.69 - 3.60 \sin \theta}$$

Substituting the above result into Eq. 1 yields

$$V = 98.1\left(\frac{1.5 \sin \theta}{2}\right) - m(98.1)(1.92 - \sqrt{3.69 - 3.60 \sin \theta}) \tag{2}$$

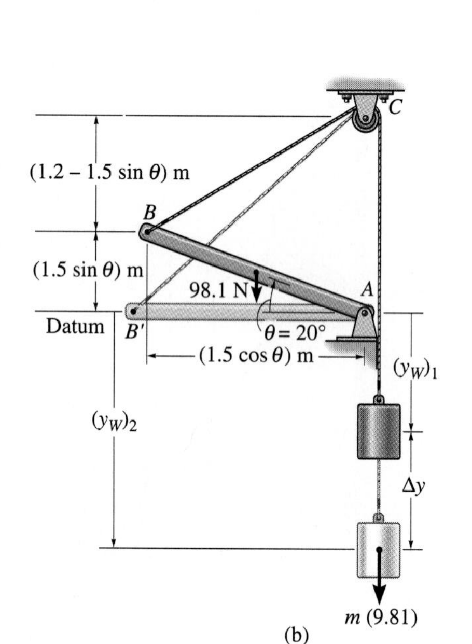

(b)

Equilibrium Position

$$\frac{dV}{d\theta} = 73.6 \cos \theta - \left[\frac{m(98.1)}{2}\right]\left(\frac{3.60 \cos \theta}{\sqrt{3.69 - 3.60 \sin \theta}}\right) = 0$$

$$\left.\frac{dV}{d\theta}\right|_{\theta=20°} = 69.16 - 10.58m = 0$$

$$m = \frac{69.16}{10.58} = 6.54 \text{ kg}$$ ***Ans.***

Stability. Taking the second derivative of Eq. 2, we obtain

$$\frac{d^2V}{d\theta^2} = -73.6 \sin \theta - \left[\frac{m(9.81)}{2}\right]\left(\frac{-1}{2}\right)\frac{(3.60 \cos \theta)^2}{(3.69 - 3.60 \sin \theta)^{3/2}} - \frac{m(9.81)}{2}\left(\frac{-3.60 \sin \theta}{\sqrt{3.69 - 3.60 \sin \theta}}\right)$$

Fig. 11–15

For the equilibrium position $\theta = 20°$, also $m = 6.54$ kg, so

$$\frac{d^2V}{d\theta^2} = 47.6 > 0 \qquad \text{(stable equilibrium at } \theta = 20°\text{)}$$ ***Ans.***

Example 11–7

The homogeneous block having a mass m rests on the top surface of the cylinder, Fig. 11–16a. Show that this is a condition of unstable equilibrium if $h > 2R$.

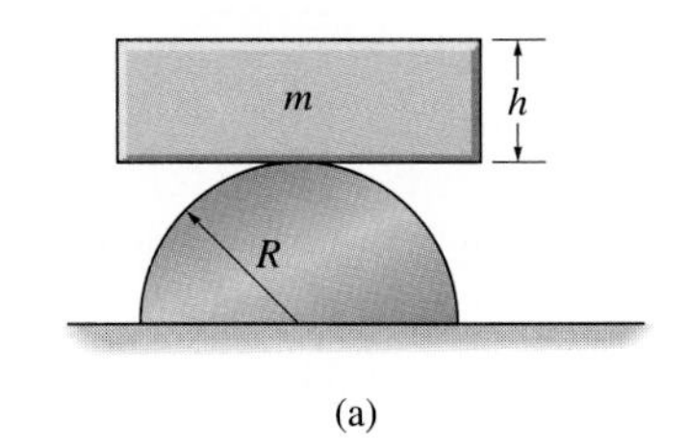

(a)

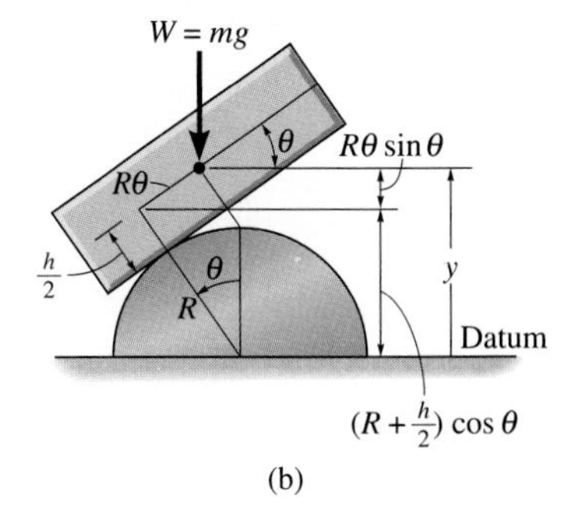

(b)

Fig. 11–16

SOLUTION

Potential Function. The datum is established at the base of the cylinder, Fig. 11–16b. If the block is displaced by an amount θ from the equilibrium position, the potential function may be written in the form

$$V = V_e + V_g$$
$$= 0 + mgy$$

From Fig. 11–16b,

$$y = \left(R + \frac{h}{2}\right)\cos\theta + R\theta\sin\theta$$

Thus,

$$V = mg\left[\left(R + \frac{h}{2}\right)\cos\theta + R\theta\sin\theta\right]$$

Equilibrium Position

$$\frac{dV}{d\theta} = mg\left[-\left(R + \frac{h}{2}\right)\sin\theta + R\sin\theta + R\theta\cos\theta\right] = 0$$
$$= mg\left(-\frac{h}{2}\sin\theta + R\theta\cos\theta\right) = 0$$

Obviously, $\theta = 0°$ is the equilibrium position that satisfies this equation.

Stability. Taking the second derivative of V yields

$$\frac{d^2V}{d\theta^2} = mg\left(-\frac{h}{2}\cos\theta + R\cos\theta - R\theta\sin\theta\right)$$

At $\theta = 0°$,

$$\left.\frac{d^2V}{d\theta^2}\right|_{\theta=0°} = -mg\left(\frac{h}{2} - R\right)$$

Since all the constants are positive, the block is in unstable equilibrium if $h > 2R$, for then $d^2V/d\theta^2 < 0$.

PROBLEMS

11–27. If the potential function for a conservative one-degree-of-freedom system is $V = (8x^3 - 2x^2 - 10)$ J, where x is given in meters, determine the positions for equilibrium and investigate the stability at each of these positions.

***11–28.** If the potential function for a conservative one-degree-of-freedom system is $V = (12 \sin 2\theta + 15 \cos \theta)$ J, where $0° < \theta < 180°$, determine the positions for equilibrium and investigate the stability at each of these positions.

11–29. If the potential function for a conservative one-degree-of-freedom system is $V = (10 \cos 2\theta + 25 \sin \theta)$ J, where $0° < \theta < 180°$, determine the positions for equilibrium and investigate the stability at each of these positions.

11–30. If the potential function for a conservative two-degree-of-freedom system is $V = (9y^2 + 18x^2)$ J, where x and y are given in meters, determine the equilibrium position and investigate the stability at this position.

11–31. Solve Prob. 11–6 using the principle of potential energy, and investigate the stability at the equilibrium position.

***11–32.** Solve Prob. 11–12 using the principle of potential energy, and investigate the stability at the equilibrium position.

11–33. Solve Prob. 11–13 using the principle of potential energy, and investigate the stability at the equilibrium position.

11–34. Solve Prob. 11–24 using the principle of potential energy.

■**11–35.** The two bars each have a weight of 8 lb. Determine the angle θ for equilibrium and investigate the stability at the equilibrium position. The spring has an unstretched length of 1 ft. Take $k = 30$ lb/ft.

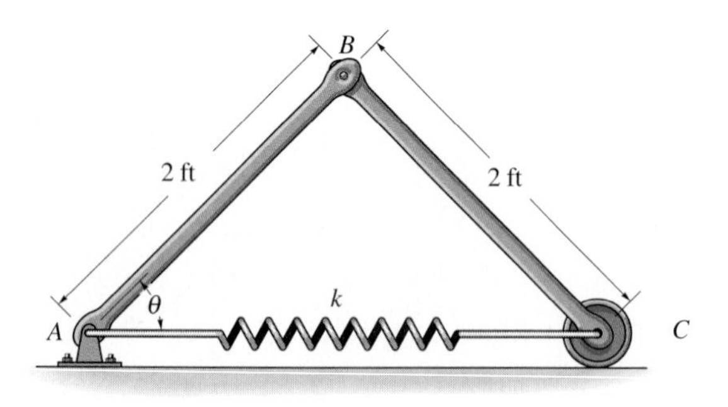

Prob. 11–35

***11–36.** The two bars each have a weight of 8 lb. Determine the required stiffness k of the spring so that the two bars are in neutral equilibrium when $\theta = 30°$. The spring has an unstretched length of 1 ft.

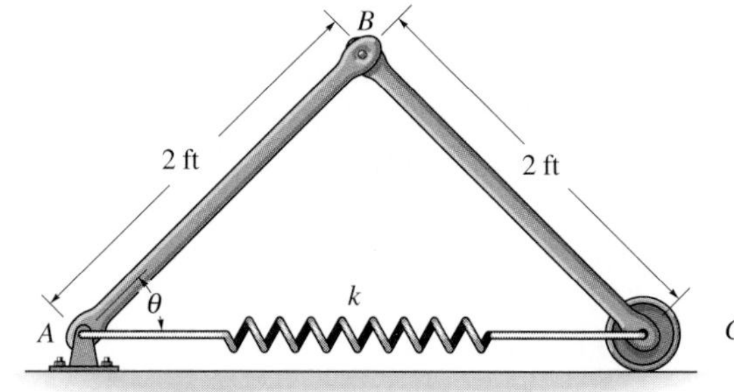

Prob. 11–36

11–37. The uniform beam has a weight W. If the contacting surfaces are smooth, determine the angle θ for equilibrium. The spring is uncompressed when $\theta = 90°$.

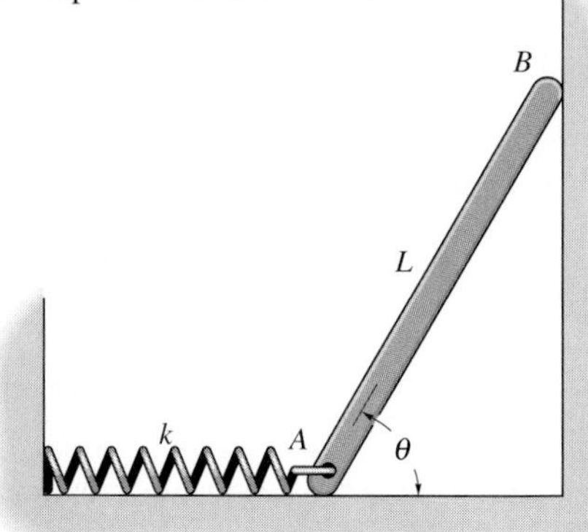

Prob. 11–37

11–38. The uniform beam has a weight W. If the contacting surfaces are smooth, determine the angle θ for equilibrium. The spring is unstretched when $\theta = 0°$.

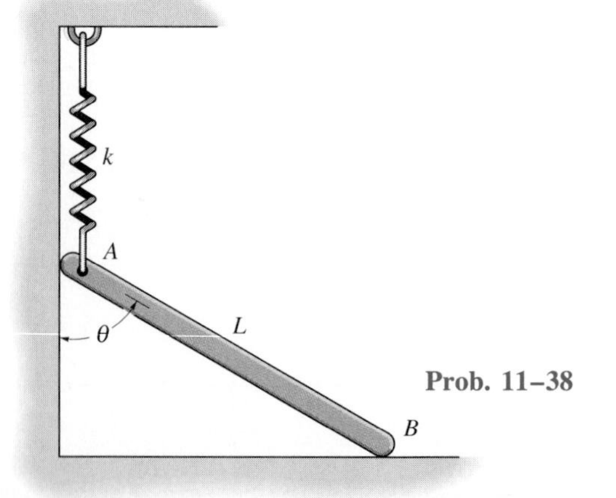

Prob. 11–38

11–39. The bar supports a weight of $W = 500$ lb at its end. If the springs are originally unstretched when the bar is vertical, and $k_1 = 300$ lb/ft, $k_2 = 500$ lb/ft, investigate the stability of the bar when it is in the vertical position.

***11–40.** The bar supports a weight of $W = 500$ lb at its end. If the springs are originally unstretched when the bar is vertical, determine the required stiffness $k_1 = k_2 = k$ of the springs so that the bar is in neutral equilibrium when it is vertical.

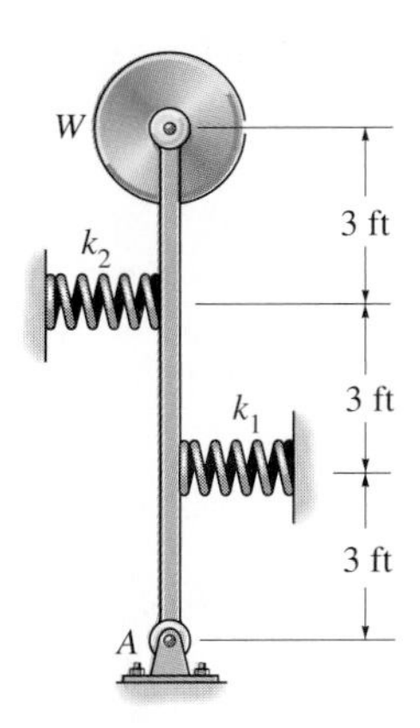

Probs. 11–39/11–40

■**11–42.** The spring has a stiffness $k = 400$ N/m and an unstretched length of 0.3 m. Determine the angle θ for equilibrium if the uniform links each have a mass of 5 kg.

11–43. The spring has an unstretched length of 0.3 m. Determine its stiffness k so that it is in equilibrium when $\theta = 20°$. Each uniform link has a mass of 5 kg. Investigate the stability at this position.

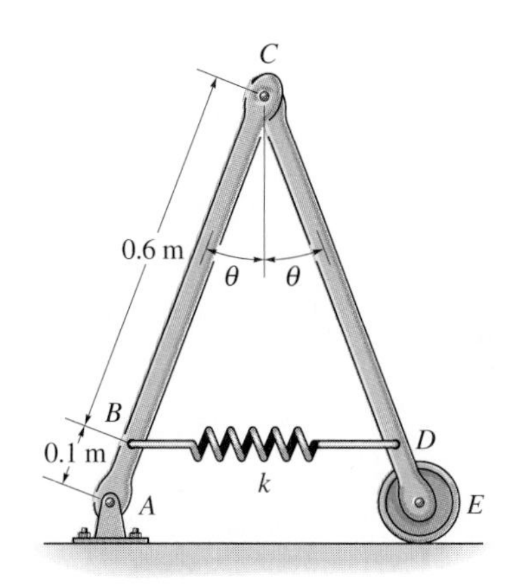

Probs. 11–42/11–43

11–41. The rod supports a disk B having a weight of 800 N. It is held in equilibrium using four springs for which $k = 400$ N/m. Determine the minimum distance d between the springs so that the rod remains in stable equilibrium when it is vertical.

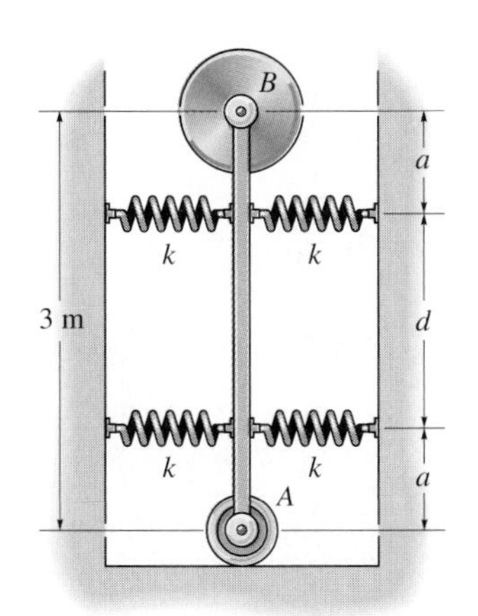

Prob. 11–41

***11–44.** Determine the angle θ for equilibrium and investigate the stability at this position. The bars each have a mass of 3 kg and the suspended block D has a mass of 7 kg. Cord DC has a total length of 1 m.

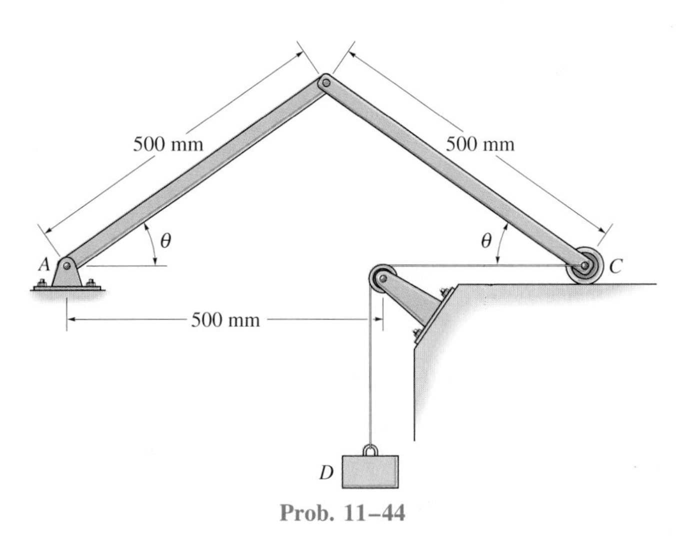

Prob. 11–44

11–45. The door has a uniform weight of 50 lb. It is hinged at A and is held open by the 30-lb weight and the pulley. Determine the angle θ for equilibrium.

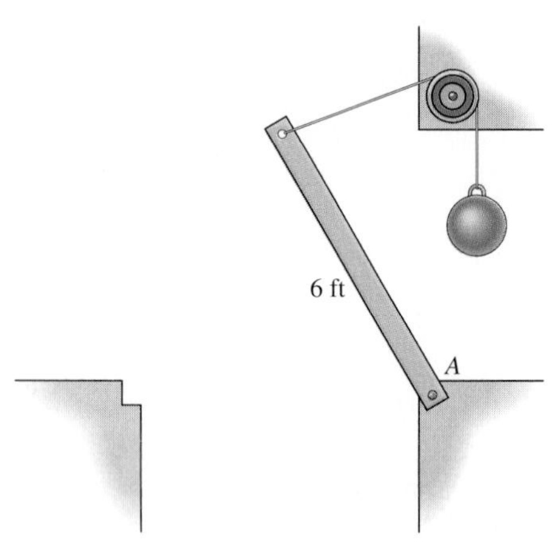

Prob. 11–45

11–46. The spring of the scale has an unstretched length a. Determine the angle θ for equilibrium when a weight W is supported on the platform. Neglect the weights of the members. What value of W would be required to keep the scale in neutral equilibrium when $\theta = 0°$?

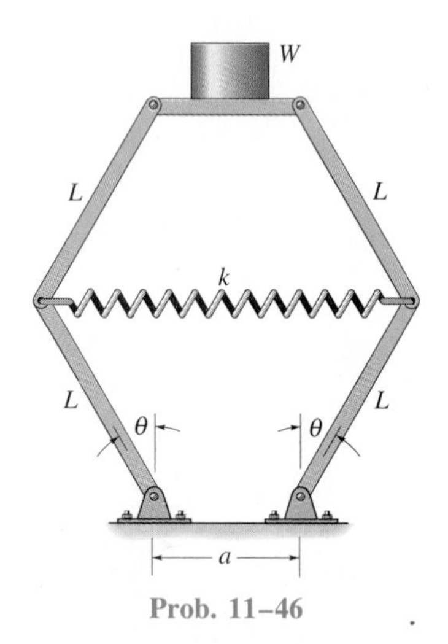

Prob. 11–46

11–47. The block weighs W and is supported by links AB and BC. Determine the necessary spring stiffness k required to hold the system in neutral equilibrium. The springs are subjected to an initial compression F_0 when the links are vertical as shown.

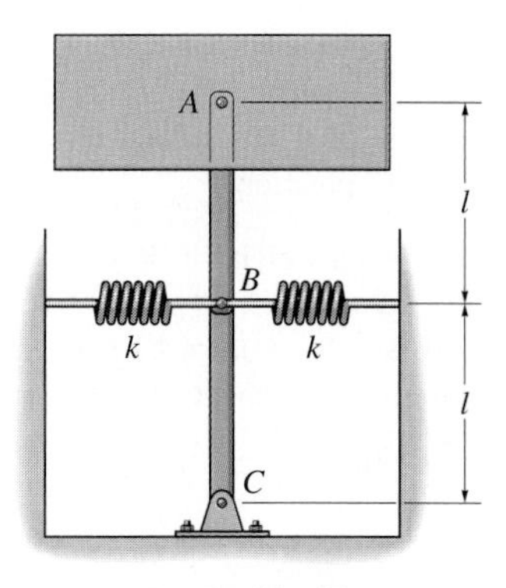

Prob. 11–47

***11–48.** The Roberval balance is in equilibrium when no weights are placed on the pans A and B. If two masses m_A and m_B are placed at *any* location a and b on the pans, show that neutral equilibrium is maintained if $m_A d_A = m_B d_B$.

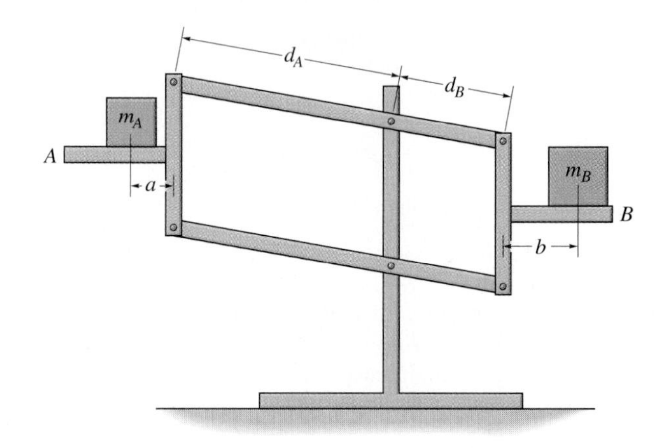

Prob. 11–48

11–49. The cup has a hemispherical bottom and a mass m. Determine the position h of the center of mass G so that the cup is in neutral equilibrium.

Prob. 11–49

11–50. The homogeneous cylinder has a conical cavity cut into its base as shown. Determine the depth d of the cavity so that the cylinder balances on the pivot and remains in neutral equilibrium.

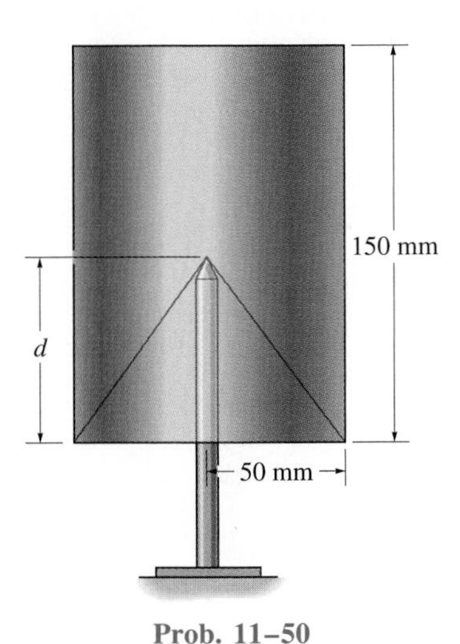

Prob. 11–50

11–51. The uniform right circular cone having a mass m is suspended from the cord as shown. Determine the angle θ at which it hangs from the wall for equilibrium. Is the cone in stable equilibrium?

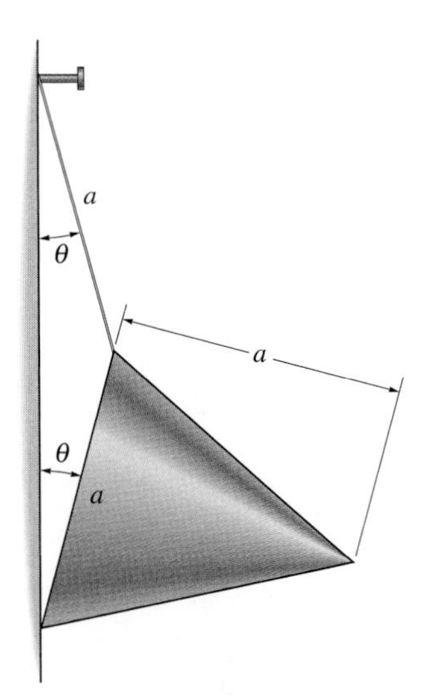

Prob. 11–51

***11–52.** The triangular block of weight W rests on the smooth corners which are a distance a apart. If the block has three equal sides of length d, determine the angle θ for equilibrium.

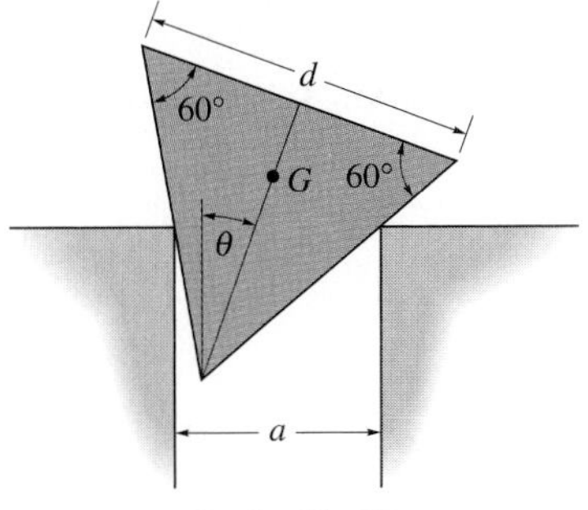

Prob. 11–52

11–53. Two uniform bars, each having a weight W, are pin-connected at their ends. If they are placed over a smooth cylindrical surface, show that the angle θ for equilibrium must satisfy the equation $\cos\theta/\sin^3\theta = a/2r$.

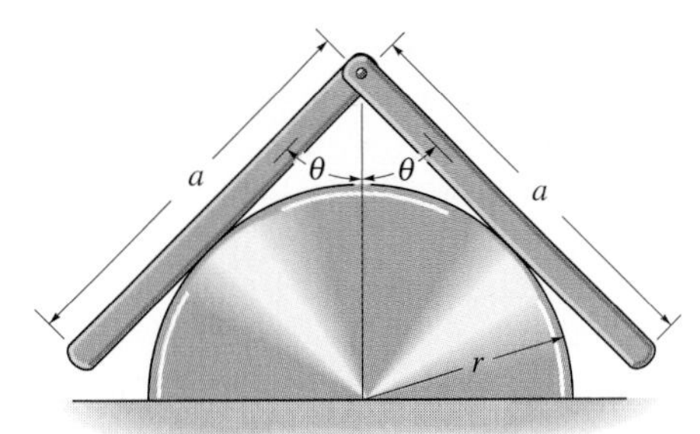

Prob. 11–53

11–54. The homogeneous block has a mass of 10 kg and rests on the smooth corners of two ledges. Determine the angle θ for placement that will cause the block to be stable.

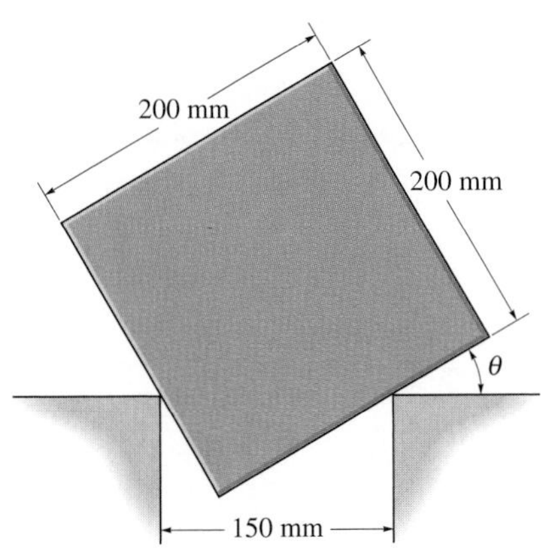

Prob. 11–54

REVIEW PROBLEMS

11–55. The three-bar mechanism of negligible weight is subjected to a couple moment $M_A = 8$ lb · ft. Determine the magnitude of the couple moment $\mathbf{M}_D$ needed to maintain the equilibrium position $\theta = 30°$, $\phi = 90°$.

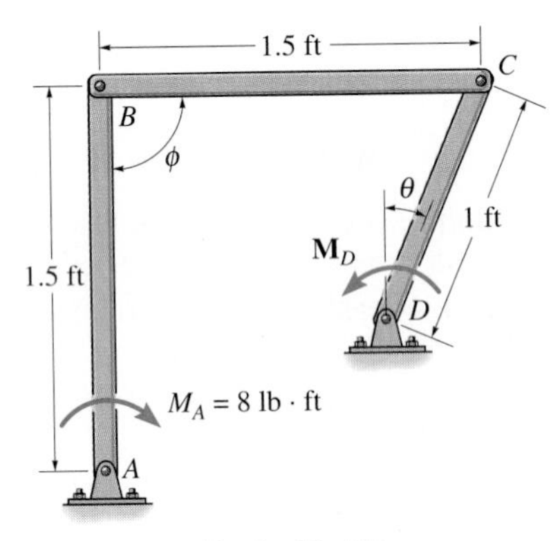

Prob. 11–55

*11–56.** Compute the force developed in the spring required to keep the 6-kg rod in equilibrium when $\theta = 30°$. The spring remains horizontal due to the roller guide.

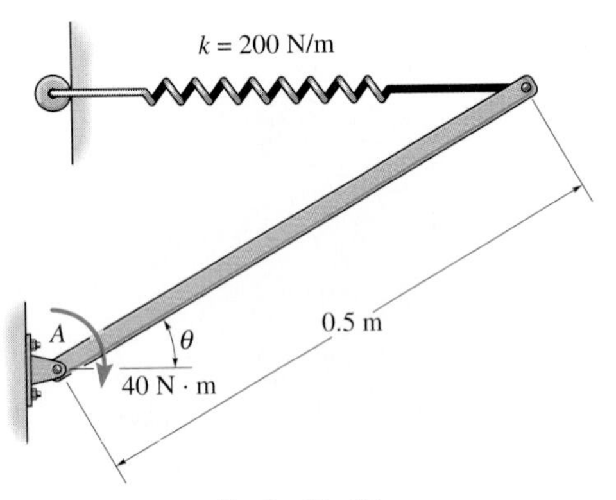

Prob. 11–56

11–57. The uniform bar AB weighs 10 lb. If the attached spring is unstretched when $\theta = 90°$, use the method of virtual work and determine the angle θ for equilibrium. Note that the spring always remains in the vertical position due to the roller guide.

11–58. Solve Prob. 11–57 using the principle of potential energy. Investigate the stability of the bar when it is in the equilibrium position.

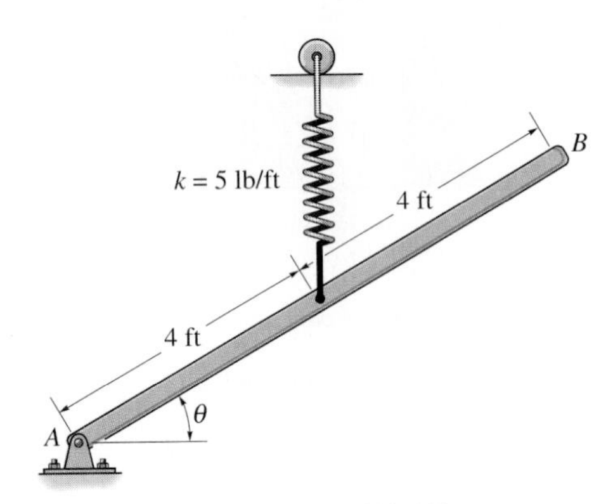

Probs. 11–57/11–58

11–59. The uniform rod AB has a weight of 10 lb. If the spring DC is unstretched when $\theta = 90°$, determine the angle θ for equilibrium using the principle of virtual work. The spring is always in the horizontal position because of the roller guide at D.

***11–60.** Solve Prob. 11–59 using the principle of potential energy. Investigate the stability of the rod when it is in the equilibrium position.

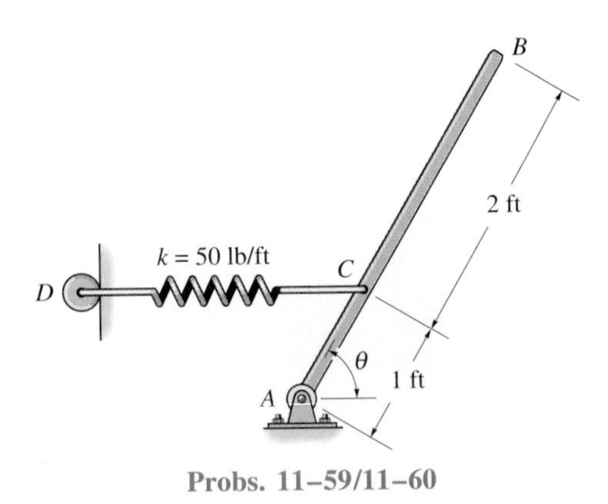

Probs. 11–59/11–60

11–61. The chain puller is used to draw two ends of a chain together in order to attach the "master link." The device is operated by turning the screw S, which pushes the bar AB downward, thereby drawing the tips C and D towards one another. If the sliding contacts at A and B are smooth, determine the force F maintained by the screw at E, which for the position shown is required to develop a drawing tension of 120 lb in the chain.

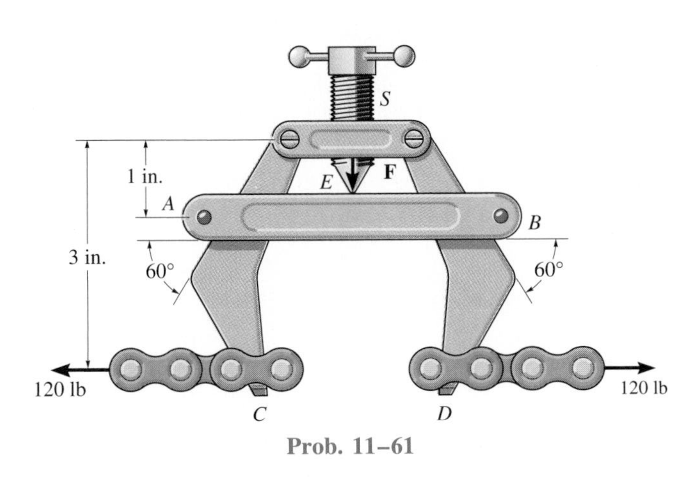

Prob. 11–61

■**11–62.** A disk of weight W is attached to the end of rod ABC. If the rod is supported by a smooth slider block at C and rod BD, determine the angle θ for equilibrium. Neglect the weight of the rods and the slider.

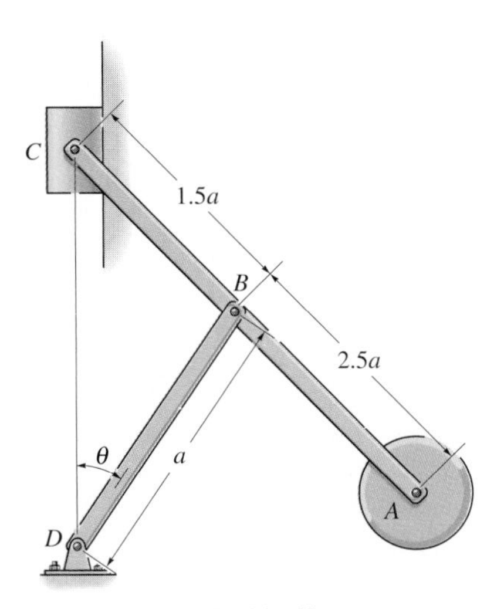

Prob. 11–62

11–63. If $P = 10$ lb, determine the angle θ for equilibrium of the pin-connected mechanism. The spring is unstretched when $\theta = 45°$. Neglect the weight of the members.

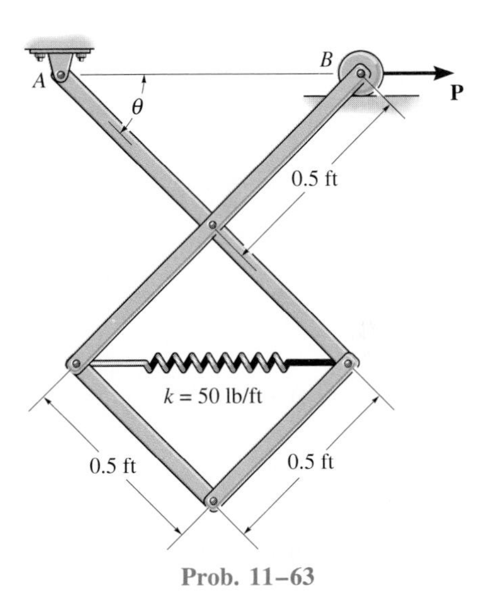

Prob. 11–63

***11–64.** Each member of the pin-connected mechanism has a mass of 8 kg. If the spring is unstretched when $\theta = 0°$, determine the angle θ for equilibrium.

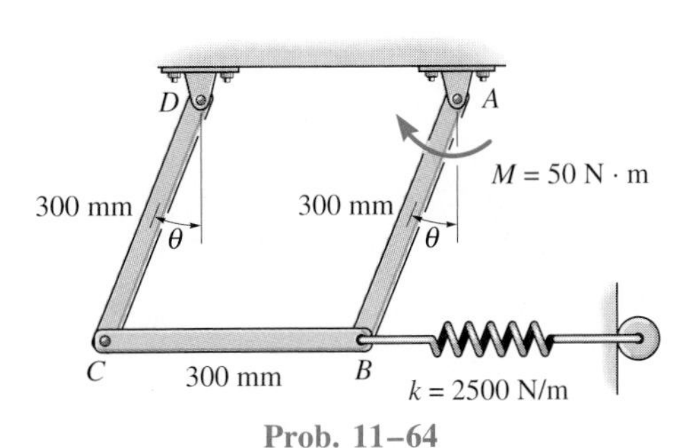

Prob. 11–64

A

Mathematical Expressions

Quadratic Formula

If $ax^2 + bx + c = 0$, then $x = \dfrac{-b \pm \sqrt{b^2 - 4ac}}{2a}$

Hyperbolic Functions

$$\sinh x = \frac{e^x - e^{-x}}{2}, \quad \cosh x = \frac{e^x + e^{-x}}{2}, \quad \tanh x = \frac{\sinh x}{\cosh x}$$

Trigonometric Identities

$$\sin\theta = \frac{A}{C}, \quad \csc\theta = \frac{C}{A}$$

$$\cos\theta = \frac{B}{C}, \quad \sec\theta = \frac{C}{B}$$

$$\tan\theta = \frac{A}{B}, \quad \cot\theta = \frac{B}{A}$$

C A θ B

$$\sin^2\theta + \cos^2\theta = 1$$

$$\sin(\theta \pm \phi) = \sin\theta\cos\phi \pm \cos\theta\sin\phi$$

$$\sin 2\theta = 2\sin\theta\cos\theta$$

$$\cos(\theta \pm \phi) = \cos\theta\cos\phi \mp \sin\theta\sin\phi$$

$$\cos 2\theta = \cos^2\theta - \sin^2\theta$$

$$\cos\theta = \pm\sqrt{\frac{1 + \cos 2\theta}{2}}, \quad \sin\theta = \pm\sqrt{\frac{1 - \cos 2\theta}{2}}$$

$$\tan\theta = \frac{\sin\theta}{\cos\theta}$$

$$1 + \tan^2\theta = \sec^2\theta \qquad 1 + \cot^2\theta = \csc^2\theta$$

Power-Series Expansions

$$\sin x = x - \frac{x^3}{3!} + \cdots, \quad \cos x = 1 - \frac{x^2}{2!} + \cdots$$

$$\sinh x = x + \frac{x^3}{3!} + \cdots, \quad \cosh x = 1 + \frac{x^2}{2!} + \cdots$$

Derivatives

$$\frac{d}{dx}(u^n) = nu^{n-1}\frac{du}{dx} \qquad \frac{d}{dx}(\sin u) = \cos u\frac{du}{dx}$$

$$\frac{d}{dx}(uv) = u\frac{dv}{dx} + v\frac{du}{dx} \qquad \frac{d}{dx}(\cos u) = -\sin u\frac{du}{dx}$$

$$\frac{d}{dx}\left(\frac{u}{v}\right) = \frac{v\dfrac{du}{dx} - u\dfrac{dv}{dx}}{v^2} \qquad \frac{d}{dx}(\tan u) = \sec^2 u\frac{du}{dx}$$

$$\frac{d}{dx}(\cot u) = -\csc^2 u\frac{du}{dx} \qquad \frac{d}{dx}(\sinh u) = \cosh u\frac{du}{dx}$$

$$\frac{d}{dx}(\sec u) = \tan u\sec u\frac{du}{dx} \qquad \frac{d}{dx}(\cosh u) = \sinh u\frac{du}{dx}$$

$$\frac{d}{dx}(\csc u) = -\csc u\cot u\frac{du}{dx}$$

Integrals

$$\int x^n\,dx = \frac{x^{n+1}}{n+1} + C,\ n \neq -1$$

$$\int \frac{dx}{a+bx} = \frac{1}{b}\ln(a+bx) + C$$

$$\int \frac{dx}{a+bx^2} = \frac{1}{2\sqrt{-ba}}\ln\left[\frac{\sqrt{a}+2\sqrt{-b}}{\sqrt{a}-x\sqrt{-b}}\right] + C,\quad a>0,\ b<0$$

$$\int \frac{x\,dx}{a+bx^2} = \frac{1}{2b}\ln(bx^2+a) + C$$

$$\int \frac{x^2\,dx}{a+bx^2} = \frac{x}{b} - \frac{a}{b\sqrt{ab}}\tan^{-1}\frac{x\sqrt{ab}}{a} + C$$

$$\int \frac{dx}{a^2-x^2} = \frac{1}{2a}\ln\left[\frac{a+x}{a-x}\right] + C,\ a^2 > x^2$$

$$\int \sqrt{a+bx}\,dx = \frac{2}{3b}\sqrt{(a+bx)^3} + C$$

$$\int x\sqrt{a+bx}\,dx = \frac{-2(2a-3bx)\sqrt{(a+bx)^3}}{15b^2} + C$$

$$\int x^2\sqrt{a+bx}\,dx = \frac{2(8a^2-12abx+15b^2x^2)\sqrt{(a+bx)^3}}{105b^3} + C$$

$$\int \sqrt{a^2-x^2}\,dx = \frac{1}{2}\left[x\sqrt{a^2-x^2} + a^2\sin^{-1}\frac{x}{a}\right] + C,\quad a>0$$

$$\int x\sqrt{a^2-x^2}\,dx = -\frac{1}{3}\sqrt{(a^2-x^2)^3} + C$$

$$\int x^2\sqrt{a^2-x^2}\,dx = -\frac{x}{4}\sqrt{(a^2-x^2)^3} + \frac{a^2}{8}\left(x\sqrt{a^2-x^2} + a^2\sin^{-1}\frac{x}{a}\right) + C,\ a>0$$

$$\int \sqrt{x^2 \pm a^2}\,dx = \frac{1}{2}\left[x\sqrt{x^2\pm a^2} \pm a^2\ln\left(x+\sqrt{x^2\pm a^2}\right)\right] + C$$

$$\int x\sqrt{x^2\pm a^2}\,dx = \frac{1}{3}\sqrt{(x^2\pm a^2)^3} + C$$

$$\int x^2\sqrt{x^2\pm a^2}\,dx = \frac{x}{4}\sqrt{(x^2\pm a^2)^3} \pm \frac{a^2}{8}x\sqrt{x^2\pm a^2} - \frac{a^4}{8}\ln\left(x+\sqrt{x^2\pm a^2}\right) + C$$

$$\int \frac{dx}{\sqrt{a+bx}} = \frac{2\sqrt{a+bx}}{b} + C$$

$$\int \frac{x\,dx}{\sqrt{x^2\pm a^2}} = \sqrt{x^2\pm a^2} + C$$

$$\int \frac{dx}{\sqrt{a+bx+cx^2}} = \frac{1}{\sqrt{c}}\ln\left[\sqrt{a+bx+cx^2} + x\sqrt{c} + \frac{b}{2\sqrt{c}}\right] + C,\ c>0$$

$$= \frac{1}{\sqrt{-c}}\sin^{-1}\left(\frac{-2cx-b}{\sqrt{b^2-4ac}}\right) + C,\ c<0$$

$$\int \sin x\,dx = -\cos x + C$$

$$\int \cos x\,dx = \sin x + C$$

$$\int x\cos(ax)\,dx = \frac{1}{a^2}\cos(ax) + \frac{x}{a}\sin(ax) + C$$

$$\int x^2\cos(ax)\,dx = \frac{2x}{a^2}\cos(ax) + \frac{a^2x^2-2}{a^3}\sin(ax) + C$$

$$\int e^{ax}\,dx = \frac{1}{a}e^{ax} + C$$

$$\int x\,e^{ax}\,dx = \frac{e^{ax}}{a^2}(ax-1) + C$$

$$\int \sinh x\,dx = \cosh x + C$$

$$\int \cosh x\,dx = \sinh x + C$$

B

Numerical and Computer Analysis

Occasionally the application of the laws of mechanics will lead to a system of equations for which a closed-form solution is difficult or impossible to obtain. When confronted with this situation, engineers will often use a numerical method which in most cases can be programmed on a microcomputer or "programmable" pocket calculator. Here we will briefly present a computer program for solving a set of linear algebraic equations and three numerical methods which can be used to solve an algebraic or transcendental equation, evaluate a definite integral, and solve an ordinary differential equation. Application of each method will be explained by example, and an associated computer program written in Microsoft BASIC, which is designed to run on most personal computers, is provided.* A text on numerical analysis should be consulted for further discussion regarding a check of the accuracy of each method and the inherent errors that can develop from the methods.

B.1 Linear Algebraic Equations

Application of the equations of static equilibrium or the equations of motion sometimes requires solving a set of linear algebraic equations. The computer program listed in Fig. B–1 can be used for this purpose. It is based on the method of a Gaussian elimination and can solve at most 10 equations with 10

*Similar types of programs can be written or purchased for programmable pocket calculators.

```
1 PRINT"Linear system of equations":PRINT
2 DIM A(10,11)
3 INPUT"Input number of equations : ",N
4 PRINT
5 PRINT"A  coefficients"
6 FOR I = 1 TO N
7 FOR J = 1 TO N
8 PRINT "A(";I;",";J;
9 INPUT")=",A(I,J)
10 NEXT J
11 NEXT I
12 PRINT
13 PRINT"B  coefficients"
14 FOR I = 1 TO N
15 PRINT "B(";I;
16 INPUT")=",A(I,N+1)
17 NEXT I
18 GOSUB 25
19 PRINT
20 PRINT"Unknowns"
21 FOR I = 1 TO N
22 PRINT "X(";I;")=";A(I,N+1)
23 NEXT I
24 END
25 REM Subroutine Guassian
26 FOR M=1 TO N
27 NP=M
28 BG=ABS(A(M,M))
29 FOR I = M TO N
30 IF ABS(A(I,M))<=BG THEN 33
31 BG=ABS(A(I,M))
32 NP=I
33 NEXT I
34 IF NP=M THEN 40
35 FOR I = M TO N+1
36 TE=A(M,I)
37 A(M,I)=A(NP,I)
38 A(NP,I)=TE
39 NEXT I
40 FOR I = M+1 TO N
41 FC=A(I,M)/A(M,M)
42 FOR J = M+1 TO N+1
43 A(I,J)=A(I,J)-FC*A(M,J)
44 NEXT J
45 NEXT I
46 NEXT M
47 A(N,N+1)=A(N,N+1)/A(N,N)
48 FOR I = N-1 TO 1 STEP -1
49 SM=0
50 FOR J=I+1 TO N
51 SM=SM+A(I,J)*A(J,N+1)
52 NEXT J
53 A(I,N+1)=(A(I,N+1)-SM)/A(I,I)
54 NEXT I
55 RETURN
```

Fig. B–1

unknowns. To do so, the equations should first be written in the following general format:

$$A_{11}x_1 + A_{12}x_2 + \cdots + A_{1n}x_n = B_1$$
$$A_{21}x_1 + A_{22}x_2 + \cdots + A_{2n}x_n = B_2$$
$$\vdots$$
$$A_{n1}x_1 + A_{n2}x_2 + \cdots + A_{nn}x_n = B_n$$

The "A" and "B" coefficients are "called" for when running the program. The output presents the unknowns $x_1, \ldots, x_n$.

Example B–1

Solve the two equations

$$3x_1 + x_2 = 4$$
$$2x_1 - x_2 = 10$$

SOLUTION

When the program begins to run, it first calls for the number of equations (2); then the A coefficients in the sequence $A_{11} = 3$, $A_{12} = 1$, $A_{21} = 2$, $A_{22} = -1$; and finally the B coefficients $B_1 = 4$, $B_2 = 10$. The output appears as

Unknowns

$X(1) = 2.8$ ***Ans.***

$X(2) = -4.4$ ***Ans.***

B.2 Simpson's Rule

Simpson's rule is a numerical method that can be used to determine the area under a curve given as a graph or as an explicit function $y = f(x)$. Likewise, it can be used to compute the value of a definite integral which involves the function $y = f(x)$. To do so, the area must be subdivided into an *even number* of strips or intervals having a width h. The curve between three consecutive ordinates is approximated by a parabola, and the entire area or definite integral is then determined from the formula

$$\int_{x_0}^{x_n} f(x)\,dx \simeq \frac{h}{3}[y_0 + 4(y_1 + y_3 + \cdots + y_{n-1}) + 2(y_2 + y_4 + \cdots + y_{n-2}) + y_n] \quad \text{(B–1)}$$

The computer program for this equation is given in Fig. B–2. For its use, we must first specify the function (on line 6 of the program). The upper and lower limits of the integral and the number of intervals are called for when the program is executed. The value of the integral is then given as the output.

```
1 PRINT"Simpson's rule":PRINT
2 PRINT" To execute this program :":PRINT
3 PRINT"    1- Modify right-hand side of the equation given below,
4 PRINT"         then press RETURN key"
5 PRINT"    2- Type  RUN 6":PRINT:EDIT 6
6 DEF FNF(X)=LOG(X)
7 PRINT:INPUT" Enter Lower Limit = ",A
8 INPUT" Enter Upper Limit = ",B
9 INPUT" Enter Number (even) of Intervals = ",N%
10 H=(B-A)/N%:AR=FNF(A):X=A+H
11 FOR J%=2 TO N%
12 K=2*(2-J%+2*INT(J%/2))
13 AR=AR+K*FNF(X)
14 X=X+H:NEXT J%
15 AR=H*(AR+FNF(B))/3
16 PRINT" Integral = ",AR
17 END
```

Fig. B–2

Example B–2

Evaluate the definite integral

$$\int_2^5 \ln x \, dx$$

SOLUTION

The interval $x_0 = 2$ to $x_6 = 5$ will be divided into six equal parts ($n = 6$), each having a width $h = (5 - 2)/6 = 0.5$. We then compute $y = f(x) = \ln x$ at each point of subdivision.

n	x_n	y_n
0	2	0.693
1	2.5	0.916
2	3	1.099
3	3.5	1.253
4	4	1.386
5	4.5	1.504
6	5	1.609

Thus, Eq. B–1 becomes

$$\int_2^5 \ln x \, dx \simeq \frac{0.5}{3}[0.693 + 4(0.916 + 1.253 + 1.504) + 2(1.099 + 1.386) + 1.609]$$

$$\simeq 3.66 \qquad \textit{Ans.}$$

This answer is equivalent to the exact answer to three significant figures. Obviously, accuracy to a greater number of significant figures can be improved by selecting a smaller interval h (or larger n).

Using the computer program, we first specify the function $\ln x$, line 6 in Fig. B–2. During execution, the program input requires the upper and lower limits 2 and 5, and the number of intervals $n = 6$. The output appears as

$$\text{Integral} = 3.66082 \qquad \textit{Ans.}$$

B.3 The Secant Method

The secant method is used to find the real roots of an algebraic or transcendental equation $f(x) = 0$. The method derives its name from the fact that the formula used is established from the slope of the secant line to the graph $y = f(x)$. This slope is $[f(x_n) - f(x_{n-1})]/(x_n - x_{n-1})$, and the secant formula is

$$x_{n+1} = x_n - f(x_n)\left[\frac{x_n - x_{n-1}}{f(x_n) - f(x_{n-1})}\right] \qquad \text{(B–2)}$$

For application it is necessary to provide two initial guesses, x_0 and x_1, and thereby evaluate x_2 from Eq. B–2 ($n = 1$). One then proceeds to reapply Eq. B–2 with x_1 and the calculated value of x_2 and obtain x_3 ($n = 2$), etc., until the value $x_{n+1} \simeq x_n$. One can see this will occur if x_n is approaching the root of the function $f(x) = 0$, since the correction term on the right of Eq. B–2 will tend toward zero. In particular, the larger the slope, the smaller the correction to x_n, and the faster the root will be found. On the other hand, if the slope is very small in the neighborhood of the root, the method leads to large corrections for x_n, and convergence to the root is slow and may even lead to a failure to find it. In such cases other numerical techniques must be used for solution.

A computer program based on Eq. B–2 is listed in Fig. B–3. We must first specify the function on line 7 of the program. When the program is executed, two initial guesses, x_0 and x_1, must be entered in order to approximate the solution. The output specifies the value of the root. If it cannot be determined, this is so stated.

```
1 PRINT"Secant method":PRINT
2 PRINT" To execute this program :":PRINT
3 PRINT"     1) Modify right hand side of the equation given below,"
4 PRINT"        then press RETURN key."
5 PRINT"     2) Type  RUN 7"
6 PRINT:EDIT 7
7 DEF FNF(X)=.5*SIN(X)-2*COS(X)+1.3
8 INPUT"Enter point #1 =",X
9 INPUT"Enter point #2 =",X1
10 IF X=X1 THEN 14
11 EP=.00001:TL=2E-20
12 FP=(FNF(X1)-FNF(X))/(X1-X)
13 IF ABS(FP)>TL THEN 15
14 PRINT"Root can not be found.":END
15 DX=FNF(X1)/FP
16 IF ABS(DX)>EP THEN 19
17 PRINT "Root = ";X1;"     Function evaluated at this root = ";FNF(X1)
18 END
19 X=X1:X1=X1-DX
20 GOTO 12
```

Fig. B–3

Example B–3

Determine the root of the equation

$$f(x) = 0.5 \sin x - 2 \cos x + 1.30 = 0$$

SOLUTION

Guesses of the initial roots will be $x_0 = 45°$ and $x_1 = 30°$. Applying Eq. B–2,

$$x_2 = 30° - (-0.1821)\frac{(30° - 45°)}{(-0.1821 - 0.2393)} = 36.48°$$

Using this value in Eq. B–2, along with $x_1 = 30°$, we have

$$x_3 = 36.48° - (-0.0108)\frac{36.48° - 30°}{(-0.0108 + 0.1821)} = 36.89°$$

Repeating the process with this value and $x_2 = 36.48°$ yields

$$x_4 = 36.89° - (0.0005)\left[\frac{36.89° - 36.48°}{(0.0005 + 0.0108)}\right] = 36.87°$$

Thus $x = 36.9°$ is appropriate to three significant figures.

If the problem is solved using the computer program, first we specify the function, line 7 in Fig. B–3. During execution, the first and second guesses must be entered in radians. Choosing these to be 0.8 rad and 0.5 rad, the result appears as

Root = 0.6435022.

Function evaluated at this root = 1.66893E–06.

This result converted from radians to degrees is therefore

$$x = 36.9° \qquad \textit{Ans.}$$

Answers

Chapter 1

1–1. (*a*) 78.5 N, (*b*) 0.392 mN, (*c*) 7.46 MN

1–2. 2.42 Mg/m^3

1–5. (*a*) 45.3 MN, (*b*) 56.8 km, (*c*) 5.63 μg

1–6. (*a*) 0.185 Mg^2, (*b*) 4 μg^2, (*c*) 0.0122 km^3

1–7. (*a*) kN · m, (*b*) Gg/m, (*c*) $\mu N/s^2$, (*d*) GN/s

1–9. 0.0209 lb/ft^2, 101 kPa

1–10. (*a*) 27.1 N · m, (*b*) 70.7 kN/m^3, (*c*) 1.27 mm/s

1–11. $16.4(10^3)$ mm^3

1–13. 238 lb

1–14. 584 kg

1–15. 4.96 μN

1–17. $44.9(10)^{-3}$ N^2, $2.79(10^3)$ s^2, 23.4 s

1–18. (*a*) 4.81 slug, (*b*) 70.2 kg, (*c*) 689 N, (*d*) 25.5 lb, (*e*) 70.2 kg

1–19. (*a*) 15.9 mm/s, (*b*) 3.69 Mm · s/kg, (*c*) 1.14 km · kg

Chapter 2

2–1. 867 N, 108°

2–2. 308 N, 91.9°

2–3. 393 lb, 353°

2–5. 45.8 lb, 162°

2–6. 605 N, 85.4°

2–7. $F_{1u} = 205$ N, $F_{1v} = 160$ N

2–9. 115 lb

2–10. $F_u = 91.9$ lb, $F_v = 80.8$ lb

2–11. $F_{\parallel} = 46.5$ lb, $F_{\perp} = 99.7$ lb

2–13. $F_{AC} = 43.9$ lb, $F_{AB} = 53.8$ lb

2–14. 79.5°, 90.8 lb

2–15. 10.8 kN, 3.16°

2–17. $F_a = 30.6$ lb, $F_b = 26.9$ lb

2–18. $F = 19.6$ lb, $F_b = 26.4$ lb

2–19. 60°

2–21. $T = 877$ lb, $F_R = 1.34$ kip

2–22. 744 lb, 23.8°

2–23. (*a*) $F_n = -14.1$ lb, $F_t = 14.1$ lb, (*b*) $F_x = 19.3$ lb, $F_y = 5.18$ lb

2–25. $F_2 = 8.35$ lb, $F_1 = 14.0$ lb

2–26. 70.5°

2–27. 19.2 N, 2.37° ↘

2–29. 53.5°, 621 lb

2–30. 38.3°

2–31. $F_A = 3.66$ kN, $F_B = 7.07$ kN

2–33. $F_x = 514$ lb, $F_y = -613$ lb

2–34. 747 N, 85.5°

2–35. 97.8 N, 46.5°

2–37. 546 N, 253°

2–38. $\mathbf{F}_1 = \{-15.0\mathbf{i} - 26.0\mathbf{j}\}$ kN, $\mathbf{F}_2 = \{-10.0\mathbf{i} + 24.0\mathbf{j}\}$ kN

2–39. 25.1 kN, 185°

2–41. 217 N, 87.0°

2–42. 867 N, 108°

2–43. 308 N, 91.9°

2–45. 10.8 kN, 3.16°

2–46. 19.2 N, 2.37° ⦨

2–47. $F_{1x} = -200$ lb, $F_{1y} = 0$, $F_{2x} = 320$ lb, $F_{2y} = -240$ lb, $F_{3x} = 180$ lb, $F_{3y} = 240$ lb, $F_{4x} = -300$ lb, $F_{4y} = 0$

2–49. 68.3 kN, 1.44° ⦩

2–50. $\mathbf{F}_1 = \{90\mathbf{i} - 120\mathbf{j}\}$ lb, $\mathbf{F}_2 = \{-275\mathbf{j}\}$ lb, $\mathbf{F}_3 = \{-37.5\mathbf{i} - 65.0\mathbf{j}\}$ lb, 463 lb

2–51. 37.0°, 889 N

2–53. $0 \leq P \leq 1.62$ kN

2–54. 54.3°, 686 N

2–55. 1.23 kN, 6.08°

2–57. 389 N, 42.7°

2–58. $\mathbf{F}_1 = \{F_1 \cos\theta\mathbf{i} + F_1 \sin\theta\mathbf{j}\}$ N, $\mathbf{F}_2 = \{350\mathbf{i}\}$ N, $\mathbf{F}_3 = \{-100\mathbf{j}\}$ N, 67.0°, $F_1 = 434$ N

2–59. 117°, 1.12 F_1

2–61. 1.03 kN, 87.9°

2–62. 5.96 kN

2–63. $\mathbf{F} = \{217\mathbf{i} + 85.5\mathbf{j} - 91.2\mathbf{k}\}$ lb

2–65. 50 N, $\alpha = 74.1°$, $\beta = 41.3°$, $\gamma = 53.1°$

2–66. 90°, $\{-30\mathbf{i} - 52.0\mathbf{k}\}$ N

2–67. $\mathbf{F}_1 = \{4.80\mathbf{i} + 6.40\mathbf{j}\}$ kN, $\mathbf{F}_2 = \{3.00\mathbf{i} - 3.00\mathbf{j} + 4.24\mathbf{k}\}$ kN, $\mathbf{F}_R = \{7.80\mathbf{i} + 3.40\mathbf{j} + 4.24\mathbf{k}\}$ kN, $F_R = 9.51$ kN, $\alpha = 34.9°$, $\beta = 69.0°$, $\gamma = 63.5°$

2–69. $\mathbf{F}_1 = \{53.1\mathbf{i} - 44.5\mathbf{j} + 40\mathbf{k}\}$ lb, $\alpha_1 = 48.4°$, $\beta_1 = 124°$, $\gamma_1 = 60°$, $\mathbf{F}_2 = \{-130\mathbf{k}\}$ lb, $\alpha_2 = 90°$, $\beta_2 = 90°$, $\gamma_2 = 180°$

2–70. $\mathbf{F}_1 = \{175\mathbf{i} + 175\mathbf{j} - 247\mathbf{k}\}$ N, $\mathbf{F}_2 = \{173\mathbf{i} - 100\mathbf{j} + 150\mathbf{k}\}$ N

2–71. 369 N, $\alpha = 19.5°$, $\beta = 78.3°$, $\gamma = 105°$

2–73. $\mathbf{F}_1 = \{176\mathbf{j} - 605\mathbf{k}\}$ lb, $\mathbf{F}_2 = \{125\mathbf{i} - 177\mathbf{j} + 125\mathbf{k}\}$ lb, $\mathbf{F}_R = \{125\mathbf{i} - 0.377\mathbf{j} - 480\mathbf{k}\}$ lb, $F_R = 496$ lb, $\alpha = 75.4°$, $\beta = 90.0°$, $\gamma = 165°$

2–74. $\alpha_1 = 45.6°$, $\beta_1 = 53.1°$, $\gamma_1 = 66.4°$

2–75. $\alpha_1 = 90°$, $\beta_1 = 53.1°$, $\gamma_1 = 66.4°$

2–77. $\alpha = 124°$, $\beta = 71.3°$, $\gamma = 140°$

2–78. $F_x = 1.28$ kN, $F_y = 2.60$ kN, $F_z = 0.776$ kN

2–79. $F = 2.02$ kN, $F_y = 0.523$ kN

2–81. $F_x = 40$ N, $F_y = 40$ N, $F_z = 56.6$ N

2–82. $F = 32.7$ N, $F_y = 16.3$ N

2–83. 32.4 lb, $\alpha_2 = 122°$, $\beta_2 = 74.5°$, $\gamma_2 = 144°$

2–85. 7 ft, $\alpha = 31.0°$, $\beta = 107°$, $\gamma = 115°$

2–86. $\{-2.35\mathbf{i} + 3.93\mathbf{j} + 3.71\mathbf{k}\}$ ft, 5.89 ft, $\alpha = 113°$, $\beta = 48.2°$, $\gamma = 51.0°$

2–87. 2.11 m

2–89. 4.42 m

2–90. 467 mm

2–91. 732 mm

2–93. 12.3 km

2–94. $r_{AD} = 1.50$ m, $r_{BD} = 1.50$ m, $r_{CD} = 1.73$ m

2–95. $\{-37.7\mathbf{i} - 15.6\mathbf{j} + 44.0\mathbf{k}\}$ lb, $\alpha = 129°$, $\beta = 105°$, $\gamma = 42.8°$

2–97. $\mathbf{F}_1 = \{38.0\mathbf{i} + 104\mathbf{j} - 101\mathbf{k}\}$ N, $\mathbf{F}_2 = \{119\mathbf{i} - 19.9\mathbf{j} - 159\mathbf{k}\}$ N

2–98. 316 N, $\alpha = 60.1°$, $\beta = 74.6°$, $\gamma = 146°$

2–99. $\mathbf{F}_1 = \{-3.79\mathbf{i} + 11.4\mathbf{k}\}$ lb, $\mathbf{F}_2 = \{-6.65\mathbf{i} - 11.8\mathbf{j} + 11.8\mathbf{k}\}$ lb

2–101. $x = 7.65$ ft, $y = 4.24$ ft, $z = 3.76$ ft

2–102. $x = 8.67$ ft, $y = 1.89$ ft

2–103. $\{-34.3\mathbf{i} + 22.9\mathbf{j} - 68.6\mathbf{k}\}$ lb

2–105. $\{98.1\mathbf{i} + 269\mathbf{j} - 201\mathbf{k}\}$ lb

2–106. $\{59.4\mathbf{i} - 88.2\mathbf{j} - 83.2\mathbf{k}\}$ lb, $\alpha = 63.9°$, $\beta = 131°$, $\gamma = 128°$

2–107. $\mathbf{F}_1 = \{-26.2\mathbf{i} - 41.9\mathbf{j} + 62.9\mathbf{k}\}$ lb, $\mathbf{F}_2 = \{13.4\mathbf{i} - 26.7\mathbf{j} - 40.1\mathbf{k}\}$ lb, $F_R = 73.5$ lb, $\alpha = 100°$, $\beta = 159°$, $\gamma = 71.9°$

2–109. 757 N, $\alpha = 149°$, $\beta = 90.0°$, $\gamma = 59.0°$

2–110. 492 mm, $\{-13.2\mathbf{i} - 18.3\mathbf{j} + 19.8\mathbf{k}\}$ N

2–111. $\mathbf{F}_{EA} = \{12\mathbf{i} - 8\mathbf{j} - 24\mathbf{k}\}$ kN, $\mathbf{F}_{EB} = \{12\mathbf{i} + 8\mathbf{j} - 24\mathbf{k}\}$ kN, $\mathbf{F}_{EC} = \{-12\mathbf{i} + 8\mathbf{j} - 24\mathbf{k}\}$ kN, $\mathbf{F}_{ED} = \{-12\mathbf{i} - 8\mathbf{j} - 24\mathbf{k}\}$ kN, $\mathbf{F}_R = \{-96\mathbf{k}\}$ kN

2–113. $\{476\mathbf{i} + 329\mathbf{j} - 159\mathbf{k}\}$ N

2–114. $x = -6.41$ m, $y = 13.4$ m

2–115. 1.50 kN, $\alpha = 77.6°$, $\beta = 90.6°$, $\gamma = 168°$

2–117. 121°

2–118. $\mathbf{r}_1 = |2.57 \text{ m}|$, $\mathbf{r}_2 = |3.60 \text{ m}|$

2–119. 109°

2–121. $F_B = 566$ N, $F_A = 293$ N, $F_{oa} = 693$ N, $F_{ob} = 773$ N

2–122. 82.0°

2–123. 2.67 m

2–125. 115°

2–126. $F_1 = 20$ N, $F_2 = 31.6$ N

2–127. $F_1 = 333$ N, $F_2 = 373$ N

2–129. 50.6 N

2–130. 97.3°
2–131. $\mathbf{F}_{\parallel} = 99.1$ N, $F_{\perp} = 592$ N
2–133. $\theta = 74.4°$, $\phi = 55.4°$
2–134. 22.7 N, 14.8 N
2–135. 70.5°
2–137. 10.5 lb
2–138. 143°
2–139. $T = 54.7$ lb, $P = 42.6$ lb
2–141. $F_u = 320$ N, $F_v = 332$ N
2–142. 428 lb, $\alpha = 88.3°$, $\beta = 20.6°$, $\gamma = 69.5°$
2–143. 250 lb, $\alpha = 87.0°$, $\beta = 143°$, $\gamma = 53.1°$
2–145. $\mathbf{F}_1 = \{43.3\mathbf{i} + 25\mathbf{j}\}$ lb, $\mathbf{F}_2 = \{-14.8\mathbf{i} - 31.7\mathbf{j}\}$ lb
2–146. 29.3 lb, 347°
2–147. $\theta = 74.0°$, $\phi = 33.9°$
2–149. $F_x = -13.2$ kN, $F_y = 18.8$ kN

Chapter 3

3–1. $F_1 = 439$ N, $F_2 = 233$ N
3–2. 16.4°, 137 lb
3–3. 31.8°, 4.94 kN
3–5. $F_1 = 1.83$ kN, $F_2 = 9.60$ kN
3–6. 4.69°, 4.31 kN
3–7. 1.32 kip
3–9. 1.13 mN
3–10. 158 N
3–11. 1.56 m
3–13. $m = 12.8$ kg
3–14. $\dfrac{1}{k_T} = \dfrac{1}{k_1} + \dfrac{1}{k_2}$
3–15. 78.7°, 127 lb
3–17. 43.1°, 20.5 kg
3–18. 11.5°
3–19. $F_{BC} = 70.7$ lb, $F_{AB} = 50$ lb, $F_{AD} = 70.7$ lb, 23 ft
3–21. 60°, 34.6 lb
3–22. 60°, 46.2 lb
3–23. $F_A = 34.6$ lb, $F_B = 57.3$ lb
3–25. 6 lb
3–26. 35.0°
3–27. 2.66 ft
3–29. $T = \dfrac{50}{\cos\theta}$
3–30. 40.8 lb
3–31. 2.46 ft
3–33. 43.0°
3–34. 88.8 lb
3–35. $T_{AB} = 340$ N, $T_{AE} = 170$ N, $T_{BD} = 490$ N, $T_{BC} = 562$ N
3–37. 40.2°
3–38. 4.98 ft
3–39. 0 and 6.59 m
3–41. $F_1 = 5.10$ kN, $F_2 = 11.8$ kN, $F_3 = 3.92$ kN
3–42. $F_1 = 800$ N, $F_2 = 147$ N, $F_3 = 564$ N
3–43. $F_1 = 5.60$ kN, $F_2 = 8.55$ kN, $F_3 = 9.44$ kN
3–45. 55.8 N
3–46. $F_{AB} = 441$ N, $F_{AC} = 515$ N, $F_{AD} = 221$ N
3–47. $F_{AB} = 348$ N, $F_{AC} = 413$ N, $F_{AD} = 174$ N
3–49. 771 lb
3–50. $F_{AD} = 1.20$ kN, $F_{AC} = 0.40$ kN, $F_{AB} = 0.80$ kN
3–51. $F_{AB} = 0.980$ kN, $F_{AC} = 0.463$ kN, $F_{AD} = 1.55$ kN
3–53. 138 N
3–54. $F_{AC} = 92.9$ lb, $F_{AD} = 364$ lb, $F_{AO} = 757$ lb
3–55. $F_{AC} = 574$ lb, $F_{AB} = 500$ lb, $F_{CD} = 1.22$ kip, $F_{CE} = F_{CF} = 416$ lb
3–57. $F_{AD} = 1.70$ kip, $F_{AC} = 0.744$ kip, $F_{AB} = 1.37$ kip
3–58. $F_{AD} = 1.42$ kip, $F_{AC} = 0.914$ kip, $F_{AB} = 1.47$ kip
3–59. $F_{AB} = 469$ lb, $F_{AC} = F_{AD} = 331$ lb
3–61. 267 lb
3–62. 1.64 ft
3–63. 120 lb
3–65. $T_B = 109$ lb, $T_C = 47.4$ lb, $T_D = 87.9$ lb
3–66. yes, yes
3–67. $l'_{AB} = 0.452$ m, $l'_{AC} = 0.658$ m
3–69. 4.11 m
3–70. $T = 25.3$ N, $F = 22.3$ N
3–71. 20°, 305 lb
3–73. 240 lb

Chapter 4

4–5. 8.00 kN · m ↻
4–6. 18.5 kN · m ↺
4–7. 1.68 kN · m ↺
4–9. 3.57 kN · m ↺
4–10. 3.15 kN · m ↺
4–11. 2.42 kip · ft ↻

4–13. (*a*) 73.9 N · m ↻, (*b*) 82.2 N ←
4–14. 37.9°, 79.8 N · m, 128°, 0
4–15. $M_B = 90.6$ lb · ft ↺, $M_C = 141$ lb · ft ↺
4–17. $m = \left(\dfrac{l}{d + l}\right)M$
4–18. 2.53 kN · m
4–19. 1.59 kN
4–21. 319 lb
4–22. (*a*) 330 lb · ft, 76.0°, (*b*) 0, 166°
4–23. 80 kN · m, 24.0 m
4–25. (*a*) 13.0 N · m, (*b*) 35.2 N
4–26. (*a*) 2594 lb · ft ↻, (*b*) 869 lb · ft ↻
4–27. 306 lb · ft, 652 lb
4–29. 23.8 lb
4–30. $M_A = 60.8$ lb · in, $M_B = 79.9$ lb · in ↻
4–31. $\{-84\mathbf{i} - 8\mathbf{j} - 39\mathbf{k}\}$ kN · m
4–33. $\{260\mathbf{i} + 180\mathbf{j} + 510\mathbf{k}\}$ N · m
4–34. $\{440\mathbf{i} + 220\mathbf{j} + 990\mathbf{k}\}$ N · m
4–35. $\{-229\mathbf{i} + 132\mathbf{j}\}$ lb · ft
4–37. $\mathbf{M}_A = \{-1.90\mathbf{i} + 6.00\mathbf{j}\}$ kN · m
4–38. $\{61.2\mathbf{i} + 81.6\mathbf{j}\}$ N · m
4–39. 176 N
4–41. $\{-37.6\mathbf{i} + 90.7\mathbf{j} - 155\mathbf{k}\}$ N · m
4–42. $\{-840\mathbf{i} + 360\mathbf{j} - 660\mathbf{k}\}$ N · m
4–43. $\{-720\mathbf{i} + 120\mathbf{j} - 660\mathbf{k}\}$ N · m
4–45. 18.6 lb
4–46. $y = 2$ m, $z = 1$ m
4–47. $y = 1$ m, $z = 3$ m, $d = 1.15$ m
4–49. $\{218\mathbf{j} + 163\mathbf{k}\}$ N · m
4–50. $\{-43.0\mathbf{i} - 39.5\mathbf{j} + 14.3\mathbf{k}\}$ kN · m
4–51. $\{26.1\mathbf{i} - 15.1\mathbf{j}\}$ lb · ft
4–53. 0.277 N · m
4–54. $\{-78.4\mathbf{j}\}$ lb · ft
4–55. 6.71 lb
4–57. 165 N · m
4–58. 226 N · m
4–59. 38.4 lb · in.
4–61. 8.50 lb
4–62. 14.8 N · m
4–63. 20.2 N
4–65. 5.66 N
4–66. 18.3 kN · m ↺
4–67. 17.6 kN · m ↺
4–69. 720 lb · ft ↺
4–70. 108 lb
4–71. 39.7 N · m
4–73. 167 lb, resultant couple can act anywhere
4–74. 348 lb · ft, resultant couple can act anywhere
4–75. 2.03 ft
4–77. (*a*) $\{126\mathbf{k}\}$ lb · ft, (*b*) 126 lb · ft
4–78. 0.909 kip
4–79. 139 lb
4–81. $\{557\mathbf{i} - 7.14\mathbf{j} + 857\mathbf{k}\}$ N · m
4–82. $\{37.5\mathbf{i} - 25\mathbf{j}\}$ N · m, 45.1 N · m
4–83. 832 N
4–85. (*a*) $\{-5\mathbf{i} + 8.75\mathbf{j}\}$ N · m, (*b*) $\{-5\mathbf{i} + 8.75\mathbf{j}\}$ N · m
4–86. 992 N
4–87. 59.9 N · m, $\alpha = 99.0°$, $\beta = 106°$, $\gamma = 18.3°$
4–89. $\{-50\mathbf{i} + 60\mathbf{j}\}$ lb · ft, 78.1 lb · ft
4–90. $\{7.01\mathbf{i} + 42.1\mathbf{j}\}$ N · m
4–91. 35.1 N
4–93. 375 N, 737 N · m ↺
4–94. 80 lb, 399 lb · ft ↻
4–95. 80 lb, 696 lb · ft ↻
4–97. 274 lb, 5.24° ⦩, 5.48 kip · ft ↺
4–98. 2.10 kN, 81.6° ∠, 10.6 kN · m ↻
4–99. 2.10 kN, 81.6° ∠, 16.8 kN · m ↻
4–101. 375 lb ↑, 2.47 ft (to the left)
4–102. $\{-1\mathbf{i} - 2\mathbf{j} - 5\mathbf{k}\}$ kN, $\{0.650\mathbf{i} + 19.75\mathbf{j} - 9.05\mathbf{k}\}$ kN · m
4–103. 798 lb, 67.9° ⦣, 7.43 ft
4–105. 1302 N, 84.5° ⦣, 7.36 m
4–106. 1302 N, 84.5°, 1.36 m (to the right)
4–107. 922 lb, 77.5° ⦣, 3.56 ft
4–109. 991 N, 1.78 m
4–110. 991 N, 2.64 m
4–111. $\mathbf{F}_R = \{-700\mathbf{j} - 2200\mathbf{k}\}$ lb,
$\mathbf{M}_{RA} = \{19.6\mathbf{i} - 21.6\mathbf{j} + 8.40\mathbf{k}\}$ kip · ft
4–113. $\{-70\mathbf{i} + 140\mathbf{j} - 408\mathbf{k}\}$ N, $\{-26\mathbf{i} + 357\mathbf{j} + 127\mathbf{k}\}$ N · m
4–114. $\{-40\mathbf{j} - 40\mathbf{k}\}$ N, $\{-12\mathbf{j} + 12\mathbf{k}\}$ N · m
4–115. $\{-28.3\mathbf{j} - 68.3\mathbf{k}\}$ N, $\{-20.5\mathbf{j} + 8.49\mathbf{k}\}$ N · m
4–117. $F_B = 163$ lb, $F_C = 223$ lb
4–118. $\{270\mathbf{k}\}$ N, $\{-2.22\mathbf{i}\}$ N · m
4–119. $\{270\mathbf{k}\}$ N, $y = -8.22$ mm, $x = 0$
4–121. 140 kN, $x = 6.43$ m, $y = 7.29$ m

4–122. $\{-180\mathbf{k}\}$ N, 1.06 m
4–123. $\{-210\mathbf{k}\}$ N, $\{-15\mathbf{i} + 225\mathbf{j}\}$ N · m
4–125. 0.20 m, $\{-40\mathbf{i}\}$ N, $\{-30\mathbf{i}\}$ N · m
4–126. 53.3 N, $\{-40\mathbf{i}\}$ N, $\{-30\mathbf{i}\}$ N · m
4–127. 990 N, 3.07 kN · m, $x = 1.16$ m, $y = 2.06$ m
4–129. 13.2 lb↓, 3.09 ft
4–130. $F_R = 0$, $M_{R_O} = 1.35$ kip · ft
4–131. 18.0 kip↓, 11.7 ft
4–133. 90 kN↓, 338 kN · m ↻
4–134. 0.525 kN, 0.171 m
4–135. 10.6 kip↓, 0.479 ft
4–137. 3.60 kN↓, 16.2 kN · m ↻
4–138. 1.35 kN, 42.0° ↙, 0.1 m
4–139. 1.35 kN, 42.0° ↙, 0.556 m
4–141. 3.90 kip↑, 11.3 ft
4–142. 1.5 m, 175 N/m
4–143. 22.4 kip, 7.50 ft
4–145. 107 kN, 2.40 m
4–146. 14.9 kN, 2.27 m
4–147. 43.6 lb, 3.27 ft
4–149. 73.3 lb, 2.22 ft
4–150. $F_R = \dfrac{w_0}{a}(e^{aL} - 1)$, $x = \dfrac{(e^{aL}aL - e^{aL} + 1)}{a(e^{aL} - 1)}$
4–151. $\{-5.39\mathbf{i} + 13.1\mathbf{j} + 11.4\mathbf{k}\}$ N · m
4–153. 13.8 kip · ft ↻
4–154. 10.1 kip · ft ↺
4–155. $\{15\mathbf{i} + 35.4\mathbf{j} + 25\mathbf{k}\}$ N · m, 45.8 N · m, $\alpha = 70.9°$, $\beta = 39.5°$, $\gamma = 56.9°$
4–157. −809 lb · ft
4–158. 19.8 lb
4–159. $\mathbf{M}_x = \{15\mathbf{i}\}$ lb · ft, $\mathbf{M}_y = \{4\mathbf{j}\}$ lb · ft, $\mathbf{M}_z = \{36\mathbf{k}\}$ lb · ft

Chapter 5

5–11. 10 lb
5–13. $B_y = 642$ N, $A_x = 192$ N, $A_y = 180$ N
5–14. $F_B = 3.76$ kip, $A_x = 3.26$ kip, $A_y = 3.12$ kip
5–15. $N_B = 2.14$ kip, $A_y = 1.49$ kip, $A_x = 1.29$ kip
5–17. $N_C = 5.77$ lb, $N_A = 23.7$ lb, $N_B = 12.2$ lb
5–18. $C_y = 586$ N, $F_A = 413$ N
5–19. $F_D = 110$ kip, $F_L = 170$ kip, 11.4 ft
5–21. $R_A = 667$ N, $R_B = 220$ N, $R_C = 440$ N
5–22. $F_{CD} = 195$ lb, $A_x = 97.4$ lb, $A_y = 31.2$ lb
5–23. $F_H = 59.4$ lb, $T_B = 67.4$ lb
5–25. $F = 22$ kN, $A_x = 30$ kN, $A_y = 16$ kN
5–26. 78.6 lb
5–27. $F_B = 6.38$ N, $A_x = 3.19$ N, $A_y = 2.48$ N
5–29. $B_x = 6.67$ kN, $A_x = 6.67$ kN, $A_y = 8.00$ kN
5–30. 14.4 kN
5–31. $A_x = 1462$ lb, $F_B = 1.66$ kip
5–33. $D_x = 0$, $D_y = 1.65$ kip, $M_D = 1.40$ kip · ft, $(M_D)_{max} = 3.00$ kip · ft
5–34. 6 ft, 267 lb/ft
5–35. 105 N
5–37. $T_{CD} = 416$ lb, $F_A = 461$ lb, 81.4° ⦣
5–38. 15.8 ft
5–41. $F_{CB} = 782$ N, $A_x = 625$ N, $A_y = 681$ N
5–42. $F_2 = 724$ lb, $F_1 = 1.45$ kip, $F_A = 1.75$ kip
5–43. $R_A = 1.06$ kN, $R_B = 1.42$ kN, $R_C = 0.501$ kN
5–45. $T_{BC} = 16.4$ kN, $T = 5$ kN, $F_A = 20.6$ kN
5–46. 41.4°
5–47. $R_A = 30.3$ kip, $R_B = 136$ kip
5–49. $F_A = 101$ lb, $R = 202$ lb
5–50. $F_A = 1.85$ kN, $F_B = 2.02$ kN, $F_C = 391$ N
5–51. $F_B = 0.3P$, $F_C = 0.6P$, $x_C = \dfrac{0.6P}{k}$
5–53. $R_C = 63.9$ lb, $R_B = 11.9$ lb, $R_A = 26.0$ lb
5–54. $\dfrac{a}{\cos^3\theta}$
5–55. $F_C = 8.50$ lb, $F_B = 16.6$ lb, $F_A = 1.90$ lb
5–57. $\tan^{-1}\dfrac{b}{a}$
5–59. 13.0 m
5–61. 568 mm
5–62. 47.5°
5–63. $A_x = 0$, $A_y = -200$ N, $A_z = 150$ N, $(M_A)_x = 100$ N · m, $(M_A)_y = 0$, $(M_A)_z = 500$ N · m
5–65. $T_B = 2.75$ kip, $T_C = 1.38$ kip, $T_A = 1.38$ kip
5–66. $T_A = 235$ lb, $T_B = 293$ lb, $T_C = 372$ lb
5–67. 750 lb, $x = 5.20$ ft, $y = 5.27$ ft
5–69. $B_z = 373$ N, $A_x = 0$, $A_y = 0$, $A_z = 333$ N, $T_{CD} = 43.5$ N
5–70. $O_x = 0$, $O_y = -84.9$ lb, $O_z = 80$ lb, $M_{O_x} = 948$ lb · ft, $M_{O_y} = 0$, $M_{O_z} = 0$
5–71. $P = 75$ lb, $A_y = 0$, $B_z = 75$ lb, $A_z = 75$ lb, $B_x = 112$ lb, $A_x = 37.5$ lb
5–73. $T_{BC} = 205$ N, $T_{ED} = 629$ N, $A_x = 32.4$ N, $A_y = 107$ N, $A_z = 1.28$ kN
5–74. $A_x = 0$, $A_y = 1.50$ kip, $A_z = 750$ lb, $T = 919$ lb

5–75. $F = 1.31$ kip, $A_x = 0$, $A_y = 1.31$ kip, $A_z = 653$ lb

5–77. $A_z = 5$ kN, $A_x = 0$, $A_y = 16.7$ kN

5–78. $T_{BC} = T_{BD} = 17$ kN, $A_y = 11.3$ kN, $A_x = 0$, $A_z = -15.7$ kN

5–79. $F_{DE} = 0.721$ kip, $F_{BC} = 2.16$ kip, $A_x = -0.309$ kip, $A_y = 1.55$ kip, $A_z = -1.21$ kip

5–81. $T = 58$ N, $C_z = 87$ N, $C_y = -28.8$ N, $D_y = -79.2$ N, $D_z = 58$ N, $D_x = 0$

5–82. $T = 58$ N, $C_z = 77.6$ N, $C_y = -24.9$ N, $D_x = 0$, $D_y = -68.5$ N, $D_z = 32.1$ N

5–83. $F_{AC} = F_{BC} = 6.13$ kN, $F_{DE} = 19.6$ kN

5–85. $T_{BC} = 131$ lb, $T_{BD} = 510$ lb, $A_x = 0$, $A_y = 0$, $A_z = 589$ lb

5–86. $B_z = 1167$ lb, $C_z = 734$ lb, $A_z = 1600$ lb

5–87. $F_{BC} = 0$, $A_y = 0$, $A_z = 800$ lb, $(M_A)_x = 4.80$ kip · ft, $(M_A)_y = 0$, $(M_A)_z) = 0$

5–89. $F_{BDC} = 62.0$ lb, $F_{CE} = 110$ lb, $A_x = 19.4$ lb, $A_y = 192$ lb, $A_z = -25.8$ lb

5–90. $A_x = 633$ lb, $A_y = -141$ lb, $B_x = -721$ lb, $B_z = 895$ lb, $C_y = 200$ lb, $C_z = -506$ lb

5–91. $F_2 = 674$ lb

5–93. 43.75 N · m

5–94. $A_z = 10.6$ lb, $D_y = -0.230$ lb, $A_y = 0.230$ lb, $D_x = 5.17$ lb, $A_x = 5.44$ lb, $M = 0.459$ lb · ft

5–95. $A_x = 0$, $F_{BD} = 208$ N, $F_{BC} = 792$ N, $A_z = 0$, $M_{Ax} = 0$, $M_{Az} = 700$ N · m

5–97. $R_A = 105$ lb, $B_x = 97.4$ lb, $B_y = 269$ lb

5–98. $kR(\cot\theta - \cos\theta)$

5–99. $A_y = 390$ N, $B_x = 0$, $B_y = 60$ N

5–101. $F_{BC} = 175$ lb, $A_x = 130$ lb, $A_y = -10$ lb, $M_{Ax} = -300$ lb · ft, $M_{Ay} = 0$, $M_{Az} = -720$ lb · ft

5–102. 105 lb

5–103. $w_2 = 137$ lb, $w_1 = 549$ lb

5–105. $A_z = 49.0$ N, $B_y = 24.5$ N, $A_y = -24.5$ N, $T_{BC} = 24.5$ N, $A_x = 24.5$ N

5–106. $A_y = 8$ kN, $B_y = 5$ kN, $B_x = 5.20$ kN

Chapter 6

6–1. $F_{AD} = 849$ lb (C), $F_{AB} = 600$ lb (T), $F_{BD} = 400$ lb (C), $F_{BC} = 600$ lb (T), $F_{DC} = 1.41$ kip (T), $F_{DE} = 1.60$ kip (C)

6–2. $F_{AD} = 1.13$ kip (C), $F_{AB} = 800$ lb (T), $F_{BD} = 0$, $F_{BC} = 800$ lb (T), $F_{DC} = 1.13$ kip (T), $F_{DE} = 1.60$ kip (C)

6–3. $F_{AD} = 9.90$ kN (C), $F_{AB} = 7$ kN (T), $F_{DB} = 4.95$ kN (T), $F_{DC} = 14.8$ kN (C), $F_{CB} = 10.5$ kN (T)

6–5. $F_{AG} = 471$ lb (C), $F_{AB} = 333$ lb (T), $F_{BG} = 0$, $F_{BC} = 333$ lb (T), $F_{DE} = 943$ lb (C), $F_{DC} = 667$ lb (T), $F_{EC} = 667$ lb (T), $F_{EG} = 667$ lb (C), $F_{CG} = 471$ lb (T)

6–6. $F_{AG} = 1179$ lb (C), $F_{AB} = 833$ lb (T), $F_{BC} = 833$ lb (T), $F_{BG} = 500$ lb (T), $F_{DE} = 1650$ lb (C), $F_{DC} = 1167$ lb (T), $F_{EC} = 1167$ lb (T), $F_{EG} = 1167$ lb (C), $F_{CG} = 471$ lb (T)

6–7. $F_{GB} = 27.5$ kN (T), $F_{AF} = 15.0$ kN (C), $F_{AB} = 28.0$ kN (C), $F_{BF} = 25.0$ kN (T), $F_{BC} = 15.0$ kN (T), $F_{FC} = 21.2$ kN (C), $F_{FE} = 0$, $F_{ED} = 0$, $F_{EC} = 15.0$ kN (T), $F_{DC} = 0$

6–9. $F_{BC} = 3$ kN (C), $F_{BA} = 8$ kN (C), $F_{AC} = 1.46$ kN (C), $F_{AF} = 4.17$ kN (T), $F_{CD} = 4.17$ kN (C), $F_{CF} = 3.12$ kN (C), $F_{EF} = 0$, $F_{ED} = 13.1$ kN (C), $F_{DF} = 5.21$ kN (T)

6–10. $F_{AB} = 330$ lb (C), $F_{AF} = 79.4$ lb (T), $F_{BF} = 233$ lb (T), $F_{BC} = 233$ lb (C), $F_{FC} = 47.1$ lb (C), $F_{FE} = 113$ lb (T), $F_{EC} = 300$ lb (T), $F_{ED} = 113$ lb (T), $F_{CD} = 377$ lb (C)

6–11. $F_{AB} = 377$ lb (C), $F_{AF} = 190$ lb (T), $F_{BF} = 267$ lb (T), $F_{BC} = 267$ lb (C), $F_{FC} = 189$ lb (T), $F_{FE} = 56.7$ lb (T), $F_{EC} = 0$, $F_{ED} = 56.7$ lb (T), $F_{CD} = 189$ lb (C)

6–13. 849 lb

6–14. 849 lb

6–15. $F_{AE} = 8.94$ kN (C), $F_{AB} = 8$ kN (T), $F_{BC} = 8$ kN (T), $F_{BE} = 8$ kN (C), $F_{EC} = 8.94$ kN (T), $F_{ED} = 17.9$ kN (C), $F_{DC} = 8$ kN (T)

6–17. $F_{AB} = 7.5$ kN (T), $F_{AE} = 4.5$ kN (C), $F_{ED} = 4.5$ kN (C), $F_{EB} = 8$ kN (T), $F_{BD} = 19.8$ kN (C), $F_{BC} = 18.5$ kN (T)

6–18. $F_{AB} = 196$ N (T), $F_{AE} = 118$ N (C), $F_{ED} = 118$ N (C), $F_{EB} = 216$ N (T), $F_{BD} = 1.04$ kN (C), $F_{BC} = 857$ N (T)

6–19. $F_{CD} = 3.61$ kN (C), $F_{CB} = 3$ kN (T), $F_{BA} = 3$ kN (T), $F_{BD} = 3$ kN (C), $F_{DA} = 2.70$ kN (T), $F_{DE} = 6.31$ kN (C)

6–21. $F_{CB} = 400$ lb (C), $F_{CD} = 693$ lb (C), $F_{BD} = 667$ lb (T), $F_{BA} = 1.13$ kip (C)

6–22. $F_{EF} = 0.667P$ (T), $F_{FD} = 1.67P$ (T), $F_{AB} = 0.471P$ (C), $F_{AE} = 1.67P$ (T), $F_{AC} = 1.49P$ (C), $F_{BF} = 1.41P$ (T), $F_{BD} = 1.49P$ (C), $F_{EC} = 1.41P$ (T), $F_{CD} = 0.471\,P$ (C)

6–23. $F_{BC} = 1.89P$ (C), $F_{CD} = 1.37P$ (T), $F_{AB} = 0.471P$ (C), $F_{BD} = 1.67P$ (T), $F_{DA} = 1.37P$ (T)

6–25. $F_{BA} = P\csc 2\theta$ (C), $F_{BC} = P\cot 2\theta$ (C), $F_{CA} = (\cot\theta\cos\theta - \sin\theta + 2\cos\theta)P$ (T), $F_{CD} = (\cot 2\theta + 1)P$ (C), $F_{DA} = (\cot 2\theta + 1)(\cos 2\theta)(P)$ (C)

6–26. 732 N

6–27. $F_{CB} = 0$, $F_{CD} = 0$, $F_{AB} = 2.40P$ (C), $F_{AF} = 2.00P$ (T), $F_{EF} = 1.86P$ (T), $F_{ED} = 0.373P$ (C), $F_{FB} = 1.86P$ (T), $F_{FD} = 0.333P$ (T), $F_{DB} = 0.373P$ (C)

6–29. $F_{DE} = 16.3$ kN (C), $F_{DC} = 8.40$ kN (T), $F_{EA} = 8.85$ kN (C), $F_{EC} = 6.20$ kN (C), $F_{AB} = 3.11$ kN (T), $F_{AF} = 6.20$ kN (T), $F_{BC} = 2.20$ kN (T), $F_{BF} = 6.20$ kN (C), $F_{CF} = 8.77$ kN (T)

6–30. $F_{DE} = 18.7$ kN (C), $F_{DC} = 9.60$ kN (T), $F_{EA} = 10.1$ kN (C), $F_{EC} = 4.80$ kN (C),

$F_{AB} = 6.79$ kN (T), $F_{AF} = 4.80$ kN (T), $F_{BC} = 4.80$ kN (T), $F_{BF} = 4.80$ kN (C), $F_{CF} = 6.79$ kN (T)

6–31. $F_{HG} = 29$ kN (C), $F_{BC} = 20.5$ kN (T), $F_{HC} = 12.0$ kN (T)

6–33. $F_{CD} = 50$ kN (T), $F_{HD} = 7.07$ kN (C), $F_{GD} = 5$ kN (T)

6–34. $F_{HI} = 35$ kN (C), $F_{BC} = 50$ kN (T), $F_{HB} = 21.2$ kN (C)

6–35. $F_{KJ} = 13.3$ kN (T), $F_{BC} = 14.9$ kN (C), $F_{CK} = 0$

6–37. $F_{KJ} = 11.2$ kip (T), $F_{CD} = 9.38$ kip (C), $F_{CJ} = 3.12$ kip (C), $F_{DJ} = 0$

6–38. $F_{JI} = 7.50$ kip (T), $F_{EI} = 2.50$ kip (C)

6–39. $F_{CD} = 2.5$ kip (T), $F_{FE} = 3.50$ kip (T), $F_{CE} = 2.50$ kip (C)

6–41. $F_{GF} = 671$ lb (C), $F_{GB} = 671$ lb (T)

6–42. $F_{BC} = 700$ lb (T), $F_{FC} = 76.9$ lb (T)

6–43. $F_{BC} = 10.4$ kN (C), $F_{HG} = 9.16$ kN (T), $F_{HC} = 2.24$ kN (T)

6–45. 2.60 kip (T)

6–46. 2.00 kip (C)

6–47. 1.00 kip (T)

6–49. $F_{AB} = F_{BC} = F_{CD} = F_{DE} = F_{HI} = F_{GI} = 0$, $F_{IC} = 5.62$ kN (C), $F_{CG} = 9.00$ kN (T)

6–50. $F_{AB} = F_{BC} = F_{CD} = F_{DE} = F_{HI} = F_{GI} = 0$, $F_{JE} = 9.38$ kN (C), $F_{GF} = 5.625$ kN (T)

6–51. $F_{EF} = P$ (C), $F_{CB} = 1.12P$ (T), $F_{BE} = 0.5P$ (T)

6–53. $F_{EH} = 3.33$ kN (T), $F_{IH} = 5.33$ kN (C), $F_{EF} = 1.33$ kN (T)

6–54. $F_{KD} = 0$, $F_{CD} = 4.00$ kN (T), $F_{KJ} = 5.33$ kN (C)

6–55. $F_{HG} = 12.7$ kN (T), $F_{BC} = 15.1$ kN (C), $F_{HC} = 1.50$ kN (T)

6–57. $F_{BC} = 9.55$ kip (C), $F_{GC} = 6.91$ kip (T), $F_{GF} = 2.92$ kip (T)

6–58. $F_{DE} = 1.87$ kip (C), $F_{JI} = 1.87$ kip (T), $F_{DO} = 1.17$ kip (C)

6–59. $F_{CD} = 2.53$ kip (C), $F_{KJ} = 2.53$ kip (T)

6–61. $F_{DC} = F_{DA} = 2.59$ kN (C), $F_{DB} = 3.85$ kN (C), $F_{BC} = F_{BA} = 0.890$ kN (T), $F_{AC} = 0.616$ kN (T)

6–62. $F_{CA} = F_{CB} = 122$ lb (C), $F_{CD} = 173$ lb (T), $F_{BD} = 86.6$ lb (T), $F_{BA} = 0$, $F_{DA} = 86.6$ lb (T)

6–63. $F_{BC} = 0$, $F_{CD} = 0$, $F_{CF} = 8$ kN (C), $F_{BD} = 0$, $F_{BA} = 6$ kN (C), $F_{AD} = 0$, $F_{DF} = 0$, $F_{DE} = 9$ kN (C), $F_{EF} = 0$, $F_{EA} = 0$, $F_{AF} = 0$

6–65. $F_{BF} = 0$, $F_{BC} = 0$, $F_{BE} = 500$ lb (T), $F_{AB} = 300$ lb (C), $F_{AC} = 972$ lb (T), $F_{AD} = 0$, $F_{AE} = 367$ lb (C), $F_{DE} = 0$, $F_{EF} = 300$ lb (C), $F_{CD} = 500$ lb (C), $F_{CF} = 300$ lb (C), $F_{DF} = 424$ lb (T)

6–66. $F_{BE} = 900$ lb (T), $F_{BC} = 0$, $F_{AB} = 600$ lb (C), $F_{DE} = F_{CD} = F_{AD} = F_{CE} = F_{AC} = F_{BC} = 0$, $F_{AE} = 671$ lb (C)

6–67. $F_{AD} = 686$ N (T), $F_{BD} = 0$, $F_{CD} = 615$ N (C), $F_{BC} = 229$ N (T), $F_{EC} = 457$ N (C), $F_{AC} = 343$ N (T)

6–69. $F_{BC} = F_{BD} = 1.34$ kN (C), $F_{AB} = 2.4$ kN (C), $F_{AG} = F_{AE} = 1.01$ kN (T), $F_{BG} = 1.80$ kN (T), $F_{BE} = 1.80$ kN (T)

6–70. $R_B = 26.7$ lb, $A_x = 0$, $A_y = 34.7$ lb

6–71. 30 lb, 130 lb

6–73. $R_E = 177$ lb, $R_A = 128$ lb

6–74. $R_B = 113$ N, $R_A = 144$ N, $R_D = 79.8$ N, $R_C = 79.8$ N

6–75. $A_x = C_x = 5$ kN, $A_y = C_y = 6.67$ kN, $B_x = 5$ kN, $B_y = 1.33$ kN, $D_x = 0$, $D_y = 8$ kN, 10 kN · m

6–77. $F_B = 907$ N, $P = 156$ N

6–78. 464 N, 25.6 N · m

6–79. $A_x = 5.61$ kN, $A_y = 1.91$ kN, $C_x = 2.39$ kN, $C_y = 1.91$ kN, $B_x = 2.39$ kN, $B_y = 6.91$ kN

6–81. $A_x = 6.43$ kN, $A_y = 2.62$ kN, $B_x = 1.57$ kN, $B_y = 2.62$ kN, $C_x = 1.57$ kN, $C_y = 2.62$ kN

6–82. $A_x = 1500$ N, $A_y = 600$ N

6–83. 743 N

6–85. $P = 21.8$ N, $R_A = 43.6$ N, $R_B = 43.6$ N, $R_C = 131$ N

6–86. $P = 81.8$ N, $R_A = 183$ N, $R_B = 183$ N, $R_C = 441$ N

6–87. $P = 25$ lb, $R_A = 25$ lb, $R_B = 60$ lb

6–89. 100 lb, 14.6°

6–90. $C_x = 75$ lb, $C_y = 100$ lb

6–91. $B_x = 75$ lb, $B_y = 300$ lb, $E_x = 225$ lb, $E_y = 600$ lb, $D_x = 300$ lb, $D_y = 300$ lb

6–93. $A_x = 240$ lb, $A_y = 20$ lb, $N_C = 50.0$ lb

6–94. $A_x = 80$ lb, $A_y = 80$ lb, $B_y = 133$ lb, $B_x = 333$ lb, $C_x = 413$ lb, $C_y = 53.3$ lb

6–95. $C_y = 34.4$ lb, $C_x = 16.7$ lb, $B_x = 66.7$ lb, $B_y = 15.6$ lb

6–97. 75 lb

6–98. 46.9 lb

6–99. 0.853 in.

6–101. $F_B = 223$ N, $F_A = 386$ N

6–102. $m = 366$ kg, $F_A = 2.93$ kN

6–103. 1.11 Mg

6–105. $A_y = 657$ N, $C_y = 229$ N, $C_x = 0$, $B_x = 0$, $B_y = 429$ N

6–106. $E_x = 300$ N, $D_x = 300$ N, $D_y = E_y = 42.9$ N

6–107. 41.7 lb, frictional force stops wheel

6–109. (*a*) $F = 205$ lb, $N_C = 380$ lb, (*b*) $F = 102$ lb, $N_C = 72.5$ lb

6–110. $A_x = 0$, $A_y = 34.0$ N, $C_y = 6.54$ N, $C_x = 0$, 292 mm, $B_x = 0$, $B_y = 1.06$ N

6–111. $N_C = 20$ lb, $B_x = 34$ lb, $B_y = 62$ lb, $A_x = 34$ lb, $A_y = 12$ lb, 336 lb · ft

6–113. 2.21 kip

6–114. $T = W$

6–115. 4.38 in.

6–117. 66.0 lb

6–118. 2.42 lb

6–119. 2.88 kip, 3.99 kip

6–121. $F_{CA} = 12.9$ kip, $F_{AB} = 11.9$ kip, $F_{AD} = 2.39$ kip

6–122. $W_1 = 3$ lb, $W_2 = 21$ lb, $W_3 = 75$ lb

6–123. 15.5°

6–125. 13.9 N

6–126. $\dfrac{kL}{2 \sin \theta \tan \theta}(2 - \csc \theta)$

6–127. $2.41P/L$

6–129. 3.86 kg

6–130. $F_{DE} = 525$ lb, $B_z = 0$, $B_x = -214$ lb, $B_y = 288$ lb

6–131. $F_{DE} = 270$ lb, $B_z = 0$, $B_x = -30$ lb, $B_y = -13.3$ lb

6–133. $P = 283$ N, $B_z = 283$ N, $D_z = 283$ N, $B_y = 283$ N, $D_y = 283$ N, $D_x = -70.7$ N, $B_x = -70.7$ N

6–134. $F_{AB} = 1.56$ kN, $M_{Ex} = 0.5$ kN · m, $M_{Ey} = 0$, $E_y = 0$, $E_x = 0$

6–135. $M_{Cx} = 0$, $C_x = 0$, $F_{BA} = 1.54$ kip, $C_z = -0.18$ kip, $C_y = -1.17$ kip, -4.14 kip · ft, $A_x = 0$, $A_y = 1.44$ kip, $A_z = 0.540$ kip

6–137. $A_x = 1.40$ kN, $A_y = 250$ N, $C_x = 500$ N, $C_y = 1.70$ kN

6–138. $F_{CB} = 3$ kN (T), $F_{CD} = 2.60$ kN (C), $F_{DE} = 2.60$ kN (C), $F_{DB} = 2$ kN (T), $F_{BE} = 2$ kN (C), $F_{BA} = 5$ kN (T)

6–139. $C_y = 5$ kN, $B_y = 15$ kN, $A_x = 0$, $A_y = 5$ kN, 30 kN · m

6–141. $R_B = R_C = 49.5$ N, $F_C = 49.5$ N

6–142. $F_{BE} = 1.53$ kip, $F_{CD} = 350$ lb

6–143 16.1°

6–145. $F_{DC} = 0.577P$ (C), $F_{DE} = 0.289P$ (T), $F_{CE} = 0.577P$ (T), $F_{CB} = 0.577P$ (C), $F_{BE} = F_{CE} = 0.577P$ (T), $F_{AB} = F_{DC} = 0.577P$ (C), $F_{AE} = F_{DE} = 0.289P$ (T)

6–146. $F_{DC} = 2.89W$ (C), $F_{DE} = 1.44W$ (T), $F_{CE} = 1.16W$ (T), $F_{CB} = 2.02W$ (C), $F_{BE} = F_{CE} = 1.16W$ (T), $F_{AB} = F_{DC} = 2.89W$ (C), $F_{AE} = F_{DE} = 1.44W$ (T)

6–147. 160 mm, 240 mm

6–149. $F_{AC} = 2.51$ kip (C), $F_{AB} = 3.08$ kip (C), $F_{AD} = 3.43$ kip (T)

Chapter 7

7–1. $T_A = 680$ N · m, $T_B = 350$ N · m, $T_C = 800$ N · m, $T_D = 0$

7–2. $T_C = 0$, $T_D = 400$ N · m, $T_E = 950$ N · m

7–3. $T_A = 550$ N · m, $T_B = 400$ N · m, $T_C = 600$ N · m, $T_D = 0$

7–5. $N_C = 0$, $V_C = 0.5$ kip, 3.6 kip · ft

7–6. -292 kN · m, $N_C = -20$ kN, $V_C = 70.6$ kN

7–7. $N_D = -800$ N, $V_D = 0$, 1.20 kN · m

7–9. 48 kip · ft, 6 kip

7–10. $N_A = 86.6$ lb, $V_A = 150$ lb, 1800 lb · in.

7–11. 43.8 lb · ft

7–13. $N_A = 0$, $V_A = 450$ lb, $M_A = 1125$ lb · ft, $N_B = 0$, $V_B = 850$ lb, $M_B = 6325$ lb · ft, $V_C = 0$, $N_C = 1200$ lb, $M_C = 8125$ lb · ft

7–14. $N_D = 1.92$ kN, $V_D = 100$ N, $M_D = 900$ N · m

7–15. $N_E = -1.92$ kN, $V_E = 800$ N, $M_E = 2.40$ kN · m

7–17. $P = 0.533$ kN, $N_C = -2$ kN, $V_C = 0.533$ kN, $M_C = 0.400$ kN · m

7–18. $V_D = 0$, $N_D = 0$, $M_D = 0$, $V_E = 533$ lb, $N_E = 1200$ lb, $M_E = 1600$ lb · ft, $V_F = 0$, $N_F = 2400$ lb, $M_F = 3200$ lb · ft

7–19. $N_C = -406$ lb, $V_C = 903$ lb, 1.35 kip · ft

7–21. $N_D = 0$, $V_D = 0$, $M_D = 9$ kip · ft, $N_E = 0$, $V_E = -7$ kip, $M_E = -12$ kip · ft

7–22. 3.75 kip, 30 kip · ft

7–23. $N_D = 0$, $V_D = 0.75$ kip, $M_D = 13.5$ kip · ft, $N_E = 0$, $V_E = -9$ kip, $M_E = -24.0$ kip · ft

7–25. $N_E = -250$ lb, $V_E = 245$ lb, $M_E = -490$ lb · ft, $N_F = 0$, $V_F = -308$ lb, $M_F = -1.23$ kip · ft

7–26. $N_B = 59.8$ lb, $V_B = -496$ lb, $M_B = -480$ lb · ft, $N_C = -495$ lb, $V_C = 70.7$ lb, $M_C = -1.59$ kip · ft

7–27. $V = 0.293rw_0$, $N = -0.707rw_0$, $M = -0.0783r^2w_0$

7–29. 1.69 kip, 10.0 kip · ft

7–30. $V_{Ax} = 0$, $N_{Ay} = 0$, $V_{Az} = 250$ lb, $M_{Ax} = 187.5$ lb · ft, $M_{Ay} = 187.5$ lb · ft, $M_{Az} = 0$, $V_{Bx} = 0$, $V_{By} = 0$, $N_{Bz} = 250$ lb, $M_{Bx} = 437.5$ lb · ft, $M_{By} = 187.5$ lb · ft, $M_{Bz} = 0$

7–31. $C_x = -150$ lb, $C_y = -350$ lb, $C_z = 700$ lb, $M_{Cx} = 1.40$ kip · ft, $M_{Cy} = -1.20$ kip · ft, $M_{Cz} = -750$ lb · ft

7–33. $D_x = -116$ kN, $D_y = 65.6$ kN, $D_z = 0$, $M_{Dx} = -49.2$ kN · m, $M_{Dy} = -87.0$ kN · m, $M_{Dz} = -26.2$ kN · m

7–34. (*a*) $V = (1 - a/L)P$, $M = (1 - a/L)Px$, $V = -(a/L)P$, $M = P(a - a/L)x$

7–35. (*a*) $V = P$, $M = Px$, $V = 0$, $M = Pa$, $V = -P$, $M = P(L - x)$, (*b*) 800 lb, $800x$ lb · ft, 0, 4000 lb · ft, -800 lb, $(9600 - 800x)$ lb · ft

7–37. 2 kN · m

7–38. (*a*) $V = \frac{w}{2}(L - 2x)$, $M = \frac{w}{2}(Lx - x^2)$,
(*b*) $(2500 - 500x)$ lb, $(2500x - 250x^2)$ lb · ft

7–39. 400 lb/ft

7–41. $V = 2.5 - 2x$, $M = 2.5x - x^2$, $V = -7.5$, $M = -7.5x + 25$

7–42. $V = 250(10 - x)$, $M = 25(100x - 5x^2 - 6)$

7–43. $V = 133.75 - 40x$, $M = 133.75x - 20x^2$, $V = 20$, $M = 20x - 370$

7–45. (*a*) $V = \frac{w}{8}(7L - 8x)$, $M = -\frac{w}{8}(4x^2 - 7Lx + 3L^2)$,
(*b*) $V = 5(10.5 - x)$, $M = -2.5(x^2 - 21x + 108)$

7–46. $V = 3 - \frac{x^2}{4}$, $M = 3x - \frac{x^3}{12}$

7–47. $0.366L$

7–49. 22.2 lb/ft

7–50. $x = \frac{L}{2}$, $P = \frac{4M_{max}}{L}$

7–51. $\frac{L}{3}$

7–53. $wr \sin\theta$, $wr(\cos\theta - 1)$, $M = wr^2(1 - \cos\theta)$

7–54. $N = \frac{P}{5}(4\cos\theta + 3\sin\theta)$, $V = \frac{P}{5}(4\sin\theta - 3\cos\theta)$,
$M = \frac{Pr}{5}(4 - 4\cos\theta - 3\sin\theta)$

7–55. See Prob. 7–35

7–57. See Prob. 7–38

7–58. See Prob. 7–42

7–59. See Prob. 7–46

7–61. $x = 0$, $V = 1200$, $M = 0$; $x = 24^-$, $V = 400$, $M = 19200$

7–62. $x = 0$, $V = 3.75$, $M = 0$; $x = 4^+$, $V = -3.25$, $M = 13$

7–63. $x = 0$, $V = 1.525$, $M = 0$; $x = 0.8^-$, $V = 0.025$, $M = 0.395$

7–65. $x = 0$, $V = -400$, $M = 0$; $x = 14.5$, $V = 0$, $M = 3912$

7–66. $x = 0$, $V = -200$, $M = 0$; $x = 2.75$, $V = 0$, $M = -46.9$

7–67. $x = 0$, $V = -14.3$, $M = 0$; $x = 3^+$, $V = -22.3$, $M = -22.9$

7–69. $x = 0$, $V = 0$, $M = 0$; $x = 6^+$, $V = 800$, $M = -1200$

7–70. 2 kip/ft

7–71. $x = 0$, $V = 52.8$, $M = 0$, $x = 9^-$, $V = -172.2$, $M = -200$

7–73. $x = 0$, $V = wL/2$, $M = -5wL^2/24$; $x = L/2$, $V = wL/4$, $M = -wL^2/24$.

7–74. $x = 0$, $V = 0$, $M = 0$; $x = 18$, $V = 0$, $M = 0$

7–75. $x = 0$, $V = 0$, $M = 0$; $x = 6^+$, $V = 9$, $M = -36$

7–77. $x = 0$, $V = 1100$, $M = -10400$; $x = 8^+$, $V = 400$, $M = -1600$

7–78. $x = 0$, $V = 7wL/18$, $M = 0$; $x = L^+$, $V = wL/2$, $M = -72wL^2/648$

7–79. $x = 0$, $V = wL/3$, $M = 0$; $x = 0.528L$, $V = 0$, $M = 0.0940wL^2$

7–81. 2.43 m, 62.9 N

7–82. $y_B = 8.67$ ft, $y_D = 7.04$ ft

7–83. 6.44 ft, 658 lb

7–85. 71.4 lb

7–86. $T_{AB} = 166$ lb, $T_{CD} = 176$ lb, $T_{BC} = 93.4$ lb, 20.2 ft

7–87. 3.98 ft

7–89. $T_A = 61.7$ kip, $T_B = F_H = 36.5$ kip, $T_C = 50.7$ kip

7–90. 51.9 lb/ft

7–91. $y = \frac{x^2}{7813}\left(75 - \frac{x^2}{200}\right)$ ft, 9.28 kip

7–93. $(38.5x^2 + 577x)(10^{-3})$ m, 5.20 kN

7–94. 0.141

7–97. 302 ft

7–98. $L = 94.3$ ft, $h = 25.3$ ft

7–99. 292 lb

7–101. 5.14 ft

7–102. $(T_{max})_B = 2.73$ kip, $(T_{max})_C = 2.99$ kip

7–103. 5.18 ft

7–105. 66.6 N

7–106. 121 N, 5.91 m

7–107. 170 lb, 150 ft

7–109. 35.8 m

7–110. 238 ft, $h = 93.8$ ft

7–111. $x = 0$, $V = 10$, $M = -30$, $x = 5$, $V = 0$, $M = -5$

7–113. $x = 0$, $V = 1022$, $M = 0$; $x = 0.6$, $V = 0$, $M = 407$

7–114. 11.1 lb, 23.5 ft

7–115. $x = 0$, $V = 9.11$, $M = 0$, $x = 9$, $V = -5.89$, $M = 0$

7–117. $x = 0$, $V = 2.5$, $M = 0$, $x = 1.25$, $V = 0$, $M = 1.56$

7–118. $V_D = M_D = 0$, $N_D = F_{CD} = 86.6$ lb, $N_E = 0$, $V_E = 28.9$ lb, $M_E = 86.6$ lb · ft

7–119. $\{16 - 4y\}$ lb, $\{-2y^2 + 16y - 40\}$ lb · ft

Chapter 8

8–1. 224 lb

8–2. $76.4 \text{ lb} \le P \le 144 \text{ lb}$

8–3. $F_C = 27.4$ lb, $N_C = 309$ lb
8–5. 140 N, 500 mm
8–6. $F_C = 30.5$ N, $N_C = 152$ N
8–7. $0.4L$
8–9. (*a*) $P = 30$ N < 39.8 N no, (*b*) $P = 70$ N > 39.8 N yes
8–10. (*a*) $P = 30$ N < 34.26 N no, (*b*) $P = 70$ N > 34.26 N yes
8–11. will *not slip*
8–13. 3.67 kip
8–14. 21.8°
8–15. 12.0 lb
8–17. 0.268
8–18. hoop slips first
8–19. 1.53 kip
8–21. 227 N, 0.662
8–22. 54.9 kg
8–23. (*a*) 318 lb, (*b*) 360 lb
8–25. $\dfrac{h}{\mu_s} + \dfrac{d}{2}$
8–26. 22.5 lb, 0.15
8–27. 30.4 lb, 0.195
8–29. 200 N
8–30. car *A will not* move
8–31. 2.70 in.
8–33. 9.63 ft
8–34. $\theta = 16.7°$, $\phi = 40.5°$
8–35. slipping at *A*
8–37. 3.35 ft
8–38. 0.962 N
8–39. 0.962 N
8–41. 14.9 lb
8–42. 100 lb, 1.50 ft
8–43. 63.4°
8–45. 33.4°
8–46. 0.176
8–47. $P = 355$ N
8–49. 77.3 N · m
8–50. 71.4 N
8–51. 13.3 lb
8–53. 28.1 N
8–54. 375 N
8–55. 589 N
8–58. $\alpha = -\theta$, $\phi = \theta$, $P = W \sin(\alpha + \phi)$
8–59. $(l^2 - a^2 - h^2)^{1/2}(l^2 - a^2)^{1/2}/ha$
8–61. any number
8–62. 1.61 kip, 1.34 ft
8–63. $P' = 4.96$ kip, $P = 8.16$ kip
8–65. 34.5 N
8–66. 5.53 kN, yes
8–67. 9.36 kN
8–69. 304 N
8–70. 32.9 mm
8–71. 25.3°
8–73. 5.69 lb · in.
8–74. 7.19 kN
8–75. self locking
8–77. 1.98 kN
8–78. 0.202 N · m
8–79. 0.0637
8–81. 31.0 N · m
8–82. 72.7 N
8–83. $T_A = 187$ lb, $T_B = 137$ lb
8–85. (*a*) 1.31 kN, (*b*) 372 N
8–86. (*a*) 4.60 kN, (*b*) 16.2 kN
8–87. approx. 2 turns (695°)
8–89. 8.28 ft
8–90. 73.3 lb
8–91. 99.2°
8–93. 12.7 lb
8–94. 15.4 lb, 5.47 lb
8–95. 42.3 N
8–97. 17.1 lb
8–99. 25.6 kg
8–101. 304 lb · in.
8–102. 0.0568
8–103. 36.3 lb · ft
8–105. 21.3 lb
8–106. 118 N
8–107. $\frac{1}{2}\mu PR$
8–109. $0.521P\mu R$
8–110. 0.442 psi, 450 lb
8–111. 2.00 N
8–113. 826 N
8–114. 814 N

8–115. 0.215, 0.211 (approx.), 6 lb
8–117. 13.8 lb
8–118. 29.0 lb
8–119. 68.2°, 0.0455 N · m
8–121. 44.6 N
8–122. 78.8 lb
8–123. 245 N
8–126. 2.25 kN
8–127. 486 lb
8–129. 0.0626
8–130. cam *cannot* support broom
8–131. 38.5 mm
8–133. 1.55 lb
8–134. $F_{BC} = 1.38$ kN (T), $F_{BD} = 1.73$ kN (C), $F_{AB} = 1.04$ kN (T), $F_{AC} = 1.73$ kN (C), $F_{CD} = 1.04$ kN (T), $F_{AD} = 1.38$ kN (T)
8–135. 3.62 N · m

Chapter 9

9–1. $\frac{4}{\pi}$, $B_x = 1$ lb, $A_x = 1$ lb, $A_y = 3.14$ lb
9–2. $0.75l$
9–3. $\frac{r \sin \alpha}{\alpha}$
9–5. 0.531 ft, $O_x = 0$, $O_y = 0.574$ lb, 0.305 lb · ft
9–6. 0.183 ft
9–7. $\bar{x} = 0$, $\bar{y} = 1.60$ m
9–9. $\bar{x} = \frac{3}{8}b$, $\bar{y} = \frac{3}{5}h$
9–10. $\bar{x} = \frac{4}{7}a$, $\bar{y} = \frac{a}{5}$
9–11. $\bar{y} = \frac{a\pi}{8}$, $\bar{x} = \frac{L}{2}$
9–13. $\bar{x} = \frac{n+1}{2(n+2)}a$, $\bar{y} = \frac{n}{2n+1}h$
9–14. $\bar{x} = \frac{3}{8}a$, $\bar{y} = \frac{3}{5}a$
9–15. $\bar{y} = \frac{4b}{3\pi}$, $\bar{x} = \frac{4a}{3\pi}$
9–17. $\bar{x} = \frac{4r}{3\pi}$, $\bar{y} = \frac{4r}{3\pi}$
9–18. $\bar{y} = 0.223r$, $\bar{x} = 0.223r$
9–19. $\bar{x} = 0$, $\bar{y} = \frac{4r}{3\pi}$
9–21. 0.649 ft
9–22. 2.04 ft
9–23. $\bar{x} = 1.26$ m, $\bar{y} = 0.143$ m, $N_B = 47.9$ kN, $A_x = 33.9$ kN, $A_y = 73.9$ kN
9–25. 0.980 ft
9–26. 0.587 m
9–27. 0.404 m
9–29. $1.11a$
9–30. $\bar{y} = \frac{3}{8}b$, $\bar{x} = \bar{z} = 0$
9–31. $\bar{z} = \frac{2}{3}a$, $\bar{x} = \bar{y} = 0$
9–33. $\bar{x} = \frac{a}{2}$, $\bar{y} = \bar{z} = 0$
9–34. $\frac{2}{3}h$
9–37. $\bar{z} = \frac{h}{4}$, $\bar{x} = \bar{y} = \frac{a}{\pi}$
9–38. $0.4a$
9–39. $\frac{h}{4}\left(\frac{R^2 + 2Rr + 3r^2}{R^2 + Rr + r^2}\right)$
9–41. 2 kg, 5 m
9–42. $\bar{x} = 1.44$ m, $\bar{y} = 2.22$ m, $\bar{z} = 2.67$ m
9–43. 67.9 mm
9–45. 3 m
9–46. $\bar{x} = 1.60$ ft, $\bar{y} = 7.04$ ft, $A_x = 0$, $A_y = 149$ lb, $M_A = 502$ lb · ft
9–47. $\bar{x} = 6.00$ in., $\bar{y} = 1.78$ in.
9–49. $\bar{x} = 3.52$ ft, $\bar{y} = 4.09$ ft, 42.8 kip
9–50. 135 mm
9–51. $\frac{h}{3}\left(\frac{2b_1 + b_2}{b_1 + b_2}\right)$
9–53. $\bar{x} = 3$ in., $\bar{y} = 2$ in.
9–54. 210 mm
9–55. 154 mm
9–57. $\bar{x} = 4.62$ in., $\bar{y} = 1.00$ in.
9–58. 37.5 mm
9–59. $\bar{x} = 432$ mm, $\bar{y} = -339$ mm
9–61. 37.8°
9–62. 2.48 ft, 38.9°
9–63. $\frac{\frac{2}{3} r \sin^3 \alpha}{\alpha - \frac{\sin 2\alpha}{2}}$
9–65. 1.625 in.
9–66. $\bar{x} = 2.81$ ft, $\bar{y} = 1.73$ ft, $N_B = 72.1$ lb, $N_A = 86.9$ lb
9–67. $\bar{x} = 1.47$ in., $\bar{y} = 2.68$ in., $\bar{z} = 2.84$ in.

9–69. 2.00 ft
9–70. 8.22 in.
9–71. 323 mm
9–73. 101 mm
9–74. 385 mm
9–75. 122 mm
9–77. 0.667 ft^2, 0.375 ft, 1.57 ft^3
9–78. 1.39 ft^2, 0.541 ft, 4.71 ft^3
9–79. 826 in^2
9–81. 3.33 ft^2, 1.2 ft, 25.1 ft^3
9–82. 3.14 m^3
9–83. 3.49 m^3
9–85. 2.25 m^2
9–86. $4.25(10^6)$ mm^3
9–87. 16.8 lb
9–89. $3.12(10^6)$ lb
9–90. 84.7 kip
9–91. 2.75 gal.
9–93. 36.3 kN
9–94. 0.357 lb
9–95. $143(10)^{-6}$ m^3
9–97. 29.9 mm
9–98. 207 m^3, 188 m^2
9–99. $6.79(10^6)$ mm^3
9–101. $119(10^3)$ mm^2
9–102. 1.41 MN, 4 m
9–103. 157 kN, 235 kN, 4.22 m
9–105. 17.2 kip, 5.22 ft, 18.8 kip
9–106. 391 kN/m
9–107. 45.9 kN, 1.41 m
9–109. $B_x = 2.20$ MN, $B_y = 0.859$ MN, $A_y = 2.51$ MN
9–110. 6.67 m
9–111. 7.73 m
9–113. $F_{R_1} = 2.45$ kN, 1.02 m, $F_{R_2} = 9.75$ kN, at center of plate B
9–114. 331 kN/m
9–115. 2.02 kip
9–117. 3.85 kN, 0.625 m
9–118. 749 lb, −1.77 ft
9–119. 4.00 kip, −6.49 ft
9–121. 1250 lb, $\bar{x} = 2.33$ ft, $\bar{y} = 4.33$ ft
9–122. $\bar{x} = 0$, $\bar{y} = 2.40$ m, $F_R = 42.7$ kN, $B_y = C_y = 12.8$ kN, $A_y = 17.1$ kN
9–123. 24 kN, $\bar{x} = 2$ m, $\bar{y} = 1.33$ m
9–125. 1.25 m^2
9–126. 39.5 kg
9–127. 4.42 in.
9–129. $\bar{x} = 1.22$ ft, $\bar{y} = 0.778$ ft, $\bar{z} = 0.778$ ft, $M_{Ax} = 16$ lb · ft, $M_{Ay} = 57.1$ lb · ft, $M_{Az} = 0$, $A_x = 0$, $A_y = 0$, $A_z = 20.6$ lb
9–130. 0.600 in.
9–131. $901(10^3)$ mm^3, $57.7(10^3)$ mm^2
9–133. $\bar{x} = 4.56$ m, $\bar{y} = 3.07$ m, $R_B = 4.66$ kN, $R_A = 5.99$ kN
9–134. $\bar{y} = 150$ mm, $\bar{x} = 33.9$ mm

Chapter 10

10–1. 973 in^4
10–2. $15.4(10)^6$ mm^4
10–3. 39.0 m^4
10–5. 10.7 in^4
10–6. 307 in^4
10–7. 3.20 m^4
10–9. $\frac{2}{15}bh^3$
10–10. $\frac{2}{7}b^3h$
10–11. (*a*) 23.8 ft^4, (*b*) 23.8 ft^4
10–13. 6.87 m^4
10–14. 5.81 m^4
10–15. 0.533 m^4
10–17. 3.20 m^4
10–18. 0.762 m^4
10–19. $\frac{\sqrt{3}a^4}{96}$
10–21. 0.935 m^4
10–22. 333 in^4
10–23. 32.4 in^4
10–25. 0.176 m^4
10–26. $2.26(10^3)$ in^4
10–27. 341 in^4
10–29. 832 in^4
10–30. 170 mm, $722(10^6)$ mm^4
10–31. $91.7(10^6)$ mm^4
10–33. 54.7 in^4
10–34. 1.29 in., 6.74 in^4
10–35. 18.3 in^4
10–37. $122(10^6)$ mm^4

10–38. 81 in^4
10–39. 91.1 in^4
10–41. 207 mm, $\bar{I}_{x'} = 222(10^6)\ \text{mm}^4$, $I_y = 115(10^6)\ \text{mm}^4$
10–42. 648 in^4
10–43. 1971 in^4
10–45. 80.7 mm, $67.6(10^6)\ \text{mm}^4$
10–46. 61.6 mm, $41.2(10^6)\ \text{mm}^4$
10–47. 503 in^4
10–49. $0.187d^4$
10–50. $251(10^3)\ \text{mm}^4$, $172(10^3)\ \text{mm}^4$, 68.4%
10–51. $\dfrac{b^3d^3}{6(b^2 + d^2)}$
10–53. 109 mm
10–54. $\frac{1}{12}a^3b \sin^3 \theta$
10–55. $\dfrac{ab \sin \theta}{12}(b^2 + a^2 \cos^2 \theta)$
10–57. $\dfrac{h^2b^2}{16}$
10–58. 8 in^4
10–59. 1.33 in^4
10–61. 48 in^4
10–62. 0.333 m^4
10–63. $\frac{3}{16}b^2h^2$
10–65. 3.30 m^4
10–66. 97.8 in^4
10–67. $\bar{x} = 85.0$ mm, $\bar{y} = 35.0$ mm, $-7.50(10^6)\ \text{mm}^4$
10–69. $-110\ \text{in}^4$
10–70. 36 in^4
10–71. 0.740 in^4
10–73. $135(10)^6\ \text{mm}^4$
10–74. $I_u = 274\ \text{in}^4$, $I_v = 571\ \text{in}^4$, $I_{uv} = -258\ \text{in}^4$
10–75. $I_u = 195\ \text{in}^4$, $I_v = 650\ \text{in}^4$, $I_{uv} = -191\ \text{in}^4$
10–77. 6.08°, $I_{\max} = 1.74(10^3)\ \text{in}^4$, $I_{\min} = 435\ \text{in}^4$
10–78. $I_u = 3.47(10^3)\ \text{in}^4$, $I_v = 3.47(10^3)\ \text{in}^4$, $I_{uv} = 2.05(10^3)\ \text{in}^4$
10–79. $I_u = 2.44(10^3)\ \text{in}^4$, $I_v = 4.49(10^3)\ \text{in}^4$, $I_{uv} = 1.77(10^3)\ \text{in}^4$
10–81. $I_{\max} = 64.1\ \text{in}^4$, $I_{\min} = 5.33\ \text{in}^4$
10–82. $I_{\max} = 4.92(10^6)\ \text{mm}^4$, $I_{\min} = 1.36(10^6)\ \text{mm}^4$
10–83. x and y principal axes, 4.67 in., $I_x = 1.47(10^3)\ \text{in}^4$, $I_y = 2.50(10^3)\ \text{in}^4$
10–85. $-22.5°$, $I_{\max} = 250\ \text{in}^4$, $I_{\min} = 20.4\ \text{in}^4$
10–86. $I_{\max} = 1.74(10^3)\ \text{in}^4$, $I_{\min} = 435\ \text{in}^4$
10–87. $I_{\max} = 64.1\ \text{in}^4$, $I_{\min} = 5.33\ \text{in}^4$
10–89. $I_{\max} = 4.92(10^6)\ \text{mm}^4$, $I_{\min} = 1.36(10^6)\ \text{mm}^4$
10–90. $I_{\max} = 250\ \text{in}^4$, $I_{\min} = 20.4\ \text{in}^4$
10–91. $\frac{1}{3}ml^2$
10–93. $\frac{3}{10}mr^2$
10–94. $\frac{2}{5}mr^2$
10–95. $\dfrac{a}{\sqrt{3}}$
10–97. $\frac{2}{5}mb^2$
10–98. 2.25 slug · ft^2
10–99. $\frac{93}{70}mb^2$
10–101. $\dfrac{m}{6}(a^2 + h^2)$
10–102. 2.17 slug · ft^2
10–103. 1.53 kg · m^2
10–105. 293 slug · ft^2
10–106. 0.888 m, 5.61 kg · m^2
10–107. 6.39 m, 53.2 kg · m^2
10–109. 0.203 m, 0.230 kg · m^2
10–110. 34.2 kg · m^2
10–111. $I_u = 5.09(10^6)\ \text{mm}^4$, $I_v = 5.09(10^6)\ \text{mm}^4$, $I_{uv} = 0$
10–113. $I_x = \dfrac{1}{3(3n + 1)}bh^3$, $I_y = \dfrac{1}{n + 3}b^3h$
10–114. 307 in^4
10–115. 10.7 in^4
10–117. 30.9 in^4
10–118. 0.305 m^4
10–119. 1.60 m^4
10–121. $I_x = 0.167\ \text{in}^4$, $I_y = 0.0333\ \text{in}^4$
10–122. 0.0625 in^4
10–123. $\theta_{P_1} = -21.6°$, $\theta_{P_2} = 68.4°$, $I_{\max} = 0.191\ \text{in}^4$, $I_{\min} = 0.00859\ \text{in}^4$

Chapter 11

11–1. 512 N
11–2. $\dfrac{500\sqrt{0.04 \cos^2 \theta + 0.6}}{(0.2 \cos \theta + \sqrt{0.04 \cos^2 \theta + 0.6}) \sin \theta}$
11–3. $\dfrac{P}{2 \tan \theta}$
11–5. 27.4°
11–6. 1.05 kN/m
11–7. $\frac{1}{2}Wl \cos \theta$

11–9. 63.2 N/m

11–10. 16.6°

11–11. 16.6°

11–13. 53.1°

11–14. 9.21°

11–15. 35.2°

11–17. 50 lb

11–18. 50 lb

11–19. $\cos^{-1}\left(\frac{a}{2L}\right)^{1/3}$

11–21. 13.9 N

11–22. 43.3 N · m

11–23. 25.0 N · m

11–25. $m_B = 2$ kg, $m_A = 1.60$ kg

11–26. 427 N

11–27. $x = 0$ unstable, $x = 0.167$ m stable

11–29. 38.7° unstable, 90° stable

11–30. (0, 0), unstable

11–31. 1.05 kN/m, stable

11–33. 53.1° stable

11–34. 100 kg

11–35. 2.55°, unstable

11–37. 90°, $\sin^{-1}\left(\frac{W}{2kL}\right)$

11–38. 0°, $\cos^{-1}\left(1 - \frac{W}{2kL}\right)$

11–39. 0°, stable

11–41. 2.45 m

11–42. 15.5°

11–43. 94.3 N/m, stable

11–45. 30°

11–46. 0°, $\cos^{-1}\left(\frac{W}{2kL}\right)$, $2kL$

11–47. $\frac{W}{L}$

11–49. 0

11–50. 87.9 mm

11–51. 9.46°, stable

11–53. $\frac{a}{2r}$

11–54. 45°

11–55. 4.62 lb · ft

11–57. 90°, 30°

11–58. 90° unstable, 30° stable

11–59. 90°, 17.5°

11–61. 312 lb

11–62. 0°, 33.1°

11–63. 24.9°

Index

Dynamics

Although the space shuttle is a very large object, from a distance its motion can be modeled as a particle. In this chapter we will analyze the geometry of its motion.

12

Kinematics of a Particle

In this chapter we will study the geometric aspects of the motion of a particle, measured with respect to both fixed and moving reference frames. The path will be described using several different types of coordinates systems, and the components of the motion along the coordinate axes will be determined. For simplicity, motion along a straight line will be considered before the more general study of motion along a curved path. Once these ideas are fully understood, the analysis of the forces that cause the motion will be presented in the following chapters.

12.1 Rectilinear Kinematics: Continuous Motion

The first part of the study of engineering mechanics is devoted to *statics,* which is concerned with the equilibrium of bodies at rest or moving with constant velocity. The second part is devoted to *dynamics,* which is concerned with bodies having accelerated motion. In this text, the subject of dynamics will be presented in two parts: *kinematics,* which treats only the geometric aspects of motion; and *kinetics,* which is the analysis of the forces causing the motion. In order to understand better the principles involved, particle dynamics will be discussed first, followed by topics in rigid-body dynamics presented in two and then three dimensions.

We will begin our study of dynamics by discussing particle kinematics. Recall that a *particle* has a mass but negligible size and shape. Therefore we must limit application to those objects that have dimensions that are of no consequence in the analysis of the motion. In most problems, one is interested in bodies of finite size, such as rockets, projectiles, or vehicles. Such objects may be considered as particles, provided motion of the body is characterized by motion of its mass center and any rotation of the body is neglected.

Rectilinear Kinematics. A particle can move along either a straight or a curved path. In order to introduce the kinematics of particle motion, we will begin with a study of *rectilinear* or straight-line motion. The kinematics of this motion is characterized by specifying, at any given instant, the particle's position, velocity, and acceleration.

Position. The straight-line path of the particle can be defined using a single coordinate axis s, Fig. 12–1a. The origin O on the path is a fixed point, and from this point the *position vector* $\mathbf{r}$ is used to specify the position of the particle P at any given instant. For *rectilinear motion,* however, the *direction* of $\mathbf{r}$ is *always* along the s axis, and so it never changes. What will change is its magnitude and its sense or arrowhead direction. For analytical work it is therefore convenient to represent $\mathbf{r}$ by an *algebraic scalar s,* representing the *position coordinate* of the particle, Fig. 12–1a. The magnitude of s (and $\mathbf{r}$) is the distance from O to P, usually measured in meters (m) or feet (ft), and the sense (or arrowhead direction of $\mathbf{r}$) is defined by the algebraic sign of s. Although the choice is arbitrary, in this case s is positive since the coordinate axis is positive to the right of the origin. Likewise, it is negative if the particle is located to the left of O.

Displacement. The *displacement* of the particle is defined as the *change* in its *position.* For example, if the particle moves from P to P', Fig. 12–1b, the displacement is $\Delta\mathbf{r} = \mathbf{r}' - \mathbf{r}$. Using algebraic scalars to represent $\Delta\mathbf{r}$, we also have

$$\Delta s = s' - s$$

Here Δs is *positive* since the particle's final position is to the *right* of its initial position, i.e., $s' > s$. Likewise, if the final position is to the *left* of its initial position, Δs is *negative.*

Since the displacement of a particle is a vector quantity, it should be distinguished from the distance the particle travels. Specifically, the *distance traveled* is a *positive scalar* which represents the total length of path traversed by the particle.

Velocity. If the particle moves through a displacement $\Delta\mathbf{r}$ from P to P' during the time interval Δt, Fig. 12–1b, the *average velocity* of the particle during this time interval is

$$\mathbf{v}_{avg} = \frac{\Delta\mathbf{r}}{\Delta t}$$

If we take smaller and smaller values of Δt, the magnitude of $\Delta\mathbf{r}$ becomes smaller and smaller. Consequently, the *instantaneous velocity* is defined as $\mathbf{v} = \lim_{\Delta t \to 0} (\Delta\mathbf{r}/\Delta t)$, or

$$\mathbf{v} = \frac{d\mathbf{r}}{dt}$$

Representing **v** as an algebraic scalar, Fig. 12–1*c*, we can also write

$$(\overset{+}{\rightarrow}) \qquad v = \frac{ds}{dt} \qquad (12\text{–}1)$$

Since Δt or dt is always positive, the sign used to define the *sense* of the velocity is the same as that of Δs (or ds). For example, if the particle is moving to the *right,* Fig. 12–1*c*, the velocity is *positive;* whereas if it is moving to the *left,* the velocity is *negative.* (This fact is emphasized here by the arrow written at the left of Eq. 12–1.) The *magnitude* of the velocity is known as the *speed,* and it is generally expressed in units of m/s or ft/s.

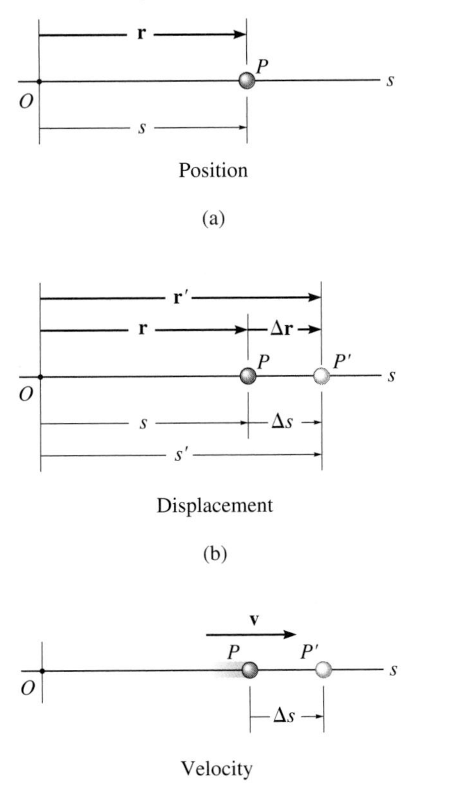

Fig. 12–1

Occasionally, the term "average speed" is used. The *average speed* is always a positive scalar and is defined as the total distance traveled by a particle, s_T, divided by the elapsed time Δt; i.e.,

$$v_{\text{sp}} = \frac{s_T}{\Delta t}$$

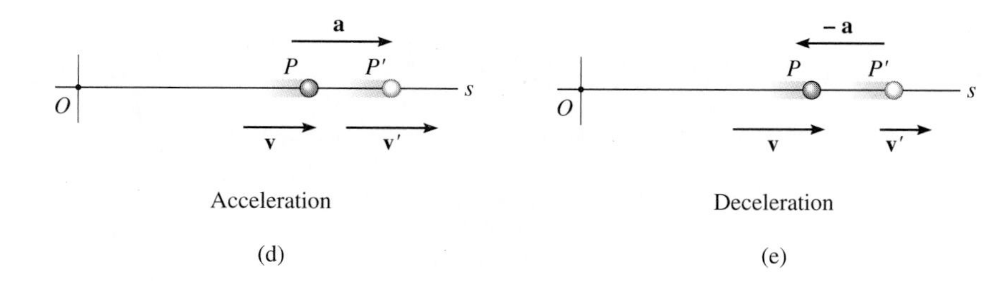

Fig. 12–1 (*cont'd*)

Acceleration. Provided the velocity of the particle is known at the two points P and P', the *average acceleration* of the particle during the time interval Δt is defined as

$$\mathbf{a}_{\text{avg}} = \frac{\Delta \mathbf{v}}{\Delta t}$$

Here $\Delta \mathbf{v}$ represents the difference in the velocity during the time interval Δt, i.e., $\Delta \mathbf{v} = \mathbf{v}' - \mathbf{v}$, Fig. 12–1*d*.

The *instantaneous acceleration* at time t is found by taking smaller and smaller values of Δt and corresponding smaller and smaller values of $\Delta \mathbf{v}$, so that $\mathbf{a} = \lim_{\Delta t \to 0} (\Delta \mathbf{v}/\Delta t)$ or, using algebraic scalars,

$$(\overset{+}{\rightarrow}) \qquad a = \frac{dv}{dt} \qquad (12\text{–}2)$$

Substituting Eq. 12–1 into this result, we can also write

$$(\overset{+}{\rightarrow}) \qquad a = \frac{d^2 s}{dt^2}$$

Both the average and instantaneous acceleration can be either positive or negative. In particular, when the particle is *slowing down,* or its speed is decreasing, it is said to be *decelerating.* In this case, v' in Fig. 12–1*e* is *less* than v and so $\Delta v = v' - v$ will be negative. Consequently, a will also be negative and therefore it will act to the *left,* in the opposite *sense* to v. Also, note that when the *velocity* is *constant,* the *acceleration is zero* since $\Delta v = v - v = 0$. Units commonly used to express the magnitude of acceleration are m/s^2 or ft/s^2.

A differential relation involving the displacement, velocity, and acceleration along the path may be obtained by eliminating the time differential dt between Eqs. 12–1 and 12–2. By doing so, realize that although we can then establish another equation, it will *not* be independent of Eqs. 12–1 and 12–2. Show that

$$(\overset{+}{\rightarrow}) \qquad a\,ds = v\,dv \qquad (12\text{–}3)$$

Constant Acceleration, $a = a_c$. When the acceleration is constant, each of the three kinematic equations $a_c = dv/dt$, $v = ds/dt$, and $a_c\, ds = v\, dv$ may be integrated to obtain formulas that relate a_c, v, s, and t.

Velocity as a Function of Time. Integrate $a_c = dv/dt$, assuming that initially $v = v_0$ when $t = 0$.

$$\int_{v_0}^{v} dv = \int_{0}^{t} a_c\, dt$$

$$v - v_0 = a_c(t - 0)$$

$$(\overset{+}{\rightarrow}) \qquad v = v_0 + a_c t \qquad (12\text{–}4)$$

Constant Acceleration

Position as a Function of Time. Integrate $v = ds/dt = v_0 + a_c t$, assuming that initially $s = s_0$ when $t = 0$.

$$\int_{s_0}^{s} ds = \int_{0}^{t} (v_0 + a_c t)\, dt$$

$$s - s_0 = v_0(t - 0) + a_c(\tfrac{1}{2}t^2 - 0)$$

$$(\overset{+}{\rightarrow}) \qquad s = s_0 + v_0 t + \tfrac{1}{2}a_c t^2 \qquad (12\text{–}5)$$

Constant Acceleration

Velocity as a Function of Position. Either solve for t in Eq. 12–4 and substitute into Eq. 12–5, or integrate $v\, dv = a_c\, ds$, assuming that initially $v = v_0$ at $s = s_0$.

$$\int_{v_0}^{v} v\, dv = \int_{s_0}^{s} a_c\, ds$$

$$\tfrac{1}{2}v^2 - \tfrac{1}{2}v_0^2 = a_c(s - s_0)$$

$$(\overset{+}{\rightarrow}) \qquad v^2 = v_0^2 + 2a_c(s - s_0) \qquad (12\text{–}6)$$

Constant Acceleration

This equation is not independent of Eqs. 12–4 and 12–5. Why?

The magnitudes and signs of s_0, v_0, and a_c, used in the above three equations, are determined from the chosen origin and positive direction of the s axis. As indicated by the arrow written at the left of each equation, we have assumed positive quantities act to the right, in accordance with the coordinate axis s shown in Fig. 12–1. Also, it is important to remember that the above equations are useful *only when the acceleration is constant and when* $t = 0$, $s = s_0$, $v = v_0$. A common example of constant accelerated motion occurs when a body falls freely toward the earth. If air resistance is neglected and the distance of fall is short, then the *downward* acceleration of the body when it is close to the earth is constant and approximately $9.81\ \text{m/s}^2$ or $32.2\ \text{ft/s}^2$. The proof of this is given in Example 13–2.

PROCEDURE FOR ANALYSIS

Coordinate System. Whenever the kinematic equations are applied, it is *very important* first to establish a position coordinate s along the path, and to specify its *fixed origin* and positive direction. Because the path is *rectilinear,* the lines of direction of the particle's position, velocity, and acceleration *never change.* Therefore these quantities can be represented as algebraic scalars. For analytical work the sense of s, v, and a can be determined from their *algebraic signs.* In the examples that follow, the positive sense for each scalar will be indicated by an arrow shown alongside each kinematic equation as it is applied.

Kinematic Equations. Often a mathematical relationship between *any two* of the four variables a, v, s, and t can be established either by observation or by experiment. When this is the case, the relationships among the remaining variables can be obtained by differentiation or integration, using the kinematic equations $a = dv/dt$, $v = ds/dt$, or $a\,ds = v\,dv$.* Since each of these equations relates *three* variables, then, if a variable is *known* as a function of another variable, a third variable can be determined by *choosing the kinematic equation which relates all three.* For example, suppose that the *acceleration* is known as a function of *position,* $a = f(s)$. The *velocity* can be determined from $a\,ds = v\,dv$ since $f(s)$ can be substituted for a to yield $f(s)\,ds = v\,dv$. Solution for v requires integration. Note that the velocity *cannot* be obtained by using $a = dv/dt$, since $f(s)\,dt = dv$ contains two variables, s and t, on the left side and so it *cannot* be integrated. Whenever integration is performed, it is important that the position and velocity be known at a given instant in order to evaluate either the constant of integration if an indefinite integral is used, or the limits of integration if a definite integral is used. Lastly, keep in mind that Eqs. 12–4 through 12–6 have only a limited use. *Never apply these equations unless it is absolutely certain that the acceleration is constant.*

*Some standard differentiation and integration formulas are given in Appendix A.

Example 12–1

During a test, the car in Fig. 12–2 moves in a straight line such that for a short time its velocity is defined by $v = (9t^2 + 2t)$ ft/s, where t is in seconds. Determine its position and acceleration when $t = 3$ s. When $t = 0$, $s = 0$.

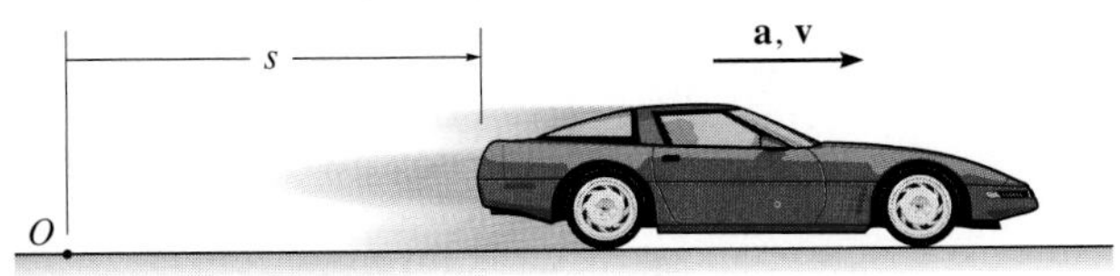

Fig. 12–2

SOLUTION

Coordinate System. The position coordinate extends from the fixed origin O to the car, positive to the right.

Position. The car's velocity is given as a function of time so that its position can be determined from $v = ds/dt$, since this equation relates v, s, and t. Noting that $s = 0$ when $t = 0$, we have*

$$(\overset{+}{\rightarrow}) \qquad v = \frac{ds}{dt} = (9t^2 + 2t)$$

$$\int_0^s ds = \int_0^t (9t^2 + 2t)\, dt$$

$$s\Big|_0^s = 3t^3 + t^2\Big|_0^t$$

$$s = 3t^3 + t^2$$

When $t = 3$ s,

$$s = 3(3)^3 + (3)^2 = 90 \text{ ft} \qquad \textit{Ans.}$$

Acceleration. Knowing the velocity as a function of time, the acceleration is determined from $a = dv/dt$, since this equation relates a, v, and t.

$$(\overset{+}{\rightarrow}) \qquad a = \frac{dv}{dt} = \frac{d}{dt}(9t^2 + 2t)$$

$$= 18t + 2$$

When $t = 3$ s,

$$a = 18(3) + 2 = 56 \text{ ft/s}^2 \rightarrow \qquad \textit{Ans.}$$

The formulas for constant acceleration *cannot* be used to solve this problem. Why?

*The *same result* can be obtained by evaluating a constant of integration C rather than using definite limits on the integral. For example, integrating $ds = (9t^2 + 2t)\, dt$ yields $s = 3t^3 + t^2 + C$. Using the condition that at $t = 0$, $s = 0$, then $C = 0$.

Example 12–2

A small projectile is fired vertically *downward* into a fluid medium with an initial velocity of 60 m/s. If the projectile experiences a deceleration which is equal to $a = (-0.4v^3)$ m/s^2, where v is measured in m/s, determine the projectile's velocity and position 4 s after it is fired.

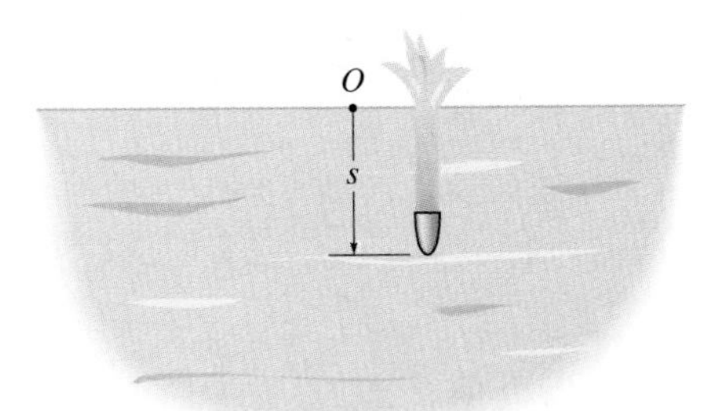

Fig. 12–3

SOLUTION

Coordinate System. Since the motion is downward, the position coordinate is positive downward, with origin located at O, Fig. 12–3.

Velocity. The acceleration is given as a function of velocity so that the velocity can be obtained from $a = dv/dt$, since this equation relates v, a, and t. (Why not use $v = v_0 + a_c t$?) Separating the variables and integrating, with $v_0 = 60$ m/s when $t = 0$, yields

$(+\downarrow)$
$$a = \frac{dv}{dt} = -0.4v^3$$

$$\int_{60}^{v} \frac{dv}{-0.4v^3} = \int_0^t dt$$

$$\frac{1}{-0.4}\left(\frac{1}{-2}\right)\frac{1}{v^2}\Bigg|_{60}^{v} = t - 0$$

$$\frac{1}{0.8}\left[\frac{1}{v^2} - \frac{1}{(60)^2}\right] = t$$

$$v = \left\{\left[\frac{1}{(60)^2} + 0.8t\right]^{-1/2}\right\} \text{ m/s}$$

Here the positive root is taken, since the projectile is moving downward. When $t = 4$ s,

$$v = 0.559 \text{ m/s} \downarrow \qquad \textit{Ans.}$$

Position. Knowing the velocity as a function of time, we can obtain the projectile's position from $v = ds/dt$, since this equation relates s, v, and t. Using the initial condition $s = 0$, when $t = 0$, we have

$(+\downarrow)$
$$v = \frac{ds}{dt} = \left[\frac{1}{(60)^2} + 0.8t\right]^{-1/2}$$

$$\int_0^s ds = \int_0^t \left[\frac{1}{(60)^2} + 0.8t\right]^{-1/2} dt$$

$$s = \frac{2}{0.8}\left[\frac{1}{(60)^2} + 0.8t\right]^{1/2}\Bigg|_0^t$$

$$s = \frac{1}{0.4}\left\{\left[\frac{1}{(60)^2} + 0.8t\right]^{1/2} - \frac{1}{60}\right\} \text{ m}$$

When $t = 4$ s,

$$s = 4.43 \text{ m} \qquad \textit{Ans.}$$

Example 12–3

During a test an elevator is traveling upward at 15 m/s and the hoisting cable is cut when it is 40 m from the ground. Determine the maximum height s_B reached by the elevator and its speed just before it hits the ground. During the entire time the elevator is in motion, it is subjected to a constant downward acceleration of 9.81 m/s^2 due to gravity. Neglect the effect of air resistance.

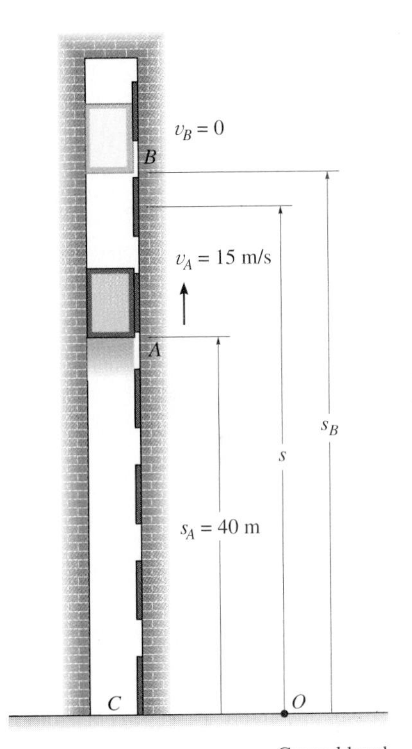

Fig. 12–4

SOLUTION

Coordinate System. The origin O for the position coordinate s is taken at ground level with positive upward, Fig. 12–4.

Maximum Height. At the maximum height $s = s_B$ the velocity $v_B = 0$. Since the elevator is traveling *upward* when $t = 0$, it is subjected to a velocity $v_A = +15$ m/s (positive since it is in the same sense as positive displacement). For the entire motion, the acceleration is *constant* such that $a_c = -9.81$ m/s^2 (negative since it acts in the *opposite* sense to positive velocity or positive displacement). Since a_c is *constant* throughout the entire motion, the elevator's position may be related to its velocity at the two points A and B on the path by using Eq. 12–6, namely,

$$(+\uparrow) \qquad v_B^2 = v_A^2 + 2a_c(s_B - s_A)$$

$$0 = (15 \text{ m/s})^2 + 2(-9.81 \text{ m/s}^2)(s_B - 40 \text{ m})$$

$$s_B = 51.5 \text{ m} \qquad \textit{Ans.}$$

Velocity. To obtain the velocity of the elevator just before it hits the ground, we can apply Eq. 12–6 between points B and C, Fig. 12–4.

$$(+\uparrow) \qquad v_C^2 = v_B^2 + 2a_c(s_C - s_B)$$

$$= 0 + 2(-9.81 \text{ m/s}^2)(0 - 51.5 \text{ m})$$

$$v_C = -31.8 \text{ m/s} = 31.8 \text{ m/s} \downarrow \qquad \textit{Ans.}$$

The negative root was chosen since the elevator is moving *downward.*

Similarly, Eq. 12–6 may also be applied between points A and C, i.e.,

$$(+\uparrow) \qquad v_C^2 = v_A^2 + 2a_c(s_C - s_A)$$

$$= (15 \text{ m/s})^2 + 2(-9.81 \text{ m/s}^2)(0 - 40 \text{ m})$$

$$v_C = -31.8 \text{ m/s} = 31.8 \text{ m/s} \downarrow$$

Note: It should be realized that the elevator is subjected to a *deceleration* from A to B of 9.81 m/s^2, and then from B to C it is *accelerated* at this rate. Furthermore, even though the elevator momentarily comes to *rest* at B ($v_B = 0$) the acceleration at B is 9.81 m/s^2 downward!

Example 12–4

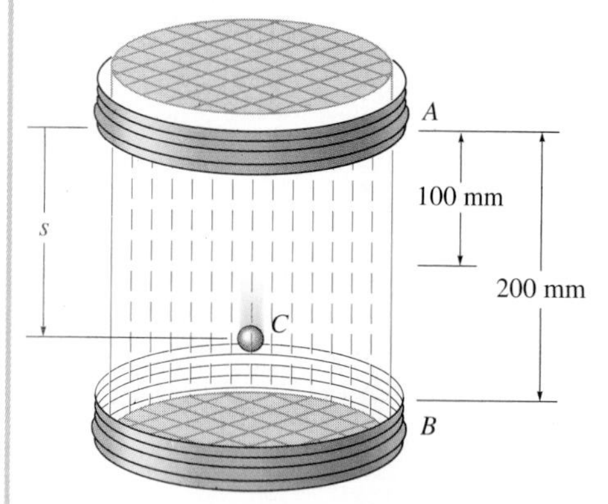

Fig. 12–5

A metallic particle is subjected to the influence of a magnetic field such that it travels downward through a fluid that extends from plate A to plate B, Fig. 12–5. If the particle is released from rest at the midpoint C, $s = 100$ mm, and the acceleration is measured as $a = (4s)$ m/s^2, where s is in meters, determine the velocity of the particle when it reaches plate B, $s = 200$ mm, and the time it needs to travel from C to B.

SOLUTION

Coordinate System. As shown in Fig. 12–5, s is taken positive downward, measured from plate A.

Velocity. Since the particle's acceleration is known as a function of position, the velocity as a function of position can be obtained by using $v\,dv = a\,ds$. Why not use the formulas for constant acceleration? Realizing that $v = 0$ at $s = 100$ mm $= 0.1$ m, we have

$(+\downarrow)$

$$v\,dv = a\,ds$$

$$\int_0^v v\,dv = \int_{0.1}^s 4s\,ds$$

$$\tfrac{1}{2}v^2\Big|_0^v = \frac{4}{2}s^2\Big|_{0.1}^s$$

$$v = 2(s^2 - 0.01)^{1/2} \tag{1}$$

At $s = 200$ mm $= 0.2$ m,

$$v_B = 0.346 \text{ m/s} = 346 \text{ mm/s} \downarrow \qquad \textit{Ans.}$$

The positive root is chosen since the particle is traveling downward, i.e., in the $+s$ direction.

Time. The time for the particle to travel from C to B can be obtained using $v = ds/dt$ and Eq. 1, where $s = 0.1$ m when $t = 0$.

$(+\downarrow)$

$$ds = v\,dt$$

$$= 2(s^2 - 0.01)^{1/2}\,dt$$

$$\int_{0.1}^s \frac{ds}{(s^2 - 0.01)^{1/2}} = \int_0^t 2\,dt$$

$$\ln\left(s + \sqrt{s^2 - 0.01}\right)\Big|_{0.1}^s = 2t\Big|_0^t$$

$$\ln\left(s + \sqrt{s^2 - 0.01}\right) + 2.30 = 2t$$

At $s = 200$ mm $= 0.2$ m,

$$t = \frac{\ln\left(0.2 + \sqrt{(0.2)^2 - 0.01}\right) + 2.30}{2} = 0.657 \text{ s} \qquad \textit{Ans.}$$

Example 12–5

A particle moves along a horizontal path such that its velocity is given by $v = (3t^2 - 6t)$ m/s, where t is the time in seconds. If it is initially located at the origin O, determine the distance traveled by the particle during the time interval $t = 0$ to $t = 3.5$ s, and the particle's average velocity and average speed during this time interval.

SOLUTION

Coordinate System. Here we will assume positive motion to the right, measured from the origin O.

Distance Traveled. Since the velocity is related to time, the position related to time may be found by integrating $v = ds/dt$ with $t = 0$, $s = 0$.

$$(\overset{+}{\rightarrow}) \qquad ds = v\,dt$$

$$= (3t^2 - 6t)\,dt$$

$$\int_0^s ds = 3\int_0^t t^2\,dt - 6\int_0^t t\,dt$$

$$s = (t^3 - 3t^2)\text{ m} \qquad (1)$$

In order to determine the distance traveled in 3.5 s, it is necessary to investigate the path of motion. The graph of the velocity function, Fig. 12–6*a*, reveals that for $0 \le t < 2$ s the velocity is *negative*, which means the particle is traveling to the *left*, and for $t > 2$ s the velocity is *positive*, and hence the particle is traveling to the *right*. Also, $v = 0$ at $t = 2$ s. The particle's position when $t = 0$, $t = 2$ s, and $t = 3.5$ s can be determined from Eq. 1. This yields

$$s|_{t=0} = 0 \qquad s|_{t=2\text{s}} = -4.0\text{ m} \qquad s|_{t=3.5\text{s}} = 6.12\text{ m}$$

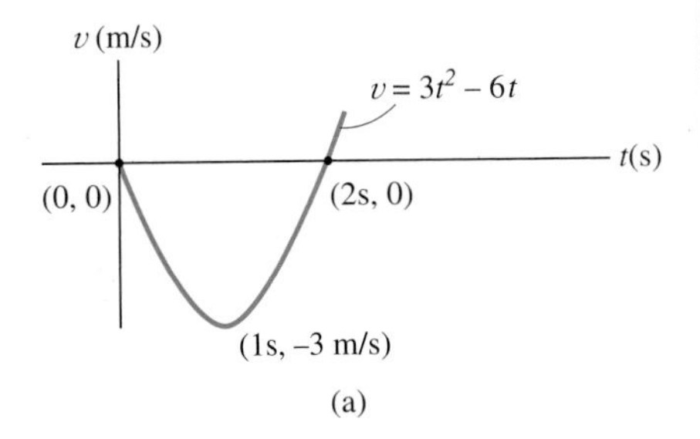

(a)

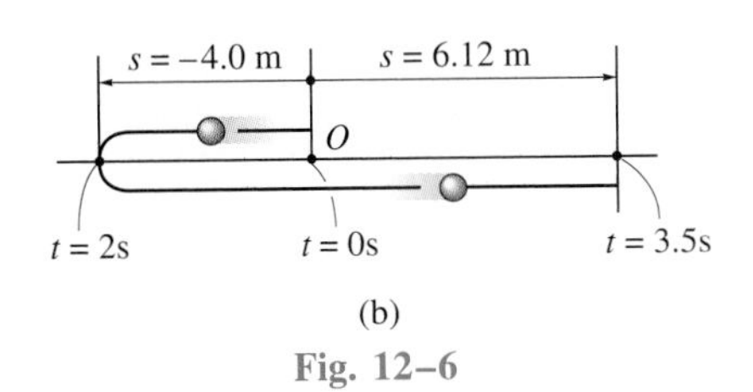

(b)

Fig. 12–6

The path is shown in Fig. 12–6*b*. Hence, the distance traveled in 3.5 s is

$$s_T = 4.0 + 4.0 + 6.12 = 14.12\text{ m} = 14.1\text{ m} \qquad \textit{Ans.}$$

Velocity. The *displacement* from $t = 0$ to $t = 3.5$ s is

$$\Delta s = s|_{t=3.5\text{s}} - s|_{t=0} = 6.12 - 0 = 6.12\text{ m}$$

so the average velocity is

$$v_{avg} = \frac{\Delta s}{\Delta t} = \frac{6.12}{3.5 - 0} = 1.75\text{ m/s} \rightarrow \qquad \textit{Ans.}$$

The average speed is defined in terms of the *distance traveled* s_T. This positive scalar is

$$(v_{sp})_{avg} = \frac{s_T}{\Delta t} = \frac{14.12}{3.5 - 0} = 4.03\text{ m/s} \qquad \textit{Ans.}$$

PROBLEMS

12–1. A bicyclist starts from rest and after traveling along a straight path a distance of 20 m reaches a speed of 30 km/h. Determine his acceleration if it is *constant*. Also, how long does it take to reach the speed of 30 km/h?

12–2. A car starts from rest and reaches a speed of 80 ft/s after traveling 500 ft along a straight road. Determine its constant acceleration and the time of travel.

12–3. A baseball is thrown downward from a 50-ft tower with an initial speed of 18 ft/s. Determine the speed at which it hits the ground and the time of travel.

***12–4.** A particle travels along a straight line such that in 2 s it moves from an initial position $s_A = +0.5$ m to a position $s_B = -1.5$ m. Then in another 4 s it moves from s_B to $s_C = +2.5$ m. Determine the particle's average velocity and average speed during the 6-s time interval.

12–5. Traveling with an initial speed of 70 km/h, a car accelerates at 6000 km/h^2 along a straight road. How long will it take to reach a speed of 120 km/h? Also, through what distance does the car travel during this time?

12–6. A freight train travels at $v = 60(1 - e^{-t})$ ft/s, where t is the elapsed time in seconds. Determine the distance traveled in three seconds, and the acceleration at this time.

Prob. 12–6

12–7. A sphere is fired downward into a medium with an initial speed of 27 m/s. If it experiences a deceleration $a = (-6t)$ m/s^2, where t is in seconds, determine the distance traveled before it stops.

***12–8.** A sphere is fired downward into a medium with an initial speed of 27 m/s. If it experiences a deceleration $a = (-6t)$ m/s^2, where t is in seconds, determine its velocity when $t = 2$ s. How far has it traveled?

12–9. The position of a particle along a straight line is given by $s = (t^3 - 9t^2 + 15t)$ ft, where t is in seconds. Determine the position when $t = 6$ s and the total distance the particle travels during the 6-s time interval. *Hint:* Plot the path to determine the total distance traveled.

12–10. The position of a particle along a straight line is given by $s = (t^3 - 9t^2 + 15t)$ ft, where t is in seconds. Determine its maximum acceleration and maximum velocity during the time interval $0 \leq t \leq 10$ s.

12–11. From approximately what floor of a building must a car be dropped from an at-rest position so that it reaches a speed of 80.7 ft/s (55 mi/h) when it hits the ground? Each floor is 12 ft higher than the one below it. (*Note:* You may want to remember this when traveling 55 mi/h.)

***12–12.** A car is to be hoisted by elevator to the fourth floor of a parking garage, which is 48 ft above the ground. If the elevator can accelerate at 0.6 ft/s^2, decelerate at 0.3 ft/s^2, and reach a maximum speed of 8 ft/s, determine the shortest time to make the lift, starting from rest and ending at rest.

12–13. A particle travels in a straight line such that for a short time $2\text{ s} \leq t \leq 6\text{ s}$ its motion is described by $v = (4/a)$ ft/s, where a is in ft/s^2. If $v = 6$ ft/s when $t = 2$ s, determine the particle's acceleration when $t = 3$ s.

12–14. The acceleration of a particle as it moves along a straight line is given by $a = (2t - 1)$ m/s^2, where t is in seconds. If $s = 1$ m and $v = 2$ m/s when $t = 0$, determine the particle's velocity and position when $t = 6$ s. Also, determine the total distance the particle travels during this time period.

12–15. Car B is traveling a distance d ahead of car A. Both cars are traveling at 60 ft/s when the driver of B suddenly applies the brakes, causing his car to decelerate at 12 ft/s^2. It takes the driver of car A 0.75 s to react (this is the normal reaction time for drivers). When he applies his brakes, he decelerates at 15 ft/s^2. Determine the minimum distance d between the cars so as to avoid a collision.

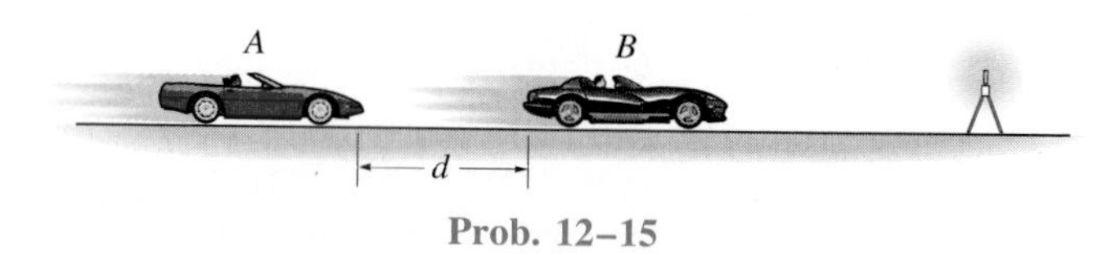

Prob. 12–15

***12–16.** The position of a particle along a straight-line path is defined by $s = (t^3 - 6t^2 - 15t + 7)$ ft, where t is in seconds. Determine the total distance traveled when $t = 10$ s. What are the particle's average velocity, average speed, and the instantaneous velocity and acceleration at this time?

12–17. When a train is traveling along a straight track at 2 m/s, it begins to accelerate at $a = (60\, v^{-4})$ m/s^2, where v is in m/s. Determine its velocity v and the position 3 s after the acceleration.

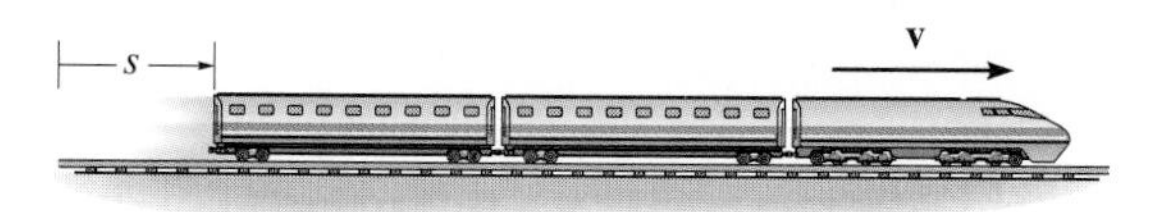

Prob. 12–17

12–18. A particle travels to the right along a straight line with a velocity $v = [5/(4 + s)]$ m/s, where s is in meters. Determine its position when $t = 6$ s if $s = 5$ m when $t = 0$.

12–19. A particle travels to the right along a straight line with a velocity $v = [5/(4 + s)]$ m/s, where s is in meters. Determine its deceleration when $s = 2$ m.

***12–20.** A stone A is dropped from rest down a well, and in 1 s another stone B is dropped from rest. Determine the distance between the stones another second later.

12–21. A stone A is dropped from rest down a well, and in 1 s another stone B is dropped from rest. Determine the time interval between the instant A strikes the water and the instant B strikes the water. Also, at what speed do they strike the water?

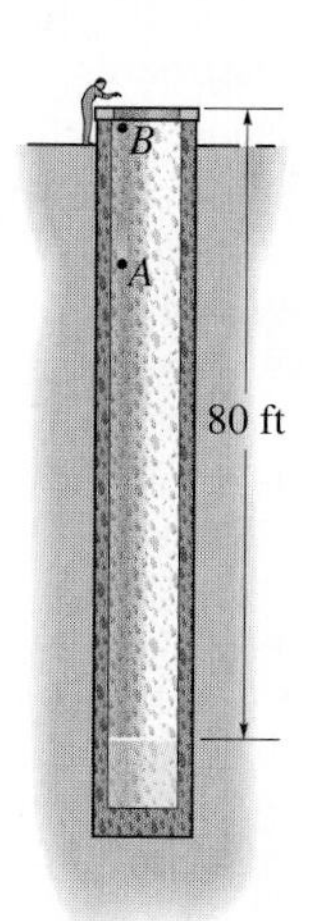

Probs. 12–20/12–21

12–22. A particle moving along a straight line is subjected to a deceleration $a = (-2v^3)$ m/s^2, where v is in m/s. If it has a velocity $v = 8$ m/s and a position $s = 10$ m when $t = 0$, determine its velocity and position when $t = 4$ s.

12–23. A particle travels in a straight line with accelerated motion such that $a = -ks$, where s is the distance from the starting point and k is a proportionality constant which is to be determined. For $s = 2$ ft the velocity is 4 ft/s, and for $s = 3.5$ ft the velocity is 10 ft/s. What is s when $v = 0$?

***12–24.** The acceleration of a rocket traveling upward is given by $a = (6 + 0.02s)$ m/s^2, where s is in meters. Determine the rocket's velocity when $s = 2$ km and the time needed to reach this altitude. Initially, $v = 0$ and $s = 0$ when $t = 0$.

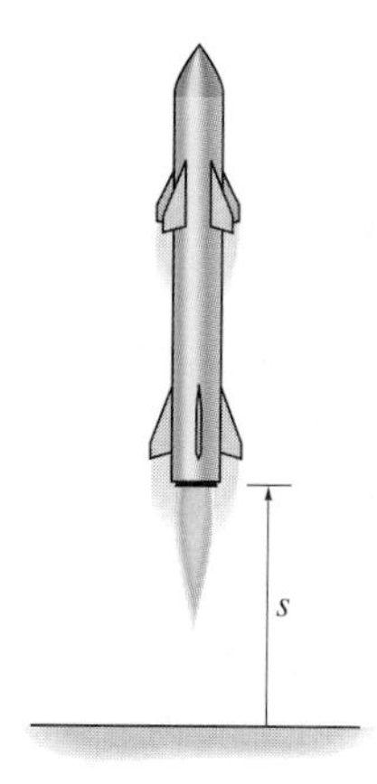

Prob. 12–24

12–25. The acceleration of a rocket traveling upward is given by $a = (6 + 0.02s)$ m/s^2, where s is in meters. Determine the time needed for the rocket to reach an altitude of $s = 100$ m. Initially, $v = 0$ and $s = 0$ when $t = 0$.

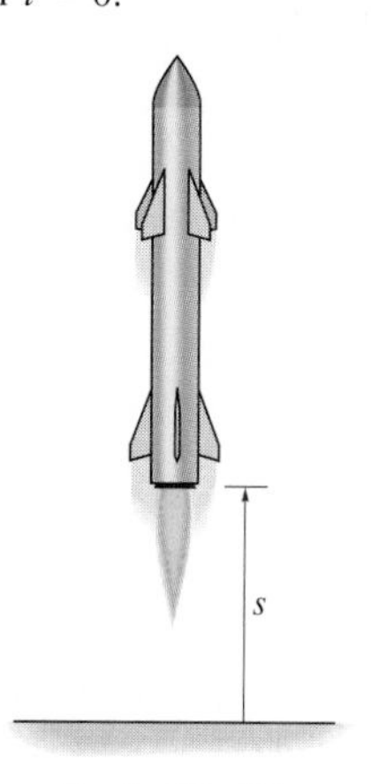

Prob. 12–25

12–26. At $t = 0$ bullet A is fired vertically with an initial (muzzle) velocity of 450 m/s. When $t = 3$ s, bullet B is fired upward with a muzzle velocity of 600 m/s. Determine the time t, after A is fired, as to when bullet B passes bullet A. At what altitude does this occur?

12–27. Ball A is released from rest at a height of 40 ft at the same time that a second ball B is thrown upward 5 ft from the ground. If the balls pass one another at a height of 20 ft, determine the speed at which ball B was thrown upward.

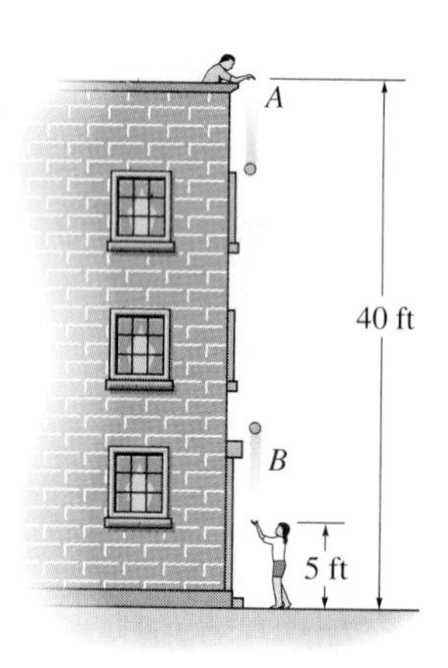

Prob. 12–27

*■**12–28.** A projectile, initially at the origin, moves vertically downward along a straight-line path through a fluid medium such that its velocity is defined as $v = 3(8e^{-t} + t)^{1/2}$ m/s, where t is in seconds. Plot the position s of the projectile during the first 2 s. Use the Runge-Kutta method to evaluate s with incremental values of $h = 0.25$ s.

12–29. The acceleration of a particle along a straight line is defined by $a = (2t - 9)$ m/s^2, where t is in seconds. At $t = 0$, $s = 1$ m and $v = 10$ m/s. When $t = 9$ s, determine (a) the particle's position, (b) the total distance traveled, and (c) the velocity.

12–30. Determine the time required for a car to travel 1 km along a road if the car starts from rest, reaches a maximum speed at some intermediate point, and then stops at the end of the road. The car can accelerate at 1.5 m/s^2 and decelerate at 2 m/s^2.

12–31. Two particles A and B start from rest at the origin $s = 0$ and move along a straight line such that $a_A = (6t - 3)$ ft/s^2 and $a_B = (12t^2 - 8)$ ft/s^2, where t is in seconds. Determine the distance between them when $t = 4$ s and the total distance each has traveled in $t = 4$ s.

***12–32.** When two cars A and B are next to one another, they are traveling in the same direction with speeds v_A and v_B, respectively. If B maintains its constant speed, while A begins to decelerate at a_A, determine the distance d between the cars at the instant A stops.

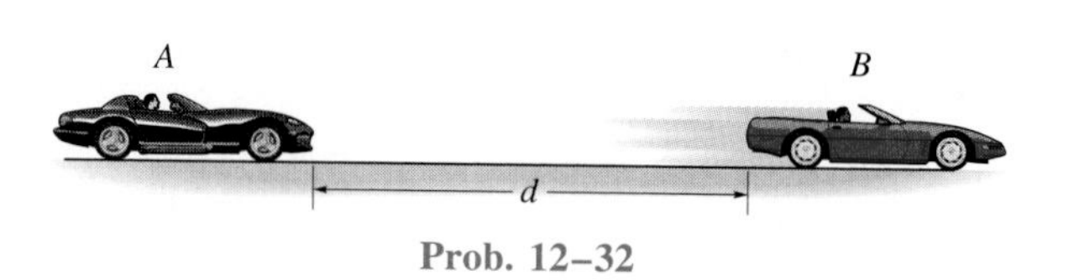

Prob. 12–32

12–33. When a particle falls through the air, its initial acceleration $a = g$ diminishes until it is zero, and thereafter it falls at a constant or terminal velocity v_f. If this variation of the acceleration can be expressed as $a = (g/v_f^2)(v_f^2 - v^2)$, determine the time needed for the velocity to become $v < v_f$. Initially the particle falls from rest.

12–34. As a body is projected to a high altitude above the earth's *surface,* the variation of the acceleration of gravity with respect to altitude y must be taken into account. Neglecting air resistance, this acceleration is determined from the formula $a = -g_o[R^2/(R + y)^2]$, where g_o is the constant gravitational acceleration at sea level, R is the radius of the earth, and the positive direction is measured upward. If $g_o = 9.81$ m/s^2 and $R = 6356$ km, determine the minimum initial velocity (escape velocity) at which a projectile should be shot vertically from the earth's surface so that it does not fall back to the earth. *Hint:* This requires that $v = 0$ as $y \to \infty$.

12–35. Accounting for the variation of gravitational acceleration a with respect to altitude y (see Prob. 12–34), derive an equation that relates the velocity of a freely falling particle to its altitude. Assume that the particle is released from rest at an altitude y_o from the earth's surface. With what velocity does the particle strike the earth if it is released from rest at an altitude $y_o = 500$ km? Use the numerical data in Prob. 12–34.

12.2 Rectangular Kinematics: Erratic Motion

When a particle's motion during a time period is erratic, it may be difficult to obtain a continuous mathematical function to describe its position, velocity, or acceleration. Instead, the motion may best be described graphically using a series of curves that can be generated experimentally from computer output. If the resulting graph describes the relationship between any two of the variables, a, v, s, t, a graph describing the relationship between the other variables can be established by using the kinematic equations $a = dv/dt$, $v = ds/dt$, $a\,ds = v\,dv$. The following situations occur frequently.

Given the *s–t* Graph, Construct the *v–t* Graph. If the position of a particle can be *determined experimentally* during a time period t, the s–t graph for the particle can be plotted, Fig. 12–7*a*. To determine the particle's velocity as a function of time, i.e., the v–t graph, we must use $v = ds/dt$ since this equation relates v, s, and t. Therefore, the v–t graph is established by measuring the *slope* (ds/dt) of the s–t graph at various times and plotting the results. For example, measurement of the slopes v_0, v_1, v_2, and v_3 at the intermediate points (0, 0), (t_1, s_1), (t_2, s_2), and (t_3, s_3) on the s–t graph, Fig. 12–7*a*, gives the corresponding points on the v–t graph shown in Fig. 12–7*b*.

It may also be possible to establish the v–t graph *mathematically*, provided curved segments of the s–t graph can be expressed in the form of equations $s = f(t)$. Corresponding equations describing the curved segments of the v–t graph are then determined by *differentiation*, since $v = ds/dt = d[f(t)]/dt$.

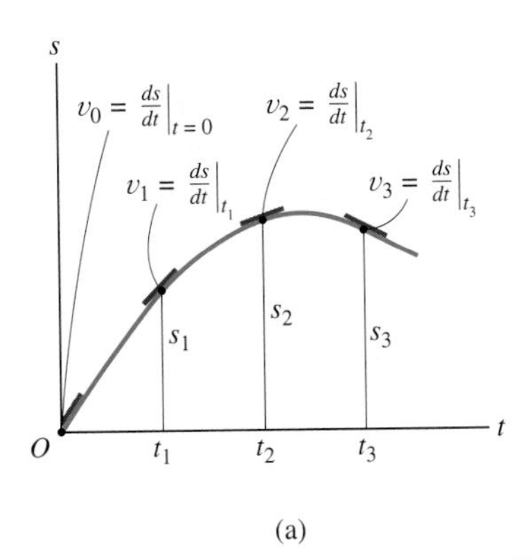

(a)

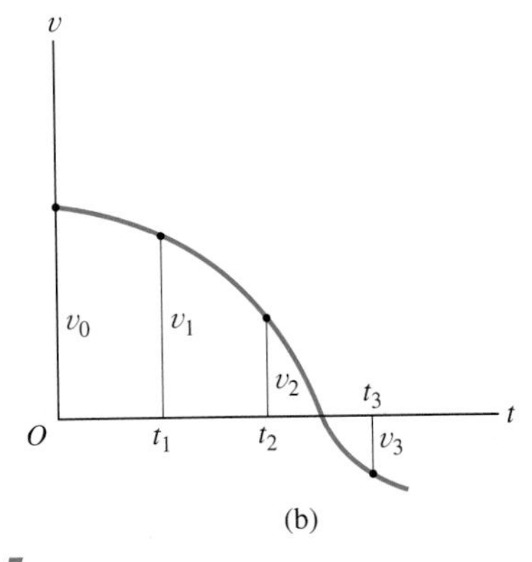

(b)

Fig. 12–7

Given the v–t Graph, Construct the a–t Graph. When the particle's v–t graph is known, as in Fig. 12–8*a*, the acceleration as a function of time, i.e., the a–t graph, can be determined using $a = dv/dt$. (Why?) Hence, the a–t graph is established by measuring the slope (dv/dt) of the v–t graph at various times and plotting the results. For example, measurement of the slopes a_0, a_1, a_2, and a_3 at the intermediate points $(0, 0)$, (t_1, v_1), (t_2, v_2), and (t_3, v_3) on the v–t graph, Fig. 12–8*a*, yields the corresponding points on the a–t graph shown in Fig. 12–8*b*.

Any curved segments of the a–t graph can also be determined *mathematically,* provided the equations of the corresponding curves of the v–t graph are known, $v = g(t)$. This is done by simply taking the *derivative* of $v = g(t)$, since $a = dv/dt = d[g(t)]/dt$.

Since differentiation reduces a polynomial of degree n to that of degree $n - 1$, then from the above explanation, if the s–t graph is parabolic (a second-degree curve), the v–t graph will be a sloping line (a first-degree curve), and the a–t graph will be a constant or a horizontal line (a zero-degree curve).

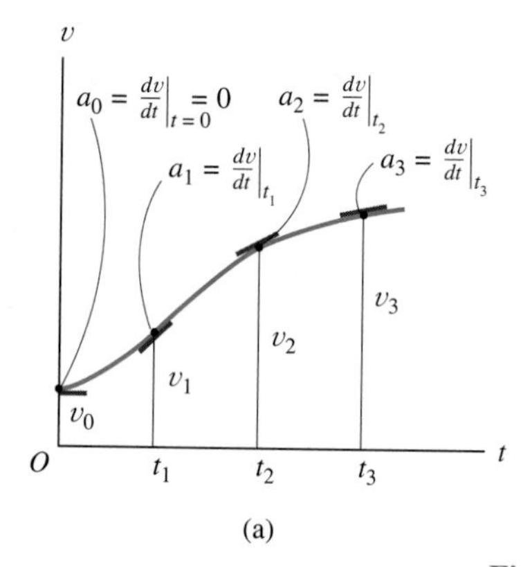

(a)

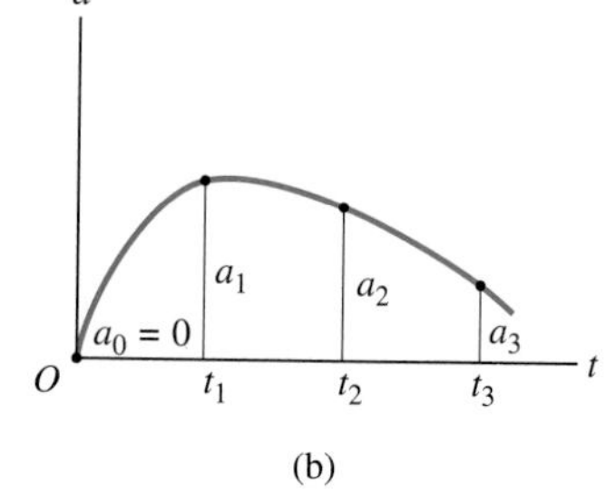

(b)

Fig. 12–8

Example 12–6

By experiment, a car moves along a straight road such that its position is described by the graph shown in Fig. 12–9*a*. Construct the v–t and a–t graphs for the time period $0 \le t \le 30$ s.

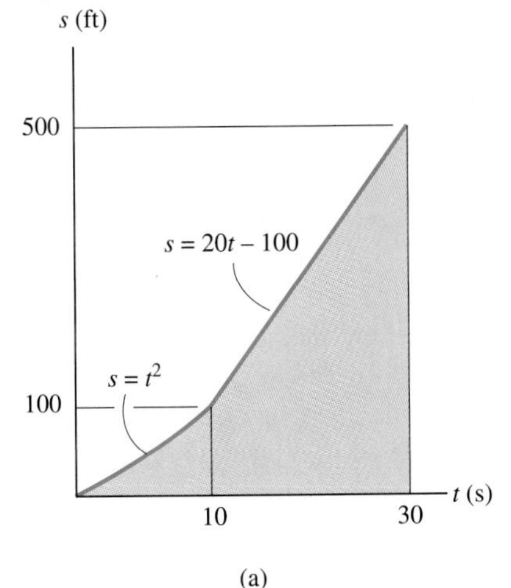

(a)

SOLUTION

***v–t* Graph.** Since $v = ds/dt$, the v–t graph can be determined by differentiating the equations defining the s–t graph, Fig. 12–9*a*. We have

$$0 \le t < 10 \text{ s}; \qquad s = t^2 \qquad v = \frac{ds}{dt} = 2t$$

$$10 \text{ s} < t \le 30 \text{ s}; \qquad s = 20t - 100 \qquad v = \frac{ds}{dt} = 20$$

The results are plotted in Fig. 12–9*b*. We can also obtain specific values of v by measuring the *slope* of the s–t graph at a given instant. For example, at $t = 20$ s, the slope of the s–t graph is determined from the straight line from 10 s to 30 s, i.e.,

$$t = 20 \text{ s}; \qquad v = \frac{\Delta s}{\Delta t} = \frac{500 - 100}{30 - 10} = 20 \text{ ft/s}$$

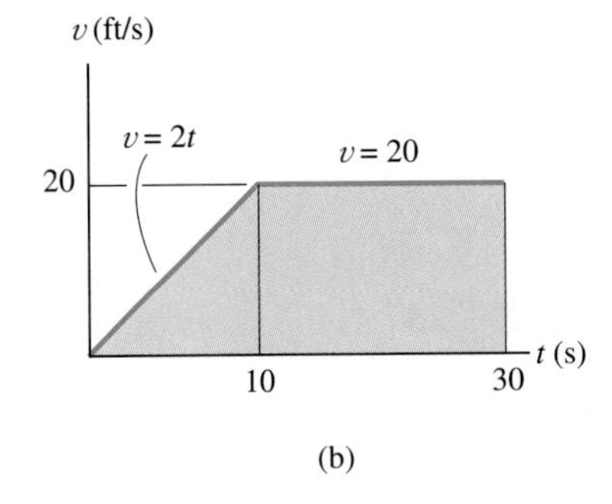

(b)

***a–t* Graph.** Since $a = dv/dt$, the a–t graph can be determined by differentiating the equations defining the lines of the v–t graph. This yields

$$0 \le t < 10 \text{ s}; \qquad v = 2t \qquad a = \frac{dv}{dt} = 2$$

$$10 < t \le 30 \text{ s}; \qquad v = 20 \qquad a = \frac{dv}{dt} = 0$$

The results are plotted in Fig. 12–9*c*. Show that $a = 2$ ft/s^2 when $t = 5$ s by measuring the slope of the v–t graph.

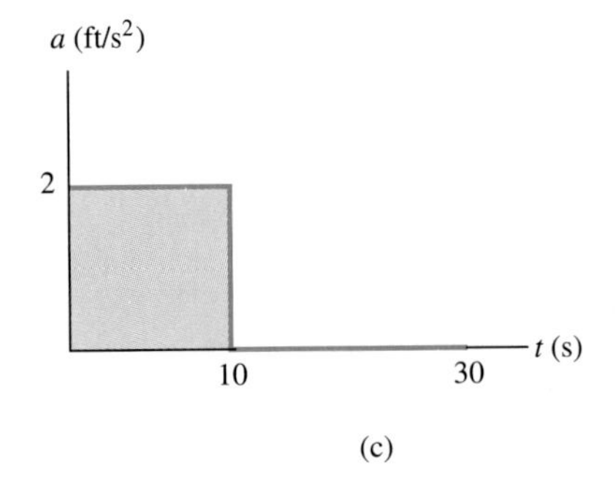

(c)

Fig. 12–9

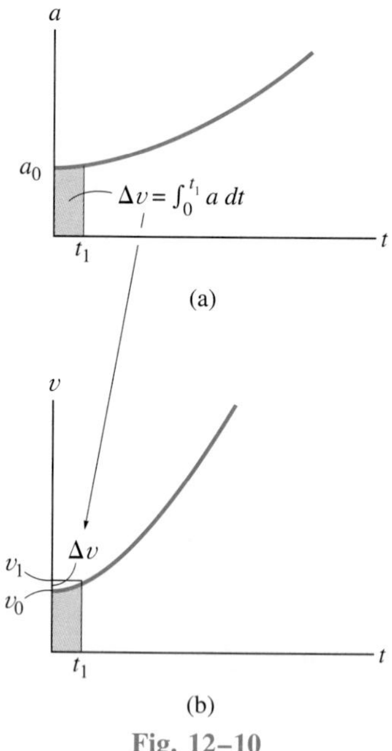

Fig. 12–10

Given the a–t Graph, Construct the v–t Graph. If the a–t graph is given, Fig. 12–10a, the v–t graph may be constructed using the equation $a = dv/dt$, written in integrated form as $\Delta v = \int a\,dt$. In this case, the change in the particle's speed during a period of time is equal to the *area* under the a–t graph during the same time period, Fig. 12–10b. Hence, to construct the v–t graph, we begin by first knowing the particle's initial velocity v_0 and then add to this small increments of area (Δv) determined from the a–t graph. In this manner, successive points, $v_1 = v_0 + \Delta v$, etc., for the v–t graph are determined. Notice that an algebraic addition of the area increments is necessary, since areas lying above the t axis correspond to an increase in v (''positive'' area), whereas those lying below the t axis indicate a decrease in v (''negative'' area).

If curved segments of the a–t graph can be described by a series of equations, then each of these equations may be *integrated* to yield equations describing the corresponding curved segments of the v–t graph. Hence, if the a–t graph is linear (a first-degree curve), the integration will yield a v–t graph that is parabolic (a second-degree curve), etc.

Given the v–t Graph, Construct the s–t Graph. When the v–t graph is given, Fig. 12–11a, it is possible to determine the s–t graph using $v = ds/dt$, written in integrated form as $\Delta s = \int v\,dt$. In this case the particle's displacement during a period of time is equal to the *area* under the v–t graph during the same time period, Fig. 12–11b. In the same manner as stated above, we begin by knowing the particle's initial position s_0 and add (algebraically) to this small area increments Δs determined from the v–t graph.

If it is possible to describe curved segments of the v–t graph by a series of equations, then each of these equations may be *integrated* to yield equations that describe corresponding curved segments of the s–t graph.

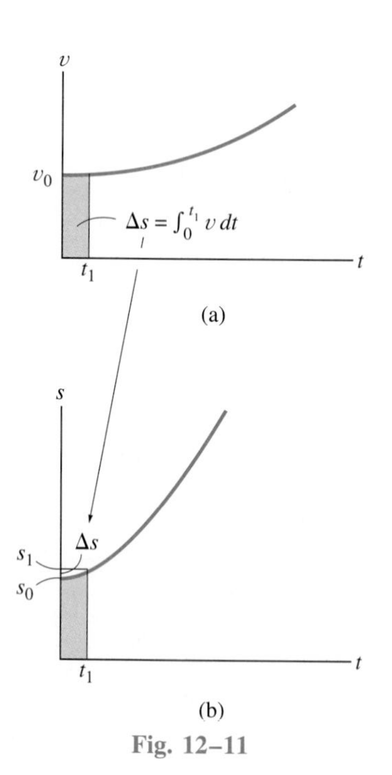

Fig. 12–11

Example 12–7

The rocket sled in Fig. 12–12*a* starts from rest and travels along a straight track such that it accelerates at a constant rate for 10 s and then decelerates at a constant rate. Draw the v–t and s–t graphs and determine the time t' needed to stop the sled. How far has the sled traveled?

SOLUTION

***v–t* Graph.** Since $dv = a\,dt$, the v–t graph is determined by integrating the straight-line segments of the a–t graph. Using the *initial condition* $v = 0$ when $t = 0$, we have

$$0 \le t < 10 \text{ s};\; a = 10; \qquad \int_0^v dv = \int_0^t 10\,dt, \qquad v = 10t$$

When $t = 10$ s, $v = 10(10) = 100$ m/s. Using this as the *initial condition* for the next time period, we have

$$10 \text{ s} < t \le t';\; a = -2; \qquad \int_{100}^v dv = \int_{10}^t -2\,dt, \qquad v = -2t + 120$$

When $t = t'$ we require $v = 0$. This yields, Fig. 12–12*b*,

$$t' = 60 \text{ s} \qquad \textit{Ans.}$$

A more direct solution for t' is possible by realizing that the area under the a–t graph is equal to the change in the sled's velocity. We require $\Delta v = 0 = A_1 + A_2$, Fig. 12–12*a*. Thus

$$0 = 101 \text{ m/s}^2(10 \text{ s}) - (-2 \text{ m/s}^2)(t' - 10 \text{ s}) = 0$$

$$t' = 60 \text{ s} \qquad \textit{Ans.}$$

***s–t* Graph.** Since $ds = v\,dt$, integrating the equations of the v–t graph yields the corresponding equations of the s–t graph. Using the *initial condition* $s = 0$ when $t = 0$, we have

$$0 \le t \le 10 \text{ s};\; v = 10t; \qquad \int_0^s ds = \int_0^t 10t\,dt, \qquad s = 5t^2$$

When $t = 10$ s, $s = 5(10)^2 = 500$ m. Using this *initial condition,*

$$10 \text{ s} \le t \le 60 \text{ s};\; v = -2t + 120; \quad \int_{500}^s ds = \int_{10}^t (-2t + 120)\,dt$$

$$s - 500 = -t^2 + 120t - [-(10)^2 + 120(10)]$$

$$s = -t^2 + 120t - 600$$

When $t' = 60$ s, the position is

$$s = -(60)^2 + 120(60) - 600 = 3000 \text{ m} \qquad \textit{Ans.}$$

The s–t graph is shown in Fig. 12–12*c*. Note that a direct solution for s when $t' = 60$ s is possible, since the *triangular area* under the v–t graph would yield the displacement $\Delta s = s - 0$ from $t = 0$ to $t' = 60$ s. Hence,

$$\Delta s = \tfrac{1}{2}(60)(100) = 3000 \text{ m} \qquad \textit{Ans.}$$

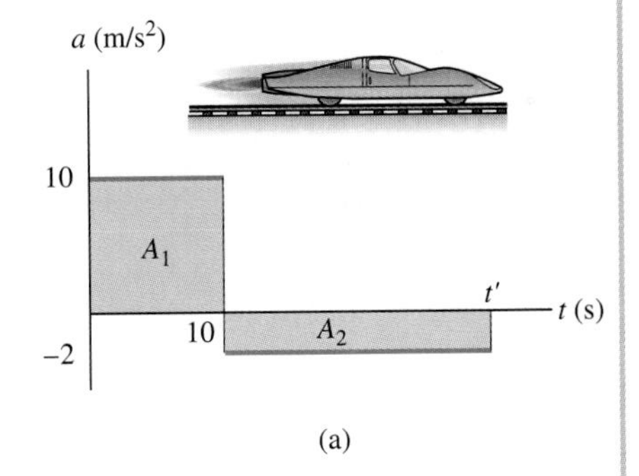

(a)

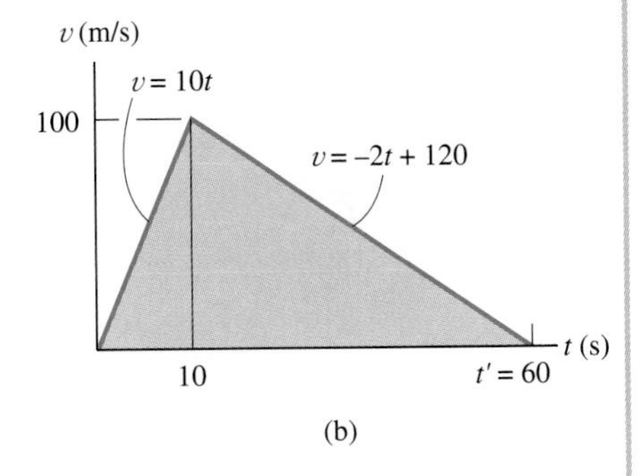

(b)

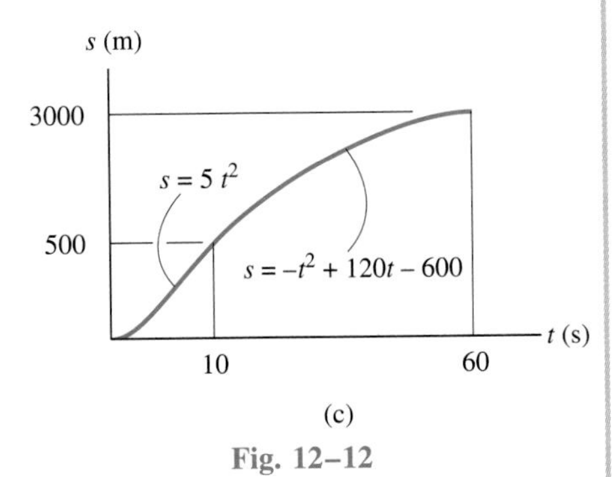

(c)

Fig. 12–12

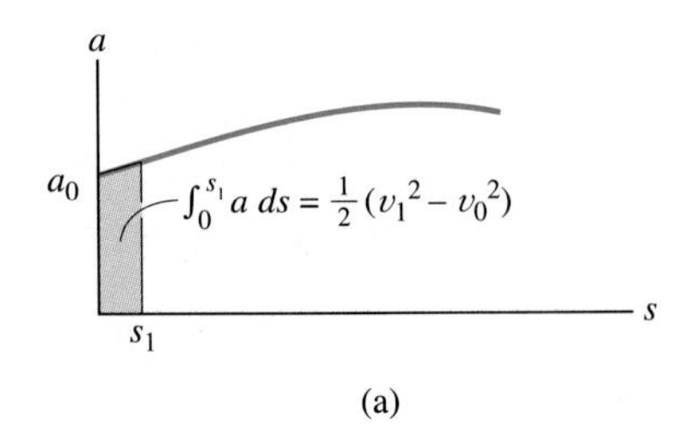

(a)

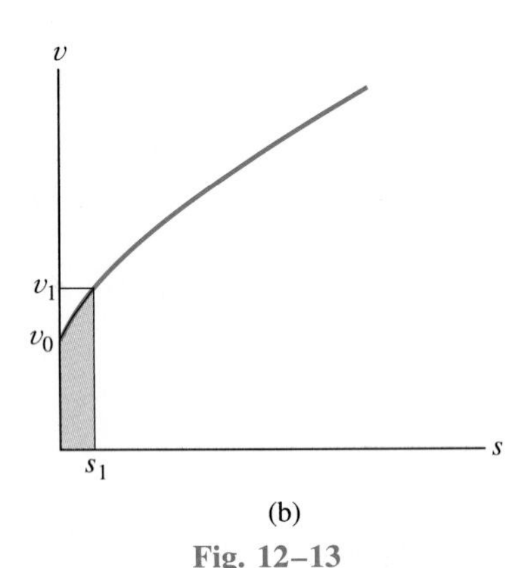

(b)

Fig. 12–13

Given the a–s Graph, Construct the v–s Graph. In some cases an a–s graph for the particle can be constructed, so that points on the v–s graph can be determined by using $v\,dv = a\,ds$. (Why?) Integrating this equation between the limits $v = v_0$ at $s = s_0$ and $v = v_1$ at $s = s_1$, we have $\frac{1}{2}(v_1^2 - v_0^2) = \int_{s_0}^{s_1} a\,ds$. Thus, the small segment of area under the a–s graph, $\int_{s_0}^{s_1} a\,ds$, shown colored in Fig. 12–13a, equals one-half the difference in the squares of the speed, $\frac{1}{2}(v_1^2 - v_0^2)$. Therefore, if the area $\int_{s_0}^{s_1} a\,ds$ is determined, it is possible to calculate the value of v_1 at s_1 if the initial value of v_0 at s_0 is known, i.e., $v_1 = (2\int_{s_0}^{s_1} a\,ds + v_0^2)^{1/2}$, Fig. 12–13$b$. Successive points on the v–s graph can be constructed in this manner starting from the initial velocity v_0.

Another way to construct the v–s graph is to first determine the equations which define the curved segments of the a–s graph. Then the corresponding equations defining the curves of the v–s graph can be obtained directly from integration, using $v\,dv = a\,ds$.

Given the v–s Graph, Construct the a–s Graph. If the v–s graph is known, the acceleration a at any position s can be determined using the following graphical procedure. At any point (s, v), Fig. 12–14a, the slope dv/ds of the v–s graph is determined. Since $a\,ds = v\,dv$ or $a = v(dv/ds)$, then since v and dv/ds are known, the value of a can be calculated, Fig. 12–14b.

We can also determine the curved segments describing the a–s graph analytically, provided the equations of the corresponding curved segments of the v–s graph are known. As above, this requires integration using $a\,ds = v\,dv$.

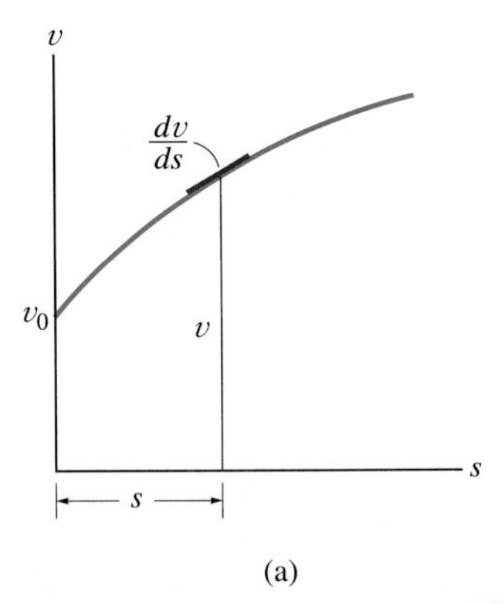

(a)

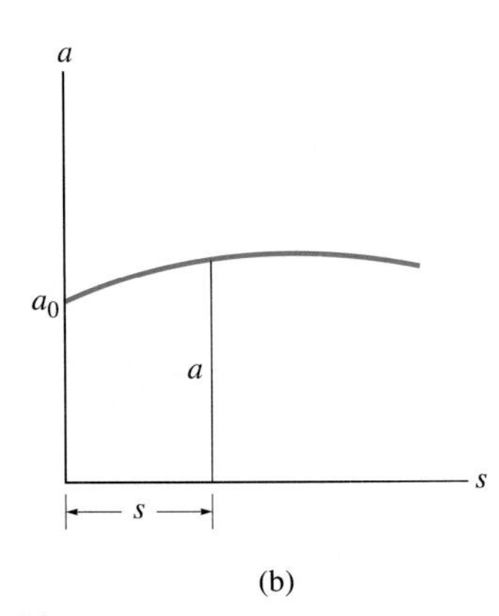

(b)

Fig. 12–14

Example 12–8

The v–s graph describing the motion of a motorcycle is shown in Fig. 12–15a. Construct the a–s graph of the motion and determine the time needed for the motorcycle to reach the position $s = 400$ ft.

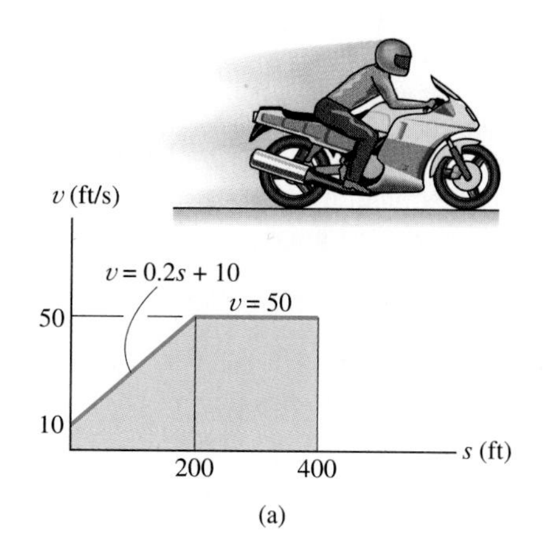

(a)

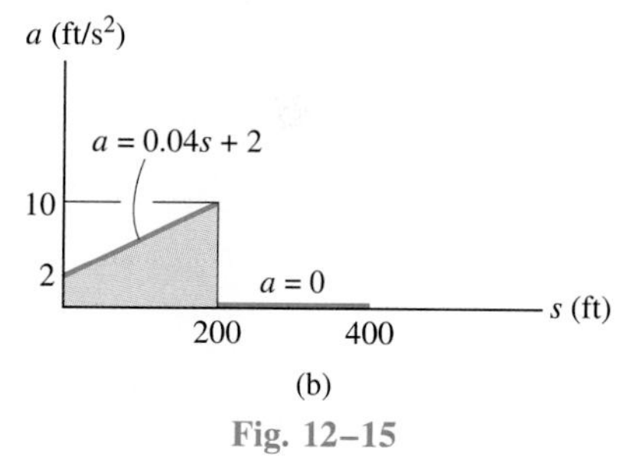

(b)

Fig. 12–15

SOLUTION

a–s Graph. Since the equations of the v–s graph are given, the a–s graph can be determined using the equation $a\,ds = v\,dv$, which yields

$0 \leq s < 200$ ft;

$$v = 0.2s + 10$$

$$a = v\frac{dv}{ds} = (0.2s + 10)\frac{d}{ds}(0.2s + 10) = 0.04s + 2$$

$200\text{ ft} < s \leq 400$ ft; $v = 50$;

$$a = v\frac{dv}{ds} = (50)\frac{d}{ds}(50) = 0$$

The results are plotted in Fig. 12–15b.

Time. The time can be obtained using the v–s graph and $v = ds/dt$, because this equation relates v, s, and t. For the first segment of motion, $s = 0$ at $t = 0$, so

$$0 \leq s < 200\text{ ft};\ v = 0.2s + 10;\ dt = \frac{ds}{v} = \frac{ds}{0.2s + 10}$$

$$\int_0^t dt = \int_0^s \frac{ds}{0.2s + 10}$$

$$t = 5\ln(0.2s + 10) - 5\ln 10$$

At $s = 200$ ft, $t = 5\ln[0.2(200) + 10] - 5\ln 10 = 8.05$ s. Therefore, for the second segment of motion,

$$200\text{ ft} < s \leq 400\text{ ft};\quad v = 50;\quad dt = \frac{ds}{v} = \frac{ds}{50}$$

$$\int_{8.05}^t dt = \int_{200}^s \frac{ds}{50}$$

$$t - 8.05 = \frac{s}{50} - 4$$

$$t = \frac{s}{50} + 4.05$$

Therefore, at $s = 400$ ft,

$$t = \frac{400}{50} + 4.05 = 12.0\text{ s}$$ *Ans.*

PROBLEMS

***12–36.** An airplane starts from rest, travels 5000 ft down a runway, and after uniform acceleration, takes off with a speed of 162 mi/h. It then climbs in a straight line with a uniform acceleration of 3 ft/s^2 until it reaches a constant speed of 220 mi/h. Draw the s-t, v-t, and a-t graphs that describe the motion.

12–37. The elevator starts from rest at the first floor of the building. It can accelerate at 5 ft/s^2 and then decelerate at 2 ft/s^2. Determine the shortest time it takes to reach a floor 40 ft above the ground. The elevator starts from rest and then stops. Draw the a-t, v-t, and s-t graphs for the motion.

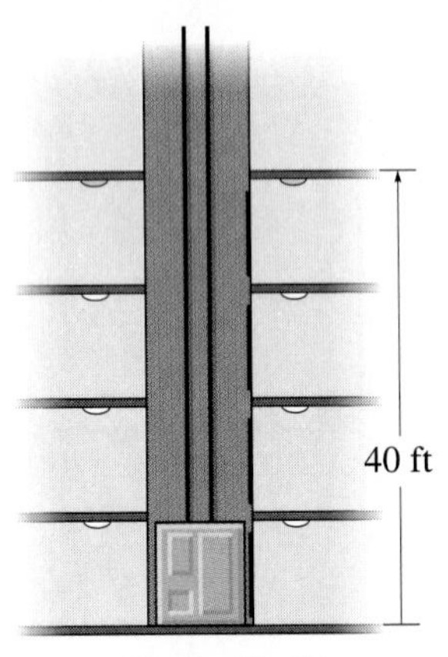

Prob. 12–37

12–38. The bobsled moves down along the straight course, such that its v-t graph is as shown. Construct the s-t and a-t graphs for the 50-s time interval. When $t = 0$, $s = 0$.

12–39. The bobsled moves down along the straight course such that its v-t graph is as shown. Determine the position and acceleration of the bobsled when $t = 15$ s and $t = 40$ s. When $t = 0$, $s = 0$.

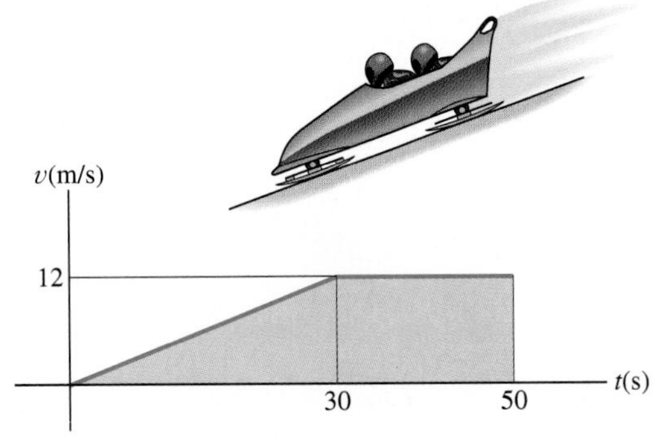

Probs. 12–38/12–39

***12–40.** The v-t graph for a particle moving through an electric field from one plate to another has the shape shown in the figure. The acceleration and deceleration that occur are constant and both have a magnitude of 4 m/s^2. If the plates are spaced 200 mm apart, determine the maximum velocity v_{max} and the time t' for the particle to travel from one plate to the other. Also draw the s-t graph. When $t = t'/2$ the particle is at $s = 100$ mm.

***12–41.** The v-t graph for a particle moving through an electric field from one plate to another has the shape shown in the figure, where $t' = 0.2$ s and $v_{max} = 10$ m/s. Draw the s-t and a-t graphs for the particle. When $t = t'/2$ the particle is at $s = 0.5$ m.

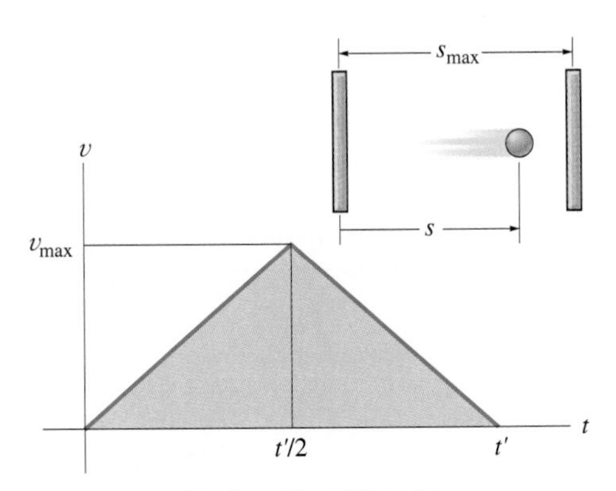

Probs. 12–40/12–41

12–42. The v-t graph for the motion of a train as it moves from station A to station B is shown. Draw the a-t graph and determine the average speed for the train and the distance between the stations.

12–43. The v-t graph for the motion of a train as it moves from station A to station B is shown. Draw the s-t graph and determine the acceleration of the train when $t = 50$ s and $t = 100$ s.

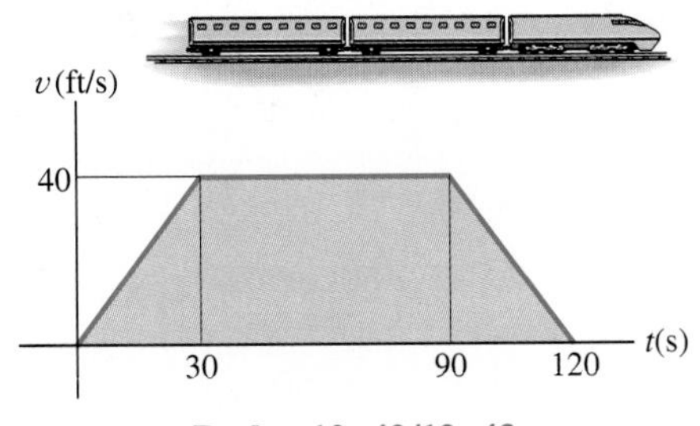

Probs. 12–42/12–43

*12–44. A motorcycle starts from rest at $s = 0$ and travels along a straight road with the speed shown by the v-t graph. Determine the total distance the motorcycle travels until it stops when $t = 15$ s. Also plot the a-t and s-t graphs.

12–45. A motorcycle starts from rest at $s = 0$ and travels along a straight road with the speed shown by the v-t graph. Determine the motorcycle's acceleration and position when $t = 8$ s and $t = 12$ s.

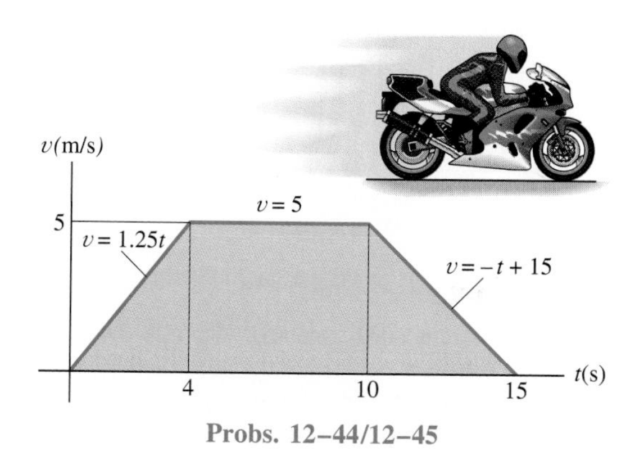

Probs. 12–44/12–45

12–47. From experimental data, the motion of a jet plane while traveling along a runway is defined by the v-t graph shown. Construct the s-t and a-t graphs for the motion. The plane starts from rest.

*12–48. From experimental data, the motion of a jet plane while traveling along a runway is defined by the v-t graph shown. Determine the plane's acceleration and position when $t = 10$ s and $t = 25$ s. The plane starts from rest.

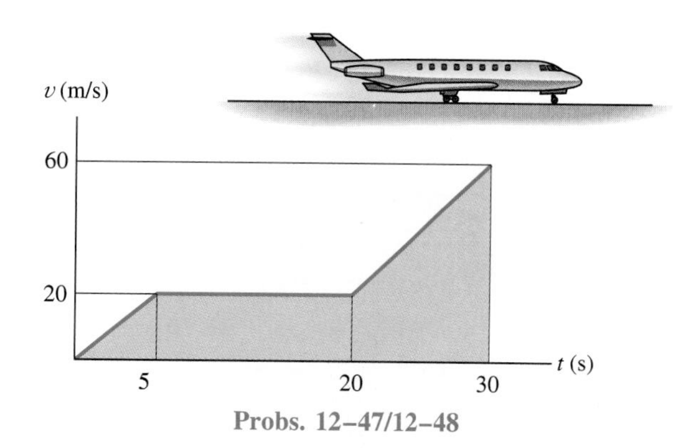

Probs. 12–47/12–48

12–46. An airplane lands on the straight runway, originally traveling at 110 ft/s when $s = 0$. If it is subjected to the decelerations shown, determine the time t' needed to stop the plane and construct the s-t graph for the motion.

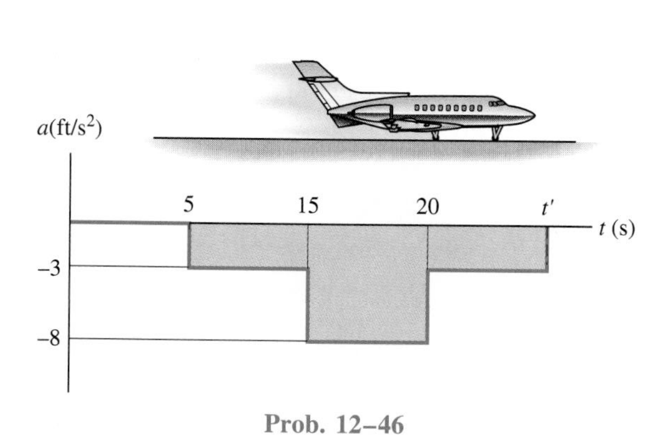

Prob. 12–46

12–49. The s-t graph for a train traveling along a straight track has been experimentally determined. From the data, construct the v-t and a-t graphs for the motion; $0 \leq t \leq 60$ s. For $0 \leq t \leq 30$ s, the curve is $s = (0.4t^2)$ m, where t is in seconds.

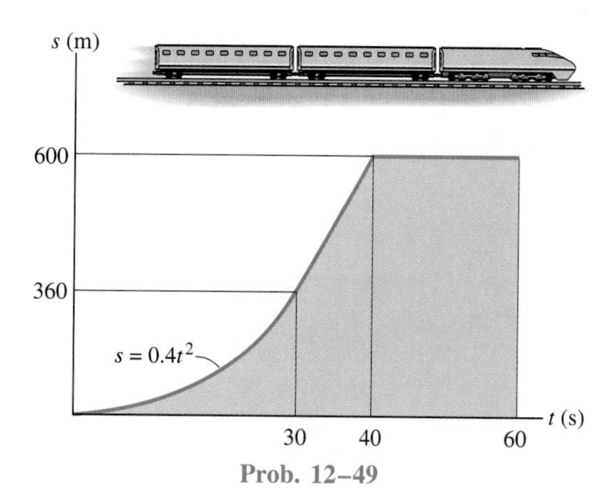

Prob. 12–49

■**12–50.** The a-s graph for a rocket moving along a straight track has been experimentally determined. If the rocket starts at $s = 0$ when $v = 0$, determine its speed when it is at $s = 75$ ft, and 125 ft, respectively. Use Simpson's rule with $n = 100$ to evaluate v at $s = 125$ ft.

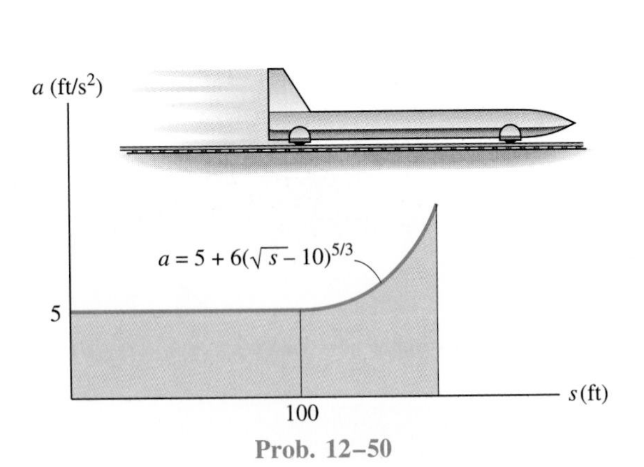

Prob. 12–50

12–51. A two-stage rocket is fired vertically from rest with the acceleration shown. After 15 s the first stage A burns out and the second stage B ignites. Plot the v-t and s-t graphs which describe the motion of the second stage for $0 \leq t \leq 40$ s.

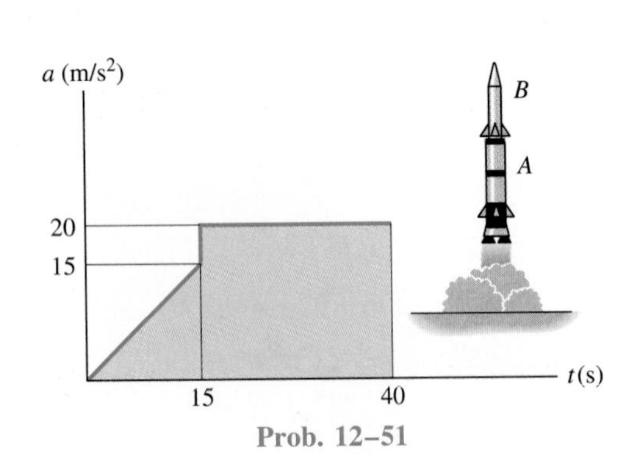

Prob. 12–51

***12–52.** A car travels along a straight road with the speed shown by the v-t graph. Determine the total distance the car travels until it stops when $t = 48$ s. Also plot the s-t and a-t graphs.

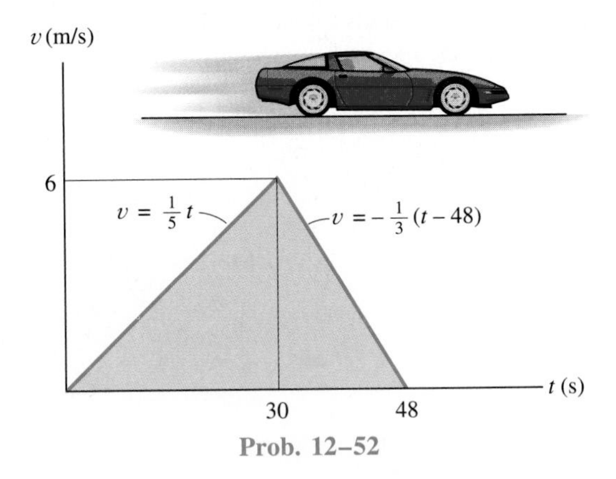

Prob. 12–52

12–53. A freight train starts from rest when $t = 0$ and travels on a straight track with a constant acceleration of 0.75 ft/s^2. When $t = t'$ it maintains a constant speed so that when $t = 160$ s the train has traveled 1500 ft. Determine the time t' and draw the v-t graph for the motion.

12–54. The car is originally traveling along a straight road at 15 ft/s when it is subjected to the motion shown by the a-t graph. Determine the car's maximum speed and the time t' when it stops.

12–55. The car is originally at rest when it is subjected to the motion shown by the a-t graph. Draw the v-t and s-t graphs for $0 \leq t \leq 8$ s.

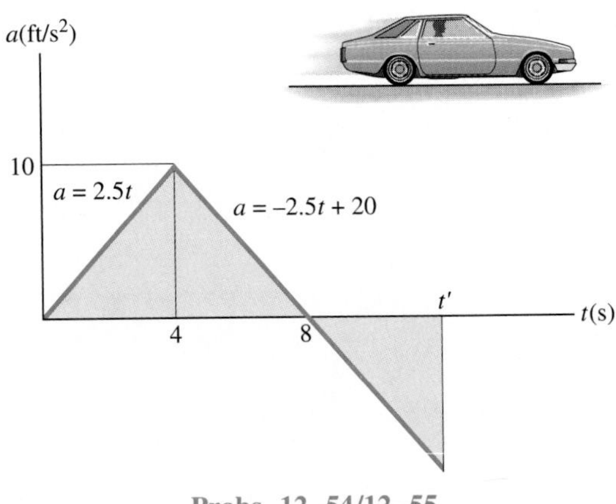

Probs. 12–54/12–55

***12–56.** Two rockets start from rest at the same elevation. Rocket A accelerates vertically at 20 m/s^2 for 12 s and then maintains a constant speed. Rocket B accelerates at 15 m/s^2 until reaching a constant speed of 150 m/s. Construct the a-t, v-t, and s-t graphs for each rocket until $t = 20$ s. What is the distance between the rockets when $t = 20$ s?

12–57. The race car starts from rest and travels along a straight road until it reaches a speed of 26 m/s in 8 s as shown on the v-t graph. The flat part of the graph is caused by shifting gears. Draw the a-t graph and determine the maximum acceleration of the car.

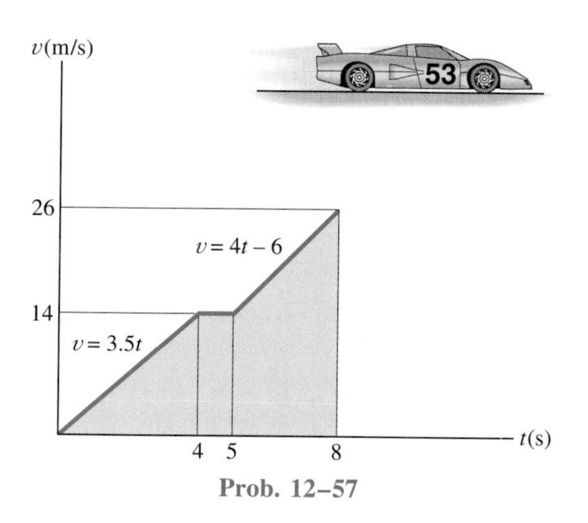

Prob. 12–57

12–58. A two-stage rocket is fired vertically from rest at $s = 0$ with an acceleration as shown. After 30 s the first stage A burns out and the second stage B ignites. Plot the v-t and s-t graphs which describe the motion of the second stage for $0 \leq t \leq 60$ s.

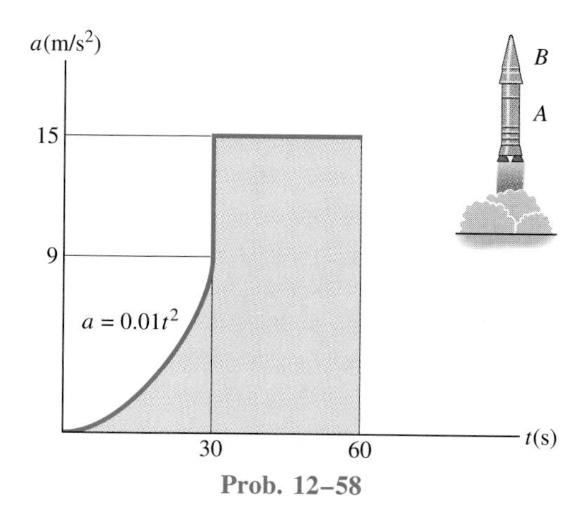

Prob. 12–58

12–59. The jet plane starts from rest at $s = 0$ and is subjected to the acceleration shown. Construct the v-s graph and determine the time needed to travel 500 ft.

***12–60.** The jet plane starts from rest at $s = 0$ and is subjected to the acceleration shown. Determine the speed of the plane when it has traveled 200 ft. Also, how much time is required for it to travel 200 ft?

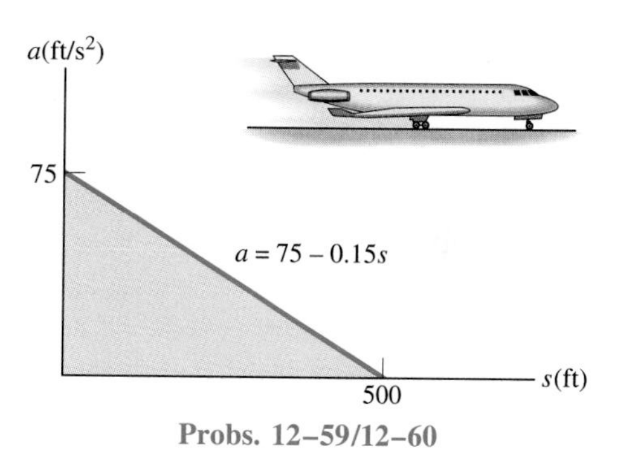

Probs. 12–59/12–60

12–61. The v-s graph for a go-cart traveling on a straight road is shown. Determine the acceleration of the go-cart at $s = 50$ m and $s = 150$ m. Draw the a-s graph.

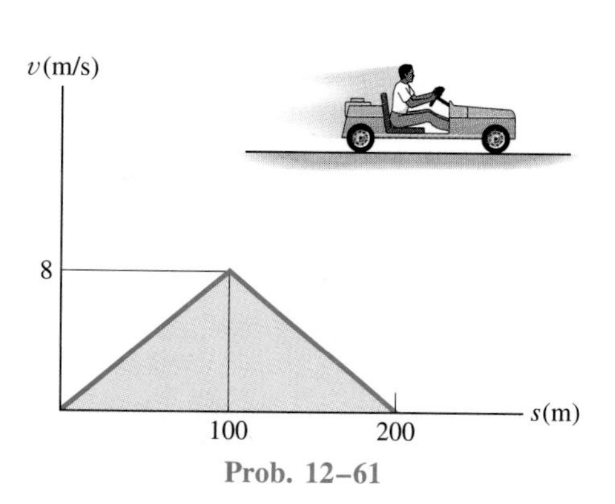

Prob. 12–61

12–62. The v-s graph for the car is given for the first 500 ft of its motion. Plot the a-s graph for $0 \leq s \leq 500$ ft. How long does it take to travel 500 ft? The car starts at $s = 0$ when $t = 0$.

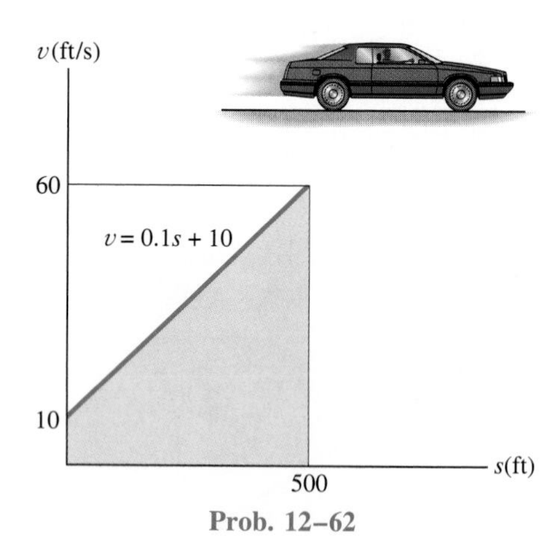

Prob. 12–62

12–63. The car starts from rest at $s = 0$ and is subjected to an acceleration shown by the a-s graph. Draw the v-s graph and determine the time needed to travel 200 ft.

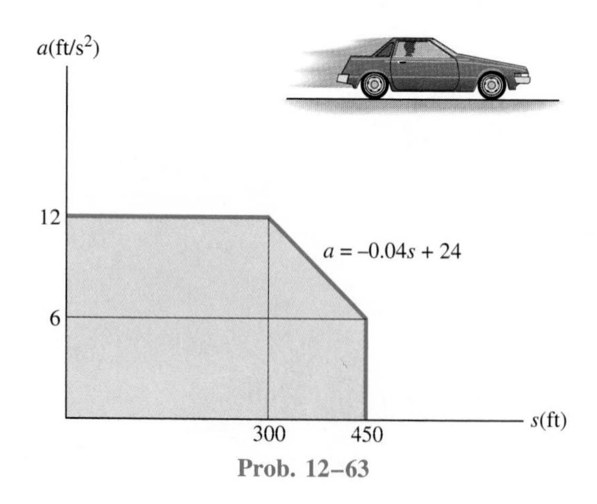

Prob. 12–63

***12–64.** The v-s graph for an airplane traveling on a straight runway is shown. Determine the acceleration of the plane at $s = 100$ m and $s = 150$ m. Draw the a-s graph.

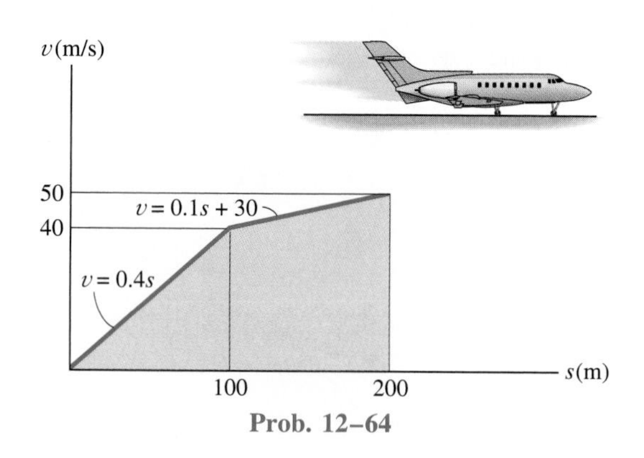

Prob. 12–64

12–65 Starting from rest at $s = 0$, a boat travels in a straight line with an acceleration as shown by the a-s graph. Determine the boat's speed when $s = 40$, 90, and 200 ft.

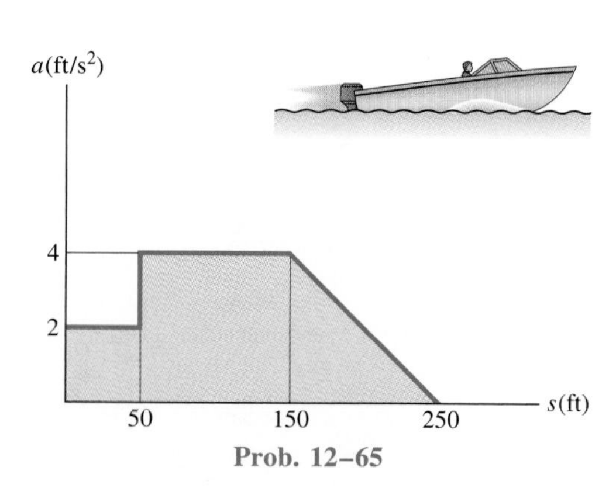

Prob. 12–65

12.3 General Curvilinear Motion

Curvilinear motion occurs when the particle moves along a curved path. Since this path is often described in three dimensions, vector analysis will be used to formulate the particle's position, velocity, and acceleration.* In this section the general aspects of curvilinear motion are discussed, and in subsequent sections three types of coordinate systems often used to analyze this motion will be introduced.

Position. Consider a particle located at point P on a space curve defined by the path function s, Fig. 12–16a. The position of the particle, measured from a fixed point O, will be designated by the *position vector* $\mathbf{r} = \mathbf{r}(t)$. This vector is a function of time since, in general, both its magnitude and direction change as the particle moves along the curve.

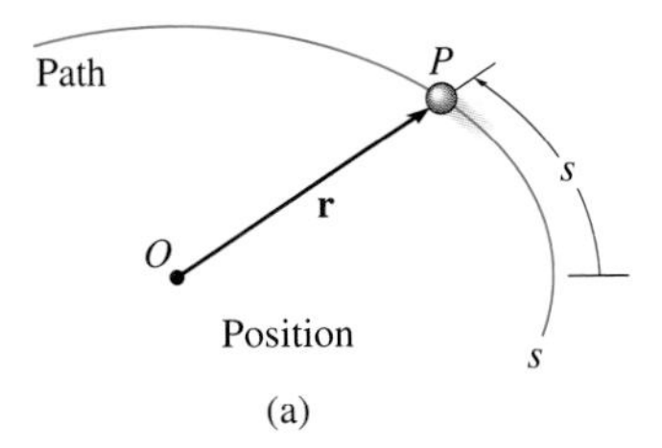

(a)

Displacement. Suppose that during a small time interval Δt the particle moves a distance Δs along the curve to a new position P', defined by $\mathbf{r}' = \mathbf{r} + \Delta\mathbf{r}$, Fig. 12–16$b$. The *displacement* $\Delta\mathbf{r}$ represents the change in the particle's position and is determined by vector subtraction; i.e., $\Delta\mathbf{r} = \mathbf{r}' - \mathbf{r}$.

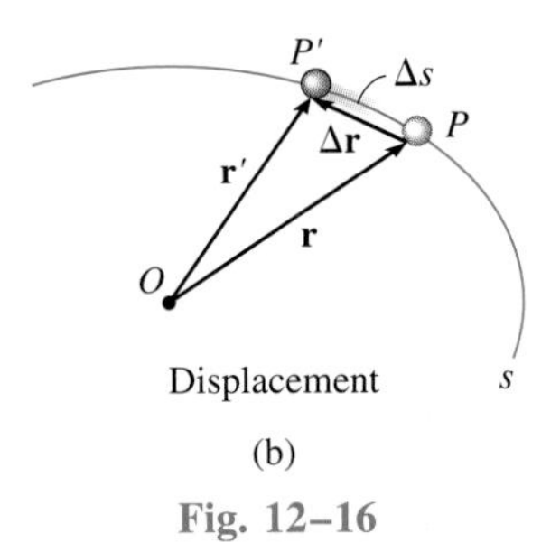

(b)

Fig. 12–16

*A summary of some of the important concepts of vector analysis is given in Appendix C.

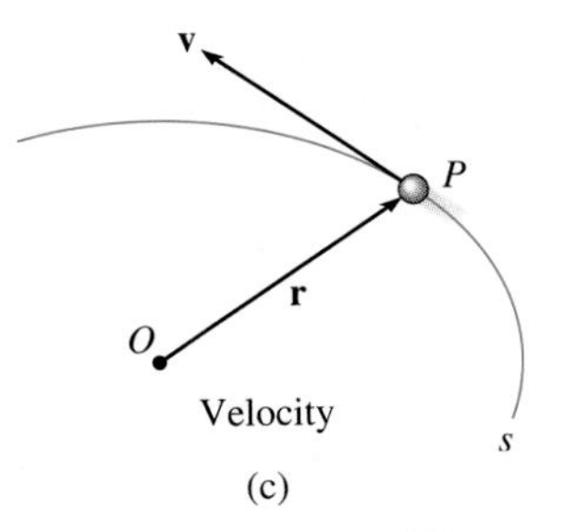

(c)

Fig. 12–16 *(cont'd)*

Velocity. During the time Δt, the *average velocity* of the particle is defined as

$$\mathbf{v}_{avg} = \frac{\Delta \mathbf{r}}{\Delta t}$$

The *instantaneous velocity* is determined from this equation by letting $\Delta t \rightarrow 0$ and consequently the direction of $\Delta \mathbf{r}$ *approaches* the *tangent* to the curve at point P. Hence, $\mathbf{v} = \lim_{\Delta t \rightarrow 0} (\Delta \mathbf{r}/\Delta t)$ or

$$\mathbf{v} = \frac{d\mathbf{r}}{dt} \tag{12-7}$$

Since $d\mathbf{r}$ will be tangent to the curve at P, the *direction* of $\mathbf{v}$ is also *tangent to the curve,* Fig. 12–16*c*. The *magnitude* of $\mathbf{v}$, which is called the *speed,* may be obtained by noting that the magnitude of the displacement $\Delta \mathbf{r}$ is the length of the straight line segment from P to P', Fig. 12–16*b*. Realizing that this length, Δr, approaches the arc length Δs as $\Delta t \rightarrow 0$, we have $v = \lim_{\Delta t \rightarrow 0} (\Delta r/\Delta t) = \lim_{\Delta t \rightarrow 0} (\Delta s/\Delta t)$, or

$$v = \frac{ds}{dt} \tag{12-8}$$

Thus, the *speed* can be obtained by differentiating the path function s with respect to time.

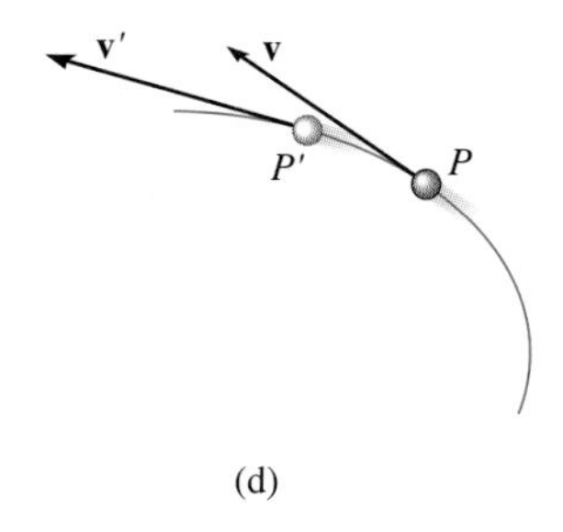

(d)

Acceleration. If the particle has a velocity $\mathbf{v}$ at time t and a velocity $\mathbf{v}' = \mathbf{v} + \Delta\mathbf{v}$ at $t + \Delta t$, Fig. 12–16*d*, then the *average acceleration* of the particle during the time interval Δt is

$$\mathbf{a}_{avg} = \frac{\Delta \mathbf{v}}{\Delta t}$$

where $\Delta\mathbf{v} = \mathbf{v}' - \mathbf{v}$. To study this time rate of change, the two velocity vectors in Fig. 12–16*d* are plotted in Fig. 12–16*e* such that their tails are located at the fixed point O' and their arrowheads touch points on the dashed curve. This curve is called a *hodograph,* and when constructed, it describes the locus of points for the arrowhead of the velocity vector in the same manner as the *path s* describes the locus of points for the arrowhead of the position vector, Fig. 12–16*a*.

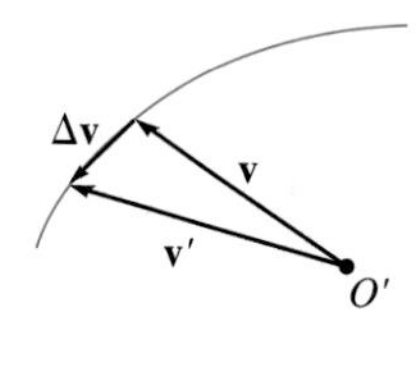

(e)

To obtain the *instantaneous acceleration,* let $\Delta t \to 0$ in the above equation. In the limit $\Delta\mathbf{v}$ will approach the *tangent to the hodograph* and so $\mathbf{a} = \lim_{\Delta t \to 0} (\Delta\mathbf{v}/\Delta t)$, or

$$\mathbf{a} = \frac{d\mathbf{v}}{dt} \qquad (12\text{–}9)$$

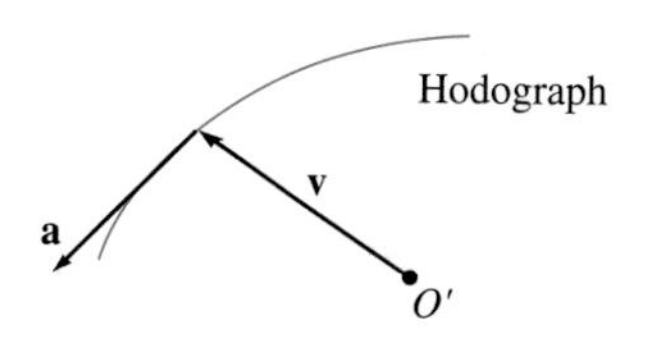

(f)

Substituting Eq. 12–7 into this result, we can also write

$$\mathbf{a} = \frac{d^2\mathbf{r}}{dt^2}$$

By definition of the derivative, $\mathbf{a}$ acts *tangent to the hodograph,* Fig. 12–16*f*, and therefore, *in general,* $\mathbf{a}$ *is not tangent to the path of motion,* Fig. 12–16*g*. To clarify this point, realize that $\Delta\mathbf{v}$ and consequently $\mathbf{a}$ must account for the change made in *both* the magnitude *and* direction of the velocity $\mathbf{v}$ as the particle moves from P to P', Fig. 12–16*d*. Just a magnitude change increases (or decreases) the "length" of $\mathbf{v}$, and this in itself would allow $\mathbf{a}$ to remain tangent to the path. However, in order for the particle to follow the path, the directional change always swings the velocity vector toward the "inside" or "concave side" of the path, and therefore $\mathbf{a}$ *cannot* remain tangent to the path. In summary then, $\mathbf{v}$ is always tangent to the *path* and $\mathbf{a}$ is always tangent to the *hodograph.*

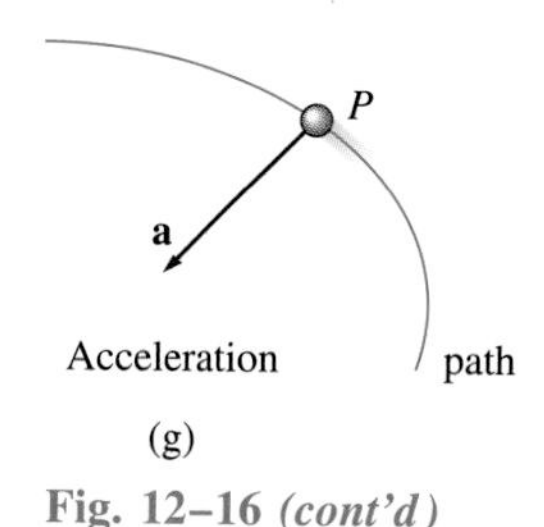

(g)

Fig. 12–16 *(cont'd)*

12.4 Curvilinear Motion: Rectangular Components

Occasionally the motion of a particle can best be described along a path that is represented using a fixed x, y, z frame of reference.

Position. If at a given instant the particle P is at point (x, y, z) on the curved path s, Fig. 12–17a, its location is then defined by the *position vector*

$$\mathbf{r} = x\mathbf{i} + y\mathbf{j} + z\mathbf{k} \tag{12–10}$$

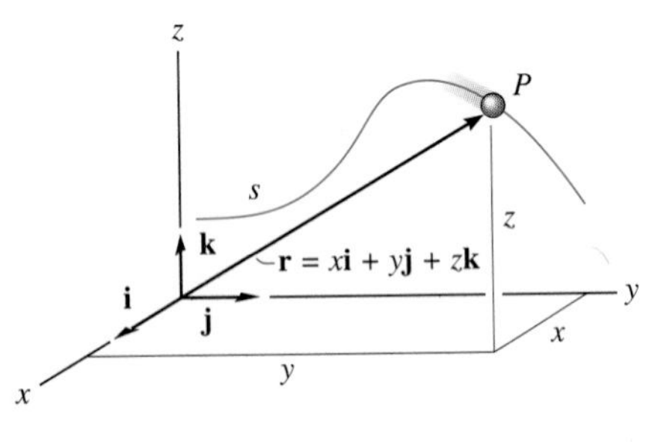

Position

(a)

Because of the particle motion and the shape of the path, the x, y, z components of $\mathbf{r}$ are generally all functions of time; i.e., $x = x(t)$, $y = y(t)$, and $z = z(t)$, so that $\mathbf{r} = \mathbf{r}(t)$.

In accordance with the discussion in Appendix C, the *magnitude* of $\mathbf{r}$ is *always positive* and defined from Eq. C–3 as

$$r = \sqrt{x^2 + y^2 + z^2}$$

The *direction* of $\mathbf{r}$ is specified by the components of the unit vector $\mathbf{u}_r = \mathbf{r}/r$.

Velocity. The first time derivative of $\mathbf{r}$ yields the velocity $\mathbf{v}$ of the particle. Hence,

$$\mathbf{v} = \frac{d\mathbf{r}}{dt} = \frac{d}{dt}(x\mathbf{i}) + \frac{d}{dt}(y\mathbf{j}) + \frac{d}{dt}(z\mathbf{k})$$

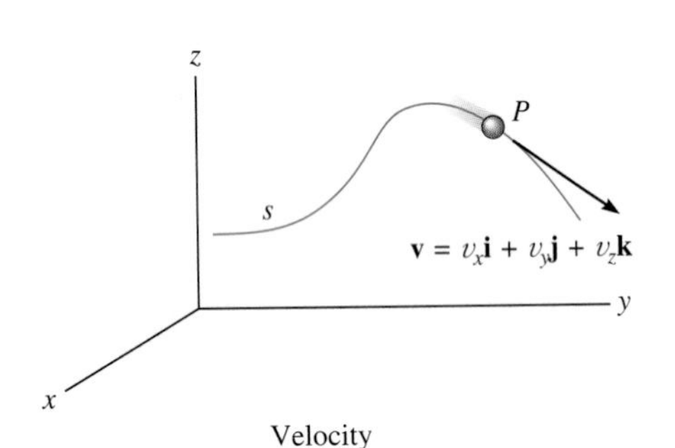

Velocity

(b)

Fig. 12–17

When taking the derivative, it is necessary to account for changes in *both* the magnitude and direction of each of the vector's components. The derivative of the $\mathbf{i}$ component of $\mathbf{v}$ is therefore

$$\frac{d}{dt}(x\mathbf{i}) = \frac{dx}{dt}\mathbf{i} + x\frac{d\mathbf{i}}{dt}$$

The second term on the right side is zero, since the x, y, z reference frame is *fixed,* and therefore the *direction* (and the *magnitude*) of $\mathbf{i}$ does not change with time. Differentiation of the $\mathbf{j}$ and $\mathbf{k}$ components may be carried out in a similar manner, which yields the final result,

$$\mathbf{v} = \frac{d\mathbf{r}}{dt} = v_x\mathbf{i} + v_y\mathbf{j} + v_z\mathbf{k} \tag{12–11}$$

where

$$\begin{aligned} v_x &= \dot{x} \\ v_y &= \dot{y} \\ v_z &= \dot{z} \end{aligned} \tag{12–12}$$

The "dot" notation $\dot{x}$, $\dot{y}$, $\dot{z}$ represents the first time derivatives of the parametric equations $x = x(t)$, $y = y(t)$, and $z = z(t)$, respectively.

The velocity has a *magnitude* defined as the positive value of

$$v = \sqrt{v_x^2 + v_y^2 + v_z^2}$$

and a *direction* that is specified by the components of the unit vector $\mathbf{u}_v = \mathbf{v}/v$. This direction is *always tangent to the path,* as shown in Fig. 12–17*b*.

Acceleration. The acceleration of the particle is obtained by taking the first time derivative of Eq. 12–11 (or the second time derivative of Eq. 12–10). Using dots to represent the derivatives of the components, we have

$$\mathbf{a} = \frac{d\mathbf{v}}{dt} = a_x\mathbf{i} + a_y\mathbf{j} + a_z\mathbf{k} \qquad (12\text{–}13)$$

where

$$\begin{aligned} a_x &= \dot{v}_x = \ddot{x} \\ a_y &= \dot{v}_y = \ddot{y} \\ a_z &= \dot{v}_z = \ddot{z} \end{aligned} \qquad (12\text{–}14)$$

Here a_x, a_y, and a_z represent, respectively, the first time derivatives of the functions $v_x = v_x(t)$, $v_y = v_y(t)$, and $v_z = v_z(t)$, or the second time derivatives of the functions $x = x(t)$, $y = y(t)$, and $z = z(t)$.

The acceleration has a *magnitude* defined as the positive value of

$$a = \sqrt{a_x^2 + a_y^2 + a_z^2}$$

and a *direction* specified by the components of the unit vector $\mathbf{u}_a = \mathbf{a}/a$. Since $\mathbf{a}$ represents the time rate of *change* in velocity, in general $\mathbf{a}$ will *not* be tangent to the path traveled by the particle, Fig. 12–17*c*.

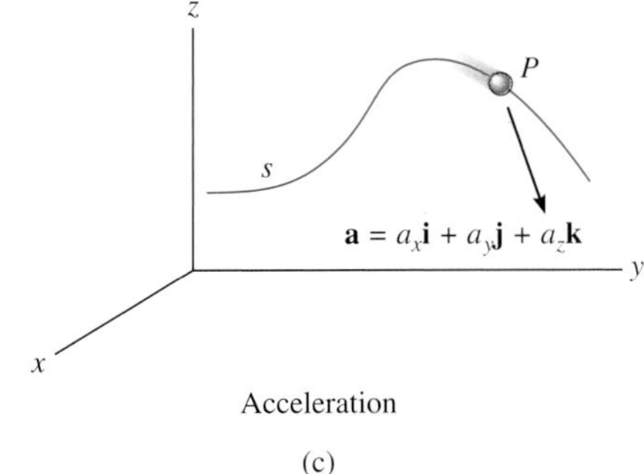

Acceleration

(c)

Fig. 12–17 ***(cont'd)***

PROCEDURE FOR ANALYSIS

Coordinate System. A rectangular coordinate system can be used to solve problems when the motion can conveniently be expressed in terms of its x, y, z components.

Kinematic Quantities. Since *rectilinear motion* occurs along each axis, a description of the motion of each component can be determined using $v = ds/dt$ and $a = dv/dt$, as outlined in Sec. 12.1 and formalized above. Also, if the motion is not expressed as a parameter of time, the equation $a\,ds = v\,dv$ can be used. Once the x, y, z components of $\mathbf{v}$ and $\mathbf{a}$ have been determined, the magnitudes of these vectors are computed from the Pythagorean theorem, Eq. C–3, and their directions from the components of their unit vectors, Eqs. C–4 and C–5.

Example 12–9

At any instant the horizontal position of the weather balloon in Fig. 12–18a is defined by $x = (30t)$ ft where t is given in seconds. If the equation of the path is $y = x^2/100$, determine *(a)* the distance of the balloon from the station at A when $t = 2$ s, *(b)* the magnitude and direction of the velocity when $t = 2$ s, and *(c)* the magnitude and direction of the acceleration when $t = 2$ s.

SOLUTION

Position. When $t = 2$ s, $x = 30(2)$ ft $= 60$ ft, and so

$$y = (60)^2/100 = 36 \text{ ft}$$

The straight-line distance from A to B is therefore

$$r = \sqrt{(60)^2 + (36)^2} = 70.0 \text{ ft} \qquad \textit{Ans.}$$

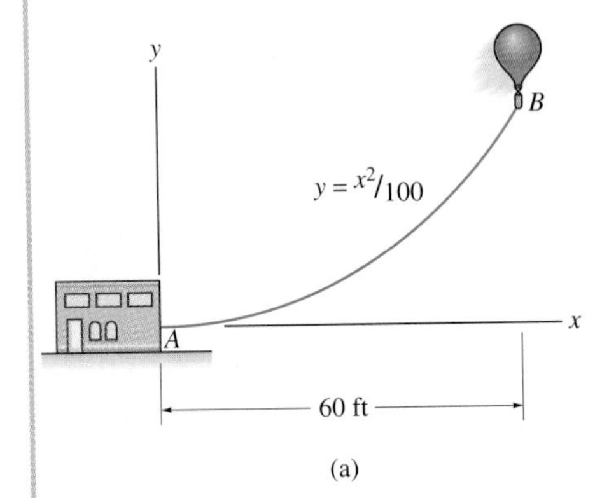

(a)

Velocity. Using Eqs. 12–12, and application of the chain rule of calculus the components of velocity when $t = 2$ s are

$$v_x = \dot{x} = \frac{d}{dt}(30t) = 30 \text{ ft/s} \rightarrow$$

$$v_y = \dot{y} = 2x\dot{x}/100 = 2(60)(30)/100 = 36 \text{ ft/s} \uparrow$$

When $t = 2$ s, the magnitude of velocity is therefore

$$v = \sqrt{(30)^2 + (36)^2} = 46.9 \text{ ft/s} \qquad \textit{Ans.}$$

The direction is tangent to the path, Fig. 12–18b, where

$$\theta_v = \tan^{-1}\frac{v_y}{v_x} = \tan^{-1}\frac{36}{30} = 50.2^\circ \qquad \textit{Ans.}$$

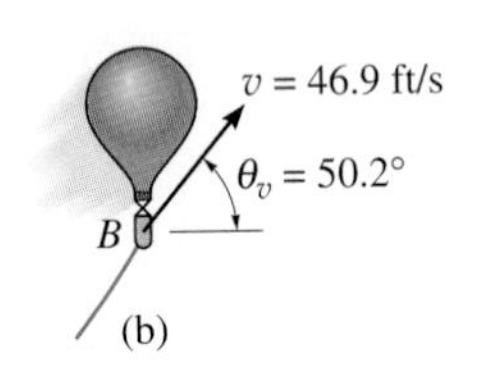

(b)

Acceleration. The components of acceleration are determined from Eqs. 12–14, and application of the chain rule. We have

$$a_x = \dot{v}_x = 0$$

$$a_y = \dot{v}_y = \frac{d}{dt}(2x\dot{x}/100) = 2(\dot{x})\dot{x}/100 + 2x(\ddot{x})/100$$

$$= 2(30)^2/100 + 2(60)(0)/100 = 18 \text{ ft/s}^2 \uparrow$$

Thus

$$a = \sqrt{(0)^2 + (18)^2} = 18 \text{ ft/s}^2 \qquad \textit{Ans.}$$

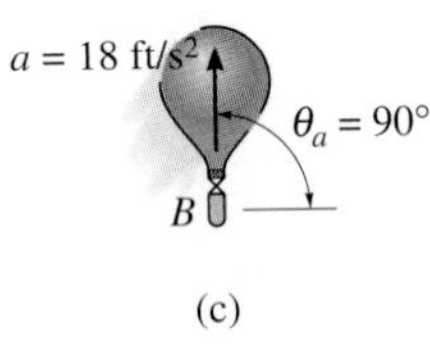

(c)

Fig. 12–18

The direction of **a**, as shown in Fig. 12–18c, is

$$\theta_a = \tan^{-1}\frac{18}{0} = 90^\circ \qquad \textit{Ans.}$$

Note: It is also possible to obtain v_y and a_y by first expressing $y = f(t) = (30t)^2/100 = 9t^2$ and then taking successive time derivatives.

Example 12–10

The motion of a box B moving along the spiral conveyor shown in Fig. 12–19 is defined by the position vector $\mathbf{r} = \{0.5 \sin (2t)\mathbf{i} + 0.5 \cos (2t)\mathbf{j} - 0.2t\mathbf{k}\}$ m, where t is given in seconds and the arguments for sine and cosine are given in radians (π rad $= 180°$). Determine the location of the box when $t = 0.75$ s and the magnitude of its velocity and acceleration at this instant.

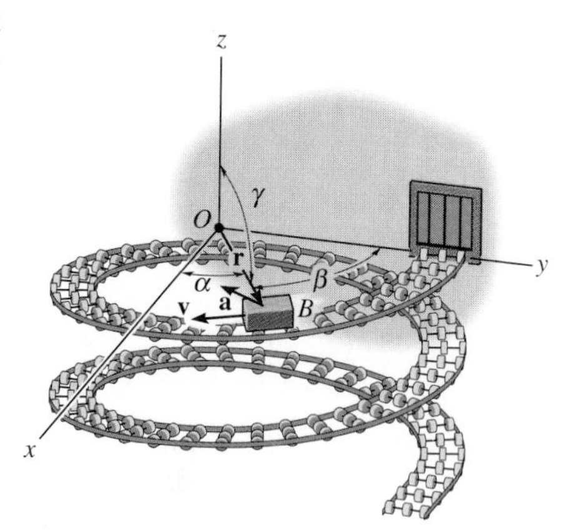

Fig. 12–19

SOLUTION

Position. Evaluating $\mathbf{r}$ when $t = 0.75$ s yields

$$\mathbf{r}|_{t=0.75\,\text{s}} = \{0.5 \sin (1.5 \text{ rad})\mathbf{i} + 0.5 \cos (1.5 \text{ rad})\mathbf{j} - 0.2(0.75)\mathbf{k}\} \text{ m}$$
$$= \{0.499\mathbf{i} + 0.035\mathbf{j} - 0.150\mathbf{k}\} \text{ m} \qquad \textit{Ans.}$$

The distance of the box from the origin O is

$$r = \sqrt{(0.499)^2 + (0.035)^2 + (-0.150)^2} = 0.522 \text{ m} \qquad \textit{Ans.}$$

The direction of $\mathbf{r}$ is obtained from the components of the unit vector,

$$\mathbf{u}_r = \frac{\mathbf{r}}{r} = \frac{0.499}{0.522}\mathbf{i} + \frac{0.035}{0.522}\mathbf{j} - \frac{0.150}{0.522}\mathbf{k}$$
$$= 0.956\mathbf{i} + 0.067\mathbf{j} - 0.287\mathbf{k}$$

Hence, the coordinate direction angles α, β, and γ, Fig. 12–19, are

$$\alpha = \cos^{-1}(0.956) = 17.1° \qquad \textit{Ans.}$$
$$\beta = \cos^{-1}(0.067) = 86.2° \qquad \textit{Ans.}$$
$$\gamma = \cos^{-1}(-0.287) = 106° \qquad \textit{Ans.}$$

Velocity. The velocity is defined by

$$\mathbf{v} = \frac{d\mathbf{r}}{dt} = \frac{d}{dt}[0.5 \sin (2t)\mathbf{i} + 0.5 \cos (2t)\mathbf{j} - 0.2t\mathbf{k}]$$
$$= \{1 \cos (2t)\mathbf{i} - 1 \sin (2t)\mathbf{j} - 0.2\mathbf{k}\} \text{ m/s}$$

Hence, when $t = 0.75$ s the magnitude of velocity, or speed, is

$$v = \sqrt{v_x^2 + v_y^2 + v_z^2}$$
$$= \sqrt{[1 \cos (1.5 \text{ rad})]^2 + [-1 \sin (1.5 \text{ rad})]^2 + (-0.2)^2}$$
$$= 1.02 \text{ m/s} \qquad \textit{Ans.}$$

The velocity is tangent to the path as shown in Fig. 12–19. Its coordinate direction angles can be determined from $\mathbf{u}_v = \mathbf{v}/v$.

Acceleration. The acceleration $\mathbf{a}$ of the box, which is shown in Fig. 12–19, is *not* tangent to the path. Show that

$$\mathbf{a} = \frac{d\mathbf{v}}{dt} = \{-2 \sin (2t)\mathbf{i} - 2 \cos (2t)\mathbf{j}\} \text{ m/s}^2 \Big|_{t=0.75\,\text{s}} = 2 \text{ m/s}^2 \qquad \textit{Ans.}$$

12.5 Motion of a Projectile

The free-flight motion of a projectile is often studied in terms of its rectangular components, since the projectile's acceleration *always* acts in the vertical direction. To illustrate the concepts involved in the kinematic analysis, consider a projectile launched at point (x_0, y_0), as shown in Fig. 12–20. The path is defined in the x–y plane such that the initial velocity is $\mathbf{v}_0$, having components $(\mathbf{v}_x)_0$ and $(\mathbf{v}_y)_0$. When air resistance is neglected, the only force acting on the projectile is its weight, which causes the projectile to have a *constant downward acceleration* of approximately $a_c = g = 9.81 \text{ m/s}^2$ or $g = 32.2 \text{ ft/s}^2$.* Hence, $a_y = -g$ and $a_x = 0$.

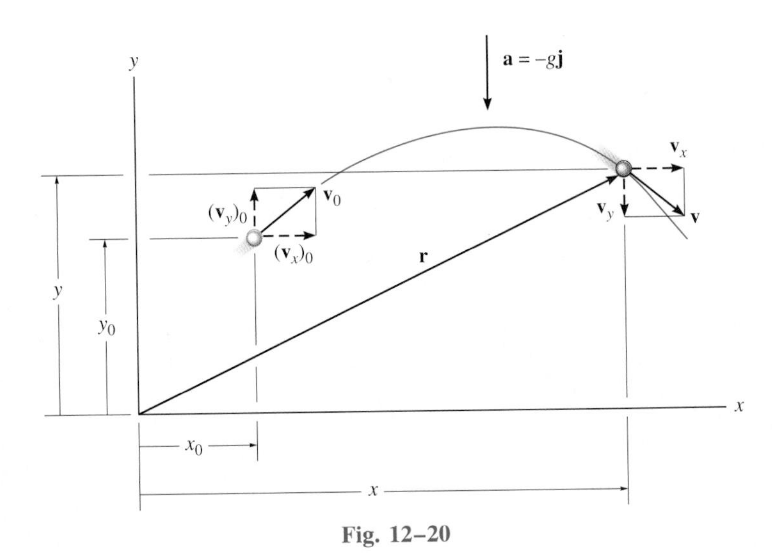

Fig. 12–20

Horizontal Motion. Since $a_x = 0$, application of the constant acceleration equations, 12–4 to 12–6, yields

$$(\overset{+}{\rightarrow})\, v = v_0 + a_c t; \qquad v_x = (v_x)_0$$

$$(\overset{+}{\rightarrow})\, x = x_0 + v_0 t + \tfrac{1}{2} a_c t^2; \qquad x = x_0 + (v_x)_0 t$$

$$(\overset{+}{\rightarrow})\, v^2 = v_0^2 + 2a_c(s - s_0); \qquad v_x = (v_x)_0$$

The first and last equations indicate that *the horizontal component of velocity always remains constant during the motion.*

Vertical Motion. Since the positive y axis is directed upward, then $a_y = -g$. Applying Eqs. 12–4 to 12–6, we get

*This assumes that the earth's gravitational field does not vary with altitude (see Example 13–2).

$$(+\uparrow)v = v_0 + a_c t; \qquad v_y = (v_y)_0 - gt$$
$$(+\uparrow)y = y_0 + v_0 t + \tfrac{1}{2}a_c t^2; \qquad y = y_0 + (v_y)_0 t - \tfrac{1}{2}gt^2$$
$$(+\uparrow)v^2 = v_0^2 + 2a_c(y - y_0); \qquad v_y^2 = (v_y)_0^2 - 2g(y - y_0)$$

Recall that the last equation can be formulated on the basis of eliminating the time t between the first two equations, and therefore *only two of the above three equations are independent of one another.*

To summarize, problems involving the motion of a projectile can have at most three unknowns since only three independent equations can be written. These equations consist of one equation in the horizontal direction and two in the vertical direction. Scalar analysis can be used here because the component motions along the x and y axes are *rectilinear.* Once $\mathbf{v}_x$ and $\mathbf{v}_y$ are obtained, realize that the resultant velocity $\mathbf{v}$, which is *always tangent* to the path, is defined by the *vector sum* as shown in Fig. 12–20.

PROCEDURE FOR ANALYSIS

Using the above results, the following procedure provides a method for solving problems concerning free-flight projectile motion.

Coordinate System. Establish the fixed x, y coordinate axes and sketch the trajectory of the particle. Between any *two points* on the path specify the given problem data and the *three unknowns.* In all cases the acceleration of gravity acts downward. The particle's initial and final velocities should be represented in terms of their x and y components. Remember that positive and negative position, velocity, and acceleration components always act in accordance with their associated coordinate directions.

Kinematic Equations. Depending upon the known data and what is to be determined, a choice should be made as to which three of the following four equations should be applied between the two points on the path to obtain the most direct solution to the problem.

Horizontal Motion. The *velocity* in the horizontal or x direction is *constant,* i.e., $(v_x) = (v_x)_0$, and

$$x = x_0 + (v_x)_0 t$$

Vertical Motion. In the vertical or y direction *only two* of the following three equations can be used for solution.

$$v_y = (v_y)_0 + a_c t$$
$$y = y_0 + (v_y)_0 t + \tfrac{1}{2}a_c t^2$$
$$v_y^2 = (v_y)_0^2 + 2a_c(y - y_0)$$

For example, if the particles' final velocity v_y is not needed, then the first and third of these equations (for y) will not be useful.

Example 12–11

A sack slides off the ramp, shown in Fig. 12–21, with a horizontal velocity of 12 m/s. If the height of the ramp is 6 m from the floor, determine the time needed for the sack to strike the floor and the range R where sacks begin to pile up.

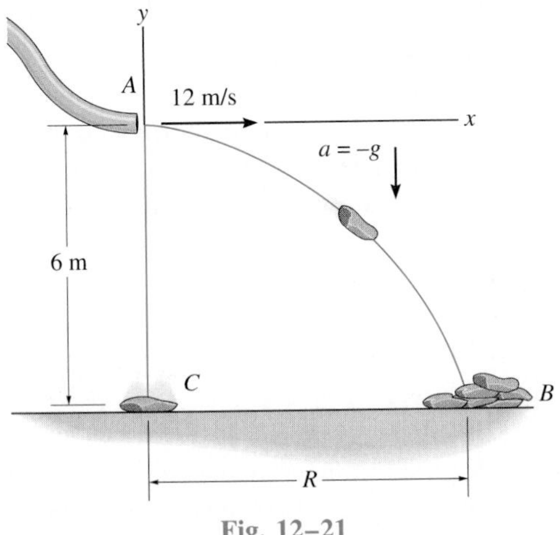

Fig. 12–21

SOLUTION

Coordinate System. The origin of coordinates is established at the beginning of the path, point A, Fig. 12–21. The initial velocity of a sack has components $(v_A)_x = 12$ m/s and $(v_A)_y = 0$. Also, between points A and B the acceleration is $a_y = -9.81$ m/s^2. Since $(v_B)_x = (v_A)_x = 12$ m/s, the three unknowns are $(v_B)_y$, R, and the time of flight t. Here we do not need to determine $(v_B)_y$.

Vertical Motion. The vertical distance from A to B is known, and therefore we can obtain a direct solution for t by using the equation

$(+\uparrow)$
$$y = y_0 + (v_y)_0 t + \tfrac{1}{2}a_c t^2$$
$$-6 \text{ m} = 0 + 0 + \tfrac{1}{2}(-9.81 \text{ m/s}^2)t^2$$
$$t = 1.11 \text{ s} \qquad \textit{Ans.}$$

This calculation also indicates that if a sack was released *from rest* at A, it would take the same amount of time to strike the floor at C, Fig. 12–21.

Horizontal Motion. Since t has been calculated, R is determined as follows:

$(\overset{+}{\rightarrow})$
$$x = x_0 + (v_x)_0 t$$
$$R = 0 + 12 \text{ m/s } (1.11 \text{ s})$$
$$R = 13.3 \text{ m} \qquad \textit{Ans.}$$

Example 12–12

The chipping machine is designed to eject wood chips at $v_O = 25$ ft/s as shown in Fig. 12–22. If the tube is oriented at 30° from the horizontal, determine how high, h, the chips strike the pile if they land on the pile 20 ft from the tube.

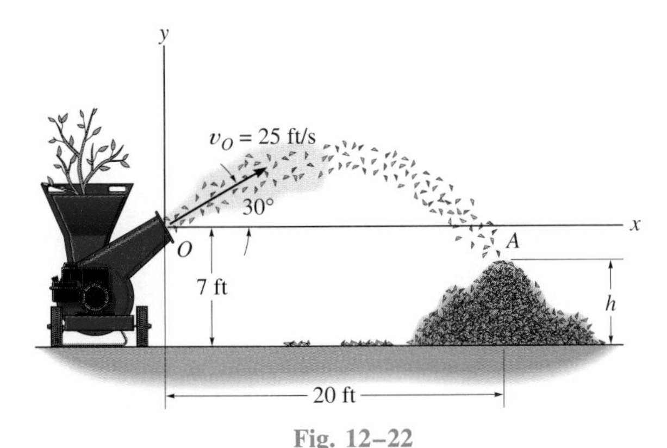

Fig. 12–22

SOLUTION

Coordinate System. When the motion is analyzed between points O and A, the three unknowns are represented as the height h, time of flight t_{OA}, and vertical component of velocity $(v_A)_y$. With the origin of coordinates at O, Fig. 12–22, the initial velocity of a chip has components of

$$(v_O)_x = (25 \cos 30°) \text{ ft/s} = 21.65 \text{ ft/s} \rightarrow$$
$$(v_O)_y = (25 \sin 30°) \text{ ft/s} = 12.5 \text{ ft/s} \uparrow$$

Also, $(v_A)_x = (v_O)_x = 21.65$ ft/s and $a_y = -32.2 \text{ ft/s}^2$. Since we do not need to determine $(v_A)_y$, we have

Horizontal Motion

$(\overset{+}{\rightarrow})$

$$x_A = x_O + (v_O)_x t_{OA}$$
$$20 \text{ ft} = 0 + (21.65 \text{ ft/s}) t_{OA}$$
$$t_{OA} = 0.9238 \text{ s}$$

Vertical Motion. Relating t_{OA} to the initial and final elevations of a chip, we have

$(+\uparrow)$

$$y_A = y_O + (v_O)_y t_{OA} + \tfrac{1}{2} a_c t_{OA}^2$$
$$(h - 7 \text{ ft}) = 0 + (12.5 \text{ ft/s})(0.9238 \text{ s}) + \tfrac{1}{2}(-32.2 \text{ ft/s}^2)(0.9238 \text{ s})^2$$
$$h = 4.81 \text{ ft}$$

Ans.

Example 12–13

Using a video camera, it is observed that when a ball is kicked from A as shown in Fig. 12–23, it just clears the top of a wall at B as it reaches its maximum height. Knowing that the distance from A to the wall is 20 m and the wall is 4 m high, determine the initial speed at which the ball was kicked. Neglect the size of the ball.

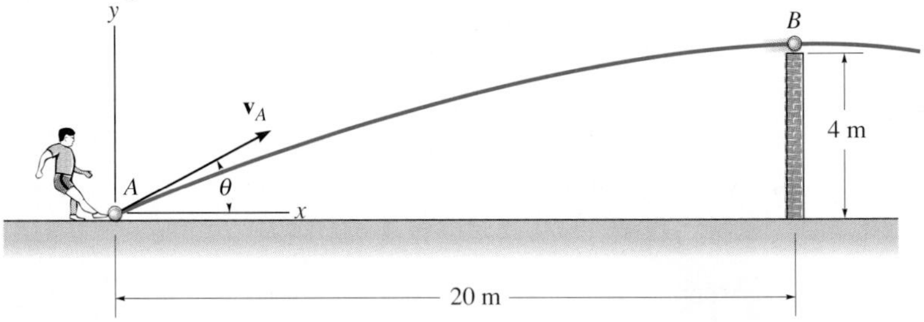

Fig. 12–23

SOLUTION

Coordinate System. The three unknowns are represented by the initial speed v_A, angle of inclination θ, and the time t_{AB} to travel from A to B, Fig. 12–23. At the highest point, B, the velocity $(v_B)_y = 0$, and $(v_B)_x = (v_A)_x = v_A \cos \theta$.

Horizontal Motion

$(\overset{+}{\rightarrow})$

$$x_B = x_A + (v_A)_x t_{AB}$$

$$20 \text{ m} = 0 + (v_A \cos \theta) t_{AB} \qquad (1)$$

Vertical Motion

$(+\uparrow)$

$$(v_B)_y = (v_A)_y + a_c t_{AB}$$

$$0 = v_A \sin \theta - 9.81 t_{AB} \qquad (2)$$

$(+\uparrow)$

$$(v_B)_y^2 = (v_A)_y^2 + 2a_c[y_B - y_A]$$

$$0 = v_A^2 \sin^2 \theta + 2(-9.81 \text{ m/s}^2)(4 \text{ m} - 0) \qquad (3)$$

To obtain v_A, eliminate t_{AB} from Eqs. 1 and 2, which yields

$$v_A^2 \sin \theta \cos \theta = 196.2 \qquad (4)$$

Solve for v_A^2 in Eq. 4 and substitute into Eq. 3, so that

$$\frac{\sin \theta}{\cos \theta} = \tan \theta = \frac{2(9.81)(4)}{196.2} = 0.4$$

$$\theta = \tan^{-1}(0.4) = 21.8^\circ$$

Then, using Eq. 4, the required initial speed is

$$v_A = \sqrt{\frac{196.2}{(\sin 21.8^\circ)(\cos 21.8^\circ)}} = 23.9 \text{ m/s} \qquad \textit{Ans.}$$

PROBLEMS

12–66. A particle moves along the path $\mathbf{r} = \{8t^2\mathbf{i} + (t^3 + 5)\,\mathbf{j}\}$ m, where t is in seconds. Determine the magnitudes of the particle's velocity and acceleration when $t = 3$ s. Also determine the equation $y = f(x)$ of the path.

12–67. Motion of particles A and B is described by the position vectors $\mathbf{r}_A = \{3t\mathbf{i} + 9t(2 - t)\mathbf{j}\}$ m and $\mathbf{r}_B = \{3(t^2 - 2t + 2)\mathbf{i} + 3(t - 2)\mathbf{j}\}$ m, respectively, where t is in seconds. Determine the point where the particles collide and their speeds just before the collision. How long does it take before the collision occurs?

***12–68.** Motion of particles A and B is described by the position vectors $\mathbf{r}_A = \{2t\mathbf{i} + (t^2 - 1)\mathbf{j}\}$ ft and $\mathbf{r}_B = \{(t + 2)\mathbf{i} + (2t^2 - 5)\mathbf{j}\}$ ft, respectively, where t is in seconds. Determine the point where the particles collide and their speeds just before the collision.

12–69. If $x = 1 - t$ and $y = t^2$, where x and y are in meters and t is in seconds, determine the x and y components of velocity and acceleration and construct the path $y = f(x)$.

12–70. If $x = 3 \sin t$ and $y = 4 \cos t$, where x and y are in meters and t is in seconds, determine the x and y components of velocity and acceleration and construct the path $y = f(x)$.

12–71. A particle, originally at rest and located at point (3 ft, 2 ft, 5 ft), is subjected to an acceleration $\mathbf{a} = \{6t\mathbf{i} + 12t^2\mathbf{k}\}$ ft/s^2. Determine the particle's position (x, y, z) when $t = 2$ s.

***12–72.** A particle travels with a constant speed v around a helix defined by the parametric equations $x = (5 \cos 2t)$ m, $y = (5 \sin 2t)$ m, $z = (3t)$ m, where t is in seconds. Determine the particle's acceleration as a function of time.

12–73. The curvilinear motion of a particle is defined by $x = 3t^2$, $y = 4t + 2$, and $z = 6t^3 - 8$, where the x, y, z position is given in meters and the time in seconds. Determine the magnitudes and coordinate direction angles α, β, γ of the particle's velocity and acceleration when $t = 2$ s.

12–74. A particle travels along the curve from A to B in 2 s. It takes 4 s for it to go from B to C and then 3 s to go from C to D. Determine its average velocity when it goes from A to D.

12–75. A particle travels along the curve from A to B in 2 s. It takes 4 s for it to go from B to C and then 3 s to go from C to D. Determine its average speed when it goes from A to D.

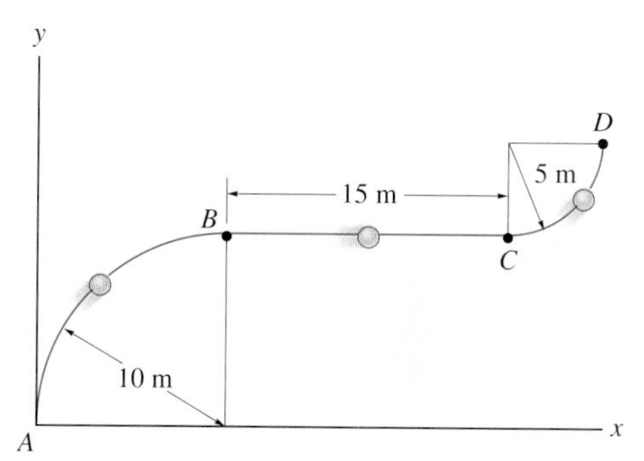

Probs. 12–74/12–75

***12–76.** A particle travels along the path $y = x^2$, where x and y are in meters. If the particle's component of velocity in the y direction is always $v_y = 3$ m/s, determine the magnitude of the particle's velocity when $t = 2$ s. When $t = 0$ the particle is at point (1 m, 1 m). At what distance is the particle from the origin when $t = 2$ s?

12–77. A car traveling along the straight portions of the road has the velocities indicated in the figure when it arrives at points A, B, and C. If it takes 3 s to go from A to B, and then 5 s to go from B to C, determine the average acceleration between points A and B and between points A and C.

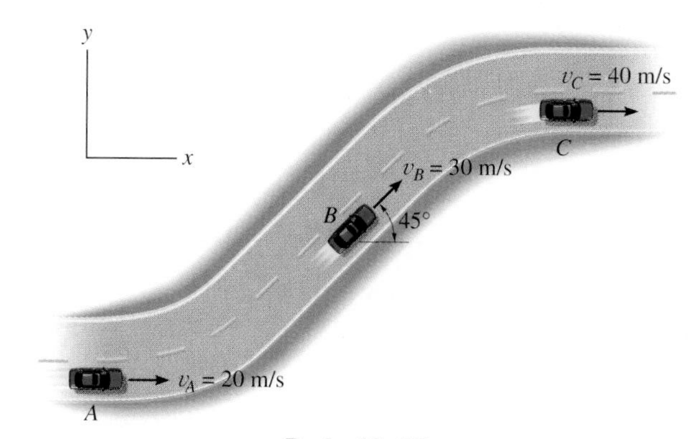

Prob. 12–77

12–78. A particle travels along a path such that its position is $\mathbf{r} = \{(2 \sin t)\mathbf{i} + 2(1 - \cos t)\mathbf{j}\}$ ft, where t is in seconds and the arguments for the sine and cosine are given in radians. Find the equation $y = f(x)$ which describes the path, and show that the magnitudes of the particle's velocity and acceleration are constant. What is the magnitude of the particle's displacement from $t = 0$ to $t = 2$ s?

12–79. The particle travels along the path defined by the parabola $y = 0.5x^2$. If the component of velocity along the x axis is $v_x = (5t)$ ft/s, where t is in seconds, determine the particle's distance from the origin O and the magnitude of acceleration when $t = 1$ s. When $t = 0$, $x = 0$, $y = 0$.

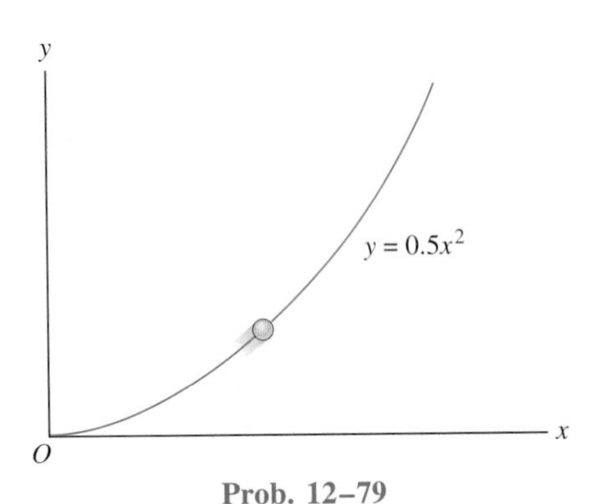

Prob. 12–79

***12–80.** The particle travels along the path defined by the parabola $y = 0.5x^2$. If the component of velocity along the y axis is $v_y = (2t^2)$ ft/s, where t is in seconds, determine the particle's distance from the origin O and the magnitude of acceleration when $t = 1$ s. When $t = 0$, $x = 0$, $y = 0$.

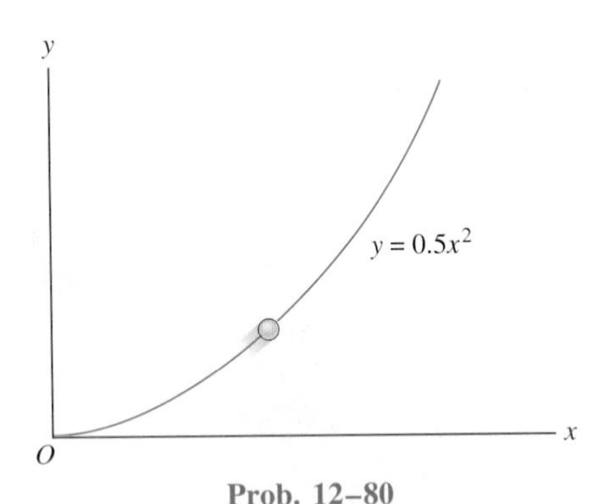

Prob. 12–80

12–81. A particle moves along a hyperbolic path $x^2/16 - y^2 = 28$. If the x component of velocity is $v_x = 4$ m/s and remains constant, determine the magnitudes of the particle's velocity and acceleration when it is at point (32 m, 6 m).

12–82. A particle moves in the positive quadrant along a hyperbolic path $x^2/16 - y^2 = 28$ $(x > 0)$. If the y component of velocity is $v_y = (2t)$ m/s, where t is in seconds, determine the magnitudes of the particle's velocity and acceleration when $t = 2$ s. When $t = 0$, $y = 0$.

■**12–83.** A particle is traveling with a velocity of $\mathbf{v} = \{3\sqrt{t}e^{-0.2t}\mathbf{i} + 4e^{-0.8t^2}\mathbf{j}\}$ m/s, where t is in seconds. Determine the magnitude of the particle's displacement from $t = 0$ to $t = 3$ s. Use Simpson's rule to evaluate the integrals. What is the magnitude of the particle's acceleration when $t = 2$ s?

***12–84.** A particle moves in the x-y plane such that its position is defined by $\mathbf{r} = \{\sin^2 \theta\mathbf{i} + (\theta^2 + \cos 2\theta)\mathbf{j}\}$ ft, where θ is in radians and r is in feet. If $\theta = (0.8t^3)$ rad, where t is in seconds, determine the particle's velocity and acceleration when $t = 3$ s.

12–85. The position of a particle is defined by $\mathbf{r} = \{\theta^3\mathbf{i} + \sin 2\theta\mathbf{j} + \cos^2 \theta\mathbf{k}\}$ ft, where θ is in radians. If $\theta = (2t^2)$ rad, where t is in seconds, determine the particle's velocity and acceleration when $t = 1$ s. Express $\mathbf{v}$ and $\mathbf{a}$ as Cartesian vectors.

12–86. A particle moves along the helical path defined by the parametric equations $x = br \cos at$, $y = r \sin at$, and $z = kt$, where a, b, k, and r are constants. Determine the magnitudes of the particle's velocity and acceleration as a function of time.

12–87. When a rocket reaches an altitude of 40 m it begins to travel along the parabolic path $(y - 40)^2 = 160x$, where the coordinates are measured in meters. If the component of velocity in the vertical direction is constant at $v_y = 180$ m/s, determine the magnitudes of the rocket's velocity and acceleration when it reaches an altitude of 80 m.

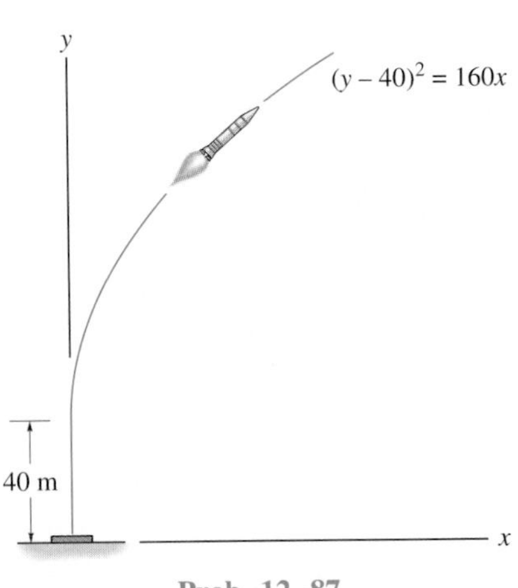

Prob. 12–87

•12–88. The girl throws the toy at an angle of 30° from point A as shown. Determine the maximum and minimum speed v_A it can have so that it lands in the pool.

12–89. The girl always throws the toys at an angle of 30° from point A as shown. Determine the time between throws so that both toys strike the edges of the pool B and C at the same instant. With what speed must she throw each toy?

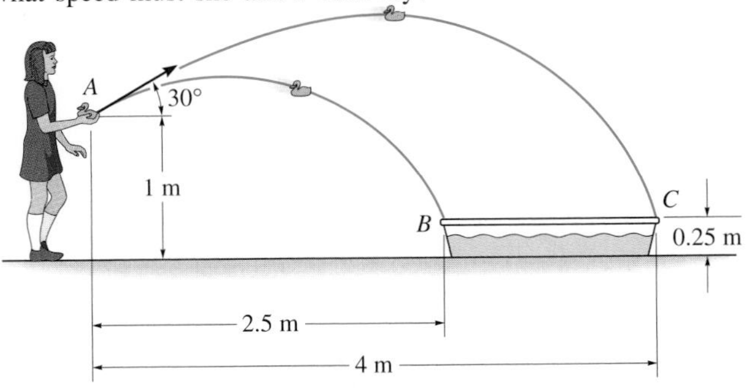

Probs. 12–88/12–89

12–90. The balloon A is ascending at the rate $v_A = 12$ km/h and is being carried horizontally by the wind at $v_w = 20$ km/h. If a ballast bag is dropped from the balloon such that it takes 8 s for it to reach the ground, determine the balloon's altitude h at the instant the bag was released. Assume that the bag was released from the balloon with the same velocity as the balloon.

12–91. The balloon A is ascending at the rate $v_A = 12$ km/h and is being carried horizontally by the wind at $v_w = 20$ km/h. If a ballast bag is dropped from the balloon at the instant $h = 50$ m, determine the time needed for it to strike the ground. Assume that the bag was released from the balloon with the same velocity as the balloon. Also, with what speed does the bag strike the ground?

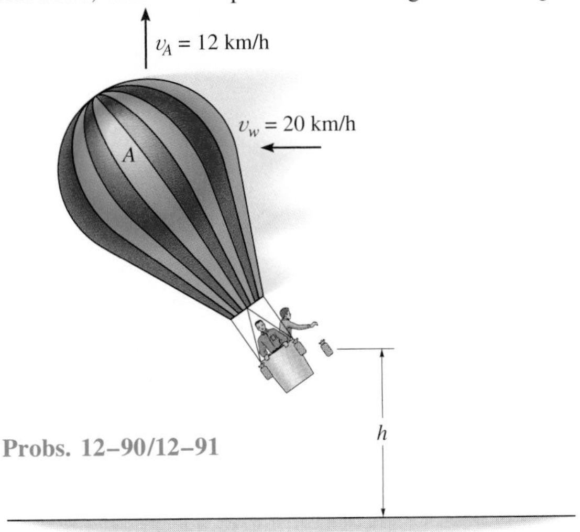

Probs. 12–90/12–91

*12–92. Determine the maximum height on the wall to which the firefighter can project water from the hose, if the speed of the water at the nozzle is $v_C = 48$ ft/s.

■12–93. Determine the smallest angle θ, measured above the horizontal, that the hose should be directed so that the water stream strikes the bottom of the wall at B. The speed of the water at the nozzle is $v_C = 48$ ft/s.

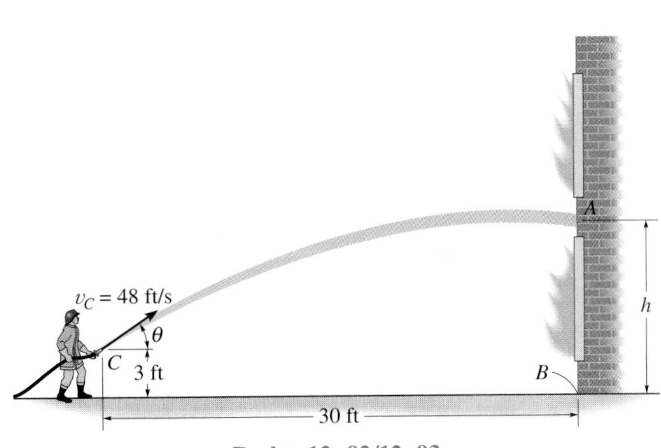

Probs. 12–92/12–93

12–94. During a race the dirt bike was observed to leap up off the small hill at A at an angle of 60° with the horizontal. If the point of landing is 20 ft away, determine the approximate speed at which the bike was traveling just before it left the ground. Neglect the size of the bike for the calculation.

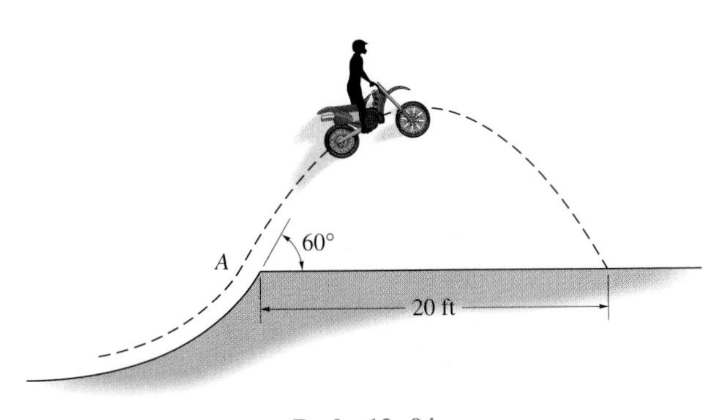

Prob. 12–94

12–95. Measurements of a shot recorded on a videotape during a basketball game are shown. The ball passed through the hoop even though it barely cleared the hands of the player B who attempted to block it. Neglecting the size of the ball, determine the magnitude v_A of its initial velocity and the height h of the ball when it passes over player B.

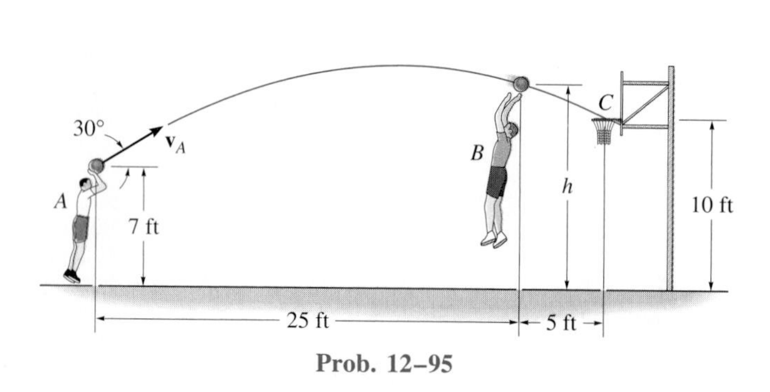

Prob. 12–95

12–97. At the same instant two boys throw balls A and B from the window with a speed v_0 and kv_0, respectively, where k is a constant. Show that the balls will collide if $k = \cos\theta_2/\cos\theta_1$.

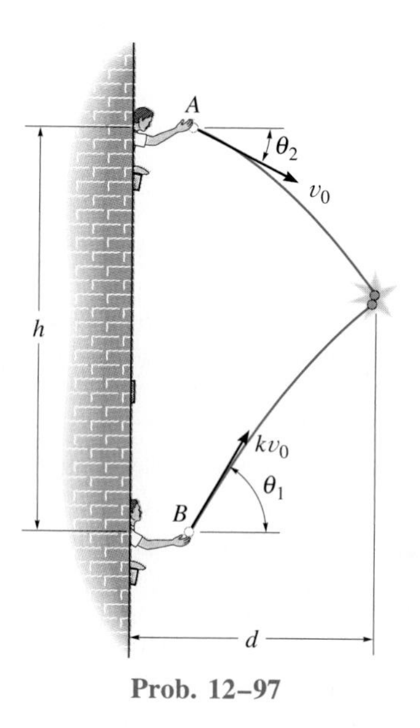

Prob. 12–97

■*12–96. The projectile is thrown in the air with a speed of 8 m/s and at an angle $\theta = 30°$ with the horizontal. Determine the distance it must travel along the path to reach its highest point B.

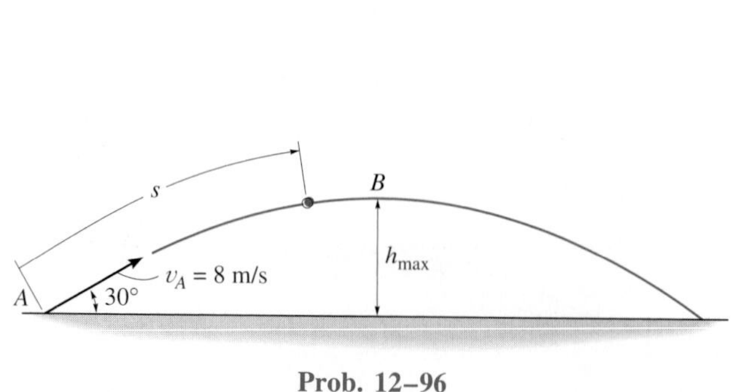

Prob. 12–96

12–98. The snowmobile is traveling at 10 m/s when it leaves the embankment at A. Determine the time of flight from A to B and the range R of the trajectory.

12–99. The snowmobile is traveling at 10 m/s when it leaves the embankment at A. Determine the speed at which it strikes the ground at B and its maximum acceleration along the trajectory AB.

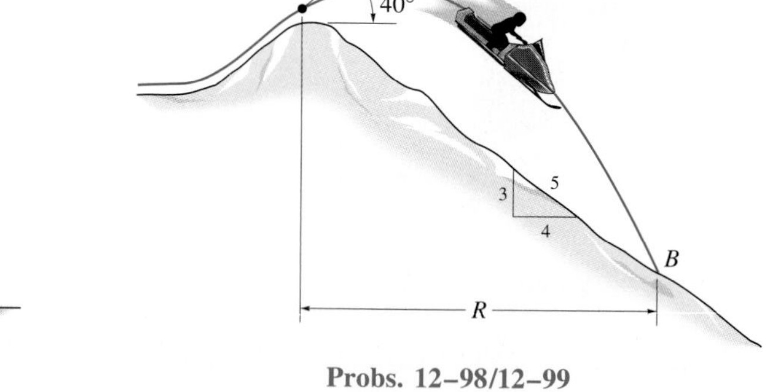

Probs. 12–98/12–99

***12–100.** A golf ball is struck with a velocity of 80 ft/s as shown. Determine the distance d to where it will land.

12–101. A golf ball is struck with a velocity of 80 ft/s as shown. Determine the speed at which it strikes the ground at B and the time of flight from A to B.

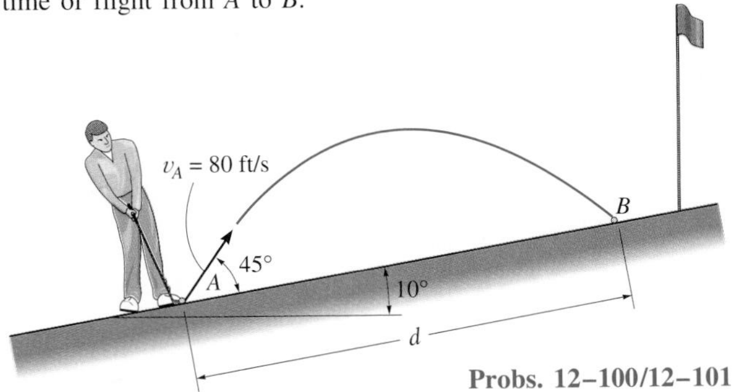

Probs. 12–100/12–101

12–102. The man stands 60 ft from the wall and throws a ball at it with a speed $v_0 = 50$ ft/s. Determine the angle θ at which he should release the ball so that it strikes the wall at the highest point possible. What is this height? The room has a ceiling height of 20 ft.

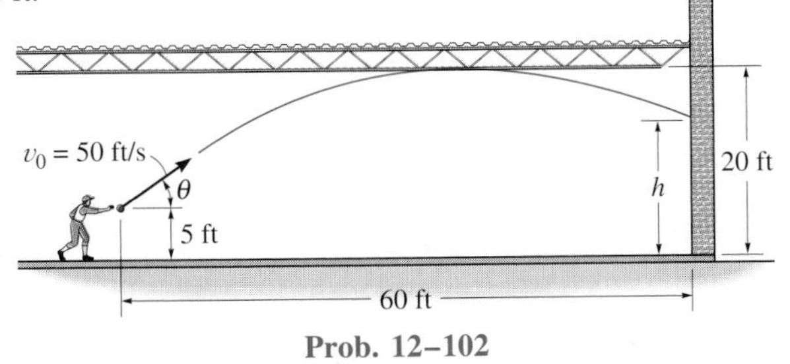

Prob. 12–102

12–103. Determine the minimum speed of the stunt rider, so that when he leaves the ramp at A he passes through the center of the hoop at B. Also, how far h should the landing ramp be from the hoop so that he lands on it safely at C? Neglect the size of the motorcycle and rider.

***12–104.** The ball at A is kicked with a speed $v_A = 80$ ft/s and at an angle $\theta_A = 30°$. Determine the point $(x, -y)$ where it strikes the ground. Assume the ground has the shape of a parabola as shown.

12–105. The ball at A is kicked such that $\theta_A = 30°$. If it strikes the ground at B having coordinates $x = 15$ ft, $y = -9$ ft, determine the speed at which it is kicked and the speed at which it strikes the ground.

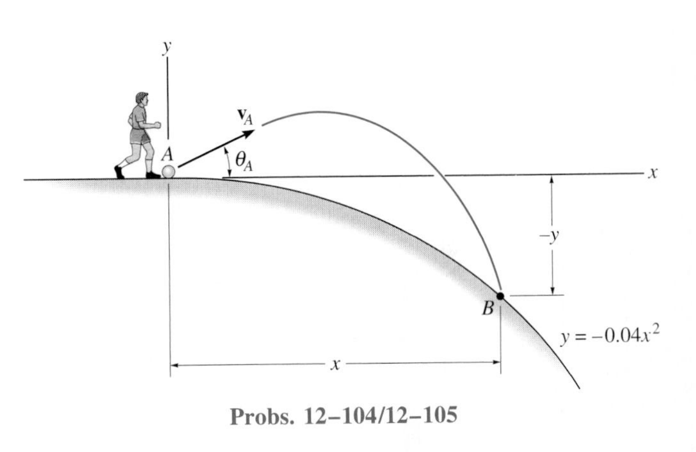

Probs. 12–104/12–105

12–106. A boy at A wishes to throw two balls to the boy at B so that they arrive at the *same time*. If each ball is thrown with a speed of 10 m/s, determine the angles θ_C and θ_D at which each ball should be thrown and the time between each throw. Note that the first ball must be thrown at $\theta_C (> \theta_D)$, then the second ball is thrown at θ_D.

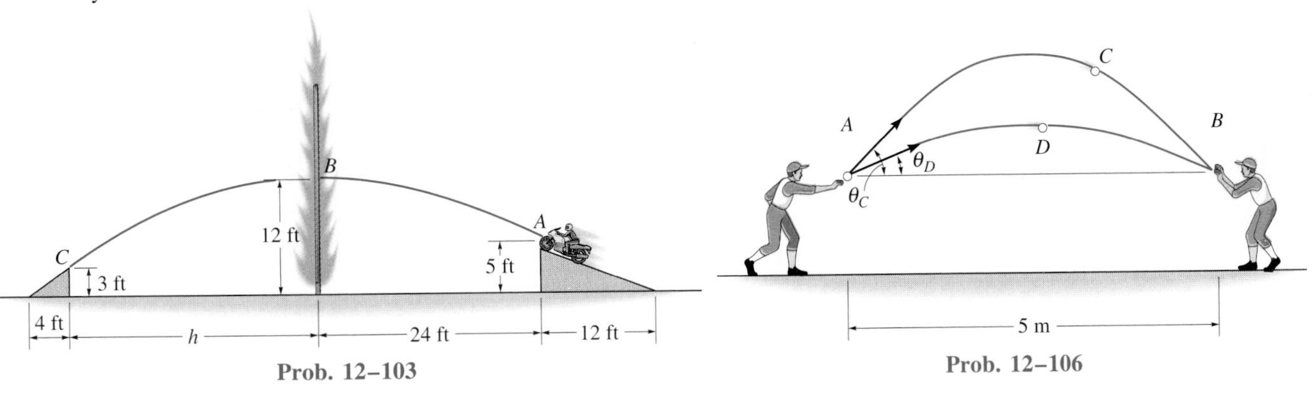

Prob. 12–103

Prob. 12–106

12–107. A boy at O throws a ball in the air with a speed v_O at an angle θ_1. If he then throws another ball at the same speed v_O at an angle $\theta_2 < \theta_1$, determine the time between the throws so the balls collide in mid air at B.

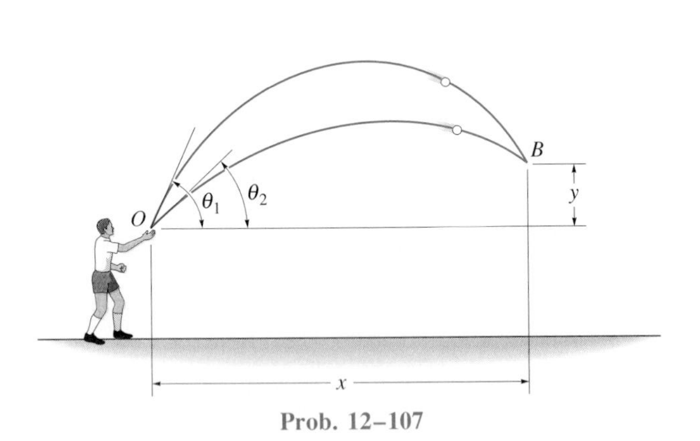

Prob. 12–107

***12–108.** The girl at A can throw a ball at $v_A = 10$ m/s. Calculate the maximum possible range $R = R_{max}$ and the associated angle θ at which it should be thrown. Assume the ball is caught at B at the same elevation from which it is thrown.

12–109. Show that the girl at A can throw the ball to the boy at B by launching it at equal angles measured up or down from a 45° inclination. If $v_A = 10$ m/s, determine the range R if this value is 15°, i.e., $\theta_1 = 45° - 15° = 30°$ and $\theta_2 = 45° + 15° = 60°$. Assume the ball is caught at the same elevation from which it is thrown.

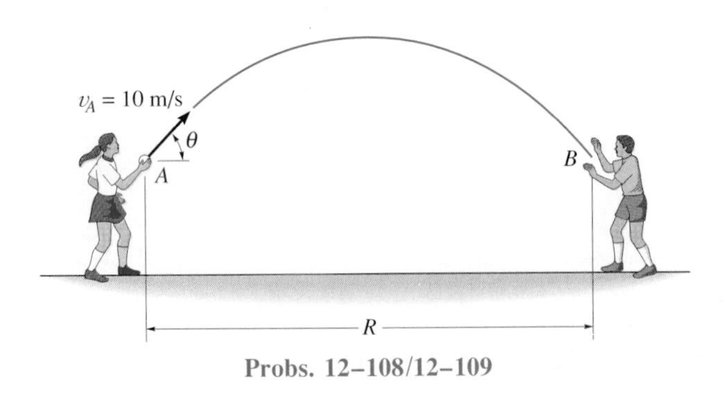

Probs. 12–108/12–109

12–110. The man at A wishes to throw two darts at the target at B so that they arrive at the *same time*. If each dart is thrown with a speed of 10 m/s, determine the angles θ_C and θ_D at which they should be thrown and the time between each throw. Note that the first dart must be thrown at $\theta_C (> \theta_D)$, then the second dart is thrown at θ_D.

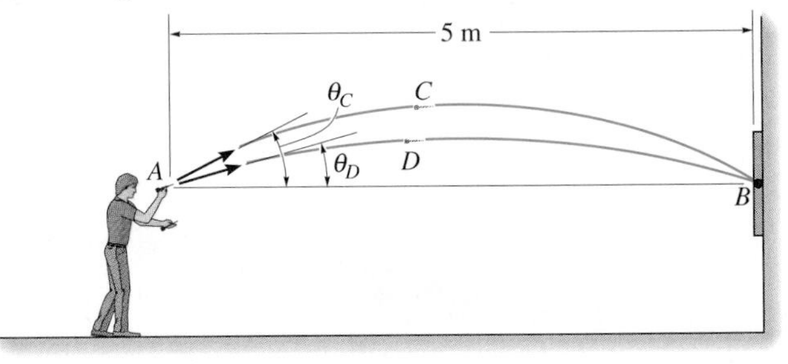

Prob. 12–110

12–111. The football player at A throws the ball in the y-z plane at a speed $v_A = 50$ ft/s and an angle $\theta_A = 60°$ with the horizontal. At the instant the ball was thrown, the player was at B and runs with constant speed along the line BC in order to catch it. Determine this speed, v_B, so that he makes the catch at the same elevation from which the ball was thrown.

***12–112.** The football player at A throws the ball in the y-z plane with a speed $v_A = 50$ ft/s and an angle $\theta_A = 60°$ with the horizontal. At the instant the ball was thrown, the player was at B and runs at a constant speed of $v_B = 23$ ft/s along the line BC. Determine if he can reach point C, which has the same elevation as A, before the ball gets there.

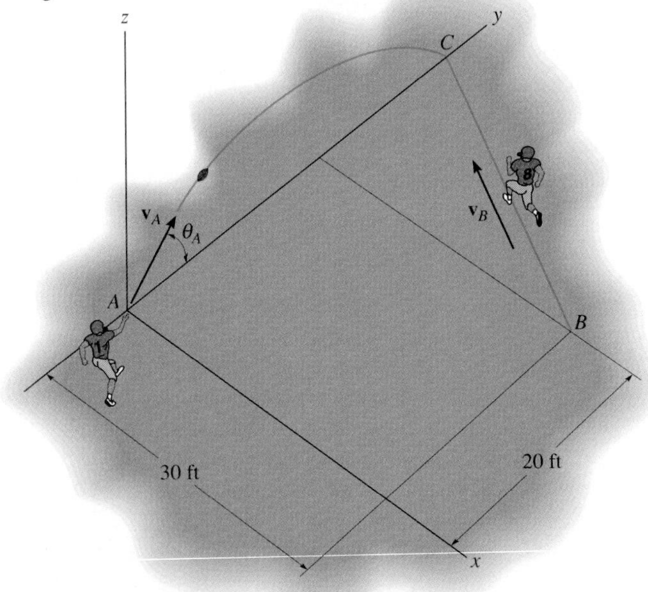

Probs. 12–111/12–112

12.6 Curvilinear Motion: Normal and Tangential Components

When the path along which a particle is moving is *known,* it is often convenient to describe the motion using n and t coordinates which act normal and tangent to the path, respectively, and at the instant considered have their *origin located at the particle.*

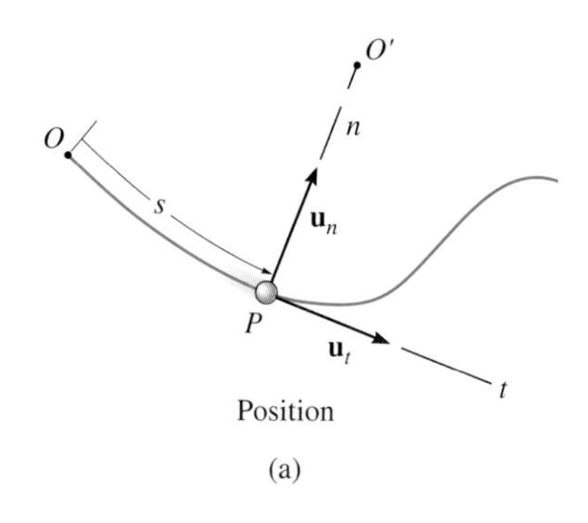

Position

(a)

Planar Motion. Consider the particle P shown in Fig. 12–24a, which is moving in a plane along a fixed curve, such that at a given instant it is located at position s, measured from point O. We will now consider a coordinate system that has its origin at a *fixed point* on the curve, and at the instant considered this origin happens to *coincide* with the location of the particle. The t axis is *tangent* to the curve at P and is positive in the direction of *increasing* s. We will designate this positive direction with the unit vector $\mathbf{u}_t$. A unique choice for the *normal axis* can be made by considering the fact that geometrically the curve is constructed from a series of differential arc segments ds. As shown in Fig. 12–24b, each segment ds is formed from the arc of an associated circle having a *radius of curvature* ρ (rho) and *center of curvature* O'. The normal axis n which will be chosen is perpendicular to the t axis and is directed from P *toward* the center of curvature O', Fig. 12–24a. This positive direction, which is *always* on the concave side of the curve, will be designated by the unit vector $\mathbf{u}_n$. The plane which contains the n and t axes is referred to as the *osculating plane,* and in this case it is fixed in the plane of motion.*

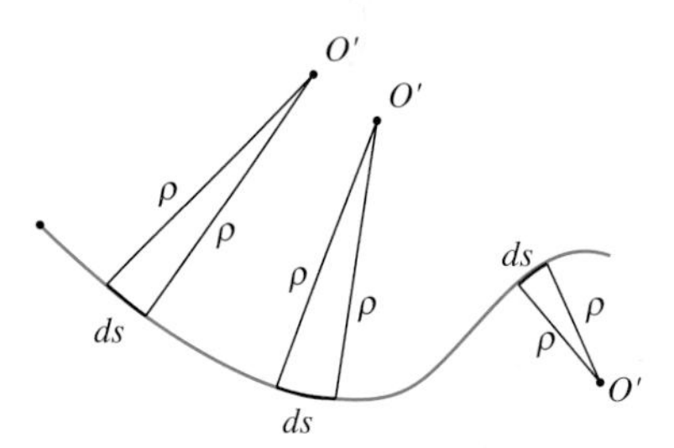

Radius of curvature

(b)

Velocity. Since the particle is moving, s is a function of time. As indicated in Sec. 12.3, the particle's velocity $\mathbf{v}$ has a *direction* that is *always tangent to the path,* Fig. 12–24c, and a *magnitude* that is determined by taking the time derivative of the path function $s = s(t)$, i.e., $v = ds/dt$ (Eq. 12–8). Hence

$$\mathbf{v} = v\mathbf{u}_t \tag{12–15}$$

where

$$v = \dot{s} \tag{12–16}$$

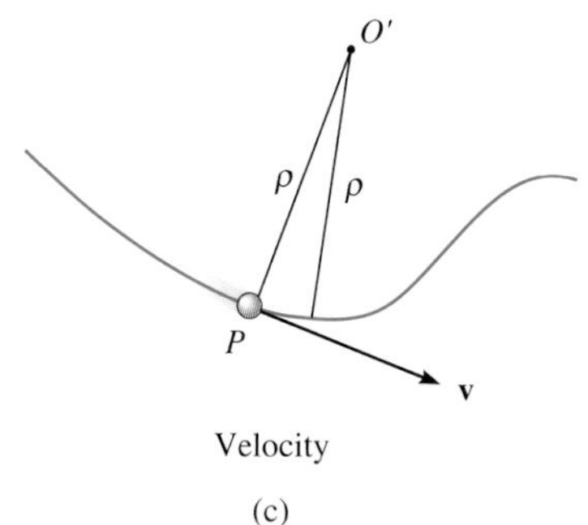

Velocity

(c)

Fig. 12–24

*The osculating plane may also be defined as that plane which has the greatest contact with the curve at a point. It is the limiting position of a plane contacting both the point and the arc segment ds. As noted above, the osculating plane is always coincident with a plane curve; however, each point on a three-dimensional curve has a unique osculating plane.

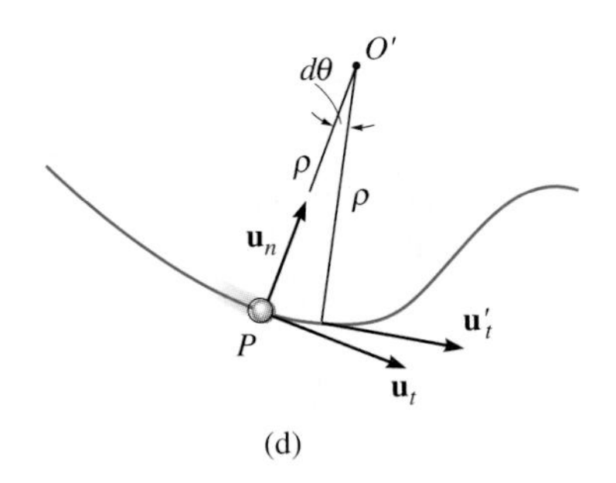

(d)

Acceleration. The acceleration of the particle is the time rate of change of the velocity. Thus,

$$\mathbf{a} = \dot{\mathbf{v}} = \dot{v}\mathbf{u}_t + v\dot{\mathbf{u}}_t \tag{12–17}$$

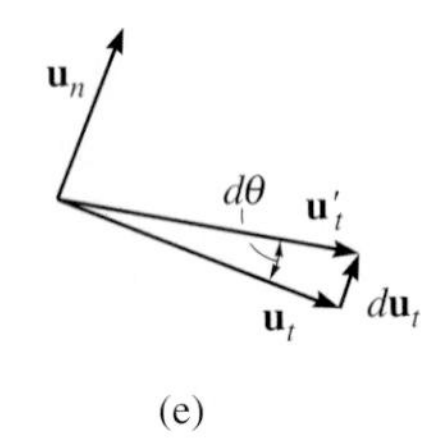

(e)

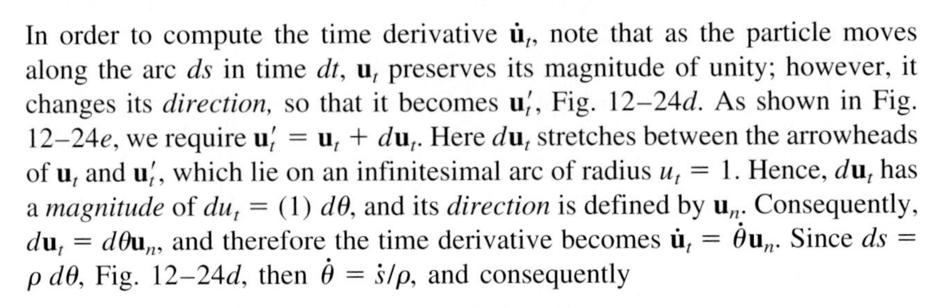

In order to compute the time derivative $\dot{\mathbf{u}}_t$, note that as the particle moves along the arc ds in time dt, $\mathbf{u}_t$ preserves its magnitude of unity; however, it changes its *direction,* so that it becomes $\mathbf{u}'_t$, Fig. 12–24*d*. As shown in Fig. 12–24*e*, we require $\mathbf{u}'_t = \mathbf{u}_t + d\mathbf{u}_t$. Here $d\mathbf{u}_t$ stretches between the arrowheads of $\mathbf{u}_t$ and $\mathbf{u}'_t$, which lie on an infinitesimal arc of radius $u_t = 1$. Hence, $d\mathbf{u}_t$ has a *magnitude* of $du_t = (1)\, d\theta$, and its *direction* is defined by $\mathbf{u}_n$. Consequently, $d\mathbf{u}_t = d\theta\mathbf{u}_n$, and therefore the time derivative becomes $\dot{\mathbf{u}}_t = \dot{\theta}\mathbf{u}_n$. Since $ds = \rho\, d\theta$, Fig. 12–24*d*, then $\dot{\theta} = \dot{s}/\rho$, and consequently

$$\dot{\mathbf{u}}_t = \dot{\theta}\mathbf{u}_n = \frac{\dot{s}}{\rho}\mathbf{u}_n = \frac{v}{\rho}\mathbf{u}_n$$

Substituting into Eq. 12–17, $\mathbf{a}$ can be written as the sum of its two components,

$$\mathbf{a} = a_t\mathbf{u}_t + a_n\mathbf{u}_n \tag{12–18}$$

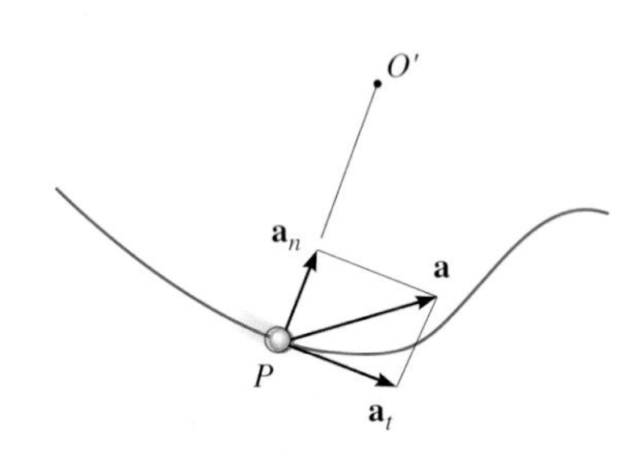

Acceleration

(f)

Fig. 12–24 *(cont'd)*

where

$$a_t = \dot{v} \quad \text{or} \quad a_t = v\frac{dv}{ds} \tag{12–19}$$

and

$$a_n = \frac{v^2}{\rho} \tag{12–20}$$

These two mutually perpendicular components are shown in Fig. 12–24*f*, in which case the *magnitude* of acceleration is the positive value of

$$a = \sqrt{a_t^2 + a_n^2} \tag{12–21}$$

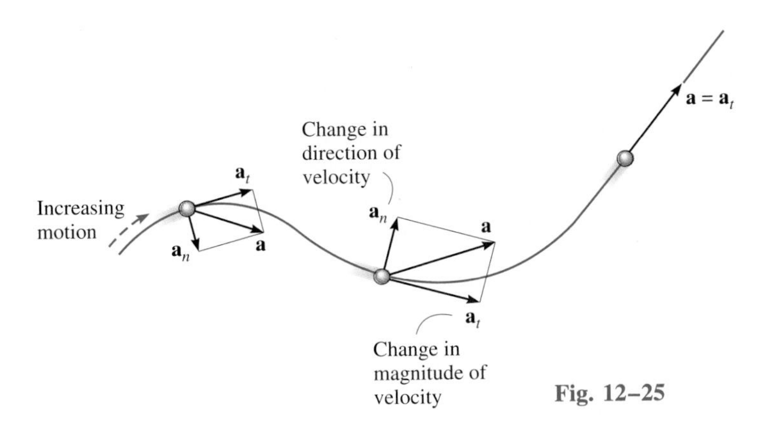

Fig. 12–25

To summarize these concepts, consider the following two special cases of motion.

1. If the particle moves along a straight line, then $\rho \rightarrow \infty$ and from Eq. 12–20, $a_n = 0$. Thus $a = a_t = \dot{v}$, and we can conclude that the *tangential component of acceleration represents the time rate of change in the magnitude of the velocity.*
2. If the particle moves along a curve with a constant speed, then $a_t = \dot{v} = 0$ and $a = a_n = v^2/\rho$. Therefore, the *normal component of acceleration represents the time rate of change in the direction of the velocity.* Since it *always* acts toward the center of curvature, this component is sometimes referred to as the *centripetal acceleration.*

As a result of these interpretations, a particle moving along the curved path shown in Fig. 12–25 will have accelerations directed as shown in the figure.

Three-Dimensional Motion. If the particle is moving along a space curve, Fig. 12–26, then at a given instant the t axis is uniquely specified; however, an infinite number of straight lines can be constructed normal to the tangent axis at P. As in the case of planar motion, we will choose the positive n axis directed from P toward the path's center of curvature O'. This axis is referred to as the *principal normal* to the curve at P. With the n and t axes so defined, Eqs. 12–15 to 12–21 can be used to determine $\mathbf{v}$ and $\mathbf{a}$. Since $\mathbf{u}_t$ and $\mathbf{u}_n$ are always perpendicular to one another, and lie in the osculating plane, for spatial motion a third unit vector, $\mathbf{u}_b$, defines a *binormal axis b* which is perpendicular to $\mathbf{u}_t$ and $\mathbf{u}_n$, Fig. 12–26.

Since the three unit vectors are related to one another by the vector cross product, e.g., $\mathbf{u}_b = \mathbf{u}_t \times \mathbf{u}_n$, Fig. 12–26, it may be possible to use this relation to establish the direction of one of the axes, if the directions of the other two are known. For example, no motion occurs in the $\mathbf{u}_b$ direction, and so if this direction and $\mathbf{u}_t$ are known, then $\mathbf{u}_n$ can be determined, where in this case $\mathbf{u}_n = \mathbf{u}_b \times \mathbf{u}_t$, Fig. 12–26. Remember, though, that $\mathbf{u}_n$ is always on the concave side of the curve.

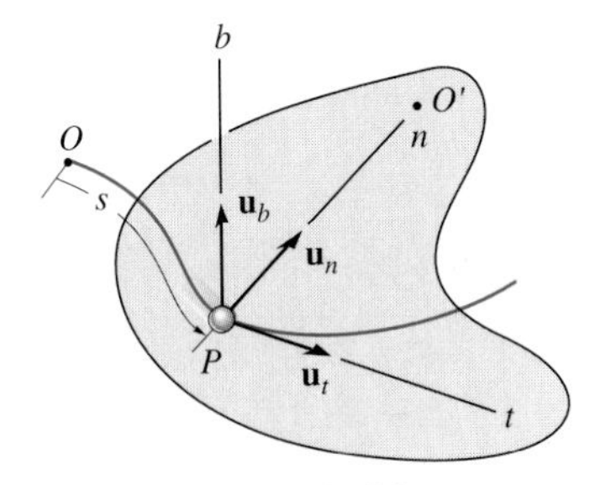

Fig. 12–26

PROCEDURE FOR ANALYSIS

Coordinate System. Provided the *path* of the particle is *known,* we can establish a set of n and t coordinates having a *fixed origin* which is coincident with the particle at the instant considered. The positive tangent axis acts in the direction of motion and the positive normal axis is directed toward the path's center of curvature. This set of axes is particularly advantageous for studying the velocity and acceleration of the particle, because the t and n components of $\mathbf{a}$ are expressed by Eqs. 12–19 and 12–20, respectively.

Velocity. The particle's *velocity* is tangent to the path. Its magnitude is found from the time derivative of the path function.

$$v = \dot{s}$$

Note that v can be positive or negative depending on whether it acts in the positive or negative s direction.

Tangential Acceleration. The *tangential component of acceleration is the result of the time rate of change in the magnitude of velocity.* This component acts in the positive s direction if the particle's speed is increasing or in the opposite direction if the speed is decreasing. The magnitudes of $\mathbf{v}$ and $\mathbf{a}_t$ are related to one another and the time t or the path function s by using the equations of rectilinear motion, namely,

$$a_t = \dot{v} \qquad a_t\, ds = v\, dv$$

If a_t is *constant,* $a_t = (a_t)_c$, the above equations, when integrated, yield

$$s = s_0 + v_0 t + \tfrac{1}{2}(a_t)_c t^2$$
$$v = v_0 + (a_t)_c t$$
$$v^2 = v_0^2 + 2(a_t)_c(s - s_0)$$

Normal Acceleration. The *normal component of acceleration is the result of the time rate of change in the direction of the particle's velocity.* This component is *always* directed toward the center of curvature of the path, i.e., along the positive n axis. Its magnitude is determined from

$$a_n = \frac{v^2}{\rho}$$

In particular, if the path is expressed as $y = f(x)$, the radius of curvature ρ at any point on the path is computed from the equation*

$$\rho = \left| \frac{[1 + (dy/dx)^2]^{3/2}}{d^2y/dx^2} \right|$$

*The derivation of this result is given in any standard calculus text.

Example 12–14

A skier travels with a constant speed of 6 m/s along the parabolic path $y = \frac{1}{20}x^2$ shown in Fig. 12–27. Determine his velocity and acceleration at the instant he arrives at A. Neglect the size of the skier in the calculation.

SOLUTION

Coordinate System. Although the path has been expressed in terms of its x and y coordinates, we can still establish the origin of the n, t axes at the fixed point A on the path and determine the components of $\mathbf{v}$ and $\mathbf{a}$ along these axes, Fig. 12–27.

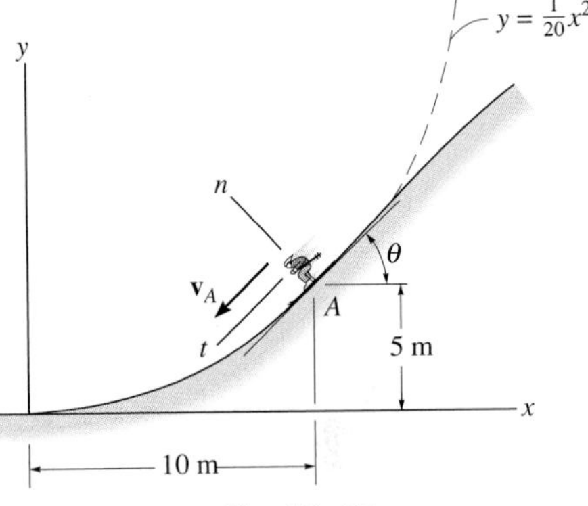

Fig. 12–27

Velocity. By definition, the velocity is always directed tangent to the path. Since $y = \frac{1}{20}x^2$, $dy/dx = \frac{1}{10}x$, then $dy/dx|_{x=10} = 1$. Hence, at A, $\mathbf{v}$ makes an angle of $\theta = \tan^{-1} 1 = 45°$ with the x axis, Fig. 12–27. Therefore,

$$v_A = 6 \text{ m/s} \quad 45° \measuredangle \mathbf{v}_A \qquad \textit{Ans.}$$

Acceleration. The acceleration is determined by using $\mathbf{a} = \dot{v}\mathbf{u}_t + (v^2/\rho)\mathbf{u}_n$. However, it is first necessary to determine the radius of curvature of the path at A (10 m, 5 m). Since $d^2y/dx^2 = \frac{1}{10}$, then

$$\rho = \left| \frac{[1 + (dy/dx)^2]^{3/2}}{d^2y/dx^2} \right| = \left| \frac{[1 + (\frac{1}{10}x)^2]^{3/2}}{\frac{1}{10}} \right|_{x=10\text{m}}$$

The acceleration becomes

$$\mathbf{a}_A = \dot{v}\mathbf{u}_t + \frac{v^2}{\rho}\mathbf{u}_n$$
$$= 0\mathbf{u}_t + \frac{(6 \text{ m/s})^2}{28.3 \text{ m}}\mathbf{u}_n$$
$$= \{1.27\mathbf{u}_n\} \text{ m/s}^2$$

Since $\mathbf{a}_A$ acts in the direction of the positive n axis, it makes an angle of $\theta + 90° = 135°$ with the positive x axis. Hence,

$$a_A = 1.27 \text{ m/s}^2 \quad \mathbf{a}_A \; 135° \qquad \textit{Ans.}$$

Note: By using n, t coordinates, we were able to readily solve this problem since the n and t components account for the *separate* changes in the magnitude and direction of $\mathbf{v}$, and each of these changes have been directly formulated as Eqs. 12–19 and 12–20, respectively.

Example 12–15

A race car C travels around the horizontal circular track that has a radius of 300 ft, Fig. 12–28. If the car increases its speed at a constant rate of 7 ft/s^2, starting from rest, determine the time for it to reach an acceleration of 8 ft/s^2. What is its speed at this instant?

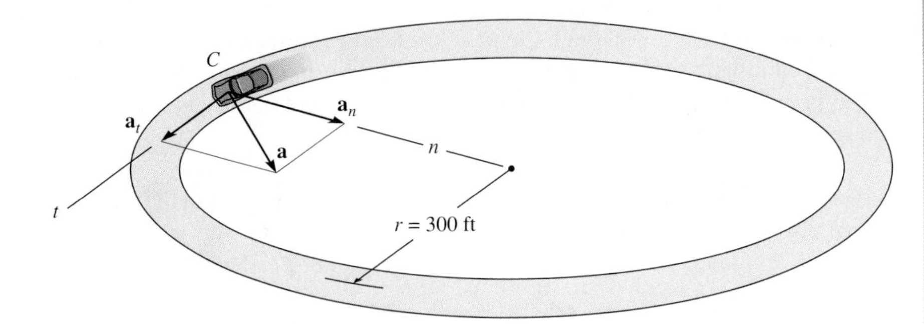

Fig. 12–28

SOLUTION

Coordinate System. The origin of the n and t axes is coincident with the car at the instant considered. The t axis is in the direction of motion, and the positive n axis is directed toward the center of the circle.

Acceleration. The magnitude of acceleration can be related to its components using $a = \sqrt{a_t^2 + a_n^2}$. Here $a_t = 7$ ft/s^2. Also $a_n = v^2/\rho$, where $\rho = 300$ ft and

$$v = v_0 + (a_t)_c t$$
$$v = 0 + 7t$$

Thus

$$a_n = \frac{v^2}{\rho} = \frac{(7t)^2}{300} = 0.163t^2 \text{ ft/s}^2$$

The time needed for the acceleration to reach 8 ft/s^2 is therefore

$$a = \sqrt{a_t^2 + a_n^2}$$
$$8 = \sqrt{(7)^2 + (0.163t^2)^2}$$

Solving for the positive value of t yields

$$0.163t^2 = \sqrt{(8)^2 - (7)^2}$$
$$t = 4.87 \text{ s} \qquad \textit{Ans.}$$

Velocity. The speed at time $t = 4.87$ s is

$$v = 7t = 7(4.87) = 34.1 \text{ ft/s} \qquad \textit{Ans.}$$

Example 12–16

A box starts from rest at point A and travels along the horizontal conveyor shown in Fig. 12–29a. During the motion, the increase in speed is $a_t = (0.2t)$ m/s^2, where t is in seconds. Determine the magnitude of its acceleration when it arrives at point B.

SOLUTION

Coordinate System. The position of the box at any instant is defined from the fixed point A using the position or path coordinate s, Fig. 12–29a. The acceleration is to be determined at B, so the origin of the n, t axes is at this point. Why are n and t coordinates selected to solve this problem?

The acceleration is calculated from its components $a_t = \dot{v}$ and $a_n = v^2/\rho$. To determine these components, however, it is first necessary to formulate v and $\dot{v}$ so that they may be evaluated at B. Since $v_A = 0$ when $t = 0$, then

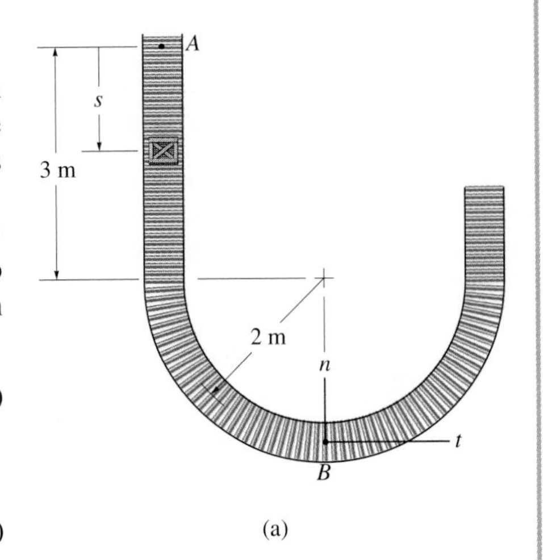

(a)

$$a_t = \dot{v} = 0.2t \qquad (1)$$

$$\int_0^v dv = \int_0^t 0.2t\,dt$$

$$v = 0.1t^2 \qquad (2)$$

The time needed for the box to reach point B can be determined by realizing that the position of B is $s_B = 3 + 2\pi(2)/4 = 6.14$ m, Fig. 12–29a, and since $s_A = 0$ when $t = 0$ we have

$$v = \frac{ds}{dt} = 0.1t^2$$

$$\int_0^{6.14} ds = \int_0^{t_B} 0.1t^2\,dt$$

$$6.14 = 0.0333t_B^3$$

$$t_B = 5.69 \text{ s}$$

Substituting into Eqs. 1 and 2 yields

$$(a_t)_B = \dot{v}_B = 0.2(5.69) = 1.14 \text{ m/s}^2$$

$$v_B = 0.1(5.69)^2 = 3.24 \text{ m/s}$$

At B, $\rho_B = 2$ m, so that

$$(a_n)_B = \frac{v_B^2}{\rho_B} = \frac{(3.24 \text{ m/s})^2}{2 \text{ m}} = 5.25 \text{ m/s}^2$$

The magnitude of $\mathbf{a}_B$, Fig. 12–29b, is therefore

$$a_B = \sqrt{(1.14)^2 + (5.25)^2} = 5.37 \text{ m/s}^2 \qquad \textit{Ans.}$$

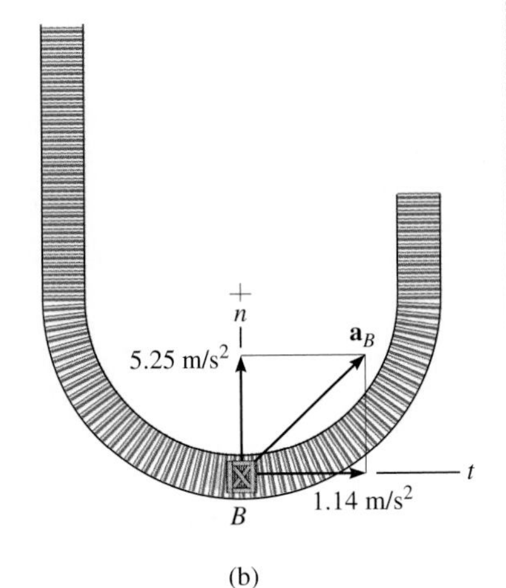

(b)

Fig. 12–29

PROBLEMS

12–113. A car is traveling along a circular curve that has a radius of 50 m. If its speed is 16 m/s and is increasing uniformly at 8 m/s^2, determine the magnitude of its acceleration at this instant.

12–114. A car moves along a circular track of radius 250 ft such that its speed for a short period of time $0 \le t \le 4$ s, is $v = 3(t + t^2)$ ft/s, where t is in seconds. Determine the magnitude of its acceleration when $t = 3$ s. How far has it traveled in $t = 3$ s?

12–115. At a given instant the jet plane has a speed of 400 ft/s and an acceleration of 70 ft/s^2 acting in the direction shown. Determine the rate of increase in the plane's speed and the radius of curvature ρ of the path.

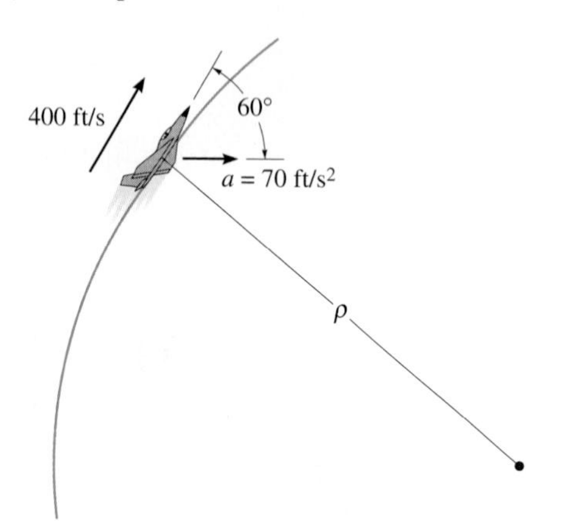

Prob. 12–115

***12–116.** A particle moves along the curve $y = 180/x^2$ such that when it is at $x = 5$ ft it has a speed of 8 ft/s, which is increasing at 12 ft/s^2. Determine the magnitude of the particle's acceleration at this instant.

12–117. The position of a particle is defined by $\mathbf{r} = \{4(t - \sin t)\mathbf{i} + (2t^2 - 3)\mathbf{j}\}$ m, where t is in seconds and the argument for the sine is in radians. Determine the speed of the particle and its normal and tangential components of acceleration when $t = 1$ s.

12–118. The position of a particle is defined by $\mathbf{r} = \{t^2\mathbf{i} + (e^t \sin t)\mathbf{j}\}$ ft, where t is in seconds and the argument for the sine is in radians. Determine the speed of the particle and its normal and tangential components of acceleration when $t = 2$ s.

12–119. A rocket follows a path such that its acceleration is defined by $\mathbf{a} = \{16\mathbf{i} + 4t\mathbf{j}\}$ ft/s^2. If it starts from rest at $\mathbf{r} = \mathbf{0}$, determine the speed of the rocket and the radius of curvature of its path when $t = 10$ s.

***12–120.** A particle travels along the path $y = a + bx + cx^2$, where a, b, c are constants. If the speed of the particle is constant, $v = v_0$, determine the x and y components of velocity and the normal component of acceleration when $x = 0$.

12–121. Starting from rest, the motorboat travels around the circular path, $\rho = 50$ m, at a speed $v = (0.8t)$ m/s, where t is in seconds. Determine the magnitudes of the boat's velocity and acceleration when it has traveled 20 m.

12–122. Starting from rest, the motorboat travels around the circular path, $\rho = 50$ m, at a speed $v = (0.2t^2)$ m/s, where t is in seconds. Determine the magnitudes of the boat's velocity and acceleration at the instant $t = 3$ s.

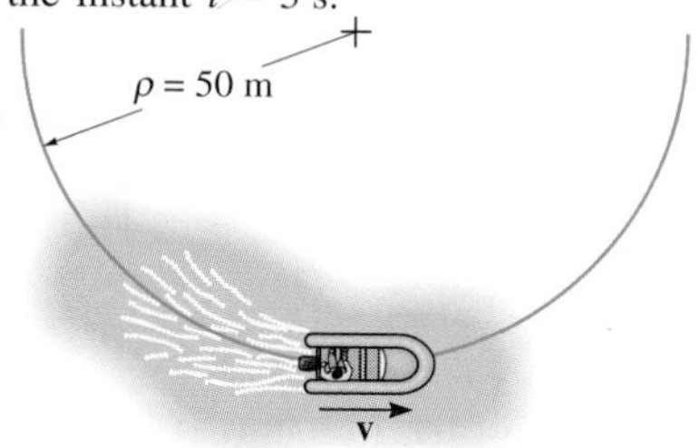

Probs. 12–121/12–122

12–123. The plane travels along the vertical parabolic path at a constant speed of 200 m/s. Determine the magnitude of acceleration of the plane when it is at point A.

***12–124.** The jet plane travels along the vertical parabolic path. When it is at point A it has a speed of 200 m/s, which is increasing at the rate of 0.8 m/s^2. Determine the magnitude of acceleration of the plane when it is at point A.

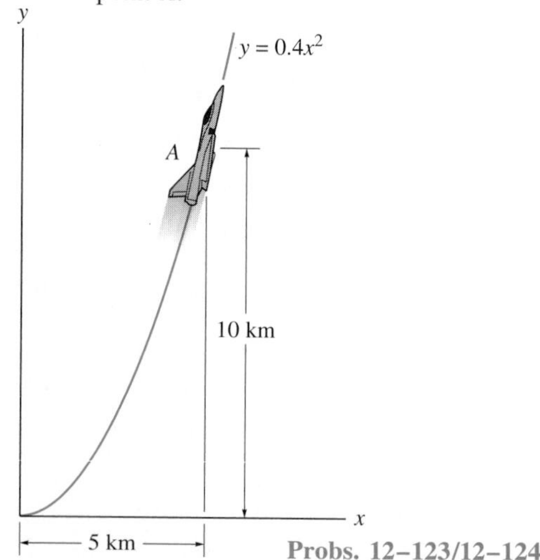

Probs. 12–123/12–124

12–125. At a given instant the train engine at E has a speed of 20 m/s and an acceleration of 14 m/s^2 acting in the direction shown. Determine the rate of increase in the train's speed and the radius of curvature ρ of the path.

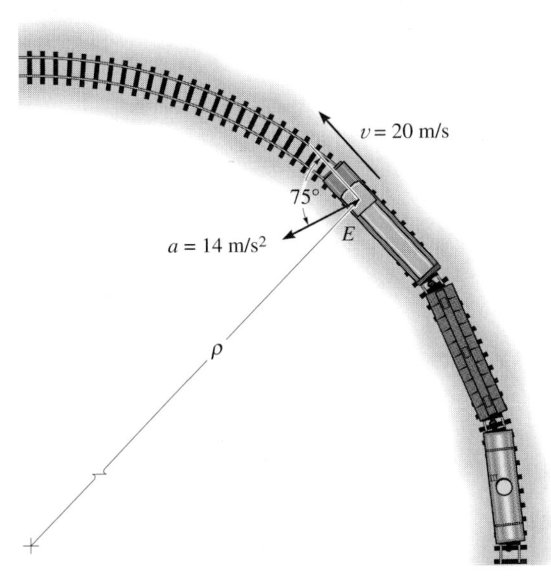

Prob. 12–125

12–126. The truck travels at a speed of 4 m/s along a circular road that has a radius of 50 m. For a short distance from $s = 0$, its speed is then increased by $\dot{v} = (0.05s)\ \text{m/s}^2$, where s is in meters. Determine its speed and the magnitude of its acceleration when it has moved $s = 10$ m.

12–127. The truck travels along a circular road that has a radius of 50 m at a speed of 4 m/s. For a short distance when $t = 0$, its speed is then increased by $\dot{v} = (0.4t)\ \text{m/s}^2$, where t is in seconds. Determine the speed and the magnitude of the truck's acceleration when $t = 4$ s.

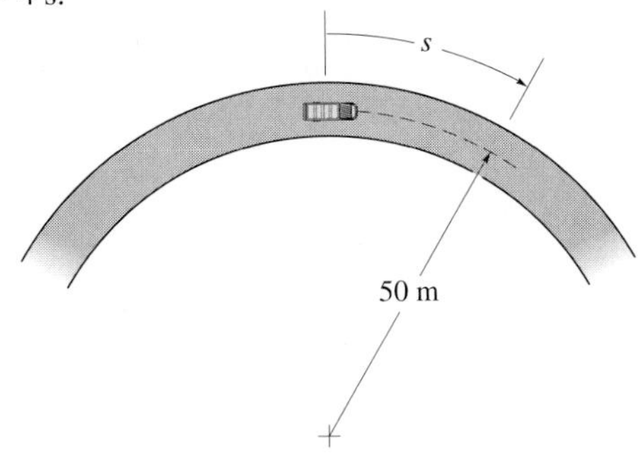

Probs. 12–126/12–127

***12–128.** The automobile is originally at rest at $s = 0$. If its speed is increased by $\dot{v} = (0.05t^2)\ \text{ft/s}^2$, where t is in seconds, determine the magnitudes of its velocity and acceleration when $t = 18$ s.

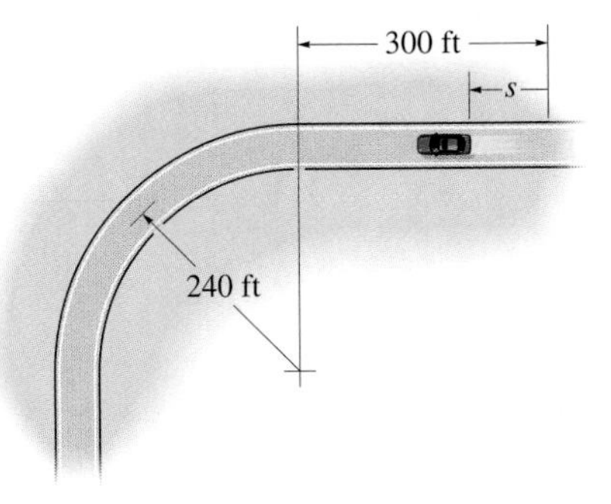

Prob. 12–128

12–129. The automobile is originally at rest at $s = 0$. If it then starts to increase its speed at $\dot{v} = (0.05t^2)\ \text{ft/s}^2$, where t is in seconds, determine the magnitudes of its velocity and acceleration at $s = 550$ ft.

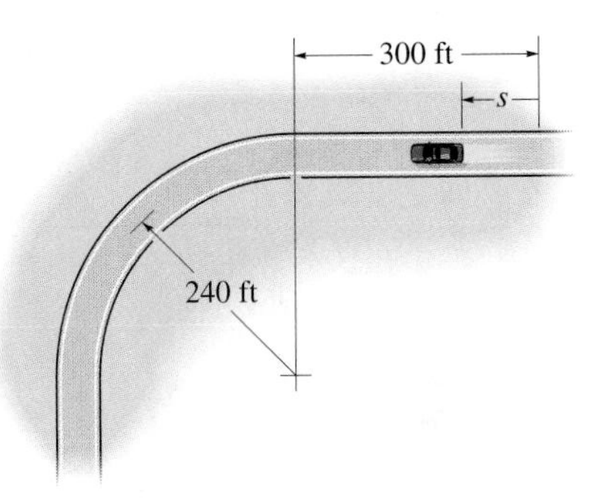

Prob. 12–129

12–130. A boat has an initial speed of 16 ft/s. If it then increases its speed along a circular path of radius $\rho = 80$ ft at the rate of $\dot{v} = (1.5s)$ ft/s, where s is in feet, determine the time needed for the boat to travel $s = 50$ ft.

12–131. A boat has an initial speed of 16 ft/s. If it then increases its speed along a circular path of radius $\rho = 80$ ft at the rate of $\dot{v} = (1.5s)$ ft/s, where s is in feet, determine the normal and tangential components of the boat's acceleration at $s = 16$ ft.

***■12–132.** A go-cart moves along a circular track of radius 100 ft such that its speed for a short period of time, $0 \le t \le 4$ s, is $v = 60(1 - e^{-t^2})$ ft/s. Determine the magnitude of its acceleration when $t = 2$ s. How far has it traveled in $t = 2$ s? Use Simpson's rule with $n = 50$ to evaluate the integral.

■**12–133.** The car B turns such that its speed is increased by $\dot{v}_B = (0.5e^t)$ m/s^2, where t is in seconds. If the car starts from rest when $\theta = 0°$, determine the magnitudes of its velocity and acceleration when the arm AB rotates $\theta = 30°$. Neglect the size of the car.

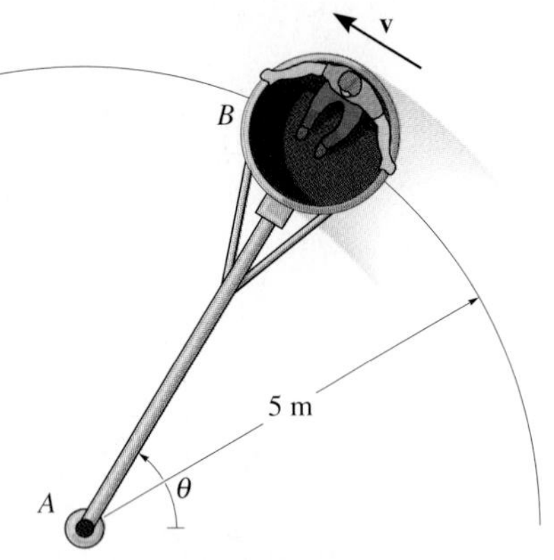

Prob. 12–133

12–134. The car B turns such that its speed is increased by $\dot{v}_B = (0.5e^t)$ m/s^2, where t is in seconds. If the car starts from rest when $\theta = 0°$, determine the magnitudes of its velocity and acceleration when $t = 2$ s. Neglect the size of the car. Also, through what angle θ has it traveled?

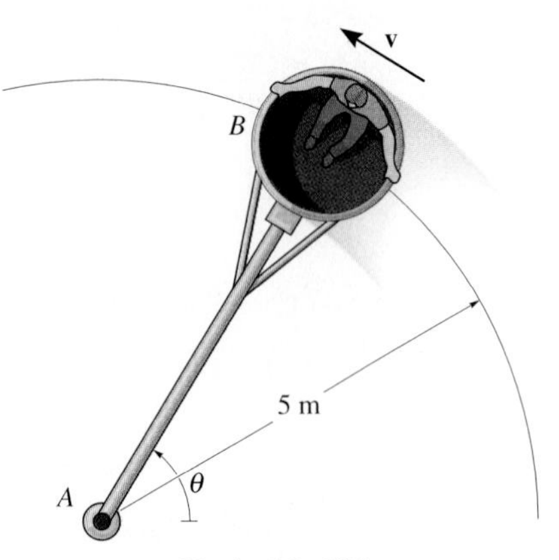

Prob. 12–134

12–135. A boy sits on a merry-go-round so that he is always located at $r = 8$ ft from the center of rotation. The merry-go-round is originally at rest, and then due to rotation the boy's speed is uniformly increased at 2 ft/s^2. Determine the time needed for his acceleration to become 4 ft/s^2.

***12–136.** A particle moves along the hyperbolic curve $y = 1/x$ with a constant speed of 6 ft/s. Determine the magnitude and direction of the particle's acceleration at the instant it reaches the point (1 ft, 1 ft).

12–137. The ball is kicked with an initial speed $v_A = 8$ m/s at an angle $\theta_A = 40°$ with the horizontal. Find the equation of the path, $y = f(x)$, and then determine the ball's velocity and the normal and tangential components of its acceleration when $t = 0.25$ s.

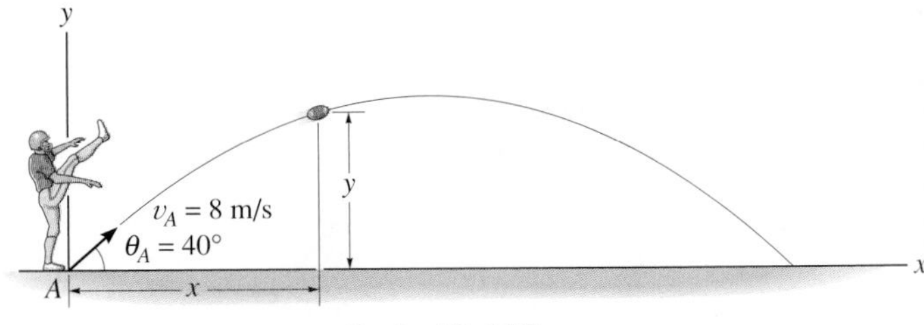

Prob. 12–137

12–138. Particles A and B are traveling counterclockwise around a circular track at a constant speed of 8 m/s. If at the instant shown the speed of A is increased by $\dot{v}_A = (4s_A)$ m/s^2, where s_A is in meters, determine the distance measured counterclockwise along the track from A to B when $t = 1$ s. What is the magnitude of the acceleration of each particle at this instant?

■**12–139.** Particles A and B are traveling around a circular track at a speed of 8 m/s at the instant shown. If the speed of B is increased by $\dot{v}_B = 4$ m/s^2, and at the same instant A has an increase in speed $\dot{v}_A = 0.8t$ m/s^2, determine how long it takes for a collision to occur. What is the magnitude of the acceleration of each particle just before the collision occurs?

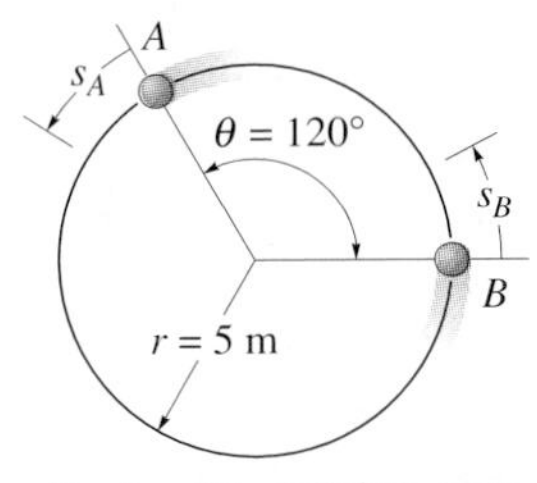

Probs. 12–138/12–139

***12–140.** A particle moves along the curve $y = \sin x$ with a constant speed $v = 2$ m/s. Determine the normal and tangential components of its velocity and acceleration at any instant.

12–141. A particle moves along the curve $y = 1 + \cos x$ with a constant speed $v = 4$ ft/s. Determine the normal and tangential components of its velocity and acceleration at any instant.

12–142. The motion of a particle is defined by the equations $x = (4t^2)$ ft and $y = (2t^3)$ ft, where t is in seconds. Determine the normal and tangential components of the particle's velocity and acceleration when $t = 2$ s.

12–143. The motion of a particle is defined by the equations $x = (2t + t^2)$ m and $y = (t^2)$ m, where t is in seconds. Determine the normal and tangential components of the particle's velocity and acceleration when $t = 2$ s.

***12–144.** The motorcycle travels along the elliptical track at a constant speed v. Determine its greatest acceleration if $a > b$.

12–145. The motorcycle travels along the elliptical track at a constant speed v. Determine its smallest acceleration if $a > b$.

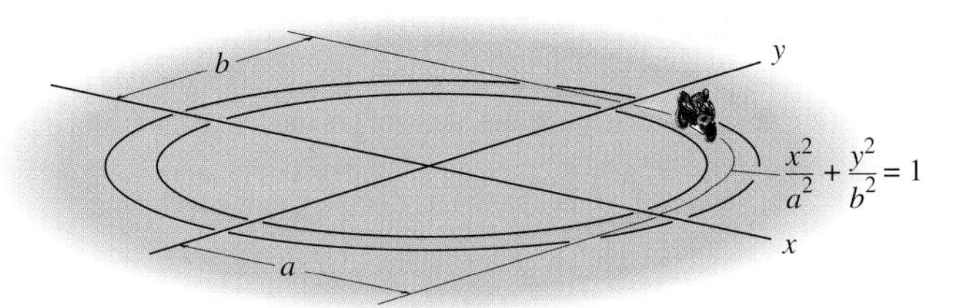

Probs. 12–144/12–145

12–146. The two particles A and B start at the origin O and travel in opposite directions along the circular path at constant speeds $v_A = 0.7$ m/s and $v_B = 1.5$ m/s, respectively. Determine in $t = 2$ s, (a) the displacement along the path of each particle, (b) the position vector to each particle, and (c) the shortest distance between the particles.

12–147. The two particles A and B start at the origin O and travel in opposite directions along the circular path at constant speeds $v_A = 0.7$ m/s and $v_B = 1.5$ m/s, respectively. Determine the time when they collide and the magnitude of the acceleration of B just before this happens.

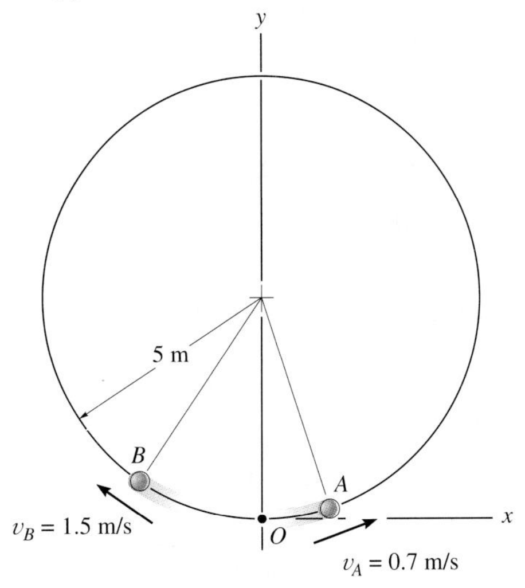

Probs. 12–146/12–147

12.7 Curvilinear Motion: Cylindrical Components

In some engineering problems it is often convenient to express the path of motion of a particle in terms of cylindrical coordinates, r, θ, and z. If motion is restricted to the plane, the polar coordinates r and θ are used.

Polar Coordinates. We can specify the location of particle P shown in Fig. 12–30a using both the *radial coordinate r,* which extends outward from the fixed origin O to the particle, and a *transverse coordinate* θ, which is the counterclockwise angle between a fixed reference line and the r axis. The angle is generally measured in degrees or radians, where 1 rad $= 180°/\pi$. The positive directions of the r and θ coordinates are defined by the unit vectors $\mathbf{u}_r$ and $\mathbf{u}_\theta$, respectively. These directions are *perpendicular* to one another. Here $\mathbf{u}_r$ or the radial direction $+r$ extends from P along increasing r, when θ is held fixed, and $\mathbf{u}_\theta$ or $+\theta$ extends from P in a direction that occurs when r is held fixed and θ is increased.

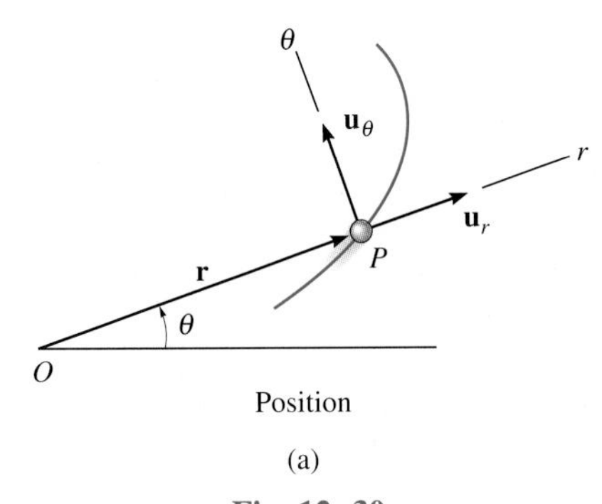

Position

(a)

Fig. 12–30

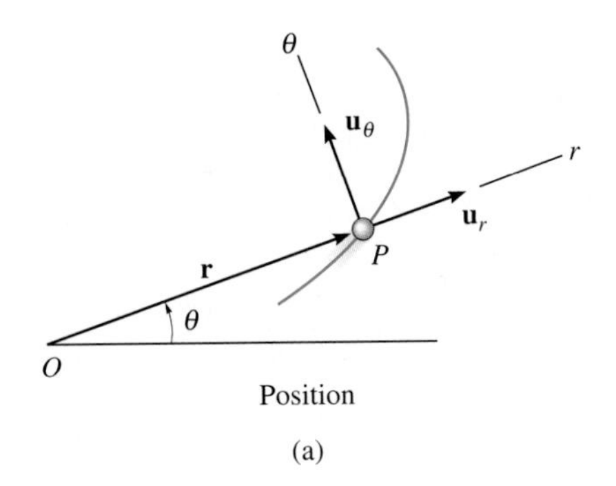

Position

(a)

Position. At any instant the position of the particle, Fig. 12–30*a*, is defined by the position vector

$$\mathbf{r} = r\,\mathbf{u}_r \tag{12–22}$$

Velocity. The instantaneous velocity $\mathbf{v}$ is obtained by taking the time derivative of $\mathbf{r}$. Using a dot to represent time differentiation, we have

$$\mathbf{v} = \dot{\mathbf{r}} = \dot{r}\mathbf{u}_r + r\,\dot{\mathbf{u}}_r$$

In order to evaluate $\dot{\mathbf{u}}_r$, notice that $\mathbf{u}_r$ changes only its direction with respect to time, since by definition the magnitude of this vector is always unity. Hence, during the time Δt, a change Δr will not cause a change in the direction of $\mathbf{u}_r$; however, a change $\Delta\theta$ will cause $\mathbf{u}_r$ to become $\mathbf{u}'_r$, where $\mathbf{u}'_r = \mathbf{u}_r + \Delta\mathbf{u}_r$, Fig. 12–30*b*. The time change in $\mathbf{u}_r$ is then $\Delta\mathbf{u}_r$. For small angles $\Delta\theta$ this vector has a magnitude $\Delta u_r \approx 1(\Delta\theta)$ and acts in the $\mathbf{u}_\theta$ direction. Therefore, $\Delta\mathbf{u}_r = \Delta\theta\mathbf{u}_\theta$, and so

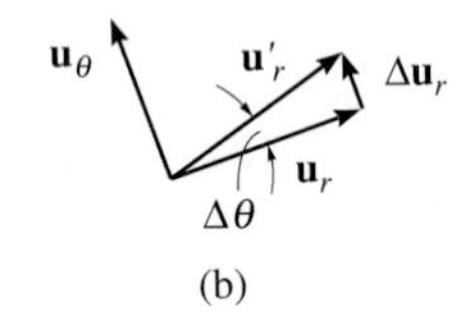

(b)

$$\dot{\mathbf{u}}_r = \lim_{\Delta t \to 0} \frac{\Delta\mathbf{u}_r}{\Delta t} = \left(\lim_{\Delta t \to 0} \frac{\Delta\theta}{\Delta t}\right)\mathbf{u}_\theta$$

$$\dot{\mathbf{u}}_r = \dot{\theta}\mathbf{u}_\theta \tag{12–23}$$

Substituting into the above equation, the velocity can be written in component form as

$$\mathbf{v} = v_r\mathbf{u}_r + v_\theta\mathbf{u}_\theta \tag{12–24}$$

where

$$\begin{aligned} v_r &= \dot{r} \\ v_\theta &= r\dot{\theta} \end{aligned} \tag{12–25}$$

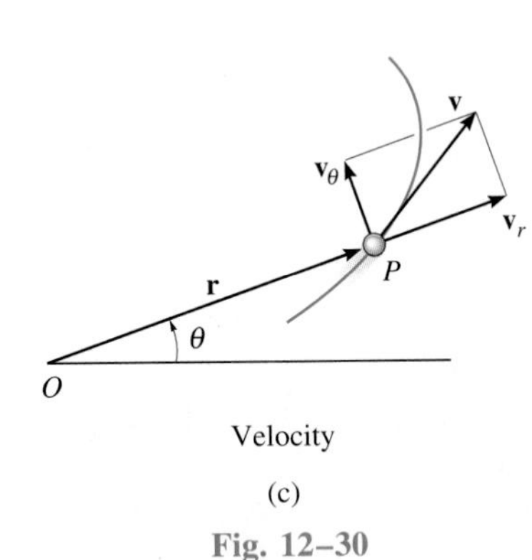

Velocity

(c)

Fig. 12–30

These components are shown graphically in Fig. 12–30*c*. Note that the *radial component* $\mathbf{v}_r$ is a measure of the rate of increase or decrease in the length of the radial coordinate, i.e., $\dot{r}$; whereas the *transverse component* $\mathbf{v}_\theta$ can be interpreted as the rate of motion along the circumference of a circle having a radius r. In particular, the term $\dot{\theta} = d\theta/dt$ is called the *angular velocity*, since it provides a measure of the time rate of change of the angle θ. Common units used for this measurement are rad/s.

Since $\mathbf{v}_r$ and $\mathbf{v}_\theta$ are mutually perpendicular, the *magnitude* of velocity or speed is simply the positive value of

$$v = \sqrt{(\dot{r})^2 + (r\dot{\theta})^2} \tag{12–26}$$

and the *direction* of $\mathbf{v}$ is, of course, tangent to the path at P, Fig. 12–30*c*.

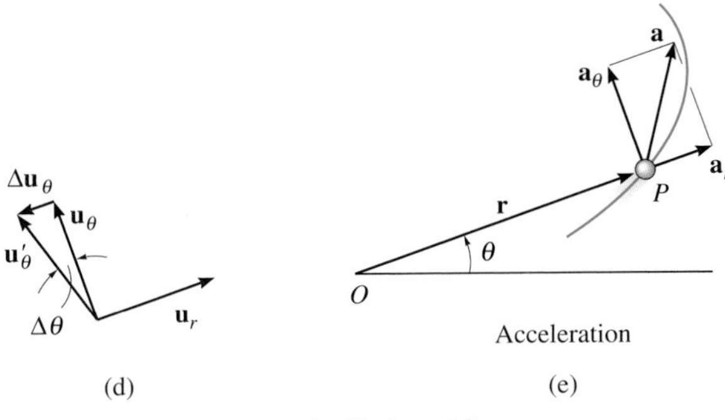

Fig. 12–30 *(cont'd)*

Acceleration. Taking the time derivatives of Eq. 12–24, using Eqs. 12–25, we obtain the particle's instantaneous acceleration,

$$\mathbf{a} = \dot{\mathbf{v}} = \ddot{r}\mathbf{u}_r + \dot{r}\dot{\mathbf{u}}_r + \dot{r}\dot{\theta}\mathbf{u}_\theta + r\ddot{\theta}\mathbf{u}_\theta + r\dot{\theta}\dot{\mathbf{u}}_\theta$$

To evaluate the term involving $\dot{\mathbf{u}}_\theta$, it is necessary only to find the change made in the direction of $\mathbf{u}_\theta$ since its magnitude is always unity. During the time Δt, a change Δr will not change the direction of $\mathbf{u}_\theta$, although a change $\Delta\theta$ will cause $\mathbf{u}_\theta$ to become $\mathbf{u}'_\theta$, where $\mathbf{u}'_\theta = \mathbf{u}_\theta + \Delta\mathbf{u}_\theta$, Fig. 12–30*d*. The time change in $\mathbf{u}_\theta$ is thus $\Delta\mathbf{u}_\theta$. For small angles this vector has a magnitude $\Delta u_\theta \approx 1(\Delta\theta)$ and acts in the $-\mathbf{u}_r$ direction; i.e., $\Delta\mathbf{u}_\theta = -\Delta\theta\mathbf{u}_r$. Thus,

$$\dot{\mathbf{u}}_\theta = \lim_{\Delta t \to 0} \frac{\Delta\mathbf{u}_\theta}{\Delta t} = -\left(\lim_{\Delta t \to 0} \frac{\Delta\theta}{\Delta t}\right)\mathbf{u}_r$$

$$\dot{\mathbf{u}}_\theta = -\dot{\theta}\mathbf{u}_r \qquad (12\text{–}27)$$

Substituting this result and Eq. 12–23 into the above equation for **a**, we can write the acceleration in component form as

$$\mathbf{a} = a_r\mathbf{u}_r + a_\theta\mathbf{u}_\theta \qquad (12\text{–}28)$$

where

$$a_r = \ddot{r} - r\dot{\theta}^2$$
$$a_\theta = r\ddot{\theta} + 2\dot{r}\dot{\theta} \qquad (12\text{–}29)$$

The term $\ddot{\theta} = d^2\theta/dt^2 = d/dt(d\theta/dt)$ is called the *angular acceleration* since it measures the change made in the rate of change of θ during an instant of time. Common units for this measurement are rad/s^2.

Since $\mathbf{a}_r$ and $\mathbf{a}_\theta$ are always perpendicular, the *magnitude* of acceleration is simply the positive value of

$$a = \sqrt{(\ddot{r} - r\dot{\theta}^2)^2 + (r\ddot{\theta} + 2\dot{r}\dot{\theta})^2} \qquad (12\text{–}30)$$

The *direction* is determined from the vector addition of its two components. In general, **a** will *not* be tangent to the path, Fig. 12–30*e*.

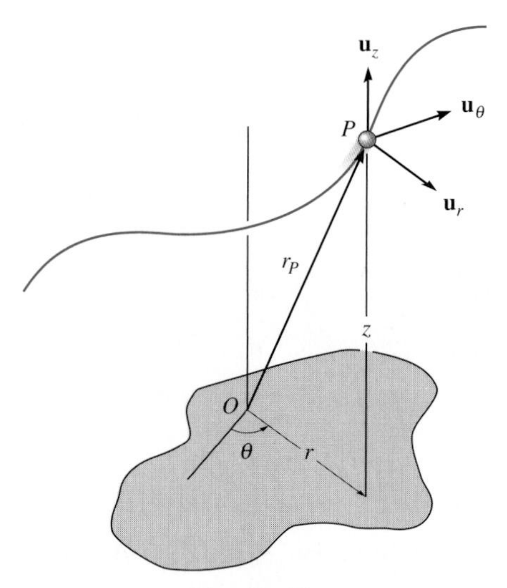

Fig. 12–31

Cylindrical Coordinates. If the particle P moves along a space curve as shown in Fig. 12–31, then its location may be specified by the three *cylindrical coordinates, r, θ, z*. The z coordinate is identical to that used for rectangular coordinates. Since the unit vector defining its direction, $\mathbf{u}_z$, is constant, the time derivatives of this vector are zero and therefore the position, velocity, and acceleration of the particle can be written in terms of its cylindrical coordinates as follows:

$$\mathbf{r}_P = r\mathbf{u}_r + z\mathbf{u}_z$$

$$\mathbf{v} = \dot{r}\mathbf{u}_r + r\dot{\theta}\mathbf{u}_\theta + \dot{z}\mathbf{u}_z \qquad (12\text{–}31)$$

$$\mathbf{a} = (\ddot{r} - r\dot{\theta}^2)\mathbf{u}_r + (r\ddot{\theta} + 2\dot{r}\dot{\theta})\mathbf{u}_\theta + \ddot{z}\mathbf{u}_z \qquad (12\text{–}32)$$

Time Derivatives. The equations of kinematics require that we obtain the time derivatives $\dot{r}$, $\ddot{r}$, $\dot{\theta}$, and $\ddot{\theta}$ in order to evaluate the r and θ components of $\mathbf{v}$ and $\mathbf{a}$. Two types of problems generally occur:

1. If the coordinates are specified as time parametric equations, $r = r(t)$ and $\theta = \theta(t)$, then the time derivatives can be found directly. For example, consider

$$r = 4t^2 \qquad \theta = (8t^3 + 6)$$

$$\dot{r} = 8t \qquad \dot{\theta} = 24\,t^2$$

$$\ddot{r} = 8 \qquad \ddot{\theta} = 48t$$

2. If the time-parametric equations are not given, then it will be necessary to specify the path $r = f(\theta)$ and compute the *relationship* between the time derivatives using the chain rule of calculus. Consider the following examples.

$$
\begin{aligned}
r &= 5\theta^2 \\
\dot{r} &= 10\theta\dot{\theta} \\
\ddot{r} &= 10[(\dot{\theta})\dot{\theta} + \theta(\ddot{\theta})] \\
&= 10\dot{\theta}^2 + 10\theta\ddot{\theta}
\end{aligned}
$$

or

$$
\begin{aligned}
r^2 &= 6\theta^3 \\
2r\dot{r} &= 18\theta^2\dot{\theta} \\
2[(\dot{r})\dot{r} + r(\ddot{r})] &= 18[(2\theta\dot{\theta})\dot{\theta} + \theta^2(\ddot{\theta})] \\
\dot{r}^2 + r\ddot{r} &= 9(2\theta\dot{\theta}^2 + \theta^2\ddot{\theta})
\end{aligned}
$$

Thus, if two of the *four* time derivatives $\dot{r}$, $\ddot{r}$, $\dot{\theta}$, and $\ddot{\theta}$ are *known,* the other two can be obtained from the equations of the first and second time derivatives. In some problems, however, two of these time derivatives may *not* be known; instead the magnitude of the particle's velocity or acceleration may be specified. If this is the case, Eqs 12–26 and 12–30 [$v^2 = \dot{r}^2 + (r\dot{\theta})^2$ and $a^2 = (\ddot{r} - r\dot{\theta}^2)^2 + (r\ddot{\theta} + 2\dot{r}\dot{\theta})^2$] may be used to obtain the necessary relationships involving $\dot{r}$, $\ddot{r}$, $\dot{\theta}$, and $\ddot{\theta}$.

PROCEDURE FOR ANALYSIS

Coordinate System. Polar coordinates are a suitable choice for solving problems for which data regarding the angular motion of the radial coordinate r is given to describe the particle's motion. Also, some paths of motion can conveniently be described in terms of these coordinates. To use cylindrical coordinates, the origin is established at a fixed point, and the radial line r is directed to the particle. Also, the transverse coordinate θ is measured counterclockwise from a fixed reference line to the radial line. If necessary, the equation of the path can then be determined by relating r to θ, $r = f(\theta)$.

Velocity and Acceleration. Once r and the time derivatives $\dot{r}$, $\ddot{r}$, $\dot{\theta}$, and $\ddot{\theta}$ have been evaluated at the instant considered, their values can be substituted into Eqs. 12–25 and 12–29 to obtain the radial and transverse components of **v** and **a**.

Motion in three dimensions requires a simple extension of the above procedure to include $\dot{z}$ and $\ddot{z}$.

Besides the examples which follow, further examples involving the calculation of a_r and a_θ can be found in the "kinematics" sections of Examples 13–10 through 13–12.

Example 12–17

The amusement park ride shown in Fig. 12–32*a* consists of a chair that is rotating in a horizontal circular path of radius r such that the arm OB has an angular velocity $\dot{\theta}$ and angular acceleration $\ddot{\theta}$. Determine the radial and transverse components of velocity and acceleration of the passenger. Neglect his size in the calculation.

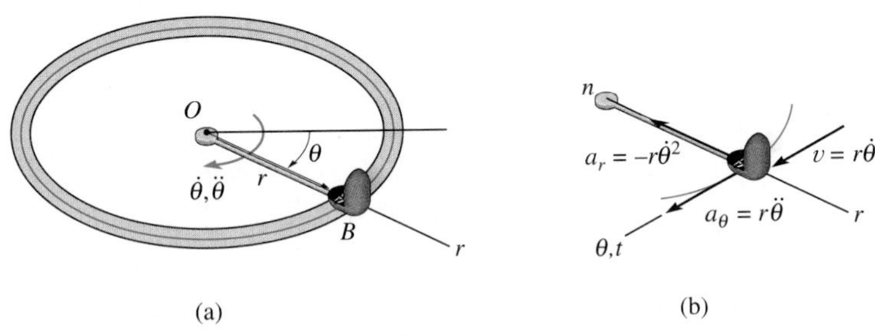

Fig. 12–32

SOLUTION

Coordinate System. Since the angular motion of the cord is reported, polar coordinates are chosen for the solution, Fig. 12–32*a*. Here θ is not related to r, since the radius is constant for all θ.

Velocity and Acceleration. Equations 12–25 and 12–29 will be used for the solution, and so it is first necessary to specify the first and second time derivatives of r and θ. Since r is *constant,* we have

$$r = r \qquad \dot{r} = 0 \qquad \ddot{r} = 0$$

Thus

$$v_r = \dot{r} = 0 \qquad \textit{Ans.}$$

$$v_\theta = r\dot{\theta} \qquad \textit{Ans.}$$

$$a_r = \ddot{r} - r\dot{\theta}^2 = -r\dot{\theta}^2 \qquad \textit{Ans.}$$

$$a_\theta = r\ddot{\theta} + 2\dot{r}\dot{\theta} = r\ddot{\theta} \qquad \textit{Ans.}$$

These results are shown in Fig. 12–32*b*. Also shown are the n, t axes, which in this special case of circular motion happen to be *collinear* with the r and θ axes, respectively. In particular note that $v = v_\theta = v_t = r\dot{\theta}$. Also,

$$-a_r = a_n = \frac{v^2}{\rho} = \frac{(r\dot{\theta})^2}{r} = r\dot{\theta}^2$$

$$a_\theta = a_t = \frac{dv}{dt} = \frac{d}{dt}(r\dot{\theta}) = \frac{dr}{dt}\dot{\theta} + r\frac{d\dot{\theta}}{dt} = 0 + r\ddot{\theta}$$

Example 12–18

The rod OA, shown in Fig. 12–33a, is rotating in the horizontal plane such that $\theta = (t^3)$ rad. At the same time, the collar B is sliding outward along OA so that $r = (100t^2)$ mm. If in both cases t is in seconds, determine the velocity and acceleration of the collar when $t = 1$ s.

SOLUTION

Coordinate System. Since time-parametric equations of the path are given, it is not necessary to relate r to θ.

Velocity and Acceleration. Computing the time derivatives, and evaluating when $t = 1$ s, we have

$$r = 100t^2\Big|_{t=1\,\text{s}} = 100 \text{ mm} \qquad \theta = t^3\Big|_{t=1\,\text{s}} = 1 \text{ rad} = 57.3^\circ$$

$$\dot{r} = 200t\Big|_{t=1\,\text{s}} = 200 \text{ mm/s} \qquad \dot{\theta} = 3t^2\Big|_{t=1\,\text{s}} = 3 \text{ rad/s}$$

$$\ddot{r} = 200\Big|_{t=1\,\text{s}} = 200 \text{ mm/s}^2 \qquad \ddot{\theta} = 6t\Big|_{t=1\,\text{s}} = 6 \text{ rad/s}^2$$

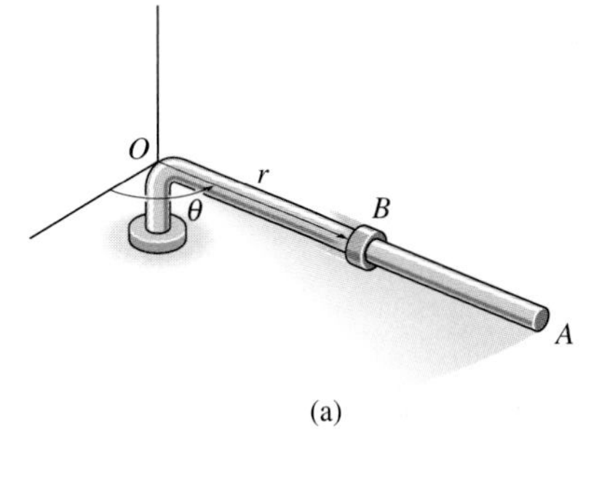

(a)

As shown in Fig. 12–33b,

$$\begin{aligned}\mathbf{v} &= \dot{r}\mathbf{u}_r + r\dot{\theta}\mathbf{u}_\theta \\ &= 200\mathbf{u}_r + 100(3)\mathbf{u}_\theta \\ &= \{200\mathbf{u}_r + 300\mathbf{u}_\theta\} \text{ mm/s}\end{aligned}$$

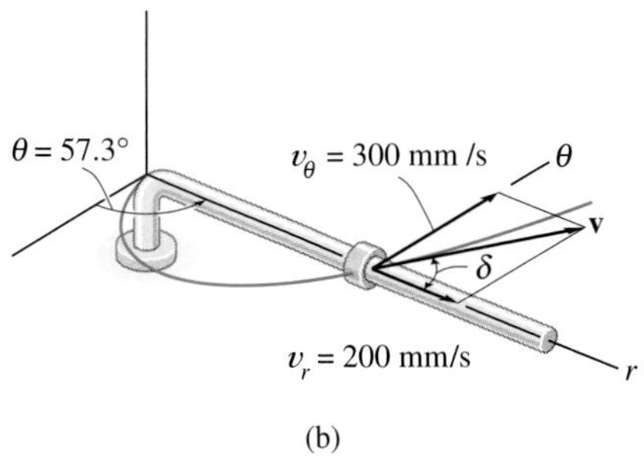

(b)

The magnitude of **v** is

$$v = \sqrt{(200)^2 + (300)^2} = 361 \text{ mm/s} \qquad \textit{Ans.}$$

$$\delta = \tan^{-1}\left(\frac{300}{200}\right) = 56.3^\circ \qquad \delta + 57.3^\circ = 114^\circ \; \mathbf{v}\; 114^\circ \qquad \textit{Ans.}$$

As shown in Fig. 12–33c,

$$\begin{aligned}\mathbf{a} &= (\ddot{r} - r\dot{\theta}^2)\mathbf{u}_r + (r\ddot{\theta} + 2\dot{r}\dot{\theta})\mathbf{u}_\theta \\ &= [200 - 100(3)^2]\mathbf{u}_r + [100(6) + 2(200)3]\mathbf{u}_\theta \\ &= \{-700\mathbf{u}_r + 1800\mathbf{u}_\theta\} \text{ mm/s}^2\end{aligned}$$

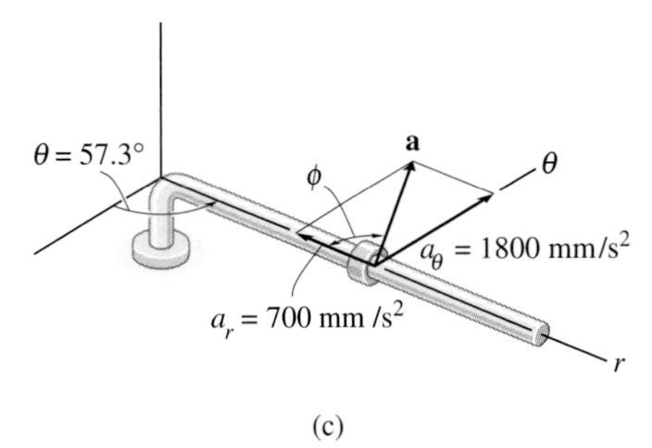

(c)

Fig. 12–33

The magnitude of **a** is

$$a = \sqrt{(700)^2 + (1800)^2} = 1930 \text{ mm/s}^2 \qquad \textit{Ans.}$$

$$\phi = \tan^{-1}\left(\frac{1800}{700}\right) = 68.7^\circ \qquad \phi - 57.3^\circ = 11.4^\circ \; 11.4^\circ \mathbf{a} \qquad \textit{Ans.}$$

Example 12–19

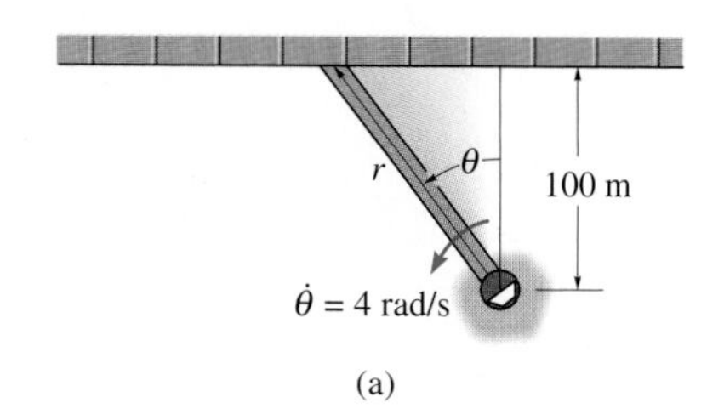

(a)

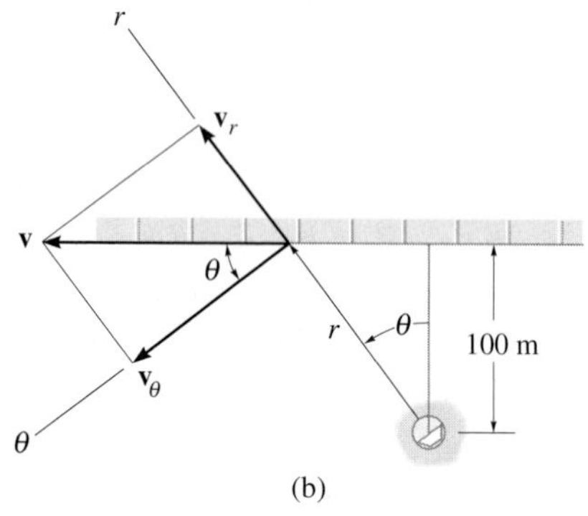

(b)

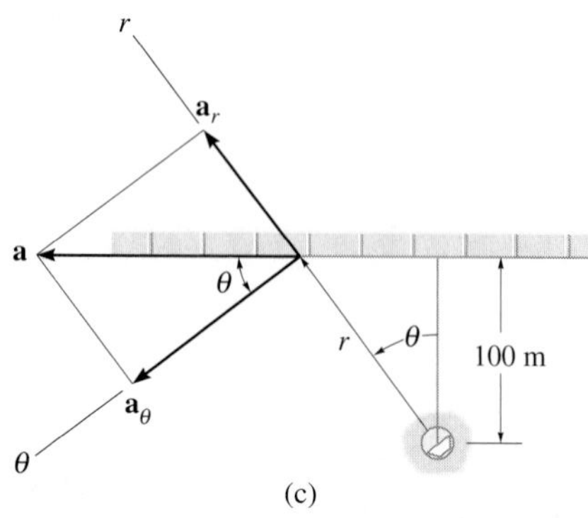

(c)

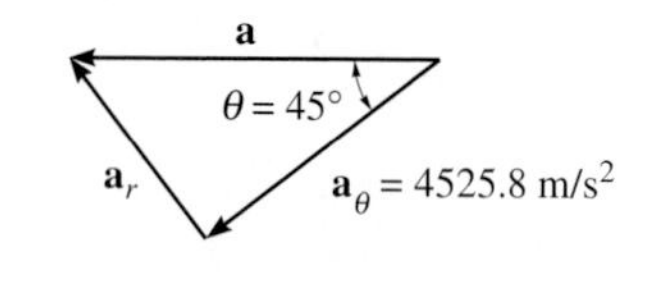

(d)

Fig. 12–34

The searchlight shown in Fig. 12–34*a* casts a spot of light along the face of a wall that is located 100 m from the searchlight. Determine the magnitude of the velocity and acceleration at which the spot appears to travel across the wall at the instant $\theta = 45°$. The searchlight is rotating at a constant rate of $\dot{\theta} = 4$ rad/s.

SOLUTION

Coordinate System. Why should polar coordinates be used to solve this problem? To compute the necessary time derivatives it is first necessary to relate r to θ. From Fig. 12–34*a*, this relation is

$$r = 100/\cos\theta = 100\sec\theta$$

Velocity and Acceleration. Using the chain rule of calculus, noting that $d(\sec\theta) = \sec\theta\tan\theta\,d\theta$, and $d(\tan\theta) = \sec^2\theta\,d\theta$, we have

$$\dot{r} = 100(\sec\theta\tan\theta)\dot{\theta}$$

$$\ddot{r} = 100(\sec\theta\tan\theta)\dot{\theta}(\tan\theta)\dot{\theta} + 100\sec\theta(\sec^2\theta)\dot{\theta}(\dot{\theta}) + 100\sec\theta\tan\theta(\ddot{\theta})$$

$$= 100\sec\theta\tan^2\theta(\dot{\theta})^2 + 100\sec^3\theta(\dot{\theta})^2 + 100(\sec\theta\tan\theta)\ddot{\theta}$$

Since $\dot{\theta} = 4$ rad/s = constant, then $\ddot{\theta} = 0$, and the above equations, when $\theta = 45°$, become

$$r = 100\sec 45° = 141.4$$

$$\dot{r} = 400\sec 45°\tan 45° = 565.7$$

$$\ddot{r} = 1600(\sec 45°\tan^2 45° + \sec^3 45°) = 6788.2$$

As shown in Fig. 12–34*b*,

$$\mathbf{v} = \dot{r}\mathbf{u}_r + r\dot{\theta}\mathbf{u}_\theta$$

$$= 565.7\mathbf{u}_r + 141.4(4)\mathbf{u}_\theta$$

$$= 565.7\mathbf{u}_r + 565.7\mathbf{u}_\theta$$

$$v = \sqrt{v_r^2 + v_\theta^2} = \sqrt{(565.7)^2 + (565.7)^2}$$

$$= 800 \text{ m/s} \qquad \textit{Ans.}$$

As shown in Fig. 12–34*c*,

$$\mathbf{a} = (\ddot{r} - r\dot{\theta}^2)\mathbf{u}_r + (r\ddot{\theta} + 2\dot{r}\dot{\theta})\mathbf{u}_\theta$$

$$= [6788.2 - 141.4(4)^2]\mathbf{u}_r + [141.4(0) + 2(565.7)4]\mathbf{u}_\theta$$

$$= 4525.8\mathbf{u}_r + 4525.8\mathbf{u}_\theta$$

$$a = \sqrt{a_r^2 + a_\theta^2} = \sqrt{(4525.8)^2 + (4525.8)^2}$$

$$= 6400 \text{ m/s}^2 \qquad \textit{Ans.}$$

Note: It is also possible to compute a, *without* having to calculate $\ddot{r}$ (or a_r). As shown in Fig. 12–34*d*, since $a_\theta = 4525.8$ m/s², then by vector resolution, $a = 4525.8/\cos 45° = 6400$ m/s².

Example 12–20

Due to the rotation of the forked rod, the cylindrical peg *A* in Fig. 12–35*a* travels around the slotted path, a portion of which is in the shape of a cardioid, $r = 0.5(1 - \cos\theta)$ ft, where θ is in radians. If the peg's velocity is $v = 4$ ft/s and its acceleration is $a = 30$ ft/s^2 at the instant $\theta = 180°$, determine the angular velocity $\dot{\theta}$ and angular acceleration $\ddot{\theta}$ of the fork.

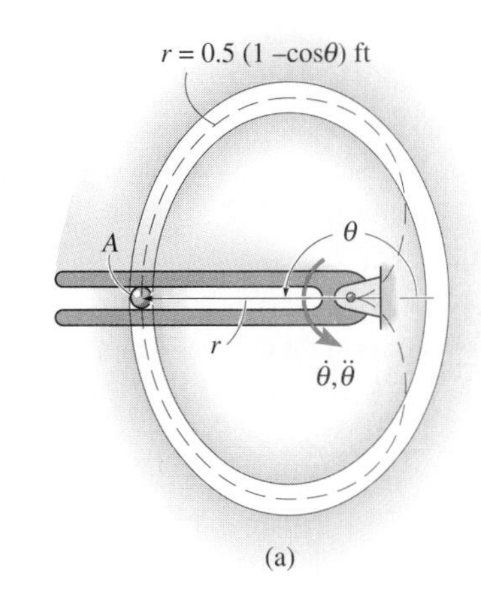

(a)

SOLUTION

Coordinate System. This path is most unusual, and mathematically it is best expressed using polar coordinates, as done here, rather than rectangular coordinates. Also, $\dot{\theta}$ and $\ddot{\theta}$ must be determined so r, θ coordinates are an obvious choice.

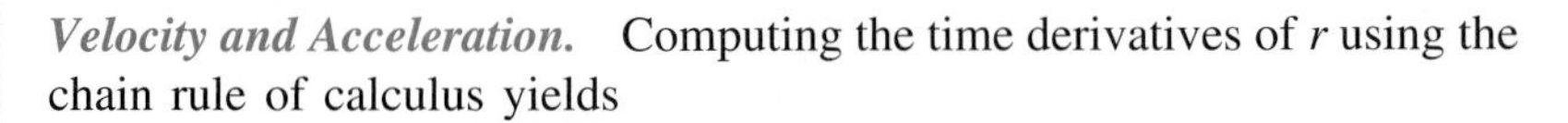

Velocity and Acceleration. Computing the time derivatives of r using the chain rule of calculus yields

$$r = 0.5(1 - \cos\theta)$$
$$\dot{r} = 0.5(\sin\theta)\dot{\theta}$$
$$\ddot{r} = 0.5(\cos\theta)\dot{\theta}(\dot{\theta}) + 0.5(\sin\theta)\ddot{\theta}$$

Evaluating these results at $\theta = 180°$, we have

$$r = 1\text{ ft} \qquad \dot{r} = 0 \qquad \ddot{r} = -0.5\dot{\theta}^2$$

Since $v = 4$ ft/s, using Eq. 12–26 to determine $\dot{\theta}$ yields

$$v = \sqrt{(\dot{r})^2 + (r\dot{\theta})^2}$$
$$4 = \sqrt{(0)^2 + (1\dot{\theta})^2}$$
$$\dot{\theta} = 4\text{ rad/s} \qquad \textit{Ans.}$$

In a similar manner, $\ddot{\theta}$ can be found using Eq. 12–30.

$$a = \sqrt{(\ddot{r} - r\dot{\theta}^2)^2 + (r\ddot{\theta} + 2\dot{r}\dot{\theta})^2}$$
$$30 = \sqrt{[-0.5(4)^2 - 1(4)^2]^2 + [1\ddot{\theta} + 2(0)(4)]^2}$$
$$(30)^2 = (-24)^2 + \ddot{\theta}^2$$
$$\ddot{\theta} = 18\text{ rad/s}^2 \qquad \textit{Ans.}$$

Vectors **a** and **v** are shown in Fig. 12–35*b*.

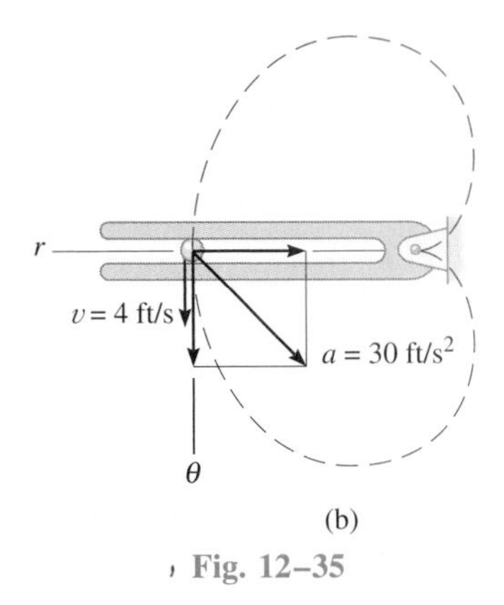

(b)

Fig. 12–35

PROBLEMS

***12–148.** The time rate of change of acceleration is referred to as a *jerk*, which is often used as a means of measuring passenger discomfort. Calculate this vector, $\dot{\mathbf{a}}$, in terms of its cylindrical components, using Eq. 12–29.

12–149. If a particle's position is described by the polar coordinates $r = 4(1 + \sin t)$ m and $\theta = (2e^{-t})$ rad, where t is in seconds and the argument for the sine is in radians, determine the radial and tangential components of the particle's velocity and acceleration when $t = 2$ s.

12–150. If a particle's position is described by the polar coordinates $r = (2 \sin 2\theta)$ m and $\theta = (4t)$ rad, where t is in seconds, determine the radial and tangential components of its velocity and acceleration when $t = 1$ s.

12–151. A particle moves in the x-y plane such that its position is defined by $\mathbf{r} = \{2t\mathbf{i} + 4t^2\mathbf{j}\}$ft, where t is in seconds. Determine the radial and tangential components of the particle's velocity and acceleration when $t = 2$ s.

***12–152.** A particle moves along a path defined by polar coordinates $r = (3t^3)$ m and $\theta = (5t^2)$ rad, where t is in seconds. Determine the components of its velocity and acceleration when $t = 2$ s.

12–153. A particle moves along a path defined by polar coordinates $r = (2e^t)$ ft and $\theta = (8t^2)$ rad, where t is in seconds. Determine the components of its velocity and acceleration when $t = 1$ s.

12–154. A particle moves along a path defined by polar coordinates $r = (3 \sin t)$ m and $\theta = (2t^3)$ rad, where t is in seconds and the argument for the sine is in radians. Determine the components of its velocity and acceleration when $t = 1$ s.

12–155. A boy sits on a merry-go-round so that he is always located at $r = 8$ ft from the center of rotation. The merry-go-round is originally at rest, and then due to rotation the boy's angular rate of rotation is increased by $\ddot{\theta} = 2$ rad/s^2. Determine his speed when his acceleration becomes 25 ft/s^2.

***12–156.** A boy sits on a merry-go-round so that he is always located at $r = 8$ ft from the center of rotation. The merry-go-round is originally at rest, and then due to rotation the boy's angular rate of rotation is increased by $\ddot{\theta} = 2$ rad/s^2. Determine the magnitude of his acceleration in $t = 2$ s. Also, through what distance does he travel?

12–157. A truck is traveling along the horizontal circular curve of radius $r = 60$ m with a constant speed $v = 20$ m/s. Determine the angular rate of rotation $\dot{\theta}$ of the radial line r and the magnitude of the truck's acceleration.

12–158. A truck is traveling along the horizontal circular curve of radius $r = 60$ m with a speed of 20 m/s which is increasing at 3 m/s^2. Determine the truck's radial and transverse components of acceleration.

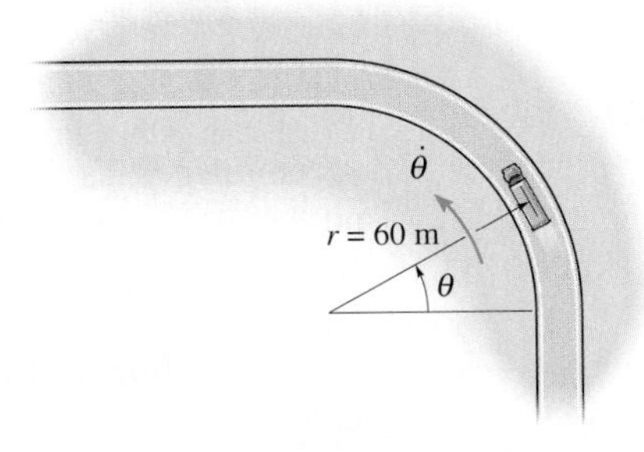

Probs. 12–157/12–158

12–159. A particle is moving along a circular path having a radius of 0.4 m. Its angular position as a function of time is given by $\theta = (2t^2)$ rad, where t is in seconds. Determine the magnitude of the particle's acceleration when $\theta = 30°$. The particle starts from rest when $\theta = 0°$.

***12–160.** At the instant shown, the man is twirling a hose over his head with an angular velocity $\dot{\theta} = 2$ rad/s and an angular acceleration $\ddot{\theta} = 3$ rad/s^2. If it is assumed that the hose lies in a horizontal plane, and water is flowing through it at a constant rate of 3 m/s, determine the magnitudes of the velocity and acceleration of a water particle as it exits the open end, $r = 1.5$ m.

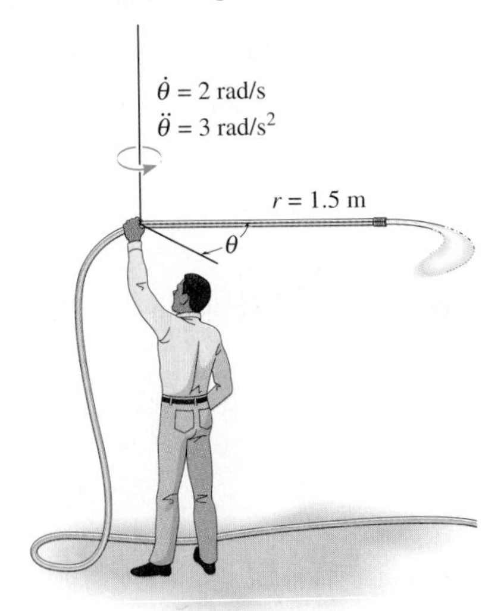

Prob. 12–160

12–161. The slotted link is pinned at O, and as a result of the constant angular velocity $\theta = 3$ rad/s it drives the peg P for a short distance along the spiral guide $r = (0.4\ \theta)$ m, where θ is in radians. Determine the radial and transverse components of the velocity and acceleration of P at the instant $\theta = \pi/3$ rad.

12–162. Solve Prob. 12–161 if the slotted link has an angular acceleration $\theta = 8$ rad/s^2 when $\theta = 3$ rad/s at $\theta = \pi/3$ rad.

12–163. The slotted link is pinned at O, and as a result of the constant angular velocity $\theta = 3$ rad/s it drives the peg P for a short distance along the spiral guide $r = (0.4\ \theta)$ m, where θ is in radians. Determine the velocity and acceleration components of the particle at the instant it leaves the slot in the link, i.e., when $r =$ 0.5 m.

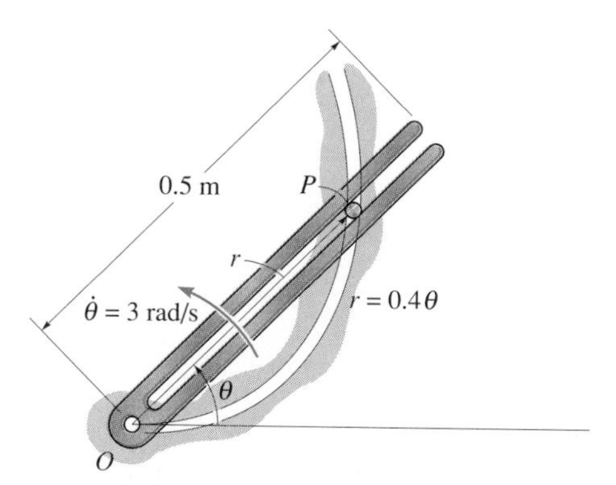

Probs. 12–161/12–162/12–163

***12–164.** The car travels around the circular track with a constant speed of 20 m/s. Determine the car's radial and transverse components of velocity and acceleration at the instant $\theta = \pi/4$ rad.

12–165. The car travels around the circular track such that its transverse component is $\theta = (0.006t^2)$ rad, where t is in seconds. Determine the car's radial and transverse components of velocity and acceleration at the instant $t = 4$ s.

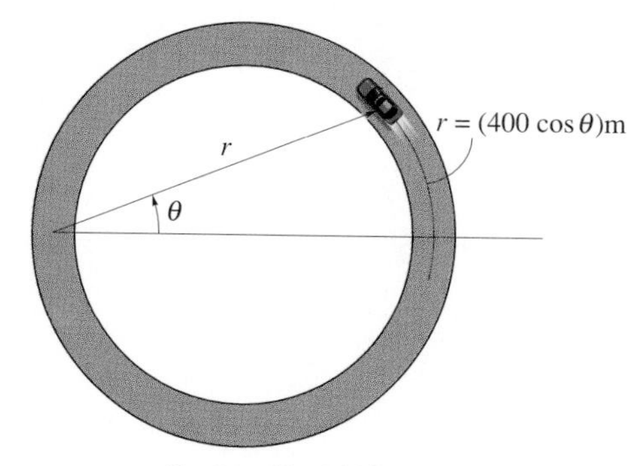

Probs. 12–164/12–165

12–166. A horse on the merry-go-round moves according to the equations $r = 8$ ft, $\theta = (0.6t)$ rad, and $z = (1.5 \sin \theta)$ ft, where t is in seconds. Determine the cylindrical components of the velocity and acceleration of the horse when $t = 4$ s.

12–167. A horse on the merry-go-round moves according to the equations $r = 8$ ft, $\theta = 2$ rad/s and $z = (1.5 \sin \theta)$ ft, where t is in seconds. Determine the maximum and minimum magnitudes of the velocity and acceleration of the horse during the motion.

Probs. 12–166/12–167

***12–168.** At the instant shown, the watersprinkler is rotating with an angular speed $\theta = 2$ rad/s and an angular acceleration $\theta =$ 3 rad/s^2. If the nozzle lies in the vertical plane and water is flowing through it at a constant rate of 3 m/s, determine the magnitudes of the velocity and acceleration of a water particle as it exits the open end, $r = 0.2$ m.

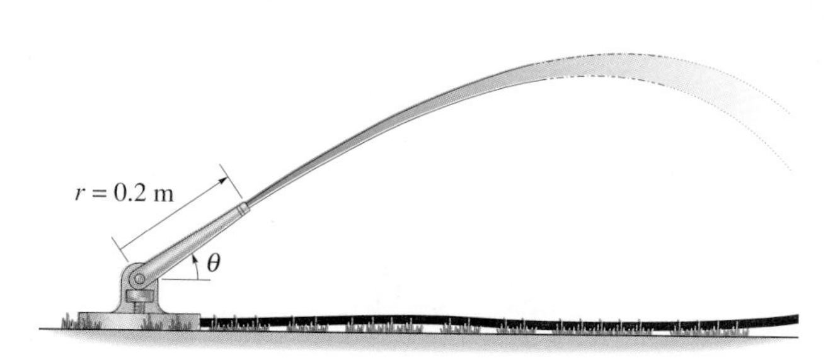

Prob. 12–168

12–169. The arm of the robot has a fixed length so that $r = 3$ ft and its grip A moves along the path $z = (3 \sin 4\theta)$ ft, where θ is in radians. If $\theta = (0.5\,t)$ rad, where t is in seconds, determine the magnitudes of the grip's velocity and acceleration when $t = 3$ s.

12–170. For a short time the arm of the robot is extending at a constant rate such that $\dot{r} = 1.5$ ft/s when $r = 3$ ft, $z = (4t^2)$ ft, and $\theta = 0.5t$ rad, where t is in seconds. Determine the magnitudes of the velocity and acceleration of the grip A when $t = 3$ s.

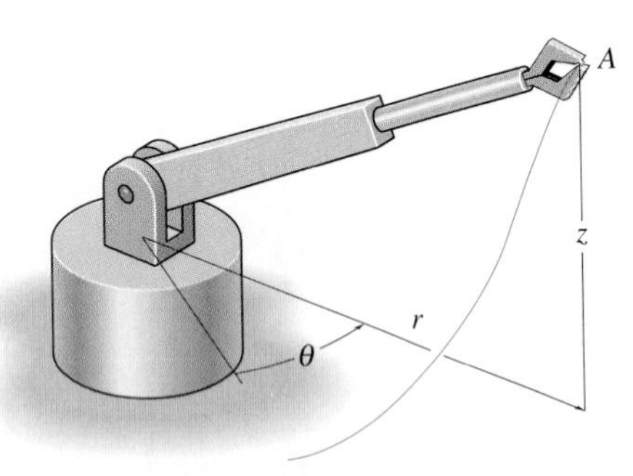

Probs. 12–169/12–170

12–171. Because of telescopic action, the end of the industrial robotic arm extends along the path of the limaçon $r = (1 + 0.5 \cos \theta)$ m. At the instant $\theta = \pi/4$, the arm has an angular rotation $\dot{\theta} = 0.6$ rad/s, which is increasing at $\ddot{\theta} = 0.25$ rad/s^2. Determine the radial and transverse components of the velocity and acceleration of the object A held in its grip at this instant.

***12–172.** Because of telescopic action, the end of the industrial robotic arm extends along the path of the limaçon $r = (1 + 0.5 \cos \theta)$ m. If design requirements restrict the maximum acceleration of the object held in the grip at A to have a magnitude of 5 m/s^2, determine the greatest constant angular velocity of the arm. Require $-30° \leq \theta \leq 30°$.

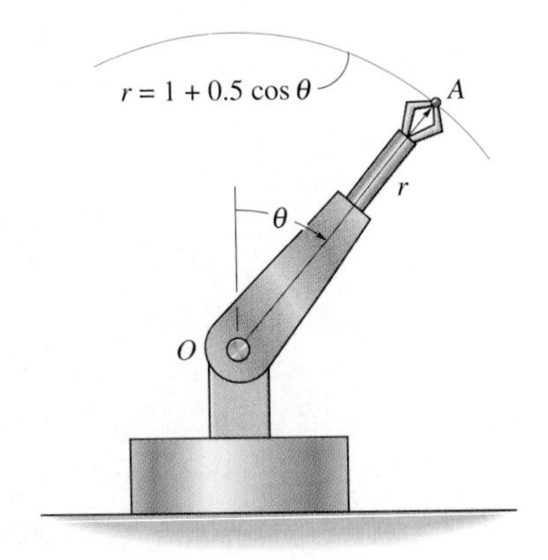

Probs. 12–171/12–172

12–173. For a short distance the train travels along a track having the shape of a spiral, $r = (1000/\theta)$ m, where θ is in radians. If it maintains a constant speed $v = 20$ m/s, determine the radial and transverse components of its velocity when $\theta = (9\pi/4)$ rad.

12–174. For a short distance the train travels along a track having the shape of a spiral, $r = (1000/\theta)$ m, where θ is in radians. If the angular rate is constant, $\dot{\theta} = 0.2$ rad/s, determine the radial and transverse components of its velocity and acceleration when $\theta = (9\pi/4)$ rad.

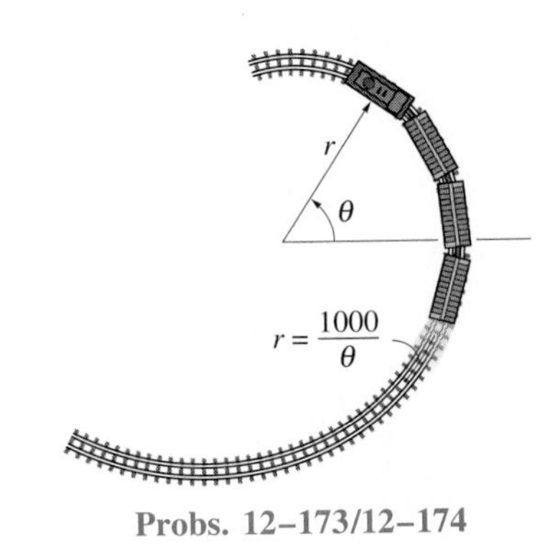

Probs. 12–173/12–174

12–175. A cameraman standing at A is following the movement of a race car, B, which is traveling around a curved track at a constant speed of 30 m/s. Determine the angular rate $\dot{\theta}$ at which the man must turn in order to keep the camera directed on the car at the instant $\theta = 30°$.

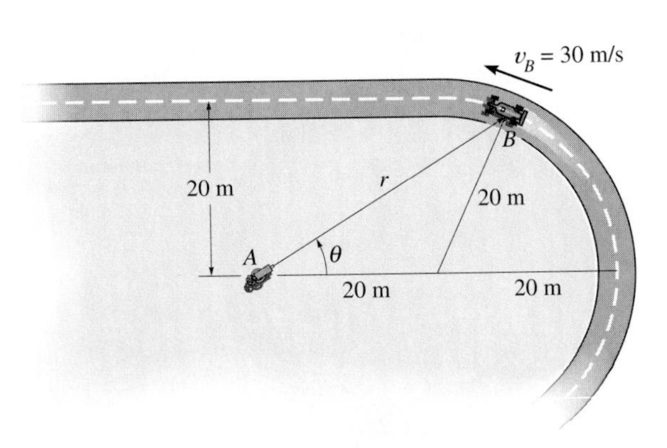

Prob. 12–175

*12–176. A particle travels along the portion of the "four-leaf rose" defined by the equation $r = (5 \cos 2\theta)$ m. If the angular velocity of the radial coordinate line is $\dot{\theta} = (3t^2)$ rad/s, where t is in seconds, determine the radial and transverse components of the particle's velocity and acceleration at the instant $\theta = 30°$. When $t = 0$, $\theta = 0$.

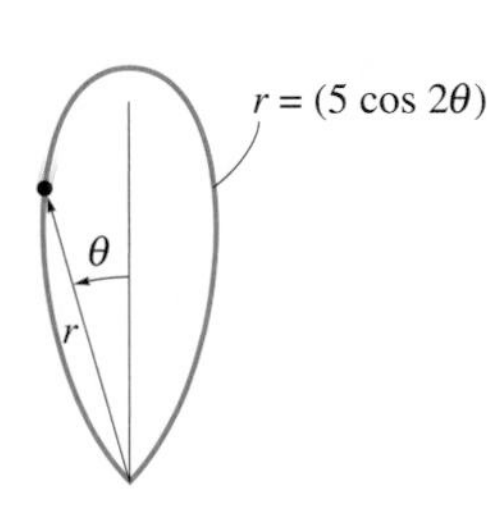

Prob. 12–176

12–177. The slotted arm AB drives the pin C through the spiral groove described by the equation $r = a\theta$. If the angular rate of rotation is constant at $\dot{\theta}$, determine the radial and transverse components of velocity and acceleration of the pin.

12–178. Solve Prob. 12–177 if the spiral path is logarithmic, i.e., $r = a\,e^{k\theta}$.

12–179. The slotted arm AB drives the pin C through the spiral groove described by the equation $r = (1.5\theta)$ ft, where θ is in radians. If the arm starts from rest when $\theta = 60°$ and is driven at an angular rate of $\dot{\theta} = (4t)$ rad/s, where t is in seconds, determine the radial and transverse components of velocity and acceleration of the pin C when $t = 1$ s.

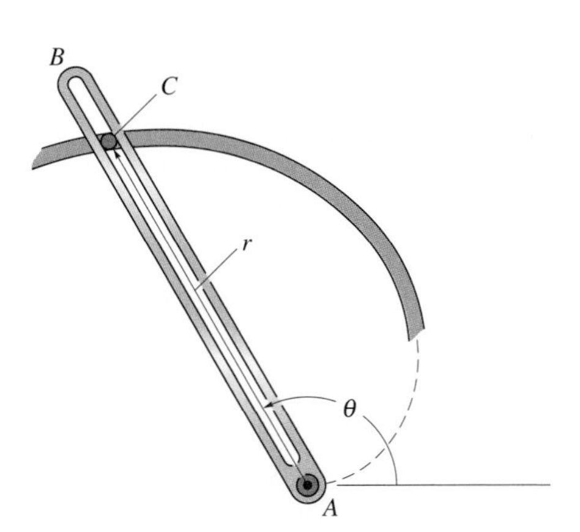

Probs. 12–177/12–178/12–179

*12–180. For a short time the jet plane moves along a path in the shape of a lemniscate, $r^2 = (2500 \cos 2\theta)$ km^2. At the instant $\theta = 30°$, the radar tracking device is rotating at $\dot{\theta} = 5(10^{-3})$ rad/s with $\ddot{\theta} = 2(10^{-3})$ rad/s^2. Determine the radial and transverse components of velocity and acceleration of the plane at this instant.

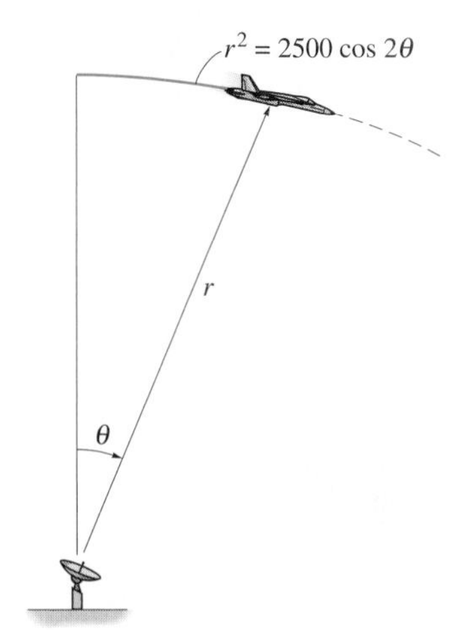

Prob. 12–180

12–181. The searchlight on the boat anchored 2000 ft from shore is turned on the automobile, which is traveling along the straight road at a constant speed of 80 ft/s. Determine the angular rate of rotation of the light when the automobile is $r = 3000$ ft from the boat.

12–182. If the car in Prob. 12–181 is accelerating at 15 ft/s^2 at the instant $r = 3000$ ft, determine the required angular acceleration $\ddot{\theta}$ of the light at this instant.

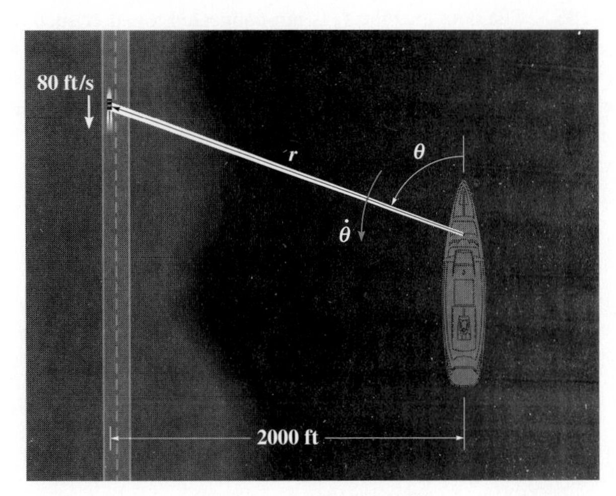

Probs. 12–181/12–182

12–183. The roller coaster is traveling down along the spiral ramp with a constant speed $v = 6$ m/s. If the track descends a distance of 10 m for every full revolution, $\theta = 2\pi$ rad, determine the magnitude of the roller coaster's acceleration as it moves along the track, $r = 5$ m. *Hint:* For part of the solution, note that the tangent to the ramp at any point is at an angle $\phi = \tan^{-1}[10/2\pi(5)] = 17.66°$ from the horizontal. Use this to determine the velocity components v_θ and v_z, which in turn are used to determine θ and z.

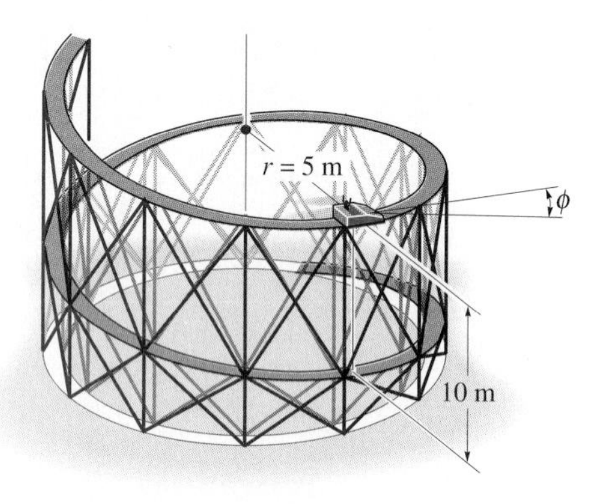

Prob. 12–183

*■**12–184.** The double collar C is pin-connected together such that one collar slides over the fixed rod and the other slides over the rotating rod AB. If the angular velocity of AB is given as $\theta = (e^{0.5t^2})$ rad/s, where t is in seconds, and the path defined by the fixed rod is $r = |(0.4 \sin\theta + 0.2)|$ m, determine the radial and transverse components of the collar's velocity and acceleration when $t = 1$ s. When $t = 0$, $\theta = 0$. Use Simpson's rule with $n = 50$ to determine θ at $t = 1$ s.

12–185. The double collar C is pin-connected together such that one collar slides over the fixed rod and the other slides over the rotating rod AB. If the mechanism is to be designed so that the largest speed given to the collar is 6 m/s, determine the required constant angular velocity θ of rod AB. The path defined by the fixed rod is $r = (0.4 \sin\theta + 0.2)$m.

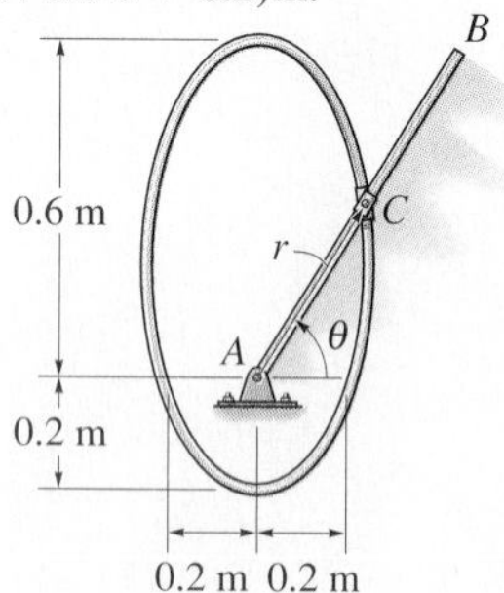

Probs. 12–184/12–185

12–186. For a short time the position of the roller-coaster car along its path is defined by the equations $r = 25$ m, $\theta = (0.3t)$ rad, and $z = (-8\cos\theta)$ m, where t is in seconds. Determine the magnitude of the car's velocity and acceleration when $t = 4$ s.

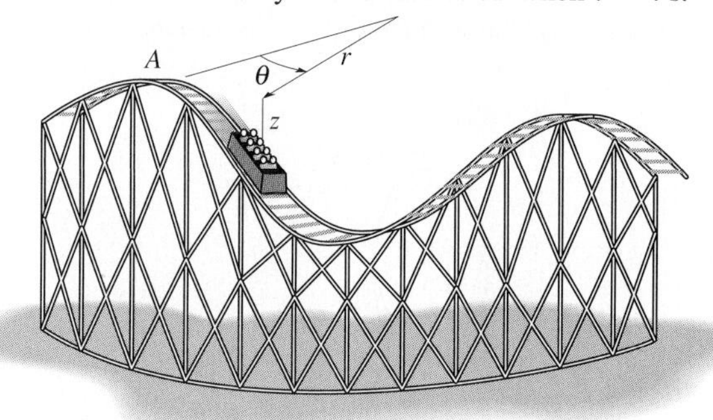

Prob. 12–186

12–187. The mechanism of a machine is constructed so that the roller at A follows the surface of the cam described by the equation $r = (0.3 + 0.2\cos\theta)$ m. If $\theta = 0.5$ rad/s and $\theta = 0$, determine the magnitudes of the roller's velocity and acceleration when $\theta = 30°$. Neglect the size of the roller. Also compute the velocity components $(\mathbf{v}_A)_x$ and $(\mathbf{v}_A)_y$ of the roller at this instant. The rod to which the roller is attached remains vertical and can slide up or down along the guides while the guides translate horizontally to the left.

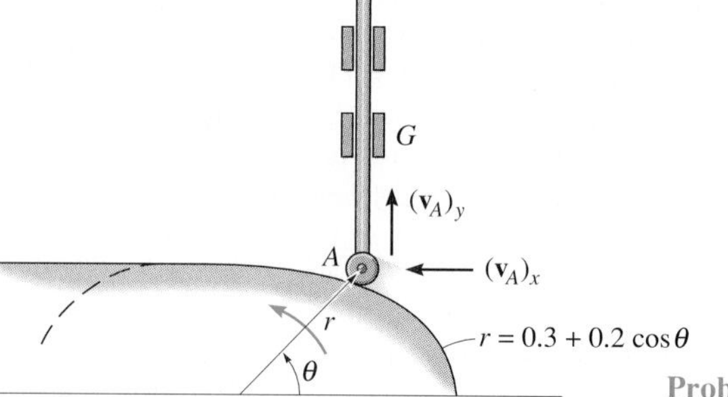

Prob. 12–187

***12–188.** For a short time a rocket travels at a constant speed of 800 m/s along the parabolic path $y = 600 - 35x^2$. Determine the radial and transverse components of velocity of the rocket at the instant $\theta = 60°$, where θ is measured counterclockwise from the x axis.

12–189. The motion of a particle along a fixed path is defined by the parametric equations $r = 1.5$ m, $\theta = (2t)$ rad, and $z = (t^2)$ m, where t is in seconds. Determine the unit vector that specifies the direction of the binormal axis to the osculating plane with respect to a set of fixed x, y, z coordinate axes when $t = 0.25$ s. *Hint:* Formulate the particle's velocity $\mathbf{v}$ and acceleration $\mathbf{a}$ in terms of their $\mathbf{i}$, $\mathbf{j}$, $\mathbf{k}$ components. Note that $x = r\cos\theta$ and $y = r\sin\theta$. The binormal is collinear to $\mathbf{v} \times \mathbf{a}$. Why?

12.8 Absolute Dependent Motion Analysis of Two Particles

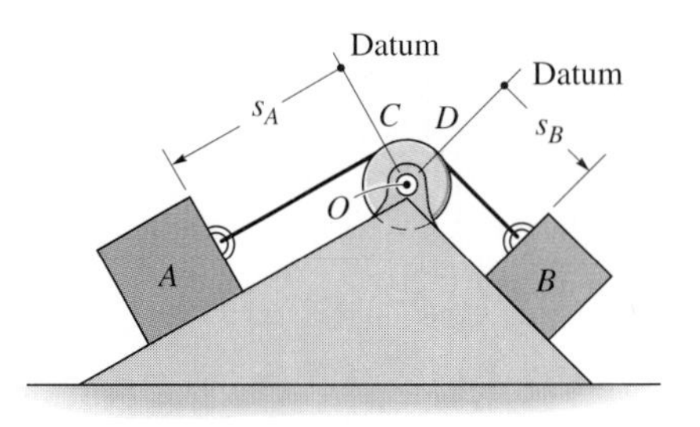

Fig. 12–36

In some types of problems the motion of one particle will *depend* on the corresponding motion of another particle. This dependency commonly occurs if the particles are interconnected by inextensible cords which are wrapped around pulleys. For example, the movement of block A downward along the inclined plane in Fig. 12–36 will cause a corresponding movement of block B up the other incline. We can show this mathematically by first specifying the location of the blocks using *position coordinates* s_A and s_B. Note that each of the coordinate axes is (1) referenced from a *fixed* point (O) or *fixed* datum line, (2) measured along each inclined plane in the direction of motion of block A and block B, and (3) has a positive sense from C to A and D to B. If the total cord length is l_T, the position coordinates are related by the equation

$$s_A + l_{CD} + s_B = l_T$$

Here l_{CD} is the length of the cord passing over arc CD. Taking the time derivative of this expression, realizing that l_{CD} and l_T remain constant, while s_A and s_B measure the lengths of the changing segments of the cord, we have

$$\frac{ds_B}{dt} + \frac{ds_A}{dt} = 0 \qquad \text{or} \qquad v_B = -v_A$$

The negative sign indicates that when block A has a velocity downward, i.e., in the direction of positive s_A, it causes a corresponding upward velocity of block B; i.e., B moves in the negative s_B direction.

In a similar manner, time differentiation of the velocities yields the relation between the accelerations, i.e.,

$$a_B = -a_A$$

A more complicated example involving dependent motion of two blocks is shown in Fig. 12–37*a*. In this case, the position of block A is specified by s_A, and the position of the *end* of the cord from which block B is suspended is defined by s_B. Here we have chosen coordinate axes which are (1) referenced from fixed points or datums, (2) measured in the direction of motion of each block, and (3) positive to the right (s_A) and positive downward (s_B). During the motion, the colored segments of the cord in Fig. 12–37*a* remain constant. If l represents the total length of cord minus these segments, then the position coordinates can be related by the equation

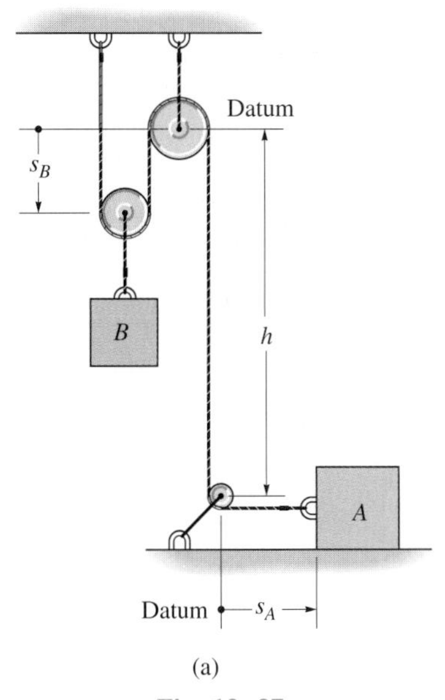

(a)
Fig. 12–37

$$2s_B + h + s_A = l$$

Since l and h are constant during the motion, the two time derivatives yield

$$2v_B = -v_A \qquad 2a_B = -a_A$$

Hence, when B moves downward ($+s_B$), A moves to the left ($-s_A$) at two times the motion.

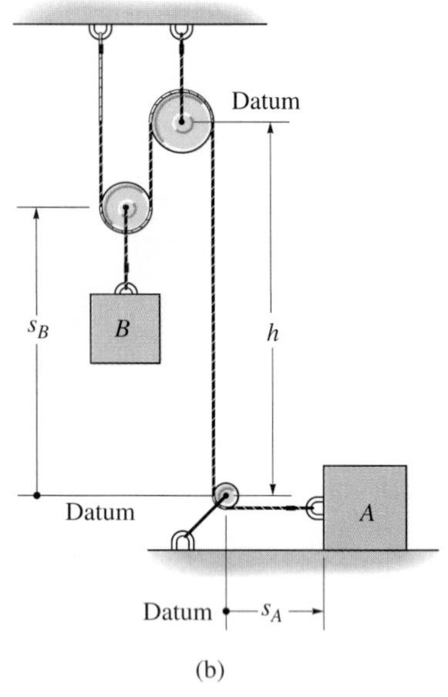

(b)

Fig. 12–37 *(cont'd)*

This example can also be worked by defining the position of block *B* from the center of the bottom pulley (a fixed point), Fig. 12–37*b*. In this case

$$2(h - s_B) + h + s_A = l$$

Time differentiation yields

$$2v_B = v_A \qquad 2a_B = a_A$$

Here the signs are the same. Why?

PROCEDURE FOR ANALYSIS

The above method of relating the dependent motion of one particle to that of another can be performed using algebraic scalars or position coordinates provided each particle moves along a rectilinear path. When this is the case, only the magnitude of the velocity and acceleration of each particle will change, not their line of direction. The following procedure is required.

Position-Coordinate Equation. Establish position coordinates which have their origin located at a *fixed* point or datum and extend from this point to a point having the same motion as each of the particles. It is *not necessary* that the *origin* be the *same* for each of these coordinates; however, it is *important* that each coordinate axis selected be directed along the *path of motion* of the particle.

Using geometry or trigonometry, relate the coordinates to the total length of the cord, l_T, or to that portion of cord, l, which *excludes* the segments that do not change length as the particles move—such as the arc segments wrapped over pulleys.

Time Derivatives. The two successive time derivatives of the position-coordinate equation yield the required velocity and acceleration equations which relate the motions of the two particles. The signs of the terms in these equations will be consistent with those that specify the positive and negative sense of the position coordinates.

If a problem involves a *system* of two or more cords wrapped around pulleys, then the motion of a point on one cord must be related to the motion of a point on another cord using the above procedure. Separate equations are written for a fixed length of each cord of the system and the positions of the two particles are then related by these equations (see Examples 12–22 and 12–23).

Example 12–21

Determine the speed of block A in Fig. 12–38 if block B has an upward speed of 6 ft/s.

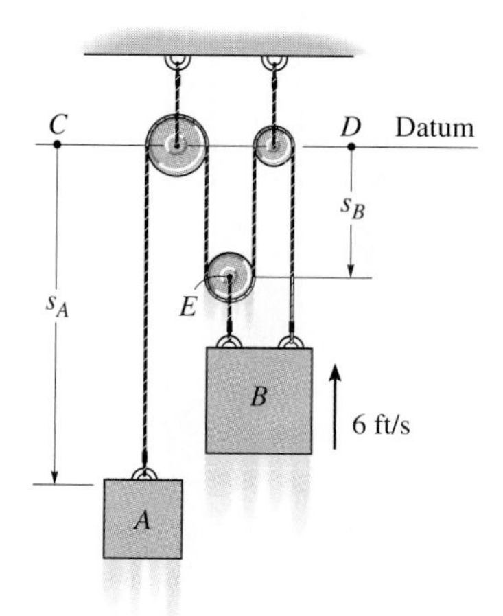

Fig. 12–38

SOLUTION

Position-Coordinate Equation. There is *one cord* in this system having segments which are changing length. Position coordinates s_A and s_B will be used since each is measured from a fixed point (C or D) and extends along each block's *path of motion*. In particular, s_B is directed to point E since motion of B and E is the *same*.

The colored segments of the cord in Fig. 12–38 remain at a constant length and do not have to be considered as the blocks move. The remaining length of cord, l, is also constant and is related to the changing position coordinates s_A and s_B by the equation

$$s_A + 3s_B = l$$

Time Derivative. Taking the time derivative yields

$$v_A + 3v_B = 0$$

so that when $v_B = -6$ ft/s (upward),

$$v_A = 18 \text{ ft/s} \downarrow \qquad \textit{Ans.}$$

Example 12–22

Determine the speed of block A in Fig. 12–39 if block B has an upward speed of 6 ft/s.

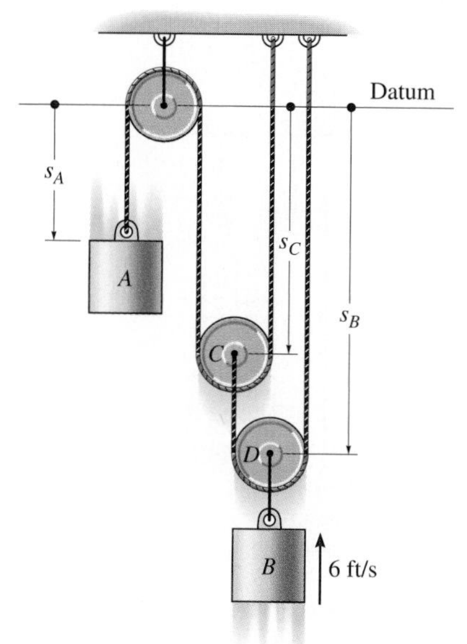

Fig. 12–39

SOLUTION

Position-Coordinate Equation. As shown, the positions of blocks A and B are defined using the coordinates s_A and s_B. Since the system has two cords which change length, it will be necessary to use a third coordinate, s_C, in order to relate s_A to s_B. In other words, the length of one of the cords can be expressed in terms of s_A and s_C, and the length of the other cord can be expressed in terms of s_B and s_C.

The colored segments of the cords in Fig. 12–39 do not have to be considered in the analysis. Why? For the remaining cord lengths, say l_1 and l_2, we have

$$s_A + 2s_C = l_1 \qquad s_B + (s_B - s_C) = l_2$$

Eliminating s_C yields an equation defining the positions of both blocks, i.e.,

$$s_A + 4s_B = 2l_2 + l_1$$

Time Derivative. The time derivative gives

$$v_A + 4v_B = 0$$

so that when $v_B = -6$ ft/s (upward),

$$v_A = +24 \text{ ft/s} = 24 \text{ ft/s} \downarrow \qquad \textit{Ans.}$$

Example 12–23

Determine the speed with which block B rises in Fig. 12–40 if the end of the cord at A is pulled down with a speed of 2 m/s.

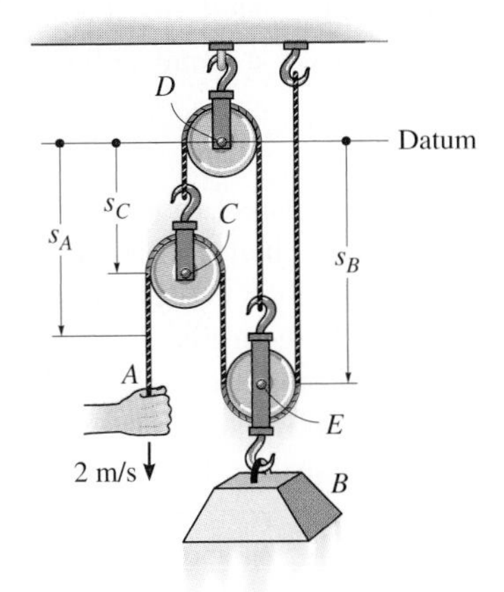

Fig. 12–40

SOLUTION

Position-Coordinate Equation. The position of point A is defined by s_A, and the position of block B is specified by s_B since point E on the pulley will have the same motion as the block. Both coordinates are measured from a horizontal datum passing through the fixed pin at pulley D. Since the system consists of two cables, the coordinates s_A and s_B cannot be related directly. Instead, by establishing a third position coordinate, s_C, we can now express the length of one of the cables in terms of s_A and s_C, and the length of the other cable in terms of s_B and s_C.

Excluding the colored segments of the cords in Fig. 12–40, the remaining constant cord lengths l_1 and l_2 (along with the hook dimensions) can be expressed as

$$(s_A - s_C) + (s_B - s_C) + s_B = l_1$$
$$s_C + s_B = l_2$$

Eliminating s_C yields

$$s_A + 4s_B = l_1 + 2l_2$$

As required, this equation relates the position s_B of block B to the position s_A of point A.

Time Derivative. The time derivative gives

$$v_A + 4v_B = 0$$

so that when $v_A = 2$ m/s (downward),

$$v_B = -0.5 \text{ m/s} = 0.5 \text{ m/s} \uparrow \qquad \textit{Ans.}$$

Example 12–24

A man at A is hoisting a safe S as shown in Fig. 12–41 by walking to the right with a constant velocity $v_A = 0.5$ m/s. Determine the velocity and acceleration of the safe when it reaches the window elevation at E. The rope is 30 m long and passes over a small pulley at D.

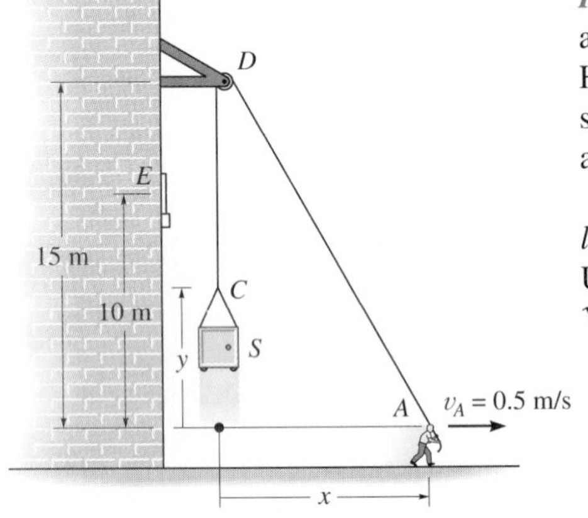

Fig. 12–41

SOLUTION

Position-Coordinate Equation. This problem is unlike the previous examples, since rope segment DA changes both direction and magnitude. However, the ends of the rope, which define the positions of S and A, are specified by means of the x and y coordinates measured from fixed points and *directed along the paths of motion* of the ends of the rope.

The x and y coordinates may be related since the rope has a fixed length $l = 30$ m, which at all times is equal to the length of segment DA plus CD. Using the Pythagorean theorem to determine l_{DA}, we have $l_{DA} = \sqrt{(15)^2 + x^2}$; also, $l_{CD} = 15 - y$. Hence,

$$l = l_{DA} + l_{CD}$$

$$30 = \sqrt{(15)^2 + x^2} + (15 - y)$$

$$y = \sqrt{225 + x^2} - 15 \qquad (1)$$

Time Derivatives. Taking the time derivative, where $v_S = dy/dt$ and $v_A = dx/dt$, yields

$$v_S = \frac{dy}{dt} = \frac{1}{2}\frac{2x}{\sqrt{225 + x^2}}\frac{dx}{dt}$$

$$= \frac{x}{\sqrt{225 + x^2}}v_A \qquad (2)$$

At $y = 10$ m, x is determined from Eq. 1, i.e., $x = 20$ m. Hence, from Eq. 2 with $v_A = 0.5$ m/s,

$$v_S = \frac{20}{\sqrt{225 + (20)^2}}(0.5) = 0.4 \text{ m/s} = 400 \text{ mm/s} \uparrow \qquad \textit{Ans.}$$

The acceleration is determined by taking the time derivative of Eq. 2. Since v_A is constant, then $a_A = dv_A/dt = 0$, and we have

$$a_S = \frac{d^2y}{dt^2} = \left[\frac{-x^2}{(225 + x^2)^{3/2}}\frac{dx}{dt} + \frac{1}{(225 + x^2)^{1/2}}\frac{dx}{dt}\right]v_A = \frac{225v_A^2}{(225 + x^2)^{3/2}}$$

At $x = 20$ m, with $v_A = 0.5$ m/s, the acceleration becomes

$$a_S = \frac{225(0.5 \text{ m/s})^2}{[225 + (20 \text{ m})^2]^{3/2}} = 0.00360 \text{ m/s}^2 = 3.60 \text{ mm/s}^2 \uparrow \qquad \textit{Ans.}$$

Note that the constant velocity at A causes the other end C of the rope to have this acceleration since $\mathbf{v}_A$ causes segment DA to change its direction.

12.9 Relative-Motion Analysis of Two Particles Using Translating Axes

Throughout this chapter the absolute motion of a particle has been determined using a single fixed reference frame for measurement. There are many cases, however, where the path of motion for a particle is complicated, so that it may be feasible to analyze the motion in parts by using two or more frames of reference. For example, the motion of a particle located at the tip of an airplane propeller, while the plane is in flight, is more easily described if one observes first the motion of the airplane from a fixed reference and then superimposes (vectorially) the circular motion of the particle measured from a reference attached to the airplane. Any type of coordinates—rectangular, cylindrical, etc.—may be chosen to describe these two different motions.

In this section only *translating frames of reference* will be considered for the analysis. Relative-motion analysis of particles using rotating frames of reference will be treated in Secs. 16.8 and 20.4, since such an analysis depends on prior knowledge of the kinematics of line segments.

Position. Consider particles A and B, which move along the arbitrary paths aa and bb, respectively, as shown in Fig. 12–42a. The *absolute position* of each particle, $\mathbf{r}_A$ and $\mathbf{r}_B$, is measured from the common origin O of the *fixed* x, y, z reference frame. The origin of a second frame of reference x', y', z' is attached to and moves with particle A. The axes of this frame do not rotate; rather, as stated above, they are *only permitted to translate* relative to the fixed frame. The *relative position* of "B with respect to A" is designated by a *relative-position vector* $\mathbf{r}_{B/A}$. Using vector addition, the three vectors shown in Fig. 12–42a can be related by the equation*

$$\mathbf{r}_B = \mathbf{r}_A + \mathbf{r}_{B/A} \tag{12–33}$$

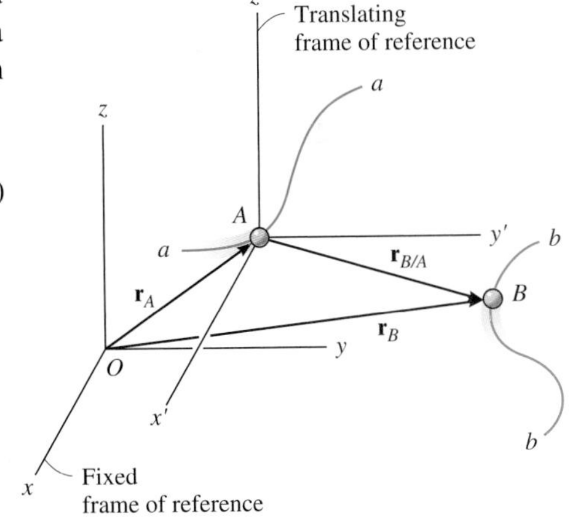

(a)

Fig. 12–42

*An easy way to remember the setup of this equation, and others like it, is to note the "cancellation" of the subscript A between the two terms, i.e., $\mathbf{r}_B = \mathbf{r}_{\not{A}} + \mathbf{r}_{B/\not{A}}$.

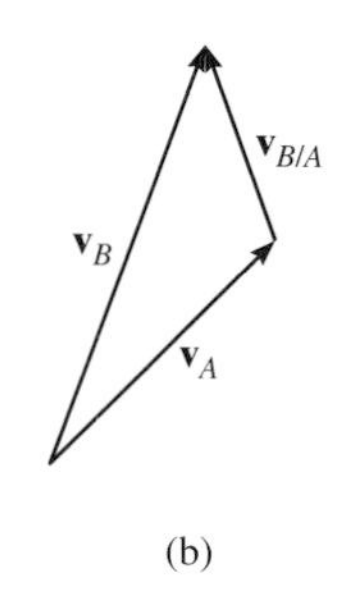

(b)

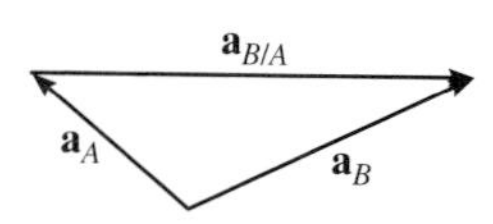

(c)

Fig. 12–42 *(cont'd)*

Velocity. An equation that relates the velocities of the particles can be determined by taking the time derivative of Eq. 12–33, i.e.,

$$\mathbf{v}_B = \mathbf{v}_A + \mathbf{v}_{B/A} \tag{12–34}$$

Here $\mathbf{v}_B = d\mathbf{r}_B/dt$ and $\mathbf{v}_A = d\mathbf{r}_A/dt$ refer to *absolute velocities,* since they are measured from the fixed frame of reference; whereas the *relative velocity* $\mathbf{v}_{B/A} = d\mathbf{r}_{B/A}/dt$ is observed from the translating reference frame. It is important to note that since the x', y', z' axes translate, the *components* of $\mathbf{r}_{B/A}$ will *not* change direction and therefore the time derivative of this vector will only have to account for the change in the vector's magnitude. The above equation therefore states that the velocity of B is equal to the velocity of A plus (vectorially) the relative velocity of "B with respect to A," as measured by a *translating observer* fixed in the x', y', z' reference, Fig. 12–42b.

Acceleration. The time derivative of Eq. 12–34 yields a similar vector relationship between the *absolute* and *relative accelerations* of particles A and B.

$$\mathbf{a}_B = \mathbf{a}_A + \mathbf{a}_{B/A} \tag{12–35}$$

Here $\mathbf{a}_{B/A}$ is the acceleration of B as seen by an observer located at A and translating with the x', y', z' reference frame. The vector addition is shown in Fig. 12–42c.

PROCEDURE FOR ANALYSIS

When applying the relative-position equation, $\mathbf{r}_B = \mathbf{r}_A + \mathbf{r}_{B/A}$, it is first necessary to specify the locations of the fixed x, y, z and translating x', y', z' axes. Usually, the origin A of the translating axes is located at a point having a *known position,* $\mathbf{r}_A$, Fig. 12–42a. A graphical representation of the vector addition $\mathbf{r}_B = \mathbf{r}_A + \mathbf{r}_{B/A}$ can be shown, and both the known and unknown quantities labeled on this sketch. Since vector addition forms a triangle, there can be at most *two unknowns,* represented by the magnitudes and/or directions of the vector quantities. These unknowns can be solved for either graphically, using trigonometry (law of sines, law of cosines), or by resolving each of the three vectors $\mathbf{r}_B$, $\mathbf{r}_A$, and $\mathbf{r}_{B/A}$ into rectangular or Cartesian components, thereby generating a set of scalar equations. The latter method is illustrated in the example problems which follow.

The relative-motion equations $\mathbf{v}_B = \mathbf{v}_A + \mathbf{v}_{B/A}$ and $\mathbf{a}_B = \mathbf{a}_A + \mathbf{a}_{B/A}$ are applied in the same manner as explained above, except in this case the origin O of the fixed x, y, z axes does not have to be specified, Figs. 12–42b and 12–42c.

Example 12–25

A train, traveling at a constant speed of 60 mi/h, crosses over a road as shown in Fig. 12–43*a*. If automobile *A* is traveling at 45 mi/h along the road, determine the relative velocity of the train with respect to the automobile.

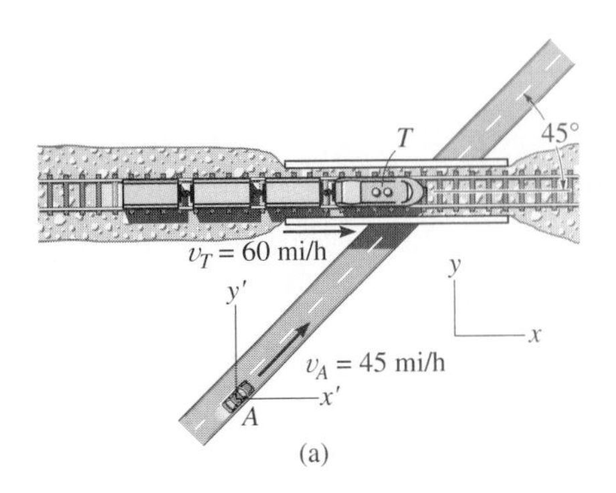

(a)

SOLUTION I

Vector Analysis. The relative velocity $\mathbf{v}_{T/A}$ is measured from the translating x', y' axes attached to the automobile, Fig. 12–43*a*. It is determined from $\mathbf{v}_T = \mathbf{v}_A + \mathbf{v}_{T/A}$. Since $\mathbf{v}_T$ and $\mathbf{v}_A$ are known in *both* magnitude and direction, the unknowns become the components of $\mathbf{v}_{T/A}$. Using the x, y axes in Fig. 12–43*a* and a Cartesian vector analysis, we have

$$\mathbf{v}_T = \mathbf{v}_A + \mathbf{v}_{T/A}$$

$$60\mathbf{i} = (45 \cos 45^\circ\, \mathbf{i} + 45 \sin 45^\circ\, \mathbf{j}) + \mathbf{v}_{T/A}$$

$$\mathbf{v}_{T/A} = \{28.2\mathbf{i} - 31.8\mathbf{j}\} \text{ mi/h} \qquad \textit{Ans.}$$

The magnitude of $\mathbf{v}_{T/A}$ is thus

$$v_{T/A} = \sqrt{(28.2)^2 + (-31.8)^2} = 42.5 \text{ mi/h} \qquad \textit{Ans.}$$

From the direction of each component, the direction of $\mathbf{v}_{T/A}$ defined from the x axis, is

$$\tan\theta = \frac{(v_{T/A})_y}{(v_{T/A})_x} = \frac{31.8}{28.2}$$

$$\theta = 48.4^\circ \measuredangle^{\theta} \qquad \textit{Ans.}$$

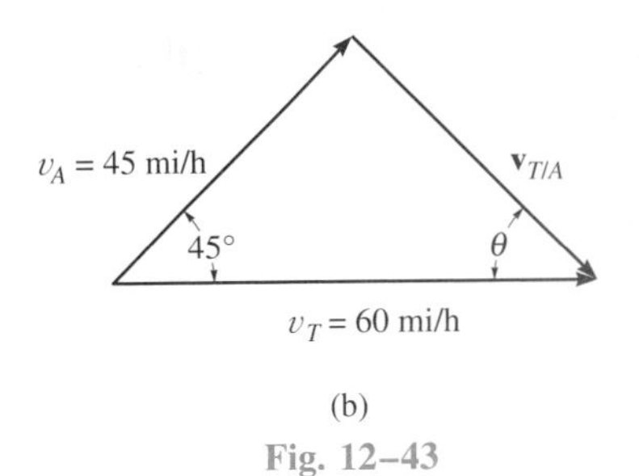

(b)

Fig. 12–43

Note that the vector addition shown in Fig. 12–43*b* indicates the correct sense for $\mathbf{v}_{T/A}$. This figure anticipates the answer and can be used to check it.

SOLUTION II

Scalar Analysis. The unknown components of $\mathbf{v}_{T/A}$ can also be determined by applying a scalar analysis. We will assume these components act in the *positive* x and y directions. Thus,

$$\mathbf{v}_T = \mathbf{v}_A + \mathbf{v}_{T/A}$$

$$\begin{bmatrix} 60 \text{ mi/h} \\ \rightarrow \end{bmatrix} = \begin{bmatrix} 45 \text{ mi/h} \\ \measuredangle 45^\circ \end{bmatrix} + \begin{bmatrix} (v_{T/A})_x \\ \rightarrow \end{bmatrix} + \begin{bmatrix} (v_{T/A})_y \\ \uparrow \end{bmatrix}$$

Resolving each vector into its x and y components yields

$(\overset{+}{\rightarrow})$ $$60 = 45 \cos 45^\circ + (v_{T/A})_x + 0$$

$(+\uparrow)$ $$0 = 45 \sin 45^\circ + 0 + (v_{T/A})_y$$

Solving, we obtain the previous results,

$$(v_{T/A})_x = 28.2 \text{ mi/h} = 28.2 \text{ mi/h} \rightarrow$$

$$(v_{T/A})_y = -31.8 \text{ mi/h} = 31.8 \text{ mi/h} \downarrow \qquad \textit{Ans.}$$

Example 12–26

Two planes are flying at the same elevation and have the motion shown in Fig. 12–44*a*. Plane *A* is flying along a straight-line path, whereas plane *B* is flying along a circular path having a radius of curvature of $\rho_B = 400$ km. Determine the velocity and acceleration of *B* as measured by the pilot of *A*.

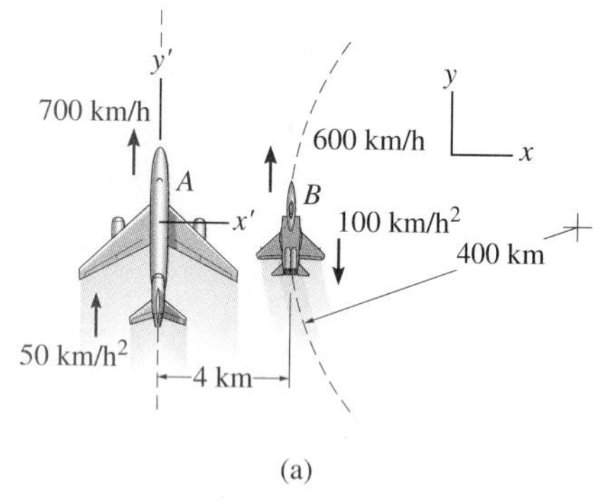

(a)

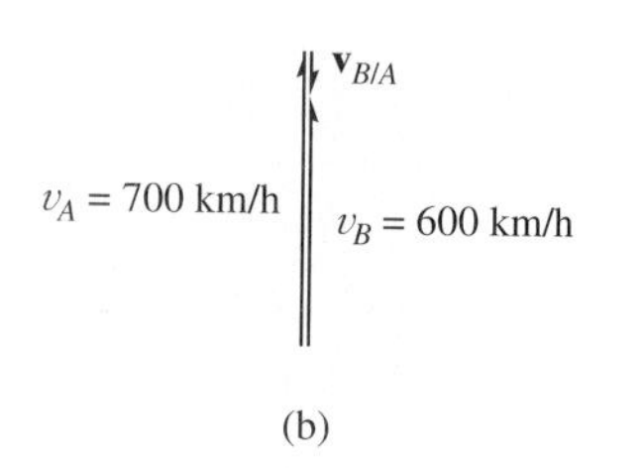

(b)

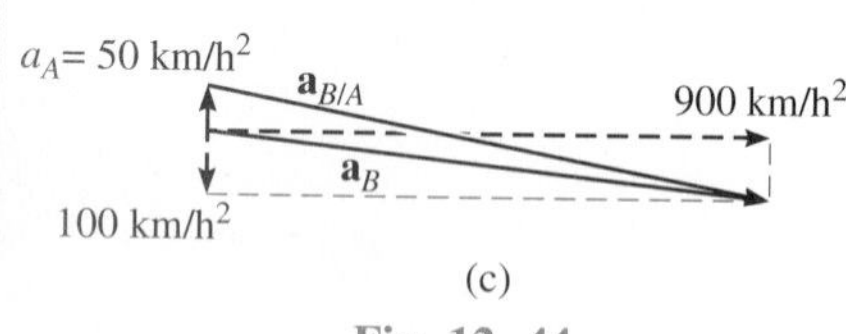

(c)

Fig. 12–44

SOLUTION

Velocity. The fixed *x*, *y* axes are located at an arbitrary point. Plane *A* is traveling with rectilinear motion and a *translating frame of reference x′, y′* will be attached to it. Applying the relative-velocity equation in scalar form since the velocity vectors of both planes are parallel at the instant shown, we have

$(+\uparrow)$

$$v_B = v_A + v_{B/A}$$

$$600 = 700 + v_{B/A}$$

$$v_{B/A} = -100 \text{ km/h} = 100 \text{ km/h} \downarrow \qquad \textit{Ans.}$$

The vector addition is shown in Fig. 12–44*b*.

Acceleration. Plane *B* has both tangential and normal components of acceleration, since it is flying along a *curved path.* From Eq. 12–20, the magnitude of the normal component is

$$(a_B)_n = \frac{v_B^2}{\rho} = \frac{(600 \text{ km/h})^2}{400 \text{ km}} = 900 \text{ km/h}^2$$

Applying the relative-acceleration equation, we have

$$\mathbf{a}_B = \mathbf{a}_A + \mathbf{a}_{B/A}$$

$$900\mathbf{i} - 100\mathbf{j} = 50\mathbf{j} + \mathbf{a}_{B/A}$$

Thus,

$$\mathbf{a}_{B/A} = \{900\mathbf{i} - 150\mathbf{j}\} \text{ km/h}^2$$

The magnitude and direction of $\mathbf{a}_{B/A}$ are therefore

$$a_{B/A} = 912 \text{ km/h}^2 \qquad \theta = \tan^{-1}\frac{150}{900} = 9.46° \measuredangle^{\theta} \qquad \textit{Ans.}$$

The vector addition is shown in Fig. 12–44*c*.

Notice that the solution to this problem is possible using a translating frame of reference, since the pilot in plane *A* is "translating." Observation of plane *A* with respect to the pilot of plane *B*, however, must be obtained using a *rotating* set of axes attached to plane *B*. (This assumes, of course, that the pilot of *B* is fixed in the rotating frame, so he does not turn his eyes to follow the motion of *A*.) The analysis for this case is given in Example 16–21.

Example 12–27

At the instant shown in Fig. 12–45 cars A and B are traveling with speeds of 8 m/s and 12 m/s, respectively. Also at this instant, A has a decrease in speed of 2 m/s^2 and B has an increase in speed of 3 m/s^2. Determine the velocity and acceleration of B with respect to A.

SOLUTION

Velocity. The fixed x, y axes are established at a point on the ground and the translating x', y' axes are attached to car A, Fig. 12–45. The relative velocity is determined from $\mathbf{v}_B = \mathbf{v}_A + \mathbf{v}_{B/A}$. What are the two unknowns? Using a Cartesian vector analysis, we have

$$\mathbf{v}_B = \mathbf{v}_A + \mathbf{v}_{B/A}$$
$$-12\mathbf{j} = (-8 \cos 60°\mathbf{i} - 8 \sin 60°\mathbf{j}) + \mathbf{v}_{B/A}$$
$$\mathbf{v}_{B/A} = \{4\mathbf{i} - 5.072\mathbf{j}\} \text{ m/s}$$

Thus,

$$v_{B/A} = \sqrt{(4)^2 + (5.072)^2} = 6.46 \text{ m/s} \qquad \textit{Ans.}$$

Noting that $\mathbf{v}_{B/A}$ has $+\mathbf{i}$ and $-\mathbf{j}$ components, its direction is

$$\tan\theta = \frac{(v_{B/A})_y}{(v_{B/A})_x} = \frac{5.072}{4}$$
$$\theta = 51.7° \qquad \textit{Ans.}$$

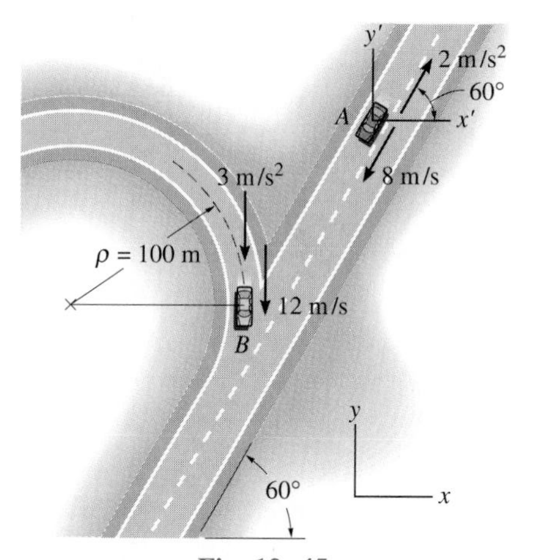

Fig. 12–45

Acceleration. Car B has both tangential and normal components of acceleration. Why? The magnitude of the normal component is

$$(a_B)_n = \frac{v_B^2}{\rho} = \frac{(12 \text{ m/s})^2}{100 \text{ m}} = 1.440 \text{ m/s}^2$$

Applying the equation for relative acceleration yields

$$\mathbf{a}_B = \mathbf{a}_A + \mathbf{a}_{B/A}$$
$$(-1.440\mathbf{i} - 3\mathbf{j}) = (2 \cos 60°\mathbf{i} + 2 \sin 60°\mathbf{j}) + \mathbf{a}_{B/A}$$
$$\mathbf{a}_{B/A} = \{-2.440\mathbf{i} - 4.732\mathbf{j}\} \text{ m/s}^2$$

Thus

$$a_{B/A} = \sqrt{(2.440)^2 + (4.732)^2} = 5.32 \text{ m/s}^2 \qquad \textit{Ans.}$$

$$\tan\phi = \frac{(a_{B/A})_y}{(a_{B/A})_x} = \frac{4.732}{2.440}$$
$$\phi = 62.7° \qquad \textit{Ans.}$$

Is it possible to obtain the relative acceleration of $\mathbf{a}_{A/B}$ using this method? Refer to the comment made at the end of Example 12–26.

PROBLEMS

12–190. The cable and pulley system is used to lift the load A. Determine the displacement of the load if the end C of the cable is pulled 4 ft upward.

12–191. The cable and pulley system is used to lift the load A. Determine the speed of A if the end C of the cable is given an upward speed of 6 ft/s.

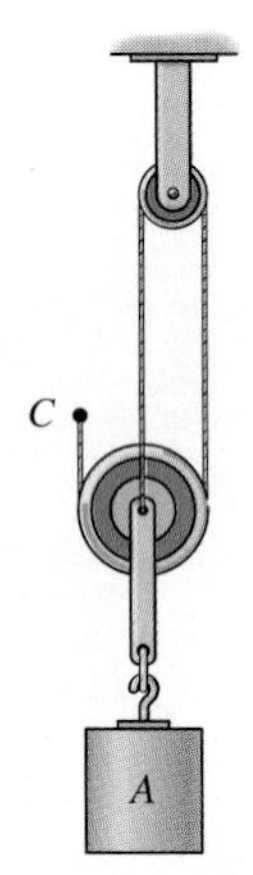

Probs. 12–190/12–191

***12–192.** Determine the constant speed at which the cable at A must be drawn in by the motor in order to hoist the load at B 15 ft in 5 s.

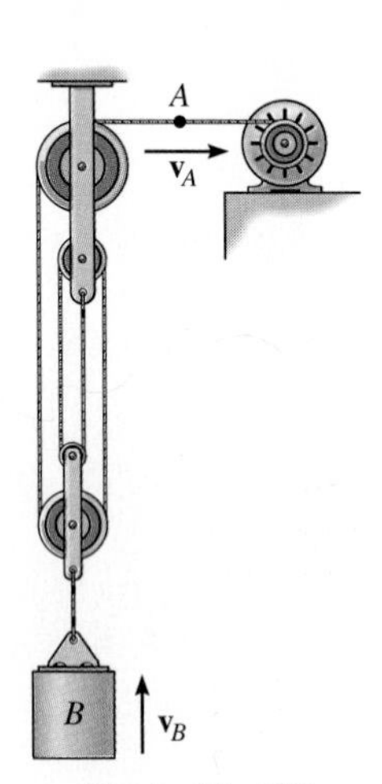

Prob. 12–192

12–193. Determine the time needed for the load at B to attain a speed of 8 m/s, starting from rest, if the cable is drawn into the motor with an acceleration of 0.2 m/s^2.

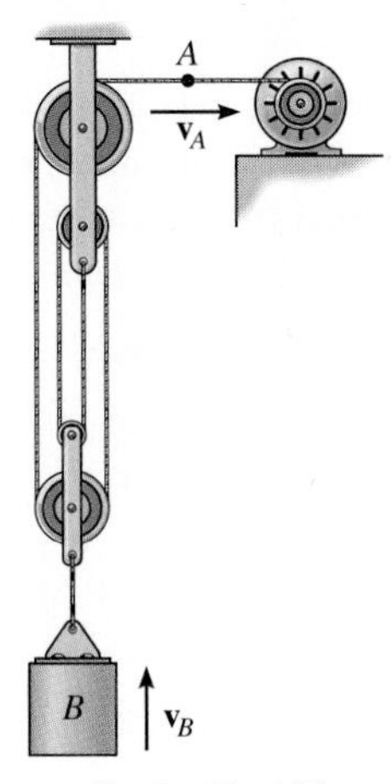

Prob. 12–193

12–194. Determine the speed of the block at B if the cord at A is pulled down with a speed of 8 ft/s. What is the relative velocity of B with respect to A?

12–195. Determine the displacement of the block at B if A is pulled down 4 ft.

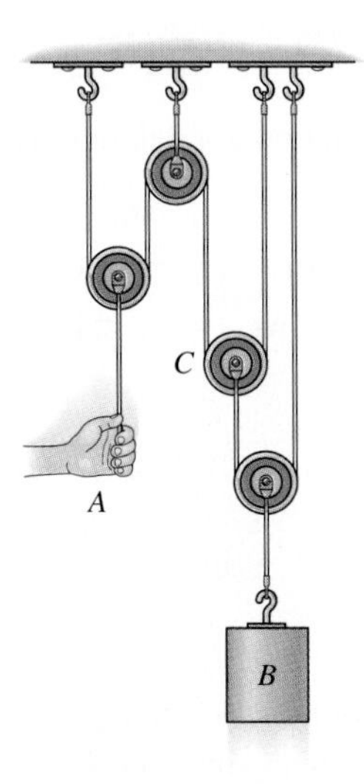

Probs. 12–194/12–195

*12–196. The pulley arrangement shown is designed for hoisting materials. If BC *remains fixed* while the plunger P is pushed downward with a speed of 4 ft/s, determine the speed of the load at A.

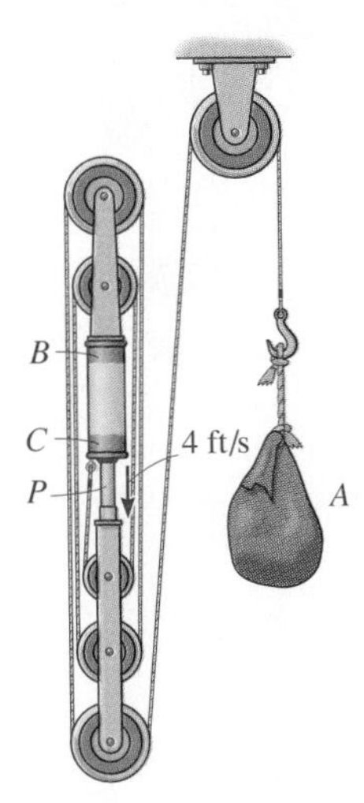

Prob. 12–196

12–197. The hook B on the oil rig is supported by a cable which is connected to the drum at A, passes over a pulley at E, down and around the hook's pulley at B, up and around another pulley at E, and is then attached to the drum at C. Determine the time needed to hoist the hook from $h = 0$ to $h = 100$ ft if (a) drum A draws the cable in at a constant speed of 4 ft/s and drum C is stationary, and (b) drum A draws in the cable at a constant speed of 3 ft/s and drum C draws in the cable at a constant speed of 2 ft/s.

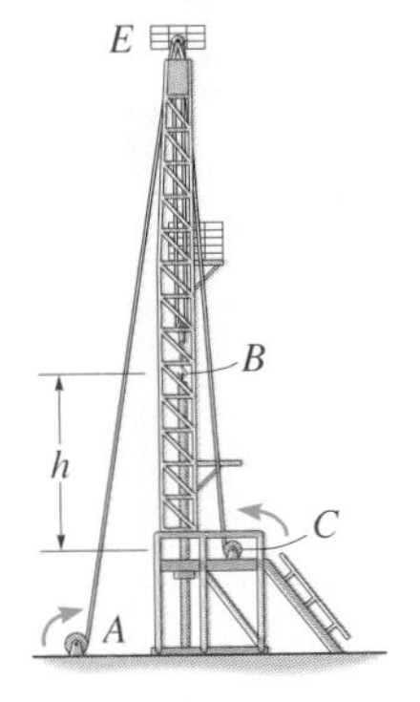

Prob. 12–197

12–198 The hook B on the oil rig is supported by a cable which is connected to the drum at A, passes over a pulley at E, down and around the hook's pulley at B, up and around another pulley at E, and is then attached to the drum at C. Gearing at drum A has been designed to draw in its cable at 5 ft/s. If it is restricted that the pipe be lifted out of the well at 8 ft/s, determine the rate at which the drum at C must draw in its cable.

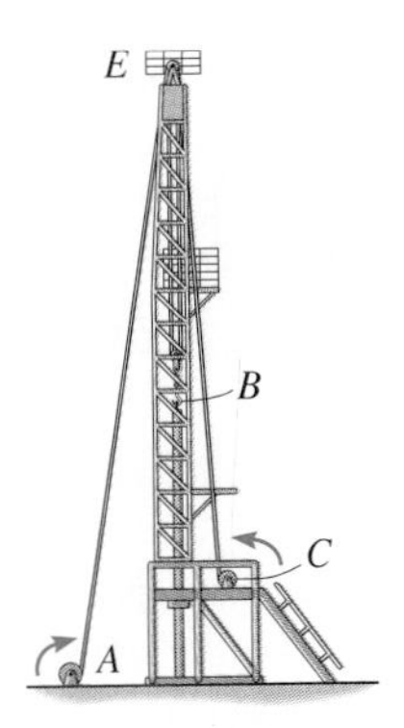

Prob. 12–198

12–199. If block A of the pulley system is moving downward with a speed of 4 ft/s while block C is moving up at 2 ft/s, determine the speed of block B.

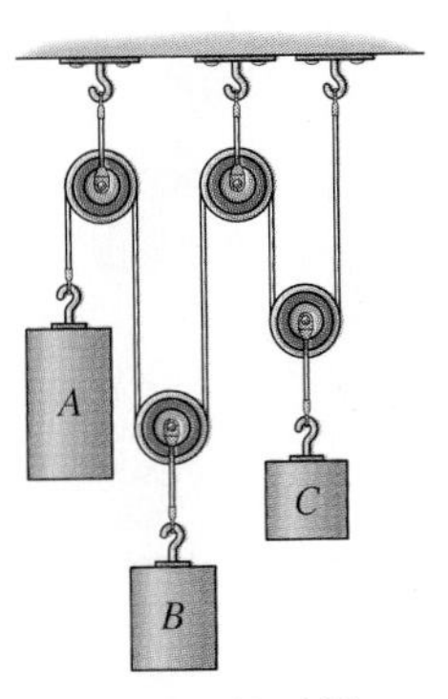

Prob. 12–199

*12–200. If block A of the pulley system is moving downward at 6 ft/s while block C is moving down at 18 ft/s, determine the relative velocity of block B with respect to C.

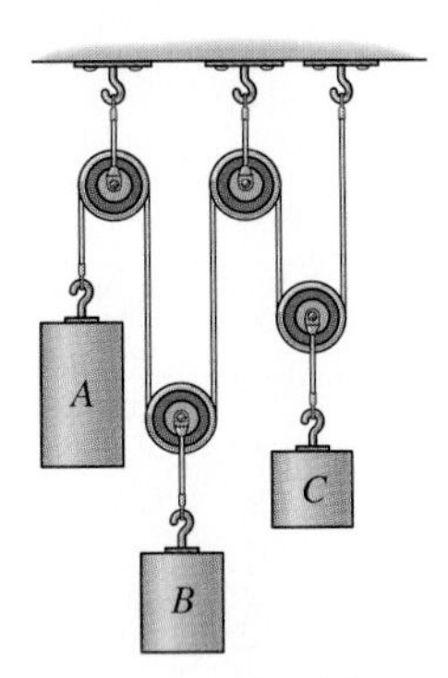

Prob. 12–200

12–201. Determine the constant speed at which the cable at A must be drawn in by the motor in order to hoist the load 6 m in 1.5 s.

12–202. Starting from rest, the cable can be wound onto the drum of the motor at a rate of $v_A = (3t^2)$ m/s, where t is in seconds. Determine the time needed to lift the load 7 m.

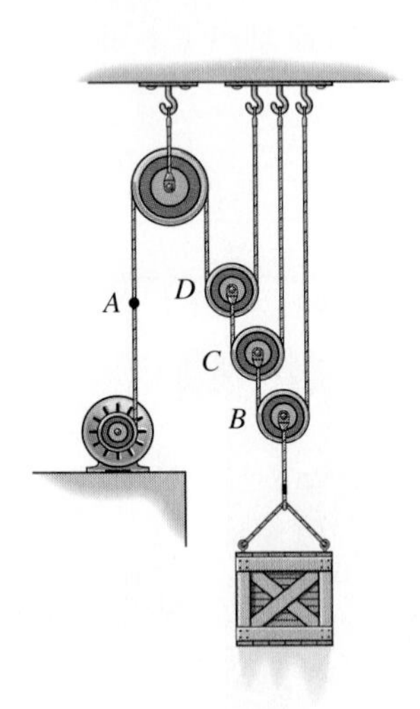

Probs. 12–201/12–202

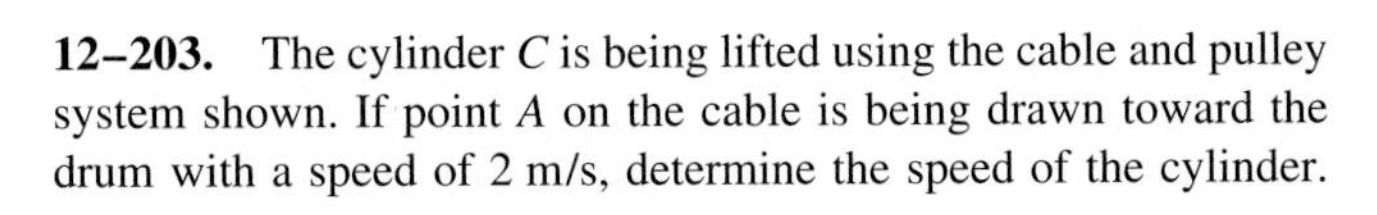

12–203. The cylinder C is being lifted using the cable and pulley system shown. If point A on the cable is being drawn toward the drum with a speed of 2 m/s, determine the speed of the cylinder.

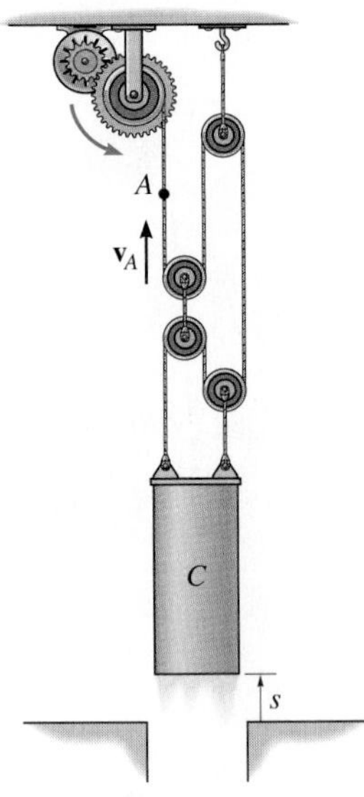

Probs. 12–203/12–204

*12–204. The cylinder C can be lifted with a maximum acceleration of $a_C = 3$ m/s^2 without causing the cables to fail. Determine the speed at which point A is moving toward the drum when $s = 4$ m if the cylinder is lifted from rest in the shortest time possible.

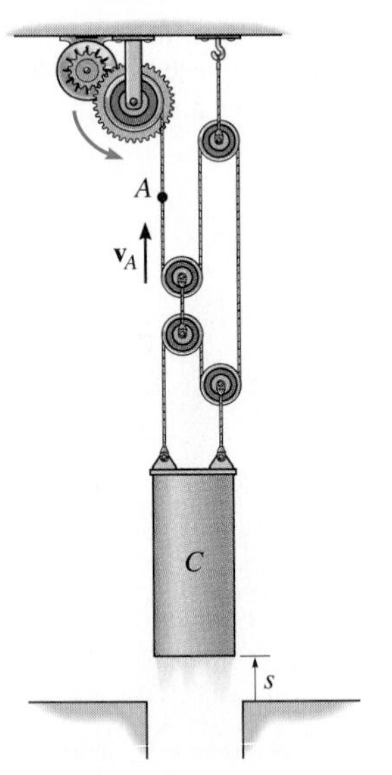

Probs. 12–203/12–204

12–205. Vertical motion of the load is produced by movement of the piston at A on the boom. If the piston or pulley C is moving to the left at 6 ft/s, determine the vertical motion of the load. The cable is attached at B, passes over the pulley at C, then D, E, F, and again around E, and is then attached at G.

12–206. Vertical motion of the load is produced by movement of the piston at A on the boom. Determine the distance the piston or pulley at C must move to the left in order to lift the load 2 ft. The cable is attached at B, passes over the pulley at C, then D, E, F, and again around E, and is attached at G.

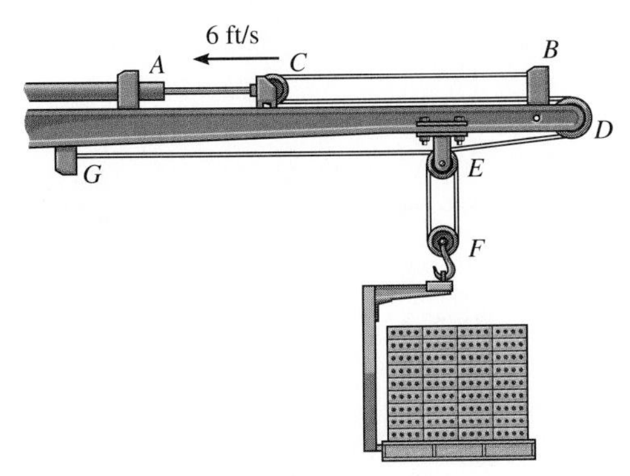

Probs. 12–205/12–206

12–207. The wireline used to lift a tool from the bottom of the oil well is controlled by the logging unit at A and the draw works at B. If the line is being wound up on the drum of the logging unit with a speed of 5 m/s, while the draw works are being lifted at 2 m/s, determine the speed at which the tool is being lifted.

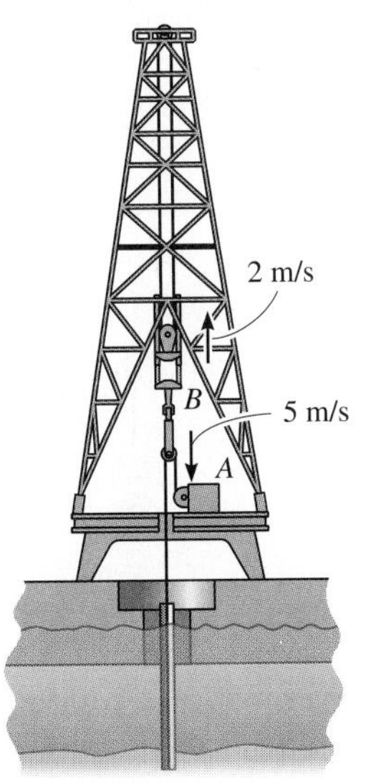

Prob. 12–207

***12–208.** The crate C is being lifted by moving the roller at A downward with a constant speed of $v_A = 2$ m/s along the guide. Determine the velocity and acceleration of the crate at the instant $s = 1$ m. When the roller is at B, the crate rests on the ground. Neglect the size of the pulley in the calculation. *Hint:* Relate the coordinates x_C and x_A using the problem geometry, then take the first and second time derivatives.

12–209. The crate C is being lifted by moving the roller at A downward with a constant speed of $v_A = 2$ m/s along the vertical guide. Find the variation of the velocity of the crate versus its position s. When the roller is at B, the crate rests on the ground. Neglect the size of the pulley in the calculation. *Hint:* Relate the coordinates x_C and x_A using the problem geometry, then take the first time derivative.

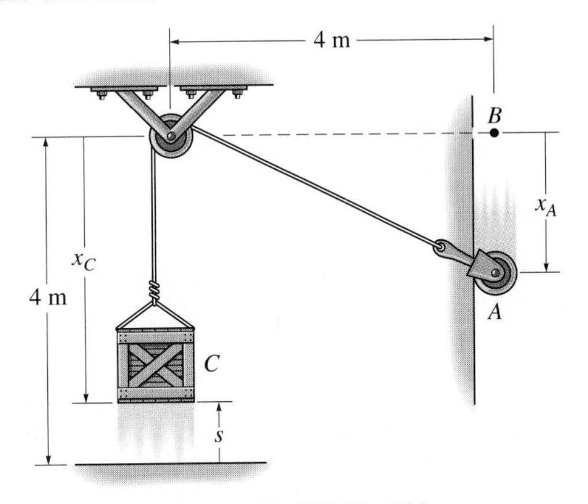

Probs. 12–208/12–209

12–210. The girl at C stands near the edge of the pier and pulls in the rope *horizontally* at a constant speed of 6 ft/s. Determine how fast the boat approaches the pier at the instant the rope length AB is 50 ft.

12–211. The girl at C stands near the edge of the pier and pulls in the rope *horizontally* at a constant speed of 6 ft/s. Determine the acceleration of the boat at the instant the rope length AB is 50 ft.

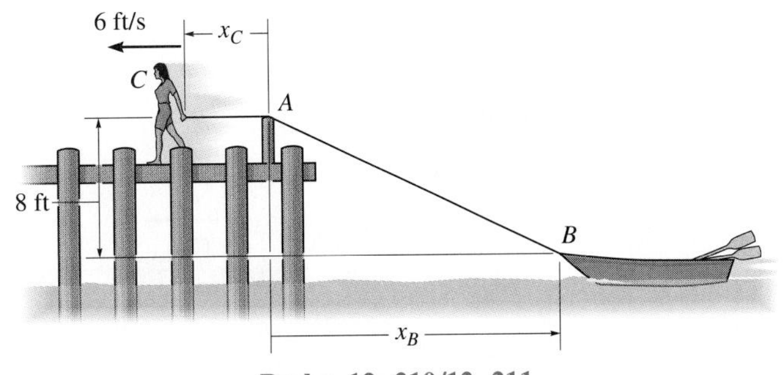

Probs. 12–210/12–211

***12–212.** If block B is moving down with a velocity v_B and has an acceleration a_B, determine the velocity and acceleration of block A in terms of the parameters shown.

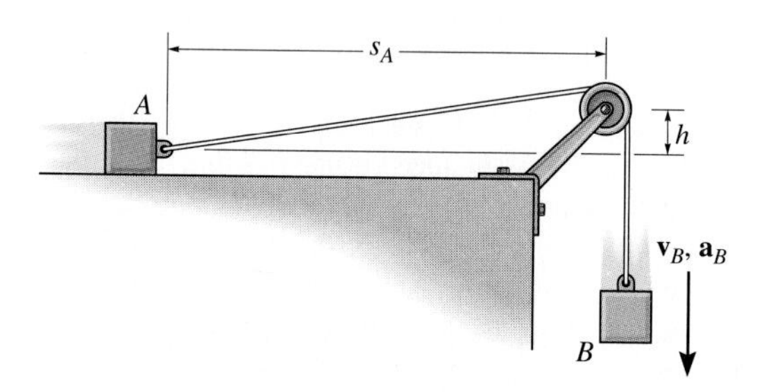

Prob. 12–212

12–213. The truck at B is used to hoist the cabinet up to the fourth floor of a building using the rope and pulley arrangement shown. If the truck is moving forward at a constant speed of $v_B = 2$ ft/s, determine the speed at which the cabinet rises at the instant $s_A = 40$ ft. Neglect the size of the pulley. When $s_B = 0$, $s_A = 0$, so that points A and B are coincident, i.e., the rope is 100 ft long.

12–214. Determine the acceleration of the cabinet in Prob. 12–213 at the instant $s_A = 40$ ft if the truck is accelerating at $a_B = 0.6$ ft/s^2 and has a speed of $v_B = 2$ ft/s.

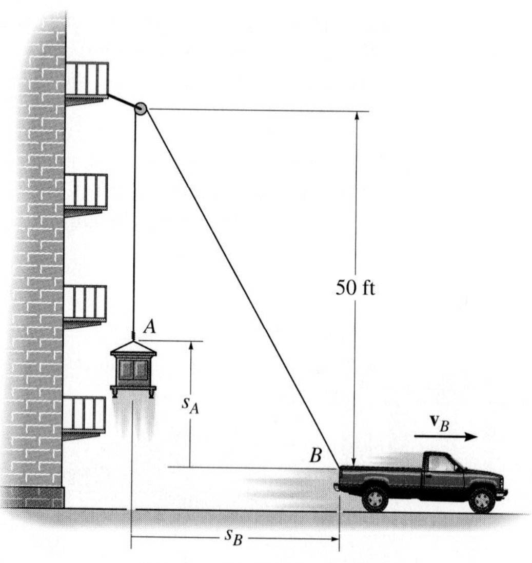

Probs. 12–213/12–214

12–215. Two planes, A and B, are flying at the same altitude. If their velocities are $v_A = 600$ km/h and $v_B = 500$ km/h such that the angle between their straight-line courses is $\theta = 75°$, determine the velocity of plane B with respect to plane A.

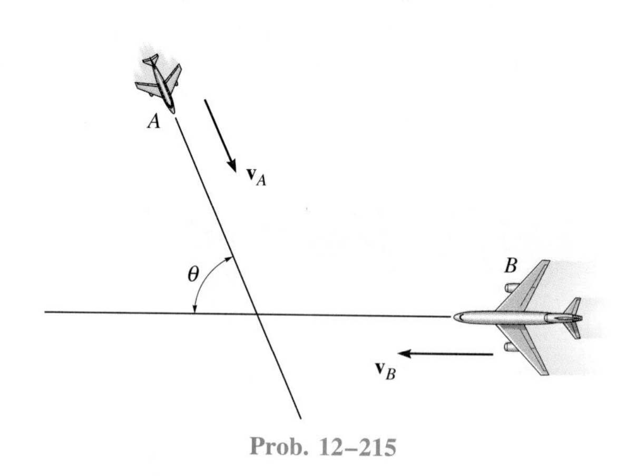

Prob. 12–215

***12–216.** At the instant shown, cars A and B are traveling at speeds of 30 mi/h and 20 mi/h, respectively. If B is increasing its speed by 1200 mi/h^2, while A maintains a constant speed, determine the velocity and acceleration of B with respect to A.

12–217. At the instant shown, cars A and B are traveling at speeds of 30 mi/h and 20 mi/h, respectively. If A is increasing its speed at 400 mi/h^2 whereas the speed of B is decreasing at 800 mi/h^2, determine the velocity and acceleration of B with respect to A.

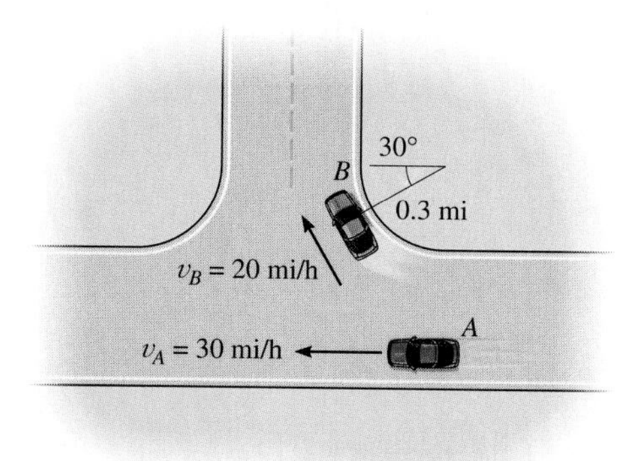

Probs. 12–216/12–217

12–218. Planes A and B are flying at the same altitude. If their velocities are $v_A = 300$ mi/h and $v_B = 250$ mi/h when the angle between their straight-line courses is 35° as shown, determine the velocity of plane A with respect to plane B.

12–219. At a given instant planes A and B are located as shown. If they maintain their straight-line courses and speeds, determine the distance between them in $t = 5$ min. Plane A is traveling at $v_A = 300$ mi/h and plane B is traveling at $v_B = 250$ mi/h.

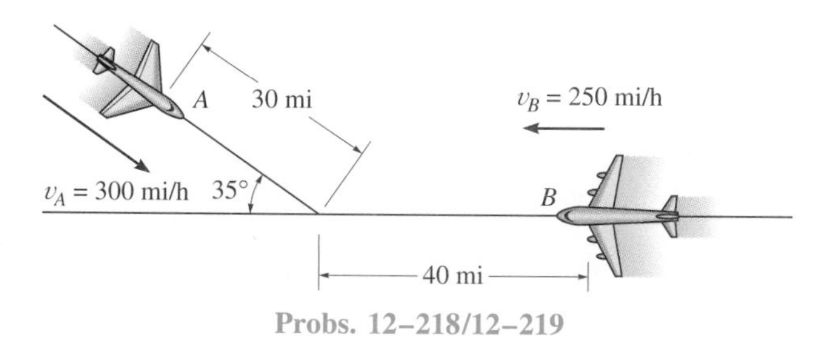

Probs. 12–218/12–219

*__12–220.__ Two boats leave the shore at the same time and travel in the directions shown. If $v_A = 20$ ft/s and $v_B = 15$ ft/s, determine the speed of boat A with respect to boat B. How long after leaving the shore will the boats be 800 ft apart?

12–221. Each boat begins from rest at point O and heads in the direction shown at the same instant. If A has an acceleration of $a_A = 2$ ft/s², and B has an acceleration of $a_B = 3$ ft/s², determine the speed of boat A with respect to boat B at the instant they become 800 ft apart. How long will this take?

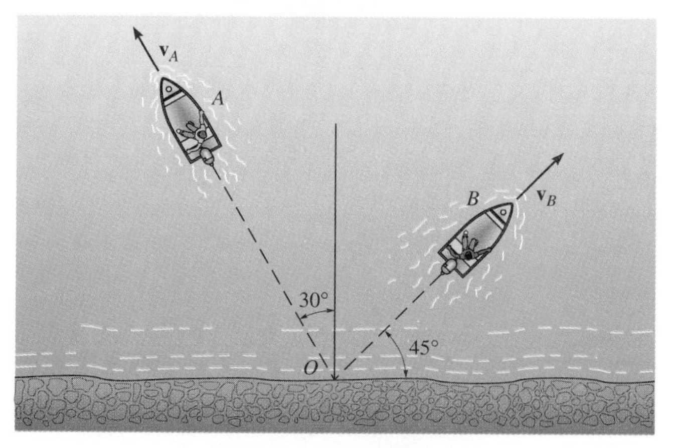

Probs. 12–220/12–221

12–222. At the instant shown, the bicyclist at A is traveling at 7 m/s around the curve on the race track while increasing his speed at 0.5 m/s². The bicyclist at B is traveling at 8.5 m/s along the straight-a-way and increasing his speed at 0.7 m/s². Determine the relative velocity and relative acceleration of A with respect to B at this instant.

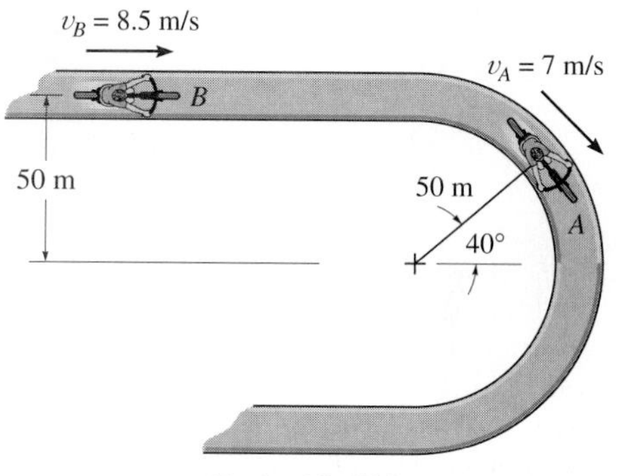

Prob. 12–222

12–223. At the instant shown, cars A and B are traveling at speeds of 55 mi/h and 40 mi/h, respectively. If B is increasing its speed by 1200 mi/h², while A maintains a constant speed, determine the velocity and acceleration of B with respect to A. Car B moves along a curve having a radius of curvature of 0.5 mi.

*__12–224.__ At the instant shown, cars A and B are traveling at speeds of 55 mi/h and 40 mi/h, respectively. If B is decreasing its speed at 1500 mi/h² while A is increasing its speed at 800 mi/h², determine the acceleration of B with respect to A. Car B moves along a curve having a radius of curvature of 0.75 mi.

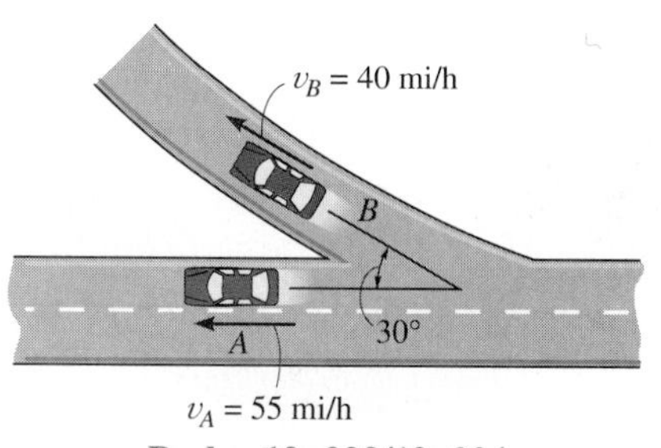

Probs. 12–223/12–224

12–225. At the instant shown, car A has a speed of 20 km/h, which is being increased at the rate of 300 km/h^2 as the car enters an expressway. At the same instant, car B is decelerating at 250 km/h^2 while traveling forward at 100 km/h. Determine the velocity and acceleration of car A with respect to car B.

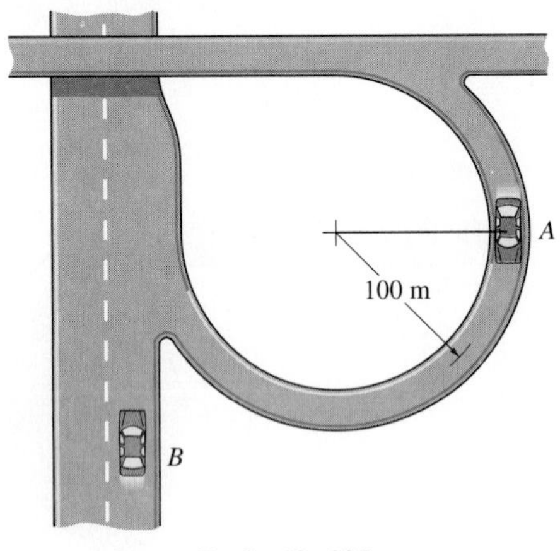

Prob. 12–225

12–226. Cars A and B are traveling around the circular race track. At the instant shown, A has a speed of 90 ft/s and is increasing its speed at the rate of 15 ft/s^2, whereas B has a speed of 105 ft/s and is decreasing its speed at 25 ft/s^2. Determine the relative velocity and relative acceleration of car A with respect to car B at this instant.

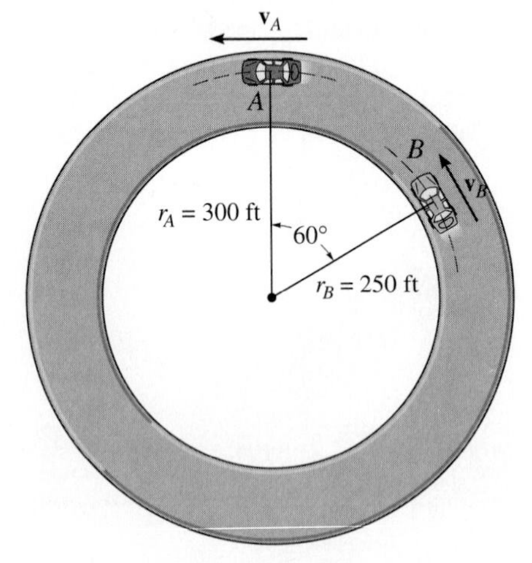

Prob. 12–226

12–227. A passenger in an automobile observes that raindrops make an angle of 30° with the horizontal as the auto travels forward with a speed of 60 km/h. Compute the terminal (constant) velocity $\mathbf{v}_r$ of the rain if it is assumed to fall vertically.

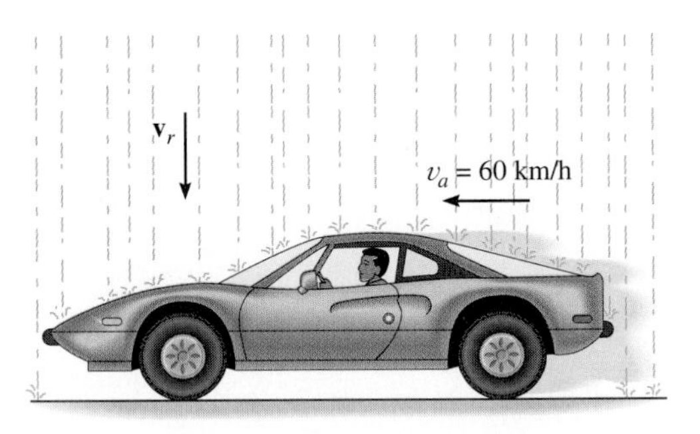

Prob. 12–227

■***12–228.** The boat can travel with a speed of 16 km/h in still water. The point of destination is located along the dashed line. If the water is moving at 4 km/h, determine the bearing angle θ at which the boat must travel to stay on course.

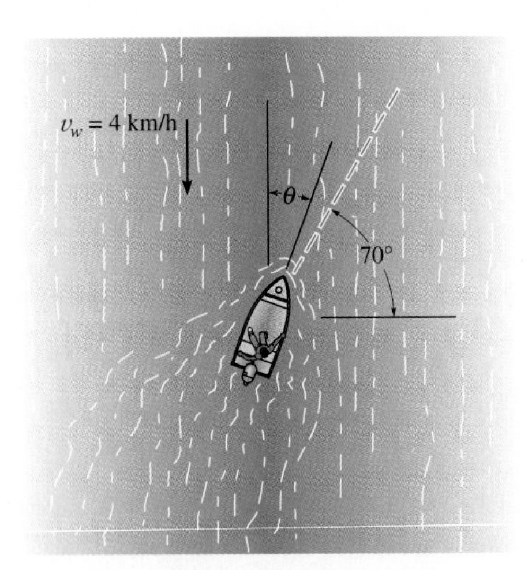

Prob. 12–228

12–229. At a given instant the football player at A throws a football C with a velocity of 20 m/s in the direction shown. Determine the constant speed at which the player at B must run so that he can catch the football at the same elevation at which it was thrown. Also calculate the relative velocity and relative acceleration of the football with respect to B at the instant the catch is made. Player B is 15 m away from A when A starts to throw the football.

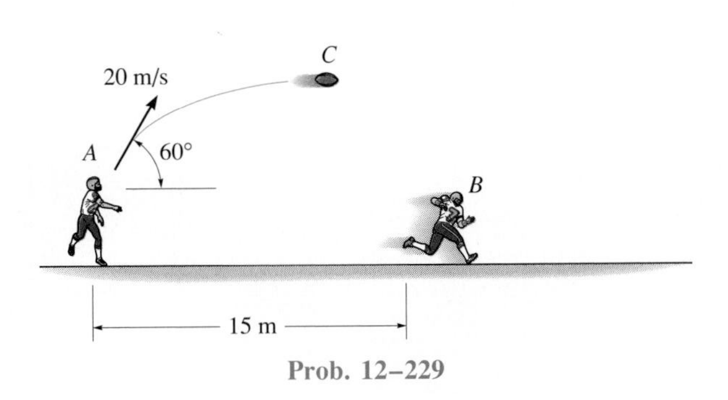

Prob. 12–229

12–230. At a given instant, each of the two particles A and B are moving with a speed of 8 m/s along the paths shown. If B is decelerating at 6 m/s^2 and the speed of A is increasing at 5 m/s^2, determine the relative acceleration of A with respect to B at this instant.

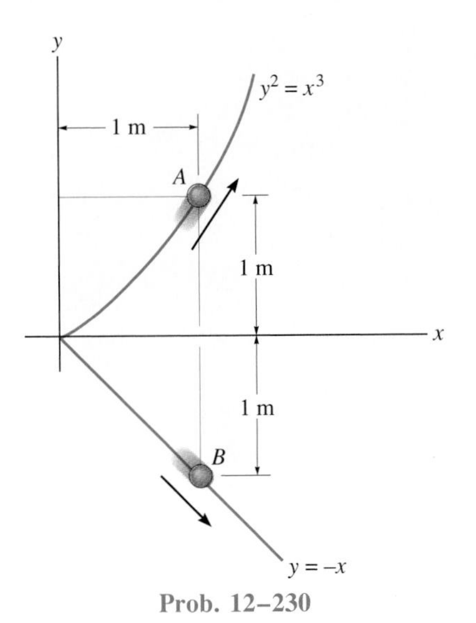

Prob. 12–230

12–231. The boy B is running down the slope at a constant speed of 12 ft/s. At the instant he reaches A, his friend throws a ball horizontally off the side of the slope. Determine the time of flight before the ball is caught by B, the launch velocity $\mathbf{v}_0$, and the relative speed at which it is caught by B.

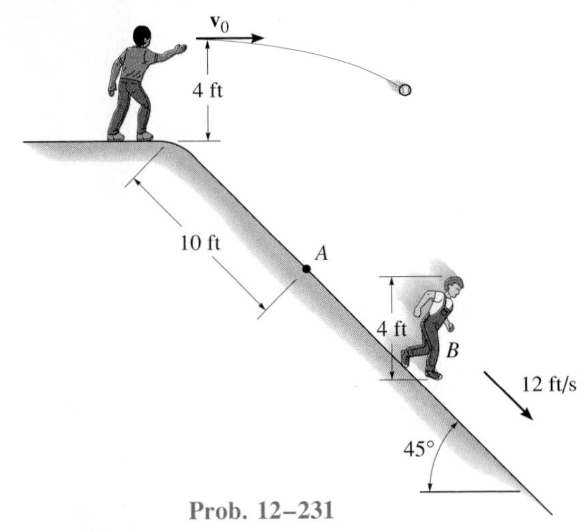

Prob. 12–231

***12–232.** The boy A is running in a straight line away from the building at a constant speed of 4 ft/s. The boy C throws the ball B horizontally when A is at $d = 10$ ft. At what speed must C throw the ball so that A can catch it? Also determine the relative speed of the ball with respect to boy A at the instant the catch is made.

12–233. The boy A is running in a straight line away from the building at a constant speed of 4 ft/s and is at $d = 0$ when the ball is thrown at $v_C = 10$ ft/s. At what horizontal distance d must he be from C in order to make the catch? with a horizontal velocity of $v_C = 10$ ft/s? Also determine the relative speed of the ball with respect to boy A at the instant the catch is made.

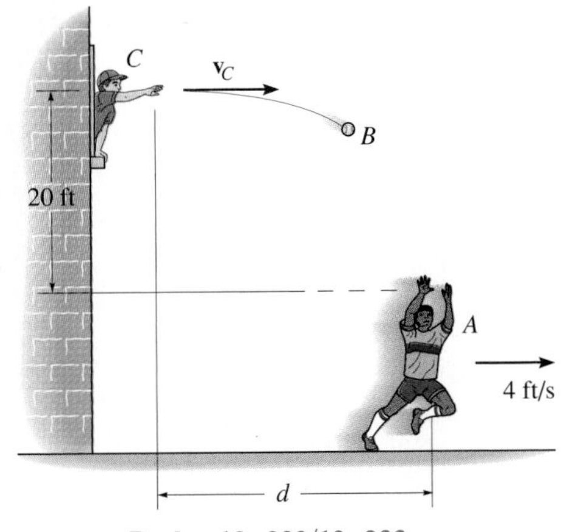

Probs. 12–232/12–233

The design of industrial conveyors requires knowledge of the forces acting on them, and the ability to predict the motion of the material they transport. The principles involved in this study are outlined in this chapter.

13

Kinetics of a Particle: Force and Acceleration

In Chapter 12 we developed the methods needed to formulate the acceleration of a particle in terms of its velocity and position. In this chapter we will use these concepts when applying Newton's second law of motion to study the effects caused by an unbalanced force acting on a particle. Depending on the geometry of the path, the analysis of problems will be performed using rectangular, normal and tangential, or cylindrical coordinates. In the last part of the chapter, Newton's second law of motion will be used to study problems involving space mechanics.

13.1 Newton's Laws of Motion

Many of the earlier notions about dynamics were dispelled after 1590 when Galileo performed experiments to study the motions of pendulums and falling bodies. The conclusions drawn from these experiments gave some insight as to the effects of forces acting on bodies in motion. The general laws of motion of a body subjected to forces were not known, however, until 1687, when Isaac Newton first presented three basic laws governing the motion of a particle. In a slightly reworded form, Newton's three laws of motion can be stated as follows:

First Law: *A particle originally at rest, or moving in a straight line with a constant velocity, will remain in this state provided the particle is not subjected to an unbalanced force.*

Second Law: *A particle acted upon by an unbalanced force* **F** *experiences an acceleration* **a** *that has the same direction as the force and a magnitude that is directly proportional to the force.**

Third Law: *The mutual forces of action and reaction between two particles are equal, opposite, and collinear.*

The first and third laws were used extensively in developing the concepts of statics. Although these laws are also considered in dynamics, Newton's second law of motion forms the basis for most of this study, since this law relates the accelerated motion of a particle to the forces that act on it. It should be noted that statics is a special case of dynamics, since Newton's second law yields the results of his first law when the resultant force is equal to zero; namely, no acceleration occurs, and therefore the particle's velocity is constant.

Measurements of force and acceleration can be recorded in a laboratory so that in accordance with the second law, if a known unbalanced force $\mathbf{F}_1$ is applied to a particle, the acceleration $\mathbf{a}_1$ of the particle may be measured. Since the force and acceleration are directly proportional, the constant of proportionality, m, may be determined from the ratio $m = F_1/a_1$. Provided the units of measurement are consistent, a different unbalanced force $\mathbf{F}_2$ applied to the particle will create an acceleration $\mathbf{a}_2$, such that $F_2/a_2 = m$.† In both cases the ratio will be the same and the acceleration and the force, both being vector quantities, will have the same direction. The positive scalar m is called the *mass* of the particle. Being constant during any acceleration, m provides a quantitative measure of the resistance of the particle to a change in its velocity.

If the mass of the particle is m, Newton's second law of motion may be written in mathematical form as

$$\mathbf{F} = m\mathbf{a}$$

This equation, which is referred to as the *equation of motion,* is one of the

*Stated another way, the unbalanced force acting on the particle is proportional to the time rate of change of the particle's linear momentum. See footnote ‡ on next page.

†Throughout this discussion, the units of force, mass, length, and time used to measure F, m, and a must be chosen such that the units for one of these quantities are defined in terms of all the others. Such is the case for the SI and FPS units, where it may be recalled that $\text{N} = \text{kg} \cdot \text{m/s}^2$ and $\text{slug} = \text{lb} \cdot \text{s}^2/\text{ft}$ (see Sec. 1.3 of *Statics*). Note, however, that if units of force, mass, length, and time are *all* selected arbitrarily, then $F = kma$ and k (a dimensionless constant) would have to be determined experimentally in order to preserve the equality.

most important formulations in mechanics.‡ As stated above, its validity has been based solely on *experimental evidence.* In 1905, however, Albert Einstein developed the theory of relativity and placed limitations on the use of Newton's second law for describing general particle motion. Through experiments it was proven that *time* is not an absolute quantity as assumed by Newton; and as a result, the equation of motion fails to predict the exact behavior of a particle, especially when the particle's speed approaches the speed of light (0.3 Gm/s). Developments of the theory of quantum mechanics by Erwin Schrödinger and others indicate further that conclusions drawn from using this equation are also invalid when particles move within an atomic distance of one another. For the most part, however, these requirements regarding particle speed and size are not encountered in engineering problems, so their effects will not be considered in this book.

Newton's Law of Gravitational Attraction. Shortly after formulating his three laws of motion, Newton postulated a law governing the mutual attraction between any two particles. In mathematical form this law can be expressed as

$$F = G\frac{m_1 m_2}{r^2} \qquad (13\text{–}1)$$

where

F = force of attraction between the two particles
G = universal constant of gravitation; according to experimental evidence $G = 66.73(10^{-12})\ \text{m}^3/(\text{kg} \cdot \text{s}^2)$
m_1, m_2 = mass of each of the two particles
r = distance between the centers of the two particles

Any two particles or bodies have a mutually attractive gravitational force acting between them. In the case of a particle located at or near the surface of the earth, however, the only gravitational force having any sizable magnitude is that between the earth and the particle. This force is termed the "weight" and, for our purpose, it will be the only gravitational force considered.

Mass and Weight. *Mass* is a property of matter by which we can compare the action of one body with that of another. As indicated above, this property manifests itself as a gravitational attraction between two bodies and provides a quantitative measure of the resistance of matter to a change in velocity. It is an *absolute* quantity since the measurement of mass can be made at any location. The weight of a body, however, is *not absolute* since it is measured in a gravitational field, and hence its magnitude depends on where the measurement is made. From Eq. 13–1, we can develop a general expres-

‡Since m is constant, we can also write $\mathbf{F} = (d/dt)(m\mathbf{v})$, where $m\mathbf{v}$ is the particle's linear momentum.

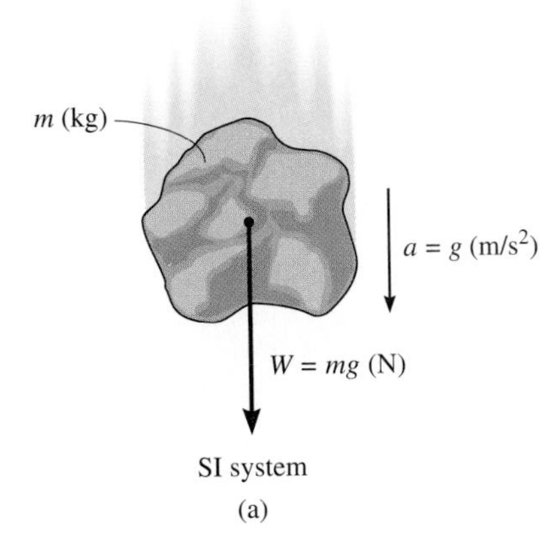

SI system
(a)

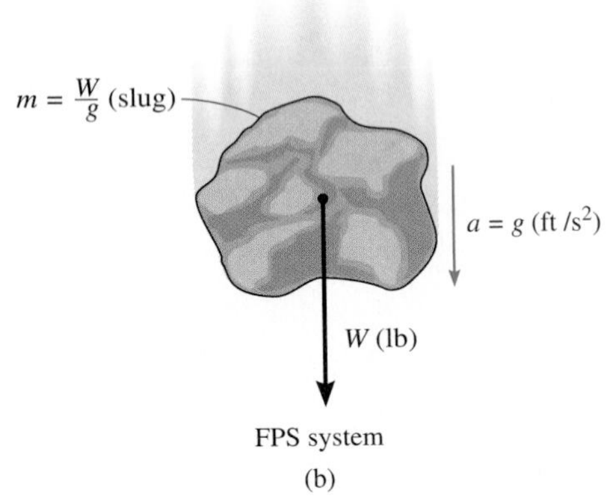

FPS system
(b)

Fig. 13–1

sion for finding the weight W of a particle having a mass $m_1 = m$. Let m_2 be the mass of the earth and r the distance between the earth's center and the particle. Then, if $g = Gm_2/r^2$, we have

$$W = mg$$

By comparison with $F = ma$, we term g the acceleration due to gravity. For most engineering calculations g is measured at a point on the surface of the earth at sea level, and at a latitude of 45°, considered the "standard location."

The mass and weight of a body are measured differently in the SI and FPS systems of units, and the method of defining these units should be thoroughly understood.

SI System of Units. In the SI system the mass of the body is specified in kilograms and the weight must be calculated using the equation of motion, $F = ma$. Hence, if a body has a mass m (kg) and is located at a point where the acceleration due to gravity is g (m/s²), then the weight is expressed in *newtons* as $W = mg$ (N), Fig. 13–1*a*. In particular, if the body is located at the "standard location," the acceleration due to gravity is $g = 9.806\,65$ m/s². For calculations, the value $g = 9.81$ m/s² will be used, so that

$$W = mg \text{ (N)} \qquad (g = 9.81 \text{ m/s}^2) \tag{13–2}$$

Therefore, a body of mass 1 kg has a weight of 9.81 N; a 2-kg body weighs 19.62 N; and so on.

FPS System of Units. In the FPS system the weight of the body is specified in pounds and the mass must be calculated from $F = ma$. Hence, if a body has a weight W (lb) and is located at a point where the acceleration due to gravity is g (ft/s²), then the mass is expressed in *slugs* as $m = W/g$ (slug), Fig. 13–1*b*. Since the acceleration of gravity at the standard location is approximately 32.2 ft/s²(= 9.81 m/s²), the mass of the body measured in slugs is

$$m = \frac{W}{g} \text{ (slug)} \qquad (g = 32.2 \text{ ft/s}^2) \tag{13–3}$$

Therefore, a body weighing 32.2 lb has a mass of 1 slug; a 64.4-lb body has a mass of 2 slugs; and so on.

13.2 The Equation of Motion

When more than one force acts on a particle, the resultant force is determined by a vector summation of all the forces; i.e., $\mathbf{F}_R = \Sigma\mathbf{F}$. For this more general case, the equation of motion may be written as

$$\Sigma\mathbf{F} = m\mathbf{a} \qquad (13\text{–}4)$$

To illustrate application of this equation, consider the particle P shown in Fig. 13–2*a*, which has a mass m and is subjected to the action of two forces, $\mathbf{F}_1$ and $\mathbf{F}_2$. We can graphically account for the magnitude and direction of each force acting on the particle by drawing the particle's *free-body diagram,* Fig. 13–2*b*. Since the *resultant* of these forces *produces* the vector $m\mathbf{a}$, its magnitude and direction can be represented graphically on the *kinetic diagram,* shown in Fig. 13–2*c*. The equal sign written between the diagrams symbolizes the *graphical* equivalency between the free-body diagram and the kinetic diagram; i.e., $\Sigma\mathbf{F} = m\mathbf{a}$.* In particular, note that if $\mathbf{F}_R = \Sigma\mathbf{F} = \mathbf{0}$, then the acceleration is also zero, so that the particle will either remain at *rest* or move along a straight-line path with *constant velocity.* Such are the conditions of *static equilibrium,* Newton's first law of motion.

Inertial Frame of Reference. Whenever the equation of motion is applied, it is required that measurements of the acceleration be made from a *Newtonian* or *inertial frame of reference. Such a coordinate system does not rotate and is either fixed or translates in a given direction with a constant velocity (zero acceleration).* This definition ensures that the particle's *acceleration* measured by observers in two different inertial frames of reference will always be the *same.* For example, consider the particle P moving with an absolute acceleration $\mathbf{a}_P$ along a straight path as shown in Fig. 13–3. If the

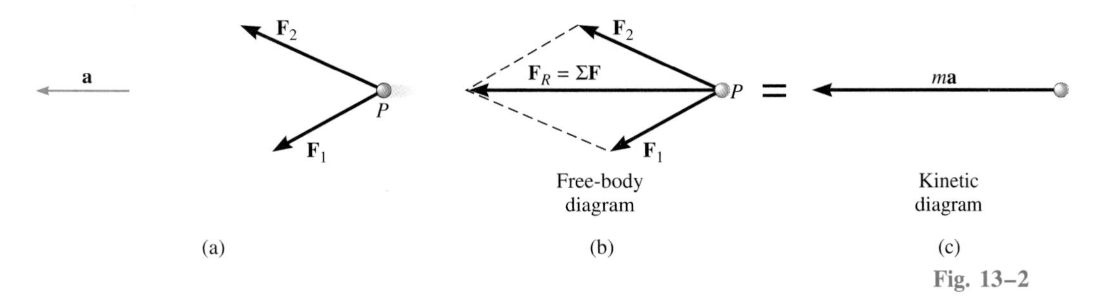

Fig. 13–2

*The equation of motion can also be rewritten in the form $\Sigma\mathbf{F} - m\mathbf{a} = \mathbf{0}$. The vector $-m\mathbf{a}$ is referred to as the *inertia force vector.* If it is treated in the same way as a "force vector," then the state of "equilibrium" created is referred to as *dynamic equilibrium.* This method for application of the equation of motion is often referred to as the *D'Alembert principle,* named after the French mathematician Jean le Rond d'Alembert.

observer is *fixed* in the inertial x, y frame of reference, this acceleration, $\mathbf{a}_P$, will be measured by the observer regardless of the direction and magnitude of the velocity $\mathbf{v}_O$ of the frame of reference. On the other hand, if the observer is *fixed* in the noninertial x', y' frame of reference, Fig. 13–3, the observer will not measure the particle's acceleration as $\mathbf{a}_P$. Instead, if the frame is *accelerating* at $\mathbf{a}_{O'}$ the particle will appear to have an acceleration of $\mathbf{a}_{P/O'} = \mathbf{a}_P - \mathbf{a}_{O'}$. Also, if the frame is *rotating,* as indicated by the curl, then the particle will appear to move along a *curved path,* in which case it will appear to have other components of acceleration (see Sec. 16.8). In any case, the measured acceleration cannot be used in Newton's law of motion to determine the forces acting on the particle.

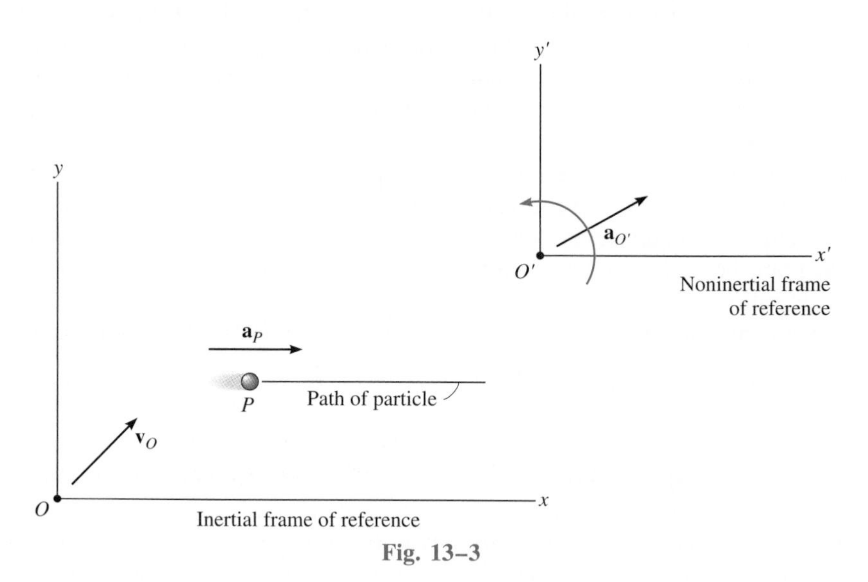

Fig. 13–3

When studying the motions of rockets and satellites it is justifiable to consider the inertial reference frame as fixed to the stars, whereas dynamics problems concerned with motions on or near the surface of the earth may be solved by using an inertial frame which is assumed fixed to the earth. Even though the earth both rotates about its own axis and revolves about the sun, the acceleration created by these rotations can be neglected in most computations.

13.3 Equation of Motion for a System of Particles

The equation of motion will now be extended to include a system of n particles isolated within an enclosed region in space, as shown in Fig. 13–4*a*. In particular, there is no restriction in the way the particles are connected, and as a result the following analysis will apply equally well to the motion of a solid, liquid, or gas system. At the instant considered, the arbitrary ith particle, having a mass m_i, is subjected to a system of internal forces and a resultant external force. The *resultant internal force,* represented symbolically as $\mathbf{f}_i = \Sigma_{j=1(j\neq i)}^{n} \mathbf{f}_{ij}$, is determined from the forces which the other particles exert on the ith particle. Usually these forces are developed by direct contact, although the summation extends over all n particles within the dashed boundary. Note, however, that it is meaningless for $j = i$ since the ith particle cannot exert a force on itself. The *resultant external force* $\mathbf{F}_i$ represents, for example, the effect of gravitational, electrical, magnetic, or contact forces between adjacent bodies or particles *not* included within the system.

The free-body and kinetic diagrams for the ith particle are shown in Fig. 13–4*b*. Applying the equation of motion yields

$$\Sigma \mathbf{F} = m\mathbf{a}; \qquad \mathbf{F}_i + \mathbf{f}_i = m_i \mathbf{a}_i$$

When the equation of motion is applied to each of the other particles of the system, similar equations will result. If all these equations are added together *vectorially,* we obtain

$$\Sigma \mathbf{F}_i + \Sigma \mathbf{f}_i = \Sigma m_i \mathbf{a}_i$$

The summation of the internal forces, if carried out, will equal zero, since the internal forces between particles all occur in equal but opposite collinear pairs. Consequently, only the sum of the external forces will remain and therefore the equation of motion, written for the system of particles, becomes

$$\Sigma \mathbf{F}_i = \Sigma m_i \mathbf{a}_i \qquad (13\text{–}5)$$

If $\mathbf{r}_G$ is a position vector which locates the *center of mass G* of the particles, Fig. 13–4*a*, then by definition of the center of mass, $m\mathbf{r}_G = \Sigma m_i \mathbf{r}_i$, where $m = \Sigma m_i$ is the total mass of all the particles. Differentiating this equation twice with respect to time, assuming that no mass is entering or leaving the system,* yields

$$m\mathbf{a}_G = \Sigma m_i \mathbf{a}_i$$

Substituting this result into the above equation, we obtain

$$\Sigma \mathbf{F} = m\mathbf{a}_G \qquad (13\text{–}6)$$

This equation states that the sum of the external forces acting on the system of particles is equal to the total mass of the particles times the acceleration of its center of mass G. Since in reality all particles must have a finite size to possess mass, Eq. 13–6 justifies application of the equation of motion to a *body* that is represented as a single particle.

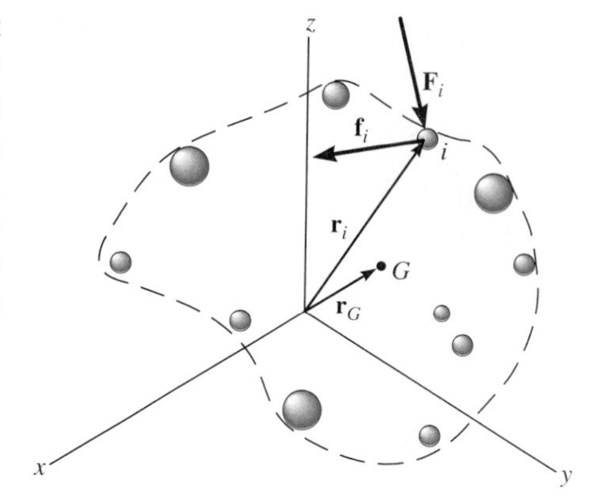

Inertial coordinate system

(a)

$\mathbf{f}_i$ $\mathbf{F}_i$ = $m_i \mathbf{a}_i$

Free-body diagram

Kinetic diagram

(b)

Fig. 13–4

*A case in which m is a function of time (variable mass) is discussed in Sec. 15.9.

13.4 Equations of Motion: Rectangular Coordinates

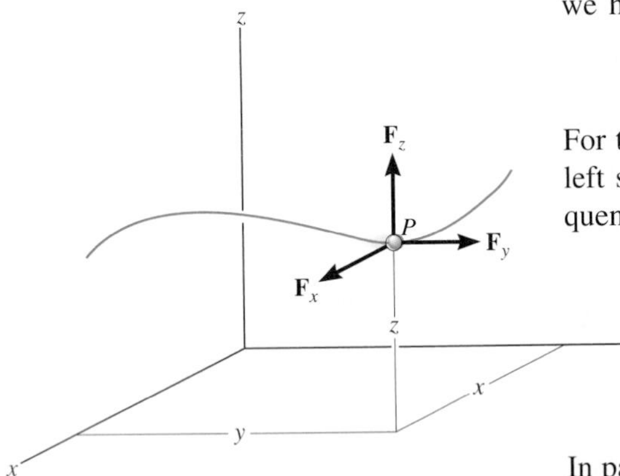

Fig. 13–5

When a particle is moving relative to an inertial x, y, z frame of reference, the forces acting on the particle, as well as its acceleration, may be expressed in terms of their **i, j, k** components, Fig. 13–5. Applying the equation of motion, we have

$$\Sigma \mathbf{F} = m\mathbf{a}$$

$$\Sigma F_x \mathbf{i} + \Sigma F_y \mathbf{j} + \Sigma F_z \mathbf{k} = m(a_x \mathbf{i} + a_y \mathbf{j} + a_z \mathbf{k})$$

For this equation to be satisfied, the respective **i, j,** and **k** components on the left side must equal the corresponding components on the right side. Consequently, we may write the following three scalar equations:

$$\begin{aligned} \Sigma F_x &= ma_x \\ \Sigma F_y &= ma_y \\ \Sigma F_z &= ma_z \end{aligned} \qquad (13\text{–}7)$$

In particular, if the particle is constrained to move in the x–y plane, then only the first two of these equations are used to specify the motion.

PROCEDURE FOR ANALYSIS

The equations of motion are used to solve problems which require a relationship between the forces acting on a particle and the accelerated motion they cause. Whenever they are applied, the unknown force and acceleration components should be identified and an equivalent number of equations should be written. If further equations are necessary for the solution, kinematics may be considered. Recall that these equations only relate the geometric properties of the motion, and therefore they become useful if the particle's position or velocity is to be related to its acceleration. Whatever the situation, the following procedure provides a general method for solving problems in kinetics.

Free-Body Diagram. Select the inertial coordinate system. Most often, rectangular or x, y, z coordinates are chosen to analyze problems for which the particle has *rectilinear motion.* If this occurs, one of the axes should extend in the direction of motion. Once the coordinates are established, draw the particle's free-body diagram. Drawing the free-body diagram is *very important* since it provides a graphical representation that accounts for *all the forces* ($\Sigma \mathbf{F}$) which act on the particle, and thereby makes it possible to resolve these forces into their x, y, z components. The direction and sense of the particle's acceleration **a** should also be established. If the sense of its components is unknown, for mathematical convenience assume that they are in the same *direction* as the *positive* inertial coordinate axes. The acceleration may be sketched on the x, y, z coordinate system, *but not on* the free-body diagram, or it may be represented as the $m\mathbf{a}$ vector on the kinetic diagram.* Once the free-body diagram has been constructed, identify the unknowns.

Equations of Motion. If the forces can be resolved directly from the free-body diagram, apply the equations of motion in their scalar component form. If the geometry of the problem appears complicated, which often occurs in three dimensions, Cartesian vector analysis can be used for the solution.

Friction. If the particle contacts a rough surface, it may be necessary to use the *frictional equation,* which relates the coefficient of kinetic friction μ_k to the magnitudes of the frictional and normal forces $\mathbf{F}_f$ and $\mathbf{N}$ acting at the surfaces of contact, i.e., $F_f = \mu_k N$. Remember that $\mathbf{F}_f$ always acts on the free-body diagram such that it opposes the motion of the particle relative to the surface it contacts.

Spring. If the particle is connected to an *elastic spring* having negligible mass, the spring force F_s can be related to the deformation of the spring by the equation $F_s = ks$. Here k is the spring's stiffness measured as a force per unit length, and s is the stretch or compression defined as the difference between the deformed length l and the undeformed length l_0, i.e., $s = l - l_0$.

Kinematics. As stated previously, if a complete solution cannot be obtained strictly from the equation of motion, the equations of kinematics may be considered. For example, if the velocity or position of the particle is to be found, it will be necessary to apply the proper kinematic equations once the particle's acceleration is determined from $\Sigma\mathbf{F} = m\mathbf{a}$.

If *acceleration is a function of time,* use $a = dv/dt$ and $v = ds/dt$ which, when integrated, yield the particle's velocity and position.

If *acceleration is a function of displacement,* integrate $a\,ds = v\,dv$ to obtain the velocity as a function of position.

If *acceleration is constant,* use $v = v_0 + a_c t$, $s = s_0 + v_0 t + \frac{1}{2}a_c t^2$, $v^2 = v_0^2 + 2a_c(s - s_0)$ to determine the velocity or position of the particle.

If the problem involves the dependent motion of several particles, use the method outlined in Sec. 12.8 to relate their accelerations.

In all cases, make sure the positive inertial coordinate directions used for writing the kinematic equations are the same as those used for writing the equations of motion; otherwise, simultaneous solution of the equations will result in errors. Also, if the solution for an unknown vector component yields a negative scalar, it indicates that the component acts in the direction opposite to that which was assumed.

The following examples illustrate numerical application of this procedure.

*It is a convention in this text always to use the kinetic diagram as a graphical aid when developing the proofs and theory. The particle's acceleration or its components will be shown as colored vectors near the free-body diagram in the examples.

Example 13–1

The 50-kg crate shown in Fig. 13–6a rests on a horizontal plane for which the coefficient of kinetic friction is $\mu_k = 0.3$. If the crate does not tip over when it is subjected to a 400-N towing force as shown, determine the velocity of the crate in 5 s starting from rest.

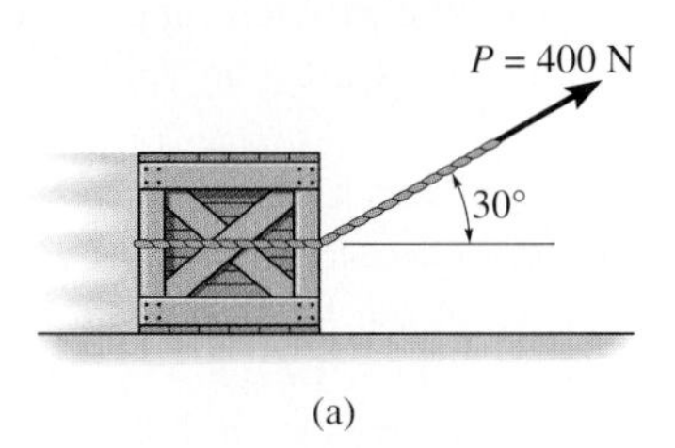

(a)

SOLUTION

This problem requires application of the equations of motion, since the crate's acceleration can be related to the force causing the motion. The crate's velocity can then be determined using kinematics.

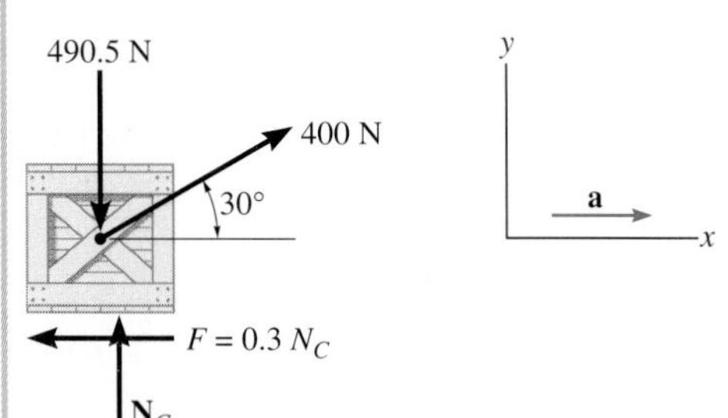

(b)

Free-Body Diagram. The weight of the crate in Fig. 13–6a is $W = mg = 50\text{ kg }(9.81\text{ m/s}^2) = 490.5\text{ N}$. As shown in Fig. 13–6$b$, the frictional force has a magnitude $F = \mu_k N_C$ and acts to the left, since it opposes the motion of the crate. The acceleration $\mathbf{a}$ is assumed to act horizontal, in the positive x direction. There are two unknowns, namely N_C and a.

Equations of Motion. Using the data shown on the free-body diagram, we have

$$\overset{+}{\rightarrow}\Sigma F_x = ma_x; \qquad 400\cos 30° - 0.3\,N_C = 50a \qquad (1)$$

$$+\uparrow \Sigma F_y = ma_y; \qquad N_C - 490.5 + 400\sin 30° = 0 \qquad (2)$$

Solving Eq. 2 for N_C, substituting the result into Eq. 1, and solving for a yields

$$N_C = 290.5\text{ N}$$

$$a = 5.19\text{ m/s}^2$$

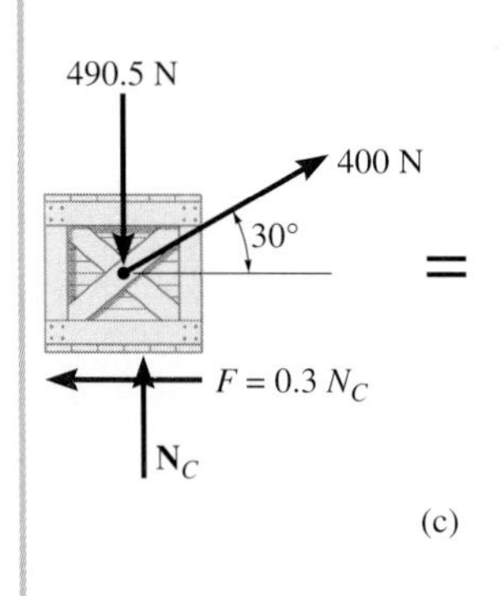

(c)

Fig. 13–6

Kinematics. Since the acceleration is *constant,* and the initial velocity is zero, the velocity of the crate in 5 s is

$$(\overset{+}{\rightarrow}) \qquad v = v_0 + a_c t$$

$$= 0 + 5.19(5)$$

$$= 26.0\text{ m/s} \rightarrow \qquad \textit{Ans.}$$

Note the alternative procedure of drawing the crate's free-body *and* kinetic diagrams, Fig. 13–6c, prior to applying the equations of motion.

Example 13–2

A 10-kg projectile is fired vertically upward from the ground, with an initial velocity of 50 m/s, Fig. 13–7*a*. Determine the maximum height to which it will travel if (*a*) atmospheric resistance is neglected; and (*b*) atmospheric resistance is measured as $F_D = (0.01v^2)$ N, where v is the speed at any instant, measured in m/s.

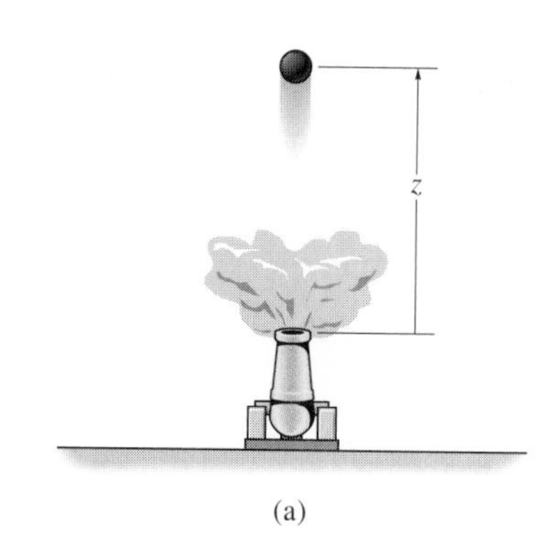

(a)

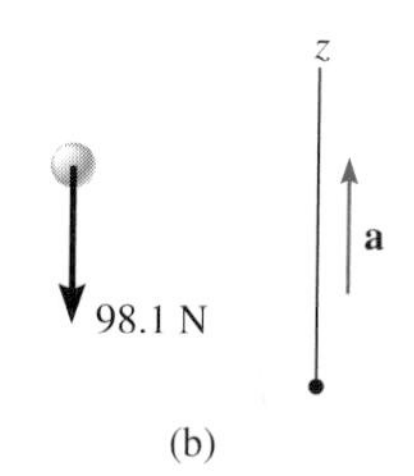

(b)

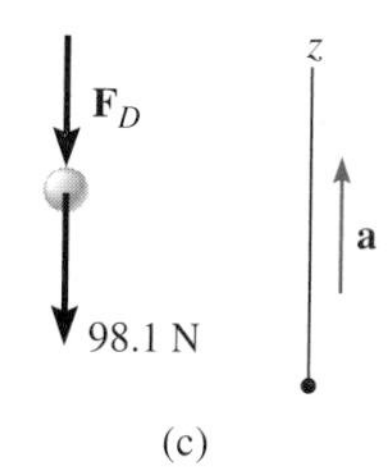

(c)

Fig. 13–7

SOLUTION

In both cases the known force on the projectile can be related to its acceleration using the equation of motion. Kinematics can then be used to relate the projectile's acceleration to its position.

Part (a) Free-Body Diagram. As shown in Fig. 13–7*b*, the projectile's weight is $W = mg = 10(9.81) = 98.1$ N. We will assume the unknown acceleration **a** acts upward in the *positive* z direction.

Equation of Motion

$+\uparrow \Sigma F_z = ma_z;$ $\qquad -98.1 = 10a, \qquad a = -9.81 \text{ m/s}^2$

The result indicates that the projectile, like every object having free-flight motion near the earth's surface, is subjected to a constant downward acceleration of 9.81 m/s^2.

Kinematics. Initially, $z_0 = 0$ and $v_0 = 50$ m/s, and at the maximum height $z = h$, $v = 0$. Since the acceleration is *constant,* then

$(+\uparrow)$
$$v^2 = v_0^2 + 2a_c(z - z_0)$$
$$0 = (50)^2 + 2(-9.81)(h - 0)$$
$$h = 127 \text{ m} \qquad \textit{Ans.}$$

Part (b) Free-Body Diagram. Since the force $F_D = (0.01v^2)$ N tends to retard the upward motion of the projectile, it acts downward as shown on the free-body diagram, Fig. 13–7*c*.

Equation of Motion*

$+\uparrow \Sigma F_z = ma_z;$ $\qquad -0.01v^2 - 98.1 = 10a, \qquad a = -0.001v^2 - 9.81$

Kinematics. Here the acceleration is *not constant;* however, it can be related to the velocity and displacement by using $a\,dz = v\,dv$.

$(+\uparrow)\ a\,dz = v\,dv; \qquad (-0.001v^2 - 9.81)dz = v\,dv$

Separating the variables and integrating, realizing that initially $z_0 = 0$, $v_0 = 50$ m/s (positive upward), and at $z = h$, $v = 0$, we have

$$\int_0^h dz = -\int_{50}^{0} \frac{v\,dv}{0.001v^2 + 9.81} = -500 \ln (v^2 + 9810)\Big|_{50}^{0}$$

$$h = 114 \text{ m} \qquad \textit{Ans.}$$

The answer indicates a lower elevation than that obtained in part (*a*). Why?

*Note that if the projectile were fired downward, with z positive downward, the equation of motion would then be $-0.01v^2 + 98.1 = 10a$.

Example 13–3

The sled with load shown in Fig. 13–8*a* has a weight of 50 lb and is acted upon by a force having a variable magnitude $P = 20t$, where P is in pounds and t is in seconds. Compute the sled's velocity 2 s after **P** has been applied. The sled's initial velocity is $v_0 = 3$ ft/s down the plane, and the coefficient of kinetic friction between the sled and the plane is $\mu_k = 0.3$.

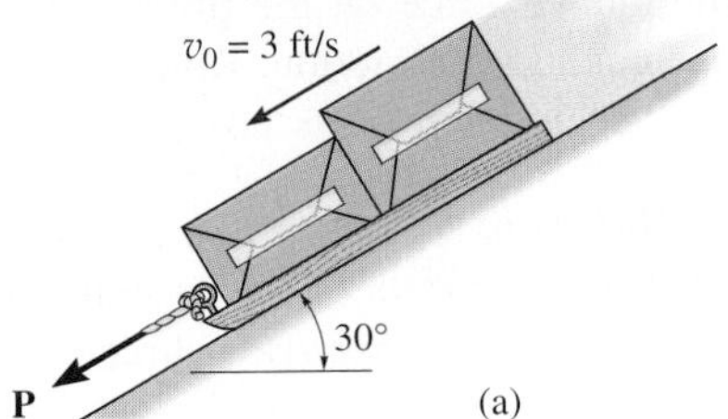

SOLUTION

Before reading further, do you see *why* it is necessary to apply both the equations of motion *and* kinematics to obtain a solution of this problem?

Free-Body Diagram. As shown in Fig. 13–8*b*, the frictional force is directed opposite to the sled's sliding motion and has a magnitude $F = \mu_k N_C = 0.3N_C$. The mass is $m = W/g = 50/32.2 = 1.55$ slug. There are two unknowns, namely N_C and a.

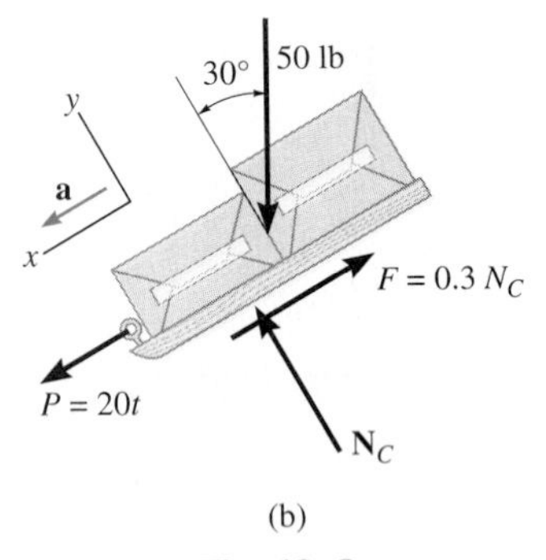

Fig. 13–8

Equations of Motion. Using the data shown on the free-body diagram, we have

$$+\swarrow \Sigma F_x = ma_x; \quad 20t - 0.3N_C + 50 \sin 30° = 1.55a \quad (1)$$

$$+\nwarrow \Sigma F_y = ma_y; \quad N_C - 50 \cos 30° = 0 \quad (2)$$

Solving for N_C in Eq. 2 ($N_C = 43.3$ lb), substituting into Eq. 1, and simplifying yields

$$a = 12.88t + 7.73 \quad (3)$$

Kinematics. Since the acceleration is a function of time, the velocity of the sled is obtained by using $a = dv/dt$ with the initial condition that $v_0 = 3$ ft/s at $t = 0$. We have

$$(+\swarrow) \qquad dv = a\,dt$$

$$\int_3^v dv = \int_0^t (12.88t + 7.73)dt$$

$$v = 6.44t^2 + 7.73t + 3$$

When $t = 2$ s,

$$v = 44.2 \text{ ft/s} \swarrow \qquad \textit{Ans.}$$

Example 13–4

A smooth 2-kg collar C, shown in Fig. 13–9a, is attached to a spring having a stiffness $k = 3$ N/m and an unstretched length of 0.75 m. If the collar is released from rest at A, determine its acceleration and the normal force of the rod on the collar at the instant $y = 1$ m.

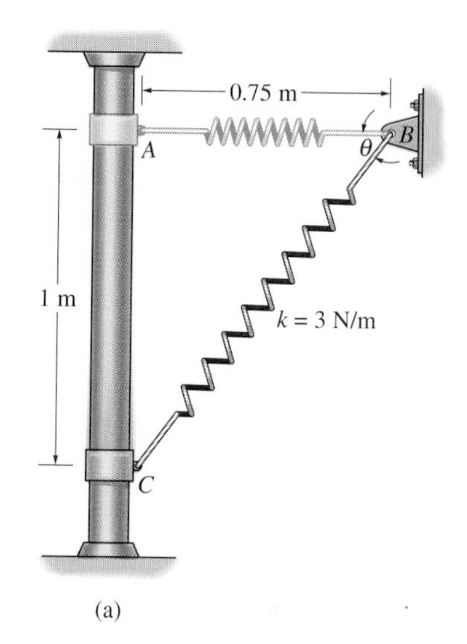

(a)

SOLUTION

Free-Body Diagram. The free-body diagram of the collar when it is located at the arbitrary position y is shown in Fig. 13–9b. Note that the weight is $W = 2(9.81) = 19.62$ N. Furthermore, the collar is *assumed* to be accelerating so that "**a**" acts downward in the *positive* y direction. There are four unknowns, namely, N_C, F_s, a, and θ.

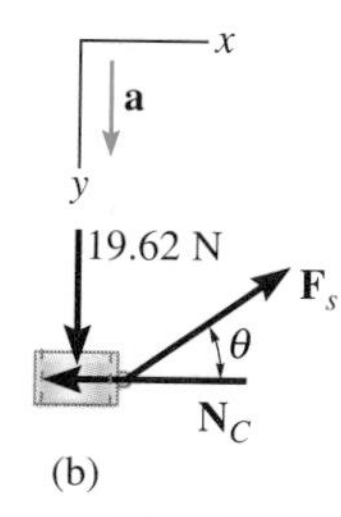

(b)

Fig. 13–9

Equations of Motion. Using the data in Fig. 13–9b,

$$\overset{+}{\rightarrow}\Sigma F_x = ma_x; \qquad -N_C + F_s \cos\theta = 0 \tag{1}$$

$$+\downarrow \Sigma F_y = ma_y; \qquad 19.62 - F_s \sin\theta = 2a \tag{2}$$

From Eq. 2 it is seen that the acceleration is not constant; rather, it depends on the magnitude and direction of the spring force. Solution is possible once F_s and θ are known.

The magnitude of the spring force is a function of the stretch s of the spring; i.e., $F_s = ks$. Here the unstretched length is $AB = 0.75$ m, Fig. 13–9a; therefore, $s = CB - AB = \sqrt{y^2 + (0.75)^2} - 0.75$. Since $k =$ 3 N/m, then

$$F_s = ks = 3\left(\sqrt{y^2 + (0.75)^2} - 0.75\right) \tag{3}$$

From Fig. 13–9a, the angle θ is related to y by trigonometry.

$$\sin\theta = \frac{y}{\sqrt{y^2 + (0.75)^2}} \tag{4}$$

Substituting $y = 1$ m into Eqs. 3 and 4 yields $F_s = 1.50$ N and $\theta = 53.1°$. Substituting these results into Eqs. 1 and 2, we obtain

$$N_C = 0.900 \text{ N} \qquad \textit{Ans.}$$

$$a = 9.21 \text{ m/s}^2 \downarrow \qquad \textit{Ans.}$$

Example 13–5

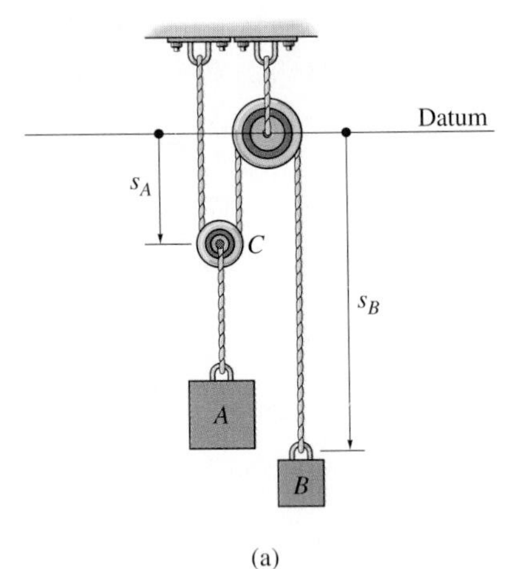

(a)

(b)

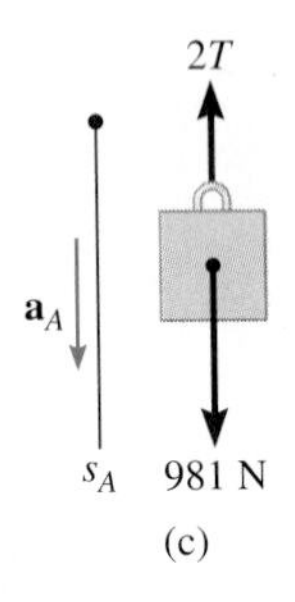

(c)

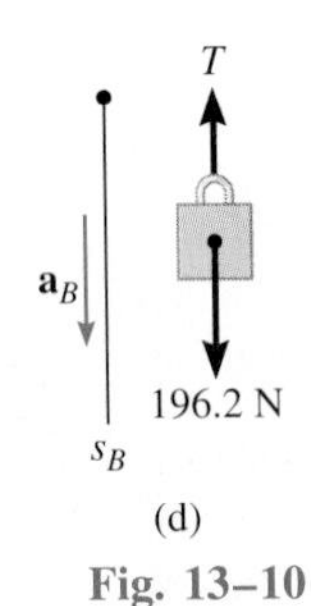

(d)

Fig. 13–10

The 100-kg block A shown in Fig. 13–10a is released from rest. If the mass of the pulleys and the cord is neglected, determine the speed of the 20-kg block B in 2 s.

SOLUTION

Motion of blocks A and B will be analyzed separately since they move along different paths.

Free-Body Diagrams. Since the mass of the pulleys is *neglected,* the effect of their inertia is zero and so we can apply the equilibrium condition for pulley C as shown in Fig. 13–10b. The free-body diagrams for blocks A and B are shown in Fig. 13–10c and d, respectively. Here we will *assume* both blocks accelerate downward, in the direction of $+s_A$ and $+s_B$. What are the three unknowns?

Equations of Motion

Block A (Fig. 13–10c):

$$+\downarrow \Sigma F_y = ma_y; \qquad 981 - 2T = 100a_A \qquad (1)$$

Block B (Fig. 13–10d):

$$+\downarrow \Sigma F_y = ma_y; \qquad 196.2 - T = 20a_B \qquad (2)$$

Kinematics. The necessary third equation is obtained by studying the kinematics of the pulley arrangement in order to relate a_A to a_B. Using the technique developed in Sec. 12.8, the coordinates s_A and s_B measure the positions of A and B from the fixed datum, Fig. 13–10a. It is seen that

$$2s_A + s_B = l$$

where l is constant and represents the total vertical length of cord. Differentiating this expression twice with respect to time yields

$$2a_A = -a_B \qquad (3)$$

Notice, however, that in writing Eqs. 1 to 3, the *positive direction was always assumed downward.* It is very important to be *consistent* in this assumption since we are seeking a simultaneous solution of equations. The solution yields

$$T = 327.0 \text{ N}$$
$$a_A = 3.27 \text{ m/s}^2$$
$$a_B = -6.54 \text{ m/s}^2$$

Hence when block A accelerates *downward,* block B accelerates *upward.* Since a_B is constant, the velocity of block B in 2 s is thus

$$(+\downarrow) \qquad v = v_0 + a_B t$$
$$= 0 + (-6.54)(2)$$
$$= -13.1 \text{ m/s} \qquad \textit{Ans.}$$

The negative sign indicates that block B is moving upward. Why?

PROBLEMS

13–1. The moon has a mass of $73.5(10^{21})$ kg, and the earth has a mass of $5.98(10^{24})$ kg. If their centers are $384(10^6)$ m apart, determine the gravitational attractive force between the two bodies.

13–2. Mars and the earth have diameters of 6775 km and 12 755 km, respectively. The mass of Mars is 0.107 times that of the earth. If a body weighs 200 N on the surface of the earth, what would its weight be on Mars? Also, what is the mass of the body and the acceleration of gravity on Mars?

13–3. Determine (only) the ''gravitational attraction'' between an 80-kg man and a 50-kg woman. The distance between their centers of mass is 0.5 m.

***13–4.** The baggage truck A has a mass of 800 kg and is used to pull each of the 300-kg cars. Determine the tension in the couplings at B and C if the tractive force **F** on the truck is $F = 480$ N. What is the speed of the truck when $t = 2$ s, starting from rest? The car wheels are free to roll. Neglect the mass of the wheels.

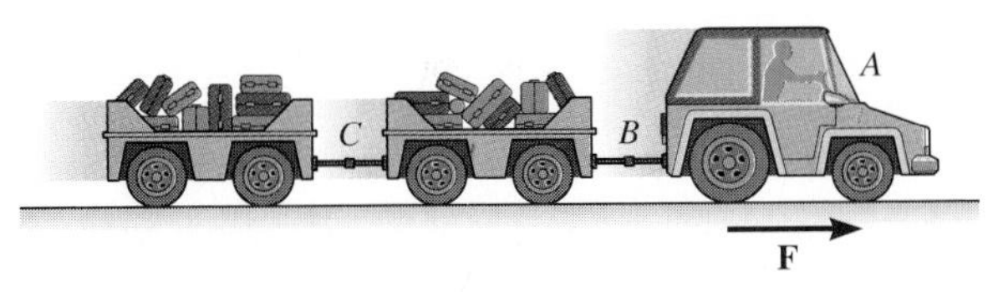

Prob. 13–4

13–5. The baggage truck A has a mass of 800 kg and is used to pull each of the 300-kg cars. If the tractive force **F** on the truck is $F = 480$ N, determine the initial acceleration of the truck. What is the acceleration of the truck if the coupling at C suddenly fails? The car wheels are free to roll. Neglect the mass of the wheels.

Prob. 13–5

13–6. By using an inclined plane to retard the motion of a falling object, and thus make the observations more accurate, Galileo was able to determine experimentally that the distance through which an object moves in free fall is proportional to the square of the time for travel. Show that this is the case, i.e., $s \propto t^2$, by determining the time t_B, t_C, and t_D needed for a block of mass m to slide from rest at A to points B, C, and D, respectively. Neglect the effects of friction.

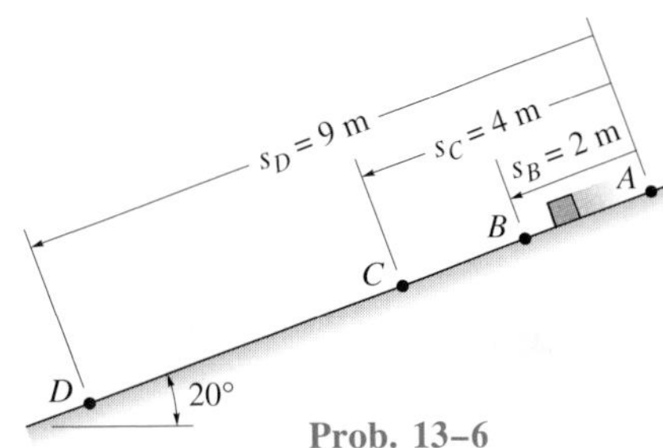

Prob. 13–6

13–7. The 500-kg fuel assembly for a nuclear reactor is being lifted out from the core of the nuclear reactor using the pulley system shown. It is hoisted upward with a constant acceleration such that $s = 0$ when $t = 0$ and $s = 2.5$ m when $t = 1.5$ s. Determine the tension in the cable at A during the motion.

***13–8.** The 500-kg fuel assembly for a nuclear reactor is being lifted out from the core of the nuclear reactor using the pulley system shown. If the allowable load in the cable cannot exceed 8 kN, determine the shortest possible time needed to lift the assembly to $s = 2.5$ m. Also, what speed does it have when $s = 2.5$ m? Originally the assembly is at rest when $s = 0$.

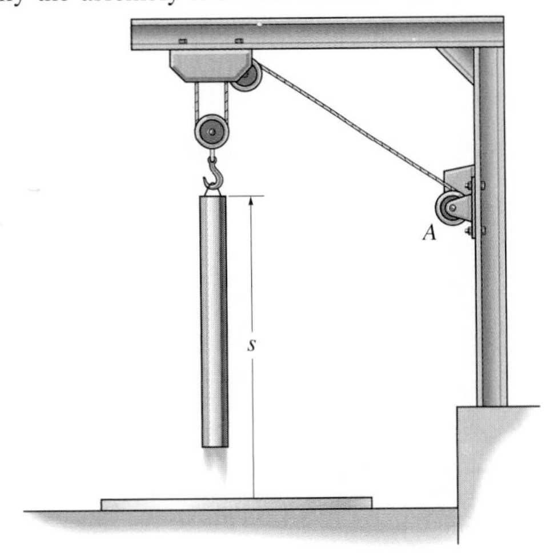

Probs. 13–7/13–8

13–9. The 4.5-Gg tanker is being towed with a constant acceleration of 0.001 m/s^2 by the tugboat using a cable that makes an angle of 15° with the stern of the tugboat as shown. Determine the force of the cable on the stern of the tanker. Neglect water resistance.

13–10. The 4.5-Gg tanker is being towed with a constant acceleration of 0.001 m/s^2 by the tugboat using a cable that makes an angle of 15° with the stern of the tugboat as shown. Determine the horizontal force of the water acting on the propeller of the tugboat, necessary to do this. The tugboat has a mass of 50 Mg. Also, what is the force of the cable on the stern of the tugboat? Neglect water resistance.

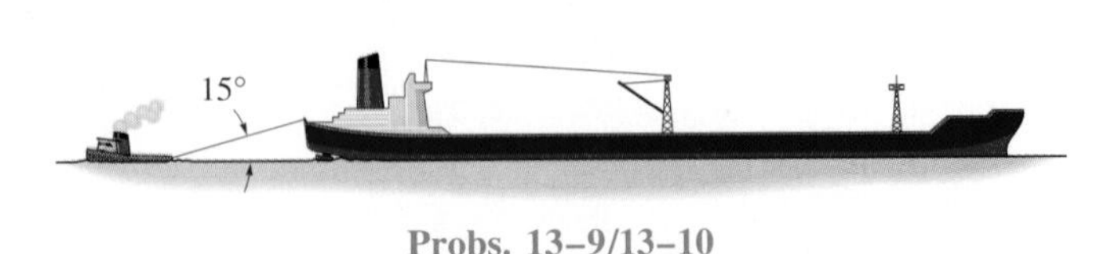

Probs. 13–9/13–10

13–11. The speed of the 3500-lb sports car is plotted over the 30-s time period. Plot the variation of the traction force F needed to cause the motion.

***13–12.** The speed of the 3500-lb sports car is plotted over the 30-s time period. Determine the traction force F acting on the car needed to cause the motion at $t = 5$ s and $t = 20$ s.

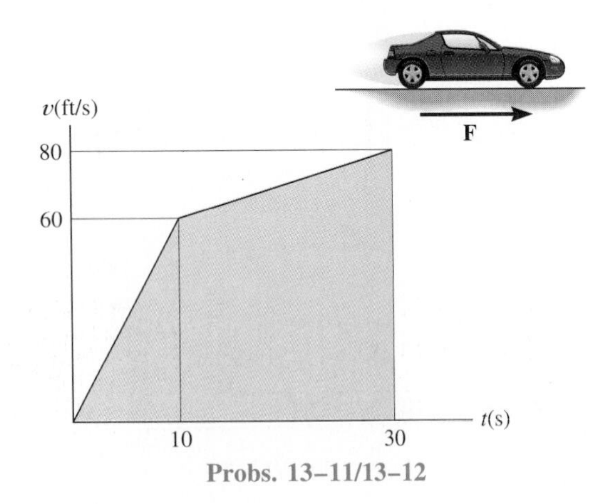

Probs. 13–11/13–12

13–13. The man weighs 180 lb and supports the barbells which have a weight of 100 lb. If he lifts them 2 ft in the air in 1.5 s starting from rest, determine the reaction of *both* of his feet on the ground during the lift. Assume the motion is with uniform acceleration.

Prob. 13–13

13–14. The boy having a weight of 80 lb hangs uniformly from the bar. Determine the force in each of his arms in $t = 2$ s if the bar is moving upward with (a) a constant velocity of 3 ft/s, and (b) a speed of $v = (4t^2)$ ft/s, where t is in seconds.

Prob. 13–14

13–15. The water-park ride consists of an 800-lb sled which slides from rest down the incline and then into the pool. If the frictional resistance on the incline is $F_r = 30$ lb, and in the pool for a short distance $F_r = 80$ lb, determine how fast the sled is traveling when $s = 5$ ft.

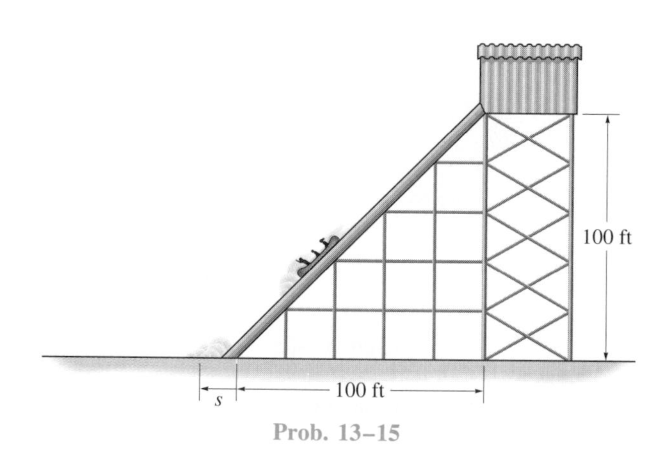

Prob. 13–15

***13–16.** Each of the two blocks has a mass m. The coefficient of kinetic friction at all surfaces of contact is μ. If a horizontal force **P** is applied to the bottom block, determine the acceleration of the bottom block in each case.

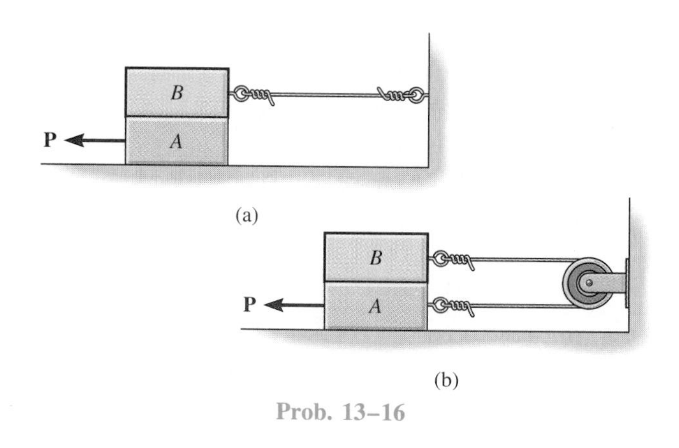

Prob. 13–16

13–17. A smooth 2-lb collar C fits loosely on the horizontal shaft. If the spring is unextended when $s = 0$, determine the velocity of the collar when $s = 1$ ft if the collar is given an initial horizontal velocity of 15 ft/s when $s = 0$.

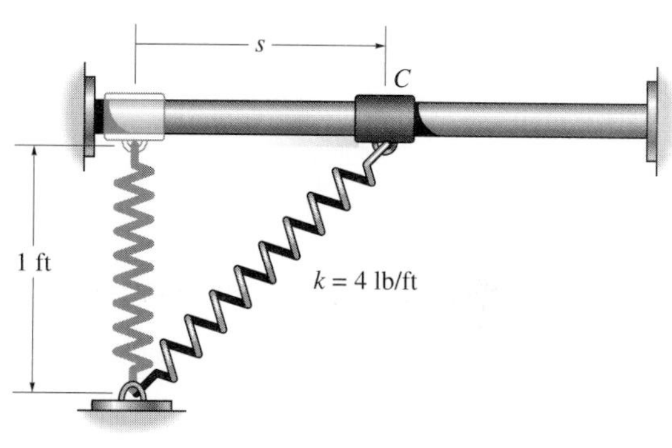

Prob. 13–17

13–18. The driver attempts to tow the crate using a rope that has a tensile strength of 200 lb. If the crate is originally at rest and has a weight of 500 lb, determine the greatest acceleration it can have if the coefficient of static friction between the crate and the road is $\mu_s = 0.4$, and the coefficient of kinetic friction is $\mu_k = 0.3$.

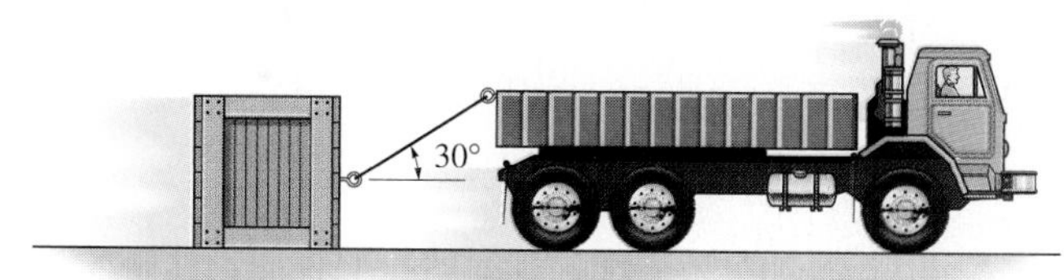

Prob. 13–18

13–19. A 40-lb suitcase slides from rest 20 ft down the smooth ramp. Determine the point where it strikes the ground at C. How long does it take to go from A to C?

***13–20.** Solve Prob. 13–19 if the suitcase has an initial velocity down the ramp of $v_A = 10$ ft/s and the coefficient of kinetic friction along AB is $\mu_k = 0.2$.

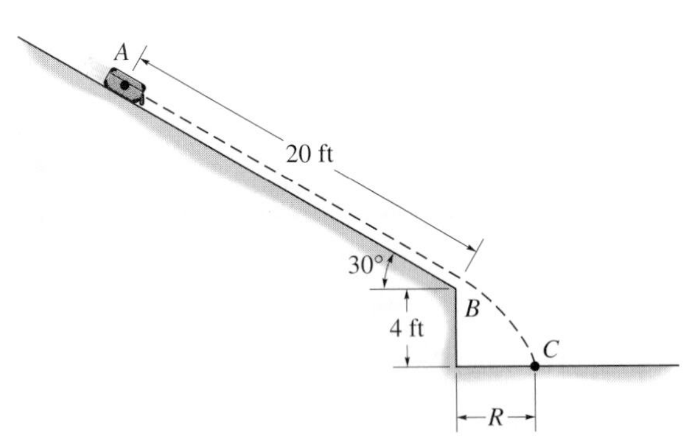

Probs. 13–19/13–20

13–21. Determine the acceleration of the 5-kg cylinder A. Neglect the mass of the pulleys and cords. The block at B has a mass of 10 kg. Assume the surface at B is smooth.

13–22. Determine the acceleration of the 5-kg cylinder A. Neglect the mass of the pulleys and cords. The block at B has a mass of 10 kg. The coefficient of kinetic friction between block B and the surface is $\mu_k = 0.1$.

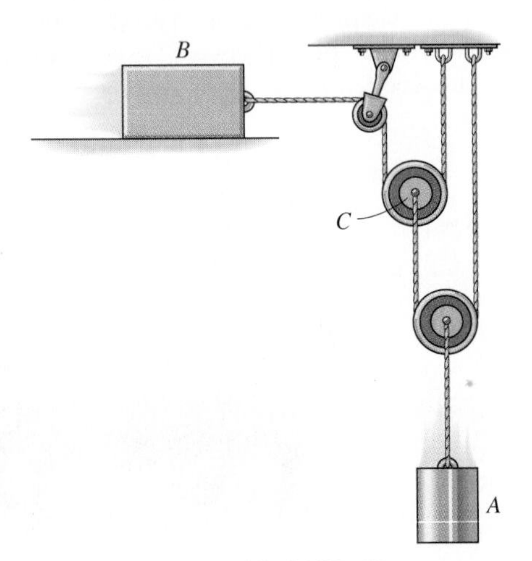

Probs. 13–21/13–22

13–23. The winding drum D is drawing in the cable at an accelerated rate of 5 m/s^2. Determine the cable tension if the suspended crate has a mass of 800 kg.

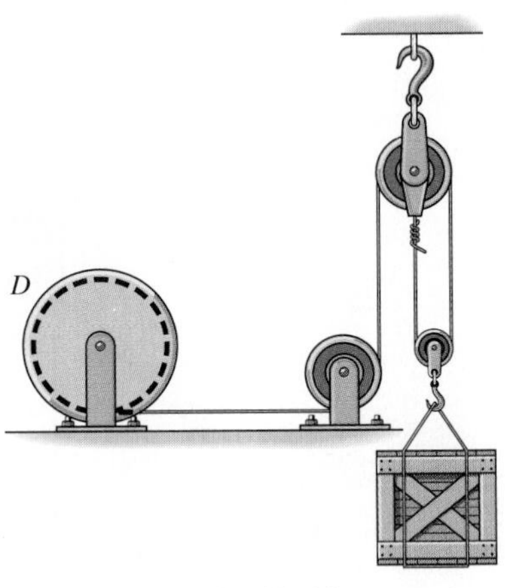

Prob. 13–23

***13–24.** A force $F = 15$ lb is applied to the cord. Determine how high the 30-lb block A rises in 2 s starting from rest. Neglect the weight of the pulleys and cord.

13–25. Determine the constant force F which must be applied to the cord in order to cause the 30-lb block A to have a speed of 12 ft/s when it has been displaced 3 ft upward starting from rest. Neglect the weight of the pulleys and cord.

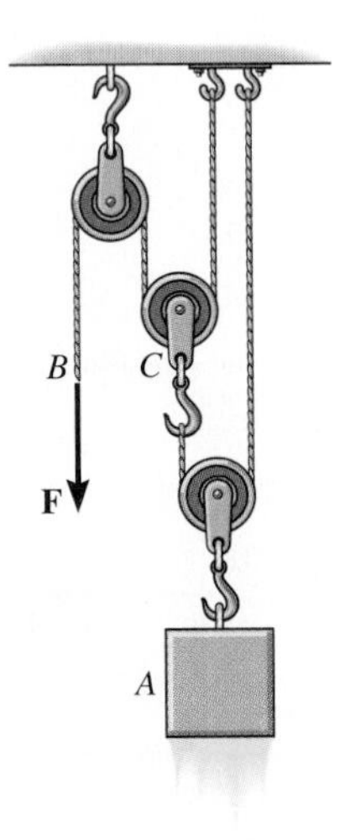

Probs. 13–24/13–25

13–26. At a given instant the 10-lb block A is moving downward with a speed of 6 ft/s. Determine its speed 2 s later. Block B has a weight of 4 lb, and the coefficient of kinetic friction between it and the horizontal plane is $\mu_k = 0.2$. Neglect the mass of the pulleys and cord.

13–27. The 5-lb block B rests on the smooth surface. Determine its acceleration when the 3-lb block A is released from rest. What would be the acceleration of B if the block at A was replaced by a 3-lb vertical force acting on the attached cord?

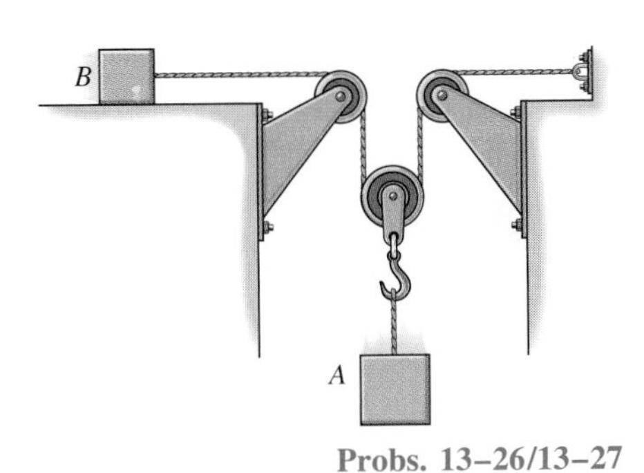

Probs. 13–26/13–27

13–30. The 2-kg collar C is free to slide along the smooth shaft AB. Determine the acceleration of collar C if (*a*) the shaft is fixed from moving, (*b*) collar A, which is fixed to shaft AB, moves downward at constant velocity along the vertical rod, and (*c*) collar A is subjected to a downward acceleration of 2 m/s^2. In all cases, the collar moves in the plane.

13–31. The 2-kg collar C is free to slide along the smooth shaft AB. Determine the acceleration of collar C if collar A is subjected to an upward acceleration of 4 m/s^2.

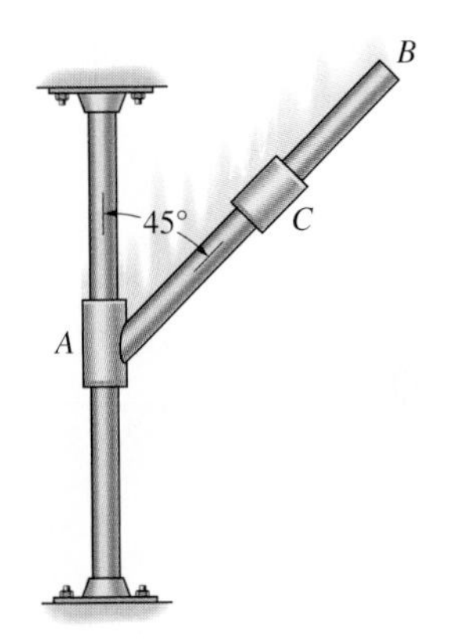

Probs. 13–30/13–31

***13–28.** The 400-kg mine car is hoisted up the incline using the cable and motor M. For a short time, the force in the cable is $F = (3200t^2)$ N, where t is in seconds. If the car has an initial velocity $v_1 = 2$ m/s when $t = 0$, determine its velocity when $t =$ 2 s.

13–29. The 400-kg mine car is hoisted up the incline using the cable and motor M. For a short time, the force in the cable is $F = (3200t^2)$ N, where t is in seconds. If the car has an initial velocity $v_1 = 2$ m/s at $s = 0$ and $t = 0$, determine the distance it moves up the plane when $t = 2$ s.

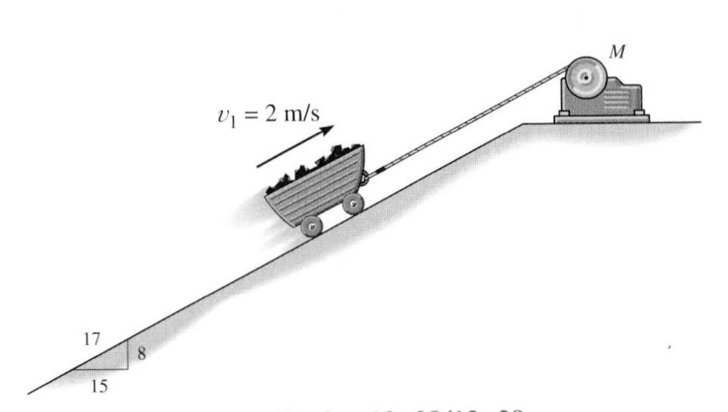

Probs. 13–28/13–29

■***13–32.** The 10-kg crate rests on the cart for which the coefficient of static friction is $\mu_s = 0.3$ between the crate and cart. Determine the largest angle θ of the plane so that a crate does not slide on the cart when the cart is given an acceleration of $a = 6$ m/s^2 down the plane.

13–33. Determine the normal force the 10-kg crate A exerts on the smooth cart if the cart is given an acceleration of $a = 2$ m/s^2 down the plane. Also, what is the acceleration of the crate? Set $\theta = 30°$.

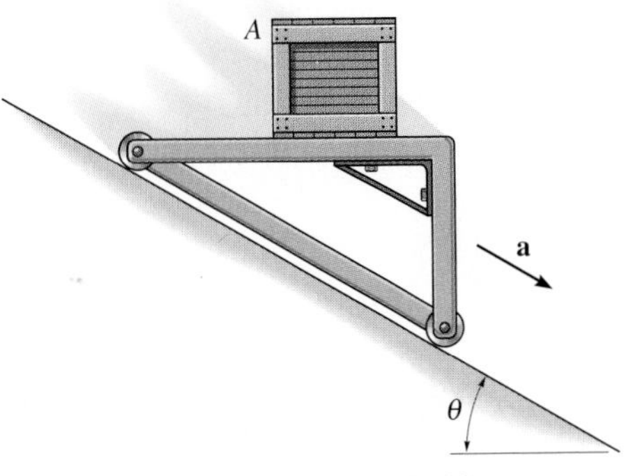

Probs. 13–32/13–33

13–34. The conveyor belt delivers each 12-kg crate to the ramp at A such that the crate's speed is $v_A = 2.5$ m/s, directed down *along* the ramp. If the coefficient of kinetic friction between each crate and the ramp is $\mu_k = 0.3$, determine the speed at which each crate slides off the ramp at B. Assume that no tipping occurs. Take $\theta = 30°$.

■**13–35.** The conveyor belt delivers each 12-kg crate to the ramp at A such that the crate's speed is $v_A = 2.5$ m/s, directed down *along* the ramp. If the coefficient of kinetic friction between each crate and the ramp is $\mu_k = 0.3$, determine the smallest incline θ of the ramp so that the crates will slide off and fall into the cart.

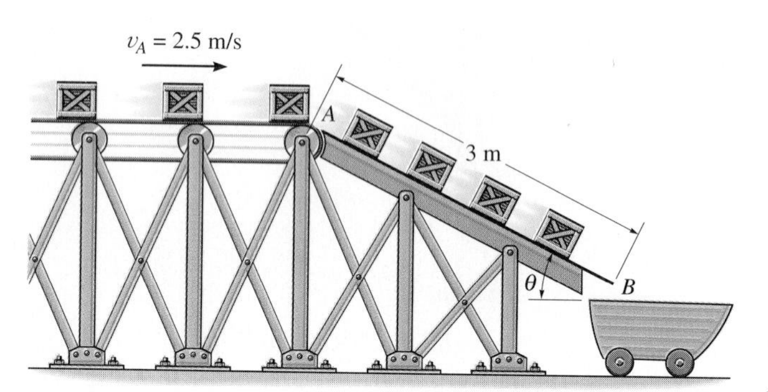

Probs. 13–34/13–35

*__13–36.__ The conveyor belt is moving at 4 m/s. If the coefficient of static friction between the conveyor and the 10-kg package B is $\mu_s = 0.2$, determine the shortest time the belt can stop so that the package does not slide on the belt.

13–37. The conveyor belt is designed to transport packages of various weights. Each 10-kg package has a coefficient of kinetic friction $\mu_k = 0.15$. If the speed of the conveyor is 5 m/s, and then it suddenly stops, determine the distance the package will slide on the belt before coming to rest.

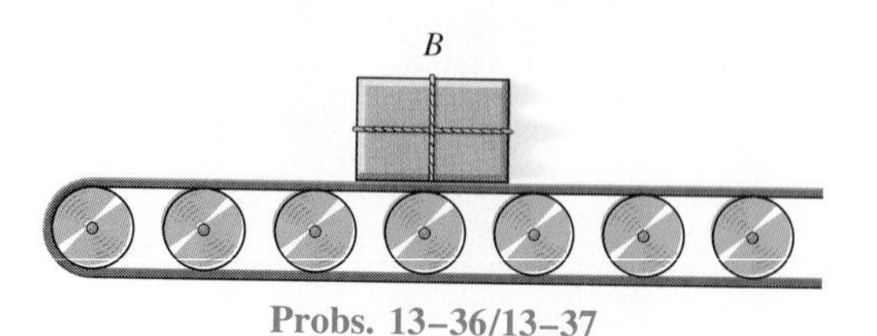

Probs. 13–36/13–37

13–38. Determine the speed of the collar in Example 13–4 at the instant $y = 1$ m. *Hint:* Combine Eqs. (2), (3), and (4) to determine $a = f(y)$. Then substitute into the equation $v\,dv = a\,dy$ and integrate.

13–39. Blocks A and B each have a mass m. Determine the largest horizontal force P which can be applied to B so that A will not move relative to B. All surfaces are smooth.

*__13–40.__ Blocks A and B each have a mass m. Determine the largest horizontal force P which can be applied to B so that A will not slip up B. The coefficient of static friction between A and B is μ. Neglect friction between B and C.

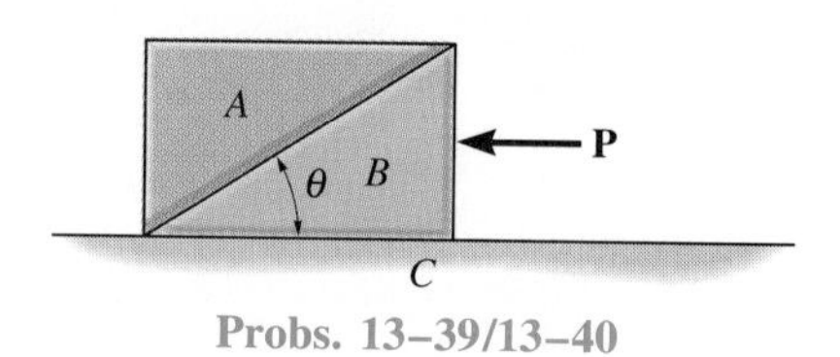

Probs. 13–39/13–40

13–41. A car of mass m is traveling at a slow velocity v_0. If it is subjected to the drag resistance of the wind, which is proportional to its velocity, i.e., $F_D = kv$, determine the distance and the time the car will travel before its velocity becomes $0.5\,v_0$. Assume no other frictional forces act on the car.

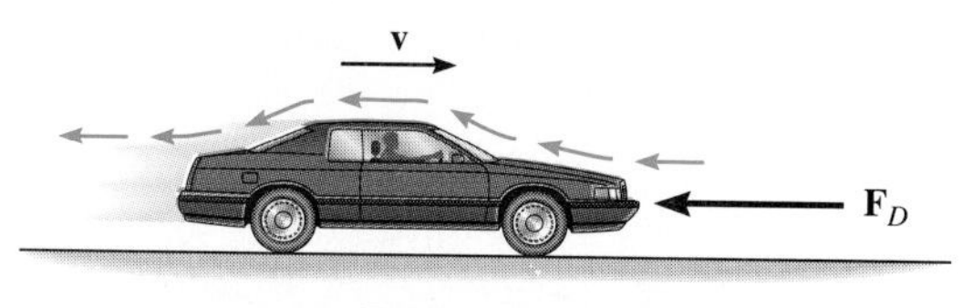

Prob. 13–41

13–42. A parachutist having a mass m is falling at v_0 when he opens his parachute at a very high altitude. If the atmospheric drag resistance is $F_D = kv^2$, where k is a constant, determine his velocity when he has fallen a distance h. What is his velocity when he lands on the ground? This velocity is referred to as the *terminal velocity,* which is found by letting the distance of fall $y \to \infty$.

13–43. A parachutist having a mass m opens his parachute from an at-rest position at a very high altitude. If the atmospheric drag resistance is $F_D = kv^2$, where k is a constant, determine his velocity when he has fallen for a time t. What is his velocity when he lands on the ground? This velocity is referred to as the *terminal velocity,* which is found by letting the time of fall $t \rightarrow \infty$.

Probs. 13–42/13–43

*13–44. A particle of mass m is fired at an angle θ_0 with a velocity $\mathbf{v}_0$ in a liquid that develops a drag resistance $F = -kv$, where k is a constant. Determine the maximum or terminal speed reached by the particle.

13–45. A projectile of mass m is fired into a liquid at an angle θ_0 with an initial velocity $\mathbf{v}_0$ as shown. If the liquid develops a frictional or drag resistance on the projectile which is proportional to its velocity, i.e., $F = kv$, where k is a constant, determine the x and y components of its position at any instant. Also, what is the maximum distance x_{max} that it travels?

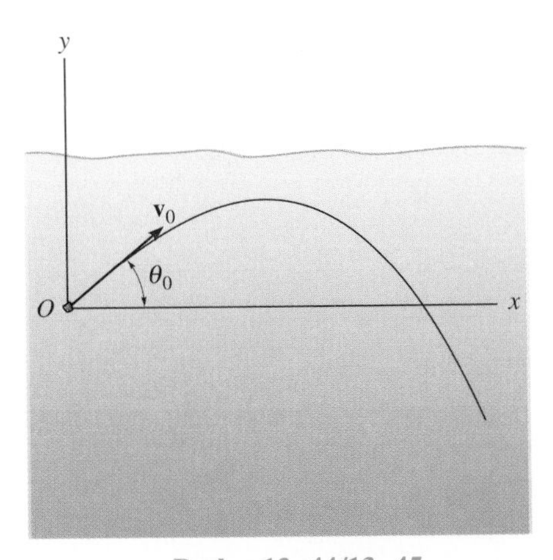

Probs. 13–44/13–45

13–46. Block A has a weight of 8 lb and block B has a weight of 6 lb. They rest on a surface for which the coefficient of kinetic friction is $\mu_k = 0.2$. If the spring has a stiffness of $k = 20$ lb/ft, and it is compressed 0.2 ft, determine the acceleration of each block just after they are released.

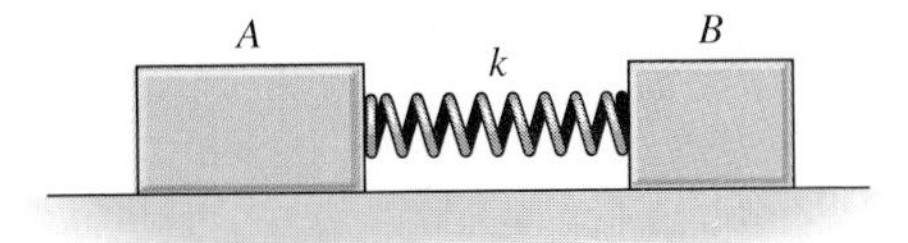

Prob. 13–46

13–47. Each of the three plates has a mass of 10 kg. If the coefficients of static and kinetic friction at each surface of contact are $\mu_s = 0.3$ and $\mu_k = 0.2$, respectively, determine the acceleration of each plate when the three horizontal forces are applied.

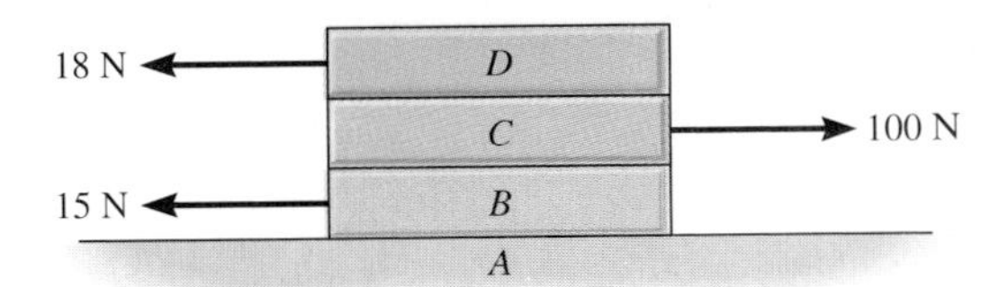

Prob. 13–47

*13–48. Determine the time needed to pull the cord at B down 4 ft starting from rest when a force of 10 lb is applied to the cord. Block A weighs 20 lb. Neglect the mass of the pulleys and cords.

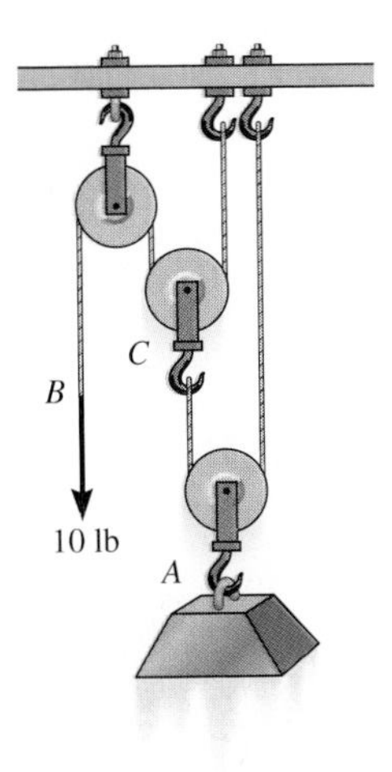

Prob. 13–48

13–49. Block B rests on a smooth surface. If the coefficients of static and kinetic friction between A and B are $\mu_s = 0.4$ and $\mu_k = 0.3$, respectively, determine the acceleration of each block if someone pushes horizontally on block A with a force of (*a*) $F = 6$ lb, and (*b*) $F = 50$ lb.

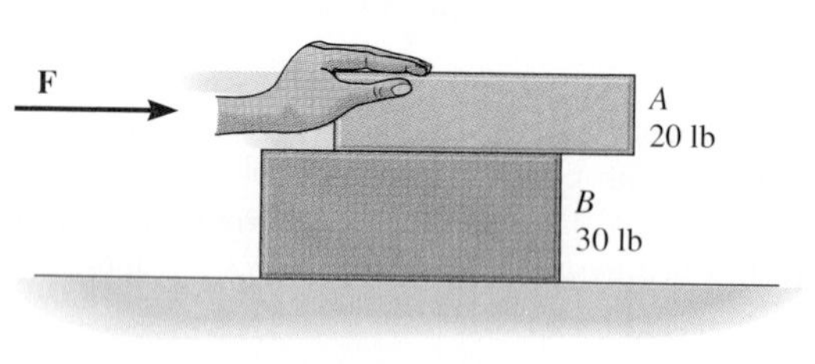

Prob. 13–49

13–50. Crate B has a mass m and is released from rest when it is on top of cart A, which has a mass $3m$. Determine the tension in cord CD needed to hold the cart from moving while B is sliding down A. Neglect friction.

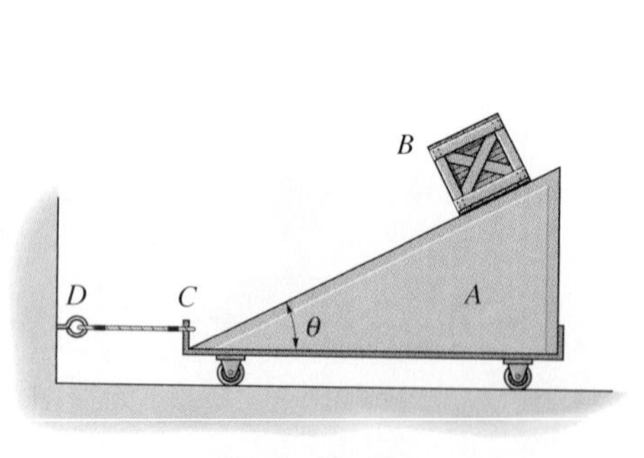

Prob. 13–50

13–51. The smooth block B of negligible size has a mass m and rests on the horizontal plane. If the board AC pushes on the block at a constant angle θ with an acceleration $\mathbf{a}_0$, determine the velocity of the block along the board and the distance s the block moves along the board as a function of time t. The board and block start from rest when $s = 0$, $t = 0$.

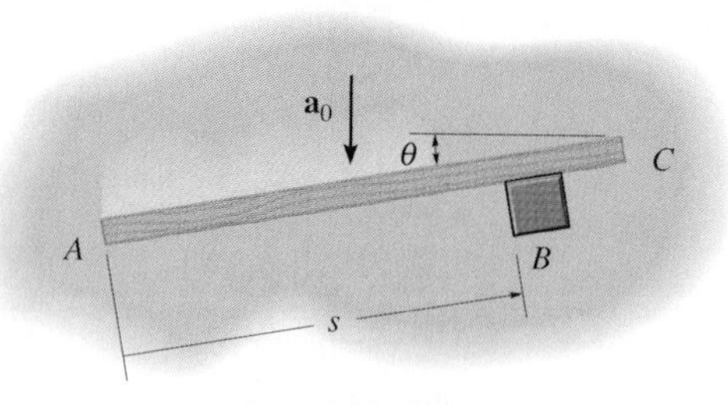

Prob. 13–51

***13–52.** Cylinder B has a mass m and is hoisted using the cord and pulley system shown. Determine the magnitude of force $\mathbf{F}$ as a function of the block's vertical position y so that when $\mathbf{F}$ is applied the block rises with a constant acceleration $\mathbf{a}_B$. Neglect the mass of the cord and pulleys.

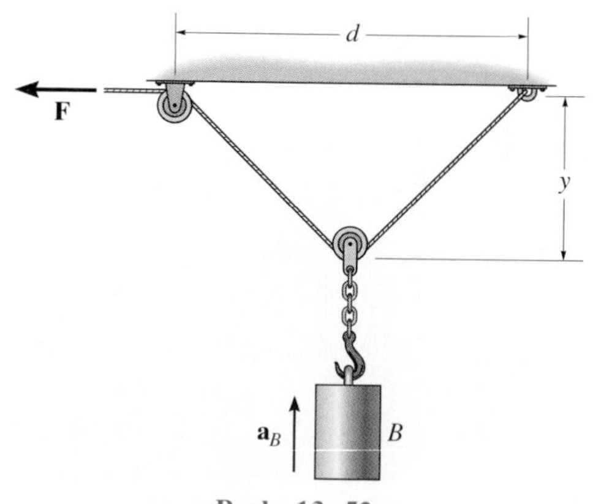

Prob. 13–52

13–53. The elevator E has a mass of 500 kg and the counterweight at A has a mass of 150 kg. If the motor supplies a constant force of 5 kN on the cable at B, determine the speed of the elevator in $t = 3$ s starting from rest. Neglect the mass of the pulleys and cable.

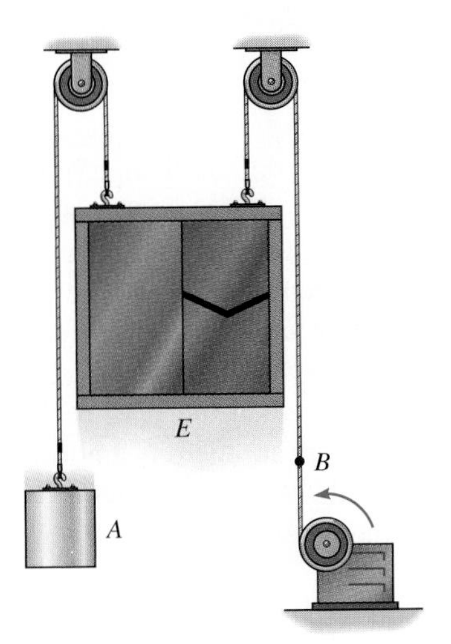

Prob. 13–53

13–54. The block A has a mass m_A and rests on the pan B, which has a mass m_B. Both are supported by a spring having a stiffness k that is attached to the bottom of the pan and to the ground. Determine the distance d the pan should be pushed down from the equilibrium position and then released from rest so that separation of the block will take place from the surface of the pan at the instant the spring becomes unstretched.

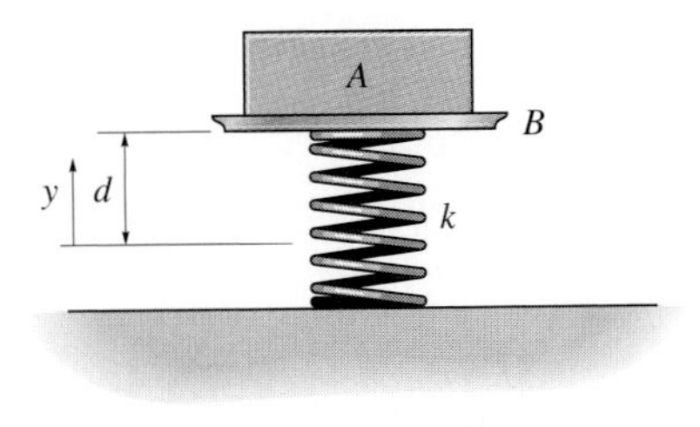

Prob. 13–54

13.5 Equations of Motion: Normal and Tangential Coordinates

When a particle moves over a curved path which is known, the equation of motion for the particle may be written in the normal and tangential directions. We have

$$\Sigma \mathbf{F} = m\mathbf{a}$$

$$\Sigma F_t \mathbf{u}_t + \Sigma F_n \mathbf{u}_n + \Sigma F_b \mathbf{u}_b = m\mathbf{a}_t + m\mathbf{a}_n$$

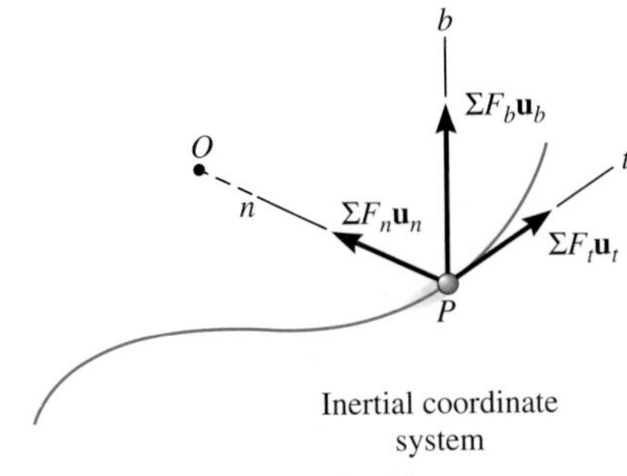

Inertial coordinate system

Fig. 13–11

Here ΣF_n, ΣF_t, and ΣF_b represent the sums of all the force components acting on the particle in the normal, tangential, and binormal directions, respectively, Fig. 13–11. Note that there is no motion of the particle in the binormal direc-

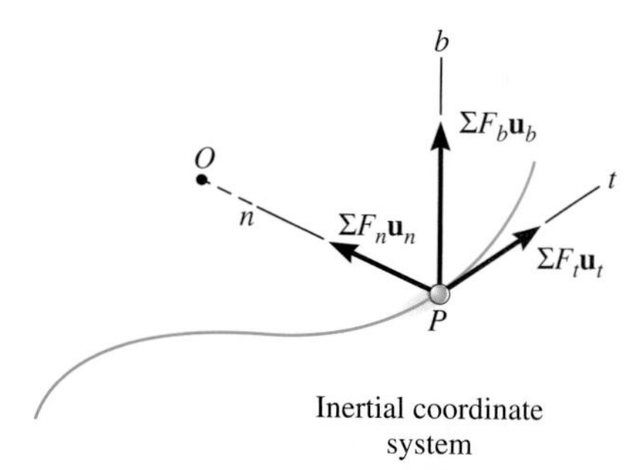

Inertial coordinate system

Fig. 13–11 (repeated)

tion, since the particle is constrained to move along the path. The above equation is satisfied provided

$$\begin{aligned} \Sigma F_t &= ma_t \\ \Sigma F_n &= ma_n \\ \Sigma F_b &= 0 \end{aligned} \qquad (13\text{–}8)$$

Recall that $a_t(= dv/dt)$ represents the time rate of change in the magnitude of velocity. Consequently, if $\Sigma\mathbf{F}_t$ acts in the direction of motion, the particle's speed will increase, whereas if it acts in the opposite direction, the particle will slow down. Likewise, a_n $(= v^2/\rho)$ represents the time rate of change in the velocity's direction. Since this vector *always* acts in the positive n direction, i.e., toward the path's center of curvature, then $\Sigma\mathbf{F}_n$, which causes $\mathbf{a}_n$, also acts in this direction. In particular, when the particle is constrained to travel in a circular path with a constant speed, there is a normal force exerted on the particle by the constraint. Since this force is always directed toward the center of the path, it is often referred to as the *centripetal force.*

PROCEDURE FOR ANALYSIS

When a problem involves the motion of a particle along a *known curved path,* normal and tangential coordinates should be considered for the analysis since the acceleration components can be readily formulated. The method for applying the equations of motion, which relate the forces to the acceleration, has been outlined in the procedure given in Sec. 13.4. Specifically, for n, t, b coordinates it may be stated as follows:

Free-Body Diagram. Establish the inertial n, t, b coordinate system at the particle and draw the particle's free-body diagram. The particle's normal acceleration $\mathbf{a}_n$ *always* acts in the positive n direction. If the tangential acceleration $\mathbf{a}_t$ is unknown, assume it acts in the positive t direction. From the diagram identify the unknowns in the problem.

Equations of Motion. Apply the equations of motion, Eqs. 13–8.

Kinematics. Formulate the tangential and normal components of acceleration; i.e., $a_t = dv/dt$ or $a_t = v\,dv/ds$ and $a_n = v^2/\rho$. If the path is defined as $y = f(x)$, the radius of curvature at the point where the particle is located can be obtained from $\rho = |[1 + (dy/dx)^2]^{3/2}/(d^2y/dx^2)|$.

The following examples illustrate application of this procedure numerically.

Example 13–6

Determine the banking angle θ of the circular track so that the wheels of the sports car shown in Fig. 13–12*a* will not have to depend on friction to prevent the car from sliding either up or down the curve. The car has a negligible size and travels at a constant speed of 100 ft/s. The radius of the track is 600 ft.

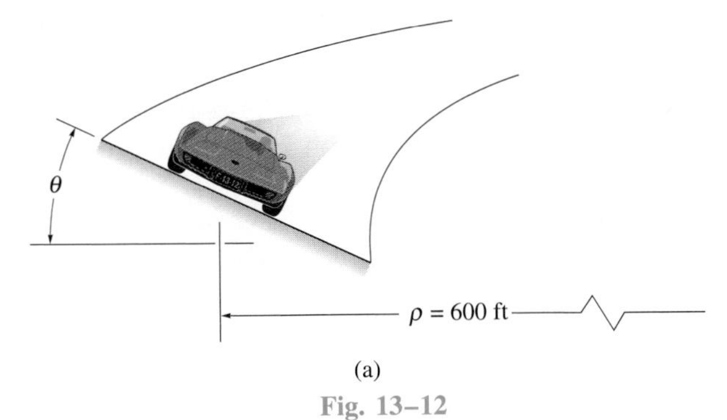

(a)

Fig. 13–12

SOLUTION

Before looking at the solution to this problem, give some thought as to why it should be solved using *n*, *t*, *b* coordinates.

Free-Body Diagram. As shown in Fig. 13–12*b*, the car is assumed to have a mass *m*. As stated in the problem, no frictional force acts on the car. Here $\mathbf{N}_C$ represents the *resultant* of all the wheels on the ground. Since a_n can be calculated, the unknowns are N_C and θ.

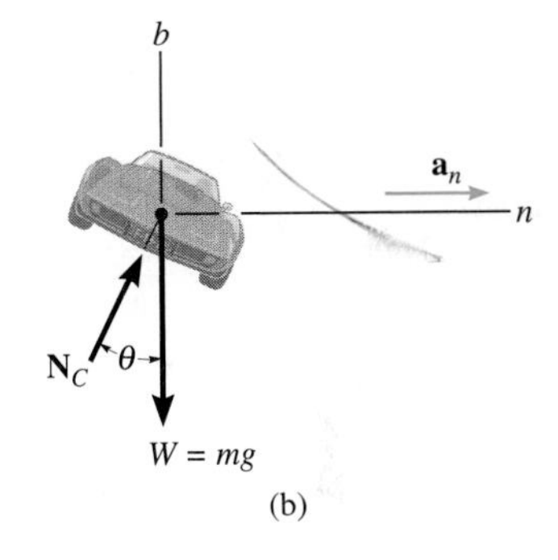

(b)

Equations of Motion. Using the *n*, *b* axes shown,

$$+\uparrow \Sigma F_b = 0; \qquad N_C \cos\theta - mg = 0 \qquad (1)$$

$$\overset{+}{\rightarrow}\Sigma F_n = ma_n; \qquad N_C \sin\theta = m\frac{v^2}{\rho} \qquad (2)$$

Eliminating N_C and m from these equations by dividing Eq. 2 by Eq. 1, we obtain

$$\tan\theta = \frac{v^2}{g\rho} = \frac{(100)^2}{32.2(600)}$$

$$\theta = \tan^{-1}(0.518)$$

$$= 27.4^\circ \qquad \textit{Ans.}$$

A force summation in the tangential direction of motion is of no consequence to the solution. If it were considered, note that $a_t = dv/dt = 0$, since the car moves with *constant speed*. A further analysis of this problem is discussed in Prob. 21–59.

Example 13–7

The 3-kg disk D is attached to the end of a cord as shown in Fig. 13–13a. The other end of the cord is attached to a ball-and-socket joint located at the center of a platform. If the platform is rotating rapidly, and the disk is placed on it and released from rest as shown, determine the time it takes for the disk to reach a speed great enough to break the cord. The maximum tension the cord can sustain is 100 N, and the coefficient of kinetic friction between the disk and the platform is $\mu_k = 0.1$.

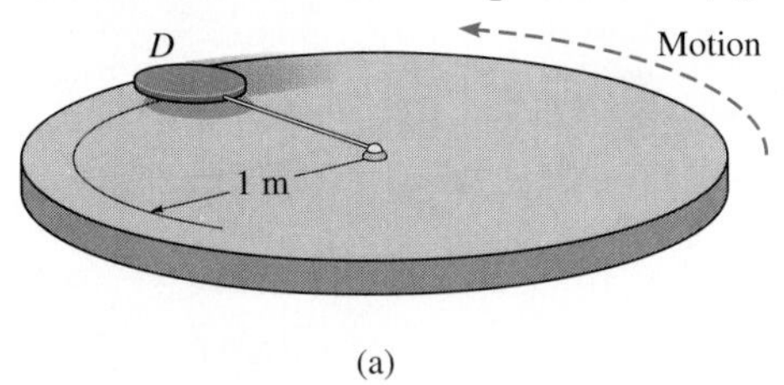

(a)

SOLUTION

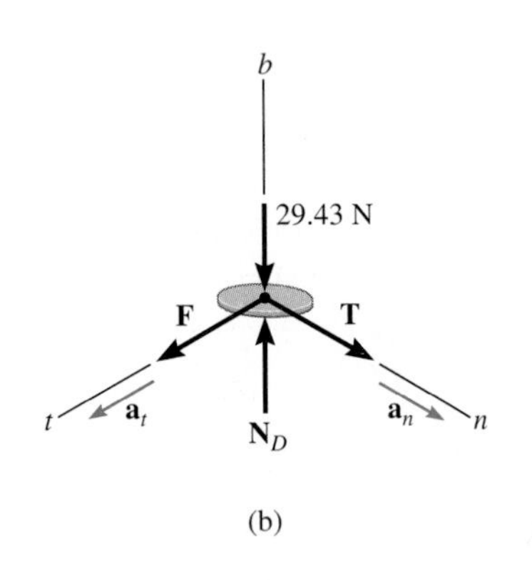

(b)

Fig. 13–13

Free-Body Diagram. As shown in Fig. 13–13b, the disk has *both* normal and tangential components of acceleration as a result of the unbalanced forces **T** and **F**. Since sliding occurs, the frictional force has a magnitude $F = \mu_k N_D = 0.1N_D$ and a sense of direction that opposes the *relative motion* of the disk with respect to the platform. The cord restricts motion of the disk in the n direction, and therefore **F** acts in the positive t direction. The weight of the disk is $W = 3(9.81) = 29.43$ N. Since a_n can be related to v, and at maximum speed $T = 100$ N, the unknowns are N_D, a_t, and v.

Equations of Motion

$\Sigma F_b = 0;$ $$N_D - 29.43 = 0 \quad (1)$$

$\Sigma F_t = ma_t;$ $$0.1N_D = 3a_t \quad (2)$$

$\Sigma F_n = ma_n;$ $$T = 3\left(\frac{v^2}{1}\right) \quad (3)$$

Setting $T = 100$ N, Eq. 3 can be solved for the critical speed v_{cr} of the disk needed to break the cord. Solving all the equations, we obtain

$$N_D = 29.43 \text{ N}$$
$$a_t = 0.981 \text{ m/s}^2$$
$$v_{cr} = 5.77 \text{ m/s}$$

Kinematics. Since a_t is *constant,* the time needed to break the cord is

$$v_{cr} = v_0 + a_t t$$
$$5.77 = 0 + (0.981)t$$
$$t = 5.89 \text{ s}$$ *Ans.*

Example 13–8

The skier in Fig. 13–14*a* descends the smooth slope, which may be approximated by a parabola. If she has a weight of 120 lb, determine the normal force she exerts on the ground at the instant she arrives at point *A*, where her velocity is 30 ft/s. Also compute her acceleration at *A*.

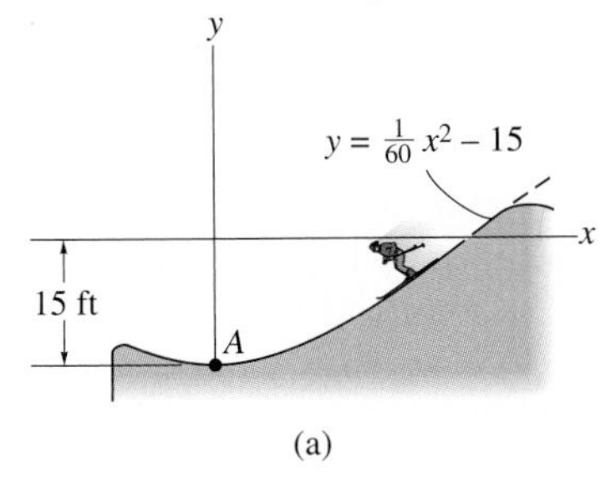

(a)

SOLUTION

Why consider using *n*, *t* coordinates to solve this problem?

Free-Body Diagram. The free-body diagram for the skier when she is at *A* is shown in Fig. 13–14*b*. Since the path is *curved,* there are two components of acceleration, $\mathbf{a}_n$ and $\mathbf{a}_t$. Since a_n can be calculated, the unknowns are a_t and N_A.

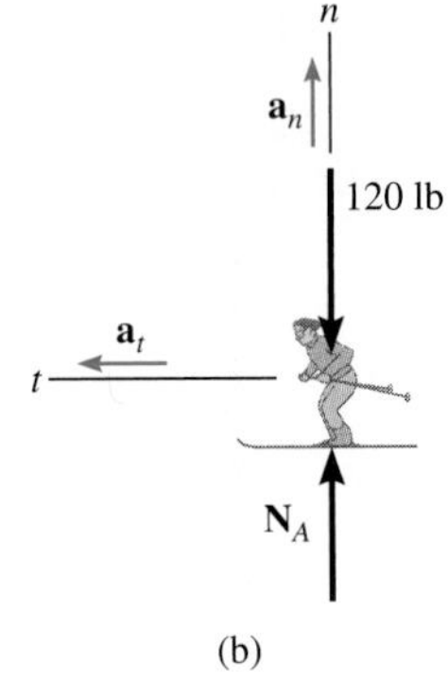

(b)

Fig. 13–14

Equations of Motion

$$+\uparrow \Sigma F_n = ma_n; \qquad N_A - 120 = \frac{120}{32.2}\left(\frac{(30)^2}{\rho}\right) \qquad (1)$$

$$\overset{+}{\leftarrow}\Sigma F_t = ma_t; \qquad 0 = \frac{120}{32.2}a_t \qquad (2)$$

The radius of curvature ρ for the path must be computed at point $A(0, -15 \text{ ft})$. Here $y = \frac{1}{60}x^2 - 15$, $dy/dx = \frac{1}{30}x$, $d^2y/dx^2 = \frac{1}{30}$, so that at $x = 0$,

$$\rho = \left|\frac{[1 + (dy/dx)^2]^{3/2}}{d^2y/dx^2}\right|_{x=0} = \left|\frac{[1 + (0)^2]^{3/2}}{\frac{1}{30}}\right| = 30 \text{ ft}$$

Substituting into Eq. 1 and solving for N_A, we have

$$N_A = 232 \text{ lb} \qquad \textit{Ans.}$$

Kinematics. From Eq. 2,

$$a_t = 0$$

Thus,

$$a_n = \frac{v^2}{\rho} = \frac{(30)^2}{30} = 30 \text{ ft/s}^2$$

$$a_A = a_n = 30 \text{ ft/s}^2 \uparrow \qquad \textit{Ans.}$$

Example 13–9

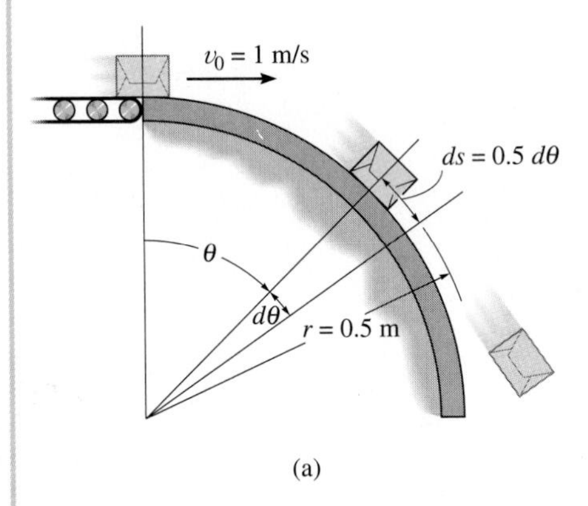

(a)

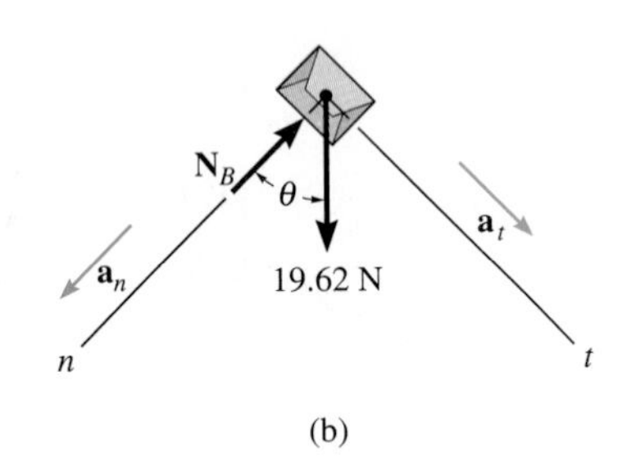

(b)

Fig. 13–15

Packages, each having a mass of 2 kg, are delivered from a conveyor to a smooth circular ramp with a velocity of $v_0 = 1$ m/s as shown in Fig. 13–15a. If the radius of the ramp is 0.5 m, determine the angle $\theta = \theta_{max}$ at which each package begins to leave the surface.

SOLUTION

Free-Body Diagram. The free-body diagram for a package, when it is located at the *general position* θ, is shown in Fig. 13–15b. The package must have a tangential acceleration $\mathbf{a}_t$, since its *speed* is always *increasing* as it slides downward. The weight is $W = 2(9.81) = 19.62$ N. Specify the three unknowns.

Equations of Motion

$$+\swarrow \Sigma F_n = ma_n; \qquad -N_B + 19.62\cos\theta = 2\frac{v^2}{0.5} \tag{1}$$

$$+\searrow \Sigma F_t = ma_t; \qquad 19.62\sin\theta = 2a_t \tag{2}$$

At the instant $\theta = \theta_{max}$, the package leaves the surface of the ramp so that $N_B = 0$. Therefore, there are three unknowns, v, a_t, and θ.

Kinematics. The third equation for the solution is obtained by noting that the magnitude of tangential acceleration a_t may be related to the speed of the package v and the angle θ. Since $a_t\,ds = v\,dv$ and $ds = r\,d\theta = 0.5\,d\theta$, Fig. 13–15$a$, we have

$$a_t = \frac{v\,dv}{0.5\,d\theta} \tag{3}$$

To solve, substitute Eq. 3 into Eq. 2 and separate the variables. This gives

$$v\,dv = 4.905\sin\theta\,d\theta$$

Integrate both sides, realizing that when $\theta = 0°$, $v_0 = 1$ m/s.

$$\int_1^v v\,dv = 4.905\int_{0°}^{\theta}\sin\theta\,d\theta$$

$$\left.\frac{v^2}{2}\right|_1^v = -4.905\cos\theta\Big|_{0°}^{\theta}$$

$$v^2 = 9.81(1 - \cos\theta) + 1$$

Substituting into Eq. 1 with $N_B = 0$ and solving for $\cos\theta_{max}$ yields

$$19.62\cos\theta_{max} = \frac{2}{0.5}[9.81(1 - \cos\theta_{max}) + 1]$$

$$\cos\theta_{max} = \frac{43.24}{58.86};$$

$$\theta_{max} = 42.7° \qquad \textit{Ans.}$$

PROBLEMS

13–55. Determine the maximum constant speed at which the pilot can travel around the vertical curve having a radius of curvature $\rho = 800$ m, so that he experiences a maximum acceleration $a_n = 8g = 78.5$ m/s^2. If he has a mass of 70 kg, determine the normal force he exerts on the seat of the airplane when the plane is traveling at this speed and is at its lowest point.

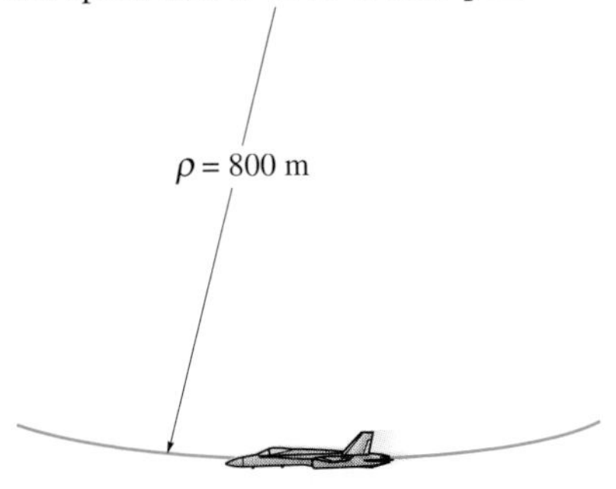

Prob. 13–55

***13–56.** At the instant $\theta = 60°$, the boy's center of mass G has a downward speed $v_G = 15$ ft/s. Determine the rate of increase in his speed and the tension in each of the two supporting cords of the swing at this instant. The boy has a weight of 60 lb. Neglect his size and the mass of the seat and cords.

13–57. At the instant $\theta = 60°$, the boy's center of mass G is momentarily at rest. Determine his speed and the tension in each of the two supporting cords of the swing when $\theta = 90°$. The boy has a weight of 60 lb. Neglect his size and the mass of the seat and cords.

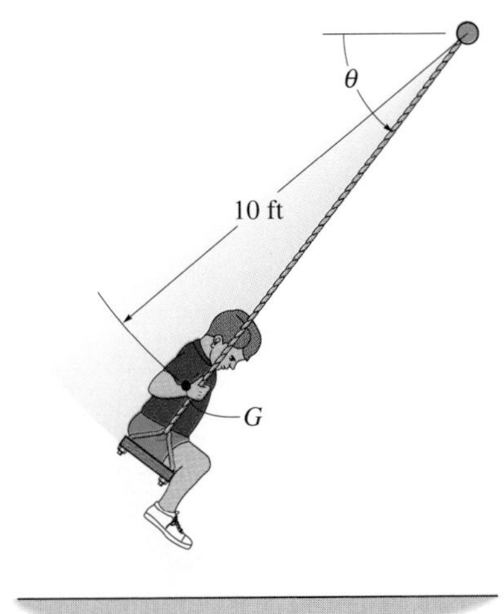

Probs. 13–56/13–57

13–58. The device shown is designed to produce the experience of weightlessness in a passenger when he reaches point A, $\theta = 90°$, along the path. If the passenger has a mass of 75 kg, determine the minimum speed he should have when he reaches A so that he does not exert a normal reaction on the seat. The chair is pin-connected to the frame BC so that he is always seated in an upright position. During the motion his speed remains constant.

13–59. The passenger has a mass of 75 kg and always sits in an upright position on the chair. At the instant $\theta = 30°$, he has a speed of 5 m/s and an increase of speed of 2 m/s^2. Determine the horizontal and vertical forces that the chair exerts on him in order to produce this motion.

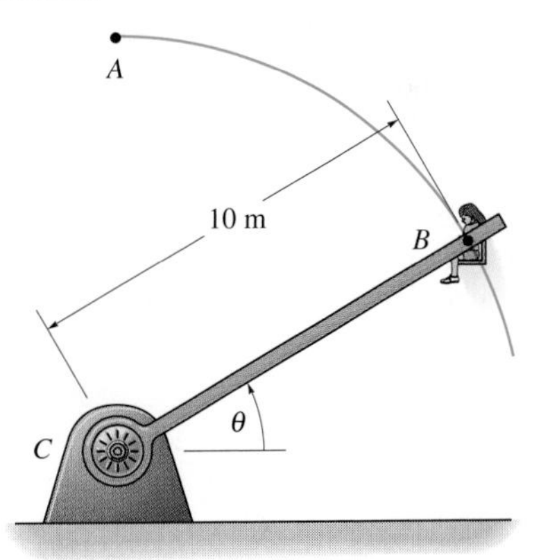

Probs. 13–58/13–59.

***13–60.** The 600-kg wrecking ball is suspended from the crane by a cable having a negligible mass. If the ball has a speed $v = 8$ m/s at the instant it is at its lowest point, $\theta = 0°$, determine the tension in the cable at this instant. Also, determine the angle θ to which the ball swings before it stops.

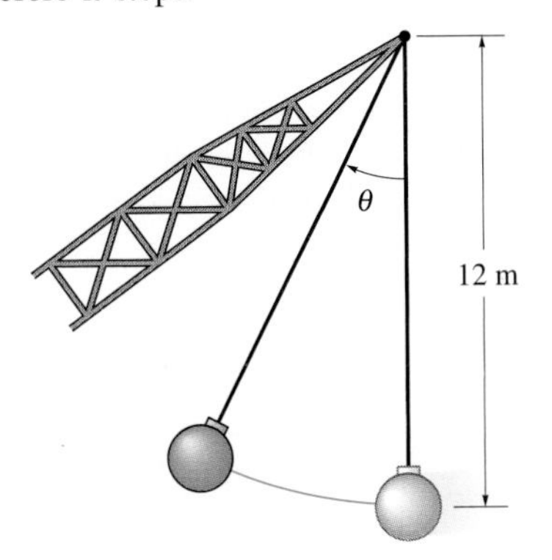

Prob. 13–60

13–61. The 600-kg wrecking ball is suspended from the crane by a cable having a negligible mass. Determine the speed of the ball when it is at its lowest point ($\theta = 0°$) if the cable is observed to swing to $\theta = 30°$ when the motion is momentarily stopped. Compute the tension in the cable at each position.

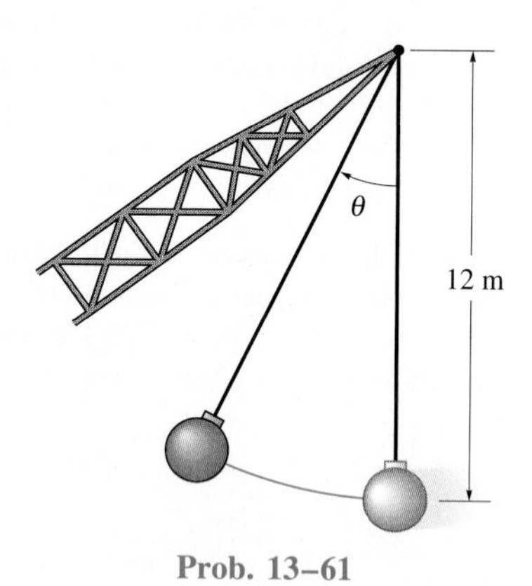

Prob. 13–61

13–62. The pendulum bob has a mass m and is released from rest when $\theta = 0°$. Determine the tension in the cord as a function of the angle of descent θ. Neglect the size of the bob.

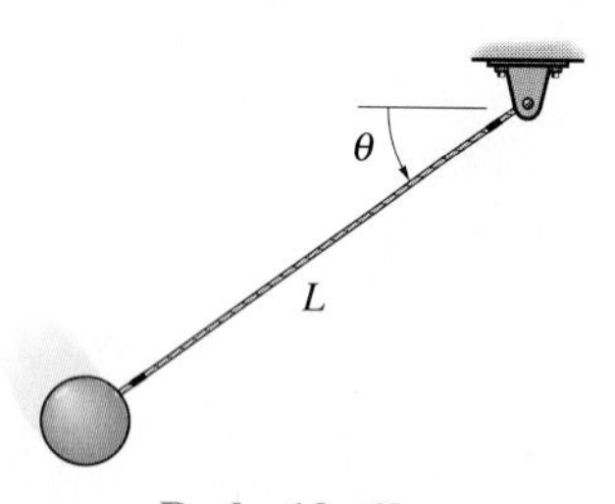

Prob. 13–62

13–63. The jet ski and rider have a total mass of 350 kg and mass center at G. If the rider undergoes a turn having a radius of curvature $\rho = 8$ m and bank angle of 30°, while traveling with a constant speed $v = 10$ m/s, determine the magnitude of the resultant force of the water on the hull. Neglect the size of the ski and rider.

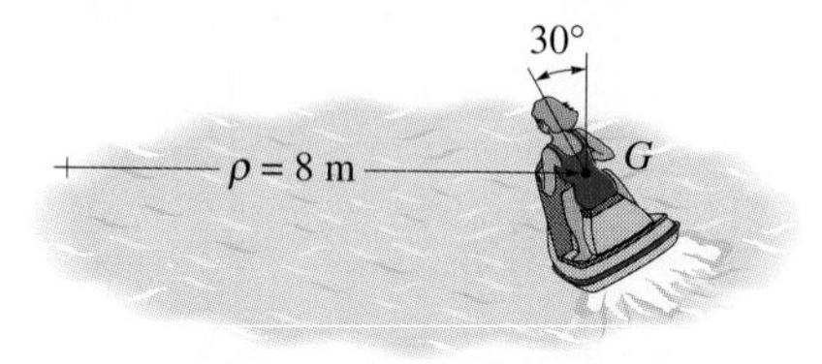

Prob. 13–63

***13–64.** If the crest of the hill has a radius of curvature $\rho = 200$ ft, determine the maximum constant speed at which the car can travel over it without leaving the surface of the road. Neglect the size of the car in the calculation. The car has a weight of 3500 lb.

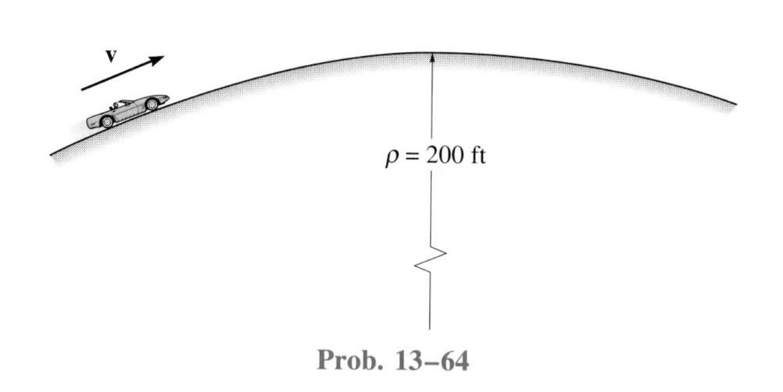

Prob. 13–64

13–65. The 150-lb man lies against the cushion for which the coefficient of static friction is $\mu_s = 0.5$. Determine the resultant normal and frictional forces the cushion exerts on him if, due to rotation about the z axis, he has a constant speed $v = 20$ ft/s. Neglect the size of the man. Take $\theta = 60°$.

13–66. The 150-lb man lies against the cushion for which the coefficient of static friction is $\mu_s = 0.5$. If he rotates about the z axis with a constant speed $v = 30$ ft/s, determine the smallest angle θ of the cushion at which he will begin to slip off.

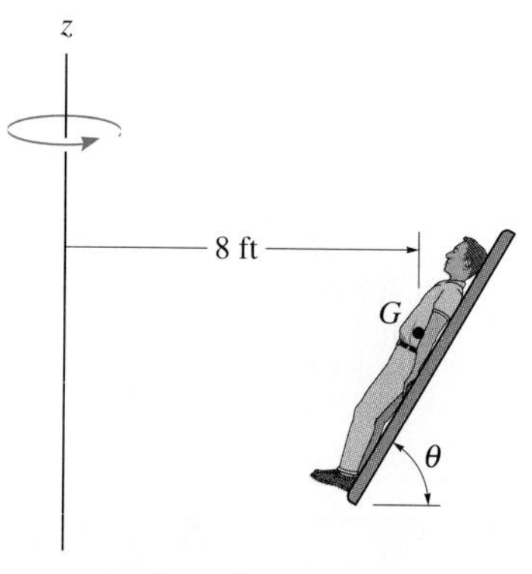

Probs. 13–65/13–66

13–67. Determine the constant speed of the passengers on the amusement-park ride if it is observed that the supporting cables are directed at $\theta = 30°$ from the vertical. Each chair including its passenger has a mass of 80 kg. Also, what are the components of force in the n, t, and b directions which the chair exerts on a 50-kg passenger during the motion?

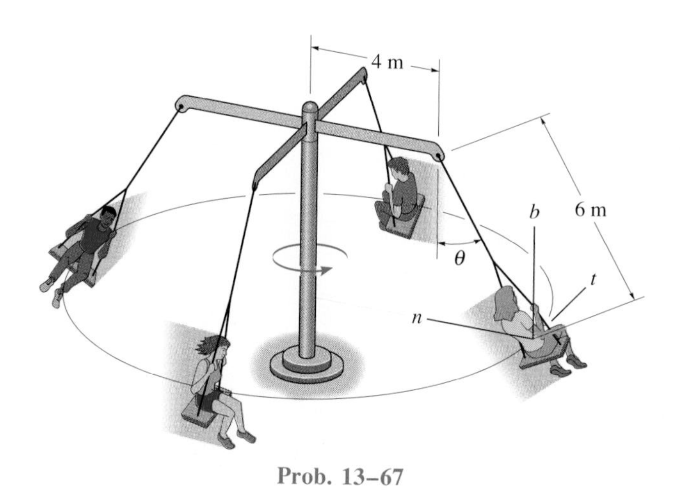

Prob. 13–67

13–69. A motorcyclist in a circus rides his motorcycle within the confines of the hollow sphere. If the coefficient of static friction between the wheels of the motorcycle and the sphere is $\mu_s = 0.4$, determine the minimum speed at which he must travel if he is to ride along the wall when $\theta = 90°$. The mass of the motorcycle and rider is 250 kg, and the radius of curvature to the center of gravity is $\rho = 20$ m. Neglect the size of the motorcycle for the calculation.

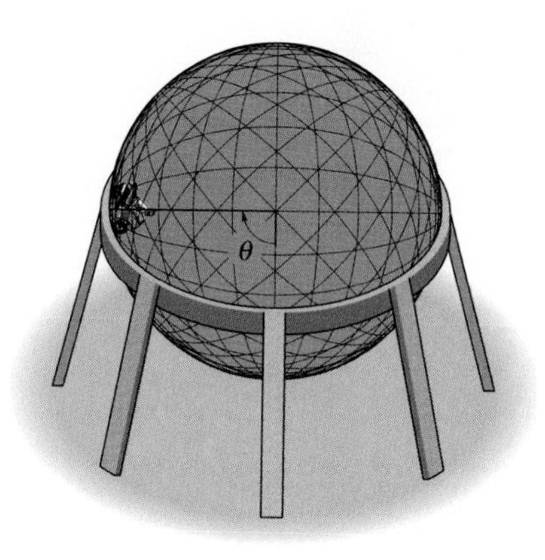

Prob. 13–69

***13–68.** The airplane, traveling at a constant speed of 50 m/s, is executing a horizontal turn. If the plane is banked at $\theta = 15°$, when the pilot experiences only a normal force on the seat of the plane, determine the radius of curvature ρ of the turn. Also, what is the normal force of the seat on the pilot if he has a mass of 70 kg.

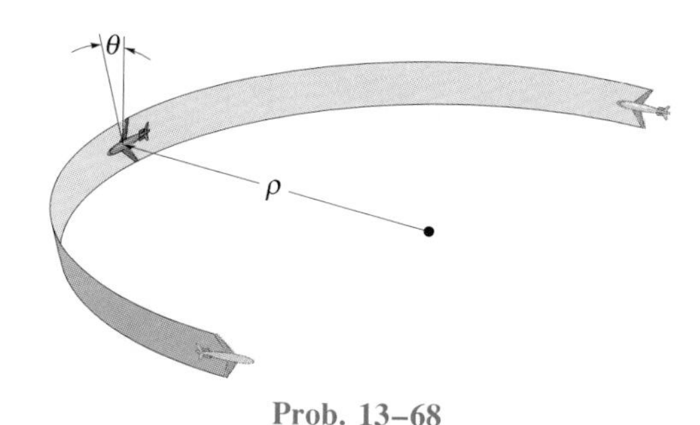

Prob. 13–68

13–70. The package has a weight of 5 lb and slides down the chute. When it reaches the curved portion AB, it is traveling at 8 ft/s ($\theta = 0°$). If the chute is smooth, determine the speed of the package when it reaches the intermediate point C ($\theta = 30°$) and when it reaches the horizontal plane ($\theta = 45°$). Also, find the normal force on the package at C.

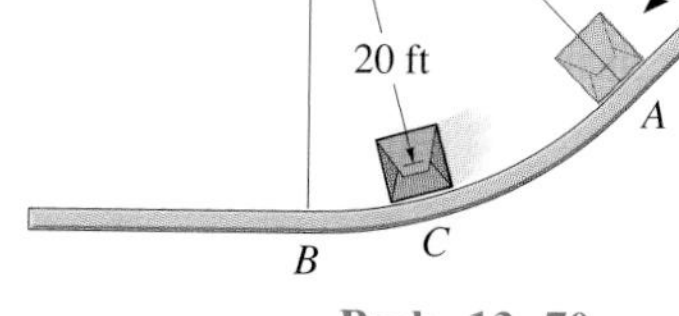

Prob. 13–70

13–71. Cartons having a mass of 5 kg are required to move along the assembly line at a constant speed of 8 m/s. Determine the smallest radius of curvature, ρ, for the conveyor so the cartons do not slip. The coefficients of static and kinetic friction between a carton and the conveyor are $\mu_s = 0.7$ and $\mu_k = 0.5$, respectively.

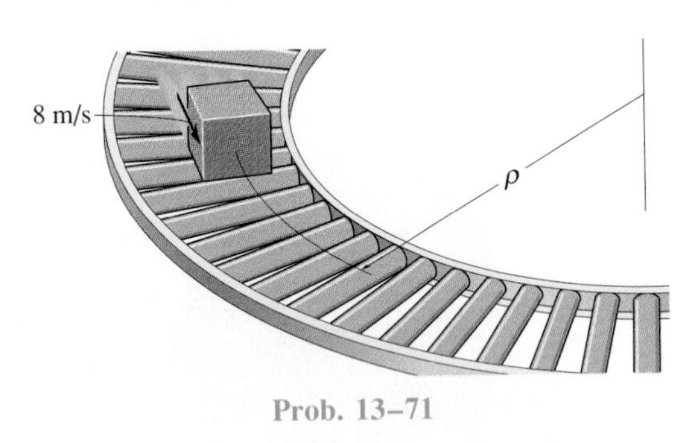

Prob. 13–71

***13–72.** When crossing an intersection, a motorcyclist encounters the slight bump or crown caused by the intersecting road. If the crest of the bump has a radius of curvature $\rho = 50$ ft, determine the maximum constant speed at which he can travel without leaving the surface of the road. Neglect the size of the motorcycle and rider in the calculation. The rider and his motorcycle have a total weight of 450 lb.

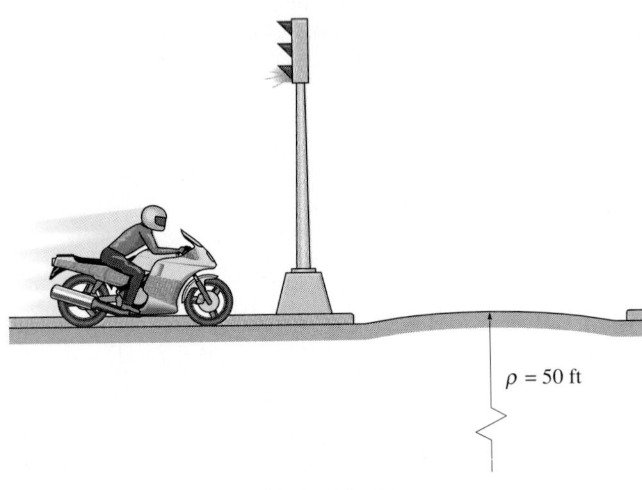

Prob. 13–72

13–73. A girl having a mass of 25 kg sits at the edge of the merry-go-round so her center of mass G is at a distance of 1.5 m from the axis of rotation. If the angular motion of the platform is *slowly* increased so that the girl's tangential component of acceleration can be neglected, determine the maximum speed which she can have before she begins to slip off the merry-go-round. The coefficient of static friction between the girl and the merry-go-round is $\mu_s = 0.3$.

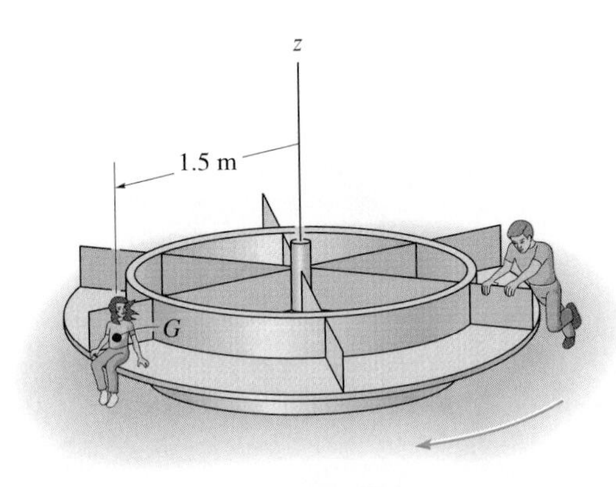

Prob. 13–73

13–74. The 2-kg spool S fits loosely on the inclined rod for which the coefficient of static friction is $\mu_s = 0.2$. If the spool is located 0.25 m from A, determine the minimum constant speed the spool can have so that it does not slip down the rod.

13–75. The 2-kg spool S fits loosely on the inclined rod for which the coefficient of static friction is $\mu_s = 0.2$. If the spool is located 0.25 m from A while the rod is rotating, determine the maximum constant speed the spool can have so that it does not slip up the rod.

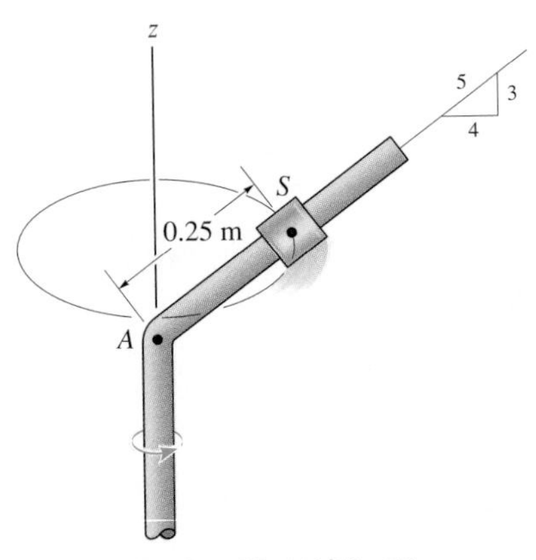

Probs. 13–74/13–75

***13–76.** The collar has a weight of 5 lb, and the attached spring has an unstretched length of 3 ft. If at the instant $\theta = 30°$ the collar has a speed $v = 4$ ft/s, determine the normal force on the collar and the magnitude of the collar's acceleration. Neglect friction.

13–77. The collar has a weight of 5 lb, and the attached spring has an unstretched length of 3 ft. If the collar is positioned on the rod so that $\theta = 30°$ and released from rest, determine the initial acceleration of the collar and the normal force on it. Neglect friction.

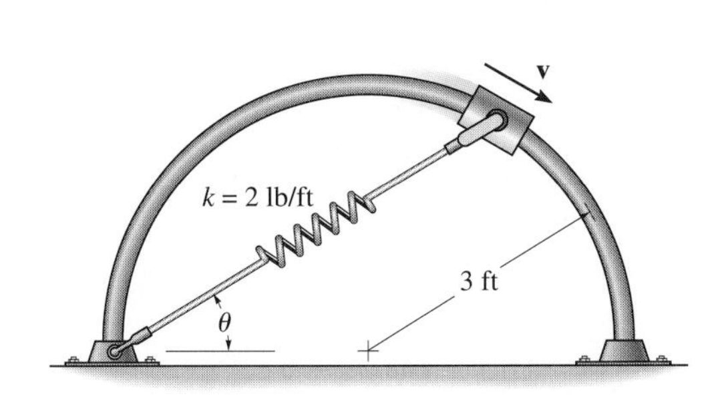

Probs. 13–76/13–77

13–78. The man has a mass of 80 kg and sits 3 m from the center of the rotating platform. Due to the rotation his speed is increased from rest by $\dot{v} = 0.4 \text{ m/s}^2$. If the coefficient of static friction between his clothes and the platform is $\mu_s = 0.3$, determine the time required to cause him to slip.

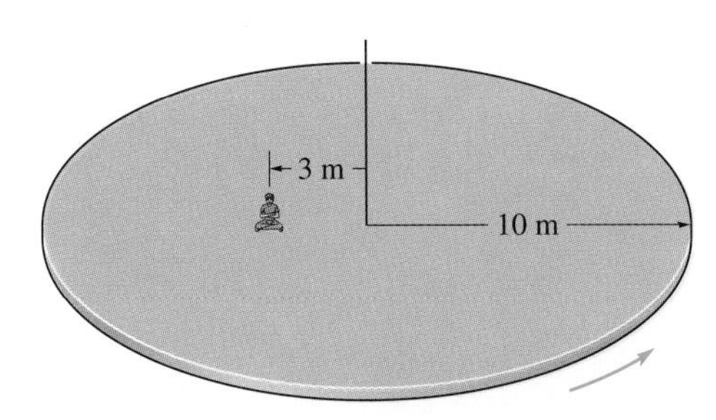

Prob. 13–78

13–79. The cylindrical plug has a weight of 2 lb and it is free to move within the confines of the smooth pipe. The spring has a stiffness $k = 14$ lb/ft and when no motion occurs the distance $d = 0.5$ ft. Determine the force of the spring on the plug when the plug is at rest with respect to the pipe. The plug is traveling with a constant speed of 15 ft/s, which is caused by the rotation of the pipe about the vertical axis.

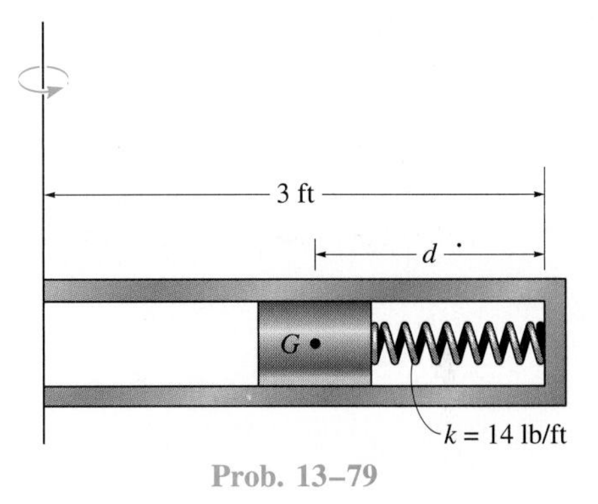

Prob. 13–79

***13–80.** An elastic cord having an unstretched length l, stiffness k, and mass per unit length m_0 is stretched around the drum of radius r $(2\pi r > l)$. Determine the speed of the cord, due to the rotation of the drum, which will allow the cord to loosen its contact with the drum.

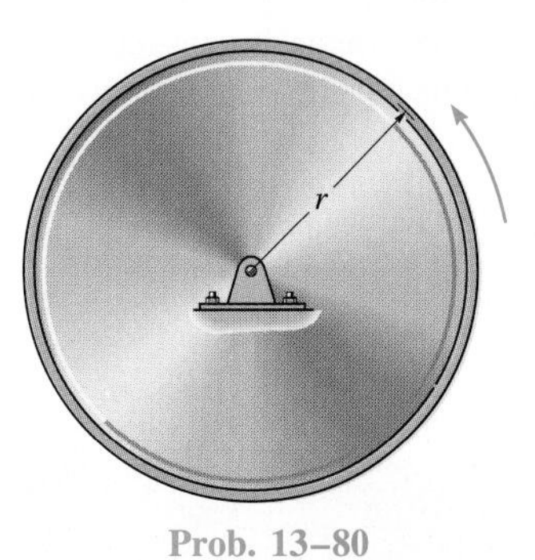

Prob. 13–80

13–81. Compute the mass of the sun, knowing that the distance from the earth to the sun is $149.6(10^6)$ km. *Hint:* use Eq. 13–1 to represent the force of gravity acting on the earth.

13–82. The block has a weight of 2 lb and it is free to move along the smooth slot in the rotating disk. The spring has a stiffness of 2.5 lb/ft and an unstretched length of 1.25 ft. Determine the force of the spring on the block and the tangential component of force which the slot exerts on the side of the block, when the block is at rest with respect to the disk and is traveling with a constant speed of 12 ft/s.

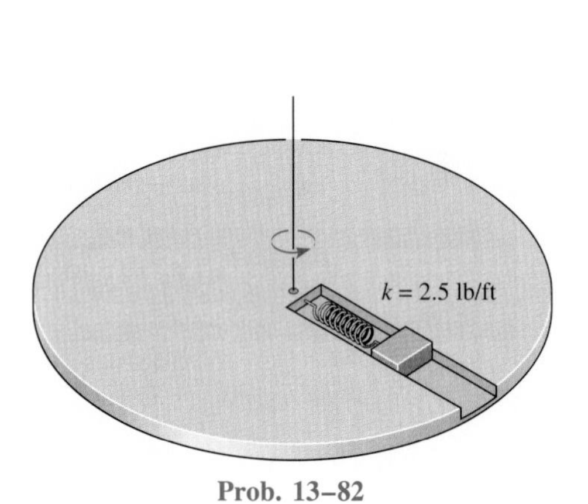

Prob. 13–82

***13–84.** If the bicycle and rider have a total weight of 180 lb, determine the resultant normal force acting on the bicycle when it is at point A while it is freely coasting at $v_A = 6$ ft/s. Also, compute the increase in the bicyclist's speed at this point. Neglect the resistance due to the wind and the size of the bicycle and rider.

Prob. 13–84

13–83. The 2-lb block is released from rest at A and slides down along the smooth cylindrical surface. If the attached spring has a stiffness $k = 2$ lb/ft, determine its unstretched length so that it does not allow the block to leave the surface until $\theta = 60°$.

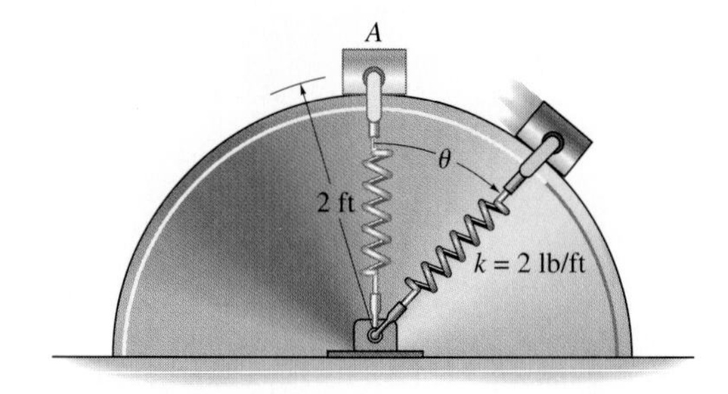

13–85. The 2-kg block B is given a velocity $v_A = 2$ m/s when it reaches point A. Determine the speed v of the block and the normal force N_B of the plane on the block as a function of θ. Plot these results as v vs. θ and N_B vs. θ and specify the angle at which the normal force is maximum. Neglect friction and the size of the block in the calculation.

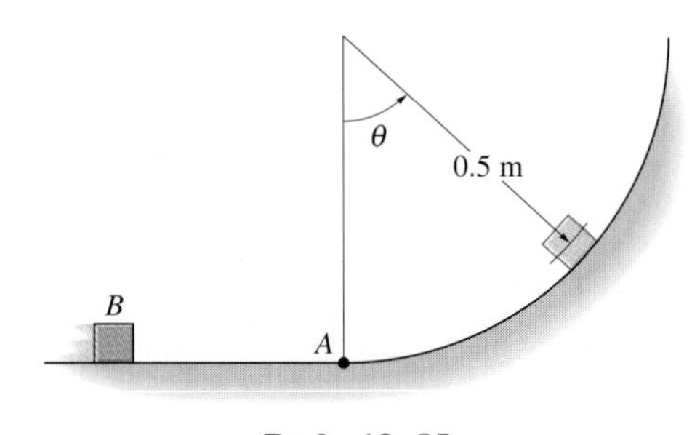

Prob. 13–85

13.6 Equations of Motion: Cylindrical Coordinates

When all the forces acting on a particle are resolved into cylindrical components, i.e., along the unit-vector directions $\mathbf{u}_r$, $\mathbf{u}_\theta$, and $\mathbf{u}_z$, Fig. 13–16, the equation of motion may be expressed as

$$\Sigma \mathbf{F} = m\mathbf{a}$$

$$\Sigma F_r \mathbf{u}_r + \Sigma F_\theta \mathbf{u}_\theta + \Sigma F_z \mathbf{u}_z = ma_r \mathbf{u}_r + ma_\theta \mathbf{u}_\theta + ma_z \mathbf{u}_z$$

To satisfy this equation, the respective $\mathbf{u}_r$, $\mathbf{u}_\theta$, and $\mathbf{u}_z$ components on the left side must equal the corresponding components on the right side. Consequently, we may write the following three scalar equations of motion:

$$\begin{aligned} \Sigma F_r &= ma_r \\ \Sigma F_\theta &= ma_\theta \\ \Sigma F_z &= ma_z \end{aligned} \tag{13–9}$$

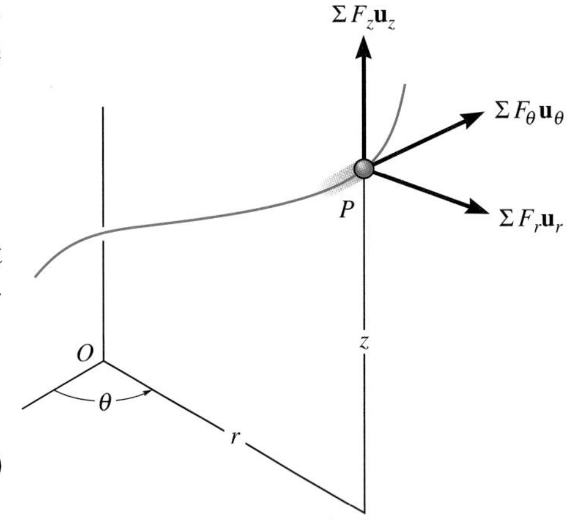

inertial coordinate system

Fig. 13–16

If the particle is constrained to move only in the r–θ plane, then only the first two of Eqs. 13–9 are used to specify the motion.

Tangential and Normal Forces. The most straightforward type of problem involving cylindrical coordinates requires the determination of the resultant force components ΣF_r, ΣF_θ, and ΣF_z causing a particle to move with a *known* acceleration. If, however, the particle's accelerated motion is not completely specified at the given instant, then some information regarding the directions or magnitudes of the forces acting on the particle must be known or computed in order to solve Eqs. 13–9. For example, the force **P** causes the particle in Fig. 13–17*a* to move along a path defined in polar coordinates as $r = f(\theta)$. The *normal force* **N** which the path exerts on the particle is always *perpendicular to the tangent of the path,* whereas the frictional force **F** always acts along the tangent in the opposite direction of motion. The *directions* of **N** and **F** can be specified relative to the radial coordinate by computing the angle ψ (psi), Fig. 13–17*b*, which is defined between the *extended* radial line $r = OP$ and the tangent to the curve. This angle can be obtained by noting that

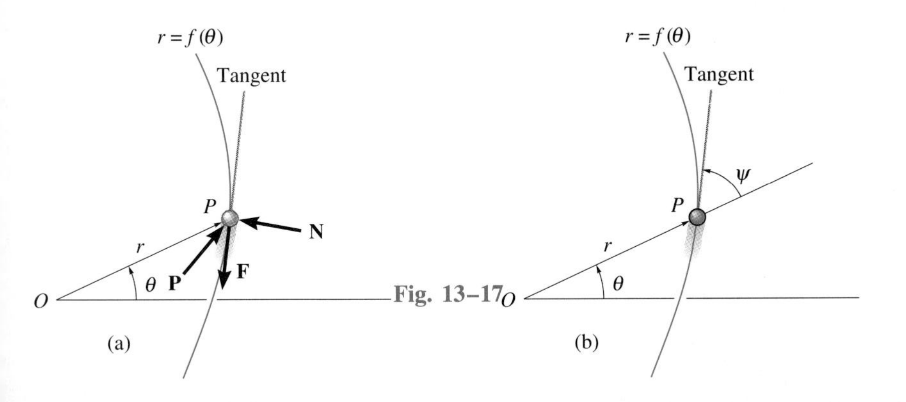

Fig. 13–17

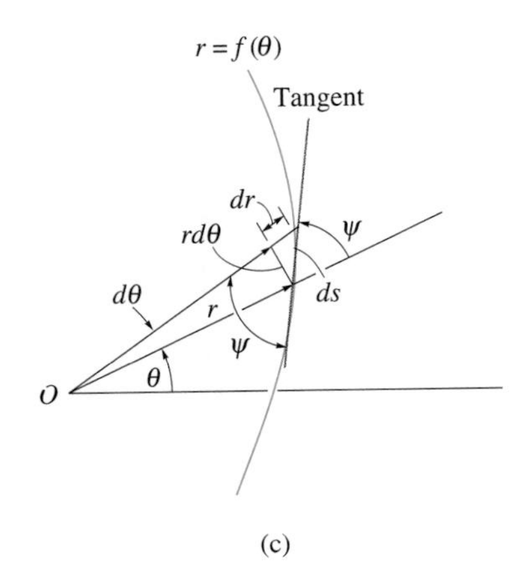

(c)

Fig. 13–17 *(cont'd)*

when the particle is displaced a distance ds along the path, Fig. 13–17c, the component of displacement in the radial direction is dr and the component of displacement in the transverse direction is $r\,d\theta$. Since these two components are mutually perpendicular, the angle ψ can be determined from $\tan\psi = r\,d\theta/dr$, or

$$\tan\psi = \frac{r}{dr/d\theta} \qquad (13\text{–}10)$$

If ψ is calculated as a positive quantity, it is measured from the *extended radial line* to the tangent in a counterclockwise sense or in the positive direction of θ. If it is negative, it is measured in the opposite direction to positive θ.

For example, consider $r = a(1 + \cos\theta)$, the cardioid shown in Fig. 13–18. Because $dr/d\theta = -a\sin\theta$, then when $\theta = 30°$, $\tan\psi = a(1 + \cos 30°)/(-a\sin 30°) = -3.732$, or $\psi = -75°$, measured clockwise, as shown in the figure.

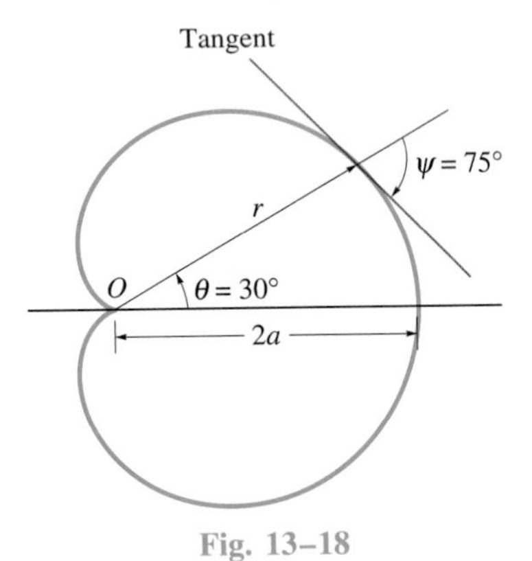

Fig. 13–18

PROCEDURE FOR ANALYSIS

Cylindrical or polar coordinates are a suitable choice for the analysis of a problem for which data regarding the angular motion of the radial line r are given, or in cases where the path can be conveniently expressed in terms of these coordinates. Once these coordinates have been established, the equations of motion can be applied in order to relate the forces acting on the particle to its acceleration components. The method for doing this has been outlined in the procedure for analysis given in Sec. 13.4. The following is a summary of this procedure.

Free-Body Diagram. Establish the r, θ, z inertial coordinate system and draw the particle's free-body diagram. Assume that $\mathbf{a}_r$, $\mathbf{a}_\theta$, and $\mathbf{a}_z$ act in the *positive directions* of r, θ, and z if they are unknown. Before applying the equations of motion, identify all the unknowns.

Equations of Motion. Apply the equations of motion, Eqs. 13–9.

Kinematics. Use the methods of Sec. 12.7 to determine r and the time derivatives $\dot{r}$, $\ddot{r}$, $\dot{\theta}$, $\ddot{\theta}$, and $\ddot{z}$, and evaluate the acceleration components $a_r = \ddot{r} - r\dot{\theta}^2$, $a_\theta = r\ddot{\theta} + 2\dot{r}\dot{\theta}$, and $a_z = \ddot{z}$. If any of these three components is computed as a negative quantity, it indicates that it acts in its negative coordinate direction.

The following examples illustrate application of this procedure numerically. Further application of Eqs. 13–9 is given in Sec. 13.7, where problems involving rockets, satellites, and planetary bodies subjected to gravitational forces are considered.

Example 13–10

The 2-lb block in Fig. 13–19a moves on a smooth horizontal track, such that its path is specified in polar coordinates by the parametric equations $r = (10t^2)$ ft and $\theta = (0.5t)$ rad, where t is in seconds. Determine the magnitude of the tangential force **F** causing the motion at the instant $t = 1$ s.

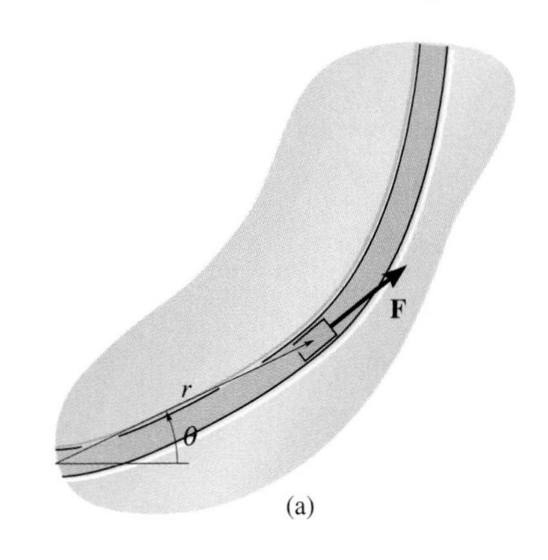

(a)

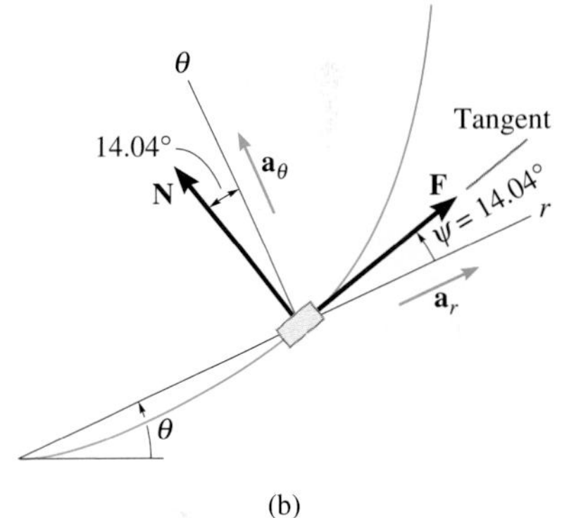

(b)

Fig. 13–19

SOLUTION

Free-Body Diagram. As shown on the block's free-body diagram, Fig. 13–19b, the normal force of the track on the block, **N,** and the tangential force **F** are located at an angle ψ from the r and θ axes. This angle can be obtained from Eq. 13–10. To do this, we must first express the path as $r = f(\theta)$ by eliminating the parameter t between r and θ. This yields $r = 40\theta^2$. Also, when $t = 1$ s, $\theta = 0.5(1\text{ s}) = 0.5$ rad. Thus

$$\tan\psi = \frac{r}{dr/d\theta} = \left.\frac{40\theta^2}{40(2\theta)}\right|_{\theta=0.5\text{ rad}} = 0.25$$

$$\psi = 14.04^\circ$$

Because ψ is a positive quantity, it is measured counterclockwise from the r axis to the tangent (the same direction as θ) as shown in Fig. 13–19b. There are presently four unknowns: F, N, a_r and a_θ.

Equations of Motion

$$+\nearrow\Sigma F_r = ma_r; \qquad F\cos 14.04^\circ - N\sin 14.04^\circ = \frac{2}{32.2}a_r \qquad (1)$$

$$\nwarrow+\Sigma F_\theta = ma_\theta; \qquad F\sin 14.04^\circ + N\cos 14.04^\circ = \frac{2}{32.2}a_\theta \qquad (2)$$

Kinematics. Since the motion is specified, the coordinates and the required time derivatives can be computed and evaluated at $t = 1$ s.

$$r = 10t^2\Big|_{t=1\text{ s}} = 10\text{ ft} \qquad \theta = 0.5t\Big|_{t=1\text{ s}} = 0.5\text{ rad}$$

$$\dot{r} = 20t\Big|_{t=1\text{ s}} = 20\text{ ft/s} \qquad \dot{\theta} = 0.5\text{ rad/s}$$

$$\ddot{r} = 20\text{ ft/s}^2 \qquad \ddot{\theta} = 0$$

$$a_r = \ddot{r} - r\dot{\theta}^2 = 20 - 10(0.5)^2 = 17.5\text{ ft/s}^2$$

$$a_\theta = r\ddot{\theta} + 2\dot{r}\dot{\theta} = 10(0) + 2(20)(0.5) = 20\text{ ft/s}^2$$

Substituting into Eqs. 1 and 2 and solving, we get

$$F = 1.36\text{ lb} \qquad \textit{Ans.}$$

$$N = 0.942\text{ lb}$$

Example 13–11

The smooth 2-kg cylinder C in Fig. 13–20a has a peg P through its center which passes through the slot in arm OA. If the arm rotates in the *vertical plane* at a constant rate $\dot{\theta} = 0.5$ rad/s, determine the force that the arm exerts on the peg at the instant $\theta = 60°$.

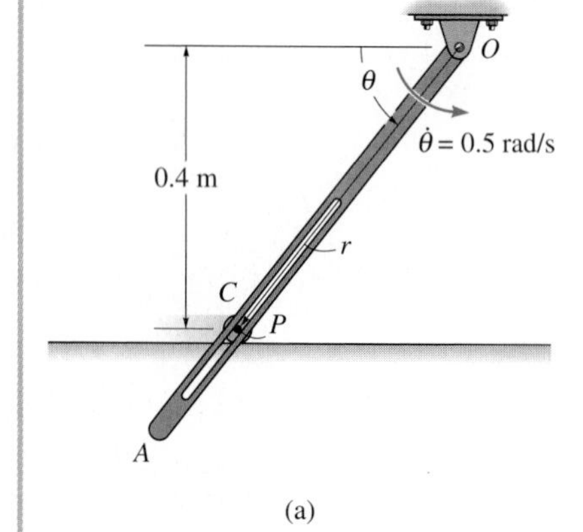

(a)

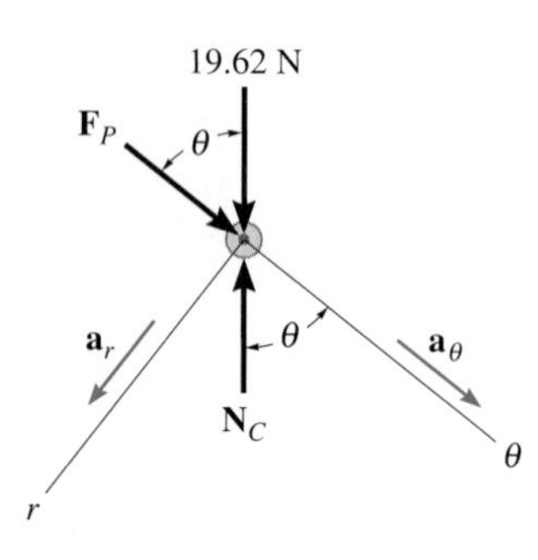

(b)

Fig. 13–20

SOLUTION

Why is it a good idea to use polar coordinates to solve this problem?

Free-Body Diagram. The free-body diagram for the cylinder is shown in Fig. 13–20b. The force of the peg, $\mathbf{F}_P$, acts perpendicular to the slot in the arm. As usual, $\mathbf{a}_r$ and $\mathbf{a}_\theta$ are assumed to act in the directions of *positive* r and θ, respectively. Identify the four unknowns.

Equations of Motion. Using the data in Fig. 13–20b, we have

$$+\swarrow \Sigma F_r = ma_r; \qquad 19.62 \sin\theta - N_C \sin\theta = 2a_r \qquad (1)$$

$$+\searrow \Sigma F_\theta = ma_\theta; \qquad 19.62 \cos\theta + F_P - N_C \cos\theta = 2a_\theta \qquad (2)$$

Kinematics. From Fig. 13–20a, r can be related to θ by the equation

$$r = \frac{0.4}{\sin\theta} = 0.4 \csc\theta$$

Since $d(\csc\theta) = -(\csc\theta \cot\theta)\, d\theta$ and $d(\cot\theta) = -(\csc^2\theta)\, d\theta$, then r and the necessary time derivatives become

$$\dot{\theta} = 0.5 \qquad r = 0.4 \csc\theta$$

$$\ddot{\theta} = 0 \qquad \dot{r} = -0.4(\csc\theta \cot\theta)\dot{\theta}$$

$$= -0.2 \csc\theta \cot\theta$$

$$\ddot{r} = -0.2(-\csc\theta \cot\theta)(\dot{\theta}) \cot\theta - 0.2 \csc\theta(-\csc^2\theta)\dot{\theta}$$

$$= 0.1 \csc\theta(\cot^2\theta + \csc^2\theta)$$

Evaluating these formulas at $\theta = 60°$, we get

$$\dot{\theta} = 0.5 \qquad r = 0.462$$

$$\ddot{\theta} = 0 \qquad \dot{r} = -0.133$$

$$\ddot{r} = 0.192$$

$$a_r = \ddot{r} - r\dot{\theta}^2 = 0.192 - 0.462(0.5)^2 = 0.0765$$

$$a_\theta = r\ddot{\theta} + 2\dot{r}\dot{\theta} = 0 + 2(-0.133)(0.5) = -0.133$$

Substituting these results into Eqs. 1 and 2 with $\theta = 60°$ and solving yields

$$N_C = 19.4 \text{ N} \qquad F_P = -0.355 \text{ N} \qquad \textit{Ans.}$$

The negative sign indicates that $\mathbf{F}_P$ acts opposite to that shown in Fig. 13–20b.

Note: Try and obtain $a_r = 0.0765$ m/s^2 from using the shorter method explained in Example 12–19.

Example 13–12

A can C, having a mass of 0.5 kg, moves along a grooved horizontal slot shown in Fig. 13–21*a*. The slot is in the form of a spiral, which is defined by the equation $r = (0.1\theta)$ m, where θ is in radians. If the arm OA is rotating at a constant rate $\dot{\theta} = 4$ rad/s in the horizontal plane, determine the force it exerts on the can at the instant $\theta = \pi$ rad. Neglect friction and the size of the can.

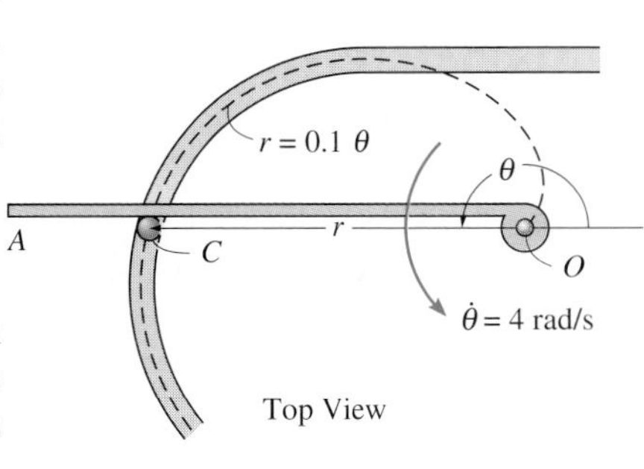

(a)

SOLUTION

Free-Body Diagram. The driving force $\mathbf{F}_C$ acts perpendicular to the arm OA, whereas the normal force of the wall of the slot on the can, $\mathbf{N}_C$, acts perpendicular to the tangent to the curve at $\theta = \pi$ rad, Fig. 13–21*b*. As usual, $\mathbf{a}_r$ and $\mathbf{a}_\theta$ are assumed to act in the *positive directions* of r and θ, respectively. Since the path is specified, the angle ψ which the extended radial line r makes with the tangent, Fig. 13–21*c*, can be determined from Eq. 13–10. We have $r = 0.1\theta$, so that $dr/d\theta = 0.1$, and therefore

$$\tan\psi = \frac{r}{dr/d\theta} = \frac{0.1\theta}{0.1} = \theta$$

When $\theta = \pi$, $\psi = \tan^{-1}\pi = 72.3°$, so that $\phi = 90° - \psi = 17.7°$, as shown in Fig. 13–21*c*. Identify the four unknowns in Fig. 13–21*b*.

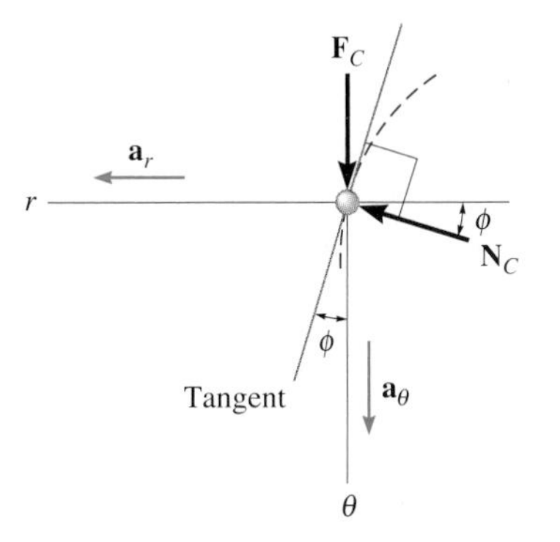

(b)

Equations of Motion. Using $\phi = 17.7°$ and the data shown in Fig. 13–21*b*, we have

$$\overset{+}{\leftarrow}\Sigma F_r = ma_r; \qquad N_C \cos 17.7° = 0.5a_r \qquad (1)$$

$$+\downarrow\Sigma F_\theta = ma_\theta; \qquad F_C - N_C \sin 17.7° = 0.5a_\theta \qquad (2)$$

Kinematics. Computing the time derivatives of r and θ gives

$$\dot{\theta} = 4 \text{ rad/s} \qquad r = 0.1\theta$$

$$\ddot{\theta} = 0 \qquad \dot{r} = 0.1\dot{\theta} = 0.1(4) = 0.4 \text{ m/s}$$

$$\ddot{r} = 0.1\ddot{\theta} = 0$$

At the instant $\theta = \pi$ rad,

$$a_r = \ddot{r} - r\dot{\theta}^2 = 0 - 0.1(\pi)(4)^2 = -5.03 \text{ m/s}^2$$

$$a_\theta = r\ddot{\theta} + 2\dot{r}\dot{\theta} = 0 + 2(0.4)(4) = 3.20 \text{ m/s}^2 \qquad \textit{Ans.}$$

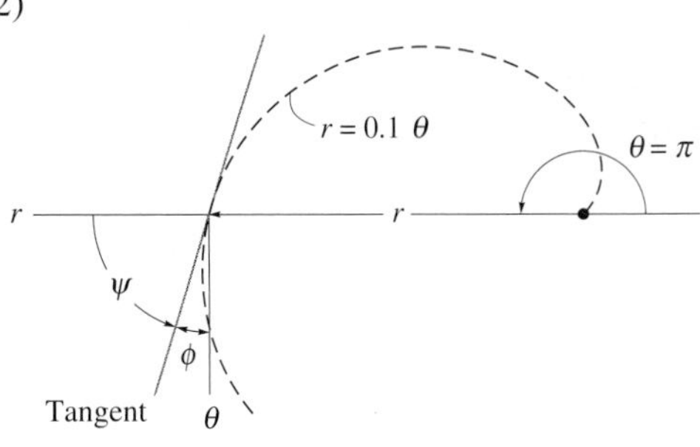

(c)

Fig. 13–21

Substituting these results into Eqs. 1 and 2 and solving yields

$$N_C = -2.64 \text{ N}$$

$$F_C = 0.797 \text{ N} \qquad \textit{Ans.}$$

What does the negative sign for N_C indicate?

PROBLEMS

13–86. A particle, having a mass of 2 kg, moves along a three-dimensional path defined by the equations $r = (10t^2 + 3t)$ m, $\theta = (0.1t^3)$ rad, and $z = (4t^2 + 15t - 6)$ m, where t is in seconds. Determine the r, θ, and z components of force which the path exerts on the particle when $t = 2$ s.

13–87. The path of motion of a 5-lb particle in the horizontal plane is described in terms of polar coordinates as $r = (2t + 10)$ ft and $\theta = (5t^2 - 6t)$ rad, where t is in seconds. Determine the magnitude of the unbalanced force acting on the particle when $t = 2$ s.

***13–88.** A particle, having a mass of 1.5 kg, moves along a three-dimensional path defined by the equations $r = (4 + 3t)$ m, $\theta = (t^2 + 2)$ rad, and $z = (6 - t^3)$ m, where t is in seconds. Determine the r, θ, and z components of force which the path exerts on the particle when $t = 2$ s.

13–89. The spring-held follower AB has a weight of 0.75 lb and moves back and forth as its end rolls on the contoured surface of the cam, where $r = 0.2$ ft and $z = (0.1 \sin \theta)$ ft. If the cam is rotating at a constant rate of 6 rad/s, determine the force at the end A of the follower when $\theta = 90°$. In this position the spring is compressed 0.4 ft. Neglect friction at the bearing C.

13–90. The spring-held follower AB has a weight of 0.75 lb and moves back and forth as its end rolls on the contoured surface of the cam, where $r = 0.2$ ft and $z = (0.1 \sin \theta)$ ft. If the cam is rotating at a constant rate of 6 rad/s, determine the maximum and minimum force the follower exerts on the cam if the spring is compressed 0.2 ft when $\theta = 90°$.

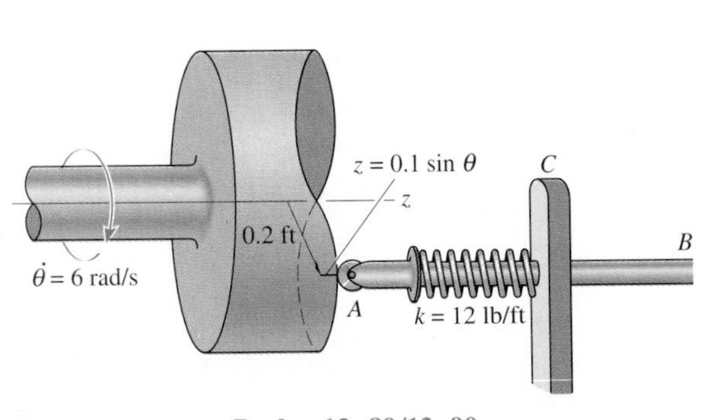

Probs. 13–89/13–90

13–91. The boy of mass 40 kg is sliding down the spiral slide at a constant speed such that his position, measured from the top of the chute, has components $r = 1.5$ m, $\theta = (0.7t)$ rad, and $z = (-0.5t)$ m, where t is in seconds. Determine the components of force $\mathbf{F}_r$, $\mathbf{F}_\theta$, and $\mathbf{F}_z$ which the slide exerts on him at the instant $t = 2$ s. Neglect the size of the boy.

***13–92.** The 40-kg boy is sliding down the smooth spiral slide such that after one revolution $z = 2$ m and his speed is 2 m/s. Determine the r, θ, z components of force the slide exerts on him at this instant. Neglect the size of the boy.

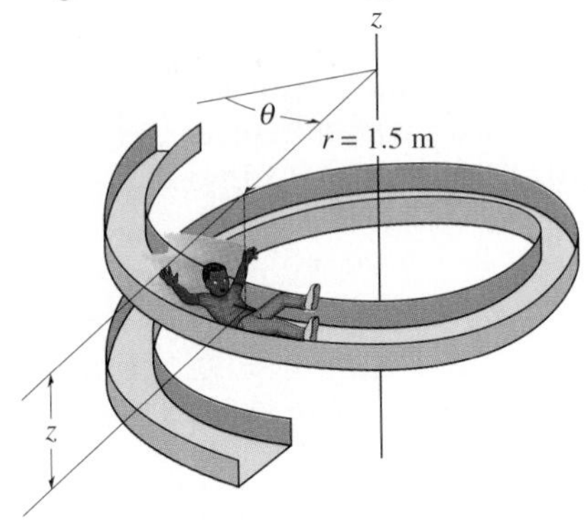

Probs. 13–91/13–92

13–93. The particle has a mass of 0.5 kg and is confined to move along the smooth horizontal slot due to the rotation of the arm OA. Determine the force of the rod on the particle and the normal force of the slot on the particle when $\theta = 30°$. The rod is rotating with a constant angular velocity $\dot{\theta} = 2$ rad/s. Assume the particle contacts only one side of the slot at any instant.

13–94. Solve Prob. 13–93 if the arm has an angular acceleration of $\ddot{\theta} = 3$ rad/s² when $\dot{\theta} = 2$ rad/s at $\theta = 30°$.

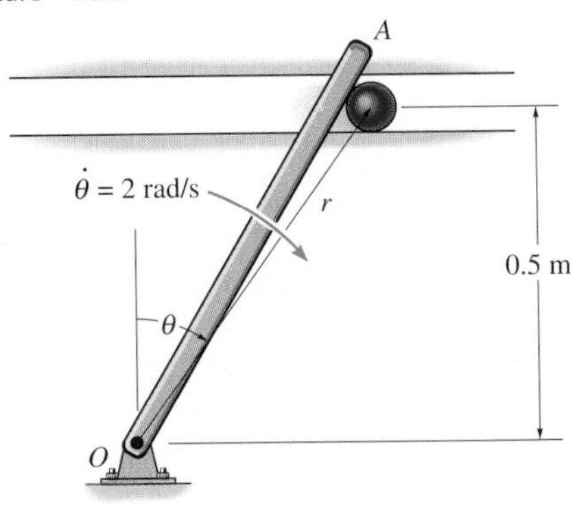

Probs. 13–93/13–94

13–95. Rod OA rotates counterclockwise at a constant angular rate $\dot{\theta} = 4$ rad/s. The double collar B is pin-connected together such that one collar slides over the rotating rod and the other collar slides over the circular rod described by the equation $r = (1.6 \cos \theta)$ m. If *both* collars have a mass of 0.5 kg, determine the force which the circular rod exerts on one of the collars and the force that OA exerts on the other collar at the instant $\theta = 45°$. Motion is in the horizontal plane.

***13–96.** Solve Prob. 13–95 if motion is in the vertical plane.

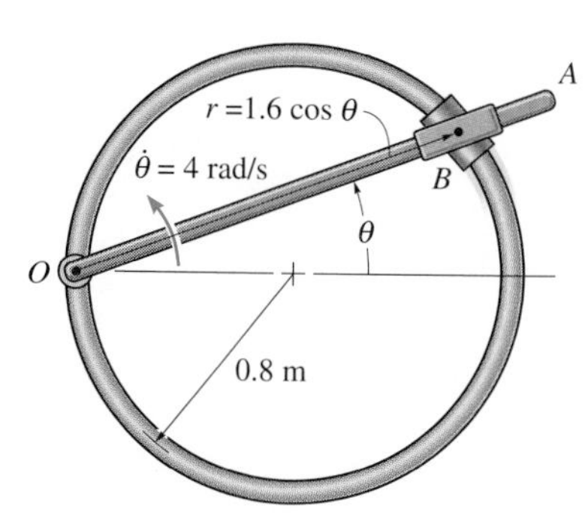

Probs. 13–95/13–96

13–97. The 0.5-lb particle is guided along the circular path using the slotted arm guide. If the arm has an angular velocity $\dot{\theta} = 4$ rad/s and an angular acceleration $\ddot{\theta} = 8\ \text{rad/s}^2$ at the instant $\theta = 30°$, determine the force of the guide on the particle. Motion occurs in the *horizontal plane.*

13–98. Solve Prob. 13–97 if motion occurs in the *vertical plane.*

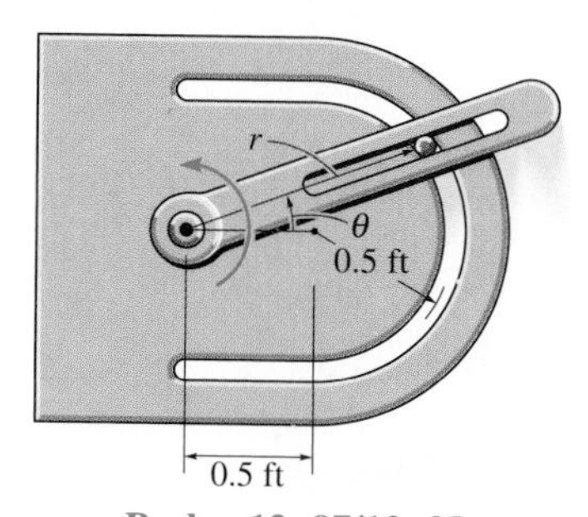

Probs. 13–97/13–98

13–99. The smooth particle has a mass of 80 g. It is attached to an elastic cord extending from O to P and due to the slotted arm guide moves along the *horizontal* circular path $r = (0.8 \sin \theta)$ m. If the cord has a stiffness $k = 30$ N/m and an unstretched length of 0.25 m, determine the force of the guide on the particle when $\theta = 60°$. The guide has a constant angular velocity $\dot{\theta} = 5$ rad/s.

***13–100.** Solve Prob. 13–99 if $\ddot{\theta} = 2\ \text{rad/s}^2$ when $\dot{\theta} = 5$ rad/s and $\theta = 60°$.

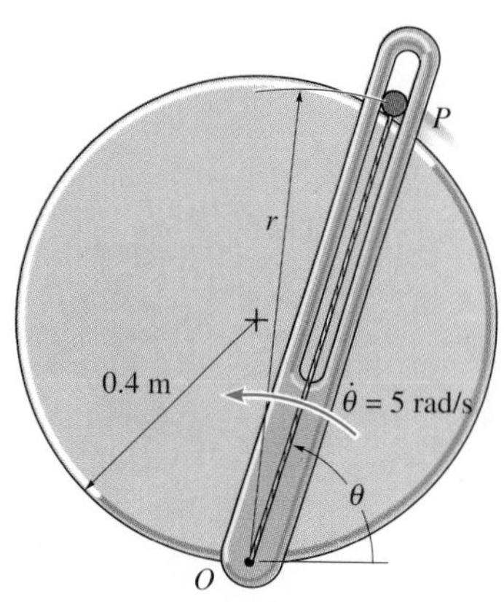

Probs. 13–99/13–100

13–101. For a short time, the 250-kg roller-coaster car is traveling along the spiral track at a constant speed such that its position measured from the top of the track has components $r = 10$ m, $\theta = (0.2t)$ rad, and $z = (-0.3t)$ m, where t is in seconds. Determine the magnitudes of the components of force which the track exerts on the car in the r, θ, and z directions at the instant $t = 2$ s. Neglect the size of the car.

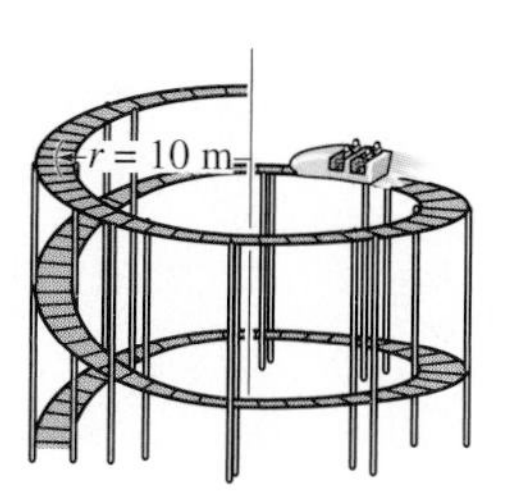

Prob. 13–101

13–102. The ball has a mass of 2 kg and a negligible size. It is originally traveling around the horizontal circular path of radius $r_0 = 0.5$ m such that the angular rate of rotation is $\dot{\theta}_0 = 1$ rad/s. If the attached cord ABC is drawn down through the hole at a constant speed of 0.2 m/s, determine the force the cord exerts on the ball at the instant $r = 0.25$ m. Also, compute the angular velocity of the ball at this instant. Neglect the effects of friction between the ball and horizontal plane. *Hint:* First show that the equation of motion in the θ direction yields $a_\theta = r\ddot{\theta} + 2\dot{r}\dot{\theta} = \dfrac{1}{r}\dfrac{d(r^2\dot{\theta})}{dt} = 0$. When integrated, $r^2\dot{\theta} = C$, where the constant C is determined from the problem data.

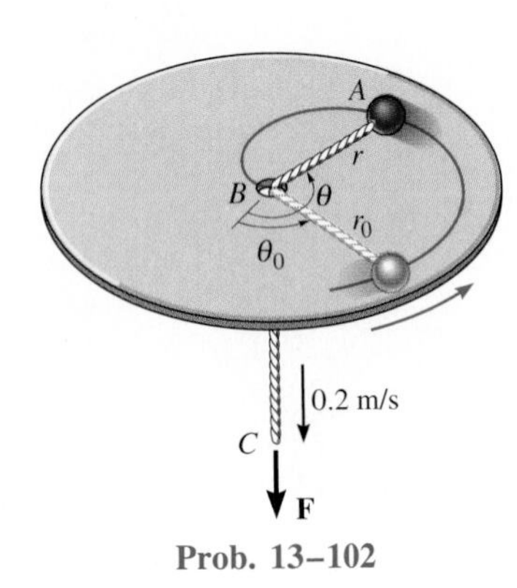

Prob. 13–102

13–103. The 0.5-lb ball is guided along the vertical circular path using the arm OA. If the arm has an angular velocity $\dot{\theta} = 0.4$ rad/s and an angular acceleration $\ddot{\theta} = 0.8$ rad/s^2 at the instant $\theta = 30°$, determine the force of the arm on the ball. Neglect friction and the size of the ball. Set $r_c = 0.4$ ft.

***13–104.** The particle of mass m is guided along the vertical circular path of radius r_c using the arm OA. If the arm has a constant angular velocity $\dot{\theta}_0$, determine the angle θ at which the particle starts to leave the surface of the semicylinder.

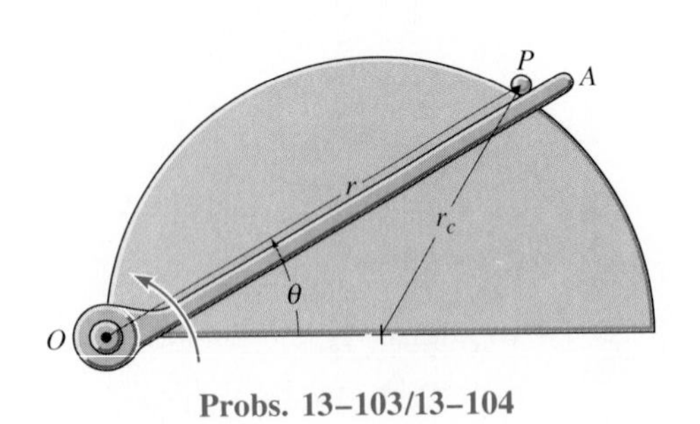

Probs. 13–103/13–104

13–105. The collar, which has a weight of 3 lb, slides along the smooth rod lying in the *horizontal plane* and having the shape of a parabola $r = 4/(1 - \cos\theta)$, where θ is in radians and r is in feet. If the collar's angular rate is constant and equals $\dot{\theta} = 4$ rad/s, determine the tangential retarding force P needed to cause the motion and the normal force that the collar exerts on the rod at the instant $\theta = 90°$.

13–106. Solve Prob. 13–105 if the parabolic path (rod) lies in the *vertical plane.*

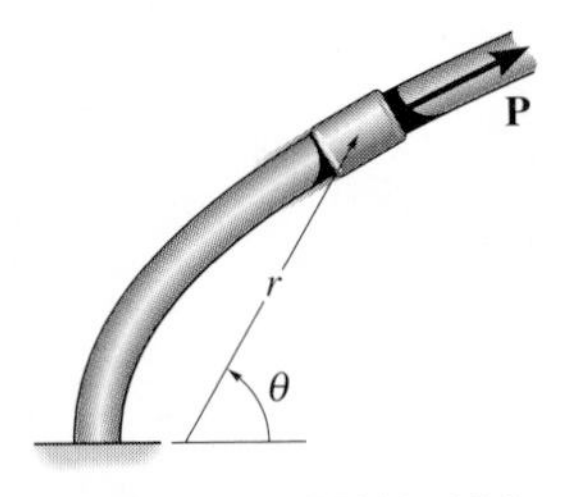

Probs. 13–105/13–106

13–107. The arm is rotating at a rate of $\dot{\theta} = 5$ rad/s when $\ddot{\theta} = 2$ rad/s^2 and $\theta = 90°$. Determine the normal force it must exert on the 0.5-kg particle if the particle is confined to move along the slotted path defined by the *horizontal* hyperbolic spiral $r\theta = 0.2$ m.

***13–108.** Solve Prob. 13–107 if the path is *vertical.*

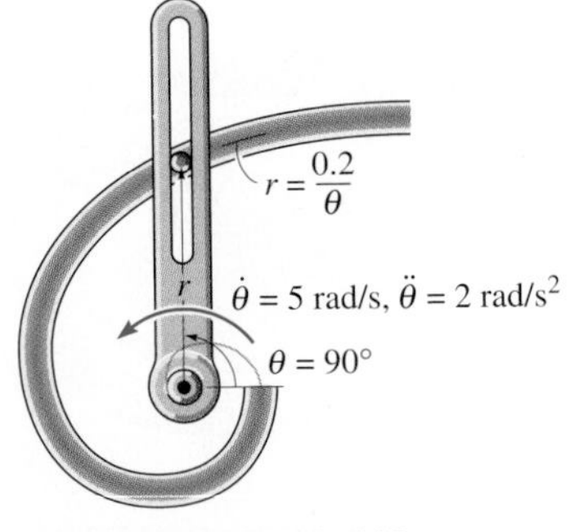

Probs. 13–107/13–108

13–109. A boy standing on firm ground spins the girl sitting on a circular "dish" or sled in a circular path of radius $r_0 = 3$ m such that her angular rate of rotation is $\dot{\theta}_0 = 0.1$ rad/s. If the attached cable OC is drawn inward at a constant speed $\dot{r} = -0.5$ m/s, determine the tension it exerts on the sled at the instant $r = 2$ m. The sled and girl have a total mass of 50 kg. Neglect the size of the girl and sled and the effects of friction between the sled and ice. *Hint:* First show that the equation of motion in the θ direction yields $a_\theta = r\ddot{\theta} + 2\dot{r}\dot{\theta} = d/dt(r^2\dot{\theta}) = 0$. When integrated, $r^2\dot{\theta} = C$, where the constant C is determined from the problem data.

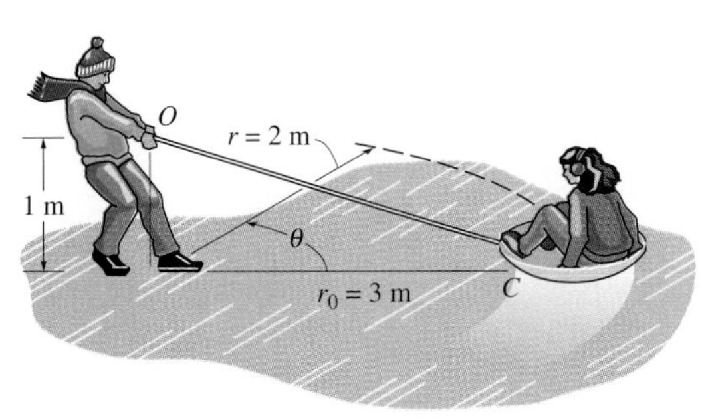

Prob. 13–109

13–110. Using air pressure, the 0.5-kg ball is forced to move through the tube lying in the *horizontal plane* and having the shape of a logarithmic spiral. If the tangential force exerted on the ball due to the air is 6 N, determine the rate of increase in the ball's speed at the instant $\theta = \pi/2$. What direction does it act in, measured from the horizontal?

13–111. Solve Prob. 13–110 if the tube lies in a *vertical plane.*

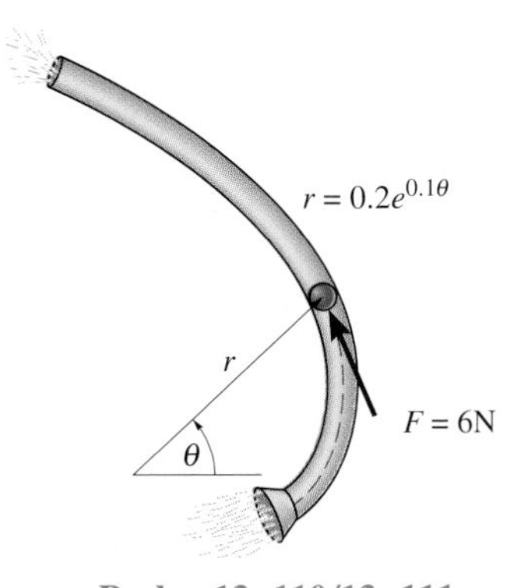

Probs. 13–110/13–111

***13–112.** The collar has a mass of 2 kg and travels along the smooth horizontal rod defined by the equiangular spiral $r = (e^\theta)$ m, where θ is in radians. Determine the tangential force F and the normal force N acting on the collar when $\theta = 45°$, if the force F maintains a constant angular motion $\dot{\theta} = 2$ rad/s.

13–113. The collar has a mass of 2 kg and travels along the smooth horizontal rod defined by the equiangular spiral $r = (e^\theta)$ m, where θ is in radians. Determine the tangential force F and the normal force N acting on the collar when $\theta = 90°$, if the force F maintains a constant angular motion $\dot{\theta} = 2$ rad/s.

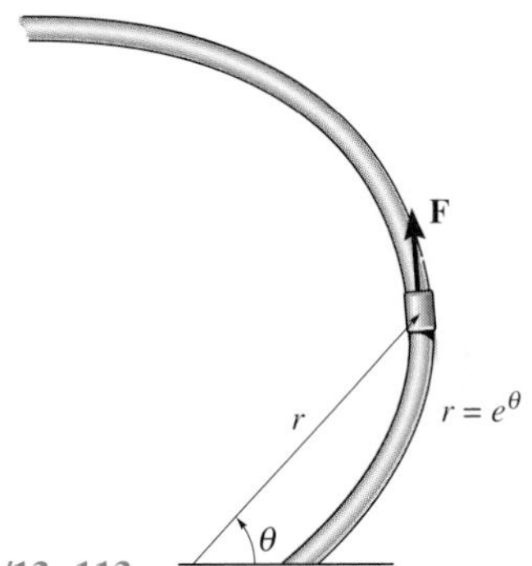

Probs. 13–112/13–113

13–114. The forked rod is used to move the smooth 2-lb particle around the horizontal path in the shape of a limaçon, $r = (2 + \cos\theta)$ ft. If at all times $\dot{\theta} = 0.5$ rad/s, determine the force which the rod exerts on the particle at the instant $\theta = 90°$. The fork and path contact the particle on only one side.

13–115. Solve Prob. 13–114 at the instant $\theta = 60°$.

***13–116.** The forked rod is used to move the smooth 2-lb particle around the horizontal path in the shape of a limaçon, $r = (2 + \cos\theta)$ ft. If $\theta = (0.5t^2)$ rad, where t is in seconds, determine the force which the rod exerts on the particle at the instant $t = 1$ s. The fork and path contact the particle on only one side.

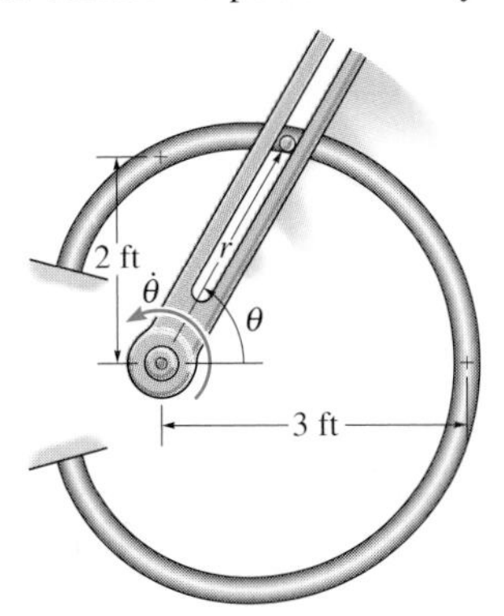

Probs. 13–114/13–115/13–116

*13.7 Central-Force Motion and Space Mechanics

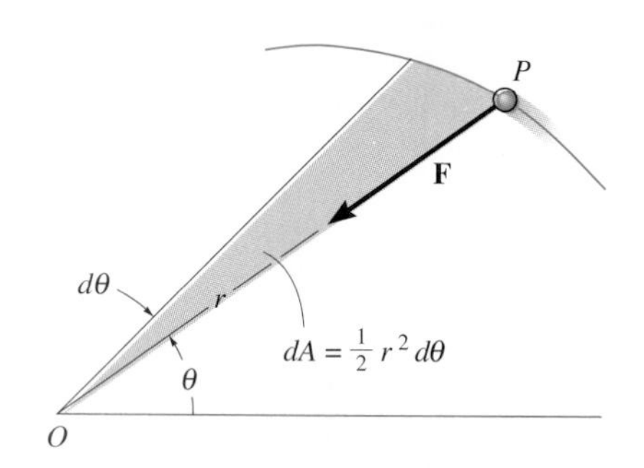

(a)

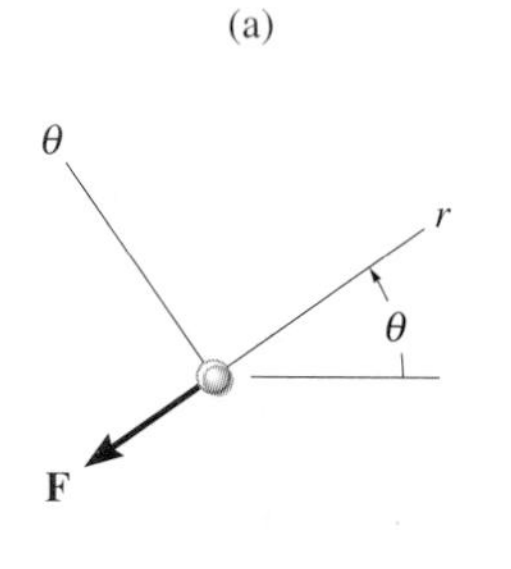

(b)

Fig. 13–22

If a particle is moving only under the influence of a force having a line of action which is always directed toward a fixed point, the motion is called *central-force motion.* This type of motion is commonly caused by electrostatic and gravitational forces.

In order to determine the motion, we will consider the particle P shown in Fig. 13–22a, which has a mass m and is acted upon only by the central force **F.** The free-body diagram for the particle is shown in Fig. 13–22b. Using polar coordinates (r, θ), the equations of motion, Eqs. 13–9, become

$$-F = m\left[\frac{d^2r}{dt^2} - r\left(\frac{d\theta}{dt}\right)^2\right]$$
$$0 = m\left(r\frac{d^2\theta}{dt^2} + 2\frac{dr}{dt}\frac{d\theta}{dt}\right) \tag{13–11}$$

The second of these equations may be written in the form

$$\frac{1}{r}\left[\frac{d}{dt}\left(r^2\frac{d\theta}{dt}\right)\right] = 0$$

so that integrating yields

$$r^2\frac{d\theta}{dt} = h \tag{13–12}$$

Here h is a constant of integration. From Fig. 13–22a notice that the shaded area described by the radius r, as r moves through an angle $d\theta$, is $dA = \frac{1}{2}r^2\,d\theta$. If the *areal velocity* is defined as

$$\frac{dA}{dt} = \frac{1}{2}r^2\frac{d\theta}{dt} = \frac{h}{2} \tag{13–13}$$

then, by comparison with Eq. 13–12, it is seen that the areal velocity for a particle subjected to central-force motion is *constant.* In other words, the particle will sweep out equal segments of area per unit of time as it travels along the path. To obtain the *path of motion,* $r = f(\theta)$, the independent variable t must be eliminated from Eqs. 13–11. Using the chain rule of calculus and Eq. 13–12, the time derivatives of Eqs. 13–11 may be replaced by

$$\frac{dr}{dt} = \frac{dr}{d\theta}\frac{d\theta}{dt} = \frac{h}{r^2}\frac{dr}{d\theta}$$
$$\frac{d^2r}{dt^2} = \frac{d}{dt}\left(\frac{h}{r^2}\frac{dr}{d\theta}\right) = \frac{d}{d\theta}\left(\frac{h}{r^2}\frac{dr}{d\theta}\right)\frac{d\theta}{dt} = \left[\frac{d}{d\theta}\left(\frac{h}{r^2}\frac{dr}{d\theta}\right)\right]\frac{h}{r^2}$$

Substituting a new dependent variable (xi) $\xi = 1/r$ into the second equation, we have

$$\frac{d^2r}{dt^2} = -h^2\xi^2\frac{d^2\xi}{d\theta^2}$$

Also, the square of Eq. 13–12 becomes

$$\left(\frac{d\theta}{dt}\right)^2 = h^2\xi^4$$

Substituting these last two equations into the first of Eqs. 13–11 yields

$$-h^2\xi^2\frac{d^2\xi}{d\theta^2} - h^2\xi^3 = -\frac{F}{m}$$

or

$$\frac{d^2\xi}{d\theta^2} + \xi = \frac{F}{mh^2\xi^2} \qquad (13\text{–}14)$$

This differential equation defines the path over which the particle travels when it is subjected to the central force* **F**.

For application, the force of gravitational attraction will be considered. Some common examples of central-force systems which depend on gravitation include the motion of the moon and artificial satellites about the earth, and the motion of the planets about the sun. As a typical problem in space mechanics, consider the trajectory of a space satellite or space vehicle launched into orbit with an initial velocity $\mathbf{v}_0$, Fig. 13–23. It will be assumed that this velocity is initially *parallel* to the tangent at the surface of the earth, as shown in the figure.† Just after the satellite is released into free flight, the only force acting on it is the gravitational force of the earth. (Gravitational attractions involving other bodies such as the moon or sun will be neglected, since for orbits close to the earth their effect is small in comparison with the earth's gravitation.) According to Newton's law of gravitation, force **F** will always act between the mass centers of the earth and the satellite, Fig. 13–23. From Eq. 13–1, this force of attraction has a magnitude of

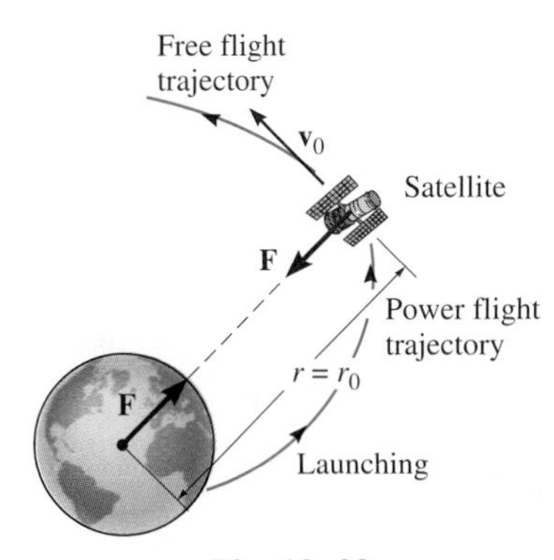

Fig. 13–23

$$F = G\frac{M_e m}{r^2}$$

where M_e and m represent the mass of the earth and the satellite, respectively, G is the gravitational constant, and r is the distance between the mass centers. Setting $\xi = 1/r$ in the above equation and substituting the result into Eq. 13–14, we obtain

*In the derivation, **F** is considered positive when it is directed toward point O. If **F** is oppositely directed, the right side of Eq. 13–14 should be negative.

†The case where $\mathbf{v}_0$ acts at some initial angle θ to the tangent is best described using the conservation of angular momentum (see Prob. 15–116).

$$\frac{d^2\xi}{d\theta^2} + \xi = \frac{GM_e}{h^2} \tag{13–15}$$

This second-order ordinary differential equation has constant coefficients and is nonhomogeneous. The solution is represented as the sum of the complementary and particular solutions. The complementary solution is obtained when the term on the right is equal to zero. It is

$$\xi_c = C \cos(\theta - \phi)$$

where C and ϕ are constants of integration. The particular solution is

$$\xi_p = \frac{GM_e}{h^2}$$

Thus, the complete solution to Eq. 13–15 is

$$\xi = \xi_c + \xi_p$$
$$= \frac{1}{r} = C\cos(\theta - \phi) + \frac{GM_e}{h^2} \tag{13–16}$$

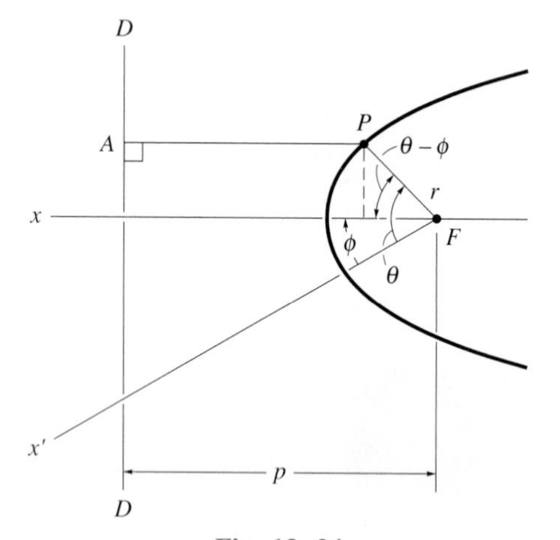

Fig. 13–24

The validity of this result may be checked by substitution into Eq. 13–15.

Equation 13–16 represents the *free-flight trajectory* of the satellite. It is the equation of a conic section expressed in terms of polar coordinates. As shown in Fig. 13–24, a *conic section* is defined as the locus of point P, which moves in a plane in such a way that the ratio of its distance from a fixed point F to its distance from a fixed line is constant. The fixed point is called the *focus*, and the fixed line DD is called the *directrix*. The constant ratio is called the *eccentricity* of the conic section and is denoted by e. Thus,

$$e = \frac{FP}{PA}$$

which may be written in the form

$$FP = r = e(PA) = e[p - r\cos(\theta - \phi)]$$

or

$$\frac{1}{r} = \frac{1}{p}\cos(\theta - \phi) + \frac{1}{ep}$$

Comparing this equation with Eq. 13–16, it is seen that the eccentricity of the conic section for the trajectory is

$$e = \frac{Ch^2}{GM_e} \tag{13–17}$$

and the fixed distance from the focus to the directrix is

$$p = \frac{1}{C} \tag{13–18}$$

Provided the polar angle θ is measured from the x axis (an axis of symmetry since it is perpendicular to the directrix DD), the angle ϕ is zero, Fig. 13–24, and therefore Eq. 13–16 reduces to

$$\frac{1}{r} = C \cos \theta + \frac{GM_e}{h^2} \tag{13–19}$$

The constants h and C are determined from the data obtained for the position and velocity of the satellite at the end of the *power-flight trajectory.* For example, if the initial height or distance to the space vehicle is r_0 (measured from the center of the earth) and its initial speed is v_0 at the beginning of its free flight, Fig. 13–25, then the constant h may be obtained from Eq. 13–12. When $\theta = \phi = 0°$, the velocity $\mathbf{v}_0$ has no radial component; therefore, from Eq. 12–25, $v_0 = r_0(d\theta/dt)$, so that

$$h = r_0^2 \frac{d\theta}{dt}$$

or

$$h = r_0 v_0 \tag{13–20}$$

To determine C, use Eq. 13–19 with $\theta = 0°$, $r = r_0$, and substitute Eq. 13–20 for h:

$$C = \frac{1}{r_0}\left(1 - \frac{GM_e}{r_0 v_0^2}\right) \tag{13–21}$$

The equation for the free-flight trajectory therefore becomes

$$\frac{1}{r} = \frac{1}{r_0}\left(1 - \frac{GM_e}{r_0 v_0^2}\right) \cos \theta + \frac{GM_e}{r_0^2 v_0^2} \tag{13–22}$$

The type of path taken by the satellite is determined from the value of the eccentricity of the conic section as given by Eq. 13–17. If

$e = 0$	free-flight trajectory is a circle	(13–23)
$e = 1$	free-flight trajectory is a parabola	
$e < 1$	free-flight trajectory is an ellipse	
$e > 1$	free-flight trajectory is a hyperbola	

Each of these trajectories is shown in Fig. 13–25. From the curves it is seen that when the satellite follows a parabolic path, it is "on the border" of never returning to its initial starting point. The initial launch velocity, $\mathbf{v}_0$, required for the satellite to follow a parabolic path is called the *escape velocity*. The speed, v_e, can be determined by using the second of Eqs. 13–23 with Eqs. 13–17, 13–20, and 13–21. It is left as an exercise to show that

$$v_e = \sqrt{\frac{2GM_e}{r_0}} \tag{13–24}$$

The speed v_c required to launch a satellite into a *circular orbit* can be found using the first of Eqs. 13–23. Since e is related to h and C, Eq. 13–17, C must be zero to satisfy this equation (from Eq. 13–20, h cannot be zero); and therefore, using Eq. 13–21, we have

$$v_c = \sqrt{\frac{GM_e}{r_0}} \tag{13–25}$$

Provided r_0 represents a minimum height for launching, in which frictional resistance from the atmosphere is neglected, speeds at launch which are less than v_c will cause the satellite to reenter the earth's atmosphere and either burn up or crash, Fig. 13–25.

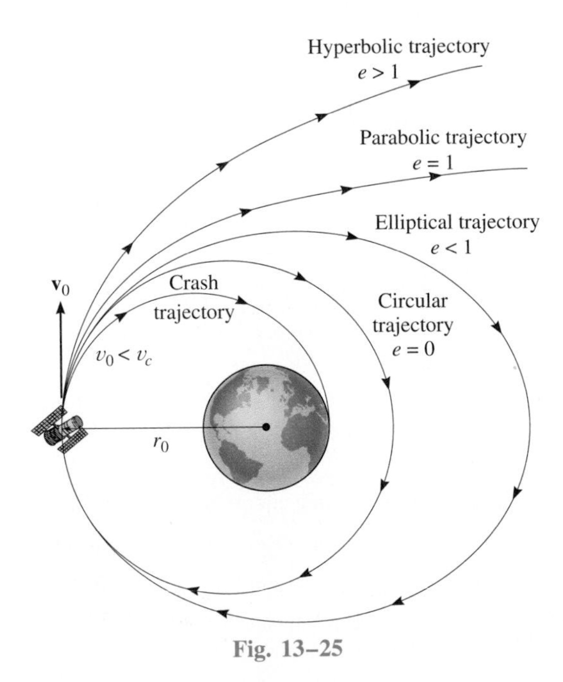

Fig. 13–25

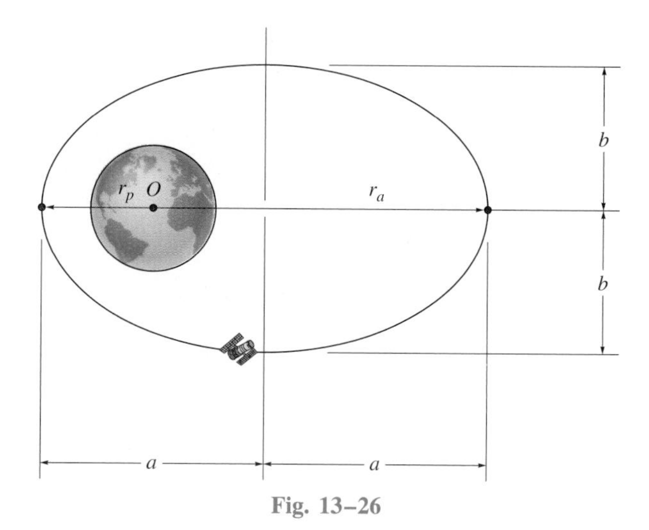

Fig. 13–26

All the trajectories attained by planets and most satellites are elliptical, Fig. 13–26. For a satellite's orbit about the earth, the *minimum distance* from the orbit to the center of the earth O (which is located at one of the foci of the ellipse) is r_p and can be found using Eq. 13–22 with $\theta = 0°$. Therefore,

$$r_p = r_0 \tag{13–26}$$

This minimum distance is called the *perigee* of the orbit. The *apogee* or maximum distance r_a can be found using Eq. 13–22 with $\theta = 180°$.* Thus,

$$r_a = \frac{r_0}{(2GM_e/r_0 v_0^2) - 1} \tag{13–27}$$

With reference to Fig. 13–26, the semimajor axis a of the ellipse is

$$a = \frac{r_p + r_a}{2} \tag{13–28}$$

Using analytical geometry, it can be shown that the minor axis b is determined from the equation

$$b = \sqrt{r_p r_a} \tag{13–29}$$

*Actually, the terminology perigee and apogee pertains only to orbits about the *earth*. If any other heavenly body is located at the focus of an elliptical orbit, the minimum and maximum distances are referred to respectively as the *periapsis* and *apoapsis* of the orbit.

Furthermore, by direct integration, the area of an ellipse is

$$A = \pi ab = \frac{\pi}{2}(r_p + r_a)\sqrt{r_p r_a} \qquad (13\text{–}30)$$

The areal velocity has been defined by Eq. 13–13, $dA/dt = h/2$. Integrating yields $A = hT/2$, where T is the *period* of time required to make one orbital revolution. From Eq. 13–30, the period is

$$T = \frac{\pi}{h}(r_p + r_a)\sqrt{r_p r_a} \qquad (13\text{–}31)$$

In addition to predicting the orbital trajectory of earth satellites, the theory developed in this section is valid, to a surprisingly close approximation, at predicting the actual motion of the planets traveling around the sun. In this case the mass of the sun, M_s, should be substituted for M_e when the appropriate formulas are used.

The fact that the planets do indeed follow elliptic orbits about the sun was discovered by the German astronomer Johannes Kepler in the early seventeenth century. His discovery was made *before* Newton had developed the laws of motion and the law of gravitation, and so at the time it provided important proof as to the validity of these laws. Kepler's laws, stipulated after 20 years of planetary observation, are summarized by the following three statements:

1. Every planet moves in its orbit such that the line joining it to the sun sweeps over equal areas in equal intervals of time, whatever the line's length.
2. The orbit of every planet is an ellipse with the sun placed at one of its foci.
3. The square of the period of any planet is directly proportional to the cube of the minor axis of its orbit.

A mathematical statement of the first and second laws is given by Eqs. 13–13 and 13–22. The third law can be shown from Eq. 13–31 using Eqs. 13–19, 13–28, and 13–29.

Example 13–13

A satellite is launched 600 km from the surface of the earth, with an initial velocity of 30 Mm/h acting parallel to the tangent at the surface of the earth, Fig. 13–27. Assuming that the radius of the earth is 6378 km and that its mass is $5.976(10^{24})$ kg, determine *(a)* the eccentricity of the orbital path, and *(b)* the velocity of the satellite at apogee.

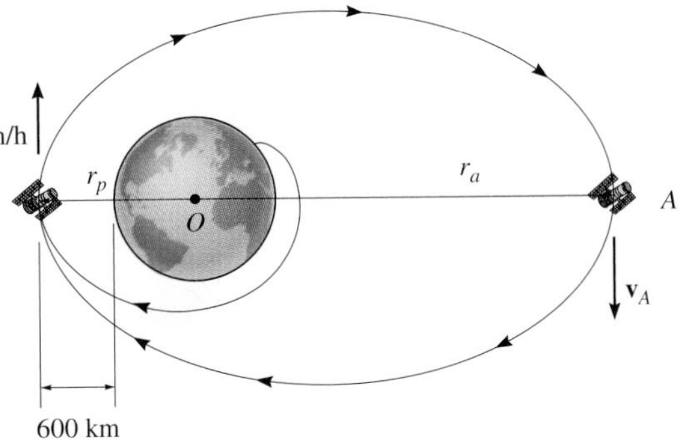

SOLUTION

Part (a). The eccentricity of the orbit is obtained using Eq. 13–17. The constants h and C are first determined from Eqs. 13–20 and 13–21. Since

$$r_p = 6378 \text{ km} + 600 \text{ km} = 6.978(10^6) \text{ m}$$
$$v_0 = 30 \text{ Mm/h} = 8333.3 \text{ m/s}$$

then

$$h = r_p v_0 = 6.978(10^6)(8333.3) = 5.815(10^{10}) \text{ m}^2\text{/s}$$

$$C = \frac{1}{r_p}\left(1 - \frac{GM_e}{r_p v_0^2}\right)$$
$$= \frac{1}{6.978(10^6)}\left\{1 - \frac{6.673(10^{-11})[5.976(10^{24})]}{6.978(10^6)(8333.3)^2}\right\} = 2.54(10^{-8}) \text{ m}^{-1}$$

Hence,

$$e = \frac{Ch^2}{GM_e} = \frac{2.54(10^{-8})[5.815(10^{10})]^2}{6.673(10^{-11})[5.976(10^{24})]} = 0.215 < 1 \qquad \textit{Ans.}$$

From Eq. 13–23, observe that the orbit is an *ellipse*.

Part (b). If the satellite were launched at the apogee A shown in Fig. 13–27, with a velocity $\mathbf{v}_A$, the same orbit would be maintained provided that

$$h = r_p v_0 = r_a v_A = 5.815(10^{10}) \text{ m}^2\text{/s}$$

Using Eq. 13–27, we have

$$r_a = \frac{r_p}{\dfrac{2GM_e}{r_p v_0^2} - 1} = \frac{6.978(10^6)}{\left\{\dfrac{2[6.673(10^{-11})][5.976(10^{24})]}{6.978(10^6)(8333.3)^2} - 1\right\}} = 10.804(10^6)$$

Thus,

$$v_A = \frac{5.815(10^{10})}{10.804(10^6)} = 5382.3 \text{ m/s} = 19.4 \text{ Mm/h} \qquad \textit{Ans.}$$

Fig. 13–27

PROBLEMS

In the following problems, assume that the radius of the earth is 6378 km, the earth's mass is $5.976(10^{24})$ kg, the mass of the sun is $1.99(10^{30})$ kg, and the gravitational constant is $G = 66.73(10^{-12})\ \text{m}^3/(\text{kg} \cdot \text{s}^2)$.

13–117. If the orbit of an asteroid has an eccentricity of $e = 0.056$ about the sun, determine the periapsis of the orbit. The orbit's apoapsis is $2.0(10^9)$ km.

13–118. The satellite is moving in an elliptical path with an eccentricity $e = 0.25$. Determine its speed when it is at its maximum distance A and minimum distance B from the earth.

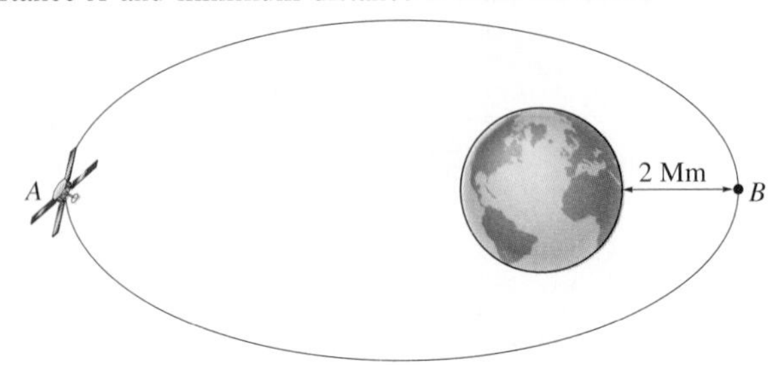

Probs. 13–117/13–118

13–119. A satellite is placed into orbit at a velocity of 6 km/s, parallel to the surface of the earth. Determine the proper altitude of the satellite above the earth's surface such that its orbit remains circular. What will happen to the satellite if its initial velocity is 5 km/s when placed tangentially into this orbit?

***13–120.** A communications satellite is to be placed into an equatorial circular orbit around the earth so that it always remains directly over a point on the earth's surface. If this requires the period to be 24 hours (approximately), determine the radius of the orbit and the satellite's velocity.

13–121. The rocket is traveling in a free-flight elliptical orbit about the earth such that $e = 0.76$ and its perigee is 9 Mm as shown. Determine its speed when it is at point B. Also determine the sudden decrease in speed the rocket must experience at A in order to travel in a circular orbit about the earth.

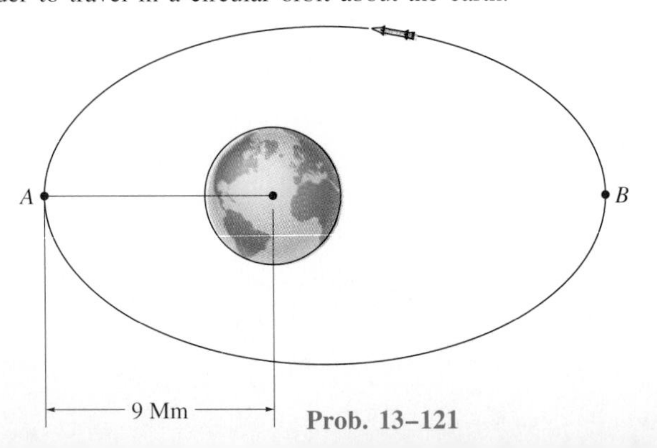

Prob. 13–121

13–122. An asteroid is in an elliptical orbit about the sun such that its periapsis is $9.30(10^9)$ km. If the eccentricity of the orbit is $e = 0.073$, determine the apoapsis of the orbit.

13–123. A satellite is to be placed into an elliptical orbit about the earth such that at the perigee of its orbit it has an *altitude* of 800 km and at apogee its *altitude* is 2400 km. Determine its required launch velocity tangent to the earth's surface at perigee and the period of its orbit.

***13–124.** Show the speed of a satellite launched into a circular orbit about the earth is given by Eq. 13–25. Determine the speed of a satellite launched parallel to the surface of the earth so that it travels in a circular orbit 800 km from the earth's surface.

13–125. The elliptical path of a satellite has an eccentricity $e = 0.130$. If it has a speed of 15 Mm/h when it is at perigee, P, determine its speed when it arrives at apogee, A. Also, how far is it from the earth's surface when it is at A?

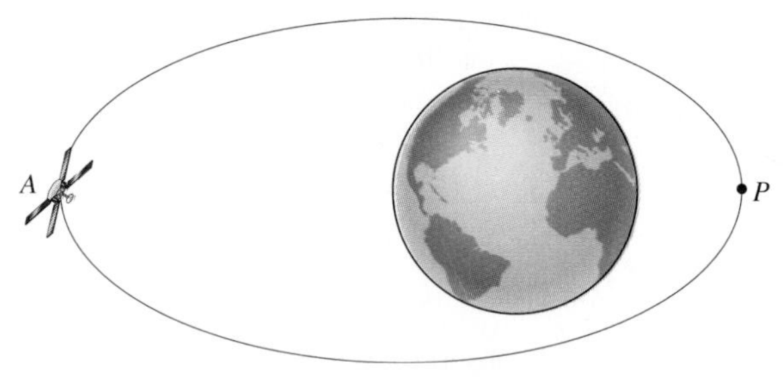

Prob. 13–125

13–126. A satellite is launched with an initial velocity $v_0 = 2500$ mi/h, parallel to the surface of the earth. Determine the required altitude (or range of altitudes) above the earth's surface for launching if the free-flight trajectory is to be (a) circular, (b) parabolic, (c) elliptical, and (d) hyperbolic. Take $G = 34.4(10^{-9})\ (\text{lb} \cdot \text{ft}^2)/\text{slug}^2$, $M_e = 409(10^{21})$ slug, the earth's radius $r_e = 3960$ mi, and 1 mi = 5280 ft.

13–127. The planet Jupiter travels around the sun in an elliptical orbit such that the eccentricity is $e = 0.048$. If the periapsis between Jupiter and the sun is $r_0 = 440(10^6)$ mi, determine (a) Jupiter's speed at periapsis and (b) the apoapsis of the orbit. Take $G = 34.4(10^{-9})\ \text{lb} \cdot \text{ft}^2/\text{slug}^2$, $M_s = 197(10^{27})$ slug, and 1 mi = 5280 ft.

***13–128.** Prove Kepler's third law of motion. *Hint:* Use Eqs. 13–19, 13–28, 13–29, and 13–31.

13–129. The rocket is traveling in free flight along an elliptical trajectory $A'A$. The planet has a mass 0.60 times that of the earth's. If the rocket has an apoapsis and periapsis as shown in the figure, determine the speed of the rocket when it is at point A.

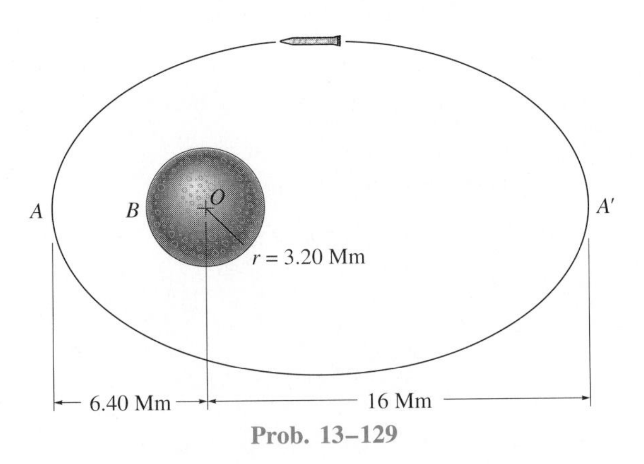

Prob. 13–129

13–130. A rocket is docked next to a satellite located 18 Mm above the earth's surface. If the satellite is traveling in a circular orbit, determine the speed which must suddenly be given to the rocket, relative to the satellite, such that it travels in free flight away from the satellite along a parabolic trajectory as shown.

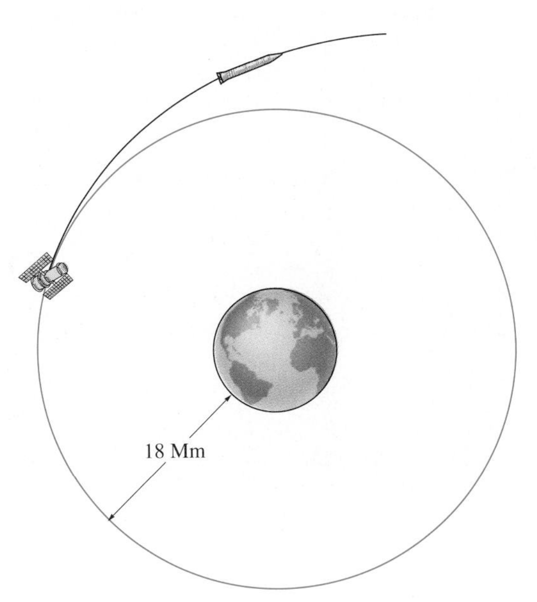

Prob. 13–130

13–131. A satellite S travels in a circular orbit around the earth. A rocket is located at the apogee of its elliptical orbit for which $e = 0.58$. Determine the sudden change in speed that must occur at A so that the rocket can enter the satellite's orbit while in free flight along the dashed elliptical trajectory. When it arrives at B, determine the sudden adjustment in speed that must be given to the rocket in order to maintain the circular orbit.

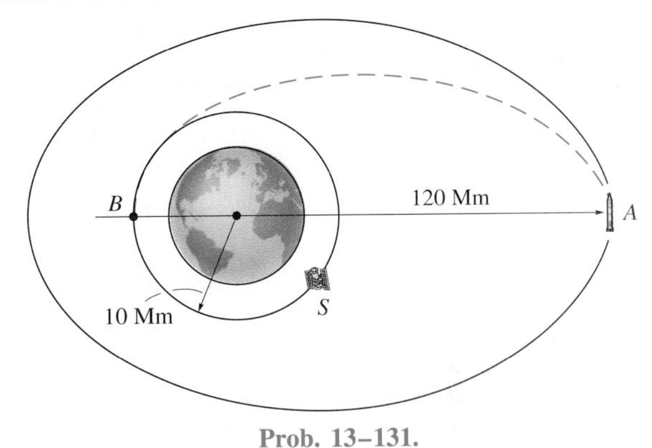

Prob. 13–131.

***13–132.** An asteroid is in an elliptical orbit about the sun such that its periapsis is $9.30(10^9)$ km. If the eccentricity of the orbit is $e = 0.073$, determine the apoapsis of the orbit.

13–133. The rocket shown is originally in a circular orbit 6 Mm above the surface of the earth. It is required that it travel in another circular orbit having an altitude of 14 Mm. To do this, the rocket is given a short pulse of power at A so that it travels in free flight along the dashed elliptical path from the first orbit to the second orbit. Determine the necessary speed it must have at A just after the power pulse, and the time required to get to the outer orbit along the path AA'. What adjustment in speed must be made at A' to maintain the second circular orbit?

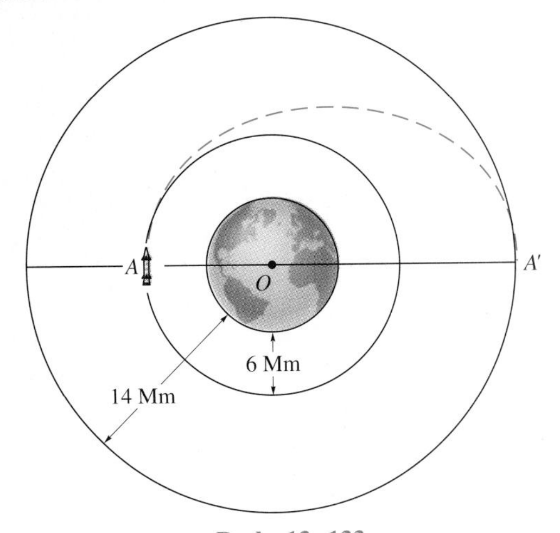

Prob. 13–133

The power requirements of this construction elevator and its motion involve work and energy principles which can be analyzed using the methods outlined in this chapter.

14

Kinetics of a Particle: Work and Energy

In this chapter we will integrate the equation of motion with respect to displacement and thereby obtain the principle of work and energy. The resulting equation is useful for solving problems which involve force, velocity, and displacement. Later in the chapter the concept of power will be discussed, and a method presented for solving kinetic problems using the theorem of conservation of energy. Before presenting these topics, however, it is first necessary to define the work done by various types of forces.

14.1 The Work of a Force

In mechanics a force $\mathbf{F}$ does *work* on a particle only when the particle undergoes a *displacement in the direction of the force.* For example, consider the force $\mathbf{F}$ acting on the particle in Fig. 14–1, which has a location on the path s that is specified by the position vector $\mathbf{r}$. If the particle moves along the path to a new position $\mathbf{r}'$, the displacement is then $d\mathbf{r} = \mathbf{r}' - \mathbf{r}$. The magnitude of $d\mathbf{r}$ is represented by ds, the differential segment along the path. If the angle between the tails of $d\mathbf{r}$ and $\mathbf{F}$ is θ, Fig. 14–1, then the work dU which is done by $\mathbf{F}$ is a *scalar quantity,* defined by

$$dU = F\,ds \cos\theta$$

By definition of the dot product (see Eq. C–14) the above equation may also be written as

$$dU = \mathbf{F} \cdot d\mathbf{r}$$

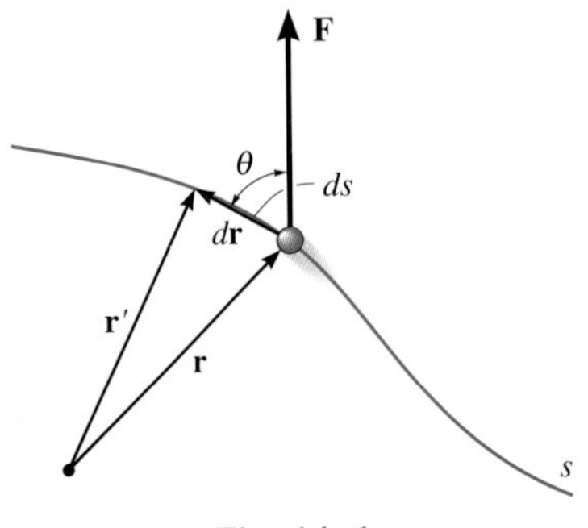

Fig. 14–1

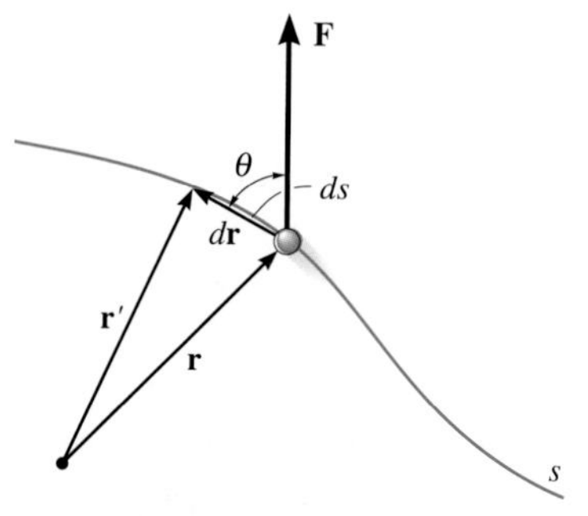

Fig. 14–1 (*repeated*)

Work as expressed by this equation, $\mathbf{U} = \mathbf{F} \cdot d\mathbf{r}$, may be interpreted in one of two ways: either as the product of F and the component of displacement in the direction of the force, i.e., $ds \cos \theta$, or as the product of ds and the component of force in the direction of displacement, i.e., $F \cos \theta$. Note that if $0° \leq \theta < 90°$, then the force component and the displacement have the *same sense* so that the work is *positive;* whereas if $90° < \theta \leq 180°$, these vectors have an *opposite sense* and therefore the work is *negative*. Also, $dU = 0$ if the force is *perpendicular* to displacement, since $\cos 90° = 0$, or if the force is applied at a *fixed point,* in which case the displacement is zero.

The basic unit for work in the SI system is called a joule (J). This unit combines the units of force and displacement. Specifically, 1 *joule* of work is done when a force of 1 newton moves 1 meter along its line of action (1 J = 1 N · m). The moment of a force has this same combination of units (N · m); however, the concepts of moment and work are in no way related. A moment is a vector quantity, whereas work is a scalar. In the FPS system work is generally defined by writing the units as ft · lb, which is distinguished from the units for a moment, written as lb · ft.

Work of a Variable Force. If the particle undergoes a finite displacement along its path from $\mathbf{r}_1$ to $\mathbf{r}_2$ or s_1 to s_2, Fig. 14–2*a*, the work is determined by integration. If $\mathbf{F}$ is expressed as a function of position, $F = F(s)$, we have

$$U_{1-2} = \int_{\mathbf{r}_1}^{\mathbf{r}_2} \mathbf{F} \cdot d\mathbf{r} = \int_{s_1}^{s_2} F \cos \theta \, ds \qquad (14\text{–}1)$$

If the working component of the force, $F \cos \theta$, is plotted versus s, Fig. 14–2*b*, the integral in this equation can be interpreted as the *area under the curve* from position s_1 to position s_2.

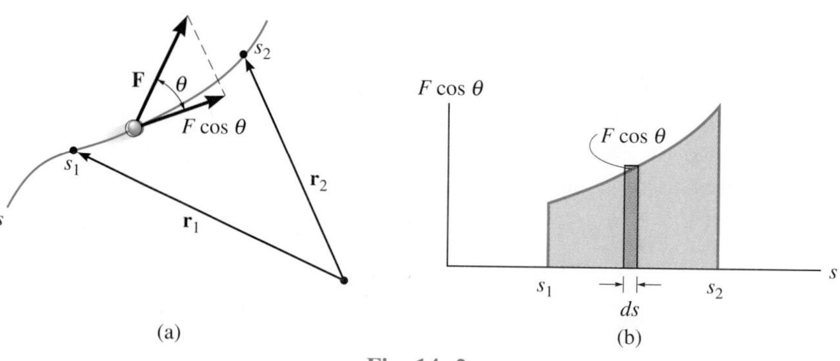

Fig. 14–2

Work of a Constant Force Moving Along a Straight Line. If the force $\mathbf{F}_c$ has a constant magnitude and acts at a constant angle θ from its straight-line path, Fig. 14–3*a*, then the component of $\mathbf{F}_c$ in the direction of displacement is $F_c \cos\theta$. The work done by $\mathbf{F}_c$ when the particle is displaced from s_1 to s_2 is determined by Eq. 14–1, in which case

$$U_{1-2} = F_c \cos\theta \int_{s_1}^{s_2} ds$$

or

$$U_{1-2} = F_c \cos\theta(s_2 - s_1) \tag{4–2}$$

Here the work of $\mathbf{F}_c$ represents the *area under the rectangle* in Fig. 14–3*b*.

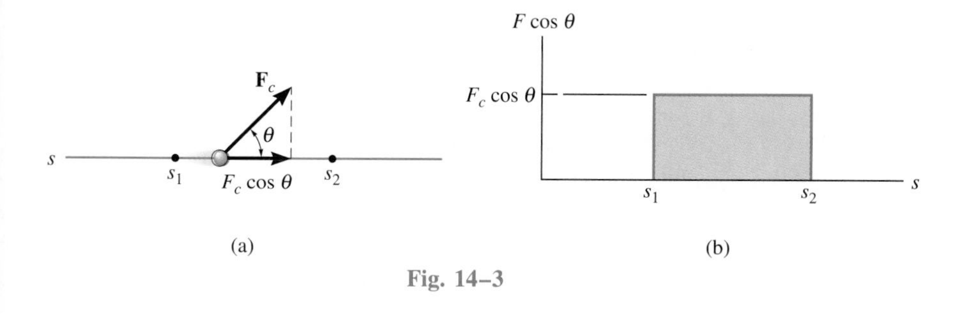

Fig. 14–3

Work of a Weight. Consider a particle which moves up along the path s shown in Fig. 14–4 from position s_1 to position s_2. At an intermediate point, the displacement $d\mathbf{r} = dx\,\mathbf{i} + dy\,\mathbf{j} + dz\,\mathbf{k}$. Since $\mathbf{W} = -W\mathbf{j}$, applying Eq. 14–1 yields

$$U_{1-2} = \int \mathbf{F} \cdot d\mathbf{r} = \int_{\mathbf{r}_1}^{\mathbf{r}_2} (-W\mathbf{j}) \cdot (dx\,\mathbf{i} + dy\,\mathbf{j} + dz\,\mathbf{k})$$

$$= \int_{y_1}^{y_2} -W\,dy = -W(y_2 - y_1)$$

or

$$U_{1-2} = -W\,\Delta y \tag{14–3}$$

Thus, the work done is equal to the magnitude of the particle's weight times its vertical displacement. In the case shown in Fig. 14–4 the work is *negative,* since W is downward and Δy is upward. Note, however, that if the particle is displaced *downward* $(-\Delta y)$, the work of the weight is *positive.* Why?

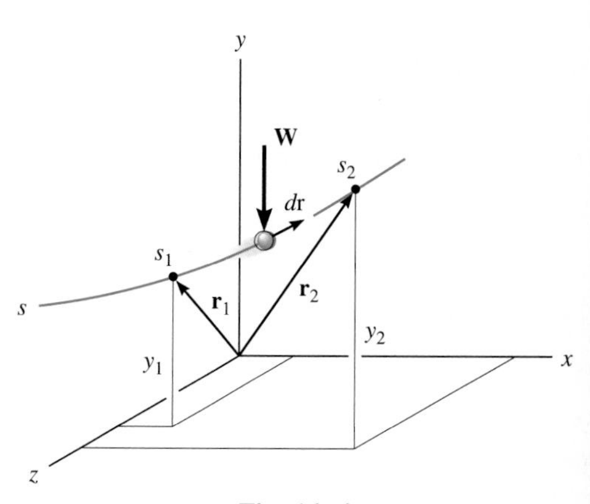

Fig. 14–4

Work of a Spring Force. The magnitude of force developed in a linear elastic spring when the spring is displaced a distance s from its unstretched position is $F_s = ks$, where k is the spring stiffness. If the spring is elongated or compressed from a position s_1 to a further position s_2, Fig. 14–5*a*, the work done *on the spring* by $\mathbf{F}_s$ is *positive,* since in each case the force and displacement are in the *same direction.* We require

$$U_{1-2} = \int_{s_1}^{s_2} F_s\, ds = \int_{s_1}^{s_2} ks\, ds$$
$$= \tfrac{1}{2}ks_2^2 - \tfrac{1}{2}ks_1^2$$

This equation represents the trapezoidal area under the line $F_s = ks$, Fig. 14–5*b*.

If a particle (or body) is attached to a spring, then the force $\mathbf{F}_s$ exerted on the particle is *opposite* to that exerted on the spring, Fig. 14–5*c*. Consequently, the force will do *negative work* on the particle when the particle is moving so as to further elongate (or compress) the spring. Hence, the above equation becomes

$$U_{1-2} = -\left(\tfrac{1}{2}ks_2^2 - \tfrac{1}{2}ks_1^2\right) \qquad (14\text{–}4)$$

When this equation is used, a mistake in sign can be eliminated if one simply notes the direction of the spring force acting on the particle and compares it with the direction of displacement of the particle—if both are in the *same direction, positive work* results; if they are *opposite* to one another, the *work is negative.*

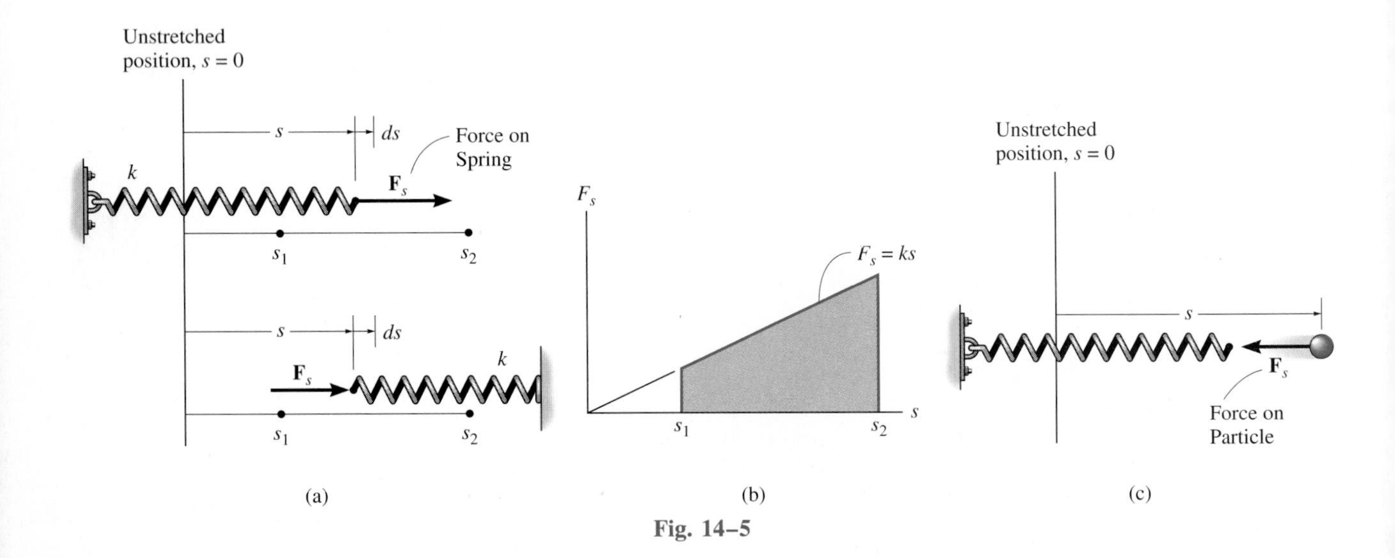

Fig. 14–5

Example 14–1

The 10-kg block shown in Fig. 14–6*a* rests on the smooth incline. If the spring is originally unstretched, determine the total work done by all the forces acting on the block when a horizontal force $P = 400$ N pushes the block up the plane $s = 2$ m.

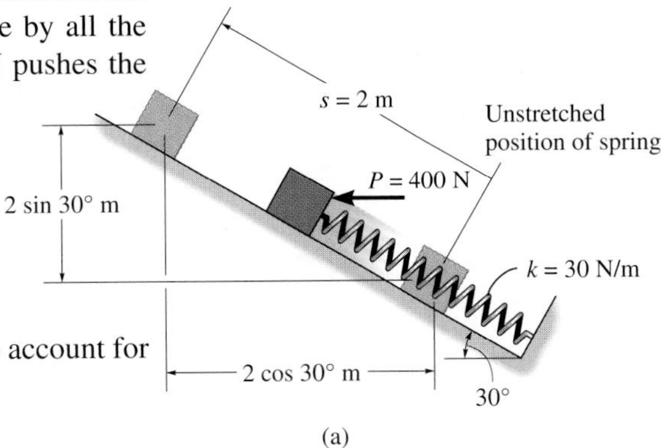

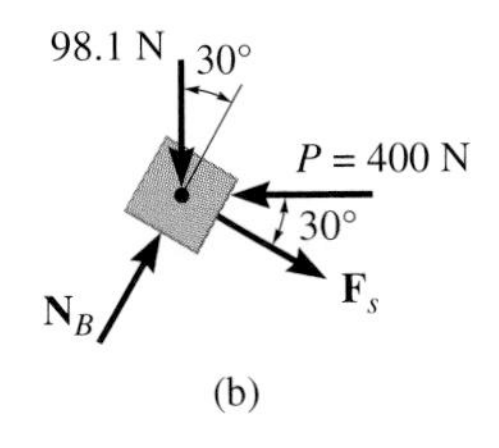

Fig. 14–6

SOLUTION

First the free-body diagram of the block is drawn in order to account for all the forces that act on the block, Fig. 14–6*b*.

***Horizontal Force* P.** Since this force is *constant,* the work is computed using Eq. 14–2. The result can be calculated as the force times the component of displacement in the direction of the force; i.e.,

$$U_P = 400 \text{ N } (2 \text{ m} \cos 30°) = 692.8 \text{ J}$$

or the displacement times the component of force in the direction of displacement, i.e.,

$$U_P = 400 \text{ N} \cos 30°(2 \text{ m}) = 692.8 \text{ J}$$

***Spring Force* $\mathbf{F}_s$.** Since the spring is originally unstretched and in the final position is stretched 2 m, the work of $\mathbf{F}_s$ is

$$U_s = -\tfrac{1}{2}(30 \text{ N/m})(2 \text{ m})^2 = -60 \text{ J}$$

Why is the work negative?

***Weight* W.** Since the weight acts in the opposite direction to its vertical displacement the work is negative; i.e.,

$$U_W = -98.1 \text{ N } (2 \text{ m} \sin 30°) = -98.1 \text{ J}$$

Note that it is also possible to consider the component of weight in the direction of displacement; i.e.,

$$U_W = -(98.1 \text{ N} \sin 30°)2 \text{ m} = -98.1 \text{ J}$$

***Normal Force* $\mathbf{N}_B$.** This force does *no work* since it is *always* perpendicular to the displacement.

Total Work. The work of all the forces when the block is displaced 2 m is thus

$$U_T = 692.8 - 60 - 98.1 = 535 \text{ J} \qquad \textit{Ans.}$$

14.2 Principle of Work and Energy

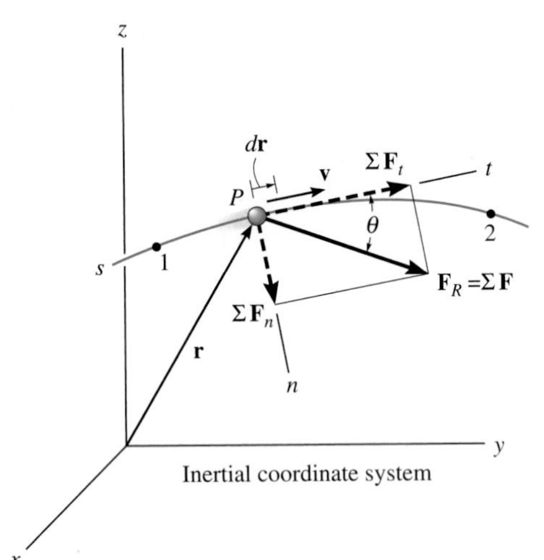

Fig. 14–7

Consider a particle P in Fig. 14–7, which at the instant considered is located at position $\mathbf{r}$ on the path as measured from an inertial coordinate system. If the particle has a mass m and is subjected to a system of external forces represented by the resultant $\mathbf{F}_R = \Sigma\mathbf{F}$, then the equation of motion for the particle is $\Sigma\mathbf{F} = m\mathbf{a}$. When the particle undergoes a displacement $d\mathbf{r}$ along the path, the work done by the forces is

$$\Sigma\mathbf{F}\cdot d\mathbf{r} = m\mathbf{a}\cdot d\mathbf{r}$$

If we establish the n and t axes at the particle, then $\Sigma\mathbf{F}$ can be resolved into its normal and tangential components, Fig. 14–7. Realizing that the magnitude of $d\mathbf{r}$ is ds, it can be seen that $\Sigma\mathbf{F}\cdot d\mathbf{r} = \Sigma F\, ds \cos\theta = \Sigma F_t\, ds$. In other words, the work of $\Sigma\mathbf{F}$ is computed *only from its tangential components.* The normal components $\Sigma F_n = \Sigma F \sin\theta$ do *no work* since the particle cannot be displaced in the normal direction. Since $\mathbf{a}\cdot d\mathbf{r} = a_t\, ds$, the above equation can also be written as

$$\Sigma\mathbf{F}\cdot d\mathbf{r} = \Sigma F_t\, ds = ma_t\, ds$$

Applying the kinematic equation $a_t\, ds = v\, dv$ and integrating both sides, assuming initially that the particle has a position $\mathbf{r} = \mathbf{r}_1$ and a speed $v = v_1$, and later $\mathbf{r} = \mathbf{r}_2$, $v = v_2$, yields

$$\Sigma\int_{\mathbf{r}_1}^{\mathbf{r}_2} \mathbf{F}\cdot d\mathbf{r} = \int_{v_1}^{v_2} mv\, dv$$

or

$$\Sigma\int_{\mathbf{r}_1}^{\mathbf{r}_2} \mathbf{F}\cdot d\mathbf{r} = \tfrac{1}{2}mv_2^2 - \tfrac{1}{2}mv_1^2 \qquad (14\text{–}5)$$

Using Eq. 14–1, the final result may be written as

$$\Sigma U_{1-2} = \tfrac{1}{2}mv_2^2 - \tfrac{1}{2}mv_1^2 \qquad (14\text{–}6)$$

This equation represents the *principle of work and energy* for the particle. The term on the left is the sum of the work done by *all* the forces acting on the particle as the particle moves from point 1 to point 2. The two terms on the right side, which are of the form $T = \frac{1}{2}mv^2$, define the particle's final and initial *kinetic energy,* respectively. These terms are *positive* scalar quantities since they do not depend on the direction of the particle's velocity. Furthermore, Eq. 14–6 must be dimensionally homogeneous so that the kinetic energy has the same units as work, e.g., joules (J) or ft · lb.

When Eq. 14–6 is applied, it is often symbolized in the form

$$T_1 + \Sigma U_{1-2} = T_2 \qquad (14\text{–}7)$$

which states that the particle's initial kinetic energy plus the work done by all the forces acting on the particle as it moves from its initial to its final position is equal to the particle's final kinetic energy.

As noted from the derivation, the principle of work and energy represents an integrated form of $\Sigma F_t = ma_t$, acquired by using the kinematic equation $a_t = v\,dv/ds$. As a result, this principle will provide a convenient *substitution* for $\Sigma F_t = ma_t$ when solving those types of kinetic problems which involve force, velocity, and displacement, since these variables are involved in the terms of Eq. 14–7. For example, if a particle's initial speed is known, and the work of all the forces acting on the particle can be computed, then Eq. 14–7 provides a *direct means* of obtaining the final speed v_2 of the particle after it undergoes a specified displacement. If instead v_2 is determined by means of the equation of motion, a two-step process is necessary; i.e., apply $\Sigma F_t = ma_t$ to obtain a_t, then integrate $a_t = v\,dv/ds$ to obtain v_2. Note that the principle of work and energy cannot be used, for example, to determine forces directed *normal* to the path of motion, since these forces do no work on the particle. Instead $\Sigma F_n = ma_n$ must be applied. For curved paths, however, the magnitude of the normal force is a function of speed. Hence, it may be easier to obtain this speed using the principle of work and energy, and then substitute this quantity into the equation of motion $\Sigma F_n = mv^2/\rho$ to obtain the normal force.

PROCEDURE FOR ANALYSIS

As stated above, the principle of work and energy is used to solve kinetic problems that involve *velocity, force,* and *displacement,* since these terms are involved in the equation. For application it is suggested that the following procedure be used.

Work (Free-Body Diagram). Establish the inertial coordinate system and draw a free-body diagram of the particle in order to account for all the forces that do work on the particle as it moves along its path.

Principle of Work and Energy. Apply the principle of work and energy, $T_1 + \Sigma U_{1-2} = T_2$. The kinetic energy at the initial and final points is always positive, since it involves the speed squared ($T = \frac{1}{2}mv^2$). The work done by each force shown on the free-body diagram is computed by using the appropriate equations developed in Sec. 14.1. Since *algebraic addition* of the work terms is required, it is important that the proper sign of each term be specified. Specifically, work is *positive* when the force component is in the *same direction* as its displacement, otherwise it is negative.

Numerical application of this procedure is illustrated in the examples following Sec. 14.3.

14.3 Principle of Work and Energy for a System of Particles

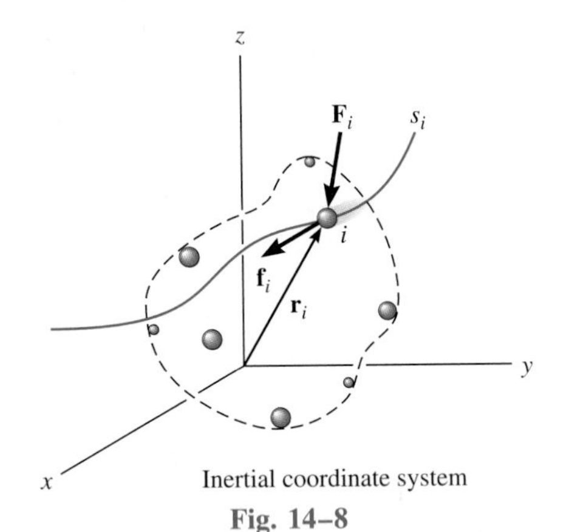

Inertial coordinate system

Fig. 14–8

The principle of work and energy can be extended to include a system of n particles isolated within an enclosed region of space as shown in Fig. 14–8. Here the arbitrary ith particle, having a mass m_i, is subjected to a resultant external force $\mathbf{F}_i$ and a resultant internal force $\mathbf{f}_i = \Sigma_{j=1(j\neq i)}^{n} \mathbf{f}_{ij}$ which each of the other particles exerts on the ith particle. Using Eq. 14–5, the principle of work and energy written for the ith particle is thus

$$\tfrac{1}{2}m_i v_{i1}^2 + \int_{\mathbf{r}_{i1}}^{\mathbf{r}_{i2}} \mathbf{F}_i \cdot d\mathbf{r}_i + \int_{\mathbf{r}_{i1}}^{\mathbf{r}_{i2}} \mathbf{f}_i \cdot d\mathbf{r}_i = \tfrac{1}{2}m_i v_{i2}^2$$

Similar equations result if the principle of work and energy is applied to each of the other particles of the system. Since both work and kinetic energy are scalars, the results may be added together algebraically, so that

$$\Sigma\tfrac{1}{2}m_i v_{i1}^2 + \Sigma\int_{\mathbf{r}_{i1}}^{\mathbf{r}_{i2}} \mathbf{F}_i \cdot d\mathbf{r}_i + \Sigma\int_{\mathbf{r}_{i1}}^{\mathbf{r}_{i2}} \mathbf{f}_i \cdot d\mathbf{r}_i = \Sigma\tfrac{1}{2}m_i v_{i2}^2$$

We can write this equation symbolically as

$$\Sigma T_1 + \Sigma U_{1-2} = \Sigma T_2 \tag{14–8}$$

This equation states that the system's initial kinetic energy (ΣT_1) plus the work done by all the external and internal forces acting on the particles of the system (ΣU_{1-2}) is equal to the system's final kinetic energy (ΣT_2). To maintain this balance of energy, strict accountability of the work done by all the forces must be made. In this regard, note that although the internal forces on adjacent particles occur in equal but opposite collinear pairs, the total work done by each of these forces will, in general, *not cancel* out since the paths over which corresponding particles travel will be *different.* There are, however, two important exceptions to this rule which often occur in practice. If the particles are contained within the boundary of a *translating rigid body,* the internal forces all undergo the same displacement, and therefore the internal work will be zero. Also, particles connected by inextensible cables make up a system that has internal forces which are displaced by an equal amount. In this case, adjacent particles exert equal but opposite internal forces that have components which undergo the same displacement, and therefore the work of these forces cancels. On the other hand, note that if the body is assumed to be *nonrigid,* the particles of the body are displaced along *different paths,* and some of the energy due to force interactions would be given off and lost as heat or stored in the body if permanent deformations occur. We will discuss these effects briefly at the end of this section and in Sec. 15–4. Throughout this text, however, the principle of work and energy will be applied to the solution of problems only when direct accountability of these energy losses does not have to be made.

The procedure for analysis outlined in Sec. 14.2 provides a method for applying Eq. 14–8; however, only one equation applies for the entire system.

If the particles are connected by cords, other equations can generally be obtained by using the kinematic principles outlined in Sec. 12.8 in order to relate the particles' speeds. When doing this, assume that the particles are displaced in the *positive coordinate directions* when writing *both* the kinematic equation and the principle of work and energy (see Example 14–6).

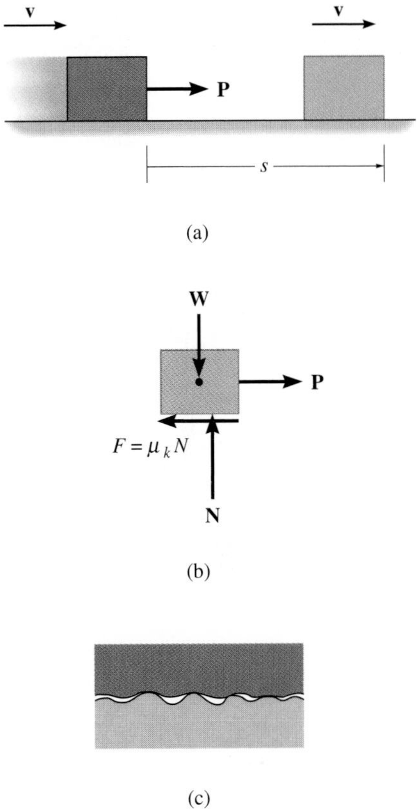

Work of Friction Caused by Sliding. A special class of problems will now be investigated which requires a careful application of Eq. 14–8. These problems all involve cases where a body is sliding over the surface of another body in the presence of friction. Consider, for example, a block which is translating a distance s over a rough surface as shown in Fig. 14–9*a*. If the applied force **P** just balances the *resultant* frictional force $\mu_k N$, Fig. 14–9*b*, then due to equilibrium a constant velocity **v** is maintained and one would expect Eq. 14–8 to be applied as follows:

$$\tfrac{1}{2}mv^2 + Ps - \mu_k Ns = \tfrac{1}{2}mv^2$$

Indeed this equation is satisfied if $P = \mu_k N$; however, as one realizes from experience, the sliding motion will *generate heat,* a form of energy which seems not to be accounted for in the work–energy equation. In order to explain this paradox and thereby more closely represent the nature of friction, we should actually model the block so that both surfaces of contact are *deformable* (nonrigid).* Recall that the rough portions at the bottom of the block act as "teeth," and when the block slides these teeth *deform slightly* and either break off or vibrate due to interlocking effects and pulling away from "teeth" at the contacting surface, Fig. 14–9*c*. As a result, frictional forces that act on the block at these points are displaced slightly, due to the localized deformations, and then they are replaced by other frictional forces as other points of contact are made. At any instant, the *resultant* **F** of all these frictional forces remains essentially constant, i.e., $\mu_k N$; however, due to the localized deformations, the actual displacement s' of $\mu_k N$ is *not* the same displacement s as the applied force **P.** Instead, s' will be *less* than s ($s' < s$), and therefore the *external work* done by the resultant frictional force will be $\mu_k Ns'$ and not $\mu_k Ns$. The remaining amount of work, $\mu_k N(s - s')$, manifests itself as an increase in *internal energy,* which in fact causes the block's temperature to rise.

In summary then, Eq. 14–8 can be applied to problems involving sliding friction; however, it should be fully realized that the work of the resultant frictional force is not represented by $\mu_k Ns$; instead, this term represents *both* the external work of friction ($\mu_k Ns'$) and internal work [$\mu_k N(s - s')$] which is converted into various forms of internal energy, such as heat.†

v v P s

(a)

W P $F = \mu_k N$ N

(b)

(c)

Fig. 14–9

*See Chapter 8 of *Engineering Mechanics: Statics.*

†See B. A. Sherwood and W. H. Bernard, "Work and Heat Transfer in the Presence of Sliding Friction," *Am. J. Phys.* 52, 1001 (1984).

Example 14–2

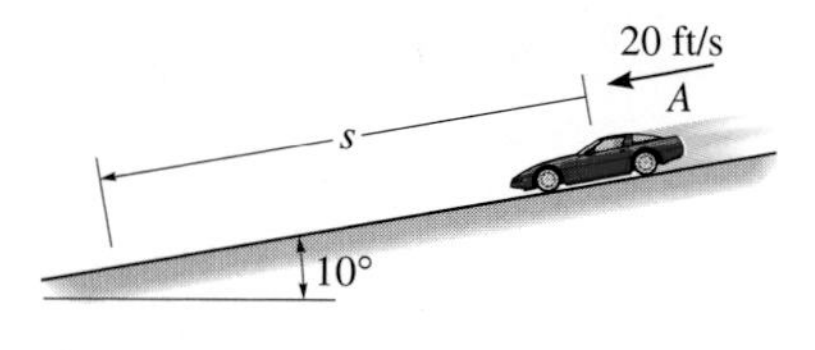

(a)

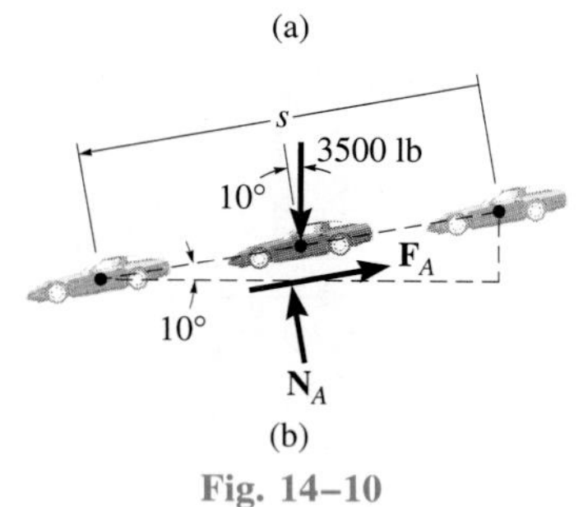

(b)

Fig. 14–10

The 3500-lb automobile shown in Fig. 14–10*a* is traveling down the 10° inclined road at a speed of 20 ft/s. If the driver wishes to stop his car, determine how far s his tires skid on the road if he jams on the brakes, causing his wheels to lock. The coefficient of kinetic friction between the wheels and the road is $\mu_k = 0.5$.

SOLUTION

This problem can be solved using the principle of work and energy, since it involves force, velocity, and displacement.

Work (Free-Body Diagram). As shown in Fig. 14–10*b*, the normal force $\mathbf{N}_A$ does no work since it never undergoes displacement along its line of action. The weight, 3500 lb, is displaced $s \sin 10°$ and does positive work. Why? The frictional force $\mathbf{F}_A$ does both external and internal work when it is *thought* to undergo a displacement s. This work is negative since it is in the opposite direction to displacement. Applying the equation of equilibrium normal to the road, we have

$$+\nwarrow \Sigma F_n = 0; \qquad N_A - 3500 \cos 10° \text{ lb} = 0 \qquad N_A = 3446.8 \text{ lb}$$

Thus,

$$F_A = 0.5 N_A = 1723.4 \text{ lb}$$

Principle of Work and Energy

$$\{T_1\} + \{\Sigma U_{1-2}\} = \{T_2\}$$

$$\left\{\frac{1}{2}\left(\frac{3500 \text{ lb}}{32.2 \text{ ft/s}^2}\right)(20 \text{ ft/s})^2\right\} + \{3500 \text{ lb}(s \sin 10°) - (1723.4 \text{ lb})\, s\} = \{0\}$$

Solving for s yields

$$s = 19.5 \text{ ft} \qquad \textit{Ans.}$$

Note that if this problem is solved by using the equation of motion, *two steps* are involved. First, from the free-body diagram, Fig. 14–10*b*, the equation of motion is applied along the incline. This yields

$$+\swarrow \Sigma F_s = ma_s; \quad 3500 \sin 10° \text{ lb} - 1723.4 \text{ lb} = \frac{3500 \text{ lb}}{32.2 \text{ ft/s}^2} a$$

$$a = -10.3 \text{ ft/s}^2$$

Then, using the integrated form of $a\,ds = v\,dv$ (kinematics), since a is constant, we have

$$(+\swarrow) \quad v^2 = v_0^2 + 2a_c(s - s_0);$$

$$(0)^2 = (-20 \text{ ft/s})^2 + 2(-10.3 \text{ ft/s}^2)(s - 0)$$

$$s = 19.5 \text{ ft} \qquad \textit{Ans.}$$

Example 14–3

A 10-kg block rests on the horizontal surface shown in Fig. 14–11*a*. The spring, which is not attached to the block, has a stiffness $k = 500$ N/m and is initially compressed 0.2 m from C to A. After the block is released from rest at A, determine its velocity when it passes point D. The coefficient of kinetic friction between the block and the plane is $\mu_k = 0.2$.

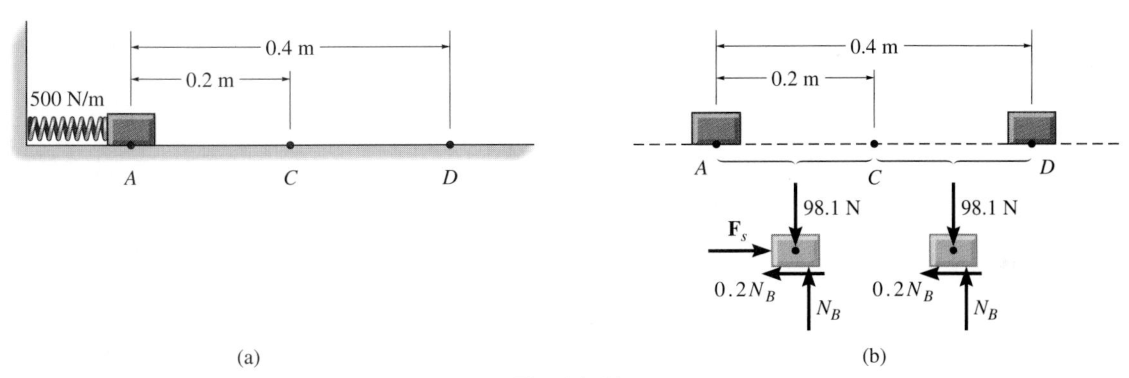

Fig. 14–11

SOLUTION

Why can this problem be solved using the principle of work and energy?

Work (Free-Body Diagrams). Two free-body diagrams for the block are shown in Fig. 14–11*b*. The block moves under the influence of the spring force $\mathbf{F}_s$ along the 0.2-m-long path AC, after which it continues to slide along the plane to point D. With reference to either free-body diagram, $\Sigma F_y = 0$; hence, $N_B = 98.1$ N. Only the spring and friction forces do work during the displacement—the spring force does positive work from A to C, whereas the frictional force not only creates heat but also does negative work. Why?

Principle of Work and Energy

$$\{T_A\} + \{\Sigma U_{A-D}\} = \{T_D\}$$

$$\{\tfrac{1}{2}m(v_A)^2\} + \{\tfrac{1}{2}ks_{AC}^2 - (0.2N_B)s_{AD}\} = \{\tfrac{1}{2}m(v_D)^2\}$$

$$\{0\} + \{\tfrac{1}{2}(500\text{ N/m})(0.2\text{ m})^2 - 0.2(98.1\text{ N})(0.4\text{ m})\} = \{\tfrac{1}{2}(10\text{ kg})(v_D)^2\}$$

Solving for v_D, we get

$$v_D = 0.656\text{ m/s} \rightarrow \qquad \textit{Ans.}$$

Example 14–4

The platform P, shown in Fig. 14–12a, has negligible mass and is tied down so that the 0.4-m-long cords keep the spring compressed 0.6 m when *nothing* is on the platform. If a 2-kg block is placed on the platform and released from rest after the platform is pushed down 0.1 m, determine the maximum height h the block rises in the air, measured from the ground.

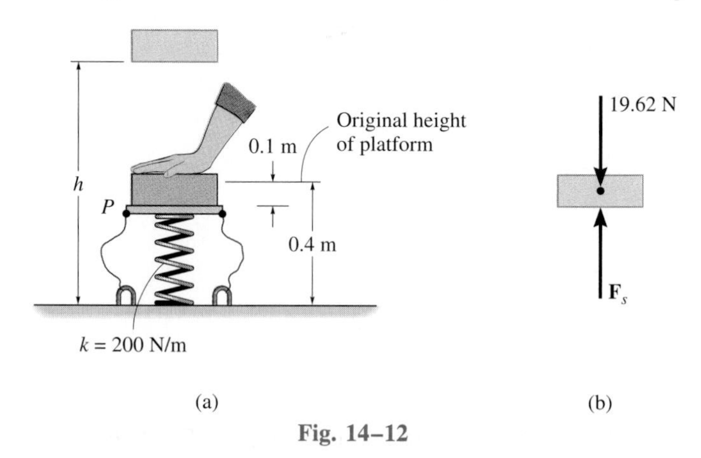

Fig. 14–12

SOLUTION

Work (Free-Body Diagram). Since the block is released from rest and later reaches its maximum height, the initial and final velocities are zero. The free-body diagram of the block when it is still in contact with the platform is shown in Fig. 14–12b. Note that the weight does negative work and the spring force does positive work. Why? In particular, the *initial compression* in the spring is $s_1 = 0.6\text{ m} + 0.1\text{ m} = 0.7\text{ m}$. Due to the cords, the spring's *final compression* is $s_2 = 0.6\text{ m}$ (after the block leaves the platform). The bottom of the block rises from a height of $(0.4\text{ m} - 0.1\text{ m}) = 0.3\text{ m}$ to a final height h.

Principle of Work and Energy

$$\{T_1\} + \{\Sigma U_{1-2}\} = \{T_2\}$$

$$\{\tfrac{1}{2}mv_1^2\} + \{-(\tfrac{1}{2}ks_2^2 - \tfrac{1}{2}ks_1^2) - W\,\Delta y\} = \{\tfrac{1}{2}mv_2^2\}$$

Note that here $s_1 > s_2$, so the work of the spring will indeed be positive. Thus,

$$\{0\} + \{-[\tfrac{1}{2}(200\text{ N/m})(0.6\text{ m})^2 - \tfrac{1}{2}(200\text{ N/m})(0.7\text{ m})^2] - (19.62\text{ N})[h - (0.3\text{ m})]\} = \{0\}$$

Solving yields

$$h = 0.963\text{ m} \qquad \textit{Ans.}$$

Example 14–5

Packages having a mass of 2 kg are delivered from a conveyor to a smooth circular ramp with a velocity of $v_0 = 1$ m/s as shown in Fig. 14–13*a*. If the radius of the ramp is 0.5 m, determine the angle $\theta = \theta_{max}$ at which each package begins to leave the surface.

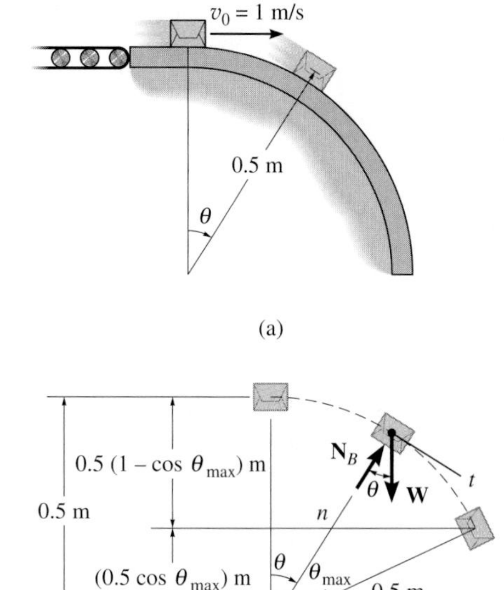

Fig. 14–13

SOLUTION

Work (Free-Body Diagram). Inspection of Fig. 14–13*b* reveals that only the weight $W = 2(9.81) = 19.62$ N does work during the displacement. Is this work positive or negative? If a package is assumed to leave the surface when $\theta = \theta_{max}$ then the weight moves through a vertical displacement of $0.5(1 - \cos\theta_{max})$ m, as shown in the figure.

Principle of Work and Energy

$$\{T_1\} + \{\Sigma U_{1-2}\} = \{T_2\}$$

$$\{\tfrac{1}{2}(2\text{ kg})(1\text{ m/s})^2\} + \{19.62\text{ N}(0.5\text{ m})(1 - \cos\theta_{max})\} = \{\tfrac{1}{2}(2\text{ kg})v_2^2\}$$

$$v_2^2 = 9.81(1 - \cos\theta_{max}) + 1 \qquad (1)$$

Equation of Motion. There are two unknowns in Eq. 1, θ_{max} and v_2. A second equation relating these two variables may be obtained by applying the equation of motion in the *normal direction* to the forces on the free-body diagram. Thus,

$$+\swarrow\Sigma F_n = ma_n; \qquad -N_B + 19.62\text{ N}\cos\theta = (2\text{ kg})\left(\frac{v^2}{0.5\text{ m}}\right)$$

When the package leaves the ramp at $\theta = \theta_{max}$, $N_B = 0$ and $v = v_2$; hence,

$$\cos\theta_{max} = \frac{v_2^2}{4.905} \qquad (2)$$

Eliminating the unknown v_2^2 between Eqs. 1 and 2 gives

$$4.905\cos\theta_{max} = 9.81(1 - \cos\theta_{max}) + 1$$

Solving, we have

$$\cos\theta_{max} = 0.735$$

$$\theta_{max} = 42.7° \qquad \textit{Ans.}$$

This problem has also been solved in Example 13–9. If the two methods of solution are compared, it will be apparent that a work–energy approach yields a more direct solution.

Example 14–6

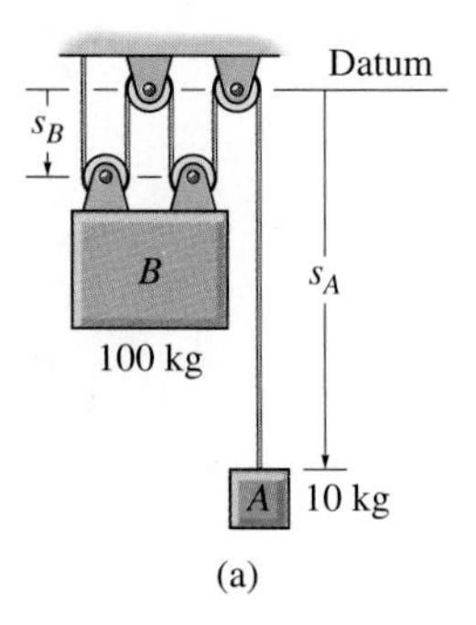

(a)

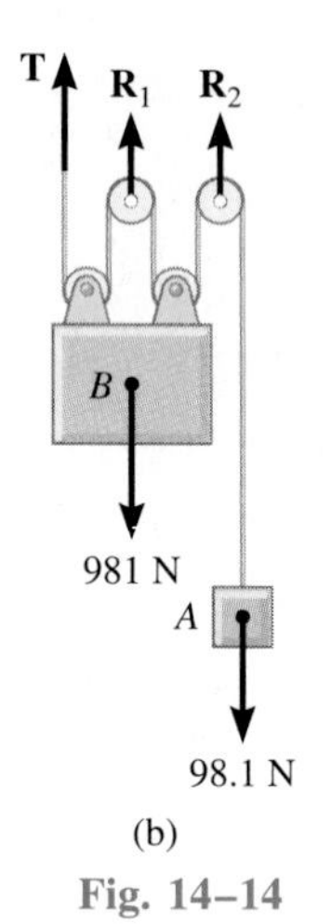

(b)

Fig. 14–14

The blocks A and B shown in Fig. 14–14*a* have a mass of 10 kg and 100 kg, respectively. Determine the distance B travels from the point where it is released from rest to the point where its speed becomes 2 m/s.

SOLUTION

This problem may be solved by considering the blocks separately and applying the principle of work and energy to each block. However, the work of the (unknown) cable tension can be eliminated from the analysis by considering blocks A and B together as a *system.* The solution will require simultaneous solution of the equations of work and energy *and* kinematics. To be consistent with our sign convention, we will assume both blocks move in the positive *downward* direction.

Work (Free-Body Diagram). As shown on the free-body diagram of the system, Fig. 14–14*b*, the cable force **T** and reactions $\mathbf{R}_1$ and $\mathbf{R}_2$ do *no work,* since these forces represent the reactions at the supports and consequently do not move while the blocks are being displaced. The weights both do positive work since, as stated above, they are both assumed to move downward.

Principle of Work and Energy. Realizing the blocks are released from rest, we have

$$\{\Sigma T_1\} + \{\Sigma U_{1-2}\} = \{\Sigma T_2\}$$

$$\{\tfrac{1}{2}m_A(v_A)_1^2 + \tfrac{1}{2}m_B(v_B)_1^2\} + \{W_B\,\Delta s_B + W_A\,\Delta s_A\} = \{\tfrac{1}{2}m_A(v_A)_2^2 + \tfrac{1}{2}m_B(v_B)_2^2\}$$

$$\{0 + 0\} + \{981\text{ N}(\Delta s_B) + 98.1\text{ N}(\Delta s_A)\} =$$

$$\{\tfrac{1}{2}(10\text{ kg})(v_A)_2^2 + \tfrac{1}{2}(100\text{ kg})(2\text{ m/s})^2\} \quad (1)$$

Kinematics. Using the methods of kinematics discussed in Sec. 12.8, it may be seen from Fig. 14–14*a* that at any given instant the total length l of all the vertical segments of cable may be expressed in terms of the position coordinates s_A and s_B as

$$s_A + 4s_B = l$$

Hence, a change in position yields the displacement equation

$$\Delta s_A + 4\,\Delta s_B = 0$$

$$\Delta s_A = -4\,\Delta s_B \quad (2)$$

As required, both of these displacements are positive downward. Taking the time derivative yields

$$v_A = -4v_B = -4(2\text{ m/s}) = -8\text{ m/s}$$

Retaining the negative sign in Eq. 2 and substituting into Eq. 1 yields

$$\Delta s_B = 0.883\text{ m}\downarrow \qquad \textit{Ans.}$$

PROBLEMS

14–1. A woman having a mass of 70 kg stands in an elevator which has a downward acceleration of 4 m/s^2 starting from rest. Determine the work done by her weight and the work of the normal force which the floor exerts on her when the elevator descends 6 m. Explain why the work of these forces is different.

14–2. A bullet moving at a speed of 1000 ft/s has its speed reduced to 900 ft/s when it passes through a board. Determine how many such boards the bullet will penetrate before it stops.

14–3. A hammer has a weight W and is moving with a velocity v when it strikes a nail. If the nail is driven a distance s into a block, determine the average force of resistance. Neglect the mass of the nail, and any energy lost during the striking, and assume the hammer remains in contact with the nail.

***14–4.** The car having a mass of 2 Mg is towed along an inclined road. If the car starts from rest and attains a speed of 5 m/s after traveling a distance of 150 m, determine the constant towing force F applied to the car. Neglect friction and the mass of the wheels.

14–5. The car having a mass of 2 Mg is originally traveling at 2 m/s. Determine the distance it must be towed by a force $F =$ 4 kN in order to attain a speed of 5 m/s. Neglect friction and the mass of the wheels.

Probs. 14–4/14–5

14–6. The "air spring" A is used to protect the support structure B and prevent damage to the conveyor-belt tensioning weight C in the event of a belt failure D. The force developed by the spring as a function of its deflection is shown in the graph. Determine the suspended height d of the 50-lb counterweight C so that if the conveyor belt fails the counterweight will deform the spring $s_{max} = 0.2$ ft. Neglect the mass of the pulley and belt.

14–7. The "air spring" A is used to protect the support structure B and prevent damage to the conveyor-belt tensioning weight C in the event of a belt failure D. The force developed by the spring as a function of its deflection is shown by the graph. If the weight is 50 lb and is suspended a height $d = 1.5$ ft above the top of the spring, determine the maximum deformation of the spring in the event the conveyor belt fails. Neglect the mass of the pulley and belt.

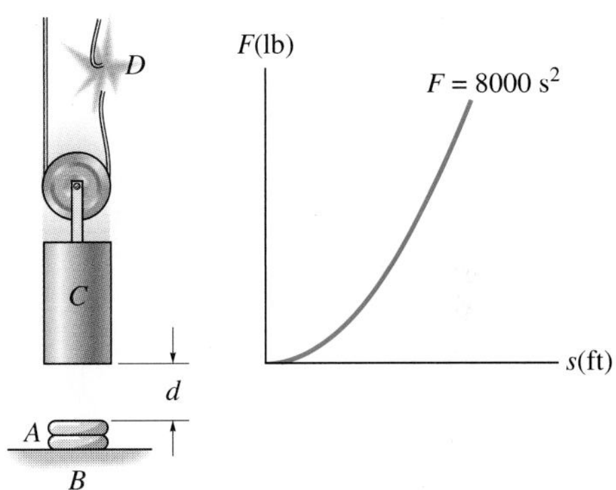

Probs. 14–6/14–7

***14–8.** The bumper B on the 5-Mg train car will deflect in accordance with the load-deflection graph shown, where $k = 15(10^6)$ N/m^2. Determine the maximum deflection s that will occur in order to stop the car if it is traveling at $v = 5$ m/s when it collides with the rigid stop. Neglect the mass of the car wheels.

14–9. Design considerations for the bumper B on the 5-Mg train train car require use of a nonlinear spring having the load-deflection characteristics shown in the graph. Select the proper value of k so that the maximum deflection of the spring is limited to 0.2 m when the car, traveling at 4 m/s, strikes the rigid stop. Neglect the mass of the car wheels.

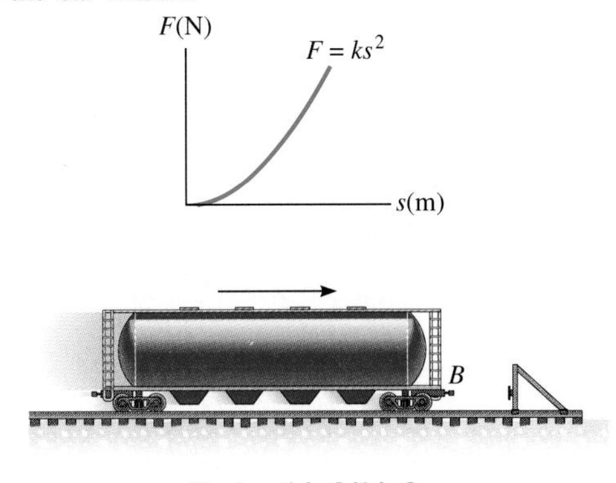

Probs. 14–8/14–9

14–10. The 0.5-kg ball of negligible size is fired up the vertical circular track using the spring plunger. The plunger keeps the spring compressed 0.08 m when $s = 0$. Determine how far s it must be pulled back and released so that the ball will begin to leave the track when $\theta = 135°$.

14–11. A 0.5-kg ball of negligible size is fired up the vertical circular track using the spring plunger. The plunger keeps the spring compressed 0.08 m when $s = 0$. If the plunger is pulled back $s = 0.2$ m and released, determine the speed of the ball when $\theta = 90°$. Also, what is the normal force of the track on the ball at this instant?

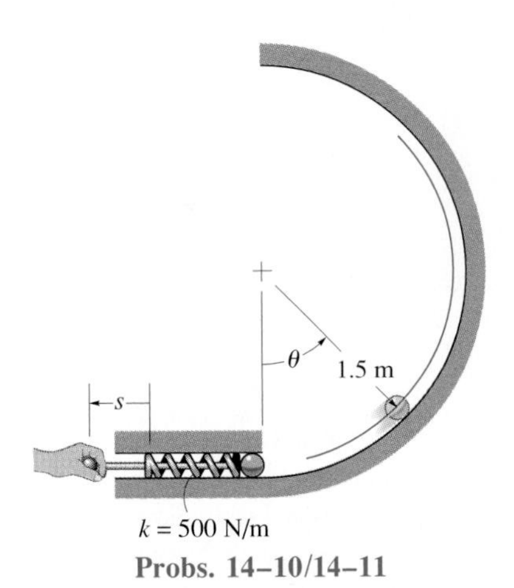

Probs. 14–10/14–11

■***14–12.** The force **F**, acting in a constant direction on the 20-kg block, has a magnitude which varies with the position s of the block. Determine how far the block slides before its velocity becomes 5 m/s. When $s = 0$ the block is moving to the right at 2 m/s. The coefficient of kinetic friction between the block and surface is $\mu_k = 0.3$. Check if it is indeed possible for the block to reach a speed of 5 m/s by finding the distance at which the block stops.

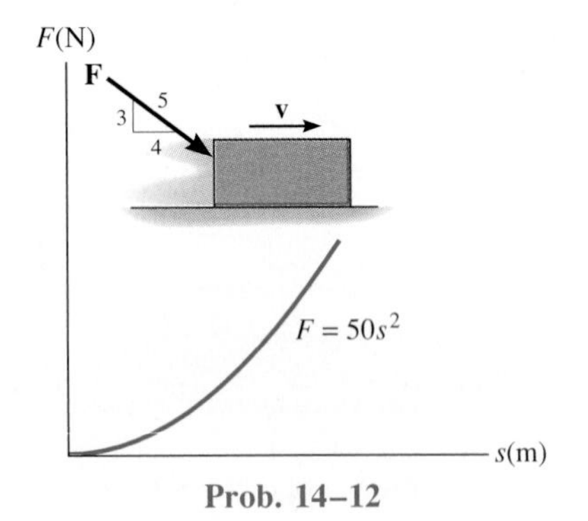

Prob. 14–12

14–13. The force **F**, acting in a constant direction on the 20-kg block, has a magnitude which varies with position s of the block. Determine the speed of the block after it slides 3 m. When $s = 0$ the block is moving to the right at 2 m/s. The coefficient of kinetic friction between the block and surface is $\mu_k = 0.3$. Check if it is indeed possible for the block to slide 3 m as intended.

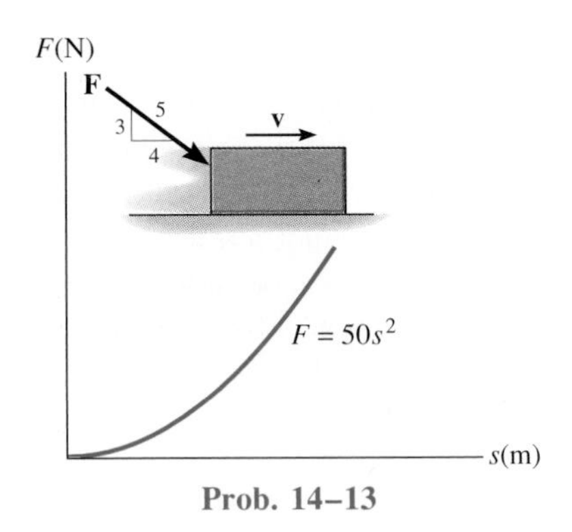

Prob. 14–13

14–14. The 200-kg roller coaster car is accelerated uniformly from rest at A until it reaches a maximum speed at B in $t = 3.5$ s, at which point it begins to travel freely along the spiral loop. Determine this maximum speed at B so that the car just makes the loop without falling off the track. Also, compute the constant horizontal force F needed to give the car the necessary acceleration from A to B. The radius of curvature at C is $\rho_C = 25$ m.

14–15. Determine the height h to the top of the incline D to which the 200-kg roller coaster car will reach, if it is launched at B with a speed just sufficient for it to round the top of the loop at C without leaving the track. The radius of curvature at C is $\rho_C = 25$ m.

Probs. 14–14/14–15

***14–16.** The 2-kg block is subjected to a force having a constant direction and a magnitude $F = (300/(1 + s))$ N, where s is in meters. When $s = 4$ m, the block is moving to the left with a speed of 8 m/s. Determine its speed when $s = 12$ m. The coefficient of kinetic friction between the block and the ground is $\mu_k = 0.25$. Check if it is indeed possible for the block to slide to $s = 12$ m as intended.

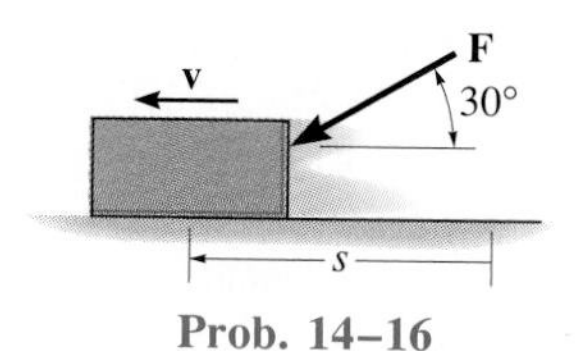

Prob. 14–16

14–17. Block A has a weight of 60 lb and block B has a weight of 10 lb. Determine the speed of block A after it moves 5 ft down the plane, starting from rest. Neglect friction and the mass of the cord and pulleys.

7.18 ft/s

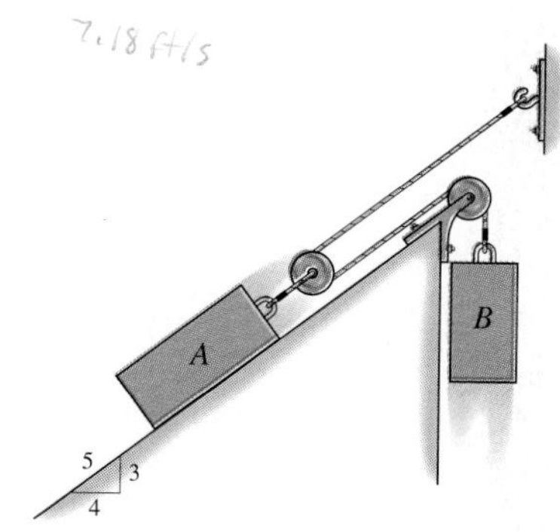

Prob. 14–17

14–18. Block A has a weight of 60 lb and block B has a weight of 10 lb. If the coefficient of kinetic friction between the incline and block A is $\mu_k = 0.2$, determine the speed of block A after it moves 3 ft down the plane, starting from rest. Neglect the mass of the cord and pulleys.

3.52 ft/s

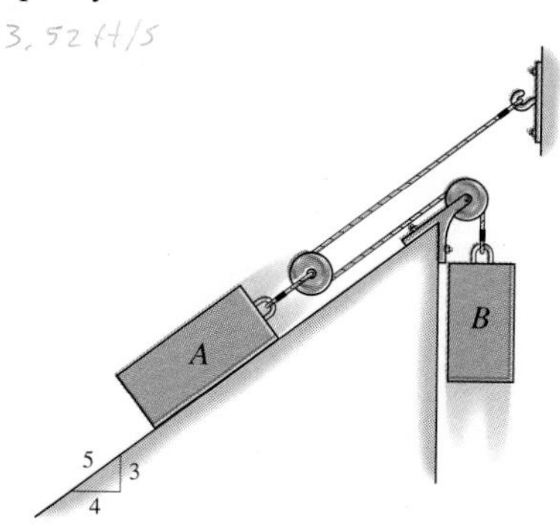

Prob. 14–18

14–19. The 2-kg smooth collar is attached to a spring that has an unstretched length of 3 m. If it is drawn to point B and released from rest, determine its speed when it arrives at point A.

***14–20.** It is required that when the 2-kg smooth collar is drawn back to point B and released from rest, the collar attains a speed of 2 m/s when it arrives at point A. Determine the necessary unstretched length of the spring if it has a stiffness $k = 3$ N/m.

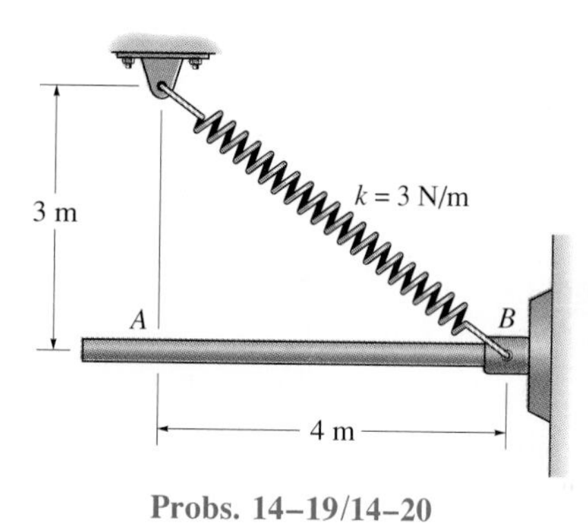

Probs. 14–19/14–20

14–21. Marbles having a mass of 5 g fall from rest at A through the glass tube and accumulate in the can at C. Determine the placement R of the can from the end of the tube and the speed at which the marbles fall into the can. Neglect the size of the can.

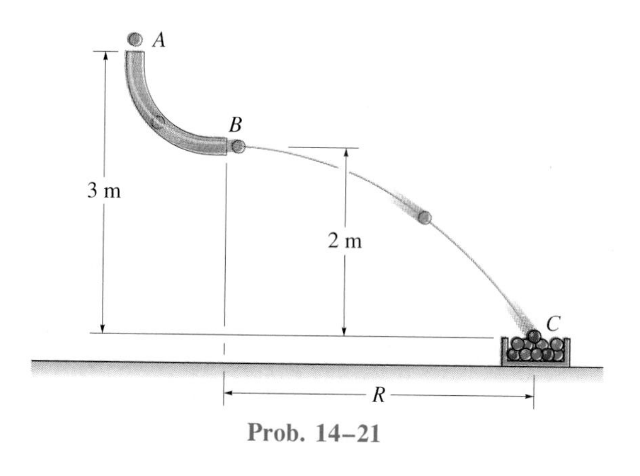

Prob. 14–21

14–22. The steel ingot has a mass of 1800 kg. It travels along the conveyor at a speed $v = 0.5$ m/s when it collides with the "nested" spring assembly. Determine the maximum deflection in each spring needed to stop the motion of the ingot. Take $k_A = 5$ kN/m, $k_B = 3$ kN/m.

14–23. The steel ingot has a mass of 1800 kg. It travels along the conveyor at a speed $v = 0.5$ m/s when it collides with the "nested" spring assembly. If the stiffness of the outer spring is $k_A = 5$ kN/m, determine the required stiffness k_B of the inner spring so that the motion of the ingot is stopped at the moment the front, C, of the ingot is 0.3 m from the wall.

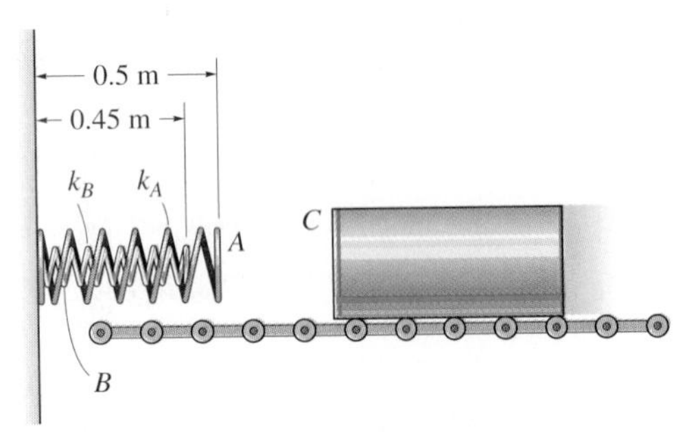

Probs. 14–22/14–23

***14–24.** Block A has a weight of 60 lb and block B has a weight of 10 lb. Determine the speed of A just after it moves downward 3 ft starting from rest. Neglect the mass of the cord and pulleys.

14–25. Block A has a weight of 60 lb and block B has a weight of 10 lb. Determine the distance A must descend from rest before it obtains a speed of 8 ft/s. Also, what is the tension in the cord supporting block A? Neglect the mass of the cord and pulleys.

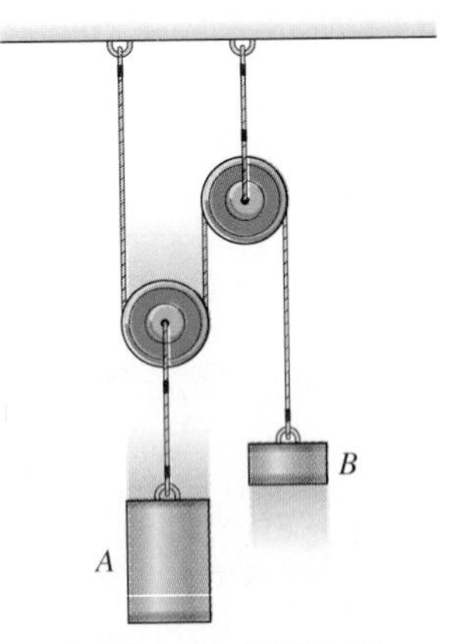

Probs. 14–24/14–25

14–26. The motion of a truck is arrested using a bed of loose stones AB and a set of crash barrels BC. If experiments show that the stones provide a rolling resistance of 160 lb per wheel and the crash barrels provide a resistance as shown in the graph, determine the distance x the 4500-lb truck penetrates the barrels if the truck is coasting at 60 ft/s when it approaches A. Take $s = 50$ ft and neglect the size of the truck.

14–27. The motion of a truck is arrested using a bed of loose stones AB and a set of crash barrels BC. If experiments show that the stones provide a rolling resistance of 160 lb per wheel and the crash barrels provide a resistance as shown in the graph, determine the proper length s of the stone path so that the 4500-lb truck coasting at 60 ft/s when it approaches A, will not penetrate the barrels more than $x = 5.5$ ft before being brought to a stop.

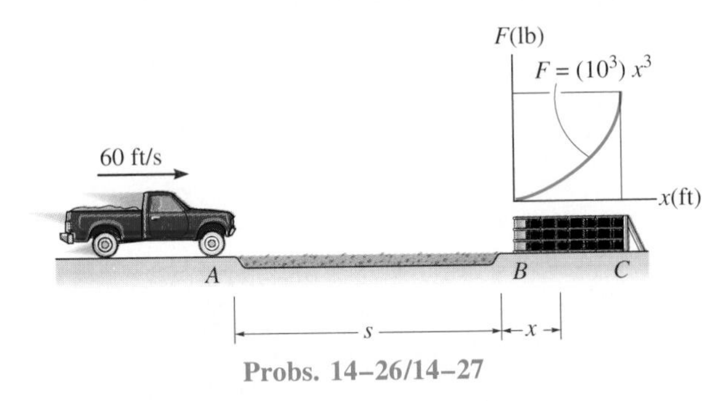

Probs. 14–26/14–27

***14–28.** The crash cushion for a highway barrier consists of a nest of barrels filled with an impact-absorbing material. The barrier stopping force is measured versus the vehicle penetration into the barrier. Determine the distance a car having a weight of 4000 lb will penetrate the barrier if it is originally traveling at 55 ft/s when it strikes the first barrel.

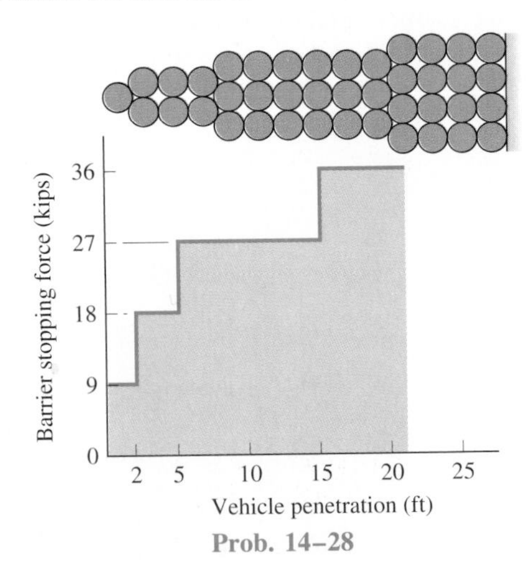

Prob. 14–28

14–29. The catapulting mechanism is used to propel the 10-kg slider A to the right along the smooth track. The propelling action is obtained by drawing the pulley attached to rod BC rapidly to the left by means of a piston P. If the piston applies a constant force $F = 20$ kN to rod BC such that it moves it 0.2 m, determine the speed attained by the slider if it was originally at rest. Neglect the mass of the pulleys, cable, piston, and rod BC.

14–30. The catapulting mechanism is used to propel the 10-kg slider A to the right along the smooth track. The propelling action is obtained by drawing the pulley attached to rod BC rapidly to the left by means of a piston P. If the piston applies a force to rod BC of $F = (30s)$ kN, where s is the displacement in meters, determine the speed attained by the slider when the piston moves 0.1 m, starting from rest. Neglect the mass of the pulleys, cable, and rod BC.

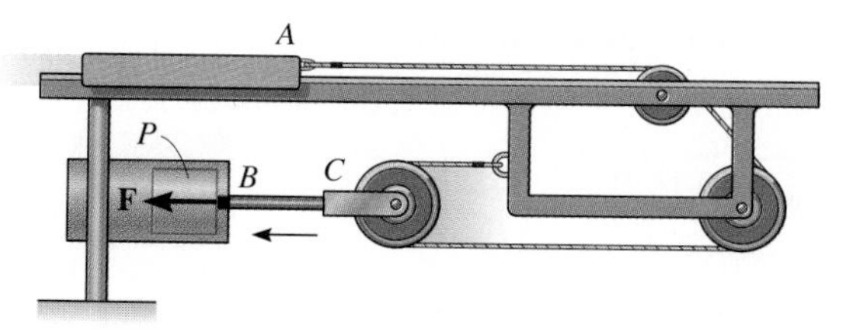

Probs. 14–29/14–30

14–31. The 5-lb cylinder is falling from A with a speed $v_A = 10$ ft/s onto the platform. Determine the maximum displacement of the platform, caused by the collision. The spring has an unstretched length of 1.75 ft and is originally kept in compression by the 1-ft long cables attached to the platform. Neglect the mass of the platform and spring and any energy lost during the collision.

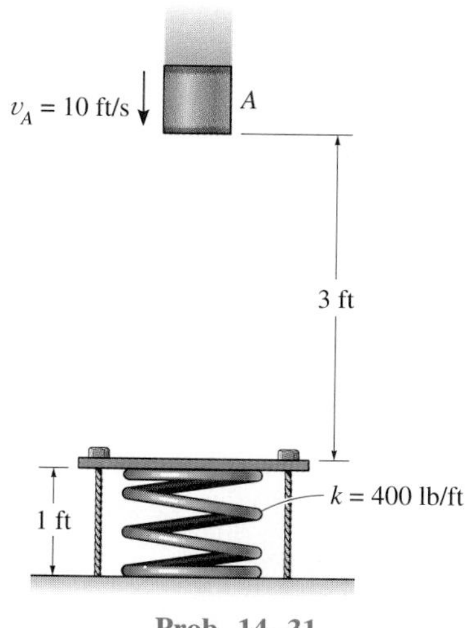

Prob. 14–31

***14–32.** The 100-kg stone is being dragged across the smooth surface by means of the truck T. If the towing cable passes over a small pulley at A, determine the amount of work that must be done by the truck in order to increase the cable angle θ from $\theta_1 = 30°$ to $\theta_2 = 45°$. The truck exerts a constant force $F = 500$ N on the cable at B. Neglect the mass of the pulley and cable.

14–33. The 100-kg stone is being dragged across the smooth surface by means of the truck T. If the towing cable passes over a small pulley at A, determine the speed of the stone when $\theta = 60°$. The stone is at rest when $\theta = 30°$, and the truck exerts a constant force $F = 500$ N on the cable at B.

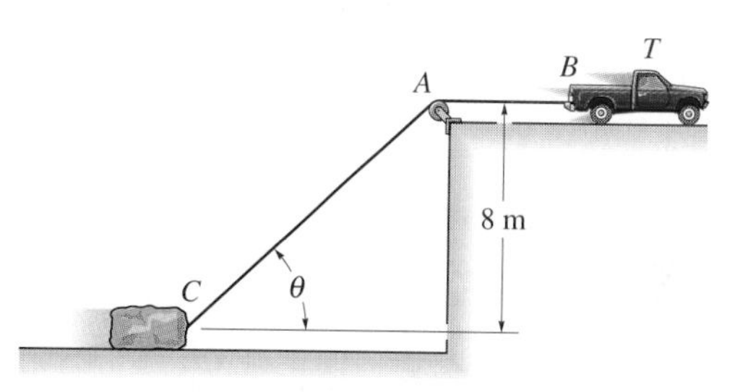

Probs. 14–32/14–33

14–34. The cyclist travels to point A, pedaling until he reaches a speed $v_A = 8$ m/s. He then coasts freely up the curved surface. Determine the normal force he exerts on the surface when he reaches point B. The total mass of the bike and man is 75 kg. Neglect friction, the mass of the wheels, and the size of the bicycle.

14–35. The cyclist travels to point A, pedaling until he reaches a speed $v_A = 4$ m/s. He then coasts freely up the curved surface. Determine how high he reaches up the surface before he comes to a stop. Also, what are the resultant normal force on the surface at this point and his acceleration? The total mass of the bike and man is 75 kg. Neglect friction, the mass of the wheels, and the size of the bicycle.

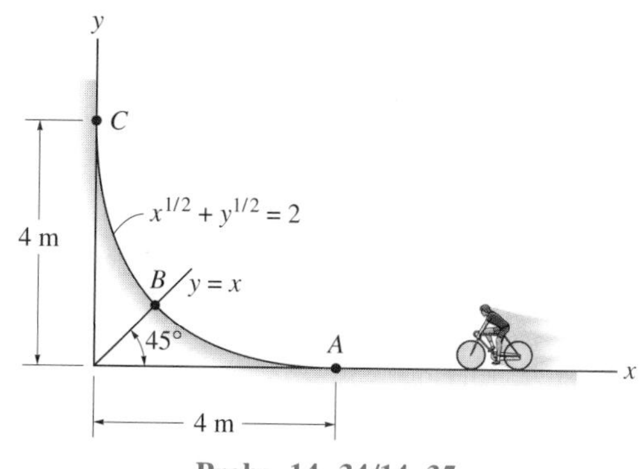

Probs. 14–34/14–35

*14–36. The man at the window A wishes to throw the 30-kg sack on the ground. To do this he allows it to swing from rest at B to point C, when he releases the cord at $\theta = 30°$. Determine the speed at which it strikes the ground and the distance R.

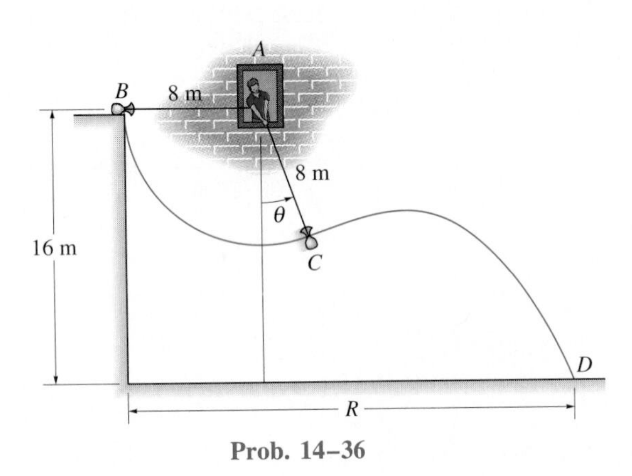

Prob. 14–36

14–37. The collar has a mass of 20 kg and slides along the smooth rod. Two springs are attached to it and the ends of the rod as shown. If each spring has an uncompressed length of 1 m and the collar has a speed of 2 m/s when $s = 0$, determine the maximum compression of each spring due to the back-and-forth (oscillating) motion of the collar.

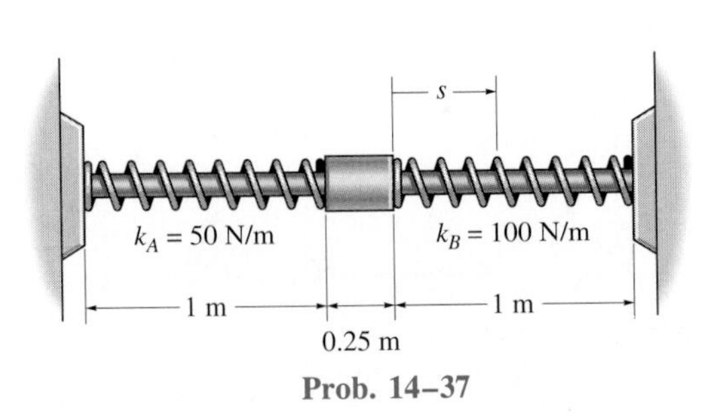

Prob. 14–37

14–38. The spring has a stiffness $k = 50$ lb/ft and an *unstretched length* of 2 ft. As shown, it is confined by the plate P and wall using cables so that its length is 1.5 ft. Determine the speed v_A of the 4-lb block when it is at A, so that it slides along the horizontal plane and strikes the plate and pushes it forward 0.25 ft before momentarily stopping. The coefficient of kinetic friction between the block and plane is $\mu_k = 0.2$. Neglect the mass of the plate and spring and the energy loss between the plate and block during the collision.

14–39. The spring bumper is used to arrest the motion of the 4-lb block, which is sliding toward it at $v = 9$ ft/s. As shown, the spring is confined by the plate P and wall using cables so that its length is 1.5 ft. If the stiffness of the spring is $k = 50$ lb/ft, determine the required unstretched length of the spring so that the plate is not displaced more than 0.2 ft after the block collides into it. Neglect friction, the mass of the plate and spring, and the energy loss between the plate and block during the collision.

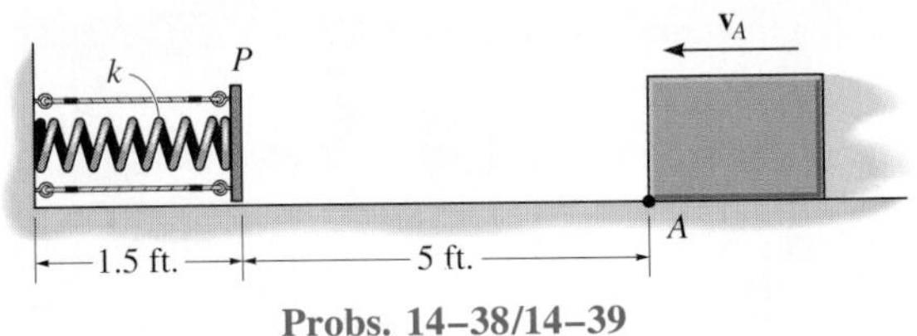

Probs. 14–38/14–39

*14–40. The collar has a mass of 20 kg and is supported on the smooth rod. The attached springs are undeformed when $d = 0.5$ m. Determine the speed of the collar after the applied force $F = 100$ N causes it to be displaced so that $d = 0.3$ m. When $d = 0.5$ m the collar is at rest.

14–41. The collar has a mass of 20 kg and is supported on the smooth rod. The attached springs are both compressed 0.4 m when $d = 0.5$ m. Determine the speed of the collar after the applied force $F = 100$ N causes it to be displaced so that $d = 0.3$ m. When $d = 0.5$ m the collar is at rest.

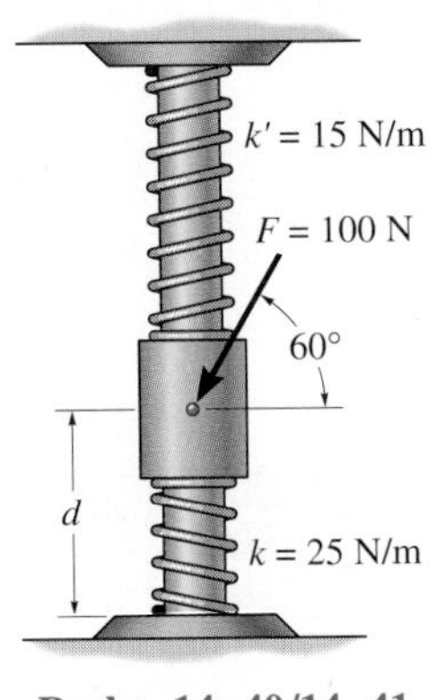

Probs. 14–40/14–41

14.4 Power and Efficiency

Power. *Power* is defined as the amount of work performed per unit of time. Hence, the *power* generated by a machine or engine that performs an amount of work dU within the time interval dt is

$$P = \frac{dU}{dt} \tag{14-9}$$

Provided the work dU is expressed by $dU = \mathbf{F} \cdot d\mathbf{r}$, then it is also possible to write

$$P = \frac{dU}{dt} = \frac{\mathbf{F} \cdot d\mathbf{r}}{dt}$$

or

$$P = \mathbf{F} \cdot \mathbf{v} \tag{14-10}$$

Hence, power is a *scalar,* where in the formulation $\mathbf{v}$ represents the velocity of the point which is acted upon by the force $\mathbf{F}$.

The basic units of power used in the SI and FPS systems are the watt (W) and horsepower (hp), respectively. These units are defined as

$$1 \text{ W} = 1 \text{ J/s} = 1 \text{ N} \cdot \text{m/s}$$
$$1 \text{ hp} = 550 \text{ ft} \cdot \text{lb/s}$$

For conversion between the two systems of units, 1 hp = 746 W.

The term ''power'' provides a useful basis for determining the type of motor or machine which is required to do a certain amount of work in a given time. For example, two pumps may each be able to empty a reservoir if given enough time; however, the pump having the larger power will complete the job sooner.

Efficiency. The *mechanical efficiency* of a machine is defined as the ratio of the output of useful power produced by the machine to the input of power supplied to the machine. Hence,

$$\epsilon = \frac{\text{power output}}{\text{power input}} \tag{14–11}$$

If energy applied to the machine occurs during the *same time interval* at which it is removed, then the efficiency may also be expressed in terms of the ratio of output energy to input energy; i.e.

$$\epsilon = \frac{\text{energy output}}{\text{energy input}} \tag{14–12}$$

Since machines consist of a series of moving parts, frictional forces will always be developed within the machine, and as a result, extra energy or power is needed to overcome these forces. Consequently, *the efficiency of a machine is always less than 1.*

PROCEDURE FOR ANALYSIS

When computing the power supplied to a body, one must first determine the unbalanced external force **F** acting on the body which causes the motion. This force is usually developed by a machine or engine placed either within or external to the body. If the body is accelerating, it may be necessary to draw its free-body diagram and apply the equation of motion ($\Sigma \mathbf{F} = m\mathbf{a}$) to determine **F.** Once **F** and the velocity **v** of the point where **F** is applied have been computed, the power is determined by multiplying the force magnitude by the component of velocity acting in the direction of **F,** i.e., $P = \mathbf{F} \cdot \mathbf{v} = Fv \cos \theta$.

In some problems the power may also be computed by calculating the work done per unit of time ($P_{avg} = \Delta U/\Delta t$, or $P = dU/dt$). Depending on the problem, this work can be done either by the external or internal force of a machine or engine, by the weight of the body, or by an elastic spring force acting on the body.

Example 14–7

The motor M of the hoist shown in Fig. 14–15a operates with an efficiency of 0.85. Determine the power that must be supplied to the motor to lift the 75-lb crate C so that point P on the cable is being drawn in with an acceleration of 4 ft/s^2, and at the instant shown its speed is 2 ft/s. Neglect the mass of the pulley and cable.

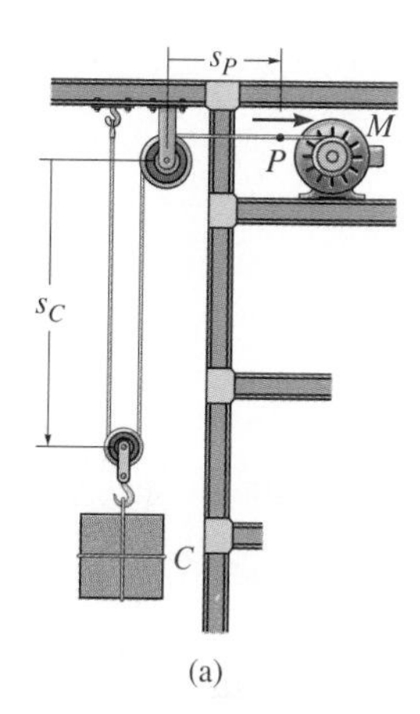

(a)

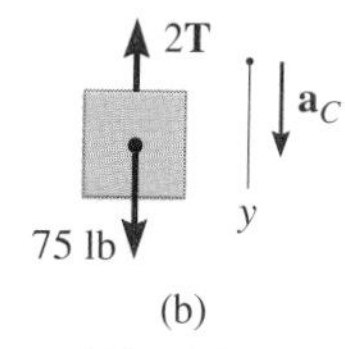

(b)

Fig. 14–15

SOLUTION

In order to compute the power output of the motor, it is first necessary to determine the tension in the cable since this force is developed by the motor.

From the free-body diagram, Fig. 14–15b, we have

$$+\downarrow \Sigma F_y = ma_y; \qquad -2T + 75 \text{ lb} = \frac{75 \text{ lb}}{32.2 \text{ ft/s}^2} a_C \qquad (1)$$

The acceleration of the crate can be obtained by using kinematics to relate the motion of the crate to the known motion of point P, Fig. 14–15a. Hence, by the methods of Sec. 12.8, the coordinates s_C and s_P in Fig. 14–15a can be related to a constant portion of cable length l which is changing in the vertical and horizontal directions. We have $2s_C + s_P = l$. Taking the second time derivative of this equation yields

$$2a_C = -a_P \qquad (2)$$

Since $a_P = +4$ ft/s^2, then $a_C = (-4 \text{ ft/s}^2)/2 = -2$ ft/s^2. What does the negative sign indicate? Substituting this result into Eq. 1 and *retaining* the negative sign since the acceleration in *both* Eqs. 1 and 2 is considered positive downward, we have

$$-2T + 75 \text{ lb} = \frac{75 \text{ lb}}{32.2 \text{ ft/s}^2}(-2 \text{ ft/s}^2)$$

$$T = 39.8 \text{ lb}$$

The power output, measured in units of horsepower, required to draw the cable in at a rate of 2 ft/s is therefore

$$P = \mathbf{T} \cdot \mathbf{v} = (39.8 \text{ lb})(2 \text{ ft/s})(1 \text{ hp}/550 \text{ lb} \cdot \text{ft/s})$$
$$= 0.145 \text{ hp}$$

This *power output* requires that the motor provide a *power input* of

$$\text{power input} = \frac{1}{\epsilon}(\text{power output})$$

$$= \frac{1}{0.85}(0.145 \text{ hp}) = 0.170 \text{ hp} \qquad \textit{Ans.}$$

Since the velocity of the crate is constantly changing, notice that this power requirement is *instantaneous*.

Example 14–8

The sports car shown in Fig. 14–16*a* has a mass of 2 Mg and an engine running efficiency of $\epsilon = 0.63$. As it moves forward, the wind creates a drag resistance on the car of $F_D = 1.2v^2$ N, where v is the velocity in m/s. If the car is traveling at a constant speed of 50 m/s, determine the maximum power supplied by the engine.

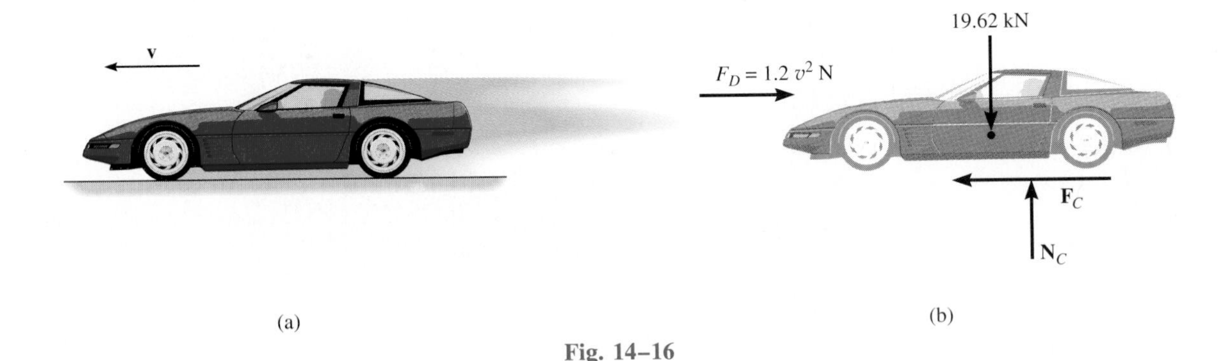

Fig. 14–16

SOLUTION

As shown on the free-body diagram, Fig. 14–16*b*, the normal force $\mathbf{N}_C$ and frictional force $\mathbf{F}_C$ represent the *resultant forces* of all four wheels. In particular, the unbalanced frictional force drives or pushes the car *forward*. This effect is, of course, created by the rotating motion of the rear wheels on the pavement and is developed by the power of the engine.

Applying the equation of motion in the x direction, we have

$$\overset{+}{\leftarrow}\Sigma F_x = ma_x; \qquad F_C - 1.2v^2\text{ N} = (2000\text{ kg})\frac{dv}{dt}$$

Since the car is traveling with *constant velocity, dv/dt* = 0. Hence, with $v = 50$ m/s,

$$F_C = 1.2(50\text{ m/s})^2 = 3000\text{ N}$$

The power output of the car is manifested by the driving (frictional) force $\mathbf{F}_C$. Thus

$$P = \mathbf{F}_C \cdot \mathbf{v} = (3000\text{ N})(50\text{ m/s}) = 150\text{ kW}$$

The power supplied by the engine (power input) is therefore

$$\text{power input} = \frac{1}{\epsilon}(\text{power output}) = \frac{1}{0.63}(150\text{ kW}) = 238\text{ kW} \qquad \textit{Ans.}$$

PROBLEMS

14–42. A spring having a stiffness of 5 kN/m is compressed 400 mm. The stored energy in the spring is used to drive a machine which requires 90 W of power. Determine how long the spring can supply energy at the required rate.

14–43. A train having a weight of $4(10^6)$ lb is traveling at 20 ft/s. If it is brought to a stop, determine how many days a 75-W light bulb must burn to expend the same amount of energy.

***14–44.** The jeep has a weight of 2500 lb and an engine which transmits a power of 100 hp to *all* the wheels. Assuming the wheels do not slip on the ground, determine the angle θ of the largest incline the jeep can climb at a constant speed $v = 30$ ft/s.

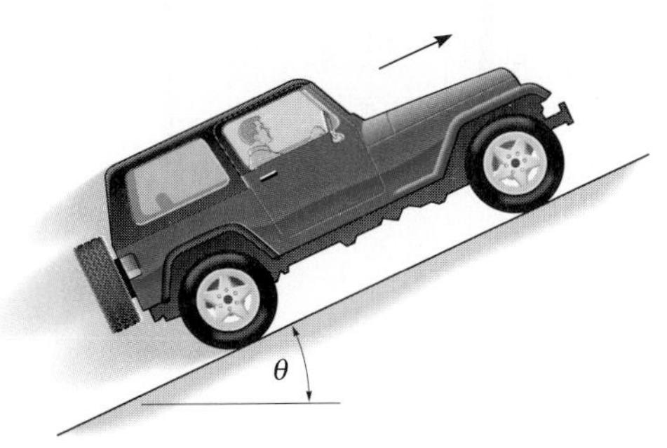

Prob. 14–44

14–45. The jeep has a weight of 2500 lb and travels up the incline, $\theta = 35°$, with a constant speed of 20 ft/s. Determine the power the engine must transmit to the wheels. Assume the wheels do not slip on the ground.

Prob. 14–45

14–46. The locomotive of a train exerts a constant pull of 400 kN on the cars, which have a total mass of 2 Gg. The cars have a total frictional resistance of 6.5 kN and are originally traveling at 2 m/s up a slope of 1°. Determine the speed of the cars after they travel 2 km. What is the power output of the locomotive when it has reached this point?

14–47. An electric streetcar has a weight of 15 000 lb and accelerates along a horizontal straight road from rest such that the power is always 100 hp. Determine how far it must travel to reach a speed of 40 ft/s.

***14–48.** The 50-lb crate is hoisted by the motor M. If the crate starts from rest and, by constant acceleration, attains a speed of 12 ft/s after rising $s = 10$ ft, determine the power that must be supplied to the motor at the instant $s = 10$ ft. The motor has an efficiency $\epsilon = 0.76$. Neglect the mass of the pulley and cable.

14–49. The 50-lb crate is given a speed of 10 ft/s in $t = 4$ s starting from rest. If the acceleration is constant, determine the power that must be supplied to the motor when $t = 2$ s. The motor has an efficiency $\epsilon = 0.76$. Neglect the mass of the pulley and cable.

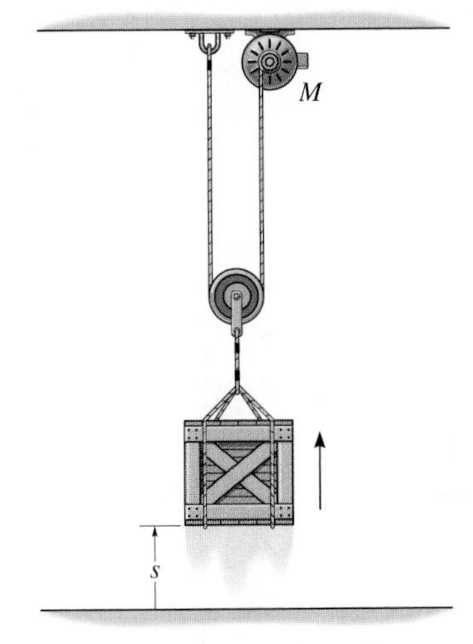

Probs. 14–48/14–49

14–50. The sports car has a mass of 2.3 Mg, and while it is traveling at 28 m/s the driver causes it to accelerate at 5 m/s². If the drag resistance on the car due to the wind is $F_D = (0.3v^2)$ N, where v is the velocity in m/s, determine the power supplied to the engine at this instant. The engine has a running efficiency of $\epsilon = 0.68$.

Prob. 14–50

14–51. The sports car has a mass of 2.3 Mg and accelerates at 6 m/s^2, starting from rest. If the drag resistance on the car due to the wind is $F_D = (10v)$ N, where v is the velocity in m/s, determine the power supplied to the engine when $t = 5$ s. The engine has a running efficiency of $\epsilon = 0.68$.

Prob. 14–51

***14–52.** To dramatize the loss of energy in an automobile, consider a car having a weight of 5 000 lb that is traveling at 35 mi/h. If the car is brought to a stop, determine how long a 100-W light bulb must burn to expend the same amount of energy. (1 mi = 5280 ft.)

14–53. The motor M is used to hoist the 500-kg elevator upward with a constant velocity $v_E = 8$ m/s. If the motor draws 60 kW of electrical power, determine the motor's efficiency. Neglect the mass of the pulleys and cable.

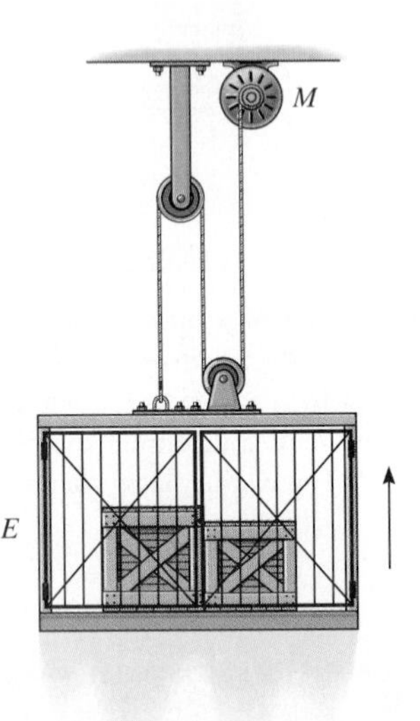

Prob. 14–53

14–54. The 500-kg elevator starts from rest and travels upward with a constant acceleration $a_c = 2$ m/s^2. Determine the power output of the motor M when $t = 3$ s. Neglect the mass of the pulleys and cable.

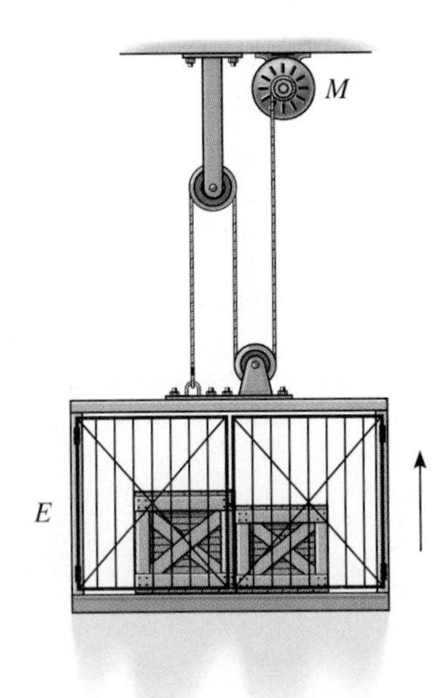

Prob. 14–54

14–55. The 10-lb collar starts from rest at A and is lifted by applying a constant horizontal force of $F = 25$ lb to the cord. If the rod is smooth, determine the power developed by the force at the instant $\theta = 60°$.

***14–56.** The 10-lb collar starts from rest at A and is lifted with a constant speed of 2 ft/s along the smooth rod. Determine the power developed by the force **F** at the instant shown.

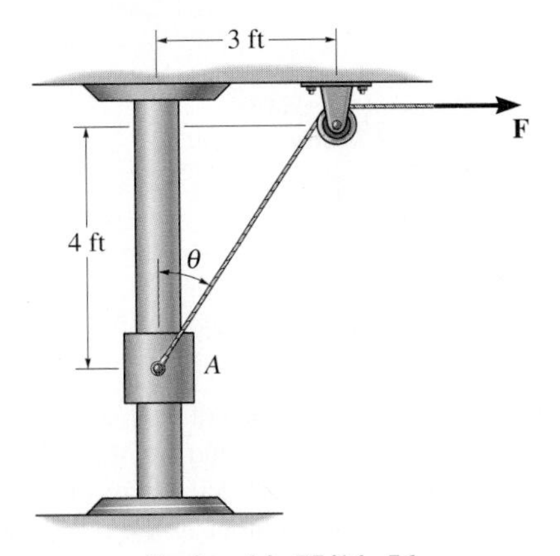

Probs. 14–55/14–56

14–57. The crate has a mass of 150 kg and rests on a surface for which the coefficients of static and kinetic friction are $\mu_s = 0.3$ and $\mu_k = 0.2$, respectively. If the motor M supplies a cable force of $F = (8t^2 + 20)$ N, where t is in seconds, determine the power output developed by the motor when $t = 5$ s.

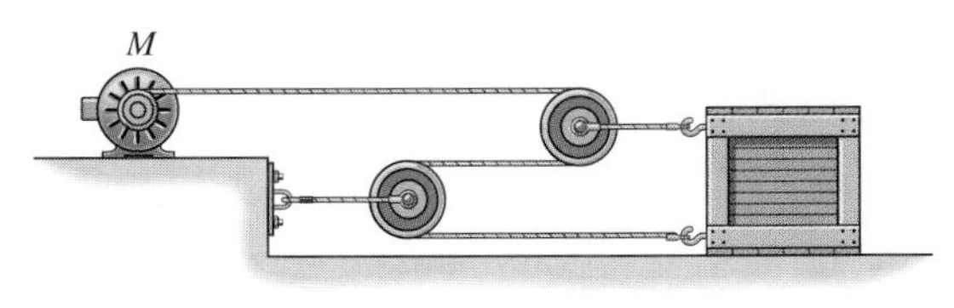

Prob. 14–57

14–58. The 50-lb load is hoisted by the pulley system and motor M. If the crate starts from rest and by constant acceleration attains a speed of 15 ft/s after rising $s = 6$ ft, determine the power that must be supplied to the motor at this instant. The motor has an efficiency $\epsilon = 0.76$. Neglect the mass of the pulleys and cable.

14–59. The 50-lb load is hoisted by the pulley system and motor M. If the motor exerts a constant force of 30 lb on the cable, determine the power that must be supplied to the motor if the load has been hoisted $s = 10$ ft starting from rest. The motor has an efficiency of $\epsilon = 0.76$.

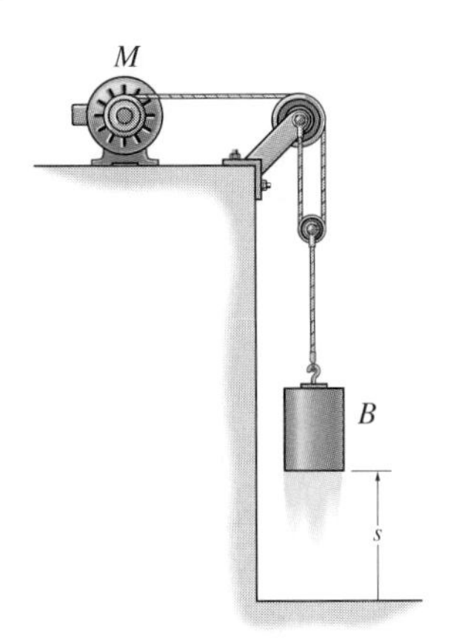

Probs. 14–58/14–59

*__14–60.__ A car has a mass m and accelerates along a horizontal straight road from rest such that the power is always a constant amount P. Determine how far it must travel to reach a speed of v.

14–61. A baseball having a mass of 0.4 kg is thrown such that the force acting on it varies with time as shown in the first graph. Also, the velocity of the ball acting in the same direction as the force varies with time as shown in the second graph. Determine the power applied as a function of time and the work done in $t = 0.3$ s.

14–62. A baseball having a mass of 0.4 kg is thrown such that the force acting on it varies with time as shown in the first graph. Also, the velocity of the ball acting in the same direction as the force varies with time as shown in the second graph. Determine the maximum power developed during the 0.3-second time period.

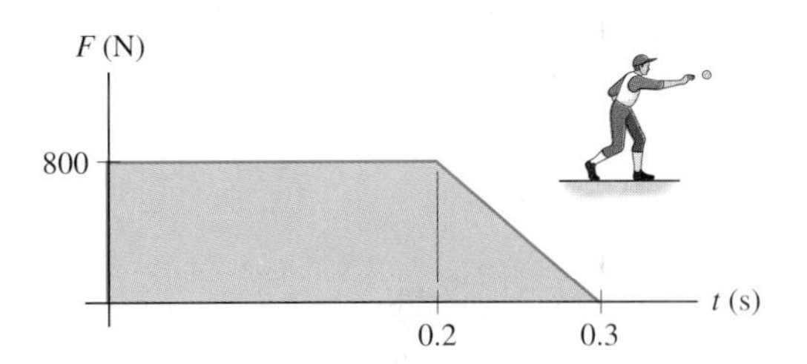

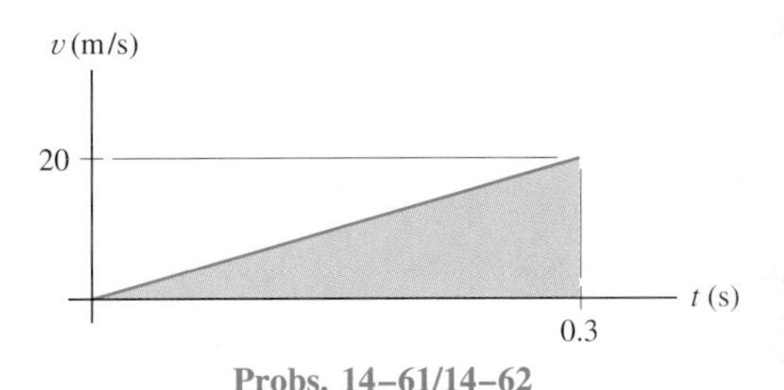

Probs. 14–61/14–62

14–63. The rocket sled has a mass of 4 Mg and travels from rest along the horizontal track for which the coefficient of kinetic friction is $\mu_k = 0.20$. If the engine provides a constant thrust $T = 150$ kN, determine the power output of the engine as a function of time. Neglect the loss of fuel mass and air resistance.

*__14–64.__ The rocket sled has a mass of 4 Mg and travels from rest along the smooth horizontal track such that it maintains a constant power output of 60 kW. Neglect the loss of fuel mass and air resistance, and determine how far it must travel to reach a speed of $v = 60$ m/s.

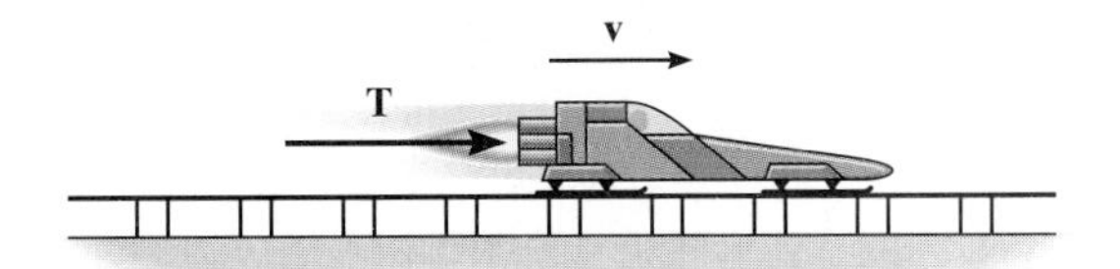

Probs. 14–63/14–64

14.5 Conservative Forces and Potential Energy

Conservative Force. A unique type of force acting on a particle is one that depends *only* on the net *change* in the particle's position and is independent of the particle's velocity and acceleration. Furthermore, if the work done by this force in moving the particle from one point to another is *independent of the path* followed by the particle, this force is called a *conservative force.* The weight of a particle and the force of an elastic spring are two examples of conservative forces often encountered in mechanics.

Weight. The work done by the weight of a particle is *independent of the path;* rather, it depends only on the particle's *vertical displacement.* If this displacement is Δy (positive upward), then from Eq. 14–3,

$$U = -W(\Delta y)$$

Elastic Spring. The work done by a spring force *acting on a particle* is *independent of the path* of the particle, but depends only on the extension or compression s of the spring. If the spring is elongated or compressed from position s_1 to a further position s_2, then from Eq. 14–4,

$$U = -(\tfrac{1}{2}ks_2^2 - \tfrac{1}{2}ks_1^2)$$

Friction. In contrast to a conservative force, consider the force of friction exerted *on a moving object* by a fixed surface. The work done by the frictional force *depends on the path*—the longer the path, the greater the work. Consequently, *frictional forces are nonconservative.* The work is dissipated from the body in the form of heat.

Potential Energy. Energy may be defined as the capacity for doing work. When energy comes from the *motion* of the particle, it is referred to as *kinetic energy.* When it comes from the *position* of the particle, measured from a fixed datum or reference plane, it is called potential energy. Thus, *potential energy* is a measure of the amount of work a conservative force will do when it moves from a given position to the datum. In mechanics, the potential energy due to both gravity (weight) and an elastic spring is important.

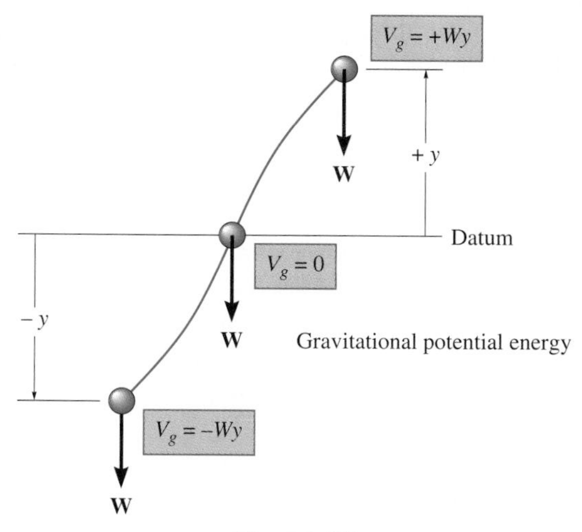

Fig. 14–17

Gravitational Potential Energy. If a particle is located a distance y *above* an arbitrarily selected datum, as shown in Fig. 14–17, the particle's weight **W** has positive *gravitational potential energy,* V_g, since **W** has the capacity of doing positive work when the particle is moved back down to the datum. Likewise, if the particle is located a distance y *below* the datum, V_g is negative since the weight does negative work when the particle is moved back up to the datum. At the datum $V_g = 0$.

In general, if y is *positive upward,* the gravitational potential energy of the particle of weight W is thus*

$$V_g = Wy \tag{14–13}$$

Elastic Potential Energy. When an elastic spring is elongated or compressed a distance s from its unstretched position, the elastic potential energy V_e due to the spring's configuration can be expressed as

$$V_e = +\tfrac{1}{2}ks^2 \tag{14–14}$$

Here V_e is *always positive* since, in the deformed position, the force of the spring has the *capacity* for always doing positive work on the particle when the spring is returned to its unstretched position, Fig. 14–18.

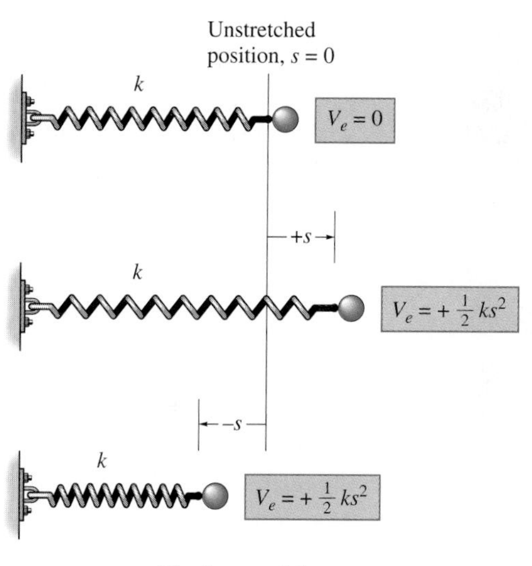

Fig. 14–18

*Here the weight is assumed to be *constant.* This assumption is suitable for small differences in elevation Δy. If the elevation change is significant, however, a variation of weight with elevation must be taken into account (see Prob. 14–92).

Potential Function. In the general case, if a particle is subjected to both gravitational and elastic forces, the particle's potential energy can be expressed as a *potential function,* which is the algebraic sum

$$V = V_g + V_e \tag{14-15}$$

Measurement of V depends on the location of the particle with respect to a selected datum in accordance with Eqs. 14–13 and 14–14.

If the particle is located at an arbitrary point (x, y, z) in space, this potential function is then $V = V(x, y, z)$. The work done by a conservative force in moving the particle from point (x_1, y_1, z_1) to point (x_2, y_2, z_2) is measured by the *difference* of this function, i.e.,

$$U_{1-2} = V_1 - V_2 \tag{14-16}$$

For example, the potential function for a particle of weight W suspended from a spring can be expressed in terms of its position, s, measured from a datum located at the unstretched length of the spring, Fig. 14–19. We have

$$\begin{aligned} V &= V_g + V_e \\ &= -Ws + \tfrac{1}{2}ks^2 \end{aligned}$$

If the particle moves from s_1 to a lower position s_2, then applying Eq. 14–16 it can be seen that the work of $\mathbf{W}$ and $\mathbf{F}_s$ is

$$\begin{aligned} U_{1-2} = V_1 - V_2 &= (-Ws_1 + \tfrac{1}{2}ks_1^2) - (-Ws_2 + \tfrac{1}{2}ks_2^2) \\ &= W(s_2 - s_1) - (\tfrac{1}{2}ks_2^2 - \tfrac{1}{2}ks_1^2) \end{aligned}$$

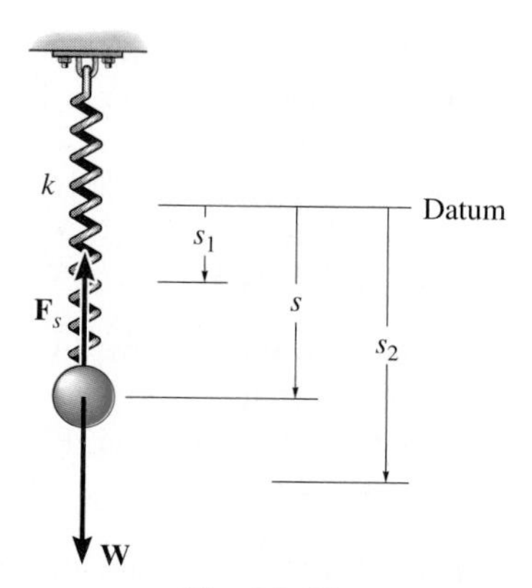

Fig. 14–19

When the displacement along the path is infinitesimal, i.e., from point (x, y, z) to $(x + dx, y + dy, z + dz)$, Eq. 14–16 becomes

$$\begin{aligned} dU &= V(x, y, z) - V(x + dx, y + dy, z + dz) \\ &= -dV(x, y, z) \end{aligned} \tag{14–17}$$

Provided both the force and displacement are defined using rectangular coordinates, then the work can also be expressed as

$$\begin{aligned} dU &= \mathbf{F} \cdot d\mathbf{r} = (F_x\mathbf{i} + F_y\mathbf{j} + F_z\mathbf{k}) \cdot (dx\mathbf{i} + dy\mathbf{j} + dz\mathbf{k}) \\ &= F_x\,dx + F_y\,dy + F_z\,dz \end{aligned}$$

Substituting this result into Eq. 14–17 and expressing the differential $dV(x, y, z)$ in terms of partial derivatives yields

$$F_x\,dx + F_y\,dy + F_z\,dz = -\left(\frac{\partial V}{\partial x}dx + \frac{\partial V}{\partial y}dy + \frac{\partial V}{\partial z}dz\right)$$

Since changes in x, y, and z are all independent of one another, this equation is satisfied provided

$$F_x = -\frac{\partial V}{\partial x}, \qquad F_y = -\frac{\partial V}{\partial y}, \qquad F_z = -\frac{\partial V}{\partial z} \tag{14–18}$$

Thus,

$$\begin{aligned} \mathbf{F} &= -\frac{\partial V}{\partial x}\mathbf{i} - \frac{\partial V}{\partial y}\mathbf{j} - \frac{\partial V}{\partial z}\mathbf{k} \\ &= -\left(\frac{\partial}{\partial x}\mathbf{i} + \frac{\partial}{\partial y}\mathbf{j} + \frac{\partial}{\partial z}\mathbf{k}\right)V \end{aligned}$$

or

$$\mathbf{F} = -\boldsymbol{\nabla}V \tag{14–19}$$

where $\boldsymbol{\nabla}$ (del) represents the vector operator $\boldsymbol{\nabla} = (\partial/\partial x)\mathbf{i} + (\partial/\partial y)\mathbf{j} + (\partial/\partial z)\mathbf{k}$.

Equation 14–19 relates a force $\mathbf{F}$ to its potential function V and thereby provides a mathematical criterion for proving that $\mathbf{F}$ is conservative. For example, the gravitational potential function for a weight located a distance y above a datum is $V_g = Wy$, Eq. 14–13. To prove that $\mathbf{W}$ is conservative, it is necessary to show that it satisfies Eq. 14–19 (or Eq. 14–18), in which case

$$F_y = -\frac{\partial V}{\partial y}; \qquad F = -\frac{\partial}{\partial y}(Wy) = -W$$

The negative sign indicates that $\mathbf{W}$ acts downward, opposite to positive y, which is upward.

14.6 Conservation of Energy

When a particle is acted upon by a system of *both* conservative and nonconservative forces, the portion of the work done by the *conservative forces* can be written in terms of the difference in their potential energies using Eq. 14–16, i.e., $(\Sigma U_{1-2})_{\text{cons.}} = V_1 - V_2$. As a result, the principle of work and energy can be written as

$$T_1 + V_1 + (\Sigma U_{1-2})_{\text{noncons.}} = T_2 + V_2 \tag{14–20}$$

Here $(\Sigma U_{1-2})_{\text{noncons.}}$ represents the work of the nonconservative forces acting on the particle. If *only conservative forces* are applied to the body, this term is zero and then we have

$$T_1 + V_1 = T_2 + V_2 \tag{14–21}$$

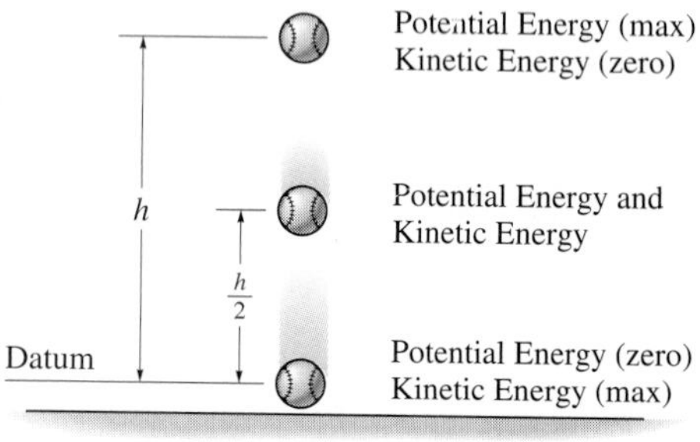

Potential Energy (max)
Kinetic Energy (zero)

Potential Energy and
Kinetic Energy

Potential Energy (zero)
Kinetic Energy (max)

h $\frac{h}{2}$ Datum

Fig. 14–20

This equation is referred to as the *conservation of mechanical energy* or simply the *conservation of energy.* It states that during the motion the sum of the particle's kinetic and potential energies remains *constant.* For this to occur, kinetic energy must be transformed into potential energy, and vice versa. For example, if a ball of weight **W** is dropped from a height h above the ground (datum), Fig. 14–20, the potential energy of the ball is maximum before it is dropped, at which time its kinetic energy is zero. The total mechanical energy of the ball in its initial position is thus

$$E = T_1 + V_1 = 0 + Wh = Wh$$

When the ball has fallen a distance $h/2$, its speed can be determined by using $v^2 = v_0^2 + 2a_c(y - y_0)$, which yields $v = \sqrt{2g(h/2)} = \sqrt{gh}$. The energy of the ball at the mid-height position is therefore

$$E = V_2 + T_2 = W\frac{h}{2} + \frac{1}{2}\frac{W}{g}(\sqrt{gh})^2 = Wh$$

Just before the ball strikes the ground, its potential energy is zero and its speed is $v = \sqrt{2gh}$. Here, again, the total energy of the ball is

$$E = V_3 + T_3 = 0 + \frac{1}{2}\frac{W}{g}(\sqrt{2gh})^2 = Wh$$

Note that when the ball comes in contact with the ground, it deforms somewhat, and provided the ground is hard enough, the ball will rebound off the surface, reaching a new height h', which will be less than the height h from which it was first released. Neglecting air friction, the difference in height accounts for an energy loss, $E_l = W(h - h')$, which occurs during the collision. Portions of this loss produce noise, localized deformation of the ball and ground, and heat.

System of Particles. If a system of particles is *subjected only to conservative forces,* then an equation similar to Eq. 14–21 can be written for the particles. Applying the ideas of the preceding discussion, Eq. 14–8 ($\Sigma T_1 + \Sigma U_{1-2} = \Sigma T_2$) becomes

$$\Sigma T_1 + \Sigma V_1 = \Sigma T_2 + \Sigma V_2 \qquad (14\text{–}22)$$

Here, the sum of the system's initial kinetic and potential energy is equal to the sum of the system's final kinetic and potential energy. In other words, the system's kinetic energy and potential energy, caused by *both* the internal and external forces, remain constant; i.e., $\Sigma T + \Sigma V = \text{const.}$

PROCEDURE FOR ANALYSIS

The conservation of energy equation is used to solve problems involving *velocity, displacement,* and *conservative force systems.* It is generally *easier to apply* than the principle of work and energy. This is because the energy equation just requires specifying the particle's kinetic and potential energy at only *two points* along the path, rather than determining the work when the particle moves through a *displacement.* For application it is suggested that the following procedure be used.

Potential Energy. Draw two diagrams showing the particle located at its initial and final points along the path. If the particle is subjected to a vertical displacement, establish the fixed horizontal datum from which to measure the particle's gravitational potential energy V_g. Although this position can be selected arbitrarily, it is best to locate the datum either at the initial or final point of the path, since at the datum $V_g = 0$. Data pertaining to the elevation y of the particle from the datum and the extension or compression s of any connecting springs can be determined from the geometry associated with the two diagrams. Recall that the potential energy $V = V_g + V_e$. Here $V_g = Wy$, where y is positive upward from the datum, and negative downward from the datum; and $V_e = \frac{1}{2}ks^2$, which is *always positive.*

Conservation of Energy. Apply the equation $T_1 + V_1 = T_2 + V_2$. When computing the kinetic energy, $T = \frac{1}{2}mv^2$, the particle's speed v must be measured from an inertial reference frame.

It is important to remember that only problems involving conservative force systems may be solved by using the conservation of energy theorem. As stated previously, friction or other drag-resistant forces, which depend upon velocity or acceleration, are nonconservative. A portion of the work done by such forces is transformed into thermal energy used to heat up the surfaces of contact, and consequently this energy dissipates into the surroundings and may not be recovered. Therefore, problems involving frictional forces can be solved by using either the principle of work and energy written in the form of Eq. 14–20, if it applies, or the equation of motion.

Example 14–9

The boy and his bicycle shown in Fig. 14–21*a* have a total weight of 125 lb and a center of mass at *G*. If he is coasting, i.e., not pedaling, with a speed of 10 ft/s at the top of the hill *A*, determine the normal force exerted on both wheels of the bicycle when he arrives at *B*, where the radius of curvature of the road is $\rho = 50$ ft. Neglect friction.

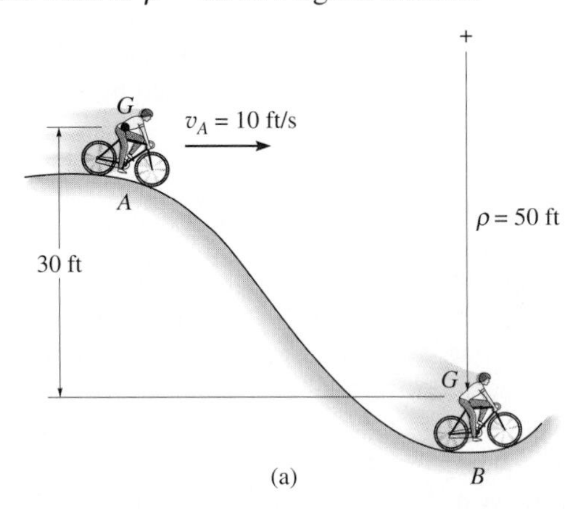

(a)

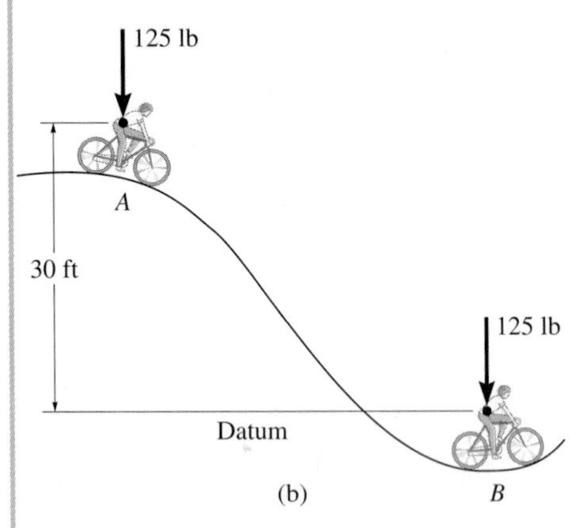

(b)

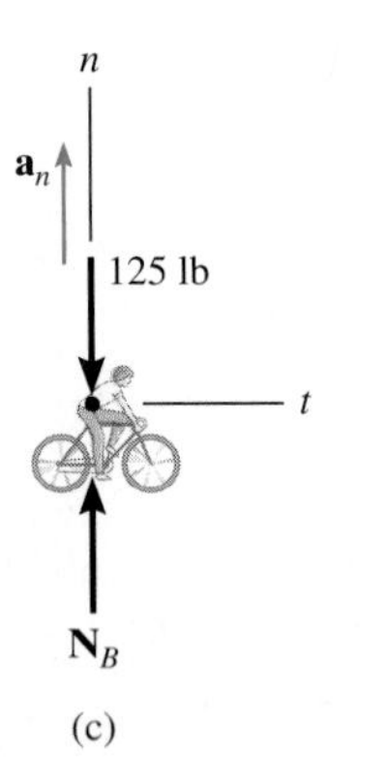

(c)

Fig. 14–21

SOLUTION

Since the normal force does *no work,* it must be obtained using the equation of motion, $\Sigma F_n = m(v^2/\rho)$. We can, however, determine the bicycle's speed at *B* using the conservation of energy equation. Why?

Potential Energy. Figure 14–21*b* shows the bicycle at points *A* and *B*. For convenience, the potential-energy datum has been established through the center of mass when the bicycle is located at *B*.

Conservation of Energy

$$\{T_A\} + \{V_A\} = \{T_B\} + \{V_B\}$$

$$\left\{\frac{1}{2}\left(\frac{125\text{ lb}}{32.2\text{ ft/s}^2}\right)(10\text{ ft/s})^2\right\} + \{(125\text{ lb})(30\text{ ft})\} = \left\{\frac{1}{2}\left(\frac{125\text{ lb}}{32.2\text{ ft/s}^2}\right)(v_B)^2\right\} + \{0\}$$

$$v_B = 45.1\text{ ft/s}$$

Equation of Motion. Using the data tabulated on the free-body diagram when the bicycle is at *B*, Fig. 14–21*c*, we have

$$+\uparrow \Sigma F_n = ma_n; \quad N_B - 125 = \frac{125\text{ lb}}{32.2\text{ ft/s}^2}\frac{(45.1\text{ ft/s})^2}{(50\text{ ft})}$$

$$N_B = 283\text{ lb}$$ ***Ans.***

Example 14–10

The ram R shown in Fig. 14–22a has a mass of 100 kg and is released from rest 0.75 m from the top of a spring, A, that has a stiffness $k_A = 12$ kN/m. If a second spring B, having a stiffness $k_B = 15$ kN/m, is "nested" in A, determine the maximum displacement of A needed to stop the downward motion of the ram. The unstretched length of each spring is indicated in the figure. Neglect the mass of the springs.

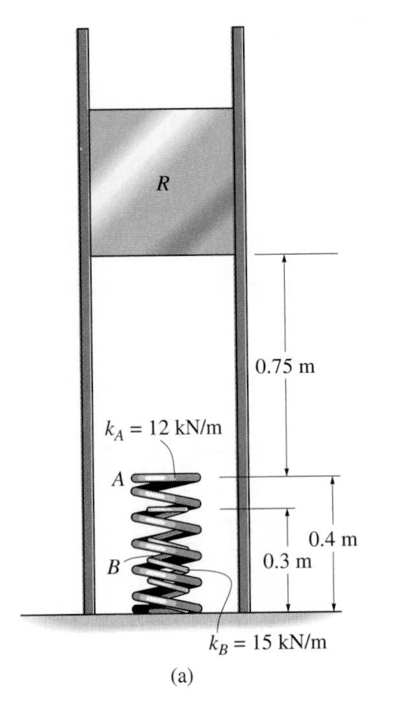

(a)

SOLUTION

Potential Energy. We will *assume* that the ram compresses *both* springs at the instant it comes to rest. The datum is located through the center of gravity of the ram at its initial position, Fig. 14–22b. When the kinetic energy is reduced to zero ($v_2 = 0$), A is compressed a distance s_A so that B compresses $s_B = s_A - 0.1$ m.

Conservation of Energy

$$T_1 + V_1 = T_2 + V_2$$

$$\{0\} + \{0\} = \{0\} + \{\tfrac{1}{2}k_A s_A^2 + \tfrac{1}{2}k_B(s_A - 0.1)^2 - Wh\}$$

$$\{0\} + \{0\} = \{0\} + \{\tfrac{1}{2}(12\,000 \text{ N/m})s_A^2 + \tfrac{1}{2}(15\,000 \text{ N/m})(s_A - 0.1 \text{ m})^2 - 981 \text{ N}(0.75 \text{ m} + s_A)\}$$

Rearranging the terms,

$$13\,500 s_A^2 - 2481 s_A - 660.75 = 0$$

Using the quadratic formula and solving for the positive root,* we have

$$s_A = 0.331 \text{ m} \qquad \textit{Ans.}$$

Since $s_B = 0.331$ m $- 0.1$ m $= 0.231$ m, which is positive, the assumption that *both* springs are compressed by the ram is correct.

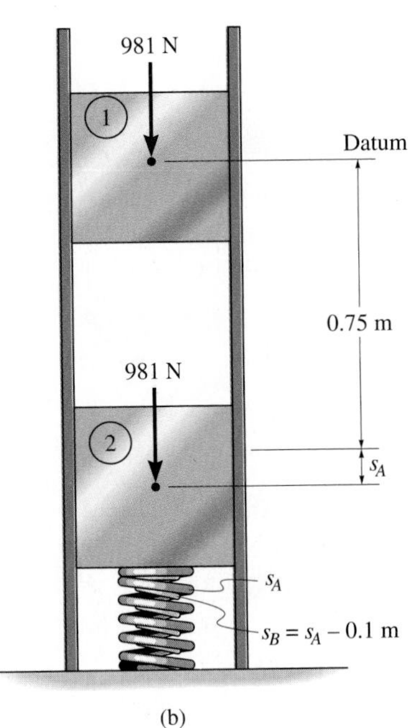

(b)

Fig. 14–22

*The second root, $s_A = -0.148$ m, does not represent the physical situation. Since positive s is measured downward, the negative sign indicates that spring A would have to be "extended" by an amount of 0.148 m to stop the ram.

Example 14–11

A smooth 2-kg collar C, shown in Fig. 14–23a, fits loosely on the vertical shaft. If the spring is unstretched when the collar is in the position A, determine the speed at which the collar is moving when $y = 1$ m, if *(a)* it is released from rest at A, and *(b)* it is released at A with an *upward* velocity $v_A = 2$ m/s.

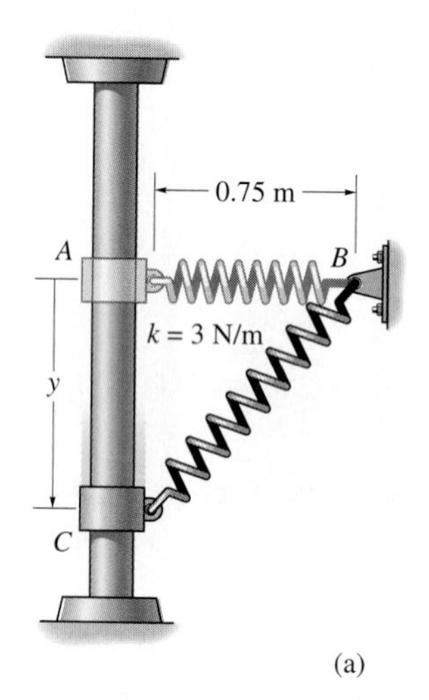

SOLUTION

Part (a)

Potential Energy. For convenience, the datum is established through AB, Fig. 14–23b. When the collar is at C, the gravitational potential energy is $-(mg)y$, since the collar is *below* the datum, and the elastic potential energy is $\frac{1}{2}ks_{CB}^2$. Here $s_{CB} = 0.5$ m, which represents the *stretch* in the spring as computed in the figure.

Conservation of Energy

$$\{T_A\} + \{V_A\} = \{T_C\} + \{V_C\}$$

$$\{0\} + \{0\} = \{\tfrac{1}{2}mv_C^2\} + \{\tfrac{1}{2}ks_{CB}^2 - mgy\}$$

$$\{0\} + \{0\} = \{\tfrac{1}{2}(2\text{ kg})v_C^2\} + \{\tfrac{1}{2}(3\text{ N/m})(0.5\text{ m})^2 - 2(9.81)\text{ N}(1\text{ m})\}$$

$$v_C = 4.39\text{ m/s}\downarrow \qquad \textit{Ans.}$$

This problem can also be solved by using the equation of motion or the principle of work and energy. Note that in *both* of these methods the variation of the magnitude and direction of the spring force must be taken into account (see Example 13–4). Here, however, the above method of solution is clearly advantageous since the calculations depend *only* on data calculated at the initial and final points of the path.

Part (b)

Conservation of Energy. If $v_A = 2$ m/s, using the data in Fig. 14–23b, we have

$$\{T_A\} + \{V_A\} = \{T_C\} + \{V_C\}$$

$$\{\tfrac{1}{2}mv_A^2\} + \{0\} = \{\tfrac{1}{2}mv_C^2\} + \{\tfrac{1}{2}ks_{CB}^2 - mgy\}$$

$$\{\tfrac{1}{2}(2\text{ kg})(2\text{ m/s})^2\} + \{0\} = \{\tfrac{1}{2}(2\text{ kg})v_C^2\} + \{\tfrac{1}{2}(3\text{ N/m})(0.5\text{ m})^2 - 2(9.81)\text{ N}(1\text{ m})\}$$

$$v_C = 4.82\text{ m/s}\downarrow \qquad \textit{Ans.}$$

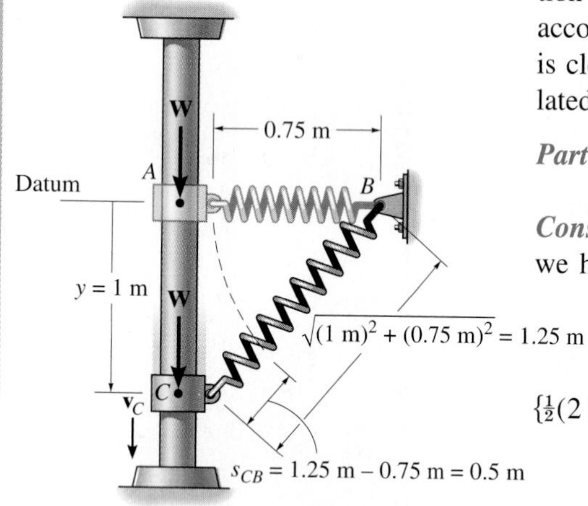

Fig. 14–23

Note that the kinetic energy of the collar depends only on the *magnitude* of velocity, and therefore it is immaterial if the collar is moving up or down at 2 m/s when released at A.

PROBLEMS

14–65. Solve Prob. 14–10 using the conservation of energy equation.

14–66. Solve Prob. 14–15 using the conservation of energy equation.

14–67. Solve Prob. 14–17 using the conservation of energy equation.

***14–68.** Solve Prob. 14–19 using the conservation of energy equation.

14–69. Solve Prob. 14–22 using the conservation of energy equation.

14–70. Solve Prob. 14–31 using the conservation of energy equation.

14–71. Two equal-length springs having stiffnesses $k_A = 300$ N/m and $k_B = 200$ N/m are ''nested'' together in order to form a shock absorber. If a 2-kg block is dropped from an at-rest position $s = 0.6$ m above the top of the springs, determine their maximum deformation required to stop the motion of the block.

***14–72.** Two equal-length springs are ''nested'' together in order to form a shock absorber. If it is designed to arrest the motion of a 2-kg mass that is dropped $s = 0.5$ m above the top of the springs from an at-rest position, and the maximum compression of the springs is to be 0.2 m, determine the required stiffness of the inner spring, k_B, if the outer spring has a stiffness $k_A = 400$ N/m.

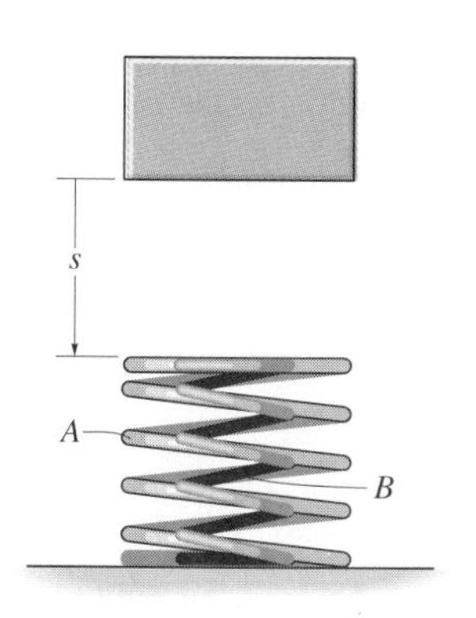

Probs. 14–71/14–72

14–73. The girl has a mass of 40 kg and center of mass at G. If she is swinging to a maximum height defined by $\theta = 60°$, determine the force developed along each of the four supporting posts such as AB at the instant $\theta = 0°$. The swing is centrally located between the posts.

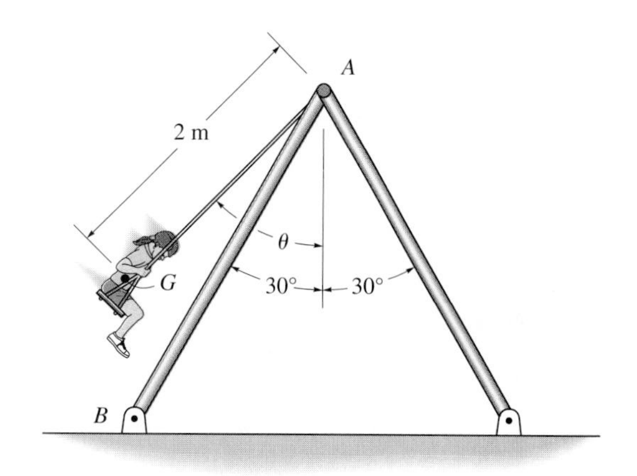

Prob. 14–73

14–74. The collar has a weight of 8 lb. If it is pushed down so as to compress the spring 2 ft and then released from rest ($h = 0$), determine its speed when it is displaced $h = 4.5$ ft. The spring is not attached to the collar. Neglect friction.

14–75. The collar has a weight of 8 lb. If it is released from rest at a height of $h = 2$ ft from the top of the uncompressed spring, determine the speed of the collar after it falls and compresses the spring 0.3 ft.

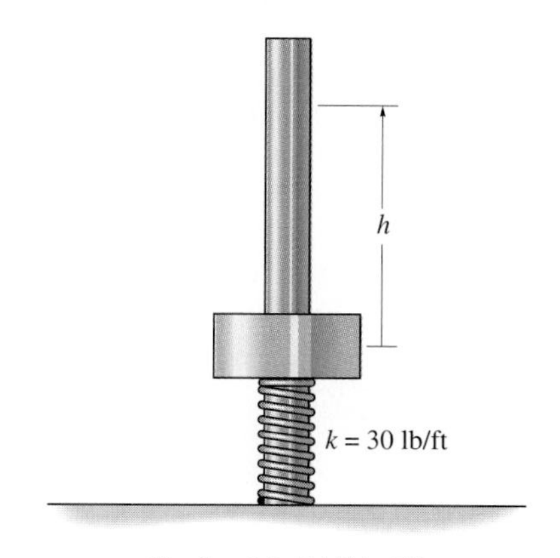

Probs. 14–74/14–75

***14–76.** The 5-lb collar is released from rest at A and travels along the smooth guide. Determine the speed of the collar just before it strikes the stop at B. The spring has an unstretched length of 12 in.

14–77. The 5-lb collar is released from rest at A and travels along the smooth guide. Determine its speed when its center reaches point C and the normal force it exerts on the rod at this point. The spring has an unstretched length of 12 in., and point C is located just before the end of the curved portion of the rod.

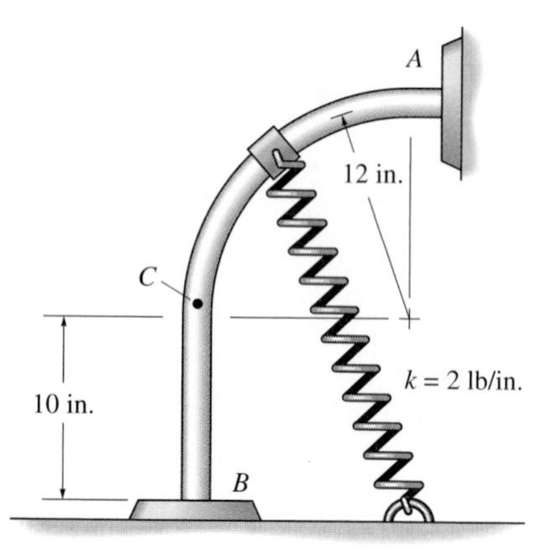

Probs. 14–76/14–77

14–78. The 20-lb collar is constrained to move on the smooth rod. It is attached to the three springs which are unstretched when $s = 0$. If the collar is displaced $s = 0.5$ ft and released from rest, determine its speed when $s = 0$.

14–79. The 20-lb collar is constrained to move on the smooth rod. It is attached to the three springs which are unstretched when $s = 0$. If the collar is given a velocity of 2 ft/s to the right when $s = 0$, determine the maximum compression of each spring as the collar travels back and forth (oscillates) on the rod.

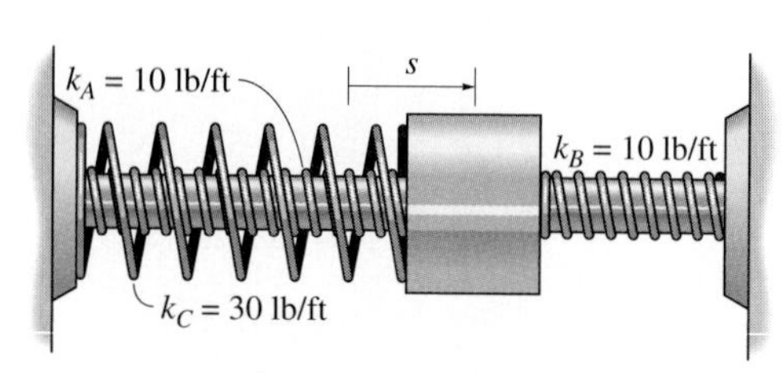

Probs. 14–78/14–79

***14–80.** Each of the two elastic rubberbands of the slingshot has an unstretched length of 200 mm. If they are pulled back to the position shown and released from rest, determine the speed of the 25-g pellet just after the rubberbands become unstretched. Neglect the mass of the rubberbands and the change in elevation of the pellet while it is constrained by the rubberbands. Each rubberband has a stiffness of $k = 50$ N/m.

14–81. Each of the two elastic rubberbands of the slingshot has an unstretched length of 200 mm. If they are pulled back to the position shown and released from rest, determine the maximum height the 25-g pellet will reach if it is fired vertically upward. Neglect the mass of the rubberbands and the change in elevation of the pellet while it is constrained by the rubberbands. Each rubberband has a stiffness $k = 50$ N/m.

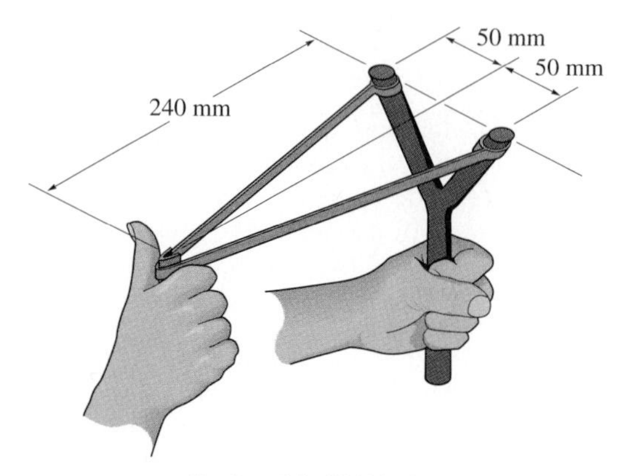

Probs. 14–80/14–81

14–82. The roller-coaster car has a mass of 800 kg, including its passenger, and starts from the top of the hill A with a speed $v_A = 3$ m/s. Determine the minimum height h of the hill crest so that the car travels around both inside loops without leaving the track. Neglect friction, the mass of the wheels, and the size of the car. What is the normal reaction on the car when the car is at B and when it is at C?

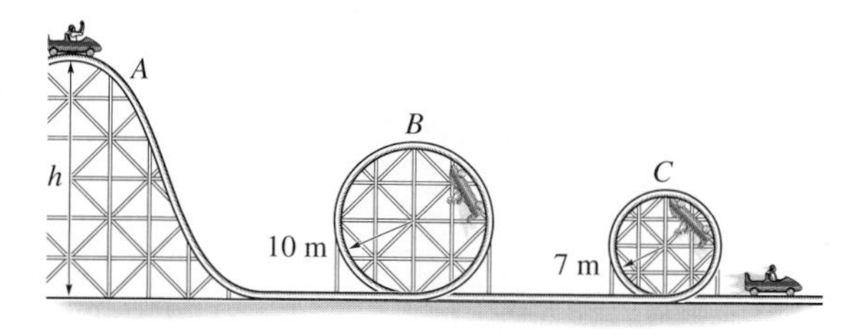

Prob. 14–82

14–83. The roller-coaster car has a mass of 800 kg, including its passenger. If it is released from rest at the top of the hill A, determine the minimum height h of the hill crest so that the car travels around both inside loops without leaving the track. Neglect friction, the mass of the wheels, and the size of the car. What is the normal reaction on the car when the car is at B and when it is at C?

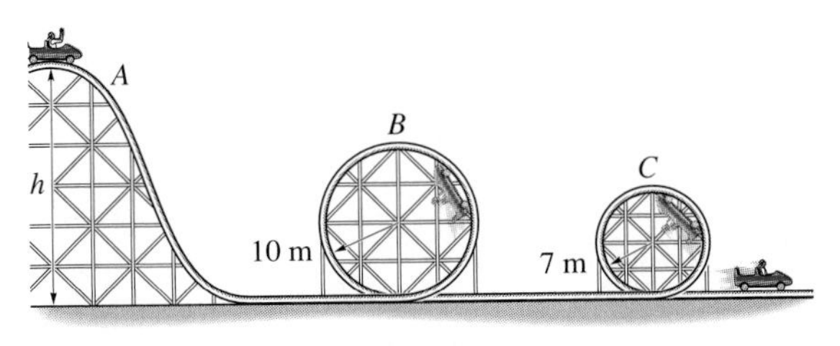

Prob. 14–83

***14–84.** The spring has a stiffness $k = 3$ lb/ft and an unstretched length of 2 ft. If it is attached to the 5-lb smooth collar and the collar is released from rest at A, determine the speed of the collar just before it strikes the end of the rod at B. Neglect the size of the collar.

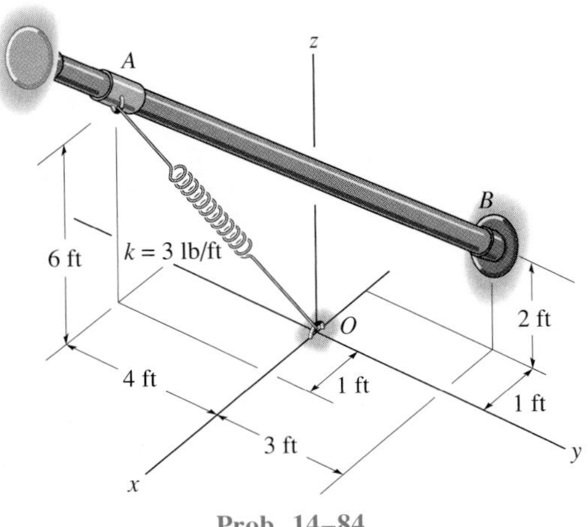

Prob. 14–84

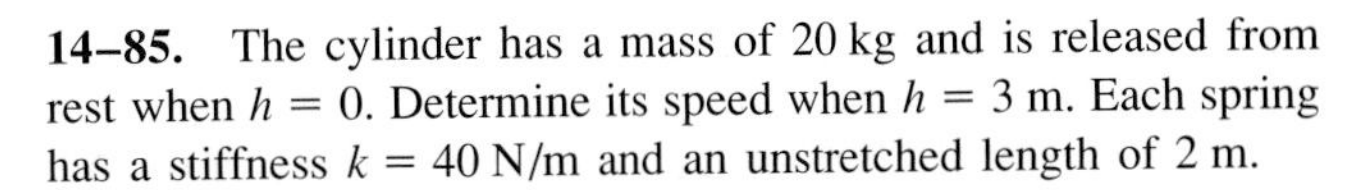

14–85. The cylinder has a mass of 20 kg and is released from rest when $h = 0$. Determine its speed when $h = 3$ m. Each spring has a stiffness $k = 40$ N/m and an unstretched length of 2 m.

14–86. If the 20-kg cylinder is released from rest at $h = 0$, determine the required stiffness k of each spring so that its motion is arrested or momentarily stops when $h = 0.5$ m. Each spring has an unstretched length of 1 m.

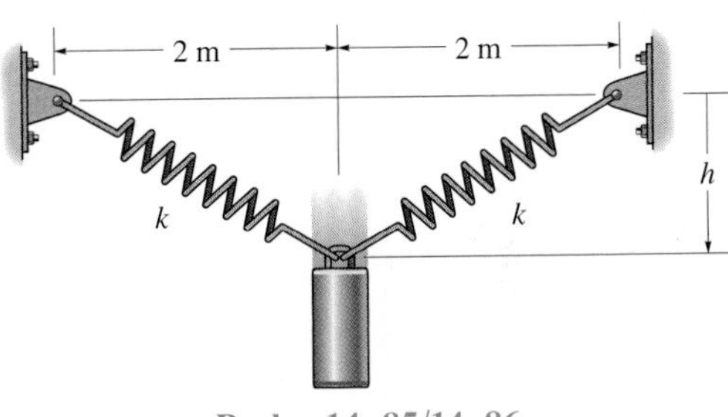

Probs. 14–85/14–86

14–87. The toboggan and passenger have a total mass of 60 kg. When it reaches point A it has a speed $v_A = 4$ m/s. Determine the angle θ at which it leaves the smooth circular curve and the distance s where it plunges in the snow. Neglect friction.

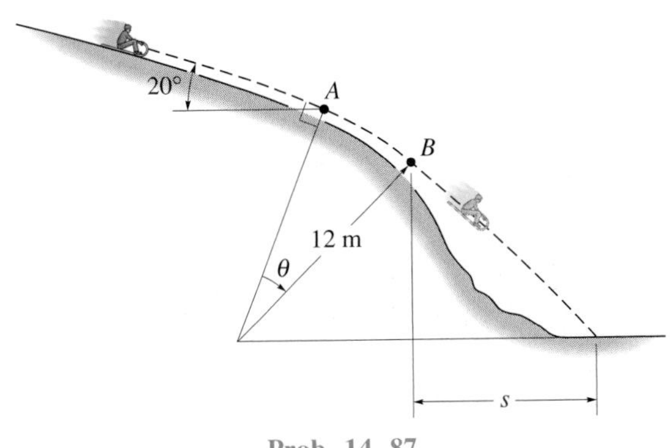

Prob. 14–87

***14–88.** Just for fun, two 150-lb engineering students A and B intend to jump off the bridge from rest using an elastic cord (bungee cord) having a stiffness $k = 80$ lb/ft. They wish to just reach the surface of the river, when A, attached to the cord, lets go of B at the instant they touch the water. Determine the proper unstretched length of the cord to do the stunt, and calculate the maximum acceleration of student A and the maximum height he reaches above the water after the rebound. From your results, comment on the feasibility of doing this stunt.

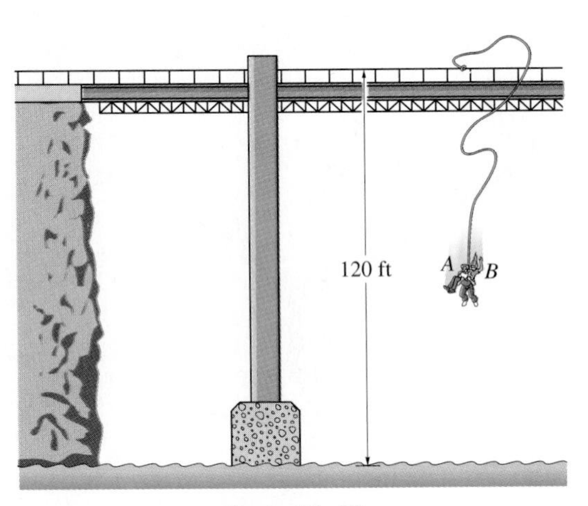

Prob. 14–88

14–89. The ride at an amusement park consists of a gondola which is lifted to a height of 120 ft at A. If it is released from rest and falls along the parabolic track, determine the speed at the instant $y = 20$ ft. Also determine the normal reaction of the tracks on the gondola at this instant. The gondola and passenger have a total weight of 500 lb. Neglect the effects of friction and the mass of the wheels.

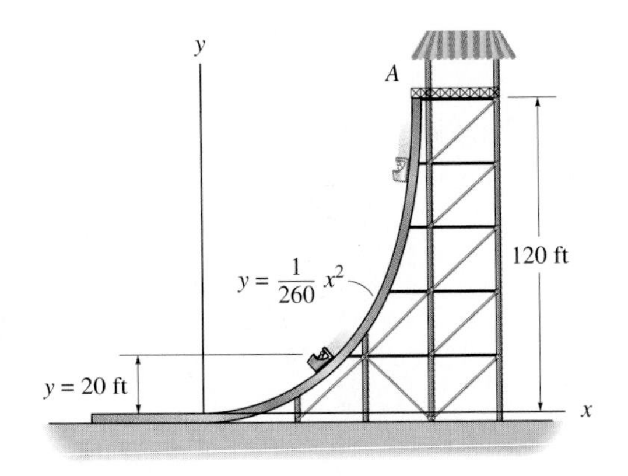

Prob. 14–89

14–90. The ball has a weight of 15 lb and is fixed to a rod having a negligible mass. If it is released from rest when $\theta = 0°$, determine the angle θ at which the compressive force in the rod becomes zero.

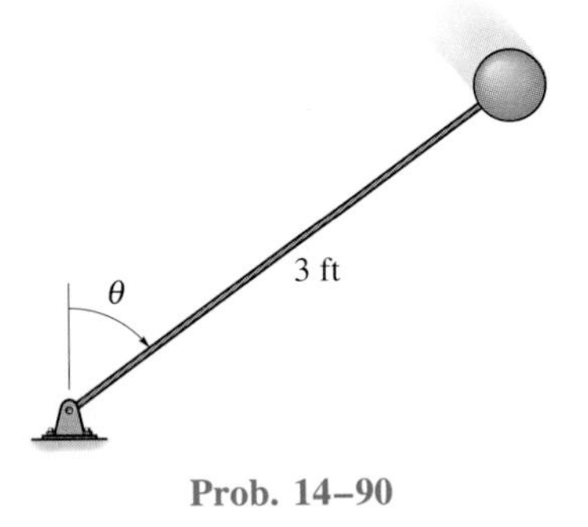

Prob. 14–90

14–91. If the roller-coaster car has a speed $v_A = 5$ ft/s when it is at A and coasts freely down the track, determine the speed v_B it attains when it reaches point B. Also determine the normal force a 150-lb passenger exerts on the car when it is at B. At this point the track follows a path defined by $y = x^2/200$. Neglect the effects of friction, the mass of the wheels, and the size of the car.

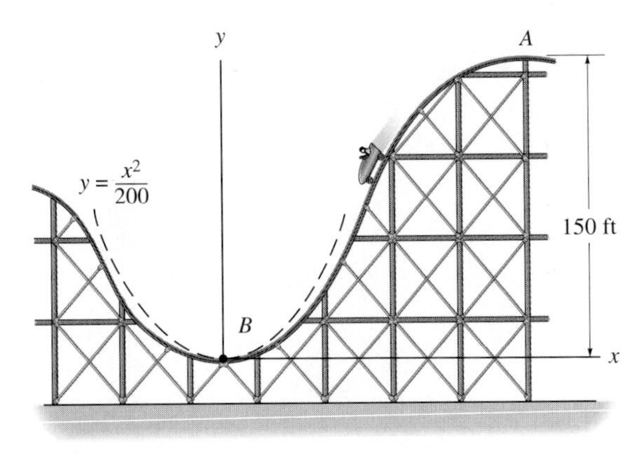

Prob. 14–91

***14–92.** If the mass of the earth is M_e, show that the gravitational potential energy of a body of mass m located a distance r from the center of the earth is $V_g = -GM_e m/r$. Recall that the gravitational force acting between the earth and the body is $F = G(M_e m/r^2)$, Eq. 13–1. For the calculation, locate the datum at $r \to \infty$. Also, prove that **F** is a conservative force.

14–93. A rocket of mass m is fired vertically from the surface of the earth, i.e., at $r = r_1$. Assuming that no mass is lost as it travels upward, determine the work it must do against gravity to reach a distance r_2. The force of gravity is $F = GM_e m/r^2$ (Eq. 13–1), where M_e is the mass of the earth and r the distance between the rocket and the center of the earth.

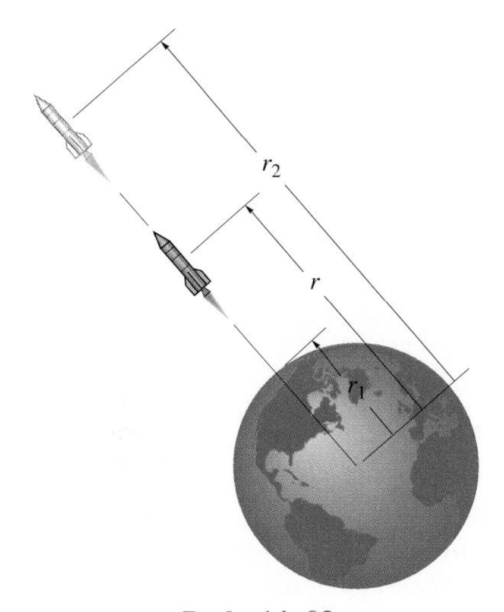

Prob. 14–93

14–94. The block has a weight W and is placed on the top of an *undeformed* spring. If it is released from this position, determine the distance Δ_{max} the spring deforms when the block is released from rest. Compare this result with the deformation Δ_{st} that occurs if the block is applied statically or gradually to the spring.

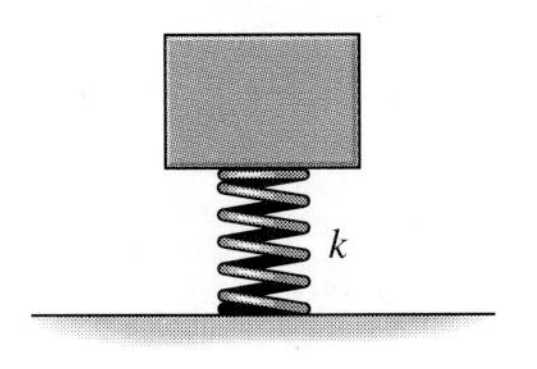

Prob. 14–94

14–95. The assembly consists of two blocks A and B which have a mass of 20 kg and 30 kg, respectively. Determine the speed of each block when B descends 1.5 m. The blocks are released from rest. Neglect the mass of the pulleys and cords.

Prob. 14–95

***14–96.** The assembly consists of two blocks A and B, which have a mass of 20 kg and 30 kg, respectively. Determine the distance B must descend in order for A to achieve a speed of 3 m/s starting from rest.

Prob. 14–96

14–97. The double-spring bumper is used to stop the 1500-lb steel billet in the rolling mill. Determine the maximum deflection of the plate A caused by the billet if it strikes the plate with a speed of 8 ft/s. Neglect the mass of the springs, rollers and the plates A and B. Take $k_1 = 3000$ lb/ft., $k_2 = 4500$ lb/ft.

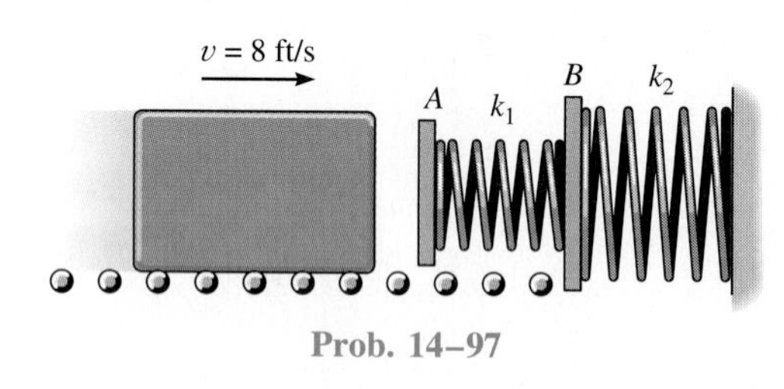

Prob. 14–97

14–98. The double-spring bumper is used to stop the 1500-lb steel billet in a rolling mill. Determine the stiffness $k = k_1 = k_2$ of each spring so that no spring is compressed more than 0.2 ft after it is struck by the billet traveling with a speed of 8 ft/s. Neglect the mass of the springs, rollers and the plates A and B.

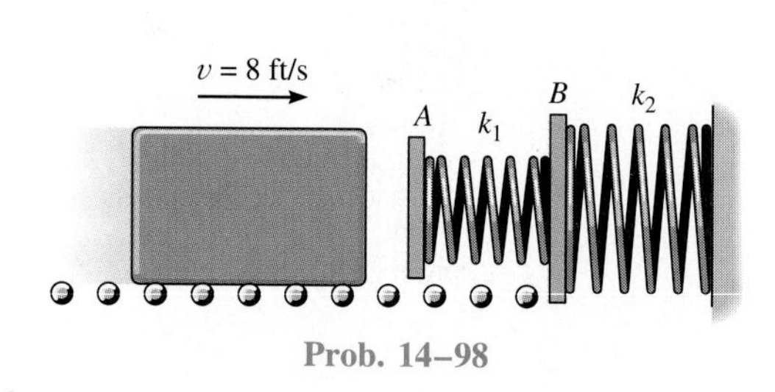

Prob. 14–98

14–99. Four inelastic cables C are attached to a plate P and hold the 20-in. spring 6 in. in compression when *no force* acts on the plate. If a block B, having a weight of 7 lb, is placed on the plate and the plate is pushed down $y = 8$ in. and released from rest, determine how high the block rises from the point where it was released. Neglect the mass of the plate. The spring has a stiffness $k = 20$ lb/ft.

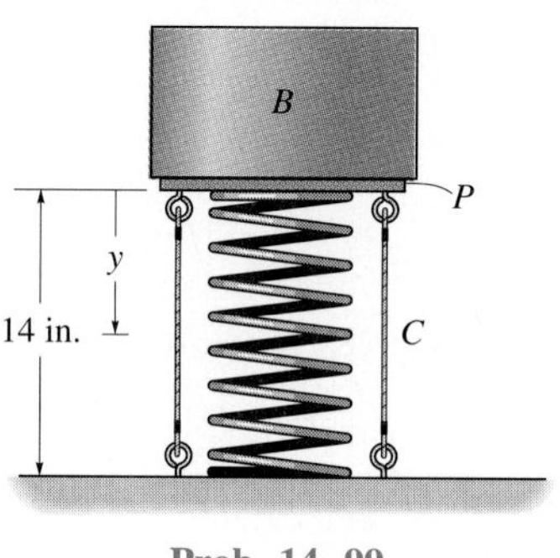

Prob. 14–99

***14–100.** The plunger is designed to fire the 3-lb block 2 ft in the air, measured from the point where the block is pushed down, $y = 5$ in., and released from rest. If the stiffness of the spring is $k = 30$ lb/ft, determine the required uncompressed length of the spring that must be used for the device. The plate P is held in place by the four inelastic cables C, each of which is 14 in. long.

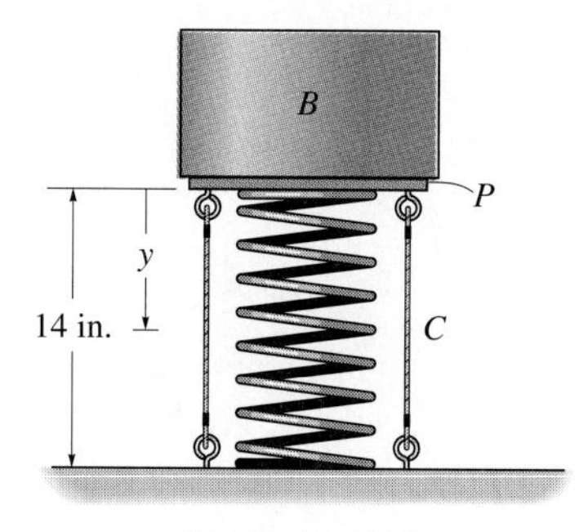

Prob. 14–100

14–101. Four inelastic cables C are attached to a plate P and hold the 1-ft-long spring 0.25 ft in compression when *no weight* is on the plate. There is also an undeformed spring nested within this compressed spring. If the block, having a weight of 10 lb, is moving downward at $v = 4$ ft/s, when it is 2 ft above the plate, determine the maximum compression in each spring after it strikes the plate. Neglect the mass of the plate and spring and any energy lost in the collision.

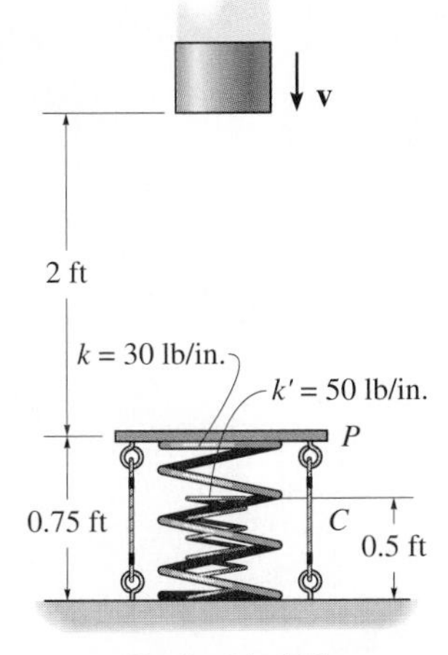

Prob. 14–101

14–102. Four inelastic cables C are attached to a plate P and hold the 1-ft-long spring 0.25 ft in compression when *no weight* is on the plate. There is also an undeformed spring nested within this compressed spring. Determine the speed v of the 10-lb block when it is 2 ft above the plate, so that after it strikes the plate, it compresses the nested spring, having a stiffness of 50 lb/in., an amount of 0.20 ft. Neglect the mass of the plate and spring and any energy lost in the collision.

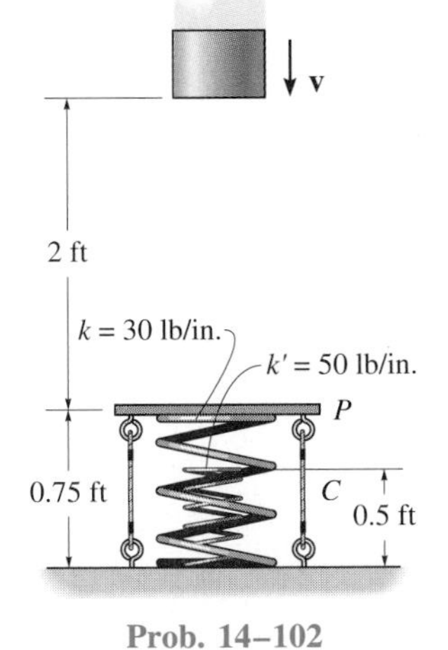

Prob. 14–102

14–103. The skier starts from rest at A and travels down the ramp. If friction and air resistance can be neglected, determine his speed v_B when he reaches B. Also, compute the distance s to where he strikes the ground at C, if he makes the jump traveling horizontally at B. Neglect the skier's size. He has a mass of 70 kg.

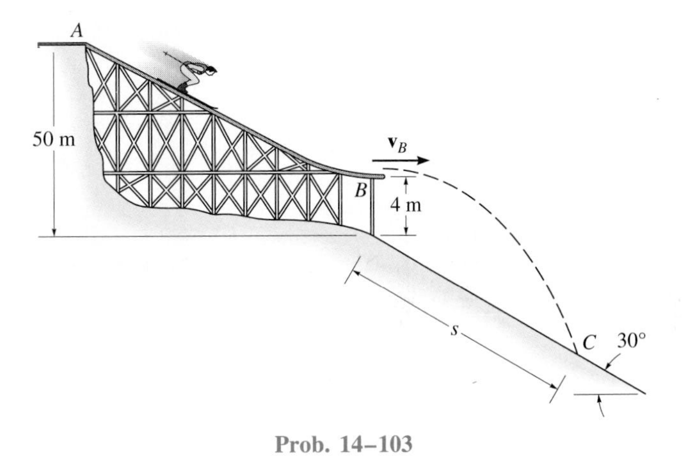

Prob. 14–103

***14–104.** The block has a mass of 20 kg and is released from rest when $s = 0.5$ m. If the mass of the bumpers A and B can be neglected, determine the maximum deformation of each spring due to the collision.

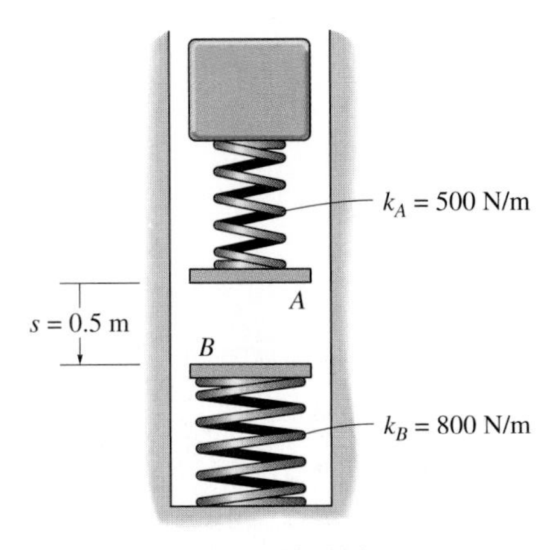

Prob. 14–104

When engineers design running shoes they must select materials that reduce the impact loading of the ground on the shoe, so that this loading does not produce discomfort within the joints of the body. The theory of impulse and momentum used for such an analysis is explained in this chapter.

15

Kinetics of a Particle: Impulse and Momentum

In this chapter we will integrate the equation of motion with respect to time and thereby obtain the principle of impulse and momentum. The resulting equation is useful for solving problems involving force, velocity, and time. It will be shown that impulse and momentum principles also provide the necessary means for analyzing problems of impact, steady fluid flow, and systems which gain or lose mass.

15.1 Principle of Linear Impulse and Momentum

The equation of motion for a particle of mass m can be written as

$$\Sigma \mathbf{F} = m\mathbf{a} = m\frac{d\mathbf{v}}{dt} \tag{15–1}$$

where $\mathbf{a}$ and $\mathbf{v}$ are both measured from an inertial frame of reference. Rearranging the terms and integrating between the limits $\mathbf{v} = \mathbf{v}_1$ at $t = t_1$ and $\mathbf{v} = \mathbf{v}_2$ at $t = t_2$, we have

$$\Sigma \int_{t_1}^{t_2} \mathbf{F}\, dt = m \int_{\mathbf{v}_1}^{\mathbf{v}_2} d\mathbf{v}$$

or

$$\Sigma \int_{t_1}^{t_2} \mathbf{F}\, dt = m\mathbf{v}_2 - m\mathbf{v}_1 \qquad (15\text{–}2)$$

This equation is referred to as the *principle of linear impulse and momentum.* From the derivation it can be seen that it is simply a time integration of the equation of motion. It provides a *direct means* of obtaining the particle's final velocity $\mathbf{v}_2$ after a specified time period when the particle's initial velocity is known and the forces acting on the particle are either constant or can be expressed as functions of time. Notice that if $\mathbf{v}_2$ is determined using the equation of motion, a two-step process is necessary; i.e., apply $\Sigma\mathbf{F} = m\mathbf{a}$ to obtain **a,** then integrate $\mathbf{a} = d\mathbf{v}/dt$ to obtain $\mathbf{v}_2$.

Linear Impulse. The integral $\mathbf{I} = \int \mathbf{F}\, dt$ in Eq. 15–2 is defined as the *linear impulse.* This term is a vector quantity which measures the effect of a force during the time the force acts. Since time is a positive scalar, the impulse vector acts in the same direction as the force, and its magnitude has units of force–time, e.g., N · s or lb · s. If the force is expressed as a function of time, the impulse may be determined by direct evaluation of the integral. In particular, if **F** acts in a *constant direction* during the time period t_1 to t_2, the magnitude of the impulse $\mathbf{I} = \int_{t_1}^{t_2} \mathbf{F}\, dt$ can be represented experimentally by the shaded area under the curve of force versus time, Fig. 15–1. However, if the force is constant in magnitude and direction, the resulting impulse becomes $\mathbf{I} = \int_{t_1}^{t_2} \mathbf{F}_c\, dt = \mathbf{F}_c(t_2 - t_1)$, which represents the shaded rectangular area shown in Fig. 15–2.

Linear Momentum. Each of the two vectors of the form $\mathbf{L} = m\mathbf{v}$ in Eq. 15–2 is defined as the *linear momentum* of the particle. Since m is a positive scalar, the linear-momentum vector has the same direction as **v,** and its magnitude mv has units of mass–velocity, e.g., kg · m/s, slug · ft/s.*

*Although the units for impulse and momentum are defined differently, one can show that Eq. 15–2 is dimensionally homogeneous.

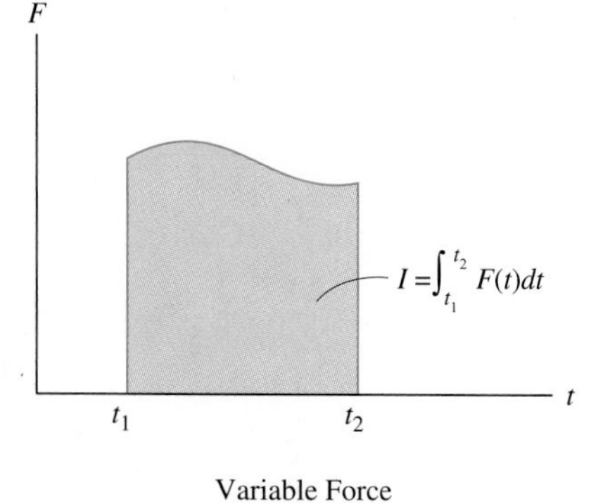

Variable Force

Fig. 15–1

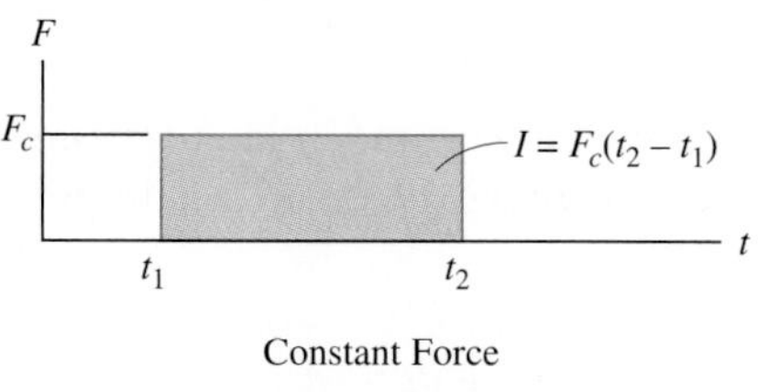

Constant Force

Fig. 15–2

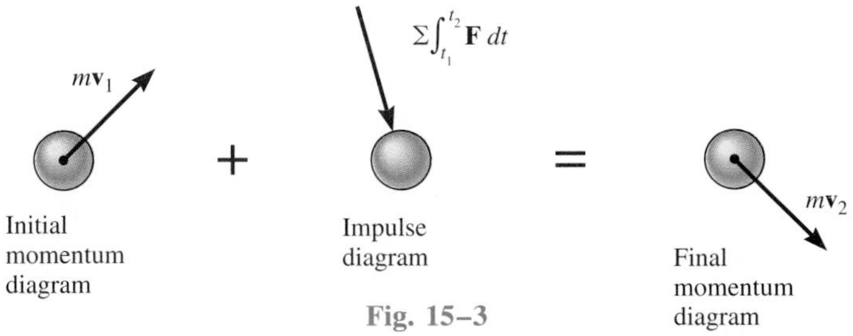

$m\mathbf{v}_1$ + $\Sigma\int_{t_1}^{t_2}\mathbf{F}\,dt$ = $m\mathbf{v}_2$

Initial momentum diagram

Impulse diagram

Final momentum diagram

Fig. 15–3

Principle of Linear Impulse and Momentum. For problem solving, Eq. 15–2 will be rewritten in the form

$$m\mathbf{v}_1 + \Sigma\int_{t_1}^{t_2}\mathbf{F}\,dt = m\mathbf{v}_2 \qquad (15\text{–}3)$$

which states that the initial momentum of the particle at t_1 plus the vector sum of all the impulses applied to the particle during the time interval t_1 to t_2 is equivalent to the final momentum of the particle at t_2. These three terms are illustrated graphically on the *impulse and momentum diagrams* shown in Fig. 15–3. Each of these diagrams accounts graphically for all the vectors in the equation $m\mathbf{v}_1 + \Sigma\int_{t_1}^{t_2}\mathbf{F}\,dt = m\mathbf{v}_2$. The two *momentum diagrams* are simply outlined shapes of the particle which indicate the direction and magnitude of the particle's initial and final momenta, $m\mathbf{v}_1$ and $m\mathbf{v}_2$, respectively, Fig. 15–3. Similar to the free-body diagram, the *impulse diagram* is an outlined shape of the particle showing all the impulses that act on the particle when it is located at some intermediate point along its path. In general, whenever the magnitude or direction of a force *varies,* the impulse of the force is determined by integration and represented on the impulse diagram as $\mathbf{I} = \int_{t_1}^{t_2}\mathbf{F}\,dt$. If the force is *constant* for the time interval $(t_2 - t_1)$, the impulse applied to the particle is $\mathbf{I} = \mathbf{F}_c(t_2 - t_1)$, acting in the same direction as $\mathbf{F}_c$.

Scalar Equations. If each of the vectors in Eq. 15–3 or on Fig. 15–3 is resolved into its x, y, and z components, we can write symbolically the following three scalar equations:

$$\begin{aligned} m(v_x)_1 + \Sigma\int_{t_1}^{t_2} F_x\,dt &= m(v_x)_2 \\ m(v_y)_1 + \Sigma\int_{t_1}^{t_2} F_y\,dt &= m(v_y)_2 \\ m(v_z)_1 + \Sigma\int_{t_1}^{t_2} F_z\,dt &= m(v_z)_2 \end{aligned} \qquad (15\text{–}4)$$

These equations represent the principle of linear impulse and momentum for the particle in the x, y, and z directions, respectively.

PROCEDURE FOR ANALYSIS

The principle of linear impulse and momentum is used to solve problems involving *force, time,* and *velocity,* since these terms are involved in the formulation. For application it is suggested that the following procedure be used.

Free-Body Diagram. Establish the x, y, z inertial frame of reference and draw the particle's free-body diagram in order to account for all the forces that produce impulses on the particle. The direction and sense of the particle's initial and final velocities should also be established. To account for these vectors, they may be sketched on the coordinate system, but *not* on the free-body diagram. If a velocity is unknown, assume that the sense of its components is in the direction of the positive inertial coordinate(s).

As an alternative procedure, draw the impulse and momentum diagrams for the particle as discussed in reference to Fig. 15–3.*

Principle of Impulse and Momentum. Apply the principle of linear impulse and momentum, $m\mathbf{v}_1 + \Sigma \int_{t_1}^{t_2} \mathbf{F}\, dt = m\mathbf{v}_2$. If motion occurs in the x–y plane, the two scalar component equations can be formulated by either resolving the vector components of $\mathbf{F}$ from the free-body diagram, or by using the data on the impulse and momentum diagrams.

If the problem involves the dependent motion of several particles, use the method outlined in Sec. 12.8 to relate their velocities. Make sure the positive coordinate directions used for writing these kinematic equations are the *same* as those used for writing the equations of impulse and momentum.

*This procedure will be followed when developing the proofs and theory in the text.

The following examples numerically illustrate application of this procedure.

Example 15–1

The 100-kg crate shown in Fig. 15–4*a* is originally at rest on the smooth horizontal surface. If a force of 200 N, acting at an angle of 45°, is applied to the crate for 10 s, determine the final velocity of the crate and the normal force which the surface exerts on the crate during the time interval.

SOLUTION

This problem can be solved using the principle of impulse and momentum since it involves force, velocity, and time.

Free-Body Diagram. See Fig. 15–4*b*. Here it has been assumed that during the motion the crate remains on the surface and after 10 s the crate moves to the right with a velocity $\mathbf{v}_2$. Since all the forces acting on the crate are *constant,* the respective impulses are simply the product of the force magnitude and 10 s $[\mathbf{I} = \mathbf{F}_c(t_2 - t_1)]$.

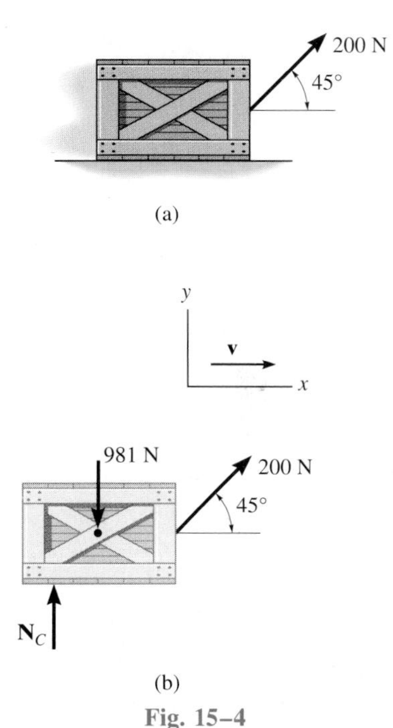

Fig. 15–4

Principle of Impulse and Momentum. Resolving the vectors in Fig. 15–4*b* along the *x, y* axes and applying Eqs. 15–4 yields

$$(\overset{+}{\rightarrow}) \qquad m(v_x)_1 + \Sigma \int_{t_1}^{t_2} F_x\,dt = m(v_x)_2$$

$$0 + 200\text{ N}(10\text{ s})\cos 45^\circ = (100\text{ kg})v_2$$

$$v_2 = 14.1\text{ m/s} \rightarrow \qquad \textit{Ans.}$$

$$(+\uparrow) \qquad m(v_y)_1 + \Sigma \int_{t_1}^{t_2} F_y\,dt = m(v_y)_2$$

$$0 + N_C(10\text{ s}) - 981\text{ N}(10\text{ s}) + 200\text{ N}(10\text{ s})\sin 45^\circ = 0$$

$$N_C = 840\text{ N} \qquad \textit{Ans.}$$

Since no motion occurs in the *y* direction, direct application of $\Sigma F_y = 0$ gives the same result for N_C.

Note the alternative procedure of drawing the crate's impulse and momentum diagrams, Fig. 15–4*c*, prior to applying Eqs. 15–4.

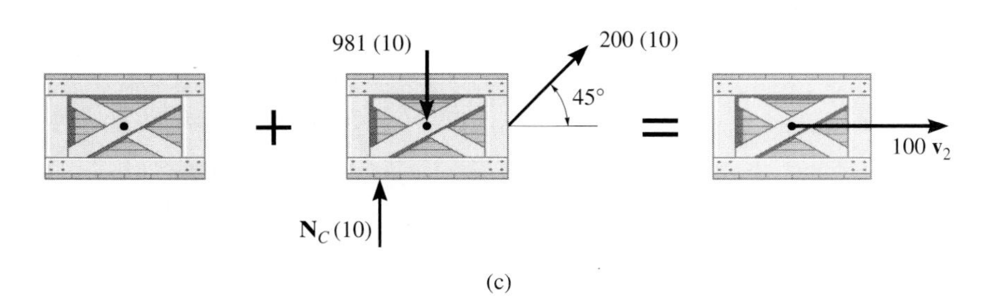

(c)

Example 15–2

The crate shown in Fig. 15–5a has a weight of 50 lb and is acted upon by a force having a variable magnitude $P = (20t)$ lb, where t is in seconds. Determine the crate's velocity 2 s after **P** has been applied. The crate has an initial velocity $v_1 = 3$ ft/s down the plane, and the coefficient of kinetic friction between the crate and the plane is $\mu_k = 0.3$.

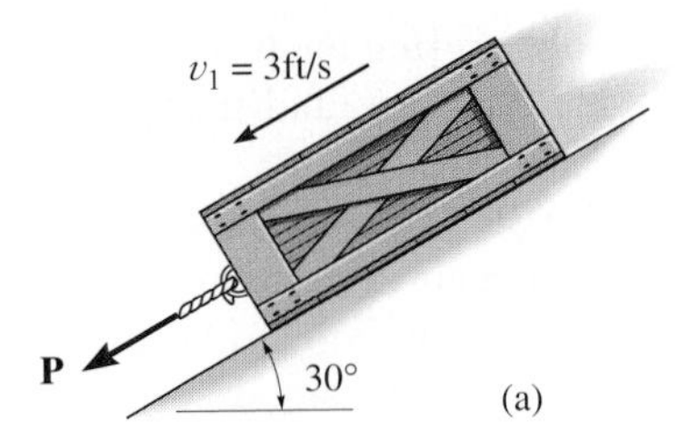

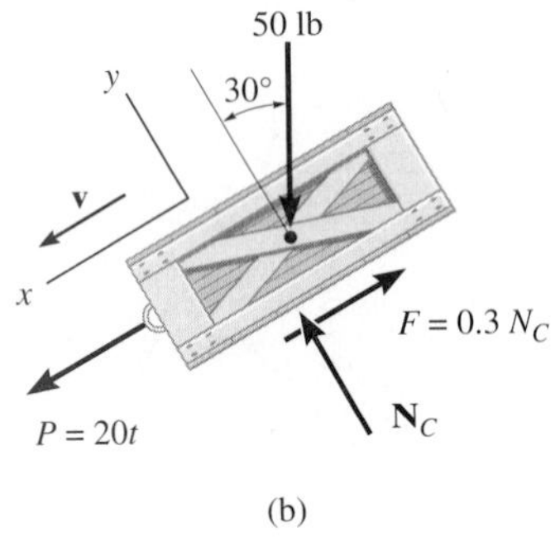

(b)

Fig. 15–5

SOLUTION

Free-Body Diagram. See Fig. 15–5b. Since the magnitude of force $P = 20t$ *varies* with time, the impulse it creates must be determined by *integrating* over the 2-s time interval. The weight, normal force, and frictional force (which acts opposite to the direction of motion) are all *constant,* so that the impulse created by each of these forces is simply the magnitude of the force times 2 s.

Principle of Impulse and Momentum. Applying Eqs. 15–4 in the x direction, we have

$$(+\swarrow) \qquad m(v_x)_1 + \Sigma\int_{t_1}^{t_2} F_x\,dt = m(v_x)_2$$

$$\frac{50\text{ lb}}{32.2\text{ ft/s}^2}(3\text{ ft/s}) + \int_0^2 20t\,dt - 0.3N_C(2\text{ s}) + (50\text{ lb})(2\text{ s})\sin 30^\circ = \frac{50\text{ lb}}{32.2\text{ ft/s}^2}v_2$$

$$4.66 + 40 - 0.6N_C + 50 = 1.55v_2$$

The equation of equilibrium can be applied in the y direction. Why?

$$+\nwarrow\Sigma F_y = 0; \qquad N_C - 50\cos 30^\circ\text{ lb} = 0$$

Solving,

$$N_C = 43.3\text{ lb}$$

$$v_2 = 44.2\text{ ft/s}\ \swarrow \qquad \textit{Ans.}$$

This problem has also been solved using the equation of motion in Example 13–3. The two methods of solution should be compared. Since *force, velocity,* and *time* are involved in the problem, application of the principle of impulse and momentum eliminates the need for using kinematics ($a = dv/dt$) and thereby yields an easier method for solution.

Example 15–3

Blocks A and B shown in Fig. 15–6a have a mass of 3 kg and 5 kg, respectively. If the system is released from rest, determine the velocity of block B in 6 s. Neglect the mass of the pulleys and cord.

(a)

SOLUTION

Free-Body Diagram. See Fig. 15–6b. Since the weight of each block is constant, the cord tensions will also be constant. Furthermore, since the mass of pulley D is neglected, the cord tension $T_A = 2T_B$, Fig. 15–6b. The blocks are both assumed to be traveling downward in the positive direction.

Principle of Impulse and Momentum

Block A:

$$(+\downarrow) \qquad m(v_A)_1 + \Sigma \int_{t_1}^{t_2} F_y \, dt = m(v_A)_2$$

$$0 - 2T_B(6 \text{ s}) + 3(9.81) \text{ N}(6 \text{ s}) = (3 \text{ kg})(v_A)_2 \qquad (1)$$

Block B:

$$(+\downarrow) \qquad m(v_B)_1 + \Sigma \int_{t_1}^{t_2} F_y \, dt = m(v_B)_2$$

$$0 + 5(9.81) \text{ N}(6 \text{ s}) - T_B(6 \text{ s}) = (5 \text{ kg})(v_B)_2 \qquad (2)$$

Kinematics. Since the blocks are subjected to dependent motion, the velocity of A may be related to that of B by using the kinematic analysis discussed in Sec. 12.8. A horizontal datum is established through the fixed point at C, Fig. 15–6a, and the changing positions of the blocks, s_A and s_B, are related to the constant total length l of the vertical segments of the cord by the equation

$$2s_A + s_B = l$$

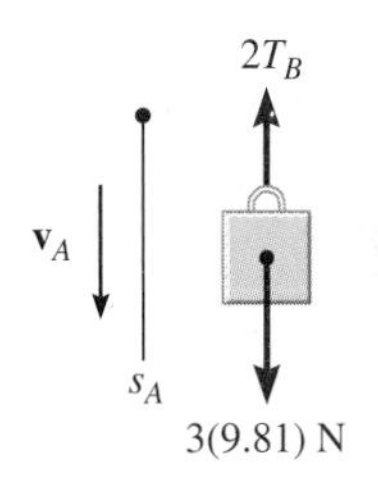

Taking the time derivative yields

$$2v_A = -v_B \qquad (3)$$

As indicated by the negative sign, when B moves downward A moves upward.* Substituting this result into Eq. 1 and solving Eqs. 1 and 2 yields

$$(v_B)_2 = 35.8 \text{ m/s} \downarrow \qquad \textit{Ans.}$$

$$T_B = 19.2 \text{ N}$$

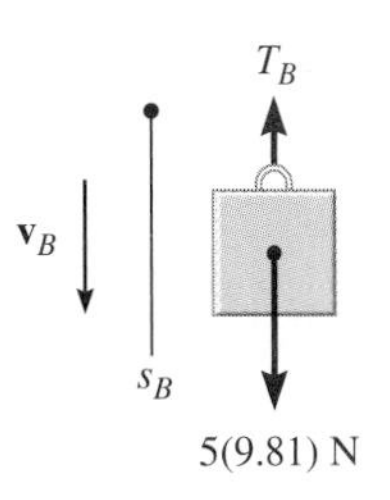

(b)

Fig. 15–6

*Note that the *positive* (downward) directions for $(\mathbf{v}_A)_2$ and $(\mathbf{v}_B)_2$ are *consistent* in Figs. 15–6a and 15–6b and in Eqs. 1 to 3. Why is this important?

PROBLEMS

15–1. A ball has a mass of 30 kg and is thrown upward with a speed of 15 m/s. Determine how long it takes for it to stop. Also, how high does it rise before stopping? Use the principle of impulse and momentum for the solution.

15–2. A 40-g golf ball is struck in 3 ms by a driver such that it is given a velocity $v_1 = 35$ m/s directed to the right, 30° from the horizontal. Determine the average impulsive force exerted on the ball and the momentum of the ball in $t = 1$ s after it leaves the ground. Neglect the impulse caused by the ball's weight while it is struck.

15–3. A 2-lb ball is thrown in the direction shown with an initial speed $v_A = 18$ ft/s. Determine the time needed for it to reach its highest point B and the speed at which it is traveling at B. Use the principle of impulse and momentum for the solution.

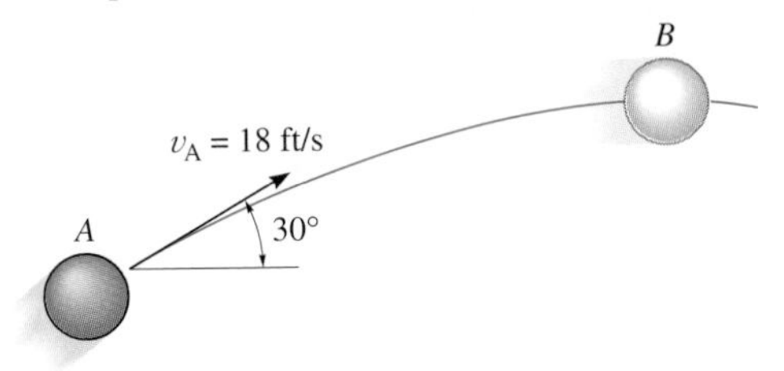

Prob. 15–3

***15–4.** The 180-lb iron worker is secured by a fall-arrest system consisting of a harness and lanyard AB, which is fixed to the beam. If the lanyard has a slack of 4 ft, determine the average impulsive force developed in the lanyard if he happens to fall 4 feet. Neglect his size in the calculation and assume the impulse takes place in 0.6 seconds.

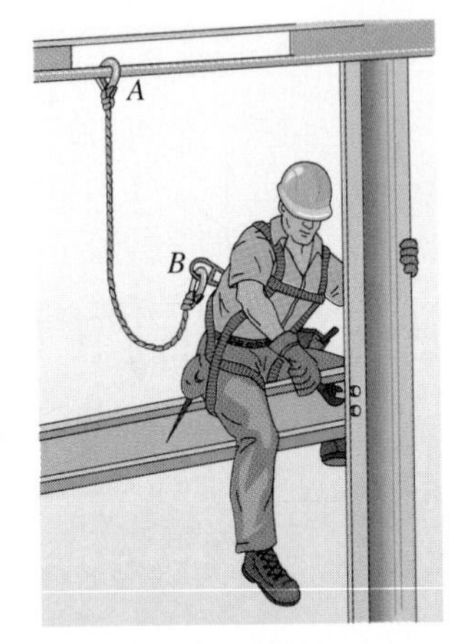

Prob. 15–4

15–5. A man hits the 50-g golf ball such that it leaves the tee at an angle of 40° with the horizontal and strikes the ground at the same elevation a distance of 20 m away. Determine the impulse of the club C on the ball. Neglect the impulse caused by the ball's weight while the club is striking the ball.

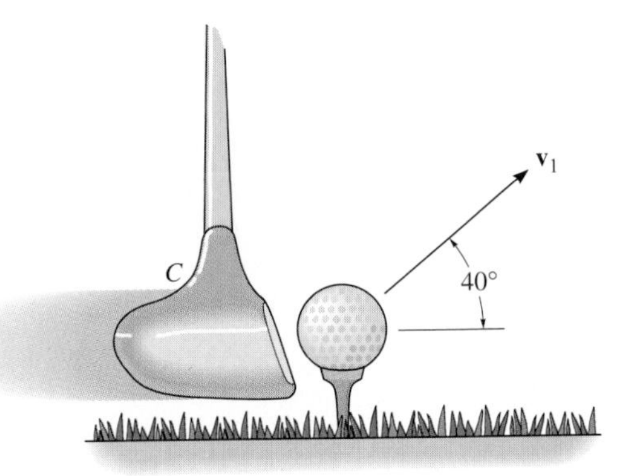

Prob. 15–5

15–6. A tankcar has a mass of 20 Mg and is freely rolling to the right with a speed of 0.75 m/s. If it strikes the barrier, determine the horizontal impulse needed to stop the car if the spring in the bumper B has a stiffness (a) $k \rightarrow \infty$ (bumper is rigid), and (b) $k = 15$ kN/m.

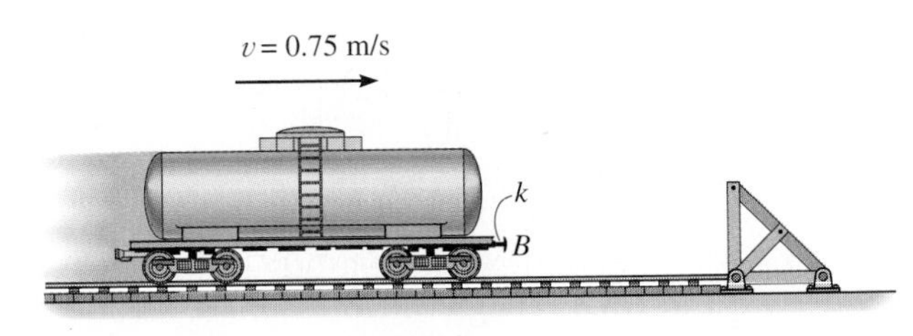

Prob. 15–6

15–7. During operation the breaker hammer develops on the concrete surface a force which is indicated in the graph. To achieve this the 2-lb spike S is fired from rest into the surface at 200 ft/s. Determine the speed of the spike just after rebounding.

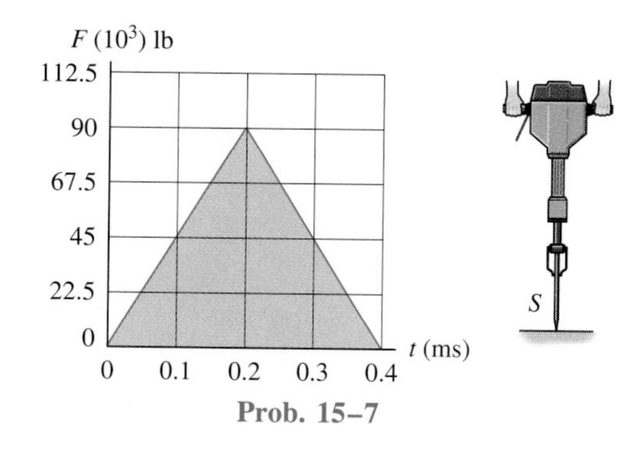

Prob. 15–7

15–9. The tennis ball has a horizontal speed of 15 m/s when it is struck by the racket. If it then travels away at an angle of 25° from the horizontal and reaches a maximum altitude of 10 m, measured from the height of the racket, determine the magnitude of the net impulse of the racket on the ball. The ball has a mass of 180 g. Neglect the weight of the ball during the time the racket strikes the ball.

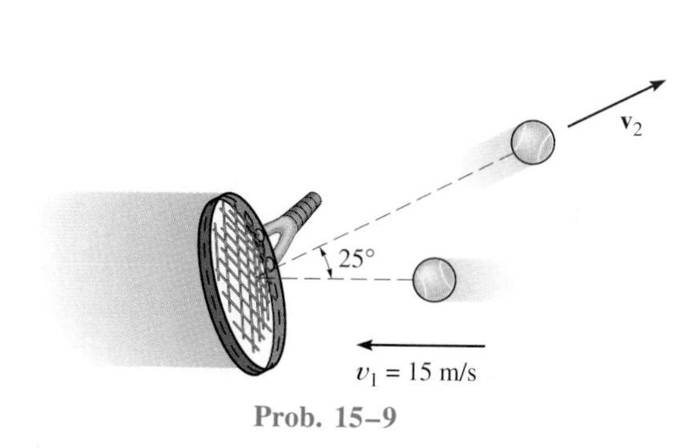

Prob. 15–9

***15–8.** A man kicks the 200-g ball such that it leaves the ground at an angle of 30° with the horizontal and strikes the ground at the same elevation a distance of 15 m away. Determine the impulse of his foot F on the ball. Neglect the impulse caused by the ball's weight while its being kicked.

Prob. 15–8

15–10. The twitch in a muscle of the arm develops a force which can be measured as a function of time as shown in the graph. If the effective contraction of the muscle lasts for a time t_0, determine the impulse developed by the muscle.

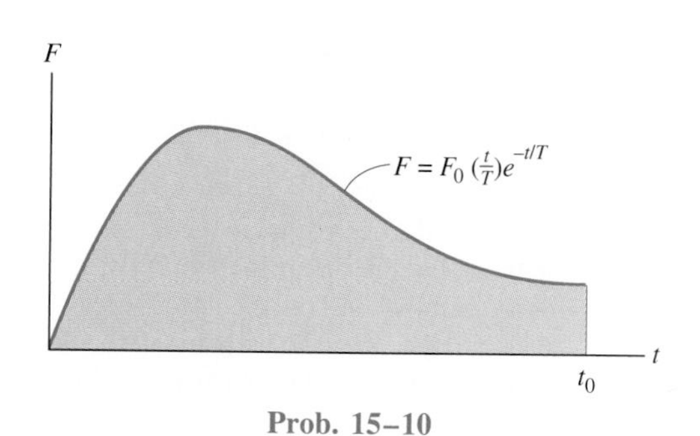

Prob. 15–10

15–11. If it takes 35 s for the 50-Mg tugboat to increase its speed uniformly to 25 km/h, starting from rest, determine the force of the rope on the tugboat. The propeller provides the propulsion force **F** which gives the tugboat forward motion, whereas the barge moves freely. Also, determine F acting on the tugboat. The barge has a mass of 75 Mg.

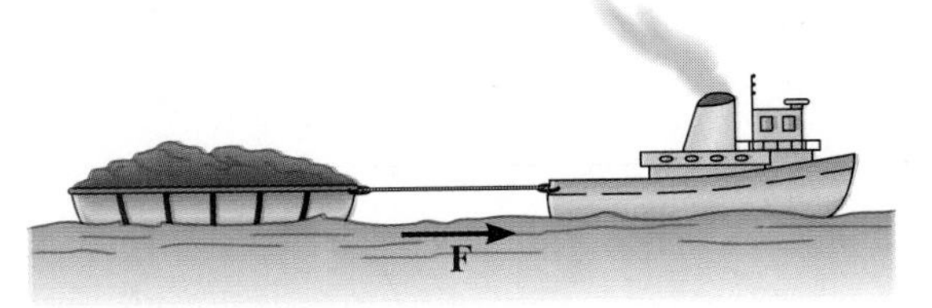

Prob. 15–11

***15–12.** The uniform beam has a weight of 5000 lb. Determine the average tension in each of the two cables AB and AC if the beam is given an upward speed of 8 ft/s in 1.5 s starting from rest. Neglect the mass of the cables.

15–13. Each of the cables can sustain a maximum tension of 5000 lb. If the uniform beam has a weight of 5000 lb, determine the shortest time possible to lift the beam with a speed of 10 ft/s starting from rest.

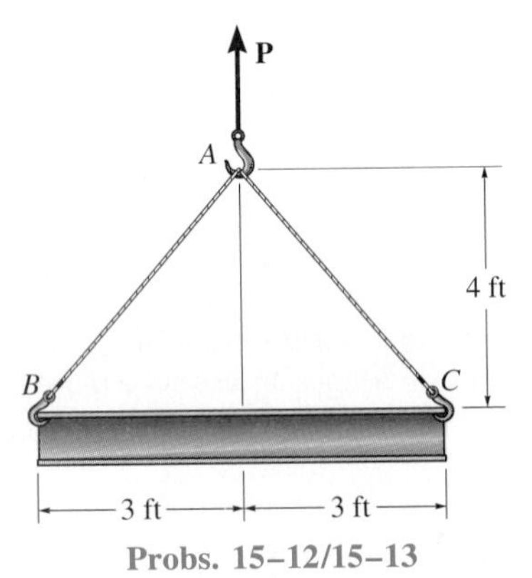

Probs. 15–12/15–13

15–14. The 5-kg block is moving downward at $v_1 = 2$ m/s when it is 8 m from the sandy surface. Determine the impulse of the sand on the block necessary to stop its motion. Neglect the distance the block dents into the sand and assume the block does not rebound. Neglect the weight of the block during the impact with the sand.

15–15. The 5-kg block is falling downward at $v_1 = 2$ m/s when it is 8 m from the sandy surface. Determine the average impulsive force acting on the block by the sand if the motion of the block is stopped in 0.9 s once the block strikes the sand. Neglect the distance the block dents into the sand and assume the block does not rebound. Neglect the weight of the block during the impact with the sand.

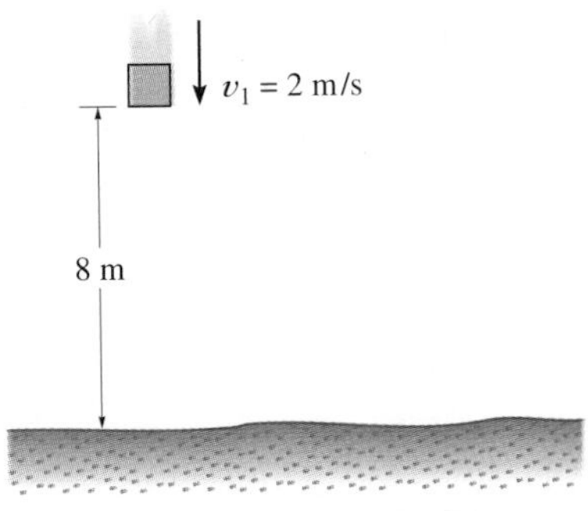

Probs. 15–14/15–15

***15–16.** A jet plane has a mass of 250 Mg and a horizontal velocity of 100 m/s when $t = 0$. If the engines provide a resultant horizontal thrust $F = (200 + 2t^2)$ kN, where t is in seconds, determine the plane's velocity in $t = 5$ s. Neglect air resistance and the loss of fuel during the motion.

15–17. A jet plane has a mass of 250 Mg and a horizontal velocity of 100 m/s when $t = 0$. If the engines provide a resultant horizontal thrust $F = (40 + 0.5t)$ kN, where t is in seconds, determine the time needed for the plane to attain a velocity of 200 m/s. Neglect air resistance and the loss of fuel during the motion.

15–18. The rocket sled has a mass of 3 Mg and starts from rest when $t = 0$. If the engines provide a horizontal thrust T which varies as shown in the graph, determine the sled's velocity in $t = 4$ s. Neglect air resistance, friction, and the loss of fuel during the motion.

15–19. Solve Prob. 15–18 if the coefficients of static and kinetic friction between the tracks and the sled are $\mu_s = 0.3$ and $\mu_k = 0.2$, respectively. Neglect air resistance and the loss of fuel during the motion.

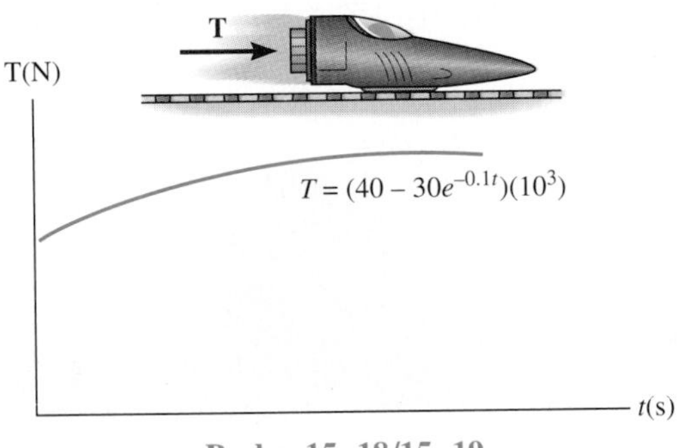

Probs. 15–18/15–19

***15–20.** As indicated by the derivation, the principle of impulse and momentum is valid for observers in *any* inertial reference frame. Show that this is so, by considering the 10-kg block which rests on the smooth surface and is subjected to a horizontal force of 6 N. If observer A is in a *fixed* frame x, determine the final speed of the block in 4 s if it has an initial speed of 5 m/s measured from the fixed frame. Compare the result with that obtained by an observer B, attached to the x' axis that moves at a constant velocity of 2 m/s relative to A.

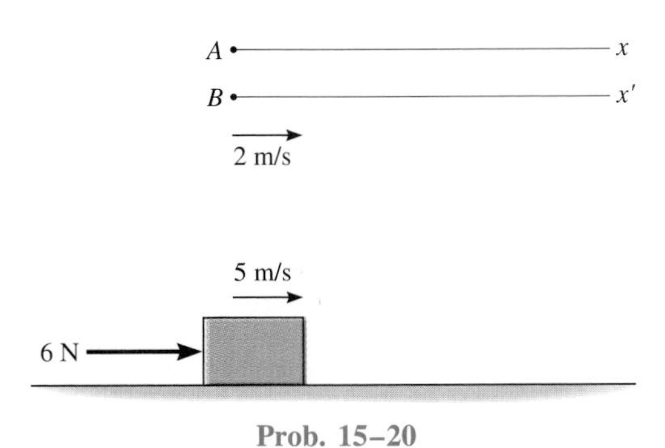

Prob. 15–20

15–21. The winch delivers a horizontal towing force F to its cable at A which varies as shown in the graph. Determine the speed of the 70-kg block B when $t = 18$ s. Originally the block is moving upward at $v_1 = 3$ m/s.

15–22. The winch delivers a horizontal towing force F to its cable at A which varies as shown in the graph. Determine the speed of the 40-kg block B when $t = 24$ s. Originally the block is resting on the ground.

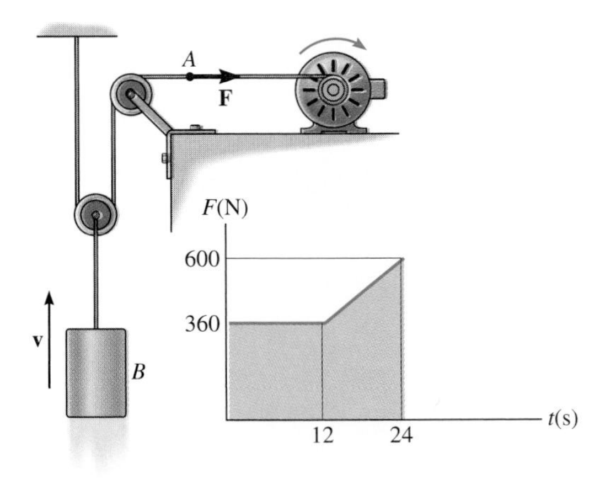

Probs. 15–21/15–22

15–23. Block A weighs 10 lb and block B weighs 3 lb. If B is moving downward with a velocity $(v_B)_1 = 3$ ft/s at $t = 0$, determine the velocity of A when $t = 1$ s. Assume that the horizontal plane is smooth. Neglect the mass of the pulleys and cord.

***15–24.** Solve Prob. 15–23 if the coefficient of kinetic friction between the horizontal plane and block A is $\mu_A = 0.15$.

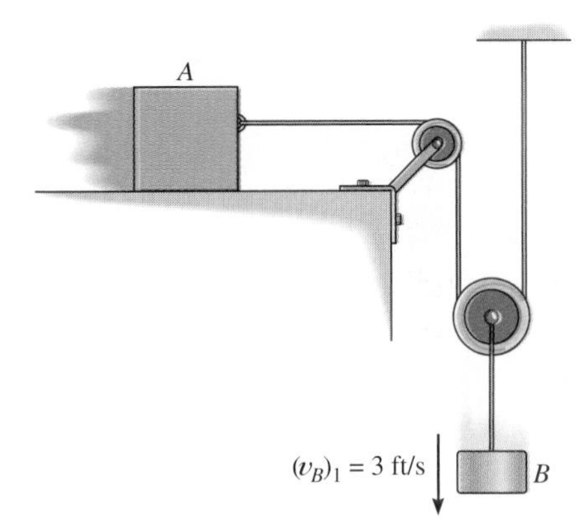

Probs. 15–23/15–24

15–25. The force acting on the 50-lb crate has a magnitude of $F = (2.4t^2 + 15t)$ lb, where t is in seconds. If the crate starts from rest, determine its speed when $t = 2$ s. Also determine the distance the crate moves in 2 s. Neglect friction.

15–26. The force acting on the 50-lb crate has a magnitude of $F = (2.4t^2)$ lb, where t is in seconds. If the crate starts from rest, determine its speed when $t = 5$ s. The coefficients of static and kinetic friction between the crate and floor are $\mu_s = 0.3$ and $\mu_k = 0.2$, respectively.

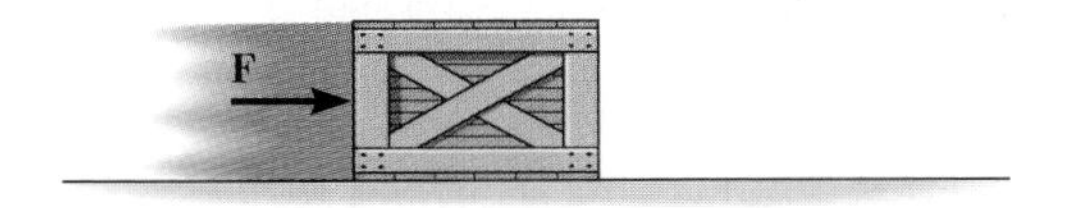

Probs. 15–25/15–26

15–27. The motor pulls on the cable at A with a force $F = (30 + t^2)$ lb, where t is in seconds. If the 34-lb crate is originally at rest at $t = 0$, determine its speed in $t = 4$ s. Neglect the mass of the cable and pulleys. *Hint:* First find the time needed to begin lifting the crate.

***15–28.** The motor pulls on the cable at A with a force $F = (e^{2t})$ lb, where t is in seconds. If the 34-lb crate is originally at rest on the ground at $t = 0$, determine the crate's velocity when $t = 2$ s. Neglect the mass of the cable and pulleys. *Hint:* First find the time needed to begin lifting the crate.

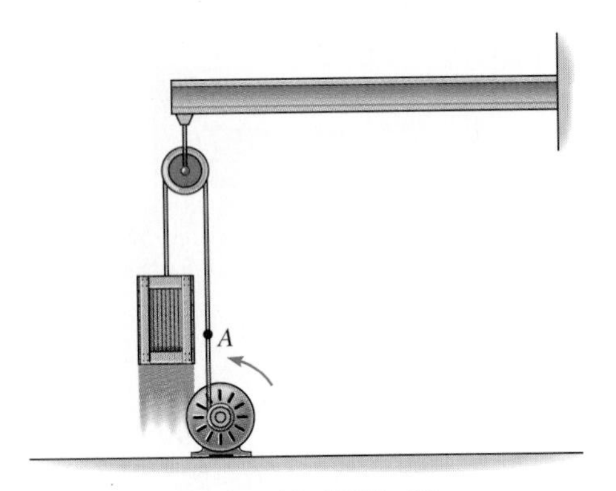

Probs. 15–27/15–28

15–29. The 50-kg block B is hoisted using the cable and motor arrangement shown. If the block is moving upward at $v_1 = 2$ m/s when $t = 0$, and the motor develops a tension in the cord of $T = (500 + 120\sqrt{t})$ N, where t is in seconds, determine the velocity of the block when $t = 2$ s. Neglect the mass of the pulleys and cable.

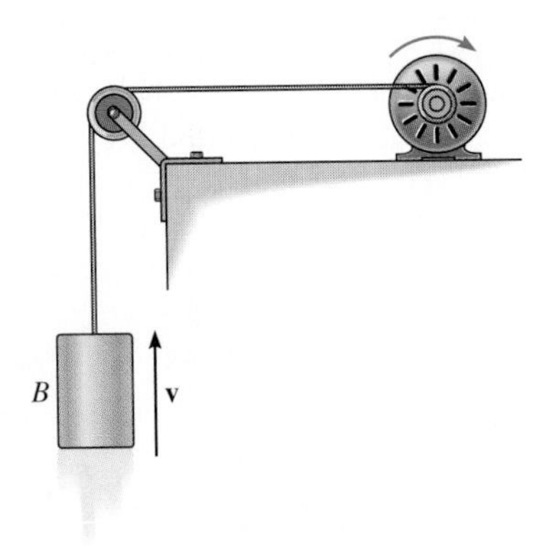

Prob. 15–29

15–30. The log has a mass of 500 kg and rests on the ground for which the coefficients of static and kinetic friction are $\mu_s = 0.5$ and $\mu_k = 0.4$, respectively. The winch delivers a horizontal towing force T to its cable at A which varies as shown in the graph. Determine the speed of the log when $t = 5$ s. Originally the tension in the cable is zero. *Hint:* First determine the force needed to begin moving the log.

15–31. The log has a mass of 500 kg and rests on the ground for which the coefficients of static and kinetic friction are $\mu_s = 0.5$ and $\mu_k = 0.4$, respectively. The winch delivers a horizontal towing force T to its cable at A which varies as shown in the graph. How much time is required to give the log a speed of 10 m/s? Originally the tension in the cable is zero. *Hint:* First determine the force needed to begin moving the log.

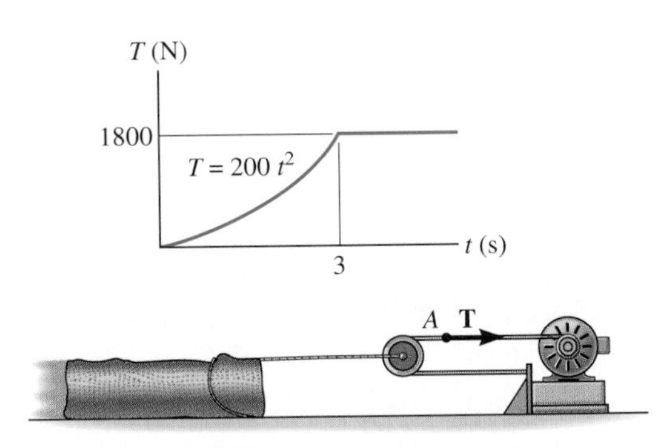

Probs. 15–30/15–31

***15–32.** The block has a mass of 5 kg and is released from rest at A. It slides down the smooth plane onto the rough surface having a coefficient of kinetic friction of $\mu_k = 0.2$. Determine the total time of travel before the block stops sliding. Also, how far s does the box slide before stopping? Neglect the size of the block.

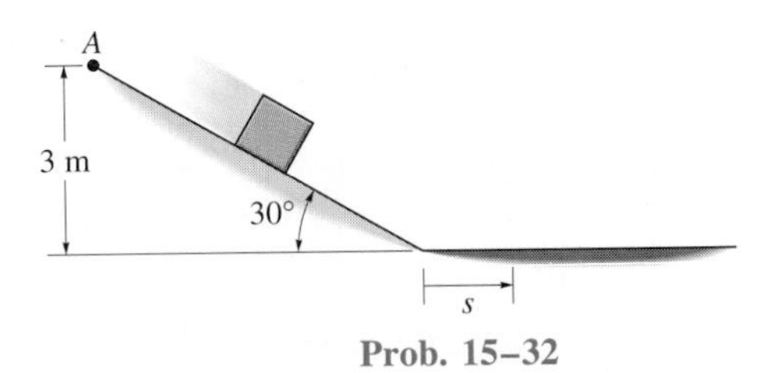

Prob. 15–32

15.2 Principle of Linear Impulse and Momentum for a System of Particles

The principle of linear impulse and momentum for a system of particles moving relative to an inertial reference, Fig. 15–7, is obtained from the equation of motion $\Sigma\mathbf{F}_i = \Sigma m_i\mathbf{a}_i$, Eq. 13–5, which may be rewritten as

$$\Sigma\mathbf{F}_i = \Sigma m_i \frac{d\mathbf{v}_i}{dt} \tag{15–5}$$

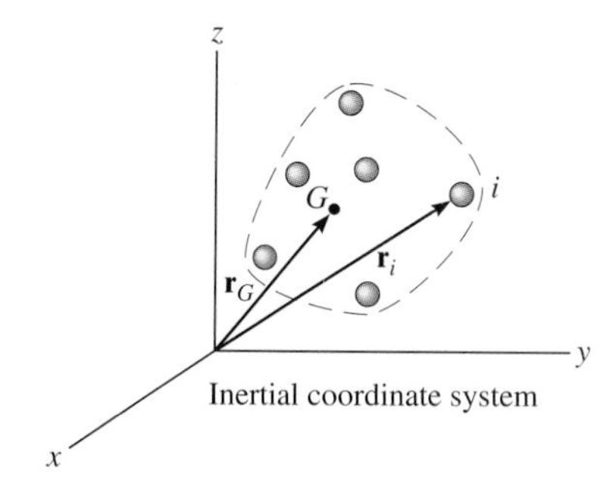

Fig. 15–7

The term on the left side represents only the sum of all the *external forces* acting on the system of particles. Recall that the internal forces $\mathbf{f}_i$ between particles do not appear with this summation, since by Newton's third law they occur in equal but opposite collinear pairs and therefore cancel out. Multiplying both sides of Eq. 15–5 by dt and integrating between the limits $t = t_1$, $\mathbf{v}_i = (\mathbf{v}_i)_1$ and $t = t_2$, $\mathbf{v}_i = (\mathbf{v}_i)_2$ yields

$$\Sigma m_i(\mathbf{v}_i)_1 + \Sigma \int_{t_1}^{t_2} \mathbf{F}_i\, dt = \Sigma m_i(\mathbf{v}_i)_2 \tag{15–6}$$

This equation states that the initial linear momenta of the system added vectorially to the impulses of all the *external forces* acting on the system during the time period t_1 to t_2 are equal to the system's final linear momenta.

By definition, the location of the mass center G of the system is determined from $m\mathbf{r}_G = \Sigma m_i\mathbf{r}_i$, where $m = \Sigma m_i$ is the total mass of all the particles, and $\mathbf{r}_G$ and $\mathbf{r}_i$ are defined in Fig. 15–7. Taking the time derivatives, we have

$$m\mathbf{v}_G = \Sigma m_i\mathbf{v}_i$$

which states that the total linear momentum of the system of particles is equivalent to the linear momentum of a "fictitious" aggregate particle of mass $m = \Sigma m_i$ moving with the velocity of the mass center G of the system. Substituting into Eq. 15–6 yields

$$m(\mathbf{v}_G)_1 + \Sigma \int_{t_1}^{t_2} \mathbf{F}\, dt = m(\mathbf{v}_G)_2 \tag{15–7}$$

This equation states that the initial linear momentum of the aggregate particle plus (vectorially) the external impulses acting on the system of particles during the time interval t_1 to t_2 is equal to the aggregate particle's final linear momentum. Since in reality all particles must have finite size to possess mass, the above equation justifies application of the principle of linear impulse and momentum to a rigid body represented as a single particle.

15.3 Conservation of Linear Momentum for a System of Particles

When the sum of the *external impulses* acting on a system of particles is *zero,* Eq. 15–6 reduces to a simplified form, namely,

$$\Sigma m_i(\mathbf{v}_i)_1 = \Sigma m_i(\mathbf{v}_i)_2 \tag{15-8}$$

This equation is referred to as the *conservation of linear momentum.* It states that the vector sum of the linear momenta for a system of particles remains constant throughout the time period t_1 to t_2. Substituting $m\mathbf{v}_G = \Sigma m_i \mathbf{v}_i$ into Eq. 15–8, we can also write

$$(\mathbf{v}_G)_1 = (\mathbf{v}_G)_2 \tag{15-9}$$

which indicates that the velocity $\mathbf{v}_G$ of the mass center for the system of particles does not change when no external impulses are applied to the system.

The conservation of linear momentum is often applied when particles collide or interact. For application, a careful study of the free-body diagram for the *entire* system of particles should be made. By doing this, one will be able to identify the forces as creating either external or internal impulses and thereby determine in what direction(s) linear momentum is conserved. As stated earlier, the *internal impulses* for the system will always cancel out, since they occur in equal but opposite collinear pairs. If the time period over which the motion is studied is *very short,* some of the external impulses may also be neglected or considered approximately equal to zero. The forces causing these negligible impulses are called *nonimpulsive forces.* By comparison, forces which are very large, act for a very short period of time, and yet produce a significant change in momentum are called *impulsive forces.* They, of course, cannot be neglected in the impulse–momentum analysis.

Impulsive forces normally occur due to an explosion or the striking of one body against another, whereas nonimpulsive forces may include the weight of a body, the force imparted by a slightly deformed spring having a relatively small stiffness, or for that matter, any force that is very small compared to other larger (impulsive) forces. When making this distinction between impulsive and nonimpulsive forces, it is important to realize that it only applies during a *specific time period.* To illustrate, consider the effect of striking a baseball with a bat. During the *very short* time of interaction, the force of the bat on the ball (or particle) is impulsive since it changes the ball's momentum drastically. By comparison, the ball's weight will have a negligible effect on the change in momentum, and therefore it is nonimpulsive. Consequently, it can be neglected from an impulse–momentum analysis during this time period. It should be pointed out, however, that if an impulse–momentum analysis is considered during the much longer time of flight after the ball–bat interaction, then the impulse of the ball's weight is important since it, along with air resistance, causes the change in the momentum of the ball.

PROCEDURE FOR ANALYSIS

Generally, the principle of linear impulse and momentum or the conservation of linear momentum is applied to a *system of particles* in order to determine the final velocities of the particles *just after* the time period considered. By applying these equations to the entire system, the internal impulses acting within the system, which may be unknown, are *eliminated* from the analysis. For application it is suggested that the following procedure be used.

Free-Body Diagram. Establish the x, y, z inertial frame of reference and draw the free-body diagram for each particle of the system in order to identify the internal and external forces. The conservation of linear momentum applies to the system in a given direction when *no external impulsive forces* act on the system in that direction. Also, establish the direction and sense of the particles' initial and final velocities. If the sense is unknown, assume it is along a positive inertial coordinate axis.

As an alternative procedure, draw the impulse and momentum diagrams for each particle of the system just before, during, and just after the impulsive forces are applied. Then investigate the impulse diagram in order to clearly distinguish the external impulsive and nonimpulsive forces from the system's internal impulses.

Momentum Equations. Apply the principle of linear impulse and momentum or the conservation of linear momentum in the appropriate directions. If the particles are subjected to dependent motion using cables and pulleys, kinematics as discussed in Sec. 12.8 can be used to relate the velocities.

If it is necessary to determine the *internal impulsive force* acting on only one particle of a system, the particle must be *isolated* (free-body diagram) and the principle of linear impulse and momentum must be applied *to the particle.* After the impulse $\int F\,dt$ is calculated, then, provided the time Δt for which the impulse acts is known, the *average impulsive force* F_{avg} can be determined from $F_{avg} = \int F\,dt/\Delta t$.

The following examples illustrate application of this procedure numerically.

Example 15–4

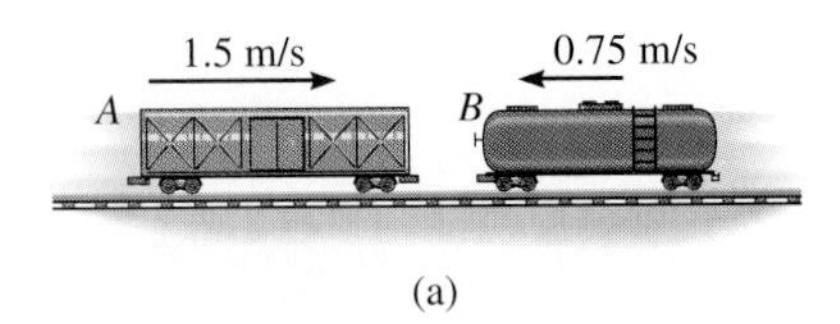

(a)

The 15-Mg boxcar A is coasting freely at 1.5 m/s on the horizontal track when it encounters a tank car B having a mass of 12 Mg and coasting at 0.75 m/s toward it as shown in Fig. 15–8*a*. If the cars meet and couple together, determine *(a)* the speed of both cars just after the coupling, and *(b)* the average force between them if the coupling takes place in 0.8 s.

SOLUTION

Part (a)

*Free-Body Diagram.** As shown in Fig. 15–8*b*, we have considered *both* cars as a single system. By inspection, momentum is conserved in the x direction since the coupling force **F** is *internal* to the system and will therefore cancel out. It is assumed both cars, when coupled, move at $\mathbf{v}_2$ in the positive x direction.

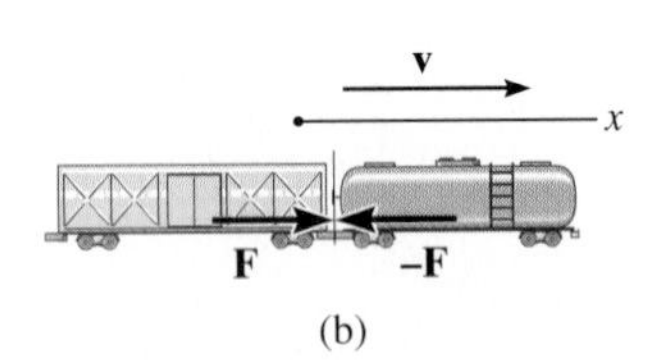

(b)

Conservation of Linear Momentum

$$(\overset{+}{\rightarrow}) \qquad m_A(v_A)_1 + m_B(v_B)_1 = (m_A + m_B)v_2$$

$$(15\,000 \text{ kg})(1.5 \text{ m/s}) - 12\,000 \text{ kg}(0.75 \text{ m/s}) = (27\,000 \text{ kg})v_2$$

$$v_2 = 0.5 \text{ m/s} \rightarrow \qquad \textit{Ans.}$$

Part (b). The average (impulsive) coupling force, $\mathbf{F}_{avg}$, can be determined by applying the principle of linear momentum to *either one* of the cars.

Free-Body Diagram. As shown in Fig. 15–8*c*, by isolating the boxcar the coupling force is *external* to the car.

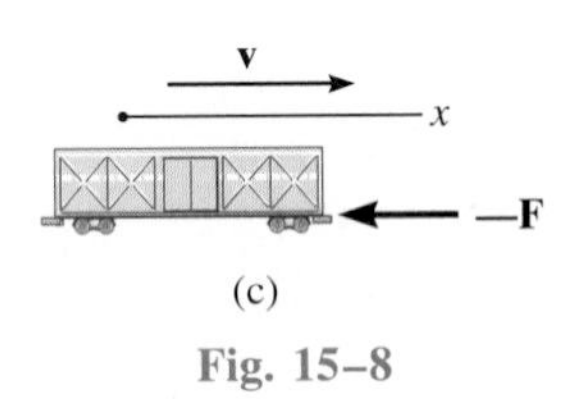

(c)

Fig. 15–8

Principle of Impulse and Momentum. Since $\int F\,dt = F_{avg}\,\Delta t = F_{avg}(0.8)$, we have

$$(\overset{+}{\rightarrow}) \qquad m_A(v_A)_1 + \Sigma \int F\,dt = m_A v_2$$

$$(15\,000 \text{ kg})(1.5 \text{ m/s}) - F_{avg}(0.8 \text{ s}) = (15\,000 \text{ kg})(0.5 \text{ m/s})$$

$$F_{avg} = 18.8 \text{ kN} \qquad \textit{Ans.}$$

Solution was possible here since the boxcar's final velocity was obtained in Part *(a)*. Try solving for F_{avg} by applying the principle of impulse and momentum to the tank car.

*Only horizontal forces are shown on the free-body diagram.

Example 15–5

The 1200-lb cannon shown in Fig. 15–9*a* fires an 8-lb projectile with a muzzle velocity of 1500 ft/s relative to the ground. If firing takes place in 0.03 s, determine *(a)* the recoil velocity of the cannon just after firing, and *(b)* the average impulsive force acting on the projectile. The cannon support is fixed to the ground and the horizontal recoil of the cannon is absorbed by two springs.

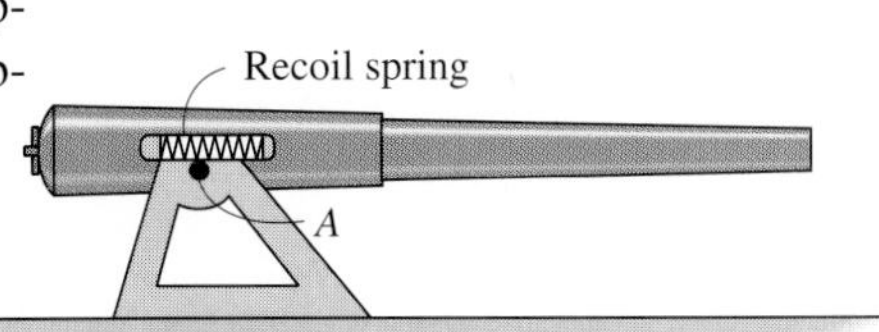

(a)

SOLUTION

Part (a)

Free-Body Diagram.* As shown in Fig. 15–9*b*, we have considered the projectile and cannon as a single system, because then the impulsive forces, **F**, between the cannon and projectile are *internal* to the system and will therefore cancel from the analysis. Furthermore, during the time $\Delta t = 0.03$ s, the two recoil springs which are attached to the support each exert a *nonimpulsive force* $\mathbf{F}_s$ on the cannon. This is because Δt is very short, so that during this time the cannon only moves through a very small distance† s. Consequently, $F_s = ks \approx 0$, where k is the spring's stiffness. Hence it may be concluded that momentum for the system is conserved in the *horizontal direction.* Here we will assume that the cannon moves to the left, while the projectile moves to the right after firing.

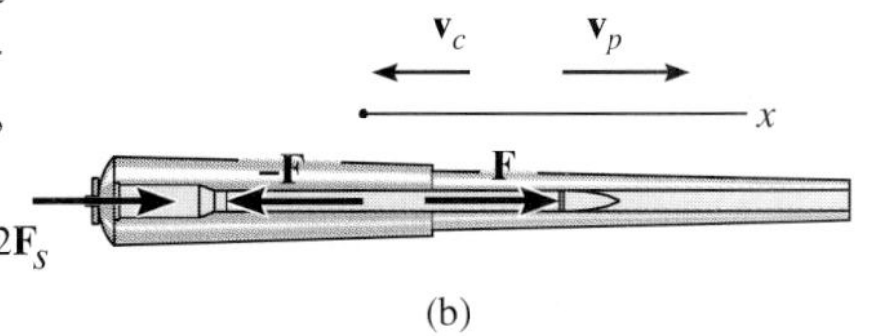

(b)

Conservation of Linear Momentum.

$$(\overset{+}{\rightarrow}) \quad m_c(v_c)_1 + m_p(v_p)_1 = -m_c(v_c)_2 + m_p(v_p)_2$$

$$0 + 0 = -\frac{1200 \text{ lb}}{32.2 \text{ ft/s}^2}(v_c)_2 + \frac{8 \text{ lb}}{32.2 \text{ ft/s}^2}(1500 \text{ ft/s})$$

$$(v_c)_2 = 10 \text{ ft/s} \leftarrow \qquad \textit{Ans.}$$

Part (b). The average impulsive force exerted by the cannon on the projectile can be determined by applying the principle of linear impulse and momentum to the projectile (or to the cannon). Why?

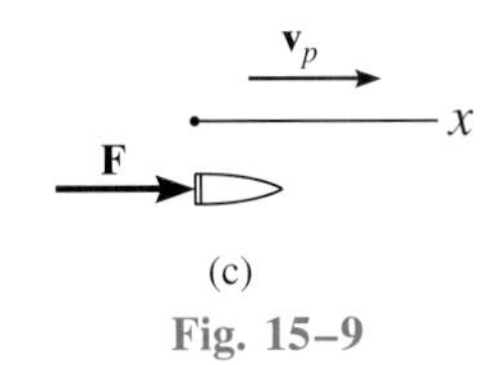

(c)

Fig. 15–9

Principle of Impulse and Momentum. Using the data on the free-body diagram, Fig. 15–9*c*, noting that $\int F\,dt = F_{avg}\,\Delta t = F_{avg}(0.03)$, we have

$$(\overset{+}{\rightarrow}) \qquad m(v_p)_1 + \Sigma \int F\,dt = m(v_p)_2$$

$$0 + F_{avg}(0.03 \text{ s}) = \frac{8 \text{ lb}}{32.2 \text{ ft/s}^2}(1500 \text{ ft/s})$$

$$F_{avg} = 12.4(10^3) \text{ lb} = 12.4 \text{ kip} \qquad \textit{Ans.}$$

*Only horizontal forces are shown on the free-body diagram.

†If the cannon is firmly fixed to its support (no springs), the reactive force of the support on the cannon must be considered as an external impulse to the system, since the support would allow no movement of the cannon.

Example 15–6

The 350-Mg tugboat T shown in Fig. 15–10a is used to pull the 50-Mg barge B with a rope R. If the barge is initially at rest, and the tugboat is coasting freely with a velocity of $(v_T)_1 = 3$ m/s while the rope is *slack*, determine the velocity of the tugboat *directly after* the rope becomes taut. Assume the rope does not stretch. Neglect the frictional effects of the water.

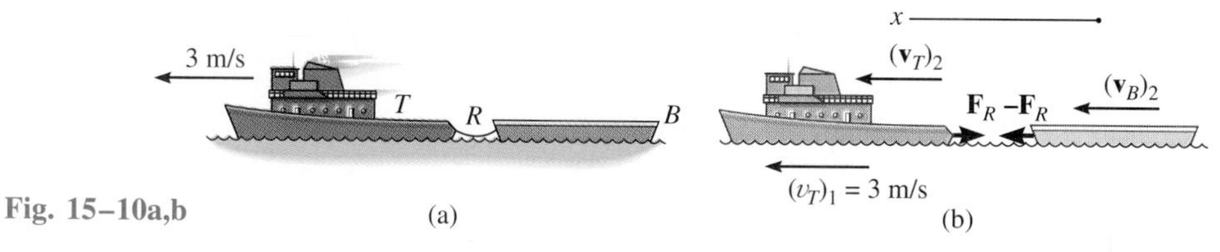

Fig. 15–10a,b

SOLUTION

Free-Body Diagram.* As shown in Fig. 15–10b, we have considered the entire system (tugboat and barge). Hence, the impulsive force created between the tugboat and the barge is *internal* to the system, and therefore momentum of the system is conserved during the instant of towing.

Conservation of Momentum. Noting that $(v_B)_2 = (v_T)_2$, we have

$$(\overset{+}{\leftarrow}) \qquad m_T(v_T)_1 + m_B(v_B)_1 = m_T(v_T)_2 + m_B(v_B)_2$$

$$350(10^3)\text{ kg}(3\text{ m/s}) + 0 = 350(10^3)\text{ kg}(v_T)_2 + 50(10^3)\text{ kg}(v_T)_2$$

Solving,

$$(v_T)_2 = 2.62\text{ m/s} \leftarrow \qquad \textit{Ans.}$$

This value represents the tugboat's velocity *just after* the towing impulse. Use this result and show that the towing *impulse* is 131 kN · s.

The alternative procedure of drawing the system's impulse and momentum diagrams is shown in Fig. 15–10c. The conservation of momentum is formulated as above by writing the vector terms directly from the diagrams.

*Only horizontal forces are shown on the free-body diagram.

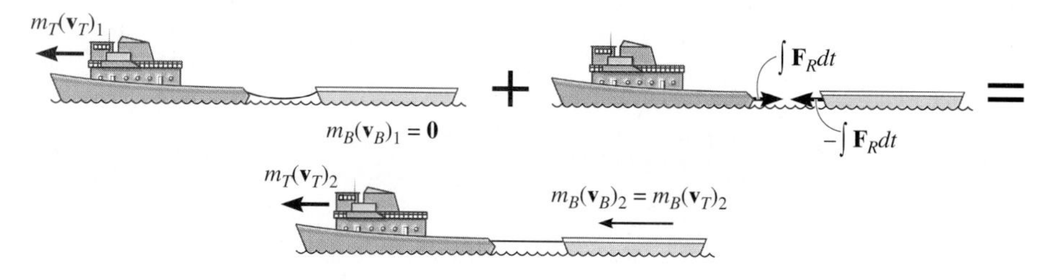

Fig. 15–10c

Example 15–7

A rigid pile P shown in Fig. 15–11*a* has a mass of 800 kg and is driven into the ground using a hammer H that has a mass of 300 kg. The hammer falls from rest from a height $y_0 = 0.5$ m and strikes the top of the pile. Determine the impulse which the hammer imparts on the pile if the pile is surrounded entirely by loose sand so that after striking the hammer does *not* rebound off the pile.

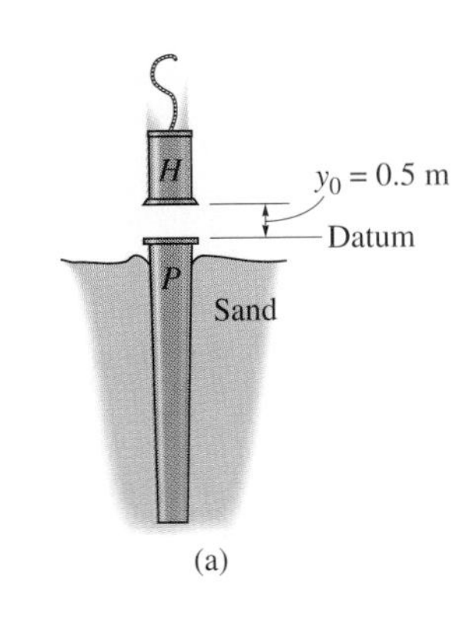

(a)

SOLUTION

Conservation of Energy. The velocity at which the hammer strikes the pile can be determined using the conservation of energy equation applied to the hammer. With the datum at the top of the pile, Fig. 15–11*a*, we have

$$T_0 + V_0 = T_1 + V_1$$

$$\tfrac{1}{2}m_H(v_H)_0^2 + W_H y_0 = \tfrac{1}{2}m_H(v_H)_1^2 + W_H y_1$$

$$0 + 300(9.81)\text{ N}(0.5\text{ m}) = \frac{1}{2}(300\text{ kg})(v_H)_1^2 + 0$$

$$(v_H)_1 = 3.13\text{ m/s}$$

Free-Body Diagram. From the physical aspects of the problem, the free-body diagram of the hammer and pile, Fig. 15–11*b*, indicates that during the *short time* occurring just before to just after the *collision* the weights of the hammer and pile and the resistance force $\mathbf{F}_s$ of the sand are all *nonimpulsive*. The impulsive force $\mathbf{R}$ is internal to the system and therefore cancels. Consequently, momentum is conserved in the vertical direction.

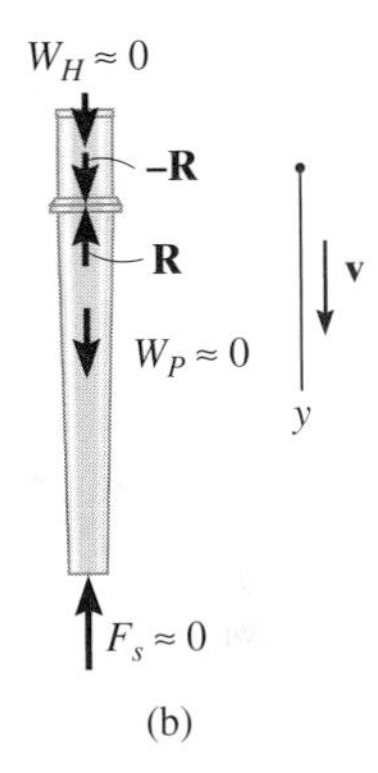

(b)

Conservation of Momentum. Since the hammer does not rebound off the pile just after collision, then $(v_H)_2 = (v_P)_2 = v_2$.

$$(+\downarrow) \qquad m_H(v_H)_1 + m_P(v_P)_1 = m_H v_2 + m_P v_2$$

$$(300\text{ kg})(3.13\text{ m/s}) + 0 = (300\text{ kg})v_2 + (800\text{ kg})v_2$$

$$v_2 = 0.854\text{ m/s}$$

Principle of Impulse and Momentum. The impulse which the hammer imparts to the pile can now be determined since $\mathbf{v}_2$ is known. From the free-body diagram for the hammer, Fig. 15–11*c*, we have

$$(+\downarrow) \qquad m_H(v_H)_1 + \Sigma\int_{t_1}^{t_2} F_y\,dt = m_H v_2$$

$$(300\text{ kg})(3.13\text{ m/s}) - \int R\,dt = (300\text{ kg})(0.854\text{ m/s})$$

$$\int R\,dt = 683\text{ N}\cdot\text{s} \qquad \textit{Ans.}$$

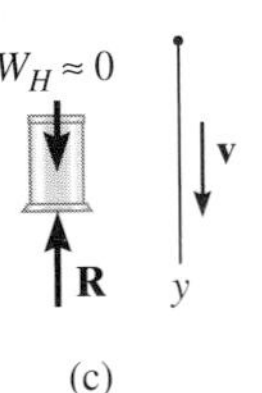

(c)

Fig. 15–11

Try finding this impulse by applying the principle of impulse and momentum to the pile.

Example 15–8

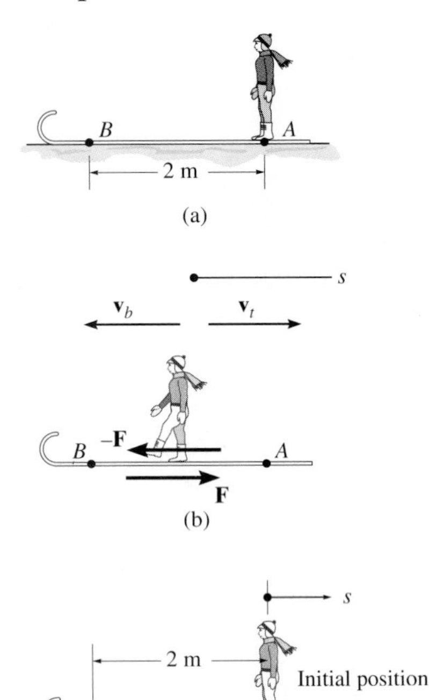

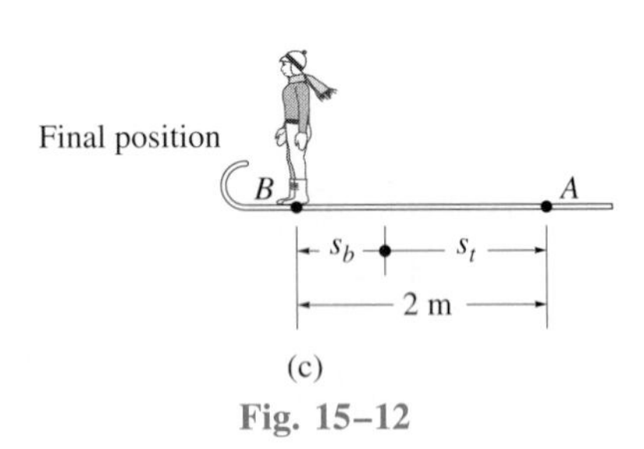

Fig. 15–12

A boy having a mass of 40 kg stands on the back of a 15-kg toboggan which is originally at rest, Fig. 15–12*a*. If he walks to the front *B* and stops, determine the distance the toboggan moves. Neglect friction between the bottom of the toboggan and the ground (ice).

SOLUTION I

Free-Body Diagram. The unknown frictional force of the boy's shoes on the bottom of the toboggan can be *excluded* from the analysis if the toboggan and boy on it are considered as a single system. In this way the frictional force **F** becomes *internal* and the conservation of momentum applies, Fig. 15–12*b*.

Conservation of Momentum. Since both the initial and final momenta of the system are zero (because the initial and final velocities are zero), the systems's momentum must also be zero when the boy is at some intermediate point between *A* and *B*. Thus

$$(\overset{+}{\rightarrow}) \qquad -m_b v_b + m_t v_t = 0 \qquad (1)$$

Here the two unknowns v_b and v_t represent the velocities of the boy moving to the left and the toboggan moving to the right. Both are measured from a *fixed inertial reference* on the ground.

At any instant the *position* of point *A* on the toboggan and the *position* of the boy must be determined by integration. Since $v = ds/dt$, then $-m_b ds_b + m_t ds_t = 0$. Assuming the initial position of point *A* to be at the origin, Fig. 15–12*c*, then at the final position we have $-m_b s_b + m_t s_t = 0$. Since $s_b + s_t = 2$ m, or $s_b = (2 - s_t)$ then

$$-m_b(2 - s_t) + m_t s_t = 0 \qquad (2)$$

$$s_t = \frac{2m_b}{m_b + m_t} = \frac{2(40)}{40 + 15} = 1.45 \text{ m} \qquad \textit{Ans.}$$

SOLUTION II

The problem may also be solved by considering the relative motion of the boy with respect to the toboggan, $\mathbf{v}_{b/t}$. This velocity is related to the velocities of the boy and toboggan by the equation $\mathbf{v}_b = \mathbf{v}_t + \mathbf{v}_{b/t}$, Eq. 12–34. Since positive motion is assumed to be to the right in Eq. 1, $\mathbf{v}_b$ and $\mathbf{v}_{b/t}$ are negative, because the boy's motion is to the left. Hence, in scalar form, $-v_b = v_t - v_{b/t}$, and Eq. 1 then becomes $m_b(v_t - v_{b/t}) + m_t v_t = 0$. Integrating gives

$$m_b(s_t - s_{b/t}) + m_t s_t = 0$$

Realizing $s_{b/t} = 2$ m, we obtain Eq. 2.

PROBLEMS

15–33. A railroad car having a mass of 15 Mg is coasting at 1.5 m/s on a horizontal track. At the same time another car having a mass of 12 Mg is coasting at 0.75 m/s in the opposite direction. If the cars meet and couple together, determine the speed of both cars just after the coupling. Compute the difference between the total kinetic energy before and after coupling has occurred, and explain qualitatively what happened to this energy.

15–34. The car A has a weight of 4500 lb and is traveling to the right at 3 ft/s. Meanwhile a 3000-lb car B is traveling at 6 ft/s to the left. If the cars crash head-on and become entangled, determine their common velocity just after the collision. Assume that the brakes are not applied during collision.

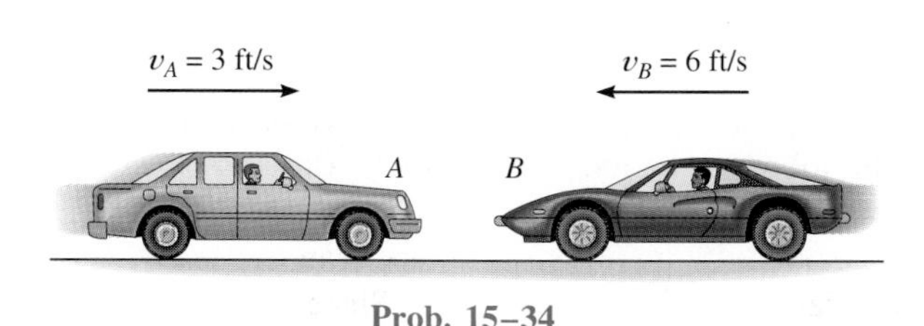

Prob. 15–34

15–35. The bus B has a weight of 15000 lb and is traveling to the right at 5 ft/s. Meanwhile a 3000-lb car A is traveling at 4 ft/s to the left. If the vehicles crash head-on and become entangled, determine their common velocity just after the collision. Assume that the vehicles are free to roll during collision.

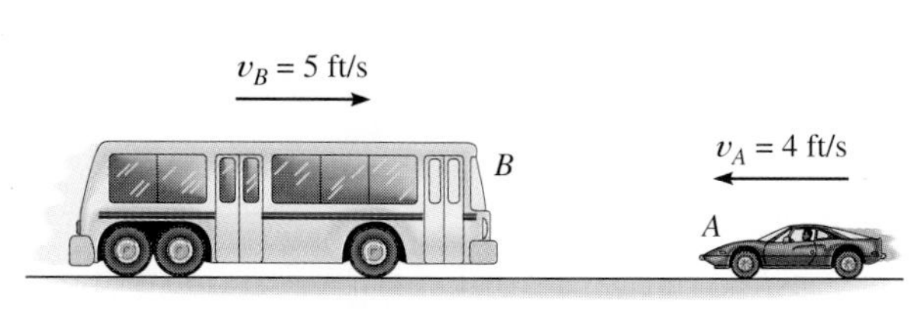

Prob. 15–35

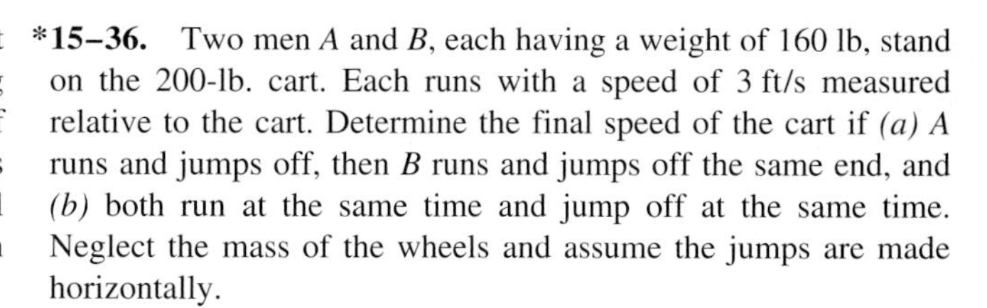

***15–36.** Two men A and B, each having a weight of 160 lb, stand on the 200-lb. cart. Each runs with a speed of 3 ft/s measured relative to the cart. Determine the final speed of the cart if *(a)* A runs and jumps off, then B runs and jumps off the same end, and *(b)* both run at the same time and jump off at the same time. Neglect the mass of the wheels and assume the jumps are made horizontally.

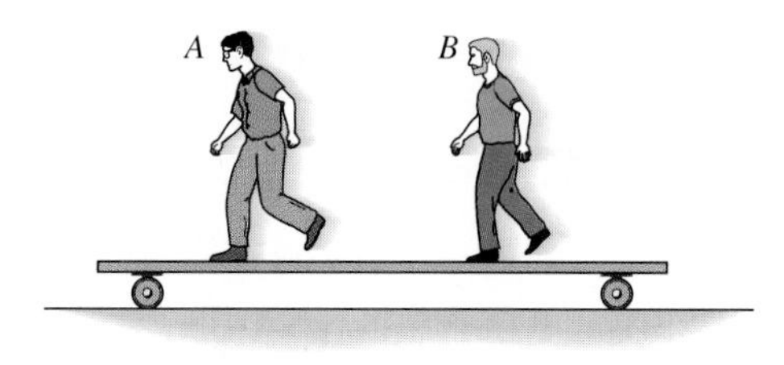

Prob. 15–36

15–37. The man M weighs 150 lb and jumps onto the boat B which has a weight of 200 lb. If he has a horizontal component of velocity *relative to the boat* of 3 ft/s, just before he enters the boat, and the boat is traveling $v_B = 2$ ft/s away from the pier when he makes the jump, determine the resulting velocity of the man and boat.

Prob. 15–37

15–38. The man M weighs 150 lb and jumps onto the boat B which is originally at rest. If he has a horizontal component of velocity of 3 ft/s just before he enters the boat, determine the weight of the boat if it has a velocity of 2 ft/s once the man enters it.

Prob. 15–38

15–39. A 20-g bullet is fired horizontally into the 300-g block which rests on the smooth surface. After the bullet becomes embedded into the block, the block moves to the right 0.3 m before momentarily coming to rest. Determine the speed $(v_B)_1$ of the bullet. The spring has a stiffness $k = 200$ N/m and is originally unstretched.

***15–40.** The 20-g bullet is fired horizontally at $(v_B)_1 = 1200$ m/s into the 300-g block which rests on the smooth surface. Determine the distance the block moves to the right before momentarily coming to rest. The spring has a stiffness $k = 200$ N/m and is originally unstretched.

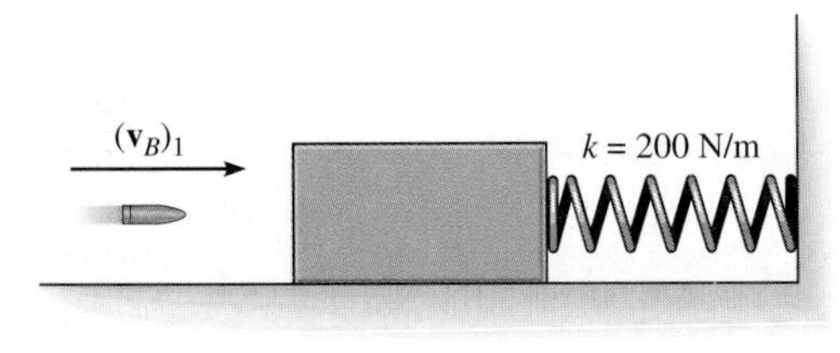

Probs. 15–39/15–40

15–41. A toboggan having a mass of 10 kg starts from rest at A and carries a girl and boy having a mass of 40 kg and 45 kg, respectively. When the toboggan reaches the bottom of the slope at B, the boy is pushed off from the back with a horizontal velocity of $v_{b/t} = 2$ m/s, measured relative to the toboggan. Determine the velocity of the toboggan afterwards. Neglect friction in the calculation.

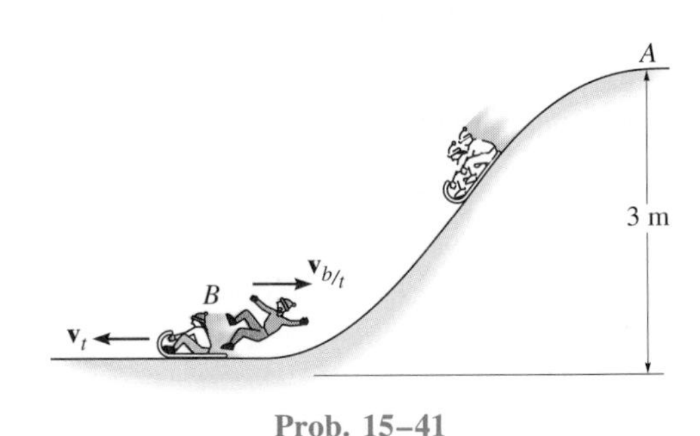

Prob. 15–41

15–42. A boy A having a weight of 80 lb and a girl B having a weight of 65 lb stand motionless at the ends of the toboggan, which has a weight of 20 lb. If they exchange positions, A going to B and then B going to A's original position, determine the final position of the toboggan just after the motion. Neglect friction.

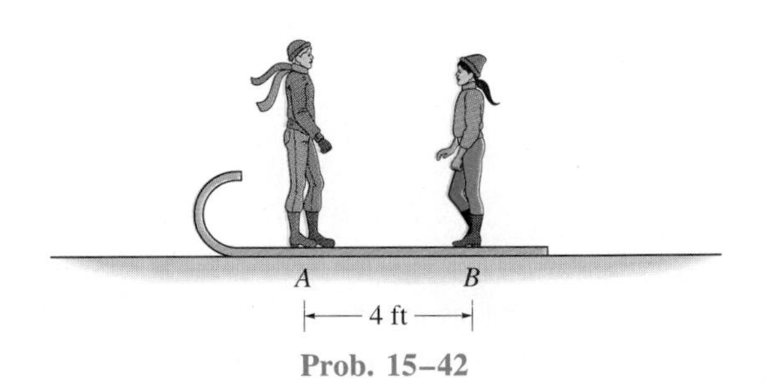

Prob. 15–42

15–43. A boy A having a weight of 80 lb and a girl B having a weight of 65 lb stand motionless at the ends of the toboggan, which has a weight of 20 lb. If A walks to B and stops, and both walk back together to the original position of A, determine the final position of the toboggan just after the motion stops. Neglect friction.

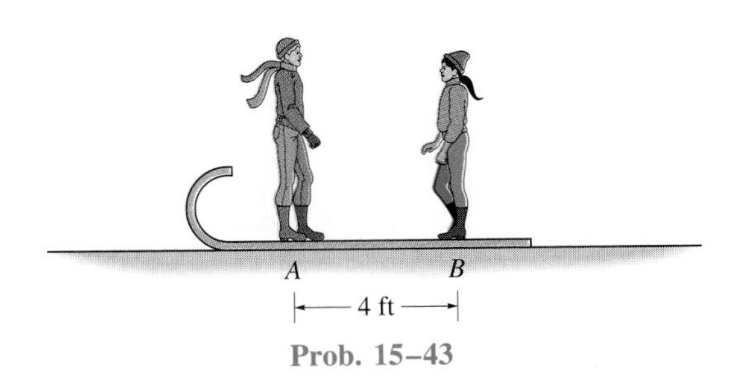

Prob. 15–43

***15–44.** The 10-lb projectile is fired from ground level with an initial velocity of $v_A = 80$ ft/s in the direction shown. When it reaches its highest point B it explodes into two 5-lb fragments. If one fragment travels vertically upward at 12 ft/s, determine the distance between the fragments after they strike the ground. Neglect the size of the gun.

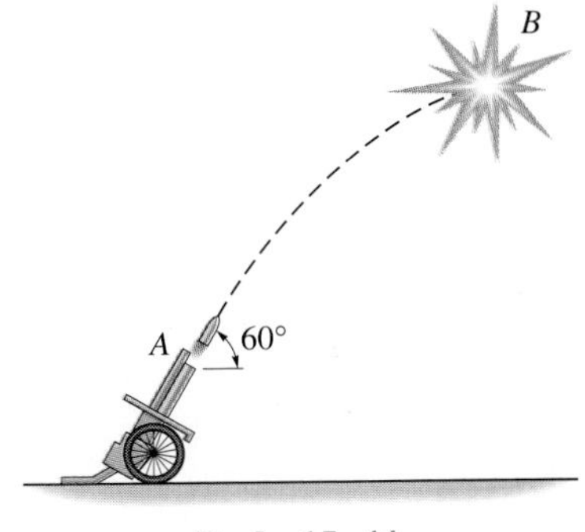

Prob. 15–44

15–45. The 10-lb projectile is fired from ground level with an initial velocity of $v_A = 80$ ft/s in the direction shown. When it reaches its highest point B it explodes into two 5-lb fragments. If one fragment is seen to travel vertically upward, and after they fall they are 150 ft apart, determine the speed of each fragment just after the explosion. Neglect the size of the gun.

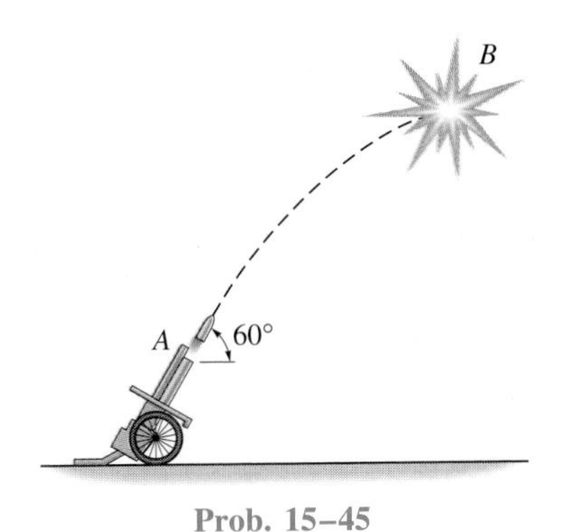

Prob. 15–45

15–46. A tugboat T having a mass of 19 Mg is tied to a barge B having a mass of 75 Mg. If the rope is ''elastic'' such that it has a stiffness $k = 600$ kN/m, determine the maximum stretch in the rope during the initial towing. Originally both the tugboat and barge are moving in the same direction with speeds $(v_T)_1 = 15$ km/h and $(v_B)_1 = 10$ km/h, respectively. Neglect the resistance of the water.

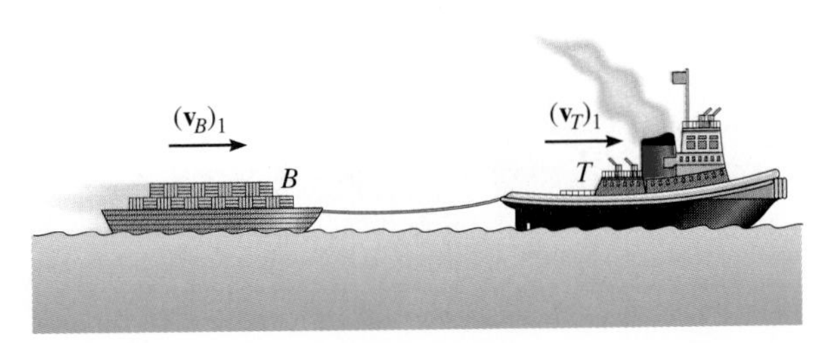

Prob. 15–46

15–47. The boy jumps off the flat car at A with a velocity of $v' = 4$ ft/s relative to the car as shown. If he lands on the second flat car B, determine the final speed of both cars after the motion. Each car has a weight of 80 lb. The boy's weight is 60 lb. Both cars are originally at rest. Neglect the mass of the car's wheels.

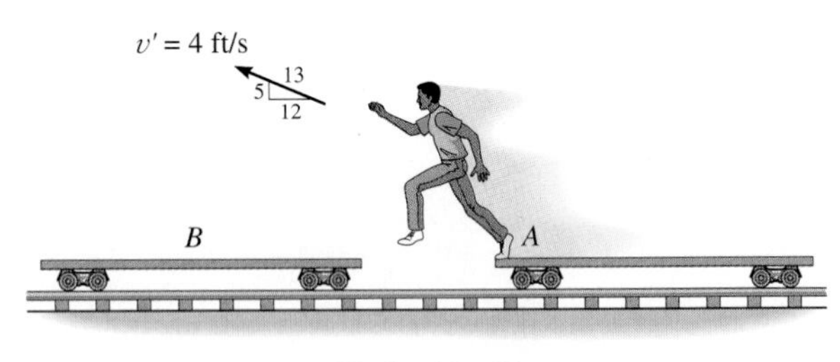

Prob. 15–47

***15–48.** The boy B jumps off the canoe at A with a velocity of 5 m/s relative to the canoe as shown. If he lands in the second canoe C, determine the final speed of both canoes after the motion. Each canoe has a mass of 40 kg. The boy's mass is 30 kg, and the girl D has a mass of 25 kg. Both canoes are originally at rest.

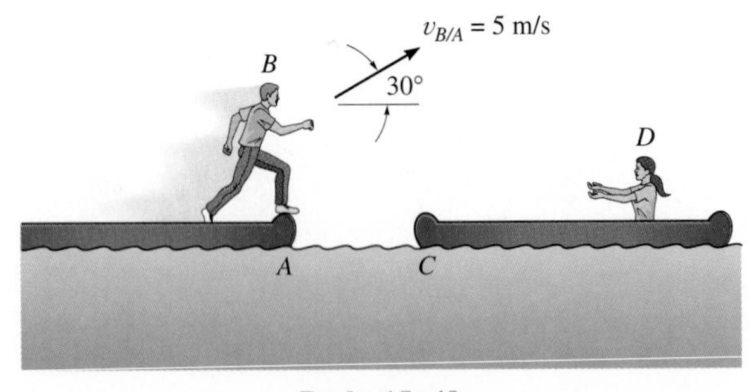

Prob. 15–48

15–49. Block A has a mass of 5 kg and is placed on the smooth triangular block B having a mass of 30 kg. If the system is released from rest, determine the distance B moves from point O when A reaches the bottom. Neglect the size of block A.

15–50. Solve Prob. 15–49 if the coefficient of kinetic friction between A and B is $\mu_k = 0.3$. Neglect friction between block B and the horizontal plane.

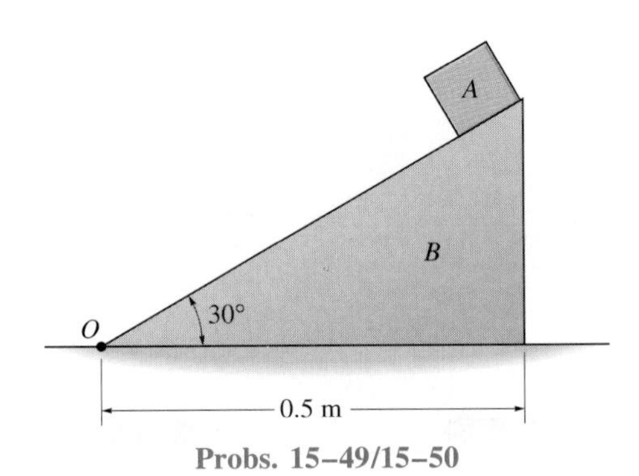

Probs. 15–49/15–50

15–51. The crate has a mass m and rests on the barge which has a mass M and is initially at rest. The coefficient of kinetic friction between the crate and barge is μ. The rope is pulled (or jerked) such that it snaps and as a result the impulse gives the barge a sudden velocity $\mathbf{v}_0$. Determine the time the crate slides on the barge before coming to rest on the barge. Also, what is the final velocity of the barge?

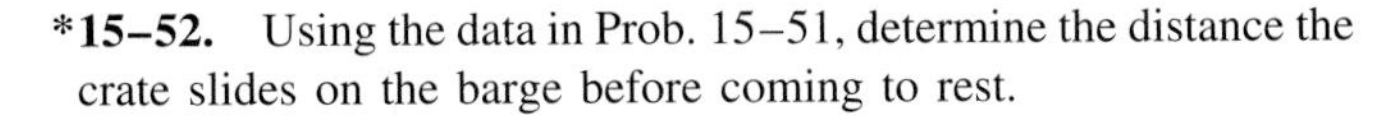

***15–52.** Using the data in Prob. 15–51, determine the distance the crate slides on the barge before coming to rest.

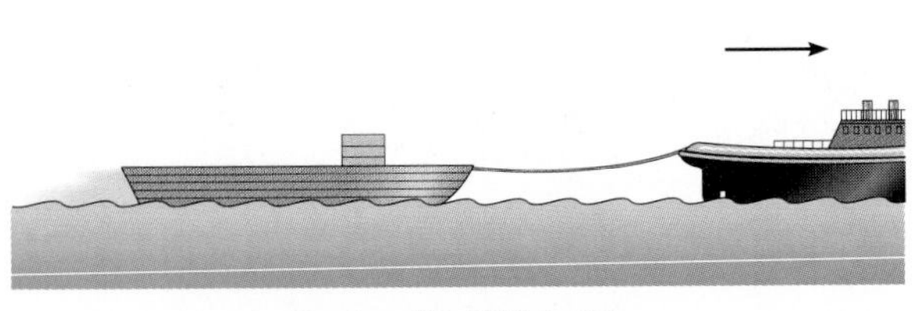
Probs. 15–51/15–52

15–53. The block has a mass of 50 kg and rests on the surface of the cart having a mass of 75 kg. If the spring which is attached to the cart and not the block is compressed 0.2 m and the system is released from rest, determine the speed of the block after the spring becomes undeformed. Neglect the mass of the cart's wheels and the spring in the calculation. Also neglect friction. Take $k = 300$ N/m.

15–54. The block has a mass of 50 kg and rests on the surface of the cart having a mass of 75 kg. If the spring which is attached to the cart and not the block is compressed 0.2 m and the system is released from rest, determine the speed of the block with respect to the cart after the spring becomes undeformed. Neglect the mass of the wheels and the spring in the calculation. Also neglect friction. Take $k = 300$ N/m.

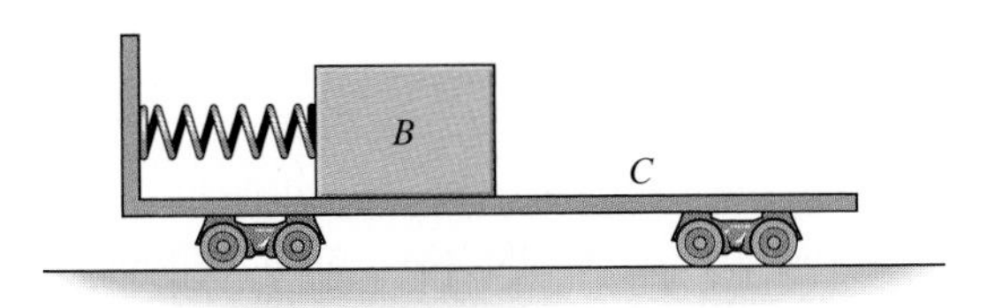

Probs. 15–53/15–54

15–55. The free-rolling ramp has a weight of 120 lb. The crate, whose weight is 80 lb, slides from rest at A, 15 ft down the ramp to B. Determine the ramp's speed when the crate reaches B. Assume that the ramp is smooth, and neglect the mass of the wheels.

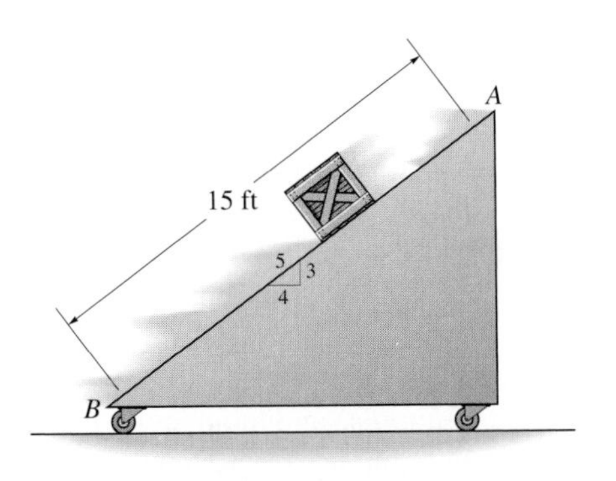

Prob. 15–55

***15–56.** The free-rolling ramp has a weight of 120 lb. If the 80 lb crate is released from rest at A, determine the distance the ramp moves when the crate slides 15 ft down the ramp and reaches the bottom B.

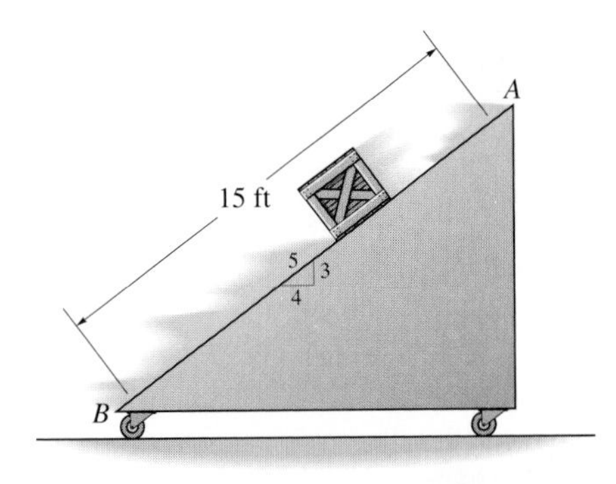

Prob. 15–56

15–57. Blocks A and B have masses of 40 kg and 60 kg, respectively. They are placed on a smooth surface and the spring connected between them is stretched 2 m. If they are released from rest, determine the speeds of both blocks the instant the spring becomes unstretched.

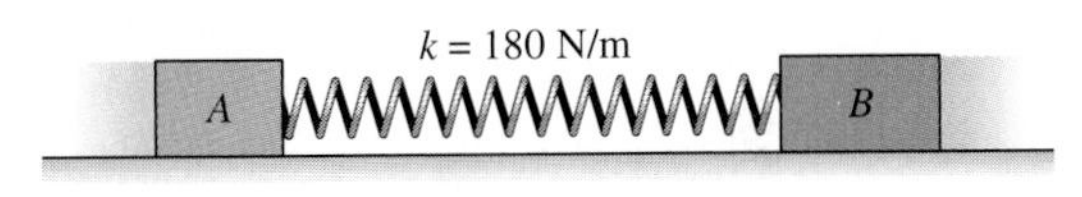

Prob. 15–57

15–58. Blocks A and B have masses of 40 kg and 60 kg, respectively. They are placed on a smooth surface and the spring connected between them is compressed 1.5 m. If they are released from rest, determine the speeds of both blocks the instant the spring becomes uncompressed.

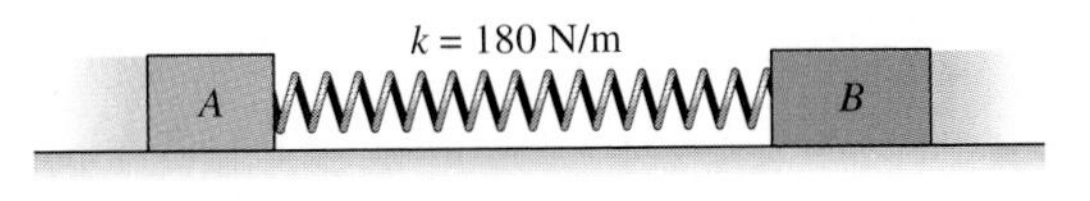

Prob. 15–58

15.4 Impact

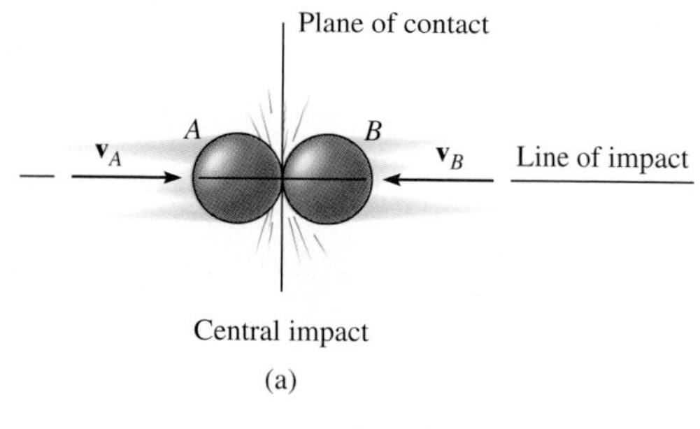

Central impact

(a)

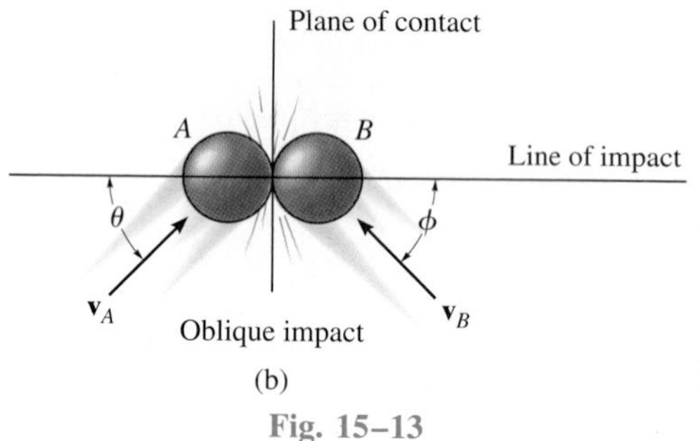

Oblique impact

(b)

Fig. 15–13

Impact occurs when two bodies collide with each other during a very *short* interval of time, causing relatively large (impulsive) forces to be exerted between the bodies. The striking of a hammer and nail, or a golf club and ball, are common examples of impact loadings.

In general, there are two types of impact. *Central impact* occurs when the direction of motion of the mass centers of the two colliding particles is along a line passing through the mass centers of the particles. This line is called the *line of impact,* Fig. 15–13*a*. When the motion of one or both of the particles is at an angle with the line of impact, Fig. 15–13*b*, the impact is said to be *oblique impact.*

Central Impact. To illustrate the method for analyzing the mechanics of impact, consider the case involving the central impact of two *smooth* particles A and B shown in Fig. 15–14.

1. The particles have the initial momenta shown in Fig. 15–14*a*. Provided $(v_A)_1 > (v_B)_1$, collision will eventually occur.
2. During the collision the material of the particles must be thought of as *deformable* or nonrigid. The particles will undergo a *period of deformation* such that they exert an equal but opposite deformation impulse $\int \mathbf{P}\,dt$ on each other, Fig. 15–14*b*.
3. Only at the instant of *maximum deformation* will both particles move with a common velocity $\mathbf{v}$, Fig. 15–14*c*.
4. Afterward a *period of restitution* occurs, in which case the material from which the particles are made will either return to its original shape or remain permanently deformed. The equal but opposite *restitution impulse* $\int \mathbf{R}\,dt$ pushes the particles apart from one another, Fig. 15–14*d*. In reality, the physical properties of any two bodies are such that the deformation impulse is *always greater* than that of restitution, i.e., $\int P\,dt > \int R\,dt$.
5. Just after separation the particles will have the final momenta shown in Fig. 15–14*e*, where $(v_B)_2 > (v_A)_2$.

In most problems the initial velocities of the particles will be *known* and it will be necessary to determine their final velocities $(v_A)_2$ and $(v_B)_2$. In this regard, *momentum* for the *system of particles* is *conserved* since during collision the internal impulses of deformation and restitution *cancel*. Hence, referring to Figs. 15–14*a* and *e* requires that

$$(\overset{+}{\rightarrow}) \qquad m_A(v_A)_1 + m_B(v_B)_1 = m_A(v_A)_2 + m_B(v_B)_2 \qquad (15\text{–}10)$$

In order to obtain a second equation, necessary to solve for $(v_A)_2$ and $(v_B)_2$, we must apply the principle of impulse and momentum to *each particle.* For example, during the deformation phase for particle A, Figs. 15–14*a*, 15–14*b*, and 15–14*c*, we have

$$(\overset{+}{\rightarrow}) \qquad m_A(v_A)_1 - \int P\,dt = m_A v$$

For the restitution phase, Figs. 15–14*c*, 15–14*d*, and 15–14*e*,

$$(\overset{+}{\rightarrow}) \qquad m_A v - \int R\,dt = m_A(v_A)_2$$

The ratio of the restitution impulse to the deformation impulse is called the *coefficient of restitution, e.* From the above equations, this value for particle *A* is

$$e = \frac{\int R\,dt}{\int P\,dt} = \frac{v - (v_A)_2}{(v_A)_1 - v}$$

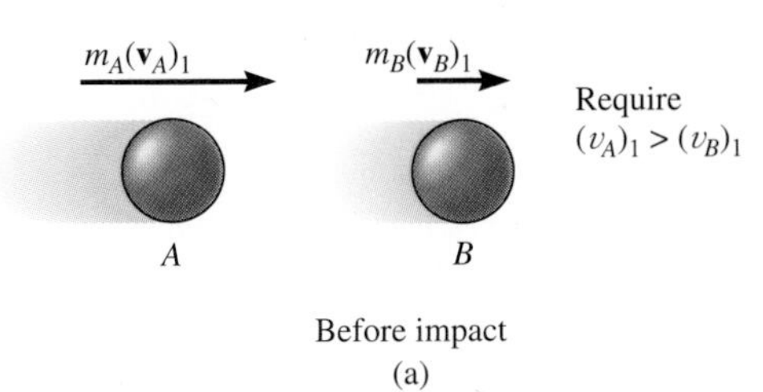

Before impact
(a)

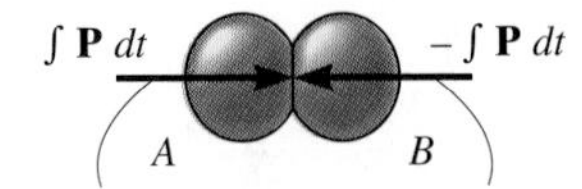

Deformation impulse
(b)

In a similar manner, we can establish *e* by considering particle *B*, Fig. 15–14. This yields

$$e = \frac{\int R\,dt}{\int P\,dt} = \frac{(v_B)_2 - v}{v - (v_B)_1}$$

If the unknown v is eliminated from the above two equations, the coefficient of restitution can be expressed in terms of the particles' initial and final velocities as

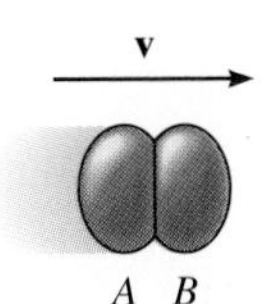

Maximum deformation
(c)

$$(\overset{+}{\rightarrow}) \qquad e = \frac{(v_B)_2 - (v_A)_2}{(v_A)_1 - (v_B)_1} \qquad (15\text{–}11)$$

Provided a value for *e* is specified, Eqs. 15–10 and 15–11 may be solved simultaneously to obtain $(v_A)_2$ and $(v_B)_2$. In doing so, however, it is important to carefully establish a sign convention for defining the positive direction for both $\mathbf{v}_B$ and $\mathbf{v}_A$, and then use it *consistently* when writing *both* equations. As noted from the application above, and indicated symbolically by the arrow in parentheses, we have defined the positive direction to the right when referring to the motions of both *A* and *B*. Consequently, if a negative value results from the solution of either $(v_A)_2$ or $(v_B)_2$, it indicates motion is to the left.

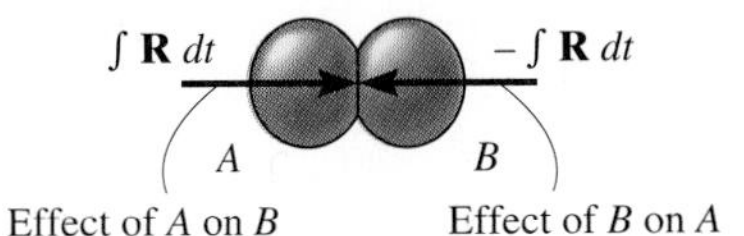

Restitution impulse
(d)

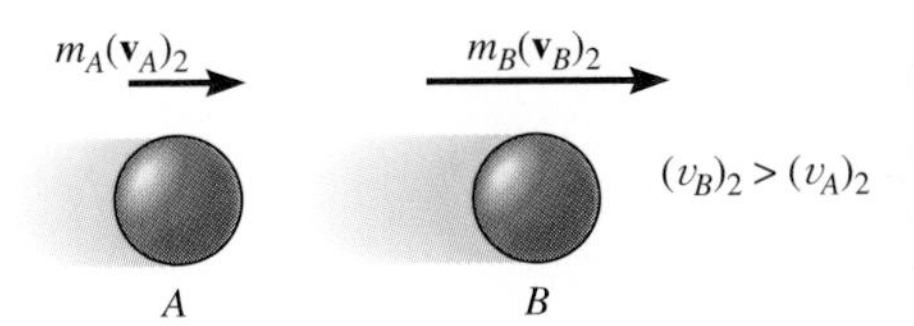

After impact
(e)

Fig. 15–14

Coefficient of Restitution. With reference to Figs. 15–14*a* and 15–14*e*, it is seen that Eq. 15–11 states that the coefficient of restitution is equal to the ratio of the relative velocity of the particles' separation *just after impact,* $(v_B)_2 - (v_A)_2$, to the relative velocity of the particles' approach *just before impact,* $(v_A)_1 - (v_B)_1$. By measuring these relative velocities experimentally, it has been found that *e* varies appreciably with impact velocity as well as with the size and shape of the colliding bodies. For these reasons the coefficient of restitution is reliable only when used under conditions which closely approximate those which were known to exist when measurements of it were made. In general e has a value between zero and one, and one should be aware of the physical meaning of these two limits.

Elastic Impact ($e = 1$): If the collision between the two particles is *perfectly elastic,* the deformation impulse ($\int \mathbf{P}\, dt$) is equal and opposite to the restitution impulse ($\int \mathbf{R}\, dt$). Although in reality this can never be achieved, $e = 1$ for an elastic collision.

Plastic Impact ($e = 0$): The impact is said to be *inelastic or plastic* when $e = 0$. In this case there is no restitution impulse given to the particles ($\int \mathbf{R}\, dt = \mathbf{0}$), so that after collision both particles couple or stick *together* and move with a common velocity.

From the above derivation it should be evident that the principle of work and energy cannot be used for the analysis of this problem since it is not possible to know how the *internal forces* of deformation and restitution vary or move during the collision. By knowing the particle's velocities before and after collision, however, the energy loss during collision can be calculated on the basis of the difference in the particle's kinetic energy. This energy loss, $\Sigma U_{1-2} = \Sigma T_2 - \Sigma T_1$, occurs because some of the initial kinetic energy of the particle is transformed into thermal energy as well as creating sound and localized deformation of the material when the collision occurs. In particular, if the impact is *perfectly elastic,* no energy is lost in the collision; whereas if the collision is *plastic,* the energy lost during collision is a maximum.

PROCEDURE FOR ANALYSIS (CENTRAL IMPACT)

In most cases the *final velocities* of two smooth particles are to be determined *just after* they are subjected to direct central impact. Provided the coefficient of restitution, the mass of each particle, and each particle's initial velocity *just before* impact are known, the solution to the problem can be obtained using the following two equations:

1. The conservation of momentum applies to the system of particles, $\Sigma m v_1 = \Sigma m v_2$.
2. The coefficient of restitution, $e = ((v_B)_2 - (v_A)_2)/((v_A)_1 - (v_B)_1)$, relates the relative velocities of the particles along the line of impact, just before and just after collision.

When applying these two equations, the sense of an unknown velocity can be assumed. If the solution yields a negative magnitude, the velocity acts in the opposite sense.

Oblique Impact. When oblique impact occurs between two smooth particles, the particles move away from each other with velocities having unknown directions as well as unknown magnitudes. Provided the initial velocities are known, four unknowns are present in the problem. As shown in Fig. 15–15*a*, these unknowns may be represented either as $(v_A)_2$, $(v_B)_2$, θ_2, and ϕ_2, or as the x and y components of the final velocities.

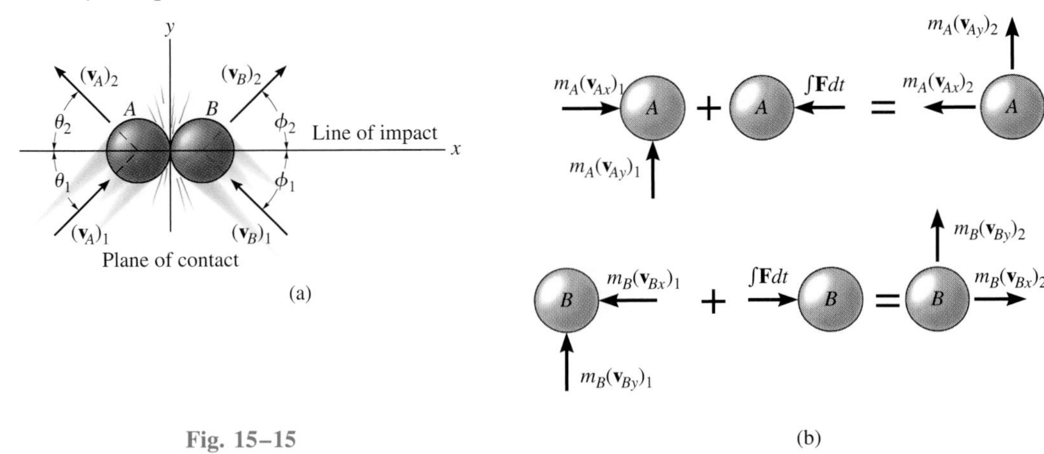

Fig. 15–15

PROCEDURE FOR ANALYSIS (OBLIQUE IMPACT)

If the y axis is established within the plane of contact and the x axis along the line of impact, the impulsive forces of deformation and restitution act *only in the x direction,* Fig. 15–15*b*. Resolving the velocity or momentum vectors into components along the x and y axes, Fig. 15–15*b*, it is possible to write four independent scalar equations in order to determine $(v_{Ax})_2$, $(v_{Ay})_2$, $(v_{Bx})_2$, and $(v_{By})_2$.

1. Momentum of the system is conserved *along the line of impact*, x axis, so that $\Sigma m(v_x)_1 = \Sigma m(v_x)_2$.
2. The coefficient of restitution, $e = ((v_{Bx})_2 - (v_{Ax})_2)/((v_{Ax})_1 - (v_{Bx})_1)$, relates the relative-velocity *components* of the particles *along the line of impact* (x axis).
3. Momentum of particle A is conserved along the y axis, perpendicular to the line of impact, since no impulse acts on particle A in this direction.
4. Momentum of particle B is conserved along the y axis, perpendicular to the line of impact, since no impulse acts on particle B in this direction.

Application of these four equations is illustrated numerically in Example 15–11.

Example 15–9

The bag A, having a weight of 6 lb, is released from rest at the position $\theta = 0°$, as shown in Fig. 15–16a. After falling $\theta = 90°$, it strikes an 18-lb box B. If the coefficient of restitution between the bag and box is $e = 0.5$, determine the velocities of the bag and box just after impact and the loss of energy during collision.

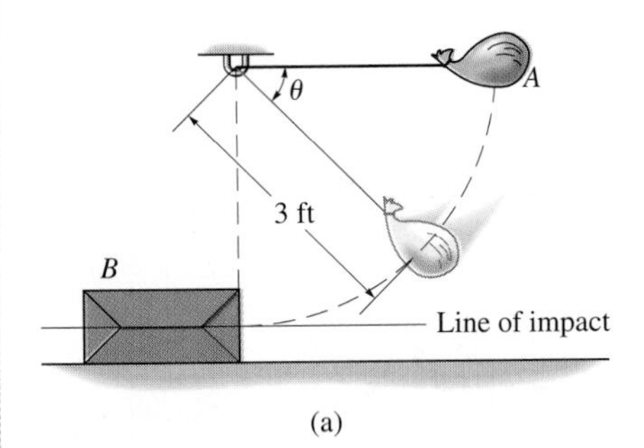

(a)

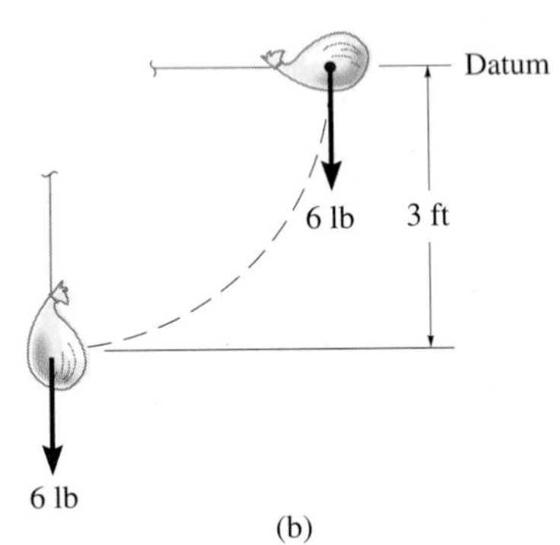

(b)

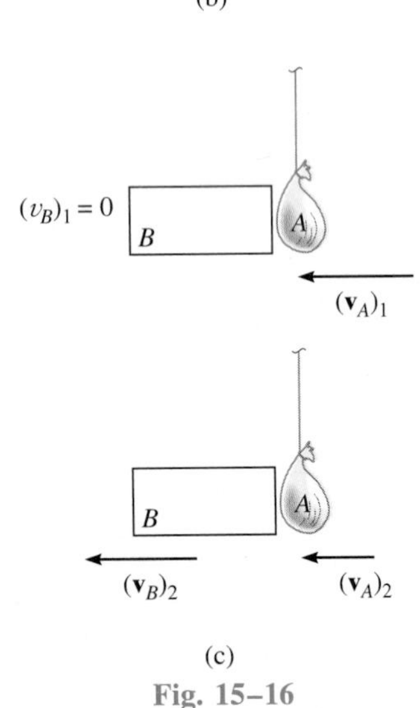

(c)

Fig. 15–16

SOLUTION

This problem involves central impact. Why? Before analyzing the mechanics of the impact, however, it is first necessary to obtain the velocity of the bag *just before* it strikes the box.

Conservation of Energy. With the datum at $\theta = 0°$, Fig. 15–16b, we have

$$T_0 + V_0 = T_1 + V_1$$

$$0 + 0 = \frac{1}{2}\left(\frac{6\text{ lb}}{32.2\text{ ft/s}^2}\right)(v_A)_1^2 - 6\text{ lb}(3\text{ ft}); \quad (v_A)_1 = 13.9\text{ ft/s}$$

Conservation of Momentum. After impact we will assume A and B travel to the left. Applying the conservation of momentum to the system, we have

$$(\overset{+}{\leftarrow}) \qquad m_B(v_B)_1 + m_A(v_A)_1 = m_B(v_B)_2 + m_A(v_A)_2$$

$$0 + \frac{6\text{ lb}}{32.2\text{ ft/s}^2}(13.9\text{ ft/s}) = \frac{18\text{ lb}}{32.2\text{ ft/s}^2}(v_B)_2 + \frac{6\text{ lb}}{32.2\text{ ft/s}^2}(v_A)_2$$

$$(v_A)_2 = 13.9 - 3(v_B)_2 \qquad (1)$$

Coefficient of Restitution. Realizing that for separation to occur after collision $(v_B)_2 > (v_A)_2$, Fig. 15–16c, we have

$$(\overset{+}{\leftarrow}) \qquad e = \frac{(v_B)_2 - (v_A)_2}{(v_A)_1 - (v_B)_1}; \qquad 0.5 = \frac{(v_B)_2 - (v_A)_2}{13.9\text{ ft/s} - 0}$$

$$(v_A)_2 = (v_B)_2 - 6.95 \qquad (2)$$

Solving Eqs. 1 and 2 simultaneously yields

$(v_A)_2 = -1.74\text{ ft/s} = 1.74\text{ ft/s} \rightarrow$ and $(v_B)_2 = 5.21\text{ ft/s} \leftarrow$ ***Ans.***

Loss of Energy. Applying the principle of work and energy just before and just after collision, we have

$$\Sigma U_{1-2} = T_2 - T_1;$$

$$\Sigma U_{1-2} = \left[\frac{1}{2}\left(\frac{18\text{ lb}}{32.2\text{ ft/s}^2}\right)(5.21\text{ ft/s})^2 + \frac{1}{2}\left(\frac{6\text{ lb}}{32.2\text{ ft/s}^2}\right)(1.74\text{ ft/s})^2\right] - \left[\frac{1}{2}\left(\frac{6\text{ lb}}{32.2\text{ ft/s}^2}\right)(13.9\text{ ft/s})^2\right]$$

$$\Sigma U_{1-2} = -10.1\text{ ft}\cdot\text{lb}$$ ***Ans.***

Why is there an energy loss?

Example 15–10

The ball B shown in Fig. 15–17a has a mass of 1.5 kg and is suspended from the ceiling by a 1-m-long elastic cord. If the cord is *stretched* downward 0.25 m and the ball is released from rest, determine how far the cord stretches after the ball rebounds from the ceiling. The stiffness of the cord is $k = 800$ N/m and the coefficient of restitution between the ball and ceiling is $e = 0.8$. The ball makes a central impact with the ceiling.

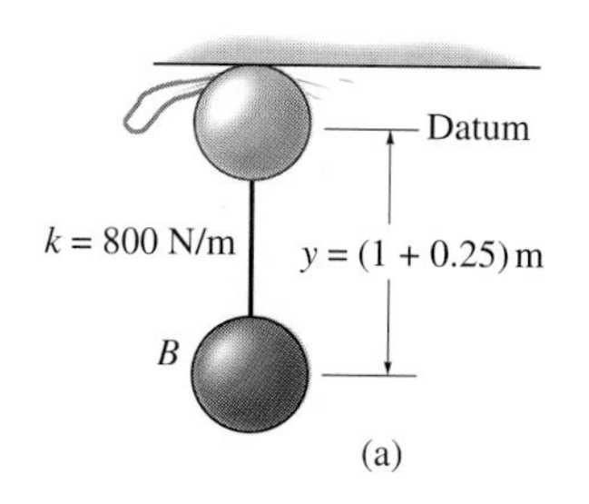

(a)

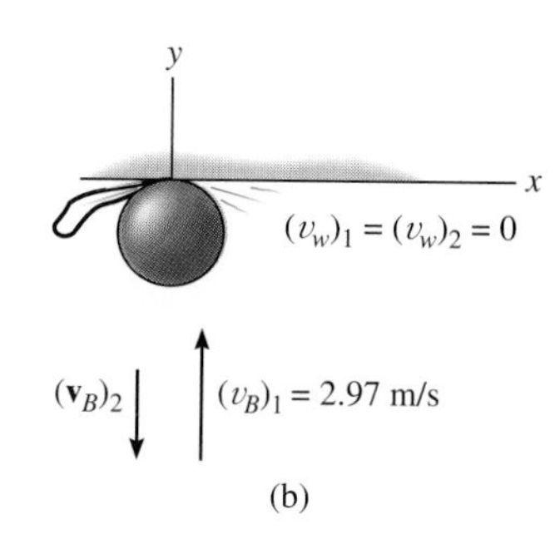

(b)

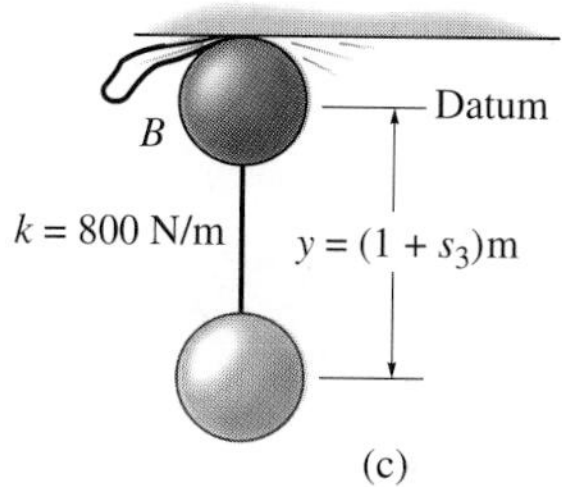

(c)

Fig. 15–17

SOLUTION

It is necessary to first obtain the initial velocity of the ball *just before* it strikes the ceiling.

Conservation of Energy. With the datum located as shown in Fig. 15–17a, realizing that initially $y = y_0 = (1 + 0.25)$ m $= 1.25$ m, we have

$$T_0 + V_0 = T_1 + V_1$$
$$\tfrac{1}{2}m(v_B)_0^2 - W_B y_0 + \tfrac{1}{2}ks^2 = \tfrac{1}{2}m(v_B)_1^2 + 0$$
$$0 - 1.5(9.81)\text{ N}(1.25\text{ m}) + \tfrac{1}{2}(800\text{ N/m})(0.25\text{ m})^2 = \tfrac{1}{2}(1.5\text{ kg})(v_B)_1^2$$
$$(v_B)_1 = 2.97\text{ m/s}\uparrow$$

The interaction of the ball with the ceiling will now be considered using the principles of impact.* Note that the y axis, Fig. 15–17b, represents the line of impact for the ball. Since an unknown portion of the mass of the ceiling is involved in the impact, the conservation of momentum for the ball–ceiling system will not be written. The "velocity" of this portion of ceiling is zero since it remains at rest *both* before and after impact.

Coefficient of Restitution.

$$(+\uparrow)\ e = \frac{(v_B)_2 - (v_A)_2}{(v_A)_1 - (v_B)_1};\qquad 0.8 = \frac{(v_B)_2 - 0}{0 - 2.97\text{ m/s}}$$
$$(v_B)_2 = -2.37\text{ m/s} = 2.37\text{ m/s}\downarrow$$

Conservation of Energy. The maximum stretch s_3 in the cord may be determined by again applying the conservation of energy equation to the ball just after collision. Assuming that $y = y_3 = (1 + s_3)$ m, Fig. 15–17c, then

$$T_2 + V_2 = T_3 + V_3$$
$$\tfrac{1}{2}m(v_B)_2^2 + 0 = \tfrac{1}{2}m(v_B)_3^2 - W_B y_3 + \tfrac{1}{2}ks_3^2$$
$$\tfrac{1}{2}(1.5\text{ kg})(2.37\text{ m/s})^2 = 0 - 9.81(1.5)\text{ N}(1\text{ m} + s_3) + \tfrac{1}{2}(800\text{ N/m})s_3^2$$
$$400s_3^2 - 14.72s_3 - 18.94 = 0$$

Solving this quadratic equation for the positive root yields

$$s_3 = 0.237\text{ m} = 237\text{ mm}\qquad\textit{Ans.}$$

*The weight of the ball is considered a nonimpulsive force.

Example 15–11

Two smooth disks A and B, having a mass of 1 and 2 kg, respectively, collide with initial velocities as shown in Fig. 15–18*a*. If the coefficient of restitution for the disks is $e = 0.75$, determine the x and y components of the final velocity of each disk after collision. Neglect friction.

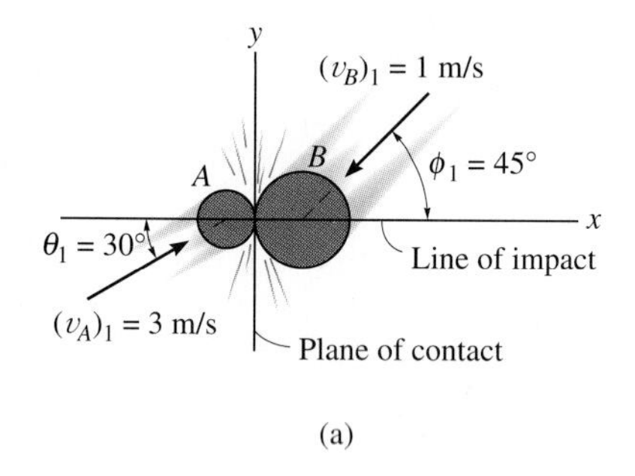

(a)

SOLUTION

The problem involves *oblique impact*. Why? In order to seek a solution, we have established the x and y axes along the line of impact and the plane of contact, respectively, Fig. 15–18*a*.

Resolving each of the initial velocities into x and y components, we have

$$(v_{Ax})_1 = 3 \cos 30° = 2.60 \text{ m/s} \qquad (v_{Ay})_1 = 3 \sin 30° = 1.50 \text{ m/s}$$

$$(v_{Bx})_1 = -1 \cos 45° = -0.707 \text{ m/s} \qquad (v_{By})_1 = -1 \sin 45° = -0.707 \text{ m/s}$$

The four unknown velocity components after collision are assumed to act in the positive directions, Fig. 15–18*b*. Since the impact occurs only in the x direction (line of impact), the conservation of momentum for *both* disks can be applied in this direction. Why?

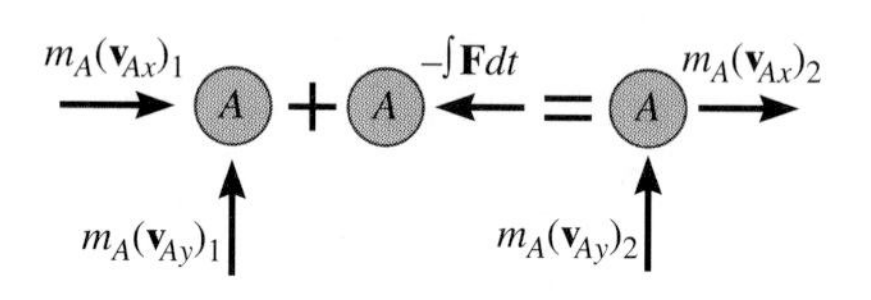

Conservation of "x" Momentum. In reference to the momentum diagrams, we have

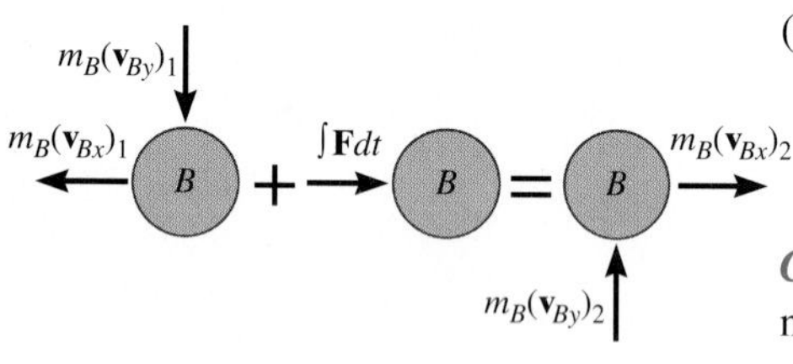

(b)

$$(\overset{+}{\rightarrow}) \qquad m_A(v_{Ax})_1 + m_B(v_{Bx})_1 = m_A(v_{Ax})_2 + m_B(v_{Bx})_2$$

$$1 \text{ kg}(2.60 \text{ m/s}) + 2 \text{ kg}(-0.707 \text{ m/s}) = 1 \text{ kg}(v_{Ax})_2 + 2 \text{ kg}(v_{Bx})_2$$

$$(v_{Ax})_2 + 2(v_{Bx})_2 = 1.18 \qquad (1)$$

Coefficient of Restitution (x). Both disks are *assumed* to have components of velocity in the $+x$ direction after collision, Fig. 15–18*b*.

$$(\overset{+}{\rightarrow}) \ e = \frac{(v_{Bx})_2 - (v_{Ax})_2}{(v_{Ax})_1 - (v_{Bx})_1}; \quad 0.75 = \frac{(v_{Bx})_2 - (v_{Ax})_2}{2.60 \text{ m/s} - (-0.707 \text{ m/s})}$$

$$(v_{Bx})_2 - (v_{Ax})_2 = 2.48 \qquad (2)$$

Solving Eqs. 1 and 2 for $(v_{Ax})_2$ and $(v_{Bx})_2$ yields

$$(v_{Ax})_2 = -1.26 \text{ m/s} = 1.26 \text{ m/s} \leftarrow \qquad (v_{Bx})_2 = 1.22 \text{ m/s} \rightarrow \qquad \textit{Ans.}$$

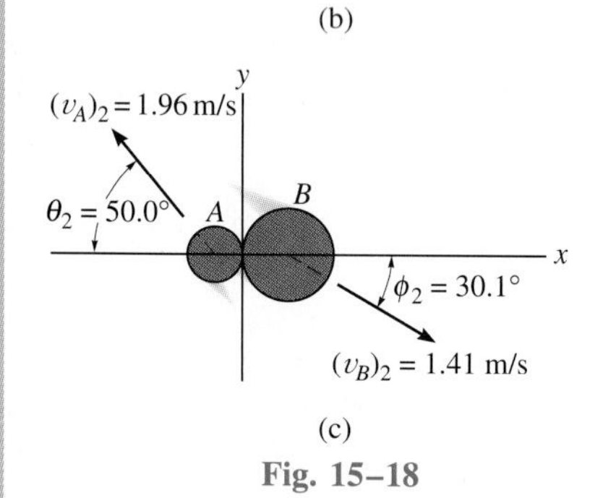

(c)

Fig. 15–18

Conservation of "y" Momentum. The momentum of *each disk* is *conserved* in the y direction (plane of contact), since the disks are smooth and therefore *no* external impulse acts in this direction. From Fig. 15–18*b*,

$$(+\uparrow) \quad m_A(v_{Ay})_1 = m_A(v_{Ay})_2 \quad (v_{Ay})_2 = 1.50 \text{ m/s} \uparrow \qquad \textit{Ans.}$$

$$(+\uparrow) \quad m_B(v_{By})_1 = m_B(v_{By})_2 \quad (v_{By})_2 = -0.707 \text{ m/s} = 0.707 \text{ m/s} \downarrow \qquad \textit{Ans.}$$

Show that when the velocity components are summed, one obtains the results shown in Fig. 15–18*c*.

PROBLEMS

15–59. Disk A has a mass of 250 g and is sliding on a *smooth* horizontal surface with an initial velocity $(v_A)_1 = 2$ m/s. It makes a direct collision with disk B, which has a mass of 175 g and is originally at rest. If both disks are of the same size and the collision is perfectly elastic ($e = 1$), determine the velocity of each disk just after collision. Show that the kinetic energy of the disks before and after collision is the same.

***15–60.** Two cars each have the same mass m. Car A is traveling with a speed of 12 ft/s when it crashes head-on with B, which is at rest. If after collision the driver of B locks his brakes, determine how far car B will slide before coming to a stop. Take $\mu_k = 0.5$ and $e = 0.6$. Neglect the size of the cars.

15–61. Block A has a mass of 3 kg and is sliding on a rough horizontal surface with a velocity $(v_A)_1 = 2$ m/s when it makes a direct collision with block B, which has a mass of 2 kg and is originally at rest. If the collision is perfectly elastic ($e = 1$), determine the velocity of each block just after collision and the distance between the blocks when they stop sliding. The coefficient of kinetic friction between the blocks and the plane is $\mu_k = 0.3$.

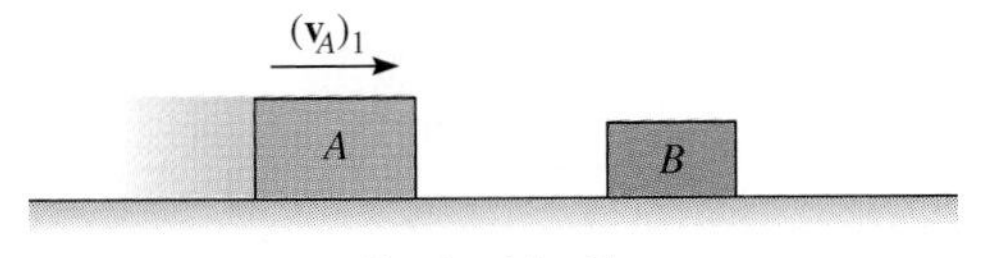

Prob. 15–61

15–62. A 0.5 kg ball is ejected from the tube at A with a horizontal velocity $v_A = 2$ m/s. Determine the horizontal distance R where it strikes the smooth inclined plane. If the coefficient of restitution is $e = 0.6$, determine the speed at which it bounces from the plane.

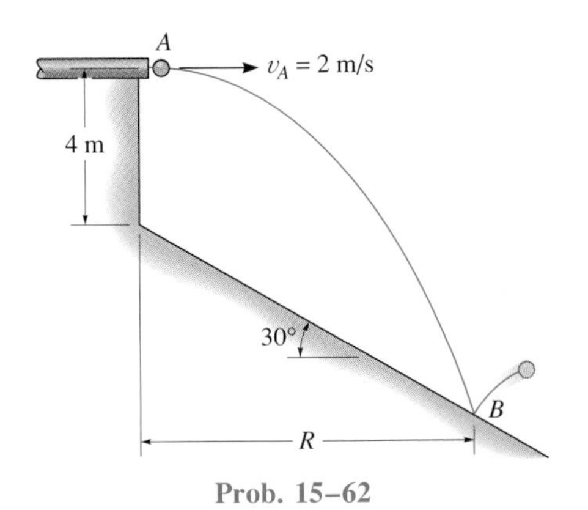

Prob. 15–62

15–63. The 1-lb ball is dropped from rest and falls a distance of 4 ft before striking the smooth plane at A. If $e = 0.8$, determine the distance d to where it again strikes the plane at B.

***15–64.** The 1-lb ball is dropped from rest and falls a distance of 4 ft before striking the smooth plane at A. If it rebounds and in $t = 1$ s again strikes the plane at B, determine the coefficient of restitution e between the ball and the plane. Also, what is the distance d?

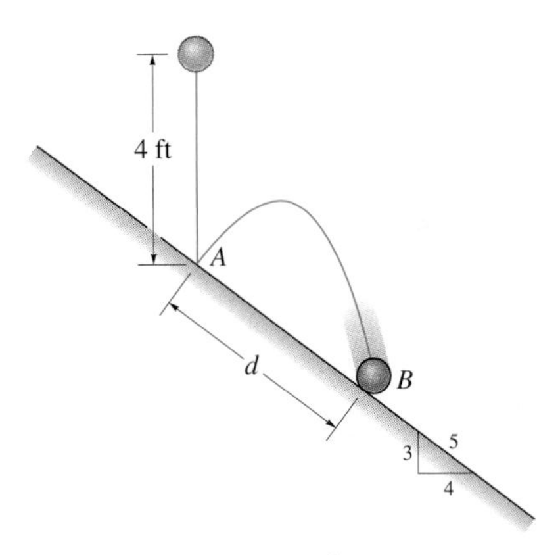

Probs. 15–63/15–64

15–65. Determine the horizontal velocity $\mathbf{v}_A$ at which the girl must throw the ball so that it bounces once on the smooth surface and then lands into the cup at C. Take $e = 0.6$ and $d = 8$ ft, and neglect the size of the cup.

15–66. If the girl throws the ball with a horizontal velocity of 8 ft/s, determine the distance d so that the ball bounces once on the smooth surface and then lands in the cup at C. Take $e = 0.8$.

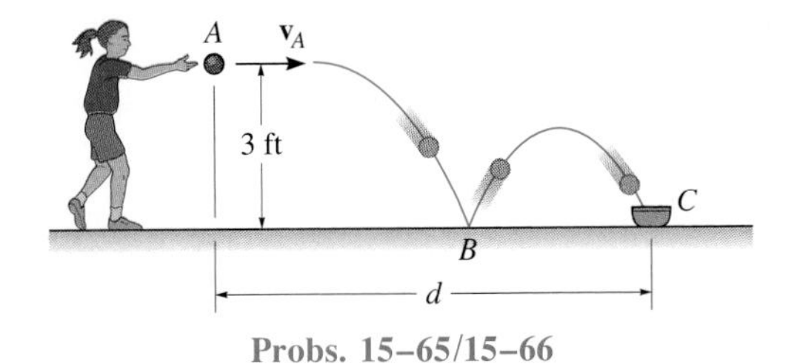

Probs. 15–65/15–66

15–67. A ball has a mass m and is dropped vertically onto a horizontal surface from a height h. If the coefficient of restitution is e between the ball and the surface, determine the time needed for the ball to stop bouncing.

*15–68. The three balls each have the same mass m. If A is released from rest at θ, determine the angle ϕ to which C rises after collision. The coefficient of restitution between each ball is e.

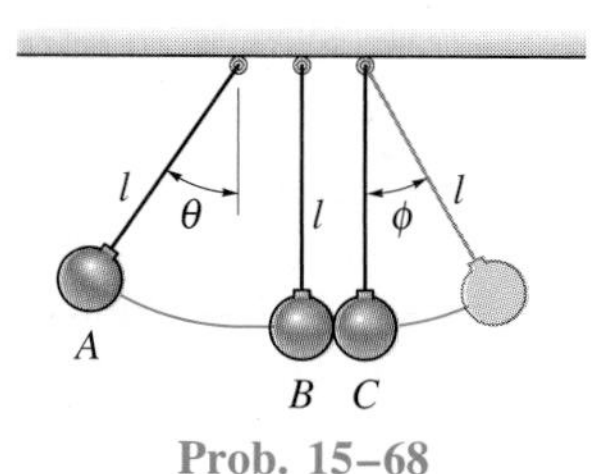

Prob. 15–68

15–69. The four smooth balls each have the same mass m. If A and B are rolling forward with velocity $\mathbf{v}$ and strike C, explain why after collision C and D each move off with velocity $\mathbf{v}$. Why doesn't D move off with velocity $2\mathbf{v}$? The collision is elastic, $e = 1$. Neglect the size of each ball.

15–70. The four balls each have the same mass m. If A and B are rolling forward with velocity $\mathbf{v}$ and strike C, determine the velocity of each ball after the first three collisions. Take $e = 0.5$ between each ball.

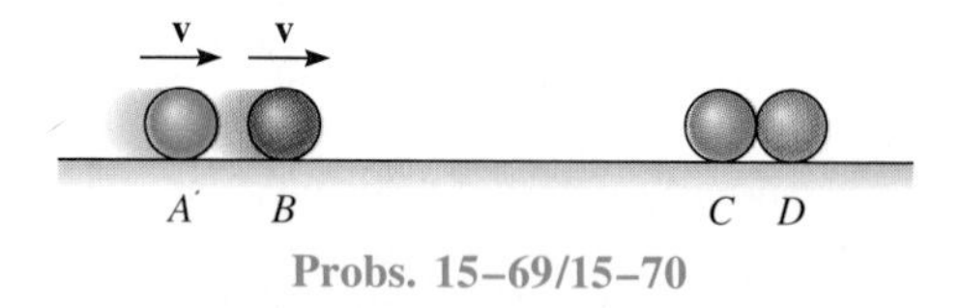

Probs. 15–69/15–70

15–71. The 20-lb suitcase A is released from rest at C. After it slides down the smooth ramp it strikes the 10-lb suitcase B, which is originally at rest. If the coefficient of restitution between the suitcases is $e = 0.3$ and the coefficient of kinetic friction between the floor DE and each suitcase is $\mu_k = 0.4$, determine (a) the velocity of A just before impact, (b) the velocities of A and B just after impact, and (c) the distance B slides before coming to rest.

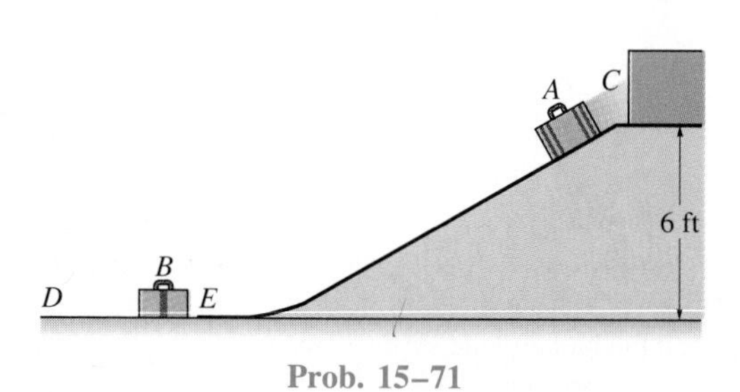

Prob. 15–71

*15–72. The man A has a weight of 175 lb and jumps from rest $h = 8$ ft onto a platform P that has a weight of 60 lb. The platform is mounted on a spring, which has a stiffness $k = 200$ lb/ft. Determine (a) the velocities of A and P just after impact and (b) the maximum compression imparted to the spring by the impact. Assume the coefficient of restitution between the man and the platform is $e = 0.6$, and the man holds himself rigid during the motion.

15–73. The man A has a weight of 100 lb and jumps from rest onto the platform P that has a weight of 60 lb. The platform is mounted on a spring, which has a stiffness $k = 200$ lb/ft. If the coefficient of restitution between the man and the platform is $e = 0.6$, and the man holds himself rigid during the motion, determine the required height h of the jump if the maximum compression of the spring becomes 2 ft.

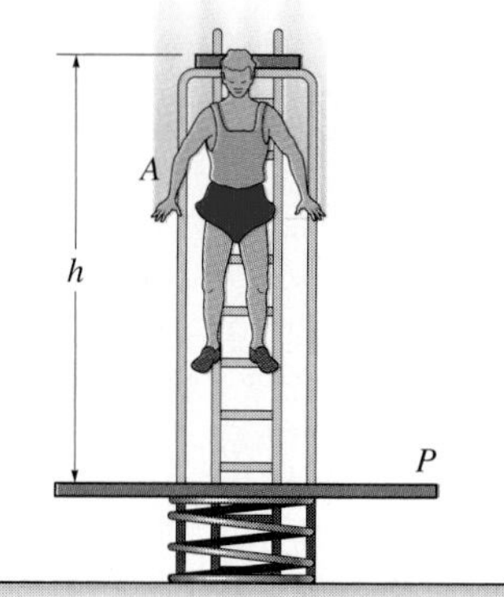

Probs. 15–72/15–73

15–74. Block A, having a mass m, is released from rest, falls a distance h and strikes the plate B having a mass $2m$. If the coefficient of restitution between A and B is e, determine the velocity of the plate just after collision. The spring has a stiffness k.

15–75. Block A, having a mass of 2 kg, is released from rest, falls a distance $h = 0.5$ m, and strikes the plate B having a mass of 3 kg. If the coefficient of restitution between A and B is $e = 0.6$, determine the velocity of the block just after collision. The spring has a stiffness $k = 30$ N/m.

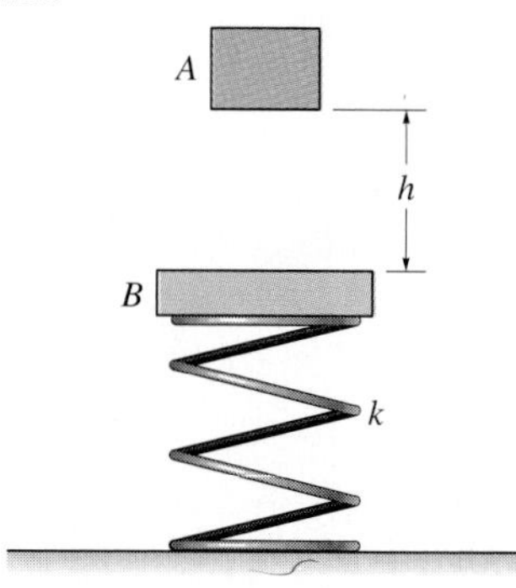

Probs. 15–74/15–75

*15–76. The ping-pong ball has a mass of 2 g. If it is struck with the velocity shown, determine how high h it rises above the end of the smooth table after the rebound. Take $e = 0.8$.

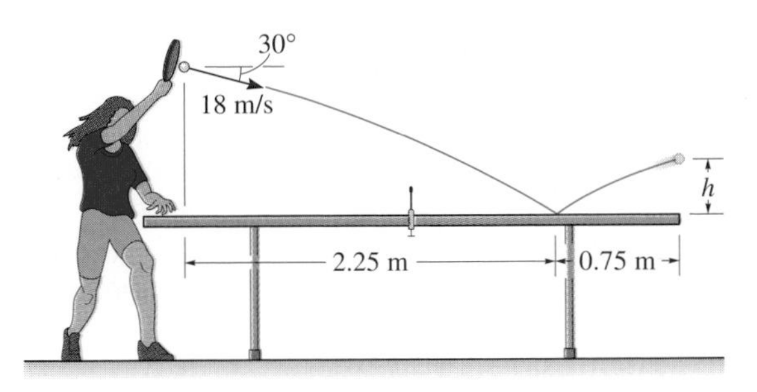

Prob. 15–76

15–77. A ball of negligible size and mass m is given a velocity of $\mathbf{v}_0$ on the center of the cart which has a mass M and is originally at rest. If the coefficient of restitution between the ball and walls A and B is e, determine the velocity of the ball and the cart just after the ball strikes A. Also, determine the total time needed for the ball to strike A, rebound, then strike B, and rebound and then return to the center of the cart. Neglect friction.

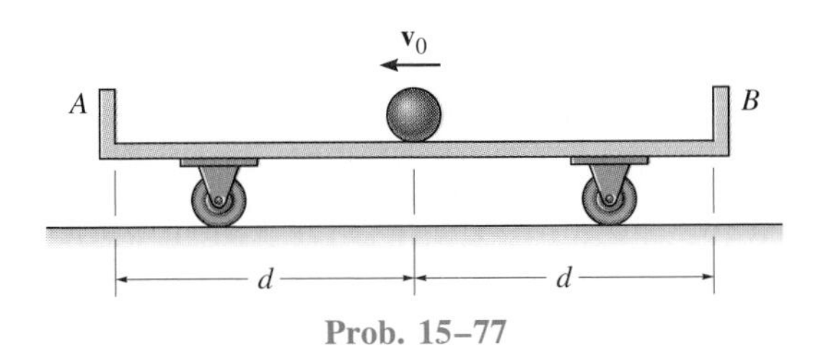

Prob. 15–77

15–78. A ball is thrown onto a rough floor at an angle θ. If it rebounds at an angle ϕ and the coefficient of kinetic friction is μ, determine the coefficient of restitution e. Neglect the size of the ball. *Hint:* Show that during impact, the average impulses in the x and y directions are related by $I_x = \mu I_y$. Since the time of impact is the same, $F_x \Delta t = \mu F_y \Delta t$ or $F_x = \mu F_y$.

15–79. A ball is thrown onto a rough floor at an angle of $\theta = 45°$. If it rebounds at the same angle $\phi = 45°$, determine the coefficient of kinetic friction between the floor and the ball. The coefficient of restitution is $e = 0.6$. *Hint:* Show that during impact, the average impulses in the x and y directions are related by $I_x = \mu I_y$. Since the time of impact is the same, $F_x \Delta t = \mu F_y \Delta t$ or $F_x = \mu F_y$.

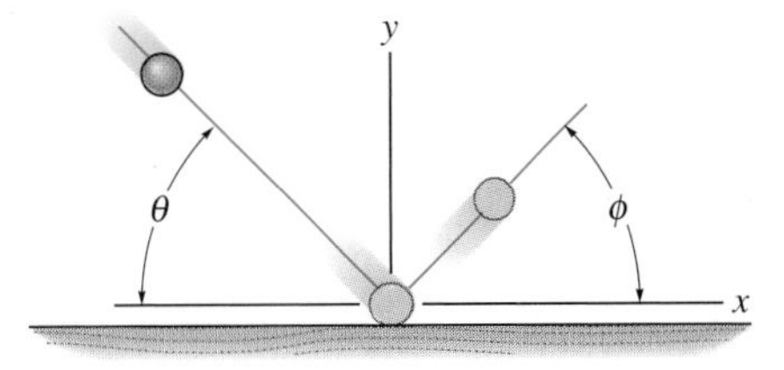

Probs. 15–78/15–79

*15–80. It was observed that a tennis ball when served horizontally 7.5 ft overhead strikes the smooth ground at B 20 ft away. Determine the initial velocity $\mathbf{v}_A$ of the ball and the velocity $\mathbf{v}_B$ (and θ) of the ball just after it strikes the court at B. Take $e = 0.7$.

15–81. The tennis ball is struck with a horizontal velocity $\mathbf{v}_A$, strikes the smooth ground at B, and bounces upward at $\theta = 30°$. Determine the initial velocity $\mathbf{v}_A$, the final velocity $\mathbf{v}_B$, and the coefficient of restitution between the ball and the ground.

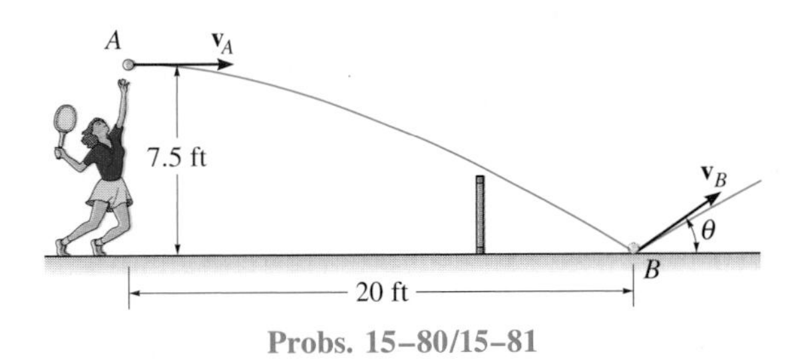

Probs. 15–80/15–81

15–82. The 20-lb box slides on the surface for which $\mu_k = 0.3$. The box has a velocity $v = 15$ ft/s when it is 2 ft from the plate. If it strikes the plate, which has a weight of 10 lb and is held in position by an unstretched spring of stiffness $k = 400$ lb/ft, determine the maximum compression imparted to the spring. Take $e = 0.8$ between the box and the plate.

15–83. The 20-lb box slides on the smooth surface at $v = 15$ ft/s when it strikes the 10 lb plate. Determine the required stiffness of the spring if it is allowed to compress a maximum of 4 in. Take $e = 0.8$ between the box and the plate.

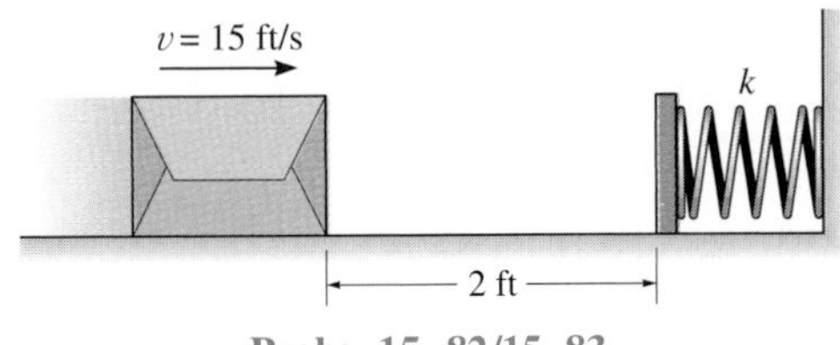

Probs. 15–82/15–83

***15–84.** Two smooth coins A and B, each having the same mass, slide on a smooth surface with the motion shown. Determine the speed of each coin after collision if they move off along the colored paths. *Hint:* Since the line of impact has not been defined, apply the conservation of momentum along the x and y axes, respectively.

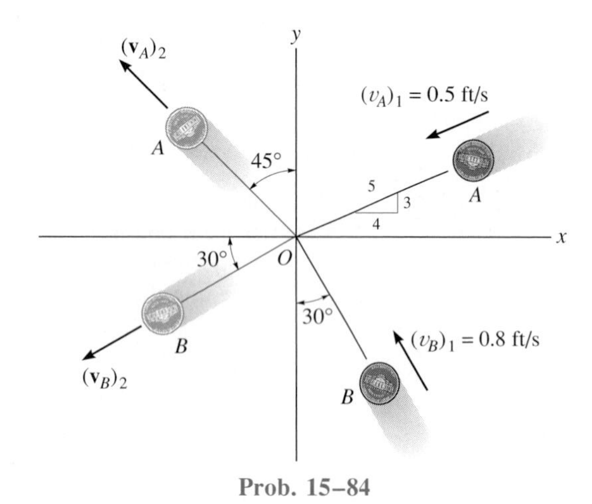

Prob. 15–84

15–85. Two smooth disks A and B each have a mass of 0.5 kg. If both disks are moving with the velocities shown when they collide, determine their final velocities just after collision. The coefficient of restitution is $e = 0.75$.

15–86. Two smooth disks A and B each have a mass of 0.5 kg. If both disks are moving with the velocities shown when they collide, determine the coefficient of restitution between the disks if after collision B travels along a line, 30° counterclockwise from the y axis.

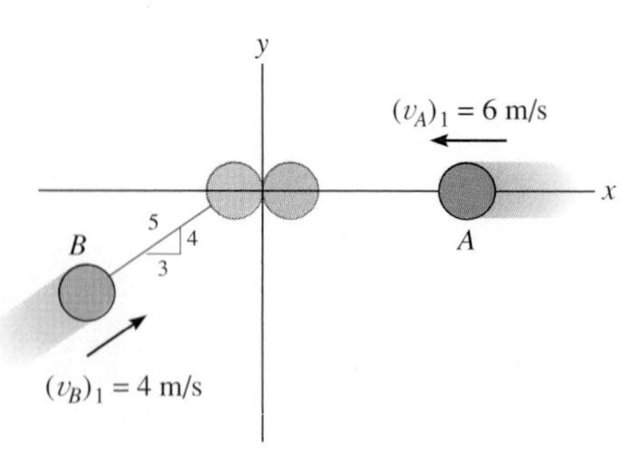

Probs. 15–85/15–86

15–87. The "stone" A used in the sport of curling slides over the ice track and strikes another "stone" B as shown. If each "stone" is smooth and has a weight of 47 lb, and the coefficient of restitution between the "stones" is $e = 0.8$, determine their speeds just after collision. Initially A has a velocity of 8 ft/s and B is at rest. Neglect friction.

***15–88.** The "stone" A used in the sport of curling slides over the ice track and strikes another "stone" B as shown. If each "stone" is smooth and has a weight of 47 lb, and the coefficient of restitution between the "stones" is $e = 0.8$, determine the time required just after collision for B to slide off the runway. This requires the horizontal component of displacement to be 3 ft.

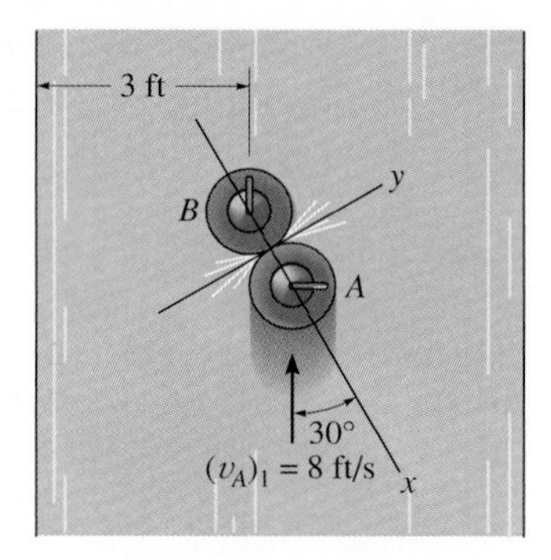

Probs. 15–87/15–88

15–89. Two smooth disks A and B have the initial velocities shown just before they collide at O. If they have masses $m_A = 8$ kg and $m_B = 6$ kg, determine their speeds just after impact. The coefficient of restitution is $e = 0.5$.

15–90. Two smooth disks A and B have the initial velocities shown just before they collide at O. If they have masses $m_A = 10$ kg and $m_B = 8$ kg, determine their speeds just after impact. The coefficient of restitution is $e = 0.4$.

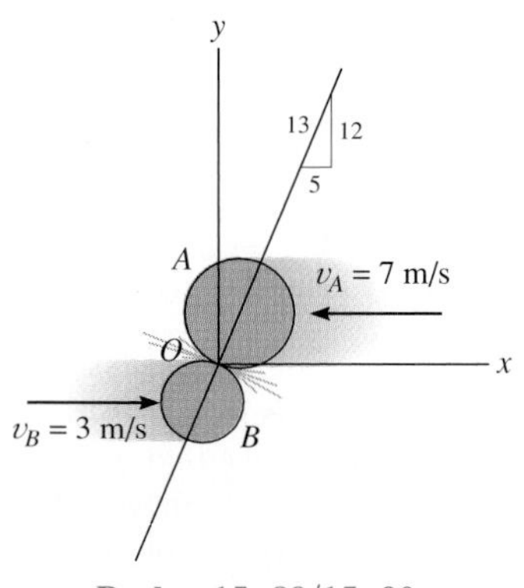

Probs. 15–89/15–90

15.5 Angular Momentum

The *angular momentum* of a particle about point O is defined as the "moment" of the particle's linear momentum about O. Since this concept is analogous to finding the moment of a force about a point, the angular momentum, $\mathbf{H}_O$, is sometimes referred to as the *moment of momentum.*

Scalar Formulation. If a particle is moving along a curve lying in the x–y plane, Fig. 15–19, the angular momentum at any instant can be computed about point O (actually the z axis) by using a scalar formulation. The *magnitude* of $\mathbf{H}_O$ is

$$(H_O)_z = (d)(mv) \tag{15–12}$$

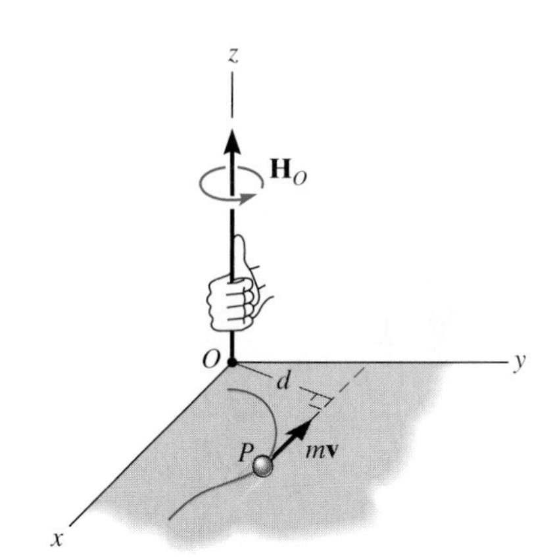

Fig. 15–19

Here d is the moment arm or perpendicular distance from O to the line of action of $m\mathbf{v}$. Common units for this magnitude are kg · m²/s or slug · ft²/s. The *direction* of $\mathbf{H}_O$ is defined by the right-hand rule. As shown in Fig. 15–19, the curl of the fingers of the right hand indicates the sense of rotation of $m\mathbf{v}$ about O, so that in this case the thumb (or $\mathbf{H}_O$) is directed perpendicular to the x–y plane along the $+z$ axis.

Vector Formulation. If the particle is moving along a space curve, Fig. 15–20, the vector cross product can be used to determine the *angular momentum* about O. In this case

$$\mathbf{H}_O = \mathbf{r} \times m\mathbf{v} \tag{15–13}$$

Here $\mathbf{r}$ denotes a position vector drawn from the moment point O to the particle P. As shown in the figure, $\mathbf{H}_O$ is *perpendicular* to the shaded plane containing $\mathbf{r}$ and $m\mathbf{v}$.

In order to evaluate the cross product, $\mathbf{r}$ and $m\mathbf{v}$ should be expressed in terms of their Cartesian components, so that the angular momentum is determined by evaluating the determinant:

$$\mathbf{H}_O = \begin{vmatrix} \mathbf{i} & \mathbf{j} & \mathbf{k} \\ r_x & r_y & r_z \\ mv_x & mv_y & mv_z \end{vmatrix} \tag{15–14}$$

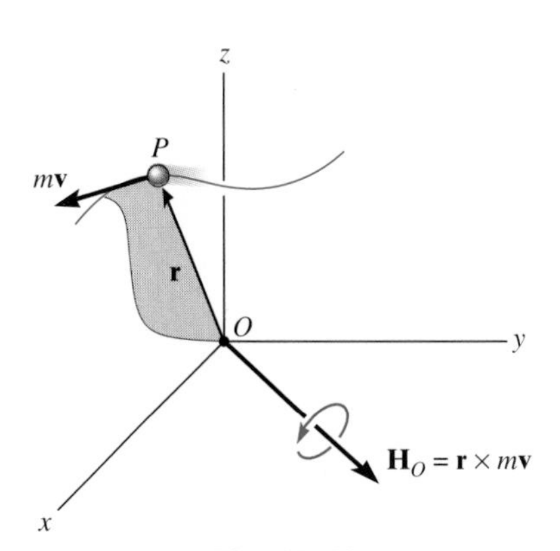

Fig. 15–20

15.6 Relation Between Moment of a Force and Angular Momentum

The moments about point O of all the forces acting on the particle in Fig. 15–21a may be related to the particle's angular momentum by using the equation of motion. If the mass of the particle is constant, we may write

$$\Sigma \mathbf{F} = m\dot{\mathbf{v}}$$

The moments of the forces about point O can be obtained by performing a cross-product multiplication of each side of this equation by the position vector $\mathbf{r}$, which is measured in the x, y, z inertial frame of reference. We have

$$\Sigma \mathbf{M}_O = \mathbf{r} \times \Sigma \mathbf{F} = \mathbf{r} \times m\dot{\mathbf{v}}$$

From Appendix C, the derivative of $\mathbf{r} \times m\mathbf{v}$ can be written as

$$\dot{\mathbf{H}}_O = \frac{d}{dt}(\mathbf{r} \times m\mathbf{v}) = \dot{\mathbf{r}} \times m\mathbf{v} + \mathbf{r} \times m\dot{\mathbf{v}}$$

The first term on the right side, $\dot{\mathbf{r}} \times m\mathbf{v} = m(\dot{\mathbf{r}} \times \dot{\mathbf{r}}) = \mathbf{0}$, since the cross product of a vector with itself is zero. Hence, the above equation becomes

$$\Sigma \mathbf{M}_O = \dot{\mathbf{H}}_O \qquad (15\text{–}15)$$

This equation states that *the resultant moment about point O of all the forces acting on the particle is equal to the time rate of change of the particle's angular momentum about point O.* This result is similar to Eq. 15–1, i.e.,

$$\Sigma \mathbf{F} = \dot{\mathbf{L}} \qquad (15\text{–}16)$$

Here $\mathbf{L} = m\mathbf{v}$, so *the resultant force acting on the particle is equal to the time rate of change of the particle's linear momentum.*

From the derivations, it is seen that Eqs. 15–15 and 15–16 are actually another way of stating Newton's second law of motion. In other sections of this book it will be shown that these equations have many practical applications when extended and applied to the solution of problems involving either a system of particles or a rigid body.

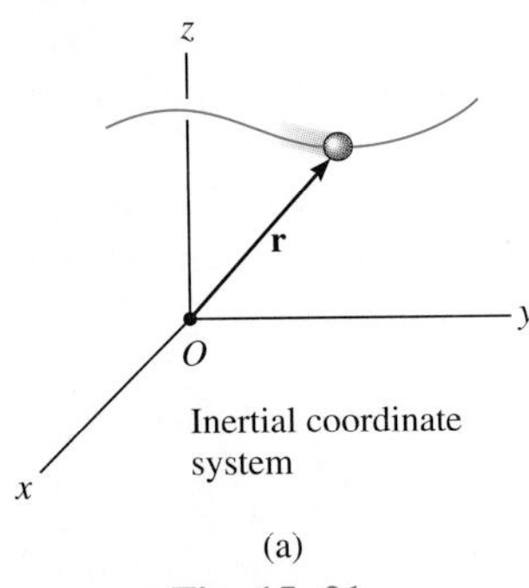

(a)

Fig. 15–21

System of Particles. An equation having the same form as Eq. 15–15 may be derived for the system of n particles shown in Fig. 15–21b. The forces acting on the arbitrary ith particle of the system consist of a resultant *external force* $\mathbf{F}_i$ and a resultant *internal force* $\mathbf{f}_i = \Sigma_{j=1(j \neq i)}^{n} \mathbf{f}_{ij}$. Expressing the moments of these forces about point O, using the form of Eq. 15–15, we have

$$(\mathbf{r}_i \times \mathbf{F}_i) + (\mathbf{r}_i \times \mathbf{f}_i) = (\dot{\mathbf{H}}_i)_O$$

Here $\mathbf{r}_i$ represents the position vector drawn from the origin O of an inertial frame of reference to the ith particle, and $(\dot{\mathbf{H}}_i)_O$ is the time rate of change in the angular momentum of the ith particle about O. Similar equations can be written for each of the other particles of the system. When the results are summed vectorially, the result is

$$\Sigma(\mathbf{r}_i \times \mathbf{F}_i) + \Sigma(\mathbf{r}_i \times \mathbf{f}_i) = \Sigma(\dot{\mathbf{H}}_i)_O$$

The second term is zero since the internal forces occur in equal but opposite pairs, and because each pair of forces has the same line of action, the moment of each pair of forces about point O is therefore zero. Hence, dropping the index notation, the above equation can be written in a simplified form as

$$\Sigma \mathbf{M}_O = \dot{\mathbf{H}}_O \tag{15–17}$$

which states that *the sum of the moments about point O of all the external forces acting on a system of particles is equal to the time rate of change of the total angular momentum of the system of particles about point O.* Although point O has been chosen here as the origin of coordinates, it actually can represent any point *fixed* in the inertial frame of reference.

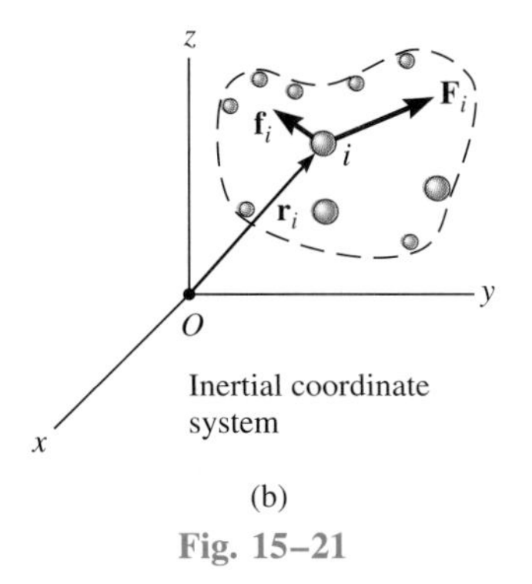

(b)

Fig. 15–21

Example 15–12

The box shown in Fig. 15–22*a* has a mass m and is traveling down the smooth circular ramp such that when it is at the angle θ it has a speed v. Compute its angular momentum about point O at this instant and the rate of increase in its speed.

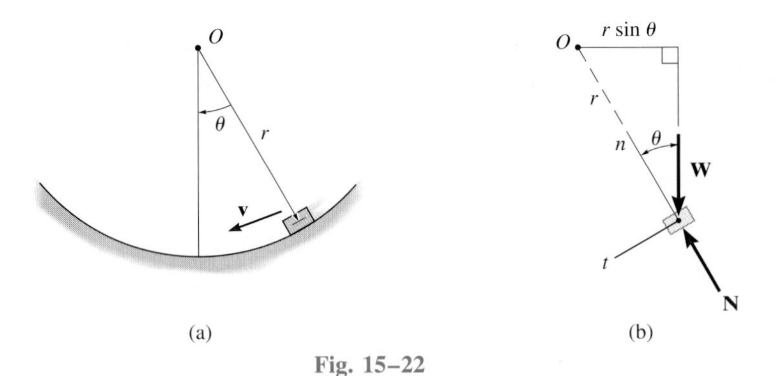

Fig. 15–22

SOLUTION

Since **v** is tangent to the path, applying Eq. 15–12 the angular momentum is

$$H_O = rmv \curvearrowright \qquad \textit{Ans.}$$

The rate of increase in its speed (dv/dt) can be found by applying Eq. 15–15. From the free-body diagram of the block, Fig. 15–22*b*, it is seen that only the weight $W = mg$ contributes a moment about point O. We have

$$\curvearrowright + \Sigma M_O = \dot{H}_O; \qquad mg(r \sin \theta) = \frac{d}{dt}(rmv)$$

Since r and m are constant,

$$mgr \sin \theta = rm\frac{dv}{dt}$$

$$\frac{dv}{dt} = g \sin \theta \qquad \textit{Ans.}$$

This same result can, of course, be obtained from the equation of motion applied in the tangential direction, Fig. 15–22*b*, i.e.,

$$+\swarrow \Sigma F_t = ma_t; \qquad mg \sin \theta = m\left(\frac{dv}{dt}\right)$$

$$\frac{dv}{dt} = g \sin \theta \qquad \textit{Ans.}$$

15.7 Angular Impulse and Momentum Principles

Principle of Angular Impulse and Momentum. If Eq. 15–15 is rewritten in the form $\Sigma \mathbf{M}_O \, dt = d\mathbf{H}_O$ and integrated, we have, assuming that at time $t = t_1$, $\mathbf{H}_O = (\mathbf{H}_O)_1$ and at time $t = t_2$, $\mathbf{H}_O = (\mathbf{H}_O)_2$,

$$\Sigma \int_{t_1}^{t_2} \mathbf{M}_O \, dt = (\mathbf{H}_O)_2 - (\mathbf{H}_O)_1$$

or

$$(\mathbf{H}_O)_1 + \Sigma \int_{t_1}^{t_2} \mathbf{M}_O \, dt = (\mathbf{H}_O)_2 \qquad (15\text{–}18)$$

This equation is referred to as the *principle of angular impulse and momentum.* The initial and final angular momenta $(\mathbf{H}_O)_1$ and $(\mathbf{H}_O)_2$ are defined as the moment of the linear momentum of the particle ($\mathbf{H}_O = \mathbf{r} \times m\mathbf{v}$) at the instants t_1 and t_2, respectively. The second term on the left side, $\Sigma \int_{t_1}^{t_2} \mathbf{M}_O \, dt$, is called the *angular impulse.* It is computed on the basis of integrating, with respect to time, the moments of all the forces acting on the particle over the time interval t_1 to t_2. Since the moment of a force about point O is defined as $\mathbf{M}_O = \mathbf{r} \times \mathbf{F}$, the angular impulse may be expressed in vector form as

$$\text{angular impulse} = \int_{t_1}^{t_2} \mathbf{M}_O \, dt = \int_{t_1}^{t_2} (\mathbf{r} \times \mathbf{F}) \, dt \qquad (15\text{–}19)$$

Here $\mathbf{r}$ is a position vector which extends from point O to any point on the line of action of $\mathbf{F}$.

In a similar manner, using Eq. 15–17, the principle of angular impulse and momentum for a system of particles may be written as

$$\Sigma(\mathbf{H}_O)_1 + \Sigma \int_{t_1}^{t_2} \mathbf{M}_O \, dt = \Sigma(\mathbf{H}_O)_2 \qquad (15\text{–}20)$$

Here the first and third terms represent the angular momenta of the system of particles $[\Sigma \mathbf{H}_O = \Sigma(\mathbf{r}_i \times m\mathbf{v}_i)]$ at the instants t_1 and t_2. The second term is the vector sum of the angular impulses given to all the particles during the time period t_1 to t_2. Recall that these impulses are created only by the moments of the external forces acting on the system where, for the ith particle, $\mathbf{M}_O = \mathbf{r}_i \times \mathbf{F}_i$.

Vector Formulation. Using impulse and momentum principles, it is therefore possible to write two vector equations which define the particle motion: namely, Eqs. 15–3 and 15–18, restated as

$$m\mathbf{v}_1 + \Sigma \int_{t_1}^{t_2} \mathbf{F}\, dt = m\mathbf{v}_2$$
$$(\mathbf{H}_O)_1 + \Sigma \int_{t_1}^{t_2} \mathbf{M}_O\, dt = (\mathbf{H}_O)_2 \qquad (15\text{–}21)$$

Scalar Formulation. In general, the above equations may be expressed in x, y, z component form, yielding a total of six independent scalar equations. If the particle is confined to move in the x–y plane, three independent scalar equations may be written to express the motion, namely,

$$m(v_x)_1 + \Sigma \int_{t_1}^{t_2} F_x\, dt = m(v_x)_2$$
$$m(v_y)_1 + \Sigma \int_{t_1}^{t_2} F_y\, dt = m(v_y)_2 \qquad (15\text{–}22)$$
$$(H_O)_1 + \Sigma \int_{t_1}^{t_2} M_O\, dt = (H_O)_2$$

The first two of these equations represent the principle of linear impulse and momentum in the x and y directions, which has been discussed in Sec. 15.1, and the third equation represents the principle of angular impulse and momentum about the z axis.

Conservation of Angular Momentum.

When the angular impulses acting on a particle are all zero during the time t_1 to t_2, Eq. 15–18 reduces to the following simplified form:

$$(\mathbf{H}_O)_1 = (\mathbf{H}_O)_2 \qquad (15\text{–}23)$$

This equation is known as the *conservation of angular momentum.* It states that from t_1 to t_2 the particle's angular momentum remains constant. Obviously, if no external impulse is applied to the particle, both linear and angular momentum will be conserved. In some cases, however, the particle's angular momentum will be conserved and linear momentum may not. An example of this occurs when the particle is subjected *only* to a *central force* (see Sec. 13.7). As shown in Fig. 15–23, the impulsive central force **F** is always directed toward point O as the particle moves along the path. Hence, the angular impulse (moment) created by **F** about the z axis passing through point O is always zero, and therefore angular momentum of the particle is conserved about this axis.

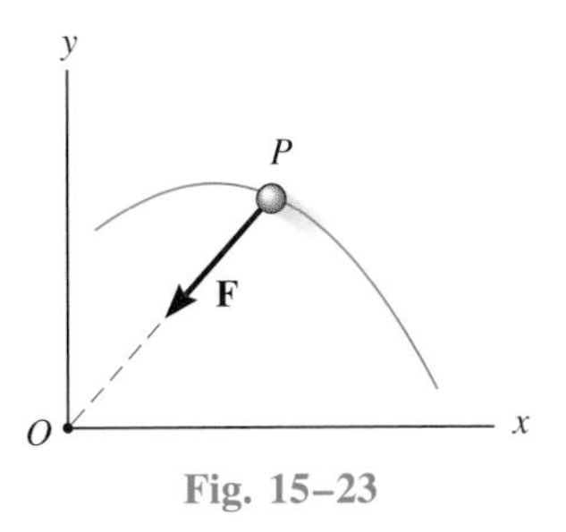

Fig. 15–23

From Eq. 15–20, we can also write the conservation of angular momentum for a system of particles, namely,

$$\Sigma(\mathbf{H}_O)_1 = \Sigma(\mathbf{H}_O)_2 \tag{15–24}$$

In this case the summation must include the angular momenta of all n particles in the system.

PROCEDURE FOR ANALYSIS

When applying the principles of angular impulse and momentum, or the conservation of angular momentum, it is suggested that the following procedure be used.

Free-Body Diagram. Draw the particle's free-body diagram in order to determine any axis about which angular momentum may be conserved. For this to occur, the moments of the forces (or impulses) about the axis must be zero throughout the time period t_1 to t_2. Apart from the free-body diagram, the direction and sense of the particle's initial and final velocities should also be established.

An alternative procedure would be to draw the impulse and momentum diagrams for the particle.

Momentum Equations. Apply the principle of angular impulse and momentum, $(\mathbf{H}_O)_1 + \Sigma \int_{t_1}^{t_2} \mathbf{M}_O \, dt = (\mathbf{H}_O)_2$, or if appropriate, the conservation of angular momentum, $(\mathbf{H}_O)_1 = (\mathbf{H}_O)_2$.

If other equations are needed for the problem solution, when appropriate, use the principle of linear impulse and momentum, the principle of work and energy, the equations of motion, or kinematics.

The following examples illustrate numerical application of this procedure.

Example 15–13

The 5-kg block of negligible size rests on the smooth horizontal plane, Fig. 15–24*a*. It is attached at A to a slender rod of negligible mass. The rod is attached to a ball-and-socket joint at B. If a moment $M = (3t)$ N · m, where t is in seconds, is applied to the rod and a horizontal force $P = 10$ N is applied to the block as shown, determine the speed of the block in 4 s starting from rest.

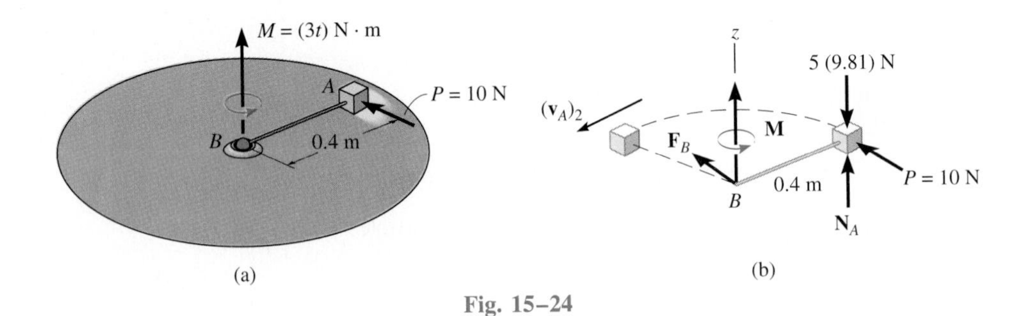

Fig. 15–24

SOLUTION

Free-Body Diagram. If we consider the system of both the rod and block, Fig. 15–24*b*, then the resultant force reaction $\mathbf{F}_B$ can be eliminated from the analysis by applying the principle of angular impulse and momentum about the z axis. Why? If this is done, the angular impulses created by the weight and normal reaction $\mathbf{N}_A$ are also eliminated, since they act parallel to the z axis and therefore create zero moment about this axis.

Principle of Angular Impulse and Momentum

$$(H_z)_1 + \Sigma \int_{t_1}^{t_2} M_z \, dt = (H_z)_2$$

$$(H_z)_1 + \int_{t_1}^{t_2} M \, dt + r_{BA} P(\Delta t) = (H_z)_2$$

$$0 + \int_0^4 3t \, dt + (0.4 \text{ m})(10 \text{ N})(4 \text{ s}) = 5 \text{ kg}(v_A)_2(0.4 \text{ m})$$

$$24 + 16 = 2(v_A)_2$$

$$(v_A)_2 = 20 \text{ m/s} \qquad \textit{Ans.}$$

Example 15–14

The ball B, shown in Fig. 15–25a, has a weight of 0.8 lb and is attached to a cord which passes through a hole at A in a smooth table. When the ball is $r_1 = 1.75$ ft from the hole, it is rotating around in a circle such that its speed is $v_1 = 4$ ft/s. If by applying a force **F** the cord is pulled downward through the hole with a constant speed $v_c = 6$ ft/s, determine *(a)* the speed of the ball at the instant it is $r_2 = 0.6$ ft from the hole, and *(b)* the amount of work done by the force **F** in shortening the radial distance r. Neglect the size of the ball.

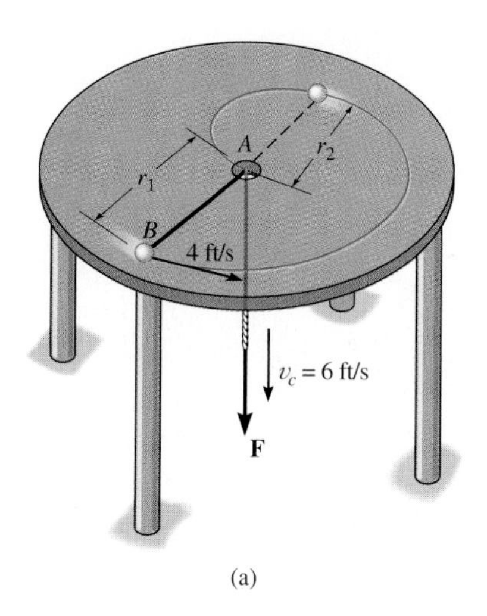

(a)

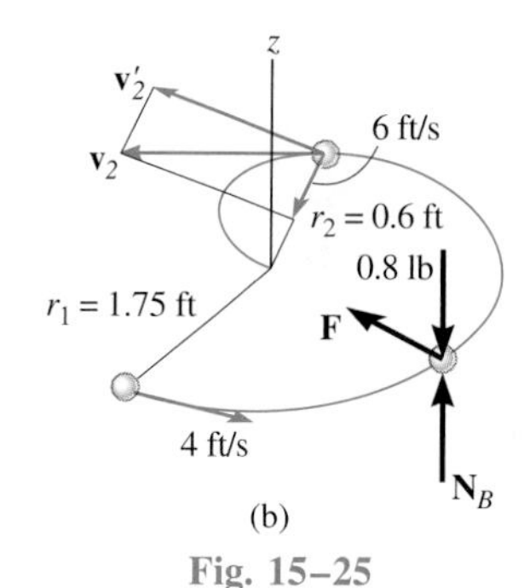

(b)

Fig. 15–25

SOLUTION

Part (a)

Free-Body Diagram. As the ball moves from radial position r_1 to r_2, Fig. 15–25b, two of the forces acting on it are unknown; however, since the effect of the cord **F** on the ball passes through the z axis and the weight and $\mathbf{N}_B$ are parallel to it, the moments, or angular impulses created by the forces, are all *zero* about this axis. Hence, the conservation of angular momentum applies about the z axis.

Conservation of Angular Momentum. As indicated in Fig. 15–25b, the ball's velocity $\mathbf{v}_2$ is resolved into two components. The radial component, 6 ft/s, is known; however, it produces zero angular momentum about the z axis. Thus,

$$\mathbf{H}_1 = \mathbf{H}_2$$

$$r_1 m_B v_1 = r_2 m_B v_2'$$

$$1.75 \text{ ft}\left(\frac{0.8 \text{ lb}}{32.2 \text{ ft/s}^2}\right)4 = 0.6 \text{ ft}\left(\frac{0.8 \text{ lb}}{32.2 \text{ ft/s}^2}\right) v_2'$$

$$v_2' = 11.67 \text{ ft/s}$$

The speed of the ball is thus

$$v_2 = \sqrt{(11.67)^2 + (6)^2}$$

$$= 13.1 \text{ ft/s} \qquad \textit{Ans.}$$

Part (b). The only force that does work on the ball is **F**. (The normal force and weight do not move vertically.) The initial and final kinetic energies of the ball can be determined so that from the principle of work and energy we have

$$T_1 + \Sigma U_{1-2} = T_2$$

$$\frac{1}{2}\left(\frac{0.8 \text{ lb}}{32.2 \text{ ft/s}^2}\right)(4 \text{ ft/s})^2 + U_F = \frac{1}{2}\left(\frac{0.8 \text{ lb}}{32.2 \text{ ft/s}^2}\right)(13.1 \text{ ft/s})^2$$

$$U_F = 1.94 \text{ ft} \cdot \text{lb} \qquad \textit{Ans.}$$

Example 15–15

The 2-kg disk shown in Fig. 15–26*a* rests on a smooth horizontal surface and is attached to an elastic cord that has a stiffness $k_c = 20$ N/m and is initially unstretched. If the disk is given a velocity $(v_D)_1 = 1.5$ m/s, perpendicular to the cord, determine the rate at which the cord is being stretched and the speed of the disk at the instant the cord is stretched 0.2 m.

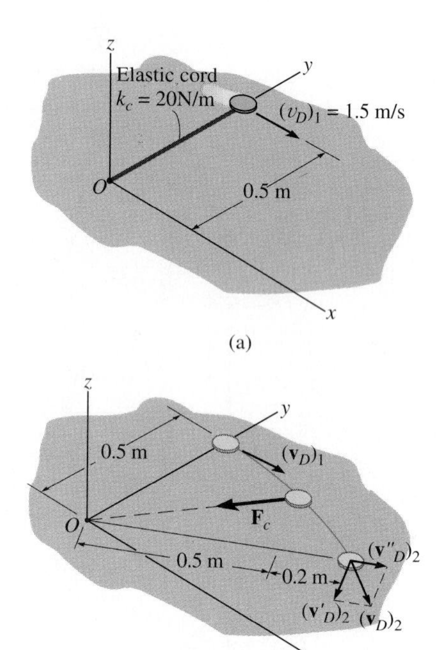

Fig. 15–26

SOLUTION

Free-Body Diagram. After the disk has been launched, it slides along the dashed path shown in Fig. 15–26*b*. By inspection, angular momentum about point O (or the z axis) is *conserved,* since the moment of the central force $\mathbf{F}_c$ about O is always zero. Also, when the distance is 0.7 m, only the component of $(\mathbf{v}'_D)_2$ is effective in producing angular momentum of the disk about O.

Conservation of Angular Momentum. The component $(\mathbf{v}'_D)_2$ can be obtained by applying the conservation of angular momentum about O (the z axis), i.e.,

$$(\mathbf{H}_O)_1 = (\mathbf{H}_O)_2$$

$$(\circlearrowleft +) \qquad r_1 m_D (v_D)_1 = r_2 m_D (v'_D)_2$$

$$0.5\text{ m}(2\text{ kg})(1.5\text{ m/s}) = 0.7\text{ m}(2\text{ kg})(v'_D)_2$$

$$(v'_D)_2 = 1.07\text{ m/s}$$

Conservation of Energy. The speed of the disk may be obtained by applying the conservation of energy equation at the point where the disk was launched and at the point where the cord is stretched 0.2 m.

$$T_1 + V_1 = T_2 + V_2$$

$$\tfrac{1}{2}(2\text{ kg})(1.5\text{ m/s})^2 + 0 = \tfrac{1}{2}(2\text{ kg})(v_D)_2^2 + \tfrac{1}{2}(20\text{ N/m})(0.2\text{ m})^2$$

Thus,

$$(v_D)_2 = 1.36\text{ m/s} \qquad \textit{Ans.}$$

Having determined $(v_D)_2$ and its component $(v'_D)_2$, the rate of stretch of the cord $(v''_D)_2$ is determined from the Pythagorean theorem,

$$(v''_D)_2 = \sqrt{(v_D)_2^2 - (v'_D)_2^2}$$

$$= \sqrt{(1.36)^2 - (1.07)^2}$$

$$= 0.838\text{ m/s} \qquad \textit{Ans.}$$

PROBLEMS

15–91. A particle has a linear momentum $\mathbf{L} = \{-8\mathbf{i} + 14\mathbf{j} - 12\mathbf{k}\}$ kg · m/s. If it is at point A (3 m, −4 m, 2 m), determine its angular momentum about point B (−1 m, −2 m, 8 m).

***15–92.** A particle has a linear momentum $\mathbf{L} = \{2\mathbf{i} + 4\mathbf{j} + 6\mathbf{k}\}$ kg · m/s. If it is at point A (2 m, −1 m, 4 m), determine its angular momentum about point B (5 m, −2 m, 6 m).

15–93. Determine the angular momentum $\mathbf{H}_O$ of each of the particles about point O. Use a scalar solution.

15–94. Determine the angular momentum $\mathbf{H}_P$ of each of the particles about point P. Use a scalar solution.

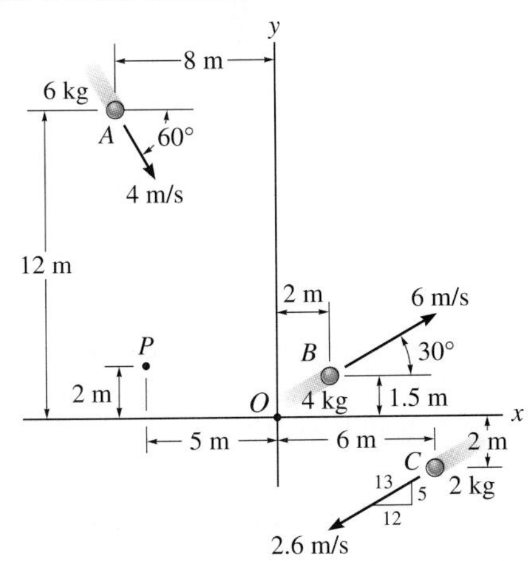

Probs. 15–93/15–94

15–95. Determine the angular momentum $\mathbf{H}_O$ of the particle about point O. Use a Cartesian vector solution.

***15–96.** Determine the angular momentum $\mathbf{H}_P$ of the particle about point P. Use a Cartesian vector solution.

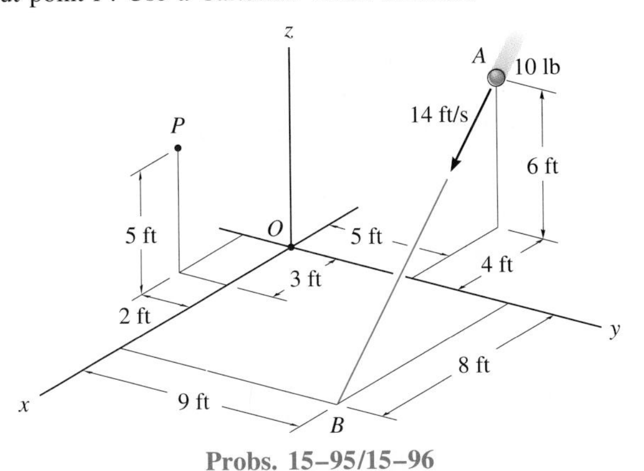

Probs. 15–95/15–96

15–97. Determine the angular momentum of the 2-lb particle A about point O. Use a Cartesian vector solution.

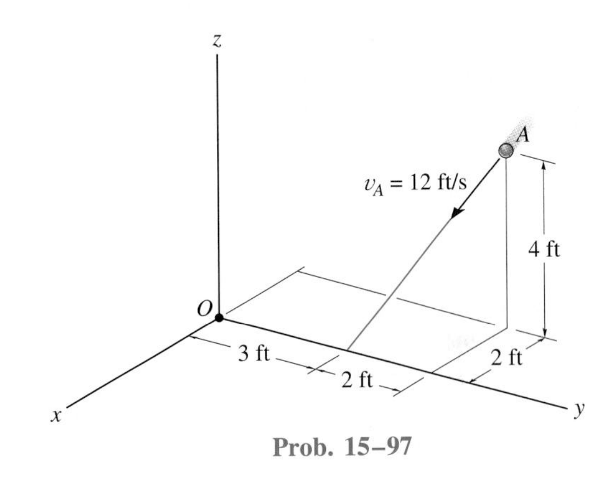

Prob. 15–97

15–98. Determine the angular momentum $\mathbf{H}_O$ of each of the two particles about point O. Use a scalar solution.

15–99. Determine the angular momentum $\mathbf{H}_P$ of each of the two particles about point P. Use a scalar solution.

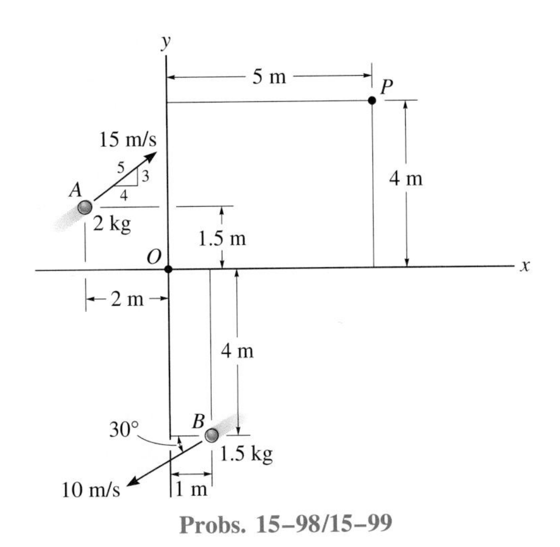

Probs. 15–98/15–99

***15–100.** A projectile having a mass of 3 kg is fired upward from a cannon with a muzzle velocity of $v_0 = 500$ m/s at an angle of 45° with the horizontal. Determine the projectile's angular momentum measured from the point of launch at the instant it is at the maximum height of its trajectory.

15–101. A projectile having a mass of 3 kg is fired upward from a cannon at an angle of 60° with the horizontal. If it strikes the horizontal 800 m away from where it was fired, at the same elevation, determine the projectile's maximum angular momentum measured from the point of launch.

15–102. The two spheres each have a mass of 3 kg and are attached to the rod of negligible mass. If a torque $M = (6e^{0.2t})$ N · m, where t is in seconds, is applied to the rod as shown, determine the speed of each of the spheres in 2 s, starting from rest.

15–103. The two spheres each have a mass of 3 kg and are attached to the rod of negligible mass. Determine the time the torque $M = (8t)$ N · m, where t is in seconds, must be applied to the rod so that each sphere attains a speed of 3 m/s starting from rest.

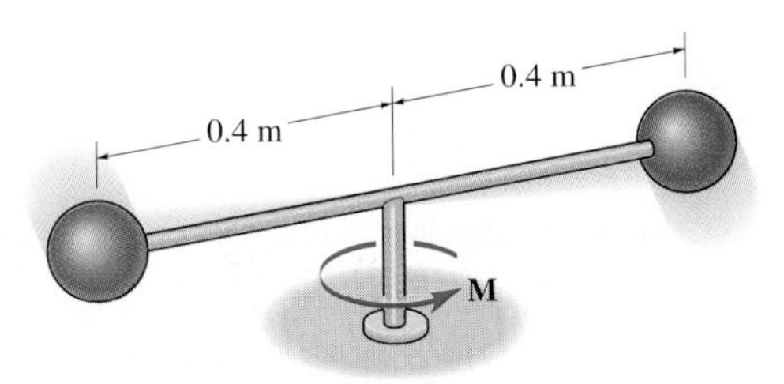

Probs. 15–102/15–103

***15–104.** The ball B has a mass of 10 kg and is attached to the end of a rod whose mass may be neglected. If the rod is subjected to a torque $M = (3t^2 + 5t + 2)$ N · m, where t is in seconds, determine the speed of the ball when $t = 2$ s. The ball has a speed $v = 2$ m/s when $t = 0$.

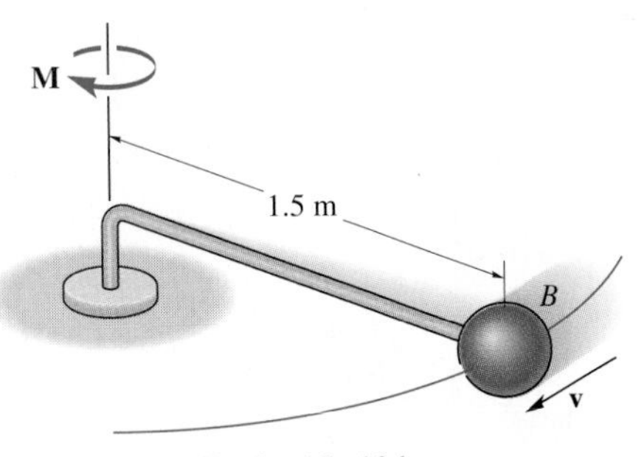

Prob. 15–104

15–105. The small cylinder C has a mass of 10 kg and is attached to the end of a rod whose mass may be neglected. If the frame is subjected to a couple $M = (8t^2 + 5)$ N · m, where t is in seconds, and the cylinder is subjected to a force of 60 N, which is always directed as shown, determine the speed of the cylinder when $t = 2$ s. The cylinder has a speed $v_0 = 2$ m/s when $t = 0$.

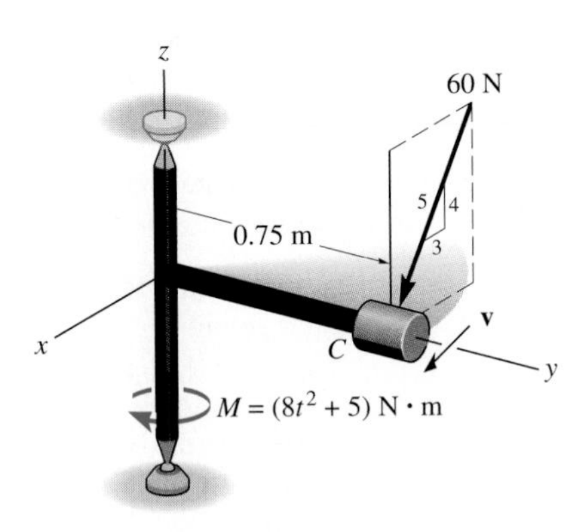

Prob. 15–105

15–106. A 4-lb ball B is traveling around in a circle of radius $r_1 = 3$ ft with a speed $(v_B)_1 = 6$ ft/s. If the attached cord is pulled down through the hole with a constant speed $v_r = 2$ ft/s, determine the ball's speed at the instant $r_2 = 2$ ft. How much work has to be done to pull down the cord? Neglect friction and the size of the ball.

15–107. A 4-lb ball B is traveling around in a circle of radius $r_1 = 3$ ft with a speed $(v_B)_1 = 6$ ft/s. If the attached cord is pulled down through the hole with a constant speed $v_r = 2$ ft/s, determine how much time is required for the ball to reach a speed of 12 ft/s. How far r_2 is the ball from the hole when this occurs? Neglect friction and the size of the ball.

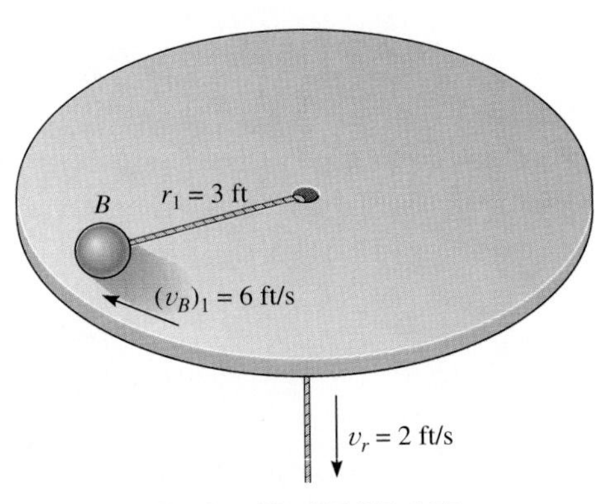

Probs. 15–106/15–107

*15–108. The 800-lb roller-coaster car starts from rest on the track having the shape of a cylindrical helix. If the helix descends 8 ft for every one revolution, determine the speed of the car in $t = 4$ s. Also, how far has the car descended in this time? Neglect friction and the size of the car.

15–109. The 800-lb roller-coaster car starts from rest on the track having the shape of a cylindrical helix. If the helix descends 8 ft for every one revolution, determine the time required for the car to attain a speed of 60 ft/s. Neglect friction and the size of the car.

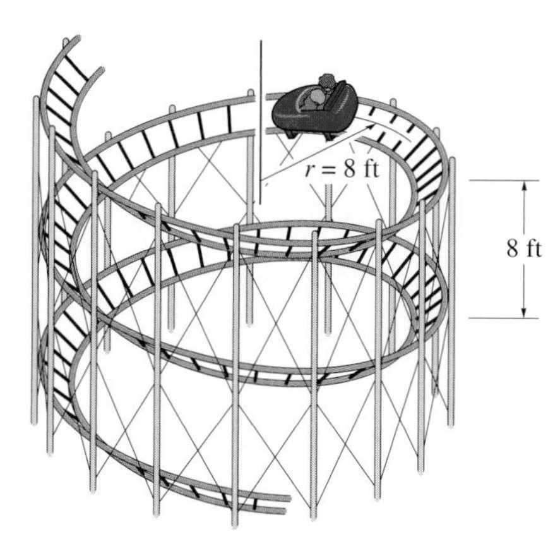

Probs. 15–108/15–109

15–110. A small block having a mass of 0.1 kg is given a horizontal velocity $v_1 = 0.4$ m/s when $r_1 = 500$ mm. It slides along the smooth conical surface. When it descends to $h = 100$ mm, determine its speed and the angle of descent θ, that is, the angle measured from the horizontal to the tangent of the path.

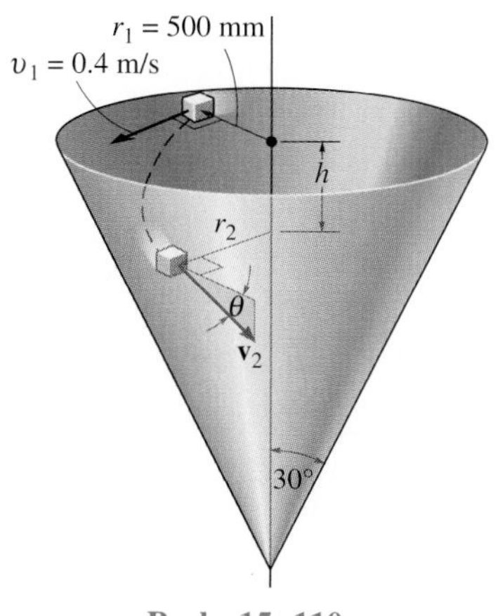

Prob. 15–110

15–111. A small block having a mass of 0.1 kg is given a horizontal velocity of $v_1 = 0.4$ m/s when $r_1 = 500$ mm. It slides along the smooth conical surface. Determine the distance h it must descend for it to reach a speed of $v_2 = 2$ m/s. Also, what is the angle of descent θ, that is, the angle measured from the horizontal to the tangent of the path?

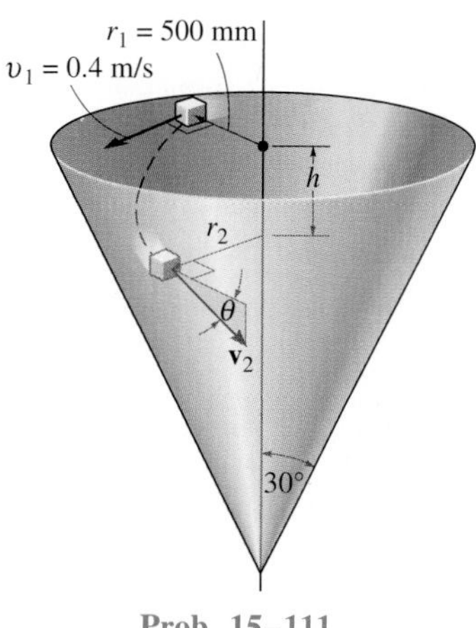

Prob. 15–111

*15–112. A child having a mass of 50 kg holds her legs up as shown as she swings downward from rest at $\theta_1 = 30°$. Her center of mass is located at point G_1. When she is at the bottom position $\theta = 0°$, she *suddenly* lets her legs come down, shifting her center of mass to position G_2. Determine her speed in the upswing due to this sudden movement and the angle θ_2 to which she swings before momentarily coming to rest. Treat the child's body as a particle.

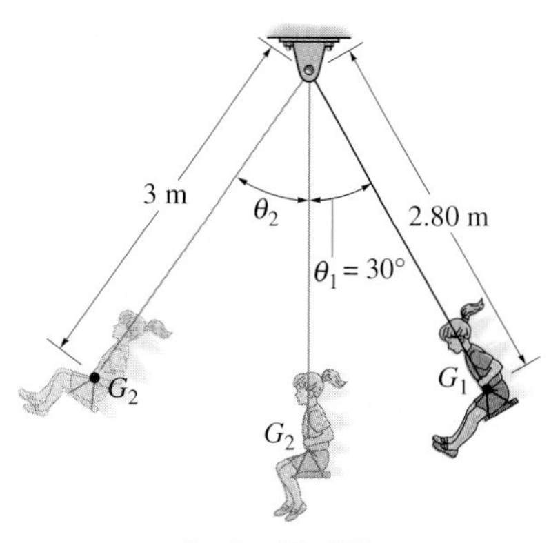

Prob. 15–112

15–113. The 10-lb block rests on a surface for which $\mu_k = 0.5$. It is acted upon by a radial force of 2 lb and a horizontal force of 7 lb, always directed at 30° from the tangent to the path as shown. If the block is initially moving in a circular path with a speed $v_1 = 2$ ft/s at the instant the forces are applied, determine the time required before the tension in cord AB becomes 20 lb. Neglect the size of the block for the calculation.

15–114. The 10-lb block is originally at rest on the smooth surface. It is acted upon by a radial force of 2 lb and a horizontal force of 7 lb, always directed at 30° from the tangent to the path as shown. Determine the time required to break the cord, which requires a tension $T = 30$ lb. What is the speed of the block when this occurs? Neglect the size of the block for the calculation.

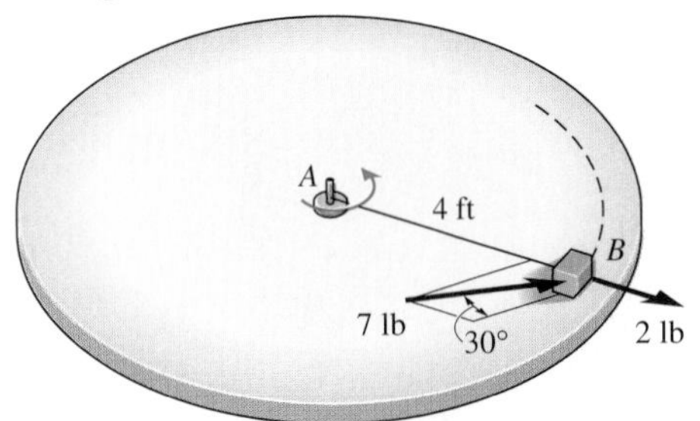

Probs. 15–113/15–114

15–115. The elastic cord has an unstretched length $l_0 = 1.5$ ft and a stiffness $k = 12$ lb/ft. It is attached to a fixed point at A and a block at B, which has a weight of 2 lb. If the block is released from rest from the position shown, determine its speed when it reaches point C after it slides along the smooth guide. After leaving the guide, it is launched onto the smooth *horizontal* plane. Determine if the cord becomes unstretched. Also, calculate the angular momentum of the block about point A, at any instant after it passes point C.

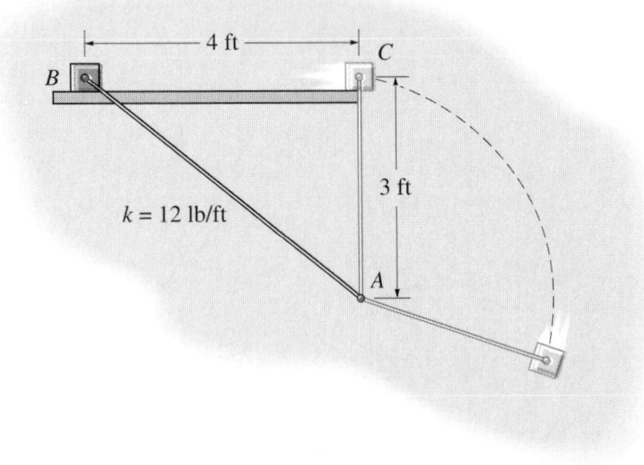

Prob. 15–115

***15–116.** An earth satellite of mass 700 kg is launched into a free-flight trajectory about the earth with an initial speed $v_A = 10$ km/s when the distance from the center of the earth is $r_A = 15$ Mm. If the launch angle at this position is $\phi_A = 70°$, determine the speed v_B of the satellite and its closest distance r_B from the center of the earth. The earth has a mass $M_e = 5.976(10^{24})$ kg. *Hint:* Under these conditions, the satellite is subjected only to the earth's gravitational force, $F = GM_e m_s/r^2$, Eq. 13-1. For part of the solution, use the conservation of energy (see Prob. 14–92).

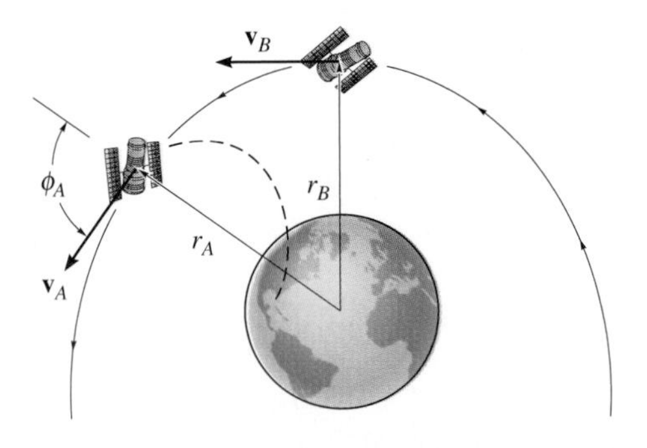

Prob. 15–116

15–117. The boy weighs 80 lb. While holding on to a ring, he runs in a circle and then lifts his feet off the ground, holding himself in the crouched position shown. If *initially* his center of gravity G is $r_A = 10$ ft from the pole and his velocity is *horizontal* such that $v_A = 8$ ft/s, determine (a) his velocity when he is at B, where $r_B = 7$ ft, $\Delta z = 2$ ft, and (b) the vertical component of his velocity, $(v_B)_z$, which is causing him to fall downward.

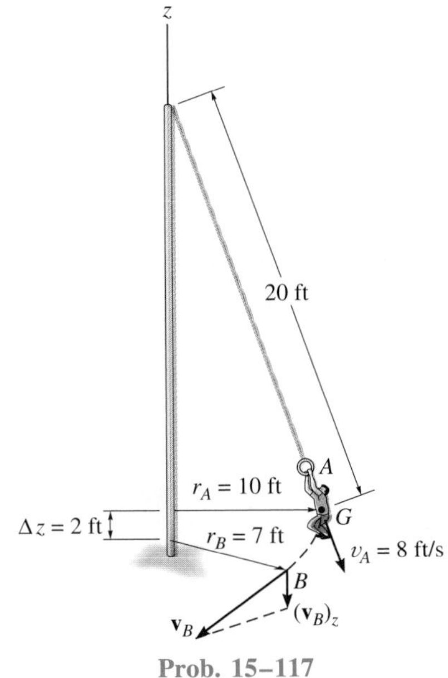

Prob. 15–117

*15.8 Steady Fluid Streams

Knowledge of the forces developed by steadily moving fluid streams is of importance in the design and analysis of turbines, pumps, blades, and fans. To illustrate how the principle of impulse and momentum may be used to determine these forces, consider the diversion of a steady stream of fluid (liquid or gas) by a fixed pipe, Fig. 15–27*a*. The fluid enters the pipe with a velocity $\mathbf{v}_A$ and exits with a velocity $\mathbf{v}_B$. Both of these velocities are measured by an observer fixed in an inertial frame of reference. The impulse and momentum diagrams for the fluid stream are shown in Fig. 15–27*b*. The force $\Sigma\mathbf{F}$, shown on the impulse diagram, represents the resultant of all the external forces acting on the fluid stream. It is this loading which gives the fluid stream an impulse whereby the original momentum of the fluid is changed in both its magnitude and direction. Since the flow is steady, $\Sigma\mathbf{F}$ will be *constant* during the time interval dt. During this time the fluid stream is in motion, and as a result a small amount of fluid, having a mass dm, is about to enter the pipe with a velocity $\mathbf{v}_A$ at time t. If this element of mass and the mass of fluid in the pipe are considered as a "closed system," then at time $t + dt$ a corresponding element of mass dm must leave the pipe with a velocity $\mathbf{v}_B$. Also, the fluid stream *within* the pipe section has a mass m and an *average velocity* $\mathbf{v}$ which is constant during the time interval dt. Applying the principle of linear impulse and momentum to the fluid stream, we have

$$dm\ \mathbf{v}_A + m\mathbf{v} + \Sigma\mathbf{F}\ dt = dm\ \mathbf{v}_B + m\mathbf{v}$$

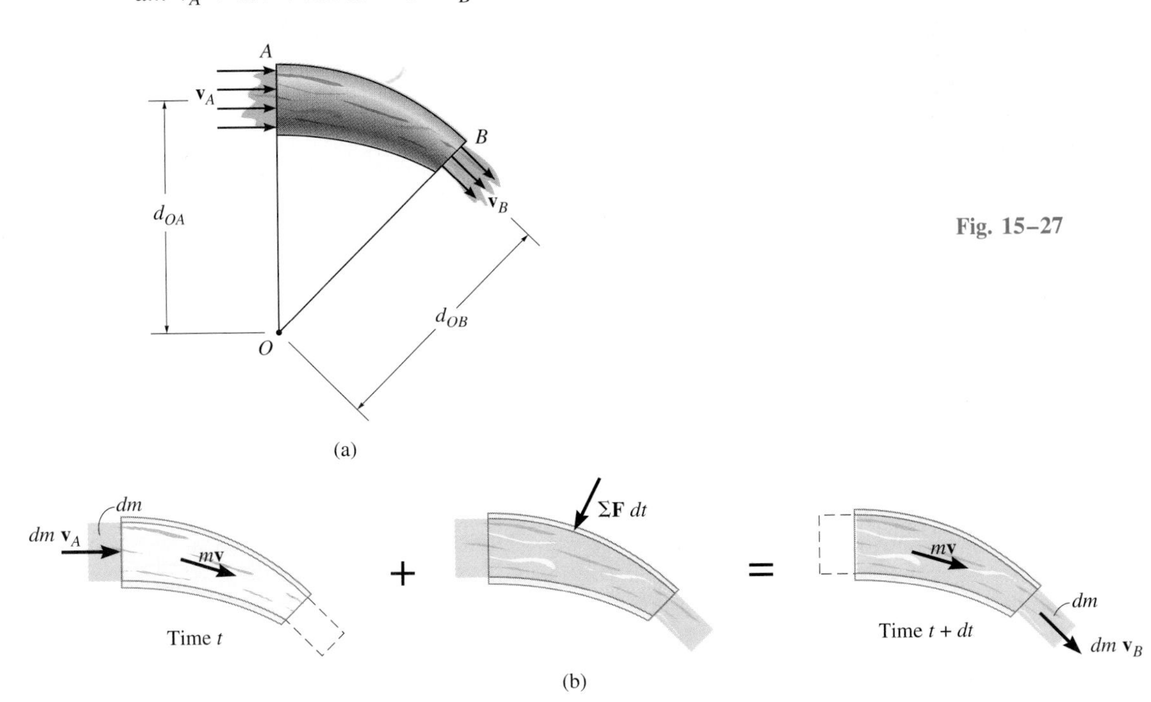

Fig. 15–27

Force Resultant. Solving for the resultant force yields

$$\Sigma \mathbf{F} = \frac{dm}{dt}(\mathbf{v}_B - \mathbf{v}_A) \tag{15–25}$$

Provided the motion of the fluid can be represented in the x–y plane, it is usually convenient to express this vector equation in the form of two scalar component equations, i.e.,

$$\Sigma F_x = \frac{dm}{dt}(v_{Bx} - v_{Ax})$$
$$\Sigma F_y = \frac{dm}{dt}(v_{By} - v_{Ay}) \tag{15–26}$$

The term dm/dt is called the *mass flow* and indicates the constant amount of fluid which flows either into or out of the pipe per unit of time. If the cross-sectional areas and densities of the fluid at the entrance A and exit B are A_A, ρ_A and A_B, ρ_B, respectively, Fig. 15–27*c*, then *continuity of mass* requires that $dm = \rho\, dV = \rho_A(ds_A\, A_A) = \rho_B(ds_B\, A_B)$. Hence, during the time dt, since $v_A = ds_A/dt$ and $v_B = ds_B/dt$, we have

$$\frac{dm}{dt} = \rho_A v_A A_A = \rho_B v_B A_B = \rho_A Q_A = \rho_B Q_B \tag{15–27}$$

Here $Q = vA$ is the volumetric *flow rate,* which measures the volume of fluid flowing per unit of time.

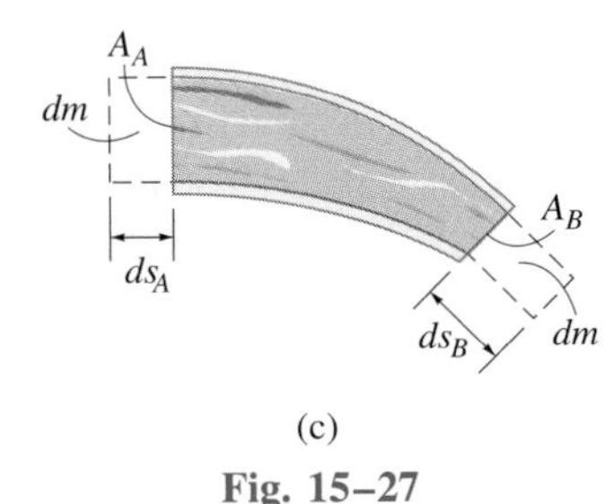

(c)

Fig. 15–27

Moment Resultant. In some cases it is necessary to obtain the support reactions on the fluid-carrying device. If Eq. 15–25 does not provide enough information to do this, the principle of angular impulse and momentum must be used. The formulation of this principle applied to fluid streams can be obtained from Eq. 15–17, $\Sigma \mathbf{M}_O = \dot{\mathbf{H}}_O$, which states that the moment of all the external forces acting on the system about point O is equal to the time rate of change of angular momentum about O. In the case of the pipe shown in Fig. 15–27a, the flow is steady in the x–y plane; hence we have

$$(\circlearrowleft +) \qquad \Sigma M_O = \frac{dm}{dt}(d_{OB}\, v_B - d_{OA}\, v_A) \qquad (15\text{–}28)$$

where the moment arms d_{OB} and d_{OA} are directed from O to the *geometric center* or *centroid* of the openings at A and B.

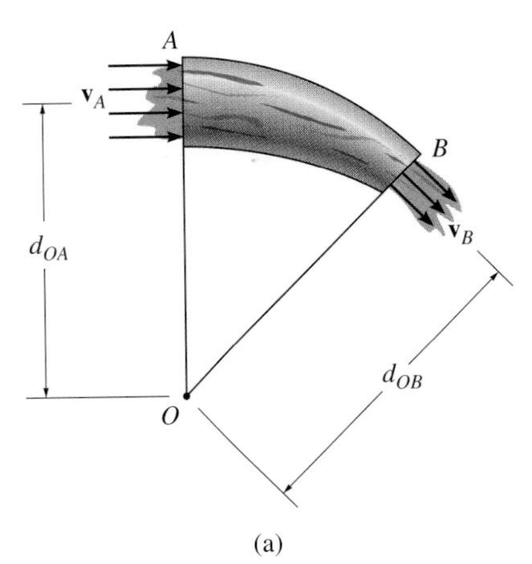

(a)
Fig. 15–27a *(Repeated)*

PROCEDURE FOR ANALYSIS

The following procedure provides a method for solving problems involving steady flow.

Kinematic Diagram. In problems where the device is *moving*, a *kinematic diagram* may be helpful for determining the entrance and exit velocities of the fluid flowing onto the device, since a *relative-motion analysis* of velocity will be involved. The *kinematic diagram* in this case is simply a graphical representation of the velocities showing the vector addition of the relative-motion components. In all cases, the measurement of velocity must be made by an observer fixed in an inertial frame of reference. Once the velocity of the fluid flowing onto the device is determined, the mass flow is calculated using Eq. 15–27.

Free-Body Diagram. Draw a free-body diagram of the device which is directing the fluid in order to establish the forces $\Sigma \mathbf{F}$ acting on it. These external forces will include the support reactions, the weight of the device and the fluid contained within it, and the static pressure forces of the fluid at the entrance and exit sections of the device.*

Equations of Steady Flow. Apply the equations of steady flow, Eqs. 15–26 and 15–28, using the appropriate components of velocity and force shown on the kinematic and free-body diagrams.

*In the SI system pressure is measured using the *pascal* (Pa), where 1 Pa = 1 N/m^2.

The following examples illustrate application of this procedure numerically.

Example 15–16

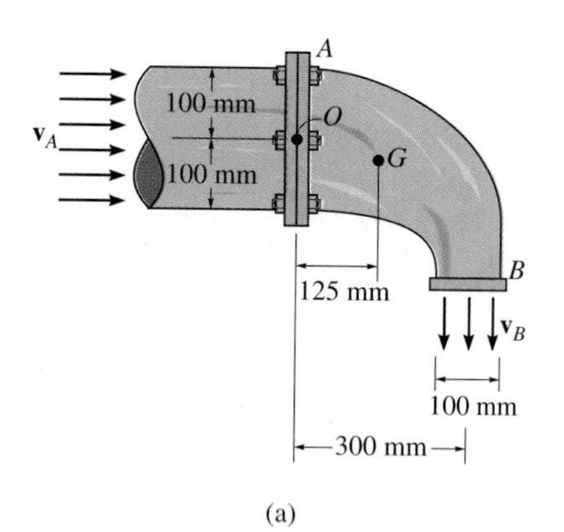

(a)

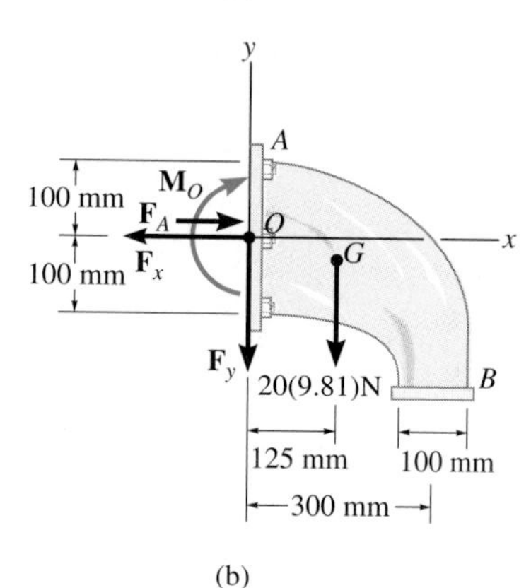

(b)

Fig. 15–28

Determine the reaction components which the pipe joint at A exerts on the elbow in Fig. 15–28a, if water flowing through the pipe is subjected to a static gauge pressure of 100 kPa at A. The discharge at B is $Q_B = 0.2$ m^3/s. Water has a density $\rho_w = 1000$ kg/m^3, and the water-filled elbow has a mass of 20 kg and center of mass at G.

SOLUTION

Using a fixed inertial coordinate system, the velocity of flow at A and B and the mass flow rate can be obtained from Eq. 15–27. Since the density of water is constant, $Q_B = Q_A = Q$. Hence,

$$\frac{dm}{dt} = \rho_w Q = (1000\ \text{kg/m}^3)(0.2\ \text{m}^3/\text{s}) = 200\ \text{kg/s}$$

$$v_B = \frac{Q}{A_B} = \frac{0.2\ \text{m}^3/\text{s}}{\pi(0.05\ \text{m})^2} = 25.46\ \text{m/s} \downarrow$$

$$v_A = \frac{Q}{A_A} = \frac{0.2\ \text{m}^3/\text{s}}{\pi(0.1\ \text{m})^2} = 6.37\ \text{m/s} \rightarrow$$

Free-Body Diagram. As shown on the free-body diagram, Fig. 15–28b, the *fixed* connection at A exerts a resultant couple $\mathbf{M}_O$ and force components $\mathbf{F}_x$ and $\mathbf{F}_y$ on the elbow. Due to the static pressure of water in the pipe, the pressure force acting on the fluid at A is $F_A = p_A A_A$. Since 1 kPa = 1000 N/m^2,

$$F_A = p_A A_A = [100(10^3)\ \text{N/m}^2][\pi(0.1\ \text{m})^2] = 3141.6\ \text{N}$$

There is no static pressure acting at B, since the water is discharged at atmospheric pressure; i.e., the pressure measured by a gauge at B is equal to zero, $p_B = 0$.

Equations of Steady Flow

$$\overset{+}{\rightarrow} \Sigma F_x = \frac{dm}{dt}(v_{Bx} - v_{Ax}); \quad -F_x + 3141.6\ \text{N} = 200\ \text{kg/s}(0 - 6.37\ \text{m/s})$$

$$F_x = 4.41\ \text{kN} \qquad \textit{Ans.}$$

$$+\uparrow \Sigma F_y = \frac{dm}{dt}(v_{By} - v_{Ay}); -F_y - 20(9.81)\ \text{N} = 200\ \text{kg/s}(-25.46\ \text{m/s} - 0)$$

$$F_y = 4.90\ \text{kN} \qquad \textit{Ans.}$$

If moments are summed about point O, Fig. 15–28b, then $\mathbf{F}_x$, $\mathbf{F}_y$, and the static pressure $\mathbf{F}_A$ are eliminated, as well as the moment of momentum of the water entering at A, Fig. 15–28a. Hence,

$$\curvearrowleft + \Sigma M_O = \frac{dm}{dt}(d_{OB}v_B - d_{OA}v_A)$$

$$M_O + 20(9.81)\ \text{N}(0.125\ \text{m}) = 200\ \text{kg/s}[(0.3\ \text{m})(25.46\ \text{m/s}) - 0]$$

$$M_O = 1.50\ \text{kN}\cdot\text{m} \qquad \textit{Ans.}$$

Example 15–17

A 2-in.-diameter water jet having a velocity of 25 ft/s impinges upon a single moving blade, Fig. 15–29*a*. If the blade is moving at 5 ft/s away from the jet, determine the horizontal and vertical components of force which the blade is exerting on the water. What power does the water generate on the blade? Water has a specific weight of $\gamma_w = 62.4\ \text{lb/ft}^3$.

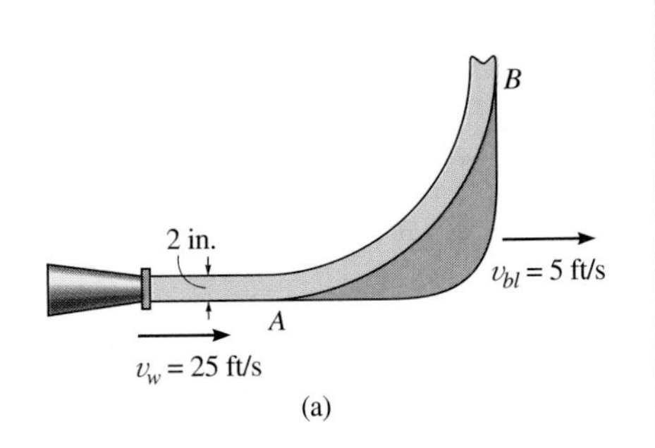

(a)

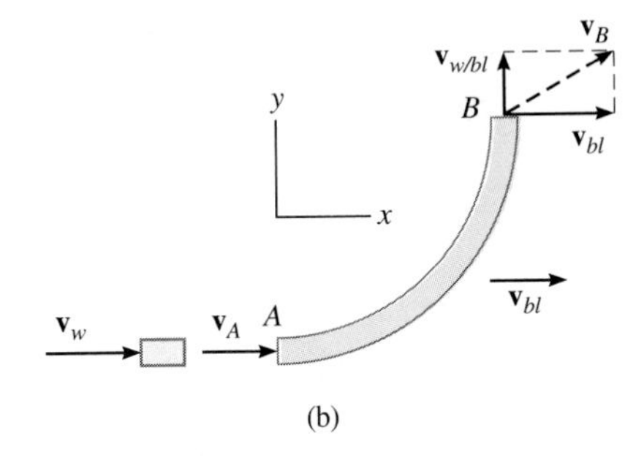

(b)

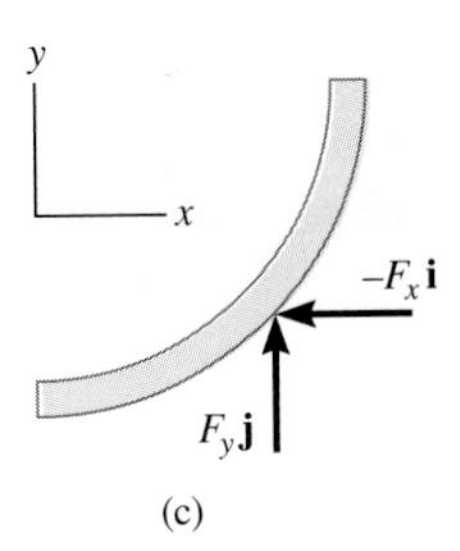

(c)

Fig. 15–29

SOLUTION

Kinematic Diagram. From a fixed inertial coordinate system, Fig. 15–29*b*, the rate at which water enters the blade is

$$\mathbf{v}_A = \{25\mathbf{i}\}\ \text{ft/s}$$

The *relative-flow velocity* of the water onto the blade is $\mathbf{v}_{w/bl} = \mathbf{v}_w - \mathbf{v}_{bl} = 25\mathbf{i} - 5\mathbf{i} = \{20\mathbf{i}\}$ ft/s. Since the blade is moving with a velocity of $\mathbf{v}_{bl} = \{5\mathbf{i}\}$ ft/s, the velocity of flow at B measured from x, y is the vector sum, shown in Fig. 15–29*b*. Here,

$$\begin{aligned}\mathbf{v}_B &= \mathbf{v}_{bl} + \mathbf{v}_{w/bl}\\ &= \{5\mathbf{i} + 20\mathbf{j}\}\ \text{ft/s}\end{aligned}$$

Thus, the mass flow of water *onto* the blade that undergoes a momentum change is

$$\frac{dm}{dt} = \rho_w(v_{w/bl})A_A = \frac{62.4}{32.2}(20)\left[\pi\left(\frac{1}{12}\right)^2\right] = 0.846\ \text{slug/s}$$

Free-Body Diagram. The free-body diagram of a section of water acting on the blade is shown in Fig. 15–29*c*. The weight of the water will be neglected in the calculation, since this force will be small compared to the reactive components $\mathbf{F}_x$ and $\mathbf{F}_y$.

Equations of Steady Flow

$$\Sigma\mathbf{F} = \frac{dm}{dt}(\mathbf{v}_B - \mathbf{v}_A)$$

$$-F_x\mathbf{i} + F_y\mathbf{j} = 0.846(5\mathbf{i} + 20\mathbf{j} - 25\mathbf{i})$$

Equating the respective **i** and **j** components gives

$$F_x = 0.846(20) = 16.9\ \text{lb} \leftarrow \qquad \textit{Ans.}$$

$$F_y = 0.846(20) = 16.9\ \text{lb} \uparrow \qquad \textit{Ans.}$$

The water exerts equal but opposite forces on the blade.

Since the water force which causes the blade to move forward horizontally with a velocity of 5 ft/s is $F_x = 16.9$ lb, then from Eq. 14–10 the power is

$$P = \mathbf{F}\cdot\mathbf{v}; \qquad P = \frac{16.9\ \text{lb}(5\ \text{ft/s})}{550\ \text{hp/(ft}\cdot\text{lb/s)}} = 0.154\ \text{hp} \qquad \textit{Ans.}$$

★15.9 Propulsion with Variable Mass

In the previous section we considered the case in which a *constant* amount of mass dm enters and leaves a *"closed system."* There are, however, two other important cases involving mass flow, which are represented by a system that is either gaining or losing mass. In this section we will discuss each of these cases separately.

A System That Loses Mass. Consider a device which at an instant of time has a mass m and is moving forward with a velocity $\mathbf{v}$, as measured from a fixed inertial reference frame, Fig. 15–30a. At this same instant the device is expelling an amount of mass m_e such that the mass flow velocity $\mathbf{v}_e$ is *constant* when the measurement is made from the inertial frame of reference. For the analysis, consider the *"closed system"* to include *both the mass m of the device and the expelled mass m_e,* as shown by the dashed line in the figure. The impulse and momentum diagrams for the system are shown in Fig. 15–30b. During the time interval dt, the velocity of the device is increased from $\mathbf{v}$ to $\mathbf{v} + d\mathbf{v}$. This increase in forward velocity, however, does not change the velocity $\mathbf{v}_e$ of the expelled mass, since this mass moves at a constant speed once it has been ejected. To increase the velocity of the device during the time dt, an amount of mass dm_e has been ejected and thereby gained in the exhaust. The impulses are created by $\Sigma\mathbf{F}_s$, which represents the resultant of all the external forces which *act on the system* in the direction of motion. This force resultant *does not include* the force which causes the device to move forward, since this force (called a *thrust*) is *internal to the system;* that is, the thrust acts with equal magnitude but opposite direction on the mass m of the device and the expelled exhaust mass m_e.* Applying the principle of impulse and momentum to the system, in reference to Fig. 15–30b, we have

$$(\overset{+}{\rightarrow}) \qquad mv - m_e v_e + \Sigma F_s\, dt = (m - dm_e)(v + dv) - (m_e + dm_e)v_e$$

or

$$\Sigma F_s\, dt = -v\, dm_e + m\, dv - dm_e\, dv - v_e\, dm_e$$

(a)

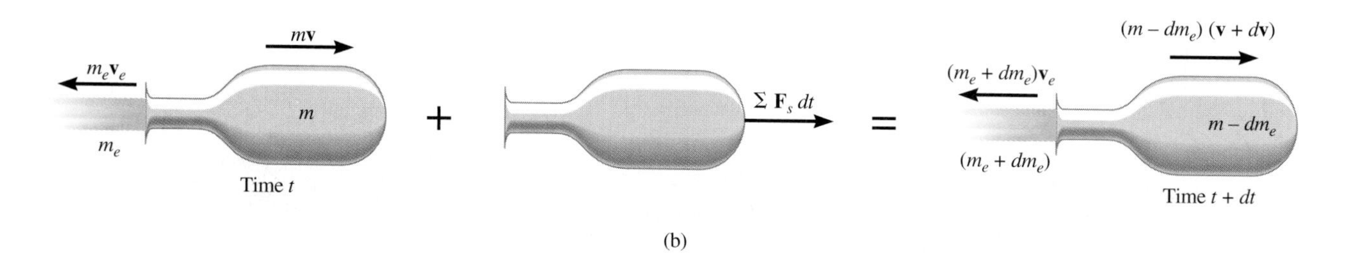

(b)

Fig. 15–30

*$\Sigma\mathbf{F}_s$ represents the external resultant force *acting on the system,* which is different from $\Sigma\mathbf{F}$, the resultant force acting only on the device.

Without loss of accuracy, the third term on the right side of this equation may be neglected since it is a "second-order" differential. Dividing by dt gives

$$\Sigma F_s = m\frac{dv}{dt} - (v + v_e)\frac{dm_e}{dt}$$

Noting that the relative velocity of the device as seen by an observer moving with the particles of the ejected mass is $v_{D/e} = (v + v_e)$, the final result can be written as

$$\Sigma F_s = m\frac{dv}{dt} - v_{D/e}\frac{dm_e}{dt} \qquad (15\text{–}29)$$

Here the term dm_e/dt represents the rate at which mass is being ejected.

To illustrate an application of Eq. 15–29, consider the rocket shown in Fig. 15–31, which has a weight $\mathbf{W}$ and is moving upward against an atmospheric drag force $\mathbf{F}_D$. The system to be considered consists of the mass of the rocket and the mass of ejected gas m_e. Applying Eq. 15–29 to this system gives

$$(+\uparrow) \qquad -F_D - W = \frac{W}{g}\frac{dv}{dt} - v_{D/e}\frac{dm_e}{dt}$$

The last term of this equation represents the *thrust* $\mathbf{T}$ which the engine exhaust exerts on the rocket, Fig. 15–31. Recognizing that $dv/dt = a$, we may therefore write

$$(+\uparrow) \qquad T - F_D - W = \frac{W}{g}a$$

If a free-body diagram of the rocket is drawn, it becomes obvious that this equation represents an application of $\Sigma\mathbf{F} = m\mathbf{a}$ for the rocket.

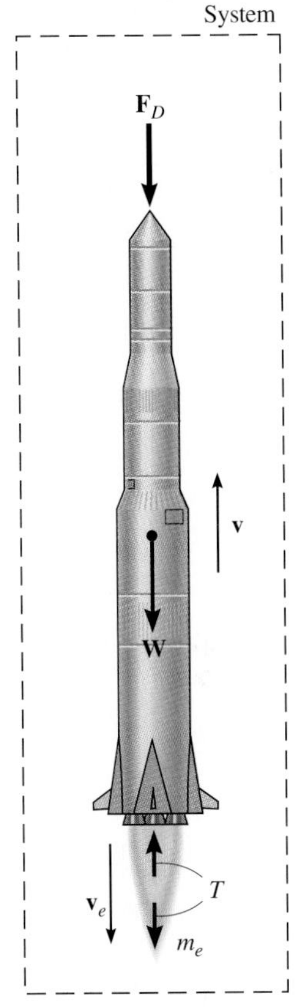

Fig. 15–31

A System That Gains Mass. A device such as a scoop or a shovel may gain mass as it moves forward. Consider, for example, the device shown in Fig. 15–32a, which at an instant of time has a mass m and is moving forward with a velocity $\mathbf{v}$, as measured by an observer fixed in an inertial frame of reference. At the same instant, the device is collecting a particle stream of mass m_i. The flow velocity $\mathbf{v}_i$ of this injected mass is constant and independent of the velocity $\mathbf{v}$. It is required that $v > v_i$. The system to be considered at this

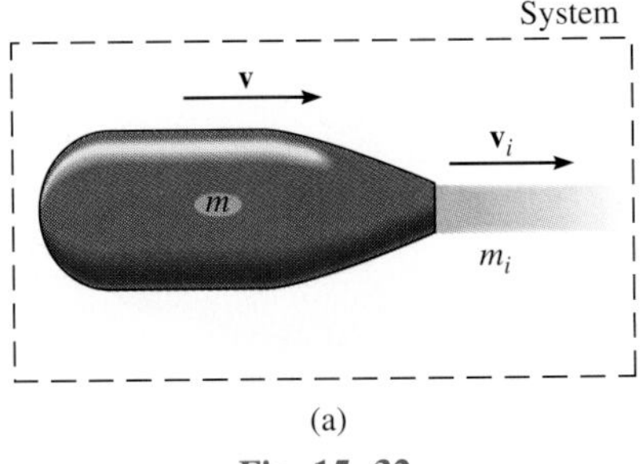

(a)

Fig. 15–32

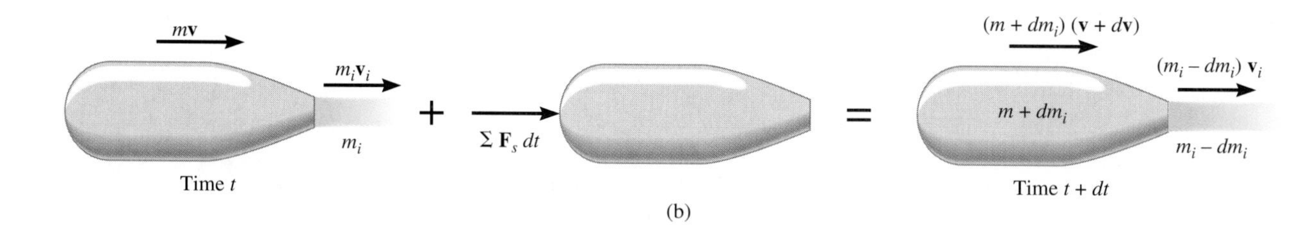

(b)

instant includes both the mass of the device and the mass of the injected particles, as shown by the dashed line in the figure. The impulse and momentum diagrams for this system are shown in Fig. 15–32*b*. With an increase in mass dm_i gained by the device, there is an assumed increase in velocity $d\mathbf{v}$ during the time interval dt. This increase is caused by the impulse created by $\Sigma\mathbf{F}_s$, the resultant of all the external forces *acting on the system* in the direction of motion. The force summation does not include the retarding force of the injected mass acting on the device. Why? Applying the principle of impulse and momentum to the system, we have

$$(\overset{+}{\rightarrow}) \qquad mv + m_i v_i + \Sigma F_s\,dt = (m + dm_i)(v + dv) + (m_i - dm_i)v_i$$

Using the same procedure as in the previous case, we may write this equation as

$$\Sigma F_s = m\frac{dv}{dt} + (v - v_i)\frac{dm_i}{dt}$$

Since the relative velocity of the device as seen by an observer moving with the particles of the injected mass is $v_{D/i} = (v - v_i)$, the final result can be written as

$$\Sigma F_s = m\frac{dv}{dt} + v_{D/i}\frac{dm_i}{dt} \qquad (15\text{–}30)$$

where dm_i/dt is the rate of mass injected into the device. The last term in this equation represents the magnitude of force **R,** which the injected mass *exerts on the device.* Since $dv/dt = a$, Eq. 15–30 becomes

$$\Sigma F_s - R = ma$$

This is the application of $\Sigma\mathbf{F} = m\mathbf{a}$, Fig. 15–32*c*.

As in the case of steady flow, problems which are solved using Eqs. 15–29 and 15–30 should be accompanied by the necessary free-body diagram. With this diagram one can then determine ΣF_s *for the system* and isolate the force exerted on the device by the particle stream.

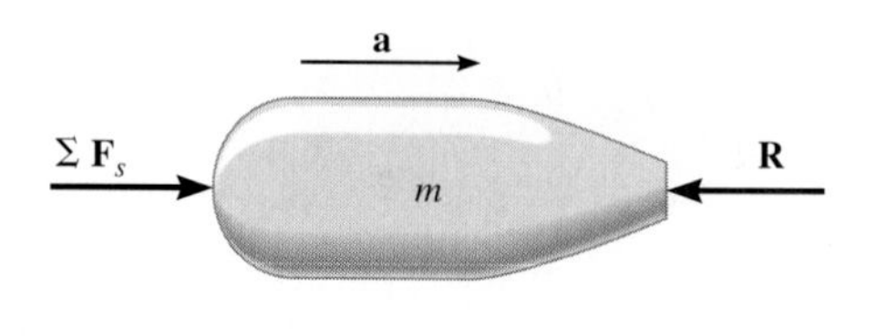

(c)

Fig. 15–32

Example 15–18

The initial combined mass of a rocket and its fuel is m_0. A total mass m_f of fuel is consumed at a constant rate of $dm_e/dt = c$ and expelled at a constant speed of u relative to the rocket. Determine the maximum velocity of the rocket, i.e., at the instant the fuel runs out. Neglect the change in the rocket's weight with altitude and the drag resistance of the air. The rocket is fired vertically from rest.

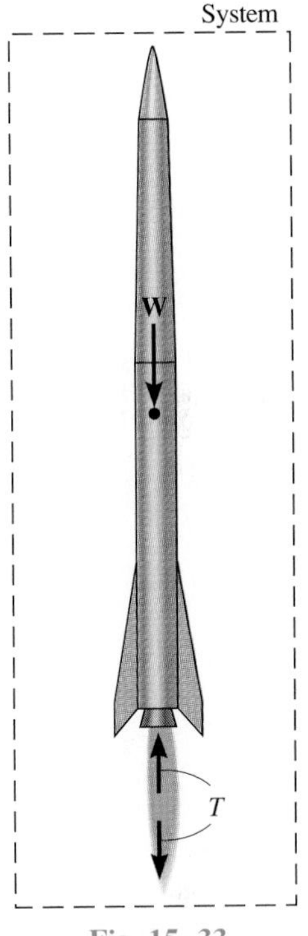

Fig. 15–33

SOLUTION

Since the rocket is losing mass as it moves upward, Eq. 15–29 can be used for the solution. The only *external force* acting on the *system* consisting of the rocket and a portion of the expelled mass is the weight **W**, Fig. 15–33. Hence,

$$+\uparrow \Sigma F_s = m\frac{dv}{dt} - v_{D/e}\frac{dm_e}{dt}; \qquad -W = m\frac{dv}{dt} - uc \tag{1}$$

The rocket's velocity is obtained by integrating this equation.

At any given instant t during the flight, the mass of the rocket can be expressed as $m = m_0 - (dm_e/dt)t = m_0 - ct$. Since $W = mg$, Eq. 1 becomes

$$-(m_0 - ct)g = (m_0 - ct)\frac{dv}{dt} - uc$$

Separating the variables and integrating, realizing that $v = 0$ at $t = 0$, we have

$$\int_0^v dv = \int_0^t \left(\frac{uc}{m_0 - ct} - g\right) dt$$

$$v = -u\ln(m_0 - ct) - gt\Big|_0^t = u\ln\left(\frac{m_0}{m_0 - ct}\right) - gt \tag{2}$$

Note that lift off requires the first term on the left to be greater than the second during the initial phase of the motion. The time t' needed to consume all the fuel is

$$m_f = \left(\frac{dm_e}{dt}\right)t' = ct'$$

Hence,

$$t' = m_f/c$$

Substituting into Eq. 2 yields

$$v_{max} = u\ln\left(\frac{m_0}{m_0 - m_f}\right) - \frac{gm_f}{c} \qquad \textit{Ans.}$$

Example 15–19

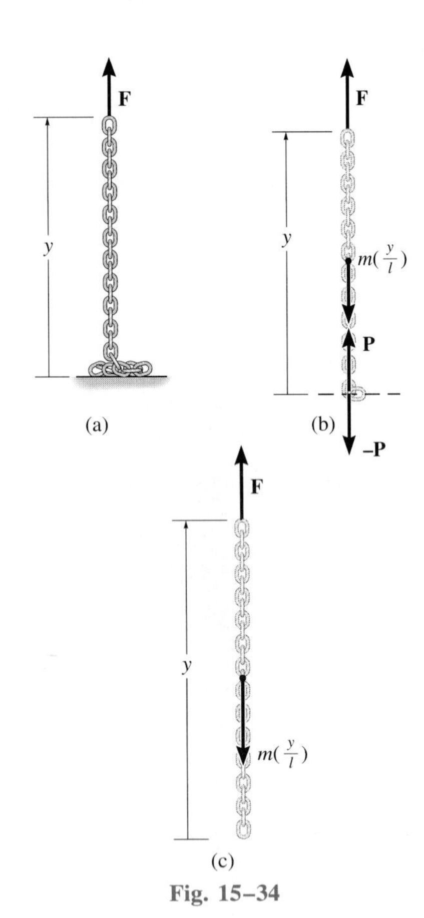

Fig. 15–34

A chain of length l, Fig. 15–34a, has a mass m. Determine the magnitude of force **F** required to *(a)* raise the chain with a constant speed v_c, starting from rest when $y = 0$; and *(b)* lower the chain with a constant speed v_c, starting from rest when $y = l$.

SOLUTION

Part (a). As the chain is raised, all the suspended links are given a sudden impulse downward by each added link which is lifted off the ground. Thus, the *suspended portion* of the chain may be considered as a device which is *gaining mass*. A free-body diagram of a portion of the chain which is located at an arbitrary height y above the ground is shown in Fig. 15–34b. The system to be considered is the length of chain y which is suspended by **F** at any instant, including the next link which is about to be added but is still at rest. The forces acting on this system *exclude* the internal forces **P** and **−P,** which act between the added link and the suspended portion of the chain. Hence, $\Sigma F_s = F - mg(y/l)$.

To apply Eq. 15–30, it is also necessary to find the rate at which mass is being added to the system. The velocity $\mathbf{v}_c$ of the chain is equivalent to $\mathbf{v}_{D/i}$. Why? Since v_c is constant, $dv_c/dt = 0$ and $dy/dt = v_c$. Integrating, using the initial condition that $y = 0$ at $t = 0$, gives $y = v_c t$. Thus, the mass of the system at any instant is $m_i = m(y/l) = m(v_c t/l)$, and therefore the *rate* at which mass is *added* to the suspended chain is

$$\frac{dm_i}{dt} = m\left(\frac{v_c}{l}\right)$$

Applying Eq. 15–30 to the system, using this data, we have

$$+\uparrow \Sigma F_s = m\frac{dv_c}{dt} + v_{D/i}\frac{dm_i}{dt}$$

$$F - mg\left(\frac{y}{l}\right) = 0 + v_c m\left(\frac{v_c}{l}\right)$$

Hence,

$$F = (m/l)(gy + v_c^2) \qquad \textit{Ans.}$$

Part (b). When the chain is being lowered, the links which are expelled (given zero velocity) *do not* impart an impulse to the *remaining* suspended links. Why? Thus, the system in Part *(a)* cannot be considered. Instead, the equation of motion will be used to obtain the solution. At time t the portion of chain still off the floor is y. The free-body diagram for a suspended portion of the chain is shown in Fig. 15–34c. Thus,

$$+\uparrow \Sigma F = ma; \qquad F - mg\left(\frac{y}{l}\right) = 0$$

$$F = mg\left(\frac{y}{l}\right) \qquad \textit{Ans.}$$

PROBLEMS

15–118. Water is discharged at 16 m/s against the fixed cone diffuser. If the opening diameter of the nozzle is 40 mm, determine the horizontal force exerted by the water on the diffuser. $\rho_w = 1\ \text{Mg/m}^3$.

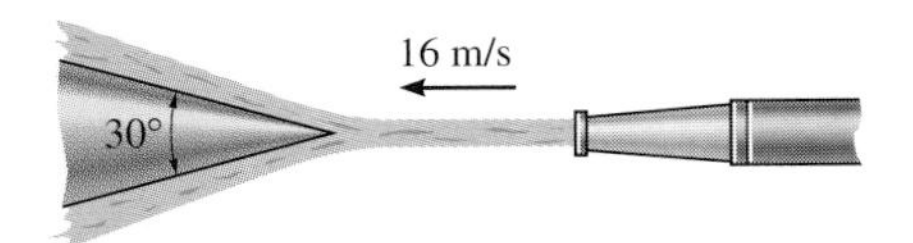

Prob. 15–118

15–119. A jet of water having a cross-sectional area of 4 in^2 strikes the fixed blade with a speed of 25 ft/s. Determine the horizontal and vertical components of force which the blade exerts on the water. $\gamma_w = 62.4\ \text{lb/ft}^3$.

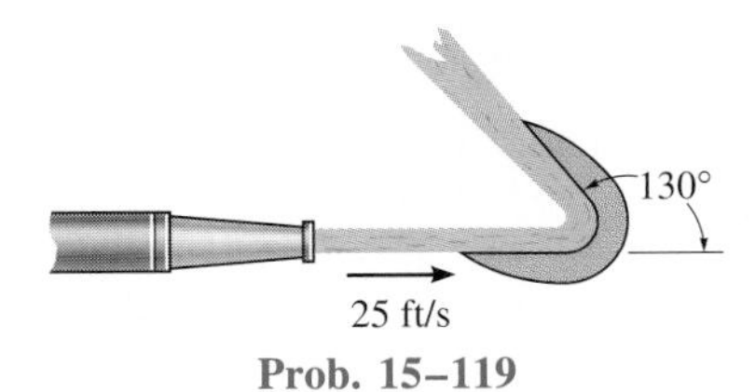

Prob. 15–119

***15–120.** The blade divides the jet of water having a diameter of 3 in. One-fourth of the water flows downward while the other three-fourths flow upward. If the flow is $Q = 0.5\ \text{ft}^3/\text{s}$, determine the horizontal and vertical components of force exerted on the blade by the jet. $\gamma_w = 62.4\ \text{lb/ft}^3$.

15–121. The blade divides the jet of water having a diameter of 3 in. and traveling to the right at 8 ft/s. The blade is moving horizontally to the left at 2 ft/s, while one-fourth of the water flows downward and the other three-fourths flow upward. Determine the horizontal and vertical components of force exerted on the blade by the jet. $\gamma_w = 62.4\ \text{lb/ft}^3$.

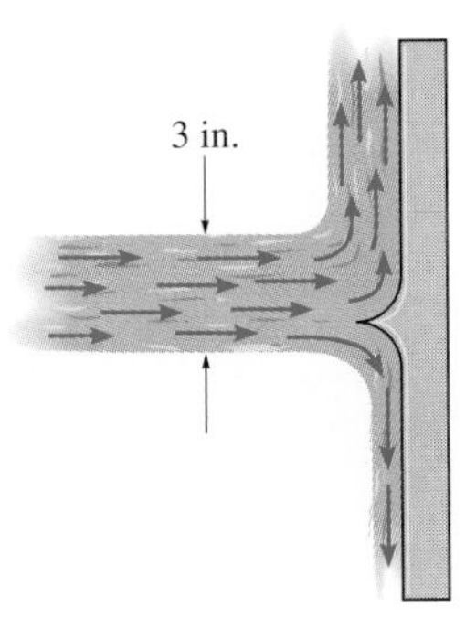

Probs. 15–120/15–121

15–122. Water is flowing from the 150-mm-diameter fire hydrant with a velocity $v_B = 15$ m/s. Determine the horizontal and vertical components of force and the moment developed at the base joint A, if the static (gauge) pressure at A is 50 kPa. The diameter of the fire hydrant at A is 200 mm. $\rho_w = 1\ \text{Mg/m}^3$.

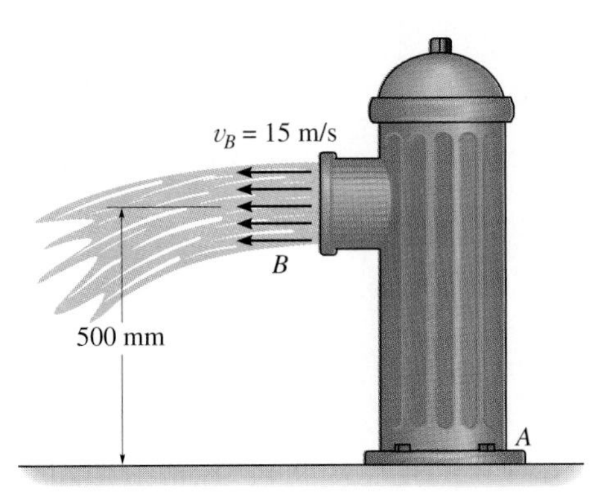

Prob. 15–122

15–123. The fan draws air through a vent with a speed of 12 ft/s. If the cross-sectional area of the vent is 2 ft^2, determine the horizontal thrust on the blade. The specific weight of the air is $\gamma_a = 0.076\ \text{lb/ft}^3$.

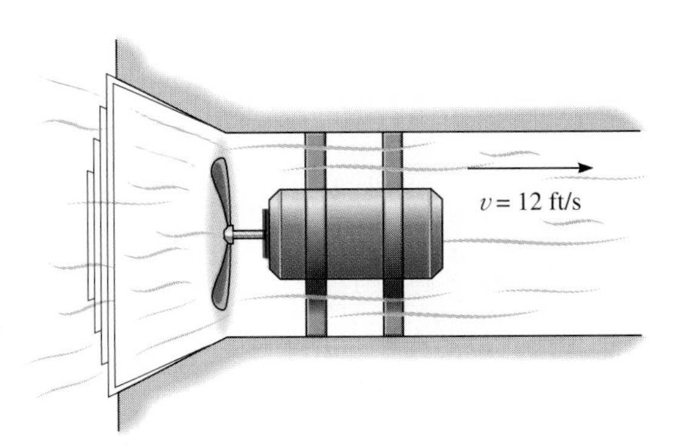

Prob. 15–123

***15–124.** The chute is used to divert the flow of water $Q = 0.6\ \text{m}^3/\text{s}$. If the water has a cross-sectional area of $0.05\ \text{m}^2$ determine the force components at the pin A and roller B necessary for equilibrium. Neglect both the weight of the chute and the weight of the water on the chute. $\rho_w = 1\ \text{Mg/m}^3$.

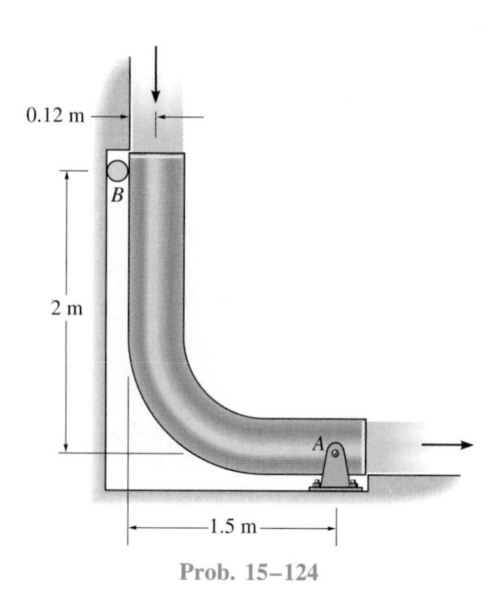

Prob. 15–124

15–125. The buckets on the *Pelton wheel* are subjected to a 2-in.-diameter jet of water, which has a velocity of 150 ft/s. If each bucket is traveling at 95 ft/s when the water strikes it, determine the power developed by the wheel. $\gamma_w = 62.4\ \text{lb/ft}^3$.

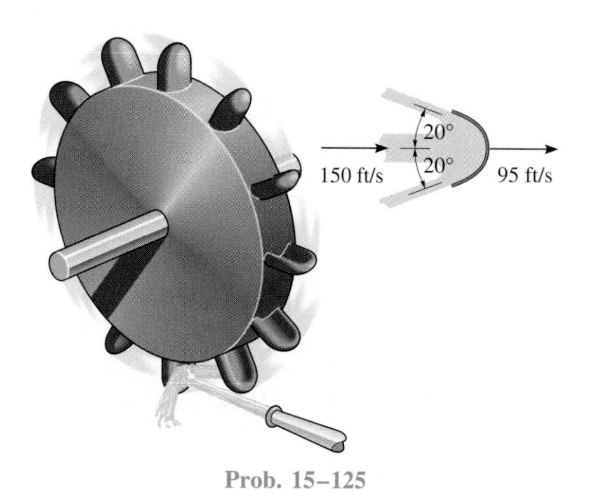

Prob. 15–125

15–126. The 200-kg boat is powered by a fan F which develops a slipstream having a diameter of 0.75 m. If the fan ejects air with a speed of 14 m/s, measured relative to the boat, determine the initial acceleration of the boat if it is initially at rest. Assume that air has a constant density $\rho_a = 1.22\ \text{kg/m}^3$ and that the entering air is essentially at rest. Neglect the drag resistance of the water.

Prob. 15–126

15–127. The boat has a mass of 180 kg and is traveling forward on a river with a constant velocity of 70 km/h, measured *relative* to the river. The river is flowing in the opposite direction at 5 km/h. If a tube is placed in the water, as shown, and it collects 40 kg of water in the boat in 80 s, determine the horizontal thrust T on the tube that is required to overcome the resistance to the water collection. $\rho_w = 1\ \text{Mg/m}^3$.

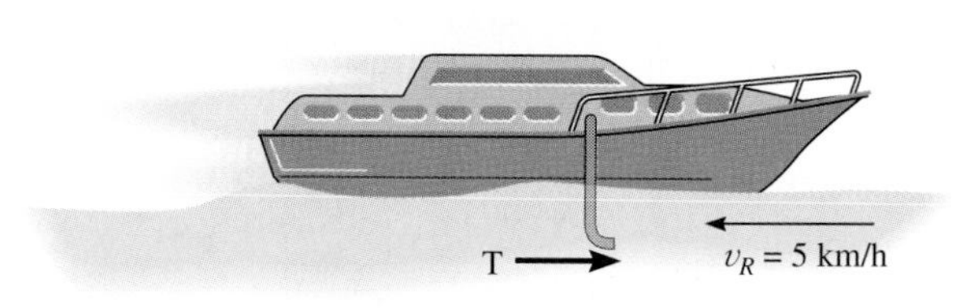

Prob. 15–127

*15–128. The static pressure of water at C is 40 lb/in^2. If water flows out of the pipe at A and B with velocities $v_A = 12$ ft/s and $v_B = 25$ ft/s, determine the horizontal and vertical components of force exerted on the elbow at C necessary to hold the pipe assembly in equilibrium. Neglect the weight of water within the pipe and the weight of the pipe. The pipe has a diameter of 0.75 in. at C, and at A and B the diameter is 0.5 in. $\gamma_w = 62.4$ lb/ft^3.

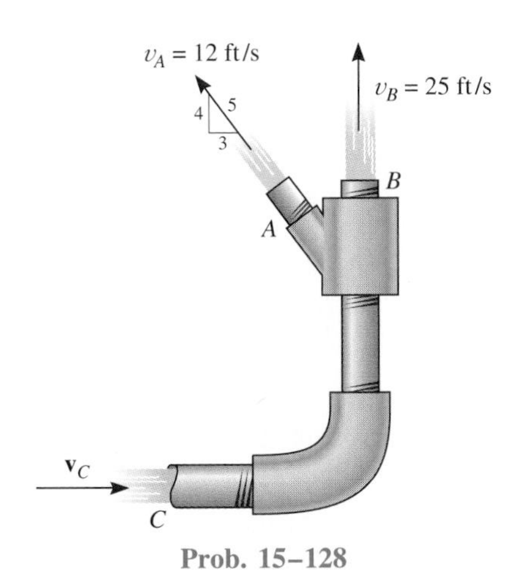

Prob. 15–128

15–129. The car is used to scoop up water that is lying in a trough at the tracks. Determine the force needed to pull the car forward at constant velocity **v** for each of the three cases. The scoop has a cross-sectional area A and the density of water is ρ_w.

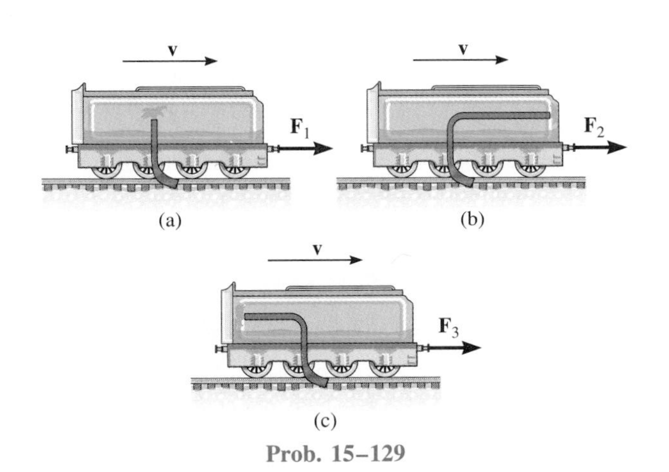

Prob. 15–129

15–130. A power lawn mower hovers very close over the ground. This is done by drawing air in at a speed of 6 m/s through an intake unit A, which has a cross-sectional area of $A_A = 0.25$ m^2, and then discharging it at the ground, B, where the cross-sectional area is $A_B = 0.35$ m^2. If air at A is subjected only to atmospheric pressure, determine the air pressure which the lawn mower exerts on the ground when the weight of the mower is freely supported and no load is placed on the handle. The mower has a mass of 15 kg with center of mass at G. Assume that air has a constant density of $\rho_a = 1.22$ kg/m^3.

Prob. 15–130

15–131. A missile weighs 40 000 lb. The constant thrust provided by a turbojet engine is $T = 15\,000$ lb. Additional thrust is provided by *two* rocket boosters B. The propellant in each booster is burned at a constant rate of 150 lb/s, with a relative exhaust velocity of 3000 ft/s. If the mass of the propellant lost by the turbojet engine can be neglected, determine the velocity of the missile after the 4-s burn time of the boosters. The initial velocity of the missile is 300 mi/h.

*15–132. A rocket has an empty weight of 500 lb and carries 300 lb of fuel. If the fuel is burned at the rate of 15 lb/s and ejected with a relative velocity of 4400 ft/s, determine the maximum speed attained by the rocket starting from rest. Neglect the effect of gravitation on the rocket.

15–133. The rocket car has a mass of 3 Mg (empty) and carries 150 kg of fuel. If the fuel is consumed at a constant rate of 4 kg/s and ejected from the car with a relative velocity of 250 m/s, determine the maximum speed attained by the car starting from rest. The drag resistance due to the atmosphere is $F_D = (60v^2)$ N, where v is the speed measured in m/s.

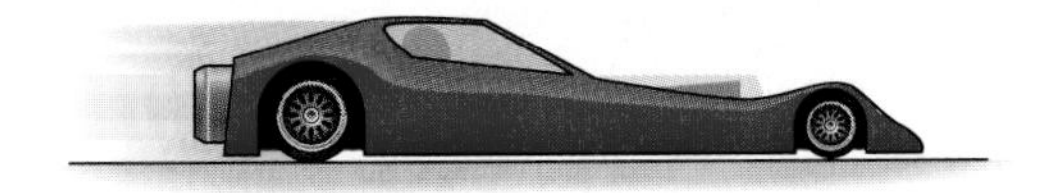
Prob. 15–133

15–134. The 12-Mg jet airplane has a constant speed of 950 km/h when it is flying along a horizontal straight-line path. Air enters the intake scoops S at the rate of 50 m^3/s. If the engine burns fuel at the rate of 0.4 kg/s and the gas (air and fuel) is exhausted relative to the plane with a speed of 450 m/s, determine the resultant drag force exerted on the plane by air resistance. Assume that air has a constant density of 1.22 kg/m^3. *Hint:* Since mass both enters and exits the plane, Eqs. 15–29 and 15–30 must be combined to yield

$$\Sigma F_s = m\frac{dv}{dt} - v_{D/e}\frac{dm_e}{dt} + v_{D/i}\frac{dm_i}{dt}$$

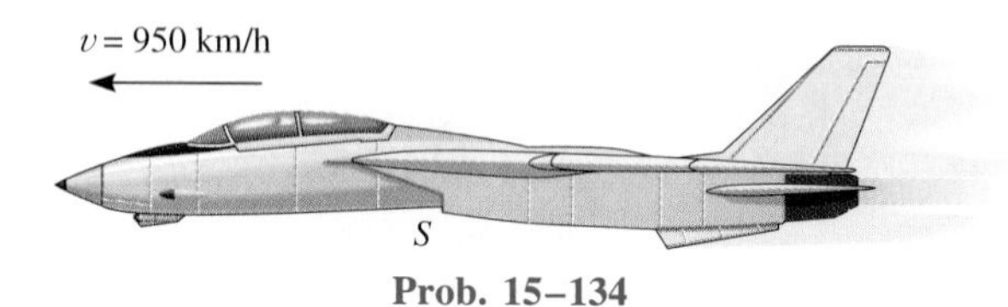

Prob. 15–134

15–135. The jet is traveling at a speed of 500 mi/h, 30° with the horizontal. If the fuel is being spent at 3 lb/s, and the engine takes in air at 400 lb/s, whereas the exhaust gas (air and fuel) has a relative speed of 32 800 ft/s, determine the acceleration of the plane at this instant. The drag resistance of the air is $F_D = (0.7v^2)$ lb, where the speed is measured in ft/s. The jet has a weight of 15 000 lb. *Hint:* See Prob. 15–134.

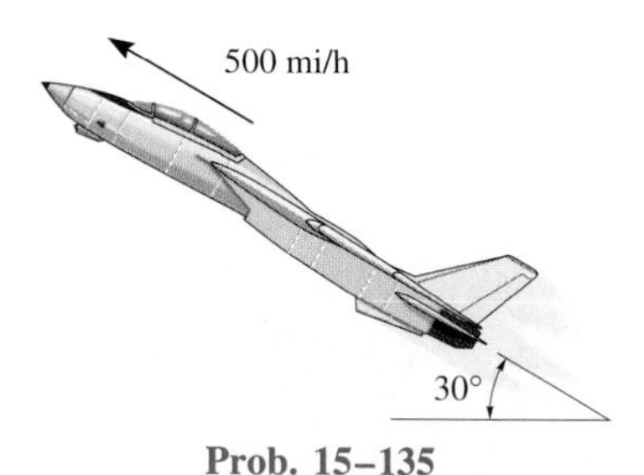

Prob. 15–135

***15–136.** The rocket has an initial mass m_0, including the fuel. For practical reasons desired for the crew, it is required that it maintain a constant upward acceleration a_0. If the fuel is expelled from the rocket at a relative speed $v_{e/r}$ determine the rate at which the fuel should be consumed to maintain the motion. Neglect air resistance, and assume that the gravitational acceleration is constant.

Prob. 15–136

15–137. The earthmover initially carries 10 m^3 of sand having a density of 1520 kg/m^3. The sand is unloaded horizontally through a 2.5-m^2 dumping port P at a rate of 900 kg/s measured relative to the port. If the earthmover maintains a constant resultant tractive force $F = 4$ kN at its front wheels to provide forward motion, determine its acceleration when half the sand is dumped. When empty, the earthmover has a mass of 30 Mg. Neglect any resistance to forward motion and the mass of the wheels. The rear wheels are free to roll.

15–138. The earthmover initially carries 10 m^3 of sand having a density of 1520 kg/m^3. The sand is unloaded horizontally through a 2.5-m^2 dumping port P at a rate of 900 kg/s measured relative to the port. Determine the resultant tractive force **F** at its front wheels if the acceleration of the earthmover is 0.1 m/s^2 when half the sand is dumped. When empty, the earthmover has a mass of 30 Mg. Neglect any resistance to forward motion and the mass of the wheels. The rear wheels are free to roll.

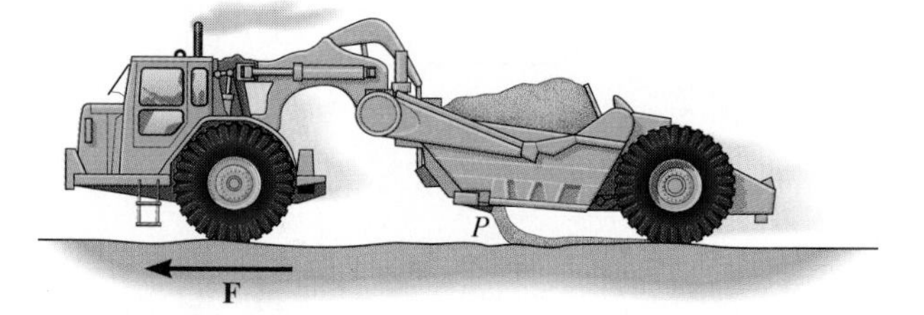

Probs. 15–137/15–138

15–139. The car has a mass m_0 and is used to tow the smooth chain having a total length l and a mass per unit of length m'. If the chain is originally piled up, determine the tractive force F that must be supplied by the rear wheels of the car, necessary to maintain a constant speed v while the chain is being drawn out.

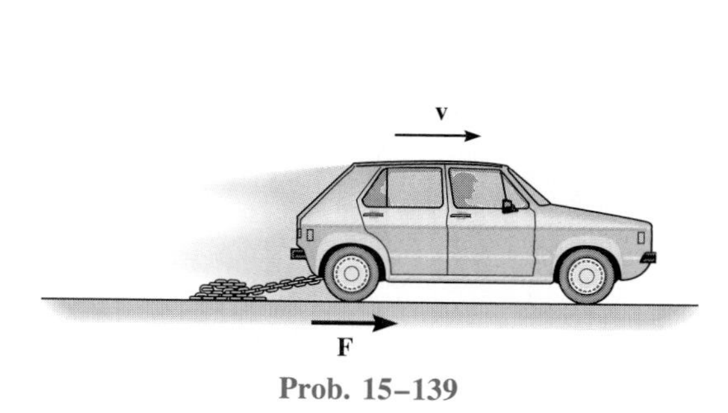

Prob. 15–139

***15–140.** Determine the magnitude of force **F** as a function of time, which must be applied to the end of the cord at A to raise the hook H with a constant speed $v = 0.4$ m/s. Initially the chain is at rest on the ground. Neglect the mass of the cord and the hook. The chain has a mass of 2 kg/m.

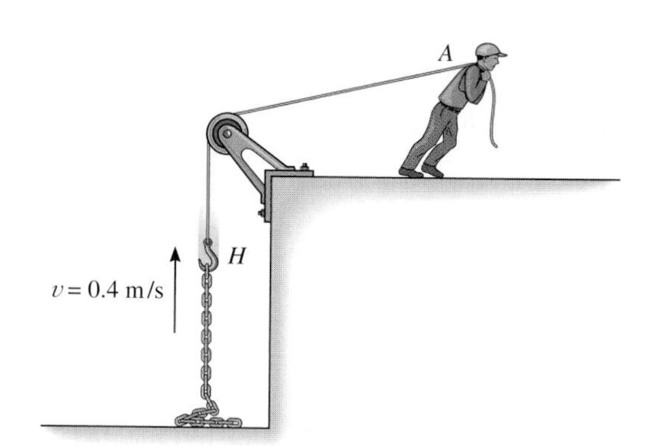

Prob. 15–140

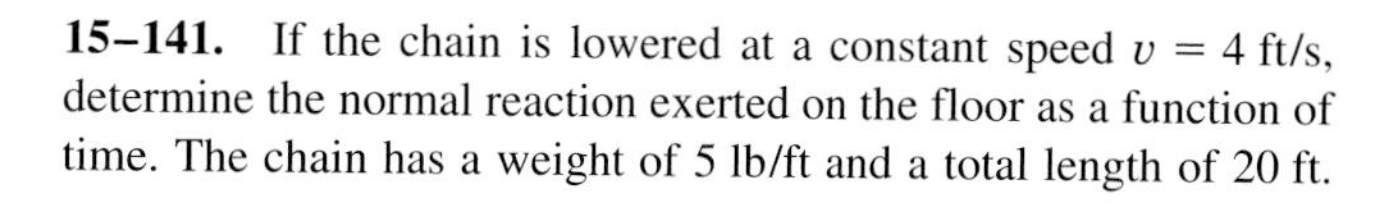
15–141. If the chain is lowered at a constant speed $v = 4$ ft/s, determine the normal reaction exerted on the floor as a function of time. The chain has a weight of 5 lb/ft and a total length of 20 ft.

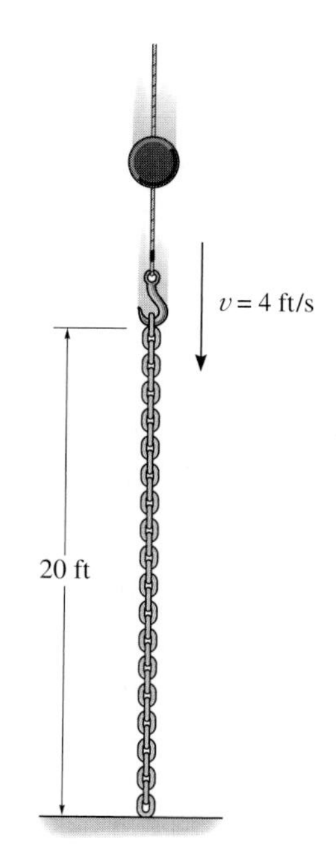

Prob. 15–141

15–142. The rope has a mass m' per unit length. If the end length $y = h$ is draped off the edge of the table, and released, determine the velocity of its end A for any position y, as the rope uncoils and begins to fall.

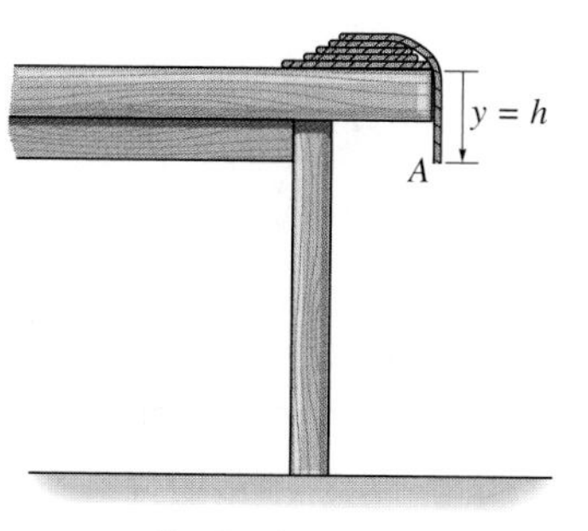

Prob. 15–142

Review 1:

Kinematics and Kinetics of a Particle

The topics and problems presented in Chapters 12 through 15 have all been *categorized* in order to provide a *clear focus* for learning the various problem-solving principles involved. In engineering practice, however, it is most important to be able to *identify* an appropriate method for the solution of a particular problem. In this regard, one must fully understand the limitations and use of the equations of dynamics, and be able to recognize which equations and principles to use for the problem's solution. For these reasons, we will now summarize the equations and principles of particle dynamics and provide the opportunity for applying them to a variety of problems.

Kinematics. Problems in kinematics require a study only of the geometry of motion, and do not account for the forces causing the motion. When the equations of kinematics are applied, one should clearly establish a fixed origin and select an appropriate coordinate system used to define the position of the particle. Once the positive direction of each coordinate axis is established, then the directions of the components of position, velocity, and acceleration can be determined from the algebraic sign of their numerical quantities.

Rectilinear Motion

Variable Acceleration. If a mathematical (or graphical) relationship is established between *any two* of the *four* variables s, v, a, and t, then a *third* variable can be determined by solving one of the following equations which relates all three variables.

$$v = \frac{ds}{dt} \qquad a = \frac{dv}{dt} \qquad a\,ds = v\,dv$$

Constant Acceleration. Be *absolutely certain* that the acceleration is constant when using the following equations:

$$s = s_0 + v_0 t + \tfrac{1}{2} a_c t^2 \qquad v = v_0 + a_c t \qquad v^2 = v_0^2 + 2a_c(s - s_0)$$

Curvilinear Motion

x, y, z Coordinates. These coordinates are often used when the motion can be resolved into horizontal and vertical components. Also they are useful for studying projectile motion since the acceleration of the projectile is *always* downward.

$$\begin{aligned} v_x &= \dot{x} & a_x &= \dot{v}_x \\ v_y &= \dot{y} & a_y &= \dot{v}_y \\ v_z &= \dot{z} & a_z &= \dot{v}_z \end{aligned}$$

n, t, b Coordinates. These coordinates are particularly advantageous for studying the particle's *acceleration* along a known path. This is because the t and n components of **a** represent the separate changes in the magnitude and direction of the velocity, respectively, and these components can be readily formulated.

$$v = \dot{s}$$

$$a_t = \dot{v} = v\frac{dv}{ds}$$

$$a_n = \frac{v^2}{\rho}$$

where

$$\rho = \left| \frac{[1 + (dy/dx)^2]^{3/2}}{d^2y/dx^2} \right|$$

when the path $y = f(x)$ is given.

r, θ, z Coordinates. These coordinates are used when data regarding the angular motion of the radial coordinate r is given to describe the particle's motion. Also, some paths of motion can conveniently be described using these coordinates.

$$\begin{aligned} v_r &= \dot{r} & a_r &= \ddot{r} - r\dot{\theta}^2 \\ v_\theta &= r\dot{\theta} & a_\theta &= r\ddot{\theta} + 2\dot{r}\dot{\theta} \\ v_z &= \dot{z} & a_z &= \ddot{z} \end{aligned}$$

Relative Motion. If the origin of a translating coordinate system is established at particle A, then for particle B,

$$\mathbf{r}_B = \mathbf{r}_A + \mathbf{r}_{B/A}$$
$$\mathbf{v}_B = \mathbf{v}_A + \mathbf{v}_{B/A}$$
$$\mathbf{a}_B = \mathbf{a}_A + \mathbf{a}_{B/A}$$

Here the relative motion is measured by an observer fixed in the translating coordinate system.

Kinetics. Problems in kinetics involve the analysis of forces which cause the motion. When applying the equations of kinetics, it is absolutely necessary that measurements of the motion be made from an *inertial coordinate system,* i.e., one that does not rotate and is either fixed or translates with constant velocity. If a problem requires *simultaneous solution* of the equations of kinetics and kinematics, then the coordinate systems selected for writing each of the equations should define the *positive directions* of the axes in the *same* manner.

Equations of Motion. These equations are used to solve for the particle's acceleration or the forces causing the motion. If they are used to determine a particle's position, velocity, or time of motion, then kinematics will also have to be considered in the solution. Before applying the equations of motion, *always draw a free-body diagram* to identify all the forces acting on the particle. Also, establish the direction of the particle's acceleration or its components. (A kinetic diagram may accompany the solution in order to graphically account for the $m\mathbf{a}$ vector.)

$$\Sigma F_x = ma_x \qquad \Sigma F_n = ma_n \qquad \Sigma F_r = ma_r$$
$$\Sigma F_y = ma_y \qquad \Sigma F_t = ma_t \qquad \Sigma F_\theta = ma_\theta$$
$$\Sigma F_z = ma_z \qquad \Sigma F_b = 0 \qquad \Sigma F_z = ma_z$$

Work and Energy. The equation of work and energy represents an integrated form of the tangential equation of motion, $\Sigma F_t = ma_t$, combined with kinematics ($a_t\,ds = v\,dv$). *It is used to solve problems involving force, velocity, and displacement.* Before applying this equation, *always draw a free-body diagram* in order to identify the forces which do work on the particle.

$$T_1 + \Sigma U_{1-2} = T_2$$

where

$$T = \tfrac{1}{2}mv^2 \qquad \text{(kinetic energy)}$$
$$U_F = \int_{s_1}^{s_2} F\cos\theta\,ds \qquad \text{(work of a variable force)}$$
$$U_{F_c} = F_c\cos\theta(s_2 - s_1) \qquad \text{(work of a constant force)}$$
$$U_W = W\,\Delta y \qquad \text{(work of a weight)}$$
$$U_s = -(\tfrac{1}{2}ks_2^2 - \tfrac{1}{2}ks_1^2) \qquad \text{(work of an elastic spring)}$$

If the forces acting on the particle are *conservative forces,* i.e., those that *do not* cause a dissipation of energy such as friction, then apply the conservation of energy equation. This equation is easier to use than the equation of work and energy since it applies only at *two points* on the path and *does not* require calculation of the work done by a force as the particle moves along the path.

$$T_1 + V_1 = T_2 + V_2$$

where

$$V_g = Wy \qquad \text{(gravitational potential energy)}$$
$$V_e = \tfrac{1}{2}ks^2 \qquad \text{(elastic potential energy)}$$

If the *power* developed by a force is to be calculated, use

$$P = \frac{dU}{dt} = \mathbf{F} \cdot \mathbf{v}$$

where $\mathbf{v}$ is the velocity of a particle acted upon by the force $\mathbf{F}$.

Impulse and Momentum. The equation of *linear impulse and momentum* is an integrated form of the equation of motion, $\Sigma\mathbf{F} = m\mathbf{a}$, combined with kinematics ($\mathbf{a} = d\mathbf{v}/dt$). *It is used to solve problems involving force, velocity, and time.* Before applying this equation, one should *always draw the free-body diagram,* in order to identify all the forces that cause impulses on the particle. From the diagram the impulsive and nonimpulsive forces should be identified. Recall that the nonimpulsive forces can be neglected in the analysis during the time of impact. Also, establish the direction of the particle's velocity just before and just after the impulses are applied. As an alternative procedure, the impulse and momentum diagrams may accompany the solution in order to graphically account for the terms in the equation.

$$m\mathbf{v}_1 + \Sigma \int_{t_1}^{t_2} \mathbf{F}\, dt = m\mathbf{v}_2$$

If several particles are involved in the problem, consider applying the *conservation of momentum* to the system in order to eliminate the internal impulses from the analysis. This can be done in a specified direction, provided no external impulses act on the particles in that direction.

$$\Sigma m\mathbf{v}_1 = \Sigma m\mathbf{v}_2$$

If the problem involves impact and the coefficient of restitution e is given, then apply the following equation.

$$e = \frac{(v_B)_2 - (v_A)_2}{(v_A)_1 - (v_B)_1} \qquad \text{(along line of impact)}$$

Remember that during impact the principle of work and energy cannot be used, since the particles deform and therefore the work due to the internal forces will be unknown. The principle of work and energy can be used, however, to determine the energy loss during the collision once the particle's initial and final velocities are determined.

The *principle of angular impulse and momentum* and the *conservation of angular momentum* may be applied about an axis in order to *eliminate* some of the unknown impulses acting on the particle during the time period when its motion is studied. Investigation of the particle's free-body diagram (or the impulse diagram) will aid in choosing the axis for application.

$$(\mathbf{H}_O)_1 + \Sigma \int_{t_1}^{t_2} \mathbf{M}_O \, dt = (\mathbf{H}_O)_2$$

$$(\mathbf{H}_O)_1 = (\mathbf{H}_O)_2$$

The following problems provide an opportunity for applying the above concepts. They are presented in *random order* so that practice may be gained in identifying the various types of problems and developing the skills necessary for their solution.

REVIEW PROBLEMS

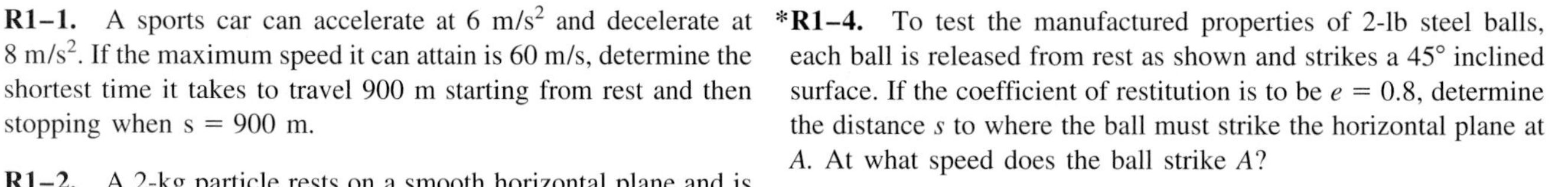

R1–1. A sports car can accelerate at 6 m/s^2 and decelerate at 8 m/s^2. If the maximum speed it can attain is 60 m/s, determine the shortest time it takes to travel 900 m starting from rest and then stopping when s = 900 m.

R1–2. A 2-kg particle rests on a smooth horizontal plane and is acted upon by forces $F_x = 0$ and $F_y = 3$ N. If $x = 0$, $y = 0$, $v_x = 6$ m/s, and $v_y = 2$ m/s when $t = 0$, determine the equation $y = f(x)$ which describes the path.

R1–3. Determine the velocity of each block 2 s after the blocks are released from rest. Neglect the mass of the pulleys and cord.

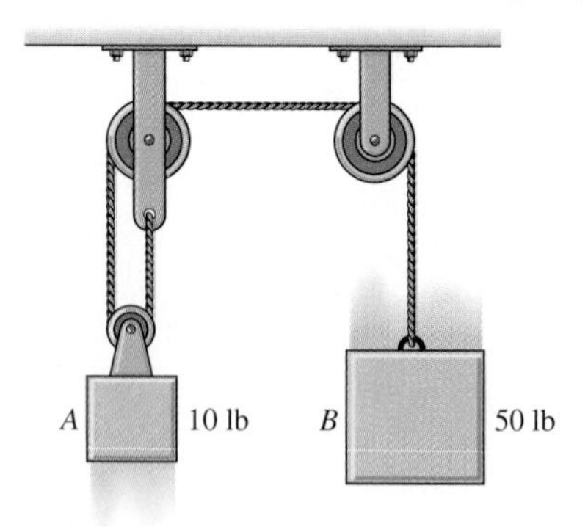

Prob. R1–3

***R1–4.** To test the manufactured properties of 2-lb steel balls, each ball is released from rest as shown and strikes a 45° inclined surface. If the coefficient of restitution is to be $e = 0.8$, determine the distance s to where the ball must strike the horizontal plane at A. At what speed does the ball strike A?

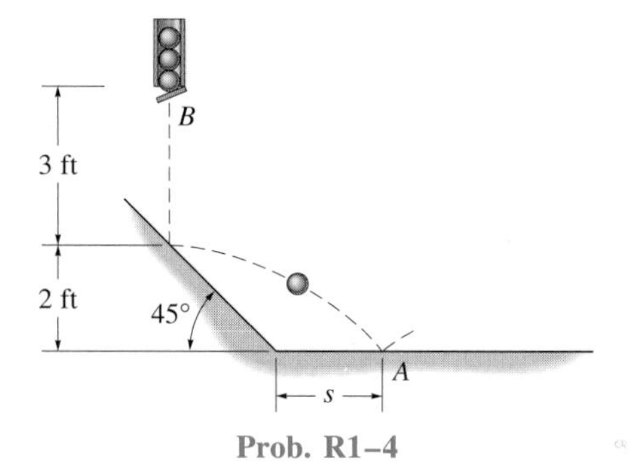

Prob. R1–4

R1–5. The 90-lb force is required to drag the 200-lb block 60 ft up the *rough* inclined plane at constant velocity. If the force is removed when the block reaches point *B*, and the block is then released from rest, determine the block's velocity when it slides back down the plane and reaches point *A*.

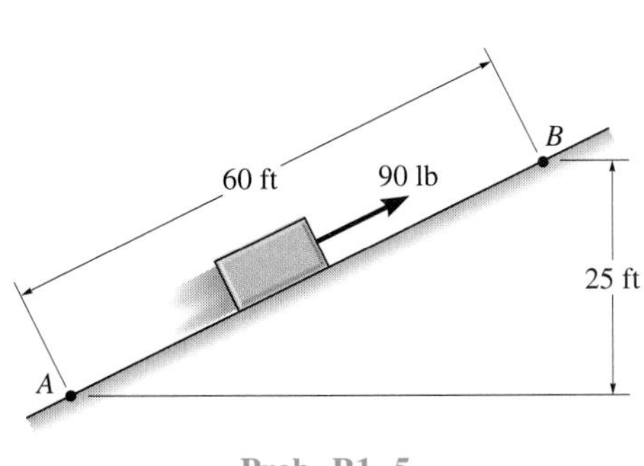

Prob. R1–5

R1–6. The motor at *C* pulls in the cable with an acceleration $a_C = (3t^2)\ \text{m/s}^2$, where *t* is in seconds. The motor at *D* draws in its cable at $a_D = 5\ \text{m/s}^2$. If both motors start at the same instant from rest when $d = 3$ m, determine (a) the time needed for $d = 0$, and (b) the relative velocity of block *A* with respect to block *B* when this occurs.

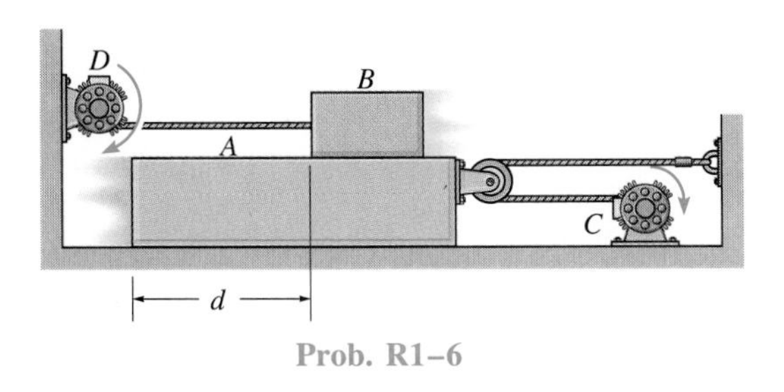

Prob. R1–6

R1–7. A spring having a stiffness of 5 kN/m is compressed 400 mm. The stored energy in the spring is used to drive a machine which requires 80 W of power. Determine how long the spring can supply energy at the required rate.

***R1–8.** The two cyclists *A* and *B* travel at the same constant speed v. Determine the velocity of *A* with respect to *B* if *A* travels along the circular track, while *B* travels along the diameter of the circle.

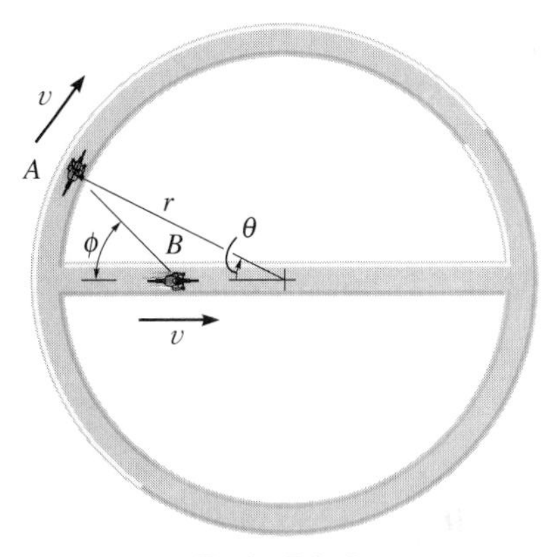

Prob. R1–8

R1–9. A 2-kg particle rests on a smooth horizontal plane and is acted upon by forces $F_x = (8x)$ N, where *x* is in meters, and $F_y = 0$. If $x = 0$, $y = 0$, $v_x = 4$ m/s, and $v_y = 6$ m/s when $t = 0$, determine the equation $y = f(x)$ which describes the path.

R1–10. Assuming that the force acting on a 2-g bullet, as it passes horizontally through the barrel of a rifle, varies with time in the manner shown, determine the maximum net force F_0, applied to the bullet when it is fired. The muzzle velocity is 500 m/s when $t = 0.75$ ms. Neglect friction between the bullet and the rifle barrel.

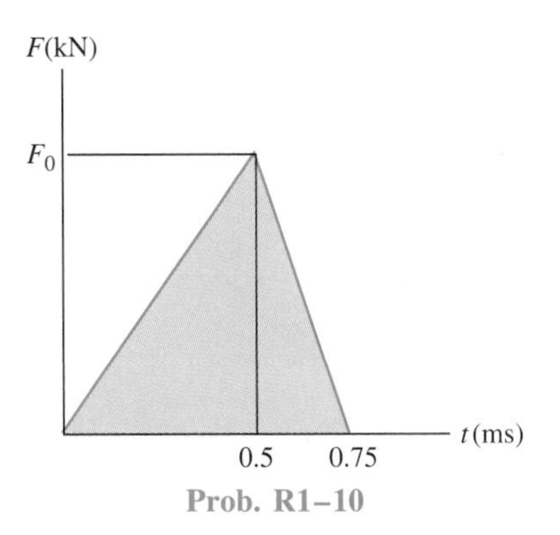

Prob. R1–10

R1–11. A 0.2-kg spool slides down along a smooth rod. If the rod has a constant angular rate of rotation $\dot{\theta} = 2$ rad/s in the vertical plane, show that the equations of motion for the spool are $\ddot{r} - 4r - 9.81 \sin \theta = 0$ and $0.8\dot{r} + N_s - 1.962 \cos \theta = 0$, where N_s is the magnitude of the normal force of the rod on the spool. Using the methods of differential equations, it can be shown that the solution of the first of these equations is $r = C_1 e^{-2t} + C_2 e^{2t} - (9.81/8) \sin 2t$. If r, $\dot{r}$, and θ are zero when $t = 0$, evaluate the constants C_1 and C_2 and determine r at the instant $\theta = \pi/4$ rad.

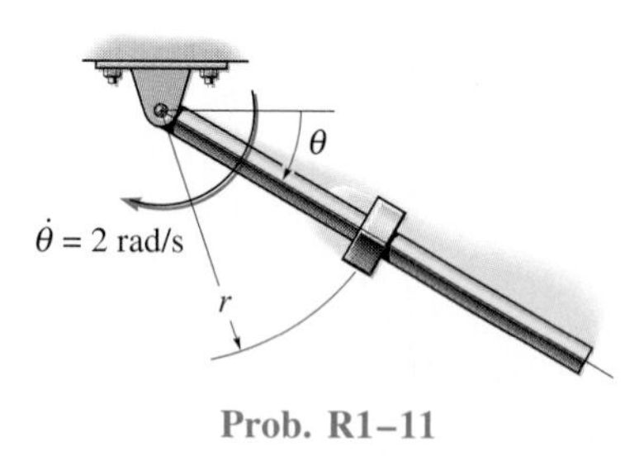

Prob. R1–11

*■**R1–12.** Packages having a mass of 2.5 kg ride on the surface of the conveyor belt. If the belt starts from rest and with constant acceleration increases to a speed of 0.75 m/s in 2 s, determine the maximum angle of tilt, θ, so that none of the packages slip on the inclined surface AB of the belt. The coefficient of static friction between the belt and each package is $\mu_s = 0.3$. At what angle ϕ do the packages first begin to slip off the surface of the belt if the belt is moving at a constant speed of 0.75 m/s?

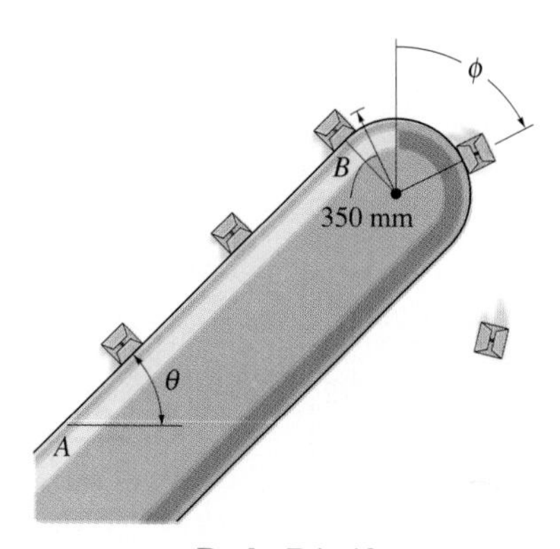

Prob. R1–12

R1–13. A projectile, initially at the origin, moves along a straight-line path through a fluid medium such that its velocity is $v = 1800(1 - e^{-0.3t})$ mm/s, where t is in seconds. Determine the displacement of the projectile during the first 3 s.

R1–14. The speed of a train during the first minute of its motion has been recorded as follows:

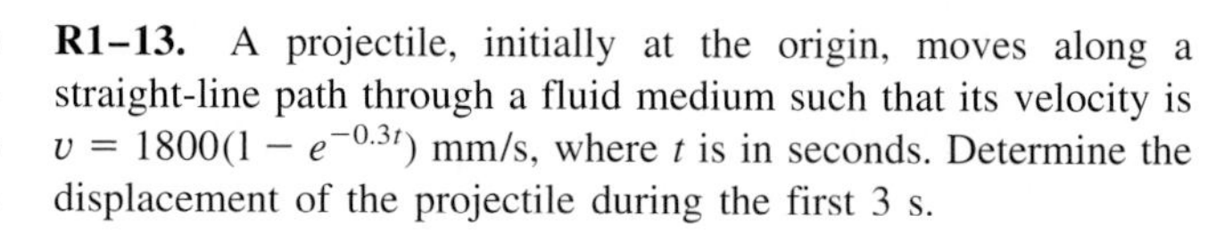

t(s)	0	20	40	60
v(m/s)	0	16	21	24

Plot the v-t graph, approximating the curve as straight line segments between the given points. Determine the total distance traveled.

R1–15. A train car, having a mass of 25 Mg, travels up a 10° incline with a constant speed of 80 km/h. Determine the power required to overcome the force of gravity.

***R1–16.** The slotted arm AB drives the pin C through the spiral groove described by the equation $r = (1.5\theta)$ ft, where θ is in radians. If the arm starts from rest when $\theta = 60°$ and is driven at an angular rate of $\dot{\theta} = (4t)$ rad/s, where t is in seconds, determine the radial and transverse components of velocity and acceleration of the pin when $t = 1$ s.

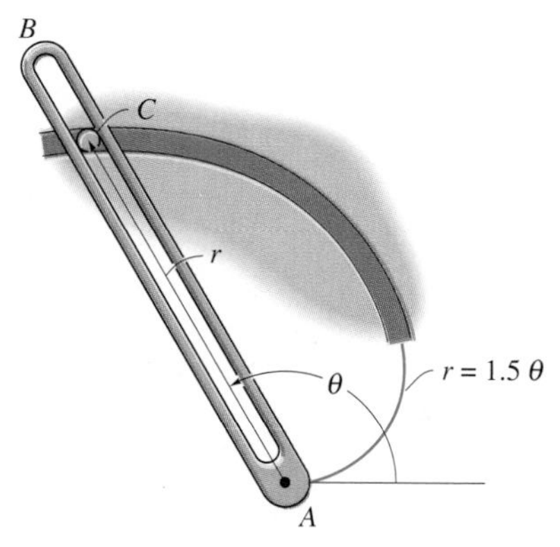

Prob. R1–16

R1–17. The chain has a mass of 3 kg/m. If the coefficient of kinetic friction between the chain and the plane is $\mu_k = 0.2$, determine the velocity at which the end A will pass point B when the chain is released from rest.

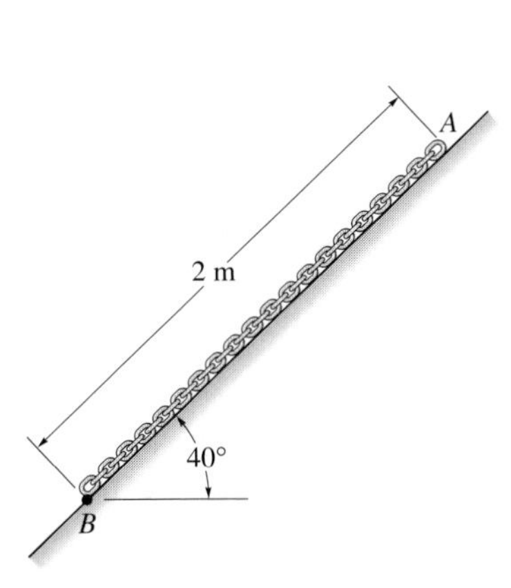

Prob. R1–17

R1–18. The 6-lb ball is fired from a tube by a spring having a stiffness $k = 20$ lb/in. Determine how far the spring must be compressed to fire the ball from the compressed position to a height of 8 ft, at which point it has a velocity of 6 ft/s.

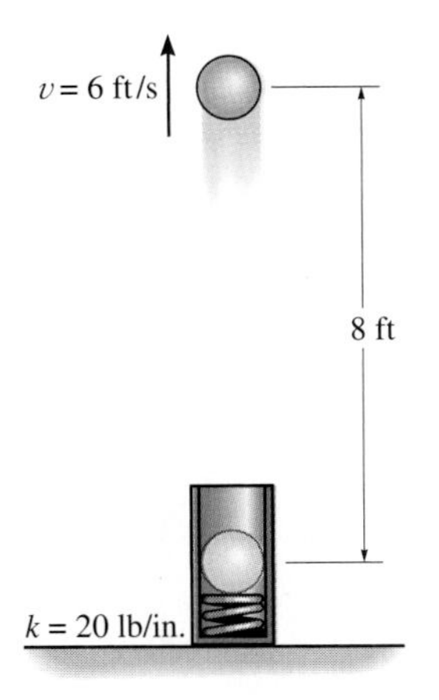

Prob. R1–18

R1–19. The collar of negligible size has a mass of 0.25 kg and is attached to a spring having an unstretched length of 100 mm. If the collar is released from rest at A and travels along the smooth guide, determine its speed just before it strikes B.

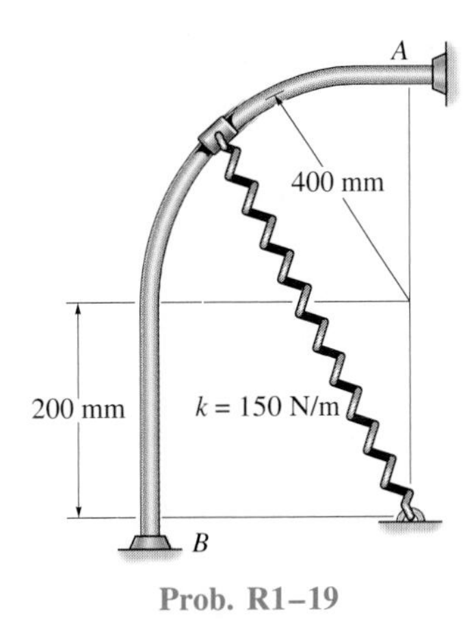

Prob. R1–19

***R1–20.** A crate has a weight of 1500 lb. If it is pulled along the ground at a constant speed for a distance of 20 ft, and the towing cable makes an angle of 15° with the horizontal, determine the tension in the cable and the work done by the towing force. The coefficient of kinetic friction between the crate and the ground is $\mu_k = 0.55$.

R1–21. Disk A weighs 2 lb and is sliding on a smooth horizontal plane with a velocity of 3 ft/s. Disk B weighs 11 lb and is initially at rest. If after the impact A has a velocity of 1 ft/s directed along the positive x axis, determine the velocity of B after impact. How much kinetic energy is lost in the collision?

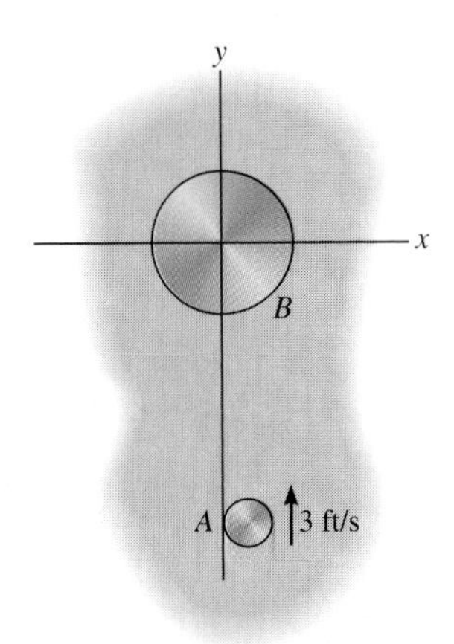

Prob. R1–21

R1–22. A particle is moving along a circular path of 2-m radius such that its position as a function of time is given by $\theta = (5t^2)$ rad, where t is in seconds. Determine the magnitude of the particle's acceleration when $\theta = 30°$. The particle starts from rest when $\theta = 0°$.

R1–23. If the end of the cable at A is pulled down with a speed of 2 m/s, determine the speed at which block B rises.

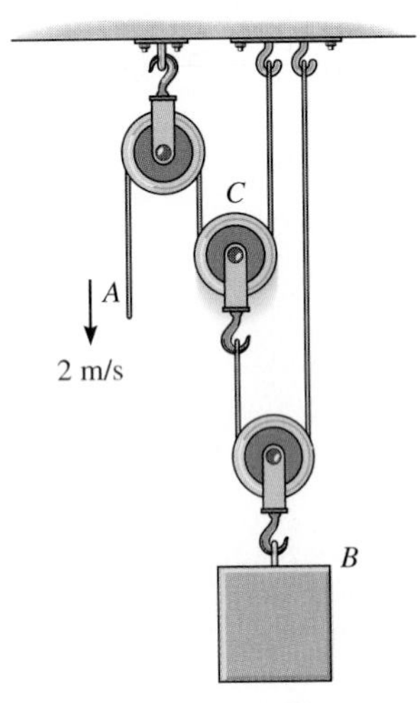

Prob. R1–23

***R1–24.** A rifle has a mass of 2.5 kg. If it is loosely gripped and a 1.5-g bullet is fired from it with a horizontal muzzle velocity of 1400 m/s, determine the recoil velocity of the rifle just after firing.

R1–25. The drinking fountain is designed such that the nozzle is located from the edge of the basin as shown. Determine the maximum and minimum speed at which water can be ejected from the nozzle so that it does not splash over the sides of the basin at B and C.

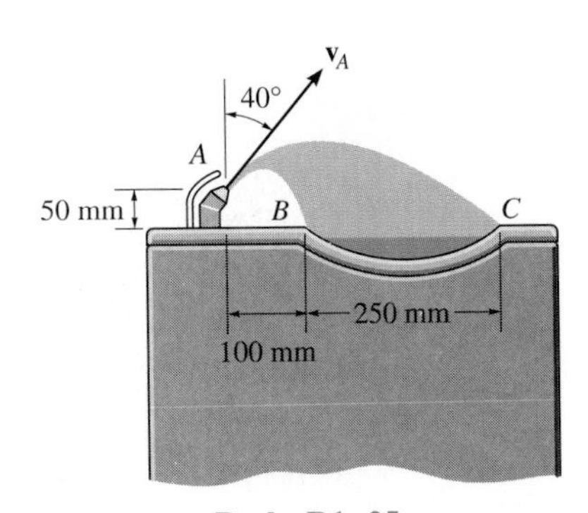

Prob. R1–25

R1–26. The 20-lb block B rests on the surface of a table for which the coefficient of kinetic friction is $\mu_k = 0.1$. Determine the speed of the 10-lb block A after it has moved downward 2 ft from rest. Neglect the mass of the pulleys and cords.

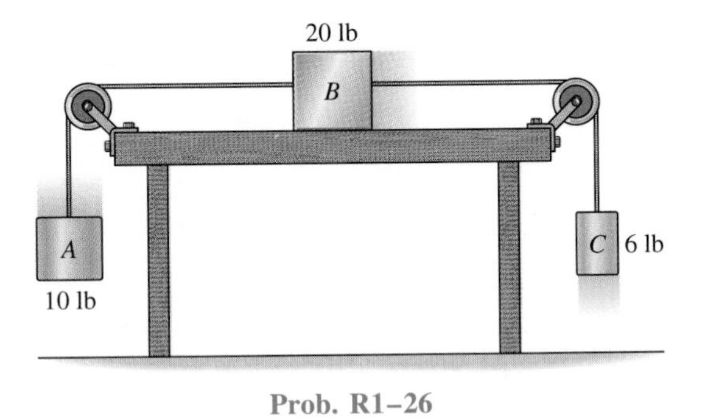

Prob. R1–26

R1–27. The 5-lb ball, attached to the cord, is struck by the boy. Determine the smallest speed he must impart to the ball so that it will swing around in a vertical circle, without causing the cord to become slack.

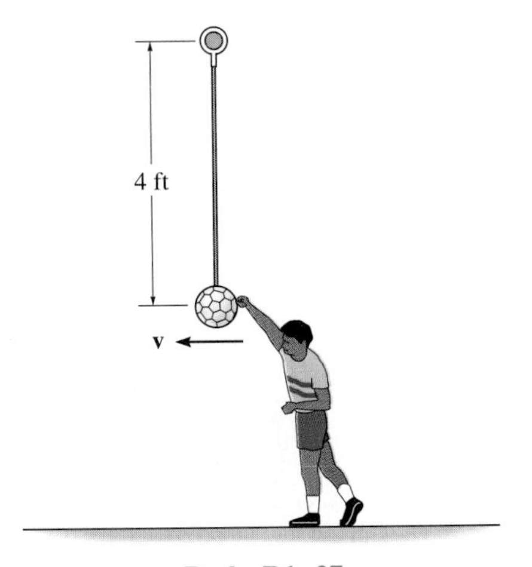

Prob. R1–27

***R1–28.** Two planes A and B are flying side by side at a constant speed of 900 km/h. Maintaining this speed, plane A begins to travel along the spiral path $r = (1500\theta)$ km, where θ is in radians, whereas plane B continues to fly in a straight line. Determine the speed of plane A with respect to plane B when $r = 750$ km.

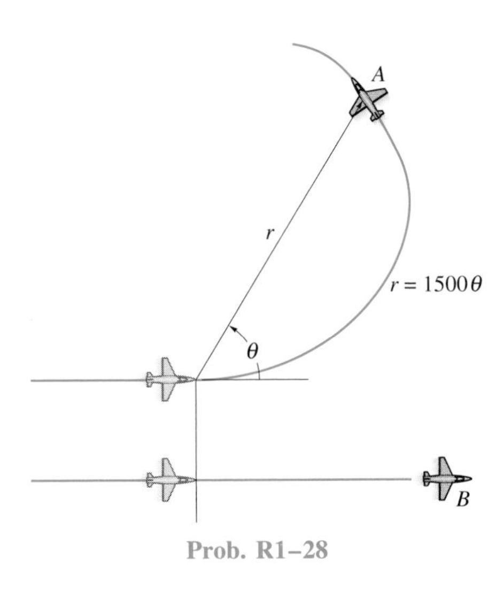

Prob. R1–28

R1–29. The particle P travels with a constant speed of 300 mm/s along the curve. Determine its acceleration when it is located at point (200 mm, 100 mm).

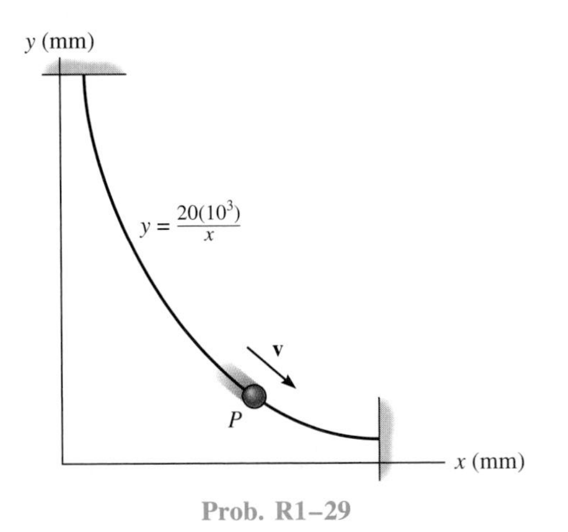

Prob. R1–29

R1–30. The block has a mass of 0.5 kg and moves within the smooth vertical slot. If the block starts from rest when the *attached* spring is in the unstretched position at A, determine the *constant* vertical force F which must be applied to the cord so that the block attains a speed $v_B = 2.5$ m/s when it reaches B; $s_B = 0.15$ m. Neglect the mass of the cord and pulley.

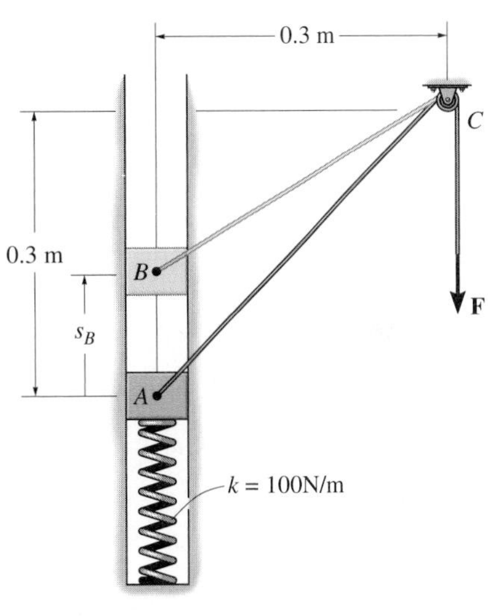

Prob. R1–30

R1–31. If a horizontal force of $P = 10$ lb is applied to block A, determine the acceleration of block B. Neglect friction. *Hint:* Show that $a_B = a_A \tan 15°$.

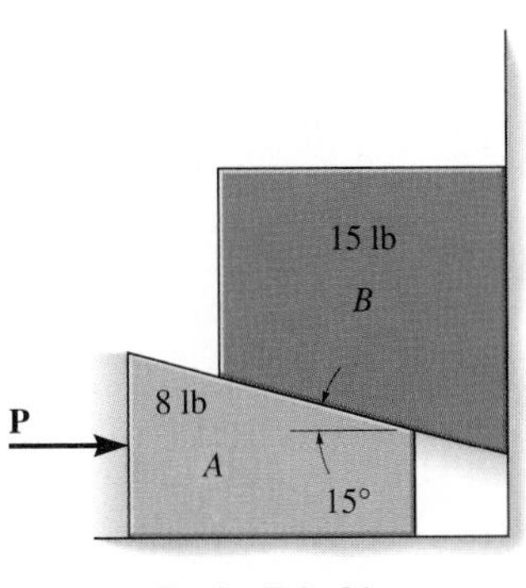

Prob. R1–31

***R1–32.** The spool, which has a mass of 4 kg, slides along the rotating rod. At the instant shown, the angular rate of rotation of the rod is $\dot{\theta} = 6$ rad/s and this rotation is increasing at $\ddot{\theta} = 2$ rad/s^2. At this same instant, the spool has a velocity of 3 m/s and an acceleration of 1 m/s^2, both measured relative to the rod and directed away from the center O when $r = 0.5$ m. Determine the radial frictional force and the normal force, both exerted by the rod on the spool at this instant.

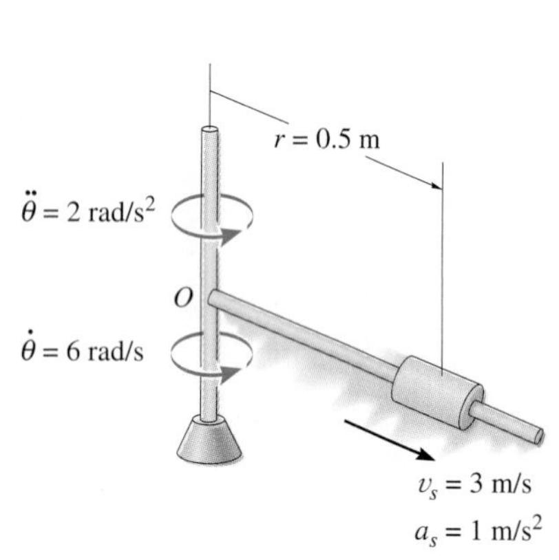

Prob. R1–32

R1–33. A skier starts from rest at A (30 ft, 0) and descends the smooth slope, which may be approximated by a parabola. If she has a weight of 120 lb, determine the normal force she exerts on the ground at the instant she arrives at point B.

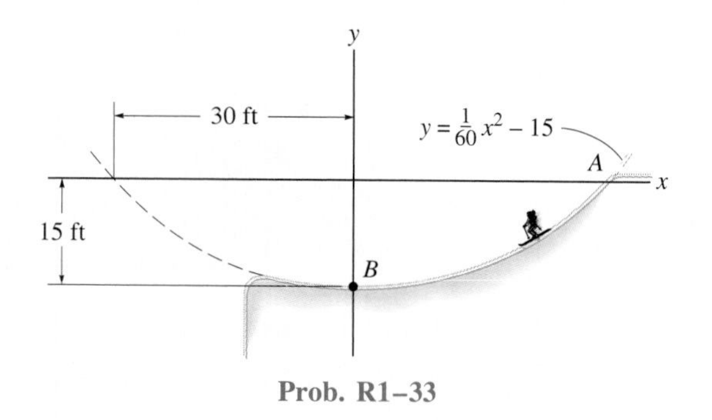

Prob. R1–33

R1–34. The small 2-lb collar starting from rest at A slides down along the smooth rod. During the motion, the collar is acted upon by a force $\mathbf{F} = \{10\mathbf{i} + 6y\mathbf{j} + 2z\mathbf{k}\}$ lb, where x, y, z are in feet. Determine the collar's speed when it strikes the wall at B.

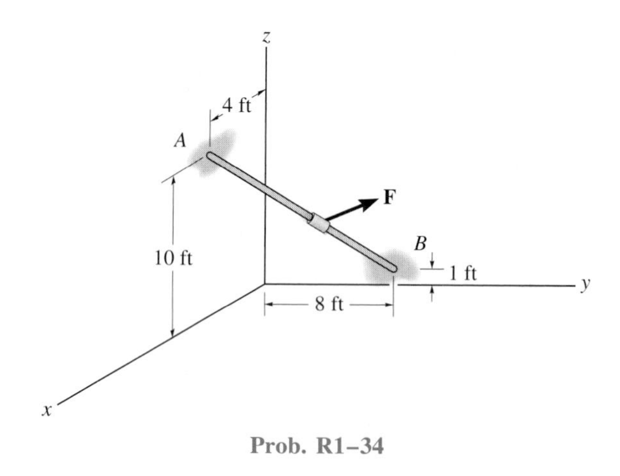

Prob. R1–34

R1–35. A ball having a mass of 200 g is released from rest at a height of 400 mm above a very large fixed metal surface. If the ball rebounds to a height of 325 mm above the surface, determine the coefficient of restitution between the ball and the surface.

***R1–36.** Packages having a mass of 6 kg slide down a smooth chute and land horizontally with a speed of 3 m/s on the surface of a conveyor belt. If the coefficient of kinetic friction between the belt and a package is $\mu_k = 0.2$, determine the time needed to bring the package to rest on the belt if the belt is moving in the same direction as the package with a speed $v = 1$ m/s.

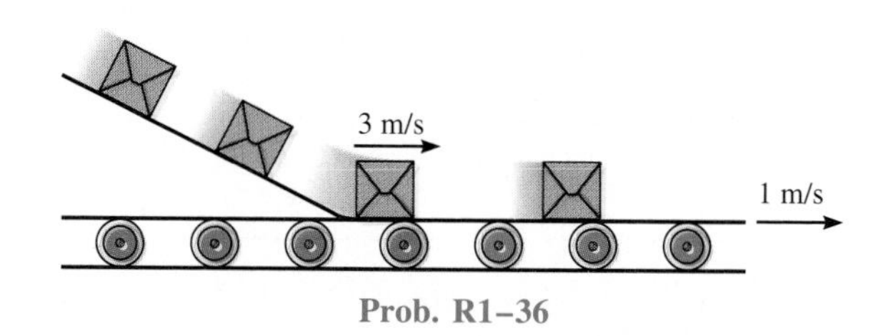

Prob. R1–36

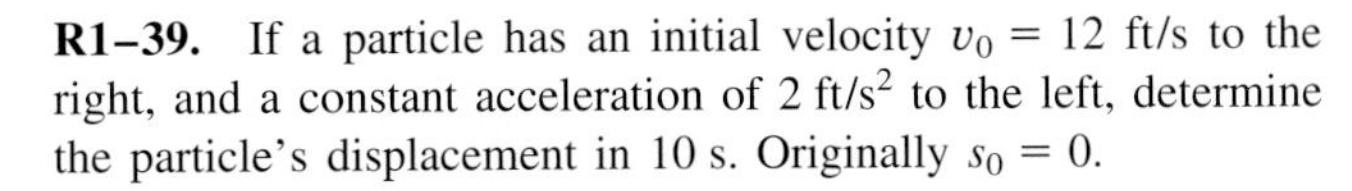

R1–37. The blocks A and B weigh 10 and 30 lb, respectively. They are connected together by a light cord and ride in the frictionless grooves. Determine the speed of each block after block A moves 6 ft up along the plane. The blocks are released from rest.

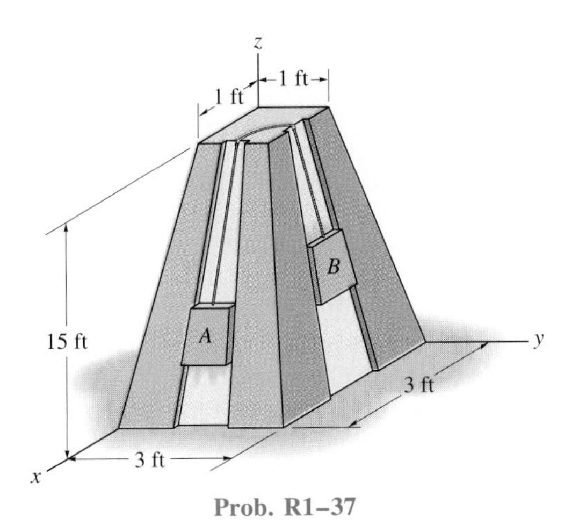

Prob. R1–37

R1–38. The motor M pulls in its attached rope with an acceleration $a_P = 6\ \text{m/s}^2$. Determine the towing force exerted by M on the rope in order to move the 50-kg crate up the inclined plane. The coefficient of kinetic friction between the crate and the plane is $\mu_k = 0.3$. Neglect the mass of the pulleys and rope.

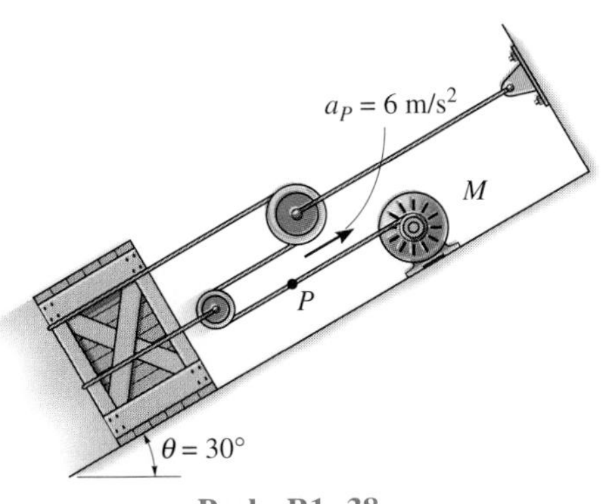

Prob. R1–38

R1–39. If a particle has an initial velocity $v_0 = 12$ ft/s to the right, and a constant acceleration of 2 ft/s^2 to the left, determine the particle's displacement in 10 s. Originally $s_0 = 0$.

***R1–40.** A 3-lb block, initially at rest at point A, slides along the smooth parabolic surface. Determine the normal force acting on the block when it reaches B. Neglect the size of the block.

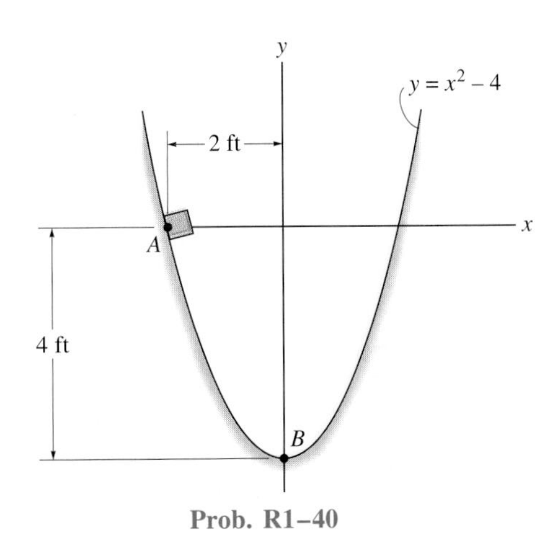

Prob. R1–40

R1–41. At a given instant the 10-lb block A is moving downward with a speed of 6 ft/s. Determine its speed 2 s later. Block B has a weight of 4 lb, and the coefficient of kinetic friction between it and the horizontal plane is $\mu_k = 0.2$. Neglect the mass of the pulleys and cord.

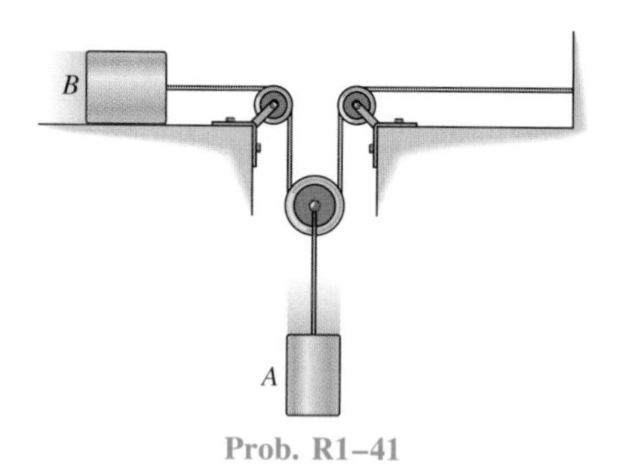

Prob. R1–41

R1–42. A freight train starts from rest and travels with a constant acceleration of 0.5 ft/s^2. After a time t' it maintains a constant speed so that when $t = 160$ s it has traveled 2000 ft. Determine the time t' and draw the v-t graph for the motion.

R1–43. The crate, having a weight of 50 lb, is hoisted by the pulley system and motor M. If the crate starts from rest and, by constant acceleration, attains a speed of 12 ft/s after rising 10 ft, determine the power that must be supplied to the motor at the instant $s = 10$ ft. The motor has an efficiency $\epsilon = 0.74$.

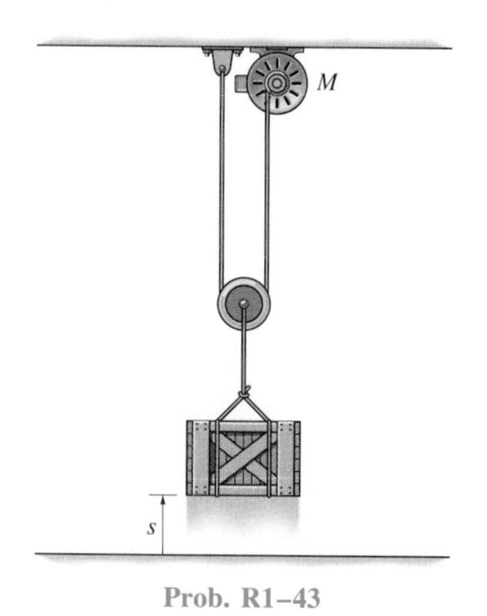

Prob. R1–43

***R1–44.** An automobile is traveling with a *constant speed* along a horizontal circular curve that has a radius $e = 750$ ft. If the magnitude of acceleration is $a = 8$ ft/s^2, determine the speed at which the automobile is traveling.

R1–45. Block B rests on a smooth surface. If the coefficient of static friction between A and B is $\mu_s = 0.4$, determine the acceleration of each block if (a) $F = 6$ lb, and (b) $F = 50$ lb.

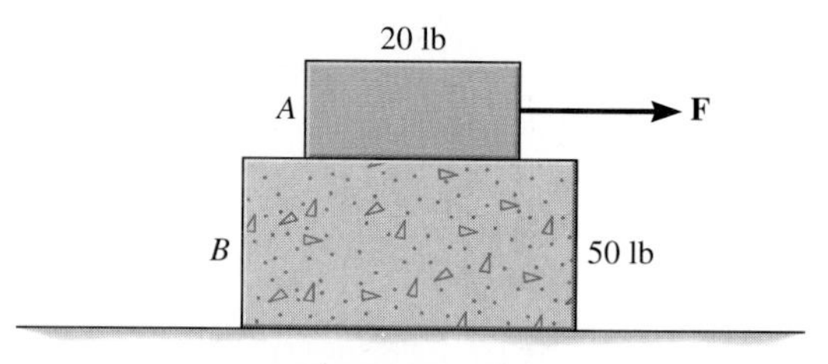

Prob. R1–45

R1–46. The 100-kg crate is subjected to the action of two forces, $F_1 = 800$ N and $F_2 = 1.5$ kN, as shown. If it is originally at rest, determine the distance it slides in order to attain a speed of 6 m/s. The coefficient of kinetic friction between the crate and the surface is $\mu_k = 0.2$.

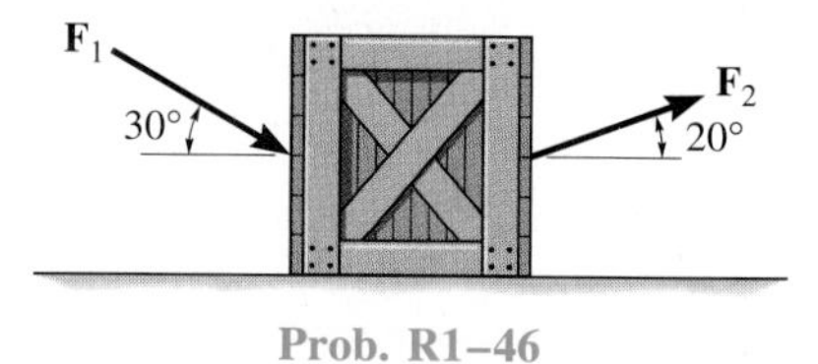

Prob. R1–46

R1–47. A 20-kg block is originally at rest on a horizontal surface for which the coefficient of static friction is $\mu_s = 0.6$ and the coefficient of kinetic friction is $\mu_k = 0.5$. If a horizontal force F is applied such that it varies with time as shown, determine the speed of the block in 10 s. *Hint:* First determine the time needed to overcome friction and start the block moving.

F

F(N)

200

5

10

t(s)

Prob. R1–47

***R1–48.** Two smooth billiard balls A and B have an equal mass of $m = 200$ g. If A strikes B with a velocity of $(v_A)_1 = 2$ m/s as shown, determine their final velocities just after collision. Ball B is originally at rest and the coefficient of restitution is $e = 0.75$.

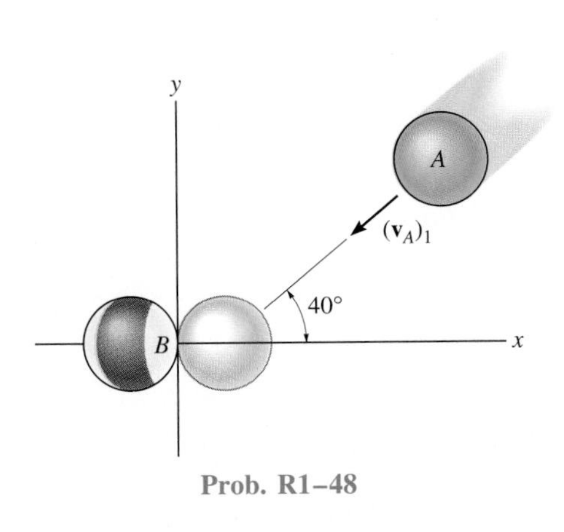

Prob. R1–48

R1–50. Determine the tension developed in the two cords and the acceleration of each block. Neglect the mass of the pulleys and cords. *Hint:* Since the system consists of *two* cords, relate the motion of block A to C, and of block B to C. Then, by elimination, relate the motion of A to B.

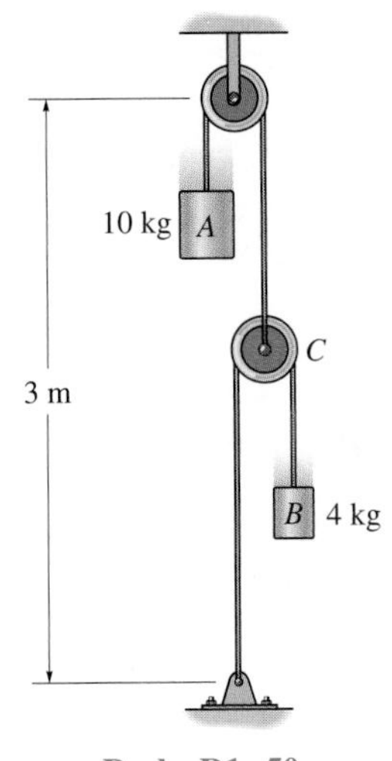

Prob. R1–50

R1–49. If a 150-lb crate is released from rest at A, determine its speed after it slides 30 ft down the plane. The coefficient of kinetic friction between the crate and plane is $\mu_k = 0.3$.

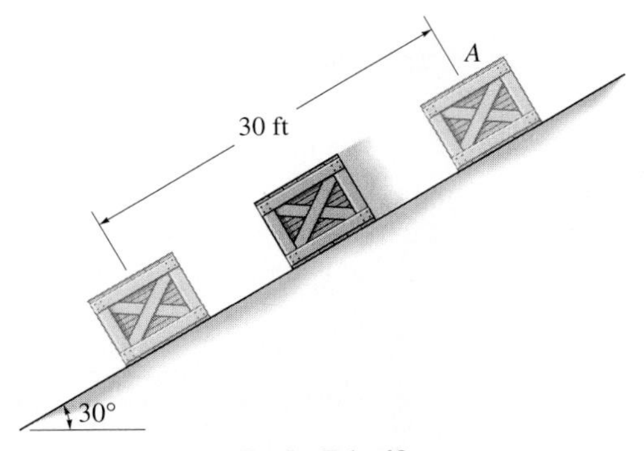

Prob. R1–49

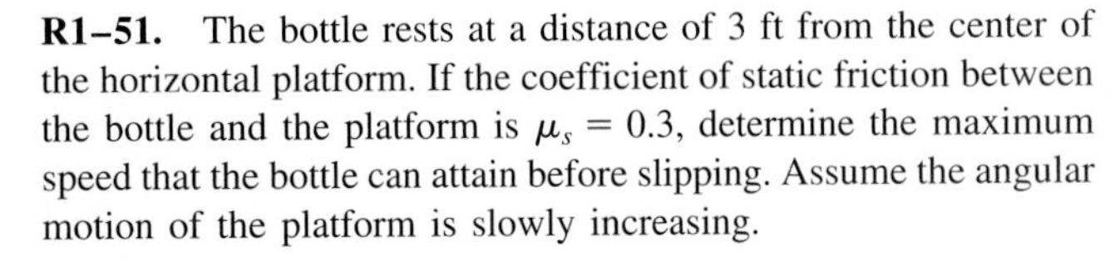

R1–51. The bottle rests at a distance of 3 ft from the center of the horizontal platform. If the coefficient of static friction between the bottle and the platform is $\mu_s = 0.3$, determine the maximum speed that the bottle can attain before slipping. Assume the angular motion of the platform is slowly increasing.

R1–52. Work Prob. R1–51 assuming that the platform starts rotating from rest so that the speed of the bottle is increased at 2 ft/s_2.

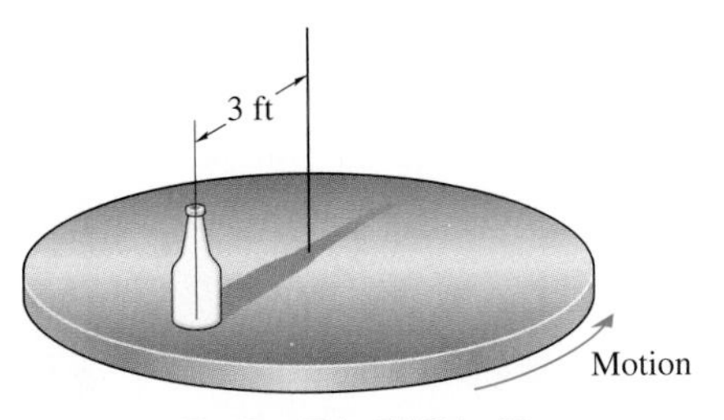

Probs. R1–51/R1–52

The angular motion of wind-turbine generators is quite variable and can be analyzed using the technique developed in this chapter.

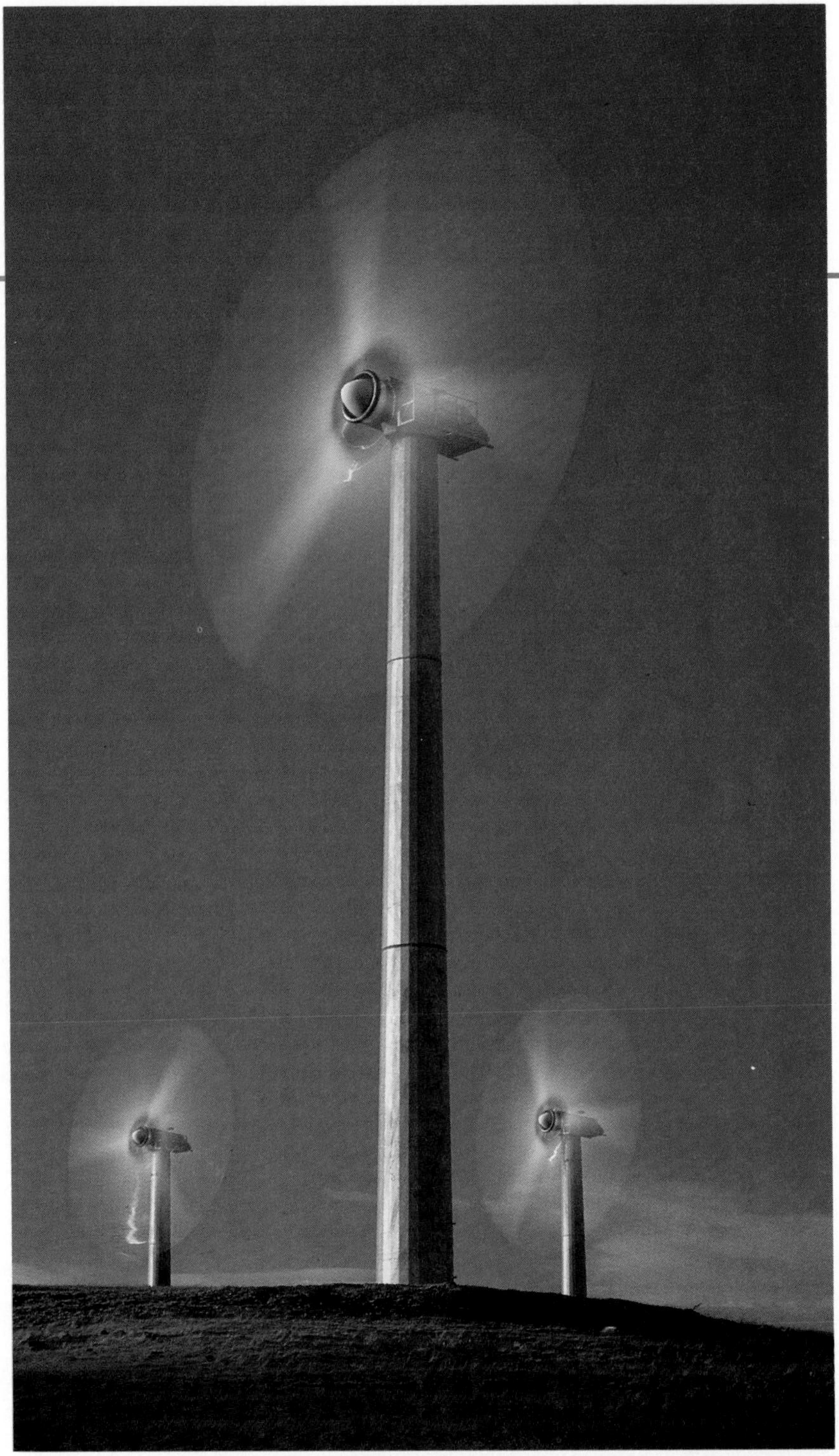

16

Planar Kinematics of a Rigid Body

In this chapter, planar kinematics, or the study of the geometry of the planar motion for a rigid body, will be discussed. This study is important for the design of gears, cams, and mechanisms used for many machine operations. Furthermore, once the kinematics of a rigid body is thoroughly understood, it will be possible to apply the equations of motion, which relate the forces on the body to the body's motion.

A rigid body can be subjected to three types of planar motion, namely, translation, rotation about a fixed axis, and general plane motion. We will first define each of these motions, then discuss the analysis of each separately. In all cases it will be shown that rigid-body planar motion is completely specified provided the motions of any two points on the body are known. The more complex study of three-dimensional kinematics of a rigid body is presented in Chapter 20.

16.1 Rigid-Body Motion

When all the particles of a rigid body move along paths which are equidistant from a fixed plane, the body is said to undergo *planar motion.* As stated previously, there are three types of planar motion; in order of increasing complexity, they are

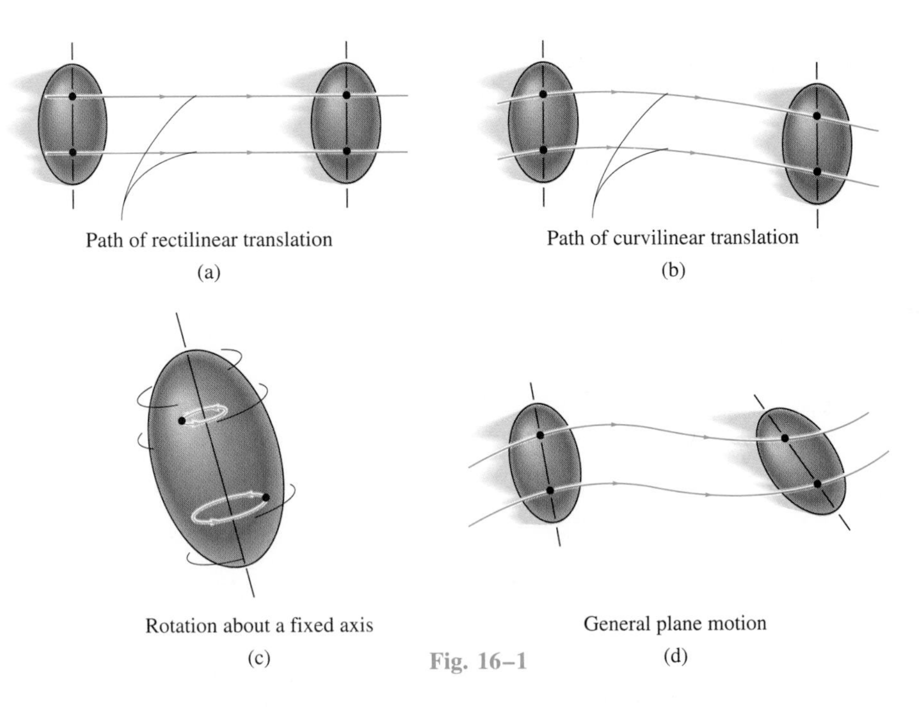

Fig. 16–1

1. *Translation.* This type of motion occurs if any line segment on the body remains parallel to its original direction during the motion. When the paths of motion for any two particles of the body are along equidistant straight lines, Fig. 16–1*a*, the motion is called *rectilinear translation.* However, if the paths of motion are along curved lines which are equidistant, Fig. 16–1*b*, the motion is called *curvilinear translation.*
2. *Rotation about a fixed axis.* When a rigid body rotates about a fixed axis, all the particles of the body, except those which lie on the axis of rotation, move along circular paths, Fig. 16–1*c*.
3. *General plane motion.* When a body is subjected to general plane motion, it undergoes a combination of translation *and* rotation, Fig. 16–1*d*. The translation occurs within a reference plane, and the rotation occurs about an axis perpendicular to the reference plane.

The above planar motions are exemplified by the moving parts of the crank mechanism shown in Fig. 16–2.

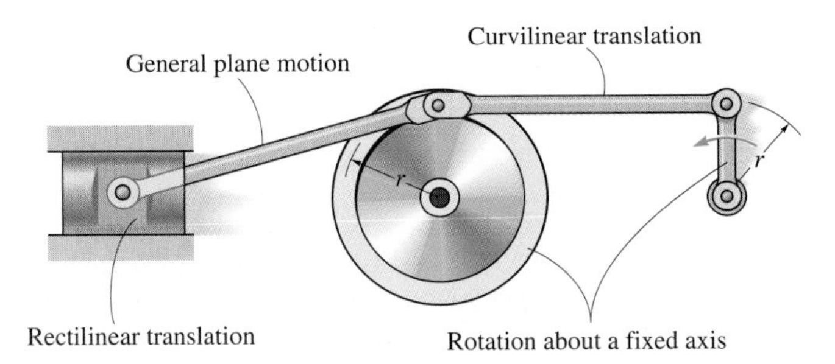

Fig. 16–2

16.2 Translation

Consider a rigid body which is subjected to either rectilinear or curvilinear translation in the x–y plane, Fig. 16–3.

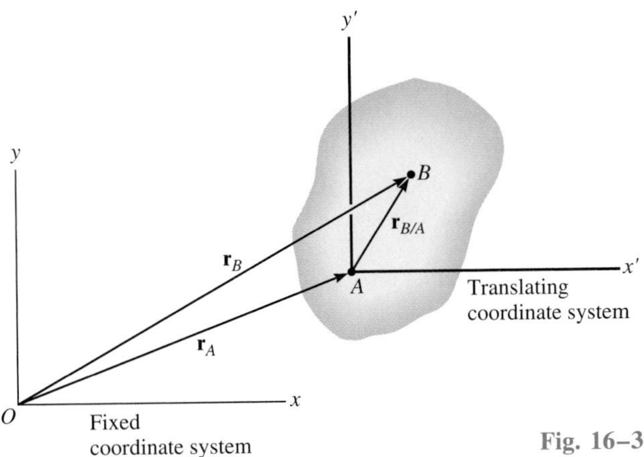

Fig. 16–3

Position. The locations of points A and B in the body are defined from the fixed x, y reference frame by using *position vectors* $\mathbf{r}_A$ and $\mathbf{r}_B$. The translating x', y' coordinate system is *fixed in the body* and has its origin located at A, hereafter referred to as the *base point.* The position of B with respect to A is denoted by the *relative-position vector* $\mathbf{r}_{B/A}$ (“$\mathbf{r}$ of B with respect to A”). By vector addition,

$$\mathbf{r}_B = \mathbf{r}_A + \mathbf{r}_{B/A}$$

Velocity. A relationship between the instantaneous velocities of A and B is obtained by taking the time derivative of the position equation, which yields $\mathbf{v}_B = \mathbf{v}_A + d\mathbf{r}_{B/A}/dt$. Here $\mathbf{v}_A$ and $\mathbf{v}_B$ denote *absolute velocities* since these vectors are measured from the x, y axes. The term $d\mathbf{r}_{B/A}/dt = \mathbf{0}$, since the *magnitude* of $\mathbf{r}_{B/A}$ is constant by definition of a rigid body, and because the body is translating the *direction* of $\mathbf{r}_{B/A}$ is constant. Therefore,

$$\mathbf{v}_B = \mathbf{v}_A$$

Acceleration. Taking the time derivative of the velocity equation yields a similar relationship between the instantaneous accelerations of A and B:

$$\mathbf{a}_B = \mathbf{a}_A$$

The above two equations indicate that *all points in a rigid body subjected to either curvilinear or rectilinear translation move with the same velocity and acceleration.* As a result, the kinematics of particle motion, discussed in Chapter 12, may also be used to specify the kinematics of points located in a translating rigid body.

16.3 Rotation About a Fixed Axis

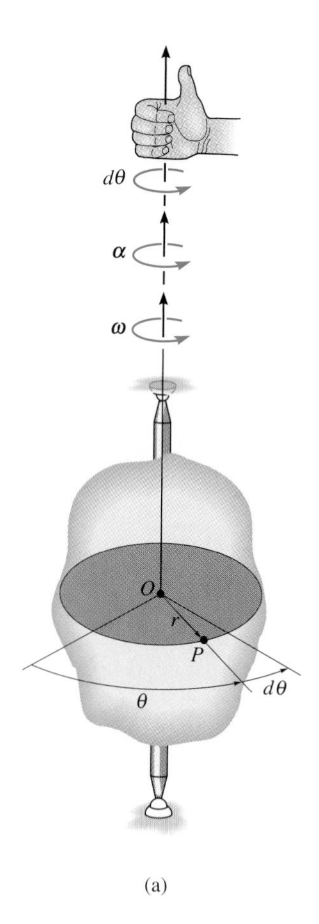

(a)

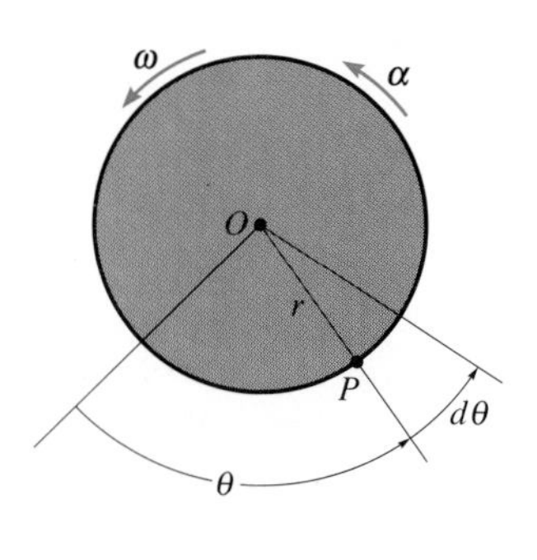

(b)

Fig. 16–4

When a body is rotating about a fixed axis, any point P located in the body travels along a *circular path.* This motion depends on the angular motion of the body about the axis. For this reason, we will first study the angular-motion properties of the body before analyzing the circular motion of P.

Angular Motion. Since a point is without dimension, it has no angular motion. *Only lines or bodies undergo angular motion.* To study these effects, we will consider the body shown in Fig. 16–4*a* and the angular motion of a radial line r located within the shaded plane.

Angular Position. At the instant shown, the *angular position* of r is defined by the angle θ, measured between a *fixed* reference line and r. Here r extends perpendicular from the axis of rotation at point O to a point P in the body.

Angular Displacement. The change in the angular position, often measured as a differential $d\boldsymbol{\theta}$, is called the *angular displacement.** This vector has a *magnitude* of $d\theta$, which is measured in degrees, radians, or revolutions, where $1 \text{ rev} = 2\pi \text{ rad}$. Since motion is about a *fixed axis,* the direction of $d\boldsymbol{\theta}$ is *always* along the axis. Specifically, the *direction of* $d\boldsymbol{\theta}$ is determined by the right-hand rule; that is, the fingers of the right hand are curled with the sense of rotation, so that in this case the thumb, or $d\boldsymbol{\theta}$, points upward, Fig. 16–4*a*. In two dimensions, as shown by the top view of the shaded plane, Fig. 16–4*b*, both θ and $d\boldsymbol{\theta}$ are directed counterclockwise, and so the thumb points outward.

Angular Velocity. The time rate of change in the angular position is called the *angular velocity* $\boldsymbol{\omega}$ (omega). Since $d\boldsymbol{\theta}$ occurs during an instant of time dt, then,

$$(\circlearrowleft +) \qquad \omega = \frac{d\theta}{dt} \qquad (16\text{–}1)$$

This vector has a *magnitude* which is often measured in rad/s. It is expressed here in scalar form, since its *direction* is always along the axis of rotation, i.e., in the same direction as $d\boldsymbol{\theta}$, Fig. 16–4*a*. When indicating the angular motion in the shaded plane, Fig. 16–4*b*, we can refer to the sense of rotation as clockwise or counterclockwise. Here we have *arbitrarily* chosen counterclockwise rotations as *positive,* and indicated this by the curl shown in parentheses next to Eq. 16–1. Realize, however, that the directional sense of $\boldsymbol{\omega}$ is actually outward.

*It is shown in Sec. 20.1 that finite rotations or finite angular displacements are *not* vector quantities, although differential rotations $d\boldsymbol{\theta}$ are vectors.

Angular Acceleration. The *angular acceleration* $\boldsymbol{\alpha}$ (alpha) measures the time rate of change of the angular velocity. Hence, the *magnitude* of this vector may be written as

$$(\circlearrowright +) \qquad \alpha = \frac{d\omega}{dt} \qquad (16\text{–}2)$$

Using Eq. 16–1, it is possible to express α as

$$(\circlearrowright +) \qquad \alpha = \frac{d^2\theta}{dt^2} \qquad (16\text{–}3)$$

The line of action of $\boldsymbol{\alpha}$ is the same as that for $\boldsymbol{\omega}$, Fig. 16–4*a*; however, its sense of *direction* depends on whether $\boldsymbol{\omega}$ is increasing or decreasing with time. In particular, if $\boldsymbol{\omega}$ is decreasing, $\boldsymbol{\alpha}$ is called an *angular deceleration* and therefore has a sense of direction which is opposite to $\boldsymbol{\omega}$.

By eliminating dt from Eqs. 16–1 and 16–2, we obtain a differential relation between the angular acceleration, angular velocity, and angular displacement, namely,

$$(\circlearrowright +) \qquad \alpha\, d\theta = \omega\, d\omega \qquad (16\text{–}4)$$

The similarity between the differential relations for angular motion and those developed for rectilinear motion of a particle ($v = ds/dt$, $a = dv/dt$, and $a\, ds = v\, dv$) should be apparent.

Constant Angular Acceleration. If the angular acceleration of the body is constant, $\boldsymbol{\alpha} = \boldsymbol{\alpha}_c$, then Eqs. 16–1, 16–2, and 16–4, when integrated, yield a set of formulas which relate the body's angular velocity, angular position, and time. These equations are similar to Eqs. 12–4 to 12–6 used for rectilinear motion. The results are

$$(\circlearrowright +) \qquad \omega = \omega_0 + \alpha_c t \qquad (16\text{–}5)$$

$$(\circlearrowright +) \qquad \theta = \theta_0 + \omega_0 t + \tfrac{1}{2}\alpha_c t^2 \qquad (16\text{–}6)$$

$$(\circlearrowright +) \qquad \omega^2 = \omega_0^2 + 2\alpha_c(\theta - \theta_0) \qquad (16\text{–}7)$$

Constant Angular Acceleration

Here θ_0 and ω_0 are the initial values of the body's angular position and angular velocity, respectively.

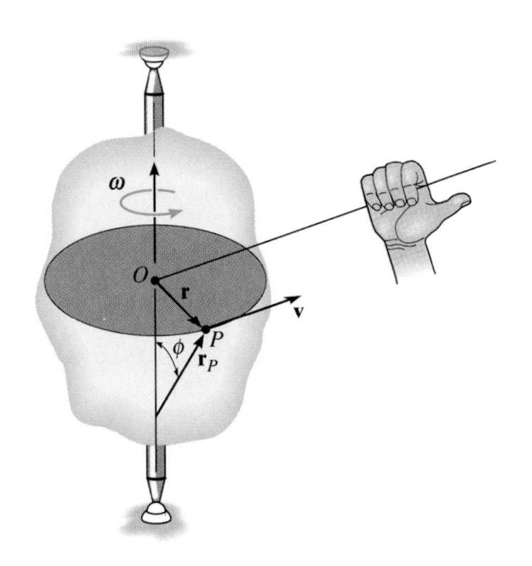

(c)

Fig. 16–4 *(cont'd)*

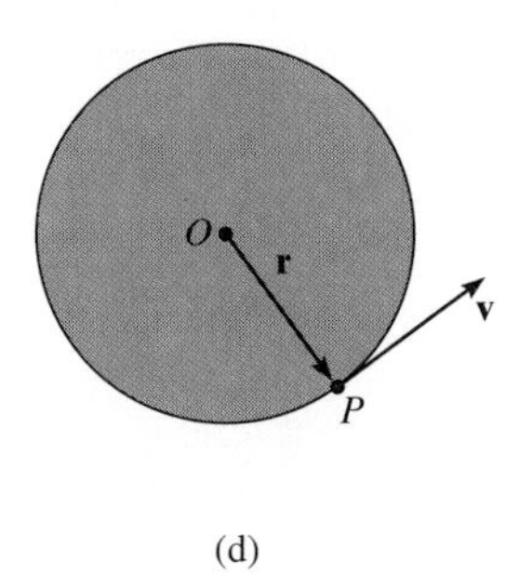

(d)

Motion of Point *P*. As the rigid body in Fig. 16–4*c* rotates, point *P* travels along a *circular path* of radius *r* and center at point *O*. This path is contained in the shaded plane shown in top view, Fig. 16–4*d*.

Position. The position of *P* is defined by the position vector **r**, which extends from *O* to *P*.

Velocity. The velocity of *P* has a magnitude which can be determined from the circular motion of *P* using its polar coordinate components $v_r = \dot{r}$ and $v_\theta = r\dot{\theta}$, Eqs. 12–25. Since *r* is constant, the radial component $v_r = \dot{r} = 0$, so that $v = v_\theta = r\dot{\theta}$. Because $\omega = \dot{\theta}$, Eq. 16–1, the result is

$$v = \omega r \tag{16–8}$$

As shown in Figs. 16–4*c* and 16–4*d*, the *direction* of **v** is *tangent* to the circular path.

Both the magnitude and direction of **v** can be accounted for by using the cross product of $\boldsymbol{\omega}$ and $\mathbf{r}_P$ (see Appendix C). Here, $\mathbf{r}_P$ is directed from *any point* on the axis of rotation to point *P*, Fig. 16–4*c*. We have

$$\mathbf{v} = \boldsymbol{\omega} \times \mathbf{r}_P \tag{16–9}$$

The order of the vectors in this formulation is important, since the cross product is not commutative, i.e., $\boldsymbol{\omega} \times \mathbf{r}_P \neq \mathbf{r}_P \times \boldsymbol{\omega}$. In this regard, notice in Fig. 16–4*c* how the correct direction of **v** is established by the right-hand rule. The fingers of the right hand are curled from $\boldsymbol{\omega}$ toward $\mathbf{r}_P$ ($\boldsymbol{\omega}$ ''cross'' $\mathbf{r}_P$). The thumb indicates the correct direction of **v**, that is, tangent to the path in the direction of motion. From Eq. C–8, the magnitude of **v** in Eq. 16–9 is $v = \omega r_P \sin\phi$. Since $r = r_P \sin\phi$, Fig. 16–4*c*, then $v = \omega r$, which agrees with Eq. 16–8. As a special case, the position vector **r** can be chosen from point *O* to point *P*, Fig. 16–4*c*. Here **r** lies in the plane of motion and again the velocity of point *P* is

$$\mathbf{v} = \boldsymbol{\omega} \times \mathbf{r} \tag{16–10}$$

Acceleration. For convenience, the acceleration of P will be expressed in terms of its normal and tangential components.* Using $a_t = dv/dt$ and $a_n = v^2/\rho$, noting that $\rho = r$, $v = \omega r$, and $\alpha = d\omega/dt$, we have

$$a_t = \alpha r \tag{16–11}$$

$$a_n = \omega^2 r \tag{16–12}$$

The *tangential component of acceleration,* Figs. 16–4*e* and 16–4*f*, represents the time rate of change in the velocity's magnitude. If the speed of P is increasing, then $\mathbf{a}_t$ acts in the same direction as $\mathbf{v}$; if the speed is decreasing, $\mathbf{a}_t$ acts in the opposite direction of $\mathbf{v}$; and finally, if the speed is constant, $\mathbf{a}_t$ is zero.

The *normal component of acceleration* represents the time rate of change in the velocity's direction. The *direction* of $\mathbf{a}_n$ is always toward O, the center of the circular path, Figs. 16–4*e* and 16–4*f*.

Like the velocity, the acceleration of point P may be expressed in terms of the vector cross product. Taking the time derivative of Eq. 16–9 yields

$$\mathbf{a} = \frac{d\mathbf{v}}{dt} = \frac{d\boldsymbol{\omega}}{dt} \times \mathbf{r}_P + \boldsymbol{\omega} \times \frac{d\mathbf{r}_P}{dt}$$

Recalling that $\boldsymbol{\alpha} = d\boldsymbol{\omega}/dt$, and using Eq. 16–9 ($d\mathbf{r}_P/dt = \mathbf{v} = \boldsymbol{\omega} \times \mathbf{r}_P$), we have

$$\mathbf{a} = \boldsymbol{\alpha} \times \mathbf{r}_P + \boldsymbol{\omega} \times (\boldsymbol{\omega} \times \mathbf{r}_P) \tag{16–13}$$

By definition of the cross product, the first term on the right has a magnitude $a_t = \alpha r_P \sin\phi = \alpha r$, and by the right-hand rule, $\boldsymbol{\alpha} \times \mathbf{r}_P$ is in the direction of $\mathbf{a}_t$, Fig. 16–4*e*. Likewise, the second term has a magnitude $a_n = \omega^2 r_P \sin\phi = \omega^2 r$, and applying the right-hand rule twice, first $\boldsymbol{\omega} \times \mathbf{r}_P$ then $\boldsymbol{\omega} \times (\boldsymbol{\omega} \times \mathbf{r}_P)$, it can be seen that this result is in the same direction as $\mathbf{a}_n$, Fig. 16–4*e*. Noting that this is also the *same* direction as $-\mathbf{r}$, which lies in the plane of motion, we can express $\mathbf{a}_n$ in a much simpler form as $\mathbf{a}_n = -\omega^2\mathbf{r}$. Hence, Eq. 16–12 can be identified by its two components as

$$\begin{aligned}\mathbf{a} &= \mathbf{a}_t + \mathbf{a}_n \\ &= \boldsymbol{\alpha} \times \mathbf{r} - \omega^2\mathbf{r}\end{aligned} \tag{16–14}$$

Since $\mathbf{a}_t$ and $\mathbf{a}_n$ are perpendicular, Figs. 16–4*e* and 16–4*f*, if needed the magnitude of acceleration can be determined from the Pythagorean theorem; namely, $a = \sqrt{a_n^2 + a_t^2}$.

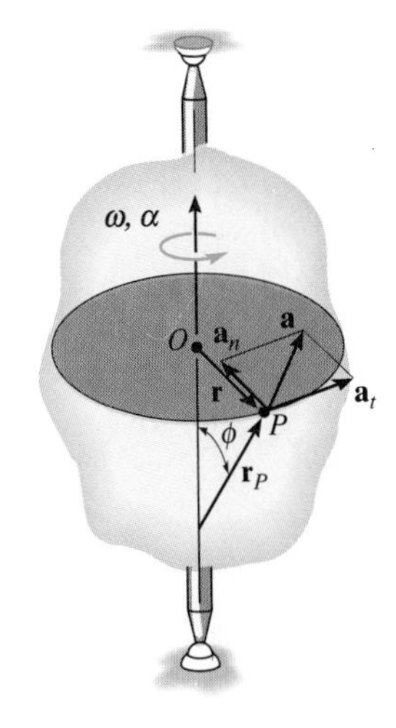

(e)

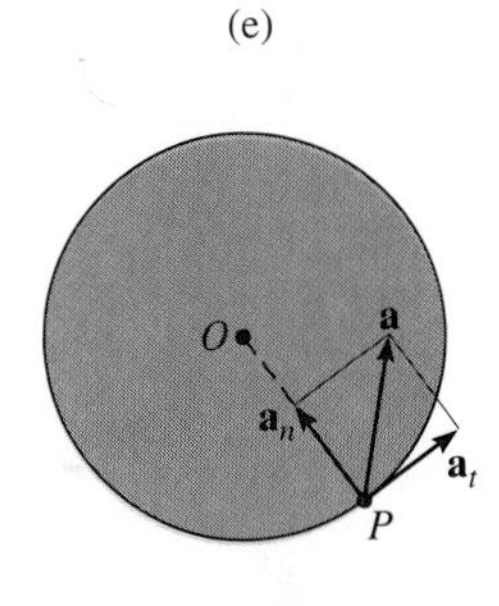

(f)

Fig. 16–4 *(cont'd)*

*Polar coordinates can also be used. Since $a_r = \ddot{r} - r\dot{\theta}^2$ and $a_\theta = r\ddot{\theta} + 2\dot{r}\dot{\theta}$, substituting $\dot{r} = \ddot{r} = 0$, $\dot{\theta} = \omega$, $\ddot{\theta} = \alpha$, we obtain Eqs. 16–11 and 16–12.

PROCEDURE FOR ANALYSIS

In order to determine the velocity and acceleration of a point located in a rigid body that is rotating about a fixed axis, it is first necessary to know the body's angular velocity and angular acceleration. If Eqs. 16–1 to 16–7 are used to obtain ω and α, it is important that a *positive sense of direction* along the axis of rotation be established. By doing so, the sense of θ, ω, and α can be determined from the algebraic signs of their numerical quantities. In the examples that follow, the positive sense will be indicated by a curl alongside each kinematic equation as it is applied.

Angular Motion. If $\boldsymbol{\alpha}$ or $\boldsymbol{\omega}$ is unknown, then the relationships between the angular motions are defined by the differential equations

$$\omega = \frac{d\theta}{dt} \qquad \alpha = \frac{d\omega}{dt} \qquad \alpha\, d\theta = \omega\, d\omega$$

If one is *absolutely certain* that the body's angular acceleration is *constant,* then the following equations can be used:

$$\omega = \omega_0 + \alpha_c t$$
$$\theta = \theta_0 + \omega_0 t + \tfrac{1}{2}\alpha_c t^2$$
$$\omega^2 = \omega_0^2 + 2\alpha_c(\theta - \theta_0)$$

Motion of P. When the motion of a point P in the body is to be determined, it is suggested that a *kinematic diagram* accompany the problem solution. This diagram is simply a graphical representation showing the motion of the point.

In most cases the velocity and the two components of acceleration can be determined from the scalar equations

$$v = \omega r$$
$$a_t = \alpha r$$
$$a_n = \omega^2 r$$

However, if the geometry of the problem is difficult to visualize, the following vector equations should be used:

$$\mathbf{v} = \boldsymbol{\omega} \times \mathbf{r}_P = \boldsymbol{\omega} \times \mathbf{r}$$
$$\mathbf{a}_t = \boldsymbol{\alpha} \times \mathbf{r}_P = \boldsymbol{\alpha} \times \mathbf{r}$$
$$\mathbf{a}_n = \boldsymbol{\omega} \times (\boldsymbol{\omega} \times \mathbf{r}_P) = -\omega^2\mathbf{r}$$

For application, $\mathbf{r}_P$ is directed from any point on the axis of rotation to point P, whereas $\mathbf{r}$ lies in the plane of motion of P. Either of these vectors, along with $\boldsymbol{\omega}$ and $\boldsymbol{\alpha}$, should be expressed in terms of its **i, j, k** components and, if necessary, the cross products computed by using a determinant expansion (see Eq. C–12).

Example 16–1

A cord is wrapped around a wheel which is initially at rest as shown in Fig. 16–5. If a force is applied to the cord and gives it an acceleration $a = (4t)$ m/s^2, where t is in seconds, determine as a function of time (*a*) the angular velocity of the wheel, and (*b*) the angular position of line *OP* in radians.

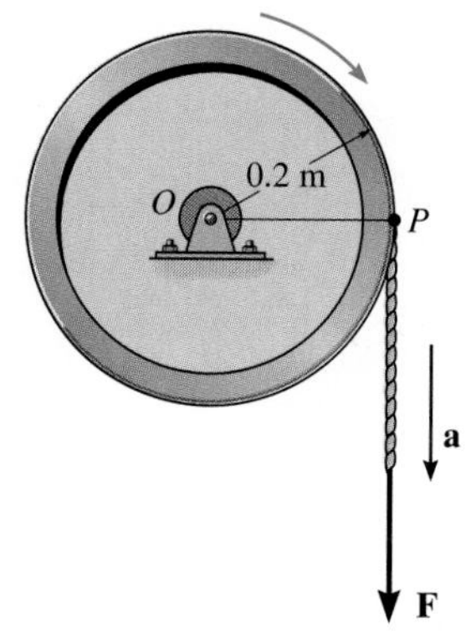

Fig. 16–5

SOLUTION

Part (a). The wheel is subjected to rotation about a fixed axis passing through point *O*. Thus, point *P* on the wheel has motion about a circular path, and therefore the acceleration of this point has *both* tangential and normal components. In particular, the tangential component is $(a_P)_t = (4t)$ m/s^2, since the cord is connected to the wheel and *tangent* to it at *P*. Hence the angular acceleration of the wheel is

$$(\circlearrowright +) \qquad (a_P)_t = \alpha r$$

$$(4t) \text{ m/s}^2 = \alpha(0.2 \text{ m})$$

$$\alpha = 20t \text{ rad/s}^2$$

Using this result, the wheel's angular velocity $\boldsymbol{\omega}$ can now be determined from $\alpha = d\omega/dt$, since this equation relates α, t, and ω. Integrating, with the initial condition that $\omega = 0$ at $t = 0$, yields

$$(\circlearrowright +) \qquad \alpha = \frac{d\omega}{dt} = (20t) \text{ rad/s}^2$$

$$\int_0^{\omega} d\omega = \int_0^t 20t\, dt$$

$$\omega = 10t^2 \text{ rad/s} \circlearrowright \qquad \textit{Ans.}$$

Why is it not possible to use Eq. 16–5 ($\omega = \omega_0 + \alpha_c t$) to obtain this result?

Part (b). Using this result, the angular position θ of the radial line *OP* can be computed from $\omega = d\theta/dt$, since this equation relates θ, ω, and t. Integrating, with the initial condition $\theta = 0$ at $t = 0$, we have

$$(\circlearrowright +) \qquad \frac{d\theta}{dt} = \omega = (10t^2) \text{ rad/s}$$

$$\int_0^{\theta} d\theta = \int_0^t 10t^2\, dt$$

$$\theta = 3.33\, t^3 \text{ rad} \qquad \textit{Ans.}$$

Example 16–2

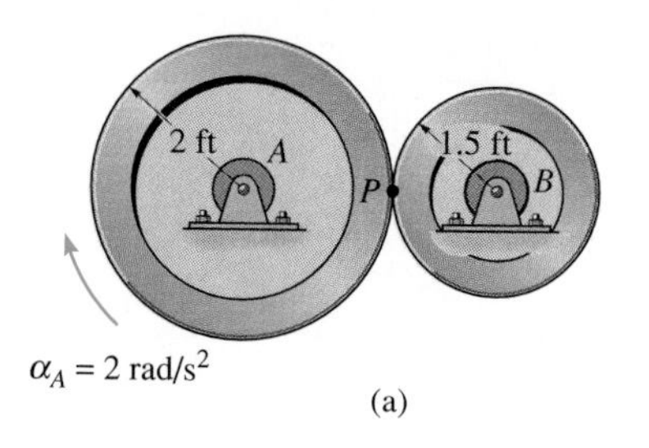

(a)

Disk A, shown in Fig. 16–6a, starts from rest and through the use of a motor begins to rotate with a constant angular acceleration of $\alpha_A = 2\ \text{rad/s}^2$. If no slipping occurs between the disks, determine the angular velocity and angular acceleration of disk B just after A turns 10 revolutions.

SOLUTION

First we will convert the 10 revolutions to radians. Since there are 2π rad to one revolution, then

$$\theta_A = 10\ \text{rev}\left(\frac{2\pi\ \text{rad}}{1\ \text{rev}}\right) = 62.83\ \text{rad}$$

Since α_A is *constant*, the angular velocity of A is then

$$(\curvearrowleft+)\qquad \omega^2 = \omega_0^2 + 2\alpha_c(\theta - \theta_0)$$
$$\omega_A^2 = 0 + 2(2\ \text{rad/s}^2)(62.83\ \text{rad} - 0)$$
$$\omega_A = 15.9\ \text{rad/s}\ \circlearrowleft$$

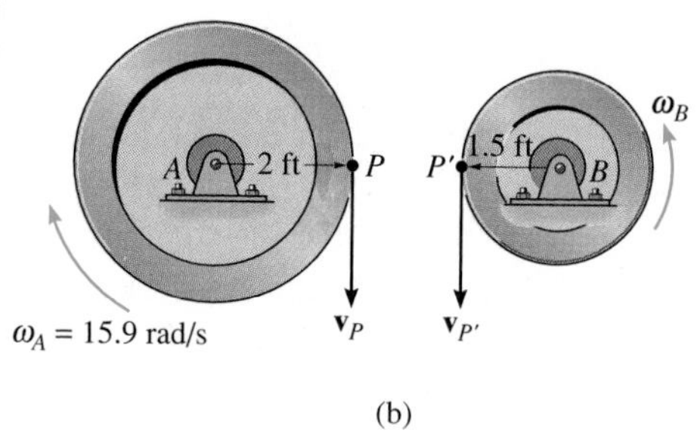

(b)

As shown in Fig. 16–6b, the speed of the contacting point P on the rim of A is

$$(+\downarrow)\qquad v_P = \omega_A r_A = (15.9\ \text{rad/s})(2\ \text{ft}) = 31.8\ \text{ft/s}\ \downarrow$$

The velocity is always tangent to the path of motion; and since no slipping occurs between the disks, the speed of point P' on B is the *same* as the speed of P on A. The angular velocity of B is therefore

$$(\circlearrowright+)\qquad \omega_B = \frac{v_{P'}}{r_B} = \frac{31.8\ \text{ft/s}}{1.5\ \text{ft}} = 21.1\ \text{rad/s}\ \curvearrowleft \qquad \textit{Ans.}$$

The *tangential components* of acceleration of both disks are also equal, since the disks are in contact with one another. Hence, from Fig. 16–6c,

$$(a_P)_t = (a_{P'})_t$$
$$\alpha_A r_A = \alpha_B r_B$$
$$\alpha_B = \alpha_A\left(\frac{r_A}{r_B}\right) = 2\left(\frac{2\ \text{ft}}{1.5\ \text{ft}}\right) = 2.67\ \text{rad/s}^2\ \curvearrowleft \qquad \textit{Ans.}$$

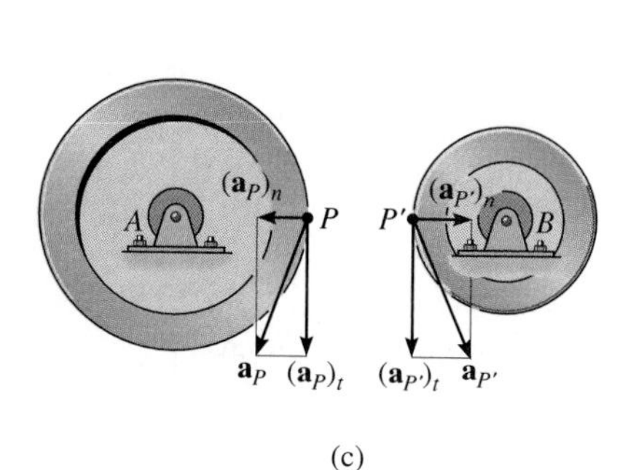

(c)

Fig. 16–6

Notice that the normal components of acceleration $(a_P)_n$ and $(a_{P'})_n$ act in *opposite directions*, since the paths of motion for both points are *different*. Furthermore, $(a_P)_n \neq (a_{P'})_n$, since the *magnitudes* of these components depend on both the radius and angular velocity of each disk, i.e., $(a_P)_n = \omega_A^2 r_A$ and $(a_{P'})_n = \omega_B^2 r_B$. Consequently, $\mathbf{a}_P \neq \mathbf{a}_{P'}$.

PROBLEMS

16–1. A wheel has an initial clockwise angular velocity of 10 rad/s and a constant angular acceleration of 3 rad/s^2. Determine the number of revolutions it must undergo to acquire a clockwise angular velocity of 15 rad/s. What time is required?

16–2. A flywheel has its angular speed increased uniformly from 15 rad/s to 60 rad/s in 80s. If the diameter of the wheel is 2 ft, determine the magnitudes of the normal and tangential components of acceleration of a point on the rim of the wheel when $t = 80\text{s}$, and the total distance the point travels during the time period.

16–3. If the angular velocity of the disk is increased uniformly from 3 rev/min when $t = 0$ to 10 rev/min when $t = 4$ s, determine the magnitudes of the velocity and acceleration of point A on the disk when $t = 4$ s.

***16–4.** The angular velocity of the disk is defined by $\omega = (5t^2 + 2)$ rad/s, where t is in seconds. Determine the magnitudes of the velocity and acceleration of point A on the disk when $t = 0.5$ s.

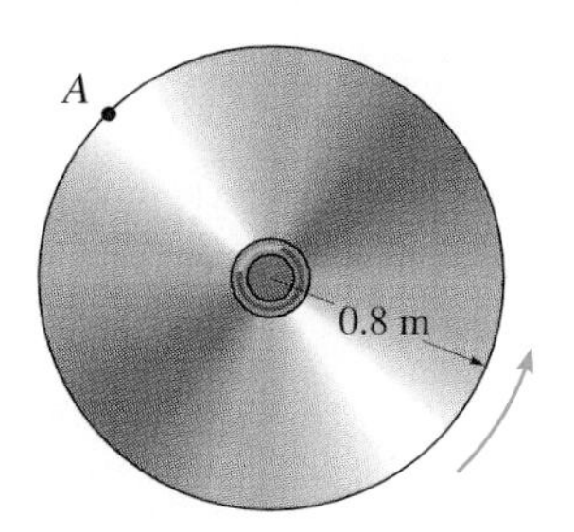

Probs. 16–3/16–4

16–5. The tub of a washing machine is rotating at 50 rad/s when the power is turned off. If it takes 15 s for the tub to come to rest, determine (a) its constant angular deceleration, and (b) the total number of revolutions it makes.

16–6. The angular position of a disk is defined by $\theta = (t + 4t^2)$ rad, where t is in minutes. Determine the number of revolutions, the angular velocity, and angular acceleration of the disk in 90 seconds.

16–7. If the motor turns gear A with an angular acceleration of $\alpha_A = 2\ \text{rad/s}^2$ when the angular velocity is $\omega_A = 20$ rad/s, determine the angular acceleration and angular velocity of gear D.

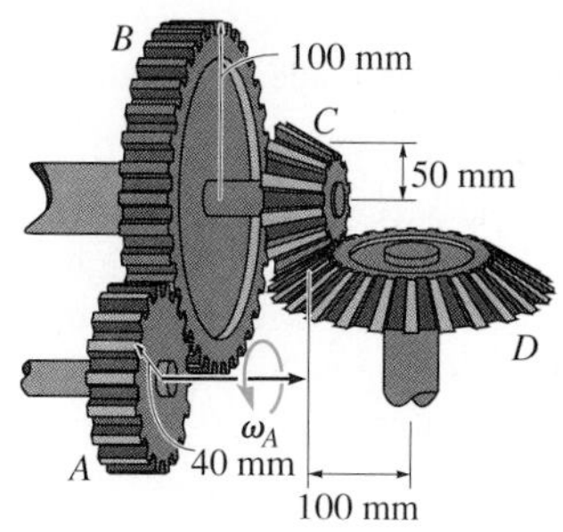

Prob. 16–7

***16–8.** The pinion gear A on the motor shaft is given a constant angular acceleration $\alpha = 3\ \text{rad/s}^2$. If the gears A and B have the dimensions shown, determine the angular velocity and angular displacement of the output shaft C, when $t = 2$ s starting from rest. The shaft is fixed to B and turns with it.

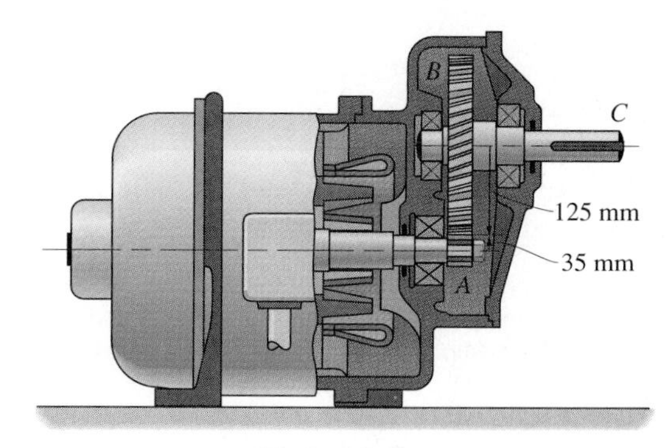

Prob. 16–8

16–9. The motor M begins rotating at $\omega = 4(1 - e^{-t})$ rad/s, where t is in seconds. If the pulleys and fan have the radii shown, determine the magnitudes of the velocity and acceleration of point P on the fan blade when $t = 0.5$ s. Also, what is the maximum speed of this point?

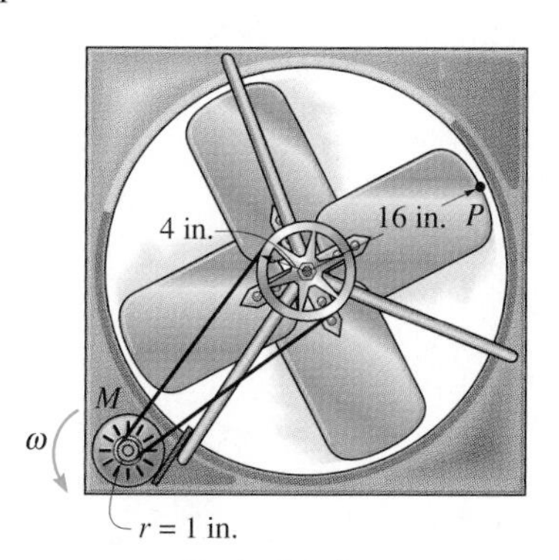

Prob. 16–9

16–10. The vertical-axis windmill consists of two blades that have a parabolic shape. If the blades are originally at rest and begin to turn with a constant angular acceleration $\alpha_c = 0.5\ \text{rad/s}^2$, determine the magnitudes of the velocity and acceleration of points A and B on a blade after the blade has rotated through two revolutions.

16–11. The vertical-axis windmill consists of two blades that have a parabolic shape. If the blades are originally at rest and begin to turn with a constant angular acceleration $\alpha_c = 0.5\ \text{rad/s}^2$, determine the magnitudes of the velocity and acceleration of points A and B on a blade when $t = 4$ s.

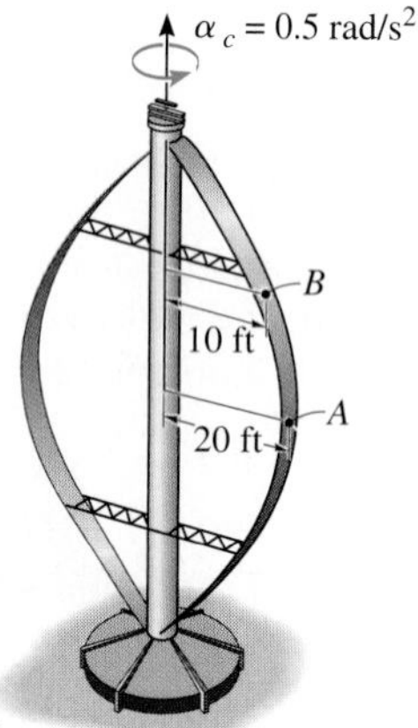

Probs. 16–10/16–11

*__16–12.__ If disk A has an initial angular velocity of $\omega_0 = 6$ rad/s and a constant angular acceleration $\alpha_A = 3\ \text{rad/s}^2$, determine the magnitudes of the velocity and acceleration of block B when $t =$ 2 s.

16–13. A motor gives disk A an angular acceleration of $\alpha_A = (0.6t^2 + 0.75)\ \text{rad/s}^2$, where t is in seconds. If the initial angular velocity of the disk is $\omega_0 = 6$ rad/s, determine the magnitudes of the velocity and acceleration of block B when $t = 2$s.

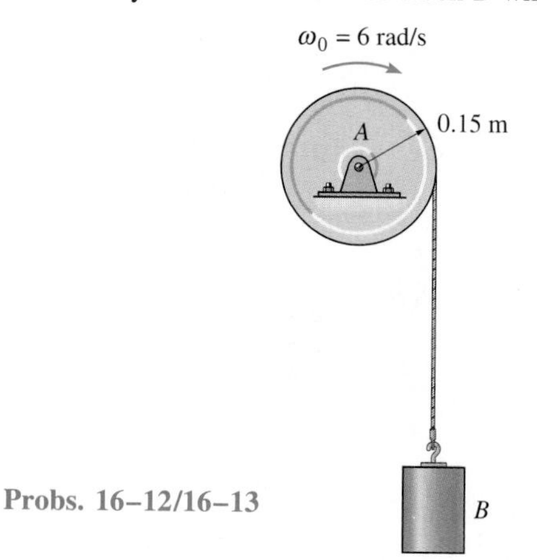

Probs. 16–12/16–13

16–14. The disk is originally rotating at $\omega_0 = 8$ rad/s. If it is subjected to a constant angular acceleration $\alpha_c = 6\ \text{rad/s}^2$, determine the magnitudes of the velocity and the n and t components of acceleration of point A at the instant $t = 3$ s.

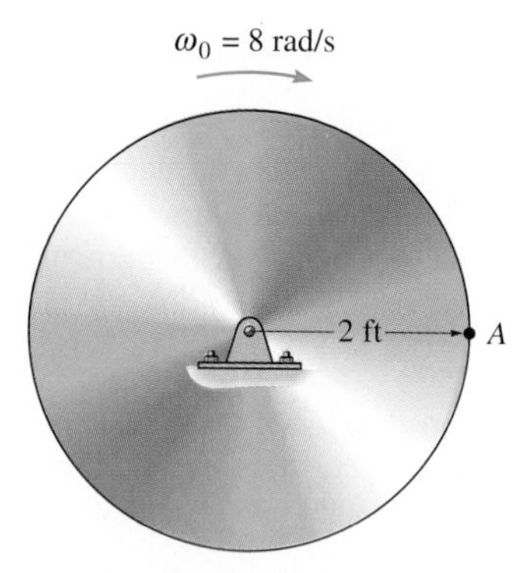

Prob. 16–14

16–15. Starting from rest when $s = 0$, pulley A is given an angular acceleration $\alpha = (6\theta)\ \text{rad/s}^2$, where θ is in radians. Determine the speed of block B when it has risen $s = 6$ m. The pulley has an inner hub D which is fixed to C and turns with it.

*__16–16.__ Starting from rest when $s = 0$, pulley A is given a constant angular acceleration $\alpha_c = 6\ \text{rad/s}^2$. Determine the speed of block B when it has risen $s = 6$ m. The pulley has an inner hub D which is fixed to C and turns with it.

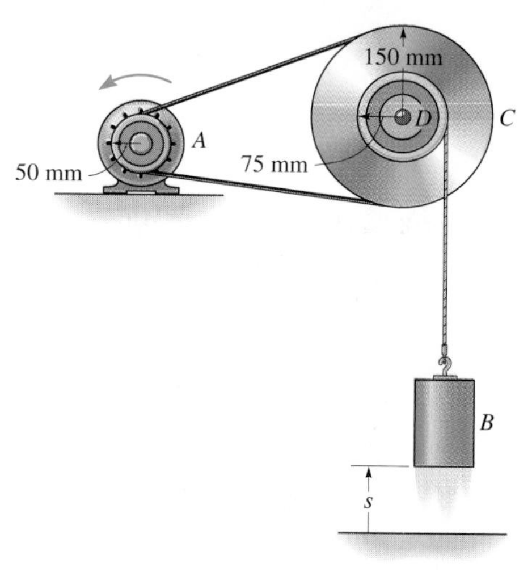

Probs. 16–15/16–16

16–17. Initially the motor on the circular saw turns its drive shaft at $\omega = (20\,t^{2/3})$ rad/s, where t is in seconds. If the radii of gears A and B are 0.25 in. and 1 in., respectively, determine the magnitudes of the velocity and acceleration of a tooth C on the saw blade after the drive shaft rotates $\theta = 5$ rad starting from rest.

Prob. 16–17

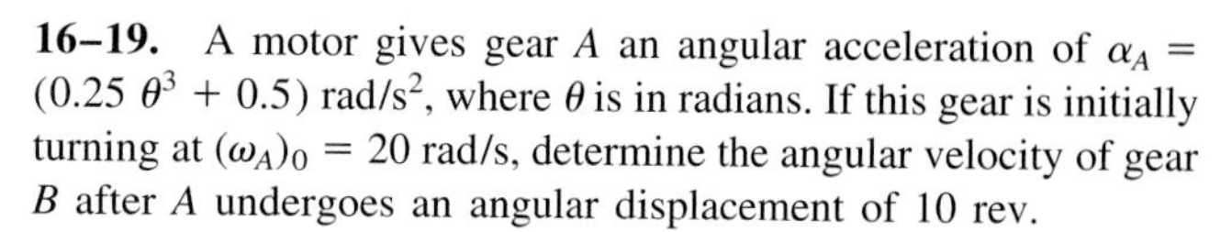

16–19. A motor gives gear A an angular acceleration of $\alpha_A = (0.25\,\theta^3 + 0.5)$ rad/s^2, where θ is in radians. If this gear is initially turning at $(\omega_A)_0 = 20$ rad/s, determine the angular velocity of gear B after A undergoes an angular displacement of 10 rev.

***16–20.** A motor gives gear A an angular acceleration of $\alpha_A = (4t^3)$ rad/s^2, where t is in seconds. If this gear is initially turning at $(\omega_A)_0 = 20$ rad/s, determine the angular velocity of gear B when $t = 2$ s.

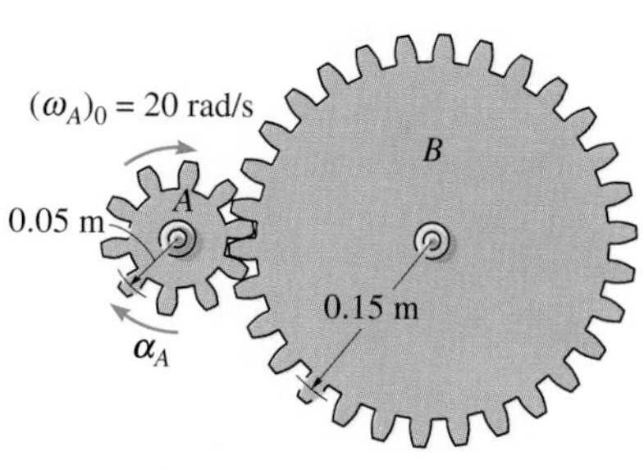

Probs. 16–19/16–20

16–18. A motor rotates its attached input shaft S at 1750 rev/min. This shaft turns gear A, which turns gear B and then gear C. Determine the required radius of gear D so its attached output shaft E turns at 50 rad/s. The radii of gears A, B, and C are $r_A = 28$ mm, $r_B = 80$ mm, and $r_C = 15$ mm, respectively.

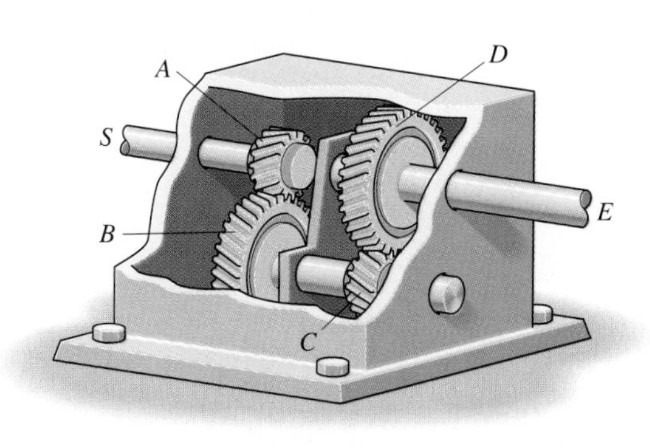

Prob. 16–18

16–21. Due to the screw at E, the actuator provides linear motion to the arm at F when the motor turns the gear at A. If the gears have the radii listed in the figure, and the screw at E has a pitch $p = 2$ mm, determine the speed at F when the motor turns A at $\omega_A = 20$ rad/s. *Hint:* The screw pitch indicates the amount of advance of the screw for each full revolution.

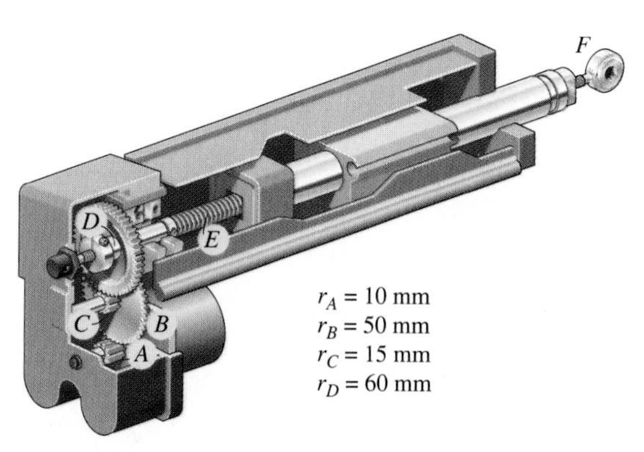

Prob. 16–21

16–22. If the angular velocity of the drum is increased uniformly from 6 rad/s when $t = 0$ to 12 rad/s when $t = 5$ s, determine the magnitudes of the velocity and acceleration of points A and B on the belt when $t = 1$ s. At this instant the points are located as shown.

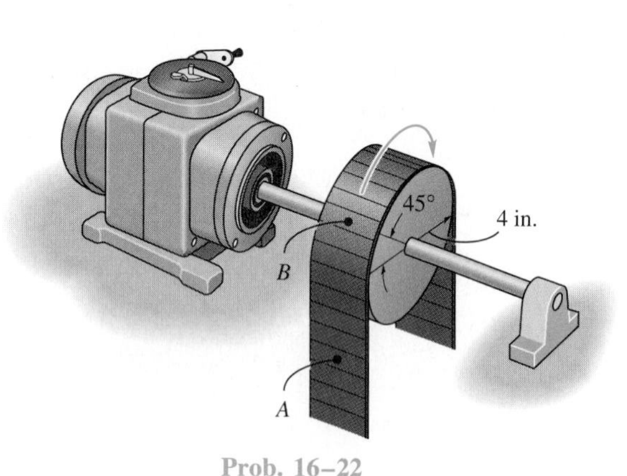

Prob. 16–22

16–23. The operation of "reverse" for a three-speed automotive transmission is illustrated schematically in the figure. If the crank shaft G is turning with an angular speed of 60 rad/s, determine the angular speed of the drive shaft H. Each of the gears rotates about a fixed axis. Note that gears A and B, C and D, E and F are in mesh. The radii of each of these gears are reported in the figure.

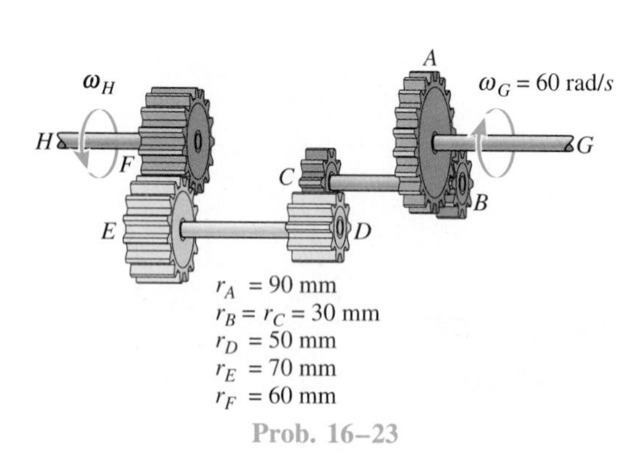

Prob. 16–23

***16–24.** If the hoisting gear A has an initial angular velocity $\omega_A = 8$ rad/s and an angular deceleration $\alpha_A = -1.5$ rad/s^2, determine the velocity and acceleration of block C in 2 s.

■**16–25.** Solve Prob. 16–24 assuming that the angular deceleration of gear A is defined by the relation $\alpha_A = (-0.5e^{-0.4t^2})$ rad/s^2, where t is in seconds. Use Simpson's rule with $n = 50$ to evaluate the integral.

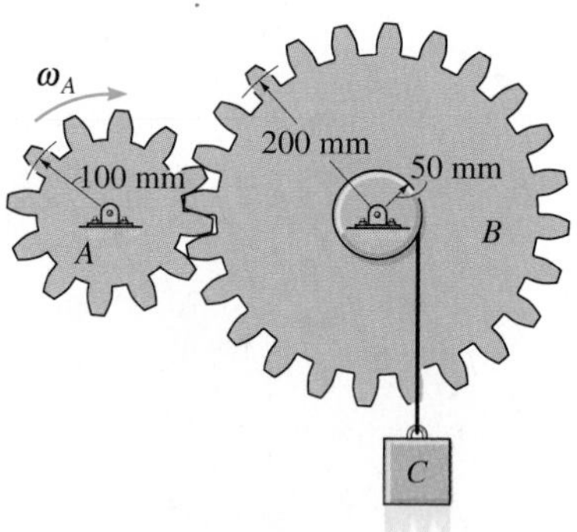

Probs. 16–24/16–25

16–26. Rotation of the robotic arm occurs due to linear movement of the hydraulic cylinders A and B. If cylinder A is extending at the constant rate of 0.5 ft/s while B is contracting at 0.5 ft/s, determine the magnitudes of velocity and acceleration of the part C held in the grips of the arm. The gear at D has a radius of 0.10 ft and $\theta = 45°$.

16–27. Rotation of the robotic arm occurs due to linear movement of the hydraulic cylinders A and B. If cylinder A is extending at the constant rate of 0.5 ft/s while B is contracting at 0.5 ft/s, determine the angle θ so that the part C held in the grip of the arm has the largest magnitude of velocity. Also, determine the angle θ for the largest magnitude of acceleration. What are these magnitudes? The gear D has a radius of 0.10 ft.

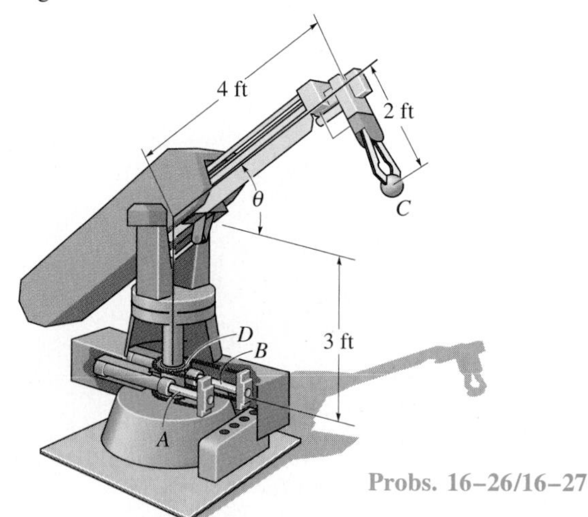

Probs. 16–26/16–27

***16–28.** The angular acceleration of the wheel is defined by $\alpha = (3 + 0.1\theta)\ \text{rad/s}^2$, where θ is in radians. Determine the magnitudes of the velocity and acceleration of a point P on its rim when $\theta = 2$ rad. How much time is needed to reach this angular position? Originally, $\theta = 0$, and $\omega = 0$ when $t = 0$.

16–29. The angular acceleration of the wheel is defined by $\alpha = 4(1 - e^{-2t})\ \text{rad/s}^2$, where t is in seconds. Determine the magnitudes of the velocity and acceleration of a point P on its rim when $t = 0.5$ s. Originally the wheel is at rest.

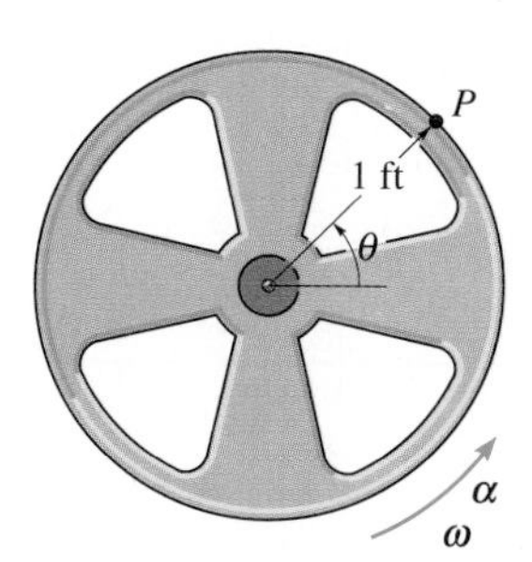

Probs. 16–28/16–29

16–30. A mill in a textile plant uses the belt-and-pulley arrangement shown to transmit power. If an electric motor turns pulley A at 5 rad/s, compute the angular velocities of pulleys B and C. The hub at D is rigidly *connected* to C and turns with it.

16–31. A mill in a textile plant uses the belt-and-pulley arrangement shown to transmit power. When $t = 0$ an electric motor is turning pulley A with an angular velocity of $\omega_A = 5$ rad/s. If this pulley is subjected to a constant angular acceleration 2 rad/s², determine the angular velocity of pulley B after B turns 6 revolutions. The hub at D is rigidly *connected* to pulley C and turns with it.

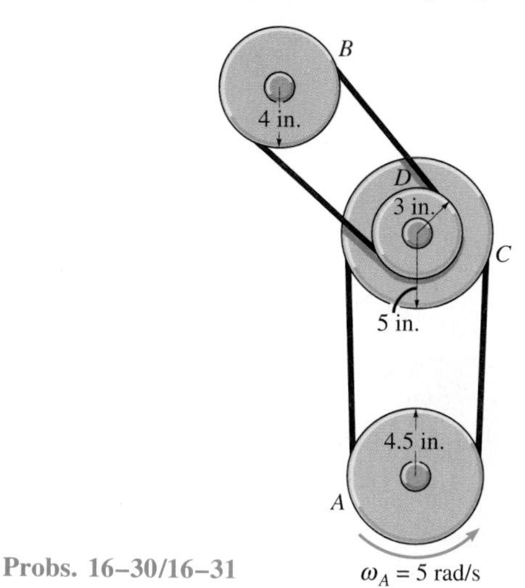

Probs. 16–30/16–31

***16–32.** A belt is wrapped around the inner hubs of pulleys A and B as shown. Each hub is fixed to its respective pulley and turns with it. If A has a constant angular acceleration $\alpha_A = 6\ \text{rad/s}^2$, determine the velocity of block C in 3 s if it starts from rest. How high does it rise in 3 s?

16–33. A belt is wrapped around the inner hubs of pulleys A and B as shown. Each hub is fixed to its respective pulley and turns with it. If A has an acceleration $\alpha_A = (0.4t)\ \text{rad/s}^2$, where t is in seconds, determine the velocity of block C when $t = 3$ s if it starts from rest. How high does the block rise in 3 s?

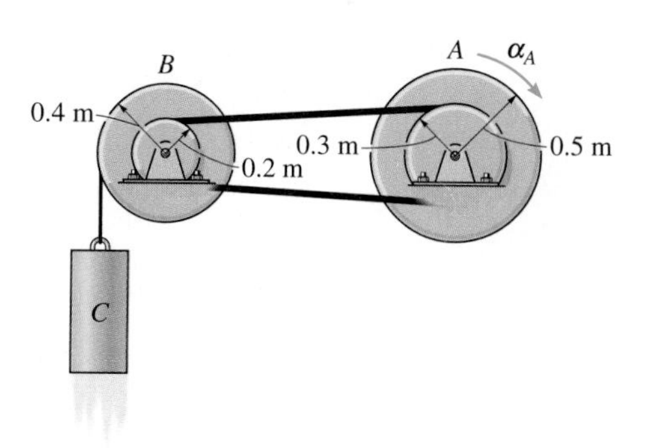

Probs. 16–32/16–33

16–34. The power of a bus engine is transmitted using the belt-and-pulley arrangement shown. If the engine turns pulley A at $\omega_A = 60$ rad/s, determine the angular velocities of the generator pulley B and the air-conditioning pulley C. The hub at D is rigidly *connected* to B and turns with it.

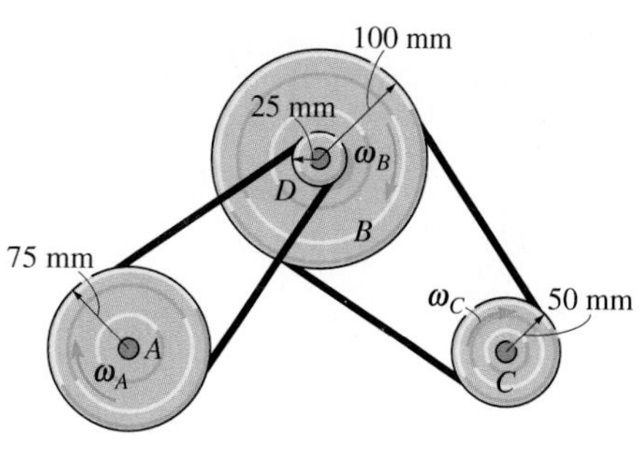

Prob. 16–34

16–35. The power of a bus engine is transmitted using the belt-and-pulley arrangement shown. If the engine turns pulley A at $\omega_A = (20t + 40)$ rad/s, where t is in seconds, determine the angular velocities of the generator pulley B and the air-conditioning pulley C when $t = 3$ s.

16–37. The rod assembly is supported by ball-and-socket joints at A and B. At the instant shown it is rotating about the y axis with an angular velocity $\omega = 5$ rad/s and has an angular acceleration $\alpha = 8$ rad/s^2. Determine the magnitudes of the velocity and acceleration of point C at this instant. Solve the problem using Cartesian vectors and Eqs. 16–9 and 16–13.

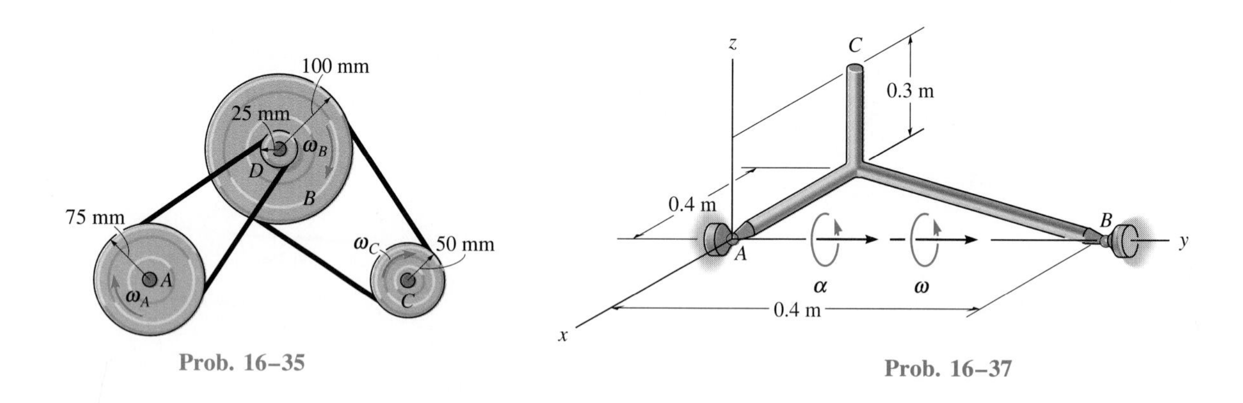

Prob. 16–35

Prob. 16–37

***16–36.** The rope of diameter d is wrapped around the tapered drum which has the dimensions shown. If the drum is rotating at a constant rate of ω, determine the upward acceleration of the block. Neglect the small horizontal displacement of the block.

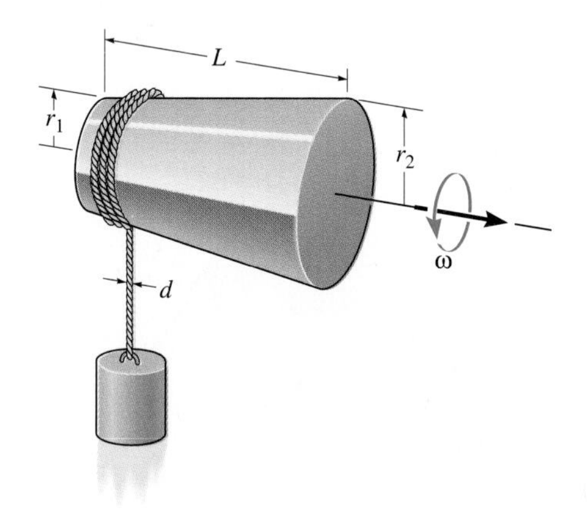

Prob. 16–36

16–38. The board rests on the surface of two drums. At the instant shown, it has an acceleration of 1 ft/s^2, while points located on the periphery of the drums have an acceleration of 3 ft/s^2. If the board does not slip on the surface of the drums, determine the velocity of the board.

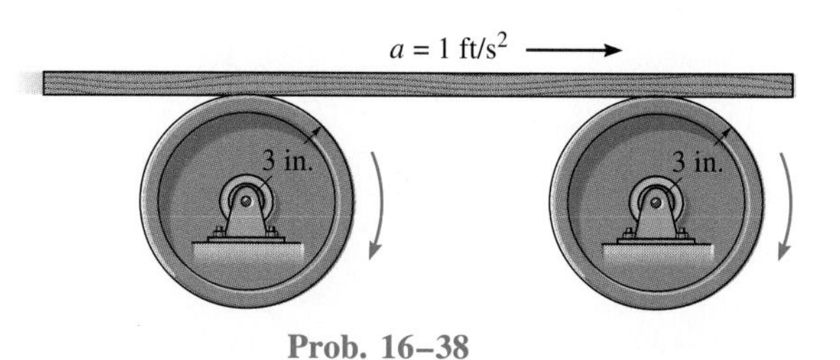

Prob. 16–38

*16.4 Absolute General Plane Motion Analysis

A body subjected to *general plane motion* undergoes a *simultaneous* translation and rotation. If the body is represented by a thin slab, the slab translates in the plane and rotates about an axis perpendicular to the plane. The analysis of this motion can be specified by knowing *both* the angular rotation of a line fixed in the body and the rectilinear motion of a point in the body. One way to define these motions is to use a position coordinate s to specify the location of the point along its path and an angular position coordinate θ to specify the orientation of the line. The two coordinates are then related using the geometry of the problem. By *direct application* of the time-differential equations $v = ds/dt$, $a = dv/dt$, $\omega = d\theta/dt$, and $\alpha = d\omega/dt$, the *motion* of the point and the *angular motion* of the line can then be related. In some cases, this procedure may also be used to relate the motions of one body to those of a connected body.

PROCEDURE FOR ANALYSIS

The following procedure provides a method for relating the velocity and acceleration of a point P undergoing rectilinear motion to the angular velocity and angular acceleration of a line contained in a body.

Position Coordinate Equation. Locate point P using a position coordinate s, which is measured from a *fixed origin* and is *directed along the straight-line path of motion* of point P. Also measure from a fixed reference line the angular position θ of a line lying in the body. From the dimensions of the body, relate s to θ, $s = f(\theta)$, using geometry and/or trigonometry.

Time Derivatives. Take the first derivative of $s = f(\theta)$ with respect to time to get a relationship between v and ω. Take the second time derivative to get a relationship between a and α.

This procedure is illustrated in the following examples.

Example 16–3

At a given instant, the cylinder of radius r, shown in Fig. 16–7, has an angular velocity $\boldsymbol{\omega}$ and angular acceleration $\boldsymbol{\alpha}$. Determine the velocity and acceleration of its center G if the cylinder rolls without slipping.

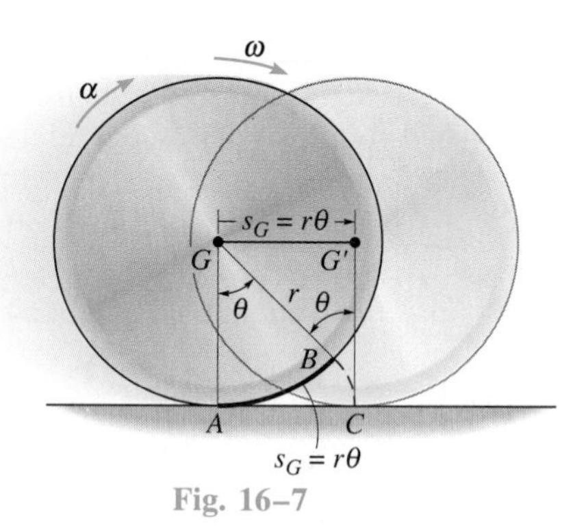

Fig. 16–7

SOLUTION

Position-Coordinate Equation. By inspection, point G moves *horizontally* to the right from G to G' as the cylinder rolls, Fig. 16–7. Consequently its new location G' will be specified by the *horizontal* position coordinate s_G, which is measured from the original position (G) of the cylinder's center. Notice also, that as the cylinder rolls (without slipping), points on its surface contact the ground such that the arc length AB of contact must be equal to the distance s_G. Consequently, motion from the colored to the dashed position requires the radial line GB to rotate θ to the position $G'C$. Since the arc $AB = r\theta$, then G travels a distance

$$s_G = r\theta$$

Time Derivatives. Taking successive time derivatives of this equation, realizing that r is constant, $\omega = d\theta/dt$, and $\alpha = d\omega/dt$, gives the necessary relationships:

$$s_G = r\theta$$

$$v_G = r\omega \qquad \textit{Ans.}$$

$$a_G = r\alpha \qquad \textit{Ans.}$$

Remember that these relationships are valid only if the cylinder (disk, wheel, ball, etc.) rolls *without* slipping.

Example 16–4

The end of rod R shown in Fig. 16–8 maintains contact with the cam by means of a spring. If the cam rotates about an axis through point O with an angular acceleration $\boldsymbol{\alpha}$ and angular velocity $\boldsymbol{\omega}$, compute the velocity and acceleration of the rod when the cam is in the arbitrary position θ.

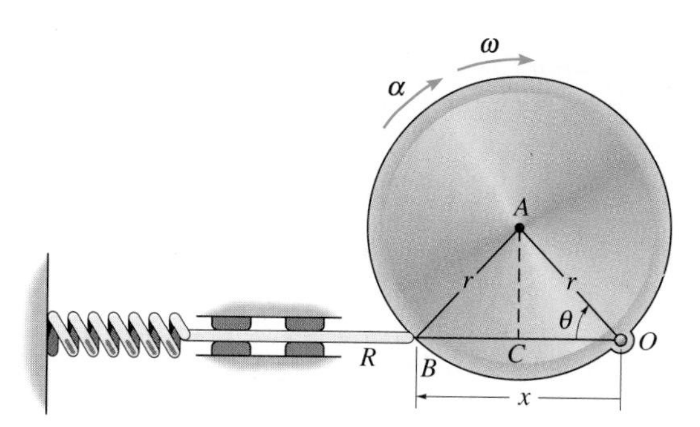

Fig. 16–8

SOLUTION

Position Coordinate Equation. Coordinates x and θ are chosen for the analysis in order to express the *rotational motion* of the cam, which is defined by the angular motion of line OA, i.e., $\omega = d\theta/dt$, and the *rectilinear motion* of the rod (or horizontal component of the motion of point B), i.e., $v = dx/dt$. These coordinates are measured from the *fixed point O* and may be related to each other using trigonometry. Since $OC = CB = r\cos\theta$, Fig. 16–8, then

$$x = 2r\cos\theta$$

Time Derivatives. Using the chain rule of calculus, we have

$$\frac{dx}{dt} = -2r\sin\theta\frac{d\theta}{dt}$$

$$v = -2r\,\omega\sin\theta \qquad \textit{Ans.}$$

$$\frac{dv}{dt} = -2r\left(\frac{d\omega}{dt}\right)\sin\theta - 2r\omega\left(\cos\theta\frac{d\theta}{dt}\right)$$

$$a = -2r(\alpha\sin\theta + \omega^2\cos\theta) \qquad \textit{Ans.}$$

The negative signs indicate that v and a are opposite to the direction of positive x.

Example 16–5

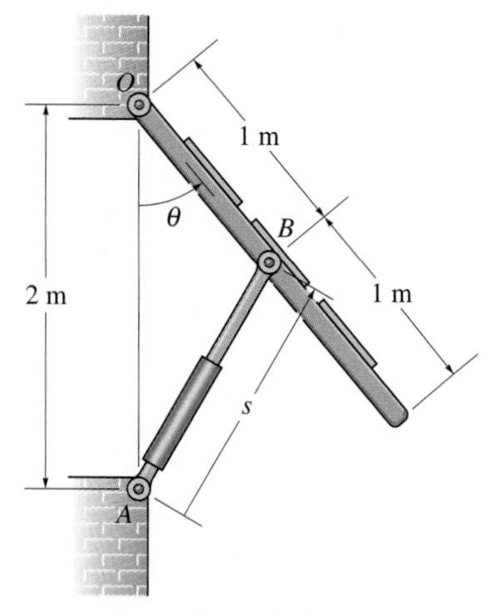

Fig. 16–9

A large window is opened using a hydraulic cylinder AB. If the cylinder extends at a constant rate of 0.5 m/s, determine the angular velocity and angular acceleration of the window at the instant $\theta = 30°$.

SOLUTION

Position Coordinates Equation. The angular motion of the window can be obtained using the coordinate θ, whereas the extension or motion *along the hydraulic cylinder* is defined using a coordinate s, which measures the length from the fixed point A to the moving point B. These coordinates can be related using the law of cosines, namely,

$$s^2 = (2\text{ m})^2 + (1\text{ m})^2 - 2(2\text{ m})(1\text{ m})\cos\theta$$

$$s^2 = 5 - 4\cos\theta \qquad (1)$$

When $\theta = 30°$,

$$s = 1.239\text{ m}$$

Time Derivatives. Taking the time derivatives of Eq. (1), we have

$$2s\frac{ds}{dt} = 0 - 4(-\sin\theta)\frac{d\theta}{dt}$$

$$s(v_s) = 2\sin\theta\omega \qquad (2)$$

Since $v_s = 0.5$ m/s, then at $\theta = 30°$,

$$(1.239\text{ m})(0.5\text{ m/s}) = 2\sin 30°\omega$$

$$\omega = 0.620\text{ rad/s} \qquad \textit{Ans.}$$

Taking the time derivative of Eq. (2) yields

$$\frac{ds}{dt}v_s + s\frac{dv_s}{dt} = 2\left(\cos\theta\frac{d\theta}{dt}\right)\omega + 2\sin\theta\frac{d\omega}{dt}$$

$$v_s^2 + sa_s = 2\cos\theta\omega^2 + 2\sin\theta\alpha$$

Since $a_s = dv_s/dt = 0$, then

$$(0.5\text{ m/s})^2 + 0 = 2\cos 30°(0.620\text{ rad/s})^2 + 2\sin 30°\alpha$$

$$\alpha = -0.415\text{ rad/s}^2 \qquad \textit{Ans.}$$

Since the result is negative, it indicates the window has an angular deceleration.

PROBLEMS

16–39. The bar DC rotates uniformly about the shaft at D with a constant angular velocity $\boldsymbol{\omega}$. Determine the velocity and acceleration of the bar AB, which is confined by the guides to move vertically.

***16–40.** Solve Prob. 16–39 if the bar also has an angular acceleration α.

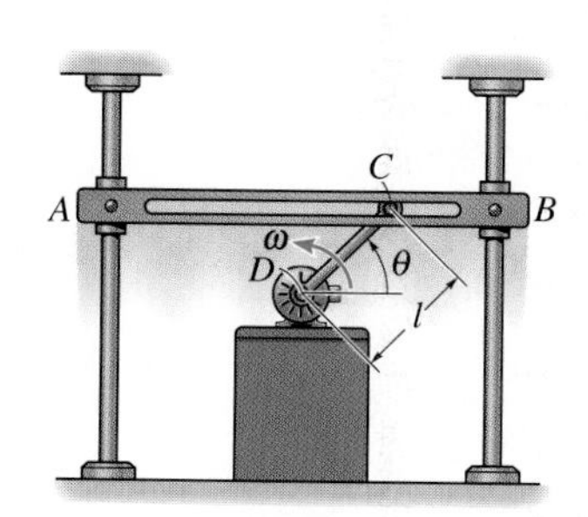

Probs. 16–39/16–40

16–41. The *Scotch yoke* is used to convert the constant circular motion of crank OA into translating motion of rod BC. If OA is rotating with a constant angular velocity $\omega = 5$ rad/s, determine the velocity and acceleration of BC for any angle θ of the crank.

16–42. The *Scotch yoke* is used to convert the constant circular motion of crank OA into translating motion of rod BC. If OA is rotating with an angular acceleration $\alpha = 2$ rad/s^2 and has an angular velocity $\omega = 5$ rad/s when $\theta = 30°$, determine the velocity and acceleration of BC at this instant.

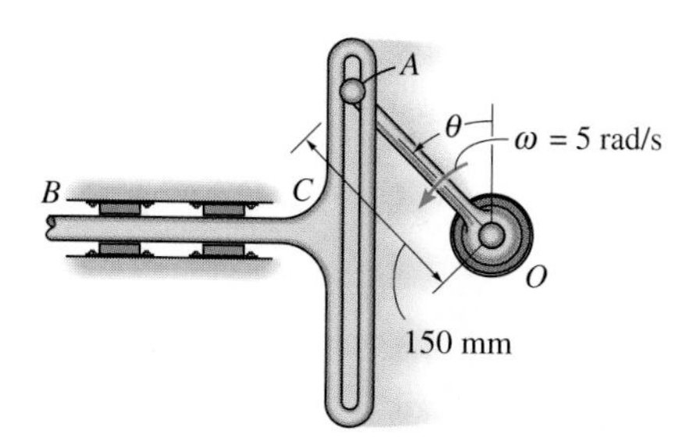

Probs. 16–41/16–42

16–43. The mechanism is used to convert the constant circular motion ω of rod AB into translating motion of rod CD. Determine the velocity and acceleration of CD for any angle θ of AB.

***16–44.** Solve Prob. 16–43 if the rod also has an angular acceleration α.

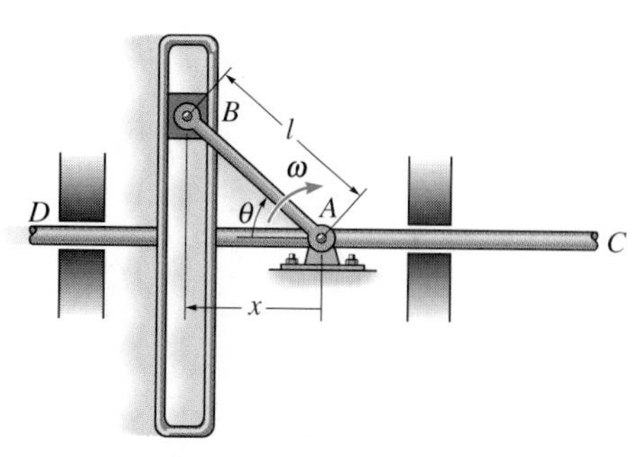

Probs. 16–43/16–44

16–45. At the instant $\theta = 50°$, the slotted guide is moving upward with an acceleration of 3 m/s^2 and a velocity of 2 m/s. Determine the angular acceleration and angular velocity of link AB at this instant. *Note:* The upward motion of the guide is in the negative y direction.

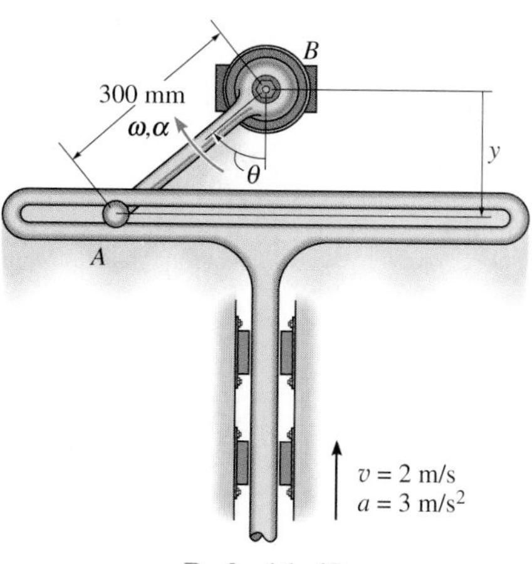

Prob. 16–45

16–46. The block moves to the left with a constant velocity $\mathbf{v}_0$. Determine the angular velocity and angular acceleration of the bar as a function of θ.

16–47. The block moves to the left with an acceleration $\mathbf{a}_0$ and velocity $\mathbf{v}$. Determine the angular velocity and angular acceleration of the bar as a function of θ.

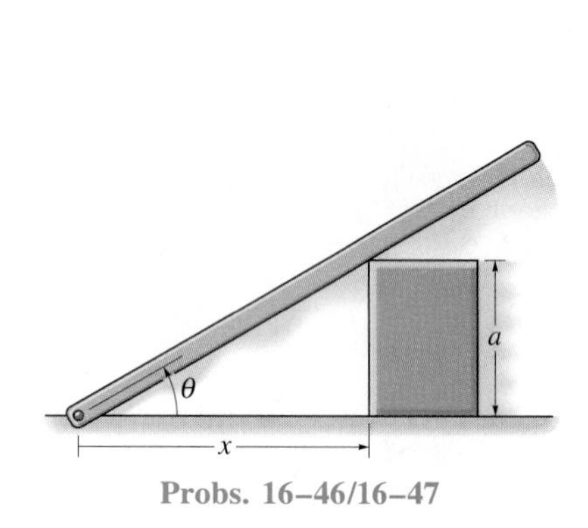

Probs. 16–46/16–47

***16–48.** The inclined plate moves to the left with a constant velocity $\mathbf{v}$. Determine the angular velocity and angular acceleration of the slender rod of length l. The rod pivots about the step at C as it slides on the plate.

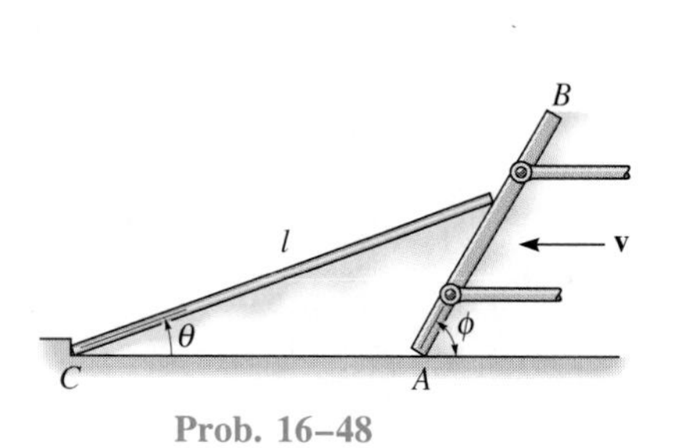

Prob. 16–48

16–49. Bar AB rotates uniformly about the fixed pin A with a constant angular velocity $\boldsymbol{\omega}$. Determine the velocity and acceleration of block C, at the instant $\theta = 60°$.

16–50. When $\theta = 60°$ bar AB is rotating uniformly about the fixed pin A with an angular acceleration $\alpha = 2$ rad/s^2 and angular velocity $\omega = 3$ rad/s. Determine the velocity and acceleration of block C at this instant.

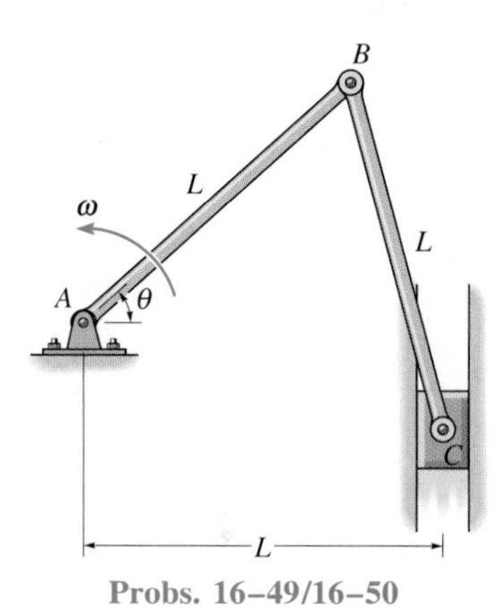

Probs. 16–49/16–50

16–51. The bar is confined to move along the vertical and inclined planes. If the velocity of the roller at A is $v_A = 6$ ft/s when $\theta = 45°$, determine the bar's angular velocity and the velocity of roller B at this instant

***16–52.** The bar is confined to move along the vertical and inclined planes. If the velocity of the roller at A is $v_A = 6$ ft/s, when $\theta = 45°$, determine the bar's angular acceleration and the acceleration of roller B at this instant.

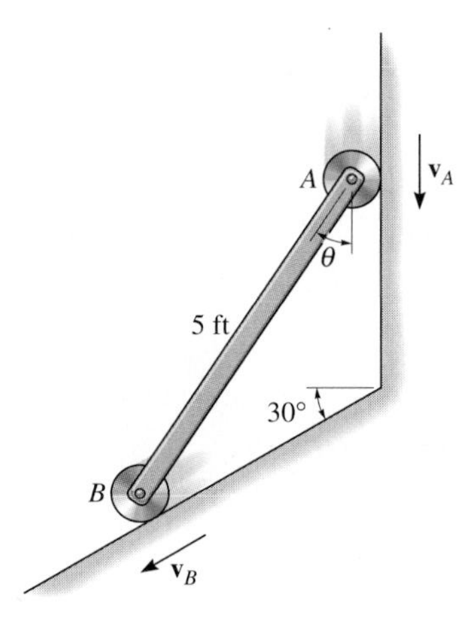

Probs. 16–51/16–52

16–53. The end A of the bar is moving to the left with a constant velocity $\mathbf{v}_A$. Determine the angular velocity ω and angular acceleration α of the bar as a function of its position x.

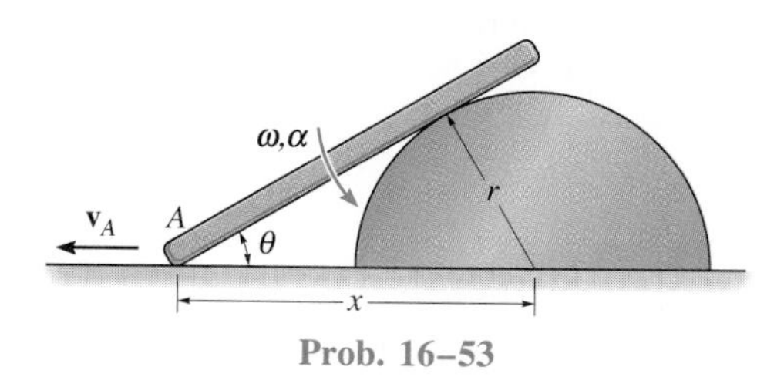

Prob. 16–53

16–54. The slotted yoke is pinned at A while end B is used to move the ram R horizontally. If the disk rotates with a constant angular velocity $\boldsymbol{\omega}$, determine the velocity and acceleration of the ram. The crank pin C is fixed to the disk and turns with it.

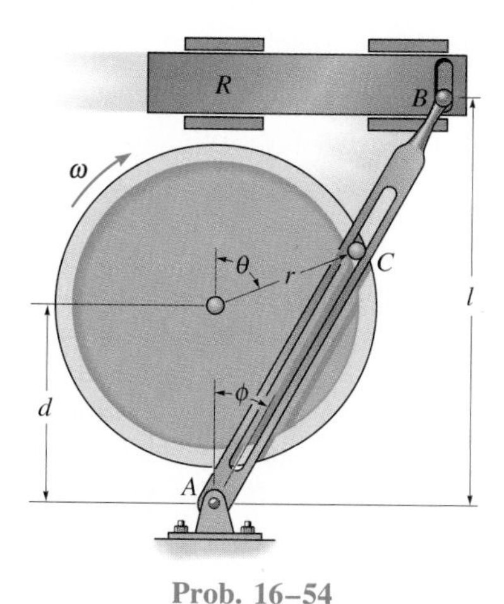

Prob. 16–54

16–55. The circular cam rotates about the fixed point O with a constant angular velocity $\boldsymbol{\omega}$. Determine the velocity v of the follower rod AB as a function of θ.

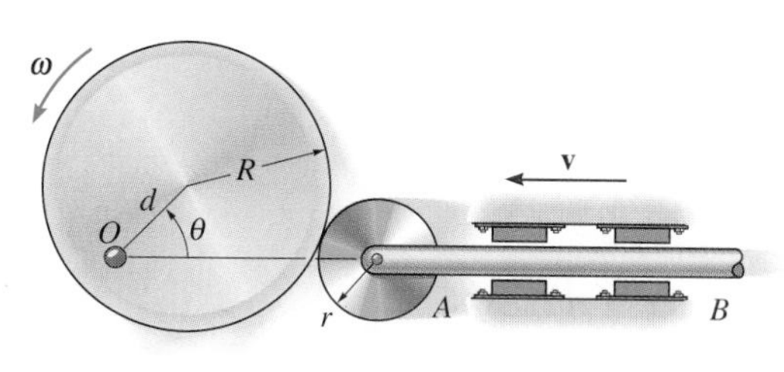

Prob. 16–55

***16–56.** The Geneva wheel A provides intermittent rotary motion $\boldsymbol{\omega}_A$ for continuous motion $\omega_D = 2$ rad/s of disk D. By choosing $d = 100\sqrt{2}$ mm, the wheel has zero angular velocity at the instant pin B enters or leaves one of the four slots. Determine the magnitude of the angular velocity $\boldsymbol{\omega}_A$ of the Geneva wheel at any angle θ for which pin B is in contact with the slot.

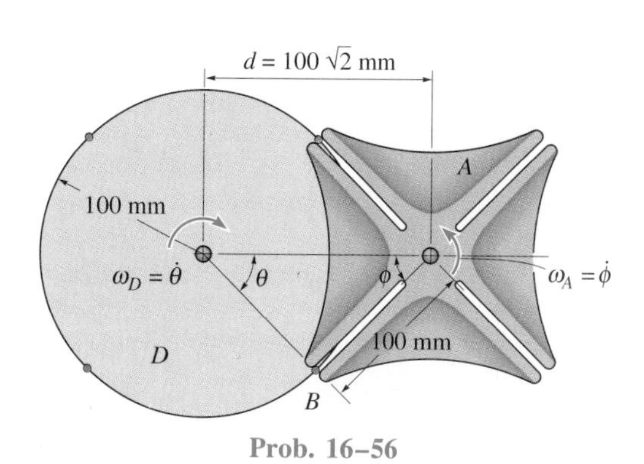

Prob. 16–56

16.5 Relative-Motion Analysis: Velocity

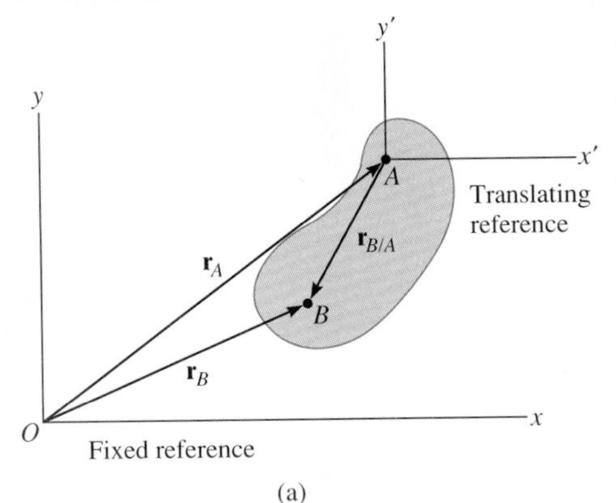

(a)

Time t

Time $t + dt$

General plane motion

(b)

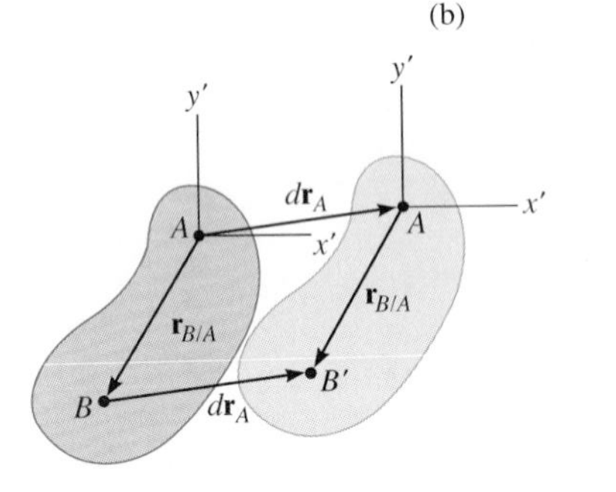

Translation

(c)

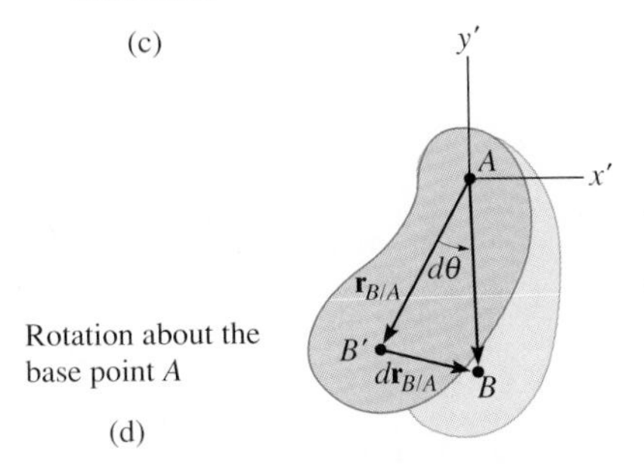

Rotation about the base point A

(d)

Fig. 16–10

It was pointed out in Sec. 16.1 that general plane motion of a rigid body consists of a *combination* of translation and rotation. Oftentimes it is convenient to view these "component" motions *separately* using a *relative-motion analysis*. Two sets of coordinate axes will be used to do this. The x, y coordinate system is fixed and will be used to measure the *absolute* positions, velocities, and accelerations of two points A and B on the body, Fig. 16–10a. The origin of the x', y' coordinate system will be attached to the selected "base point" A, which generally has a *known* motion. The axes of this coordinate system are *not fixed to the body;* rather they will be allowed to *translate* with respect to the fixed frame.

Position. The position vector $\mathbf{r}_A$ in Fig. 16–10a specifies the location of the "base point" A, whereas the relative-position vector $\mathbf{r}_{B/A}$ locates point B on the body with respect to A. By vector addition, the *position* of B can be determined from the equation

$$\mathbf{r}_B = \mathbf{r}_A + \mathbf{r}_{B/A}$$

Displacement. During an instant of time dt, points A and B undergo displacements $d\mathbf{r}_A$ and $d\mathbf{r}_B$ as shown in Fig. 16–10b. As stated above, here we will consider the motion by its component parts. For example, the *entire body* first *translates* by an amount $d\mathbf{r}_A$ so that A, the base point, moves to its *final position* and B moves to B', Fig. 16–10c. The body is then rotated by an amount $d\theta$ about A, so that B' undergoes a *relative displacement* $d\mathbf{r}_{B/A}$ and thus moves to its final position B, Fig. 16–10d. Note that since the body is *rigid*, $\mathbf{r}_{B/A}$ has a fixed magnitude and thus $d\mathbf{r}_{B/A}$ accounts only for a change in the *direction* of $\mathbf{r}_{B/A}$. Due to the rotation about A the magnitude of the relative displacement is $dr_{B/A} = r_{B/A}\,d\theta$ and the displacement of B in Fig. 16–10b is thus

$$d\mathbf{r}_B = d\mathbf{r}_A + d\mathbf{r}_{B/A}$$

- $d\mathbf{r}_{B/A}$: due to rotation about A
- $d\mathbf{r}_A$: due to translation
- $d\mathbf{r}_B$: due to translation and rotation about A

Velocity. To determine the relationship between the velocities of points A and B, it is necessary to take the time derivative of the position equation, or simply divide the displacement equation by dt. This yields

$$\frac{d\mathbf{r}_B}{dt} = \frac{d\mathbf{r}_A}{dt} + \frac{d\mathbf{r}_{B/A}}{dt}$$

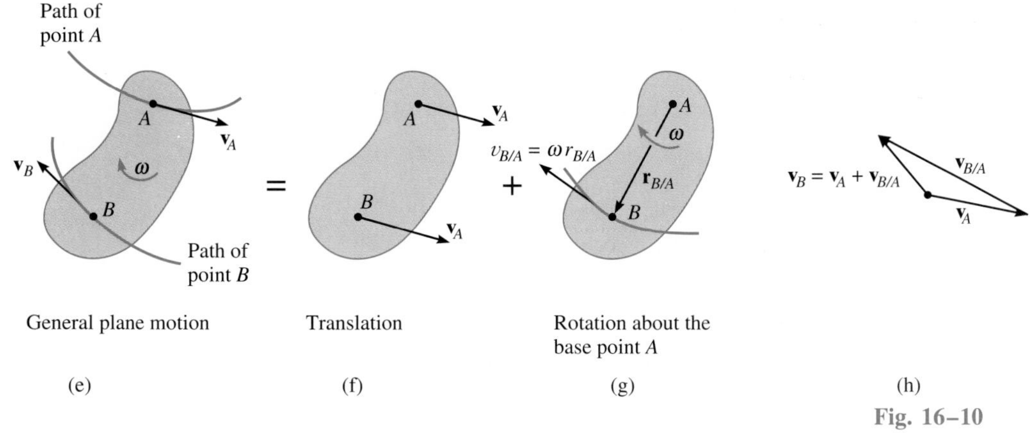

Fig. 16–10

The terms $d\mathbf{r}_B/dt = \mathbf{v}_B$ and $d\mathbf{r}_A/dt = \mathbf{v}_A$ are measured from the fixed x, y axes and represent the *absolute velocities* of points A and B, respectively. The magnitude of the third term is $d(r_{B/A}\theta)/dt = r_{B/A}\dot{\theta} = r_{B/A}\omega$, where ω is the angular velocity of the body at the instant considered. Note that $\boldsymbol{\omega}$ is *absolute,* since it is the *same* when measured from either the fixed x, y or translating x', y' axes. We will denote this third term as the *relative velocity* $\mathbf{v}_{B/A}$, since it represents the velocity of B with respect to A as measured by an observer fixed to the translating x', y' axes. Since the body is rigid, it is important to realize that this observer sees point B move along a *circular arc* that has a radius of curvature $r_{B/A}$. In other words, *the body appears to move as if it were rotating with an angular velocity* $\boldsymbol{\omega}$ *about the* z' *axis passing through* A. Consequently, $\mathbf{v}_{B/A}$ has a magnitude of $v_{B/A} = \omega r_{B/A}$ and a *direction* which is perpendicular to $r_{B/A}$. We therefore have

$$\mathbf{v}_B = \mathbf{v}_A + \mathbf{v}_{B/A} \tag{16–15}$$

where

$\mathbf{v}_B$ = velocity of point B

$\mathbf{v}_A$ = velocity of the base point A

$\mathbf{v}_{B/A}$ = relative velocity of "B with respect to A."
Since the relative motion is *circular,* the *magnitude* of $\mathbf{v}_{B/A}$ is $v_{B/A} = \omega r_{B/A}$ and the *direction* is perpendicular to $\mathbf{r}_{B/A}$.

Each of the three terms in Eq. 16–15 is represented graphically on the *kinematic diagrams* in Figs. 16–10*e*, 16–10*f*, and 16–10*g*. Here it is seen that at a given instant the velocity of B, Fig. 16–10*e*, is determined by considering the entire body to translate with a velocity of $\mathbf{v}_A$, Fig. 16–10*f*, and then rotate about the base point A with an angular velocity $\boldsymbol{\omega}$**,** Fig. 16–10*g*. Vector addition of these two effects, applied to B, yields $\mathbf{v}_B$, as shown in Fig. 16–10*h*.

Since the relative velocity $\mathbf{v}_{B/A}$ represents the effect of *circular motion,* observed from translating axes having their origin at the base point A, this term can be expressed by the cross product $\mathbf{v}_{B/A} = \boldsymbol{\omega} \times \mathbf{r}_{B/A}$, Eq. 16–9. Hence, for application, we can also write Eq. 16–15 as

$$\mathbf{v}_B = \mathbf{v}_A + \boldsymbol{\omega} \times \mathbf{r}_{B/A} \tag{16–16}$$

where

$\mathbf{v}_B$ = velocity of B

$\mathbf{v}_A$ = velocity of the base point A

$\boldsymbol{\omega}$ = angular velocity of the body

$\mathbf{r}_{B/A}$ = relative-position vector drawn from A to B

The velocity equation 16–15 or 16–16 may be used in a practical manner to study the planar motion of a rigid body which is either pin-connected to or in contact with other moving bodies. To obtain the necessary data when applying this equation, points A and B should generally be selected at joints which are pin-connected, or at points in contact with adjacent bodies which have a *known motion.* For example, both points A and B on link AB, Fig. 16–11*a*, have circular paths of motion. Hence, at the instant shown, the magnitudes of their velocities are determined by the angular motion of the wheel and link BC so that $v_A = \omega_A r$ and $v_B = \omega_{BC} l$. The *directions* of $\mathbf{v}_A$ and $\mathbf{v}_B$ are always *tangent* to their paths of motion, Fig. 16–11*b*. In the case of the wheel in Fig. 16–12, which rolls without slipping, point A can be selected at the ground. Here A momentarily has zero velocity since the ground does not move. Furthermore, the center of the wheel, B, moves along a horizontal path so that $\mathbf{v}_B$ is horizontal.

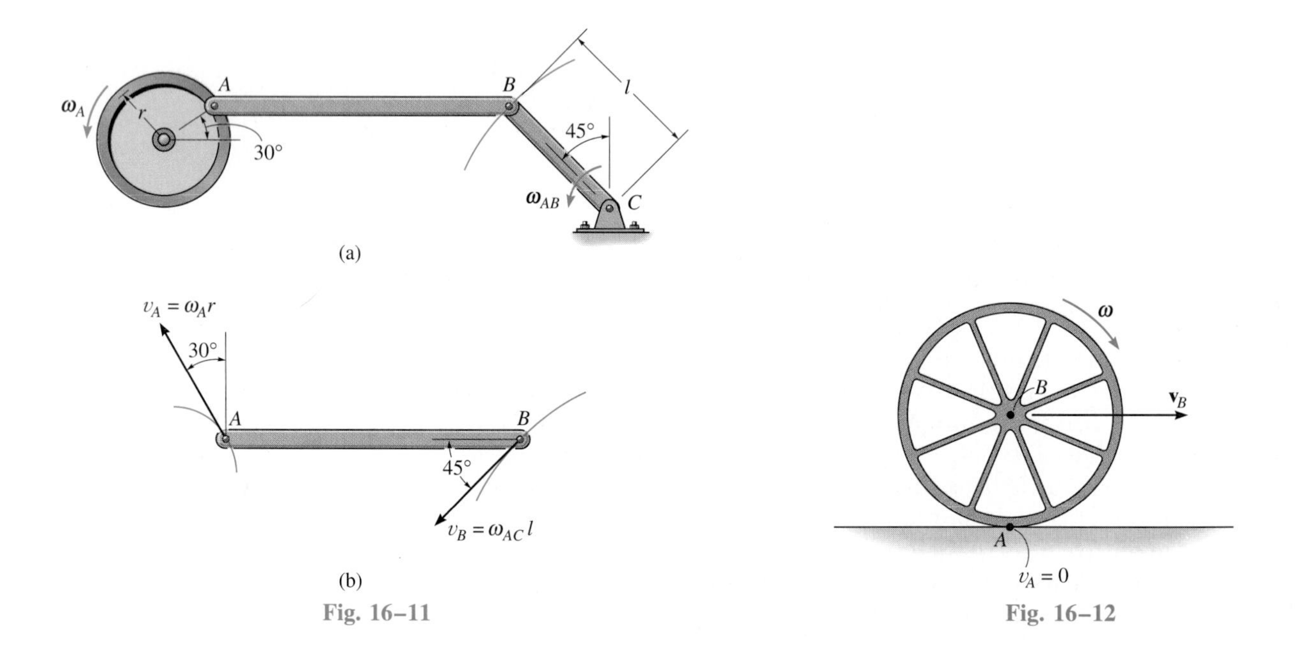

Fig. 16–11

Fig. 16–12

PROCEDURE FOR ANALYSIS

Equation 16–15 relates the velocities of any two points A and B located on the *same* rigid body. It can be applied either by using Cartesian vector analysis, in the form of Eq. 16–16, or by writing the x and y scalar component equations directly. For a solution, there can be at most two unknowns. For application, it is suggested that the following procedure be used.

Kinematic Diagram. Establish the directions of the fixed x, y coordinates, and draw a kinematic diagram of the body. Indicate on it the velocities $\mathbf{v}_A$, $\mathbf{v}_B$ of points A and B, the angular velocity $\boldsymbol{\omega}$, and the relative-position vector $\mathbf{r}_{B/A}$. Normally the base point A is selected as a point having a known velocity. Identify the *two* unknowns on the diagram. If the magnitudes of $\mathbf{v}_A$, $\mathbf{v}_B$, or $\boldsymbol{\omega}$ are unknown, the sense of direction of these vectors may be assumed.

If the velocity equation is to be applied *without* using a Cartesian vector analysis, then the magnitude and direction of the relative velocity $\mathbf{v}_{B/A}$ must be established. This can be done by drawing a kinematic diagram such as shown in Fig. 16–10g, which shows the relative motion. Since the body is considered to be "pinned" momentarily at the base point A, the *magnitude* of $\mathbf{v}_{B/A}$ is $v_{B/A} = \omega r_{B/A}$. The *sense of direction* is established from the diagram, such that $\mathbf{v}_{B/A}$ acts perpendicular to $\mathbf{r}_{B/A}$ in accordance with the rotational motion $\boldsymbol{\omega}$ of the body.*

Velocity Equation. To apply Eq. 16–16, express the vectors in Cartesian vector form and substitute them into the equation. Evaluate the cross product and then equate the respective $\mathbf{i}$ and $\mathbf{j}$ components to obtain two scalar equations. To obtain these scalar equations *directly,* write Eq. 16–15 in symbolic form, $\mathbf{v}_B = \mathbf{v}_A + \mathbf{v}_{B/A}$, and underneath each of the terms represent the vectors by their magnitudes and directions. To do this, use the data tabulated on the kinematic diagrams. The scalar equations are determined from the x and y components of each of the vectors. Solve these equations for the two unknowns. If the solution yields a *negative* answer for an *unknown,* it indicates the sense of direction of the vector is opposite to that shown on the kinematic diagram.

Realize that once the velocity of a single point on the body and the body's angular velocity are *known,* the velocity of any other point on the body can then be determined by applying the relative-velocity equation between the two points.

*The notation $\mathbf{v}_B = \mathbf{v}_A + \mathbf{v}_{B/A(\text{pin})}$ may be helpful in recalling that A is "pinned."

The following example problems illustrate both methods of application numerically.

Example 16–6

The link shown in Fig. 16–13*a* is guided by two blocks at *A* and *B*, which move in the fixed slots. If the velocity of *A* is 2 m/s downward, determine the velocity of *B* at the instant $\theta = 45°$.

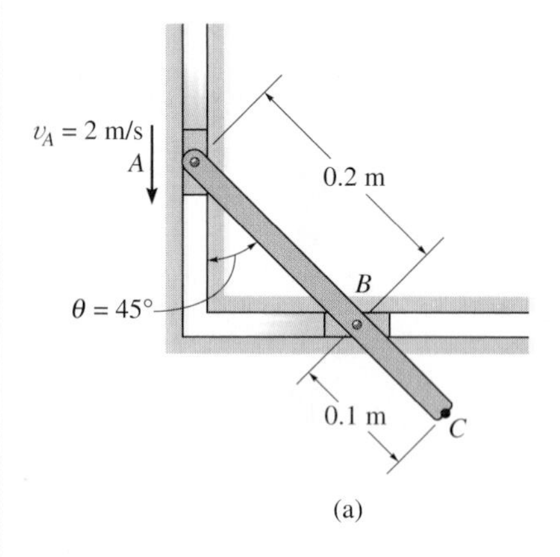

(a)

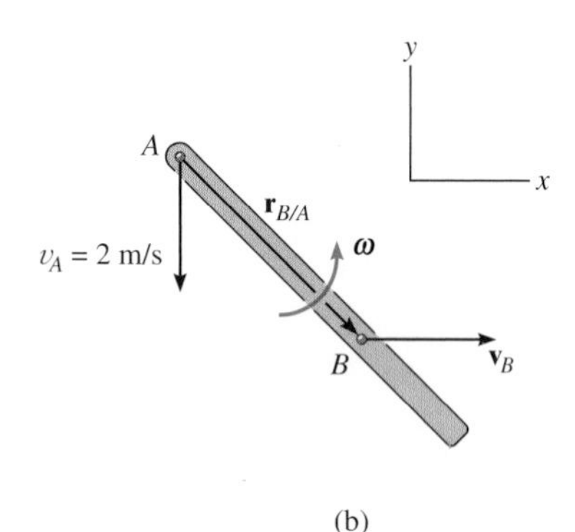

(b)

Fig. 16–13

SOLUTION *(VECTOR ANALYSIS)*

Kinematic Diagram. Since points *A* and *B* are restricted to move along the fixed slots and $\mathbf{v}_A$ is directed downward, the velocity $\mathbf{v}_B$ must be directed horizontally to the right, Fig. 16–13*b*. This motion causes the link to rotate counterclockwise; that is, by the right-hand rule the angular velocity $\boldsymbol{\omega}$ is directed outward, perpendicular to the plane of motion. Knowing the magnitude and direction of $\mathbf{v}_A$ and the lines of action of $\mathbf{v}_B$ and $\boldsymbol{\omega}$, it is possible to apply the velocity equation $\mathbf{v}_B = \mathbf{v}_A + \boldsymbol{\omega} \times \mathbf{r}_{B/A}$ to points *A* and *B* in order to solve for the two unknown magnitudes v_B and ω. Since $\mathbf{r}_{B/A}$ is needed, it is also shown in Fig. 16–13*b*.

Velocity Equation. Expressing each of the vectors in Fig. 16–13*b* in terms of their **i, j,** and **k** components and applying Eq. 16–16 to *A*, the base point, and *B*, we have

$$\mathbf{v}_B = \mathbf{v}_A + \boldsymbol{\omega} \times \mathbf{r}_{B/A}$$

$$v_B\mathbf{i} = -2\mathbf{j} + [\omega\mathbf{k} \times (0.2 \sin 45°\mathbf{i} - 0.2 \cos 45°\mathbf{j})]$$

or

$$v_B\mathbf{i} = -2\mathbf{j} + 0.2\omega \sin 45°\mathbf{j} + 0.2\omega \cos 45°\mathbf{i}$$

Equating the **i** and **j** components gives

$$v_B = 0.2\omega \cos 45° \qquad 0 = -2 + 0.2\omega \sin 45°$$

Thus,

$$\omega = 14.1 \text{ rad/s} \circlearrowleft$$

$$v_B = 2 \text{ m/s} \rightarrow \qquad \textit{Ans.}$$

Since both results are *positive,* the *directions* of $\mathbf{v}_B$ and $\boldsymbol{\omega}$ are indeed *correct* as shown in Fig. 16–13*b*. It should be emphasized that these results are *valid only* at the instant $\theta = 45°$. A recalculation for $\theta = 44°$ yields $v_B = 2.07$ m/s and $\omega = 14.4$ rad/s; whereas when $\theta = 46°$, $v_B = 1.93$ m/s and $\omega = 13.9$ rad/s, etc.

Now that the velocity of a point (*A*) on the link and the angular velocity are *known,* the velocity of any other point on the link can be determined. As an exercise, see if you can apply Eq. 16–16 to points *A* and *C* and show that $v_C = 3.16$ m/s, directed at $\theta = 18.4°$ up from the horizontal.

Example 16–7

The cylinder shown in Fig. 16–14*a* rolls freely on the surface of a conveyor belt which is moving at 2 ft/s. Assuming that no slipping occurs between the cylinder and the belt, determine the velocity of point *A*. The cylinder has a clockwise angular velocity $\omega = 15$ rad/s at the instant shown.

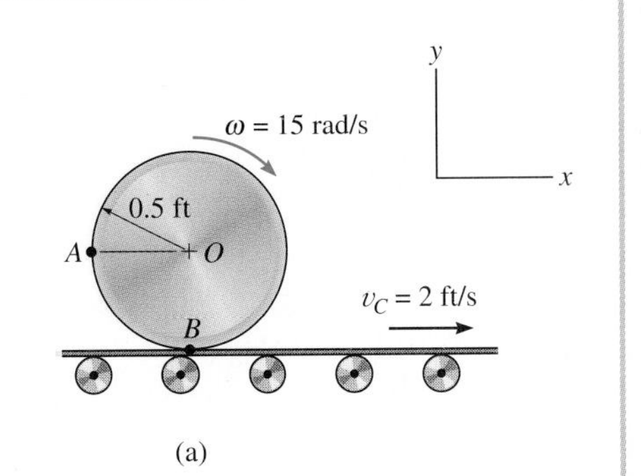

(a)

SOLUTION I (*VECTOR ANALYSIS*)

Kinematic Diagram. Since no slipping occurs, point *B* on the cylinder has the same velocity as the conveyor, Fig. 16–14*b*. Also, the angular velocity of the cylinder is known, so we can apply the velocity equation to *B*, the base point, and *A* to determine $\mathbf{v}_A$.

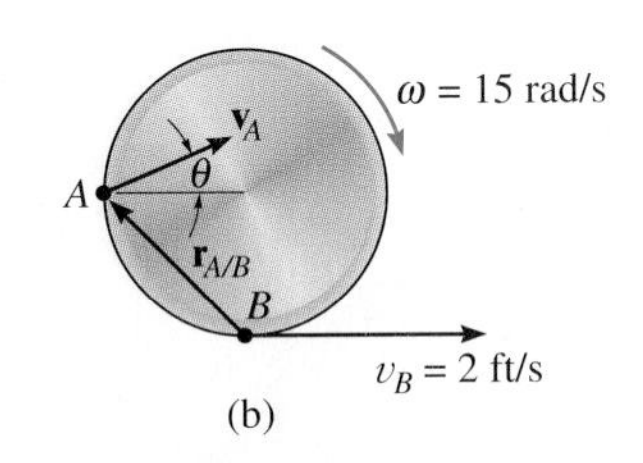

(b)

Velocity Equation

$$\mathbf{v}_A = \mathbf{v}_B + \boldsymbol{\omega} \times \mathbf{r}_{A/B}$$
$$(v_A)_x\mathbf{i} + (v_A)_y\mathbf{j} = 2\mathbf{i} + (-15\mathbf{k}) \times (-0.5\mathbf{i} + 0.5\mathbf{j})$$
$$(v_A)_x\mathbf{i} + (v_A)_y\mathbf{j} = 2\mathbf{i} + 7.50\mathbf{j} + 7.50\mathbf{i}$$

so that

$$(v_A)_x = 2 + 7.50 = 9.50 \text{ ft/s} \quad (1)$$
$$(v_A)_y = 7.50 \text{ ft/s} \quad (2)$$

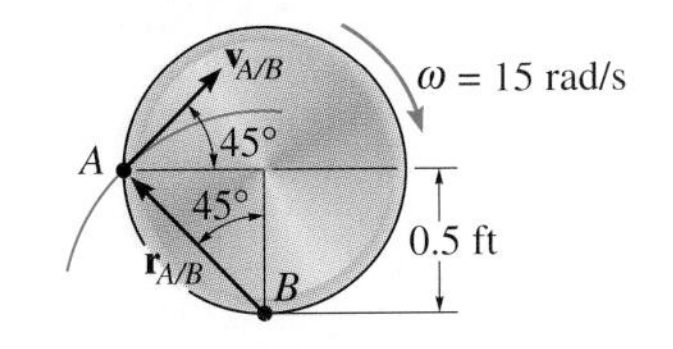

Relative motion

(c)

Fig. 16–14

Thus,

$$v_A = \sqrt{(9.50)^2 + (7.50)^2} = 12.1 \text{ ft/s} \qquad \textit{Ans.}$$
$$\theta = \tan^{-1}\frac{7.50}{9.50} = 38.3^\circ \ \angle^{\mathbf{v}_A}\theta \qquad \textit{Ans.}$$

SOLUTION II (*SCALAR COMPONENTS*)

As an alternative procedure, the scalar components of $\mathbf{v}_A = \mathbf{v}_B + \mathbf{v}_{A/B}$ can be obtained directly. From the kinematic diagram showing the relative "circular" motion $\mathbf{v}_{A/B}$, Fig. 16–14*c*, we have

$$v_{A/B} = \omega r_{A/B} = (15 \text{ rad/s})\left(\frac{0.5 \text{ ft}}{\cos 45^\circ}\right) = 10.6 \text{ ft/s} \ \angle^{45^\circ}$$

Thus

$$v_A = v_B + v_{A/B}$$
$$\begin{bmatrix}(v_A)_x\\ \rightarrow\end{bmatrix} + \begin{bmatrix}(v_A)_y\\ \uparrow\end{bmatrix} = \begin{bmatrix}2 \text{ ft/s}\\ \rightarrow\end{bmatrix} + \begin{bmatrix}10.6 \text{ ft/s}\\ \angle^{45^\circ}\end{bmatrix}$$

Equating the *x* and *y* components gives the same results as before, namely,

$(\overset{+}{\rightarrow})$ $\qquad (v_A)_x = 2 + 10.6 \cos 45^\circ = 9.50$ ft/s

$(+\uparrow)$ $\qquad (v_A)_y = 0 + 10.6 \sin 45^\circ = 7.50$ ft/s

Example 16–8

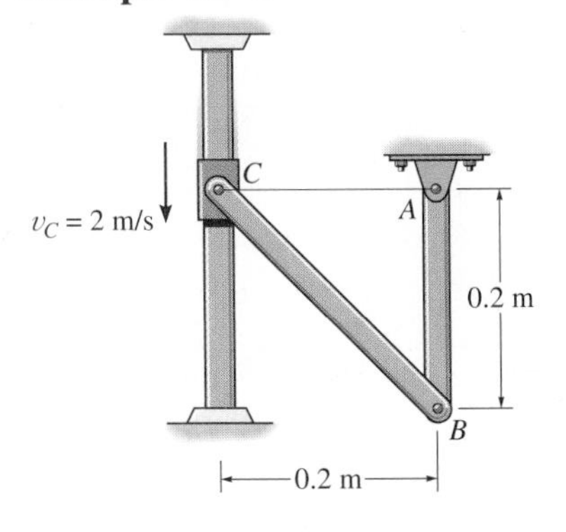

(a)

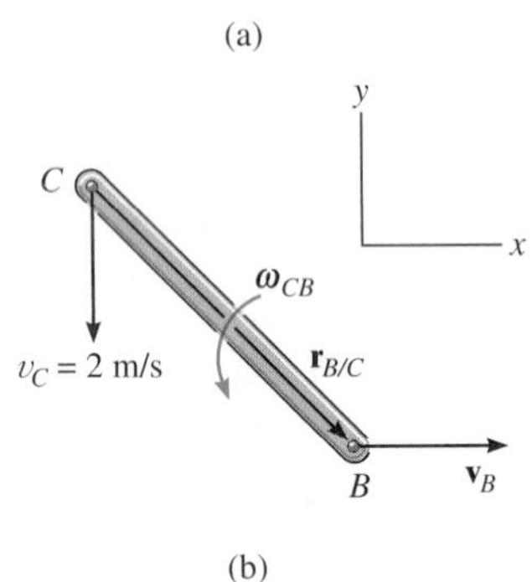

(b)

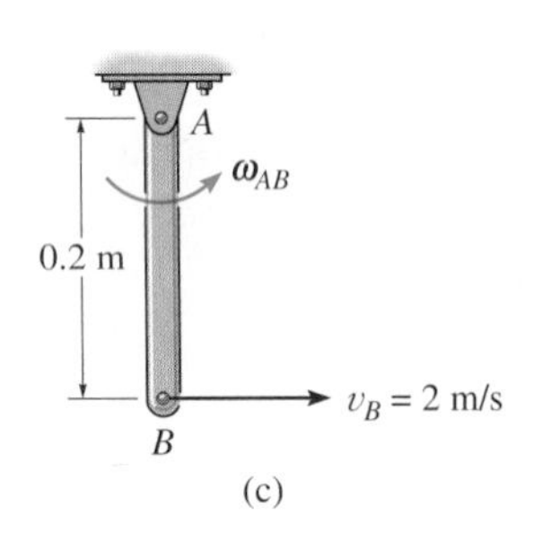

(c)

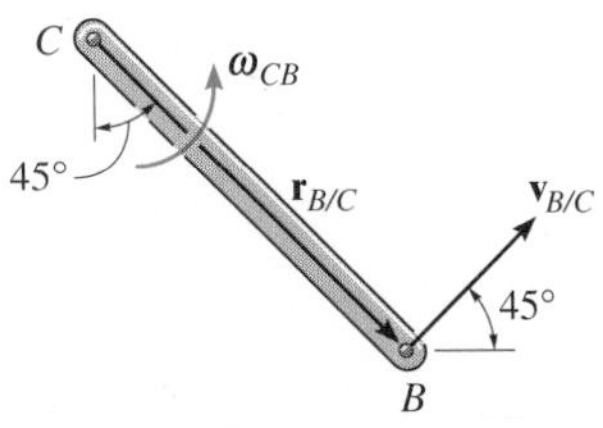

Relative motion

(d)

Fig. 16–15

The collar C in Fig. 16–15a is moving downward with a velocity of 2 m/s. Determine the angular velocity of CB and AB at this instant.

SOLUTION I (*VECTOR ANALYSIS*)

Kinematic Diagram. The downward motion of C causes B to move to the right. Also, CB and AB rotate counterclockwise. To solve, we will write the appropriate kinematic equation for each link.

Velocity Equation

Link CB (general plane motion): See Fig. 16–15b.

$$\mathbf{v}_B = \mathbf{v}_C + \boldsymbol{\omega}_{CB} \times \mathbf{r}_{B/C}$$
$$v_B\mathbf{i} = -2\mathbf{j} + \omega_{CB}\mathbf{k} \times (0.2\mathbf{i} - 0.2\mathbf{j})$$
$$v_B\mathbf{i} = -2\mathbf{j} + 0.2\omega_{CB}\mathbf{j} + 0.2\omega_{CB}\mathbf{i}$$

$$v_B = 0.2\omega_{CB} \qquad (1)$$
$$0 = -2 + 0.2\omega_{CB} \qquad (2)$$

$$\omega_{CB} = 10 \text{ rad/s} \circlearrowleft \qquad \textit{Ans.}$$
$$v_B = 2 \text{ m/s} \rightarrow$$

Link AB (rotation about a fixed axis): See Fig. 16–15c.

$$\mathbf{v}_B = \boldsymbol{\omega}_{AB} \times \mathbf{r}_B$$
$$2\mathbf{i} = \omega_{AB}\mathbf{k} \times (-0.2\mathbf{j})$$
$$2 = 0.2\omega_{AB}$$
$$\omega_{AB} = 10 \text{ rad/s} \circlearrowleft \qquad \textit{Ans.}$$

SOLUTION II (*SCALAR COMPONENTS*)

The scalar component equations of $\mathbf{v}_B = \mathbf{v}_C + \mathbf{v}_{B/C}$ can be obtained directly. The kinematic diagram in Fig. 16–15d shows the relative "circular" motion $\mathbf{v}_{B/C}$. We have

$$\mathbf{v}_B = \mathbf{v}_C + \mathbf{v}_{B/C}$$
$$\begin{bmatrix} v_B \\ \rightarrow \end{bmatrix} = \begin{bmatrix} 2 \text{ m/s} \\ \downarrow \end{bmatrix} + \begin{bmatrix} \omega_{CB}(0.2\sqrt{2} \text{ m}) \\ \measuredangle 45^\circ \end{bmatrix}$$

Resolving these vectors in the x and y directions yields

$$(\overset{+}{\rightarrow}) \qquad v_B = 0 + \omega_{CB}(0.2\sqrt{2}\cos 45^\circ)$$
$$(+\uparrow) \qquad 0 = -2 + \omega_{CB}(0.2\sqrt{2}\sin 45^\circ)$$

which is the same as Eqs. 1 and 2.

Example 16–9

The bar AB of the linkage shown in Fig. 16–16a has a clockwise angular velocity of 30 rad/s when $\theta = 60°$. Compute the angular velocities of member BC and the wheel at this instant.

SOLUTION *(VECTOR ANALYSIS)*

Kinematic Diagrams. By inspection, the velocities of points B and C are defined by the rotation of link AB and the wheel about their fixed axes. The position vectors and the angular velocity of each member are shown on the kinematic diagram in Fig. 16–16b. To solve, we will write the appropriate kinematic equation for each member.

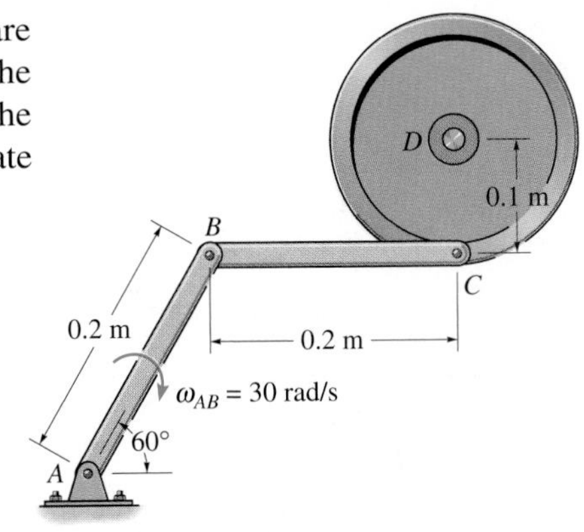

(a)

Velocity Equation

Link AB (rotation about a fixed axis):

$$\mathbf{v}_B = \boldsymbol{\omega}_{AB} \times \mathbf{r}_B$$
$$= (-30\mathbf{k}) \times (0.2 \cos 60°\mathbf{i} + 0.2 \sin 60°\mathbf{j})$$
$$= \{5.20\mathbf{i} - 3.0\mathbf{j}\} \text{ m/s}$$

Link BC (general plane motion):

$$\mathbf{v}_C = \mathbf{v}_B + \boldsymbol{\omega}_{BC} \times \mathbf{r}_{C/B}$$
$$v_C\mathbf{i} = 5.20\mathbf{i} - 3.0\mathbf{j} + (\omega_{BC}\mathbf{k}) \times (0.2\mathbf{i})$$
$$v_C\mathbf{i} = 5.20\mathbf{i} + (0.2\omega_{BC} - 3.0)\mathbf{j}$$
$$v_C = 5.20 \text{ m/s}$$
$$0 = 0.2\omega_{BC} - 3.0$$
$$\omega_{BC} = 15 \text{ rad/s} \circlearrowleft \qquad \textit{Ans.}$$

Wheel (rotation about a fixed axis):

$$\mathbf{v}_C = \boldsymbol{\omega}_D \times \mathbf{r}_C$$
$$5.20\mathbf{i} = (\omega_D\mathbf{k}) \times (-0.1\mathbf{j})$$
$$5.20 = 0.1\omega_D$$
$$\omega_D = 52 \text{ rad/s} \circlearrowleft \qquad \textit{Ans.}$$

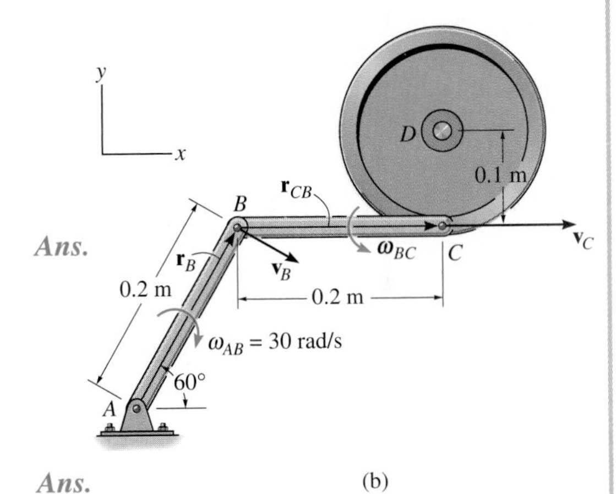

(b)

Fig. 16–16

Note that, by inspection, Fig. 16–16a, $v_B = (0.2)(30) = 6$ m/s, $\measuredangle^{30°}$ and $\mathbf{v}_C$ is directed to the right. As an exercise, use this information and try to obtain ω_{BC} by applying $\mathbf{v}_C = \mathbf{v}_B + \mathbf{v}_{C/B}$ using scalar components.

PROBLEMS

16–57. At the instant shown the boomerang has an angular velocity $\omega = 4$ rad/s, and its mass center G has a velocity $v_G = 6$ in./s. Determine the velocity of point A at this instant.

16–58. At the instant shown the boomerang has an angular velocity $\omega = 4$ rad/s, and its mass center G has a velocity $v_G = 6$ in./s. Determine the velocity of point B at this instant.

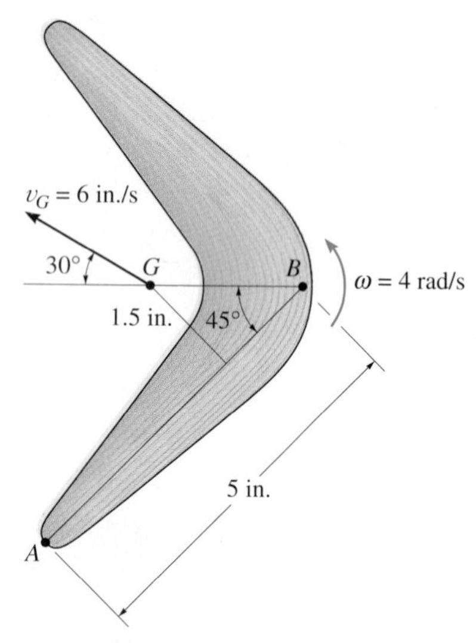

Probs. 16–57/16–58

16–59. The wheel is rotating with an angular velocity $\omega = 8$ rad/s. Determine the velocity of the collar A at this instant.

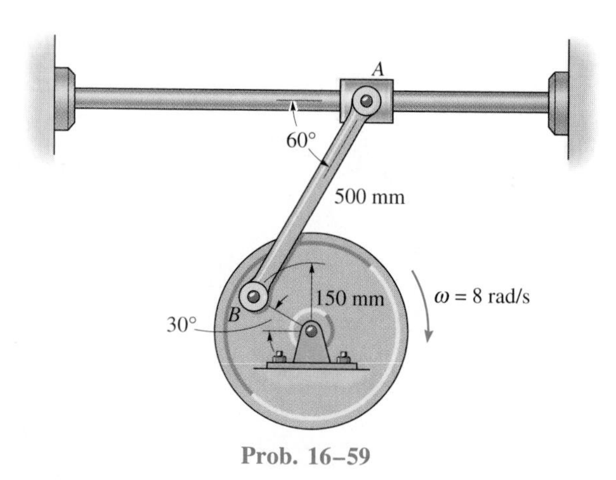

Prob. 16–59

***16–60.** The crankshaft AB is rotating at 500 rad/s about a fixed axis passing through A. Determine the speed of the piston P at the instant it is in the position shown.

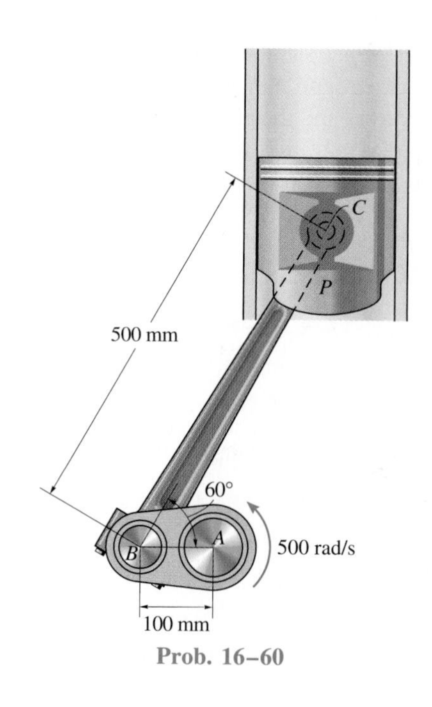

Prob. 16–60

16–61. The velocity of the slider block C is 4 ft/s up the inclined groove. Determine the angular velocity of links AB and BC and the velocity of point B at the instant shown.

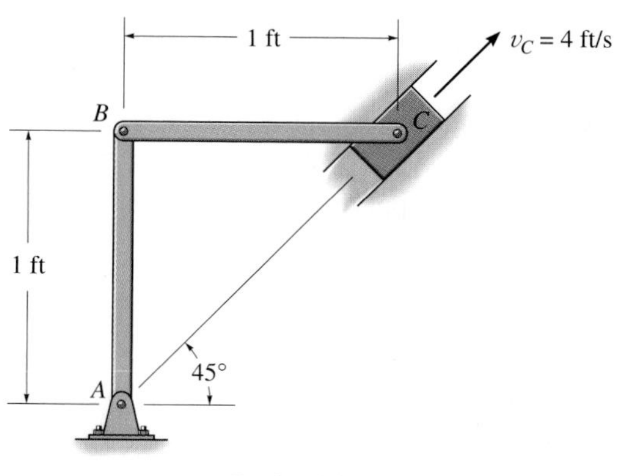

Prob. 16–61

16–62. The shaper mechanism is designed to give a slow cutting stroke and a quick return to a blade attached to the slider at C. Determine the velocity of the slider block C at the instant $\theta = 60°$, if link AB is rotating at 4 rad/s.

16–63. Determine the velocity of the slider block at C at the instant $\theta = 45°$, if link AB is rotating at 4 rad/s.

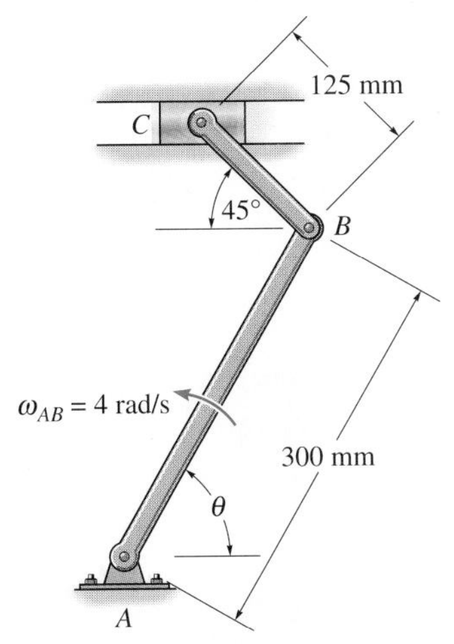

Probs. 16–62/16–63

16–66. At the instant shown, the truck is traveling to the right at 3 m/s, while the pipe is rolling counterclockwise at $\omega = 8$ rad/s without slipping at B. Determine the velocity of the pipe's center G.

16–67. At the instant shown, the truck is traveling to the right at 8 m/s. If the spool does not slip at B, determine its angular velocity so that its mass center G appears to an observer on the ground to remain stationary.

Probs. 16–66/16–67

***16–64.** If link AB is rotating at $\omega_{AB} = 3$ rad/s, determine the angular velocity of link CD at the instant shown.

16–65. If link CD is rotating at $\omega_{CD} = 5$ rad/s, determine the angular velocity of link AB at the instant shown.

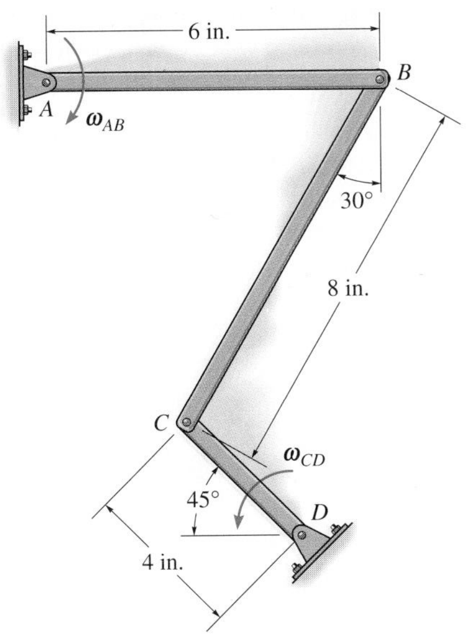

Probs. 16–64/16–65

***16–68.** If the link AB is rotating about the pin at A with an angular velocity $\omega_{AB} = 5$ rad/s, determine the velocities of blocks C and E at the instant shown.

16–69. If the block at C has a velocity of 8 ft/s to the left at the instant shown, determine the angular velocity ω_{AB} of link AB and the velocity of block E at this instant.

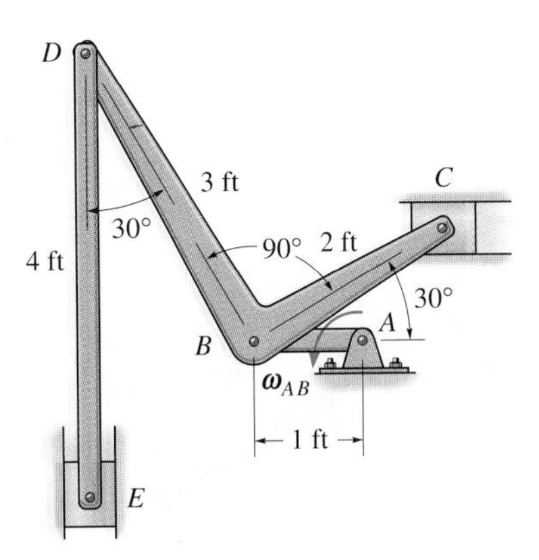

Probs. 16–68/16–69

16–70. If disk A has a constant angular velocity $\omega_A = 4$ rad/s, determine the angular velocity of disk D at the instant $\theta = 45°$.

16–71. If disk D has a constant angular velocity $\omega_D = 2$ rad/s, determine the angular velocity of disk A at the instant $\theta = 60°$.

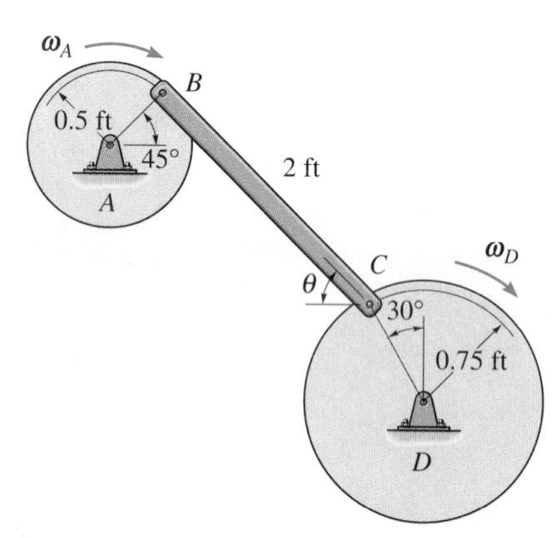

Probs. 16–70/16–71

16–74. If the end of the cord is pulled downward with a speed $v_C = 120$ mm/s, determine the angular velocities of pulleys A and B and the speed of block D. Assume that the cord does not slip on the pulleys.

16–75. Determine the speed v_C at which the cord C should be pulled downward so that the block at D has an upward speed of 3 m/s.

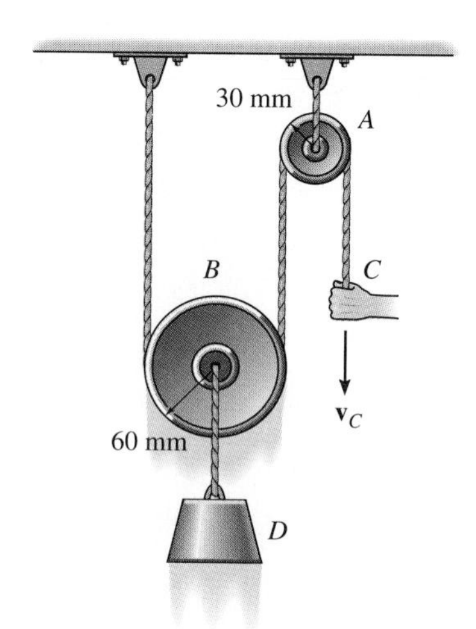

Probs. 16–74/16–75

***16–72.** If rod AB is rotating with an angular velocity $\omega_{AB} = 12$ rad/s, determine the angular velocities of rods BC and CD at the instant shown.

16–73. If rod CD is rotating with an angular velocity $\omega_{DC} = 8$ rad/s, determine the angular velocities of rods AB and CB at the instant shown.

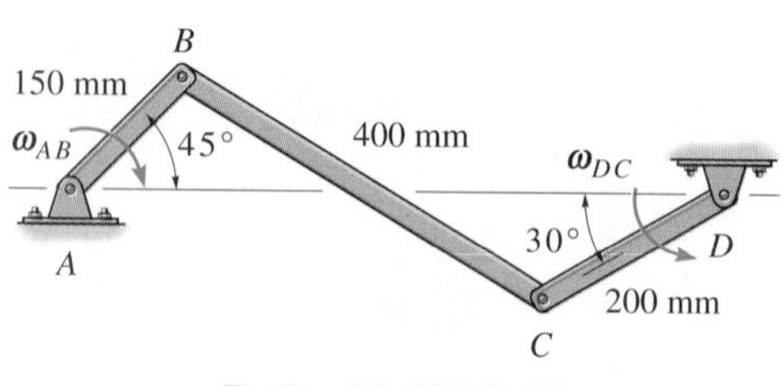

Probs. 16–72/16–73

***16–76.** The top view of an automatic service window at a fast-food restaurant is shown in the figure. During operation, a motor drives the pin-connected link CB with an angular velocity $\omega_{CB} = 0.5$ rad/s. Determine the velocity at the instant shown of the end A, which moves along the slotted guide.

16–77. The top view of an automatic service window at a fast-food restaurant is shown in the figure. During operation, a motor drives the pin-connected link CB with an angular velocity $\omega_{CB} = 0.5$ rad/s. Determine the velocity of the end D at the instant shown. End A moves along the slotted guide.

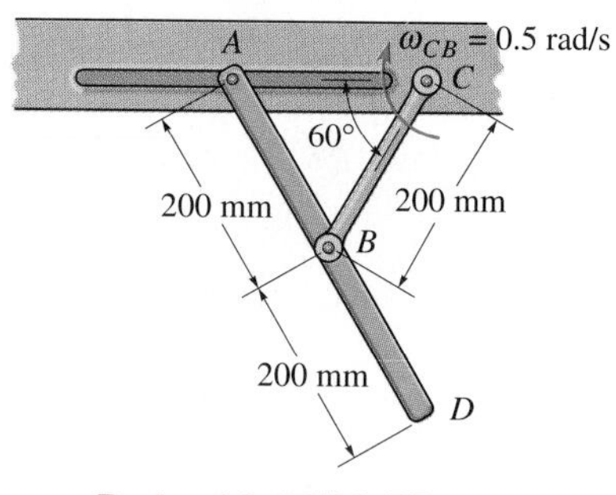

Probs. 16–76/16–77

16–78. Part of an automatic transmission consists of a *fixed* ring gear R, three equal planet gears P, the sun gear S, and the planet carrier C, which is shaded. If the sun gear is rotating at $\omega_S = 6$ rad/s, determine the angular velocity ω_C of the *planet carrier*. Note that C is pin-connected to the center of each of the planet gears.

16–79. Part of an automatic transmission consists of a fixed ring gear R, three equal planet gears P, the sun gear S, and the planet carrier C, which is shaded. If the planet gears are rotating clockwise at $\omega_P = 6$ rad/s, determine the angular velocity ω_S of the sun gear. Note that C is pin-connected to the center of each of the planet gears.

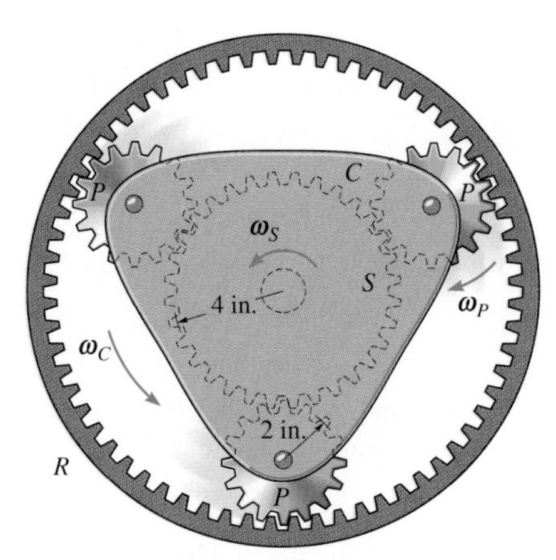

Probs. 16–78/16–79

*16–80. The pinion gear A rolls on the fixed gear rack B with an angular velocity $\omega = 4$ rad/s. Determine the velocity of the gear rack C.

16–81. The pinion gear rolls on the gear racks. If B is moving to the right at 8 ft/s and C is moving to the left at 4 ft/s, determine the angular velocity of the pinion gear and the velocity of its center A.

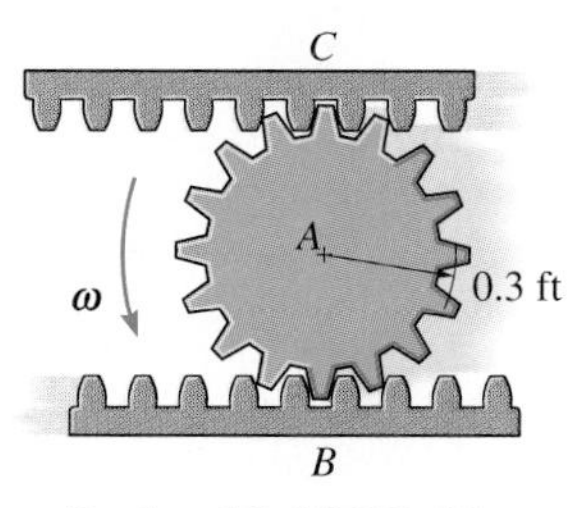

Probs. 16–80/16–81

16–82. The similar links AB and CD rotate about the fixed pins at A and C. If AB has an angular velocity $\omega_{AB} = 8$ rad/s, determine the angular velocity of BDP and the velocity of point P.

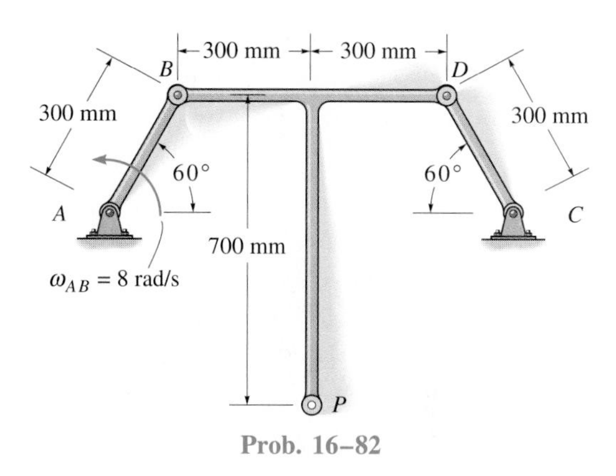

Prob. 16–82

16–83. Knowing that angular velocity of link AB is $\omega_{AB} = 4$ rad/s, determine the velocity of the collar at C and the angular velocity of link CB at the instant shown. Link CB is horizontal at this instant.

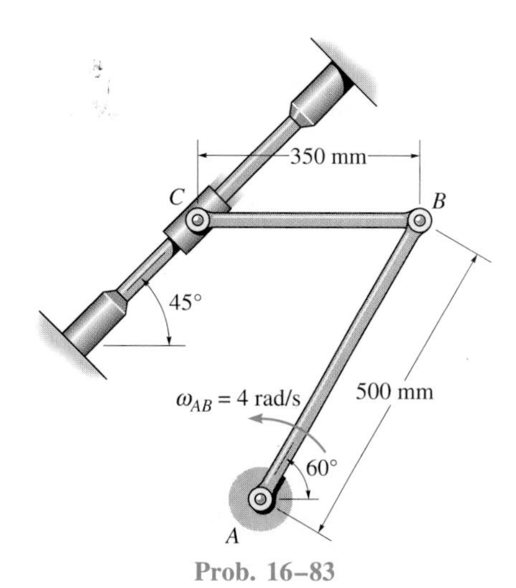

Prob. 16–83

*16–84. In an automobile transmission the planet pinions A and B rotate on shafts that are mounted on the planet-pinion carrier CD. As shown, CD is attached to a shaft at E which is aligned with the center of the *fixed* sun-gear S. This shaft is not attached to the sun gear. If CD is rotating at $\omega_{CD} = 8$ rad/s, determine the angular velocity of the ring gear R.

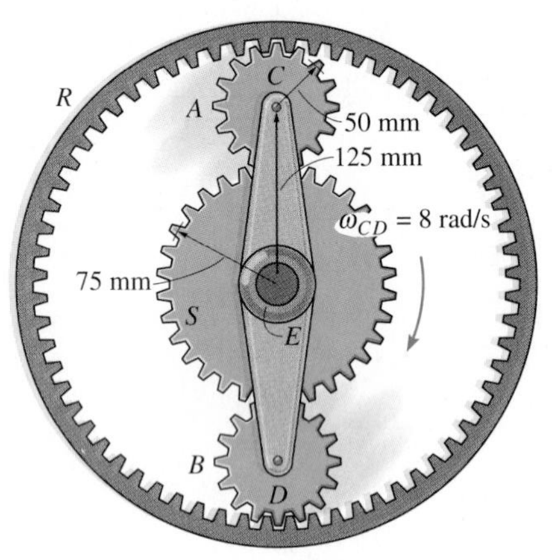

Prob. 16–84

16–85. If the hub gear H and ring gear R have angular velocities $\omega_H = 5$ rad/s and $\omega_R = 20$ rad/s, respectively, determine the angular velocity ω_S of the spur gear S and the angular velocity of its attached arm OA.

16–86. If the hub gear H has an angular velocity $\omega_H = 5$ rad/s, determine the angular velocity of the ring gear R so that the arm OA attached to the spur gear S remains stationary ($\omega_{OA} = 0$). What is the angular velocity of the spur gear?

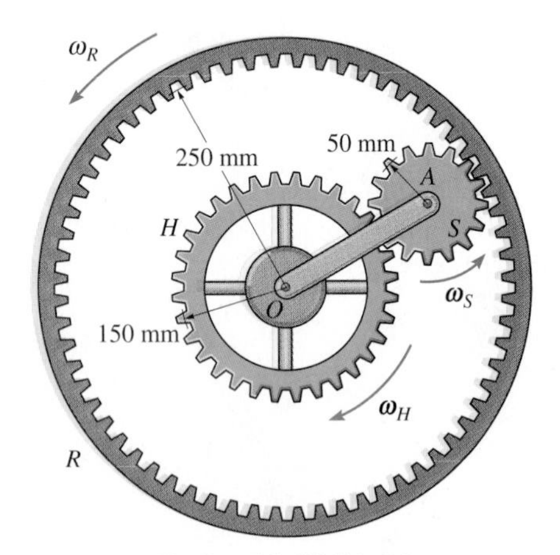

Probs. 16–85/16–86

16–87. If the slider block A is moving downward at $v_A = 4$ m/s, determine the velocities of blocks B and C at the instant shown.

*16–88. If the slider block A is moving downward at $v_A = 4$ m/s, determine the velocity of point E at the instant shown.

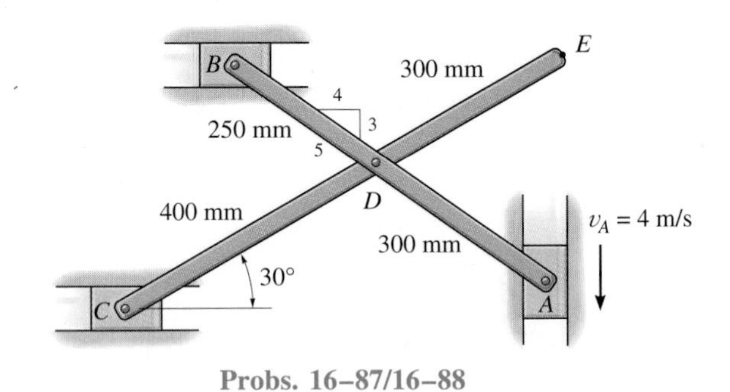

Probs. 16–87/16–88

16–89. The mechanism is used on a machine for the manufacturing of a wire product. Because of the rotational motion of link AB and the sliding of block F, the segmental gear lever DE undergoes general plane motion. If AB is rotating at $\omega_{AB} = 5$ rad/s, determine the velocity of point E at the instant shown.

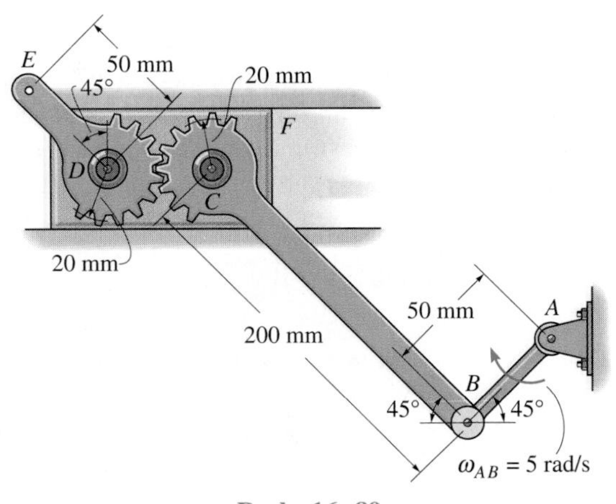

Prob. 16–89

16–90. The two-cylinder engine is designed so that the pistons are connected to the crankshaft *BE* using a master rod *ABC* and articulated rod *AD*. If the crankshaft is rotating at $\omega = 30$ rad/s, determine the velocities of the pistons *C* and *D* at the instant shown.

16–91. Mechanical toy animals often use a walking mechanism as shown idealized in the figure. If the driving crank *AB* is propelled by a spring motor such that $\omega_{AB} = 5$ rad/s, determine the velocity of the rear foot *E* at the instant shown. Although not part of this problem, the upper end of the foreleg has a slotted guide which is constrained by the fixed pin at *G*.

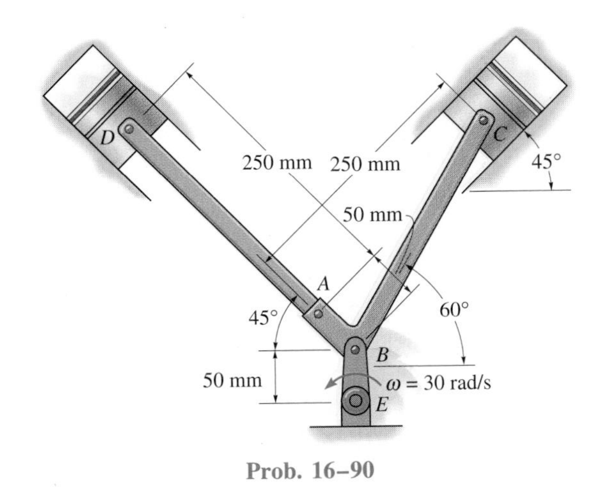

Prob. 16–90

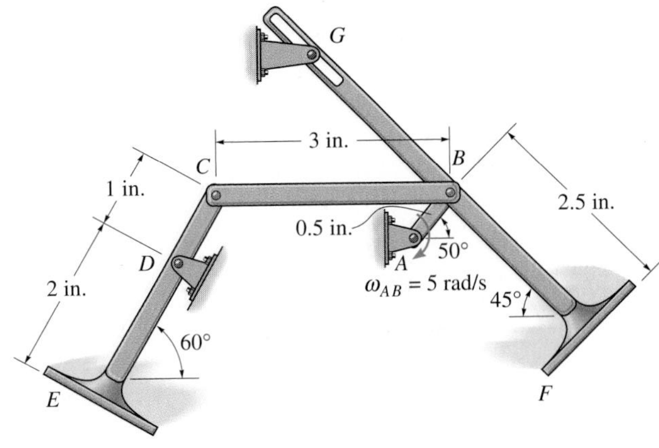

Prob. 16–91

16.6 Instantaneous Center of Zero Velocity

The velocity of any point *B* located on a rigid body can be obtained in a simple way if one chooses the base point *A* to be a point that has *zero velocity* at the instant considered. In this case, $\mathbf{v}_A = \mathbf{0}$, and therefore the velocity equation, $\mathbf{v}_B = \mathbf{v}_A + \boldsymbol{\omega} \times \mathbf{r}_{B/A}$, becomes $\mathbf{v}_B = \boldsymbol{\omega} \times \mathbf{r}_{B/A}$. For a body having general plane motion, point *A* so chosen is called the *instantaneous center of zero velocity (IC)* and it lies on the *instantaneous axis of zero velocity*. This axis is always perpendicular to the plane used to represent the motion, and the intersection of the axis with this plane defines the location of the *IC*. Since point *A* is coincident with the *IC*, then $\mathbf{v}_B = \boldsymbol{\omega} \times \mathbf{r}_{B/IC}$ and so point *B* moves momentarily about the *IC* in a *circular path;* in other words, the body appears to rotate about the instantaneous axis. If the relative-position vector $\mathbf{r}_{B/IC}$ is established from the *IC* to point *B*, then the *magnitude* of $\mathbf{v}_B$ is simply $v_B = \omega r_{B/IC}$, where ω is the angular velocity of the body. Due to the circular motion, the *direction* of $\mathbf{v}_B$ must always be *perpendicular* to $\mathbf{r}_{B/IC}$.

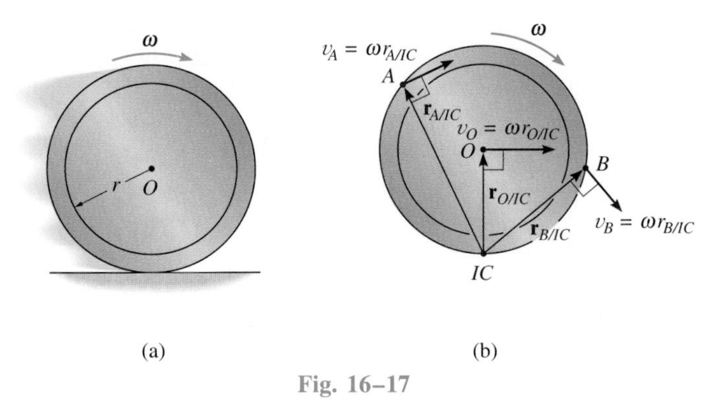

Fig. 16–17

For example, consider a wheel which rolls *without slipping,* Fig. 16–17*a*. In this case the point of *contact* with the ground has *zero velocity.* Hence this point represents the *IC* for the wheel, Fig. 16–17*b*. If it is imagined that the wheel is momentarily pinned at this point, the velocities of points *A, B, O,* and so on, can be found using $v = \omega r$. Here the radial distances $r_{A/IC}$, $r_{B/IC}$, and $r_{O/IC}$, shown in Fig. 16–17*b*, must be determined from the geometry of the wheel.

When a body is subjected to general plane motion, the point determined as the instantaneous center of zero velocity for the body can only be used for an *instant of time.* Since the body changes its position from one instant to the next, then for each position of the body a unique instantaneous center must be determined. The locus of points which define the *IC* during the body's motion is called a *centrode,* Fig. 16–18*a*. Thus, each point on the centrode acts as the *IC* for the body only for an instant of time.

Although the *IC* may be conveniently used to determine the velocity of any point in a body, it generally *does not have zero acceleration* and therefore it *should not* be used for finding the accelerations of points in a body.

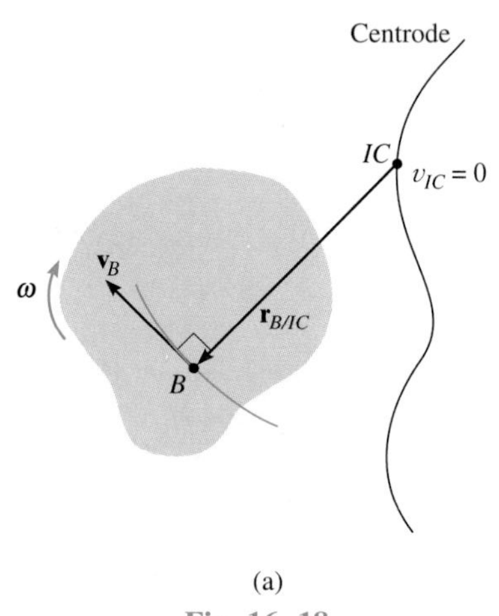

Fig. 16–18

Location of the IC. If the location of the *IC* is unknown, it may be determined by using the fact that the *relative-position vector* extending from the *IC* to a point is *always perpendicular* to the *velocity* of the point. Several possibilities exist:

1. *Given the velocity of a point on the body and the angular velocity of the body.* In this case, if $\mathbf{v}_B$ and $\boldsymbol{\omega}$ are known, the *IC* is located along the line drawn perpendicular to $\mathbf{v}_B$ at *B*, such that the distance from *B* to the *IC* is $r_{B/IC} = v_B/\omega$, Fig. 16–18*a*. Note that the *IC* lies on that side of *B* which causes rotation about the *IC*, which is consistent with the direction of motion caused by $\boldsymbol{\omega}$ and $\mathbf{v}_B$.

2. *Given the lines of action of two nonparallel velocities.* Consider the body in Fig. 16–18*b*, where the lines of action of the velocities $\mathbf{v}_A$ and $\mathbf{v}_B$ are known. From each of these lines construct at points A and B line segments that are perpendicular to v_A and v_B and therefore define the lines of action of $\mathbf{r}_{A/IC}$ and $\mathbf{r}_{B/IC}$, respectively. Extending these perpendiculars to their *point of intersection* as shown locates the IC at the instant considered. The magnitudes of $\mathbf{r}_{A/IC}$ and $\mathbf{r}_{B/IC}$ are generally determined from the geometry of the body and trigonometry. Furthermore, if the magnitude and sense of $\mathbf{v}_A$ are known, then the angular velocity of the body is determined from $\omega = v_A/r_{A/IC}$. Once computed, $\boldsymbol{\omega}$ can then be used to determine $v_B = \omega r_{B/IC}$.
3. *Given the magnitude and direction of two parallel velocities.* When the velocities of points A and B are parallel and have *known* magnitudes v_A and v_B, then the location of the IC is determined by proportional triangles. Examples are shown in Fig. 16–18*c* and *d*. In both cases $r_{A/IC} = v_A/\omega$ and $r_{B/IC} = v_B/\omega$. If d is a known distance between points A and B, then in Fig. 16–18*c*, $r_{A/IC} + r_{B/IC} = d$ and in Fig. 16–18*d*, $r_{B/IC} - r_{A/IC} = d$. As a special case, note that if the body is *translating*, $\mathbf{v}_A = \mathbf{v}_B$ and the IC would be located at infinity, in which case $r_{A/IC} = r_{B/IC} \to \infty$. This being the case, $\omega = v_A/\infty = v_B/\infty = 0$, as expected.

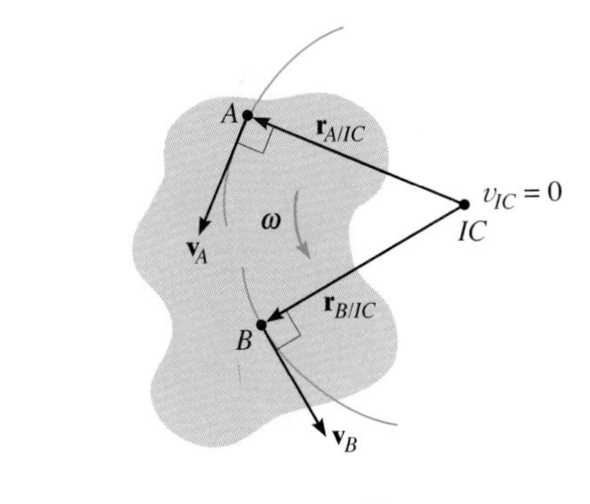

(b)

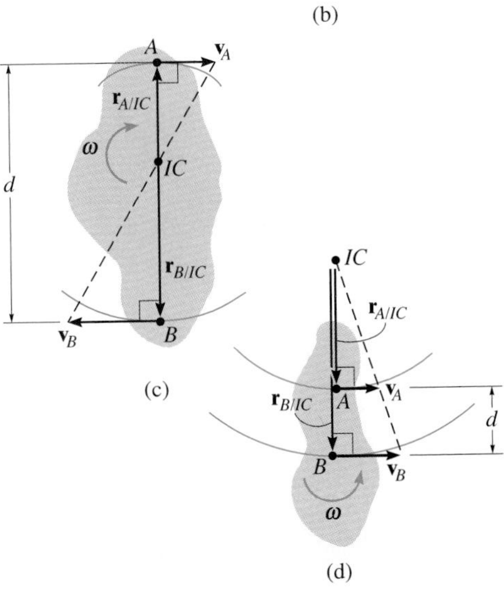

(c)

(d)

Fig. 16–18 *(cont'd)*

PROCEDURE FOR ANALYSIS

The velocity of a point on a body which is subjected to general plane motion can be determined with reference to its instantaneous center of zero velocity, provided the location of the IC is first established. This is done by using one of the three methods described above. As shown on the kinematic diagram in Fig. 16–19, the body is imagined as "extended and pinned" at the IC such that, at the instant considered, it rotates about this pin with its angular velocity $\boldsymbol{\omega}$. The *magnitude* of velocity for the arbitrary points A, B, and C in the body can then be determined by using the equation $v = \omega r$, where r is the radial line drawn from the IC to the point. The line of action of each velocity vector is *perpendicular* to its associated radial line, and the velocity has a *sense of direction* which tends to move the point in a manner consistent with the angular rotation of the radial line, Fig. 16–19.

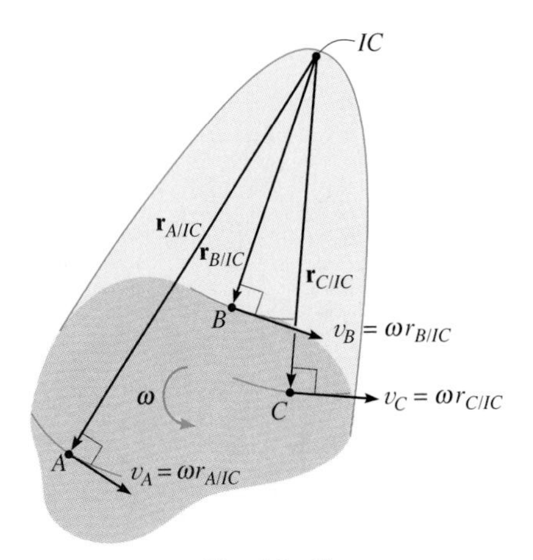

Fig. 16–19

Example 16–10

Show how to determine the location of the instantaneous center of zero velocity for (*a*) the crankshaft *BC* shown in Fig. 16–20*a*; and (*b*) the link *CB* shown in Fig. 16–20*b*.

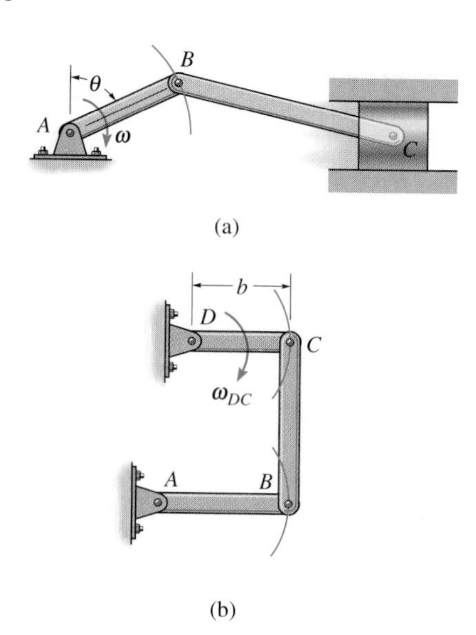

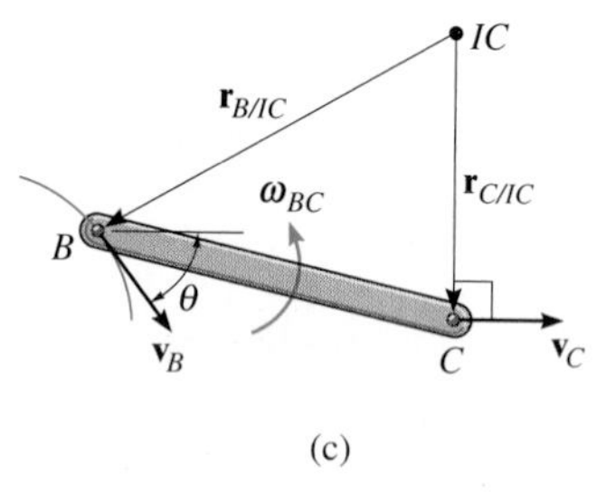

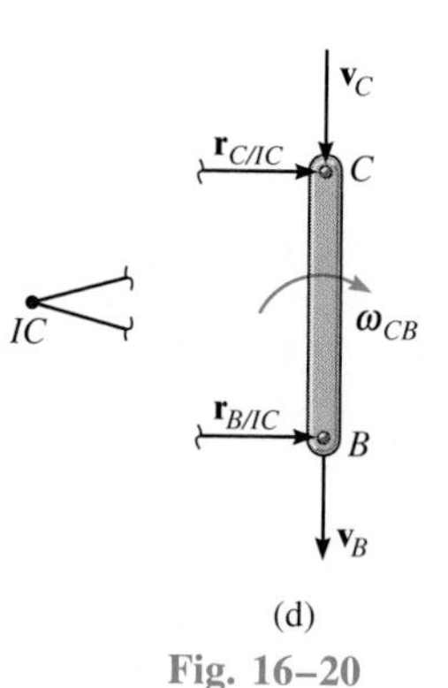

Fig. 16–20

SOLUTION

Part (a). As shown in Fig. 16–20*a*, point *B* has a velocity $\mathbf{v}_B$, which is caused by the clockwise rotation of link *AB*. Point *B* moves in a circular path such that $\mathbf{v}_B$ is perpendicular to *AB*, and so it acts at an angle θ from the horizontal as shown in Fig. 16–20*c*. The motion of point *B* causes the piston to move forward *horizontally* with a velocity $\mathbf{v}_C$. Consequently, point *C* on the crankshaft moves horizontally with this same velocity. When lines are drawn perpendicular to $\mathbf{v}_B$ and $\mathbf{v}_C$, Fig. 16–20*c*, they intersect at the *IC*.

Part (b). Points *B* and *C* follow circular paths of motion since rods *AB* and *DC* are each subjected to rotation about a fixed axis, Fig. 16–20*b*. Since the velocity is always tangent to the path, at the instant considered, $\mathbf{v}_C$ on rod *DC* and $\mathbf{v}_B$ on rod *AB* are both directed vertically downward, along the axis of link *CB*, Fig. 16–20*d*. Furthermore, since *CB* is *rigid*, no relative displacement occurs between points *B* and *C*, so that $\mathbf{v}_B = \mathbf{v}_C$. Radial lines drawn perpendicular to these two velocities form parallel lines which intersect at ''infinity;'' i.e., $r_{C/IC} \to \infty$ and $r_{B/IC} \to \infty$. Thus, $\omega_{CB} = v_C/r_{C/IC} = \omega_{DC}(b)/\infty = 0$. As a result, rod *CB* momentarily *translates*. An instant later, however, *CB* will move to a new position, causing the instantaneous center to move to some finite location.

Example 16–11

Block D shown in Fig. 16–21a moves with a speed of 3 m/s. Determine the angular velocities of links BD and AB, and the velocity of point B at the instant shown.

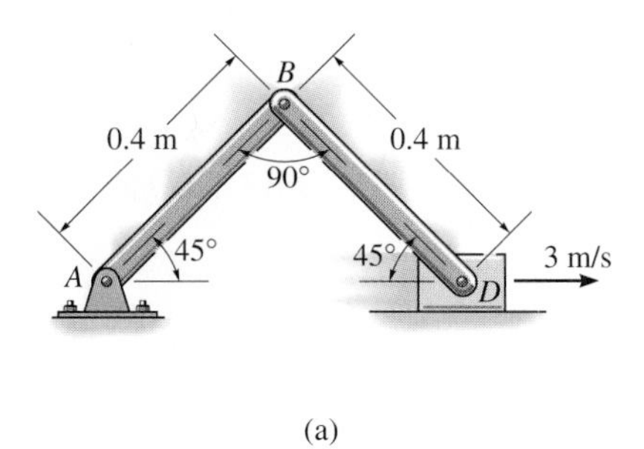

(a)

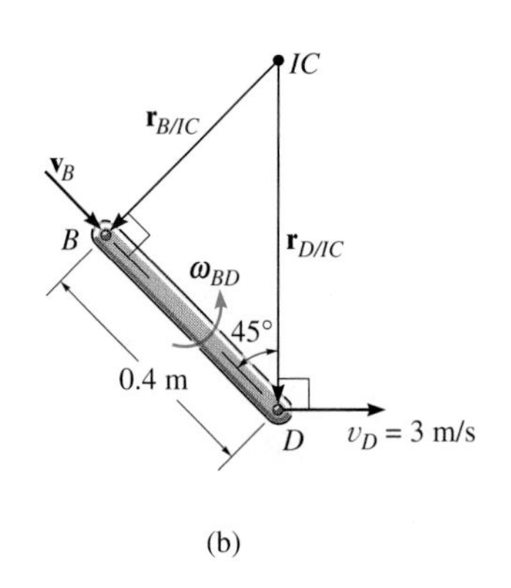

(b)

SOLUTION

Since D moves to the right at 3 m/s, it causes arm AB to rotate about point A in a clockwise direction. Hence, $\mathbf{v}_B$ is directed perpendicular to AB as shown in Fig. 16–21b. The instantaneous center of zero velocity for BD is located at the intersection of the line segments drawn perpendicular to $\mathbf{v}_B$ and $\mathbf{v}_D$, Fig. 16–21b. From the geometry,

$$r_{B/IC} = 0.4 \tan 45° \text{ m} = 0.4 \text{ m}$$

$$r_{D/IC} = \frac{0.4 \text{ m}}{\cos 45°} = 0.566 \text{ m}$$

Since the magnitude of $\mathbf{v}_D$ is known, the angular velocity of link BD is

$$\omega_{BD} = \frac{v_D}{r_{D/IC}} = \frac{3 \text{ m/s}}{0.566 \text{ m}} = 5.30 \text{ rad/s} \circlearrowleft \qquad \textit{Ans.}$$

The velocity of B is therefore

$$v_B = \omega_{BD}(r_{B/IC}) = 5.30 \text{ rad/s}(0.4 \text{ m}) = 2.12 \text{ m/s} \searrow 45° \qquad \textit{Ans.}$$

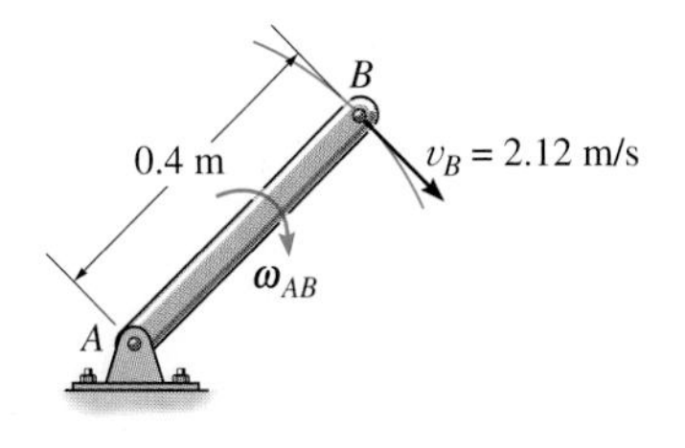

(c)

Fig. 16–21

Link AB is subjected to rotation about a fixed axis passing through A, Fig. 16–21c, and since v_B is known, the angular velocity of AB is

$$\omega_{AB} = \frac{v_B}{r_{B/A}} = \frac{2.12 \text{ m/s}}{0.4 \text{ m}} = 5.30 \text{ rad/s} \circlearrowright \qquad \textit{Ans.}$$

Example 16–12

The cylinder shown in Fig. 16–22*a* rolls without slipping between the two moving plates *E* and *D*. Determine the angular velocity of the cylinder and the velocity of its center *C* at the instant shown.

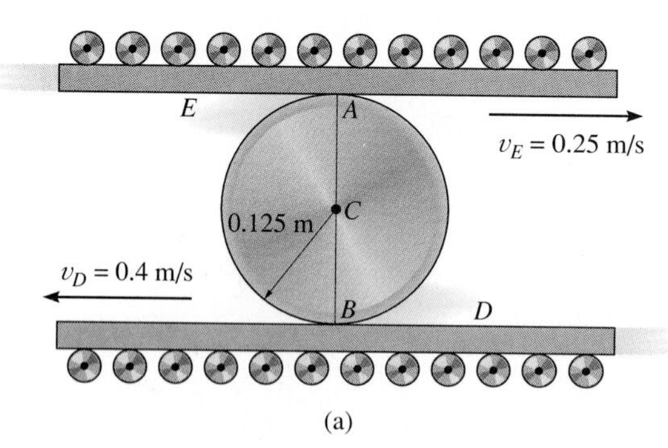

Fig. 16–22

SOLUTION

Since no slipping occurs, the contact points *A* and *B* on the cylinder have the same velocity as the plates *E* and *D*. Furthermore, the velocities $\mathbf{v}_A$ and $\mathbf{v}_B$ are *parallel,* so that by the proportionality of right triangles the *IC* is located at a point on line *AB*, Fig. 16–22*b*. Assuming this point to be a distance x from *B*, we have

$$v_B = \omega x; \qquad 0.4 \text{ m/s} = \omega x$$

$$v_A = \omega(0.25 \text{ m} - x); \quad 0.25 \text{ m/s} = \omega(0.25 \text{ m} - x)$$

Dividing one of these equations into the other eliminates ω and yields

$$0.4(0.25 - x) = 0.25x$$

$$x = \frac{0.1}{0.65} = 0.154 \text{ m}$$

Hence, the angular velocity of the cylinder is

$$\omega = \frac{v_B}{x} = \frac{0.4 \text{ m/s}}{0.154 \text{ m}} = 2.60 \text{ rad/s} \circlearrowright \qquad \textit{Ans.}$$

The velocity of point *C* is therefore

$$v_C = \omega r_{C/IC} = 2.60 \text{ rad/s}(0.154 \text{ m} - 0.125 \text{ m})$$
$$= 0.0750 \text{ m/s} \leftarrow \qquad \textit{Ans.}$$

PROBLEMS

***16–92.** Solve Prob. 16–60 using the method of instantaneous center of zero velocity.

16–93. Solve Prob. 16–62 using the method of instantaneous center of zero velocity.

16–94. Solve Prob. 16–66 using the method of instantaneous center of zero velocity.

16–95. Solve Prob. 16–67 using the method of instantaneous center of zero velocity.

***16–96.** Solve Prob. 16–71 using the method of instantaneous center of zero velocity.

16–97. Solve Prob. 16–74 using the method of instantaneous center of zero velocity.

16–98. Solve Prob. 16–82 using the method of instantaneous center of zero velocity.

16–99. Solve Prob. 16–84 using the method of instantaneous center of zero velocity.

***16–100.** In each case show graphically how to locate the instantaneous center of zero velocity of link AB. Assume the geometry is known.

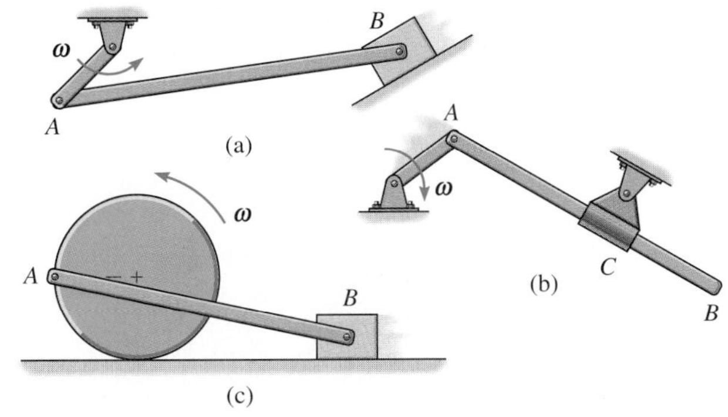

Prob. 16–100

16–101. At the instant shown, the disk is rotating at $\omega = 4$ rad/s. Determine the velocities of points A, B, and C.

16–102. At the instant shown, the disk is rotating at $\omega = 4$ rad/s. Determine the velocities of points A, O, and E.

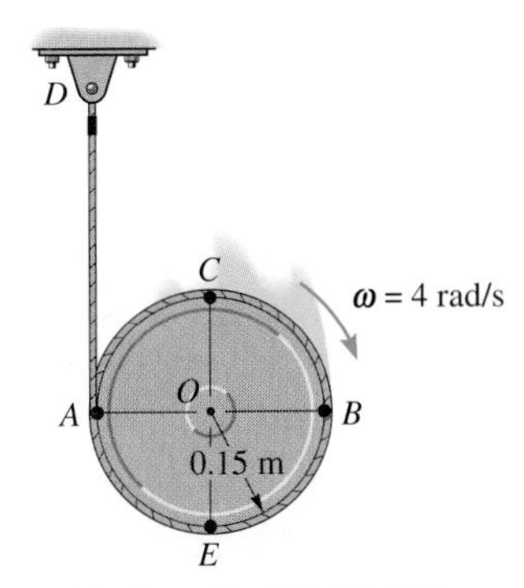

Probs. 16–101/16–102

16–103. The conveyor belt is moving to the right at $v = 8$ ft/s, and at the same instant the cylinder is rolling counterclockwise at $\omega = 2$ rad/s without slipping. Determine the velocities of the cylinder's center C and point B at this instant.

***16–104.** The conveyor belt is moving to the right at $v = 12$ ft/s, and at the same instant the cylinder is rolling counterclockwise at $\omega = 6$ rad/s while its center has a velocity of 4 ft/s to the left. Determine the velocities of points A and B on the disk at this instant. Does the cylinder slip on the conveyor?

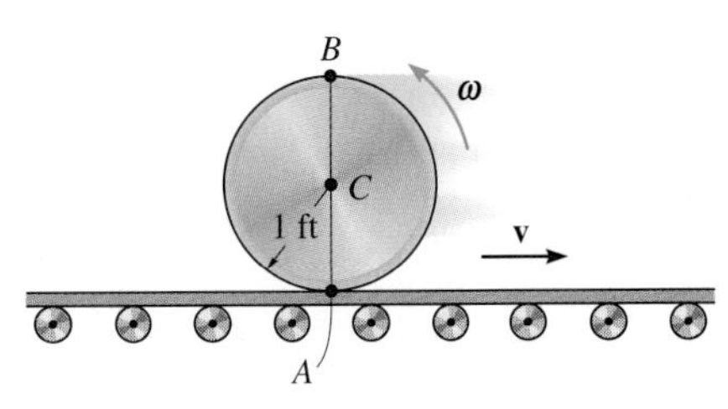

Probs. 16–103/16–104

16–105. If link AB is rotating at $\omega_{AB} = 6$ rad/s, determine the angular velocities of links BC and CD at the instant $\theta = 60°$.

16–106. If link AB is rotating at $\omega_{AB} = 6$ rad/s, determine the angular velocities of links BC and CD at the instant $\theta = 45°$.

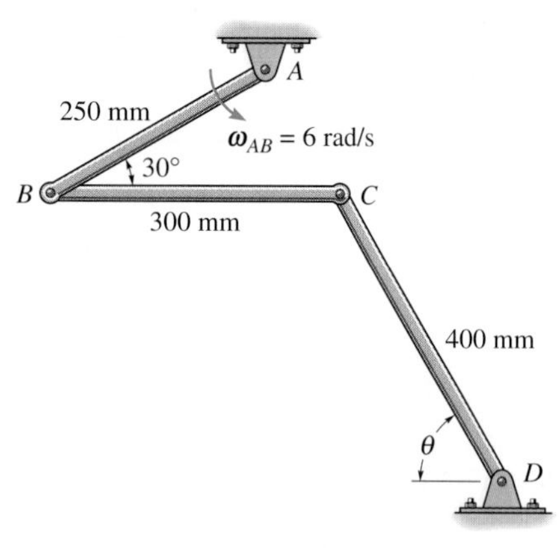

Probs. 16–105/16–106

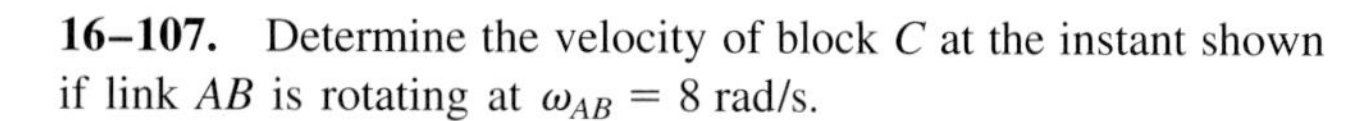

16–107. Determine the velocity of block C at the instant shown if link AB is rotating at $\omega_{AB} = 8$ rad/s.

***16–108.** Determine the angular velocity of link AB at the instant shown if block C is moving upward at 12 in./s.

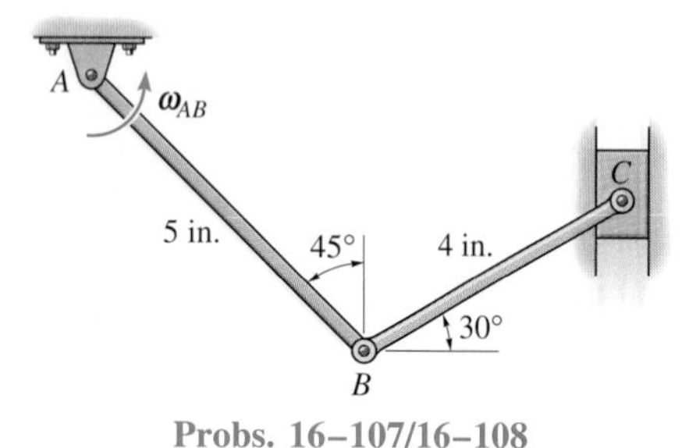

Probs. 16–107/16–108

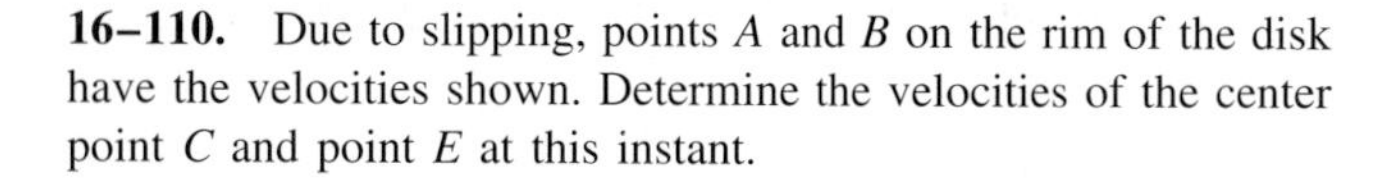

16–109. Due to slipping, points A and B on the rim of the disk have the velocities shown. Determine the velocities of the center point C and point D at this instant.

16–110. Due to slipping, points A and B on the rim of the disk have the velocities shown. Determine the velocities of the center point C and point E at this instant.

16–111. Due to slipping, points A and B on the rim of the disk have the velocities shown. Determine the velocities of the center point C and point F at this instant.

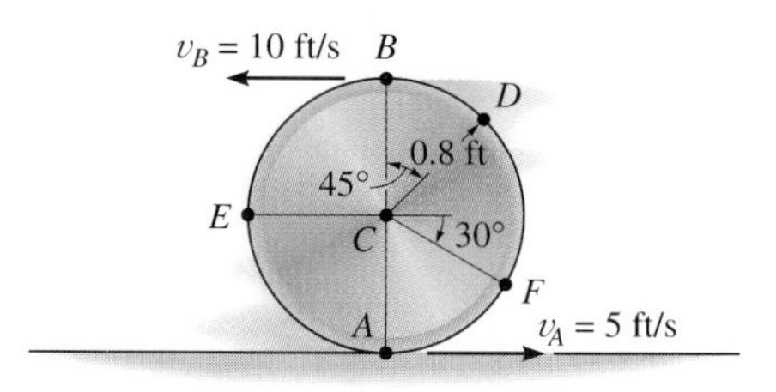

Probs. 16–109/16–110/16–111

***16–112.** The disk of radius r is confined to roll without slipping at A and B. If the plates have the velocities shown, determine the angular velocity of the disk.

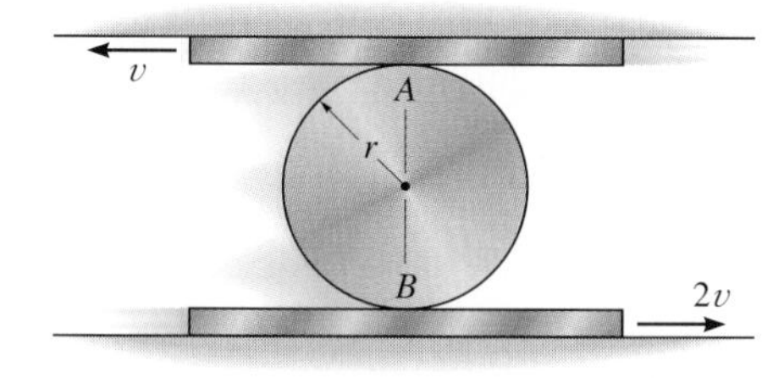

Prob. 16–112

16–113. As the car travels forward at 80 ft/s on a wet road, due to slipping, the rear wheels have an angular velocity $\omega = 100$ rad/s. Determine the speeds of points A, B, and C caused by the motion.

Prob. 16–113

16–114. Knowing that the angular velocity of link AB is $\omega_{AB} = 4$ rad/s, determine the velocity of the collar at C and the angular velocity of link CB at the instant shown. Link CB is horizontal at this instant.

16–115. If the collar at C is moving downward to the left at $v_C = 8$ m/s, determine the angular velocity of link AB at the instant shown.

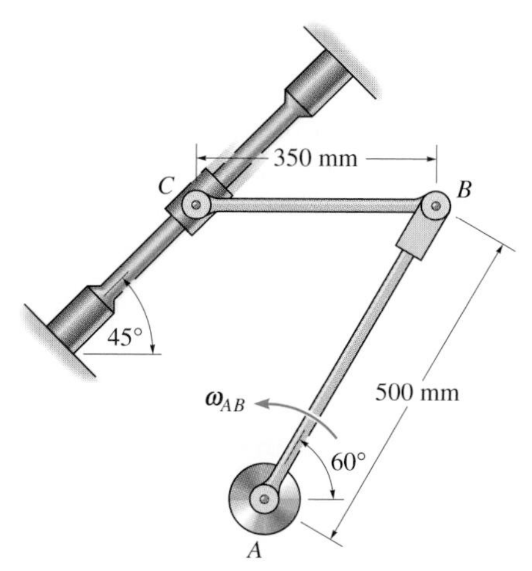

Probs. 16–114/16–115

16–117. The epicyclic gear train is driven by the rotating link DE, which has an angular velocity $\omega_{DE} = 5$ rad/s. If the ring gear F is fixed, determine the angular velocities of gears A, B, and C.

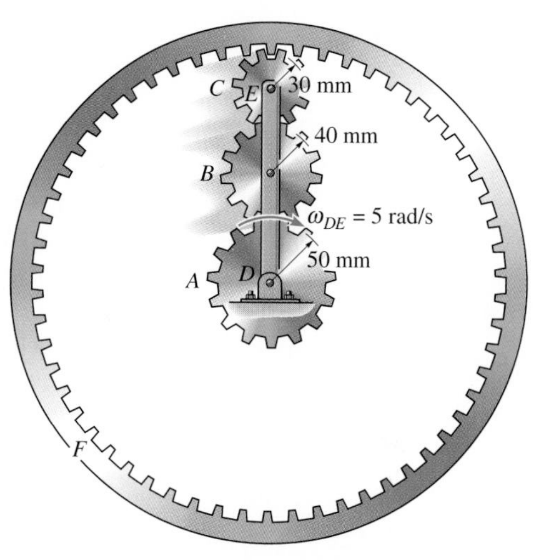

Prob. 16–117

16–118. As the crankshaft AB rotates at $\omega_{AB} = 50$ rad/s about the fixed axis through point A, the wrist pins C and D rock between their bearing supports E and F. If C and D are allowed to turn freely about their centers, while they are held from slipping at E or F, determine the angular velocity of the connecting rod CBD at the instant shown.

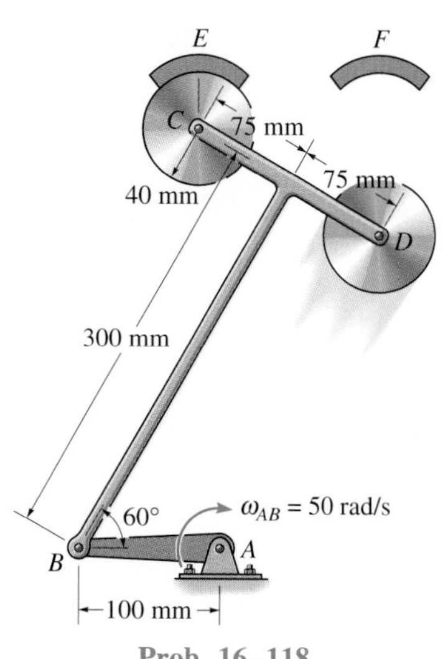

Prob. 16–118

***16–116.** The cylinder B rolls on the fixed cylinder A without slipping. If the connected bar CD is rotating with an angular velocity $\omega_{CD} = 5$ rad/s, determine the angular velocity of cylinder B. Point C is a fixed point.

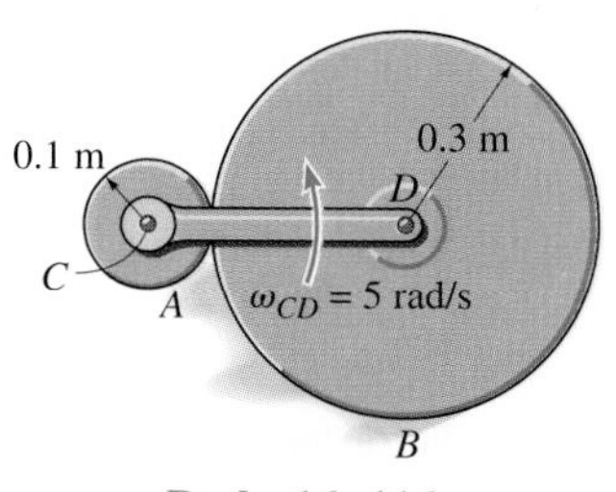

Prob. 16–116

16.7 Relative-Motion Analysis: Acceleration

An equation that relates the accelerations of two points on a rigid body subjected to general plane motion may be determined by differentiating the velocity equation $\mathbf{v}_B = \mathbf{v}_A + \mathbf{v}_{B/A}$ with respect to time. Thus,

$$\frac{d\mathbf{v}_B}{dt} = \frac{d\mathbf{v}_A}{dt} + \frac{d\mathbf{v}_{B/A}}{dt}$$

The terms $d\mathbf{v}_B/dt = \mathbf{a}_B$ and $d\mathbf{v}_A/dt = \mathbf{a}_A$ are measured from a set of *fixed x, y axes* and represent the *absolute accelerations* of points B and A. The last term represents the acceleration of B with respect to A as measured by an observer fixed to the translating x', y' axes having their origin at the base point A. In Sec. 16.5 it was shown that to this observer point B appears to move along a *circular arc* that has a radius of curvature $r_{B/A}$. In other words, *the body appears to move as if it were rotating about the z' axis passing through point A.* Consequently, $\mathbf{a}_{B/A}$ can be expressed in terms of its tangential and normal components of motion; i.e., $\mathbf{a}_{B/A} = (\mathbf{a}_{B/A})_t + (\mathbf{a}_{B/A})_n$, where $(a_{B/A})_t = \alpha r_{B/A}$ and $(a_{B/A})_n = \omega^2 r_{B/A}$. Hence, the relative-acceleration equation can be written in the form

$$\mathbf{a}_B = \mathbf{a}_A + (\mathbf{a}_{B/A})_t + (\mathbf{a}_{B/A})_n \qquad (16\text{–}17)$$

where

$\mathbf{a}_B$ = acceleration of point B

$\mathbf{a}_A$ = acceleration of point A

$(\mathbf{a}_{B/A})_t$ = relative tangential acceleration component of "B with respect to A." Since the relative motion is circular, the *magnitude* of $(\mathbf{a}_{B/A})_t$ is $(a_{B/A})_t = \alpha r_{B/A}$ and the *direction* is perpendicular to $\mathbf{r}_{B/A}$

$(\mathbf{a}_{B/A})_n$ = relative normal acceleration component of "B with respect to A." Since the relative motion is circular, the *magnitude* of $(\mathbf{a}_{B/A})_n$ is $(a_{B/A})_n = \omega^2 r_{B/A}$ and the *direction* is always B toward A

Each of the four terms in Eq. 16–17 is represented graphically on the *kinematic diagrams* shown in Fig. 16–23. Here it is seen that at a given instant the acceleration of B, Fig. 16–23*a*, is determined by considering the body to translate with an acceleration $\mathbf{a}_A$, Fig. 16–23*b*, and simultaneously rotate about the base point A with an instantaneous angular velocity $\boldsymbol{\omega}$ and angular acceleration $\boldsymbol{\alpha}$, Fig. 16–23*c*. Vector addition of these two effects, applied to B, yields $\mathbf{a}_B$, as shown in Fig. 16–23*d*. It should be noted from Fig. 16–23*a* that points A and B move along *curved paths,* and as a result the accelerations of these points have *both tangential and normal components.* (Recall that the acceleration of a point is *tangent to the path only* when the path is *rectilinear* or when it is an inflection point.)

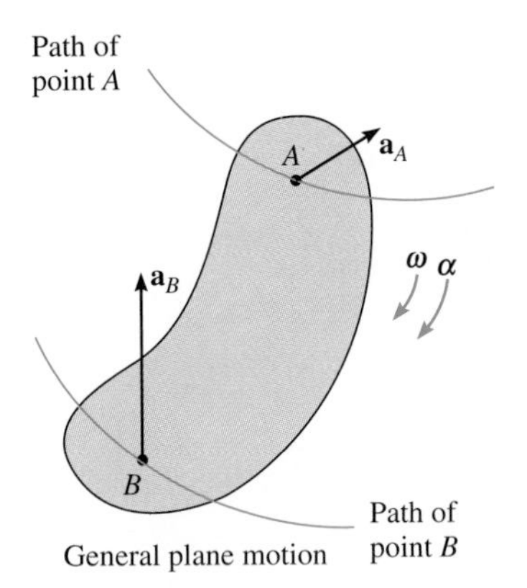

(a)

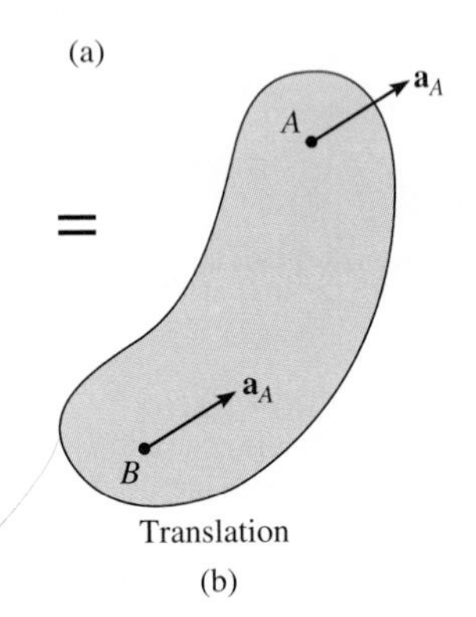

(b)

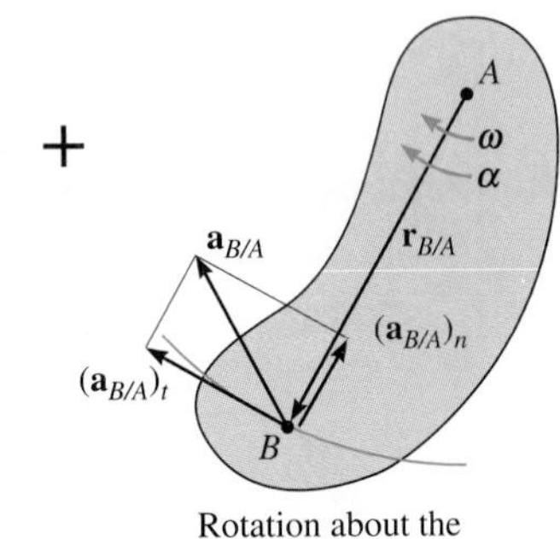

(c)

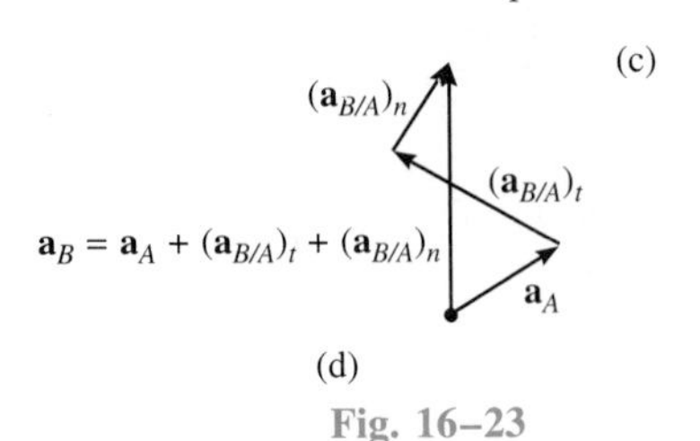

(d)

Fig. 16–23

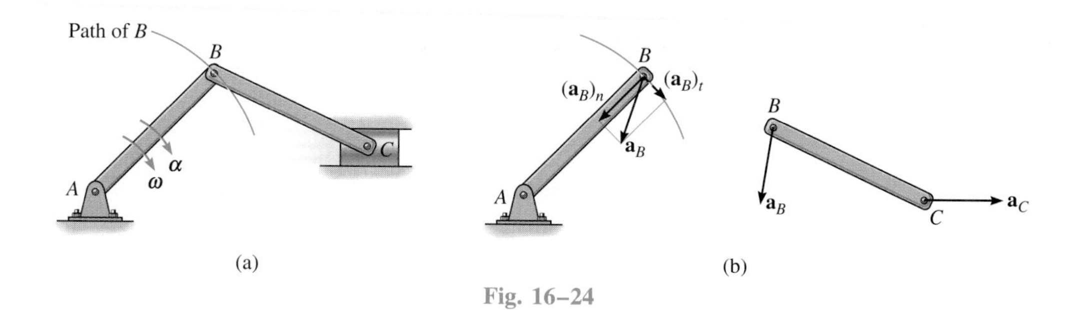

Fig. 16–24

Since the relative-acceleration components represent the effect of *circular motion* observed from translating axes having their origin at the base point A, these terms can be expressed as $(\mathbf{a}_{B/A})_t = \boldsymbol{\alpha} \times \mathbf{r}_{B/A}$ and $(\mathbf{a}_{B/A})_n = -\omega^2\mathbf{r}_{B/A}$, Eq. 16–14. Hence, Eq. 16–17 becomes

$$\mathbf{a}_B = \mathbf{a}_A + \boldsymbol{\alpha} \times \mathbf{r}_{B/A} - \omega^2\mathbf{r}_{B/A} \tag{16–18}$$

where

$\mathbf{a}_B$ = acceleration of point B
$\mathbf{a}_A$ = acceleration of the base point A
$\boldsymbol{\alpha}$ = angular acceleration of the body
$\boldsymbol{\omega}$ = angular velocity of the body
$\mathbf{r}_{B/A}$ = relative-position vector drawn from A to B

If Eq. 16–17 or 16–18 is applied in a practical manner to study the accelerated motion of a rigid body which is pin-connected to two other bodies, it should be realized that points which are *coincident at the pin* move with the *same acceleration,* since the path of motion over which they travel is the *same.* For example, point B lying on either rod AB or BC of the crank mechanism shown in Fig. 16–24a has the same acceleration, since the rods are pin-connected at B. Here the motion of B is along a *curved path,* so that $\mathbf{a}_B$ is calculated on the basis of its tangential, $(a_B)_t = \alpha r_{B/A}$, and normal component, $(a_B)_n = \omega^2 r_{B/A}$, which are defined by the angular motion of AB. At the other end of rod BC, however, point C moves along a *rectilinear path,* which is defined by the piston. Hence, in this case, the acceleration $\mathbf{a}_C$ is directed along the path, Fig. 16–24b.

If two bodies contact one another *without slipping,* and the *points in contact* move along *different paths,* the *tangential components* of acceleration of the points will be the *same;* however, the *normal components* will *not* be the same. For example, consider the two meshed gears in Fig. 16–25a. Point A is located on gear B and a coincident point A' is located on gear C. Due to the rotational motion, $(\mathbf{a}_A)_t = (\mathbf{a}_{A'})_t$; however, since both points follow different curved paths, $(\mathbf{a}_A)_n \neq (\mathbf{a}_{A'})_n$ and therefore $\mathbf{a}_A \neq \mathbf{a}_{A'}$, Fig. 16–25$b$.

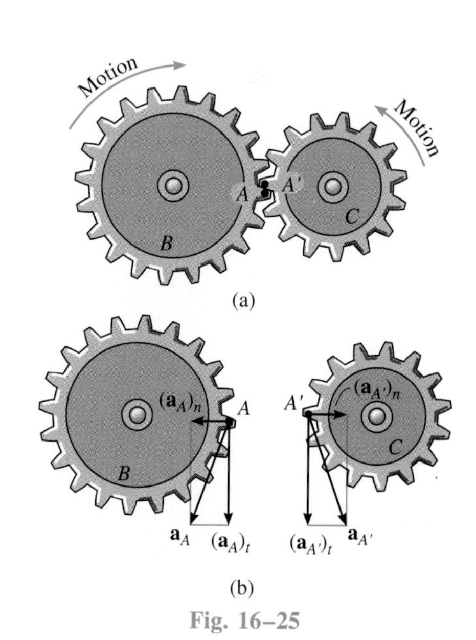

Fig. 16–25

PROCEDURE FOR ANALYSIS

Equation 16–17 relates the accelerations of any two points A and B located on the *same* rigid body. Like the velocity equation, Eq. 16–15, it can be applied either by using a Cartesian vector analysis, in the form of Eq. 16–18, or by writing the x and y scalar component equations directly. For application, it is suggested that the following procedure be used.

Velocity Analysis. If the angular velocity $\boldsymbol{\omega}$ of the body is unknown, determine it by using a velocity analysis as discussed in Sec. 16.5 or 16.6. Determine also the velocities $\mathbf{v}_A$ and $\mathbf{v}_B$ of points A and B *if these points move along curved paths.*

Kinematic Diagrams. Establish the directions of the fixed x, y coordinates, and draw the kinematic diagram of the body. Indicate on it the accelerations of points A and B, $\mathbf{a}_A$ and $\mathbf{a}_B$, the angular velocity $\boldsymbol{\omega}$, the angular acceleration $\boldsymbol{\alpha}$, and the relative position vector $\mathbf{r}_{B/A}$. Normally the base point A is selected as a point having a known acceleration. In particular, if points A and B move along *curved paths,* their accelerations should be expressed in terms of their tangential and normal components, i.e., $\mathbf{a}_A = (\mathbf{a}_A)_t + (\mathbf{a}_A)_n$ and $\mathbf{a}_B = (\mathbf{a}_B)_t + (\mathbf{a}_B)_n$. Here $(a_A)_n = (v_A)^2/\rho_A$ and $(a_B)_n = (v_B)^2/\rho_B$, where ρ_A and ρ_B define the radii of curvature of the paths of points A and B, respectively. Identify the two unknowns on the diagram. If the magnitudes of $(\mathbf{a}_A)_t$, $(\mathbf{a}_B)_t$, or $\boldsymbol{\alpha}$ are unknown, the sense of direction of these vectors may be assumed.

If the acceleration equation is applied *without* using a Cartesian vector analysis, then the magnitude and direction of the relative acceleration components $(\mathbf{a}_{B/A})_t$ and $(\mathbf{a}_{B/A})_n$ must be established. This can be done by drawing a kinematic diagram such as shown in Fig. 16–23*c*. Since the body is considered to be momentarily "pinned" at the base point A, these components have *magnitudes* of $(a_{B/A})_t = \alpha r_{B/A}$ and $(a_{B/A})_n = \omega^2 r_{B/A}$. Their *sense of direction* is established from the diagram such that $(\mathbf{a}_{B/A})_t$ acts perpendicular to $\mathbf{r}_{B/A}$, in accordance with the rotational motion $\boldsymbol{\alpha}$ of the body, and $(\mathbf{a}_{B/A})_n$ is directed from B toward A.*

Acceleration Equation. To apply Eq. 16–18, express the vectors in Cartesian vector form and substitute them into the equation. Evaluate the cross product and then equate the respective $\mathbf{i}$ and $\mathbf{j}$ components to obtain two scalar equations. To obtain these scalar equations *directly* write Eq. 16–17 in a symbolic form, $\mathbf{a}_B = \mathbf{a}_A + (\mathbf{a}_{B/A})_t + (\mathbf{a}_{B/A})_n$, and underneath each of the terms represent the vectors by their magnitudes and directions. To do this, use the data tabulated on the kinematic diagrams. The scalar equations are determined from the x and y components of these vectors. Solve these equations for the two unknowns. If the solution yields a *negative* answer for an *unknown,* it indicates the sense of direction of the vector is opposite to that shown on the kinematic diagram.

*Perhaps the notation $\mathbf{a}_B = \mathbf{a}_A + (\mathbf{a}_{B/A(\text{pin})})_t + (\mathbf{a}_{B/A(\text{pin})})_n$ may be helpful in recalling that A is pinned.

Example 16–13

The rod AB shown in Fig. 16–26a is confined to move along the inclined planes at A and B. If point A has an acceleration of 3 m/s^2 and a velocity of 2 m/s, both directed down the plane at the instant the rod becomes horizontal, determine the angular acceleration of the rod at this instant.

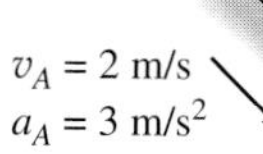

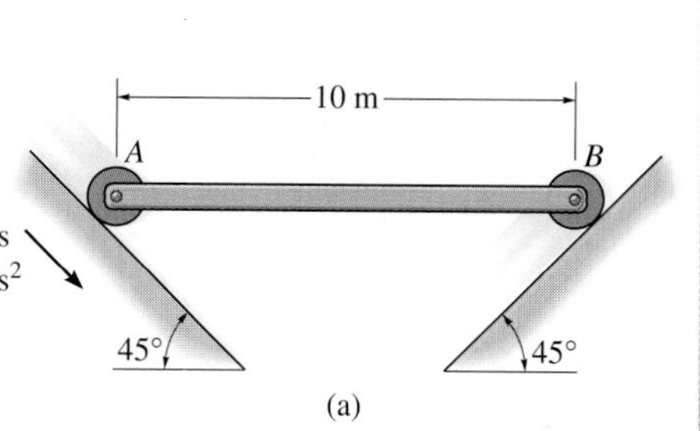

(a)

SOLUTION I (*VECTOR ANALYSIS*)

We will apply the acceleration equation to points A and B on the rod. To do so it is first necessary to determine the angular velocity of the rod. Show that it is $\omega = 0.283$ rad/s ↰ using either the velocity equation or the method of instantaneous centers.

Kinematic Diagram. The x, y axes are established in Fig. 16–26b. Since points A and B both move along straight-line paths, they have *no* components of acceleration normal to the paths. There are two unknowns in Fig. 16–26b, namely, a_B and α.

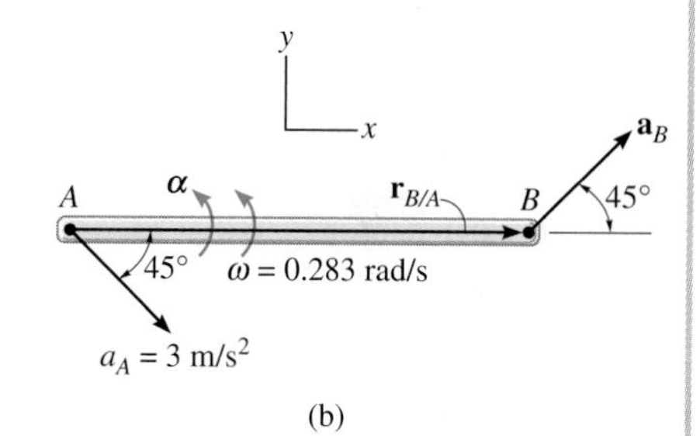

(b)

Acceleration Equation. Applying Eq. 16–18 to points A and B on the rod, and expressing each of the vectors in Cartesian vector form, we have

$$\mathbf{a}_B = \mathbf{a}_A + \boldsymbol{\alpha} \times \mathbf{r}_{B/A} - \omega^2 \mathbf{r}_{B/A}$$

$$a_B \cos 45^\circ \mathbf{i} + a_B \sin 45^\circ \mathbf{j}$$
$$= 3 \cos 45^\circ \mathbf{i} - 3 \sin 45^\circ \mathbf{j} + (\alpha \mathbf{k}) \times (10\mathbf{i})$$

Carrying out the cross product and equating the **i** and **j** components yields

$$a_B \cos 45^\circ = 3 \cos 45^\circ - (0.283)^2(10) \qquad (1)$$
$$a_B \sin 45^\circ = -3 \sin 45^\circ + \alpha(10) \qquad (2)$$

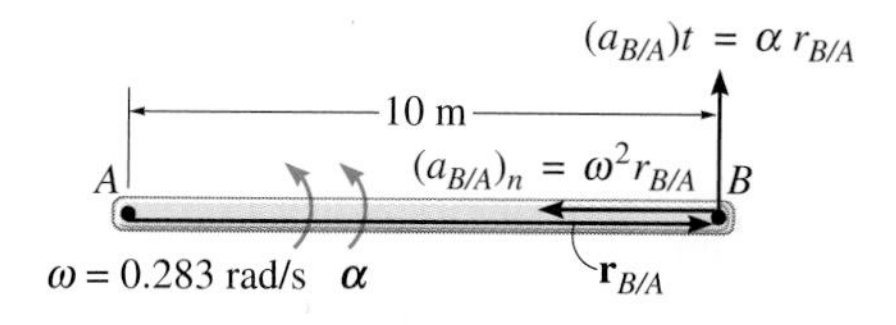

(c)

Fig. 16–26

Solving, we have

$$a_B = 1.87 \text{ m/s}^2 \ \angle 45^\circ$$
$$\alpha = 0.344 \text{ rad/s}^2 \ ↰ \qquad \textit{Ans.}$$

SOLUTION II (*SCALAR COMPONENTS*)

As an alternative procedure, the scalar component equations 1 and 2 can be obtained directly. From the kinematic diagram, showing the relative acceleration components $(\mathbf{a}_{B/A})_t$ and $(\mathbf{a}_{B/A})_n$, Fig. 16–26c, we have

$$\mathbf{a}_B = \mathbf{a}_A + (\mathbf{a}_{B/A})_t + (\mathbf{a}_{B/A})_n$$

$$\begin{bmatrix} a_B \\ \angle 45^\circ \end{bmatrix} = \begin{bmatrix} 3 \text{ m/s}^2 \\ \searrow 45^\circ \end{bmatrix} + \begin{bmatrix} \alpha(10 \text{ m}) \\ \uparrow \end{bmatrix} + \begin{bmatrix} (0.283 \text{ rad/s})^2(10 \text{ m}) \\ \leftarrow \end{bmatrix}$$

Equating the x and y components yields Eqs. 1 and 2, and the solution proceeds as before.

Example 16–14

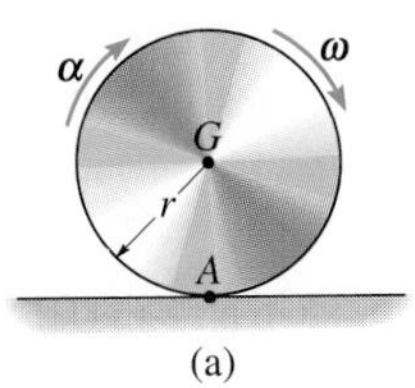

(a)

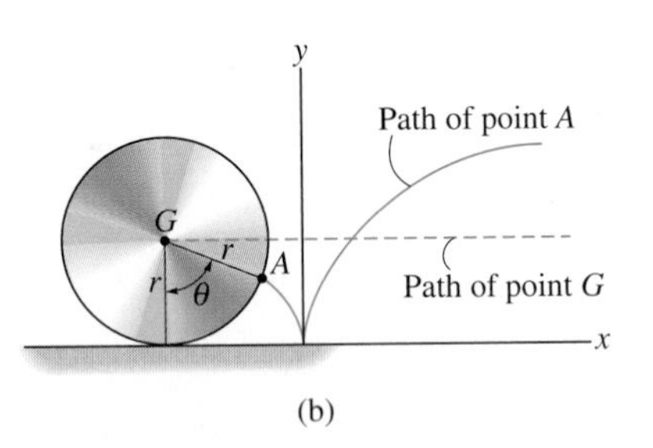

(b)

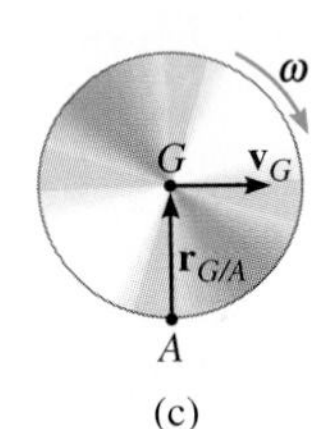

(c)

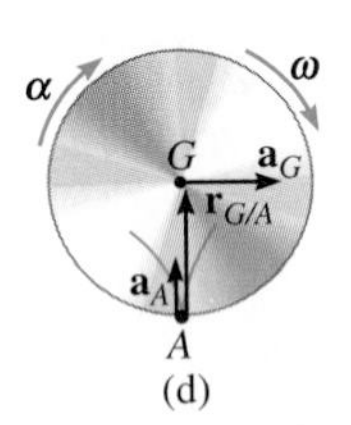

(d)

Fig. 16–27

At a given instant, the cylinder of radius r, shown in Fig. 16–27a, has an angular velocity $\boldsymbol{\omega}$ and angular acceleration $\boldsymbol{\alpha}$. Determine the velocity and acceleration of its center G if it rolls without slipping.

SOLUTION

As the cylinder rolls, point G moves along a straight line, and point A, located on the rim of the cylinder, moves along a curved path called a *cycloid,** Fig. 16–27b. We will apply the velocity and acceleration equations to these two points.

Velocity Analysis. Since no slipping occurs, at the instant A contacts the ground, $\mathbf{v}_A = \mathbf{0}$. Thus, from the kinematic diagram in Fig. 16–27c we have

$$\mathbf{v}_G = \mathbf{v}_A + \boldsymbol{\omega} \times \mathbf{r}_{G/A}$$

$$v_G\mathbf{i} = \mathbf{0} + (-\omega\mathbf{k}) \times (r\mathbf{j})$$

$$v_G = \omega r \qquad (1) \quad \textit{Ans.}$$

This same result can also be obtained directly by noting that point A represents the instantaneous center of zero velocity.

Kinematic Diagram. The acceleration of point G is horizontal since it moves along a *straight-line path. Just before* point A touches the ground, its velocity is directed downward along the y axis, Fig. 16–27b, and just after contact, its velocity is directed *upward.*† For this reason, point A begins to accelerate upward when it leaves the ground at A, Fig. 16–27d. The magnitudes of $\mathbf{a}_A$ and $\mathbf{a}_G$ are unknown.

Acceleration Equation

$$\mathbf{a}_G = \mathbf{a}_A + \boldsymbol{\alpha} \times \mathbf{r}_{G/A} - \omega^2\mathbf{r}_{G/A}$$

$$a_G\mathbf{i} = a_A\mathbf{j} + (-\alpha\mathbf{k}) \times (r\mathbf{j}) - \omega^2(r\mathbf{j})$$

Evaluating the cross product and equating the **i** and **j** components yields

$$a_G = \alpha r \qquad (2) \quad \textit{Ans.}$$

$$a_A = \omega^2 r \qquad (3)$$

These important results, that $v_G = \omega r$ and $a_G = \alpha r$, were also obtained in Example 16–3. They apply to any circular object, such as a ball, pulley, disk, etc., that rolls *without* slipping. Also, the fact that $a_A = \omega^2 r$ indicates that the instantaneous center of zero velocity, point A, is *not* a point of zero acceleration.

*Although it is not necessary here, one can show that the equation of the path (cycloid) can be written in terms of θ as $x = r(\theta - \sin\theta)$ and $y = r(1 - \cos\theta)$.

†The slope of the path, dy/dx, is along the y axis at $\theta = 90°$.

Example 16–15

The ball rolls without slipping and has the angular motion shown in Fig. 16–28*a*. Determine the acceleration of point *B* and point *A* at this instant.

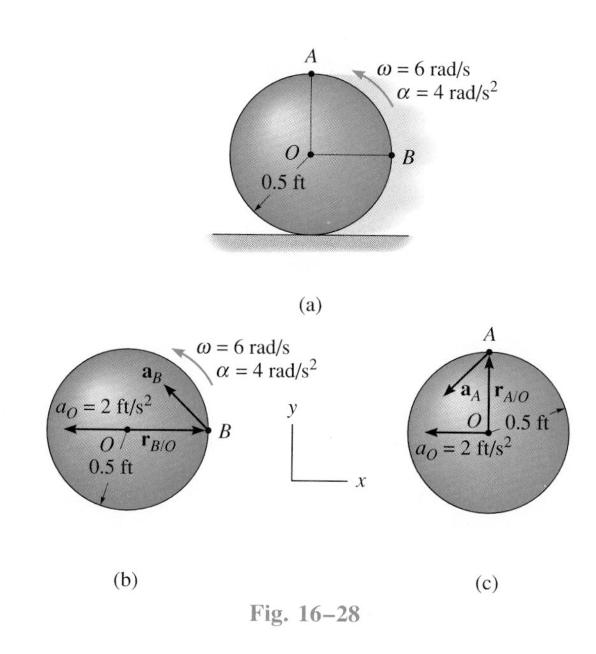

Fig. 16–28

SOLUTION *(VECTOR ANALYSIS)*

Kinematic Diagram. Using the results of Example 16–3, the center has an acceleration of $a_O = \alpha r = (4 \text{ rad/s}^2)(0.5 \text{ ft}) = 2 \text{ ft/s}^2$. We will apply the acceleration equation to points *O* and *B* and points *O* and *A*.

Acceleration Equation

For point *B*

$$\mathbf{a}_B = \mathbf{a}_O + \boldsymbol{\alpha} \times \mathbf{r}_{B/O} - \omega^2 \mathbf{r}_{B/O}$$
$$\mathbf{a}_B = -2\mathbf{i} + (4\mathbf{k}) \times (0.5\mathbf{i}) - (6)^2(0.5\mathbf{i})$$
$$\mathbf{a}_B = \{-20\mathbf{i} + 2\mathbf{j}\} \text{ ft/s}^2 \qquad \textit{Ans.}$$

For point *A*

$$\mathbf{a}_A = \mathbf{a}_O + \boldsymbol{\alpha} \times \mathbf{r}_{A/O} - \omega^2 \mathbf{r}_{A/O}$$
$$\mathbf{a}_A = -2\mathbf{i} + (4\mathbf{k}) \times (0.5\mathbf{j}) - (6)^2(0.5\mathbf{j})$$
$$\mathbf{a}_A = \{-4\mathbf{i} - 18\mathbf{j}\} \text{ ft/s}^2 \qquad \textit{Ans.}$$

Example 16–16

The spool shown in Fig. 16–29*a* unravels from the cord, such that at the instant shown it has an angular velocity of 3 rad/s and an angular acceleration of 4 rad/s^2. Determine the acceleration of point B.

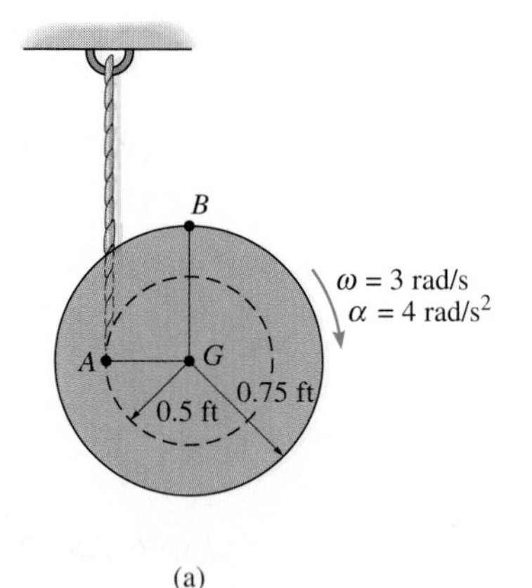

(a)

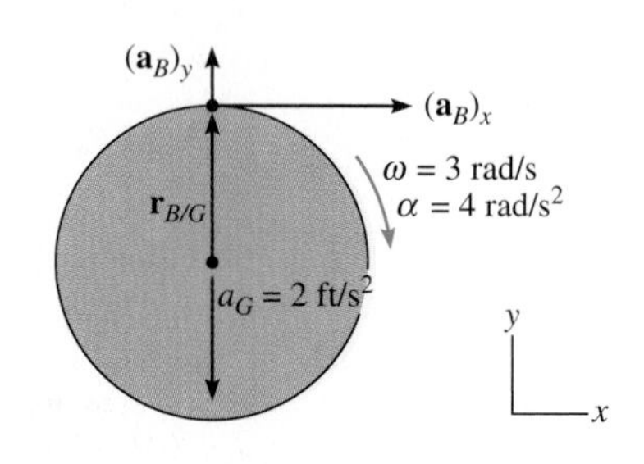

(b)

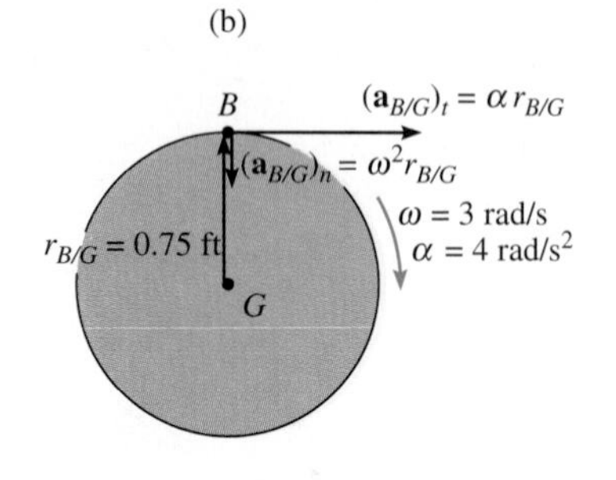

(c)

Fig. 16–29

SOLUTION I (*VECTOR ANALYSIS*)

The spool "appears" to be rolling downward without slipping at point A. Therefore, we can use the results of Example 16–3 to determine the acceleration of point G, i.e.,

$$a_G = \alpha r = 4(0.5) = 2 \text{ ft/s}^2$$

We will therefore apply the acceleration equation at points G and B.

Kinematic Diagram. Point B moves along a *curved path* having an *unknown* radius of curvature.* Its acceleration will be represented by its unknown x and y components as shown in Fig. 16–29*b*.

Acceleration Equation

$$\mathbf{a}_B = \mathbf{a}_G + \boldsymbol{\alpha} \times \mathbf{r}_{B/G} - \omega^2 \mathbf{r}_{B/G}$$

$$(a_B)_x\mathbf{i} + (a_B)_y\mathbf{j} = -2\mathbf{j} + (-4\mathbf{k}) \times (0.75\mathbf{j}) - (3)^2(0.75\mathbf{j})$$

Therefore, the component equations are

$$(a_B)_x = 4(0.75) = 3 \text{ ft/s}^2 \rightarrow \qquad (1)$$

$$(a_B)_y = -2 - 6.75 = -8.75 \text{ ft/s}^2 = 8.75 \text{ ft/s}^2 \downarrow \qquad (2)$$

The magnitude and direction of $\mathbf{a}_B$ are therefore

$$a_B = \sqrt{(3)^2 + (8.75)^2} = 9.25 \text{ ft/s}^2 \qquad \textit{Ans.}$$

$$\theta = \tan^{-1}\frac{8.75}{3} = 71.1^\circ \qquad \textit{Ans.}$$

SOLUTION II (*SCALAR COMPONENTS*)

This problem may be solved by writing the scalar component equations directly. The kinematic diagram in Fig. 16–29*c* shows the relative acceleration components $(\mathbf{a}_{B/G})_t$ and $(\mathbf{a}_{B/G})_n$. Thus,

$$\mathbf{a}_B = \mathbf{a}_G + (\mathbf{a}_{B/G})_t + (\mathbf{a}_{B/G})_n$$

$$\begin{bmatrix}(a_B)_x \\ \rightarrow\end{bmatrix} + \begin{bmatrix}(a_B)_y \\ \uparrow\end{bmatrix} = \begin{bmatrix}2 \text{ ft/s}^2 \\ \downarrow\end{bmatrix} + \begin{bmatrix}4 \text{ rad/s}^2(0.75 \text{ ft}) \\ \rightarrow\end{bmatrix} + \begin{bmatrix}(3 \text{ rad/s})^2(0.75 \text{ ft}) \\ \downarrow\end{bmatrix}$$

The x and y components yield Eqs. 1 and 2 above.

*Realize that the path's radius of curvature ρ is *not* equal to the radius of the cylinder since the cylinder is *not* rotating about point G. Furthermore, ρ is *not* defined as the distance from A (*IC*) to B, since the location of the *IC* depends only on the velocity of a point and *not* the geometry of its path.

Example 16–17

The collar C in Fig. 16–30a is moving downward with an acceleration of 1 m/s^2. At the instant shown, it has a speed of 2 m/s which gives links CB and AB an angular velocity $\omega_{AB} = \omega_{CB} = 10$ rad/s. (See Example 16–8.) Determine the angular accelerations of CB and AB at this instant.

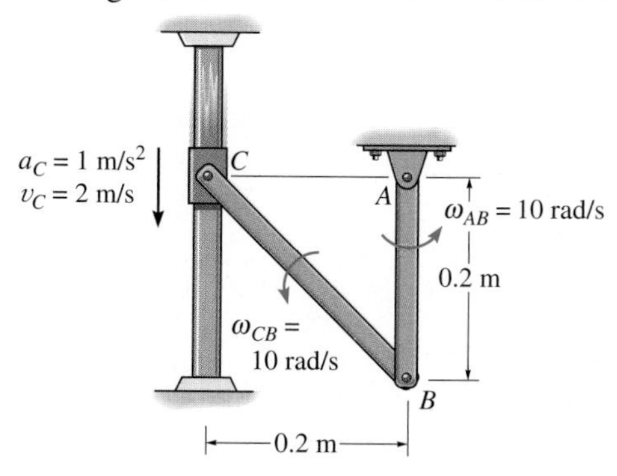

(a)

SOLUTION (*VECTOR ANALYSIS*)

Kinematic Diagram. The kinematic diagrams of *both* links AB and CB are shown in Fig. 16–30b. To solve, we will apply the appropriate kinematic equation to each link.

Acceleration Equation

Link AB (rotation about a fixed axis):

$$\mathbf{a}_B = \boldsymbol{\alpha}_{AB} \times \mathbf{r}_B - \omega_{AB}^2 \mathbf{r}_B$$
$$\mathbf{a}_B = (\alpha_{AB}\mathbf{k}) \times (-0.2\mathbf{j}) - (10)^2(-0.2\mathbf{j})$$
$$\mathbf{a}_B = 0.2\alpha_{AB}\mathbf{i} + 20\mathbf{j}$$

Note that $\mathbf{a}_B$ has two components since it moves along a *curved path.*

Link BC (general plane motion): Using the result for $\mathbf{a}_B$ and applying Eq. 16–18, we have

$$\mathbf{a}_B = \mathbf{a}_C + \boldsymbol{\alpha}_{CB} \times \mathbf{r}_{B/C} - \omega_{CB}^2 \mathbf{r}_{B/C}$$
$$0.2\alpha_{AB}\mathbf{i} + 20\mathbf{j} = -1\mathbf{j} + (\alpha_{CB}\mathbf{k}) \times (0.2\mathbf{i} - 0.2\mathbf{j}) - (10)^2(0.2\mathbf{i} - 0.2\mathbf{j})$$
$$0.2\alpha_{AB}\mathbf{i} + 20\mathbf{j} = -1\mathbf{j} + 0.2\alpha_{CB}\mathbf{j} + 0.2\alpha_{CB}\mathbf{i} - 20\mathbf{i} + 20\mathbf{j}$$

Thus,

$$0.2\alpha_{AB} = 0.2\alpha_{CB} - 20$$
$$20 = -1 + 0.2\alpha_{CB} + 20$$

Solving,

$$\alpha_{CB} = 5 \text{ rad/s}^2 \circlearrowleft \qquad \textit{Ans.}$$
$$\alpha_{AB} = -95 \text{ rad/s}^2 = 95 \text{ rad/s}^2 \circlearrowright \qquad \textit{Ans.}$$

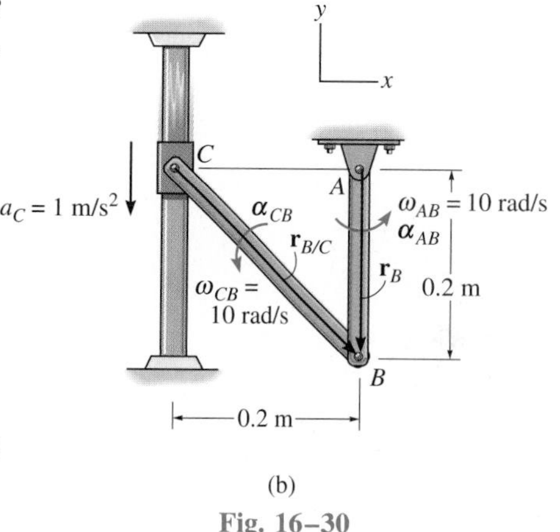

(b)

Fig. 16–30

Example 16–18

The crankshaft AB of an engine turns with a clockwise angular acceleration of 20 rad/s^2, Fig. 16–31a. Determine the acceleration of the piston at the instant AB is in the position shown. At this instant $\omega_{AB} = 10$ rad/s and $\omega_{BC} = 2.43$ rad/s.

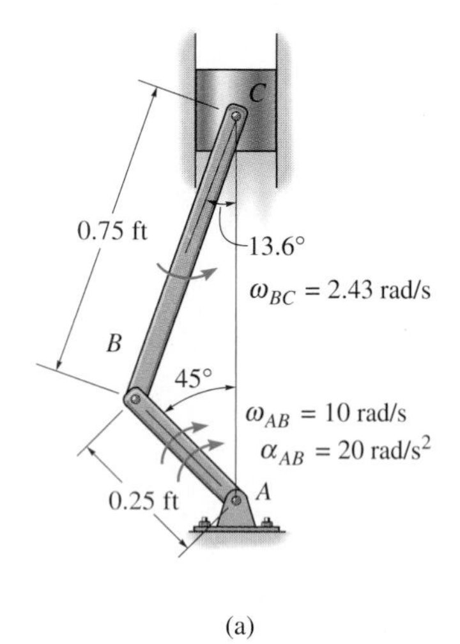

(a)

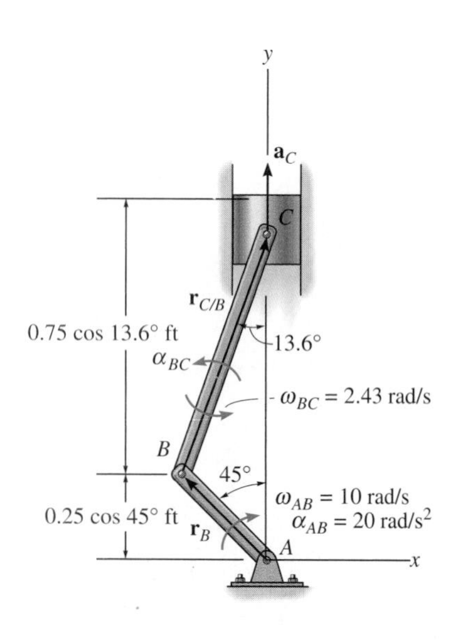

(b)

Fig. 16–31

SOLUTION *(VECTOR ANALYSIS)*

Kinematic Diagram. The kinematic diagrams for both AB and BC are shown in Fig. 16–31b. Here $\mathbf{a}_C$ is vertical since C moves along a straight-line path.

Acceleration Equation. Expressing each of the position vectors in Cartesian vector form

$$\mathbf{r}_{B/A} = \{-0.25 \sin 45°\mathbf{i} + 0.25 \cos 45°\mathbf{j}\} \text{ ft} = \{-0.177\mathbf{i} + 0.177\mathbf{j}\} \text{ ft}$$
$$\mathbf{r}_{C/B} = \{0.75 \sin 13.6°\mathbf{i} + 0.75 \cos 13.6°\mathbf{j}\} \text{ ft} = \{0.176\mathbf{i} + 0.729\mathbf{j}\} \text{ ft}$$

Crankshaft rod AB (rotation about a fixed axis):

$$\begin{aligned}\mathbf{a}_B &= \boldsymbol{\alpha}_{AB} \times \mathbf{r}_{B/A} - \omega_{AB}^2\mathbf{r}_{B/A}\\ &= (-20\mathbf{k}) \times (-0.177\mathbf{i} + 0.177\mathbf{j}) - (10)^2(-0.177\mathbf{i} + 0.177\mathbf{j})\\ &= \{21.24\mathbf{i} - 14.16\mathbf{j}\} \text{ ft/s}^2\end{aligned}$$

Connecting rod BC (general plane motion): Using the result for $\mathbf{a}_B$ and noting that a_C is in the vertical direction, we have

$$\mathbf{a}_C = \mathbf{a}_B + \boldsymbol{\alpha}_{BC} \times \mathbf{r}_{C/B} - \omega_{BC}^2\mathbf{r}_{C/B}$$
$$a_C\mathbf{j} = 21.24\mathbf{i} - 14.16\mathbf{j} + (\alpha_{BC}\mathbf{k}) \times (0.176\mathbf{i} + 0.729\mathbf{j}) - (2.43)^2(0.176\mathbf{i} + 0.729\mathbf{j})$$
$$a_C\mathbf{j} = 21.24\mathbf{i} - 14.16\mathbf{j} + 0.176\alpha_{BC}\mathbf{j} - 0.729\alpha_{BC}\mathbf{i} - 1.04\mathbf{i} - 4.30\mathbf{j}$$
$$0 = 20.20 - 0.729\alpha_{BC}$$
$$a_C = 0.176\alpha_{BC} - 18.46$$

Solving yields

$$\alpha_{BC} = 27.7 \text{ rad/s}^2 \circlearrowleft$$
$$a_C = -13.6 \text{ ft/s}^2 \qquad \textit{Ans.}$$

Since the piston is moving upward, the negative sign for a_C indicates that the piston is decelerating, i.e., $\mathbf{a}_C = \{-13.6\mathbf{j}\}$ ft/s^2. This causes the speed of the piston to decrease until the crankshaft AB becomes vertical, at which time the piston is momentarily at rest.

PROBLEMS

16–119. At a given instant the top end A of the bar has the velocity and acceleration shown. Determine the acceleration of the bottom B and the bar's angular acceleration at this instant.

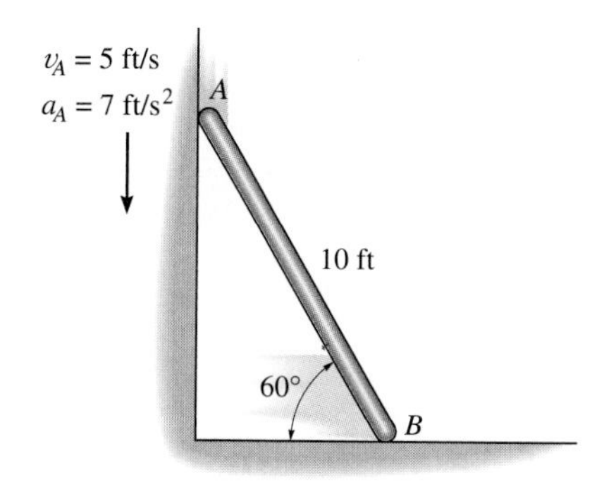

Prob. 16–119

***16–120.** Determine the acceleration of piston P in Prob. 16–60 at the instant shown. The crankshaft is rotating at a constant rate of $\omega = 500$ rad/s as shown.

16–121. At a given instant the bottom A of the ladder has an acceleration $a_A = 4$ ft/s^2 and velocity $v_A = 6$ ft/s, both acting to the left. Determine the acceleration of the top of the ladder, B, and the ladder's angular acceleration at this same instant.

16–122. At a given instant the top B of the ladder has an acceleration $a_B = 2$ ft/s^2 and a velocity of $v_B = 4$ ft/s, both acting downward. Determine the acceleration of the bottom A of the ladder, and the ladder's angular acceleration at this instant.

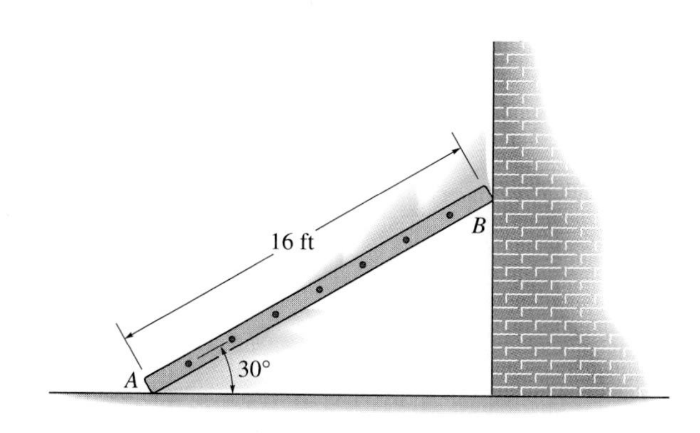

Probs. 16–121/16–122

16–123. At a given instant, link AB has an angular acceleration $\alpha_{AB} = 12$ rad/s^2 and an angular velocity $\omega_{AB} = 4$ rad/s. Determine the angular velocity and angular acceleration of link CD at this instant.

***16–124.** At a given instant, link CD has an angular acceleration $\alpha_{CD} = 5$ rad/s^2 and angular velocity $\omega_{CD} = 2$ rad/s. Determine the angular velocity and angular acceleration of link AB at this instant.

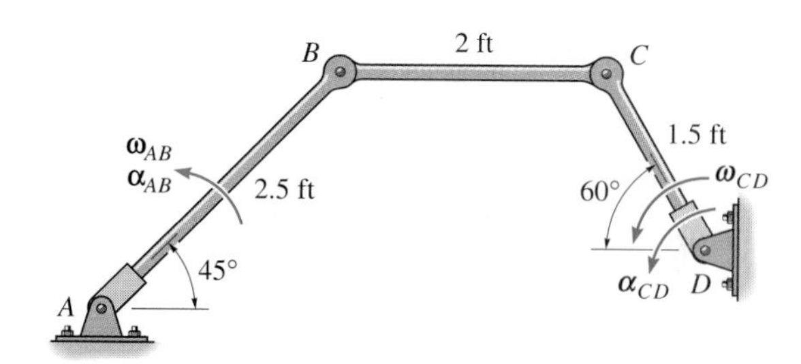

Probs. 16–123/16–124

16–125. The wheel is moving to the right such that it has an angular velocity $\omega = 2$ rad/s and angular acceleration $\alpha = 4$ rad/s^2 at the instant shown. If it does not slip at A, determine the acceleration of point B.

16–126. The wheel is moving to the right such that it has an angular velocity $\omega = 2$ rad/s and angular acceleration $\alpha = 4$ rad/s^2 at the instant shown. If it does not slip at A, determine the acceleration of point D.

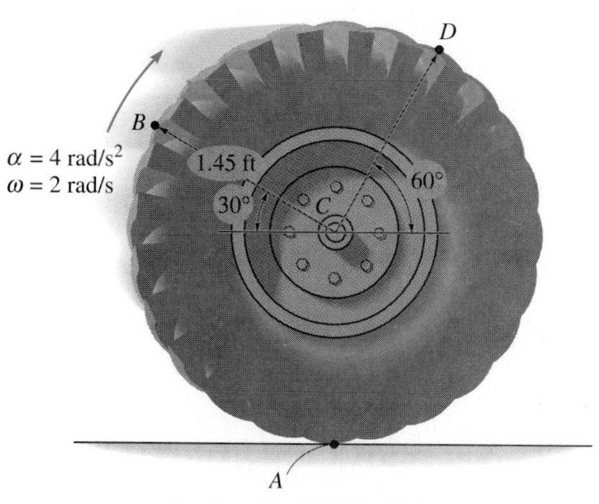

Probs. 16–125/16–126

16–127. At a given instant the wheel is rotating with the angular motions shown. Determine the acceleration of the collar at A at this instant.

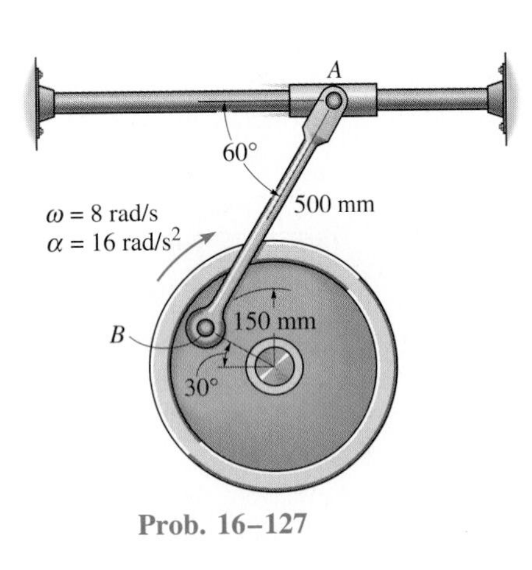

Prob. 16–127

***16–128.** At a given instant the wheel is rotating with the angular motions shown. Determine the acceleration of the collar at A at this instant.

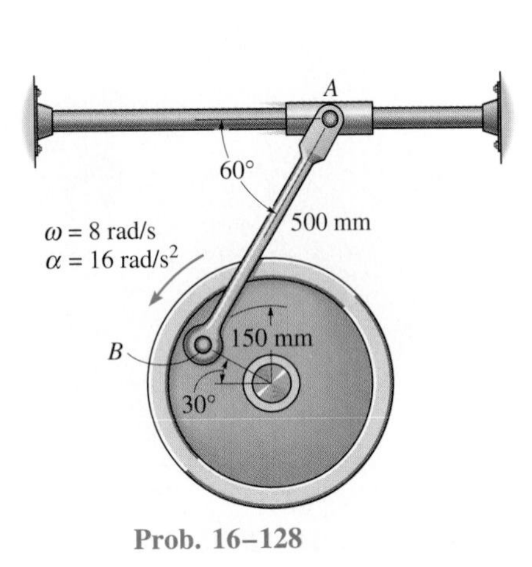

Prob. 16–128

16–129. The flywheel rotates with an angular velocity $\omega = 2$ rad/s and an angular acceleration $\alpha = 6$ rad/s^2. Determine the angular acceleration of links AB and BC at this instant.

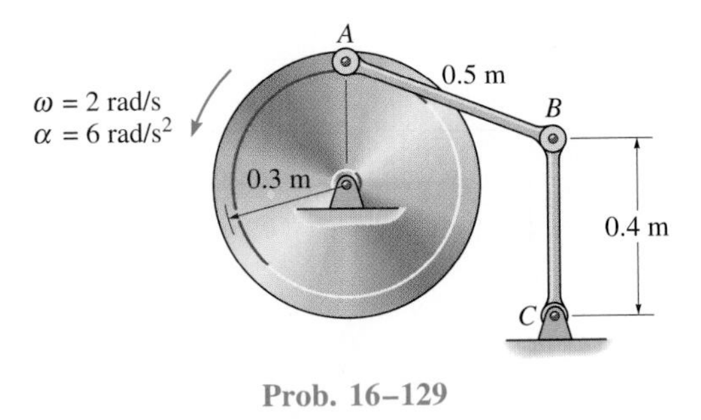

Prob. 16–129

16–130. The ends of the bar AB are confined to move along the paths shown. At a given instant, A has a velocity of 8 ft/s and an acceleration of 3 ft/s^2. Determine the angular velocity and angular acceleration of AB at this instant.

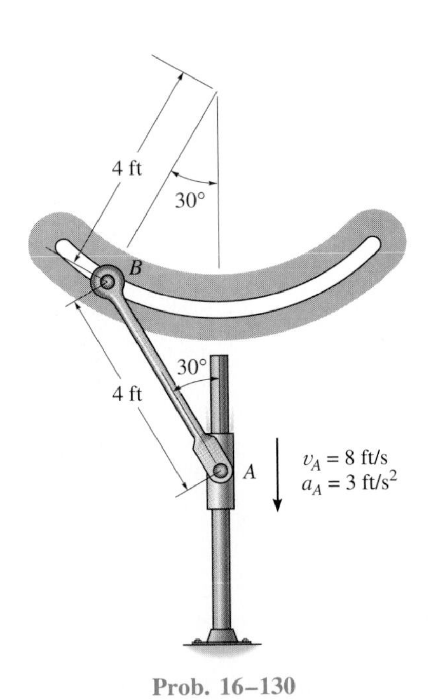

Prob. 16–130

16–131. The disk rotates with an angular velocity $\omega = 5$ rad/s and an angular acceleration $\alpha = 6$ rad/s^2. Determine the angular acceleration of link CB at this instant.

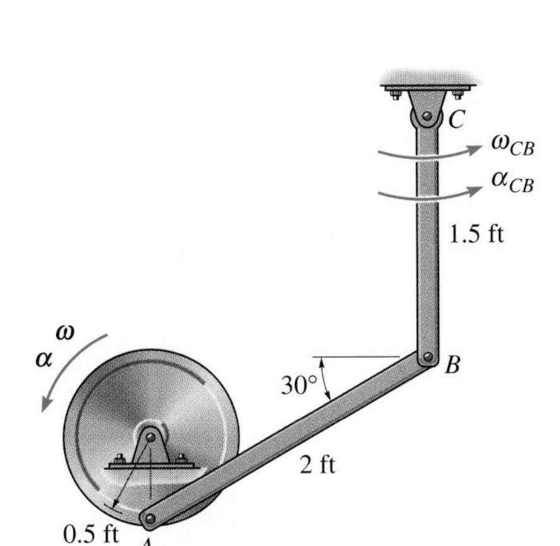

Prob. 16–131

*16–132. Link CB rotates with an angular velocity $\omega_{CB} = 2$ rad/s and angular acceleration $\alpha_{CB} = 4$ rad/s^2. Determine the angular velocity and angular acceleration of the disk at this instant.

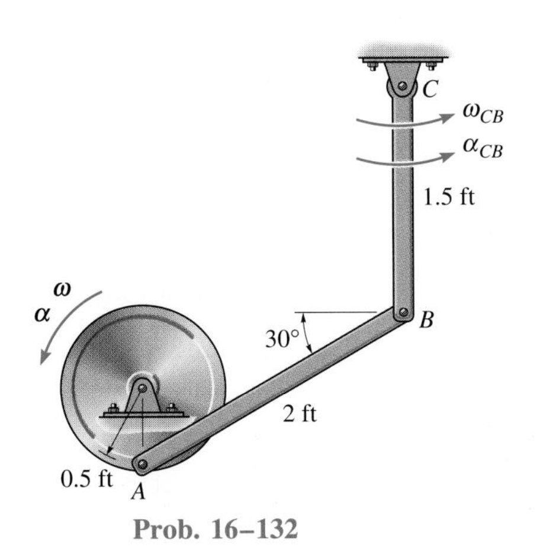

Prob. 16–132

16–133. The tied crank and gear mechanism gives rocking motion to crank AC, necessary for the operation of a printing press. If link DE has the angular motion shown, determine the respective angular velocities and angular accelerations of gear F and crank AC at this instant.

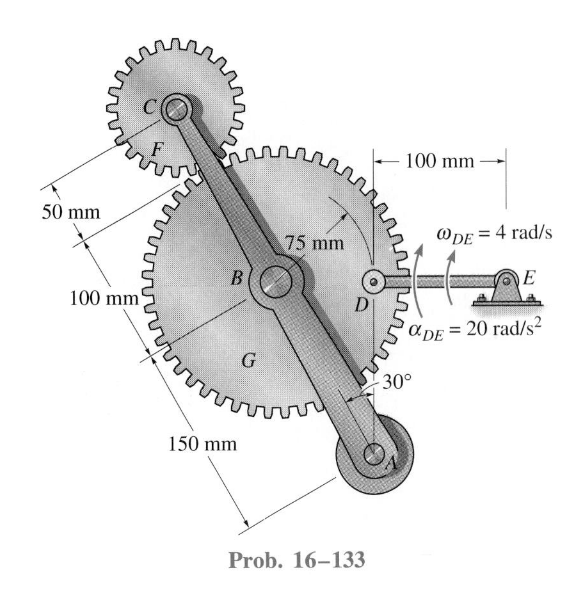

Prob. 16–133

16–134. At a given instant the wheel is rotating with the angular velocity and angular acceleration shown. Determine the acceleration of block B at this instant.

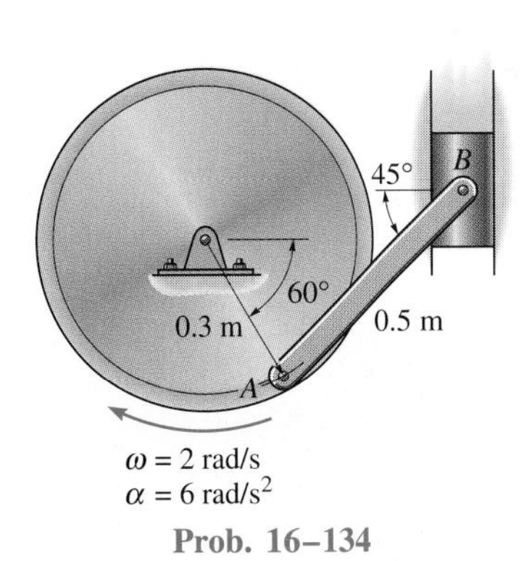

Prob. 16–134

16–135. At a given instant gears A and B have the angular motions shown. Determine the angular acceleration of gear C and the acceleration of its center point D at this instant. Note that the inner hub of gear C is in mesh with gear A and its outer rim is in mesh with gear B.

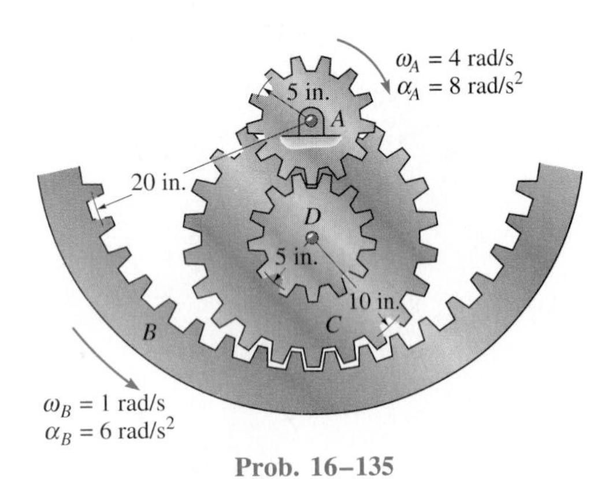

Prob. 16–135

16–138. The hoop is cast on the rough surface such that it has an angular velocity $\omega = 4$ rad/s and an angular deceleration $\alpha = 5$ rad/s^2. Also, its center has a velocity $v_O = 5$ m/s and a deceleration $a_O = 2$ m/s^2. Determine the acceleration of point A at this instant.

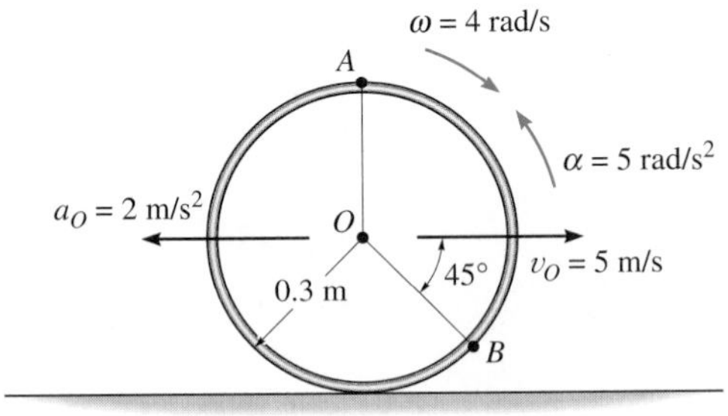

Prob. 16–138

***16–136.** The wheel rolls without slipping such that at the instant shown it has an angular velocity $\boldsymbol{\omega}$ and angular acceleration $\boldsymbol{\alpha}$. Determine the angular velocity and angular acceleration of the rod at this instant.

16–137. The wheel rolls without slipping such that at the instant shown it has an angular velocity $\boldsymbol{\omega}$ and angular acceleration $\boldsymbol{\alpha}$. Determine the velocity and acceleration of point B on the rod at this instant.

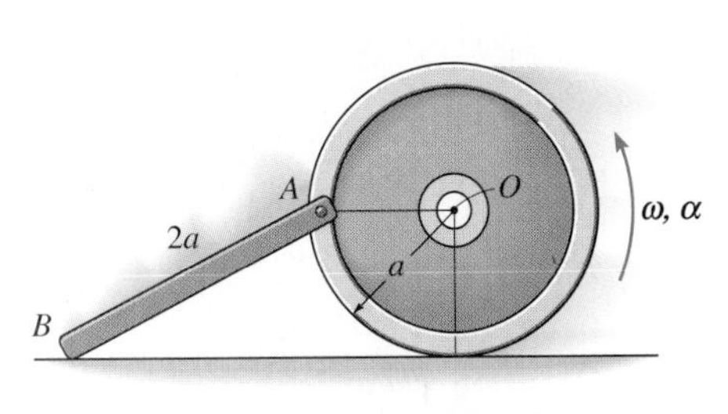

Probs. 16–136/16–137

16–139. The hoop is cast on the rough surface such that it has an angular velocity $\omega = 4$ rad/s and an angular deceleration $\alpha = 5$ rad/s^2. Also, its center has a velocity of $v_O = 5$ m/s and a deceleration $a_O = 2$ m/s^2. Determine the acceleration of point B at this instant.

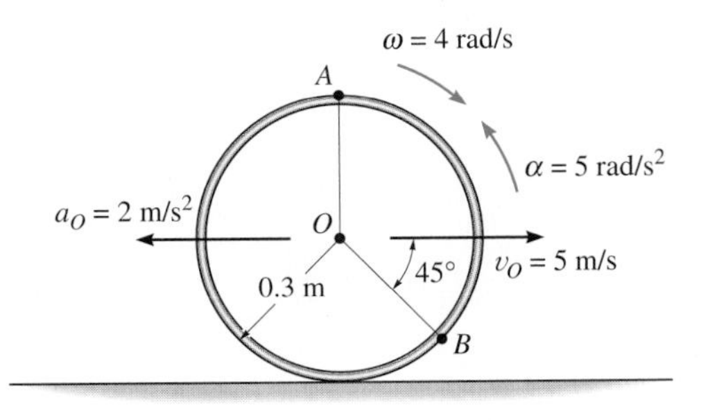

Prob. 16–139

***16–140.** The slider block B is moving to the right with an acceleration of 2 ft/s^2. At the instant shown, its velocity is 6 ft/s. Determine the angular acceleration of link AB and the acceleration of point A at this instant.

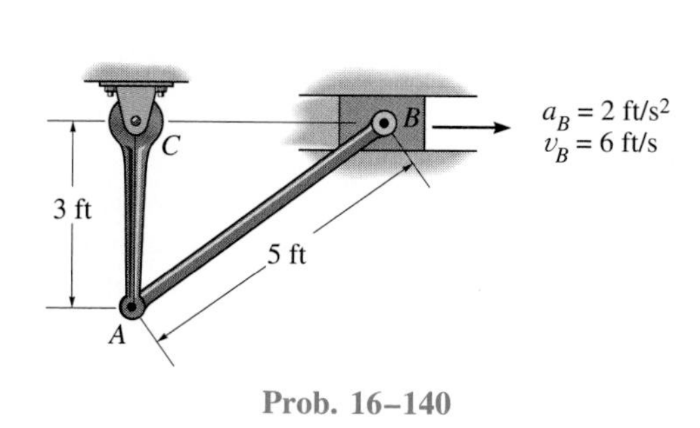

Prob. 16–140

16–141. If the disk turns with a constant angular velocity of 3 rad/s, determine the acceleration of point A at the instant shown.

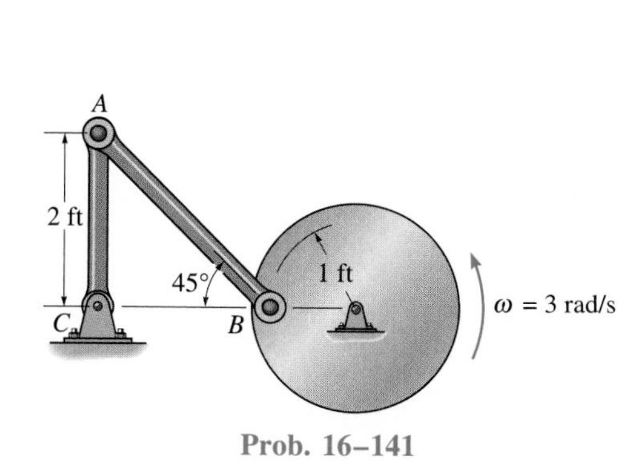

Prob. 16–141

16–142. The ends of the bar AB are confined to move along the paths shown. At a given instant, A has a velocity of $v_A = 4$ ft/s and an acceleration of $a_A = 7$ ft/s^2. Determine the angular velocity and angular acceleration of AB at this instant.

Prob. 16–142

16–143. At a given instant, the cables supporting the pipe have the motions shown. Determine the angular velocity and angular acceleration of the pipe and the velocity and acceleration of point B located on the pipe.

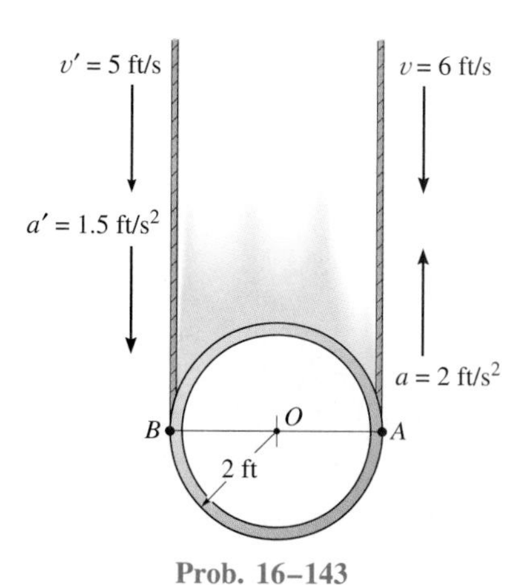

Prob. 16–143

16.8 Relative-Motion Analysis Using Rotating Axes

In the previous sections the relative-motion analysis for velocity and acceleration was described using a translating coordinate system. This type of analysis is useful for determining the motion of points on the *same* rigid body, or the motion of points located on several pin-connected rigid bodies. In some problems, however, rigid bodies (mechanisms) are constructed such that *sliding* will occur at their connections. The kinematic analysis for such cases is best performed if the motion is analyzed using a coordinate system which both *translates* and *rotates*. Furthermore, this frame of reference is useful for analyzing the motions of two points on a mechanism which are *not* located in the *same* rigid body and for specifying the kinematics of particle motion when the particle is moving along a rotating path.

In the following analysis two equations are developed which relate the velocity and acceleration of two points, one of which is the origin of a moving frame of reference subjected to both a translation and a rotation in the plane.* Due to the generality in the derivation which follows, these two points may represent either two particles moving independently of one another or two points located on the same (or different) rigid bodies.

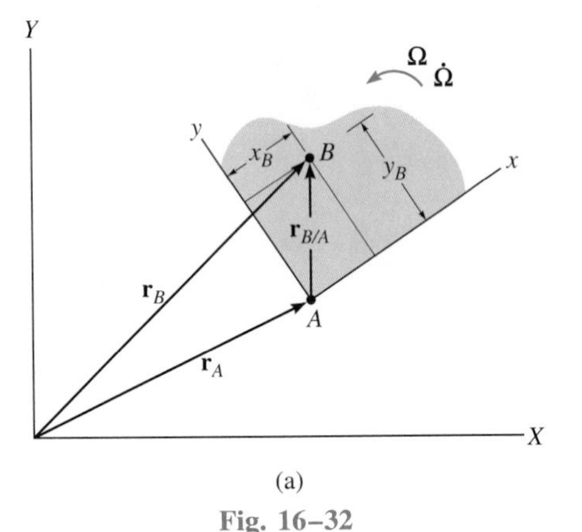

Fig. 16–32

Position. Consider the two points A and B shown in Fig. 16–32*a*. Their location is specified by the position vectors $\mathbf{r}_A$ and $\mathbf{r}_B$, which are measured from the fixed X, Y, Z coordinate system. As shown in the figure, the "base point" A represents the origin of the x, y, z coordinate system, which is assumed to be both translating and rotating with respect to the X, Y, Z system. The position of B with respect to A is specified by the relative-position vector $\mathbf{r}_{B/A}$. The components of this vector may be expressed either in terms of unit vectors along the X, Y axes, i.e., $\mathbf{I}$ and $\mathbf{J}$, or by unit vectors along the x, y axes, i.e., $\mathbf{i}$ and $\mathbf{j}$. Although the magnitude of $\mathbf{r}_{B/A}$ is the same when measured in both coordinate systems, the direction of this vector will be measured differently if the x, y axes are not parallel to the X, Y axes. For the proof which follows, $\mathbf{r}_{B/A}$ will be measured relative to the moving x, y frame of reference. Thus, if B has coordinates (x_B, y_B), Fig. 16–32*a*, then

$$\mathbf{r}_{B/A} = x_B\mathbf{i} + y_B\mathbf{j}$$

Using vector addition, the three position vectors in Fig. 16–32*a* are related by the equation

$$\mathbf{r}_B = \mathbf{r}_A + \mathbf{r}_{B/A} \tag{16–19}$$

*The more general, three-dimensional motion of the points is developed in Sec. 20.4.

At the instant considered, point A has a velocity $\mathbf{v}_A$ and an acceleration $\mathbf{a}_A$, while the angular velocity and angular acceleration of the x, y axes are $\mathbf{\Omega}$ (omega) and $\dot{\mathbf{\Omega}} = d\mathbf{\Omega}/dt$, respectively. All these vectors are measured from the X, Y, Z frame of reference, although they may be expressed in terms of either **I, J, K** or **i, j, k** components. Since planar motion is specified, then by the right-hand rule $\mathbf{\Omega}$ and $\dot{\mathbf{\Omega}}$ are always directed *perpendicular* to the reference plane of motion, whereas $\mathbf{v}_A$ and $\mathbf{a}_A$ lie in this plane.

Velocity. The velocity of point B is determined by taking the time derivative of Eq. 16–19, which yields

$$\mathbf{v}_B = \mathbf{v}_A + \frac{d\mathbf{r}_{B/A}}{dt} \qquad (16\text{–}20)$$

The last term in this equation is evaluated as follows:

$$\begin{aligned}\frac{d\mathbf{r}_{B/A}}{dt} &= \frac{d}{dt}(x_B\mathbf{i} + y_B\mathbf{j})\\ &= \frac{dx_B}{dt}\mathbf{i} + x_B\frac{d\mathbf{i}}{dt} + \frac{dy_B}{dt}\mathbf{j} + y_B\frac{d\mathbf{j}}{dt}\\ &= \left(\frac{dx_B}{dt}\mathbf{i} + \frac{dy_B}{dt}\mathbf{j}\right) + \left(x_B\frac{d\mathbf{i}}{dt} + y_B\frac{d\mathbf{j}}{dt}\right) \qquad (16\text{–}21)\end{aligned}$$

The two terms in the first set of parentheses represent the components of velocity of point B as measured by an observer attached to the moving x, y, z coordinate system. These terms will be denoted by vector $(\mathbf{v}_{B/A})_{xyz}$. In the second set of parentheses the instantaneous time rate of change of the unit vectors **i** and **j** is measured by an observer located in the fixed X, Y, Z coordinate system. These changes, $d\mathbf{i}$ and $d\mathbf{j}$, are due *only* to the instantaneous *rotation* $d\theta$ of the x, y, z axes, causing **i** to become $\mathbf{i}' = \mathbf{i} + d\mathbf{i}$ and **j** to become $\mathbf{j}' = \mathbf{j} + d\mathbf{j}$, Fig. 16–32*b*. As shown, the *magnitudes* of both $d\mathbf{i}$ and $d\mathbf{j}$ equal 1 $(d\theta)$, since $i = i' = j = j' = 1$. The *direction* of $d\mathbf{i}$ is defined by $+\mathbf{j}$, since $d\mathbf{i}$ is tangent to the path described by the arrowhead of **i** in the limit as $\Delta t \rightarrow dt$. Likewise, $d\mathbf{j}$ acts in the $-\mathbf{i}$ direction, Fig. 16–32*b*. Hence,

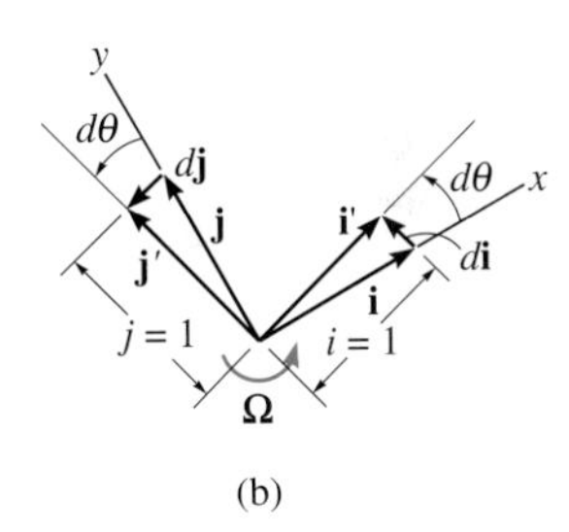

(b)

$$\frac{d\mathbf{i}}{dt} = \frac{d\theta}{dt}(\mathbf{j}) = \Omega\mathbf{j} \qquad \frac{d\mathbf{j}}{dt} = \frac{d\theta}{dt}(-\mathbf{i}) = -\Omega\mathbf{i}$$

Viewing the axes in three dimensions, Fig. 16–32*c*, and noting that $\mathbf{\Omega} = \Omega\mathbf{k}$, we can express the above derivatives in terms of the cross product as

$$\frac{d\mathbf{i}}{dt} = \mathbf{\Omega} \times \mathbf{i} \qquad \frac{d\mathbf{j}}{dt} = \mathbf{\Omega} \times \mathbf{j} \qquad (16\text{–}22)$$

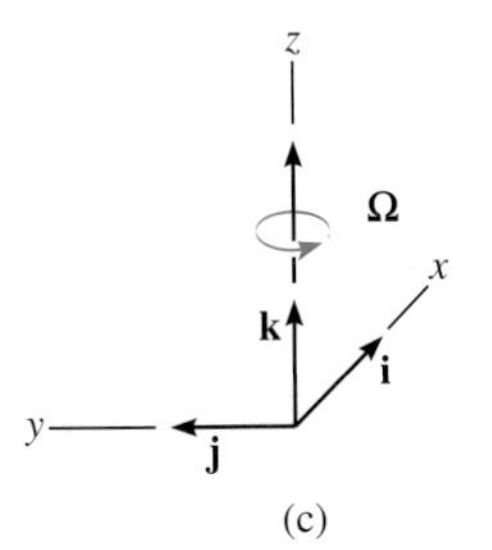

(c)

Fig. 16–32 *(cont'd)*

Substituting these results into Eq. 16–21 and using the distributive property of the vector cross product, we obtain

$$\frac{d\mathbf{r}_{B/A}}{dt} = (\mathbf{v}_{B/A})_{xyz} + \mathbf{\Omega} \times (x_B\mathbf{i} + y_B\mathbf{j}) = (\mathbf{v}_{B/A})_{xyz} + \mathbf{\Omega} \times \mathbf{r}_{B/A} \qquad (16\text{–}23)$$

Hence, Eq. 16–20 becomes

$$\mathbf{v}_B = \mathbf{v}_A + \mathbf{\Omega} \times \mathbf{r}_{B/A} + (\mathbf{v}_{B/A})_{xyz} \tag{16–24}$$

where

$\mathbf{v}_B$ = velocity of B, measured from the X, Y, Z reference

$\mathbf{v}_A$ = velocity of the origin A of the x, y, z reference, measured from the X, Y, Z reference

$(\mathbf{v}_{B/A})_{xyz}$ = relative velocity of "B with respect to A," as measured by an observer attached to the *rotating* x, y, z reference

$\mathbf{\Omega}$ = angular velocity of the x, y, z reference, measured from the X, Y, Z reference

$\mathbf{r}_{B/A}$ = relative position of "B with respect to A"

Comparing Eq. 16–24 with Eq. 16–16 ($\mathbf{v}_B = \mathbf{v}_A + \mathbf{\Omega} \times \mathbf{r}_{B/A}$), which is valid for a translating frame of reference, it can be seen that the only difference between the equations is represented by the term $(\mathbf{v}_{B/A})_{xyz}$.

When applying Eq. 16–24 it is often useful to understand what each of the terms represents. In order of appearance, they are as follows:

$\mathbf{v}_B$ {absolute velocity of B

(equals)

$\mathbf{v}_A$ {absolute velocity of origin of x, y, z frame

(plus)

$\mathbf{\Omega} \times \mathbf{r}_{B/A}$ {angular velocity effect caused by rotation of x, y, z frame

} motion of x, y, z frame observed from the X, Y, Z frame

(plus)

$(\mathbf{v}_{B/A})_{xyz}$ {relative velocity of B with respect to A } motion of B observed from the x, y, z frame

Acceleration. The acceleration of point B, observed from the X, Y, Z coordinate system, may be expressed in terms of its motion measured with respect to the rotating or moving system of coordinates by taking the time derivative of Eq. 16–24, i.e.,

$$\frac{d\mathbf{v}_B}{dt} = \frac{d\mathbf{v}_A}{dt} + \frac{d\mathbf{\Omega}}{dt} \times \mathbf{r}_{B/A} + \mathbf{\Omega} \times \frac{d\mathbf{r}_{B/A}}{dt} + \frac{d(\mathbf{v}_{B/A})_{xyz}}{dt}$$

$$\mathbf{a}_B = \mathbf{a}_A + \dot{\mathbf{\Omega}} \times \mathbf{r}_{B/A} + \mathbf{\Omega} \times \frac{d\mathbf{r}_{B/A}}{dt} + \frac{d(\mathbf{v}_{B/A})_{xyz}}{dt} \qquad (16\text{–}25)$$

Here $\dot{\mathbf{\Omega}} = d\mathbf{\Omega}/dt$ is the angular acceleration of the x, y, z coordinate system. For planar motion $\mathbf{\Omega}$ is always perpendicular to the plane of motion, and therefore $\dot{\mathbf{\Omega}}$ measures *only the change in magnitude* of $\mathbf{\Omega}$. The derivative $d\mathbf{r}_{B/A}/dt$ in Eq. 16–25 is defined by Eq. 16–23, so that

$$\mathbf{\Omega} \times \frac{d\mathbf{r}_{B/A}}{dt} = \mathbf{\Omega} \times (\mathbf{v}_{B/A})_{xyz} + \mathbf{\Omega} \times (\mathbf{\Omega} \times \mathbf{r}_{B/A}) \qquad (16\text{–}26)$$

Computing the time derivative of $(\mathbf{v}_{B/A})_{xyz} = (v_{B/A})_x\mathbf{i} + (v_{B/A})_y\mathbf{j}$, we have

$$\frac{d(\mathbf{v}_{B/A})_{xyz}}{dt} = \left[\frac{d(v_{B/A})_x}{dt}\mathbf{i} + \frac{d(v_{B/A})_y}{dt}\mathbf{j}\right] + \left[(v_{B/A})_x\frac{d\mathbf{i}}{dt} + (v_{B/A})_y\frac{d\mathbf{j}}{dt}\right]$$

The two terms in the first set of brackets represent the components of acceleration of point B as measured by an observer attached to the moving coordinate system. These terms will be denoted by vector $(\mathbf{a}_{B/A})_{xyz}$. The terms in the second set of brackets can be simplified using Eqs. 16–22. Hence,

$$\frac{d(\mathbf{v}_{B/A})_{xyz}}{dt} = (\mathbf{a}_{B/A})_{xyz} + \mathbf{\Omega} \times (\mathbf{v}_{B/A})_{xyz}$$

Substituting this and Eq. 16–26 into Eq. 16–25 and rearranging terms yields

$$\mathbf{a}_B = \mathbf{a}_A + \dot{\mathbf{\Omega}} \times \mathbf{r}_{B/A} + \mathbf{\Omega} \times (\mathbf{\Omega} \times \mathbf{r}_{B/A}) + 2\mathbf{\Omega} \times (\mathbf{v}_{B/A})_{xyz} + (\mathbf{a}_{B/A})_{xyz} \qquad (16\text{–}27)$$

where

$\mathbf{a}_B$ = acceleration of B, measured from the X, Y, Z reference

$\mathbf{a}_A$ = acceleration of the origin A of the x, y, z reference, measured from the X, Y, Z reference

$(\mathbf{a}_{B/A})_{xyz}$, $(\mathbf{v}_{B/A})_{xyz}$ = relative acceleration and relative velocity of "B with respect to A," as measured by an observer attached to the *rotating* x, y, z reference

$\dot{\mathbf{\Omega}}, \mathbf{\Omega}$ = angular acceleration and angular velocity of the x, y, z reference, measured from the X, Y, Z reference

$\mathbf{r}_{B/A}$ = relative position of "B with respect to A"

If the motions of points A and B are along *curved paths,* it is often convenient to express the accelerations $\mathbf{a}_B$, $\mathbf{a}_A$, and $(\mathbf{a}_{B/A})_{xyz}$ in Eq. 16–27 in terms of their normal and tangential components. If Eq. 16–27 is compared with Eq. 16–18, written in the form $\mathbf{a}_B = \mathbf{a}_A + \dot{\mathbf{\Omega}} \times \mathbf{r}_{B/A} + \mathbf{\Omega} \times (\mathbf{\Omega} \times \mathbf{r}_{B/A})$, which is valid for a translating frame of reference, it can be seen that the difference between the equations is represented by the terms $2\mathbf{\Omega} \times (\mathbf{v}_{B/A})_{xyz}$ and $(\mathbf{a}_{B/A})_{xyz}$. In particular, $2\mathbf{\Omega} \times (\mathbf{v}_{B/A})_{xyz}$ is called the *Coriolis acceleration,* named after the French engineer G. C. Coriolis, who was the first to determine it. This term represents the difference in the acceleration of B as measured from nonrotating and rotating x, y, z axes. As indicated by the vector cross product, the Coriolis acceleration will *always* be perpendicular to both $\mathbf{\Omega}$ and $(\mathbf{v}_{B/A})_{xyz}$. It is an important component of the acceleration which must be considered whenever rotating reference frames are used. This often occurs, for example, when studying the accelerations and forces which act on rockets, long-range projectiles, or other bodies having motions that are largely affected by the rotation of the earth.

The following interpretation of the terms in Eq. 16–27 may be useful when applying this equation to the solution of problems.

Term	Interpretation	
$\mathbf{a}_B$	absolute acceleration of B	
	(equals)	
$\mathbf{a}_A$	absolute acceleration of origin of x, y, z frame	motion of x, y, z frame observed from the X, Y, Z frame
	(plus)	
$\dot{\mathbf{\Omega}} \times \mathbf{r}_{B/A}$	angular acceleration effect caused by rotation of x, y, z frame	
	(plus)	
$\mathbf{\Omega} \times (\mathbf{\Omega} \times \mathbf{r}_{B/A})$	angular velocity effect caused by rotation of x, y, z frame	
	(plus)	
$2\mathbf{\Omega} \times (\mathbf{v}_{B/A})_{xyz}$	combined effect of B moving relative to x, y, z coordinates and rotation of x, y, z frame	interacting motion
	(plus)	
$(\mathbf{a}_{B/A})_{xyz}$	relative acceleration of B with respect to A	motion of B observed from the x, y, z frame

PROCEDURE FOR ANALYSIS

The following procedure provides a method for applying Eqs. 16–24 and 16–27 to the solution of problems involving the planar motion of particles or rigid bodies.

Coordinate Axes. Choose an appropriate location for the origin and proper orientation of the axes for both the X, Y, Z and moving x, y, z reference frames. Most often solutions are easily obtained if at the instant considered: (1) the origins are coincident; (2) the axes are collinear; and/or (3) the axes are parallel. The moving frame should be selected fixed to the body or device where the relative motion occurs.

Kinematic Equations. After defining the origin A of the moving reference and specifying the moving point B, Eqs. 16–24 and 16–27 should be written in symbolic form

$$\mathbf{v}_B = \mathbf{v}_A + \mathbf{\Omega} \times \mathbf{r}_{B/A} + (\mathbf{v}_{B/A})_{xyz}$$
$$\mathbf{a}_B = \mathbf{a}_A + \dot{\mathbf{\Omega}} \times \mathbf{r}_{B/A} + \mathbf{\Omega} \times (\mathbf{\Omega} \times \mathbf{r}_{B/A}) + 2\mathbf{\Omega} \times (\mathbf{v}_{B/A})_{xyz} + (\mathbf{a}_{B/A})_{xyz}$$

Each of the vectors in these equations should be defined from the problem data and expressed in Cartesian vector form. This essentially requires a determination of (1) the motion of the moving reference, i.e., $\mathbf{v}_A$, $\mathbf{a}_A$, $\mathbf{\Omega}$, and $\dot{\mathbf{\Omega}}$; and (2) the motion of B measured with respect to the moving reference, i.e., $\mathbf{r}_{B/A}$, $(\mathbf{v}_{B/A})_{xyz}$, and $(\mathbf{a}_{B/A})_{xyz}$. The components of all these vectors may be selected along either the X, Y, Z axes or the x, y, z axes. The choice is arbitrary provided a consistent set of unit vectors is used.

Finally, substitute the data into the kinematic equations and perform the vector operations.

The following examples illustrate this procedure numerically.

Example 16–19

At the instant $\theta = 60°$, the rod in Fig. 16–33 has an angular velocity of 3 rad/s and an angular acceleration of 2 rad/s^2. At this same instant, the collar C is traveling outward along the rod such that when $x = 0.2$ m the velocity is 2 m/s and the acceleration is 3 m/s^2, both measured relative to the rod. Determine the Coriolis acceleration and the velocity and acceleration of the collar at this instant.

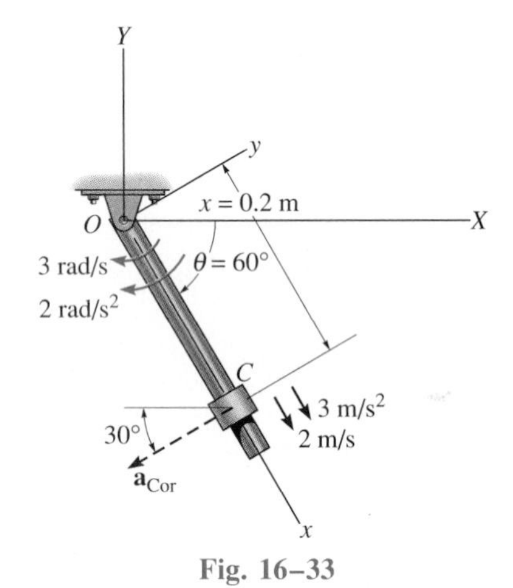

Fig. 16–33

SOLUTION

Coordinate Axes. The origin of both coordinate systems is located at point O, Fig. 16–33. Since motion of the collar is reported relative to the rod, the moving x, y, z frame of reference is *attached* to the rod.

Kinematic Equations

$$\mathbf{v}_C = \mathbf{v}_O + \mathbf{\Omega} \times \mathbf{r}_{C/O} + (\mathbf{v}_{C/O})_{xyz} \quad (1)$$

$$\mathbf{a}_C = \mathbf{a}_O + \mathbf{\Omega} \times \mathbf{r}_{C/O} + \mathbf{\Omega} \times (\mathbf{\Omega} \times \mathbf{r}_{C/O}) + 2\mathbf{\Omega} \times (\mathbf{v}_{C/O})_{xyz} + (\mathbf{a}_{C/O})_{xyz} \quad (2)$$

It will be simpler to express the data in terms of **i, j, k** component vectors rather than **I, J, K** components. Hence,

Motion of moving reference	*Motion of C with respect to moving reference*
$\mathbf{v}_O = \mathbf{0}$	$\mathbf{r}_{C/O} = \{0.2\mathbf{i}\}$ m
$\mathbf{a}_O = \mathbf{0}$	$(\mathbf{v}_{C/O})_{rel} = \{2\mathbf{i}\}$ m/s
$\mathbf{\Omega} = \{-3\mathbf{k}\}$ rad/s	$(\mathbf{a}_{C/O})_{rel} = \{3\mathbf{i}\}$ m/s^2
$\mathbf{\Omega} = \{-2\mathbf{k}\}$ rad/s^2	

From Eq. 2, the Coriolis acceleration is defined as

$$\mathbf{a}_{Cor} = 2\mathbf{\Omega} \times (\mathbf{v}_{C/O})_{xyz} = 2(-3\mathbf{k}) \times (2\mathbf{i}) = \{-12\mathbf{j}\} \text{ m/s}^2 \qquad \textit{Ans.}$$

This vector is shown dashed in Fig. 16–33. If desired, it may be resolved into **I, J** components acting along the X and Y axes, respectively.

The velocity and acceleration of the collar are determined by substituting the data into Eqs. 1 and 2 and evaluating the cross products, which yields

$$\begin{aligned}\mathbf{v}_C &= \mathbf{v}_O + \mathbf{\Omega} \times \mathbf{r}_{C/O} + (\mathbf{v}_{C/O})_{xyz} \\ &= \mathbf{0} + (-3\mathbf{k}) \times (0.2\mathbf{i}) + 2\mathbf{i} \\ &= \{2\mathbf{i} - 0.6\mathbf{j}\} \text{ m/s} \qquad \textit{Ans.}\end{aligned}$$

$$\begin{aligned}\mathbf{a}_C &= \mathbf{a}_O + \mathbf{\Omega} \times \mathbf{r}_{C/O} + \mathbf{\Omega} \times (\mathbf{\Omega} \times \mathbf{r}_{C/O}) + 2\mathbf{\Omega} \times (\mathbf{v}_{C/O})_{xyz} + (\mathbf{a}_{C/O})_{xyz} \\ &= \mathbf{0} + (-2\mathbf{k}) \times (0.2\mathbf{i}) + (-3\mathbf{k}) \times [(-3\mathbf{k}) \times (0.2\mathbf{i})] + 2(-3\mathbf{k}) \times (2\mathbf{i}) + 3\mathbf{i} \\ &= \mathbf{0} - 0.4\mathbf{j} - 1.80\mathbf{i} - 12\mathbf{j} + 3\mathbf{i} \\ &= \{1.20\mathbf{i} - 12.4\mathbf{j}\} \text{ m/s}^2 \qquad \textit{Ans.}\end{aligned}$$

Example 16–20

The rod AB, shown in Fig. 16–34, rotates clockwise such that it has an angular velocity $\omega_{AB} = 2$ rad/s and angular acceleration $\alpha_{AB} = 4$ rad/s^2 when $\theta = 45°$. Determine the angular motion of rod DE at this instant. The collar at C is pin-connected to AB and slides over rod DE.

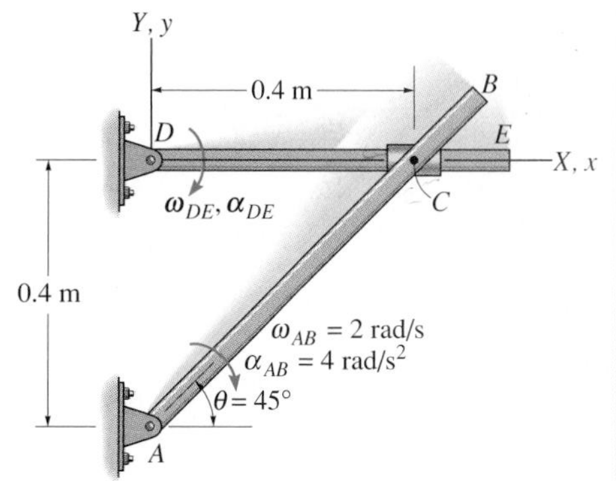

Fig. 16–34

SOLUTION

Coordinate Axes. The origins of both the fixed and moving frames of reference are located at D, Fig. 16–34. Furthermore, the x, y, z reference is attached to and rotates with rod DE.

Kinematic Equations

$$\mathbf{v}_C = \mathbf{v}_D + \mathbf{\Omega} \times \mathbf{r}_{C/D} + (\mathbf{v}_{C/D})_{xyz} \tag{1}$$

$$\mathbf{a}_C = \mathbf{a}_D + \mathbf{\Omega} \times \mathbf{r}_{C/D} + \mathbf{\Omega} \times (\mathbf{\Omega} \times \mathbf{r}_{C/D}) + 2\mathbf{\Omega} \times (\mathbf{v}_{C/D})_{xyz} + (\mathbf{a}_{C/D})_{xyz} \tag{2}$$

All vectors will be expressed in terms of **i, j, k** components.

Motion of moving reference	*Motion of C with respect to moving reference*
$\mathbf{v}_D = \mathbf{0}$	$\mathbf{r}_{C/D} = \{0.4\mathbf{i}\}$ m
$\mathbf{a}_D = \mathbf{0}$	$(\mathbf{v}_{C/D})_{xyz} = (v_{C/D})_{xyz}\mathbf{i}$
$\mathbf{\Omega} = -\omega_{DE}\mathbf{k}$	$(\mathbf{a}_{C/D})_{xyz} = (a_{C/D})_{xyz}\mathbf{i}$
$\dot{\mathbf{\Omega}} = -\alpha_{DE}\mathbf{k}$	

Motion of C: Since the collar moves along a *circular path,* its velocity and acceleration can be determined using Eqs. 16–9 and 16–14.

$$\mathbf{v}_C = \boldsymbol{\omega}_{AB} \times \mathbf{r}_{C/A} = (-2\mathbf{k}) \times (0.4\mathbf{i} + 0.4\mathbf{j}) = \{0.8\mathbf{i} - 0.8\mathbf{j}\} \text{ m/s}$$

$$\mathbf{a}_C = \boldsymbol{\alpha}_{AB} \times \mathbf{r}_{C/A} - \omega_{AB}^2\mathbf{r}_{C/A}$$
$$= (-4\mathbf{k}) \times (0.4\mathbf{i} + 0.4\mathbf{j}) - (2)^2(0.4\mathbf{i} + 0.4\mathbf{j}) = \{-3.2\mathbf{j}\} \text{ m/s}^2$$

Substituting the data into Eqs. 1 and 2, we have

$$\mathbf{v}_C = \mathbf{v}_D + \mathbf{\Omega} \times \mathbf{r}_{C/D} + (\mathbf{v}_{C/D})_{xyz}$$
$$0.8\mathbf{i} - 0.8\mathbf{j} = \mathbf{0} + (-\omega_{DE}\mathbf{k}) \times (0.4\mathbf{i}) + (v_{C/D})_{xyz}\mathbf{i}$$
$$0.8\mathbf{i} - 0.8\mathbf{j} = \mathbf{0} - 0.4\omega_{DE}\mathbf{j} + (v_{C/D})_{xyz}\mathbf{i}$$
$$(v_{C/D})_{xyz} = 0.8 \text{ m/s}$$
$$\omega_{DE} = 2 \text{ rad/s} \qquad \textit{Ans.}$$

$$\mathbf{a}_C = \mathbf{a}_D + \dot{\mathbf{\Omega}} \times \mathbf{r}_{C/D} + \mathbf{\Omega} \times (\mathbf{\Omega} \times \mathbf{r}_{C/D}) + 2\mathbf{\Omega} \times (\mathbf{v}_{C/D})_{xyz} + (\mathbf{a}_{C/D})_{xyz}$$
$$-3.2\mathbf{j} = \mathbf{0} + (-\alpha_{DE}\mathbf{k}) \times (0.4\mathbf{i}) + (-2\mathbf{k}) \times [(-2\mathbf{k}) \times (0.4\mathbf{i})]$$
$$+ 2(-2\mathbf{k}) \times (0.8\mathbf{i}) + (a_{C/D})_{xyz}\mathbf{i}$$
$$-3.2\mathbf{j} = -0.4\alpha_{DE}\mathbf{j} - 1.6\mathbf{i} - 3.2\mathbf{j} + (a_{C/D})_{xyz}\mathbf{i}$$
$$(a_{C/D})_{xyz} = 1.6 \text{ m/s}^2$$
$$\alpha_{DE} = 0 \qquad \textit{Ans.}$$

Example 16–21

Two planes A and B are flying at the same elevation and have the motions shown in Fig. 16–35. Determine the velocity and acceleration of A as measured by the pilot of B.

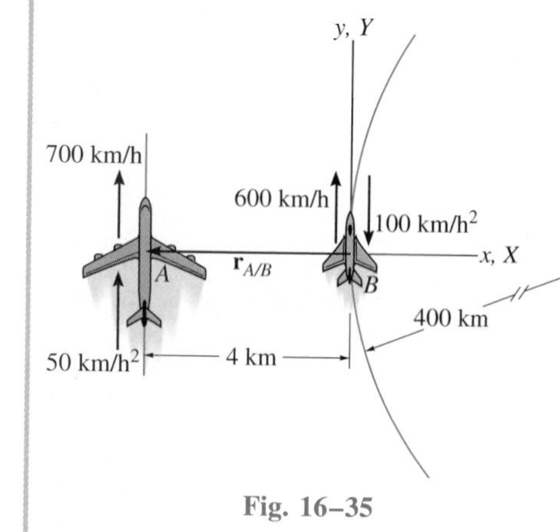

Fig. 16–35

SOLUTION

Coordinate Axes. Since the relative motion of A with respect to the pilot in B is being sought, the x, y, z axes are attached to plane B, Fig. 16–35. At the *instant* considered, the origin B coincides with the origin of the fixed X, Y, Z frame.

Kinematic Equations

$$\mathbf{v}_A = \mathbf{v}_B + \boldsymbol{\Omega} \times \mathbf{r}_{A/B} + (\mathbf{v}_{A/B})_{xyz} \quad (1)$$

$$\mathbf{a}_A = \mathbf{a}_B + \dot{\boldsymbol{\Omega}} \times \mathbf{r}_{A/B} + \boldsymbol{\Omega} \times (\boldsymbol{\Omega} \times \mathbf{r}_{A/B}) + 2\boldsymbol{\Omega} \times (\mathbf{v}_{A/B})_{xyz} + (\mathbf{a}_{A/B})_{xyz} \quad (2)$$

Motion of moving reference:

$$\mathbf{v}_B = \{600\mathbf{j}\} \text{ km/h}$$

$$\mathbf{a}_B = (\mathbf{a}_B)_n + (\mathbf{a}_B)_t = \{900\mathbf{i} - 100\mathbf{j}\} \text{ km/h}^2$$

$$(a_B)_n = \frac{v_B^2}{\rho} = \frac{(600)^2}{400} = 900 \text{ km/h}^2$$

$$\Omega = \frac{v_B}{\rho} = \frac{600 \text{ km/h}}{400 \text{ km}} = 1.5 \text{ rad/h} \curvearrowright \qquad \boldsymbol{\Omega} = \{-1.5\mathbf{k}\} \text{ rad/h}$$

$$\dot{\Omega} = \frac{(a_B)_t}{\rho} = \frac{100 \text{ km/h}^2}{400 \text{ km}} = 0.25 \text{ rad/h}^2 \curvearrowleft \qquad \dot{\boldsymbol{\Omega}} = \{0.25\mathbf{k}\} \text{ rad/h}^2$$

Motion of A with respect to moving reference:

$$\mathbf{r}_{A/B} = \{-4\mathbf{i}\} \text{ km} \qquad (\mathbf{v}_{A/B})_{xyz} = ? \qquad (\mathbf{a}_{A/B})_{xyz} = ?$$

Substituting the data into Eqs. 1 and 2, realizing that $\mathbf{v}_A = \{700\mathbf{j}\}$ km/h and $\mathbf{a}_A = \{50\mathbf{j}\}$ km/h^2, we have

$$\mathbf{v}_A = \mathbf{v}_B + \boldsymbol{\Omega} \times \mathbf{r}_{A/B} + (\mathbf{v}_{A/B})_{xyz}$$

$$700\mathbf{j} = 600\mathbf{j} + (-1.5\mathbf{k}) \times (-4\mathbf{i}) + (\mathbf{v}_{A/B})_{xyz}$$

$$(\mathbf{v}_{A/B})_{xyz} = \{94\mathbf{j}\} \text{ km/h} \qquad \textit{Ans.}$$

$$\mathbf{a}_A = \mathbf{a}_B + \dot{\boldsymbol{\Omega}} \times \mathbf{r}_{A/B} + \boldsymbol{\Omega} \times (\boldsymbol{\Omega} \times \mathbf{r}_{A/B}) + 2\boldsymbol{\Omega} \times (\mathbf{v}_{A/B})_{xyz} + (\mathbf{a}_{A/B})_{xyz}$$

$$50\mathbf{j} = (900\mathbf{i} - 100\mathbf{j}) + (0.25\mathbf{k}) \times (-4\mathbf{i})$$

$$+ (-1.5\mathbf{k}) \times [(-1.5\mathbf{k}) \times (-4\mathbf{i})] + 2(-1.5\mathbf{k}) \times (94\mathbf{j}) + (\mathbf{a}_{A/B})_{xyz}$$

$$(\mathbf{a}_{A/B})_{xyz} = \{-1190\mathbf{i} + 151\mathbf{j}\} \text{ km/h}^2 \qquad \textit{Ans.}$$

The solution of this problem should be compared with that of Example 12–26.

PROBLEMS

***16–144.** Block A, which is attached to a cord, moves along the slot of a horizontal forked rod. If the cord is pulled down through the hole at O at a constant rate of 2 m/s, determine the acceleration of the block at the instant shown. The rod rotates about O with a constant angular velocity $\omega = 4$ rad/s.

16–145. Block A, which is attached to a cord, moves along the slot of a horizontal forked rod. At the instant shown, the cord is pulled down through the hole at O with an acceleration of 4 m/s^2 and its velocity is 2 m/s. Determine the acceleration of the block at this instant. The rod rotates about O with a constant angular velocity $\omega = 4$ rad/s.

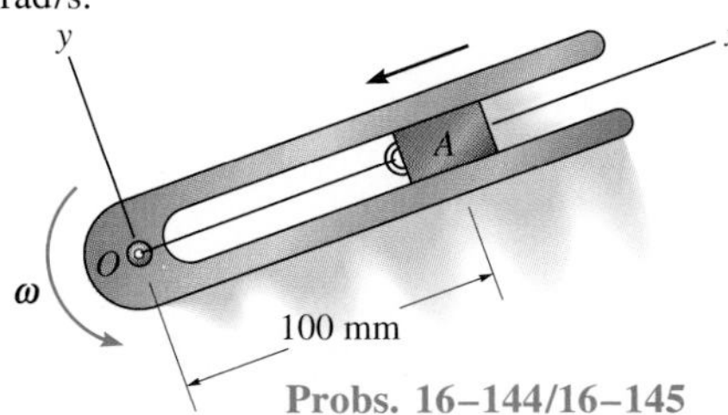

Probs. 16–144/16–145

16–146. Block B moves along the slot in the platform with a constant speed of 2 ft/s, measured relative to the platform in the direction shown. If the platform is rotating at a constant rate of $\omega = 5$ rad/s, determine the velocity and acceleration of the block at the instant $\theta = 90°$.

16–147. Solve Prob. 16–146 if when $\theta = 90°$ the block has a velocity of 2 ft/s and an acceleration of 4 ft/s^2, measured relative to the platform in the direction shown, and at this same instant the platform has an angular velocity $\omega = 5$ rad/s and an angular acceleration $\alpha = 8$ rad/s^2.

***16–148.** Block B moves along the slot in the platform with a constant speed of 2 ft/s, measured relative to the platform in the direction shown. If the platform is rotating at a constant rate of $\omega = 5$ rad/s, determine the velocity and acceleration of the block at the instant $\theta = 60°$.

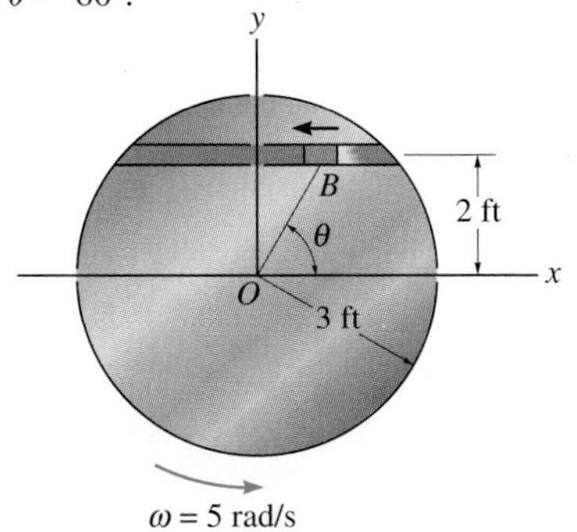

Probs. 16–146/16–147/16–148

16–149. While the swing bridge is opening with a constant rotation of 0.5 rad/s, a man runs along the roadway at a constant speed of 5 ft/s relative to the roadway. Determine his velocity and acceleration at the instant $d = 15$ ft.

16–150. While the swing bridge is opening with a constant rotation of 0.5 rad/s, a man runs along the roadway such that when $d = 10$ ft he is running outward from the center at 5 ft/s with an acceleration of 2 ft/s^2, both measured relative to the roadway. Determine his velocity and acceleration at this instant.

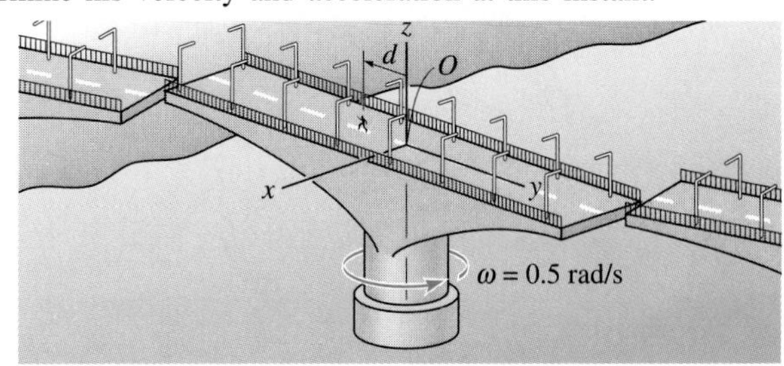

Probs. 16–149/16–150

16–151. A girl stands at A on a platform which is rotating with a constant angular velocity $\omega = 0.5$ rad/s. If she walks at a constant speed of $v = 0.75$ m/s measured relative to the platform, determine her acceleration (a) when she reaches point D in going along the path ADC, $d = 1$ m; and (b) when she reaches point B if she follows the path ABC, $r = 3$ m.

***16–152.** A girl stands at A on a platform which is rotating with an angular acceleration $\alpha = 0.2$ rad/s^2 and at the instant shown has an angular velocity $\omega = 0.5$ rad/s. If she walks at a constant speed $v = 0.75$ m/s measured relative to the platform, determine her acceleration (a) when she reaches point D in going along the path ADC, $d = 1$ m; and (b) when she reaches point B if she follows the path ABC, $r = 3$ m.

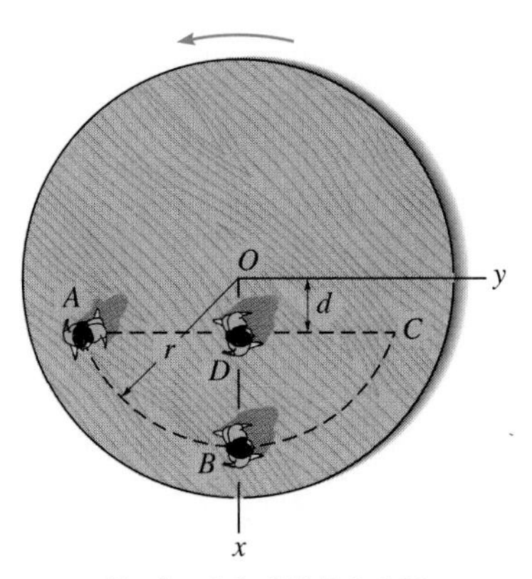

Probs. 16–151/16–152

16–153. Rod AB rotates counterclockwise with a constant angular velocity $\omega = 3$ rad/s. Determine the velocity of point C located on the double collar when $\theta = 30°$. The collar consists of two pin-connected slider blocks which are constrained to move along the circular path and the rod AB.

16–154. Rod AB rotates counterclockwise with a constant angular velocity $\omega = 3$ rad/s. Determine the velocity and acceleration of point C located on the double collar when $\theta = 45°$. The collar consists of two pin-connected slider blocks which are constrained to move along the circular path and the rod AB.

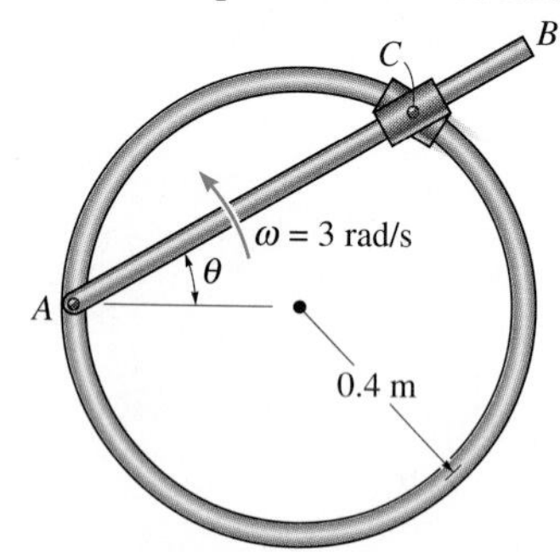

Probs. 16–153/16–154

16–155. A ride in an amusement park consists of a rotating platform P, having a constant angular velocity $\omega_P = 1.5$ rad/s, and four cars, mounted on the platform, which have constant angular velocities $\omega_{C/P} = 2$ rad/s measured relative to the platform. Determine the velocity and acceleration of the passenger at A at the instant shown.

***16–156.** A ride in an amusement park consists of a rotating platform P, having a constant angular velocity $\omega_P = 1.5$ rad/s, and four cars, C, mounted on the platform, which have constant angular velocities $\omega_{C/P} = 2$ rad/s measured relative to the platform. Determine the velocity and acceleration of the passenger at B at the instant shown.

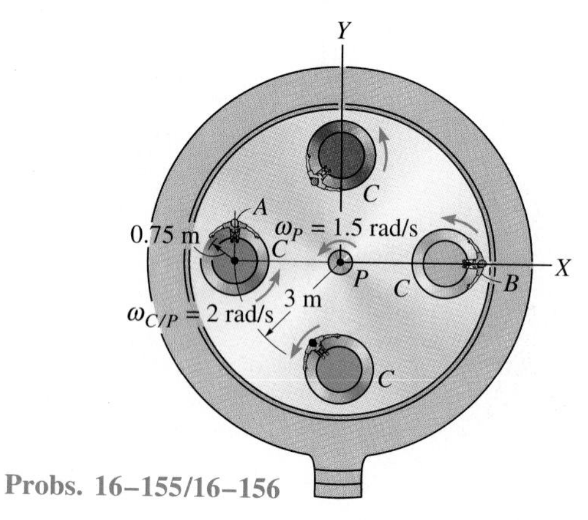

Probs. 16–155/16–156

16–157. The cars on the amusement-park ride rotate around the axle at A with a constant angular velocity $\omega_{A/f} = 2$ rad/s, measured relative to the frame AB. At the same time the frame rotates around the main axle support at B with a constant angular velocity $\omega_f = 1$ rad/s. Determine the velocity and acceleration of the passenger at C at the instant shown.

16–158. The cars on the amusement-park ride rotate around the axle at A with a constant angular velocity $\omega_{A/f} = 2$ rad/s, measured relative to the frame AB. At the same time the frame rotates around the main axle support at B with a constant angular velocity $\omega_f = 1$ rad/s. Determine the velocity and acceleration of the passenger at D at the instant shown.

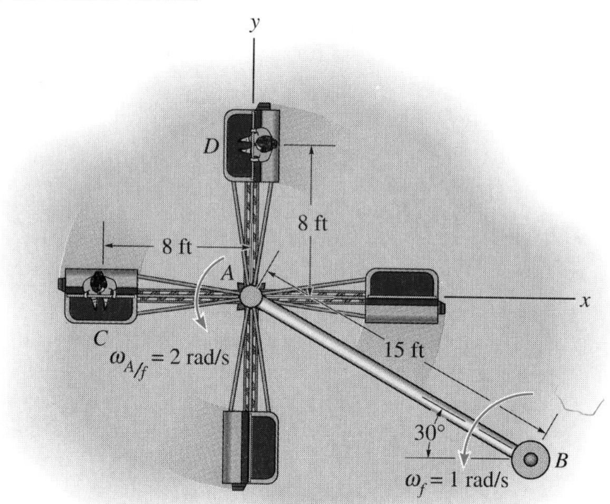

Probs. 16–157/16–158

16–159. At the instant $\theta = 60°$ rod AB has an angular velocity $\omega_{AB} = 3$ rad/s and an angular acceleration $\alpha_{AB} = 5$ rad/s^2. Determine the angular velocity and angular acceleration of rod CD at this instant. The collar at C is pin-connected to CD and slides over AB.

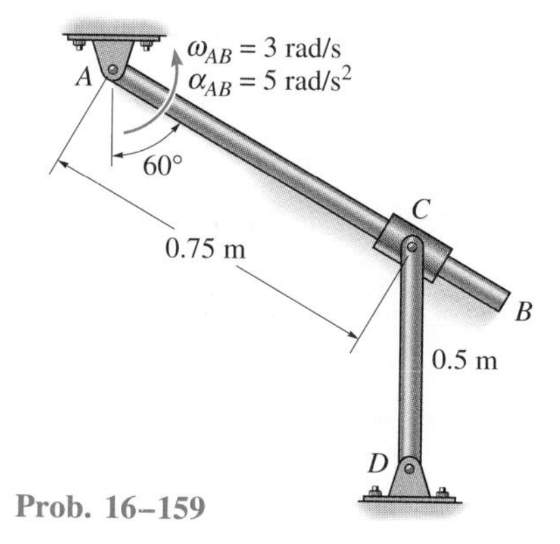

Prob. 16–159

*16–160. The "quick return" mechanism consists of a crank AB, slider block B, and slotted link CD. If the crank has the angular motions shown, determine the angular velocity and angular acceleration of CD at this instant.

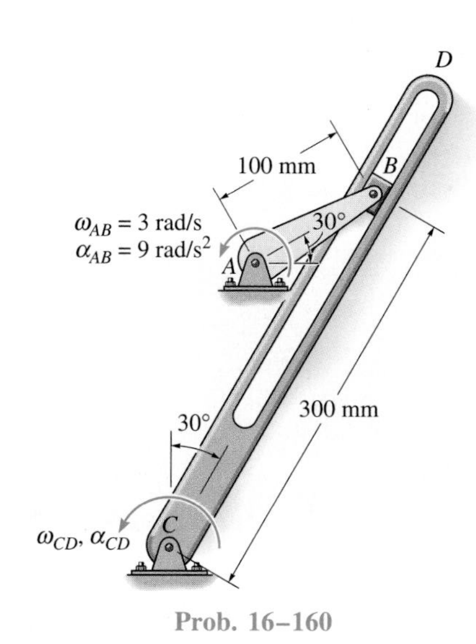

Prob. 16–160

16–161. The gear has the angular motion shown. Determine the angular velocity and angular acceleration of the slotted link BC at this instant. The peg at A is fixed to the gear.

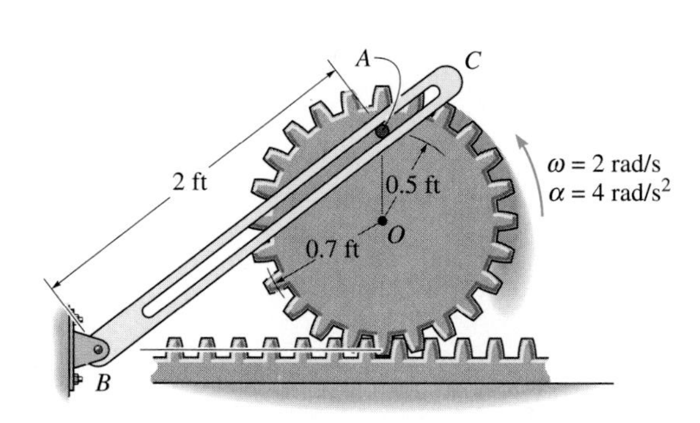

Prob. 16–161

16–162. A ride in an amusement park consists of a rotating arm AB having a constant angular velocity $\omega_{AB} = 2$ rad/s about point A and a car mounted at the end of the arm which has a constant angular velocity $\boldsymbol{\omega}' = \{-0.5\mathbf{k}\}$ rad/s, measured relative to the arm. At the instant shown, determine the velocity and acceleration of the passenger at C.

16–163. Solve Prob. 16–162 if at the instant shown the arm has an angular acceleration $\alpha_{AB} = 1$ rad/s^2 when $\omega_{AB} = 2$ rad/s and the car has a relative angular acceleration $\boldsymbol{\alpha}' = \{-0.6\mathbf{k}\}$ rad/s^2 when $\boldsymbol{\omega}' = \{-0.5\mathbf{k}\}$ rad/s.

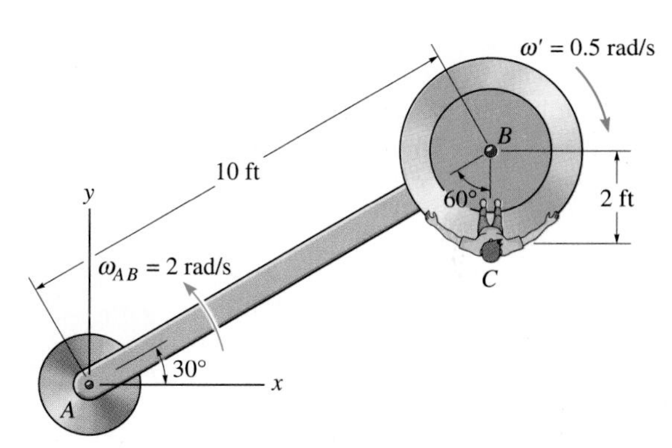

Probs. 16–162/16–163

*16–164. At the instant shown, the robotic arm AB is rotating counterclockwise at $\omega = 5$ rad/s and has an angular acceleration $\alpha = 2$ rad/s^2. Simultaneously, the grip BC is rotating counterclockwise at $\omega' = 6$ rad/s and $\alpha' = 2$ rad/s^2, both measured relative to a *fixed* reference. Determine the velocity and acceleration of the object held at the grip C.

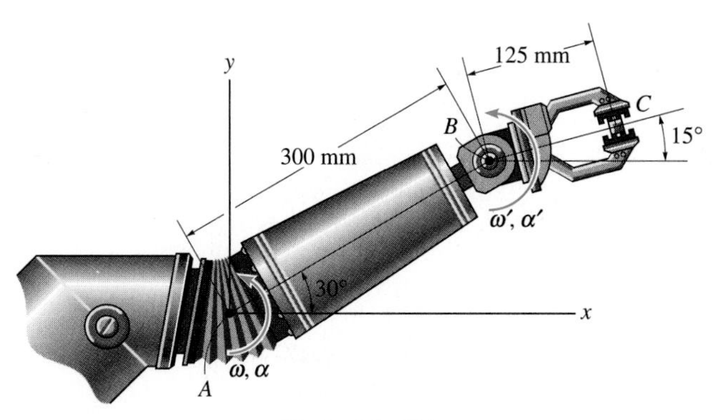

Prob. 16–164

The forces acting on this plane as it is in the process of landing are quite severe and must be accounted for in the design of its structure. In this chapter we will discuss the dynamic principles involved in this analysis.

17

Planar Kinetics of a Rigid Body: Force and Acceleration

In the previous chapter, the planar kinematics of rigid-body motion was presented in order of increasing complexity, that is, translation, rotation about a fixed axis, and general plane motion. The study of rigid-body kinetics in this chapter will be presented in somewhat the same order. The chapter begins by introducing a property of a body called the moment of inertia. Afterwards, a derivation of the equations of general plane motion for a symmetric rigid body is given. These equations are then applied to specific problems of rigid-body translation, rotation about a fixed axis, and finally general plane motion. A kinetic study of these motions is referred to as the kinetics of planar motions or simply *planar kinetics*. The more complex study of three-dimensional rigid-body kinetics, which includes planar motion of unsymmetrical rigid bodies, is presented in Chapter 21.

17.1 Moment of Inertia

Since a body has a definite size and shape, an applied nonconcurrent force system may cause the body to both translate and rotate. The translational aspects of the motion were studied in Chapter 13 and are governed by the equation $\mathbf{F} = m\mathbf{a}$. It will be shown in Sec. 17.2 that the rotational aspects, caused by the moment $\mathbf{M}$, are governed by an equation of the form $\mathbf{M} = I\boldsymbol{\alpha}$. The symbol I in this equation is termed the moment of inertia. By comparison, the *moment of inertia* is a measure of the resistance of a body to *angular acceleration* ($\mathbf{M} = I\boldsymbol{\alpha}$) in the same way that *mass* is a measure of the body's resistance to *acceleration* ($\mathbf{F} = m\mathbf{a}$).

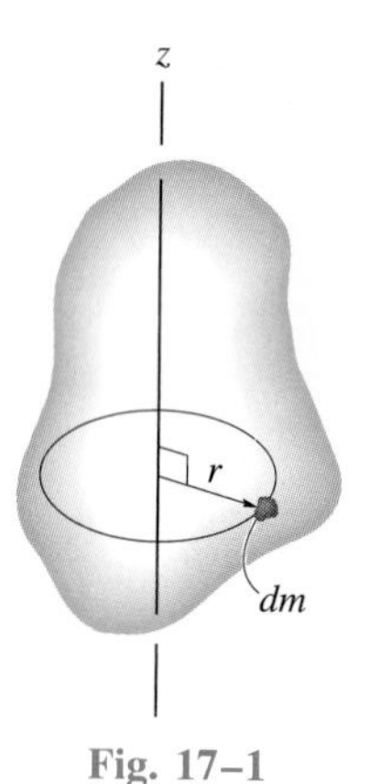

Fig. 17–1

We define the *moment of inertia* as the integral of the "second moment" about an axis of all the elements of mass dm which compose the body.* For example, consider the rigid body shown in Fig. 17–1. The body's moment of inertia about the z axis is

$$I = \int_m r^2\,dm \tag{17–1}$$

Here the "moment arm" r is the perpendicular distance from the axis to the arbitrary element dm. Since the formulation involves r, the value of I is different for each axis about which it is computed. For example, if the axis coincides with the longitudinal axis for a slender rod, I will be small, since r is small for each element of the rod. If the axis is perpendicular to the rod, I will be large, because the rod will have a larger mass distributed farther from the axis. In the study of planar kinetics, the axis which is generally chosen for analysis passes through the body's mass center G and is always perpendicular to the plane of motion. The moment of inertia computed about this axis will be denoted as I_G. Realize that because r is squared in Eq. 17–1, the mass moment of inertia is always a positive quantity. Common units used for its measurement are $\text{kg} \cdot \text{m}^2$ or $\text{slug} \cdot \text{ft}^2$.

*Another property of the body, which measures the symmetry of the body's mass with respect to a coordinate system, is the product of inertia. This property applies to the three-dimensional motion of a body and will be discussed in Chapter 21.

PROCEDURE FOR ANALYSIS

In this treatment, we will consider only symmetric bodies having surfaces which are generated by revolving a curve about an axis. An example of such a body of which the volume of revolution is generated about the z axis is shown in Fig. 17–2. If the body consists of material having a variable density, $\rho = \rho\,(x, y, z)$, the elemental mass dm of the body may be expressed in terms of its density and volume as $dm = \rho\, dV$. Substituting dm into Eq. 17–1, the body's moment of inertia is then computed using *volume elements* for integration; i.e.,

$$I = \int_V r^2 \rho\, dV \qquad (17\text{–}2)$$

In the special case of ρ being a *constant,* this term may be factored out of the integral and the integration is then purely a function of geometry,

$$I = \rho \int_V r^2\, dV \qquad (17\text{–}3)$$

When the elemental volume chosen for integration has infinitesimal dimensions in all three directions, e.g., $dV = dx\,dy\,dz$, Fig. 17–2*a*, the moment of inertia of the body must be computed using "triple integration." The integration process can, however, be simplified to a *single integration* provided the chosen elemental volume has a differential size or thickness in only *one direction.* Shell or disk elements are often used for this purpose.

Shell Element. If a *shell element* having a height z, radius y, and thickness dy is chosen for integration, Fig. 17–2*b*, then the volume is $dV = (2\pi y)(z)\,dy$. This element may be used in Eq. 17–2 or 17–3 for computing the moment of inertia I_z of the body about the z axis, since the *entire element,* due to its "thinness," lies at the *same* perpendicular distance $r = y$ from the z axis (see Example 17–1).

Disk Element. If a disk element having a radius y and a thickness dz is chosen for integration, Fig. 17–2*c*, then the volume $dV = (\pi y^2)\,dz$. In this case, however, the element is *finite* in the radial direction, and consequently its parts *do not* all lie at the *same radial distance* r from the z axis. As a result, Eq. 17–2 or 17–3 *cannot* be used to determine I_z directly. Instead, to perform the integration using this element, it is first necessary to determine the moment of inertia *of the element* about the z axis and then integrate this result (see Example 17–2).

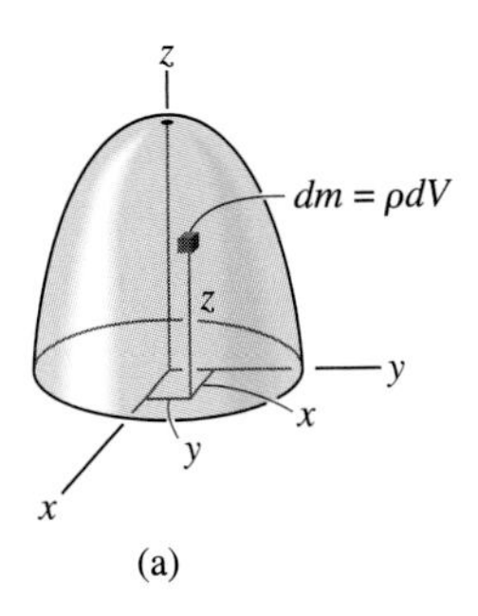

(a)

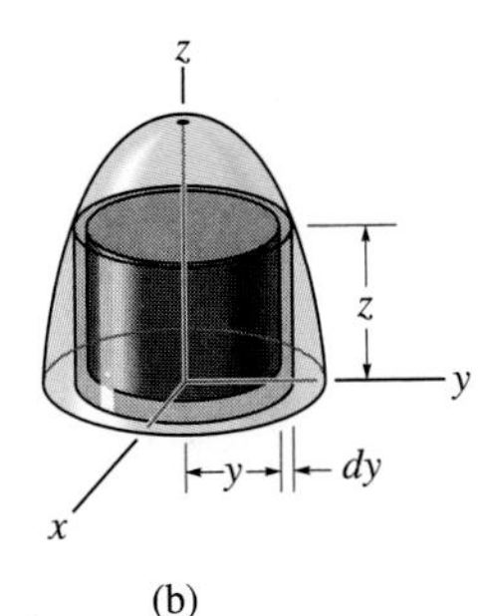

(b)

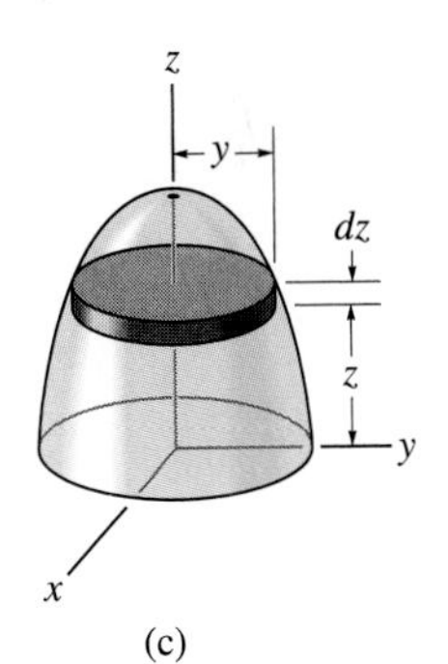

(c)

Fig. 17–2

Example 17–1

Determine the moment of inertia of the cylinder shown in Fig. 17–3*a* about the z axis. The density ρ of the material is constant.

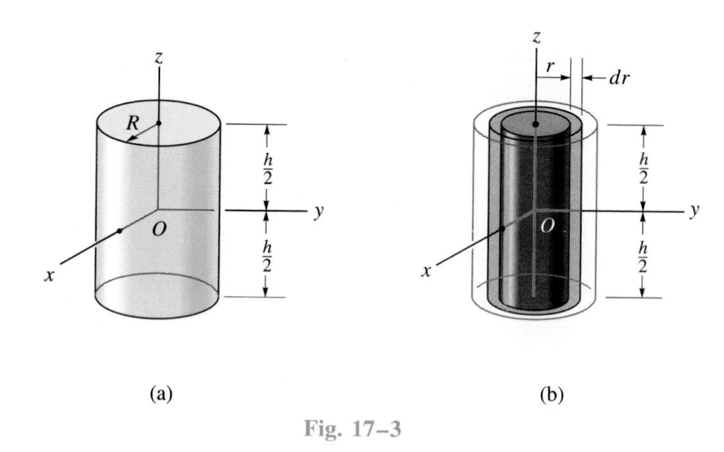

Fig. 17–3

SOLUTION

Shell Element. This problem may be solved using the *shell element* in Fig. 17–3*b* and single integration. The volume of the element is $dV = (2\pi r)(h)\,dr$, so that its mass is $dm = \rho\,dV = \rho(2\pi hr\,dr)$. Since the *entire element* lies at the same distance r from the z axis, the moment of inertia *of the element* is

$$dI_z = r^2\,dm = \rho 2\pi h r^3\,dr$$

Integrating over the entire region of the cylinder yields

$$I_z = \int_m r^2\,dm = \rho 2\pi h \int_0^R r^3\,dr = \frac{\rho\pi}{2}R^4 h$$

The mass of the cylinder is

$$m = \int_m dm = \rho 2\pi h \int_0^R r\,dr = \rho\pi h R^2$$

so that

$$I_z = \frac{1}{2}mR^2 \qquad \textit{Ans.}$$

Example 17–2

A solid is formed by revolving the shaded area shown in Fig. 17–4*a* about the y axis. If the density of the material is 5 slug/ft^3, determine the moment of inertia about the y axis.

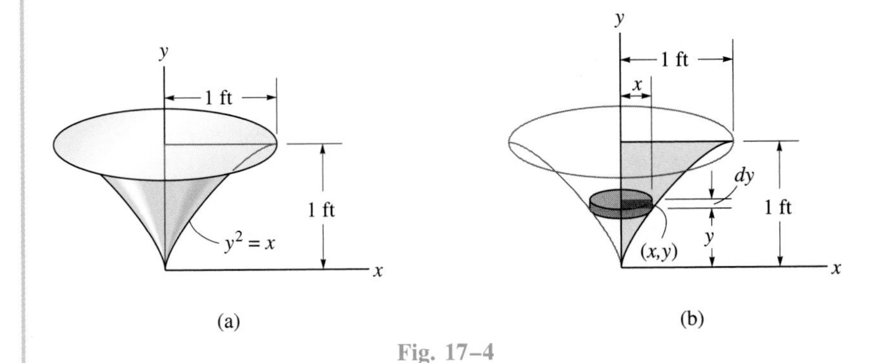

Fig. 17–4

SOLUTION

Disk Element. The moment of inertia will be computed using a *disk element,* as shown in Fig. 17–4*b*. Here the element intersects the curve at the arbitrary point (x, y) and has a mass

$$dm = \rho\, dV = \rho(\pi x^2)\, dy$$

Although all portions of the element are *not* located at the same distance from the y axis, it is still possible to determine the moment of inertia dI_y *of the element* about the y axis. In the preceding example it was shown that the moment of inertia of a cylinder about its longitudinal axis is $I = \frac{1}{2}mR^2$, where m and R are the mass and radius of the cylinder. Since the height of the cylinder is not involved in this formula, we can also use it for a disk. Thus, for the disk element in Fig. 17–4*b*, we have

$$dI_y = \tfrac{1}{2}(dm)x^2 = \tfrac{1}{2}[\rho(\pi x^2)\, dy]x^2$$

Substituting $x = y^2$, $\rho = 5$ slug/ft^3, and integrating with respect to y, from $y = 0$ to $y = 1$ ft, yields the moment of inertia for the entire solid.

$$I_y = \frac{\pi(5)}{2}\int_0^1 x^4\, dy = \frac{\pi(5)}{2}\int_0^1 y^8\, dy = 0.873 \text{ slug}\cdot\text{ft}^2 \qquad \textit{Ans.}$$

Parallel-Axis Theorem. If the moment of inertia of the body about an axis passing through the body's mass center is known, then the moment of inertia about any other *parallel axis* may be determined by using the *parallel-axis theorem.* This theorem can be derived by considering the body shown in Fig. 17–5. The z' axis passes through the mass center G, whereas the corresponding *parallel z axis* lies at a constant distance d away. Selecting the differential element of mass dm, which is located at point (x', y'), and using the Pythagorean theorem, $r^2 = (d + x')^2 + y'^2$, we can express the moment of inertia of the body about the z axis as

$$I = \int_m r^2\,dm = \int_m [(d + x')^2 + y'^2]\,dm$$

$$= \int_m (x'^2 + y'^2)\,dm + 2d\int_m x'\,dm + d^2\int_m dm$$

Since $r'^2 = x'^2 + y'^2$, the first integral represents I_G. The second integral equals *zero,* since the z' axis passes through the body's mass center, i.e., $\int x'\,dm = \bar{x}\int dm = 0$ since $\bar{x} = 0$. Finally, the third integral represents the total mass m of the body. Hence, the moment of inertia about the z axis can be written as

$$I = I_G + md^2 \tag{17–4}$$

where

I_G = moment of inertia about the z' axis passing through the mass center G
m = mass of the body
d = perpendicular distance between the parallel axes

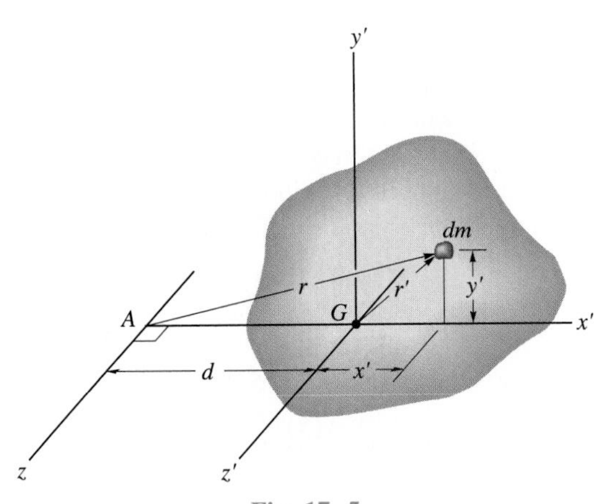

Fig. 17–5

Radius of Gyration. Occasionally, the moment of inertia of a body about a specified axis is reported in handbooks using the *radius of gyration, k.* This value has units of length, and when it and the body's mass m are known, the body's moment of inertia is determined from the equation

$$I = mk^2 \quad \text{or} \quad k = \sqrt{\frac{I}{m}} \tag{17-5}$$

Note the *similarity* between the definition of k in this formula and r in the equation $dI = r^2\,dm$, which defines the moment of inertia of an elemental mass dm of the body about an axis.

Composite Bodies. If a body is constructed of a number of simple shapes such as disks, spheres, and rods, the moment of inertia of the body about any axis z can be determined by adding algebraically the moments of inertia of all the composite shapes computed about the z axis. Algebraic addition is necessary since a composite part must be considered as a negative quantity if it has already been counted as part of another part—for example a ''hole'' subtracted from a solid plate. The parallel-axis theorem is needed for the calculations if the center of mass of each composite part does not lie on the z axis. For the calculation, then, $I = \Sigma(I_G + md^2)$. Here I_G for each of the composite parts is computed by integration or can be determined from a table, such as the one given on the inside back cover.

Example 17–3

If the plate shown in Fig. 17–6*a* has a density of 8000 kg/m^3 and a thickness of 10 mm, compute its moment of inertia about an axis directed perpendicular to the page and passing through point O.

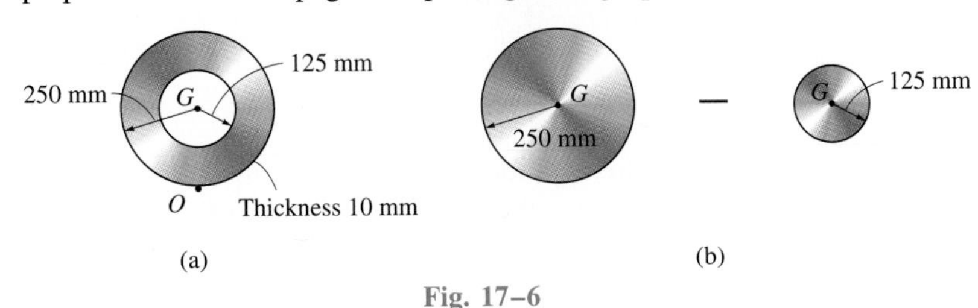

Fig. 17–6

SOLUTION

The plate consists of two composite parts, the 250-mm-radius disk *minus* a 125-mm-radius disk, Fig. 17–6*b*. The moment of inertia about O can be determined by computing the moment of inertia of each of these parts about O and then adding the results *algebraically*. The computations are performed by using the parallel-axis theorem in conjunction with the data listed in the table on the inside back cover.

Disk. The moment of inertia of a disk about the centroidal axis perpendicular to the plane of the disk is $I_G = \frac{1}{2}mr^2$. The mass center of the disk is located at a distance of 0.25 m from point O. Thus,

$$m_d = \rho_d V_d = 8000\text{ kg/m}^3[\pi(0.25\text{ m})^2(0.01\text{ m})] = 15.71\text{ kg}$$

$$(I_d)_O = \tfrac{1}{2}m_d r_d^2 + m_d d^2$$

$$= \frac{1}{2}(15.71\text{ kg})(0.25\text{ m})^2 + (15.71\text{ kg})(0.25\text{ m})^2$$

$$= 1.473\text{ kg}\cdot\text{m}^2$$

Hole. For the 125-mm-radius disk (hole), we have

$$m_h = \rho_h V_h = 8000\text{ kg/m}^3[\pi(0.125\text{ m})^2(0.01\text{ m})] = 3.93\text{ kg}$$

$$(I_h)_O = \tfrac{1}{2}m_h r_h^2 + m_h d^2$$

$$= \frac{1}{2}(3.93\text{ kg})(0.125\text{ m})^2 + (3.93\text{ kg})(0.25\text{ m})^2$$

$$= 0.276\text{ kg}\cdot\text{m}^2$$

The moment of inertia of the plate about point O is therefore

$$I_O = (I_d)_O - (I_h)_O$$

$$= 1.473 - 0.276$$

$$= 1.20\text{ kg}\cdot\text{m}^2$$

Ans.

Example 17–4

The pendulum consists of two thin rods, each having a weight of 10 lb and suspended from point O as shown in Fig. 17–7. Compute the pendulum's moment of inertia about an axis passing through *(a)* the pin at O, and *(b)* the mass center G of the pendulum.

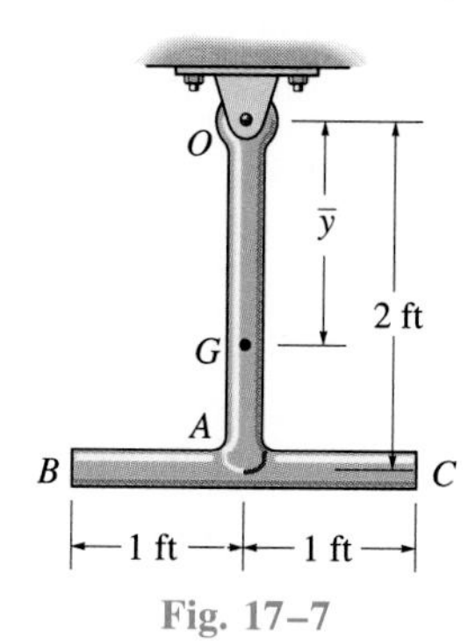

Fig. 17–7

SOLUTION

Part (a). Using the table on the inside back cover, the moment of inertia of rod OA about an axis perpendicular to the page and passing through the end point O of the rod is $I_O = \frac{1}{3}ml^2$. Hence,

$$(I_{OA})_O = \frac{1}{3}ml^2 = \frac{1}{3}\left(\frac{10 \text{ lb}}{32.2 \text{ ft/s}}\right)(2 \text{ ft})^2 = 0.414 \text{ slug} \cdot \text{ft}^2$$

The same value is obtained using $I_G = \frac{1}{12}ml^2$ and the parallel-axis theorem.

$$(I_{OA})_O = \frac{1}{12}ml^2 + md^2 = \frac{1}{12}\left(\frac{10 \text{ lb}}{32.2 \text{ ft/s}^2}\right)(2 \text{ ft})^2 + \left(\frac{10 \text{ lb}}{32.2 \text{ ft/s}^2}\right)(1 \text{ ft})^2$$
$$= 0.414 \text{ slug} \cdot \text{ft}^2$$

For rod BC we have

$$(I_{BC})_O = \frac{1}{12}ml^2 + md^2 = \frac{1}{12}\left(\frac{10 \text{ lb}}{32.2 \text{ ft/s}^2}\right)(2 \text{ ft})^2 + \left(\frac{10 \text{ lb}}{32.2 \text{ ft/s}^2}\right)(2 \text{ ft})^2$$
$$= 1.346 \text{ slug} \cdot \text{ft}^2$$

The moment of inertia of the pendulum about O is therefore

$$I_O = 0.414 + 1.346 = 1.76 \text{ slug} \cdot \text{ft}^2 \qquad \textit{Ans.}$$

Part (b). The mass center G will be located relative to the pin at O. Assuming this distance to be $\bar{y}$, Fig. 17–7, and using the formula for determining the mass center, we have

$$\bar{y} = \frac{\Sigma \tilde{y}m}{\Sigma m} = \frac{1(10/32.2) + 2(10/32.2)}{(10/32.2) + (10/32.2)} = 1.50 \text{ ft}$$

The moment of inertia I_G may be computed in the same manner as I_O, which requires successive applications of the parallel-axis theorem to transfer the moments of inertia of rods OA and BC to G. A more direct solution, however, involves using the result for I_O, i.e.,

$$I_O = I_G + md^2; \qquad 1.76 \text{ slug} \cdot \text{ft}^2 = I_G + \left(\frac{20 \text{ lb}}{32.2 \text{ ft/s}^2}\right)(1.50 \text{ ft})^2$$
$$I_G = 0.362 \text{ slug} \cdot \text{ft}^2 \qquad \textit{Ans.}$$

PROBLEMS

17–1. Determine the moment of inertia I_y for the slender rod. The rod's density ρ and cross-sectional area A are constant. Express the result in terms of the rod's total mass m.

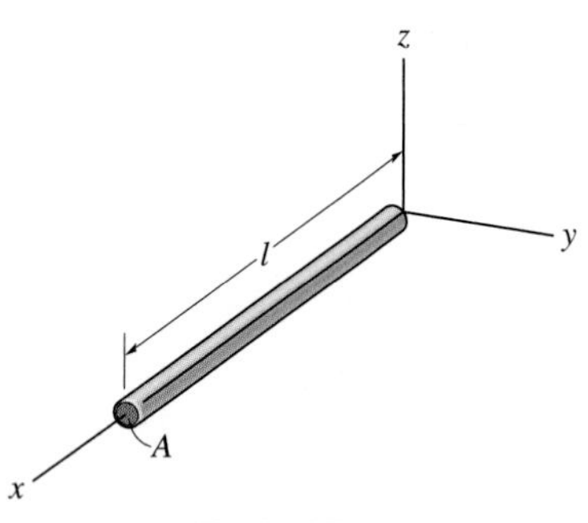

Prob. 17–1

17–2. The paraboloid is formed by revolving the shaded area around the x axis. Determine the radius of gyration k_x. The material has a constant density ρ.

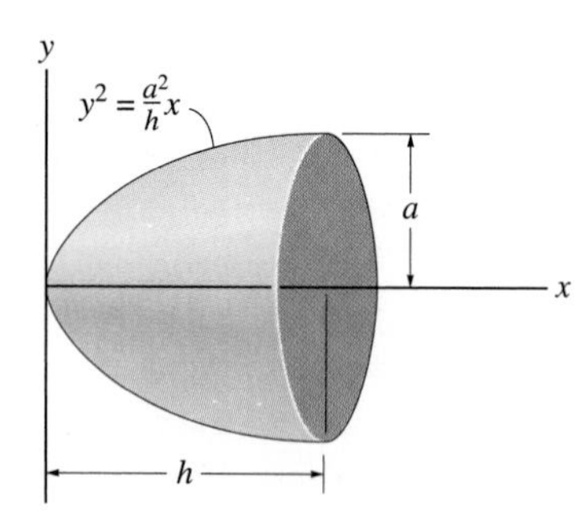

Prob. 17–2

17–3. Determine the moment of inertia I_x and express the result in terms of the total mass m of the sphere. The sphere has a constant density ρ.

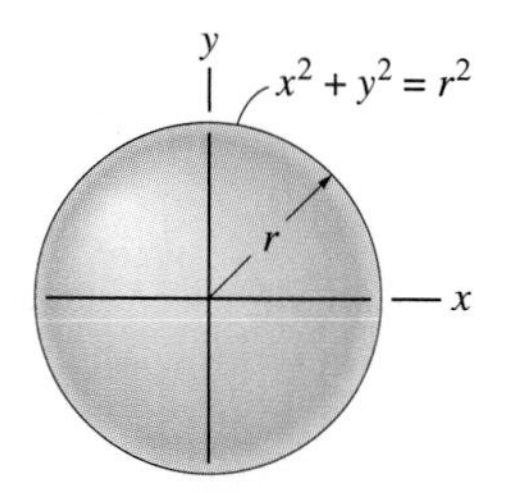

Prob. 17–3

***17–4.** Determine the moment of inertia of the thin ring about the z axis. The ring has a mass m.

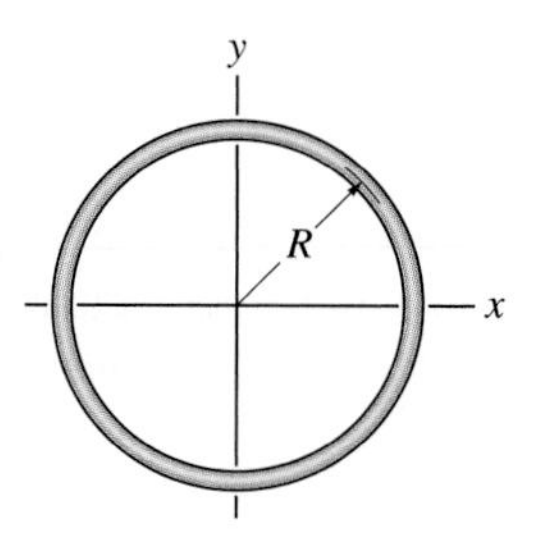

Prob. 17–4

17–5. A semiellipsoid is formed by rotating the shaded area about the x axis. Determine the moment of inertia of this solid with respect to the x axis and express the result in terms of the mass m of the solid. The material has a constant density ρ.

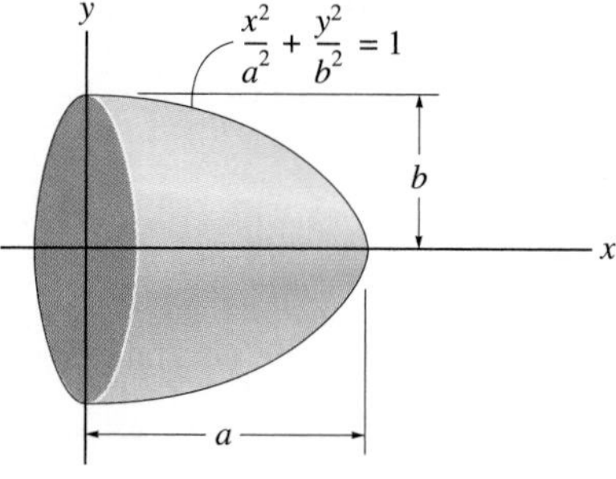

Prob. 17–5

17–6. Determine the moment of inertia I_z of the torus. The mass of the torus is m and the density ρ is constant. *Suggestion:* Use a shell element.

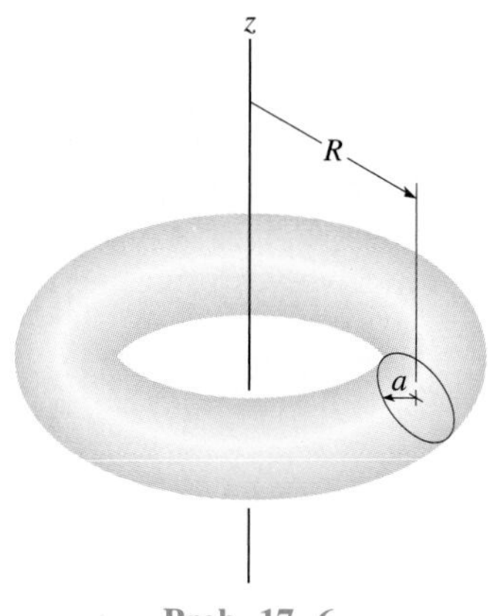

Prob. 17–6

17–7. Determine the moment of inertia of the homogeneous triangular prism with respect to the y axis. Express the result in terms of the mass m of the prism. *Hint:* For integration, use thin plate elements parallel to the x-y plane and having a thickness dz.

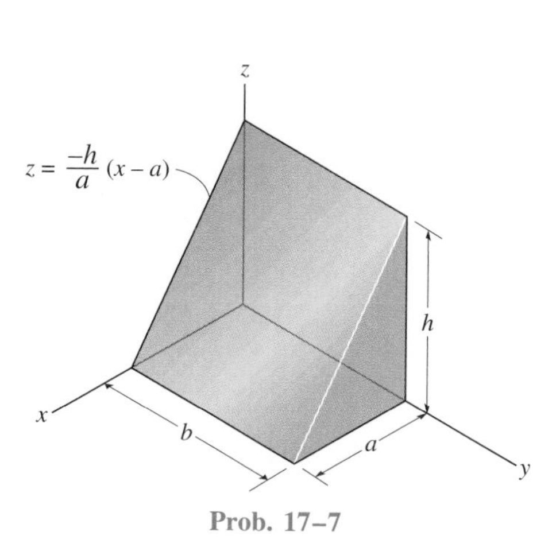

Prob. 17–7

***17–8.** Determine the moment of inertia of the homogeneous pyramid of mass m with respect to the z axis. The density of the material is ρ. *Suggestion:* Use a rectangular plate element having a volume $dV = (2x)(2y)dz$.

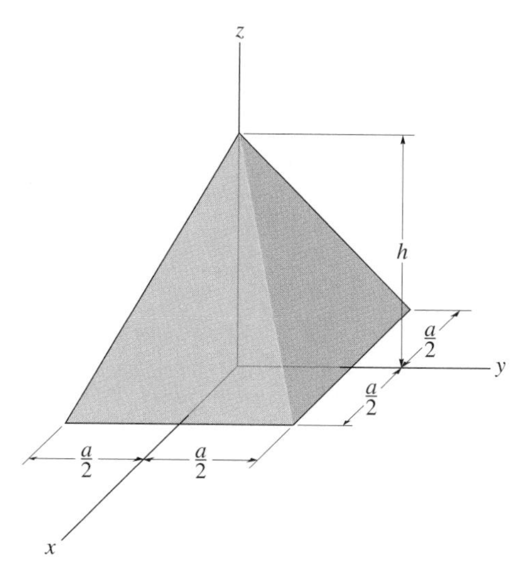

Prob. 17–8

17–9. The solid cylinder has an outer radius R, height h, and is made from a material having a density that varies from its center as $\rho = k + ar^2$, where k and a are constants. Determine the mass of the cylinder and its moment of inertia about the z axis.

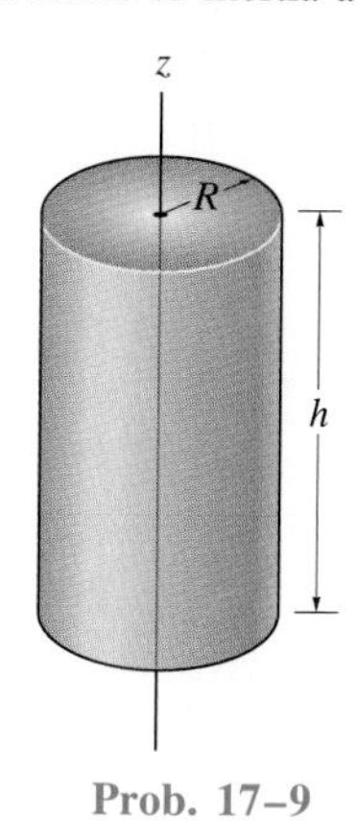

Prob. 17–9

17–10. The pendulum consists of a disk having a mass of 6 kg and slender rods AB and DC which have a mass of 2 kg/m. Determine the length L of DC so that the center of mass is at the bearing O. What is the moment of inertia of the assembly about an axis perpendicular to the page and passing through O?

17–11. The pendulum consists of a disk having a mass of 6 kg and slender rods AB and DC which have a mass of 2 kg/m. If $L = 0.75$ m, determine the moment of inertia of the assembly about an axis perpendicular to the page and passing through O.

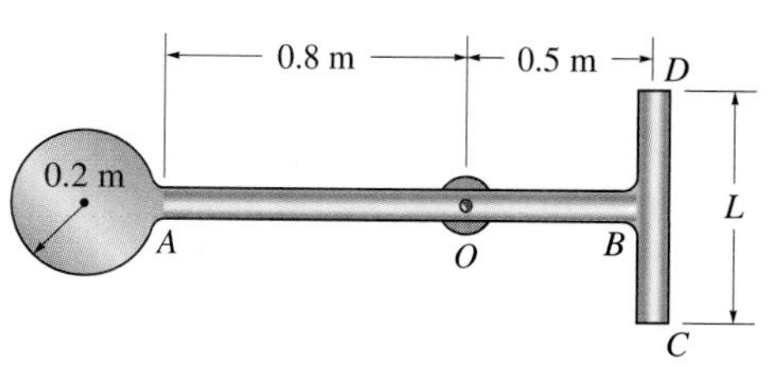

Probs. 17–10/17–11

***17–12.** The slender rods have a weight of 3 lb/ft. Determine the moment of inertia of the assembly about an axis perpendicular to the page and passing through the pin at A.

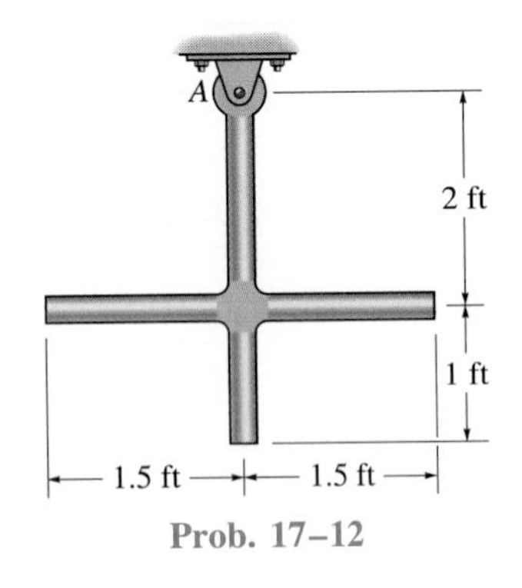

Prob. 17–12

17–13. The slender rods have a weight of 3 lb/ft. Determine the moment of inertia of the assembly about an axis perpendicular to the page and passing through the assembly's center of gravity G.

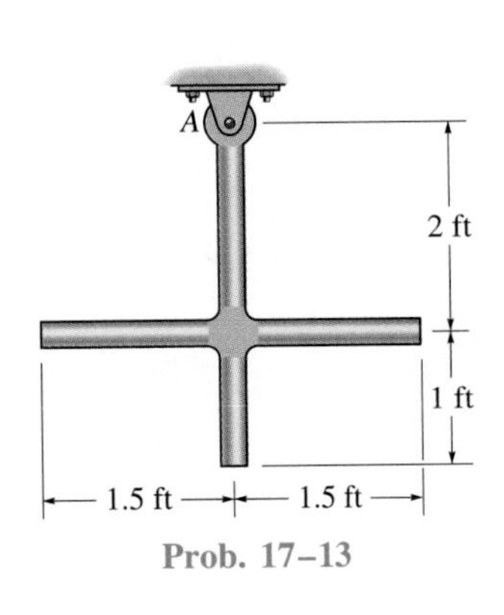

Prob. 17–13

17–14. The wheel consists of a thin ring having a mass of 10 kg and *four* spokes made from slender rods and each having a mass of 2 kg. Determine the wheel's moment of inertia about an axis perpendicular to the page and passing through point A.

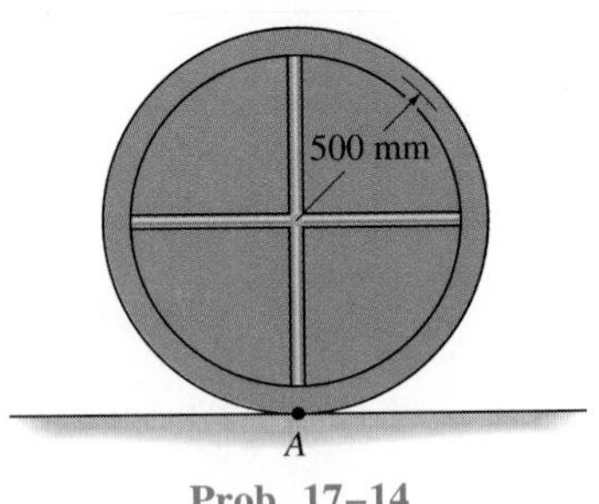

Prob. 17–14

17–15. Determine the moment of inertia about an axis perpendicular to the page and passing through the pin at O. The thin plate has a hole in its center. Its thickness is 50 mm, and the material has a density $\rho = 50\ \text{kg/m}^3$.

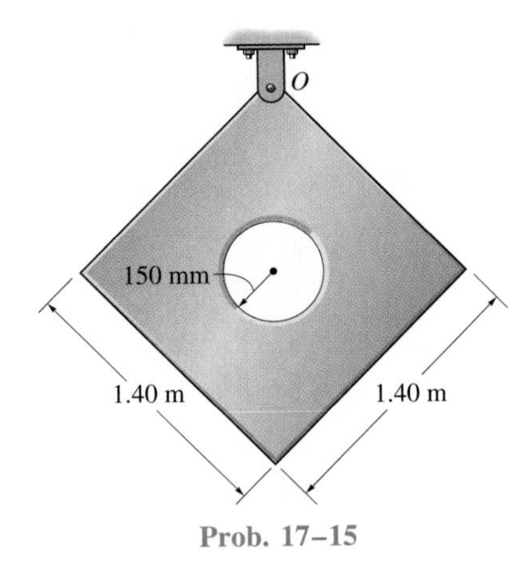

Prob. 17–15

***17–16.** The pendulum consists of two slender rods AB and OC which have a mass of 3 kg/m. The thin plate has a mass of 12 kg/m^2. Determine the location $\bar{y}$ of the center of mass G of the pendulum, then calculate the moment of inertia of the pendulum about an axis perpendicular to the page and passing through G.

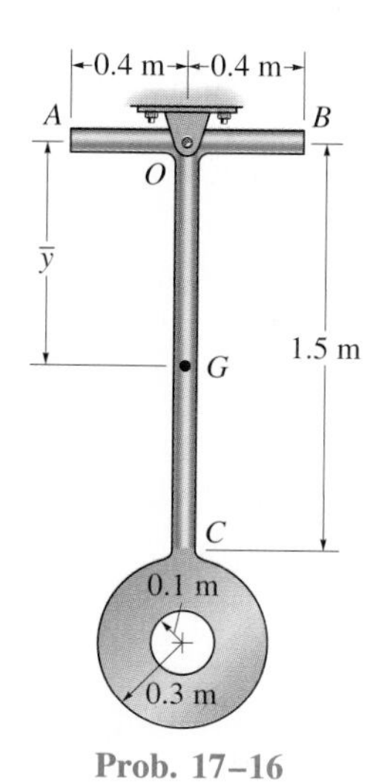

Prob. 17–16

17–17. The pendulum consists of two slender rods AB and OC which have a mass of 3 kg/m. The thin plate has a mass of 12 kg/m^2. Determine the moment of inertia of the pendulum about an axis perpendicular to the page and passing through the pin at O.

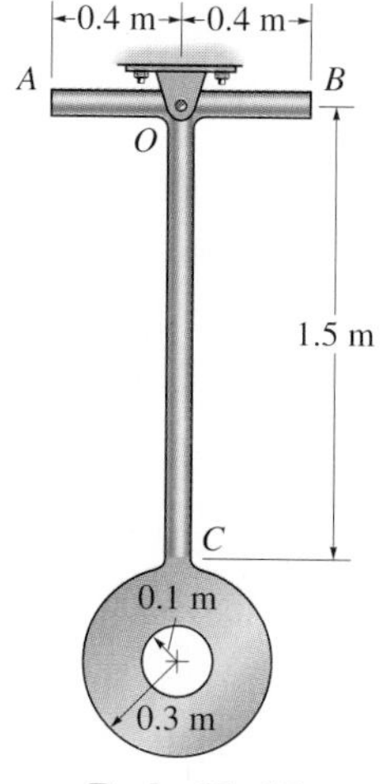

Prob. 17–17

17–18. Determine the moment of inertia of the wheel about an axis which is perpendicular to the page and passes through the center of mass G. The material has a specific weight $\gamma = 90$ lb/ft^3.

17–19. Determine the moment of inertia of the wheel about an axis which is perpendicular to the page and passes through point O. The material has a specific weight $\gamma = 90$ lb/ft^3.

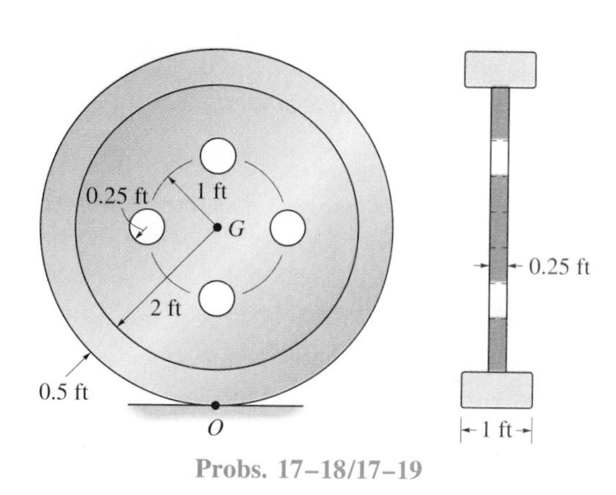

Probs. 17–18/17–19

***17–20.** Determine the moment of inertia I_z of the frustum of the cone which has a conical depression. The material has a density $\rho = 200$ kg/m^3.

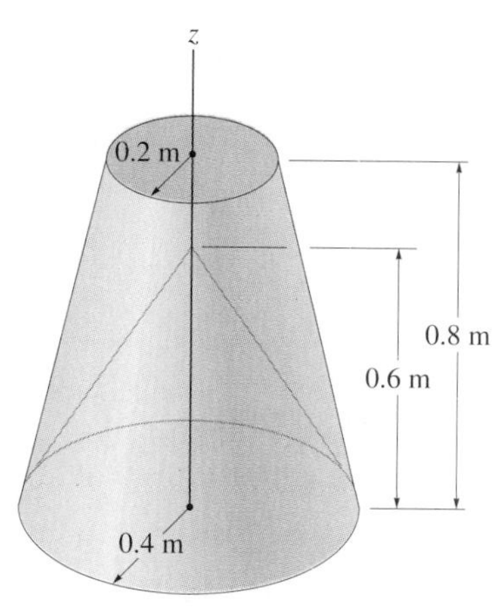

Prob. 17–20

17–21. Determine the moment of inertia of the solid steel assembly about the xx axis. Steel has a specific weight $\gamma = 490$ lb/ft^3.

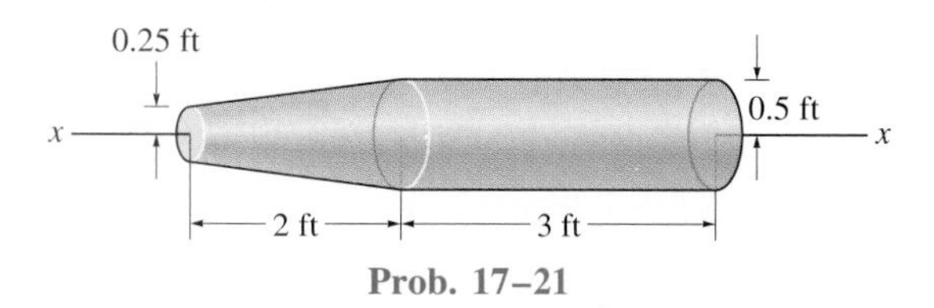

Prob. 17–21

17–22. Determine the location $\bar{y}$ of the center of mass G of the assembly and then calculate the moment of inertia about an axis perpendicular to the page and passing through G. The block has a mass of 3 kg and the semicylinder has a mass of 5 kg.

17–23. Determine the moment of inertia of the assembly about an axis perpendicular to the page and passing through point O. The block has a mass of 3 kg, and the semicylinder has a mass of 5 kg.

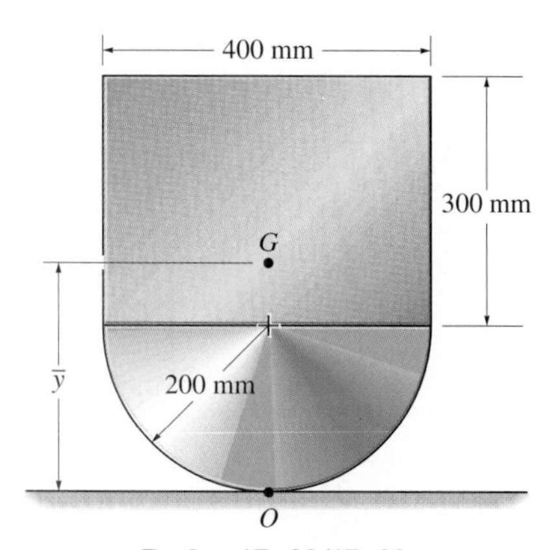

Probs. 17–22/17–23

***17–24.** The pendulum consists of the 3-kg slender rod and the 5-kg thin plate. Determine the location $\bar{y}$ of the center of mass G of the pendulum, then calculate the moment of inertia of the pendulum about an axis perpendicular to the page and passing through G.

17–25. The pendulum consists of the 3-kg slender rod and the 5-kg thin plate. Determine the moment of inertia of the pendulum about an axis perpendicular to the page and passing through the pin at O.

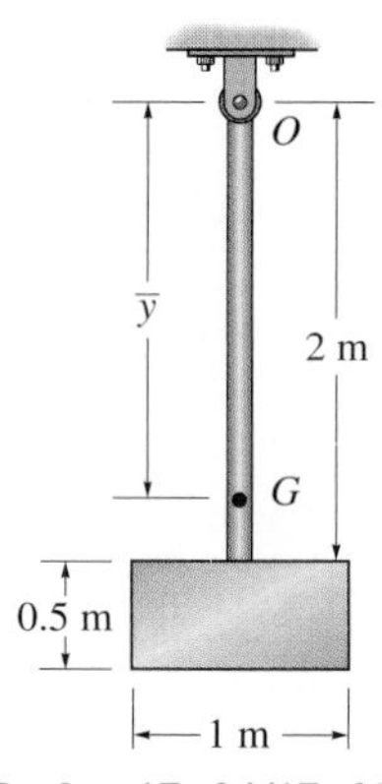

Probs. 17–24/17–25

17–26. Each of the three rods has a mass m. Determine the moment of inertia of the assembly about an axis which is perpendicular to the page and passes through the center point O.

17–27. Each of the three rods has a mass m. Determine the moment of inertia of the assembly about an axis which is perpendicular to the page and passes through point A.

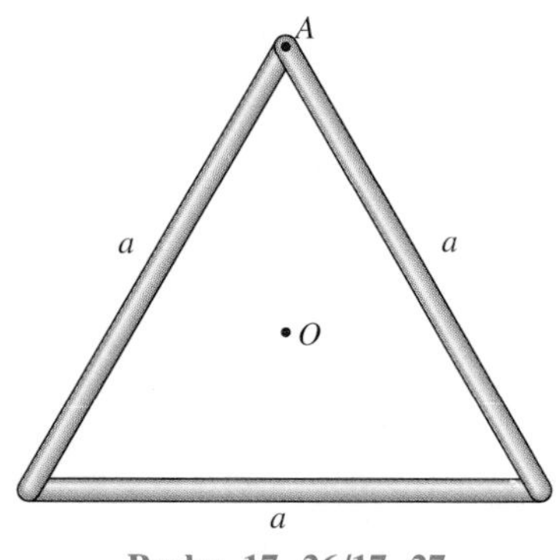

Probs. 17–26/17–27

17.2 Planar Kinetic Equations of Motion

In the following analysis we will limit our study of planar kinetics to rigid bodies which, along with their loadings, are considered to be *symmetrical* with respect to a fixed reference plane.* The plane motion of such a rigid body may be analyzed in a fixed reference plane because the path of motion of each particle of the body is a plane curve parallel to the reference plane. Thus, for the analysis, the motion of the body may be viewed within the reference plane, and all the forces (and couple moments) acting on the body can then be projected onto the plane. An example of an arbitrary body of this type is shown in Fig. 17–8*a*. Here the *inertial frame of reference x, y, z* has its origin *coincident* with the arbitrary point P in the body. By definition, *these axes do not rotate and are either fixed or translate with constant velocity.*

Equation of Translational Motion. The external forces shown on the body in Fig. 17–8*a* symbolically represent the effect of gravitational, electrical, magnetic, or contact forces between adjacent bodies. Since this force system has been considered previously in Sec. 13.3, for the analysis of a system of particles, the results may be used here, in which case the particles are contained within the boundary of the body. Hence, if the equation of motion is applied to each of the particles of the body, and the results added vectorially, it may be concluded that

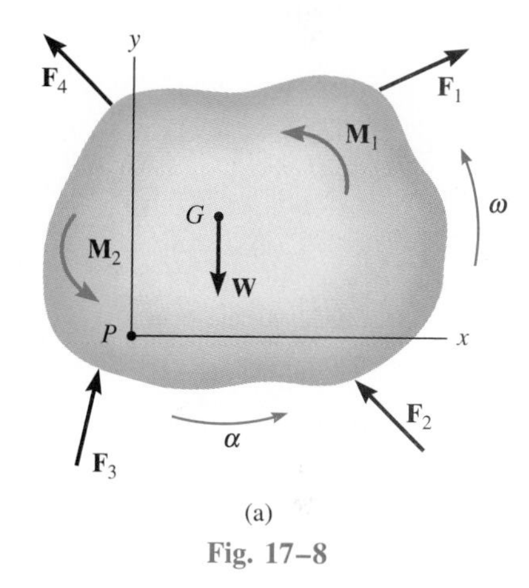

(a)
Fig. 17–8

$$\Sigma \mathbf{F} = m\mathbf{a}_G$$

This equation is referred to as the *translational equation of motion* for the mass center of a rigid body. It states that *the sum of all the external forces acting on the body is equal to the body's mass times the acceleration of its mass center G.*

For motion of the body in the x–y plane, the equation of motion for G may be written in the form of two independent scalar equations, namely,

$$\Sigma F_x = m(a_G)_x$$
$$\Sigma F_y = m(a_G)_y$$

*By doing this, the rotational equation of motion reduces to a rather simplified form. The more general case of body shape and loading is considered in Chapter 21.

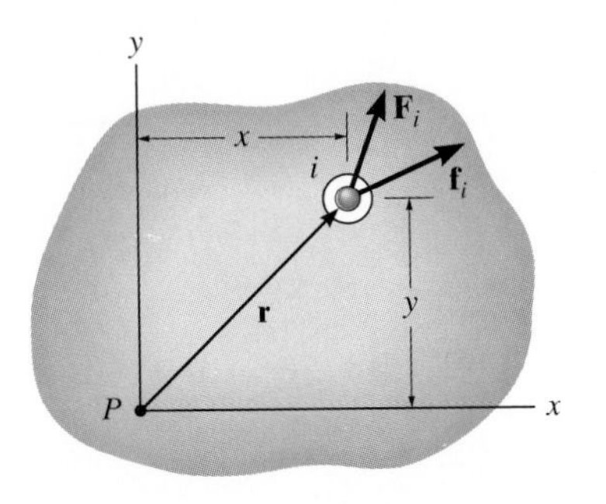

Particle free-body diagram

(b)

=

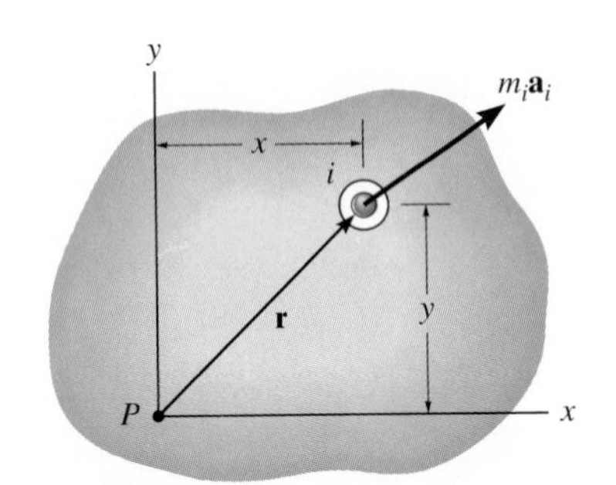

Particle kinetic diagram

(c)

Fig. 17–8 ***(cont'd)***

Equation of Rotational Motion. We will now determine the effects caused by the moments of the external force system computed about an axis perpendicular to the plane of motion (the z axis) and passing through point P. As shown on the free-body diagram of the ith particle, Fig. 17–8b, $\mathbf{F}_i$ represents the *resultant external force* acting on the particle, and $\mathbf{f}_i$ is the *resultant of the internal forces* caused by interactions with adjacent particles. If the particle has a mass m_i and at the instant considered its acceleration is $\mathbf{a}_i$, then the kinetic diagram is constructed as shown in Fig. 17–8c. If moments of the forces acting on the particle are summed about point P, we require

$$\mathbf{r} \times \mathbf{F}_i + \mathbf{r} \times \mathbf{f}_i = \mathbf{r} \times m_i\mathbf{a}_i$$

or

$$(\mathbf{M}_P)_i = \mathbf{r} \times m_i\mathbf{a}_i$$

We will now write this equation in terms of the acceleration $\mathbf{a}_P$ of point P. If the body has an angular acceleration $\boldsymbol{\alpha}$ and angular velocity $\boldsymbol{\omega}$, Fig. 17–8d, then using Eq. 16–17 we have

$$\begin{aligned}(\mathbf{M}_P)_i &= m_i\mathbf{r} \times (\mathbf{a}_P + \boldsymbol{\alpha} \times \mathbf{r} - \omega^2\mathbf{r}) \\ &= m_i[\mathbf{r} \times \mathbf{a}_P + \mathbf{r} \times (\boldsymbol{\alpha} \times \mathbf{r}) - \omega^2(\mathbf{r} \times \mathbf{r})]\end{aligned}$$

The last term is zero, since $\mathbf{r} \times \mathbf{r} = \mathbf{0}$. Expressing the vectors with Cartesian components and carrying out the cross-product operations yields

$$\begin{aligned}(M_P)_i\,\mathbf{k} &= m_i\{(x\mathbf{i} + y\mathbf{j}) \times [(a_P)_x\mathbf{i} + (a_P)_y\mathbf{j}] \\ &\quad + (x\mathbf{i} + y\mathbf{j}) \times [\alpha\mathbf{k} \times (x\mathbf{i} + y\mathbf{j})]\} \\ (M_P)_i\mathbf{k} &= m_i[-y(a_P)_x + x(a_P)_y + \alpha x^2 + \alpha y^2]\mathbf{k} \\ \circlearrowleft (M_P)_i &= m_i[-y(a_P)_x + x(a_P)_y + \alpha r^2]\end{aligned}$$

Letting $m_i \rightarrow dm$, and integrating with respect to the entire mass m of the body, we obtain the resultant moment equation

$$\circlearrowleft \Sigma M_P = -\left(\int_m y\,dm\right)(a_P)_x + \left(\int_m x\,dm\right)(a_P)_y + \left(\int_m r^2\,dm\right)\alpha$$

Here ΣM_P represents only the moment of the *external forces* acting on the body about point P. The resultant moment of the internal forces is zero, since for the entire body these forces occur in equal and opposite collinear pairs and thus the moment of each pair of forces about point P cancels. The integrals in the first and second terms on the right are used to locate the body's center of mass G with respect to P, since $\bar{y}m = \int y\,dm$ and $\bar{x}m = \int x\,dm$, Fig. 17–8d. Also, the last integral represents the body's moment of inertia computed about the z axis, i.e., $I_P = \int r^2\,dm$. Thus,

$$\circlearrowleft \Sigma M_P = -\bar{y}m(a_P)_x + \bar{x}m(a_P)_y + I_P\alpha \qquad (17\text{–}6)$$

It is possible to reduce this equation to a simpler form if point P coincides with the mass center G for the body. If this is the case, then $\bar{x} = \bar{y} = 0$, and therefore*

$$\Sigma M_G = I_G \alpha \tag{17–7}$$

This rotational equation of motion states that the sum of the moments of all the external forces computed about the body's mass center G is equal to the product of the moment of inertia of the body about an axis passing through G and the body's angular acceleration.

Equation 17–6 can also be rewritten in terms of the x and y components of $\mathbf{a}_G$ and the body's moment of inertia I_G. If point G is located at point $(\bar{x}, \bar{y})$, Fig. 17–8d, then by the parallel-axis theorem, $I_P = I_G + m(\bar{x}^2 + \bar{y}^2)$. Substituting into Eq. 17–6 and rearranging terms, we get

$$\circlearrowleft \Sigma M_P = \bar{y}m[-(a_P)_x + \bar{y}\alpha] + \bar{x}m[(a_P)_y + \bar{x}\alpha] + I_G\alpha \tag{17–8}$$

From the kinematic diagram of Fig. 17–8d, $\mathbf{a}_P$ can be expressed in terms of $\mathbf{a}_G$ as

$$\mathbf{a}_G = \mathbf{a}_P + \boldsymbol{\alpha} \times \bar{\mathbf{r}} - \omega^2 \bar{\mathbf{r}}$$

$$(a_G)_x\mathbf{i} + (a_G)_y\mathbf{j} = (a_P)_x\mathbf{i} + (a_P)_y\mathbf{j} + \alpha\mathbf{k} \times (\bar{x}\mathbf{i} + \bar{y}\mathbf{j}) - \omega^2(\bar{x}\mathbf{i} + \bar{y}\mathbf{j})$$

Carrying out the cross product and equating the respective **i** and **j** components yields the two scalar equations.

$$(a_G)_x = (a_P)_x - \bar{y}\alpha - \bar{x}\omega^2$$

$$(a_G)_y = (a_P)_y + \bar{x}\alpha - \bar{y}\omega^2$$

From these equations, $[-(a_P)_x + \bar{y}\alpha] = [-(a_G)_x - \bar{x}\omega^2]$ and $[(a_P)_y + \bar{x}\alpha] = [(a_G)_y + \bar{y}\omega^2]$. Substituting these results into Eq. 17–8 and simplifying gives

$$\circlearrowleft \Sigma M_P = -\bar{y}m(a_G)_x + \bar{x}m(a_G)_y + I_G\alpha \tag{17–9}$$

This important result indicates that when moments of the external forces shown on the free-body diagram are summed about point P, Fig. 17–8e, they are equivalent to the sum of the "kinetic moments" of the components of $m\mathbf{a}_G$ *about P plus the "kinetic moment" of* $I_G\boldsymbol{\alpha}$, *Fig. 17–8f.* In other words, when the "kinetic moments," $\Sigma(\mathcal{M}_k)_P$, are computed, Fig. 17–8f, the vectors $m(\mathbf{a}_G)_x$ and $m(\mathbf{a}_G)_y$ are treated as sliding vectors; that is, they can act at *any point along their line of action.* In a similar manner, $I_G\boldsymbol{\alpha}$ can be treated as a free vector and can therefore act at *any point. It is important to keep in mind that* $m\mathbf{a}_G$ *and* $I_G\boldsymbol{\alpha}$ *are not the same as a force or a couple moment. Instead, they are caused by the external effects of forces and couple moments acting on the body.* With this in mind we can therefore write Eq. 17–9 in a more general form as

$$\Sigma M_P = \Sigma(\mathcal{M}_k)_P \tag{17–10}$$

*It also reduces to this same simple form $\Sigma M_P = I_P\alpha$ if point P is a *fixed point* (see Eq. 17–16) or the acceleration of point P is directed along the line PG.

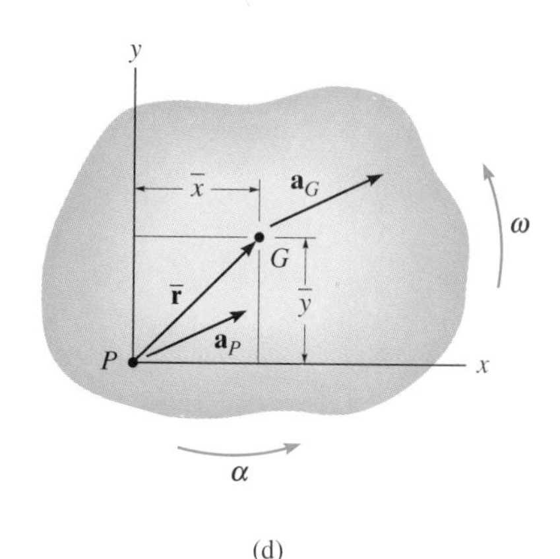

(d)

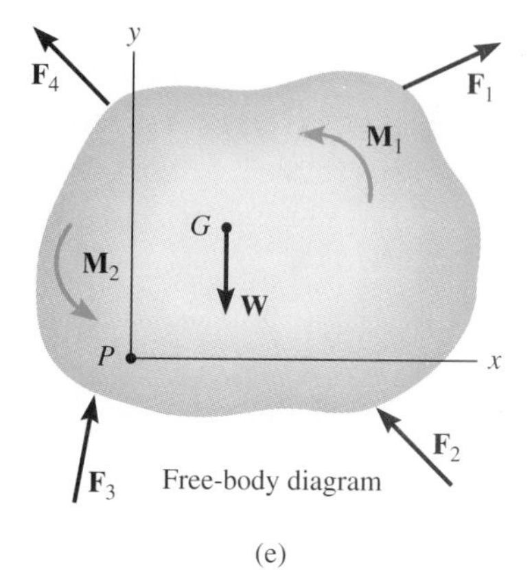

Free-body diagram

(e)

=

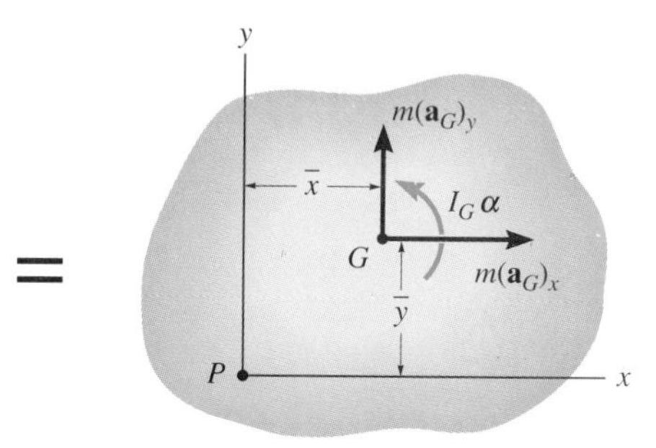

Kinetic diagram

(f)

Fig. 17–8 *(cont'd)*

General Application of the Equations of Motion. To summarize this analysis, *three* independent scalar equations may be written to describe the general plane motion of a symmetrical rigid body.

$$\Sigma F_x = m(a_G)_x$$
$$\Sigma F_y = m(a_G)_y \qquad (17\text{–}11)$$
$$\Sigma M_G = I_G\alpha \quad \text{or} \quad \Sigma M_P = \Sigma(\mathcal{M}_k)_P$$

When applying these equations, one should *always* draw a free-body diagram, Fig. 17–8*e*, in order to account for the terms involved in ΣF_x, ΣF_y, ΣM_G, or ΣM_P. In some problems it may also be helpful to draw the *kinetic diagram* for the body. This diagram accounts graphically for the terms $m(\mathbf{a}_G)_x$, $m(\mathbf{a}_G)_y$, and $I_G\boldsymbol{\alpha}$, and it is especially convenient when used to determine the components of $m\mathbf{a}_G$ and the moment terms in $\Sigma(\mathcal{M}_k)_P$.*

17.3 Equations of Motion: Translation

When a rigid body undergoes a *translation,* Fig. 17–9*a*, all the particles of the body have the *same acceleration,* so that $\mathbf{a}_G = \mathbf{a}$. Furthermore, $\boldsymbol{\alpha} = \mathbf{0}$, in which case the rotational equation of motion applied at point G reduces to a simplified form, namely, $\Sigma M_G = 0$. Application of this and the translational equations of motion will now be discussed for each of the two types of translation presented in Chapter 16.

Rectilinear Translation. When a body is subjected to *rectilinear translation,* all the particles of the body (slab) travel along parallel straight-line paths. The free-body and kinetic diagrams are shown in Fig. 17–9*b*. Since $I_G\boldsymbol{\alpha} = \mathbf{0}$, only $m\mathbf{a}_G$ is shown on the kinetic diagram. Hence, the equations of motion which apply in this case become

$$\Sigma F_x = m(a_G)_x$$
$$\Sigma F_y = m(a_G)_y \qquad (17\text{–}12)$$
$$\Sigma M_G = 0$$

The last equation requires that the sum of the moments of all the external forces (and couple moments) computed about the body's center of mass be equal to zero. It is possible, of course, to sum moments about other points on or off the body, in which case the moment of $m\mathbf{a}_G$ must be taken into account.

*For this reason, the kinetic diagram will be used in the solution of an example problem whenever $\Sigma M_P = \Sigma(\mathcal{M}_k)_P$ is applied.

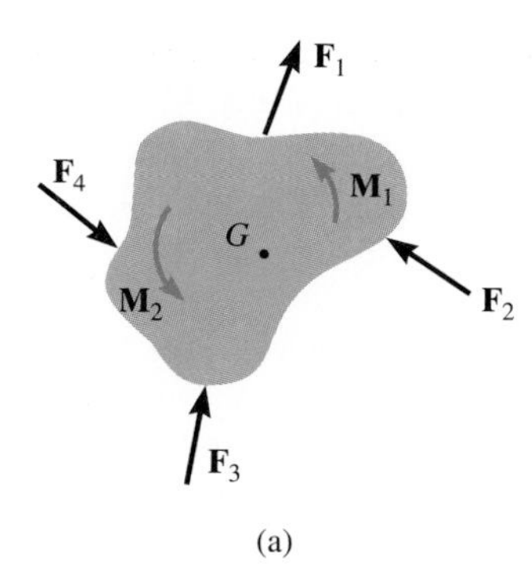

(a)

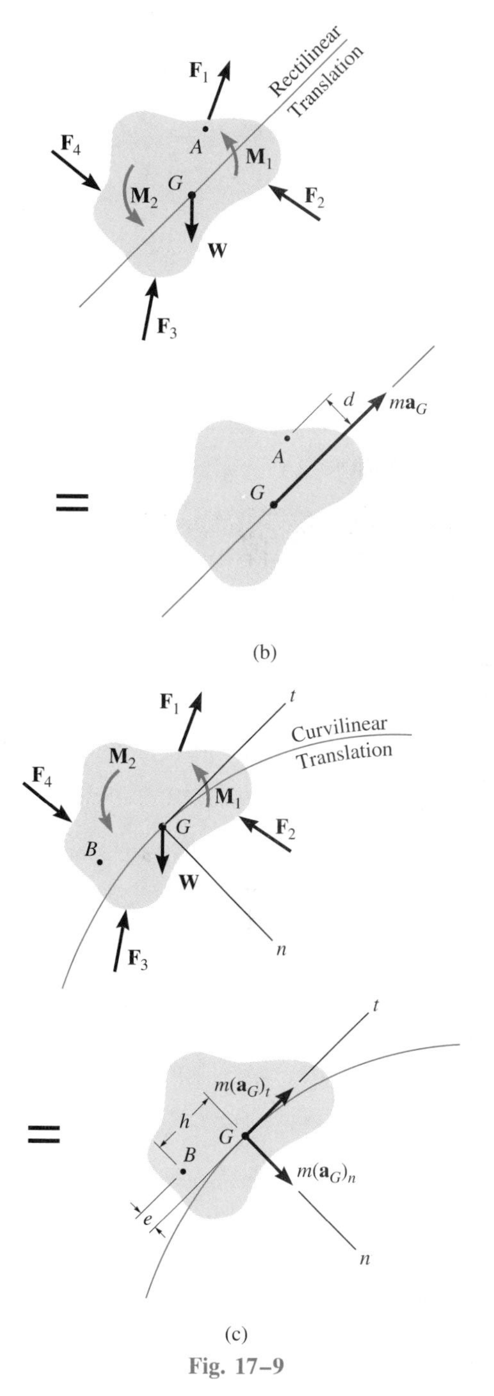

Fig. 17–9

For example, if point A is chosen, which lies at a perpendicular distance d from the line of action of $m\mathbf{a}_G$, the following moment equation applies:

$$\circlearrowright\!+\Sigma M_A = \Sigma(\mathcal{M}_k)_A; \qquad \Sigma M_A = (ma_G)d$$

Here the sum of moments of the external forces and couple moments about A (ΣM_A, free-body diagram) equals the moment of $m\mathbf{a}_G$ about A ($\Sigma(\mathcal{M}_k)_A$, kinetic diagram).

Curvilinear Translation. When a rigid body is subjected to *curvilinear translation,* all the particles of the body travel along *parallel curved paths.* For analysis, it is often convenient to use an inertial coordinate system having an origin which is coincident with the body's mass center at the instant considered, and axes which are oriented in the normal and tangential directions of the path of motion, Fig. 17–9c. The three scalar equations of motion are then

$$\begin{aligned} \Sigma F_n &= m(a_G)_n \\ \Sigma F_t &= m(a_G)_t \\ \Sigma M_G &= 0 \end{aligned} \qquad (17\text{–}13)$$

where $(a_G)_t$ and $(a_G)_n$ represent, respectively, the magnitudes of the tangential and normal components of acceleration of point G.

If the moment equation $\Sigma M_G = 0$ is replaced by a moment summation about the arbitrary point B, Fig. 17–9c, it is necessary to account for the moments, $\Sigma(\mathcal{M}_k)_B$, of the two components $m(\mathbf{a}_G)_n$ and $m(\mathbf{a}_G)_t$ about this point. From the kinetic diagram, h and e represent the perpendicular distances (or "moment arms") from B to the lines of action of the components. The required moment equation therefore becomes

$$\circlearrowright\!+\Sigma M_B = \Sigma(\mathcal{M}_k)_B; \quad \Sigma M_B = e[m(a_G)_t] - h[m(a_G)_n]$$

PROCEDURE FOR ANALYSIS

The following procedure provides a method for solving kinetic problems involving rigid-body *translation.*

Free-Body Diagram. Establish the x, y or n, t inertial coordinate system and draw the free-body diagram in order to account for all the external forces and couple moments that act on the body. The direction and sense of the acceleration of the body's mass center $\mathbf{a}_G$ should also be established. If the sense of its components *cannot* be determined, assume they are in the direction of the positive inertial coordinate axes. The acceleration may be sketched on the coordinate system, but not on the free-body diagram. Identify the unknowns in the problem. If it is decided that the rotational equation of motion $\Sigma M_P = \Sigma(\mathcal{M}_k)_P$ is to be used in the solution, then it may be beneficial to draw the kinetic diagram, since it accounts graphically for the components $m(\mathbf{a}_G)_x$, $m(\mathbf{a}_G)_y$ or $m(\mathbf{a}_G)_t$, $m(\mathbf{a}_G)_n$ and is therefore convenient for "visualizing" the terms needed in the moment sum $\Sigma(\mathcal{M}_k)_P$.

Equations of Motion. Apply the three equations of motion, Eqs. 17–12 or Eqs. 17–13. To simplify the analysis, the moment equation $\Sigma M_G = 0$ can be replaced by the more general equation $\Sigma M_P = \Sigma(\mathcal{M}_k)_P$, where point P is usually located at the intersection of the lines of action of as many unknown forces as possible.

If the body is in contact with a *rough surface* and slipping occurs, use the frictional equation $F = \mu_k N$ to relate the normal force $\mathbf{N}$ to its associated frictional force $\mathbf{F}$. Remember, $\mathbf{F}$ always acts on the body so as to oppose the motion of the body relative to the surface it contacts.

Kinematics. Use kinematics if the velocity and position of the body are to be determined. For *rectilinear translation* with *variable acceleration* use

$$a_G = \frac{dv_G}{dt} \qquad a_G\, ds_G = v_G\, dv_G \qquad v_G = \frac{ds_G}{dt}$$

For *rectilinear translation* with *constant acceleration,* use

$$v_G = (v_G)_0 + a_G t$$
$$v_G^2 = (v_G)_0^2 + 2a_G[s_G - (s_G)_0]$$
$$s_G = (s_G)_0 + (v_G)_0 t + \tfrac{1}{2}a_G t^2$$

For *curvilinear translation,* use

$$(a_G)_n = \frac{v_G^2}{\rho} = \omega^2\rho$$

$$(a_G)_t = \frac{dv_G}{dt} \qquad (a_G)_t\, ds_G = v_G\, dv_G \qquad (a_G)_t = \alpha\rho$$

The following examples illustrate application of this procedure.

Example 17–5

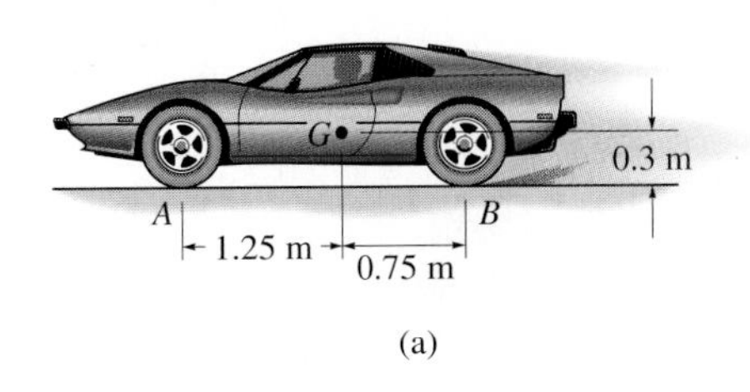

(a)

The car shown in Fig. 17–10*a* has a mass of 2 Mg and a center of mass at *G*. Determine the car's acceleration if the "driving" wheels in the back are always slipping, whereas the front wheels freely rotate. Neglect the mass of the wheels. The coefficient of kinetic friction between the wheels and the road is $\mu_k = 0.25$.

SOLUTION I

Free-Body Diagram. As shown in Fig. 17–10*b*, the rear-wheel frictional force $\mathbf{F}_B$ pushes the car forward, and since *slipping occurs,* the magnitude of this force is related to the magnitude of its associated normal force $\mathbf{N}_B$ by $F_B = 0.25N_B$. The frictional forces acting on the *front wheels* are *zero,* since these wheels have negligible mass.* There are three unknowns in the problem, N_A, N_B, and a_G. Here we will sum moments about the mass center. The car (point *G*) is assumed to accelerate to the left, i.e., in the negative *x* direction, Fig. 17–10*b*.

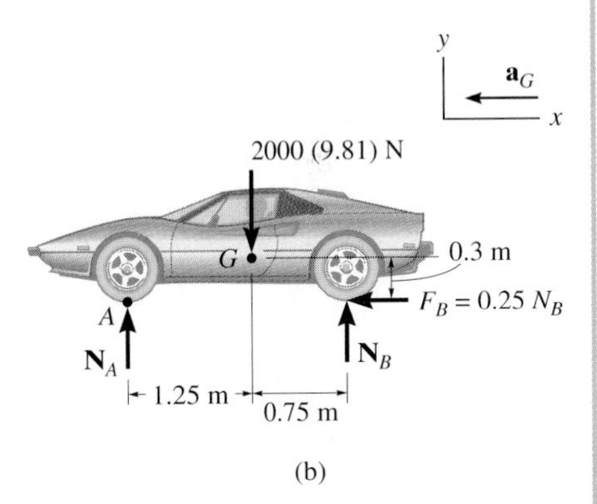

(b)

Equations of Motion

$$\stackrel{+}{\rightarrow}\Sigma F_x = m(a_G)_x; \qquad -0.25N_B = -(2000\text{ kg})a_G \qquad (1)$$

$$+\uparrow \Sigma F_y = m(a_G)_y; \quad N_A + N_B - 2000(9.81)\text{ N} = 0 \qquad (2)$$

$$\circlearrowright+\Sigma M_G = 0; \quad -N_A(1.25\text{ m}) - 0.25N_B(0.3\text{ m}) + N_B(0.75\text{ m}) = 0 \quad (3)$$

Solving,

$$a_G = 1.59\text{ m/s}^2 \leftarrow \qquad \textit{Ans.}$$

$$N_A = 6.88\text{ kN}$$

$$N_B = 12.7\text{ kN}$$

SOLUTION II

Free-Body and Kinetic Diagrams. If the "moment" equation is applied about point *A*, then the unknown N_A will be eliminated from the equation. To "visualize" the moment of $m\mathbf{a}_G$ about *A*, we will include the kinetic diagram as part of the analysis, Fig. 17–10*c*.

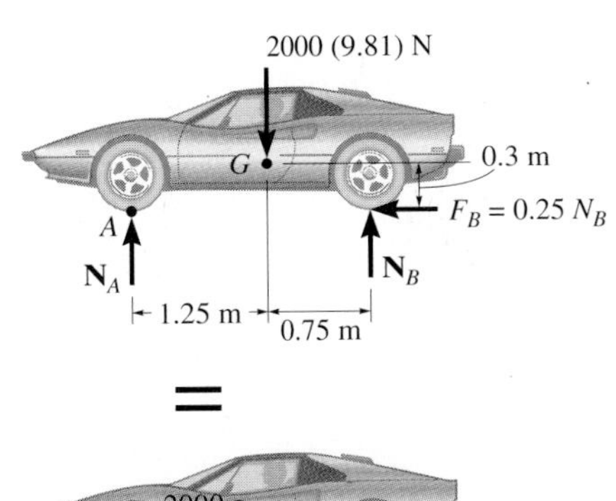

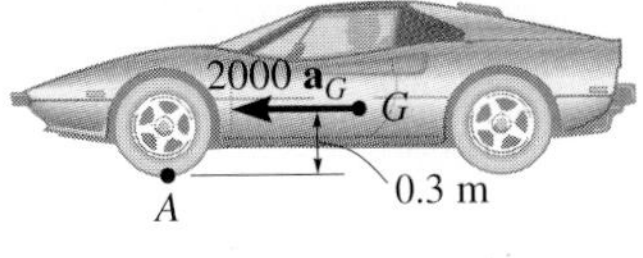

(c)

Fig. 17–10

Equation of Motion. We require

$$\circlearrowright+\Sigma M_A = \Sigma(\mathcal{M}_k)_A; \qquad N_B(2\text{ m}) - 2000(9.81)\text{ N}(1.25\text{ m}) = (2000\text{ kg})a_G(0.3\text{ m})$$

Solving this and Eq. 1 for a_G leads to a simpler solution than that obtained from Eqs. 1 to 3.

*If the mass of the front wheels were to be included in the analysis, the frictional force acting at *A* would be *directed to the right* to create the necessary counterclockwise rotation of the wheels. The problem solution for this case would be more involved since a general-plane-motion analysis of the wheels would have to be considered (see Sec. 17.5).

Example 17–6

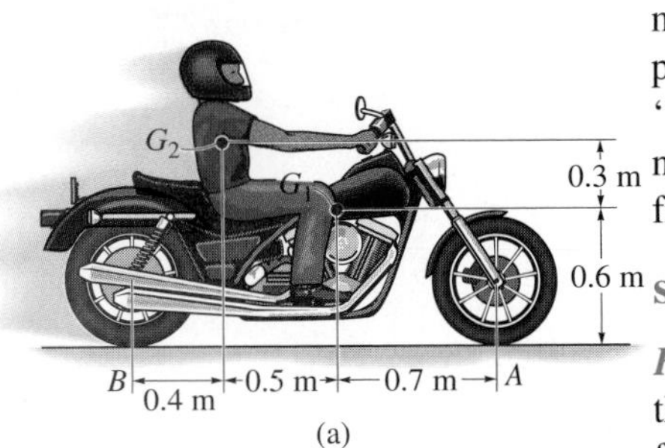

(a)

The motorcycle shown in Fig. 17–11*a* has a mass of 125 kg and a center of mass at G_1, while the rider has a mass of 75 kg and a center of mass at G_2. If the coefficient of static friction between the wheels and the pavement is $\mu_s = 0.8$, determine if it is possible for the rider to do a "wheely," i.e., lift the front wheel off the ground. What acceleration is necessary to do this? Neglect the mass of the wheels and assume that the front wheel is free to roll.

SOLUTION

Free-Body and Kinetic Diagrams. In this problem we will consider both the motorcycle and the rider as the "system" to be analyzed. It is possible first to determine the location of the center of mass for this "system" by using the equations $\bar{x} = \Sigma\tilde{x}m/\Sigma m$ and $\bar{y} = \Sigma\tilde{y}m/\Sigma m$. Here, however, we will consider the separate weight and mass of each of its *component parts* as shown on the free-body and kinetic diagrams, Fig. 17–11*b*. Both parts move with the *same* acceleration and we have assumed that the front wheel is about to leave the ground, so that the normal reaction $N_A \approx 0$. In order for this to happen it is required that the friction force $F_B \leq 0.8N_B$, otherwise slipping will occur. The three unknowns in the problem are N_B, F_B, and a_G.

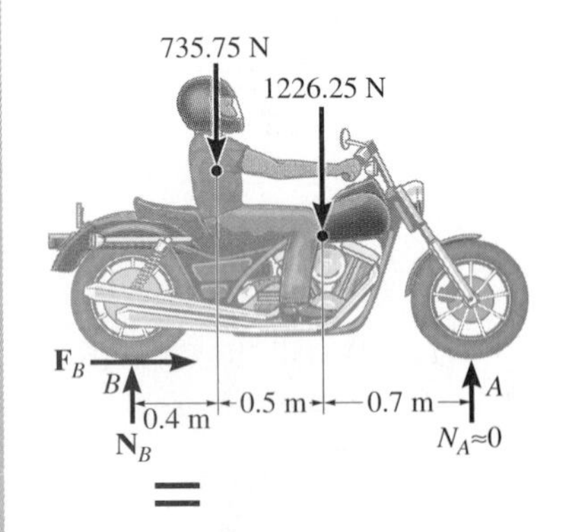

=

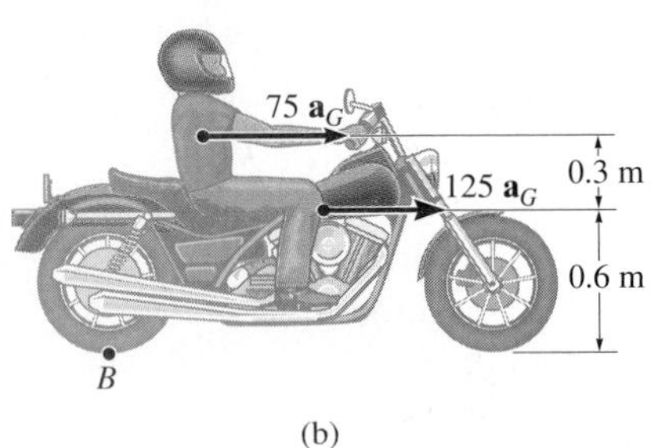

(b)

Fig. 17–11

Equations of Motion

$$\overset{+}{\rightarrow}\Sigma F_x = m(a_G)_x; \quad F_B = (75 \text{ kg} + 125 \text{ kg})a_G \tag{1}$$

$$+\uparrow \Sigma F_y = m(a_G)_y; \; N_B - 735.75 \text{ N} - 1226.25 \text{ N} = 0 \tag{2}$$

$$\circlearrowright + \Sigma M_B = \Sigma(\mathcal{M}_k)_B; \quad -(735.75 \text{ N})(0.4 \text{ m}) - (1226.25 \text{ N})(0.9 \text{ m}) = -(75 \text{ kg } a_G)(0.9 \text{ m}) - (125 \text{ kg } a_G)(0.6 \text{ m})$$

Solving,

$$a_G = 9.81 \text{ m/s}^2 \rightarrow \qquad \textit{Ans.}$$

$$N_B = 1.96 \text{ kN}$$

$$F_B = 1.96 \text{ kN}$$

The maximum frictional force that can be developed at B is

$$(F_B)_{max} = \mu_s N_B = 0.8(1.96 \text{ kN}) = 1.57 \text{ kN}$$

Since this force is less than that required ($F_B = 1.96$ kN), it is *not possible* to lift the front wheel off the ground.

Note: If the maximum acceleration of the motorcycle is to be obtained, then $F_B = 0.8N_B$ and $N_A \neq 0$. The problem would have to be solved in a manner similar to that of Example 17–5. Try it, and show that $(a_G)_{max} = 6.76 \text{ m/s}^2$.

Example 17–7

A uniform 50-kg crate rests on a horizontal surface for which the coefficient of kinetic friction is $\mu_k = 0.2$. Determine the crate's acceleration if a force of $P = 600$ N is applied to the crate as shown in Fig. 17–12*a*.

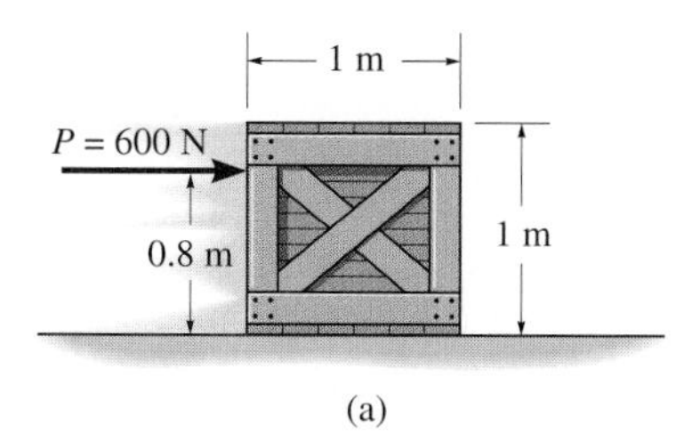

(a)

SOLUTION

Free-Body Diagram. The force **P** can cause the crate either to slide or to tip over. As shown in Fig. 17–12*b*, it is assumed that the crate slides, so that $F = \mu_k N_C = 0.2N_C$. Also, the resultant normal force $\mathbf{N}_C$ acts at O, a distance x (where $0 < x \leq 0.5$ m) from the crate's center line.* The three unknowns are N_C, x, and a_G.

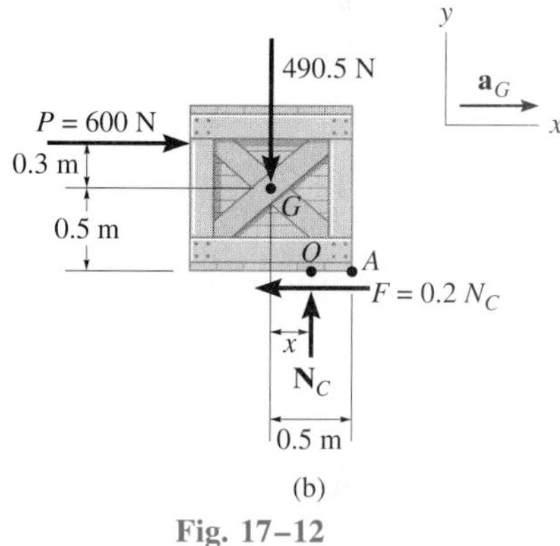

(b)

Fig. 17–12

Equations of Motion

$$\overset{+}{\rightarrow}\Sigma F_x = m(a_G)_x; \qquad 600\text{ N} - 0.2N_C = (50\text{ kg})\,a_G \qquad (1)$$

$$+\uparrow \Sigma F_y = m(a_G)_y; \qquad N_C - 490.5\text{ N} = 0 \qquad (2)$$

$$\circlearrowright\!+\Sigma M_G = 0; \quad -600\text{ N}(0.3\text{ m}) + N_C(x) - 0.2N_C(0.5\text{ m}) = 0 \qquad (3)$$

Solving, we obtain

$$N_C = 490\text{ N}$$

$$x = 0.467\text{ m}$$

$$a_G = 10.0\text{ m/s}^2 \rightarrow \qquad \textit{Ans.}$$

Since $x = 0.467$ m < 0.5 m, indeed the crate slides as originally assumed. If the solution had given a value of $x > 0.5$ m, the problem would have to be reworked with the assumption that tipping occurred. If this were the case, $\mathbf{N}_C$ would act at the *corner point A* and $F \leq 0.2N_C$.

*The line of action of $\mathbf{N}_C$ does not necessarily pass through the mass center G ($x = 0$), since $\mathbf{N}_C$ must counteract the tendency for tipping caused by **P.** See Sec. 8.1 of *Engineering Mechanics: Statics.*

Example 17–8

The 100-kg beam BD shown in Fig. 17–13a is supported by two rods having negligible mass. Determine the force created in each rod if at the instant $\theta = 30°$ the rods are both rotating with an angular velocity of $\omega = 6$ rad/s.

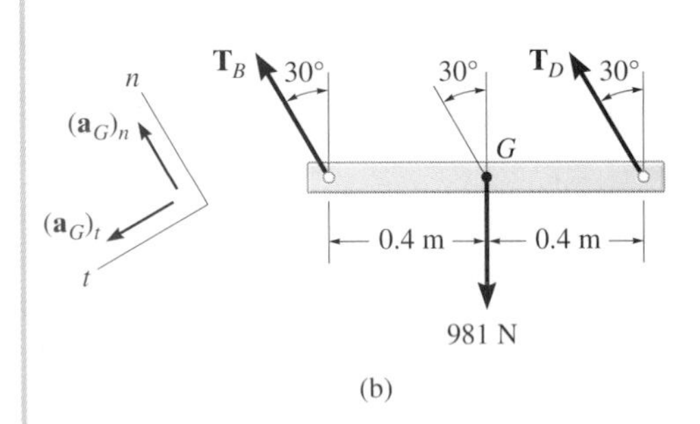

(b)

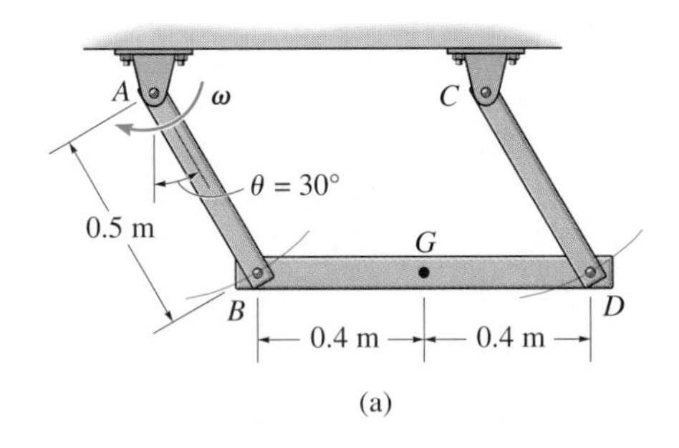

(a)

(c)

Fig. 17–13

SOLUTION

Free-Body Diagram. The beam moves with *curvilinear translation* since points B and D and the center of mass G all move along circular paths, each path having the same radius of 0.5 m. Using normal and tangential coordinates, the free-body diagram for the beam is shown in Fig. 17–13b. Because of the *translation,* G has the *same* motion as the pin at B, which is connected to both the rod and the beam. By studying the angular motion of rod AB, Fig. 17–13c, note that the tangential component of acceleration acts downward to the left due to the clockwise direction of $\boldsymbol{\alpha}$. Furthermore, the normal component of acceleration is *always* directed toward the center of curvature (toward point A for rod AB). Since the angular velocity of AB is 6 rad/s, then

$$(a_G)_n = \omega^2 r = (6 \text{ rad/s})^2(0.5 \text{ m}) = 18 \text{ m/s}^2$$

The three unknowns are T_B, T_D, and $(a_G)_t$. The directions of $(\mathbf{a}_G)_n$ and $(\mathbf{a}_G)_t$ have been established, and are indicated on the coordinate axes.

Equations of Motion

$$+\nwarrow \Sigma F_n = m(a_G)_n; \qquad T_B + T_D - 981 \cos 30° \text{ N} = 100 \text{ kg}(18 \text{ m/s}^2) \quad (1)$$

$$+\swarrow \Sigma F_t = m(a_G)_t; \qquad 981 \sin 30° = 100 \text{ kg}(a_G)_t \quad (2)$$

$$\circlearrowright + \Sigma M_G = 0; \quad -(T_B \cos 30°)(0.4 \text{ m}) + (T_D \cos 30°)(0.4 \text{ m}) = 0 \quad (3)$$

Simultaneous solution of these three equations gives

$$T_B = T_D = 1.33 \text{ kN} \; \nwarrow\!\!{}^{30°} \qquad \textit{Ans.}$$

$$(a_G)_t = 4.90 \text{ m/s}^2$$

PROBLEMS

***17–28.** The 2-lb bottle rests on the check-out conveyor at a grocery store. If the coefficient of static friction is $\mu_s = 0.2$, determine the largest acceleration the conveyor can have without causing the bottle to slip or tip. The center of gravity is at G.

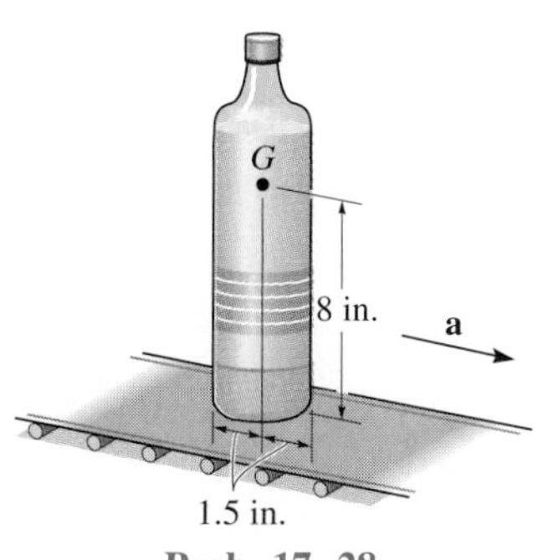

Prob. 17–28

17–29. The door has a weight of 200 lb and a center of gravity at G. Determine how far the door moves in 2 s, starting from rest, if a man pushes on it at C with a horizontal force $F = 30$ lb. Also, find the vertical reactions at the rollers A and B.

17–30. The door has a weight of 200 lb and a center of gravity at G. Determine the constant force F that must be applied to the door to push it open 12 ft to the right in 5 s, starting from rest. Also, find the vertical reactions at the rollers A and B.

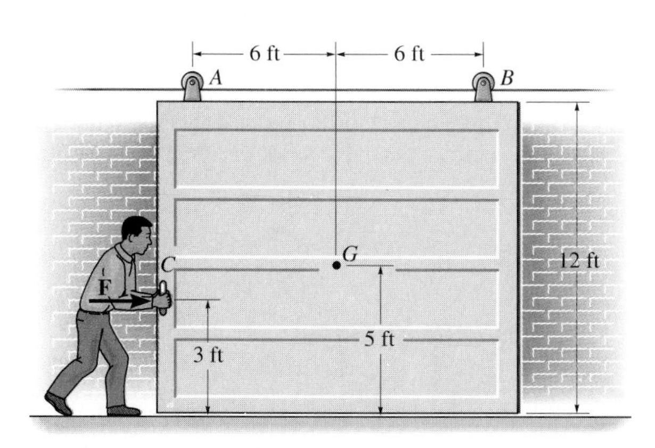

Probs. 17–29/17–30

17–31. The jet aircraft has a total mass of 22 Mg and a center of mass at G. Initially at take-off the engines provide a thrust $2T = 4$ kN and $T' = 1.5$ kN. Determine the acceleration of the plane and the normal reactions on the nose wheel and each of the *two* wing wheels located at B. Neglect the mass of the wheels and, due to low velocity, neglect any lift caused by the wings.

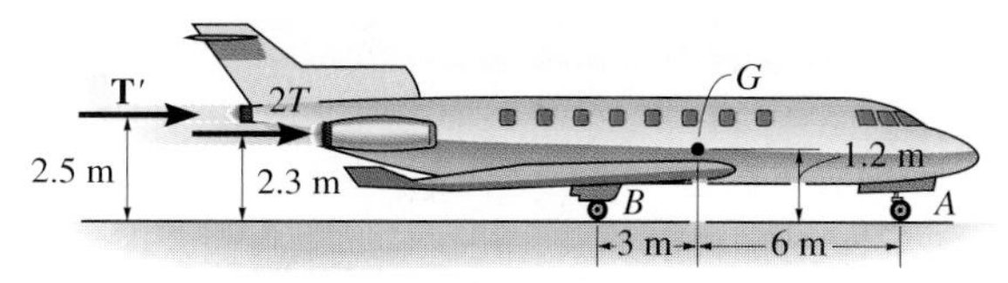

Prob. 17–31

***17–32.** The assembly has a mass of 8 Mg and is hoisted using the boom and pulley system. If the winch at B draws in the cable with an acceleration of 2 m/s^2, determine the compressive force in the hydraulic cylinder needed to support the boom. The boom has a mass of 2 Mg and mass center at G.

17–33. The assembly has a mass of 4 Mg and is hoisted using the winch at B. Determine the greatest acceleration of the assembly so that the compressive force in the hydraulic cylinder supporting the boom does not exceed 180 kN. What is the tension in the supporting cable? The boom has a mass of 2 Mg and mass center at G.

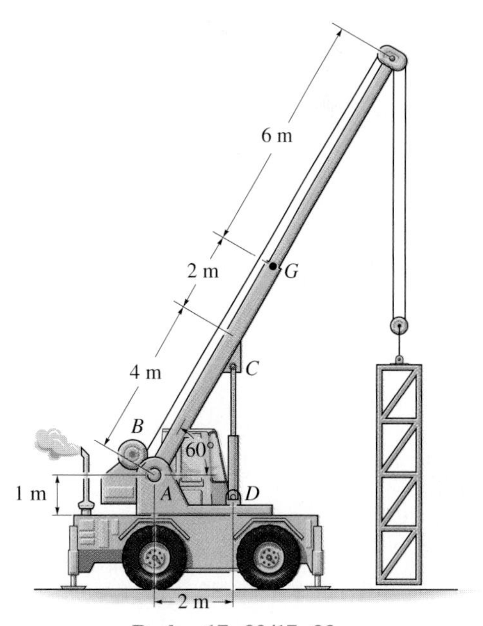

Probs. 17–32/17–33

17–34. The uniform pipe has a weight of 500 lb/ft and diameter of 2 ft. If it is hoisted as shown with an acceleration of 0.5 ft/s^2, determine the internal moment at the center A of the pipe due to the lift.

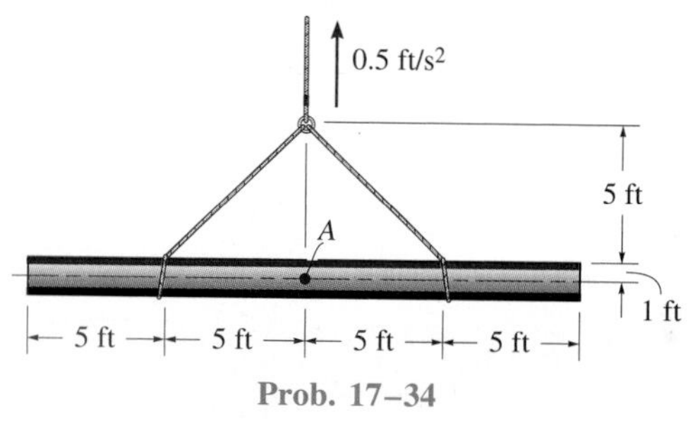

Prob. 17–34

17–35. The bar has a weight per length w and is supported by the smooth collar. If it is released from rest, determine the internal normal force, shear force, and bending moment in the bar as a function of x.

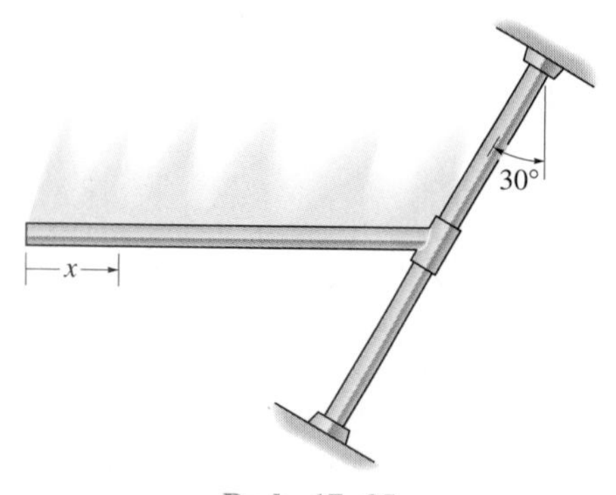

Prob. 17–35

***17–36.** The uniform girder AB has a mass of 8 Mg. Determine the internal axial, shear, and bending-moment loadings at the center of the girder if a crane gives it an upward acceleration of 3 m/s^2.

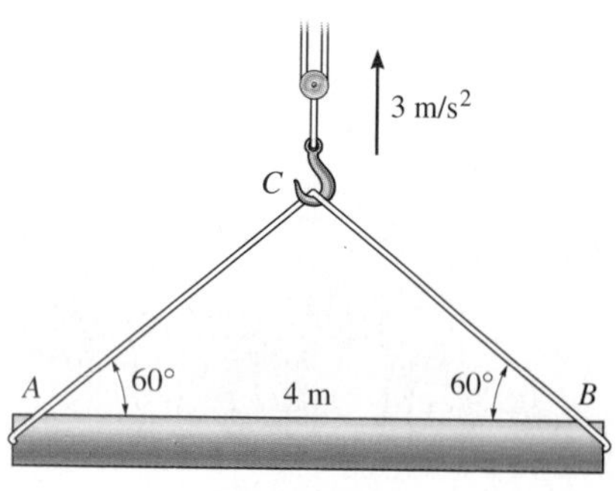

Prob. 17–36

17–37. The top truck has a mass of 1.75 Mg and a center of mass at G. It is tied to the transport using a chain DE. Determine the tension developed in the chain and the normal reactions on each of the truck's four wheels if the transport accelerates at 2 m/s^2.

17–38. The top truck has a mass of 1.75 Mg and a center of mass at G. The deck on which it is supported has negligible mass and is pin-connected at H and supported by the hydraulic cylinder IJ. Determine the horizontal and vertical components of force developed at the pin and the force in the hydraulic cylinder if the transport accelerates at 2 m/s^2. Assume a similar type of support exists on the other side of the transport.

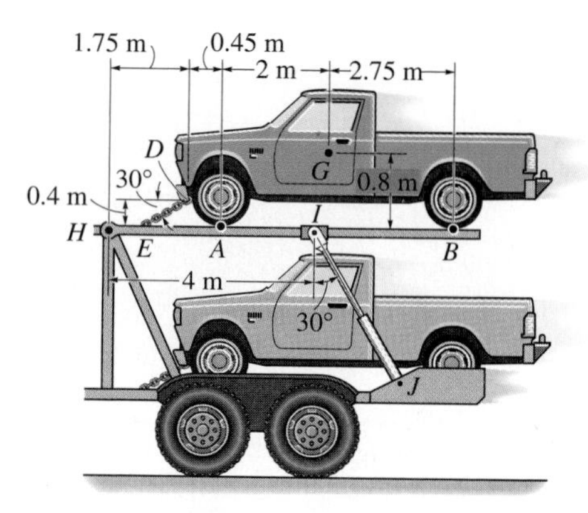

Probs. 17–37/17–38

17–39. The drop gate at the end of the trailer has a mass of 1.25 Mg and mass center at G. If it is supported by the cable AB and hinge at C, determine the tension in the cable when the truck begins to accelerate at 5 m/s^2. Also, what are the horizontal and vertical components of reaction at the hinge C?

***17–40.** The drop gate at the end of the trailer has a mass of 1.25 Mg and mass center at G. If it is supported by the cable AB and hinge at C, determine the maximum deceleration of the truck so that the gate does not begin to rotate forward. What are the horizontal and vertical components of reaction at the hinge C?

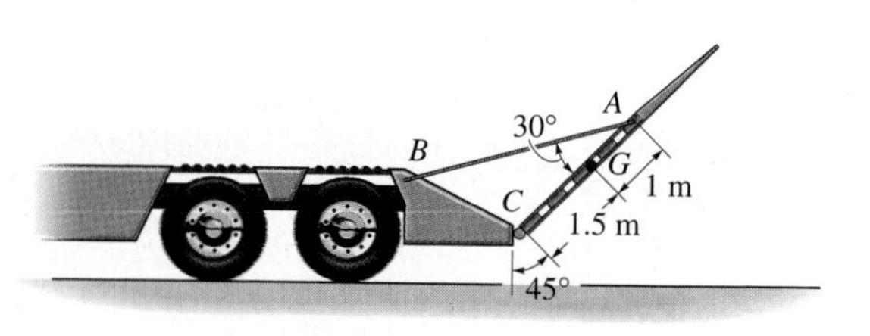

Probs. 17–39/17–40

17–41. The sports car has a weight of 4500 lb and center of gravity at G. If it starts from rest it causes the rear wheels to slip as it accelerates. Determine how long it takes for it to reach a speed of 10 ft/s. Also, what are the normal reactions at *each* of the four wheels on the road? The coefficients of static and kinetic friction at the road are $\mu_s = 0.5$ and $\mu_k = 0.3$, respectively. Neglect the mass of the wheels.

17–42. Determine the shortest possible time for the car in Prob. 17–41 to obtain a speed of 10 ft/s.

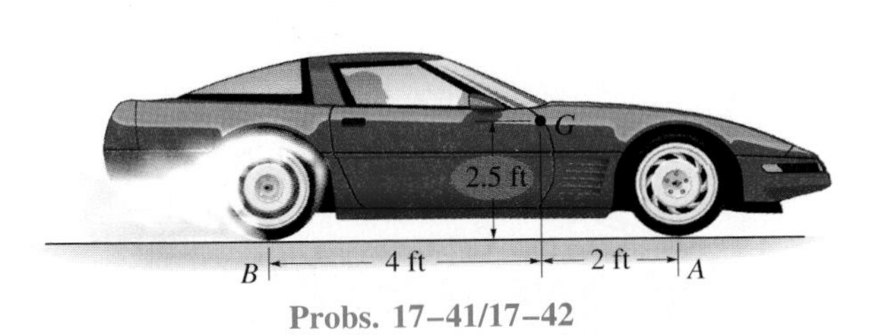

Probs. 17–41/17–42

***17–44.** The drum truck supports the 600-lb drum that has a center of gravity at G. If the operator pushes it forward with a horizontal force of 20 lb, determine the acceleration of the truck and the normal reactions at each of the four wheels. Neglect the mass of the wheels.

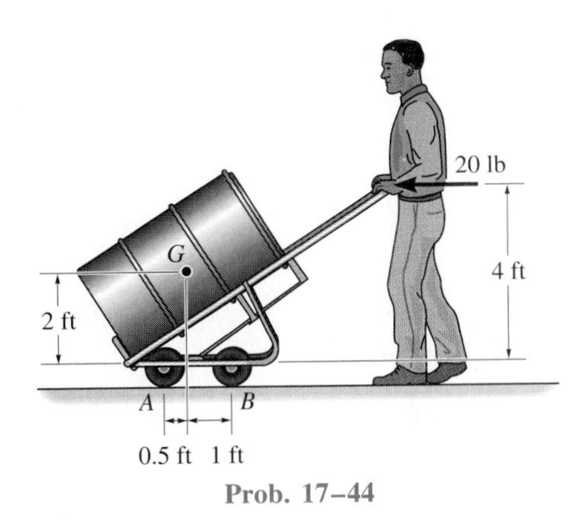

Prob. 17–44

17–43. The car accelerates uniformly from rest to 88 ft/s in 15 seconds. If it has a weight of 3800 lb and a center of gravity at G, determine the normal reaction of *each wheel* on the pavement during the motion. Power is developed at the front wheels, whereas the rear wheels are free to roll. Neglect the mass of the wheels and take the coefficients of static and kinetic friction to be $\mu_s = 0.4$ and $\mu_k = 0.2$, respectively.

Prob. 17–43

17–45. The arched pipe has a mass of 80 kg and rests on the surface of the platform. As it is hoisted from one level to the next, $\alpha = 0.25$ rad/s^2 and $\omega = 0.5$ rad/s at the instant $\theta = 30°$. If it does not slip, determine the normal reactions of the arch on the platform at this instant.

17–46. The arched pipe has a mass of 80 kg and rests on the surface of the platform for which the coefficient of static friction is $\mu_s = 0.3$. Determine the greatest angular acceleration α of the platform, starting from rest when $\theta = 45°$, without causing the pipe to slip on the platform.

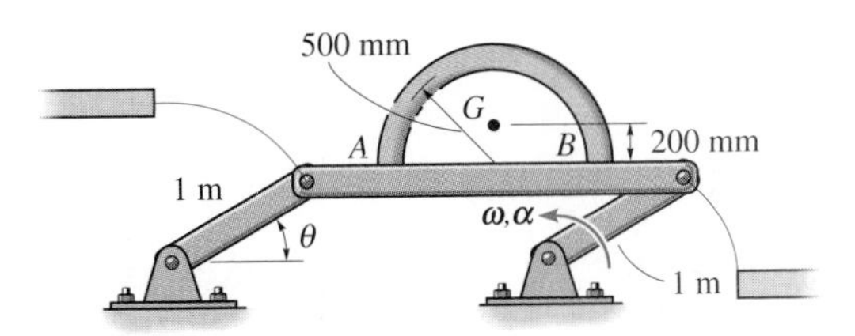

Probs. 17–45/17–46

17–47. A car having a weight of 4000 lb begins to skid and turn with the brakes applied to all four wheels. If the coefficient of kinetic friction between the wheels and the road is $\mu_k = 0.8$, determine the maximum critical height h of the center of gravity G such that the car does not overturn. Tipping will begin to occur after the car rotates 90° from its original direction of motion and, as shown in the figure, undergoes *translation* while skidding. *Hint:* Draw a free-body diagram of the car viewed from the front. When tipping occurs, the normal reactions of the wheels on the right side (or passenger side) are zero.

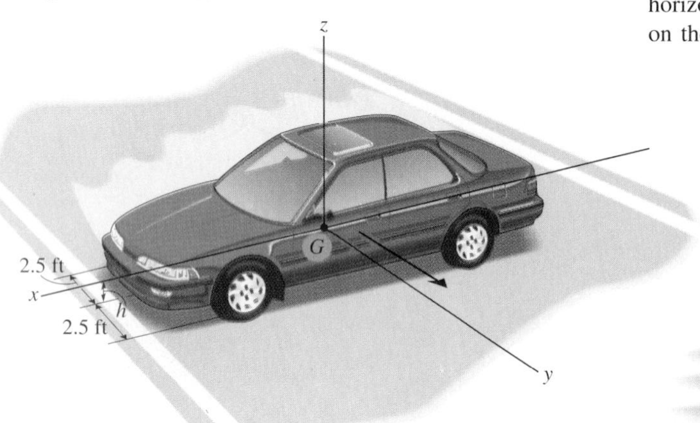

Prob. 17–47

***17–48.** The desk has a weight of 75 lb and a center of gravity at G. Determine its initial acceleration if a man pushes on it at C with a force $F = 60$ lb. The coefficient of kinetic friction at A and B is $\mu_k = 0.2$.

17–49. The desk has a weight of 75 lb and a center of gravity at G. Determine the initial acceleration of the desk when the man applies enough force F to overcome the static friction at A and B. Also, find the vertical reactions on each of the two legs at A and at B. The coefficients of static and kinetic friction at A and B are $\mu_s = 0.5$ and $\mu_k = 0.2$, respectively.

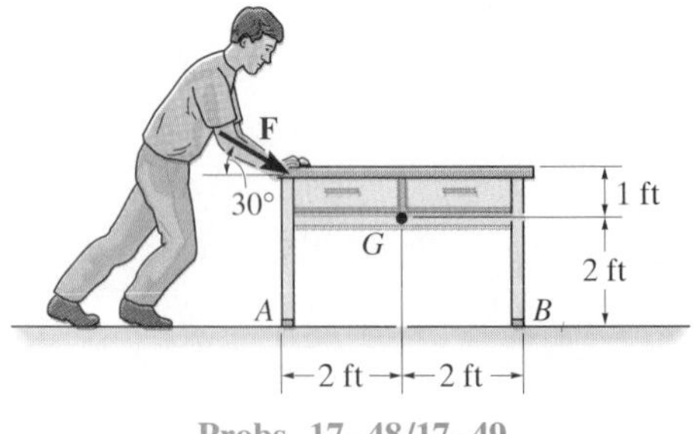

Probs. 17–48/17–49

17–50. The smooth 180-lb pipe has a length of 20 ft and a negligible diameter. It is carried on a truck as shown. Determine the maximum acceleration which the truck can have without causing the normal reaction at A to be zero. Also determine the horizontal and vertical components of force which the truck exerts on the pipe at B.

17–51. The smooth 180-lb pipe has a length of 20 ft and a negligible diameter. It is carried on a truck as shown. If the truck accelerates at $a = 5 \text{ ft/s}^2$, determine the normal reaction at A and the horizontal and vertical components of force which the truck exerts on the pipe at B.

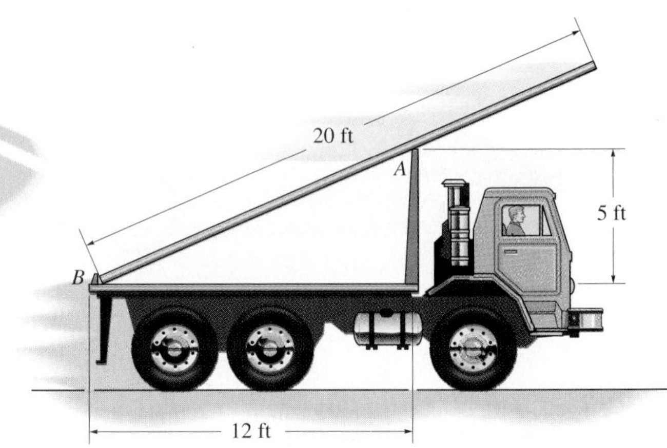

Probs. 17–50/17–51

***17–52.** The van has a weight of 4500 lb and center of gravity at G_v. It carries a fixed 800-lb load which has a center of gravity at G_l. If the van is traveling at 40 ft/s, determine the distance it skids before stopping. The brakes cause *all* the wheels to lock or skid. The coefficient of kinetic friction between the wheels and the pavement is $\mu_k = 0.3$. Compare this distance with that of the van being empty. Neglect the mass of the wheels.

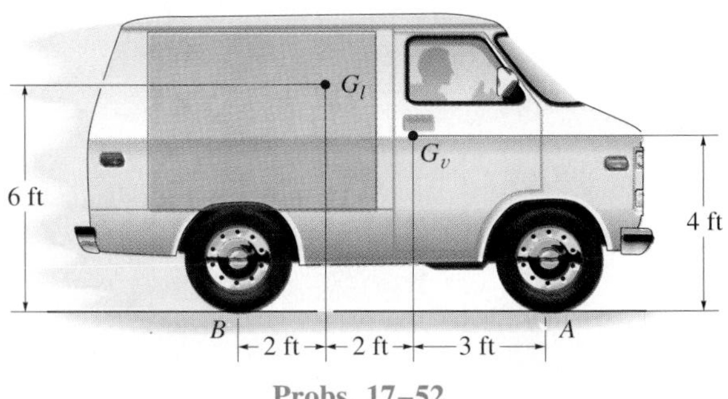

Probs. 17–52

17–53. The van has a weight of 4500 lb and center of gravity at G_v. It carries a fixed 800-lb load which has a center of gravity at G_l. If power is applied only to the rear wheels B, causing them to slip on the pavement, determine the acceleration of the van. The coefficient of kinetic friction between the wheels and the pavement is $\mu_k = 0.3$. Neglect the mass of the wheels. The front wheels are free to roll.

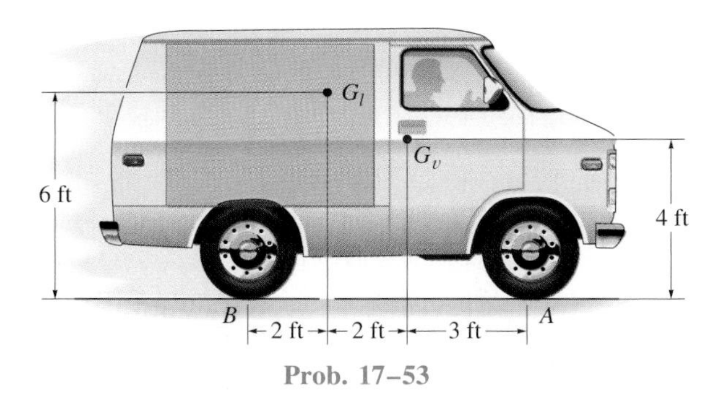

Prob. 17–53

17–54. The forks on the tractor support the pallet that carries a mass of 400 kg. If the coefficient of static friction between the pallet and the forks is $\mu_s = 0.4$, determine the greatest initial angular acceleration of the linkage for which the load can be *lowered* before it begins to slip on the pallet. Note that the tractor linkage subjects the load to curvilinear translation having a radius of 3 m. Initially the load is at rest.

17–55. Solve Prob. 17–54 if the load is *raised*.

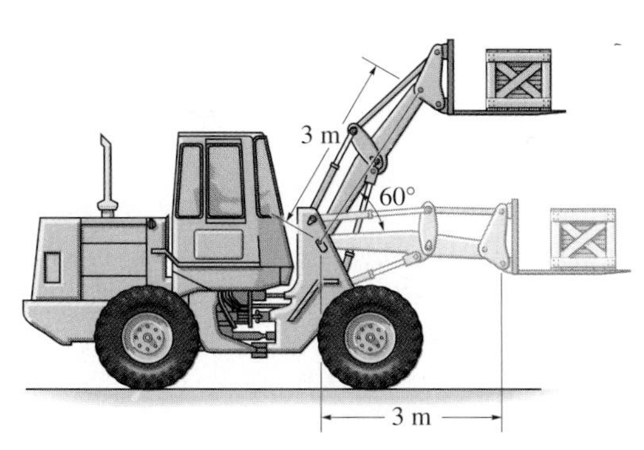

Probs. 17–54/17–55

***17–56.** The crate C has a weight of 150 lb and rests on the truck elevator for which the coefficient of static friction is $\mu_s = 0.4$. Determine the largest initial angular acceleration α, starting from rest, which the parallel links AB and DE can have without causing the crate to slip. No tipping occurs.

17–57. The crate C has a weight of 150 lb and rests on the truck elevator. Determine the initial friction and normal force of the elevator on the crate if the parallel links are given an angular acceleration $\alpha = 2$ rad/s^2 starting from rest.

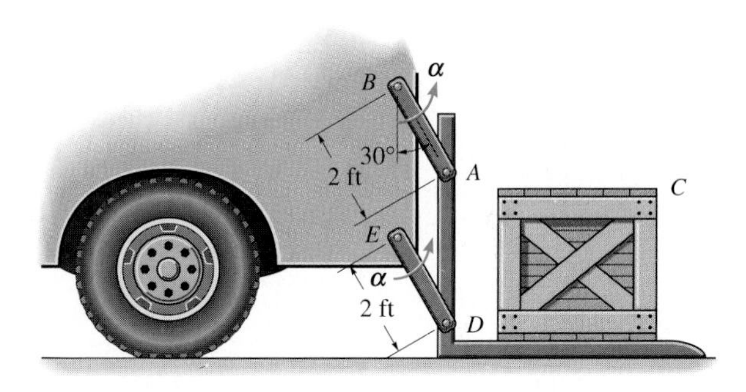

Probs. 17–56/17–57

17–58. The uniform bar BC has a weight of 40 lb and is pin-connected to the two links which have negligible mass. Determine the moment **M** that must be applied to link DC so that DC has the angular motion shown at the instant $\theta = 30°$. Also, what is the force developed in member AB?

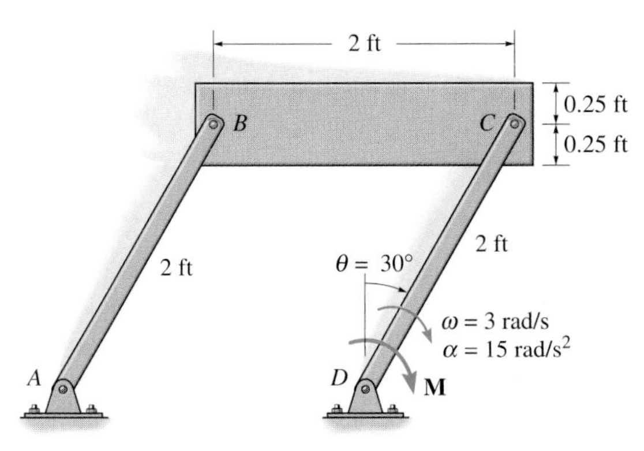

Prob. 17–58

17.4 Equations of Motion: Rotation About a Fixed Axis

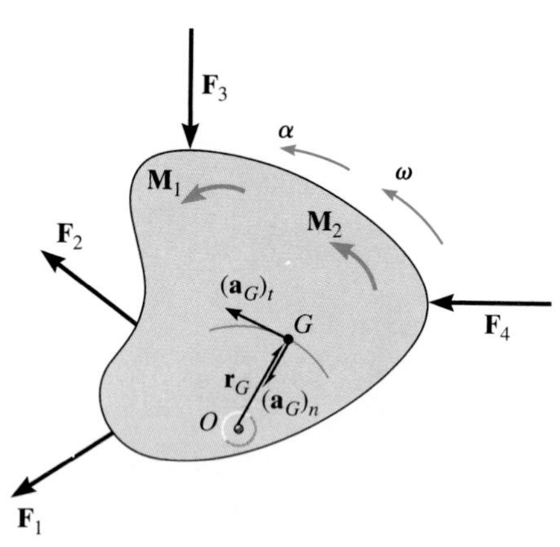

(a)

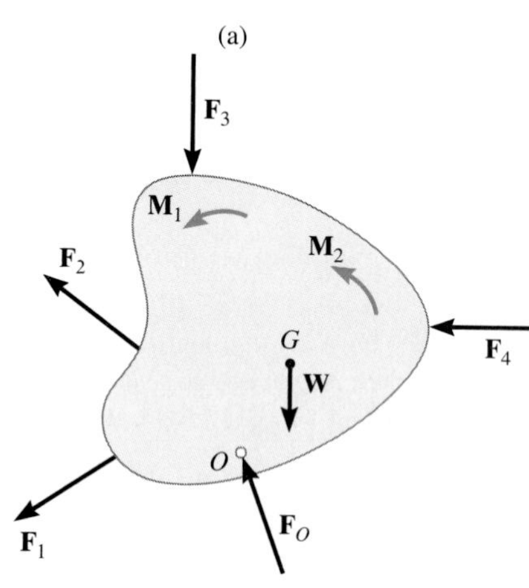

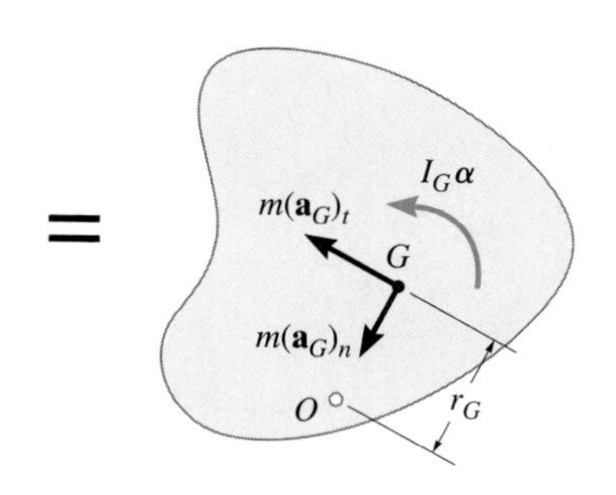

(b)

Fig. 17–14

Consider the rigid body (or slab) shown in Fig. 17–14*a*, which is constrained to rotate in the vertical plane about a fixed axis perpendicular to the page and passing through the pin at O. The angular velocity and angular acceleration are caused by the external force and couple moment system acting on the body. Because the body's center of mass G moves in a *circular path,* the acceleration of this point is represented by its tangential and normal components. The *tangential component of acceleration* has a *magnitude* of $(a_G)_t = \alpha r_G$ and must act in a *direction* which is *consistent* with the body's angular acceleration $\boldsymbol{\alpha}$. The *magnitude* of the *normal component of acceleration* is $(a_G)_n = \omega^2 r_G$. This component is *always directed* from point G to O, regardless of the direction of $\boldsymbol{\omega}$.

The free-body and kinetic diagrams for the body are shown in Fig. 17–14*b*. The weight of the body, $W = mg$, and the pin reaction $\mathbf{F}_O$ are included on the free-body diagram since they represent external forces acting on the body. The two components $m(\mathbf{a}_G)_t$ and $m(\mathbf{a}_G)_n$, shown on the kinetic diagram, are associated with the tangential and normal acceleration components of the body's mass center. These vectors act in the same *direction* as the acceleration components and have *magnitudes* of $m(a_G)_t$ and $m(a_G)_n$. The $I_G\boldsymbol{\alpha}$ vector acts in the same *direction* as $\boldsymbol{\alpha}$ and has a *magnitude* of $I_G\alpha$, where I_G is the body's moment of inertia calculated about an axis which is perpendicular to the page and passing through G. From the derivation given in Sec. 17.2, the equations of motion which apply to the body may be written in the form

$$\begin{aligned} \Sigma F_n &= m(a_G)_n = m\omega^2 r_G \\ \Sigma F_t &= m(a_G)_t = m\alpha r_G \\ \Sigma M_G &= I_G\alpha \end{aligned} \qquad (17\text{–}14)$$

The moment equation may be replaced by a moment summation about any arbitrary point P on or off the body provided one accounts for the moments $\Sigma(\mathcal{M}_k)_P$ produced by $I_G\boldsymbol{\alpha}$, $m(\mathbf{a}_G)_t$, and $m(\mathbf{a}_G)_n$ about the point. In many problems it is convenient to sum moments about the pin at O in order to eliminate the *unknown* force $\mathbf{F}_O$. From the kinetic diagram, Fig. 17–14*b*, this requires

$$\curvearrowright\!+\Sigma M_O = \Sigma(\mathcal{M}_k)_O; \qquad \Sigma M_O = r_G m(a_G)_t + I_G\alpha \qquad (17\text{–}15)$$

Note that the moment of $m(\mathbf{a}_G)_n$ is not included in the summation since the line of action of this vector passes through O. Substituting $(a_G)_t = r_G\alpha$, we may rewrite the above equation as $\curvearrowright\!+\Sigma M_O = (I_G + mr_G^2)\alpha$. From the parallel-axis theorem, $I_O = I_G + md^2$, and therefore the term in parentheses represents the *moment of inertia of the body about the fixed axis of rotation passing through O.** Consequently, we can write the three equations of motion for the body as

*The result $\Sigma M_O = I_O\alpha$ can also be obtained *directly* from Eq. 17–6 by selecting point P to coincide with O, realizing that $(a_P)_x = (a_P)_y = 0$.

$$\Sigma F_n = m(a_G)_n = m\omega^2 r_G$$
$$\Sigma F_t = m(a_G)_t = m\alpha r_G \qquad (17\text{–}16)$$
$$\Sigma M_O = I_O\alpha$$

For applications, one should remember that "$I_O\alpha$" accounts for the "moment" of *both* $m(\mathbf{a}_G)_t$ *and* $I_G\boldsymbol{\alpha}$ about point O, Fig. 17–14*b*. In other words, $\Sigma M_O = \Sigma(\mathcal{M}_k)_O = I_O\alpha$, as indicated by Eqs. 17–15 and 17–16.

PROCEDURE FOR ANALYSIS

The following procedure provides a method for solving kinetic problems which involve the rotation of a body about a fixed axis.

Free-Body Diagram. Establish the inertial x, y or n, t coordinate system, and specify the direction and sense of the accelerations $(\mathbf{a}_G)_n$ and $(\mathbf{a}_G)_t$, and the angular acceleration $\boldsymbol{\alpha}$ of the body. Recall that $(\mathbf{a}_G)_t$ must act in a direction which is in accordance with $\boldsymbol{\alpha}$, whereas $(\mathbf{a}_G)_n$ always acts toward the axis of rotation, point O. If the sense of $(\mathbf{a}_G)_t$ or $\boldsymbol{\alpha}$ *cannot* be established, assume they are in the direction of the positive coordinate axes. It may be convenient to show these vectors on the coordinate system. Draw the free-body diagram to account for all the external forces that act on the body and compute the moment of inertia I_G or I_O. Identify the unknowns in the problem. If it is decided that the rotational equation of motion $\Sigma M_P = \Sigma(\mathcal{M}_k)_P$ is to be used, i.e., moments will *not* be summed about point G or O, then consider drawing the kinetic diagram in order to help "visualize" the "moment" developed by the components $m(\mathbf{a}_G)_n$, $m(\mathbf{a}_G)_t$, and $I_G\boldsymbol{\alpha}$ when writing the terms for the moment sum $\Sigma(\mathcal{M}_k)_P$.

Equations of Motion. Apply the three equations of motion, Eqs. 17–14 or 17–16.

Kinematics. Use kinematics if a complete solution cannot be obtained strictly from the equations of motion. In this regard, if the *angular acceleration is variable,* use

$$\alpha = \frac{d\omega}{dt} \qquad \alpha\, d\theta = \omega\, d\omega \qquad \omega = \frac{d\theta}{dt}$$

If the *angular acceleration is constant,* use

$$\omega = \omega_0 + \alpha_c t$$
$$\theta = \theta_0 + \omega_0 t + \tfrac{1}{2}\alpha_c t^2$$
$$\omega^2 = \omega_0^2 + 2\alpha_c(\theta - \theta_0)$$

The following examples illustrate application of this procedure.

Example 17–9

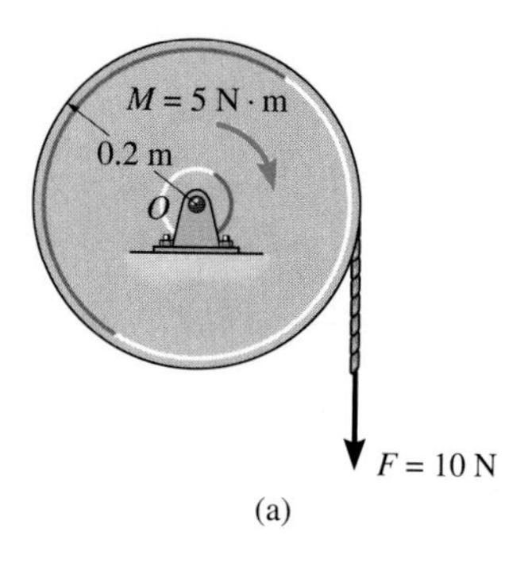

(a)

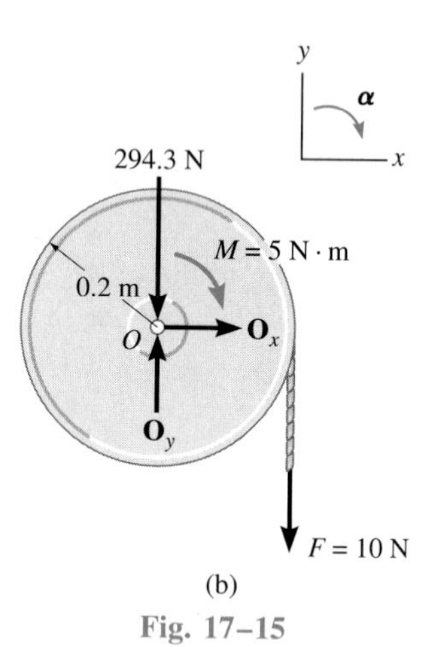

(b)

Fig. 17–15

The 30-kg disk shown in Fig. 17–15*a* is pin-supported at its center. If it starts from rest, determine the number of revolutions it must make to attain an angular velocity of 20 rad/s. Also, what are the reactions at the pin? The disk is acted upon by a constant force $F = 10$ N, which is applied to a cord wrapped around its periphery, and a constant couple moment $M = 5$ N · m. Neglect the mass of the cord in the calculation.

SOLUTION

Free-Body Diagram. Fig. 17–15*b*. Note that the mass center is not subjected to an acceleration; however, the disk has a clockwise angular acceleration.

The moment of inertia of the disk about the pin is

$$I_O = \tfrac{1}{2}mr^2 = \frac{1}{2}(30\text{ kg})(0.2\text{ m})^2 = 0.6\text{ kg}\cdot\text{m}^2$$

The three unknowns are O_x, O_y, and α.

Equations of Motion

$$\overset{+}{\rightarrow}\Sigma F_x = m(a_G)_x; \qquad O_x = 0 \qquad \textit{Ans.}$$

$$+\uparrow \Sigma F_y = m(a_G)_y; \qquad O_y - 294.3\text{ N} - 10\text{ N} = 0$$

$$O_y = 304\text{ N} \qquad \textit{Ans.}$$

$$\circlearrowright + \Sigma M_O = I_O\alpha; \qquad -10\text{ N}(0.2\text{ m}) - 5\text{ N}\cdot\text{m} = -(0.6\text{ kg}\cdot\text{m}^2)\alpha$$

$$\alpha = 11.7\text{ rad/s}^2 \circlearrowright$$

Kinematics. Since α is constant and rotates clockwise as assumed, the number of radians the disk must turn to obtain a clockwise angular velocity of 20 rad/s is

$$\circlearrowright + \qquad \omega^2 = \omega_0^2 + 2\alpha_c(\theta - \theta_0)$$

$$(-20\text{ rad/s})^2 = 0 + 2(-11.7\text{ rad/s}^2)(\theta - 0)$$

$$\theta = -17.1\text{ rad} = 17.1\text{ rad} \circlearrowright$$

Hence,

$$\theta = 17.1\text{ rad}\left(\frac{1\text{ rev}}{2\pi\text{ rad}}\right) = 2.73\text{ rev} \circlearrowright \qquad \textit{Ans.}$$

Example 17–10

The 20-kg slender rod shown in Fig. 17–16*a* is rotating in the vertical plane, and at the instant shown it has an angular velocity of $\omega = 5$ rad/s. Determine the rod's angular acceleration and the horizontal and vertical components of reaction at the pin at this instant.

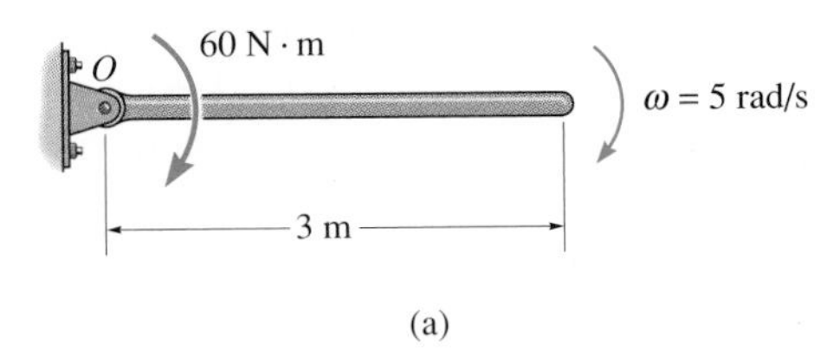

(a)

(b)

Fig. 17–16

SOLUTION

Free-Body and Kinetic Diagrams. Fig. 17–16*b*. As shown on the kinetic diagram, point *G* moves in a circular path and so has two components of acceleration. It is important that the tangential component $a_t = \alpha r_G$ act downward since it must be in accordance with the angular acceleration $\boldsymbol{\alpha}$ of the rod. The three unknowns are O_n, O_t, and α.

Equations of Motion

$$\overset{+}{\leftarrow}\Sigma F_n = m\omega^2 r_G; \qquad O_n = (20\text{ kg})(5\text{ rad/s})^2(1.5\text{ m})$$

$$+\downarrow\Sigma F_t = m\alpha r_G; \qquad -O_t + 20(9.81)\text{ N} = (20\text{ kg})(\alpha)(1.5\text{ m})$$

$$\curvearrowright+\Sigma M_G = I_G\alpha; \qquad O_t(1.5\text{ m}) + 60\text{ N}\cdot\text{m} = [\tfrac{1}{12}(20\text{ kg})(3\text{ m})^2]\alpha$$

Solving

$$O_n = 750\text{ N} \qquad O_t = 19.0\text{ N} \qquad \alpha = 5.90\text{ rad/s}^2 \qquad \textit{Ans.}$$

A more direct solution to this problem would be to sum moments about point *O* to eliminate $\mathbf{O}_n$ and $\mathbf{O}_t$ and obtain a *direct solution* for α. Hence,

$$\curvearrowright+\Sigma M_O = \Sigma(\mathcal{M}_k)_O; \quad 60\text{ N}\cdot\text{m} + 20(9.81)\text{ N}(1.5\text{ m}) = [\tfrac{1}{12}(20\text{ kg})(3\text{ m})^2]\alpha + [20\text{ kg}(\alpha)(1.5\text{ m})](1.5\text{ m})$$

$$\alpha = 5.90\text{ rad/s}^2 \qquad \textit{Ans.}$$

Also, since $I_O = \frac{1}{3}ml^2$ for a slender rod, we can apply

$$\curvearrowright+\Sigma M_O = I_O\alpha; \quad 60\text{ N}\cdot\text{m} + 20(9.81)\text{ N}(1.5\text{ m}) = [\ (20\text{ kg})(3\text{ m})^2]\alpha$$

$$\alpha = 5.90\text{ rad/s}^2 \qquad \textit{Ans.}$$

By comparison, the last equation provides the simplest solution to the problem and *does not* require use of the kinetic diagram.

Example 17–11

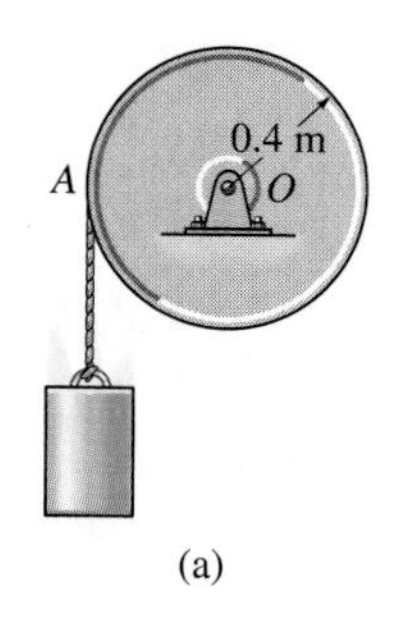

(a)

The drum shown in Fig. 17–17*a* has a mass of 60 kg and a radius of gyration $k_O = 0.25$ m. A cord of negligible mass is wrapped around the periphery of the drum and attached to a block having a mass of 20 kg. If the crate is released, determine the drum's angular acceleration.

SOLUTION I

Free-Body Diagram. Here we will consider the drum and block separately, Fig. 17–17*b*. Assuming the block accelerates *downward* at **a**, it creates a *counterclockwise* angular acceleration $\boldsymbol{\alpha}$ of the drum.

The moment of inertia of the drum is

$$I_O = mk_O^2 = (60\text{ kg})(0.25\text{ m})^2 = 3.75\text{ kg}\cdot\text{m}^2$$

There are five unknowns, namely O_x, O_y, T, a, and α.

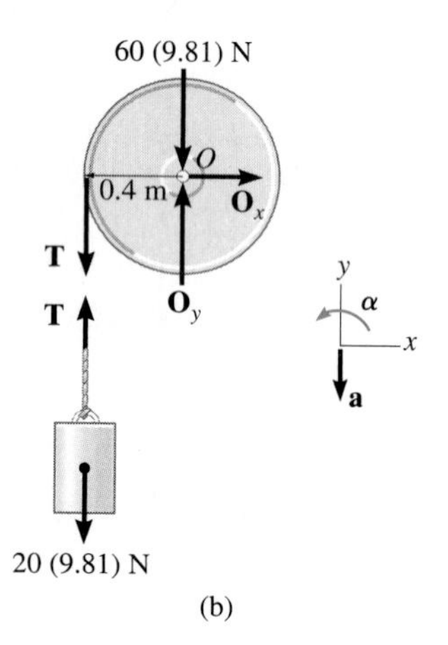

(b)

Equations of Motion. Applying the translational equations of motion $\Sigma F_x = m(a_G)_x$ and $\Sigma F_y = m(a_G)_y$ to the drum is of no consequence to the solution, since these equations involve the unknowns O_x and O_y. Thus, for the drum and block, respectively,

$$\circlearrowleft + \Sigma M_O = I_O\alpha; \qquad T(0.4\text{ m}) = (3.75\text{ kg}\cdot\text{m}^2)\alpha \qquad (1)$$

$$+\uparrow \Sigma F_y = m(a_G)_y; \qquad -20(9.81)\text{ N} + T = -20a \qquad (2)$$

Kinematics. Since the point of contact A between the cord and drum has a tangential component of acceleration **a**, Fig. 17–17*a*, then

$$\circlearrowleft + a = \alpha r; \qquad a = \alpha(0.4) \qquad (3)$$

Solving the above equations,

$$T = 106\text{ N}$$
$$a = 4.52\text{ m/s}^2$$
$$\alpha = 11.3\text{ rad/s}^2 \circlearrowleft \qquad \textit{Ans.}$$

SOLUTION II

Free-Body and Kinetic Diagrams. The cable tension T can be eliminated from the analysis by considering the drum and block as a *single system*, Fig. 17–17*c*. The kinetic diagram is shown since moments will be summed about point O.

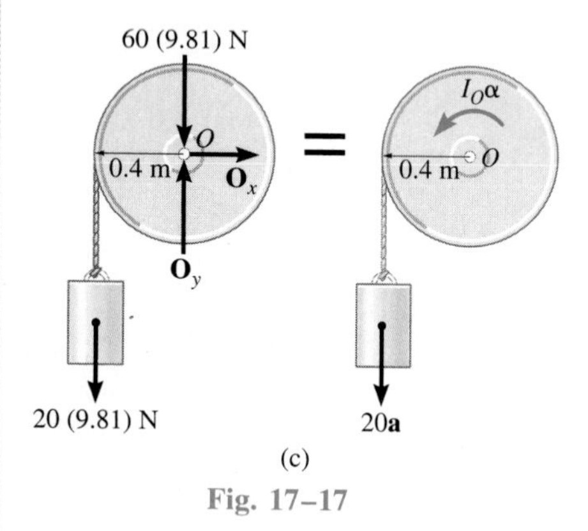

(c)

Fig. 17–17

Equations of Motion. Using Eq. 3 and applying the moment equation about O to eliminate the unknowns O_x and O_y, we have

$$\circlearrowleft + \Sigma M_O = \Sigma(\mathcal{M}_k)_O; \qquad 20(9.81)\text{ N}(0.4\text{ m}) =$$
$$(3.75\text{ kg}\cdot\text{m}^2)\alpha + [20\text{ kg}(0.4\text{ m }\alpha)](0.4\text{ m})$$
$$\alpha = 11.3\text{ rad/s}^2 \circlearrowleft \qquad \textit{Ans.}$$

Note: If the block was *removed* and a force of 20(9.81) N was applied to the cord, show that $\alpha = 20.9\text{ rad/s}^2$ and explain the reason for the difference in the results.

Example 17–12

The unbalanced 50-lb flywheel shown in Fig. 17–18*a* has a radius of gyration of $k_G = 0.6$ ft about an axis passing through its mass center *G*. If it has a clockwise angular velocity of 8 rad/s at the instant shown, determine the horizontal and vertical components of reaction at the pin *O*.

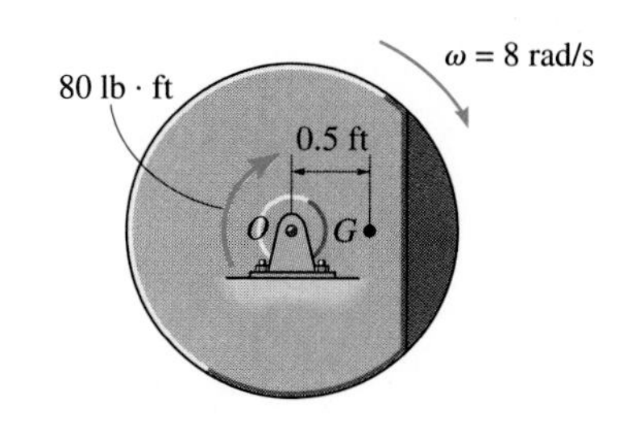

(a)

SOLUTION

Free-Body and Kinetic Diagrams. Since *G* moves in a circular path, it will have both normal and tangential components of acceleration. Also, since α, which is caused by the flywheel's weight, acts clockwise, the tangential component of acceleration will act downward. Why? The vectors $m(a_G)_t = m\alpha r_G$, $m(a_G)_n = m\omega^2 r_G$, and $I_G\alpha$ are shown on the kinematic diagram in Fig. 17–18*b*. Here, the moment of inertia of the flywheel about its mass center is determined from the radius of gyration and the flywheel's mass; i.e., $I_G = mk_G^2 = (50 \text{ lb}/32.2 \text{ ft/s}^2)(0.6 \text{ ft})^2 = 0.559 \text{ slug} \cdot \text{ft}^2$.

The three unknowns are O_x, O_y, and α.

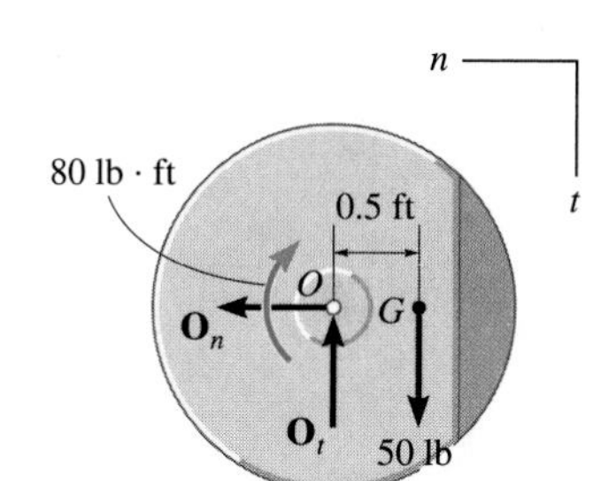

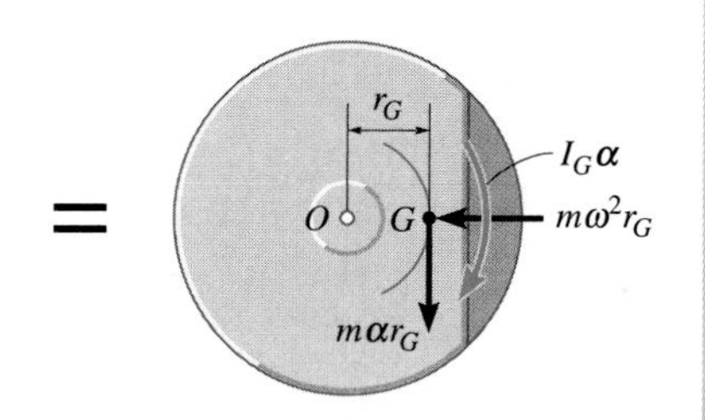

(b)

Fig. 17–18

Equations of Motion

$$\overset{+}{\leftarrow}\Sigma F_n = m\omega^2 r_G; \quad O_n = \left(\frac{50 \text{ lb}}{32.2 \text{ ft/s}^2}\right)(8 \text{ rad/s})^2(0.5 \text{ ft}) \qquad (1)$$

$$+\downarrow \Sigma F_t = m\alpha r_G; \quad -O_t + 50 \text{ lb} = \left(\frac{50 \text{ lb}}{32.2 \text{ ft/s}^2}\right)(\alpha)(0.5 \text{ ft}) \qquad (2)$$

$$\curvearrowright\!+\Sigma M_G = I_G\alpha; \quad 80 \text{ lb} \cdot \text{ft} + O_t(0.5 \text{ ft}) = (0.559 \text{ slug} \cdot \text{ft}^2)\alpha \qquad (3)$$

Solving, $\alpha = 111 \text{ rad/s}^2$ $\quad O_n = 49.7 \text{ lb} \quad O_t = -36.1 \text{ lb}$ *Ans.*

Moments can also be summed about point *O* in order to eliminate $\mathbf{O}_n$ and $\mathbf{O}_t$ and thereby obtain a *direct solution* for $\boldsymbol{\alpha}$, Fig. 17–18*b*. This can be done in one of *two* ways, i.e., by using either $\Sigma M_O = \Sigma(\mathcal{M}_k)_O$ or $\Sigma M_O = I_O\alpha$. If the first of these equations is applied, we have

$$\curvearrowright\!+\Sigma M_O = \Sigma(\mathcal{M}_k)_O; \quad 80 \text{ lb} \cdot \text{ft} + 50 \text{ lb}(0.5 \text{ ft}) = (0.559 \text{ slug} \cdot \text{ft}^2)\alpha + \left[\left(\frac{50 \text{ lb}}{32.2 \text{ ft/s}^2}\right)\alpha(0.5 \text{ ft})\right](0.5 \text{ ft})$$

$$105 = 0.947\alpha \qquad (4)$$

If $\Sigma M_O = I_O\alpha$ is applied, then by the parallel-axis theorem the moment of inertia of the flywheel about *O* is

$$I_O = I_G + mr_G^2 = 0.559 + \left(\frac{50}{32.2}\right)(0.5)^2 = 0.947 \text{ slug} \cdot \text{ft}^2$$

Hence, from the free-body diagram, Fig. 17–18*b*, we require

$$\curvearrowright\!+\Sigma M_O = I_O\alpha; \quad 80 \text{ lb} \cdot \text{ft} + 50 \text{ lb}(0.5 \text{ ft}) = (0.947 \text{ slug} \cdot \text{ft}^2)\alpha$$

which is the same as Eq. 4. Solving for α and substituting into Eq. 2 yields the answer for O_t obtained previously.

Example 17–13

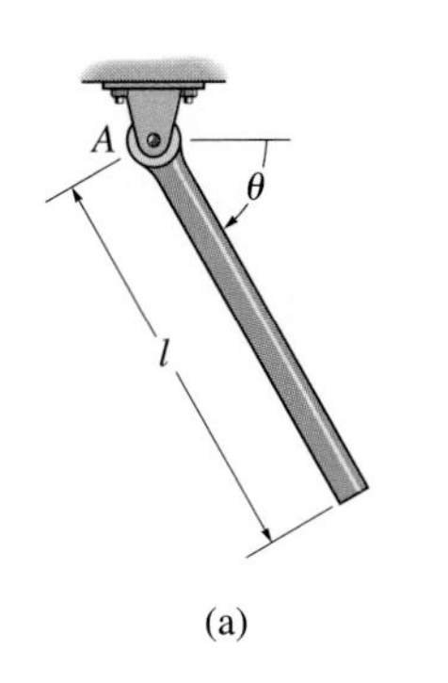

(a)

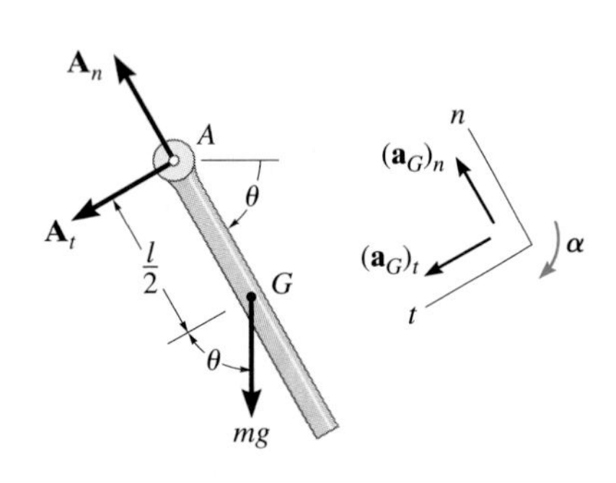

(b)

Fig. 17–19

The slender rod shown in Fig. 17–19*a* has a mass m and length l and is released from rest when $\theta = 0°$. Determine the horizontal and vertical components of force which the pin at A exerts on the rod at the instant $\theta = 90°$.

SOLUTION

Free-Body Diagram. The free-body diagram for the rod is shown when the rod is in the general position θ, Fig. 17–19*b*. For convenience, the force components at A are shown acting in the n and t directions. Note that $\boldsymbol{\alpha}$ acts clockwise.

The moment of inertia of the rod about point A is $I_A = \frac{1}{3}ml^2$.

Equations of Motion. Moments will be summed about A in order to eliminate the reactive forces there.*

$$+\nwarrow \Sigma F_n = m\omega^2 r_G; \qquad A_n - mg\sin\theta = m\omega^2(l/2) \qquad (1)$$

$$+\swarrow \Sigma F_t = m\alpha r_G; \qquad A_t + mg\cos\theta = m\alpha(l/2) \qquad (2)$$

$$\curvearrowright + \Sigma M_A = I_A\alpha; \qquad mg\cos\theta(l/2) = (\tfrac{1}{3}ml^2)\alpha \qquad (3)$$

Kinematics. For a given angle θ there are four unknowns in the above three equations: A_n, A_t, ω, and α. As shown by Eq. 3, α is *not constant;* rather, it depends on the position θ of the rod. The necessary fourth equation is obtained using kinematics, where α and ω can be related to θ by the equation

$$(\curvearrowright +) \qquad \omega\, d\omega = \alpha\, d\theta \qquad (4)$$

Note that the positive direction for this equation *agrees* with that of Eq. 3. This is important since we are seeking a simultaneous solution.

In order to solve for ω at $\theta = 90°$, eliminate α from Eqs. 3 and 4, which yields

$$\omega\, d\omega = (1.5\, g/l)\cos\theta\, d\theta$$

Since $\omega = 0$ at $\theta = 0°$, we have

$$\int_0^{\omega} \omega\, d\omega = (1.5\, g/l)\int_{0°}^{90°} \cos\theta\, d\theta$$

$$\omega^2 = 3\, g/l$$

Substituting this value into Eq. 1 with $\theta = 90°$ and solving Eqs. 1 to 3 yields

$$\alpha = 0 \qquad A_t = 0 \qquad A_n = 2.5mg \qquad \textit{Ans.}$$

*If $\Sigma M_A = \Sigma(\mathcal{M}_k)_A$ is used, one must account for the moments of $I_G\boldsymbol{\alpha}$ and $m(\mathbf{a}_G)_t$ about A. Here, however, we have used $\Sigma M_A = I_A\alpha$. (Refer to Eq. 17–15.)

PROBLEMS

17–59. The 80-kg disk is supported by a pin at A. If it is released from rest from the position shown, determine the initial horizontal and vertical components of reaction at the pin.

***17–60.** The 80-kg disk is supported by a pin at A. If it is rotating clockwise at $\omega = 0.5$ rad/s when it is in the position shown, determine the horizontal and vertical components of reaction at the pin at this instant.

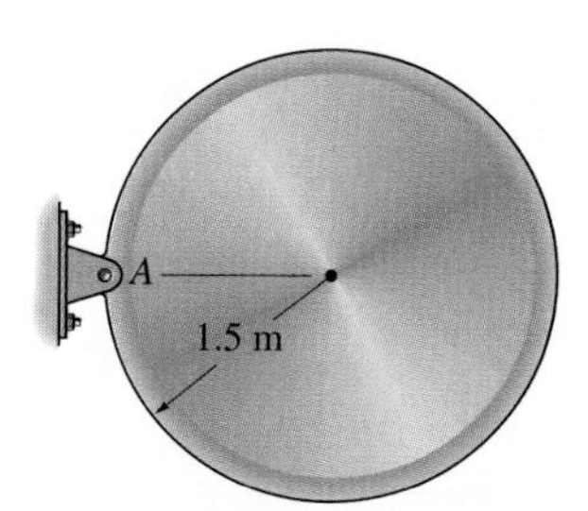

Probs. 17–59/17–60

17–61. The 10-kg wheel has a radius of gyration $k_A = 200$ mm. If the wheel is subjected to a moment $M = (5t)$ N · m, where t is in seconds, determine its angular velocity when $t = 3$ s starting from rest. Also, compute the reactions which the fixed pin A exerts on the wheel during the motion.

17–62. The 10-kg wheel has a radius of gyration $k_A = 200$ mm. If the wheel is subjected to a moment $M = (4\theta)$ N · m, where θ is in radians, determine its angular velocity when it undergoes two revolutions starting from rest. Also, compute the reactions which the fixed pin A exerts on the wheel during the motion.

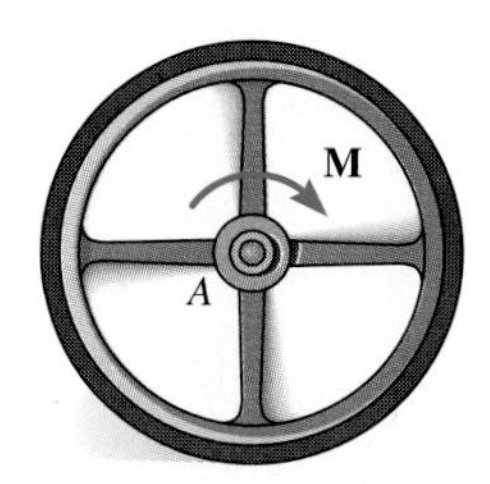

Probs. 17–61/17–62

17–63. The drum has a weight of 80 lb and a radius of gyration $k_O = 0.4$ ft. If the cable, which is wrapped around the drum, is subjected to a vertical force $P = 15$ lb, determine the time needed to increase the drum's angular velocity from $\omega_1 = 5$ rad/s to $\omega_2 = 25$ rad/s. Neglect the mass of the cable.

***17–64.** The drum has a weight of 80 lb and a radius of gyration $k_O = 0.4$ ft. If the cable, which is wrapped around the drum, is subjected to a vertical force $P = (3t^2 + 5)$ lb, where t is in seconds, determine the angular velocity of the drum in $t = 2$ s if at $t = 0$, $\omega_0 = 5$ rad/s. Neglect the mass of the cable.

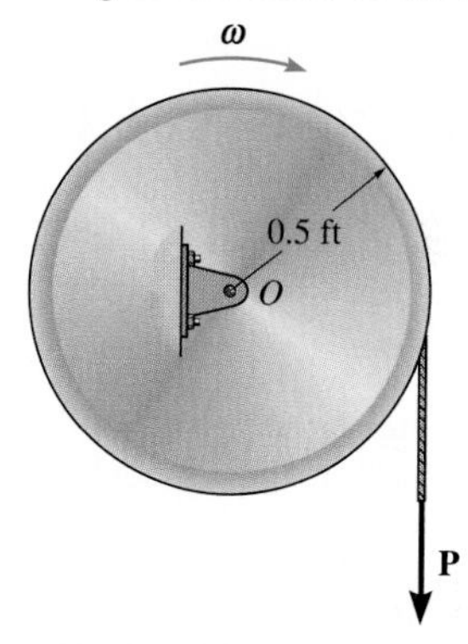

Probs. 17–63/17–64

17–65. The drum has a weight of 20 lb and a radius of gyration about its mass center of 0.8 ft. If the block has a weight of 12 lb, determine the angular acceleration α_D of the drum if the block is allowed to fall freely. Compare this value of $\boldsymbol{\alpha}_D$ with that determined by removing the block and applying a force of 12 lb to the cord. Explain the reason for the difference in the results.

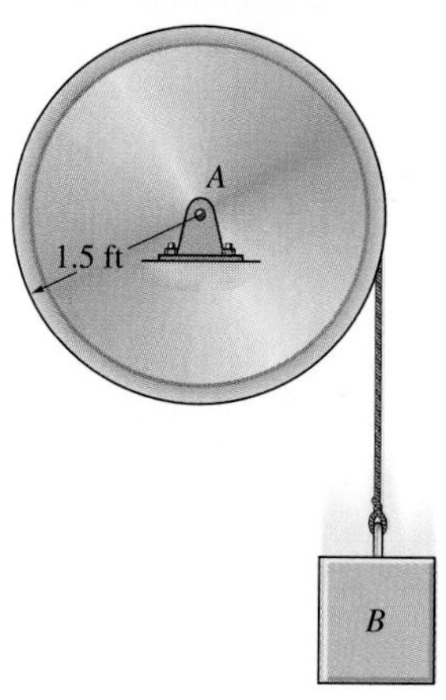

Prob. 17–65

17–66. The drum has a weight of 20 lb and can be treated as a solid cylinder. If the block has a weight of 8 lb, determine the angular acceleration α_D of the drum if the block is allowed to fall freely. Compare this value of $\boldsymbol{\alpha}_D$ with that determined by removing the block and applying a force of 8 lb to the cord. Explain the reason for the difference in the results.

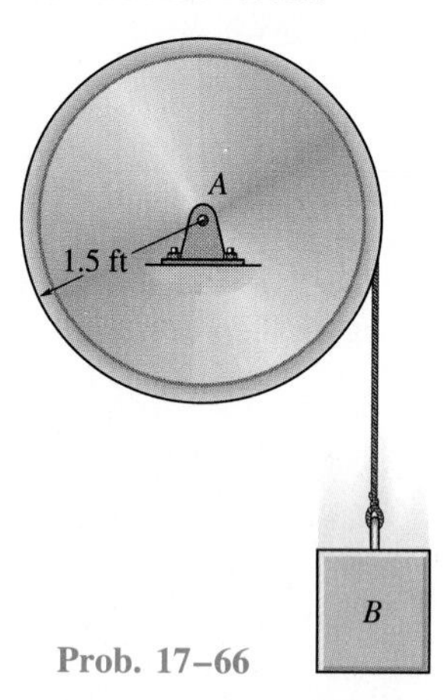

Prob. 17–66

17–67. The spool is supported on small rollers at A and B. Determine the angular acceleration of the spool and the normal forces at A and B if a vertical force $P = 80$ N is applied to the cable. The spool has a mass of 60 kg and a radius of gyration $k_O = 0.65$ m. For the calculation neglect the mass of the cable and the mass of the rollers at A and B.

***17–68.** The spool is supported on small rollers at A and B. Determine the constant force P that must be applied to the cable in order to unwind 8 m of cable in 4 s starting from rest. Also calculate the normal forces at A and B during this time. The spool has a mass of 60 kg and a radius of gyration $k_O = 0.65$ m. For the calculation neglect the mass of the cable and the mass of the rollers at A and B.

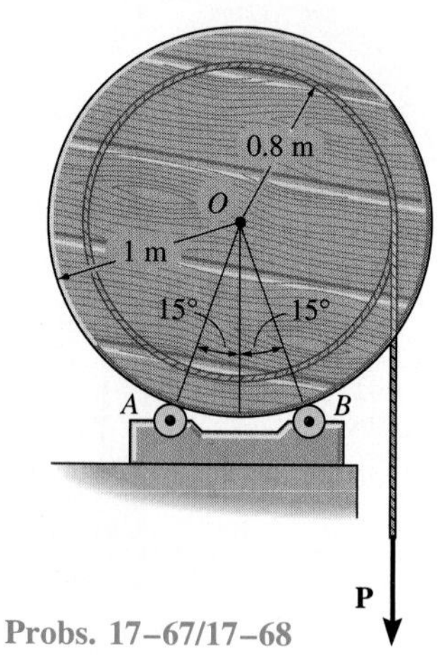

Probs. 17–67/17–68

17–69. The 10-lb bar is pinned at its center O and connected to a torsional spring. The spring has a stiffness $k = 5$ lb · ft/rad, so that the torque developed is $M = (5\theta)$ lb · ft, where θ is in radians. If the bar is released from rest when it is vertical at $\theta = 90°$, determine its angular velocity at the instant $\theta = 0°$.

17–70. The 10-lb bar is pinned at its center O and connected to a torsional spring. The spring has a stiffness $k = 5$ lb · ft/rad, so that the torque developed is $M = (5\theta)$ lb · ft, where θ is in radians. If the bar is released from rest when it is vertical at $\theta = 90°$, determine its angular velocity at the instant $\theta = 45°$.

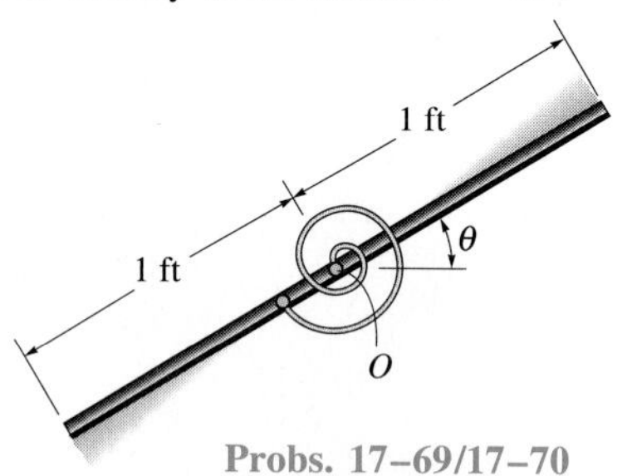

Probs. 17–69/17–70

17–71. The 20-kg roll of paper has a radius of gyration $k_A = 90$ mm about an axis passing through point A. It is pin-supported at both ends by two brackets AB. If the roll rests against a wall for which the coefficient of kinetic friction is $\mu_k = 0.2$ and a vertical force $F = 30$ N is applied to the end of the paper, determine the angular acceleration of the roll as the paper unrolls.

***17–72.** The 20-kg roll of paper has a radius of gyration $k_A = 90$ mm about an axis passing through point A. It is pin-supported at both ends by two brackets AB. If the roll rests against a wall for which the coefficient of kinetic friction is $\mu_k = 0.2$, determine the constant vertical force F that must be applied to the roll to pull off 1 m of paper in $t = 3$ s starting from rest. Neglect the mass of paper that is removed.

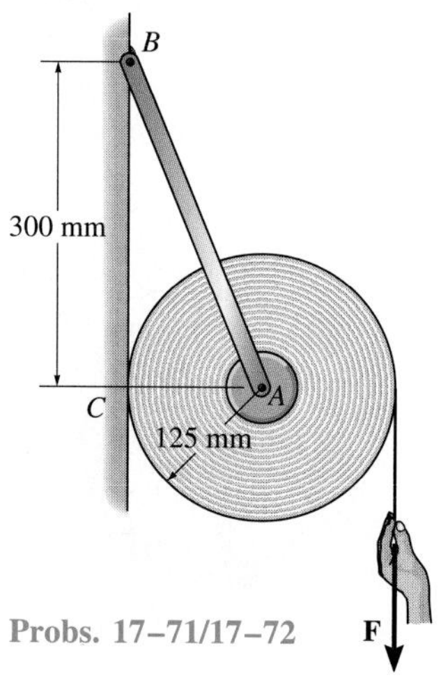

Probs. 17–71/17–72

17–73. The door will close automatically using torsional springs mounted on the hinges. Each spring has a stiffness $k = 50\ \text{N}\cdot\text{m/rad}$ so that the torque on each hinge is $M = (50\theta)\ \text{N}\cdot\text{m}$, where θ is measured in radians. If the door is released from rest when it is open at $\theta = 90^\circ$, determine its angular velocity at the instant $\theta = 0^\circ$. For the calculation, treat the door as a thin plate having a mass of 70 kg.

17–74. The door will close automatically using torsional springs mounted on the hinges. If the torque on each hinge is $M = k\theta$, where θ is measured in radians, determine the required torsional stiffness k so that the door will close ($\theta = 0^\circ$) with an angular velocity $\omega = 2$ rad/s when it is released from rest at $\theta = 90^\circ$. For the calculation, treat the door as a thin plate having a mass of 70 kg.

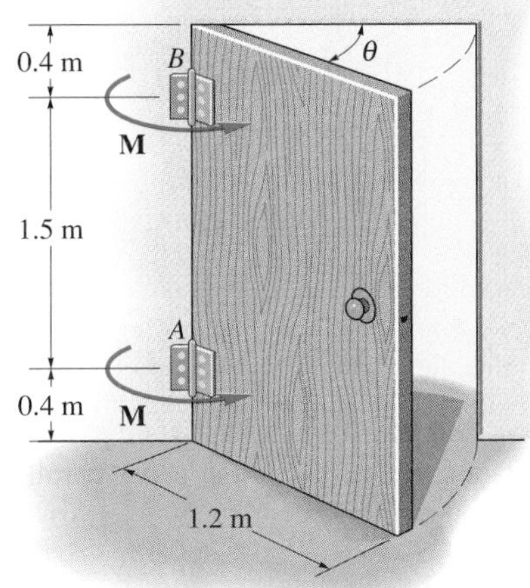

Probs. 17–73/17–74

17–75. Cable is unwound from a spool supported on small rollers at A and B by exerting a force $T = 300$ N on the cable. Compute the time needed to unravel 5 m of cable from the spool if the spool and cable have a total mass of 600 kg and a radius of gyration of $k_O = 1.2$ m. For the calculation, neglect the mass of the cable being unwound and the mass of the rollers at A and B. The rollers turn with no friction.

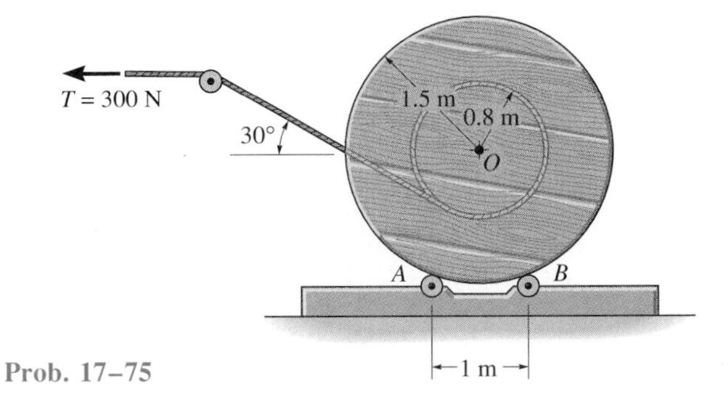

Prob. 17–75

***17–76.** The operation of the doorbell requires the use of the electromagnet, which attracts the iron clapper AB, pinned at end A and consisting of a 0.2-kg slender rod to which is attached the 0.04-kg steel ball having a radius of 6 mm. If the attractive force of the magnet at C is 0.5 N on the center of the ball when the button is pushed, determine the initial angular acceleration of the clapper. The spring is originally stretched 20 mm. *Hint:* The normal force at B becomes zero.

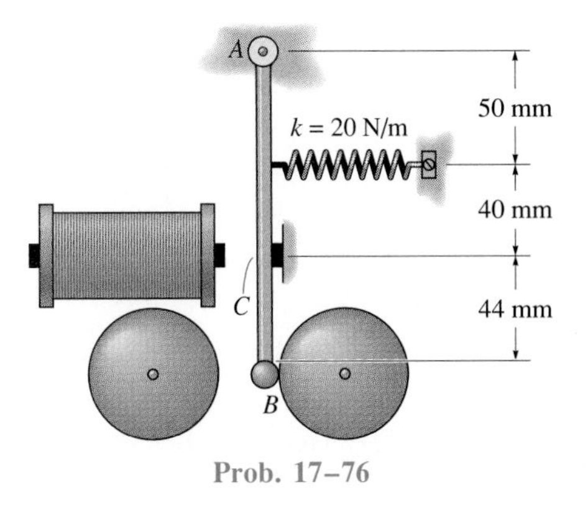

Prob. 17–76

17–77. The disk has a mass M and a radius R. If a block of mass m is attached to the cord, determine the angular acceleration of the disk when the block is released from rest. Also, what is the distance the block falls from rest in the time t?

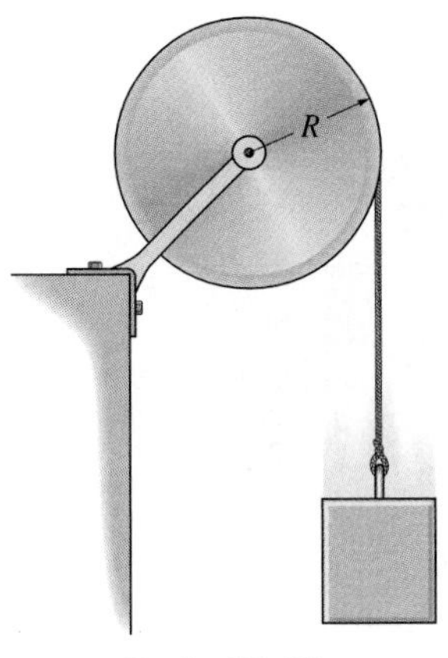

Prob. 17–77

17–78. The disk has a mass M and a radius R. If a block of mass m is attached to the cord, determine the angular acceleration of the disk when the block is released from rest. Also, what is the velocity of the block after it falls a distance $2R$ starting from rest?

Prob. 17–78

17–79. Determine the angular acceleration of the 25-kg diving board and the horizontal and vertical components of reaction at the pin A the instant the man jumps off. Assume that the board is uniform and rigid, and that at the instant he jumps off the spring is compressed a maximum amount of 200 mm, $\omega = 0$, and the board is horizontal. Take $k = 7$ kN/m.

***17–80.** The spring is displaced 300 mm vertically when the 70-kg man jumps off the end of the 25-kg diving board. Assuming the board is uniform and rigid, and that the instant he jumps off the board $\omega = 0$ and the board is horizontal, determine the initial angular acceleration of the board. Take $k = 10$ kN/m.

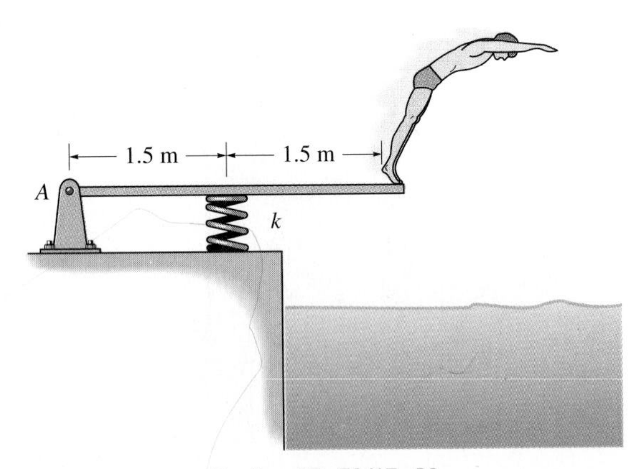

Probs. 17–79/17–80

17–81. The armature (slender rod) AB has a mass of 0.2 kg and can pivot about the pin at A. Movement is controlled by the electromagnet E, which exerts a horizontal attractive force on the armature at B of $F_B = (0.2(10^{-3})l^{-2})$ N, where l in meters is the gap between the armature and the magnet at any instant. If the armature lies in the horizontal plane, and is originally at rest, determine the speed of the contact at B the instant $l = 0.01$ m. Originally $l = 0.02$ m.

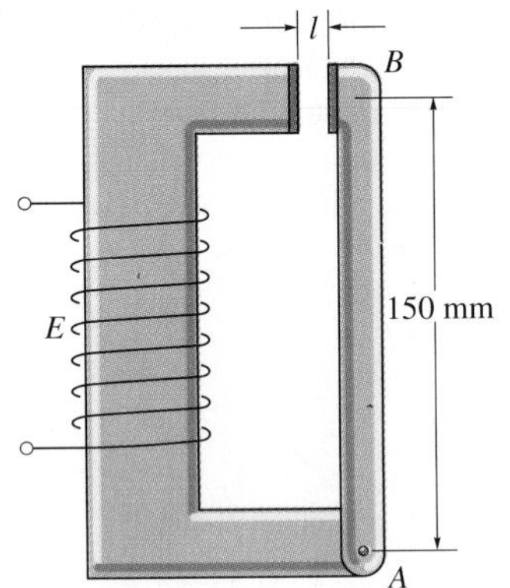

Prob. 17–81

17–82. The two blocks A and B have a mass m_A and m_B, respectively, where $m_B > m_A$. If the pulley can be treated as a disk of mass M, determine the acceleration of block A. Neglect the mass of the cord and any slipping on the pulley.

17–83. The two blocks A and B have a mass of 10 kg and 15 kg, respectively. If the pulley can be treated as a 4-kg disk, determine the acceleration of block A. Take $r = 200$ mm. Neglect the mass of the cord and any slipping on the pulley.

***17–84.** The two blocks A and B have a weight of 10 lb and 15 lb, respectively. If the pulley can be treated as a 5-lb disk having a radius of 0.75 ft, determine the acceleration of block A. Neglect the mass of the cord and any slipping on the pulley. Also, what would be the acceleration of block A if block B was replaced with a force of 15 lb?

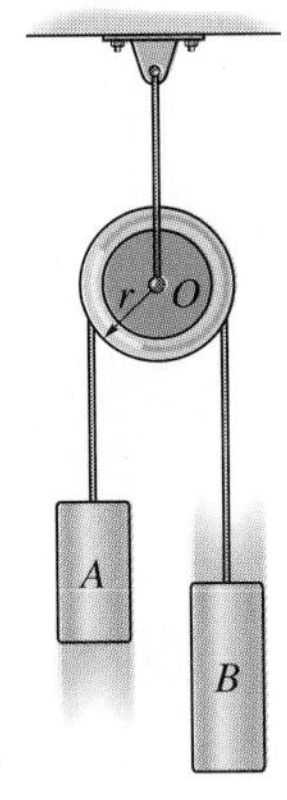

Probs. 17–82/17–83/17–84

17–85. The cord is wrapped around the inner core of the spool. If a 5-lb weight B is suspended from the cord and released from rest, determine the spool's angular velocity in 3 s. Neglect the mass of the cord. The spool has a weight of 180 lb and its radius of gyration about the axle A is $k_A = 1.25$ ft. Solve the problem in two ways. First by considering the "system" consisting of the block and spool, and then by considering the block and spool separately.

17–86. The cord is wrapped around the inner core of the spool. If a 5-lb weight B is suspended from the cord and released from rest, determine the spool's angular velocity after 2 ft of cord have been unraveled. Neglect the mass of the cord. The spool has a weight of 180 lb and its radius of gyration about the axle A is $k_A = 1.25$ ft. Solve the problem in two ways. First by considering the "system" consisting of the block and spool, and then by considering the block and spool separately.

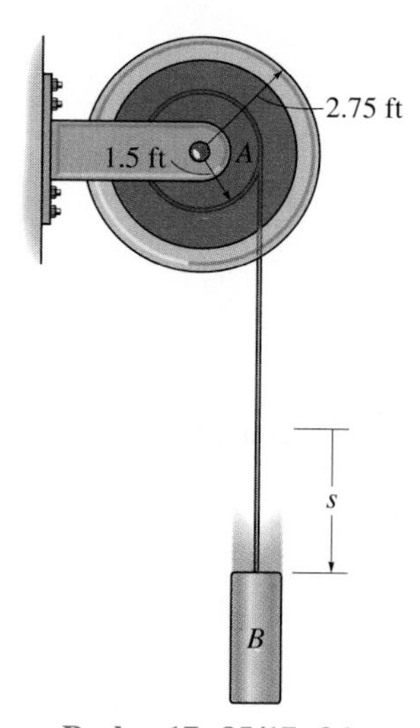

Probs. 17–85/17–86

17–87. The two-bar assembly is released from rest in the position shown. Determine the initial bending moment at the fixed joint B. Each bar has a mass m and length l.

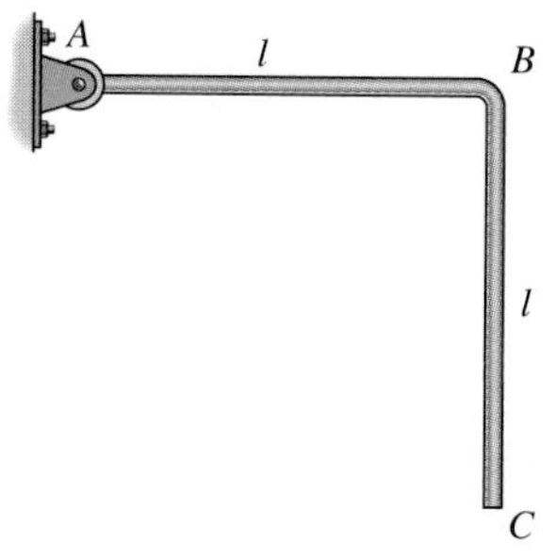

Prob. 17–87

***17–88.** The 5-Mg pipe is supported on a set of two conical rollers located at each of its two ends. If each roller has a mass of 8 kg and a radius of gyration about its central horizontal axis of $k_G = 0.12$ m, determine the initial acceleration of the pipe and the angular acceleration of each roller if a horizontal force $P = 600$ N is applied to the pipe. Assume the pipe does not slip on the rollers.

17–89. The 5-Mg pipe is supported on a set of two conical rollers located at each of its two ends. If each roller has a mass of 8 kg and a radius of gyration about its central horizontal axis of $k_G = 0.12$ m, determine the greatest axial force P that can be applied to the pipe so that initially it does not slip on the rollers. The coefficient of static friction between the pipe and rollers is $\mu_s = 0.2$.

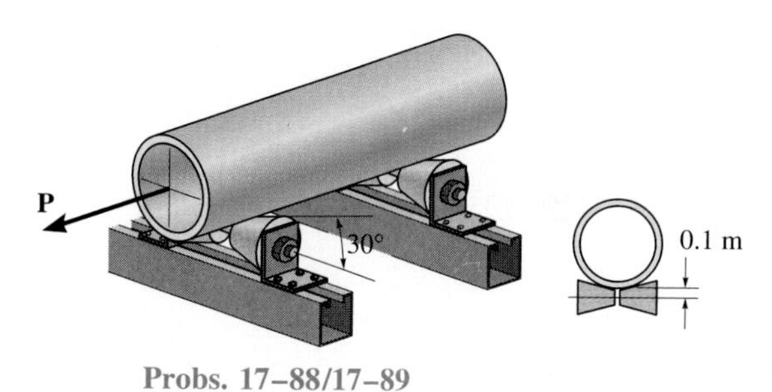

Probs. 17–88/17–89

17–90. The wheel has a mass of 25 kg and a radius of gyration $k_B = 0.15$ m. It is originally spinning at $\omega_1 = 40$ rad/s. If it is placed on the ground, for which the coefficient of kinetic friction is $\mu_C = 0.5$, determine the time required for the motion to stop. What are the horizontal and vertical components of reaction which the pin at A exerts on AB during this time? Neglect the mass of AB.

17–91. The wheel has a mass of 25 kg and a radius of gyration $k_B = 0.15$ m. It is originally spinning at $\omega_1 = 40$ rad/s. If it is placed on the ground, for which the coefficient of kinetic friction is $\mu_C = 0.5$, determine the angular displacement of the wheel required to stop the motion. What are the horizontal and vertical components of reaction that the pin at A exerts on AB during this time? Neglect the mass of AB.

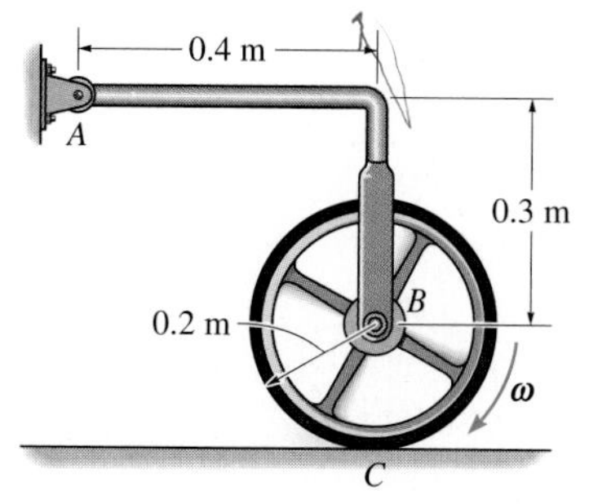

Probs. 17–90/17–91

*■**17–92.** A 40-kg boy sits on top of the large wheel which has a mass of 400 kg and a radius of gyration $k_G = 5.5$ m. If the boy essentially starts from rest at $\theta = 0°$, and the wheel begins to rotate freely, determine the angle at which the boy begins to slip. The coefficient of static friction between the wheel and the boy is $\mu_s = 0.5$. Neglect the size of the boy in the calculation.

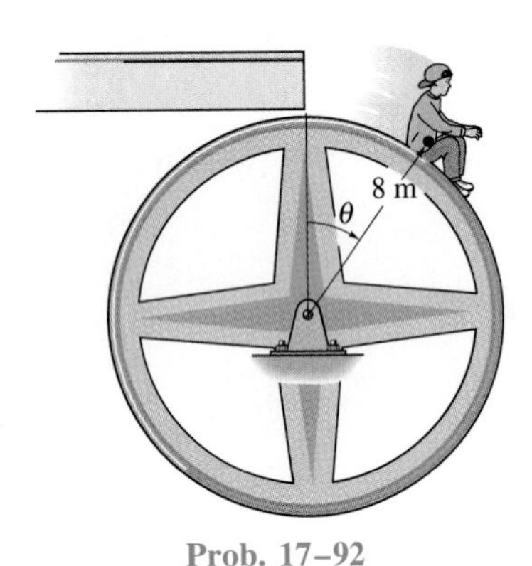

Prob. 17–92

17–93. The slender rod of length L and mass m is released from rest when $\theta = 0°$. Determine as a function of θ the normal and frictional forces which are exerted on the ledge at A as it falls downward. At what angle θ does it begin to slip if the coefficient of static friction at A is μ?

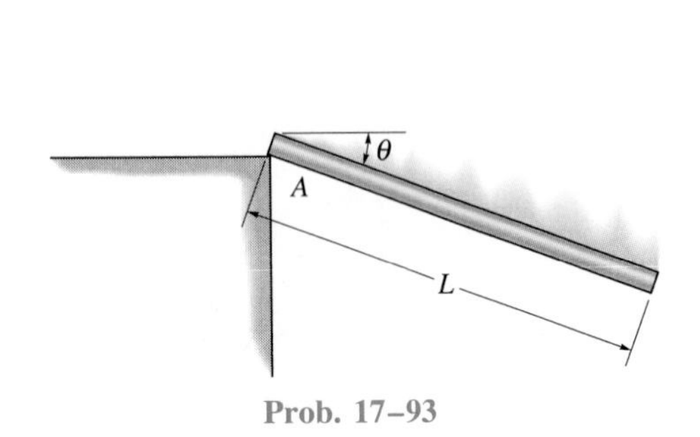

Prob. 17–93

17–94. A force $F = 2$ lb is applied perpendicular to the axis of the 5-lb rod and moves from O to A at a constant rate of 4 ft/s. If the rod is at rest when $\theta = 0°$ and **F** is at O when $t = 0$, determine the rod's angular velocity at the instant the force is at A. Through what angle has the rod rotated when this occurs? The rod rotates in the *horizontal plane.*

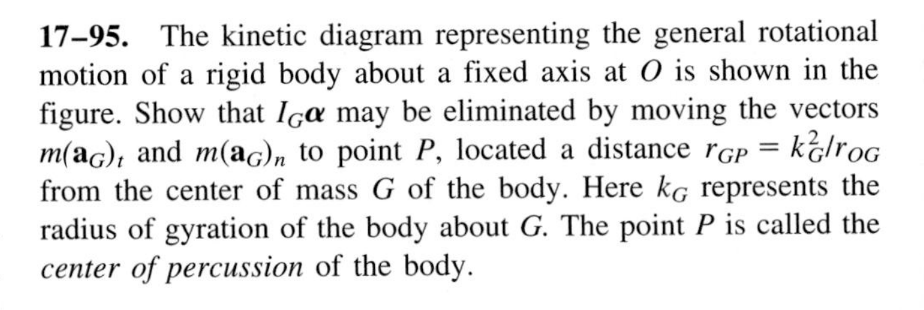

Prob. 17–94

17–95. The kinetic diagram representing the general rotational motion of a rigid body about a fixed axis at O is shown in the figure. Show that $I_G\boldsymbol{\alpha}$ may be eliminated by moving the vectors $m(\mathbf{a}_G)_t$ and $m(\mathbf{a}_G)_n$ to point P, located a distance $r_{GP} = k_G^2/r_{OG}$ from the center of mass G of the body. Here k_G represents the radius of gyration of the body about G. The point P is called the *center of percussion* of the body.

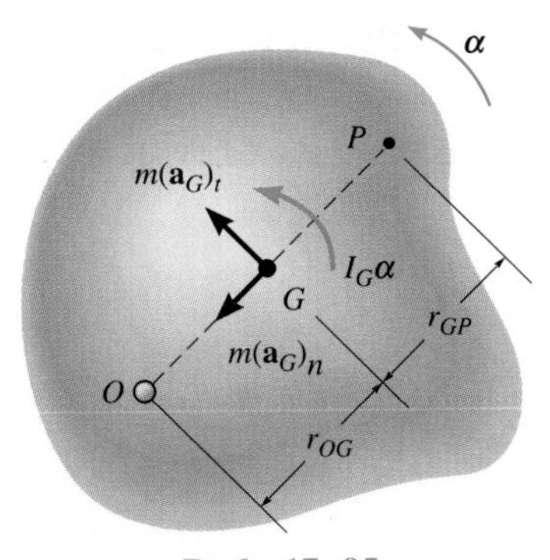

Prob. 17–95

***17–96.** Determine the position r_P of the center of percussion P of the 10-lb slender bar. (See Prob. 17–95.) What is the horizontal force A_x at the pin when the bar is struck at P with a force $F =$ 20 lb?

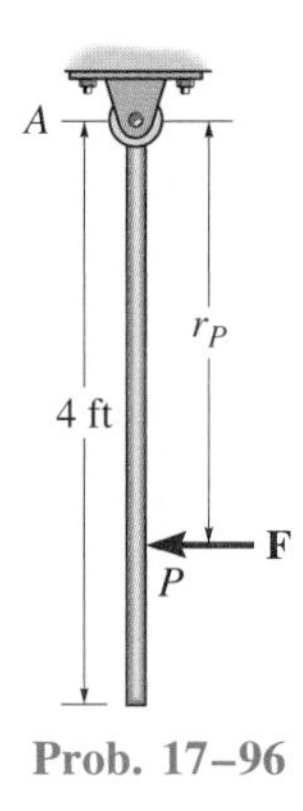

Prob. 17–96

17–97. If the support at B is suddenly removed, determine the initial horizontal and vertical components of reaction which the pin A exerts on the rod ACB. Segments AC and CB each have a weight of 10 lb.

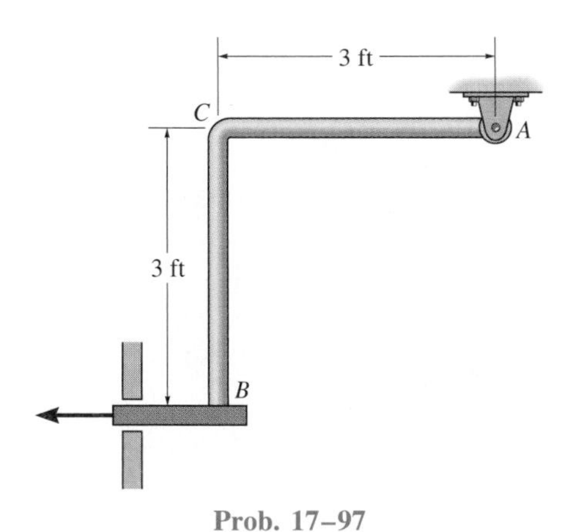

Prob. 17–97

17–98. The ladder is released from rest when $\theta = 0°$. Determine as a function of θ the normal and frictional forces which are exerted on the ladder by the ground as it falls downward. Is it possible for the ladder to slip as it falls? Why? For the calculation, treat the ladder as a slender rod of length l and mass m.

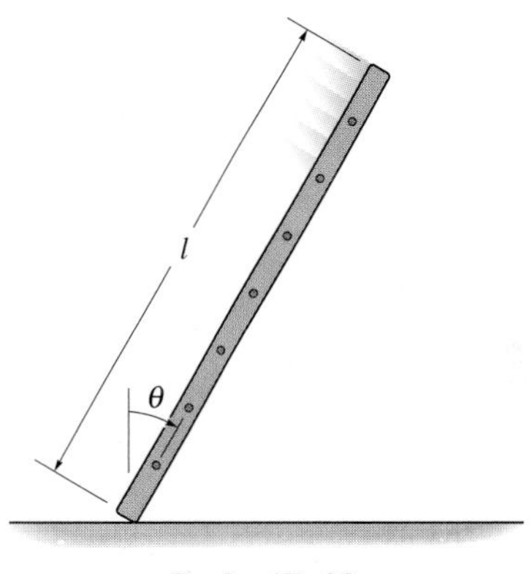

Prob. 17–98

17–99. The bar has a weight per length w and rotates in the vertical plane at constant angular velocity $\boldsymbol{\omega}$. Determine the internal normal force, shear force, and bending moment as a function of x and the bar's position θ.

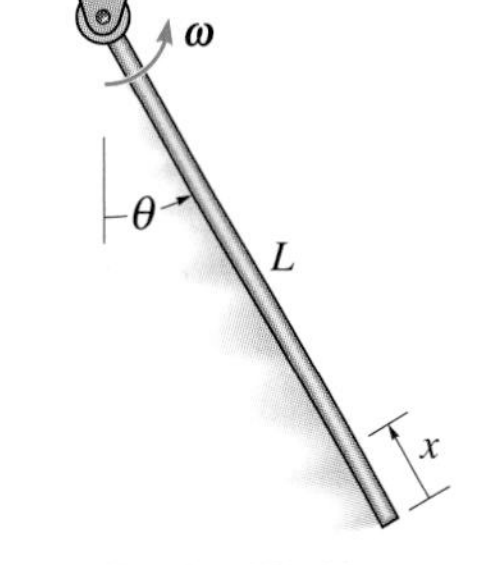

Prob. 17–99

17.5 Equations of Motion: General Plane Motion

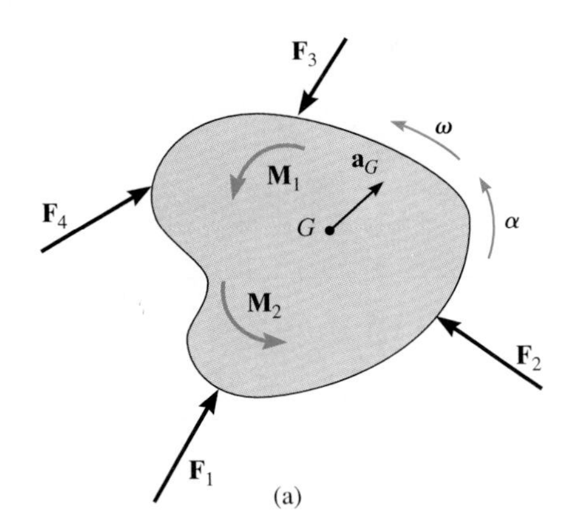

(a)

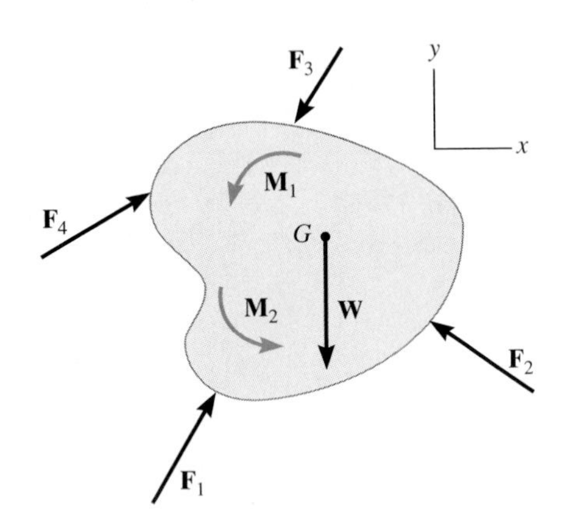

=

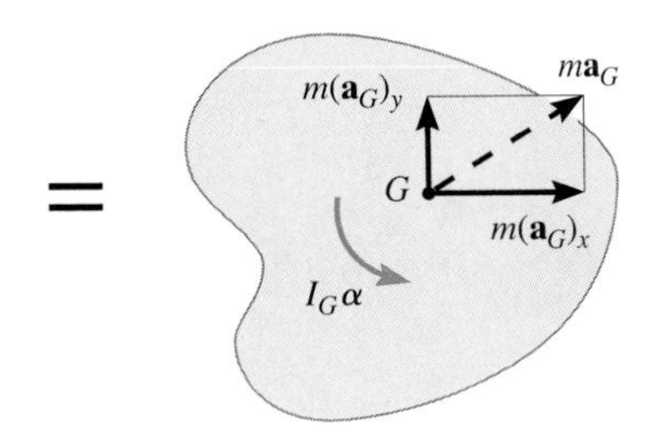

(b)

Fig. 17–20

The rigid body (or slab) shown in Fig. 17–20*a* is subjected to general plane motion caused by the externally applied force and couple-moment system. The free-body and kinetic diagrams for the body are shown in Fig. 17–20*b*. The vector $m\mathbf{a}_G$ (shown dashed) has the same *direction* as the acceleration of the body's mass center, and $I_G\boldsymbol{\alpha}$ acts in the same *direction* as the angular acceleration. If an x and y inertial coordinate system is chosen as shown, the three equations of motion may be written as

$$\Sigma F_x = m(a_G)_x$$
$$\Sigma F_y = m(a_G)_y \qquad (17\text{–}17)$$
$$\Sigma M_G = I_G\alpha$$

In some problems it may be convenient to sum moments about some point P other than G. This is usually done in order to eliminate unknown forces from the moment summation. When used in this more general sense, the three equations of motion become

$$\Sigma F_x = m(a_G)_x$$
$$\Sigma F_y = m(a_G)_y \qquad (17\text{–}18)$$
$$\Sigma M_P = \Sigma(\mathcal{M}_k)_P$$

Here $\Sigma(\mathcal{M}_k)_P$ represents the moment sum of $I_G\boldsymbol{\alpha}$ and $m\mathbf{a}_G$ (or its components) about P as determined by the data on the kinetic diagram.

PROCEDURE FOR ANALYSIS

The following procedure provides a method for solving kinetic problems involving general plane motion of a rigid body.

Free-Body Diagram. Establish the x, y inertial coordinate system and draw the free-body diagram for the body. Specify the direction and sense of the acceleration of the mass center, $\mathbf{a}_G$, and the angular acceleration $\boldsymbol{\alpha}$ of the body. It may be convenient to show these vectors on the coordinate system, but not on the free-body diagram. Also, compute the moment of inertia I_G. Identify the unknowns in the problem. If it is decided that the rotational equation of motion $\Sigma M_P = \Sigma(\mathcal{M}_k)_P$ is to be used, then consider drawing the kinetic diagram in order to help "visualize" the "moments" developed by the components $m(\mathbf{a}_G)_x$, $m(\mathbf{a}_G)_y$, and $I_G\boldsymbol{\alpha}$ when writing the terms in the moment sum $\Sigma(\mathcal{M}_k)_P$.

Equations of Motion. Apply the three equations of motion, Eqs. 17–17 or 17–18.

Kinematics. Use kinematics if a complete solution cannot be obtained strictly from the equations of motion. In particular, if the body's motion is *constrained* due to its supports, additional equations may be obtained by using $\mathbf{a}_B = \mathbf{a}_A + \mathbf{a}_{B/A}$, which relate the accelerations of any two points A and B on the body (see Example 17–17).

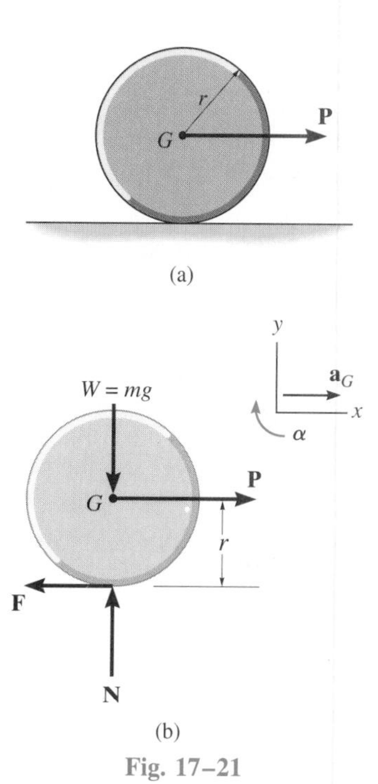

Fig. 17–21

Frictional Rolling Problems. There is a class of planar kinetics problems which deserves special mention. These problems involve wheels, cylinders, or bodies of similar shape, which roll on a *rough* plane surface. Because of the applied loadings, it may not be known if the body *rolls without slipping,* or if it *slides as it rolls.* For example, consider the homogeneous disk shown in Fig. 17–21*a*, which has a mass m and is subjected to a known horizontal force **P.** Following the procedure outlined above, the free-body diagram is shown in Fig. 17–21*b*. Since $\mathbf{a}_G$ is directed to the right and $\boldsymbol{\alpha}$ is clockwise, we have

$$\overset{+}{\rightarrow}\Sigma F_x = m(a_G)_x; \qquad P - F = ma_G \qquad (17\text{–}19)$$

$$+\uparrow \Sigma F_y = m(a_G)_y; \qquad N - mg = 0 \qquad (17\text{–}20)$$

$$\circlearrowright+\Sigma M_G = I_G\alpha; \qquad Fr = I_G\alpha \qquad (17\text{–}21)$$

A fourth equation is needed since these *three equations* contain *four unknowns: F, N,* α, and a_G.

No Slipping. If the frictional force **F** is great enough to allow the disk to roll *without slipping,* then a_G may be related to α by the *kinematic equation,**

$$(\circlearrowright+) \qquad a_G = \alpha r \qquad (17\text{–}22)$$

The four unknowns are determined by *solving simultaneously* Eqs. 17–19 to 17–22. When the solution is obtained, the assumption of no slipping must be *checked.* Recall that no slipping occurs provided $F \leq \mu_s N$, where μ_s is the coefficient of static friction. If the inequality is satisfied, the problem is solved. However, if $F > \mu_s N$, the problem must be *reworked,* since then the disk slips as it rolls.

Slipping. In the case of slipping, α and a_G are *independent of one another* so that Eq. 17–22 does not apply. Instead, the magnitude of the frictional force is related to the magnitude of the normal force using the coefficient of kinetic friction μ_k, i.e.,

$$F = \mu_k N \qquad (17\text{–}23)$$

In this case Eqs. 17–19 to 17–21 and 17–23 are used for the solution. It is important to keep in mind that whenever Eq. 17–22 or 17–23 is applied, it is necessary that there is a consistency in the directional sense of the vectors. In the case of Eq. 17–22, $\mathbf{a}_G$ must be directed to the right when $\boldsymbol{\alpha}$ is clockwise, since the rolling motion requires it. And in Eq. 17–23, **F** must be directed to the left to prevent the assumed slipping motion to the right, Fig. 17–21*b*. On the other hand, if these equations are *not used* for the solution, these vectors can have *any* assumed directional sense. Then if the calculated numerical value of these quantities is negative, the vectors act in their opposite sense of direction. Examples 17–15 and 17–16 illustrate these concepts numerically.

*See Example 16–3 or 16–14.

Example 17–14

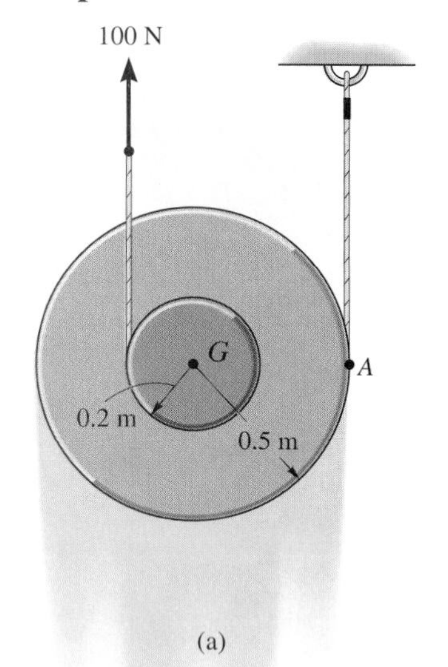

(a)

The spool in Fig. 17–22*a* has a mass of 8 kg and a radius of gyration of $k_G = 0.35$ m. If cords of negligible mass are wrapped around its inner hub and outer rim as shown, determine the spool's angular acceleration.

SOLUTION I

The angular velocity is obtained by first finding the spool's angular acceleration using the equations of motion.

Free-Body Diagram. Fig. 17–22*b*. The 100-N force causes $\mathbf{a}_G$ to act upward. Also, $\boldsymbol{\alpha}$ acts clockwise, since the spool winds around the cord at A.

There are three unknowns, T, a_G, and α. The moment of inertia of the spool about its mass center is

$$I_G = mk_G^2 = 8(0.35)^2 = 0.980 \text{ kg} \cdot \text{m}^2$$

Equations of Motion

$$+\uparrow \Sigma F_y = m(a_G)_y; \qquad T + 100 \text{ N} - 78.48 \text{ N} = (8 \text{ kg})a_G \qquad (1)$$

$$\circlearrowright+\Sigma M_G = I_G\alpha; \qquad 100 \text{ N}(0.2 \text{ m}) - T(0.5 \text{ m}) = (0.980 \text{ kg} \cdot \text{m}^2)\alpha \qquad (2)$$

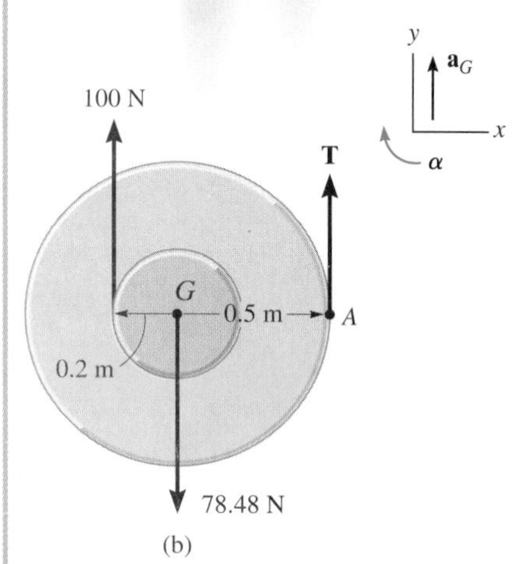

(b)

Kinematics. A complete solution is obtained if kinematics is used to relate a_G to α. In this case the spool "rolls without slipping" on the cord at A. Hence, we can use the results of Example 16–3 or 16–14, so that

$$(\circlearrowright+)a_G = \alpha r; \qquad a_G = 0.5\alpha \qquad (3)$$

Solving Eqs. 1 to 3, we have

$$\alpha = 10.3 \text{ rad/s}^2 \qquad \textit{Ans.}$$

$$a_G = 5.16 \text{ m/s}^2$$

$$T = 19.8 \text{ N}$$

SOLUTION II

Equations of Motion. We can eliminate the unknown T by summing moments about point A. From the free-body and kinetic diagrams Figs. 17–22*b* and 17–22*c*, we have

$$\circlearrowright+\Sigma M_A = \Sigma(M_k)_A; \; 100 \text{ N}(0.7 \text{ m}) - 78.48 \text{ N}(0.5 \text{ m})$$
$$= (0.980 \text{ kg} \cdot \text{m}^2)\alpha + [(8 \text{ kg})a_G](0.5 \text{ m})$$

Using Eq. (3),

$$\alpha = 10.3 \text{ rad/s}^2 \qquad \textit{Ans.}$$

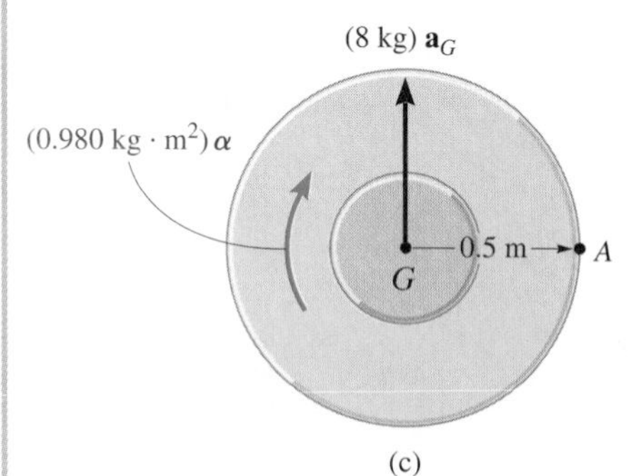

(c)

Fig. 17–22

Example 17–15

The 50-lb wheel shown in Fig. 17–23a has a radius of gyration $k_G = 0.70$ ft. If a 35-lb · ft couple moment is applied to the wheel, determine the acceleration of its mass center G. The coefficients of static and kinetic friction between the wheel and the plane at A are $\mu_s = 0.3$ and $\mu_k = 0.25$, respectively.

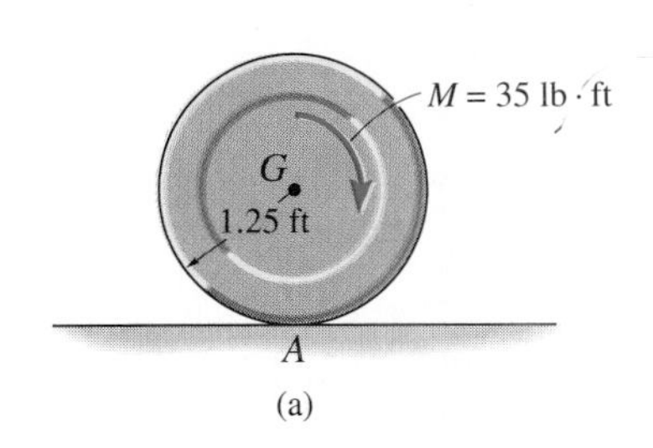

(a)

SOLUTION

Free-Body Diagram. By inspection of Fig. 17–23b, it is seen that the couple moment causes the wheel to have a clockwise angular acceleration of $\boldsymbol{\alpha}$. As a result, the acceleration of the mass center, $\mathbf{a}_G$, is directed to the right. The moment of inertia is

$$I_G = mk_G^2 = \frac{50 \text{ lb}}{32.2 \text{ ft/s}^2}(0.70 \text{ ft})^2 = 0.761 \text{ slug} \cdot \text{ft}^2$$

The unknowns are N_A, F_A, a_G, and α.

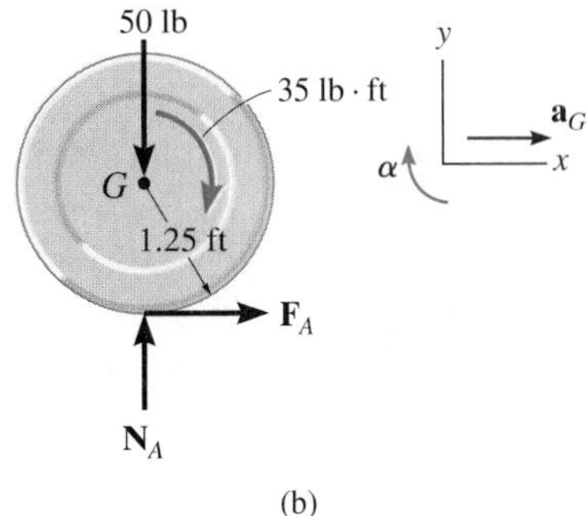

(b)

Fig. 17–23

Equations of Motion

$$\overset{+}{\rightarrow}\Sigma F_x = m(a_G)_x; \qquad F_A = \frac{50 \text{ lb}}{32.2 \text{ ft/s}^2} a_G \qquad (1)$$

$$+\uparrow \Sigma F_y = m(a_G)_y; \qquad N_A - 50 \text{ lb} = 0 \qquad (2)$$

$$\circlearrowright\!+\Sigma M_G = I_G\alpha; \qquad 35 \text{ lb} \cdot \text{ft} - 1.25 \text{ ft}(F_A) = (0.761 \text{ slug} \cdot \text{ft}^2)\alpha \qquad (3)$$

A fourth equation is needed for a complete solution.

Kinematics (No Slipping). If this assumption is made, then

$$(\circlearrowright\!+) \qquad a_G = (1.25 \text{ ft})\alpha \qquad (4)$$

Solving Eqs. 1 to 4,

$$N_A = 50.0 \text{ lb} \qquad F_A = 21.3 \text{ lb}$$
$$\alpha = 11.0 \text{ rad/s}^2 \qquad a_G = 13.7 \text{ ft/s}^2$$

The original assumption of no slipping requires $F_A \le \mu_s N_A$. However, since $21.3 \text{ lb} > 0.3(50 \text{ lb}) = 15$ lb, the wheel slips as it rolls.

(Slipping.) Equation 4 is not valid. Instead, it is necessary that $\mathbf{F}_A$ acts to the right. Why? Also, $F_A = \mu_k N_A$, or

$$F_A = 0.25 N_A \qquad (5)$$

Solving Eqs. 1 to 3 and 5 yields

$$N_A = 50.0 \text{ lb} \qquad F_A = 12.5 \text{ lb}$$
$$\alpha = 25.5 \text{ rad/s}^2$$
$$a_G = 8.05 \text{ ft/s}^2 \rightarrow \qquad \textit{Ans.}$$

Example 17–16

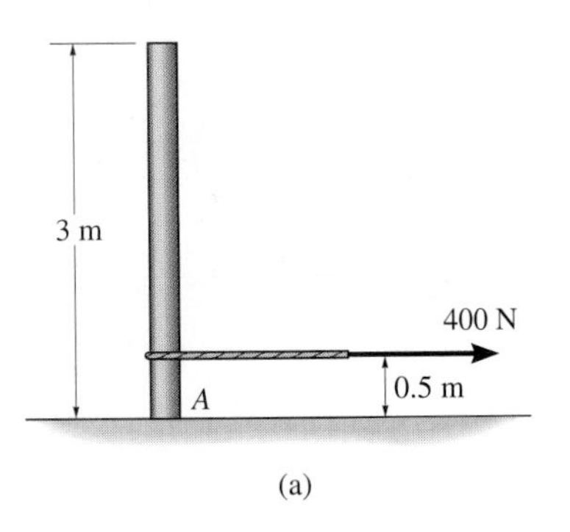

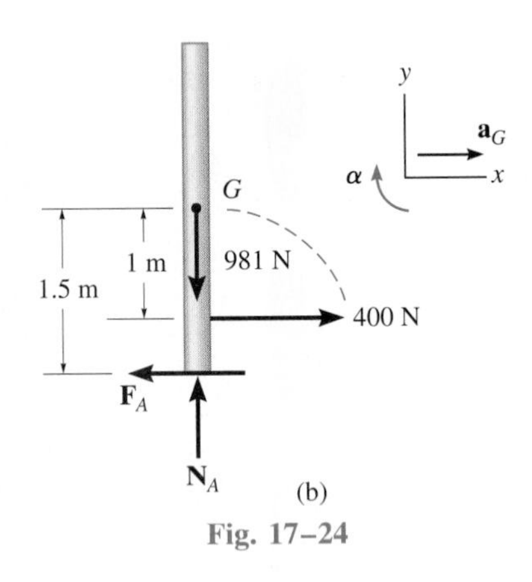

Fig. 17–24

The uniform slender pole shown in Fig. 17–24*a* has a mass of 100 kg and a moment of inertia $I_G = 75\ \text{kg}\cdot\text{m}^2$. If the coefficients of static and kinetic friction between the end of the pole and the surface are $\mu_s = 0.3$ and $\mu_k = 0.25$, respectively, determine the pole's angular acceleration at the instant the 400-N horizontal force is applied. The pole is originally at rest.

SOLUTION

Free-Body Diagram. Figure 17–24*b*. The path of motion of the mass center *G* will be along an unknown (dashed) curved path having a radius of curvature ρ which is initially parallel to the *y* axis. There is no normal or *y* component of acceleration since the pole is originally at rest, i.e., $\mathbf{v}_G = \mathbf{0}$, so that $(a_G)_y = v_G^2/\rho = 0$. We will assume the mass center accelerates to the right, and the pole has a clockwise angular acceleration of $\boldsymbol{\alpha}$. The unknowns are N_A, F_A, a_G, and α.

Equations of Motion

$$\overset{+}{\rightarrow}\Sigma F_x = m(a_G)_x; \qquad 400\ \text{N} - F_A = (100\ \text{kg})a_G \qquad (1)$$

$$+\uparrow \Sigma F_y = m(a_G)_y; \qquad N_A - 981\ \text{N} = 0 \qquad (2)$$

$$\curvearrowright +\Sigma M_G = I_G\alpha; \quad F_A(1.5\ \text{m}) - 400\ \text{N}(1\ \text{m}) = (75\ \text{kg}\cdot\text{m}^2)\alpha \qquad (3)$$

A fourth equation is needed for a complete solution.

Kinematics (No Slipping). In this case point *A* acts as a ''pivot'' so that indeed, if α is clockwise, then a_G is directed to the right.

$$\curvearrowright + a_G = \alpha r_{AG}; \qquad a_G = (1.5\ \text{m})\alpha \qquad (4)$$

Solving Eqs. 1 to 4 yields

$$N_A = 981\ \text{N} \qquad F_A = 300\ \text{N}$$
$$a_G = 1\ \text{m/s}^2 \qquad \alpha = 0.667\ \text{rad/s}^2$$

Testing the original assumption of no slipping requires $F_A \le \mu_s N_A$. However, $300\ \text{N} > 0.3(981\ \text{N}) = 294\ \text{N}$. (Slips at *A*.)

(Slipping.) For this case Eq. 4 does *not* apply. Instead, $\mathbf{F}_A$ must act to the left and the frictional equation $F_A = \mu_k N_A$ is used. Hence,

$$F_A = 0.25N_A \qquad (5)$$

Solving Eqs. 1 to 3 and 5 simultaneously yields

$$N_A = 981\ \text{N} \qquad F_A = 245\ \text{N} \qquad a_G = 1.55\ \text{m/s}^2$$
$$\alpha = -0.428\ \text{rad/s}^2 = 0.428\ \text{rad/s}^2 \curvearrowleft \qquad \textbf{\textit{Ans.}}$$

Example 17–17

The 30-kg wheel shown in Fig. 17–25a has a mass center at G and a radius of gyration $k_G = 0.15$ m. If the wheel is originally at rest and released from the position shown, determine its angular acceleration. No slipping occurs.

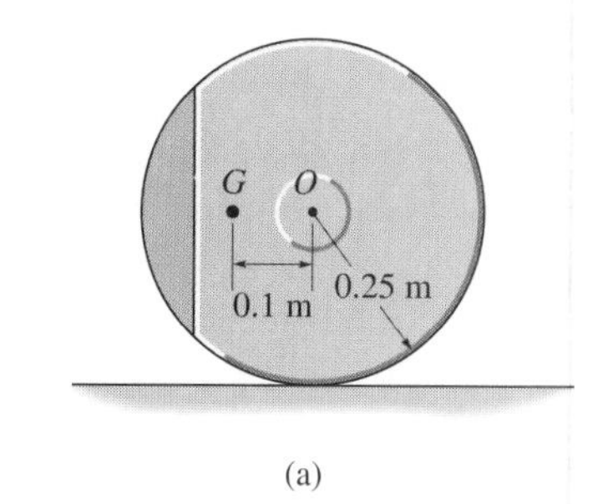

(a)

SOLUTION

Free-Body and Kinetic Diagrams. The two unknowns $\mathbf{F}_A$ and $\mathbf{N}_A$ shown on the free-body diagram, Fig. 17–25b, can be eliminated from the analysis by summing moments about point A. The kinetic diagram accompanies the solution in order to illustrate application of $\Sigma(\mathcal{M}_k)_A$. Since the *path of motion* of G is *unknown,* the two components $m(\mathbf{a}_G)_x$ and $m(\mathbf{a}_G)_y$ must be shown on the kinetic diagram, Fig. 17–25b.

The moment of inertia is

$$I_G = mk_G^2 = 30(0.15)^2 = 0.675 \text{ kg} \cdot \text{m}^2$$

There are five unknowns, N_A, F_A, $(a_G)_x$, $(a_G)_y$, and α.

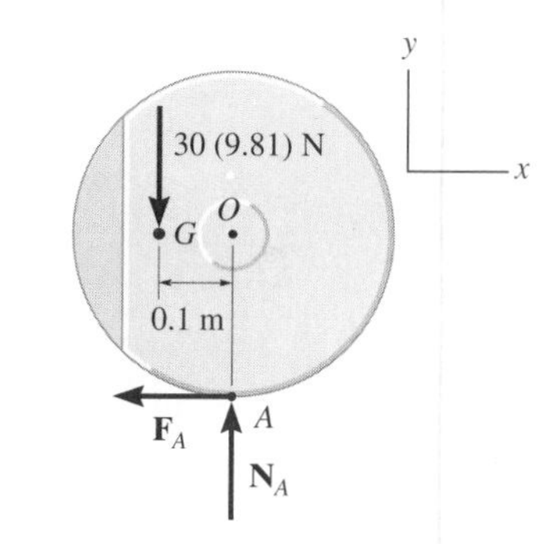

Equation of Motion. Applying the rotational equation of motion about point A, to eliminate N_A, and F_A, we have

$$\circlearrowleft + \Sigma M_A = \Sigma(\mathcal{M}_k)_A; \qquad 30(9.81)\text{ N}(0.1\text{ m}) =$$
$$(0.675 \text{ kg} \cdot \text{m}^2)\alpha + (30\text{ kg})(a_G)_x(0.25\text{ m}) + (30\text{ kg})(a_G)_y(0.1\text{ m}) \quad (1)$$

There are three unknowns in this equation: $(a_G)_x$, $(a_G)_y$, and α.

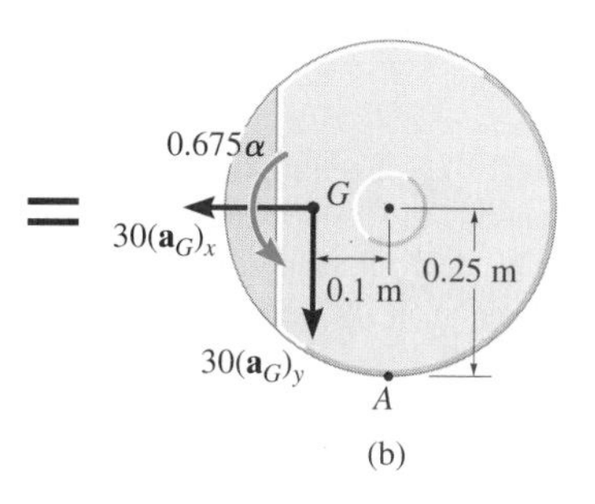

(b)

Kinematics. Using kinematics, $(a_G)_x$, $(a_G)_y$ will be related to α. As shown in Fig. 17–25c, these vectors must have the same directional sense as the corresponding vectors on the kinetic diagram since we are seeking a simultaneous solution with Eq. 1. Since no slipping occurs, $a_O = \alpha r = \alpha(0.25\text{ m})$, directed to the left, Fig. 17–25c. Also, $\omega = 0$, since the wheel is originally at rest. Applying the acceleration equation to point O (base point) and point G, we have

$$\mathbf{a}_G = \mathbf{a}_O + \boldsymbol{\alpha} \times \mathbf{r}_{G/O} - \omega^2\mathbf{r}_{G/O}$$
$$-(a_G)_x\mathbf{i} - (a_G)_y\mathbf{j} = -\alpha(0.25)\mathbf{i} + (-\alpha\mathbf{k}) \times (-0.1\mathbf{i}) - \mathbf{0}$$

Expanding and equating the respective $\mathbf{i}$ and $\mathbf{j}$ components, we have

$$(a_G)_x = \alpha(0.25) \quad (2)$$
$$(a_G)_y = \alpha(0.1) \quad (3)$$

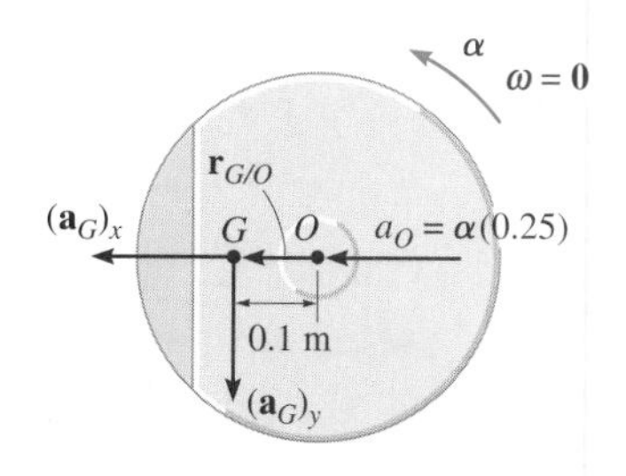

(c)

Fig. 17–25

Solving Eqs. 1 to 3 yields

$$\alpha = 10.3 \text{ rad/s}^2 \circlearrowleft \qquad \textit{Ans.}$$
$$(a_G)_x = 2.58 \text{ m/s}^2$$
$$(a_G)_y = 1.03 \text{ m/s}^2$$

As an exercise, show that $F_A = 77.4$ N and $N_A = 263$ N.

PROBLEMS

***17–100.** If the disk *rolls without slipping*, show that when moments are summed about the instantaneous center of zero velocity, *IC*, it is possible to use the moment equation $\Sigma M_{IC} = I_{IC}\alpha$, where I_{IC} represents the moment of inertia of the disk calculated about the instantaneous axis of zero velocity.

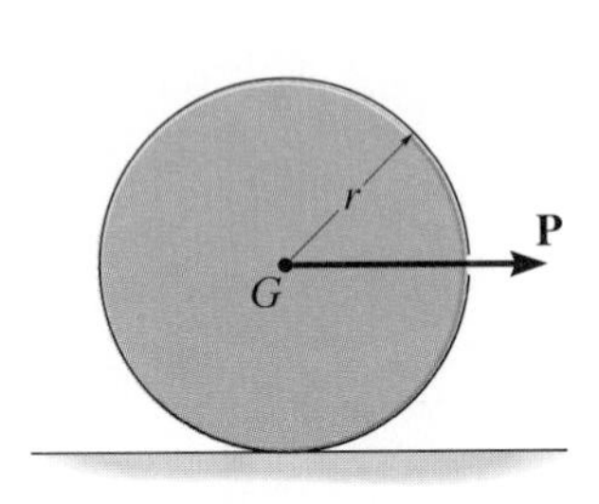

Prob. 17–100

17–101. The 20-kg punching bag has a radius of gyration about its center of mass G of $k_G = 0.4$ m. If it is subjected to a horizontal force $F = 30$ N, determine the initial angular acceleration of the bag and the tension in the supporting cable AB.

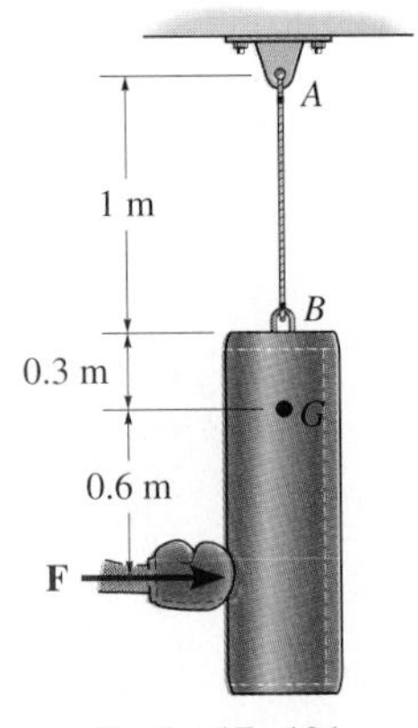

Prob. 17–101

17–102. The rocket has a weight of 20 000 lb, mass center at G, and radius of gyration about the mass center of $k_G = 21$ ft when it is fired. Each of its two engines provides a thrust $T = 50\,000$ lb. At a given instant, engine A suddenly fails to operate. Determine the angular acceleration of the rocket and the acceleration of its nose B.

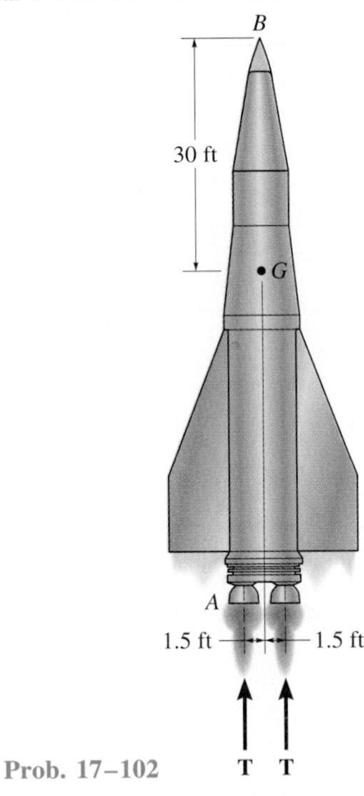

Prob. 17–102

17–103. The uniform 50-lb board is suspended from cords at C and D. If these cords are subjected to constant forces of 30 lb and 45 lb, respectively, determine the acceleration of the board's center and the board's angular acceleration. Assume the board is a thin plate. Neglect the mass of the pulleys at E and F.

***17–104.** Solve Prob. 17–103 if blocks having a weight $W_A = 30$ lb and $W_B = 45$ lb are suspended from the cords at A and B.

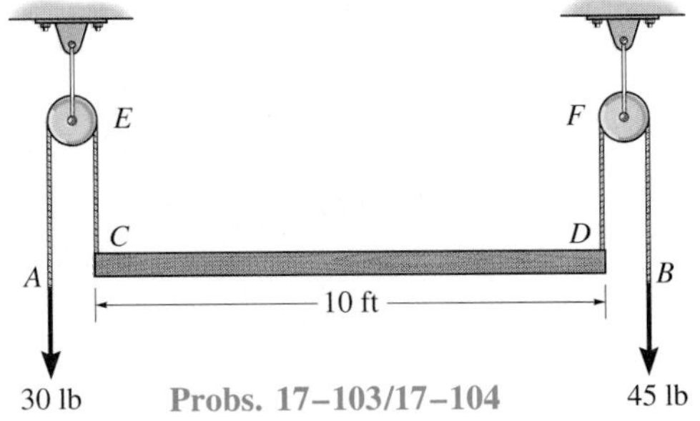

Probs. 17–103/17–104

17–105. The spool has a mass of 500 kg and a radius of gyration $k_G = 1.30$ m. It rests on the surface of a conveyor belt for which the coefficient of kinetic friction is $\mu_k = 0.4$. If the conveyor accelerates at $a_C = 1\ \text{m/s}^2$, determine the initial tension in the wire and the angular acceleration of the spool. The spool is originally at rest.

17–106. The spool has a mass of 500 kg and a radius of gyration $k_G = 1.30$ m. It rests on the surface of a conveyor belt for which the coefficient of static friction is $\mu_s = 0.5$. Determine the greatest acceleration a_C of the conveyor so that the spool will not slip. Also, what are the initial tension in the wire and the angular acceleration of the spool? The spool is originally at rest.

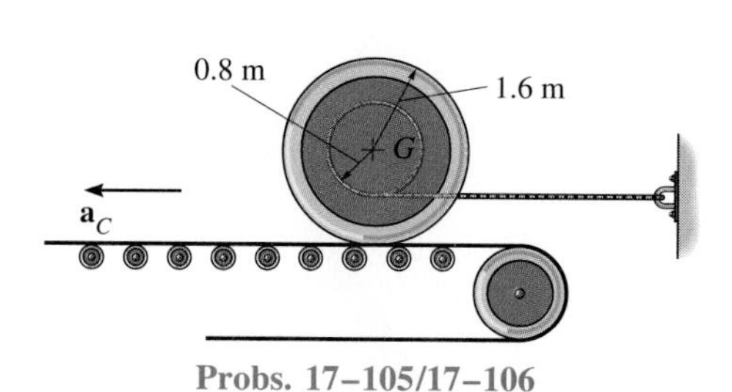

Probs. 17–105/17–106

***17–108.** The 2-kg slender bar is supported by cord BC and then released from rest at A. Determine the initial angular acceleration of the bar and the tension in the cord.

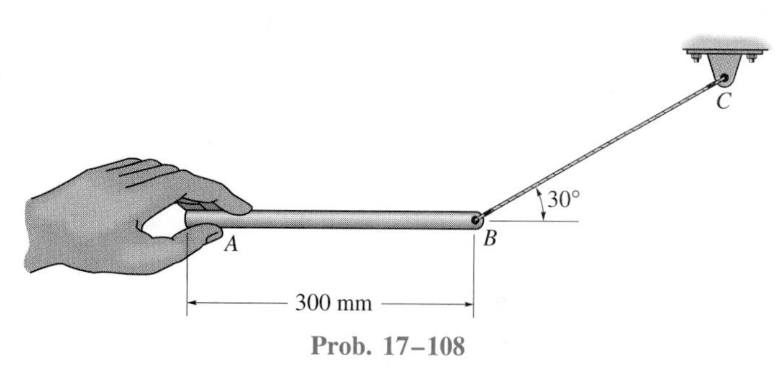

Prob. 17–108

17–107. The upper body of the crash dummy has a mass of 75 lb, a center of gravity at G, and a radius of gyration about G of $k_G = 0.7$ ft. By means of the seat belt this body segment is assumed to be pin-connected to the seat of the car at A. If a crash causes the car to decelerate at 50 ft/s^2, determine the angular velocity of the body when it has rotated to $\theta = 30°$.

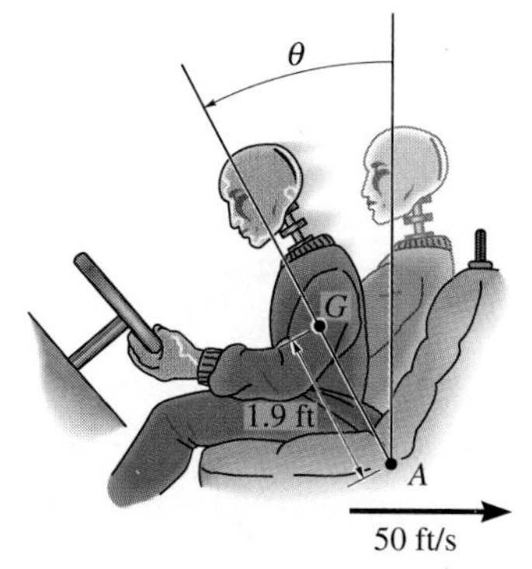

Prob. 17–107

17–109. The semicircular disk has a mass of 10 kg. If it is rotating at $\omega = 4$ rad/s at the instant $\theta = 60°$, determine the normal and frictional forces it exerts on the ground at this instant. Assume the disk does not slip as it rolls.

17–110. The semicircular disk having a mass of 10 kg is rotating at $\omega = 4$ rad/s at the instant $\theta = 60°$. If the coefficient of static friction at A is $\mu_s = 0.5$, determine if the disk slips at this instant.

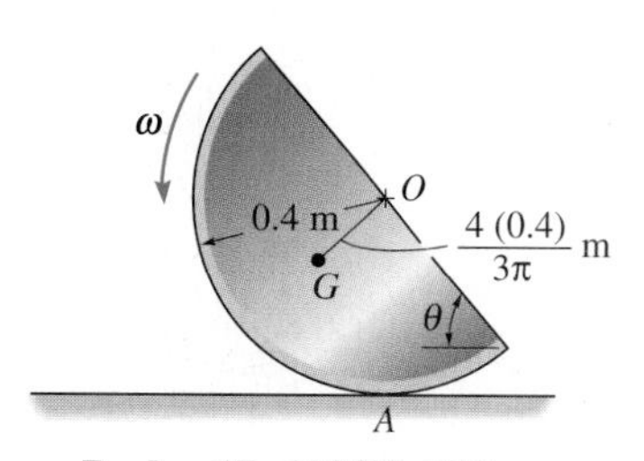

Probs. 17–109/17–110

17–111. A biomechanical model of the head and neck of a restrained passenger in an automobile is shown in the figure. Measurements indicate that the head plus half the neck, assumed pinned at A, have a mass of 0.28 slug and a moment of inertia about an axis through the center of gravity of $I_G = 0.024 \text{ slug} \cdot \text{ft}^2$. If an automobile and the passenger's torso are subjected to a horizontal deceleration $a_A = 15g = 483 \text{ ft/s}^2$, when the head is in the position shown, determine the horizontal and vertical components of force which the neck exerts on the head at A at this instant. The restraining moment $\mathbf{M}_A$ is required to hold the head in the *nonrotating* equilibrium position shown, prior to the deceleration. Assume this moment remains constant during the deceleration.

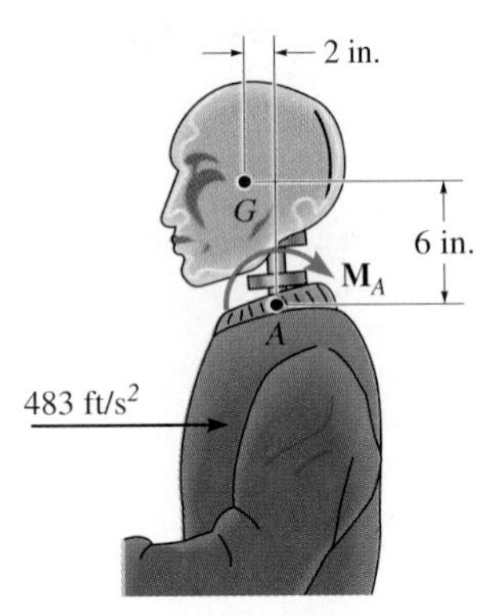

Prob. 17–111

*__17–112.__ The uniform bar of mass m and length L is balanced in the vertical position when the horizontal force $\mathbf{P}$ is applied to the roller at A. Determine the bar's initial angular acceleration and the acceleration of its top point B.

17–113. Solve Prob. 17–112 if the roller is removed and the coefficient of kinetic friction at the ground is μ_k.

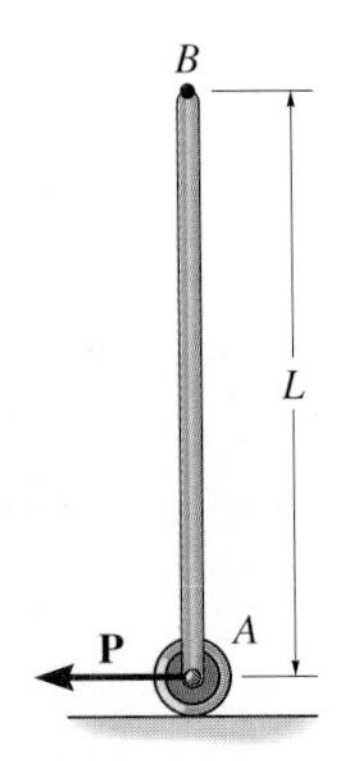

Probs. 17–112/17–113

17–114. The ladder has a weight W and rests against the smooth wall and ground. Determine its angular acceleration as a function of θ when it is released and allowed to slide downward. For the calculation, treat the ladder as a slender rod.

17–115. The ladder has a weight W and rests against the smooth wall and ground. Determine its angular velocity as a function of θ when it is released from rest when $\theta = 0$ and allowed to slide downward. For the calculation, treat the ladder as a slender rod.

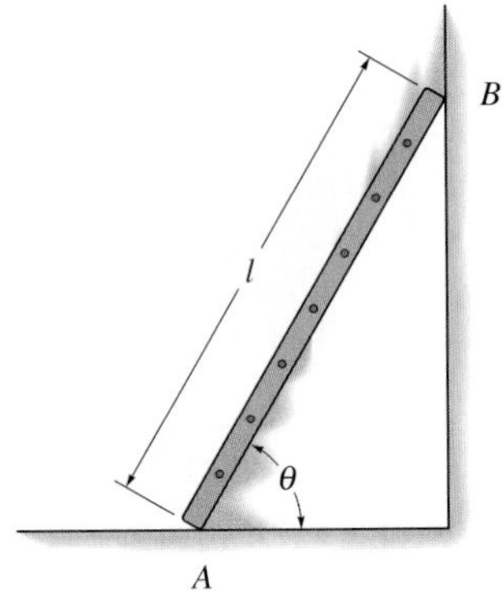

Probs. 17–114/17–115

*__17–116.__ The 10-lb hoop or thin ring is given an initial angular velocity of 6 rad/s when it is placed on the surface. If the coefficient of kinetic friction between the hoop and the surface is $\mu_k = 0.3$, determine the distance the hoop moves before it stops slipping.

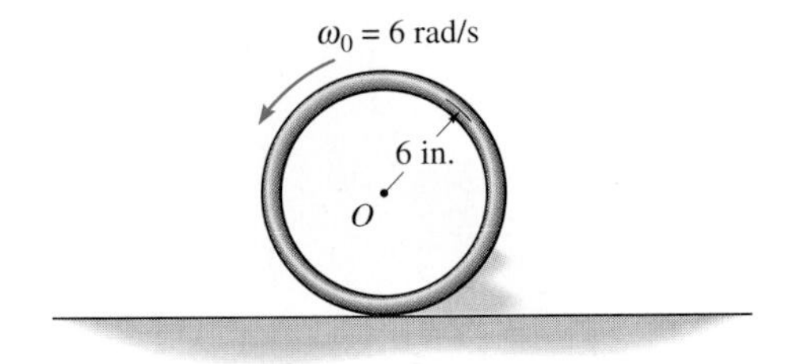

Prob. 17–116

17–117. The 15-lb circular plate is suspended from a pin at A. If the pin is connected to a track which is given an acceleration $a_A = 3\ \text{ft/s}^2$, determine the horizontal and vertical components of reaction at A and the acceleration of the plate's mass center G. The plate is originally at rest.

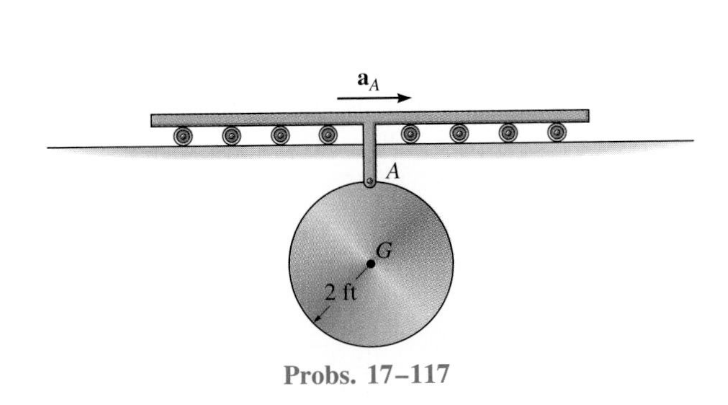

Probs. 17–117

17–118. The 15-lb circular plate is suspended from a pin at A. If the pin is connected to a track which is given an acceleration $a_A = 5\ \text{ft/s}^2$, determine the horizontal and vertical components of reaction at A and the angular acceleration of the plate. The plate is originally at rest.

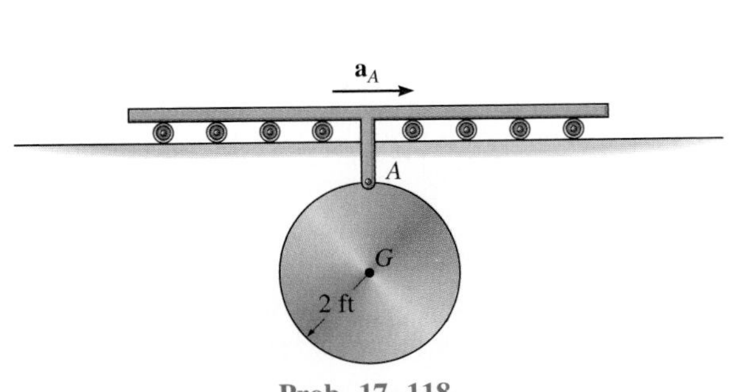

Prob. 17–118

17–119. A cord is wrapped around each of the two 10-kg disks. If they are released from rest determine the angular acceleration of each disk and the tension in the cord C. Neglect the mass of the cord.

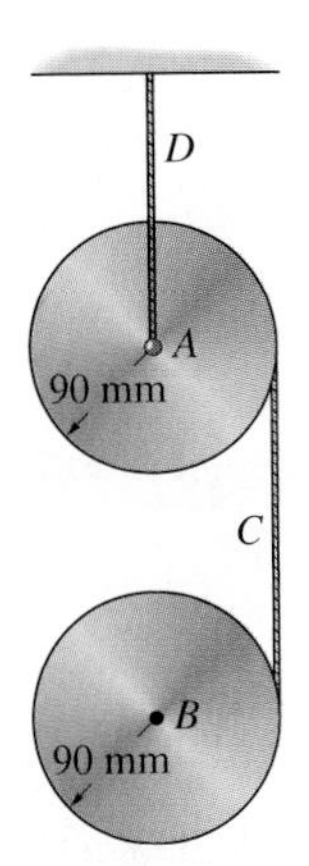

Prob. 17–119

***17–120.** The 10-lb disks are connected together using a bar having a negligible mass. If they roll on the inclined planes without slipping, determine the initial angular acceleration of each disk if the system is released from rest from the position shown.

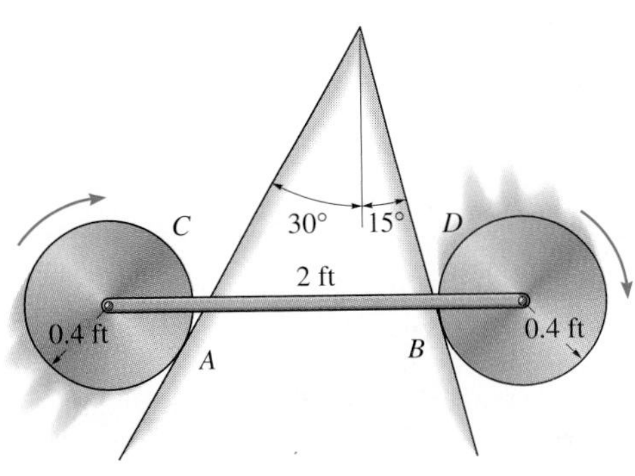

Prob. 17–120

17–121. The 500-lb beam is supported at A and B when it is subjected to a force of 1000 lb as shown. If the pin support at A suddenly fails, determine the beam's initial angular acceleration and the force of the roller support on the beam. For the calculation, assume that the beam is a slender rod so that its thickness can be neglected.

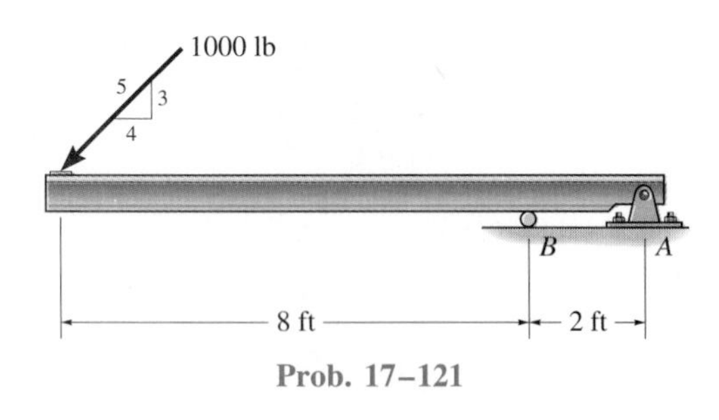

Prob. 17–121

17–122. The 15-lb disk rests on the 5-lb plate. A cord is wrapped around the periphery of the disk and attached to the wall at B. If a torque $M = 40$ lb · ft is applied to the disk, determine the angular acceleration of the disk and the time needed for the end C of the plate to travel 3 ft and strike the wall. The disk does not slip on the plate and the surface at D is smooth. Neglect the mass of the cord.

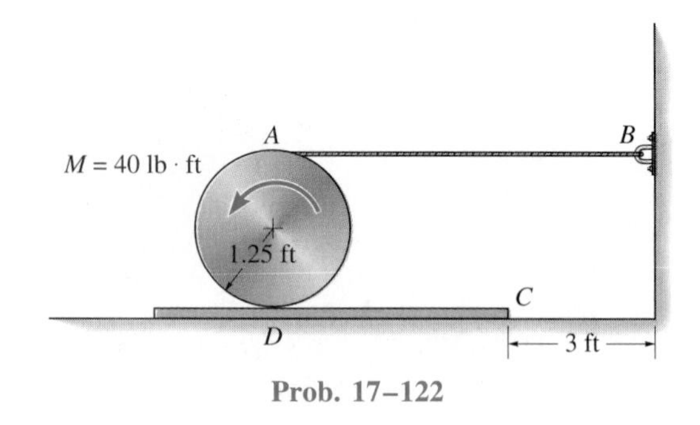

Prob. 17–122

17–123. The 15-lb disk rests on the 5-lb plate. A cord is wrapped around the periphery of the disk and attached to the wall at B. If a torque $M = 40$ lb · ft is applied to the disk, determine the angular acceleration of the disk and the time needed for the end C of the plate to travel 3 ft and strike the wall. Assume the disk does not slip on the plate and the plate rests on the surface at D having a coefficient of kinetic friction of $\mu_k = 0.2$. Neglect the mass of the cord.

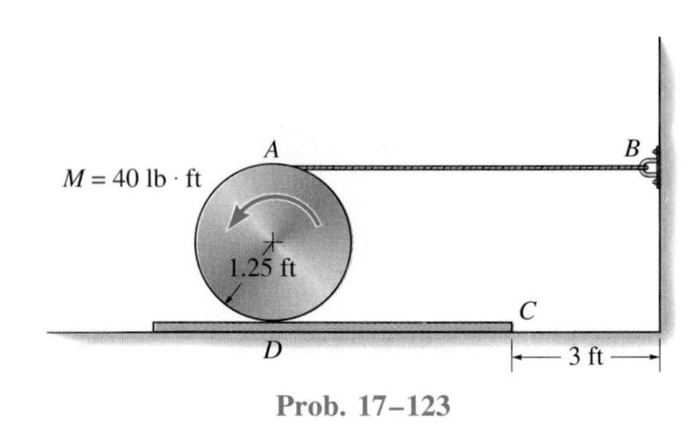

Prob. 17–123

***17–124.** A long strip of paper is wrapped into two rolls each having a mass of 10 kg. The rolls are placed on a *smooth surface.* The core of roll A is subjected to a horizontal force of 60 N. Determine the initial tension in the paper between the rolls and the angular acceleration of each roll. For the calculation, assume the rolls to be approximated by cylinders. Neglect the mass of the paper between the rolls.

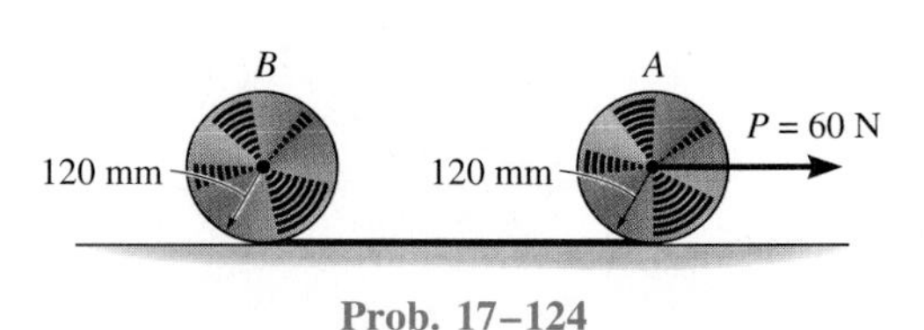

Prob. 17–124

17–125. The 10-lb disk and 4-lb block are released from rest. Determine the velocity of the block when $t = 3$ s. The coefficient of static friction at A is $\mu_A = 0.2$. Neglect the mass of the cord and pulleys.

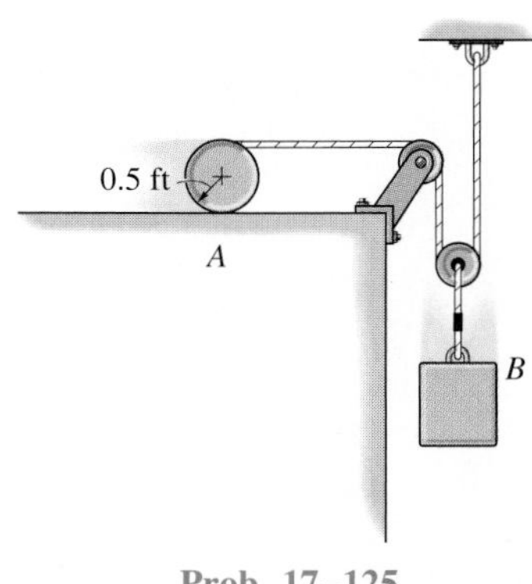

Prob. 17–125

17–126. The 10-lb disk and 4-lb block are released from rest. Determine the velocity of the block when it has descended 1 ft. The coefficient of static friction at A is $\mu_A = 0.2$. Neglect the mass of the cord and pulleys.

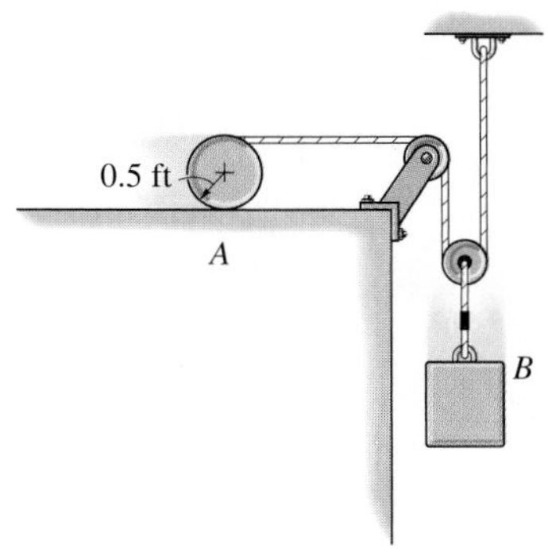

Prob. 17–127

17–127. A concrete pipe (thin ring) has a mass of 500 kg and radius of 0.5 m and rolls without slipping down a 300 kg ramp. If the ramp is free to move, determine its acceleration. Neglect the size of the wheels at A and B.

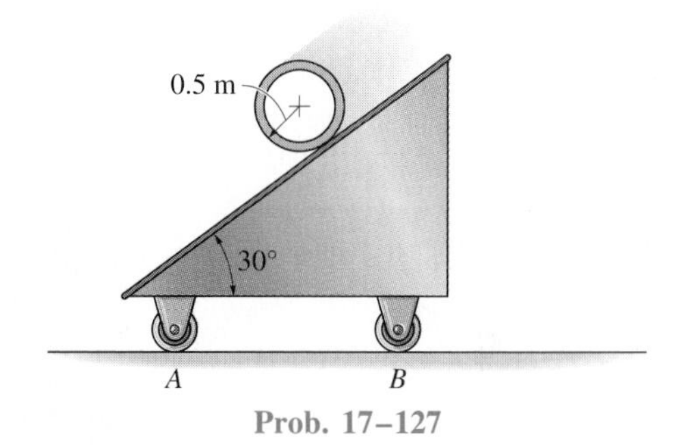

Prob. 17–127

***17–128.** The 25-lb slender rod has a length of 6 ft. Using a collar of negligible mass, its end A is confined to move along the smooth circular bar of radius $3\sqrt{2}$ ft. End B rests on the floor, for which the coefficient of kinetic friction is $\mu_B = 0.4$. If the bar is released from rest when $\theta = 30°$, determine the angular acceleration of the bar at this instant.

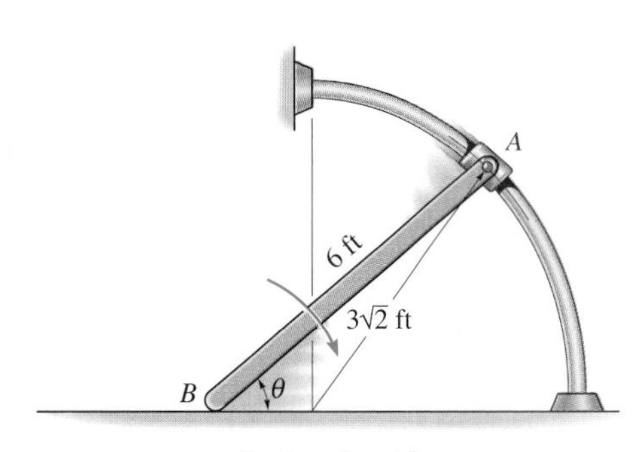

Prob. 17–128

Work and energy principles play an important role in the motion of the draw works used to lift pipe on this drilling rig. The fundamental concepts involved are explained in this chapter.

18

Planar Kinetics of a Rigid Body: Work and Energy

It was shown in Chapter 14 that problems which involve force, velocity, and displacement can conveniently be solved by using the principle of work and energy, or if the force system is "conservative," the conservation of energy theorem. In this chapter we will apply work and energy methods to solve problems involving the planar motion of a rigid body. The more general discussion as applied to the three-dimensional motion of a rigid body is presented in Chapter 21.

Before discussing the principle of work and energy for a body, however, the methods for obtaining the body's kinetic energy when it is subjected to translation, rotation about a fixed axis, or general plane motion will be developed.

18.1 Kinetic Energy

Consider the rigid body shown in Fig. 18–1, which is represented here by a *slab* moving in the inertial x–y reference plane. An arbitrary ith particle of the body, having a mass dm, is located at r from the arbitrary point P. If at the *instant* shown the particle has a velocity $\mathbf{v}_i$, then the particle's kinetic energy is $T_i = \frac{1}{2}dm\, v_i^2$. The kinetic energy of the entire body is determined by writing similar expressions for each particle of the body and integrating the results,

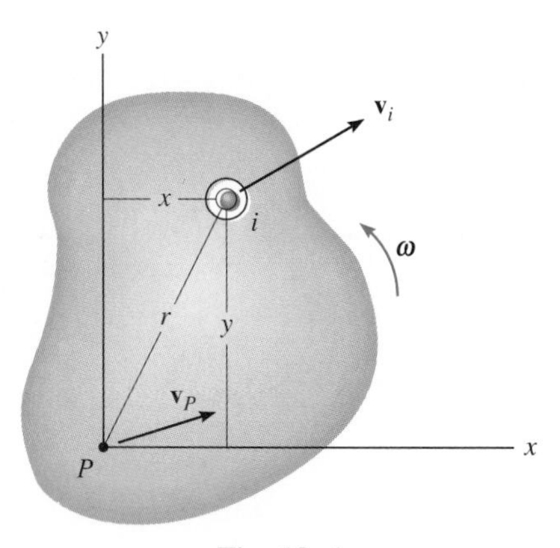

Fig. 18–1

i.e.,

$$T = \frac{1}{2}\int_m dm\, v_i^2$$

This equation may be written in another manner by expressing $\mathbf{v}_i$ in terms of the velocity of point P, i.e., $\mathbf{v}_P$. If the body has an angular velocity $\boldsymbol{\omega}$, then from Fig. 18–1 we have

$$\begin{aligned}\mathbf{v}_i &= \mathbf{v}_P + \mathbf{v}_{i/P} \\ &= (v_P)_x\mathbf{i} + (v_P)_y\mathbf{j} + \omega\mathbf{k} \times (x\mathbf{i} + y\mathbf{j}) \\ &= [(v_P)_x - \omega y]\mathbf{i} + [(v_P)_y + \omega x]\mathbf{j}\end{aligned}$$

The square of the magnitude of $\mathbf{v}_i$ is thus

$$\begin{aligned}\mathbf{v}_i \cdot \mathbf{v}_i = v_i^2 &= [(v_P)_x - \omega y]^2 + [(v_P)_y + \omega x]^2 \\ &= (v_P^2)_x - 2(v_P)_x\omega y + \omega^2 y^2 + (v_P^2)_y + 2(v_P)_y\omega x + \omega^2 x^2 \\ &= v_P^2 - 2(v_P)_x\omega y + 2(v_P)_y\omega x + \omega^2 r^2\end{aligned}$$

Substituting into the equation of kinetic energy yields

$$T = \tfrac{1}{2}\left(\int_m dm\right)v_P^2 - (v_P)_x\omega\left(\int_m y\, dm\right) + (v_P)_y\omega\left(\int_m x\, dm\right) + \tfrac{1}{2}\omega^2\left(\int_m r^2\, dm\right)$$

The first integral on the right represents the entire mass m of the body. Since $\bar{x}m = \int x\, dm$ and $\bar{y}m = \int y\, dm$, the second and third integrals locate the body's center of mass G with respect to P. The last integral represents the body's moment of inertia I_P, computed about the z axis passing through point P. Thus,

$$T = \tfrac{1}{2}mv_P^2 - (v_P)_x\, \omega\bar{y}\, m + (v_P)_y\, \omega\bar{x}\, m + \tfrac{1}{2}I_P\, \omega^2 \qquad (18\text{–}1)$$

This equation reduces to a simpler form if point P coincides with the mass center G for the body, in which case $\bar{x} = \bar{y} = 0$, and therefore

$$T = \tfrac{1}{2}mv_G^2 + \tfrac{1}{2}I_G\omega^2 \qquad (18\text{–}2)$$

Here I_G is the moment of inertia of the body about an axis which is perpendicular to the plane of motion and passes through the mass center. Both terms on the right side are always *positive,* since the velocities are squared. Furthermore, it may be verified that these terms have units of length times force, common units being m · N or ft · lb. Recall, however, that in the SI system the unit of energy is the joule (J), where 1 J = 1 m · N.

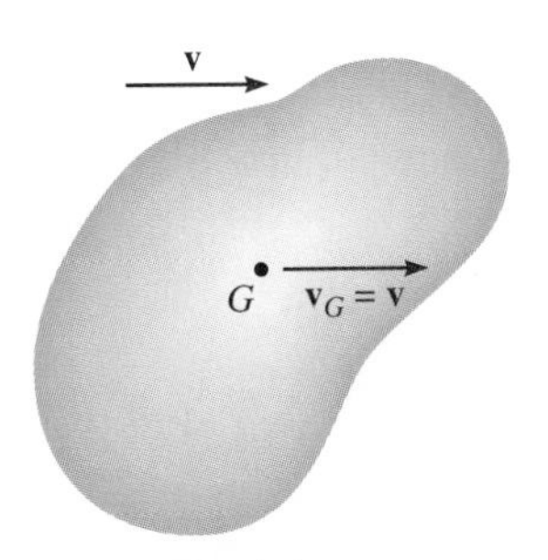

Translation
Fig. 18–2

Translation. When a rigid body of mass m is subjected to either rectilinear or curvilinear *translation,* the kinetic energy due to rotation is zero, since $\boldsymbol{\omega} = \mathbf{0}$. From Eq. 18–2, the kinetic energy of the body is therefore

$$T = \tfrac{1}{2}mv_G^2 \tag{18–3}$$

where v_G is the magnitude of the translational velocity at the instant considered, Fig. 18–2.

Rotation About a Fixed Axis. When a rigid body is *rotating about a fixed axis* passing through point O, Fig. 18–3, the body has both *translational* and *rotational* kinetic energy as defined by Eq. 18–2, i.e.,

$$T = \tfrac{1}{2}mv_G^2 + \tfrac{1}{2}I_G\omega^2 \tag{18–4}$$

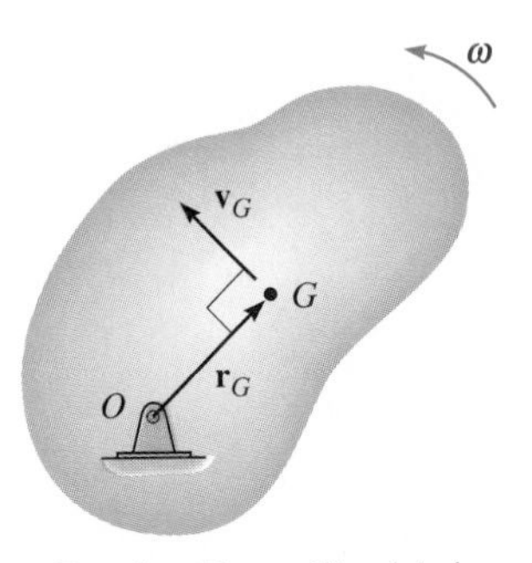

Rotation About a Fixed Axis
Fig. 18–3

The body's kinetic energy may be formulated in another manner by noting that $v_G = r_G\omega$, in which case $T = \tfrac{1}{2}(I_G + mr_G^2)\omega^2$. By the parallel-axis theorem, the terms inside the parentheses represent the moment of inertia I_O of the body about an axis perpendicular to the plane of motion and passing through point O. Hence,*

$$T = \tfrac{1}{2}I_O\omega^2 \tag{18–5}$$

From the derivation, this equation may be substituted for Eq. 18–4, since it accounts for *both* the translational kinetic energy of the body's mass center and the rotational kinetic energy of the body computed about the mass center.

General Plane Motion. When a rigid body is subjected to general plane motion, Fig. 18–4, it has an angular velocity $\boldsymbol{\omega}$ and its mass center has a velocity $\mathbf{v}_G$. Hence, the kinetic energy is defined by Eq. 18–2, i.e.,

$$T = \tfrac{1}{2}mv_G^2 + \tfrac{1}{2}I_G\omega^2 \tag{18–6}$$

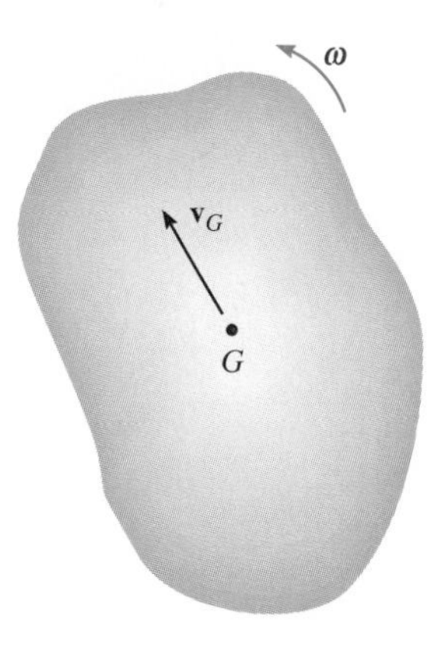

General plane motion
Fig. 18–4

Here it is seen that the total kinetic energy of the body consists of the *scalar* sum of the body's *translational* kinetic energy, $\tfrac{1}{2}mv_G^2$, and *rotational* kinetic energy about its mass center, $\tfrac{1}{2}I_G\omega^2$.

Because energy is a scalar quantity, the total kinetic energy for a system of *connected* rigid bodies is the sum of the kinetic energies of all its moving parts. Depending on the type of motion, the kinetic energy of *each body* is found by applying Eq. 18–2 or the alternative forms mentioned above.

*The similarity between this derivation and that of $\Sigma M_O = I_O\alpha$, Eq. 17–16, should be noted. Also note that the same result can be obtained from Eq. 18–1 by selecting point P at O, realizing that $\mathbf{v}_O = \mathbf{0}$.

Example 18–1

The system of three elements shown in Fig. 18–5a consists of a 6-kg block B, a 10-kg disk D, and a 12-kg cylinder C. A continuous cord of negligible mass is wrapped around the cylinder, passes over the disk, and is then attached to the block. If the block is moving downward with a speed of 0.8 m/s and the cylinder rolls without slipping, determine the total kinetic energy of the system at this instant.

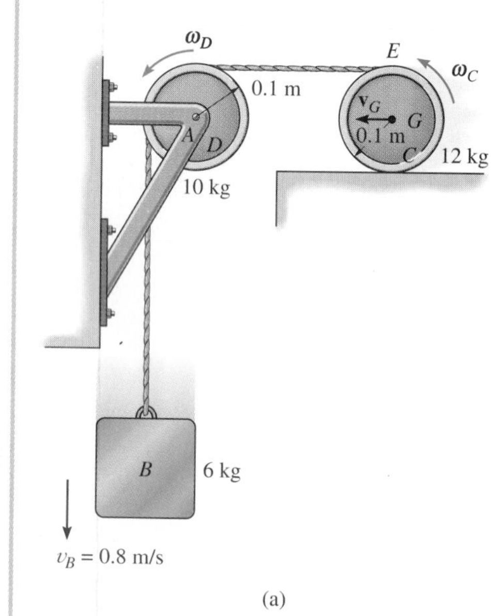

(a)

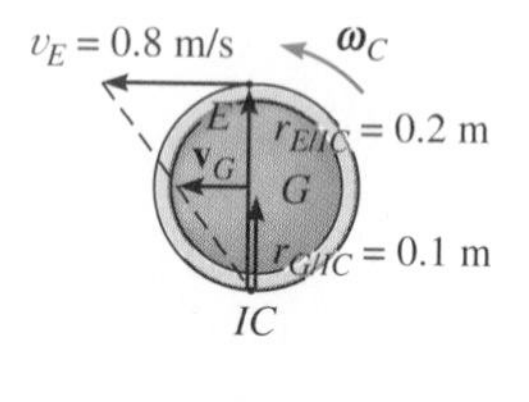

(b)

Fig. 18–5

SOLUTION

By inspection, the block is translating, the disk rotates about a fixed axis, and the cylinder has general plane motion. Hence, in order to compute the kinetic energy of the disk and cylinder, it is first necessary to determine ω_D, ω_C, and v_G, Fig. 18–5a. From the *kinematics* of the disk,

$$v_B = r_D\omega_D; \qquad 0.8 \text{ m/s} = (0.1 \text{ m})\omega_D \qquad \omega_D = 8 \text{ rad/s}$$

Since the cylinder rolls without slipping, the instantaneous center of zero velocity is at the point of contact with the ground, Fig. 18–5b, hence,

$$v_E = r_{E/IC}\omega_C; \qquad 0.8 \text{ m/s} = (0.2 \text{ m})\omega_C \qquad \omega_C = 4 \text{ rad/s}$$

$$v_G = r_{G/IC}\omega_C; \qquad v_G = (0.1 \text{ m})(4 \text{ rad/s}) = 0.4 \text{ m/s}$$

Block

$$T_B = \tfrac{1}{2}m_B v_B^2 = \tfrac{1}{2}(6 \text{ kg})(0.8 \text{ m/s})^2 = 1.92 \text{ J}$$

Disk

$$T_D = \tfrac{1}{2}I_D\omega_D^2 = \tfrac{1}{2}(\tfrac{1}{2}m_D r_D^2)\omega_D^2$$
$$= \tfrac{1}{2}[\tfrac{1}{2}(10 \text{ kg})(0.1 \text{ m})^2](8 \text{ rad/s})^2 = 1.60 \text{ J}$$

Cylinder

$$T_C = \tfrac{1}{2}mv_G^2 + \tfrac{1}{2}I_G\omega_C^2 = \tfrac{1}{2}mv_G^2 + \tfrac{1}{2}(\tfrac{1}{2}m_C r_C^2)\omega_C^2$$
$$= \tfrac{1}{2}(12 \text{ kg})(0.4 \text{ m/s})^2 + \tfrac{1}{2}[\tfrac{1}{2}(12 \text{ kg})(0.1 \text{ m})^2](4 \text{ rad/s})^2 = 1.44 \text{ J}$$

The total kinetic energy of the system is therefore

$$T = T_B + T_D + T_C$$
$$= 1.92 \text{ J} + 1.60 \text{ J} + 1.44 \text{ J} = 4.96 \text{ J} \qquad \textit{Ans.}$$

18.2 The Work of a Force

Several types of forces are often encountered in planar kinetics problems involving a rigid body. The work of each of these forces has been presented in Sec. 14.1 and is listed below as a summary.

Work of a Variable Force. If an external force **F** acts on a rigid body the work done by the force when it moves along the path s, Fig. 18–6, is defined as

$$U_F = \int_s F \cos \theta \, ds \qquad (18\text{–}7)$$

Here θ is the angle between the "tails" of the force vector and the differential displacement. In general, the integration must account for the variation of the force's direction and magnitude.

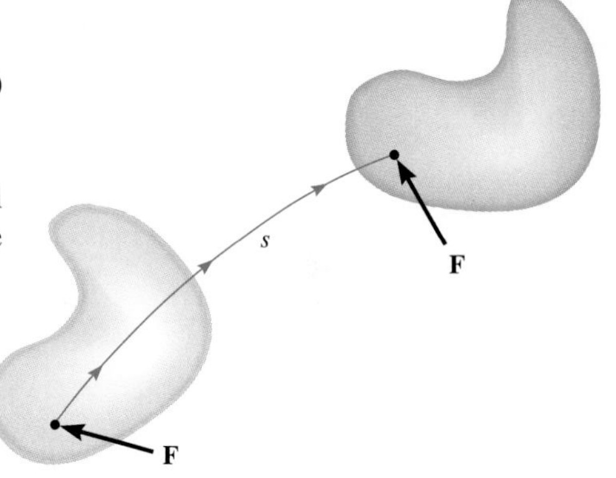

Fig. 18–6

Work of a Constant Force. If an external force $\mathbf{F}_c$ acts on a rigid body, Fig. 18–7, and maintains a constant magnitude F_c and constant direction θ, while the body undergoes a translation s, Eq. 18–7 can be integrated so that the work becomes

$$U_{F_c} = (F_c \cos \theta)s \qquad (18\text{–}8)$$

Here $F_c \cos \theta$ represents the magnitude of the component of force in the direction of displacement.

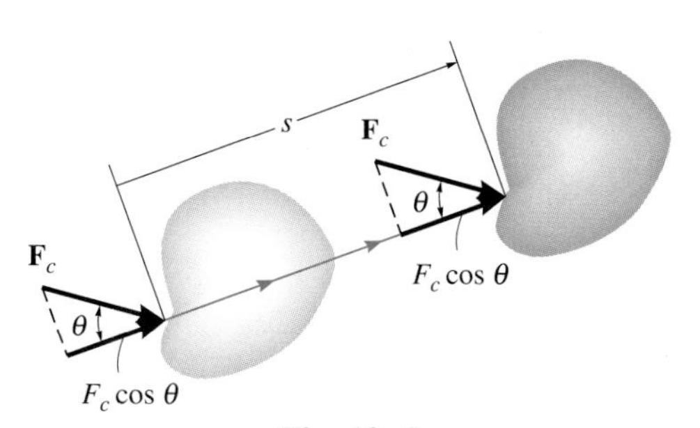

Fig. 18–7

Work of a Weight. The weight of a body does work only when the body's center of mass G undergoes a *vertical displacement* Δy. If this displacement is *upward,* Fig. 18–8, the work is negative, since the weight and displacement are in opposite directions.

$$U_W = -W\,\Delta y \tag{18–9}$$

Likewise, if the displacement is *downward* $(-\Delta y)$ the work becomes *positive.* Here the elevation change is considered to be small so that **W,** which is caused by gravitation, is constant.

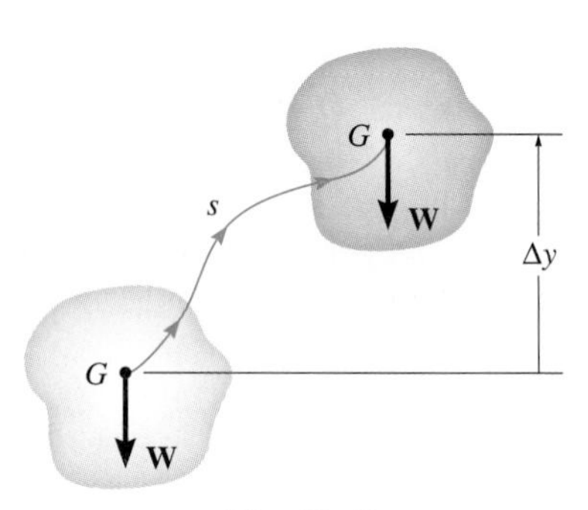

Fig. 18–8

Work of a Spring Force. If a linear elastic spring is attached to a body, the spring force $F_s = ks$ *acting on the body* does work when the spring either stretches or compresses from s_1 to a *further* position s_2. In both cases the work will be *negative* since the *displacement of the body* is always in the opposite direction to the force, Fig. 18–9. The work done is

$$U_s = -\left(\tfrac{1}{2}ks_2^2 - \tfrac{1}{2}ks_1^2\right) \tag{18–10}$$

where $|s_2| > |s_1|$.

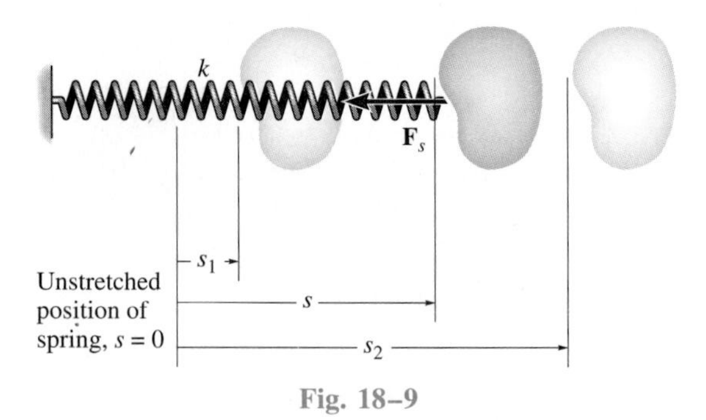

Fig. 18–9

Forces That Do No Work. There are some external forces that do no work when the body is displaced. These forces can act either at *fixed points* on the body or they can have a direction *perpendicular to their displacement.* Examples include the reactions at a pin support about which a body rotates, the normal reaction acting on a body that moves along a fixed surface, and the weight of a body when the center of gravity of the body moves in a *horizontal plane,* Fig. 18–10. A rolling resistance force $\mathbf{F}_r$ acting on a body as it *rolls without slipping* over a rough surface also does no work, Fig. 18–10.* This is because, during any *instant of time dt,* $\mathbf{F}_r$ acts at a point on the body which has *zero velocity* (instantaneous center, *IC*), and so the work done by the force on the point is zero. In other words, the point is not displaced in the direction of the force during this instant. Refer to Fig. 16–27*b*, where it can be seen that at the instant point *A* is at the origin of the *x, y, z* coordinate system, its displacement will be vertical, not horizontal. Since $\mathbf{F}_r$ contacts each successive particle for only an instant, the work of $\mathbf{F}_r$ will be zero.

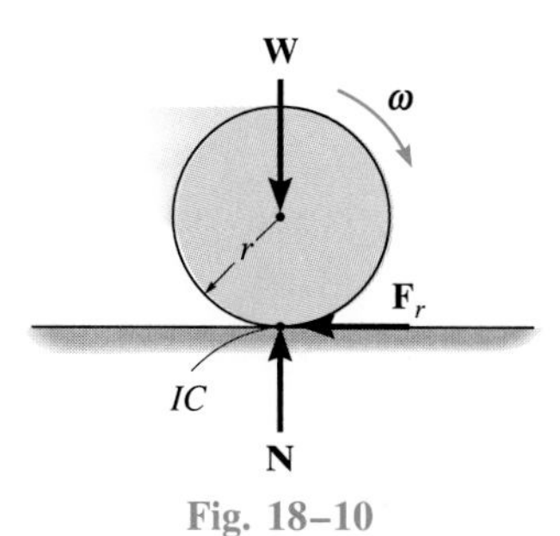

Fig. 18–10

18.3 The Work of a Couple

Recall that a *couple* consists of a pair of noncollinear forces which have equal magnitudes and opposite directions. When a body subjected to a couple undergoes general plane motion, the two forces do work *only* when the body undergoes a *rotation.* To show this, consider the body in Fig. 18–11*a*, which is subjected to a couple having a magnitude of $M = Fr$. Any general differential displacement of the body can be considered as a separate translation and rotation. When the body *translates* such that the *component of displacement* along the line of action of the forces is ds_t, Fig. 18–11*b*, clearly the "positive" work of one force *cancels* the "negative" work of the other. Consider now a differential rotation $d\theta$ of the body about an axis which is perpendicular to the plane of the couple and intersects the plane at point *O*, Fig. 18–11*c*. (For the derivation any other point in the plane may also be considered.) As shown, each force undergoes a displacement $ds_\theta = (r/2)\,d\theta$ in the direction of the force; hence, the total work done is

$$dU_M = F\left(\frac{r}{2}d\theta\right) + F\left(\frac{r}{2}d\theta\right) = (Fr)\,d\theta$$
$$= M\,d\theta$$

Here the line of action of $d\theta$ is parallel to the line of action of *M*. This is *always the case for general plane motion,* since $\mathbf{M}$ and $d\boldsymbol{\theta}$ are perpendicular to the plane of motion. Furthermore, the resultant work is *positive* when $\mathbf{M}$ and $d\boldsymbol{\theta}$ have the *same sense of direction* and *negative* if these vectors have an *opposite sense of direction.*

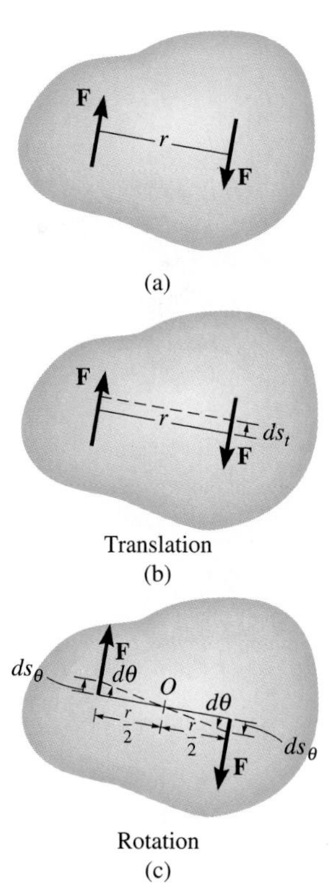

(a)

Translation
(b)

Rotation
(c)

Fig. 18–11

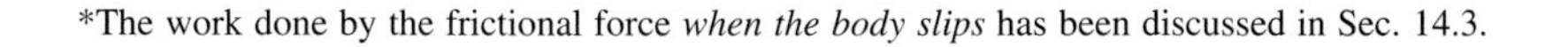
*The work done by the frictional force *when the body slips* has been discussed in Sec. 14.3.

When the body rotates in the plane through a finite angle θ measured in radians, from θ_1 to θ_2, the work of a couple is

$$U_M = \int_{\theta_1}^{\theta_2} M\,d\theta \tag{18–11}$$

If the couple moment **M** has a *constant magnitude,* then

$$U_M = M(\theta_2 - \theta_1) \tag{18–12}$$

Here the work is *positive* provided **M** and $(\boldsymbol{\theta}_2 - \boldsymbol{\theta}_1)$ are in the same direction.

18.4 Principle of Work and Energy

In Sec. 14.2 the principle of work and energy was developed for a particle. By applying this principle to each of the particles of a rigid body and adding the results algebraically, since energy is a scalar, the principle of work and energy for a rigid body may be developed. In this regard, the body's initial and final kinetic energies have been defined by the formulations in Sec. 18.1, and the work done by the *external* forces and couple moments is defined by formulations in Secs. 18.2 and 18.3. Note that the work of the body's *internal forces* does not have to be considered since the body is rigid. These forces occur in equal but opposite collinear pairs, so that when the body moves, the work of one force cancels that of its counterpart. Furthermore, since the body is rigid, *no relative movement* between these forces occurs, so that no internal work is done. Thus the principle of work and energy for a rigid body may be written as

$$T_1 + \Sigma U_{1-2} = T_2 \tag{18–13}$$

This equation states that the body's initial translational *and* rotational kinetic energy plus the work done by all the external forces and couple moments acting on the body as the body moves from its initial to its final position is equal to the body's final translational *and* rotational kinetic energy.

When several rigid bodies are pin-connected, connected by inextensible cables, or in mesh with one another, this equation may be applied to the entire system of connected bodies. In all these cases the internal forces, which hold the various members together, do no work and hence are eliminated from the analysis.

PROCEDURE FOR ANALYSIS

The principle of work and energy is used to solve kinetics problems that involve *velocity, force,* and *displacement,* since these terms are involved in the formulation. For application, it is suggested that the following procedure be used.

Kinetic Energy (Kinematic Diagrams). Determine the kinetic-energy terms T_1 and T_2 by applying the equation $T = \frac{1}{2}mv_G^2 + \frac{1}{2}I_G\omega^2$ or an appropriate form of this equation developed in Sec. 18.1. In this regard, *kinematic diagrams* for velocity may be useful for determining v_G and ω, or for establishing a *relationship* between v_G and ω.*

Work (Free-Body Diagram). Draw a free-body diagram of the body when it is located at an intermediate point along the path, in order to account for all the forces and couple moments which do work on the body as it moves along the path. The work of each force and couple moment can be computed using the appropriate formulations outlined in Secs. 18.2 and 18.3. Since *algebraic addition* of the work terms is required, it is important that the proper sign of each term be specified. Specifically, work is *positive* when the force (couple moment) is in the *same direction* as its displacement (rotation); otherwise, it is negative.

Principle of Work and Energy. Apply the principle of work and energy, $T_1 + \Sigma U_{1-2} = T_2$. Since this is a scalar equation, it can be used to solve for only one unknown when it is applied to a single rigid body. This is in contrast to the three scalar equations of motion, Eqs. 17–11, which may be written for the same body.

*A brief review of Secs. 16.5 to 16.7 may prove helpful in solving problems, since computations for kinetic energy require a kinematic analysis of velocity.

Example 18–2

The bar shown in Fig. 18–12*a* has a mass of 10 kg and is subjected to a couple moment of 50 N · m and a force of $P = 80$ N, which is always applied perpendicular to the end of the bar. Also, the spring has an unstretched length of 0.5 m and remains in the vertical position due to the roller guide at *B*. Determine the total work done by all the forces acting on the bar when it has rotated downward from $\theta = 0°$ to $\theta = 90°$.

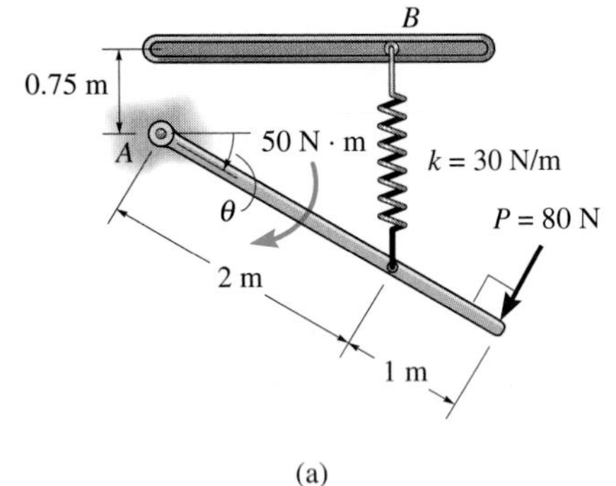

(a)

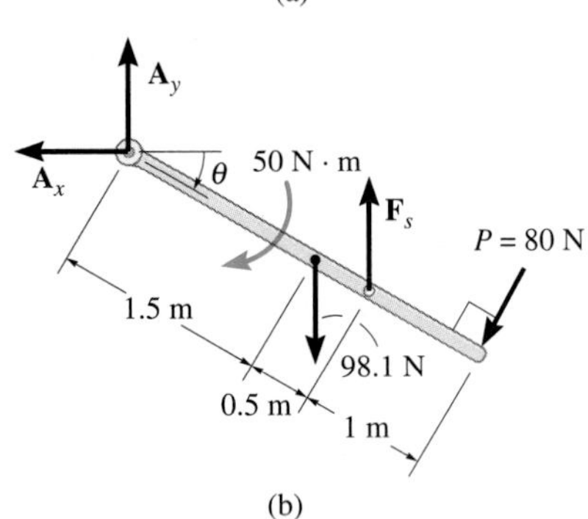

(b)

Fig. 18–12

SOLUTION

First the free-body diagram of the bar is drawn in order to account for all the forces that act on it, Fig. 18–12*b*.

Weight W. Since the weight 10(9.81) N = 98.1 N is displaced downward 1.5 m, the work is

$$U_W = 98.1 \text{ N}(1.5 \text{ m}) = 147.2 \text{ J}$$

Why is the work positive?

Couple Moment M. The couple moment rotates through an angle of $\theta = \pi/2$ rad. Hence

$$U_M = 50 \text{ N} \cdot \text{m}(\pi/2) = 78.5 \text{ J}$$

Spring Force F_s. When $\theta = 0°$ the spring is stretched (0.75 m − 0.5 m) = 0.25 m, and when $\theta = 90°$, the stretch is (2 m + 0.75 m) − 0.5 m = 2.25 m. Thus

$$U_s = -[\tfrac{1}{2}(30 \text{ N/m})(2.25 \text{ m})^2 - \tfrac{1}{2}(30 \text{ N/m})(0.25 \text{ m})^2] = -75.0 \text{ J}$$

By inspection the spring does negative work on the bar since F_s acts in the opposite direction to displacement.

Force P. As the bar moves downward, the force is displaced through a distance of $\pi/2(3 \text{ m}) = 4.712$ m. The work is positive. Why?

$$U_P = 80 \text{ N}(4.712 \text{ m}) = 377.0 \text{ J}$$

Pin Reactions. Forces $\mathbf{A}_x$ and $\mathbf{A}_y$ do no work since they are not displaced.

Total Work. The work of all the forces when the bar is displaced is thus

$$U = 147.2 + 78.5 - 75.0 + 377.0 = 528 \text{ J} \qquad \textit{Ans.}$$

Example 18–3

The 30-kg disk shown in Fig. 18–13a is pin-supported at its center. Determine the number of revolutions it must make to attain an angular velocity of 20 rad/s starting from rest. It is acted upon by a constant force $F = 10$ N, which is applied to a cord wrapped around its periphery, and a constant couple moment $M = 5$ N · m. Neglect the mass of the cord in the calculation.

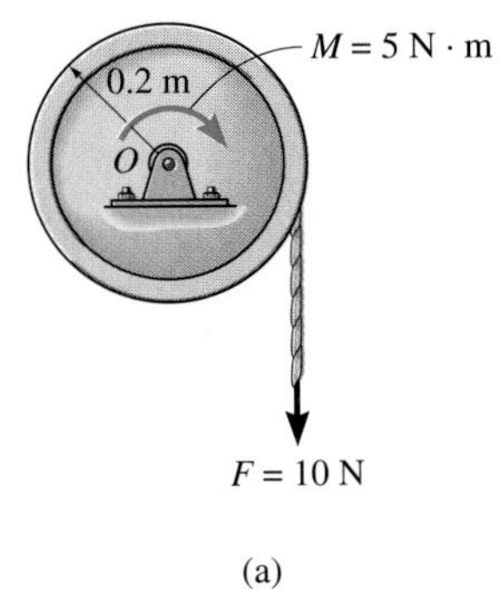

(a)

SOLUTION

Kinetic Energy. Since the disk rotates about a fixed axis, the kinetic energy can be computed using $T = \frac{1}{2} I_O \omega^2$, where the moment of inertia is $I_O = \frac{1}{2} m r^2$. Initially, the disk is at rest, so that

$$T_1 = 0$$
$$T_2 = \tfrac{1}{2} I_O \omega_2^2 = \tfrac{1}{2}[\tfrac{1}{2}(30 \text{ kg})(0.2 \text{ m})^2](20 \text{ rad/s})^2 = 120 \text{ J}$$

Work (Free-Body Diagram). As shown in Fig. 18–13b, the pin reactions $\mathbf{O}_x$ and $\mathbf{O}_y$ and the weight (294.3 N) do no work, since they are not displaced. The *couple moment,* having a constant magnitude, does positive work $U_M = M\theta$ as the disk *rotates* through a clockwise angle of θ rad, and the *constant force* $\mathbf{F}$ does positive work $U_{F_c} = Fs$ as the cord *moves* downward $s = \theta r = \theta(0.2 \text{ m})$.

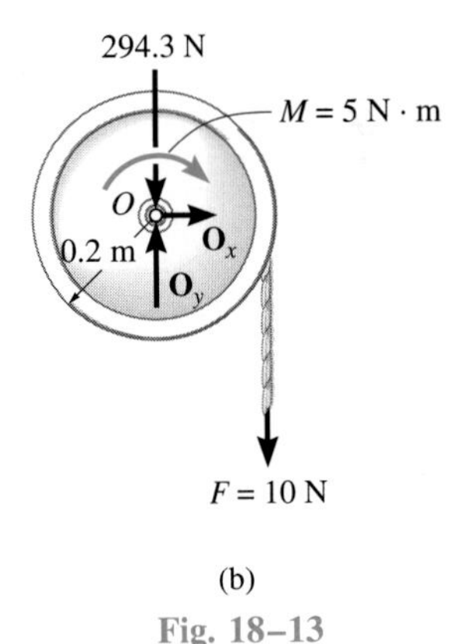

(b)

Fig. 18–13

Principle of Work and Energy

$$\{T_1\} + \{\Sigma U_{1-2}\} = \{T_2\}$$
$$\{T_1\} + \{M\theta + Fs\} = \{T_2\}$$
$$\{0\} + \{(5 \text{ N} \cdot \text{m})\theta + (10 \text{ N})\theta(0.2 \text{ m})\} = \{120 \text{ J}\}$$
$$\theta = 17.1 \text{ rad} = 17.1 \text{ rad}\left(\frac{1 \text{ rev}}{2\pi \text{ rad}}\right) = 2.73 \text{ rev} \qquad \textit{Ans.}$$

This problem has also been solved in Example 17–9. Compare the two methods of solution and note that since force, velocity, and displacement θ are involved, a work–energy approach yields a more direct solution.

Example 18–4

The uniform 5-kg bar shown in Fig. 18–14*a* is acted upon by the 30-N force which always acts perpendicular to the bar as shown. If the bar has an initial clockwise angular velocity $\omega_1 = 10$ rad/s when $\theta = 0°$, determine its angular velocity at the instant $\theta = 90°$.

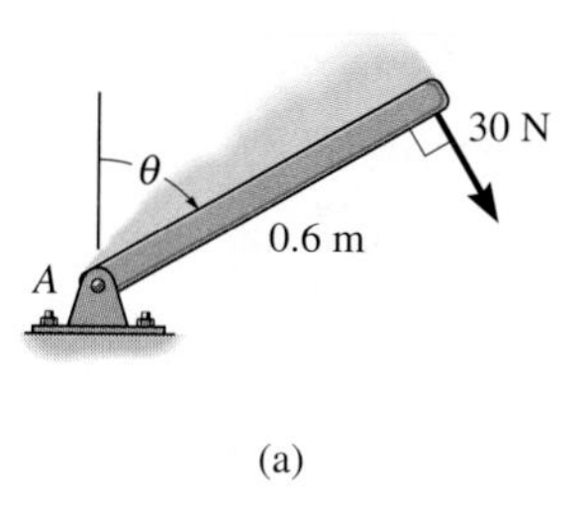

(a)

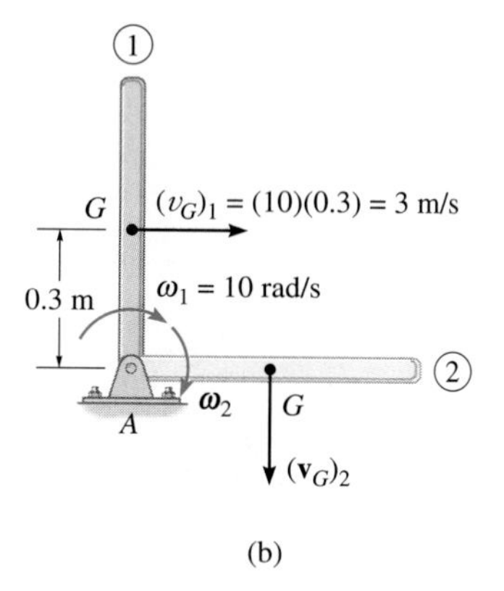

(b)

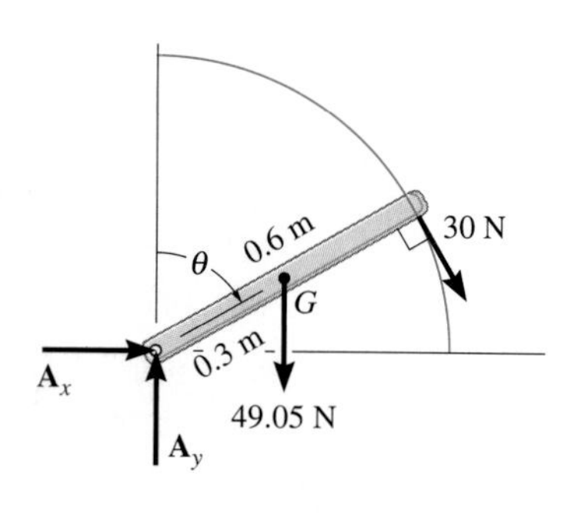

(c)

Fig. 18–14

SOLUTION

Kinetic Energy (Kinematic Diagrams). Two kinematic diagrams of the bar when $\theta = 0°$ (position 1) and $\theta = 90°$ (position 2) are shown in Fig. 18–14*b*. The initial kinetic energy may be computed with reference to either the fixed point of rotation *A* or the center of mass *G*. If *A* is considered, then

$$T_1 = \tfrac{1}{2} I_A \omega_1^2 = \tfrac{1}{2}[\tfrac{1}{3}(5\text{ kg})(0.6\text{ m})^2](10\text{ rad/s})^2 = 30\text{ J}$$

If point *G* is considered, then

$$T_1 = \tfrac{1}{2}m(v_G)_1^2 + \tfrac{1}{2} I_G \omega_1^2$$
$$= \tfrac{1}{2}(5\text{ kg})(3\text{ m/s})^2 + \tfrac{1}{2}[\tfrac{1}{12}(5\text{ kg})(0.6\text{ m})^2](10\text{ rad/s})^2 = 30\text{ J}$$

In the final position,

$$T_2 = \tfrac{1}{2}[\tfrac{1}{3}(5\text{ kg})(0.6\text{ m})^2]\omega_2^2 = 0.3\omega_2^2$$

Work (Free-Body Diagram). Fig. 18–14*c*. The reactions $\mathbf{A}_x$ and $\mathbf{A}_y$ do no work, since these forces do not move. The 49.05-N weight, centered at *G*, moves downward through a vertical distance $\Delta y = 0.3$ m. The 30-N force moves tangent to its path of length $\tfrac{1}{2}\pi(0.6)$ m.

Principle of Work and Energy

$$\{T_1\} + \{\Sigma U_{1-2}\} = \{T_2\}$$
$$\{30\text{ J}\} + \{49.05\text{ N}(0.3\text{ m}) + 30\text{ N}[\tfrac{1}{2}\pi(0.6\text{ m})]\} = \{0.3\omega_2^2\text{ J}\}$$
$$\omega_2 = 15.6\text{ rad/s} \curvearrowright$$ ***Ans.***

Example 18–5

The wheel shown in Fig. 18–15a weighs 40 lb and has a radius of gyration $k_G = 0.6$ ft about its mass center G. If it is subjected to a clockwise couple moment of 15 lb · ft and rolls from rest without slipping, determine its angular velocity after its center G moves 0.5 ft. The spring has a stiffness $k = 10$ lb/ft and is initially unstretched when the couple moment is applied.

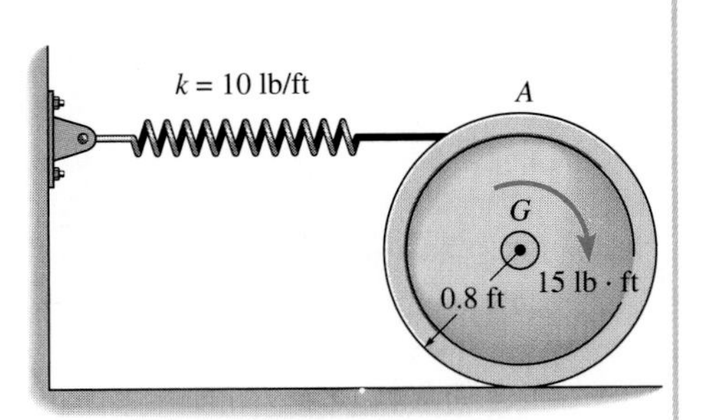

(a)

SOLUTION

Kinetic Energy (Kinematic Diagram). Since the wheel is initially at rest,

$$T_1 = 0$$

The kinematic diagram of the wheel when it is in the final position is shown in Fig. 18–15b. The velocity of the mass center $(\mathbf{v}_G)_2$ can be related to the angular velocity $\boldsymbol{\omega}_2$ from the instantaneous center of zero velocity (IC), i.e., $(v_G)_2 = 0.8\omega_2$. Hence, the final kinetic energy is

$$\begin{aligned} T_2 &= \tfrac{1}{2}m(v_G)_2^2 + \tfrac{1}{2}I_G(\omega_2)^2 \\ &= \frac{1}{2}\left(\frac{40 \text{ lb}}{32.2 \text{ ft/s}^2}\right)((0.8 \text{ ft})\,\omega_2)^2 + \frac{1}{2}\left[\frac{40 \text{ lb}}{32.2 \text{ ft/s}^2}(0.6 \text{ ft})^2\right](\omega_2)^2 \\ &= 0.621(\omega_2)^2 \end{aligned}$$

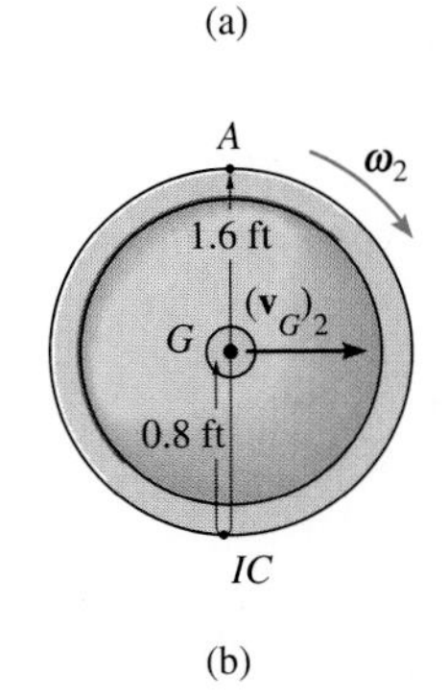

(b)

Work (Free-Body Diagram). As shown in Fig. 18–15c, only the spring force $\mathbf{F}_s$ and the couple moment do work. The normal force does not move along its line of action and the frictional force does *no work,* since the wheel does not slip as it rolls.

The work of $\mathbf{F}_s$ may be computed using $U_s = -\frac{1}{2}ks^2$. Here the work is negative since $\mathbf{F}_s$ is in the opposite direction to displacement. Since the wheel does not slip when the center G moves 0.5 ft, then the wheel rotates $\theta = s_G/r_{G/IC} = 0.5 \text{ ft}/0.8 \text{ ft} = 0.625$ rad, Fig. 18–15b. Hence, the spring stretches $s_A = \theta r_{A/IC} = 0.625 \text{ rad}(1.6 \text{ ft}) = 1$ ft.

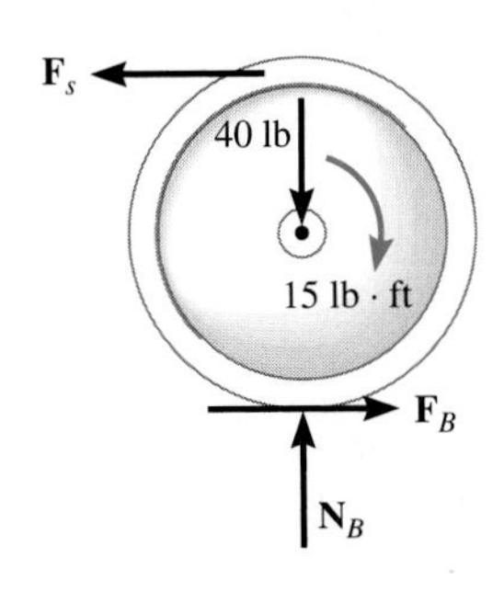

(c)

Fig. 18–15

Principle of Work and Energy

$$\{T_1\} + \{\Sigma U_{1-2}\} = \{T_2\}$$

$$\{T_1\} + \{M\theta - \tfrac{1}{2}ks^2\} = \{T_2\}$$

$$\{0\} + \{15 \text{ lb} \cdot \text{ft}(0.625 \text{ rad}) - \frac{1}{2}(10 \text{ lb/ft})(1 \text{ ft})^2\} = \{0.621(\omega_2)^2 \text{ ft} \cdot \text{lb}\}$$

$$\omega_2 = 2.65 \text{ rad/s} \circlearrowright \qquad \textit{Ans.}$$

Example 18–6

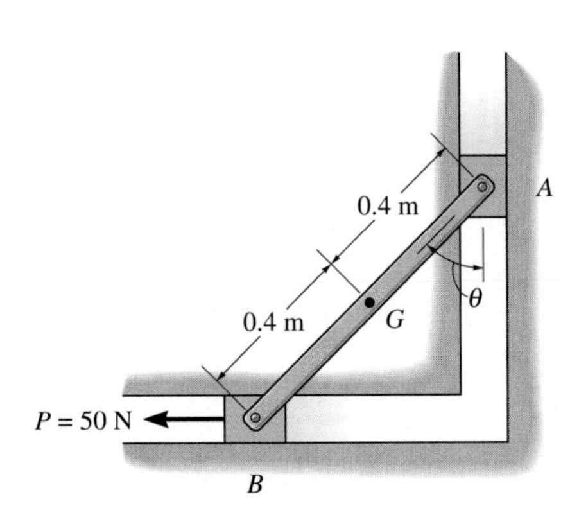

(a)

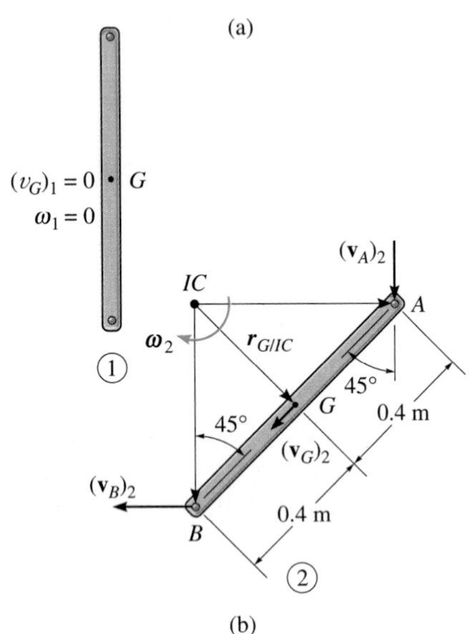

(b)

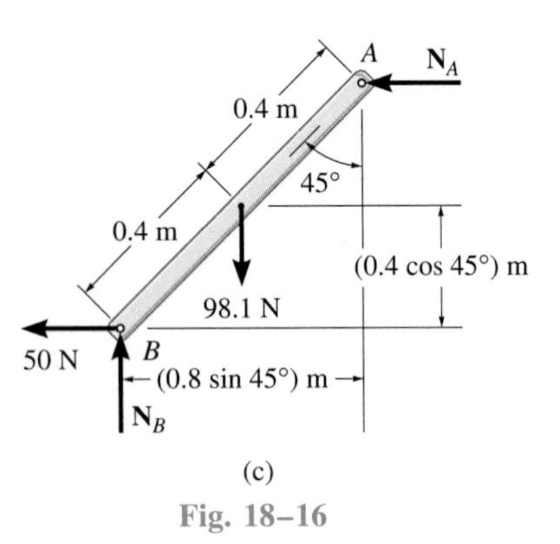

(c)

Fig. 18–16

The 10-kg rod shown in Fig. 18–16*a* is constrained so that its ends move along the grooved slots. The rod is initially at rest when $\theta = 0°$. If the slider block at B is acted upon by a horizontal force $P = 50$ N, determine the angular velocity of the rod at the instant $\theta = 45°$. Neglect the mass of blocks A and B.

SOLUTION

Why can the principle of work and energy be used to solve this problem?

Kinetic Energy (Kinematic Diagrams). Two kinematic diagrams of the rod, when it is in the initial position 1 and final position 2, are shown in Fig. 18–16*b*. When the rod is in position 1, $T_1 = 0$ since $(\mathbf{v}_G)_1 = \boldsymbol{\omega}_1 = \mathbf{0}$. In position 2 the angular velocity is $\boldsymbol{\omega}_2$ and the velocity of the mass center is $(\mathbf{v}_G)_2$. Hence, the kinetic energy is

$$\begin{aligned} T_2 &= \tfrac{1}{2}m(v_G)_2^2 + \tfrac{1}{2}I_G(\omega_2)^2 \\ &= \tfrac{1}{2}(10\text{ kg})(v_G)_2^2 + \tfrac{1}{2}[\tfrac{1}{12}(10\text{ kg})(0.8\text{ m})^2](\omega_2)^2 \\ &= 5(v_G)_2^2 + 0.267(\omega_2)^2 \end{aligned} \tag{1}$$

The two unknowns $(v_G)_2$ and ω_2 may be related from the instantaneous center of zero velocity for the rod, Fig. 18–16*b*. It is seen that as A moves downward with a velocity $(\mathbf{v}_A)_2$, B moves horizontally to the left with a velocity $(\mathbf{v}_B)_2$. Knowing these directions, the IC may be determined as shown in the figure. Hence,

$$\begin{aligned} (v_G)_2 &= r_{G/IC}\omega_2 = (0.4 \tan 45°\text{ m})\omega_2 \\ &= 0.4\omega_2 \end{aligned}$$

Substituting into Eq. 1, we have

$$T_2 = 5(0.4\omega_2)^2 + 0.267(\omega_2)^2 = 1.067(\omega_2)^2$$

Work (Free-Body Diagram). Fig. 18–16*c*. The normal forces $\mathbf{N}_A$ and $\mathbf{N}_B$ do no work as the rod is displaced. Why? The 98.1-N weight is displaced a vertical distance of $\Delta y = (0.4 - 0.4 \cos 45°)$ m; whereas the 50-N force moves a horizontal distance of $s = (0.8 \sin 45°)$ m. Both of these forces do positive work. Why?

Principle of Work and Energy

$$\{T_1\} + \{\Sigma U_{1-2}\} = \{T_2\}$$

$$\{T_1\} + \{W\Delta y + Ps\} = \{T_2\}$$

$$\{0\} + \{98.1\text{ N}(0.4\text{ m} - 0.4 \cos 45°\text{ m}) + 50\text{ N}(0.8 \sin 45°\text{ m})\} = \{1.067(\omega_2)^2\text{ J}\}$$

Solving for ω_2 gives

$$\omega_2 = 6.11\text{ rad/s} \circlearrowright \qquad \textit{Ans.}$$

PROBLEMS

18–1. At a given instant the body of mass m has an angular velocity $\boldsymbol{\omega}$ and its mass center has a velocity $\mathbf{v}_G$. Show that its kinetic energy can be represented as $T = \frac{1}{2}I_{IC}\omega^2$, where I_{IC} is the moment of inertia of the body computed about the instantaneous axis of zero velocity, located a distance $r_{G/IC}$ from the mass center as shown.

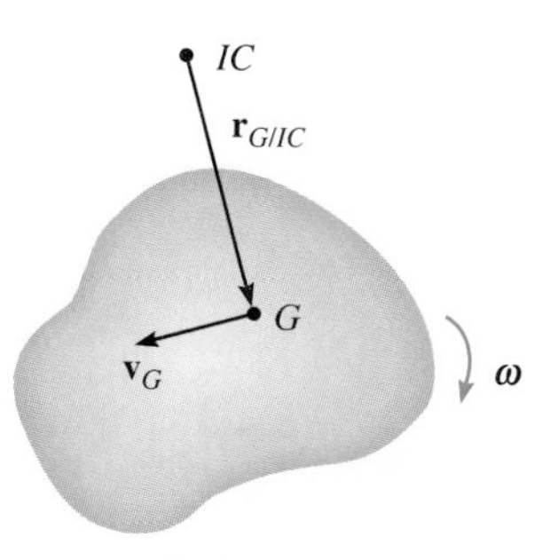

Prob. 18–1

18–2. Solve Prob. 17–69 using the principle of work and energy.

18–3. Solve Prob. 17–73 using the principle of work and energy.

***18–4.** The wheel is made from a 5-kg thin ring and two 2-kg slender rods. If the torsional spring attached to the wheel's center has a stiffness $k = 2\ \text{N}\cdot\text{m/rad}$, so that the torque on the center of the wheel is $M = (2\theta)\ \text{N}\cdot\text{m}$, where θ is in radians, determine the maximum angular velocity of the wheel if it is rotated two revolutions and then released from rest.

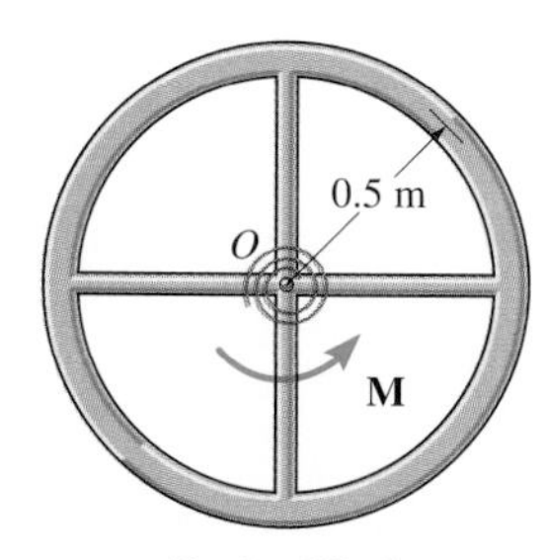

Prob. 18–4

18–5. At the instant shown, the 30-lb disk has a counterclockwise angular velocity of 5 rad/s when its center has a velocity of 20 ft/s. Determine the kinetic energy of the disk at this instant.

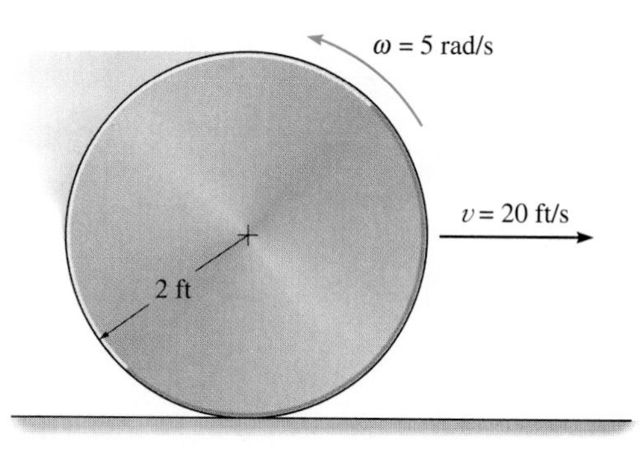

Prob. 18–5

18–6. The uniform slender rod has a mass m and is pinned at its end. If at the instant θ the angular velocity of the rod is $\boldsymbol{\omega}$, show that the kinetic energy is the same when computed with respect to its mass center G or point A.

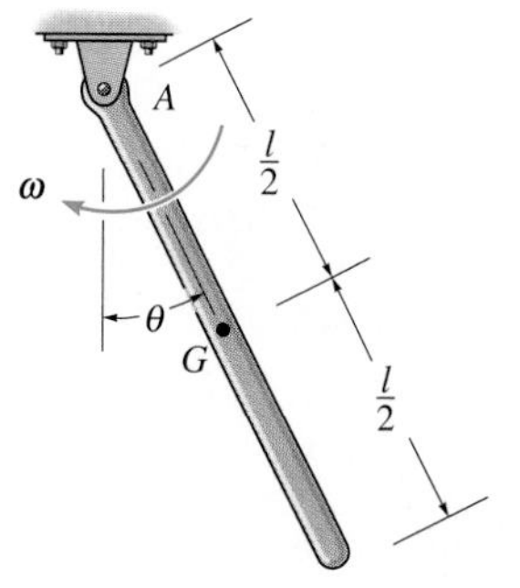

Prob. 18–6

18–7. At the instant shown, link AB has an angular velocity $\omega_{AB} = 2$ rad/s. If each link is considered as a uniform slender bar with a weight of 0.5 lb/in., determine the total kinetic energy of the system.

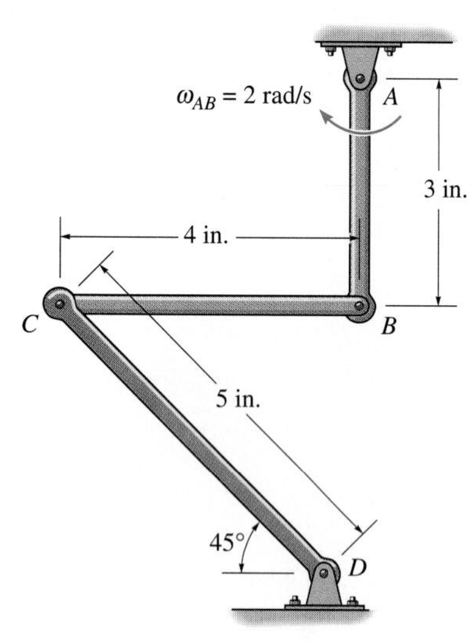

Prob. 18–7

***18–8.** The tub of the mixer has a weight of 70 lb and a radius of gyration $k_G = 1.3$ ft about its center of gravity. If a constant torque $M = 60$ lb · ft is applied to the dumping wheel, determine the angular velocity of the tub when it has rotated $\theta = 90°$. Originally the tub is at rest when $\theta = 0°$.

18–9. Solve Prob. 18–8 if the applied torque is $M = (50\,\theta)$ lb · ft, where θ is in radians.

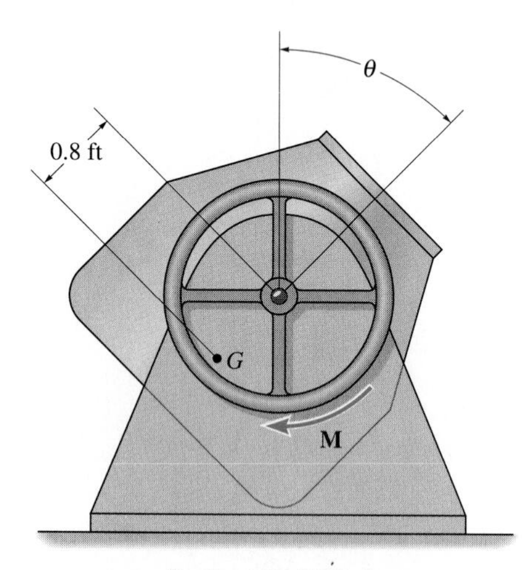

Probs. 18–8/18–9

18–10. A yo-yo has a weight of 0.3 lb and a radius of gyration $k_O = 0.06$ ft. If it is released from rest, determine how far it must descend in order to attain an angular velocity $\omega = 70$ rad/s. Neglect the mass of the string and assume that the string is wound around the central peg such that the mean radius at which it unravels is $r = 0.02$ ft.

18–11. A yo-yo has a weight of 0.3 lb and a radius of gyration $k_O = 0.06$ ft. If it is released from rest, determine its angular velocity when it has descended $s = 3$ ft. Neglect the mass of the string and assume that the string is wound around the central peg such that the mean radius at which it unravels is $r = 0.02$ ft.

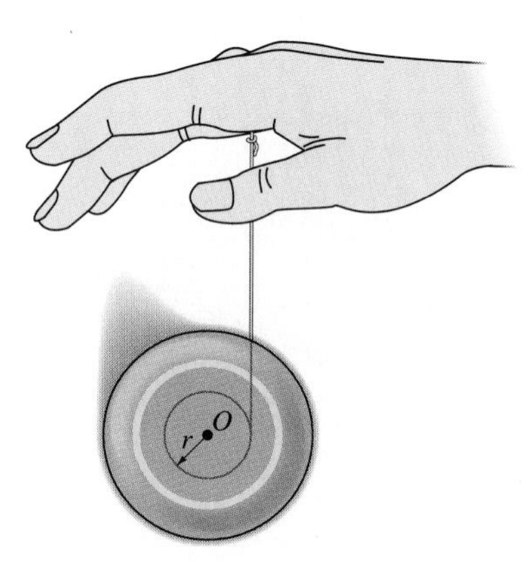

Probs. 18–10/18–11

***18–12.** The soap-box car has a weight of 110 lb, including the passenger but *excluding* its four wheels. Each wheel has a weight of 5 lb, radius of 0.5 ft, and a radius of gyration $k = 0.3$ ft, computed about an axis passing through the wheel's axle. Determine the car's speed after it has traveled 100 ft starting from rest. The wheels roll without slipping. Neglect air resistance.

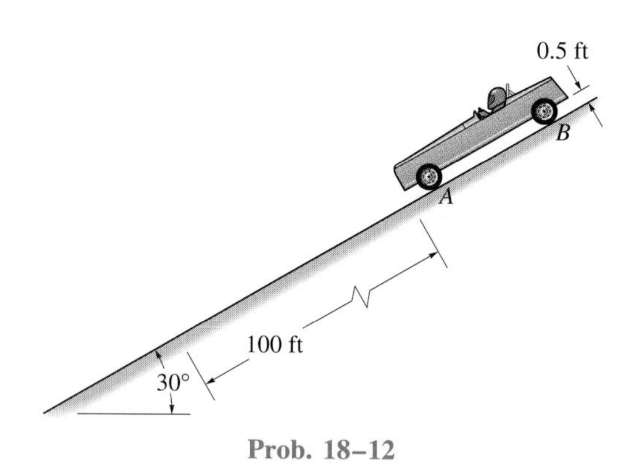

Prob. 18–12

18–13. The hand winch is used to lift the 50-kg load. Determine the work required to rotate the handle five revolutions. The gear at A has a radius of 20 mm.

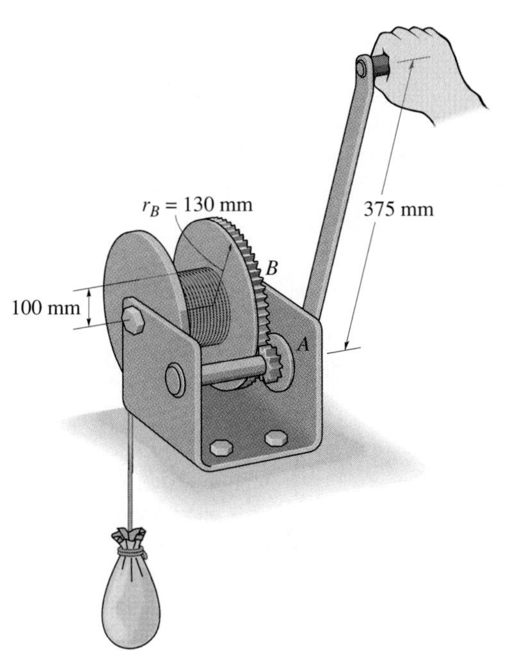

Prob. 18–13

18–14. The link is rotating about point O in the vertical plane with an angular velocity $\omega = 5$ rad/s when $\theta = 0°$. If it has a mass of 20 kg, a mass center at G, and a radius of gyration $k_G =$ 300 mm, determine its angular velocity at the instant $\theta = 90°$.

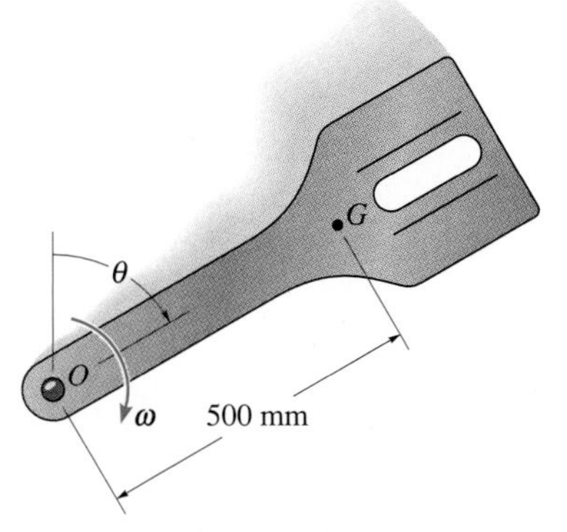

Prob. 18–14

18–15. The 4-kg slender rod is subjected to the force and couple moment. When it is in the position shown it has an angular velocity $\omega_1 = 6$ rad/s. Determine its angular velocity at the instant it has rotated downward 90°. The force is always applied perpendicular to the axis of the rod. Motion occurs in the vertical plane.

***18–16.** The 4-kg slender rod is subjected to the force and couple moment. When the rod is in the position shown it has a angular velocity $\omega_1 = 6$ rad/s. Determine its angular velocity at the instant it has rotated 360°. The force is always applied perpendicular to the axis of the rod and motion occurs in the vertical plane.

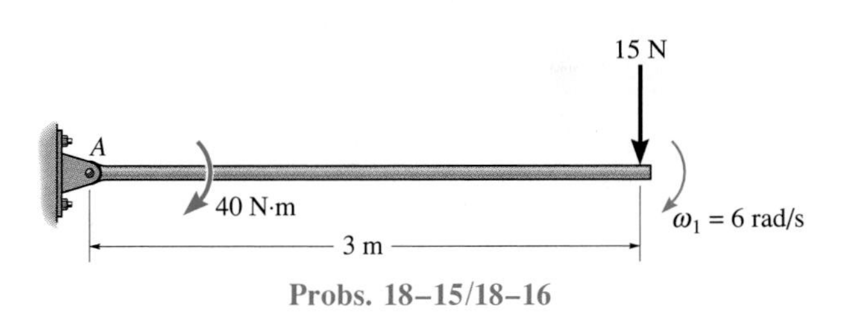

Probs. 18–15/18–16

18–17. The pendulum consists of two slender rods each having a mass of 4 kg/m. If it is acted upon by a moment $M = 50$ N · m and released from the position shown, determine its angular velocity when it has rotated (a) 90° and (b) 180°. Motion occurs in the vertical plane.

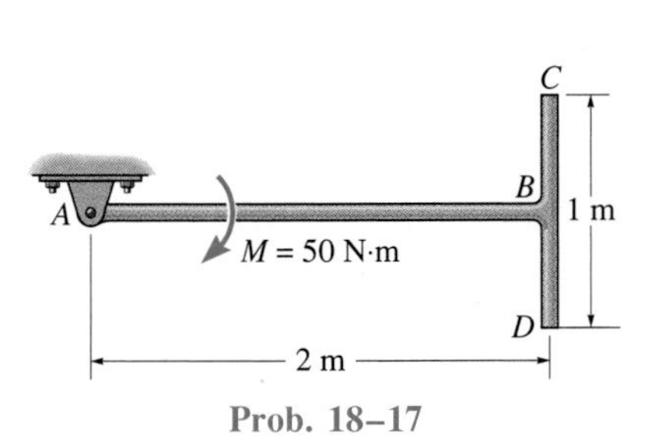

Prob. 18–17

18–18. The uniform pipe has a mass of 16 Mg and radius of gyration about the z axis of $k_G = 2.7$ m. If the worker pushes on it with a horizontal force of 50 N, applied perpendicular to the pipe, determine the pipes angular velocity when it has rotated 90° about the z axis, starting from rest. Assume the pipe does not swing.

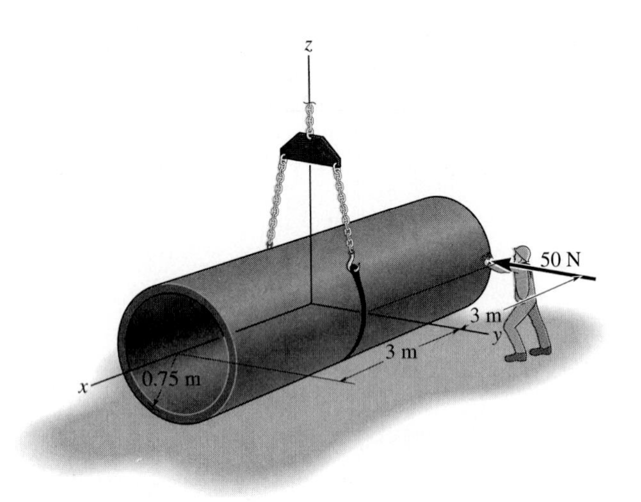

Prob. 18–18

18–19. A motor supplies a constant torque $M = 6$ kN · m to the winding drum that operates the elevator. If the elevator has a mass of 900 kg, the counterweight C has a mass of 200 kg, and the winding drum has a mass of 600 kg and radius of gyration about its axis of $k = 0.6$ m, determine the speed of the elevator after it rises 5 m starting from rest. Neglect the mass of the pulleys.

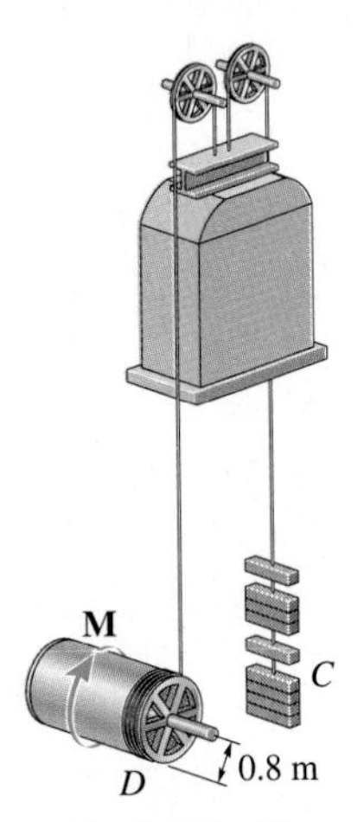

Prob. 18–19

***18–20.** The linkage consists of two 8-lb rods AB and CD and a 10-lb bar AD. When $\theta = 0°$, rod AB is rotating with an angular velocity $\omega_{AB} = 2$ rad/s. If rod CD is subjected to a couple moment $M = 15$ lb · ft and bar AD is subjected to a horizontal force $P = 20$ lb as shown, determine ω_{AB} at the instant $\theta = 90°$.

18–21. The linkage consists of two 8-lb rods AB and CD and a 10-lb bar AD. When $\theta = 0°$, rod AB is rotating with an angular velocity $\omega_{AB} = 2$ rad/s. If rod CD is subjected to a couple moment $M = 15$ lb · ft and bar AD is subjected to a horizontal force $P = 20$ lb as shown, determine ω_{AB} at the instant $\theta = 45°$.

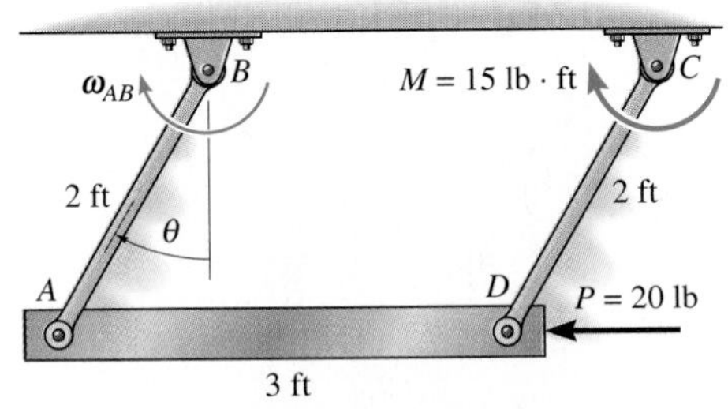

Probs. 18–20/18–21

18–22. The girl having a weight of 110 lb is sitting on a swing such that when she is at A, $\theta = 30°$, she is momentarily at rest and her center of gravity is at G. If she keeps this same fixed (rigid) position as she swings downward, determine the angular velocity of the swing when she reaches the lowest point at B ($\theta = 0°$). Her radius of gyration about an axis passing through G is 1.8 ft.

18–23. The girl having a weight of 110 lb is sitting on a swing such that when she is at A, $\theta = 45°$, she is momentarily at rest and her center of gravity is at G. If she keeps this same fixed (rigid) position as she swings downward, determine the angular velocity of the swing when she reaches the position $\theta = 15°$. Her radius of gyration about an axis passing through G is 1.8 ft.

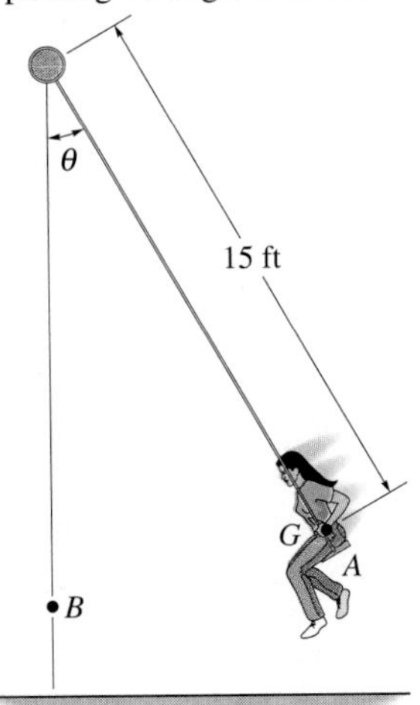

Probs. 18–22/18–23

***18–24.** The uniform bar has a mass m and length l. If it is released from rest when $\theta = 0°$, determine the angle θ at which it first begins to slip. The coefficient of static friction at O is $\mu_s = 0.3$.

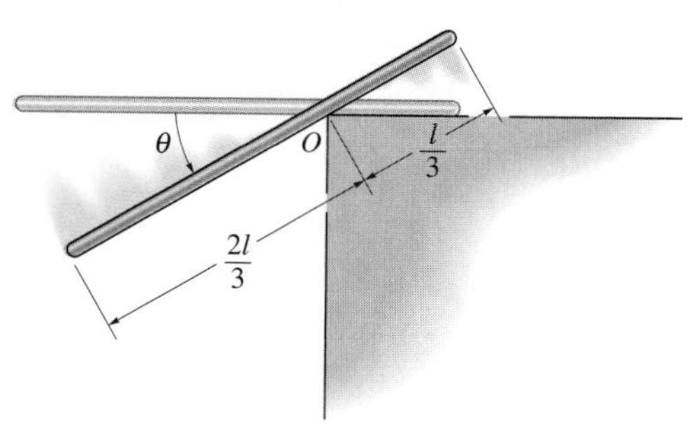

Prob. 18–24

18–25. The system consists of a 20-lb disk A, 4-lb slender rod BC, and a 1-lb smooth collar C. If the disk rolls without slipping, determine the velocity of the collar at the instant the rod becomes horizontal, i.e., $\theta = 0°$. The system is released from rest when $\theta = 45°$.

18–26. The system consists of a 20-lb disk A, 4-lb slender rod BC, and a 1-lb smooth collar C. If the disk rolls without slipping, determine the velocity of the collar at the instant $\theta = 30°$. The system is released from rest when $\theta = 45°$.

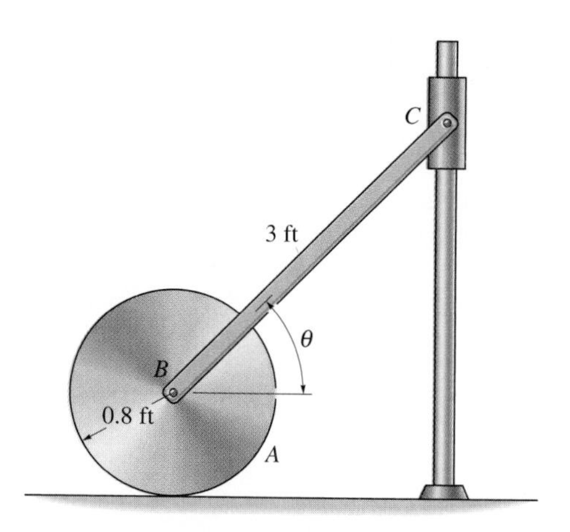

Probs. 18–25/18–26

18–27. The 20-kg disk is originally at rest and the spring holds it in equilibrium. A couple moment $M = 30\text{ N}\cdot\text{m}$ is then applied to the disk as shown. Determine the disk's angular velocity at the instant its mass center G has moved $s = 0.8$ m down along the inclined plane. The disk rolls without slipping.

***18–28.** The 20-kg disk is originally at rest and the spring holds it in equilibrium. A couple moment $M = 30\text{ N}\cdot\text{m}$ is then applied to the disk as shown. Determine how far s the center of mass of the disk travels down along the incline, measured from the equilibrium position, before it stops. The disk rolls without slipping.

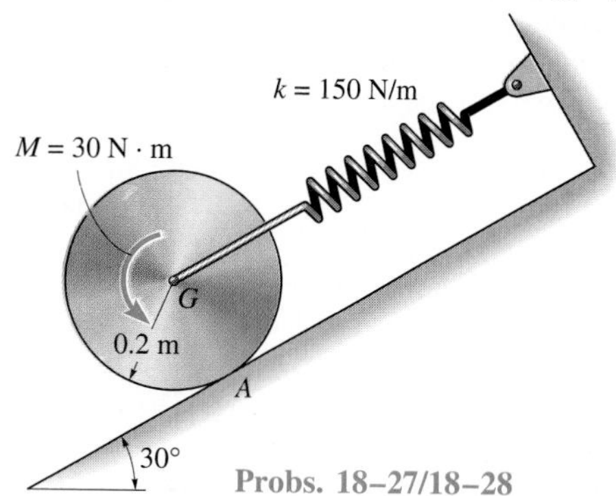

Probs. 18–27/18–28

18–29. The uniform door has a mass of 20 kg and can be treated as a thin plate having the dimensions shown. If it is connected to a torsional spring at A, which has an initial twist of 2 rad, when the door is closed ($\theta = 0°$), determine the minimum work required to open it, $\theta = 90°$. The spring has a stiffness $k = 80\text{ N}\cdot\text{m/rad}$. *Hint:* For a torsional spring $M = k\theta$, where k is the stiffness and θ is the angle of twist.

18–30. What is the angular velocity of the door in Prob. 18–29 if it is released from rest in the open position ($\theta = 90°$) just before it strikes the jamb, $\theta = 0°$?

18–31. The uniform door has a mass of 20 kg and can be treated as a thin plate having the dimensions shown. If it is connected to a torsional spring at A, which has a stiffness of $k = 80\text{ N}\cdot\text{m/rad}$, determine the required initial twist of the spring in radians so that the door has an angular velocity of 12 rad/s when it closes at $\theta = 0°$ after being opened at $\theta = 90°$ and released from rest. *Hint:* For a torsional spring $M = k\theta$, when k is the stiffness and θ is the angle of twist.

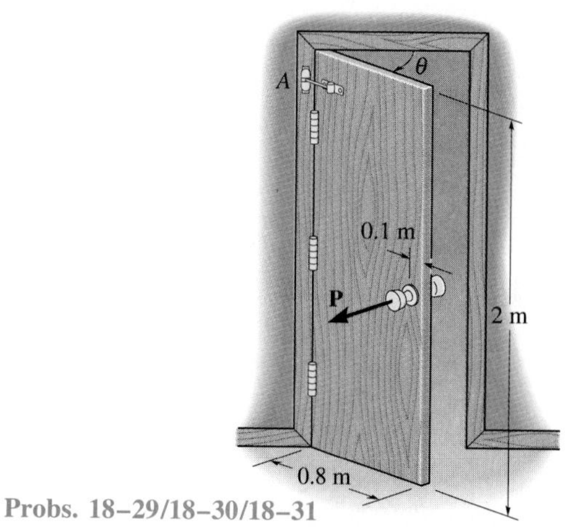

Probs. 18–29/18–30/18–31

***18–32.** The drum has a mass of 50 kg and a radius of gyration about the pin at O of $k_O = 0.23$ m. Starting from rest, the suspended 15-kg block B is allowed to fall 3 m without applying the brake ACD. Determine the speed of the block at this instant. If the coefficient of kinetic friction at the brake pad C is $\mu_k = 0.5$, determine the constant force P that must be applied at the brake arm, which will then stop the block after it descends *another* 3 m. Neglect the thickness of the arm.

18–33. The drum has a mass of 50 kg and a radius of gyration about the pin at O of $k_O = 0.23$ m. If the 15-kg block is moving downward at 3 m/s, and a force $P = 100$ N is applied to the brake arm, determine how far the block descends from the instant the brake is applied until it stops. Neglect the thickness of the arm. The coefficient of kinetic friction at the brake pad is $\mu_k = 0.5$.

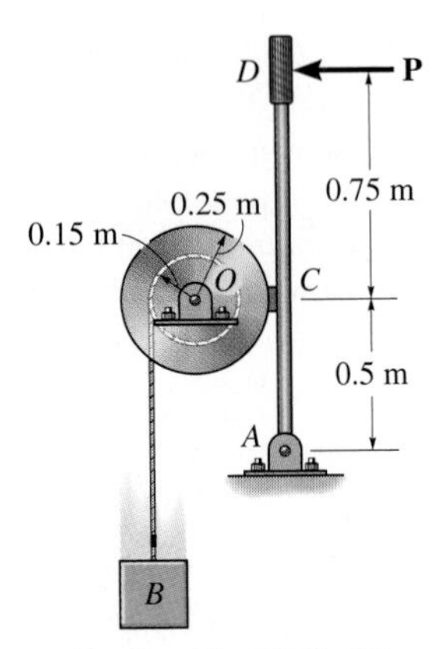

Probs. 18–32/18–33

18–34. The uniform slender bar has a mass m and a length L. It is subjected to a uniform distributed load w_0 which is always directed perpendicular to the axis of the bar. If it is released from rest from the position shown, determine its angular velocity at the instant it has rotated 90°. Solve the problem for rotation in (a) the horizontal plane, and (b) the vertical plane.

18–35. Solve Prob. 18–34 if the distributed load is triangular, which varies from zero at the pin O to w_0 at its other end.

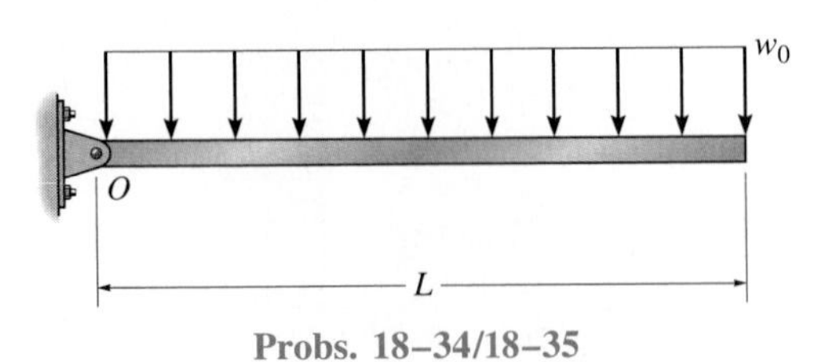

Probs. 18–34/18–35

***18–36.** The beam has a weight of 1500 lb and is being raised to a vertical position by pulling very slowly on its bottom end A. If the cord fails when $\theta = 60°$ and the beam is essentially at rest, determine the speed of A at the instant cord BC becomes vertical. Neglect friction and the mass of the cords, and treat the beam as a slender rod.

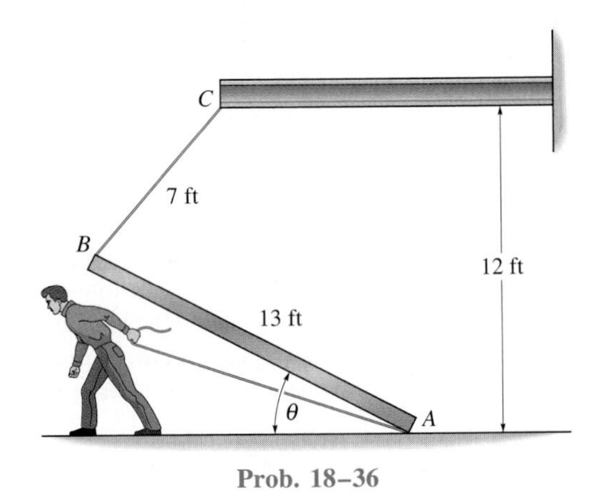

Prob. 18–36

18–37. A ball of mass m and radius r is cast onto the horizontal surface such that it rolls without slipping. Determine the minimum speed v_G of its mass center G so that it rolls completely around the loop of radius $R + r$ without leaving the track.

18–38. The ball has a mass of 10 kg and radius $r = 100$ mm and rolls without slipping on the horizontal surface at $v_G = 5$ m/s. Determine its angular velocity and the normal force it exerts on the track when it reaches the position $\theta = 90°$. Take $R = 500$ mm.

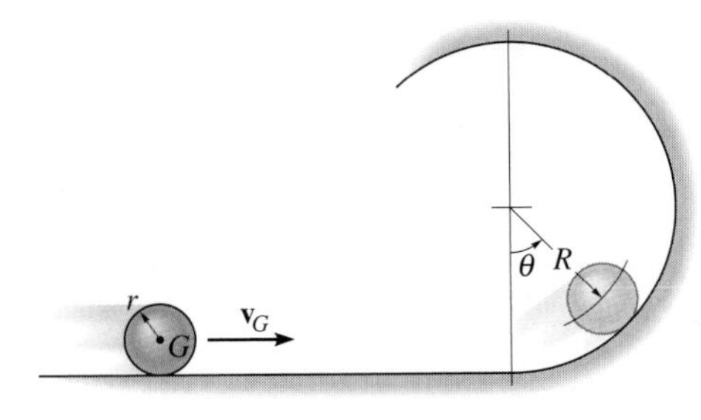

Probs. 18–37/18–38

18.5 Conservation of Energy

When a force system acting on a rigid body consists only of *conservative forces,* the conservation of energy theorem may be used to solve a problem which otherwise would be solved using the principle of work and energy. This theorem is often easier to apply since the work of a conservative force is *independent of the path* and depends on only the initial and final positions of the body. It was shown in Sec. 14.5 that the work of a conservative force may be expressed as the difference in the body's potential energy measured from an arbitrarily selected reference or datum.

Gravitational Potential Energy. Since the total weight of a body can be considered concentrated at its center of gravity, the *gravitational potential energy* of the body is determined by knowing the height of the body's center of gravity above or below a horizontal datum. Measuring y_G as *positive upward,* the gravitational potential energy of the body is thus

$$V_g = Wy_G \tag{18–14}$$

Here the potential energy is *positive* when y_G is positive, since the weight has the ability to do *positive work* when the body is moved back to the datum, Fig. 18–17. Likewise, if the body is located *below* the datum ($-y_G$), the gravitational potential energy is *negative,* since the weight does *negative work* when the body is moved back to the datum.

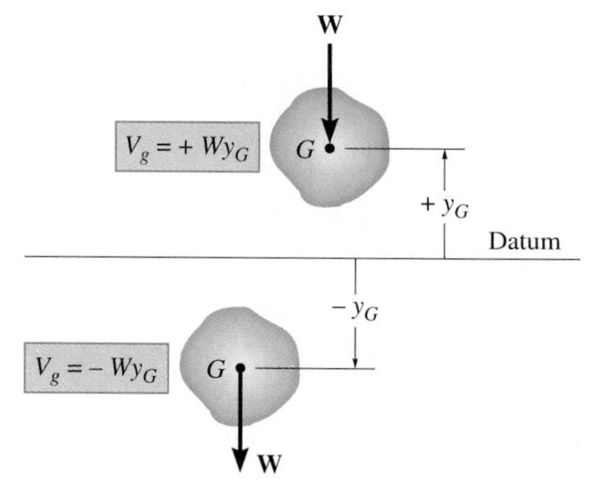

Gravitational potential energy

Fig. 18–17

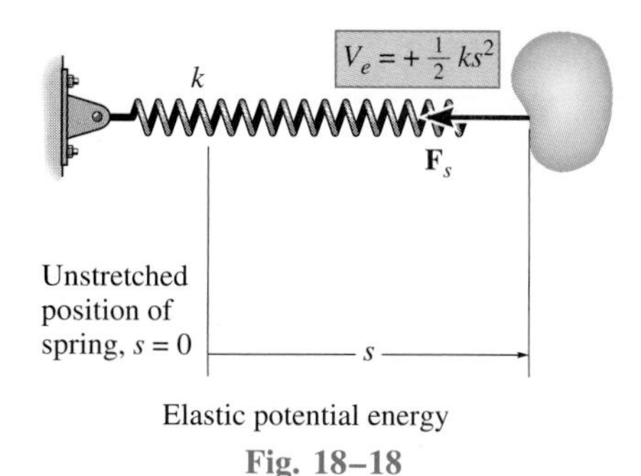

Elastic potential energy

Fig. 18–18

Elastic Potential Energy. The force developed by an elastic spring is also a conservative force. The *elastic potential energy* which a spring imparts to an attached body when the spring is elongated or compressed from an initial undeformed position ($s = 0$) to a final position s, Fig. 18–18, is

$$V_e = +\tfrac{1}{2}ks^2 \tag{18–15}$$

In the deformed position, the spring force acting *on the body* always has the capacity for doing positive work when the spring is returned back to its original undeformed position (see Sec. 14.5).

Conservation of Energy. In general, if a body is subjected to both gravitational and elastic forces, the total *potential energy* is expressed as a potential function V represented as the algebraic sum

$$V = V_g + V_e \tag{18–16}$$

Here measurement of V depends on the location of the body with respect to a selected datum in accordance with Eqs. 18–14 and 18–15.

Realizing that the work of conservative forces can be written as a difference in their potential energies, i.e., $(\Sigma U_{1-2})_{\text{cons}} = V_1 - V_2$, Eq. 14–16, we can rewrite the principle of work and energy for a rigid body as

$$T_1 + V_1 + (\Sigma U_{1-2})_{\text{noncons}} = T_2 + V_2 \tag{18–17}$$

Here $(\Sigma U_{1-2})_{\text{noncons}}$ represents the work of the nonconservative forces, such as friction, acting on the body. If this term is zero, then

$$T_1 + V_1 = T_2 + V_2 \tag{18–18}$$

This is the equation of the conservation of mechanical energy for the body. It states that the *sum* of the potential and kinetic energies of the body remains *constant* when the body moves from one position to another. It also applies to a system of smooth, pin-connected rigid bodies, bodies connected by inextensible cords, and bodies in mesh with other bodies. In all these cases the forces acting at the points of contact are *eliminated* from the analysis, since they occur in equal and opposite collinear pairs and each pair of forces moves through an equal distance when the system undergoes a displacement.

PROCEDURE FOR ANALYSIS

The conservation of energy equation is used to solve problems involving *velocity, displacement,* and *conservative force systems.* For application it is suggested that the following procedure be used.

Potential Energy. Draw two diagrams showing the body located at its initial and final positions along the path. If the center of gravity of the body, G, is subjected to a *vertical displacement,* determine where to establish the fixed horizontal datum from which to measure the body's gravitational potential energy V_g. Although the location of the datum is arbitrary, it is advantageous to place it through G when the body is either at its initial or final position, since at the datum $V_g = 0$. Data pertaining to the elevation y of the body's center of gravity from the datum and the extension or compression of any connecting springs can be determined from the geometry associated with the two diagrams. Recall that the potential energy $V = V_g + V_e$. Here $V_g = Wy_G$, where y_G is positive upward from the datum, and $V_e = \frac{1}{2}ks^2$, which is always positive.

Kinetic Energy. The kinetic-energy terms T_1 and T_2 are determined from $T = \frac{1}{2}mv_G^2 + \frac{1}{2}I_G\omega^2$ or an appropriate form of this equation as developed in Sec. 18.1. In this regard, kinematic diagrams for velocity may be useful for determining v_G and ω, or for establishing a *relationship* between these quantities.

Conservation of Energy. Apply the conservation of energy equation $T_1 + V_1 = T_2 + V_2$.

It is important to remember that *only problems involving conservative force systems may be solved by using this equation.* As stated in Sec. 14.5, friction or other drag-resistant forces, which depend on velocity or acceleration, are nonconservative. The work of such forces is transformed into thermal energy used to heat up the surfaces of contact, and consequently this energy is dissipated into the surroundings and may not be recovered. Therefore, problems involving frictional forces can be solved by using either the principle of work and energy written in the form of Eq. 18–17, if it applies, or the equations of motion.

The following example problems illustrate application of the above procedure.

Example 18–7

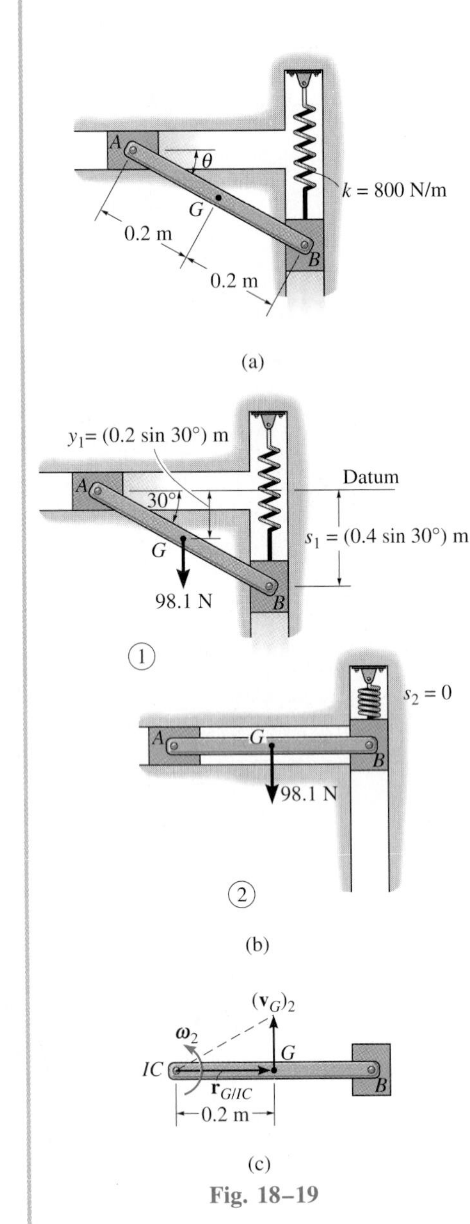

Fig. 18–19

The 10-kg rod AB shown in Fig. 18–19a is confined so that its ends move in the horizontal and vertical slots. The spring has a stiffness of $k = 800$ N/m and is unstretched when $\theta = 0°$. Determine the angular velocity of AB when $\theta = 0°$, if AB is released from rest when $\theta = 30°$. Neglect the mass of the slider blocks.

SOLUTION

Potential Energy. The two diagrams of the rod, when it is located at its initial and final positions, are shown in Fig. 18–19b. The datum, used to measure the gravitational potential energy, is placed in line with the rod when $\theta = 0°$.

When the rod is in position 1, the center of gravity G is located *below the datum* so that the gravitational potential energy is *negative*. Furthermore, (positive) elastic potential energy is stored in the spring, since it is stretched a distance of $s_1 = (0.4 \sin 30°)$ m. Thus,

$$V_1 = -Wy_1 + \tfrac{1}{2}ks_1^2$$
$$= -98.1\text{ N}(0.2 \sin 30°\text{ m}) + \tfrac{1}{2}(800\text{ N/m})(0.4 \sin 30°\text{ m})^2 = 6.19\text{ J}$$

When the rod is in position 2, the potential energy of the rod is zero, since the spring is unstretched, $s_2 = 0$, and the center of gravity G is located at the datum. Thus,

$$V_2 = 0$$

Kinetic Energy. The rod is released from rest from position 1, thus $(\mathbf{v}_G)_1 = \mathbf{0}$ and $\boldsymbol{\omega}_1 = \mathbf{0}$, and

$$T_1 = 0$$

In position 2 the angular velocity is $\boldsymbol{\omega}_2$ and the rod's mass center has a velocity of $(\mathbf{v}_G)_2$. Using *kinematics*, $(\mathbf{v}_G)_2$ can be related to $\boldsymbol{\omega}_2$ as shown in Fig. 18–19c. At the instant considered, the instantaneous center of zero velocity (IC) for the rod is at point A; hence, $(v_G)_2 = (r_{G/IC})\omega_2 = (0.2)\omega_2$. Thus,

$$T_2 = \tfrac{1}{2}m(v_G)_2^2 + \tfrac{1}{2}I_G(\omega_2)^2$$
$$= \tfrac{1}{2}(10\text{ kg})((0.2\text{ m})\omega_2)^2 + \tfrac{1}{2}[\tfrac{1}{12}(10\text{ kg})(0.4\text{ m})^2](\omega_2)^2 = 0.267\omega_2^2$$

Conservation of Energy

$$\{T_1\} + \{V_1\} = \{T_2\} + \{V_2\}$$
$$\{0\} + \{6.19\} = \{0.267\omega_2^2\} + \{0\}$$
$$\omega_2 = 4.82\text{ rad/s} \circlearrowleft \qquad \textit{Ans.}$$

Example 18–8

The disk shown in Fig. 18–20*a* has a weight of 30 lb and is attached to a spring which has a stiffness $k = 2$ lb/ft and an unstretched length of 1 ft. If the disk is released from rest in the position shown and rolls without slipping, determine its angular velocity at the instant it is displaced 3 ft.

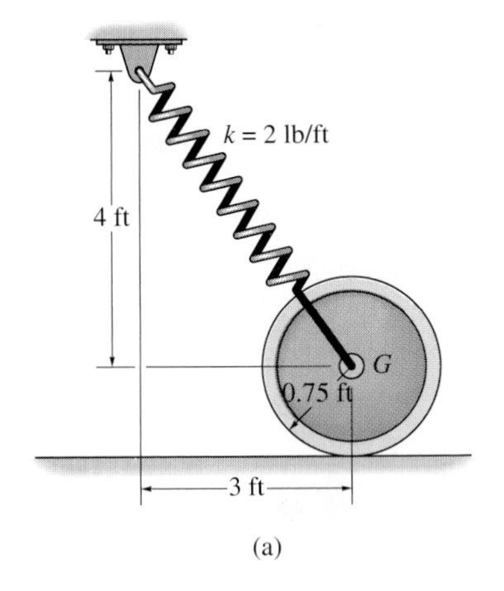

(a)

SOLUTION

Potential Energy. Two diagrams of the disk, when it is located in its initial and final positions, are shown in Fig. 18–20*b*. A gravitational datum is not needed here since the weight is not displaced vertically. From the problem geometry the spring is stretched $s_1 = (\sqrt{3^2 + 4^2} - 1) = 4$ ft and $s_2 = (4 - 1) = 3$ ft in the initial and final positions, respectively. Hence,

$$V_1 = \tfrac{1}{2}ks_1^2 = \tfrac{1}{2}(2\text{ lb/ft})(4\text{ ft})^2 = 16\text{ J}$$
$$V_2 = \tfrac{1}{2}ks_2^2 = \tfrac{1}{2}(2\text{ lb/ft})(3\text{ ft})^2 = 9\text{ J}$$

Kinetic Energy. The disk is released from rest so that $(\mathbf{v}_G)_1 = \mathbf{0}$, $\boldsymbol{\omega}_1 = \mathbf{0}$, and

$$T_1 = 0$$

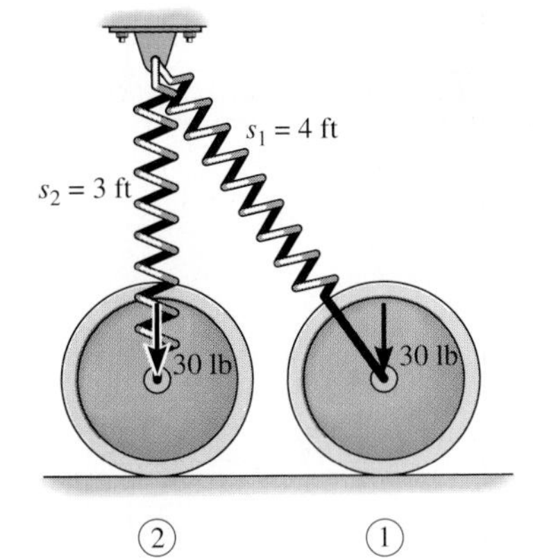

(b)

Since the disk rolls without slipping, $(\mathbf{v}_G)_2$ can be related to $\boldsymbol{\omega}_2$ from the instantaneous center of zero velocity, Fig. 18–20*c*. Hence, $(v_G)_2 = (0.75\text{ ft})\omega_2$. Thus,

$$T_2 = \frac{1}{2}m(v_G)_2^2 + \frac{1}{2}I_G(\omega_2)^2$$
$$= \frac{1}{2}\left(\frac{30\text{ lb}}{32.2\text{ ft/s}^2}\right)((0.75\text{ ft})\omega_2)^2 + \frac{1}{2}\left[\frac{1}{2}\left(\frac{30\text{ lb}}{32.2\text{ ft/s}^2}\right)(0.75\text{ ft})^2\right](\omega_2)^2$$
$$= 0.393(\omega_2)^2$$

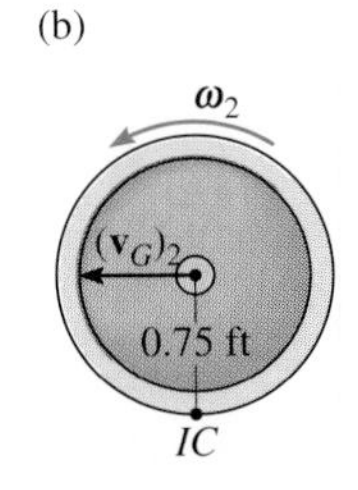

(c)

Fig. 18–20

Conservation of Energy

$$\{T_1\} + \{V_1\} = \{T_2\} + \{V_2\}$$
$$\{0\} + \{16\} = \{0.393(\omega_2)^2\} + \{9\}$$
$$\omega_2 = 4.22\text{ rad/s} \circlearrowleft \qquad \textit{Ans.}$$

Example 18–9

The 10-kg homogeneous disk shown in Fig. 18–21*a* is attached to a uniform 5-kg rod *AB*. If the assembly is released from rest when $\theta = 60°$, determine the angular velocity of the rod when $\theta = 0°$. Assume that the disk rolls without slipping. Neglect friction along the guide and the mass of the collar at *B*.

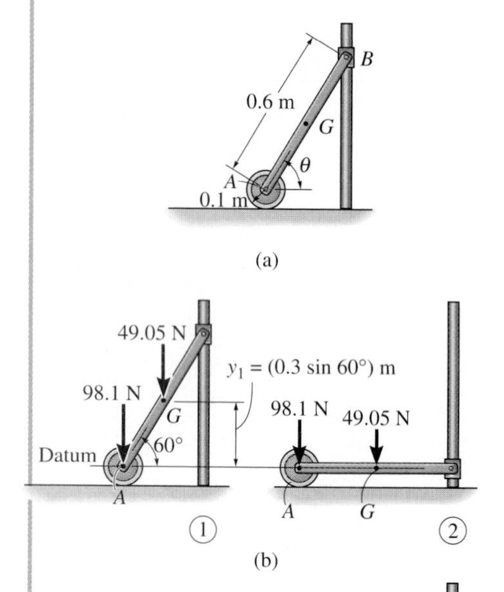

Fig. 18–21

SOLUTION

Potential Energy. Two diagrams for the rod and disk, when they are located at their initial and final positions, are shown in Fig. 18–21*b*. For convenience the datum, which is horizontally fixed, passes through point *A*.

When the system is in position 1, the rod's weight has positive potential energy. Thus,

$$V_1 = W_R y_1 = 49.05 \text{ N}(0.3 \sin 60° \text{ m}) = 12.74 \text{ J}$$

When the system is in position 2, both the weight of the rod and the weight of the disk have zero potential energy. Why? Thus,

$$V_2 = 0$$

Kinetic Energy. Since the entire system is at rest in the initial position,

$$T_1 = 0$$

In the final position the rod has an angular velocity $(\boldsymbol{\omega}_R)_2$ and its mass center has a velocity $(\mathbf{v}_G)_2$, Fig. 18–21*c*. Since the rod is *fully extended* in this position, the disk is momentarily at rest, so $(\boldsymbol{\omega}_D)_2 = \mathbf{0}$ and $(\mathbf{v}_A)_2 = \mathbf{0}$. For the rod $(\mathbf{v}_G)_2$ can be related to $(\boldsymbol{\omega}_R)_2$ from the instantaneous center of zero velocity, which is located at point *A*, Fig. 18–21*c*. Hence, $(v_G)_2 = r_{G/IC}(\omega_R)_2$ or $(v_G)_2 = 0.3(\omega_R)_2$. Thus,

$$T_2 = \frac{1}{2}m_R(v_G)_2^2 + \frac{1}{2}I_G(\omega_R)_2^2 + \frac{1}{2}m_D(v_A)_2^2 + \frac{1}{2}I_A(\omega_D)_2^2$$

$$= \frac{1}{2}(5 \text{ kg})(0.3 \text{ m}(\omega_R)_2)^2 + \frac{1}{2}\left[\frac{1}{12}(5 \text{ kg})(0.6 \text{ m})^2\right](\omega_R)_2^2 + 0 + 0$$

$$= 0.3(\omega_R)_2^2$$

Conservation of Energy

$$\{T_1\} + \{V_1\} = \{T_2\} + \{V_2\}$$

$$\{0\} + \{12.74\} = \{0.3(\omega_R)_2^2\} + \{0\}$$

$$(\omega_R)_2 = 6.52 \text{ rad/s} \circlearrowright \qquad \textit{Ans.}$$

PROBLEMS

18–39. Solve Prob. 18–10 using the conservation of energy equation.

*__18–40.__ Solve Prob. 18–12 using the conservation of energy equation.

18–41. Solve Prob. 18–14 using the conservation of energy equation.

18–42. Solve Prob. 18–22 using the conservation of energy equation.

18–43. Solve Prob. 18–25 using the conservation of energy equation.

*__18–44.__ Solve Prob. 18–36 using the conservation of energy equation.

18–45. Solve Prob. 18–37 using the conservation of energy equation.

18–46. The spool has a mass of 50 kg and a radius of gyration $k_O = 0.280$ m. If the 20-kg block A is released from rest, determine its velocity just after it descends 2 m. Neglect the mass of the cord.

18–47. The spool has a mass of 50 kg and a radius of gyration $k_O = 0.280$ m. If the 20-kg block A is released from rest, determine the distance the block must fall in order for the spool to have an angular velocity $\omega = 5$ rad/s. Also, what is the tension in the cord while the block is in motion? Neglect the mass of the cord.

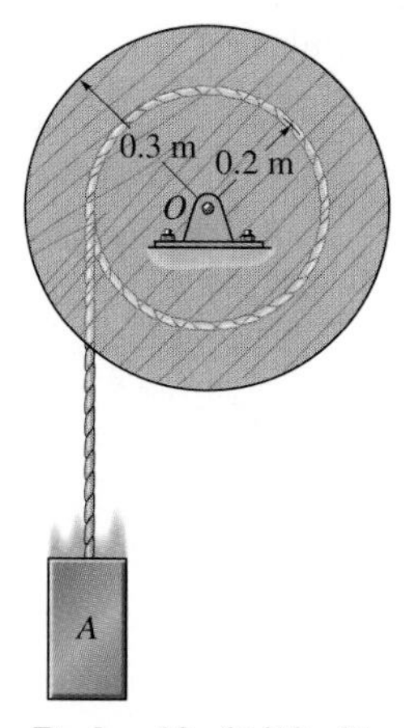

Probs. 18–46/18–47

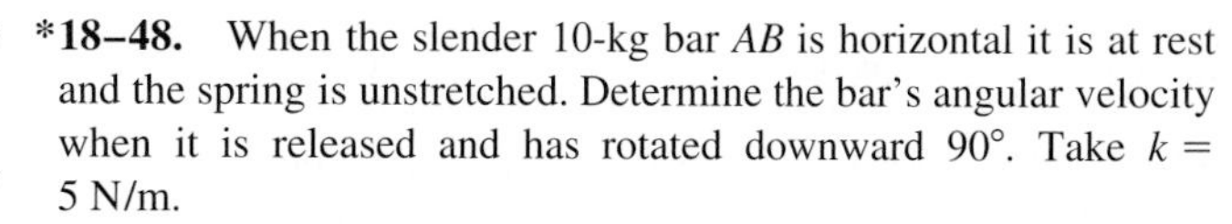

*__18–48.__ When the slender 10-kg bar AB is horizontal it is at rest and the spring is unstretched. Determine the bar's angular velocity when it is released and has rotated downward 90°. Take $k = 5$ N/m.

18–49. When the slender 10-kg bar AB is horizontal it is at rest and the spring is unstretched. Determine the stiffness k of the spring so that the motion of the bar is momentarily stopped when it has rotated downward 90°.

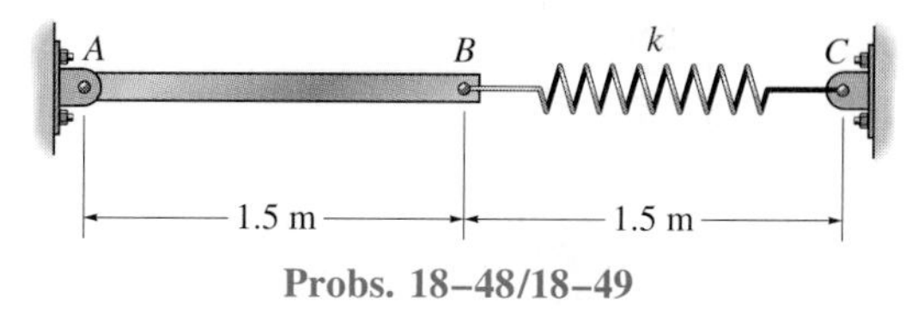

Probs. 18–48/18–49

18–50. The 50-lb wheel has a radius of gyration about its center of gravity G of $k_G = 0.7$ ft. If it rolls without slipping, determine its angular velocity when it has rotated clockwise 90° from the position shown. The spring AB has a stiffness $k = 1.20$ lb/ft and an unstretched length of 0.5 ft. The wheel is released from rest.

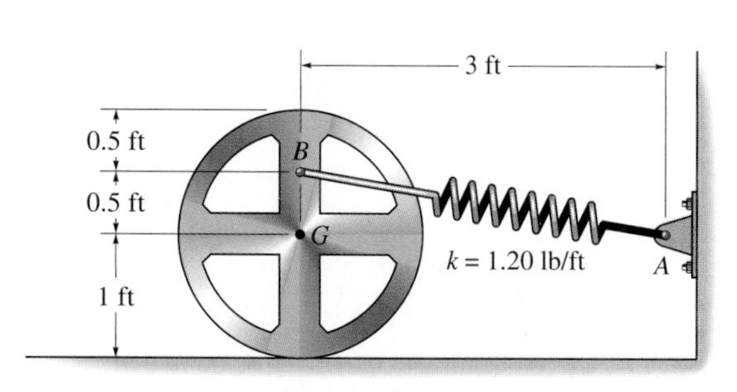

Prob. 18–50

18–51. The door is made from one piece, whose ends move along the horizontal and vertical tracks. If the door is in the open position, $\theta = 0°$, and then released, determine the speed at which its end A strikes the stop at C. Assume the door is a 180-lb thin plate having a width of 10 ft.

***18–52.** The door is made from one piece, whose ends move along the horizontal and vertical tracks. If the door is in the open position, $\theta = 0°$, and then released, determine its angular velocity at the instant $\theta = 30°$. Assume the door is a 180-lb thin plate having a width of 10 ft.

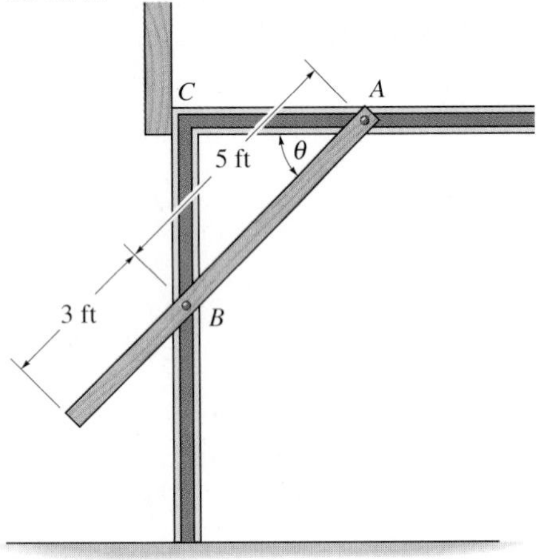

Probs. 18–51/18–52

18–53. The compound disk pulley consists of a hub and attached outer rim. If it has a mass of 3 kg and a radius of gyration $k_G = 45$ mm, determine the speed of block A after A descends 0.2 m from rest. Blocks A and B each have a mass of 2 kg. Neglect the mass of the cords.

18–54. The compound disk pulley consists of a hub and attached outer rim. If it has a mass of 3 kg and a radius of gyration $k_G = 45$ mm, determine the distance block A must descend in order to give the pulley an angular velocity of 4 rad/s. Blocks A and B each have a mass of 2 kg. Neglect the mass of the cords.

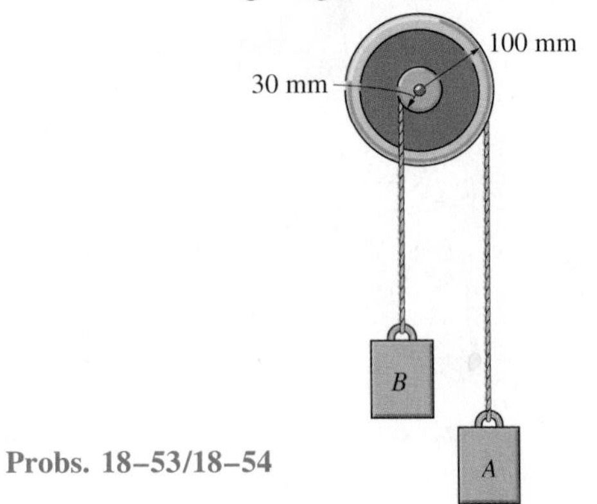

Probs. 18–53/18–54

18–55. The assembly consists of a 3-kg pulley A and 10-kg pulley B. If a 2-kg block is suspended from the cord, determine the block's speed after it descends 0.5 m starting from rest. Neglect the mass of the cord and treat the pulleys as thin disks. No slipping occurs.

***18–56.** The assembly consists of a 3-kg pulley A and 10-kg pulley B. If a 2-kg block is suspended from the cord, determine the distance the block must descend, starting from rest, in order to cause B to have an angular velocity of 6 rad/s. Neglect the mass of the cord and treat the pulleys as thin disks. No slipping ocurs.

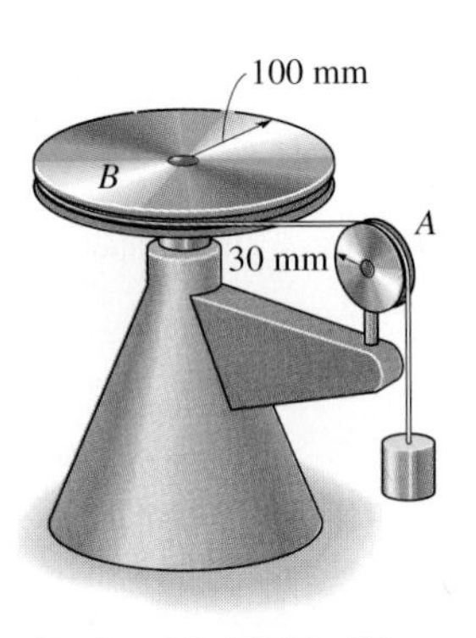

Probs. 18–55/18–56

18–57. The 25-lb slender rod AB is attached to a spring BC which has an unstretched length of 4 ft. If the rod is released from rest when $\theta = 30°$, determine its angular velocity at the instant $\theta = 90°$.

18–58. The 25-lb slender rod AB is attached to a spring BC which has an unstretched length of 4 ft. If the rod is released from rest when $\theta = 30°$, determine the angular velocity of the rod the instant the spring becomes unstretched.

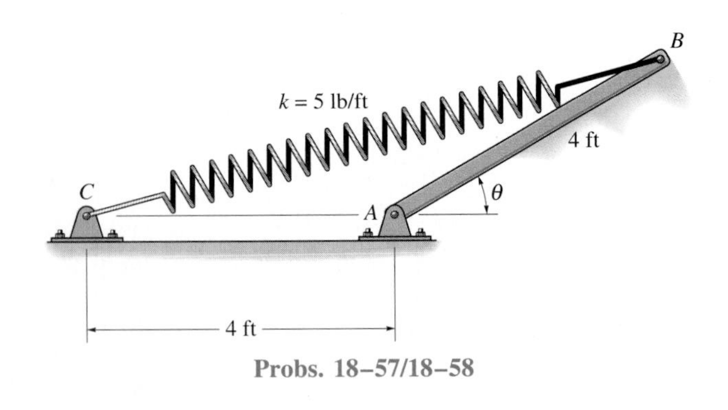

Probs. 18–57/18–58

18–59. The assembly consists of two 8-lb bars which are pin-connected to the two 10-lb disks. If the bars are released from rest when $\theta = 60°$, determine their angular velocities at the instant $\theta = 0°$. Assume the disks roll without slipping.

***18–60.** The assembly consists of two 8-lb bars which are pin-connected to the two 10-lb disks. If the bars are released from rest when $\theta = 60°$, determine their angular velocities at the instant $\theta = 30°$. Assume the disks roll without slipping.

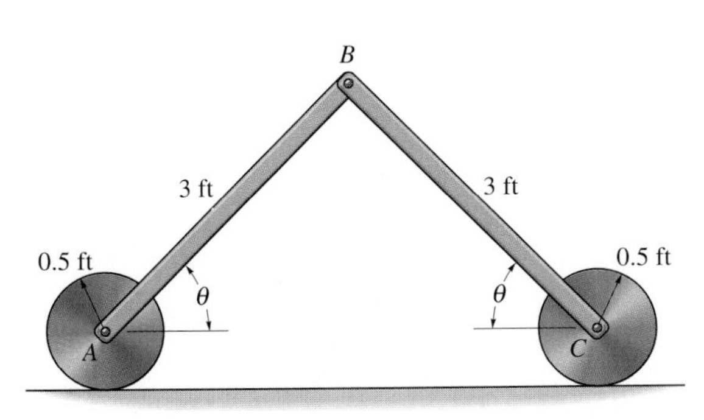

Probs. 18–59/18–60

18–61. The disk A is pinned at O and weighs 15 lb. A 1-ft rod weighing 2 lb and a 1-ft-diameter sphere weighing 10 lb are welded to the disk, as shown. If the spring is originally stretched 1 ft and the sphere is released from the position shown, determine the angular velocity of the disk when it has rotated 90°.

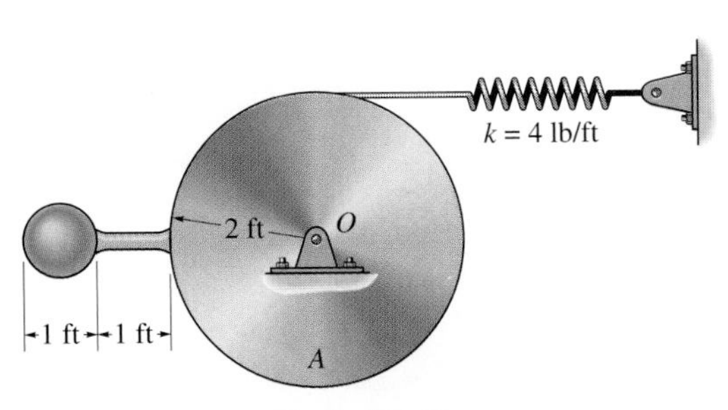

Prob. 18–61

18–62. The 20-lb disk rolls on the curved surface without slipping. The link AB weighs 5 lb. If the disk is released from rest when $\theta = 0°$, determine its angular velocity when $\theta = 90°$.

18–63. The 20-lb disk rolls on the curved surface without slipping. The link AB weighs 5 lb. If the disk is released from rest when $\theta = 0°$, determine its angular velocity when $\theta = 45°$.

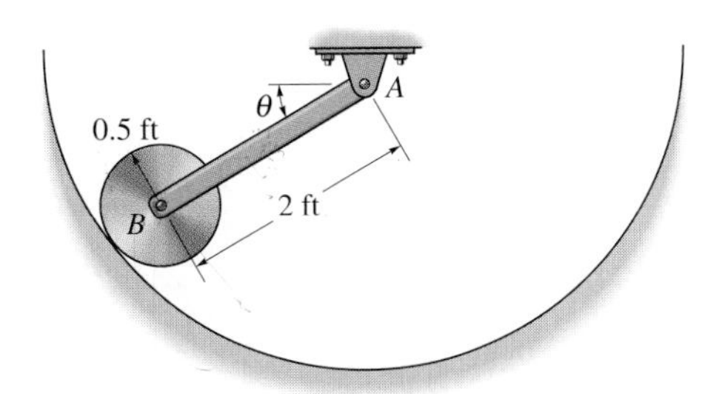

Probs. 18–62/18–63

***18–64.** The elevator E has a mass of 500 kg, the counterweight C has a mass of 200 kg, and the pulley or traction sheave has a mass of 150 kg and radius of gyration about its axle of $k_A = 0.5$ m. Starting from rest, determine the speed of the elevator after it falls freely and descends 4 m. Neglect the mass of the cables.

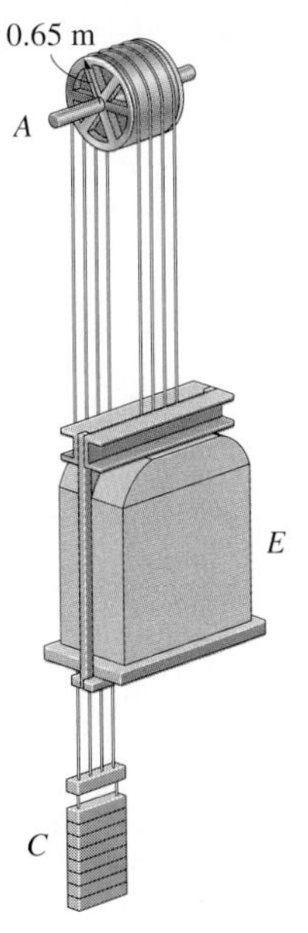

Prob. 18–64

18–65. The assembly consists of a 5-lb slender rod AC to which is pin-connected a 12-lb disk and spring BD. If the rod is brought into the horizontal position $\theta = 0°$, and the disk is given a counterclockwise rotation of 3 rad/s when the rod is released from rest, determine the angular velocity of the rod at the instant $\theta = 90°$. The spring has an unstretched length of 1 ft.

18–66. Solve Prob. 18–65 if the disk is originally at rest and is fixed from rotating about its pin-connection at C.

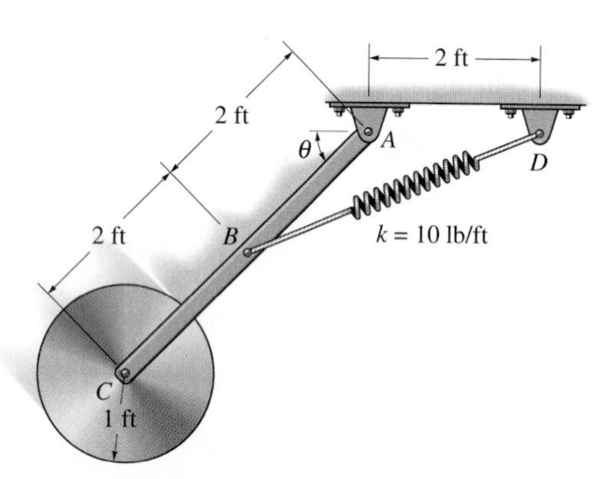

Probs. 18–65/18–66

***18–68.** The spool at A has a mass of 12 kg and a radius of gyration $k_C = 225$ mm. Pulley B has a mass of 7 kg and may be considered as a disk. If the system is released from rest and the cord does not slip, determine the angular velocity of the pulley after the spool descends 0.5 m. Neglect the mass of the cord.

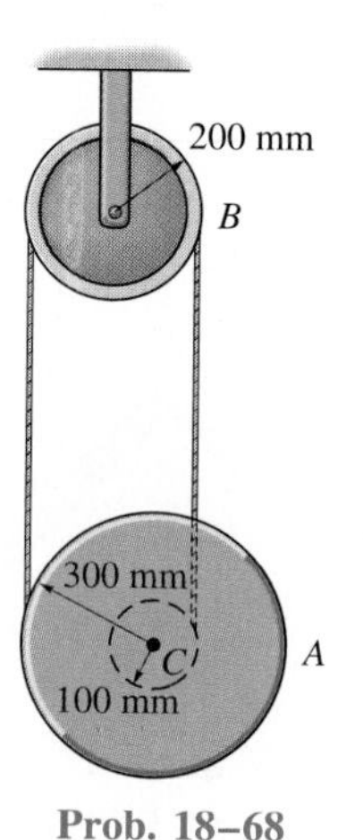

Prob. 18–68

18–67. The end A of the garage door AB travels along the horizontal track, and the end of member BC is attached to a spring at C. If the spring is originally unstretched, determine the stiffness k so that when the door falls downward from rest in the position shown, it will have zero angular velocity the moment it closes, i.e., when it and BC become vertical. Neglect the mass of member BC and assume the door is a thin plate having a weight of 200 lb and a width and height of 12 ft. There is a similar connection and spring on the other side of the door.

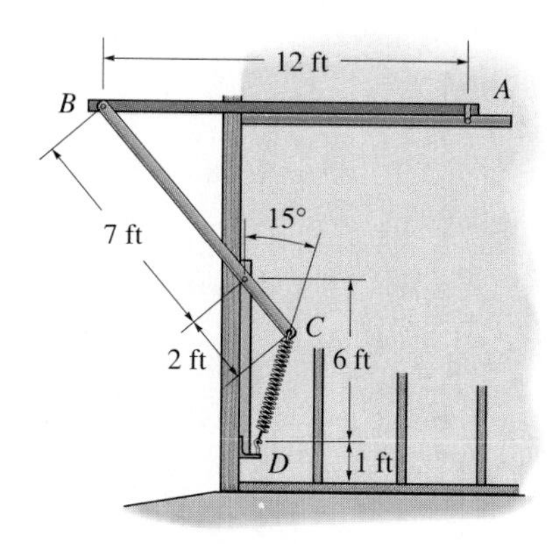

Prob. 18–67

18–69. The spool at A has a mass of 12 kg and a radius of gyration $k_C = 225$ mm. Pulley B has a mass of 7 kg and may be considered as a disk. If the system is released from rest and the cord does not slip, determine the distance the spool must descend in order for the pulley to attain an angular velocity of 4 rad/s.

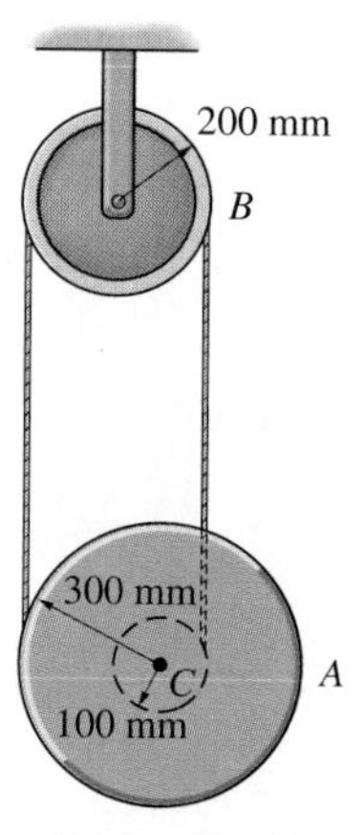

Prob. 18–69

18–70. The uniform 80-lb garage door is guided at its ends on the track. Determine the required initial stretch in the spring when the door is open, $\theta = 0°$, so that when it falls freely it comes to rest when it just reaches the fully closed position, $\theta = 90°$. Assume the door can be treated as a thin plate, and there is a spring and pulley system on each of the two sides of the door.

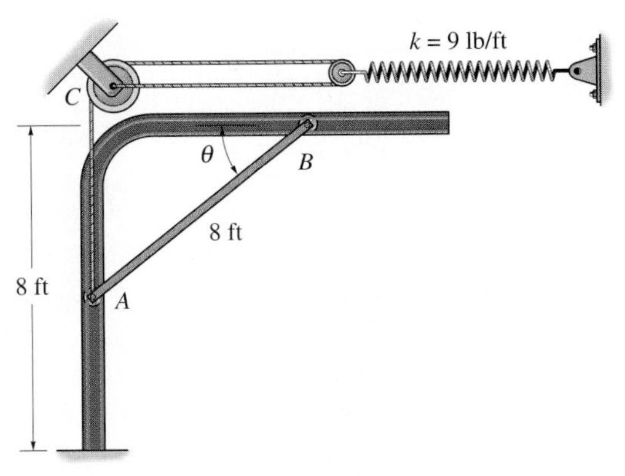

Prob. 18–70

18–71. The uniform 80-lb garage door is guided at its ends on the track. If it is released from rest at $\theta = 0°$, determine the door's angular velocity at the instant $\theta = 30°$. The spring is originally stretched 1 ft when the door is held open, $\theta = 0°$. Assume the door can be treated as a thin plate, and there is a spring and pulley system on each of two sides of the door.

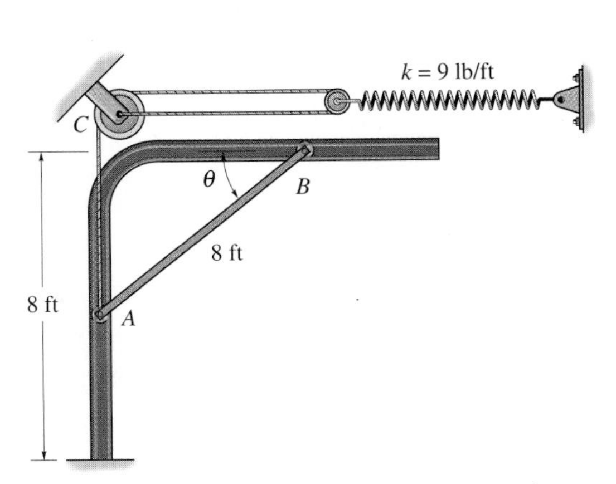

Prob. 18–71

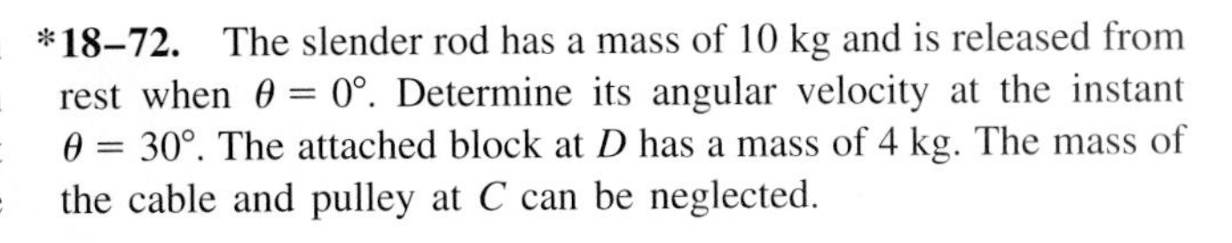

***18–72.** The slender rod has a mass of 10 kg and is released from rest when $\theta = 0°$. Determine its angular velocity at the instant $\theta = 30°$. The attached block at D has a mass of 4 kg. The mass of the cable and pulley at C can be neglected.

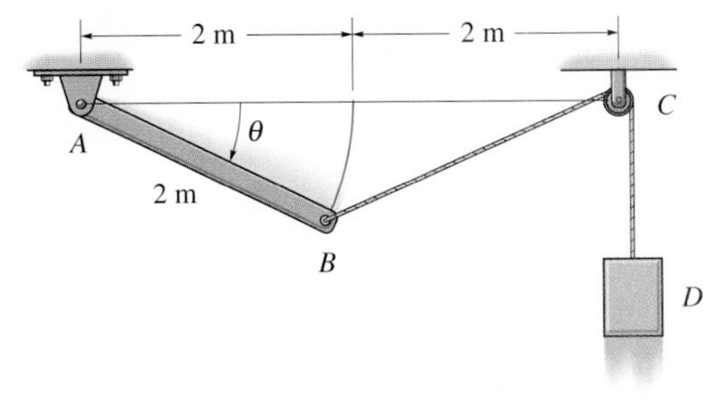

Prob. 18–72

18–73. The slender 15-kg bar is initially at rest and standing in the vertical position when the bottom end A is displaced slightly to the right. If the track in which it moves is smooth, determine the speed at which end A strikes the corner D. The bar is constrained to move in the vertical plane. Neglect the mass of the cord BC.

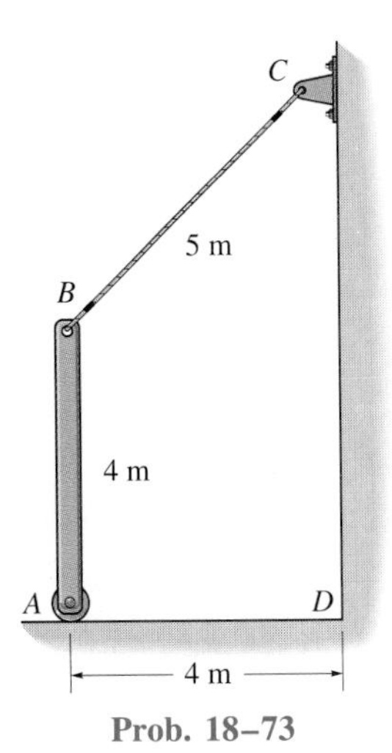

Prob. 18–73

The launching of this weather satellite requires application of impulse and momentum principles to accurately predict its angular motion and proper orientation once it is in orbit. We will discuss these principles in this chapter.

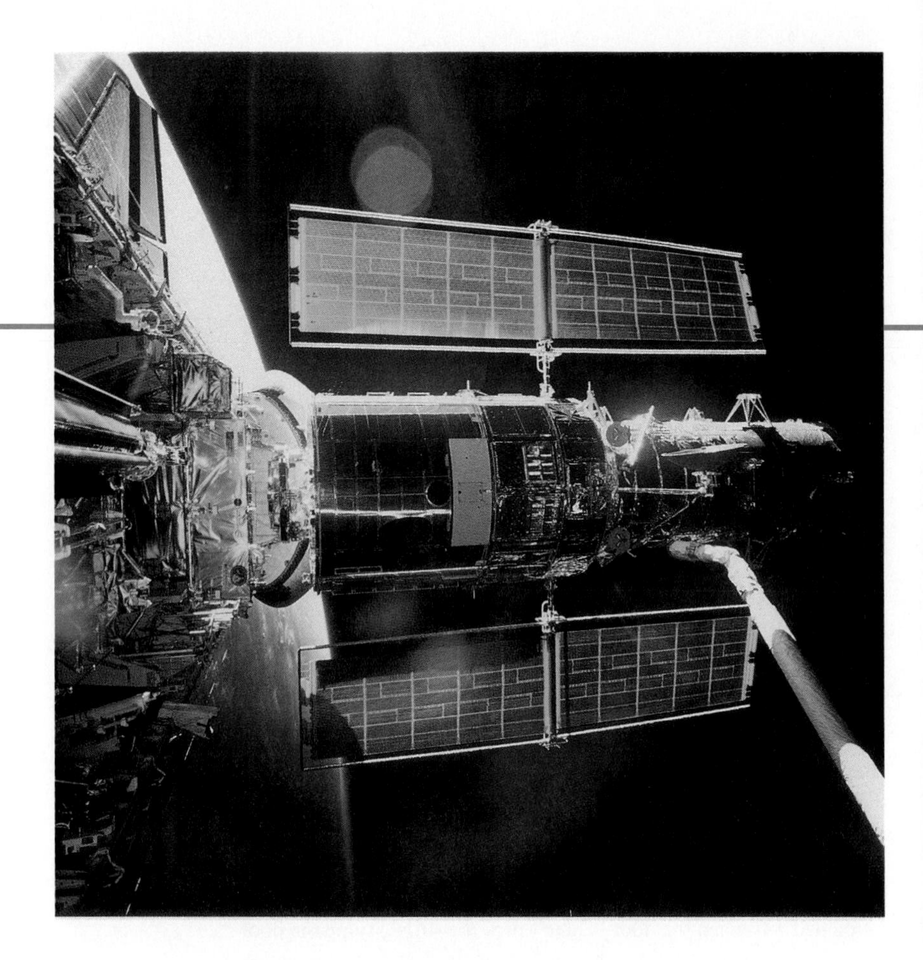

19

Planar Kinetics of a Rigid Body: Impulse and Momentum

In Chapter 15 it was shown that problems which involve force, velocity, and time can conveniently be solved by using the principle of linear or angular impulse and momentum. In this chapter we will extend these concepts somewhat in order to determine the impulse and momentum relationships that apply to a rigid body. The three planar motions which will be considered are translation, rotation about a fixed axis, and general plane motion. The more general discussion of momentum principles applied to the three-dimensional motion of a rigid body is presented in Chapter 21.

19.1 Linear and Angular Momentum

In this section we will develop formulations for the linear and angular momentum of a rigid body which is symmetric with respect to an inertial x–y reference plane.

Linear Momentum. The linear momentum of a rigid body is determined by summing vectorially the linear momenta of all the particles of the body, i.e., $\mathbf{L} = \Sigma m_i \mathbf{v}_i$. We can simplify this expression by noting that $\Sigma m_i \mathbf{v}_i = m\mathbf{v}_G$ (see Sec. 15.2) so that

$$\mathbf{L} = m\mathbf{v}_G \tag{19–1}$$

This equation states that the body's linear momentum is a vector quantity having a *magnitude* mv_G, which is commonly measured in units of kg · m/s or slug · ft/s, and a *direction* defined by $\mathbf{v}_G$, the velocity of the body's mass center.

Angular Momentum. Consider the body in Fig. 19–1*a*, which is subjected to general plane motion. At the instant shown, the arbitrary point *P* has a velocity $\mathbf{v}_P$ and the body has an angular velocity $\boldsymbol{\omega}$. If the velocity of the *i*th particle of the body is to be determined, Fig. 19–1*a*, then

$$\mathbf{v}_i = \mathbf{v}_P + \mathbf{v}_{i/P} = \mathbf{v}_P + \boldsymbol{\omega} \times \mathbf{r}$$

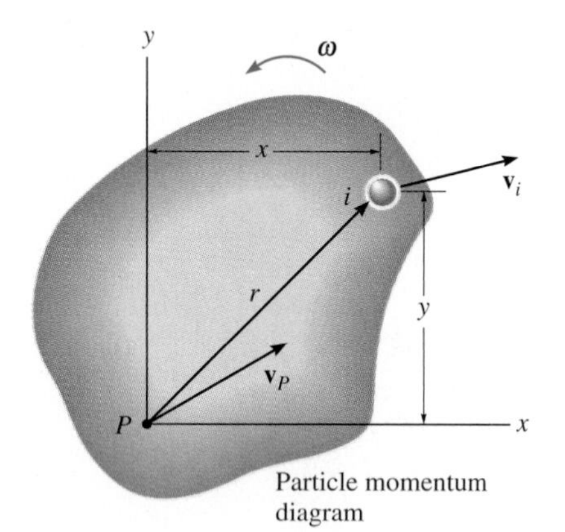

(a)

Fig. 19–1

The angular momentum of particle *i* about point *P* is equal to the ''moment'' of the particle's linear momentum about *P*, Fig. 19–1*a*. Thus,

$$(\mathbf{H}_P)_i = \mathbf{r} \times m_i\, \mathbf{v}_i$$

Expressing $\mathbf{v}_i$ in terms of $\mathbf{v}_P$ and using Cartesian vectors, we have

$$(H_P)_i\, \mathbf{k} = m_i(x\mathbf{i} + y\mathbf{j}) \times [(v_P)_x\mathbf{i} + (v_P)_y\mathbf{j} + \omega\mathbf{k} \times (x\mathbf{i} + y\mathbf{j})]$$

$$(H_P)_i = -m_i\, y(v_P)_x + m_i\, x(v_P)_y + m_i\, \omega r^2$$

Letting $m_i \rightarrow dm$ and integrating over the entire mass *m* of the body, we obtain

$$H_P = -\left(\int_m y\, dm\right)(v_P)_x + \left(\int_m x\, dm\right)(v_P)_y + \left(\int_m r^2\, dm\right)\omega$$

Here H_P represents the angular momentum of the body about an axis (the *z* axis) perpendicular to the plane of motion and passing through point *P*. Since $\bar{x}m = \int x\, dm$ and $\bar{y}m = \int y\, dm$, the integrals for the first and second terms on the right are used to locate the body's center of mass *G* with respect to *P*. Also, the last integral represents the body's moment of inertia computed about the *z* axis, i.e., $I_P = \int r^2\, dm$. Thus,

$$H_P = -\bar{y}m(v_P)_x + \bar{x}m(v_P)_y + I_P\omega \tag{19–2}$$

This equation reduces to a simpler form if point *P* coincides with the mass center *G* for the body,* in which case $\bar{x} = \bar{y} = 0$. Hence, the angular momentum of the body, computed about an axis passing through point *G*, is therefore

*It *also* reduces to the same simple form, $H_P = I_P\omega$, if point *P* is a *fixed point* (see Eq. 19–9) or the velocity of point *P* is directed along the line *PG*.

$$H_G = I_G\omega \tag{19–3}$$

This equation states that the angular momentum of the body computed about point G is equal to the product of the moment of inertia of the body about an axis passing through G and the body's angular velocity. It is important to realize that $\mathbf{H}_G$ is a vector quantity having a *magnitude* $I_G\omega$, which is commonly measured in units of kg · m²/s or slug · ft²/s, and a *direction* defined by $\boldsymbol{\omega}$, which is always perpendicular to the plane of motion.

Equation 19–2 can also be rewritten in terms of the x and y components of the velocity of the body's mass center, $(v_G)_x$ and $(v_G)_y$, and the body's moment of inertia I_G. If point G is located at coordinates $(\bar{x}, \bar{y})$, Fig. 19–1*b*, then, by the parallel-axis theorem, $I_P = I_G + m(\bar{x}^2 + \bar{y}^2)$. Substituting into Eq. 19–2 and rearranging terms, we have

$$H_P = \bar{y}m[-(v_P)_x + \bar{y}\omega] + \bar{x}m[(v_P)_y + \bar{x}\omega] + I_G\omega \tag{19–4}$$

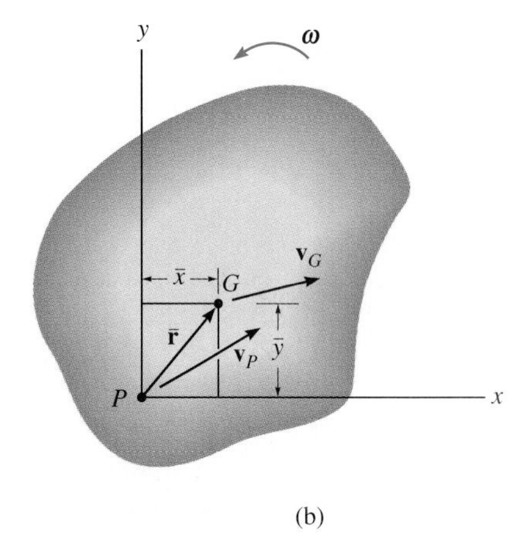

(b)

From the kinematic diagram of Fig. 19–1*b*, $\mathbf{v}_G$ can be expressed in terms of $\mathbf{v}_P$ as

$$\mathbf{v}_G = \mathbf{v}_P + \boldsymbol{\omega} \times \bar{\mathbf{r}}$$
$$(v_G)_x\mathbf{i} + (v_G)_y\mathbf{j} = (v_P)_x\mathbf{i} + (v_P)_y\mathbf{j} + \omega\mathbf{k} \times (\bar{x}\mathbf{i} + \bar{y}\mathbf{j})$$

Carrying out the cross product and equating the respective **i** and **j** components yields the two scalar equations

$$(v_G)_x = (v_P)_x - \bar{y}\omega$$
$$(v_G)_y = (v_P)_y + \bar{x}\omega$$

Substituting these results into Eq. 19–4 yields

$$H_P = -\bar{y}m(v_G)_x + \bar{x}(v_G)_y + I_G\omega \tag{19–5}$$

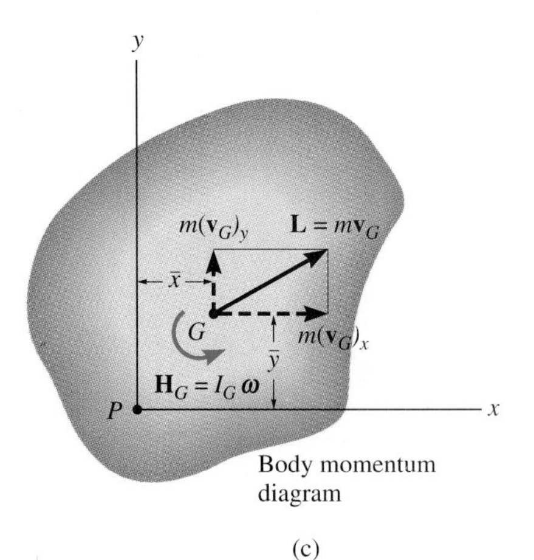

Body momentum diagram

(c)

Fig. 19–1

With reference to Fig. 19–1*c*, *this result indicates that when the angular momentum of the body is computed about point P, it is equivalent to the moment of the linear momentum* $m\mathbf{v}_G$ *or its components* $m(\mathbf{v}_G)_x$ *and* $m(\mathbf{v}_G)_y$ *about P plus the angular momentum* $I_G\boldsymbol{\omega}$. Note that since $\boldsymbol{\omega}$ is a free vector, $\mathbf{H}_G$ *can act at any point on the body* provided it preserves its same magnitude and direction. Furthermore, since angular momentum is equal to the "moment" of the linear momentum, the *line of action of* **L** *must pass through the body's mass center G* in order to preserve the correct magnitude of $\mathbf{H}_P$ when "moments" are computed about P, Fig. 19–1*c*. As a result of this analysis, we will now consider three types of motion.

Translation. When a rigid body of mass m is subjected to either rectilinear or curvilinear *translation,* Fig. 19–2a, its mass center has a velocity of $\mathbf{v}_G = \mathbf{v}$ and $\boldsymbol{\omega} = \mathbf{0}$ for the body. Hence, the linear momentum and the angular momentum computed about G become

$$\begin{aligned} L &= mv_G \\ H_G &= 0 \end{aligned} \tag{19–6}$$

If the angular momentum is computed about any other point A on or off the body, Fig. 19–2a, the "moment" of the linear momentum $\mathbf{L}$ must be computed about the point. Since d is the "moment arm" as shown in the figure, then in accordance with Eq. 19–5, $H_A = (d)(mv_G) \circlearrowleft$.

Rotation About a Fixed Axis. When a rigid body is *rotating about a fixed axis* passing through point O, Fig. 19–2b, the linear momentum and the angular momentum computed about G are

$$\begin{aligned} L &= mv_G \\ H_G &= I_G\omega \end{aligned} \tag{19–7}$$

It is sometimes convenient to compute the angular momentum of the body about point O. In this case it is necessary to account for the "moments" of *both* $\mathbf{L}$ and $\mathbf{H}_G$ about O. Noting that $\mathbf{L}$ (or $\mathbf{v}_G$) is always *perpendicular to* $\mathbf{r}_G$, we have

$$\circlearrowright + \qquad H_O = I_G\omega + r_G(mv_G) \tag{19–8}$$

This equation may be *simplified* by first substituting $v_G = r_G\omega$, in which case $H_O = (I_G + mr_G^2)\omega$, and, by the parallel-axis theorem, noting that the terms inside the parentheses represent the moment of inertia I_O of the body about an axis perpendicular to the plane of motion and passing through point O. Hence,*

$$H_O = I_O\omega \tag{19–9}$$

For the computation, then, either Eq. 19–8 or 19–9 can be used.

*The similarity between this derivation and that of Eq. 17–16 ($\Sigma M_O = I_O\alpha$) and Eq. 18–5 ($T = \frac{1}{2}I_O\omega^2$) should be noted. Also note that the same result can be obtained from Eq. 19–2 by selecting point P at O, realizing that $(v_O)_x = (v_O)_y = 0$.

General Plane Motion. When a rigid body is subjected to general plane motion, Fig. 19–2*c*, the linear momentum and the angular momentum computed about G become

$$\begin{aligned} L &= mv_G \\ H_G &= I_G\omega \end{aligned} \qquad (19\text{–}10)$$

If the angular momentum is computed about a point A located either on or off the body, Fig. 19–2*c*, it is necessary to compute the moments of *both* $\mathbf{L}$ and $\mathbf{H}_G$ about this point. In this case,

$$\circlearrowleft + \qquad H_A = I_G\omega + (d)(mv_G)$$

Here d is the moment arm, as shown in the figure.

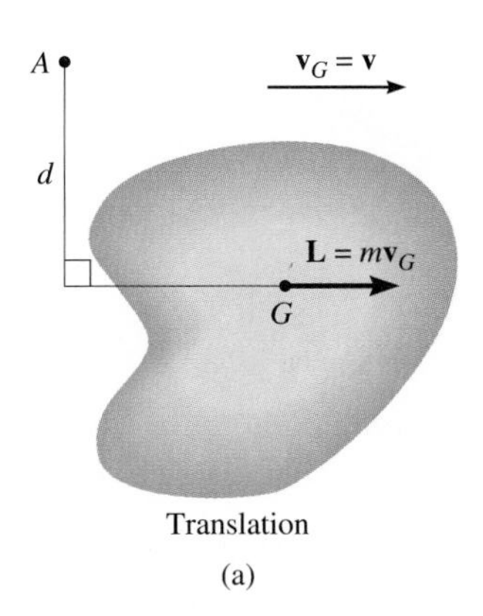

Translation

(a)

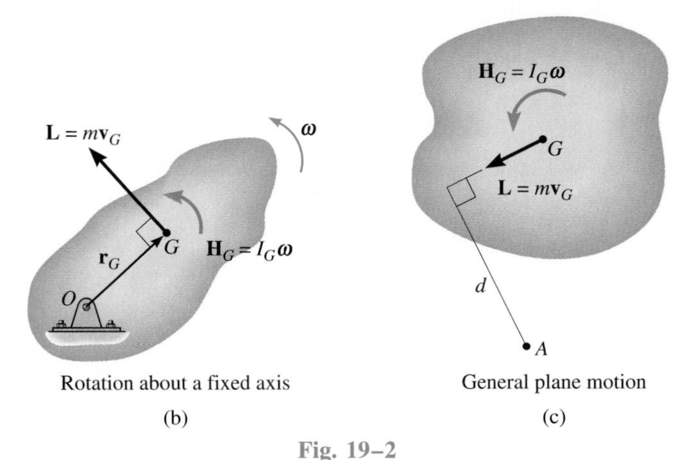

Rotation about a fixed axis

(b)

General plane motion

(c)

Fig. 19–2

Example 19–1

At a given instant the 10-kg disk and 5-kg bar have the motion shown in Fig. 19–3a. Determine their angular momentum about point G and about point B for the disk, and about point G and the IC for the bar.

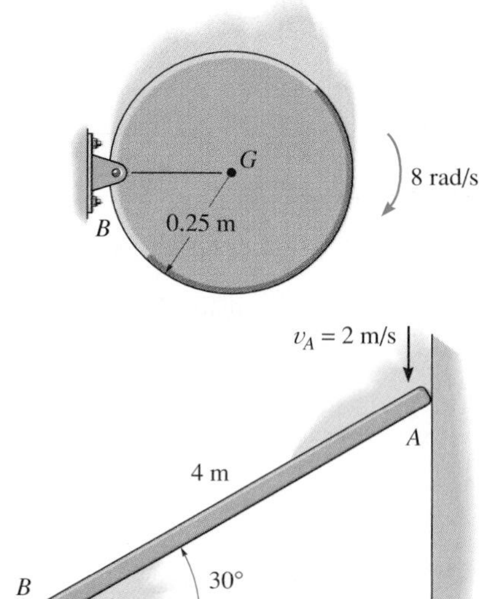

(a)

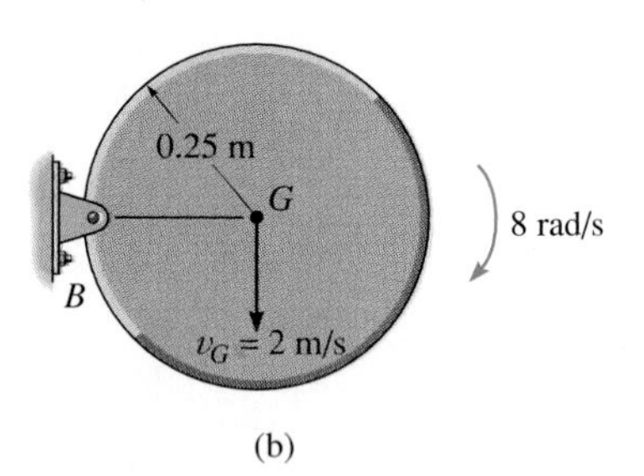

(b)

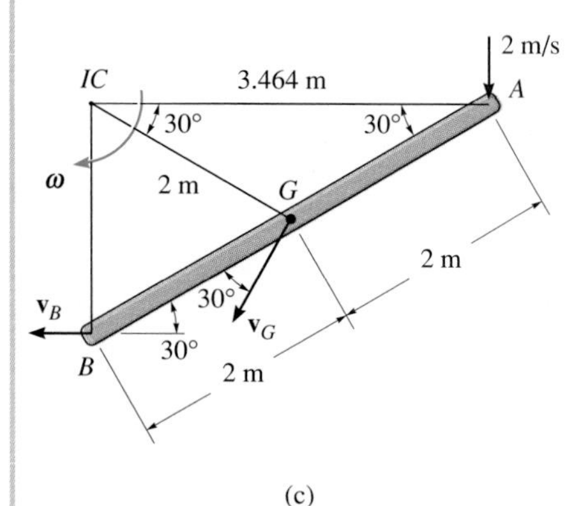

(c)

Fig. 19–3

SOLUTION

Disk. Since the disk is *rotating about a fixed axis* (through point B), then $v_G = (8 \text{ rad/s})(0.25 \text{ m}) = 2 \text{ m/s}$, Fig. 19–3$b$. Hence

$$\circlearrowright + H_G = I_G\omega = [\tfrac{1}{2}(10 \text{ kg})(0.25 \text{ m})^2](8 \text{ rad/s}) = 2.50 \text{ kg}\cdot\text{m}^2/\text{s} \qquad \textit{Ans.}$$

$$\circlearrowright + H_B = I_G\omega + (mv_G)r_G = 2.50 \text{ kg}\cdot\text{m}^2/\text{s} + (10 \text{ kg})(2 \text{ m/s})(0.25 \text{ m})$$
$$= 7.50 \text{ kg}\cdot\text{m}^2/\text{s} \circlearrowright \qquad \textit{Ans.}$$

Also, since $I_B = (3/2)mr^2$, then

$$\circlearrowright + H_B = I_B\omega = [\tfrac{3}{2}(10 \text{ kg})(0.25 \text{ m})^2](8 \text{ rad/s}) = 7.50 \text{ kg}\cdot\text{m}^2/\text{s} \circlearrowright \qquad \textit{Ans.}$$

Bar. The bar undergoes *general plane motion*. The IC is established in Fig. 19–3c, so that $\omega = (2 \text{ m/s})/(3.464 \text{ m}) = 0.5774 \text{ rad/s}$ and $v_G = (0.5774 \text{ rad/s})(2 \text{ m}) = 1.155 \text{ m/s}$. Thus,

$$\circlearrowright + H_G = I_G\omega = [\tfrac{1}{12}(5 \text{ kg})(4 \text{ m})^2](0.5774 \text{ rad/s}) = 3.85 \text{ kg}\cdot\text{m}^2/\text{s} \circlearrowright \qquad \textit{Ans.}$$

Moments of $I_G\omega$ and mv_G about the IC yield

$$\circlearrowright + H_{IC} = I_G\omega + d(mv_G) = 3.85 \text{ kg}\cdot\text{m}^2/\text{s} + (2 \text{ m})(5 \text{ kg})(1.155 \text{ m/s})$$
$$= 15.4 \text{ kg}\cdot\text{m}^2/\text{s} \circlearrowright \qquad \textit{Ans.}$$

19.2 Principle of Impulse and Momentum

Like the case for particle motion, the principle of impulse and momentum for a rigid body is developed by *combining* the equation of motion with kinematics. The resulting equation will allow a *direct solution to problems involving force, velocity, and time.*

Principle of Linear Impulse and Momentum. The equation of translational motion for a rigid body can be written as $\Sigma \mathbf{F} = m\mathbf{a}_G = m\,(d\mathbf{v}_G/dt)$. Since the mass of the body is constant,

$$\Sigma \mathbf{F} = \frac{d}{dt}(m\mathbf{v}_G)$$

Multiplying both sides by dt and integrating from $t = t_1$, $\mathbf{v}_G = (\mathbf{v}_G)_1$ to $t = t_2$, $\mathbf{v}_G = (\mathbf{v}_G)_2$ yields

$$\Sigma \int_{t_1}^{t_2} \mathbf{F}\,dt = m(\mathbf{v}_G)_2 - m(\mathbf{v}_G)_1 \qquad (19\text{–}11)$$

This equation is referred to as the *principle of linear impulse and momentum.* It states that the sum of all the impulses created by the *external force system* which acts on the body during the time interval t_1 to t_2 is equal to the change in the linear momentum of the body during the time interval, Fig. 19-4.

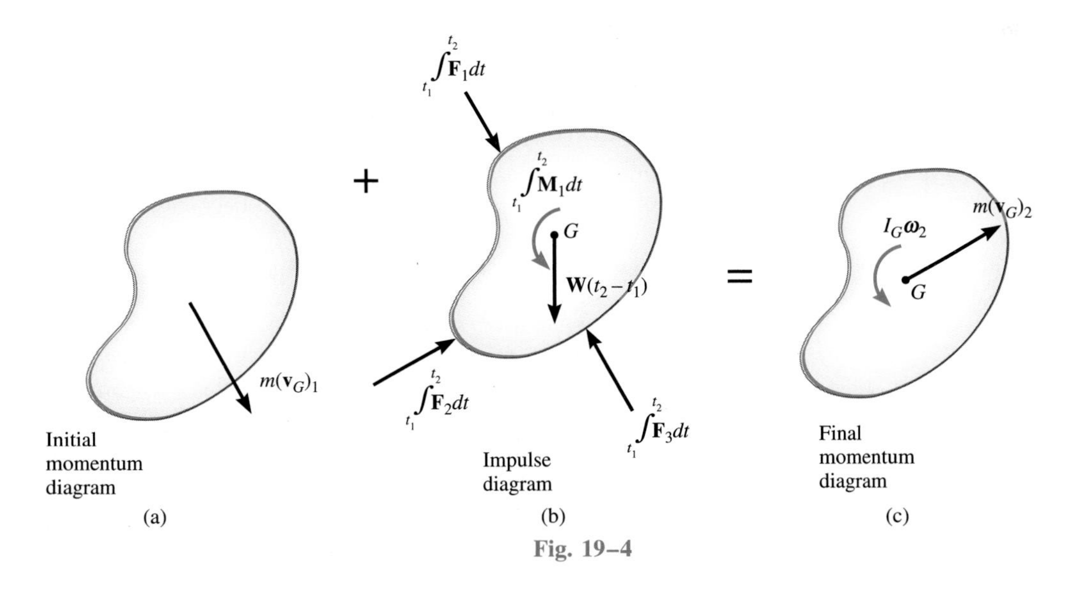

Fig. 19–4

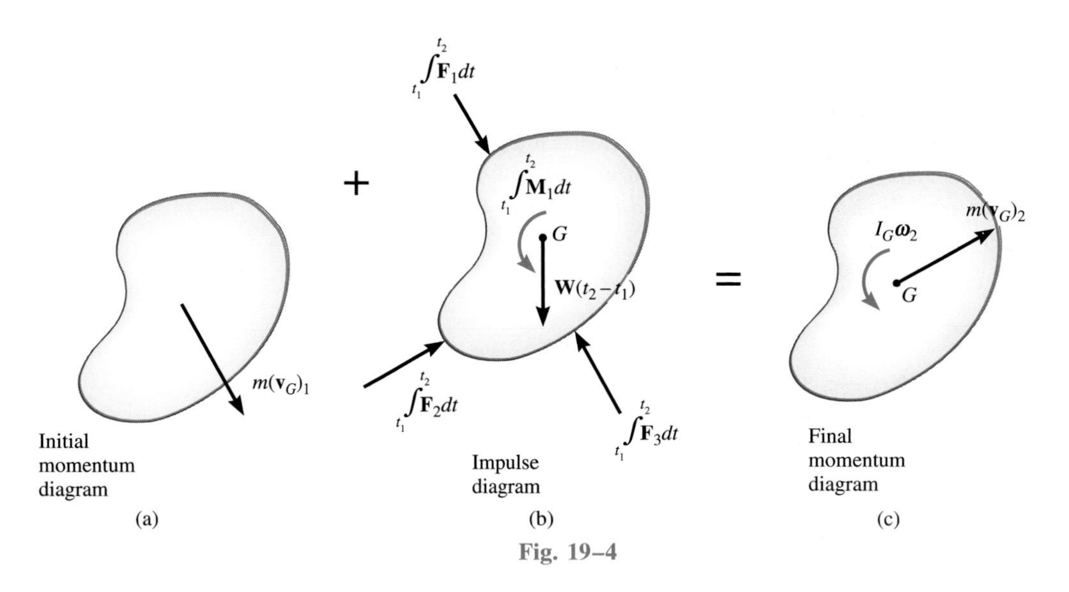

Fig. 19–4

Principle of Angular Impulse and Momentum. If the body has *general plane motion* we can write $\Sigma M_G = I_G\alpha = I_G(d\omega/dt)$. Since the moment of inertia is constant,

$$\Sigma M_G = \frac{d}{dt}(I_G\omega)$$

Multiplying both sides by dt and integrating from $t = t_1$, $\omega = \omega_1$ to $t = t_2$, $\omega = \omega_2$ gives

$$\Sigma \int_{t_1}^{t_2} M_G \, dt = I_G\omega_2 - I_G\omega_1 \qquad (19\text{–}12)$$

In a similar manner, for *rotation about a fixed axis* passing through point O, Eq. 17–16 ($\Sigma M_O = I_O\alpha$) when integrated becomes

$$\Sigma \int_{t_1}^{t_2} M_O \, dt = I_O\omega_2 - I_O\omega_1 \qquad (19\text{–}13)$$

Equations 19–12 and 19–13 are referred to as the *principle of angular impulse and momentum.* Both equations state that the sum of the angular impulses acting on the body during the time interval t_1 to t_2 is equal to the change in the body's angular momentum during this time interval. In particular, the angular impulse considered is determined by integrating the moments about point G or O of all the external forces and couple moments applied to the body.

To summarize the preceding concepts, if motion is occurring in the x–y plane, using impulse and momentum principles the following *three scalar equations* may be written which describe the *planar motion* of the body:

$$
\begin{aligned}
m(v_{Gx})_1 + \Sigma \int_{t_1}^{t_2} F_x \, dt &= m(v_{Gx})_2 \\
m(v_{Gy})_1 + \Sigma \int_{t_1}^{t_2} F_y \, dt &= m(v_{Gy})_2 \\
I_G \omega_1 + \Sigma \int_{t_1}^{t_2} M_G \, dt &= I_G \omega_2
\end{aligned}
\qquad (19\text{–}14)
$$

The first two of these equations represent the principle of linear impulse and momentum in the x–y plane, Eq. 19–11, and the third equation represents the principle of angular impulse and momentum about the z axis, which passes through the body's mass center G, Eq. 19–12.

The terms in Eqs. 19–14 can be graphically accounted for by drawing a set of impulse and momentum diagrams for the body, Fig. 19–4. Note that the linear momenta $m\mathbf{v}_G$ are applied at the body's mass center, Figs. 19–4*a* and 19–4*c*; whereas the angular momenta $I_G\boldsymbol{\omega}$ are free vectors, and therefore, like a couple moment, they may be applied at any point on the body. When the impulse diagram is constructed, Fig. 19–4*b*, vectors **F** and **M,** which vary with time, are indicated by the integrals. However, if **F** and **M** are *constant* from t_1 to t_2, integration of the impulses yields $\mathbf{F}(t_2 - t_1)$ and $\mathbf{M}(t_2 - t_1)$, respectively. Such is the case for the body's weight **W,** Fig. 19–4*b*.

Equations 19–14 may also be applied to an entire system of connected bodies rather than to each body separately. Doing this eliminates the need to include reactive impulses which occur at the connections since they are *internal* to the system. The resultant equations may be written in symbolic form as

$$
\begin{aligned}
\left(\Sigma \begin{matrix}\text{syst. linear}\\ \text{momentum}\end{matrix}\right)_{x1} + \left(\Sigma \begin{matrix}\text{syst. linear}\\ \text{impulse}\end{matrix}\right)_{x(1-2)} &= \left(\Sigma \begin{matrix}\text{syst. linear}\\ \text{momentum}\end{matrix}\right)_{x2} \\
\left(\Sigma \begin{matrix}\text{syst. linear}\\ \text{momentum}\end{matrix}\right)_{y1} + \left(\Sigma \begin{matrix}\text{syst. linear}\\ \text{impulse}\end{matrix}\right)_{y(1-2)} &= \left(\Sigma \begin{matrix}\text{syst. linear}\\ \text{momentum}\end{matrix}\right)_{y2} \\
\left(\Sigma \begin{matrix}\text{syst. angular}\\ \text{momentum}\end{matrix}\right)_{O1} + \left(\Sigma \begin{matrix}\text{syst. angular}\\ \text{impulse}\end{matrix}\right)_{O(1-2)} &= \left(\Sigma \begin{matrix}\text{syst. angular}\\ \text{momentum}\end{matrix}\right)_{O2}
\end{aligned}
$$

(19–15)

As indicated, the system's angular momentum and angular impulse must be computed with respect to the *same fixed reference point O* for all the bodies of the system.

PROCEDURE FOR ANALYSIS

Impulse and momentum principles are used to solve kinetics problems that involve *velocity, force,* and *time* since these terms are involved in the formulation. For application it is suggested that the following procedure be used.

Free-Body Diagram. Establish the x, y, z inertial frame of reference and draw the free-body diagram in order to account for all the forces that produce impulses on the body. The direction and sense of the initial and final velocity of the body's mass center, $\mathbf{v}_G$, and the body's angular velocity $\boldsymbol{\omega}$ should also be established. If any of these motions is unknown, assume that the sense of its components is in the direction of the positive inertial coordinates. To account for these vectors, they may be sketched on the x, y, z coordinate system, but not on the free-body diagram. Also compute the moment of inertia I_G or I_O.

As an alternative procedure, draw the impulse and momentum diagrams for the body or system of bodies. Each of these diagrams represents an outlined shape of the body which accounts graphically for the data required for each of the three terms in Eqs. 19–14 or 19–15, Fig. 19–4. Note that these diagrams are particularly helpful in order to visualize the "moment" terms used in the principle of angular impulse and momentum, when it has been decided that application of this equation is to be about a point other than the body's mass center G or a fixed point O.

Principle of Impulse and Momentum. Apply the three scalar equations 19–14 (or 19–15). In cases where the body is rotating about a fixed axis, Eq. 19–13 may be substituted for the third of Eqs. 19–14.

Kinematics. If more than three equations are needed for a complete solution, it may be possible to relate the velocity of the body's mass center to the body's angular velocity using *kinematics.* If the motion appears to be complicated, kinematic (velocity) diagrams may be helpful in obtaining the necessary relation.

In general, a method to be used for the solution of a particular type of problem should be decided upon *before* attempting to solve the problem. As stated above, the *principle of impulse and momentum* is most suitable for solving problems which involve *velocity, force,* and *time.* For some problems, however, a combination of the equation of motion and its two integrated forms, the principle of work and energy and the principle of impulse and momentum, will yield the most direct solution to the problem.

Example 19–2

The disk shown in Fig. 19–5*a* weighs 20 lb and is pin-supported at its center. If it is acted upon by a constant couple moment of 4 lb · ft and a force of 10 lb which is applied to a cord wrapped around its periphery, determine the angular velocity of the disk two seconds after starting from rest. Also, what are the force components of reaction at the pin?

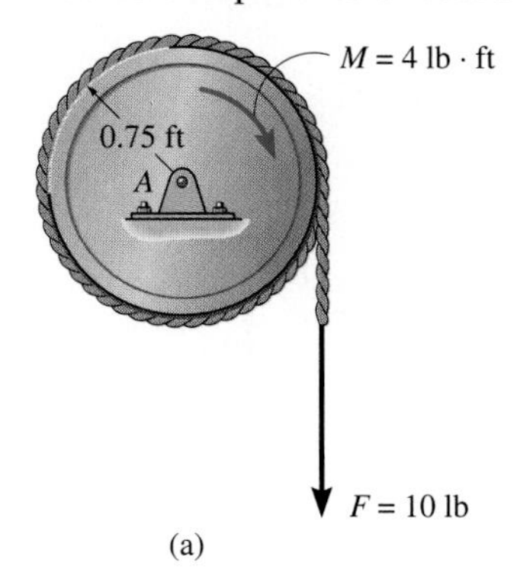

(a)

SOLUTION

Free-Body Diagram Fig. 19–5*b*. The disk's mass center does not move; however, the loading causes the disk to rotate clockwise.

The moment of inertia of the disk about its fixed axis of rotation is

$$I_A = \frac{1}{2}mr^2 = \frac{1}{2}\left(\frac{20 \text{ lb}}{32.2 \text{ ft/s}^2}\right)(0.75 \text{ ft})^2 = 0.175 \text{ slug} \cdot \text{ft}^2$$

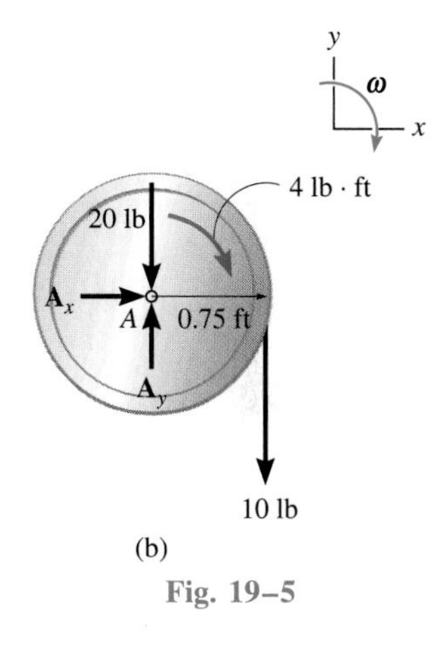

(b)

Fig. 19–5

Principle of Impulse and Momentum

$(\overset{+}{\rightarrow})$
$$m(v_{Ax})_1 + \Sigma\int_{t_1}^{t_2} F_x\,dt = m(v_{Ax})_2$$
$$0 + A_x(2 \text{ s}) = 0$$

$(+\uparrow)$
$$m(v_{Ay})_1 + \Sigma\int_{t_1}^{t_2} F_y\,dt = m(v_{Ay})_2$$
$$0 + A_y(2 \text{ s}) - 20 \text{ lb}(2 \text{ s}) - 10 \text{ lb}(2 \text{ s}) = 0$$

$(\curvearrowright +)$
$$I_A\omega_1 + \Sigma\int_{t_1}^{t_2} M_A\,dt = I_A\omega_2$$
$$0 + 4 \text{ lb} \cdot \text{ft}(2 \text{ s}) + [10 \text{ lb}(2 \text{ s})](0.75 \text{ ft}) = 0.175\omega_2$$

Solving these equations yields

$A_x = 0$ *Ans.*

$A_y = 30 \text{ lb}$ *Ans.*

$\omega_2 = 132 \text{ rad/s} \curvearrowright$ *Ans.*

Example 19–3

The block shown in Fig. 19–6*a* has a mass of 6 kg. It is attached to a cord which is wrapped around the periphery of a 20-kg disk that has a moment of inertia $I_A = 0.40 \text{ kg} \cdot \text{m}^2$. If the block is initially moving downward with a speed of 2 m/s, determine its speed in 3 s. Neglect the mass of the cord in the calculation.

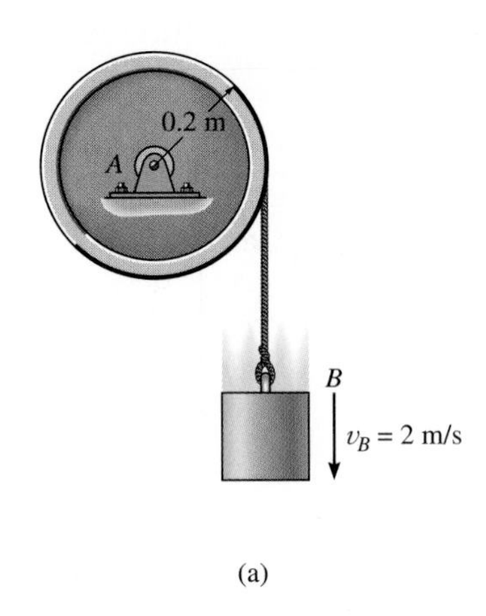

(a)

SOLUTION I

Free-Body Diagram. The free-body diagrams of the block and disk are shown in Fig. 19–6*b*. All the forces are *constant* since the weight of the block causes the motion. The downward motion of the block, $\mathbf{v}_B$, causes $\boldsymbol{\omega}$ of the disk to be clockwise.

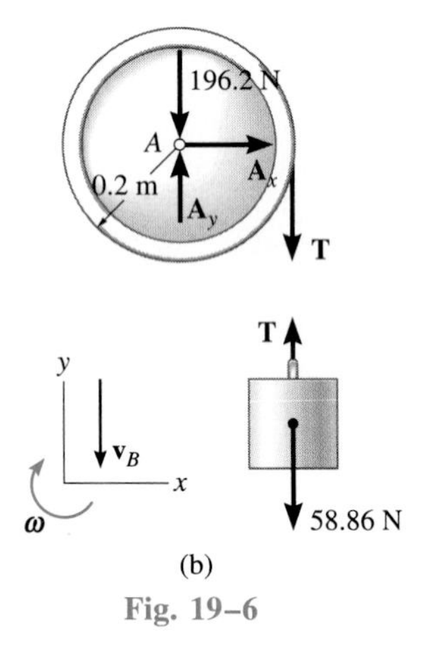

(b)

Fig. 19–6

Principle of Impulse and Momentum. We can eliminate $\mathbf{A}_x$ and $\mathbf{A}_y$ from the analysis by applying the principle of angular momentum about point A. Hence

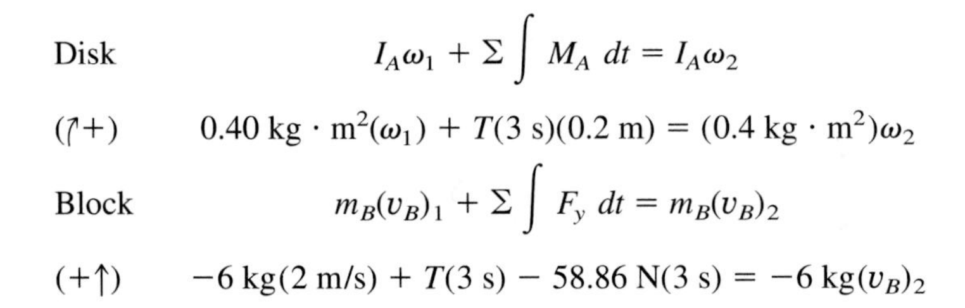

Disk $\qquad I_A\omega_1 + \Sigma \int M_A \, dt = I_A\omega_2$

$(\curvearrowleft +) \qquad 0.40 \text{ kg} \cdot \text{m}^2(\omega_1) + T(3 \text{ s})(0.2 \text{ m}) = (0.4 \text{ kg} \cdot \text{m}^2)\omega_2$

Block $\qquad m_B(v_B)_1 + \Sigma \int F_y \, dt = m_B(v_B)_2$

$(+\uparrow) \qquad -6 \text{ kg}(2 \text{ m/s}) + T(3 \text{ s}) - 58.86 \text{ N}(3 \text{ s}) = -6 \text{ kg}(v_B)_2$

Kinematics. Since $\omega = v_B/r$, then $\omega_1 = (2 \text{ m/s})/(0.2 \text{ m}) = 10$ rad/s and $\omega_2 = (v_B)_2/0.2 \text{ m} = 5(v_B)_2$. Substituting and solving the equations simultaneously for $(v_B)_2$ yields

$$(v_B)_2 = 13.0 \text{ m/s} \downarrow \qquad \textit{Ans.}$$

SOLUTION II

Impulse and Momentum Diagrams. We can obtain $(v_B)_2$ *directly* by considering the *system* consisting of the block, the cord, and the disk. The impulse and momentum diagrams have been drawn to clarify application of the principle of angular impulse and momentum about point A, Fig. 19–6c.

Principle of Angular Impulse and Momentum. Realizing that $\omega_1 = 10$ rad/s and $\omega_2 = 5(v_B)_2$, we have

$$\left(\Sigma \begin{matrix}\text{syst. angular}\\ \text{momentum}\end{matrix}\right)_{A1} + \left(\Sigma \begin{matrix}\text{syst. angular}\\ \text{impulse}\end{matrix}\right)_{A(1-2)} = \left(\Sigma \begin{matrix}\text{syst. angular}\\ \text{momentum}\end{matrix}\right)_{A2}$$

$$(\circlearrowleft +) \quad 6\text{ kg}(2\text{ m/s})(0.2\text{ m}) + 0.4\text{ kg}\cdot\text{m}^2(10\text{ rad/s}) + 58.86\text{ N}(3\text{ s})(0.2\text{ m})$$
$$= 6\text{ kg}(v_B)_2(0.2\text{ m}) + 0.40\text{ kg}\cdot\text{m}^2[(v_B)_2(0.2\text{ m})]$$
$$(v_B)_2 = 13.0\text{ m/s}\downarrow \qquad \textit{Ans.}$$

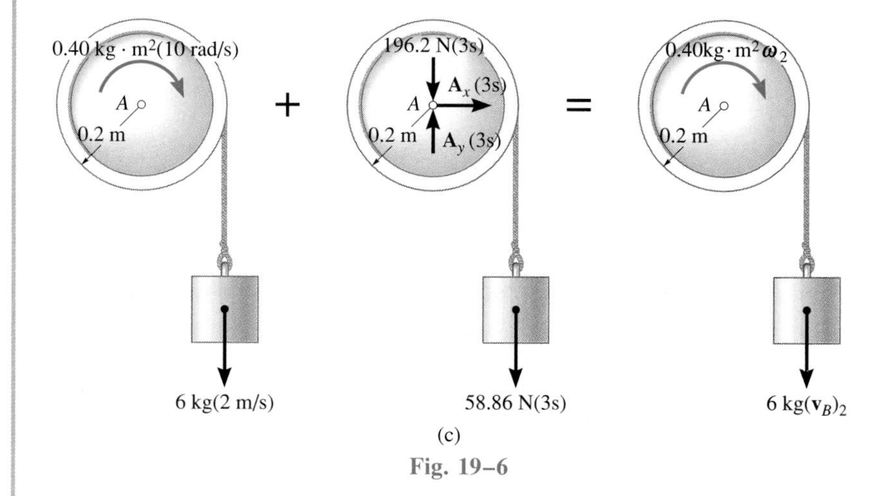

(c)

Fig. 19–6

Example 19–4

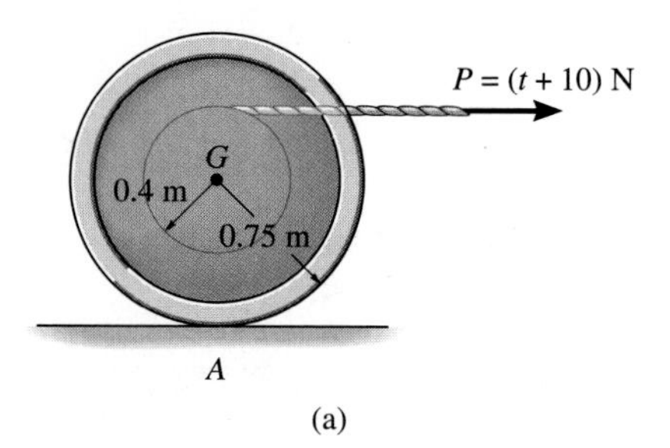

(a)

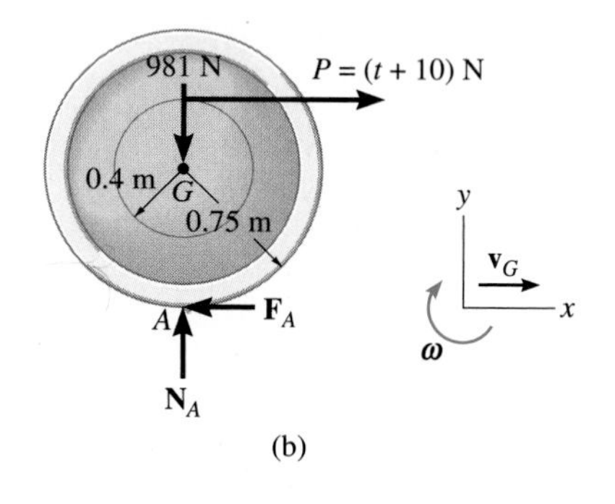

(b)

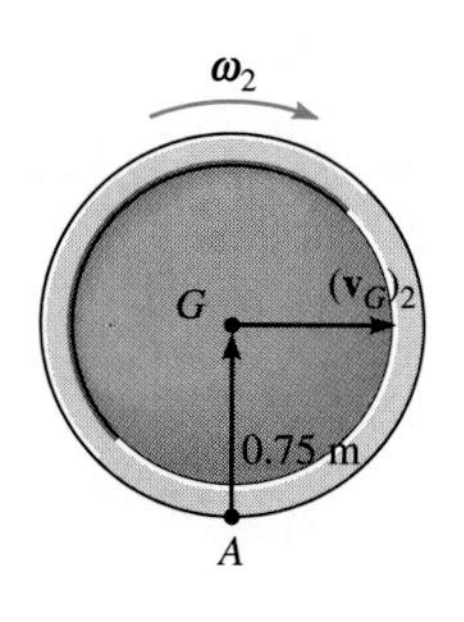

(c)

Fig. 19–7

The 100-kg spool shown in Fig. 19–7*a* has a radius of gyration $k_G = 0.35$ m. A cable is wrapped around the central hub of the spool and a horizontal force having a variable magnitude of $P = (t + 10)$ N is applied, where t is measured in seconds. If the spool is initially at rest, determine its angular velocity in 5 s. Assume that the spool rolls without slipping at A.

SOLUTION

Free-Body Diagram. By inspection of the free-body diagram, Fig. 19–7*b*, the *variable* force P will cause the friction force F_A to be variable, and thus the impulses created by both P and F_A must be determined by integration. The force $\mathbf{P}$ causes the mass center to have a velocity $\mathbf{v}_G$ to the right, and the spool has a clockwise angular velocity $\boldsymbol{\omega}$.

The moment of inertia of the spool about its mass center is

$$I_G = mk_G^2 = 100(0.35)^2 = 12.25 \text{ kg}\cdot\text{m}^2$$

Principle of Impulse and Momentum

$$(\overset{+}{\rightarrow}) \qquad m(v_G)_1 + \Sigma\int F_x\,dt = m(v_G)_2$$

$$0 + \int_0^{5\text{ s}} (t+10)\text{ N}\,dt - \int F_A\,dt = 100\text{ kg}(v_G)_2$$

$$62.5 - \int F_A\,dt = 100(v_G)_2 \qquad (1)$$

$$(\curvearrowright +) \qquad I_G\omega_1 + \Sigma\int M_G\,dt = I_G\omega_2$$

$$0 + \left[\int_0^{5\text{ s}} (t+10)\text{ N}\,dt\right](0.4\text{ m}) + \left(\int F_A\,dt\right)(0.75\text{ m}) = (12.25\text{ kg}\cdot\text{m}^2)\,\omega_2$$

$$25 + \left(\int F_A\,dt\right)(0.75) = 12.25\,\omega_2 \qquad (2)$$

Kinematics. Since the spool does not slip, the instantaneous center of zero velocity is at point A, Fig. 19–7*c*. Hence, the velocity of G can be expressed in terms of the spool's angular velocity as $(v_G)_2 = (0.75\text{ m})\omega_2$. Substituting this into Eq. 1 and eliminating the unknown impulse $\int F_A\,dt$ between Eqs. 1 and 2, we obtain

$$\omega_2 = 1.05 \text{ rad/s} \curvearrowright \qquad \textit{Ans.}$$

Note: A more direct solution can be obtained by applying the principle of angular impulse and momentum about point A. As an exercise, do this and show that one obtains the same result.

Example 19–5

The Charpy impact test is used in materials testing to determine the energy absorption characteristics of a material during impact. The test is performed using the pendulum shown in Fig. 19–8*a*, which has a mass m, mass center at G, and a radius of gyration k_G about G. Determine the distance r_P from the pin at A to the point P where the impact with the specimen S should occur so that the horizontal force at the pin is essentially zero during the impact. For the computation, assume the specimen absorbs all the pendulum's kinetic energy during the time it falls and thereby stops the pendulum from swinging when $\theta = 0°$.

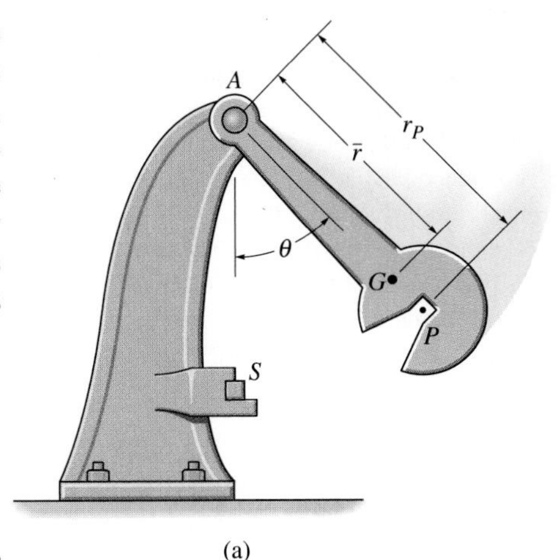

(a)

SOLUTION

Free-Body Diagram. As shown on the free-body diagram, Fig. 19–8*b*, the conditions of the problem require the horizontal impulse at A to be zero. Just before impact, the pendulum has a clockwise angular velocity $\boldsymbol{\omega}_1$ and the mass center of the pendulum is moving to the left at $(v_G)_1 = \bar{r}\omega_1$.

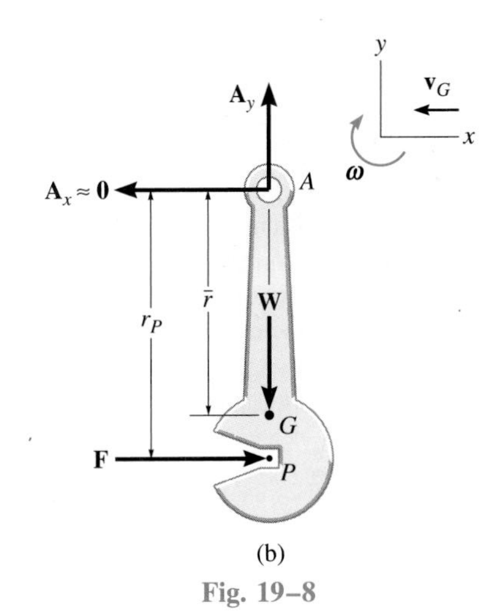

(b)

Fig. 19–8

Principle of Impulse and Momentum. We will apply the principle of angular impulse and momentum about point A. Thus,

$$(\curvearrowright +) \qquad I_A\omega_1 - \left(\int F\,dt\right)r_P = 0$$

$$(\overset{+}{\leftarrow}) \qquad -m(\bar{r}\omega_1) + \int F\,dt = 0$$

Eliminating the impulse $\int F\,dt$ and substituting $I_A = mk_G^2 + m\bar{r}^2$ yields

$$[mk_G^2 + m\bar{r}^2]\omega_1 - m(\bar{r}\omega_1)r_P = 0$$

Factoring out $m\omega_1$ and solving for r_P, we obtain

$$r_P = \bar{r} + \frac{k_G^2}{\bar{r}} \qquad \textit{Ans.}$$

The point P, so defined, is called the *center of percussion*. By placing the striking point at P, the force developed at the pin will be minimized. Many sports rackets, clubs, etc. are designed so that the object being struck occurs at the center of percussion. As a consequence, no "sting" or little sensation occurs in the hand of the player. (Also see Probs. 17–95 and 19–1.)

PROBLEMS

19–1. The rigid body (slab) has a mass m and is rotating with an angular velocity $\boldsymbol{\omega}$ about an axis passing through the fixed point O. Show that the momenta of all the particles composing the body can be represented by a single vector having a magnitude mv_G and acting through point P, called the *center of percussion*, which lies at a distance $r_{P/G} = k_G^2/r_{G/O}$ from the mass center G. Here k_G is the radius of gyration of the body, computed about an axis perpendicular to the plane of motion and passing through G.

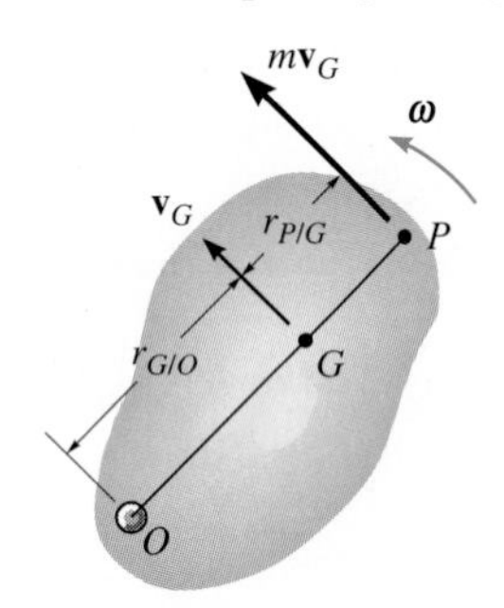

Prob. 19–1

19–2. At a given instant, the body has a linear momentum $\mathbf{L} = m\mathbf{v}_G$ and an angular momentum $\mathbf{H}_G = I_G\boldsymbol{\omega}$ computed about its mass center. Show that the angular momentum of the body computed about the instantaneous center of zero velocity IC can be expressed as $\mathbf{H}_{IC} = I_{IC}\boldsymbol{\omega}$, where I_{IC} represents the body's moment of inertia computed about the instantaneous axis of zero velocity. As shown, the IC is located at a distance $r_{G/IC}$ away from the mass center G.

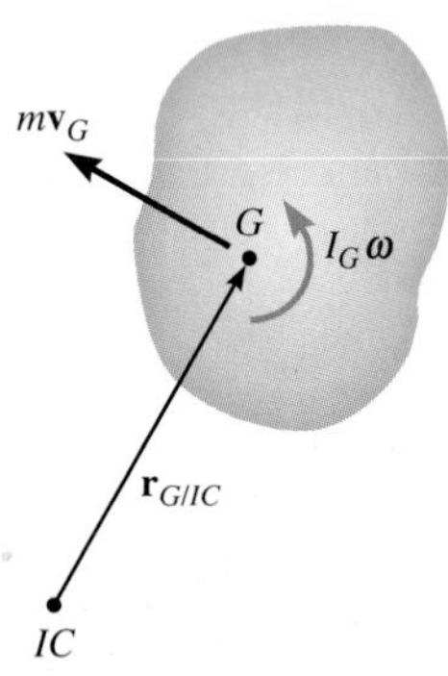

Prob. 19–2

19–3. Show that if a slab is rotating about a fixed axis perpendicular to the slab and passing through its mass center G, the angular momentum is the same when computed about any other point P on the slab.

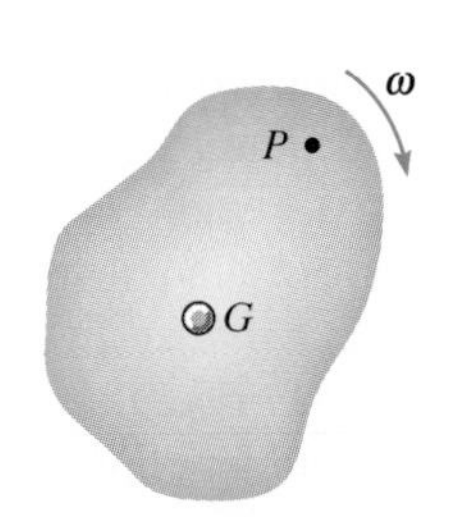

Prob. 19–3

***19–4.** Solve Prob. 17–64 using the principle of impulse and momentum.

19–5. Solve Prob. 17–61 using the principle of impulse and momentum.

19–6. The space capsule has a mass of 1200 kg and a moment of inertia $I_G = 900 \text{ kg} \cdot \text{m}^2$ about an axis passing through G and directed perpendicular to the page. If it is traveling forward with a speed $v_G = 800$ m/s and executes a turn by means of two jets, which provide a constant thrust of 400 N for 0.3 s, determine the capsule's angular velocity just after the jets are turned off.

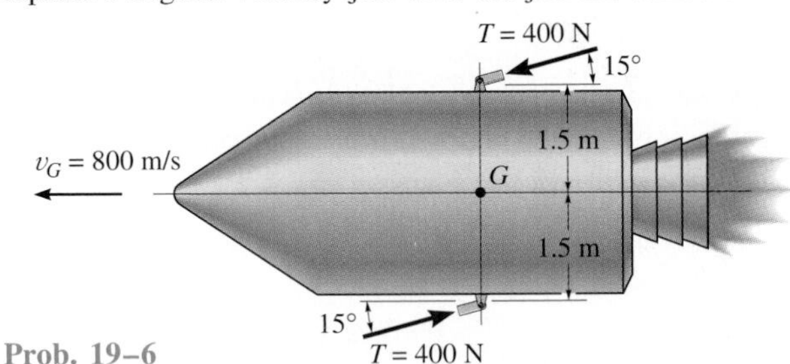

Prob. 19–6

19–7. The 25-lb disk has an angular velocity of 30 rad/s. It is suddenly brought into contact with the horizontal surface which has a coefficient of kinetic friction of $\mu_k = 0.2$. Determine how long it takes for the disk to stop spinning. What is the force acting along bar AB? Neglect the weight of AB.

***19–8.** Solve Prob. 19–7 if the disk is rotating counterclockwise at $\omega = 30$ rad/s.

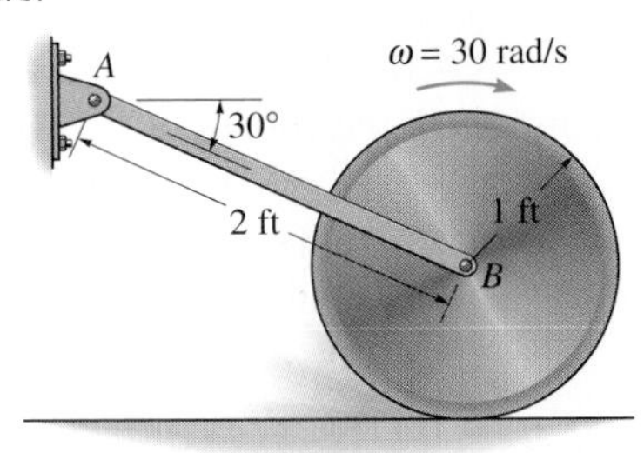

Probs. 19–7/19–8

19–9. The inner hub of the wheel rests on the inclined track. If it does not slip at A, determine its angular velocity 2 s after it is released from rest. The wheel has a weight of 30 lb and a radius of gyration $k_G = 1.30$ ft.

19–10. The inner hub of the wheel rests on the inclined track. If it does not slip at A determine the time required for its angular velocity to become $\omega = 5$ rad/s, starting from rest. The wheel has a weight of 30 lb and a radius of gyration $k_G = 1.30$ ft.

19–11. The inner hub of the wheel rests on the inclined track. If the coefficients of static and kinetic friction at A are $\mu_s = 0.3$ and $\mu_k = 0.2$, respectively, determine the wheel's angular velocity 2 s after it is released from rest. The wheel has a weight of 30 lb and a radius of gyration $k_G = 1.30$ ft.

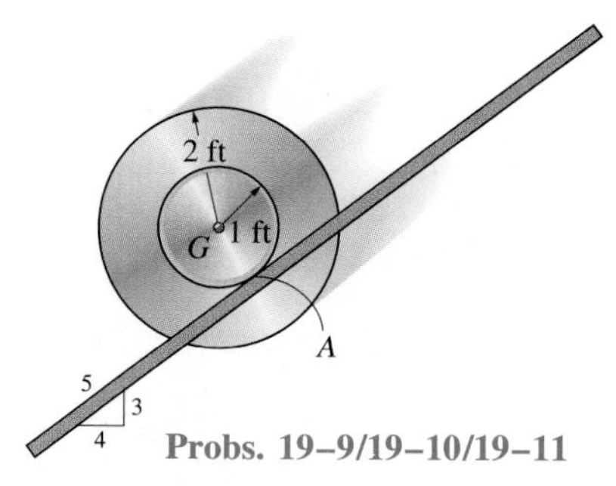

Probs. 19–9/19–10/19–11

***19–12.** The spool has a mass of 30 kg and a radius of gyration $k_O = 0.25$ m. Block A has a mass of 25 kg and block B has a mass of 10 kg. If they are released from rest, determine their speeds in 2 s starting from rest. Neglect the mass of the ropes.

19–13. The spool has a mass of 30 kg and a radius of gyration $k_O = 0.25$ m. Block A has a mass of 25 kg, and block B has a mass of 10 kg. If they are released from rest, determine the time required for block A to attain a speed of 2 m/s. Neglect the mass of the ropes.

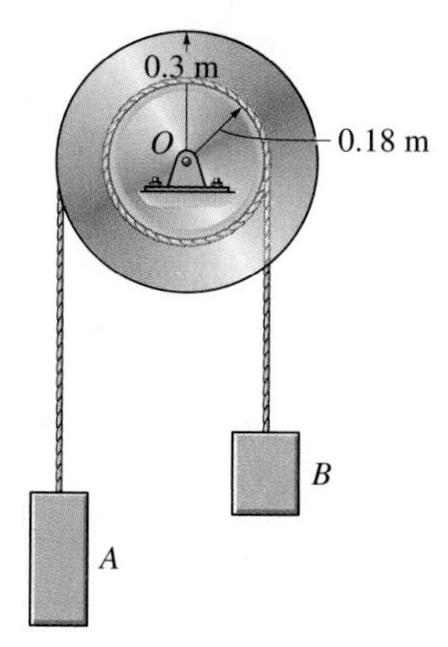

Probs. 19–12/19–13

19–14. The man pulls the rope off the reel with a constant force of 8 lb in the direction shown. If the reel has a weight of 250 lb and radius of gyration $k_G = 0.8$ ft about the trunnion (pin) at A, determine the angular velocity of the reel in 3 s starting from rest. Neglect friction and the weight of rope that is removed.

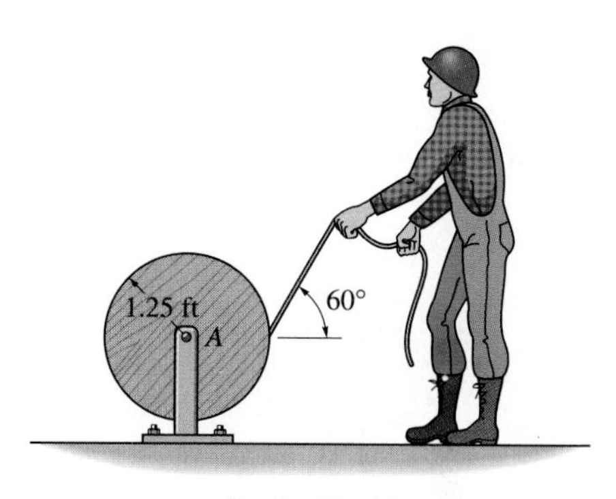

Prob. 19–14

19–15. Angular motion is transmitted from a driver wheel A to the driven wheel B by friction between the wheels at C. If A always rotates at a constant rate of 16 rad/s, and the coefficient of kinetic friction between the wheels is $\mu_k = 0.2$, determine the time required for B to reach a constant angular velocity once the wheels make contact with a normal force of 50 N. What is the final angular velocity of wheel B? Wheel B has a mass of 90 kg and a radius of gyration about its axis of rotation of $k_G = 120$ mm.

***19–16.** Solve Prob. 19–15 if the driver wheel A contacts wheel B at D. Does wheel B rotate in the same direction when contact is made at C or D?

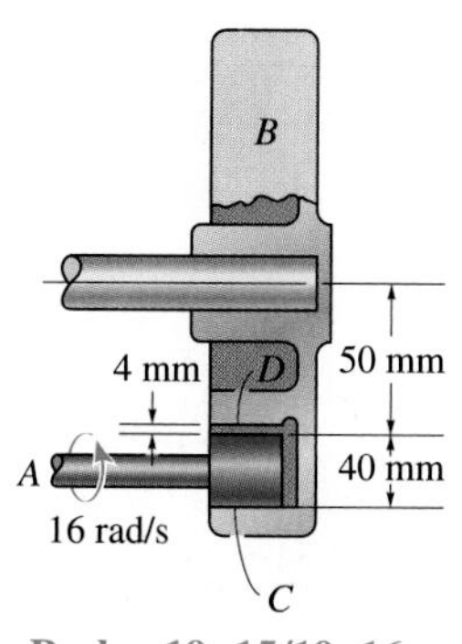

Probs. 19–15/19–16

19–17. The drum of mass m, radius r, and radius of gyration k_O rolls along an inclined plane for which the coefficient of static friction is μ. If the drum is released from rest, determine the maximum angle θ for the incline so that it rolls without slipping at A.

19–18. The drum has a mass of 70 kg, a radius of 300 mm, and radius of gyration $k_O = 125$ mm. If the coefficients of static and kinetic friction at A are $\mu_s = 0.4$ and $\mu_k = 0.3$, respectively, determine the drum's angular velocity 2 s after it is released from rest. Take $\theta = 30°$.

19–19. Solve Prob. 19–18 if $\theta = 75°$.

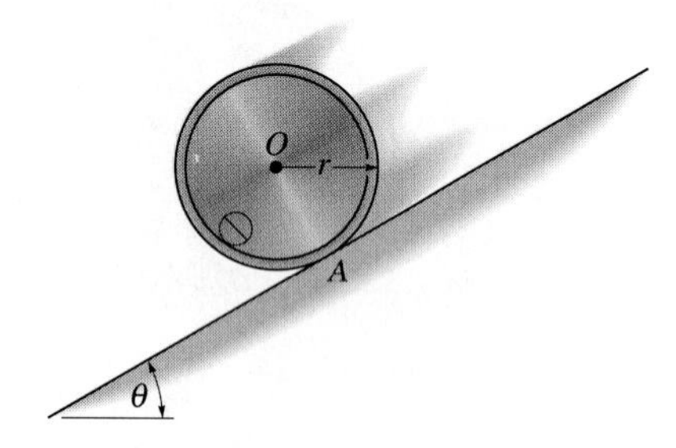

Probs. 19–17/19–18/19–19

***19–20.** The double pulley consists of two wheels which are attached to one another and turn at the same rate. The pulley has a mass of 15 kg and a radius of gyration $k_O = 110$ mm. If the block at A has a mass of 40 kg, determine the speed of the block in 3 s after a constant force $F = 2$ kN is applied to the rope wrapped around the inner hub of the pulley. The block is originally at rest. Neglect the mass of the rope.

19–21. The double pulley consists of two wheels which are attached to one another and turn at the same rate. The pulley has a mass of 15 kg and a radius of gyration $k_O = 110$ mm. If the block at A has a mass of 40 kg, determine the constant force F that should be applied to the rope in order to give the block a speed of 4 m/s in 3 s. The block is originally at rest. Neglect the mass of the rope.

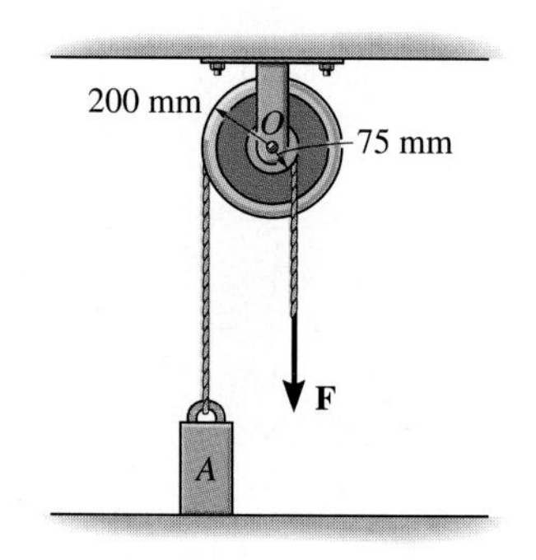

Probs. 19–20/19–21

19–22. The wheel weighs 100 lb and has a radius of gyration $k_G = 1.5$ ft. A cord is wrapped around the inner hub and is subjected to a force $P = (2t + 3)$ lb, where t is in seconds. If the wheel rolls without slipping, determine the speed of its center 5 s after the force is applied. The wheel starts from rest.

19–23. The wheel weighs 100 lb and has a radius of gyration $k_G = 1.5$ ft. A cord is wrapped around the inner hub and is subjected to a force $P = 8$ lb. If the wheel rolls without slipping, determine the speed of its center 5 s after the force is applied. The wheel starts from rest.

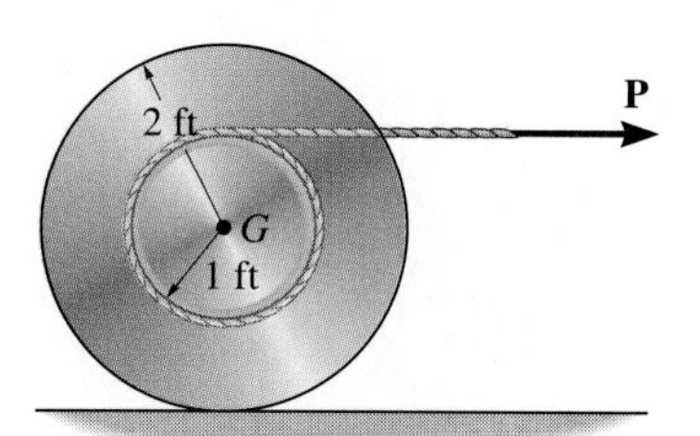

Probs. 19–22/19–23

***19–24.** The spool has a weight of 30 lb and a radius of gyration $k_O = 0.45$ ft. A cord is wrapped around its inner hub and the end subjected to a horizontal force $P = 5$ lb. Determine the spool's angular velocity in 4 s starting from rest. Assume the spool rolls without slipping.

19–25. Solve Prob. 19–24 if the coefficient of static friction is $\mu_s = 0.3$ and the coefficient of kinetic friction is $\mu_k = 0.25$.

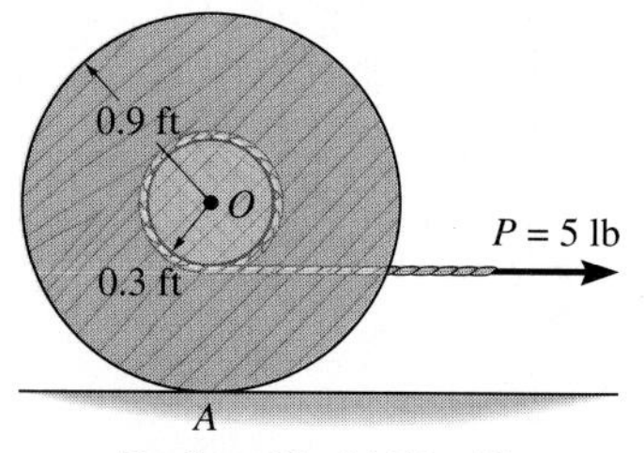

Probs. 19–24/19–25

19–26. The spool has a weight of 75 lb and a radius of gyration $k_O = 1.20$ ft. If the block B weighs 60 lb, and a force $P = 25$ lb is applied to the cord, determine the speed of the block in 5 s starting from rest. Neglect the mass of the cord.

19–27. The spool has a weight of 75 lb and a radius of gyration $k_O = 1.20$ ft. If the block B weighs 60 lb, determine the horizontal force P which must be applied to the cord in order to give the block an upward speed of 10 ft/s in 4 s, starting from rest. Neglect the mass of the cord.

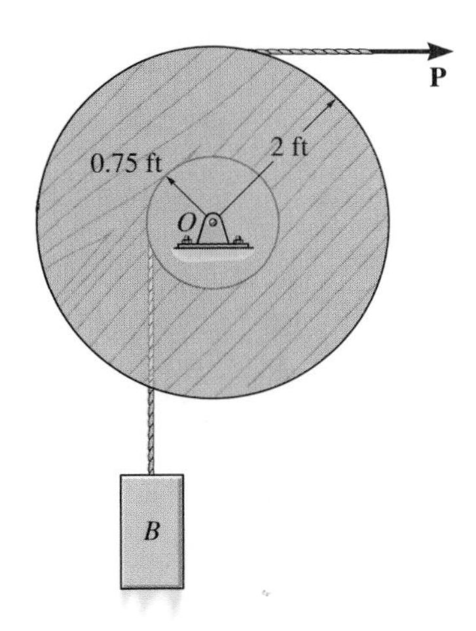

Probs. 19–26/19–27

***19–28.** The inner hub of the wheel rests on the horizontal track. If it does not slip at A, determine the speed of the 10-lb block in 2 s after the block is released from rest. The wheel has a weight of 30 lb and a radius of gyration $k_G = 1.30$ ft. Neglect the mass of the pulley and cord.

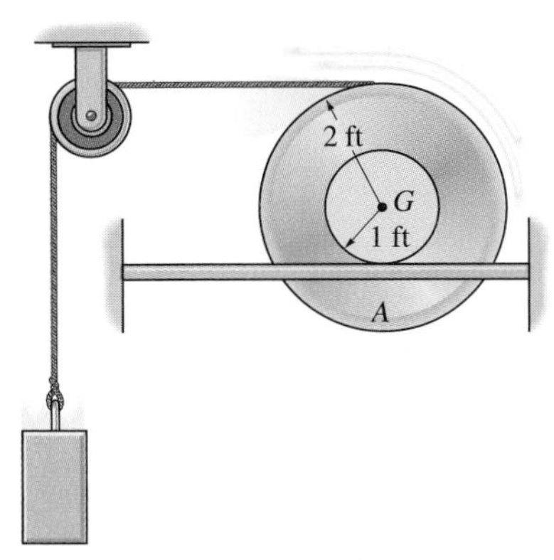

Prob. 19–28

19–29. The 10-lb rectangular plate is at rest on a smooth *horizontal* floor. If it is given the horizontal impulses shown, determine its angular velocity and the velocity of the mass center.

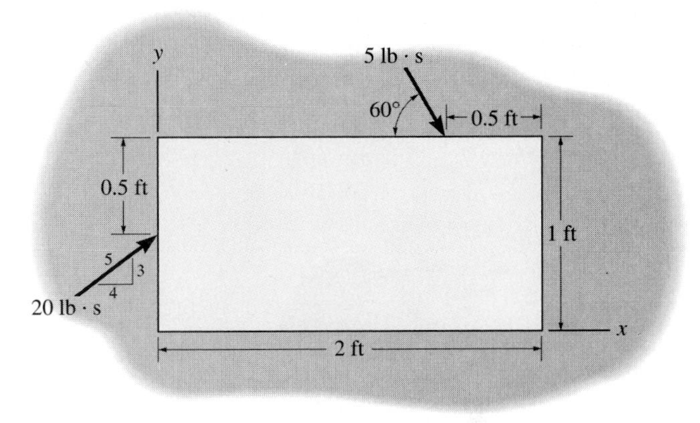

Prob. 19–29

19–30. A constant torque $M = 0.8$ N · m is applied to gear A. If A is originally rotating at 5 rad/s, determine its angular velocity in 2 s. The two smaller *equivalent* gears B and C are pinned at their centers. The mass and centroidal radii of gyration of the gears about axes perpendicular to the page and passing through their centers are given in the figure.

19–31. A torque $M = (0.05e^{0.5t})$ N · m, where t is in seconds, is applied to gear A. If A is originally rotating at 5 rad/s, determine its angular velocity when $t = 2$ s. The two smaller *equivalent* gears B and C are pinned at their centers. The mass and centroidal radii of gyration of the gears about axes perpendicular to the page and passing through their centers are given in the figure.

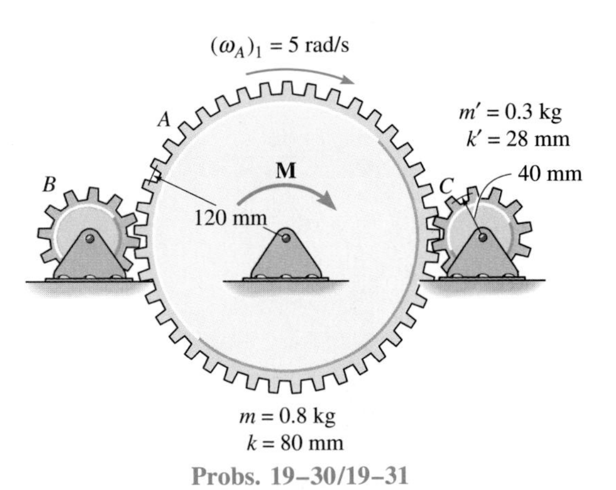

Probs. 19–30/19–31

***19–32.** The ball of mass m and radius r rolls along an inclined plane for which the coefficient of static friction is μ. If the ball is released from rest, determine the maximum angle θ for the incline so that it rolls without slipping at A.

19–33. The 5-kg ball has a radius $r = 100$ mm and rolls along an inclined plane for which the coefficients of static and kinetic friction are $\mu_s = 0.4$ and $\mu_k = 0.3$, respectively. If the ball is released from rest, determine its angular velocity in 2 s. Take $\theta = 30°$.

19–34. Solve Prob. 19–33 if $\theta = 60°$.

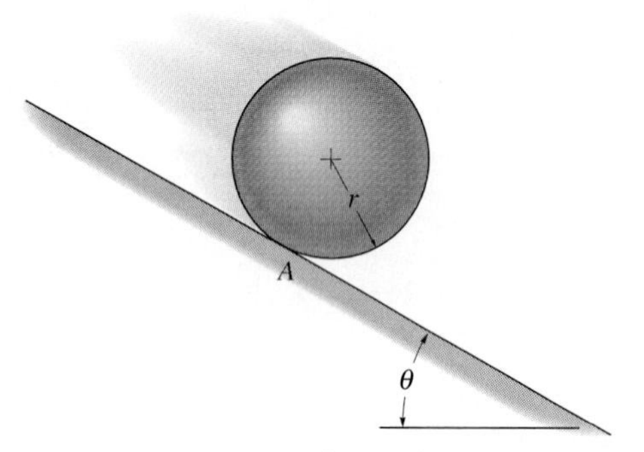

Probs. 19–32/19–33/19–34

19–35. The hoop (thin ring) has a mass of 5 kg and is released down the inclined plane such that it has a backspin $\omega = 8$ rad/s and its center has a velocity $v_G = 3$ m/s as shown. If the coefficient of kinetic friction between the hoop and the plane is $\mu_k = 0.6$, determine how long the hoop rolls before it stops slipping.

***19–36.** The hoop (thin ring) has a mass of 5 kg and is released down the inclined plane such that it has a backspin $\omega = 8$ rad/s and its center has a velocity $v_G = 3$ m/s as shown. If the coefficient of kinetic friction between the hoop and the plane is $\mu_k = 0.6$, determine the hoop's angular velocity in 1 s.

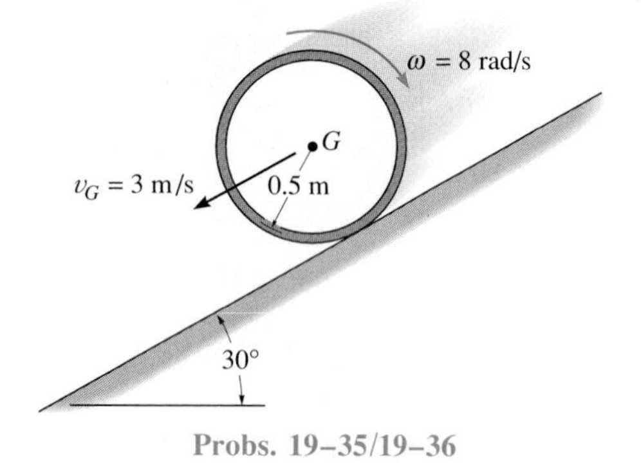

Probs. 19–35/19–36

19–37. The slender rod of mass m and length L is released from rest when it is in the vertical position. If it falls and strikes the soft ledge at A without rebounding, determine the impulse which the ledge exerts on the rod.

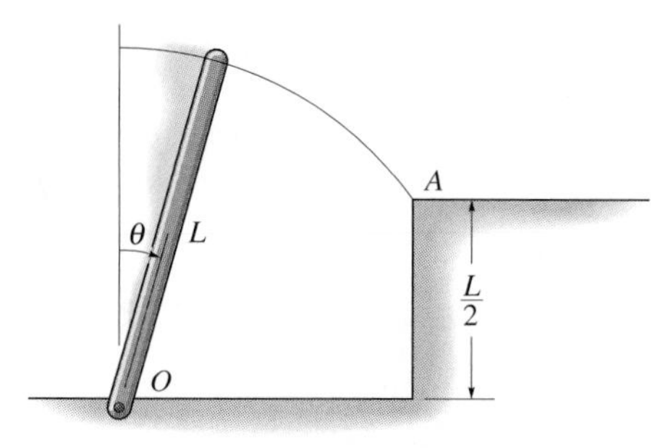

Prob. 19–37

19–38. The square plate has a mass m and is suspended at its corner A by a cord. If it receives a horizontal impulse **I** at corner B, determine the location y of the point P about which the plate appears to rotate during the impact.

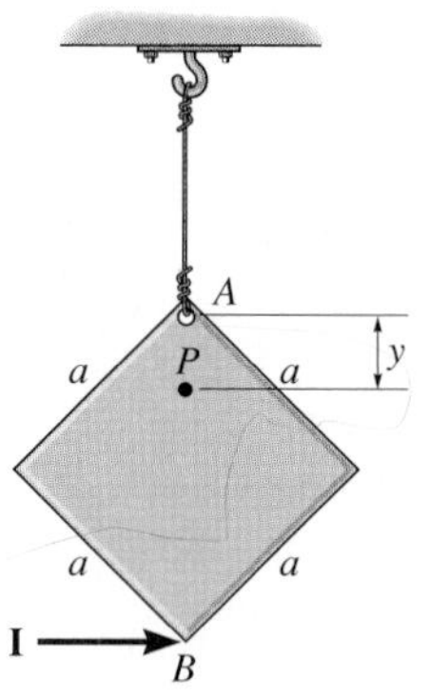

Prob. 19–38

19–39. The rod of length L and mass m lies on a smooth horizontal surface and is subjected to a force **P** at its end A as shown. Determine the location d of the point about which the rod begins to turn, i.e., the point that has zero velocity.

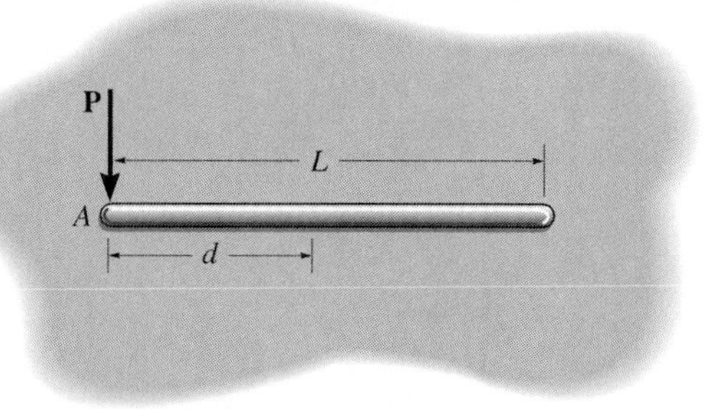

Prob. 19–39

***19–40.** A cord is wrapped around the rim of each 10-lb disk. If disk B is released from rest, determine the angular velocity of disk A in 2 s. Neglect the mass of the cord.

19–41. A cord is wrapped around the rim of each 10-lb disk. If disk B is released from rest, determine how much time t is required before A is given an angular velocity $\omega_A = 5$ rad/s.

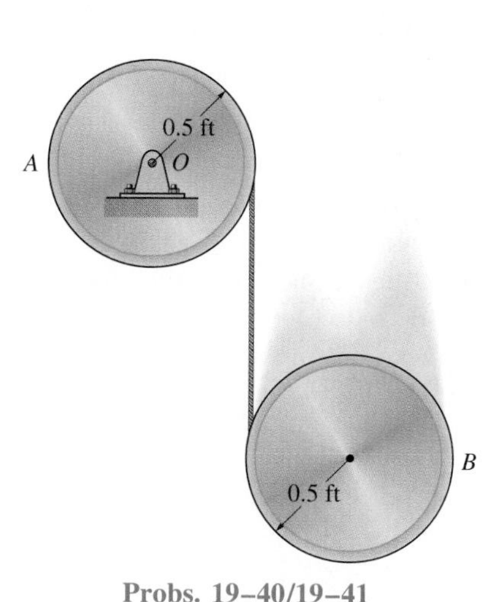

Probs. 19–40/19–41

19–42. The crate has a mass m_c. Determine the constant speed v_0 it acquires as it moves down the conveyor. The rollers each have a radius of r, mass m, and are spaced d apart. Note that friction causes each roller to rotate when the crate comes in contact with it.

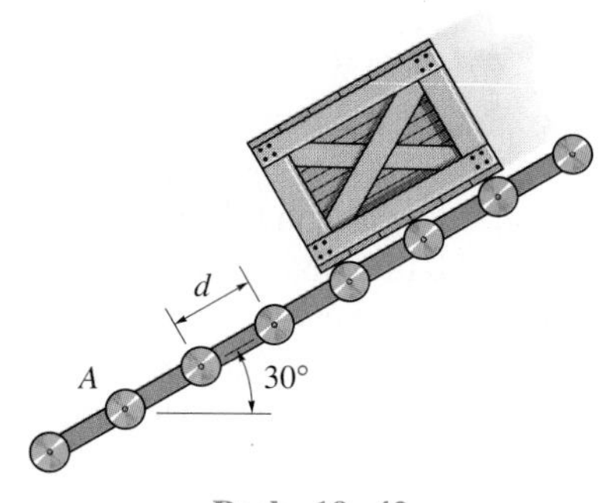

Prob. 19–42

19–43. The double pulley consists of two wheels which are attached to one another and turn at the same rate. The pulley has a mass of 30 kg and a radius of gyration $k_O = 250$ mm. If two men A and B grab the suspended ropes and step off the ledges at the same time, determine their speeds in 4 s starting from rest. The men A and B have a mass of 60 kg and 70 kg, respectively. Assume they do not move relative to the rope during the motion. Neglect the mass of the rope.

Prob. 19–43

***19–44.** The 10-lb cylinder rests on the 20-lb dolly. If the system is released from rest, determine the angular velocity of the cylinder in 2 s. The cylinder does not slip on the dolly. Neglect the mass of the wheels on the dolly.

19–45. Solve Prob. 19–44 if the coefficients of static and kinetic friction between the cylinder and the dolly are $\mu_s = 0.3$ and $\mu_k = 0.2$, respectively.

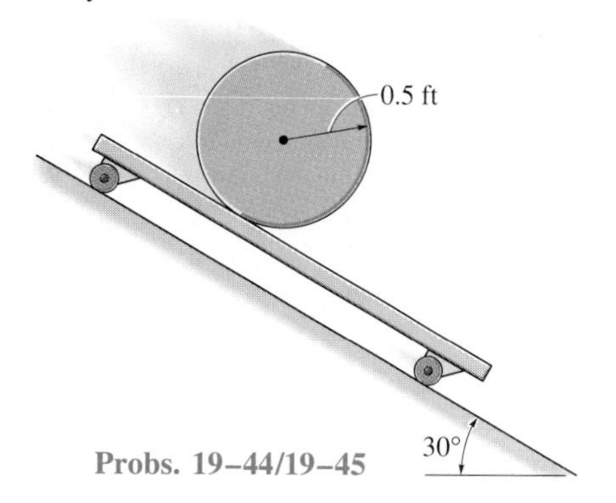

Probs. 19–44/19–45

19.3 Conservation of Momentum

Conservation of Linear Momentum. If the sum of all the *linear impulses* acting on a system of connected rigid bodies is *zero,* the linear momentum of the system is constant or conserved. Consequently, the first two of Eqs. 19–15 reduce to the form

$$\left(\sum \begin{matrix}\text{syst. linear}\\ \text{momentum}\end{matrix}\right)_1 = \left(\sum \begin{matrix}\text{syst. linear}\\ \text{momentum}\end{matrix}\right)_2 \tag{19–16}$$

This equation is referred to as the *conservation of linear momentum.*

Without inducing appreciable errors in the computations, it may be possible to apply Eq. 19–16 in a specified direction for which the linear impulses are small or *nonimpulsive.* Specifically, nonimpulsive forces occur when small forces act over very short periods of time. For example, the impulse created by the force of a tennis racket hitting a ball during a very short time interval Δt is large, whereas the impulse of the weight of the ball during this time is small by comparison and may therefore be neglected in the motion analysis of the ball during Δt.

Conservation of Angular Momentum. The angular momentum of a system of connected rigid bodies is conserved about the system's center of mass G, or a fixed point O, when the sum of all the angular impulses created by the external forces acting on the system is zero or appreciably small (nonimpulsive) when computed about these points. The third of Eqs. 19–15 then becomes

$$\left(\sum \begin{matrix}\text{syst. angular}\\ \text{momentum}\end{matrix}\right)_{O1} = \left(\sum \begin{matrix}\text{syst. angular}\\ \text{momentum}\end{matrix}\right)_{O2} \tag{19–17}$$

This equation is referred to as the *conservation of angular momentum.* In the case of a single rigid body, Eq. 19–17 applied to point G becomes $(I_G\omega)_1 = (I_G\omega)_2$. To illustrate an application of this equation, consider a swimmer who executes a somersault after jumping off a diving board. By tucking his arms and legs in close to his chest, he *decreases* his body's moment of inertia and thus *increases* his angular velocity ($I_G\omega$ must be constant). If he straightens out just before entering the water, his body's moment of inertia is *increased* and his angular velocity *decreases.* Since the weight of his body creates a linear impulse during the time of motion, this example also illustrates that the angular momentum of a body is conserved and yet the linear momentum is *not.* Such cases occur whenever the external forces creating the linear impulse pass through either the center of mass of the body or a fixed axis of rotation.

PROCEDURE FOR ANALYSIS

Provided the initial linear or angular velocity of the body is known, the conservation of linear or angular momentum is used to determine the respective final linear or angular velocity of the body *just after* the time period considered. Furthermore, by applying these equations to a *system* of bodies, the internal impulses acting within the system, which may be unknown, are eliminated from the analysis, since they occur in equal but opposite collinear pairs. For application it is suggested that the following procedure be used.

Free-Body Diagram. Establish the x, y inertial frame of reference and draw the free-body diagram for the body or system of bodies during the time of impact. From this diagram classify each of the applied forces as being either "impulsive" or "nonimpulsive." In general, for short times, "nonimpulsive forces" consist of the weight of a body, the force of a slightly deformed spring, or any force that is *known to be small* when compared to an impulsive force. From the free-body diagram, it will then be possible to tell if the conservation of linear or angular momentum can be applied. Specifically, the *conservation of linear momentum* applies in a given direction when *no* external impulsive forces act on the body or system in that direction; whereas the *conservation of angular momentum* applies about a fixed point O or at the mass center G of a body or system of bodies when all the external impulsive forces acting on the body or system create zero moment (or zero angular impulse) about O or G.

As an alternative procedure, consider drawing the impulse and momentum diagrams for the body or system of bodies. These diagrams are particularly helpful in order to visualize the "moment" terms used in the conservation of angular momentum equation, when it has been decided that angular momenta are to be computed about a point other than the body's mass center G.

Conservation of Momentum. Apply the conservation of linear or angular momentum in the appropriate directions.

Kinematics. Use *kinematics* if further equations are necessary for the solution of a problem. If the motion appears to be complicated, kinematic (velocity) diagrams may be helpful in obtaining the necessary kinematic relations.

If it is necessary to determine an *internal impulsive force* acting on only one body of a system of connected bodies, the body must be *isolated* (free-body diagram) and the principle of linear or angular impulse and momentum must be applied *to the body*. After the impulse $\int F\,dt$ is calculated, then, provided the time Δt for which the impulse acts is known, the *average impulsive force* F_{avg} can be determined from $F_{avg} = \int F\,dt/\Delta t$.

The following examples illustrate application of this procedure.

Example 19–6

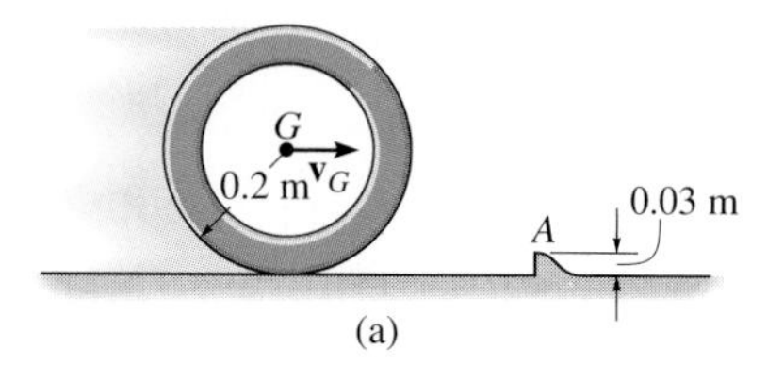

The 10-kg wheel shown in Fig. 19–9*a* has a moment of inertia $I_G = 0.156 \text{ kg} \cdot \text{m}^2$. Assuming that the wheel does not slip or rebound, determine the minimum velocity $\mathbf{v}_G$ it must have to just roll over the obstruction at *A*.

SOLUTION

Impulse and Momentum Diagrams. Since no slipping or rebounding occurs, the wheel essentially *pivots* about point *A* during contact. This condition is shown in Fig. 19–9*b*, which indicates, respectively, the momentum of the wheel *just before impact,* the impulses given to the wheel *during impact,* and the momentum of the wheel *just after impact.* Only two impulses (forces) act on the wheel. By comparison, the impulse at *A* is much greater than that caused by the weight, and since the time of impact is very short, the weight can be considered nonimpulsive. The impulsive force **F** at *A* has both an unknown magnitude and an unknown direction θ. To eliminate this force from the analysis, note that angular momentum about *A* is essentially *conserved* since $(98.1\Delta t)d \approx 0$.

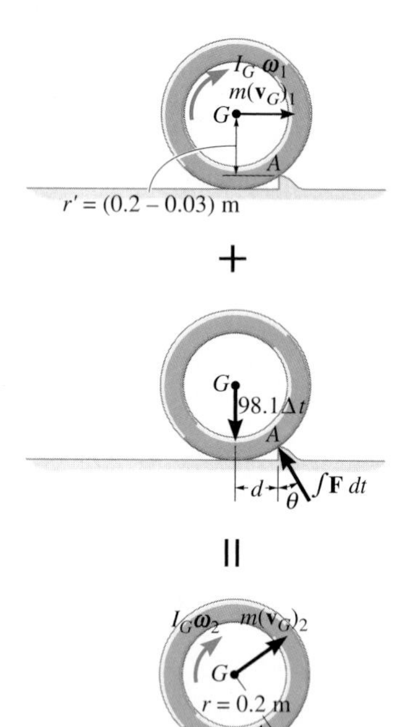

Conservation of Angular Momentum. With reference to Fig. 19–9*b*,

$$(\circlearrowleft +) \qquad (\mathbf{H}_A)_1 = (\mathbf{H}_A)_2$$

$$r'm(v_G)_1 + I_G\omega_1 = rm(v_G)_2 + I_G\omega_2$$

$$(0.2 \text{ m} - 0.03 \text{ m})(10 \text{ kg})(v_G)_1 + (0.156 \text{ kg} \cdot \text{m}^2)(\omega_1) = (0.2 \text{ m})(10 \text{ kg})(v_G)_2 + (0.156 \text{ kg} \cdot \text{m}^2)(\omega_2)$$

Kinematics. Since no slipping occurs, in general $\omega = v_G/r = v_G/0.2 \text{ m} = 5v_G$. Substituting this into the above equation and simplifying yields

$$(v_G)_2 = 0.892(v_G)_1 \qquad (1)$$

Conservation of Energy.* In order to roll over the obstruction, the wheel must pass the dashed position 3 shown in Fig. 19–9*c*. Hence, if $(v_G)_2$ [or $(v_G)_1$] is to be a minimum, it is necessary that the kinetic energy of the wheel at position 2 be equal to the potential energy at position 3. Constructing the datum through the center of gravity, as shown in the figure, and applying the conservation of energy equation, we have

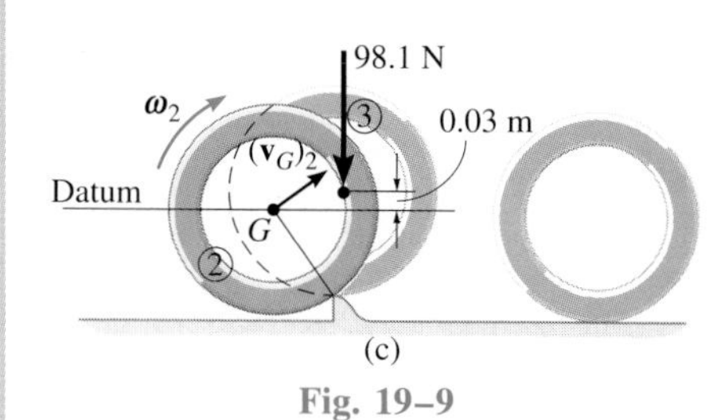

Fig. 19–9

$$\{T_2\} + \{V_2\} = \{T_3\} + \{V_3\}$$

$$\{\tfrac{1}{2}(10 \text{ kg})(v_G)_2^2 + \tfrac{1}{2}(0.156 \text{ kg} \cdot \text{m}^2)(\omega_2)^2\} + \{0\} = \{0\} + \{(98.1 \text{ N})(0.03 \text{ m})\}$$

Substituting $\omega_2 = 5(v_G)_2$ and Eq. 1 into this equation, and solving,

$$(v_G)_1 = 0.729 \text{ m/s} \rightarrow \qquad \textit{Ans.}$$

*This principle *does not* apply during *impact,* since energy is *lost* during the collision; however, just after impact it can be used.

Example 19–7

The 5-kg slender rod shown in Fig. 19–10*a* is pinned at O and is initially at rest. If a 4-g bullet is fired into the rod with a velocity of 400 m/s, as shown in the figure, determine the angular velocity of the rod just after the bullet becomes embedded in it.

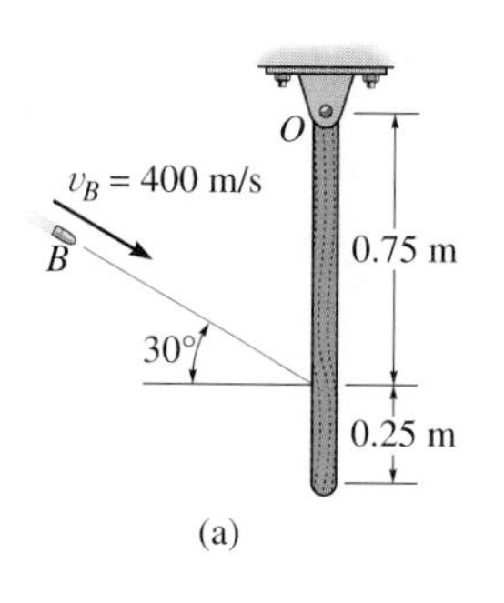

(a)

SOLUTION

Impulse and Momentum Diagrams. The impulse which the bullet exerts on the rod can be eliminated from the analysis and the angular velocity of the rod just after impact can be determined by considering the bullet and rod as a single system. To clarify the principles involved, the impulse and momentum diagrams are shown in Fig. 19–10*b*. The momentum diagrams are drawn *just before and just after impact.* During impact, the bullet and rod exchange equal but opposite *internal impulses* at A. As shown on the impulse diagram, the impulses that are external to the system are due to the reactions at O and the weights of the bullet and rod. Since the time of impact, Δt, is very short, the rod moves only a slight amount and so the "moments" of these impulses about point O are essentially zero, and therefore angular momentum is conserved about this point.

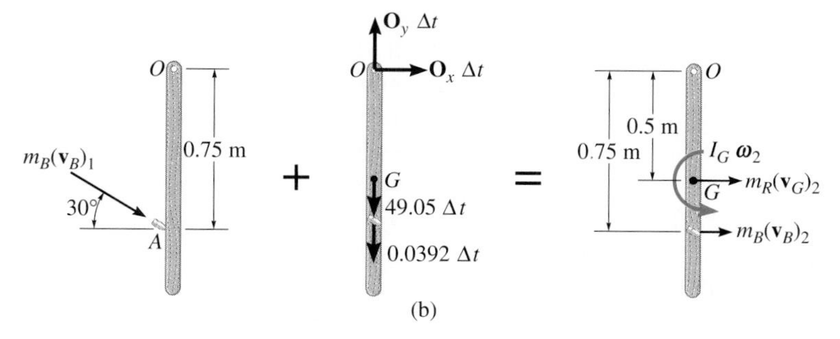

(b)

Conservation of Angular Momentum. From Fig. 19–10*b*, we have

$$(\circlearrowright+) \qquad \Sigma(H_O)_1 = \Sigma(H_O)_2$$

$$m_B(v_B)_1 \cos 30°(0.75\text{ m}) = m_B(v_B)_2(0.75\text{ m}) + m_R(v_G)_2(0.5\text{ m}) + I_G\omega_2$$

$$(0.004\text{ kg})(400 \cos 30°\text{ m/s})(0.75\text{ m}) =$$
$$(0.004\text{ kg})(v_B)_2(0.75\text{ m}) + (5\text{ kg})(v_G)_2(0.5\text{ m}) + [\tfrac{1}{12}(5\text{ kg})(1\text{ m})^2]\omega_2$$

or

$$1.039 = 0.003(v_B)_2 + 2.50(v_G)_2 + 0.417\omega_2 \qquad (1)$$

Kinematics. Since the rod is pinned at O, from Fig. 19–10*c* we have

$$(v_G)_2 = (0.5\text{ m})\omega_2 \qquad (v_B)_2 = (0.75\text{ m})\omega_2$$

Substituting into Eq. 1 and solving yields

$$\omega_2 = 0.622\text{ rad/s} \circlearrowleft \qquad \textit{Ans.}$$

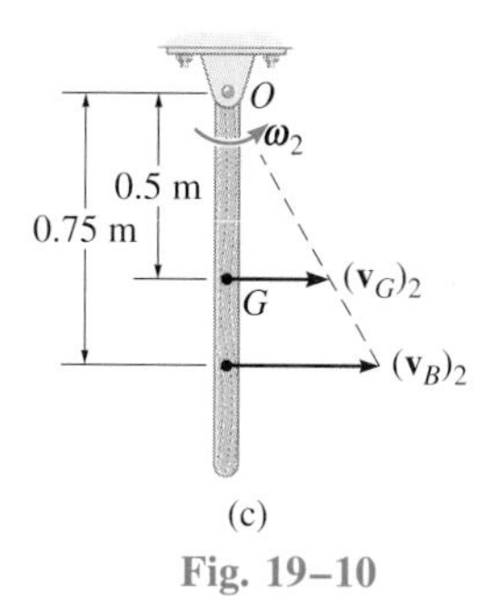

(c)

Fig. 19–10

*19.4 Eccentric Impact

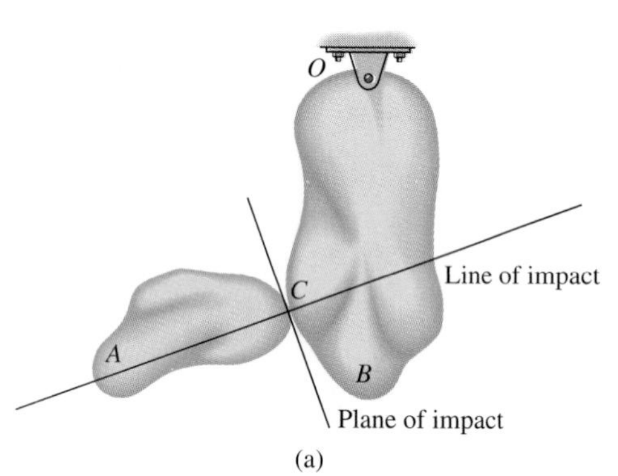

(a)

Fig. 19–11

The concepts involving central and oblique impact of particles have been presented in Sec. 15.4. We will now expand this treatment and discuss the eccentric impact of two bodies. *Eccentric impact* occurs when the line connecting the *mass centers* of the two bodies *does not* coincide with the line of impact.* This type of impact often occurs when one or both of the bodies is constrained to rotate about a fixed axis. Consider, for example, the collision between the two bodies A and B, shown in Fig. 19–11*a*, which collide at point C. Body B rotates about an axis passing through point O, whereas body A both rotates and translates. It is assumed that just before collision B is rotating counterclockwise with an angular velocity $(\boldsymbol{\omega}_B)_1$, and the velocity of the contact point C located on A is $(\mathbf{u}_A)_1$. Kinematic diagrams for both bodies just before collision are shown in Fig. 19–11*b*. Provided the bodies are smooth at point C, the *impulsive forces* they exert on each other *are directed along the line of impact.* Hence, the component of velocity of point C on body B, which is directed along the line of impact, is $(v_B)_1 = (\omega_B)_1 r$, Fig. 19–11*b*. Likewise, on body A the component of velocity $(\mathbf{u}_A)_1$ along the line of impact is $(\mathbf{v}_A)_1$. In order for a collision to occur, $(v_A)_1 > (v_B)_1$.

During the impact an equal but opposite impulsive force $\mathbf{P}$ is exerted between the bodies which *deforms* their shapes at the point of contact. The resulting impulse is shown on the impulse diagrams for both bodies, Fig. 19–11*c*. Note that the impulsive force created at point C, on the rotating body B, creates impulsive pin reactions at the supporting pin O. On these diagrams it is assumed the impact creates forces which are much larger than the weights of the bodies. Hence, the impulses created by the weights are not shown since they are negligible compared to $\int \mathbf{P}\,dt$ and the reactive impulses at O. When the deformation of point C is a maximum, C on both the bodies moves with a common velocity $\mathbf{v}$ along the line of impact, Fig. 19–11*d*. A period of *restitution* then occurs in which the bodies tend to regain their original shapes. The restitution phase creates an equal but opposite impulsive force $\mathbf{R}$ acting between the bodies as shown on the impulse diagram, Fig. 19–11*e*. After restitution the bodies move apart such that point C on body B has a velocity $(\mathbf{v}_B)_2$ and point C on body A has a velocity $(\mathbf{u}_A)_2$, Fig. 19–11*f*, where $(v_B)_2 > (v_A)_2$.

In general, a problem involving the impact of two bodies requires determining the *two unknowns* $(v_A)_2$ and $(v_B)_2$, assuming $(v_A)_1$ and $(v_B)_1$ are known (or can be determined using kinematics, energy methods, the equations of motion, etc.) To solve this problem two equations must be written. The *first equation* generally involves application of *the conservation of angular momentum to the two bodies.* In the case of both bodies A and B, we can state that angular momentum is conserved about point O since the impulses at C are internal to the system and the impulses at O create zero moment (or zero

*When these lines coincide, central impact occurs and the problem can be analyzed as discussed in Sec. 15.4.

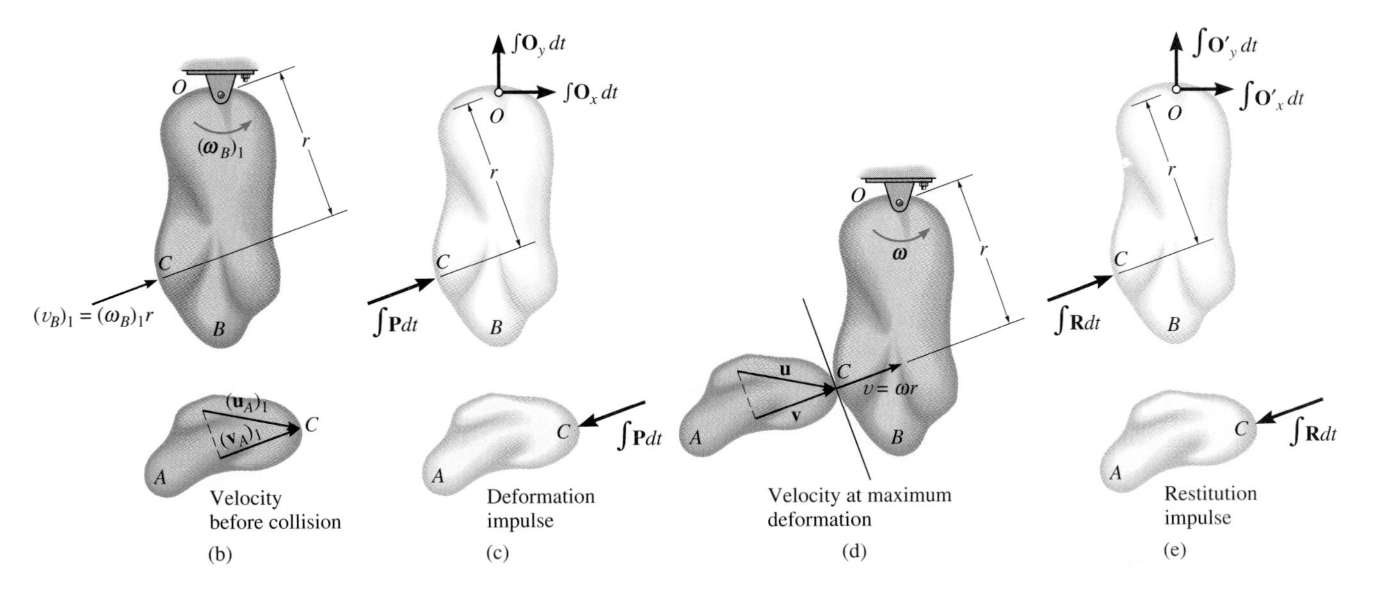

angular impulse) about point O. The *second equation* is obtained using the definition of the *coefficient of restitution, e,* which is a ratio of the restitution impulse to the deformation impulse. To establish a useful form of this equation we must first apply the principle of angular impulse and momentum about point O to bodies B and A separately. Combining the results, we then obtain the necessary equation. Proceeding in this manner, the principle of impulse and momentum applied to body B from the time just before the collision to the instant of maximum deformation, Figs. 19–11b, 19–11c, and 19–11d, becomes

$$(\circlearrowright +) \qquad I_O(\omega_B)_1 + r\int P\,dt = I_O\omega \qquad (19\text{–}18)$$

Here I_O is the moment of inertia of body B about point O. Similarly, applying the principle of angular impulse and momentum from the instant of maximum deformation to the time just after the impact, Figs. 19–11d, 19–11e, and 19–11f, yields

$$(\circlearrowright +) \qquad I_O\omega + r\int R\,dt = I_O(\omega_B)_2 \qquad (19\text{–}19)$$

Solving Eqs. 19–18 and 19–19 for $\int P\,dt$ and $\int R\,dt$, respectively, and formulating e, we have

$$e = \frac{\int R\,dt}{\int P\,dt} = \frac{r(\omega_B)_2 - r\omega}{r\omega - r(\omega_B)_1} = \frac{(v_B)_2 - v}{v - (v_B)_1}$$

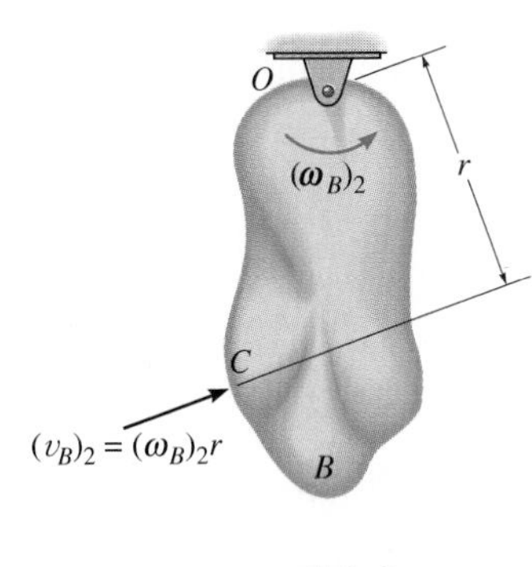

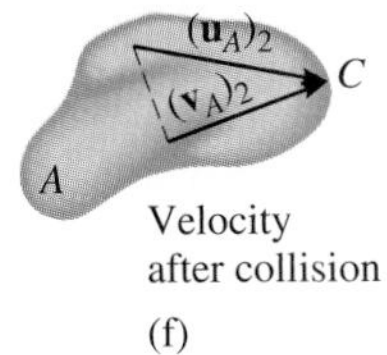

Fig. 19–11 ***(cont'd)***

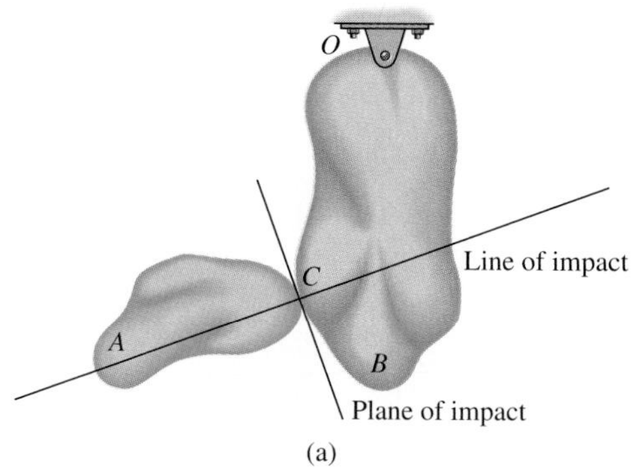

(a)

Fig. 19–11

In the same manner, we may write an equation which relates the magnitudes of velocity $(v_A)_1$ and $(v_A)_2$ of body A. The result is

$$e = \frac{(v_A)_2 - v}{v - (v_A)_1}$$

Combining the above equations by eliminating the common velocity v yields the desired result, i.e.,

$$(+\nearrow) \qquad e = \frac{(v_B)_2 - (v_A)_2}{(v_A)_1 - (v_B)_1} \qquad (19\text{–}20)$$

This equation is identical to Eq. 15–11, which was derived for the central impact occurring between two particles. Equation 19–20, however, states that the coefficient of restitution, measured at the points of contact (C) between two colliding bodies, is equal to the ratio of the relative velocity of *separation* of the points *just after impact* to the relative velocity at which the points *approach* one another *just* before impact. In deriving Eq. 19–20, we assumed that the points of contact for both bodies move up and to the right *both* before and after impact. If motion of any one of the contacting points occurs down and to the left, the velocity of this point is considered a negative quantity in Eq. 19–20.

As stated previously, when Eq. 19–20 is used in conjunction with the conservation of angular momentum for the bodies, it provides a useful means of obtaining the velocities of two colliding bodies just after collision.

Example 19–8

The 10-lb slender rod is suspended from the pin at A, Fig. 19–12a. If a 2-lb ball B is thrown at the rod and strikes its center with a horizontal velocity of 30 ft/s, determine the angular velocity of the rod just after impact. The coefficient of restitution is $e = 0.4$.

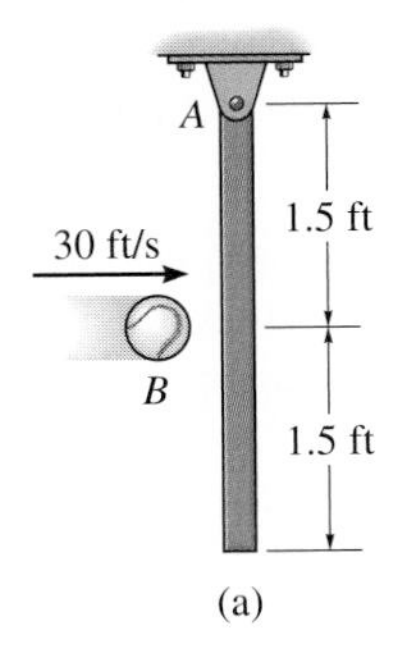

(a)

SOLUTION

Conservation of Angular Momentum. Consider the ball and rod as a system, Fig. 19–12b. Angular momentum is conserved about point A since the impulsive force between the rod and ball is *internal*. Also, the *weights* of the ball and rod are *nonimpulsive*. Noting the directions of the velocities of the ball and rod just after impact as shown on the kinematic diagram, Fig. 19–12c, we require

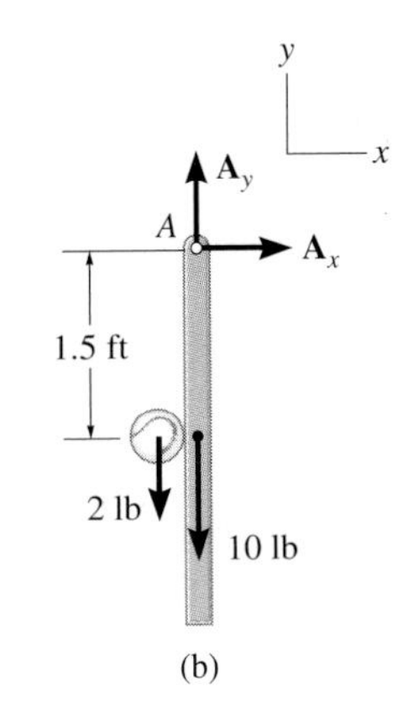

(b)

$$(\circlearrowright +) \qquad (H_A)_1 = (H_A)_2$$

$$m_B(v_B)_1(1.5 \text{ ft}) = m_B(v_B)_2(1.5 \text{ ft}) + m_R(v_G)_2(1.5 \text{ ft}) + I_G\omega_2$$

$$\left(\frac{2 \text{ lb}}{32.2 \text{ ft/s}^2}\right)(30 \text{ ft/s})(1.5 \text{ ft}) = \left(\frac{2 \text{ lb}}{32.2 \text{ ft/s}^2}\right)(v_B)_2(1.5 \text{ ft}) + \left(\frac{10 \text{ lb}}{32.2 \text{ ft/s}^2}\right)(v_G)_2(1.5 \text{ ft}) + \left[\frac{1}{12}\frac{10 \text{ lb}}{32.2 \text{ ft/s}^2}(3 \text{ ft})^2\right]\omega_2$$

Since $(v_G)_2 = 1.5\omega_2$ then

$$2.795 = 0.09317(v_B)_2 + 0.9317\omega_2 \qquad (1)$$

Coefficient of Restitution. With reference to Fig. 19–12c, we have

$$(\overset{+}{\rightarrow}) \qquad e = \frac{(v_G)_2 - (v_B)_2}{(v_B)_1 - (v_G)_1} \qquad 0.4 = \frac{(1.5 \text{ ft})\omega_2 - (v_B)_2}{30 \text{ ft/s} - 0}$$

$$12.0 = 1.5\omega_2 - (v_B)_2$$

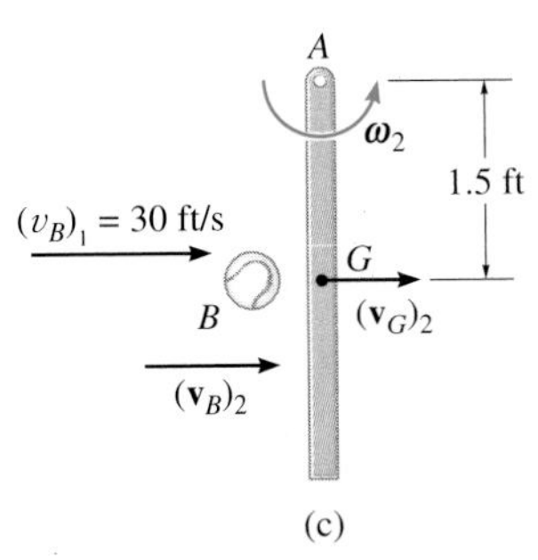

(c)

Fig. 19–12

Solving,

$$(v_B)_2 = -6.52 \text{ ft/s} = 6.52 \text{ ft/s} \leftarrow$$

$$\omega_2 = 3.65 \text{ rad/s} \circlearrowleft \qquad \textit{Ans.}$$

PROBLEMS

19–46. The turntable T of a record player has a mass of 0.75 kg and a radius of gyration $k_z = 125$ mm. It is *turning freely* at $\omega_T = 2$ rad/s when a 50-g record (thin disk) falls on it. Determine the final angular velocity of the turntable just after the record stops slipping on the turntable.

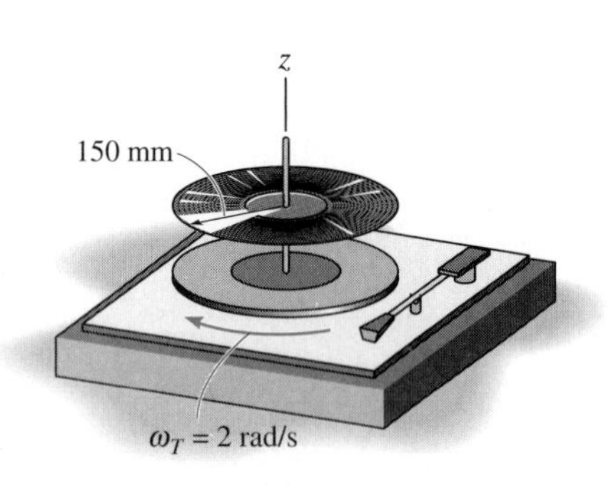

Prob. 19–46

19–47. Each of the two slender rods and the disk have the same mass m. Also, the length of each rod is equal to the diameter d of the disk. If the assembly is rotating with an angular velocity $\boldsymbol{\omega}_1$ when the rods are directed outward, determine the angular velocity of the assembly if by internal means the rods are brought to an upright vertical position.

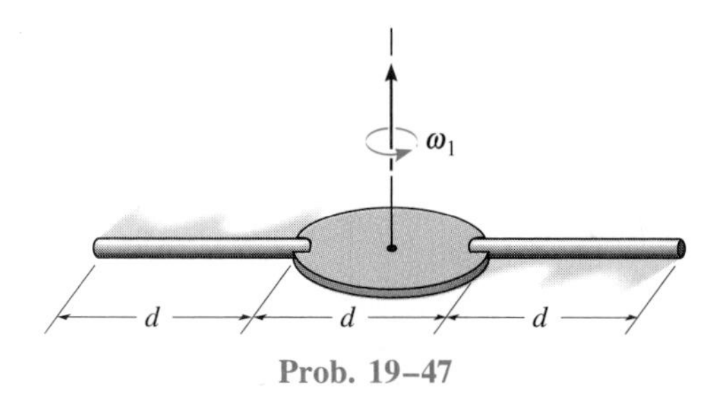

Prob. 19–47

***19–48.** Two wheels A and B have a mass m_A and m_B, and radii of gyration about their central vertical axes of k_A and k_B, respectively. If they are freely rotating in the same direction at $\boldsymbol{\omega}_A$ and $\boldsymbol{\omega}_B$ about the same vertical axis, determine their common angular velocity after they are brought into contact and slipping between them stops.

19–49. The 2-kg rod ACB supports the two 4-kg disks at its ends. If both disks are given a clockwise angular velocity $(\omega_A)_1 = (\omega_B)_1 = 5$ rad/s while the rod is held stationary and then released, determine the angular velocity of the rod after both disks have stopped spinning relative to the rod due to frictional resistance at the pins A and B. Motion is in the *horizontal plane*. Neglect friction at pin C.

19–50. Solve Prob. 19–49 if disk A is given an angular velocity $(\omega_A)_1 = 5$ rad/s clockwise, and disk B has an angular velocity $(\omega_B)_1 = 3$ rad/s counterclockwise.

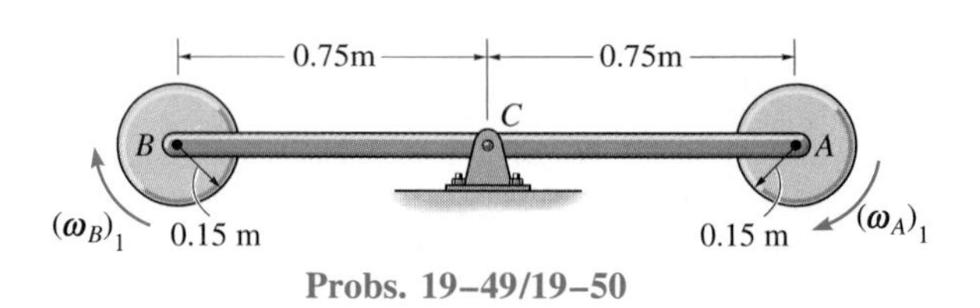

Probs. 19–49/19–50

19–51. The space satellite has a mass of 125 kg and a moment of inertia $I_z = 0.940$ kg · m^2, excluding the four solar panels A, B, C, and D. Each solar panel has a mass of 20 kg and can be approximated as a thin plate. If the satellite is originally spinning about the z axis at a constant rate $\omega_z = 0.5$ rad/s when $\theta = 90°$, determine the rate of spin if all the panels are raised and reach the upward position, $\theta = 0°$, at the same instant.

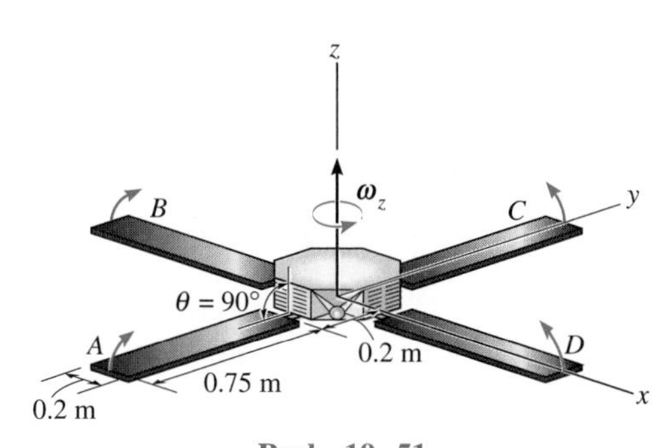

Prob. 19–51

***19–52.** The man sits on the swivel chair holding two 5-lb weights with his arms outstretched. If he is turning at 3 rad/s in this position, determine his angular velocity when the weights are drawn in 0.3 ft from the axis of rotation. Assume he weighs 160 lb and has a radius of gyration $k_z = 0.55$ ft about the z axis. Neglect the mass of his arms and the size of the weights for the calculation.

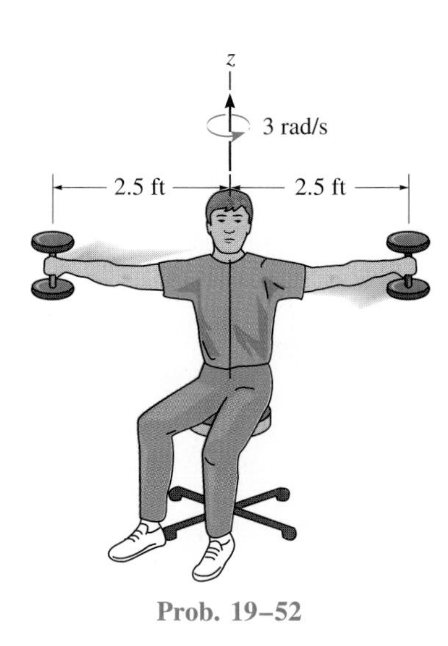

Prob. 19–52

19–53. The rod has a length L and mass m. A smooth collar having a negligible size and one-fourth the mass of the rod is placed on the rod at its midpoint. If the rod is freely rotating at $\boldsymbol{\omega}$ about its end and the collar is released, determine the rod's angular velocity just before the collar flies off the rod. Also, what is the speed of the collar as it leaves the rod?

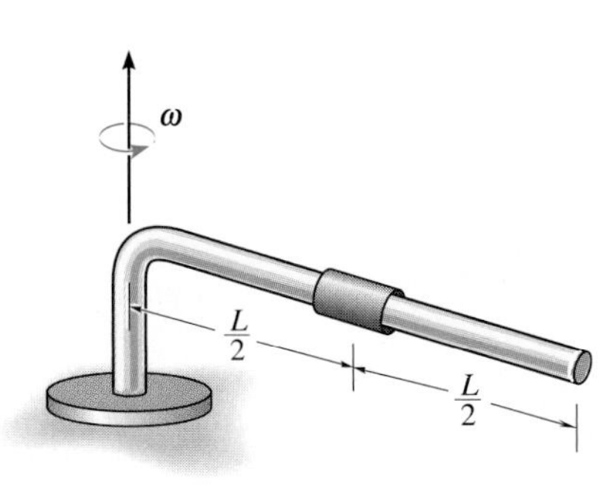

Prob. 19–53

19–54. The 5-lb rod AB supports the 3-lb disk at its end. If the disk is given an angular velocity $\omega_D = 8$ rad/s while the rod is held stationary and then released, determine the angular velocity of the rod after the disk has stopped spinning relative to the rod due to frictional resistance at the bearing A. Motion is in the *horizontal plane.* Neglect friction at the fixed bearing B.

Prob. 19–54

19–55. Two children A and B, each having a mass of 30 kg, sit at the edge of the merry-go-round which is rotating at $\omega = 2$ rad/s. Excluding the children, the merry-go-round has a mass of 180 kg and a radius of gyration $k_z = 0.6$ m. Determine the angular velocity of the merry-go-round if A jumps off horizontally in the $-n$ direction with a speed of 2 m/s, measured with respect to the merry-go-round. What is the merry-go-round's angular velocity if B then jumps off horizontally in the $+t$ direction with a speed of 2 m/s, measured with respect to the merry-go-round? Neglect friction and the size of each child.

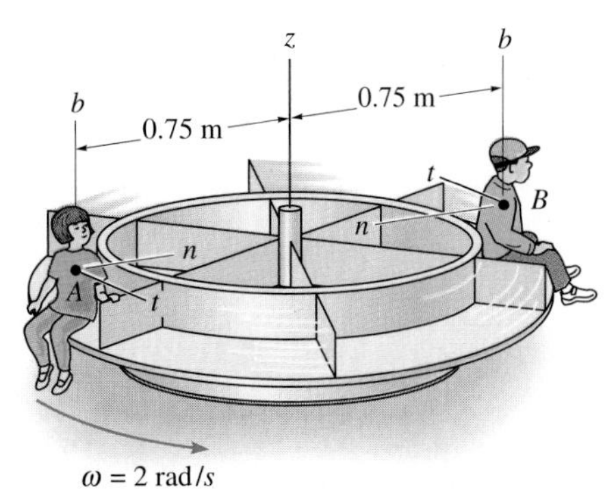

Prob. 19–55

***19–56.** A thin disk of mass m has an angular velocity $\boldsymbol{\omega}_1$ while rotating on a smooth surface. Determine its new angular velocity just after the hook at its edge strikes the peg P and the disk starts to rotate about P without rebounding.

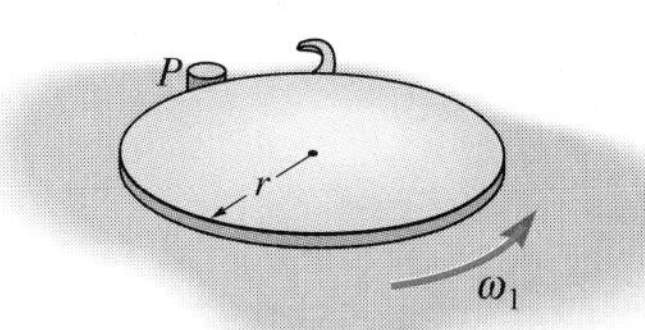

Prob. 19–56

19–57. The rod of mass m and length L is released from rest without rotating. When it falls a distance L, the end A strikes the hook S, which provides a permanent connection. Determine the angular velocity ω of the rod after it has rotated 90°. Treat the rod's weight during impact as a nonimpulsive force.

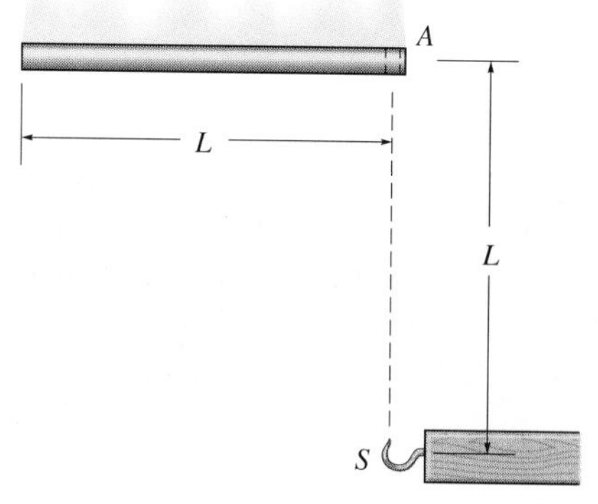

Prob. 19–57

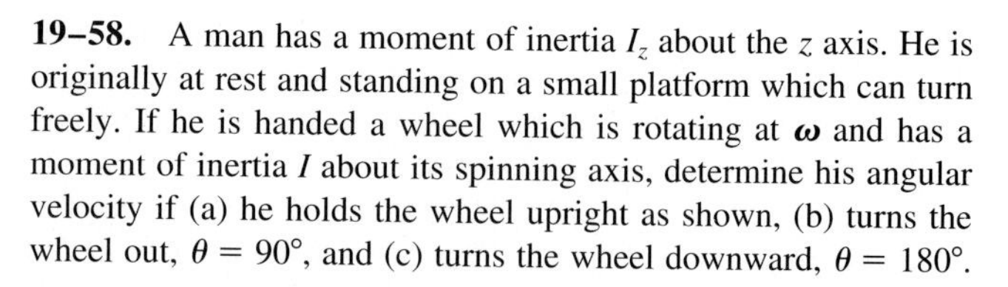

19–58. A man has a moment of inertia I_z about the z axis. He is originally at rest and standing on a small platform which can turn freely. If he is handed a wheel which is rotating at $\boldsymbol{\omega}$ and has a moment of inertia I about its spinning axis, determine his angular velocity if (a) he holds the wheel upright as shown, (b) turns the wheel out, $\theta = 90°$, and (c) turns the wheel downward, $\theta = 180°$.

19–59. If the man in Prob. 19–58 is given the wheel when it is at *rest* and he starts it spinning with an angular velocity $\boldsymbol{\omega}$, determine his angular velocity if (a) he holds the wheel upright as shown, (b) turns the wheel out, $\theta = 90°$, and (c) turns the wheel downward, $\theta = 180°$.

Probs. 19–58/19–59

***19–60.** Determine the height h at which a billiard ball of mass m must be struck so that no frictional force develops between it and the table at A. Assume that the cue C only exerts a horizontal force **P** on the ball.

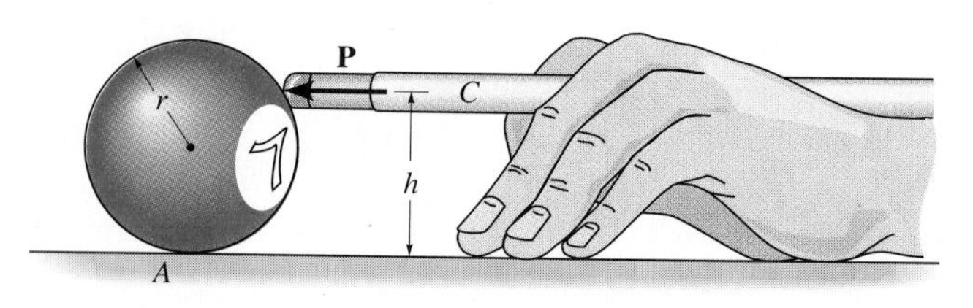

Prob. 19–60

19–61. The solid ball has a mass m and rolls without slipping along a horizontal plane with an angular velocity $\boldsymbol{\omega}_1$. Provided it does not slip or rebound, determine its angular velocity as it just starts to roll up the inclined plane.

19–62. Solve Prob. 19–61 if the body is a disk of mass m and radius r.

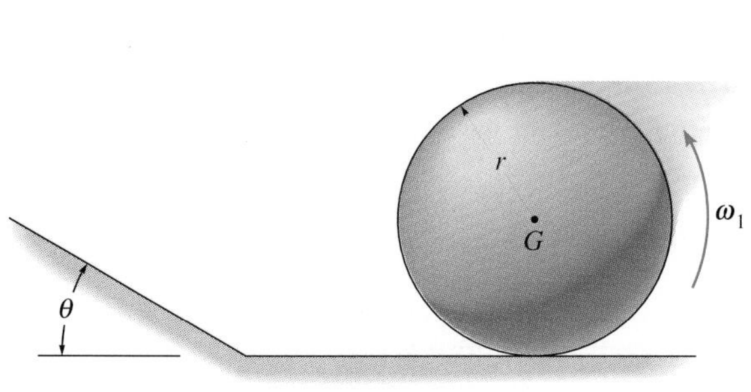

Probs. 19–61/19–62

***19–64.** The disk has a mass m and radius r. If it strikes the rough step having a height $\frac{1}{8}r$ as shown, determine the smallest angular velocity ω_1 the disk can have and not rebound off the step when it strikes it.

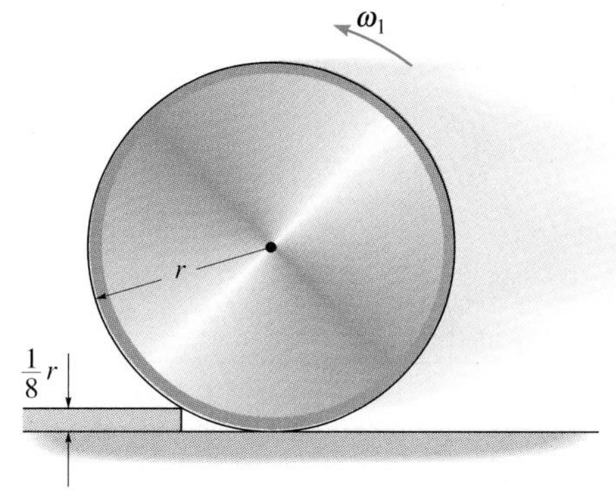

Prob. 19–64

19–63. Disk A has a weight of 20 lb. An inextensible cable is attached to the 10-lb weight and wrapped around the disk. The weight is dropped 2 ft before the slack is taken up. If the impact is perfectly elastic, i.e., $e = 1$, determine the angular velocity of the disk just after impact.

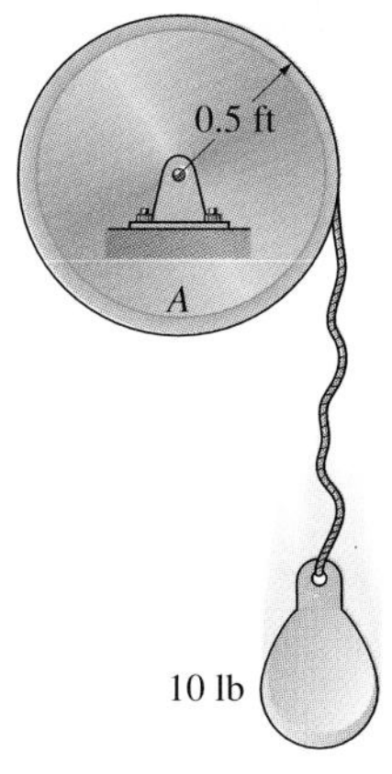

Prob. 19–63

19–65. The plank has a weight of 30 lb, center of gravity at G, and it rests on the two sawhorses at A and B. If the end D is raised 2 ft above the top of the sawhorses and is released from rest, determine how high end C will rise from the top of the sawhorses after the plank falls so that it rotates clockwise about A, strikes and pivots on the sawhorse at B, and rotates clockwise off the sawhorse at A.

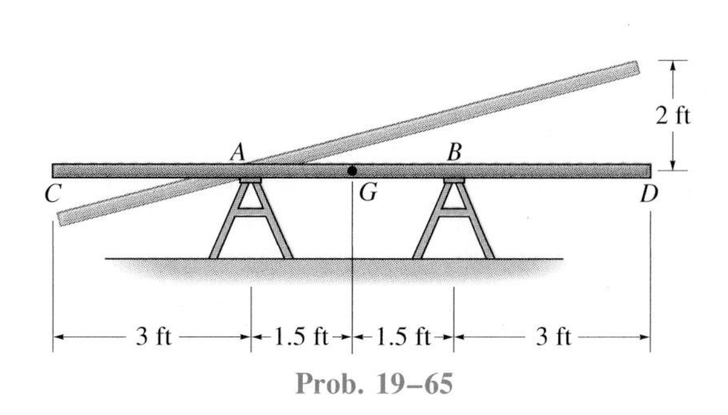

Prob. 19–65

19–66. A 7-g bullet having a velocity of 800 m/s is fired into the edge of the 5-kg disk as shown. Determine the angular velocity of the disk just after the bullet becomes embedded in it. Also, calculate how far θ the disk will swing until it stops. The disk is originally at rest.

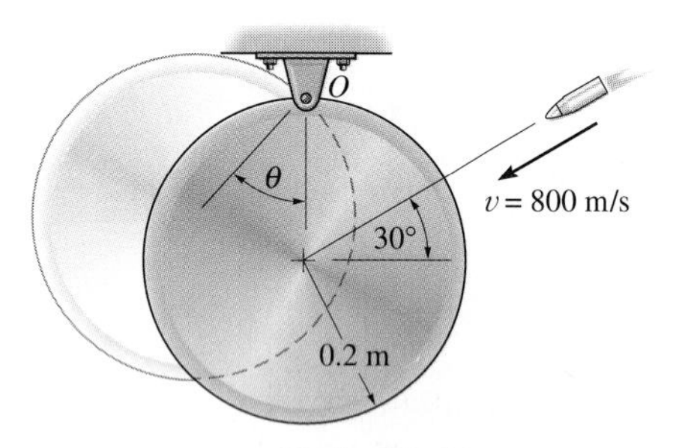

Prob. 19–66

19–67. The two disks each weigh 10 lb. If they are released from rest when $\theta = 30°$, determine θ after they collide and rebound from each other. The coefficient of restitution is $e = 0.75$. When $\theta = 0°$, the disks hang so that they just touch one another.

***19–68.** The two disks each weigh 10 lb. When they are released from rest at $\theta = 45°$, they collide and rebound from each other, coming momentarily to rest when $\theta = 30°$. Determine the coefficient of restitution between the disks. When $\theta = 0°$, the disks hang so that they just touch one another.

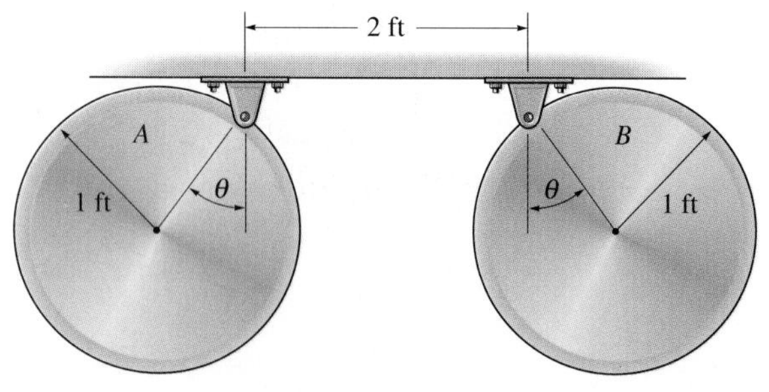

Probs. 19–67/19–68

19–69. The 6-lb slender rod AB is released from rest when it is in the *horizontal position* so that it begins to rotate clockwise. A 1-lb ball is thrown at the rod with a velocity $v = 50$ ft/s. The ball strikes the rod at C at the instant the rod is in the vertical position as shown. Determine the angular velocity of the rod just after the impact. Take $e = 0.7$ and $d = 2$ ft.

19–70. The 6-lb slender rod AB is originally at rest, suspended in the vertical position. A 1-lb ball is thrown at the rod with a velocity $v = 50$ ft/s and strikes the rod at C. Determine the angular velocity of the rod just after the impact. Take $e = 0.7$ and $d = 2$ ft.

19–71. The 6-lb rod is originally at rest, suspended in the vertical position. Determine the distance d where the 1-lb ball, traveling at $v = 50$ ft/s, should strike it so that it does not create a horizontal impulse at A. What is the rod's angular velocity just after the impact? Take $e = 0.5$.

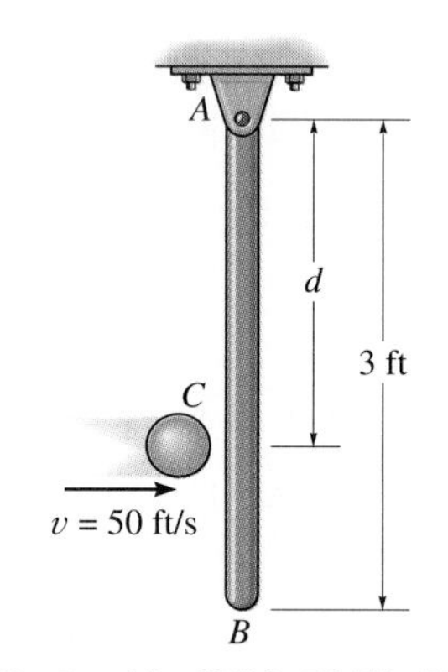

Probs. 19–69/19–70/19–71

***19–72.** The pendulum consists of a 10-lb solid ball and 4-lb rod. If it is released from rest when $\theta_1 = 0°$, determine the angle θ_2 after the ball strikes the wall, rebounds, and the pendulum swings up to the point of momentary rest. Take $e = 0.6$.

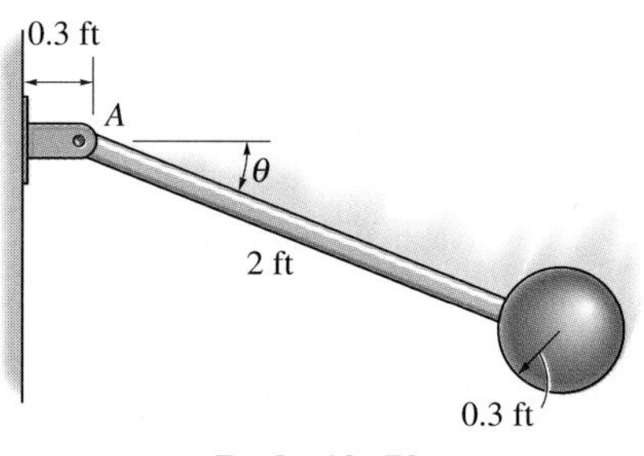

Prob. 19–72

19–73. The 15-lb rod AB is released from rest in the vertical position. If the coefficient of restitution between the floor and the cushion at B is $e = 0.7$, determine how high the end of the rod rebounds after impact with the floor.

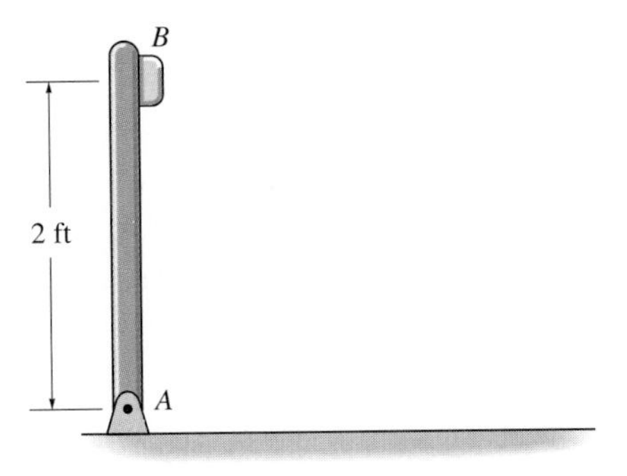

Prob. 19–73

19–74. A solid ball with a mass m is thrown on the ground such that at the instant of contact it has an angular velocity $\boldsymbol{\omega}_1$ and velocity components $(\mathbf{v}_G)_{x1}$ and $(\mathbf{v}_G)_{y1}$ as shown. If the ground is rough so no slipping occurs, determine the components of the velocity of its mass center just after impact. The coefficient of restitution is e.

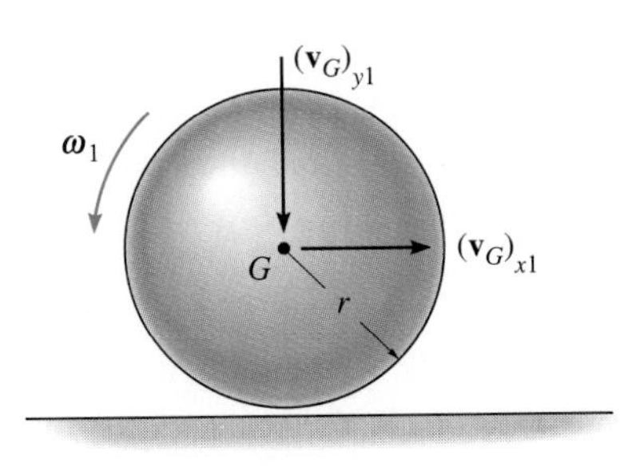

Prob. 19–74

19–75. A solid 10-kg ball having a radius of 0.1 m is thrown on the ground such that at the instant of contact the velocity components of its mass center are $(v_G)_{x1} = 2$ m/s and $(v_G)_{y1} = 3$ m/s as shown. If the ground is rough so no slipping occurs, determine the required backspin ω_1 so that the ball rebounds vertically from the surface. The coefficient of restitution is $e = 0.8$. How high does it bounce?

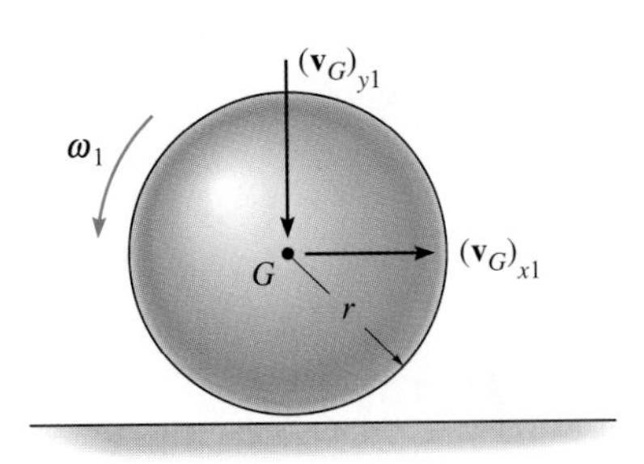

Prob. 19–75

***19–76.** The solid ball of mass m is dropped with a velocity $\mathbf{v}_1$ onto the edge of the rough step. If it rebounds horizontally off the step with a velocity $\mathbf{v}_2$, determine the angle θ at which contact occurs. Assume no slipping when the ball strikes the step. The coefficient of restitution is e.

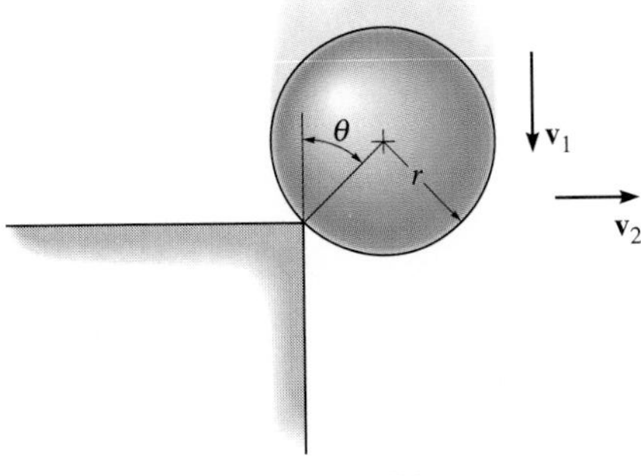

Prob. 19–76

Review 2: Planar Kinematics and Kinetics of a Rigid Body

Having presented the various topics in planar kinematics and kinetics in Chapters 16 through 19, we will now summarize these principles and provide an opportunity for applying them to the solution of various types of problems.

Kinematics. Here we are interested in studying the geometry of motion, without concern for the forces which cause the motion. Before solving a planar kinematics problem, it is *first* necessary to *classify the motion* as being either rectilinear or curvilinear translation, rotation about a fixed axis, or general plane motion. In particular, problems involving general plane motion can be solved either with reference to a fixed axis (absolute motion analysis) or using a translating or rotating frame of reference (relative motion analysis). The choice generally depends upon the type of constraints and the problem geometry. In all cases, application of the necessary equations may be clarified by drawing a kinematic diagram. Remember that the *velocity* of a point is always *tangent* to its path of motion, and the *acceleration* of a point can have *components* in the *n–t* directions when the path is *curved.*

Translation. When the body moves with rectilinear or curvilinear translation, *all* the points on the body have the *same motion.*

$$\mathbf{v}_B = \mathbf{v}_A \qquad \mathbf{a}_B = \mathbf{a}_A$$

Rotation About a Fixed Axis

Angular Motion

Variable Angular Acceleration. Provided a mathematical relationship is given between *any two* of the *four* variables θ, ω, α, and t, then a *third* variable can be determined by solving one of the following equations which relate all three variables.

$$\omega = \frac{d\theta}{dt} \qquad \alpha = \frac{d\omega}{dt} \qquad \alpha\, d\theta = \omega\, d\omega$$

Constant Angular Acceleration. The following equations apply when it is *absolutely certain* that the angular acceleration is constant.

$$\theta = \theta_0 + \omega_0 t + \tfrac{1}{2}\alpha_c t^2 \qquad \omega = \omega_0 + \alpha_c t \qquad \omega^2 = \omega_0^2 + 2\alpha_c(\theta - \theta_0)$$

Motion of Point P. Once $\boldsymbol{\omega}$ and $\boldsymbol{\alpha}$ have been determined, then the circular motion of point P can be specified using the following scalar or vector equations.

$$v = \omega r \qquad \mathbf{v} = \boldsymbol{\omega} \times \mathbf{r}$$

$$a_t = \alpha r \quad a_n = \omega^2 r \qquad \mathbf{a} = \boldsymbol{\alpha} \times \mathbf{r} + \boldsymbol{\omega} \times (\boldsymbol{\omega} \times \mathbf{r})$$

General Plane Motion—Relative-Motion Analysis. Recall that when *translating axes* are placed at the ''base point'' A, the *relative motion* of point B with respect to A is simply *circular motion of B about A.* The following equations apply to two points A and B located on the *same* rigid body.

$$\mathbf{v}_B = \mathbf{v}_A + \mathbf{v}_{B/A} = \mathbf{v}_A + (\boldsymbol{\omega} \times \mathbf{r}_{B/A})$$

$$\mathbf{a}_B = \mathbf{a}_A + \mathbf{a}_{B/A} = \mathbf{a}_A + \boldsymbol{\alpha} \times \mathbf{r}_{B/A} + \boldsymbol{\omega} \times (\boldsymbol{\omega} \times \mathbf{r}_{B/A})$$

Rotating and translating axes are often used to analyze the motion of rigid bodies which are connected together by collars or slider blocks.

$$\mathbf{v}_B = \mathbf{v}_A + \boldsymbol{\Omega} \times \mathbf{r}_{B/A} + (\mathbf{v}_{B/A})_{xyz}$$

$$\mathbf{a}_B = \mathbf{a}_A + \dot{\boldsymbol{\Omega}} \times \mathbf{r}_{B/A} + \boldsymbol{\Omega} \times (\boldsymbol{\Omega} \times \mathbf{r}_{B/A}) + 2\boldsymbol{\Omega} \times (\mathbf{v}_{B/A})_{xyz} + (\mathbf{a}_{B/A})_{xyz}$$

Kinetics. To analyze the forces which cause the motion we must use the principles of kinetics. When applying the necessary equations, it is important to first establish the inertial coordinate system and define the positive directions of the axes. The *directions* should be the *same* as those selected when writing any equations of kinematics provided *simultaneous solution* of equations becomes necessary.

Equations of Motion. These equations are used to determine accelerated motions or forces causing the motion. If used to determine position, velocity, or time of motion, then kinematics will have to be considered for part of the solution. Before applying the equations of motion, *always draw a free-body diagram* in order to identify all the forces acting on the body. Also, establish the directions of the acceleration of the mass center and the angular acceleration of the body. (A kinetic diagram may also be drawn in order to represent $m\mathbf{a}_G$ and $I_G\boldsymbol{\alpha}$ graphically. This diagram is particularly convenient for resolving $m\mathbf{a}_G$ into components and for identifying the terms in the moment sum $\Sigma(\mathcal{M}_k)_P$.)

The three equations of motion are

$$\Sigma F_x = m(a_G)_x$$
$$\Sigma F_y = m(a_G)_y$$
$$\Sigma M_G = I_G\alpha \qquad \text{or} \qquad \Sigma M_P = \Sigma(\mathcal{M}_k)_P$$

In particular, if the body is *rotating about a fixed axis,* moments may also be summed about point O on the axis, in which case

$$\Sigma M_O = \Sigma(\mathcal{M}_k)_O = I_o\alpha$$

Work and Energy. *The equation of work and energy is used to solve problems involving force, velocity, and displacement.* Before applying this equation, *always draw a free-body diagram* of the body in order to identify the forces which do work. Recall that the kinetic energy of the body consists of that due to translational motion of the mass center, $\mathbf{v}_G$, *and* rotational motion of the body, $\boldsymbol{\omega}$.

$$T_1 + \Sigma U_{1-2} = T_2$$

where

$$T = \tfrac{1}{2}mv_G^2 + \tfrac{1}{2}I_G\omega^2$$
$$U_F = \int F\cos\theta\, ds \qquad \text{(variable force)}$$
$$U_{F_c} = F_c\cos\theta\,(s_2 - s_1) \qquad \text{(constant force)}$$
$$U_W = W\,\Delta y \qquad \text{(weight)}$$
$$U_s = -(\tfrac{1}{2}ks_2^2 - \tfrac{1}{2}ks_1^2) \qquad \text{(spring)}$$
$$U_M = M\theta \qquad \text{(constant couple moment)}$$

If the forces acting on the body are conservative forces, then apply the conservation of energy equation. This equation is easier to use than the equation of work and energy, since it applies only at *two points* on the path and *does not* require calculation of the work done by a force as the body moves along the path.

$$T_1 + V_1 = T_2 + V_2$$

where

$$V_g = Wy \qquad \text{(gravitational potential energy)}$$
$$V_e = \tfrac{1}{2}ks^2 \qquad \text{(elastic potential energy)}$$

Impulse and Momentum. *The principles of linear and angular impulse and momentum are used to solve problems involving force, velocity, and time.* Before applying the equations, *draw a free-body diagram* in order to identify all the forces which cause linear and angular impulses on the body. Also, establish the directions of the velocity of the mass center and the angular velocity of the body just before and just after the impulses are applied. (As an alternative procedure, the impulse and momentum diagrams may accompany the solution in order to graphically account for the terms in the equations. These diagrams are particularly advantageous when computing the angular impulses and angular momenta about a point other than the body's mass center.)

$$m(\mathbf{v}_G)_1 + \Sigma \int \mathbf{F}\, dt = m(\mathbf{v}_G)_2$$

$$(\mathbf{H}_G)_1 + \Sigma \int \mathbf{M}_G\, dt = (\mathbf{H}_G)_2$$

or

$$(\mathbf{H}_O)_1 + \Sigma \int \mathbf{M}_O\, dt = (\mathbf{H}_O)_2$$

Conservation of Momentum. If nonimpulsive forces or no impulsive forces act on the body in a particular direction, or the motions of several bodies are involved in the problem, then consider applying the conservation of linear or angular momentum for the solution. Investigation of the free-body diagram (or the impulse diagram) will aid in determining the directions for which the impulsive forces are zero, or axes about which the impulsive forces cause zero angular momentum. When this occurs, then

$$m(\mathbf{v}_G)_1 = m(\mathbf{v}_G)_2$$
$$(\mathbf{H}_O)_1 = (\mathbf{H}_O)_2$$

The problems that follow involve application of all the above concepts. They are presented in *random order* so that practice may be gained at identifying the various types of problems and developing the skills necessary for their solution.

REVIEW PROBLEMS

R2–1. The 10-lb block is sliding on the smooth surface when the corner D hits a stop block S. Determine the minimum velocity v the block should have which would allow it to tip over on its side and land in the position shown. Neglect the size of S. *Hint:* During impact consider the weight of the block to be nonimpulsive.

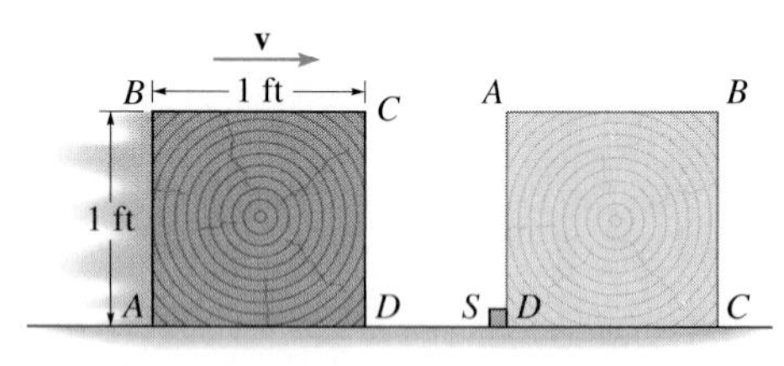

Prob. R2–1

R2–2. The hoisting gear A has an initial angular velocity of 60 rad/s and a constant deceleration of 1 rad/s^2. Determine the velocity and deceleration of the block which is being hoisted by the hub on gear B when $t = 3$ s.

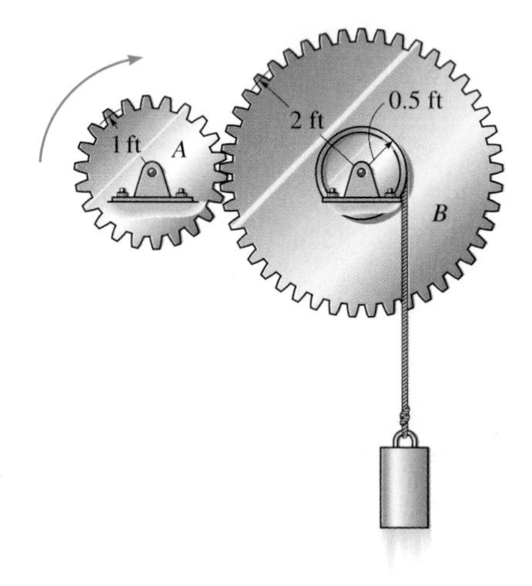

Prob. R2–2

R2–3. The rod is bent into the shape of a sine curve and is forced to rotate about the y axis by connecting the spindle S to a motor. If the rod starts from rest in the position shown and a motor drives it for a short time with an angular acceleration $\alpha = (1.5e^t)$ rad/s^2, where t is in seconds, determine the magnitudes of the angular velocity and angular displacement of the rod when $t = 3$ s. Locate the point on the rod which has the greatest velocity and acceleration, and compute the magnitudes of the velocity and acceleration of this point when $t = 3$ s. The curve defining the rod is $z = 0.25 \sin(\pi y)$, where the argument for the sine is given in radians when y is in meters.

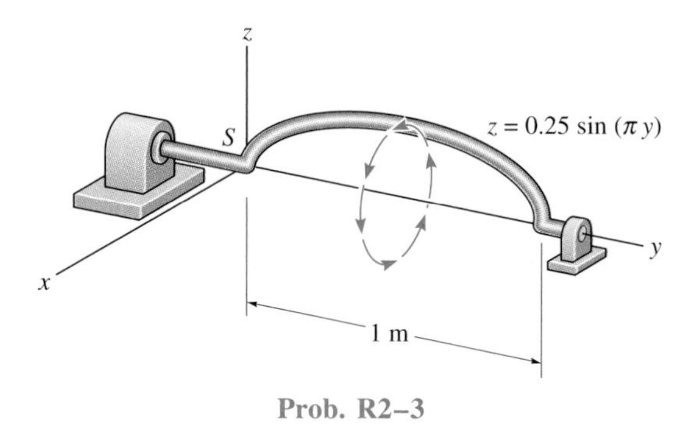

Prob. R2–3

***R2–4.** A cord is wrapped around the inner spool of the gear. If it is pulled with a constant velocity **v**, determine the velocity and acceleration of points A and B. The gear rolls on the fixed gear rack.

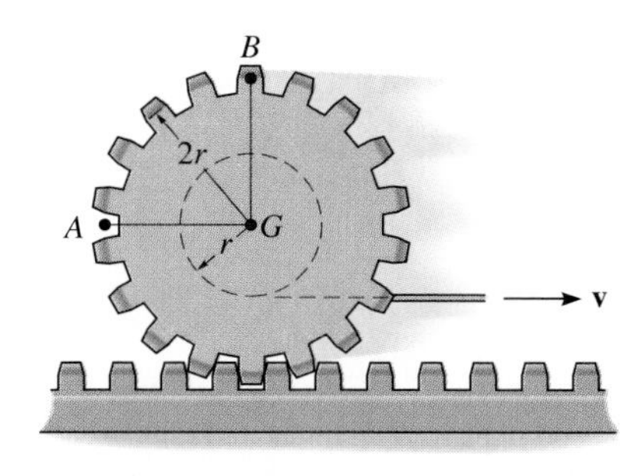

Prob. R2–4

R2–5. A 7-kg automobile tire is released from rest at A on the incline and rolls without slipping to point B, where it then travels in free flight. Determine the maximum height h the tire attains. The radius of gyration of the tire about its mass center is $k_G = 0.3$ m.

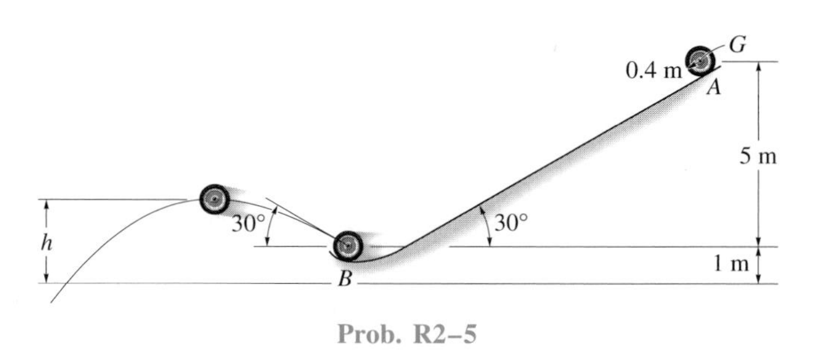

Prob. R2–5

R2–6. The link OA is pinned at O and rotates because of the sliding action of rod R along the horizontal groove. If R starts from rest when $\theta = 0°$ and has a constant acceleration $a_R = 60$ mm/s^2 to the right, determine the angular velocity and angular acceleration of OA when $t = 2$ s.

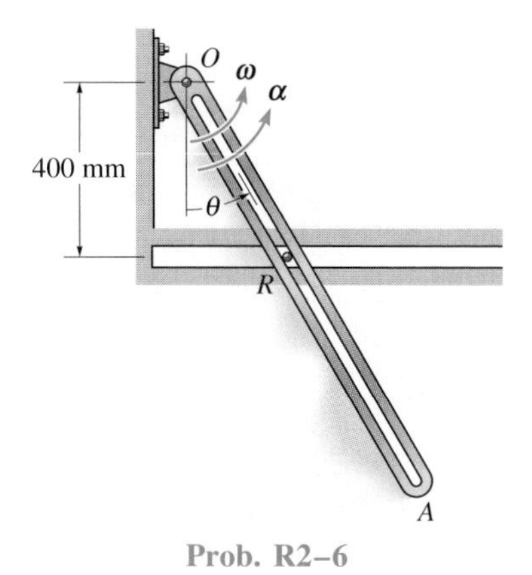

Prob. R2–6

R2–7. The uniform connecting rod BC has a mass of 3 kg and is pin-connected at its end points. Determine the vertical forces which the pins exert on the ends B and C of the rod at the instant (a) $\theta = 0°$, and (b) $\theta = 90°$. The crank AB is turning with a constant angular velocity $\omega_{AB} = 5$ rad/s.

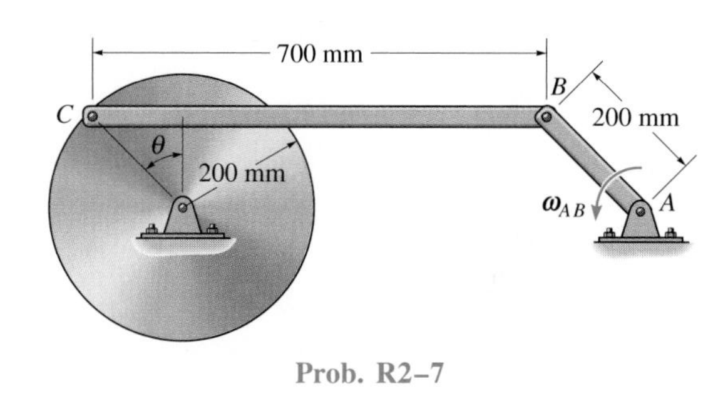

Prob. R2–7

***R2–8.** The tire has a mass of 9 kg and a radius of gyration $k_O =$ 225 mm. If it is released from rest and rolls down the plane without slipping, determine the speed of its center O when $t = 3$ s.

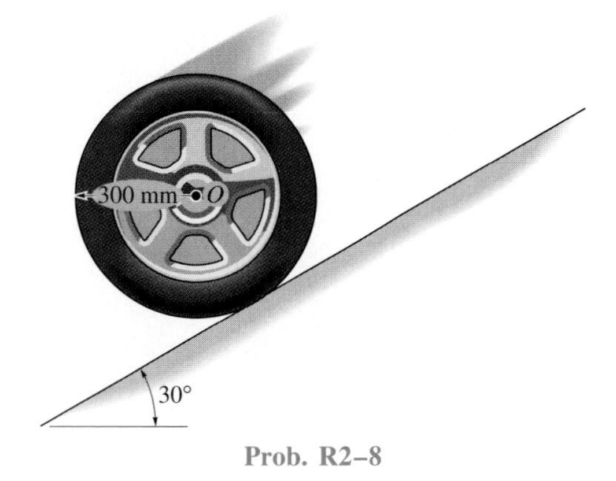

Prob. R2–8

R2–9. The double pendulum consists of two rods. Rod AB has a constant angular velocity of 3 rad/s, and rod BC has a constant angular velocity of 2 rad/s. Both of these absolute motions are measured counterclockwise. Determine the velocity and acceleration of point C at the instant shown.

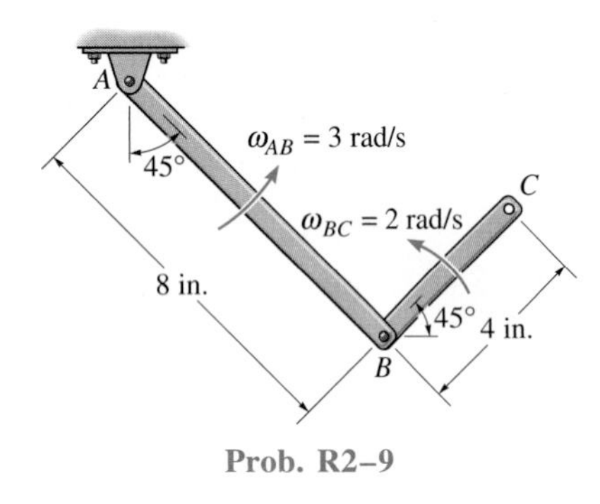

Prob. R2–9

R2–10. The spool and wire wrapped around its core have a mass of 20 kg and a centroidal radius of gyration $k_G = 250$ mm. If the coefficient of kinetic friction at the ground is $\mu_B = 0.1$, determine the angular acceleration of the spool when the 30-N · m couple moment is applied.

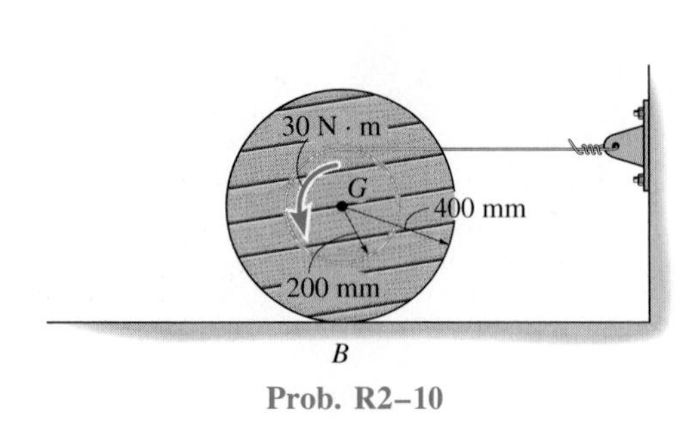

Prob. R2–10

R2–11. If the ball has a weight of 15 lb and is thrown onto a *rough surface* so that its center has a velocity of 6 ft/s parallel to the surface, determine the amount of backspin, $\boldsymbol{\omega}$, the ball must be given so that it stops spinning at the same instant that its forward velocity is zero. It is not necessary to know the coefficient of kinetic friction at A for the calculation.

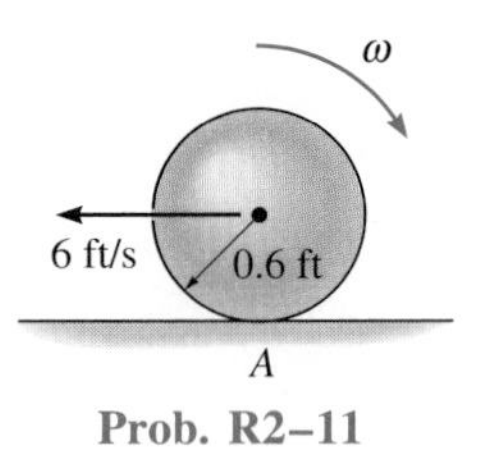

Prob. R2–11

***R2–12.** Blocks A and B weigh 50 and 10 lb, respectively. If P = 100 lb, determine the normal force exerted by block A on block B. Neglect friction and the weights of the pulleys, cord, and bars of the triangular frame.

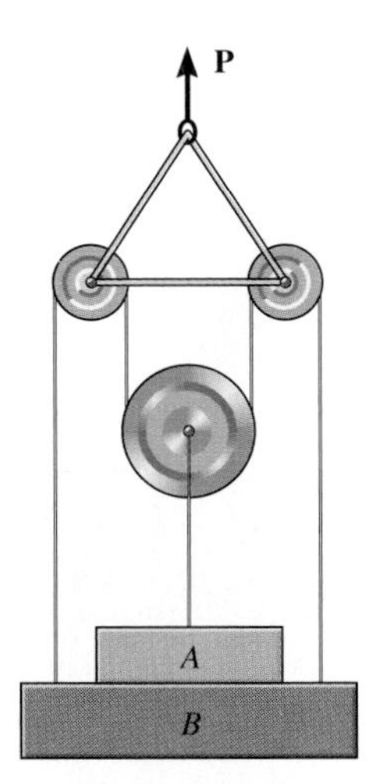

Prob. R2–12

R2–13. Determine the velocity and acceleration of rod R for any angle θ of cam C if the cam rotates with a constant angular velocity $\boldsymbol{\omega}$. The pin connection at O does not cause an interference with the motion of A on C.

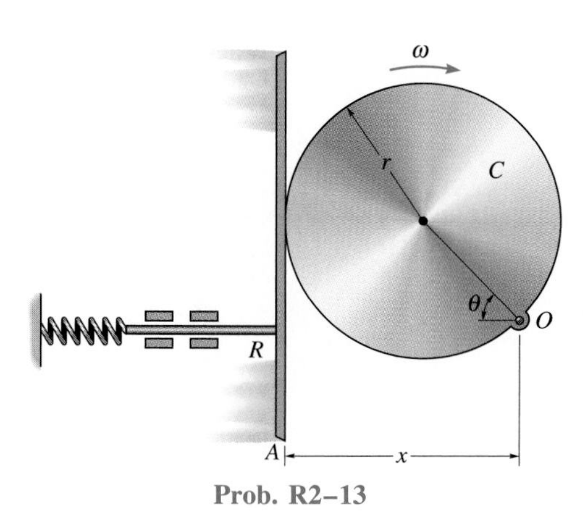

Prob. R2–13

R2–15. A tape having a thickness s wraps around the wheel which is turning at a constant rate $\boldsymbol{\omega}$. Assuming the unwrapped portion of tape remains horizontal, determine the acceleration of point P on the tape when the radius is r. *Hint:* Since $v_P = \omega r$, take the time derivative and note that $dr/dt = \omega(s/2\pi)$.

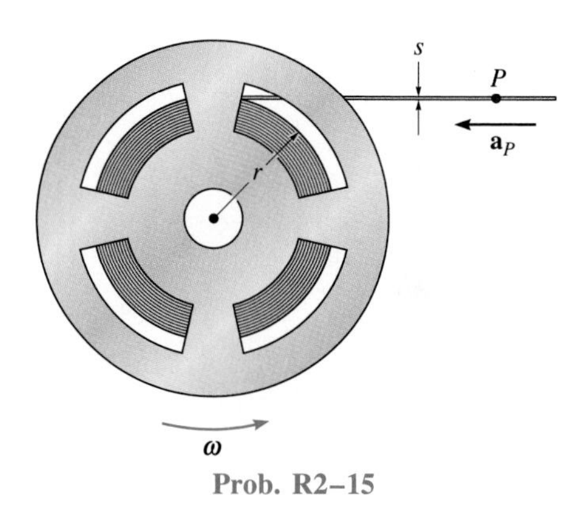

Prob. R2–15

R2–14. The uniform plate weighs 40 lb and is supported by a roller at A. If a horizontal force $F = 70$ lb is suddenly applied to the roller, determine the acceleration of the center of the roller at the instant the force is applied. The plate has a moment of inertia about its center of mass of $I_G = 0.414 \text{ slug} \cdot \text{ft}^2$. Neglect the weight of the roller.

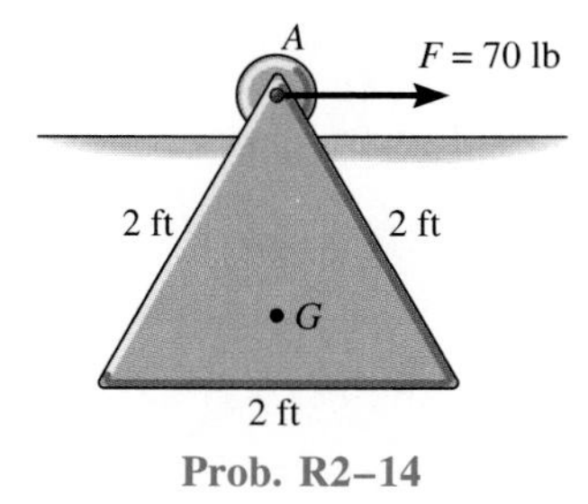

Prob. R2–14

***R2–16.** The 15-lb cylinder is initially at rest on a 5-lb plate. If a couple moment $M = 40 \text{ lb} \cdot \text{ft}$ is applied to the cylinder, determine the angular acceleration of the cylinder and the time needed for the end B of the plate to travel 3 ft and strike the wall. Assume the cylinder does not slip on the plate, and neglect the mass of the rollers under the plate.

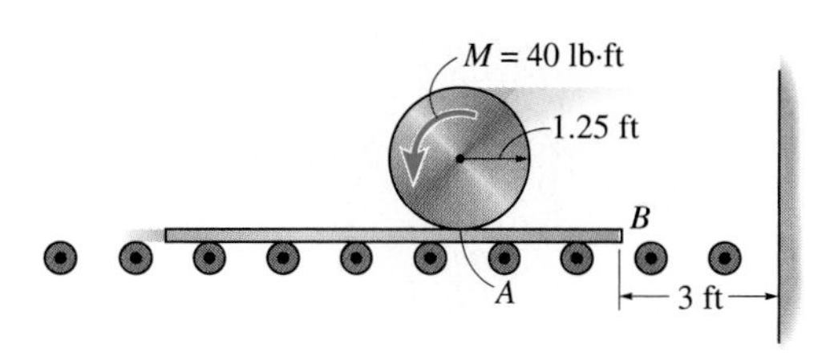

Prob. R2–16

R2–17. The wheelbarrow and its contents have a mass of 40 kg and a mass center at G, excluding the wheel. The wheel has a mass of 4 kg and a radius of gyration $k_O = 0.120$ m. If the wheelbarrow is released from rest from the position shown, determine its speed after it travels 4 m down the incline. The coefficient of kinetic friction between the incline and A is $\mu_A = 0.3$. The wheels roll without slipping at B.

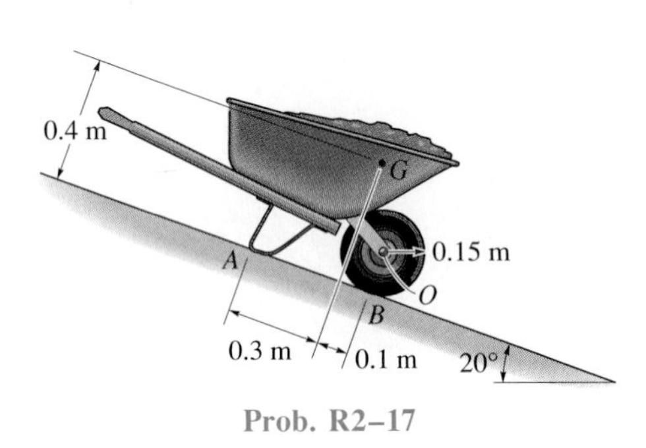

Prob. R2–17

R2–18. The drum of mass m, radius r, and radius of gyration k_O rolls along an inclined plane for which the coefficient of static friction is μ. If the drum is released from rest, determine the maximum angle θ for the incline so that it rolls without slipping.

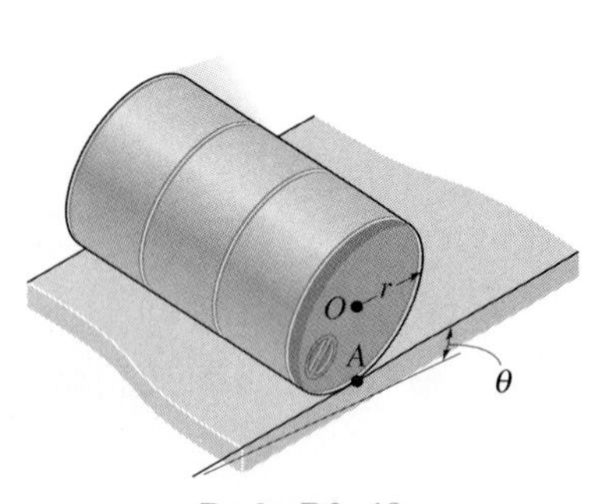

Prob. R2–18

R2–19. The 20-lb solid ball is cast on the floor such that it has a backspin $\omega = 15$ rad/s and its center has an initial horizontal velocity $v_G = 20$ ft/s. If the coefficient of kinetic friction between the floor and the ball is $\mu_A = 0.3$, determine the distance it travels before it stops spinning.

***R2–20.** Determine the backspin $\boldsymbol{\omega}$ which should be given to the 20-lb ball so that when its center is given an initial horizontal velocity $v_G = 20$ ft/s it stops spinning and translating at the same instant. The coefficient of kinetic friction is $\mu_A = 0.3$.

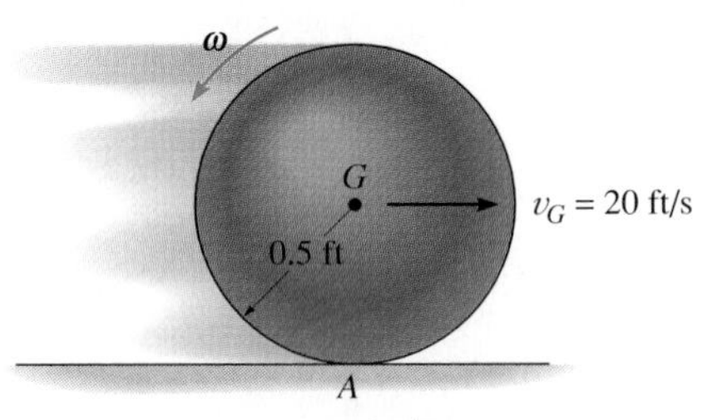

Probs. R2–19/R2–20

R2–21. A 20-kg roll of paper, originally at rest, is pin-supported at its ends to bracket AB. The roll rests against a wall for which the coefficient of kinetic friction at C is $\mu_C = 0.3$. If a force of 40 N is applied uniformly to the end of the sheet, determine the initial angular acceleration of the roll and the tension in the bracket as the paper unwraps. For the calculation, treat the roll as a cylinder.

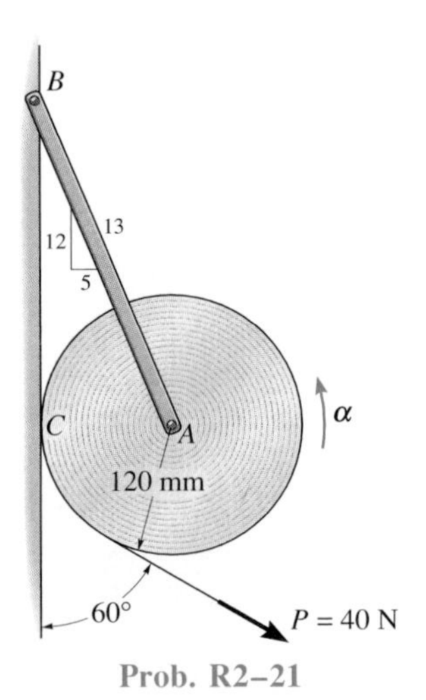

Prob. R2–21

R2–22. Compute the velocity of rod R for any angle θ of the cam C if the cam rotates with a constant angular velocity $\boldsymbol{\omega}$. The pin connection at O does not cause an interference with the motion of A on C.

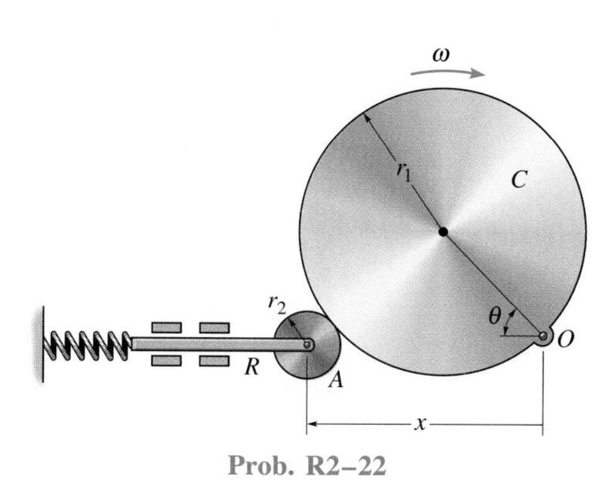

Prob. R2–22

R2–23. The assembly weighs 10 lb and has a radius of gyration $k_G = 0.6$ ft about its center of mass G. The kinetic energy of the assembly is 31 ft-lb when it is in the position shown. If it is rolling counterclockwise on the surface without slipping, determine its linear momentum at this instant.

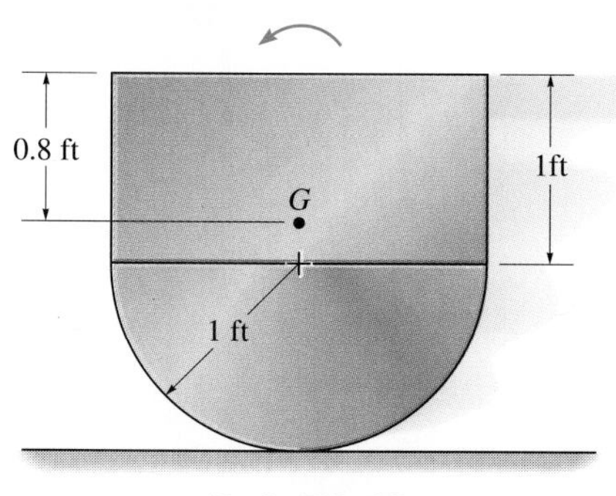

Prob. R2–23

***R2–24.** The pendulum consists of a 30-lb sphere and a 10-lb slender rod. Compute the reaction at the pin O just after the cord AB is cut.

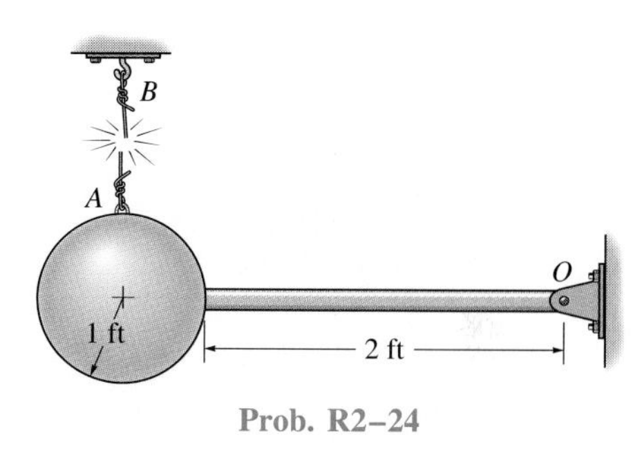

Prob. R2–24

R2–25. The board rests on the surface of two drums. At the instant shown, it has an acceleration of 0.5 m/s^2 to the right, while at the same instant points on the outer rim of each drum have an acceleration with a magnitude of 3 m/s^2. If the board does not slip on the drums, determine its speed due to the motion.

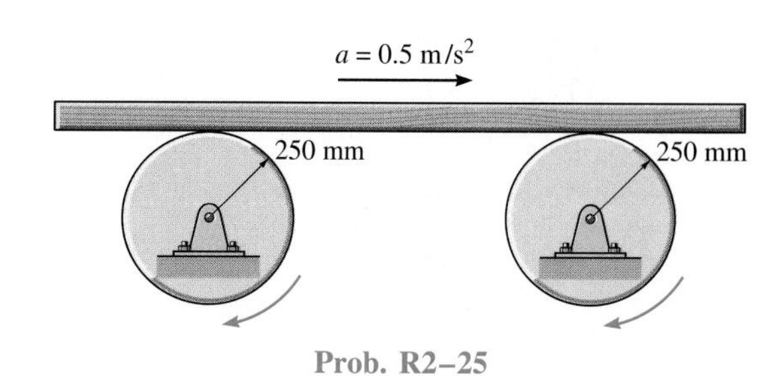

Prob. R2–25

R2–26. The center of the pulley is being lifted vertically with an acceleration of 4 m/s^2 at the instant it has a velocity of 2 m/s. If the cable does not slip on the pulley's surface, determine the accelerations of the cylinder B and point C on the pulley.

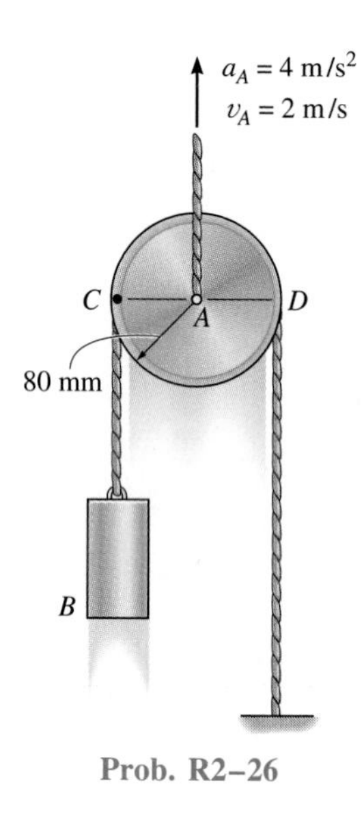

Prob. R2–26

R2–27. At the instant shown, two forces act on the 30-lb slender rod which is pinned at O. Determine the magnitude of force **F** and the initial angular acceleration of the rod so that the horizontal reaction which the *pin exerts on the rod* is 5 lb directed to the right.

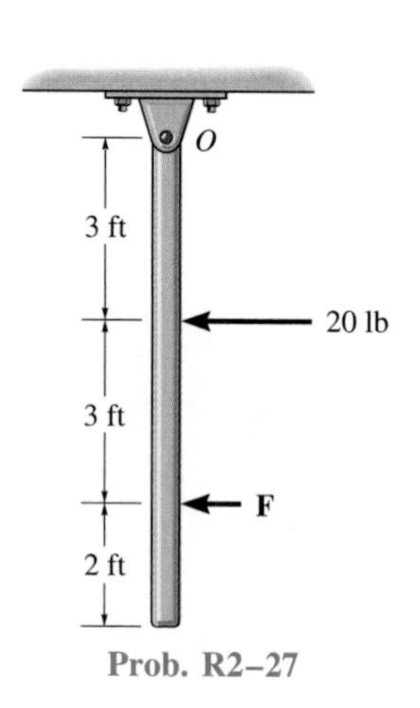

Prob. R2–27

***R2–28.** A chain that has a negligible mass is draped over a sprocket which has a mass of 2 kg and a radius of gyration $k_O = 50$ mm. If the 4-kg block A is released from rest in the position $s = 1$ m, determine the angular velocity which the chain imparts to the sprocket when $s = 2$ m.

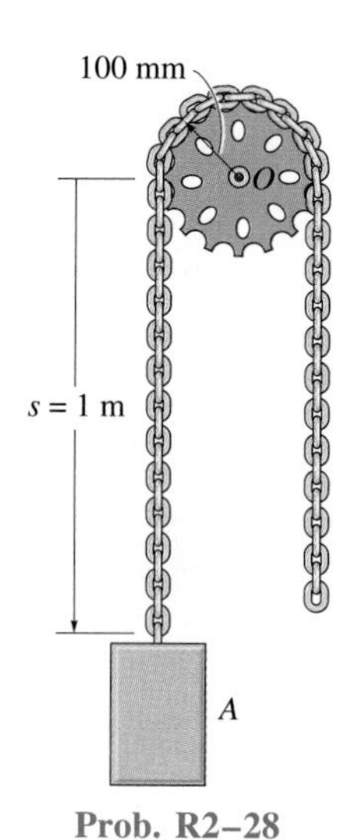

Prob. R2–28

R2–29. The spool has a weight of 30 lb and a radius of gyration $k_O = 0.65$ ft. If a force of 40 lb is applied to the cord at A, determine the angular velocity of the spool in $t = 3$ s starting from rest. Neglect the mass of the pulley and cord.

R2–30. Solve Prob. R2–29 if a 40-lb block is suspended from the cord at A, rather than applying the 40-lb force.

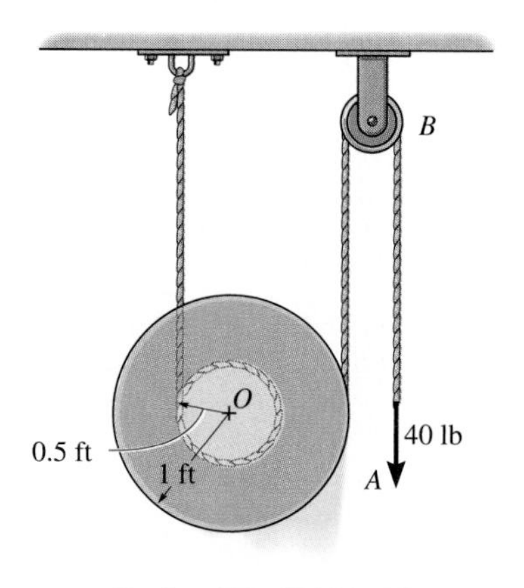

Probs. R2–29/R2–30

R2–31. The dresser has a weight of 80 lb and is pushed along the floor. If the coefficient of static friction at A and B is $\mu_s = 0.3$ and the coefficient of kinetic friction is $\mu_k = 0.2$, determine the smallest horizontal force P needed to cause motion. If this force is increased slightly, determine the acceleration of the dresser. Also, what are the normal reactions at A and B when it begins to move?

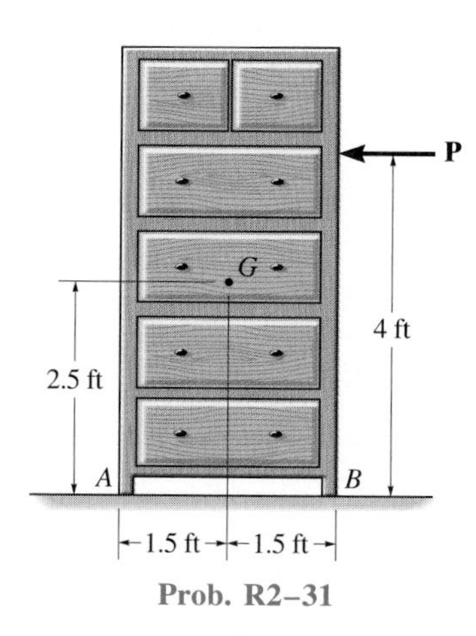

Prob. R2–31

***R2–32.** When the crank on the Chinese windlass is turning, the rope on shaft A unwinds while that on shaft B winds up. Determine the speed at which the block lowers if the crank is turning with an angular velocity $\omega = 4$ rad/s. What is the angular velocity of the pulley at C? The rope segments on each side of the pulley are both parallel and vertical, and the rope does not slip on the pulley.

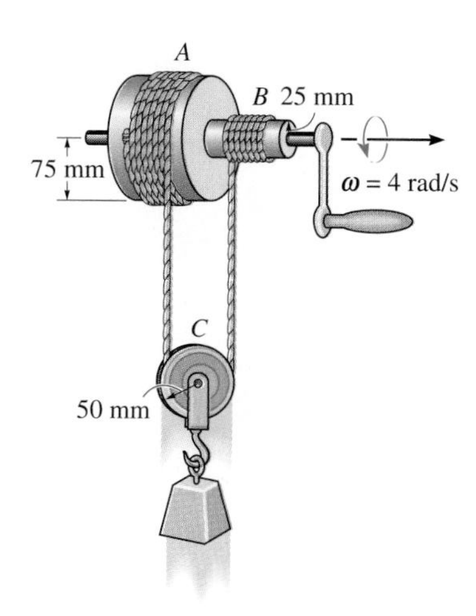

Prob. R2–32

R2–33. The semicircular disk has a mass of 50 kg and is released from rest from the position shown. The coefficients of static and kinetic friction between the disk and the beam are $\mu_s = 0.5$ and $\mu_k = 0.3$, respectively. Determine the initial reactions at the pin A and roller B, used to support the beam. Neglect the mass of the beam for the calculation.

R2–34. The semicircular disk has a mass of 50 kg and is released from rest from the position shown. The coefficients of static and kinetic friction between the disk and the beam are $\mu_s = 0.2$ and $\mu_k = 0.1$, respectively. Determine the initial reactions at the pin A and roller B used to support the beam. Neglect the mass of the beam for the calculation.

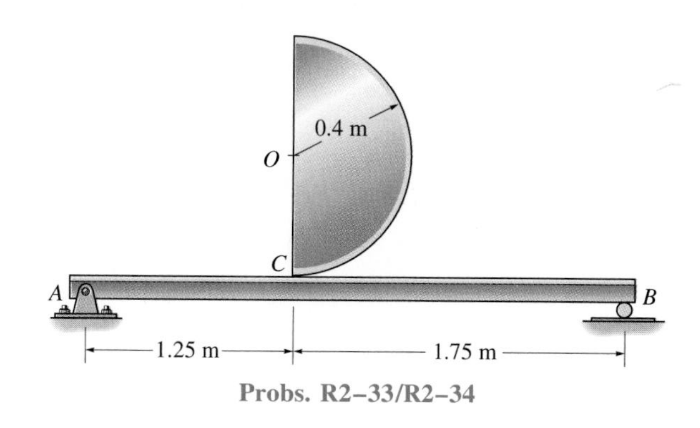

Probs. R2–33/R2–34

R2–35. The cylinder having a mass of 5 kg is initially at rest when it is placed in contact with the wall B and the rotor at A. If the rotor always maintains a constant clockwise angular velocity $\omega = 6$ rad/s, determine the initial angular acceleration of the cylinder. The coefficient of kinetic friction at the contacting surfaces B and C is $\mu_k = 0.2$.

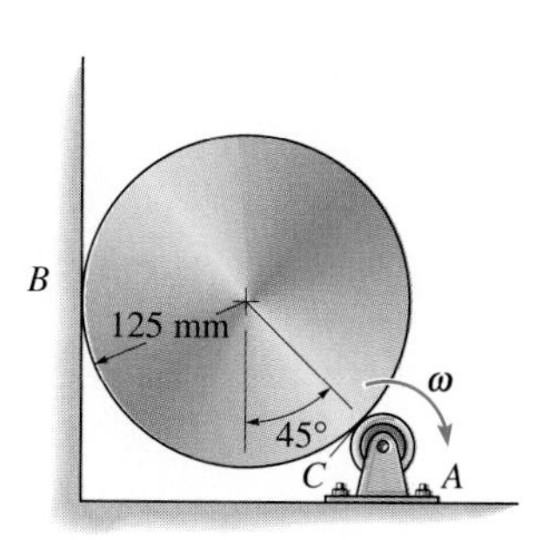

Prob. R2–35

***R2–36.** The truck carries the 800-lb crate which has a center of gravity at G_c. Determine the largest acceleration of the truck so that the crate will not slip or tip on the truck bed. The coefficient of static friction between the crate and the truck is $\mu_s = 0.6$.

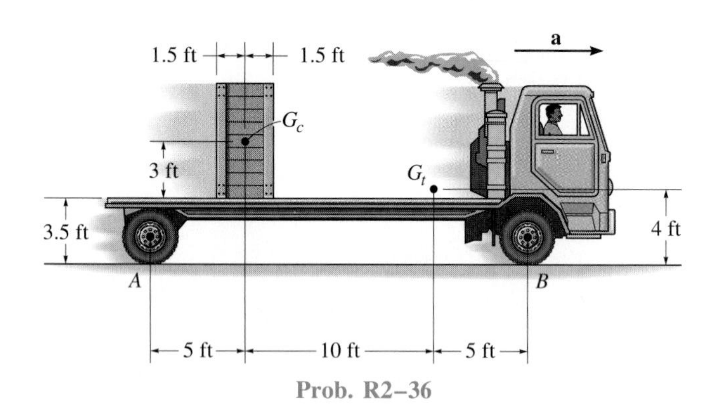

Prob. R2–36

R2–37. The truck has a weight of 8000 lb and center of gravity at G_t. It carries the 800-lb crate, which has a center of gravity at G_c. Determine the normal reaction at *each* of its four tires if it accelerates at $a = 0.5$ ft/s^2. Also, what is the frictional force acting between the crate and the truck, and between *each* of the rear tires and the road? Assume that power is delivered only to the rear tires. The front tires are free to roll. Neglect the mass of the tires. The crate does not slip or tip on the truck.

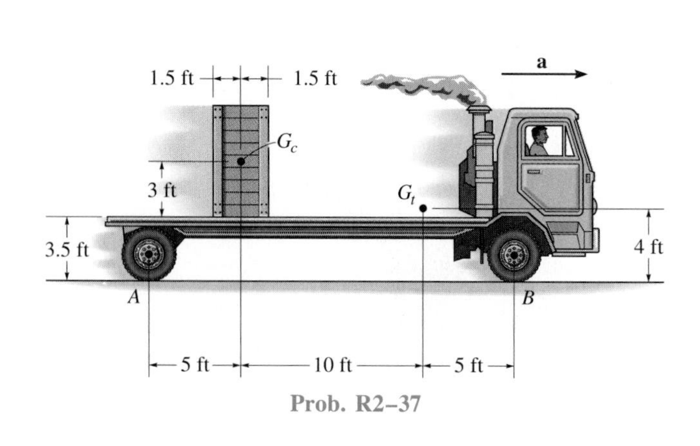

Prob. R2–37

R2–38. Spool B is at rest and spool A is rotating at 6 rad/s when the slack in the cord connecting them is taken up. Determine the angular velocity of each spool immediately after the cord is jerked tight. The spools A and B have weights and radii of gyration $W_A = 30$ lb, $k_A = 0.8$ ft and $W_B = 15$ lb, $k_B = 0.6$ ft, respectively.

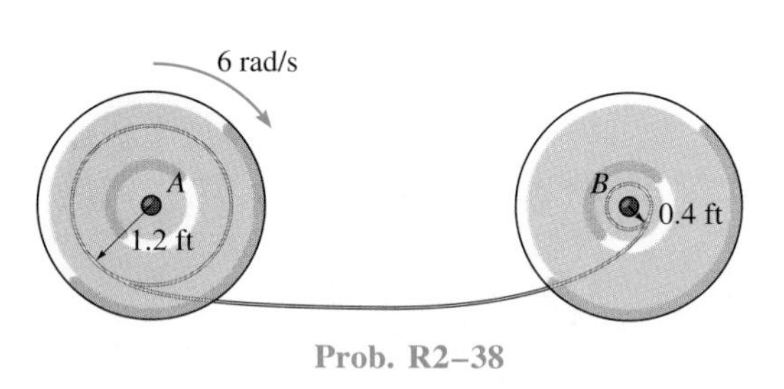

Prob. R2–38

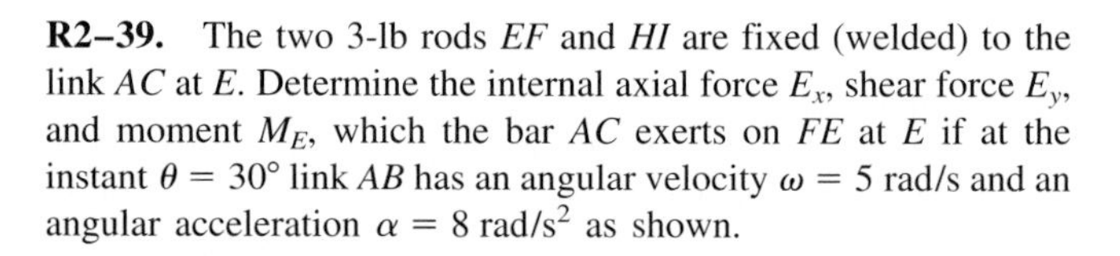

R2–39. The two 3-lb rods EF and HI are fixed (welded) to the link AC at E. Determine the internal axial force E_x, shear force E_y, and moment M_E, which the bar AC exerts on FE at E if at the instant $\theta = 30°$ link AB has an angular velocity $\omega = 5$ rad/s and an angular acceleration $\alpha = 8$ rad/s^2 as shown.

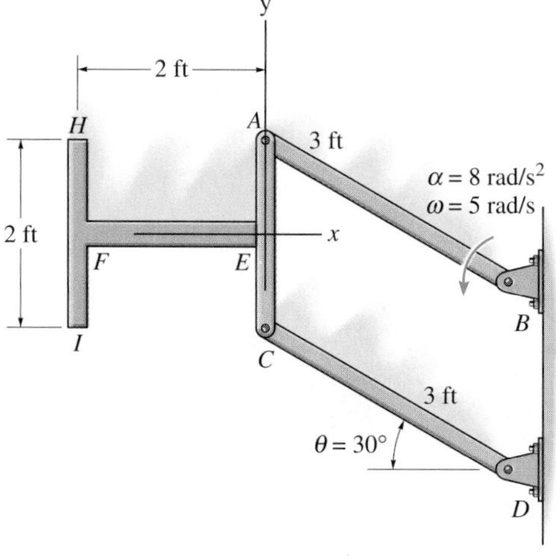

Prob. R2–39

***R2–40.** The dragster has a mass of 1500 kg and a center of mass at G. If the coefficient of kinetic friction between the rear wheels and the pavement is $\mu_k = 0.6$, determine if it is possible for the driver to lift the front wheels, A, off the ground while the rear wheels are slipping. If so, what acceleration is necessary to do this? Neglect the mass of the wheels and assume that the front wheels are free to roll.

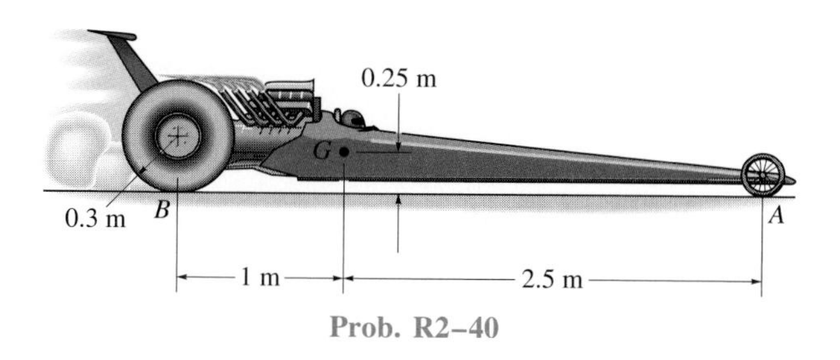

Prob. R2–40

R2–41. The dragster has a mass of 1500 kg and a center of mass at G. If no slipping occurs, determine the friction force F_B which must be applied to *each* of the rear wheels B in order to develop an acceleration $a = 6\text{ m/s}^2$. What are the normal reactions of *each* wheel on the ground? Neglect the mass of the wheels and assume that the front wheels are free to roll.

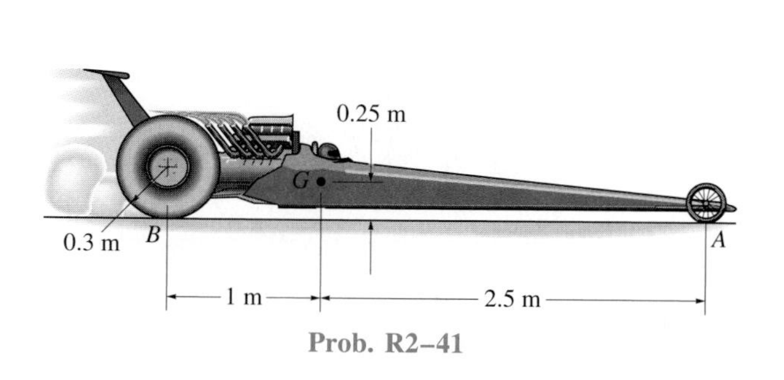

Prob. R2–41

R2–42. The 1.6-Mg car shown has been ''raked'' by increasing the height $h = 0.2$ m of its center of mass. This was done by raising the springs on the rear axle. If the coefficient of kinetic friction between the rear wheels and the ground is $\mu_k = 0.3$, show that the car can accelerate slightly faster than its counterpart for which $h = 0$. Neglect the mass of the wheels and driver and assume the front wheels at B are free to roll while the rear wheels slip.

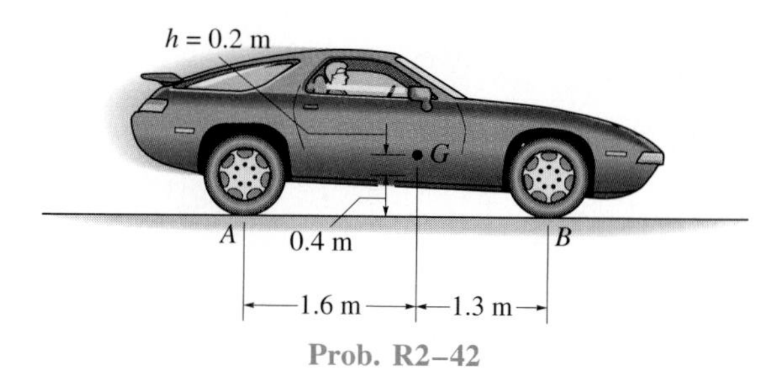

Prob. R2–42

R2–43. The handcart has a mass of 200 kg and center of mass at G. Determine the normal reactions at *each* of the wheels at A and B if a force $P = 50$ N is applied to the handle. Neglect the mass and rolling resistance of the wheels.

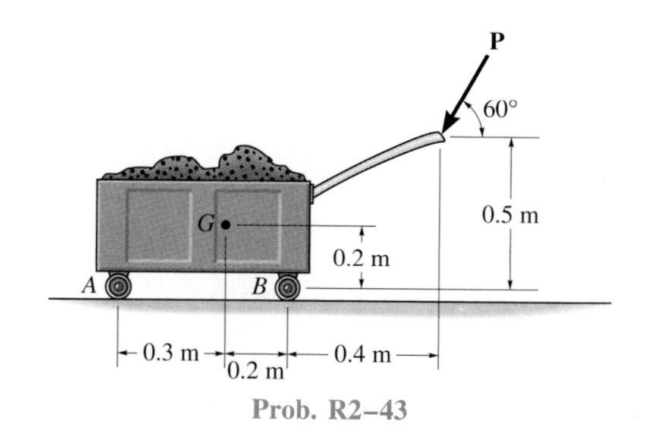

Prob. R2–43

***R2–44.** If bar AB has an angular velocity $\omega_{AB} = 6$ rad/s, determine the velocity of the slider block C at the instant shown.

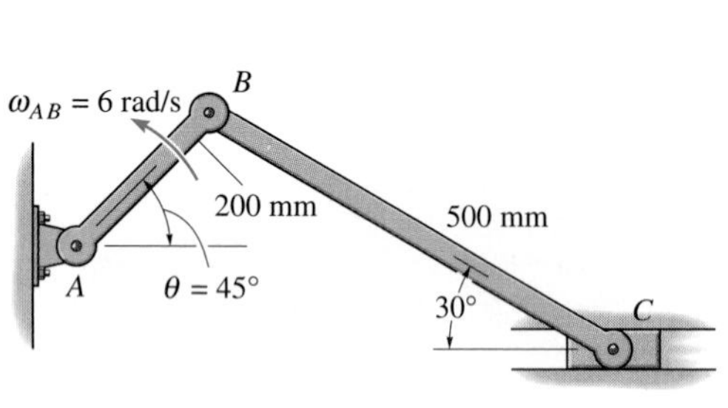

Prob. R2–44

R2–45. The disk is rotating at a constant rate $\omega = 4$ rad/s, and as it falls freely, its center has an acceleration of 32.2 ft/s^2. Determine the acceleration of points A and B on the rim of the disk at the instant shown.

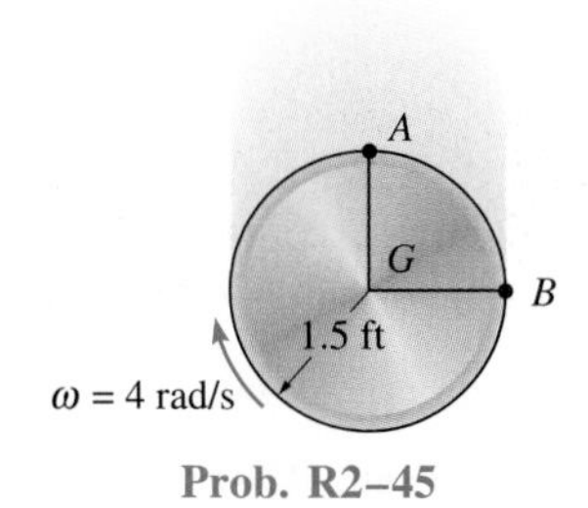

Prob. R2–45

R2–46. The 80-lb cylinder is attached to the 10-lb slender rod which is pinned from point A. At the instant $\theta = 30°$ it has an angular velocity $\omega_1 = 1$ rad/s as shown. Determine the largest angle θ to which the rod swings before it momentarily stops.

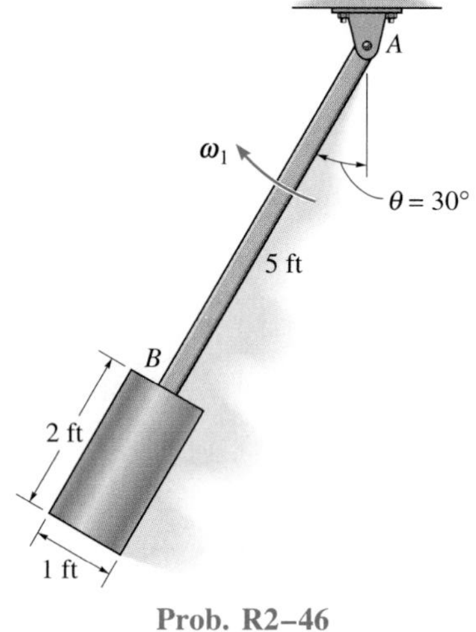

Prob. R2–46

R2–47. The bicycle and rider have a mass of 80 kg with center of mass located at G. If the coefficient of kinetic friction at the rear tire is $\mu_B = 0.8$, determine the normal reactions at the tires A and B, and the deceleration of the rider, when the rear wheel locks for braking. What is the normal reaction at the rear wheel when the bicycle is traveling at constant velocity and the brakes are not applied? Neglect the mass of the wheels.

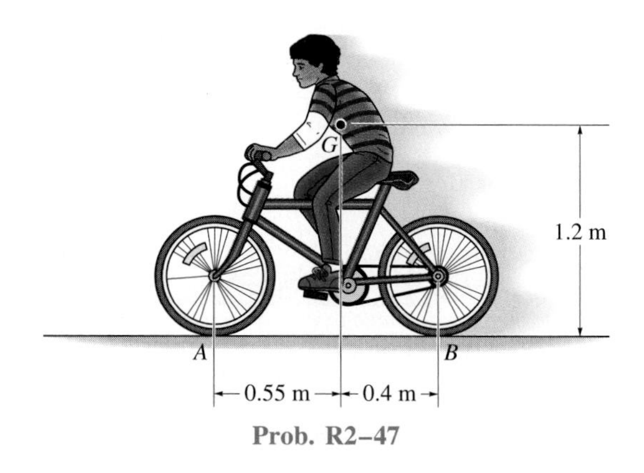

Prob. R2–47

***R2–48.** At the instant shown, link AB has an angular velocity $\omega_{AB} = 2$ rad/s and an angular acceleration $\alpha_{AB} = 6$ rad/s^2. Determine the acceleration of the pin at C and the angular acceleration of link CB at this instant, when $\theta = 60°$.

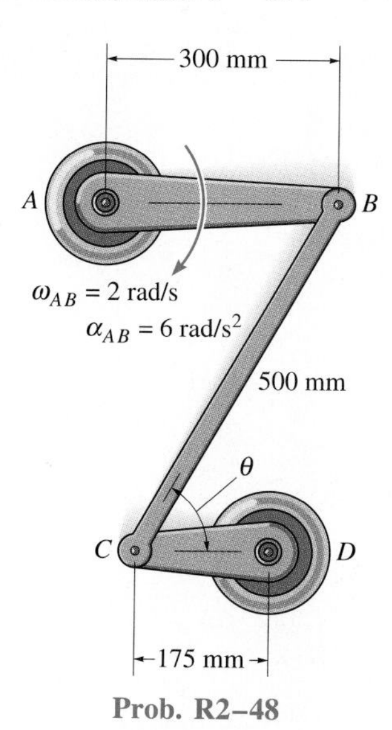

Prob. R2–48

R2–49. The spool has a mass of 60 kg and a radius of gyration $k_G = 0.3$ m. If it is released from rest, determine how far it descends down the smooth plane before it attains an angular velocity $\omega = 6$ rad/s. Neglect friction and the mass of the cord which is wound around the central core.

R2–50. Solve Prob. R2–49 if the plane is rough, such that the coefficient of kinetic friction at A is $\mu_A = 0.2$.

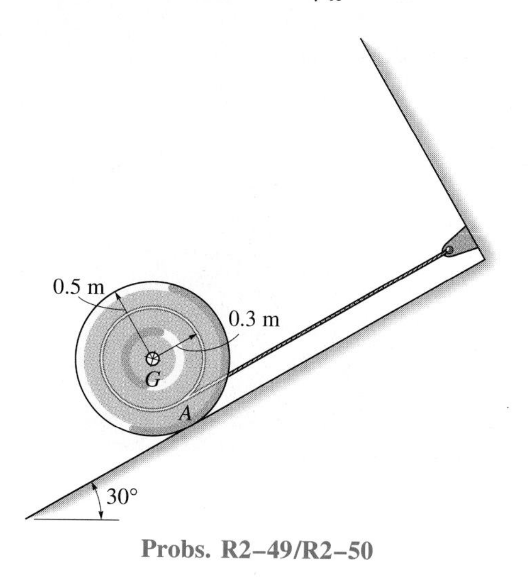

Probs. R2–49/R2–50

R2–51. The gear rack has a mass of 6 kg, and the gears each have a mass of 4 kg and a radius of gyration $k = 30$ mm at their centers. If the rack is originally moving downward at 2 m/s, when $s = 0$, determine the speed of the rack when $s = 600$ mm. The gears are free to turn about their centers, A and B.

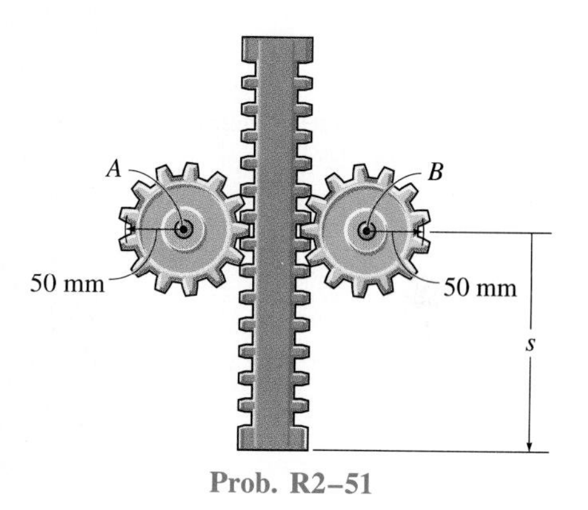

Prob. R2–51

***R2–52.** The car has a mass of 1.50 Mg and a mass center at G. Determine the maximum acceleration it can have if (a) power is supplied only to the rear wheels, (b) power is supplied only to the front wheels. Neglect the mass of the wheels in the calculation, and assume that the wheels that do not receive power are free to roll. Also, assume that slipping of the powered wheels occurs, where the coefficient of kinetic friction is $\mu_k = 0.3$.

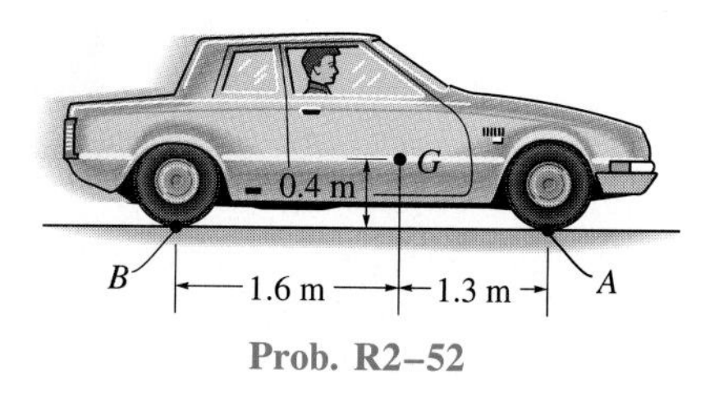

Prob. R2–52

Telescopic robotic arms are often used in the manufacturing industry. The principles outlined in this chapter can be used to determine the kinematics of their three-dimensional motion.

20

Three-Dimensional Kinematics of a Rigid Body

Three types of planar rigid-body motion have been presented in Chapter 16: translation, rotation about a fixed axis, and general plane motion. In this chapter the three-dimensional motion of a rigid body, which consists of rotation about a fixed point and general motion, will be discussed. This is followed by a general study of the motion of particles and rigid bodies using coordinate systems that both translate and rotate. The analysis of these motions is more complex than planar-motion analysis, since the angular acceleration of the body will measure a change in both the *magnitude* and *direction* of the body's angular velocity. In order to simplify the motion's three-dimensional aspects, throughout the chapter we will make use of vector analysis.*

⋆20.1 Rotation About a Fixed Point

When a rigid body rotates about a fixed point, the distance r from the point to a particle P located on the body is the *same* for *any position* of the body. Thus, the path of motion for the particle lies on the *surface of a sphere* having a radius r and centered at the fixed point. Since motion along this path occurs only from a series of rotations made during a finite time interval, we will first develop a familiarity with some of the properties of rotational displacements.

*A brief review of vector analysis is given in Appendix C.

Euler's Theorem. Euler's theorem states that two "component" rotations about different axes passing through a point are equivalent to a single resultant rotation about an axis passing through the point. If more than two rotations are applied, they can be combined into pairs, and each pair can be further reduced to combine into one rotation.

Finite Rotations. If the component rotations used in Euler's theorem are *finite,* it is important that the *order* in which they are applied be maintained. This is because finite rotations do *not* obey the law of vector addition, and hence they cannot be classified as vector quantities. To show this, consider the two finite rotations $\boldsymbol{\theta}_1 + \boldsymbol{\theta}_2$ applied to the block in Fig. 20–1*a*. Each rotation has a magnitude of 90° and a direction defined by the right-hand rule, as indicated by the arrow. The resultant orientation of the block is shown at the right. When these two rotations are applied in the order $\boldsymbol{\theta}_2 + \boldsymbol{\theta}_1$, as shown in Fig. 20–1*b*, the resultant position of the block is *not* the same as it is in Fig. 20–1*a*. Consequently, *finite rotations* do not obey the commutative law of addition ($\boldsymbol{\theta}_1 + \boldsymbol{\theta}_2 \neq \boldsymbol{\theta}_2 + \boldsymbol{\theta}_1$), and therefore *they cannot be classified as vectors.* If smaller, yet finite, rotations had been used to illustrate this point, e.g., 10° instead of 90°, the *resultant* orientation of the block after each combination of rotations would also be different; however, in this case, only by a small amount.

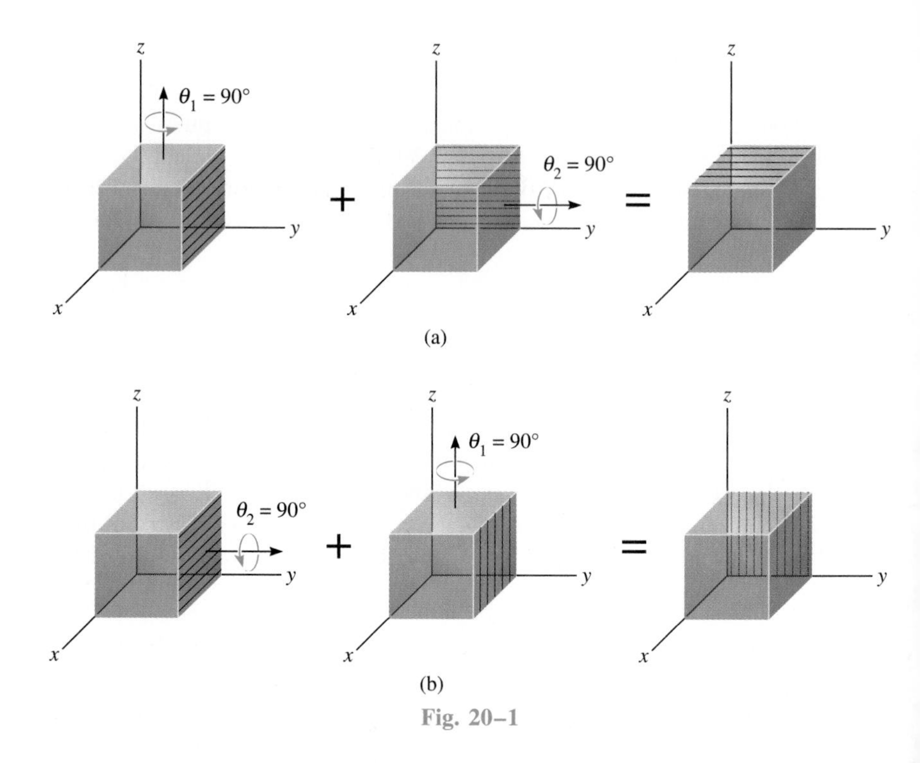

Fig. 20–1

Infinitesimal Rotations. When defining the angular motions of a body subjected to three-dimensional motion, only rotations which are *infinitesimally small* will be considered. *Such rotations may be classified as vectors, since they can be added vectorially in any manner.* To show this, let us for purposes of simplicity consider the rigid body itself to be a sphere which is allowed to rotate about its central fixed point O, Fig. 20–2a. If we impose two infinitesimal rotations $d\boldsymbol{\theta}_1 + d\boldsymbol{\theta}_2$ on the body, it is seen that point P moves along the path $d\boldsymbol{\theta}_1 \times \mathbf{r} + d\boldsymbol{\theta}_2 \times \mathbf{r}$ and ends up at P'. Had the two successive rotations occurred in the order $d\boldsymbol{\theta}_2 + d\boldsymbol{\theta}_1$, then the resultant displacements of P would have been $d\boldsymbol{\theta}_2 \times \mathbf{r} + d\boldsymbol{\theta}_1 \times \mathbf{r}$. Since the vector cross product obeys the distributive law, by comparison $(d\boldsymbol{\theta}_1 + d\boldsymbol{\theta}_2) \times \mathbf{r} = (d\boldsymbol{\theta}_2 + d\boldsymbol{\theta}_1) \times \mathbf{r}$. Hence infinitesimal rotations $d\boldsymbol{\theta}$ are vectors, since these quantities have both a magnitude and direction for which the order of (vector) addition is not important, i.e., $d\boldsymbol{\theta}_1 + d\boldsymbol{\theta}_2 = d\boldsymbol{\theta}_2 + d\boldsymbol{\theta}_1$. Furthermore, as shown in Fig. 20–2a, the two "component" rotations $d\boldsymbol{\theta}_1$ and $d\boldsymbol{\theta}_2$ are equivalent to a single resultant rotation $d\boldsymbol{\theta} = d\boldsymbol{\theta}_1 + d\boldsymbol{\theta}_2$, a consequence of Euler's theorem.

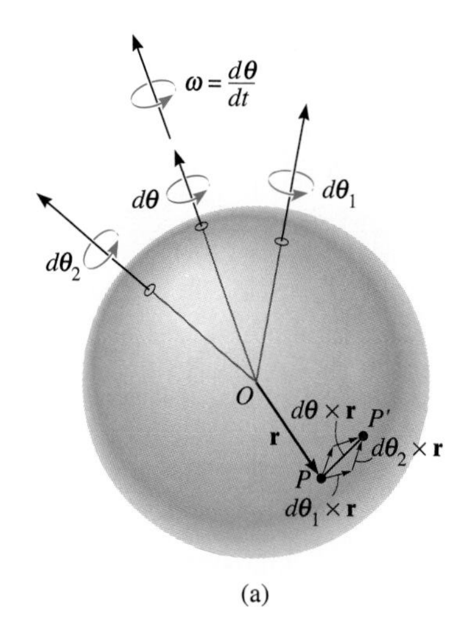

(a)

Angular Velocity. If the body is subjected to an angular rotation $d\boldsymbol{\theta}$ about a fixed point, the angular velocity of the body is defined by the time derivative,

$$\boldsymbol{\omega} = \dot{\boldsymbol{\theta}} \qquad (20\text{–}1)$$

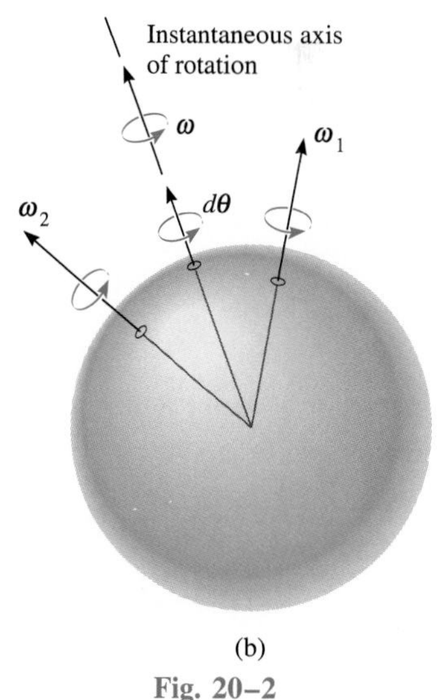

(b)
Fig. 20–2

The line specifying the direction of $\boldsymbol{\omega}$, which is collinear with $d\boldsymbol{\theta}$, is referred to as the *instantaneous axis of rotation,* Fig. 20–2b. In general, this axis changes direction during each instant of time. Since $d\boldsymbol{\theta}$ is a vector quantity, so too is $\boldsymbol{\omega}$, and it follows from vector addition that if the body is subjected to two component angular motions, $\boldsymbol{\omega}_1 = \dot{\boldsymbol{\theta}}_1$ and $\boldsymbol{\omega}_2 = \dot{\boldsymbol{\theta}}_2$, the resultant angular velocity is $\boldsymbol{\omega} = \boldsymbol{\omega}_1 + \boldsymbol{\omega}_2$.

Angular Acceleration. The body's angular acceleration is determined from the time derivative of the angular velocity, i.e.,

$$\boldsymbol{\alpha} = \dot{\boldsymbol{\omega}} \qquad (20\text{–}2)$$

For motion about a fixed point, $\boldsymbol{\alpha}$ must account for a change in *both* the magnitude and direction of $\boldsymbol{\omega}$, so that, in general, $\boldsymbol{\alpha}$ is not directed along the instantaneous axis of rotation, Fig. 20–3.

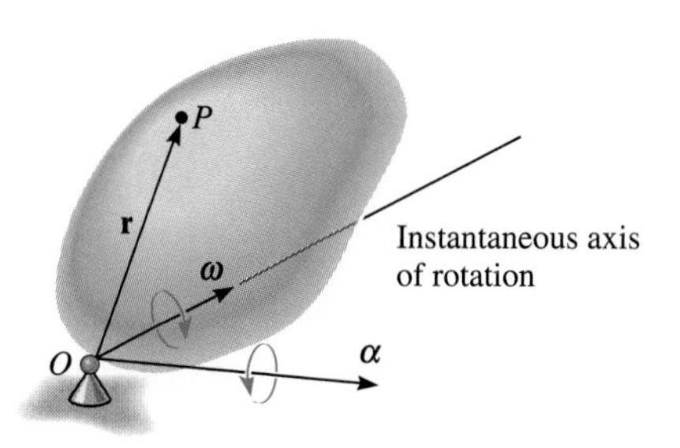

Fig. 20–3

As the direction of the instantaneous axis of rotation (or the line of action of $\boldsymbol{\omega}$) changes in space, the locus of points defined by the axis generates a fixed *space cone.* If the change in this axis is viewed with respect to the

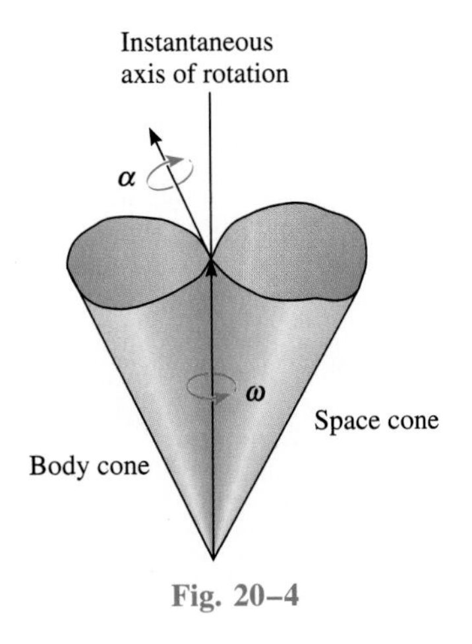

Fig. 20–4

rotating body, the locus of the axis generates a *body cone,* Fig. 20–4. At any given instant, these cones are tangent along the instantaneous axis of rotation, and when the body is in motion, the body cone appears to roll either on the inside or the outside surface of the fixed space cone. Provided the paths defined by the open ends of the cones are described by the head of the $\boldsymbol{\omega}$ vector, $\boldsymbol{\alpha}$ must act tangent to these paths at any given instant, since the time rate of change of $\boldsymbol{\omega}$ is equal to $\boldsymbol{\alpha}$, Fig. 20–4.

Velocity. Once $\boldsymbol{\omega}$ is specified, the velocity of any point P on a body rotating about a fixed point can be determined using the same methods as for a body rotating about a fixed axis (Sec. 16.3). Hence, by the cross product,

$$\mathbf{v} = \boldsymbol{\omega} \times \mathbf{r} \tag{20–3}$$

Here $\mathbf{r}$ defines the position of P measured from the fixed point O, Fig. 20–3.

Acceleration. If $\boldsymbol{\omega}$ and $\boldsymbol{\alpha}$ are known at a given instant, the acceleration of any point P on the body can be obtained by time differentiation of Eq. 20–3, which yields

$$\mathbf{a} = \boldsymbol{\alpha} \times \mathbf{r} + \boldsymbol{\omega} \times (\boldsymbol{\omega} \times \mathbf{r}) \tag{20–4}$$

The form of this equation is the same as that developed in Sec. 16.3, which defines the acceleration of a point located on a body subjected to rotation about a fixed axis.

*20.2 The Time Derivative of a Vector Measured from a Fixed and Translating-Rotating System

In many types of problems involving the motion of a body about a fixed point, the angular velocity $\boldsymbol{\omega}$ is specified in terms of its component angular motions. For example, the disk in Fig. 20–5 spins about the horizontal y axis at $\boldsymbol{\omega}_s$ while it rotates or precesses about the vertical z axis at $\boldsymbol{\omega}_p$. Therefore, its resultant angular velocity is $\boldsymbol{\omega} = \boldsymbol{\omega}_s + \boldsymbol{\omega}_p$. If the angular acceleration $\boldsymbol{\alpha}$ of such a body is to be determined, it is sometimes easier to compute the time derivative of $\boldsymbol{\omega}$, Eq. 20–2, by using a coordinate system which has a *rotation* defined by one or more of the components of $\boldsymbol{\omega}$.* For this reason, and for other uses later, an equation will presently be derived that relates the time derivative of any vector $\mathbf{A}$ defined from a translating-rotating reference to its derivative defined from a fixed reference.

*In the case of the spinning disk, Fig. 20–5, the x, y, z axes may be given an angular velocity of $\boldsymbol{\omega}_p$.

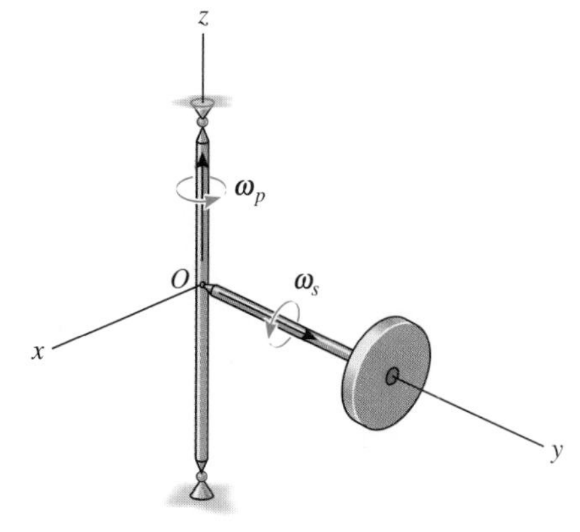

Fig. 20–5

Consider the x, y, z axes of the moving frame of reference to have an angular velocity $\mathbf{\Omega}$ which is measured from the fixed X, Y, Z axes, Fig. 20–6*a*. In the following discussion, it will be convenient to express vector $\mathbf{A}$ in terms of its $\mathbf{i}$, $\mathbf{j}$, $\mathbf{k}$ components, which define the directions of the moving axes. Hence,

$$\mathbf{A} = A_x\mathbf{i} + A_y\mathbf{j} + A_z\mathbf{k}$$

In general, the time derivative of $\mathbf{A}$ must account for the change in both the vector's magnitude and direction. However, if this derivative is taken *with respect to the moving frame of reference,* only a change in the magnitudes of the components of $\mathbf{A}$ must be accounted for, since the directions of the components do not change with respect to the moving reference. Hence,

$$(\dot{\mathbf{A}})_{xyz} = \dot{A}_x\mathbf{i} + \dot{A}_y\mathbf{j} + \dot{A}_z\mathbf{k} \qquad (20\text{–}5)$$

When the time derivative of $\mathbf{A}$ is taken *with respect to the fixed frame of reference,* the *directions* of $\mathbf{i}$, $\mathbf{j}$, and $\mathbf{k}$ change only on account of the *rotation* $\mathbf{\Omega}$ of the axes and not their translation. Hence, in general,

$$\dot{\mathbf{A}} = \dot{A}_x\mathbf{i} + \dot{A}_y\mathbf{j} + \dot{A}_z\mathbf{k} + A_x\dot{\mathbf{i}} + A_y\dot{\mathbf{j}} + A_z\dot{\mathbf{k}}$$

The time derivatives of the unit vectors will now be considered. For example, $\dot{\mathbf{i}} = d\mathbf{i}/dt$ represents only a change in the *direction* of $\mathbf{i}$ with respect to time, since $\mathbf{i}$ has a fixed magnitude of 1 unit. As shown in Fig. 20–6*b*, the change, $d\mathbf{i}$, is *tangent to the path* described by the arrowhead of $\mathbf{i}$ as $\mathbf{i}$ moves due to the rotation $\mathbf{\Omega}$. Accounting for both the magnitude and direction of $d\mathbf{i}$, we can therefore define $\dot{\mathbf{i}}$ using the cross product, $\dot{\mathbf{i}} = \mathbf{\Omega} \times \mathbf{i}$. In general,

$$\dot{\mathbf{i}} = \mathbf{\Omega} \times \mathbf{i} \qquad \dot{\mathbf{j}} = \mathbf{\Omega} \times \mathbf{j} \qquad \dot{\mathbf{k}} = \mathbf{\Omega} \times \mathbf{k}$$

These formulations were also developed in Sec. 16.8, regarding planar motion of the axes. Substituting the results into the above equation and using Eq. 20–5 yields

$$\dot{\mathbf{A}} = (\dot{\mathbf{A}})_{xyz} + \mathbf{\Omega} \times \mathbf{A} \qquad (20\text{–}6)$$

This result is rather important and it will be used throughout Sec. 20.4 and Chapter 21. In words, it states that the time derivative of *any vector* $\mathbf{A}$ as observed from the fixed X, Y, Z frame of reference is equal to the time rate of change of $\mathbf{A}$ as observed from the x, y, z translating-rotating frame of reference, Eq. 20–5, plus $\mathbf{\Omega} \times \mathbf{A}$, the change of $\mathbf{A}$ caused by the rotation of the x, y, z frame. In other words, $(\dot{\mathbf{A}})_{xyz}$ indicates the time rate of change of the *magnitude* of $\mathbf{A}$, and $\mathbf{\Omega} \times \mathbf{A}$ indicates the time rate of change in the *direction* of $\mathbf{A}$. As a result, Eq. 20–6 should always be used whenever $\mathbf{\Omega}$ produces a change in the direction of $\mathbf{A}$ as seen from the X, Y, Z reference. If this change does not occur, i.e., $\mathbf{\Omega} = \mathbf{0}$, then $\dot{\mathbf{A}} = (\dot{\mathbf{A}})_{xyz}$, and so the time rate of change of $\mathbf{A}$ as observed from both coordinate systems will be the *same.*

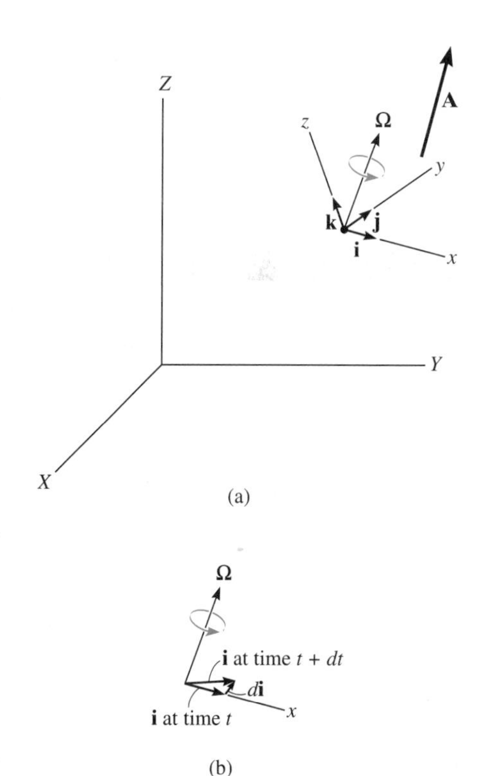

Fig. 20–6

Example 20–1

The disk shown in Fig. 20–7*a* is spinning about its horizontal axis with a constant angular velocity $\omega_s = 3$ rad/s, while the horizontal platform on which the disk is mounted is rotating about the vertical axis at a constant rate $\omega_p = 1$ rad/s. Determine the angular acceleration of the disk and the velocity and acceleration of point *A* on the disk when it is in the position shown.

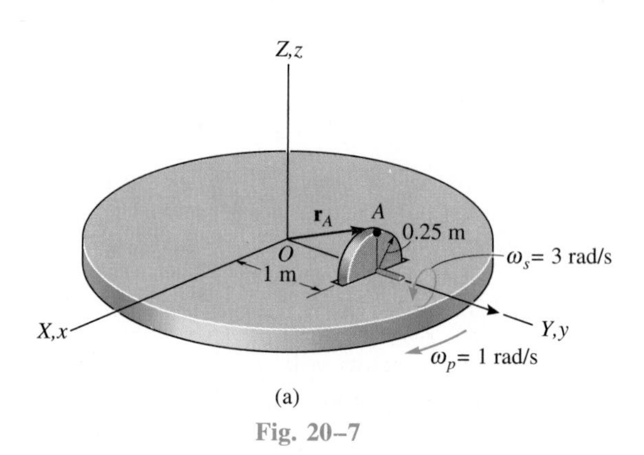

(a)

Fig. 20–7

SOLUTION

Point *O* represents a fixed point of rotation for the disk if one considers a hypothetical extension of the disk to this point. To determine the velocity and acceleration of point *A*, it is first necessary to determine the resultant angular velocity $\boldsymbol{\omega}$ and angular acceleration $\boldsymbol{\alpha}$ of the disk, since these vectors are used in Eqs. 20–3 and 20–4.

Angular Velocity. The angular velocity, which is measured from *X*, *Y*, *Z*, is simply the vector addition of the two component motions. Thus,

$$\boldsymbol{\omega} = \boldsymbol{\omega}_s + \boldsymbol{\omega}_p = \{3\mathbf{j} - 1\mathbf{k}\} \text{ rad/s}$$

At first glance, it may not appear that the disk is actually rotating with this angular velocity, since it is generally more difficult to imagine the resultant of angular motions in comparison with linear motions. To further understand the angular motion, consider the disk as being replaced by a cone (a body cone), which is rolling over the stationary space cone, Fig. 20–7*b*. The instantaneous axis of rotation is along the line of contact of the cones. This axis defines the direction of the resultant $\boldsymbol{\omega}$, having components $\boldsymbol{\omega}_s$ and $\boldsymbol{\omega}_p$.

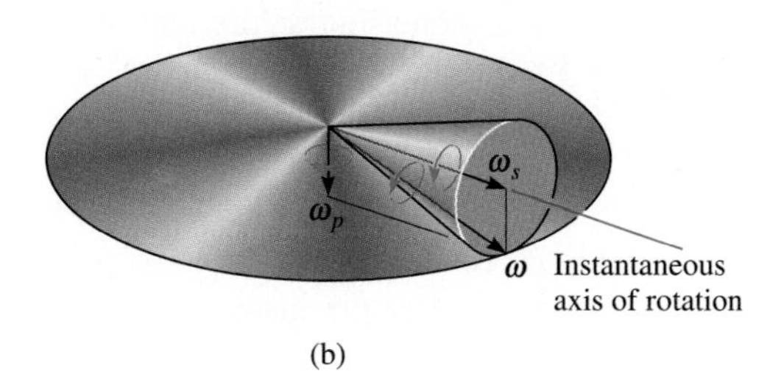

(b)

Angular Acceleration. Since the magnitude of $\boldsymbol{\omega}$ is constant, only a change in its direction, as seen from a fixed reference, creates the angular acceleration $\boldsymbol{\alpha}$ of the disk. One way to obtain $\boldsymbol{\alpha}$ is to compute the time derivative of *each of the two components* of $\boldsymbol{\omega}$ using Eq. 20–6. At the instant shown in Fig. 20–7*a*, imagine the fixed *X*, *Y*, *Z* and a rotating *x*, *y*, *z* frame to be coincident. If the rotating *x*, *y*, *z* frame is chosen to have an angular velocity of $\boldsymbol{\Omega} = \boldsymbol{\omega}_p = \{1\mathbf{k}\}$ rad/s, then $\boldsymbol{\omega}_s$ will *always* be directed along the *y* (not *Y*) axis, and the time rate of change of $\boldsymbol{\omega}_s$ as seen from *x*, *y*, *z* is *zero;* i.e., $(\dot{\boldsymbol{\omega}}_s)_{xyz} = \mathbf{0}$ (the magnitude of $\boldsymbol{\omega}_s$ is constant). However, $\boldsymbol{\Omega} = \boldsymbol{\omega}_p$ does change the *direction* of $\boldsymbol{\omega}_s$. Thus,

$$\dot{\boldsymbol{\omega}}_s = (\dot{\boldsymbol{\omega}}_s)_{xyz} + \boldsymbol{\omega}_p \times \boldsymbol{\omega}_s = \mathbf{0} + (1\mathbf{k}) \times (3\mathbf{j}) = \{-3\mathbf{i}\}\ \text{rad/s}^2$$

By the same choice of axes rotation, $\boldsymbol{\Omega} = \boldsymbol{\omega}_p$, or even with $\boldsymbol{\Omega} = \mathbf{0}$, the time derivative $(\dot{\boldsymbol{\omega}}_p)_{xyz} = \mathbf{0}$, since $\boldsymbol{\omega}_p$ is *always* directed along the *z* (or *Z*) axis and has a constant magnitude.

$$\dot{\boldsymbol{\omega}}_p = (\dot{\boldsymbol{\omega}}_p)_{xyz} + \boldsymbol{\omega}_p \times \boldsymbol{\omega}_p = \mathbf{0} + \mathbf{0} = \mathbf{0}$$

The angular acceleration of the disk is therefore

$$\boldsymbol{\alpha} = \dot{\boldsymbol{\omega}} = \dot{\boldsymbol{\omega}}_s + \dot{\boldsymbol{\omega}}_p = \{-3\mathbf{i}\}\ \text{rad/s}^2 \qquad \textit{Ans.}$$

Velocity and Acceleration. Since $\boldsymbol{\omega}$ and $\boldsymbol{\alpha}$ have been determined, the velocity and acceleration of point *A* can be computed using Eqs. 20–3 and 20–4. Realizing that $\mathbf{r}_A = \{1\mathbf{j} + 0.25\mathbf{k}\}$ m, Fig. 20–7, we have

$$\mathbf{v}_A = \boldsymbol{\omega} \times \mathbf{r}_A = (3\mathbf{j} - 1\mathbf{k}) \times (1\mathbf{j} + 0.25\mathbf{k}) = \{-0.25\mathbf{i}\}\ \text{m/s} \qquad \textit{Ans.}$$

$$\begin{aligned}\mathbf{a}_A &= \boldsymbol{\alpha} \times \mathbf{r}_A + \boldsymbol{\omega} \times (\boldsymbol{\omega} \times \mathbf{r}_A)\\ &= (-3\mathbf{i}) \times (1\mathbf{j} + 0.25\mathbf{k}) + (3\mathbf{j} - 1\mathbf{k}) \times [(3\mathbf{j} - 1\mathbf{k}) \times (1\mathbf{j} + 0.25\mathbf{k})]\\ &= \{-1\mathbf{j} - 8.25\mathbf{k}\}\ \text{m/s}^2 \qquad \textit{Ans.}\end{aligned}$$

Example 20–2

At the instant $\theta = 60°$, the gyrotop in Fig. 20–8 has three components of angular motion directed as shown and having magnitudes defined as:

spin: $\omega_s = 10$ rad/s, increasing at the rate of 6 rad/s^2
nutation: $\omega_n = 3$ rad/s, increasing at the rate of 2 rad/s^2
precession: $\omega_p = 5$ rad/s, increasing at the rate of 4 rad/s^2

Determine the angular velocity and angular acceleration of the top.

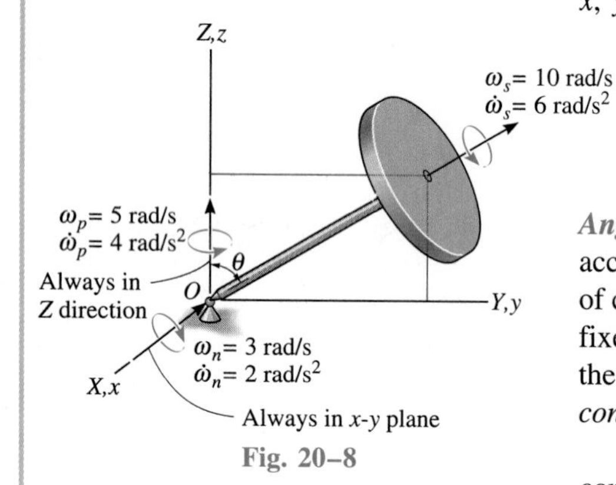

Fig. 20–8

SOLUTION

Angular Velocity. The top is rotating about the fixed point O. If the fixed and rotating frames are coincident at the instant shown, then the angular velocity can be expressed in terms of **i, j, k** components, appropriate to the x, y, z frame; i.e.,

$$\begin{aligned}\boldsymbol{\omega} &= -\omega_n\mathbf{i} + \omega_s \sin\theta\mathbf{j} + (\omega_p + \omega_s\cos\theta)\mathbf{k}\\ &= -3\mathbf{i} + 10\sin 60°\mathbf{j} + (5 + 10\cos 60°)\mathbf{k}\\ &= \{-3\mathbf{i} + 8.66\mathbf{j} + 10\mathbf{k}\}\text{ rad/s}\end{aligned}$$ *Ans.*

Angular Acceleration. As in the solution of Example 20–1, the angular acceleration $\boldsymbol{\alpha}$ will be determined by investigating separately the time rate of change of *each of the angular velocity components* as observed from the fixed X, Y, Z reference. We will choose an $\boldsymbol{\Omega}$ for the x, y, z reference so that the component of $\boldsymbol{\omega}$ which is being considered is viewed as having a *constant direction* when observed from x, y, z.

Careful examination of the motion of the top reveals that $\boldsymbol{\omega}_s$ has a *constant direction* relative to x, y, z if these axes rotate at $\boldsymbol{\Omega} = \boldsymbol{\omega}_n + \boldsymbol{\omega}_p$. Since $\boldsymbol{\omega}_n$ *always* lies in the fixed X–Y plane, this vector has a *constant direction* if the motion is viewed from axes x, y, z having a rotation of $\boldsymbol{\Omega} = \boldsymbol{\omega}_p$ (not $\boldsymbol{\Omega} = \boldsymbol{\omega}_s + \boldsymbol{\omega}_p$). Finally, the component $\boldsymbol{\omega}_p$ is *always directed* along the Z axis so that here it is not necessary to think of x, y, z as rotating, i.e., $\boldsymbol{\Omega} = \mathbf{0}$. Expressing the data in terms of the **i, j, k** components, we therefore have

For $\boldsymbol{\omega}_s$, $\boldsymbol{\Omega} = \boldsymbol{\omega}_n + \boldsymbol{\omega}_p$ (this rotation changes the direction of $\boldsymbol{\omega}_s$ relative to X, Y, Z):

$$\begin{aligned}\dot{\boldsymbol{\omega}}_s &= (\dot{\boldsymbol{\omega}}_s)_{xyz} + (\boldsymbol{\omega}_n + \boldsymbol{\omega}_p) \times \boldsymbol{\omega}_s\\ &= (6\sin 60°\mathbf{j} + 6\cos 60°\mathbf{k}) + (-3\mathbf{i} + 5\mathbf{k}) \times (10\sin 60°\mathbf{j} + 10\cos 60°\mathbf{k})\\ &= \{-43.30\mathbf{i} + 20.20\mathbf{j} - 22.98\mathbf{k}\}\text{ rad/s}^2\end{aligned}$$

For $\boldsymbol{\omega}_n$, $\boldsymbol{\Omega} = \boldsymbol{\omega}_p$ (this rotation changes the direction of $\boldsymbol{\omega}_n$ relative to X, Y, Z):

$$\dot{\boldsymbol{\omega}}_n = (\dot{\boldsymbol{\omega}}_n)_{xyz} + \boldsymbol{\omega}_p \times \boldsymbol{\omega}_n = -2\mathbf{i} + (5\mathbf{k}) \times (-3\mathbf{i}) = \{-2\mathbf{i} - 15\mathbf{j}\}\text{ rad/s}^2$$

For $\boldsymbol{\omega}_p$, $\boldsymbol{\Omega} = \mathbf{0}$ (since the rotation is zero, the direction of $\boldsymbol{\omega}_p$ is not changed relative to X, Y, Z):

$$\dot{\boldsymbol{\omega}}_p = (\dot{\boldsymbol{\omega}}_p)_{xyz} = \{4\mathbf{k}\}\text{ rad/s}^2$$

Thus, the angular acceleration of the top is

$$\boldsymbol{\alpha} = \dot{\boldsymbol{\omega}}_s + \dot{\boldsymbol{\omega}}_n + \dot{\boldsymbol{\omega}}_p = \{-45.3\mathbf{i} + 5.20\mathbf{j} - 19.0\mathbf{k}\}\text{ rad/s}^2$$ *Ans.*

*20.3 General Motion

Shown in Fig. 20–9 is a rigid body subjected to general motion in three dimensions for which the angular velocity is $\boldsymbol{\omega}$ and the angular acceleration is $\boldsymbol{\alpha}$. If point A has a known motion of $\mathbf{v}_A$ and $\mathbf{a}_A$, the motion of any other point B may be determined by using a relative-motion analysis. In this section a *translating coordinate system* will be used to define the relative motion, and in the next section a reference that is both rotating and translating will be considered.

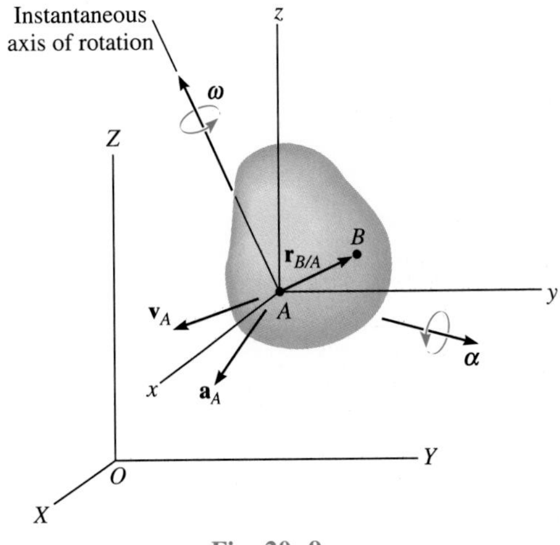

Fig. 20–9

If the origin of the translating coordinate system x, y, z ($\boldsymbol{\Omega} = \mathbf{0}$) is located at the "base point" A, then, at the instant shown, the motion of the body may be regarded as the sum of an instantaneous translation of the body having a motion of $\mathbf{v}_A$ and $\mathbf{a}_A$ and a rotation of the body about an instantaneous axis passing through the base point. Since the body is rigid, the motion of point B measured by an observer located at A is the same as *motion of the body about a fixed point.* This relative motion occurs about the instantaneous axis of rotation and is defined by $\mathbf{v}_{B/A} = \boldsymbol{\omega} \times \mathbf{r}_{B/A}$, Eq. 20–3, and $\mathbf{a}_{B/A} = \boldsymbol{\alpha} \times \mathbf{r}_{B/A} + \boldsymbol{\omega} \times (\boldsymbol{\omega} \times \mathbf{r}_{B/A})$, Eq. 20–4. For translating axes the relative motions are related to absolute motions by $\mathbf{v}_B = \mathbf{v}_A + \mathbf{v}_{B/A}$ and $\mathbf{a}_B = \mathbf{a}_A + \mathbf{a}_{B/A}$, Eqs. 16–14 and 16–16, so that the absolute velocity and acceleration of point B can be determined from the equations

$$\mathbf{v}_B = \mathbf{v}_A + \boldsymbol{\omega} \times \mathbf{r}_{B/A} \tag{20–7}$$

and

$$\mathbf{a}_B = \mathbf{a}_A + \boldsymbol{\alpha} \times \mathbf{r}_{B/A} + \boldsymbol{\omega} \times (\boldsymbol{\omega} \times \mathbf{r}_{B/A}) \tag{20–8}$$

These two equations are identical to those describing the general plane motion of a rigid body, Eqs. 16–15 and 16–17. However, difficulty in application arises for three-dimensional motion, because $\boldsymbol{\alpha}$ measures the change in *both* the magnitude and direction of $\boldsymbol{\omega}$. (Recall that, for general plane motion, $\boldsymbol{\alpha}$ and $\boldsymbol{\omega}$ are always parallel or perpendicular to the plane of motion, and therefore $\boldsymbol{\alpha}$ measures only a change in the magnitude of $\boldsymbol{\omega}$.) In some problems the constraints or connections of a body will require that the directions of the angular motions or displacement paths of points on the body be defined. As illustrated in the following example, this information is useful for obtaining some of the terms in the above equations.

Example 20–3

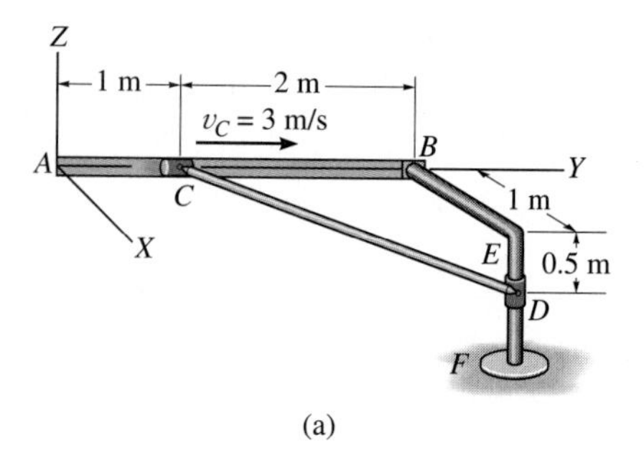

(a)

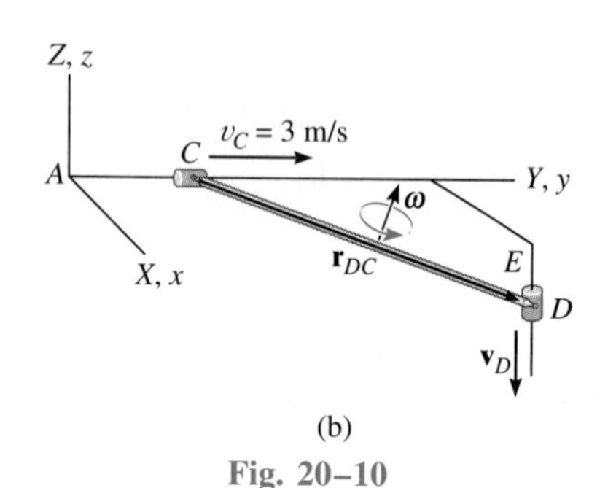

(b)

Fig. 20–10

One end of the rigid bar CD shown in Fig. 20–10a slides along the grooved wall slot AB, and the other end slides along the vertical member EF. If the collar at C is moving toward B at a speed of 3 m/s, determine the velocity of the collar at D and the angular velocity of the bar at the instant shown. The bar is connected to the collars at its end points by ball-and-socket joints.

SOLUTION

Bar CD is subjected to general motion. Why? The velocity of point D on the bar may be related to the velocity of point C by the equation

$$\mathbf{v}_D = \mathbf{v}_C + \boldsymbol{\omega} \times \mathbf{r}_{D/C}$$

The fixed and translating frames of reference are assumed to coincide at the instant considered, Fig. 20–10b. We have

$$\mathbf{v}_D = -v_D\mathbf{k} \qquad \mathbf{v}_C = \{3\mathbf{j}\} \text{ m/s}$$
$$\mathbf{r}_{D/C} = \{1\mathbf{i} + 2\mathbf{j} - 0.5\mathbf{k}\} \text{ m} \qquad \boldsymbol{\omega} = \omega_x\mathbf{i} + \omega_y\mathbf{j} + \omega_z\mathbf{k}$$

Substituting these quantities into the above equation gives

$$-v_D\mathbf{k} = 3\mathbf{j} + \begin{vmatrix} \mathbf{i} & \mathbf{j} & \mathbf{k} \\ \omega_x & \omega_y & \omega_z \\ 1 & 2 & -0.5 \end{vmatrix}$$

Expanding and equating the respective **i, j, k** components yields

$$-0.5\omega_y - 2\omega_z = 0 \tag{1}$$
$$0.5\omega_x + 1\omega_z + 3 = 0 \tag{2}$$
$$2\omega_x - 1\omega_y + v_D = 0 \tag{3}$$

These equations contain four unknowns.* A fourth equation can be written if the direction of $\boldsymbol{\omega}$ is specified. In particular, any component of $\boldsymbol{\omega}$ acting along the bar's axis has no effect on moving the collars. This is because the bar is *free to rotate* about its axis. Therefore, if $\boldsymbol{\omega}$ is specified as acting *perpendicular* to the axis of the bar, then $\boldsymbol{\omega}$ must have a unique magnitude to satisfy the above equations. Perpendicularity is guaranteed provided the dot product of $\boldsymbol{\omega}$ and $\mathbf{r}_{D/C}$ is zero (see Eq. C–14). Hence,

$$\boldsymbol{\omega} \cdot \mathbf{r}_{D/C} = (\omega_x\mathbf{i} + \omega_y\mathbf{j} + \omega_z\mathbf{k}) \cdot (1\mathbf{i} + 2\mathbf{j} - 0.5\mathbf{k}) = 0$$
$$1\omega_x + 2\omega_y - 0.5\omega_z = 0 \tag{4}$$

Solving Eqs. 1 through 4 simultaneously yields

$$\omega_x = -4.86 \text{ rad/s} \qquad \omega_y = 2.29 \text{ rad/s} \qquad \omega_z = -0.571 \text{ rad/s} \qquad \textit{Ans.}$$
$$v_D = 12.0 \text{ m/s} \downarrow \qquad \textit{Ans.}$$

*Although this is the case the magnitude of $\mathbf{v}_D$ can be obtained. For example, solve Eqs. 1 and 2 for ω_y and ω_x in terms of ω_z and substitute into Eq. 3. It will be noted that ω_z will *cancel out,* which will allow a solution for v_D.

PROBLEMS

20–1. The ladder of the fire truck rotates around the z axis with an angular velocity $\omega_1 = 0.15$ rad/s, which is increasing at 0.8 rad/s^2. At the same instant it is rotating upward at a constant rate $\omega_2 = 0.6$ rad/s. Determine the velocity and acceleration of point A located at the top of the ladder at this instant.

20–2. The ladder of the fire truck rotates around the z axis with an angular velocity of $\omega_1 = 0.15$ rad/s, which is increasing at 0.2 rad/s^2. At the same instant it is rotating upwards at $\omega_2 = 0.6$ rad/s while increasing at 0.4 rad/s^2. Determine the velocity and acceleration of point A located at the top of the ladder at this instant.

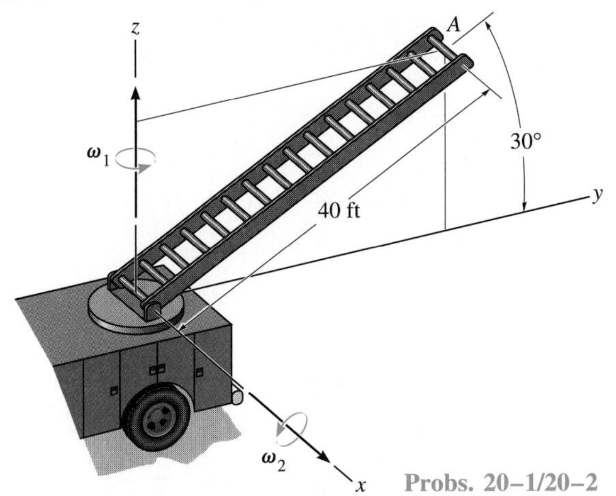

Probs. 20–1/20–2

20–3. An automobile maintains a constant speed of 40 ft/s. Assuming that the 2.2-ft-diameter tires do not slip on the pavement, determine the magnitudes of the angular velocity and angular acceleration of the tires if the automobile enters a horizontal curve having a 100-ft radius.

***20–4.** Gear A is *fixed* while gear B is free to rotate on the shaft S. If the shaft is turning about the z axis at $\omega_z = 5$ rad/s, while increasing at 2 rad/s^2, determine the velocity and acceleration of point P at the instant shown. The face of gear B lies in a vertical plane.

Prob. 20–4

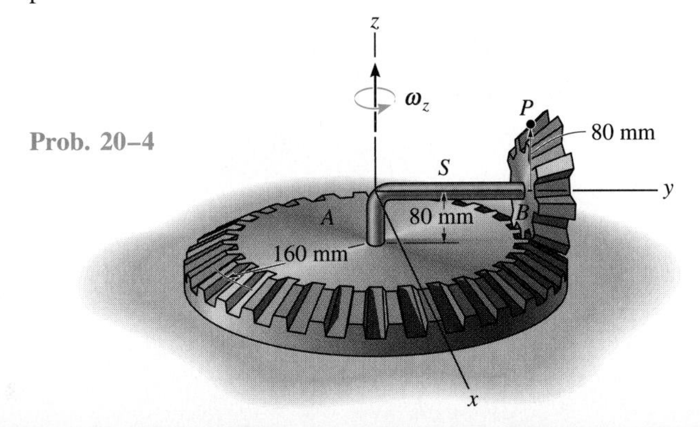

20–5. Gear A is fixed while gear B is free to rotate on the shaft S. If the shaft is turning about the z axis at $\omega_z = 5$ rad/s, while increasing at 2 rad/s^2, determine the velocity and acceleration of point C at the instant shown. The face of gear B lies in a vertical plane.

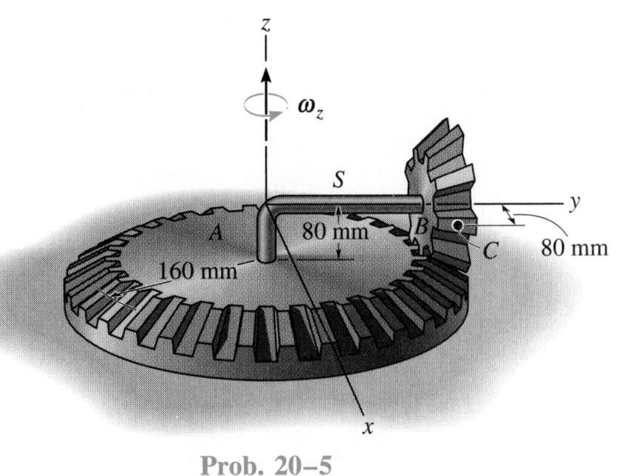

Prob. 20–5

20–6. The drill pipe P turns at a constant angular rate $\omega_P = 4$ rad/s. Determine the angular velocity and angular acceleration of the conical rock bit, which rolls without slipping. Also, what are the velocity and acceleration of point A?

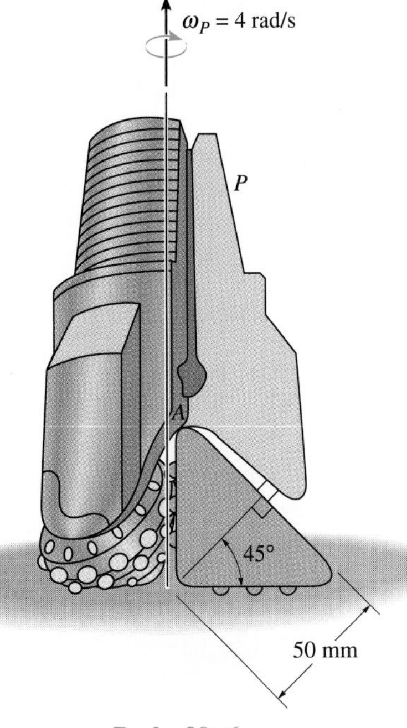

Prob. 20–6

20–7. The bevel gear A rolls on the fixed gear B. If at the instant shown the shaft to which A is attached is rotating at 2 rad/s and has an angular acceleration of 4 rad/s^2, determine the angular velocity and angular acceleration of gear A.

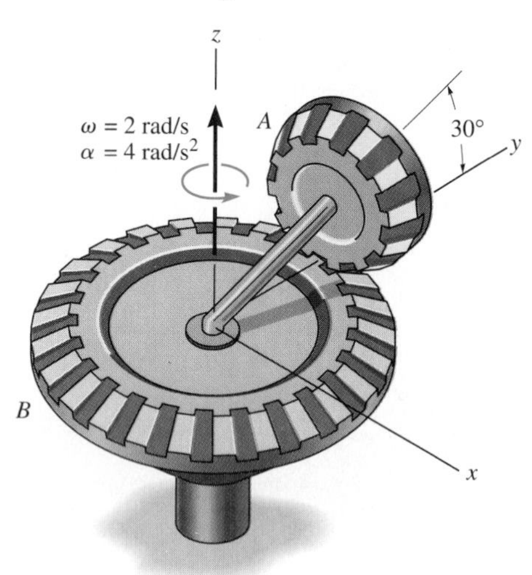

Prob. 20–7

***20–8.** The electric fan is mounted on a swivel support such that the fan rotates about the z axis at a constant rate of $\omega_z = 1$ rad/s and the fan blade is spinning at a constant rate $\omega_s = 60$ rad/s. If $\phi = 45°$ for the motion, determine the angular velocity and the angular acceleration of the blade.

20–9. The electric fan is mounted on a swivel support such that the fan rotates about the z axis at a constant rate of $\omega_z = 1$ rad/s and the fan blade is spinning at a constant rate $\omega_s = 60$ rad/s. If at the instant $\phi = 45°$, $\dot{\phi} = 2$ rad/s for the motion, determine the angular velocity and the angular acceleration of the blade.

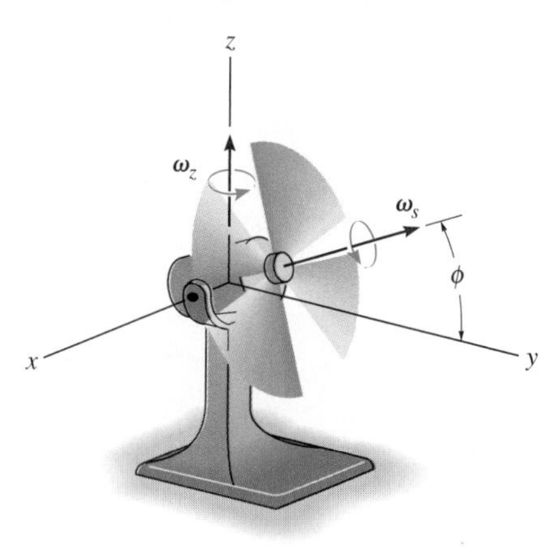

Probs. 20–8/20–9

20–10. At a given instant, the antenna has an angular motion $\omega_1 = 3$ rad/s and $\dot{\omega}_1 = 2$ rad/s^2 about the z axis. At this same instant $\theta = 30°$, the angular motion about the x axis is $\omega_2 = 1.5$ rad/s, and $\dot{\omega}_2 = 4$ rad/s^2. Determine the velocity and acceleration of the signal horn A at this instant. The distance from O to A is $d = 3$ ft.

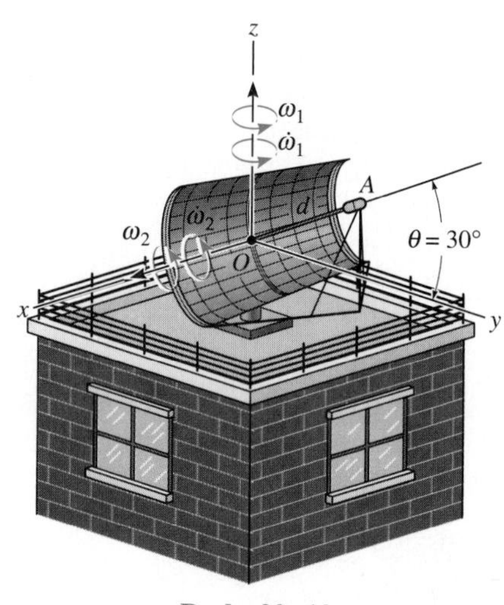

Prob. 20–10

20–11. The cone rolls without slipping such that at the instant shown $\omega_z = 4$ rad/s and $\dot{\omega}_z = 3$ rad/s^2. Determine the velocity and acceleration of point A at this instant.

***20–12.** The cone rolls without slipping such that at the instant shown $\omega_z = 4$ rad/s and $\dot{\omega}_z = 3$ rad/s^2. Determine the velocity and acceleration of point B at this instant.

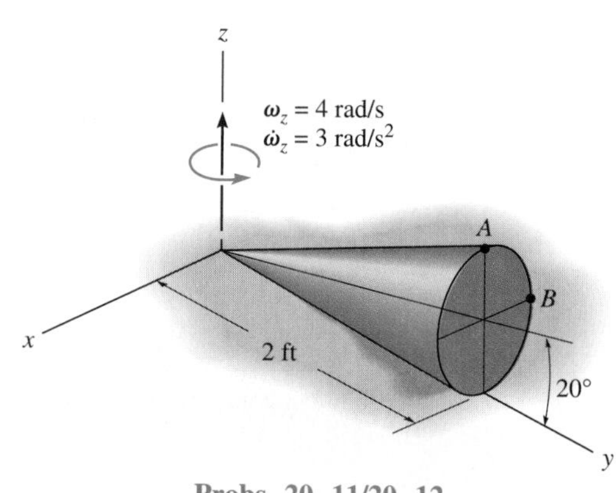

Probs. 20–11/20–12

20–13. Gear A is fixed to the crankshaft S, while gear C is fixed and B are free to rotate. The crankshaft is turning at 80 rad/s about its axis. Determine the magnitudes of the angular velocity of the propeller and the angular acceleration of gear B.

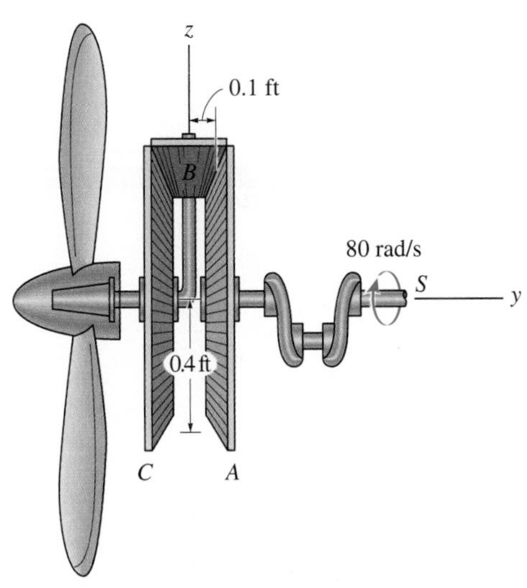

Prob. 20–13

20–14. Shaft BD is connected to a ball-and-socket joint at B, and a beveled gear A is attached to its other end. The gear is in mesh with a fixed gear C. If the shaft and gear A are *spinning* with a constant angular velocity $\omega_1 = 8$ rad/s, determine the angular velocity and angular acceleration of gear A.

20–15. Solve Prob. 20–14 if shaft BD has $\omega_1 = 8$ rad/s and $\dot{\omega}_1 = 4$ rad/s^2.

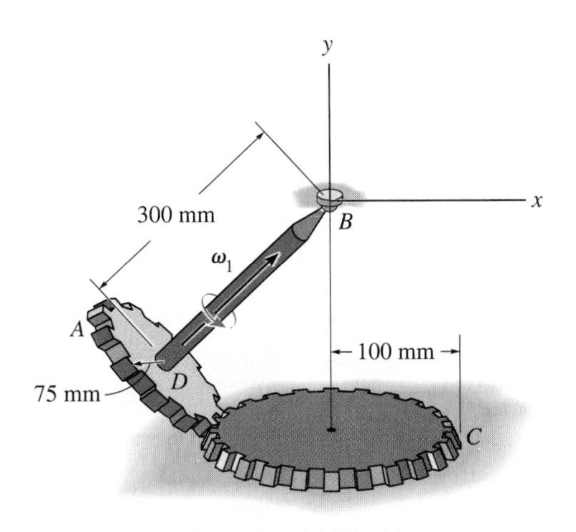

Probs. 20–14/20–15

***20–16.** The truncated cone rotates about the z axis at a constant rate $\omega_z = 0.4$ rad/s without slipping on the horizontal plane. Determine the velocity and acceleration of point A on the cone.

20–17. The truncated cone rotates about the z axis at $\omega_z = 0.4$ rad/s without slipping on the horizontal plane. If at this same instant ω_z is increasing at $\dot{\omega}_z = 0.5$ rad/s^2, determine the velocity and acceleration of point A on the cone.

Probs. 20–16/20–17

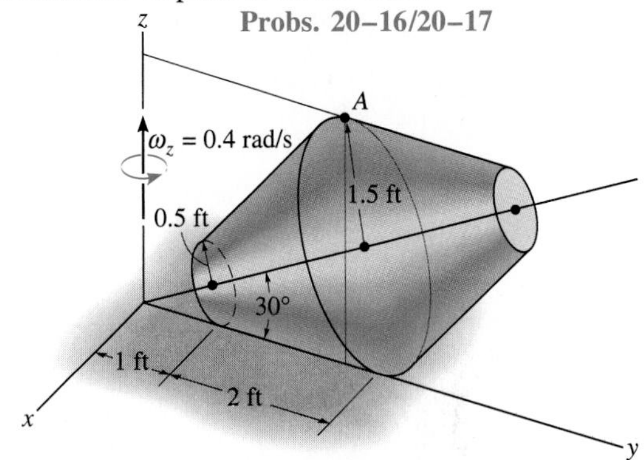

20–18. Rod AB is attached to the rotating arm using ball-and-socket joints. If AC is rotating with a constant angular velocity of 8 rad/s about the pin at C, determine the angular velocity of link BD at the instant shown.

20–19. Rod AB is attached to the rotating arm using ball-and-socket joints. If AC is rotating about point C with an angular velocity of 8 rad/s and has an angular acceleration of 6 rad/s^2 at the instant shown, determine the angular velocity and angular acceleration of link BD at this instant.

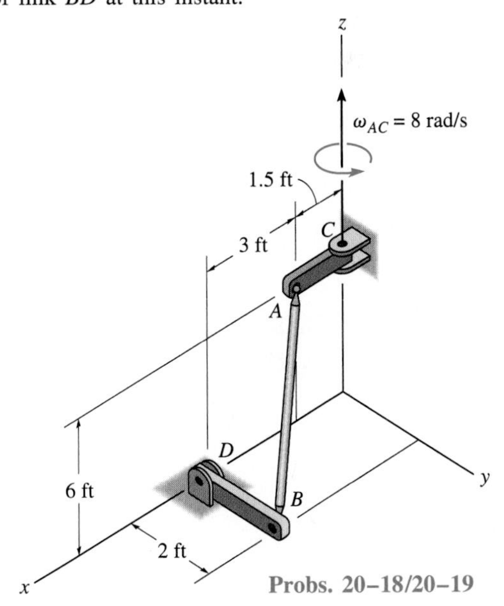

Probs. 20–18/20–19

***20–20.** If the rod is attached with ball-and-socket joints to smooth collars A and B at its end points, determine the speed of B at the instant shown if A is moving downward at a constant speed of $v_A = 8$ ft/s. Also, determine the angular velocity of the rod if it is directed perpendicular to the axis of the rod.

20–21. If the collar at A in Prob. 20–20 is moving downward with an acceleration $\mathbf{a}_A = \{-5\mathbf{k}\}$ ft/s^2, at the instant its speed is $v_A = 8$ ft/s, determine the acceleration of the collar at B at this instant.

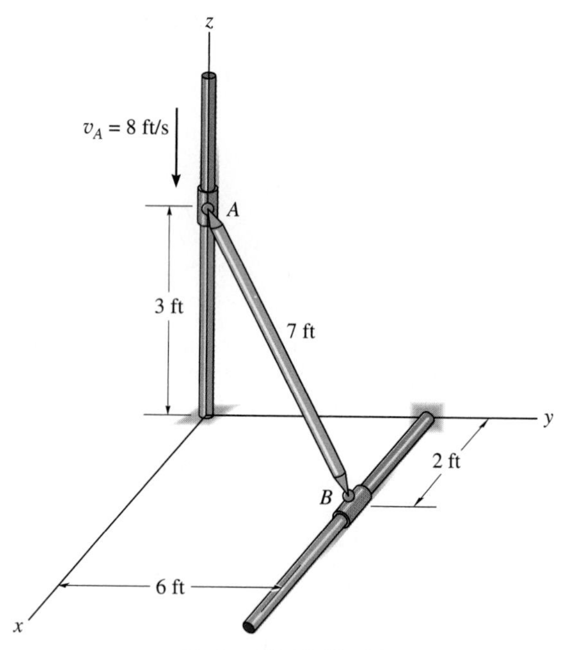

Probs. 20–20/20–21

20–22. The rod AB is attached to collars at its ends by ball-and-socket joints. If collar A has a speed $v_A = 20$ ft/s, determine the speed of collar B at the instant shown.

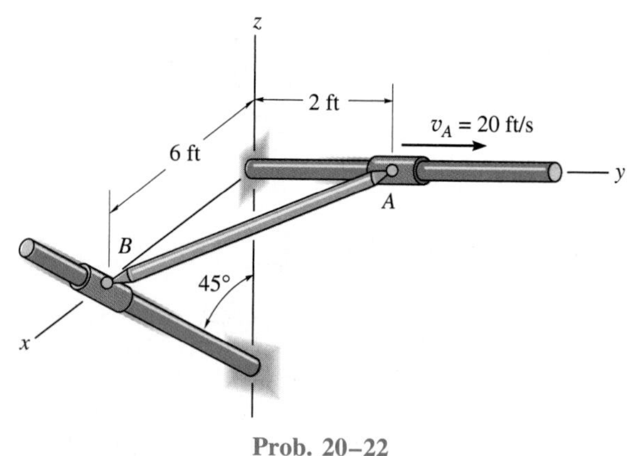

Prob. 20–22

20–23. If the collar at A in Prob. 20–22 has an acceleration of $\mathbf{a}_A = \{12\mathbf{j}\}$ ft/s^2 at the instant its speed is $v_A = 20$ ft/s, determine the acceleration of the collar at B at this instant.

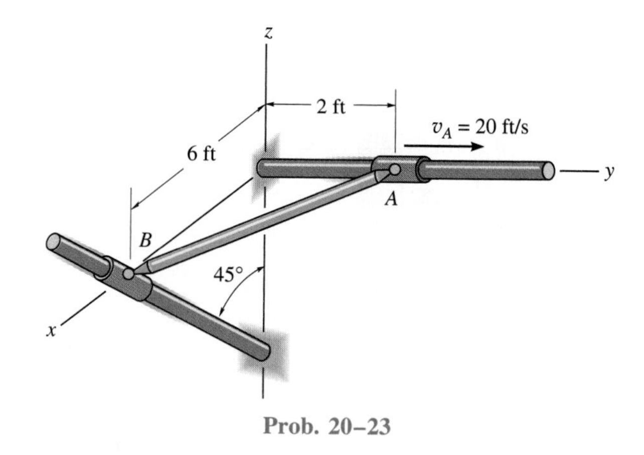

Prob. 20–23

***20–24.** The rod is attached to smooth collars A and B at its ends using ball-and-socket joints. Determine the speed of B at the instant shown if A is moving at $v_A = 6$ m/s. Also, determine the angular velocity of the rod if it is directed perpendicular to the axis of the rod.

20–25. If the collar at A in Prob. 20–24 has a deceleration of $\mathbf{a}_A = \{-5\mathbf{i}\}$ m/s^2, at the instant shown, determine the acceleration of collar B at this instant.

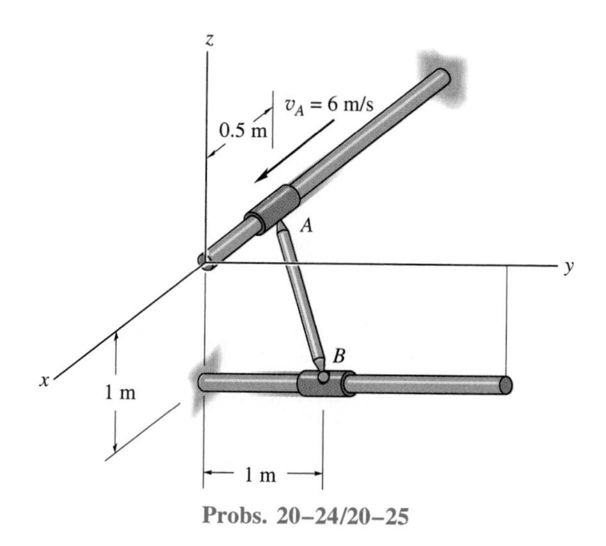

Probs. 20–24/20–25

20–26. Rod AB is attached to collars at its ends by using ball-and-socket joints. If collar A moves along the fixed rod at $v_A = 8$ ft/s, determine the angular velocity of the rod and the velocity of collar B at the instant shown. Assume that the rod's angular velocity is directed perpendicular to the axis of the rod.

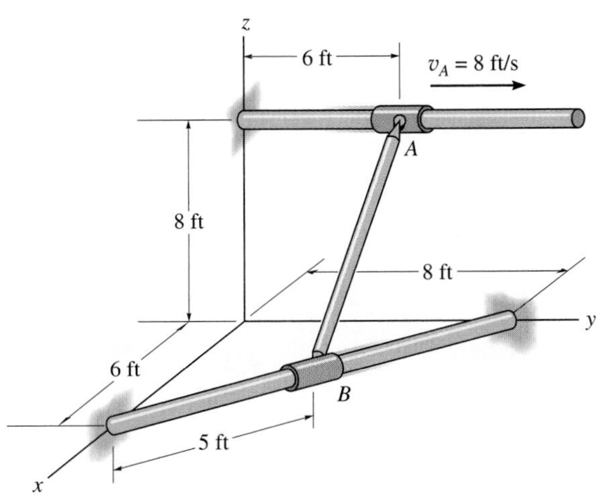

Prob. 20–26

20–27. Rod AB is attached to collars at its ends by using ball-and-socket joints. If collar A moves along the fixed rod with a velocity $v_A = 8$ ft/s and has an acceleration $a_A = 4$ ft/s^2 at the instant shown, determine the angular acceleration of the rod and the acceleration of collar B at this instant. Assume that the rod's angular velocity and angular acceleration are directed perpendicular to the axis of the rod.

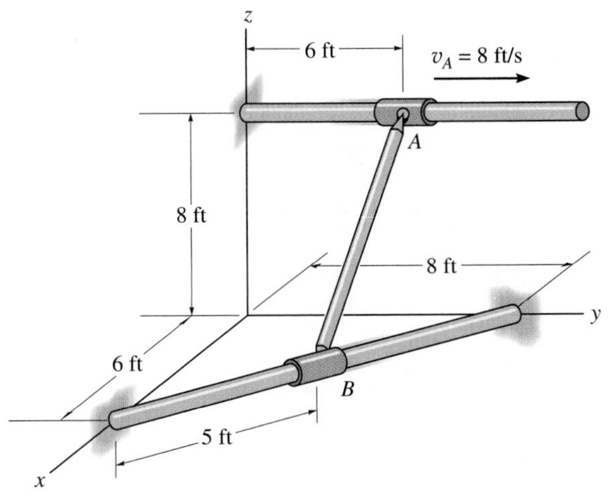

Prob. 20–27

***20–28.** The rod AB is attached to collars at its ends by ball-and-socket joints. If collar A has a velocity of $v_A = 5$ ft/s, determine the angular velocity of the rod and the velocity of collar B at the instant shown. Assume the angular velocity of the rod is directed perpendicular to the rod.

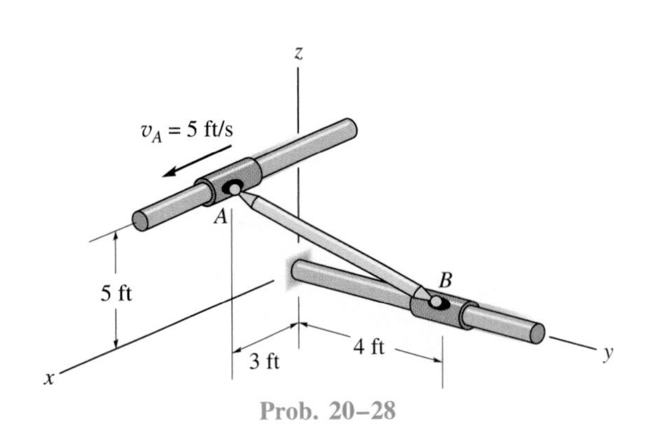

Prob. 20–28

20–29. The rod AB is attached to the collars at its ends by ball-and-socket joints. If collar A has an acceleration $\mathbf{a}_A = \{6\mathbf{i}\}$ ft/s^2, and a velocity $\boldsymbol{v}_A = \{5\mathbf{i}\}$ ft/s, determine the angular acceleration of the rod and the acceleration of collar B at the instant shown. Assume the angular velocity and angular acceleration of the rod are both directed perpendicular to the rod.

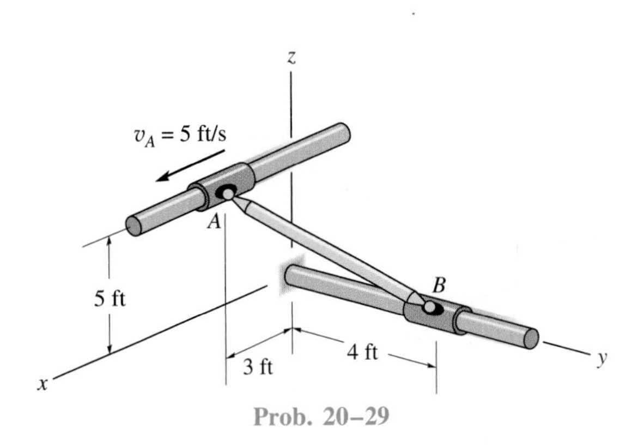

Prob. 20–29

20–30. Rod AB is attached to a disk and a collar by ball-and-socket joints. If the disk is rotating at a constant angular velocity $\omega = \{2\mathbf{i}\}$ rad/s, determine the velocity and acceleration of the collar at A at the instant shown. Assume the angular velocity is directed perpendicular to the rod.

20–31. Rod AB is attached to a disk and a collar by ball-and-socket joints. If the disk is rotating with an angular acceleration $\boldsymbol{\alpha} = \{4\mathbf{i}\}$ rad/s^2, and at the instant shown has an angular velocity $\boldsymbol{\omega} = \{2\mathbf{i}\}$ rad/s, determine the velocity and acceleration of the collar at A at the instant shown.

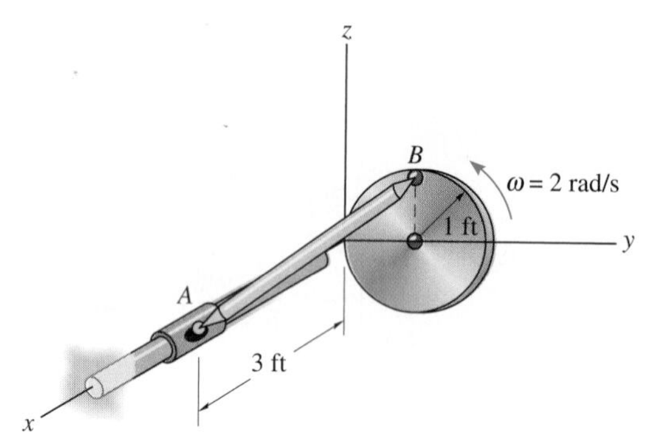

Probs. 20–30/20–31

***20–32.** Disk A is rotating at a constant angular velocity of 10 rad/s. If rod BC is joined to the disk and a collar by ball-and-socket joints, determine the velocity of collar B at the instant shown. Also, what is the rod's angular velocity ω_{BC} if it is directed perpendicular to the axis of the rod?

20–33 If the disk in Prob. 20–32 has an angular acceleration $\alpha = 5$ rad/s^2, determine the acceleration of the collar B at the instant shown, when $\omega = 10$ rad/s. Also, what is the rod's angular acceleration $\boldsymbol{\alpha}_{BC}$ if it is directed perpendicular to the axis of the rod?

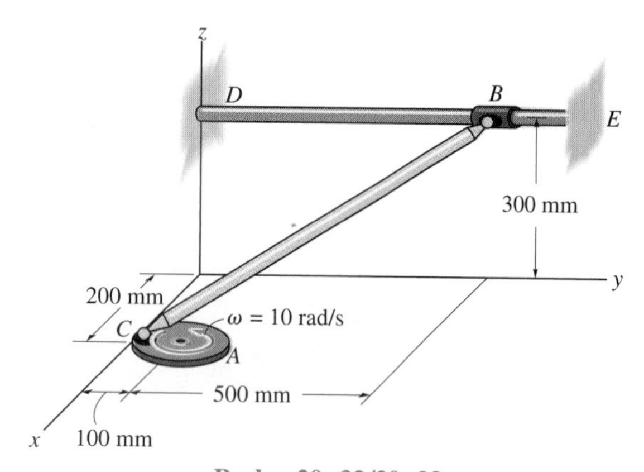

Probs. 20–32/20–33

20–34. Solve Prob. 20–32 if the connection at B consists of a pin as shown in the figure below, rather than a ball-and-socket joint. *Hint:* The constraint allows rotation of the rod both along bar DE ($\mathbf{j}$ direction) and along the axis of the pin ($\mathbf{n}$ direction). Since there is no rotational component in the $\mathbf{u}$ direction, i.e., perpendicular to $\mathbf{n}$ and $\mathbf{j}$ where $\mathbf{u} = \mathbf{j} \times \mathbf{n}$, an additional equation for solution can be obtained from $\boldsymbol{\omega} \cdot \mathbf{u} = 0$. The vector $\mathbf{n}$ is in the same direction as $\mathbf{r}_{B/C} \times \mathbf{r}_{D/C}$.

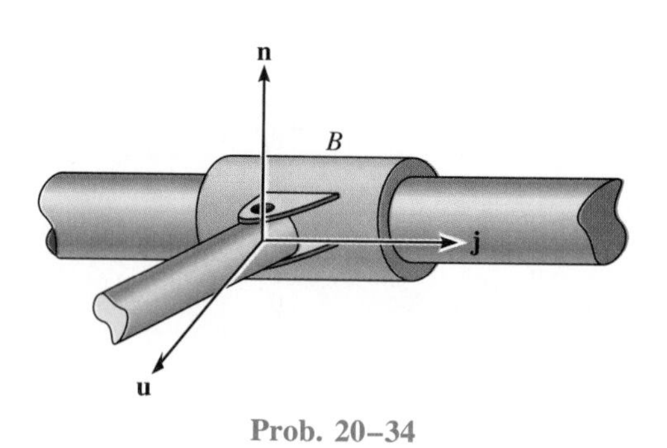

Prob. 20–34

20–35. The rod assembly is supported at B by a ball-and-socket joint and at A by a clevis. If the collar at B moves in the x-z plane with a speed $v_B = 5$ ft/s, determine the velocity of points A and C on the rod assembly at the instant shown. *Hint:* See Prob. 20–34.

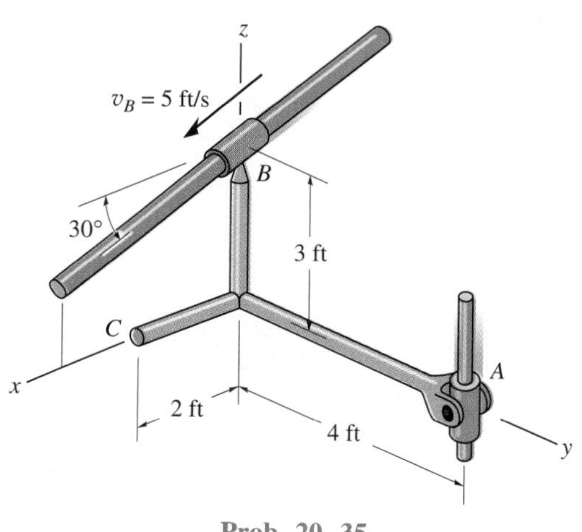

Prob. 20–35

***20–36.** Rod AB is supported at B by a ball-and-socket joint and at A by a clevis. If the collar at B has a speed of $v_B = 6$ ft/s, determine the velocity of the clevis at A at the instant shown. *Hint:* See Prob. 20–34.

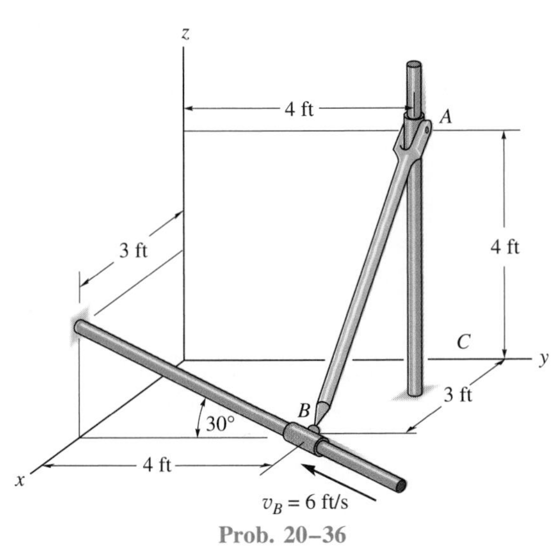

Prob. 20–36

20–37. The triangular plate ABC is supported at A by a ball-and-socket joint and at C by the x-z plane. The side AB lies in the x-y plane. At the instant $\theta = 60°$. $\dot{\theta} = 2$ rad/s and point C has the coordinates shown. Determine the angular velocity of the plate and the velocity of point C at this instant.

20–38. The triangular plate ABC is supported at A by a ball-and-socket joint and at C by the x-z plane. The side AB lies in the x-y plane. At the instant $\theta = 60°$, $\dot{\theta} = 2$ rad/s, $\ddot{\theta} = 3$ rad/s^2, and point C has the coordinates shown. Determine the angular acceleration of the plate and the acceleration of point C at this instant.

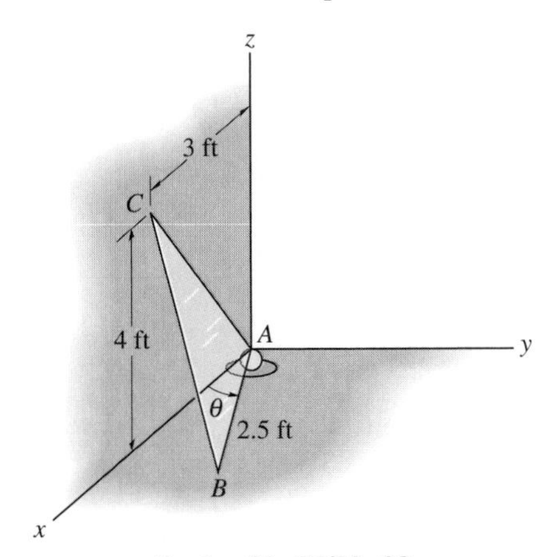

Probs. 20–37/20–38

20–39. The rod ABC is supported at A by a ball-and-socket joint and at C by the y-z plane. The segment AB lies in the x-y plane. At the instant $\theta = 60°$, $\dot{\theta} = 2$ rad/s, and point C has the coordinates shown. Determine the angular velocity of the rod and the velocity of point C at this instant.

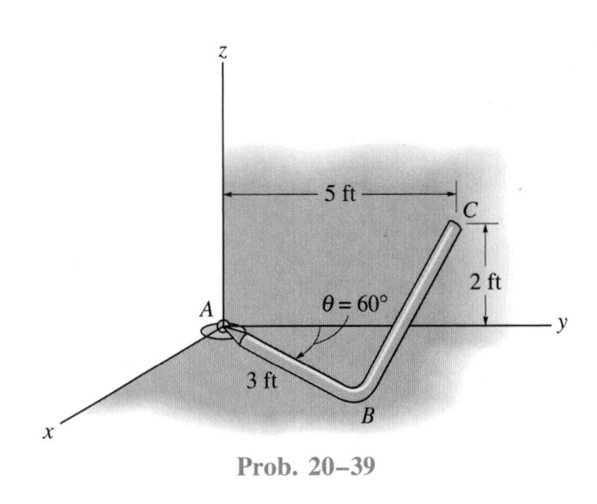

Prob. 20–39

***20–40.** The end C of the plate rests on the horizontal plane, while end points A and B are restricted to move along the grooved slots. If at the instant shown A is moving downward with a constant velocity of $v_A = 4$ ft/s, determine the angular velocity of the plate and the velocities of points B and C.

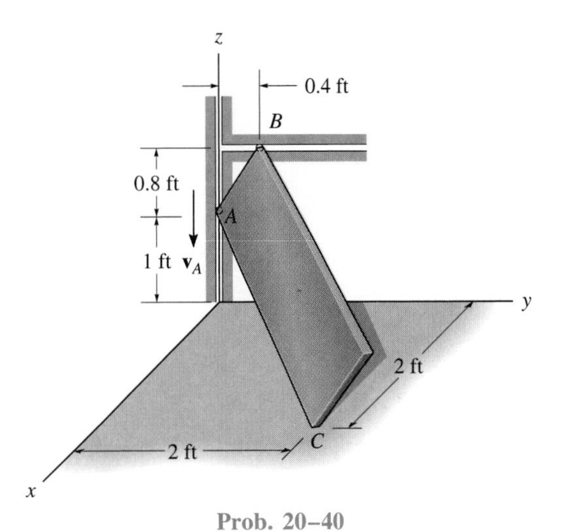

Prob. 20–40

*20.4 Relative-Motion Analysis Using Translating and Rotating Axes

The most general way to analyze the three-dimensional motion of a rigid body requires the use of a system of x, y, z axes that both translates and rotates relative to a second frame X, Y, Z. This analysis also provides a means for determining the motions of two points located on separate members of a mechanism, and for determining the relative motion of one particle with respect to another when one or both particles are moving along *rotating paths.* In this section two equations will be developed which relate the velocities and accelerations of two points A and B, of which one point moves relative to a frame of reference subjected to both translation and rotation. Because of the generality in the derivation, A and B may represent either two particles moving independently of one another or two points located in the same or different rigid bodies.

As shown in Fig. 20–11, the locations of points A and B are specified relative to the X, Y, Z frame of reference by position vectors $\mathbf{r}_A$ and $\mathbf{r}_B$. The base point A represents the origin of the x, y, z coordinate system, which is translating and rotating with respect to X, Y, Z. At the instant considered, the velocity and acceleration of point A are $\mathbf{v}_A$ and $\mathbf{a}_A$, respectively, and the angular velocity and angular acceleration of the x, y, z axes are $\mathbf{\Omega}$ and $\dot{\mathbf{\Omega}} = d\mathbf{\Omega}/dt$, respectively. All these vectors are *measured* with respect to the X, Y, Z frame of reference, although they may be expressed in Cartesian component form along either set of axes.

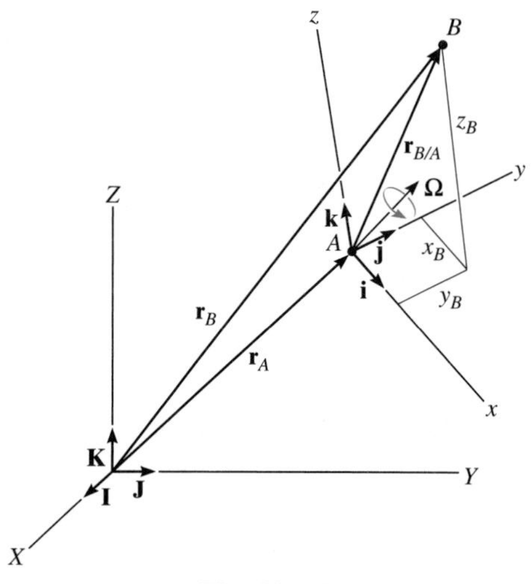

Fig. 20–11

Position. If the position of "B with respect to A" is specified by the *relative-position vector* $\mathbf{r}_{B/A}$, Fig. 20–11, then, by vector addition,

$$\mathbf{r}_B = \mathbf{r}_A + \mathbf{r}_{B/A} \tag{20–9}$$

where

$\mathbf{r}_B$ = position of B
$\mathbf{r}_A$ = position of the origin A
$\mathbf{r}_{B/A}$ = relative position of "B with respect to A"

Velocity. The velocity of point B measured from X, Y, Z is determined by taking the time derivative of Eq. 20–9, which yields

$$\dot{\mathbf{r}}_B = \dot{\mathbf{r}}_A + \dot{\mathbf{r}}_{B/A}$$

The first two terms represent $\mathbf{v}_B$ and $\mathbf{v}_A$. The last term is evaluated by applying Eq. 20–6, since $\mathbf{r}_{B/A}$ is measured between two points in a rotating reference. Hence,

$$\dot{\mathbf{r}}_{B/A} = (\dot{\mathbf{r}}_{B/A})_{xyz} + \boldsymbol{\Omega} \times \mathbf{r}_{B/A} = (\mathbf{v}_{B/A})_{xyz} + \boldsymbol{\Omega} \times \mathbf{r}_{B/A} \tag{20–10}$$

Here $(\mathbf{v}_{B/A})_{xyz}$ is the relative velocity of B with respect to A measured from x, y, z. Thus,

$$\mathbf{v}_B = \mathbf{v}_A + \boldsymbol{\Omega} \times \mathbf{r}_{B/A} + (\mathbf{v}_{B/A})_{xyz} \tag{20–11}$$

where

$\mathbf{v}_B$ = velocity of B
$\mathbf{v}_A$ = velocity of the origin A of the x, y, z frame of reference
$(\mathbf{v}_{B/A})_{xyz}$ = relative velocity of "B with respect to A" as measured by an observer attached to the rotating x, y, z frame of reference
$\boldsymbol{\Omega}$ = angular velocity of the x, y, z frame of reference
$\mathbf{r}_{B/A}$ = relative position of "B with respect to A"

Acceleration. The acceleration of point B measured from X, Y, Z is determined by taking the time derivative of Eq. 20–11, which yields

$$\dot{\mathbf{v}}_B = \dot{\mathbf{v}}_A + \dot{\mathbf{\Omega}} \times \mathbf{r}_{B/A} + \mathbf{\Omega} \times \dot{\mathbf{r}}_{B/A} + \dot{\mathbf{v}}_{B/A}$$

The time derivatives defined in the first and second terms represent $\mathbf{a}_B$ and $\mathbf{a}_A$ respectively. The fourth term is evaluated using Eq. 20–10, and the last term is evaluated by applying Eq. 20–6, which yields

$$\begin{aligned}\dot{\mathbf{v}}_{B/A} &= (\dot{\mathbf{v}}_{B/A})_{xyz} + \mathbf{\Omega} \times (\mathbf{v}_{B/A})_{xyz} \\ &= (\mathbf{a}_{B/A})_{xyz} + \mathbf{\Omega} \times (\mathbf{v}_{B/A})_{xyz}\end{aligned}$$

Here $(\mathbf{a}_{B/A})_{xyz}$ is the relative acceleration of B with respect to A measured from x, y, z. Substituting this result and Eq. 20–10 into the above equation and simplifying, we have

$$\mathbf{a}_B = \mathbf{a}_A + \dot{\mathbf{\Omega}} \times \mathbf{r}_{B/A} + \mathbf{\Omega} \times (\mathbf{\Omega} \times \mathbf{r}_{B/A}) + 2\mathbf{\Omega} \times (\mathbf{v}_{B/A})_{xyz} + (\mathbf{a}_{B/A})_{xyz} \qquad (20\text{–}12)$$

where

$\mathbf{a}_B$ = acceleration of B

$\mathbf{a}_A$ = acceleration of the origin A of the x, y, z frame of reference

$(\mathbf{a}_{B/A})_{xyz}$, $(\mathbf{v}_{B/A})_{xyz}$ = relative acceleration and relative velocity of "B with respect to A" as measured by an observer attached to the rotating x, y, z frame of reference

$\dot{\mathbf{\Omega}}$, $\mathbf{\Omega}$ = angular acceleration and angular velocity of the x, y, z frame of reference

$\mathbf{r}_{B/A}$ = relative position of "B with respect to A"

Equations 20–11 and 20–12 are identical to those used in Sec. 16.8 for analyzing relative plane motion.* In that case, however, application was simplified since $\mathbf{\Omega}$ and $\dot{\mathbf{\Omega}}$ have a *constant direction* which is always perpendicular to the plane of motion. For three-dimensional motion, $\dot{\mathbf{\Omega}}$ must be computed by using Eq. 20–6, since $\dot{\mathbf{\Omega}}$ depends on the change in both the magnitude and direction of $\mathbf{\Omega}$.

*Refer to Sec. 16.8 for an interpretation of the terms.

PROCEDURE FOR ANALYSIS

The following procedure provides a method for applying Eqs. 20–11 and 20–12 to solve problems involving the three-dimensional motion of particles or rigid bodies.

Coordinate Axes. Define the location and orientation of the X, Y, Z and x, y, z coordinate axes. Most often solutions are easily obtained if at the instant considered: (1) the origins are *coincident,* (2) the axes are collinear, and/or (3) the axes are parallel. Since several components of angular velocity may be involved in a problem, the calculations will be reduced if the x, y, z axes are selected such that only one component of angular velocity is observed in this frame ($\mathbf{\Omega}_{xyz}$) and the frame rotates with $\mathbf{\Omega}$ defined by the other components of angular velocity.

Kinematic Equations. After the origin of the moving reference, A, is defined and the moving point B is specified, Eqs. 20–11 and 20–12 should be written in symbolic form as

$$\mathbf{v}_B = \mathbf{v}_A + \mathbf{\Omega} \times \mathbf{r}_{B/A} + (\mathbf{v}_{B/A})_{xyz}$$

$$\mathbf{a}_B = \mathbf{a}_A + \dot{\mathbf{\Omega}} \times \mathbf{r}_{B/A} + \mathbf{\Omega} \times (\mathbf{\Omega} \times \mathbf{r}_{B/A}) + 2\mathbf{\Omega} \times (\mathbf{v}_{B/A})_{xyz} + (\mathbf{a}_{B/A})_{xyz}$$

Vectors $\mathbf{\Omega}$, $\mathbf{r}_A$, and $\mathbf{r}_{B/A}$ should be defined from the problem data and represented in Cartesian form.

Motion of the *moving reference* requires specifying $\mathbf{v}_A$, $\mathbf{a}_A$, and $\dot{\mathbf{\Omega}}$. Equation 20–6 must be used to calculate these vectors if $\mathbf{r}_A$ and $\mathbf{\Omega}$ appear to *change direction* when observed from the fixed X, Y, Z reference. In a similar manner, if $\mathbf{r}_{B/A}$ and $\mathbf{\Omega}_{xyz}$ appear to *change direction* as observed from the rotating reference x, y, z, then Eq. 20–6 must be used to calculate $(v_{B/A})_{xyz}$, $(a_{B/A})_{xyz}$, and $\dot{\Omega}_{xyz}$. In all cases, the kinematic quantities of position, velocity, and angular velocity that are used to compute the necessary time derivatives must be treated as *variables.*

After the final forms of $\dot{\mathbf{\Omega}}$, $\mathbf{v}_A$, $\mathbf{a}_A$, $(\dot{\mathbf{\Omega}})_{xyz}$, $(\mathbf{v}_{B/A})_{xyz}$, and $(\mathbf{a}_{B/A})_{xyz}$ are obtained, numerical problem data may be substituted and the kinematic terms evaluated. The components of all these vectors may be selected either along the X, Y, Z axes or along x, y, z. The choice is arbitrary, provided a consistent set of unit vectors is used.

Finally, substitute the data into the kinematic equations and perform the vector operations.

The following two examples illustrate this procedure.

Example 20–4

A motor M and attached rod AB have the angular motion shown in Fig. 20–12. A collar C on the rod is located 0.25 m from A, and is moving downward along the rod with a velocity of 3 m/s and an acceleration of 2 m/s^2. Determine the velocity and acceleration of C at this instant.

SOLUTION

Coordinate Axes. The origin of the fixed X, Y, Z reference is chosen at the center of the platform, and the origin of the moving, x, y, z frame at point A, Fig. 20–12. Since the collar is subjected to two components of angular motion, $\boldsymbol{\omega}_p$ and $\boldsymbol{\omega}_M$, it is viewed as having an angular velocity of $\boldsymbol{\Omega}_{xyz} = \boldsymbol{\omega}_M$ in x, y, z and the x, y, z axes are attached to the platform so that $\boldsymbol{\Omega} = \boldsymbol{\omega}_p$.

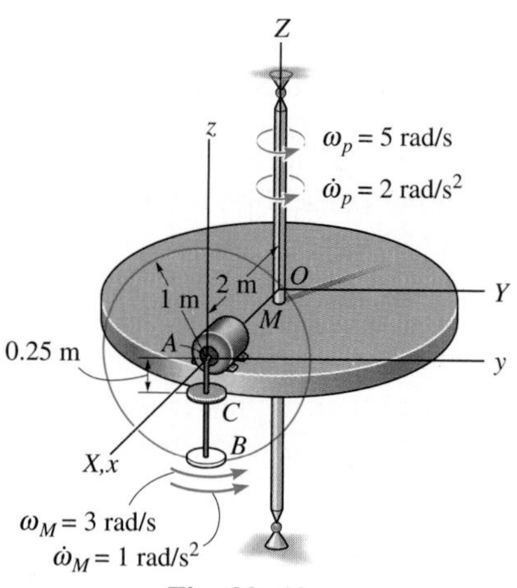

Fig. 20–12

Kinematic Equations. Equations 20–11 and 20–12, applied to points C and A, become

$$\mathbf{v}_C = \mathbf{v}_A + \mathbf{\Omega} \times \mathbf{r}_{C/A} + (\mathbf{v}_{C/A})_{xyz} \quad (1)$$

$$\mathbf{a}_C = \mathbf{a}_A + \dot{\mathbf{\Omega}} \times \mathbf{r}_{C/A} + \mathbf{\Omega} \times (\mathbf{\Omega} \times \mathbf{r}_{C/A}) + 2\mathbf{\Omega} \times (\mathbf{v}_{C/A})_{xyz} + (\mathbf{a}_{C/A})_{xyz} \quad (2)$$

Motion of Moving Reference

$\mathbf{\Omega} = \boldsymbol{\omega}_p = \{5\mathbf{k}\}$ rad/s $\quad$ $\mathbf{\Omega}$ does not appear to change direction relative to X, Y, Z.

$\dot{\mathbf{\Omega}} = \dot{\boldsymbol{\omega}}_p = \{2\mathbf{k}\}\text{ rad/s}^2$

$\mathbf{r}_A = \{2\mathbf{i}\}$ m $\quad$ $\mathbf{r}_A$ appears to change direction relative to X, Y, Z. Thus Eq. 20–6 has to be applied.

$$\mathbf{v}_A = \dot{\mathbf{r}}_A = (\dot{\mathbf{r}}_A)_{xyz} + \boldsymbol{\omega}_p \times \mathbf{r}_A = \mathbf{0} + 5\mathbf{k} \times 2\mathbf{i} = \{10\mathbf{j}\}\text{ m/s}$$

$$\mathbf{a}_A = \ddot{\mathbf{r}}_A = [(\ddot{\mathbf{r}}_A)_{xyz} + \boldsymbol{\omega}_p \times (\dot{\mathbf{r}}_A)_{xyz}] + \dot{\boldsymbol{\omega}}_p \times \mathbf{r}_A + \boldsymbol{\omega}_p \times \dot{\mathbf{r}}_A$$
$$= \mathbf{0} + \mathbf{0} + 2\mathbf{k} \times 2\mathbf{i} + 5\mathbf{k} \times 10\mathbf{j} = \{-50\mathbf{i} + 4\mathbf{j}\}\text{ m/s}^2$$

Since point A moves in a circular path lying in the X–Y plane we can also apply Eqs. 16–15 and 16–17 to obtain these same results.

Motion of C with Respect to Moving Reference

$\mathbf{\Omega}_{xyz} = \boldsymbol{\omega}_M = \{3\mathbf{i}\}$ rad/s $\quad$ $\dot{\mathbf{\Omega}}_{xyz} = \dot{\boldsymbol{\omega}}_M = \{1\mathbf{i}\}\text{ rad/s}^2$

Note that $\mathbf{\Omega}_{xyz}$ does not appear to change direction relative to x, y, z.

$\mathbf{r}_{C/A} = \{-0.25\mathbf{k}\}$ m

$\mathbf{r}_{C/A}$ appears to change direction relative to x, y, z, so that Eq. 20–6 has to be applied.

$$(\mathbf{v}_{C/A})_{xyz} = \dot{\mathbf{r}}_{C/A} = (\dot{\mathbf{r}}_{C/A})_{xyz} + \boldsymbol{\omega}_M \times \mathbf{r}_{C/A}$$
$$= -3\mathbf{k} + [3\mathbf{i} \times (-0.25\mathbf{k})] = \{0.75\mathbf{j} - 3\mathbf{k}\}\text{ m/s}$$

$$(\mathbf{a}_{C/A})_{xyz} = \ddot{\mathbf{r}}_{C/A} = [(\ddot{\mathbf{r}}_{C/A})_{xyz} + \boldsymbol{\omega}_M \times (\dot{\mathbf{r}}_{C/A})_{xyz}] + \dot{\boldsymbol{\omega}}_M \times \mathbf{r}_{C/A} + \boldsymbol{\omega}_M \times \dot{\mathbf{r}}_{C/A}$$
$$= [-2\mathbf{k} + 3\mathbf{i} \times (-3\mathbf{k})] + [(1\mathbf{i}) \times (-0.25\mathbf{k})] + [(3\mathbf{i}) \times (0.75\mathbf{j} - 3\mathbf{k})]$$
$$= \{18.25\mathbf{j} + 0.25\mathbf{k}\}\text{ m/s}^2$$

Substituting the data into Eqs. 1 and 2 yields

$$\mathbf{v}_C = \mathbf{v}_A + \mathbf{\Omega} \times \mathbf{r}_{C/A} + (\mathbf{v}_{C/A})_{xyz}$$
$$= 10\mathbf{j} + [5\mathbf{k} \times (-0.25\mathbf{k})] + (0.75\mathbf{j} - 3\mathbf{k})$$
$$= \{10.8\mathbf{j} - 3\mathbf{k}\}\text{ m/s} \qquad \textit{Ans.}$$

$$\mathbf{a}_C = \mathbf{a}_A + \dot{\mathbf{\Omega}} \times \mathbf{r}_{C/A} + \mathbf{\Omega} \times (\mathbf{\Omega} \times \mathbf{r}_{C/A}) + 2\mathbf{\Omega} \times (\mathbf{v}_{C/A})_{xyz} + (\mathbf{a}_{C/A})_{xyz}$$
$$= (-50\mathbf{i} + 4\mathbf{j}) + [2\mathbf{k} \times (-0.25\mathbf{k})] + 5\mathbf{k} \times [5\mathbf{k} \times (-0.25\mathbf{k})]$$
$$+ 2[5\mathbf{k} \times (0.75\mathbf{j} - 3\mathbf{k})] + (18.25\mathbf{j} + 0.25\mathbf{k})$$
$$= \{-57.5\mathbf{i} + 22.2\mathbf{j} + 0.25\mathbf{k}\}\text{ m/s}^2 \qquad \textit{Ans.}$$

Example 20–5

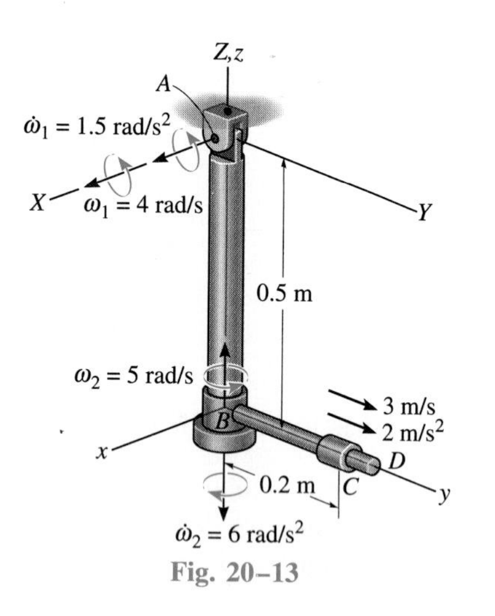

Fig. 20–13

The pendulum shown in Fig. 20–13 consists of two rods. AB is pin-supported at A and swings only in the Y–Z plane, whereas a bearing at B allows the attached rod BD to spin about rod AB. At a given instant, the rods have the angular motions shown. If a collar C, located 0.2 m from B, has a velocity of 3 m/s and an acceleration of 2 m/s² along the rod, determine the velocity and acceleration of the collar at this instant.

SOLUTION I $(\mathbf{\Omega} = \boldsymbol{\omega}_1,\ \mathbf{\Omega}_{xyz} = \boldsymbol{\omega}_2)$

Coordinate Axes. Since the rod is rotating about the fixed point A, the origin of the X, Y, Z frame will be chosen at A. Motion of the collar is conveniently observed from B, so the origin of the x, y, z frame is located at this point.

Kinematic Equations

$$\mathbf{v}_C = \mathbf{v}_B + \mathbf{\Omega} \times \mathbf{r}_{C/B} + (\mathbf{v}_{C/B})_{xyz} \quad (1)$$

$$\mathbf{a}_C = \mathbf{a}_B + \dot{\mathbf{\Omega}} \times \mathbf{r}_{C/B} + \mathbf{\Omega} \times (\mathbf{\Omega} \times \mathbf{r}_{C/B}) + 2\mathbf{\Omega} \times (\mathbf{v}_{C/B})_{xyz} + (\mathbf{a}_{C/B})_{xyz} \quad (2)$$

Motion of Moving Reference

$\mathbf{\Omega} = \boldsymbol{\omega}_1 = \{4\mathbf{i}\}$ rad/s $\quad \dot{\mathbf{\Omega}} = \dot{\boldsymbol{\omega}}_1 = \{1.5\mathbf{i}\}$ rad/s² $\quad \mathbf{\Omega}$ does not appear to change direction relative to X, Y, Z.

$\mathbf{r}_B = \{-0.5\mathbf{k}\}$ m $\quad \mathbf{r}_B$ appears to change direction relative to X, Y, Z.

$$\mathbf{v}_B = \dot{\mathbf{r}}_B = (\dot{\mathbf{r}}_B)_{xyz} + \boldsymbol{\omega}_1 \times \mathbf{r}_B = \mathbf{0} + 4\mathbf{i} \times (-0.5\mathbf{k}) = \{2\mathbf{j}\} \text{ m/s}$$

$$\mathbf{a}_B = \ddot{\mathbf{r}}_B = [(\ddot{\mathbf{r}}_B)_{xyz} + \boldsymbol{\omega}_1 \times (\dot{\mathbf{r}}_B)_{xyz}] + \dot{\boldsymbol{\omega}}_1 \times \mathbf{r}_B + \boldsymbol{\omega}_1 \times \dot{\mathbf{r}}_B$$
$$= (\mathbf{0} + \mathbf{0}) + [1.5\mathbf{i} \times (-0.5\mathbf{k})] + 4\mathbf{i} \times 2\mathbf{j} = \{0.75\mathbf{j} + 8\mathbf{k}\} \text{ m/s}^2$$

This is the same as applying Eqs. 16–15 and 16–17, since B has circular motion in the X–Y plane.

Motion of C with Respect to Moving Reference

$\mathbf{\Omega}_{xyz} = \boldsymbol{\omega}_2 = \{5\mathbf{k}\}$ rad/s $\quad \dot{\mathbf{\Omega}}_{xyz} = \dot{\boldsymbol{\omega}}_2 = \{-6\mathbf{k}\}$ rad/s² $\quad \boldsymbol{\omega}_2$ does not appear to change direction relative to x, y, z.

$\mathbf{r}_{C/B} = 0.2\mathbf{j}\}$ $\quad \mathbf{r}_{C/B}$ appears to change direction relative to x, y, z.

$$(\mathbf{v}_{C/B})_{xyz} = \dot{\mathbf{r}}_{C/B} = (\dot{\mathbf{r}}_{C/B})_{xyz} + \boldsymbol{\omega}_2 \times \mathbf{r}_{C/B} = 3\mathbf{j} + 5\mathbf{k} \times 0.2\mathbf{j} = \{-1\mathbf{i} + 3\mathbf{j}\} \text{ m/s}$$

$$(\mathbf{a}_{C/B})_{xyz} = \ddot{\mathbf{r}}_{C/B} = [(\ddot{\mathbf{r}}_{C/B})_{xyz} + \boldsymbol{\omega}_2 \times (\dot{\mathbf{r}}_{C/B})_{xyz}] + \dot{\boldsymbol{\omega}}_2 \times \mathbf{r}_{C/B} + \boldsymbol{\omega}_2 \times \dot{\mathbf{r}}_{C/B}$$
$$(2\mathbf{j} + 5\mathbf{k} \times 3\mathbf{j}) + (-6\mathbf{k} \times 0.2\mathbf{j}) + [5\mathbf{k} \times (-1\mathbf{i} + 3\mathbf{j})]$$
$$= \{-28.8\mathbf{i} - 3\mathbf{j}\} \text{ m/s}^2$$

Substituting the data into Eqs. 1 and 2 yields

$$\mathbf{v}_C = \mathbf{v}_B + \mathbf{\Omega} \times \mathbf{r}_{C/B} + (\mathbf{v}_{C/B})_{xyz} = 2\mathbf{j} + 4\mathbf{i} \times 0.2\mathbf{j} + (-1\mathbf{i} + 3\mathbf{j})$$
$$= \{-1\mathbf{i} + 5\mathbf{j} + 0.8\mathbf{k}\} \text{ m/s} \qquad \textit{Ans.}$$

$$\mathbf{a}_C = \mathbf{a}_B + \dot{\mathbf{\Omega}} \times \mathbf{r}_{C/B} + \mathbf{\Omega} \times (\mathbf{\Omega} \times \mathbf{r}_{C/B}) + 2\mathbf{\Omega} \times (\mathbf{v}_{C/B})_{xyz} + (\mathbf{a}_{C/B})_{xyz}$$
$$= (0.75\mathbf{j} + 8\mathbf{k}) + (1.5\mathbf{i} \times 0.2\mathbf{j}) + [4\mathbf{i} \times (4\mathbf{i} \times 0.2\mathbf{j})]$$
$$+ 2[4\mathbf{i} \times (-1\mathbf{i} + 3\mathbf{j})] + (-28.8\mathbf{i} - 3\mathbf{j})$$
$$= \{-28.8\mathbf{i} - 5.45\mathbf{j} + 32.3\mathbf{k}\} \text{ m/s}^2 \qquad \textit{Ans.}$$

SOLUTION II $(\mathbf{\Omega} = \boldsymbol{\omega}_1 + \boldsymbol{\omega}_2, \mathbf{\Omega}_{xyz} = \mathbf{0})$

Motion of Moving Reference

$$\mathbf{\Omega} = \boldsymbol{\omega}_1 + \boldsymbol{\omega}_2 = \{4\mathbf{i} + 5\mathbf{k}\} \text{ rad/s}$$

From the constraints of the problem $\boldsymbol{\omega}_1$ does not change direction relative to X, Y, Z; however, the direction of $\boldsymbol{\omega}_2$ is changed by $\boldsymbol{\omega}_1$. Thus, a simple way of obtaining $\dot{\mathbf{\Omega}}$ is to consider x', y', z' axes coincident with the X, Y, Z axes at A, such that the primed axes have an angular velocity $\boldsymbol{\omega}_1$. Then

$$\begin{aligned}\dot{\mathbf{\Omega}} = \dot{\boldsymbol{\omega}}_1 + \dot{\boldsymbol{\omega}}_2 &= [(\dot{\boldsymbol{\omega}}_1)_{x'y'z'} + \boldsymbol{\omega}_1 \times \boldsymbol{\omega}_1] + [(\dot{\boldsymbol{\omega}}_2)_{x'y'z'} + \boldsymbol{\omega}_1 \times \boldsymbol{\omega}_2] \\ &= (1.5\mathbf{i} + \mathbf{0}) + (-6\mathbf{k} + 4\mathbf{i} \times 5\mathbf{k}) = \{1.5\mathbf{i} - 20\mathbf{j} - 6\mathbf{k}\} \text{ rad/s}^2\end{aligned}$$

Also,

$$\mathbf{r}_B = \{-0.5\mathbf{k}\} \text{ m}$$

Here again only $\boldsymbol{\omega}_1$ changes the direction of $\mathbf{r}_B$ so that the time derivatives of $\mathbf{r}_B$ can be computed using the primed axes defined above, which are rotating at $\boldsymbol{\omega}_1$. Hence,

$$\begin{aligned}\mathbf{v}_B = \dot{\mathbf{r}}_B &= (\dot{\mathbf{r}}_B)_{x'y'z'} + \boldsymbol{\omega}_1 \times \mathbf{r}_B \\ &= \mathbf{0} + 4\mathbf{i} \times (-0.5\mathbf{k}) = \{2\mathbf{j}\} \text{ m/s} \\ \mathbf{a}_B = \ddot{\mathbf{r}}_B &= [(\ddot{\mathbf{r}}_B)_{x'y'z'} + \boldsymbol{\omega}_1 \times (\dot{\mathbf{r}}_B)_{x'y'z'}] + \dot{\boldsymbol{\omega}}_1 \times \mathbf{r}_B + \boldsymbol{\omega}_1 \times \dot{\mathbf{r}}_B \\ &= [\mathbf{0} + \mathbf{0}] + [1.5\mathbf{i} \times (-0.5\mathbf{k})] + 4\mathbf{i} \times 2\mathbf{j} = \{0.75\mathbf{j} + 8\mathbf{k}\} \text{ m/s}^2\end{aligned}$$

Motion of C with Respect to Moving Reference

$$\begin{aligned}\mathbf{\Omega}_{xyz} &= \mathbf{0} \\ \dot{\mathbf{\Omega}}_{xyz} &= \mathbf{0} \\ \mathbf{r}_{C/B} &= \{0.2\mathbf{j}\} \text{ m} \qquad \mathbf{r}_{C/B} \text{ does not appear to change direction relative to } x, y, z. \\ (\mathbf{v}_{C/B})_{xyz} &= \dot{\mathbf{r}}_{C/B} = \{3\mathbf{j}\} \text{ m/s} \\ (\mathbf{a}_{C/B})_{xyz} &= \ddot{\mathbf{r}}_{C/B} = \{2\mathbf{j}\} \text{ m/s}^2\end{aligned}$$

Substituting the data into Eqs. 1 and 2 yields

$$\begin{aligned}\mathbf{v}_C &= \mathbf{v}_B + \mathbf{\Omega} \times \mathbf{r}_{C/B} + (\mathbf{v}_{C/B})_{xyz} \\ &= 2\mathbf{j} + [(4\mathbf{i} + 5\mathbf{k}) \times (0.2\mathbf{j})] + 3\mathbf{j} \\ &= \{-1\mathbf{i} + 5\mathbf{j} + 0.8\mathbf{k}\} \text{ m/s}\end{aligned}$$ ***Ans.***

$$\begin{aligned}\mathbf{a}_C &= \mathbf{a}_B + \dot{\mathbf{\Omega}} \times \mathbf{r}_{C/B} + \mathbf{\Omega} \times (\mathbf{\Omega} \times \mathbf{r}_{C/B}) + 2\mathbf{\Omega} \times (\mathbf{v}_{C/B})_{xyz} + (\mathbf{a}_{C/B})_{xyz} \\ &= (0.75\mathbf{j} + 8\mathbf{k}) + [(1.5\mathbf{i} - 20\mathbf{j} - 6\mathbf{k}) \times (0.2\mathbf{j})] \\ &\quad + (4\mathbf{i} + 5\mathbf{k}) \times [(4\mathbf{i} + 5\mathbf{k}) \times 0.2\mathbf{j}] + 2[(4\mathbf{i} + 5\mathbf{k}) \times 3\mathbf{j}] + 2\mathbf{j} \\ &= \{-28.8\mathbf{i} - 5.45\mathbf{j} + 32.3\mathbf{k}\} \text{ m/s}^2\end{aligned}$$ ***Ans.***

PROBLEMS

20–41. Solve Example 20–5 such that the x, y, z axes move with curvilinear translation, $\mathbf{\Omega} = \mathbf{0}$, in which case the collar appears to have both an angular velocity $\mathbf{\Omega}_{C/B} = \boldsymbol{\omega}_1 + \boldsymbol{\omega}_2$ and radial motion.

20–42. At a given instant, rod BD is rotating about the y axis with an angular velocity $\omega_{BD} = 2$ rad/s and an angular acceleration $\dot{\omega}_{BD} = 5$ rad/s^2. Also, when $\theta = 45°$ link AC is rotating downward such that $\dot{\theta} = 4$ rad/s and $\ddot{\theta} = 2$ rad/s^2. Determine the velocity and acceleration of point A on the link at this instant.

20–43. At a given instant, rod BD is rotating about the y axis with an angular velocity $\omega_{BD} = 2$ rad/s and an angular acceleration $\dot{\omega}_{BD} = 5$ rad/s^2. Also, when $\theta = 60°$ link AC is rotating downward such that $\dot{\theta} = 2$ rad/s and $\ddot{\theta} = 8$ rad/s^2. Determine the velocity and acceleration of point A on the link at this instant.

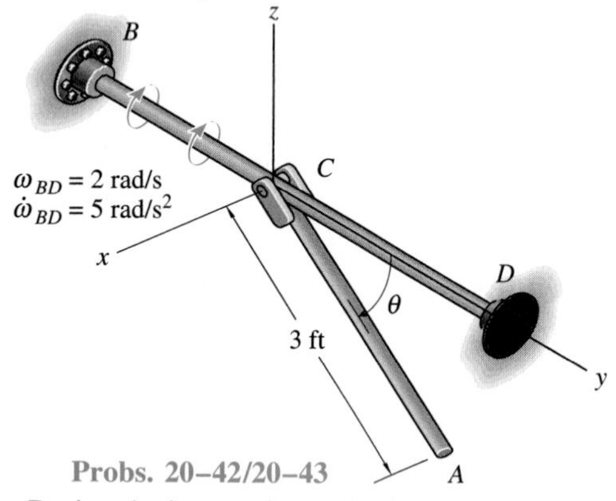

Probs. 20–42/20–43

***20–44.** During the instant shown the frame of the X-ray camera is rotating about the vertical axis at $\omega_z = 5$ rad/s and $\dot{\omega}_z = 2$ rad/s^2. Relative to the frame the arm is rotating at $\omega_{rel} = 2$ rad/s and $\dot{\omega}_{rel} = 1$ rad/s^2. Determine the velocity and acceleration of the center of the camera C at this instant.

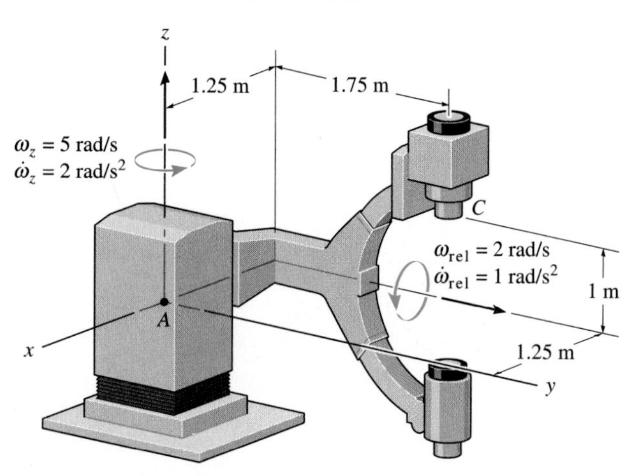

Prob. 20–44

20–45. At the instant shown, the backhoe is traveling forward at a constant speed $v_O = 2$ ft/s, and the boom ABC is rotating about the z axis with an angular velocity $\omega_1 = 0.8$ rad/s and an angular acceleration $\dot{\omega}_1 = 1.30$ rad/s^2. At this same instant the boom is rotating with $\omega_2 = 3$ rad/s when $\dot{\omega}_2 = 2$ rad/s^2, both measured relative to the frame. Determine the velocity and acceleration of point P on the bucket at this instant.

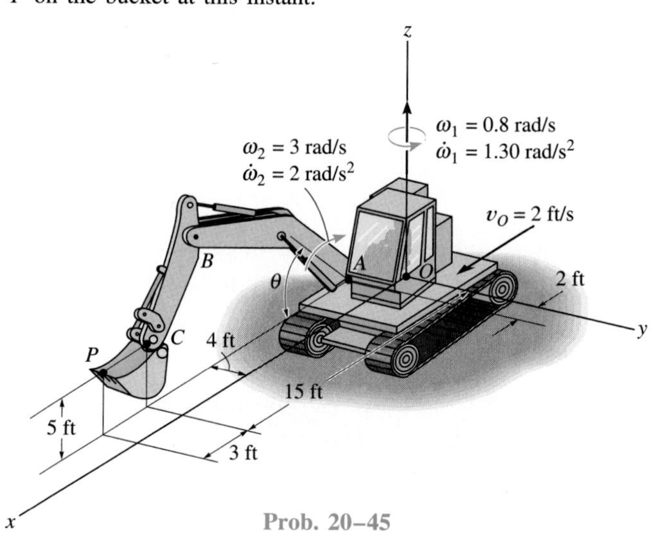

Prob. 20–45

20–46. At the given instant, the rod is spinning about the z axis with an angular velocity $\omega_1 = 3$ rad/s and angular acceleration $\dot{\omega}_1 = 4$ rad/s^2. At this same instant, the disk is spinning with $\omega_2 = 2$ rad/s when $\dot{\omega}_2 = 1$ rad/s^2, both measured *relative* to the rod. Determine the velocity and acceleration of point P on the disk at this instant.

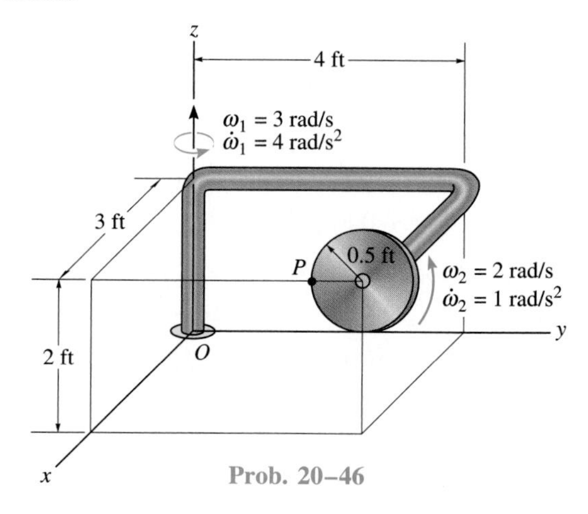

Prob. 20–46

20–47. The boom AB of the crane is rotating about the z axis with an angular velocity $\omega_z = 0.75$ rad/s, which is increasing at $\dot{\omega}_z = 2$ rad/s^2. At the same instant, $\theta = 60°$ and the boom is rotating upward at a constant rate $\dot{\theta} = 0.5$ rad/s^2. Determine the velocity and acceleration of the tip B of the boom at this instant.

***20–48.** The boom AB of the crane is rotating about the z axis with an angular velocity of $\omega_z = 0.75$ rad/s, which is increasing at $\dot{\omega}_z = 2$ rad/s^2. At the same instant, $\theta = 60°$ and the boom is rotating upward at $\dot{\theta} = 0.5$ rad/s^2, which is increasing at $\ddot{\theta} = 0.75$ rad/s^2. Determine the velocity and acceleration of the tip B of the boom at this instant.

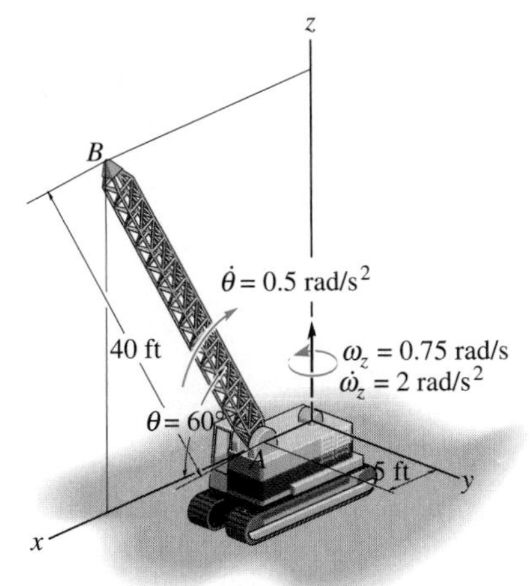

Probs. 20–47/20–48

20–49. At the instant shown, the arm AB is rotating about the fixed bearing with an angular velocity $\omega_1 = 2$ rad/s and angular acceleration $\dot{\omega}_1 = 6$ rad/s^2. At the same instant, rod BD is rotating relative to rod AB at $\omega_2 = 7$ rad/s, which is increasing at $\dot{\omega}_2 = 1$ rad/s^2. Also, the collar C is moving along rod BD with a velocity $\dot{r} = 2$ ft/s and a *deceleration* $\ddot{r} = -0.5$ ft/s^2, both measured relative to the rod. Determine the velocity and acceleration of the collar at this instant.

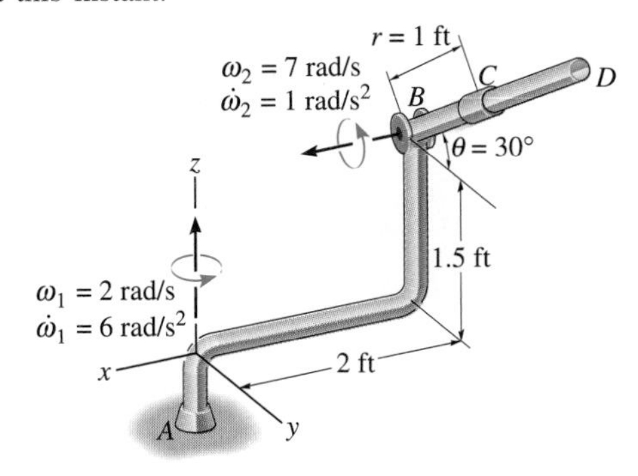

Prob. 20–49

20–50. The crane is rotating about the z axis with a constant rate $\omega_1 = 0.25$ rad/s, while the boom OA is rotating downward with a constant rate $\omega_2 = 0.4$ rad/s. Compute the velocity and acceleration of point A located at the top of the boom at the instant shown.

20–51. Solve Prob. 20–50 if the angular motions are increasing at $\dot{\omega}_1 = 0.4$ rad/s^2 and $\dot{\omega}_2 = 0.8$ rad/s^2 at the instant shown.

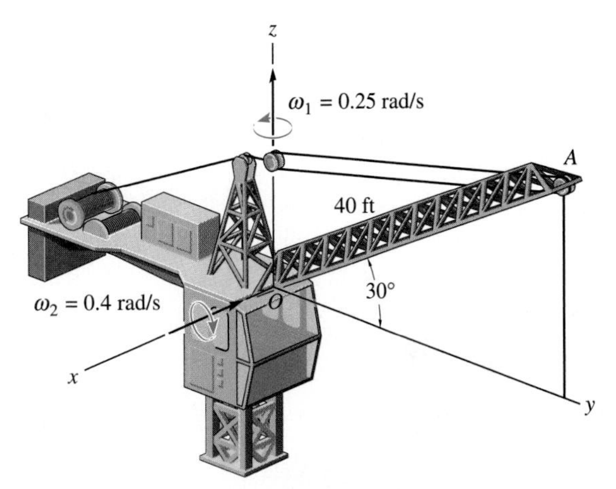

Probs. 20–50/20–51

***20–52.** The boom AB of the locomotive crane is rotating about the Z axis with an angular velocity $\omega_1 = 0.5$ rad/s, which is increasing at $\dot{\omega}_1 = 3$ rad/s^2. At this same instant, $\theta = 30°$ and the boom is rotating upward at a constant rate of $\dot{\theta} = 3$ rad/s. Determine the velocity and acceleration of the tip B of the boom at this instant.

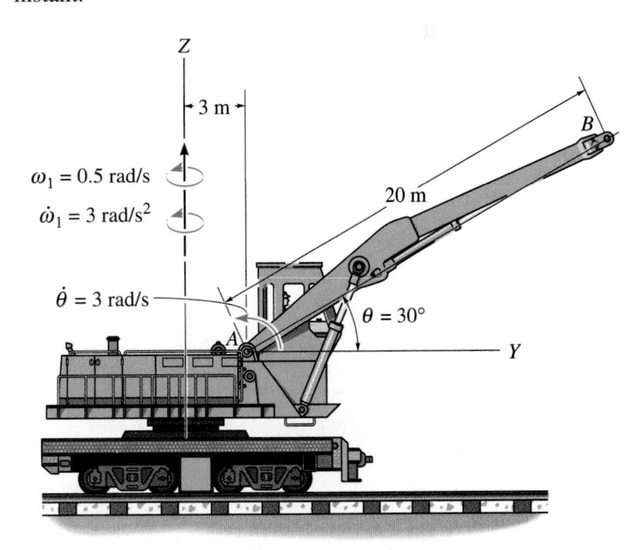

Prob. 20–52

20–53. The locomotive crane is traveling to the right at 2 m/s and has an acceleration of 1.5 m/s^2, while the boom is rotating about the Z axis with an angular velocity $\omega_1 = 0.5$ rad/s, which is increasing at $\dot{\omega}_1 = 3$ rad/s^2. At this same instant, $\theta = 30°$ and the boom is rotating upward at a constant rate $\dot{\theta} = 3$ rad/s. Determine the velocity and acceleration of the tip B of the boom at this instant.

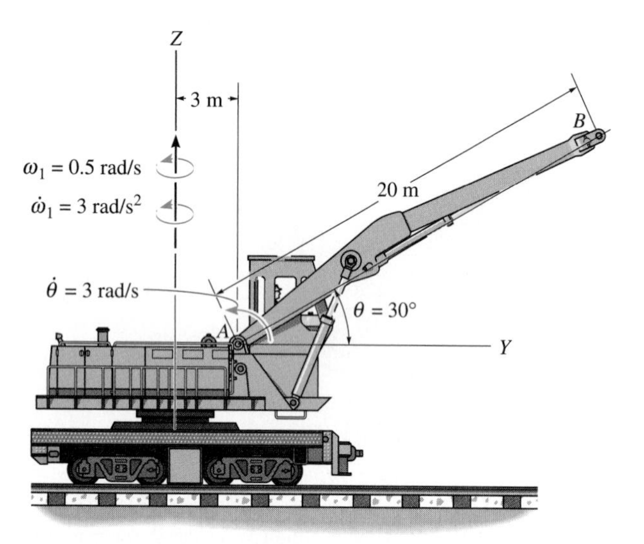

Prob. 20–53

20–54. At the instant shown, the helicopter is moving upward with a velocity $v_H = 4$ ft/s and has an acceleration $a_H = 2$ ft/s^2. At the same instant the frame H, *not* the horizontal blade, is rotating about a vertical axis with a constant angular velocity $\omega_H = 0.9$ rad/s. If the tail blade B is rotating with a constant angular velocity $\omega_{B/H} = 180$ rad/s, measured relative to H, determine the velocity and acceleration of point P, located on the tip of the blade, at the instant the blade is in the vertical position.

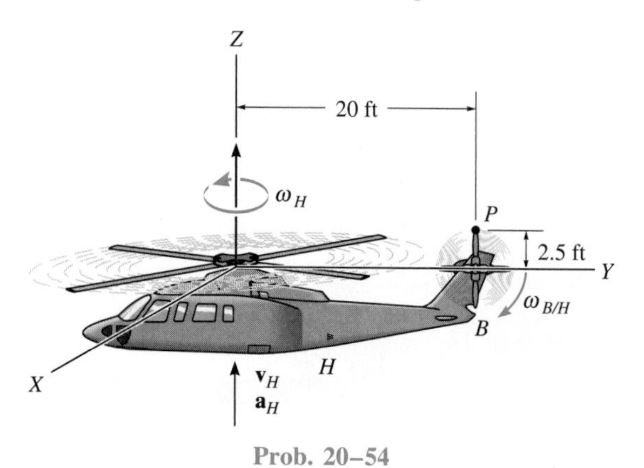

Prob. 20–54

20–55. At the instant shown, the arm OA of the conveyor belt is rotating about the z axis with a constant angular velocity $\omega_1 = 6$ rad/s, while at the same instant the arm is rotating upward at a constant rate $\omega_2 = 4$ rad/s. If the conveyor is running at a constant rate $\dot{r} = 5$ ft/s, determine the velocity and acceleration of the package P at the instant shown. Neglect the size of the package.

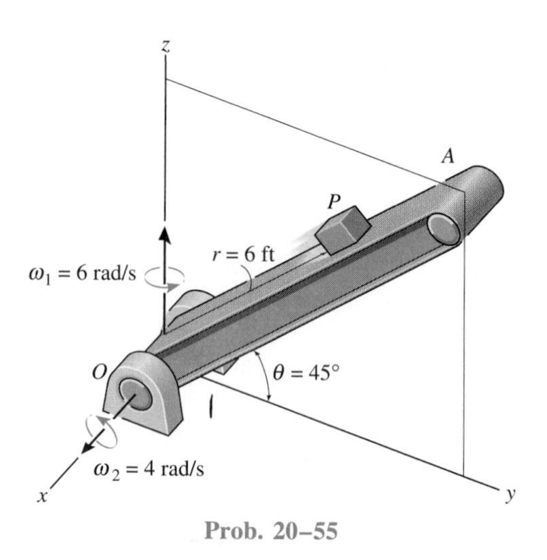

Prob. 20–55

***20–56.** At the instant shown, the arm OA of the conveyor belt is rotating about the z axis with a constant angular velocity $\omega_1 = 6$ rad/s, while at the same instant the arm is rotating upward at a constant rate $\omega_2 = 4$ rad/s. If the conveyor is running at a rate $\dot{r} = 5$ ft/s, which is increasing at $\ddot{r} = 8$ ft/s^2, determine the velocity and acceleration of the package P at the instant shown. Neglect the size of the package.

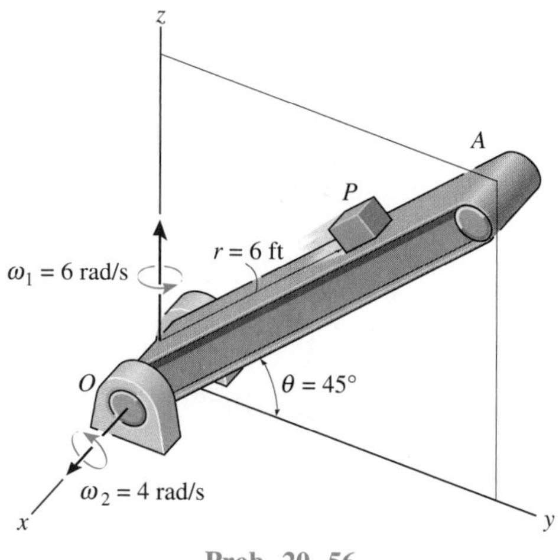

Prob. 20–56

20–57. At the instant shown, the rod AB is rotating about the z axis with an angular velocity $\omega_1 = 4$ rad/s and an angular acceleration $\dot{\omega}_1 = 3$ rad/s^2. At this same instant, the circular rod has an angular motion relative to the rod as shown. If the collar C is moving down around the circular rod with a speed of 3 in./s, which is increasing at 8 in./s^2, both measured relative to the rod, determine the collar's velocity and acceleration at this instant.

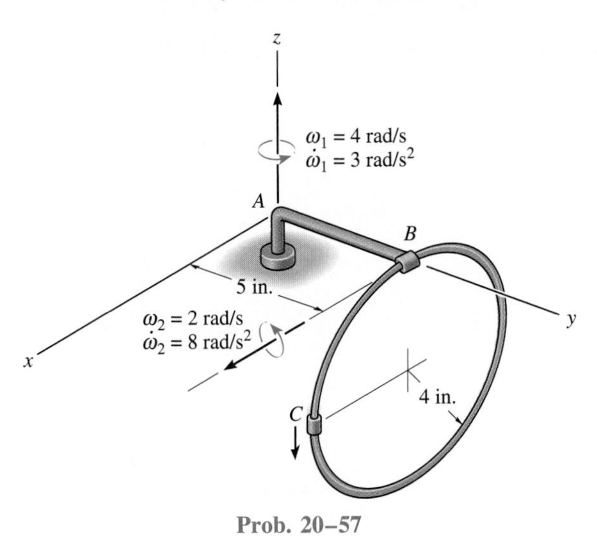

Prob. 20–57

20–58. At the instant shown, the arm AB is rotating about the fixed pin A with an angular velocity $\omega_1 = 4$ rad/s and angular acceleration $\dot{\omega}_1 = 3$ rad/s^2. At this same instant, rod BD is rotating relative to rod AB with an angular velocity $\omega_2 = 5$ rad/s, which is increasing at $\dot{\omega}_2 = 7$ rad/s^2. Also, the collar C is moving along rod BD with a velocity of 3 m/s and an acceleration of 2 m/s^2, both measured relative to the rod. Determine the velocity and acceleration of the collar at this instant.

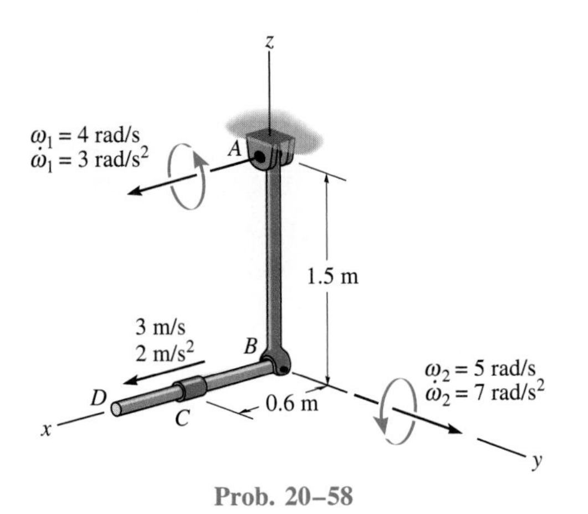

Prob. 20–58

20–59. At a given instant, the rod has the angular motions shown, while the collar C is moving down *relative* to the rod with a velocity of 6 ft/s and an acceleration of 2 ft/s^2. Determine the collar's velocity and acceleration at this instant.

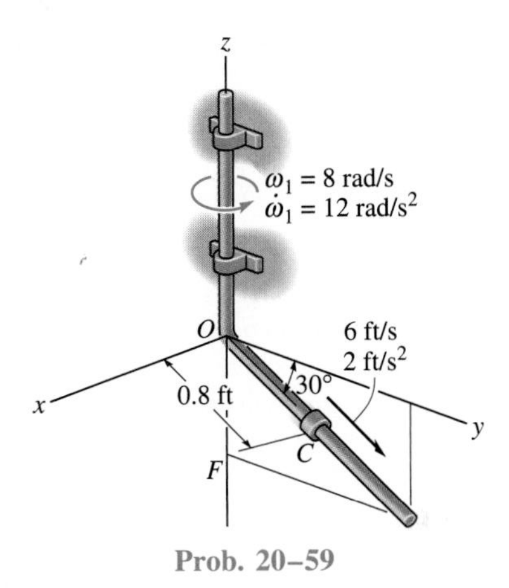

Prob. 20–59

***20–60.** At the instant shown, the industrial manipulator is rotating about the z axis at $\omega_1 = 5$ rad/s, and about joint B at $\omega_2 = 2$ rad/s. Determine the velocity and acceleration of the grip A at this instant, when $\phi = 30°$, $\theta = 45°$, and $r = 1.6$ m.

20–61. At the instant shown, the industrial manipulator is rotating about the z axis at $\omega_1 = 5$ rad/s and $\dot{\omega}_1 = 2$ rad/s^2; and about joint B at $\omega_2 = 2$ rad/s and $\dot{\omega}_2 = 3$ rad/s^2. Determine the velocity and acceleration of the grip A at this instant, when $\phi = 30°$, $\theta = 45°$, and $r = 1.6$ m.

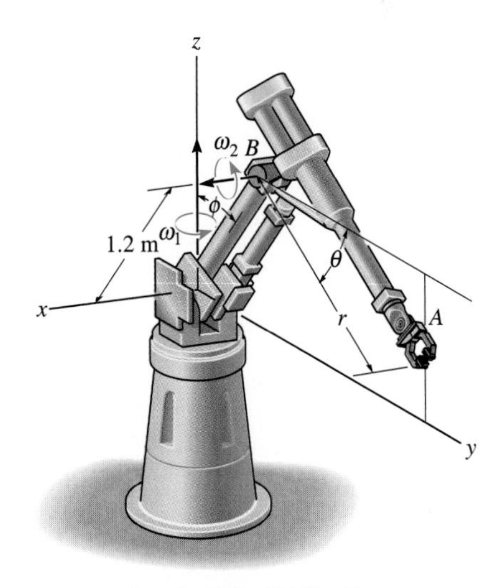

Probs. 20–60/20–61

The design of amusement-park rides requires a force analysis that depends on their three-dimensional motion. The principles required to do this will be explained in this chapter.

21

Three-Dimensional Kinetics of a Rigid Body

In general, as a body moves through space it has a simultaneous translation and rotation at a given instant. The kinetic aspects of the *translation* have been discussed in Chapter 17, where it was shown that a system of external forces acting on the body may be related to the acceleration of the body's mass center by the equation $\Sigma \mathbf{F} = m\mathbf{a}_G$. In this chapter emphasis is placed primarily on the *rotational* aspects of rigid-body motion, since motion of the body's mass center, defined by $\Sigma \mathbf{F} = m\mathbf{a}_G$, is treated in the same manner as particle motion.

The rotational equations of motion relate the body's components of angular motion to the moment components created by the external forces about some point located either on or off the body. To apply these equations, it is first necessary to formulate the moments and products of inertia of the body and to compute the body's angular momentum. Afterward, the principles of impulse and momentum and work and energy will be at our disposal for solving problems in three dimensions. The rotational equations of motion will then be developed, and special topics involving the motion of an unsymmetrical body about a fixed axis, motion of a gyroscope, and torque-free motion will be discussed.

*21.1 Moments and Products of Inertia

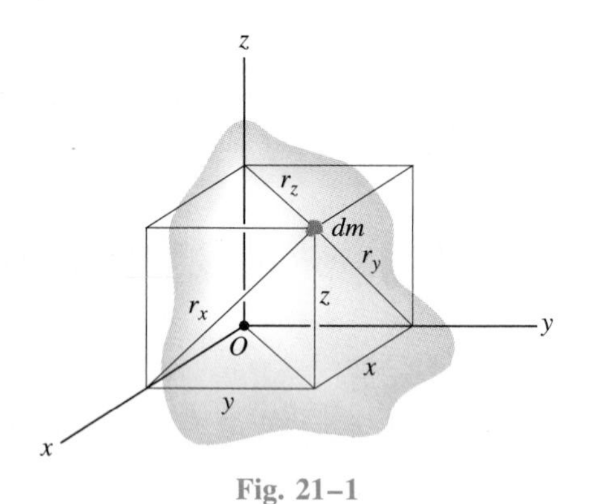

Fig. 21–1

When studying the planar kinetics of a body, it was necessary to introduce the moment of inertia I_G, which was computed about an axis perpendicular to the plane of motion and passing through the mass center G. For the kinetic analysis of three-dimensional motion it will sometimes be necessary to calculate six inertial quantities. These terms, called the moments and products of inertia, describe in a particular way the distribution of mass for a body relative to a given coordinate system that has a specified orientation and point of origin.

Moment of Inertia. Consider the rigid body shown in Fig. 21–1. The *moment of inertia* for a differential element dm of the body about any one of the three coordinate axes is defined as the product of the mass of the element and the square of the shortest distance from the axis to the element. For example, as noted in the figure, $r_x = \sqrt{y^2 + z^2}$, so that the mass moment of inertia of dm about the x axis is

$$dI_{xx} = r_x^2\,dm = (y^2 + z^2)\,dm$$

The moment of inertia I_{xx} for the body is determined by integrating this expression over the entire mass of the body. Hence, for each of the axes, we may write

$$\begin{aligned} I_{xx} &= \int_m r_x^2\,dm = \int_m (y^2 + z^2)\,dm \\ I_{yy} &= \int_m r_y^2\,dm = \int_m (x^2 + z^2)\,dm \\ I_{zz} &= \int_m r_z^2\,dm = \int_m (x^2 + y^2)\,dm \end{aligned} \tag{21–1}$$

Here it is seen that the moment of inertia is *always a positive quantity,* since it is the summation of the product of the mass dm, which is always positive, and distances squared.

Product of Inertia. The *product of inertia* for a differential element dm is defined with respect to a set of *two orthogonal planes* as the product of the mass of the element and the perpendicular (or shortest) distances from the planes to the element. For example, with respect to the y–z and x–z planes, the product of inertia dI_{xy} for the element dm shown in Fig. 21–1 is

$$dI_{xy} = xy\,dm$$

Note also that $dI_{yx} = dI_{xy}$. By integrating over the entire mass, the product of inertia of the body for each combination of planes may be expressed as

$$
\begin{aligned}
I_{xy} &= I_{yx} = \int_m xy\,dm \\
I_{yz} &= I_{zy} = \int_m yz\,dm \\
I_{xz} &= I_{zx} = \int_m xz\,dm
\end{aligned}
\tag{21–2}
$$

Unlike the moment of inertia, which is always positive, the product of inertia may be positive, negative, or zero. The result depends on the signs of the two defining coordinates, which vary independently from one another. In particular, if either one or both of the orthogonal planes are *planes of symmetry* for the mass, the *product of inertia* with respect to these planes will be *zero*. In such cases, elements of mass will occur in *pairs,* located on each side of the plane of symmetry. On one side of the plane the product of inertia for the element will be positive, while on the other side the product of inertia for the corresponding element will be negative, the sum therefore yielding zero. Examples of this are shown in Fig. 21–2. In the first case, Fig. 21–2*a*, the y–z plane is a plane of symmetry, and hence $I_{xz} = I_{xy} = 0$. Computation for I_{yz} will yield a *positive* result, since all elements of mass are located using only positive y and z coordinates. For the cylinder, with the coordinate axes located as shown in Fig. 21–2*b*, the x–z and y–z planes are both planes of symmetry. Thus, $I_{zx} = I_{yz} = I_{xy} = 0$.

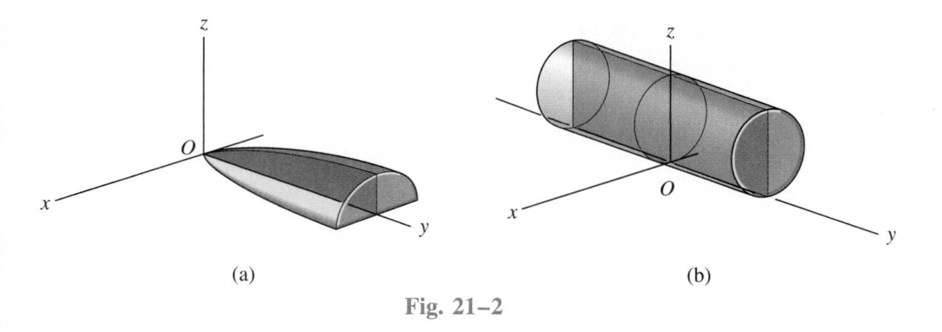

Fig. 21–2

Parallel-Axis and Parallel-Plane Theorems. The techniques of integration which are used to determine the moment of inertia of a body were described in Sec. 17.1. Also discussed were methods to determine the moment of inertia of a composite body, i.e., a body that is composed of simpler segments, as tabulated on the inside back cover. In both of these cases the *parallel-axis theorem* is often used for the calculations. This theorem, which was developed in Sec. 17.1, is used to transfer the moment of inertia of a body from an axis passing through its mass center G to a parallel axis passing through some other point. In this regard, if G has coordinates x_G, y_G, z_G defined from the x, y, z axes, Fig. 21–3, then the parallel-axis equations used to calculate the moments of inertia about the x, y, z axes are

$$\begin{aligned} I_{xx} &= (I_{x'x'})_G + m(y_G^2 + z_G^2) \\ I_{yy} &= (I_{y'y'})_G + m(x_G^2 + z_G^2) \\ I_{zz} &= (I_{z'z'})_G + m(x_G^2 + y_G^2) \end{aligned} \tag{21–3}$$

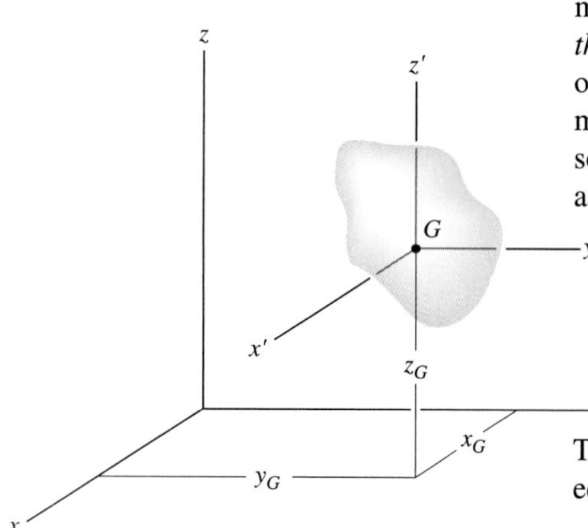

Fig. 21–3

The products of inertia of a body or a composite are computed in the same manner as the body's moments of inertia. Here, however, the *parallel-plane theorem* is important. This theorem is used to transfer the products of inertia of the body from a set of three orthogonal planes passing through the body's mass center to a corresponding set of three parallel planes passing through some other point O. Defining the perpendicular distances between the planes as x_G, y_G, and z_G, Fig. 21–3, the parallel-plane equations can be written as

$$\begin{aligned} I_{xy} &= (I_{x'y'})_G + m x_G y_G \\ I_{yz} &= (I_{y'z'})_G + m y_G z_G \\ I_{zx} &= (I_{z'x'})_G + m z_G x_G \end{aligned} \tag{21–4}$$

The derivation of these formulas is similar to that given for the parallel-axis equation, Sec. 17.1.

Inertia Tensor. The inertial properties of a body are completely characterized by nine terms, six of which are independent of one another. This set of terms is defined using Eqs. 21–1 and 21–2 and can be written as

$$\begin{pmatrix} I_{xx} & -I_{xy} & -I_{xz} \\ -I_{yx} & I_{yy} & -I_{yz} \\ -I_{zx} & -I_{zy} & I_{zz} \end{pmatrix}$$

This array is called an *inertia tensor.* It has a unique set of values for a body when it is computed for each location of the origin O and orientation of the coordinate axes.

For point O we can specify a unique axes inclination for which the products of inertia for the body are zero when computed with respect to these axes. When this is done, the inertia tensor is said to be ''diagonalized'' and may be written in the simplified form

$$\begin{pmatrix} I_x & 0 & 0 \\ 0 & I_y & 0 \\ 0 & 0 & I_z \end{pmatrix}$$

Here $I_x = I_{xx}$, $I_y = I_{yy}$, and $I_z = I_{zz}$ are termed the *principal moments of inertia* for the body, which are computed from the *principal axes of inertia.* Of these three principal moments of inertia, one will be a maximum and another a minimum of the body's moment of inertia.

Mathematical determination of the directions of principal axes of inertia will not be discussed here (see Prob. 21–25). There are many cases, however, in which the principal axes may be determined by inspection. From the previous discussion it was noted that if the coordinate axes are oriented such that *two* of the three orthogonal planes containing the axes are planes of *symmetry* for the body, then all the products of inertia for the body are zero with respect to the coordinate planes, and hence the coordinate axes are principal axes of inertia. For example, the x, y, z axes shown in Fig. 21–2*b* represent the principal axes of inertia for the cylinder at point O.

Moment of Inertia About an Arbitrary Axis. Consider the body shown in Fig. 21–4, where the nine elements of the inertia tensor have been computed for the x, y, z axes having an origin at O. Here we wish to compute the moment of inertia of the body about the Oa axis, for which the direction is defined by the unit vector $\mathbf{u}_a$. By definition $I_{Oa} = \int b^2\,dm$, where b is the *perpendicular distance* from dm to Oa. If the position of dm is located using $\mathbf{r}$, then $b = r\sin\theta$, which represents the *magnitude* of the cross product $\mathbf{u}_a \times \mathbf{r}$. Hence, the moment of inertia can be expressed as

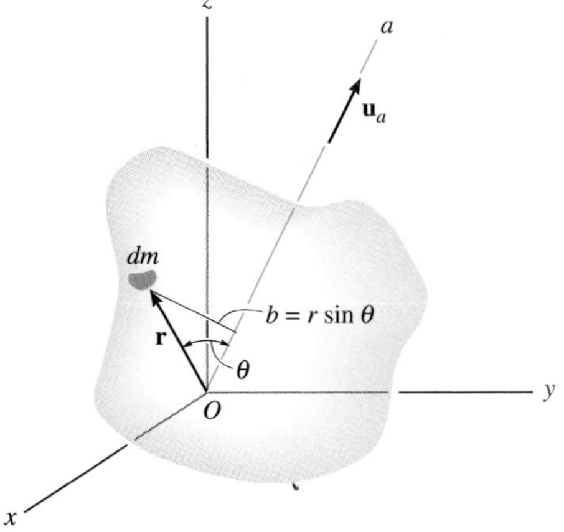

Fig. 21–4

$$I_{Oa} = \int_m |(\mathbf{u}_a \times \mathbf{r})|^2\,dm = \int_m (\mathbf{u}_a \times \mathbf{r}) \cdot (\mathbf{u}_a \times \mathbf{r})\,dm$$

Provided $\mathbf{u}_a = u_x\mathbf{i} + u_y\mathbf{j} + u_z\mathbf{k}$ and $\mathbf{r} = x\mathbf{i} + y\mathbf{j} + z\mathbf{k}$, so that $\mathbf{u}_a \times \mathbf{r} = (u_y z - u_z y)\mathbf{i} + (u_z x - u_x z)\mathbf{j} + (u_x y - u_y x)\mathbf{k}$, then, after substituting and performing the dot-product operation, we can write the moment of inertia as

$$\begin{aligned} I_{Oa} &= \int_m [(u_y z - u_z y)^2 + (u_z x - u_x z)^2 + (u_x y - u_y x)^2]\,dm \\ &= u_x^2 \int_m (y^2 + z^2)\,dm + u_y^2 \int_m (z^2 + x^2)\,dm + u_z^2 \int_m (x^2 + y^2)\,dm \\ &\quad - 2u_x u_y \int_m xy\,dm - 2u_y u_z \int_m yz\,dm - 2u_z u_x \int_m zx\,dm \end{aligned}$$

Recognizing the integrals to be the moments and products of inertia of the body, Eqs. 21–1 and 21–2, we have

$$I_{Oa} = I_{xx}u_x^2 + I_{yy}u_y^2 + I_{zz}u_z^2 - 2I_{xy}u_x u_y - 2I_{yz}u_y u_z - 2I_{zx}u_z u_x \qquad (21\text{–}5)$$

Thus, if the inertia tensor is specified for the x, y, z axes, the moment of inertia of the body about the inclined Oa axis can be computed by using Eq. 21–5. For the calculation the direction cosines u_x, u_y, u_z of the axes must be determined. These terms specify the cosines of the coordinate direction angles α, β, γ made between the Oa axis and the x, y, z axes, respectively (see Appendix C).

Example 21–1

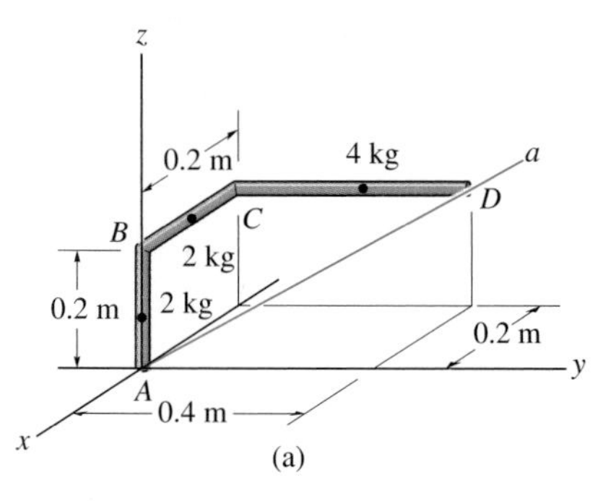

(a)

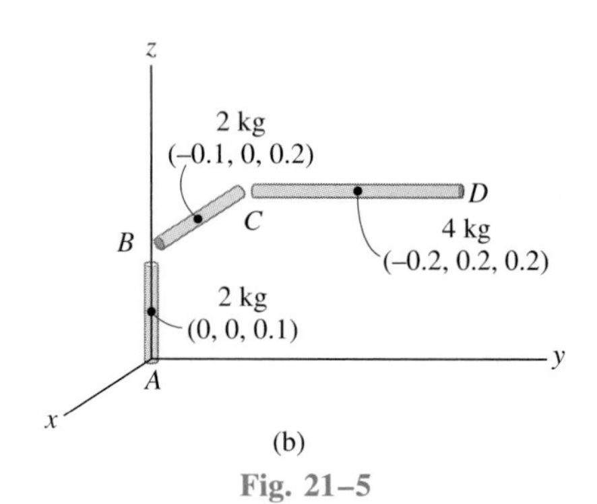

(b)

Fig. 21–5

Determine the moment of inertia of the bent rod shown in Fig. 21–5*a* about the *Aa* axis. The mass of each of the three segments is shown in the figure.

SOLUTION

The moment of inertia I_{Aa} can be computed by using Eq. 21–5. It is first necessary, however, to determine the moments and products of inertia of the rod about the *x*, *y*, *z* axes. This is done using the formula for the moment of inertia of a slender rod, $I = \frac{1}{12}ml^2$, and the parallel-axis and parallel-plane theorems, Eqs. 21–3 and 21–4. Dividing the rod into three parts and locating the mass center of each segment, Fig. 21–5*b*, we have

$$I_{xx} = [\tfrac{1}{12}(2)(0.2)^2 + 2(0.1)^2] + [0 + 2(0.2)^2] + [\tfrac{1}{12}(4)(0.4)^2 + 4((0.2)^2 + (0.2)^2)] = 0.480 \text{ kg} \cdot \text{m}^2$$

$$I_{yy} = [\tfrac{1}{12}(2)(0.2)^2 + 2(0.1)^2] + [\tfrac{1}{12}(2)(0.2)^2 + 2((-0.1)^2 + (0.2)^2)] + [0 + 4((-0.2)^2 + (0.2)^2)] = 0.453 \text{ kg} \cdot \text{m}^2$$

$$I_{zz} = [0 + 0] + [\tfrac{1}{12}(2)(0.2)^2 + 2(0.1)^2] + [\tfrac{1}{12}(4)(0.4)^2 + 4((-0.2)^2 + (0.2)^2)] = 0.400 \text{ kg} \cdot \text{m}^2$$

$$I_{xy} = [0 + 0] + [0 + 0] + [0 + 4(-0.2)(0.2)] = -0.160 \text{ kg} \cdot \text{m}^2$$

$$I_{yz} = [0 + 0] + [0 + 0] + [0 + 4(0.2)(0.2)] = 0.160 \text{ kg} \cdot \text{m}^2$$

$$I_{zx} = [0 + 0] + [0 + 2(0.2)(-0.1)] + [0 + 4(0.2)(-0.2)] = -0.200 \text{ kg} \cdot \text{m}^2$$

The *Aa* axis is defined by the unit vector

$$\mathbf{u}_{Aa} = \frac{\mathbf{r}_D}{r_D} = \frac{-0.2\mathbf{i} + 0.4\mathbf{j} + 0.2\mathbf{k}}{\sqrt{(-0.2)^2 + (0.4)^2 + (0.2)^2}} = -0.408\mathbf{i} + 0.816\mathbf{j} + 0.408\mathbf{k}$$

Thus,

$$u_x = -0.408 \qquad u_y = 0.816 \qquad u_z = 0.408$$

Substituting the computed data into Eq. 21–5 yields

$$\begin{aligned} I_{Aa} &= I_{xx}u_x^2 + I_{yy}u_y^2 + I_{zz}u_z^2 - 2I_{xy}u_xu_y - 2I_{yz}u_yu_z - 2I_{zx}u_zu_x \\ &= 0.480(-0.408)^2 + (0.453)(0.816)^2 + 0.400(0.408)^2 \\ &\quad - 2(-0.160)(-0.408)(0.816) - 2(0.160)(0.816)(0.408) \\ &\quad - 2(-0.200)(0.408)(-0.408) \\ &= 0.168 \text{ kg} \cdot \text{m}^2 \end{aligned}$$

Ans.

PROBLEMS

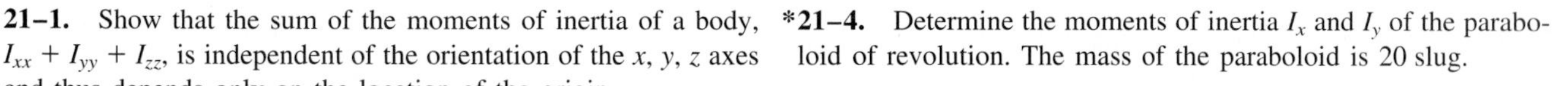

21–1. Show that the sum of the moments of inertia of a body, $I_{xx} + I_{yy} + I_{zz}$, is independent of the orientation of the x, y, z axes and thus depends only on the location of the origin.

21–2. Determine the moment of inertia of the cylinder with respect to the a–a axis of the cylinder. The cylinder has a mass m.

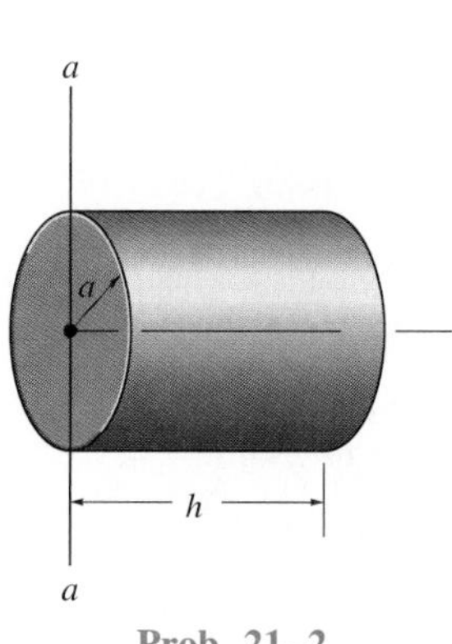

Prob. 21–2

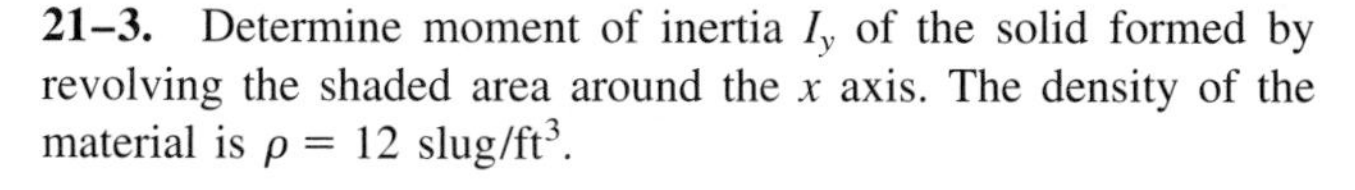

21–3. Determine moment of inertia I_y of the solid formed by revolving the shaded area around the x axis. The density of the material is $\rho = 12$ slug/ft^3.

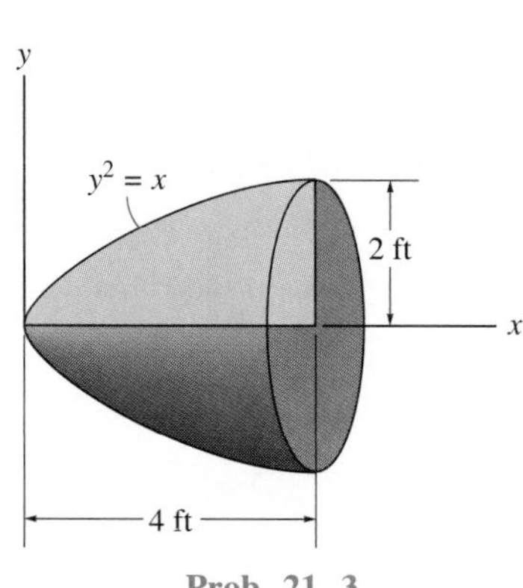

Prob. 21–3

***21–4.** Determine the moments of inertia I_x and I_y of the paraboloid of revolution. The mass of the paraboloid is 20 slug.

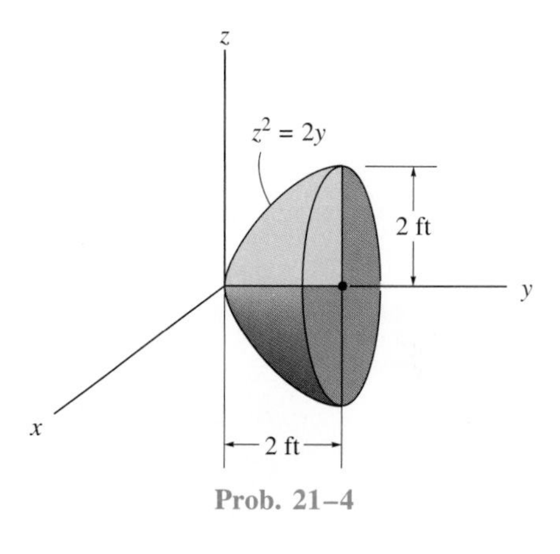

Prob. 21–4

21–5. Determine the radii of gyration k_x and k_y for the solid formed by revolving the shaded area about the y axis. The density of the material is ρ.

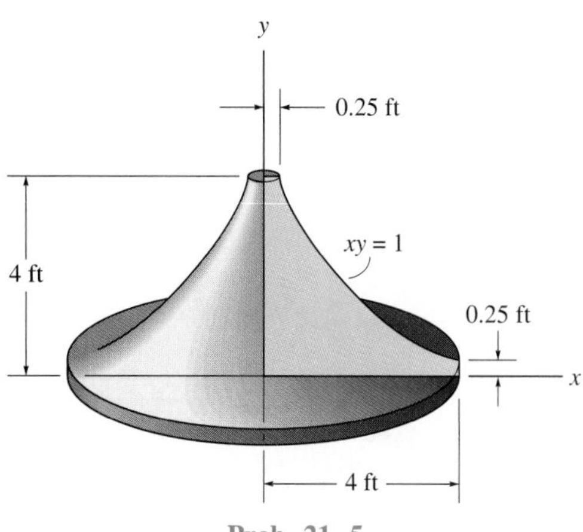

Prob. 21–5

21–6. Determine the product of inertia I_{xy} of the body formed by revolving the shaded area about the line $x = 5$ ft. Express the result in terms of the density of the material, ρ

21–7. Determine the moment of inertia I_y of the body formed by revolving the shaded area about the line $x = 5$ ft. Express the result in terms of the density of the material, ρ.

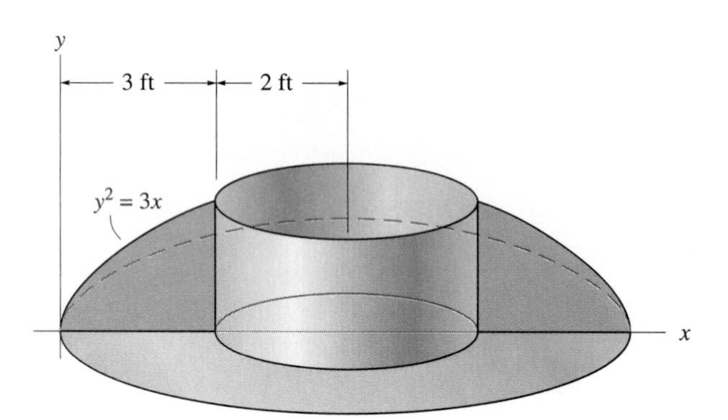

Probs. 21–6/21–7

*21–8. Determine the product of inertia I_{xy} for the homogeneous prism. The density of the material is ρ. Express the result in terms of the mass m of the prism.

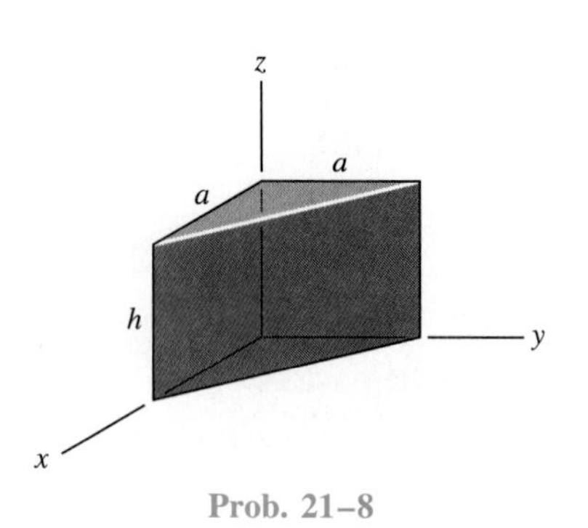

Prob. 21–8

21–9. Determine the product of inertia I_{yz} for the homogeneous tetrahedron. The mass density of the material is ρ. Express the result in terms of the total mass m of the solid.

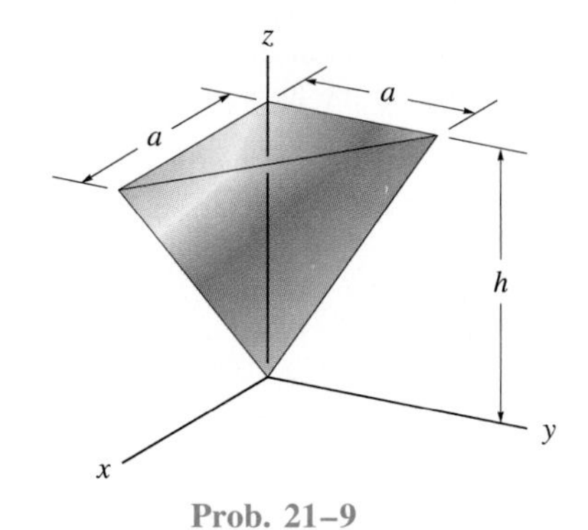

Prob. 21–9

21–10. Determine the product of inertia I_{xy} for the bent rod. The rod has a mass of 2 kg/m.

21–11. Determine the moments of inertia I_{xx}, I_{yy}, I_{zz} for the bent rod. The rod has a mass of 2 kg/m.

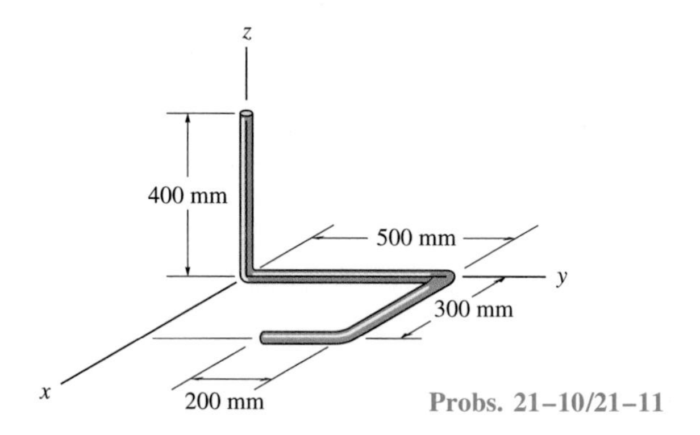

Probs. 21–10/21–11

***21–12.** Determine the elements of the inertia tensor for the cube with respect to the x,y,z coordinate system. The mass of the cube is m.

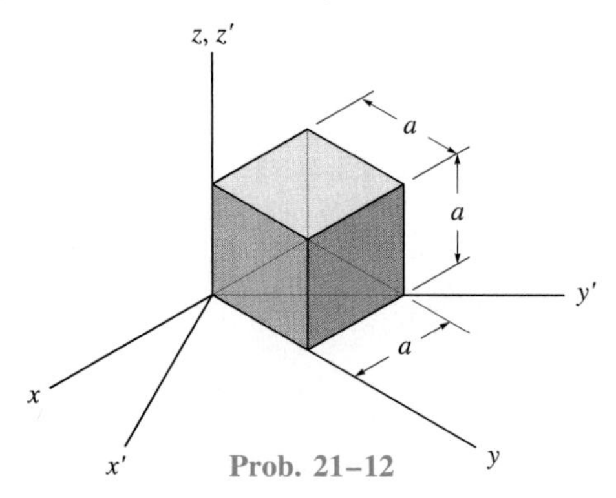

Prob. 21–12

21–13. Determine the moment of inertia of the cone about the z' axis. The weight of the cone is 15 lb, the *height* is $h = 1.5$ ft, and the radius is $r = 0.5$ ft.

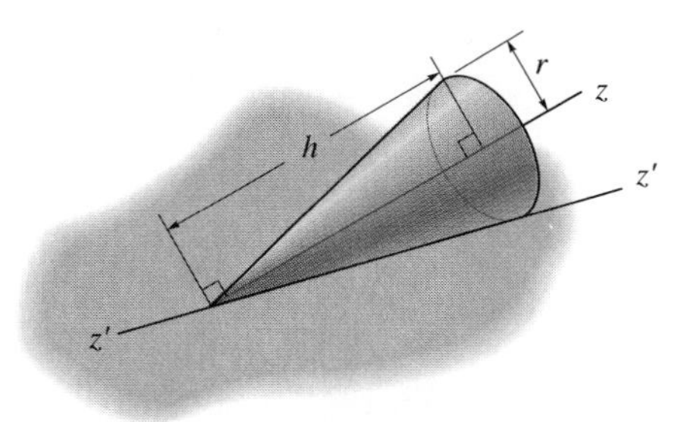

Prob. 21–13

21–14. Determine the moment of inertia of the composite body about the aa axis. The cylinder weighs 20 lb, and each hemisphere weighs 10 lb.

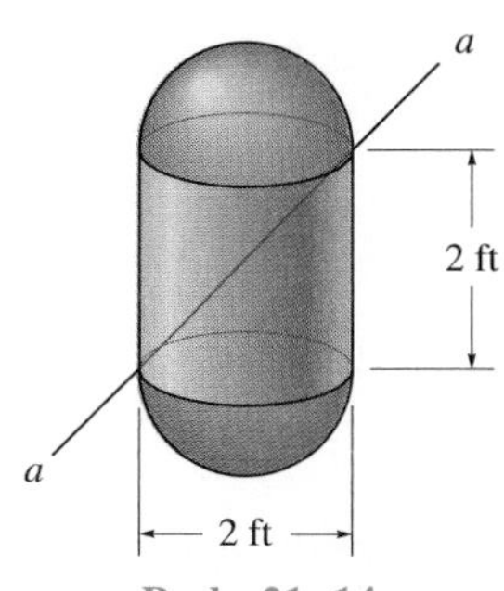

Prob. 21–14

21–15. Rod AB has a weight of 6 lb and each sphere has a weight of 8 lb. Determine the moment of inertia of this assembly about the x axis.

***21–16.** Rod AB has a weight of 6 lb and each sphere has a weight of 8 lb. Determine the moment of inertia of this assembly about the y axis.

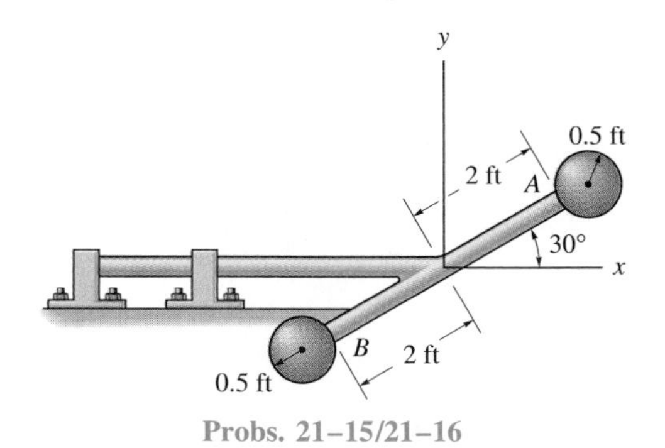

Probs. 21–15/21–16

21–17. Determine the moment of inertia of the 4-kg circular plate about the axis of rod OA.

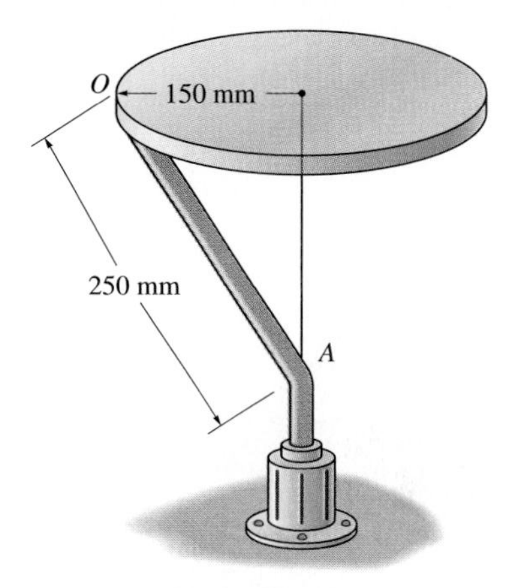

Prob. 21–17

21–18 Compute the moment of inertia of the rod-and-thin-ring assembly about the z axis. The rods and ring have a mass of 2 kg/m.

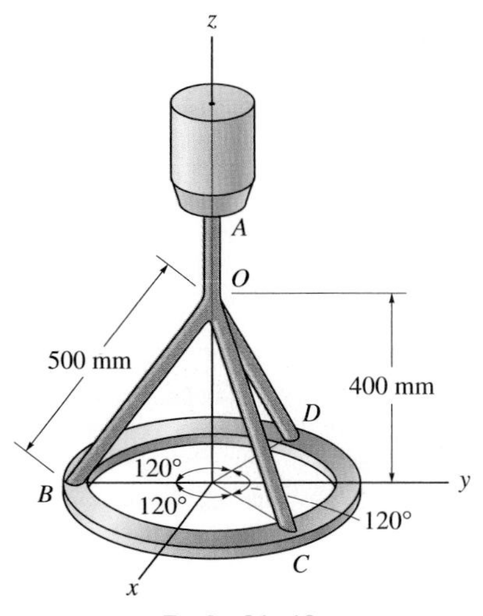

Prob. 21–18

21–19. Determine the moment of inertia I_x of the composite plate assembly. The plates have a specific weight of 6 lb/ft^2.

***21–20.** Determine the product of inertia I_{yz} of the composite plate assembly. The plates have a weight of 6 lb/ft^2.

21–21. Determine the product of inertia I_{xy} of the composite plate assembly. The plates have a weight of 6 lb/ft^2.

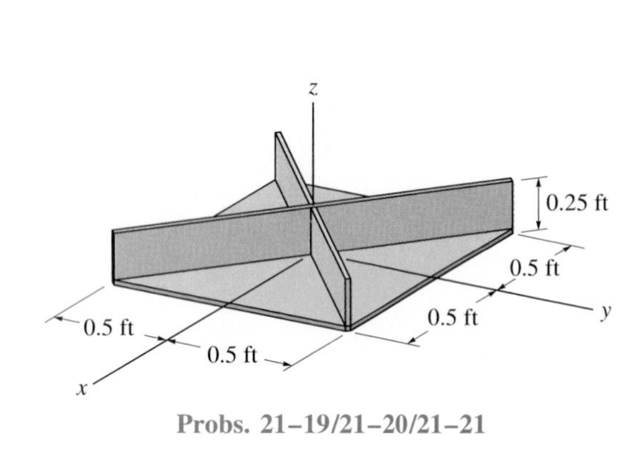

Probs. 21–19/21–20/21–21

21–23. The assembly consists of a 15-lb plate A, 40-lb plate B, and four 7-lb rods. Determine the moments of inertia of the assembly with respect to the principal x, y, z axes.

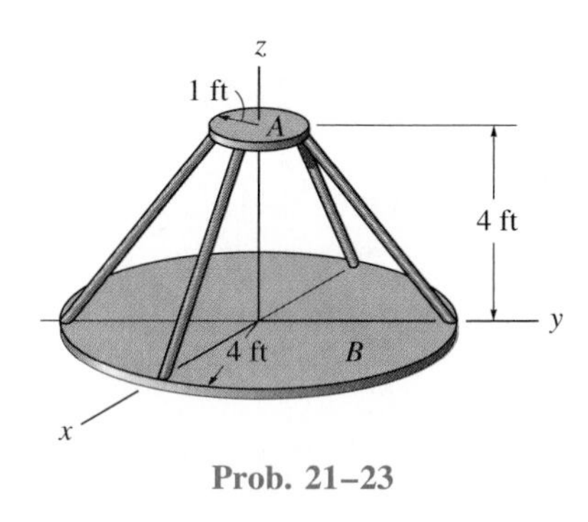

Prob. 21–23

21–22. The bent rod has a weight of 1.5 lb/ft. Locate the center of gravity $G(\bar{x}, \bar{y})$ and determine the principal moments of inertia $I_{x'}$, $I_{y'}$, and $I_{z'}$ of the rod with respect to the x', y', z' axes.

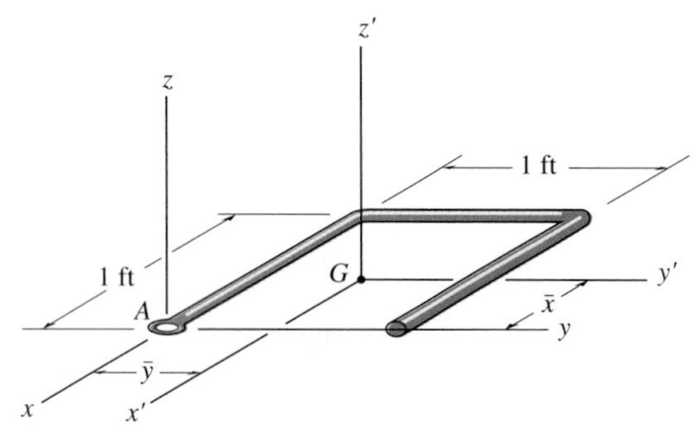

Prob. 21–22

***21–24.** The assembly consists of two square plates A and B which have a mass of 3 kg each and a rectangular plate C which has a mass of 4.5 kg. Determine the moments of inertia I_x, I_y and I_z.

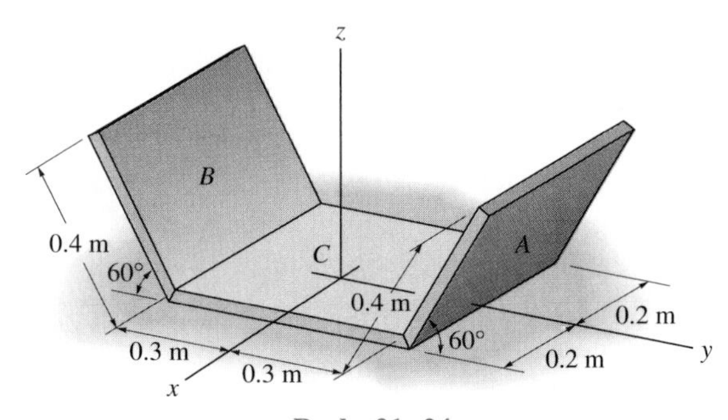

Prob. 21–24

*21.2 Angular Momentum

In this section we will develop the necessary equations used to determine the angular momentum of a rigid body about an arbitrary point. This formulation will provide the necessary means for developing both the principle of impulse and momentum and the equations of rotational motion for a rigid body.

Consider the rigid body in Fig. 21–6, which has a total mass m and center of mass located at G. The X, Y, Z coordinate system represents an inertial frame of reference, and hence, its axes are fixed or translate with a constant velocity. The angular momentum as measured from this reference will be computed relative to the arbitrary point A. The position vectors $\mathbf{r}_A$ and $\boldsymbol{\rho}_A$ are drawn from the origin of coordinates to point A and from A to the ith particle of the body. If the particle's mass is m_i, the angular momentum about point A is

$$(\mathbf{H}_A)_i = \boldsymbol{\rho}_A \times m_i \mathbf{v}_i$$

where $\mathbf{v}_i$ represents the particle's velocity as measured from the X, Y, Z coordinate system. If the body has an angular velocity $\boldsymbol{\omega}$ at the instant considered, $\mathbf{v}_i$ may be related to the velocity of A by applying Eq. 20–7, i.e.,

$$\mathbf{v}_i = \mathbf{v}_A + \boldsymbol{\omega} \times \boldsymbol{\rho}_A$$

Thus,

$$\begin{aligned}(\mathbf{H}_A)_i &= \boldsymbol{\rho}_A \times m_i(\mathbf{v}_A + \boldsymbol{\omega} \times \boldsymbol{\rho}_A) \\ &= (\boldsymbol{\rho}_A\, m_i) \times \mathbf{v}_A + \boldsymbol{\rho}_A \times (\boldsymbol{\omega} \times \boldsymbol{\rho}_A)\, m_i\end{aligned}$$

For the entire body, summing all the particles of the body requires an integration, i.e., $m_i \rightarrow dm$ so that

$$\mathbf{H}_A = \left(\int_m \boldsymbol{\rho}_A\, dm\right) \times \mathbf{v}_A + \int_m \boldsymbol{\rho}_A \times (\boldsymbol{\omega} \times \boldsymbol{\rho}_A)\, dm \qquad (21\text{–}6)$$

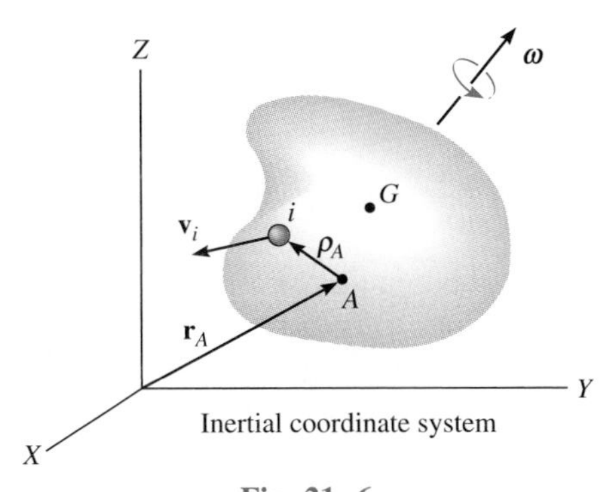

Fig. 21–6

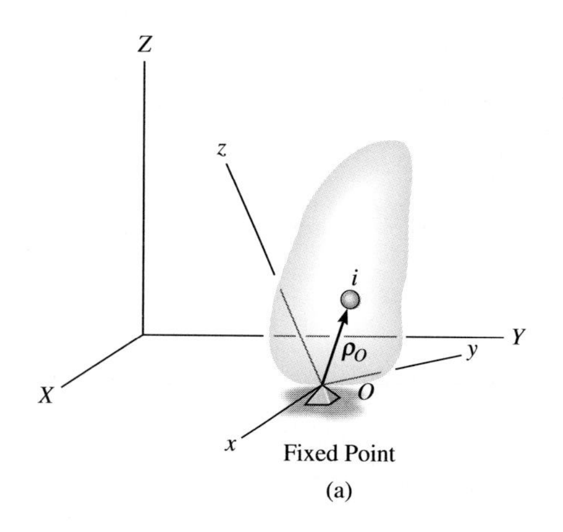

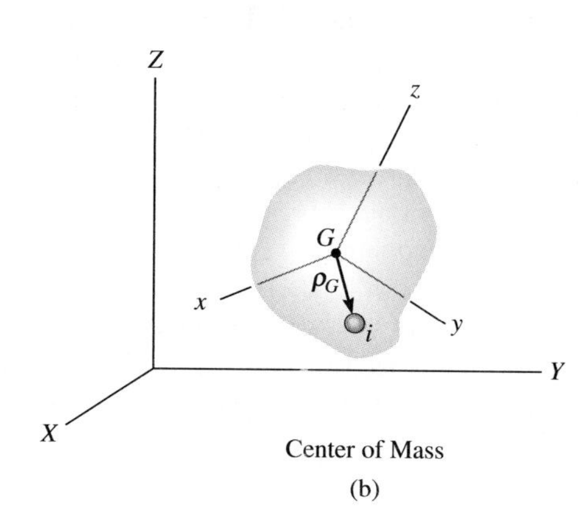

Fixed Point O. If A becomes a *fixed point O* in the body, Fig. 21–7a, then $\mathbf{v}_A = \mathbf{0}$ and Eq. 21–6 reduces to

$$\mathbf{H}_O = \int_m \boldsymbol{\rho}_O \times (\boldsymbol{\omega} \times \boldsymbol{\rho}_O)\, dm \qquad (21\text{–}7)$$

Center of Mass G. If A is located at the *center of mass G* of the body, Fig. 21–7b, then $\int_m \boldsymbol{\rho}_A\, dm = \mathbf{0}$ and

$$\mathbf{H}_G = \int_m \boldsymbol{\rho}_G \times (\boldsymbol{\omega} \times \boldsymbol{\rho}_G)\, dm \qquad (21\text{–}8)$$

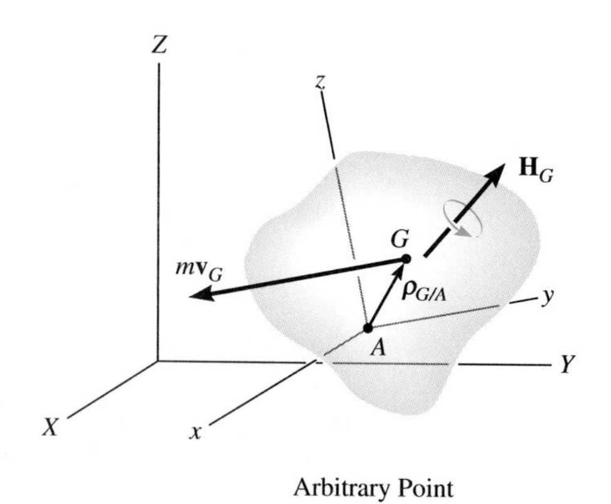

Fig. 21–7

Arbitrary Point A. In general, A may be some point other than O or G, Fig. 21–7c, in which case Eq. 21–6 may nevertheless be simplified to the following form (see Prob. 21–21).

$$\mathbf{H}_A = \boldsymbol{\rho}_{G/A} \times m\mathbf{v}_G + \mathbf{H}_G \qquad (21\text{–}9)$$

Here it can be seen that the angular momentum consists of two parts—the moment of the linear momentum $m\mathbf{v}_G$ of the body about point A added (vectorially) to the angular momentum $\mathbf{H}_G$. Equation 21–9 may also be used for computing the angular momentum of the body about a fixed point O; the results, of course, will be the same as those computed using the more convenient Eq. 21–7.

Rectangular Components of H. To make practical use of Eqs. 21–7 through 21–9, the angular momentum must be expressed in terms of its scalar components. For this purpose, it is convenient to choose a second set of x, y, z axes having an arbitrary orientation relative to the X, Y, Z axes, Fig. 21–7, and for a general explanation, note that Eqs. 21–7 to 21–9 all contain the form

$$\mathbf{H} = \int_m \boldsymbol{\rho} \times (\boldsymbol{\omega} \times \boldsymbol{\rho})\, dm$$

Expressing $\mathbf{H}$, $\boldsymbol{\rho}$, and $\boldsymbol{\omega}$ in terms of x, y, and z components, we have

$$H_x\mathbf{i} + H_y\mathbf{j} + H_z\mathbf{k} = \int_m (x\mathbf{i} + y\mathbf{j} + z\mathbf{k}) \times [(\omega_x\mathbf{i} + \omega_y\mathbf{j} + \omega_z\mathbf{k}) \times (x\mathbf{i} + y\mathbf{j} + z\mathbf{k})]\, dm$$

Expanding the cross products and combining terms yields

$$\begin{aligned} H_x\mathbf{i} + H_y\mathbf{j} + H_z\mathbf{k} = {} & \left[\omega_x \int_m (y^2 + z^2)\, dm - \omega_y \int_m xy\, dm - \omega_z \int_m xz\, dm\right]\mathbf{i} \\ & + \left[-\omega_x \int_m xy\, dm + \omega_y \int_m (x^2 + z^2)\, dm - \omega_z \int_m yz\, dm\right]\mathbf{j} \\ & + \left[-\omega_x \int_m zx\, dm - \omega_y \int_m yz\, dm + \omega_z \int_m (x^2 + y^2)\, dm\right]\mathbf{k} \end{aligned}$$

Equating the respective **i, j, k** components and recognizing that the integrals represent the moments and products of inertia, we obtain

$$\begin{aligned} H_x &= I_{xx}\omega_x - I_{xy}\omega_y - I_{xz}\omega_z \\ H_y &= -I_{yx}\omega_x + I_{yy}\omega_y - I_{yz}\omega_z \\ H_z &= -I_{zx}\omega_x - I_{zy}\omega_y + I_{zz}\omega_z \end{aligned} \qquad (21\text{–}10)$$

These three equations represent the scalar form of the **i, j,** and **k** components of $\mathbf{H}_O$ or $\mathbf{H}_G$ (given in vector form by Eqs. 21–7 and 21–8). The angular momentum of the body about the arbitrary point A, other than the fixed point O or the center of mass G, may also be expressed in scalar form. Here it is necessary to use Eq. 21–9 and to represent $\boldsymbol{\rho}_{G/A}$ and $\mathbf{v}_G$ as Cartesian vectors, carry out the cross-product operation, and substitute the components, Eqs. 21–10, for $\mathbf{H}_G$.

Equations 21–10 may be simplified further if the x, y, z coordinate axes are oriented such that they become *principal axes of inertia* for the body at the point. When these axes are used, the products of inertia $I_{xy} = I_{yz} = I_{zx} = 0$, and if the principal moments of inertia about the x, y, z axes are represented as $I_x = I_{xx}$, $I_y = I_{yy}$, and $I_z = I_{zz}$, the three components of angular momentum become

$$H_x = I_x\omega_x \qquad H_y = I_y\omega_y \qquad H_z = I_z\omega_z \qquad (21\text{–}11)$$

Principle of Impulse and Momentum. Now that the means for computing the angular momentum for a body have been presented, the *principle of impulse and momentum,* as discussed in Sec. 19.2, may be used to solve kinetics problems which involve *force, velocity, and time.* For this case, the following two vector equations are available:

$$m(\mathbf{v}_G)_1 + \Sigma \int_{t_1}^{t_2} \mathbf{F}\, dt = m(\mathbf{v}_G)_2 \qquad (21\text{–}12)$$

$$\mathbf{H}_1 + \Sigma \int_{t_1}^{t_2} \mathbf{M}_O\, dt = \mathbf{H}_2 \qquad (21\text{–}13)$$

In three dimensions each vector term can be represented by three scalar components, and therefore a total of *six scalar equations* can be written. Three equations relate the linear impulse and momentum in the x, y, z directions, and three equations relate the body's angular impulse and momentum about the x, y, z axes. Before applying Eqs. 21–12 and 21–13 to the solution of problems, the material in Secs. 19.2 and 19.3 should be reviewed.

*21.3 Kinetic Energy

In order to apply the principle of work and energy to the solution of problems involving general rigid-body motion, it is first necessary to formulate expressions for the kinetic energy of the body. In this regard, consider the rigid body shown in Fig. 21–8, which has a total mass m and center of mass located at G. The kinetic energy of the ith particle of the body having a mass m_i and velocity $\mathbf{v}_i$, measured relative to the inertial X, Y, Z frame of reference, is

$$T_i = \tfrac{1}{2} m_i v_i^2 = \tfrac{1}{2} m_i (\mathbf{v}_i \cdot \mathbf{v}_i)$$

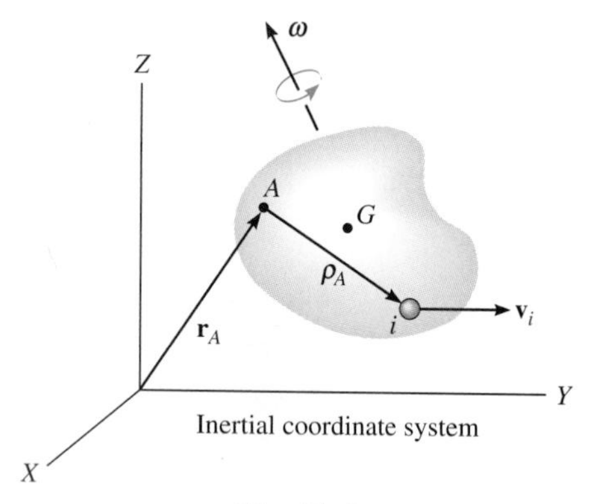

Fig. 21–8

Provided the velocity of an arbitrary point A in the body is known, $\mathbf{v}_i$ may be related to $\mathbf{v}_A$ by the equation $\mathbf{v}_i = \mathbf{v}_A + \boldsymbol{\omega} \times \boldsymbol{\rho}_A$, where $\boldsymbol{\omega}$ is the instantaneous angular velocity of the body, measured from the X, Y, Z coordinate system, and $\boldsymbol{\rho}_A$ is a position vector drawn from A to i. Using this expression for $\mathbf{v}_i$, the kinetic energy for the particle may be written as

$$\begin{aligned} T_i &= \tfrac{1}{2} m_i (\mathbf{v}_A + \boldsymbol{\omega} \times \boldsymbol{\rho}_A) \cdot (\mathbf{v}_A + \boldsymbol{\omega} \times \boldsymbol{\rho}_A) \\ &= \tfrac{1}{2}(\mathbf{v}_A \cdot \mathbf{v}_A)\, m_i + \mathbf{v}_A \cdot (\boldsymbol{\omega} \times \boldsymbol{\rho}_A)\, m_i + \tfrac{1}{2}(\boldsymbol{\omega} \times \boldsymbol{\rho}_A) \cdot (\boldsymbol{\omega} \times \boldsymbol{\rho}_A)\, m_i \end{aligned}$$

The kinetic energy for the entire body is obtained by summing the kinetic energies of all the particles of the body. This requires an integration, since $m_i \rightarrow dm$, so that

$$T = \tfrac{1}{2} m(\mathbf{v}_A \cdot \mathbf{v}_A) + \mathbf{v}_A \cdot \left(\boldsymbol{\omega} \times \int_m \boldsymbol{\rho}_A\, dm \right) + \tfrac{1}{2} \int_m (\boldsymbol{\omega} \times \boldsymbol{\rho}_A) \cdot (\boldsymbol{\omega} \times \boldsymbol{\rho}_A)\, dm$$

The last term on the right may be rewritten using the vector identity $\mathbf{a} \times \mathbf{b} \cdot \mathbf{c} = \mathbf{a} \cdot \mathbf{b} \times \mathbf{c}$, where $\mathbf{a} = \boldsymbol{\omega}$, $\mathbf{b} = \boldsymbol{\rho}_A$, and $\mathbf{c} = \boldsymbol{\omega} \times \boldsymbol{\rho}_A$. The final result is

$$T = \tfrac{1}{2}m(\mathbf{v}_A \cdot \mathbf{v}_A) + \mathbf{v}_A \cdot \left(\boldsymbol{\omega} \times \int_m \boldsymbol{\rho}_A \, dm\right)$$

$$+ \tfrac{1}{2}\boldsymbol{\omega} \cdot \int_m \boldsymbol{\rho}_A \times (\boldsymbol{\omega} \times \boldsymbol{\rho}_A) \, dm \qquad (21\text{–}14)$$

This equation is rarely used because of the computations involving the integrals. Simplification occurs, however, if the reference point A is either a fixed point O or the center of mass G.

Fixed Point O. If A is a *fixed point* O in the body, Fig. 21–7*a*, then $\mathbf{v}_A = \mathbf{0}$, and using Eq. 21–7, we can express Eq. 21–14 as

$$T = \tfrac{1}{2}\boldsymbol{\omega} \cdot \mathbf{H}_O$$

If the x, y, z axes represent the principal axes of inertia for the body, then $\boldsymbol{\omega} = \omega_x\mathbf{i} + \omega_y\mathbf{j} + \omega_z\mathbf{k}$ and $\mathbf{H}_O = I_x\omega_x\mathbf{i} + I_y\omega_y\mathbf{j} + I_z\omega_z\mathbf{k}$. Substituting into the above equation and performing the dot product operations yields

$$T = \tfrac{1}{2}I_x\omega_x^2 + \tfrac{1}{2}I_y\omega_y^2 + \tfrac{1}{2}I_z\omega_z^2 \qquad (21\text{–}15)$$

Center of Mass G. If A is located at the *center of mass* G of the body, Fig. 21–7*b*, then $\int_m \boldsymbol{\rho}_A \, dm = \mathbf{0}$ and, using Eq. 21–8, we can write Eq. 21–14 as

$$T = \tfrac{1}{2}mv_G^2 + \tfrac{1}{2}\boldsymbol{\omega} \cdot \mathbf{H}_G$$

In a manner similar to that for a fixed point, the last term on the right side may be represented in scalar form, in which case

$$T = \tfrac{1}{2}mv_G^2 + \tfrac{1}{2}I_x\omega_x^2 + \tfrac{1}{2}I_y\omega_y^2 + \tfrac{1}{2}I_z\omega_z^2 \qquad (21\text{–}16)$$

Here it is seen that the kinetic energy consists of two parts; namely, the translational kinetic energy of the mass center, $\tfrac{1}{2}mv_G^2$, and the body's rotational kinetic energy.

Principle of Work and Energy. Using one of the above expressions for computing the kinetic energy of a body, the *principle of work and energy* may be applied to solve kinetics problems which involve *force, velocity, and displacement.* For this case only one scalar equation can be written for each body, i.e.,

$$T_1 + \Sigma U_{1-2} = T_2 \qquad (21\text{–}17)$$

Before applying this equation to the solution of problems, the material in Chapter 18 should be reviewed.

Example 21–2

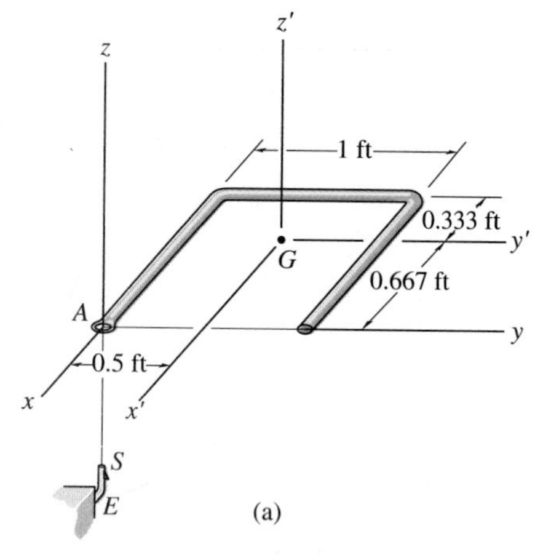

The rod in Fig. 21–9a has a weight of 1.5 lb/ft. Determine its angular velocity just after the end A falls onto the hook at E. The hook provides a permanent connection for the rod due to the spring-lock mechanism S. Just before striking the hook the rod is falling downward with a speed $(v_G)_1 =$ 10 ft/s.

SOLUTION

The principle of impulse and momentum will be used since impact occurs.

Impulse and Momentum Diagrams. Fig. 21–9b. During the short time Δt, the impulsive force $\mathbf{F}$ acting at A changes the momentum of the rod. (The impulse created by the rod's weight $\mathbf{W}$ during this time is small compared to $\int \mathbf{F}\,dt$, so that it is neglected, i.e., the weight is a nonimpulsive force.) Hence, the angular momentum of the rod is *conserved* about point A since the moment of $\int \mathbf{F}\,dt$ about A is zero.

Conservation of Angular Momentum. Equation 21–9 must be used for computing the angular momentum of the rod, since A does not become a *fixed point* until *after* the impulsive interaction with the hook. Thus, with reference to Fig. 21–9b, $(\mathbf{H}_A)_1 = (\mathbf{H}_A)_2$, or

$$\mathbf{r}_{G/A} \times m(\mathbf{v}_G)_1 = \mathbf{r}_{G/A} \times m(\mathbf{v}_G)_2 + (\mathbf{H}_G)_2 \qquad (1)$$

From Fig. 21–9a, $\mathbf{r}_{G/A} = \{-0.667\mathbf{i} + 0.5\mathbf{j}\}$ ft. Furthermore, the primed axes are principal axes of inertia for the rod because $I_{x'y'} = I_{x'z'} = I_{z'y'} =$ 0. Hence, from Eqs. 21–11, $(\mathbf{H}_G)_2 = I_{x'}\omega_x\mathbf{i} + I_{y'}\omega_y\mathbf{j} + I_{z'}\omega_z\mathbf{k}.$ The principal moments of inertia are $I_{x'} = I_{y'} = 0.0155 \text{ slug}\cdot\text{ft}^2$, $I_{z'} =$ $0.0401 \text{ slug}\cdot\text{ft}^2$ (see Prob. 21–11). Substituting into Eq. 1, we have

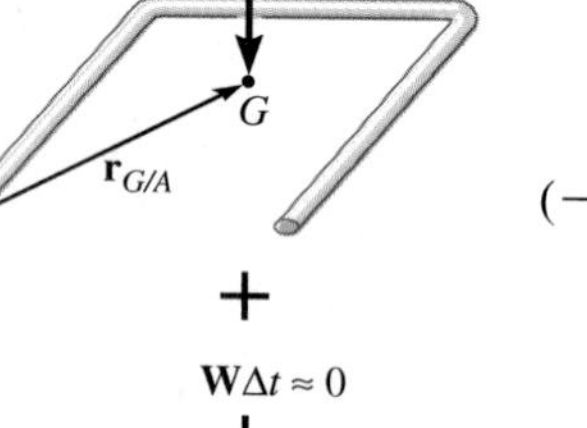

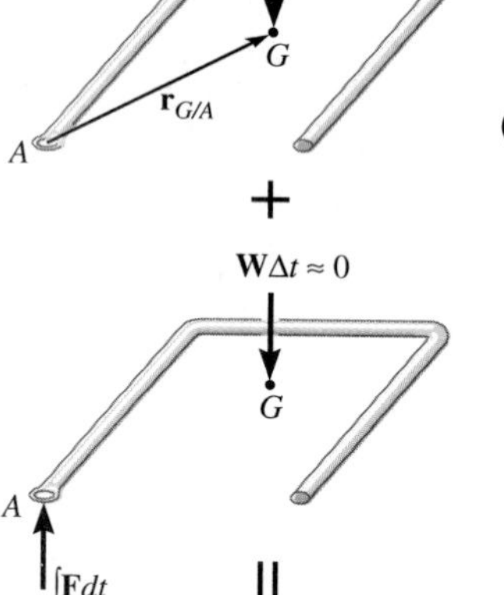

Fig. 21–9

$$(-0.667\mathbf{i} + 0.5\mathbf{j}) \times \left[\left(\frac{4.5}{32.2}\right)(-10\mathbf{k})\right] = (-0.667\mathbf{i} + 0.5\mathbf{j}) \times \left[\left(\frac{4.5}{32.2}\right)(-v_G)_2\mathbf{k}\right]$$
$$+ 0.0155\omega_x\mathbf{i} + 0.0155\omega_y\mathbf{j} + 0.0401\omega_z\mathbf{k}$$

Expanding and equating the respective **i**, **j**, and **k** components yields

$$-0.699 = -0.0699(v_G)_2 + 0.0155\omega_x \qquad (2)$$
$$-0.932 = -0.0932(v_G)_2 + 0.0155\omega_y \qquad (3)$$
$$0 = 0.0401\omega_z \qquad (4)$$

Kinematics. There are four unknowns in the above equations; however, another equation may be obtained by relating $\boldsymbol{\omega}$ to $(\mathbf{v}_G)_2$ using *kinematics.* Since $\omega_z = 0$ (Eq. 4) and after impact the rod rotates about the fixed point A, Eq. 20–3 may be applied, in which case $(\mathbf{v}_G)_2 = \boldsymbol{\omega} \times \mathbf{r}_{G/A}$, or

$$-(v_G)_2\mathbf{k} = (\omega_x\mathbf{i} + \omega_y\mathbf{j}) \times (-0.667\mathbf{i} + 0.5\mathbf{j})$$
$$-(v_G)_2 = 0.5\omega_x + 0.667\omega_y \qquad (5)$$

Solving Eqs. 2 through 5 simultaneously yields

$$(\mathbf{v}_G)_2 = \{-8.62\mathbf{k}\} \text{ ft/s} \qquad \boldsymbol{\omega} = \{-6.21\mathbf{i} - 8.29\mathbf{j}\} \text{ rad/s} \qquad \textit{Ans.}$$

Example 21–3

A 5-N · m torque is applied to the vertical shaft CD shown in Fig. 21–10a, which allows the 10-kg gear A to turn freely about CE. Assuming that gear A starts from rest, determine the angular velocity of CD after it has turned two revolutions. Neglect the mass of shaft CD and axle CE and assume that gear A can be approximated by a thin disk. Gear B is fixed.

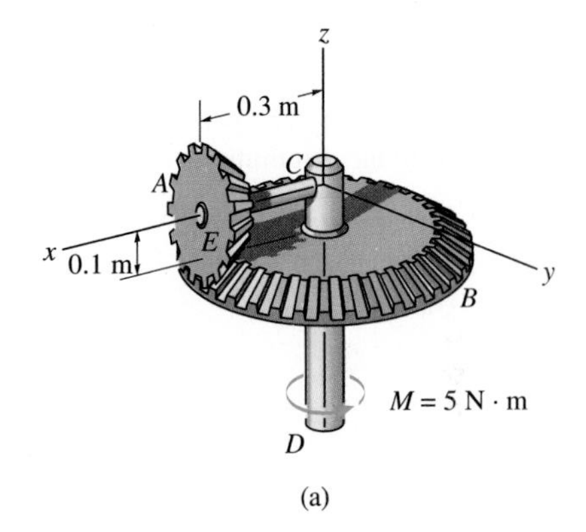

(a)

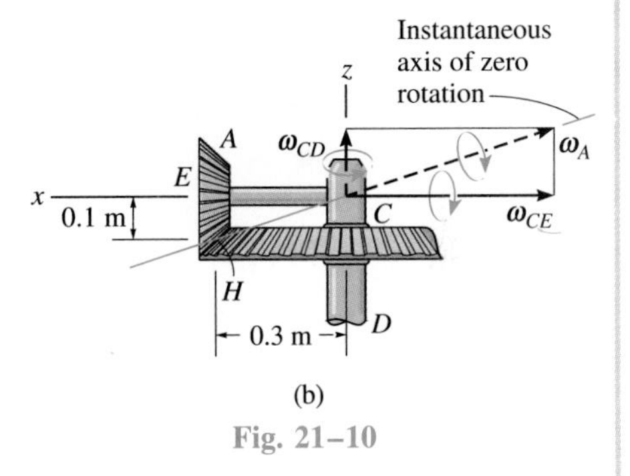

(b)

Fig. 21–10

SOLUTION

The principle of work and energy may be used for the solution. Why?

Work. If shaft CD, the axle CE, and gear A are considered as a system of connected bodies, only the applied torque $\mathbf{M}$ does work. For two revolutions of CD, this work is $\Sigma U_{1-2} = (5 \text{ N} \cdot \text{m})(4\pi \text{ rad}) = 62.83$ J.

Kinetic Energy. Since the gear is initially at rest, its initial kinetic energy is zero. A kinematic diagram for the gear is shown in Fig. 21–10b. If the angular velocity of CD is taken as $\boldsymbol{\omega}_{CD}$, then the angular velocity of gear A is $\boldsymbol{\omega}_A = \boldsymbol{\omega}_{CD} + \boldsymbol{\omega}_{CE}$. The gear may be imagined as a portion of a massless extended body which is rotating about the *fixed point C.* The instantaneous axis of rotation for this body is along line CH, because both points C and H on the body (gear) have zero velocity and must therefore lie on this axis. This requires that the components $\boldsymbol{\omega}_{CD}$ and $\boldsymbol{\omega}_{CE}$ be related by the equation $\omega_{CD}/0.1 \text{ m} = \omega_{CE}/0.3 \text{ m}$ or $\omega_{CE} = 3\omega_{CD}$. Thus,

$$\boldsymbol{\omega}_A = -\omega_{CE}\mathbf{i} + \omega_{CD}\mathbf{k} = -3\omega_{CD}\mathbf{i} + \omega_{CD}\mathbf{k} \qquad (1)$$

The x, y, z axes in Fig. 21–10a represent *principal axes of inertia* at C for the gear. Since point C is a fixed point of rotation, Eq. 21–15 may be applied to determine the kinetic energy, i.e.,

$$T = \tfrac{1}{2}I_x\omega_x^2 + \tfrac{1}{2}I_y\omega_y^2 + \tfrac{1}{2}I_z\omega_z^2 \qquad (2)$$

Using the parallel-axis theorem, the moments of inertia of the gear about point C are as follows:

$$I_x = \tfrac{1}{2}(10 \text{ kg})(0.1 \text{ m})^2 = 0.05 \text{ kg} \cdot \text{m}^2$$
$$I_y = I_z = \tfrac{1}{4}(10 \text{ kg})(0.1 \text{ m})^2 + 10 \text{ kg}(0.3 \text{ m})^2 = 0.925 \text{ kg} \cdot \text{m}^2$$

Since $\omega_x = -3\omega_{CD}$, $\omega_y = 0$, $\omega_z = \omega_{CD}$, Eq. 2 becomes

$$T_A = \tfrac{1}{2}(0.05)(-3\omega_{CD})^2 + 0 + \tfrac{1}{2}(0.925)(\omega_{CD})^2 = 0.6875\omega_{CD}^2$$

Principle of Work and Energy. Applying the principle of work and energy, we obtain

$$T_1 + \Sigma U_{1-2} = T_2 \qquad 0 + 62.83 = 0.6875\omega_{CD}^2$$
$$\omega_{CD} = 9.56 \text{ rad/s} \qquad \textit{Ans.}$$

PROBLEMS

21–25. If a body contains *no planes of symmetry*, the principal moments of inertia can be computed mathematically. To show how this is done, consider the rigid body which is spinning with an angular velocity $\boldsymbol{\omega}$, directed along one of its principal axes of inertia. If the principal moment of inertia about this axis is I, the angular momentum can be expressed as $\mathbf{H} = I\boldsymbol{\omega} = I\omega_x\mathbf{i} + I\omega_y\mathbf{j} + I\omega_z\mathbf{k}$. The components of $\mathbf{H}$ may also be expressed by Eqs. 21–10, where the inertia tensor is assumed to be known. Equate the $\mathbf{i}$, $\mathbf{j}$, and $\mathbf{k}$ components of both expressions for $\mathbf{H}$ and consider ω_x, ω_y, and ω_z to be unknown. The solution of these three equations is obtained provided the determinant of the coefficients is zero. Show that this determinant, when expanded, yields the cubic equation

$$I^3 - (I_{xx} + I_{yy} + I_{zz})I^2 + (I_{xx}I_{yy} + I_{yy}I_{zz} + I_{zz}I_{xx} - I_{xy}^2 - I_{yz}^2 - I_{zx}^2)I - (I_{xx}I_{yy}I_{zz} - 2I_{xy}I_{yz}I_{zx} - I_{xx}I_{yz}^2 - I_{yy}I_{zx}^2 - I_{zz}I_{xy}^2) = 0$$

The three positive roots of I, obtained from the solution of this equation, represent the principal moments of inertia I_x, I_y, and I_z.

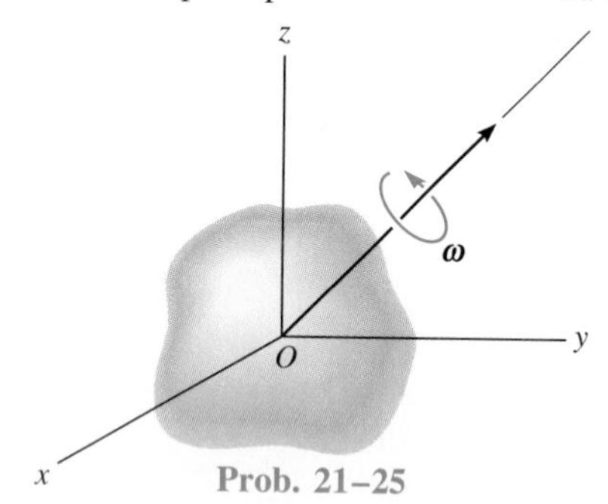

Prob. 21–25

21–26. Show that if the angular momentum of a body is computed with respect to an arbitrary point A, then $\mathbf{H}_A$ can be expressed by Eq. 21–9. This requires substituting $\boldsymbol{\rho}_A = \boldsymbol{\rho}_G + \boldsymbol{\rho}_{G/A}$ into Eq. 21–6 and expanding, noting that $\int \boldsymbol{\rho}_G \, dm = \mathbf{0}$ by definition of the mass center and $\mathbf{v}_G = \mathbf{v}_A + \boldsymbol{\omega} \times \boldsymbol{\rho}_{G/A}$.

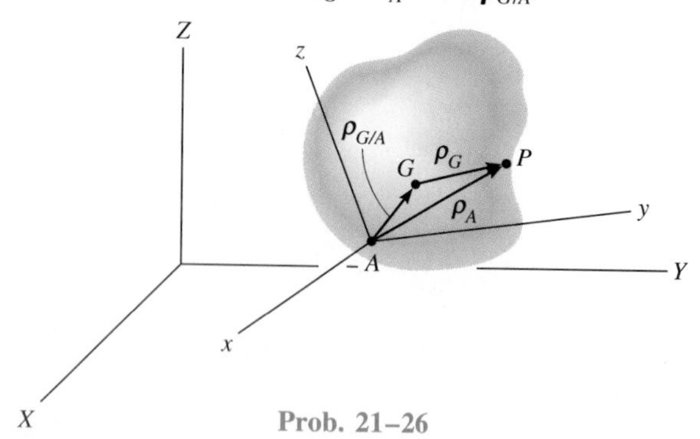

Prob. 21–26

21–27. Determine the kinetic energy of the 7-kg disk and 1.5-kg rod when the assembly is rotating about the z axis at $\omega = 5$ rad/s.

***21–28.** Determine the angular momentum $\mathbf{H}_z$ of the 7-kg disk and 1.5-kg rod when the assembly is rotating about the z axis at $\omega = 5$ rad/s.

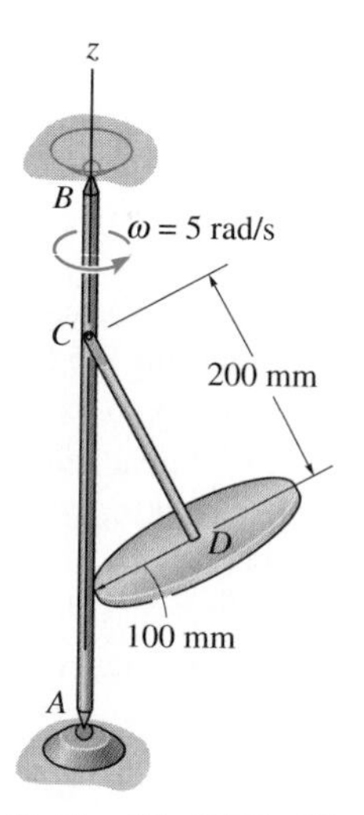

Probs. 21–27/21–28

21–29. The 2-kg gear A rolls on the fixed plate gear C. Determine the angular velocity of rod OB about the z axis after it rotates one revolution about the z axis, starting from rest. The rod is acted upon by the constant moment $M = 5$ N · m. Neglect the mass of rod OB. Assume that gear A is a uniform disk having a radius of 100 mm.

21–30. Solve Prob. 21–29 if the rod OB has a mass of 0.5 kg.

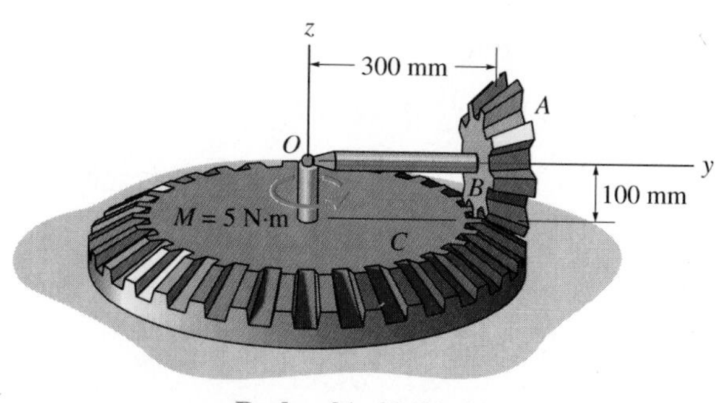

Probs. 21–29/21–30

21–31. The 5-kg thin plate is suspended at O using a ball-and-socket joint. It is rotating with a constant angular velocity $\boldsymbol{\omega} = \{2\mathbf{k}\}$ rad/s when the corner A strikes the hook at S, which provides a permanent connection. Determine the angular velocity of the plate immediately after impact.

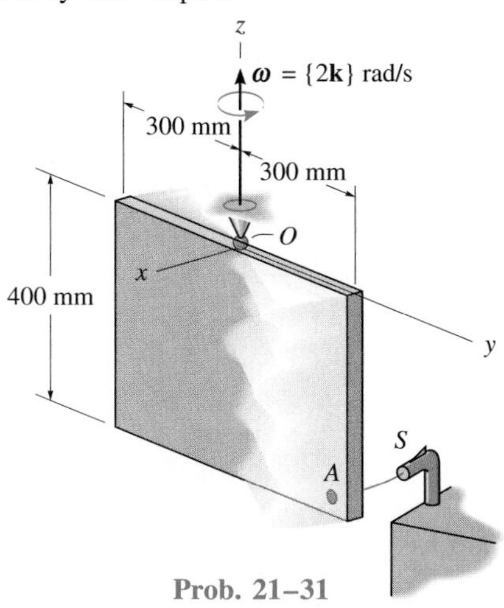

Prob. 21–31

***21–32.** Rod AB has a weight of 6 lb and is attached to two smooth collars at its end points by ball-and-socket joints. If collar A is moving downward at a speed of 8 ft/s, determine the kinetic energy of the rod at the instant shown. Assume that at this instant the angular velocity of the rod is directed perpendicular to the rod's axis.

21–33. At the instant shown the collar at A on the 6-lb rod AB has a velocity of $v_A = 8$ ft/s. Determine the kinetic energy of the rod after the collar has descended 3 ft. Neglect friction and the thickness of the rod. Neglect the mass of the collar and the collar are attached to the rod using ball-and-socket joints.

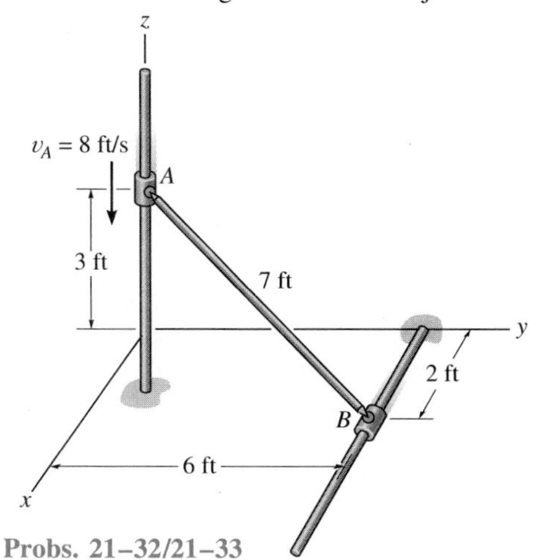

Probs. 21–32/21–33

21–34. The rod weighs 3 lb/ft and is suspended from parallel cords at A and B. If the rod has an angular velocity of 2 rad/s about the z axis at the instant shown, determine how high the center of the rod rises at the instant the rod momentarily stops swinging.

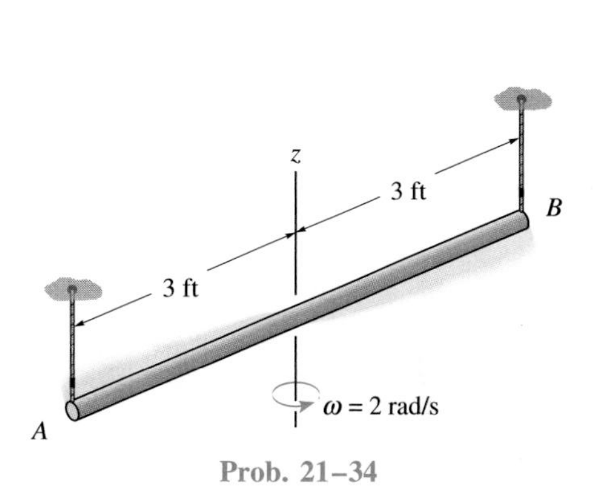

Prob. 21–34

21–35. The 4-lb rod AB is attached to the 1-lb collar at A and a 2-lb link BC using ball-and-socket joints. If the rod is released from rest in the position shown, determine the angular velocity of the link after it has rotated 180°.

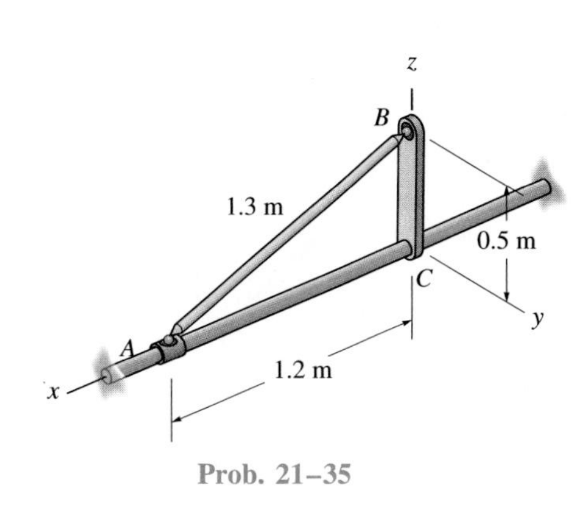

Prob. 21–35

***21–36.** The circular plate has a weight of 19 lb and a diameter of 1.5 ft. If it is released from rest and falls horizontally 2.5 ft onto the hook at S, which provides a permanent connection, determine the velocity of the mass center of the plate just after the connection with the hook is made.

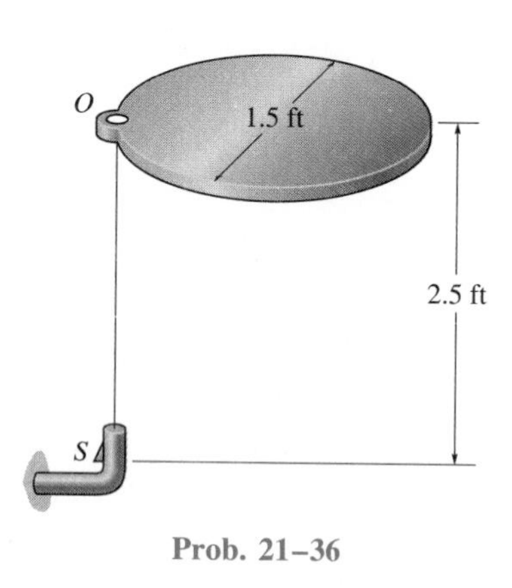

Prob. 21–36

21–37. Rod AB has a weight of 6 lb and is attached to two smooth collars at its ends by ball-and-socket joints. If collar A is moving downward with a speed of 8 ft/s when $z = 3$ ft, determine the speed of A at the instant $z = 0$. The spring has an unstretched length of 2 ft. Neglect the mass of the collars. Assume the angular velocity of rod AB is perpendicular to its axis.

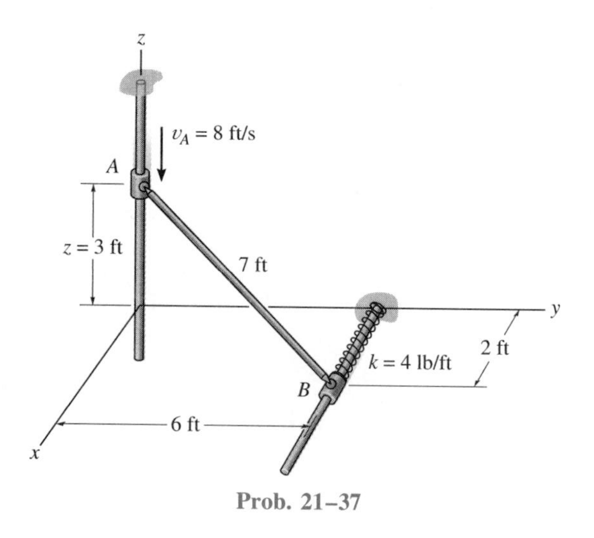

Prob. 21–37

21–38. The circular disk has a weight of 15 lb and is mounted on the shaft AB at an angle of 45° with the horizontal. Determine the angular velocity of the shaft when $t = 3$ s if a constant torque $M = 2$ lb · ft is applied to the shaft. The shaft is originally spinning at $\omega_1 = 8$ rad/s when the torque is applied.

21–39. The circular disk has a weight of 15 lb and is mounted on the shaft AB at an angle of 45° with the horizontal. Determine the angular velocity of the shaft when $t = 2$ s if a torque $M = (4e^{0.1t})$ lb · ft, where t is in seconds, is applied to the shaft. The shaft is originally spinning at $\omega_1 = 8$ rad/s when the torque is applied.

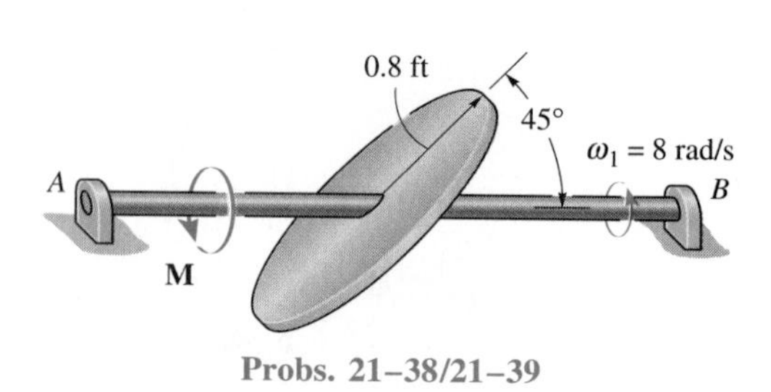

Probs. 21–38/21–39

***21–40.** The 15-kg rectangular plate is free to rotate about the y axis because of the bearing supports at A and B. When the plate is balanced in the vertical plane, a 3-g bullet is fired into it, perpendicular to its surface, with a velocity $\mathbf{v} = \{-2000\mathbf{i}\}$ m/s. Compute the angular velocity of the plate at the instant it has rotated 180°. If the bullet strikes corner D with the same velocity $\mathbf{v}$, instead of at C, does the angular velocity remain the same? Why or why not?

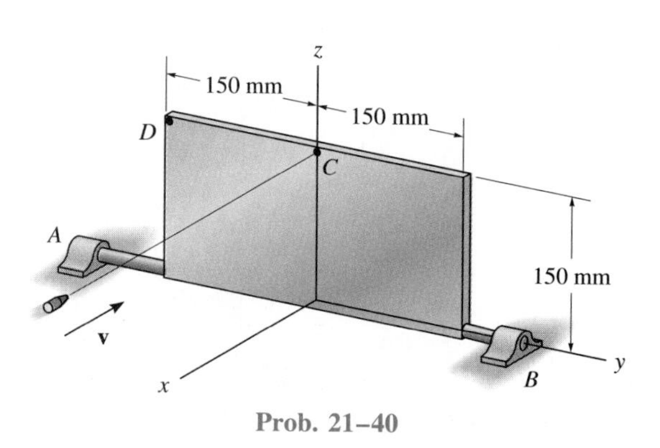

Prob. 21–40

21–41. The assembly consists of a 4-kg rod AB which is connected to link OA and the collar at B by ball-and-socket joints. When $\theta = 0°$, $y = 600$ mm, the system is at rest, the spring is unstretched, and a couple moment $M = 7$ N · m is applied to the link at O. Determine the angular velocity of the link at the instant $\theta = 90°$. Neglect the mass of the link.

21–42. The assembly consists of a 4-kg rod AB which is connected to link OA and the collar at B by ball-and-socket joints. When $\theta = 0°$, $y = 600$ mm, the system is at rest, the spring is unstretched, and a couple moment $M = (4\theta + 2)$ N · m, where θ is in radians, is applied to the link at O. Determine the angular velocity of the link at the instant $\theta = 90°$. Neglect the mass of the link.

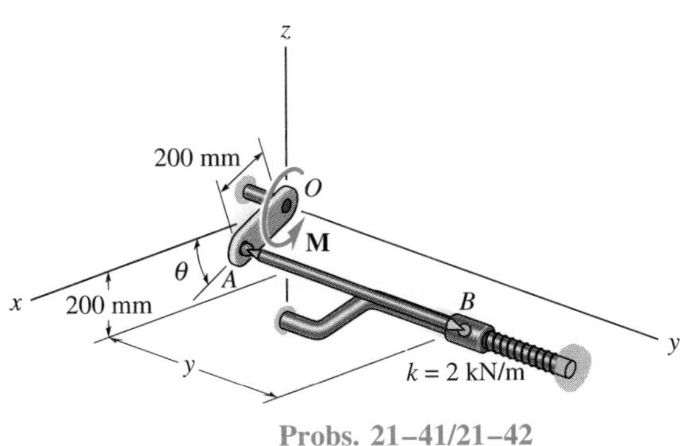

Probs. 21–41/21–42

21–43. The space capsule has a mass of 3.5 Mg and the radii of gyration are $k_x = k_z = 0.8$ m and $k_y = 0.5$ m. If it is traveling with a velocity $\mathbf{v}_G = \{600\mathbf{j}\}$ m/s, compute its angular velocity just after it is struck by a meteoroid having a mass of 0.60 kg and a velocity $\mathbf{v}_m = \{-200\mathbf{i} - 400\mathbf{j} + 200\mathbf{k}\}$ m/s. Assume that the meteoroid embeds itself into the capsule at point A and that the capsule initially has no angular velocity.

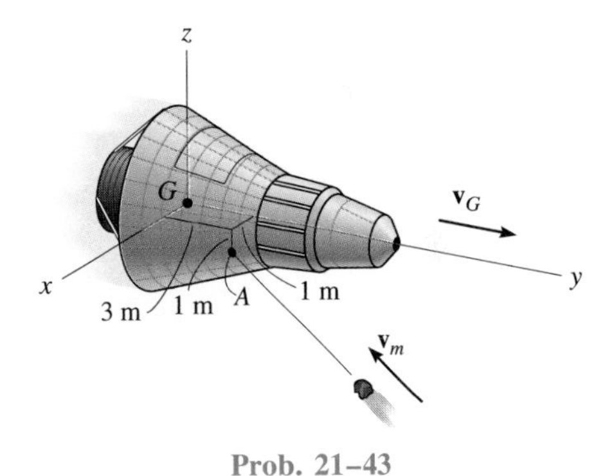

Prob. 21–43

*21.4 Equations of Motion

Having become familiar with the techniques used to describe both the inertial properties and the angular momentum of a body, we can now write the equations which describe the motion of the body in their most useful forms.

Equations of Translational Motion. The *translational motion* of a body is defined in terms of the acceleration of the body's mass center, which is measured from an inertial X, Y, Z reference. The equation of translational motion for the body can be written in vector form as

$$\Sigma\mathbf{F} = m\mathbf{a}_G \tag{21–18}$$

or by the three scalar equations

$$\begin{aligned}\Sigma F_x &= m(a_G)_x\\ \Sigma F_y &= m(a_G)_y\\ \Sigma F_z &= m(a_G)_z\end{aligned} \tag{21–19}$$

Here, $\Sigma\mathbf{F} = \Sigma F_x\mathbf{i} + \Sigma F_y\mathbf{j} + \Sigma F_z\mathbf{k}$ represents the sum of all the external forces acting on the body.

Equations of Rotational Motion. In Sec. 15.6, we developed Eq. 15–17, namely,

$$\Sigma \mathbf{M}_O = \dot{\mathbf{H}}_O \tag{21–20}$$

which states that the sum of the moments about a fixed point O of all the external forces acting on a system of particles (contained in a rigid body) is equal to the time rate of change of the total angular momentum of the body about point O. When moments of the external forces acting on the particles are summed about the system's *mass center G,* one again obtains the same simple form of Eq. 21–20, relating the moment summation $\Sigma \mathbf{M}_G$ to the angular momentum $\mathbf{H}_G$. To show this, consider the system of particles in Fig. 21–11, where x, y, z represents an inertial frame of reference and the x', y', z' axes, with origin at G, *translate* with respect to this frame. In general, G is *accelerating,* so by definition the translating frame is *not* an inertial reference. The angular momentum of the ith particle with respect to this frame is, however,

$$(\mathbf{H}_i)_G = \mathbf{r}_{i/G} \times m_i \mathbf{v}_{i/G}$$

where $\mathbf{r}_{i/G}$ and $\mathbf{v}_{i/G}$ represent the relative position and relative velocity of the ith particle with respect to G. Taking the time derivative gives

$$(\dot{\mathbf{H}}_i)_G = \dot{\mathbf{r}}_{i/G} \times m_i \mathbf{v}_{i/G} + \mathbf{r}_{i/G} \times m_i \dot{\mathbf{v}}_{i/G}$$

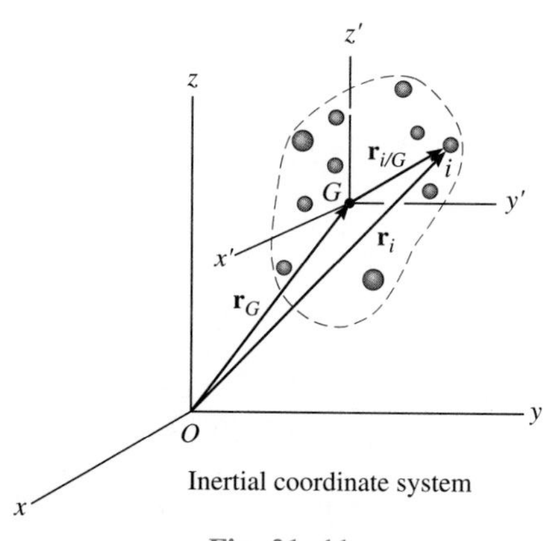

Fig. 21–11

Here $\dot{\mathbf{H}}_G$ is the time rate of change of the total angular momentum of the body computed relative to point G.

The relative acceleration for the ith particle is defined by the equation $\mathbf{a}_{i/G} = \mathbf{a}_i - \mathbf{a}_G$, where $\mathbf{a}_i$ and $\mathbf{a}_G$ represent, respectively, the accelerations of the ith particle and point G measured with respect to the *inertial frame of reference.* Substituting and expanding, using the distributive property of the vector cross product, yields

$$\dot{\mathbf{H}}_G = \Sigma(\mathbf{r}_{i/G} \times m_i\mathbf{a}_i) - (\Sigma m_i\mathbf{r}_{i/G}) \times \mathbf{a}_G$$

By definition of the mass center, the sum $(\Sigma m_i\mathbf{r}_{i/G}) = (\Sigma m_i)\bar{\mathbf{r}}$ is equal to zero, since the position vector $\bar{\mathbf{r}}$ relative to G is zero. Hence, the last term in the above equation is zero. Using the equation of motion, the product $m_i\mathbf{a}_i$ may be replaced by the resultant *external force* $\mathbf{F}_i$ acting on the ith particle. Denoting $\Sigma\mathbf{M}_G = \Sigma\mathbf{r}_{i/G} \times \mathbf{F}_i$, the final result may be written as

$$\Sigma\mathbf{M}_G = \dot{\mathbf{H}}_G \qquad (21\text{–}21)$$

The rotational equation of motion for the body will now be developed from either Eq. 21–20 or 21–21. In this regard, the scalar components of the angular momentum $\mathbf{H}_O$ or $\mathbf{H}_G$ are defined by Eqs. 21–10 or, if principal axes of inertia are used either at point O or G, by Eqs. 21–11. If these components are computed about x, y, z axes that are *rotating* with an angular velocity $\mathbf{\Omega}$, which may be *different* from the body's angular velocity $\boldsymbol{\omega}$, then the time derivative $\dot{\mathbf{H}} = d\mathbf{H}/dt$, as used in Eqs. 21–20 and 21–21, must account for the rotation of the x, y, z axes as measured from the inertial X, Y, Z axes. Hence, the time derivative of $\mathbf{H}$ must be determined from Eq. 20–6, in which case Eqs. 21–20 and 21–21 become

$$\Sigma\mathbf{M}_O = (\dot{\mathbf{H}}_O)_{xyz} + \mathbf{\Omega} \times \mathbf{H}_O$$
$$\Sigma\mathbf{M}_G = (\dot{\mathbf{H}}_G)_{xyz} + \mathbf{\Omega} \times \mathbf{H}_G$$

Here $(\dot{\mathbf{H}})_{xyz}$ is the time rate of change of $\mathbf{H}$ measured from the x, y, z reference.

There are three ways in which one can define the motion of the x, y, z axes. Obviously, motion of this reference should be chosen to yield the simplest set of moment equations for the solution of a particular problem.

x, y, z Axes Having Motion $\mathbf{\Omega} = \mathbf{0}$. If the body has general motion, the x, y, z axes may be chosen with origin at G, such that the axes only *translate* relative to the inertial X, Y, Z frame of reference. Doing this would certainly simplify Eq. 21–22, since $\mathbf{\Omega} = \mathbf{0}$. However, the body may have a rotation $\boldsymbol{\omega}$ about these axes, and therefore the moments and products of inertia of the body would have to be expressed as *functions of time.* In most cases this would be a difficult task, so that such a choice of axes has restricted value.

x, y, z Axes Having Motion $\boldsymbol{\Omega} = \boldsymbol{\omega}$. The x, y, z axes may be chosen such that they are *fixed in and move with the body.* The moments and products of inertia of the body relative to these axes will be *constant* during the motion. Since $\boldsymbol{\Omega} = \boldsymbol{\omega}$, Eqs. 21–22 become

$$
\begin{aligned}
\Sigma \mathbf{M}_O &= (\dot{\mathbf{H}}_O)_{xyz} + \boldsymbol{\omega} \times \mathbf{H}_O \\
\Sigma \mathbf{M}_G &= (\dot{\mathbf{H}}_G)_{xyz} + \boldsymbol{\omega} \times \mathbf{H}_G
\end{aligned}
\tag{21–23}
$$

We may express each of these vector equations as three scalar equations using Eqs. 21–10. Neglecting the subscripts O and G yields

$$
\begin{aligned}
\Sigma M_x &= I_{xx}\,\dot{\omega}_x - (I_{yy} - I_{zz})\omega_y\omega_z - I_{xy}(\dot{\omega}_y - \omega_z\omega_x) - I_{yz}(\omega_y^2 - \omega_z^2) \\
&\qquad - I_{zx}\,(\dot{\omega}_z + \omega_x\omega_y) \\
\Sigma M_y &= I_{yy}\,\dot{\omega}_y - (I_{zz} - I_{xx})\omega_z\omega_x - I_{yz}\,(\dot{\omega}_z - \omega_x\omega_y) - I_{zx}(\omega_z^2 - \omega_x^2) \\
&\qquad - I_{xy}\,(\dot{\omega}_x + \omega_y\omega_z) \\
\Sigma M_z &= I_{zz}\,\dot{\omega}_z - (I_{xx} - I_{yy})\omega_x\omega_y - I_{zx}\,(\dot{\omega}_x - \omega_y\omega_z) - I_{xy}(\omega_x^2 - \omega_y^2) \\
&\qquad - I_{yz}(\dot{\omega}_y + \omega_z\omega_x)
\end{aligned}
\tag{21–24}
$$

Notice that for a rigid body symmetric with respect to the xy reference plane, and undergoing general plane motion in this plane, $I_{xz} = I_{yz} = 0$, and $\omega_x = \omega_y = d\omega_x/dt = d\omega_y/dt = 0$. Equations 21–24 reduce to the form $\Sigma M_x = \Sigma M_y = 0$, and $\Sigma M_z = I_{zz}\alpha_z$ (where $\alpha_z = \dot{\omega}_z$), which is essentially the third of Eqs. 17–11 or 17–16 depending on the choice of point G or O for summing moments.

If the x, y, and z axes are chosen as *principal axes of inertia,* the products of inertia are zero, $I_{xx} = I_x$, etc., and Eqs. 21–24 reduce to the form

$$
\begin{aligned}
\Sigma M_x &= I_x\dot{\omega}_x - (I_y - I_z)\omega_y\omega_z \\
\Sigma M_y &= I_y\dot{\omega}_y - (I_z - I_x)\omega_z\omega_x \\
\Sigma M_z &= I_z\dot{\omega}_z - (I_x - I_y)\omega_x\omega_y
\end{aligned}
\tag{21–25}
$$

This set of equations is known historically as the *Euler equations of motion,* named after the Swiss mathematician Leonhard Euler, who first developed them. They apply *only* for moments summed about either point O or G.

When applying these equations it should be realized that $\dot{\omega}_x$, $\dot{\omega}_y$, and $\dot{\omega}_z$ represent the time derivatives of the magnitudes of the x, y, z components of $\boldsymbol{\omega}$ as observed from x, y, z. Since the x, y, z axes are rotating at $\boldsymbol{\Omega} = \boldsymbol{\omega}$, then, from Eq. 20–6, it may be noted that $\dot{\boldsymbol{\omega}} = (\dot{\boldsymbol{\omega}})_{xyz} + \boldsymbol{\omega} \times \boldsymbol{\omega}$. Since $\boldsymbol{\omega} \times \boldsymbol{\omega} = \mathbf{0}$, $\dot{\boldsymbol{\omega}} = (\dot{\boldsymbol{\omega}})_{xyz}$. This important result indicates that the required time derivative of $\boldsymbol{\omega}$ can be obtained either by first finding the components of $\boldsymbol{\omega}$ along the x, y, z axes and then taking the time derivative of the magnitudes of these components, i.e., $(\dot{\boldsymbol{\omega}})_{xyz}$, or by finding the time derivative of $\boldsymbol{\omega}$ with respect to the X, Y, Z axes, i.e., $\dot{\boldsymbol{\omega}}$, and then determining the components $\dot{\omega}_x$, $\dot{\omega}_y$, and $\dot{\omega}_z$. In practice, it is generally easier to compute $\dot{\omega}_x$, $\dot{\omega}_y$, and $\dot{\omega}_z$ on the basis of finding $\dot{\boldsymbol{\omega}}$. See Example 21–5.

x, y, z Axes Having Motion $\boldsymbol{\Omega} \neq \boldsymbol{\omega}$. To simplify the calculations for the time derivative of $\boldsymbol{\omega}$, it is often convenient to choose the x, y, z axes having an angular velocity $\boldsymbol{\Omega}$ which is different from the angular velocity $\boldsymbol{\omega}$ of the body. This is particularly suitable for the analysis of spinning tops and gyroscopes which are *symmetrical* about their spinning axis.* When this is the case, the moments and products of inertia remain constant during the motion.

Equations 21–22 are applicable for such a set of chosen axes. Each of these two vector equations may be reduced to a set of three scalar equations which are derived in a manner similar to Eqs. 21–25,† i.e.,

$$
\begin{aligned}
\Sigma M_x &= I_x\dot{\omega}_x - I_y\Omega_z\omega_y + I_z\Omega_y\omega_z \\
\Sigma M_y &= I_y\dot{\omega}_y - I_z\Omega_x\omega_z + I_x\Omega_z\omega_x \\
\Sigma M_z &= I_z\dot{\omega}_z - I_x\Omega_y\omega_x + I_y\Omega_x\omega_y
\end{aligned}
\qquad (21\text{–}26)
$$

Here Ω_x, Ω_y, Ω_z represent the x, y, z components of $\boldsymbol{\Omega}$, measured from the inertial frame of reference.

Any one of these sets of moment equations, Eqs. 21–24, 21–25, or 21–26, represents a series of three first-order nonlinear differential equations. These equations are "coupled," since the angular-velocity components are present in all the terms. Success in determining the solution for a particular problem therefore depends upon what is unknown in these equations. Difficulty certainly arises when one attempts to solve for the unknown components of $\boldsymbol{\omega}$, given the external moments as functions of time. Further complications can arise if the moment equations are coupled to the three scalar equations of translation, Eqs. 21–19. This can happen because of the existence of kinematic constraints which relate the rotation of the body to the translation of its mass center, as in the case of a hoop which rolls without slipping. Problems necessitating the simultaneous solution of differential equations generally require application of numerical methods with the aid of a computer. In many engineering problems, however, one is required to determine the applied moments acting on the body, given information about the motion of the body. Fortunately, many of these types of problems have direct solutions, so that there is no need to resort to computer techniques.

*A detailed discussion of such devices is given in Sec. 21.5.
†See Prob. 21–37.

PROCEDURE FOR ANALYSIS

The following procedure provides a method for solving problems involving the three-dimensional motion of a rigid body.

Free-Body Diagram. Draw a *free-body diagram* of the body at the instant considered and specify the x, y, z coordinate system. The origin of this reference must be located either at the body's mass center G, or at point O, considered fixed in an inertial reference frame and located either in the body or on a massless extension of the body. Depending on the nature of the problem, a decision should be made as to what type of rotational motion $\mathbf{\Omega}$ this coordinate system should have, i.e., $\mathbf{\Omega} = \mathbf{0}$, $\mathbf{\Omega} = \boldsymbol{\omega}$, or $\mathbf{\Omega} \neq \boldsymbol{\omega}$. When deciding, one should keep in mind that the moment equations are simplified when the axes move in such a manner that they represent principal axes of inertia for the body at all times. Also, compute the necessary moments and products of inertia for the body relative to the x, y, z axes.

Kinematics. Determine the components of the body's angular velocity and angular acceleration. In some cases a *kinematic diagram* may be helpful, since it provides a graphical aid for determining the acceleration of the body's mass center and the body's angular acceleration. The angular velocity can usually be determined from the constraints of the body or from the given problem data. The components of $\boldsymbol{\omega}$ are observed from the inertial X, Y, Z axes. Also compute the time derivative of $\boldsymbol{\omega}$ as observed from the x, y, z coordinates. In particular, note that if $\mathbf{\Omega} = \boldsymbol{\omega}$, then, from Eq. 20–6, this derivative can also be computed from the inertial X, Y, Z frame of reference, since $\dot{\boldsymbol{\omega}} = (\dot{\boldsymbol{\omega}})_{xyz}$.

Equations of Motion. Apply either the two vector equations 21–18 and 21–22, or the six scalar component equations appropriate for the x, y, z coordinate axes chosen for the problem.

The following example problems illustrate application of this procedure.

Example 21–4

The gear shown in Fig. 21–12*a* has a mass of 10 kg and is mounted at an angle of 10° with a rotating shaft having negligible mass. If $I_z = 0.1 \text{ kg}\cdot\text{m}^2$, $I_x = I_y = 0.05 \text{ kg}\cdot\text{m}^2$, and the shaft is rotating with a constant angular velocity of $\omega_{AB} = 30$ rad/s, determine the reactions that the bearing supports A and B exert on the shaft at the instant shown.

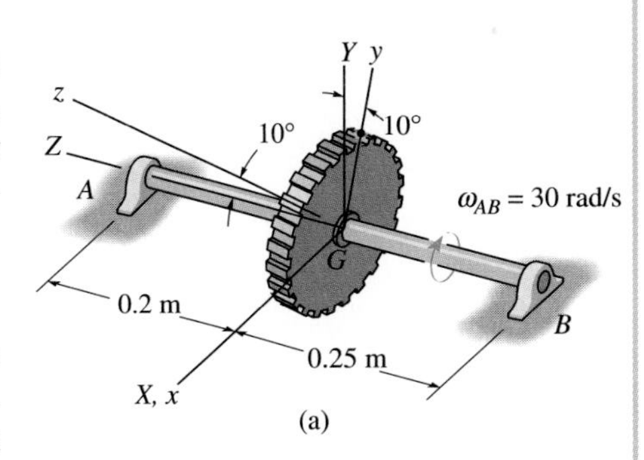

(a)

SOLUTION

Free-Body Diagram Fig. 21–12*b*. The origin of the x, y, z coordinate system is located at the gear's center of mass G, which is also a fixed point. The axes are fixed in and rotate with the gear, since these axes will then always represent the principal axes of inertia for the gear. Hence $\boldsymbol{\Omega} = \boldsymbol{\omega}$.

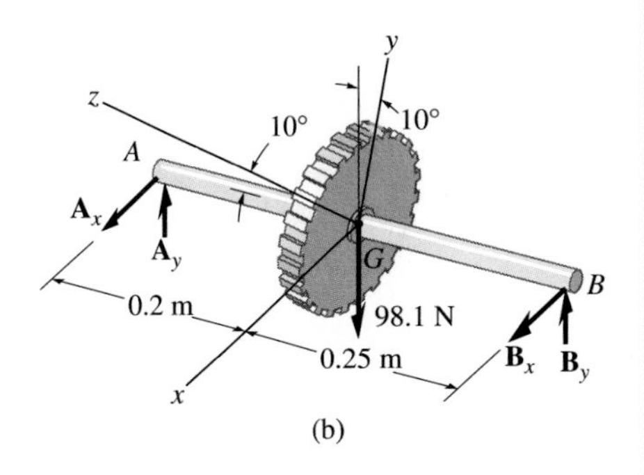

(b)

Kinematics. As shown in Fig. 21–12*c*, the angular velocity $\boldsymbol{\omega}$ of the gear is constant in magnitude and is always directed along the axis of the shaft AB. Since this vector is measured from the X, Y, Z inertial frame of reference, for any position of the x, y, z axes $\boldsymbol{\omega}$ has x, y, z components of

$$\omega_x = 0 \qquad \omega_y = -30 \sin 10° \qquad \omega_z = 30 \cos 10°$$

We will compute $(\dot{\omega})_{xyz}$ from the X, Y, Z axes since $\Omega = \omega$, so that $\dot{\omega} = (\dot{\omega})_{xyz}$. To do this we observe that $\boldsymbol{\omega}$ has a constant magnitude and direction when observed from X, Y, Z, so that $\dot{\omega} = (\dot{\omega})_{xyz} = 0$. Hence, $\dot{\omega}_x = \dot{\omega}_y = \dot{\omega}_z = 0$. Also, since G is a fixed point, $(a_G)_x = (a_G)_y = (a_G)_z = 0$.

Equations of Motion. Applying Eqs. 21–25 ($\boldsymbol{\Omega} = \boldsymbol{\omega}$) yields

$$\Sigma M_x = I_x\dot{\omega}_x - (I_y - I_z)\omega_y\omega_z$$
$$-(A_Y)(0.2) + (B_Y)(0.25) = 0 - (0.05 - 0.1)(-30 \sin 10°)(30 \cos 10°)$$
$$-0.2A_Y + 0.25B_Y = -7.70 \qquad (1)$$
$$\Sigma M_y = I_y\dot{\omega}_y - (I_z - I_x)\omega_z\omega_x$$
$$A_X(0.2)\cos 10° - B_X(0.25)\cos 10° = 0 + 0$$
$$A_X = 1.25B_X \qquad (2)$$
$$\Sigma M_z = I_z\dot{\omega}_z - (I_x - I_y)\omega_x\omega_y$$
$$A_X(0.2)\sin 10° - B_X(0.25)\sin 10° = 0$$
$$A_X = 1.25B_X$$

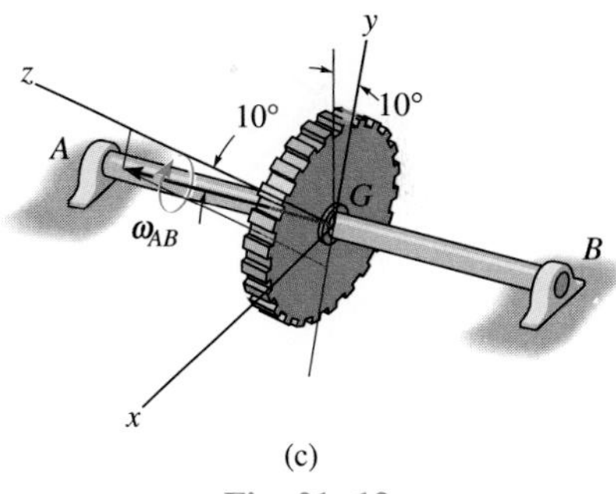

(c)

Fig. 21–12

Applying Eqs. 21–19, we have

$$\Sigma F_X = m(a_G)_X; \qquad A_X + B_X = 0 \qquad (3)$$
$$\Sigma F_Y = m(a_G)_Y; \qquad A_Y + B_Y - 98.1 = 0 \qquad (4)$$
$$\Sigma F_Z = m(a_G)_Z; \qquad 0 = 0$$

Solving Eqs. 1 through 4 simultaneously gives

$$A_X = B_X = 0 \qquad A_Y = 71.6 \text{ N} \qquad B_Y = 26.4 \text{ N} \qquad \textit{Ans.}$$

Keep in mind that these reactions will change as the gear rotates, since they depend on the components of angular velocity along the x, y, z axes, which are fixed to and rotate with the gear.

Example 21–5

The airplane shown in Fig. 21–13*a* is in the process of making a steady *horizontal* turn at the rate of ω_p. During this motion, the airplane's propeller is spinning at the rate of ω_s. If the propeller has two blades, determine the moments which the propeller shaft exerts on the propeller when the blades are in the vertical position. For simplicity, assume the blades to be a uniform slender bar having a moment of inertia I about an axis perpendicular to the blades and passing through their center, and having zero moment of inertia about a longitudinal axis.

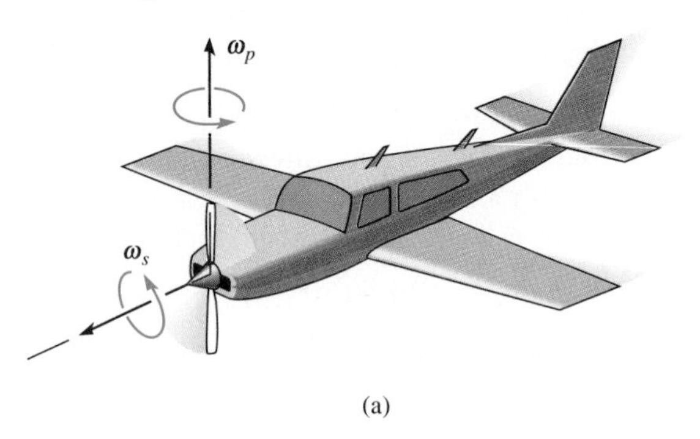

(a)

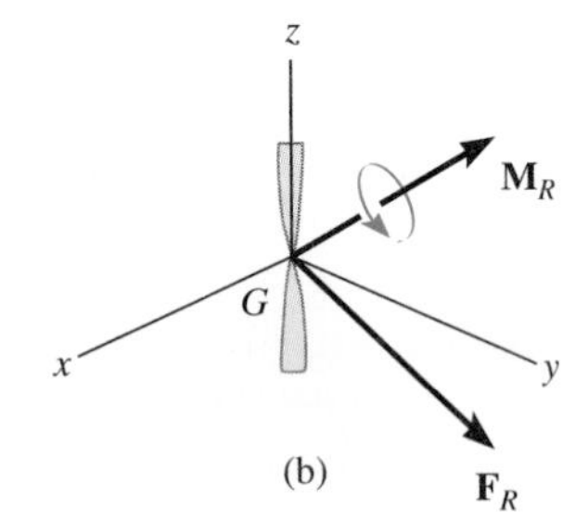

(b)

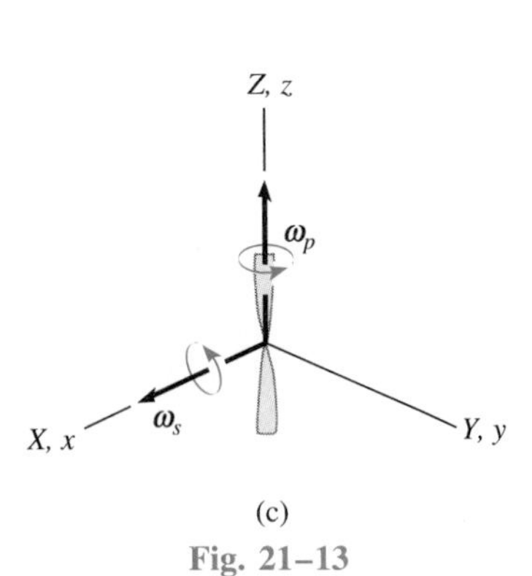

(c)

Fig. 21–13

SOLUTION

Free-Body Diagram. Fig. 21–13*b*. The effect of the connecting shaft on the propeller is indicated by the resultants $\mathbf{F}_R$ and $\mathbf{M}_R$. (The propeller's weight is assumed to be negligible.) The x, y, z axes will be taken fixed to the propeller, since these axes always represent the principal axes of inertia for the propeller. Thus, $\mathbf{\Omega} = \boldsymbol{\omega}$. The moments of inertia I_x and I_y are equal ($I_x = I_y = I$) and $I_z = 0$.

Kinematics. The angular velocity of the x, y, z axes observed from the X, Y, Z axes, coincident with the x, y, z axes, Fig. 21–13*c*, is $\boldsymbol{\omega} = \boldsymbol{\omega}_s + \boldsymbol{\omega}_p = \omega_s\mathbf{i} + \omega_p\mathbf{k}$, so that the x, y, z components of $\boldsymbol{\omega}$ are

$$\omega_x = \omega_s \qquad \omega_y = 0 \qquad \omega_z = \omega_p$$

Since $\mathbf{\Omega} = \boldsymbol{\omega}$, then $\dot{\omega} = (\dot{\omega})_{xyz}$. Hence, like Example 21–4, the time derivative of $\boldsymbol{\omega}$ will be computed with respect to the fixed X, Y, Z axes and then $\dot{\boldsymbol{\omega}}$ will be resolved into components along the moving x, y, z axes to obtain $(\dot{\omega})_{xyz}$. To do this, Eq. 20–6 must be used since $\boldsymbol{\omega}$ is changing direction relative to X, Y, Z. (Note that this was unnecessary for the case in Example 21–4.) Since $\boldsymbol{\omega} = \boldsymbol{\omega}_s + \boldsymbol{\omega}_p$, then $\dot{\boldsymbol{\omega}} = \dot{\boldsymbol{\omega}}_s + \dot{\boldsymbol{\omega}}_p$. Similar to Example 20–1, the time rate of change of each of these components relative to the X, Y, Z axes can be obtained by using a third coordinate system x', y', z', which has an angular velocity $\mathbf{\Omega}' = \boldsymbol{\omega}_p$ and is coincident with the X, Y, Z axes at the instant shown. Thus

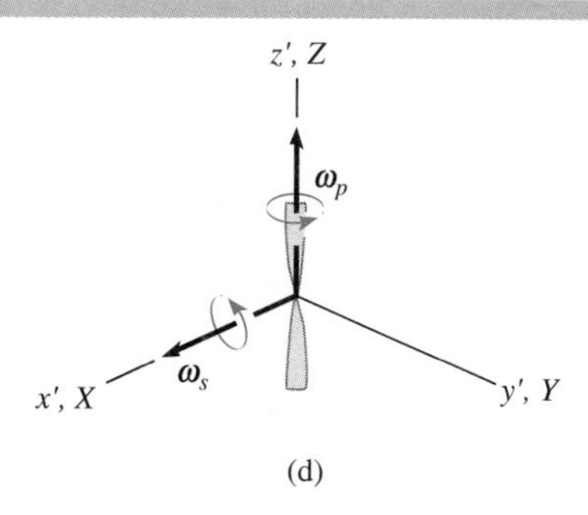

(d)

$$\begin{aligned}\dot{\boldsymbol{\omega}} &= (\dot{\boldsymbol{\omega}})_{x'y'z'} + \boldsymbol{\Omega}' \times \boldsymbol{\omega} \\ &= (\dot{\boldsymbol{\omega}}_s)_{x'y'z'} + (\dot{\boldsymbol{\omega}}_p)_{x'y'z'} + \boldsymbol{\omega}_p \times (\boldsymbol{\omega}_s + \boldsymbol{\omega}_p) \\ &= \mathbf{0} + \mathbf{0} + \boldsymbol{\omega}_p \times \boldsymbol{\omega}_s + \boldsymbol{\omega}_p \times \boldsymbol{\omega}_p \\ &= \mathbf{0} + \mathbf{0} + \omega_p \mathbf{k} \times \omega_s \mathbf{i} + \mathbf{0} = \omega_p \omega_s \mathbf{j}\end{aligned}$$

Since the X, Y, Z axes are also coincident with the x, y, z axes at the instant shown, Fig. 21–13*d*, the components of $\dot{\boldsymbol{\omega}}$ along these axes are

$$\dot{\omega}_x = 0 \qquad \dot{\omega}_y = \omega_p \omega_s \qquad \dot{\omega}_z = 0$$

These same results can, of course, also be determined by direct calculation of $(\dot{\omega})_{xyz}$. To do this, it will be necessary to view the propeller in some *general position* such as shown in Fig. 21–13*e*. Here the plane has turned through an angle ϕ and the propeller has turned through an angle ψ relative to the plane. Notice that $\boldsymbol{\omega}_p$ is always directed along the fixed Z axis and $\boldsymbol{\omega}_s$ follows the x axis. Thus the components of $\boldsymbol{\omega}$ are

$$\omega_x = \omega_s \qquad \omega_y = -\omega_p \sin\psi \qquad \omega_z = \omega_p \cos\psi$$

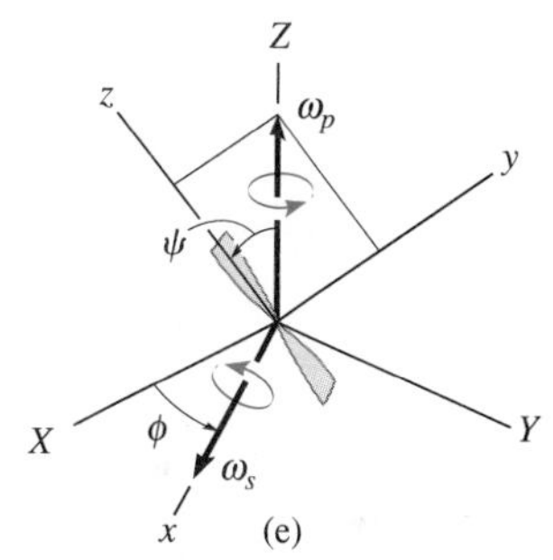

(e)

Fig. 21–13 *(cont'd)*

Since ω_s and ω_p are constant, the time derivatives of these components become

$$\dot{\omega}_x = 0 \qquad \dot{\omega}_y = \omega_p \cos\psi\,\dot{\psi} \qquad \omega_z = \omega_p \sin\psi\,\dot{\psi}$$

but $\psi = 0°$ and $\dot{\psi} = \omega_s$ at the instant considered. Thus,

$$\dot{\omega}_x = 0 \qquad \dot{\omega}_y = \omega_p \omega_s \qquad \dot{\omega}_z = 0$$

which are the same results as those computed above.

Equations of Motion. Using Eqs. 21–25, we have

$$\Sigma M_x = I_x \dot{\omega}_x - (I_y - I_z)\omega_y \omega_z = I(0) - (I - 0)(0)\omega_p$$

$$M_x = 0 \qquad \textit{Ans.}$$

$$\Sigma M_y = I_y \dot{\omega}_y - (I_z - I_x)\omega_z \omega_x = I(\omega_p \omega_s) - (0 - I)\omega_p \omega_s$$

$$M_y = 2I\omega_p \omega_s \qquad \textit{Ans.}$$

$$\Sigma M_z = I_z \dot{\omega}_z - (I_x - I_y)\omega_x \omega_y = 0(0) - (I - I)\omega_s(0)$$

$$M_z = 0 \qquad \textit{Ans.}$$

PROBLEMS

***21–44.** Derive the scalar form of the rotational equation of motion along the x axis, when $\mathbf{\Omega} \neq \boldsymbol{\omega}$ and the moments and products of inertia of the body are *not constant* with respect to time.

21–45. Derive the scalar form of the rotational equation of motion along the x axis, when $\mathbf{\Omega} \neq \boldsymbol{\omega}$ and the moments and products of inertia of the body are *constant* with respect to time.

21–46. Derive the Euler equations of motions for $\mathbf{\Omega} \neq \boldsymbol{\omega}$, i.e., Eqs. 21–26.

21–47. The 40-kg flywheel (disk) is mounted 20 mm off its true center at G. If the shaft is rotating at a constant speed $\omega = 8$ rad/s, determine the maximum reactions exerted on the journal bearings at A and B.

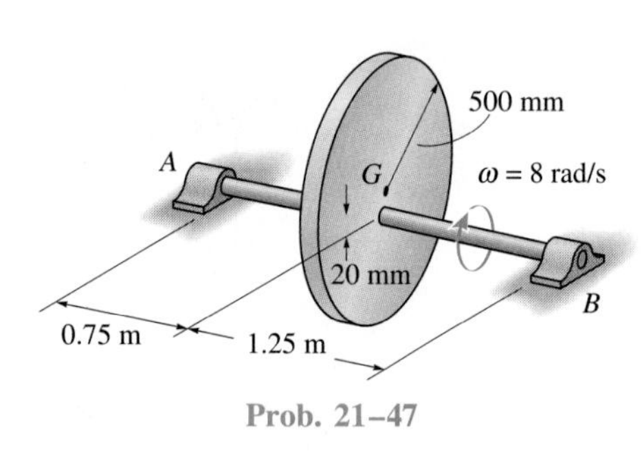

Prob. 21–47

***21–48.** The 40-kg flywheel (disk) is mounted 20 mm off its true center at G. If the shaft is rotating at a constant speed $\omega = 8$ rad/s, determine the minimum reactions exerted on the journal bearings at A and B during the motion.

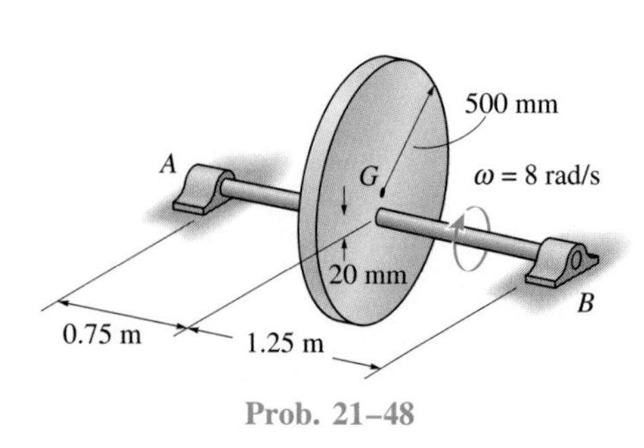

Prob. 21–48

21–49. The 4-lb bar rests along the smooth corners of an open box. At the instant shown, the box has a velocity $\mathbf{v} = \{5\mathbf{k}\}$ ft/s and an acceleration $\mathbf{a} = \{2\mathbf{k}\}$ ft/s^2. Determine the x, y, z components of force which the corners exert on the bar.

21–50. The 4-lb bar rests along the smooth corners of an open box. At the instant shown, the box has a velocity $\mathbf{v} = \{3\mathbf{j}\}$ ft/s and an acceleration $\mathbf{a} = \{-6\mathbf{j}\}$ ft/s^2. Determine the x, y, z components of force which the corners exert on the bar.

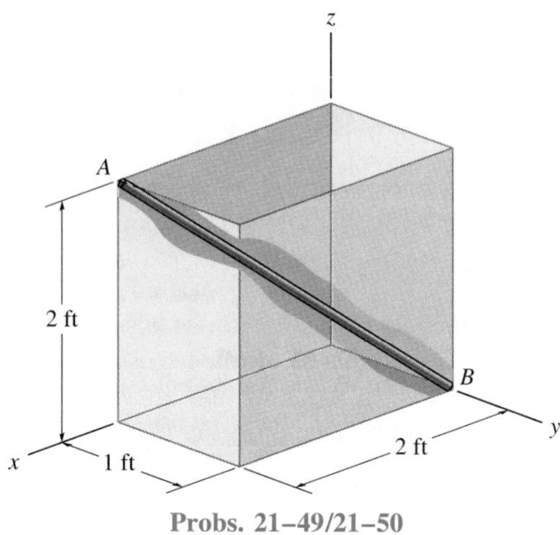

Probs. 21–49/21–50

21–51. Solve Example 21–5 if the airplane is equipped with a four-bladed propeller rather than one with two blades.

***21–52.** The 20-lb plate is mounted on the shaft AB so that the plane of the plate makes an angle $\theta = 30°$ with the vertical. If the shaft is turning in the direction shown with an angular velocity of 25 rad/s, determine the vertical reactions at the bearing supports A and B when the plate is in the position shown.

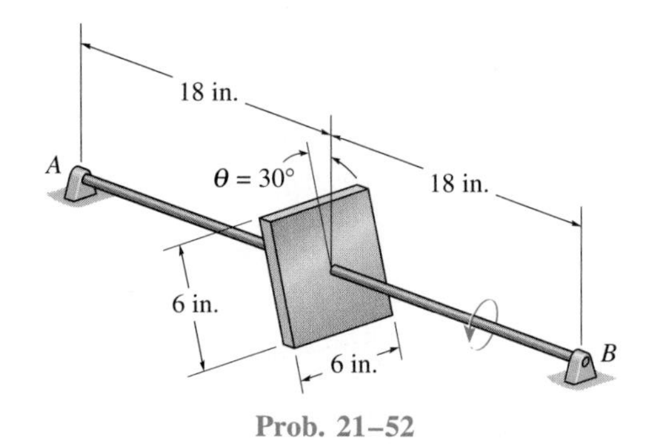

Prob. 21–52

21–53. Solve Prob. 21–52 if the shaft AB is subjected to a couple moment $M = 8\ \text{lb} \cdot \text{ft}$ acting in the direction indicated by the curl.

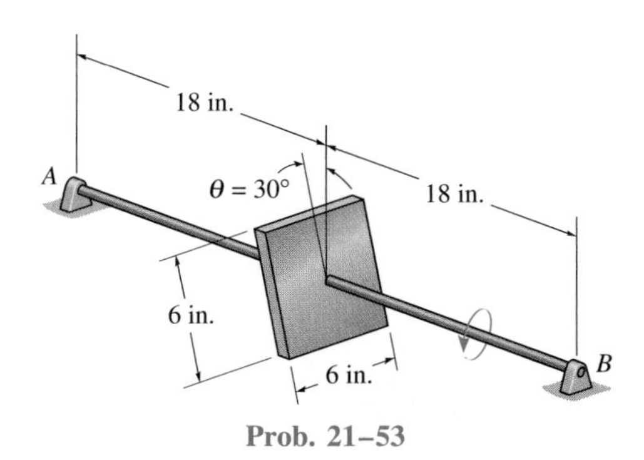

Prob. 21–53

21–54. The rod assembly is supported by a ball-and-socket joint at C and a journal bearing at D, which develops only x and y force reactions. The rods have a mass of 0.75 kg/m. Determine the angular acceleration of the rods and the components of reaction at the supports at the instant $\omega = 8$ rad/s as shown.

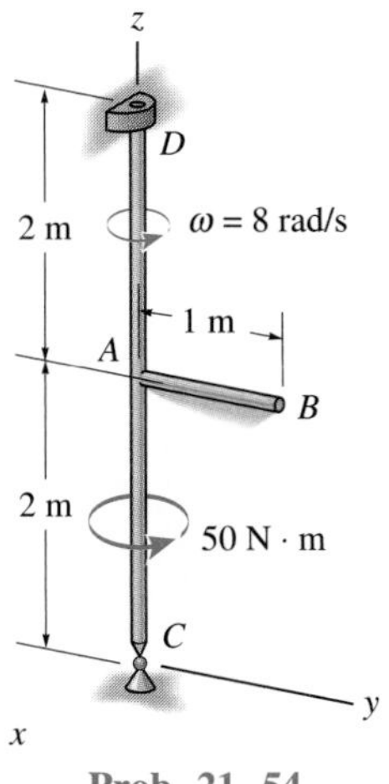

Prob. 21–54

21–55. The rod assembly is supported by journal bearings at A and B, which develops only x and y force reactions on the shaft. If the shaft AB is rotating in the direction shown at $\boldsymbol{\omega} = \{-5\mathbf{j}\}$ rad/s, determine the reactions at the bearings when the assembly is in the position shown. Also, what is the shaft's angular acceleration? The mass of each rod is 1.5 kg/m.

***21–56.** The rod assembly is supported by journal bearings at A and B, which develops only x and z force reactions on the shaft. If the shaft AB is subjected to a couple moment $\mathbf{M} = \{8\mathbf{j}\}\ \text{N} \cdot \text{m}$, and at the instant shown the shaft has an angular velocity of $\boldsymbol{\omega} = \{-5\mathbf{j}\}$ rad/s, determine the reactions at the bearings of the assembly at this instant. Also, what is the shaft's angular acceleration? The mass of each rod is 1.5 kg/m.

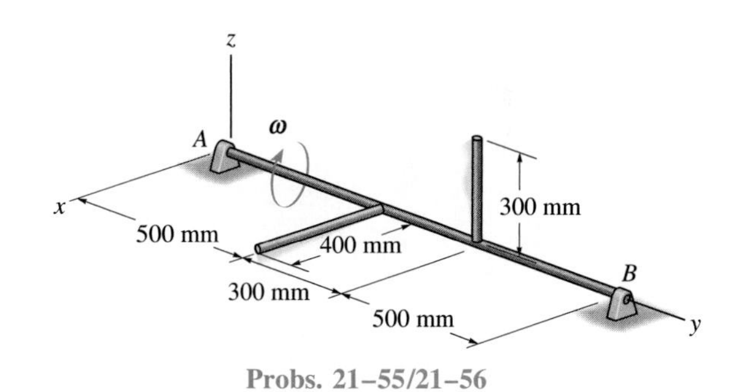

Probs. 21–55/21–56

21–57. The 5-kg rod AB is supported by a rotating shaft. The support at A is a journal bearing, which develops reactions normal to the rod. The support at B is a thrust bearing, which develops reactions both normal to the rod and perpendicular to the axis of the rod. Neglecting friction, determine the x, y, z components of reaction at these supports when the rod rotates with a constant angular velocity $\omega = 10$ rad/s.

21–58 Solve Prob. 21–57 if the shaft is given an angular acceleration $\alpha = 3\ \text{rad/s}^2$ when its angular velocity is $\omega = 10$ rad/s.

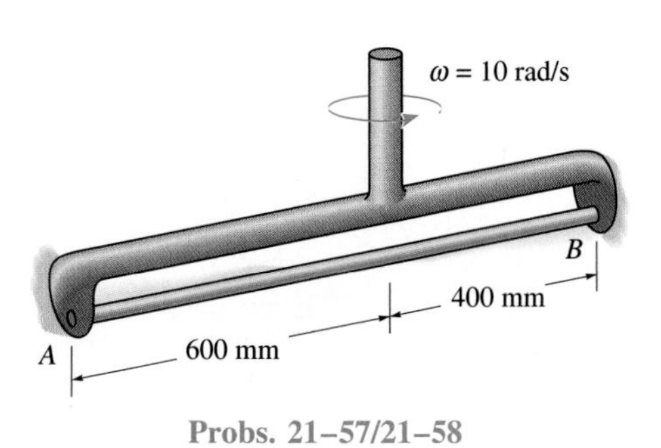

Probs. 21–57/21–58

21–59. The car is traveling around the curved road of radius e such that its mass center has a constant speed v_G. Write the equations of rotational motion with respect to the x, y, z axes. Assume that the car's six moments and products of inertia with respect to these axes are known.

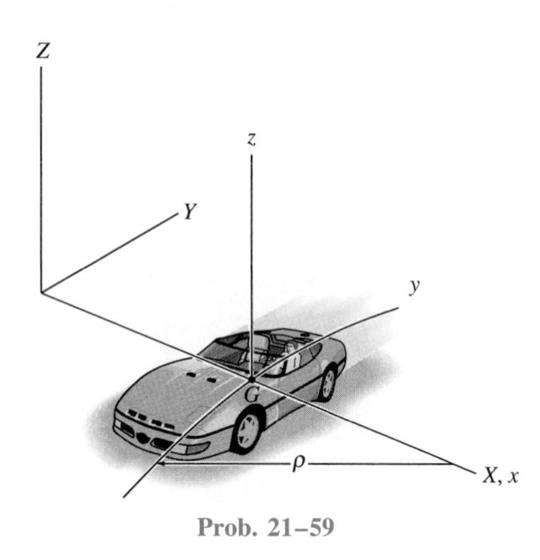

Prob. 21–59

21–61. The rod AB supports the 10-lb sphere. If the rod is pinned at A to the vertical shaft which is rotating at a constant rate $\boldsymbol{\omega} = \{7\mathbf{k}\}$ rad/s, determine the angle θ of the rod during the motion. Neglect the mass of the rod in the calculation.

21–62. The rod AB supports the 10-lb sphere. If the rod is pinned at A to the vertical shaft which is rotating with an angular acceleration $\boldsymbol{\alpha} = \{2\mathbf{k}\}$ rad/s^2, and at the instant shown the shaft has an angular velocity $\boldsymbol{\omega} = \{7\mathbf{k}\}$ rad/s, determine the angle θ of the rod at this instant. Neglect the mass of the rod in the calculation.

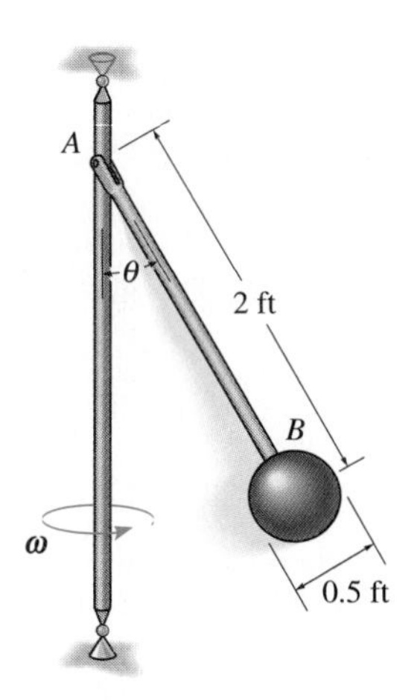

Probs. 21–61/21–62

***21–60.** The *thin rod* has a mass of 0.8 kg and a total length of 150 mm. It is rotating about its midpoint at a constant rate $\dot{\theta} = 6$ rad/s, while the table to which its axle A is fastened is rotating at 2 rad/s. Determine the x, y, z moment components which the axle exerts on the rod when the rod is in any position θ.

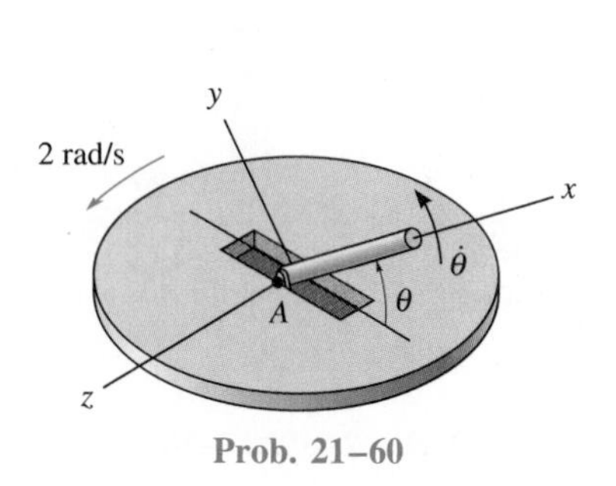

Prob. 21–60

21–63. The cylinder has a mass of 30 kg and is mounted on an axle that is supported by bearings at A and B. If the axle is turning at $\boldsymbol{\omega} = \{-40\mathbf{j}\}$ rad/s, determine the vertical components of force acting at the bearings at this instant.

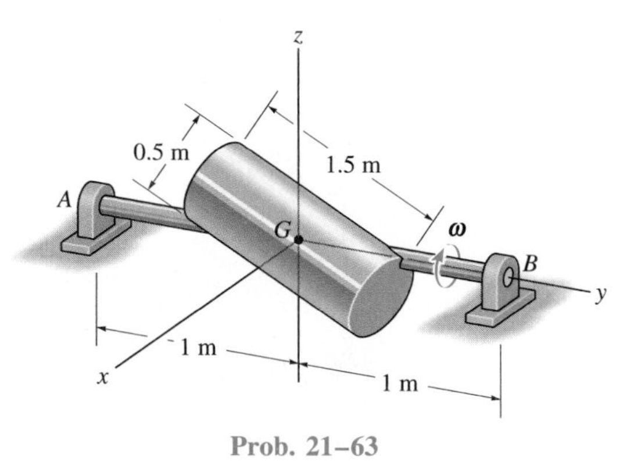

Prob. 21–63

***21–64.** The cylinder has a mass of 30 kg and is mounted on an axle which is supported by bearings at A and B. If the axle is subjected to a couple moment $\mathbf{M} = \{-30\mathbf{j}\}$ N · m, and at the instant shown has an angular velocity $\boldsymbol{\omega} = \{-40\mathbf{j}\}$ rad/s, determine the vertical components of force acting at the bearings at this instant.

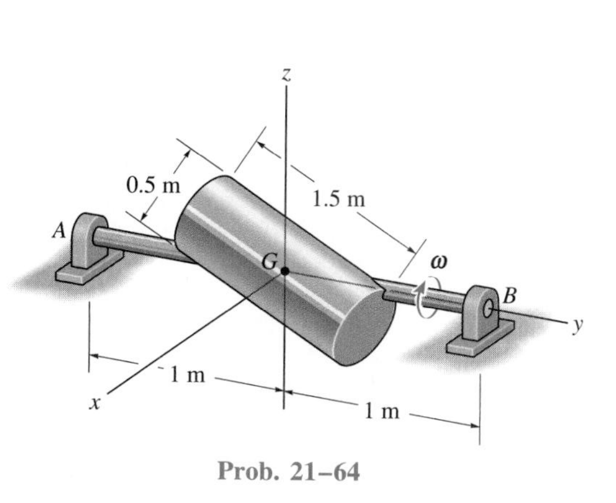

Prob. 21–64

21–65. The uniform hatch door, having a mass of 15 kg and a mass center at G, is supported in the horizontal plane by bearings at A and B. If a vertical force $F = 300$ N is applied to the door as shown, determine the components of reaction at the bearings and the angular acceleration of the door. The bearing at A will resist a component of force in the y direction, whereas the bearing at B will not. For the calculation, assume the door to be a thin plate and neglect the size of each bearing. The door is originally at rest.

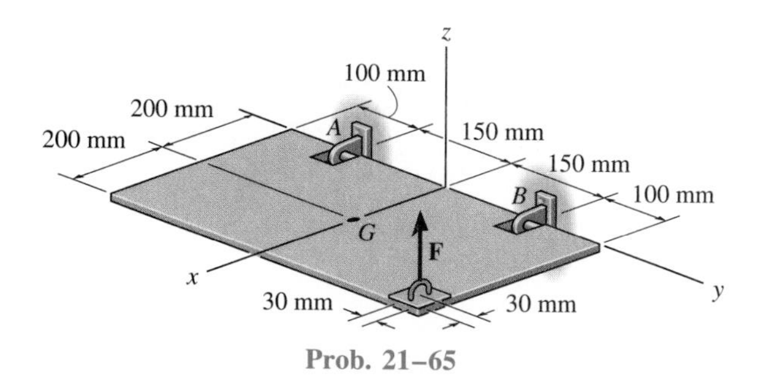

Prob. 21–65

21–66. The man sits on a swivel chair which is rotating with a constant angular velocity of 3 rad/s. He holds the uniform 5-lb rod AB horizontal. He suddenly gives it an angular acceleration of 2 rad/s², measured relative to him, as shown. Determine the required force and moment components at the grip, A, necessary to do this. Establish axes at the rod's center of mass G, with $+z$ upward, and $+y$ directed along the axis of the rod towards A.

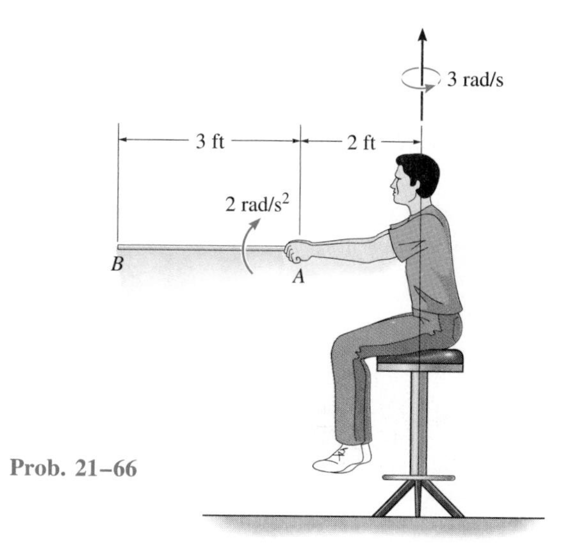

Prob. 21–66

21–67. The bent uniform rod ACD has a weight of 5 lb/ft and is supported at A by a pin and at B by a cord. If the vertical shaft rotates with a constant angular velocity $\omega = 20$ rad/s, determine the x, y, z components of force and moment developed at A and the tension in the cord.

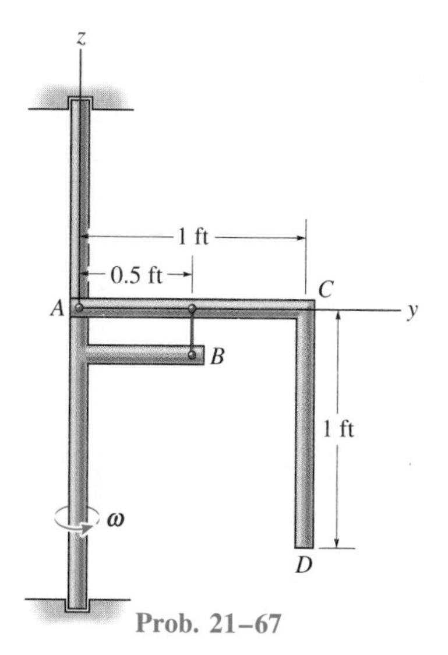

Prob. 21–67

★21.5 Gyroscopic Motion

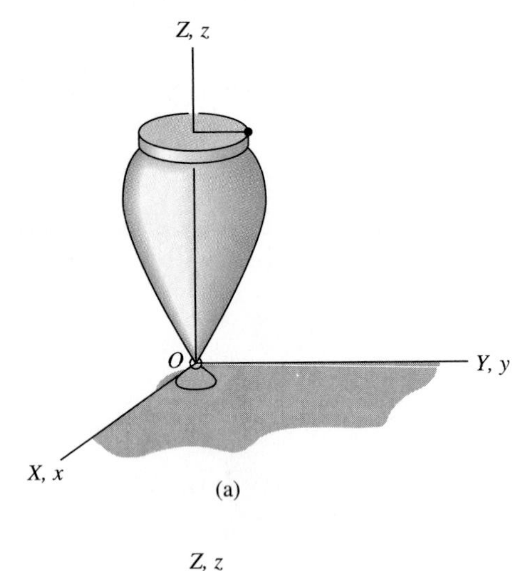

(a)

In this section the equations used for analyzing the motion of a body (or top) which is symmetrical with respect to an axis and moving about a fixed point lying on the axis will be developed. These equations will then be applied to study the motion of a particularly interesting device, the gyroscope.

The body's motion will be analyzed using *Euler angles* ϕ, θ, ψ (phi, theta, psi). To illustrate how these angles define the position of a body, reference is made to the top shown in Fig. 21–14*a*. The top is attached to point O and has an orientation relative to the fixed X, Y, Z axes at some instant of time as shown in Fig. 21–14*d*. To define this final position, a second set of x, y, z axes will be needed. For purposes of discussion, assume that this reference is fixed in the top. Starting with the X, Y, Z and x, y, z axes in coincidence, Fig. 21–14*a*, the final position of the top is determined using the following three steps:

1. Rotate the top about the Z (or z) axis through an angle ϕ ($0 \leq \phi < 2\pi$), Fig. 21–14*b*.
2. Rotate the top about the x axis through an angle θ ($0 \leq \theta \leq \pi$), Fig. 21–14*c*.
3. Rotate the top about the z axis through an angle ψ ($0 \leq \psi < 2\pi$) to obtain the final position, Fig. 20–14*d*.

The sequence of these three angles, ϕ, θ, then ψ, must be maintained, since finite rotations are *not vectors* (see Fig. 20–1). Although this is the case, the infinitesimal rotations $d\boldsymbol{\phi}$, $d\boldsymbol{\theta}$, and $d\boldsymbol{\psi}$ are vectors, and thus the angular velocity $\boldsymbol{\omega}$ of the top can be expressed in terms of the time derivatives of the Euler angles. The angular-velocity components $\dot{\phi}$, $\dot{\theta}$, and $\dot{\psi}$ are known as the *pre-*

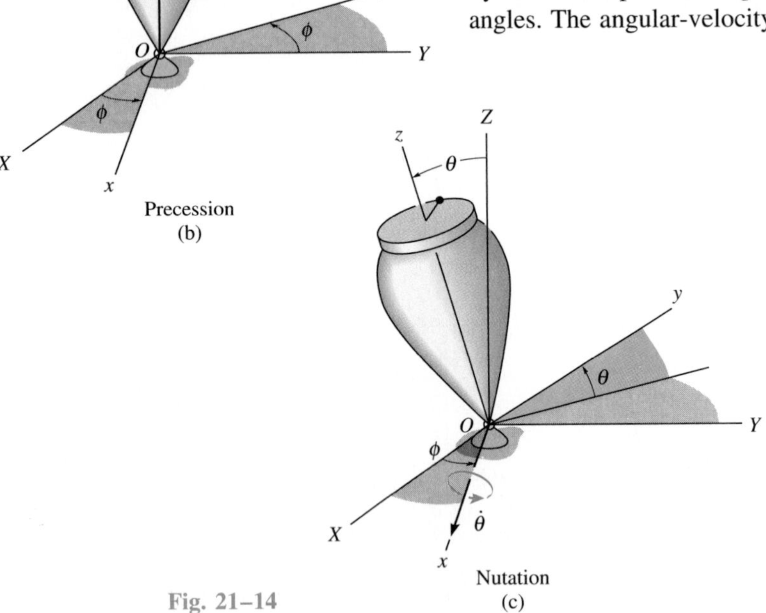

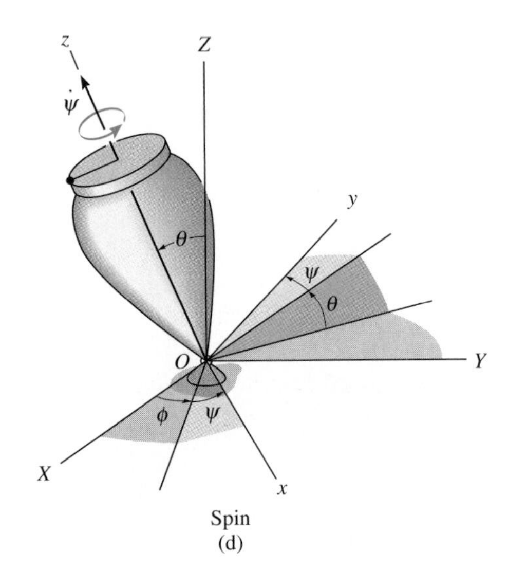

Fig. 21–14

cession, nutation, and *spin,* respectively. Their positive directions are shown in Fig. 21–14. It is seen that these vectors are not all perpendicular to one another; however, $\boldsymbol{\omega}$ of the top can still be expressed in terms of these three components.

In our case the body (top) is symmetric with respect to the z or spin axis. If we consider the top in Fig. 21–15, for which the x axis is oriented such that at the instant considered the spin angle $\psi = 0$ and *the x, y, z axes follow the motion of the body only in nutation and precession,* i.e., $\boldsymbol{\Omega} = \boldsymbol{\omega}_p + \boldsymbol{\omega}_n$, then the nutation and spin are always directed along the x and z axes, respectively. Hence, the angular velocity of the body is specified only in terms of the Euler angle θ, i.e.,

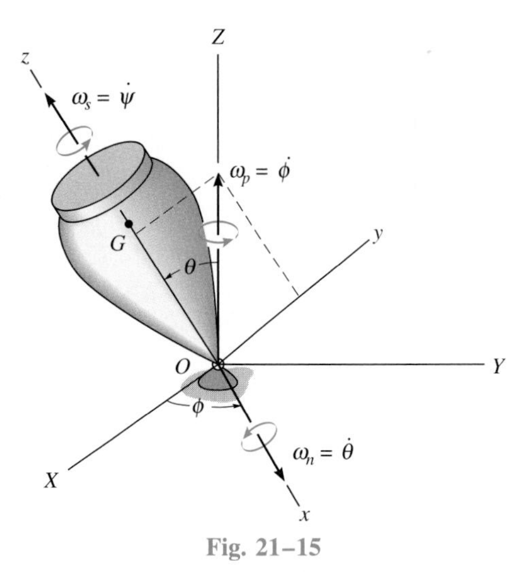

Fig. 21–15

$$\begin{aligned}\boldsymbol{\omega} &= \omega_x\mathbf{i} + \omega_y\mathbf{j} + \omega_z\mathbf{k} \\ &= \dot{\theta}\mathbf{i} + (\dot{\phi}\sin\theta)\mathbf{j} + (\dot{\phi}\cos\theta + \dot{\psi})\mathbf{k}\end{aligned} \tag{21–27}$$

Since motion of the axes is not affected by the spin component,

$$\begin{aligned}\boldsymbol{\Omega} &= \Omega_x\mathbf{i} + \Omega_y\mathbf{j} + \Omega_z\mathbf{k} \\ &= \dot{\theta}\mathbf{i} + (\dot{\phi}\sin\theta)\mathbf{j} + (\dot{\phi}\cos\theta)\mathbf{k}\end{aligned} \tag{21–28}$$

The x, y, z axes in Fig. 21–15 represent *principal axes of inertia* of the body for *any* spin of the body about these axes. Hence, the moments of inertia are constant and will be represented as $I_{xx} = I_{yy} = I$ and $I_{zz} = I_z$. Since $\boldsymbol{\Omega} \neq \boldsymbol{\omega}$, the Euler Eqs. 21–26 are used to establish the rotational equations of motion. Substituting into these equations the respective angular-velocity components defined by Eqs. 21–27 and 21–28, their corresponding time derivatives, and the moment of inertia components yields

$$\begin{aligned}\Sigma M_x &= I(\ddot{\theta} - \dot{\phi}^2\sin\theta\cos\theta) + I_z\dot{\phi}\sin\theta(\dot{\phi}\cos\theta + \dot{\psi}) \\ \Sigma M_y &= I(\ddot{\phi}\sin\theta + 2\dot{\phi}\dot{\theta}\cos\theta) - I_z\dot{\theta}(\dot{\phi}\cos\theta + \dot{\psi}) \\ \Sigma M_z &= I_z(\ddot{\psi} + \ddot{\phi}\cos\theta - \dot{\phi}\dot{\theta}\sin\theta)\end{aligned} \tag{21–29}$$

Each moment summation applies only at the fixed point O or the center of mass G of the body. Since the equations represent a coupled set of nonlinear second-order differential equations, in general a closed-form solution may not be obtained. Instead, the Euler angles ϕ, θ, and ψ may be obtained graphically as functions of time using numerical analysis and computer techniques.

A special case, however, does exist for which simplification of Eqs. 21–29 is possible. Commonly referred to as *steady precession,* it occurs when the nutation angle θ, precession $\dot{\phi}$, and spin $\dot{\psi}$ all remain *constant.* Equations 21–29 then reduce to the form

$$\Sigma M_x = -I\dot{\phi}^2\sin\theta\cos\theta + I_z\dot{\phi}\sin\theta(\dot{\phi}\cos\theta + \dot{\psi}) \tag{21–30}$$

$$\begin{aligned}\Sigma M_y &= 0 \\ \Sigma M_z &= 0\end{aligned}$$

Equation 21–30 may be further simplified by noting that, from Eq. 21–27, $\omega_z = \dot{\phi}\cos\theta + \dot{\psi}$, so that

$$\Sigma M_x = -I\dot{\phi}^2 \sin\theta\cos\theta + I_z\dot{\phi}(\sin\theta)\omega_z$$

or

$$\Sigma M_x = \dot{\phi}\sin\theta(I_z\omega_z - I\dot{\phi}\cos\theta) \qquad (21\text{–}31)$$

It is interesting to note what effects the spin $\dot{\psi}$ has on the moment about the x axis. In this regard consider the spinning rotor shown in Fig. 21–16. Here $\theta = 90°$, in which case Eq. 21–30 reduces to the form

$$\Sigma M_x = I_z\dot{\phi}\dot{\psi}$$

or

$$\Sigma M_x = I_z\Omega_y\omega_z \qquad (21\text{–}32)$$

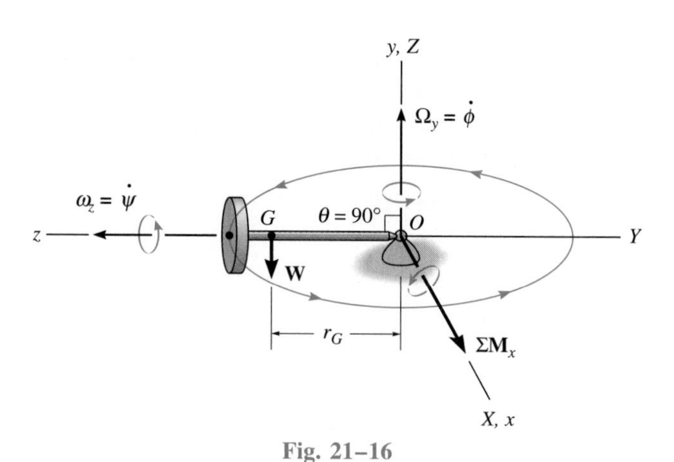

Fig. 21–16

From the figure it is seen that vectors $\Sigma\mathbf{M}_x$, $\mathbf{\Omega}_y$, and $\boldsymbol{\omega}_z$ all act along their respective *positive axes* and therefore are mutually perpendicular. Instinctively, one would expect the rotor to fall down under the influence of gravity! However, this is not the case at all, provided the product $I_z\Omega_y\omega_z$ is correctly chosen to counterbalance the moment $\Sigma M_x = Wr_G$ of the rotor's weight about O. This unusual phenomenon of rigid-body motion is often referred to as the *gyroscopic effect.*

Perhaps a more intriguing demonstration of the gyroscopic effect comes from studying the action of a *gyroscope*, frequently referred to as a *gyro*. A gyro is a rotor which spins at a very high rate about its axis of symmetry. This rate of spin is considerably greater than its precessional rate of rotation about the vertical axis. Hence, for all practical purposes, the angular momentum of the gyro can be assumed directed along its axis of spin. Thus, for the gyro rotor shown in Fig. 21–17, $\omega_z \gg \Omega_y$, and the magnitude of the angular momentum about point O, as computed by Eqs. 21–11, reduces to the form $H_O = I_z\omega_z$. Since both the magnitude and direction of $\mathbf{H}_O$ are constant as observed from x, y, z, direct application of Eq. 21–22 yields

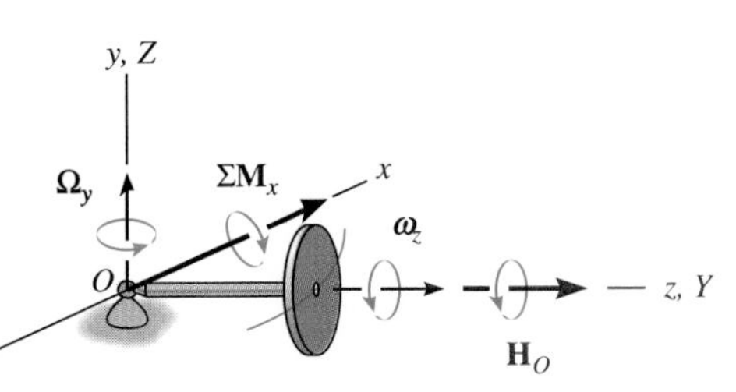

Fig. 21–17

$$\Sigma \mathbf{M}_x = \mathbf{\Omega}_y \times \mathbf{H}_O \tag{21–33}$$

Using the right-hand rule applied to the cross product, it is seen that $\mathbf{\Omega}_y$ always swings $\mathbf{H}_O$ (or $\boldsymbol{\omega}_z$) toward the sense of $\Sigma\mathbf{M}_x$. In effect, the *change in direction* of the gyro's angular momentum, $d\mathbf{H}_O$, is equivalent to the angular impulse caused by the gyro's weight about O, i.e., $d\mathbf{H}_O = \Sigma\mathbf{M}_x\, dt$, Eq. 21–20. Also, since $H_O = I_z\omega_z$ and ΣM_x, Ω_y, and H_O are mutually perpendicular, Eq. 21–33 reduces to Eq. 21–32.

When a gyro is mounted in gimbal rings, Fig. 21–18, it becomes *free* of external moments applied to its base. Thus, in theory, its angular momentum **H** will never precess but, instead, maintain its same fixed orientation along the axis of spin when the base is rotated. This type of gyroscope is called a *free gyro* and is useful as a gyrocompass when the spin axis of the gyro is directed north. In reality, the gimbal mechanism is never completely free of friction, so such a device is useful only for the local navigation of ships and aircraft. The gyroscopic effect is also useful as a means of stabilizing both the rolling motion of ships at sea and the trajectories of missiles and projectiles. Furthermore, this effect is of significant importance in the design of shafts and bearings for rotors which are subjected to forced precessions.

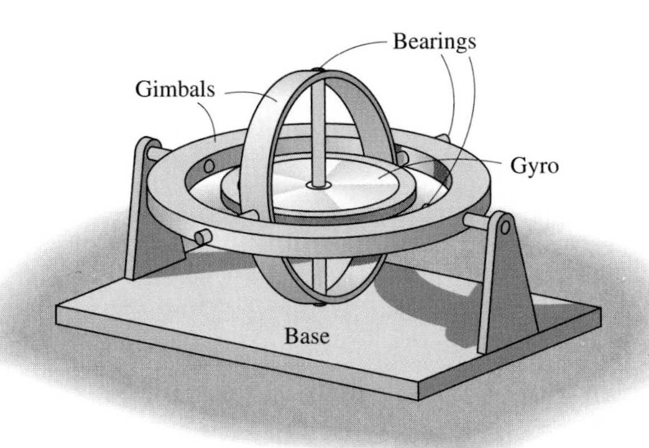

Fig. 21–18

Example 21–6

The top shown in Fig. 21–19*a* has a mass of 0.5 kg and is precessing about the vertical axis at a constant angle of $\theta = 60°$. If it spins with an angular velocity $\omega_s = 100$ rad/s, determine the precessional velocity $\boldsymbol{\omega}_p$. Assume that the axial and transverse moments of inertia of the top are $4.5(10^{-4})$ kg · m² and $12.0(10^{-4})$ kg · m², respectively, measured with respect to the fixed point O.

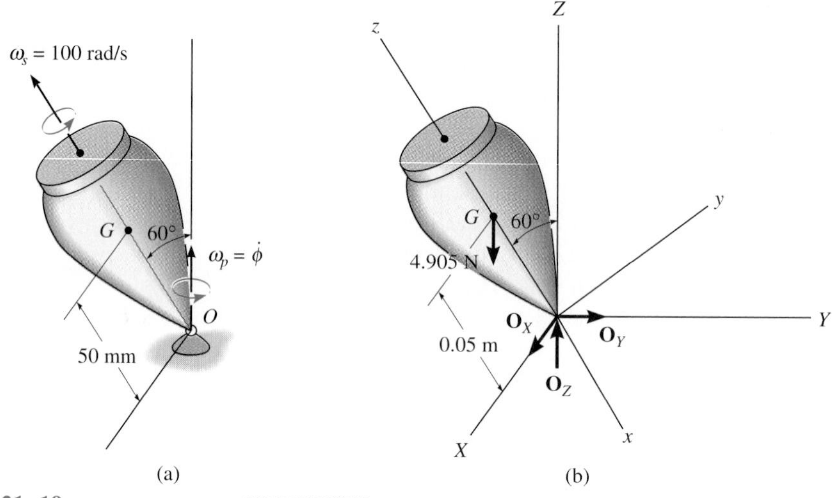

Fig. 21–19

SOLUTION

Equation 21–30 will be used for the solution since the motion is a *steady precession.* As shown on the free-body diagram, Fig. 21–19*b*, the coordinate axes are established in the usual manner, that is, with the positive z axis in the direction of spin, the positive Z axis in the direction of precession, and the positive x axis in the direction of the moment $\Sigma \mathbf{M}_x$ (refer to Fig. 21–15). Thus,

$$\Sigma M_x = -I\dot{\phi}^2 \sin\theta\cos\theta + I_z\dot{\phi}\sin\theta(\dot{\phi}\cos\theta + \dot{\psi})$$

$$4.905\text{ N}(0.05\text{ m})\sin 60° = -[12.0(10^{-4})\text{kg}\cdot\text{m}^2\,\dot{\phi}^2]\sin 60°\cos 60° + [4.5(10^{-4})\text{kg}\cdot\text{m}^2]\,\dot{\phi}\sin 60°(\dot{\phi}\cos 60° + 100\text{ rad/s})$$

or

$$\dot{\phi}^2 - 120.0\dot{\phi} + 654.0 = 0 \qquad (1)$$

Solving this quadratic equation for the precession gives

$$\dot{\phi} = 114\text{ rad/s} \qquad \text{(high precession)} \qquad \textit{Ans.}$$

and

$$\dot{\phi} = 5.72\text{ rad/s} \qquad \text{(low precession)} \qquad \textit{Ans.}$$

In reality, low precession of the top would generally be observed, since high precession would require a larger kinetic energy.

Example 21–7

The 1-kg disk shown in Fig. 21–20*a* is spinning about its axis with a constant angular velocity $\omega_D = 70$ rad/s. The block at *B* has a mass of 2 kg, and by adjusting its position *s* one can change the precession of the disk about its supporting pivot at *O*. Compute the position *s* which will enable the disk to have a constant precessional velocity $\omega_p = 0.5$ rad/s about the pivot. Neglect the weight of the shaft.

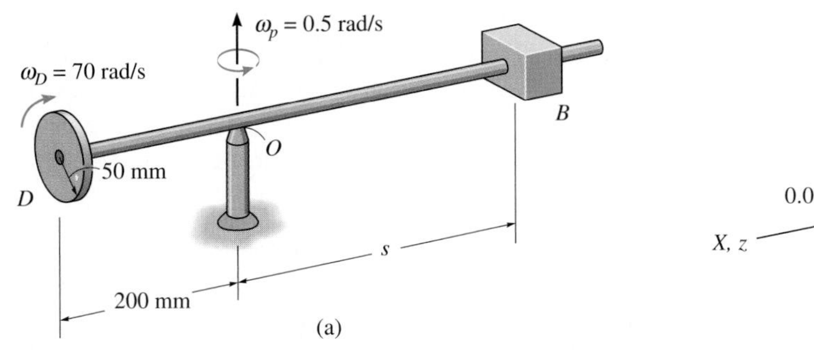

(a)

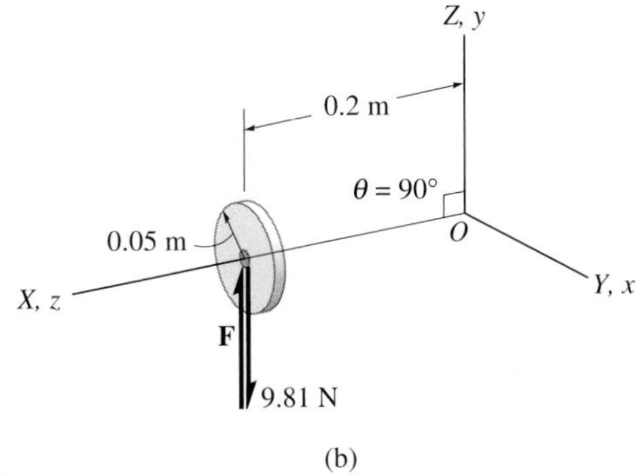

(b)

Fig. 21–20

SOLUTION

The free-body diagram of the disk is shown in Fig. 21–20*b*, where **F** represents the force reaction of the shaft on the disk. The origin for both the *x*, *y*, *z* and *X*, *Y*, *Z* coordinate systems is located at point *O*, which represents a *fixed point* for the disk. (Although point *O* does not lie on the disk, imagine a massless extension of the disk to this point.) In the conventional sense, the *Z* axis is chosen along the axis of precession, and the *z* axis is along the axis of spin, so that $\theta = 90°$. Since the precession is *steady*, Eq. 21–31 may be used for the solution. This equation reduces to

$$\Sigma M_x = \dot{\phi} I_z \omega_z$$

which is the same as Eq. 21–32. Substituting the required data gives

$$9.81 \text{ N}(0.2 \text{ m}) - F(0.2 \text{ m}) = 0.5 \text{ rad/s}[\tfrac{1}{2}(1 \text{ kg})(0.05 \text{ m})^2](-70 \text{ rad/s})$$

$$F = 10.0 \text{ N}$$

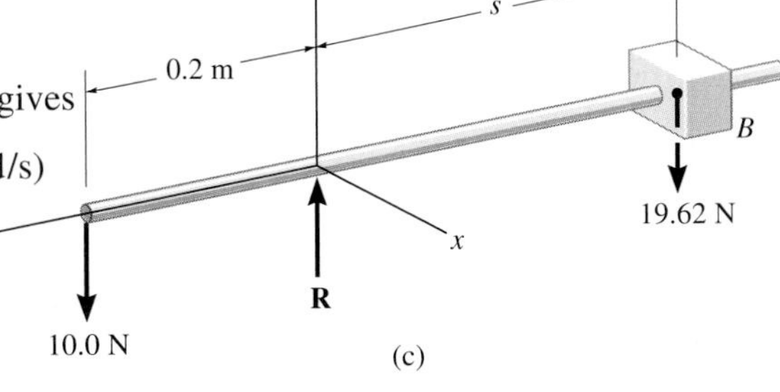

(c)

As shown on the free-body diagram of the shaft and block *B*, Fig. 21–20*c*, summing moments about the *x* axis requires

$$(19.62 \text{ N})s = (10.0 \text{ N})(0.2 \text{ m})$$

$$= 0.102 \text{ m} = 102 \text{ mm} \qquad \textit{Ans.}$$

★21.6 Torque-Free Motion

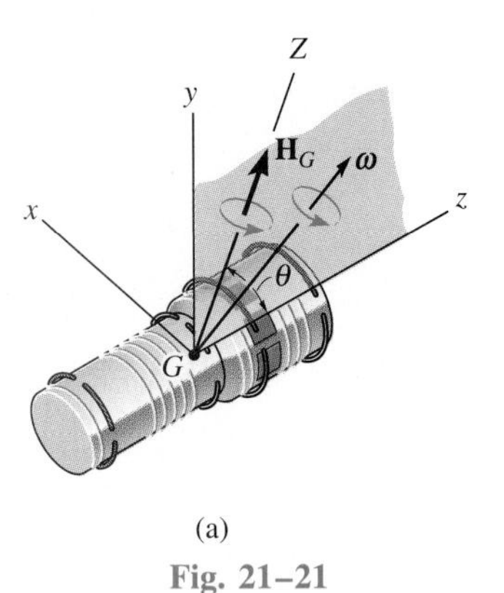

(a)

Fig. 21–21

When the only external force acting on a body is caused by gravitation, the general motion of the body is referred to as *torque-free motion.* This type of motion is characteristic of planets, artificial satellites, and projectiles—provided the effects of air friction are neglected.

In order to describe the characteristics of this motion, the distribution of the body's mass will be assumed *axisymmetric.* The satellite shown in Fig. 21–21*a* is an example of such a body, where the z axis represents an axis of symmetry. The origin of the x, y, z coordinates is located at the mass center G, such that $I_{zz} = I_z$ and $I_{xx} = I_{yy} = I$ for the body. If gravitation is the only external force present, the summation of moments about the mass center is zero. From Eq. 21–21, this requires the angular momentum of the body to be constant, i.e.,

$$\mathbf{H}_G = \text{const}$$

At the instant considered, it will be assumed that the inertial frame of reference is oriented such that the positive Z axis is directed along $\mathbf{H}_G$ and the y axis lies in the plane formed by the z and Z axes, Fig. 21–21*a*. The Euler angle formed between Z and z is θ, and therefore, with this choice of axes the angular momentum may be expressed as

$$\mathbf{H}_G = H_G \sin \theta \mathbf{j} + H_G \cos \theta \mathbf{k}$$

Furthermore, using Eqs. 21–11, we have

$$\mathbf{H}_G = I\omega_x \mathbf{i} + I\omega_y \mathbf{j} + I_z \omega_z \mathbf{k}$$

where ω_x, ω_y, ω_z represent the x, y, z components of the body's angular velocity. Equating the respective **i, j,** and **k** components of the above two equations yields

$$\omega_x = 0 \qquad \omega_y = \frac{H_G \sin \theta}{I} \qquad \omega_z = \frac{H_G \cos \theta}{I_z} \tag{21–34}$$

or

$$\boldsymbol{\omega} = \frac{H_G \sin \theta}{I} \mathbf{j} + \frac{H_G \cos \theta}{I_z} \mathbf{k} \tag{21–35}$$

In a similar manner, equating the respective **i, j,** and **k** components of Eq. 21–27 to those of Eq. 21–34, we obtain

$$\dot{\theta} = 0$$

$$\dot{\phi} \sin \theta = \frac{H_G \sin \theta}{I}$$

$$\dot{\phi} \cos \theta + \dot{\psi} = \frac{H_G \cos \theta}{I_z}$$

Solving, we get

$$\theta = \text{const}$$
$$\dot{\phi} = \frac{H_G}{I} \tag{21–36}$$
$$\dot{\psi} = \frac{I - I_z}{I I_z} H_G \cos\theta$$

Thus, for torque-free motion of an axisymmetrical body, the angle θ formed between the angular-momentum vector and the spin of the body remains constant. Furthermore, the angular momentum $\mathbf{H}_G$, precession $\dot{\phi}$, and spin $\dot{\psi}$ for the body remain constant at all times during the motion. Eliminating H_G from the second and third of Eqs. 21–36 yields the following relationship between the spin and precession:

$$\dot{\psi} = \frac{I - I_z}{I_z} \dot{\phi} \cos\theta \tag{21–37}$$

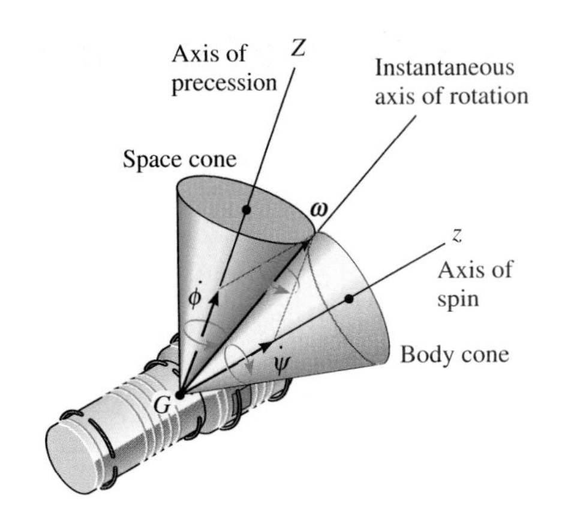

(b)
Fig. 21–21 *(cont'd)*

As shown in Fig. 21–21*b*, the body precesses about the Z axis, which is fixed in direction, while it spins about the z axis. These two components of angular motion may be studied by using a simple cone model, introduced in Sec. 20.1. The *space cone* defining the precession is fixed from rotating, since the precession has a fixed direction, while the *body cone* rotates around the space cone's outer surface without slipping. On this basis, an attempt should be made to imagine the motion. The interior angle of each cone is chosen such that the resultant angular velocity of the body is directed along the line of contact of the two cones. This line of contact represents the instantaneous axis of rotation for the body cone, and hence the angular velocity of both the body cone and the body must be directed along this line. Since the spin is a function of the moments of inertia I and I_z of the body, Eq. 21–36, the cone model in Fig. 21–21*b* is satisfactory for describing the motion, provided $I > I_z$. Torque-free motion which meets these requirements is called *regular precession.* If $I < I_z$, the spin is negative and the precession positive. This motion is represented by the satellite motion shown in Fig. 21–22 ($I < I_z$). The cone model may again be used to represent the motion; however, to preserve the correct vector addition of spin and precession to obtain the angular velocity $\boldsymbol{\omega}$, the inside surface of the body cone must roll on the outside surface of the (fixed) space cone. This motion is referred to as *retrograde precession.*

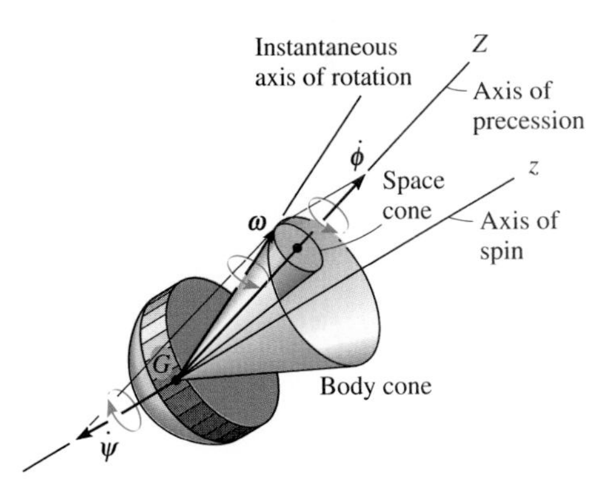

Fig. 21–22

Example 21–8

The motion of a football is observed using a slow-motion projector. From the film, the spin of the football is seen to be directed 30° from the horizontal, as shown in Fig. 21–23*a*. Also, the football is precessing about the vertical axis at a rate $\dot{\phi} = 3$ rad/s. If the ratio of the axial to transverse moments of inertia of the football is $\frac{1}{3}$, measured with respect to the center of mass, determine the magnitude of the football's spin and its angular velocity. Neglect the effect of air resistance.

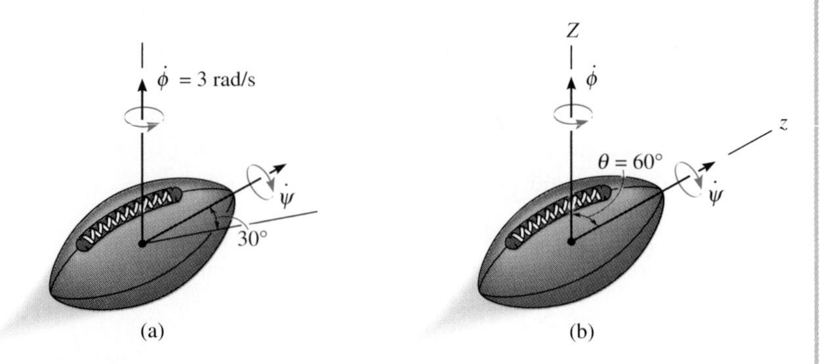

Fig. 21–23

SOLUTION

Since the weight of the football is the only force acting, the motion is torque-free. In the conventional sense, if the z axis is established along the axis of spin and the Z axis along the precession axis, as shown in Fig. 21–23*b*, then the angle $\theta = 60°$. Applying Eq. 21–37, the spin is

$$\dot{\psi} = \frac{I - I_z}{I_z}\dot{\phi}\cos\theta = \frac{I - \frac{1}{3}I}{\frac{1}{3}I}(3)\cos 60°$$

$$= 3 \text{ rad/s} \qquad \textit{Ans.}$$

Using Eqs. 21–34, where $H_G = \dot{\phi}I$ (Eq. 21–36), we have

$$\omega_x = 0$$

$$\omega_y = \frac{H_G \sin\theta}{I} = \frac{3I\sin 60°}{I} = 2.60 \text{ rad/s}$$

$$\omega_z = \frac{H_G \cos\theta}{I_z} = \frac{3I\cos 60°}{\frac{1}{3}I} = 4.50 \text{ rad/s}$$

Thus,

$$\omega = \sqrt{(\omega_x)^2 + (\omega_y)^2 + (\omega_z)^2}$$

$$= \sqrt{(0)^2 + (2.60)^2 + (4.50)^2}$$

$$= 5.20 \text{ rad/s} \qquad \textit{Ans.}$$

PROBLEMS

*21–68. An airplane descends at a steep angle and then levels off horizontally to land. If the propeller is turning clockwise when observed from the rear of the plane, determine the direction in which the plane tends to turn as caused by the gyroscopic effect as it levels off.

21–69. A thin rod is initially coincident with the Z axis when it is given three rotations defined by the Euler angles $\theta = 45°$, $\phi = 30°$, and $\psi = 60°$. If these rotations are given in the order stated, determine the coordinate direction angles α, β, γ of the axis of the rod with respect to the X, Y, and Z axes. Is this direction the same for any order of the rotations? Why?

21–70. Show that the angular velocity of a body, in terms of Euler angles ϕ, θ, and ψ, may be expressed as $\boldsymbol{\omega} = (\dot{\phi}\sin\theta\sin\psi + \dot{\theta}\cos\psi)\mathbf{i} + (\dot{\phi}\sin\theta\cos\psi - \dot{\theta}\sin\psi)\mathbf{j} + (\dot{\phi}\cos\theta + \dot{\psi})\mathbf{k}$, where **i, j,** and **k** are directed along the x, y, z axes as shown in Fig. 21–14d.

21–71. The turbine on a ship has a mass of 400 kg and is mounted on bearings A and B as shown. Its center of mass is at G, its radius of gyration is $k_z = 0.3$ m, and $k_x = k_y = 0.5$ m. If it is spinning at 200 rad/s, determine the vertical reactions at the bearings when the ship undergoes each of the following motions: (a) rolling, $\omega_1 = 0.2$ rad/s, (b) turning, $\omega_2 = 0.8$ rad/s, (c) pitching, $\omega_3 = 1.4$ rad/s.

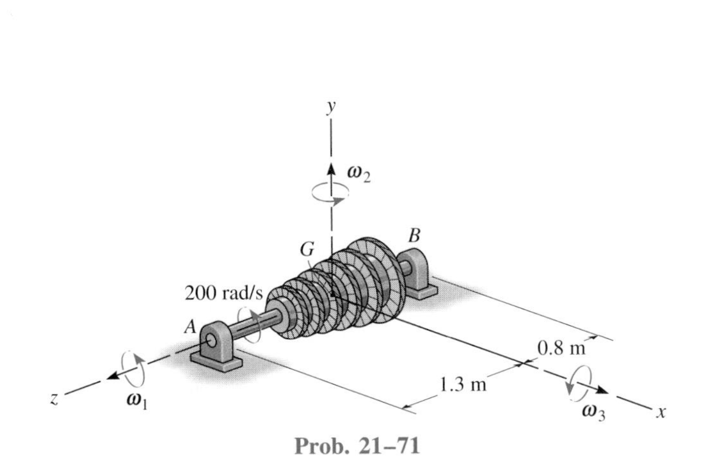

Prob. 21–71

*21–72. A wheel of mass m and radius r rolls with constant spin $\boldsymbol{\omega}$ about a circular path having a radius a. If the angle of inclination is θ, determine the rate of precession. Treat the wheel as a thin ring. No slipping occurs.

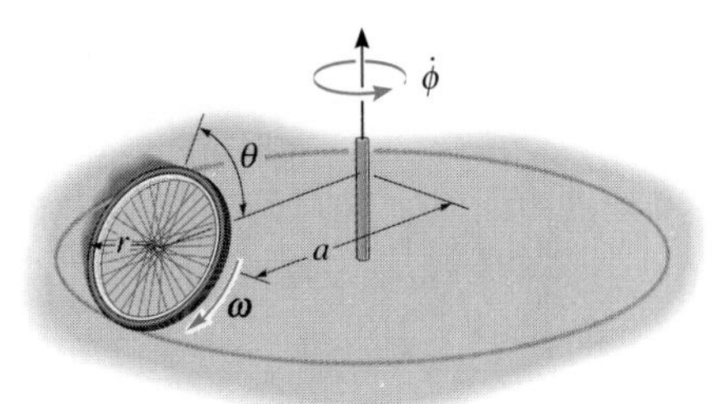

Prob. 21–72

21–73. The toy gyroscope consists of a rotor R which is attached to the frame of negligible mass. If the rotor is spinning about its axle with an angular velocity $\omega_R = 150$ rad/s, determine the constant angular velocity ω_p at which the frame is precessing about the pivot point at O. The stem OA moves in the horizontal plane. The rotor has a mass of 200 g and a radius of gyration $k_{OA} = 20$ mm about OA.

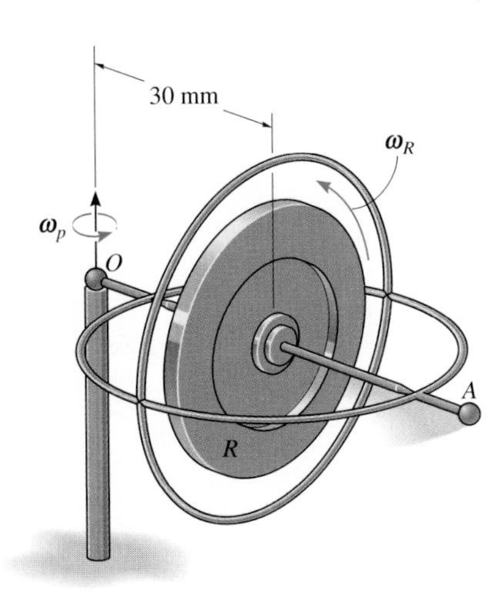

Prob. 21–73

21–74. The toy gyroscope consists of a rotor R which is attached to the frame of negligible mass. If it is observed that the frame is precessing about the pivot point O at $\omega_p = 2$ rad/s, determine the angular velocity ω_R of the rotor. The stem OA moves in the horizontal plane. The rotor has a mass of 200 g and a radius of gyration $k_{OA} = 20$ mm about OA.

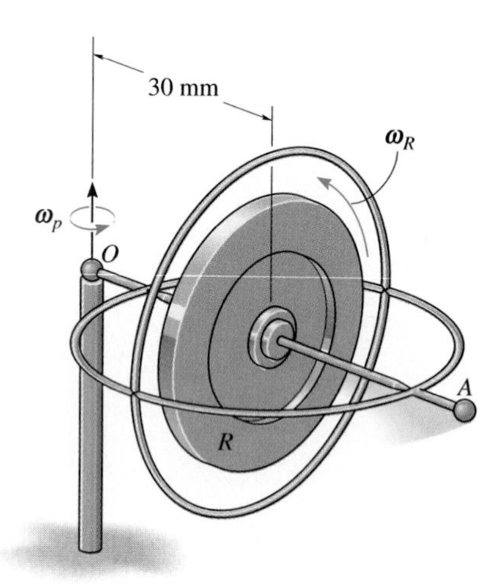

Prob. 21–74

21–75. The motor weighs 50 lb and has a radius of gyration of 0.2 ft about the z axis. The shaft of the motor is supported by bearings at A and B, and is turning at a constant rate $\boldsymbol{\omega}_s = \{100\mathbf{k}\}$ rad/s, while the frame has an angular velocity $\boldsymbol{\omega}_y = \{2\mathbf{j}\}$ rad/s. Determine the moment which the bearing forces at A and B exert on the shaft due to this motion.

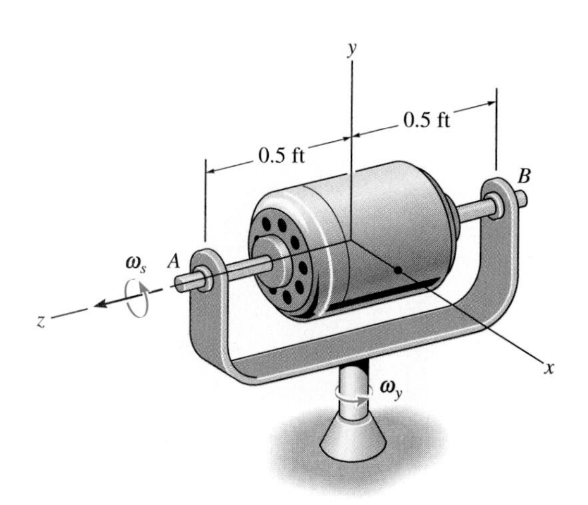

Prob. 21–75

***21–76.** The propeller on a single-engine airplane has a mass of 15 kg and a centroidal radius of gyration of 0.3 m computed about the axis of spin. When viewed from the front of the airplane, the propeller is turning clockwise at 350 rad/s about the spin axis. If the airplane enters a vertical curve having a radius of 80 m and is traveling at 200 km/h, determine the gyroscopic bending moment which the propeller exerts on the bearings of the engine when the airplane is in its lowest position.

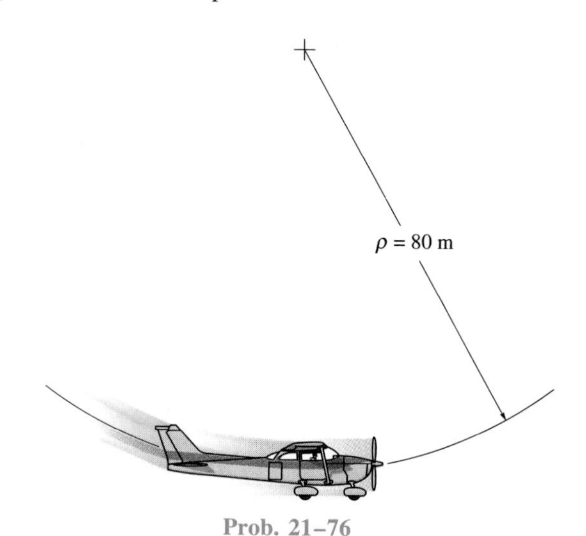

Prob. 21–76

21–77. The car is traveling at $v_C = 100$ km/h around the horizontal curve having a radius of 80 m. If each wheel has a mass of 16 kg, a radius of gyration $k_G = 300$ mm about its spinning axis, and a diameter of 400 mm, determine the difference between the normal forces of the rear wheels, caused by the gyroscopic effect. The distance between the wheels is 1.30 m.

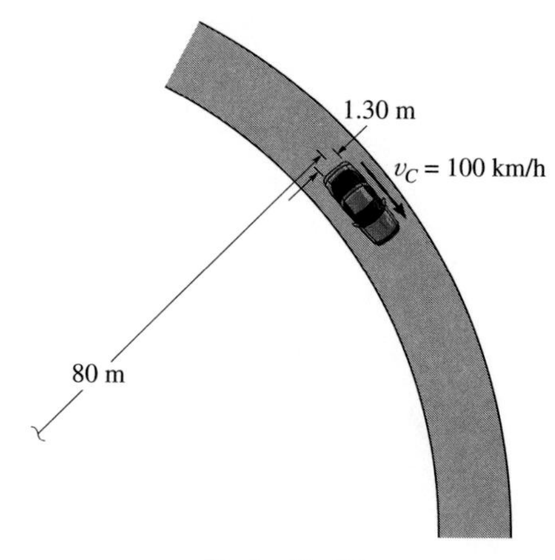

Prob. 21–77

21–78. The conical top has a mass of 0.8 kg, and the moments of inertia are $I_x = I_y = 3.5(10^{-3})\ \text{kg}\cdot\text{m}^2$ and $I_z = 0.8(10^{-3})\ \text{kg}\cdot\text{m}^2$. If it spins freely in the ball-and-socket joint at A with an angular velocity $\omega_s = 750$ rad/s, compute the precession of the top about the axis of the shaft AB.

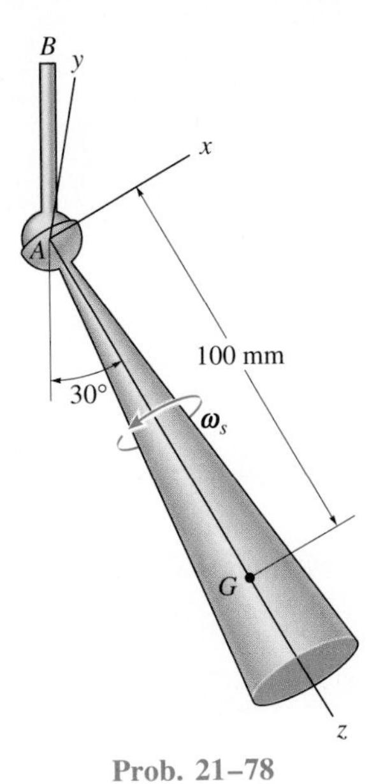

Prob. 21–78

21–79. The 4-kg disk is thrown with a spin $\omega_z = 6$ rad/s. If the angle θ is measured as 160°, determine the precision about the Z axis.

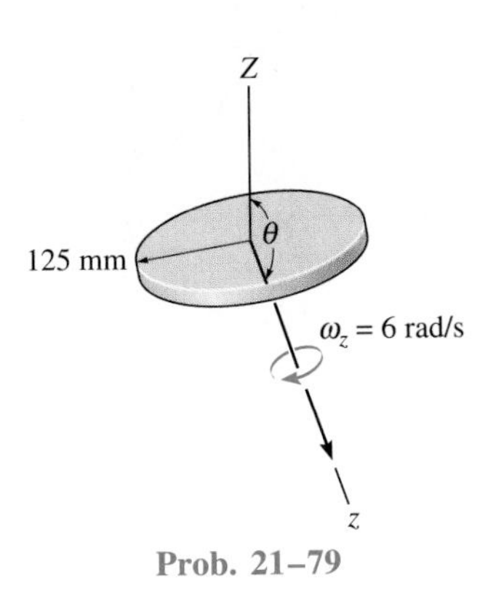

Prob. 21–79

***21–80.** The projectile shown is subjected to torque-free motion. The transverse and axial moments of inertia are I and I_z, respectively. If θ represents the angle between the precessional axis Z and the axis of symmetry z, and β is the angle between the angular velocity $\boldsymbol{\omega}$ and the z axis, show that β and θ are related by the equation $\tan\theta = (I/I_z)\tan\beta$.

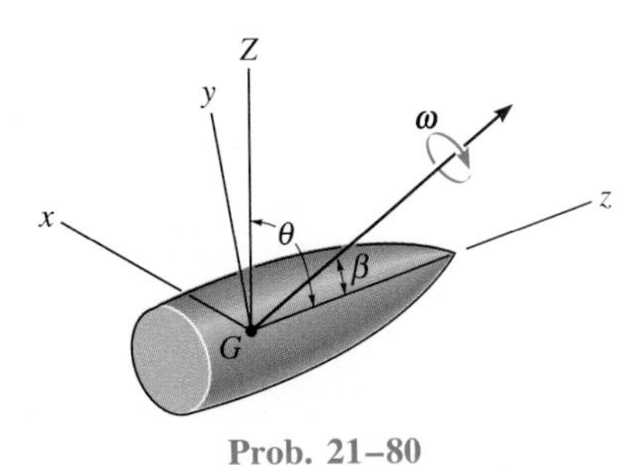

Prob. 21–80

21–81. While the rocket is in free flight, it has a spin of 3 rad/s and precesses about an axis measured 10° from the axis of spin. If the ratio of the axial to transverse moments of inertia of the rocket is 1/15, computed about axes which pass through the mass center G, determine the angle which the resultant angular velocity makes with the spin axis. Construct the body and space cones used to describe the motion. Is the precession regular or retrograde?

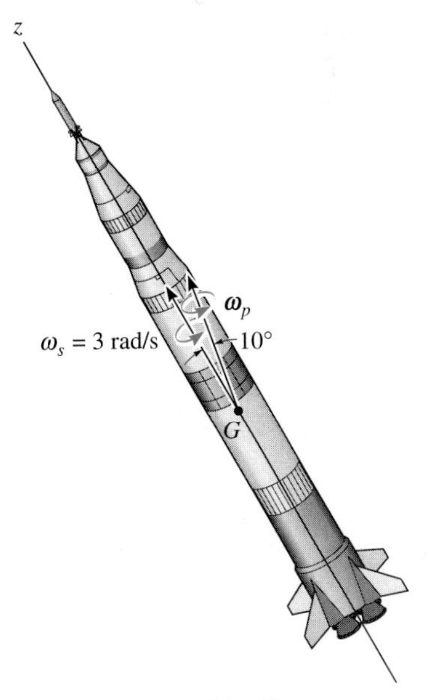

Prob. 21–81

The analysis of vibrations plays an important role in the study of the behavior of structures subjected to earthquakes. In this chapter we will discuss some of the important aspects of vibrational motion.

22

Vibrations

A *vibration* is the periodic motion of a body or system of connected bodies displaced from a position of equilibrium. In general, there are two types of vibration, free and forced. *Free vibration* occurs when the motion is maintained by gravitational or elastic restoring forces, such as the swinging motion of a pendulum or the vibration of an elastic rod. *Forced vibration* is caused by an external periodic or intermittent force applied to the system. Both of these types of vibration may be either damped or undamped. *Undamped* vibrations can continue indefinitely because frictional effects are neglected in the analysis. Since in reality both internal and external frictional forces are present, the motion of all vibrating bodies is actually *damped.*

In this chapter we will study the characteristics of the above types of vibrating motion. The analysis will apply to those bodies which are constrained to move only in one direction. These single-degree-of-freedom systems require only one coordinate to specify completely the position of the system at any time. The analysis of multiple-degree-of-freedom systems is based on this simplified case and is thoroughly treated in textbooks devoted to vibrational theory.

★22.1 Undamped Free Vibration

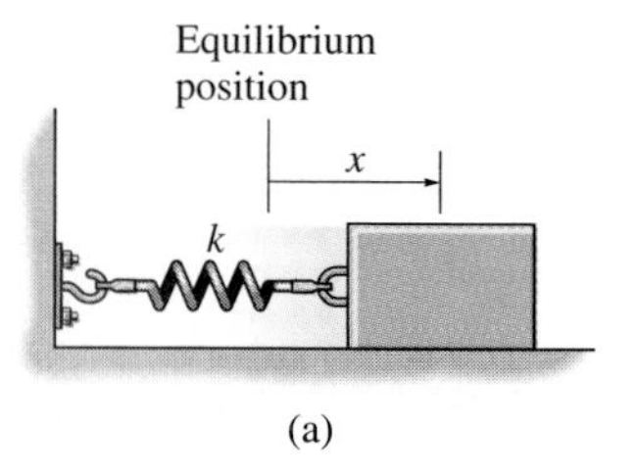

(a)

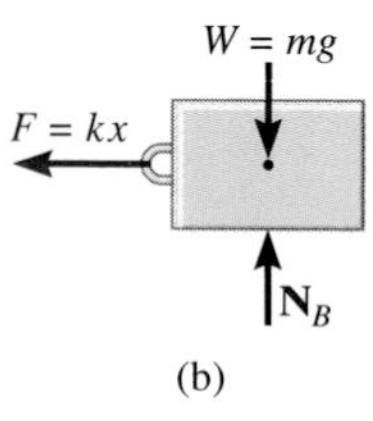

(b)

Fig. 22–1

The simplest type of vibrating motion is undamped free vibration, represented by the model shown in Fig. 22–1*a*. The block has a mass *m* and is attached to a spring having a stiffness *k*. Vibrating motion occurs when the block is released from a displaced position *x* so that the spring pulls on the block. When this occurs, the block will attain a velocity such that it will proceed to move out of equilibrium when $x = 0$. Provided the supporting surface is smooth, oscillation will continue indefinitely.

The time-dependent path of motion of the block may be determined by applying the equation of motion to the block when it is in the displaced position *x*. The free-body diagram is shown in Fig. 22–1*b*. The elastic restoring force $F = kx$ is always directed toward the equilibrium position, whereas the acceleration **a** is assumed to act in the direction of *positive displacement.* Noting that $a = d^2x/dt^2 = \ddot{x}$, we have

$$\overset{+}{\rightarrow}\Sigma F_x = ma_x; \qquad -kx = m\ddot{x}$$

Here it is seen that the acceleration is proportional to the block's position. Motion described in this manner is called *simple harmonic motion.* Rearranging the terms into a "standard form" gives

$$\ddot{x} + p^2x = 0 \qquad (22\text{–}1)$$

The constant *p* is called the *circular frequency,* expressed in rad/s, and in this case

$$p = \sqrt{\frac{k}{m}} \qquad (22\text{–}2)$$

Equation 22–1 may also be obtained by considering the block to be suspended, and measuring the displacement *y* from the block's *equilibrium position,* Fig. 22–2*a*. When the block is in equilibrium, the spring exerts an upward force of $F = W = mg$ on the block. Hence, when the block is displaced a distance *y* downward from this position, the magnitude of the spring force is $F = W + ky$, Fig. 22–2*b*. Applying the equation of motion gives

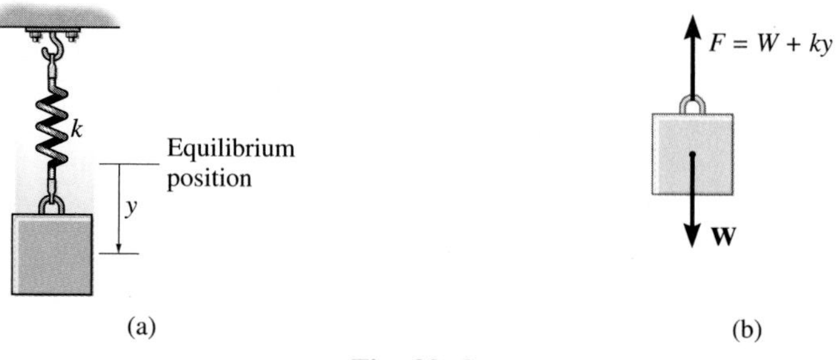

Fig. 22–2

$+\downarrow \Sigma F_y = ma_y; \qquad -W - ky + W = m\ddot{y}$

or

$$\ddot{y} + p^2 y = 0$$

which is the same form as Eq. 22–1, where p is defined by Eq. 22–2.

Equation 22–1 is a homogeneous, second-order, linear, differential equation with constant coefficients. It can be shown, using the methods of differential equations, that the general solution of this equation is

$$x = A \sin pt + B \cos pt \tag{22–3}$$

where A and B represent two constants of integration. The block's velocity and acceleration are determined by taking successive time derivatives, which yields

$$v = \dot{x} = Ap \cos pt - Bp \sin pt \tag{22–4}$$

$$a = \ddot{x} = -Ap^2 \sin pt - Bp^2 \cos pt \tag{22–5}$$

When Eqs. 22–3 and 22–5 are substituted into Eq. 22–1, the differential equation is indeed satisfied, and therefore Eq. 22–3 represents the true solution to Eq. 22–1.

The integration constants A and B in Eq. 22–3 are generally determined from the initial conditions of the problem. For example, suppose that the block in Fig. 22–1*a* has been displaced a distance x_1 to the right from its equilibrium position and given an initial (positive) velocity $\mathbf{v}_1$ directed to the right. Substituting $x = x_1$ at $t = 0$ into Eq. 22–3 yields $B = x_1$. Since $v = v_1$ at $t = 0$, using Eq. 22–4 we obtain $A = v_1/p$. If these values are substituted into Eq. 22–3, the equation describing the motion becomes

$$x = \frac{v_1}{p} \sin pt + x_1 \cos pt \tag{22–6}$$

Equation 22–3 may also be expressed in terms of simple sinusoidal motion. Let

$$A = C \cos \phi \tag{22–7}$$

and

$$B = C \sin \phi \tag{22–8}$$

where C and ϕ are new constants to be determined in place of A and B. Substituting into Eq. 22–3 yields

$$x = C \cos \phi \sin pt + C \sin \phi \cos pt$$

Since $\sin(\theta + \phi) = \sin \theta \cos \phi + \cos \theta \sin \phi$, then

$$x = C \sin(pt + \phi) \tag{22–9}$$

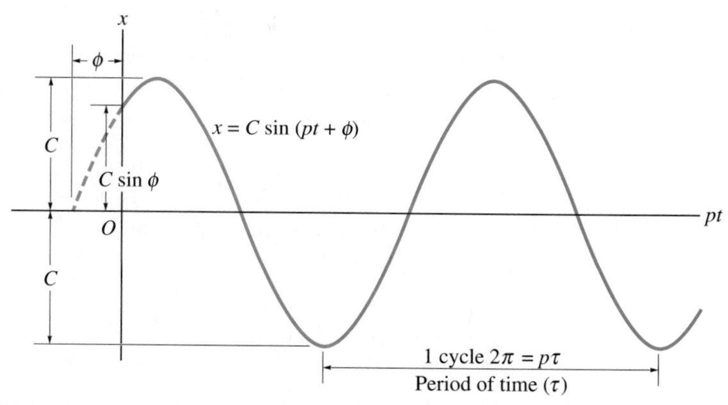

Fig. 22–3

If this equation is plotted on an x-versus-pt axis, the graph shown in Fig. 22–3 is obtained. The maximum displacement of the block from its equilibrium position is defined as the *amplitude* of vibration. From either the figure or Eq. 22–9 the amplitude is C. The angle ϕ is called the *phase angle* since it represents the amount by which the curve is displaced from the origin when $t = 0$. The constants C and ϕ are related to A and B by Eqs. 22–7 and 22–8. Squaring and adding these two equations, the amplitude becomes

$$C = \sqrt{A^2 + B^2} \tag{22–10}$$

If Eq. 22–8 is divided by Eq. 22–7, the phase angle is

$$\phi = \tan^{-1}\frac{B}{A} \tag{22–11}$$

Note that the sine curve, Eq. 22–9, completes one *cycle* in time $t = \tau$ (tau) when $p\tau = 2\pi$, or

$$\tau = \frac{2\pi}{p} \tag{22–12}$$

This length of time is called a *period,* Fig. 22–3. Using Eq. 22–2, the period may also be represented as

$$\tau = 2\pi\sqrt{\frac{m}{k}} \tag{22–13}$$

The *frequency* f is defined as the number of cycles completed per unit of time, which is the reciprocal of the period:

$$f = \frac{1}{\tau} = \frac{p}{2\pi} \tag{22–14}$$

or

$$f = \frac{1}{2\pi}\sqrt{\frac{k}{m}} \tag{22–15}$$

The frequency is expressed in cycles/s. This ratio of units is called a *hertz* (Hz), where 1 Hz = 1 cycle/s = 2π rad/s.

When a body or system of connected bodies is given an initial displacement from its equilibrium position and released, it will vibrate with a definite frequency known as the *natural frequency.* This type of vibration is called *free vibration,* provided no external forces except gravitational or elastic forces act on the body during the motion. Also, if the *amplitude* of vibration remains *constant,* the motion is said to be *undamped.* The undamped free vibration of a body having a single degree of freedom has the same characteristics as simple harmonic motion of the block and spring. Consequently, the body's motion is described by a differential equation of the *same form* as Eq. 22–1, i.e.,

$$\ddot{x} + p^2 x = 0 \tag{22–16}$$

Hence, if the circular frequency p of the body is known, the period of vibration τ, natural frequency f, and other vibrating characteristics of the body can be established using Eqs. 22–3 through 22–15.

PROCEDURE FOR ANALYSIS

As in the case of the block and spring, the circular frequency p of a rigid body or system of connected rigid bodies having a single degree of freedom can be determined using the following procedure:

Free-Body Diagram. Draw the free-body diagram of the body when the body is displaced by a *small amount* from its equilibrium position. Locate the body with respect to its equilibrium position by using an appropriate *inertial coordinate* q. The acceleration of the body's mass center $\mathbf{a}_G$ or the body's angular acceleration $\boldsymbol{\alpha}$ should have a sense which is in the positive direction of the position coordinate. If it is decided that the rotational equation of motion $\Sigma M_P = \Sigma(\mathcal{M}_k)_P$ is to be used, then it may be beneficial to also draw the kinetic diagram since it graphically accounts for the components $m(\mathbf{a}_G)_x$, $m(\mathbf{a}_G)_y$, and $I_G\boldsymbol{\alpha}$, and thereby makes it convenient for visualizing the terms needed in the moment sum $\Sigma(\mathcal{M}_k)_P$.

Equation of Motion. Apply the equation of motion to relate the elastic or gravitational *restoring* forces and couple moments acting on the body to the body's accelerated motion.

Kinematics. Using kinematics, express the body's accelerated motion in terms of the second time derivative of the position coordinate, $\ddot{q}$. Substitute this result into the equation of motion and determine p by rearranging the terms so that the resulting equation is of the form $\ddot{q} + p^2 q = 0$.

The following examples illustrate this procedure.

Example 22–1

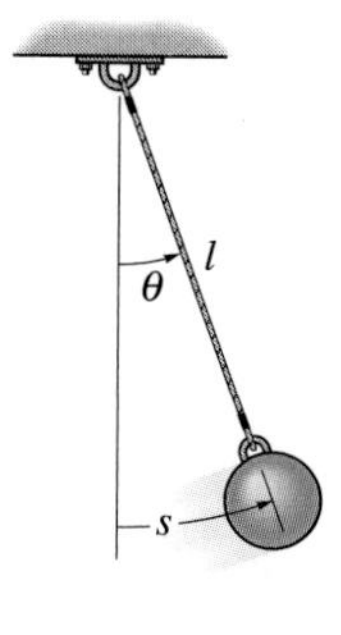

(a)

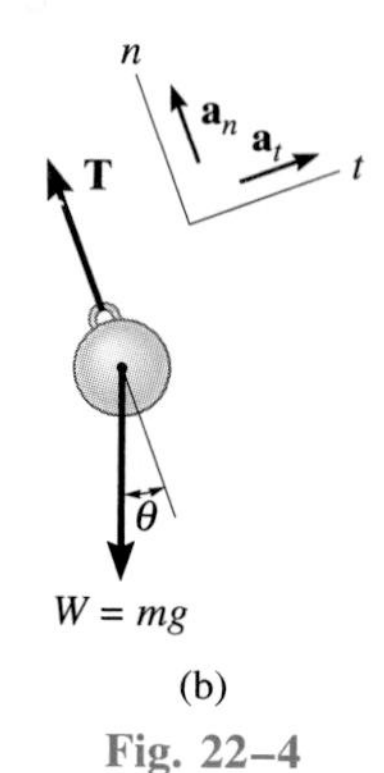

(b)

Fig. 22–4

Determine the period of vibration for the simple pendulum shown in Fig. 22–4*a*. The bob has a mass m and is attached to a cord of length l. Neglect the size of the bob.

SOLUTION

Free-Body Diagram. Motion of the system will be related to the position coordinate $(q =)\theta$, Fig. 22–4*b*. When the bob is displaced by an angle θ, the *restoring force* acting on the bob is created by the *weight component* $mg \sin \theta$. Furthermore, $\mathbf{a}_t$ acts in the direction of *increasing* s (or θ).

Equation of Motion. Applying the equation of motion in the *tangential direction,* since it involves the restoring force, yields

$$+\nearrow \Sigma F_t = ma_t; \qquad -mg \sin \theta = ma_t \qquad (1)$$

Kinematics. $a_t = d^2s/dt^2 = \ddot{s}$. Furthermore, s may be related to θ by the equation $s = l\theta$, so that $a_t = l\ddot{\theta}$. Hence, Eq. 1 reduces to the form

$$\ddot{\theta} + \frac{g}{l} \sin \theta = 0 \qquad (2)$$

The solution of this equation involves the use of an elliptic integral. For *small displacements,* however, $\sin \theta \approx \theta$, in which case

$$\ddot{\theta} + \frac{g}{l}\theta = 0 \qquad (3)$$

Comparing this equation with Eq. 22–16 ($\ddot{x} + p^2x = 0$), which is the "standard form" for simple harmonic motion, it is seen that $p = \sqrt{g/l}$. From Eq. 22–12, the period of time required for the bob to make one complete swing is therefore

$$\tau = \frac{2\pi}{p} = 2\pi\sqrt{\frac{l}{g}} \qquad \textit{Ans.}$$

This interesting result, originally discovered by Galileo Galilei through experiment, indicates that the period depends only on the length of the cord and not on the mass of the pendulum bob or the angle θ.

The solution of Eq. 3 is given by Eq. 22–3, where $p = \sqrt{g/l}$ and θ is substituted for x. Like the block and spring, the constants A and B in this problem may be determined if, for example, one knows the displacement and velocity of the bob at a given instant.

Example 22–2

The 10-kg rectangular plate shown in Fig. 22–5*a* is suspended at its center from a rod having a torsional stiffness $k = 1.5 \text{ N} \cdot \text{m/rad}$. Determine the natural period of vibration of the plate when it is given a small angular displacement θ in the plane of the plate.

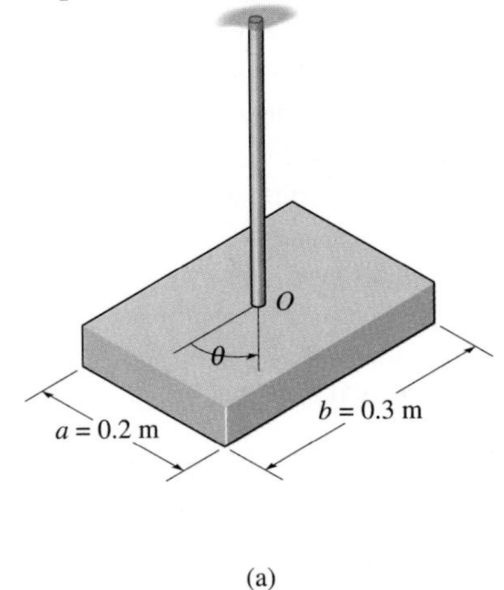

(a)

Fig. 22–5

SOLUTION

Free-Body Diagram. Fig. 22–5*b*. Since the plate is displaced in its own plane, the torsional *restoring* moment created by the rod is $M = k\theta$. This moment acts in the direction opposite to the angular displacement θ. The angular acceleration $\ddot{\theta}$ acts in the direction of *positive* θ.

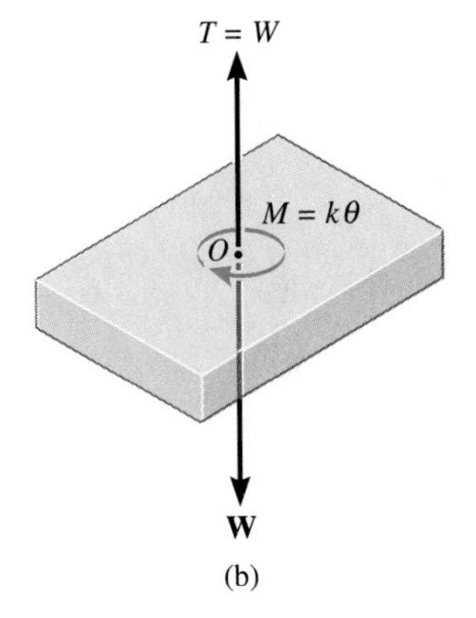

(b)

Equation of Motion

$\Sigma M_O = I_O \alpha;$ $\qquad -k\theta = I_O \ddot{\theta}$

or

$$\ddot{\theta} + \frac{k}{I_O}\theta = 0$$

Since this equation is in the "standard form," the circular frequency is $p = \sqrt{k/I_O}$.

The moment of inertia of the plate about an axis coincident with the rod is $I_O = \frac{1}{12}m(a^2 + b^2)$. Hence,

$$I_O = \frac{1}{12}(10 \text{ kg})[(0.2 \text{ m})^2 + (0.3 \text{ m})^2] = 0.108 \text{ kg} \cdot \text{m}^2$$

The natural period of vibration is, therefore,

$$\tau = \frac{2\pi}{p} = 2\pi\sqrt{\frac{I_O}{k}} = 2\pi\sqrt{\frac{0.108}{1.5}} = 1.69 \text{ s} \qquad \textit{Ans.}$$

Example 22–3

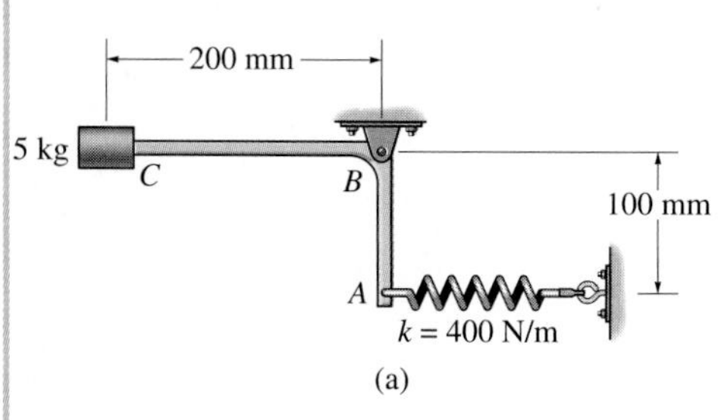

(a)

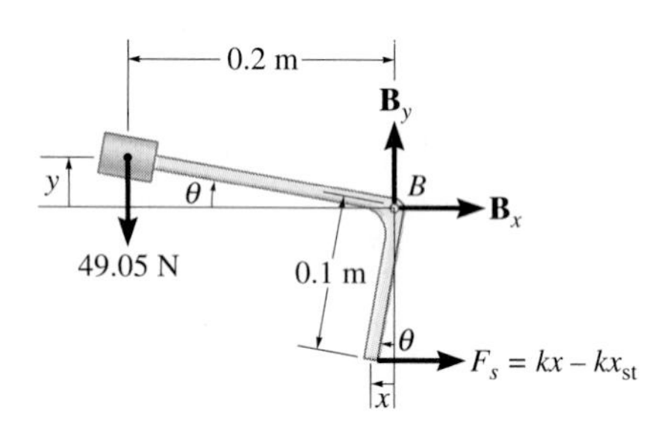

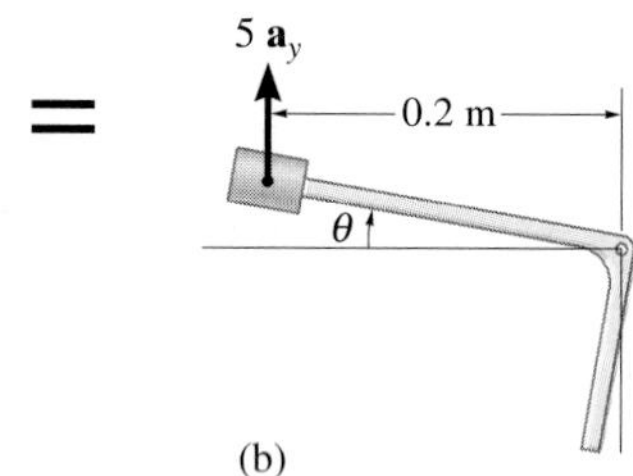

(b)

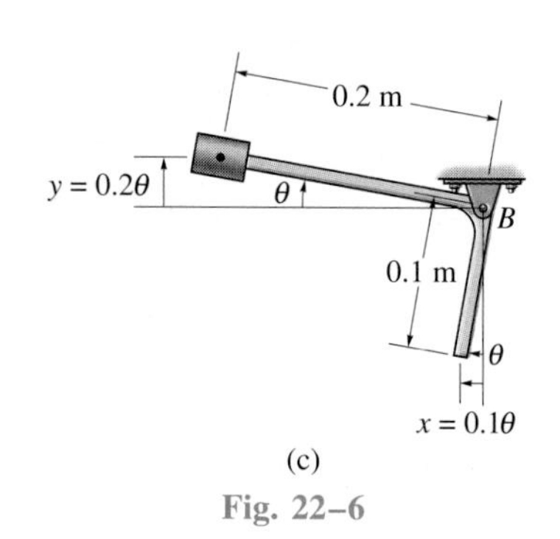

(c)

Fig. 22–6

The bent rod shown in Fig. 22–6*a* has a negligible mass and supports a 5-kg collar at its end. Determine the natural period of vibration for the system.

SOLUTION

Free-Body and Kinetic Diagrams. Fig. 22–6*b*. Here the rod is displaced by a small amount θ from the equilibrium position. Since the spring is subjected to an initial compression of x_{st} for equilibrium, then when the displacement $x > x_{st}$ the spring exerts a force of $F_s = kx - kx_{st}$ on the rod. To obtain the "standard form," Eq. 22–16, $5\mathbf{a}_y$ acts *upward,* which is in accordance with positive θ displacement.

Equation of Motion. Moments will be summed about point B to eliminate the unknown reaction at this point. Since θ is small,

$$\circlearrowright + \Sigma M_B = \Sigma(\mathcal{M}_k)_B;$$
$$kx(0.1\text{ m}) - kx_{st}(0.1\text{ m}) + 49.05\text{ N}(0.2\text{ m}) = -(5\text{ kg})a_y(0.2\text{ m})$$

The second term on the left side, $-kx_{st}(0.1\text{ m})$, represents the moment created by the spring force which is necessary to hold the collar in *equilibrium,* i.e., at $x = 0$. Since this moment is equal and opposite to the moment $49.05(0.2)$ created by the weight of the collar, these two terms cancel in the above equation, so that

$$kx(0.1) = -5a_y(0.2) \tag{1}$$

Kinematics. The positions of the spring and the collar may be related to the angle θ, Fig. 22–6*c*. Since θ is small, $x = (0.1\text{ m})\theta$ and $y = (0.2\text{ m})\theta$. Therefore, $a_y = \ddot{y} = 0.2\ddot{\theta}$. Substituting into Eq. 1 yields

$$400(0.1\theta)0.1 = -5(0.2\ddot{\theta})0.2$$

Rewriting this equation in standard form gives

$$\ddot{\theta} + 20\theta = 0$$

Compared with $\ddot{x} + p^2x = 0$ (Eq. 22–16), we have

$$p^2 = 20 \qquad p = 4.47\text{ rad/s}$$

The natural period of vibration is therefore

$$\tau = \frac{2\pi}{p} = \frac{2\pi}{4.47} = 1.40\text{ s} \qquad \textit{Ans.}$$

Example 22–4

A 10-lb block is suspended from a cord that passes over a 15-lb disk, as shown in Fig. 22–7*a*. The spring has a stiffness $k = 200$ lb/ft. Determine the natural period of vibration for the system.

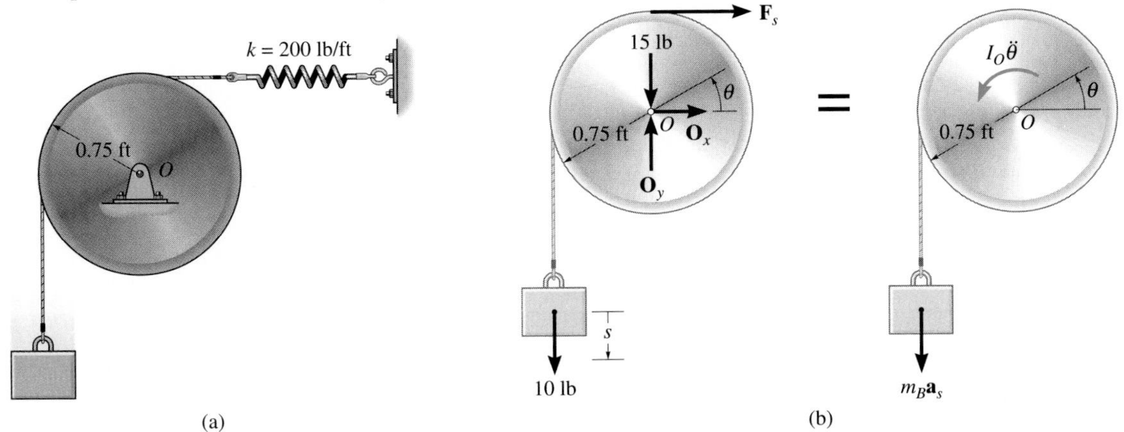

SOLUTION

Free-Body and Kinetic Diagrams. Fig. 22–7*b*. The *system* consists of the disk, which undergoes a rotation defined by the angle θ, and the block, which translates by an amount s. The vector $I_O\ddot{\theta}$ acts in the direction of *positive* θ, and consequently $m_B\mathbf{a}_s$ acts downward in the direction of *positive* s.

Equation of Motion. Summing moments about point O to eliminate the reactions $\mathbf{O}_x$ and $\mathbf{O}_y$, realizing that $I_O = \frac{1}{2}mr^2$, yields

$$\circlearrowright +\Sigma M_O = \Sigma(\mathcal{M}_k)_O;$$

$$10\text{ lb}(0.75\text{ ft}) - F_s(0.75\text{ ft}) = \frac{1}{2}\left(\frac{15\text{ lb}}{32.2\text{ ft/s}}\right)(0.75\text{ ft})^2\ddot{\theta} + \left(\frac{10\text{ lb}}{32.2\text{ ft/s}}\right)a_s(0.75\text{ ft}) \quad (1)$$

Kinematics. As shown on the kinematic diagram in Fig. 22–7*c*, a small positive displacement θ of the disk causes the block to lower by an amount $s = 0.75\theta$; hence, $a_s = \ddot{s} = 0.75\ddot{\theta}$. When $\theta = 0°$, the spring force required for *equilibrium* of the disk is 10 lb, acting to the right. For position θ, the spring force is $F_s = (200\text{ lb/ft})(0.75\theta\text{ ft}) + 10$ lb. Substituting these results into Eq. 1 and simplifying yields

$$\ddot{\theta} + 368\theta = 0$$

Hence,

$$p^2 = 368 \qquad p = 19.2\text{ rad/s}$$

Therefore, the natural period of vibration is

$$\tau = \frac{2\pi}{p} = \frac{2\pi}{19.2} = 0.328\text{ s} \qquad \textit{Ans.}$$

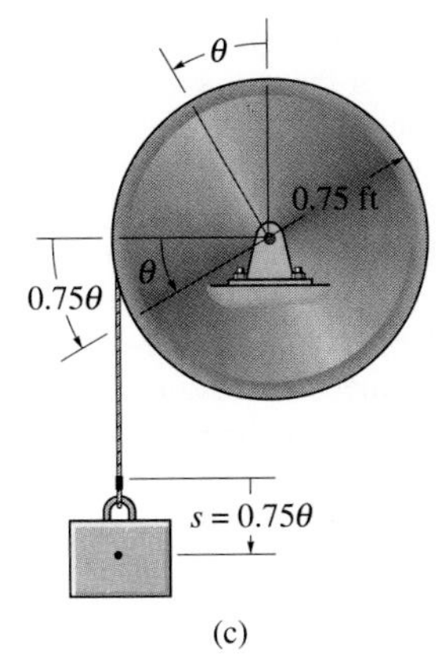

Fig. 22–7

PROBLEMS

22–1. When a 20-lb weight is suspended from a spring, the spring is stretched a distance of 4 in. Determine the natural frequency and the period of vibration for a 10-lb weight attached to the same spring.

22–2. A spring has a stiffness of 600 N/m. If a 4-kg block is attached to the spring, pushed 50 mm above its equilibrium position, and released from rest, determine the equation which describes the block's motion. Assume that positive displacement is measured downward.

22–3. When a 3-kg block is suspended from a spring, the spring is stretched a distance of 60 mm. Determine the natural frequency and the period of vibration for a 0.2-kg block attached to the same spring.

***22–4.** An 8-kg block is suspended from a spring having a stiffness $k = 80$ N/m. If the block is given an upward velocity of 0.4 m/s when it is 90 mm above its equilibrium position, determine the equation which describes the motion and the maximum upward displacement of the block measured from the equilibrium position. Assume that positive displacement is measured downward.

22–5. A 2-lb weight is suspended from a spring having a stiffness $k = 2$ lb/in. If the weight is pushed 1 in. upward from its equilibrium position and then released from rest, determine the equation which describes the motion. What is the amplitude and the natural frequency of the vibration?

22–6. A 6-lb weight is suspended from a spring having a stiffness $k = 3$ lb/in. If the weight is given an upward velocity of 20 ft/s when it is 2 in. above its equilibrium position, determine the equation which describes the motion and the maximum upward displacement of the weight, measured from the equilibrium position. Assume positive displacement is downward.

■**22–7.** A spring is stretched 175 mm by an 8-kg block. If the block is displaced 100 mm downward from its equilibrium position and given a downward velocity of 1.50 m/s, determine the differential equation which describes the motion. Assume that positive displacement is measured downward. Use the Runge Kutta method to determine the position of the block, measured from its unstretched position, when $t = 0.22$ s. Use a time increment of $\Delta t = 0.02$ s.

***22–8.** If the block in Prob. 22–7 is given an upward velocity of 4 m/s when it is displaced downward a distance of 60 mm from its equilibrium position, determine the equation which describes the motion. What is the amplitude of the motion? Assume that positive displacement is measured downward.

22–9. A pendulum has a 0.4-m-long cord and is given a tangential velocity of 0.2 m/s toward the vertical from a position $\theta = 0.3$ rad. Determine the equation which describes the angular motion.

22–10. Determine to the nearest degree the maximum angular displacement of the bob in Prob. 22–9 if it is initially displaced $\theta = 0.2$ rad from the vertical and given a tangential velocity of 0.4 m/s away from the vertical.

22–11. Determine the frequency of vibration for the block. The springs are originally compressed Δ.

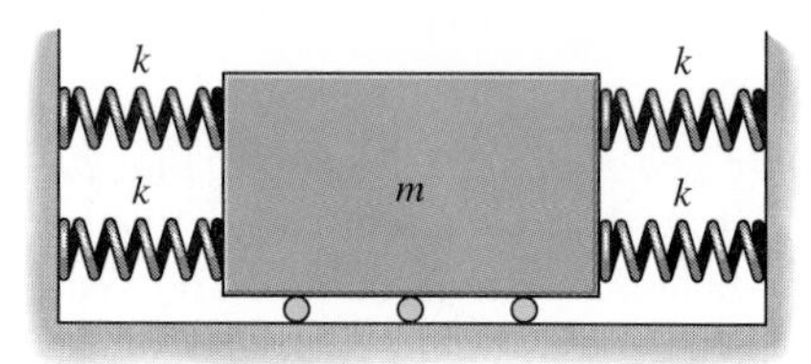

Prob. 22–11

***22–12.** The square plate has a mass m and is suspended at its corner by the pin O. Determine the natural period of oscillation if it is displaced a small amount and released.

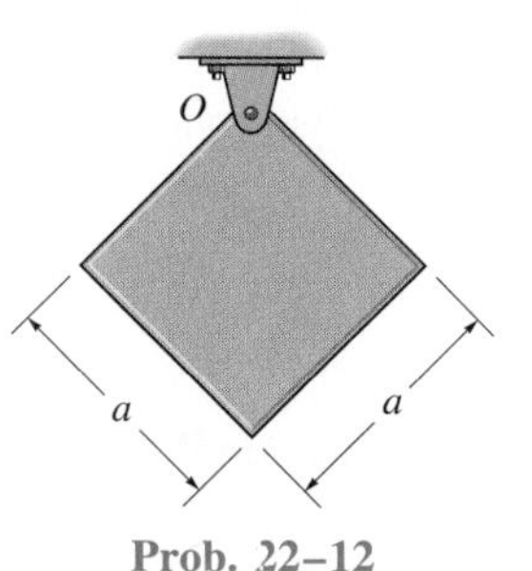

Prob. 22–12

22–13. The uniform beam is supported at its ends by two springs A and B, each having the same stiffness k. When nothing is supported on the beam, it has a period of vertical vibration of 0.83 s. If a 50-kg mass is placed at its center, the period of vertical vibration is 1.52 s. Compute the stiffness of each spring and the mass of the beam.

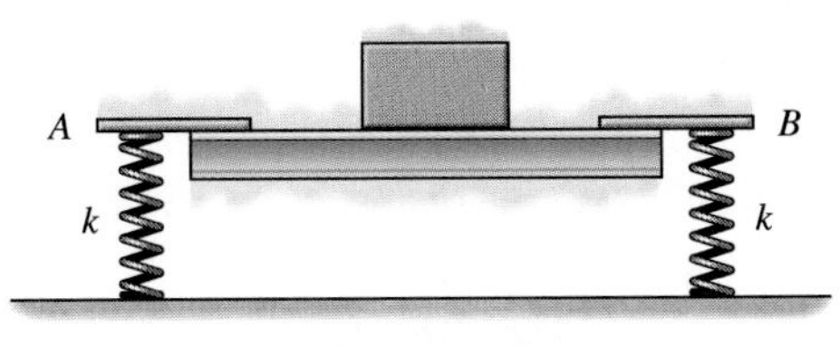

Prob. 22–13

22–14. The semicircular disk weighs 20 lb. Determine the natural period of vibration if it is displaced a small amount and released.

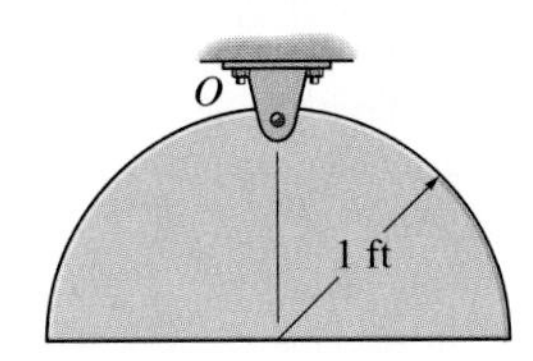

Prob. 22–14

22–15. The body of arbitrary shape has a mass m, mass center at G, and a radius of gyration about G of k_G. If it is displaced a slight amount θ from its equilibrium position and released, determine the natural period of vibration.

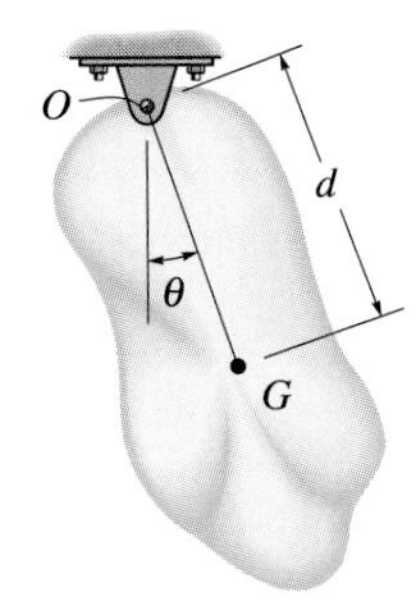

Prob. 22–15

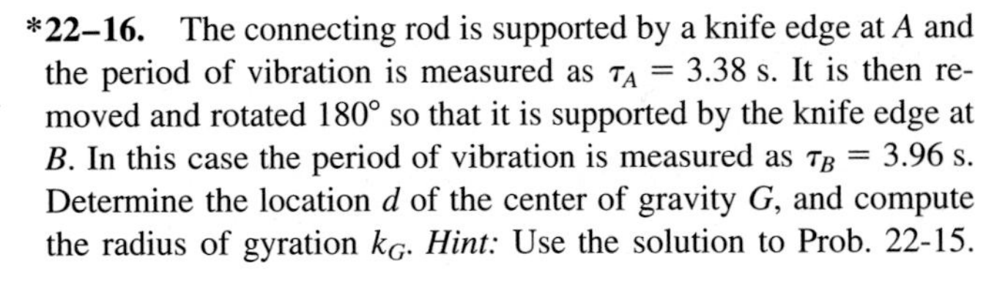

***22–16.** The connecting rod is supported by a knife edge at A and the period of vibration is measured as $\tau_A = 3.38$ s. It is then removed and rotated 180° so that it is supported by the knife edge at B. In this case the period of vibration is measured as $\tau_B = 3.96$ s. Determine the location d of the center of gravity G, and compute the radius of gyration k_G. *Hint:* Use the solution to Prob. 22-15.

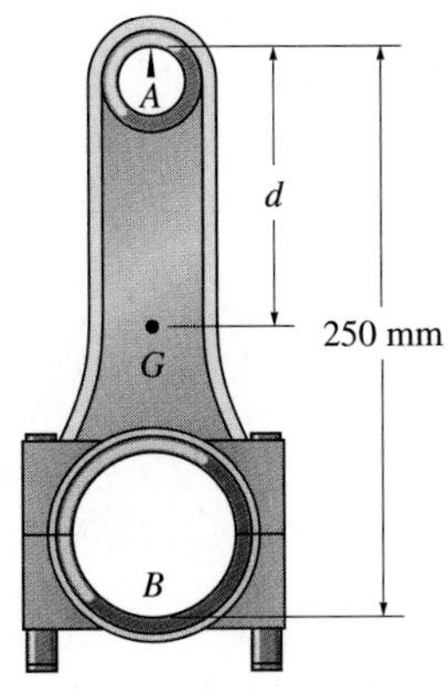

Prob. 22–16

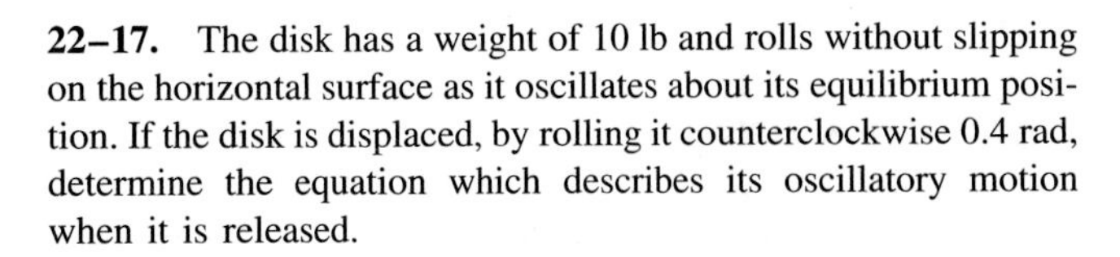

22–17. The disk has a weight of 10 lb and rolls without slipping on the horizontal surface as it oscillates about its equilibrium position. If the disk is displaced, by rolling it counterclockwise 0.4 rad, determine the equation which describes its oscillatory motion when it is released.

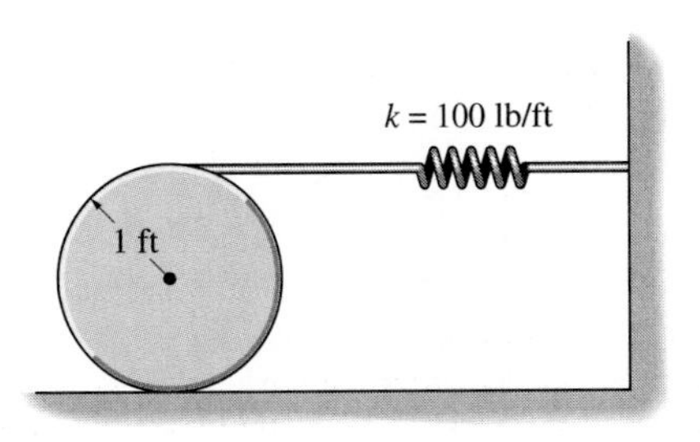

Prob. 22–17

22–18. The plate of mass m is supported by three symmetrically placed cords of length l as shown. If the plate is given a slight rotation about a vertical axis through its center and released, determine the natural period of oscillation.

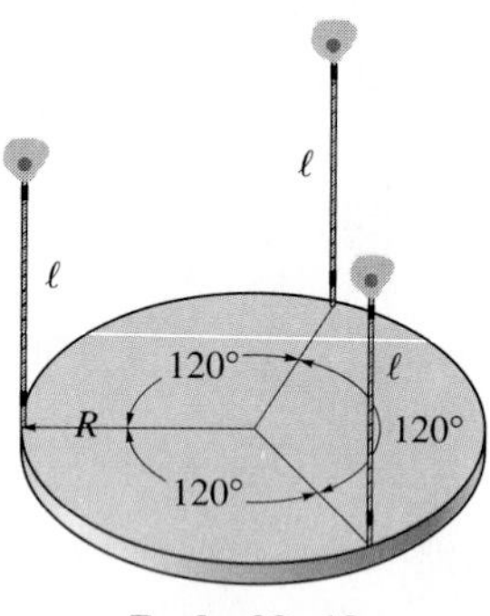

Prob. 22–18

22–19. While standing in an elevator, the man holds a pendulum which consists of an 18-in. cord and a 0.5-lb bob. If the elevator is descending with an acceleration $a = 4\ \text{ft/s}^2$, determine the natural period of vibration for small amplitudes of swing.

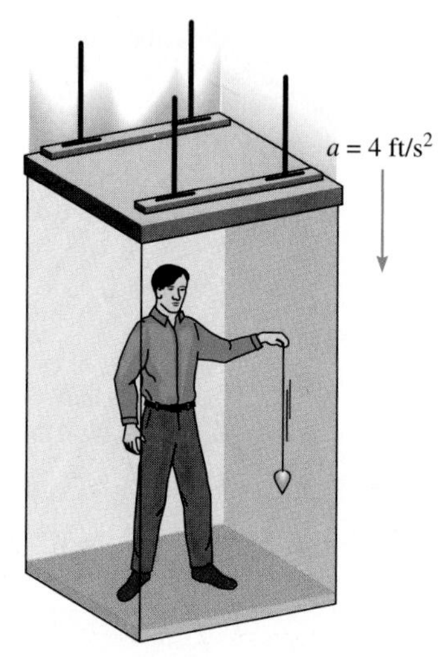

Prob. 22–19

***22–20.** The platform AB when empty has a mass of 400 kg, center of mass at G_1, and natural period of oscillation $\tau_1 = 2.38$ s. If a car, having a mass of 1.2 Mg and center of mass at G_2, is placed on the platform, the natural period of oscillation becomes $\tau_2 = 3.16$ s. Determine the moment of inertia of the car about an axis passing through G_2.

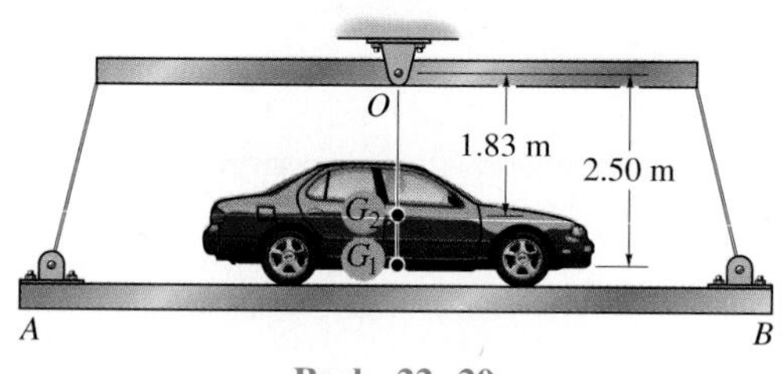

Prob. 22–20

22–21. The pointer on a metronome supports a 0.4-lb slider A, which is positioned at a fixed distance from the pivot O of the pointer. When the pointer is displaced, a torsional spring at O exerts a restoring torque on the pointer having a magnitude $M = (1.2\theta)$ lb · ft, where θ represents the angle of displacement from the vertical, measured in radians. Determine the natural period of vibration when the pointer is displaced a small amount θ and released. Neglect the mass of the pointer.

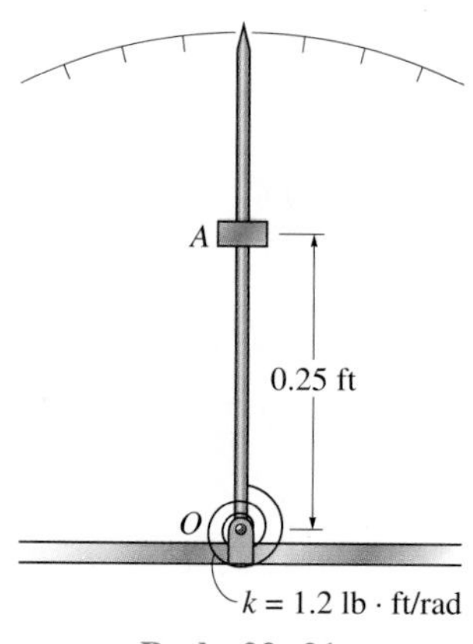

Prob. 22–21

22–22. The 50-lb spool is attached to two springs. If the spool is displaced a small amount and released, determine the natural period of vibration. The radius of gyration of the spool is $k_G = 1.5$ ft. The spool rolls without slipping.

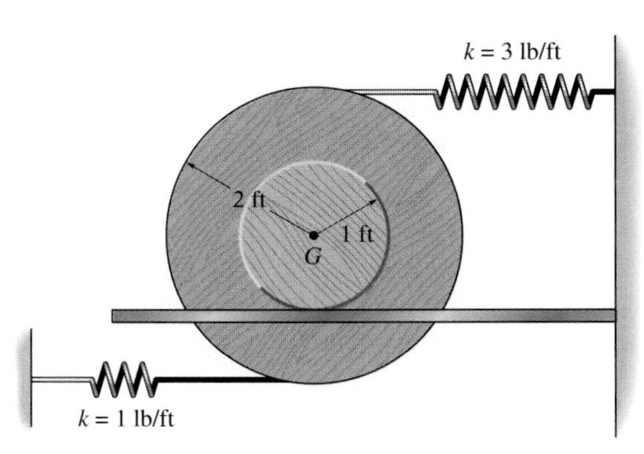

Prob. 22–22

22–23. The 20-lb rectangular plate has a natural period of vibration $\tau = 0.3$ s, as it oscillates around the axis of rod AB. Determine the torsional stiffness k, measured in lb · ft/rad, of the rod. Neglect the mass of the rod.

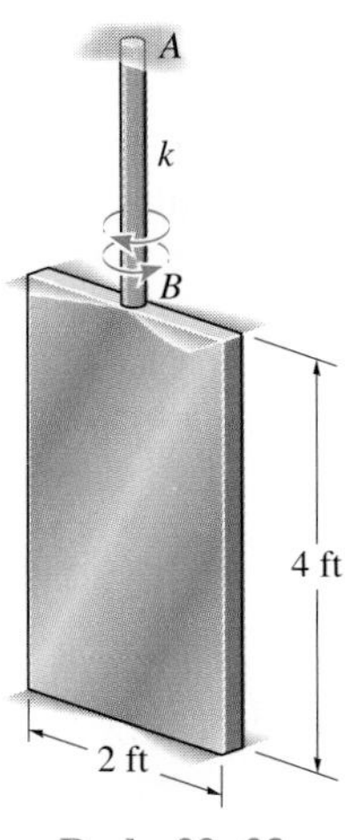

Prob. 22–23

***22–24.** The 25-lb weight is fixed to the end of the rod assembly. If both springs are unstretched when the assembly is in the position shown, determine the natural period of vibration for the weight when it is displaced slightly and released. Neglect the size of the block and the mass of the rods.

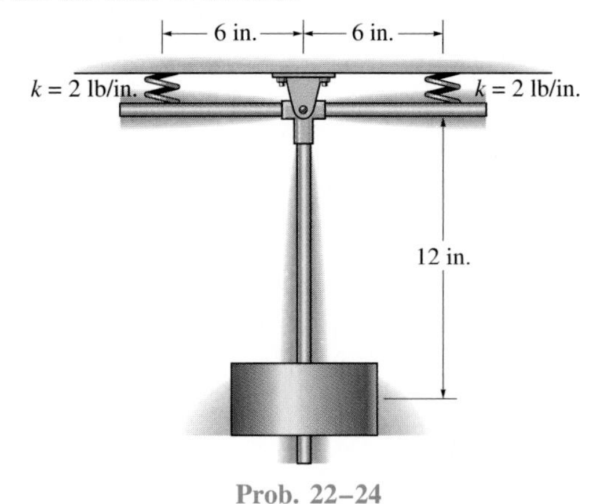

Prob. 22–24

22–25. The uniform rod has a mass m and is supported by the pin O. If the rod is given a small displacement and released, determine the natural period of vibration. The springs are unstretched when the rod is in the position shown.

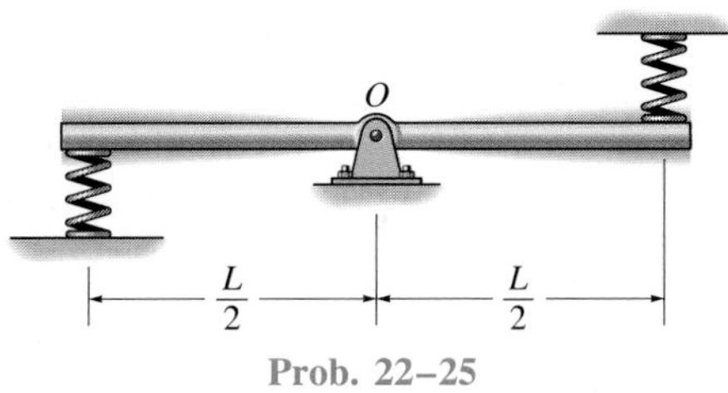

Prob. 22–25

22–26. If the wire AB is subjected to a tension of 20 lb, determine the equation which describes the motion when the 5-lb weight is displaced 2 in. horizontally and released from rest.

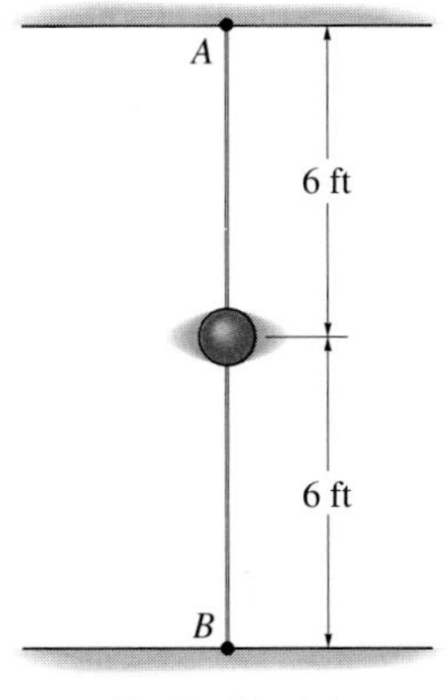

Prob. 22–26

*22.2 Energy Methods

The simple harmonic motion of a body, discussed in Sec. 22.1, is due only to gravitational and elastic restoring forces acting on the body. Since these types of forces are *conservative,* it is possible to use the conservation of energy equation to obtain the body's natural frequency or period of vibration. To show how to do this, consider the block and spring in Fig. 22–8. When the block is displaced an arbitrary amount x from the equilibrium position, the kinetic energy is $T = \frac{1}{2}mv^2 = \frac{1}{2}m\dot{x}^2$ and the potential energy is $V = \frac{1}{2}kx^2$. By the conservation of energy equation, Eq. 14–21, it is necessary that

$$T + V = \text{const}$$

$$\frac{1}{2}m\dot{x}^2 + \frac{1}{2}kx^2 = \text{const} \qquad (22\text{–}17)$$

The differential equation describing the *accelerated motion* of the block can be obtained by *differentiating* this equation with respect to time; i.e.,

$$m\dot{x}\ddot{x} + kx\dot{x} = 0$$

$$\dot{x}(m\ddot{x} + kx) = 0$$

Since the velocity $\dot{x}$ is not *always* zero in a vibrating system,

$$\ddot{x} + p^2x = 0 \qquad p = \sqrt{k/m}$$

which is the same as Eq. 22–1.

If the energy equation is written for a *system of connected bodies,* the natural frequency or the equation of motion can also be determined by time differentiation. Here it is *not necessary* to dismember the system to account for reactive and connective forces which do no work. Also, by this method, the circular frequency p may be obtained *directly.* For example, consider the total mechanical energy of the block and spring in Fig. 22–8 when the block is at its *maximum displacement.* In this position the block is temporarily at rest, so that the kinetic energy is zero and the potential energy, stored in the spring, is a maximum. Therefore, Eq. 22–17 becomes $\frac{1}{2}kx_{\max}^2 = \text{const}$. At the instant the block passes the equilibrium position, the kinetic energy of the block is a maximum and the potential energy of the spring is zero. Hence, Eq. 22–17 becomes $\frac{1}{2}m(\dot{x})_{\max}^2 = \text{const}$. Since the vibrating motion of the block is *har-*

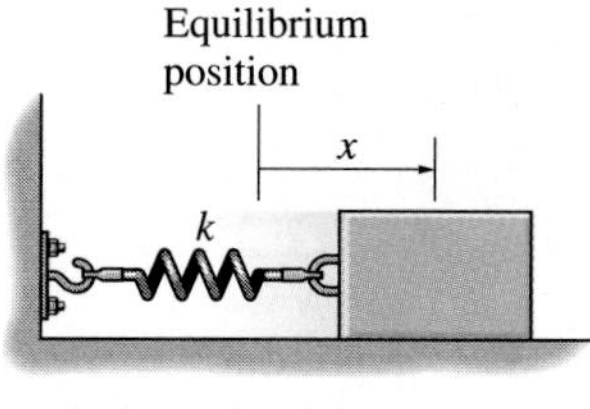

Fig. 22–8

monic, the *solution* for the displacement and velocity may be written in the form of Eq. 22–9 and its time derivative, i.e.,

$$x = C \sin(pt + \phi) \qquad \dot{x} = Cp \cos(pt + \phi)$$

so that

$$x_{\max} = C \qquad \dot{x}_{\max} = Cp$$

Applying the conservation of energy equation ($T + V = \text{const}$) yields

$$V_{\max} = T_{\max}; \qquad \tfrac{1}{2}kx_{\max}^2 = \tfrac{1}{2}m\dot{x}_{\max}^2 = \text{const}$$

or

$$kC^2 = mC^2p^2$$

Solving for p yields

$$p = \sqrt{\frac{k}{m}}$$

which is identical to Eq. 22–2.

PROCEDURE FOR ANALYSIS

The following procedure provides a method for determining the circular frequency p of a body or system of connected bodies using the conservation of energy equation.

Energy Equation. Draw the body when it is displaced by a *small amount* from its equilibrium position and define the location of the body from its equilibrium position by an appropriate position coordinate q. Formulate the equation of energy for the body, $T + V = \text{const}$, in terms of the position coordinate.* Recall, in general, the kinetic energy must account for both the body's translational and rotational motion, $T = \frac{1}{2}mv_G^2 + \frac{1}{2}I_G\omega^2$, Eq. 18–2, and the potential energy is the sum of the gravitational and elastic potential energies of the body, $V = V_g + V_e$, Eq. 18–15. In particular, V_g should be measured from a datum for which $q = 0$ (equilibrium position).

Time Derivative. Take the time derivative of the energy equation and factor out the common terms. The resultant differential equation represents the equation of motion for the system. The value of p is obtained after rearranging the terms in the standard form $\ddot{q} + p^2q = 0$.

*It is suggested that the material in Sec. 18.5 be reviewed.

The following examples illustrate this procedure.

Example 22–5

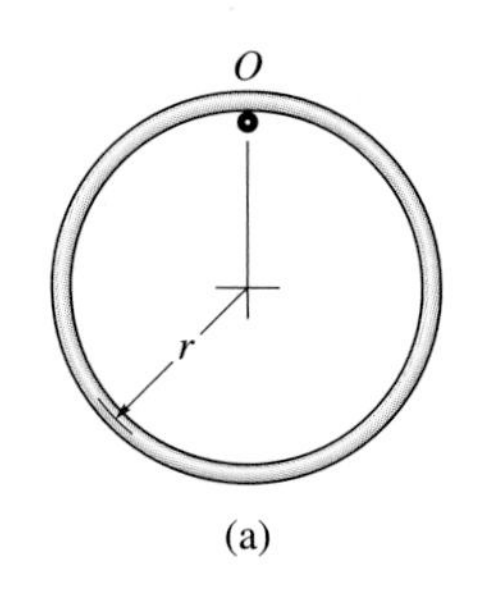

(a)

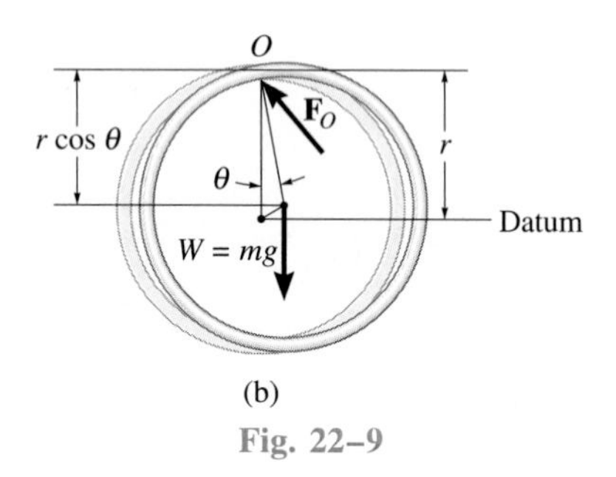

(b)

Fig. 22–9

The thin hoop shown in Fig. 22–9*a* is supported by a peg at *O*. Determine the period of oscillation for small amplitudes of swing. The hoop has a mass *m*.

SOLUTION

Energy Equation. A diagram of the hoop when it is displaced a small amount $(q =)\theta$ from the equilibrium position is shown in Fig. 22–9*b*. Using the table on the inside back cover and the parallel-axis theorem to determine I_O, we can express the kinetic energy as

$$T = \tfrac{1}{2}I_O\omega^2 = \tfrac{1}{2}[mr^2 + mr^2]\dot{\theta}^2 = mr^2\dot{\theta}^2$$

If a horizontal datum is placed through the center of gravity of the hoop when $\theta = 0$, then the center of gravity moves upward $r(1 - \cos\theta)$ in the displaced position. For *small angles*, $\cos\theta$ may be replaced by the first two terms of its series expansion, $\cos\theta = 1 - \theta^2/2 + \cdots$. Therefore, the potential energy is

$$V = mgr\left[1 - \left(1 - \frac{\theta^2}{2}\right)\right] = mgr\frac{\theta^2}{2}$$

The total energy in the system is

$$T + V = mr^2\dot{\theta}^2 + mgr\frac{\theta^2}{2}$$

Time Derivative.

$$mr^2 2\dot{\theta}\ddot{\theta} + mgr\theta\dot{\theta} = 0$$

$$mr\dot{\theta}(2r\ddot{\theta} + g\theta) = 0$$

Since $\dot{\theta}$ is not always equal to zero, from the terms in parentheses,

$$\ddot{\theta} + \frac{g}{2r}\theta = 0$$

Hence,

$$p = \sqrt{\frac{g}{2r}}$$

so that

$$\tau = \frac{2\pi}{p} = 2\pi\sqrt{\frac{2r}{g}} \qquad \textit{Ans.}$$

Example 22–6

A 10-kg block is suspended from a cord wrapped around a 5-kg disk, as shown in Fig. 22–10*a*. If the spring has a stiffness $k = 200$ N/m, determine the natural period of vibration for the system.

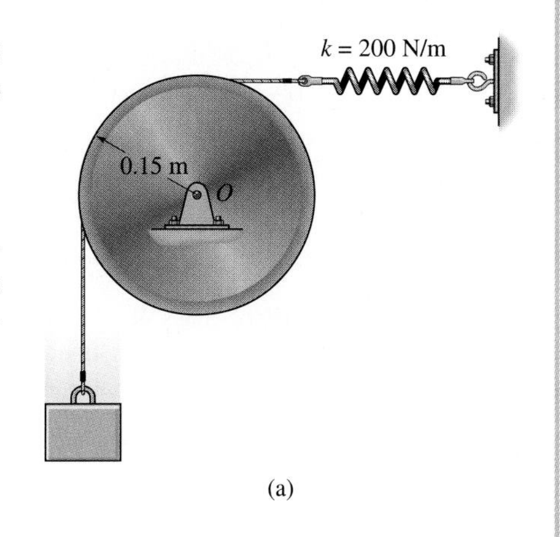

(a)

SOLUTION

Energy Equation. A diagram of the block and disk when they are displaced by respective amounts s and θ from the equilibrium position is shown in Fig. 22–10*b*. Since $s = (0.15\text{ m})\theta$, the kinetic energy of the system is

$$\begin{aligned} T &= \tfrac{1}{2}m_b v_b^2 + \tfrac{1}{2}I_O\omega_d^2 \\ &= \tfrac{1}{2}(10\text{ kg})[(0.15\text{ m})\dot{\theta}]^2 + \tfrac{1}{2}[\tfrac{1}{2}(5\text{ kg})(0.15\text{ m})^2](\dot{\theta})^2 \\ &= 0.141(\dot{\theta})^2 \end{aligned}$$

Establishing the datum at the equilibrium position of the block and realizing that the spring stretches s_{st} for equilibrium, we can write the potential energy as

$$\begin{aligned} V &= \tfrac{1}{2}k(s_{st} + s)^2 - Ws \\ &= \tfrac{1}{2}(200\text{ N/m})[s_{st} + (0.15\text{ m})\theta]^2 - 98.1\text{ N}[(0.15\text{ m})\theta] \end{aligned}$$

The total energy for the system is, therefore,

$$T + V = 0.141(\dot{\theta})^2 + 100(s_{st} + 0.15\theta)^2 - 14.72\theta$$

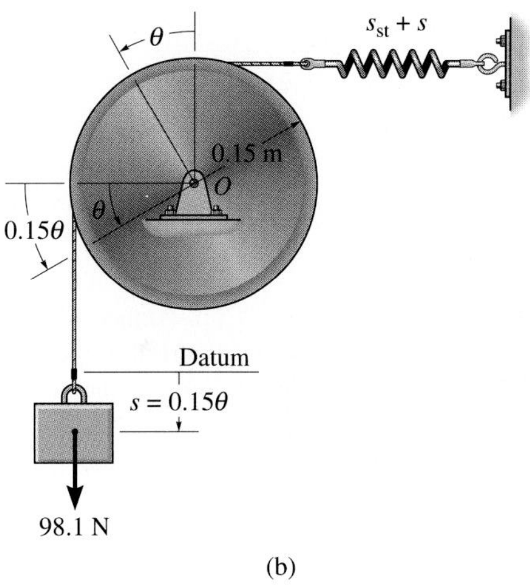

(b)

Fig. 22–10

Time Derivative.

$$0.282(\dot{\theta})\ddot{\theta} + 200(s_{st} + 0.15\theta)0.15\dot{\theta} - 14.72\dot{\theta} = 0$$

Since $s_{st} = 98.1/200 = 0.4905$ m, the above equation reduces to the standard form

$$\ddot{\theta} + 16\theta = 0$$

so that

$$p = \sqrt{16} = 4\text{ rad/s}$$

Thus,

$$\tau = \frac{2\pi}{p} = \frac{2\pi}{4} = 1.57\text{ s} \qquad \textit{Ans.}$$

PROBLEMS

22–27. Solve Prob. 22–11 using energy methods.

***22–28.** Solve Prob. 22–12 using energy methods.

22–29. Solve Prob. 22–14 using energy methods.

22–30. Solve Prob. 22–15 using energy methods.

22–31. The uniform rod of mass m is supported by a pin at A and a spring at B. If the end B is given a small downward displacement and released, determine the natural period of vibration.

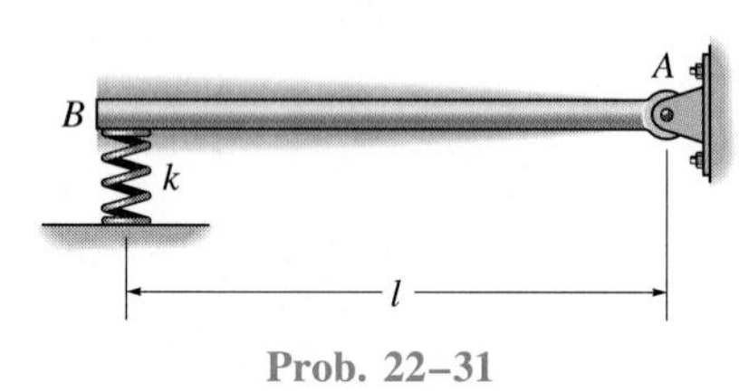

Prob. 22–31

***22–32.** The 7-kg disk is pin-connected at its midpoint. Determine the natural period of vibration of the disk if the springs have sufficient tension in them to prevent the cord from slipping on the disk as it oscillates. *Hint:* Assume that the initial stretch in each spring is δ_O. This term will cancel out after taking the time derivative of the energy equation.

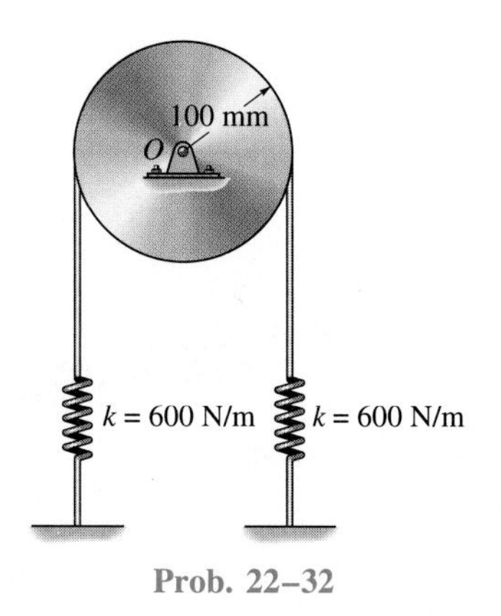

Prob. 22–32

22–33. The machine has a mass m and is uniformly supported by four springs each having a stiffness k. Determine the natural period of vertical vibration.

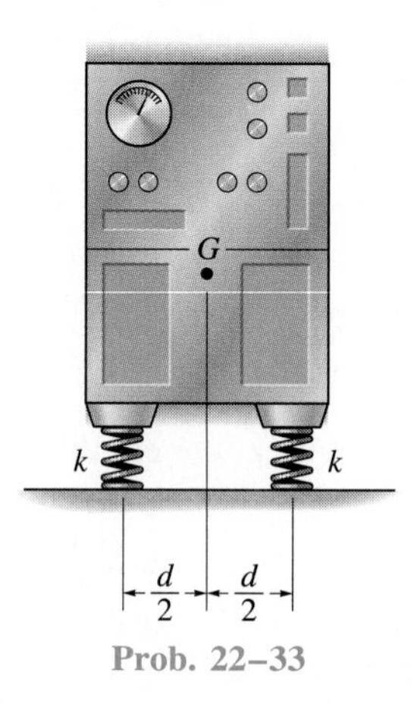

Prob. 22–33

22–34. Determine the natural period of vibration of the 10-lb semicircular disk.

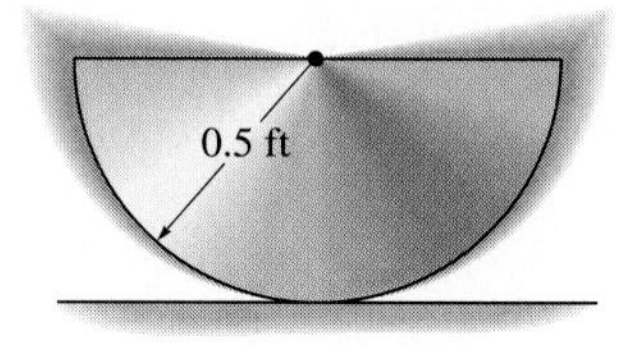

Prob. 22–34

22–35. Determine the natural period of vibration of the 3-kg sphere. Neglect the mass of the rod and the size of the sphere.

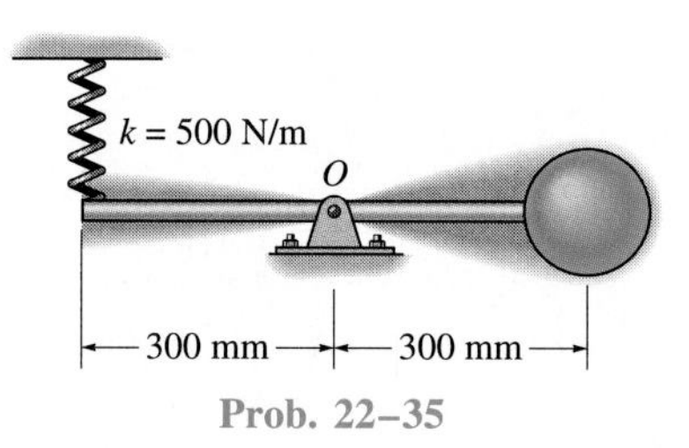

Prob. 22–35

*22–36. The 5-lb sphere is attached to a rod of negligible mass. Determine the natural frequency of vibration of the sphere. Neglect the size of the sphere.

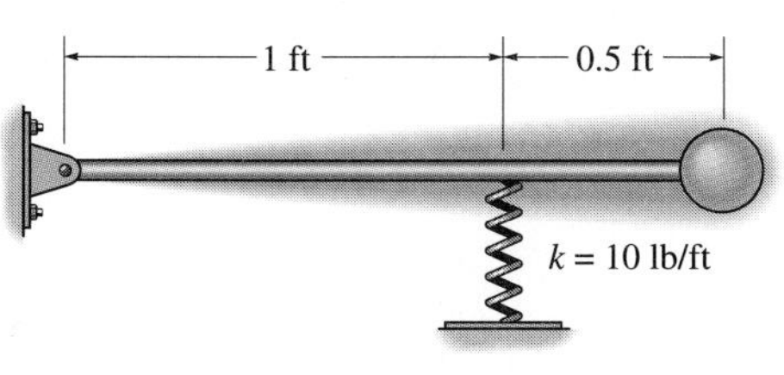

Prob. 22–36

22–37. Determine the natural frequency of vibration of the 20-lb disk. Assume the disk does not slip on the inclined surface.

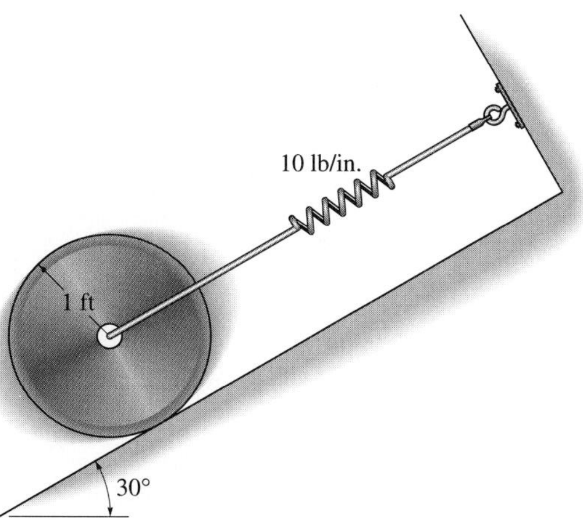

Prob. 22–37

22–38. Determine the natural period of vibration of the pendulum. Consider the two rods to be slender, each having a weight of 8 lb/ft.

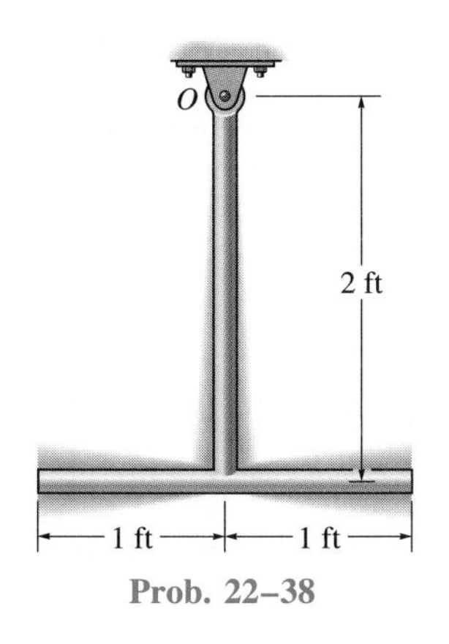

Prob. 22–38

22–39. If the disk has a mass of 8 kg, determine the natural frequency of vibration. The springs are originally unstretched.

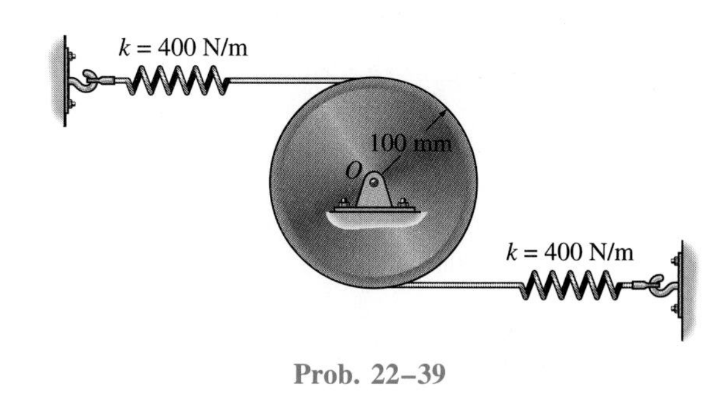

Prob. 22–39

*22–40. The slender rod has a weight of 4 lb/ft. If it is supported in the horizontal plane by a ball-and-socket joint at A and a cable at B, determine the natural frequency of vibration when the end B is given a small horizontal displacement and then released.

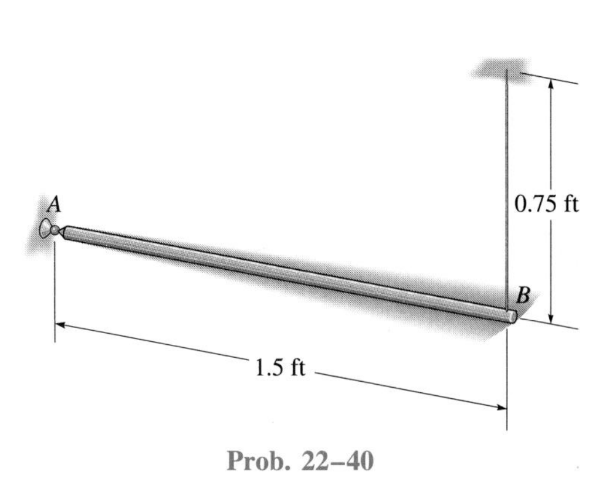

Prob. 22–40

22–41. The bar has a mass of 8 kg and is suspended from two springs such that when it is in equilibrium, the springs make an angle of 45° with the horizontal as shown. Determine the natural period of vibration if the bar is pulled down a short distance and released. Each spring has a stiffness of $k = 40$ N/m.

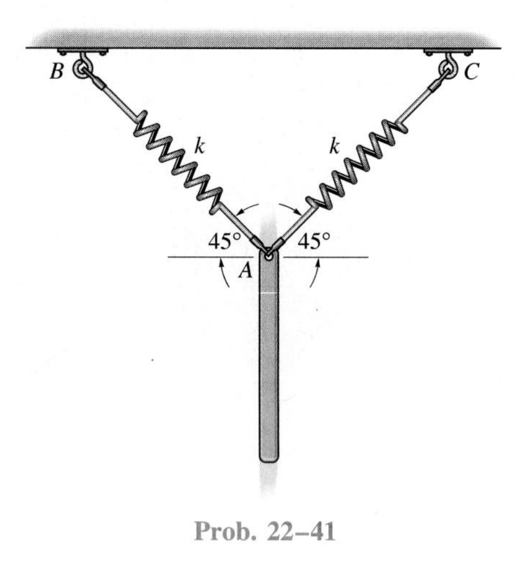

Prob. 22–41

22–42. Determine the differential equation of motion of the 3-kg spool. Assume that it does not slip at the surface of contact as it oscillates. The radius of gyration of the spool about its center of mass is $k_G = 125$ mm.

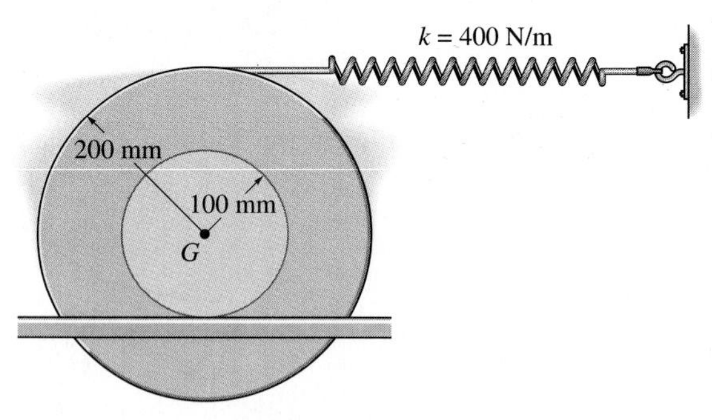

Prob. 22–42

*22.3 Undamped Forced Vibration

Undamped forced vibration is considered to be one of the most important types of vibrating motion in engineering work. The principles which describe the nature of this motion may be used to analyze the forces which cause vibration in many types of machines and structures.

Periodic Force. The block and spring shown in Fig. 22–11*a* provide a convenient ''model'' which represents the vibrational characteristics of a system subjected to a periodic force $F = F_O \sin \omega t$. This force has a maximum magnitude of F_O and a *forcing frequency* ω. The free-body diagram for the block when it is displaced a distance x is shown in Fig. 22–11*b*. Applying the equation of motion yields

$$\overset{+}{\rightarrow}\Sigma F_x = ma_x; \qquad F_O \sin \omega t - kx = m\ddot{x}$$

or

$$\ddot{x} + \frac{k}{m}x = \frac{F_O}{m}\sin \omega t \tag{22–18}$$

This equation is referred to as a nonhomogeneous second-order differential equation. The general solution consists of a complementary solution, x_c, *plus* a particular solution, x_p.

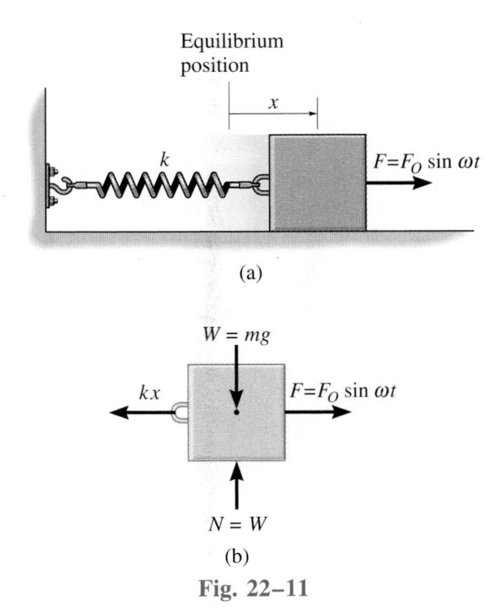

Fig. 22–11

The *complementary solution* is determined by setting the term on the right side of Eq. 22–18 equal to zero and solving the resulting homogeneous equation, which is equivalent to Eq. 22–1. The solution is defined by Eq. 22–3, i.e.,

$$x_c = A \sin pt + B \cos pt \tag{22–19}$$

where p is the circular frequency, $p = \sqrt{k/m}$, Eq. 22–2.

Since the motion is periodic, the *particular solution* of Eq. 22–18 may be determined by assuming a solution of the form

$$x_p = C \sin \omega t \tag{22–20}$$

where C is a constant. Taking the second time derivative and substituting into Eq. 22–18 yields

$$-C\omega^2 \sin \omega t + \frac{k}{m}(C \sin \omega t) = \frac{F_O}{m} \sin \omega t$$

Factoring out $\sin \omega t$ and solving for C gives

$$C = \frac{F_O/m}{(k/m) - \omega^2} = \frac{F_O/k}{1 - (\omega/p)^2} \tag{22–21}$$

Substituting into Eq. 22–20, we obtain the particular solution

$$x_p = \frac{F_O/k}{1 - (\omega/p)^2} \sin \omega t \tag{22–22}$$

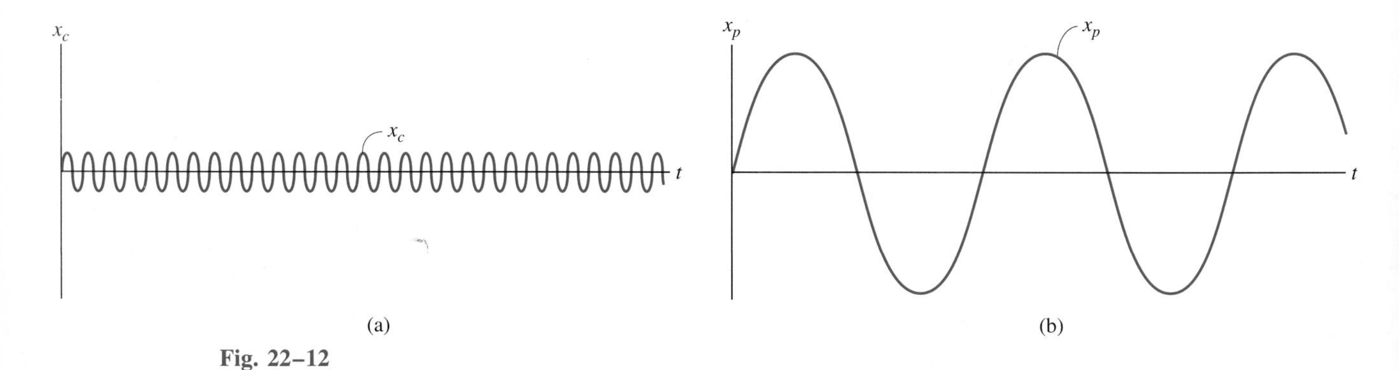

Fig. 22–12

The *general solution* is therefore

$$x = x_c + x_p = A \sin pt + B \cos pt + \frac{F_O/k}{1 - (\omega/p)^2} \sin \omega t \quad (22\text{–}23)$$

Here x describes two types of vibrating motion of the block. The *complementary solution* x_c defines the *free vibration,* which depends on the circular frequency $p = \sqrt{k/m}$ and the constants A and B, Fig. 22–12*a*. Specific values for A and B are obtained by evaluating Eq. 22–23 at a given instant when the displacement and velocity are known. The *particular solution* x_p describes the *forced vibration* of the block caused by the applied force $F = F_O \sin \omega t$, Fig. 22–12*b*. The resultant vibration x is shown in Fig. 22–12*c*. Since all vibrating systems are subject to *friction,* the free vibration, x_c, will in time dampen out. For this reason the free vibration is referred to as *transient,* and the forced vibration is called *steady state,* since it is the only vibration that remains, Fig. 22–12*d*.

From Eq. 22–21 it is seen that the *amplitude* of forced vibration depends on the *frequency ratio* ω/p. If the *magnification factor* MF is defined as the ratio of the amplitude of steady-state vibration, $(x_p)_{\max}$, to the static deflection F_O/k, which is caused by the amplitude of the periodic force F_O, then, from Eq. 22–22,

$$\text{MF} = \frac{(x_p)_{\max}}{F_O/k} = \frac{1}{1 - (\omega/p)^2} \quad (22\text{–}24)$$

This equation is graphed in Fig. 22–13, where it is seen that for $\omega \approx 0$, the MF ≈ 1. In this case, because of the very low frequency $\omega \ll p$, the magnitude of the force **F** changes slowly and so the vibration of the block will be in phase with the applied force **F.** If the force or displacement is applied with a frequency close to the natural frequency of the system, i.e., $\omega/p \approx 1$, the amplitude of vibration of the block becomes extremely large. This occurs because the force **F** is applied to the block so that it always follows the motion of the block. This condition is called *resonance,* and in practice, resonating vibra-

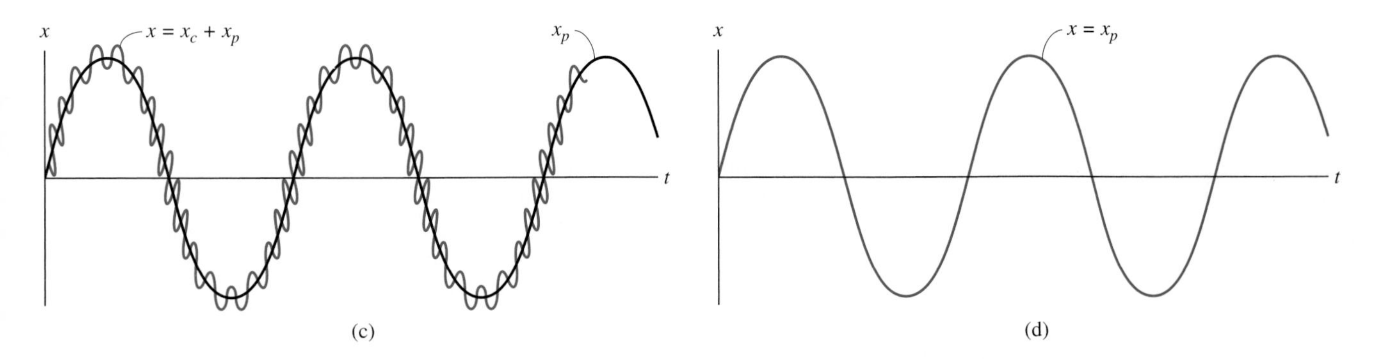

tions can cause tremendous stress and rapid failure of parts. When the cyclic force $F_O \sin \omega t$ is applied at high frequencies ($\omega > p$), the value of the MF becomes negative, indicating that the motion of the block is out of phase with the force. Under these conditions, as the block is displaced to the right, the force acts to the left, and vice versa. For extremely high frequencies ($\omega \gg p$) the block remains almost stationary, and hence the MF is approximately zero.

Fig. 22–13

Periodic Support Displacement. Forced vibrations can also arise from the periodic excitation of the support of a system. The model shown in Fig. 22–14*a* represents the periodic vibration of a block which is caused by harmonic movement $\delta = \delta_0 \sin \omega t$ of the support. The free-body diagram for the block in this case is shown in Fig. 22–14*b*. The coordinate x is measured from the point of zero displacement of the support, i.e., when the radial line OA coincides with OB, Fig. 22–14*a*. Therefore, general displacement of the spring is $(x - \delta_0 \sin \omega t)$. Applying the equation of motion yields

$$\xrightarrow{+}\Sigma F_x = ma_x; \qquad -k(x - \delta_0 \sin \omega t) = m\ddot{x}$$

or

$$\ddot{x} + \frac{k}{m}x = \frac{k\delta_0}{m}\sin \omega t \qquad (22\text{–}25)$$

By comparison, this equation is identical to the form of Eq. 22–18, *provided* F_O *is replaced* by $k\delta_0$. If this substitution is made into the solutions defined by Eqs. 22–21 to 22–23, the results are appropriate for describing the motion of the block when subjected to the support displacement $\delta = \delta_0 \sin \omega t$.

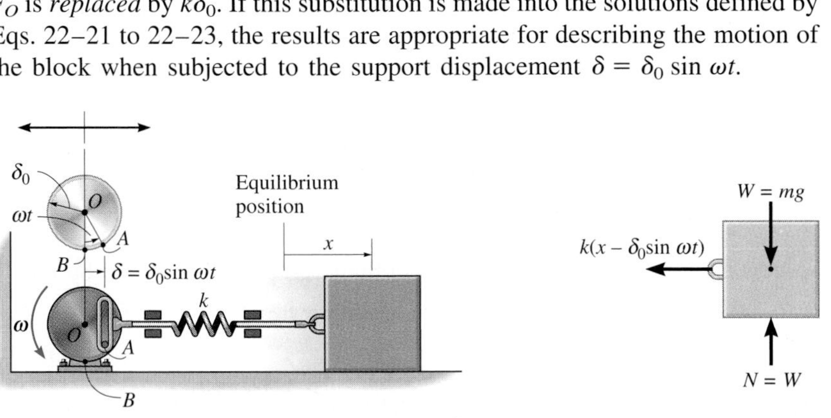

Fig. 22–14

Example 22–7

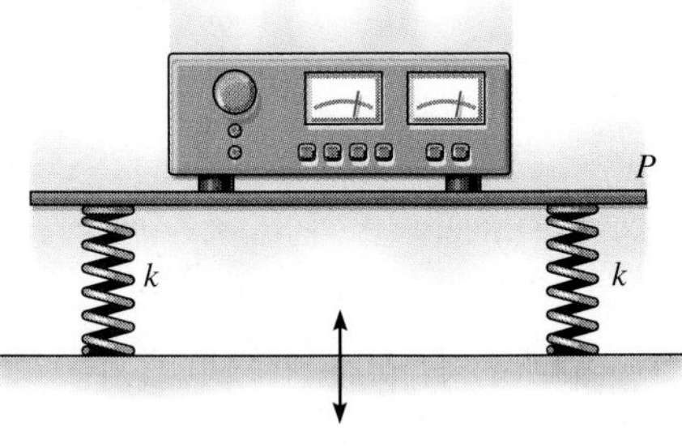

Fig. 22–15

The instrument shown in Fig. 22–15 is rigidly attached to a platform P, which in turn is supported by *four* springs, each having a stiffness $k = 800$ N/m. Initially the platform is at rest when the floor is subjected to a displacement $\delta = 10\sin(8t)$ mm, where t is in seconds. If the instrument is constrained to move vertically, and the total mass of the instrument and platform is 20 kg, determine the vertical displacement y of the platform, measured from the equilibrium position, as a function of time. What floor vibration is required to cause resonance?

SOLUTION

Since the induced vibration is caused by the displacement of the supports, the motion is described by Eq. 22–23, with F_O replaced by $k\delta_O$, i.e.,

$$y = A\sin pt + B\cos pt + \frac{\delta_O}{1 - (\omega/p)^2}\sin\omega t \qquad (1)$$

Here $\delta = \delta_O \sin\omega t = 10\sin(8t)$ mm, so that

$$\delta_O = 10 \text{ mm} \qquad \omega = 8 \text{ rad/s}$$

$$p = \sqrt{\frac{k}{m}} = \sqrt{\frac{4(800 \text{ N/m})}{20 \text{ kg}}} = 12.6 \text{ rad/s}$$

From Eq. 22–22, with $k\delta_O$ replacing F_O, the amplitude of vibration caused by the floor displacement is

$$(y_p)_{\max} = \frac{\delta_O}{1 - (\omega/p)^2} = \frac{10}{1 - (8 \text{ rad/s}/12.6 \text{ rad/s})^2} = 16.7 \text{ mm} \qquad (2)$$

Hence, Eq. 1 and its time derivative become

$$y = A\sin(12.6t) + B\cos(12.6t) + 16.7\sin(8t)$$

$$\dot{y} = A(12.6)\cos(12.6t) - B(12.6)\sin(12.6t) + 133.3\cos(8t)$$

The constants A and B are evaluated from these equations. Since $y = 0$ and $\dot{y} = 0$ at $t = 0$, then

$$0 = 0 + B + 0 \qquad B = 0$$

$$0 = A(12.6) - 0 + 133.3 \qquad A = -10.6$$

The vibrating motion is therefore described by the equation

$$y = -10.6\sin(12.6t) + 16.7\sin(8t) \qquad \textit{Ans.}$$

Resonance will occur when the amplitude of vibration caused by the floor displacement approaches infinity. From Eq. 2, this requires

$$\omega = p = 12.6 \text{ rad/s} \qquad \textit{Ans.}$$

*22.4 Viscous Damped Free Vibration

The vibration analysis considered thus far has not included the effects of friction or damping in the system, and as a result, the solutions obtained are only in close agreement with the actual motion. Since all vibrations die out in time, the presence of damping forces should be included in the analysis.

In many cases damping is attributed to the resistance created by the substance, such as water, oil, or air, in which the system vibrates. Provided the body moves slowly through this substance, the resistance to motion is directly proportional to the body's speed. The type of force developed under these conditions is called a *viscous damping force*. The magnitude of this force is expressed by an equation of the form

$$F = c\dot{x} \tag{22-26}$$

where the constant c is called the *coefficient of viscous damping* and has units of N · s/m or lb · s/ft.

The vibrating motion of a body or system having viscous damping may be characterized by the block and spring shown in Fig. 22–16*a*. The effect of damping is provided by the *dashpot* connected to the block on the right side. Damping occurs when the piston P moves to the right or left within the enclosed cylinder. The cylinder contains a fluid, and the motion of the piston is retarded since the fluid must flow around or through a small hole in the piston. The dashpot is assumed to have a coefficient of viscous damping c.

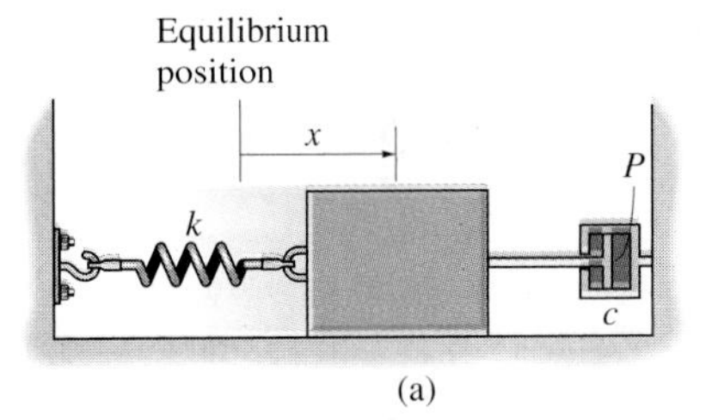

(a)

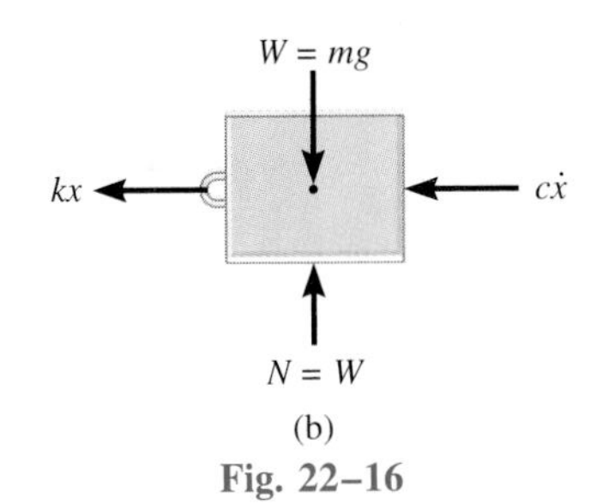

(b)

Fig. 22–16

If the block is displaced a distance x from its equilibrium position, the resulting free-body diagram is shown in Fig. 22–16*b*. Both the spring force kx and the damping force $c\dot{x}$ oppose the forward motion of the block, so that applying the equation of motion yields

$$\overset{+}{\rightarrow}\Sigma F_x = ma_x; \qquad -kx - c\dot{x} = m\ddot{x}$$

or

$$m\ddot{x} + c\dot{x} + kx = 0 \tag{22-27}$$

This linear, second-order, homogeneous, differential equation has solutions of the form

$$x = e^{\lambda t}$$

where e is the base of the natural logarithm and λ (lambda) is a constant. The value of λ may be obtained by substituting this solution into Eq. 22–27, which yields

$$m\lambda^2 e^{\lambda t} + c\lambda e^{\lambda t} + ke^{\lambda t} = 0$$

or

$$e^{\lambda t}(m\lambda^2 + c\lambda + k) = 0$$

Since $e^{\lambda t}$ is never zero, a solution is possible provided

$$m\lambda^2 + c\lambda + k = 0$$

Hence, by the quadratic formula, the two values of λ are

$$\lambda_1 = -\frac{c}{2m} + \sqrt{\left(\frac{c}{2m}\right)^2 - \frac{k}{m}}$$
$$\lambda_2 = -\frac{c}{2m} - \sqrt{\left(\frac{c}{2m}\right)^2 - \frac{k}{m}} \tag{22-28}$$

The general solution of Eq. 22–27 is therefore a linear combination of exponentials which involves both of these roots. There are three possible combinations of λ_1 and λ_2 which must be considered for the general solution. Before discussing these combinations, however, let us first define the *critical damping coefficient* c_c as the value of c which makes the radical in Eqs. 22–28 equal to zero; i.e.,

$$\left(\frac{c_c}{2m}\right)^2 - \frac{k}{m} = 0$$

or

$$c_c = 2m\sqrt{\frac{k}{m}} = 2mp \tag{22-29}$$

Here the value of p is the circular frequency $p = \sqrt{k/m}$, Eq. 22–2.

Overdamped System. When $c > c_c$, the roots λ_1 and λ_2 are both real. The general solution of Eq. 22–27 may then be written as

$$x = Ae^{\lambda_1 t} + Be^{\lambda_2 t} \tag{22-30}$$

Motion corresponding to this solution is *nonvibrating*. The effect of damping is so strong that when the block is displaced and released, it simply creeps back to its original position without oscillating. The system is said to be *overdamped.*

Critically Damped System. If $c = c_c$, then $\lambda_1 = \lambda_2 = -c_c/2m = -p$. This situation is known as *critical damping,* since it represents a condition where c has the smallest value necessary to cause the system to be nonvibrating. Using the methods of differential equations, it may be shown that the solution to Eq. 22–27 for critical damping is

$$x = (A + Bt)e^{-pt} \tag{22-31}$$

Underdamped System. Most often $c < c_c$, in which case the system is referred to as *underdamped.* In this case the roots λ_1 and λ_2 are complex numbers and it may be shown that the general solution of Eq. 22–27 can be written as

$$x = D[e^{-(c/2m)t} \sin(p_d t + \phi)] \tag{22–32}$$

where D and ϕ are constants generally determined from the initial conditions of the problem. The constant p_d is called the *damped natural frequency* of the system. It has a value of

$$p_d = \sqrt{\frac{k}{m} - \left(\frac{c}{2m}\right)^2} = p\sqrt{1 - \left(\frac{c}{c_c}\right)^2} \tag{22–33}$$

where the ratio c/c_c is called the *damping factor.*

The graph of Eq. 22–32 is shown in Fig. 22–17. The initial limit of motion, D, diminishes with each cycle of vibration, since motion is confined within the bounds of the exponential curve. Using the damped natural frequency p_d, the period of damped vibration may be written as

$$\tau_d = \frac{2\pi}{p_d} \tag{22–34}$$

Since $p_d < p$, Eq. 22–33, the period of damped vibration, τ_d, will be greater than that of free vibration, $\tau = 2\pi/p$.

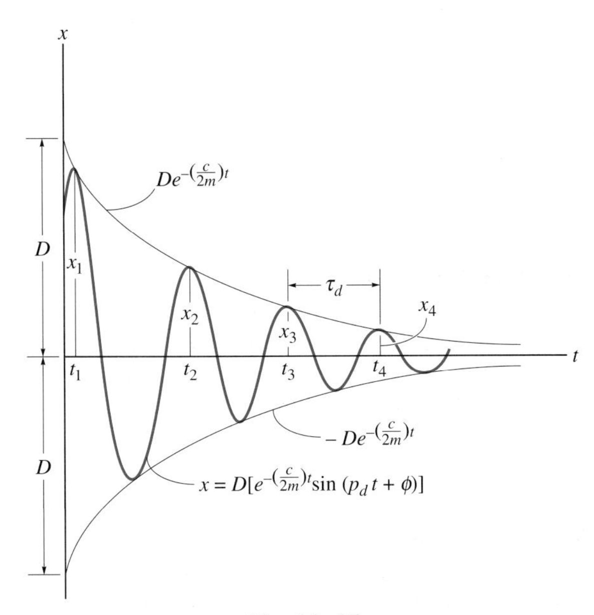

Fig. 22–17

★22.5 Viscous Damped Forced Vibration

The most general case of single-degree-of-freedom vibrating motion occurs when the system includes the effects of forced motion and induced damping. The analysis of this particular type of vibration is of practical value when applied to systems having significant damping characteristics.

If a dashpot is attached to the block and spring shown in Fig. 22–11*a*, the differential equation which describes the motion becomes

$$m\ddot{x} + c\dot{x} + kx = F_O \sin \omega t \qquad (22\text{–}35)$$

A similar equation may be written for a block and spring having a periodic support displacement, Fig. 22–14*a*, which includes the effects of damping. In that case, however, F_O is replaced by $k\delta_O$. Since Eq. 22–35 is nonhomogeneous, the general solution is the sum of a complementary solution, x_c, and a particular solution, x_p. The complementary solution is determined by setting the right side of Eq. 22–35 equal to zero and solving the homogeneous equation, which is equivalent to Eq. 22–27. The solution is therefore given by Eq. 22–30, 22–31, or 22–32, depending on the values of λ_1 and λ_2. Because all systems contain friction, however, this solution will dampen out with time. Only the particular solution, which describes the *steady-state vibration* of the system, will remain. Since the applied forcing function is harmonic, the steady-state motion will also be harmonic. Consequently, the particular solution will be of the form

$$x_p = A' \sin \omega t + B' \cos \omega t \qquad (22\text{–}36)$$

The constants A' and B' are determined by taking the necessary time derivatives and substituting them into Eq. 22–35, which after simplification yields

$$(-A'm\omega^2 - cB'\omega + kA') \sin \omega t + (-B'm\omega^2 + cA'\omega + kB') \cos \omega t = F_O \sin \omega t$$

Since this equation holds for all time, the constant coefficients of $\sin \omega t$ and $\cos \omega t$ may be equated; i.e.,

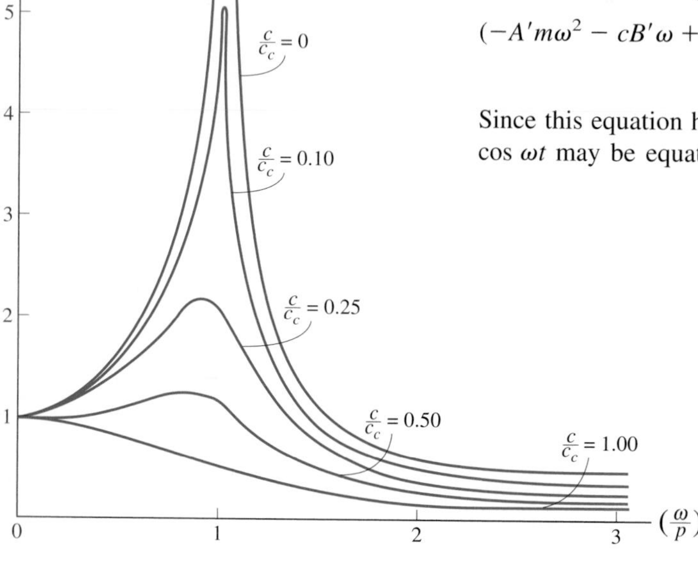

Fig. 22–18

$$-A'm\omega^2 - cB'\omega + kA' = F_O$$
$$-B'm\omega^2 + cA'\omega + kB' = 0$$

Solving for A' and B', realizing that $p^2 = k/m$, yields

$$A' = \frac{(F_O/m)(p^2 - \omega^2)}{(p^2 - \omega^2)^2 + (c\omega/m)^2}$$
$$B' = \frac{-F_O(c\omega/m^2)}{(p^2 - \omega^2)^2 + (c\omega/m)^2} \qquad (22\text{–}37)$$

It is also possible to express Eq. 22–36 in a form similar to Eq. 22–9,

$$x_p = C' \sin(\omega t - \phi') \qquad (22\text{–}38)$$

in which case the constants C' and ϕ' are

$$C' = \frac{F_O/k}{\sqrt{[1 - (\omega/p)^2]^2 + [2(c/c_c)(\omega/p)]^2}}$$
$$\phi' = \tan^{-1}\left[\frac{2(c/c_c)(\omega/p)}{1 - (\omega/p)^2}\right] \qquad (22\text{–}39)$$

The angle ϕ' represents the phase difference between the applied force and the resulting steady-state vibration of the damped system.

The *magnification factor* MF has been defined in Sec. 22.3 as the ratio of the amplitude of deflection caused by the forced vibration to the deflection caused by a static force $\mathbf{F}_O$. From Eq. 22–38, the forced vibration has an amplitude of C'; thus,

$$\text{MF} = \frac{C'}{F_O/k} = \frac{1}{\sqrt{[1 - (\omega/p)^2]^2 + [2(c/c_c)(\omega/p)]^2}} \qquad (22\text{–}40)$$

The MF is plotted in Fig. 22–18 versus the frequency ratio ω/p for various values of the damping factor c/c_c. It can be seen from this graph that the magnification of the amplitude increases as the damping factor decreases. Resonance obviously occurs only when the damping factor is zero and the frequency ratio equals 1.

Example 22–8

The 30-kg electric motor shown in Fig. 22–19 is supported by *four* springs, each spring having a stiffness of 200 N/m. If the rotor R is unbalanced such that its effect is equivalent to a 4-kg mass located 60 mm from the axis of rotation, determine the amplitude of vibration when the rotor is turning at $\omega = 10$ rad/s. The damping factor is $c/c_c = 0.15$.

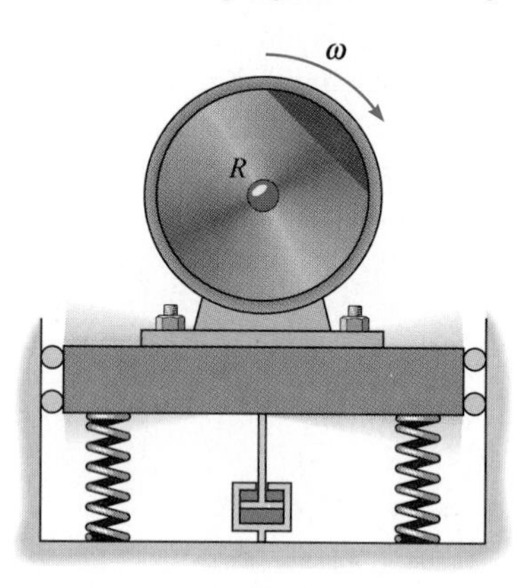

Fig. 22–19

SOLUTION

The periodic force which causes the motor to vibrate is the centrifugal force due to the unbalanced rotor. This force has a constant magnitude of

$$F_O = ma_n = mr\omega^2 = 4\text{ kg}(0.06\text{ m})(10\text{ rad/s})^2 = 24\text{ N}$$

Oscillation in the vertical direction may be expressed in the periodic form $F = F_O \sin \omega t$, where $\omega = 10$ rad/s. Thus,

$$F = 24 \sin 10t$$

The stiffness of the entire system of four springs is $k = 4(200\text{ N/m}) = 800\text{ N/m}$. Therefore, the circular frequency of vibration is

$$p = \sqrt{\frac{k}{m}} = \sqrt{\frac{800\text{ N/m}}{30\text{ kg}}} = 5.16\text{ rad/s}$$

Since the damping factor is known, the steady-state amplitude may be determined from the first of Eqs. 22–39, i.e.,

$$C' = \frac{F_O/k}{\sqrt{[1 - (\omega/p)^2]^2 + [2(c/c_c)(\omega/p)]^2}}$$

$$= \frac{24/800}{\sqrt{[1 - (10/5.16)^2]^2 + [2(0.15)(10/5.16)]^2}}$$

$$= 0.0107\text{ m} = 10.7\text{ mm}$$ ***Ans.***

*22.6 Electrical Circuit Analogs

The characteristics of a vibrating mechanical system may be represented by an electric circuit. Consider the circuit shown in Fig. 22–20*a*, which consists of an inductor L, a resistor R, and a capacitor C. When a voltage $E(t)$ is applied, it causes a current of magnitude i to flow through the circuit. As the current flows past the inductor the voltage drop is $L(di/dt)$, when it flows across the resistor the drop is Ri, and when it arrives at the capacitor the drop is $(1/C)\int i\,dt$. Since current cannot flow past a capacitor, it is only possible to measure the charge q acting on the capacitor. The charge may, however, be related to the current by the equation $i = dq/dt$. Thus, the voltage drops which occur across the inductor, resistor, and capacitor may be written as $L\,d^2q/dt^2$, $R\,dq/dt$, and q/C, respectively. According to Kirchhoff's voltage law, the applied voltage balances the sum of the voltage drops around the circuit. Therefore,

$$L\frac{d^2q}{dt^2} + R\frac{dq}{dt} + \frac{1}{C}q = E(t) \qquad (22\text{–}41)$$

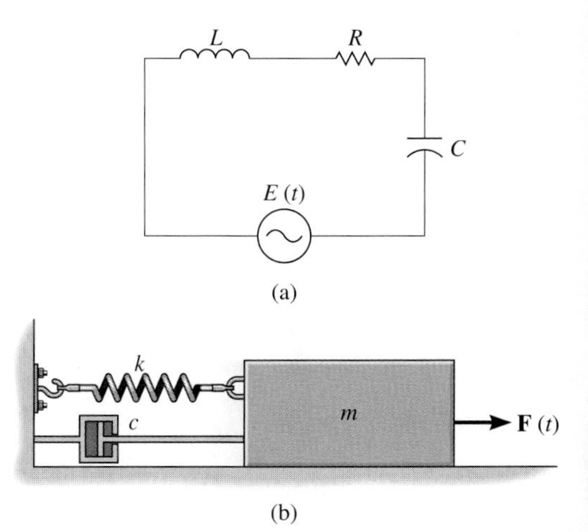

Fig. 22–20

Consider now the model of a single-degree-of-freedom mechanical system, Fig. 22–20*b*, which is subjected to both a general forcing function $F(t)$ and damping. The equation of motion for this system was established in the previous section and can be written as

$$m\frac{d^2x}{dt^2} + c\frac{dx}{dt} + kx = F(t) \qquad (22\text{–}42)$$

By comparison, it is seen that Eqs. 22–41 and 22–42 have the same form, hence mathematically the problem of analyzing an electric circuit is the same as that of analyzing a vibrating mechanical system. The analogs between the two equations are given in Table 22–1.

This analogy has important application to experimental work, for it is much easier to simulate the vibration of a complex mechanical system using an electric circuit, which can be constructed on an analog computer, than to make an equivalent mechanical spring-and-dashpot model.

Table 22–1 Electrical-Mechanical Analogs

Electrical		*Mechanical*	
Electric charge	q	Displacement	x
Electric current	i	Velocity	dx/dt
Voltage	$E(t)$	Applied force	$F(t)$
Inductance	L	Mass	m
Resistance	R	Viscous damping coefficient	c
Reciprocal of capacitance	$1/C$	Spring stiffness	k

PROBLEMS

22–43. Use a block-and-spring model like that shown in Fig. 22-14a but suspended from a vertical position and subjected to a periodic support displacement of $\delta = \delta_0 \cos \omega t$, determine the equation of motion for the system, and obtain its general solution. Define the displacement y measured from the static equilibrium position of the block when $t = 0$.

***22–44.** A 4-lb weight is attached to a spring having a stiffness $k = 10$ lb/ft. The weight is drawn downward a distance of 4 in. and released from rest. If the support moves with a vertical displacement $\delta = (0.5 \sin 4t)$ in., where t is in seconds, determine the equation which describes the position of the weight as a function of time.

22–45. A 7-kg block is suspended from a spring that has a stiffness $k = 350$ N/m. The block is drawn downward 70 mm from the equilibrium position and released from rest at $t = 0$. If the support moves with an impressed displacement $\delta = (20 \sin 4t)$ mm, where t is in seconds, determine the equation which describes the vertical motion of the block. Assume that positive displacement is measured downward.

22–46. Use a block-and-spring model like that shown in Fig. 22–14a, but suspended from a vertical position and subjected to a periodic support displacement $\delta = \delta_0 \sin \omega t$, determine the equation of motion for the system, and obtain its general solution. Define the displacement y measured from the static equilibrium position of the block when $t = 0$.

22–47. A block which has a mass m is suspended from a spring having a stiffness k. If an impressed downward vertical force $F = F_O$ acts on the weight, determine the equation which describes the position of the block as a function of time.

***22–48.** A 5-lb weight is suspended from a vertical spring having a stiffness of 50 lb/ft. An impressed force $F = (0.25 \sin 8t)$ lb, where t is in seconds, is acting on the weight. Determine the equation of motion of the weight when it is pulled down 3 in. from the equilibrium position and released from rest.

22–49. The 20-lb block is attached to a spring having a stiffness of 20 lb/ft. A force $F = (6 \cos 2t)$ lb, where t is in seconds, is applied to the block. Determine the maximum speed of the block after frictional forces cause the free vibrations to dampen out.

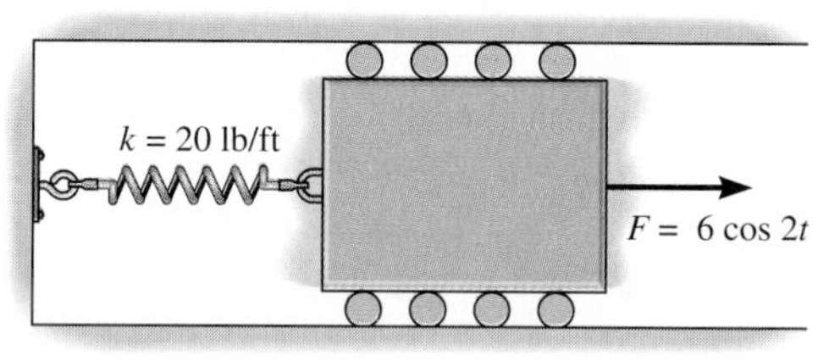

Prob. 22–49

22–50. The light elastic rod supports a 4-kg sphere. When an 18-N vertical force is applied to the sphere, the rod deflects 14 mm. If the wall oscillates with harmonic frequency of 2 Hz and has an amplitude of 15 mm, determine the amplitude of vibration for the sphere.

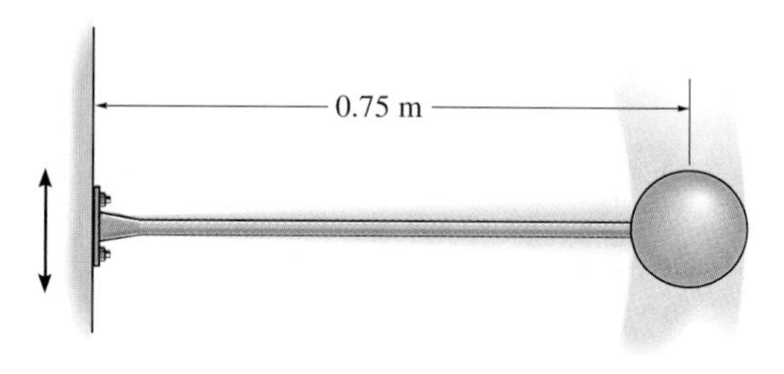

Prob. 22–50

22–51. A spring has a stiffness $k = 2$ lb/in. If a 5-lb weight is attached to the spring and is given an upward velocity of 2 ft/s from the equilibrium position, determine the position of the weight as a function of time. Assume that motion takes place in a medium which furnishes a retarding force F (pounds) having a magnitude numerically equal to four times the speed v of the weight, where v is in ft/s.

*22–52. The electric motor has a mass of 50 kg and is supported by *four springs,* each spring having a stiffness of 100 N/m. If the motor turns a disk D which is mounted eccentrically, 20 mm from the disk's center, determine the angular rotation ω at which resonance occurs. Assume that the motor only vibrates in the vertical direction.

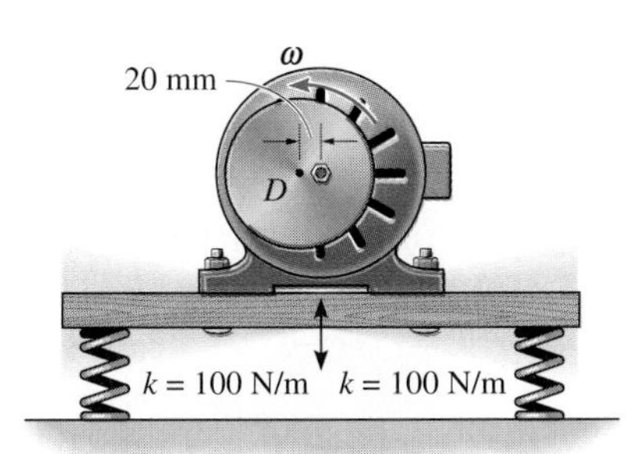

Prob. 22–52

22–53. The instrument is centered uniformly on a platform P, which in turn is supported by *four* springs, each spring having a stiffness $k = 130$ N/m. If the floor is subjected to a vibration $\omega = 7$ Hz, having a vertical displacement amplitude $\delta_O = 0.17$ ft, determine the vertical displacement amplitude of the platform and instrument. The instrument and the platform have a total weight of 18 lb.

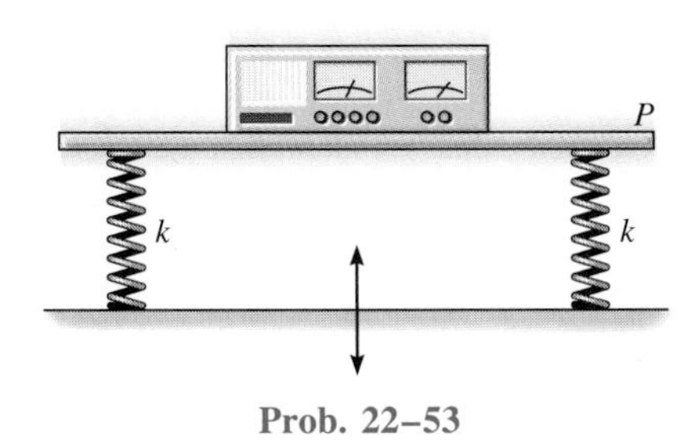

Prob. 22–53

22–54. The engine is mounted on a foundation block which is spring-supported. Describe the steady-state vibration of the system if the block and engine have a total weight of 1500 lb and the engine, when running, creates an impressed force $F = (50 \sin 2t)$ lb, where t is in seconds. Assume that the system vibrates only in the vertical direction, with the positive displacement measured downward, and that the total stiffness of the springs can be represented as $k = 2000$ lb/ft.

22–55. Determine the rotational speed ω of the engine in Prob. 22–54 which will cause resonance.

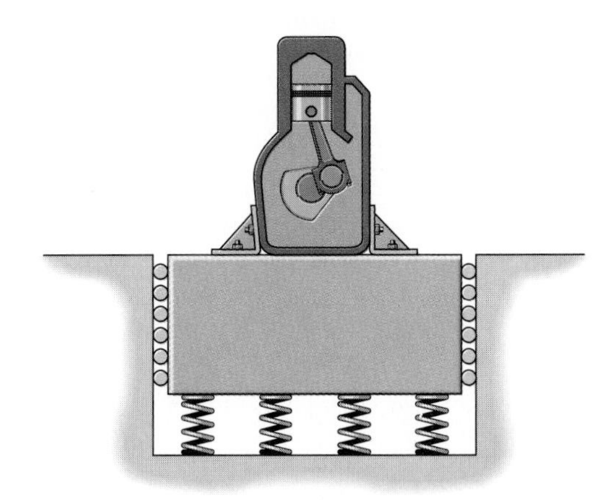

Probs. 22–54/22–55

*22–56. The fan has a mass of 25 kg and is fixed to the end of a horizontal beam that has a negligible mass. The fan blade is mounted eccentrically on the shaft such that it is equivalent to an unbalanced 3.5-kg mass located 100 mm from the axis of rotation. If the static deflection of the beam is 50 mm as a result of the weight of the fan, determine the angular speed of the fan at which resonance will occur. *Hint:* See the first part of Example 22–8.

22–57. What is the amplitude of steady-state vibration of the fan in Prob. 22–56 when the angular velocity of the fan is 10 rad/s? *Hint:* See the first part of Example 22–8.

22–58. What will be the amplitude of steady-state vibration of the fan in Prob. 22–56 if the angular velocity of the fan is 18 rad/s? *Hint:* See the first part of Example 22–8.

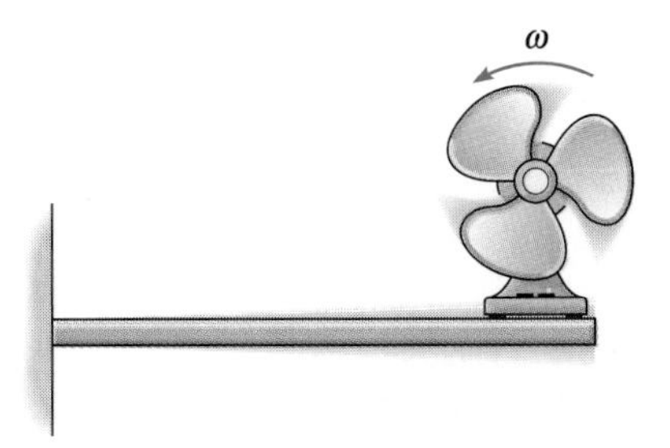

Probs. 22–56/22–57/22–58

22–59. The electric motor turns an eccentric flywheel which is equivalent to an unbalanced 0.25-lb weight located 10 in. from the axis of rotation. If the static deflection of the beam is 1 in. because of the weight of the motor, determine the angular velocity of the flywheel at which resonance will occur. The motor weighs 150 lb. Neglect the mass of the beam.

***22–60.** What will be the amplitude of steady-state vibration of the motor in Prob. 22–59 if the angular velocity of the flywheel is 20 rad/s?

22–61. Determine the angular velocity of the motor in Prob. 22–59 which will produce an amplitude of vibration of 0.25 in.

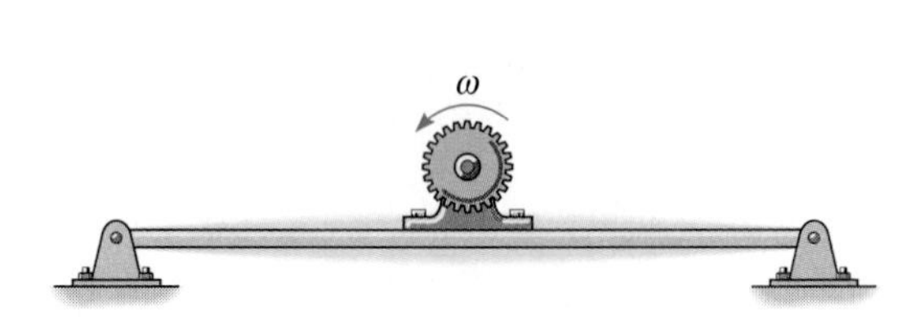

Probs. 22–59/22–60/22–61

22–62. A 5-kg block is suspended from a spring having a stiffness of 300 N/m. If the block is acted upon by a vertical force $F = (7 \sin 8t)$ N, where t is in seconds, determine the equation which describes the motion of the block when it is pulled down 100 mm from the equilibrium position and released from rest at $t = 0$. Assume that positive displacement is measured downward.

22–63. The 80-lb block is attached to a spring, the end of which is subjected to a periodic support displacement $\delta_A = (0.5 \sin 8t)$ ft, where t is in seconds. Determine the amplitude of the steady-state motion.

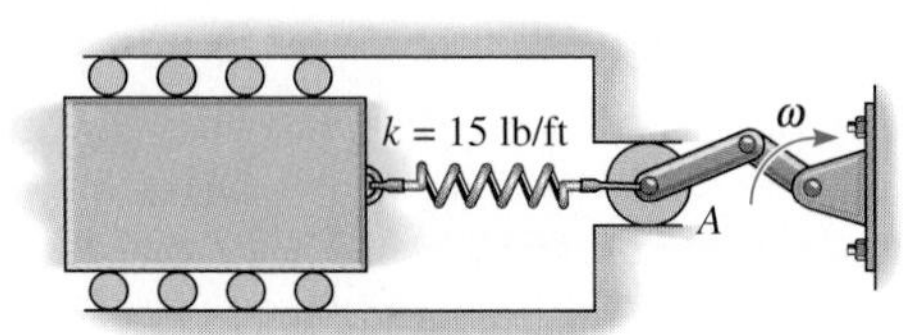

Prob. 22–63

***22–64.** The 450-kg trailer is pulled with a constant speed over the surface of a bumpy road, which may be approximated by a cosine curve having an amplitude of 50 mm and wave length of 4 m. If the two springs s which support the trailer each have a stiffness of 800 N/m, determine the speed v which will cause the greatest vibration (resonance) of the trailer. Neglect the weight of the wheels.

22–65. Determine the amplitude of vibration of the trailer in Prob. 22–64 if the speed $v = 15$ km/h.

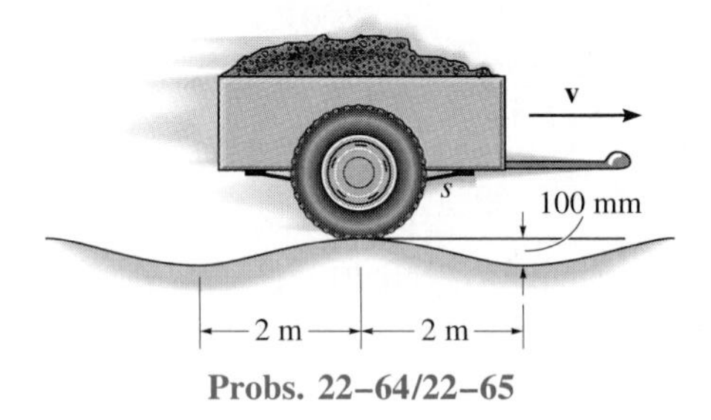

Probs. 22–64/22–65

22–66. The damping factor, c/c_c, may be determined experimentally by measuring the successive amplitudes of vibrating motion of a system. If two of these maximum displacements can be approximated by x_1 and x_2, as shown in Fig. 22–17, show that the ratio $\ln x_1/x_2 = 2\pi(c/c_c)/\sqrt{1-(c/c_c)^2}$. The quantity $\ln x_1/x_2$ is called the *logarithmic decrement.*

22–67. The 200-lb electric motor is fastened to the midpoint of the simply supported beam. It is found that the beam deflects 2 in. when the motor is not running. The motor turns an eccentric flywheel which is equivalent to an unbalanced weight of 1 lb located 5 in. from the axis of rotation. If the motor is turning at 100 rpm, determine the amplitude of steady-state vibration. The damping factor is $c/c_c = 0.20$. Neglect the mass of the beam.

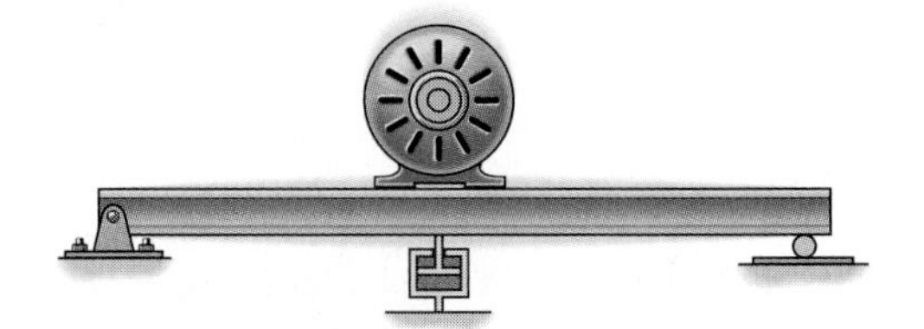

Prob. 22–67

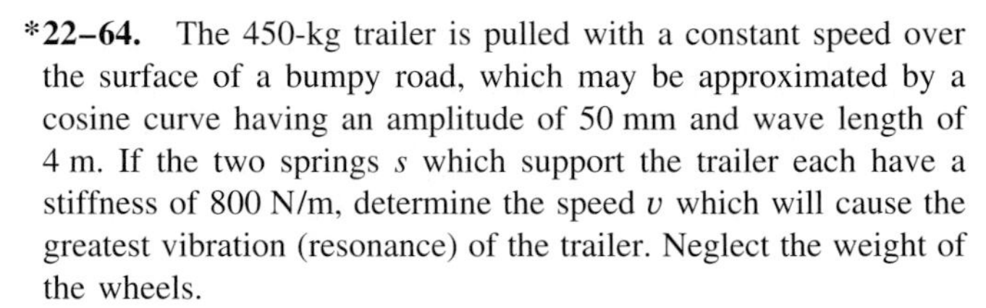

*22–68. The bell-crank mechanism consists of a bent rod, having a negligible mass, and an attached 5-lb weight. Determine the critical damping coefficient c_c and the damping natural frequency for small vibrations about the equilibrium position. Neglect the size of the weight.

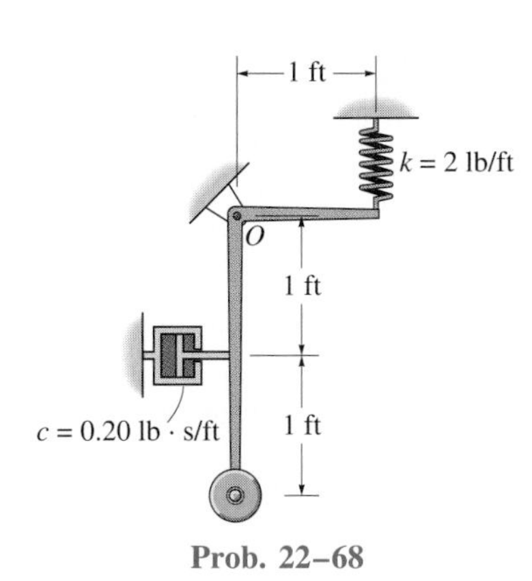

Prob. 22–68

22–69. A block having a weight of 7 lb is suspended from a spring having a stiffness $k = 75$ lb/ft. The support, to which the spring is attached is given simple harmonic motion, which may be damping factor is $c/c_c = 0.8$, determine the phase angle ϕ of forced vibration.

22–70. Determine the magnification factor of the block, spring, and dashpot combination in Prob. 22–69.

22–71. A block having a mass of 0.5 slug is suspended from a spring having a stiffness $k = 0.6$ lb/in. If a dashpot provides a damping force of 0.2 lb on the block when the speed of the block is $v = 1$ ft/s, determine the period of free vibration.

*22–72. The 4-kg circular disk is attached to three springs, each spring having a stiffness $k = 180$ N/m. If the disk is immersed in a fluid and given a downward velocity of 0.3 m/s at the equilibrium position, determine the equation which describes the motion. Assume that positive displacement is measured downward, and that fluid resistance acting on the disk furnishes a damping force having a magnitude $F = (60|v|)$ N, where v is in m/s.

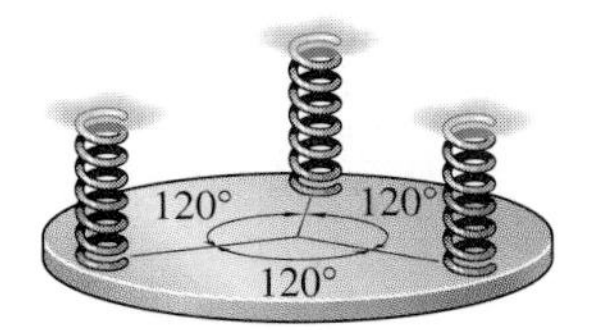

Prob. 22–72

22–73. The 20-kg block is subjected to the action of the harmonic force $F = (90 \cos 6t)$ N, where t is in seconds. Write the equation which describes the steady-state motion.

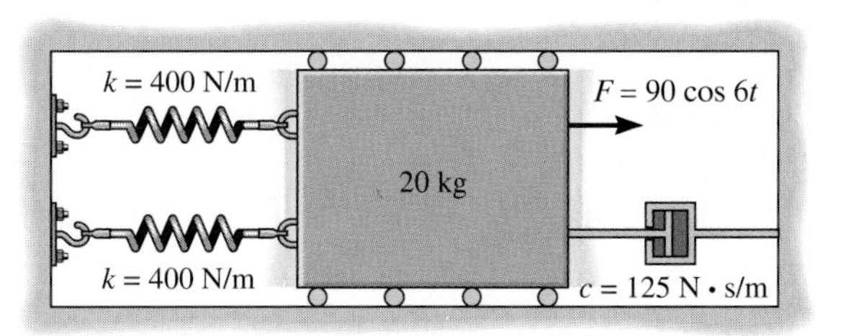

Prob. 22–73

22–74. A block having a weight of 7 lb is suspended from a spring having a stiffness $k = 75$ lb/ft. The support to which the spring is attached is given simple harmonic motion that may be expressed by $\delta = (0.15 \sin 2t)$ ft, where t is in seconds. If the damping factor is $c/c_c = 0.8$, determine the phase angle ϕ of forced vibration.

22–75. Determine the magnification factor of the block, spring, and dashpot combination in Prob. 22–74.

*22–76. The bar has a weight of 6 lb. If the stiffness of the spring is $k = 8$ lb/ft and the dashpot has a damping coefficient $c = 60$ lb · s/ft, determine the differential equation which describes the motion in terms of the angle θ of the bar's rotation. Also, what should be the damping coefficient of the dashpot if the bar is to be critically damped?

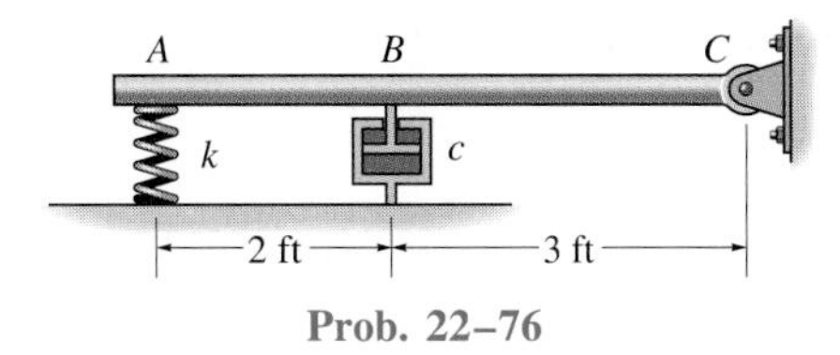

Prob. 22–76

22–77. The barrel of a cannon has a mass of 700 kg, and after firing it recoils a distance of 0.64 m. If it returns to its original position by means of a single recuperator having a damping coefficient of 2 kN · s/m, determine the required stiffness of each of the two springs fixed to the base and attached to the barrel so that the barrel recuperates without vibration.

22–78. The 10-kg block-spring-damper system is continually damped. If the block is displaced to $x = 50$ mm and released from rest, determine the time required for it to return to the position $x = 2$ mm.

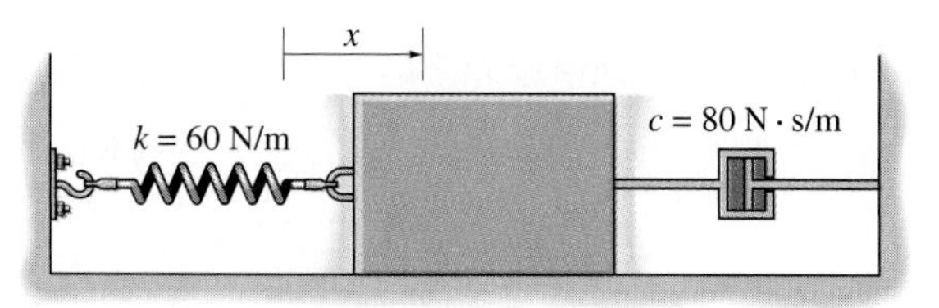

Prob. 22–78

22–79. Draw the electrical circuit that is equivalent to the mechanical system shown. What is the differential equation which describes the motion of the current in the circuit?

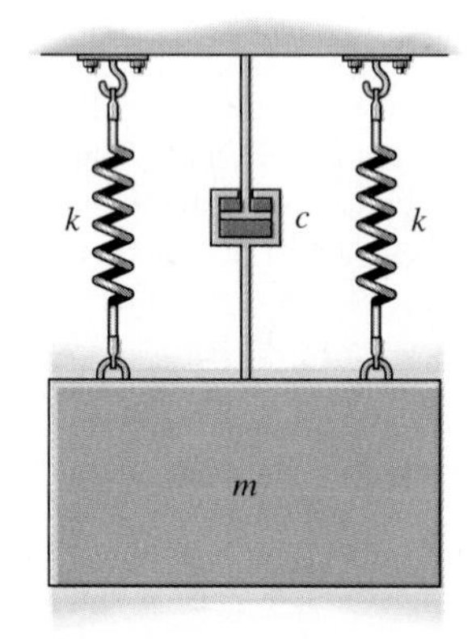

Prob. 22–79

***22–80.** Determine the differential equation of motion for the damped vibratory system shown. What type of motion occurs?

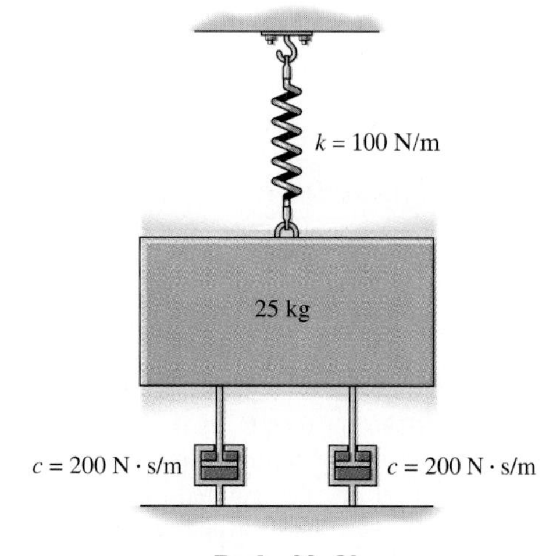

Prob. 22–80

22–81. Draw the electrical circuit that is equivalent to the mechanical system shown. Determine the differential equation which describes the motion of the current in the circuit.

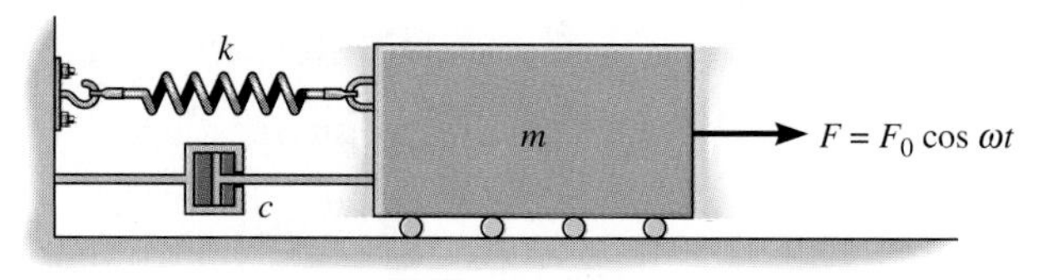

Prob. 22–81

22–82. Draw the electrical circuit that is equivalent to the mechanical system shown. Determine the differential equation which describes the motion of the current in the circuit.

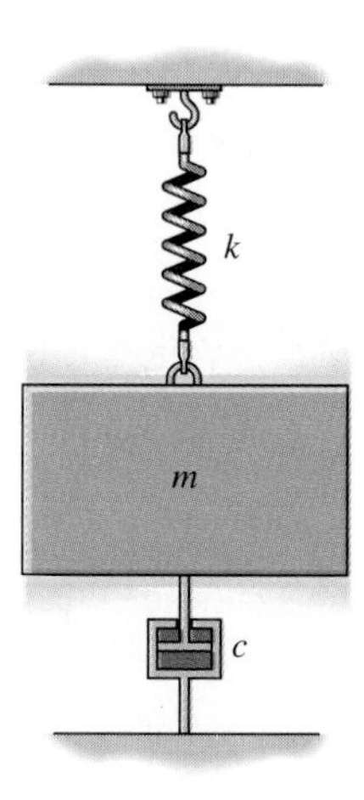

Prob. 22–82

22–83. Determine the mechanical analog for the electrical circuit. What differential equations describe the mechanical and electrical systems?

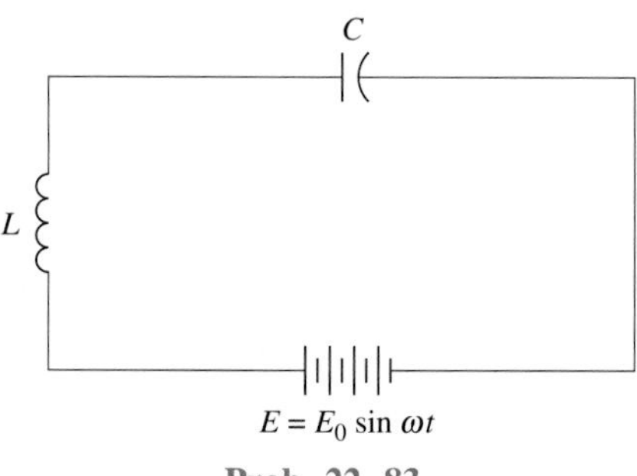

Prob. 22–83

A

Mathematical Expressions

Quadratic Formula

If $ax^2 + bx + c = 0$, then $x = \dfrac{-b \pm \sqrt{b^2 - 4ac}}{2a}$

Hyperbolic Functions

$$\sinh x = \frac{e^x - e^{-x}}{2}, \quad \cosh x = \frac{e^x + e^{-x}}{2}, \quad \tanh x = \frac{\sinh x}{\cosh x}$$

Trigonometric Identities

$$\sin \theta = \frac{A}{C}, \quad \csc \theta = \frac{C}{A}$$

$$\cos \theta = \frac{B}{C}, \quad \sec \theta = \frac{C}{B}$$

$$\tan \theta = \frac{A}{B}, \quad \cot \theta = \frac{B}{A}$$

C

A

θ

B

$$\sin^2 \theta + \cos^2 \theta = 1$$

$$\sin (\theta \pm \phi) = \sin \theta \cos \phi \pm \cos \theta \sin \phi$$

$$\sin 2\theta = 2 \sin \theta \cos \theta$$

$$\cos (\theta \pm \phi) = \cos \theta \cos \phi \mp \sin \theta \sin \phi$$

$$\cos 2\theta = \cos^2 \theta - \sin^2 \theta$$

$$\cos \theta = \pm \sqrt{\frac{1 + \cos 2\theta}{2}}, \quad \sin \theta = \pm \sqrt{\frac{1 - \cos 2\theta}{2}}$$

$$\tan \theta = \frac{\sin \theta}{\cos \theta}$$

$$1 + \tan^2 \theta = \sec^2 \theta \qquad 1 + \cot^2 \theta = \csc^2 \theta$$

Power-Series Expansions

$$\sin x = x - \frac{x^3}{3!} + \cdots$$

$$\cos x = 1 - \frac{x^2}{2!} + \cdots$$

$$\sinh x = x + \frac{x^3}{3!} + \cdots$$

$$\cosh x = 1 + \frac{x^2}{2!} + \cdots$$

Derivatives

$$\frac{d}{dx}(u^n) = nu^{n-1}\frac{du}{dx}$$

$$\frac{d}{dx}(uv) = u\frac{dv}{dx} + v\frac{du}{dx}$$

$$\frac{d}{dx}\left(\frac{u}{v}\right) = \frac{v\frac{du}{dx} - u\frac{dv}{dx}}{v^2}$$

$$\frac{d}{dx}(\cot u) = -\csc^2 u\frac{du}{dx}$$

$$\frac{d}{dx}(\sec u) = \tan u \sec u\frac{du}{dx}$$

$$\frac{d}{dx}(\csc u) = -\csc u \cot u\frac{du}{dx}$$

$$\frac{d}{dx}(\sin u) = \cos u\frac{du}{dx}$$

$$\frac{d}{dx}(\cos u) = -\sin u\frac{du}{dx}$$

$$\frac{d}{dx}(\tan u) = \sec^2 u\frac{du}{dx}$$

$$\frac{d}{dx}(\sinh u) = \cosh u\frac{du}{dx}$$

$$\frac{d}{dx}(\cosh u) = \sinh u\frac{du}{dx}$$

Integrals

$$\int x^n\,dx = \frac{x^{n+1}}{n+1} + C,\ n \neq -1$$

$$\int \frac{dx}{a+bx} = \frac{1}{b}\ln(a+bx) + C$$

$$\int \frac{dx}{a+bx^2} = \frac{1}{2\sqrt{-ba}}\ln\left[\frac{\sqrt{a} + 2\sqrt{-b}}{\sqrt{a} - x\sqrt{-b}}\right] + C, \quad a > 0,\ b < 0$$

$$\int \frac{x\,dx}{a+bx^2} = \frac{1}{2b}\ln(bx^2 + a) + C$$

$$\int \frac{x^2\,dx}{a+bx^2} = \frac{x}{b} - \frac{a}{b\sqrt{ab}}\tan^{-1}\frac{x\sqrt{ab}}{a} + C$$

$$\int \frac{dx}{a^2 - x^2} = \frac{1}{2a}\ln\left[\frac{a+x}{a-x}\right] + C,\ a^2 > x^2$$

$$\int \sqrt{a+bx}\,dx = \frac{2}{3b}\sqrt{(a+bx)^3} + C$$

$$\int x\sqrt{a+bx}\,dx = \frac{-2(2a - 3bx)\sqrt{(a+bx)^3}}{15b^2} + C$$

$$\int x^2\sqrt{a+bx}\,dx = \frac{2(8a^2 - 12abx + 15b^2x^2)\sqrt{(a+bx)^3}}{105b^3} + C$$

$$\int \sqrt{a^2 - x^2}\,dx = \frac{1}{2}\left[x\sqrt{a^2 - x^2} + a^2\sin^{-1}\frac{x}{a}\right] + C, \quad a > 0$$

$$\int x\sqrt{a^2 - x^2}\,dx = -\frac{1}{3}\sqrt{(a^2 - x^2)^3} + C$$

$$\int x^2\sqrt{a^2 - x^2}\,dx = -\frac{x}{4}\sqrt{(a^2 - x^2)^3} + \frac{a^2}{8}\left(x\sqrt{a^2 - x^2} + a^2\sin^{-1}\frac{x}{a}\right) + C,\ a > 0$$

$$\int \sqrt{x^2 \pm a^2}\, dx = \frac{1}{2}\left[x\sqrt{x^2 \pm a^2} \pm a^2 \ln\left(x + \sqrt{x^2 \pm a^2}\right)\right] + C$$

$$\int x\sqrt{x^2 \pm a^2}\, dx = \frac{1}{3}\sqrt{(x^2 \pm a^2)^3} + C$$

$$\int x^2\sqrt{x^2 \pm a^2}\, dx = \frac{x}{4}\sqrt{(x^2 \pm a^2)^3} \pm \frac{a^2}{8}x\sqrt{x^2 \pm a^2} - \frac{a^4}{8}\ln\left(x + \sqrt{x^2 \pm a^2}\right) + C$$

$$\int \frac{dx}{\sqrt{a + bx}} = \frac{2\sqrt{a + bx}}{b} + C$$

$$\int \frac{x\, dx}{\sqrt{x^2 \pm a^2}} = \sqrt{x^2 \pm a^2} + C$$

$$\int \frac{dx}{\sqrt{a + bx + cx^2}} = \frac{1}{\sqrt{c}}\ln\left[\sqrt{a + bx + cx^2} + x\sqrt{c} + \frac{b}{2\sqrt{c}}\right] + C,\ c > 0$$

$$= \frac{1}{\sqrt{-c}}\sin^{-1}\left(\frac{-2cx - b}{\sqrt{b^2 - 4ac}}\right) + C,\ c < 0$$

$$\int \sin x\, dx = -\cos x + C$$

$$\int \cos x\, dx = \sin x + C$$

$$\int x\cos(ax)\, dx = \frac{1}{a^2}\cos(ax) + \frac{x}{a}\sin(ax) + C$$

$$\int x^2\cos(ax)\, dx = \frac{2x}{a^2}\cos(ax) + \frac{a^2x^2 - 2}{a^3}\sin(ax) + C$$

$$\int e^{ax}\, dx = \frac{1}{a}e^{ax} + C$$

$$\int x\, e^{ax}\, dx = \frac{e^{ax}}{a^2}(ax - 1) + C$$

$$\int \sinh x\, dx = \cosh x + C$$

$$\int \cosh x\, dx = \sinh x + C$$

B

Numerical and Computer Analysis

Occasionally the application of the laws of mechanics will lead to a system of equations for which a closed-form solution is difficult or impossible to obtain. When confronted with this situation, engineers will often use a numerical method which in most cases can be programmed on a microcomputer or ''programmable'' pocket calculator. Here we will briefly present a computer program for solving a set of linear algebraic equations and three numerical methods which can be used to solve an algebraic or transcendental equation, evaluate a definite integral, and solve an ordinary differential equation. Application of each method will be explained by example, and an associated computer program written in Microsoft BASIC, which is designed to run on most personal computers, is provided.* A text on numerical analysis should be consulted for further discussion regarding a check of the accuracy of each method and the inherent errors that can develop from the methods.

B.1 Linear Algebraic Equations

Application of the equations of static equilibrium or the equations of motion sometimes requires solving a set of linear algebraic equations. The computer program listed in Fig. B–1 can be used for this purpose. It is based on the method of a Gaussian elimination and can solve at most 10 equations with 10

*Similar types of programs can be written or purchased for programmable pocket calculators.

```
1 PRINT"Linear system of equations":PRINT
2 DIM A(10,11)
3 INPUT"Input number of equations : ",N
4 PRINT
5 PRINT"A  coefficients"
6 FOR I = 1 TO N
7 FOR J = 1 TO N
8 PRINT "A(";I;",";J;
9 INPUT")=",A(I,J)
10 NEXT J
11 NEXT I
12 PRINT
13 PRINT"B  coefficients"
14 FOR I = 1 TO N
15 PRINT "B(";I;
16 INPUT")=",A(I,N+1)
17 NEXT I
18 GOSUB 25
19 PRINT
20 PRINT"Unknowns"
21 FOR I = 1 TO N
22 PRINT "X(";I;")=";A(I,N+1)
23 NEXT I
24 END
25 REM Subroutine Guassian
26 FOR M=1 TO N
27 NP=M
28 BG=ABS(A(M,M))
29 FOR I = M TO N
30 IF ABS(A(I,M))<=BG THEN 33
31 BG=ABS(A(I,M))
32 NP=I
33 NEXT I
34 IF NP=M THEN 40
35 FOR I = M TO N+1
36 TE=A(M,I)
37 A(M,I)=A(NP,I)
38 A(NP,I)=TE
39 NEXT I
40 FOR I = M+1 TO N
41 FC=A(I,M)/A(M,M)
42 FOR J = M+1 TO N+1
43 A(I,J)=A(I,J)-FC*A(M,J)
44 NEXT J
45 NEXT I
46 NEXT M
47 A(N,N+1)=A(N,N+1)/A(N,N)
48 FOR I = N-1 TO 1 STEP -1
49 SM=0
50 FOR J=I+1 TO N
51 SM=SM+A(I,J)*A(J,N+1)
52 NEXT J
53 A(I,N+1)=(A(I,N+1)-SM)/A(I,I)
54 NEXT I
55 RETURN
```

Fig. B–1

unknowns. To do so, the equations should first be written in the following general format:

$$A_{11}x_1 + A_{12}x_2 + \cdots + A_{1n}x_n = B_1$$
$$A_{21}x_1 + A_{22}x_2 + \cdots + A_{2n}x_n = B_2$$
$$\vdots$$
$$A_{n1}x_1 + A_{n2}x_2 + \cdots + A_{nn}x_n = B_n$$

The "A" and "B" coefficients are "called" for when running the program. The output presents the unknowns $x_1, \ldots, x_n$.

Example B–1

Solve the two equations

$$3x_1 + x_2 = 4$$
$$2x_1 - x_2 = 10$$

SOLUTION

When the program begins to run, it first calls for the number of equations (2); then the A coefficients in the sequence $A_{11} = 3$, $A_{12} = 1$, $A_{21} = 2$, $A_{22} = -1$; and finally the B coefficients $B_1 = 4$, $B_2 = 10$. The output appears as

Unknowns

$X(1) = 2.8$ *Ans.*

$X(2) = -4.4$ *Ans.*

B.2 Simpson's Rule

Simpson's rule is a numerical method that can be used to determine the area under a curve given as a graph or as an explicit function $y = f(x)$. Likewise, it can be used to compute the value of a definite integral which involves the function $y = f(x)$. To do so, the area must be subdivided into an *even number* of strips or intervals having a width h. The curve between three consecutive ordinates is approximated by a parabola, and the entire area or definite integral is then determined from the formula

$$\int_{x_0}^{x_n} f(x)\, dx \simeq \frac{h}{3}[y_0 + 4(y_1 + y_3 + \cdots + y_{n-1}) + 2(y_2 + y_4 + \cdots + y_{n-2}) + y_n] \quad \text{(B-1)}$$

The computer program for this equation is given in Fig. B–2. For its use, we must first specify the function (on line 6 of the program). The upper and lower limits of the integral and the number of intervals are called for when the program is executed. The value of the integral is then given as the output.

```
1 PRINT"Simpson's rule":PRINT
2 PRINT" To execute this program :":PRINT
3 PRINT"   1- Modify right-hand side of the equation given below,
4 PRINT"        then press RETURN key"
5 PRINT"   2- Type  RUN 6":PRINT:EDIT 6
6 DEF FNF(X)=LOG(X)
7 PRINT:INPUT" Enter Lower Limit = ",A
8 INPUT" Enter Upper Limit = ",B
9 INPUT" Enter Number (even) of Intervals = ",N%
10 H=(B-A)/N%:AR=FNF(A):X=A+H
11 FOR J%=2 TO N%
12 K=2*(2-J%+2*INT(J%/2))
13 AR=AR+K*FNF(X)
14 X=X+H:NEXT J%
15 AR=H*(AR+FNF(B))/3
16 PRINT" Integral = ",AR
17 END
```

Fig. B–2

Example B–2

Evaluate the definite integral

$$\int_2^5 \ln x \, dx$$

SOLUTION

The interval $x_0 = 2$ to $x_6 = 5$ will be divided into six equal parts ($n = 6$), each having a width $h = (5 - 2)/6 = 0.5$. We then compute $y = f(x) = \ln x$ at each point of subdivision.

n	x_n	y_n
0	2	0.693
1	2.5	0.916
2	3	1.099
3	3.5	1.253
4	4	1.386
5	4.5	1.504
6	5	1.609

Thus, Eq. B–1 becomes

$$\int_2^5 \ln x \, dx \simeq \frac{0.5}{3}[0.693 + 4(0.916 + 1.253 + 1.504) + 2(1.099 + 1.386) + 1.609]$$

$$\simeq 3.66 \qquad \textit{Ans.}$$

This answer is equivalent to the exact answer to three significant figures. Obviously, accuracy to a greater number of significant figures can be improved by selecting a smaller interval h (or larger n).

Using the computer program, we first specify the function $\ln x$, line 6 in Fig. B–2. During execution, the program input requires the upper and lower limits 2 and 5, and the number of intervals $n = 6$. The output appears as

Integral = 3.66082 *Ans.*

B.3 The Secant Method

The secant method is used to find the real roots of an algebraic or transcendental equation $f(x) = 0$. The method derives its name from the fact that the formula used is established from the slope of the secant line to the graph $y = f(x)$. This slope is $[f(x_n) - f(x_{n-1})]/(x_n - x_{n-1})$, and the secant formula is

$$x_{n+1} = x_n - f(x_n)\left[\frac{x_n - x_{n-1}}{f(x_n) - f(x_{n-1})}\right] \tag{B-2}$$

For application it is necessary to provide two initial guesses, x_0 and x_1, and thereby evaluate x_2 from Eq. B–2 ($n = 1$). One then proceeds to reapply Eq. B–2 with x_1 and the calculated value of x_2 and obtain x_3 ($n = 2$), etc., until the value $x_{n+1} \simeq x_n$. One can see this will occur if x_n is approaching the root of the function $f(x) = 0$, since the correction term on the right of Eq. B–2 will tend toward zero. In particular, the larger the slope, the smaller the correction to x_n, and the faster the root will be found. On the other hand, if the slope is very small in the neighborhood of the root, the method leads to large corrections for x_n, and convergence to the root is slow and may even lead to a failure to find it. In such cases other numerical techniques must be used for solution.

A computer program based on Eq. B–2 is listed in Fig. B–3. We must first specify the function on line 7 of the program. When the program is executed, two initial guesses, x_0 and x_1, must be entered in order to approximate the solution. The output specifies the value of the root. If it cannot be determined, this is so stated.

```
1 PRINT"Secant method":PRINT
2 PRINT" To execute this program :":PRINT
3 PRINT"     1) Modify right hand side of the equation given below,"
4 PRINT"        then press RETURN key."
5 PRINT"     2) Type  RUN 7"
6 PRINT:EDIT 7
7 DEF FNF(X)=.5*SIN(X)-2*COS(X)+1.3
8 INPUT"Enter point #1 =",X
9 INPUT"Enter point #2 =",X1
10 IF X=X1 THEN 14
11 EP=.00001:TL=2E-20
12 FP=(FNF(X1)-FNF(X))/(X1-X)
13 IF ABS(FP)>TL THEN 15
14 PRINT"Root can not be found.":END
15 DX=FNF(X1)/FP
16 IF ABS(DX)>EP THEN 19
17 PRINT "Root = ";X1;"      Function evaluated at this root = ";FNF(X1)
18 END
19 X=X1:X1=X1-DX
20 GOTO 12
```

Fig. B–3

Example B–3

Determine the root of the equation

$$f(x) = 0.5 \sin x - 2 \cos x + 1.30 = 0$$

SOLUTION

Guesses of the initial roots will be $x_0 = 45°$ and $x_1 = 30°$. Applying Eq. B–2,

$$x_2 = 30° - (-0.1821)\frac{(30° - 45°)}{(-0.1821 - 0.2393)} = 36.48°$$

Using this value in Eq. B–2, along with $x_1 = 30°$, we have

$$x_3 = 36.48° - (-0.0108)\frac{36.48° - 30°}{(-0.0108 + 0.1821)} = 36.89°$$

Repeating the process with this value and $x_2 = 36.48°$ yields

$$x_4 = 36.89° - (0.0005)\left[\frac{36.89° - 36.48°}{(0.0005 + 0.0108)}\right] = 36.87°$$

Thus $x = 36.9°$ is appropriate to three significant figures.

If the problem is solved using the computer program, first we specify the function, line 7 in Fig. B–3. During execution, the first and second guesses must be entered in radians. Choosing these to be 0.8 rad and 0.5 rad, the result appears as

Root = 0.6435022.

Function evaluated at this root = 1.66893E–06.

This result converted from radians to degrees is therefore

$$x = 36.9°$$ ***Ans.***

B.4 Runge-Kutta Method.

The Runge-Kutta method is used to solve an ordinary differential equation. It consists of applying a set of formulas which are used to find specific values of y for corresponding incremental values h in x. The formulas given in general form are as follows:

First-Order Equation. To integrate $\dot{x} = f(t, x)$ step by step, use

$$x_{i+1} = x_i + \frac{1}{6}(k_1 + 2k_2 + 2k_3 + k_4) \tag{B–3}$$

where

$$\begin{aligned} k_1 &= hf(t_i, x_i) \\ k_2 &= hf\left(t_i + \frac{h}{2}, x_i + \frac{k_1}{2}\right) \\ k_3 &= hf\left(t_i + \frac{h}{2}, x_i + \frac{k_2}{2}\right) \\ k_4 &= hf(t_i + h, x_i + k_3) \end{aligned} \tag{B–4}$$

Second-Order Equation. To integrate $\ddot{x} = f(t, x, \dot{x})$, use

$$\begin{aligned} x_{i+1} &= x_i + h\left[\dot{x}_i + \frac{1}{6}(k_1 + k_2 + k_3)\right] \\ \dot{x}_{i+1} &= \dot{x}_i + \frac{1}{6}(k_1 + 2k_2 + 2k_3 + k_4) \end{aligned} \tag{B–5}$$

where

$$\begin{aligned} k_1 &= hf(t_i, x_i, \dot{x}_i) \\ k_2 &= hf\left(t_i + \frac{h}{2}, x_i + \frac{h}{2}\dot{x}_i, \dot{x}_i + \frac{k_1}{2}\right) \\ k_3 &= hf\left(t_i + \frac{h}{2}, x_i + \frac{h}{2}\dot{x}_i + \frac{h}{4}k_1, \dot{x}_i + \frac{k_2}{2}\right) \\ k_4 &= hf\left(t_i + h, x_i + h\dot{x}_i + \frac{h}{2}k_2, \dot{x}_i + k_3\right) \end{aligned} \tag{B–6}$$

To apply these equations, one starts with initial values $t_i = t_0$, $x_i = x_0$ and $\dot{x}_i = \dot{x}_0$ (for the second-order equation). Choosing an increment h for t_0, the four constants k are computed and these results are substituted into Eq. B–3 or B–5 in order to compute $x_{i+1} = x_1$, $\dot{x}_{i+1} = x_1$, corresponding to $t_{i+1} = t_1 = t_0 + h$. Repeating this process using t_1, x_1, $\dot{x}_1$ and h, values for x_2, $\dot{x}_2$ and $t_2 = t_1 + h$ are then computed, etc.

Computer programs which solve first- and second-order differential equations by this method are listed in Figs. B–4 and B–5, respectively. In order to use these programs, the operator specifies the function $\dot{x} = f(t, x)$ or $\ddot{x} = f(t, x, \dot{x})$ (line 7), the initial values t_0, x_0, $\dot{x}_0$ (for second-order equation), the final time t_n, and the step size h. The output gives the values of t, x, and $\dot{x}$ for each time increment until t_n is reached.

```
1 PRINT"Runge-Kutta Method for 1-st order Differential Equation":PRINT
2 PRINT" To execute this program :":PRINT
3 PRINT"   1) Modify right hand side of the equation given below,"
4 PRINT"      then Press RETURN key"
5 PRINT"   2) Type  RUN 7"
6 PRINT:EDIT 7
7 DEF FNF(T,X)=5*T+X
8 CLS:PRINT" Initial Conditions":PRINT
9 INPUT"Input  t    = ",T
10 INPUT"       x    = ",X
11 INPUT"Final  t    = ",T1
12 INPUT"step size  = ",H:PRINT
13 PRINT"        t            x"
14 IF T>=T1+H THEN 23
15 PRINT USING"######.#####";T;X
16 K1=H*FNF(T,X)
17 K2=H*FNF(T+.5*H,X+.5*K1)
18 K3=H*FNF(T+.5*H,X+.5*K2)
19 K4=H*FNF(T+H,X+K3)
20 T=T+H
21 X=X+(K1+K2+K2+K3+K3+K4)/6
22 GOTO 14
23 END
```

Fig. B–4

```
1 PRINT"Runge-Kutta Method for 2-nd order Differential Equation":PRINT
2 PRINT" To execute this program :":PRINT
3 PRINT"   1) Modify right hand side of the equation given below,"
4 PRINT"      then Press RETURN key"
5 PRINT"   2) Type  RUN 7"
6 PRINT:EDIT 7
7 DEF FNF(T,X,XD)=
8 INPUT"Input  t    = ",T
9 INPUT"       x    = ",X
10 INPUT"      dx/dt = ",XD
11 INPUT"Final  t    = ",T1
12 INPUT"step size  = ",H:PRINT
13 PRINT"        t            x          dx/dt"
14 IF T>=T1+H THEN 24
15 PRINT USING"######.#####";T;X;XD
16 K1=H*FNF(T,X,XD)
17 K2=H*FNF(T+.5*H,X+.5*H*XD,XD+.5*K1)
18 K3=H*FNF(T+.5*H,X+(.5*H)*(XD+.5*K1),XD+.5*K2)
19 K4=H*FNF(T+H,X+H*XD+.5*H*K2,XD+K3)
20 T=T+H
21 X=X+H*XD+H*(K1+K2+K3)/6
22 XD=XD+(K1+K2+K2+K3+K3+K4)/6
23 GOTO 14
24 END
```

Fig. B–5

Example B–4

Solve the differential equations $\dot{x} = 5t + x$. Obtain the results for two steps using time increments of $h = 0.02$ s. At $t_0 = 0$, $x_0 = 0$.

SOLUTION

This is a first-order equation, so Eqs. B–3 and B–4 apply. Thus, for $t_0 = 0$, $x_0 = 0$, $h = 0.02$, we have

$$k_1 = 0.02(0 + 0) = 0$$
$$k_2 = 0.02[5(0.1) + 0] = 0.001$$
$$k_3 = 0.02[5(0.01) + 0.0005] = 0.00101$$
$$k_4 = 0.02[5(0.02) + 0.00101] = 0.00202$$
$$x_1 = 0 + \tfrac{1}{6}[0 + 2(0.001) + 2(0.00101) + 0.00202] = 0.00101$$

Using the values $t_1 = 0 + 0.02 = 0.02$ and $x_1 = 0.00101$ with $h = 0.02$, the value for x_2 is now computed from Eqs. B–3 and B–4.

$$k_1 = 0.02[5(0.02) + 0.00101] = 0.00202$$
$$k_2 = 0.02[5(0.03) + 0.00202] = 0.00304$$
$$k_3 = 0.02[5(0.03) + 0.00253] = 0.00305$$
$$k_4 = 0.02[5(0.04) + 0.00406] = 0.00408$$
$$x_2 = 0.001 + \tfrac{1}{6}[0.00202 + 2(0.00304) + 2(0.00305) + 0.00408]$$
$$= 0.00405 \qquad \textit{Ans.}$$

To solve this problem using the computer program in Fig. B–4, the function is first entered on line 7, then the data $t_0 = 0$, $x_0 = 0$, $t_n = 0.04$, and $h = 0.02$ is specified. The results appear as

t	x
0.00000	0.00000
0.02000	0.00101
0.04000	0.00405

Ans.

C

Vector Analysis

The following discussion provides a brief review of the vector analysis. A more detailed treatment of these topics is given in *Engineering Mechanics: Statics.*

Vector. A vector, **A,** is a quantity which has a magnitude and direction, and adds according to the parallelogram law. As shown in Fig. C–1, **A** = **B** + **C,** where **A** is the *resultant vector* and **B** and **C** are *component vectors.*

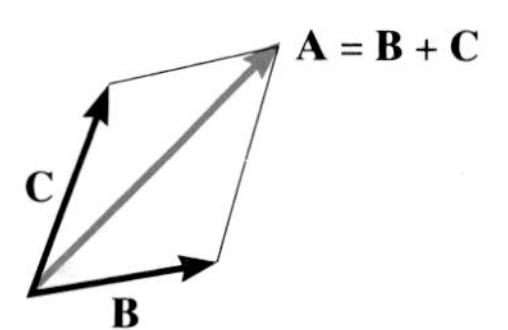

Fig. C–1

Unit Vector. A unit vector, $\mathbf{u}_A$, has a magnitude of one "dimensionless" unit and acts in the same direction as **A.** It is determined by dividing **A** by its magnitude A, i.e.,

$$\mathbf{u}_A = \frac{\mathbf{A}}{A} \qquad \text{(C–1)}$$

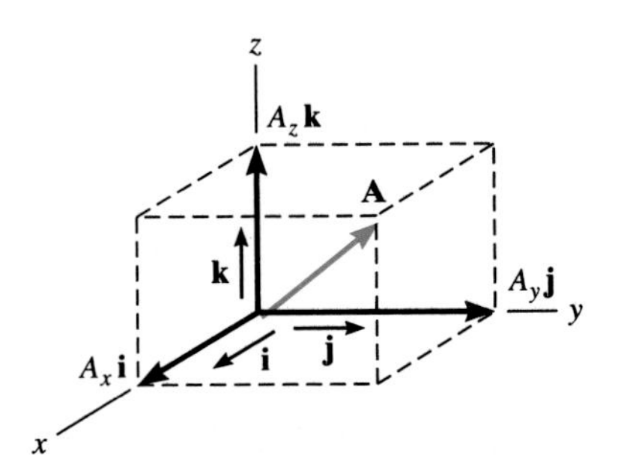

Fig. C–2

Cartesian Vector Notation. The directions of the positive x, y, z axes are defined by the Cartesian unit vectors, **i, j, k,** respectively.

As shown in Fig. C–2, vector **A** is formulated by the addition of its x, y, z components as

$$\mathbf{A} = A_x\mathbf{i} + A_y\mathbf{j} + A_z\mathbf{k} \tag{C–2}$$

The *magnitude* of **A** is determined from

$$A = \sqrt{A_x^2 + A_y^2 + A_z^2} \tag{C–3}$$

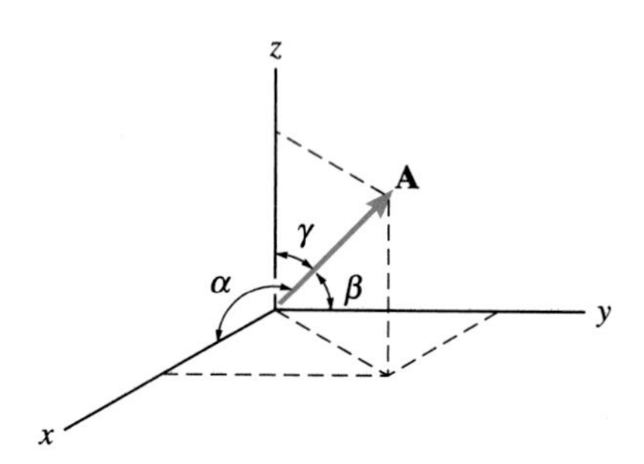

Fig. C–3

The *direction* of **A** is defined in terms of its *coordinate direction angles*, α, β, γ, measured from the *tail* of **A** to the *positive* x, y, z axes, Fig. C–3. These angles are determined from the *direction cosines* which represent the **i, j, k** components of the unit vector $\mathbf{u}_A$; i.e., from Eqs. C–1 and C–2,

$$\mathbf{u}_A = \frac{A_x}{A}\mathbf{i} + \frac{A_y}{A}\mathbf{j} + \frac{A_z}{A}\mathbf{k} \tag{C–4}$$

so that the direction cosines are

$$\cos\alpha = \frac{A_x}{A} \qquad \cos\beta = \frac{A_y}{A} \qquad \cos\gamma = \frac{A_z}{A} \tag{C–5}$$

Hence, $\mathbf{u}_A = \cos\alpha\mathbf{i} + \cos\beta\mathbf{j} + \cos\gamma\mathbf{k}$, and using Eq. C–3, it is seen that

$$\cos^2\alpha + \cos^2\beta + \cos^2\gamma = 1 \tag{C–6}$$

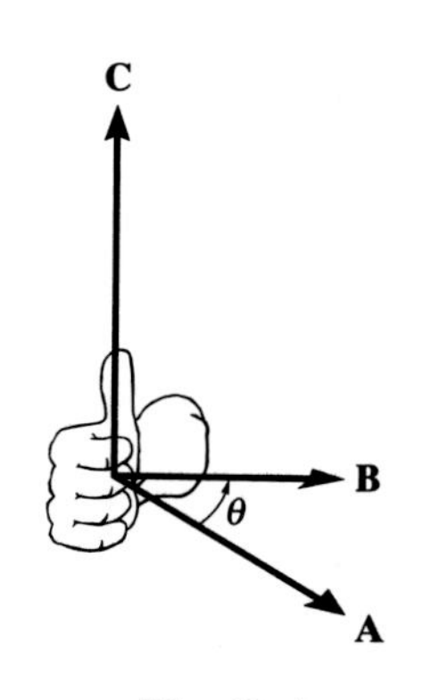

Fig. C–4

The Cross Product. The cross product of two vectors **A** and **B,** which yields the resultant vector **C,** is written as

$$\mathbf{C} = \mathbf{A} \times \mathbf{B} \tag{C–7}$$

and reads **C** equals **A** "cross" **B.** The *magnitude* of **C** is

$$C = AB\sin\theta \tag{C–8}$$

where θ is the angle made between the *tails* of **A** and **B** ($0° \le \theta \le 180°$). The *direction* of **C** is determined by the right-hand rule, whereby the fingers of the right hand are curled *from* **A** *to* **B** and the thumb points in the direction of **C,** Fig. C–4. This vector is perpendicular to the plane containing vectors **A** and **B.**

The vector cross product is *not* commutative, i.e., $\mathbf{A} \times \mathbf{B} \neq \mathbf{B} \times \mathbf{A}$. Rather,

$$\mathbf{A} \times \mathbf{B} = -\mathbf{B} \times \mathbf{A} \tag{C–9}$$

The distributive law is valid; i.e.,

$$\mathbf{A} \times (\mathbf{B} + \mathbf{D}) = \mathbf{A} \times \mathbf{B} + \mathbf{A} \times \mathbf{D} \qquad \text{(C–10)}$$

And the cross product may be multiplied by a scalar m in any manner; i.e.,

$$m(\mathbf{A} \times \mathbf{B}) = (m\mathbf{A}) \times \mathbf{B} = \mathbf{A} \times (m\mathbf{B}) = (\mathbf{A} \times \mathbf{B})m \qquad \text{(C–11)}$$

Equation C–7 can be used to find the cross product of any pair of Cartesian unit vectors. For example, to find $\mathbf{i} \times \mathbf{j}$, the magnitude is $(i)(j) \sin 90° = (1)(1)(1) = 1$, and its direction $+\mathbf{k}$ is determined from the right-hand rule, applied to $\mathbf{i} \times \mathbf{j}$, Fig. C–2. A simple scheme shown in Fig. C–5 may be helpful in obtaining this and other results when the need arises. If the circle is constructed as shown, then "crossing" two of the unit vectors in a *counterclockwise* fashion around the circle yields a *positive* third unit vector, e.g., $\mathbf{k} \times \mathbf{i} = \mathbf{j}$. Moving *clockwise*, a *negative* unit vector is obtained, e.g., $\mathbf{i} \times \mathbf{k} = -\mathbf{j}$.

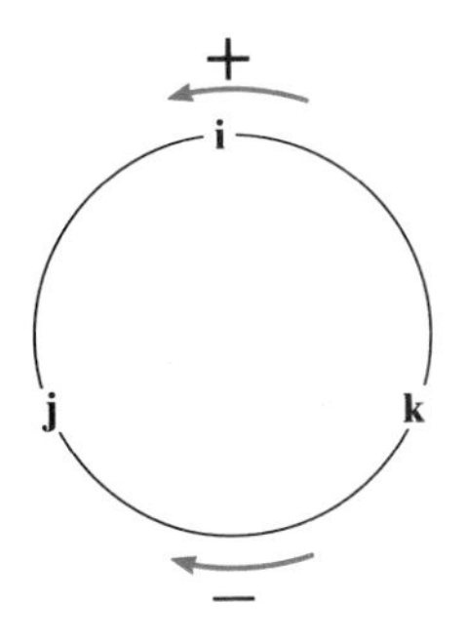

Fig. C–5

If $\mathbf{A}$ and $\mathbf{B}$ are expressed in Cartesian component form, then the cross product, Eq. C–7, may be evaluated by expanding the determinant

$$\mathbf{C} = \mathbf{A} \times \mathbf{B} = \begin{vmatrix} \mathbf{i} & \mathbf{j} & \mathbf{k} \\ A_x & A_y & A_z \\ B_x & B_y & B_z \end{vmatrix} \qquad \text{(C–12)}$$

which yields

$$\mathbf{C} = (A_y B_z - A_z B_y)\mathbf{i} - (A_x B_z - A_z B_x)\mathbf{j} + (A_x B_y - A_y B_x)\mathbf{k}$$

Recall that the cross product is used in statics to define the moment of a force $\mathbf{F}$ about point O, in which case

$$\mathbf{M}_O = \mathbf{r} \times \mathbf{F} \qquad \text{(C–13)}$$

where $\mathbf{r}$ is a position vector directed from point O to *any point* on the line of action of $\mathbf{F}$.

The Dot Product. The dot product of two vectors **A** and **B**, which yields a scalar, is defined as

$$\mathbf{A} \cdot \mathbf{B} = AB \cos \theta \qquad \text{(C–14)}$$

and reads **A** "dot" **B**. The angle θ is formed between the *tails* of **A** and **B** $(0° \leq \theta \leq 180°)$.

The dot product is commutative; i.e.,

$$\mathbf{A} \cdot \mathbf{B} = \mathbf{B} \cdot \mathbf{A} \qquad \text{(C–15)}$$

The distributive law is valid; i.e.,

$$\mathbf{A} \cdot (\mathbf{B} + \mathbf{D}) = \mathbf{A} \cdot \mathbf{B} + \mathbf{A} \cdot \mathbf{D} \qquad \text{(C–16)}$$

And scalar multiplication can be performed in any manner; i.e.,

$$m(\mathbf{A} \cdot \mathbf{B}) = (m\mathbf{A}) \cdot \mathbf{B} = \mathbf{A} \cdot (m\mathbf{B}) = (\mathbf{A} \cdot \mathbf{B})m \qquad \text{(C–17)}$$

Using Eq. C–14, the dot product between any two Cartesian vectors can be determined. For example, $\mathbf{i} \cdot \mathbf{i} = (1)(1) \cos 0° = 1$ and $\mathbf{i} \cdot \mathbf{j} = (1)(1) \cos 90° = 0$.

If **A** and **B** are expressed in Cartesian component form, then the dot product, Eq. C–14, can be determined from

$$\mathbf{A} \cdot \mathbf{B} = A_x B_x + A_y B_y + A_z B_z \qquad \text{(C–18)}$$

The dot product may be used to determine the *angle θ formed between two vectors.* From Eq. C–14,

$$\theta = \cos^{-1}\left(\frac{\mathbf{A} \cdot \mathbf{B}}{AB}\right) \qquad \text{(C–19)}$$

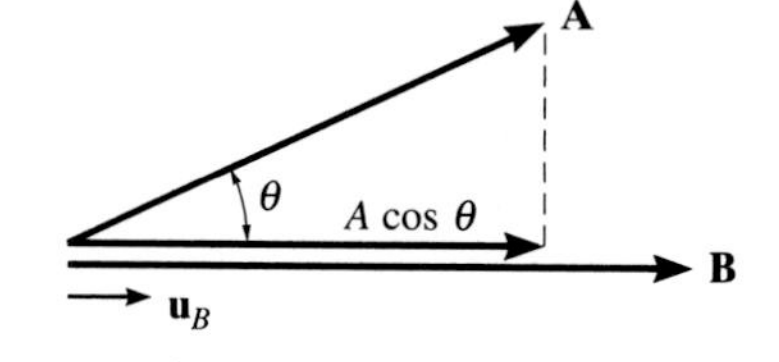

Fig. C–6

It is also possible to find the *component of a vector in a given direction* using the dot product. For example, the magnitude of the component (or projection) of vector **A** in the direction of **B**, Fig. C–6, is defined by $A \cos \theta$. From Eq. C–14, this magnitude is

$$A \cos \theta = \mathbf{A} \cdot \frac{\mathbf{B}}{B} = \mathbf{A} \cdot \mathbf{u}_B \qquad \text{(C–20)}$$

where $\mathbf{u}_B$ represents a unit vector acting in the direction of **B**, Fig. C–6.

Differentiation and Integration of Vector Functions. The rules for differentiation and integration of the sums and products of scalar functions apply as well to vector functions. Consider, for example, the two vector functions $\mathbf{A}(s)$ and $\mathbf{B}(s)$. Provided these functions are smooth and continuous for all s, then

$$\frac{d}{ds}(\mathbf{A} + \mathbf{B}) = \frac{d\mathbf{A}}{ds} + \frac{d\mathbf{B}}{ds} \qquad \text{(C–21)}$$

$$\int (\mathbf{A} + \mathbf{B})\, ds = \int \mathbf{A}\, ds + \int \mathbf{B}\, ds \qquad \text{(C–22)}$$

For the cross product,

$$\frac{d}{ds}(\mathbf{A} \times \mathbf{B}) = \left(\frac{d\mathbf{A}}{ds} \times \mathbf{B}\right) + \left(\mathbf{A} \times \frac{d\mathbf{B}}{ds}\right) \qquad \text{(C–23)}$$

Similarly, for the dot product,

$$\frac{d}{ds}(\mathbf{A} \cdot \mathbf{B}) = \frac{d\mathbf{A}}{ds} \cdot \mathbf{B} + \mathbf{A} \cdot \frac{d\mathbf{B}}{ds} \qquad \text{(C–24)}$$

Answers

Chapter 12

12–1. 1.74 m/s^2, 4.80 s **12–2.** 6.40 ft/s^2, 12.5 s

12–3. 59.5 ft/s, 1.29 s **12–5.** 30 s, 792 m

12–6. 123 ft, 2.99 ft/s^2 **12–7.** 54.0 m

12–9. $s = -18$ ft, $d = 46$ ft **12–10.** 42 ft/s^2, 135 ft/s

12–11. 9th floor **12–13.** 0.603 ft/s^2

12–14. $v = 32$ m/s, $s = 67$ m, $d = 66$ m

12–15. 16.9 ft **12–17.** 3.93 m/s, 9.98 m

12–18. 7.87 m **12–19.** -0.116 m/s^2

12–21. 1 s, 71.8 ft/s **12–22.** 11.9 m, 0.250 m/s

12–23. -10.2 s^{-2}, 1.56 ft **12–25.** 5.62 s

12–26. 10.3 s, 4.11 km **12–27.** 31.4 ft/s

12–29. (*a*) -30.5 m, (*b*) 56.0 m, (*c*) 10 m/s

12–30. 48.3 s

12–31. $\Delta s = 152$ ft, $(s_A)_{\text{total}} = 41$ ft, $(s_B)_{\text{total}} = 200$ ft

12–33. $t = \dfrac{v_f}{2g} \ln\left(\dfrac{v_f + v}{v_f - v}\right)$ **12–34.** 11.2 km/s

12–35. $v = R\sqrt{\dfrac{2g_0(y_0 - y)}{(R + y)(R + y_0)}}$, $v = 3.02$ km/s

12–37. 7.48 s

12–38. $t = 50$ s, $a = 0$, $s = 420$ m

12–39. 0.4 m/s^2, 0, 45 m, 300 m

12–41. $t = 0.2$ s, $s = 1$ m **12–42.** 3600 ft, 30 ft/s

12–43. 0, -1.33 ft/s^2

12–45. 0, 30 m, -1 m/s^2, 48 m

12–46. 33.3 s

12–47. $t = 5$ s, $s = 50$ m, $t = 30$ s, $s = 750$ m

12–49. $t^{-} = 30$ s, $v = 24$ m/s, $a = 0.8$ m/s^2

12–50. 27.4 ft/s, 37.4 ft/s

12–51. $t = 40$ s, $v = 612.5$ m/s, $s = 9625$ m

12–53. 13.0 s **12–54.** 55 ft/s, 14.6 s

12–55. $t = 8$ s, $v = 40$ ft/s, $s = 160$ ft

12–57. 4.00 m/s^2

12–58. $t = 60$ s, $v = 540$ m/s, $s = 10\,125$ m

12–59. 4.06 s

12–61. 0.32 m/s^2, -0.32 m/s^2

12–62. 17.9 s **12–63.** 5.77 s

12–65. 12.6 ft/s, 22.8 ft/s, 36.1 ft/s

12–66. $y = (\frac{1}{8}x)^{3/2} + 5$, $v = 55.1$ m/s, $a = 24.1$ m/s^2

12–67. $t = 2$ s, (6 m, 0), $v_A = 18.3$ m/s, $v_B = 6.71$ m/s

12–69. -1 m/s, $\{2t\}$ m/s, 0, 2 m/s^2, $y = (1 - x)^2$

12–70. $v_x = \{3 \cos t\}$ m/s, $v_y = \{-4 \sin t\}$ m/s, $a_x = \{-3 \sin t\}$ m/s^2, $a_y = \{-4 \cos t\}$ m/s^2, $\dfrac{y^2}{16} + \dfrac{x^2}{9} = 1$

12–71. $\mathbf{r} = \{11\mathbf{i} + 2\mathbf{j} + 21\mathbf{k}\}$ ft

12–73. $v = 73.1$ m/s, $\alpha = 80.6°$, $\beta = 86.9°$, $\gamma = 9.96°$, $a = 72.2$ m/s^2, $\alpha = 85.2°$, $\beta = 90.0°$, $\gamma = 4.76°$

12–74. $\mathbf{v}_{AD} = \{3.33\mathbf{i} + 1.67\mathbf{j}\}$ m/s

12–75. 4.28 m/s

12–77. $\mathbf{a}_{AB} = \{0.404\mathbf{i} + 7.07\mathbf{j}\}$ m/s^2, $\mathbf{a}_{AC} = \{2.50\mathbf{i}\}$ m/s^2

12–78. $y^2 + x^2 - 4y = 0$, $v = 2$ ft/s, $a = 2$ ft/s^2, $r = 3.37$ ft

12–79. 4.00 ft, 37.8 ft/s^2 **12–81.** 4.22 m/s, 0.130 m/s^2

12–82. 10.4 m/s, 11.1 m/s^2 **12–83.** 8.34 m, 0.541 m/s^2

12–85. $\mathbf{v} = \{48\mathbf{i} - 5.23\mathbf{j} + 3.03\mathbf{k}\}$ ft/s,
$\mathbf{a} = \{240\mathbf{i} + 43.2\mathbf{j} + 23.9\mathbf{k}\}$ ft/s^2

12–86. $v = ar\sqrt{b^2 \sin^2 at + \cos^2 at + \dfrac{k^2}{a^2r^2}}$,
$a = a^2r\sqrt{b^2 \cos^2 at + \sin^2 at}$

12–87. 201 m/s, 405 m/s^2

12–89. $(v_A)_B = 4.32$ m/s, $(v_A)_C = 5.85$ m/s, 0.121 s

12–90. 287 m **12–91.** 3.55 s, 32.0 m/s

12–93. 6.41° **12–94.** 27.3 ft/s

12–95. 36.7 ft/s, 11.5 ft **12–98.** 19.0 m, 2.48 s

12–99. -9.81 m/s^2, 19.5 m/s **12–101.** 3.57 s, 67.4 ft/s

12–102. 38.4°, 14.8 ft **12–103.** 60.2 ft/s, 41.6 ft

12–105. $v_A = 16.5$ ft/s, $v_B = 29.2$ ft/s

12–106. $\theta_D = 14.7°$, $\theta_C = 75.3°$, 1.45 s

12–107. $\Delta t = \dfrac{2v_0}{g}\left[\dfrac{\sin(\theta_1 - \theta_2)}{\cos\theta_2 + \cos\theta_1}\right]$

12–109. 8.83 m

12–110. $\theta_D = 14.7°$, $\theta_C = 75.3°$, 1.45 s

12–111. 20.8 ft/s **12–113.** 9.50 m/s^2

12–114. 21.6 ft/s^2, 40.5 ft

12–115. 35.0 ft/s^2, 2.64(10^3) ft

12–117. 4.40 m/s, $a_t = 5.04$ m/s^2, $a_n = 1.39$ m/s^2

12–118. 5.41 ft/s, $a_n = 5.89$ ft/s^2, $a_t = 2.66$ ft/s^2

12–119. 256 ft/s, 5.25(10^3) ft **12–121.** 5.66 m/s, 1.02 m/s^2

12–122. 1.8 m/s, 1.20 m/s^2 **12–123.** 0.457 m/s^2

12–125. 3.62 m/s^2, 29.6 m **12–126.** 4.58 m/s, 0.653 m/s^2

12–127. 7.20 m/s, 1.91 m/s^2 **12–129.** 115 ft/s, 58.4 ft/s^2

12–130. 1.68 s

12–131. $a_n = 8.00$ ft/s^2, $a_t = 24$ ft/s^2

12–133. 3.68 m/s, 4.98 m/s^2

12–134. 3.19 m/s, 4.22 m/s^2, 25.1° **12–135.** 2.63 s

12–137. $y = 0.8391x - 0.1306x^2$, $a_n = 8.98$ m/s^2,
$a_t = 3.94$ m/s^2

12–138. 14.4 m, $a_B = 12.8$ m/s^2, $a_A = 190$ m/s^2

12–139. 2.51 s, $a_B = 65.2$ m/s^2, $a_A = 22.2$ m/s^2

12–141. $v_t = 4$ ft/s, $v_n = 0$, $a_t = 0$, $a_n = \dfrac{16\cos x}{(1 + \sin^2 x)^{3/2}}$

12–142. $v_t = 28.8$ ft/s, $v_n = 0$, $a_n = 6.66$ ft/s^2, $a_t = 24.4$ ft/s^2

12–143. $v_t = 7.21$ m/s, $v_n = 0$, $a_n = 0.555$ m/s^2, $a_t = 2.77$ m/s^2

12–145. $a_{\min} = a_n = \dfrac{v^2 b}{a^2}$

12–146. (*a*) $s_A = 1.40$ m, $s_B = 3$ m,
(*b*) $\mathbf{r}_A = \{1.38\mathbf{i} + 0.195\mathbf{j}\}$ m,
$\mathbf{r}_B = \{-2.82\mathbf{i} + 0.873\mathbf{j}\}$ m, (*c*) 4.26 m

12–147. 14.3 s, 0.45 m/s^2

12–149. $v_r = -1.66$ m/s, $v_\theta = -2.07$ m/s, $a_r = -4.20$ m/s^2,
$a_\theta = 2.97$ m/s^2

12–150. $v_r = -2.33$ m/s, $v_\theta = 7.91$ m/s, $a_r = -158$ m/s^2,
$a_\theta = -18.6$ m/s^2

12–151. $v_r = 16.0$ ft/s, $v_\theta = 1.94$ ft/s, $a_r = 7.76$ ft/s^2,
$a_\theta = 1.94$ ft/s^2

12–153. $v_r = 5.44$ ft/s, $v_\theta = 87.0$ ft/s, $a_r = -1386$ ft/s^2,
$a_\theta = 261$ ft/s^2

12–154. $v_r = 1.62$ m/s, $v_\theta = 15.1$ m/s, $a_r = -93.4$ m/s^2,
$a_\theta = 49.7$ m/s^2

12–155. 12.4 ft/s **12–157.** 0.333 rad/s, 6.67 m/s^2

12–158. $a_r = -6.67$ m/s^2, $a_\theta = 3$ m/s^2 **12–159.** 2.32 m/s^2

12–161. $v_r = 1.20$ m/s, $v_\theta = 1.26$ m/s, $a_r = -3.77$ m/s^2,
$a_\theta = 7.20$ m/s^2

12–162. $v_r = 1.20$ m/s, $v_\theta = 1.26$ m/s, $a_r = -0.570$ m/s^2,
$a_\theta = 10.6$ m/s^2

12–163. $v_r = 1.20$ m/s, $v_\theta = 1.50$ m/s, $a_r = -4.50$ m/s^2,
$a_\theta = 7.20$ m/s^2

12–165. $v_r = -1.84$ m/s, $v_\theta = 19.1$ m/s, $a_r = -2.29$ m/s^2,
$a_\theta = 4.60$ m/s^2

12–166. $v_r = 0$, $v_\theta = 4.80$ ft/s, $v_z = -0.664$ ft/s,
$a_r = -2.88$ ft/s^2, $a_\theta = 0$, $a_z = -0.365$ ft/s^2

12–167. $v_{\max} = 16.3$ ft/s, $v_{\min} = 16$ ft/s, $a_{\max} = 32.6$ ft/s^2,
$a_{\min} = 32$ ft/s^2

12–169. 5.95 ft/s, 3.44 ft/s^2

12–170. $v = 24.1$ ft/s, $a = 8.17$ ft/s^2

12–171. $v_r = -0.212$ m/s, $v_\theta = 0.812$ m/s, $a_r = -0.703$ m/s^2,
$a_\theta = 0.0838$ m/s^2

12–173. $v_r = -2.80$ m/s, $v_\theta = 19.8$ m/s

12–174. $v_r = -4.00$ m/s, $v_\theta = 28.3$ m/s, $a_r = -5.43$ m/s^2,
$a_\theta = -1.60$ m/s^2

12–175. 0.75 rad/s

12–177. $v_r = a\dot\theta$, $v_\theta = a\theta\dot\theta$, $a_r = -a\theta\dot\theta^2$, $a_\theta = 2a\dot\theta^2$

12–178. $v_r = ake^{k\theta}\,\dot\theta$, $v_\theta = ae^{k\theta}\,\dot\theta$, $a_r = ae^{k\theta}\,\dot\theta^2(k^2 - 1)$,
$a_\theta = 2ake^{k\theta}\,\dot\theta^2$

12–179. $v_r = 6$ ft/s, $v_\theta = 18.3$ ft/s, $a_r = -67.1$ ft/s^2,
$a_\theta = 66.3$ ft/s^2

12–181. 0.0178 rad/s **12–182.** 0.00404 rad/s^2

12–183. 6.54 m/s^2 **12–185.** 10.0 rad/s

12–186. 7.83 m/s, 2.27 m/s^2

12–187. $v = 0.242$ m/s, $a = 0.169$ m/s^2, $v_x = 0.162$ m/s, $v_y = 0.180$ m/s

12–189. $\mathbf{u}_b = \{-0.349\mathbf{i} - 0.0127\mathbf{j} - 0.937\mathbf{k}\}$, $\mathbf{u}_b = \{0.349\mathbf{i} + 0.0127\mathbf{j} + 0.937\mathbf{k}\}$

12–190. 1.33 ft ↑ **12–191.** 2 ft/s ↑

12–193. 160 s

12–194. $v_B = 4$ ft/s ↑, $v_{B/A} = 12$ ft/s ↑

12–195. 2 ft ↑ **12–197.** (*a*) 50 s, (*b*) 40 s

12–198. 11 ft/s **12–199.** 0

12–201. 32 m/s ↓ **12–202.** 3.83 s

12–203. 0.667 m/s ↑ **12–205.** 6 ft/s ↑

12–206. 2 ft **12–207.** 9 m/s ↑

12–209. $v_C = \dfrac{-2\sqrt{(s+4)^2 - 16}}{(s+4)}$

12–210. 6.08 ft/s ← **12–211.** 0.0192 ft/s^2

12–213. 1.66 ft/s **12–214.** 0.513 ft/s^2

12–215. 875 km/h, 41.5° θ ◺

12–217. 1955 mi/h^2, 0.767° ◸ θ

12–218. 525 mi/h, 19.1° ◸ θ

12–219. 23.4 mi

12–221. 22.6 s, 70.9 ft/s, 7.11° θ ◹

12–222. 6.69 m/s, 53.3° θ ◹, 1.52 m/s^2, 41.9° θ ◹

12–223. 28.5 mi/h, 44.5° ∠θ, 3418 mi/h^2, 80.6° ∠θ

12–225. 120 km/h ↓, 4000 km/h^2, 0.716° θ ◹

12–226. 98.4 ft/s, 67.6° θ ◹, 19.8 ft/s^2, 57.4° ∠θ

12–227. 34.6 km/h

12–229. $v_B = 5.75$ m/s, $v_{C/B} = 17.8$ m/s, 76.2° ◸θ, 9.81 m/s^2 ↓

12–230. 4.47 m/s^2, 87.4° ∠θ

12–231. 0.977 s, $v_0 = 15.7$ ft/s, $v_{b/B} = 24.1$ ft/s, $v_D = 15.7$ ft/s →

12–233. 11.1 ft, 36.4 ft/s

Chapter 13

13–1. 199(10^{18}) N **13–2.** 76.1 N, 20.4 kg, 3.73 m/s^2

13–3. 1.07 μN **13–5.** 0.343 m/s^2, 0.436 m/s^2

13–6. 1.09 s, 1.54 s, 2.32 s **13–7.** 3.01 kN

13–9. 4.66 kN **13–10.** 4.55 kN

13–11. $t = 5$ s, $F = 652$ lb, $t = 20$ s, $F = 109$ lb

13–13. 286 lb

13–14. (*a*) 40 lb, (*b*) 59.9 lb

13–15. 77.9 ft/s **13–17.** 14.6 ft/s

13–18. 3.43 ft/s^2 **13–19.** 5.30 ft, 1.82 s

13–21. 0.297 m/s^2 **13–22.** 0.0595 m/s^2

13–23. 4.92 kN **13–25.** 13.1 lb

13–26. 26.8 ft/s **13–27.** 8.40 ft/s^2, 9.66 ft/s^2

13–29. 5.43 m

13–30. (*a*), (*b*) 6.94 m/s^2, (*c*) 7.08 m/s^2, 56.5° θ ◹

13–31. 7.49 m/s^2 **13–33.** 88.1 N, 1 m/s^2

13–34. 4.52 m/s **13–35.** 22.6°

13–37. 8.49 m **13–38.** 4.39 m/s

13–39. $P = 2\,mg \tan\theta$

13–41. $t = 0.693\dfrac{m}{k}$, $x = 0.5\dfrac{mv_0}{k}$

13–42. $v = \left[\sqrt{\dfrac{mg}{k}} - \left(\dfrac{mg}{k} - v_0^2\right)e^{-(2kh/m)}\right]^{1/2}$, $v = \sqrt{\dfrac{mg}{k}}$

13–43. $v = \sqrt{\dfrac{mg}{k}}\left[\dfrac{e^{2t\sqrt{mg/k}} - 1}{e^{2t\sqrt{mg/k}} + 1}\right]$, $v_t = \sqrt{\dfrac{mg}{k}}$

13–45. $x = \dfrac{mv_0}{k}\cos\theta_0(1 - e^{-(k/m)t})$,

$y = -\dfrac{mgt}{k} + \dfrac{m}{k}\left(v_0 \sin\theta_0 + \dfrac{mg}{k}\right)\left(1 - e^{-(k/m)t}\right)$,

$x_{\max} = \dfrac{mv_0\cos\theta_0}{k}$

13–46. $a_A = 9.66$ ft/s^2 ←, $a_B = 15.0$ ft/s^2 →

13–47. $a_B = 0$, $a_C = 4.11$ m/s^2 →, $a_D = 0.162$ m/s^2 →

13–49. (*a*) 3.86 ft/s^2, (*b*) $a_A = 70.8$ ft/s^2, $a_B = 6.44$ ft/s^2

13–50. $T = \dfrac{mg}{2}\sin 2\theta$

13–51. $v_{B/AC} = a_0 \sin\theta t$, $s = \frac{1}{2}a_0 \sin\theta t^2$

13–53. 7.23 m/s **13–54.** $d = \dfrac{(m_A + m_B)g}{k}$

13–55. 251 m/s, 6.18 kN **13–57.** 38.0 lb

13–58. 9.90 m/s

13–59. $F_x = 237$ N, $F_y = 772$ N

13–61. $T = 5.10$ kN, $v = 5.62$ m/s, $T = 7.46$ kN

13–62. $T = 3\,mg \sin\theta$ **13–63.** 5.56 kN

13–65. $N = 277$ lb, $F = 13.4$ lb

13–66. 47.5°

13–67. 6.30 m/s, $F_n = 283$ N, $F_t = 0$, $F_b = 490$ N

13–69. 22.1 m/s

13–70. $v_C = 19.9$ ft/s, $N_C = 7.91$ lb, $v_B = 21.0$ ft/s

13–71. 9.32 m **13–73.** 2.10 m/s

13–74. 0.969 m/s **13–75.** 1.48 m/s

13–77. 1.96 ft/s^2, 8.13 lb **13–78.** 7.39 s
13–79. 4.90 lb **13–81.** 1.99(10^{30}) kg
13–82. $F_t = 0$, $F_s = 3.42$ lb **13–83.** 1.5 ft
13–85. $v = \sqrt{v_0^2 - 2gr(1 - \cos\theta)}$,
$N = 58.86\cos\theta - 23.24$, 0°
13–86. $F_r = -92.5$ N, $F_\theta = 317$ N, $F_z = 35.6$ N
13–87. 427 lb **13–89.** 4.72 lb
13–90. $(F_A)_{max} = 2.32$ lb, $(F_A)_{min} = 0.0839$ lb
13–91. $F_r = -29.4$ N, $F_\theta = 0$, $F_z = 392$ N
13–93. $N = 5.79$ N, $F = 1.78$ N
13–94. $N = 6.37$ N, $F = 2.93$ N
13–95. $N = 25.6$ N, $F = 0$
13–97. 0.143 lb **13–98.** 0.432 lb
13–99. 7.67 N
13–101. $F_r = -100$ N, $F_\theta = 0$, $F_z = 2.45$ kN
13–102. 4.00 rad/s, 8 N **13–103.** 0.300 lb
13–105. $P = 12.6$ lb, $N = 4.22$ lb
13–106. $P = 14.8$ lb, $N = 2.09$ lb
13–107. $F = -1.66$ N **13–109.** 5.66 N
13–110. 12 m/s^2, 5.71° θ⦣
13–111. 11.0 m/s^2, 5.71° θ⦣
13–113. $N_C = 0$, $F = 77.0$ N
13–114. −0.0155 lb **13–115.** −0.0108 lb
13–117. 1.79(10^9) km
13–118. $v_B = 7.71$ km/s, $v_A = 4.63$ km/s
13–119. 4.70 Mm, crash
13–121. $v_B = 1.20$ km/s, $\Delta v = 2.17$ km/s
13–122. 10.8(10^9) km **13–123.** 7.82 km/s, 1.97 h
13–125. 11.6 Mm/h, 27.3 Mm
13–126. (*a*) 194(10^3) mi, (*b*) 392(10^3) mi,
(*c*) 194(10^3) mi $< r <$ 392(10^3) mi,
(*d*) $r >$ 392(10^3) mi
13–127. 55.3(10^3) ft/s, 484(10^6) mi
13–129. 7.31 km/s **13–130.** 1.68 km/s
13–131. $\Delta v_A = 466$ m/s, $\Delta v_B = 2.27$ km/s
13–133. 2.90 h, 578 m/s

Chapter 14

14–1. $U_w = 4.12$ kJ, $U_{N_f} = -2.44$ kJ, difference accounts for change in kinetic energy
14–2. 6 **14–3.** $F_{avg} = \dfrac{W(v^2 + 2gs)}{2gs}$
14–5. 59.7 m **14–6.** 2.72 in.
14–7. 3.90 in. **14–9.** 15.0 MN/m^2
14–10. 179 mm **14–11.** 6.52 m/s, 14.2 N
14–13. 3.77 m/s, block actually stops when $s = 0.755$ m.
14–14. $v_B = 30.5$ m/s, 1.74 kN
14–15. $h = 47.5$ m **14–17.** 7.18 ft/s
14–18. 3.52 ft/s **14–19.** 2.45 m/s
14–21. 2.83 m, 7.67 m/s
14–22. $s_A = 0.255$ m, $s_B = 0.205$ m
14–23. 11.1 kN/m **14–25.** 2.48 ft, 36.0 lb
14–26. 5.44 ft **14–27.** 35.6 ft
14–29. 28.3 m/s **14–30.** 5.48 m/s
14–31. 0.0735 ft **14–33.** 8.22 m/s
14–34. 1.70 kN
14–35. 0.815 m, 568 N, 6.23 m/s^2
14–37. $s = 0.730$ m **14–38.** 13.9 ft/s
14–39. 1.90 ft **14–41.** 2.34 m/s
14–42. 4.44 s **14–43.** 5.20 days
14–45. 52.1 hp **14–46.** 10.3 m/s, 4.12 MW
14–47. 181 ft **14–49.** 0.645 hp
14–50. 483 kW **14–51.** 622 kW
14–53. 0.654 **14–54.** 35.4 kW
14–55. 0.229 hp **14–57.** 1.13 kW
14–58. 2.84 hp **14–59.** 1.63 hp
14–61. $P = 53.3t$ kW, $P = (160t - 533t^2)$ kW, $u = 1.69$ kJ
14–62. 10.7 kW **14–63.** $5.33t$ MW
14–65. 179 mm **14–66.** 47.5 m
14–67. 7.18 ft/s
14–69. $s_A = 0.255$ m, $s_B = 0.205$ m
14–70. 0.0735 ft **14–71.** 0.260 m
14–73. 227 N **14–74.** 13.9 ft/s
14–75. 11.7 ft/s **14–77.** 12.6 ft/s, 18.9 lb
14–78. 4.49 ft/s **14–79.** 2.67 in.
14–81. 416 mm
14–82. 24.5 m, $N_B = 0$, $N_C = 16.8$ kN
14–83. 25.0 m, $N_B = 0$, $N_C = 16.8$ kN
14–85. 6.97 m/s **14–86.** 773 N/m
14–87. 27.8°, 4.63 m **14–89.** 80.2 ft/s, 952 lb
14–90. 48.2° **14–91.** 98.4 ft/s, 601 lb
14–93. $U_{1-2} = GM_e m\left(\dfrac{1}{r_2} - \dfrac{1}{r_1}\right)$
14–94. $\Delta_{max} = 2W/k$, $\Delta_{max} = 2\,\Delta_{st}$

14–95. $v_A = 1.54$ m/s, $v_B = 4.62$ m/s
14–97. 1.29 ft **14–98.** 3.11 kip/in.
14–99. 1.59 ft **14–101.** 0.195 ft
14–102. 20.4 ft/s **14–103.** 30.0 m/s, 130 m

Chapter 15

15–1. 1.53 s, 11.5 m
15–2. 467 N, $30° \measuredangle \theta$, 1.25 kg · m/s, $14.2° \measuredangle \theta$
15–3. 0.280 s, 15.6 ft/s **15–5.** 0.706 N · s, $40° \measuredangle \theta$
15–6. 15 kN · s **15–7.** 89.8 ft/s
15–9. 8.49 N · s, $17.3° \measuredangle \theta$
15–10. $I = TF_0\left[1 - e^{-t_0/T}\left(1 + \frac{t_0}{T}\right)\right]$
15–11. $F = 24.8$ kN, $T = 14.9$ kN
15–13. 0.518 s **15–14.** 63.4 N · s
15–15. 70.5 N **15–17.** 246 s
15–18. 20.4 m/s **15–19.** 12.5 m/s
15–21. 21.8 m/s **15–22.** 269 m/s
15–23. 10.5 ft/s **15–25.** 23.4 ft/s, 14.9 ft
15–26. 40.2 ft/s **15–27.** 10.1 ft/s
15–29. 6.91 m/s **15–30.** 7.65 m/s
15–31. 5.72 s **15–33.** 0.5 m/s, 16.9 kJ
15–34. 0.600 ft/s **15–35.** 3.5 ft/s
15–37. 3.29 ft/s **15–38.** 75 lb
15–39. 120 m/s **15–41.** 8.62 m/s
15–42. 0.364 ft **15–43.** 1.58 ft
15–45. $v_{F_1} = 80.6$ ft/s, $v_{F_2} = 9.56$ ft/s
15–46. 0.221 m
15–47. $v_A = 1.58$ ft/s, $v = 0.904$ ft/s
15–49. 71.4 mm **15–50.** 71.4 mm
15–51. $v = \frac{Mv_0}{m + M}$, $t = \frac{v_0}{\mu g\left(1 + \frac{m}{M}\right)}$
15–53. 0.379 m/s
15–54. $v_b = 0.379$ m/s, $v_{b/c} = 0.632$ m/s
15–55. 8.93 ft/s
15–57. $v_A = 3.29$ m/s, $v_B = 2.19$ m/s
15–58. $v_A = 2.46$ m/s, $v_B = 1.64$ m/s
15–59. $(v_A)_2 = 0.353$ m/s, $(v_B)_2 = 2.35$ m/s
15–61. $(v_A)_2 = 0.400$ m/s, $(v_B)_2 = 2.40$ m/s, 0.951 m
15–62. 2.06 m, 8.21 m/s **15–63.** 13.8 ft
15–65. 8.42 ft/s **15–66.** 8.98 ft
15–67. $t = \sqrt{\frac{2h}{g}}\left(\frac{1 + e}{1 - e}\right)$
15–69. $v_C = 0$, $v_D = v$, $v_B = v$, $v_A = 0$, $v_C = v$, $v_B = 0$
15–70. $v_D = 0.5625v \rightarrow$, $v_B = 0.8125v$, $v_A = 0.4375v \rightarrow$
15–71. $(v_A)_1 = 19.7$ ft/s, $(v_A)_2 = 11.1$ ft/s, $(v_B)_2 = 17.0$ ft/s, 11.3 ft
15–73. 4.82 ft
15–74. $(v_B)_2 = \frac{1}{3}\sqrt{2gh}(1 + e)$
15–75. 0.125 m/s
15–77. $v_c = \frac{v_0(1 + e)m}{(m + M)}$, $v_b = v_0\left(\frac{m - eM}{m + M}\right)$, $t = \frac{d}{v_0}\left(1 + \frac{1}{e}\right)^2$
15–78. $e = \frac{\sin\phi}{\sin\theta}\left(\frac{\cos\theta - \mu\sin\theta}{\mu\sin\phi + \cos\phi}\right)$
15–79. 0.25
15–81. 29.3 ft/s, 33.8 ft/s, 0.770
15–82. 0.456 ft **15–83.** 906 lb/ft
15–85. $v_A = 1.35$ m/s $\rightarrow$, $v_B = 5.89$ m/s, $\theta \measuredangle 32.9°$
15–86. 0.0113
15–87. $(v_A)_2 = 4.06$ ft/s, $(v_B)_2 = 6.24$ ft/s
15–89. $v_B = 3.50$ m/s, $v_A = 6.47$ m/s
15–90. $v_B = 3.32$ m/s, $v_A = 6.47$ m/s
15–91. $\mathbf{H}_B = \{108\mathbf{i} + 96\mathbf{j} + 40\mathbf{k}\}$ kg · m²/s
15–93. $(H_O)_A = 22.3$ kg · m²/s, $(H_O)_B = -7.18$ kg · m²/s, $(H_O)_C = -21.6$ kg · m²/s
15–94. $(H_P)_A = -57.6$ kg · m²/s, $(H_P)_B = 94.4$ kg · m²/s, $(H_P)_C = -41.2$ kg · m²/s
15–95. $\mathbf{H}_O = \{-16.8\mathbf{i} + 14.9\mathbf{j} - 23.6\mathbf{k}\}$ slug · ft²/s
15–97. $(\mathbf{H}_A)_0 = \{-1.83\mathbf{i} - 0.913\mathbf{k}\}$ slug · ft²/s
15–98. $\circlearrowright + (H_A)_0 = 72.0$ kg · m²/s $\circlearrowright$, $\circlearrowright + (H_B)_0 = 59.5$ kg · m²/s $\circlearrowright$
15–99. $\circlearrowright + (H_A)_P = 66.0$ kg · m²/s $\circlearrowright$, $\circlearrowright + (H_B)_P = 73.9$ kg · m²/s $\circlearrowright$
15–101. 198 Mg · m²/s **15–102.** 6.15 m/s
15–103. 1.34 s **15–105.** 13.4 m/s
15–106. 9.22 ft/s, 3.04 ft · lb **15–107.** 1.52 ft, 0.739 s
15–109. 11.9 s **15–110.** 1.46 m/s, 71.9°
15–111. 196 mm, 75.0° **15–113.** 3.41 s
15–114. 0.910 s, 17.8 ft/s
15–115. 44.0 ft/s, 8.19 slug · ft²/s, cord *cannot* become unstretched
15–117. $v_B = 13.9$ ft/s, $(v_B)_z = 7.89$ ft/s

15–118. 11.0 N
15–119. $F_x = 55.3$ lb, $F_y = 25.8$ lb
15–121. $F_x = 14.1$ lb, $F_y = 7.06$ lb
15–122. $A_x = 3.98$ kN, $A_y = 3.81$ kN, 1.99 kN · m
15–123. 0.680 lb **15–125.** 4.69 hp
15–126. 0.528 m/s^2 **15–127.** 10.4 N
15–129. $F = v^2 \rho A$ **15–130.** 452 Pa
15–131. 580 ft/s **15–133.** 16.1 m/s
15–134. 11.5 kN **15–135.** 37.5 ft/s^2
15–137. 112 mm/s^2 **15–138.** 3.49 kN
15–139. $F = m'v^2$ **15–141.** $R = (20t + 2.48)$ lb
15–142. $v = \sqrt{\frac{2}{3} g \left(\frac{y^3 - h^3}{y^2} \right)}$

Review Problems 1

R1–1. 23.8 s
R1–2. $y = 0.0208x^2 + 0.333x$ (parabola)
R1–3. $v_B = 55.2$ ft/s, $v_A = 27.6$ ft/s
R1–5. 38.5 ft/s **R1–6.** 1.07 s, 5.93 m/s →
R1–7. 5 s
R1–9. $y = 3 \ln \left(\frac{\sqrt{x^2 + 4} + x}{2} \right)$
R1–10. 2.67 kN **R1–11.** 0.198 m
R1–13. 1.84 m **R1–14.** 980 m
R1–15. 946 kW **R1–17.** 4.38 m/s
R1–18. 7.85 in. **R1–19.** 10.4 m/s
R1–21. 0.575 ft/s, 71.5° ⦩, −0.192 ft · lb
R1–22. 29.0 m/s^2 **R1–23.** 0.5 m/s
R1–25. 0.838 m/s, 1.76 m/s **R1–26.** 2.68 ft/s
R1–27. 25.4 ft/s **R1–29.** 322 mm/s^2 ⦪ 26.6°
R1–30. 38.5 N **R1–31.** 5.68 ft/s^2
R1–33. 240 lb **R1–34.** 47.8 ft/s
R1–35. 0.901 **R1–37.** 13.8 ft/s
R1–38. 158 N **R1–39.** 20.0 ft
R1–41. 26.8 ft/s ↓ **R1–42.** 27.3 s
R1–43. 1.80 hp
R1–45. (*a*) 2.76 ft/s^2, (*b*) $a_A = 67.6$ ft/s^2, $a_B = 5.15$ ft/s^2
R1–46. 0.933 m **R1–47.** 31.7 m/s
R1–49. 21.5 ft/s
R1–50. $a_A = 0.755$ m/s^2, $a_B = 1.51$ m/s^2, $T_A = 90.6$ N, $T_B = 45.3$ N
R1–51. 5.38 ft/s

Chapter 16

16–1. 3.32 rev, 1.67 s
16–2. $a_t = 0.562$ ft/s^2, $a_n = 3600$ ft/s^2, 3000 ft
16–3. 0.838 m/s, 0.890 m/s^2
16–5. (*a*) −3.33 rad/s^2, (*b*) 59.7 rev
16–6. 1.67 rev, 13.0 rad/min, 8.00 rad/min^2
16–7. 0.4 rad/s^2, 4 rad/s
16–9. 6.30 in./s, 10.0 in./s^2, 16.0 in./s
16–10. $v_A = 70.9$ ft/s, $v_B = 35.4$ ft/s, $a_A = 252$ ft/s^2, $a_B = 126$ ft/s^2
16–11. $v_A = 40$ ft/s, $v_B = 20$ ft/s, $a_A = 80.6$ ft/s^2, $a_B = 40.3$ ft/s^2
16–13. 1.37 m/s, 0.472 m/s^2
16–14. 52.0 ft/s, $a_t = 12.0$ ft/s^2, $a_n = 1.35(10^3)$ ft/s^2
16–15. 14.7 m/s **16–17.** 8.81 in./s, 32.6 in./s^2
16–18. 19.2 mm **16–19.** 465 rad/s
16–21. 0.318 mm/s
16–22. 2.4 ft/s, $a_A = 0.4$ ft/s^2, $a_B = 17.3$ ft/s^2
16–23. 126 rad/s
16–25. 184 mm/s, 2.52 mm/s^2
16–26. 21.2 ft/s, 106 ft/s^2
16–27. 26.6°, 22.4 ft/s, 112 ft/s^2
16–29. 0.736 ft/s, 2.59 ft/s^2
16–30. $\omega_C = 4.50$ rad/s, $\omega_B = 3.38$ rad/s
16–31. 10.6 rad/s **16–33.** 1.08 m/s, 1.08 m
16–34. $\omega_B = 180$ rad/s, $\omega_C = 360$ rad/s
16–35. $\omega_B = 300$ rad/s, $\omega_C = 600$ rad/s
16–37. 2.50 m/s, 13.1 m/s^2 **16–38.** 0.841 ft/s
16–39. $v_{AB} = \omega l \cos \theta$, $a_{AB} = -\omega^2 l \sin \theta$
16–41. $v_{BC} = (0.75 \cos \theta)$ m/s, $a_{BC} = (-3.75 \sin \theta)$ m/s^2
16–42. 0.650 m/s, −1.62 m/s^2
16–43. $v_{CD} = -\omega l \sin \theta$, $a_{CD} = -\omega^2 l \cos \theta$
16–45. 8.70 rad/s, −50.5 rad/s^2
16–46. $\omega = \frac{v_0}{a} \sin^2 \theta$, $\alpha = \left(\frac{v_0}{a} \right)^2 \sin 2\theta \sin^2 \theta$
16–47. $\omega = \frac{v}{a} \sin^2 \theta$, $\alpha = \frac{\sin^2 \theta}{a} \left[\frac{v^2 \sin 2\theta}{a} - a_0 \right]$
16–49. $v_C = L\omega \uparrow$, $a_C = 0.577 L\omega^2 \uparrow$
16–50. $v_C = 3L \uparrow$, $a_C = 7.20L \uparrow$
16–51. $\omega = 1.08$ rad/s, $v_B = 4.39$ ft/s
16–53. $\omega = \frac{-v_A r}{x\sqrt{x^2 - r^2}}$, $\alpha = \frac{v_A^2 r(2x^2 - r^2)}{x^2(x^2 - r^2)^{3/2}}$

16–54. $\dot{x} = \dfrac{lr(r + d\cos\theta)}{(d + r\cos\theta)^2}\omega,$

$\ddot{x} = \dfrac{lr\sin\theta(2r^2 - d^2 + rd\cos\theta)}{(d + r\cos\theta)^3}\omega^2$

16–55. $v = \omega d\left(\sin\theta + \dfrac{d\sin 2\theta}{2\sqrt{(R + r)^2 - d^2\sin^2\theta}}\right)$

16–57. 9.33 in./s **16–58.** 12.6 in./s

16–59. 2.40 m/s

16–61. 2.83 ft/s, $\omega_{BC} = \omega_{AB} = 2.83$ rad/s

16–62. 1.64 m/s **16–63.** 1.70 m/s

16–65. 3.72 rad/s **16–66.** 9 m/s

16–67. 5.33 rad/s **16–69.** 13.9 rad/s, 25.9 ft/s

16–70. 10.3 rad/s **16–71.** 0

16–73. $\omega_{BC} = 1.07$ rad/s, $\omega_{AB} = 9.62$ rad/s

16–74. $\omega_A = 4$ rad/s, $\omega_B = 1$ rad/s, $v_D = 60$ mm/s

16–75. 6 m/s **16–77.** 0.1 m/s

16–78. 2 rad/s **16–79.** 6 rad/s

16–81. 20 rad/s, 2 ft/s **16–82.** 4 rad/s, 4.88 m/s

16–83. 2.45 m/s, 7.81 rad/s

16–85. $\omega_S = 57.5$ rad/s, $\omega_{OA} = 10.6$ rad/s

16–86. $\omega_S = 15.0$ rad/s, $\omega_R = 3.00$ rad/s

16–87. $v_B = 3.00$ m/s, $v_C = 0.587$ m/s

16–89. 312 mm/s, 8.13°

16–90. $v_C = 0.776$ m/s, $v_D = 1.06$ m/s

16–91. 4.42 in./s **16–93.** 1.64 m/s

16–94. 9 m/s **16–95.** 5.33 rad/s

16–97. $\omega_A = 4$ rad/s, $\omega_B = 1$ rad/s, 60 mm/s

16–98. 4 rad/s, 4.88 m/s **16–99.** 11.4 rad/s

16–101. $v_A = 0$, $v_B = 1.2$ m/s, $v_C = 0.849$ m/s

16–102. $v_A = 0$, $v_O = 0.6$ m/s, $v_E = 0.849$ m/s

16–103. $v_C = 6.00$ ft/s, $v_B = 4.00$ ft/s

16–105. $\omega_{BC} = 5.77$ rad/s, $\omega_{CD} = 2.17$ rad/s

16–106. $\omega_{BC} = 6.83$ rad/s, $\omega_{CD} = 2.65$ rad/s

16–107. 77.3 in./s

16–109. $v_C = 2.50$ ft/s, $v_D = 9.43$ ft/s

16–110. $v_C = 2.50$ ft/s, $v_E = 7.91$ ft/s

16–111. $v_C = 2.50$ ft/s, $v_F = 6.61$ ft/s

16–113. $v_A = 60.0$ ft/s, $v_C = 220$ ft/s, $v_B = 161$ ft/s ⦣ 60.3°

16–114. 7.81 rad/s, 2.45 m/s **16–115.** 13.1 rad/s

16–117. $\omega_C = 26.7$ rad/s, $\omega_B = 28.75$ rad/s, $\omega_A = 14.0$ rad/s

16–118. 58.8 rad/s

16–119. 0.332 rad/s^2, 7.88 ft/s^2

16–121. 1.47 rad/s^2, 24.9 ft/s^2

16–122. 0.0962 rad/s^2, 0.385 ft/s^2

16–123. $\omega_{CD} = 5.44$ rad/s, $\alpha_{CD} = 103$ rad/s^2

16–125. 13.9 ft/s^2, 8.80° ⦣ **16–126.** 11.2 ft/s^2

16–127. 12.5 m/s^2

16–129. $\alpha_{AB} = 0.75$ rad/s^2, $\alpha_{BC} = 3.94$ rad/s^2

16–130. 2 rad/s, 7.68 rad/s^2

16–131. $\alpha_{AB} = 4.81$ rad/s^2, $\alpha_{BC} = 5.21$ rad/s^2

16–133. $\omega_{AC} = 0$, $\alpha_{AC} = 28.7$ rad/s^2

16–134. 3.55 m/s^2

16–135. 10.67 rad/s^2, 14.1 in./s^2

16–137. $v_B = 1.58\omega a$, $a_B = 1.577\alpha a - 1.77\omega^2 a$

16–138. 5.94 m/s^2 **16–139.** 6.21 m/s^2

16–141. 14.2 ft/s^2, 18.4° ⦣

16–142. 4.73 rad/s, 131 rad/s^2

16–143. 0.25 rad/s, 5.00 ft/s, 0.875 rad/s^2, 1.51 ft/s^2, 85.2° ⦣

16–145. $\mathbf{a}_A = \{-5.60\mathbf{i} - 16\mathbf{j}\}$ m/s^2

16–146. $\mathbf{v}_B = \{-12.0\mathbf{i}\}$ ft/s, $\mathbf{a}_B = \{-70.0\mathbf{j}\}$ ft/s^2

16–147. $\mathbf{v}_B = \{-12.0\mathbf{i}\}$ ft/s, $\mathbf{a}_B = \{-20\mathbf{i} - 70\mathbf{j}\}$ ft/s^2

16–149. $\mathbf{v}_m = \{7.5\mathbf{i} - 5\mathbf{j}\}$ ft/s, $\mathbf{a}_m = \{5\mathbf{i} + 3.75\mathbf{j}\}$ ft/s^2

16–150. $\mathbf{v}_m = \{5\mathbf{i} - 5\mathbf{j}\}$ ft/s, $\mathbf{a}_m = \{5\mathbf{i} + 0.5\mathbf{j}\}$ ft/s^2

16–151. (*a*) $\mathbf{a}_B = \{-1\mathbf{i}\}$ m/s^2, (*b*) $\mathbf{a}_B = \{-1.69\mathbf{i}\}$ m/s^2

16–153. 2.40 m/s

16–154. 2.40 m/s, $\mathbf{a}_C = \{-14.4\mathbf{j}\}$ m/s^2

16–155. $\mathbf{v}_A = \{-2.63\mathbf{i} - 4.50\mathbf{j}\}$ m/s, $\mathbf{a}_A = \{6.75\mathbf{i} - 9.19\mathbf{j}\}$ m/s^2

16–157. $\mathbf{v}_C = \{-7.5\mathbf{i} - 37.0\mathbf{j}\}$ ft/s, $\mathbf{a}_C = \{85.0\mathbf{i} - 7.5\mathbf{j}\}$ ft/s^2

16–158. $\mathbf{v}_D = \{-31.5\mathbf{i} - 13.0\mathbf{j}\}$ ft/s, $\mathbf{a}_D = \{13.0\mathbf{i} - 79.5\mathbf{j}\}$ ft/s^2

16–159. $\omega_{CD} = 9.00$ rad/s, $\alpha_{CD} = 249$ rad/s^2

16–161. 0.720 rad/s, 2.02 rad/s^2

16–162. $\mathbf{v}_C = \{-7.00\mathbf{i} + 17.3\mathbf{j}\}$ ft/s,
$\mathbf{a}_C = \{-34.6\mathbf{i} - 15.5\mathbf{j}\}$ ft/s^2

16–163. $\mathbf{v}_C = \{-7.00\mathbf{i} + 17.3\mathbf{j}\}$ ft/s,
$\mathbf{a}_C = \{-38.8\mathbf{i} - 6.84\mathbf{j}\}$ ft/s^2

Chapter 17

17–1. $I_y = \frac{1}{3}ml^2$ **17–2.** $k_x = \dfrac{a}{\sqrt{3}}$

17–3. $I_x = \frac{2}{5}mr^2$ **17–5.** $I_x = \frac{2}{5}mb^2$

17–6. $I_z = m(R^2 + \frac{3}{4}a^2)$ **17–7.** $I_y = \dfrac{m}{6}(a^2 + h^2)$

17–9. $m = \pi hR^2\left(k + \dfrac{aR^2}{2}\right)$, $I_z = \dfrac{\pi hR^4}{2}\left[k + \dfrac{2aR^2}{3}\right]$

17–10. 6.39 m, 53.2 kg · m^2 **17–11.** 6.99 kg · m^2
17–13. 0.454 slug · ft^2 **17–14.** 7.67 kg · m^2
17–15. 6.23 kg · m^2 **17–17.** 13.4 kg · m^2
17–18. 118 slug · ft^2 **17–19.** 293 slug · ft^2
17–21. 5.64 slug · ft^2
17–22. 0.203 m, 0.230 kg · m^2
17–23. 0.560 kg · m^2 **17–25.** 29.8 kg · m^2
17–26. $I_o = \frac{1}{2}ma^2$ **17–27.** $I_A = 1.5ma^2$
17–29. $N_B = 95.0$ lb, $N_A = 105$ lb, 9.66 ft
17–30. $F = 5.96$ lb, $N_B = 99.0$ lb, $N_A = 101$ lb
17–31. $A_y = 72.6$ kN, $B_y = 71.6$ kN, $a_G = 0.250$ m/s^2
17–33. 25.1 kN, 2.74 m/s^2 **17–34.** $5.08(10^3)$ lb · ft
17–35. $N = 0.433wx$, $V = 0.25wx$, $M = 0.125wx^2$
17–37. $T = 4.04$ kN, $N_B = 3.67$ kN, $N_A = 5.93$ kN
17–38. $F_{IJ} = 10.8$ kN, $H_x = 3.66$ kN, $H_y = 779$ N
17–39. $T = 15.7$ kN, $C_x = 8.92$ kN, $C_y = 16.3$ kN
17–41. $N_A = 1393$ lb, $N_B = 857$ lb, 2.72 s
17–42. 1.48 s
17–43. $N'_A = 962$ lb, $N'_B = 938$ lb
17–45. $N_B = 402$ N, $N_A = 391$ N
17–46. 5.95 rad/s^2 **17–47.** 3.12 ft
17–49. 13.6 ft/s^2, $N'_A = 22.1$ lb, $N'_B = 30.6$ lb
17–50. 77.3 ft/s^2, $B_x = 432$ lb, $B_y = 180$ lb
17–51. $N_A = 120$ lb, $B_x = 73.9$ lb, $B_y = 69.7$ lb
17–53. 5.59 ft/s^2 **17–54.** 3.19 kN, 1.23 rad/s^2
17–55. 5.10 kN, 1.96 rad/s^2
17–57. $N_C = 159$ lb, $F_C = 16.1$ lb
17–58. 34.5 lb · ft **17–59.** $A_x = 0$, $A_y = 262$ N
17–61. 56.2 rad/s, $A_x = 0$, $A_y = 98.1$ N
17–62. 19.9 rad/s, $A_x = 0$, $A_y = 98.1$ N
17–63. 1.06 s
17–65. 14.6 rad/s^2, 45.3 rad/s^2
17–66. 9.54 rad/s^2, 17.2 rad/s^2
17–67. 346 N, 2.52 rad/s^2
17–69. 10.9 rad/s **17–70.** 9.45 rad/s
17–71. 7.28 rad/s^2 **17–73.** 2.71 rad/s
17–74. 27.2 N · m/rad **17–75.** 6.71 s
17–77. $\alpha = \dfrac{2mg}{R(M + 2m)}$, $h = \dfrac{mg}{M + 2m}t^2$
17–78. $\alpha = \dfrac{2mg}{R(M + 2m)}$, $v = \sqrt{\dfrac{8mgR}{(M + 2m)}}$
17–79. $A_x = 0$, $A_y = 289$ N, $\alpha = 23.1$ rad/s^2
17–81. 0.548 m/s
17–82. $a = \dfrac{g(M_B - M_A)}{(\frac{1}{2}M + M_B + M_A)}$
17–83. 1.82 m/s^2 **17–85.** 2.48 rad/s
17–86. 1.48 rad/s **17–87.** $M = 0.3\,gml$
17–89. 1.24 MN
17–90. $A_x = 89.2$ N, $A_y = 66.9$ N, 1.26 s
17–91. $A_x = 89.2$ N, $A_y = 66.9$ N, 25.2 rad
17–93. $F = 2.5\,mg \sin\theta$, $N = 0.25\,mg\cos\theta$, $\theta = \tan^{-1}(0.1\,\mu)$
17–94. 4.83 rad/s, 92.2°
17–97. $A_x = 4.50$ lb, $A_y = 6.50$ lb
17–98. $F = \dfrac{3}{2}mg\sin\theta\left(\dfrac{3}{2}\cos\theta - 1\right)$, $N = \dfrac{mg}{4}(1 - 3\cos\theta)^2$
17–99. $N = w\left(x\cos\theta + \dfrac{w^2x}{2g}(2L - x)\right)$, $V = wx\sin\theta$,
$M = -\dfrac{wx^2}{2}\sin\theta$
17–101. 5.62 rad/s^2, 196 N
17–102. 0.274 rad/s^2, 49.0 ft/s^2
17–103. 16.1 ft/s^2, 5.80 rad/s^2
17–105. 1.25 rad/s, 2.32 kN **17–106.** 1.35 m/s^2
17–107. 5.21 rad/s
17–109. $N_A = 91.3$ N, $F_A = 20.1$ N
17–110. disk does not slip
17–111. $A_x = 42.2$ lb, $A_y = 22.0$ lb
17–113. $\alpha = \dfrac{6(P - \mu_k mg)}{mL}$, $a_B = \dfrac{2(P - \mu_k mg)}{m}$
17–114. $\alpha = \dfrac{3g}{2l}\cos\theta$ **17–115.** $\omega = \left(\dfrac{3g}{l}\sin\theta\right)^{1/2}$
17–117. $A_y = 15.0$ lb, $A_x = 0.466$ lb, $a_G = 1.00$ ft/s^2
17–118. $A_y = 15.0$ lb, $A_x = 0.776$ lb, 1.67 rad/s^2
17–119. $\alpha_A = 43.6$ rad/s^2, $\alpha_B = 43.6$ rad/s^2, 19.6 N
17–121. 23.4 rad/s^2, 9.62 lb
17–122. 0.352 s **17–123.** 0.406 s
17–125. 20.3 ft/s **17–126.** 3.68 ft/s
17–127. 1.73 m/s^2

Chapter 18

18–2. 10.9 rad/s **18–3.** 2.71 rad/s
18–5. 210 ft · lb **18–6.** $T = \frac{1}{6}ml^2\omega^2$

18–7. 0.0188 ft · lb **18–9.** 1.50 rad/s
18–10. 0.304 ft **18–11.** 220 rad/s
18–13. 237 J **18–14.** 7.34 rad/s
18–15. 8.25 rad/s
18–17. (*a*) 4.18 rad/s, (*b*) 3.41 rad/s
18–18. 0.0636 rad/s **18–19.** 2.10 m/s
18–21. 5.92 rad/s **18–22.** 0.753 rad/s
18–23. 1.05 rad/s **18–25.** 13.3 ft/s
18–26. 3.07 m/s **18–27.** 11.0 rad/s
18–29. 350 J **18–30.** 12.8 rad/s
18–31. 1.66 rad **18–33.** 9.75 m
18–34. (*a*) $\omega = \sqrt{\frac{3\pi}{2}\left(\frac{w_0}{m}\right)}$, (*b*) $\omega = \sqrt{\frac{3\pi}{2}\frac{w_0}{m} + \frac{3g}{L}}$
18–35. (*a*) $\omega = \sqrt{\frac{\pi w_0}{m}}$, (*b*) $\omega = \sqrt{\frac{\pi w_0}{m} + \frac{3g}{L}}$
18–37. $v_G = 3\sqrt{\frac{3}{7}gR}$ **18–38.** 42.4 rad/s, 360 N
18–39. 0.304 ft **18–41.** 7.34 rad/s
18–42. 0.753 rad/s **18–43.** 13.3 ft/s
18–45. $(v_G)_1 = 3\sqrt{\frac{3}{7}gR}$ **18–46.** 2.58 m/s
18–47. 0.301 m, 163 N **18–49.** 42.8 N/m
18–50. 1.80 rad/s **18–51.** 31.9 ft/s
18–53. 1.40 m/s **18–54.** 16.2 mm
18–55. 1.52 m/s **18–57.** 1.18 rad/s
18–58. 2.82 rad/s **18–59.** 5.28 rad/s
18–61. 1.73 rad/s **18–62.** 19.1 rad/s
18–63. 16.1 rad/s **18–65.** 4.89 rad/s
18–66. 4.97 rad/s **18–67.** 100 lb/ft
18–69. 59.0 mm **18–70.** 2.44 ft
18–71. 1.82 rad/s **18–73.** 4.00 m/s
18–73. 4.00 m/s

Chapter 19

19–5. 56.2 rad/s, 0, 98.1 N **19–6.** 0.386 rad/s
19–7. 5.18 lb, 2.60 s **19–9.** 14.4 rad/s
19–10. 0.696 s
19–11. $\omega = 14.4$ rad/s, $\omega_2 = 6.10$ rad/s
19–13. 0.530 s **19–14.** 6.04 rad/s
19–15. 3.56 rad/s, 5.12 s
19–17. $\theta = \tan^{-1}\left[\left(1 + \frac{r^2}{k_0^2}\right)\mu\right]$
19–18. 27.9 rad/s **19–19.** 29.2 rad/s
19–21. 1.20 kN **19–22.** 12.4 ft/s
19–23. 12.4 ft/s **19–25.** 12.7 rad/s
19–26. 4.26 ft/s **19–27.** 29.8 lb
19–29. 119 rad/s, 64.5 ft/s **19–30.** 176 rad/s
19–31. 23.4 rad/s **19–33.** 70.1 rad/s
19–34. 73.6 rad/s **19–35.** 1.32 s
19–37. $I = m\sqrt{\frac{gL}{6}}$ **19–38.** $y = \frac{\sqrt{2}}{3}a$
19–39. $\frac{2}{3}L$ **19–41.** 0.194 s
19–42. $v_0 = \sqrt{(2g\sin\theta d)\left(\frac{m_c}{m}\right)}$
19–43. $v_A = 1.66$ m/s, $v_B = 1.30$ m/s
19–45. 0 **19–46.** 1.91 rad/s
19–47. $\frac{11}{3}\omega_1$ **19–49.** 0.0906 rad/s
19–50. 0.0181 rad/s **19–51.** 3.56 rad/s
19–53. $\omega' = \frac{19}{28}\omega$, $v'' = 0.985\ \omega L$
19–54. 0.0708 rad/s
19–55. $\omega_1 = 2.41$ rad/s, $\omega_2 = 1.86$ rad/s
19–57. $\omega = \sqrt{7.5\frac{g}{L}}$
19–58. (*a*) 0, (*b*) $\omega_M = \frac{I}{I_z}\omega$, (*c*) $\omega_M = \frac{2I}{I_z}\omega$
19–59. (*a*) $\omega_M = \frac{I}{I_z}\omega$, (*b*) 0, (*c*) $\omega_M = \frac{I}{I_z}\omega$
19–61. $\omega_2 = \frac{1}{7}(2 + 5\cos\theta)\omega_1$
19–62. $\omega_2 = \frac{1}{3}(1 + 2\cos\theta)\omega_1$
19–63. 22.7 rad/s **19–65.** 0.500 ft
19–66. 3.23 rad/s, 32.8° **19–67.** 22.4°
19–69. 3.81 rad/s **19–70.** 7.73 rad/s
19–71. 2 m, 6.82 rad/s **19–73.** 0.980 ft
19–74. $(v_G)_{y_2} = e(v_G)_{y_1}\uparrow$, $(v_G)_{x_2} = \frac{5}{7}((v_G)_{x_1} - \frac{2}{5}\omega_1 r)$
19–75. 50 rad/s, 294 mm

Review Problems 2

R2–1. 5.96 ft/s **R2–2.** 14.2 ft/s, 0.25 ft/s^2
R2–3. 28.6 rad/s, 24.1 rad, 7.16 m/s, 205 m/s^2
R2–5. 1.80 m
R2–6. 0.275 rad/s, 0.0922 rad/s^2
R2–7. *a*) $C_y = B_y = 7.22$ N, *b*) $C_y = B_y = 14.7$ N
R2–9. 25.3 in./s, 63.4° ∠, 73.8 in./s^2, 32.5° ∠
R2–10. 8.89 rad/s^2 **R2–11.** 25.0 rad/s

R2–13. $v = -r\omega \sin\theta$, $a = -r\omega^2 \cos\theta$

R2–14. 282 ft/s^2 **R2–15.** $a = \dfrac{s}{2\pi}\omega^2$

R2–17. 4.78 m/s

R2–18. $\theta = \tan^{-1}\left[\dfrac{\mu(k_O^2 + r^2)}{k_O^2}\right]$

R2–19. 5.75 ft **R2–21.** 218 N, 21.0 rad/s^2

R2–22. $v = -r_1\omega \sin\theta - \dfrac{r_1^2\omega \sin 2\theta}{2\sqrt{(r_1 + r_2)^2 - (r_1 \sin\theta)^2}}$

R2–23. 3.92 lb · s **R2–25.** 0.860 m/s

R2–26. $a_C = 50.6$ m/s^2, $\theta = 9.09°$ ⦣, $a_B = 8.00$ m/s^2 ↑

R2–27. 12.1 rad/s^2, 30.0 lb

R2–29. 215 rad/s **R2–30.** 39.5 rad/s

R2–31. $P = 24$ lb, $a_G = 3.22$ ft/s^2, $N_B = 14.8$ lb, $N_A = 65.3$ lb

R2–33. $B_y = 180$ N, $A_y = 252$ N, $A_x = 139$ N

R2–34. $B_y = 143$ N, $A_y = 200$ N, $A_x = 34.3$ N

R2–35. 14.2 rad/s^2

R2–37. $F_C = 12.4$ lb, $N'_B = 3.09$ kip, $N'_A = 1.31$ kip, $F'_A = 68.3$ lb

R2–38. $\omega_A = 1.70$ rad/s, $\omega_B = 5.10$ rad/s

R2–39. $E_x = 9.87$ lb, $E_y = 4.86$ lb, $M_E = 7.29$ lb · ft

R2–41. $F'_B = 4.50$ kN, $N'_A = 1.78$ kN, $N'_B = 5.58$ kN

R2–42. 1.41 m/s^2, 1.38 m/s^2

R2–43. $N'_A = 383$ N, $N'_B = 620$ N

R2–45. $a_A = 56.2$ ft/s^2 ↓, $a_B = 40.2$ ft/s^2, 53.3° ⦨

R2–46. 39.3°

R2–47. 2.26 m/s^2, $N_B = 226$ N, $N_A = 559$ N, $N_B = 454$ N

R2–49. 0.661 m **R2–50.** 0.859 m

R2–51. 3.46 m/s

Chapter 20

20–1. $\mathbf{v}_A = \{-5.20\mathbf{i} - 12\mathbf{j} + 20.8\mathbf{k}\}$ ft/s, $\mathbf{a}_A = \{-24.1\mathbf{i} - 13.3\mathbf{j} - 7.20\mathbf{k}\}$ ft/s^2

20–2. $\mathbf{v}_A = \{-5.20\mathbf{i} - 12\mathbf{j} + 20.8\mathbf{k}\}$ ft/s, $\mathbf{a}_A = \{-3.33\mathbf{i} - 21.3\mathbf{j} + 6.66\mathbf{k}\}$ ft/s^2

20–3. 36.4 rad/s, 14.5 rad/s^2

20–5. $\mathbf{v}_C = \{-0.800\mathbf{i} + 0.400\mathbf{j} + 0.800\mathbf{k}\}$ m/s, $\mathbf{a}_C = \{-10.3\mathbf{i} - 3.84\mathbf{j} + 0.320\mathbf{k}\}$ m/s^2

20–6. $\boldsymbol{\omega} = \{-4.00\mathbf{j}\}$ rad/s, $\boldsymbol{\alpha} = \{16.0\mathbf{i}\}$ rad/s^2, $\mathbf{v}_A = \{-0.283\mathbf{i}\}$ m/s, $\mathbf{a}_A = \{-1.13\mathbf{j} - 1.13\mathbf{k}\}$ m/s^2

20–7. $\boldsymbol{\omega} = \{-3.46\mathbf{j}\}$ rad/s, $\dot{\boldsymbol{\omega}} = \{6.93\mathbf{i} - 6.93\mathbf{j}\}$ rad/s^2

20–9. $\boldsymbol{\omega} = \{2\mathbf{i} + 42.4\mathbf{j} + 43.4\mathbf{k}\}$ rad/s, $\dot{\boldsymbol{\omega}} = \{-42.4\mathbf{i} - 82.9\mathbf{j} + 84.9\mathbf{k}\}$ rad/s^2

20–10. $\mathbf{v}_A = \{-7.79\mathbf{i} - 2.25\mathbf{j} + 3.90\mathbf{k}\}$ ft/s, $\mathbf{a}_A = \{8.30\mathbf{i} - 35.2\mathbf{j} + 7.02\mathbf{k}\}$ ft/s^2

20–11. $\mathbf{v}_A = \{-14.1\mathbf{i}\}$ ft/s, $\mathbf{a}_A = \{-10.6\mathbf{i} - 56.5\mathbf{j} - 87.9\mathbf{k}\}$ ft/s^2

20–13. $\boldsymbol{\omega}_P = \{-40\mathbf{j}\}$ rad/s, $\boldsymbol{\alpha} = \{-6400\mathbf{i}\}$ rad/s^2

20–14. $\boldsymbol{\omega} = \{4.35\mathbf{i} + 12.7\mathbf{j}\}$ rad/s, $\mathbf{a} = \{-26.1\mathbf{k}\}$ rad/s^2

20–15. $\boldsymbol{\omega} = \{4.35\mathbf{i} + 12.7\mathbf{j}\}$ rad/s, $\dot{\boldsymbol{\omega}} = \{2.17\mathbf{i} + 6.36\mathbf{j} - 26.1\mathbf{k}\}$ rad/s^2

20–17. $\mathbf{v}_A = \{-1.80\mathbf{i}\}$ ft/s, $\mathbf{a}_A = \{-0.750\mathbf{i} - 0.720\mathbf{j} - 0.831\mathbf{k}\}$ ft/s^2

20–18. $\boldsymbol{\omega}_{BD} = \{-2.00\mathbf{i}\}$ rad/s

20–19. $\boldsymbol{\omega}_{BD} = \{-2.00\mathbf{i}\}$ rad/s, $\boldsymbol{\alpha}_{BD} = \{34.5\mathbf{i}\}$ rad/s^2

20–21. $\mathbf{a}_B = \{-96.5\mathbf{i}\}$ ft/s^2

20–22. 9.43 ft/s **20–23.** 121 ft/s^2

20–25. $\mathbf{a}_B = \{-47.5\mathbf{j}\}$ m/s^2

20–26. $\boldsymbol{\omega} = \{-0.440\mathbf{i} + 0.293\mathbf{j} - 0.238\mathbf{k}\}$ rad/s, $\mathbf{v}_B = \{-2.82\mathbf{i} + 3.76\mathbf{j}\}$ ft/s

20–27. $\boldsymbol{\alpha} = \{0.413\mathbf{i} + 0.622\mathbf{j} - 0.412(10^{-3})\mathbf{k}\}$ rad/s^2, $\mathbf{a}_B = \{-5.98\mathbf{i} + 7.98\mathbf{j}\}$ ft/s^2

20–29. $\boldsymbol{\alpha} = \{-1.43\mathbf{i} + 0.600\mathbf{j} + 1.34\mathbf{k}\}$ rad/s^2, $\mathbf{a}_B = \{-14.3\mathbf{j}\}$ ft/s^2

20–30. $\mathbf{v}_A = \{0.667\mathbf{i}\}$ ft/s, $\mathbf{a}_A = \{-0.148\mathbf{i}\}$ ft/s^2

20–31. $\mathbf{a}_A = \{1.19\mathbf{i}\}$ ft/s^2

20–33. $\boldsymbol{\alpha}_{BC} = \{1.24\mathbf{i} - 0.306\mathbf{j} + 1.44\mathbf{k}\}$ rad/s^2, $\mathbf{a}_B = \{7.98\mathbf{j}\}$ m/s^2

20–34. $\boldsymbol{\omega}_{BC} = \{0.769\mathbf{i} - 2.31\mathbf{j} + 0.513\mathbf{k}\}$ rad/s, $\mathbf{v}_B = \{-0.333\}$ m/s

20–35. $\mathbf{v}_A = \{-2.50\mathbf{k}\}$ ft/s, $\mathbf{v}_C = \{4.33\mathbf{i} + 2.17\mathbf{j} - 2.50\mathbf{k}\}$ ft/s

20–37. $\boldsymbol{\omega} = \{1.50\mathbf{i} + 2.60\mathbf{j} + 2.00\mathbf{k}\}$ rad/s, $\mathbf{v}_C = \{10.4\mathbf{i} - 7.79\mathbf{k}\}$ ft/s

20–38. $\mathbf{a}_C = \{99.6\mathbf{i} - 117\mathbf{k}\}$ ft/s^2, $\boldsymbol{\alpha} = \{10.4\mathbf{i} + 30.0\mathbf{j} + 3\mathbf{k}\}$ rad/s^2

20–39. $\boldsymbol{\omega} = \{-8.66\mathbf{i} - 5.00\mathbf{j} - 2.00\mathbf{k}\}$ rad/s, $\mathbf{v}_C = \{17.3\mathbf{j} - 43.3\mathbf{k}\}$ ft/s

20–41. $\mathbf{v}_C = \{-1.00\mathbf{i} + 5.00\mathbf{j} + 0.800\mathbf{k}\}$ m/s, $\mathbf{a}_C = \{-28.8\mathbf{i} - 5.45\mathbf{j} + 32.3\mathbf{k}\}$ m/s^2

20–42. $\mathbf{v}_A = \{4.24\mathbf{i} - 8.49\mathbf{j} - 8.49\mathbf{k}\}$ ft/s, $\mathbf{a}_A = \{44.5\mathbf{i} - 38.2\mathbf{j} + 38.2\mathbf{k}\}$ ft/s^2

20–43. $\mathbf{v}_A = \{5.20\mathbf{i} - 5.20\mathbf{j} - 3.00\mathbf{k}\}$ ft/s, $\mathbf{a}_A = \{25\mathbf{i} - 26.8\mathbf{j} + 8.78\mathbf{k}\}$ ft/s^2

20–45. $\mathbf{v}_P = \{-9.80\mathbf{i} + 14.4\mathbf{j} + 48.0\mathbf{k}\}$ ft/s, $\mathbf{a}_P = \{-160\mathbf{i} + 5.16\mathbf{j} - 13\mathbf{k}\}$ ft/s^2

20–46. $\mathbf{v}_P = \{-10.5\mathbf{i} + 9.00\mathbf{j} - 1.00\mathbf{k}\}$ ft/s, $\mathbf{a}_P = \{-41.0\mathbf{i} - 17.5\mathbf{j} - 0.500\mathbf{k}\}$ ft/s^2

20–47. $\mathbf{v}_B = \{-17.3\mathbf{i} + 18.8\mathbf{j} + 10.0\mathbf{k}\}$ ft/s, $\mathbf{a}_B = \{-19.1\mathbf{i} + 24.0\mathbf{j} - 8.66\mathbf{k}\}$ ft/s^2

20–49. $\mathbf{v}_C = \{-1.73\mathbf{i} - 5.77\mathbf{j} + 7.06\mathbf{k}\}$ ft/s, $\mathbf{a}_C = \{9.88\mathbf{i} - 72.8\mathbf{j} + 0.365\mathbf{k}\}$ ft/s^2

20–50. $\mathbf{v}_A = \{-8.66\mathbf{i} + 8\mathbf{j} - 13.9\mathbf{k}\}$ ft/s, $\mathbf{a}_A = \{-4\mathbf{i} - 7.71\mathbf{j} - 3.20\mathbf{k}\}$ ft/s^2

20–51. $\mathbf{v}_A = \{-8.66\mathbf{i} + 8\mathbf{j} - 13.9\mathbf{k}\}$ ft/s, $\mathbf{a}_A = \{-17.9\mathbf{i} + 8.29\mathbf{j} - 30.9\mathbf{k}\}$ ft/s^2

20–53. $\mathbf{v}_B = \{-10.2\mathbf{i} - 28\mathbf{j} + 52.0\mathbf{k}\}$ m/s, $\mathbf{a}_B = \{-33.0\mathbf{i} - 159\mathbf{j} - 90\mathbf{k}\}$ m/s^2

20–54. $\mathbf{v}_P = \{-18\mathbf{i} + 450\mathbf{j} + 4\mathbf{k}\}$ ft/s, $\mathbf{a}_P = \{-810\mathbf{i} - 16.2\mathbf{j} - 81000\mathbf{k}\}$ ft/s^2

20–55. $\mathbf{v}_P = \{-25.5\mathbf{i} - 13.4\mathbf{j} + 20.5\mathbf{k}\}$ ft/s, $\mathbf{a}_P = \{161\mathbf{i} - 249\mathbf{j} - 39.6\mathbf{k}\}$ ft/s^2

20–57. $\mathbf{v}_C = \{-20\mathbf{i} + 24\mathbf{j} - 3\mathbf{k}\}$ in./s, $\mathbf{a}_C = \{-145\mathbf{i} - 24\mathbf{j} + 8\mathbf{k}\}$ in./s^2

20–58. $\mathbf{v}_C = \{3\mathbf{i} + 6\mathbf{j} - 3\mathbf{k}\}$ m/s, $\mathbf{a}_C = \{-13.0\mathbf{i} + 28.5\mathbf{j} - 10.2\mathbf{k}\}$ m/s^2

20–59. $\mathbf{v}_C = \{-5.54\mathbf{i} + 5.20\mathbf{j} - 3.00\mathbf{k}\}$ ft/s, $\mathbf{a}_C = \{-91.5\mathbf{i} - 42.6\mathbf{j} - 1.00\mathbf{k}\}$ ft/s^2

20–61. $\mathbf{v}_A = \{-8.66\mathbf{i} + 2.26\mathbf{j} + 2.26\mathbf{k}\}$ m/s, $\mathbf{a}_A = \{-26.1\mathbf{i} - 44.4\mathbf{j} + 7.92\mathbf{k}\}$ m/s^2

Chapter 21

21–2. $I_{zz} = \dfrac{m}{12}(3a^2 + 4h^2)$ **21–3.** 2614 slug · ft^2

21–5. $k_y = 2.35$ ft, $k_x = 1.80$ ft **21–6.** 636ρ

21–7. $4.48(10^3)\rho$ **21–9.** $I_{yz} = \dfrac{m}{5}ah$

21–10. 0.0930 kg · m^2

21–11. $I_x = 0.341$ kg · m^2, $I_y = 0.0967$ kg · m^2, $I_z = 0.353$ kg · m^2

21–13. 0.0961 slug · ft^2 **21–14.** 1.13 slug · ft^2

21–15. 0.888 slug · ft^2 **21–17.** 0.0945 kg · m^2

21–18. 0.429 kg · m^2 **21–19.** 0.0292 slug · ft^2

21–21. 0

21–22. $\bar{y} = 0.5$ ft, $\bar{x} = -0.667$ ft, $I_{x'} = 0.0272$ slug · ft^2, $I_{y'} = 0.0155$ slug · ft^2, $I_{z'} = 0.0427$ slug · ft^2

21–23. $I_x = I_y = 20.1$ slug · ft^2, $I_z = 16.3$ slug · ft^2

21–27. 1.14 J **21–29.** 15.1 rad/s

21–30. 14.7 rad/s

21–31. $\boldsymbol{\omega} = \{-0.750\mathbf{j} + 1.00\mathbf{k}\}$ rad/s

21–33. 15.5 ft · lb **21–34.** 2.24 in.

21–35. 19.7 rad/s **21–37.** 18.2 ft/s

21–38. 61.7 rad/s **21–39.** 87.2 rad/s

21–41. 20.2 rad/s **21–42.** 17.3 rad/s

21–43. $\boldsymbol{\omega} = \{0.0536\mathbf{i} + 0.0536\mathbf{k}\}$ rad/s

21–45. $\Sigma M_x = (I_x\dot{\omega}_x - I_{xy}\dot{\omega}_y - I_{xz}\dot{\omega}_z) - \Omega_z(I_y\omega_y - I_{yz}\omega_z - I_{yx}\omega_x) + \Omega_y(I_z\omega_z - I_{zx}\omega_x - I_{zy}\omega_y)$

21–46. $\Sigma M_x = I_x\dot{\omega}_x - I_y\Omega_z\omega_y + I_z\Omega_y\omega_z$

21–47. $F_A = 277$ N, $F_B = 166$ N

21–49. $B_z = 4.25$ lb, $A_x = -2.12$ lb, $A_y = 1.06$ lb, $B_x = 2.12$ lb, $B_y = -1.06$ lb

21–50. $B_z = 3.25$ lb, $A_x = -1.63$ lb, $A_y = 0.814$ lb, $B_x = 1.63$ lb, $B_y = -0.814$ lb

21–51. $M_x = 0$, $M_y = 2I\omega_p\omega_s$, $M_z = 0$

21–53. $F_A = 8.83$ lb, $F_B = 11.2$ lb

21–54. $D_y = -12.9$ N, $\dot{\omega}_z = 200$ rad/s^2, $D_x = -37.5$ N, $C_x = -37.5$ N, $C_y = -11.1$ N, $C_z = 36.8$ N

21–55. 25.9 rad/s^2, $B_x = -0.0791$ N, $B_z = 12.3$ N, $A_x = -1.17$ N, $A_z = 12.3$ N

21–57. $B_x = 50$ N, $A_y = B_y = 0$, $A_z = B_z = 24.5$ N

21–58. $B_x = 50$ N, $A_y = 2.00$ N, $B_y = -0.500$ N, $A_z = B_z = 24.5$ N

21–59. $\Sigma M_x = \dfrac{I_{yz}}{\rho^2}v_G^2$, $\Sigma M_y = -\dfrac{I_{zx}}{\rho^2}v_G^2$, $\Sigma M_z = 0$

21–61. 70.8° **21–62.** 70.8°

21–63. $A_X = B_X = 0$, $A_Y = -1.09$ kN, $B_Y = 1.38$ kN

21–65. -102 rad/s^2, $A_x = B_x = 0$, $A_y = 0$, $A_z = 297$ N, $B_z = -143$ N

21–66. $A_x = 0$, $A_y = 4.89$ lb, $A_z = 5.47$ lb, $M_x = -8.43$ lb · ft, $M_y = 0$, $M_z = 0$

21–67. $T_B = 47.1$ lb, $M_y = 0$, $M_z = 0$, $A_x = 0$, $A_y = -93.2$ lb, $A_z = 57.1$ lb

21–69. $\alpha = 52.2°$, $\beta = 69.3°$, $\gamma = 45.0°$

21–71. (*a*) $A_y = 1.49$ kN, $B_y = 2.43$ kN, (*b*) $A_y = -1.24$ kN, $B_y = 5.17$ kN (*c*) $A_y = 1.49$ kN, $B_y = 2.43$ kN

21–73. 4.91 rad/s **21–74.** 368 rad/s

21–75. $M_x = 12.4$ lb · ft, $M_y = M_z = 0$

21–77. 107 N

21–78. 1.31 rad/s (low precession), 255 rad/s (high precession)

21–79. 12.8 rad/s

21–81. 0.673°, regular precession

Chapter 22

22–1. 0.452 s, 2.21 Hz

22–2. $x = -0.05 \cos(12.2t)$ m

22–3. 7.88 Hz, 0.127 s

22–5. 3.13 Hz, 1 in., $y = (0.0833 \cos 19.7t)$ ft

22–6. $y = (-1.44 \sin (13.9t) - 0.167 \cos (13.9t))$ ft, 1.45 ft

22–7. $\ddot{y} + 56.1y = 0$, 192 mm

22–9. $\theta = [-0.101 \sin (4.95t) + 0.300 \cos (4.95t)]$ rad

22–10. 16.3° **22–11.** $f = \dfrac{1}{\pi}\sqrt{\dfrac{k}{m}}$

22–13. 21.2 kg, 609 N/m **22–14.** 1.18 s

22–15. $\tau = 2\pi\sqrt{\dfrac{k_G^2 + d^2}{gd}}$ **22–17.** $\theta = 0.4 \cos (20.7t)$

22–18. $\tau = 2\pi\sqrt{\dfrac{l}{2g}}$ **22–19.** 1.28 s

22–21. 0.167 s **22–22.** 2.67 s

22–23. 90.8 lb · ft/rad **22–25.** $\tau = \pi\sqrt{\dfrac{2m}{3k}}$

22–26. $x = 0.167 \cos (6.55t)$ **22–27.** $f = \dfrac{1}{\pi}\sqrt{\dfrac{k}{m}}$

22–29. 1.18 s **22–30.** $\tau = 2\pi\sqrt{\dfrac{k_G^2 + d^2}{gd}}$

22–31. $f = 2\pi\sqrt{\dfrac{m}{3k}}$ **22–33.** $\tau = \pi\sqrt{\dfrac{m}{k}}$

22–34. 0.970 s **22–35.** 0.487 s

22–37. 1.81 Hz **22–38.** 1.52 s

22–39. 2.25 Hz **22–41.** 2.81 s

22–42. $\ddot{\theta} + 468\theta = 0$

22–43. $y = A \sin pt + B \cos pt + \dfrac{\dfrac{F_0}{m}}{\left(\dfrac{k}{m} - \omega^2\right)} \cos \omega t$

22–45. $y = [-16.6 \sin 7.07t + 70 \cos 7.07t + 29.4 \sin 4t]$ mm

22–46. $\ddot{y} + p^2 y = \dfrac{k\delta_0}{m} \sin \omega t$,

$$y = A \sin pt + B \cos pt + \frac{\delta_0}{1 - \left(\frac{\omega}{p}\right)^2} \sin \omega t$$

22–47. $y = A \sin pt + B \cos pt + \dfrac{F_0}{k}$

22–49. 0.685 ft/s **22–50.** 29.5 mm

22–51. $y = (-0.297e^{-9.51t} + 0.297e^{-16.2t})$ ft

22–53. 1.89 in.

22–54. $x_p = (0.0276 \sin 2t)$ ft

22–55. 6.55 rad/s **22–57.** 14.6 mm

22–58. 35.5 mm **22–59.** 19.7 rad/s

22–61. 19.0 rad/s

22–62. $y = (361 \sin 7.75t + 100 \cos 7.75t - 350 \sin 8t)$ mm

22–63. 0.625 in. **22–65.** 4.53 mm

22–67. 0.0269 in. **22–69.** 9.89°

22–70. 0.997 **22–71.** 1.66 s

22–73. $x = 0.119 \cos (6t - 83.9°)$ m

22–74. 9.89° **22–75.** 0.997

22–77. $k \leq 714$ N/m **22–78.** 3.99 s

22–79. $L\ddot{q} + R\dot{q} + \left(\dfrac{2}{C}\right)q = 0$

22–81. $L\ddot{q} + R\dot{q} + \left(\dfrac{1}{C}\right)q = E_0 \cos \omega t$

22–82. $L\ddot{q} + R\dot{q} + \dfrac{1}{C}q = 0$

22–83. $m\ddot{x} + kx = F_0 \sin \omega t$, $L\ddot{q} + \dfrac{1}{C}q = E_0 \sin \omega t$

Index

Geometric Properties of Line and Area Elements

Centroid Location	Centroid Location	Area Moment of Inertia

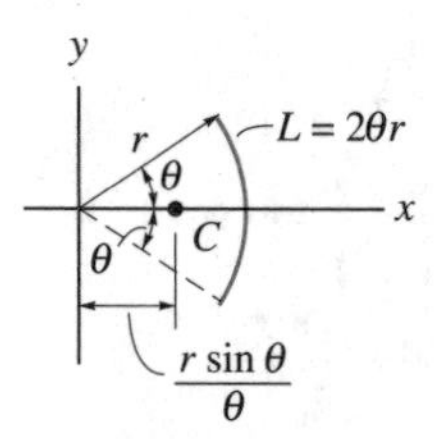

Circular arc segment

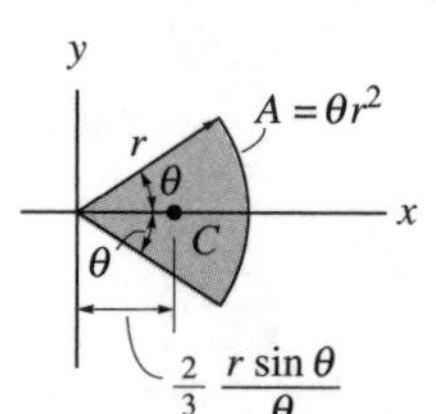

Circular sector area

$$I_x = \tfrac{1}{4}r^4(\theta - \tfrac{1}{2}\sin 2\theta)$$

$$I_y = \tfrac{1}{4}r^4(\theta - \tfrac{1}{2}\sin 2\theta)$$

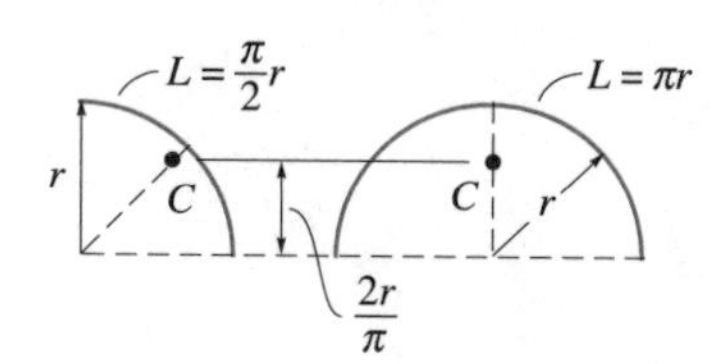

Quarter and semicircular area

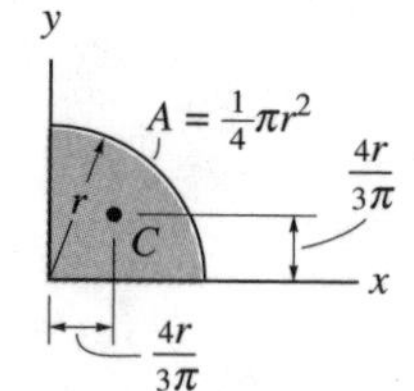

Quarter circular area

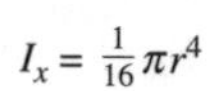

$$I_x = \tfrac{1}{16}\pi r^4$$

$$I_y = \tfrac{1}{16}\pi r^4$$

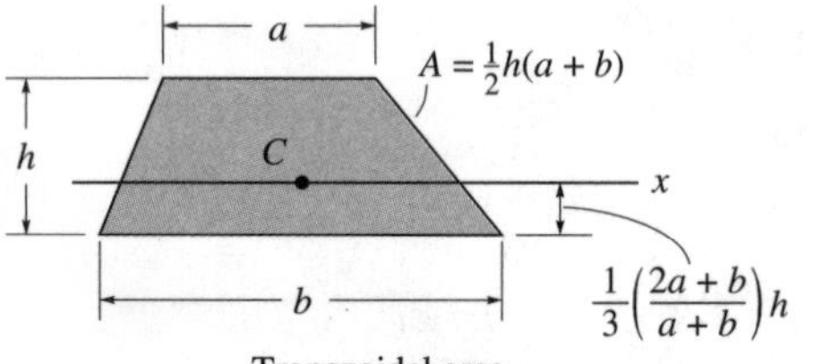

Trapezoidal area

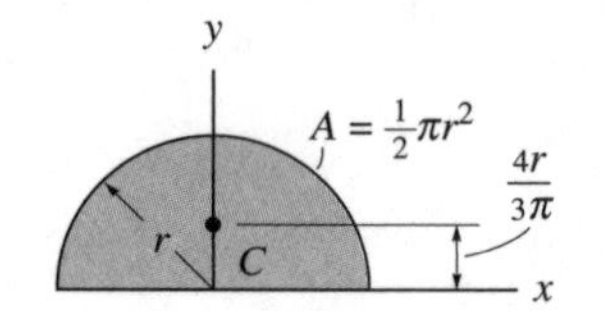

Semicircular area

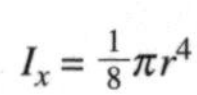

$$I_x = \tfrac{1}{8}\pi r^4$$

$$I_y = \tfrac{1}{8}\pi r^4$$

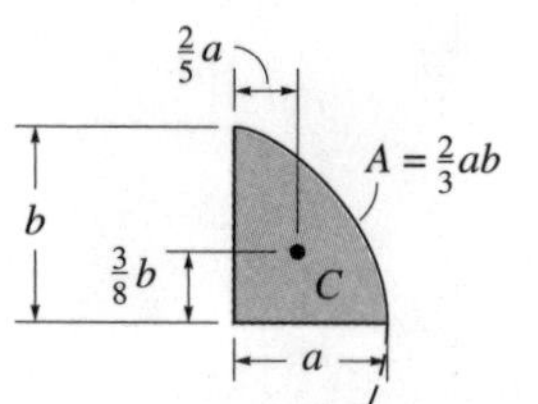

Semiparabolic area

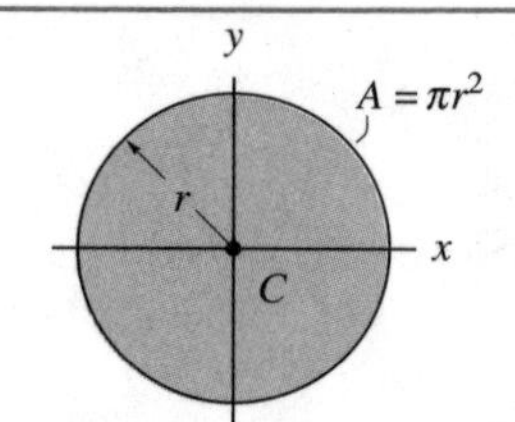

Circular area

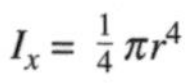

$$I_x = \tfrac{1}{4}\pi r^4$$

$$I_y = \tfrac{1}{4}\pi r^4$$

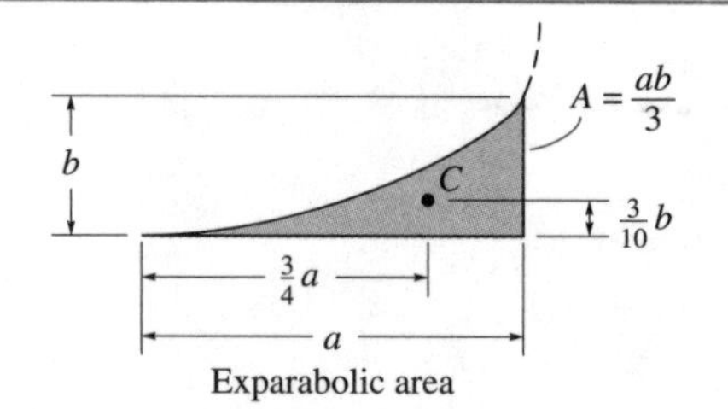

Exparabolic area

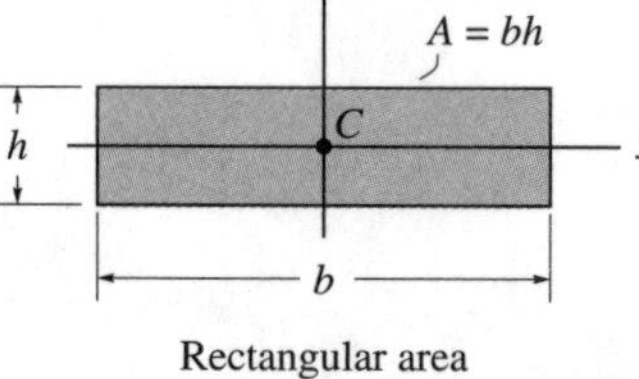
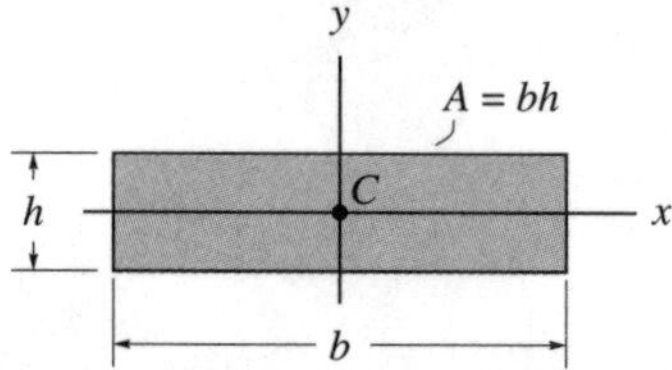

Rectangular area

$$I_x = \tfrac{1}{12}bh^3$$

$$I_y = \tfrac{1}{12}hb^3$$

a

b

C

$A = \frac{4}{3}ab$

$\frac{2}{5}a$

Parabolic area

$A = \frac{1}{2}bh$

h

C

x

b

$\frac{1}{3}h$

Triangular area

$$I_x = \tfrac{1}{36}bh^3$$